FLORA EUROPAEA

FLORA EUROPAEA

VOLUME 1

PSILOTACEAE TO PLATANACEAE

SECOND EDITION

EDITED BY

T. G. TUTIN N. A. BURGES A. O. CHATER J. R. EDMONDSON

V. H. HEYWOOD D. M. MOORE

D. H. VALENTINE S. M. WALTERS D. A. WEBB

ASSISTED BY

J. R. AKEROYD AND M. E. NEWTON

APPENDICES
EDITED BY

R. R. MILL

CAMBRIDGE
UNIVERSITY PRESS

CAMBRIDGE UNIVERSITY PRESS
Cambridge, New York, Melbourne, Madrid, Cape Town, Singapore,
São Paulo, Delhi, Dubai, Tokyo, Mexico City

Cambridge University Press
The Edinburgh Building, Cambridge CB2 8RU, UK

Published in the United States of America by Cambridge University Press, New York

www.cambridge.org
Information on this title: www.cambridge.org/9780521153669

First published 1964
Second edition 1993
Third printing 2007
First paperback printing 2010

A catalogue record for this publication is available from the British Library

Library of Congress Cataloguing in Publication data
Flora Europaea/edited by T. G. Tutin... [et al.]; assisted by J. R. Akeroyd
and M. E. Newton. - 2nd ed.
p. cm.
Includes bibliographical references and index.
Contents: v. 1. Psilotaceae to Platanaceae.
ISBN 0 521 41007 X hardback
1. Botany-Europe. 2. Botany-Europe-Classification. I. Tutin, Thomas Gaskall, 1908-.
QK281.F59 1992
581.94 – dc20 92–33771 CIP

ISBN 978-0-521-41007-6 Hardback
ISBN 978-0-521-15366-9 Paperback

CONTENTS

THE FLORA EUROPAEA ORGANIZATION

The membership shown below is as it was constituted during the preparation of the first edition of Volume 1, which was completed in January 1963.

REGIONAL ADVISERS

Albania	F. Markgraf, Zürich
Austria	E. Janchen, Wien
	K. H. Rechinger, Wien
Belgium	A. Lawalrée, Bruxelles
Bulgaria	N. Stojanov, Sofija
Czechoslovakia	D. Dostál, Olomouc
Denmark	T. W. Böcher, København
Finland	J. Jalas, Helsinki
France	H. Gaussen, Toulouse
	P. Jovet, Paris
Germany	H. Merxmüller, München
	†W. Rothmaler, Greifswald
Greece	K. H. Rechinger, Wien
Hungary	R. De Soó, Budapest
	†S. Jávorka, Budapest
Iceland	I. Óskarsson, Reykjavík
Italy	R. E. G. Pichi-Sermolli, Genova
Jugoslavia	E. Mayer, Ljubljana
Netherlands	S. J. Van Ooststroom, Leiden
Norway	R. Nordhagen, Oslo
Poland	B. Pawłowski, Kraków
Portugal	A. R. Pinto Da Silva, Sacavém
Romania	A. Borza, Cluj
	E. I. Nyárády, Cluj
Spain	E. Guinea, Madrid
	Oriol De Bolòs, Barcelona
Sweden	N. Hylander, Uppsala
Switzerland	E. Landolt, Zürich
Turkey	P. H. Davis, Edinburgh
U.S.S.R.	†B. K. Schischkin, Leningrad
	Andrey A. Fedorov, Leningrad

TECHNICAL CONSULTANTS

Á. Löve, Montreal

†W. Rothmaler, Greifswald

GEOGRAPHICAL ADVISER

H. Meusel, Halle

EDITORIAL COMMITTEE AS CONSTITUTED IN 1990

N. A. BURGES, Coleraine; *Chairman*

J. R. EDMONDSON, National Museums & Galleries on Merseyside; *Secretary*

J. R. AKEROYD, Hindolveston, Norfolk

F. A. BISBY, University of Southampton

A. O. CHATER, Natural History Museum, London

V. H. HEYWOOD, Richmond

S. L. JURY, University of Reading

D. M. MOORE, University of Reading

S. M. WALTERS, Cambridge

M. E. NEWTON, National Museums & Galleries on Merseyside; *Flora Europaea Research Officer*

LIST OF CONTRIBUTORS TO VOLUME 1, EDITIONS 1 AND 2

The following is a list of authors who have contributed accounts of genera in whole or in part.

†P. Aellen, Basel
J. R. Akeroyd, Hindolveston
P. W. Ball, Toronto
Y. I. Barkoudah, Damascus
†J. P. M. Brenan, Kew
R. K. Brummitt, Kew
N. A. Burges, Coleraine
A. O. Chater, London
†A. R. Clapham, Bentham
C. D. K. Cook, Zürich
J. A. Crabbe, London
J. Cullen, Cambridge
Å. E. Dahl, Göteborg
R. Domac, Zagreb
T. R. Dudley, Jamaica Plain, Mass.
C. Favarger, Neuchâtel
J. Do Amaral Franco, Lisboa
C. R. Fraser-Jenkins, Oxford
H. Freitag, Kassel
H. P. Fuchs, Wassenaar
†H. Gaussen, Toulouse
D. V. Geltman, St Petersburg
†E. Guinea, Madrid
G. Halliday, Lancaster
H. 't Hart, Utrecht
R. Hendrych, Praha
V. H. Heywood, Richmond
R. N. Higton, Oxford
D. J. N. Hind, Kew
J. Jalas, Helsinki
N. Jardine, Cambridge
A. C. Jermy, London
B. M. G. Jones, London
B. Jonsell, Stockholm
J. W. Kadereit, Heidelberg
A. Lawalrée, Bruxelles
E. A. Leadlay, London
M. Lidén, Göteborg
J. D. Lovis, Christchurch
M. Maier-Stolte, Kassel
K. Marhold, Bratislava
†F. Markgraf, Zürich
J. B. Martinez-Laborde, Madrid
R. D. Meikle, Minehead
†H. Merxmüller, München
R. R. Mill, Edinburgh
P. Monnier, Montpellier
A. B. Mowat, Inverness
E. Nardi, Firenze
†E. I. Nyárády, Cluj
C. N. Page, Edinburgh
J. Parnell, Dublin
A. M. Paul, London
S. Pawłowska, Kraków
†B. Pawłowski, Kraków
A. R. Pinto Da Silva, Sacavém
C. D. Preston, Abbots Ripton
J. A. Ratter, Edinburgh
K. H. Rechinger, Wien
T. C. G. Rich, Peterborough
†W. Rothmaler, Greifswald
†O. Schwarz, Jena
P. D. Sell, Cambridge
W. T. Stearn, Kew
†T. G. Tutin, Leicester
†D. H. Valentine, Manchester
†J. De Carvalho e Vasconcellos, Lisboa
R. Viane, Gent
T. G. Walker, Leeds
S. M. Walters, Cambridge
D. A. Webb, Dublin
†F. H. Whitehead, London
F. Wrigley, Westerham
M. B. Wyse Jackson, Dublin
P. S. Wyse Jackson, Kew
P. F. Yeo, Cambridge
F. Zésiger, Neuchâtel

PREFACE

On 4 September 1977, a meeting was held in Cambridge of many of those associated with the *Flora Europaea* project to mark the completion of the work and to hand formally to the Cambridge University Press the manuscript of the final volume. During this meeting there were discussions on the need for revisions, for corrigenda and addenda, for a possible second edition and on the future of the Flora Europaea Organization as a whole. There was a strong feeling, particularly amongst our continental colleagues, that some way of continuing the links which had been built up through the Flora Europaea Organization, should be maintained. It was generally agreed that neither the Editorial Committee nor the Flora Europaea Organization should be disbanded but, for the time being, would go into abeyance.

After the proofs of Volume 5 had been checked and the other responsibilities completed, the Editorial Committee arranged for the papers associated with the Flora Europaea Organization and with the preparation of the manuscripts to be deposited in the Library of the University of Reading under the care of the University Archivist. Arrangements had already been made for the royalties from the sale of the volumes of *Flora Europaea* and associated publications to be paid into a special Trust Fund administered by the Linnean Society of London. The Editorial Committee would like to record its thanks to the Linnean Society for so ably administering the Trust Fund and for the continuing support which the Society has given in so many ways.

Since Volume 5 was published, many informal discussions have taken place between members of the Editorial Committee regarding the production of indices and taxonomic publications closely associated with *Flora Europaea* and, particularly, about the continually accumulating *addenda* and *corrigenda*, which were being assembled by Professors Heywood and Moore at the University of Reading. In October 1981, the Editorial Committee reassembled on a formal basis and has continued to meet at regular intervals. There was a general feeling among the members of the Committee that, if any revision were to be attempted, it would be particularly necessary for Volume 1. This was, in part, because it was the most out of date, but also because the Committee believed it to contain more errors than did later volumes. Many of these arose from the fact that the technique of making the critical synthesis, at which the Flora aimed, took some time to learn, and also from the necessity of meeting an acceptable time-schedule, which meant that some parts of the volume were written or edited in more haste than the Editorial Committee would have wished.

At a meeting in June 1983, the Editorial Committee discussed detailed papers prepared by members of the Committee on proposals for the revision of Volume 1. It was agreed that the Committee should undertake the responsibility itself and an outline of the procedures to be followed was agreed. The accumulated royalties in the Flora Europaea Trust Fund made it possible for the Committee to appoint an experienced full-time research worker to carry out the revision under the direction of the Committee. Dr J. R. Akeroyd was appointed, taking up his duties on 1 October 1983. His previous experience with the taxonomy and evolution of British and Mediterranean plants was, in the view of the Editorial Committee, particularly appropriate. Later, a further appointment was made, and Dr M. E. Newton began work on the Flora in February 1990.

Since the publication of Volume 5, Mr A. O. Chater of the Natural History Museum, London, and Dr J. R. Edmondson of the National Museums & Galleries on Merseyside, Liverpool, have joined the Editorial Committee. It was with great regret that the members of the Organization learnt of the deaths of Professor D. H. Valentine on 10 April 1987 and Professor T. G. Tutin, F.R.S. on 7 October 1987. These colleagues both played major rôles in the inception, development and subsequent successful conclusion of the Flora Europaea project.

Following the death of Professor Tutin, Dr N. A. Burges was appointed Chairman of the Committee, whilst Professor V. H. Heywood, who had so very ably carried out the duties of Secretary since the formation of the Committee in January 1956, was succeeded by Professor D. M. Moore in November 1986 and he, in October 1989, by Dr J. R. Edmondson. Professors Heywood and Moore both remain as members of the Committee.

With the limited resources available it was impossible to continue the formal continent-wide consultation and the circulation of manuscripts in various stages of completion, valued almost as much by our European colleagues as by ourselves in the preparation of the first edition. However, many colleagues throughout Europe have significantly assisted in the revision of this volume. Throughout the revision, the Committee for Mapping the Flora of Europe, in Helsinki, has also made available a great deal of manuscript information which has proved invaluable.

Virtually every genus has been critically examined, and the accounts of most genera have been substantially revised. The relative amount of work undertaken by the authors involved in the revisions varies greatly and it has not been easy to indicate this succinctly in the text. The compromise decided on by the Editorial Committee has been to limit the attribution of authorship to three categories: (i) where the original author is deceased, inaccessible or did not wish to be associated with the revised account, the account is shown as 'original author (edition 1) revised by new contributor (edition 2)'; (ii) where a major revision has been undertaken by a new contributor with the approval of the original author, it is shown as 'original author revised by new contributor'; and (iii) where the revision has been made in collaboration with the original author, the names of both contributors are given with no reference to edition.

The Editorial Committee realizes that the restriction to only three categories may lead to an over-simplification of the variation in the amount of work undertaken by different contributors, but it believes this is preferable to attempting to provide a more detailed quantitative assessment which would be difficult to do with any precision. There have been instances where the Editorial Committee has found it necessary to take firm editorial action to achieve the consistency of treatment which has been such a fundamental feature of *Flora Europaea* since its inception and which experience has shown has been valued by the users of the Flora. As throughout the Flora, the Editorial Committee accepts responsibility for the accounts as published.

Since publication of the first edition of Volume 1 in 1964, some 250 species and 150 subspecies belonging to families in Volume 1 have been described from Europe as new to science. In addition to these, 25 species, known elsewhere, have been recorded as occurring as natives in Europe, while around 40 additional alien species have been reported as being naturalized in Europe. A critical assessment of these in accordance with the principles used throughout *Flora Europaea* has led us to include some 350 extra taxa in the new edition. At the same time, further information has enabled us to clarify many of the doubts raised in the first edition and has resulted in the deletion of about 20 taxa. Changes in taxonomy and nomenclature have been made only when there seemed to be general agreement as to their correctness by those who had studied the problem. When reasonable doubt remained, we thought it better to make no change. Although the main burden of the work in revising Volume 1 has fallen on Dr Akeroyd, the Editorial Committee has maintained a close supervision throughout.

In the preparation of the second edition of Volume 1 we have attempted to meet one of the most widely voiced criticisms of the first edition; that text synonymy was too often inadequate, many synonyms being restricted to the index alone. All synonyms are now cited in the text.

In the first edition, separate lists were given of what were regarded as Basic Floras and Standard Floras. The inclusion of synonyms in the text and the appearance of newer works has made the distinction no longer of value. Accordingly, there is now a single list of Standard Floras which gives an indication of the source of synonyms and of the most up-to-date Floras of the various areas covered by *Flora Europaea*.

The publication of a revised Volume 1 naturally raises the issue as to what revisions, if any, should be made of the other volumes. At the moment no decision has been taken but the matter is under active review by the Editorial Committee, the Secretariat of which is now based in the Department of Botany at Liverpool Museum, part of the National Museums & Galleries on Merseyside.

June 1991

ACKNOWLEDGEMENTS

The Editorial Committee gratefully acknowledges the support of the Linnean Society of London, whose Flora Europaea Trust Fund provided £120,000 towards the cost of revising this volume. It has financed the employment of two full-time Research Officers, J. R. Akeroyd (1983–1988) and M. E. Newton (1990–), and has met other expenses of the Secretariat. This was helped over a difficult period by a loan from the Royal Society's scientific publications fund. Financial assistance in the form of a legacy from the estate of the late Miss M. S. Campbell is also greatly appreciated, having provided an opportunity to commission illustrations for a new series of Notulae.

In undertaking the revision of Volume 1 of *Flora Europaea* the Committee has enjoyed continued assistance and encouragement from many colleagues throughout Europe. We wish to thank them for their willing and generous co-operation. Their help has been given in many ways, some of them botanical and others technical, but all of them vital to the success of our project.

We are deeply appreciative of the assistance of R. R. Mill, who compiled and checked the Appendices, and Mrs L. Walters, who transferred synonymy to the text and drew up the index. Our special thanks are due to Mrs M. Hart for typing the manuscript so diligently.

In addition to the authors of individual accounts, we have received advice on particular groups from many people: J. Cullen (*Alyssum*, *Aurinia*), C. R. Fraser-Jenkins (*Polypodium*), H. Freitag (Chenopodiaceae), I. C. Hedge (Chenopodiaceae), M. V. Kasakova (*Schivereckia*), R. D. Meikle (Salicaceae), C. N. Page (*Equisetum*), J. R. Press (*Rumex*, Subgen. *Platypodium*), C. D. Preston (Aizoaceae), K. H. Rechinger (*Rumex*), T. C. G. Rich (Cruciferae), L.-P. Ronse Decraene (*Polygonum*), F. Rose (*Salicornia*), G. D. Rowley (Aizoaceae, Crassulaceae), D. Tzanoudakis (*Paeonia*), P. Uotila (*Chenopodium*) and S. D. Webster (*Ranunculus*, Subgen. *Batrachium*). P. W. Ball, as well as editing Cruciferae, critically read many of the manuscripts, and A. C. Jermy and Miss A. M. Paul carefully checked Pteridophyta.

More general comment and advice for which we are grateful was received from: H. J. M. Bowen, R. K. Brummitt, G. Buzas, E. J. Clement, D. V. Geltman, A. Hansen, J. Jalas, C. E. Jarvis, S. L. Jury, D. H. Kent, Miss C. Leon, Mrs B. Molesworth Allen, D. Nicolson, Miss E. Powell, R. W. Rutherford, P. D. Sell, C. A. Stace, J. Suominen and M. F. Watson. The Editorial Committee is also most grateful to members of the Committee for Mapping the Flora of Europe for their generosity in allowing us to consult unpublished typescripts from the *Atlas Florae Europaeae* project.

The Committee acknowledges a debt of gratitude to the Directors of many Museums and Herbaria. We thank them all but wish particularly to mention the University of Cambridge; Trinity College, Dublin; the Royal Botanic Garden, Edinburgh; the National Museums & Galleries on Merseyside, Liverpool; the Natural History Museum, London; and the University of Reading, the institutions whose facilities we have enjoyed for the greater part of our work.

June 1991

PREFACE TO THE FIRST EDITION

Europe, where scientific taxonomy was developed, and whose taxonomists have participated, for more than a century, in the preparation of Floras for other continents, does not possess a Flora of its own. Many hundreds of national and regional Floras exist, together covering almost every part of Europe, but there is no modern work which deals with the area as a whole.

There are obvious historical and political reasons which explain the lack of such a Flora. It is, however, becoming increasingly important, as in all other fields of science, that a work synthesizing the accumulated data from the various regions should be produced. For one thing, the channels of distribution of information are being choked by the accumulation of unassimilated or neglected data and if these channels can be cleared, and the data sifted, collected and arranged, botanical science will be stimulated. Equally important, good taxonomy is the basis for many kinds of scientific research, but this taxonomy must be broadly based and well considered and must transcend national boundaries. To the biosystematist and the phytogeographer, for example, local taxonomy can be misleading, whereas a wider treatment, on a continental scale, can correct misconceptions and reveal problems that had not been previously suspected. A new, synthetic Flora of Europe will be useful, not only to specialists, but also to biologists of many kinds. It should make it easier for them to give a plant an accurate name, which we regard as one of the primary aims of a Flora, and it should also direct their attention to literature in which further information can be found.

The problems which face those attempting to write a Flora of Europe are formidable. The materials available are of various kinds: there are some outdated catalogues of European plants (such as Nyman's *Conspectus Florae Europaeae*), some important regional Floras, such as those by Hegi and Komarov (see p. xxix), and many excellent national Floras. In addition, there is a vast accumulation of literature, published in many places and in a variety of languages, much of which is relatively inaccessible and virtually unknown outside its country of origin. The writers of a European Flora must be prepared to seek out this literature and review it critically, in addition to studying for themselves in the herbarium and the field the plants about which they propose to write.

This, then, is the background against which any decision to write a Flora of Europe has to be taken. Mention must be made of the *Flora Europaea* projected by Werner Rothmaler in the 1940s which was to cover not only Europe but also the Caucasus, Transjordan and parts of North Africa. Owing to the conditions of wartime this scheme was abandoned. The question was reopened in a discussion on 'Progress of Work in the European Flora' at a session of the 8th International Botanical Congress held at Paris in 1954. Although no formal decision to write a Flora was taken, further discussions between a number of European botanists followed, and eventually an informal committee came into being, which held its first meeting at Leicester in January 1956. This committee (the constitution of which is given on p. vii), has remained in being, and, with few modifications, has become the executive committee in charge of the project. It was realized from the start that, while it was obviously advantageous to have a central organization in one part of Europe (the British Isles), the project could only succeed with continuous and substantial advice and help from every part of Europe. Accordingly, invitations to a number of distinguished European botanists to act as Advisory Editors (p. vii) were issued; the prompt acceptance of these invitations and the approval and support received encouraged the committee to proceed.

The next step was to plan the Flora in detail. As a result of much discussion a booklet entitled *The Presentation of Taxonomic Information: A Short Guide for Contributors to Flora Europaea* was prepared and published in 1958, and it then became possible to begin the writing of the Flora. Editors for each family were appointed from the Editorial Committee, and accounts of individual families or genera were obtained from botanists in all parts of Europe. The list of contributors to volume I is given on p. xi. Details of the style and form of the Flora are given below (p. xxi).

It should be emphasized that it has been, and remains, the intention of the Editorial Committee to produce a concise and complete Flora in the shortest possible time. Consequently, the principle has been adopted that publication of the Flora cannot be delayed for an indefinite period to allow the lengthy and detailed research required for a complete solution of all the problems that arise during its preparation. The committee believes that it is more valuable to have a complete Flora, representing a synthesis of available information, than a series of detailed monographs where completion could not be foreseen. It is planned to complete the Flora in four volumes over the next eight years.

It is important to emphasize that full collaboration with taxonomists in every part of Europe has been sought and obtained at every stage during the writing of the Flora. A panel of Regional Advisers was organized (the full list is given above). All manuscripts were circulated to these Advisers, and their comments on them, based on their specialized knowledge of both taxonomy and local distributions, have been invaluable and constitute an integral part of the organization of the Flora.

Arising out of the contacts established in this way, Symposia have been held, at Vienna in 1959, and at Genoa in 1961, to which all the Advisers of the Flora were invited, and at which topics of common interest were discussed. These Symposia have been important in establishing international collaboration on a firm basis, and in making it possible to plan the future progress of the Flora with confidence. A further Symposium was held in 1963 in Romania.

The names of the authors primarily responsible for writing the accounts of the various families and genera are given in footnotes in the text. In preparing the first volume of the Flora for the press, much trouble has been taken by the editors to produce a self-consistent manuscript, in a fairly standardized form, incorporating such revision and alteration of taxonomy, nomenclature and distribution as has been necessary. It should therefore be made clear that the Editorial Committee takes full responsibility for the form in which the text is published.

Acknowledgement is made below to the many people and organizations who have given help, financial and otherwise, in the preparation of the Flora. We should, however, record here our special gratitude to the Department of Scientific and Industrial Research in London which, by granting the Committee large sums of money over a period of several years, has made it possible to set up and maintain a permanent secretariat at the University of Liverpool, and to undertake the many ancillary activities which have arisen. It is in the prospect of their continuing support that we are now proceeding with the second volume.

October 1963

ACKNOWLEDGEMENTS TO THE FIRST EDITION

The Flora Europaea project received in its early stages several valuable *ad hoc* grants without which no substantial progress could have been made; these were from the Royal Society, the Botanical Research Fund, the Royal Horticultural Society, and Miss M. S. Campbell. A loan from Mr and Mrs J. E. Raven tided the Committee over a difficult period. In 1959 a successful application was made by the Organizing Committee to the Department of Scientific and Industrial Research for a three-year grant which has now been extended for a further three years. The total amount of the grant for the period 1959–1965 is £32,000. This has permitted adequate finance of the Secretariat and its services, and the appointment of a number of full-time Research Assistants: Miss A. B. Mowat (1959–1960), Dr P. W. Ball (1959 –) and Mr A. O. Chater (1960–). Dr Ball previously held a Leverhulme Research Fellowship at the University of Liverpool to work on the project. The Committee gratefully acknowledges all this generous support.

In addition to this main grant from British sources, the project received direct and indirect financial help from several countries, among which should be mentioned a grant for secretarial printing from the Instituto de Alta Cultura, Lisbon, and grants from the Ministries of Education in Austria and Italy, and from the International Union of Biological Sciences, towards the Symposia held in Vienna and Genoa. The third Symposium which was held in Romania was generously subsidized by the Academy of the Romanian People's Republic. We gratefully acknowledge the sponsorship of our project by the Linnean Society of London.

Acknowledgement is also due to the Universities of Cambridge, Dublin, Durham, Leicester and Liverpool for their support in making facilities available to the members of the Editorial Committee and their assistants. In particular the University of Liverpool has provided accommodation in the Hartley Botanical Laboratories for the Secretariat since 1957.

We are also grateful to the Directors of the various Museums and Herbaria who have provided the Editorial Committee and Regional Advisers with facilities to study their collections and utilize their libraries. Amongst these must be mentioned the Herbarium and Library, Royal Botanic Gardens, Kew; the Department of Botany, British Museum (Natural History), London; the Naturhistorisches Museum, Vienna; and the Herbarium of the Istituto Botanico, Florence. The Trinity College Dublin Trust has made a substantial grant towards the provision of microfilm and a reader.

A very large number of individual botanists have assisted us in various ways. The following deserve special mention: P. Aellen (Basel), Miss P. Edwards (London), A. Hansen (København), E. K. Horwood (Leicester), A. C. Jermy (London), N. Y. Sandwith (Kew). P. D. Sell (Cambridge), as well as assisting in several other ways, undertook the laborious task of preparing the index for the press.

The staff of the Secretariat, Mrs R. Seddon, Mrs J. Beck and Mrs T. Donnelly, deserve our special gratitude for their great efficiency and loyalty even during times of almost overwhelming pressure.

October 1963

INTRODUCTION

The aim of the Flora is in general diagnostic, and the descriptions, while brief, are as far as possible comparable for related species. The Standard Floras listed on p. xxviii, and the monographs or revisions given when appropriate after the descriptions of families and genera, may assist the reader in obtaining more detailed information. Other references to published work are occasionally given in cases of special taxonomic difficulty.

All available evidence, morphological, geographical, ecological and cytogenetical, has been taken into consideration in delimiting species and subspecies, but they are in all cases definable in morphological terms. (Taxa below the rank of subspecies are not normally included.)

The delimitation of genera is often controversial and the solution adopted in the Flora may on occasion be a somewhat arbitrary choice between conflicting opinions. We have endeavoured to weigh as fairly as possible the various opinions available, but there has been no consistent policy of 'lumping' or 'splitting' genera (or, for that matter, species). The order and circumscription of the Dicotyledones in the second edition follows that of the first edition, which was largely based on that of Engler-Diels, *Syllabus der Pflanzenfamilien* ed. 11 (1936). It therefore differs in certain points from that of Melchior in Engler, *Syllabus der Pflanzenfamilien* ed. 12 (1964). The order of the Pteridophyta follows Derrick, Jermy and Paul, *Sommerfeltia* **6**: 1–94 (1987).

All descriptions of taxa refer only to their representatives in Europe. In practice, we have relaxed this rule slightly for families and genera to avoid giving taxonomically misleading information, particularly in those cases where a large family or genus has only one or few, perhaps atypical, members in Europe. In such cases we have occasionally added 'in European members' or a similar phrase to emphasize the atypical representation. In no case, however, should it be assumed that the description is valid for all the non-European representatives of these taxa.

Often a short note follows the description of a species or, occasionally, a genus, containing additional information of various kinds.

(*a*) *Choice of language*

Much thought was given to the question of the best language for the Flora. There were many reasons in favour of English, but the alternative of Latin also had much to commend it. After consulting the Regional Advisers, we decided in favour of English. Opinion in Europe was fairly evenly divided, but it seemed likely that whereas in some European countries the professional taxonomist might prefer Latin, the general botanist and other biologists would prefer English. It was also clear that outside Europe English would be preferred. As we were anxious to make the Flora available to as wide a circle of readers as possible, we felt that these considerations were decisive.

We have tried, however, to bear in mind throughout the writing and editing of the work that it will be used by many readers whose knowledge of English is very imperfect. For this reason we have restricted ourselves to a much smaller technical vocabulary than is usual, believing that the gain in ease of understanding would offset the small loss of precision sometimes involved. For the same reason, in most cases where there are two English words with the same meaning, we have preferred the one derived from a Latin root. A glossary has been added, whose purpose is not the usual one of explaining technical terms in simple language (which is impracticable on an international basis), but of giving a Latin equivalent of those English words which are important for an understanding of the text, and which differ substantially from their Latin equivalents. We have also provided a short list of English definitions of those terms which experience has shown to be open to misconstruction.

Place-names used in the summary of geographical distribution have been given in their English form when they refer to independent States or to such geographical features of Europe as transcend national boundaries. All other place-names are given in the language of the country concerned. Thus we write *Sweden*, *Ukraine*, *Danube*, *Alps*, *Mediterranean*, but *Corse*, *Kriti*, *Rodopi Planina*, *Ahvenanmaa*. *Macedonia* is written in its English form if reference is intended to the whole of the region that has traditionally been known by that name; but the Jugoslav and Greek administrative units are referred to as *Makedonija* and *Makedhonia* respectively.

In transliteration from Cyrillic characters we have, as far as place-names and titles of publications are concerned, followed the ISO system recommended in the UNESCO *Bulletin for Libraries* **10**: 136–137 (1956). This is almost identical with the system recommended by Paclt in *Taxon* **2**: 159 (1953). Both are based essentially on the conventions of Serbo-Croat and, although the number of diacritical signs which they require is an undoubted disadvantage, they are the only systems which have any real claim to be considered international, all others in current use being based frankly on the acceptance of German or English conventions of pronunciation. With personal names, however, we have been influenced by the fact that Russian botanists, in transliterating their own names, usually follow German conventions; we have therefore followed the list of transliterations given in the index-volume (1962) to *Not. Syst.* (*Leningrad*)/*Botaničeskie Materialy*, and have transliterated personal names which do not occur in this list according to the same conventions.

In transliterating place-names from Greek characters we have, except for omitting the accents, followed *The Times Atlas of the World, Mid-Century Edition*, vol. 4 (London, 1956).

(*b*) *Delimitation of Europe*

We have tried as far as possible to interpret 'Europe' for the purposes of this Flora in its traditional sense. There is no doubt that phytogeographically a more natural unit would be a 'Greater Europe', bounded by the Sahara desert and the great deserts of Asia, and including, therefore, North Africa and much of south-west Asia. But to have attempted to cover this area would have more than doubled our task and would have meant the inclusion of a disproportionately large number of plants whose status and identity are very uncertain. Europe in its traditional sense, however, is completely covered by Floras which, although they vary widely in competence and modernity, reduce the unsolved problems of taxonomy and distribution to a more reasonable compass.

In framing a precise definition of Europe in this traditional sense three questions arose: where the boundary with Asia was to be drawn in eastern Russia and Kazakhstan; where it was to be drawn in the Aegean region; and which islands in the Mediterranean Sea and the Atlantic and Arctic Oceans were to be included.

Some confusion may be caused by apparent differences in geographical distribution in the former territories of the U.S.S.R. between our data and those of the invaluable *Flora Partis europaeae URSS*. In a few cases they arise from differences of taxonomic opinion, or of judgements as to the status of a plant (casual, naturalized or native), but in most cases they can be attributed to the fact that the divisions used in that Flora are, though very similar to ours, not quite identical. This is because Fedorov, in the map published in the first volume of his Flora, very sensibly made his boundaries follow the lines of administrative divisions, whereas we had perforce to follow the not very precisely indicated boundaries of Komarov's divisions, which were ecologically and floristically based and paid no attention to administrative boundaries. The most important divergences are in northern Ukraine, southern Ural, and the region north-east of St Petersburg. These are included by Fedorov in his western (Zapad), eastern (Vostok) and northern (Sever) divisions respectively, whereas for us they all fall into Rs(C). There is also a divergence in the line delimiting Europe in the cis-Caucasian region: Fedorov includes a large part of the Stavropol plateau, which we exclude from Europe.

It will be seen that, in the Arctic, Franz Joseph Land (Zemlja Franca Josifa) and Novaja Zemlja are excluded, but the islands of Kolguev and Vajgač are included; that the boundary is then defined by the crest of the Ural Mountains (with a small deviation eastwards near Sverdlovsk) and the Ural River to the Caspian Sea (thus including in Europe a small portion of W. Kazakhstan); and that the whole of the Caucasus is excluded.

In the Aegean region we have drawn our boundary along the deep-water channel which separates the Sporadhes and the islands lying close to the mainland of Anatolia from those of the C. and W. Aegean. The only large islands under Greek sovereignty which are excluded from Europe are the Sporadhes, Khios, Lesvos and Rodhos. The only island under Turkish sovereignty which is included in Europe is Imroz; Bozcaada (Tenedos) and all the islands in the Sea of Marmara are excluded. Cyprus is also excluded.

All other islands in the Mediterranean (except those administered from Africa) are included in Europe, as are also the Açores, Færöer, Iceland, Jan Mayen, Björnöya (Bear Island) and Svalbard (Spitsbergen).

(*c*) *Keys*

Artificial dichotomous keys are provided within families and genera, and, where necessary, within species. A general key to families is also given (p. xxxiii), which includes all the families in the European flora, not only those described in this volume. The keys have been designed to be practical, and do not in general use characters which are difficult to observe, even if important taxonomically; they have also been kept as brief as possible. We have tried to give characters useful both in the herbarium and in the field, though this has often been difficult because of the lack of information about field characters in certain areas, notably S.E. Europe.

The choice of the indented key was made only after thorough discussion of the advantages and disadvantages of the many possible types of key. Briefly, the advantage of the indented key is that it may also act to some extent as a conspectus; in *Papaver*, for example, it separates the scapose species of the section *Meconella* from the rest of the genus. In this way similar species are often grouped together in the key, and identification of an unknown plant is facilitated. On the other hand, we have not hesitated to key out more than once certain taxa in which the relevant characters are variable.

(*d*) *Citation and synonymy*

The use of abbreviations for author and place of publication has been standardized as far as possible, and lists of these abbreviations are given in Appendices I, II and III.

It is important to note that no attempt has been made to give a complete synonymy. Even at the binomial level, the number of names for European plants is four or five times the number of accepted species, and to include all these would be impracticable. Synonyms, whether full or partial, are given only when they are used or included in one of the Standard Floras or when they are necessary to prevent confusion. Synonyms (or the basionym) are also usually given when the combination has not previously been used in a Flora or monograph, or when the nomenclature is unfamiliar or in need of explanation. Where the name of a familiar species has been changed, an explanation has usually been published as a Notula (see Section (*j*)).

(*e*) *Species descriptions*

In order to save space and facilitate identification, descriptions may take the form of a comparison with another description. The conventional way of setting this out is, to give an example (p. 212):

131. Silene dioica (L.) Clairv... Like **129** but...

This implies that the description with which it is being compared (in this example **129**) applies to this taxon but for the differences noted. It does not necessarily mean that the two taxa are similar in appearance or that they are closely related. Additional descriptive information is sometimes also given, but in separate sentences.

Where dimensions are given, a measurement without qualification refers to length. Two measurements connected by × indicate length followed by width. Further measurements in parentheses indicate exceptional cases outside the normal ranges.

The *diploid chromosome number* ($2n =$) is given where it has been possible to verify that the count was made on material of known wild European origin. Details of those included in the first edition were provided by Moore, *Flora Europaea Check-list and Chromosome Index*, 1982, but subsequently published data have also been included in this revision.

Ecological information has been given very sparingly, and only where the ecological characteristics of a species are clearly and concisely definable for its total European range. There is an inevitable irregularity of treatment in this respect, but here, as with chromosome number, the Editorial Committee has thought it best to include only well-verified statements.

The description of each species is followed by an indication of its *distribution within Europe*. This falls into two parts: (1) a summary in a short phrase; (2) a list of abbreviations of 'territories' in which the species occurs. The summary phrase makes use of everyday geographical phrases and concepts such as 'W. Europe', 'the Mediterranean region', 'the Balkan peninsula', etc. Maps IV and V and the lists accompanying them indicate the interpretation which is to be put on these phrases. We would emphasize that they are to be interpreted in a simple geographical sense, and do not attempt in any way to divide Europe phytogeographically.

Species or subspecies believed to be endemic to Europe are distinguished by a symbol (●) before the summary of geographical distribution.

A more precise indication of distribution is given by the enumeration of the 'territories' (indicated by two-letter abbreviations from Latin) in which the plant is believed to occur. The limits of these territories follow, with very few exceptions, existing political boundaries. An alternative plan, which has been suggested by several colleagues, of dividing Europe into floristic regions and ignoring political boundaries would seem at first sight to have much to commend it. It is, however, impracticable for two reasons. First, although there is some agreement as to the broad primary units into which Europe might be divided floristically, there is no agreement on where exactly their boundaries are to be drawn, or on how they should be subdivided into smaller units of convenient size. Secondly, the information is more readily available in political than in phytogeographical terms. For rare or localized species it would have been possible to translate the information from one system to the other; but for wide-ranging species, which are described as occurring, for example, *dans une grande partie de la France mais très rare dans la région méditerranéenne*, the assignment to phytogeographical areas would have meant a search of local Floras and herbaria quite beyond the powers of authors or editors in the time available.

The political divisions have, therefore, been accepted. Each territory represents a sovereign State, with the following exceptions:

1. Certain islands or island-groups have, on account of their size, isolation, or floristic peculiarities, been treated as separate territories. These comprise Kriti, Corse, Sardegna, Sicilia, Islas Baleares, Færöer, Svalbard, Açores.
2. Malta is grouped with Sicilia, and the Channel Islands are grouped with France.
3. Ireland and Jugoslavia are each treated as a single territory.
4. The former U.S.S.R. has been subdivided, not according to its constituent republics (though these are often referred to by name in the geographical summary), but into six divisions formed by grouping together the floristic regions used in *Flora URSS*. A small adjustment has been made in the N. and E. Ukraine, so as to bring the boundaries as far as possible into line with those which delimit the regions used in *Flora RSS Ucr*.

The territories, of course, vary greatly in size, and Ga, Hs or Ju gives much less precise information than does Fa, Rs(K) or Tu. In all cases, however, the lists provide a statement as to which national Floras should be searched for further detailed information, whether on distribution or on taxonomy.

Occasionally the list of territories is followed by a brief indication, in parentheses, of extra-European distribution. We should, perhaps, explain why we have not done this for all species – a policy that was urged on us by many of our advisers. It is simply because we found that in far too many cases the information was not available. It is one thing to note that a plant has been recorded from China, from tropical Africa, or from North America under the same name as a European species; it is a very different matter to find out whether, in the opinion of a competent taxonomist, the two should really be judged as conspecific. We therefore thought it better to say nothing than to give further currency to data which have not been critically sifted and which contain many erroneous records. The only exceptions to this rule are for plants not native in Europe and for plants of which the European range is only a small fraction of their total area. If the only European record for a plant is from a limited area of S. Spain the reader may well wish to know whether it is strictly endemic there, or is found also across the straits in Morocco, or is a wide-ranging plant of North Africa that has established a small bridgehead in Europe. The three cases would be distinguished thus:

● *S. Spain*. Hs.	indicates endemic to S. Spain,
S. Spain. Hs.	indicates found also outside Europe,
S. Spain. Hs. (*N. Africa*.)	indicates widespread also in North Africa; not primarily a European plant.

(*f*) *Infraspecific taxa*

In general the only infraspecific taxa described and keyed in the Flora are subspecies. Any formal treatment of variation below the level of subspecies would have been impossible in a Flora of this kind; most of the known variation of taxa is, however, covered in the descriptions. Taxa below the level of subspecies may be mentioned when, in the opinion of the author, they seem to have some claim to recognition at higher rank, but require further investigation. Conversely, many taxa published as species or subspecies, but whose claim to the rank seems doubtful, are described or briefly mentioned in supplementary notes under the species to which they seem most closely related.

No 'experimental' categories, such as ecotypes, are used in the Flora in a formal systematic sense, though they are sometimes mentioned in notes.

(*g*) *Treatment of critical groups*

In certain cases where it is difficult to distinguish between a number of closely similar species, an *ad hoc* 'group' has been made, and these groups, not the individual species, are keyed out in the main species-key. They will serve for at least a partial identification. Following the description of a group in the text, a key to the component species is given, and they are then numbered and described so that a more detailed study, or the use of more adequate material, may enable the user to take the identification further. For example, in *Stellaria* there is the *S. media* group, which comprises the species *S. media* (L.) Vill., *S. neglecta* Weihe and *S. pallida* (Dumort.) Piré. Such groups have no taxonomic or nomenclatural status.

For the inbreeding and apomictic groups, other *ad hoc* treatments have been devised. The general approach to such groups is set out and discussed in detail in the Report of the first Flora Europaea Symposium in *Feddes Repert.* **63**: 107–228 (1961).

(*h*) *Treatment of hybrids*

Only those few hybrids which reproduce vegetatively and are frequent over a reasonably large area (e.g. *Sagina* × *normaniana*) are described and keyed as for species. Other common hybrids may be mentioned individually in notes (e.g. *Silene dioica* × *latifolia*), or collectively for the whole genus (e.g. *Quercus*).

(*i*) *Alien species*

The question of which alien species should be included is a difficult one for any Flora, and the difficulty is even greater on a continental scale. We have attempted to include the following categories:

(i) Aliens which are effectively naturalized. These include garden plants which have escaped to situations not immediately adjacent to those in which they are cultivated, as well as weeds and other plants which have been accidentally introduced; provided, in both cases, that the plant has been established in at least one station for at least 25 years, or is reported as naturalized in a number of widely separated localities.

(ii) Trees or crop-plants which are planted or cultivated in continuous stands on a fairly extensive scale.

Casual aliens, i.e. those which do not persist without constant fresh introductions, are not included unless they have often been mistaken for a native or established species, or are for any other reason of special interest.

In assessing the status of a species in any part of Europe we have, however, been dependent almost entirely on the information contained in the national Floras, and it is clear that the criteria used by different authors vary widely. There are some genera, for example *Chenopodium*, in which every transitional state from a casual to a well-established weed can be observed. All data on native, established or casual status relating to weeds or other 'palaeosynanthropic' plants must, therefore, be regarded only as approximate.

(*j*) *Publication of* Novitates

In the process of writing the accounts, new material requiring publication has naturally been brought together. For the publication of much of this material, the Committee made an arrangement with the Editorial Board of *Feddes Repertorium*, by which taxonomic and nomenclatural notes were published as part of a series entitled *Notulae Systematicae ad Floram Europaeam spectantes*. The first of these appeared in 1961. From 1971 such *Notulae* were published in the *Botanical Journal of the Linnean Society*. This continued until 1979, with the completion of the Flora in 1980. Beginning in 1987, a new series of *Notulae* relating to the revision of *Flora Europaea* vol. 1 has also been published in the *Botanical Journal of the Linnean Society*. By these arrangements, publication of new names or combinations in the Flora itself has been avoided.

LIST OF STANDARD FLORAS

More detailed information on authorship and dates of publication as well as details of other floristic works can be found in Frodin, D. G., *Guide to standard floras of the world*, Cambridge, 1984.

AESCHIMANN, D. & BURDET, H. *Flore de la Suisse et des territoires limitrophes*. Neuchâtel, 1989.

BARCELÓ Y COMBIS, F. *Flora de las Islas Baleares*. Palma [de Mallorca], 1879–1881.

BECK VON MANNAGETTA, G. *et al.* (ed.). *Flore Bosne, Hercegovine i Novipazarskog Sandžaka*. Vols. 1–4. Beograd & Sarajevo, 1903–1974.

BOISSIER, E. *Flora orientalis*. Vols. 1–5. Genève, Bâle & Lyon, 1867–1884. *Supplementum*. 1888.

BOLÒS, O. DE & VIGO, J. *Flora dels països Catalans*. Vols. 1– . Barcelona, 1984– .

BORNMÜLLER, J. *Beiträge zur Flora Mazedoniens* (In *Bot. Jahrb.* **59** (Beibl. 136), **60** (Beibl. 140), **61**). Leipzig, 1925–1928.

BRIQUET, J. *Prodrome de la Flore corse*. Vols. 1–3. Genève, Bâle, Lyon & Paris, 1910–1955.

CASTROVIEJO, S. *et al.* (ed.). *Flora ibérica*. Vols. 1– . Madrid, 1986– .

CLAPHAM, A. R., TUTIN, T. G. & WARBURG, E. F. *Flora of the British Isles*, ed. 3 by A. R. Clapham, T. G. Tutin & D. M. Moore. Cambridge, 1987.

COSTE, H. *Flore descriptive et illustrée de la France, de la Corse et des Contrées limitrophes*. Vols. 1–3. Paris, 1900–1906. Supplément 1–6, by P. Jovet & R. de Vilmorin. Paris, 1972–1985.

COUTINHO, A. X. PEREIRA. *Flora de Portugal*, ed. 2 by R. T. Palhinha. Lisboa, 1939.

DAVIS, P. H. (ed.). *Flora of Turkey and the East Aegean islands*. Vols. 1–10. Edinburgh, 1965–1988.

DEGEN, A. VON. *Flora velebitica*. Vols. 1–4. Budapest, 1936–1938.

DE LANGHE, J.-E. *et al. Nouvelle Flore de la Belgique, du Grand-Duché de Luxembourg, du Nord de la France et des Régions voisines*, ed. 3. Bruxelles, 1983.

DEMIRI, M. *Flora ekskursioniste e Shqipërisë*. Tiranë, 1983.

DIAPOULIS, K. A. *Ellenike Khloris*. Vols. 1–3. Athenai, 1939–1949.

DOMAC, R. *Flora za odredivanje i upoznavanje bilja*. Zagreb, 1950.

DOSTÁL, J. *Květena ČSR*. Praha, 1948–1950.

DOSTÁL, J. *Nová Květena ČSSR*. Vols. 1– . Praha, 1989– .

EHRENDORFER, F. (ed.). *Liste der Gefässpflanzen Mitteleuropas*, ed. 2 by W. Gutermann. Stuttgart, 1973.

FEDOROV, AN. A. (ed.). *Flora Partis europaeae URSS*. Vols. 1–6. Leningrad, 1974–1987; Vols. 8– , ed. by N. N. Tzvelev, 1989– .

FOMIN, A. V. *et al.* (ed.). *Flora Reipublicae Sovieticae Socialisticae Ucrainicae*, ed. 1. Vols. 1–12. Kijiv, 1936–1965. Ed. 2. Vol. 1. Kijiv, 1938.

FOURNIER, P. *Les quatre Flores de la France, Corse comprise*. Poinson-les-Grancey, 1934–1940. (Reprints with additions and corrections, Paris, 1946 and 1961; ed. 2, Paris, 1977, unrevised.)

FRANCO, J. DO AMARAL. *Nova flora de Portugal*. Vols. 1– . Lisboa, 1971– .

FRITSCH, K. *Exkursionsflora für Österreich und die ehemals österreichischen Nachbargebiete*, ed. 3. Wien & Leipzig, 1922.

GAMISANS, J. *Catalogue des Plantes vasculaires de la Corse*. Ajaccio, 1985.

Geideman, T. S. *Opredelitel' vysšikh Rastenij Moldavskoj SSR*, ed. 2. Kišinev, 1975.

Gorodkov, B. N. & Pojarkova, A. I. (ed.). *Flora Murmanskoj Oblasti*. Vols. 1–5. Moskva & Leningrad, 1953–1966.

Greuter, W., Burdet, H. M. & Long, G. (ed.). *Med-Checklist*. Vols. 1– . Genève, 1984– .

Halácsy, E. von. *Conspectus Florae graecae*. Vols. 1–3. Leipzig, 1900–1904. *Supplementum* 1. Leipzig, 1908. *Supplementum* 2 (In *Magyar Bot. Lapok* **11**). Budapest, 1912.

Hämet-Ahti, L. *et al.* (ed.). *Retkeilykasvio*, ed. 3. Helsinki, 1986.

Hansen, K. (ed.). *Dansk Feltflora*. Copenhagen, 1981.

Haslam, S. M., Sell, P. D. & Wolseley, P. A. *A flora of the Maltese Islands*. Msida, 1977.

Hayek, A. von. *Prodromus Florae Peninsulae balcanicae* (In *Feddes Repert.* (*Beih.*) **30**). Vols. 1–3. Berlin-Dahlem, 1924–1933.

Hegi, G. *Illustrierte Flora von Mittel-Europa*, ed. 1. Vols. 1–7. München, 1906–1931. Ed. 2. Vols. 1 — . München, 1936. Ed. 3. Vols. 1 — . München, 1966– .

Hess, H. E., Landolt, E. & Hirzel, R. *Flora der Schweiz*. Vols. 1–3. Basel & Stuttgart, 1967–1972.

Heukels, H. *Flora van Nederland*, ed. 21 by R. van der Meijden *et al.* Groningen, 1990.

Horvátić, S. (ed.). *Analitiške Flora Jugoslavije*. Vols. 1 — . Zagreb, 1967 — .

Hultén, E. *Atlas of the distribution of vascular plants in Northwestern Europe*, ed. 2. Stockholm, 1971.

Hylander, N. *Nordisk Kärlväxtflora*. Vols. 1–2. Stockholm, 1953–1966.

Jalas, J. & Suominen, J. (ed.). *Atlas Florae europaeae*. Vols. 1– . Helsinki, 1972– .

Janchen, E. *Catalogus Florae Austriae*. Vol. 1. Wien, 1956–1960. *Ergänzungsheft*. Wien, 1963. *Zweites Ergänzungsheft*. Wien, 1964. *Drittes Ergänzungsheft*. Wien, 1966. *Viertes Ergänzungsheft*. Wien & New York, 1968.

Jordanov, D. (ed.). *Flora na Narodna Republika Bălgarija*. Vol. 1– . Sofija, 1963–.

Josifović, M. (ed.). *Flora de la Republique socialiste de Serbie*. Vols. 1–8. Beograd, 1970–1976. *Dodatak*. Beograd, 1977– .

Knoche, H. *Flora balearica*. Vols. 1–4. Montpellier, 1921–1923.

Komarov, V. L. *et al.* (ed.). *Flora URSS*. Vols. 1–30. Leningrad & Moskva, 1934–1964. *Additamenta et Corrigenda*. Leningrad, 1973.

Krok, T. O. B. N. & Almquist, S. *Svensk Flora: Fanerogamer och ormbunksvaxter*, ed. 27 by L. Jonsell & B. Jonsell. Uppsala, 1984.

Lid, J. *Norsk, svensk, finsk Flora*, ed. 5 by O. Gjærevoll. Oslo, 1985.

Lid, J. *The Flora of Jan Mayen*. Oslo, 1964.

Löve, Á. *The Flora of Iceland*. Reykjavík, 1983.

Maevskij, P. F. *Flora srednej Polosy evropejskoj Časti SSSR*, ed. 9 by B. K. Schischkin. Leningrad, 1964.

Martinčič, A. & Sušnik, F. *Mala Flora Slovenije*. Ljubljana, 1969.

Merino y Román, P. B. *Flora descriptiva é ilustrada de Galicia*. Vols. 1–3. Santiago [de Compostela], 1905–1909.

Nyman, C. F. *Conspectus Florae europaeae*. Örebro, 1878–1882. *Supplementum* 1, 1882–1884. *Additamenta*, 1886. *Supplementum* 2, 1889–1890.

Ostenfeld, C. E. H. & Gröntved, J. *The Flora of Iceland and the Faeroes*. København, 1934.

Palhinha, R. T. *Catálogo das Plantas vasculares dos Açores*. Lisboa, 1966.

Pignatti, S. *Flora d'Italia*. Vols. 1–3. Bologna, 1982.

Raciborski, M., Szafer, W. & Pawłowski, B. (ed.). *Flora polska*. Vols. 1–14. Kraków & Warszawa, 1919–1980.

RASMUSSEN, R. *Föroya Flora*, ed. 2. Tórshavn, 1952.

RECHINGER, K. H. *Flora aegaea* (In *Denkschr. Akad. Wiss. Math.-Nat. Kl.* (*Wien*) **105**(**1**)). Wien, 1943. *Supplementum* (In *Phyton* (*Austria*) **1**). Horn, 1949.

ROBYNS, W. (ed.). *Flore générale de Belgique*. Spermatophytes. Vols. 1– . Bruxelles, 1952– .

ROHLENA, J. *Conspectus Florae montenegrinae* (In *Preslia* **20** and **21**). Praha, 1942.

RØNNING, O. I. *Svalbards Flora*. Oslo, 1964.

ROTHMALER, W. *Exkursionsflora für die Gebiete der DDR und der BRD*. 2: *Gefässpflanzen*, ed. 12 by H. Meusel & R. Schubert. Berlin, 1984. 4: *Kritischer Band*, ed. 6 by R. Schubert & W. Vent. Berlin, 1986. 3: *Atlas der Gefässpflanzen*, ed. 7 by R. Schubert, E. Jäger & K. Werner. Berlin, 1988.

ROUY, G. C. C. *et al. Flore de France*. Vols. 1–14. Asnières, Paris & Rochefort, 1893–1913.

RUBTSOV, N. I. (ed.). *Opredelitel' vysšikh Rastenij Kryma*. Leningrad, 1972.

SĂVULESCU, T. (ed.). *Flora Republicii populare Române*. Vols. 1–12. Bucureşti, 1952–1972.

SCHMEIL, O. & FITSCHEN, J. *Flora von Deutschland*, ed. 87 by W. Rauh & K. Senghas. Heidelberg, 1982.

STANKOV, S. S. & TALIEV, V. I. *Opredelitel' vysšikh Rastenij evropejskoj Časti SSSR*, ed. 2. Moskva, 1957.

STEFÁNSSON, S. *Flóra Íslands*, ed. 3. Akureyri, 1948.

STOJANOV, N., STEFANOV, B. & KITANOV, B. *Flora na Bălgarija*, ed. 4. Vols. 1–2. Sofija, 1966–1967.

STRID, A. (ed.). *Mountain Flora of Greece*. Vols. 1–2. Cambridge & Edinburgh, 1984–1991.

SZAFER, W., KULCZIŃSKI, S. & PAWŁOWSKI, B. *Rósliny polskie*. Warszawa, 1953.

TOLMATCHEV, A. I. *Arktičeskaja Flora SSSR*. Vols. 1–10. Moskva & Leningrad, 1960–1987.

TUTIN, T. G. *et al.* (ed.). *Flora europaea*. Vols. 1–5. Cambridge, 1964–1980.

VALDÉS, B., TALAVERA, S. & FERNANDEZ-GALIANO, E. (ed.). *Flora vascular de Andalucía occidental*. Vols. 1–3. Barcelona, 1987.

WEBB, D. A. *An Irish Flora*, ed. 6. Dundalk, 1977.

WEBB, D. A. *The Flora of European Turkey* (In *Proc. Roy. Irish Acad.* **65B**). Dublin, 1966.

WEEVERS, T. *et al.* (ed.). *Flora neerlandica*. Vols. 1– . Amsterdam, 1948– .

WILLKOMM, H. M. *Supplementum Prodromi Florae hispanicae*. Stuttgart, 1893.

WILLKOMM, H. M. & LANGE, J. *Prodromus Florae hispanicae*. Vols. 1–3. Stuttgart, 1861–1880.

ZEROV, D. K. *et al.* (ed.). *Vyznačnyk Roslyn Ukrajiny*, ed. 2. Kijiv, 1965.

SYNOPSIS OF FAMILIES IN VOLUME 1

KEYS TO MAJOR TAXA

1 Plant reproducing by spores; always herbaceous **Pteridophyta**
1 Plant reproducing by seeds; often woody
2 Ovules not enclosed in an ovary, borne either on the upper surface of scales arranged in cones, or solitary and terminal on lateral stems; perianth absent; trees or shrubs **Gymnospermae**
2 Ovules completely enclosed in an ovary; perianth usually present; herbs, trees or shrubs **Angiospermae**

KEY TO PTERIDOPHYTA

1 Plant without true roots or leaves **1. Psilotaceae**
1 Plant with roots and leaves
2 Stems jointed; leaves forming a sheath at the nodes **5. Equisetaceae**
2 Stems not jointed; leaves not fused into a sheath
3 Plants usually free-floating on water
4 Leaves 10–40 cm, 2- or 3-pinnatifid **8. Parkeriaceae**
4 Leaves less than 10 cm, not pinnatifid
5 Leaves in whorls of 3, two entire, floating, the third submerged and root-like **24. Salviniaceae**
5 Leaves small, 2-ranked, imbricate **25. Azollaceae**
3 Plant epiphytic or rooted to the ground, aquatic or terrestrial
6 Leaves not differentiated into lamina and petiole
7 Leaves forming a basal rosette **4. Isoetaceae**
7 Leaves not forming a basal rosette
8 Aquatic; leaves filiform **23. Marsileaceae**
8 Terrestrial; leaves not filiform
9 Plant homosporous; leaves without ligule **2. Lycopodiaceae**
9 Plant heterosporous; leaves with ligule **3. Selaginellaceae**
6 Leaves with distinct lamina and petiole
10 Leaves long-petiolate, with 4 leaflets **23. Marsileaceae**
10 Leaves not with 4 leaflets
11 Sporangia without an annulus
12 Plant 60–150 cm; vernation circinate **7. Osmundaceae**
12 Plant rarely more than 60 cm; vernation not circinate **6. Ophioglossaceae**
11 Sporangia with an annulus
13 Fertile and sterile leaves dissimilar
14 Leaves entire **20. Lomariopsidaceae**
14 Leaves not entire
15 Leaves 1-pinnate; pinnae entire **22. Blechnaceae**
15 Leaves 2- to 4-pinnate (or if 1-pinnate, pinnae pinnatifid)
16 Sori completely enclosed by thickened, inrolled margin of lamina **18. Woodsiaceae**
16 Sori exposed, or if enclosed then by herbaceous or membranous margin of lamina
17 Leaves often proliferous, the veins anastomosing; plants aquatic or in swamps **8. Parkeriaceae**
17 Leaves not proliferous, the veins free; plants terrestrial, in well-drained areas **9. Adiantaceae**
13 Fertile and sterile leaves similar (or differing only in size)
18 Leaves mostly not more than 1 cell thick **11. Hymenophyllaceae**
18 Leaves more than 1 cell thick
19 Leaves entire, or pinnatifid, or palmately lobed, or dichotomously forked 1–3 times
20 Leaves not pinnatifid
21 Sori orbicular, without indusia **13. Grammitidaceae**
21 Sori linear, with indusia **17. Aspleniaceae**
20 Leaves pinnatifid
22 Leaves not covered with scales **12. Polypodiaceae**
22 Leaves covered with scales on lower surface **17. Aspleniaceae**
19 Leaves pinnate
23 Sori covered by deflexed margin of leaf
24 Leaf-segments long, linear-lanceolate **10. Pteridaceae**
24 Leaf-segments not linear-lanceolate
25 Rhizome short, superficial **9. Adiantaceae**
25 Rhizome long, subterranean **15. Hypolepidaceae**
23 Sori not covered by deflexed margin of leaf
26 Indusium absent
27 Leaves not more than 8 cm **17. Aspleniaceae**
27 Leaves more than 8 cm
28 Pinnae entire **12. Polypodiaceae**
28 Pinnae divided
29 Leaves forming a crown **18. Woodsiaceae**
29 Leaves solitary
30 Groove of rhachis interrupted to admit grooves of costae; lamina tripartite **19. Dryopteridaceae**
30 Groove of rhachis not interrupted to admit grooves of costae; lamina not tripartite **16. Thelypteridaceae**
26 Indusium present
31 Leaves up to 250 cm, producing young plants vegetatively on the distal part **22. Blechnaceae**
31 Leaves not more than 200 cm, not producing plants vegetatively
32 Rhizome very hairy, without scales **14. Dicksoniaceae**
32 Rhizome not hairy, but with scales

33 Indusium cup-shaped **21. Davalliaceae**
33 Indusium not cup-shaped
34 Indusium a circumbasal ring of hairy scales **18. Woodsiaceae**
34 Indusium not a ring of scales
35 Indusium hood-like, attached at basiscopic side of sorus **18. Woodsiaceae**
35 Indusium not hood-like
36 Indusium peltate **19. Dryopteridaceae**
36 Indusium not peltate
37 Sori orbicular
38 Sori marginal; indusium lying along vein **16. Thelypteridaceae**
38 Sori not marginal; indusium lying across vein **19. Dryopteridaceae**
37 Sori ovate or linear
39 Leaves solitary on rhizome **18. Woodsiaceae**
39 Leaves forming a crown or apical tuft on rhizome
40 Sori linear or ovate; lower margin of indusium straight **17. Aspleniaceae**
40 Sori ovate; lower margin of indusium bent in the middle **18. Woodsiaceae**

KEY TO GYMNOSPERMAE

1 Leaves all scale-like and usually brownish; internodes long **30. Ephedraceae**
1 Most leaves green; internodes short
2 Female flowers solitary; seed surrounded by a fleshy aril **29. Taxaceae**
2 Female flowers in cones; seeds without a fleshy aril
3 Leaves opposite or whorled **28. Cupressaceae**
3 Leaves alternate or 2–5 on short shoots
4 Bracts and cone-scales distinct from one another **26. Pinaceae**
4 Bracts and cone-scales wholly or partially united **27. Taxodiaceae**

KEY TO ANGIOSPERMAE

This key covers all the families of Angiospermae in vols. 1 (arabic numerals) and 2–5 (roman numerals).

1 Plant free-floating on or below the surface of the water, not rooted in mud
2 Plant without obvious differentiation into stems and leaves **CXCVI. Lemnaceae**
2 Plant with obvious stems and leaves
3 Leaves divided into numerous filiform segments
4 Plant with small bladders on leaves or on apparently leafless stems **CLXI. Lentibulariaceae**
4 Plant without small bladders
5 Leaves dichotomously divided, the segments often again divided **60. Ceratophyllaceae**
5 Leaves pinnately divided, the segments simple **CXXIV. Haloragaceae**
3 Leaves not divided into numerous filiform segments
6 Leaves with a cuneate basal part, 4–6 lateral setaceous segments and a terminal orbicular lobe **71. Droseraceae**
6 Leaves not as above
7 Perianth entirely petaloid; basal part of petioles inflated **CLXXXVII. Pontederiaceae**
7 Perianth with a distinct calyx and corolla; petioles not inflated
8 Sepals, petals and stamens 4 **CXX. Trapaceae**
8 Sepals and petals 3; stamens 9–12 **CLXXII. Hydrocharitaceae**
1 Land plant or aquatic rooted in mud
9 Bifid to quadrifid coloured staminodes present; leaves often fasciculate **53. Molluginaceae**
9 Not as above
10 Perianth absent, or of a single whorl or of 2 or more whorls all ± similar in shape, size, colour and texture
11 Perianth petaloid
12 Plant without chlorophyll
13 Flowers mostly unisexual; stamen 1 **46. Balanophoraceae**
13 Flowers hermaphrodite; stamens 6–16
14 Filaments free **CXXXI. Pyrolaceae**
14 Filaments connate into a column **45. Rafflesiaceae**
12 Plant with chlorophyll
15 Perianth-segment 1, bract-like **CLXXIV. Aponogetonaceae**
15 Perianth-segments more than 1, or perianth tubular
16 Stems succulent, leafless but with groups of spines **CXVIII. Cactaceae**
16 Not as above
17 Stamens more than 12
18 Leaves pinnate **61. Ranunculaceae**
18 Leaves not pinnate
19 Herb **61. Ranunculaceae**
19 Tree **64. Magnoliaceae**
17 Stamens not more than 12
20 Flowers in ovoid capitula; involucre absent **LXXX. Rosaceae**
20 Flowers not in capitula, or in capitula surrounded by an involucre of bracts
21 Ovary superior
22 Perianth-segments 4
23 Flowers zygomorphic **41. Proteaceae**
23 Flowers actinomorphic
24 Perianth with a long tube **CVII. Thymelaeaceae**
24 Perianth-segments free
25 Herb **CLXXXIII. Liliaceae**
25 Shrub **47. Polygonaceae**
22 Perianth-segments more than 4
26 Carpels more than 1, free or nearly so

27 Leaves triquetrous, all basal **CLXXI. Butomaceae**
27 Leaves flat, cauline **51. Phytolaccaceae**
26 Carpel 1 or carpels obviously united
28 Perianth-segments 6
29 Stem stout, woody; leaves crowded, rigid, very fibrous **CLXXXIV. Agavaceae**
29 Not as above
30 Flowers actinomorphic **CLXXXIII. Liliaceae**
30 Flowers zygomorphic **CLXXXVII. Pontederiaceae**
28 Perianth-segments 5
31 Stigmas 2 or 3; stipules sheathing, scarious **47. Polygonaceae**
31 Stigma 1; stipules absent
32 Ovules numerous; perianth divided almost to base **CXXXV. Primulaceae**
32 Ovule 1; perianth with a long tube **50. Nyctaginaceae**
21 Ovary inferior, or flowers male
33 Leaves at least partly in whorls of 4 or more **CXLIV. Rubiaceae**
33 Leaves not in whorls of 4 or more
34 Flowers sessile in capitula surrounded by an involucre of bracts
35 Anthers cohering in a tube round the style, or flowers unisexual **CLXIX. Compositae**
35 Anthers free; flowers hermaphrodite **CLXVII. Dipsacaceae**
34 Flowers pedicellate, though pedicels sometimes short and flowers in compact umbels or cymes
36 Ovules 1 or 2
37 Leaves opposite **CLXVI. Valerianaceae**
37 Leaves alternate
38 Flowers in simple cymes or solitary **42. Santalaceae**
38 Flowers in umbels or superposed whorls **CXXIX. Umbelliferae**
36 Ovules numerous
39 Perianth-segments 3 or perianth tubular with an entire, unilateral limb **44. Aristolochiaceae**
39 Perianth-segments 6 or 8
40 Perianth-segments in 2 whorls of 4 **CXXIII. Onagraceae**
40 Perianth-segments in 2 whorls of 3
41 Stamens 3 **CLXXXVIII. Iridaceae**
41 Stamens 6
42 Scapose plant with a bulb **CLXXXV. Amaryllidaceae**
42 Stem leafy; rhizomatous **CLXXXIV. Agavaceae**
11 Perianth dry and scarious (though sometimes brightly coloured) or sepaloid or absent
43 Tree or shrub, sometimes small
44 Parasitic on branches of other trees or shrubs **43. Loranthaceae**
44 Not parasitic
45 Stems creeping or climbing with adventitious roots; evergreen **CXXVIII. Araliaceae**
45 Not as above
46 Most leaves opposite or subopposite
47 Young stems and leaves fleshy **48. Chenopodiaceae**
47 Neither stems nor leaves fleshy
48 Style 1
49 Leaves pinnate **CXXXIX. Oleaceae**
49 Leaves simple
50 Flowers in catkins **31. Salicaceae**
50 Flowers not in catkins **CIII. Rhamnaceae**
48 Styles 2 or more
51 Stamens 5; flowers hermaphrodite **CIII. Rhamnaceae**
51 Stamens 4 or 8; flowers often unisexual
52 Stamens 4; plant evergreen **CII. Buxaceae**
52 Stamens 8; plant usually deciduous **XCV. Aceraceae**
46 Most leaves alternate
53 Leaves pinnate
54 Male flowers in catkins; styles 2; pith septate **33. Juglandaceae**
54 Flowers not in catkins; styles 3 or 1; pith not septate
55 Style 1; fruit a compressed legume with several seeds **LXXXI. Leguminosae**
55 Styles 3; fruit a dry, 1-seeded 'drupe' **XCIV. Anacardiaceae**
53 Leaves simple
56 Leaves not more than 2 mm wide, linear or oblong
57 Stigma 1 **CVII. Thymelaeaceae**
57 Stigmas 2–9
58 Stamens 3 **CXXXIII. Empetraceae**
58 Stamens 5 **48. Chenopodiaceae**
56 Leaves more than 2 mm wide
59 Petiole-base enclosing the bud **79. Platanaceae**
59 Petiole-base not enclosing the bud
60 Anthers opening by transverse valves **65. Lauraceae**
60 Anthers opening by longitudinal slits
61 Flowers not in catkins or dense heads
62 Inflorescence of several male flowers, each of 1 stamen, and a female flower, appearing as a stalked ovary, all surrounded by 4 or 5(–8) conspicuous glands; latex present **LXXXVII. Euphorbiaceae**
62 Inflorescence not as above; latex absent
63 Flowers unisexual
64 Peltate, scale-like silvery or ferrugineous hairs present beneath the leaves and often elsewhere; ovary 1-locular; fruit fleshy **CVIII. Elaeagnaceae**
64 Peltate hairs absent; ovary 3-locular; fruit dry **LXXXVII. Euphorbiaceae**
63 Flowers hermaphrodite
65 Tree; perianth-tube short, with stamens inserted near its base **37. Ulmaceae**

65 Shrub; perianth-tube long, with stamens inserted near its apex **CVII. Thymelaeaceae**

61 At least the male flowers in catkins or dense heads

66 Latex present; fruit or false fruit fleshy **38. Moraceae**

66 Latex absent; fruit dry

67 Usually dioecious; perianth absent

68 Bracts (catkin-scales) fimbriate or lobed at apex; flowers with a cup-like disc **31. Salicaceae**

68 Bracts (catkin-scales) entire; disc absent

69 Leaves with pellucid glands; stamens with short filaments; ovule 1 **32. Myricaceae**

69 Leaves without pellucid glands; stamens with long filaments; ovules numerous **31. Salicaceae**

67 Monoecious; perianth present in male or female flowers or in both

70 Styles 3 or more; perianth present in flowers of both sexes **36. Fagaceae**

70 Styles 2; perianth present in flowers of 1 sex only

71 Male flowers 3 to each bract, with perianth **34. Betulaceae**

71 Male flowers 1 to each bract, without perianth **35. Corylaceae**

43 Herb

72 Perianth absent or represented by scales or bristles, minute at anthesis; flowers in the axils of bracts, one or more forming a spikelet; leaves usually linear, grass-like, sheathing below

73 Flowers usually with a bract above and below; sheaths usually open; stems usually with hollow internodes, not triquetrous **CXCIII. Gramineae**

73 Flowers with a bract below only; sheaths usually closed; stems usually with solid internodes, often triquetrous **CXCIX. Cyperaceae**

72 Perianth present, or perianth absent and flowers not arranged in spikelets

74 Aquatic plant with submerged or floating leaves; inflorescence sometimes emergent

75 Leaves divided into numerous filiform segments

76 Leaves pinnately divided; flowers in a terminal spike **CXXIV. Haloragaceae**

76 Leaves dichotomously divided; flowers solitary, axillary **60. Ceratophyllaceae**

75 Leaves entire or minutely toothed

77 Fruitlets with stalks several times their own length **CLXXVIII. Ruppiaceae**

77 Fruitlets sessile

78 Flowers in spikes

79 Rhizome densely covered with rigid fibres; inflorescence subtended by several leaf-like bracts (marine) **CLXXIX. Posidoniaceae**

79 Rhizome without rigid fibres; inflorescence not subtended by several leaf-like bracts

80 Flowers on one side of a flat rhachis (marine) **CLXXX. Zosteraceae**

80 Flowers all round or on 2 sides of a terete rhachis (fresh, rarely brackish water) **CLXXVII. Potamogetonaceae**

78 Flowers not in spikes

81 Flowers in heads or in branched inflorescences

82 Flowers hermaphrodite **CLXXXIX. Juncaceae**

82 Flowers unisexual

83 Leaves all basal; heads solitary, on long scapes **CXCII. Eriocaulaceae**

83 At least some leaves cauline; inflorescence with female heads below and male heads above **CXCVII. Sparganiaceae**

81 Flowers solitary or few, sessile or shortly pedicellate, usually in leaf-axils

84 Leaves in whorls of 8 or more **CXXVI. Hippuridaceae**

84 Leaves not in whorls of 8 or more

85 Carpels 2 or more, free **CLXXXI. Zannichelliaceae**

85 Carpels connate or solitary

86 Perianth-segments 4 or 6; stamens 4 or more; leaves ovate to obovate

87 Perianth-segments 4; ovary inferior **CXXIII. Onagraceae**

87 Perianth-segments 6; ovary superior **CXIX. Lythraceae**

86 Perianth-segments fewer than 4, or perianth absent; stamens 1–3; leaves linear to lanceolate

88 Perianth-segments 3; ovary inferior; stamens 2 or 3 **CLXXII. Hydrocharitaceae**

88 Perianth 2-lipped, cupular, or absent; ovary superior; stamens 1(2)

89 Leaves without a sheathing base; ovary compressed, deeply 4-lobed **CL. Callitrichaceae**

89 Leaves with a sheathing base; ovary terete, not lobed

90 Leaves entire **CLXXXI. Zannichelliaceae**

90 Leaves dentate or denticulate **CLXXXII. Najadaceae**

74 Land plant or aquatic with emergent stems or leaves

91 Climbing plant with unisexual flowers

92 Leaves opposite; perianth-segments 5 **39. Cannabaceae**

92 Leaves alternate; perianth-segments 6 **CLXXXVI. Dioscoreaceae**

91 Not climbing, or rarely climber with hermaphrodite flowers

93 Leaves linear

94 Flowers unisexual

95 Female flowers solitary; male flowers solitary or in short cymes **48. Chenopodiaceae**

95 Male and female flowers numerous, in dense heads or spikes

96 Male and female (and some hermaphrodite)

flowers mixed together in the same spike; stamen 1 **CLXXVI. Lilaeaceae**

96 Male and female flowers separate in the inflorescence; stamens 2 or more

97 Male and female flowers in separate globose heads **CXCVII. Sparganiaceae**

97 Flowers in a dense cylindrical spike, male above, female below, sometimes with a gap between them **CXCVIII. Typhaceae**

94 Flowers hermaphrodite

98 Plant densely pubescent **48. Chenopodiaceae**

98 Plant glabrous to sparsely pubescent

99 Flowers in a dense spike which is apparently lateral on a flattened leaf-like stem **CXCV. Araceae**

99 Not as above

100 Carpel 1

101 Leaves not subverticillate; stipules absent **48. Chenopodiaceae**

101 Leaves subverticillate; small stipules present **57. Caryophyllaceae**

100 Carpels more than 1

102 Carpels free (except at base); leaves with a conspicuous pore at apex **CLXXIII. Scheuchzeriaceae**

102 Carpels ± completely united; leaves without a conspicuous pore at apex

103 Perianth-segments 5

104 Leaves opposite, not glaucous **57. Caryophyllaceae**

104 Leaves in whorls, glaucous **53. Molluginaceae**

103 Perianth-segments 6

105 Flowers in unbranched racemes; styles short or absent **CLXXV. Juncaginaceae**

105 Flowers in cymes, usually in a branched inflorescence; styles 3, distinct **CLXXXIX. Juncaceae**

93 Leaves lanceolate or wider, or small and scale-like but not linear

106 Leaves compound

107 Flowers in compound umbels **CXXIX. Umbelliferae**

107 Flowers not in compound umbels

108 Flowers in capitula

109 Leaves pinnate; styles 1 or 2 **LXXX. Rosaceae**

109 Leaves ternate; styles 3–5 **CLXV. Adoxaceae**

108 Flowers not in capitula

110 Ovary inferior; styles 3, bifid **CXVI. Datiscaceae**

110 Ovary superior; styles 1, 4 or 5

111 Stamens numerous **51. Ranunculaceae**

111 Stamens 4 or 5(–10)

112 Epicalyx present **LXXX. Rosaceae**

112 Epicalyx absent **LXXXIII. Geraniaceae**

106 Leaves simple or apparently absent

113 Flowers small, usually numerous, arranged on an axis (spadix) subtended and often ± enclosed by a conspicuous bract (spathe) **CXCV. Araceae**

113 Not as above

114 Inflorescence of several male flowers, each of 1 stamen, and a female flower, appearing as a stalked ovary, all surrounded by 4 or 5(8) conspicuous glands; latex present **LXXXVII. Euphorbiaceae**

114 Not as above

115 Leaves apparently absent; stem green and succulent **48. Chenopodiaceae**

115 Leaves obvious; stem not succulent

116 Lower leaves opposite, upper alternate; monoecious; male flowers with 2-partite perianth, the female with tubular perianth **CXXV. Theligonaceae**

116 Not as above

117 Plant densely clothed with stellate or peltate hairs; ovary 3-locular with 1 ovule in each loculus **LXXXVII. Euphorbiaceae**

117 Not as above

118 Plant densely papillose

119 Leaves oblong-lanceolate, not hastate; fruit dehiscing by 5 valves **52. Aizoaceae**

119 Leaves ovate-rhombic, often hastate; fruit indehiscent **54. Tetragoniaceae**

118 Plant not densely papillose

120 Leaves in whorls

121 Stigma 1; stems hollow **CXXVI. Hippuridaceae**

121 Stigmas 3; stems solid **53. Molluginaceae**

120 Leaves not in whorls

122 Leaves opposite (rarely a few of the upper apparently alternate)

123 Leaves toothed or lobed

124 Flowers hermaphrodite

125 Ovary inferior or semi-inferior; stigmas 2 **73. Saxifragaceae**

125 Ovary superior; stigmas 5 **LXXXIII. Geraniaceae**

124 Flowers unisexual

126 Perianth-segments 4 or 2; style 1 **40. Urticaceae**

126 Perianth-segments 3; styles 2 **LXXXVII. Euphorbiaceae**

123 Leaves entire

127 Perianth absent; ovary strongly compressed, deeply 4-lobed **CL. Callitrichaceae**

127 Perianth present; ovary not compressed and 4-lobed

128 Perianth-segments 3 **47. Polygonaceae**

128 Perianth-segments 4 or more

129 Ovary inferior **CXXIII. Onagraceae**

129 Ovary superior

130 Perianth-segments 6 or 12; style and stigma 1 **CXIX. Lythraceae**
130 Perianth-segments 4 or 5; styles and stigmas 2 or more
131 Leaves with a long spinose apex; perianth-segments transversely winged in fruit **48. Chenopodiaceae**
131 Leaves without a long spinose apex; perianth-segments not winged in fruit **57. Caryophyllaceae**
122 Leaves alternate or all basal (rarely the lower opposite)
132 Stamens numerous; carpels free, except sometimes at the base **61. Ranunculaceae**
132 Stamens 12 or fewer; carpels usually solitary or united
133 Carpels attached to a central axis, otherwise free **51. Phytolaccaceae**
133 Carpel 1, or carpels obviously united
134 Stamens 12 **44. Aristolochiaceae**
134 Stamens 10 or fewer
135 Stipules united into a sheath **47. Polygonaceae**
135 Stipules free or absent
136 Leaves very large, palmately lobed, all basal; inflorescence of dense, many-flowered spikes much shorter than the leaves **CXXIV. Haloragaceae**
136 Not as above
137 Epicalyx present; stipules leaf-like **LXXX. Rosaceae**
137 Epicalyx absent; stipules small or absent
138 Ovary inferior
139 Leaves reniform, cordate **73. Saxifragaceae**
139 Leaves subulate to linear-lanceolate **42. Santalaceae**
138 Ovary superior
140 Perianth tubular below
141 Stamens 1–5; ovule basal **48. Chenopodiaceae**
141 Stamens 8; ovule pendent **CVII. Thymelaeaceae**
140 Perianth-segments free or nearly so
142 Perianth-segments 4
143 Flowers in ebracteate racemes **68. Cruciferae**
143 Flowers in axillary clusters **40. Urticaceae**
142 Perianth-segments 5, at least in male flowers
144 Perianth herbaceous, or absent in female flowers **48. Chenopodiaceae**
144 Perianth scarious **49. Amaranthaceae**
10 Perianth of 2 (rarely more) whorls differing markedly from each other in shape, size, colour or texture
145 Petals all united at base into a longer or shorter tube
146 Ovary inferior
147 Stamens 8–10, or 4 or 5 with filaments divided to base
148 Herb; anthers opening by longitudinal slits; leaves ternate **CLXV. Adoxaceae**
148 Woody; anthers opening by apical pores; leaves simple **CXXXII. Ericaceae**
147 Stamens 5 or fewer; filaments not divided
149 Leaves in whorls of 4 or more **CXLIV. Rubiaceae**
149 Leaves not in whorls
150 Stamens opposite the corolla-lobes **CXXXV. Primulaceae**
150 Stamens alternating with the corolla-lobes
151 Stipules interpetiolar **CXLIV. Rubiaceae**
151 Stipules absent or not interpetiolar
152 Flowers in capitula surrounded by an involucre of more than 2 bracts
153 Anthers coherent in a ring round the style
154 Ovule 1; calyx represented by hairs, scales, a corona or auricle **CLXIX. Compositae**
154 Ovules numerous; calyx-lobes conspicuous, usually green **CLXVIII. Campanulaceae**
153 Anthers free
155 Ovule 1; corolla-lobes usually much shorter than tube **CLXVII. Dipsacaceae**
155 Ovules numerous; corolla-lobes longer than tube **CLXVIII. Campanulaceae**
152 Flowers not in capitula, or, if in capitula with an involucre of 2 bracts
156 Anthers coherent in a tube round the style **CLXVIII. Campanulaceae**
156 Anthers free
157 Anthers sessile; pollen-grains coherent in pollinia **CCIII. Orchidaceae**
157 Anthers with filaments; pollen-grains not in pollinia
158 Leaves more than 100 cm **CC. Musaceae**
158 Leaves not more than *c.* 50 cm
159 Stamens 1–3(4)
160 Corolla 4- or 5-merous **CLXVI. Valerianaceae**
160 Corolla 3-merous
161 Sepals connate into a tube **CCI. Zingiberaceae**
161 Sepals free **CCII. Cannaceae**
159 Stamens 4–8

162 Shrub (sometimes small and creeping) or woody climber **CLXIV. Caprifoliaceae**
162 Herb
163 Tendrils present **CXVII. Cucurbitaceae**
163 Tendrils absent
164 Leaves pinnate **CLXIV. Caprifoliaceae**
164 Leaves not pinnate
165 Flowers hermaphrodite; fruit a capsule **CLXVIII. Campanulaceae**
165 Flowers unisexual; fruit fleshy **CXVII. Cucurbitaceae**
146 Ovary superior
166 Flowers papilionaceous
167 Sepals free; stamens 8 **XCII. Polygalaceae**
167 Sepals connate; stamens 10 **LXXXI. Leguminosae**
166 Flowers not papilionaceous
168 Stamens at least twice as many as corolla-lobes
169 Herb with succulent leaves **72. Crassulaceae**
169 Shrub or tree
170 Flowers unisexual **CXXXVII. Ebenaceae**
170 Flowers hermaphrodite
171 Anthers opening by apical pores; hairs simple or scale-like **CXXXII. Ericaceae**
171 Anthers opening by longitudinal slits; hairs stellate **CXXXVIII. Styracaceae**
168 Stamens not more numerous than corolla-lobes
172 Plant without chlorophyll; leaves scale-like
173 Flowers actinomorphic; stem slender, twining **CXLVI. Convolvulaceae**
173 Flowers $\pm$zygomorphic; stem stout, erect
174 Leaves fleshy at anthesis; corolla with cylindrical tube and 2-lipped limb; upper lip entire, the lower entire or shortly 3-lobed **CLIV. Scrophulariaceae**
174 Leaves not fleshy at anthesis; corolla 5-lobed, 2-lipped or almost regular **CLX. Orobanchaceae**
172 Green plant
175 Ovary deeply (2–)4-lobed with 1 ovule in each lobe; fruit separating into nutlets when mature
176 Leaves alternate **CXLVIII. Boraginaceae**
176 Leaves opposite
177 Style gynobasic **CLI. Labiatae**
177 Style apical **CXLIX. Verbenaceae**
175 Ovary not deeply (2–)4-lobed
178 Flowers distinctly zygomorphic
179 Anthers opening by apical pores **CXXXII. Ericaceae**
179 Anthers opening by longitudinal slits
180 Calyx with patent spines and erect, membranous, usually dark-spotted lobes **CXXXV. Primulaceae**
180 Calyx not as above
181 Flowers small, crowded in capitula **CLV. Globulariaceae**
181 Flowers not in capitula
182 Ovary 1-locular; carnivorous plants **CLXI. Lentibulariaceae**
182 Ovary 2-locular; not carnivorous plants
183 Ovules 4
184 Bracts and bracteoles shorter than calyx **CXLIX. Verbenaceae**
184 Bracts or bracteoles longer than calyx **CLVI. Acanthaceae**
183 Ovules numerous
185 Leaves all basal **CLIX. Gesneriaceae**
185 Cauline leaves present
186 Capsule not more than twice as long as wide **CLIV. Scrophulariaceae**
186 Capsule many times as long as wide
187 Capsule with a short beak **CLVII. Pedaliaceae**
187 Capsule with a horn 8–20 cm **CLVIII. Martyniaceae**
178 Flowers actinomorphic or nearly so
188 Sepals 2
189 Petals 2; leaves all basal **CXCII. Eriocaulaceae**
189 Petals 5; leaves not all basal **55. Portulacaceae**
188 Sepals more than 2
190 Carpels free
191 Carpels 4 or more; latex absent **72. Crassulaceae**
191 Carpels 2; latex present
192 Corolla with a corona; styles 2, free but united by the stigma **CXLIII. Asclepiadaceae**
192 Corolla without a corona; styles 2, united except at the very base **CXLII. Apocynaceae**
190 Carpels united
193 Stamens fewer than corolla-lobes
194 Herb **CLIV. Scrophulariaceae**
194 Shrub or tree
195 Leaves opposite **CXXXIX. Oleaceae**
195 Leaves alternate
196 Leaves with numerous pellucid glands **CLXII. Myoporaceae**
196 Leaves without pellucid glands
197 Corolla yellow **CXXXIX. Oleaceae**
197 Corolla not yellow **CLIV. Scrophulariaceae**
193 Stamens as many as corolla-lobes
198 Stamens opposite corolla-lobes
199 Styles or stigmas more than 1; ovule 1 **CXXXVI. Plumbaginaceae**
199 Style and stigma 1; ovules numerous
200 Herb **CXXXV. Primulaceae**
200 Shrub **CXXXIV. Myrsinaceae**
198 Stamens alternating with corolla-lobes
201 Leaves opposite or verticillate
202 Corolla entirely scarious **CLXIII. Plantaginaceae**
202 Corolla not scarious
203 Herb
204 Aquatic plant; leaves petiolate **CXLI. Menyanthaceae**
204 Land plant; leaves sessile **CXL. Gentianaceae**

203 Shrub
205 Plant small, procumbent; leaves evergreen, coriaceous **CXXXII. Ericaceae**
205 Plant large, erect; leaves deciduous, herbaceous
206 Leaves digitate **CXLIX. Verbenaceae**
206 Leaves simple
207 Flowers in long panicles; fruit a capsule **CLIII. Buddlejaceae**
207 Flowers in corymbs; fruit a drupe **CXLIX. Verbenaceae**
201 Leaves alternate or all basal
208 Corolla-lobes and stamens 4
209 Corolla scarious **CLXIII. Plantaginaceae**
209 Corolla not scarious **CLIX. Gesneriaceae**
208 Corolla-lobes and stamens 5
210 Ovary 3-locular; stigmas 3, or 1 but distinctly 3-lobed
211 Leaves herbaceous; corolla not white **CXLV. Polemoniaceae**
211 Leaves coriaceous; corolla white **CXXX. Diapensiaceae**
210 Ovary 2-locular; stigmas 2 or 1
212 Ovules 4 or fewer
213 Flowers numerous, in scorpioid cymes **CXLVIII. Boraginaceae**
213 Flowers usually solitary or few, rarely in congested racemes **CXLVI. Convolvulaceae**
212 Ovules numerous
214 Aquatic or bog-plant; corolla-lobes fimbriate **CXLI. Menyanthaceae**
214 Land plant; corolla-lobes not fimbriate
215 Leaves all basal **CLIX. Gesneriaceae**
215 Some leaves cauline
216 Style deeply divided **CXLVII. Hydrophyllaceae**
216 Style undivided
217 Corolla-lobes imbricate in bud; internal phloem absent **CLIV. Scrophulariaceae**
217 Corolla-lobes plicate or valvate in bud; internal phloem present **CLII. Solanaceae**
145 Petals not all united into a tube at base, very rarely cohering at apex
218 Ovary inferior or partly so
219 Petals numerous
220 Aquatic plant; leaves not succulent **58. Nymphaeaceae**
220 Land plant; leaves succulent **55. Portulacaceae**
219 Petals 5 or fewer
221 Petals and sepals 3
222 Flowers zygomorphic
223 Style and filaments obvious **CLXXXVIII. Iridaceae**
223 Stigma and anthers sessile **CCIII. Orchidaceae**
222 Flowers actinomorphic
224 Leaves and bracts spinose-dentate **CXC. Bromeliaceae**
224 Leaves and bracts not spinose-dentate
225 Outer perianth-whorl sepaloid **CLXXII. Hydrocharitaceae**
225 Both perianth-whorls petaloid
226 Stamens 6 **CLXXXV. Amaryllidaceae**
226 Stamens 3 **CLXXXVIII. Iridaceae**
221 Petals and sepals 2, 4 or 5
227 Stamens numerous
228 Leaves opposite, with pellucid glands **CXXI. Myrtaceae**
228 Leaves alternate, without pellucid glands
229 Leaves entire; seeds covered with pulp **CXXII. Punicaceae**
229 Leaves serrulate; seeds dry
230 Styles free; fruit fleshy **LXXX. Rosaceae**
230 Styles connate, except at apex; fruit dry **75. Hydrangeaceae**
227 Stamens 10 or fewer
231 Aquatic; leaves pinnate, with filiform segments; flowers in spikes **CXXIV. Haloragaceae**
231 Not as above
232 Herb
233 Petals 5
234 Stamens 5 **CXXIX. Umbelliferae**
234 Stamens 10 **73. Saxifragaceae**
233 Petals 4 or 2
235 Flowers in umbels surrounded by 4 conspicuous white bracts **CXXVII. Cornaceae**
235 Flowers not in umbels; no conspicuous white bracts **CXXIII. Onagraceae**
232 Shrub or woody climber
236 Flowers in umbels
237 Climber **CXXVIII. Araliaceae**
237 Erect shrub
238 Evergreen; umbels flat **CXXIX. Umbelliferae**
238 Deciduous; umbels globose **CXXVII. Cornaceae**
236 Flowers not in umbels
239 Leaves palmately lobed **77. Grossulariaceae**
239 Leaves not lobed
240 Both perianth-whorls petaloid **CXXIII. Onagraceae**
240 Outer perianth-whorl sepaloid
241 Calyx-teeth very small; ovules 1 in each carpel; fruit a drupe **CXXVII. Cornaceae**
241 Calyx-teeth large; ovules numerous; fruit a capsule
242 Stamens 10 **75. Hydrangeaceae**
242 Stamens 5 **76. Escalloniaceae**
218 Ovary superior
243 Carpels 2 or more, free or united at the base only

244 Sepals and petals 3

245 Carpels more than 3

246 Leaves lobed **61. Ranunculaceae**

246 Leaves entire **CLXX. Alismataceae**

245 Carpels 3

247 Leaves palmately or pinnately divided; petioles spiny **CXCIV. Palmae**

247 Leaves simple, sessile **72. Crassulaceae**

244 Sepals or petals more than 3

248 Flowers zygomorphic

249 Petals deeply divided **69. Resedaceae**

249 Petals entire **61. Ranunculaceae**

248 Flowers actinomorphic

250 Stamens more than twice as many as petals

251 Shrub or herb with stipulate leaves; flowers perigynous **LXXX. Rosaceae**

251 Herb; stipules absent, though leaf-bases sometimes sheathing; flowers hypogynous

252 Fruit a head of achenes; sepals deciduous **61. Ranunculaceae**

252 Fruit of 2–5 follicles; sepals persistent **62. Paeoniaceae**

250 Stamens not more than twice as many as petals

253 Leaves 3-foliolate **LXXX. Rosaceae**

253 Leaves simple

254 Tree with palmately lobed leaves; flowers in globose capitula **79. Platanaceae**

254 Herb or shrub; leaves not palmately lobed; flowers not in globose capitula

255 Carpels spirally arranged on an elongated receptacle **61. Ranunculaceae**

255 Carpels in 1 whorl

256 Herb or dwarf shrub; leaves ±succulent **72. Crassulaceae**

256 Shrub with angular stems; leaves not succulent **XCIII. Coriariaceae**

243 Carpels obviously united for at least $\frac{1}{2}$ their length, or carpel solitary

257 Flowers zygomorphic

258 One or more perianth-segments saccate or spurred at base

259 Sepals 2, small **66. Papaveraceae**

259 Sepals 3 or 5

260 Sepals 3, very unequal, 1 spurred; petals 3, not spurred **XCVIII. Balsaminaceae**

260 Sepals and petals 5

261 Leaves peltate **LXXXIV. Tropaeolaceae**

261 Leaves not peltate

262 Leaves opposite **LXXXIII. Geraniaceae**

262 Leaves alternate

263 Outer whorl of perianth-segments sepaloid **CX. Violaceae**

263 Both whorls of perianth-segments petaloid **61. Ranunculaceae**

258 Perianth not saccate or spurred at base

264 All, or all but one of the stamens with their filaments connate into a tube **LXXXI. Leguminosae**

264 All stamens free

265 Tree or shrub

266 Leaves compound

267 Leaves 3-foliolate or pinnate **LXXXI. Leguminosae**

267 Leaves digitate, with more than 3 leaflets **XCVII. Hippocastanaceae**

266 Leaves simple

268 Ovary on a long gynophore **67. Capparaceae**

268 Ovary sessile or subsessile

269 Petals 4 **68. Cruciferae**

269 Petals 5 **LXXXI. Leguminosae**

265 Herb

270 Ovary and fruit deeply 5-lobed

271 Flowers in umbellate cymes; fruit with a long beak **LXXXIII. Geraniaceae**

271 Flowers in racemes; fruit not beaked **LXXXVIII. Rutaceae**

270 Ovary and fruit not deeply 5-lobed

272 Petals fimbriate or lobed **69. Resedaceae**

272 Petals entire or emarginate

273 Stamens 10 **LXXXI. Leguminosae**

273 Stamens not more than 6

274 Sepals inserted on a cup-like hypanthium **57. Caryophyllaceae**

274 Sepals not inserted on a cup-like hypanthium

275 Ovary 2-locular; gynophore short or absent **68. Cruciferae**

275 Ovary 1-locular; gynophore long **67. Capparaceae**

257 Flowers actinomorphic

276 Inner whorl of perianth-segments spurred **61. Ranunculaceae**

276 Both whorls of perianth-segments without spurs

277 Corona of long filaments present inside the petals **CXI. Passifloraceae**

277 Flowers without a corona

278 Petals more than 10

279 Aquatic herb with petiolate leaves

280 Leaves floating, usually with a deep basal sinus **58. Nymphaeaceae**

280 Leaves not floating, peltate **59. Nelumbonaceae**

279 Terrestrial herb or shrub with sessile or subsessile leaves

281 Stamens 4–6 **63. Berberidaceae**

281 Stamens numerous **52. Aizoaceae**

278 Petals fewer than 10

282 Stamens more than twice as many as petals

283 Stamens with their filaments connate into a tube **CVI. Malvaceae**

283 Stamens free or connate in separate bundles

284 Perianth-segments persistent in fruit, 2 large and 2 small **47. Polygonaceae**

284 Perianth-segments not as above

285 Ovary on a long gynophore **67. Capparaceae**

285 Ovary sessile or subsessile

286 Ovary surrounded by a cup-shaped hypanthium; ovule 1 **LXXX. Rosaceae**

286 Flowers without a cup-shaped hypanthium; ovules 2 or more

287 Flowers small, in dense spikes or globose clusters, arranged in racemes or panicles **LXXXI. Leguminosae**
287 Flowers not as above
288 Carpel 1; leaves 2-ternate, the lower leaflets stalked **61. Ranunculaceae**
288 Carpels 2 or more; leaves not as above
289 Large tree; inflorescence with a conspicuous bract partly adnate to peduncle **CV. Tiliaceae**
289 Not as above
290 Styles more than 1, usually free
291 Leaves opposite or verticillate **CIX. Guttiferae**
291 Most leaves alternate
292 Leaves with pellucid glands **LXXXVIII. Rutaceae**
292 Leaves without pellucid glands **61. Ranunculaceae**
290 Style 1 or absent
293 Petals 4 **66. Papaveraceae**
293 Petals 5
294 Ovary 1-locular or septate at base only; stamens numerous **CXII. Cistaceae**
294 Ovary 3-locular; stamens 15 **LXXXV. Zygophyllaceae**
282 Stamens not more than twice as many as petals
295 Tree, shrub or woody climber
296 Flowers on tough, leaf-life cladodes; leaves scale-like, brownish **CLXXXIII. Liliaceae**
296 Not as above
297 Leaves small, scale-like or ericoid
298 Perianth-segments in 2 whorls of 3; stamens 3 **CXXXIII. Empetraceae**
298 Perianth-segments and stamens more than 3 in a whorl
299 Leaves opposite **CXIV. Frankeniaceae**
299 Leaves alternate **CXIII. Tamaricaceae**
297 Leaves not scale-like or ericoid
300 Peduncles adnate to petioles; ovary on a short gynophore **LXXXIX. Cneoraceae**
300 Not as above
301 All leaves opposite
302 Leaves pinnate
303 Fruit of 2 single-seeded samaras **XCV. Aceraceae**
303 Fruit a capsule or drupe
304 Fruit a capsule **CI. Staphyleaceae**
304 Fruit a black, subglobose drupe **LXXXVIII. Rutaceae**
302 Leaves entire or palmately lobed
305 Fruit of 2 single-seeded samaras; leaves usually palmately lobed **XCV. Aceraceae**
305 Fruit a fleshy capsule; leaves entire **C. Celastraceae**

301 At least some leaves alternate
306 Stamens 8 **XCVI. Sapindaceae**
306 Stamens 4, 5, 6, 10 or 12
307 Stamens 10 or 12
308 Leaves entire **CXXXII. Ericaceae**
308 Leaves pinnate
309 Spiny tree **LXXXI. Leguminosae**
309 Unarmed shrub or tree
310 Stamens with connate filaments **XCI. Meliaceae**
310 Stamens free
311 Shrub or small tree; carpel 1; fruit a small drupe **XCIV. Anacardiaceae**
311 Large tree; carpels 5 or 6, ±free; fruit a group of samaras **XC. Simaroubaceae**
307 Stamens not more than 6
312 Stamens 6 **68. Cruciferae**
312 Stamens 4 or 5
313 Stamens opposite petals
314 Woody climber; petals longer than sepals **CIV. Vitaceae**
314 Shrub or small tree
315 Petals longer than sepals **LXXXVIII. Rutaceae**
315 Petals shorter than sepals **CIII. Rhamnaceae**
313 Stamens alternating with petals
316 Very spiny shrub **C. Celastraceae**
316 Unarmed shrub or small tree
317 Bark resinous; ovule 1 **XCIV. Anacardiaceae**
317 Bark not resinous; ovules several
318 Leaf-margin usually spiny; fruit a bright red drupe **XCIX. Aquifoliaceae**
318 Leaf-margin not spiny; fruit a capsule **78. Pittosporaceae**
295 Herb, sometimes ±woody at base
319 Sepals 2; petals 5
320 Stems erect or procumbent, not twining **55. Portulacaceae**
320 Stems twining **56. Basellaceae**
319 Sepals as many as petals
321 Flowers 3-merous **CXCI. Commelinaceae**
321 Flowers 4- or more-merous
322 Leaves forming long pitchers; stigma very large, peltate **70. Sarraceniaceae**
322 Not as above
323 Flowers strongly perigynous, with a tubular or campanulate hypanthium **CXIX. Lythraceae**
323 Flowers hypogynous, or perigynous with a flat or weakly concave hypanthium
324 Cauline leaves opposite or verticillate
325 Leaves divided or serrate
326 Petals 4 **68. Cruciferae**

326 Petals 5
327 Stamens without scales on the inner side of the filaments **LXXXIII. Geraniaceae**
327 Stamens with scales on the inner side of the filaments **LXXXV. Zygophyllaceae**
325 Leaves undivided and entire
328 Leaves in 1 whorl; flower solitary, terminal **CLXXXIII. Liliaceae**
328 Not as above
329 Stipules present
330 Stipules scarious; land-plant **57. Caryophyllaceae**
330 Stipules not scarious; usually submerged aquatic **CXV. Elatinaceae**
329 Stipules absent
331 Sepals connate for more than $\frac{1}{2}$ their length
332 Styles connate; placentation parietal **CXIV. Frankeniaceae**
332 Styles free; placentation free-central **57. Caryophyllaceae**
331 Sepals free or connate at the base only
333 Ovary 1-locular; placentation free-central **57. Caryophyllaceae**
333 Ovary 4- or 5-locular; placentation axile **LXXXVI. Linaceae**
324 Leaves alternate or all basal, rarely absent
334 Herbaceous climber; tendrils present **XCVI. Sapindaceae**
334 Not climbing; tendrils absent
335 Leaves 3- or 4-foliolate **LXXXII. Oxalidaceae**
335 Leaves not 3- or 4-foliolate
336 Sepals and petals 2 or 3 **47. Polygonaceae**
336 Sepals and petals 4 or 5
337 Sepals and petals 4; stamens 4 or 6
338 Stipules absent; stamens usually 6 **68. Cruciferae**
338 Stipules present; stamens 4 **57. Caryophyllaceae**
337 Sepals and petals 5; stamens 5 or 10
339 Leaves with conspicuous, red, viscid glandular hairs **71. Droseraceae**
339 Not as above
340 Leaves with numerous pellucid glands, smelling strongly when crushed **LXXXVIII. Rutaceae**
340 Leaves without pellucid glands
341 Style 1; stigma entire or shallowly lobed; anthers opening by apical pores **CXXXI. Pyrolaceae**
341 Styles or stigmas more than 1; anthers opening by longitudinal slits
342 Stigmas 5
343 Leaves lobed or pinnate **LXXXIII. Geraniaceae**
343 Leaves entire or absent
344 Sepals connate; leaves basal or absent **CXXXVI. Plumbaginaceae**
344 Sepals free; leaves cauline **LXXXVI. Linaceae**
342 Stigmas 2–4
345 Flowers with conspicuous glandular-fimbriate staminodes **74. Parnassiaceae**
345 Glandular-fimbriate staminodes absent
346 Stamens 5 **57. Caryophyllaceae**
346 Stamens 10 **73. Saxifragaceae**

Non-flowering specimens of marine angiosperms ('sea-grasses') may be identified by the following key:

1 Rhizome with abundant rigid fibres **CLXXIX. Posidoniaceae**
1 Rhizome with few or no fibres
2 Leaves with a distinct petiole **CLXXII. Hydrocharitaceae** (*Halophila*)
2 Leaves without a petiole
3 Rhizome filiform **CLXXXI. Zannichelliaceae** (*Althenia*)
3 Rhizome at least 0·5 mm in diameter
4 Rhizome with 2 or more unbranched roots at a node **CLXXX. Zosteraceae**
4 Rhizome with 1 branched root at a node **CLXXXI. Zannichelliaceae** (*Cymodocea*)

Ruppia, *Zannichellia* and *Potamogeton pectinatus* can also occur in brackish water or in sheltered bays, though not in the open sea; flowers or fruits are almost always present in these plants.

EXPLANATORY NOTES ON THE TEXT

Signs and abbreviations

c.	*circa*, approximately
C.	central
cm	centimetre(s)
cv.	cultivar
E.	eastern, east
incl.	including
km	kilometre(s)
loc. cit.	*loco citato*, on the same page in the work cited above
m	metre(s)
mm	millimetre(s)
N.	northern, north
nm.	nothomorph
2*n*	the somatic chromosome number
op. cit.	*opere citato*, in the work cited above
prov.(s.)	province(s)
S.	southern, south
Sect.	Sectio
sp.	species (singular)
spp.	species (plural)
Subfam.	Subfamilia
Subgen.	Subgenus
Subsect.	Subsectio
subsp.	subspecies (singular)
subspp.	subspecies (plural)
var.	varietas
W.	western, west
±	more or less
0	absent
●	endemic to Europe
[ ]	not native
*	status doubtful; possibly native
?	(before a two-letter geographical abbreviation) occurrence doubtful
†	extinct; (before contributor's name) deceased

Abbreviations of geographical territories

(*For precise definitions of these territories, see map I*)

Al	Albania
Au	Austria
Az	Açores
Be	Belgium and Luxembourg
Bl	Islas Baleares
Br	Britain
Bu	Bulgaria
Co	Corse
Cr	Kriti
Cz	Czechoslovakia
Da	Denmark
Fa	Færöer
Fe	Finland
Ga	France
Ge	Germany
Gr	Greece
Hb	Ireland
He	Switzerland
Ho	Netherlands
Hs	Spain
Hu	Hungary
Is	Iceland
It	Italy
Ju	Jugoslavia*
Lu	Portugal
No	Norway
Po	Poland
Rm	Romania
Rs	European part of the former territory of the U.S.S.R., subdivided thus: (N) Northern division; (B) Baltic division; (C) Central division; (W) South-western division; (K) Krym (Crimea); (E) South-eastern division
Sa	Sardegna
Sb	Svalbard (Spitsbergen)
Si	Sicilia
Su	Sweden
Tu	Turkey (European part)

* At the time of going to press, the future political status of the constituent republics of Jugoslavia was still uncertain; we have therefore continued to treat the territory as a unit.

General notes

The sequence of families is that of Engler-Diels, *Syllabus der Pflanzenfamilien* ed. 11 (1936), except that the Monocotyledons are placed after the Dicotyledons. A few of the families in the *Syllabus* have been subdivided in accordance with modern practice.

Descriptions of taxa refer only to the European populations of the taxon in question. When extra-European representatives differ substantially, an explanatory note is sometimes added.

Groups of species have been used in some genera where the species are very difficult to separate. These groups have no formal nomenclatural status and are simply a device to enable a partial identification to be made.

Taxa below the rank of subspecies are neither keyed nor described, and varieties are mentioned only when there are special reasons.

Aliens are included only when they appear to be effectively naturalized or when planted in continuous stands on a fairly large scale.

Hybrids are mentioned only when they occur frequently.

A measurement given without qualification refers to length. Two measurements connected by × indicate length followed by width. Further measurements in parentheses indicate exceptional cases outside the normal range.

Synonyms given in the text are principally those names under which the species or subspecies is described in any of the Standard Floras (p. xxviii) or in well-known monographs.

Chromosome numbers are given only when the editors are satisfied that the count has been made on correctly identified material known to be of wild European origin.

Ecological information is provided only when the habitat-preference of a species is sufficiently uniform over its European range to permit it to be summed up in a short phrase.

Geographical terms such as 'W. Europe', 'Mediterranean region', etc., are to be interpreted as shown on maps IV and V. The statement that a plant occurs in one or more of these regions does not necessarily imply that it occurs throughout the region.

Extra-European distribution is indicated only for those plants whose European range is small and whose range outside Europe is considerably greater, or for species which are not native in Europe.

PTERIDOPHYTA

Plants with an alternation of free-living generations. Sporophytes with vascular tissue, usually perennial and herbaceous, reproducing by spores which give rise to small filamentous or thalloid gametophytes (prothalli) bearing archegonia and antheridia on the same or different prothalli.

PSILOTOPSIDA

1. PSILOTACEAE

EDIT. D. M. MOORE

Perennials without true roots or leaves. Sporangia fused in triplets to form synangia, borne at base of 2-lobed sporophylls; homosporous, the spores monolete. Prothallus small, cylindrical, colourless, with mycorrhiza.

Mainly tropical and sub-tropical.

1. **Psilotum** Swartz

N. A. BURGES

Stem branched dichotomously. Leaf-like appendages small, inconspicuous.

1. **P. nudum** (L.) Beauv., *Prodr. Aethéog.* 112 (1805) (*P. triquetrum* Swartz). Rhizome short, creeping, wiry, much-branched. Stem 10–25 cm × 1·5–2 mm, unbranched below, with dense, dichotomous branching above, triquetrous or flattened. Leaf-like appendages 1–1·5 mm, narrowly triangular. Synangia 1–1·5 mm, exindusiate, capsule-like, composed of (2)3(4) globose sporangia subtended by a forked bract. *Crevices of sandstone cliffs. S.W. Spain (near Algeciras).* Hs. (*Tropics and subtropics.*)

The Spanish plants have been described as var. ***molesworthae*** Iranzo, Prada & Salvo on account of their small size and thick stems, but similar plants are found throughout the range of the species.

LYCOPSIDA

2. LYCOPODIACEAE

EDIT. †D. H. VALENTINE AND D. M. MOORE

Herbaceous or suffruticose perennials, with more or less elongated branches bearing small, spirally arranged leaves which are without veins or with a midrib only; ligule absent. Homosporous; sporangia at the base of scale- or leaf-like sporophylls, frequently in cones. Prothallus subterranean, mostly saprophytic and mycorrhizal.

1 Stems ascending, regularly divided dichotomously into branches of equal length; sporophylls not in terminal cones **1. Huperzia**
1 Stems creeping, with short lateral branches; sporophylls in terminal cones
 2 Leaves opposite and decussate, somewhat scale-like; branches dorsiventral **4. Diphasiastrum**
 2 Leaves spirally arranged or in whorls; branches radial
  3 Leaves linear-subulate, angular, curved upwards; sporophylls similar to leaves but somewhat broader and toothed at base **2. Lycopodiella**
  3 Leaves flat, lanceolate, appressed or deflexed, not usually subulate; sporophylls ovate to broadly lanceolate, with scarious, toothed margins **3. Lycopodium**

1. **Huperzia** Bernh.

†W. ROTHMALER

Stems short, ascending and divided dichotomously into branches of equal length. Leaves and sporophylls similar, in alternating zones in the upper branches; sporangia reniform, pedunculate; spores foveolate-punctate. Prothallus holosaprophytic, subterranean, large and cylindrical.

1. **H. selago** (L.) Bernh. ex Schrank & C. F. P. Mart., *Hort. Monac.* 3 (1829) (*Lycopodium selago* L.). Stem 5–30 cm, erect. Leaves imbricate, in many rows on the stem, appressed or patent, linear to ovate-lanceolate, usually entire, often bearing in their axils bud-like gemmae. *Europe southwards to N. Spain and N. Greece; mainly on mountains except in the north.* All except Bl Cr Gr Lu Rs(K, E) Sa Si Tu.

(**a**) Subsp. **selago**: Stem 10–30 cm, 6–12 mm wide. Leaves green, patent, linear-lanceolate. Sporangia usually well-developed. $2n = 264$, *c.* 272. *Somewhat calcifuge. Throughout the range of the species.*

(**b**) Subsp. **arctica** (Grossh. ex Tolm.) Á. & D. Löve, *Bot. Not.* **114**: 35 (1961) (*H. arctica* (Grossh. ex Tolm.) Sipliv., *Lycopodium appressum* (Desv.) Petrov., *L. recurvum* Kit. ex Willd.): Stem 5–10 cm, 5–6 mm wide. Leaves yellow-green, imbricate, appressed, ovate-lanceolate. Sporangia absent or less frequently developed than in subsp. (**a**). $2n = 90$. *N. Europe, rock outcrops, dry hummocks in bogs; mossy tundra. Iceland, Fennoscandia and N. Russia extending S. to Scotland.*

Plants from Açores, with numerous fine teeth on the leaf-margins, and without gemmae, have been called **H. dentata** (Herter) J. Holub, *Folia Geobot. Phytotax.* (*Praha*) **20**: 72 (1985) (*H. selago* subsp. *dentata* (Herter) Valentine).

2. **Lycopodiella** J. Holub

†W. ROTHMALER (EDITION 1) REVISED BY A. C. JERMY (EDITION 2)

Leaves subulate, curved upwards; sporophylls similar, but somewhat broadened and toothed at the base, forming a terminal cone. Sporangia opening at the base; spores reticulate. Prothallus hemisaprophytic, napiform, with green foliaceous appendages at the apex. (*Lepidotis* Beauv., *Palhinhaea* Franco & Vasc.)

Erect stems more than 15 cm, woody, much-branched; spikes 0·5–1 cm, ovoid, clustered on the smallest branches **1. cernua**
Erect stems to 15 cm, herbaceous, simple; spikes at least 1·5 cm, oblong, solitary **2. inundata**

1. L. cernua (L.) Pichi Serm., *Webbia* **22**: 166 (1968) (*Lycopodium cernuum* L., *Lepidotis cernua* (L.) Beauv.). Main stem subterranean, creeping, woody, with erect, branched shoots up to 35 cm, the lower branches often rooting at tips. Fertile spikes 0·5–1 cm, ovoid, yellowish, at the end of short branches, ultimately nodding, with appressed, dentate sporophylls. *Açores*. *Az [Lu Si]. (*Tropics*.)

2. L. inundata (L.) J. Holub, *Preslia* **36**: 21 (1964) (*Lycopodium inundatum* L., *Lepidotis inundata* (L.) Opiz). Stem above ground, short, creeping, herbaceous, usually with few or no branches, producing a single, erect, simple stem with curved, acute leaves, 4–6 mm, bearing a terminal cone 1.5–3(–5) cm; sporophylls subappressed. $2n = 156$. *Moors, wet heaths and dunes. Most of Europe, except the Mediterranean region and much of* E. *Russia*. Au Az Br Bu Cz Da Fe Ga Ge Hb He Ho Hs It Ju Lu No Po Rm Rs(N, B, C, W, E) Su.

3. Lycopodium L.

†W. ROTHMALER (EDITION 1) REVISED BY A. C. JERMY (EDITION 2)

Leaves flat, linear to lanceolate, appressed or deflexed. Sporophylls different from the other leaves, broadly ovate to broadly lanceolate, with scarious, toothed margins. Sporangia opening at the apex; spores markedly muricate-reticulate. Prothallus holosaprophytic, disciform or tuberous, without appendages.

Cones sessile; leaves and sporophylls without a long, white hair at the apex **1. annotinum**
Cones pedunculate; at least the young leaves and sporophylls with a long, white hair at the apex **2. clavatum**

1. L. annotinum L., *Sp. Pl.* 1103 (1753). Stems procumbent, widely creeping, branched; branches erect to ascending. Leaves 3–10 mm, linear-lanceolate to subulate, appressed to deflexed. Cones terminal on erect branches, sessile, solitary; sporophylls acute. $2n = 68$. *Heaths, mountain grassland and dwarf heath on mountains. N. Europe, extending southwards in the mountains to the Pyrenees, N. Appennini and S. Carpathians*. Au Be Br Bu Cz Da Fe Ga Ge He Ho Hs Hu It No Po Rm Rs(N, B, C, W, ?E) Su.

(a) Subsp. **annotinum**: Branches 10–15 mm wide, ascending. Leaves 5–10 mm, linear-lanceolate or linear, patent or deflexed, thin, dull green, coarsely toothed, the apex acute. Cones 1·5–3 cm. *Throughout the range of the species.*

(b) Subsp. **alpestre** (Hartman) Á. & D. Löve, *Nucleus (Calcutta)* **1 (1)**: 7 (1958) (*L. dubium* auct., non Zoega, *L. pungens* La Pylaie ex Komarov): Branches 3–7 mm wide, erect. Leaves 3–7 mm, linear-subulate, patent to appressed, thick, nearly entire, the apex acuminate. Cones 0·5–1·5 cm. *Arctic and subarctic Europe southwards to* 60° *N*.

2. L. clavatum L., *Sp. Pl.* 1101 (1753). Stems long, procumbent; branches ascending. Leaves 3–5 mm, bright green, linear, acute, subappressed, prolonged apically as a hyaline hair 2–3 mm (sometimes deciduous in older leaves). Cones 1–4, 2·5–6 cm; peduncles 1·5–15 cm, with remote, yellowish, bract-like leaves; sporophylls ovate, with a long, hyaline hair at the apex. $2n = 68$. *Heaths, moors and mountain grassland. N. & C. Europe, extending locally southwards to C. Portugal, C. Italy and Bulgaria*. Au Be Br Bu Cz Da Fe Ga Ge Hb He Ho Hs Hu Is It Ju Lu No Po Rm Rs(N, B, C, W, E) Su.

Subsp. **monostachyon** (Grev. & Hooker) Selander, *Acta Phytogeogr. Suec.* **28**: 7 (1950), a variant smaller in all its parts, leaves yellow-green, cones 1(2), 1–2·5 cm long, peduncles less than 3 cm, is common in the north of Fennoscandia, S. to 65° N.

4. Diphasiastrum J. Holub

†W. ROTHMALER (EDITION 1) REVISED BY A. C. JERMY (EDITION 2)

Stems long, creeping, dorsiventral; branches forked several times and usually flabellate. Leaves 4-ranked, opposite and decussate, somewhat scale-like, the lateral leaves carinate. Cones terminal, the sporophylls different from the leaves; sporangia opening at the apex; spores markedly muricate-reticulate. Prothallus holosaprophytic. (*Diphasium* auct., non C. Presl)

Literature: J. Dostál in G. Hegi, *Ill. Fl. Mitteleur.* ed. 3, **1(1)**: 28–42 (1984). I. Kukkonen, *Ann. Bot. Fenn.* **4**: 441–470 (1967).

1 Main stems subterranean; branches not or only slightly flattened; all leaves ± equal in width **2. tristachyum**
1 Main stems usually above ground; branches flattened; lateral leaves keeled, wider than dorsal and ventral leaves
2 Ventral leaves sessile, subulate to triangular-lanceolate; cones pedunculate **1. complanatum**
2 Ventral leaves petiolate, trullate or ovate; cones sessile **3. alpinum**

1. D. complanatum (L.) J. Holub, *Preslia* **47**: 108 (1975) (*Lycopodium complanatum* L., *Diphasium complanatum* (L.) Rothm., *L. anceps* Wallr.). Main stems far-creeping, subterranean, bearing erect, much-branched shoots; finer branches 2–3 mm wide, obviously flattened. Lateral leaves wider than the dorsal, usually appressed; ventral leaves 0·5–1·5 mm on the finer sterile branches, $\frac{1}{5}$ as wide as the branches. Cones 1–4(5), pedunculate; sporophylls broadly ovate, as long as wide, acuminate, the acumen $\frac{1}{4}$–$\frac{1}{3}$ as long as the sporophyll; spores *c.* 30 μm in diameter. $2n = 46$, 48. *N. & C. Europe, extending locally southwards to Spain, N. Italy, Bulgaria and S. Ural*; *Açores*. Au Az Be Bu Cz Da Fe Ga Ge He Ho Hs Hu It Ju No Po Rm Rs(N, B, C, W, E) Su.

Besides *D. complanatum* sensu stricto three closely related taxa, undoubtedly formed through hybridization with **2** and **3** and sometimes treated as hybrids, are included here as subspecies.

1 Ventral leaves $\frac{1}{5}$ as wide as branches; free apex of lateral leaves *c.* $\frac{1}{6}$ as long as the leaf (a) subsp. **complanatum**
1 Ventral leaves $\frac{1}{4}$–$\frac{1}{3}$ as wide as branches; free apex of lateral leaves *c.* $\frac{1}{2}$ as long as the leaf
2 Width of ventral leaf $\frac{1}{3}$ width of branch (b) subsp. **issleri**
2 Width of ventral leaf $\frac{1}{5}$–$\frac{1}{4}$ width of branch
3 At least some main stems subterranean; peduncles usually more than 5 cm; cones several (c) subsp. **zeilleri**
3 Main stems, if present, above ground; peduncles less than 5 cm; cone usually solitary (d) subsp. **montellii**

(a) Subsp. **complanatum**: Main stem above ground. Ventral leaves 0·5–1 mm, $\frac{1}{5}$ as wide as branches, subulate-triangular, appressed, the apex acute; free apex of lateral leaves *c.* $\frac{1}{6}$ as long as the leaf, squarrose. Cones 1–3(4); peduncles 5–18 cm; spores *c.* 30 μm. *Throughout the range of the species.*

(**b**) Subsp. **issleri** (Rouy) Jermy, *Fern Gaz.* **13**: 260 (1989) (*D.* × *issleri* (Rouy) J. Holub): Main stem above ground. Ventral leaves 1–1·5 mm, $\frac{1}{3}$ as wide as branches, lanceolate, the apex obtuse, more or less appressed; free apex of lateral leaves *c.* $\frac{1}{2}$ as long as the leaf, falcate. Cones 1(–3); peduncles 5–10 cm; spores 32–38 μm. *C. Europe extending westwards to C. France and Britain.*

(**c**) Subsp. **zeilleri** (Rouy) Kukkonen, *Ann. Bot. Fenn.* **23**: 265 (1986) (*D.* × *zeilleri* (Rouy) J. Holub): Main stem at least in part subterranean. Ventral leaves 1–1·5 mm, $\frac{1}{3}$ as wide as branches, lanceolate-triangular, the apex acute, more or less appressed; free apex of lateral leaves *c.* $\frac{1}{2}$ as long as the leaf, falcate. Cones 2–5(6); peduncles 5–18 cm; spores 32–38 μm. *E. C. & N.E. Europe; C. France.*

(**d**) Subsp. **montellii** (Kukkonen) Kukkonen, *Ann. Bot. Fenn.* **21**: 210 (1984). Like subsp. (**c**) but with main stem usually above ground, cones usually solitary and peduncles less than 5 cm. *N. Fennoscandia and N. Russia.*

Plants from Açores similar in morphology to subsp. (**b**) have been called **D. madeirense** (Wilce) J. Holub, *Preslia* **47**: 108 (1975) (*Lycopodium madeirense* Wilce, *Diphasium madeirense* (Wilce) Rothm.) and require further investigation.

2. D. tristachyum (Pursh) J. Holub, *Preslia* **47**: 108 (1975) (*Diphasium tristachyum* (Pursh) Rothm., *D. complanatum* subsp. *chamaecyparissus* (A. Braun ex Mutel) Čelak., *Lycopodium chamaecyparissus* A. Braun ex Mutel). Stems deeply subterranean; erect shoots repeatedly branched and caespitose; finer branches 1·2–1·5(–2) mm wide, somewhat flattened. Lateral leaves as wide as dorsal; ventral leaves on the finer sterile branches 1–1·5 mm long, $\frac{1}{3}$–$\frac{1}{4}$ as wide as the branches. Cone pedunculate; sporophylls broadly ovate, acuminate; spores *c.* 30 μm. $2n = 46$. *C. Europe, extending to S. Fennoscandia, C. Russia, Romania and C. Italy.* Au Be Cz Da Fe Ga Ge He Ho Hu It Ju No Po Rm Rs(B, C) Su.

3. D. alpinum (L.) J. Holub, *op. cit.* 107 (1975) (*Lycopodium alpinum* L., *Diphasium alpinum* (L.) Rothm.). Stems above ground, procumbent, elongated, with ascending, densely caespitose, glaucous branches; finer branches slightly flattened or cylindrical. Ventral leaves of the sterile branches 2–2·5 × 0·5 mm, petiolate, lanceolate. Cone 1·5 cm, sessile; sporophylls twice as long as the sporangia, lanceolate, acute; spores *c.* 45 μm in diameter. $2n = 46$, 48–50. *Moors and heaths. N. & C. Europe, extending to N.W. Spain, N. Appennini and S.W. Bulgaria.* Au †Be Br Bu Cz Da Fa Fe Ga Ge Hb He Hs Is It Ju No Po Rm Rs(N, C, W) Su.

3. SELAGINELLACEAE

EDIT. †D. H. VALENTINE AND D. M. MOORE

Stems erect or creeping. Leaves numerous, simple, small, 1-veined, with a ligule on the adaxial surface. Heterosporous; sporangia grouped in strobili; individual sporangia solitary at the base of the sporophylls, unilocular; spores tetrahedral.

1. Selaginella Beauv.

A. LAWALRÉE

Leaves all similar and spirally arranged, or dimorphic and 4-ranked, the lower patent, the upper appressed and pointing forwards. Strobili terminal; megasporangia at the base and microsporangia at the apex of the strobilus.

The only genus; sometimes divided into several genera.

1 Leafy stems radially symmetrical; leaves all similar — **1. selaginoides**
1 Leafy stems dorsiventral; leaves of 2 sizes
  2 Stems usually at least 30 cm, appearing articulated where branches occur — **5. kraussiana**
  2 Stems less than 30 cm, not appearing articulated
    3 Strobili on erect peduncles; median leaves of upper plane acute — **2. helvetica**
    3 Strobili sessile; median leaves of upper plane cuspidate
      4 Leaves slightly asymmetrical; stems much-branched — **3. denticulata**
      4 Leaves markedly asymmetrical; stems little-branched — **4. apoda**

1. S. selaginoides (L.) Beauv. ex Schrank & C. F. P. Mart., *Hort. Monac.* 3 (1829) (*S. spinulosa* A. Braun, *S. albarracinensis* Pau, *S. spinosa* Beauv.). Sterile branches horizontal and short, radially symmetrical; fertile branches 3–15 cm, erect. Leaves 1–3 mm, all similar, spirally arranged, with 1–5 large teeth on each side. Strobili 1–5 cm, not well differentiated from the rest of the branch, stout, yellowish. $2n = 18$. *N. Europe, southwards to 52° N. in Ireland; C. & S. Europe from Czechoslovakia to N. Spain and S.W. Bulgaria.* Al Au Br Bu Cz Da Fa Fe Ga Ge Hb He Hs Is It Ju No Po Rm Rs(N, B, C, W) Su.

2. S. helvetica (L.) Spring, *Flora (Regensb.)* **21(1)**: 149 (1838). Stem 3–10 cm, creeping, flattened and dorsiventral, not articulated. Leaves up to 3 mm, dimorphic, 4-ranked, weakly denticulate or subentire. Strobili up to 5 cm, solitary or paired, on erect peduncles, sharply differentiated from the rest of the branch. $2n = 18$. *C. & S.E. Europe.* Al Au Bu Cz Ga Ge Gr He Hu It Ju Po Rm Rs(W) Tu.

3. S. denticulata (L.) Spring, *loc. cit.* (1838). Stem 4–10 cm, creeping, flattened and dorsiventral, not articulated, very slender, much-branched. Leaves up to 2·5 mm, dimorphic, 4-ranked, the lower larger, ovate, acuminate, markedly dentate and slightly asymmetrical. Strobili not more than 2 cm, sessile, not sharply defined at the base. $2n = 18$. *Mediterranean region; Portugal.* Al Bl Co Cr Ga Gr Hs It Ju Lu Sa Si.

4. S. apoda (L.) Spring in C. F. P. Mart., *Fl. Brasil.* **1(2)**: 119 (1840) (*S. apus* Spring). Stem less than 30 cm, dorsiventral, creeping, not articulated, little-branched. Leaves up to 3 mm, dimorphic, 4-ranked, denticulate, markedly asymmetrical. Strobili 0·5–2(–3) cm, sessile. *Naturalized in Germany (Berlin).* [Ge.] (*North America.*)

5. S. kraussiana (G. Kunze) A. Braun, *Ind. Sem. Horti Berol.*

1859, app. 22 (1860) (*S. azorica* Baker). Stem 25–100 cm, creeping, flattened and dorsiventral, with false articulations where branches occur. Leaves up to 4 mm, dimorphic, 4-ranked, finely denticulate, asymmetrical. Strobili not more than 2 cm, sessile. *Açores; naturalized in S. & W. Europe.* $2n = 20$. *Az [Be Br Co Ga Hb Hs It Lu Si]. (*Tropical and southern Africa.*)

4. ISOETACEAE

EDIT. †D. H. VALENTINE AND D. M. MOORE

Stem short, 2- or 3-lobed. Leaves in a rosette; leaf-base spathulate, with a membranous margin, with a delicate ligule on the adaxial surface at the point where the leaf narrows. Heterosporous; megaspores tetrahedral; microspores bilateral, in separate sporangia.

1. Isoetes L.

A. C. JERMY AND J. R. AKEROYD

Aquatic or terrestrial plants. Leaves terete or trigonous, subulate, occasionally flat and linear; stomata usually absent in aquatic species. Sporangia solitary, sessile at the base of leaf, naked or covered with a thin tissue (velum).

Literature: P. Berthet & M. Lecocq, *Pollen et Spores* **19**: 341–359 (1977). C. Prada, *Acta Bot. Malac.* **8**: 73–99 (1983).

1 Leaf-bases dark, shiny and persistent; plant terrestrial (if submerged then only for a short period in winter)
  2 Megaspores 320–560 µm, tuberculate — **10. histrix**
  2 Megaspores 600–800 µm, reticulate — **11. duriei**
1 Leaf-bases not dark, rarely persistent; plant normally submerged
  3 Mature sporangium not covered by velum
    4 Megaspores minutely papillose, with inconspicuous ridges; microspores smooth — **5. setacea**
    4 Megaspores verrucose, with prominent ridges; microspores spinose
      5 Leaves 10–25 cm; plant amphibious — **6. heldreichii**
      5 Leaves 30–100 cm; plant aquatic — **7. malinverniana**
  3 At least $\frac{1}{3}$ of mature sporangium covered by velum
    6 Megaspores reticulate; ligule long, subulate — **3. azorica**
    6 Megaspores not reticulate; ligule short, triangular
      7 Megaspores rugose — **1. lacustris**
      7 Megaspores spinose or tuberculate
        8 Megaspores with sharp tubercles or spines, at least on proximal face; stems 2-lobed — **2. echinospora**
        8 Megaspores with papillae or rounded tubercles; stem 3-lobed (rarely 4-lobed)
          9 Megaspores with numerous regular papillae on lower face; microspores smooth or minutely papillose — **4. boryana**
          9 Megaspores with scattered tubercles of various sizes on lower face; microspores ± spinose
            10 Aquatic; leaves up to 80 cm; megaspores 580–650 µm — **8. longissima**
            10 Amphibious; leaves not more than 60 cm; megaspores 325–480 µm — **9. velata**

1. I. lacustris L., *Sp. Pl.* 1100 (1753). Aquatic; stem 2-lobed. Leaves 8–25(–40) cm × 3–5 mm, erect or rarely patent and recurved, stout, subterete, tapering abruptly at apex; membranous margin wide at base, abruptly narrowing above sporangium; ligule short, triangular. Sporangium only partially covered by velum; megaspores 530–700 µm, rugose; microspores yellowish-brown, smooth or with fine, irregular furrows. $2n = 110$. *Up to 8 m deep in lakes and pools. N. Europe, extending locally southwards to N. Spain and White Russia.* Br Cz Da Fa Fe Ga Ge Hb He Ho Hs Is ?It No Po Rs(N, B, C) Su.

Plants from high mountain lakes in the E. Pyrenees, described as **I. brochonii** Motelay, *Actes Soc. Linn. Bordeaux* **45**: 45, t. 2 (1892), are possibly only a variant of **1** and have been confused with an undescribed alloploid (duodecaploid) species resulting from **1** × **2**. The latter has tuberculate megaspores 500–700 µm, the tubercles fusing in short ridges on the lower face.

2. I. echinospora Durieu, *Bull. Soc. Bot. Fr.* **8**: 164 (1861) (*I. tenella* Léman ex Desv., *I. setacea* auct., non Lam.; incl. *I. braunii* Durieu, non Unger). Like **1** but the leaves 5–20 cm, subulate, compressed, gradually tapered, flaccid, more often patent or recurved; megaspores 360–460 µm, with long, narrow spines. $2n = 22$. *In more oligotrophic waters than* **1**. *N. Europe, extending very locally southwards to N.E. Spain and C. Greece.* Be Br Bu Cz Da Fa Fe Ga Ge Gr Hb He Ho Hs Is It No Po Rs(N, B, C) Su.

3. I. azorica Durieu ex Milde, *Fil. Eur.* 278 (1867). Aquatic; stem 2-lobed. Leaves 8–30 cm, slender, flexuous; membranous margin narrow at base; ligule long, subulate. Sporangium partially covered by velum; megaspores 350–490 µm, reticulate; microspores brown, more or less spinose. *In pools and small lakes.* ● *Açores.* Az.

4. I. boryana Durieu, *Bull. Soc. Bot. Fr.* **8**: 164 (1861). Aquatic; stem 3-lobed. Leaves 10–20 cm, loosely inserted, but otherwise very like those of **1**; ligule deltate. Velum almost or completely covering sporangium; megaspores 375–600 µm, granulate with occasional large papillae on upper faces and numerous regular papillae on lower face; microspores reddish-brown, smooth or very minutely scabrid. *Shallow lakes.* ● *Coast of S.W. France.* Ga.

5. I. setacea Lam., *Encycl. Méth. Bot.* **3**: 314 (1789) (*I. delilei* Rothm.). Aquatic; stem 3-lobed. Leaves 12–40 cm × 1–2 mm, often numerous, firm, tapering to apex; membranous margin up to 5 mm wide at base, narrowing abruptly just above sporangium and continuing up leaf for *c.* $\frac{1}{4}$ its length; ligule ovate. Sporangium without velum; megaspores 440–580 µm, minutely papillose, the ridges narrow and inconspicuous; microspores minutely scabrid. *Ponds and small lakes.* ● *W. Europe.* ?Co Ga Hs Lu.

6. I. heldreichii Wettst., *Verh. Zool.-Bot. Ges. Wien* **36**: 239, t. 8 (1886). Amphibious; stem 3-lobed. Leaves 10–25 cm × 0·5–1·5 mm, few, flexuous; basal membranous margin narrow, extending up leaf for $\frac{1}{6}$ its length; ligule obovate-acute. Sporangium without velum; megaspores *c.* 660 µm, tuberculate;

microspores spinose, narrowly winged. *Schistose soil of lake margin.* ● *C. Greece (S. end of Pindhos range, near Palaiokastron).* Gr.

7. I. malinverniana Cesati & De Not., *Ind. Sem. Horti Genuensis* **1858**: 36 (1858). Aquatic; stem 3-lobed. Leaves 30–100 cm, numerous, subulate at apex; stomata absent or few; base with wide membranous margin extending up leaf for $\frac{1}{6}$ its length; ligule triangular. Sporangium without velum; megaspores 540–610 μm, with large, cylindrical or conical tubercles, the ridges prominent; microspores minutely scabrid. $2n = 44$. *Rapidly flowing water of irrigation channels.* ● *N.W. Italy; local.* It.

8. I. longissima Bory, *Compt. Rend. Acad. (Paris)* **18**: 1165 (1844) (*I. velata* subsp. *longissima* (Bory) Greuter & Burdet). Aquatic; stem 3-lobed. Leaves 40–80 cm, flaccid, green; membranous margin at base extending 3–4 cm above sporangium. Sporangium almost or completely covered by velum; megaspores 580–650 μm, with large and small tubercles; microspores spinulose. $2n = 44$. *Rapidly flowing water.* ● *N.W. Spain (Galicia).* Hs.

9. I. velata A. Braun in Bory & Durieu, *Expl. Sci. Algér. Bot.* **1**: 19, t. 37, f. 1 (1850) (*I. variabilis* Legrand, *I. dubia* Genn.). Amphibious; stem 3-lobed. Leaves 5–40(–60) cm, erect, firm, tapering to apex; membranous margin at base usually at least 1 mm wide, extending up leaf for $\frac{1}{4}$–$\frac{1}{3}$ its length; leaf-bases usually persisting, forming a 'bulb' of brown, papery scales; ligule triangular-ovate. Sporangium usually almost or completely covered by velum; megaspores 300–480 μm, with scattered large and small tubercles, which sometimes fuse to form a more or less prominent wart on each of the upper faces; microspores reddish-brown, scabrid or with short spines. *Shallow lake-margins and seasonal pools. W. Mediterranean region; Portugal.* Bl Co Ga Hs It Lu Sa Si.

A very variable species in which a number of subspecies can be recognized.

1 Megaspores with few, often indistinct, tubercles (b) subsp. **asturicense**
1 Megaspores distinctly tuberculate (although tubercles scattered)
2 Basal margin of leaf less than 1 mm wide (leaves very slender); sporangium about $\frac{2}{3}$ covered by velum (c) subsp. **tegulensis**
2 Basal margin of leaf at least 1 mm wide; sporangium almost or completely covered by velum
3 Leaves often more than 18 cm, ±robust; megaspores more than 380 μm (a) subsp. **velata**
3 Leaves up to 18 cm, slender; megaspores *c.* 300 μm (d) subsp. **tenuissima**

(a) Subsp. **velata** (*I. baetica* Willk.): Leaves 5–30 cm, yellowish-green. Sporangium almost or completely covered by velum; megaspores 380–480 μm, tuberculate. $2n = 22$. *Throughout the range of the species.*

(b) Subsp. **asturicense** (Laínz) Rivas Martínez & Prada, *Lazaroa* **2**: 237 (1980): Leaves 7–15 cm, green; stomata few. Sporangium almost or completely covered by velum; megaspores 325–410 μm. $2n = 22$. ● *N.W. Spain.*

(c) Subsp. **tegulensis** Batt. & Trabut, *Fl. Algér. Tunisie* 407 (1905) (*I. tegulensis* Genn., nom. illeg., *I. tiguliana* Genn.): Leaves 5–30 cm, very slender, green; basal margin less than 1 mm wide. Sporangium about $\frac{2}{3}$ covered by velum; megaspores 400–440 μm. *Sardegna; Sicilia. (N. Africa.)*

(d) Subsp. **tenuissima** (Boreau) O. Bolòs & Vigo, *Butll. Inst. Catalana Hist. Nat.* **38**: 64 (1974) (*I. tenuissima* Boreau, *I. viollaei* Hy): Leaves 5–18 cm, slender, green. Sporangium almost or completely covered by velum; megaspores *c.* 300 μm. ● *S. France.*

10. I. histrix Bory, *Compt. Rend. Acad. (Paris)* **18**: 1167 (1844) (*I. delalandei* Lloyd, *I. phrygia* Hausskn.; incl. *I. chaetureti* Mendes). Terrestrial; stem 3-lobed. Leaves 5–20 cm × 0·5–1 mm, more or less flat, linear; stomata numerous; membranous margin wide at base, narrowing abruptly just above sporangium but continuing up the leaf for 1–2 cm; leaf-bases persistent over several years as black, shiny, horny scales with central, short, broad lobe and 2 longer, lateral, often spine-like lobes. Sporangium completely covered by velum; megaspores 360–560 μm, with small tubercles which often become confluent, especially on the basal surface; microspores brown, spinulose-fibrillose. $2n = 20$. *Sandy places, where water lies in winter. Mediterranean region and W. Europe, northwards to 50° N. in S.W. England.* Br Co Cr Ga Gr Hs It Ju Lu Sa Si.

11. I. duriei Bory, *op. cit.* 1166 (1844). Like **11** but the leaves 8–15 cm, firm, often recurved; megaspores 600–800 μm, reticulate with prominent ridges; microspores minutely papillose. *In similar, but often damper, habitats to* **10**. *Mediterranean region (mainly in the west); Portugal.* Bl Co Ga Gr Hs It Lu Sa Si.

SPHENOPSIDA

5. EQUISETACEAE

EDIT. †D. H. VALENTINE AND D. M. MOORE

Rhizomes subterranean, far-creeping. Stems grooved and regularly jointed. Leaves in whorls, united at the base into a sheath. Homosporous; cones composed of sporangiophores; sporangia borne beneath the peltate heads of the sporangiophores; spores spherical with 4 elaters, green. Prothallus green.

1. Equisetum L.

†T. G. TUTIN

Stems erect, usually with a hollow in the centre surrounded by 2 rings of alternating smaller cavities. Free portions of leaves small, scarious. Cones borne at the ends of the main stems and rarely of the branches also. (Incl. *Hippochaete* Milde.)

The only genus. Several hybrids are widespread and sometimes locally common, and can occur in the absence of one or both parents.

1 Cones on pale brown stems, appearing before green stems
2 Sheaths with 20–30 teeth; cone 4–8 cm **10. telmateia**
2 Sheaths with 6–12 teeth; cone 1–4 cm **9. arvense**

1 Cones appearing at the same time as, or present on, green stems

3 Branches whorled

4 Main stem 10–20 mm in diameter, branched to apex, whitish, almost smooth, with 20–40 fine ribs **10. telmateia**

4 Main stem rarely more than 7 mm in diameter, green; ribs well marked, less than 20

5 Sheaths of main stem with broad, subacute teeth, some being laterally joined and appearing fewer than the grooves; branches with secondary branches **7. sylvaticum**

5 Sheaths of main stem with distinct, subulate teeth clearly equal in number to the grooves; branches simple

6 Central hollow at least $\frac{4}{5}$ the diameter of the stem; teeth not keeled **5. fluviatile**

6 Central hollow less than $\frac{2}{3}$ the diameter of the stem; teeth with 1 or 2 keels

7 Lowest internode of branches much shorter than sheath on main stem; branches hollow

8 Stem usually with 5–10 grooves; central hollow less than $\frac{1}{2}$ the diameter of stem; cone obtuse **6. palustre**

8 Stem with 8–20 grooves; central hollow more than $\frac{1}{2}$ the diameter of stem; cone apiculate **2. ramosissimum**

7 Lowest internode of branches longer than sheath on main stem; branches solid

9 Central hollow *c.* $\frac{1}{4}$ the diameter of main stem; branches mostly 4-angled, ascending or suberect **9. arvense**

9 Central hollow *c.* $\frac{1}{2}$ the diameter of main stem; branches mostly 3-angled, horizontal or recurved **8. pratense**

3 Branches absent or not in obvious whorls

10 Cone apiculate; stems usually persistent; stomata below level of other epidermal cells

11 Teeth of leaf-sheaths caducous **1. hyemale**

11 Teeth of leaf-sheaths persistent

12 Stem freely branched; central hollow *c.* $\frac{1}{2}$ the diameter of main stem **2. ramosissimum**

12 Stem simple or branched only from base; central hollow *c.* $\frac{1}{4}$ the diameter of main stem, or absent

13 Main stem with central hollow *c.* $\frac{1}{4}$ the diameter of main stem; sheaths with 6–8 teeth **3. variegatum**

13 Main stem solid; sheaths with 3 or 4 teeth **4. scirpoides**

10 Cone obtuse; stems not persistent; stomata superficial

14 Central hollow at least $\frac{4}{5}$ the diameter of main stem; teeth without ribs **5. fluviatile**

14 Central hollow less than $\frac{2}{3}$ the diameter of main stem; teeth with ribs

15 Outermost cavities about the same size as the central hollow; branches hollow, their lowest internodes much shorter than sheath on main stem **6. palustre**

15 Outermost cavities smaller than the central hollow; branches solid, their lowest internodes longer than sheath on main stem **9. arvense**

Subgen. **Hippochaete** (Milde) Baker. Stomata sunk below level of other epidermal cells. Cones apiculate. Stems all alike, hard, usually persistent.

1. E. hyemale L., *Sp. Pl.* 1062 (1753) (*Hippochaete hyemalis* (L.) Bruhin). Stem 30–100 cm × 4–6 mm, persistent, simple; ridges with 2 rows of tubercles; grooves 10–30; sheaths about as long as broad, whitish with a black band at top and bottom; teeth very soon caducous; central hollow $\frac{2}{3}$ or more the diameter of the stem. Cone 8–15 mm. $2n = c.\ 216$. *Damp or shady places. Most of Europe, but rare in the Mediterranean region.* Al Au Be Br Bu ?Co Cz Da Fa Fe Ga Ge Gr Hb He Ho Hs Hu Is It Ju No Po Rm Rs(N, B, C, W, E) Su Tu.

The following hybrids, both of which have abortive spores, occur locally, sometimes in the absence of the parents. **E. × moorei** Newman, *Phytologist* (*Newman*) **5**: 19 (1854) (*E. occidentale* (Hy) Coste, *Hippochaete* × *moorei* (Newman) H. P. Fuchs) is like **1** but has more slender, not persistent, stems, ridges with irregular tubercles or cross-bands, and the teeth of leaf-sheaths more persistent. It is probably **1** × **2**. **E. × trachyodon** A. Braun, *Flora* (*Regensb.*) **22**: 305 (1839) (*Hippochaete* × *trachyodon* (A. Braun) Börner) is intermediate between **1** and **3**, and is probably a hybrid.

2. E. ramosissimum Desf., *Fl. Atl.* **2**: 398 (1799) (*Hippochaete ramosissima* (Desf.) Börner; incl. *E. campanulatum* Poiret). Stem 10–100 cm × 3–9 mm, usually dying in autumn, freely branched, with scattered tubercles; grooves 8–20; sheaths green, becoming brown with a dark band at the bottom; teeth dark, with a narrow, scarious border; central hollow $\frac{1}{2}$–$\frac{2}{3}$ the diameter of the stem. Branches hollow; lowest internode much shorter than sheath on main stem. Cone 6–12 mm. $2n = c.\ 216$. *C. & S. Europe, extending locally northwards to the Netherlands and C. Ural.* Al Au Az Bl Bu Co Cr Cz Ga Ge Gr He Ho Hs Hu It Ju Lu Po Rm Rs(*B, C, W, K, E) Sa Si Tu [Br].

3. E. variegatum Schleicher, *Ann. Bot.* (*Usteri*) **21**: 124 (1797) (*Hippochaete variegata* (Schleicher) Bruhin). Stem 10–30(–50) cm, up to 3 mm in diameter, persistent, simple, or branched from the base; ridges finely sulcate, with 2 rows of minute tubercles; grooves 4–10; sheaths green with a black band at the apex; teeth broadly scarious with a dark centre; central hollow *c.* $\frac{1}{4}$ the diameter of the stem. Cone 5–7 mm. $2n = c.\ 216$. *N., W. & C. Europe, extending to S. Romania and C. Ukraine.* Au Be Br Cz Da Fa Fe Ga Ge Gr Hb He Ho Hs Hu Is It Ju No Po Rm Rs(N, B, C, W) Sb Su.

4. E. scirpoides Michx, *Fl. Bor.-Amer.* **2**: 281 (1803). Like **3** but stem up to 20 cm × 1–1·5 mm, with 3 or 4 wide, deeply sulcate ridges, so that it appears to be 6- to 8-angled; tubercles large, in 1 row on each angle; central hollow absent. $2n = c.\ 216$. *N.E. Europe, extending westwards to Norway and locally southwards to S.E. Russia.* Fe Is No Rs(N, B, C, E) Sb Su.

Subgen. **Equisetum**. Stomata not sunk below the other epidermal cells. Cones obtuse. Fertile and sterile stems sometimes dissimilar, never persistent.

5. E. fluviatile L., *Sp. Pl.* 1062 (1753) (*E. limosum* L., *E. heleocharis* Ehrh.). Rhizome glabrous. Stem 30–150 cm × 2–12 mm, simple or with irregularly whorled branches; ridges very narrow, smooth; grooves 10–30; sheaths green, closely applied to stem; teeth black, at least at the apex, not ribbed; central hollow at least $\frac{4}{5}$ the diameter of the stem. Cone 10–20 mm. $2n = c.\ 216$. *Marshes and shallow water. Most of Europe, but rare in much of the south.* Al Au Be Br Bu Cz Da Fa Fe Ga Ge Gr Hb He Ho Hs Hu Is It Ju No Po Rm Rs(N, B, C, W, E) Su.

E. × litorale Kühlew. ex Rupr., *Beitr. Pfl. Russ. Reich.* **4**: 91 (1845) (**5** × **9**) is like **5** but with stems more deeply grooved, the central hollow ⅓ the diameter of the stem, more branches, and loose sheaths with teeth black at the apex. It is recorded from numerous localities in N. & C. Europe.

6. E. palustre L., *Sp. Pl.* 1061 (1753). Rhizome glabrous. Stem 10–60 cm × 1–3 mm, simple or more often branched, the branches with rounded ridges; ridges stout, rough; grooves (4–)6–10(–12); sheaths green, rather loose; teeth green, with blackish apex and wide, scarious margin, 1-ribbed; central hollow less than ½ the diameter of the stem, about as large as the outer cavities. Cone 10–30 mm. $2n = c.\ 216$. *Marshes, bogs and other wet places. Almost throughout Europe, but rare in parts of the south.* All except Az Bl Cr Rs(K) Sa Sb Si.

7. E. sylvaticum L., *Sp. Pl.* 1061 (1753). Rhizome glabrous. Sterile stems 10–80 cm × 1–4 mm, regularly and abundantly branched, branches recurved, usually again branched; ridges nearly smooth; grooves 10–18; sheaths green below; teeth united into 3–6 wide, obtuse lobes; central hollow *c.* ½ the diameter of the main stem. Fertile stems usually shorter and less branched. Cone 15–25 mm. $2n = c.\ 216$. *Much of Europe but very rare in the Mediterranean region.* Au Be Br Bu Cz Da Fe Ga Ge Gr Hb He Ho Hs Hu Is It Ju No Po Rm Rs(N, B, C, W, E) Su.

8. E. pratense Ehrh., *Hannover. Mag.* **19**: 138 (1784). Rhizome glabrous. Sterile stems 10–60 cm × 1–2 mm, regularly and abundantly branched, branches patent or recurved, simple; ridges rough; grooves 8–20; sheaths mostly green; central hollow at least ½ the diameter of the main stem. Fertile stems usually shorter, developing short branches after the spores are ripe; sheaths numerous, yellowish-white, with 10–20 pale teeth. Cone 15–40 mm. $2n = c.\ 216$. *Woods and streamsides. N., C. & E. Europe, southwards to S.E. Jugoslavia.* Au ?Be Br Cz Da Fa Fe ?Ga Ge Hb He Is It Ju No Po Rm Rs(N, B, C, W, ?K, E) Su.

9. E. arvense L., *Sp. Pl.* 1061 (1753). Rhizome pubescent. Sterile stems 10–80 cm × (1–)3–5 mm, usually regularly and abundantly branched, branches ascending or suberect, simple, with sharp, acute ridges; grooves (4–)6–19; sheaths green; central hollow less than ½ the diameter of the main stem. Fertile stems appearing before the sterile, usually shorter, simple, pale brown, dying after the spores are shed; sheaths 4–6, pale brown, with darker teeth. Cone (4–)10–40 mm. $2n = c.\ 216$. *Often in drier habitats than the other species; a serious weed in some areas. Throughout Europe, but somewhat local in the south and east.* All territories.

In small arctic plants, the fertile stems become green, at least below, and branch irregularly after the spores are shed.

10. E. telmateia Ehrh., *Hannover. Mag.* **18**: 287 (1783) (*E. maximum* auct., *E. majus* Gars.). Rhizome pubescent. Sterile stems (15–)50–200 cm × 10–20 mm, regularly and abundantly branched, ivory-white; branches green, patent, simple; ridges smooth; grooves 20–40, fine; sheaths pale below, dark at top; central hollow at least ⅔ the diameter of the main stem. Fertile stems appearing before the sterile, shorter, simple, pale brown, dying after the spores are shed; sheaths numerous, pale brown, with dark teeth. Cone 40–80 mm. $2n = c.\ 216$. *W., C. & S. Europe, northwards to N. Scotland, and extending locally eastwards to E. Ukraine.* All except Fa Fe Is No Rs(N, C, E) Sb.

FILICOPSIDA

Measurements given for leaf-length in the Filicopsida refer to the whole leaf, including the petiole.

6. OPHIOGLOSSACEAE

EDIT. †D. H. VALENTINE AND D. M. MOORE

Rhizome short, fleshy, without scales. Leaves not circinate in bud, consisting of a sterile lamina and a fertile spike or panicle of spikes. Homosporous; sporangia large, thick-walled, sessile, without an annulus and opening by a transverse slit. Prothallus subterranean, tuber-like, saprophytic, with mycorrhiza.

Literature: R. Clausen, *Mem. Torrey Bot. Club* **19(2)**: 1–177 (1938).

Lamina linear to ovate, entire, with reticulate venation; fertile spike with sunken and coalescent sporangia **1. Ophioglossum**
Sterile and fertile parts of leaf both ±compound, the veins dichotomous, free; fertile spike with subsessile free sporangia **2. Botrychium**

1. Ophioglossum L.

†W. ROTHMALER (EDITION 1) REVISED BY J. R. AKEROYD (EDITION 2)

Lamina undivided, somewhat fleshy, with reticulate veins. Fertile spikes linear, with 2 rows of sunken and coalescent sporangia.

1 Leaves usually single; lamina at least 2 cm wide **4. vulgatum**
1 Leaves often 2 or 3 together; lamina usually less than 3 × 2 cm
2 Lamina attenuate at base; veins without free endings **1. lusitanicum**
2 Lamina cuneate at base; veins in the larger meshes with free endings
3 Leaves (1)2 or 3; lamina not apiculate **2. azoricum**
3 Leaves 1 or 2(3); lamina apiculate **3. polyphyllum**

1. O. lusitanicum L., *Sp. Pl.* 1063 (1753). Leaves 1–3 together; lamina 1–3 × 0·2–0·7 cm, erect, sessile or stalked, attenuate at the base, lanceolate to linear-lanceolate, obtuse, net-veined, with no fine secondary meshes and no included free vein-endings. Fertile spike with 5–10 sporangia on each side; spores smooth. $2n = 250–260$. *Mainly coastal. Mediterranean region and W. Europe.* Az Bl Br Co Cr Ga Gr Hs It Ju Lu Sa Si Tu.

2. O. azoricum C. Presl, *Tent. Pteridogr. Suppl.* 49 (1845) (*O. vulgatum* subsp. *ambiguum* (Cosson & Germ.) E. F. Warburg,

O. vulgatum subsp. *polyphyllum* auct., non A. Braun, *O. vulgatum* var. *islandicum* Á. & D. Löve, *O. vulgatum* var. *minus* Ostenf. & Gröntved, *O. sabulicolum* Sauzé & Maillard). Leaves (1)2 or 3 together; lamina 1·5–3(–5) × 0·5–2(–3) cm, deflexed, shortly stalked, cuneate at the base, lanceolate to ovate, acute, net-veined, with fine, oblong secondary meshes and free vein-endings. Fertile spike with 6–15(–20) sporangia on each side; spores tuberculate. $2n = 720$. ?● *W. Europe; very local in C. Europe.* Az Br Co Cz Ga Hb Hs Is It Po Sa.

3. O. polyphyllum A. Braun in Seub., *Fl. Azor.* 17 (1844). Like **2** but leaves usually 1 or 2; stipular sheaths persistent at base of petiole; lamina 2–6 × 0·5–1·8 cm, lanceolate to narrowly elliptical, apiculate; fertile spike with 15–30 sporangia on each side. *N. Portugal (Douro valley)* ?It Lu. (*Africa, S.W. Asia.*)

4. O. vulgatum L., *Sp. Pl.* 1062 (1753) (*O. alpinum* Rouy). Leaves usually single, rarely in pairs; lamina 3–15 × 2–6 cm, erect, sessile, with truncate or cordate, rarely cuneate base, broadly ovate to ovate-acuminate, net-veined, with fine, ovate or roundish secondary meshes and free vein-endings. Fertile spike with 12–45 sporangia on each side; spores tuberculate. $2n = 480$, *c.* 496, 500–520. *Usually in damp grassland. Europe, except the extreme north, but local in the Mediterranean region and much of the east.* All except Az Bl Cr Fa Is Sb Tu.

2. Botrychium Swartz

†W. ROTHMALER

Lamina usually compound, 1- to 4-pinnate, rarely simple, oblong to triangular, somewhat fleshy, with dichotomous free veins. Fertile spikes always compound, with subsessile free sporangia in 2 rows.

Most species occur in fairly dry, often acid, grassland, less often in open woodland.

1 Lamina broadly triangular, wider than long, hairy at least when young
2 Lamina fleshy, petiolate, 2- or 3-pinnate; segments obtuse **6. multifidum**
2 Lamina membranous, sessile, 3- or 4-pinnate; segments acute **7. virginianum**
1 Lamina oblong, ovate or triangular-ovate, longer than wide, almost glabrous
3 Lamina simple, 3-lobed or pinnate; pinnae with dichotomous veins and without midrib
4 Lamina petiolate, simple, 3-lobed or pinnate, with 2(–4) pairs of pinnae **1. simplex**
4 Lamina sessile, always pinnate, with (2)3–9 pairs of pinnae **2. lunaria**
3 Lamina 2-pinnate; segments with ± conspicuous midrib
5 Segments about as long as wide, triangular to rhomboid or ovate **3. boreale**
5 Segments longer than wide, lanceolate to oblong (if wider, then pinnatifid)
6 Segments obtuse, ovate to oblong, with somewhat inconspicuous midrib **4. matricariifolium**
6 Segments acute, lanceolate to linear-lanceolate, with pronounced midrib **5. lanceolatum**

1. B. simplex E. Hitchc., *Amer. Jour. Sci. Arts* **6**: 103 (1823). Leaves 2–10(–15) cm; lamina with a long petiole, simple, 3-lobed, trifoliolate or pinnate, with 2 or rarely 3 or 4 pairs of obovate or roundish, entire pinnae with rounded (rarely lobed) apex. *N. & N.C. Europe northwards to 66° N., extending very locally southwards to Corse; one station in N.W. Greece.* Au Co Cz Da Fe Ga Ge Gr He Is It Ju No Po Rs(N, B, C) Su.

2. B. lunaria (L.) Swartz in Schrader, *Jour. für die Bot.* **1800** **(2)**: 110 (1802). Leaves 5–30 cm; lamina sessile, oblong, always pinnate, with (2)3–9 pairs of trapezoid or flabellate pinnae; pinnae with semilunar-cuneate base and often semicircular, rounded, entire or rarely incised apex, without midrib. $2n = 90, 96$. *Most of Europe, but local in much of the south and east.* All except Bl Lu Sb Tu.

3. B. boreale Milde, *Nova Acta Acad. Leop.-Carol.* **26(2)**: 627, 757 (1858). Leaves 10–30 cm; lamina nearly sessile, triangular-ovate, somewhat longer than broad, 2-pinnate, with triangular, ovate to rhombic, acute, crenate pinnae tapering to the apex, and with midrib. $2n = 90$. *Arctic and subarctic Europe, extending southwards to 59° N. in Sweden.* Fe Is No Rs(N) Su.

4. B. matricariifolium (Retz.) A. Braun ex Koch, *Syn. Fl. Germ.* ed. 2, 972 (1845) (*B. ramosum* (Roth) Ascherson). Leaves 4–20 cm; lamina nearly sessile, oblong, ovate or deltate-ovate, 2-pinnate; pinnae and pinnules oblong to ovate, obtuse (but tapering to the apex), entire or crenate, with often not very pronounced midrib. $2n = c.\ 180$. *N., E. & C. Europe, extending locally southwards to Corse and N. Albania.* Al Au Co Cz Da Fe Ga Ge He †Ho Hu It Ju No Po †Rm Rs(N, B, C, W) Su.

5. B. lanceolatum (S. G. Gmelin) Ångström, *Bot. Not.* **1854**: 68 (1854). Leaves 5–25 cm; lamina nearly sessile, triangular-ovate to oblong, 2-pinnate; pinnae lanceolate to linear-lanceolate or oblong, acutely serrate or pinnatifid, with pronounced midrib. $2n = 90$. *N. Europe, southwards to 55° N. in White Russia; Alps; W. Carpathians.* Au Fe †Ga He Is It No Po Rs(N, B, C) Su.

6. B. multifidum (S. G. Gmelin) Rupr., *Beitr. Pfl. Russ. Reich.* **11**: 40 (1859) (*B. matricariae* (Schrank) Sprengel, *B. palmatum* C. Presl, *B. rutaceum* Swartz, *B. rutaefolium* A. Braun, ?*B. ternatum* (Thunb.) Swartz). Leaves 5–20 cm; lamina deflexed, triangular, wider than long, petiolate, 2- or 3-pinnate, always sparsely hairy, coriaceous; pinnae roundish or ovate, obtuse, imbricate, entire to crenate, with pronounced midrib. *N., C. & E. Europe; one station in N. Appennini.* Au Cz Da Fe Ga Ge He Hu It Ju No Po Rm Rs(N, B, C, W) Su.

7. B. virginianum (L.) Swartz in Schrader, *Jour. für die Bot.* **1800** **(2)**: 111 (1802). Leaves 20–80 cm; lamina deflexed, triangular, wider than long, sessile, membranous, sparsely hairy, 3- or 4-pinnate; pinnae and pinnules oblong to ovate, acuminate, acutely toothed and decurrent, with pronounced midrib. *N.E. Europe, extending locally to Switzerland, Romania and C. Ukraine.* Au Cz Fe He Hu Ju Po Rm Rs(N, B, C, W) Su.

7. OSMUNDACEAE

EDIT. †D. H. VALENTINE AND D. M. MOORE

Stems erect, without scales. Leaves large, pinnately divided. Homosporous; sporangia not in definite sori, with walls 1 cell thick, without an annulus and opening by a longitudinal slit. Prothallus green.

1. **Osmunda** L.

D. A. WEBB

Stems short, stout, covered by persistent leaf-bases. Leaves in a dense crown, not persistent, 2-pinnate. Sporangia globular or pyriform, massed in dense clusters on certain pinnae of the fertile fronds, where they entirely replace the photosynthetic tissue.

1. O. regalis L., *Sp. Pl.* 1065 (1753). Leaves 30–150 cm, densely tufted, only the upper (inner) ones of each year's growth being fertile; sterile pinnules 3–5 × 1–1·5 cm, oblong, obtuse, sessile, sometimes pinnatifid at the base, otherwise entire, or rarely serrulate, with conspicuous, dichotomous, free lateral veins. Sporangia confined to a few terminal pinnae of the fertile leaves, not occupying more than ¼ of the leaf; fertile pinnules *c.* 12 × 3 mm, pale green, rapidly turning brown. $2n = 44$. *Damp places; often on peat and somewhat calcifuge. Widespread in W. Europe, extending very locally eastwards to E. Sweden, Poland and Turkey; much reduced by drainage in many areas.* Al Az Be Br Bu Co Cr *Cz Da Ga Ge Gr Hb He Ho Hs Hu It Ju Lu No Po Sa Si Su Tu.

8–22

These families are sometimes placed in a single family, Polypodiaceae. They have the following characters in common. Stems usually with hairs or scales. Leaves usually relatively large. Homosporous; sporangia with walls 1 cell thick; annulus present. Prothallus green.

8. PARKERIACEAE

EDIT. D. M. MOORE

Aquatic plants. Stem dictyostelic. Leaves dimorphic, pinnatifid. Sori along the veins, occupying most of the lower surface of the fertile leaves. Indusium absent.

1. **Ceratopteris** Brongn.

J. R. AKEROYD

Plant floating or sometimes rooted in mud. Stem short, erect, with sparse, translucent, brown scales. Leaves in a rosette; margins of fertile leaves reflexed and covering sori; veins anastomosing. Annulus irregular, of many somewhat indurated cells; spores large.

Literature: R. M. Lloyd, *Brittonia* **26**: 139–160 (1974).

1. C. thalictroides (L.) Brongn., *Bull. Sci. Soc. Philom. Paris* ser. 3, **8**: 186 (1821) (*Acrostichum thalictroides* L.). Leaves 10–40 cm, 2- or 3-pinnatifid, triangular, often proliferous, bearing plantlets in the axils of the pinnae; petioles inflated, with a network of air-filled spaces. Fertile leaves more deeply dissected with linear segments. *Cultivated in aquaria; naturalized in warm springs in S. Austria (Villach). (Tropics and subtropics.)* [Au ?Ju.]

9. ADIANTACEAE

(Incl. *Cryptogrammaceae*, *Gymnogrammaceae*, *Sinopteridaceae*)

EDIT. D. M. MOORE

Rhizome solenostelic or rarely dictyostelic, with opaque scales. Petiole usually dark and shining, with 1 or 2 vascular strands at the base which divide distally. Sori usually submarginal and often borne on or covered by the deflexed leaf-margin which forms a pseudoindusium; spores tetrahedral or globose.

Species of *Pityrogramma*, a genus with leaves 1- to 3-pinnatisect, triangular-lanceolate and farinose beneath, are cultivated for ornament. At least one of these, **P. calomelanos** (L.) Link, *Handb.* **3**: 20 (1833), from tropical America and Africa, has been recorded as naturalized in Açores.

1 Fertile and sterile leaves distinct
  2 Fertile leaf-margins deflexed; sori ± marginal **7. Cryptogramma**
  2 Fertile leaf-margins not deflexed; sori superficial **5. Anogramma**
1 Fertile and sterile leaves similar
  3 Leaf-segments flabellate, ± toothed **6. Adiantum**
  3 Leaf-segments not flabellate (sometimes ± triangular), entire
    4 Leaves glabrous **3. Pellaea**
    4 Leaves hairy, scaly or glandular, at least below

5 Leaves elliptical, tapered towards base; upper surface of lamina with filiform scales (woolly hairs); spores tetrahedral **4. Cosentinia**

5 Leaves lanceolate to deltate, truncate at base; upper surface of lamina glabrous; spores globose

6 Lower surface of leaf glabrous or with filiform scales or glandular hairs **1. Cheilanthes**

6 Lower surface of leaf with flat, triangular scales **2. Notholaena**

1. Cheilanthes Swartz

A. C. JERMY AND H. P. FUCHS (EDITION 1) REVISED BY A. C. JERMY AND A. M. PAUL (EDITION 2)

Rhizome more or less erect, often branched. Leaves tufted, more or less erect; petiole rigid, scaly, at least at the base; lamina finely divided, 2-pinnate, the ultimate segments with no obvious midrib; veins free; lamina glabrous or with sparse indumentum beneath. Margin of fertile segments deflexed and modified to form a continuous or interrupted, usually membranous pseudoindusium (sometimes only seen in young growth); sori *c.* 0·5 mm, marginal on ends of veins; spores globose.

Most species grow in crevices of dry, siliceous rocks and walls.

1 Lamina with reddish, stalked glands or glandular hairs beneath

2 Glandular hairs 8- to 14-celled, densely covering lamina beneath; lamina deltate **5. hispanica**

2 Glandular hairs 2- to 5-celled, sparse; lamina ovate-lanceolate **6. tinaei**

1 Lamina without reddish glands or hairs beneath

3 Ultimate segments ovate to orbicular

4 Pseudoindusium entire; rhachis and costae ±glabrous **1. maderensis**

4 Pseudoindusium with ciliate margin; rhachis and costae scaly

5 Pseudoindusium broad, irregular with short cilia **4. acrostica**

5 Pseudoindusium narrow, densely long-ciliate **7. persica**

3 Ultimate segments linear to lanceolate

6 Sori well-spaced, on rounded segment-lobes **1. maderensis**

6 Sori ±continuous along margin of segment

7 Ultimate segments linear-oblong, crenate; pseudoindusium interrupted **2. guanchica**

7 Ultimate segments linear, entire; pseudoindusium continuous **3. pulchella**

1. C. maderensis Lowe, *Trans. Cambr. Philos. Soc.* **6**: 528 (1838) (*C. fragrans* (L.) Swartz, *C. pteridioides* (Reichard) C. Chr.). Rhizome-scales brown. Leaves up to 18 cm, smelling of coumarin; petiole usually as long as or slightly shorter than the lamina, reddish-brown, shiny, with scattered scales; lamina 1–3 cm wide, subtriangular to linear-lanceolate, 2-pinnate for most of its length, with sessile glands beneath; lower pinnules pinnatifid with ultimate segments oblong or suborbicular, lobed. Sori well spaced; pseudoindusium narrow, entire with rounded lobes; spores (33–)36–42(–45) μm, finely papillose. $2n = 60$. *S. Europe.* Az Co Cr Ga Gr Hs It Lu Sa Si.

2. C. guanchica C. Bolle, *Bonplandia* **7**: 107 (1859) (*C. fragrans* auct. pro parte, non (L.) Swartz). Like **1** but rhizome-scales reddish-brown; leaves up to 30 cm, not smelling of coumarin; petiole with scales mostly near the base; lamina up to 6 cm wide, deltate-lanceolate, eglandular, glabrous; ultimate segments of pinnules linear-oblong, lobed below, crenate above; sori more or less continuous; pseudoindusium broad, subrectangular, somewhat toothed; spores (31–)42–45(–52) μm, reticulately ridged. $2n = 120$. *Mediterranean region; Portugal.* Co ?Ga Gr Hs It Lu Sa.

3. C. pulchella Bory ex Willd., *Sp. Pl.* **5(1)**: 456 (1810). Like **1** but rhizome bearing narrow, dark brown scales. Leaves up to 40 cm, not smelling of coumarin; petiole as long as or slightly longer than the lamina, with scattered scales towards the base; lamina up to 10 cm wide, ovate-lanceolate, glabrous, eglandular; ultimate segments linear, entire or sometimes lobed at the base, the apical segment usually longer than all acroscopic pinnules; pseudoindusium broad, continuous along sides of ultimate segments, the margin dentate to short-ciliate; spores (36–)39–42(–45) μm, minutely papillose. *Spain.* ?Hs. (*Islas Canarias.*)

4. C. acrostica (Balbis) Tod., *Gior. Sci. Nat. Econ. Palermo* **1**: 215 (1866) (*C. fragrans* auct. pro parte, non (L.) Swartz, *C. odora* Swartz). Rhizome-scales reddish-brown. Leaves up to 25 cm, not smelling of coumarin; petiole usually as long as or slightly shorter than the lamina, with abundant scales throughout its length, about as long as the lamina; lamina up to 4 cm wide, ovate to lanceolate, 2-pinnate for most of its length, eglandular, with scales on the rhachis and costae; ultimate segments of pinnules suborbicular. Pseudoindusium broad, irregularly lobed, ciliate; spores (45–)48–51(–54) μm, granulose. $2n = 120$. *Mediterranean region; Portugal, Açores.* Al Az Bl Co Cr Ga Gr Hs It Ju Lu Sa Si.

5. C. hispanica Mett., *Abh. Senckenb. Naturf. Ges.* **3**: 74 (1859). Like **4** but leaves up to 26 cm, the petiole 2–4 times as long as the lamina, with tuft of dark scales at base; lamina up to 6 cm wide, deltate, with dense 8- to 14-celled glandular hairs beneath; scales absent; ultimate segments ovate to suborbicular, lobed; pseudoindusium narrow, with rounded lobes entire, more or less lacking a membranous margin; spores (33–)36–39(–42) μm, striate. $2n = 60$. *S.W. Europe.* Ga Hs Lu Sa.

6. C. tinaei Tod., *Gior. Sci. Nat. Econ. Palermo* **1**: 217 (1866) (*C. corsica* Reichstein & Vida). Like **4** but leaves up to 30 cm, smelling of coumarin, the petiole 1–2 times as long as the lamina, with scattered scales; lamina up to 5 cm wide, ovate-deltate or lanceolate-deltate, with sparse 2- to 5-celled glandular hairs beneath; pinnules ovate-lanceolate, lobed; pseudoindusium with rounded lobes and narrow, entire membranous margin; spores (45–)48–51(–57) μm. $2n = 120$. *W. Mediterranean region; Portugal.* Co Ga ?Gr Hs It Lu Sa Si.

7. C. persica (Bory) Mett. ex Kuhn, *Bot. Zeit.* **26**: 234 (1868) (*C. szovitsii* Fischer & C. A. Meyer). Rhizome covered with narrow, dark brown scales. Leaves up to 25 cm, not smelling of coumarin; petiole about as long as the lamina, with many narrow, brown scales; lamina up to 4 cm wide, oblong or linear-lanceolate, 2-pinnate for most of its length, eglandular, with scales on the rhachis and costae; ultimate segments oblong to suborbicular, lobed. Pseudoindusium narrow, with many long marginal hairs which cover the lower surface of the segment; spores (33–)36–42(–45) μm, minutely papillose. $2n = 60$. *S.E. Europe.* Al Bu Cr Gr It Ju Rs(K).

2. **Notholaena** R. Br.

A. C. JERMY AND A. M. PAUL

Rhizome short-creeping and densely scaly. Leaves crowded, pinnate-pinnatisect to 2-pinnate, the ultimate segments with well-defined midribs; veins free; lamina densely covered beneath with scales or hairs; petiole rigid and scaly. Margin of fertile segments slightly deflexed with a narrow, membranous edge, forming a rudimentary pseudoindusium; sori not discrete, the sporangia submarginal, spreading inwards along veins; spores globose.

1. N. marantae (L.) Desv., *Jour. Bot. Appl.* **1**: 92 (1813) (*Cheilanthes marantae* (L.) Domin, *Paraceterach marantae* (L.) R. M. Tryon jun.). Rhizome densely covered with light brown scales. Leaves 10–35 cm, arising in groups, erect; petiole dark reddish-brown, dull, with scattered linear-filiform scales; lamina 2–5 cm wide, linear-lanceolate, pinnate-pinnatisect or, more rarely, 2-pinnate, glabrous above, densely clothed with light brown or colourless, attenuate-triangular scales beneath. Spores *c.* 48 μm, spherical, lophoreticulate. $2n = 58$. *S. Europe, extending northwards to S. Czechoslovakia.* Al Au Bu Co Cz Ga Gr He Hs Hu It Ju Rm Rs(K).

3. **Pellaea** Link

†D. H. VALENTINE (EDITION 1) REVISED BY J. R. AKEROYD AND A. M. PAUL (EDITION 2)

Rhizome short. Leaves ovate-lanceolate, 2- or 3-pinnate, glabrous, coriaceous; petiole blackish-brown, longer than the lamina. Sori in contact laterally but not confluent, forming an almost continuous marginal line, protected by a continuous pseudoindusium; spores tetrahedral or globose.

Leaves 2-pinnate; pinnules stalked, less than twice as long as wide **1. calomelanos**

Leaves 2- or 3-pinnate; pinnules sessile or subsessile, at least twice as long as wide **2. viridis**

1. P. calomelanos (Swartz) Link, *Fil. Sp.* 61 (1841). Leaves 10–25 cm; pinnules stalked, shallowly cordate, triangular, 3- or 5-angled, shaped like the leaf of *Hedera helix*. *Siliceous rocks. Açores; N.E. Spain (Prov. Gerona).* *Az Hs. (*S. & E. Africa, Asia.*)

2. P. viridis (Forskål) Prantl, *Bot. Jahrb.* **3**: 420 (1882). Leaves 15–60 cm, 2- or 3-pinnate; pinnules sessile or subsessile, lanceolate to ovate, cuneate to somewhat truncate at the base. *Naturalized in Açores.* [Az.] (*Old World Tropics.*)

4. **Cosentinia** Tod.

A. C. JERMY AND A. M. PAUL

Rhizome short-creeping to erect, densely scaly. Leaves tufted, lanuginose above, lamina 2-pinnate, the ultimate segments with no obvious midrib; veins free; lamina densely covered beneath, more sparsely above, with a woolly indumentum of white to brown multicellular hairs; petiole rigid, scaly at base. Margin of fertile segments usually deflexed, but not membranous or modified; sori not discrete, the sporangia mostly submarginal, spreading inwards along veins; spores tetrahedral.

1. C. vellea (Aiton) Tod., *Gior. Sci. Nat. Econ. Palermo* **1**: 220 (1866) (*Notholaena vellea* (Aiton) Desv., *N. lanuginosa* (Desf.) Desv. ex Poiret, *Cheilanthes vellea* (Aiton) F. Mueller, *C. catanensis* (Cosent.) H. P. Fuchs). Rhizome short; apex covered with long, narrow, pale brown scales. Leaves 8–35 cm, suberect or spreading, forming a crown; petiole usually very short (up to 3 cm), yellowish-brown; lamina 2–4 cm wide, linear-lanceolate, lower pinnae widely spaced and greatly reduced; pinnules crenately lobed, oblong or sub-orbicular; indumentum may be lost from the upper surface as the frond ages. Spores *c.* 62 μm, finely reticulate. *Mediterranean region; Portugal.* Bl Co Cr Ga Gr Hs It Lu Sa Si.

5. **Anogramma** Link

†T. G. TUTIN

Sporophyte annual; gametophyte said to be perennial. Stock very short; scales few. Leaves somewhat dimorphic, 2- or 3-pinnate, the margins flat; veins free. Sori spread along the veins; indusium absent; spores tetrahedral.

1. A. leptophylla (L.) Link, *Fil. Sp.* 137 (1841) (*Gymnogramma leptophylla* (L.) Desv., *Grammitis leptophylla* (L.) Swartz). Stock with a few scales when young. Leaves slightly hairy when young, soon becoming glabrous; outer leaves sterile, much shorter than the fertile leaves and with thin, obovate-cuneate pinnules, lobed or dentate at the apex; fertile leaves 3–20 cm, with narrowly obovate-cuneate pinnules. Sori appearing confluent when mature. $2n = 52$. *Damp shady places. S. & W. Europe, northwards to the Channel Islands.* Az Bl Co Cr Ga Gr He Hs It Ju Lu Rs(K) Sa Si.

6. **Adiantum** L.

A. LAWALRÉE (EDITION 1) REVISED BY J. R. AKEROYD AND A. M. PAUL (EDITION 2)

Rhizome short-creeping, with narrow, brown scales. Petiole usually with scales at the base only; lamina wide, pinnately compound; veins free. Sori in groups of 2–10 on the deflexed margins of the lower side of the pinnules (pseudoindusia), along and between the parallel veins; spores tetrahedral.

1 Leaves pedately branched, subcoriaceous, hairy **3. hispidulum**

1 Leaves flabellately branched, thin, glabrous

2 Veins of sterile segments ending in teeth; pseudo-indusium ±oblong, between sinuses **1. capillus-veneris**

2 Veins of sterile segments ending in sinuses; pseudo-indusium reniform, around a sinus **2. raddianum**

1. A. capillus-veneris L., *Sp. Pl.* 1096 (1753). Leaves 10–60 cm, 2- or 3-pinnate, flabellately branched, ovate to ovate-lanceolate; petiole up to 25 cm, black and shining; lamina bright green, of thin, delicate texture, glabrous. Pinnules (0·5–)1–3 cm wide, cuneate, flabellate, on fine capillary stalks, obtusely lobed, variable in depth of lobing; veins of sterile segments ending in teeth. Pseudoindusia more or less oblong, between sinuses. $2n = 60$. *Damp rocks, often on calcareous tufa. W. & S. Europe.* Al Az Bl Br Bu Co Cr Ga Gr Hb He Hs It Ju Lu Rs(K) Sa Si Tu [Be Hu].

2. A. raddianum C. Presl, *Tent. Pteridogr.* 158 (1836) (*A. cuneatum* Langsd. & Fischer). Like **1** but the leaves 3- or 4-pinnate; pinnules 0·5–1 cm wide, usually less deeply lobed; veins of sterile segments ending in sinuses; pseudoindusia reniform, around sinus. *Cultivated for ornament, and sometimes naturalized in milder regions.* [Az Lu.] (*E. South America.*)

3. A. hispidulum Swartz in Schrader, *Jour. für die Bot.* **1800**: 82 (1801). Leaves 15–60 cm, pedately branched, each branch with numerous alternate, obliquely oblong segments 0·8–2 × 0·3–0·6 cm, sharply dentate, subcoriaceous, hairy, especially below; petiole dark brown, somewhat shining; veins of sterile segments ending in teeth. Pseudoindusia reniform and bearing stiff, brown hairs. Rhachis tomentose to hispid. *Naturalized in Açores.* [Az.] (*Tropics and austral subtropics.*)

7. Cryptogramma R. Br. ex Richardson

A. LAWALRÉE

Rhizome creeping or ascending. Leaves 2- or 3-pinnate, with free veins, dimorphic, the sterile with wide and the fertile with narrower divisions; sori submarginal, becoming confluent, borne on ends of veins and covered by the continuous deflexed membranous leaf-margins (pseudoindusia). Spores tetrahedral, translucent.

Onychium japonicum (Thunb.) G. Kunze, *Bot. Zeit.* **1848**: 507 (1848), with dimorphic, 4-pinnate leaves 30–80 cm, is superficially similar to *Cryptogramma*. It is perhaps becoming naturalized in Açores.

Leaves tufted, 3-pinnate **1. crispa**
Leaves arising singly, 2-pinnate **2. stelleri**

1. C. crispa (L.) R. Br. ex Hooker, *Gen. Fil.* t. 115B (1842) (*Allosorus crispus* (L.) Röhling). Leaves 15–30 cm, up to 7 cm wide, tufted, ovate to triangular-ovate, 3-pinnate, bright green, dying down in winter; petiole twice as long as the lamina; ultimate segments of sterile leaves with cuneate base, narrowly ovate and coarsely and irregularly toothed; ultimate segments of fertile leaves 1–2 mm wide, linear. $2n = 120$. *Mountain rocks and screes; calcifuge. Much of Europe but rare in the east; only in mountain districts except in the north.* Al Au Be Br Bu Co Cz Fe Ga Ge Hb He Hs Is It Ju Lu No Po ?Rm Rs(N) Su.

2. C. stelleri (S. G. Gmelin) Prantl, *Bot. Jahrb.* **3**: 413 (1882). Leaves arising singly from a slender, creeping rhizome, 8–20 × 2–5 cm wide, 2-pinnate, ovate to ovate-elongate; petiole equalling or rather longer than the lamina; ultimate segments of sterile leaves similar to those of **1**, those of the fertile leaves 2–3 mm wide, linear-lanceolate. *Rock-crevices in shady places. N.E. Russia.* Rs(N, C). (*Arctic and subarctic Asia; N. America.*)

10. PTERIDACEAE

EDIT. †D. H. VALENTINE AND D. M. MOORE

Rhizome solenostelic or dictyostelic, with opaque scales. Petiole with a single U-shaped strand. Sori submarginal, forming coenosori which are borne on the veinlet connecting the vein ends, and which are covered by the scarious, deflexed leaf-margins; spores tetrahedral, opaque.

1. Pteris L.

T. G. WALKER

Rhizome short. Leaves tufted, pinnate or 2-pinnate. Sori with paraphyses.

1 Pinnae pinnatisect **1. incompleta**
1 Pinnae simple
 2 Not more than 7 pairs of pinnae; lamina ovate **2. cretica**
 2 10 or more pairs of pinnae; lamina lanceolate **3. vittata**

1. P. incompleta Cav., *Anal. Ci. Nat.* **2**: 107 (1801) (*P. serrulata* Forskål, non L. fil., *P. arguta* Aiton). Rhizome creeping; scales long, narrow, dark brown. Petiole up to *c.* 50 cm, brown; lamina up to *c.* 100 × 40 cm, ovate; pinnae 7–9(–11) pairs, pinnatisect, the lowest pair often forked, all pinnatisect; segments oblong-lanceolate, falcate, serrate in distal $\frac{1}{2}$, acute. Sori occupying lower $\frac{1}{2}$–$\frac{2}{3}$ of the segments, not extending to the base of the sinus; indusium prominent, thick. *Açores; two stations in S.W. Spain.* Az Hs [Lu].

P. multifida Poiret in Lam., *Encycl. Méth. Bot.* **5**: 714 (1804) (*P. serrulata* L. fil., non Forskål), from China and Japan, with oblong-linear leaf-segments, which are serrate with sori occupying the entire margin except for the apex, and **P. tremula** R. Br., *Prodr. Fl. Nov. Holl.* 154 (1810), from Australia, New Zealand and Fiji, with 2- to 4-pinnate leaves up to 90 × 60 cm, are reported from Açores as escapes from cultivation, and may be becoming naturalized.

2. P. cretica L., *Mantissa* 130 (1767). Rhizome short-creeping; scales small, dark brown. Petiole up to *c.* 30 cm, light brown; lamina up to *c.* 40 cm, ovate; pinnae 1–7 pairs, linear, simple (except for lowest 1 or 2 pairs which may be forked), serrate at the apex; sterile pinnae serrulate, slightly wider than fertile. Sori occupying entire margin except for the apex; indusium reflexed and inconspicuous when mature. $2n = 58$. *C. Mediterranean region and N. Italy; one station in S. Greece; often cultivated for ornament, and locally naturalized.* Co ?Cr Ga Gr He It *Sa Si [Az Br Ho *Hs Hu].

3. P. vittata L., *Sp. Pl.* 1074 (1753) (*P. longifolia* auct., non L., *P. ensifolia* Poiret). Like **2** but petiole not more than 10 cm; lamina up to 60 cm, lanceolate, with 10 or more pairs of pinnae; pinnae always simple, cordate at the base. *Mediterranean region.* ?Bl Cr Gr Hs It Si [Az Ga *He Hu].

11. HYMENOPHYLLACEAE

EDIT. †D. H. VALENTINE AND D. M. MOORE

Plants small. Rhizome slender, extensively creeping. Lamina dark, translucent green, only 1 cell thick except on the veins. Sori gradate, marginal; indusium tubular or 2-valved, symmetrical; spores tetrahedral.

A predominantly austral and tropical family. Delimitation of species is difficult, and the plants from Europe are held by some authors to be conspecific with those in the tropics or the southern hemisphere; as here delimited, they are believed to extend outside Europe only to Madeira and Islas Canarias.

Literature: E. B. Copeland, *Philipp. Jour. Sci.* (*Bot.*) **67**: 1–110 (1938).

Rhizome filiform; indusium of 2 distinct valves; receptacle included **1. Hymenophyllum**
Rhizome 2–4 mm in diameter; indusium tubular; receptacle exserted **2. Trichomanes**

1. Hymenophyllum Sm.

D. A. WEBB

Plants with the habit of bryophytes, and often growing with them. Rhizomes filiform, less than 1 mm in diameter, smooth. Leaves more or less procumbent, persistent after withering, forming a dense mat of foliage, deeply pinnatisect, the segments asymmetrical, subentire on their proximal side, deeply lobed on their distal side with 2–15 oblong, obtuse, dentate lobes. Sori solitary, more or less globose, situated near the base of the distal segments of the lamina; indusium of 2 ovate or suborbicular valves, united only at the base.

Literature: E. B. Copeland, *Philipp. Jour. Sci.* (*Bot.*) **64**: 1–188 (1937).

Valves of indusium dentate; leaves bluish-green **1. tunbrigense**
Valves of indusium entire; leaves somewhat olive-green **2. wilsonii**

1. H. tunbrigense (L.) Sm. in Sowerby, *Engl. Bot.* **3**: t. 162 (1794). Lamina usually 22–45 × *c*. 15 mm, flat, somewhat glossy, slightly bluish-green. Indusium-valves suborbicular, fimbriate-dentate in the distal $\frac{1}{2}$, nearly flat. $2n = 26$. *Woods and other damp, shady, sheltered places. Locally in W. Europe from N. Spain to Scotland; Açores; a few isolated stations eastwards to N. Italy.* Az Be Br †Cz Ga Ge Hb Hs It †Ju.

2. H. wilsonii Hooker, *Brit. Fl.* 450 (1830) (*H. unilaterale* auct., vix Bory). Like **1** but with usually narrower leaves, which are deflexed laterally and therefore appear even narrower, darker, somewhat olive-green; indusium-valves ovate, entire, convex. $2n = 36$. *Similar situations to* **1** *but more tolerant of exposure. N.W. Europe; Açores.* Az Br Fa Ga Hb No.

2. Trichomanes L.

D. A. WEBB

Like *Hymenophyllum*, but with stouter, more hairy rhizomes, larger, erect, more compoundly pinnatisect leaves, and elongate, cylindrical sori, of which the basal part is protected by tubular, slightly 2-lipped indusium, the apical part eventually protruding and persisting in the form of a bristle-like receptacle after the sporangia have fallen off.

Copeland, *Philipp. Jour. Sci.* (*Bot.*) **64**: 1–188 (1937), has divided this large genus into numerous smaller genera; in this treatment the European species is assigned to the genus *Vandenboschia* Copel.

1. T. speciosum Willd., *Sp. Pl.* **5(1)**: 514 (1810) (*T. radicans* auct., non Swartz). Rhizome 2–4 mm in diameter, covered with black hairs. Leaves 20–35(–50) cm, the winged petiole about equalling the ovate, deeply 3- or 4-pinnatisect lamina; ultimate segments oblong, obtuse, entire, disposed symmetrically, not unilaterally as in *Hymenophyllum*. $2n = 144$. *By waterfalls, at mouths of caves, and in similar damp, dark situations. Extreme west of Europe, eastwards to the Pyrenees.* Az Br Ga Hb Hs It [Lu].

12. POLYPODIACEAE

EDIT. †D. H. VALENTINE AND D. M. MOORE

Rhizome polystelic or solenostelic, with opaque scales. Leaves 2-ranked on the upper side of the rhizome and articulated to it; petiole with 1–3 major vascular strands. Sori superficial, without indusium; spores bilateral.

1. Polypodium L.

†D. H. VALENTINE (EDITION 1) REVISED BY J. R. AKEROYD AND A. C. JERMY (EDITION 2)

Rhizome creeping. Leaves uniform; lamina pinnatifid or pinnate, glabrous, without scales; pinnae entire or dentate, the veins free. Sori orbicular or elliptical, without indusium.

Literature: M. G. Shivas, *Jour. Linn. Soc. London* (*Bot.*) **58**: 13–38 (1961). R. H. Roberts, *Fern Gaz.* **12**: 69–74 (1980).

In Europe the genus consists of a polyploid complex in which four species can be recognized. These all have deeply pinnatifid leaves up to 50 cm or more in length; the lamina is longer than the petiole, and the pinnae are usually alternate, with 5–28 on each side. The sori are arranged in 1 row on each side of the midrib of the pinna. All the species grow on rocks and walls, or, in the wetter parts of Europe, as epiphytes on trees.

1 Some rhizome-scales at least 10 mm; sori with paraphyses
2 Sori with some large, branched paraphyses; pinnae doubly serrate **1. cambricum**
2 Sori with few small, unbranched or short-branched paraphyses (often difficult to observe); pinnae simply serrate **2. macaronesicum**
1 Longest rhizome-scales not more than 10 mm; sori without paraphyses
3 Annulus with (7–)11–14(–18) indurated cells; ripe annulus reddish-brown **3. vulgare**
3 Annulus with (4–)7–10(–13) indurated cells; ripe annulus usually colourless or pale brown **4. interjectum**

1. P. cambricum L., *Sp. Pl.* 1086 (1753) (*P. australe* Fée, *P. serratum* (Willd.) Sauter, non Aublet). Rhizome-scales 5–15 mm, linear-lanceolate to lanceolate. New leaves produced in autumn; leaves ovate to triangular-ovate; pinnae narrow, acute, doubly serrate, the basal pair projecting forward; secondary veins with 3 or 4(–6) bifurcations. Paraphyses (400–)600–1200(–1800) μm, branched; annulus with 4–10 indurated cells. $2n = 74$. *Calcicole. S. & W. Europe.* Al Bl Br Bu Co Cr Ga Gr Hb He Hs It Ju Lu Rs(K) Sa Si Tu.

2. P. macaronesicum Bobrov, *Bot. Žur.* **49**: 540 (1964) (*P. cambricum* subsp. *macaronesicum* (Bobrov) Fraser-Jenkins). Like **1** but rhizome-scales often distinctly toothed; pinnae simply serrate; paraphyses 200–400(–800) μm, simple or subsimple, often difficult to observe; annulus with (7–)9–20 indurated cells. *Often an epiphyte on trees, and on walls. S.W. Spain; Açores.* Az Hs. (*Madeira, Islas Canarias.*)

The plants from Açores have thicker, very broad, often almost suborbicular leaves, and have been treated as a separate species, **P. azoricum** (Vasc.) R. Fernandes, *Bol. Soc. Brot.* ser. 2, **42**: 242 (1968) (*P. cambricum* subsp. *azoricum* (Vasc.) Nardi).

3. P. vulgare L., *Sp. Pl.* 1085 (1753). Rhizome-scales 3–6 mm, ovate, acuminate. New leaves produced in early summer; leaves lanceolate; pinnae obtuse or somewhat acute, the basal pair not projecting forward; secondary veins with 1–3(4) bifurcations. Sori orbicular; paraphyses absent; annulus with 10–15(-20) indurated cells, reddish-brown. $2n = 148$. *Most of Europe, but local in the east.* All except Bl Cr Sb Si.

4. P. interjectum Shivas, *Jour. Linn. Soc. London (Bot.)* **58**: 28 (1961) (*P. vulgare* subsp. *prionodes* (Ascherson) Rothm.). Rhizome-scales 3–6(–10) mm, ovate-lanceolate. New leaves produced in summer; leaves ovate to ovate-lanceolate; pinnae acute, the basal pair projecting forward; secondary veins with 3(4) bifurcations. Sori elliptical; paraphyses absent; annulus with 7–10 indurated cells, usually colourless or pale brown. $2n = 222$. *W. & C. Europe, northwards to Scotland and eastwards to the Baltic region.* Au Be Bl Br Co Cz Da Ga Ge Hb He Ho Hs Hu It Ju Lu Rs(B, ?C, W, ?K) Sa ?Tu.

Some of the records from E.C. & E. Europe may refer to variants of **1** with unusually short paraphyses.

13. GRAMMITIDACEAE

EDIT. D. M. MOORE

Description as for genus.

A predominantly austral and tropical family.

1. Grammitis Swartz

J. R. AKEROYD

Epiphyte. Rhizome dictyostelic. Leaves simple, entire; veins free, branched. Sori discrete, in a line along the costa, without indusium.

1. G. jungermannioides (Klotzsch) R.-C. Ching, *Bull. Fan Mem. Inst. Biol. Bot. (Peking)* **10**: 240 (1941). Rhizome erect, without scales. Leaves 3–5 cm, narrowly oblong to narrowly obovate, obtuse, somewhat coriaceous, with soft, spreading hairs on both surfaces. Sori orbicular. *Mossy forest. One station in Açores (Pico).* *Az. (*S. Central America and Caribbean region.*)

14. DICKSONIACEAE

(Incl. *Cyatheaceae*)

EDIT. †D. H. VALENTINE AND D. M. MOORE

Stems solenostelic, stout, densely covered with scales or hairs. Petiole with a single, U-shaped vascular strand which soon divides into several strands. Sori marginal, covered by a 2-valved indusium; spores tetrahedral.

Dicksonia antarctica Labill., *Nov. Holl. Pl.* **2**: 100 (1806), from S.E. Australia, with an erect stem up to 2 m and large, 2- or 3-pinnate leaves, is cultivated for ornament in S.W. England and S.W. Ireland. It reproduces freely by spores and persists in abandoned gardens, but does not appear to be truly naturalized.

Another tree-fern, **Cyathea cooperi** (Hooker & F. Mueller) Domin, *Pterid.* 262 (1929), from Australia, is perhaps becoming naturalized in Açores.

1. Culcita C. Presl

†D. H. VALENTINE

Leaves pinnately divided, uniform. Sori 2–4 mm, more or less spherical; sporangia maturing basipetally; annulus slightly oblique.

Literature: M. C. de Rezende-Pinto, *Bol. Soc. Brot.* ser. 2, **17**: 93–140 (1943).

1. C. macrocarpa C. Presl, *Tent. Pteridogr.* 135 (1836) (*Dicksonia culcita* L'Hér.). Rhizome with setiform, ferruginous scales. Leaves 30–100(–200) cm, as long as or rather longer than wide, in a lax tuft; lamina triangular, 4- or 5-pinnate, coriaceous, shining, about as long as the petiole; ultimate segments inciso-serrate. Indusium reniform. $2n = 132–136$. *Açores; W. Spain and N.W. Portugal, very local.* Az Hs *Lu.

15. HYPOLEPIDACEAE

(Incl. *Dennstaedtiaceae*)

EDIT. D. M. MOORE

Rhizome with a perforated solenostele. Petiole with several vascular strands which fuse to form a single, U-shaped strand. Sori marginal, borne on the connecting veinlet, covered by both the deflexed margin of the leaf and by an inner indusium; spores tetrahedral.

1. Pteridium Gled. ex Scop.

†D. H. VALENTINE (EDITION 1) REVISED BY C. N. PAGE (EDITION 2)

Rhizome subterranean, far-creeping. Leaves distinct; petiole long; lamina up to 3-pinnate; pinnae and pinnules with nectaries at the base; ultimate segments numerous, ovate to linear. Sori contiguous, on the marginal vein, covered both by the revolute leaf-margins and by the indusia.

Literature: C. N. Page, *Bot. Jour. Linn. Soc.* **73**: 1–34 (1976); *Watsonia* **17**: 429–434 (1989).

1. P. aquilinum (L.) Kuhn in Kersten, *Reise Ost. Afr. Bot.* **3(3)**:

11 (1879) (*P. tauricum* (C. Presl) V. Krecz., *Pteris aquilina* L.). Leaves 40–400 cm, 2-pinnate or more usually 3-pinnate-pinnatifid, more or less woolly to glabrous on the lower surface, glabrous or subglabrous on the upper, the young unfurling fronds (croziers) with filiform, white (colourless) scales and usually with red hairs; petiole tomentose at the base, remaining so or becoming glabrous throughout; pinnae orientated horizontally; pinnules shortly acuminate to obtuse. Indusium ciliate. $2n = 104$. *Usually calcifuge. Almost throughout Europe but local in the east.* All except Fa Is Sb.

A cosmopolitan species with three or possibly more subspecies in Europe.

1 Croziers without red hairs; rhachis ±erect (c) subsp. **atlanticum**
1 Croziers with red hairs; rhachis inclined
2 Petiole fleshy, green; croziers with few red hairs (a) subsp. **aquilinum**
2 Petiole wiry, reddish; croziers with dense red hairs (b) subsp. **latiusculum**

(a) Subsp. **aquilinum**: Croziers with white filiform scales and few red hairs. Petiole fleshy, mucilaginous, green; rhachis inclined slightly further at each successive pair of pinnae; lamina oblong-lanceolate to ovate-triangular; pinnae arising at angle of *c.* 75°. *Calcifuge. Throughout the range of the species.*

(b) Subsp. **latiusculum** (Desv.) C. N. Page, *Watsonia* **17**: 429 (1989) (*P. aquilinum* var. *latiusculum* (Desv.) Underw. ex A. A. Heller): Croziers with dense red hairs and few, white, filiform scales. Petiole wiry, rigid, reddish; rhachis strongly inclined at position of first pair of pinnae; lamina broadly triangular; pinnae arising at angle of *c.* 60°. *Calcifuge; native conifer woodland. N.C. & N. Europe.*

(c) Subsp. **atlanticum** C. N. Page, *Watsonia* **17**: 431 (1989): Croziers with dense white, filiform scales, without red hairs. Petiole succulent, mucilaginous, pale green; rhachis more or less vertical; lamina ovate-triangular; pinnae spreading nearly at right-angles. *Calcicole; limestone grassland near coasts. W. Scotland.*

Subsp. **brevipes** (Tausch) Wulf, *Fl. Kryma* **1**: 20 (1927), with tough, stiff, pale-green petioles, coriaceous lamina, thickly lanate beneath, few pinnae, and with croziers lacking red hairs, has been described from Kriti and Krym. Its identity is uncertain.

16. THELYPTERIDACEAE

EDIT. †D. H. VALENTINE AND D. M. MOORE

Rhizome dictyostelic, with hairs, or with papillate or hairy scales. Hairs on leaves and rhizomes always unicellular, acicular. Petiole with U-shaped vascular strand; rhachis grooved but not interrupted to admit grooves of costae. Sori superficial, median or submarginal; indusium usually absent or caducous; spores bilateral.

Literature: R. E. Holttum, *Acta Bot. Malac.* **8**: 47–58 (1983).

1 At least the lowermost veins from adjacent leaf-segments uniting some way below the sinus **5. Christella**
1 All veins free, or the lowermost veins from adjacent leaf-segments just meeting at the sinus
2 Lamina covered in yellow glands beneath, smelling of lemon; lower pinnae markedly decrescent **2. Oreopteris**
2 Lamina not very glandular; lower pinnae not decrescent
3 Lamina not, or very sparsely, hairy above **1. Thelypteris**
3 Lamina hairy above
4 Lamina deltate-ovate, the lowermost pinnae deflexed; costules not furrowed on upper side **3. Phegopteris**
4 Lamina lanceolate, the lowermost pinnae not deflexed; costules grooved above **4. Stegnogramma**

1. Thelypteris Schmidel

A. C. JERMY AND A. M. PAUL

Rhizome creeping or ascending. Leaves 1-pinnate with pinnatifid pinnae, or 2-pinnatisect, with few, sparse hairs beneath; veins free. Sori orbicular, more or less discrete, near the margin of the segment; indusium often caducous; spores echinate.

1. T. palustris Schott, *Gen. Fil.* t. 10 (1834) (*Dryopteris thelypteris* (L.) A. Gray, *Lastrea thelypteris* (L.) C. Presl, *Nephrodium thelypteris* (L.) Strempel, *Polystichum thelypteris* (L.) Roth). Rhizome slender, far-creeping, hairy, with few, ovate, papillose scales when young. Leaves (15–)25–100(–120) cm, usually solitary, erect, pinnate; petiole as long as, to slightly longer than, the lamina, greenish-yellow, blackish at the base, glabrous except for occasional colourless hairs; lamina lanceolate, thin, pale green; pinnae pinnatisect, the longest 5–12 cm, linear-lanceolate; segments linear, entire, obtuse; sterile segments *c.* 5 mm wide, flat, the fertile narrower, with inrolled margins. Sori more or less confluent; indusium glandular. $2n = 70$. *Fens and marshes. Most of Europe except the extreme north.* Al Au Az Be Br Bu Co Cz Da Fe Ga Ge Gr Hb He Ho Hs Hu It Ju Lu No Po Rm Rs(N, B, C, W, K, E) Su.

2. Oreopteris J. Holub

A. C. JERMY AND A. M. PAUL

Rhizome ascending. Leaves 1-pinnate, with deeply pinnatisect pinnae and abundant yellow glands beneath; veins free, reaching margin. Sori orbicular, near the margin of the segment; indusium caducous; spores winged, with raised reticulum.

1. O. limbosperma (Bellardi ex All.) J. Holub, *Folia Geobot. Phytotax.* (*Praha*) **4**: 46 (1969) (*Thelypteris limbosperma* (Bellardi ex All.) H. P. Fuchs, *T. oreopteris* (Ehrh.) Slosson, *Dryopteris oreopteris* (Ehrh.) Maxon, *D. montana* (Vogler) O. Kuntze, *Lastrea oreopteris* (Ehrh.) Bory, *L. limbosperma* (Bellardi ex All.) J. Holub & Pouzar, *Nephrodium oreopteris* (Ehrh.) Desv., *Polystichum oreopteris* (Ehrh.) Bernh., *P. montanum* (Vogler) Roth). Rhizome stout, the apex covered with white, becoming brown, ovate-lanceolate, papillate scales. Leaves 30–100 cm, forming a crown at the apex of the rhizome, suberect; petiole $\frac{1}{8}$–$\frac{1}{5}$ as long as the lamina, with scattered brown, ovate to ovate-lanceolate scales; lamina lanceolate or oblanceolate, firm, yellowish-green, lemon-scented when crushed; pinnae linear-lanceolate, the longest 5–12 cm, becoming shorter towards the base; veins hairy beneath. Indusium inconspicuous, irregularly toothed, glandular. $2n = 68$. *Woods, especially by streams, and in relatively open situations, as on screes. C. & N.W. Europe, extending to S. Sweden, W. Ukraine, C. Italy and the Pyrenees.* Au Az Be Br Co Cz Da Ga Ge Hb He Ho Hs Hu It Ju No Po Rm Rs(W) Su.

3. **Phegopteris** (C. Presl) Fée

A. C. JERMY AND A. M. PAUL

Rhizome creeping. Leaves 1-pinnate; pinnae pinnatisect, the lowest pair deflexed; veins free, mostly not reaching margin. Sori orbicular, often coalescing; indusium absent; spores rugulose-granulate.

1. P. connectilis (Michx) Watt, *Canad. Nat. Quart. Jour. Sci.* ser. 2, **3**: 159 (1867) (*Thelypteris phegopteris* (L.) Slosson, *Dryopteris phegopteris* (L.) C. Chr., *Lastrea phegopteris* (L.) Bory, *Nephrodium phegopteris* (L.) Prantl, *Phegopteris vulgaris* Mett., *P. polypodioides* Fée, *Polypodium phegopteris* L.). Rhizome slender, with light brown, hairy, lanceolate scales when young. Leaves 10–50 cm, solitary; petiole up to twice as long as the lamina, greenish-yellow, sparsely covered with colourless hairs and occasional brown scales; lamina 5–15 cm wide, triangular-ovate, thin, yellowish-green, bent back almost at right-angles to the petiole; pinnae lanceolate, attenuate; segments linear-oblong, entire or crenate, rounded at the apex, hairy on both surfaces, with narrow, lanceolate scales on the veins. $2n = 90$. *Shady places. Most of Europe, but rare in the south.* Au Be Br Bu Co Cz Da Fa Fe Ga Ge Gr Hb He Ho Hs Hu Is It Ju No Po Rm Rs(N, B, C, W, E) Su.

4. **Stegnogramma** Blume

A. C. JERMY AND A. M. PAUL

Rhizome creeping. Leaves 1-pinnate; veins free. Sori linear; indusium absent; spores echinate.

1. S. pozoi (Lag.) Iwatsuki, *Acta Phytotax. Geobot.* (*Kyoto*) **19**: 124 (1963) (*Thelypteris pozoi* (Lag.) C. V. Morton, *Dryopteris africana* (Desv.) C. Chr., *Gymnogramma totta* (Willd.) Schlecht., *Hemionitis pozoi* Lag., *Ceterach hispanicum* sensu Willk., pro parte). Rhizome with ovate-lanceolate, hairy, brown scales, at least on younger parts. Leaves 25–60(–75) × 6–15 cm, solitary, erect; petiole as long as or rather shorter than the lamina, brownish-yellow, hairy; lamina lanceolate, firm, light green; rhachis densely hairy; pinnae regularly pinnatifid; segments entire, obtuse, divided about half-way to the midrib, hairy on both surfaces. Sporangia setose, spread out along lateral vein of segment. *Damp, shady places. Pyrenees; Açores.* Az Ga Hs.

5. **Christella** Léveillé

A. C. JERMY AND A. M. PAUL

Rhizome creeping. Leaves 1-pinnate, with pinnatifid pinnae, hairy; veins free except for the lowermost pair of each segment, which unite with those from the adjacent segments to form a single vein, excurrent in the sinus. Sori orbicular, medial; indusium reniform, densely hairy; spores papillose, the papillae often confluent to form ridges. (*Cyclosorus* Link.)

1. C. dentata (Forskål) Brownsey & Jermy, *Brit. Fern Gaz.* **10**: 338 (1973) (*Cyclosorus dentatus* (Forskål) R.-C. Ching, *Dryopteris dentata* (Forskål) C. Chr., *D. mollis* (Jacq.) Hieron., *Aspidium molle* (Jacq.) Swartz). Rhizome stout, the apex covered with shiny brown, lanceolate, hairy scales. Leaves 30–100 cm, forming a crown at the apex of the rhizome, pinnate; petiole less than $\frac{1}{3}$ as long as the lamina, light brown, hairy; lamina lanceolate to elliptical, tapering abruptly at the apex; lower pinnae often very reduced in size; longest pinnae 10–15 × 1–2 cm, pinnatifid; segments parallel-sided, truncate at the apex, sparsely hairy above, densely hairy beneath. Sporangial stalks glandular. *Damp places beside streams. Açores; Mediterranean region, very local.* Az Cr Hs ?It Si. (*Old World tropics.*)

17. ASPLENIACEAE

EDIT. †D. H. VALENTINE AND D. M. MOORE

Rhizome dictyostelic, with clathrate scales. Petiole with 2 vascular strands, often fusing to form a single X-shaped strand. Sori superficial, borne along one or both sides of the fertile veins; spores bilateral.

1. **Asplenium** L.

J. A. CRABBE, A. C. JERMY AND J. D. LOVIS (EDITION 1) REVISED BY R. VIANE, A. C. JERMY AND J. D. LOVIS (EDITION 2)

Rhizome short, erect or occasionally creeping, with dark, linear-triangular to linear-lanceolate, sometimes filiform scales. Leaves in apical tufts; petiole dark, at least at the base, often glabrous; lamina entire or variously dissected; veins free or rarely marginally anastomosing. Sori elliptical to linear; indusium resembling the sorus in shape and size, attached along the vein, sometimes rudimentary or absent.

Spore measurements given are those of the inner wall (exospore) and do not include the ornamented perispore. Many interspecific hybrids have been recorded but, except **5(a)** × **20**, are not frequent enough to be mentioned individually. (Incl. *Ceterach* DC., *Phyllitis* Hill, *Pleurosorus* Fée.)

1 Indusium absent; plant covered with pale hairs *c.* 1 mm **28. hispanicum**
1 Indusium present (occasionally reduced or absent); plant ± glabrous or with dense scales or unicellular glands
2 Underside of leaf brown, completely covered with scales **27. ceterach**
2 Underside of leaf green, sparsely covered with scales or subglabrous
3 Leaves simple or palmately lobed
4 Leaves palmately lobed; sori not paired (not opening to each other) **1. hemionitis**
4 Leaves simple, linear or hastate; sori paired, opening to each other
5 Leaves cordate at the base **29. scolopendrium**
5 Leaves auricled to hastate at the base **30. sagittatum**
3 Leaves pinnatifid, pinnate, ternate or irregularly divided into narrow segments
6 Leaves pinnatifid at the base, the upper segments confluent into an entire apex **31. hybridum**
6 Leaves pinnate, forked or ternate
7 Leaves forked or ternate
8 Leaves 3-partite (rarely simple), with narrowly elliptical to rhombic segments **21. seelosii**
8 Leaves with 2–4 (or more) linear segments **20. septentrionale**

7 Leaves pinnately divided, not forked or 3-partite
9 Abaxial side of rhachis entirely green or stramineous
10 Leaves 1-pinnate, the pinnae entire to pinnatifid
11 Pinnae stalked **9. trichomanes-ramosum**
11 Pinnae adnate to the rhachis **11. jahandiezii**
10 Leaves 2-pinnate or more
12 Basal pinnae much reduced **13. fontanum**
12 Basal pinnae the largest
13 Rhizome-scales uniformly clathrate (occasionally with some opaque cells)
14 Sori linear; sporangia more medial on vein, covering segment of lamina at maturity **22. ruta-muraria**
14 Sori oblong; sporangia more basal, not covering whole segment **23. lepidum**
13 Rhizome-scales with central part black and opaque, the margin clathrate
15 Ultimate leaf-segments narrowly linear; base of petiole becoming shiny reddish-brown to black **24. fissum**
15 Ultimate leaf-segments rhombic to wedge-shaped; base of petiole occasionally becoming brown
16 Rhizome-scales *c.* 1 mm; lamina triangular-ovate, with numerous unicellular glands **25. aegaeum**
16 Rhizome-scales *c.* 3 mm; lamina triangular-lanceolate, with few cylindrical glands **26. creticum**
9 Abaxial side of rhachis brown or blackish-brown, or with brown streaks, green or brown above
17 Leaves 1-pinnate; pinnae entire to pinnatifid
18 Scales clathrate, occasionally dark brown but without opaque central zone
19 Middle pinnae adnate to the rhachis **10. bourgaei**
19 Middle pinnae stalked
20 Lamina less than 4 cm wide; pinnae pinnatifid or deeply cut **12. majoricum**
20 Lamina usually more than 4 cm wide; pinnae entire, the margins crenate **2. marinum**
18 Scales on the petiole base and on the rhizome with at least a central band of dark, opaque cells
21 Petiole and lamina with dense, unicellular glands **3. petrarchae**
21 Petiole and lamina without unicellular glands
22 Rhachis with 3 wings **6. anceps**
22 Rhachis with 2 wings
23 Middle pinnae more than twice as long as wide
24 Pinnae usually with 1 single sorus, along the basiscopic entire margin **4. monanthes**
24 Pinnae usually with several sori, the basiscopic margin crenate **7. azoricum**
23 Middle pinnae usually less than twice as long as wide
25 Rhachis green in the distal $\frac{1}{3}$ **8. adulterinum**
25 Rhachis brown throughout (sometimes green near the apex when young) **5. trichomanes**
17 Leaves 2- to 4-pinnate
26 Leaves 3- or 4-pinnate; plants usually more than 10 cm **(17–19). adiantum-nigrum** group
26 Leaves 2-pinnate, rarely 3-pinnate at the base
27 Lamina narrowly triangular to deltate; petiole swollen at the base
28 Lamina triangular to deltate, usually with more than 7 pairs of pinnae; teeth usually narrow and acute, or obtuse **(17–19). adiantum-nigrum** group
28 Lamina narrowly triangular to triangular, usually with fewer than 7 pairs of pinnae; teeth broad and mucronate **16. balearicum**
27 Lamina ovate to ovate-lanceolate or elliptical; petiole not distinctly swollen at the base
29 Pinnules and ultimate segments crenate, with obtuse (obscurely mucronate) teeth **15. obovatum**
29 Pinnules dentate to pinnatifid, with acute and distinctly mucronate teeth or lobes
30 Lamina lanceolate to narrowly elliptical, with several pairs of distinctly reduced basal pinnae **13. fontanum**
30 Lamina lanceolate to narrowly ovate or triangular; basal pinnae little or not reduced
31 Pinnae with only the basal segments free, the distal segments adnate to the costa and confluent into a rounded apex
32 Basal pinnae often deflected towards petiole and somewhat reduced; sori proximal on their vein **14. foreziense**
32 Basal pinnae rarely deflected or reduced; sori medial on the vein **16. balearicum**
31 Pinnae deeply divided for most of their length, the apex acute
33 Sori distal; rhizome-scales more than 5 mm **15. obovatum**
33 Sori medial; rhizome-scales less than 5 mm **16. balearicum**

Subgen. **Asplenium.** Leaves simple or compound; lamina usually subglabrous; veins free. Sorus single on a vein; indusium present.

1. A. hemionitis L., *Sp. Pl.* 1078 (1753) (*A. palmatum* Lam.). Rhizome-scales with purplish sheen. Leaves 20–35 cm; petiole about twice as long as the lamina, reddish-brown with a few filiform scales at the base; lamina 10–20 cm wide, light green, glabrous, usually palmately 5-lobed; lobes acute, the middle lobe longer than the others. Sori linear, often running the full length of the vein. $2n = 72$. *Non-calcareous rocks and forest floors. Portugal, Açores.* Az Lu. (*Madeira, Islas Canarias.*)

2. A. marinum L., *Sp. Pl.* 1081 (1753). Plant robust. Rhizome with filiform scales at the apex. Leaves 10–30(–40) cm; petiole about $\frac{1}{2}$ as long as the lamina, reddish-brown; lamina linear-lanceolate, pinnate, coriaceous, glabrous; rhachis with green wings; pinnae 1–4 cm, oblong, more developed on the upper side, crenate-serrate, the base truncate or broadly cuneate, the apex rounded. $2n = 72$. *Rocks and walls exposed to sea-spray. W. Europe, extending eastwards very locally to S. Italy.* Az Bl Br Co Ga Hb Hs It Lu No Sa Si.

3. A. petrarchae (Guérin) DC. in Lam. & DC., *Fl. Fr.* ed. 3, **5**: 238 (1815) (*A. glandulosum* Loisel.). Rhizome caespitose; scales filiform, consisting of black, opaque cells, except occasionally for a narrow border of clathrate cells. Leaves 4–12 cm; petiole and rhachis reddish-black with green apex, dull, densely covered with glandular hairs, persistent when dead and forming wiry tufts; lamina $\frac{2}{3}$–$\frac{3}{4}$ as long as the leaf, narrowly elliptical, pinnate, thin, bright green, covered above and beneath with unicellular, capitate, glandular hairs; pinnae 0·5–1 cm, ovate, truncate or broadly cuneate at the base, inciso-crenate to lobed. Exospores (33–)36–51(–57) μm. $2n = 144$. *Calcareous rocks.* ● *Mediterranean region; Portugal.* Al Bl Ga Gr Hs It Ju Lu Si.

Diploid plants ($2n = 72$) from Spain have been named subsp. **bivalens** D. E. Meyer, Lovis & Reichstein, *Ber. Schweiz. Bot. Ges.* **79**: 336 (1969) (*A. glandulosoides* A. & D. Löve). They differ principally in having smaller exospores (33–)36–39(–45) μm.

4. A. monanthes L., *Mantissa* 130 (1767). Rhizome-scales narrow, attenuate, with dark central stripe. Leaves 20–30(–45) cm, occasionally with proliferating buds at the base of the rhachis; petiole $\frac{1}{4}$ as long as the lamina, dark reddish-brown, glabrous; lamina linear-elliptical, light green, pinnate; pinnae 0·5–1 cm, oblong to rhombic, entire to crenulate-dentate, the base truncate, the apex rounded. Sori oblong, usually borne singly on the pinna. $2n = 108$. *Açores.* Az. (*Africa; Central and South America; Madeira.*)

5. A. trichomanes L., *Sp. Pl.* 1080 (1753). Rhizome caespitose; scales up to 5 mm, with dark central stripe. Leaves 4–20(–35) cm; petiole *c.* $\frac{1}{4}$ as long as the lamina, dark reddish-brown; lamina pinnate, thicker and darker green than in **3**, with occasional multicellular glandular hairs beneath or glabrous; rhachis dark reddish-brown to the apex, with 2 narrow, pale brown wings on the upper side; pinnae not more than twice as long as wide, entire or crenate-serrate, the base truncate or asymmetrically cuneate, the apex rounded. Sori small, linear, but the sporangia usually covering the pinna at maturity. *Throughout Europe.* All except Sb.

1 Rhizome-scales not more than 3·5 mm; exospore typically 27–32 μm
 2 Pinnae suborbicular, auriculate and obliquely inserted above; terminal pinna 2–4 mm wide (a) subsp. **trichomanes**
 2 Pinnae square or oblong, squarely inserted; terminal pinna 4–7 mm wide (c) subsp. **inexpectans**
1 Rhizome-scales up to 5 mm; exospore typically 32–39 μm
 3 Leaves erect; pinnae oblong, rarely imbricate (b) subsp. **quadrivalens**
 3 Leaves procumbent; pinnae auriculate or biauriculate
  4 Pinnae often imbricate, the margins usually crenulate (d) subsp. **pachyrachis**
  4 Pinnae not imbricate, the margins irregularly undulate or lobed, mostly strongly revolute (e) subsp. **coriaceifolium**

(a) Subsp. **trichomanes**: Rhizome-scales not more than 3·5 mm, lanceolate. Pinnae 3–8 mm, suborbicular or more rarely oblong, more or less distant, especially near the apex of the lamina, thin; insertion of upper pinnae oblique; terminal pinna 2–4 mm wide; mean stomata guard-cell length 32–40 μm; exospores (23–)28–32(–34) μm. $2n = 72$. *Non-calcareous rocks, usually in the mountains. Throughout the range of the species.*

(b) Subsp. **quadrivalens** D. E. Meyer emend. Lovis, *Brit. Fern Gaz.* **9**: 152 (1964): Rhizome-scales up to 5 mm, lanceolate or linear-lanceolate. Pinnae 4–12 mm, oblong, more or less crowded and usually more robust than in subsp. (**a**); insertion of upper pinnae at right-angles or slightly oblique; terminal pinna 2–5 mm wide; mean stomata guard-cell length 40–48 μm; exospores (26–)32–38(–47) μm. $2n = 144$. *Calcareous rocks and mortared walls. Throughout the range of the species.*

(c) Subsp. **inexpectans** Lovis, *Brit. Fern Gaz.* **9**: 155 (1964). Rhizome-scales not more than 3·5 mm, lanceolate. Leaves procumbent, not tapered towards the apex; pinnae 4–10 mm, square or oblong, slightly distant, thin; insertion of upper pinnae square or slightly oblique; terminal pinna 4–7 mm wide; mean stomata guard-cell length 32–42 μm; exospores (22–)27–31(–36) μm. $2n = 72$. *Limestone rocks.* ● *E., C. & S.E. Europe; Sweden.*

(d) Subsp. **pachyrachis** (Christ) Lovis & Reichstein, *Willdenowia* **10**: 18 (1980) (*A. csikii* Kümmerle & Andrasovszky). Like subsp. (**c**) but rhizome-scales often linear-lanceolate; lamina more coriaceous; pinnae broadly lanceolate to oblong-lanceolate, auriculate, markedly so on the acroscopic side, the margins usually crenulate; lower pinnae often biauriculate; insertion of upper pinnae more or less oblique; terminal pinna small, irregularly cut; mean stomata guard-cell length 40–49 μm; exospores (26–)32–38(–50) μm. $2n = 144$. *Calcareous rocks, often under overhangs. S. & W. Europe.*

(e) Subsp. **coriaceifolium** H. & K. Rasbach, Reichstein & Bennert, *Willdenowia* **19**: 471 (1990). Like subsp. (**d**) but stipe and rhachis stiff and brittle; lamina strongly coriaceous (less so when grown in shade); pinnae slightly asymmetrical, auriculate on the acroscopic side; margins irregularly undulate or lobed, with few incisions or small teeth, mostly strongly revolute; terminal pinna deltate to elliptical, dark green, with bluish tinge, glossy when fresh; mean stomata guard-cell length 45–58 μm; exospores (33–)36–39(–43) μm. $2n = 144$. *Limestone rocks and walls.* ● *Spain, Baleares.*

6. A. anceps Lowe ex Hooker & Grev., *Icon. Fil.* **2**: t. 195 (1831). Like **5(b)** but the rhachis carinate, with a third wing below; lamina dark green and shining on the upper surface; pinnae symmetrically cuneate at the base; exospores (21–)27–30(–32) μm. $2n = 72$. *Non-calcareous rocks and forest floors. Açores (Pico).* Az. (*Madeira, Islas Canarias.*)

7. A. azoricum Lovis, H. Rasbach & Reichstein, *Amer. Fern Jour.* **67**: 88 (1977) (*A. trichomanes* var. *anceps* (Lowe ex Hooker & Grev.) Milde f. *azoricum* Milde). Like **5(b)** but rhachis black; lamina dark green and shining on the upper surface; pinnae up to 1·5 cm, 2–3 times as long as wide, ovate-oblong, with conspicuous costa, distinctly crenate-serrate, glabrous on both surfaces; lower pinnae biauriculate; exospores (27–)33–36(–39) μm. $2n = 144$. *Shady rocks and walls.* ● *Açores.* Az.

8. A. adulterinum Milde, *Höheren Sporenpfl. Deutschl. Schweiz* 40 (1865). Intermediate between **5** and **9**. At least some rhizome-scales with dark central stripe. Pinnae intermediate in shape and texture between those of **5** and **9**, with scattered glandular hairs beneath; rhachis reddish-brown, becoming green towards the apex. $2n = 144$. *Ledges and crevices of serpentine or rarely basic rocks.* ● *C. Europe, extending to Norway and N.W. Greece.* Au Cz Fe Ge Gr He It Ju No Po Rm Su.

(a) Subsp. **adulterinum**: Fronds erect; pinnae usually not touching, the margins distinctly (more than 0·2 mm deep) crenate; terminal pinna 0·5–2 mm wide. *On ultrabasic rocks (serpentine, magnesite). Throughout the range of the species.*

(b) Subsp. **presolanense** Mokry, H. Rasbach & Reichstein, *Bot. Helvet.* **96**: 8 (1986). Fronds drooping or procumbent; pinnae usually touching or imbricate, the margins entire or slightly crenate; terminal pinna 1–5 mm wide. *On limestone or mica-schist. N. Italy, S. Switzerland.*

9. A. trichomanes-ramosum L., *Sp. Pl.* 1082 (1753) (*A. viride* Hudson). Rhizome caespitose; scales without dark central stripe. Leaves 5–15(–20) cm; petiole *c.* $\frac{1}{3}$ as long as the lamina, reddish-brown at the base, abruptly becoming green; lamina linear to linear-lanceolate, light green, pinnate, thin, glabrous; rhachis green, not winged, with occasional glandular hairs; pinnae *c.* 5 mm, often convex, especially when young, semi-circular to suborbicular, crenate to incised, broadly cuneate at the base. Sori small, elliptical, but the sporangia often covering the entire surface of the lamina at maturity. $2n = 72$. *Mainly calcareous rock-crevices and ledges. Most of Europe, mainly in the mountains.* Al Au Be Br Bu Co Cr Cz Fe Ga Ge Gr Hb He Ho Hs Hu Is It Ju No Po Rm Rs(N, C, W, K) Su.

10. A. bourgaei Boiss. ex Milde, *Bot. Zeit.* **24**: 384 (1866). Rhizome caespitose. Leaves 4–28 cm; petiole up to $\frac{1}{3}$ as long as the lamina, green above, blackish below; lamina linear-lanceolate, 1-pinnate; pinnae 8–20 opposite pairs, the acroscopic base adnate, forming a wing on the rhachis, pinnatifid, the segments crenulate towards the apex. $2n = 72$. *Shady crevices of limestone rocks. Karpathos.* Cr. (*E. Mediterranean region.*)

11. A. jahandiezii (Litard.) Rouy, *Fl. Fr.* **14**: 437 (1913). Leaves 3–12 cm; petiole $\frac{1}{6}$–$\frac{1}{4}$ as long as the lamina, green, with occasional narrow scales; lamina linear-elliptical, pinnate, with occasional scales on the rhachis and costules; pinnae 0·5–1 cm, adnate to the rhachis, the lower margin decurrent in a narrow wing, oblong, serrate, occasionally incised at the base. Sori lying near the midrib. $2n = 72$. *Limestone cliffs.* ● *S.E. France (Gorges du Verdon, Var).* Ga.

12. A. majoricum Litard., *Bull. Géogr. Bot. (Le Mans)* **19**: 28 (1911) (*A. lanceolatum* var. *majoricum* (Litard.) Sennen & Pau). Rhizome erect; scales without dark central stripe. Leaves 8–12 cm, the petiole *c.* $\frac{1}{3}$ as long as the lamina, dark at the base, shiny black beneath, greenish above; lamina elliptical to oblanceolate, pinnate-pinnatifid; rhachis green above, black below for $\frac{1}{2}$ its length; median pinnae 8–10 mm, the lower progressively smaller and distant, pinnatifid or deeply lobed, the segments with wide, obtuse teeth without a mucro. Sori small, lying near the midrib of the pinna; indusium persistent, deeply and conspicuously lobed; exospores (39–)42–48(–51) μm, irregularly echinulate, sometimes with cristate wings. $2n = 144$. *Dry stone walls.* ● *Islas Baleares.* Bl.

13. A. fontanum (L.) Bernh. in Schrader, *Jour. für die Bot.* **1799(1)**: 314 (1799) (*A. halleri* DC.). Rhizome erect; scales lanceolate, dark brown. Leaves 5–25 cm; petiole $\frac{1}{3}$–$\frac{1}{2}$ as long as the lamina, reddish-brown at the base, soon becoming straw-coloured or greenish; lamina lanceolate to narrowly elliptical, thin, light green, 2-pinnate, the lower pinnae decreasing in length towards the base of the leaf; rhachis with occasional glandular hairs; pinnae 0·5–1·5 cm, with 3–8 pairs of pinnules, the terminal segment $\frac{1}{4}$ as long as the pinna or less; pinnules pinnatifid, with mucronate lobes. Sori lying near the midrib. $2n = 72$. *Calcareous rocks. S. & C. Europe.* ?Au †Be Bl Co Cr ?Cz Ga Ge Gr He Ho Hs †Hu It.

14. A. foreziense Legrand, *Bull. Soc. Dauph. Ech. Pl.* **12**: 501 (1885) (*A. forisiense* Legrand, *A. foresiacum* (Legrand) Christ; incl. *A. macedonicum* Kümmerle, *A. bornmuelleri* Kümmerle). Like **13** but rhizome with filiform scales at the apex; leaves 10–20 cm; petiole reddish-brown, becoming green at the base of the rhachis; lamina lanceolate, acuminate, thicker, the lower pinnae not decreasing in length to any marked extent towards the base of the leaf; pinnae with 1–3 pairs of toothed, rounded pinnules, the terminal segment about $\frac{1}{2}$ as long as the pinna, not deeply cut. $2n = 144$. *Siliceous rocks.* ● *S.W. Europe, extending locally northwards to the Netherlands and eastwards to S. Jugoslavia.* †Be Co Ga Ge He Ho Hs It Ju Sa.

15. A. obovatum Viv., *Fl. Lib.* 68 (1824). Rhizome-scales narrow, becoming filiform at the apex of the rhizome. Leaves 15–30 cm; petiole up to as long as the lamina, but not less than $\frac{1}{2}$ its length, not swollen below, light reddish-brown, with few filiform scales; lamina ovate-lanceolate, acuminate, bright green, glabrous, 2-pinnate, the rhachis and midrib with occasional filiform scales; pinnae 2–4 cm, ovate-triangular, pinnately divided only in proximal $\frac{1}{2}$, pinnatifid distally; pinnules suborbicular, with obtuse, mucronate teeth. Sori ovate, nearer margin than midrib. *Ledges, rocks and walls, often near the sea.* ● *W. Europe and Mediterranean region.* Az Be Br Co Cr Ga Gr Hb He Hs He It ?Ju Lu Sa Si Tu.

(**a**) Subsp. **obovatum**: Pinnae 2–4 cm, ovate-triangular, pinnately divided only in the proximal $\frac{1}{2}$, pinnatifid distally; pinnules suborbicular, almost crenate with obtuse or rounded teeth only faintly mucronate. $2n = 72$. *Mediterranean region; Portugal and one station in N.W. France.*

(**b**) Subsp. **lanceolatum** (Fiori) P. Silva, *Agron. Lusit.* **20**: 217 (1959) (*A. lanceolatum* Hudson, non Forskål, *A. billotii* F. W. Schultz, *A. obovatum* var. *billotii* (F. W. Schultz) Becherer): Pinnae 3–7 cm, linear-oblong, pinnate for most of their length, the longer pinnae often recurved; pinnules oblong, with acute, mucronate teeth. $2n = 144$. *W. Europe, extending eastwards to Switzerland and S. Italy.*

Diploid plants with the morphology of subsp. (**b**) have been found in S. Spain and need further investigation.

16. A. balearicum Shivas, *Brit. Fern Gaz.* **10**: 75 (1969). Rhizome short, erect; scales *c.* 4 mm, subulate, brown. Leaves *c.* 18 × 5 cm; petiole shorter than the lamina, not swollen below, brown, with a few subulate scales; lamina triangular, 2-pinnate or 2-pinnate-pinnatifid; pinnae up to 7 pairs, triangular, obliquely inserted; pinnules ovate to elliptical, the lowermost with mucronate lobes. Sori ovate, between the midrib and margin of the pinnule. $2n = 144$. *Non-calcareous rocks and forest floors.* ● *W. Mediterranean region.* Bl Co It Sa Si.

(17–19). A. adiantum-nigrum group. Rhizome ascending or short-creeping; scales attenuate, with some filiform scales at apex of rhizome. Leaves 10–50 cm; petiole about as long as the lamina, swollen at junction with rhizome, reddish-brown, with narrow scales at the base; lamina deltate, dark green, coriaceous, glabrous, 2- or 3-pinnate; pinnae triangular-ovate to lanceolate, the longest 2·5–7 cm; pinnules pinnatisect. Sori adjacent to the midrib.

A group of three closely related species.

1 Leaves and pinnae distinctly caudate; mean exospore-length 25–30 μm **18. onopteris**
1 Leaves and pinnae rarely caudate; mean exospore-length greater than 30 μm
  2 Ultimate leaf-segments lanceolate to linear, with acute teeth **17. adiantum-nigrum**
  2 Ultimate leaf-segments flabellate, with obtuse teeth **19. cuneifolium**

17. A. adiantum-nigrum L., *Sp. Pl.* 1081 (1753). Lamina not caudate, dark green, usually longer than the petiole. Ultimate segments lanceolate, obtuse at the base, with acute teeth; mean stomata guard-cell length 50–62 μm. Exospores (30–)31–37(–45) μm. $2n = 144$. *Rocks, walls and hedge-banks. S.,*

W. & C. Europe, extending to S. Sweden. All except Bl ?Cr Fe Is Rs(N, B, C, E) Sb ?Si.

Plants from serpentine in the British Isles, Corse, Italy and S.W. Europe are similar to **19** but have a chromosome number $2n = 144$. They probably represent an ecotypic variant of **17**.

18. A. onopteris L., *loc. cit.* (1753) (*A. adiantum-nigrum* subsp. *onopteris* (L.) Luerssen). Lamina caudate, usually shorter than the petiole; pinnae and pinnules caudate, usually curved towards the apex of the leaf and pinna respectively; ultimate segments narrowly lanceolate to linear, acute at the base with acute teeth; mean stomata guard-cell length 45–55 μm. Exospores (21–)25–30(–38) μm. $2n = 72$. *Usually on base-poor rocks. S. and W. Europe; one station in S.W. Poland.* Al Az ?Be Bl Bu Co Cr Ga Gr Hb He Hs It Ju Lu Po Rm Sa Si Tu.

19. A. cuneifolium Viv., *Fl. Ital. Fragm.* **1**: 16 (1808) (*A. forsteri* Sadler, *A. serpentini* Tausch, *A. adiantum-nigrum* subsp. *serpentini* (Tausch) Koch). Lamina not caudate; ultimate leaf-segments flabellate, with obtuse teeth; mean stomata guard-cell length 43–53 μm. Exospores (23–)30–34(–44) μm. $2n = 72$. *Usually on serpentine. S. & C. Europe.* Al Au ?Bu Co Cz Ga Ge Gr He ?Hs Hu It Ju ?Lu Po Rm Rs(W).

20. A. septentrionale (L.) Hoffm., *Deutschl. Fl. Crypt.* 12 (1795). Rhizome caespitose, sometimes short-creeping. Leaves 5–15 cm; petiole 2–3 times as long as the lamina, dark reddish-brown only at the base, abruptly becoming green; lamina much reduced, thick, dark green, glabrous, dichotomously forked 1–3 times; segments *c.* 1 mm wide, linear, decurrent along the rhachis, minutely forked again at the apex. Sori elongate, running the whole length of the segment. $2n = 144$. *Siliceous rocks and walls. Most of Europe but rarer in the east.* All except Az Bl Cr Fa Ho Is Rs(B, E) Sb.

Hybrids between **5(a)** and **20** (*A.* × *alternifolium* Wulfen, *A.* × *germanicum* auct., *A.* × *breynii* auct.) are scattered throughout Europe with the parents. They are intermediate in morphology between the parents, and differ from **20** in having the petiole 1–2 times as long as the lamina, almost entirely dark reddish-brown, and the lamina pinnate with pinnae *c.* 2 mm wide, the lowest forked. They have a chromosome number of $2n = 108$.

21. A. seelosii Leybold, *Flora (Regensb.)* **38**: 348 (1855). Rhizome caespitose, sometimes short-creeping. Leaves 3–10 cm; petiole dark reddish-brown at the base, abruptly green above; lamina much reduced, thick, dark green, ternate; segments *c.* 1 mm wide, rhombic, decurrent along the rhachis, toothed at apex, often incised, the terminal segment often again 3-lobed. $2n = 72$. *Calcareous rocks. C. & E. Alps; Pyrenees and N. Spain.* Au Ga Ge Hs It Ju.

(a) Subsp. **seelosii**: Leaves glandular-hairy, the ultimate segments incised; petiole 2–3 times as long as the lamina. $2n = 72$. ● *C. & E. Alps.*

(b) Subsp. **glabrum** (Litard. & Maire) Rothm. in Cadevall & Font Quer, *Fl. Catalunya* **6**: 339 (1937) (*A. celtibericum* Rivas Martínez): Leaves glabrous, the ultimate segments incised-dentate; petiole 3–5 times as long as the lamina. $2n = 72$. *Pyrenees and N. Spain.*

Outside Europe, subsp. (**b**) is known only from Morocco.

22. A. ruta-muraria L., *Sp. Pl.* 1081 (1753). Rhizome branched, often creeping; scales very narrow, filiform at the apex of the rhizome. Leaves 4–15 cm; petiole about as long as the lamina, dark only at very base, otherwise green, with occasional glandular hairs; lamina 1–3 cm wide, ovate-lanceolate, glabrous, 2-pinnate; pinnae of 3–5 segments, the segments often again similarly divided; ultimate segments 2–3 mm, broadly flabellate to rhombic, the apex serrate to incised. Sori linear and medial, the sporangia covering the whole segment when mature; spores cristate. *Walls and base-rich rocks. Almost throughout Europe.* All except Az Fa Is Sb.

(a) Subsp. **ruta-muraria**: Fronds thick and robust; pinnules usually without a pellucid margin. Mean exospore-length 40–50 μm. $2n = 144$. *Throughout the range of the species.*

(b) Subsp. **dolomiticum** Lovis & Reichstein, *Brit. Fern Gaz.* **9**: 143 (1964) (incl. *A. eberlei* D. E. Meyer): Leaves somewhat thinner and more delicate; pinnules usually smaller, with a distinct pellucid margin. Mean exospore-length 35–42 μm. $2n = 72$. *S. & S.C. Europe.*

23. A. lepidum C. Presl, *Verh. Ges. Vaterl. Mus. Böhm.* **1836**: 63 (1836). Like **22** but rhizome caespitose; scales very dark; filiform scales at the apex of the rhizome very few or absent; petiole more slender, wiry, persistent; lamina light green, of thinner texture; ultimate segments narrower, the apex crenately incised; sori oblong, usually basal on segment, the sporangia not covering the segment at maturity. *S.E. & S.C. Europe, extending locally westwards to N.E. Spain.* Al Au Bu Cr Ga Gr Hs Hu It Ju Rm Si.

Plants with extremely dissected leaves (var. *fissoides* Ritter & Schumacher) from Greece and Jugoslavia may be confused with **24**.

(a) Subsp. **lepidum**: Rhizome-scales 2–4 mm. Leaves glandular. $2n = 144$. *Throughout the range of the species.*

(b) Subsp. **haussknechtii** (Godet & Reuter) Brownsey, *Bot. Jour. Linn. Soc.* **72**: 261 (1976) (*A. haussknechtii* Godet & Reuter): Rhizome-scales 1–3 mm. Leaves glabrous except when young. $2n = 144$. *Mountains of C. & E. Kriti; N.E. Spain. (S.W. Asia.)*

24. A. fissum Kit. ex Willd., *Sp. Pl.* **5(1)**: 348 (1810). Rhizome branched and short-creeping; scales with central stripe of black cells. Leaves 15–22(–32) cm; petiole 1–2 times as long as the lamina, blackish-red, darkening with age, green distally; lamina very much reduced, lanceolate, glabrous (rarely with short glands), 3-pinnate; pinnules palmately dissected; ultimate segments *c.* 1 mm wide. Sori oblong. $2n = 72$. *Fissures in calcareous rocks.* ● *Mountains of S.C. & S.E. Europe, from Austria to the Maritime Alps and N. Greece.* Al Au Bu ?Cr Ga Ge Gr †He It Ju.

25. A. aegaeum Lovis, Reichstein & Greuter, *Ann. Mus. Goulandris* **1**: 141 (1973). Rhizome short, erect; scales *c.* 1 mm, with dark central stripe. Leaves 3–8 cm, densely covered with cylindrical glands less than 50 μm; petiole usually shorter than the lamina, brown at the base, green above; lamina ovate to ovate-triangular, 2-pinnate below; pinnae 3–5 pairs, the lowest pair larger than the others; ultimate segments 0·5–1 mm wide. Sori oblong, often confluent at maturity. $2n = 72$. *Limestone mountain rocks. S. Greece, Kriti.* Cr Gr. (*Anatolia.*)

26. A. creticum Lovis, Reichstein & Zaffran, *Ann. Mus. Goulandris* **1**: 145 (1973). Like **25** but rhizome-scales *c.* 3 mm; leaves up to 15 cm, glabrous, with a few cylindrical glands more than 50 μm; lamina triangular-lanceolate, 1- or 2-pinnate; pinnae 4–10 pairs, the lower shortly stalked, the upper sessile; ultimate segments wider, with short teeth. $2n = 144$. *Limestone mountain rocks.* ● *Kriti.* Cr.

Subgen. **Ceterach** (Willd.) Vida ex Bir, Fraser-Jenkins & Lovis. Leaves pinnatisect, lamina densely scaly, veins marginally anastomosing. Sorus single on a vein, indusium reduced or absent.

27. A. ceterach L., *Sp. Pl.* 1080 (1753) (*Ceterach officinarum* DC.). Leaves 3–25 cm, persistent, oblong; petiole short, $\frac{1}{6}$–$\frac{1}{4}$ as long as the lamina, covered with scales; lamina pinnatisect, greyish to dark green above, covered with light brown, overlapping scales beneath; leaf-segments alternate, ovate or oblong, 9–12 on each side, entire or crenate, rounded at the apex. Sori *c.* 2 mm. Mean exospore-length 35–41 μm. $2n = 72, 144$. *Dry rocks and walls. W., S. & C. Europe, northwards to Scotland and S.E. Sweden, eastwards to Krym.* Al Be Bl Br Bu Co Cr Cz Ga Ge Gr Hb He Ho Hs Hu It Ju Lu Rm Rs(K) Sa Si Su Tu [Au Po].

Diploid plants ($2n = 72$) from E.C. & S. Europe have been described as subsp. **bivalens** (D. E. Meyer) Greuter & Burdet, *Willdenowia* **10**: 17 (1980) (*A. javorkeanum* Vida). They differ principally by the smaller mean exospore-length, 28–35 μm.

Subgen. **Pleurosorus** (Fée) Salvo, Díaz & Prada. Leaves pinnate to bipinnate; lamina hairy; veins free. Sorus single on a vein; indusium absent.

28. A. hispanicum (Cosson) Greuter & Burdet, *Willdenowia* **10**: 17 (1980) (*Ceterach hispanicum* (Cosson) Mett., *Pleurosorus hispanicus* (Cosson) C. V. Morton, *P. pozoi* auct.). Leaves 3–8 cm, tufted; petiole and lamina about equal in length, covered with stiff, eglandular hairs; pinnae 1 cm or less, rather distant, shortly stalked or subsessile, cuneate at the base, pinnatifid or lobed; each fertile pinna with 3–5 small, ovate or elliptical sori. $2n = 72$. *Shady places on calcareous mountains. S. Spain.* Hs. (*N. Africa.*)

Subgen. **Phyllitis** (Hill) Jermy & Viane. Leaves entire; lamina subglabrous; veins free. Sori paired on 2 neighbouring veins; indusium present.

29. A. scolopendrium L., *Sp. Pl.* 1079 (1753) (*Phyllitis scolopendrium* (L.) Newman, *Scolopendrium officinale* Sm.). Leaves 5–60 cm, persistent; petiole up to $\frac{1}{2}$ as long as the lamina; lamina 3–6(–7) cm wide, linear-lanceolate, cordate at the base, the margin slightly undulate, entire. Sori 8–18 mm wide, linear, parallel, usually occupying more than $\frac{1}{2}$ the width of the lamina. *Shady places. S., W. & C. Europe, extending to S.E. Sweden and N. Ukraine.* All except Fa Fe Is Rs(N, B, C, E) Sb.

Many variants of this species, some with curled or cleft fronds, are known both in nature and in cultivation.

(a) Subsp. **scolopendrium**: Leaves up to 60 cm, the margins undulate, entire. Sori not extending to the margin of the leaf. $2n = 72$. *Throughout the range of the species.*

(b) Subsp. **antri-jovis** (Kümmerle) Brownsey & Jermy, *Brit. Fern Gaz.* **10**: 346 (1973) (*Biropteris antri-jovis* Kümmerle): Leaves not more than 20 cm, the margins indented. Sori extending to the indentations of the margin of the leaf. $2n = 72$. *Mountains of Kriti.* (*Palestine.*)

30. A. sagittatum (DC.) Bange, *Bull. Soc. Linn. Lyon* **21**: 84 (1952) (*Phyllitis sagittata* (DC.) Guinea & Heywood, *P. hemionitis* (Swartz) O. Kuntze, *Scolopendrium hemionitis* (Swartz) Lag.). Leaves 12–30 cm, tufted, persistent; petiole $\frac{1}{2}$ as long as or equalling the lamina; lamina oblong-lanceolate, deeply cordate and auricled at the base; auricles of young leaves rounded, those of mature leaves triangular, more or less lobed, sometimes projecting horizontally for 3–4 cm and leaves then hastate. Sori elliptical, shorter and wider than in **29**. $2n = 72$. *Moist, shady calcareous rocks. Mediterranean region.* Bl Co Ga Gr Hs It Ju Sa Si.

Subgen. **Phyllitopsis** (Reichstein) Jermy & Viane. Leaves simple to pinnatifid; lamina sparsely scaly; veins free or rarely anastomosing at the margins. Sorus single or paired on a vein; indusium present.

31. A. hybridum (Milde) Bange, *loc. cit.* (1952) (*Phyllitis hybrida* (Milde) C. Chr., *Phyllitopsis hybrida* (Milde) Reichstein). Leaves up to 12 × 2 cm, the petiole about $\frac{1}{2}$ as long as the lamina; lamina oblong-lanceolate, obtuse, irregularly pinnatifid and often with rounded auricles at the base, not deeply cordate. Sori narrowly elliptical to broadly ovate. $2n = 144$. *Calcareous rocks and walls.* ● *Islands of N.W. Jugoslavia.* Ju.

18. WOODSIACEAE

(*Athyriaceae*)

EDIT. †D. H. VALENTINE AND D. M. MOORE

Rhizome dictyostelic, with opaque scales. Petiole grooved, with 2 vascular strands which unite distally. Sori superficial, situated on the veins; spores bilateral.

1 Fertile and sterile leaves dissimilar
 2 Sterile leaves up to 170 cm; veins free — **7. Matteuccia**
 2 Sterile leaves up to 50 cm; veins anastomosing — **8. Onoclea**
1 Fertile and sterile leaves similar
 3 Indusium absent
  4 Rhizome thick, erect; leaves forming a rosette — **1. Athyrium**
  4 Rhizome slender, creeping; leaves solitary — **6. Gymnocarpium**
 3 Indusium present
  5 Indusium hood-like or cup-shaped, attached at the basiscopic side of sorus
   6 Indusium entire or shallowly toothed — **4. Cystopteris**
   6 Indusium with a wide fringe of narrow, hair-like teeth — **5. Woodsia**
  5 Indusium flap-like, linear or oblong, attached laterally
   7 Leaves forming a crown at apex of erect rhizome; indusium usually curved, crossing the vein — **1. Athyrium**
   7 Leaves solitary on a creeping rhizome; indusium straight, not or rarely crossing the vein
    8 Groove of frond-axis open at junction of costa or costule — **2. Diplazium**
    8 Groove of frond-axis not open at junction of costa or costule — **3. Deparia**

1. **Athyrium** Roth

A. C. JERMY

Rhizome branched, covered with broadly lanceolate, acuminate scales. Leaves uniform, forming a crown at apex of rhizome, 2-pinnate, with pinnatifid or pinnatisect pinnules; lamina broadly lanceolate with pinnae decreasing towards base; veins free. Sori oblong or orbicular; indusium rudimentary and caducous, or persistent, oblong or linear, fimbriate at margin, attached laterally, opening towards the midrib.

Sori oblong; indusium persistent; spores not winged **1. filix-femina**
Sori orbicular; indusium rudimentary or absent; spores winged **2. distentifolium**

1. A. filix-femina (L.) Roth, *Tent. Fl. Germ.* **3(1)**: 65 (1799) (*Polypodium rhaeticum* L., *P. axillare* Aiton, *Asplenium filix-femina* (L.) Bernh.). Rhizome short, suberect. Leaves 20–150 cm, suberect or spreading; petiole pale yellowish-green or purplish-red, $\frac{1}{4}$–$\frac{1}{2}$ as long as the lamina; lamina thin, flaccid, light green, lanceolate; longest pinnae 3–25 cm, decreasing towards base of leaf, linear-lanceolate, tapered at apex; pinnules 3–20 mm, sessile, oblong or oblong-lanceolate, pinnatifid or pinnatisect. Sori *c.* 1 mm, more or less oblong; indusium persistent, flap-like, the lower hooked, the upper nearly straight; spores minutely papillose. $2n = 80$. *Shady places. Throughout Europe but local in the south and east.* All except Rs(K) Sb.

This species has many variants, differing mainly in the form and dissection of the leaf, which bear no relation to habitat.

2. A. distentifolium Tausch ex Opiz, *Tent. Fl. Crypt. Boem.* **1**: 14 (1820) (*A. alpestre* (Hoppe) Rylands ex T. Moore, non Clairv., *Polypodium rhaeticum* auct., non L.). Like **1** but segments of pinnules usually blunt and often tricuspidate at apex; sori less than 1 mm, orbicular, near edge of lamina; indusium rudimentary or caducous at a very early stage; spores reticulate, with narrow wings. $2n = 80$. *Screes and rocky outcrops in exposed places. N. Europe; mountains of C. & S. Europe.* Au Br Bu Co Cz Fa Fe Ga Ge He Hs Is It Ju No Po Rm Rs(N, C, W) Su.

Var. *flexile* (Newman) Jermy (*A. flexile* (Newman) Druce), known only from Scotland, has leaves 15–25 cm and very small sori, often of 3–5 sporangia only.

2. **Diplazium** Swartz

A. C. JERMY

Rhizome creeping or erect, sparsely covered with ovate, dark brown scales. Leaves erect, distant, usually 2-pinnate, with pinnules pinnatisect, glabrous above; lamina deltate or ovate-lanceolate; articulate hairs usually absent; groove of frond-axis open at junction of costa and costule; veins free. Sori oblong or linear, often paired; indusium oblong or linear, fimbriate at margin, attached laterally along the vein.

Pinnules acuminate, glabrous beneath **1. caudatum**
Pinnules rounded at apex; main veins hairy beneath **2. sibiricum**

1. D. caudatum (Cav.) Jermy, *Brit. Fern Gaz.* **9**: 161 (1964) (*Tectaria caudata* Cav., *Asplenium umbrosum* auct. non Sm.). Rhizome up to 40 cm, creeping. Leaves 70–150(–200) cm; petiole glabrous, shorter than the lamina; lamina 25–50 cm wide, ovate or ovate-lanceolate; pinnae lanceolate, acuminate or caudate; pinnules lanceolate, acute, deeply pinnatisect, the basal pairs not markedly shorter than the upper; segments serrate-dentate, glabrous beneath. Sori oblong. *Woods and shady places. Açores; one station in S.W. Spain.* Az Hs. (*Islas Canarias.*)

2. D. sibiricum (Turcz. ex G. Kunze) Kurata in Namegata & Kurata, *Coll. Cult. Ferns Fern Allies* 340 (1961) (*Athyrium crenatum* (Sommerf.) Rupr.). Rhizome slender, often much-branched and far-creeping. Leaves 25–60 cm; petiole glabrous or sparsely scaly, as long as the lamina; lamina 20–30 cm wide, deltate; pinnae lanceolate, the lowest inflexed; pinnules obtuse or rounded, decreasing in size towards base, deeply pinnatisect, the segments crenate-dentate, oblong, the main veins covered beneath with multicellular, crispate hairs. Sori small; indusium obscure, often caducous. $2n = 82$. Picea-*woodland. N.E. Europe, southwards to 53° N.; one small region of C. Norway.* Fe No Rs(N, C) Su.

D. esculentum (Retz.) Swartz in Schrader, *Jour. für die Bot.* **1800**: 312 (1803), from tropical Asia, with a stout, erect rhizome, pinnules shallowly pinnatifid to dentate and linear sori, is cultivated for ornament. It is perhaps becoming naturalized in Açores.

3. **Deparia** Hooker & Grev.

A. C. JERMY

Like *Diplazium* but groove of frond-axis not open at junction of costa or costule; articulated hairs usually present.

1. D. petersenii (G. Kunze) Kato, *Bot. Mag. (Tokyo)* **90**: 37 (1977) (*Asplenium petersenii* G. Kunze, *Diplazium allorgei* Tardieu-Blot, *Lunathyrium japonicum* auct., non (Thunb.) Kurata). Leaves 70–100 cm; petiole hispid, dark brown, shorter than the lamina; lamina 25–50 cm wide, ovate or ovate-lanceolate; pinnae lanceolate, acuminate or caudate; pinnules crenate-serrate, obtuse or truncate at apex, glabrous beneath. Sori oblong to linear. *Naturalized in Açores.* [Az.] (*Tropical and subtropical Asia.*)

4. **Cystopteris** Bernh.

J. A. CRABBE (EDITION 1) REVISED BY A. C. JERMY (EDITION 2)

Rhizome blackish. Leaves uniform, delicate, slender-stalked, dying down in winter; veins free. Sori orbicular; indusium hood-like, pale, attached across the vein at basiscopic side of sorus, becoming deflexed and shrivelled as the sporangia mature.

1 Rhizome creeping, slender; leaves solitary; lamina deltate, usually shorter than petiole
2 Proximal basiscopic pinnule conspicuously the longest **5. montana**
2 Proximal basiscopic pinnule not the longest **6. sudetica**
1 Rhizome short, stout; leaves tufted; lamina lanceolate, usually longer than petiole
3 Spores rugose **4. dickieana**
3 Spores echinate
4 Spores densely echinate, the spines almost touching at base **3. diaphana**
4 Spores ±sparsely echinate, the spines at least their own width apart
5 Ultimate segments of pinnae ovate to oblong **1. fragilis**
5 Ultimate segments of pinnae linear-oblong **2. alpina**

1. C. fragilis (L.) Bernh. in Schrader, *Neues Jour. Bot.* **1(2)**: 27 (1805) (*C. filix-fragilis* (L.) Borbás). Rhizome *c.* 5 mm in diameter, short, much-branched, covered with old leaf-bases, and with lanceolate, acuminate scales at the apex. Leaves 5–45 cm, tufted, 2-

or 3-pinnate; petiole $\frac{1}{3}$–$\frac{2}{3}$ as long as the lamina; lamina dull green, glabrous or rarely glandular, lanceolate, acute; pinnae up to *c.* 5 cm, subopposite, subsessile, ovate-lanceolate, becoming increasingly distant below, the second pair from the base usually the longest; pinnules up to *c.* 15 mm, ovate to oblong, entire and dentate to pinnate; teeth acute to obtuse, sometimes bidentate at the apex, often dentate on the margins; veins ending in the apices of the teeth. Indusium ovate-lanceolate, acuminate; spores echinate, the spines at least their own width apart. $2n = 168, 252$. *Mostly on shady, basic rocks. Throughout Europe; mainly on mountains in the Mediterranean region.* All territories.

2. C. alpina (Lam.) Desv., *Mém. Soc. Linn. Paris* **6**: 264 (1827) (*C. regia* auct. plur. pro parte). Like **1** but pinnae more deeply dissected, the ultimate segments linear-oblong. *On calcareous rocks.* ● *Throughout Europe but detailed distribution needs clarification.* $2n = 252$. Al Bu Co ?Cz ?Da ?Fe Ga ?Ge Gr ?He Hs It Ju ?No ?Po ?Rm Rs(W) Si ?Su.

3. C. diaphana (Bory) Blasdell, *Mem. Torrey Bot. Club* **21**: 47 (1963) (*C. viridula* (Desv.) Desv.). Like **1** but lamina less tapered at the base, 2-pinnate, the ultimate segments often wide and contiguous; veins ending in the sinus of a retuse segment apex or between pinnule teeth. Spores densely echinate, their spines short, almost touching at the base. $2n = 252$. *On shady, mildly basic rocks. S. & S.W. Europe, Açores.* Az Co Ga Hs It Lu Si. (*Africa.*)

4. C. dickieana R. Sim, *Gard. Farm. Jour.* **2(20)**: 308 (1848) (*C. baenitzii* Dörfler). Like **1** but leaves 2-pinnate; indusium lanceolate; spores rugose. $2n = 168$. *Widely distributed in Europe, but the distribution is imperfectly known.* Br Co Fe Ga Ge Gr He Hs Is It Lu No Rs(N) Sa Si Su.

Originally described from Britain, where the plants represent a variant in which the pinnae are deflexed and horizontally overlapping and inclined, and the pinnules wide, obtuse, more or less overlapping, and crenate to crenately lobed.

Species **1–4** form a widespread, polymorphic, polyploid complex in which it is difficult to distinguish the variants taxonomically. Tetraploid and hexaploid populations of **1** cannot as yet be distinguished with certainty either from one another or from diploid populations which have been recorded from North America. Sterile hybrids occur interspecifically and between cytotypes of **1**.

5. C. montana (Lam.) Desv., *Mém. Soc. Linn. Paris* **6**: 264 (1827). Rhizome *c.* 2 mm in diameter, creeping, little-branched, with a few scattered scales on the younger parts. Leaves 10–40 cm, solitary, 3-pinnate; petiole 1–2 times as long as the lamina; lamina deep green, more or less thinly glandular, deltate-pentagonal; pinnae up to *c.* 10 cm, subopposite, subsessile, obliquely triangular-ovate, acute, the lowest pair the longest; pinnules up to *c.* 5 cm, the proximal basiscopic pair the longest; ultimate segments up to *c.* 10 mm, ovate to oblong, more or less acute, pinnately divided. Indusium suborbicular, often glandular, caducous; spores muricate. $2n = 168$. *Woods, usually on basic rocks. N. & N.E. Europe; mountains of C. & S. Europe from the Pyrenees to the E. Carpathians.* Au Br Cz Fe Ga Ge He Hs It Ju No Po Rm Rs(N, C, W, E) Su.

6. C. sudetica A. Braun & Milde, *Jahresb. Schles. Ges. Vaterl. Kult.* **33**: 92 (1855). Like **5** but lamina ovate-deltate, yellowish-green; pinnae oblong-lanceolate, the proximal basiscopic pinnules not the longest; ultimate segments more or less obtuse. $2n = 168$. *Mountain woods. N. Europe, from Norway to Ural; mountains of E.C. Europe.* †Au Cz Ge No Po Rm Rs(N, B, C, W).

5. Woodsia R. Br.

A. O. CHATER

Rhizome short, covered by persistent leaf-bases. Leaves tufted, pinnate, with lobed to pinnatisect pinnae; petiole articulated above the base; veins free. Sori orbicular; indusium divided into a fringe of hair-like scales surrounding the base of the sorus.

Literature: J. Poelt, *Mitt. Bot. Staatssamm. (München)* **1(5)**: 167–174 (1952). R. E. G. Pichi Sermolli, *Webbia* **12**: 179–216 (1955).

1 Petiole pale, greenish to straw-coloured above the black base; petiole and rhachis glabrous or with very few scales or hairs **3. glabella**
1 Petiole dark, brownish above the black base; petiole and rhachis sparsely to densely covered with scales and hairs
 2 Rhachis and lower surface of pinnae ±densely covered with scales and hairs; longest pinnae with 4–8 lobes on each side **1. ilvensis**
 2 Rhachis sparsely clothed with scales and hairs; lower surface of pinnae with few scales and hairs, or glabrous; longest pinnae with 1–4 lobes on each side **2. alpina**

1. W. ilvensis (L.) R. Br., *Trans. Linn. Soc. London* **11**: 173 (1815) (*W. rufidula* (Michx) G. Beck, *W. ilvensis* subsp. *rufidula* (Michx) Ascherson). Leaves 4–20 cm, dull or brownish-green; petiole densely covered with lanceolate scales near the base, with subulate scales above, and with flexuous hairs throughout; lamina oblong-lanceolate, with 7–18 pinnatifid or rarely pinnatisect pinnae on each side; longest pinnae 7–20 mm, $1\frac{1}{2}$–2 times as long as wide, with 4–8 obtuse lobes on each side; rhachis and lower surface of pinnae densely covered with subulate scales (2–3 mm) and flexuous hairs. $2n = 82$. *Crevices of siliceous rocks. N. & C. Europe; C. & S. Ural; C. Ukraine (near Žitomir).* Au Br Cz Fe Ga Ge He Hu Is It Ju No Po Rm Rs(N, B, C, W) Su.

2. W. alpina (Bolton) S. F. Gray, *Nat. Arr. Brit. Pl.* **2**: 17 (1821) (*W. hyperborea* (Liljeblad) R. Br., *W. ilvensis* subsp. *alpina* (Bolton) Ascherson). Usually smaller than **1**. Leaves pale or yellowish-green; petiole with few scales or hairs, glabrescent; lamina linear-oblong, with 7–14 pinnately lobed or pinnatifid pinnae on each side; longest pinnae 5–15 mm, 1–$1\frac{1}{2}$ times as long as wide, with 1–4 very obtuse lobes on each side; rhachis sparsely clothed with subulate, rarely lanceolate scales (1–2 mm) and with hairs; lower surface of pinnae with very few scales or hairs, or glabrous. $2n = 164$. *Rock-crevices, rarely on walls. N. Europe, and on mountains southwards to C. Italy and the Pyrenees.* Au Br Cz Fe Ga Ge He Hs Hu Is It Ju No Po Rm Rs(N, C, W) Su.

3. W. glabella R. Br. ex Richardson in Franklin, *Narr. Journey* 26 (1823). Leaves 1·5–12 cm, usually yellowish-green and more or less translucent; petiole with very few scales or hairs at the base, glabrous above; lamina broadly or narrowly lanceolate or linear; pinnae 5–16, pinnately lobed or pinnatifid; longest pinnae 3–7(–10) mm with 1–7 obtuse to subacute lobes on each side; rhachis and lower surface of pinnae usually completely lacking scales or hairs. $2n = 78$. *Rock-crevices; calcicole. Arctic and subarctic Europe; Alps; isolated stations in Pyrenees and Carpathians.* Au Fe ?Ga Ge He Hs It Ju No Rm Rs(N, C) Sb Su.

(a) Subsp. **glabella**: Leaves not more than 8 cm. Pinnae usually crowded and overlapping, the longest 3–5(–6) mm, the lowest 2–4 on each side orbicular, the others ovate to lanceolate. $2n = c.$ 80, 164. *Arctic and subarctic Europe; S. Carpathians.*

(b) Subsp. **pulchella** (Bertol.) Á. & D. Löve, *Bot. Not.* **114**: 49 (1961) (*W. pulchella* Bertol., *W. glabella* auct. eur. centr., non R. Br.): Leaves up to 12 cm. Pinnae not or scarcely overlapping, the longest 4–7(–10) mm, the lowest 1 or 2 on each side orbicular, the others ovate-lanceolate to lanceolate. $2n = 78$. ● *E. & C. Alps; Pyrenees.*

6. Gymnocarpium Newman

A. C. JERMY

Rhizome slender, creeping; scales on younger part only, light brown, broadly ovate, fringed with papillae. Leaves solitary, in 2 ranks, erect; petiole $1\frac{1}{2}$–3 times as long as the lamina, blackish-brown below, brownish-yellow above; lamina deltate; veins free. Sori submarginal, round or elongated, often confluent; indusium absent.

1 Leaves glabrous, eglandular or with a few glands, bright green; rhachis glabrous or with sparse glands only **1. dryopteris**
1 Leaves with short glandular hairs, dull green; rhachis densely glandular
  2 Basal acroscopic pinnule of lowest pinna conspicuously narrower than the opposite basiscopic pinnule **2. robertianum**
  2 Basal acroscopic pinnule of lowest pinna ± the same width as the opposite basiscopic pinnule **3. jessoense**

1. G. dryopteris (L.) Newman, *Phytologist (Newman)* **4**: 371 (1851) (*Dryopteris linnaeana* C. Chr., *D. disjuncta* (Rupr.) C. V. Morton, *Phegopteris dryopteris* (L.) Fée, *Polypodium dryopteris* L., *Lastrea dryopteris* (L.) Bory, *Nephrodium dryopteris* (L.) Michx, *Thelypteris dryopteris* (L.) Slosson). Rhizome far-creeping. Leaves 10–40 cm, rolled up when young to form 3 small balls; petiole glabrous, with few scales near the base; lamina up to 30 cm wide, markedly deflexed, light green, glabrous; pinnae linear-lanceolate, opposite, pinnately lobed; each of lowest pair of pinnae often as large as the rest of the lamina, triangular, pinnate, the basiscopic pinnule larger than the acroscopic; ultimate segments more or less oblong, entire, toothed or lobed. $2n = 160$. *Woods, and shady places amongst rocks. Most of Europe, but rare in the south.* Al Au Be Br Bu Co Cz Da Fa Fe Ga Ge Gr Hb He Ho Hs Hu Is It Ju No Po Rm Rs(N, B, C, W, E, K) Su.

2. G. robertianum (Hoffm.) Newman, *loc. cit.* (1851) (*Dryopteris robertiana* (Hoffm.) C. Chr., *D. obtusifolia* (Schrank) O. Schwarz, *Phegopteris robertiana* (Hoffm.) A. Braun, *Polypodium robertianum* Hoffm., *Lastrea robertiana* (Hoffm.) Newman, *Nephrodium robertianum* (Hoffm.) Prantl, *Thelypteris robertiana* (Hoffm.) Slosson). Like **1** but the leaves usually 15–60 cm, rolled up when young to form a single ball; petiole sparsely glandular below but increasingly glandular towards the apex; lamina not obviously deflexed, dull green, with shortly-stalked glands on rhachis, veins and undersurface; each of lowest pair of pinnae smaller than the rest of the lamina. $2n = 160$–168. *Limestone rocks and screes, in less shaded situations than* **1**. *Frequent in C. Europe; local elsewhere.* Al Au Be Br Bu Cz Fe Ga Ge Gr Hb He Ho Hs Hu It Ju No Po Rm Rs(N, B, C, W, K) Su.

3. G. jessoense (Koidz.) Koidz., *Acta Phytotax. Geobot. (Kyoto)* **5**: 40 (1936). Like **2** but the basal acroscopic pinnule only slightly narrower than the opposite basiscopic pinnule; lobes or pinnules of the basal pinna inserted obliquely. $2n = 80$. *Calcareous screes in* Picea-*woodland. N.E. Europe.* Fe No Rs(N) Su. (*N. Asia, North America.*)

The European plants belong to subsp. **parvula** Sarvela, *Acta Bot. Fenn.* **15**: 103 (1978) (*G. continentale* (Petrov) Pojark.). They are distinguished by having a bipinnate lamina less than 12 cm.

7. Matteuccia Tod.

A. LAWALRÉE

Rhizome stout. Leaves tufted, pinnate with lobed or pinnatifid pinnae, dimorphic, with free veins; fertile leaves smaller, less dissected and with longer petioles than the sterile; petiole spathulate at the base. Sori round, 1 or 2 at base of each lobe, covered by the inrolled apex of the leaf; indusium caducous.

1. M. struthiopteris (L.) Tod., *Gior. Sci. Nat. Econ. Palermo* **1**: 235 (1866) (*Struthiopteris filicastrum* All., *S. germanica* Willd., *Onoclea struthiopteris* (L.) Roth). Rhizome short, erect, producing underground stolons. Sterile leaves up to 170 × 35 cm, not persistent; petiole very short; lamina oblong-lanceolate, soft, bright green; pinnae 30–70, alternate, narrowly lanceolate, pinnatifid. Fertile leaves up to 60 × 6 cm, persistent, eventually dark brown; pinnae linear, obtuse. $2n = c.$ 80. *N., E. & C. Europe, extending to Belgium, S.W. Alps and C. Jugoslavia.* Au Be Cz Da Fe Ge He Hu It Ju No Po Rm Rs(N, B, C, W, E) ?Si Su [Br Ga Hb].

8. Onoclea L.

†D. H. VALENTINE

Rhizome creeping. Leaves dimorphic, the sterile glabrous, pinnately divided, herbaceous, with anastomosing veins, the fertile 2-pinnate without laminar expansion, the pinnules lobed. Sori 1 on each lobe, the lobes recurved and collectively forming a globose structure enclosing a group of sori; indusium caducous.

1. O. sensibilis L., *Sp. Pl.* 1062 (1753). Sterile leaves up to 50 cm, dying down in winter, triangular, deeply pinnatifid, with winged rhachis and 2–16 pairs of sinuate or coarsely pinnatifid pinnae. Fertile leaves equalling the sterile, lanceolate, stiffly erect, becoming dark brown or black. *Damp places. Naturalized in a few localities in Britain and Germany.* [Br Ge.] (*North America and E. Asia.*)

19. DRYOPTERIDACEAE

(Aspidiaceae)

EDIT. †D. H. VALENTINE AND D. M. MOORE

Rhizome dictyostelic, with opaque scales. Petiole with 5–7 vascular strands; rhachis grooved, interrupted to admit grooves of costae. Sori superficial; spores bilateral.

1 Indusium reniform **3. Dryopteris**
1 Indusium peltate
 2 Veins free; pinnae not more than 1·5 cm wide **1. Polystichum**
 2 Veins anastomosing; pinnae 1·5–4 cm wide **2. Cyrtomium**

1. Polystichum Roth

†D. H. VALENTINE

Rhizome short, ascending. Leaves tufted; lamina 1- to 3-pinnate, the ultimate segments with acuminate, bristle-pointed teeth; veins mostly free. Sori orbicular; indusium peltate.

1 Leaves pinnate; pinnae undivided **1. lonchitis**
1 Leaves usually 2- or 3-pinnate; pinnae divided
 2 Pinnules distinctly stalked, not decurrent **3. setiferum**
 2 Pinnules sessile or subsessile, decurrent
  3 Leaves rigid; pinnules glabrous above **2. aculeatum**
  3 Leaves soft; pinnules hairy above **4. braunii**

1. P. lonchitis (L.) Roth, *Tent. Fl. Germ.* **3(1)**: 71 (1799) (*Aspidium lonchitis* (L.) Swartz, *Dryopteris lonchitis* (L.) O. Kuntze). Rhizome short. Leaves 15–60 cm, short, rigid, persistent; lamina 2–6 cm wide, coriaceous, linear-lanceolate, pinnate; pinnae 15–50 on each side, sometimes overlapping, slightly curved, very shortly stalked, auricled on the upper side at the base, undivided; margin serrate. Indusium irregularly dentate. $2n = 82$. *In open, rocky and often montane habitats. Much of Europe, but very local in the former U.S.S.R.* Al Au Be Br Bu Co Cr Cz Fa Fe Ga Ge Gr Hb He *Ho Hs Hu Is It Ju No Po Rm Rs(N, B, W, K) Su.

The hybrid between **1** and **2** (*P.* × *illyricum* (Borbás) Hahne) is widely distributed in W., S. & C. Europe.

P. munitum (Kaulfuss) C. Presl, *Tent. Pteridogr.* 83 (1836), from W. North America, like **1** but the lamina 6–18 cm wide, lanceolate, and the pinnae narrower, with forward-directed or incurved teeth, is cultivated for ornament and is occasionally naturalized in W. Europe.

2. P. aculeatum (L.) Roth, *Tent. Fl. Germ.* **3(1)**: 79 (1799) (*P. lobatum* (Hudson) Bast., *P. aculeatum* subsp. *lobatum* (Hudson) Vollmann, *Dryopteris lobata* (Hudson) Schinz & Thell., *Aspidium lobatum* (Hudson) Swartz). Rhizome thick, woody. Leaves 30–90 cm, rigid, usually persistent; lamina 5–22 cm wide, lanceolate, pinnate or 2-pinnate; pinnae up to 50 on each side, pinnate or pinnatifid; pinnules serrate, sessile or subsessile, obliquely decurrent; the proximal acroscopic pinnule of each pinna longer than the rest, the proximal side straight, the distal acutely auricled, the 2 sides forming an acute angle at the base. Indusium thick. $2n = 164$. *Much of Europe, except the east and extreme north.* Al Au Be Br Bu Co Cz †Da †Fe Ga Ge Gr Hb He Ho Hs Hu It Ju No Po Rm Rs(B, C, W, K) Su ?Tu.

3. P. setiferum (Forskål) Woynar, *Mitt. Naturw. Ver. Steierm.* **49**: 181 (1913) (*P. aculeatum* auct., non (L.) Roth, *P. aculeatum* subsp. *angulare* (Kit. ex Willd.) Vollmann, *D. setifera* (Forskål) Woynar, *P. angulare* (Kit. ex Willd.) C. Presl, *Aspidium aculeatum* sensu Swartz). Rhizome thick, woody. Leaves 30–120 cm, soft, usually not persistent; lamina 10–25 cm wide, lanceolate, 2-pinnate; pinnae up to 40 on each side; pinnules serrate, distinctly stalked, not decurrent; proximal acroscopic pinnule scarcely longer than the rest, the proximal side rounded below, the distal obtusely auricled, the 2 sides forming an obtuse angle at the base. Indusium thin. $2n = 82$. *S., W. & C. Europe.* Al Au Az Be Bl Br Bu Co Cr Ga Ge Gr Hb He ?Ho Hs Hu It Ju Lu Rm Rs(K) Sa Si Tu.

The hybrid between **2** and **3** (*P.* × *bicknellii* (Christ) Hahne) is found with the parents in many places in W. & C. Europe; the plants are often vigorous and infertile.

4. P. braunii (Spenner) Fée, *Mém. Fam. Foug.* **5**: 278 (1852) (*P. paleaceum* auct., non (Borkh.) O. Schwarz, *Dryopteris braunii* (Spenner) Underw., *Aspidium braunii* Spenner). Leaves up to 80 cm, soft, usually dying in autumn; lamina up to 20 cm wide, oblong-lanceolate, generally 2-pinnate; pinnae up to 40 on each side; pinnules up to 15 on each side, softly hairy above, the margins serrate; proximal acroscopic pinnule not larger than the rest; pinnules sessile or subsessile, obliquely decurrent, not or scarcely auricled, acutely angled at the base. $2n = 164$. *Woodland and shady places. S. & C. Norway; C. Europe, extending locally to C. Spain and E. Russia.* Au Cz Ga Ge He Hs Hu It Ju No Po Rm Rs(B, C, W, E) Su.

The hybrid between **2** and **4** (*P.* × *luerssenii* (Dörfler) Hahne) is fairly widely distributed in C. Europe.

2. Cyrtomium C. Presl

J. R. AKEROYD

Like *Polystichum* but leaves 1-pinnate, the pinnae large, often falcate, usually entire; veins anastomosing; ultimate segments without bristle-pointed teeth.

1. C. falcatum (L. fil.) C. Presl, *Tent. Pteridogr.* 86 (1836) (*Polystichum falcatum* (L. fil.) Diels). Stems with large, dark scales. Leaves 20–80 cm, coriaceous, the upper surface shining; petiole $\frac{1}{4}$–$\frac{1}{2}$ as long as the lamina. Pinnae 5–12 × (1·5–)2–4 cm, ovate-lanceolate, falcate, acuminate, shortly stalked. *Cultivated for ornament, and locally naturalized in W. Europe, especially by the sea.* [Az Be Br Ga Hb Ho.] (*E. Asia.*)

C. fortunei J. Sm., *Ferns Brit. Foreign* 286 (1866) (*P. falcatum* auct., non (L. fil.) Diels), from Japan, like **1** but with smaller and narrower, finely toothed pinnae, has been reported to be naturalized in Açores and in S. Europe.

3. Dryopteris Adanson

V. H. HEYWOOD (EDITION 1) REVISED BY C. R. FRASER-JENKINS (EDITION 2)

Rhizome short, erect or ascending, densely covered with broad, soft, often fimbriate scales. Leaves tufted, 1- to 4-pinnate; veins free. Sori orbicular; indusium reniform.

Literature: C. R. Fraser-Jenkins, *Bol. Soc. Brot.* ser. 2, **15**: 176–236 (1982). C. R. Fraser-Jenkins & T. Reichstein in G. Hegi, *Ill. Fl. Mittel-Eur.* ed. 3, **1(1)**: 136–169 (1984).

1 Leaves 1(2)-pinnate
2 Leaves eglandular
3 Sterile and fertile leaves similar in size; pinnae at least 20 on each side **(1–3). filix-mas** group
3 Fertile leaves longer and more erect than sterile; pinnae 10–20 on each side **19. cristata**
2 Leaves with at least a few glands
4 Leaves usually not more than 20 cm, narrowly lanceolate **18. fragrans**
4 Leaves more than 20 cm, lanceolate to oblong
5 Leaves with scattered glands; pinnule-lobes ± obtuse **(1–3). filix-mas** group
5 Leaves densely glandular; pinnule-teeth acute **5. tyrrhena**
1 Leaves 2- to 4-pinnate
6 Leaves glandular on both surfaces
7 Pinnule-teeth with setaceous apex
8 Pinnules not crowded or crispate **10. expansa**
8 Pinnules distinctly crowded, crispate **15. crispifolia**
7 Pinnule-teeth acute, but without setaceous apex
9 Lamina 3- or 4-pinnate, smelling of coumarin when crushed **14. aemula**
9 Lamina 2- or 3-pinnate, not smelling of coumarin when crushed
10 Lamina narrowly lanceolate or narrowly triangular **7. villarii**
10 Lamina triangular-lanceolate to deltate
11 Lamina 2-pinnate, densely glandular above; leaves not persisting in winter **6. submontana**
11 Lamina 2- or 3-pinnate, slightly glandular above; leaves persisting in winter **8. pallida**
6 Leaves eglandular or glands restricted to the lower surface of leaves
12 Pinnule-teeth with setaceous apex
13 Petiole-scales concolorous
14 Lamina narrowly lanceolate or ovate-lanceolate; pinnules ± crowded **17. carthusiana**
14 Lamina broadly triangular-lanceolate; pinnules not crowded
15 Lamina thin, pale or yellowish-green **10. expansa**
15 Lamina thick, usually dark green **16. guanchica**
13 Petiole-scales with darker base or central stripe
16 Lamina 2-pinnate; indusium eglandular **11. remota**
16 Lamina 3-pinnate; indusium usually with glands
17 Spores rugulose-cristate, spinulose **16. guanchica**
17 Spores tuberculate
18 Proximal basiscopic pinnule of basal pinna at least ½ as long as pinna; spores with acute tubercles, pale brown **10. expansa**
18 Proximal basiscopic pinnule of basal pinna usually less than ½ as long as pinna; spores with obtuse tubercles, dark brown
19 Petiole-scales ovate-lanceolate; spores 44–50 μm **9. dilatata**
19 Petiole-scales lanceolate to linear-lanceolate; spores 34–40 μm **12. azorica**
12 Pinnule-teeth acute but without setaceous apex
20 Leaves (1)2-pinnate, the fertile longer and more erect than the sterile **19. cristata**
20 Leaves all at least 2-pinnate, the fertile and sterile similar in size
21 Lamina ovate-lanceolate, pale green **4. caucasica**
21 Lamina triangular-lanceolate to deltate, pale or dark green
22 Lamina broadly triangular or deltate, pale green **8. pallida**
22 Lamina narrowly triangular-lanceolate, ± dark green
23 Petiole-scales dense throughout, brown or pale brown with dark base and usually a dark central stripe **11. remota**
23 Petiole-scales dense below, scattered above, reddish-brown, concolorous **13. corleyi**

(1–3). D. filix-mas group. Rhizome erect. Leaves (15–)30–120(–150) cm; fertile leaves similar to sterile leaves, pinnate with deeply pinnatifid pinnae, or 2-pinnate; pinnae 20–35 on each side; teeth not mucronate.

1 Point of insertion of secondary rhachis blackish; pinnules or ultimate segments ± parallel-sided **2. affinis**
1 Point of insertion of secondary rhachis not blackish; pinnules or ultimate segments with somewhat curved sides, denticulate, crenate, serrate or lobed all round
2 Pinnule-teeth acute; indusium not inflexed **1. filix-mas**
2 Pinnule-teeth obtuse; indusium distinctly inflexed **3. oreades**

1. D. filix-mas (L.) Schott, *Gen. Fil.* t. 9 (1834) (*Nephrodium filix-mas* (L.) L. C. M. Richard, *Polystichum filix-mas* (L.) Roth, *Polypodium filix-mas* L.). Rhizome stout; crowns 1 to many. Leaves (15–)30–130(–150) × 5–25 cm, not persisting in winter; petiole ¼–½ as long as the lamina, pale brown, somewhat scaly; scales pale brown; point of insertion of secondary rhachis not blackish; lamina oblong- to elliptical-lanceolate, flat or recurved at apex, mid to light green, not shining; pinnae linear- to oblong-lanceolate, sessile, deeply pinnatifid to pinnate, flat or recurved at margin; pinnules or their ultimate segments oblong to oblong-lanceolate, acutely denticulate, crenate-serrate or lobed all round, rounded at apex; teeth acute. Sori 1·5 mm in diameter, (2)3–6 on each pinnule; indusium not glandular, not inflexed or embracing the sporangia when young. $2n = 164$. *Almost throughout Europe.* All except Az Bl Cr Sb.

2. D. affinis (Lowe) Fraser-Jenkins, *Fern Gaz.* **12**: 56 (1979) (*D. paleacea* auct., *D. filix-mas* auct., pro parte). Like **1** but leaves often persisting in winter; petiole and rhachis densely covered with reddish to brown or pale brown scales; point of insertion of secondary rhachis usually blackish; lamina usually dark green, shining; pinnules or their ultimate segments parallel-sided, obliquely truncate, acute or rounded at apex, denticulate only at apex; indusium usually inflexed and embracing the sporangia when young. *Much of Europe, but absent from most of the north and east.* Au Az Be Br Bu Co Cz Da Ga Ge Hb He Ho Hs Hu It Ju Lu No Po Rm Rs(W, K) Sa Si Tu.

1 Pinnule-teeth obtuse; indusium thick, ± persistent
2 Leaves persisting in winter; axes of lamina not glandular **(a)** subsp. **affinis**
2 Leaves not persisting in winter; axes of lamina glandular, at least when young **(c)** subsp. **cambrensis**
1 Pinnule-teeth acute; indusium thin, deciduous

3 Pinnule lobes oblong (**b**) subsp. **borreri**
3 Pinnule lobes rounded to acute (**d**) subsp. **persica**

(**a**) Subsp. **affinis** (*D. pseudomas* sensu J. Holub & Pouzar pro parte): Leaves persisting in winter. Scales dense, narrow, glossy, usually with a dark base. Lamina very glossy. All pinnae symmetrical. Pinnules rounded to rounded-truncate, usually with few obtuse teeth, these sometimes prominent. Lowest basiscopic pinnule of lowest pinna $\frac{1}{2}$–$\frac{2}{3}$ adnate to costa. Indusium inflexed, thick and persistent. $2n = 82$. *S.W. & W.C. Europe.*

(**b**) Subsp. **borreri** (Newman) Fraser-Jenkins, *Willdenowia* **10**: 110 (1980) (*D. borreri* (Newman) Newman ex Oberholzer & Tavel, *D. mediterranea* Fomin, *D. paleacea* (D. Don) Hand.-Mazz. pro parte, non (Swartz) C. Chr.): Leaves mostly not persisting in winter. Scales more or less dense, narrow and wide intermixed, pale or with a dark base. Lamina paler and less coriaceous and shining than that of subsp. (**a**). Pinnules truncate to subacute, with prominent, acute teeth. Lowest basiscopic pinnule of lowest pinna usually longer, with oblong side-lobes, stalked. Indusium thinner and less inflexed than subsp. (**a**), shrivelling rapidly. $2n = 123$. *Almost throughout the range of the species.*

(**c**) Subsp. **cambrensis** Fraser-Jenkins, *Sommerfeltia* **6**: XI (1987) (incl. subsp. *stilluppensis* sensu Fraser-Jenkins, non Sabr.): Leaves not persisting in winter. Scales dense, lanceolate, reddish- or yellowish-brown, without a dark base. Lamina glossy, glandular on the axis, at least when young. All pinnae more or less symmetrical, tapering. Pinnules rounded, with obtuse teeth. Indusium highly inflexed but shrivelling later. $2n = 123$. *Calcifuge. Throughout the range of the species except parts of C. & S. Europe.*

(**d**) Subsp. **persica** (Lowe) Fraser-Jenkins, *Willdenowia* **10**: 113 (1980): Scales wider, pale. Lamina pale green, slightly glossy. Pinnules narrow, rounded to acute, with long, acute teeth; point of insertion of secondary rhachis only faintly blackish. Indusium curved down but not inflexed, soon shrivelling. *Mainly in E. Europe.*

3. D. oreades Fomin, *Monit. Jard. Bot. Tiflis* **18**: 20 (1910) (*D. abbreviata* auct.). Like **1** but crowns usually many, crowded; leaves 25–60(–120) cm; lamina with scattered, minute glands beneath, pale to greyish-green, not shining; tips of pinnae curving upwards; apical teeth blunt, flabellate; margin of indusium inflexed, often glandular. $2n = 82$. *Calcifuge. W. & W.C. Europe, extending to N.C. Italy.* Br Co Ga *Ge Hb Hs It Lu Sa.

4. D. caucasica (A. Braun) Fraser-Jenkins & Corley, *Fern Gaz.* **10**: 22 (1971). Rhizome erect, unbranched; plant with more delicate habit than **1–3**. Leaves up to 100 cm, eglandular, not persistent in winter; petiole $\frac{1}{4}$–$\frac{1}{2}$ as long as the lamina; scales sparse, lanceolate to linear-lanceolate, brown, concolorous. Lamina ovate-lanceolate to elliptical, 2-pinnate, pale green; pinnules narrowly lanceolate, with acute teeth. Sori 1–2 mm in diameter; indusium lacerate, glandular. $2n = 82$. *Montane forests. From Turkey-in-Europe to Krym; local.* ?Rm Rs(W, K) Tu. (*Anatolia, Caucasus.*)

5. D. tyrrhena Fraser-Jenkins & Reichstein, *Fern Gaz.* **11**: 177 (1975). Rhizome stout. Leaves 30–60 cm, persisting in winter; petiole densely glandular; scales ovate-lanceolate, pale or reddish; lamina lanceolate, pinnate, coriaceous, greyish-green, densely glandular, scented. Pinnae not usually crowded; pinnules oblong-lanceolate, shallowly to deeply lobed, with long, acute teeth. Sori usually not more than 2 on each pinnule, usually less than 1 mm; indusium glandular, persistent. $2n = 164$. *Siliceous rocks and screes. Mountains of W. Mediterranean region; local.* Co Ga Hs It Sa.

6. D. submontana (Fraser-Jenkins & Jermy) Fraser-Jenkins, *Candollea* **32**: 311 (1977) (*D. villarii* auct., non (Bellardi) Woynar ex Schinz & Thell. pro parte). Rhizome procumbent or ascending. Leaves 15–40 × 5–15 cm, tufted, forming 1 or a few crowns; petiole long, pale yellow to greenish-yellow above the black base; scales concolorous, reddish-brown; lamina triangular-lanceolate or deltate, dull green, densely covered with yellowish glandular hairs, fragrant, 2-pinnate; pinnae 15–25 on each side, ovate- to oblong-lanceolate, sessile, pinnate; pinnules pinnatilobed to subpinnate, often with long, acute teeth near the tips. Sori in 2 rows on each fertile pinnule; indusium glandular. $2n = 164$. *Calcareous rocks and rock-crevices. S. & S.C. Europe; N. Britain.* Al Br Ga Gr Hs It Ju Rm.

7. D. villarii (Bellardi) Woynar ex Schinz & Thell., *Viert. Naturf. Ges. Zürich* **60**: 339 (1915) (*D. rigida* (Swartz) A. Gray, *Nephrodium villarii* (Bellardi) G. Beck, *Polystichum rigidum* (Swartz) DC.). Like **6** but fronds narrowly lanceolate or narrowly triangular; petiole short; pinnule-teeth long, acute. $2n = 82$. *Mountains of C. Europe and Balkan peninsula.* Al Au Bu Ga Ge Gr He It Ju ?Po.

8. D. pallida (Bory) C. Chr. ex Maire & Petitmengin, *Étude Pl. Vasc. Réc. Grèce* **2**: 238 (1908) (*D. villarii* subsp. *pallida* (Bory) Heywood, *D. pallida* (Bory) Fomin, *Polystichum rigidum* var. *australe* Willk. non Ten.). Like **6** but leaves more broadly triangular or deltate, pale green, not or only slightly glandular above, almost 3-pinnate, persisting in winter; pinnule-teeth wider. *Calcicole. Mediterranean region, mainly in the east.* Al Bl †Co Cr Gr It Ju Sa Si.

(**a**) Subsp. **pallida**: Leaves usually more than 25 cm; lamina triangular-lanceolate; petiole shorter than or equalling the lamina. $2n = 82$. *C. & E. Mediterranean region.*

(**b**) Subsp. **balearica** (Litard.) Fraser-Jenkins, *Candollea* **32**: 314 (1977): Leaves up to 25 cm; lamina deltate; petiole equalling or longer than the lamina. $2n = 82$. ● *Islas Baleares (Mallorca).*

9. D. dilatata (Hoffm.) A. Gray, *Man. Bot.* 631 (1848) (*D. austriaca* auct., *Polystichum dilatatum* (Hoffm.) Schumacher). Rhizome erect or ascending. Leaves 10–150(–180) × 4–40 cm; petiole from $\frac{1}{2}$ as long as to as long as the lamina; scales ovate-lanceolate, bicolorous, dark brown or blackish in the centre, pale brown at the margin; lamina triangular-ovate to -lanceolate, 3-pinnate, usually dark green with a few glands beneath; pinnae 15–25 on each side, stalked, triangular-ovate or -lanceolate, pinnate; pinnules oblong-ovate to -lanceolate, pinnate; the proximal basiscopic pinnule of the basal pinna nearly always less than $\frac{1}{2}$ as long as the pinna; third-order pinnules ovate to oblong, toothed, sometimes pinnately lobed, the apex setaceous. Sori 0·5–1 mm in diameter; indusium with stalked glands; spores 44–50 μm, dark brown, with dense, obtuse tubercles. $2n = 164$. *Most of Europe, but rare in the Mediterranean region and the south-east.* All except ?Al Bl Cr Is Rs(N, E) Sa Sb Si.

10. D. expansa (C. Presl) Fraser-Jenkins & Jermy, *Fern Gaz.* **11**: 338 (1977) (*D. assimilis* S. Walker). Like **9** but lamina thin, pale green; proximal basiscopic pinnule of the basal pinna at least $\frac{1}{2}$ as long as the pinna; lower third-order pinnules falcate, decurrent, less setaceous; spores pale brown, with less dense, acute tubercles. $2n = 82$. ● *N. Europe; mountains of C. & S. Europe southwards to C. Portugal, Corse and C. Greece.* Au Be Br Bu Co Cz Da Fa Fe Ga Ge Gr He Ho Hs Hu Is It Ju Lu No Po Rm Rs(N, B, C, W, K) Sb Su.

11. D. remota (A. Braun ex Döll) Druce, *List Br. Pl.* 87 (1908). Like **9** but scales narrowly lanceolate; lamina narrowly triangular-lanceolate, 2-pinnate, the pinnules deeply pinnatifid below, more or less eglandular; apex of third-order pinnules acute or setaceous; indusium eglandular; some spores abortive. 2*n* = 123. *Moist, shady places in the mountains. W. & C. Europe, extending eastwards to Macedonia.* Au †Br Cz Ga Ge Hb He Hs Hu It Ju Po Rm Rs(W).

12. D. azorica (Christ) Alston, *Bol. Soc. Brot.* **30**: 14 (1956). Like **9** but scales lanceolate to linear-lanceolate; lamina dark green; pinnules with prominent, spinulose teeth, the scales beneath more conspicuous; proximal basiscopic pinnule of the basal pinna shorter; third-order pinnules oblong; spores 34–40 µm. 2*n* = 82. ● *Açores.* Az.

13. D. corleyi Fraser-Jenkins, *Bol. Soc. Brot.* ser. 2, **55**: 248 (1982). Leaves 30–60 cm; petiole $\frac{1}{3}$–$\frac{1}{2}$ as long as the lamina, densely scaly below; scales lanceolate, reddish-brown, concolorous. Lamina 2-pinnate (3-pinnate below), narrowly triangular-lanceolate, with stalked glands beneath; pinnules adnate to the rhachis, the lobes obtuse, obscurely and remotely dentate. Indusium thick, dark reddish-brown, the margin curved down, eglandular. 2*n* = 82. ● *N. Spain (Oviedo prov.).* Hs.

14. D. aemula (Aiton) O. Kuntze, *Revis. Gen.* **2**: 812 (1891) (*Lastrea aemula* (Aiton) Brackenr.). Rhizome erect or ascending. Leaves 15–60 cm, tufted; petiole about as long as the lamina, purplish-black at least below; scales narrowly lanceolate, concolorous, reddish-brown; lamina triangular-ovate to -lanceolate, bright green, with minute sessile glands on both surfaces, fragrant, 3- or 4-pinnate; pinnae 15–20 on each side, stalked, triangular-ovate to -lanceolate, pinnate; pinnules triangular-ovate to lanceolate or oblong, the basal very shortly stalked, the others sessile, pinnate; pinnules of the third or fourth order somewhat concave with upturned margins, pinnately toothed. Sori 0·5–1 mm in diameter; indusium with sessile glands. 2*n* = 82. *W. Europe, from Açores and N.W. Spain to N. Scotland.* Az Br Ga Hb Hs.

15. D. crispifolia Rasbach, Reichstein & Vida, *Bot. Jour. Linn. Soc.* **74**: 266 (1977). Like **14** but leaves strongly glandular-viscid; scales with dark central stripe or dark brown; pinnae crowded; pinnules sessile, usually crispate. 2*n* = 123. ● *Açores.* Az.

16. D. guanchica Gibby & Jermy, *Bot. Jour. Linn. Soc.* **74**: 256 (1977). Like **14** but leaves 3-pinnate; petiole purplish-black below only; scales lanceolate, brown or somewhat reddish-brown, sometimes with darker central stripe; lamina broadly triangular-lanceolate, with few glands, dark green; pinnae crowded; pinnules with rather long stalks. 2*n* = 123. *Calcifuge. S. Spain, W. Portugal.* Hs Lu.

17. D. carthusiana (Vill.) H. P. Fuchs, *Bull. Soc. Bot. Fr.* **105**: 339 (1958) (*D. spinulosa* (O. F. Mueller) O. Kuntze, *Polystichum spinulosum* (O. F. Mueller) DC., *Polypodium spinulosum* O. F. Mueller). Rhizome procumbent or creeping. Leaves 12–60 × 5–25 cm, stiffly erect; petiole about as long as the lamina; scales pale brown, concolorous; lamina narrowly lanceolate or ovate-lanceolate, light to yellowish-green, with minute glandular hairs beneath, 3-pinnate or almost so, with crowded lobes; pinnae 15–25 on each side, triangular-ovate or -lanceolate, shortly stalked, pinnate; pinnules pinnate or almost so, more or less flat, the teeth setaceous. Sori 0·5–1 mm in diameter, in 2 rows on each fertile pinnule; indusium eglandular. 2*n* = 164. *Most of Europe, but rare in the Mediterranean region.* Al Au Be Br Bu Co Cz Da Fe Ga Ge Hb He Ho Hs Hu It Ju No Po Rm Rs(N, B, C, W, K, E) Su Tu.

18. D. fragrans (L.) Schott, *Gen. Fil.* t. 9 (1834). Rhizome short and thick, more or less vertical. Leaves (5–)10–20(–40) × 2–4 cm, tufted; petiole up to $\frac{1}{3}$ as long as the lamina; scales concolorous, shining, brown or reddish; lamina narrowly lanceolate, green, covered with minute glands, scented, pinnate; pinnae 15–40 on each side, lanceolate, deeply pinnatifid, often inrolled at the margin. 2*n* = 82. *Rocks and screes. Arctic Finland (Utsjoki); N. Ural.* Fe Rs(N). (*N. Asia; Japan; North America.*)

19. D. cristata (L.) A. Gray, *Man. Bot.* 631 (1848) (*Polystichum cristatum* (L.) Roth). Rhizome procumbent or creeping. Leaves of 2 kinds, the sterile 15–30 × 3–8·5 cm, patent, the fertile 30–60 × 14–20 cm, erect; petiole of sterile leaves $\frac{1}{2}$ as long as the lamina, of fertile leaves almost equalling the lamina; scales pale brown; lamina linear- to oblong-lanceolate, 1- or 2-pinnate, glabrous; pinnae 10–20 on each side, triangular-ovate or oblong to lanceolate, shortly stalked, deeply pinnatifid or pinnate; pinnules oblong, obtuse, pinnately lobed or serrate. Sori in 2 rows on each fertile pinnule; indusium eglandular. 2*n* = 164. *Mesotrophic mires. From subarctic Finland southwards to E. Spain and C. Romania.* Au Be Br Cz Da Fe Ga Ge He Ho Hs Hu It Ju No Po Rm Rs(N, B, C, W, K, E) Su.

20. LOMARIOPSIDACEAE

(*Elaphoglossaceae*)

EDIT. †D. H. VALENTINE AND D. M. MOORE

Rhizome polystelic, with dark, clathrate scales. Leaves dimorphic; petiole with 5–7 vascular strands. Sporangia occupying the whole fertile surface, and not in distinct sori; spores bilateral.

A predominantly tropical family.

1. Elaphoglossum Schott ex Sm.

†T. G. TUTIN

Epiphytic. Sterile leaves simple and entire, coriaceous, with scales on both surfaces; veins free; fertile leaves smaller and narrower than the sterile; petioles longer.

1. E. semicylindricum (Bowdich) Benl, *Bot. Macar.* **6**: 59 (1980) (*E. hirtum* auct., non Swartz). Leaves 10–40 × 1–3 cm, oblong, narrowed at the base and at the subacute to rounded apex; petiole 2–10 cm, covered, like the lamina, with brown, fimbriate scales. *Epiphytic on* Erica scoparia *subsp.* azorica. *Açores.* Az. (*Madeira.*)

21. DAVALLIACEAE

EDIT. †D. H. VALENTINE AND D. M. MOORE

Rhizome polystelic, with broad dorsal and ventral meristeles, and with clathrate scales. Leaves distant, in 2 ranks, articulated to the rhizome; petiole with several vascular strands. Sori submarginal, with indusium; spores bilateral.

1. Davallia Sm.

A. LAWALRÉE

Rhizome rather stout, with ciliate scales. Lamina deltate to narrowly ovate, rather finely divided, coriaceous, glabrous; veins free. Indusium attached at the base and sides, cup-shaped, almost reaching the margin of the ultimate segments.

1. **D. canariensis** (L.) Sm., *Mém. Acad. Sci. (Turin)* **5**: 414 (1793). Rhizome with bright chestnut-coloured scales. Leaves 10–60 cm, solitary; petiole with scales at the base only, about equalling the lamina; lamina glabrous, deltate, 3- or 4-pinnate; ultimate segments lanceolate or ovate-oblong, mostly bidentate. *On rocks, walls and trees. Portugal and W. Spain; Açores.* Az Hs Lu.

22. BLECHNACEAE

EDIT. †D. H. VALENTINE AND D. M. MOORE

Rhizome dictyostelic, with clathrate scales. Petiole with 2 vascular strands; veinlets, branching and anastomosing, forming a row of areolae on each side of the costae. Sori discrete in areola, or forming a coenosorus along the costae; indusium present; spores bilateral.

Doodia caudata (Cav.) R. Br., *Prodr. Fl. Nov. Holl.* 151 (1810), from Australia and New Zealand, resembling *Blechnum spicant* in habit, but with finely dentate-serrate pinnae and discrete sori, has been reported from Açores as an escape from cultivation, and is becoming naturalized.

Leaf pinnate, not producing young plants vegetatively **1. Blechnum**

Leaf 2-pinnatifid, producing young plants vegetatively on distal part **2. Woodwardia**

1. Blechnum L.

A. LAWALRÉE

Rhizome short, oblique. Leaves dimorphic; lamina pinnate, coriaceous, glabrous, the pinnules entire or crenate; sterile leaves with free veins. Sori long and linear, forming a coenosorus which occupies practically the entire length of the pinnae of the fertile leaves; indusium firm, linear.

1. **B. spicant** (L.) Roth, *Ann. Bot. (Usteri)* **10**: 56 (1794) (*B. homophyllum* (Merino) Samp., *Lomaria spicant* (L.) Desv.). Leaves 8–70 × 3–7 cm, oblong-lanceolate, attenuate at the base, with 20–60 pinnae on each side. Sterile leaves patent, with pinnae 3–5 mm wide; fertile leaves erect, with pinnae 1–2 mm wide. $2n = 68$. *Usually calcifuge. Europe, eastwards to Latvia and the E. Carpathians; local in the Mediterranean region.* All except ?Bl Rs(N, C, K, E) Sb.

B. occidentale L., *Sp. Pl.* ed. 2, 1524 (1763), from tropical America, with non-dimorphic, lanceolate, pinnatisect leaves, is becoming naturalized in Açores.

2. Woodwardia Sm.

A. LAWALRÉE

Plant large. Rhizome ascending. Leaves all similar; lamina coriaceous, 2-pinnatifid; veins anastomosing to form areolae but free towards the margin of the lamina. Sori arranged in 1 or more rows on the outer side of the areolae, parallel and close to the midrib.

1. **W. radicans** (L.) Sm., *Mém. Acad. Sci. (Turin)* **5**: 412 (1793). Leaves up to 2·5 m, arcuate, producing young plants vegetatively from buds on the distal part; lamina equalling petiole, deltate or ovate-lanceolate, pinnate; pinnae up to 30 cm, deeply and regularly pinnatifid; segments falcate, acuminate, serrulate. Sori oblong, in 2 rows along either side of the midrib of the pinnules; indusium coriaceous. *Damp, shady places. S. Europe, from Açores to Kriti.* Az Co Cr Hs It Lu Si.

23. MARSILEACEAE

EDIT. †D. H. VALENTINE AND D. M. MOORE

Semi-aquatic plants. Rhizome slender, creeping, hairy, bearing leaves and roots at each node. Leaves alternate, 2-ranked, filiform or cruciformly 4-foliolate. Sporocarps tough and stony, borne singly in the leaf-axils or on stalks adnate to the petiole, and containing sori of megasporangia and microsporangia, each sorus surrounded by an indusium; annulus lacking or rudimentary. Female prothallus many-celled; male prothallus minute.

Leaves 4-foliolate **1. Marsilea**

Leaves simple, filiform **2. Pilularia**

1. Marsilea L.

J. A. CRABBE (EDITION 1) REVISED BY J. R. AKEROYD (EDITION 2)

Leaves long-petiolate; lamina cruciform, consisting of 2 contiguous pairs of opposite, sessile, obdeltate to cuneate leaflets; veins flabellate, anastomosing. Sporocarps brown to blackish with brown hairs, 2-chambered, dehiscent along the ventral suture into 2 valves, stalked or sessile; each chamber with several sori within a delicate indusium.

Plants of this genus are rarely seen and seem to be becoming rarer. All five species occur in habitats subject to periodic shallow inundation such as water-meadows and rice-fields. Submerged, floating and terrestrial forms, usually sterile, are sometimes found.

1 Sporocarps ±sessile; nodes crowded **3. strigosa**
1 Sporocarps pedicellate; nodes ±distinct
 2 Sporocarp not sulcate; pedicels 10–20 mm, 2- to 4-branched, rarely simple **1. quadrifolia**
 2 Sporocarp sulcate; pedicels usually less than 10 mm, simple
  3 Leaflets 8–17 mm
   4 Sporocarps 2–3 mm, cuboid **2. aegyptiaca**
   4 Sporocarps 3–4·5 mm, elliptic-rectangular **4. azorica**
  3 Leaflets 3–8(–11) mm
   5 Leaflets deeply crenate or lobed at apex **2. aegyptiaca**
   5 Leaflets entire or shallowly crenate at apex **5. batardae**

1. M. quadrifolia L., *Sp. Pl.* 1099 (1753) (*M. quadrifoliata* L.). Usually far-creeping, with internodes more than 1 cm. Leaves glabrous; petiole 7–20 cm; leaflets (5–)10–20 mm, obdeltate, with entire to slightly undulate, rounded apex. Sporocarps 1 or 2 (3) near the base of the petiole, 3–5 mm, ellipsoidal, with small basal teeth, pedicellate; stalks 1–2 cm, erect, 2- to 4-branched, rarely simple, usually partly connate, adnate to the petiole. *S. Europe; northwards to* c. 48° *N. in France and E. Russia.* Al Au Bu Cz Ga Ge He Hs Hu It Ju Lu †Po Rm Rs(W, E).

2. M. aegyptiaca Willd., *Sp. Pl.* **5(1)**: 540 (1810). Usually subcaespitose, with internodes less than 1 cm. Leaves sparsely hairy or glabrous, dimorphic; some leaves with leaflets 2–5 mm, narrowly cuneate, with deeply crenate to lobed, rounded apex, on petioles 1–4 cm; other leaves with leaflets 8–15 mm, cuneate to obdeltate, with crenate to entire, rounded apex, on petioles 4–6(–9) cm. Sporocarps 2–3 mm, axillary, cuboid, with a strigose-hairy sulcus in the apex, base and sides, with prominent upper basal teeth and no lower basal teeth; pedicels *c.* 5 mm, simple, erect. *Lower Volga.* Rs(E). (*Egypt and S.W. Asia.*)

3. M. strigosa Willd., *Sp. Pl.* **5(1)**: 539 (1810) (*M. pubescens* Ten., ?*M. aegyptiaca* var. *lusitanica* Coutinho, *M. fabri* Dunal). Usually caespitose, with crowded nodes and copious brown-sericeous buds. Leaves sparsely hairy or glabrous; petiole 2·5–30 cm; leaflets 5–15 mm, cuneate to obdeltate, with entire to crenulate, rounded apex. Sporocarps 3–5 mm, axillary, obovoid, laterally flattened, with conspicuous raphe and obscure basal teeth, more or less sessile, crowded, often imbricate, brown-sericeous. *S. Europe; very local.* Bl Ga It Rs(E) Sa.

4. M. azorica Launert & Paiva in A. & R. Fernandes, *Icon. Select. Fl. Azor.* **1(2)**: 159 (1983). Subcaespitose, with internodes up to 3·5 cm. Leaves sparsely hairy when young, later becoming glabrous; petiole 2·5–12 cm; leaflets 8–17 mm, obdeltate, with rounded, entire apex. Sporocarps 2 or 3 together, 3·5–4·5 mm, elliptic-rectangular, somewhat flattened, with prominent basal teeth, densely appressed-sericeous; pedicels robust, erect, curved, simple. ● *Açores (Terceira).* Az.

5. M. batardae Launert, *Bol. Soc. Brot.* ser. 2, **56**: 101 (1983) (*M. strigosa* auct., non Willd.). Far-creeping, with internodes 0·5–2 cm. Leaves more or less glabrous; petioles 1·5–10 cm; leaflets 3–8(–11) mm, flabellate, with entire or irregularly crenate apex. Sporocarps 2 or 3 together, 3–4·5 mm, subglobose, flattened, often with basal teeth obscure; pedicels 2–8 mm, erect, simple. ● *W. Spain; C. Portugal.* Hs Lu.

2. Pilularia L.

J. A. CRABBE

Leaves filiform, subulate, entire. Sporocarps globose, blackish, with brown hairs, 2- or 4-chambered, dehiscent into as many valves as chambers.

Commonly semi-aquatic, but submerged and terrestrial forms, usually sterile, also occur.

Sporocarps *c.* 3 mm in diameter, erect, subsessile, 4-chambered; leaves usually at least 50 × 1 mm **1. globulifera**
Sporocarps *c.* 0·75 mm in diameter, deflexed, long-stalked, 2-chambered; leaves usually not more than 40 × 0·3 mm **2. minuta**

1. P. globulifera L., *Sp. Pl.* 1100 (1753). Caespitose or creeping, with internodes up to 4 cm. Leaves usually at least 5 cm, 1 mm wide. Sporocarps 4-chambered, *c.* 3 mm in diameter, erect, subsessile; megaspores ovoid, constricted above the middle, 15–20 in each chamber. *Shallow water, marshy ground, pools and lake margins.* ● *W. Europe, northwards to* 61° *N., extending very locally eastwards to Finland, Poland, Czechoslovakia and S.E. Italy.* Be Br Cz Da Fe Ga Ge Hb He Ho Hs It Ju Lu No Po Rs(?C, E) Su.

2. P. minuta Durieu ex A. Braun, *Monatsber. Koenigl. Akad. (Berlin)* **1863**: 435 (1864). Caespitose or creeping, with internodes up to 1 cm. Leaves usually not more than 4 cm, 0·3 mm wide. Sporocarps 2-chambered, *c.* 0·75 mm in diameter, deflexed, on stalks 2–3 times their length; megaspores globose, not constricted, solitary in each chamber. *In seasonally wet hollows and at margins of ditches.* ● *W. Mediterranean region, S. Portugal; very local.* Bl Co Ga Lu Sa Si.

24. SALVINIACEAE

EDIT. †D. H. VALENTINE AND D. M. MOORE

Stems slender, branched, floating. Leaves in whorls of 3, very hairy; 2 of the leaves entire, floating, the third submerged. Heterosporous; sori at the base of the submerged leaves, stalked, surrounded by a thin-walled, hairy indusium, and consisting of either numerous microsporangia or a few megasporangia, or occasionally mixed.

1. Salvinia Séguier

A. LAWALRÉE

Aquatic plants. Floating leaves covered with stellate hairs; submerged leaves finely divided and root-like. Microsporangia producing 64 microspores; megasporangia with 1 megaspore; prothalli developing inside the floating spores.

The only genus.

Literature: T. K. J. Herzog, *Hedwigia* **74**: 257–284 (1935).

1. S. natans (L.) All., *Fl. Pedem.* **2**: 289 (1785). Annual, with slender, branching stems. Floating leaves 10–14 × 6–9 mm, with 3–5 lateral veins on each side, papillae 0·2–0·8 mm; submerged leaves 2–7 cm. Sori up to 3 mm, in groups of 3–8; those with megasporangia spherical. $2n = 18$. *C. & S.E. Europe, extending westwards to N.E. Spain and northwards to 58° N. in E.C. Russia.* †Be Bu Cz †Ga Ge Gr †Ho Hs Hu It Ju Po Rm Rs(B, C, W, K, E) Si Tu.

Plants with leaf-papillae 2–3 mm and leaves 15 × 15–20 mm. recorded from Spain can be referred to **S. molesta** D. Mitch., *Brit. Fern Gaz.* **10**: 251 (1972).

25. AZOLLACEAE

EDIT. †D. H. VALENTINE AND D. M. MOORE

Stems slender, branched, floating; roots simple. Leaves small, 2-ranked, imbricate, bilobed. Heterosporous; sori in groups of 2 or 4 on the lower lobe of the first leaf of each branch, surrounded by an indusium; larger sori with several microsporangia, smaller sori with 1 megasporangium.

1. Azolla Lam.

A. LAWALRÉE (EDITION 1) REVISED BY A. C. JERMY (EDITION 2)

Small aquatic plants. Leaves always harbouring the blue-green alga *Anabaena*, the upper lobe herbaceous, the lower larger and hyaline. Microsporangia producing several groups of microspores (massulae), often covered with specialized hairs (glochidia); megasporangia producing a single megaspore with 3 floats, which germinates at the water surface to produce a small prothallus.

The only genus.

Literature: A. H. Pieterse, L. de Lange & J. P. van Vliet, *Acta Bot. Neerl.* **26(6)**: 433–449 (1977); M. Follieri, *Webbia* **31(1)**: 97–104 (1977).

Leaf-trichomes 1-celled; glochidia non-septate **1. filiculoides**
Leaf-trichomes 2-celled; glochidia septate **2. mexicana**

1. A. filiculoides Lam., *Encycl. Méth. Bot.* **1**: 343 (1783) (*A. caroliniana* Willd., non auct.). Plant 1–10 cm in diameter, elliptical to suborbicular, subglaucous, green, becoming crimson-red in autumn. Upper lobe of leaf 1·5–2·5 × 0·7–1·4 mm, subacute or obtuse, with a membranous margin; leaf-trichomes 1-celled. Microsporangial sori 2 mm; massulae 5–8; glochidia non-septate. *Naturalized in W., C. & S. Europe.* [Be Br Bu Cz Ga Ge Gr Hb Ho Hs Hu It Lu Rm Sa.] (*North and South America.*)

2. A. mexicana C. Presl, *Abh. Böhm. Ges. Wiss.* **(5)3**: 150 (1845) (*A. caroliniana* auct., non Willd.). Like **1** but leaves less glaucous, olive-green becoming brownish-red in autumn; leaf-trichomes 2-celled; massulae 3–6; glochidia septate. *Naturalized in W., C. & S. Europe.* [Cz Ga Ge Ho Hs Hu It Ju Lu Rm Rs(W).] (*North and Central America.*)

SPERMATOPHYTA

GYMNOSPERMAE

Usually evergreen trees or shrubs. Leaves acicular or scale-like, more rarely broader, and then ovate to lanceolate. Xylem without vessels (except in *Gnetales*). Flowers unisexual. Ovules not enclosed in an ovary.

Literature: G. Krüssmann, *Manual of Cultivated Conifers*. London, 1985.

CONIFEROPSIDA

CONIFERALES

26. PINACEAE

EDIT. D. M. MOORE

Monoecious; resiniferous trees, rarely shrubs, with spirally arranged leaves. Flowers (cones) made up of numerous spirally arranged scales. Scales of male cones (microsporophylls) bearing 2 pollen-sacs on the lower surface. Scales of female cones (megasporophylls) made up of 2 parts – an upper ovuliferous scale (cone-scale) bearing 2 ovules on the upper surface, and a lower, subtending scale (bract). Fruit a usually woody cone. Seeds usually winged.

1 Short shoots absent; leaves all solitary
 2 Leaves borne on persistent, peg-like projections or cushions
  3 Leaves petiolate; resin-canal 1, median **3. Tsuga**
  3 Leaves sessile; resin-canals 0–2, lateral **4. Picea**
 2 Leaves not borne on peg-like projections, but leaving disc-like scars
  4 Leaf-scars circular, not at all projecting; female cone erect **1. Abies**
  4 Leaf-scars elliptical, slightly projecting; female cone deflexed or pendent **2. Pseudotsuga**
1 Short shoots present, bearing fascicles of 2 or more leaves
 5 Leaves on long shoots scale-like, not green; short shoots much reduced, bearing 2–5 green leaves **7. Pinus**
 5 Leaves on long shoots similar to those on short shoots; short shoots prominent, bearing many densely clustered leaves
  6 Leaves deciduous; bracts exceeding ovuliferous scales in flower; female cones not more than 4 cm **5. Larix**
  6 Leaves evergreen; bracts minute or absent; female cones more than 4 cm **6. Cedrus**

1. **Abies** Miller

A. O. CHATER

Evergreen trees with regularly whorled branches. Short shoots absent. Leaves spirally arranged, linear, flattened, grooved above, rarely 4-sided; resin-canals 2, longitudinal, either marginal (in the hypodermis), or median (in the mesophyll); leaf-scars circular, not at all projecting. Male cones in leaf-axils on under-surface of twigs. Female cones ripening in the first year, erect, cylindrical to ovoid; scales falling from the persistent axis when ripe.

The descriptions of twigs and leaves apply only to vegetative, lateral twigs and to the leaves on them.

In addition to the species described below, **A. balsamea** Miller, *Gard. Dict.* ed. 8, no. 3 (1768), from N.E. North America, and **A. homolepis** Siebold & Zucc., *Fl. Jap.* **2**: 17 (1842), from Japan, are occasionally grown on a small scale in experimental plantations. Both resemble **6** in the bilateral arrangements of leaves on the twigs: *A. homolepis* may be distinguished by the glabrous young twigs and resinous buds; *A. balsamea* by the resinous buds, sparsely pubescent young twigs and median resin-canals.

Most of the species described, and several others, are cultivated for ornament in parks and gardens.

Literature: J. Mattfeld, *Mitt. Deutsch. Dendrol. Ges. (Jahrb.)* **1925**: 1–37 (1925). J. do Amaral Franco, *Abetos*. Lisboa, 1950.

1 Leaves rigid, acute
 2 Leaves 15–35 mm; resin-canals marginal; bracts exserted from female cone **8. cephalonica**
 2 Leaves 10–15 mm; resin-canals median; bracts included in female cone **9. pinsapo**
1 Leaves flexible, emarginate or obtuse
 3 Leaves glaucous above (with several longitudinal rows of stomata); twigs with reddish pubescence
  4 Twigs brown; resin-canals marginal; female cone 12–20 cm; bracts long-exserted from cone, deflexed **1. procera**
  4 Twigs ashy-grey; resin-canals median; female cone 5–10 cm; bracts included in cone **3. lasiocarpa**
 3 Leaves dark or bright green above (with longitudinal rows of stomata only at apex, or absent); twigs with whitish or brownish, but not reddish, pubescence, or ± glabrous
  5 Leaves less than 15 mm (Sicilia) **7. nebrodensis**
  5 Leaves at least 15 mm
   6 Buds resinous, small; female cone 5–10 cm; bracts included in cone
    7 Twigs olive-green; leaves 1·5–2·5 mm wide; resin-canals marginal **2. grandis**
    7 Twigs silvery-grey; leaves 1–1·3 mm wide; resin-canals median **4. sibirica**
   6 Buds not resinous; female cone 10–20 cm; bracts exserted from cone
    8 Crown compact; twigs sparsely pubescent; each stomatiferous band on the under-surface of leaves composed of 8–10 rows of stomata; female cone 4–5 cm wide **5. nordmanniana**
    8 Crown wide; twigs densely pubescent; each stomatiferous band on the under-surface of leaves composed of 6–8 rows of stomata; cone 3–4 cm wide **6. alba**

1. A. procera Rehder, *Rhodora* **42**: 522 (1940) (*A. nobilis* Lindley, non A. Dietr.). Up to 80 m; trunk stout. Young twigs densely covered with reddish pubescence; buds resinous only at the apex. Leaves on the upper surface of the twigs appressed and curving upwards, not leaving a parting. Leaves 10–35 × 1–1·5 mm, obtuse; resin-canals marginal. Female cone 12–20 × 5–8 cm; bracts long-exserted and deflexed, almost completely concealing the cone.

Planted for timber in N. & *W. Europe.* [Br Da Ge Hb No.] (*W. North America.*)

2. A. grandis (D. Don) Lindley, *Penny Cycl.* **1**: 30 (1833). Up to 75 m; trunk fairly stout. Young twigs sparsely puberulent; buds resinous. Leaves on the upper surface of the twigs pointing horizontally outwards, leaving a distinct parting. Leaves 20–60 × 1·5–2·5 mm, emarginate. Female cone 5–10 × 3–4 cm; bracts included. *Planted for timber in* N. & C. *Europe.* [Br Cz Da Ge Hb He It No.] (*W. North America.*)

3. A. lasiocarpa (Hooker) Nutt., *N. Amer. Sylva* **3**: 138 (1849). Up to 48 m; crown narrowly pyramidal; trunk slender. Young twigs with reddish pubescence; buds resinous. Leaves on upper surface of the twigs pointing forwards and curving upwards, not leaving a parting. Leaves 15–40 × 2 mm, flexible, obtuse or slightly emarginate; resin-canals median. Female cone 5–10 × 3·5 cm; bracts included. *Occasionally planted for timber in* N. *Europe.* [Is Su.] (*W. North America.*)

4. A. sibirica Ledeb., *Fl. Altaica* **4**: 202 (1833). Up to 30 m; crown pyramidal; trunk slender. Young twigs sparsely pubescent; buds very resinous. Leaves on the sides of the twigs spreading horizontally; those on the upper surface pointing forwards and upwards, and not leaving a parting. Leaves 15–30 × 1–1·3 mm, thin, flexible, rounded at the apex or slightly emarginate; resin-canals median. Female cone 6–8 × 3 cm; bracts ½ as long as cone-scales, included. *Forming extensive forests in* N.E. *Russia, westwards to* c. 41° *E. and southwards to* c. 55° *N. Planted for timber in other parts of* N. *Europe.* Rs(N, C) [Da Fe Is]. (*N. Asia.*)

5. A. nordmanniana (Steven) Spach, *Hist. Vég.* (*Phan.*) **11**: 418 (1841). Up to 70 m; trunk stout. Young twigs sparsely pubescent; buds not resinous. Leaves on the upper surface of the twigs spreading forwards and upwards, not leaving a parting. Leaves 15–35 × 1·5–2 mm, emarginate; resin-canals marginal. Female cone 12–18 × 4–5 cm; bracts exserted and deflexed. *Planted for timber in* C. *Europe and occasionally elsewhere.* [Au Cz Da It Lu Rm Su Tu.] (*W. Caucasus, N. Anatolia.*)

6. A. alba Miller, *Gard. Dict.* ed. 8, no. 1 (1768) (*A. pectinata* (Lam.) DC.). Up to 50 m; crown pyramidal; trunk stout. Young twigs densely pubescent; buds not resinous. Leaves on the sides of the twigs spreading horizontally; those on the upper surface pointing outwards and upwards, leaving a distinct parting. Leaves 15–30 × 1·5–2 mm, thick, flexible, emarginate at the apex; resin-canals marginal. Female cone 10–20 × 3–4 cm; bracts exserted, deflexed. $2n = 24$. • *C. & S. Europe, mainly in the mountains, extending eastwards to White Russia; doubtfully native in N.W. France. Planted for timber in* N. & *W. Europe.* Al Au Bu Co Cz Ga Ge Gr He Hs Hu It Ju Po Rm Rs(C, W) [Be Br Da Lu No Su].

A. pardei Gaussen, *Bull. Soc. Hist. Nat. Toulouse* **57**: 357 (1929), is cultivated in France, and may perhaps occur wild in Calabria. It is very like **6** but has median resin-canals.

7. A. nebrodensis (Lojac.) Mattei, *Boll. Orto Bot. Palermo* **7**: 64 (1908) (*A. alba* subsp. *nebrodensis* (Lojac.) Nitz.). Up to 15 m; crown broad, somewhat flattened; trunk stout. Young twigs more or less glabrous; buds resinous. Leaves all patent, dense, spreading evenly from the sides and upper surfaces of the twigs, often leaving a parting on the underside. Leaves 8–13 × 2 mm, thick, rigid, obtuse, rarely apiculate; resin-canals marginal. Female cone *c.* 20 × 5 cm; bracts exserted, deflexed. • *About* 20 *native trees in* N. *Sicilia* (*Le Madonie*). *Formerly forming extensive forests on the mountains of* N. *Sicilia, and now being replanted.* Si.

Some authors regard this taxon as only a variant of **6**, but in the present state of knowledge it seems best to treat it as a species. Trees intermediate between **6** and **7** have been recorded from S.W. Italy (Calabria).

8. A. cephalonica Loudon, *Arbor. Fruticet. Brit.* **4**: 2325 (1838). Up to 30 m; crown pyramidal; trunk fairly stout. Young twigs glabrous; buds very resinous. Leaves all patent, dense, spreading evenly from the sides and upper surfaces of the twigs, less dense and often leaving a parting on the underside. Leaves 15–35 × 2–2·5 mm, thick, rigid, acute and pungent, rarely subobtuse; resin-canals marginal. Female cone 12–16 × 4–5 cm; bracts exserted, deflexed. $2n = 24$. • *Mountains of Greece,* 750–1700 *m; planted for timber in Italy.* ?Al Gr [It].

A. borisii-regis Mattf., *Notizbl. Bot. Gart. Berlin* **9**: 235 (1925), is the name applied to trees in the Balkan peninsula from S. Bulgaria to S. Greece. They are of varying form, but always more or less intermediate between **6** and **8**. They have a chromosome number of $2n = 24$. The young twigs are densely pubescent, the buds are resinous, and the leaves are usually like those of **8** in shape and in arrangement. They may be of hybrid origin; cones are very rarely produced.

9. A. pinsapo Boiss., *Biblioth. Univ. Genève* ser. 2, **13**: 402, 406 (1838). Up to 30 m; crown pyramidal; trunk stout. Young twigs glabrous; buds resinous. Leaves patent, spreading evenly all round the twigs. Leaves 10–15 mm, thick, rigid, acute; resin-canals median. Female cone 10–16 × 3–4 cm; bracts less than ½ as long as the cone-scales, included. *North-facing slopes on limestone mountains,* 1000–2000 *m.* • *S.W. Spain* (*near Ronda*)*; occasionally planted for timber elsewhere.* Hs [Au Lu].

A. × insignis Carrière ex Bailly, *Revue Hort.* (*Paris*) **1890**: 230 (1890), a hybrid between **9** and **5**, first raised in France in 1872, is very common as an ornamental tree in parks throughout Europe. It is like **9** but the young twigs are pubescent, the leaves are 10–33 × 1·5–2·5 mm, acute or obtuse, less dense, and often leaving a parting on the underside of the twigs, the female cone is 11–20 × 3·5–5 cm, and at least the lower bracts are slightly exserted.

2. Pseudotsuga Carrière

J. DO AMARAL FRANCO

Like *Abies* but branches irregularly whorled; resin-canals 2, marginal; leaf-scars elliptical, slightly projecting; female cones deflexed, pendent, ovoid or cylindrical; scales persistent; bracts exserted, deeply trifid.

1. P. menziesii (Mirbel) Franco, *Conif. Duar. Nom.* 4 (1950) (*P. douglasii* (Lindley) Carrière, *P. taxifolia* Britton). Tree up to 100 m, with ridged, dark reddish-brown bark on adult trees. Twigs pubescent. Leaves 20–35 mm, entire at the apex, fragrant when bruised. Female cone 5–10 × 2–3·5 cm, ovoid. *Extensively planted for timber in many parts of Europe.* [Au Be Br Bu Cz Da Fe Ga Ge Gr Hb He It Ju Lu Po Rm Su.] (*W. North America.*)

3. Tsuga (Antoine) Carrière

J. DO AMARAL FRANCO

Evergreen trees with irregularly whorled branches. Short shoots absent. Leaves spirally arranged, linear, flattened or tetragonal, distinctly petiolate; resin-canal 1; leaves borne on prominent, persistent cushions. Male cones in leaf-axils of previous year's twigs. Female cones ripening in the first year, pendent, ovoid or cylindrical; scales persistent; bracts rarely exserted.

Buds globose; leaves with white stomatal bands occupying almost the whole of lower surface — **1. heterophylla**
Buds ovoid; leaves with narrow, white stomatal bands on lower surface — **2. canadensis**

1. T. heterophylla (Rafin.) Sarg., *Silva N. Amer.* **12**: 73 (1899) (*T. mertensiana* auct.). Tree up to 70 m, with a narrow, pyramidal crown. Buds globose. Leaves 6–20 mm, flattened, with 2 broad, white stomatal bands, occupying almost the whole of the lower surface, obtuse. Female cone 20–25 mm, pale or reddish-brown. *Occasionally planted for timber in N.W. Europe.* [Br Da Ga Ge Hb No.] (*W. North America.*)

2. T. canadensis (L.) Carrière, *Traité Gén. Conif.* 189 (1855) (*T. americana* (Miller) Farwell). Tree up to 30 m, with a broad, pyramidal crown. Buds ovoid. Leaves 8–18 mm, with 2 narrow, white stomatal bands on lower surface. Female cone 15–20 mm, pale chestnut-brown. *Occasionally planted for timber.* [Ge.] (*North America.*)

4. Picea A. Dietr.

J. DO AMARAL FRANCO

Evergreen trees with regularly whorled branches. Short shoots absent. Leaves spirally arranged, linear, flattened, or tetragonal; resin-canals 0–2; leaves borne on persistent, peg-like projections. Male cones in leaf-axils of previous year's twigs. Female cones ripening in the first year, ovoid or cylindrical; scales persistent; bracts minute, not exserted.

1 Leaves flattened, with 2 white stomatal bands only on the upper surface
2 Leaves 15–25 mm, pungent; female cone 6–10 cm — **6. sitchensis**
2 Leaves 8–18 mm, obtuse and mucronulate; female cone 3–6 cm — **7. omorika**
1 Leaves tetragonal, with stomatal bands on all 4 sides
3 Twigs glabrous or with scattered, minute pubescence
4 Female cone 3·5–5 cm; leaves with a strong, unpleasant smell when crushed — **3. glauca**
4 Female cone 6–18 cm; leaves not smelling unpleasantly when crushed
5 Female cone 10–18 cm; leaves dark green — **1. abies**
5 Female cone 6–10 cm; leaves bluish-green — **5. pungens**
3 Twigs densely and shortly pubescent
6 Leaves 6–10 mm, obtuse — **2. orientalis**
6 Leaves 10–25 mm, acute
7 Twigs orange-brown; leaves 10–18 mm; scales rhombic-obovate or obovate, with rounded and entire or emarginate apex — **1. abies**
7 Twigs pale yellowish-brown; leaves 15–25 mm; scales rhombic-oblong, narrowed, truncate and erose at apex — **4. engelmannii**

1. P. abies (L.) Karsten, *Deutsche Fl.* 324 (1881) (*Pinus abies* L.). Up to 60 m. Twigs usually orange-brown. Leaves tetragonal, with stomatal bands on all 4 sides, pungent, dark green. *N. Europe, extending southwards in the mountains to N. Greece and Bulgaria; widely planted elsewhere for timber.* Al Au Bu Cz Fe Ga Ge Gr He Hu It Ju No Po Rm Rs(N, B, C, W) Su [Be Br Da Hb Ho Hs].

(a) Subsp. **abies** (*Abies excelsa* (Lam.) Poiret, *Picea excelsa* (Lam.) Link, *P. vulgaris* Link): Up to 60 m. Twigs glabrous or with scattered, minute hairs. Leaves 10–25 mm. Female cone 10–18 × 2·5–4 cm, cylindrical; scales rhombic-obovate, with truncate, and erose or emarginate apex. $2n = 22, 24$. ● *Throughout the range of the species in Europe; cultivated as a forest tree in N., C. & W. Europe.*

(b) Subsp. **obovata** (Ledeb.) Hultén, *Svensk Bot. Tidskr.* **43**: 388 (1949) (*P. obovata* Ledeb., *P. abies* subsp. *europaea* (Tepl.) Hyl.): Up to 30(–50) m. Twigs densely pubescent with short hairs. Leaves 10–18 mm. Female cone 6–8 cm, cylindric-ovoid; scales obovate, with rounded, sometimes emarginate apex. $2n = 24$. *N.E. Europe; locally in the Alps.*

Plants somewhat intermediate between subspp. (a) and (b) (subsp. *alpestris* (Brügger) Domin, subsp. *fennica* (Regel) Parfenov) occur in the Alps and N.E. Europe.

2. P. orientalis (L.) Link, *Linnaea* **20**: 294 (1847). Up to 40(–60) m. Twigs densely pubescent, pale brown. Leaves 6–10 mm, tetragonal, with stomatal bands on all 4 sides, obtuse, dark green, shining, densely crowded and appressed. Female cone 6–9 cm, cylindric-ellipsoid; scales obovate, rounded and entire at the apex. *Occasionally planted for timber on a small scale.* [Au Be It Tu.] (*S.W. Asia.*)

3. P. glauca (Moench) Voss, *Mitt. Deutsch. Dendrol. Ges.* **16**: 93 (1908). Up to 30 m. Twigs glabrous, greyish or pale brown. Leaves 8–18 mm, tetragonal, with stomatal bands on all 4 sides, acute or subacute, bluish-green, with a strong unpleasant smell when crushed. Female cone 3·5–5 cm, cylindric-oblong; scales suborbicular, rounded and entire at the apex. *Planted for timber in N. Europe.* [Au Be Da Fe Is No.] (*N. North America.*)

4. P. engelmannii Parry ex Engelm., *Trans. Acad. Sci. St. Louis* **2**: 212 (1863). Up to 50 m. Twigs minutely glandular-pubescent, pale yellowish-brown. Leaves 15–25 cm, tetragonal, with stomatal bands on all 4 sides, acute, usually bluish-green, with an unpleasant smell when crushed. Female cone 3·5–7·5 cm, cylindric-oblong; scales rhombic-oblong, narrowed, truncate and erose at the apex. *Planted for timber, mainly in N. Europe.* [Au Be Da Fe Is No.] (*W. North America.*)

P. mariana (Miller) Britton, E. E. Sterns & Poggenb., *Prelim. Cat.* 71 (1888), from N. North America, a tree up to 30 m, with densely pubescent twigs, leaves 6–18 mm, glaucous, and female cone 2–3·5 cm, has been planted experimentally for timber in Europe.

P. rubens Sarg., *Silva N. Amer.* **12**: 33 (1899), from N.E. North America, a tree up to 30 m, with densely pubescent twigs, leaves 10–15 mm, dark or yellowish-green, and female cone 3–4 cm, has also been planted experimentally.

5. P. pungens Engelm., *Gard. Chron.* nov. ser., **11**: 334 (1879). Up to 30(–50) m. Twigs glabrous, becoming yellowish-brown. Leaves 20–30 mm, tetragonal, with stomatal bands on all 4 sides, rigid, pungent, bluish-green. Female cone 6–10 cm, cylindric-oblong; scales rhombic-oblong, narrowed and erose at the apex. *Planted for timber, mainly in N. & C. Europe.* [Au Cz Ga He Is It No.] (*W.C. North America.*)

P. asperata Masters, *Jour. Linn. Soc. London (Bot.)* **37**: 419 (1906), from W. China, intermediate in most respects between **1** and **5**, has been planted experimentally for timber in Europe.

6. P. sitchensis (Bong.) Carrière, *Traité Gén. Conif.* 260 (1855) (*P. falcata* (Rafin.) Valck.-Suringar, *P. menziesii* (Douglas) Carrière). Up to 40(–60) m. Leaves 15–25 mm, flattened, with 2 broad, white stomatal bands above, dark green beneath, pungent. Female cone 6–10 × 2·5–3 cm, cylindric-oblong; scales rhombic-oblong. *Cultivated for timber in N.W. & C. Europe.* [Au Br Ga Ge Hb Ho Is No.] (*W. North America.*)

7. P. omorika (Pančić) Purkyně, *Österr. Monatschr. Forstwes.* **27**: 446 (1877) (*Pinus omorika* Pančić). Up to 30 m, of very slender habit. Leaves 8–18 mm, flattened, with 2 broad, white stomatal bands above, dark green beneath, obtuse and mucronulate. Female cone 3–6 × 1·5–2·5 cm, ovoid-oblong; scales suborbicular, denticulate. ● *C. Jugoslavia (Drina basin); cultivated for timber in Britain and some Scandinavian countries.* Ju [Br Da Su].

5. Larix Miller

J. DO AMARAL FRANCO

Deciduous trees with irregularly whorled branches. Twigs of 2 kinds: long shoots bearing scattered leaves; and persistent spur-like shoots bearing tufts of leaves. All leaves linear, usually flattened and thin. Resin-canals 2. Female cones ripening in the first year, subglobose to oblong; scales rounded, persistent; bracts exceeding ovuliferous scales in flower, hidden or exserted at maturity.

In addition to the species described below, **L. laricina** (Duroi) C. Koch, *Dendrologie* **2(2)**: 262 (1873), from E. North America, is cultivated on a small scale in Europe. It is like **4** but with somewhat shorter leaves, and the young branches usually glaucous.

1 Young shoots glaucous; leaves with 2 distinct, white stomatal bands beneath; cone-scales recurved at the apex **1. kaempferi**
1 Young shoots not glaucous; leaves without white stomatal bands beneath
2 Female cone-scales glabrous outside **4. gmelinii**
2 Female cone-scales puberulent or shortly tomentose outside
3 Female cone-scales straight, about twice as long as the bracts **2. decidua**
3 Female cone-scales slightly incurved at apex, convex, about 3 times as long as the bracts **3. sibirica**

1. L. kaempferi (Lamb.) Carrière, *Fl. Serres Jard. Eur.* **11**: 97 (1856) sec. Franco, *Anais Inst. Sup. Agron. (Lisboa)* **19**: 18 (1952) (*L. leptolepis* (Siebold & Zucc.) Endl., *L. japonica* Carrière). Up to 30 m. Twigs yellowish- or reddish-brown, glaucous. Leaves 15–25 mm, with 2 conspicuous, white stomatal bands beneath. Female cones 1·5–3·5 cm, ovoid; scales numerous, the upper edge recurved, slightly pubescent. *Planted for timber, mainly in N.W. Europe.* [Au Be Br Da Fe Ga Ge Hb Ho Rs(B) Su.] (*Japan.*)

2. L. decidua Miller, *Gard. Dict.* ed. 8, no. 1 (1768) (*L. europaea* DC., *L. larix* (L.) Karsten, *Pinus larix* L.). Up to 35 m. Twigs yellowish, glabrous. Leaves 20–30 mm, thin and soft, with 2 greenish stomatal bands beneath. Female cone 2–3·5 cm, ovoid; scales 40–50, rounded, loosely appressed, softly pubescent. ● *Alps; W. Carpathians; planted elsewhere for timber and locally naturalized.* Au Cz Ga Ge He *Hu It Po Rm Rs(W) [Be Br Fe Hb *Ju No Rs(B) Su].

Plants with smaller cones and more concave, rounded scales, from Poland and N.W. Ukraine, have sometimes been treated as subsp. **polonica** (Racib.) Domin, *Pl. Čechosl. Enum.* 12 (1936) (*L. polonica* Racib.). Other variants have been described in detail by K. Šiman in Klika *et al.*, *Jehličnaté* 95–126 (1953).

A hybrid between **2** and **1** (*L. × eurolepis* A. Henry) is planted for timber in parts of Europe.

3. L. sibirica Ledeb., *Fl. Altaica* **4**: 204 (1833) (*L. russica* (Endl.) Sabine ex Trautv.). Up to 40 m. Twigs light yellowish-grey, more or less hairy. Leaves up to 35 mm, thin and soft with 2 greenish stomatal bands beneath. Female cone *c.* 3·5 cm; scales *c.* 30, becoming erecto-patent, finely tomentose outside. *N.E. Russia, westwards to* c. 37° E. *and southwards to* c. 52° N.; *planted for timber in N. Sweden and Finland.* Rs(N, C) [Fe Su].

4. L. gmelinii (Rupr.) Kuzen., *Trav. Mus. Bot. Acad. Pétersb.* **18**: 41 (1920). Up to 30 m. Twigs yellowish or reddish, often pubescent. Leaves *c.* 30 mm, bright green, with 2 greenish stomatal bands beneath. Female cone 2–2·5 cm, ovoid; scales *c.* 20, glabrous outside. *Planted for timber in N. Europe.* [Da Fe No.] (*E. Asia.*)

6. Cedrus Trew

J. DO AMARAL FRANCO

Like *Larix* but leaves evergreen, usually 3-sided; female cones ripening in the second year, the scales falling from the persistent axis when ripe; bracts minute.

Leading shoot pendent; leaves up to 50 mm; female cone rounded at apex **1. deodara**
Leading shoot stiff, erect; leaves not more than 30 mm; female cone truncate or umbilicate at apex **2. atlantica**

1. C. deodara (D. Don) G. Don fil. in Loudon, *Hort. Brit.* 388 (1830) (*Pinus deodara* D. Don). Up to 60 m or more, with pendent leading shoot. Twigs densely hairy. Leaves 20–50 mm, usually green. Male cone 5–12 cm. Female cone 7–14 × 5–9 cm, broadly ellipsoidal, rounded at the apex. *Planted for timber in S. Europe.* [Ga Gr It Lu.] (*Afghanistan and N.W. Himalaya.*)

2. C. atlantica (Endl.) Carrière, *Traité Gén. Conif.* 285 (1855) (*Pinus cedrus* L.). Up to 40 m, with stiff leading shoot. Young twigs downy. Leaves 10–30 mm, green or glaucous. Male cone 3–5 cm. Female cone 5–8 × 3–5 cm, subcylindrical, truncate or umbilicate at the apex. *Planted for timber in S. Europe.* [Ga Gr It Lu.] (*Algeria, Morocco.*)

C. libani A. Richard in Bory, *Dict. Class. Hist. Nat.* **3**: 399 (1823), from S. Anatolia and Lebanon, is frequently planted as an ornamental tree, but not on a large scale. It differs from **2** in its glabrous young twigs, dark green leaves, and female cone 9–15 cm. It is perhaps only subspecifically distinct from **2**.

7. Pinus L.

†H. GAUSSEN, V. H. HEYWOOD AND A. O. CHATER

Evergreen trees or shrubs with regularly whorled branches. Twigs of 2 kinds: long shoots bearing scale-like leaves, and deciduous short shoots bearing acicular leaves in clusters of 2–5(–8). Resin-canals 2 or more, varying in position. Male cones catkin-like, clustered at the base of the young twigs. Female cones (referred to below as 'cones') ripening in the second or third year, eventually falling in their entirety, cylindrical to ovoid; ovuliferous scales woody; bracts minute; the exposed part of the scales (apophysis) with a prominent protuberance (umbo), usually ending in a spine or prickle.

In addition to the species described below, many others are grown for ornament in parks and gardens.

Literature: H. Gaussen, *Les Gymnospermes*, Fasc. VI, ch. XI, *Pinus* (1960). J. Klika *et al.*, *Jehličnaté*. Praha, 1953. N. T. Mirov, *The Genus Pinus*. New York, 1967.

1 Leaves in groups of 5
2 Seeds winged; leaf with 2–4 marginal resin-canals
3 Young shoots with fine, reddish-brown hairs; apophysis not swollen **18. strobus**
3 Young shoots glabrous or subglabrous; apophysis swollen
4 Leaves not rigid, pendent; cone 15–25 cm, straight **16. wallichiana**
4 Leaves rigid, directed forwards to apex of branches; cone 8–15 cm, curved **17. peuce**
2 Seeds not winged; leaf with 2 or 3 median resin-canals
5 Bud-scales dull; cone less than $1\frac{1}{2}$ times as long as wide; seeds with hard husk **14. cembra**
5 Bud-scales shiny; cone more than $1\frac{1}{2}$ times as long as wide; seeds with fragile husk **15. sibirica**
1 Leaves in groups of 2 or 3
6 Leaves in groups of 3, rarely mixed with some in pairs
7 Winter buds cylindrical, resinous; leaves stout; umbo with strong, persistent mucro **5. ponderosa**
7 Winter buds ovoid, not resinous; leaves slender; umbo with small, caducous mucro, or not mucronate
8 Leaves not more than 15 cm; bud-scales closely appressed; cone 7–14 cm, strongly asymmetrical **3. radiata**
8 Leaves 20–30 cm; bud-scales free, deflexed; cone 10–20 cm, symmetrical **12. canariensis**
6 Leaves in pairs
9 Seeds wingless or with wing less than 1 mm **13. pinea**
9 Seeds with well-developed wing more than 1 mm
10 Twigs grey in first year
11 Leaves more than 1 mm wide; resin-canals median **10. heldreichii**
11 Leaves less than 1 mm wide; resin-canals submarginal, rarely median **11. halepensis**
10 Twigs not grey in first year
12 Buds not resinous; scales recurved at apex
13 Resin-canals marginal or submarginal; cone 5–12 × 4 cm **11. halepensis**
13 Resin-canals median; cone 8–20 × 5–8 cm **2. pinaster**
12 Buds resinous; scales not recurved at apex
14 Resin-canals median
15 Leaves 30–70 × 1 mm, strongly twisted; cone asymmetrical **1. contorta**
15 Leaves usually more than 80 × 1·5 mm, not or scarcely twisted; cone ± symmetrical **6. nigra**
14 Resin-canals marginal or submarginal
16 Cone strongly curved; leaves dark green **4. banksiana**
16 Cone not curved; leaves bright green or glaucous
17 Leaves glaucous; cone dull, acute, pendent **7. sylvestris**
17 Leaves bright green; cone shining, obtuse, patent
18 Erect tree up to 25 m; cone 5–7 cm; apophysis very prominent, recurved and hooked **9. uncinata**
18 Shrub up to 3 m; cone 2–5 cm; apophysis flat or concavo-convex, not recurved and hooked **8. mugo**

Subgen. **Pinus** (subgen. *Diploxylon* (Koehne) Pilger). Leaves mostly in groups of 2 or 3 (rarely 5–8), with 2 fibro-vascular bundles; scale-leaves decurrent; umbo dorsal.

1. P. contorta Douglas ex Loudon, *Arbor. Fruticet. Brit.* **4**: 2292 (1838). Tree 6–10 m; bark dark brown, with numerous scales; branches short, contorted. Twigs glabrous, clear green in first year, then orange or brown. Leaves 2·5–7 cm × 0·9–1·5 mm, acute, in pairs, twisted; resin-canals median. Cone 2–6 × 2–3 cm, narrowly ovoid to ovoid-conical, symmetrical, pale brownish-yellow, shining; apophyses near base of cone pyramidal; umbo with slender, fragile mucro. Seeds 4–5 mm; wing *c.* 8 mm. *Cultivated for timber in N. Europe and occasionally elsewhere.* [Be Br Cz Da Fe Ge Hb Is It No Rm Su.] (*W. North America.*)

2. P. pinaster Aiton, *Hort. Kew.* **3**: 367 (1789) (*Pinus maritima* Lam.). Tree up to 40 m; bark becoming deeply fissured, reddish-brown. Buds oblong, fusiform, not resinous. Twigs glabrous, reddish-brown. Leaves 10–25 cm × 2 mm, in pairs, green, rigid and spiny; resin-canals 2–9, median. Cone 8–22 × 5–8 cm, ovoid-conical, symmetrical or almost so, light brown, shining; apophysis rhomboidal, keeled, with prominent, prickly umbo. Seeds 7–8 mm; wing up to 3 mm. $2n = 24$. *S.W. Europe, extending eastwards to C. Italy; sometimes planted elsewhere for timber, sand-binding or shelter.* Co Ga Hs It *Lu Sa Si [Be Br Gr Ju Tu].

The following two subspecies have been recognized but the distinguishing characters of subsp. (**b**) may be partly the result of phenotypic modification due to cultivation of the Mediterranean plant on Atlantic coasts.

(**a**) Subsp. **pinaster** (*P. mesogeensis* Fieschi & Gaussen, *P. hamiltonii* Ten.): Leaves 18–25 cm; resin-canals more than 2 under the sheath. Cone 14–22 cm. *C. & W. Mediterranean region.*

(**b**) Subsp. **atlantica** H. del Villar, *Bol. Soc. Esp. Hist. Nat.* **33**: 427 (1933): Leaves 10–20 cm; resin-canals 2 under the sheath. Cone 9–18 cm. *S. France; Atlantic coast of Spain, Portugal.*

3. P. radiata D. Don, *Trans. Linn. Soc. London (Bot.)* **17**: 442 (1836) (*P. insignis* Douglas). Tree up to 40 m; bark becoming thick, fissured, dark brown. Buds ovoid-cylindrical, resinous. Twigs glabrous, reddish-brown. Basal sheath of short shoots less than 15 mm. Leaves 10–15 cm, in groups of 3, slender, acute, densely crowded; resin-canals median. Cone 7–14 × 5–8 cm, ovoid-conical, very asymmetrical, sessile or shortly stalked, deflexed; apophyses on upper side of cone rounded; umbo with small, caducous mucro. Seeds *c.* 7 mm; wing *c.* 20 mm. *Planted for timber or shelter in W. Europe.* [Az Br Ga Hb Hs It Lu.] (*S. California.*)

P. rigida Miller, *Gard. Dict.* ed. 8, no. 10 (1768), from E. North America, a tree up to 25 m with numerous epicormic shoots, stiff leaves 7–14 cm, and cone 3–7 cm, has been planted in several countries on a small scale.

4. P. banksiana Lamb., *Descr. Gen. Pinus* **1**: 7, t. 3 (1803)

(*P. divaricata* (Aiton) Sudworth). Tree up to 25 m; bark becoming reddish-brown and forming thick scales; branches usually irregularly arranged. Buds ovoid, very resinous. Twigs glabrous, green in first year, then brownish. Leaves 2–4(–5) cm, in pairs, rigid, acute or obtuse, twisted; resin-canals marginal. Cone 2–5 × 1·5–2·5 cm, long-conical, very curved, erect, yellowish, shining, persistent; apophysis flat or convex; umbo without mucro. Seeds 3–6 mm; wing 8–12 mm. *Planted for timber in C. Europe and rarely elsewhere.* [Au Cz Da Ge Rm.] (*N.C. & N.E. North America.*)

5. P. ponderosa Douglas ex P. & C. Lawson, *Agric. Man.* 354 (1836). Tree up to 75 m; bark becoming very thick, yellowish or dark reddish-brown. Buds cylindrical, resinous. Twigs glabrous, at first brownish or greenish, becoming blackish. Basal sheath of short shoots more than 15 mm. Leaves 10–25 cm × 1·5 mm, in groups of 3 (rarely 2 or 5), rigid, curved, densely crowded, deep yellowish-green, very aromatic; resin-canals median. Cone 8–15 × 3·5–5 cm, ovoid, symmetrical, subsessile, patent or slightly deflexed; apophysis concave or plane; umbo with strong, erect, persistent mucro. Seeds *c.* 8 mm; wing *c.* 18 mm. *Occasionally planted for timber in C. & S. Europe.* [Au Ge Gr It Rm.] (*W. North America.*)

P. jeffreyi Grev. & Balf. in A. Murray, *Bot. Exped. Oreg.* **8**: 2 (1853), from W. North America, a tree up to 60 m, with non-resinous buds, glaucous twigs, greyish- to bluish-green leaves 12–20 cm, cone 15–25 cm, and the umbo with a recurved mucro, is also occasionally cultivated, sometimes under the name of *P. ponderosa*.

P. sabiniana Douglas, *Trans. Linn. Soc. London (Bot.)* **16**: 749 (1833), from California, a tree up to 25 m, with lax habit, sparse, pale green leaves 20–30 cm, and cone 15–25 cm, is cultivated for timber on a small scale in Europe.

6. P. nigra Arnold, *Reise Mariazell* 8 (1785) (*P. maritima* Miller). Tree up to 50 m; bark grey to dark brown. Buds ovoid or cylindrical-ovoid, acute, slightly resinous. Twigs glabrous, pale brown to orange-brown. Leaves 4–19 cm × 1–2 mm, in pairs, light or dark green, somewhat rigid, flexuous, straight or curved; 1–5 layers of hypodermal cells; resin-canals 3–17, median. Cone 3–8 × 2–4 cm, yellowish- or light brown, shining, subsessile; apophysis slightly or obtusely keeled; umbo mucronate. $2n = 24$. *S. Europe, extending northwards to Austria and S. Carpathians; extensively planted in N. Europe (especially subsp.* (c)) *for timber or shelter.* Al Au Bu Co Ga Gr Hs It Ju Rm Rs(K) Si Tu [Be Br Cz Da Ge Hb He Ho Su].

Very variable, with many geographical variants which are often not clearly separable. Much of the variation follows a clinal pattern (*vide* P. Fukarek, *Rad. Polj. Šum. Fakult. Sarajevu* 3(2): 1–92 (1958) for a detailed account). The following subspecies, which are sometimes regarded as distinct species, deserve recognition.

1 Leaves ± flexible; 1 or 2(3) rows of slightly thickened hypodermal cells
2 Leaves 1·2–1·9 mm wide, somewhat pungent; 1 or 2(3) rows of hypodermal cells **(c) subsp. laricio**
2 Leaves 1–1·2 mm wide, not pungent; 1 row of hypodermal cells **(b) subsp. salzmannii**
1 Leaves ± rigid; 2–5 rows of strongly thickened hypodermal cells
3 Leaves 4–7 cm **(d) subsp. dalmatica**
3 Leaves 7–18 cm
4 Leaves straight or incurved; leaf-sheath 10–16 mm **(a) subsp. nigra**
4 Leaves twisted or irregularly curved; leaf-sheath (15–)18–26 mm **(e) subsp. pallasiana**

(a) Subsp. **nigra** (*P. nigricans* Host): Tall tree with pyramidal crown. Leaves 8–16 cm × 1·5–2 mm, rigid, pungent; 2 or 3 rows of hypodermal cells. Cone 5–8 cm; apophysis somewhat keeled. *From Austria to C. Italy and Greece.*

(b) Subsp. **salzmannii** (Dunal) Franco, *Dendrologia Florestal* 56 (1943) (*P. pyrenaica* sensu Willk., *P. clusiana* Clemente, *P. salzmannii* Dunal): Medium-sized tree with narrow, pyramidal or cylindrical crown. Leaves 8–16 cm × 1–2 mm, not pungent; 1 row of hypodermal cells. Cone 4–6 cm; apophysis somewhat keeled. *Cevennes, Pyrenees, C. & E. Spain.*

(c) Subsp. **laricio** (Poiret) Maire, *Bull. Soc. Hist. Nat. Afr. Nord* **19**: 66 (1928) (*P. laricio* Poiret): Tall tree with narrow, ovoid-elongate crown. Leaves 8–16 cm × 1·2–2 mm, slightly pungent, twisted; 1(2) rows of hypodermal cells. Cone 6–8 cm, yellowish-brown; apophysis obtusely keeled. ● *Corse; Calabria, Sicilia.*

(d) Subsp. **dalmatica** (Vis.) Franco, *Dendrologia Florestal* 55 (1943): Small tree with broadly pyramidal crown. Leaves 4–7 cm × 1·5–1·8 mm, quite rigid; 3–5 rows of hypodermal cells. Cone 3·5–4·5 cm. ● *Coastal region and islands of W. Jugoslavia.*

(e) Subsp. **pallasiana** (Lamb.) Holmboe, *Bergens Mus. Skr.* ser. 2, **1(2)**: 29 (1914) (*P. pallasiana* Lamb.): Medium-sized tree with broadly ovoid crown. Leaves 12–18 cm × 1·6–2·2 mm; 2–5 rows of hypodermal cells. Cone 5–12 cm. *Balkan peninsula; S. Carpathians; Krym.*

Trees of subsp. (c) from Calabria and Sicilia are characterized by the more lax leaves with 2(3) rows of hypodermal cells, brown cone and only slightly keeled apophysis.

Plants from N.W. Romania are sometimes regarded as a distinct species, **P. banatica** (Georgescu & Ionescu) Georgescu & Ionescu, *Feddes Repert.* **41**: 183 (1936).

P. resinosa Aiton, *Hort. Kew.* **3**: 367 (1789), from E. North America, a tree up to 50 m, with reddish-brown bark, leaves 12–17 cm, marginal resin-canals, and cone 4–6 cm, is cultivated for timber on a small scale in Europe.

7. P. sylvestris L., *Sp. Pl.* 1000 (1753). Tree up to 40 m; bark dark brown on lower part of trunk, pale ochreous-red and flaking on upper part of trunk. Buds acute, more or less resinous. Twigs glabrous, yellowish-green at first, becoming greyish-brown. Leaves 3–7 cm × 2 mm, in pairs, glaucous, twisted; resin-canals submarginal. Cone 3–6 × 2–3·5 cm, dull yellowish-brown, acute, pendent, caducous; apophysis flat, or shortly pyramidal on back of cone, weakly keeled; umbo muticous. Seeds 3–5 mm; wing *c.* 10 mm. $2n = 24$. *N. & C. Europe, extending southwards in the mountains to S. Spain, Macedonia and Krym.* Al Au Br Bu Cz Fe Ga Ge Gr He *Ho Hs Hu It Ju *Lu No Po Rm Rs(N, B, C, W, K, E) Su [Be Da *Hb Ho Is Tu].

More than 150 variants have been described. Some have been regarded as subspecies, but much of the variation follows clinal patterns and it is not possible to give a consistent formal treatment of it. The variants may, however, be conveniently arranged in the following geographical groups:

Group I (var. *lapponica* Fries). Crown narrow, branched; bark with small, thin scales. *N. Fennoscandia.*

Group II (var. *rigensis* (Desf.) Ascherson & Graebner; var.

septentrionalis Schott). Tall tree with conical crown, slender, straight trunk and thin bark. *Baltic coast.*

Group III (var. *scotica* (Willd.) Schott). Crown long remaining pyramidal, rounded only in old trees; bark thin at least above. *Scotland.*

Group IV (var. *aquitana* Schott; var. *brigantiaca* Gaussen; var. *catalaunica* Gaussen; var. *hercynica* Münch; var. *iberica* Svob.; var. *pyrenaica* Svob.; var. *vindelica* Schott). Crown conical; trunk straight; branches at right-angles to trunk; bark thin, with large scales. *Mountains of W. Europe from C. Spain to C. Germany and the W. Alps.*

Group V (var. *batava* Schott; var. *borussica* Schott; var. *carpatica* Klika; var. *engadinensis* Heer; var. *haguenensis* Loudon; var. *illyrica* Svob.; var. *nevadensis* Christ; var. *pannonica* Schott; var. *rhodopaea* Svob.; var. *romanica* Svob.; var. *sarmatica* Zapał.; var. *vocontiana* Guinier & Gaussen). Trunk crooked; branches at an acute angle to the trunk; crown broad, rounded; bark thick, deeply fissured. *Lowlands, and foothills of mountain ranges of C. Europe, extending eastwards to Russia and southwards to the N. Appennini; Sierra Nevada.*

Var. *uralensis* Fischer, from Ural, and var. *cretacea* (Kalenicz.) Fomin, from Ukraine, are eastern variants of uncertain affinities.

P. kochiana Klotzsch ex C. Koch, *Linnaea* **22**: 296 (1849) is like **7** but has leaves 1 mm wide and a slightly shining cone with strongly keeled apophyses and a shortly mucronate umbo. It occurs in Ukraine (where it has been called *P. fominii* Kondrat. or *P. hamata* (Steven) D. Sosn.). The status of this plant is uncertain; it is perhaps not clearly separable from **P. armena** C. Koch, *Linnaea* **22**: 297 (1849), from Caucasus.

8. P. mugo Turra, *Gior. Ital. Sci. Nat. Agric. Arti Commerc.* **1**: 152 (1764) (*P. montana* Miller, *P. mughus* Scop.; incl. *P. pumilio* Haenke). Shrub with thick, decumbent, ascending or erect branches up to 3·5 m. Twigs green at first, becoming brown. Buds ovoid-cylindrical, very resinous. Leaves 3–8 cm × 1·5–2 mm, in pairs, bright green. Cone 2–5 × 1·5–2·5 cm, shining, subsessile, symmetrical, obtuse; apophysis flat, or convex above and concave below; umbo central or below the middle, with a small mucro. ● *Mountains of C. Europe and the Balkan peninsula; C. Appennini. Planted for sand-binding and shelter in N. Europe.* Al Au Bu Cz Ga Ge He It Ju Po Rm Rs(W) [Da].

9. P. uncinata Miller ex Mirbel in Buffon, *Hist. Nat. Pl.* **10**: 213 (1806) (incl. *P. rotundata* Link). Like **8** but usually a tree with an erect trunk up to 85 m; cone 5–7 × 2–3 cm; apophysis very prominent, recurved and hooked, or rounded and hooded; umbo excentric. ● *C. & S.W. Alps; Pyrenees; C. Spain.* Au Ga Ge He Hs It.

There has been widespread introgressive hybridization between **8** and **9**, giving rise on the one hand to a variant of **8** with erect branches, cone 4–5 × 2·5 mm, apophyses convex above and concave below, and an excentric umbo, sometimes recognized as *P. mugo* var. *pumilio* (Haenke) Zenari; and on the other hand to a variant of **9** with shrubby habit and rounded or hooded apophyses, sometimes recognized as *P. uncinata* var. *rotundata* (Link) Antoine.

10. P. heldreichii Christ, *Verh. Naturf. Ges. Basel* **3**: 549 (1863). Tree up to 20 m; crown rounded-pyramidal; bark ashy-grey, flaking to leave yellowish patches. Twigs glabrous, grey only in the first year, then brown. Buds not resinous. Leaves 6–9 cm × 1·5 mm, in pairs, rigid, more or less pungent, with 2–6 layers of hypodermal cells; resin-canals 2–11, median. Cone 7–8 × 2·5 cm, slightly shining; apophysis somewhat thickened, pyramidal, with conspicuous ridges and a very short, straight mucro. Seeds *c.* 7 mm; wing *c.* 25 mm. ● *Mountains of C. & W. part of Balkan peninsula; S. Italy.* Al Gr Ju.

Var. *heldreichii*, which has a flat apophysis, and 2 or 3 layers of hypodermal cells in the leaf, occurs only in N. Greece (Olimbos). The more widespread variant is var. *leucodermis* (Antoine) Markgraf ex Fitschen (*P. leucodermis* Antoine).

11. P. halepensis Miller, *Gard. Dict.* ed. 8, no. 8 (1768). Tree up to 20 m; crown rounded; bark silvery-grey, becoming reddish-brown and deeply fissured. Twigs glabrous. Buds not resinous. Leaves in pairs, with 1–3 layers of hypodermal cells; resin-canals 3–8, submarginal (sometimes with 1 or 2 median). Cone 5–12 × 4 cm, shining, brown; apophysis convex. Seeds 7–8 mm; wing *c.* 20 mm. *Mediterranean region; Krym.* Al Bl Co Cr Ga Gr Hs It Ju Rs(K) Sa Si [Lu].

(**a**) Subsp. **halepensis**: Trunk and branches often crooked. Twigs light grey for many years. Leaves 6–10(–15) cm × 0·7 mm, slender, clear green. Cones pendent; peduncle 1–2 cm. $2n = 24$. *Mediterranean region eastwards to Macedonia; naturalized in Portugal.*

(**b**) Subsp. **brutia** (Ten.) Holmboe, *Bergens Mus. Skr.* ser. 2, **1(2)**: 29 (1914) (*P. brutia* Ten., *P. pityusa* Steven): Trunk and branches straight. Twigs reddish-yellow or greenish. Leaves 12–18 cm × 1–1·5 mm, thick, rigid, dark green. Cones erect or ascending; peduncle very short or absent. *S.E. Europe, from Kriti to Krym.*

P. stankewiczii (Suk.) Fomin, *Monit. Jard. Bot. Tiflis* **34**: 21 (1914), from Krym, is similar to subsp. (**b**) but differs in a number of minor features such as solitary cones and wing of seed 25–27 mm. It is probably best treated as a variety of subsp. (**b**).

12. P. canariensis Sweet ex Sprengel, *Syst. Veg.* **3**: 887 (1826). Tree up to 30 m; bark becoming thick, slightly fissured, reddish-brown. Twigs glabrous, yellow. Buds ovoid, not resinous; scales broadly white-fringed. Leaves 20–30 cm × 1 mm, in groups of 3, slender, acute, densely crowded; resin-canals submarginal. Cone 10–20 × 4–7 cm, ovoid-conical, shortly stalked, deflexed; apophysis depressed-pyramidal; umbo not mucronate. Seeds *c.* 12 mm; wing *c.* 20 mm. *Planted for timber in Italy and to a smaller extent elsewhere in the Mediterranean region.* [It.] (*Islas Canarias.*)

P. taeda L., *Sp. Pl.* 1000 (1753), from S. & S.E. North America, a tree up to 50 m, with glaucous twigs, rigid leaves 12–25 cm, submarginal resin-canals, and cone 6–10 cm, is cultivated for timber on a small scale in Europe.

13. P. pinea L., *Sp. Pl.* 1000 (1753). Tree up to 30 m, parasol-shaped; bark greyish-brown, flaking to leave reddish-orange patches. Twigs glabrous, greyish-green, becoming brown. Buds not resinous; scales white-fringed. Leaves 10–20 cm × 1·5–2 mm, in pairs, acute; resin-canals submarginal. Cone 8–14 × 10 cm, shining, brown; apophysis weakly pyramidal. Seeds 15–20 × *c.* 7–11 mm; wing less than 1 mm, caducous; husk hard. *Mediterranean region and Portugal.* Al Bl Co Ga *Gr Hs *It *Ju Lu Sa Si *Tu.

The seeds, extracted from the husk, are the pine-nuts of commerce.

Subgen. **Haploxylon** (Koehne) Pilger. Leaves in groups of 5, with 1 fibro-vascular bundle; scale-leaves not decurrent; umbo terminal.

14. P. cembra L., *Sp. Pl.* 1000 (1753). Tree up to 25 m; bark reddish-grey, becoming scaly on old trees. Twigs strongly pubescent with orange-brown tomentum. Bud-scales dull, with white, membranous margin. Leaves 5–8 cm × 1 mm. Cone 5–8 × 3·5–5 cm, less than 1½ times as long as wide. Seeds 12(–14) mm; wing absent, or less than 1 mm and caducous; husk hard. ● *Alps and Carpathians; occasionally planted for timber in N. Europe.* Au Cz Ga Ge He It Po Rm Rs(W) [Fe Is No Su].

15. P. sibirica Du Tour, *Nouv. Dict. Hist. Nat.* **18**: 18 (1803) (*P. cembra* subsp. *sibirica* (Du Tour) Krylov). Like **14** but up to 40 m; bud-scales shining, brown; leaves 6–13 cm × 0·8–1·2 mm; cone 6–13 × 5–8 cm, more than 1½ times as long as wide; seed with fragile husk. *N.E. Russia, westwards to* 50° E. Rs(N, C). (*N. Asia.*)

16. P. wallichiana A. B. Jackson, *Kew Bull.* **1938**: 85 (1938) (*P. excelsa* Wall., non Lam., *P. chylla* O. Schwarz, *P. griffithii* McClell.). Tree up to 50 m; bark greyish-brown and smooth on young trees, becoming shallowly fissured on old trees. Twigs glabrous, yellowish-green at first, becoming brownish-yellow. Leaves 8–20 cm × 0·7 mm, flexible, pendent. Cone 15–25 × 3 cm, cylindrical, straight, becoming pendent. Seeds 8–9 × 5–6 mm; wing 10–20 mm. *Planted for timber in Italy.* [It.] (*Himalaya.*)

17. P. peuce Griseb., *Spicil. Fl. Rumel.* **2**: 349 (1845). Tree up to 30 m; crown narrowly and densely pyramidal; bark grey, fissured, with small scales. Twigs glabrous. Leaves 7–12 cm × 0·7 mm, rigid, acute, densely crowded, forwardly directed. Cone 8–15 × 2·5 cm, more or less curved. Seeds 8 × 5 mm; wing 15 mm. ● *Mountains of Balkan peninsula from* c. 41° *to* 43° *N.* Al Bu Gr Ju.

18. P. strobus L., *Sp. Pl.* 1001 (1753). Tree up to 50 m; crown pyramidal at first, later becoming broader; bark greyish-green and smooth on young trees, becoming brown and fissured on old trees. Twigs becoming glabrous, or remaining pubescent below leaf-traces. Leaves 5–14 cm, slender, flexible, patent. Cone 8–20 × 3–4 cm, cylindrical, often curved near the apex, pendent. Seeds 5–8 mm; wing 18–25 mm. *Planted for timber, especially in C. Europe.* [Au Be Bu Cz Da Ga Ge He It Po Rm Rs(C) Su Tu.] (*E. & C. North America.*)

P. aristata Engelm., *Amer. Jour. Sci. Arts* ser. 2, **34**: 331 (1862), from S.W. United States, a procumbent shrub or small tree up to 15 m, with leaves 20–40 mm, covered with conspicuous resinous exudate, and cone 4–9 cm, is cultivated for timber on a small scale in Europe.

27. TAXODIACEAE

EDIT. D. M. MOORE

Monoecious; resiniferous trees, rarely shrubs, with spirally arranged leaves. Male and female cones made up of numerous spirally arranged scales. Scales of male cones (microsporophylls) bearing 2–8 pollen-sacs on the lower surface. Scales of female cones (megasporophylls) bearing 2–12 ovules on the upper surface; subtending scale (bract) indistinct, wholly or partially united with the cone-scale. Female cone more or less woody in fruit. Seeds winged.

1 Leaves of lateral twigs linear to linear-oblong, flat, distichous
2 Evergreen; leaves dark green — **1. Sequoia**
2 Deciduous; leaves pale green — **3. Taxodium**
1 All leaves subulate
3 Leaves radially arranged, appressed or slightly patent at the apex, triangular in section; female cone 3–8 cm, ovoid — **2. Sequoiadendron**
3 Leaves in 5 ranks, patent, incurved, quadrangular in section; female cone 1·5–3 cm, subglobose — **4. Cryptomeria**

1. Sequoia Endl.

V. H. HEYWOOD

Evergreen trees with irregularly whorled branches. Winter-buds scaly. Leaves of 2 kinds: those on the leading and cone-bearing twigs spirally arranged, subulate, scale-like, keeled above, appressed or slightly patent; those on the lateral twigs distichous, linear to linear-oblong, flat, with 2 white stomatal bands beneath. Female cones persistent, ripening in the second year; cone-scales 15–20.

1. S. sempervirens (D. Don) Endl., *Syn. Conif.* 198 (1847). Up to 100 m, often suckering freely from the base; trunk 3–4·5 m in diameter; outer bark thick, exfoliating in dark brown plates, revealing the inner, bright red bark. Leaves of lateral twigs 6–25 × 2–3 mm, acute, often falcate, dark green above; leaves of leading and cone-bearing twigs 6 mm, incurved and with a sharp, callose point at the apex, decurrent at the base. Female cone 18–25 × 12 mm, ovoid; scales expanded into a rhomboidal disc, rarely with a mucro. Seeds 3–5 on each scale. *Planted for ornament and occasionally for timber.* [Ba Ga It.] (*W. North America.*)

2. Sequoiadendron Buchholz

V. H. HEYWOOD

Like *Sequoia*, but winter-buds naked; leaves of 1 kind only, radially arranged, narrowly lanceolate, appressed or slightly patent at the apex; female cone-scales 25–40.

1. S. giganteum (Lindley) Buchholz, *Amer. Jour. Bot.* **26**: 536 (1939) (*Sequoia gigantea* Decne, non Endl., *S. wellingtonia* Seem.). Up to 80 m, with narrow pyramidal crown; trunk up to 7 m in diameter above the thickened, buttressed base; bark thick and spongy, reddish-brown. Leaves *c.* 1 cm, becoming longer on older twigs, ovate to lanceolate, flat above, convex beneath, broadly decurrent at the base. Female cone 3–5(–8) × 2–3·5(–4·5) cm, terminal, ovoid; scales expanded into a rhomboidal disc, with a depression in the middle often bearing a slender mucro. Seeds 3–7 on each scale. *Planted for ornament and occasionally for timber.* [Au Ga.] (*California.*)

3. Taxodium L. C. M. Richard

V. H. HEYWOOD

Deciduous trees. Winter-buds scaly. Lateral twigs deciduous with the leaves. Leaves flat, linear; those of the persistent terminal twigs spirally arranged; those of the deciduous lateral twigs distichous and somewhat longer. Male cones in terminal panicles. Female cones globose, ripening in the first year.

1. T. distichum (L.) L. C. M. Richard, *Ann. Mus. Hist. Nat. (Paris)* **16**: 298 (1810). Tree up to 50 m, with a rather narrow outline, often with conspicuous buttresses at the base of the trunk. Bark reddish-brown, fibrous, peeling. Leaves up to 20 mm, acute, channelled above. Male cones purplish, in slender panicles. Female cone 15–30 mm in diameter; some of the scales sterile, others bearing 2 irregularly trigonous seeds. *Planted for timber in S. and S.C. Europe, especially on alluvial land.* [Bu Ga Ge It Rm.] (*S.E. United States.*)

4. Cryptomeria D. Don

J. DO AMARAL FRANCO

Evergreen pyramidal trees with irregularly whorled, rigid branches. Leaves spirally arranged, subulate, incurved, decurrent. Female cones persistent, ripening in the first year; cone-scales 20–30.

1. C. japonica (L. fil.) D. Don, *Trans. Linn. Soc. London (Bot.)* **18**: 167 (1841). Up to 40 m, with short ascending branches forming a narrow, dense crown; bark reddish-brown, exfoliating in long shreds. Leaves 5–20 mm, in 5 ranks, bright green, gradually shortening towards the apex of the twig. Female cone 1·5–3 cm, subglobose; scales thick, cuneate, persistent, furnished with the recurved point of the adnate bract on the back and with pointed processes at the apex. Seeds 2–5 on each scale. *Planted for ornament and occasionally for timber; naturalized in Açores.* [Az.] (*C. & S. Japan.*)

In Açores this species sometimes produces suckers from the base.

28. CUPRESSACEAE

EDIT. D. M. MOORE

Monoecious or dioecious; resiniferous trees or shrubs with opposite or whorled, usually scale-like, leaves. Male and female cones made up of opposite or whorled scales. Scales of male cones (microsporophylls) bearing 3–5 pollen-sacs on the lower surface. Scales of female cones (megasporophylls) bearing 2 to many erect ovules on the upper surface, and completely adnate to the subtending bract. Fruit a more or less woody cone, or the scales becoming fleshy and the fruit berry-like, indehiscent. Seeds winged or not.

1 Fruit indehiscent, berry-like, with fleshy, coalescent scales **5. Juniperus**
1 Fruit a dehiscent cone with ±woody scales
2 Female cone-scales 4, in a single whorl **4. Tetraclinis**
2 Female cone-scales in 3–8 pairs
3 Ripe female cone-scales flat, oblong, imbricate **3. Thuja**
3 Ripe female cone-scales peltate, valvate
4 Twigs terete or 4-angled; seeds narrowly winged, 6–20 on each scale **1. Cupressus**
4 Twigs flattened; seeds broadly winged, up to 5 on each scale **2. Chamaecyparis**

1. Cupressus L.

J. DO AMARAL FRANCO

Monoecious trees (rarely shrubs). Twigs terete or 4-angled. Leaves opposite, decussate, scale-like, appressed and imbricate on adult plants, acicular and patent on young plants and rarely on leading shoots. Female cones solitary, terminal on young twigs, ripening in the second year, globose to ellipsoid-globose; cone-scales in 3–7 pairs, peltate, woody. Seeds flattened, 6–20 on each scale, narrowly winged.

In addition to the species described below, **C. torulosa** D. Don, *Prodr. Fl. Nepal* 155 (1825), has been planted for timber on a small scale in Italy. It has obtuse, appressed leaves, and a cone *c.* 12 mm in diameter, with 4–6 pairs of scales.

1 Leaves obtuse, closely appressed; female cone green when young; cone-scales 8–14
2 Leaves on lateral twigs 0·5–1 mm; male cone 4–8 mm; ripe female cone yellowish-grey **1. sempervirens**
2 Leaves on lateral twigs 1–2 mm; male cone 3–5 mm; ripe female cone brown **2. macrocarpa**
1 Leaves acute to acuminate, free at the apex; female cone glaucous when young; cone-scales 6–8
3 Leaves greyish when young, emitting a disagreeable odour when rubbed; female cone up to 30 mm **4. arizonica**
3 Leaves green to glaucous, not emitting a disagreeable odour; female cone 10–15 mm **3. lusitanica**

1. C. sempervirens L., *Sp. Pl.* 1002 (1753) (*C. horizontalis* Miller). Tree up to 30 m. Leaves 0·5–1 mm, dark green, obtuse. Male cone 4–8 mm; female 25–40 mm, ellipsoid-oblong (rarely globose), green when young and shining yellowish-grey when ripe, with 8–14 shortly and obtusely mucronate scales. Seeds 8–20 on each scale. *Native in Kriti and perhaps elsewhere in the E. Mediterranean; long planted throughout S. Europe, and naturalized in many districts.* Cr *Gr [*Al Bl Bu Co Ga He Hs It *Ju Lu Rs(K) Sa Si Tu].

There are two well-known varieties: var. *sempervirens*, with upright branches forming a fastigiate crown, is the most commonly cultivated; var. *horizontalis* (Miller) Aiton, with patent-ascending branches forming a pyramidal crown, is the wild plant on Kriti and is much less common in cultivation.

2. C. macrocarpa Hartweg, *Jour. Hort. Soc. London* **2**: 187 (1847). Tree up to 25 m, with erecto-patent branches, forming a pyramidal crown when young, broad and flat-topped in old trees. Leaves 1–2 mm. Male cone 3–5 mm; female cone 20–35 mm, globose to ellipsoidal, shining brown when ripe, with 8–14 shortly and obtusely mucronate scales. Seeds 8–20 on each scale. *Planted for shelter and ornament, and occasionally for timber, in W. & S. Europe, especially near the sea.* [Br Ga Hb Hs It Lu Si.] (*S. California.*)

3. C. lusitanica Miller, *Gard. Dict.* ed. 8, no. 3 (1768) (*C. glauca* Lam.). Tree up to 30 m, with spreading branches and more or less pendent leading shoot and twigs, forming a pyramidal crown on young trees, usually flat-topped on old ones. Leaves 1·5–2 mm, green to glaucous, acute to acuminate. Male cone 4–6 mm; female cone 10–15 mm, globose, glaucous when young but shining brown when fully ripe, with 6–8 usually prominently mucronate scales. Seeds 8–10 on each scale. *Planted for timber and ornament in S. Europe.* [Ga Hs It Lu.] (*Mexico and Guatemala.*)

A very polymorphic species, varying greatly in habit and leaf-colour.

4. C. arizonica E. L. Greene, *Bull. Torrey Bot. Club* **9**: 64 (1882). Tree up to 30 m, with a narrow, pyramidal crown on young trees, becoming broad and open when older. Leaves greyish when young, becoming glaucous later, resinous-glandular, emitting a disagreeable odour when crushed. Female cone up to 30 mm. *Planted for timber in Italy, and on a smaller scale elsewhere in S. Europe.* [It.] (*N. Mexico and S.W. United States.*)

2. Chamaecyparis Spach

†T. G. TUTIN

Like *Cupressus* but with flattened twigs; female cones ripening in the first year or rarely early in the second, globose; seeds up to 5 on each scale. Seedlings and occasional branches of older plants of all species produce acicular, patent leaves.

1. C. lawsoniana (A. Murray) Parl., *Ann. Mus. Firenze* nov. ser., **1**: 181 (1866) (*Cupressus lawsoniana* A. Murray). Tree up to 65 m. Leaves opposite and decussate, the lateral ones larger than the others, all closely appressed, acute, with translucent glands, and with white markings beneath. Male cone pink or red; female cone *c.* 8 mm in diameter, globose, yellowish-brown when ripe; scales 8, each with a central depression in which is a projecting ridge. Seeds 2–4(5) on each scale, ovoid, with conspicuous resinous tubercles, winged. *Widely planted for timber and shelter, and locally naturalized.* [Au Br Da Ga Ge Hb It Lu No Rm Tu.] (*W. United States.*)

Other commonly cultivated species are **C. obtusa** (Siebold & Zucc.) Siebold & Zucc. in Endl., *Syn. Conif.* 63 (1847), from Japan, with obtuse, minutely glandular leaves with white X- or Y-shaped markings beneath; **C. pisifera** (Siebold & Zucc.) Siebold & Zucc. in Endl., *op. cit.* 64 (1847), from Japan, with acute, obscurely glandular leaves with white markings beneath, and cone 5–6 mm in diameter; and **C. nootkatensis** (D. Don) Spach, *Hist. Vég.* (*Phan.*) **2**: 333 (1834), from W. North America, with dull green, eglandular leaves without white markings, and with cones which ripen early in the second year.

3. Thuja L.

†T. G. TUTIN

Like *Cupressus* but twigs flattened; female cones oblong or conical, ripening in the first year; scales thin and flexible; seeds up to 5 on each scale.

1. T. plicata D. Don in Lamb., *Descr. Gen. Pinus* **2**: 19 (1824). Tree up to 65 m. Trunk not branched from the base. Leaves opposite and decussate, ovate, acute or obtuse, appressed, eglandular or with an obscure gland, usually with faint white markings beneath. Female cone *c.* 12 mm, conical, brown when ripe; scales 10–12, thin, with a thickened process on the inner side, projecting beyond the thin, triangular, deflexed apex. Seeds 2 or 3 per scale, elliptical, winged. *Planted for timber, and locally naturalized.* [Au Br Da Ge It Lu No.] (*W. North America.*)

T. occidentalis L., *Sp. Pl.* 1002 (1753), from E. North America, is also sometimes planted, mainly in C. Europe. It has conspicuously glandular leaves which are yellowish- or bluish-green and without white markings beneath; the female cones are oblong and bright brown when ripe.

T. orientalis L., *loc. cit.* (1753), from China, is occasionally planted, mainly for ornament. It usually has several basal branches as long as the main stem, entirely green leaves, the female cone with 6(–8) thick scales, each with a stout hooked or recurved apical boss, and seeds without a wing. It has a chromosome number of $2n = 22$.

4. Tetraclinis Masters

†T. G. TUTIN

Monoecious tree. Buds concealed by the leaves. Twigs flattened. Leaves in groups of 4, the lateral ones larger than the others, all adnate to the twig except at the scale-like apex. Female cones solitary; scales 4, in a single whorl, valvate, woody. Seeds with 2 wide wings.

The only species.

1. T. articulata (Vahl) Masters, *Jour. Roy. Hort. Soc.* **14**: 250 (1892). Pyramidal tree 12–15 m. Cone 8–12 mm in diameter; scales glaucous, triangular, 2 obtuse and 2 acute at the apex, deeply grooved, with a small spine near the apex. *S.E. Spain* (*near Cartagena*); *Malta.* Hs Si. (*N.W. Africa.*)

5. Juniperus L.

J. DO AMARAL FRANCO

Monoecious or dioecious trees or shrubs. Leaves opposite or in whorls of 3, acicular or scale-like, always acicular in young plants; frequently all scale-like in mature plants, or all acicular, or a combination of both. Female cones axillary or terminal, ripening in the first, second or third year, usually ovoid to globose, indehiscent, berry-like; cone-scales fleshy, coalescent. Seeds ovoid or oblong, angled or terete, 1–12 in each cone, never winged.

1 All leaves in whorls of 3, acicular, rigid, jointed at the base; cones axillary
2 Male cones in axillary fascicles; female cone 20–25 mm; seeds united to form a stone **1. drupacea**
2 Male flowers solitary, axillary; female cone 6–15 mm, seeds free
3 Leaves with 1 broad stomatal band above (which is rarely divided by a faint midrib in the basal $\frac{1}{3}$); cone 6–9 mm, black when ripe **2. communis**
3 Leaves with 2 stomatal bands above; cone 6–15 mm, reddish-brown to dark purple when ripe
4 Leaves loosely set, patent **3. oxycedrus**
4 Leaves closely set, strongly incurved and subimbricate **4. brevifolia**
1 Leaves of 2 kinds: in juvenile plants acicular and patent, in adult plants scale-like and imbricate (rarely acicular and then not jointed at the base); cones terminal
5 Acicular leaves usually in whorls of 3; scale-like leaves obtuse, with a narrow but distinct scarious border; ripe female cone dark red **5. phoenicea**
5 Acicular leaves decussate (in whorls of 3 on leading shoots); scale-like leaves without a scarious border; ripe female cone dark purple or blackish
6 Ripe female cone 4–6 mm
7 Scale-like leaves appressed, not free at the apex, obtuse to subacute; female cone subglobose, on short, recurved stalk; shrub **9. sabina**

7 Scale-like leaves free at the apex, acute to acuminate; female cone ovoid, on erect or patent stalk; tree up to 30 m **10. virginiana**
6 Ripe female cone 7–12 mm
8 Twigs 0·6–0·8 mm, wide, terete; scale-like leaves 1–1·5 mm, acute; seeds 4–6 **8. excelsa**
8 Twigs 1 mm wide, quadrangular; scale-like leaves 1·5–2 mm, acute or acuminate; seeds 1–4
9 Twigs distichously arranged; female cone dark purple, pruinose; seeds 2–4 **6. thurifera**
9 Twigs irregularly arranged; female cone almost black; seeds 1 or 2, rarely 3 **7. foetidissima**

1. J. drupacea Labill., *Icon. Pl. Syr.* **2**: 14, t. 8 (1791) (*Arceuthos drupacea* (Labill.) Antoine & Kotschy). Dioecious tree up to 20 m, with a broad, pyramidal crown (but assuming a columnar habit in cultivation). Leaves 10–25 × 2–4 mm, linear-lanceolate, acuminate, patent, with 2 white bands above. Female cone 20–25 mm, ripening in the second year, ovoid to subglobose, brown or bluish-black, pruinose. Seeds 3, minute, united to form a single large stone. *S. Greece* (*on and north of Parnon Oros*). Gr. (*Anatolia and Syria*.)

2. J. communis L., *Sp. Pl.* 1040 (1753). Dioecious tree or shrub. Leaves with a single white stomatal band above, which is rarely divided by a faint midrib in the basal ½. Female cone 6–9 mm, ripening in the second or third year, ovoid to globose, green in the first year, then pruinose, and black when fully ripe. Seeds free, usually 3. *Throughout Europe, but mainly on mountains in the south*. All except Az Bl Cr Sb ?Tu.

1 Leaves 10–15 mm, closely set, upturned or upcurved, acute to obtuse, mucronate, with a broad, white stomatal band (c) subsp. **alpina**
1 Leaves up to 20 mm, loosely or closely set, acuminate-subulate, patent
2 Leaves 1–1·5 mm wide, loosely set, linear, with a glaucous stomatal band (a) subsp. **communis**
2 Leaves 1·3–2 mm wide, closely set, linear-oblong, with a broad, white stomatal band (b) subsp. **hemisphaerica**

(a) Subsp. **communis**: Tree up to 15 m or shrub. Leaves up to 20 × 1–1·5 mm, loosely set, linear, with a glaucous stomatal band. $2n = 22$. *Almost throughout the range of the species*.

(b) Subsp. **hemisphaerica** (J. & C. Presl) Nyman, *Consp.* 676 (1881) (*J. depressa* Steven): Shrub. Leaves up to 20 × 1·3–2 mm, closely set, linear-oblong, with a broad, white stomatal band. *S. Europe, mainly on mountains*.

(c) Subsp. **alpina** (Suter) Čelak., *Prodr. Fl. Böhm.* 17 (1867) (subsp. *nana* Syme, *J. alpina* S. F. Gray, *J. sibirica* Burgsd., *J. nana* Willd., nom. illeg.): Shrub, usually spreading and mat-like. Leaves 10–15 × 1–2 mm, closely set, upturned and upcurved, acute to obtuse, mucronate, with a broad, white stomatal band. $2n = 22$. *Higher mountains of Europe; at low altitudes in the North, including coastal rocks and cliffs on Atlantic coasts*.

A population from S.W. England (Lizard peninsula) is intermediate between subspp. (a) and (c).

3. J. oxycedrus L., *Sp. Pl.* 1038 (1753). Dioecious shrub or tree up to 15 m. Leaves patent, with 2 glaucous bands above. Female cone ripening in the second year, globose to pyriform, reddish, yellowish when unripe. Seeds free, usually 3. *S. Europe*. Al Bl Bu Co Cr Ga Gr Hs It Ju Lu Rs(K) Sa Si Tu.

1 Leaves usually acute; tree up to 15 m (b) subsp. **badia**
1 Leaves ±spinous at apex; shrub up to 4 m
2 Leaves not more than 2 mm wide; female cone 8–10 mm, shining (sometimes scarcely pruinose when young) (a) subsp. **oxycedrus**
2 Leaves up to 2·5 mm wide; female cone 12–15 mm, dull (pruinose when young) (c) subsp. **macrocarpa**

(a) Subsp. **oxycedrus** (*J. rufescens* Link, nom. illeg.): Erect shrub up to 4 m; branches upright, sometimes pendulous. Leaves 8–15 × 1–1·5 mm, acuminate-subulate. More or less spinous at apex. Female cone 8–10 mm, reddish, not pruinose. $2n = 22$. *Dry hills or plains, scrub or clearings in Mediterranean woods, up to* 1900 *m. Almost throughout the range of the species*.

(b) Subsp. **badia** (H. Gay) Debeaux, *Rev. Bot. Bull. Mens. Soc. Fr. Bot.* (*Toulouse*) **11**: 412 (1893): Tree up to 15 m, pyramidal, with patent branches, pendulous at the tip. Leaves 12–20 × 1·2–2 mm, usually acute, sometimes shorter, 8–12 mm, and subobtuse, mostly on male plants. Female cone 10–13 mm, more or less pruinose when young, purplish-brown when ripe. *Dry continental woods, rarely in semi-evergreen woods. Inland regions of Spain, N.E. and C.E. Portugal.*

(c) Subsp. **macrocarpa** (Sm.) Ball, *Jour. Linn. Soc. London* (*Bot.*) **16**: 670 (1878) (*J. macrocarpa* Sm., *J. umbilicata* Godron): Like subsp. (a) but small, erect to prostrate shrub; leaves up to 2·5 mm wide; ripe female cone 12–15 mm, dull (pruinose when young). *Maritime sands and rocky places. Mediterranean region and Bulgaria; rather local.*

4. J. navicularis Gand., *Bull. Soc. Bot. Fr.* **57**: 55 (1910) (*J. oxycedrus* subsp. *rufescens* auct. lusit., subsp. *transtagana* Franco). Small, upright, dense shrub up to 2 m. Leaves 4–12 × 1–1·5 mm, patent, acute to obtuse, dense, with 2 white stomatal bands above. Female cone 7–10 mm, ripening in the second year, globose, reddish or yellowish, not pruinose when young, coral-red when ripe. *Maritime sands and evergreen scrub.* ● *S.W. Portugal, S. Spain.* Hs Lu.

5. J. brevifolia (Seub.) Antoine, *Kupress.-Gatt.* 16 (1857). Dioecious erect shrub or small tree with a broad, pyramidal crown. Leaves 3–10 × 1–2 mm, closely set, strongly incurved and subimbricate, linear-lanceolate to ovate-linear, with 2 broad, white stomatal bands above. Female cone 8–10 × 6–8 mm, ripening in the second year, subglobose, green and pruinose when young, dark reddish-brown when ripe. Seeds 3, free, ovoid, triquetrous. *Mountain slopes up to* 1500 *m.* ● *Açores.* Az.

6. J. phoenicea L., *Sp. Pl.* 1040 (1753) (*J. oophora* G. Kunze). Monoecious small tree up to 8 m, or shrub (procumbent on seashores). Twigs 1 mm wide, terete, scaly. Juvenile leaves 5–14 × 0·5–1 mm, acicular, patent, acute, mucronate, with 2 stomatal bands above and beneath; adult leaves 0·7–1 mm, scale-like, ovate-rhombic, closely appressed, obtuse to somewhat acute, with an oblong, furrowed gland on the back, and a distinct scarious border. Female cone 8–14 mm, ripening in the second year, globose to ovoid, blackish when very young, later green or yellowish and not or slightly pruinose, dark red when ripe. Seeds 3–9, free. *Mediterranean region and Portugal; mainly near the coast.* Al Bl Co Cr Ga Gr Hs It Ju Lu Sa Si.

(a) Subsp. **phoenicea**: Scale-leaves obtuse or subacute. Female cone 8–12 mm, usually globose. *Throughout most of the range of the species.*

(b) Subsp. **turbinata** (Guss.) Nyman, *Consp.* 676 (1881): Scale-leaves acute. Female cone 12–14 mm, ovoid. *Coastal sands of W. Mediterranean region.*

7. J. thurifera L., *Sp. Pl.* 1039 (1753). Dioecious pyramidal tree up to 20 m. Twigs 1 mm wide, quadrangular, scaly, distichously arranged. Leaves all decussate; adult leaves 1·5–2 mm, appressed but free at their incurved, acuminate or acute apices, with an oblong, furrowed gland on the back, and an entire or slightly toothed, but not scarious, border. Female cone 7–8 mm, ripening in the second year, pruinose when young, dark purple when fully ripe. Seeds 2–4, free. *Mountains of S.W. Europe.* Co Ga Hs.

This species has been cited by several authors as native in Portugal and Greece. The former records should be referred to **5**, the latter to **7**.

8. J. foetidissima Willd., *Sp. Pl.* **4(2)**: 853 (1806) (*J. sabinoides* Griseb.). Monoecious or dioecious tree up to 17 m, with a straight stem, and a narrow, conical crown. Twigs 1 mm wide, distinctly quadrangular, scaly, irregularly arranged. Leaves foetid when crushed, all decussate; adult leaves 1·5 mm, ovate-rhombic, free at their acuminate apices, mostly eglandular, the border entire, not scarious. Female cone 7–12 mm, ripening in the second year, globose, pruinose when young, dark reddish-brown to nearly black when ripe. Seeds 1 or 2(3), free. *Mountains of S. and W. parts of Balkan peninsula; Krym.* Al Gr Ju Rs(K).

9. J. excelsa Bieb., *Beschr. Länd. Terek Casp.* 204 (1800) (*J. isophyllos* C. Koch). Monoecious or dioecious tree up to 20 m, with a conical crown when young, later broad and open. Twigs 0·6–0·8 mm wide, terete, scaly. Juvenile leaves 5–6 mm, very few; scale-like leaves 1–1·5 mm, ovate-rhombic, closely appressed, acute, with a central, ovate or linear gland on the back, entire, the border not scarious. Female cone 8 mm, ripening in the second year, globose, slightly pruinose, dark purplish-brown when ripe. Seeds 4–6, free. *Balkan peninsula; Krym.* Al Bu Gr Ju Rs(K).

10. J. sabina L., *Sp. Pl.* 1039 (1753). Dioecious shrub, usually low and spreading. Twigs 0·6–0·8 mm wide, quadrangular. Leaves strongly foetid, all decussate; adult leaves 1–1·25 mm, ovate, appressed, obtuse to subacute, with an elliptical gland on the back, entire, the border not scarious. Female cone 4–6 mm, ripening in the autumn of the first year or in the following spring, usually depressed-globose, bluish-black, pruinose; stalk short, recurved, scaly. Seeds usually 2, free. $2n = 22$. ● *Mountains of S. & C. Europe; also at low altitudes in S.E. Russia.* Al Au Bu Cz Ga Ge Gr He Hs It Ju Po Rm Rs(C, W, K, E).

11. J. virginiana L., *Sp. Pl.* 1039 (1753). Dioecious pyramidal tree up to 30 m. Twigs 0·6–0·8 mm in diameter, very slender. Leaves all decussate; the juvenile 5–6 mm, often present on adult trees, long, spiny-pointed; adult leaves 0·5–1·5 mm, appressed but free at their acute or acuminate apices, often with a small gland on the back, entire and without a scarious border. Female cone 4–6 mm, ripening in the first year, ovoid, glaucous, brownish-violet when ripe; stalk erect or patent. Seeds 1 or 2, free. *Planted for timber in parts of C. & S. Europe.* [Ga Hu It Rm.] (*North America.*)

TAXOPSIDA

TAXALES

29. TAXACEAE

EDIT. D. M. MOORE

Usually dioecious, evergreen trees or shrubs with spirally arranged, linear, flattened leaves; not resiniferous. Male flowers axillary, solitary or in small spikes. Female flowers axillary, solitary or 2, with 1 or more sterile scales; ovuliferous scale terminal, bearing a solitary, erect ovule. Seeds partly or wholly surrounded by an aril.

1. Taxus L.

J. DO AMARAL FRANCO

Twigs irregularly alternate. Male flowers solitary, each consisting of 6–14 peltate anthers with 4–8 pollen-sacs. Female flower consisting of several imbricate scales, the uppermost fertile. Seeds ovoid, partly surrounded by a fleshy, usually scarlet, aril, ripening the first year.

1. T. baccata L., *Sp. Pl.* 1040 (1753). Shrub or tree up to 20 m, with a wide, pyramidal crown. Leaves 10–30 mm, the margins recurved, dark, glossy green above, with 2 pale green stomatal bands beneath. Seeds 6–7 mm. $2n = 24$. *Europe, except the east and extreme north, extending northwards to* c. 63° *N. in Norway and eastwards to Estonia and Krym; only on mountains in the Mediterranean region.* All except Cr Fa Is Rs(N, E) Sb; doubtfully native in Ho.

Cultivated for centuries; the many cultivars differ in habit, leaf-characters and aril-colour. A key to and descriptions of these are given in G. Krüssman, *Manual of Cultivated Conifers*, 281–288 (London, 1985).

GNETOPSIDA

GNETALES

30. EPHEDRACEAE

EDIT. J. R. EDMONDSON

Shrubs with opposite or verticillate leaves, usually reduced to short, more or less coalescent sheaths, 1–2(–3) mm; not resiniferous. Inflorescences axillary; strobili sessile or on short shoots. Fruit a globose or ovoid syncarp, the bracts fleshy at maturity, red, or rarely yellow.

1. Ephedra L.

†F. MARKGRAF (EDITION 1) REVISED BY H. FREITAG AND M. MAIER-STOLTE (EDITION 2)

Twigs green. Male strobilus with 2–8(–12) pairs of free bracts, each with an axillary flower of 2 united scales surrounding a column bearing a few 2- or 3-celled microsporangia, the microsporangia dehiscing by apical, horizontal slits. Female strobilus of 2 or 3 pairs of more or less connate bracts surrounding 1 or 2 naked apical flowers; ovules orthotropous, with integument prolonged into a slender tube.

Literature: O. Stapf, *Denkschr. Akad. Wiss. Math.-Nat. Kl.* (*Wien*) **56(2)**: 1–112 (1889). F. Widder, *Phyton* (*Austria*) **1**: 71–75 (1948). H. Riedl, *Scient. Pharmac.* **35**: 225–228 (1967). H. Freitag & M. Maier-Stolte, *Taxon* **38**: 545–556 (1989).

1 Often gynodioecious; pith white **1. foeminea**
1 Dioecious; pith brown
 2 Twigs often fragile when dry; leaves and bracts ciliolate; bracts of female strobilus connate almost to apex **2. fragilis**
 2 Twigs usually not fragile; leaves and bracts glabrous; bracts of female strobilus connate to $\frac{1}{3}$ or $\frac{1}{2}$
  3 Twigs 0·7–1·0 mm wide; male strobilus globose, with 2–4 pairs of flowers; fruit 1-seeded **3. major**
  3 Twigs 1·0–1·5 mm wide; male strobilus ovoid or cylindrical, with 4–8(–12) pairs of flowers; fruit 2-seeded **4. distachya**

1. E. foeminea Forskål, *Fl. Aegypt.* 170 (1775) (*E. fragilis* subsp. *campylopoda* (C. A. Meyer) Ascherson & Graebner). Often gynodioecious; climbing up to 3(–5) m, or procumbent; twigs usually pendent. Predominantly male strobilus subsessile, ovoid, with 4–8 pairs of flowers, the 2 terminal ones often female; male flowers with 4–6 microsporangia. Female strobilus stipitate; peduncles 2–10(–20) mm, usually reflexed; innermost pair of bracts connate almost to apex, the margins not ciliolate. Tube of integument 0·5–1 mm, straight. Fruit 2-seeded, the seed concealed by bracts. *E. Mediterranean region.* Al Bu Cr Gr It Ju Tu.

2. E. fragilis Desf., *Fl. Atl.* **2**: 372 (1799) (*E. altissima* sensu Willk., non Desf.). Dioecious; more or less erect shrub to 1·5(–2) m; main branches spreading, rarely scandent; twigs readily disarticulating at nodes. Male strobilus subsessile, ovoid, with 4–8 pairs of flowers, the flowers with 4–6 microsporangia. Female strobilus stipitate, 1- or 2-flowered; peduncles 2–4 mm; innermost pair of bracts connate almost to apex. Tube of integument 0·5–1·5(–2) mm, straight. Seed or seeds concealed by bracts. *Mediterranean region; S. Portugal.* Bl Hs Lu Si.

3. E. major Host, *Fl. Austr.* **2**: 671 (1831) (*E. nebrodensis* Tineo ex Guss., *E. scoparia* Lange). Dioecious; shrub up to 2 m, with 1 trunk-like stem; twigs 0·7–1(–1·2) mm wide. Male strobilus sessile, globose, with 2–4 pairs of flowers; flowers with 6–8 microsporangia. Female strobilus shortly stipitate, 1-flowered. Tube of integument 0·6–1 mm, straight. Fruit 1-seeded, the seed emergent. *Mediterranean region.* Al Ga Gr Hs It Ju Sa Si Tu.

(a) Subsp. **major**: Twigs rough. Female strobilus with innermost bracts connate for *c.* $\frac{1}{3}$ their length. $2n = 14$. *Almost throughout the range of the species.*

(b) Subsp. **procera** (Fischer & C. A. Meyer) Bornm., *Bot. Jahrb.* **140**: Beibl. 185 (1928) (*E. procera* Fischer & C. A. Meyer): Twigs smooth. Female strobilus with innermost bracts connate for *c.* $\frac{1}{2}$ their length. *Mountains of Greece and S. Jugoslavia.*

4. E. distachya L., *Sp. Pl.* 1040 (1753) (*E. vulgaris* L. C. M. Richard). Dioecious; shrub up to 0·5 m, with numerous stems from subterranean rhizome; twigs (1–)1·2–1·5 mm wide, usually somewhat papillose-scabrid. Male strobilus usually stalked, ovoid to cylindrical, with 4–8 pairs of flowers; flowers with 6–8 microsporangia. Female strobilus stipitate, 2-flowered, the innermost bracts connate for *c.* $\frac{1}{2}$ their length. Seeds emergent. *S. Europe, extending northwards to N.W. France, S. Slovakia, C. Ukraine and* c. 56° *N. in E. Russia.* Al Bu Co Cz Ga Gr He Hs Hu It Ju Rm Rs(C, W, K, E) Sa Si Tu.

(a) Subsp. **distachya**: Microsporangia mostly stipitate. Tube of integument 0·7–1·2 mm, straight. $2n = 24, 28$. *Almost throughout the range of the species.*

(b) Subsp. **helvetica** (C. A. Meyer) Ascherson & Graebner, *Syn. Mitteleur. Fl.* **1**: 260 (1897): Microsporangia mostly sessile. Tube of integument (1–)1·2–2 mm, twisted, rarely straight. *Rocks.* ● *Alps.*

Plants from E. Europe with smoother stems and somewhat smaller seeds have been referred to subsp. **monostachya** (L.) H. Riedl, *Scient. Pharmac.* **35**: 228 (1967).

SPERMATOPHYTA

ANGIOSPERMAE

DICOTYLEDONES

Trees, shrubs or herbs, rarely parasitic or saprophytic. Xylem with vessels. Flowers hermaphrodite or unisexual. Ovules enclosed in an ovary. Seeds with 2 cotyledons, rarely 1 by abortion.

SALICALES

31. SALICACEAE

EDIT. J. R. EDMONDSON

Dioecious. Trees or shrubs. Leaves alternate, very rarely opposite; stipules present. Flowers subtended by bracts and arranged in catkins; perianth absent; glands present, sometimes nectariferous. Male flowers with 2 to many stamens; filaments filiform, free or more or less united. Female flowers with 1 unilocular carpel with 2–4 parietal placentae; ovules numerous, ascending, anatropous. Fruit a capsule dehiscing by 2–4 valves. Seeds numerous, very small, each with a tuft of long hairs.

Literature: R. D. Meikle, *Willows and Poplars of Great Britain and Ireland*. London, 1984.

Buds with 1 outer scale; bracts entire **1. Salix**
Buds with several outer scales; bracts dentate or ciliate-fimbriate **2. Populus**

1. Salix L.

K. H. RECHINGER (EDITION 1) REVISED BY J. R. AKEROYD (EDITION 2)

Buds with 1 outer scale. Leaves variously shaped. Flowers primarily entomophilous, appearing before or after the leaves, each flower with 1 or 2 small nectaries and subtended by an entire bract. Stamens usually 2, 3 or 5. Capsule 2-valved.

Although sections have been described in this genus they will not be indicated in the text, since most of them are represented only by a single species or species-group in Europe (with the exception of Sect. *Capreae*, species **38–41**).

Hybridization plays a very important role in *Salix*. Most hybrids are highly fertile, and in some regions hybrid swarms obscure the limits of the species. Complex hybrids are also known. On the other hand certain taxa of evidently hybrid but uncertain origin are fairly well stabilized. These have not been included in the key but are mentioned after the species to which they show most resemblance. It is not possible to provide a separate key for non-flowering plants as there is too much parallel variation in most species in indumentum, leaf-shape, leaf-margin, etc.

The key is therefore based on the rather unrealistic assumption that the plant to be named is represented by branches bearing male catkins, female catkins, and leaves, and that data on its habit and habitat are also available.

Leaf-characters refer to *adult* leaves unless otherwise stated. Catkin-characters refer to the *female* catkins unless otherwise stated.

Literature: N. J. Andersson, *Kungl. Svenska Vet.-Akad. Handl.* n.s. **6(1)**: 1–180 (1867). N. J. Andersson in DC., *Prodr.* **16(2)**: 191–323 (1868). R. Buser, *Ber. Schweiz. Bot. Ges.* **50**: 567–788 (1940). A. & E. G. Camus, *Jour. Bot. (Paris)* **18**: 177–213, 245–296, 367–372 (1904); **19**: 1–68, 87–144 (1905); **20**: 1–116 (1906). B. Floderus in Holmberg, *Skand. Fl.* **1(b, 1)**: 6–160 (1931). E. F. Linton, *Jour. Bot. (London)* **51**: suppl. (1913). A. K. Skvortsov, *Proc. Study Fauna Fl. U.S.S.R., N.S. Sect. Bot.* **15 (XXIII)**: 1–262 (1968). A. Toepffer, *Ber. Bayer. Bot. Ges.* **15**: 17–233 (1915). F. Wimmer, *Salices Europaeae* 1866.

1 Low shrub of the arctic or high mountains; catkins from terminal or subterminal buds (if catkins absent, see also **17–23** and **54–55**)
2 Woody stem subterranean, with slender, herbaceous leaf- and flower-bearing aerial stems
3 Leaves orbicular or reniform, usually emarginate at apex, crenate-serrate; ovary glabrous **7. herbacea**
3 Leaves broadly elliptical; margin entire or shallowly sinuate near the base; ovary tomentose **8. polaris**
2 Woody stem above ground
4 Leaves with 2–5 pairs of lateral veins, usually orbicular
5 Petiole 5–15 mm; capsule tomentose **6. reticulata**
5 Petiole 2–3 mm; capsule glabrous **11. nummularia**
4 Leaves with 7–8 pairs of lateral veins, usually ovate, obovate or narrower
6 Ovary villous
7 Bracts dark; leaves with prominent lateral veins
8 Stipules much shorter than the petioles, often inconspicuous; leaves ovate or obovate, ±entire **15. arctica**
8 Stipules much longer than the very short petioles, especially on the terminal leaves; leaves narrowly lanceolate, densely glandular-serrate **16. pulchra**
7 Bracts uniformly pale; lateral veins of leaves not prominent
9 Stipules absent; petiole 3–5 mm; bracts ferruginous, hairy on both surfaces **19. pyrenaica**
9 Stipules as long as the petiole; petiole less than 3 mm or almost absent; bracts almost black in their upper part, ciliate **20. reptans**
6 Ovary not villous
10 Leaves hairy, at least when young; catkins large, dark purple; ovary with short, crisped hairs **(12–14). myrsinites** group
10 Leaves usually glabrous, even when young; catkins small, greenish or yellowish; ovary glabrous
11 Plant procumbent to ascending; leaves 0·5–1·1 cm wide; catkins with more than 10 flowers **9. retusa**
11 Plant extremely compact; leaves 0·2–0·4 cm wide; catkins with 3–8 flowers **10. serpillifolia**
1 Trees and shrubs mainly of lower latitude or altitude; catkins from lateral buds
12 Usually trees; bracts uniformly pale; flowers with 2 or more nectaries, or male flowers with 2 and female with 1
13 Twigs long, slender, pendent **4. babylonica**
13 Twigs not pendent
14 Mature leaves ±densely covered with persistent, silky hairs **3. alba**
14 Mature leaves glabrous
15 Leaves finely and evenly serrate with yellow glands, very glossy above; stamens (4)5(–12) **1. pentandra**

15 Leaves rather coarsely and unevenly serrate with white teeth, not very glossy above; stamens 2 or 3

16 Stipules usually caducous; bark of old stems not flaking off; twigs easily breaking off **2. fragilis**

16 Stipules persistent; bark of old stems flaking off; twigs not easily breaking off **5. triandra**

12 Usually shrubs; bracts often with dark apex and paler base; flowers with 1 nectary

17 Filaments connate, at least in part

18 Low shrubs not more than 1 m; leaves suborbicular or up to twice as long as wide; catkins less than 2 cm

19 Leaves elliptical or obovate; ovary sessile or subsessile **62. caesia**

19 Leaves suborbicular; pedicel *c.* 2 mm **64. tarraconensis**

18 Tall shrubs up to 6 m; leaves 3–15 or more times as long as wide; catkins usually at least 2 cm

20 Leaves lanate beneath at maturity; ovary glabrous **57. elaeagnos**

20 Leaves ±glabrous beneath at maturity; ovary hairy

21 Young twigs and leaves ±densely sericeous **61. wilhelmsiana**

21 Young twigs (and usually leaves) ±glabrous

22 Leaves with prominent veins beneath **60. caspica**

22 Leaves without prominent veins beneath

23 Leaves narrowed, cuneate or narrowly rounded at base, often alternate and often petiolate **58. purpurea**

23 Leaves semi-amplexicaul or truncate at base, most or all opposite and sessile or subsessile **59. amplexicaulis**

17 Filaments free (rarely connate in abnormal flowers)

24 Leaves usually at least 4 times as long as wide, glabrous, or lower surface with silky, appressed hairs all longitudinally arranged

25 Young twigs with a bluish waxy bloom; mature leaves glabrous, serrate; ovary glabrous

26 Leaves oblong-lanceolate to -ovate, acute or shortly acuminate; lateral veins 8–12 pairs; bracts, including hairs, about as long as ovary **64. daphnoides**

26 Leaves lanceolate to linear-lanceolate, long-acuminate; lateral veins 15 or more pairs; bracts, including hairs, *c.* $\frac{1}{2}$ as long as ovary **65. acutifolia**

25 Young twigs without a waxy bloom; leaves sericeous beneath, entire; ovary sericeous

27 Usually low shrubs of swamps, bogs or damp sand; style short; catkins rarely up to 2·5 cm **48. rosmarinifolia**

27 Tall riverside shrubs or small trees; style ±long; catkins more than 2·5 cm **56. viminalis**

24 Leaves usually less than 4 times as long as wide, rarely 5 times, glabrous, or with dull grey tomentum, or white lanate tomentum of short, irregularly arranged hairs, or somewhat sericeous beneath

28 Ovary glabrous

29 Branches persistently lanate (21–23). **lanata** group

29 Branches not lanate

30 Pedicel less than $\frac{1}{2}$ as long as ovary, often ±absent

31 Leaves with thick, waxy bloom over the whole lower surface, including apex **31. glabra**

31 Leaves with waxy bloom beneath, disappearing towards apex, or entirely lacking

32 Twigs with a bluish, waxy bloom **64. daphnoides**

32 Twigs without a bluish, waxy bloom

33 Petiole 10–30 mm

34 Leaves about twice as long as wide, sericeous beneath when young **32. crataegifolia**

34 Leaves scarcely longer than wide, glabrous **53. pyrolifolia**

33 Petiole not more than 10 mm

35 Procumbent shrub; stipules absent **47. repens**

35 Ascending or erect shrub or tree, 1–4 m; stipules often well developed

36 Young twigs usually whitish-pubescent; catkins lax, subsessile, or peduncle to 15 mm (27–30). **myrsinifolia** group

36 Young twigs not pubescent; catkins densely flowered, peduncle to 30 mm **52. hastata**

30 Pedicel at least $\frac{1}{2}$ as long as ovary

37 At least the main stems creeping; leaves rarely more than 2 cm, usually entire

38 Leaves usually rounded at both ends, finally glabrous; veins not or scarcely prominent above **46. myrtilloides**

38 Leaves usually narrowed at both ends, ±persistently sericeous, at least beneath; veins fine but prominent above **47. repens**

37 Main stems erect; leaves more than 2 cm, never entire

39 Twigs and underside of leaves persistently pubescent; inflorescence-axis densely hairy **33. pedicellata**

39 Young twigs and leaves glabrous or subglabrous; inflorescence-axis with scattered, short hairs **34. silesiaca**

28 Ovary not glabrous

40 Stipules absent or small and caducous

41 Twigs glabrous or glabrescent

42 Leaves with a dull bloom on both surfaces **62. caesia**

42 Leaves ±shiny above, ±glaucous beneath

43 Robust, erect shrubs, usually more than 2 m (24–26). **phylicifolia** group

43 Low alpine shrubs or subshrubs, not more than 2 m

44 Leaves ±serrate, glabrous or with sparse hairs (49–51). **arbuscula** group

44 Leaves entire, with numerous long, appressed hairs beneath

45 Bracts uniformly pale, ferruginous **19. pyrenaica**

45 Bracts black towards apex **47. repens**

41 Twigs pubescent or lanate

46 Leaves with sericeous hairs

47 Leaves ±obtuse; capsule 10–12 mm **17. glauca**

47 Leaves subacute; capsule 6–7(–10) mm **18. glaucosericea**

46 Leaves tomentose on 1 or both surfaces

48 Shrub or tree to 6 m; pedicel well developed; filaments hairy **45. xerophila**

48 Shrub, 0·5–4 m; ovary sessile or subsessile; filaments glabrous
49 Leaves green above, white beneath; catkins becoming pedunculate, elongated and ± lax at maturity **54. helvetica**
49 Leaves grey or greyish on both surfaces; catkins sessile, remaining dense at maturity **55. lapponum**
40 Stipules well developed, caducous or not
50 Leaves with sericeous hairs, at least beneath, or glabrescent
51 Catkins with long, leafy peduncles
52 Procumbent shrubs up to 50 cm; leaves glabrescent **20. reptans**
52 Erect shrubs 1–3 m; leaves with ± sericeous, persistent hairs
53 Leaves ± obtuse or rounded; lateral veins 5–7 pairs **17. glauca**
53 Leaves subacute; lateral veins 7–9(–14) pairs **18. glaucosericea**
51 Catkins sessile or subsessile, or without leafy peduncles
54 Shrub or small tree 2–3 m **43. cantabrica**
54 Low shrub to 1(–2) m
55 Leaves slightly pubescent when young, soon glabrous; bracts greenish-yellow **44. starkeana**
55 Leaves persistently and densely sericeous beneath; bracts dark-tipped **47. repens**
50 Leaf indumentum not sericeous
56 Branches persistently lanate
57 Filaments glabrous; ovary pubescent only when young **(21–23). lanata** group
57 Filaments with long hairs towards base; ovary with persistent, dense hairs **37. laggeri**
56 Branches not persistently lanate, sometimes pubescent
58 Leaves 3–5 times as long as wide; stipules patent, with deflexed apex **42. salviifolia**
58 Leaves normally up to 3 times as long as wide; stipules never patent with deflexed apex
59 2- to 4-year-old twigs with numerous very prominent ridges beneath the bark
60 Buds and twigs of the previous year pubescent
61 Ridges beneath the bark short, discontinuous; leaves finally ± glabrous above, and with sparse, grey, persistent indumentum beneath **35. aegyptiaca**
61 Ridges beneath the bark long, continuous; leaves persistently but sparsely pubescent above, and with dense, grey, felted, persistent indumentum beneath **38. cinerea**
60 Buds and twigs of the previous year ± glabrous
62 Catkins 2–5 cm; leaves at least 3 cm, not rugose, sparsely pubescent beneath; hairs, especially along the veins, rust-coloured **39. atrocinerea**
62 Catkins 1–2·5 cm; leaves *c*. 2–3 cm, rugose, grey, ± tomentose beneath **40. aurita**
59 2- to 4-year-old twigs without or with few, scarcely prominent ridges beneath the bark
63 Leaves ± glabrous beneath; shrub less than 1 m
64 Twigs and leaves ± glabrous; petiole up to 5 mm; filaments glabrous **44. starkeana**
64 Twigs and leaves hairy; petiole 6–8 mm; filaments hairy **45. xerophila**
63 Leaves persistently hairy beneath; tree, or shrub more than 1 m
65 Leaves 1½–2 times as long as wide, ovate or obovate to ovate-oblong; no ridges beneath the bark **41. caprea**
65 Leaves usually more than twice as long as wide, obovate to oblanceolate or narrowly elliptical; scattered ± indistinct ridges usually present beneath the bark
66 Stipules usually absent **45. xerophila**
66 Stipules well developed
67 Twigs greyish-brown or grey, with flat leaf-scars; mature leaves rather thick, the veins closely reticulate, impressed above, prominent beneath **36. appendiculata**
67 Twigs blackish with protruding leaf-scars; mature leaves thin, the veins widely reticulate, ± prominent above, scarcely prominent beneath **37. laggeri**

Subgen. **Salix**. Trees or tall shrubs. Leaves lanceolate, gradually acuminate. Catkins on leafy peduncles arising from lateral buds of the previous year. Bracts concolorous, yellowish. Stamens 2 or more, free. Male flowers with 2 nectaries; female flowers with 1 or 2 nectaries.

1. S. pentandra L., *Sp. Pl.* 1016 (1753). Shrub or small tree up to 7 m. Twigs glabrous, shining, not fragile. Leaves 5–12 cm, ovate or elliptical, 2–4 times as long as wide, acuminate, rounded or broadly cuneate at base, finely and very regularly serrate, with yellow marginal glands, glabrous, dark green and glossy above, paler beneath, somewhat coriaceous when mature, sticky and fragrant when young; petiole with 1–3 pairs of glands near the top. Stipules small, caducous. Catkins 2–5 × 1–1·5 cm, appearing with the leaves, cylindrical. Stamens (4)5–8(–12). Female flowers with 2 nectaries. *Wet places. Most of Europe, but rare in the south.* Al Au Br Bu Cz Da Fe Ga Ge ?Gr Hb He Ho Hs Hu It Ju No Po Rm Rs(N, B, C, W, E) Su [Be].

2. S. fragilis L., *Sp. Pl.* 1017 (1753). Tree up to 25 m, often pollarded, sometimes shrubby. Bark greyish, becoming fissured but not flaking off. Twigs glabrous and olive, very fragile at the junctions. Leaves 6–15 × 1·5–4 cm, lanceolate, usually 4½–9 times as long as wide, long-acuminate, usually asymmetrical at apex, cuneate at base, rather coarsely serrate, glabrous and shiny above, with a glaucous bloom or, less often, paler green beneath; petiole with usually 2 glands at the top. Stipules usually caducous. Catkins 3–7 × *c*. 1 cm, appearing with the leaves, drooping, rather dense, cylindrical. Stamens 2(3), free; anthers yellow. Female flowers usually with 2 nectaries. $2n = 76$. *Wet places. Most of Europe except the Arctic, but rare in the Mediterranean region and probably introduced in much of the north.* Al Au Be Bu Co Cz Ga Ge Gr He Ho Hs Hu It Ju Po Rm Rs(B, C, W, K, E) Sa Si Tu [Az Bl *Br Da Fe Hb Lu No Rs(N) Su].

The hybrid between **2** and **3** (*S.* × *rubens* Schrank) is probably the most frequent of all *Salix* hybrids. It often resembles **2**, from which it differs, *inter alia*, in the sericeous pubescence of the young leaves, and the narrower, more gradually tapering leaves. It is commoner and more widespread than pure *S. fragilis* and more frequently cultivated.

Several derivatives of this hybrid, some of them of garden origin, are naturalized in Britain. They are distinguished from each other chiefly by leaf-characters and by the length of the catkins and ovaries. One of them, *S.* × *rubens* nothovar. *basfordiana* (Scaling ex Salter) Meikle, is readily identified by its glossy, orange-yellow or sometimes reddish twigs, bright green tapering leaves 9–15 × 1·5–2 cm, with finely but sharply serrate margins, long, usually drooping catkins 5–12 cm, and ovary subsessile or shortly pedicellate, exceeding the sparsely hairy bract.

S. excelsa S. G. Gmelin, *Reise Russl.* **3**: 308 (1774), is very like **2** × **3** but has stouter catkins, and wider, often slightly brownish, unbearded bracts.

3. S. alba L., *Sp. Pl.* 1021 (1753). Tree up to 25 m, often pollarded. Branches ascending, appearing silvery-grey in life. Bark greyish, not flaking. Twigs sericeous when young, later glabrous and brown or olive, not easily breaking off. Leaves 5–10(–12) cm, usually $5\frac{1}{2}$–$7\frac{1}{2}$ times as long as wide, acuminate, not or slightly asymmetrical at apex, cuneate at base, finely serrate, covered with white, silky, appressed hairs on both surfaces; petiole usually with a few glands. Stipules caducous. Catkins 3–5 cm, appearing with the leaves, dense, cylindrical. Bracts uniformly yellowish. Stamens 2, free; anthers yellow. Female flowers usually with 1 nectary. *Most of Europe but mainly as an introduction in the north; often planted.* Al Au Be *Br Bu Co *Cr Cz Ga Ge Gr He Ho Hs *Hu It Ju Lu Po Rm Rs(*N, B, C, W, K, E) Sa Si Tu [Da Fe Hb No Su].

1 Twigs bright yellow or orange — (c) subsp. **vitellina**
1 Twigs brown or grey
 2 Leaves dull, bluish-green beneath, soon becoming glabrous — (b) subsp. **caerulea**
 2 Leaves not bluish-green, covered with ± persistent, silky, appressed hairs
  3 Leaves dark green, somewhat glabrous above; ovary subsessile — (a) subsp. **alba**
  3 Leaves silky on both surfaces, densely covered with long, appressed, silky hairs; ovary on a short but distinct pedicel — (d) subsp. **micans**

(a) Subsp. **alba**: Twigs brown or yellowish-brown, erect. Leaves covered on both surfaces with appressed, silky hairs, the upper surface more or less glabrescent. Ovary subsessile. *Throughout most of the range of the species.*

(b) Subsp. **caerulea** (Sm.) Rech. fil., *Österr. Bot. Zeitschr.* **110**: 338 (1963) (*S. caerulea* Sm.): Twigs brown or grey, erect. Leaves bluish-green, somewhat larger, soon becoming more glabrous, more conspicuously toothed than in subsp. (a). *Frequently planted in Britain (where it is used in the manufacture of cricket-bats) and perhaps elsewhere, and occasionally naturalized.*

(c) Subsp. **vitellina** (L.) Arcangeli, *Comp. Fl. Ital.* 626 (1882): Twigs bright yellow or orange, thin, sometimes pendent. Leaves paler green and indumentum less dense than in subsp. (a). Catkin-scales somewhat longer than in subsp. (a). *Frequently planted.*

(d) Subsp. **micans** (N. J. Andersson) Rech. fil., *loc. cit.* (1963) (*S. micans* N. J. Andersson): Like subsp. (a) but leaves with very long, persistent hairs on both surfaces, therefore very shiny; ovary distinctly pedicellate. *Greece.* (*Caucasus.*)

4. S. babylonica L., *Sp. Pl.* 1017 (1753) (*S. elegantissima* C. Koch). Tree up to 20 m. Twigs long, pendent, drooping almost to the ground. Leaves 8–16 × 0·8–1·5 cm, narrowly lanceolate or linear-lanceolate, acuminate, serrulate, glabrous at maturity; petiole 3–5 mm. Catkins up to 2 × 0·3–0·4 cm, subsessile, curved. Bracts concolorous, glabrescent. Female flowers with 1 nectary. Ovary glabrous, sessile; stigma emarginate. *Widely planted and sometimes more or less naturalized. Present occurrences uncertain.* (*China.*)

Several hybrids are now more commonly cultivated and are more hardy.

The hybrid between **3(c)** and **4** (*S. chrysocoma* Dode) is frequently planted and combines the sericeous indumentum and the yellow twigs of the former with the 'weeping' growth of the latter. It is more hardy than pure *S. babylonica*. The hybrid between **4** and **3** (*S.* × *sepulcralis* Simonkai) is frequently cultivated; **2** × **4** (*S.* × *pendulina* Wenderoth) is also cultivated, but less frequently.

5. S. triandra L., *Sp. Pl.* 1016 (1753) (*S. amygdalina* L.). Shrub or small tree up to 10 m. Bark smooth, greyish, flaking off in patches. Twigs glabrous, greenish- or reddish-brown. Leaves 4–11(–15) cm, usually $3\frac{1}{2}$–$7\frac{1}{2}$ times as long as wide, oblong-ovate or -lanceolate, acute or shortly acuminate (not asymmetrical), rounded or cuneate at base, serrate, glabrous, dark green and somewhat shining above, glaucous or pale green beneath; petiole with 2 or 3 small glands at the top. Stipules large, persistent. Catkins appearing with the leaves, erect, cylindrical; male 3–5(–7) cm, slender; female rather shorter and denser than the male. Stamens 3, free. Female flowers with 1 nectary. *River-banks. Most of Europe, except the Arctic; local in the Mediterranean region.* Al Au Be Br Bu Cz Fe Ga Ge Gr Hb He Ho Hs Hu It Ju Lu No Po Rm Rs(N, B, C, W, K, E) Su.

(a) Subsp. **triandra** (*S. ligustrina* Host, *S. triandra* subsp. *concolor* (Koch) A. Neumann ex Rech. fil.): Leaves pale green beneath; petiole $\frac{1}{8}$–$\frac{1}{2}$ as long as the leaf. $2n = 38$. *Common in the lowlands of N.W. Europe.*

(b) Subsp. **discolor** (Koch) Arcangeli, *Comp. Fl. Ital.* 626 (1882): Leaves glaucous or whitish beneath; petiole $\frac{1}{6}$–$\frac{1}{5}$ as long as the leaf. *Common in the south and east of the area of the species and in the valleys of the Alps.*

Subsp. (a) more frequently forms hybrids than subsp. (b). The hybrid with **56** is one of the most frequent in N.W. Europe as a planted shrub, but is extremely rare elsewhere.

The native distribution of the two subspecies is difficult to define precisely because they are frequently cultivated. Var. *villarsiana* (Flügge) Rouy is characterized by short, shiny branches and small, ovate or elliptical leaves, rounded at base, white beneath. It seems to be common in some valleys of the W. Alps and, though related to subsp. (b), perhaps represents another geographical subspecies.

Subgen. **Chamaetia** Dumort. Creeping or procumbent shrubs. Leaves orbicular or ovate. Catkins usually on leafless peduncles arising from terminal buds of the previous year. Bracts concolorous. Stamens 2, free. Flowers with 1 or 2 nectaries.

6. S. reticulata L., *Sp. Pl.* 1018 (1753). Dwarf shrub with a creeping and rooting stem, the twigs up to 20 cm with few leaves. Leaves 1–3(–5) cm, less than twice as long as wide, rounded to retuse at apex, cordate (rarely cuneate) at base, entire, coriaceous, dark green, subglabrous and rugose with impressed veins above, whitish, subglabrous and prominently reticulate-veined beneath; lateral veins 2–5 pairs; petiole 5–15 mm. Stipules absent. Catkins 1·2–3·5 cm, appearing with the leaves, on peduncles up to 2·5 cm. Bracts obovate, light brown, villous within. Capsule ovoid, sessile or subsessile, tomentose; style short. $2n = 38$. *Somewhat calcicole. Arctic and subarctic Europe and in the high mountains southwards to the Pyrenees, Maritime Alps and Makedonija.* Al Au Br Bu Cz Fe Ga Ge He Hs It Ju No Po Rm Rs(N, C) Sb Su.

7. S. herbacea L., *loc. cit.* (1753). Dwarf shrub with long, creeping, branched underground stems. Aerial twigs usually 2–3 cm, with 2–5 leaves. Leaves 0·5–2 cm, orbicular or reniform, less than $1\frac{1}{2}$ times as long as wide and sometimes wider than long, rounded or emarginate, cordate or rounded at base, crenate-serrate, glabrous, bright green and shining, the veins prominent on both surfaces; petioles up to 5 mm. Stipules usually absent. Catkins 0·5–1·5 cm, 2- to 12-flowered, appearing with the leaves. Bracts more or less obovate, yellowish-green, usually glabrous. Female flowers with 1 or 2 nectaries; style short. $2n = 38$. *Arctic and subarctic Europe, extending southwards in the mountains to C. Spain, C. Appennini and Macedonia.* Al Au Br Bu Cz Fa Fe Ga Ge Hb He Hs Is It Ju No Po Rm Rs(N, W) Sb Su.

Hybridizes frequently with several species in Scandinavia, Iceland and Scotland but very rarely in other areas. The influence of **7** in the hybrids is always very evident in habit and leaf-shape.

8. S. polaris Wahlenb., *Fl. Lapp.* 261 (1812). Dwarf shrub; stems underground; aerial twigs very slender. Leaves 1·5 × 1·2 cm, broadly elliptical, dark green, very shiny above, paler green beneath, entire or shallowly sinuate near base; lateral veins 3 or 4 pairs, slightly reticulate; petiole relatively long. Stipules absent. Catkins *c.* 15-flowered. Bracts firm, rather large, more or less urceolate, purplish-black, with rather dense, long, straight hairs. Filaments purple, smooth. Ovary subsessile, hairy; style long. *Arctic Europe, extending southwards to 60° N. in Norway and Ural.* Fe No Rs(N, C) Sb Su.

Hybridizes freely with **7** in N. Fennoscandia.

9. S. retusa L., *Syst. Nat.* ed. 10, **2**: 1287 (1759) (incl. *S. kitaibeliana* Willd.). Stem procumbent to ascending, glabrous, dark. Leaves 0·8–3·5 × 0·5–1·1 cm, rounded, more or less truncate or emarginate at apex, glabrous, green on both surfaces, almost entire; petiole 1–2 mm. Catkins up to 2–4 × 1 cm, with more than 10 flowers, cylindrical, developing with the leaves. Style $\frac{1}{8}$–$\frac{1}{6}$ as long as the glabrous ovary. $2n = 114, 152$. ● *Alps, Pyrenees, Appennini, Carpathians, Balkan peninsula.* Al Au Bu Cz Ga Ge ?Gr He Hs It Ju Po Rm Rs(W).

Hybrids of **9** are locally frequent and always similar in habit and in leaf-characters to this species. The influence of the other parent is evident mainly in the terminal leaves of elongated shoots.

10. S. serpillifolia Scop., *Fl. Carn.* ed. 2, **2**: 255 (1772). Like **9** but more compact, completely appressed to the soil; leaves 0·4–1 × 0·2–0·4 cm, imbricate, acute or emarginate; petiole *c.* 0·5 cm; catkins 0·5 cm, globose, with 3–8 flowers, developed after the leaves; style *c.* $\frac{1}{4}$ as long as the ovary. $2n = 38$. *Usually calcicole.* ● *Alps.* Au Ga Ge He It Ju.

11. S. nummularia N. J. Andersson in DC., *Prodr.* **16**: 298 (1868) (*S. rotundifolia* auct., non Trautv.). Dwarf, spreading shrub; stems and twigs above the surface, often very long, not rooting. Vegetative stems usually with *c.* 4 leaves. Leaves 0·5–1·5 cm, orbicular, with a subcordate base and emarginate apex, entire or finely serrate in the lower part, sometimes hairy when young, later entirely glabrous, light green, prominently reticulate above; lateral veins 4 or 5; petiole 2–3 mm. Stipules usually absent. Catkins *c.* 3-flowered. Bracts thin, pale, with scattered, short, hooked hairs. Filaments glabrous. Female flowers with 1 nectary. Ovary subsessile, glabrous; style long. *Arctic Russia, extending southwards to C. Ural.* Rs(N).

(12–14). S. myrsinites group. Branches up to 40 cm, procumbent or ascending, thick, crooked, knotty. Leaves green and shiny on both surfaces, with reticulate veins slightly prominent on both surfaces, and long, straight hairs more or less evanescent but persisting mainly on the margins. Catkins large, dense, dark purple, appearing with the leaves. Ovary covered with short, somewhat crooked, evanescent hairs.

The three closely related species are geographically isolated, except that **13** and **14** overlap in parts of the E. Alps and slight intergradation may be observed locally (*vide* Samuelsson, *Viert. Naturf. Ges. Zürich* **67**: 247–250 (1922)).

1 Leaf-margins entire; catkins slender; stems procumbent **14. alpina**

1 Leaf-margins glandular-serrate; catkins stout; stems ascending

2 Dead leaves deciduous **13. breviserrata**

2 Dead leaves persistent until the end of the following growing season **12. myrsinites**

12. S. myrsinites L., *Sp. Pl.* 1018 (1753). More or less ascending. Leaves shiny on both surfaces, dark green, the margin glandular-serrate, dead leaves persisting until the end of the following growing season. Ovary sparsely hairy, becoming glabrous. $2n = 152, 190$. *Fennoscandia and N. Russia; Scotland.* Br Fe No Rs(N) Su.

This species hybridizes freely with **17** and **27**. These hybrids usually inherit the shiny green lower surface of the leaves and the purplish flowers of **12**.

13. S. breviserrata B. Flod., *Ark. Bot.* **29(18)**: 44 (1940) (*S. myrsinites* subsp. *serrata* (Neilr.) Schinz & Thell.). More or less ascending. Dead leaves deciduous; margin of leaves densely glandular-serrate. Catkins stout. Capsule broad; style short. ● *Alps; a few localities in N. Spain and N. & C. Italy.* Au Ga Ge He Hs It.

14. S. alpina Scop., *Fl. Carn.* ed. 2, **2**: 255 (1772) (*S. jacquinii* Host). More or less procumbent. Dead leaves deciduous; margin of leaves entire. Catkins slender. Capsule narrow; style long. ● *E. Alps; Carpathians; one station in S. Jugoslavia.* Au Cz Ge It Ju Po Rm Rs(W).

15. S. arctica Pallas, *Fl. Ross.* **1(2)**: 86 (1788). Twigs glabrous, shiny. Leaves 3–4 × 2–3 cm, ovate or obovate, rounded or broadly cuneate at the base, thick, leathery, hairy only when young, with more or less reticulate veins slightly prominent beneath and slightly impressed above. Stipules shorter than the petiole, often conspicuous. Catkins *c.* 5 × 1 cm; peduncles leafy, long. Bracts brownish, rounded at apex. Ovary densely covered with long, white hairs; style long, filiform, more or less cleft; stigma bifid. *Iceland and Færoër; arctic and subarctic Russia; Ural.* Fa Is Rs(N, C).

16. S. pulchra Cham., *Linnaea* **6**: 543 (1831). Low shrub; twigs procumbent or more or less ascending, dark purplish-brown, shiny. Leaves narrowly lanceolate, glandular-serrate (sometimes indistinctly), glabrous on both surfaces, finely reticulate; petiole short. Stipules 3–10 mm, glandular-serrate, narrow, acute, tapering towards base, much longer than the petioles, especially in terminal well-developed leaves. Catkins 4–5 × 1·5 cm; peduncles leafless. Bracts dark, acute. Ovary covered with long, silky hairs; style long; stigma narrow, long. *N.E. Russia.* Rs(N). (*Arctic Asia and America.*)

Subgen. **Caprisalix** Dumort. Shrubs or small trees. Catkins sessile or on leafy peduncles, from lateral buds of the previous year. Bracts often with dark apex. Stamens 2, free or united. Male and female flowers with a single nectary.

17. S. glauca L., *Sp. Pl.* 1019 (1753). Leaves obovate or oblanceolate to ovate, more or less obtuse, thin, rather bright green, decolorizing rather easily on drying; indumentum moderately dense, of somewhat sericeous, slightly intricate, hairs; lateral veins 5 or 6(7) pairs; petiole short. Stipules usually absent. Pedicel distinct, as long as the nectary. Capsule 10–12 mm. *Arctic and subarctic Europe, extending southwards to S. Norway and S. Ural.* Fe Is No Rs(N, C) Sb Su.

1 Stipules present, lanceolate (c) subsp. **stipulifera**
1 Stipules usually absent
 2 Shrub or small tree; leaves obovate (a) subsp. **glauca**
 2 Prostrate shrub; leaves oblong or ovate (b) subsp. **callicarpaea**

(a) Subsp. **glauca**: Shrub or small tree. Leaves obovate, more or less obtuse; lateral veins 5 or 6(7) pairs. Stipules usually absent. Style usually only divided at apex. *N. & W. Fennoscandia; Ural.*

(b) Subsp. **callicarpaea** (Trautv.) Böcher in Böcher, Holmen & Jakobsen, *Grønlands Fl.* 138 (1957) (*S. callicarpaea* Trautv., *S. cordifolia* Pursh subsp. *callicarpaea* (Trautv.) Á. Löve): Usually prostrate shrub. Leaves oblong or ovate, more or less obtuse; lateral veins 5–7 pairs. Stipules usually absent. Style often divided to base. $2n = 114, 190$. *Iceland, Svalbard.*

(c) Subsp. **stipulifera** (B. Flod. ex Häyrén) Hiitonen, *Suomen Kasvio* 272 (1933) (*S. stipulifera* B. Flod. ex Häyrén): Shrub. Leaves obovate or oblanceolate, rounded at apex; lateral veins 5–7 pairs. Lanceolate stipules present. Capsule 8–10(–12) mm. $2n = 152$. *N. Fennoscandia, arctic Russia.*

18. S. glaucosericea B. Flod., *Svensk Bot. Tidskr.* **37**: 169 (1943) (*S. glauca* auct. alp., *S. sericea* Vill., non Marshall). Leaves 5·5–7·5 × 1·5–2·2 cm, oblanceolate, subacute, pale green and somewhat shiny above, glaucous beneath, sericeous with appressed hairs on both surfaces; lateral veins 7–9(–14) pairs. Stipules usually absent. Capsule 6–7(–10) mm. $2n = 190$. • *Alps.* Au Ga He It.

19. S. pyrenaica Gouan, *Obs. Bot.* 77, no. 8 (1773) (*S. ciliata* DC.). Low shrub up to 50 cm, with ascending, reddish-brown glabrescent twigs. Leaves ovate-elliptical, rarely obovate, with more or less rounded base, slightly shiny and with few, long hairs above, somewhat glaucous and with numerous, long, appressed hairs beneath, glabrescent but hairs persistent at the entire margin; lateral veins 8 pairs; petiole 3–5 mm, pubescent. Stipules absent. Catkins lax-flowered, terminal on leafy, more or less elongated peduncles. Bracts 2 mm, obovate, obtuse, ferruginous, long-hairy on both surfaces. Capsule ovoid, obtuse, tomentose, somewhat glabrescent near the base; pedicel 0·5 mm; style bifid. • *Pyrenees.* Ga Hs.

20. S. reptans Rupr., *Beitr. Pfl. Russ. Reich.* **2**: 54 (1845). Twigs procumbent, glabrescent, green, suffused with red. Leaves ovate to lanceolate, entire, sometimes almost cordate at base, glabrescent and glaucous beneath. Stipules as long as the very short petiole, entire, elliptical, with straight apex. Catkins dense-flowered, on leafy peduncles. Bracts almost black in the upper part, obovate or ovate, ciliate. Capsule tomentose, large; style short, bifid. *Arctic and subarctic Russia.* Rs(N). (*Arctic Asia.*)

(21–23). S. lanata group. Shrubs up to 3 m. Twigs, persistently lanate. Stipules persistent. Leaves with whitish-grey tomentum, becoming glabrous, not blackening on drying. Catkins with short peduncles. Ovary glabrous (sometimes pubescent when young).

1 Stipules entire; leaf-margin undulate, eglandular **21. lanata**
1 Stipules toothed; leaf-margin flat, densely glandular
 2 Ovary glabrous **22. glandulifera**
 2 Ovary pubescent when young **23. recurvigemmis**

21. S. lanata L., *Sp. Pl.* 1019 (1753) (*S. chrysanthos* Vahl). Leaves 3·5–7 cm, thick, broadly ovate or obovate to suborbicular, more or less cordate at base and with a short, deflexed apex, bright yellow-hairy when young; margin entire, eglandular, undulate; lateral veins 5 or 6 pairs. Stipules wide, subobtuse, entire. Peduncles with scale-like leaves. Bracts ovate, obtuse, becoming blackish-brown, golden-hairy. Nectaries long, narrow, entire or shallowly cleft. Stamens yellow. Stigma-lobes long, filiform, entire, yellow or brownish. *N. Europe, from arctic Russia to C. Scotland.* Br Fe Is No Rs(N, C) Su.

Hybridizes freely with *S. hastata* and *S. caprea.*

22. S. glandulifera B. Flod., *Bot. Not.* **1930**: 338 (1930). Like **21** but leaves usually oblanceolate, with 6–8 pairs of lateral veins; margin flat, densely glandular; stipules long, acute, with sharp glandular teeth. $2n = 38$. *Arctic Europe.* Fe No Rs(N) Su.

23. S. recurvigemmis A. Skvortsov, *Not. Syst.* (*Leningrad*) **18**: 37 (1957). Like **21** but a procumbent shrub 1·5–2 m; bud-scales persistent; ovary pubescent when young. *N.E. Russia.* Rs(N, C).

(24–26). S. phylicifolia group. Shrub up to 5 m. Twigs dark, shiny, glabrous or becoming so with age. Leaves glabrous at maturity, more or less shiny above, more or less glaucous beneath, not blackening on drying. Catkins subsessile or shortly pedunculate, appearing with the leaves, rather lax in fruit. Ovary sericeous; pedicel 2–4 times as long as the nectary.

This is one of the most difficult groups in *Salix*. It is the more complicated because of the strong tendency to hybridize with **17** and **27** as well as with species **38–41**.

1 Leaves sericeous when young, becoming glabrous later; decorticated wood with short distinct ridges **25. bicolor**
1 Leaves almost glabrous even when young; decorticated wood without ridges
 2 Leaves with short glandular teeth, shiny above **24. phylicifolia**
 2 Leaves entire or with very indistinct crenulate-glandular teeth **26. hegetschweileri**

24. S. phylicifolia L., *Sp. Pl.* 1016 (1753) (incl. *S. hibernica* Rech. fil.). Ridges on decorticated wood absent. Buds narrow, acute, dark brown. Leaves obovate-lanceolate to ovate or narrowly elliptical, tapering at both ends, shiny above, with short glandular teeth. Catkins 1·5–4(–10) × 1–1·5 cm. Filaments entirely glabrous. Capsule up to 7 mm. $2n = 114$. *N. Europe, southwards to England and S. Ural*. Br ?Bu Fa Fe Hb Is No Rs(N, B, C) Su.

25. S. bicolor Willd., *Berlin. Baumz.* 339 (1796) (*S. phylicifolia* auct., *S. schraderiana* Willd., *S. weigeliana* Willd., *S. cantabrica* Rech. fil.). Ridges on decorticated wood short but distinct. Buds short, yellowish or orange; scale persistent after opening of bud. Leaves obtuse, sericeous on both surfaces when young, punctate, less shiny above than in **24**. Catkins and capsule shorter than in **24**. Filaments often sparsely hairy near the base. ● *From N. Germany to N. Spain and Bulgaria; mainly in the mountains but absent from the Alps*. Cz Ga Ge Hs Po Rm Rs(W).

26. S. hegetschweileri Heer in Hegetschw., *Fl. Schweiz* 963 (1840) (*S. bicolor* subsp. *rhaetica* (N. J. Andersson) B. Flod., *S. hastata-weigeliana* Wimmer pro parte, *S. phylicifolia* var. *rhaetica* A. Kerner, *S. phylicifolia* sensu Sm., non L., *S. phylicifolia-hastata* Wimmer). Ridges on decorticated wood absent. Buds 6(–8) mm, chestnut-brown, rarely yellowish. Young twigs more or less felted, later becoming glabrous. Leaves obovate or ovate, often tapering towards base, with very indistinct, scattered glandular teeth (rarely entire or with more numerous glands), completely glabrous; petiole 3–6 mm. Catkins 1·5–2·4(–4) × 1·8 cm. Filaments with crispate hairs in lower ½. Capsule *c.* 4 mm. ● *Alps*. Au Ga He It.

(27–30). S. myrsinifolia group. Shrub or small tree 1–4 m. Twigs pubescent, often glabrescent, blackish- to brownish-green. Leaves green and rather dull above, paler green or glaucous, pubescent or almost glabrous beneath, usually turning black on drying. Stipules usually rather large, semi-cordate. Catkins rather lax; peduncles short, leafy. Ovary glabrous.

This is probably the most polymorphic group in the genus as to size, outline, serration and pubescence of leaves. The circumscription of the group here follows Enander, *Salic. Scand. Exsicc.* **3**: 9 (1910), and excludes all plants with a hairy ovary. Plants showing this character usually show one or more correlated characters pointing to hybridization with **24** and *S. myrsinites* in Scandinavia, with *S. atrocinerea* and *S. phylicifolia* in Britain, or with *S. cinerea* or other members of the *Capreae* in other parts of Europe. One of the more reliable vegetative characters of **27** is the waxy bloom fading towards the apex on the lower surface of leaves.

1 Adult leaves ±glaucous beneath (but always green at apex), going black on drying; twigs slender; catkins appearing before the leaves; style and stigma long, slender **27. myrsinifolia**
1 Adult leaves ±green beneath, going black less easily on drying; twigs thicker and more nodose; catkins appearing with the leaves; style and stigma often shorter and thicker
2 Indumentum of short hairs, soon ±evanescent; peduncle short, with few leaves **30. mielichhoferi**
2 Indumentum of young twigs and young leaves of rather dense, whitish hairs; peduncle long, with many leaves
3 Leaves ±elliptical, green or somewhat glaucous beneath; often a tree **28. borealis**
3 Leaves obovate or oblanceolate, bluish and dull beneath; shrub **29. apennina**

27. S. myrsinifolia Salisb., *Prodr.* 394 (1796) (*S. nigricans* Sm.). Often a shrub. Twigs slender. Young twigs whitish-pubescent, often glabrescent. Leaves orbicular-ovate to lanceolate, often glaucous beneath (but the apex always green), serrate or subentire, going black on drying; veins slender and not prominent. Catkins shortly pedunculate or subsessile, appearing before the leaves. Style 0·7–1·5 mm; stigma 0·3–0·6 mm; both slender. $2n = 114$. *Europe southwards to Pyrenees, Corse and W. Bulgaria*. Au Br Bu ?Co Cz *Da Fe Ga Ge He †Hu †It Ju No Po Rs(N, B, C, W, E) Su.

The following hybrids occur frequently:

S. cinerea × *myrsinifolia*. Often like **27** but more hairy, decorticated wood with ridges, and ovary more or less hairy.

S. myrsinifolia × *phylicifolia*. Often like **27** but less hairy, waxy bloom of lower surface of the leaf more pronounced, and ovary more or less hairy.

S. myrsinifolia × *myrsinites*. Leaves more or less green and shiny on both surfaces; catkins purplish.

S. atrocinerea × *myrsinifolia*. Like **27** or intermediate; decorticated wood with ridges; ovary hairy; indumentum sometimes tending to be ferruginous on the lower surface of the leaf.

28. S. borealis Fries, *Bot. Not.* **1840**: 193 (1840). Like **27** but often a tree; twigs thicker and more nodose; young twigs with whiter and denser indumentum; leaves more coriaceous, larger, often more or less elliptical, often deeply serrate with large glands, going black less easily on drying; less glaucous beneath, with longer and whiter hairs; veins thicker, reticulation more prominent; petiole thicker, with long hairs; stipules larger; catkins with longer, thicker, more woolly peduncles, appearing with the leaves; style and stigma often shorter and thicker. *N. Fennoscandia and N. Russia*. Fe No Rs(N) Su.

29. S. apennina A. Skvortsov, *Nov. Syst. Pl. Vasc. (Leningrad)* **1965**: 90 (1965) (*S. nigricans* auct. ital., non Sm.). Like **27** but always a shrub; leaves obovate or oblanceolate, bluish and dull on the lower surface; peduncles of catkins usually 3–15 mm (rarely subsessile); style 0·2–0·9 mm; stigma 0·2–0·4 mm. $2n = 114$. *Damp woods and marshes*. ● *Appennini; Sicilia*. It Si.

30. S. mielichhoferi Sauter, *Flora (Regensb.)* **32**: 662 (1849). Shrub. Twigs thicker and more nodose than in **27**. Indumentum of young leaves and young twigs consisting of short hairs which soon disappear. Leaves lanceolate to obovate, almost entire, green beneath, with thick, prominent reticulation, scarcely going black on drying. Catkins with a short, few-leaved peduncle, appearing with the leaves. ● *E. Alps*. Au It.

31. S. glabra Scop., *Fl. Carn.* ed. 2, **2**: 255 (1772). Erect shrub up to 1·5 m. Twigs dark brown, entirely glabrous even when young. Leaves broadly elliptical or obovate, sometimes oblanceolate, crenulate-serrate to coarsely dentate, coriaceous, glossy, with thick, waxy bloom over whole lower surface, glabrous even when young, reticulation fine but prominent on both surfaces. Stipules rarely developed, semi-cordate, serrate. Catkins 7 × 1 cm; peduncles up to 20 mm, with small obovate, serrate leaves. Ovary entirely glabrous; pedicel *c.* ⅓ as long as the ovary. $2n = 114$. ● *E. Alps*. Au Ge He It Ju.

The hybrid with **27** (*S.* × *subglabra* A. Kerner) is not infrequent and can be recognized by the young and adult leaves which are more or less hairy, at least along the midrib, and by the waxy bloom of the lower surface of the leaf which fades out below the apex.

32. S. crataegifolia Bertol. in Desv., *Jour. Bot. Appl.* **2**: 76

(1813). Procumbent or erect shrub up to 1 m. Twigs covered at first with long hairs, becoming glabrous, dark purple. Leaves 6·5–11·5 × 3·5–4 cm, elliptical, bright green above with silky hairs beneath, ultimately glabrous; lateral veins 8–12(–20) pairs. Female catkins 5(–12) × 0·8–1 cm; axis almost glabrous. Bracts reddish-brown, with long, white, silky hairs near apex. Ovary 5 mm, glabrous; pedicel short; style 1·5(–2) mm. *Limestone and metamorphic rocks.* ● *N. Italy* (*Alpi Apuane*). It.

33. S. pedicellata Desf., *Fl. Atl.* **2**: 362 (1799) (*S. aurita* var. *pedicellata* (Desf.) Fiori, *S. cinerea* var. *pedicellata* (Desf.) Moris, *S. nigricans* var. *pedicellata* (Desf.) Bertol.). Tall shrub or tree up to 10 m. Decorticated wood with numerous prominent ridges. Young twigs grey-tomentose, the older ones gradually becoming glabrous. Leaves oblong or obovate-lanceolate, dentate, crenate or nearly entire, thinly pubescent beneath, glabrescent above; lateral veins at least 10–12 pairs, connected by a fine reticulation, prominent beneath, impressed above. Stipules large, semi-cordate, dentate, deciduous. Bracts with short hairs. Catkins 3–6 × 1–1·5 cm. Ovary glabrous; pedicel 3–4 mm; style short or moderately long. *W. Mediterranean region.* Co Hs ?It Sa Si.

Plants similar in vegetative characters but with a hairy ovary are probably of hybrid origin.

Plants from N.E. Sicilia with lanceolate leaves more densely pubescent beneath, reniform stipules and distinctly stipitate capsules have recently been described as **S. gussonei** Brullo & Spampinato, *Willdenowia* **18**: 5 (1988).

34. S. silesiaca Willd., *Sp. Pl.* **4(2)**: 660 (1806). Shrub up to 3 m. Twigs slightly pubescent when young, or almost glabrous from the beginning. Decorticated wood sometimes with scattered ridges 5(–10) mm. Leaves oblanceolate, more or less broadly cuneate towards the base, almost glabrous and green on both surfaces or bluish-green beneath; reticulation prominent beneath; apex flat, triangular; margin serrate, crenate or nearly entire. Stipules reniform or semi-cordate, irregularly glandular-sinuate, more or less equalling the petiole. Catkins 1·5–3·5 × 0·5–1 cm; peduncles short. Bracts dark reddish-brown, bearded with long, white hairs. Ovary glabrous; pedicel equalling the bract, later elongating slightly. *Riversides and damp clearings.* ● *Sudety, Carpathians, Balkan peninsula.* Bu Cz Ju Po Rm Rs(W).

Absent from the Alps; records from Steiermark are referable to *S. appendiculata* × *caprea*. Hybridizes freely with *S. aurita*, *S. caprea* and locally with *S. lapponum*.

35. S. aegyptiaca L., *Cent. Pl.* **1**: 32 (1755) (*S. medemii* Boiss.). Tall shrub up to 4 m, occasionally a tree. Decorticated wood with numerous short, prominent ridges. Young twigs reddish, with dense grey tomentum, persistent on older branches. Leaves lanceolate, cuneate or rounded at base, thick, crispate-serrate; lamina with short, sparse, grey, persistent indumentum and prominent reticulation beneath, glabrescent above; lateral veins *c.* 15 pairs; petiole 5–10(–15) mm, thick, tomentose. Stipules semi-hastate, serrate, acute, deciduous. Catkins 5–6 × 1(–1·5) cm. Ovary tomentose; pedicel *c.* 2 mm; style 0·5(–1) mm. *N.E. Greece* (*near Komotini*). *Gr. (*Mountains of Asia.*)

36. S. appendiculata Vill., *Hist. Pl. Dauph.* **3(2)**: 775 (1789) (*S. grandifolia* Ser., *S. aurita* var. *grandiflora* (Ser.) Fiori). Tall shrub or small tree with short divaricate branches. Decorticated wood with few, sometimes indistinct, elevated ridges. Twigs pubescent with short hairs, more or less glabrescent. Leaves variable in size and shape, obovate to oblanceolate, more or less glabrescent above and permanently pubescent beneath; margin roughly erose-dentate to entire; veins deeply impressed above, and very prominent beneath; petiole *c.* 10 mm. Stipules well developed, semi-cordate, coarsely serrate. Catkins up to 3 × 1 cm, lax-flowered. Filaments with few, long hairs near the base. Pedicel as long as or longer than the grey-pubescent ovary. ● *Mountains of Europe from S.W. Czechoslovakia to the Pyrenees, S. Italy and S.W. Bulgaria.* ?Al Au Bu Cz Ga Ge He It Ju.

Absent from Sudety and Carpathians; records of **36** from here are referable to **34** or its hybrids.

S. appendiculata hybridizes freely, especially with *S. caprea* (to give *S.* × *macrophylla* A. Kerner) but also with many subalpine species. The influence of *S. appendiculata* is usually shown in the rugose leaves, broadest above the middle, and the dense, very prominent reticulation of the lower surface.

37. S. laggeri Wimmer, *Flora* (*Regensb.*) **37**: 162 (1854) (*S. albicans* Bonjean, *S. pubescens* Schleicher). Shrub 2–3 m. Twigs divergent, thick, knotty, dark brown to blackish, woolly-felted with long, white hairs when young, glabrescent only in the second or third year. Leaves narrowly elliptical to oblong-lanceolate, tending to turn black on drying, lanate-pubescent above, deeply green and almost shiny when mature, woolly-felted with greyish-white hairs beneath; petiole 6–20 mm. Stipules semi-sagittate, cuneate-serrate. Catkins 2–3·8 × 1·2–1·7 cm. Filaments with long hairs in lower ½ or near the base only. Ovary densely covered with whitish hairs. ● *Alps.* Au Ga He It.

38. S. cinerea L., *Sp. Pl.* 1021 (1753) (*S. aurita* var. *cinerea* (L.) Fiori). Shrub up to 6 m. Twigs rather stout, shortly and persistently pubescent, grey. Decorticated wood with narrow continuous ridges. Leaves up to 11 × 4 cm, 2–3½ times as long as wide, lanceolate or oblanceolate, apiculate or rounded, cuneate at base, persistently pubescent beneath and grey or somewhat glaucous above; margin often somewhat undulate, distantly crenate-serrate or subentire; petiole up to 15 mm. Stipules usually well developed, semi-cordate, toothed. Catkins appearing before the leaves, dense, subsessile, finally 3·5–5 cm. $2n = 76$. *Most of Europe except the extreme north, but local in the west and the Mediterranean region.* Al Au Be Br Bu Cz Da Fe Ga Ge Gr Hb Ho Hs Hu It Ju No Po Rm Rs(N, B, C, W, E) Su Tu.

Very local in most of W. Europe, where records are mostly referable to **39** or its hybrids.

39. S. atrocinerea Brot., *Fl. Lusit.* **1**: 31 (1804) (*S. cinerea* auct. eur. occid., *S. oleifolia* Sm., *S. acuminata* Thuill., non Sm., *S. aurita* auct. lusit., non L., *S. caprea* auct. lusit., non L., *S. cinerascens* Link, *S. incerta* Lapeyr., *S. rufinervis* DC.). Tall shrub or small tree up to 10 m. Young twigs somewhat pubescent, glabrescent, shining in the second year. Decorticated wood with ridges. Leaves obovate or oblong-oblanceolate, slightly denticulate to almost entire, soon glabrescent above and somewhat shiny, glaucous, with very short, curved, ferruginous hairs beneath; lateral veins 8–15 pairs. Catkins 2–5 cm, appearing before the leaves, cylindrical, dense, sessile or subsessile; axis grey-hairy. Capsule tomentose. $2n = 76$. *W. Europe.* Be Br Co Ga Hb Ho Hs Lu Sa [Az].

Subsp. **catalaunica** R. Görz, *Cavanillesia* **2**: 142 (1930) from E. Spain (Cataluña) is said to have a petiole 3–5(–8) mm, leaves with rounded base and sinuate-dentate margin, very short style, and very indistinct stigma; it somewhat resembles *S. aurita*.

S. atrocinerea hybridizes freely with **40** and **41** and evidently also with **38** wherever they meet.

40. S. aurita L., *Sp. Pl.* 1019 (1753). Shrub 1–2(–3) m, with numerous patent branches. Twigs rather slender, soon glabrous, brown, usually angular and with wide-angled branching. Decorticated wood with ridges. Leaves 2–3 cm, *c.* 1½–2½ times as long as wide, obovate, shortly cuspidate with the cusp often obliquely recurved, more or less cuneate at base, undulate, toothed to subentire, distinctly rugose, dull grey-green and more or less pubescent above, more or less grey-tomentose beneath. Stipules large and conspicuous, more or less reniform, persistent. Catkins 1–2·5 cm, appearing before the leaves, subsessile, cylindrical. $2n = 76$. *Europe, except the Arctic and most of the Mediterranean region.* Al Au Be Br Bu Co Cz Da Fe Ga Ge ?Gr Hb He Ho Hs Hu ?It Ju No Po Rm Rs(N, B, C, W, E) Su.

41. S. caprea L., *Sp. Pl.* 1020 (1753) (*S. coaetanea* (Hartman) B. Flod.). Shrub or small tree 3–10 m; bark coarsely fissured. Twigs rather stout, those of the previous year glabrous or subglabrous. Decorticated wood without ridges. Leaves 5–10 cm, 1½–2 times as long as wide, usually broadly ovate to ovate-oblong, rounded at base, more or less obtuse or shortly acuminate, sinuate-dentate or almost serrate, dark green and finally almost glabrous above, persistently softly and densely grey-tomentose beneath; lateral veins 6–9 pairs, forming almost a right-angle with the midrib, very prominent on the lower surface; petiole long. Stipules usually large, semi-cordate. Catkins appearing before the leaves, dense, subsessile. Male catkins 2–3·5 × 1·5–2 cm, oblong-ovoid. Female catkins finally 3–7 cm, lax. Bracts blackish at the apex. Both male and female flowers with 1 nectary. $2n = 38$. *Most of Europe.* Al Au Be Br Bu Co Cz Da Fe Ga Ge Gr Hb He Ho Hs Hu It Ju No Po Rm Rs(N, B, C, W, K, E) Si Su Tu.

42. S. salviifolia Brot., *Fl. Lusit.* **1**: 29 (1804) (*S. oleifolia* auct. hisp.). Tall shrub or tree up to 6 m. Young twigs grey-tomentose, indumentum persistent on the old ones. Decorticated wood with prominent ridges. Leaves linear-lanceolate or -oblong, 3–5 times as long as wide, gradually attenuate towards base, shortly acuminate and almost rounded towards apex, white tomentose on both surfaces, tomentum dense and persistent beneath, thinner above; margin slightly serrate or almost entire; lateral veins numerous; petiole up to 6 mm. Stipules semi-cordate, serrate or crenate, patent, with deflexed apex. Fruiting catkins 3–4 × 1 cm. Capsule tomentose; pedicel short; style very short. ● *Spain and Portugal.* Hs Lu.

The capsule is hairy according to the original description and in the majority of specimens seen. Similar plants but with glabrous capsules have been seen from the Sierra Morena and the Guadalquivir valley.

43. S. cantabrica Rech. fil., *Österr. Bot. Zeitschr.* **109**: 374 (1962). Shrub 2–3 m. Young twigs covered with shiny hairs, the older twigs glabrescent. Decorticated wood with scattered but distinct ridges. Leaves broadly lanceolate, rounded or rarely cuneate at the base, acute at apex, more or less entire, with indumentum of silky, appressed hairs, denser beneath; midrib prominent; veins scarcely impressed above and scarcely prominent beneath; petiole short. Stipules lanceolate, acute, shorter than the petiole. Ovary sericeous-tomentose; pedicel shorter than the bract. *Along mountain rivers.* ● *N. Spain (Cordillera Cantábrica).* Hs.

Only female flowers are known.

44. S. starkeana Willd., *Sp. Pl.* **4(2)**: 677 (1806) (*S. livida* Wahlenb.). Low shrub, rarely up to 1 m, with ascending branches. Twigs slender, pale and glabrous in the first year. Buds appressed, greenish-orange-yellow, with flattened apex. Leaves broadly lanceolate to orbicular-obovate, slightly pubescent and reddish when young, soon glabrous, bright green, more or less shiny above, pale green or greyish with very prominent veins beneath; margin glandular-serrate; lateral veins 5–7 pairs. Stipules well developed, broadly elliptical or semi-reniform, with glandular margin. Catkins 1·5–3 × 0·5–1 cm, lax; pedicel almost as long as the ovary, densely covered with grey-white hairs. Bracts lanceolate, usually greenish-yellow. Filaments glabrous. Stigma erect, ovoid, more or less urceolate. *N.E. Europe, extending locally to Norway, S.W. Germany and N. Romania.* Cz Fe Ge No Po Rm Rs(N, B, C, W) Su.

45. S. xerophila B. Flod., *Bot. Not.* **1930**: 334 (1930) (*S. cinerascens* N. J. Andersson, *S. livida* var. *cinerascens* Wahlenb., *S. sphacelata* Sommerf.). Like **44** but sometimes a tree, up to *c.* 6 m, with erect branches; twigs dull, more or less woolly; buds somewhat patent, conical, reddish-brown; leaves oblanceolate, grey-woolly with appressed, somewhat curved hairs, the margin entire and lateral veins 7 or 8 pairs; stipules usually absent; bracts obovate, brown; filaments hairy; stigma lobes divergent, cylindrical, usually deeply cleft, longer. *N. Fennoscandia and N. Russia southwards to S. Ural.* Fe No Rs(N, C) Su.

46. S. myrtilloides L., *Sp. Pl.* 1019 (1753). Low shrub with subterranean creeping stem. Twigs 30–50 cm, erect, brown, with short, evanescent hairs. Leaves orbicular to narrowly elliptical with short, sometimes plicate apex, usually more or less rounded at both ends, dull green above, paler beneath; veins not prominent; margin entire and more or less reflexed; petiole up to 5 mm. Catkins *c.* 2 cm. Nectaries ½ as long as the bracts. Ovary glabrous; style short. *Swamps and peat-bogs. N., C. & E. Europe southwards to C. Romania.* Au Cz Fe Ge He No Po Rm Rs(N, B, C, W) Su.

The hybrid with **40** (*S.* × *rugulosa* N. J. Andersson) is frequent and can be recognized by the more or less crenate, rugose leaves with more prominent reticulation and more or less pubescent ovary. **46 × 47** (*S.* × *finnmarchica* Willd.) is also frequent.

47. S. repens L., *Sp. Pl.* 1020 (1753). Low shrub with creeping stem. Branches procumbent or ascending, sometimes erect (var. *fusca* Wimmer & Grab.). Previous year's twigs slender, pale, glabrous. Leaves ovate-elliptical or obovate, finally glabrous at least above, sericeous beneath; margin more or less entire, flat and glandular. Stipules present or absent. Catkins up to 2·5 cm, cylindrical. Bracts flat, with dark apex. Style relatively long. Pedicel, capsule and filaments glabrous or hairy. $2n = 38$. ● *N., W. & C. Europe extending to E. Jugoslavia.* Au Be Br Cz Da Fe Ga Ge Hb He Ho Hs ?It Ju Lu No Rs(N, B, C) Su.

(a) Subsp. **repens**: Plant procumbent. Leaves ovate-elliptical, more or less glabrous above, sericeous beneath; lateral veins 4–6 pairs. Stipules absent. Pedicel, filaments and capsule glabrous. $2n = 38$. *Swamps, heaths and sand-dunes. Throughout the range of the species.*

(b) Subsp. **arenaria** (L.) Hiitonen, *Suomen Kasvio* 267 (1933) (*S. arenaria* L.): Plant ascending to erect. Leaves obovate, densely sericeous beneath or on both surfaces; lateral veins 5–8 pairs.

Stipules often present. Pedicel, filaments and capsule usually hairy. *Usually on sand-dunes. Atlantic coasts of Europe.*

40 × **47** (*S.* × *ambigua* Ehrh.) is common wherever the parents meet.

48. S. rosmarinifolia L., *Sp. Pl.* 1020 (1753). Previous year's twigs slender, sparsely covered with short hairs, often with crowded basal shoots. Leaves more or less erect, thin, almost linear, dark green, finally more or less glabrous above, covered with silky hairs beneath; margin entire or slightly sinuate, finely glandular, flat; lateral veins 10–12 pairs. Stipules absent. Catkins globose. Bracts hairy. Filaments glabrous. Capsule hairy. $2n = 38$. *Bogs and swamps. C. & E. Europe extending to N. France, Denmark and N. Italy.* Au †Bu Cz Da Fe Ga Ge Hu It Ju Po Rm Rs(N, B, C, W, E) Su.

(49–51). S. arbuscula group. Shrubs up to 2 m. Twigs glabrous. Leaves shiny above, glaucous beneath, never turning black on drying. Stipules minute or absent. Ovary hairy. Pedicel shorter than the nectary or absent.

The pedicel-character is useful for distinguishing these species from the *S. phylicifolia* group, which are rather similar in leaf-characters but have the pedicel 2–4 times as long as the nectary.

1 Leaves 3–5 cm, obovate to elliptical, indistinctly and remotely crenate-serrate, sometimes subentire; peduncle long **51. waldsteiniana**
1 Leaves usually not more than 3 cm, lanceolate or elliptic-lanceolate, densely and regularly serrate; peduncle short
 2 Leaves subacutely serrate with small glands; lateral veins 7–12 pairs **49. arbuscula**
 2 Leaves acutely serrate, with large, white glands; lateral veins 5–10 pairs **50. foetida**

49. S. arbuscula L., *Sp. Pl.* 1018 (1753). Leaves 0·5–3(–5) cm, elliptic-lanceolate, acute, densely and regularly serrate, the teeth with small glands; lateral veins 7–12 pairs. Catkins 1·3–1·8 × 0·4–0·5 cm, *c.* 3 times as long as wide, appearing with the leaves; peduncle short. *Fennoscandia and N. Russia; Scotland.* Br Fe No Rs(N) Su.

50. S. foetida Schleicher in Lam. & DC., *Fl. Fr.* ed. 3, 3: 296 (1805) (*S. myrsinites* var. *arbuscula* (L.) Fiori). Leaves 0·5–3 cm, elliptic-lanceolate, small, densely and acutely serrate, with large, white glands; lateral veins 5–10 pairs. Catkins 1·3–1·8 × 0·4–0·5 cm, $2\frac{1}{2}$–3 times as long as wide; peduncle short. ● *Alps; Pyrenees; ?C. Appennini.* Au Ga He It.

51. S. waldsteiniana Willd., *Sp. Pl.* **4(2)**: 679 (1806) (*S. prunifolia* Sm.). Leaves 3–5 cm, with cuneate base, obovate to elliptical, indistinctly and remotely crenate-serrate, sometimes subentire. Catkins 1·4–3·5 × 0·5–0·8 cm, 4–5 times as long as wide; peduncle long. ● *C. & E. Alps, N. part of Balkan peninsula.* Al Au Bu Ge He It Ju.

52. S. hastata L., *Sp. Pl.* 1017 (1753). Ascending or erect shrub up to 2·5 m. Branches greenish or brownish, shiny, glabrous. Leaves very variable in size and outline, rather thin, not shiny, pale and dull green above, paler but without a bloom beneath, somewhat hairy when young, soon glabrous; venation finely reticulate but scarcely prominent; margin entire or more or less finely serrate; petiole up to 1 cm. Stipules often well developed, obliquely ovate, serrate. Female catkins 6(–10) × 1(–2) cm, dense-flowered; peduncle up to 3 cm, leafy. Bracts with long, white hairs. Pedicel $\frac{1}{3}$ as long as the ovary. Nectaries $\frac{1}{2}$ as long as the pedicel. $2n = c.$ 110. *N. & C. Europe, locally on mountains in the south.* Al Au Bu Cz Da Fe Ga Ge He Hs It Ju No Po Rm Rs(N, C, ?W) Su.

A polymorphic species of wide distribution, divisible into at least three subspecies, treated as varieties by some authors.

1 Petiole 2–3 mm; leaves orbicular **(b)** subsp. **vegeta**
1 Petiole more than 3 mm; leaves broadly elliptical, obovate or narrower
 2 Leaves broadly elliptical or obovate, finely serrate **(a)** subsp. **hastata**
 2 Leaves lanceolate, entire **(c)** subsp. **subintegrifolia**

(a) Subsp. **hastata** (incl. *S. hastatella* Rech. fil.): Shrub 0·5–1 m. Leaves broadly elliptical or obovate, finely serrate; petiole 4–10 mm. *Throughout most of the range of the species.*

(b) Subsp. **vegeta** N. J. Andersson, *Kungl. Svenska Vet.-Akad. Handl.* **6(1)**: 172 (1867): Shrub 1–2 m. Leaves orbicular, finely serrate; petiole 2–3 mm. *Lowlands of S. Fennoscandia and parts of C. Europe.*

(c) Subsp. **subintegrifolia** (B. Flod.) B. Flod. in Holmberg, *Skand. Fl.* **1(b, 1)**: 120 (1931): Shrub 2–2·5 m. Leaves lanceolate, entire; petiole 4–10 mm. *North part of the range of the species.*

53. S. pyrolifolia Ledeb., *Fl. Altaica* **4**: 270 (1833). Tall shrub or tree up to 5 m. Branches rather thick, reddish-brown, smooth and shiny, or with scattered hairs. Leaves *c.* 5 × 4 cm, thin, orbicular-ovate or -elliptical, whitish and glabrous with prominent venation beneath; lateral veins forming an angle of 60–85° with the midrib; petiole up to 3 cm, slender, often brown or pink beneath like the midrib. Stipules up to 1–1·5 cm wide, orbicular-reniform. Catkins small, *c.* 100-flowered; peduncle much reduced and often almost leafless. Capsule 3–4 mm, yellowish-brown, glabrous; style *c.* 1 mm, stigma short; pedicel usually hairy. *N. Finland, N. Russia and Ural.* Fe Rs(N, C).

54. S. helvetica Vill., *Hist. Pl. Dauph.* **3(2)**: 783 (1789). Shrub 0·5–4 m. Twigs greyish-brown, rather slender, becoming glabrous and chestnut-brown. Indumentum dense, white. Leaves not crowded at ends of twigs, often asymmetrical, obovate or obovate-oblanceolate, shortly acuminate or rounded at apex, glabrescent (especially along midrib), and greenish and shiny on upper surface, white-tomentose beneath, therefore distinctly discolorous, the base rounded, rarely cordate. Stipules small, caducous. Catkins $1\frac{1}{2}$–3 times as long as wide, somewhat slender and lax-flowered, very elongate when ripe, peduncle short, leafless. Ovary ovoid, obtuse, sparsely tomentose, rather dull; style 0·8–1·5 mm, $\frac{1}{3}$–$\frac{1}{2}$ as long as the ovary; stigma distinctly quadrifid. ● *Alps; W. Carpathians.* Au Cz Ga He It Po.

55. S. lapponum L., *Sp. Pl.* 1019 (1753) (incl. *S. marrubifolia* Tausch ex N. J. Andersson). Like **54** but indumentum, especially of leaves and ovary, less dense, grey; leaves crowded at ends of the twigs, elliptical or broadly lanceolate, tapering gradually or abruptly at the base, moderately hairy above, and therefore the upper and lower surfaces almost concolorous; catkins stout, dense-flowered, scarcely becoming laxer when ripe, sessile, with caducous leaves at the base; ovary very long, sericeous with crisped and patent hairs; style 1·2–2·5 mm, filiform, $\frac{1}{2}$–$\frac{2}{3}$ as long as the ovary; stigma distinctly quadrifid. *N. Europe, extending southwards locally in the mountains to the Pyrenees and Bulgaria.* Br Bu Cz Fe Ga Ge Hs No Po Rm Rs(N, B, C, W) Su.

56. S. viminalis L., *Sp. Pl.* 1021 (1753) (*S. veriviminalis* Nasarow, *S. rossica* Nasarow, *S. gmelinii* auct., *S. linearis* Turcz.). Shrub or small tree 3–5(–10) m. Twigs long, straight, flexible, very leafy; young twigs grey-pubescent. Buds acute or obtuse. Leaves up to 1·5 cm wide when mature, narrow, linear to oblanceolate with very narrow, cuneate or rounded base, entire, green above, sericeous beneath; margin reflexed; lateral veins usually (15–)20–35 pairs. Catkins appearing before or with the leaves. Bracts hairy, pale brown, darker at the apex, not concealing the ovary. Flowers with 1 nectary. Style to about $\frac{1}{2}$ as long as the ovary. *C. & E. Europe, extending westwards to France and southwards to S. Jugoslavia; widely naturalized elsewhere.* Au Be Cz Ga Ge *Gr He Ho Hu Ju Po *Rm Rs(N, B, C, W, E) *Tu [Br Da Fe Hb Hs It Lu No Su].

S. pseudolinearis Nasarow in Komarov, *Fl. U.R.S.S.* **5**: 137 (1936), from C. & E. Russia, is doubtfully distinct from **56**. It has leaves 0·2–0·4(–0·6) cm, narrowly linear-lanceolate, sericeous on both surfaces, glabrescent above, with wide, reflexed margins and short, almost entirely black bracts.

Plants showing intermediate characters between **56** and **38–41** occur quite frequently, mainly in W. & N. Europe. They have long leaves, more or less sericeous beneath, with many lateral veins, all characters inherited from *S. viminalis*. Three taxa are often referred to by binary names because their parents are not known with certainty, and must be mentioned here, since they behave like species (*vide* R. D. Meikle, *Watsonia* **2**: 243–248 (1952)).

1 Leaves 9–11 × 2·5–4 cm, broadly lanceolate-ovate; stigma shorter than style **S. × calodendron**
1 Leaves 10–13 × 1·5–3 cm, narrowly lanceolate; stigma as long as or longer than style
 2 Bracts light brown, acute; leaves 1·5–2·5 cm wide, silky-pubescent beneath **S. × stipularis**
 2 Bracts dark brown, subobtuse; leaves 2–3 cm wide, thinly pubescent or subglabrous beneath **S. × dasyclados**

S. × calodendron Wimmer, *Salices Eur.* 187 (1866) (*S. acuminata* Sm., non Miller). *Often cultivated and escaping.* [Br ?Da Ge Hb.] (Possibly **38 × 41 × 56.**)

S. × stipularis Sm. in Sowerby, *Engl. Bot.* **17**: t. 1214 (1803) (*S. × smithiana* Willd.). *Cultivated and occasionally escaping.* [Br Ge Rs(W).] (Possibly **39 × 56.**)

S. × dasyclados Wimmer, *Flora* (*Regensb.*) **32**: 25 (1849). *Cultivated and occasionally escaping.* [Au Cz Da Fe Ge Ho Po Rs(N, C, E) Su.] (Possibly **38 × 41 × 56.**)

57. S. elaeagnos Scop., *Fl. Carn.* ed. 2, **2**: 257 (1772) (*S. incana* Schrank). Shrub up to 6 m, or rarely a tree up to 16 m. Twigs slender, yellowish- to reddish-brown, thinly covered with whitish hairs when young. Leaves erect, linear-lanceolate to narrowly linear, tapering at both ends, densely lanate on both surfaces when young, more or less glabrescent above, remaining lanate beneath, not shiny; margin deflexed, finely glandular-serrate, mainly in the apical part; veins more or less impressed above, slightly prominent beneath, often not visible because of the dense indumentum; petiole up to 5 mm. Stipules usually absent. Catkins up to 6 × 0·8 cm; peduncle up to 1 cm, with small, lanceolate leaves. Bracts about $\frac{1}{3}$ as long as the ovary. Filaments united at base or up to the middle, hairy near base. Ovary glabrous; pedicel about $\frac{1}{4}$ as long as the ovary. *Damp, sandy or gravelly soils. S. Europe extending northwards to S. Poland; naturalized further north.* Al Au Bu Co Cz Ga Ge Gr He Hs Hu It Ju Po Rm Rs(W) [Ho].

(**a**) Subsp. **elaeagnos**: Leaves up to 12 × 2 cm, linear-lanceolate or linear. $2n = 38$. *Throughout most of the range of the species.*

(**b**) Subsp. **angustifolia** (Cariot) Rech. fil., *Österr. Bot. Zeitschr.* **104**: 314 (1957). Leaves *c.* 12 × 0·5–1 cm, very narrowly linear. *S. France and Spain; frequently cultivated elsewhere.*

Hybrids of **57** are not rare, especially along some of the rivers coming from the Alps; the most prominent characters inherited from **57** are the long, narrow, curved catkins and the leaves with more or less parallel margins, lanate on the lower surface.

58. S. purpurea L., *Sp. Pl.* 1017 (1753). Rather slender shrub up to 5 m. Bark bitter. Twigs slender, straight, glabrous, shining, usually purplish at first, becoming greenish- or yellowish-grey. Leaves 3–12 cm, 3–15 times as long as wide, sometimes opposite, at least near the ends of twigs, obovate-oblong to oblanceolate-linear, acute or acuminate, very finely serrate, glabrous, dull and slightly bluish-green above, paler and often glaucous beneath. Stipules small, caducous. Catkins 1·5–4·5 cm, appearing before the leaves, dense, cylindrical, suberect to spreading, subsessile. Bracts blackish at the apex. Stamens 2, completely united and appearing as 1. Ovary pubescent. *Wet places. Most of Europe, except Fennoscandia and the N. part of the U.S.S.R.* Al Au Be Br Bu Co Cz Ga Ge Gr Hb He Ho Hs Hu It Ju Po Rm Rs(B, C, W, K) Sa Si [Da No Su].

(**a**) Subsp. **purpurea**: Leaves alternate throughout (exceptionally 1 or 2 pairs opposite at the base of the terminal shoots), narrowly lanceolate-lingulate, cuneate at base, serrate only above the middle, *c.* 10 times as long as wide on long twigs, *c.* 5 times as long as wide on short twigs; petiole *c.* $\frac{1}{6}$ the length of the leaf. *Widespread, particularly in the mountains.*

(**b**) Subsp. **lambertiana** (Sm.) A. Neumann ex Rech. fil., *Österr. Bot. Zeitschr.* **110**: 341 (1963): At least some leaves opposite, more or less broadly lingulate, narrowly rounded at base, serrate for almost their whole length, 4–8 times as long as wide on long twigs, *c.* 4 times as long as wide on short twigs; petiole $\frac{1}{20}$–$\frac{1}{15}$ the length of the leaf. *Widespread, mainly in the lowlands.*

56 × 58 (*S. × rubra* Hudson) is a frequent hybrid.

S. × forbyana Sm., an exceptionally vigorous bush, is probably *S. atrocinerea × purpurea × viminalis*; it has yellowish twigs and dark green, lustrous leaves (resembling the leaves of *S. triandra*). The female catkins are like those of *S. purpurea*; the male plant is extremely rare.

S. vinogradovii A. Skvortsov, *Nov. Syst. Pl. Vasc.* (*Leningrad*) **1966**: 55 (1966), which apparently replaces **58** from Ukraine eastwards, is a shrub 1–3 m, with leaves all alternate, shorter and more ovoid buds, and catkins sessile or on pedicels not more than 6 mm. It may represent a third subspecies of **58**.

59. S. amplexicaulis Bory, *Expéd. Sci. Morée* **3**: 277 (1832). Like **58(b)** but most or all leaves opposite, often smaller, sessile or subsessile, truncate or semi-amplexicaul at the base. *Balkan peninsula; S.W. Italy.* Al Bu Gr It Ju Tu.

60. S. caspica Pallas, *Fl. Ross.* **1(2)**: 74 (1788) (*S. volgensis* N. J. Andersson). Twigs brown when young, usually becoming yellowish-white, shiny, glabrous. Leaves 5–8(–12) × 0·4–1 cm, narrowly linear-lanceolate, very acute, strict, serrate (rarely almost entire), 8–10 times as long as wide, glabrous at maturity; veins prominent beneath. Stipules absent. Catkins *c.* 2·5 cm. Ovary whitish-sericeous, finally somewhat glabrous, subsessile or with a very short pedicel. *S.E. Russia.* Rs(E). (*W. & C. Asia.*)

61. S. wilhelmsiana Bieb., *Fl. Taur.-Cauc.* **3**: 627 (1819) (*S. angustifolia* Willd.). Young twigs more or less densely silky, rarely almost glabrous, very thin, flexible, often forming almost a right-angle with the stem. Leaves 2–6 × 0·4–1 cm, *c.* 5 times as long as wide, linear, entire or finely glandular-serrate, densely covered with silky hairs when young, shiny, finally more or less glabrous; petiole very short. Stipules usually absent. Catkins 2–3 cm, slender. Bracts very acute. Ovary densely sericeous; stigma sessile. *S.E. Russia (near the mouth of Terek R.).* Rs(E). (*S.C. Asia.*)

62. S. caesia Vill., *Hist. Pl. Dauph.* **3(2)**: 768 (1789) (*S. myrtilloides* Willd., non L.). Shrub up to 1 m. Twigs entirely glabrous, brown and dull when young. Leaves elliptical or obovate, twice as long as wide, shortly acuminate, sometimes plicate at the apex, rounded or subcordate at the base, entire or with very few remote marginal glands, rigid, with a dull bloom on both surfaces; petiole 2–3 mm. Stipules minute. Catkins 1–1·5(–2) cm; nectary about as long as the bract. Stamens free or more or less united. Ovary woolly or more or less appressed-sericeous; sessile or subsessile. *Alps.* Au Ga He It.

63. S. tarraconensis Pau in Font Quer, *Treb. Inst. Catalana Hist. Nat.* **1915**: 7 (1915). Shrub up to 1 m. Young twigs grey-hairy, becoming glabrous in second year. Leaves 0·5–3 cm, suborbicular or elliptic-obovate, up to $1\frac{1}{2}$ times as long as wide, with a small, oblique, deflexed apex, rounded or cordate at base, glabrous above except along midrib, greyish or glaucescent and subglabrous or with only very short hairs beneath; margin irregularly sinuate-crenate; lateral veins 4 or 5(6) pairs; petiole 0·5–2(–7) mm. Stipules present only on elongate shoots. Catkins 0·5–1 cm. Nectary much shorter than the bract. Stamens united for at least $\frac{1}{2}$ their length; filaments glabrous. Ovary grey-hairy; pedicel *c.* 2 mm; style 0·3–0·4 mm. *Calcareous rocks, 900–1400 m.* ● *Spain (S.E. Cataluña).* Hs.

64. S. daphnoides Vill., *Prosp. Pl. Dauph.* 51 (1779). Tall shrub or tree 7–10 m. Twigs with a bluish, waxy bloom. Leaves 5–10 cm, $2\frac{1}{2}$–4 times as long as wide, oblong-lanceolate or oblong-ovate, acute or acuminate, glandular-serrulate, soon glabrous, dark green and shining above, glaucous beneath; lateral veins 8–12 pairs; petiole 2–4 mm. Stipules large, semi-cordate. Catkins 3–4 cm, appearing before the leaves, subsessile, cylindrical, dense. Bracts blackish at apex, obovate, hairy. Ovary ovoid-conical, glabrous, subsessile; style long, slender. $2n = 38$. ● *From S. Norway southwards to S. France and S. Romania.* Au Cz Ga Ge He Hs It Ju No *Po Rm Rs(N, B, C) Su.

65. S. acutifolia Willd., *Sp. Pl.* **4(2)**: 668 (1806) (*S. daphnoides* subsp. *acutifolia* (Willd.) Blytt & O. C. Dahl). Tall shrub, rarely a tree. Leaves 6–15 cm, more than 5 times as long as wide, lanceolate to linear-lanceolate, long and sharply acuminate, serrate, dark green and shiny above, pale to grey-green and dull beneath, only the very young ones with a faint, sericeous indumentum; lateral veins 15 or more pairs; petiole up to 15 mm. Stipules lanceolate, acuminate, serrate. Catkins appearing before the leaves, covered with long, whitish, shiny hairs before flowering; peduncles short. Bracts *c.* $\frac{1}{2}$ as long as the ovary. Pedicel *c.* $\frac{1}{3}$ as long as the ovary. $2n = 38$. *Most of European Russia, extending westwards to Lithuania and S.W. Ukraine; Finland.* Fe Rs(N, C, W, E).

2. Populus L.

J. DO AMARAL FRANCO

Deciduous trees. Buds with several unequal scales. Leaves usually ovate or triangular, entire, dentate or lobed. Petiole usually long, terete or flattened. Flowers anemophilous, appearing before the leaves, in pendent, stalked catkins, each flower with a stalked, cup-shaped disc and subtended by a toothed or laciniate bract. Stamens 4 to many. Capsule 2- to 4-valved. Seeds numerous, minute.

Many hybrids, most of them of recent origin, are now planted for pulp-wood or ornament in Europe.

1 Pistillate flowers long-pedicellate, with caducous disc; leaves very variable, some, particularly towards end of long shoots, lanceolate or linear-lanceolate **11. euphratica**
1 Pistillate flowers subsessile, with persistent disc; leaves variable but never lanceolate
2 Leaves of long shoots densely tomentose beneath, those of short shoots less tomentose or subglabrous
3 Leaves of long shoots palmately lobed, white-tomentose beneath; bracts dentate **1. alba**
3 Leaves of long shoots with wide, irregularly serrate teeth, grey-tomentose beneath; bracts laciniate **2. × canescens**
2 Leaves not tomentose beneath (sometimes thinly tomentose when very young)
4 Leaves without a translucent margin
5 Petiole strongly compressed
6 Leaves cuneate or truncate at base, coarsely dentate; buds grey-tomentose **3. grandidentata**
6 Leaves rounded or subcordate at base, sinuately crenate-dentate; buds glabrous, slightly viscid **4. tremula**
5 Petiole terete
7 Leaves rhombic-elliptical to obovate, cuneate at base; petiole 0·5–2·5 cm; twigs glabrous **5. simonii**
7 Leaves deltate-ovate, usually cordate; petiole 3–6 cm; twigs hairy **6. candicans**
4 Leaves with a translucent margin, sometimes very narrow
8 Leaves with a very narrow, translucent margin; petiole terete, though sometimes grooved on upper side **7. × berolinensis**
8 Leaves with an obvious translucent margin; petiole laterally flattened
9 Leaves 10–18 cm, deltate-ovate to ovate, densely ciliate; stamens 30–60 **10. deltoides**
9 Leaves 7–10 cm, rhombic or deltate, ciliate or not; stamens 15–30
10 Lamina not ciliate, without glands at base; trunk usually with large bosses **8. nigra**
10 Lamina shortly ciliate, usually with 1 or 2 glands at base; trunk without bosses **9. × canadensis**

Sect. POPULUS (Sect. *Leuce* Duby). Bark smooth, becoming rough on old trunks; buds tomentose, or glabrous and viscid; terminal bud present; leaves tomentose or glabrous, without a translucent margin, more or less dimorphic; bracts ciliate, tardily deciduous.

1. P. alba L., *Sp. Pl.* 1034 (1753) (*P. bolleana* Lauche, *P. nivea* A. Wesmael). Up to 30(–40) m, with a broad crown, suckering freely. Bark white on young stems. Twigs and buds white-tomentose. Leaves of long shoots 6–12 cm, ovate, with 3–5 coarsely toothed lobes, dark green above, white-tomentose beneath; leaves of short shoots ovate to elliptic-oblong, sinuate-dentate, usually greyish beneath. Stamens 5–10. Fruiting catkins 8–10 cm. $2n = 38$. *S., C. & E. Europe; often planted elsewhere.* Al Au Bu Co Cz Ga Ge Gr Hs Hu It Ju Po Rm Rs(C, W, E) Sa Si Tu [Az Be Br Da He Ho Lu].

A fastigiate variant (cv. 'Roumi' = f. *pyramidalis* (Bunge) Dippel) is commonly planted as a street tree.

2. P. × canescens (Aiton) Sm., *Fl. Brit.* **3**: 1080 (1804) (*P. alba × tremula*, *P. alba-tremula* E. H. L. Krause, *P. bachofenii* Wierzb. ex Rochel, *P. hybrida* Reichenb.). Like **1** but often taller; leaves of long shoots deltate-ovate, cordate, grey-tomentose beneath; leaves of short shoots suborbicular to ovate, glabrescent. Stamens 8–15. $2n = 38$. *C. Europe, extending to S. Italy and the Netherlands; widely planted and naturalized elsewhere.* Al Au *Br Bu Cz Ga Ge He Ho Hu It Ju Po Rm Rs(C, W, E) Tu [Be Da Gr Hb Hs].

3. P. grandidentata Michx, *Fl. Bor.-Amer.* **2**: 243 (1803). Up to 20 m, with a rather narrow crown. Young twigs and buds grey-tomentose. Leaves of long shoots 7–10 cm, ovate, acuminate, truncate to cuneate at base, coarsely dentate, grey-tomentose beneath when very young, then glabrous; leaves of short shoots elliptical. Stamens 6–12. Fruiting catkins 3·5–6 cm. *Extensively planted in Austria.* [Au.] (*North America.*)

4. P. tremula L., *Sp. Pl.* 1034 (1753). Up to 20 m, short-lived, suckering freely. Bark greyish. Buds glabrous, slightly viscid. Leaves 2–8 cm, suborbicular, sinuately crenate-dentate, thin, glaucescent beneath; petiole up to 6 cm, strongly flattened; leaves on young plants and suckers up to 15 cm, ovate. Stamens 5–12(–15). Fruiting catkins up to 12 cm. *Most of Europe, but in the south only on mountains.* All except Az Bl Cr Fa Lu Sb Si.

The hybrid between **2** and **4** (*P. × hybrida* Bieb.) occurs occasionally in the areas where the parents are found together, and is also sometimes found in cultivation.

The hybrid between **P. tremuloides** Michx, *Fl. Bor.-Amer.* **2**: 243 (1803), from North America, and **4** is widely planted in Finland and Sweden and to a lesser extent in Austria.

Sect. TACAMAHACA Spach. Bark rough and furrowed; buds large, very viscid, with a strong balsam-like odour; terminal bud present; leaves whitish (but not tomentose) beneath, without a translucent margin, often dimorphic.

5. P. simonii Carrière, *Revue Hort. (Paris)* **1867**: 360 (1867) (*P. przewalskii* Maxim.). Up to 15 m, with a narrow crown. Twigs slender, terete, glabrous, reddish-brown. Leaves 4–12 × 3–8 cm, rhombic-elliptical to obovate, abruptly acuminate, cuneate at base, crenulate, glabrous; petiole 0·5–2·5 cm, reddish. Stamens 8. *Planted for timber in C. Europe; occasionally for ornament elsewhere.* [Au Cz Ga Ge Ho Hu Ju Rm Rs.] (*N. China.*)

6. P. candicans Aiton, *Hort. Kew.* **3**: 406 (1789) (*P. × gileadensis* Rouleau, *P. tacamahaca* auct., *P. balsamifera* auct., non L.). Up to 25 m or more, with a broad, open crown, often suckering. Twigs stout, angled, hairy, brown. Leaves 6–16 cm, deltate-ovate, acuminate, usually cordate at base, crenate-serrate, ciliate, densely pubescent on the veins beneath; petiole 3–6 cm, terete. Fruiting catkins 7–16 cm. *Planted for timber, and occasionally naturalized.* [Br Da Ge He Ho Hu Po Rs Su.]

Of unknown origin; perhaps a hybrid between **P. balsamifera** L., *Sp. Pl.* 1034 (1757), from North America and temperate Asia, and **10**. It has been known in cultivation since before 1755, but only as female plants.

P. trichocarpa Torrey & A. Gray ex Hooker, *Hook. Ic.* t. 878 (1852), from W. North America, is also planted. It is like **6** but has truncate to subcordate leaf-bases and acutely angled twigs.

Sect. AIGEIROS Duby. Bark on adult trunks furrowed; buds viscid, the terminal present; leaves deltate, ovate or rhombic-ovate, green on both surfaces, with a well-defined translucent margin; bracts glabrous, soon deciduous.

7. P. × berolinensis C. Koch, *Wochenschr. Gartn. Pflanzenk.* **8**: 239 (1865) (*P. laurifolia × nigra* cv. 'Italica'). Columnar tree up to 25 m. Twigs slightly angular, somewhat pubescent, yellowish-brown, later yellowish-grey. Leaves 7–12 × 4–7 cm, ovate or rhombic-ovate, long-acuminate, crenate-serrate. Catkins 4–7 cm. Stamens *c.* 15. *Of garden origin; planted for shelter in C. & E. Europe.*

8. P. nigra L., *Sp. Pl.* 1034 (1753) (*P. thevestina* Dode). Up to 30 m, with a broad, uneven crown, rarely suckering; trunk usually with large bosses. Twigs terete, first yellowish, later greyish. Leaves not ciliate and without glands at base; those on long shoots 5–10 × 4–8 cm, rhombic-ovate, long-acuminate, minutely crenate-serrate; those on short shoots smaller and broader, often more deltate. Stamens 20–30. Fruiting catkins 10–15 cm. Capsule 2-valved. $2n = 38$. *Usually beside rivers or lakes. Much of Europe but absent from a large part of the north and rare in the south-west; often planted and sometimes naturalized.* Al Au Br Bu Co Cz Ga Ge Gr Hb Ho Hs Hu It Ju Po Rm Rs(N, C, W, K, E) Sa Si Tu [Az Be Da Lu Rs(B)].

1 Twigs and leaves glabrous — (**a**) subsp. **nigra**
1 Twigs and leaves hairy
  2 Twigs and young leaves pubescent; leaves of short shoots not caudate — (**b**) subsp. **betulifolia**
  2 Twigs and young leaves hispid; leaves of short shoots caudate — (**c**) subsp. **caudina**

(**a**) Subsp. **nigra**: Tree with broad, uneven crown; twigs, leaves and catkin-axis glabrous; leaves of short shoots acute. *C. & E. Europe; cultivated elsewhere throughout the range of the species, and sometimes naturalized.*

(**b**) Subsp. **betulifolia** (Pursh) W. Wettst., *Schriftenreihe Österr. Ges. Holzforsch.* 5 (1952) (*P. betulifolia* Pursh, *P. nigra* var. *betulifolia* (Pursh) Torrey): Like subsp. (**a**) but twigs, young leaves and catkin-axis pubescent. ● *W. Europe; sometimes naturalized.*

(**c**) Subsp. **caudina** (Ten.) Bug., *Arboret. Kórnickie* **12**: 146 (1967): Like subsp. (**a**) but twigs, leaves and catkin-axis hispid; leaves of short shoots caudate. *Mediterranean region.*

Cv. 'Italica' (*P. italica* (Duroi) Moench, *P. pyramidalis* Rozier) (Lombardy Poplar), is widely planted, often in avenues. It is characterized by a narrow columnar habit (wider in pistillate trees), and somewhat smaller leaves than subsp. (**a**).

Cv. 'Plantierensis' (*P. fastigiata* var. *plantierensis* Simon-Louis, *P. nigra* var. *elegans* Bailey) is like subsp. (**a**) cv. 'Italica' in habit but differs in the pubescence of twigs, leaves and catkin-axis. *Planted as a street tree and along rivers.*

9. P. × canadensis Moench, *Verz. Ausl. Bäume Weissenst.* 81 (1785) (*P. deltoides × nigra*, *P. bachelieri* Solemacher, *P. monilifera* auct., non Aiton, *P. regenerata* A. Henry, *P. robusta* C. K. Schneider, *P. serotina* Hartig, *P. marilandica* Bosc ex Poiret, *P. virginiana* auct., *P. fremontii* auct. germ., *P. gelrica* (Houtzagers) Houtzagers). Like **8**

but much quicker-growing; trunk without bosses; twigs almost terete or slightly angled; leaves crenate-serrate, shortly ciliate; stamens 15–25. *Originated probably in France* c. 1750; *planted almost throughout Europe for shelter and for timber.*

Several clones are known in cultivation, which apparently arose independently in different places; the most important are:

(i) cv. 'Serotina' (var. *serotina* (Hartig) Rehder): Staminate, up to 40 m tall, with broad crown, glabrous twigs and deltate leaves. *The oldest cultivar; extensively planted in W. & C. Europe.*

(ii) cv. 'Serotina de Selys': Like (i) but fastigiate. *Arose in Belgium before* 1818.

(iii) cv. 'Regenerata' (var. *regenerata* (A. Henry) Rehder): Pistillate, very similar to (i) but coming into leaf 2 weeks earlier. *Arose in France* c. 1814; *extensively planted in C. Europe.*

(iv) cv. 'Marilandica' (var. *marilandica* (Bosc ex Poiret) Rehder): Pistillate, differing from (i) in its more spreading branches; leaves rhombic-ovate, coming into leaf earlier. *Much cultivated in C. Europe.*

(v) cv. 'Gelrica': Staminate, very quick-growing tree with reddish shoots, coming into leaf between (i) and (iv). *Extensively planted in C. Europe and the Netherlands.*

(vi) cv. 'Italia 214': Pistillate, with erecto-patent branches; young shoots red, glabrous; leaves large, deltate. *Extensively planted in S. Europe.*

(vii) cv. 'Robusta' (P. × *robusta* C. K. Schneider): Like (i) but with pubescent twigs and petioles. *Extensively planted in C. & N.W. Europe.*

(viii) cv. 'Campeador': Pistillate, with broad, obovoid crown; twigs glabrous, light brown; leaves large, caudate-deltate. *Extensively planted in Spain and Portugal.*

A closely related species, **P. monilifera** Aiton, *Hort. Kew.* **3**: 406 (1789) (*P. canadensis* Michx fil., non Moench, *P. deltoides* var. *monilifera* (Aiton) A. Henry), from Canada and N.E. United States, is like **9** but twigs slightly angled or almost terete, leaves 7–12 cm, broadly ovate, wider than long, cuspidate, shallowly cordate at base, and thinner. It is very rarely cultivated in Europe, though its name is commonly applied erroneously to both **9** and **10**.

10. P. deltoides Marshall, *Arbust. Amer.* 106 (1785) (*P. angulata* Aiton, *P. canadensis* auct., non Moench, *P. carolinensis* Moench, *P. deltoides* var. *missouriensis* (A. Henry) A. Henry, *P. monilifera* auct., non Aiton). A quick-growing tree up to 30 m, with erecto-patent branches forming a broad crown. Twigs often strongly angled, first greenish, later greyish-brown. Leaves 10–18 cm, deltate-ovate to ovate but longer than wide, acute, usually truncate at base, firm, densely ciliate, crenately glandular-serrate. Stamens 30–60. Fruiting catkins 15–20 cm. Capsule 3- or 4-valved. *Planted for timber and along roadsides, and naturalized in several places.* [Au Az Be Br Bu Ga Ge Ho Hs Hu It Ju Lu.] (*S.E. United States.*)

Sect. TURANGA Bunge. Terminal bud absent; leaves very variable; stamens 8–12; pistillate flowers with a laciniate disc.

11. P. euphratica Olivier, *Voy. Emp. Othoman* ed. min., **6**: 319 (1807) (*P. illicitana* Dode). Up to 15 m, with slender branches. Bark greyish. Leaves leathery, glabrous, glaucous-green, entire or coarsely dentate, more or less dimorphic, the 'juvenile' linear to linear-lanceolate, the 'adult' deltate-rhombic to reniform. Female flowers with long pedicels, in very lax catkins. Capsule 2- or 3-valved. *Saline soils. Naturalized in S.E. Spain (near Elche, Alicante prov.).* [Hs.] (*N. Africa; S.W. & C. Asia.*)

MYRICALES

32. MYRICACEAE

EDIT. N. A. BURGES

Usually dioecious shrubs or small trees. Flowers in catkins, solitary in the axils of bracts. Male flowers usually without bracteoles; stamens 2–16. Female flowers with 2 or more bracteoles; ovary superior, 1-celled; style short.

1. Myrica L.

N. A. BURGES

Leaves alternate, simple, often gland-dotted, with strong aromatic scent. Fruit a drupe or nut.

The only genus.

1 Catkins borne on leafless upper part of stems, below which the current year's growth occurs — **1. gale**
1 Catkins borne among the leaves or in the axils of fallen leaves
2 Evergreen; leaves oblanceolate; catkins branched — **2. faya**
2 Deciduous; leaves elliptical to obovate; catkins unbranched — **3. caroliniensis**

1. M. gale L., *Sp. Pl.* 1024 (1753). Usually dioecious, but plants may change sex from year to year. Deciduous shrub up to 2·5 m. Twigs with scattered, yellowish glands. Leaves 2–6 cm, oblanceolate, cuneate at base, more or less serrate near apex, pubescent beneath, with shining, yellow, fragrant glands on both surfaces. Catkins unbranched, borne on leafless branches of upper part of stems which subsequently die. Fruit dry, compressed. $2n = 48$. *Bogs and fens. N.W. Europe, extending to S. Portugal, Poland and N.W. Russia.* Be Br Da Fe Ga Ge Hb Ho Hs Lu No Po Rs(N, B) Su.

2. M. faya Aiton, *Hort. Kew.* **3**: 397 (1789). Evergreen shrub or small tree up to 8 m. Twigs with small, ferruginous, peltate hairs. Leaves 4–11 cm, oblanceolate, cuneate at base, entire, with revolute margins, glabrous, without conspicuous glands. Catkins axillary on the leafy part of the stem, more or less branched. Fruit a drupe, but only slightly fleshy. *Açores; naturalized in W. Portugal.* Az [Lu]. (*Madeira, Islas Canarias.*)

3. M. caroliniensis Miller, *Gard. Dict.* ed. 8, no. 3 (1768) (*M. pennsylvanica* Loisel.). Deciduous shrub up to 3 m. Twigs grey-pubescent and glandular. Leaves 3–10 cm, elliptical to obovate, cuneate at base, usually crenate-dentate towards apex, pubescent and with shining, yellow glands on both surfaces. Catkins in the axils of leaves that have fallen, unbranched. Fruit a subglobose drupe. *Naturalized in S. England and the Netherlands.* [Br Ho.] (*E. North America.*)

JUGLANDALES

33. JUGLANDACEAE

EDIT. D. M. MOORE

Monoecious; deciduous trees with alternate, imparipinnate leaves. Male flowers in catkins borne on twigs of the previous year; perianth present or absent; stamens 3–40. Female flowers usually few, borne on twigs of the current year; perianth 3- to 5-lobed. Ovary inferior, unilocular or incompletely 2- or 4-celled; ovule 1. Fruit a drupe or nut.

1 Twigs with continuous pith; catkins 3 or more together; perianth of male flowers absent or very small **2. Carya**
1 Twigs with septate pith; catkins solitary; perianth of male flowers present
2 Buds sessile, with scales; leaves aromatic; fruit a large drupe **1. Juglans**
2 Buds stipitate, usually naked; leaves not aromatic; fruit a winged nut **3. Pterocarya**

1. Juglans L.

†T. G. TUTIN

Twigs with septate pith. Buds with scales, sessile. Leaves aromatic. Male catkins solitary, many-flowered, pendent. Female flowers in few-flowered terminal racemes. Fruit a large drupe; mesocarp indehiscent.

1 Leaflets usually 7–9, almost entire **1. regia**
1 Leaflets usually 11–23, serrate
2 Leaf-scars without a prominent pubescent band on their upper edge; fruit finely pubescent; stone 4-celled at base **2. nigra**
2 Leaf-scars with a prominent pubescent band on their upper edge; fruit viscid and pubescent; stone 2-celled at base **3. cinerea**

1. J. regia L., *Sp. Pl.* 997 (1753). Tree up to 30 m; bark grey, smooth, eventually becoming fissured but not scaling. Leaflets 7–9, 6–15 cm, obovate or elliptical, acute or acuminate, glabrescent. Male catkins 5–15 cm. Fruit 4–5 cm, subglobose, green, gland-dotted, glabrous; stone ovoid, acute, wrinkled, easily splitting. *Possibly native in Greece and elsewhere in the Balkan peninsula; widely cultivated for its fruits (Walnuts) and timber and locally naturalized in S. & W. Europe.* Al *Au Bu Cr Gr *It Ju Rm *Si [Br Co Ga He Hs Hu Lu Rs(W, K, E) Sa ?Tu].

2. J. nigra L., *Sp. Pl.* 997 (1753). Tree up to 50 m; bark brown, fissured. Leaf-scars without a prominent pubescent band on their upper edge. Leaflets 15–23, 6–12 cm, ovate-oblong to ovate-lanceolate, acuminate, irregularly serrate, glabrescent above, pubescent and glandular beneath. Male catkins 5–15 cm. Fruit 3·5–5 cm, globose or slightly obovoid, pubescent; stone ovoid, acute, strongly ridged, not splitting. *Extensively planted for timber in parts of C. & E. Europe.* [Au Cz Da Ge It Rm.] (*E. North America.*)

3. J. cinerea L., *Syst. Nat.* ed. 10, **2**: 1272 (1759). Tree up to 30 m; bark grey, deeply fissured. Leaf-scars with a prominent pubescent band on their upper edge. Leaflets 11–19, 6–12 cm, oblong-lanceolate, acuminate, appressed-serrate, finely pubescent above, pubescent and glandular beneath. Male catkins 5–8 cm. Fruit 4–6·5 cm, ovoid-oblong, pubescent, viscid; stone ovoid-oblong with 4 prominent and 4 less-prominent, sharp ridges and many broken ridges between them, not splitting. *Occasionally planted for timber.* [Da Rm.] (*E. North America.*)

2. Carya Nutt.

†T. G. TUTIN

Twigs with continuous pith. Buds with scales, sessile. Leaves not aromatic. Male catkins 3 or more together, many-flowered, pendent. Female flowers in 2- to 10-flowered terminal racemes. Fruit a large drupe; mesocarp dehiscing more or less completely into 4 valves.

1 Bud-scales 4–6, valvate, bright yellow; fruit 4-winged in upper half **1. cordiformis**
1 Bud-scales 6–12, imbricate, not bright yellow; fruit unwinged or slightly winged near apex
2 Bark not scaling; buds 8–12 mm; fruit slightly winged near apex **2. glabra**
2 Bark scaling; buds 13–25 mm; fruit unwinged **3. ovata**

1. C. cordiformis (Wangenh.) C. Koch, *Dendrologie* **1**: 597 (1869) (*C. amara* (Michx fil.) Nutt.). Tree up to 30 m; bark light brown, scaly. Bud-scales 4–6, valvate, bright yellow. Leaflets 5–9, 8–15 cm, ovate-lanceolate to lanceolate, acuminate, serrate, pubescent beneath when young. Fruit 2–3·5 cm, obovoid to subglobose, 4-winged in upper $\frac{1}{2}$; pericarp thin, splitting to below the middle; stone grey, almost smooth. *Planted for timber in Germany.* [Ge.] (*E. North America.*)

2. C. glabra (Miller) Sweet, *Hort. Brit.* 97 (1826) (*C. porcina* (Michx fil.) Nutt.). Tree up to 40 m; bark grey, fissured, not scaly. Buds 8–12 mm; scales more than 6, imbricate, not yellow. Leaflets usually 5, 8–15 cm, oblong to oblong-lanceolate, acuminate, serrate, almost glabrous. Fruit *c.* 2·5 × 2 cm, usually obovoid, slightly winged near apex; pericarp usually splitting to the middle only; stone pale brown, nearly smooth. *Planted for timber in Germany.* [Ge.] (*E. North America.*)

3. C. ovata (Miller) C. Koch, *Dendrologie* **1**: 598 (1869) (*C. alba* (L.) Nutt.). Tree up to 40 m; bark grey, splitting into long scales. Twigs light reddish-brown, glabrescent. Buds 13–25 mm; scales 10–12, imbricate, dark; inner scales becoming 6–8 × 2·5–4 cm, yellowish or purplish when the buds open. Leaflets 5–7, 10–20 cm, the 3 upper much larger than the lower, all elliptical to oblong-lanceolate, acuminate, serrate, densely ciliate, pubescent and glandular beneath when young, later glabrous or subglabrous.

Fruit 3·5–6 cm, subglobose to broadly ovoid, not winged; pericarp splitting to base; stone white, slightly angled. *Planted for timber in C. Europe.* [Cz Ge Rm.] (*E. North America.*)

C. tomentosa (Poiret) Nutt., *Gen. N. Amer. Pl.* **2**: 221 (1818) and **C. laciniosa** (Michx fil.) Loudon, *Hort. Brit.* **1**: 384 (1830), from E. North America, have been planted on an experimental scale in Germany. They are like **3** in having large buds and angled stones, but in *C. tomentosa* the bark is not scaly and the twigs are tomentose for most of the summer, while *C. laciniosa* has pale orange twigs and 7–9 leaflets.

3. Pterocarya Kunth

†T. G. TUTIN

Twigs with septate pith. Buds naked, stipitate. Leaves not aromatic. Male catkins solitary, many-flowered, pendent. Female flowers numerous, in pendent catkins. Fruit a winged nut.

1. P. fraxinifolia (Poiret) Spach, *Hist. Vég.* (*Phan.*) **2**: 180 (1834). Tree up to 30 m, freely suckering, with deeply fissured bark. Leaflets 11–20 pairs, sharply serrate. Fruit with semi-orbicular wings. *Widely planted, though never on a large scale.* (*S.W. Asia.*)

FAGALES

34. BETULACEAE

EDIT. S. M. WALTERS

Monoecious; deciduous trees or shrubs. Leaves alternate, simple. Stipules caducous. Male flowers 3 in the axil of each bract, in pendent catkins; perianth present. Female flowers 2 or 3 in the axil of each bract, in erect catkins; perianth absent; ovary bilocular; styles 2. Fruit a flattened nutlet, usually winged, borne in a dense, cylindrical or cone-like catkin of scales formed from the accrescent, fused bracts and bracteoles.

Fruiting catkin cylindrical or narrowly ovoid; scales 3-lobed, falling with the fruit; stamens 2, bifid below the anthers **1. Betula**
Fruiting catkin ovoid, cone-like; scales 5-lobed, woody and persistent; stamens 4, entire (but with shortly forked connective) **2. Alnus**

1. Betula L.

S. M. WALTERS

Catkins with 2 bracteoles to each group of flowers in the axil of a bract. Male flowers with 2 stamens, bifid from below the anthers; perianth minute. Female flowers 3 to each bract. Fruiting catkins cylindrical or narrowly ovoid; scales 3-lobed, falling with fruit.

The taxonomy of European *Betula* is much disputed, and the following treatment does no more than provide a basic framework. For more detailed study the bibliography provided by Natho is very useful.

Several other species are grown for ornament, especially **B. papyrifera** Marshall, *Arbust. Amer.* 19 (1785), from North America, a tree up to 30 m with ovate, subglabrous leaves 4–10 cm, and **B. utilis** D. Don, *Prodr. Fl. Nepal.* 58 (1825) (*B. jacquemontii* Spach), from E. Asia, a tree up to 20 m, often with very white, peeling bark, and ovate, hairy leaves 4–12 cm.

Literature: P. Ascherson & K. O. P. P. Graebner, *Syn. Mitteleur. Fl.* **4**: 386–412 (1910). K. B. Ashburner & S. M. Walters in S. M. Walters *et al.*, eds., *Eur. Garden Fl.* **3**: 49–55 (1989). G. Natho, *Feddes Repert.* **61**: 211–273 (1959). K. H. Rechinger in Hegi, *Ill. Fl. Mitteleur.* ed. 2, **3(1)**: 141–173 (1957). H. Winkler in Engler, *Pflanzenreich* **19(IV. 61)**: 56–101 (1904).

1 Trees or large shrubs, usually more than 3 m; young male catkins in winter pendent and unprotected
2 Young twigs glabrous, with numerous peltate resin-glands; leaves usually sharply biserrate with prominent primary teeth; nutlet glabrous **1. pendula**
2 Young twigs usually without resin-glands, often hairy; leaves irregularly serrate, without prominent primary teeth; nutlet puberulent at apex **2. pubescens**
1 Small shrubs not more than 3 m; young male catkins in winter erect, protected by bud-scales
3 Leaves longer than broad, acute or subacute **3. humilis**
3 Leaves ± orbicular, obtuse or truncate **4. nana**

1. B. pendula Roth, *Tent. Fl. Germ.* **1**: 405 (1788) (*B. alba* sensu Coste, non L., *B. verrucosa* Ehrh.; incl. *B. aetnensis* Rafin.). Tree up to 30 m; bark smooth and silvery-white except towards the base, where it is usually dark and more or less dissected into rectangular bosses. Young twigs glabrous, with numerous appressed, peltate resin-glands; twigs usually very slender and pendent. Leaves 2–7 cm, ovate-deltate, acuminate, sharply biserrate with prominent primary teeth, subglabrous when mature, thin. Male catkins 3–6 cm. Fruiting catkin 1·5–3·5 × 1 cm; scales with cuneate base, and with wide, more or less recurved lateral lobes and a deltate, obtuse median lobe. Wing of fruit 2–3 times as wide as the narrowly ovoid glabrous nutlet. $2n = 28$. *Mainly on sandy or peaty soils. Most of Europe, but rare in much of the south.* All except Az Bl Cr Fa Is Sa Sb ?Tu.

Rather variable, especially in leaf-shape and habit. Lindquist (*Svensk Bot. Tidskr.* **41**: 62 (1947)) distinguishes from the typical plant a northern var. *lapponica* with smooth bark (even in old trees), less pendent twigs and somewhat larger scales and fruits, which may merit subspecific distinction.

B. oycoviensis Besser, *Prim. Fl. Galic.* **2**: 289 (1809), described from S.E. Poland, resembles **1** and has the same chromosome number, but is of shrubby habit, has 3 or 4 (not about 2) leaves on the flowering twigs, and leaves with almost equal teeth. Other shrubby variants have been described from Scandinavia and elsewhere; some at least seem to be of hybrid origin (see **2**).

2. B. pubescens Ehrh., *Beitr. Naturk.* **6**: 98 (1791). Like **1** but small tree or shrub 1–20 m; bark brownish or greyish, not dissected into rectangular bosses; young twigs without resin-glands, not or only slightly pendent; leaves coarsely serrate, usually pubescent at least in axils of veins beneath; scales with slightly ascending lateral lobes and narrowly oblong or triangular-lanceolate median lobe; fruit with wing 1–1½ times as wide as the ovoid nutlet; apex of nutlet minutely puberulent. *Mainly on peaty soils of moorland and mountains. Most of Europe, but rare in the Mediterranean region and the south-east.* Au Be Br Cz Da Fe Ga Ge Hb He Ho Hs Hu Is It Ju Lu No Po Rm Rs(N, B, C, W, E) Su.

Extremely variable. The following subspecies have often been treated as species.

1 Young twigs glabrous or subglabrous (c) subsp. **carpatica**
1 Young twigs distinctly puberulent or glandular
2 Shrub up to 12 m; wing of fruit about as wide as nutlet (d) subsp. **tortuosa**
2 Tree or tall shrub more than 5 m; wing of fruit 1–1½ times as wide as nutlet
3 Young twigs puberulent, not glandular (a) subsp. **pubescens**
3 Young twigs glabrous or puberulent, glandular (b) subsp. **celtiberica**

(**a**) Subsp. **pubescens** (*B. concinna* Gunnarsson, *B. pubescens* subsp. *suecica* Gunnarsson, ?*B. subarctica* Orlova): Tree or tall shrub, usually more than 5 m. Young twigs distinctly puberulent. Leaves 3–5 cm. Wing of fruit 1–1½ times as wide as the nutlet. $2n = 56$. *Usually lowland; northwards to* c. 70° *N. in Finland and N. Russia.*

(**b**) Subsp. **celtiberica** (Rothm. & Vasc.) Rivas Martínez, *Trab. Dep. Bot. Fisiol. Veg.* **3**: 78 (1971) (*B. celtiberica* Rothm. & Vasc.): Small tree or tall shrub up to 10 m. Young twigs glabrous or puberulent, glandular. Leaves 3–5 cm. Wing of fruit *c.* 1½ times as wide as the nutlet. $2n = 56$. ● *Mountains of N. & C. Spain and N. Portugal.*

(**c**) Subsp. **carpatica** (Willd.) Ascherson & Graebner, *Fl. Nordostd. Flachl.* 253 (1898) (*B. carpatica* Waldst. & Kit. ex Willd., *B. pubescens* subsp. *odorata* (Bechst.) E. F. Warburg; incl. *B. odorata* Bechst., *B. coriacea* Gunnarsson, *B. murithii* Gaudin): Small tree or shrub up to 8 m. Young twigs and leaves glabrescent. Leaves often less than 3 cm. Wing of fruit about as wide as the nutlet. $2n = 56$. *Arctic Europe, and southwards, mainly in the mountains, to the Pyrenees and Carpathians.*

(**d**) Subsp. **tortuosa** (Ledeb.) Nyman, *Consp.* 672 (1881) (*B. tortuosa* Ledeb.; incl. *B. kusmisscheffii* (Regel) Suk.): Shrub up to 12 m, with many interlacing branches. Young twigs and leaves distinctly puberulent. Leaves less than 3 cm. Wing of fruit about as wide as the nutlet. $2n = 56$. *Arctic Europe; mountains of Scandinavia; Iceland.*

Plants intermediate between subsp. (**b**) and **1** occur in the W. Pyrenees.

B. borysthenica Klokov, *Jour. Bot. Acad. Sci. Ukr.* **3**: 17 (1946), seems to fall within subsp. (**c**). It grows on alluvial river-sand in C. Ukraine and S.E. Russia.

Plants very similar to those of subsp. (**d**) from the Arctic occur in Scotland and in the Alps, but require further investigation.

Hybrids between **1** and **2** (*B.* × *aurata* Borkh.) have been recorded frequently, but cytotaxonomic investigation has shown that the first-generation triploid hybrid, which is highly sterile, occurs rarely, and that most of the putative hybrids have the chromosome number of the tetraploid species. It seems likely that such plants are the complex products of hybridization; they are particularly common in regions such as Britain where most natural forest has been destroyed.

B. obscura A. Kotula, *Jahresb. Schles. Ges. Vaterl. Kult.* **65**: 314 (1888) (*B. atrata* Domin), from Poland, Czechoslovakia and W. Ukraine, is probably of such hybrid origin, resembling **1** in most characters except the dark-coloured bark.

B. callosa Notø, *Tromsø Mus. Aarshefter* **24**: 23 (1901), described originally from Norway, and reported from other parts of Fennoscandia and Iceland, differs from **2** principally in its ovoid, subsessile female catkins (cylindrical and pedunculate in **2**) and the long, narrow lobes of the catkin-scales. Its relationships are not clear.

3. B. humilis Schrank, *Baier. Fl.* **1**: 421 (1789). Small, much-branched shrub, rarely more than 2 m; twigs and leaves variably puberulent, sometimes subglabrous. Leaves 1–3 × 0·5–2·5 cm, ovate or ovate-orbicular, acute or subacute at apex, rather thick in texture; margin coarsely serrate or crenate. Fruiting catkin 8–15 mm, erect; scales with 3 subequal lobes, the 2 lateral ascending and often slightly shorter than the median lobe; wing of fruit narrow, ⅓–½ as wide as the nutlet. $2n = 28$. *Bogs and fens. C. & E. Europe, from N. Germany and N. Russia to N.E. Switzerland and C. Romania.* Au †Cz Ge He Po Rm Rs(N, B, C, W, E).

4. B. nana L., *Sp. Pl.* 953 (1753). Dwarf shrub rarely more than 1 m, with spreading or procumbent branches; twigs pubescent; mature leaves glabrous. Leaves 5–15 mm, more or less orbicular, deeply crenate, thick in texture. Fruiting catkin 5–10 mm, erect; scales with 3 equal, erect lobes at apex; wing of fruit very narrow, *c.* ⅙–¼ as wide as the nutlet. $2n = 28$. *Moorland and bogs. N. & C. Europe, from the Arctic southwards to S.C. France, E. Carpathians and C. Russia; local and mainly on mountains in the southern part of its range.* Au Br Cz Fe Ga Ge He Is †It No Po Rm Rs(N, B, C) Sb Su.

Hybrids of **3** and **4** *inter se* and with **1** and **2** are all recorded. **B. sukaczewii** Soczava, *Oč. Fitosoc. Fitogeog.* 393 (1929), described from N. Ural, is apparently **4** × **2**(**c**).

2. Alnus Miller

P. W. BALL

Catkins with 2 bracteoles to each group of flowers in the axil of each bract. Male flowers with 4 stamens and 4(5)-partite perianth; anther-lobes somewhat separated by the shortly forked connective. Female flowers 2 in axil of each bract. Fruiting catkin cone-like, ovoid or ellipsoidal, the scales 5-lobed, thick and woody, long-persistent.

1 Buds sessile; catkins appearing with the leaves, in inflorescences with 2 or 3 leaves at base **1. viridis**
1 Buds short-stalked; catkins appearing before the leaves, without leaves at the base of the inflorescence
2 Inflorescence with 1–3 female catkins; leaves crenate-dentate **5. cordata**
2 Inflorescence with 3–8 female catkins; leaves serrate or laciniate
3 Leaves serrulate, with reddish-brown hairs especially in axils of veins **3. rugosa**
3 Leaves distinctly biserrate, glabrous or with yellow or grey hairs
4 Leaves obtuse or retuse; young twigs viscid; bark dark brown, fissured; female catkins distinctly pedunculate **2. glutinosa**
4 Leaves acuminate to subacute; young twigs not viscid;

bark grey or yellowish, smooth; female catkins ±sessile **4. incana**

1. A. viridis (Chaix) DC. in Lam. & DC., *Fl. Fr.* ed. 3, **3**: 304 (1805) (*A. alnobetula* (Ehrh.) Hartig). Shrub 0·5–2·5(–4) m; buds sessile. Twigs glabrous or puberulent, greenish- or reddish-brown. Leaves elliptical to suborbicular, cuneate to subcordate at base, biserrate, viscid when young. Catkins appearing with the leaves, in inflorescences with 2 or 3 leaves at base. Fruiting catkins 8–15 × 4–8 mm, usually 3–5 in a slender, pedunculate raceme. Nutlet with broad, membranous wing. *Mountain slopes, especially near streams. C. Europe, Corse and Balkan peninsula; N.E. Russia.* Au Bu Co Cz Ga Ge He Hu It Ju Po Rm Rs(N, C, W).

This species shows considerable variation in the shape and size of the leaf, and in indumentum. At least 3 subspecies can be recognized in Europe.

1 Leaves broadly ovate or suborbicular, obtuse or with a very short acumen; lower surface glabrous or with tufts of hairs in axils of veins (c) subsp. **suaveolens**
1 Leaves elliptical or ovate, acute or acuminate; lower surface usually pubescent, at least on the veins
2 Leaves 1–5(–9) cm, usually ±cuneate at base, with 4–8 pairs of lateral veins (a) subsp. **viridis**
2 Leaves 3–8(–12) cm, cuneate to subcordate at base, with 7–10 pairs of lateral veins (b) subsp. **fruticosa**

(a) Subsp. **viridis**: Leaves 1–5(–9) cm, elliptical or ovate, acute or acuminate, usually more or less cuneate at base, with 4–8 pairs of lateral veins; lower surface usually pubescent, at least on the veins. $2n = 28$. ● *Mountains of C. Europe and Balkan peninsula.*

(b) Subsp. **fruticosa** (Rupr.) Nyman, *Consp.* 672 (1881): Leaves 3–8(–12) cm, elliptical or ovate, acute or acuminate, cuneate to subcordate at base, with 7–10 pairs of lateral veins; lower surface pubescent, at least on the veins. *N.E. Russia.* (*N. Asia and North America.*)

(c) Subsp. **suaveolens** (Req.) P. W. Ball, *Feddes Repert.* **68**: 186 (1963): Leaves 1–4·5 cm, broadly ovate or suborbicular, obtuse or with a very short acumen, cuneate or sometimes subcordate at base, with 6–8 pairs of lateral veins; lower surface glabrous or with tufts of hairs in the axils of the veins. ● *Corse.*

2. A. glutinosa (L.) Gaertner, *Fruct. Sem. Pl.* **2**: 54 (1790). Tree or shrub up to 20(–35) m, with dark brown, fissured bark; buds short-stalked. Young twigs viscid, usually glabrous. Leaves (3–)4–10 cm, obovate-elliptical to suborbicular, obtuse or retuse, cuneate or rounded at base, biserrate, green on lower surface, usually glabrous except for tufts of yellowish hairs in the axils of the veins; lateral veins 5–8 pairs. Fruiting catkins 10–30 mm, ovoid, pedunculate, 3–5 in a raceme. Nutlet narrowly winged. $2n = 28$. *Usually beside water or on wet soils. Europe, except the extreme north and south.* All except Bl Cr Fa Is Sb; naturalized alien in Az.

2 × 4 (*A. × pubescens* Tausch) is not uncommon in the areas where the species grow together. It usually has the young twigs pubescent, leaves obtuse to shortly acuminate, usually pubescent at least on the veins, and the female catkins shortly pedunculate.

3. A. rugosa (Duroi) Sprengel, *Syst. Veg.* ed. 16, **3**: 848 (1826) (*A. serrulata* (Aiton) Willd.). Shrub or small tree up to 10 m. Young twigs reddish-brown, pubescent, soon becoming glabrous, glutinous. Leaves 5–10 cm, elliptical or obovate, acute or obtuse, rounded or subcordate at base, serrulate, with reddish-brown hairs especially in the axils of the veins; lateral veins 8–13 pairs. Fruiting catkins 10–15 × 8–10 mm, ovoid, 4–10 in a raceme, the upper sessile, the lower shortly pedunculate. Nutlet narrowly winged. *Frequently planted and naturalized, mainly in C. Europe.* [Cz Da Ge Po Su.] (*E. North America.*)

4. A. incana (L.) Moench, *Meth.* 424 (1794). Tree or shrub up to 30 m, with smooth bark; buds shortly stalked. Young twigs pubescent or tomentose. Leaves 3–8(–13) cm, ovate-lanceolate to ovate-orbicular, cuneate, biserrate, grey-green on lower surface, puberulent or tomentose at least when young; lateral veins 7–12 pairs. Fruiting catkins 11–17 × 9–12 mm, ovoid or suborbicular, sessile, 3–5 in a raceme. *N.E. Europe and Fennoscandia; C. Europe; locally in the mountains of S. Europe. Planted and widely naturalized in N.W. Europe.* Al Au Bu Cz Fe Ga Ge ?Gr He Hu It Ju No Po Rm Rs(N, B, C, W, E) Su [Az *Be Br Da Hb *Ho Hs].

(a) Subsp. **incana**: Bark grey, opaque. Leaves acuminate, usually tomentose on the lower surface. Wing of nutlet *c.* 0·5 mm wide. $2n = 28$. *Throughout the range of the species.*

(b) Subsp. **kolaensis** (Orlova) Á. & D. Löve, *Bot. Not.* **114**: 51 (1961) (*A. kolaensis* Orlova): Bark yellow, translucent. Leaves obtuse or shortly acute, glabrous or pubescent on the veins on the lower surface. Wing of nutlet *c.* 1 mm wide. ● *N. & W. Fennoscandia.*

5. A. cordata (Loisel.) Loisel., *Fl. Gall.* ed. 2, **2**: 317 (1828) (*A. cordifolia* Ten.). Tree up to 15 m; buds shortly stalked. Young twigs viscid, usually glabrous. Leaves 2–12 cm, ovate or suborbicular, obtuse or shortly acuminate, truncate or cordate at base, crenate-dentate, usually glabrous or with tufts of hairs in the axils of the veins; lateral veins 5–8 pairs. Fruiting catkins 15–30 mm, oblong-ovoid, solitary or up to 3 in a raceme. Nutlet narrowly winged. ● *Corse and S. Italy; N.W. Albania; sometimes planted in S.W. Europe and elsewhere.* Al Co It [Az Hs].

This species shows some variation in shape and size of leaf. Plants from Italy and Albania usually have acute or acuminate leaves 6–12 cm, while those from Corse have obtuse leaves 3–7 cm.

35. CORYLACEAE

EDIT. †T. G. TUTIN AND S. M. WALTERS

Deciduous trees or shrubs. Leaves alternate, simple; stipules caducous. Flowers monoecious, inconspicuous, anemophilous. Male flowers 1 in the axil of each bract in pendent catkins; bracteoles 2, united with the bract, or 0; perianth absent. Female flowers 2 in the axil of each bract; bracteoles present; perianth small, irregularly lobed; ovary 2-locular, inferior; styles 2. Fruit a nut, subtended or surrounded by an involucre formed from the accrescent bract and bracteoles.

1 Buds ovoid, obtuse; leaves usually with fewer than 8 pairs of veins; fruits in clusters of 1–4 **3. Corylus**

1 Buds fusiform, acute; leaves with 9 or more pairs of veins; fruits numerous, in pendent spikes
2 Bark grey, smooth; male catkins appearing in spring; nut subtended by a 3-lobed or serrate involucre **1. Carpinus**
2 Bark brown, scaly, rough; male catkins visible throughout the winter; nut enclosed in an entire, apiculate involucre **2. Ostrya**

1. Carpinus L.

†T. G. TUTIN

Buds fusiform, acute. Leaves with 9 or more pairs of veins. Male flowers without bracteoles. Infructescence lax. Nut small, subtended by the lobed or serrate, strongly veined involucre.

Leaves 4–10 cm; involucre 3-lobed **1. betulus**
Leaves 2·5–6 cm; involucre not lobed **2. orientalis**

1. C. betulus L., *Sp. Pl.* 998 (1753). Tree up to 25 m, with fluted trunk and smooth, grey bark. Leaves 4–10 cm, ovate, acuminate, biserrate, rounded to subcordate at base, pubescent on the veins beneath. Catkins up to 5 cm, pendent. Infructescence 5–14 cm; involucre 3-lobed, the mid-lobe *c.* 3·5 cm, much longer than the laterals, oblong; margin entire or serrate. $2n = 64$. *C. & S.E. Europe, extending to S. Italy, N.E. Spain, S. England, S. Sweden and C. Ukraine.* Al Au Be Br Bu Cz Da Ga Ge Gr He Ho Hs Hu It Ju Po Rm Rs(B, C, W, E) Su Tu.

2. C. orientalis Miller, *Gard. Dict.* ed. 8, no. 3 (1768) (*C. duinensis* Scop.). Shrub or small tree. Leaves 2·5–6 cm, ovate or elliptical, acute, biserrate, cuneate to rounded at base, sparsely pubescent on the veins beneath. Infructescence 3–5 cm; involucre *c.* 2 cm, triangular-ovate, serrate but not lobed. *S.E. Europe, extending westwards to Sicilia.* Al Bu Gr Hu It Ju Rm Rs(K) Si Tu.

2. Ostrya Scop.

†T. G. TUTIN

Buds fusiform, acute. Leaves with 9 or more pairs of veins. Male flowers without bracteoles. Infructescence compact. Fruit a small nut, enclosed in the entire, apiculate involucre.

1. O. carpinifolia Scop., *Fl. Carn.* ed. 2, **2**: 244 (1772) (*O. virginiana* (Miller) C. Koch. subsp. *carpinifolia* (Scop.) Briq.). Small tree with brownish, fissured bark. Leaves 5–8 cm, ovate, acuminate, very sharply biserrate, cuneate to subcordate at base, hairy when young, glabrescent. Catkins up to 10 cm, pendent. Infructescence 3–4·5 cm, subcylindrical, resembling that of *Humulus*; involucre elliptical, with puberulent mucro. *S. Europe, from S.E. France to Thrace, extending northwards to S. Austria.* Al Au Bu Co Ga Gr He †Hu It Ju Sa Si Tu.

3. Corylus L.

†T. G. TUTIN

Buds ovoid, obtuse. Leaves usually with fewer than 8 pairs of veins. Male flowers with 2 bracteoles. Female inflorescence short, bud-like; stigmas red. Fruit a large nut, surrounded by a more or less tubular involucre which is dentate or laciniate above.

1 Involucre constricted at apex **3. maxima**
1 Involucre not constricted at apex
2 Stipules obtuse; involucre about as long as nut, divided into ±ovate, irregularly dentate or laciniate lobes **1. avellana**
2 Stipules acuminate; involucre much longer than nut, divided into long-acuminate, serrate lobes **2. colurna**

1. C. avellana L., *Sp. Pl.* 998 (1753). Shrub up to 6 m, with smooth, brown bark. Leaves suborbicular, acuminate, biserrate, often shallowly lobed, cordate at base; stipules oblong, obtuse. Catkins up to 8 cm, pendent. Infructescence of 1–4 nuts; involucre about as large as the nut, divided to half-way or a little more, into more or less ovate, usually irregularly dentate or laciniate lobes about as long as the nut; nut 1·5–2 cm, brown, with a hard, woody shell. $2n = 22$. *Europe, except some islands and the extreme north and north-east.* All except Bl Fa Is Sb; doubtfully native in Cr; naturalized in Az.

2. C. colurna L., *Sp. Pl.* 999 (1753). Like **1** but usually a tree, up to 20 m; stipules lanceolate, acuminate; catkins up to 12 cm; involucre much longer than the nut, divided almost to base into many long-acuminate, serrate lobes. *Balkan peninsula, S.W. Romania.* Al Bu Gr Ju Rm Tu.

3. C. maxima Miller, *Gard. Dict.* ed. 8, no. 2 (1768) (*C. tubulosa* Willd.). Like **1** but a shrub or small tree; involucre about twice as long as the nut, contracted above the nut and dentate at apex. *W. Jugoslavia; doubtfully native in Greece; cultivated for its nuts (Hazelnuts or Filberts) elsewhere.* *Gr Ju [Au Br Cz].

Regarded by some authors as a variant or cultivar of **1** or one of its hybrids.

36. FAGACEAE

EDIT. S. M. WALTERS

Monoecious; trees or shrubs. Leaves simple, alternate. Male flowers in catkins or heads; perianth 4- to 7-lobed; stamens 8–20(–40), usually twice as many as perianth-lobes. Female flowers 1–3, surrounded by an involucre of scales (cupule); perianth 4- to 6-lobed; styles 3 or 6. Fruit a 1-seeded nut, in groups of 1–3, surrounded by the accrescent cupule.

1 Male flowers in pendent heads or 1–3 in axils of leaves; buds fusiform; nut triquetrous
2 Male flowers numerous in pendent heads; ripe cupule *c.* 2·5 cm **1. Fagus**
2 Male flowers usually solitary; ripe cupule *c.* 1 cm **2. Nothofagus**
1 Male flowers in long, erect or pendent catkins; buds ovoid; nut oblong to subglobose
3 Male catkins erect, with female flowers in lower part; cupule completely enclosing nuts **3. Castanea**
3 Male catkins pendent; female flowers in separate inflorescences; cupule enclosing only lower $\frac{1}{2}$ of nut **4. Quercus**

1. Fagus L.

†T. G. TUTIN (EDITION 1) REVISED BY J. R. AKEROYD (EDITION 2)

Deciduous trees. Buds fusiform. Flowers anemophilous. Male flowers in long-pedunculate, pendent heads; perianth 4- to 7-lobed; stamens 8–16. Female flowers usually in pairs in each stipitate, 4-partite cupule; perianth 4- or 5-lobed; styles 3, elongate. Nuts 1 or 2, triquetrous.

1. F. sylvatica L., *Sp. Pl.* 998 (1753). Tree up to 40 m; bark grey, smooth. Buds acute, reddish-brown. Leaves 4–9 cm, ovate-elliptic, acute, ciliate and silky, at least on the veins. Male flowers numerous; peduncles 5–6 cm. Nut 12–18 mm, brown; cupule *c.* 2·5 cm. *Forming woods, usually on well-drained soils, often on mountains in the south; widely planted elsewhere. S.W. & C. Europe, extending to S. Sweden.* Al Au Be Br Bu Co Cz Da Ga Ge Gr He Ho Hs Hu It Ju No Po Rm Rs(B, W, K, E) Si Su Tu [Hb].

(a) Subsp. **sylvatica**: Tree up to 30 m. Leaves 4–9 cm, with 5–8 pairs of veins. Perianth of male flowers divided almost to base. Cupule with all scales subulate. $2n = 24$. • *Almost throughout the range of the species.*

(b) Subsp. **orientalis** (Lipsky) Greuter & Burdet, *Willdenowia* **11**: 279 (1981) (*F. orientalis* Lipsky); Tree up to 40 m. Leaves 9–12(–14) cm, with 8–12 pairs of lateral veins. Perianth of male flowers divided for not more than $\frac{1}{3}$ of its length. Cupule with linear-oblong scales above and spathulate scales below. *At lower elevations and in more sheltered sites than subsp.* (**a**). *S.E. Europe from C. Greece to S.E. Russia.*

Plants intermediate between subspp. (**a**) and (**b**) that occur in E., E.C. and S.E. Europe have been described as species under the names **F. taurica** Popl., *Österr. Bot. Zeitschr.* **77**: 41 (1928), and **F. moesiaca** (K. Malý) Czecz., *Annu. Soc. Dendrol. Pologne* **5**: 52 (1933).

2. Nothofagus Blume

J. R. AKEROYD

Deciduous or evergreen trees. Buds fusiform. Flowers anemophilous. Male flowers axillary, solitary or in 2- or 3-flowered clusters, sessile or subsessile; perianth campanulate; stamens 8–40. Female flowers usually 3 in each 4-partite cupule; styles 3, short. Nuts (1–)3, mostly triquetrous.

A genus restricted to Australasia and temperate South America; several species are grown in Europe for ornament or timber; all are somewhat calcifuge.

Leaves with 8–11 pairs of lateral veins; cupule-scales narrowly triangular, with single terminal gland — **1. obliqua**
Leaves with 14–18 pairs of lateral veins; cupule-scales dendritic, glandular — **2. procera**

1. N. obliqua (Mirbel) Örsted, *Vidensk. Selsk. Skr.* **5**: 354 (1873). Tree up to 30 m; bark greyish. Leaves 3–8 cm, ovate to oblong, obtuse, irregularly serrate, with 8–9(–11) pairs of lateral veins, subglabrous, dark green above, paler beneath; petiole 3–6 mm, puberulent. Male flowers usually solitary; stamens 30–40. Cupule 8–10 mm in fruit, puberulent; scales narrowly triangular, with a single terminal gland. Nuts slightly shorter than the cupule, the 2 outer triquetrous, the central flattened. *Planted for timber in Britain and Germany, and sometimes naturalized.* [Br Ge.] (*Chile.*)

2. N. procera (Poeppig & Endl.) Örsted, *loc. cit.* (1873). Like **1** but usually not more than 20 m; leaves 4–10 cm, somewhat narrower, shallowly dentate, with 14–18 pairs of lateral veins, puberulent, pale or yellowish-green above; petiole pubescent; cupule-scales dendritic, glandular. *Planted for timber in Britain and Germany, and sometimes naturalized.* [Br Ge.] (*Chile.*)

3. Castanea Miller

†T. G. TUTIN

Deciduous trees or shrubs. Buds ovoid. Flowers entomophilous, scented, in erect catkins, male in the upper, female in the lower part of the same catkin. Male flowers with 6-partite perianth; stamens 10–20. Female flowers usually 3 in each cupule; ovary 6-locular; styles 7–9. Nuts large, brown, coriaceous, 1–3 together in a spiny cupule which dehisces irregularly by 2–4 valves.

Usually a tree; leaves 10–25 cm, coarsely and very sharply serrate — **1. sativa**
Usually a shrub; leaves 8–16 cm, crenate-serrate — **2. crenata**

1. C. sativa Miller, *Gard. Dict.* ed. 8, no. 1 (1768) (*C. vulgaris* Lam.). Tree up to 30 m; bark brownish-grey with longitudinal, often spirally curved, fissures. Buds obtuse, the terminal absent. Leaves 10–25 cm, oblong-lanceolate, acute or acuminate, glabrescent but lepidote beneath, at least near the veins. Nut 2–3·5 cm, brown with a paler base, shining; cupule green, covered with long, branched, sparsely pubescent spines. *In woods on well-drained soils, often on mountain slopes; usually calcifuge. Probably native in parts of the Balkan peninsula and more doubtfully in Italy and S.C. Europe; widely cultivated for its nuts (Chestnuts) and timber, and naturalized in W. Europe and locally elsewhere.* Al *Au *Co *Cr *Cz Gr *Hu *It Ju *Sa *Si Tu [Be Br Bu Da Ga Ge Hb He Ho Hs Lu Rm Rs(W, K) Su].

2. C. crenata Siebold & Zucc., *Abh. Akad. Wiss.* (*München*) **4(3)**: 224 (1846). Like **1** but usually a shrub up to *c.* 10 m; leaves elliptical or oblong-lanceolate, acuminate, crenate-serrate; spines of the cupule almost glabrous. *Planted for timber in S. Europe.* [Hs It Lu.] (*Japan.*)

4. Quercus L.

†O. SCHWARZ

Trees, rarely shrubs. Buds ovoid. Leaves usually dentate, sinuate or pinnately lobed, rarely entire; evergreen (persisting for more than a year), semi-evergreen (persisting green through the winter, but falling in spring), or deciduous (withering in autumn, though sometimes not falling until spring). Flowers anemophilous. Male catkins slender, pendent; bracts minute, caducous; flowers numerous, with usually 6-lobed perianth and 6–12 stamens. Female catkins with few flowers, each solitary in an involucre. Nut oblong or ellipsoidal, the base enclosed by the enlarged cupule.

In the descriptions and key, 'lateral veins' indicates those which run to the apices of the teeth or lobes; in some species intercalary veins, which run to the sinuses or do not reach the margin, may also be present.

In many species leaf-shape is very variable, and the leaves of leading shoots or of young trees may differ considerably from the descriptions given below.

Within each subgenus most species are interfertile. Hybrids are therefore common in regions where related species grow together, and much of the intraspecific variation is due to introgressive hybridization.

Literature: A. Camus, *Les Chênes: Monographie du genre* Quercus **1–3**. Paris, 1936–54. O. Schwarz, *Cavanillesia* **8**: 65–100 (1936). O. Schwarz, *Feddes Repert.* (Sonderbeih. D.) 1–200 (1936–9). J. de Carvalho e Vasconcellos & J. do Amaral Franco, *Anais Inst. Sup. Agron.* (*Lisboa*) **21**: 1–135 (1954). C. Vicioso, *Revisión de Género* Quercus *en España*. Madrid, 1950.

1 Leaves evergreen, coriaceous
2 Mature leaves glabrous; veins prominent above but not beneath; petiole 1–4 mm **3. coccifera**
2 Mature leaves tomentose beneath; veins ±immersed above, prominent beneath; petiole 6–15 mm
3 Midrib of leaf straight; bark not thick and corky **4. ilex**
3 Midrib of leaf somewhat sinuous; bark very thick, corky **5. suber**
1 Leaves deciduous or semi-evergreen, seldom very coriaceous
4 Fruits ripening in the second year (situated on leafless part of twig)
5 Involucral scales all closely appressed
6 Leaves 10–15 cm wide, lobed not more than half-way to midrib; involucre 18–25 mm wide **1. rubra**
6 Leaves 5–10 cm wide, lobed more than half-way to midrib; involucre 10–15 mm wide **2. palustris**
5 At least some involucral scales patent or deflexed
7 Mature leaves glabrous and shining; petiole 2–5 mm **6. trojana**
7 Mature leaves pubescent or scabrid, dull; petiole usually at least 10 mm
8 Teeth or lobes of leaf aristate; upper surface of leaf smooth **7. macrolepis**
8 Teeth or lobes of leaf obtuse or slightly mucronate; upper surface of leaf rough **8. cerris**
4 Fruits ripening in the first year (situated among the leaves)
9 Scales of fruiting involucre concrescent except for their small, triangular apices; fruits usually distant, on a fairly long peduncle
10 Leaves with 8–14 pairs of straight, parallel lateral veins; intercalary veins absent **13. hartwissiana**
10 Leaves with 5–9 pairs of often rather irregular lateral veins; intercalary veins usually present
11 Leaves glabrous; lateral veins mostly straight **14. robur**
11 Leaves greyish-puberulent beneath; lateral veins arising at an acute angle with the midrib, but then curving outwards towards margin **15. pedunculiflora**
9 Scales of fruiting involucre distinct, not concrescent; fruits usually crowded; peduncle short or almost absent
12 Leaves semi-evergreen
13 Low shrub, seldom more than 50 cm; leaves subsessile **22. lusitanica**
13 Tree or tall shrub; petiole at least 4 mm
14 Young twigs and leaves with abundant, floccose, caducous hairs **20. canariensis**
14 Indumentum of young twigs and leaves variable, but not floccose and caducous **21. faginea**
12 Leaves deciduous
15 Young twigs tomentose; petiole not grooved
16 Leaves with usually more than 8 pairs of parallel lateral veins; intercalary veins ±absent; indumentum brownish **16. frainetto**
16 Leaves with usually fewer than 8 pairs of lateral veins, which are not strictly parallel, and often mixed with intercalary veins; indumentum white or grey
17 Scales of involucre ovate-lanceolate, ±acute, appressed **19. pubescens**
17 Scales of involucre narrowly lanceolate, obtuse, not appressed
18 Leaves densely white-pubescent beneath **17. pyrenaica**
18 Leaves greyish-green beneath **18. congesta**
15 Young twigs glabrous, or sericeous; petiole grooved
19 Leaves with *c.* 10 pairs of lateral veins; intercalary veins absent; fruiting peduncles with appressed hairs **9. mas**
19 Leaves with 6–8 pairs of lateral veins; intercalary veins usually present, at least near base of leaf; fruiting peduncles glabrous
20 Scales of involucre not tuberculate; mature leaves pubescent beneath, at least in the vein-axils **11. petraea**
20 Scales of involucre strongly tuberculate; mature leaves glabrous beneath
21 Leaves somewhat coriaceous, regularly sinuate; involucre brownish **10. polycarpa**
21 Leaves thin, deeply and irregularly lobed; involucre greyish **12. dalechampii**

Subgen. **Erythrobalanus** (Spach) Örsted. Leaves deciduous (in species described below). Fruit ripening in second year; endocarp tomentose.

1. Q. rubra L., *Sp. Pl.* 996 (1753) (*Q. borealis* Michx). Deciduous tree up to 25 m; twigs glabrous, not pendent, dark red; buds 6 mm, reddish-brown. Leaves 12–20 × 10–15 cm, ovate to obovate, lobed about half-way to midrib, with 7–11 pairs of lobes, each with 1–3 aristate teeth, glabrous except for slight pubescence in vein-axils beneath; petiole 25–50 mm. Involucre of fruit shallow, 18–25 mm wide; scales thin, ovate, closely appressed, finely pubescent. *Planted for timber and shelter, especially in C. Europe, sometimes naturalized.* [Au Be Cz Da Ga Ge Gr He Ho Hs Hu It Lu Po Rm Rs(W, K) Su.] (*E. North America.*)

2. Q. palustris Muenchh., *Hausv.* **5**: 253 (1770). Like **1** but the twigs somewhat pendent; buds 3 mm, pale brown; leaves 10–15 × 5–10 cm, obovate, more deeply lobed, with jaggedly toothed lobes; tufts of hairs in vein-axils conspicuous; involucre 10–15 mm wide. *Planted for timber, mainly in E.C. Europe, but much less commonly than* **1**. [Au Da Ge Hu Rm.] (*E. North America.*)

Subgen. **Sclerophyllodrys** O. Schwarz. Leaves evergreen, thick in texture. Fruit ripening in first or second year; endocarp tomentose.

3. Q. coccifera L., *Sp. Pl.* 995 (1753) (*Q. mesto* Boiss., *Q. pseudococcifera* Webb; incl. *Q. calliprinos* Webb). Evergreen shrub up to 3 m (rarely small tree); young twigs puberulent. Leaves 1·5–4 cm, thin but rigid, broadly ovate or oblong, with cordate or rounded base, spinose-dentate, undulate, dark green with prominent veins above, pale green and smooth beneath, glabrous at maturity; petiole 1–4 mm. Fruit ripening in the second year; scales of involucre rigid, subspinous, usually patent. $2n = 24$. *Mediterranean region (but absent from much of Italy); Portugal.* Al Bl Bu Cr Ga Gr Hs It Ju Lu Sa Si Tu.

4. Q. ilex L., *Sp. Pl.* 995 (1753) (*Q. avellaniformis* Colmeiro & E. Boutelou, *Q. gracilis* Lange, *Q. smilax* L.). Evergreen tree up to 25 m, or shrub; twigs and buds grey-tomentose. Leaves 3–7 cm, thick but not rigid, with cuneate or rounded base, entire, mucronate-dentate or spinose-serrate, glabrescent and smooth above, grey-tomentose and with prominent veins beneath; midrib straight; petiole 6–15 mm. Stipules subulate. Perianth-lobes lanceolate; anthers mucronate. Fruit ripening in the first year; involucral scales appressed, flat. *Mediterranean region, extending to N. Spain and W. France. Planted elsewhere in W. & S. Europe, and sometimes naturalized.* Al Bl Co Cr Ga Gr Hs It Ju Lu Sa Si [Br He Rs(K)].

(a) Subsp. **ilex**: Leaves oblong-ovate to lanceolate, dark green above, with 7–11 pairs of lateral veins. Stipules thick, densely hairy. Perianth-lobes subacute; fruit bitter. $2n = 24$. *Throughout the range of the species except Portugal and most of Spain.*

(b) Subsp. **rotundifolia** (Lam.) T. Morais, *Bol. Soc. Brot.*, ser. 2, **14**: 122 (1940) (*Q. rotundifolia* Lam., *Q. ballota* Desf.): Leaves suborbicular or broadly ovate, greyish-glaucous above, with 5–8 pairs of lateral veins. Stipules membranous, wider, glabrescent. Perianth-lobes obtuse; fruit sweet; involucral scales shorter and thicker. $2n = 24$. *Spain and Portugal.*

Subgen. **Cerris** (Spach) Örsted. Leaves evergreen or deciduous. Fruit usually ripening in second year; endocarp glabrous.

5. Q. suber L., *Sp. Pl.* 995 (1753) (incl. *Q. occidentalis* Gay). Evergreen tree up to 20 m, with thick, corky bark; twigs tomentose. Leaves 3–7 cm, ovate-oblong, sinuate-dentate, dark green above, grey-tomentose beneath; midrib sinuous; petioles 8–15 mm. Fruit ripening in the first year in spring-flowering trees, but some trees flower in autumn and ripen their fruits late in the following summer. Upper involucral scales long and patent, the lower usually shorter and more appressed. $2n = 24$. *Usually calcifuge. S. Europe, from Portugal to S.E. Italy.* Co Ga Hs It Lu Sa Si.

The bark is the cork of commerce.

6. Q. trojana Webb in Loudon, *Gard. Mag.* (*Loudon*) **15**: 590 (1839) (*Q. macedonica* DC.). Semi-evergreen or deciduous tree up to 15 m; young twigs puberulent. Leaves 3–7(–10) cm, obovate-oblong, sinuate-dentate with 8–14 pairs of mucronate or subaristate teeth, shining and subglabrous on both surfaces when mature; petiole 2–5 mm. Fruit ripening in the second year; involucre with lowest scales appressed, the middle ones deflexed and the upper erect or incurved. *Balkan peninsula (mainly in the west); S.E. Italy.* Al Bu Gr It Ju Tu.

7. Q. macrolepis Kotschy, *Eichen* t. 16 (1862) (*Q. aegilops* auct.). Semi-evergreen tree; twigs tomentose. Leaves 6–10 cm, tomentose beneath, smooth above, acute, subcordate, with 3–7 pairs of large, triangular, aristate lobes, often secondarily lobed; petiole 15–40 mm. Fruit ripening in the second year; involucre large and woody, with wide, thick, flat, usually patent scales. *S. part of Balkan peninsula; Aegean region; S.E. Italy.* Al *Cr Gr It Tu.

Formerly widely cultivated for use in tanning, which has obscured the native distribution.

8. Q. cerris L., *Sp. Pl.* 997 (1753). Deciduous tree up to 35 m; twigs rough, more or less hairy; buds surrounded by linear, persistent stipules. Leaves oblong to obovate, rounded or subcordate at the base, with 4–7 pairs of lobes or teeth, subglabrous but dull and slightly scabrid above, pubescent or subglabrous beneath; petiole 8–15 mm. Fruit ripening in the second year; scales of involucre subulate, patent. *S. & S.C. Europe westwards to S.E. France; naturalized in N.W. Europe.* Al Au Bu Cz Ga *Ge Gr He Hu It Ju Rm Si Tu [Br].

Q. crenata Lam., *Encycl. Méth. Bot.* **1**: 724 (1785) (*Q. pseudosuber* G. Santi), which is found occasionally in S. Europe from S.E. France to N.W. Jugoslavia and Sicilia, is probably a hybrid between **8** and **5**. It is like **7** but with smaller, mucronate lobes to the leaf, and smaller involucre, with narrower and softer scales.

Subgen. **Quercus** (Subgen. *Lepidobalanus* (Endl.) Örsted). Leaves deciduous or semi-evergreen. Fruit ripening in the first year; endocarp glabrous.

9. Q. mas Thore, *Essai Chlor. Land.* 381 (1803). Deciduous tree; twigs, buds and lower surface of leaves at first silky-hairy, later more or less glabrous. Leaves 8–18 cm, oblong-ovate, cuneate at the base, sinuately lobed, with 8–10 pairs of rather narrow, forwardly-directed lobes; no intercalary veins; petiole 15–25 mm. Fruiting peduncle silky-hairy; scales of involucre broadly ovate, obtuse, asymmetrically tuberculate. ● *N. Spain.* ?Ga Hs.

10. Q. polycarpa Schur, *Verh. Mitt. Siebenb.* **1851**: 170 (1851) (*Q. dshorochensis* auct., non C. Koch). Small deciduous tree; twigs, buds and mature leaves glabrous. Leaves 6–10 cm, somewhat coriaceous, elliptical, sinuate, with 7–10 pairs of equal, shallow obtuse lobes; a few intercalary veins sometimes present near the base; petiole 15–35 mm. Scales of involucre broadly ovate, acute, strongly tuberculate, puberulent, brownish. *S.E. Europe, extending northwards to C. Czechoslovakia.* Au Bu Cz Gr Hu Ju Rm Tu.

11. Q. petraea (Mattuschka) Liebl., *Fl. Fuld.* 403 (1784) (*Q. sessiliflora* Salisb., *Q. sessilis* Ehrh., *Q. dshorochensis* C. Koch). Deciduous tree up to 40 m; twigs glabrous. Leaves 7–12 cm, obovate, cuneate at the base, very finely appressed-hairy beneath, and with persistent tufts of brownish hairs in the vein-axils, sinuately lobed with 5–8 pairs of rounded lobes; intercalary veins often present; petiole 18–25 mm. Scales of involucre ovate-lanceolate, thin, not tuberculate, pubescent. $2n = 24$. *Forming often pure woods, usually on somewhat poor soils. W., C. & S.E. Europe, southwards to C. Spain and S. Albania, northwards to 60° N. in Norway.* Al Au Be Br Bu Co Cz Da Ga Ge Hb He Ho Hs Hu It Ju No Po Rm Rs(B, C, W, K, E) Si Su Tu.

12. Q. dalechampii Ten., *Ind. Sem. Horti Neap.* 15 (1830). Deciduous tree; twigs and mature leaves glabrous. Leaves thin, oblong to ovate-lanceolate, pinnately lobed with 4–7 pairs of rather narrow, subacute lobes; petiole 15–30 mm. Scales of involucre subrhombic, tuberculate, shortly pubescent, greyish. *S.E. Europe, extending to S.E. Czechoslovakia and W. Italy.* Au Bu Cz Gr Hu It Ju Rm ?Si.

13. Q. hartwissiana Steven, *Bull. Soc. Nat. Moscou* **30**(1): 387 (1857). Small deciduous tree up to 10 m; twigs glabrous, reddish. Leaves broadly obovate to oblong, with asymmetrically subcordate base, sinuate-dentate with 7–10 pairs of teeth, dark, shining green and glabrous above, sparsely brownish-pubescent beneath; lateral veins straight and parallel, without intercalary veins; petiole 13–22 mm. Peduncle long, with distant fruits; involucre not more than 12 × 12 mm, thin-walled; scales thin, concrescent except for the scarious apices. *S.E. Bulgaria and Turkey.* Bu Tu. (*N. Anatolia and W. Caucasus.*)

14. Q. robur L., *Sp. Pl.* 996 (1753) (*Q. pedunculata* Ehrh.). Deciduous tree up to 45 m, glabrous except sometimes for pubescence on young twigs and lower surface of young leaves. Leaves usually obovate, with 5–7 pairs of lobes; lateral veins making a variable angle with the midrib, with several intercalary veins; petiole 5 mm or less. Peduncle usually long; scales of involucre flat, puberulent, concrescent except for the apices. *Native and dominant in woods on a variety of soils, especially 'brown earth' soils. Most of Europe, except much of the north-east and parts of the Mediterranean region.* All except Az Bl Cr Fa Is Sa Sb Tu.

(a) Subsp. **robur** (incl. *Q. robur* subsp. *broteroana* O. Schwarz, *Q. estremadurensis* O. Schwarz): Leaves thin, usually glabrous, but occasionally puberulent beneath when young; lobes usually broad and deep. Involucre variable in size, usually *c.* 12 mm wide; scales grey-green, concrescent except for a very small apex. $2n = 24$. *Throughout the range of the species.*

(b) Subsp. **brutia** (Ten.) O. Schwarz, *Notizbl. Bot. Gart. Berlin* **13**: 13 (1936) (*Q. brutia* Ten.): Twigs and lower surface of leaves pubescent when young. Leaves rather coriaceous, with long lobes and deep, narrow sinuses. Involucre up to 23 mm wide, thick and woody; scales largely concrescent, but with patent, acute apices. ● *W. part of Balkan peninsula; S. Italy.*

In Britain and Ireland, and perhaps elsewhere in N.W. Europe, there has been extensive introgressive hybridization between **14(a)** and **11**.

15. Q. pedunculiflora C. Koch, *Linnaea* **22**: 324 (1849) (*Q. haas* auct. eur., non Kotschy). Like **14(b)** but leaves somewhat glaucous above and with persistent yellow-grey tomentum beneath; petiole up to 20 mm; involucral scales verrucose, more concrescent, and pubescent with yellowish hairs. *S.E. Europe.* Bu Gr Ju Rm ?Rs(K) Tu. (*E. Anatolia, S.W. Caucasus.*)

16. Q. frainetto Ten., *Prodr. Fl. Nap.* Suppl. 2: lxxii (1813) (*Q. farnetto* Ten., *Q. conferta* Kit.). Deciduous tree up to 30 m; twigs tomentose; buds large, surrounded by persistent stipules. Leaves 10–20 cm, crowded towards the apex of twig, obovate, tapered to the auricled base, deeply pinnatifid with 7–9 pairs of oblong, often lobed segments, pubescent beneath with grey or brownish hairs; lateral veins parallel, with few intercalary veins; petiole 2–6 mm. Involucre 6–12 × 12–15 mm; scales oblong, obtuse, pubescent, loosely imbricate. *Balkan peninsula, extending northwards to N.W. Romania; S. & C. Italy.* Al Bu *Cz Gr *Hu It Ju Rm Tu.

17. Q. pyrenaica Willd., *Sp. Pl.* **4(1)**: 451 (1805) (*Q. toza* Bast.). Deciduous tree up to 20 m, with numerous suckers; twigs tomentose, pendent. Leaves 8–16 cm, obovate or broadly oblong, deeply pinnatifid with 4–8 pairs of narrow, acute segments, more or less glabrous above, densely white-pubescent beneath; petiole up to 22 mm. Involucre usually not more than 15 × 14 mm; scales narrowly lanceolate, obtuse, loosely imbricate. $2n = 24$. *W. Europe northwards to N.W. France.* Ga Hs Lu.

18. Q. congesta C. Presl in J. & C. Presl, *Del. Prag.* 32 (1822). Deciduous tree or tall shrub; twigs tomentose. Leaves up to 14 cm, ovate to obovate-oblong, sinuate-lobed or pinnatifid, with 6–8 pairs of lobes, grey-green beneath with variably persistent pubescence; intercalary veins present. Peduncles up to 4 cm. Scales of involucre linear-lanceolate, erect but not appressed. ● *S.W. Italy and Sicilia; Sardegna.* It Sa Si.

19. Q. pubescens Willd., *Berlin. Baumz.* 279 (1796) (*Q. lanuginosa* Thuill.; incl. *Q. infectoria* Olivier, *Q. virgiliana* (Ten.) Ten., *Q. brachyphylla* Kotschy, *Q. apennina* auct.). Deciduous tree up to 25 m, or shrub; twigs densely tomentose or pubescent. Leaves 4–12(–16) cm, sinuate to pinnatifid, densely tomentose beneath when young, sometimes glabrescent; petiole 12–15 mm. Fruits subsessile or shortly stalked; involucre up to 15(–20) × 14 mm; scales lanceolate, pubescent, appressed. *W., C. & S. Europe, extending eastwards to Krym.* Al Au Be Bu Co Cr Cz Ga Ge Gr He Hs Hu It Ju *Po Rm Rs(W, K) Sa Si Tu.

A very variable species in need of further study. The following subspecies can be recognized.

1 Leaves very variable in size and toothing; involucre ± obconical (b) subsp. **anatolica**
1 Leaves ± uniform; involucre hemispherical
 2 Leaves usually at least 6 cm; scales of involucre subequal (a) subsp. **pubescens**
 2 Leaves rarely more than 6 cm; scales of involucre unequal (c) subsp. **palensis**

(a) Subsp. **pubescens**: Tree up to 25 m, or shrub 3–4 m. Leaves 6–12(–16) cm, pinnately lobed. Involucre hemispherical; scales subequal, more or less obtuse. $2n = 24$. *Throughout the range of the species, except Spain and Pyrenees.*

(b) Subsp. **anatolica** O. Schwarz, *Feddes Repert.* **33**: 336 (1934) (*Q. crispata* Steven): Shrub or small tree up to 6 m. Leaves 3–6 cm, variable in outline from subentire to deeply pinnatifid, glabrescent. Involucre more or less obconical; scales subequal, attenuate, apex acute. *E. part of Balkan peninsula; Krym.*

(c) Subsp. **palensis** (Palassou) O. Schwarz, *Notizbl. Bot. Gart. Berlin* **13**: 16 (1936): Shrub or tree. Leaves 4–7 cm, sinuate-dentate or shallowly lobed, persistently tomentose beneath. Involucre hemispherical; scales unequal, the lower short, thickened and more or less fused; the upper longer, free, cuspidate. ● *Pyrenees; N.E. Spain.*

Q. sicula Borzi in Lojac., *Fl. Sic.* **2(2)**: 374 (1907) (*Q. borzii* A. Camus), from Sicilia, is said to have a larger cupule and fruit 30–35 × 22–28 mm, but is otherwise similar to **19**. It has not been reported recently.

Q. cerrioides Willk. & Costa, *Linnaea* **30**: 123 (1859), from E. Spain, with toothed, ovate leaves usually more than 6 cm, is probably derived from **19** by hybridization with **21**.

20. Q. canariensis Willd., *Enum. Pl. Horti Berol.* 975 (1809). Semi-evergreen tree up to 30 m; twigs and young leaves densely floccose-tomentose, eventually glabrescent. Leaves 6–18 cm, elliptic-ovate, with 9–14 pairs of subacute teeth, glaucous beneath; lateral veins parallel; petiole 8–30 mm. Peduncle short; lower scales of involucre ovate-lanceolate, tuberculate, the upper much smaller, loosely appressed. *Spain and S. Portugal.* Hs Lu.

In N. Spain extensively hybridized with **19**.

21. Q. faginea Lam., *Encycl. Méth. Bot.* **1**: 725 (1785) (*Q. lusitanica* auct., non Lam.; incl. *Q. valentina* Cav., *Q. alpestris* Boiss.). Semi-evergreen shrub or tree up to 20 m. Leaves 4–10 cm, ovate, elliptical or obovate-oblong, sinuate-dentate with 5–12 pairs of more or less forwardly-directed triangular teeth, glabrescent and somewhat shining above, usually more or less tomentose beneath; petiole 4–20 mm. Involucre up to 25 × 12 mm; scales all broadly lanceolate or ovate. $2n = 24$. *Spain and Portugal.* Hs Lu.

22. Q. lusitanica Lam., *Encycl. Méth. Bot.* **1**: 719 (1785)

(*Q. fruticosa* Brot., *Q. humilis* Lam.). Semi-evergreen, stoloniferous shrub up to 2 m, but usually much less; twigs purplish-brown, shining. Leaves 3–5 cm, coriaceous, glabrescent, obovate-oblong with cuneate, entire base and 4–6 pairs of forwardly-directed teeth in the distal part; petiole very short. Involucre 6–12 × 8–12 mm. $2n = 24$. *Portugal and S.W. Spain.* Hs Lu.

URTICALES

37. ULMACEAE

EDIT. †T. G. TUTIN AND S. M. WALTERS

Deciduous trees, rarely shrubs, without latex. Leaves alternate, simple, usually asymmetrical at base. Flowers all hermaphrodite or male and hermaphrodite. Perianth herbaceous; stamens erect in bud. Fruit a drupe or samara.

1 Bark fissured; leaves 2-serrate; flowers all hermaphrodite; fruit a samara **1. Ulmus**
1 Bark not fissured; leaves simply serrate; flowers male and hermaphrodite; fruit a drupe
 2 Bark scaling; perianth-segments connate; drupe dry **2. Zelkova**
 2 Bark not scaling; perianth-segments free; drupe fleshy **3. Celtis**

1. Ulmus L.

†T. G. TUTIN

Bark fissured. Flowers all hermaphrodite, appearing before the leaves on the previous year's growth. Perianth-segments connate. Anthers purplish-red. Fruit a samara, winged all round but the wing emarginate at the apex.

There is considerable intraspecific variation in habit and leaf-shape which, combined with frequent vegetative propagation in **3** and **4**, often gives rise to more or less distinctive local populations, such as the taxa known as *U. angustifolia* (Weston) Weston, *U. coritana* Melville, and *U. plotii* Druce. Some of these have been given specific rank but intermediates between them, presumably due to hybridization, occur freely, and they are probably best treated as varieties. In addition, trees presumed to be hybrids between **1** and **4** (*U. × hollandica* Miller) are commonly found. Species and putative hybrids, both often represented by selected clones, are frequently planted as ornamental trees.

During recent years populations of *Ulmus* in W. Europe have been devastated by a disease caused by the fungus *Ophiostoma ulmi* (*Ceratocystis ulmi*).

Literature: R. H. Richens, *Elm*. Cambridge, 1983.

1 Pedicels 3–6 times as long as flowers; fruits ciliate, pendulous on long pedicels **6. laevis**
1 Pedicels shorter than flowers; fruits not ciliate, subsessile
 2 Base of longer side of leaf forming a rounded lobe ± overlapping and concealing the short petiole; upper surface of lamina scabrid; seed central in the fruit
  3 Fruit glabrous **1. glabra**
  3 Fruit pubescent in the middle **2. elliptica**
 2 Base of longer side of leaf not overlapping or concealing the petiole; seed distinctly above the middle of the fruit
  4 Leaves suborbicular, scabrid to glabrescent above, with base of long side rounded; fruit orbicular **3. procera**
  4 Leaves obovate to oblanceolate, usually smooth above, with base of long side making a 90° turn into the petiole; fruit narrowly to broadly obovate
   5 Young twigs glabrous or sparsely pubescent; leaves ± glabrous beneath, serrate **4. minor**
   5 Young twigs densely white-pubescent; leaves densely grey-pubescent beneath, crenate-serrate **5. canescens**

1. U. glabra Hudson, *Fl. Angl.* 95 (1762) (*U. montana* With., *U. scabra* Miller). Tree up to 40 m, without suckers. Twigs stout, hispid when young. Leaves suborbicular or broadly obovate to elliptical; lateral veins 12–18 pairs. Fruit 15–20 mm; seed central. $2n = 28$. *Most of Europe, but rare in the north-east and much of the Mediterranean region.* Al Au Be Br Bu Co Cz Da Fe Ga Ge Gr Hb He Ho Hs Hu It Ju No Po Rm Rs(N, B, C, W, K, E) Su.

2. U. elliptica C. Koch, *Linnaea* **22**: 599 (1849). Like **1** but leaves broader, ciliate; bud-scales with ferruginous hairs; fruit pubescent in the middle. *Ukraine, Krym; planted elsewhere.* Rs(W, K).

3. U. procera Salisb., *Prodr.* 391 (1796). Tree up to *c.* 30 m, suckering. Twigs rather stout, persistently pubescent. Leaves suborbicular to ovate; lateral veins 10–12 pairs. Fruit 10–17 mm; seed above the middle. ● *W. & S. Europe (distribution imperfectly known).* *Br Bu Ga Gr Hs Hu Ju Rm [Az Hb].

4. U. minor Miller, *Gard. Dict.* ed. 8, no. 6 (1768) (*U. carpinifolia* G. Suckow, *U. campestris* auct., non L., *U. diversifolia* Melville, *U. foliacea* sensu Hayek, *U. stricta* (Aiton) Lindley, *U. glabra* Miller, non Hudson). Tree up to 30(–40) m, suckering. Twigs usually slender, glabrous. Leaves obovate, ovate or oblanceolate, serrate; lateral veins 7–12 pairs. Fruit 7–17 mm; seed above the middle. $2n = 28$. *Most of Europe, from England, S. Sweden and C. Ural southwards.* Al Au Be *Br Bu Co Cr Cz Da Ga Ge Gr He Ho Hs Hu It Ju Lu Po Rm Rs(B, C, W, K, E) Su Tu [Bl].

5. U. canescens Melville, *Kew Bull.* **1957**: 499 (1958) (*U. minor* subsp. *canescens* (Melville) Browicz & Zieliński). Like **4** but the twigs in the first year densely and softly white-pubescent; leaves ovate-elliptical, crenate-serrate, densely greyish-pubescent; lateral veins 12–16(–18) pairs. *C. & E. Mediterranean region.* Cr Gr It Ju Sa Si [Rm].

6. U. laevis Pallas, *Fl. Ross.* **1(1)**: 75 (1784) (*U. effusa* Willd., *U. pedunculata* Foug.). Tree up to 35 m. Twigs softly pubescent or glabrous. Leaves suborbicular to ovate, glabrous or softly pubescent beneath; lateral veins 12–19 pairs. Fruit 10–12 mm; seed central. $2n = 28$. *C., E. & S.E. Europe, extending westwards to C. France and northwards to S.E. Sweden (Öland).* Al Au Be Bu Co Cz Fe Ga Ge ?Gr Hu Ju Po Rm Rs(N, B, C, W, E) Su Tu [He It].

2. Zelkova Spach

†T. G. TUTIN

Bark smooth, scaling. Flowers male and hermaphrodite, appearing with the leaves on the young growth. Perianth-segments connate. Anthers yellow. Drupe dry.

1. Z. abelicea (Lam.) Boiss., *Fl. Or.* **4**: 1159 (1879) (*Z. cretica* (Sm.) Spach). Shrub 3–5 m. Twigs slender, hispid. Leaves up to *c.* 2·5 cm, subsessile, ovate, obtuse, coarsely 7- or 9-dentate. Flowers white, scented. Drupe pubescent. *Rocky mountain slopes.* ● *Kriti.* Cr.

3. Celtis L.

†T. G. TUTIN

Bark smooth, not scaling. Flowers male and hermaphrodite, appearing with the leaves on the young growth. Perianth-segments free. Anthers yellow. Drupe fleshy.

1 Leaves entirely glabrous, even when young; endocarp weakly reticulate **3. glabrata**
1 Leaves pubescent beneath, at least when young
 2 Leaves crenate or with broad, subobtuse teeth; endocarp with 4 ridges **4. tournefortii**
 2 Leaves sharply serrate; endocarp reticulate
  3 Leaves rounded or cordate at base; endocarp strongly reticulate-rugose **1. australis**
  3 Leaves cuneate at base; endocarp weakly reticulate **2. caucasica**

1. C. australis L., *Sp. Pl.* 1043 (1753). Tree up to 25 m, with grey bark. Leaves 4–15 × 1·5–6 cm, lanceolate to ovate-lanceolate, serrate, long-acuminate, rounded or cordate at base, scabrid above, pubescent beneath. Peduncle up to 3·5 cm. Drupe 9–12 mm in diameter, globose, glabrous, brownish-black when ripe; endocarp strongly reticulate-rugose. *S. Europe.* Al Bl Bu Co Ga Gr Hs It Ju Lu Rm Sa Si [He].

2. C. caucasica Willd., *Sp. Pl.* **4(2)**: 994 (1806). Like **1** but usually smaller; leaves rhombic-ovate, cuneate at base, smooth above; peduncle up to 1·7 cm; drupe reddish-brown when ripe; endocarp weakly reticulate. *E. Bulgaria; one station in S. Jugoslavia.* Bu Ju. (*W. Asia.*)

3. C. glabrata Steven ex Planchon, *Ann. Sci. Nat.* ser. 3, **10**: 285 (1848). Small tree or shrub 3–4 m. Leaves 4–6 × *c.* 4 cm, rhombic-ovate, cuspidate, serrate, cuneate at base, entirely glabrous, yellowish-green. Drupe globose; endocarp weakly reticulate. *S. Krym.* Rs(K).

4. C. tournefortii Lam., *Encycl. Méth. Bot.* **4**: 138 (1797) (incl. *C. aspera* (Ledeb.) Steven, *C. aetnensis* Strobl). Shrub or small tree 1–6 m. Leaves 5–7 × 2·5–4 cm, ovate, crenate or with broad subobtuse teeth, subcordate at base, scabrid to almost smooth on the upper surface, pubescent beneath, dark green. Drupe obovoid, brownish-yellow when ripe; endocarp with 4 ridges. *S. Europe from Sicilia to Krym.* Cr Gr Ju Rs(K) Si.

38. MORACEAE

EDIT. †T. G. TUTIN AND S. M. WALTERS

Trees or shrubs with latex. Monoecious or dioecious. Leaves alternate (rarely opposite); stipules caducous. Flowers usually densely spicate, capitate or enclosed in a fleshy, urceolate receptacle. Flowers with 3- or 4-partite perianth. Male flowers with 3 or 4 stamens opposite the perianth-segments. Female flowers with ovary 1- or 2-locular, superior; styles 1(2). Ovule anatropous or campylotropous. Fruit a drupelet often surrounded by the fleshy perianth; drupelets often crowded into syncarps.

1 Leaves entire; branches spiny **3. Maclura**
1 Leaves dentate or lobed; branches not spiny
 2 Stipule-scar encircling the stem; stamens straight in bud; fruits borne on the inside of a fleshy pyriform structure **4. Ficus**
 2 Stipule-scar not encircling the stem; stamens inflexed in bud; fruits borne on a short axis
  3 Buds with 3–6 scales; syncarp cylindrical or ovoid **2. Morus**
  3 Buds with 2 or 3 scales; syncarp globose **1. Broussonetia**

1. Broussonetia L'Hér. ex Vent.

†T. G. TUTIN

Not spiny; leaves toothed or lobed; buds with 2 or 3 scales. Dioecious. Male flowers in catkin-like inflorescences; perianth 4-partite; stamens 4, inflexed in bud. Female flowers in dense, globose, tomentose heads; perianth denticulate, with 4 very small teeth, forming a layer of pulp in fruit.

1. B. papyrifera (L.) Vent., *Tabl. Règne Végét.* **3**: 548 (1799). Small tree; young twigs villous. Leaves 7–20 cm, ovate, serrate, sometimes lobed, scabrid above, grey-tomentose beneath. Syncarps *c.* 2 cm in diameter, orange, with red drupelets. *Planted in S. Europe and sometimes naturalized.* [Ga Hs It Rm Sa Si.] (*E. Asia.*)

2. Morus L.

†T. G. TUTIN

Not spiny; leaves dentate or lobed; buds with 3–6 scales. Monoecious or dioecious. Flowers of both sexes in short, dense spikes. Perianth of male flowers 4-partite; stamens 4, inflexed in bud. Perianth of female flowers with 4 almost or quite free segments, becoming fleshy in fruit.

Leaves pubescent beneath; syncarp subsessile **1. nigra**
Leaves almost glabrous beneath; peduncle about as long as syncarp **2. alba**

1. M. nigra L., *Sp. Pl.* 986 (1753). Tree up to 10 m, with stout, rough branches. Leaves 6–20 cm, broadly ovate-cordate, dentate or lobed, scabrid above, pubescent beneath. Syncarp 2–2·5 cm, subsessile, dark purple, very acid until completely ripe. *Widely cultivated for its fruit (Mulberry), and locally naturalized in S. Europe.* [Al Bu Cr Gr Hs It Rm.] (*C. Asia.*)

M. rubra L., *loc. cit.* (1753), from E. & C. North America, like **1** but up to 20 m, the leaves truncate or subcordate at base, the peduncle about ½ as long as the slightly larger syncarp, is cultivated in Turkey-in-Europe.

2. M. alba L., *loc. cit.* (1753). Tree up to 15 m, with slender, smooth branches. Leaves 6–18 cm, ovate, rounded or cordate at base, dentate or lobed, usually smooth above, glabrous, or pubescent on the veins only beneath. Peduncle about as long as the syncarp. Syncarp 1–2·5 cm, white, pinkish or purplish, edible long before it is ripe. *Cultivated throughout S. Europe as food for silkworms and as a roadside tree, and frequently naturalized in S.E. Europe and occasionally elsewhere.* [Al Au Bu Cr Gr Hs Rm Rs(W, K, E).] (*China.*)

3. Maclura Nutt.

†T. G. TUTIN

Spiny; leaves entire. Dioecious. Perianth 4-partite. Male flowers pedicellate, cymose-paniculate; stamens 4, inflexed in bud. Female flowers in subglobose or somewhat elongated heads.

1. M. pomifera (Rafin.) C. K. Schneider, *Ill. Handb. Laubholzk.* 1: 806 (1906) (*M. aurantiaca* Nutt.). Tree. Leaves 5–12 cm, ovate, acuminate. Syncarp 10–14 cm in diameter, subglobose, orange. *Cultivated as a hedge plant and for ornament, and locally naturalized in S. Europe.* [It Rm Rs(K).] (*North America.*)

4. Ficus L.

†T. G. TUTIN

Not spiny; leaves usually lobed; buds enclosed in the connate stipules. Stipules caducous, leaving a circular scar. Monoecious. Flowers enclosed within a fleshy urceolate structure. Stamens 4, straight in bud.

Literature: I. J. Condit & J. Enderud, *Hilgardia* **25**: 1–663 (1956).

1. F. carica L., *Sp. Pl.* 1059 (1753). Spreading, deciduous shrub or small tree. Leaves 10–20 cm long and wide, usually palmately lobed, usually cordate at base, scabrid, sparsely hispid beneath. Syncarp 5–8 cm, pyriform, greenish- or brownish-violet. $2n = 26$. *Extensively cultivated and widely naturalized in S. Europe; perhaps native in the southern parts of the C. and E. Mediterranean region.* *Al *Bl *Co Cr *Ga Gr *Hs It *Ju Sa Si *Tu [Au Az Br Cz He Hu Lu Rs(K)]. (*S.W. Asia.*)

Many cultivars exist and the wild plant (var. *caprificus* Risso) also shows considerable variability. The syncarps (Figs) are eaten both fresh and dried.

F. elastica Roxb. ex Hornem., *Hort. Hafn. Suppl.* 7 (1819), with entire, glossy, evergreen leaves, is commonly grown as a house-plant and, in the Mediterranean region, as a street tree.

F. benghalensis L., *Sp. Pl.* 1059 (1753), with massive, spreading branches, an extensive system of prop-roots, and ovate, entire leaves, is grown as a shade tree in warmer parts of the Mediterranean region.

39. CANNABACEAE

EDIT. †T. G. TUTIN AND J. R. EDMONDSON

Herbs without latex. Usually dioecious. Male flowers pedicellate; perianth 5-partite; stamens erect in bud. Female flowers sessile; perianth undivided; fruit an achene enclosed in the persistent perianth.

Climber; female inflorescence cone-like — **1. Humulus**
Erect herb; female inflorescence not cone-like — **2. Cannabis**

1. Humulus L.

†T. G. TUTIN (EDITION 1) REVISED BY J. R. AKEROYD (EDITION 2)

Climbers. Inflorescences pendent, glandular; the male much-branched; the female cone-like.

Leaves 3- to 5-lobed; inflorescence distinctly enlarged in fruit — **1. lupulus**
Leaves 5- to 7-lobed; inflorescence scarcely enlarged in fruit — **2. scandens**

1. H. lupulus L., *Sp. Pl.* 1028 (1753). Stems up to 6 m, rough with deflexed hairs. Leaves opposite, broadly ovate-cordate, usually deeply 3- to 5-lobed and coarsely dentate; lobes acuminate, petiole shorter than to about as long as the lamina. Male flowers *c.* 5 mm in diameter. Female inflorescence 15–20 mm, enlarging in fruit; flowers subtended by persistent ovate, acute, pale green bracts. Cone-like infructescence *c.* 30 mm. $2n = 20$. *Most of Europe.* All except Az Bl Cr Fa Is Sb. Not native in Hb.

The infructescences (Hops) are widely used in brewing and the plant is consequently cultivated and often naturalized, so that its natural distribution is now obscured.

2. H. scandens (Loureiro) Merr., *Trans. Amer. Philos. Soc., nov. ser.*, **24**: 138 (1935) (*H. japonicus* Siebold & Zucc.). Like **1** but annual; stems rougher; leaves deeply and palmately 5- to 7-lobed, petiole longer than blade; inflorescence scarcely enlarging in fruit; bracts ovate-orbicular, cordate, cuspidate, green, *Cultivated for ornament, and naturalized in N. Italy, W. Hungary and perhaps elsewhere.* [Hu It.] (*E. Asia.*)

2. Cannabis L.

†T. G. TUTIN (EDITION 1) REVISED BY J. R. AKEROYD (EDITION 2)

Erect herb. Inflorescences erect, glandular; the male much-branched; the female racemose.

Literature: E. Small & A. Cronquist, *Taxon* **25**: 405–435 (1976).

1. C. sativa L., *Sp. Pl.* 1027 (1753). Stems up to 2·5 m. Leaves usually alternate, except the lower, 3- to 9-palmatisect; lobes lanceolate, acute, serrate. Nut smooth, not articulated at the base. $2n = 20$. *Native on alluvial sands in S.E. Russia; widely cultivated for its fibre (Hemp), oil and narcotic resin (Marijuana), and naturalized in a large part of Europe; a frequent casual elsewhere.* Rs(E) [Au Be Bu Co Cz Ga Ge Gr He Hs Hu It Ju Po Rm Rs(C) Sa Si]. (*S. & W. Asia.*)

A variable species, with a long history of domestication and selection by man. Var. *spontanea* Vavilov (*C. ruderalis* Janisch.), which is found as a weed of cultivated fields in S.E. Russia and Romania, differs in its smaller nut, with a marbled surface, distinctly articulated at the base and easily detached.

40. URTICACEAE

EDIT. J. R. EDMONDSON

Herbs, small shrubs or rarely climbers, without latex. Leaves opposite or alternate, simple, usually stipulate. Flowers usually unisexual. Perianth 4- or 5-merous, often persistent. Male flowers with (1–)4 or 5 stamens opposite the perianth-segments and inflexed in bud, often with rudimentary ovary. Female flowers often with small staminodes; ovary superior, 1-locular, sometimes adnate to perianth; style simple; ovule 1, orthotropous. Fruit an achene. Seeds usually with endosperm.

Boehmeria nivea (L.) Gaud.-Beaup. in Freyc., *Voy. Bot.* 499 (1830), from China, a herb or shrub 50–200 cm, with leaves 6–30 cm, alternate, stipulate and white-tomentose beneath, is cultivated in S. Europe for fibre and ornament, and is perhaps locally naturalized.

1 Leaves densely white-pubescent beneath; female flowers without perianth; male flowers with 1 stamen and 3- to 5-toothed cylindrical perianth **5. Forsskalea**
1 Leaves green beneath; female flowers with perianth enclosing the achene, male flowers with 4 stamens and deeply lobed perianth
2 Leaves opposite, variously toothed or lobed, rarely entire
3 Leaves 10 mm or more, usually serrate and with stinging hairs **1. Urtica**
3 Leaves up to 5 mm, entire, without stinging hairs **2. Pilea**
2 Leaves alternate, entire
4 Flowers solitary; stems very slender, creeping and rooting at the nodes **4. Soleirolia**
4 Flowers clustered, stems not creeping or rooting at the nodes **3. Parietaria**

1. Urtica L.

P. W. BALL (EDITION 1) REVISED BY D. V. GELTMAN (EDITION 2)

Annual or perennial herbs, with stinging hairs. Leaves opposite, usually variously toothed or deeply lobed; stipules 4, free or connate in pairs. Inflorescence axillary, spike-like, with clustered ultimate branches. Flowers unisexual, plants monoecious or dioecious. Perianth 4-merous. Female flowers with more or less unequal perianth-segments, the 2 larger enclosing the achene.

Literature: D. V. Geltman, *Bot. Žurn.* **71**: 1480–1490 (1986).

1 Leaves 3- to 5-lobed; lobes serrate **7. cannabina**
1 Leaves entire to serrate, but not lobed
2 Female flowers in long-pedunculate, globose lateral heads **11. pilulifera**
2 All flowers in spike-like lateral racemes (rarely in few-flowered clusters)
3 Plants with male and female flowers in the same raceme; petiole of lower leaves usually at least ½ as long as the lamina
4 Perennial; racemes 1–6 cm; perianth-segments of the female flowers subequal in fruit **1. atrovirens**
4 Annual; racemes not more than 2 cm; outer perianth-segments of the female flowers shorter than the 2 inner in fruit **9. urens**
3 Plants usually dioecious, or with male and female flowers in separate racemes; petiole of lower leaves not more than ½ as long as the lamina
5 Stipules connate in pairs at least to ¾; raceme-axis usually not branched, male with the flowers usually inserted unilaterally on inflated peduncles **10. membranacea**
5 Stipules free or connate at base only, raceme-axis usually branched; peduncles of the male racemes not inflated
6 Female racemes shorter than petiole of subtending leaf; patent or pendent; male racemes erecto-patent; perianth glabrous in fruit; leaves cuneate to sub-truncate at base **8. rupestris**
6 All racemes pendent or patent, longer than the petiole; perianth hispid in fruit; leaves truncate, rounded to subcordate at base
7 Fruit narrowly elliptical, 1·6–2 mm; stipules broadly triangular, upper connate at base **2. kioviensis**
7 Fruit ovate or ovate-elliptical, 1·1–1·3 mm; stipules oblong, always free **(3–6). dioica** group

1. U. atrovirens Req. ex Loisel., *Mém. Soc. Linn. Paris* **6**: 432 (1827). Perennial 30–100 cm, dull green; monoecious. Leaves 2–7 cm, lanceolate, ovate or suborbicular, cordate or cuneate at base, serrate; petiole about as long as the lamina; stipules 4 at each node. Racemes longer than the petiole, with both male and female flowers, patent or pendent in fruit. Female flowers with perianth-segments subequal, pubescent. $2n = 26$. ● *W. Mediterranean region.* Bl Co It Sa.

2. U. kioviensis Rogow., *Bull. Soc. Nat. Moscou* **16**: 324 (1843). Monoecious; leaves and stems bright green, glabrescent; stem up to 200 cm, often branched and rooting at the lower nodes; stipules green, broadly triangular, upper connate at base; perianth of female flowers more or less glabrous, the 2 outer segments ½–⅔ as long as the 2 inner. $2n$ = ?22, 26. *In aquatic vegetation or in very damp woods. C. & E. Europe.* Au Bu Cz Ge Hu Rm Rs(C, W, E).

(3–6). U. dioica group. Perennial 30–150(–250) cm, with long rhizome, usually dioecious. Leaves (1–)3–12(–15) cm, ovate to lanceolate, acuminate, rounded, obtuse or cordate at base, serrate; petiole not more than ½ as long as the lamina; stipules 4 at each node. Racemes up to 10 cm, usually exceeding the subtending petiole, patent or pendent in fruit. Female flowers pubescent or hispid, with outer perianth-segments less than ½ as long as the inner.

Four taxa probably meriting specific rank can be recognized, although sometimes intermediate forms may be found.

1 Lamina with stinging hairs, often also ±pilose **6. dioica**
1 Lamina without stinging hairs
2 Lamina and stem-internodes glabrous or nearly so **3. sondenii**
2 Lamina (at least beneath) and stem-internodes pubescent
3 Lamina sparsely pilose, often only on nerves beneath lowest node of inflorescence at the 7th–14th whorl **6. dioica**
3 Lamina (at least beneath) densely pubescent; lowest node of inflorescence at the 13th–20th whorl
4 Indumentum consisting of hairs 0·4–0·7 mm long;

upper internodes 1·4–2 times longer than subtending petiole, stipules 1–2 mm wide **4. galeopsifolia**

4 Indumentum consisting of hairs 0·25–0·3 mm long, upper internodes shorter to slightly longer than subtending petiole, stipules 3–4 mm wide **5. pubescens**

3. U. sondenii (Simmons) Avr. ex Geltman, *Nov. Syst. Pl. Vasc. (Leningrad)* **25**: 76 (1988) (*U. gracilis* auct. non Aiton). Lamina and stem-internodes glabrous, bright green to yellow-green, lanceolate to ovate-lanceolate, cuneate, rounded, more or less cordate at base. Stipules 1–2 mm wide. $2n = 26$. *By rivers and in damp woods. N. & E. Europe.* Fe No Rs(N, C, W, E) Su.

4. U. galeopsifolia Wierzb. ex Opiz, *Naturalientausch* **9**: 107 (1825). Lamina without stinging hairs, densely pubescent at least beneath, dark green, ovate to ovate-lanceolate, rounded, cordate or cuneate at base. Stipules 1–2 mm wide. Lowest node of inflorescence at the 13th–20th whorl. $2n = 26, 52$. *Damp habitats, especially* Alnus *woodlands, swampy river valleys. W., C. & E. Europe.* ?Au Br Bu Cz ?Ge ?Hb Ho Hu ?Po Rm Rs(C, B, W, E) ?Su.

5. U. pubescens Ledeb., *Fl. Altaica* **4**: 240 (1833). Like **4** but indumentum shorter, 0·25–0·3 mm long; upper internodes shorter or only slightly longer than the subtending petioles, marginate teeth deeper; stipules 3–4 mm wide. $2n = 26$. *Damp vegetation in Volga delta and lower Dnepr.* Rs(W, E).

6. U. dioica L., *Sp. Pl.* 984 (1753). Lamina usually with stinging hairs, often also more or less pilose, if not, only sparsely pilose, mainly on nerves, ovate to ovate-lanceolate, rounded, cuneate or cordate at base. Stipules 1–2(–3) mm wide. Lowest node of inflorescence at the 7th–14th whorl. $2n = 48, 52$, rarely 26. *Nitrophilous habitats, ruderal places, damp woods. Throughout Europe, but only as an introduced weed in some districts.* All except Az Bl Cr Sb.

7. U. cannabina L., *Sp. Pl.* 984 (1753). Stem up to 200 cm, with stinging hairs confined to the inflorescence; leaves 3- to 5-lobed, with serrate lobes; racemes erecto-patent. *Doubtfully naturalized in Russia and Ukraine; sometimes planted elsewhere.* [Rs(C, W, E).] (*C. & N. Asia.*)

8. U. rupestris Guss., *Cat. Pl. Boccad.* 83 (1821). Perennial 30–100 cm; usually dioecious. Leaves 2–10 cm, ovate-acuminate, cuneate or truncate at base, serrate; petiole not more than ½ as long as the lamina; stipules 4 at each node. Racemes unisexual; female not more than 2 cm, shorter than the subtending petiole, patent or pendent in fruit; male 2–8 cm, erecto-patent. Female flowers with subglabrous perianth-segments. ● *Sicilia.* Si.

U. morifolia Poiret in Lam., *Encycl. Méth. Bot.* **12**: 223 (1816), from Madeira and Islas Canarias, a woody perennial with small, long-pedunculate racemes, may be becoming naturalized in Açores.

9. U. urens L., *Sp. Pl.* 984 (1753). Annual 5–60(–80) cm, clear green; monoecious. Leaves 1–4(–6) cm, ovate, rounded to cuneate at base, usually deeply serrate; petiole as long as the lamina; stipules 4 at each node. Racemes not more than 2 cm, usually not longer than the subtending petiole, appearing as few-flowered axillary clusters, with numerous female and few male flowers, erecto-patent in fruit. Perianth-segments of female flowers ciliate on the margin, glabrous or sparsely hispid on the back. $2n = 24, 26, 52$. *Mainly on cultivated or disturbed soil. Throughout Europe, but doubtfully native in much of the north.* All except Az Fa Sb.

10. U. membranacea Poiret in Lam., *Encycl. Méth. Bot.* **4**: 638 (1798) (*U. dubia* Forskål, nom. illegit., *U. caudata* Vahl, non Burm. fil.). Annual 15–80 cm; monoecious. Leaves 2–6(–10) cm, ovate, subcordate at base, serrate; petiole almost as long as the lamina; stipules connate in pairs. Racemes unisexual, the lower female, shorter than the petiole, the upper male, longer than the petiole, erecto-patent; flowers inserted unilaterally on an inflated axis. $2n = 26$. *Mediterranean region and W. Europe, northwards to N. France; Açores.* Az Bl Co Cr Ga Gr Hs It Ju Lu Sa Si Tu.

11. U. pilulifera L., *Sp. Pl.* 983 (1753) (*U. dodartii* L.). Annual 30–100 cm; monoecious. Leaves 2–6 cm, ovate, truncate to subcordate at base, serrate or entire; petiole almost as long as the lamina; stipules 4 at each node. Racemes unisexual; female long-pedunculate with flowers in globose heads; male spicate. Female flowers with inflated perianth. $2n = 24$. *S. Europe; formerly a frequent casual further north, but now very rare.* Al Bl Bu Co Cr Ga Gr Hs It Ju Lu Rs(W, K) Sa Si Tu [Ge].

2. Pilea Lindley

P. W. BALL

Annual herbs without stinging hairs. Leaves opposite, entire, fleshy; stipules apparently absent. Inflorescence axillary, globose. Flowers uni- or bisexual; female flowers with 3-lobed perianth, 1 lobe larger, cucullate, enclosing the achene.

1. P. microphylla (L.) Liebm., *Danske Vid. Selsk.* ser. 5, **2**: 296 (1851). Stems up to 40 cm, procumbent, fleshy, much-branched, glabrous. Leaves 1·5–5 mm, obovate to elliptical, obtuse. Inflorescence contracted, globose, shorter than the subtending leaves; flowers minute. *Naturalized in the Balkan peninsula.* [Gr Ju.] (*Tropical America.*)

3. Parietaria L.

P. W. BALL

Annual or perennial herbs, sometimes woody at the base, without stinging hairs. Leaves alternate, entire; stipules absent. Flowers hermaphrodite or unisexual, in axillary, bracteate, 3- to many-flowered cymes, 1 or more in each leaf-axil. Perianth green, cylindrical and 4-toothed in female flowers, 4-partite in hermaphrodite and male flowers. Achenes enclosed in the perianth and sometimes in the bracts.

1 Bracts becoming brown, hard and connate, forming a 5-lobed involucre around the achene; leaves rarely more than 1 cm **7. cretica**

1 Bracts remaining herbaceous, free or slightly connate at the base; leaves often more than 1 cm

2 Bracts shorter than the perianth in fruit

3 Annual; stem sparsely pubescent; petiole of lower leaves usually equalling or longer than lamina; achenes brown or olive **3. mauritanica**

3 Perennial; stem usually densely pubescent; petiole of lower leaves shorter than lamina; achenes black

4 Erect, usually more than 30 cm; leaves 3–12 cm; bracts completely free **1. officinalis**

4 Procumbent or ascending, rarely more than 40 cm; leaves not more than 5 cm; bracts shortly connate at base **2. judaica**

2 Bracts equalling or exceeding the perianth in fruit

5 Perianth-lobes incurved and connivent at the apex in fruit; achenes symmetrical at the apex **4. lusitanica**
5 Perianth-lobes patent or erect, not connivent in fruit; achenes asymmetrically apiculate
6 Leaves ovate to ovate-orbicular; lobes of perianth ±patent **5. debilis**
6 Leaves lanceolate; lobes of perianth erect **6. pensylvanica**

1. P. officinalis L., *Sp. Pl.* 1052 (1753) (*P. erecta* Mert. & Koch). Perennial; stem 30–100 cm, erect, simple or slightly branched, densely pubescent. Leaves 3–12 cm, ovate-lanceolate or elliptical, long-acuminate; petiole shorter than the lamina. Bracts free, shorter than the perianth. Achenes black. $2n = 14$. *Damp, shady rocks and banks. C. & S.E. Europe, extending to Corse and C. France.* Al Au Co Cz Ga Ge Gr He *Ho Hu It Ju Rm Rs(W, K) Sa Si Tu [Be Da Po Su].

2. P. judaica L., *Fl. Palaest.* 32 (1756) (*P. officinalis* auct., non L., *P. diffusa* Mert. & Koch, *P. ramiflora* Moench). Like **1** but the stem usually not more than 40 cm, procumbent or ascending, much-branched; leaves up to 5 cm, ovate-acuminate; bracts shortly connate at base. $2n = 26$. *S.W. & W.C. Europe.* Al *Au Az Be Bl Br Bu Co Cr Ga Ge Gr Hb He Ho Hs It Ju Lu Rm Rs(K) Sa Si Tu.

3. P. mauritanica Durieu in Duchartre, *Rev. Bot.* **2**: 427 (1847). Annual, erect or diffuse; stem 5–40 cm, sparsely pubescent. Leaves up to 5 cm, ovate-acuminate; petioles of the lower leaves usually equalling or longer than the lamina. Bracts shorter than the perianth; perianth 2–3 mm in fruit. Achenes brown or olive. *S. half of Iberian peninsula; Islas Baleares; Sicilia.* Bl Hs Lu Si.

4. P. lusitanica L., *Sp. Pl.* 1052 (1753). Slender, diffuse annual 5–30 cm. Leaves up to 4 cm, broadly ovate-acuminate to ovate-orbicular. Bracts equalling or longer than the perianth; perianth not more than 1·5 mm, the lobes incurved at apex and connivent in fruit. Achenes brown or olive, the apex symmetrical. *Shady rocks and walls. S. & S.E. Europe.* Bl Bu Co Cr Ga Gr Hs It Ju Lu Rm Rs(W, K) Sa Si.

1 Leaves usually 2–4 cm; bracts scarcely accrescent (**b**) subsp. **serbica**
1 Leaves usually less than 2 cm; bracts conspicuously accrescent
2 Petiole shorter than the lamina; bracts linear (**a**) subsp. **lusitanica**
2 Petiole longer than the lamina; bracts ovate or ovate-lanceolate (**c**) subsp. **chersonensis**

(**a**) Subsp. **lusitanica**: Leaves not more than 2 cm; petiole shorter than the lamina. Bracts linear, conspicuously accrescent. $2n = 16$. *S. Europe.*

(**b**) Subsp. **serbica** (Pančić) P. W. Ball, *Feddes Repert.* **68**: 186 (1963) (*P. serbica* Pančić). Leaves (1–)2–4 cm; petiole about as long as the lamina. Bracts lanceolate, scarcely accrescent. ● *N. part of Balkan peninsula and Romania.*

(**c**) Subsp. **chersonensis** (A. F. Láng & Szov.) Chrtek in Rech. fil., *Fl. Iran.* **105**: 10 (1974) (*P. chersonensis* A. F. Láng & Szov.): Leaves usually less than 2 cm; petiole longer than the lamina. Bracts ovate to ovate-lanceolate, conspicuously accrescent. *Ukraine.*

5. P. debilis G. Forster, *Fl. Ins. Austral. Prodr.* 73 (1786) (*P. micrantha* Ledeb.). Like **4** but stem up to 50 cm, usually stouter; leaves up to 6 cm; perianth 1–2 mm in fruit, the lobes patent, not connivent; achenes asymmetrically apiculate. *S. Ural; naturalized in Açores.* Rs(C) [Az]. (*Widespread in the S. Hemisphere and in parts of Asia, N. Africa and Central America.*)

6. P. pensylvanica Muhl. ex Willd., *Sp. Pl.* **4(2)**: 955 (1806). Annual; stems 20–50 cm, erect, usually sparingly branched. Leaves 2–6 cm, lanceolate, acuminate. Bracts longer than the perianth; perianth 1·8–2·5 mm in fruit, the lobes erect, not connivent. Achenes pale brown, asymmetrically apiculate. *Shady places. Naturalized in Germany (Berlin).* [Ge.] (*North America.*)

7. P. cretica L., *Sp. Pl.* 1052 (1753). A diffuse or ascending annual or perennial up to 50 cm. Leaves up to 1 cm, ovate or elliptical. Bracts of each cyme becoming brown and connate, forming a 5-lobed involucre around the fruit. Achenes brown. *Greece and Aegean region; also on small islands W. & S. of Sicilia.* Cr Gr Si.

4. Soleirolia Gaud.-Beaup.

P. W. BALL

Like *Parietaria* but creeping and rooting at the nodes; flowers solitary, surrounded by an involucre of 1 bract and 2 bracteoles; achene enclosed by the perianth and involucre. (*Helxine* Req., non L.)

1. S. soleirolii (Req.) Dandy, *Feddes Repert.* **70**: 1 (1964) (*Helxine soleirolii* Req., *Parietaria soleirolii* (Req.) Sprengel, *S. corsica* Gaud.-Beaup.). Pubescent, very slender perennial 5–20 cm. Leaves 2–6 mm, suborbicular, 3-veined. Flowers unisexual, the lower female, the upper male. $2n = 20$. ● *Islands of W. Mediterranean region; often cultivated and naturalized in W. Europe.* *Bl Co It Sa [Az Br Ga Hb Ho Lu].

5. Forsskalea L.

P. W. BALL

Perennial herbs, woody at the base. Leaves alternate, crenate-dentate; stipules free. Inflorescence axillary, cymose, bracteate; flowers unisexual. Male flowers with tubular, 3- to 5-toothed perianth; stamen 1. Female flowers without perianth; ovary woolly.

1. F. tenacissima L., *Opobalsam. Decl.* 18 (1764) (*F. cossoniana* Webb). Hispid, ascending or diffuse, up to 60 cm. Leaves up to 15 mm, obovate to obovate-orbicular, crenate-dentate, cuneate at the base, densely white-pubescent beneath; stipules persistent, scarious. Cymes 1- to 5-flowered; bracts (2)3(4), up to 10 mm, longer than the cymes and forming an involucre, densely covered with long, silky hairs. *Sandy ground. S.E. Spain.* Hs. (*N. Africa.*)

PROTEALES

41. PROTEACEAE

EDIT. N. A. BURGES

Leaves simple, exstipulate. Flowers hermaphrodite, usually in bracteate racemes, spikes or clusters. Perianth 4-merous, in a single petaloid whorl; segments united in bud, separating wholly or partly at anthesis. Stamens 4, opposite the perianth-segments, with which their filaments are usually connate. Ovary superior, unilocular; style and stigma 1. Seeds without endosperm.

A large family, predominantly of the southern hemisphere; not native to Europe, but including many species cultivated in gardens.

1. **Hakea** Schrader

P. W. BALL (EDITION 1) REVISED BY N. A. BURGES (EDITION 2)

Shrubs or small trees. Flowers zygomorphic, pedicellate, in dense axillary racemes or clusters; bracts deciduous. Perianth coiled in bud, partly straightening after anthesis; segments eventually free proximally but sometimes remaining connate distally, each consisting of a narrow claw and an orbicular-spathulate, concave limb, on the inner side of which is an apparently sessile anther. Stigma large, oblique. Fruit a woody, 2-valved follicle, tardily dehiscent; seeds 2, winged.

Leaves *c.* 1 mm in diameter, terete, pungent **1. sericea**
Leaves 5–20 mm wide, flat, not pungent **2. salicifolia**

1. H. sericea Schrader, *Sert. Hannov.* 27 (1797) (*H. acicularis* (Sm. ex Vent.) Knight). Shrub up to 3 m. Leaves 2–7 × 0·1 cm, stiff, terete, pungent, glabrous when mature. Flowers in axillary clusters; perianth 4–5 mm, white or pale pink. Fruit 2–3 cm, rough. *Planted for reclamation of arid land in Spain and Portugal, and locally naturalized.* [Hs Lu.] (*E. Australia.*)

2. H. salicifolia (Vent.) B. L. Burtt, *Kew Bull.* **1941**: 33 (1942) (*H. saligna* (Andrews) Knight). Tall shrub up to 4 m. Leaves 5–10 × 0·5–2 cm, flat, narrowly elliptical or lanceolate to linear-oblong. Perianth white. Fruit 3 cm, obliquely pyriform, rough. *Planted for reclamation of arid land in Spain and Portugal, and locally naturalized.* [Hs Lu.] (*S.E. Australia, Tasmania.*)

SANTALALES

42. SANTALACEAE

EDIT. S. M. WALTERS

Herbaceous or woody plants. Leaves alternate, simple, linear, oblong or lanceolate, exstipulate, deciduous or evergreen. Flowers small, hermaphrodite or unisexual. Perianth 3- to 5-lobed or -partite, tubular to campanulate-rotate. Stamens 3–5, opposite lobes of the perianth. Ovary inferior, 1-celled. Style 1; stigma capitate or lobed. Fruit a small green nut or a drupe. Seed solitary, with abundant endosperm. Hemiparasites, growing on the roots of herbs or shrubs.

1 Herbs; perianth-segments united; fruit a small green nut **3. Thesium**
1 Woody plants; perianth-segments free; fruit a drupe
2 Plant with a woody rhizome; stigma capitate; perianth 4- or 5-partite; flowers hermaphrodite **1. Comandra**
2 Shrubs without a rhizome; stigma 3- or 4-lobed; perianth 3- or 4-partite; dioecious **2. Osyris**

1. **Comandra** Nutt.

R. HENDRYCH

Rhizome woody. Flowers hermaphrodite, with a solitary bract. Perianth 4- or 5-partite, campanulate. Stamens 4 or 5. Stigma capitate. Fruit a globose drupe surmounted by the persistent, dry perianth.

Literature: M. A. Piehl, *Mem. Torrey Bot. Club* **22**: 1–97 (1965).

1. C. elegans (Rochel ex Reichenb.) Reichenb. fil., *Icon. Fl. Germ.* **11**: 11 (1849) (*Thesium elegans* Rochel ex Reichenb.). Plant with a persistent woody rhizome; stems erect, simple or branched. Leaves oblong. Inflorescence cymose, terminal, few-flowered; flowers cream-coloured or greenish. *Balkan peninsula; Romania.* Bu Gr Ju Rm Tu.

2. **Osyris** L.

R. HENDRYCH

Dioecious; shrubs, without a rhizome. Leaves linear-lanceolate to lanceolate, evergreen. Perianth 3- or 4-partite. Male flowers in lateral few-flowered cymes; stamens 3 or 4. Female flowers solitary and terminal on short branches; stigma 3-lobed. Fruit a globose drupe.

Bracts foliaceous, persistent; drupe 5–7 mm **1. alba**
Bracts small, caducous; drupe 7–10 mm **2. quadripartita**

1. O. alba L., *Sp. Pl.* 1022 (1753) (?*O. mediterranea* Bubani). Stems up to 130 cm, with many slender, spreading branches. Leaves coriaceous, with a single mid-vein; lateral veins obscure or absent. Bracts similar to upper cauline leaves, persistent. Flowers yellowish; perianth 3-partite. Drupe 5–7 mm, red. *S. Europe.* Al Bl Bu Co Cr Ga Gr Hs It Ju Lu Sa Si Tu.

2. O. quadripartita Salzm. ex Decne, *Ann. Sci. Nat.* ser. 2, **6**:

65 (1836) (*O. lanceolata* Steudel & Hochst. ex A. DC.). Like **1** but stems up to 300 cm; leaves very coriaceous, with pinnate venation; bracts small, caducous; perianth 3- or 4-partite; drupe 7–10 mm. $2n = 40$. *S. half of Iberian peninsula; Islas Baleares.* Bl Hs Lu.

3. Thesium L.

R. HENDRYCH

Perennials, rarely annuals, without a rhizome, but often with a woody stock and sometimes producing slender stolons; stems procumbent to erect, simple or branched. Leaves linear to lanceolate, mostly entire, 1- to 5-veined, green or yellowish-green. Inflorescence a panicle with few-flowered branches, or a raceme. Flowers hermaphrodite, on short branches, with a bract and usually with 2 bracteoles. Perianth tubular to campanulate-rotate, 4- or 5-lobed, usually white inside and yellowish-green or green outside. Stamens 4 or 5. Stigma capitate. Fruit a small green nut surmounted by the persistent, dry perianth.

Literature: R. Hendrych, *Acta Univ. Carol.* (*Biol.*) *Suppl.* **1970**: 293–358 (1972); *Preslia* **48**: 107–112 (1976).

1 Bracteoles 0
2 Stock slender, stoloniferous; persistent perianth shorter than nut **1. ebracteatum**
2 Stock thick, not stoloniferous; persistent perianth 2–3 times as long as nut **25. rostratum**
1 Bracteoles 2
3 Plant puberulent-hispid **18. macedonicum**
3 Plant glabrous, rarely minutely papillose-puberulent
4 Persistent perianth at least ½ as long as nut
5 Bract ±equalling flower **16. hispanicum**
5 Bract at least twice as long as flower
6 Persistent perianth 2–3 times as long as nut
7 Perianth usually 5-lobed; flowers not secund; nut ovoid-ellipsoidal or ellipsoidal **22. pyrenaicum**
7 Perianth usually 4-lobed; flowers ±secund; nut subglobose **23. alpinum**
6 Persistent perianth ½ as long to as long as nut
8 Persistent perianth usually ±equalling nut; inflorescence spreading **22. pyrenaicum**
8 Persistent perianth about ½ as long as nut; inflorescence subsecund **24. corsalpinum**
4 Persistent perianth not more than ½ as long as nut
9 Small but prominent extra lobes of the disc present between persistent perianth lobes
10 Plant not stoloniferous; inflorescence at least 5 cm, many-flowered; leaves 3·5–5 cm **19. auriculatum**
10 Plant stoloniferous; inflorescence 2–3 cm, few-flowered; leaves usually 2 cm or less **20. kernerianum**
9 Small but prominent extra lobes of the disc not present between persistent perianth-lobes
11 Bract ±equalling flower
12 Stem less than 10 cm; plant stoloniferous **16. hispanicum**
12 Stem usually more than 10 cm, if less, then plant not stoloniferous
13 Annual; inflorescence a raceme **10. dollineri**
13 Perennial; inflorescence paniculate
14 Leaves linear, 1-veined **7. humifusum**
14 Leaves narrowly elliptical or lanceolate, at least some 3-veined
15 Leaves 1- to 3-veined, often stiff and yellowish-green; plant usually stoloniferous **2. linophyllon**
15 Leaves distinctly 3- to 5-veined, always flaccid and dark green; plant usually not stoloniferous **3. bavarum**
11 Bract distinctly longer than, rarely shorter than, flower
16 Nut reticulately veined (sometimes indistinctly so)
17 Annual; fruiting stems ascending to erect, densely covered with nuts **17. humile**
17 Perennial; fruiting stems procumbent to ascending, sparsely covered with nuts
18 Perianth ±infundibuliform; nut distinctly reticulately veined **5. bergeri**
18 Perianth campanulate-rotate; nut indistinctly reticulately veined
19 Stems usually more than 10 cm; bracteoles ±equalling flower or fruit **4. procumbens**
19 Stems less than 10 cm; bracteoles much shorter than flower or fruit **12. vlachorum**
16 Nut longitudinally veined only
20 Plant stoloniferous
21 Stems robust, rigid; inflorescence usually a panicle; leaves mostly 3-veined **2. linophyllon**
21 Stems slender and rather weak; inflorescence a raceme; leaves 1-veined
22 Flower shortly pedicellate; stems mostly 10–20 cm **8. italicum**
22 Flower subsessile; stems mostly less than 10 cm
23 Fruiting branches (at least in the lower part of the inflorescence) mostly 3 mm or longer **13. parnassi**
23 Fruiting branches very short or absent **15. kyrnosum**
20 Plant not stoloniferous
24 Leaves lanceolate, distinctly 3- to 5-veined **3. bavarum**
24 Leaves mostly linear-oblong, rarely narrowly sublanceolate to linear, 1-veined (rarely indistinctly 3-veined)
25 Inflorescence a raceme; branches 1-flowered
26 Fruiting pedicel thickened **11. brachyphyllum**
26 Fruiting pedicel not thickened
27 Annual or perennial; stems ±robust, rigid; perianth infundibuliform; bracteoles equalling or conspicuously longer than nut **10. dollineri**
27 Always perennial; stems thin, slender; perianth infundibuliform-campanulate; bracteoles mostly shorter than nut **14. sommieri**
25 Inflorescence a panicle; branches several- to many-flowered, at least in lower part of inflorescence
28 Perianth infundibuliform; panicle-branches racemose
29 Bract usually 3–4 times as long as flower **9. arvense**
29 Bract twice as long as flower **21. refractum**
28 Perianth broadly campanulate; panicle-branches often paniculate
30 Stems usually more than 20 cm, ascending to

erect, robust, rigid; bracts, bracteoles and ultimate branches not scabrid **6. divaricatum**

30 Stems usually less than 20 cm, decumbent to somewhat ascending, slender, not rigid; bracts, bracteoles and ultimate branches scabrid **7. humifusum**

1. T. ebracteatum Hayne in Schrader, *Jour. für die Bot.* **1800** (**1**): 33 (1800). Perennial; stock slender, stoloniferous. Stems 10–15(–25) cm, ascending to erect, simple or rarely with a single branch below the inflorescence. Leaves linear-oblong, 1-veined. Inflorescence racemose. Bract about 3 times as long as flower or fruit; bracteoles absent. Perianth broadly campanulate, 5-lobed. Nut ellipsoidal, twice as long as the persistent perianth. $2n = 24$. *E. & E.C. Europe, extending to N.W. Germany, E. Denmark and Estonia.* Au Cz Da Ge ?It Po Rm Rs(B, C, W, E).

2. T. linophyllon L., *Sp. Pl.* 207 (1753) (*T. linifolium* Schrank, *T. intermedium* Schrader, *T. linophyllon* var. *linifolium* (Schrank) Fiori). Perennial; usually stoloniferous. Stems 10–30 cm, ascending to erect, usually simple. Leaves narrowly elliptical, 1- to 3-veined, often rather stiff and yellowish-green. Inflorescence lax, paniculate. Bract more or less equalling to twice as long as flower or fruit; bracteoles shorter than flower or fruit. Nut ellipsoidal, 4–6 times as long as the persistent perianth. $2n = 24$. ● *From E. France to Ukraine, northwards to Latvia and southwards to S. Italy and C. Greece.* Al Au Bu Cz Ga Ge Gr He Hu It Ju Po Rm Rs(B, C, W).

3. T. bavarum Schrank, *Baier. Reise* 129 (1786) (*T. montanum* Ehrh. ex Hoffm., *T. linophyllon* var. *bavarum* (Schrank) Fiori). Perennial; usually not stoloniferous. Stems *c.* 25–60 cm, erect, usually simple. Leaves lanceolate, distinctly 3- to 5-veined, always flaccid and dark green. Inflorescence, bract and bracteoles more or less as in **2**. Nut broadly ellipsoidal, 4–5 times as long as the persistent perianth. $2n = 24$. *C. Europe, Balkan peninsula, Romania, N. & C. Italy.* Al Au Bu Cz Ga Ge Gr He Hu It Ju Rm.

Plants intermediate between **2** and **3** occur on mountains in N. & C. Greece, and the two species are perhaps only subspecifically distinct.

4. T. procumbens C. A. Meyer, *Verz. Pfl. Cauc.* 40 (1831). Perennial. Stems 10–25 cm, more or less procumbent, usually simple. Leaves linear, 1-veined. Bract longer than flower or fruit; bracteoles more or less equalling flower or fruit. Perianth campanulate-rotate, 5-lobed. Nut ovoid, indistinctly reticulately veined, 3–4 times as long as the persistent perianth. *Ukraine and S.C. Russia.* Rs(C, W).

5. T. bergeri Zucc., *Abh. Bayer. Akad. Wiss.* **2**: 324 (1837). Perennial. Stems 5–15 cm, decumbent, procumbent or ascending, simple or branched. Leaves linear to oblong-linear, not fleshy, entire, 1-veined. Inflorescence a raceme; flowers often subsessile. Bract 4 or more times as long as flower or fruit; bracteoles more or less equalling flower or fruit. Perianth more or less infundibuliform, 5-lobed. Nut subglobose, distinctly reticulately veined, 5 times as long as the persistent perianth. $2n = 16$. *S. part of Balkan peninsula; Aegean region.* Al Cr Gr Ju Tu.

6. T. divaricatum Jan ex Mert. & Koch in Röhling, *Deutschl. Fl.* ed. 3, **2**: 285 (1826) (*T. divaricatum* subsp. *vandasii* Rohlena, *T. linophyllon* var. *divaricatum* (Jan ex Mert. & Koch) Fiori; incl. *T. nevadense* Willk.). Perennial; not stoloniferous. Stems 15–25(–35) cm, ascending to erect, robust, rigid, much-branched. Leaves narrowly linear, 1-veined. Inflorescence paniculate. Bract and bracteoles usually shorter than flower or fruit. Perianth broadly campanulate, 5-lobed. Nut ovoid to subglobose, 5 times as long as the persistent perianth. $2n = 16$. *S. Europe, extending northwards to 49° N. in France.* Al Bl Bu Ga Gr ?He Hs It Ju Lu Sa Tu.

7. T. humifusum DC. in Lam. & DC., *Fl. Fr.*, ed. 3, **5**: 366 (1815) (*T. divaricatum* auct., *T. divaricatum* var. *humifusum* Duby). Perennial; not stoloniferous. Stems 10–20 cm, decumbent or long-ascending, slender, weak, branched. Leaves linear, 1-veined. Inflorescence paniculate. Bract usually equalling flower or fruit; bracteoles shorter, distinctly scabrid. Perianth broadly campanulate, 5-lobed. Nut almost subglobose, 4–5 times as long as the persistent perianth. $2n = 16$. *France and S.E. England; locally elsewhere in W. Europe.* ● Be Br ?Co Ga †Ho Hs.

8. T. italicum A. DC. in DC., *Prodr.* **14**: 644 (1857) (*T. humifusum* subsp. *italicum* (A. DC.) Rouy, *T. linophyllum* var. *italicum* (A. DC.) Fiori). Perennial; stoloniferous. Stems 10–20 cm, ascending to erect, simple or slightly branched. Leaves *c.* 3 cm, linear, 1-veined. Inflorescence a long raceme. Bract 5–6 times as long as flower or fruit; bracteoles only slightly longer than flower or fruit. Perianth infundibuliform, 5-lobed. Nut more or less ovoid, 3 times as long as the persistent perianth. ● *Sardegna.* Sa.

9. T. arvense Horvátovszky, *Fl. Tyrnav.* 27 (1774) (*T. ramosum* Hayne, *T. linophyllon* var. *ramosum* (Hayne) Fiori; incl. *T. bulgaricum* Velen.). Perennial; not stoloniferous. Stems (10–)15–25(–30) cm, ascending to erect. Leaves linear-oblong to linear, 1(–3)-veined, often subsecund. Inflorescence usually a panicle with long, racemose branches. Bract 3–4 times as long as flower or fruit; bracteoles equalling flower or fruit. Perianth infundibuliform, (4–)5-lobed. Nut ellipsoidal, 3 times as long as the persistent perianth. *E., S.E. & E.C. Europe.* Al Au Bu Cz Gr Hu It Ju Rm Rs(C, W, K, E) Tu.

10. T. dollineri Murb., *Lunds Univ. Årsskr.* **27**: 43 (1891) (*T. humile* auct. eur. med., non Vahl). Annual to perennial; not stoloniferous. Stems 6–15 cm, decumbent, ascending or erect, simple or branched. Leaves linear to oblong-linear, 1-veined, sometimes secund. Inflorescence few- or many-flowered, flowers often subsessile. Perianth infundibuliform, 5-lobed. Nut more or less ovoid-globose, 4 times as long as the persistent perianth. ● *E.C. Europe, extending to Krym and S. Bulgaria.* Au Bu Cz Hu Ju Rm Rs(W, K).

1 Plant annual **(a)** subsp. **dollineri**

1 Plant perennial

2 Leaves thin, distinctly 1-veined **(b)** subsp. **simplex**

2 Leaves rigid, thickened, indistinctly 1-veined **(c)** subsp. **moesiacum**

(a) Subsp. **dollineri**: Annual. Stem single, horizontally branched from near middle. Bract and bracteoles more or less equalling flower or fruit. *From Czechoslovakia to N.E. Bulgaria.*

(b) Subsp. **simplex** (Velen.) Stoj. & Stefanov, *Fl. Bălg.*, ed. 2, 312 (1933) (*T. simplex* Velen.). Perennial. Stems several, unbranched. Leaves thin, distinctly 1-veined. Bract and bracteoles much longer than flower or fruit. *Almost throughout the range of the species.*

(c) Subsp. **moesiacum** (Velen.) Stoj. & Stefanov, *loc. cit.* (1933) (*T. moesiacum* Velen.). Perennial. Stems several or many, unbranched. Leaves rigid, thickened, indistinctly 1-veined. Bract and bracteoles much longer than flower or fruit. *Bulgaria.*

11. T. brachyphyllum Boiss., *Diagn. Pl. Or. Nov.* **1(5)**: 48 (1844). Perennial; not stoloniferous. Stems 5–15 cm, decumbent to ascending, simple. Leaves *c.* 1 cm, linear-oblong, 1-veined, minutely papillose. Inflorescence a raceme with subsessile or shortly stalked flowers. Bract 5–8 times as long as flower or fruit; bracteoles more or less equalling the flower or fruit. Perianth broadly campanulate, 5-lobed. Nut ellipsoidal, only longitudinally, not reticulately, veined, 5–6 times as long as the persistent perianth. *Mountains of Krym, Kuban, and C. & N. Greece.* Gr Rs(K, E).

12. T. vlachorum Aldén, *Nordic Jour. Bot.* **1**: 709 (1982). Like **11** but shortly stoloniferous; stems 2–6 cm, procumbent to ascending, with scattered, short leaves; leaves linear-oblong to elliptic-oblong; inflorescence racemose or subspicate; bracts up to $2\frac{1}{2}$ times as long as fruit; bracteoles much shorter than fruit; nut with rather broad, longitudinal veins, somewhat anastomosing, 4–5 times as long as the persistent perianth. *Serpentine rocks and screes.* ● *N.W. Greece (Grammos Oros).* Gr.

Known only from the type collection.

13. T. parnassi A. DC. in DC., *Prodr.* **14**: 643 (1857) (*T. linophyllon* var. *parnassi* (A. DC.) Fiori, *T. linophyllon* var. *ramosum* (Hayne) Fiori). Perennial; stoloniferous. Stems 5–10(–15) cm, ascending to erect, simple, weak, slender. Leaves *c.* 1 cm, oblong-linear to linear, 1-veined. Inflorescence a short, rather few-flowered raceme, often with subsessile flowers. Bract usually twice as long as flower or fruit; bracteoles almost equalling flower or fruit. Perianth infundibuliform, 5-lobed. Nut subglobose, 4 times as long as the persistent perianth. $2n = 14$. ● *Mountains of W. part of Balkan peninsula, C. & S. Italy and Sicilia.* Al Gr It Ju Si.

14. T. sommieri Hendrych, *Preslia* **36**: 118 (1964). Perennial; not stoloniferous. Stems 10–20 cm, shortly ascending to suberect, simple, weak. Leaves 1–2·5 cm, oblong to sublanceolate, 1-veined. Inflorescence simple, racemose, rather few-flowered; branches 1-flowered. Bract twice as long as flower; bracteoles shorter than flower. Perianth infundibuliform-campanulate, 5-lobed. Nut ellipsoidal, 4 times as long as the persistent perianth. ● *Alpi Apuane, N. Appennini.* It.

15. T. kyrnosum Hendrych, *Novit. Bot. Inst. Bot. Univ. Carol. Prag.* **1964**: 21 (1964). Perennial. Stems 5–10 cm, ascending, simple. Leaves 1·5–2·5 cm, linear, 1-veined. Inflorescence racemose to subspicate, few-flowered. Bract 5–7 times as long as flower; bracteoles subequalling flower. Perianth campanulate-infundibuliform, 5-lobed. Nut ellipsoidal to globose-ellipsoidal, 4 times as long as the persistent perianth. ● *Corse.* Co.

16. T. hispanicum Hendrych, *Folia Geobot. Phytotax. (Praha)* **1**: 74 (1966). Perennial; with long stolons. Stems up to 7 cm, shortly ascending, simple. Leaves up to 1 cm, narrowly oblong, 1-veined. Inflorescence simple, racemose, short, few-flowered; branches 1-flowered. Bract as long as flower; bracteoles shorter than flower. Perianth infundibuliform-campanulate, 5-lobed. Nut unknown. ● *N. Spain (Picos de Europa).* Hs.

17. T. humile Vahl, *Symb. Bot.* **3**: 43 (1794). Annual. Stems 5–20 cm, ascending to erect, usually branched from base with long, simple branches. Leaves linear or narrowly linear, rather fleshy, denticulate, 1-veined. Inflorescence often subspicate, sometimes paniculate. Bract 2–5 times as long as flower or fruit; bracteoles usually equalling flower or fruit. Perianth infundibuliform, 5-lobed. Nut subglobose, reticulately veined, 5–6 times as long as the persistent perianth. $2n = 18$. *S. Europe.* Bl Co Cr Gr Hs It Ju Lu Sa Si.

18. T. macedonicum Hendrych, *Biol. Glasn. (Zagreb)* **17**: 13 (1964). Perennial; not stoloniferous; whole plant puberulent-hispid. Stems 7–15 cm, prostrate to suberect, unbranched. Leaves linear, 1–2·5 cm, 1-veined. Inflorescence racemose to subspicate; branches 1- or 2-flowered. Bract 5–7 times as long as flower; bracteoles equalling flower. Perianth subcampanulate, 5-lobed. Nut ellipsoidal, longitudinally and indistinctly subreticulately veined, 5–6 times as long as the persistent perianth. ● *S. Jugoslavia (Nidže Planina).* Ju.

19. T. auriculatum Vandas, *Sitz.-Ber. Böhm. Ges. Wiss.* **1890/1891**: 279 (1890). Perennial; not stoloniferous. Stems *c.* 25–30 cm, ascending to erect, simple, or branched above. Leaves narrowly linear, 1-veined. Inflorescence paniculate with long, racemose or subspicate branches; flowers often subsessile and often more or less secund. Bract and bracteoles equalling flowers; bract enlarging in fruit to 6 or more times as long as fruit. Perianth broadly campanulate, 5-lobed. Nut ellipsoidal, 4–5 times as long as the persistent perianth, which has small but prominent extra lobes of the disc between the 5 lobes. ● *Mountains of Jugoslavia and N. Albania.* Al Ju.

20. T. kernerianum Simonkai, *Enum. Fl. Transs.* 478 (1887). Perennial; stoloniferous. Stems up to 15 cm, ascending, simple. Leaves up to 1·5 cm, linear-oblong, 1-veined. Inflorescence short, few-flowered, subspicate. Bract at least twice as long as flower or fruit; bracteoles shorter than flower or fruit. Perianth campanulate. Nut and perianth-lobes as in **19**. ● *S. & E. Carpathians; local.* Rm.

21. T. refractum C. A. Meyer in Bong. & C. A. Meyer, *Verz. Saisang-Nor* 58 (1841). Perennial. Stems 10–50 cm, suberect, much-branched in the upper $\frac{1}{2}$. Leaves 4–7 cm, linear-lanceolate, 1-veined (rarely indistinctly 3-veined). Inflorescence paniculate. Bract twice as long as flower; bracteoles shorter than flowers or equalling them. Perianth infundibuliform, 5-lobed. Persistent perianth 3–4 times shorter than the ellipsoidal nut. *C. & S. Ural.* Rs(C). (*Siberia and C. Asia.*)

22. T. pyrenaicum Pourret, *Mém. Acad. Sci. Toulouse* **3**: 331 (1788) (*T. pratense* Ehrh. ex Schrader, *T. alpinum* var. *pyrenaicum* (Pourret) Fiori). Perennial. Stems 10–20(–30) cm, ascending to erect, simple or branched. Leaves more or less linear, 1-veined. Inflorescence racemose or paniculate; flowers not secund. Bract usually at least twice as long as flower or fruit; bracteoles equalling flower or fruit. Perianth tubular-campanulate. Nut ovoid-ellipsoidal or ellipsoidal. ● *W. & C. Europe; C. Italy.* Au Be Cz Ga Ge He †Ho Hs It Ju Lu Po.

(a) Subsp. **pyrenaicum**: Perianth 3–4 mm, mostly 5-lobed (rarely 4-lobed in isolated flowers); persistent perianth equalling the nut. $2n = 14$. *Throughout the range of the species.*

(b) Subsp. **grandiflorum** (Richter) Hendrych, *Novit. Bot. Inst. Bot. Univ. Carol. Prag.* **1964**: 9 (1964) (*T. pyrenaicum* subsp. *alpestre* O. Schwarz, *T. grandiflorum* (A. DC.) Hand.-Mazz.): Perianth at least 5 mm, mostly 5-lobed, but often 4-lobed; persistent perianth about twice as long as the nut. *E. Alps and mountains of N.W. Jugoslavia.*

23. T. alpinum L., *Sp. Pl.* 207 (1753) (incl. *T. tenuifolium* Sauter ex Koch). Perennial. Stems 10–20(–30) cm, ascending to erect, simple or slightly branched. Leaves linear to linear-oblong, 1-veined. Inflorescence racemose, rarely paniculate; flowers often subsessile, usually more or less secund. Bract at least 2–3 times as long as flower or fruit; bracteoles equalling flower or fruit. Perianth subtubular, 4(5)-lobed. Persistent perianth 2–3 times as long as the subglobose nut. $2n = 12$. *C. & S. Europe; local in N. Europe from Sweden to N.W. Russia.* Al Au Bu Cz †Da Ga Ge He Hs It Ju Po Rm Rs(B, C, W) Su.

24. T. corsalpinum Hendrych, *Novit. Bot. Inst. Bot. Univ. Carol. Prag.* **1964**: 21 (1964). Perennial. Stems 8–15(–20) cm, ascending, little-branched. Leaves linear, 1-veined. Inflorescence a simple, subsecund raceme. Bract 4–8 times as long as flower; bracteoles equalling flower. Perianth infundibuliform, (4)5-lobed. Persistent perianth $\frac{1}{2}$ as long as the subglobose to ellipsoidal nut. ● *Corse.* Co.

25. T. rostratum Mert. & Koch in Röhling, *Deutschl. Fl.* ed. 3, **2**: 287 (1826). Perennial; stock stout, woody, not stoloniferous. Stems 10–30 cm, ascending to erect, with some non-flowering branches. Leaves linear, 1-veined. Inflorescence racemose. Bract up to 4 times as long as flower or fruit; bracteoles absent. Perianth tubular, 5-lobed. Persistent perianth 2(–3) times as long as the ovoid-globose fruit. $2n = 26$. ● *C. Europe.* Au Cz Ge He It Ju.

43. LORANTHACEAE

EDIT. V. H. HEYWOOD AND J. R. EDMONDSON

Small shrubs, hemiparasitic on gymnosperms and woody angiosperms. Leaves opposite or whorled, entire, sometimes scale-like. Flowers actinomorphic, unisexual (in European species), usually 4-merous; perianth in 1 or 2 whorls, the inner more or less petaloid, with the parts free or connate. Stamens epipetalous, opposite the petals. Ovary inferior; ovules usually not differentiated from the placenta; style simple or 0. Fruit a berry, rarely dry and dehiscent. Seed solitary, without testa; embryos 1–3.

1 Leaves scale-like, connate in pairs, forming a sheath round the stem; perianth in 1 whorl — **3. Arceuthobium**
1 Leaves not scale-like, not connate; perianth in 2 whorls
 2 Inflorescence racemose or spicate — **1. Loranthus**
 2 Inflorescence cymose, sessile or very shortly pedunculate — **2. Viscum**

1. Loranthus L.

P. W. BALL

Leaves opposite, pinnately- or slightly parallel-veined, coriaceous. Inflorescence racemose or spicate. Perianth in 2 whorls. Calyx-teeth small; petals 4–6. Stamens inserted at the base of the petals; anthers dehiscing longitudinally. Fruit a yellowish, viscid berry.

A predominantly tropical and subtropical genus with only one species in Europe.

1. L. europaeus Jacq., *Enum. Stirp. Vindob.* 55 (1762). Stem up to 50 cm, dull brown. Leaves 1–5 cm, obovate-oblong, obtuse, dull green. Berry up to 10 mm, pyriform-globose, yellow. $2n = 18$. *On members of the* Fagaceae. *S.E. & E.C. Europe; Italy and Sicilia.* Al Au Cz Ge Gr Hu It Ju ?Po Rm Rs(W, K) Si Tu.

2. Viscum L.

P. W. BALL

Leaves opposite or whorled, with 3–7 parallel veins, coriaceous. Inflorescence cymose, the flowers crowded. Perianth in 2 whorls. Calyx 4-toothed in female flowers, absent in male flowers. Petals usually 4. Stamens almost completely connate with the petals; anthers opening by pores. Fruit a white, yellow or red, viscid berry.

Literature: K. F. von Tubeuf, *Monographie der Mistel.* München. 1923.

Inflorescence shortly pedunculate; berry red — **1. cruciatum**
Inflorescence sessile; berry white or yellow — **2. album**

1. V. cruciatum Sieber ex Boiss., *Voy. Bot. Midi Esp.* **2**: 274 (1840). Stem up to 60 cm, yellowish-green. Leaves 2–4 cm, often whorled, obovate-oblong, obtuse, yellowish-green. Cymes shortly pedunculate. Berry 6–10 mm, red. *On various dicotyledonous trees and shrubs. S. Spain; one station in E. Portugal.* Hs Lu. (*N. Africa.*)

2. V. album L., *Sp. Pl.* 1023 (1753) (incl. *V. laxum* Boiss. & Reuter). Stem up to 100 cm, yellowish-green. Leaves 2–8 cm, opposite, rarely whorled, obovate-oblong, obtuse, yellowish-green. Cymes sessile. Berry 6–10 mm, globose or pyriform, white or occasionally yellow. $2n = 20$. *On various trees. Most of Europe, northwards to S. Sweden and eastwards to* 41° *E. in S. Russia.* All except Az Fa Fe Hb Is Rs(N) Sa Sb.

The following subspecies are variously treated as forms, varieties, subspecies or species in Floras, and there is no general agreement as to their status and relationships.

1 Edges of the seed straight; embryos usually 2 or 3 (on dicotyledons) — (**a**) subsp. **album**
1 Edges of the seed convex; embryos 1(2) (on conifers)
 2 Leaves up to 8 cm, not more than 3 times as long as wide — (**b**) subsp. **abietis**
 2 Leaves not more than 4(–6) cm, up to 5 times as long as wide — (**c**) subsp. **austriacum**

(**a**) Subsp. **album**: Leaves variable. Berry usually white, globose; edges of the seed straight; embryos 2 or 3. *On dicotyledonous trees. Throughout the range of the species.*

(**b**) Subsp. **abietis** (Wiesb.) Abromeit in Wünsche, *Pfl. Deutschl.* ed. 12, 182 (1928) (*V. laxum* var. *abietis* (Wiesb.) Hayek): Leaves up to 8 cm, not more than 3 times as long as broad. Berry usually white, pyriform; edges of the seed convex; embryos 1(2). *On* Abies *spp. C. & S. Europe.*

(**c**) Subsp. **austriacum** (Wiesb.) Vollmann, *Fl. Bayern* 212 (1914) (*V. laxum* Boiss. & Reuter var. *pini* (Wiesb.) Hayek): Like subsp. (**b**) but leaves 2–4(–6) cm, up to 6 times as long as wide; berry usually yellow. *On* Pinus *and* Larix *spp. C. Europe and locally in S. Europe.*

3. Arceuthobium Bieb.

P. W. BALL AND J. R. AKEROYD

Leaves opposite, scale-like, connate in pairs, forming a sheath around the stem. Perianth in 1 whorl. Male flowers solitary, sessile. Perianth 2- to 5-lobed. Stamens sessile; anthers dehiscing transversely. Female flowers solitary or in pairs, pedicellate; perianth of 2 short teeth. Fruit green, dry, dehiscing explosively. (*Razoumofskya* Hoffm.)

Literature: F. G. Hawksworth & D. Wiens, *Kew Bull.* **31**: 71–80 (1976).

Stems 1–3 mm wide at base; male flowers mostly 3-merous **1. oxycedri**
Stems 5–9 mm wide at base; male flowers mostly 4-merous **2. azoricum**

1. A. oxycedri (DC.) Bieb., *Fl. Taur.-Cauc.* **3**: 629 (1819) (*Razoumofskya oxycedri* (DC.) F. W. Schultz). Stems up to 20 cm, 1–3 mm wide at the base, yellowish-green, articulate. Leaves up to 5 mm, triangular. Male flowers on short terminal branches, *c.* 2 mm, mostly 3-merous; female flowers terminal and axillary. Fruit ovate-oblong, shortly pedicellate. *Usually on* Juniperus *spp. S. Europe.* Al Bu Ga Gr Hs Ju Rs(K) Tu.

2. A. azoricum Wiens & F. G. Hawksworth, *Kew Bull.* **31**: 73 (1976) (*A. oxycedri* auct., non (DC.) Bieb.). Like **1** but stems 5–9 mm wide at the base; male flowers *c.* 2·5 mm, mostly 4-merous; female flowers in whorls. *On* Juniperus brevifolia. • *Açores.* Az.

ARISTOLOCHIALES

44. ARISTOLOCHIACEAE

EDIT. J. R. EDMONDSON

Herbs or woody climbers. Leaves alternate, entire, exstipulate. Flowers solitary or in axillary clusters, hermaphrodite, actinomorphic or zygomorphic, usually 3-merous. Perianth in 1 whorl, more or less petaloid, forming a tube, 1- or 3-lobed at the apex; stamens 6 or 12 in 1 or 2 whorls, free or connate with the stylar column. Ovary inferior, 6-locular; placentation axillary; styles usually 6, free or connate to form a column with a 6-lobed stigma; ovules numerous. Fruit a capsule.

Perianth actinomorphic, with short tube, persistent in fruit; flowers terminal **1. Asarum**
Perianth zygomorphic, with long tube, deciduous; flowers axillary **2. Aristolochia**

1. Asarum L.

P. W. BALL AND J. R. AKEROYD

Herbs with creeping rhizome. Flowers solitary and terminal. Perianth actinomorphic, 3-lobed, persistent in fruit. Stamens 12, in 2 whorls, free or almost so; filaments short. Fruit subglobose, dehiscing irregularly.

Literature: I. Kukkonen & P. Uotila, *Ann. Bot. Fenn.* **14**: 131–142 (1977).

1. A. europaeum L., *Sp. Pl.* 442 (1753). Stems 2–10 cm, pubescent, with 2–5 leaves and brown ovate scales 1–2 cm. Leaves 2·5–10 cm, reniform to cordate; petiole longer than the lamina. Perianth 10–15 mm, brownish-purple, pubescent on the outside; lobes 5–8 mm, more or less deltate, acuminate. $2n = 26, 40$. *From S. Finland southwards to S. France, Italy and Makedonija.* Al Au Be Bu Cz Fe Ga Ge He Hu It Ju Po Rm Rs(N, B, C, W, K, E) Su [Br Da Ho No Su].

1 Leaves wider than long, obtuse; lower surface pubescent (a) subsp. **europaeum**
1 Leaves about as wide as long, subacute; lower surface sparsely pubescent to glabrous
2 Leaves sparsely hairy beneath; epidermis of upper surface not or very weakly papillose (b) subsp. **italicum**
2 Leaves subglabrous or glabrous beneath; epidermis of upper surface papillose (c) subsp. **caucasicum**

(a) Subsp. **europaeum**: Leaves reniform, sometimes subcordate, obtuse; lower surface pubescent; stomata present on upper surface; epidermis of upper surface usually without papillae. $2n = 26$. *Almost throughout the range of the species.*

(b) Subsp. **italicum** Kukkonen & Uotila, *Ann. Bot. Fenn.* **14**: 139 (1977): Leaves cordate, subacute; lower surface sparsely pubescent; stomata absent on upper surface; epidermis of upper surface without or with very weak papillae. • *C. & N. Italy; Crna Gora.*

(c) Subsp. **caucasicum** (Duchartre) Soó, *Acta Bot. Acad. Sci. Hung.* **12**: 11 (1966) (*A. europaeum* var. *caucasicum* Duchartre, *A. europaeum* var. *intermedium* C. A. Meyer, *A. ibericum* Steven ex Woronow): Leaves cordate, subacute, lower surface glabrous or subglabrous; stomata absent on upper surface; epidermis of upper surface papillose. *S.W. Alps.*

2. Aristolochia L.

P. W. BALL (EDITION 1) REVISED BY E. NARDI AND J. R. AKEROYD (EDITION 2)

Perennial herbs or woody climbers. Flowers axillary. Perianth zygomorphic, deciduous, the base swollen (utricle), the upper part narrower, more or less cylindrical (tube), with single, entire lobe (limb), rarely absent or 3-lobed. Stamens usually 6, in 1 whorl, connate with stylar column. Fruit a septicidal capsule.

A. macrophylla Lam., *Encycl. Méth. Bot.* **1**: 252 (1783) (*A. durior* Hill, *A. sipho* L'Hér.), from North America, a woody climber with cordate leaves up to 30 cm and perianth brownish with a 3-lobed limb, is cultivated for ornament, and is reported to be naturalized in Germany.

Literature: P. H. Davis & M. S. Khan, *Notes Roy. Bot. Gard. Edinb.* **23**: 515–546 (1961). E. Nardi, *Webbia* **38**: 221–300 (1984); **45**: 31–69 (1991). E. Nardi & C. N. Nardi, *Bot. Helvet.* **97**: 155–165 (1987).

1 Flowers 2 or more together in the axil of a leaf **5. clematitis**
1 Flowers solitary
 2 Stem usually climbing, woody; leaves evergreen, coriaceous
  3 Ovary and pedicels pubescent **1. sempervirens**
  3 Ovary and pedicels glabrous **2. baetica**
 2 Stem not climbing, herbaceous; leaves not coriaceous
  4 Perianth without conspicuous limb, the apex clavate, with a small pore **18. microstoma**
  4 Perianth with large conspicuous limb at least $\frac{1}{5}$ as long as tube
   5 Perianth-tube strongly curved, 10–20 mm in diameter; utricle oblong-cylindrical
    6 Leaves not more than 7·5 × 8 cm; limb of perianth ovate, with basal auricles 2–5 mm wide **3. cretica**
    6 Leaves up to 16 × 13 cm; limb of perianth obovate, with basal auricles up to 20 mm wide **4. hirta**
   5 Perianth-tube usually ± straight, less than 5 mm in diameter; utricle usually ± globose
    7 Leaves with cartilaginous teeth or papillae on the margin and lower surface; tubers numerous, fasciculate, cylindrical **6. pistolochia**
    7 Leaves without cartilaginous teeth or papillae (rarely with the margin undulate or spinulose); tuber solitary, globose or elongate
     8 Leaves sessile or with petiole not more than 5 mm, sometimes amplexicaul
      9 Plant usually robust, up to 100 cm; flowers (2–)3–6 cm; perianth-tube not striped **7. rotunda**
      9 Plant slender, up to 30 cm; flowers 1–3 cm; perianth-tube striped
       10 Flowers brownish-yellow with brownish stripes; limb reddish **8. bianorii**
       10 Flowers yellowish-green with faint violet stripes; limb brownish-violet **16. merxmuelleri**
     8 Leaves with petioles mostly more than 5 mm, not amplexicaul
      11 At least some leaves acute; flowers yellow **19. sicula**
      11 Leaves obtuse to subacute; flowers variously coloured
       12 Tuber globose; capsule ovoid-oblong to pyriform
        13 Stems at least 15 cm; flowers (2–)2·5–7·5 cm **(13–15). pallida** group
        13 Stems less than 15 cm; flowers 2–3 cm **16. merxmuelleri**
       12 Tuber elongate; capsule globose or pyriform
        14 Pedicels about equalling petioles; capsule 1·5–3 cm
         15 Leaves ovate-triangular; perianth-tube widest near apex **(9–11). paucinervis** group
         15 Leaves ovate-reniform; perianth-tube widest in middle part **(13–15). pallida** group
        14 Pedicels shorter than petioles; capsule usually less than 1·5 cm
         16 Flowers brownish to purplish; limb greenish with reddish or violet-brown stripes **12. parvifolia**
         16 Flowers greyish-green; limb with reddish-brown stripes **17. tyrrhena**

1. A. sempervirens L., *Sp. Pl.* 961 (1753) (incl. *A. altissima* Desf.). Evergreen; stems up to 5 m, climbing or rarely procumbent. Leaves up to 10 × 6 cm, triangular-ovate, cordate, glabrous, coriaceous, the basal sinus shallow, the lobes not more than $\frac{1}{7}$ as long as the leaf. Flowers 2–5 cm, yellow, striped with purple, the limb purple or dull brownish-purple, tube strongly curved; ovary and pedicels shortly pubescent. Capsule 1–4 cm, ovoid or oblong. $2n = 14$. *S. & W. Greece, Kriti; Sicilia; naturalized elsewhere in S. Europe.* Cr Gr Si [Ga It Sa].

2. A. baetica L., *Sp. Pl.* 961 (1753). Like **1** but leaves glaucous, the basal sinus deep and usually narrow, the lobes up to $\frac{1}{4}$ as long as the leaf; flowers brownish-purple to reddish; ovary and pedicels glabrous. *S. & E. Spain; S. Portugal.* Hs Lu.

3. A. cretica Lam., *Encycl. Méth. Bot.* **1**: 255 (1783). Stems 30–60 cm, erect or decumbent, pubescent. Leaves 1–7·5 × 1–8 cm, reniform to triangular-ovate, about as long as wide; petioles not more than 5 cm. Flowers 5–18 cm, dull purple; utricle oblong, strongly inflated; tube 10–20 mm in diameter, strongly curved; limb ovate, with basal auricles narrow, emarginate to 2-lobed. Capsule 3–6 cm, ovate-oblong. $2n = 10$. ● *Kriti and Karpathos.* Cr.

4. A. hirta L., *Sp. Pl.* 961 (1753) (*A. bodamae* Dingler, *A. pontica* auct., non Lam.). Like **3** but leaves up to 12–16 × 8–13 cm, longer than wide, subdeltate to ovate; petiole up to 3–5 cm; flowers blackish-purple, the limb obovate, with broad, obtuse to bilobed basal auricles. *E. Thrace; Samothraki.* Gr Tu.

5. A. clematitis L., *Sp. Pl.* 962 (1753). Stems up to 100 cm, subsimple, glabrous; stock creeping, much divided. Leaves 3–15 cm, deltate to broadly ovate, obtuse; petiole 1·5–5 cm. Flowers 2–3 cm, 2–8(–12) together in the axils of the leaves, yellow, the limb brownish; pedicels very short. Capsule 2–5 cm, ovoid to pyriform. $2n = 14$. *Formerly cultivated as a medicinal herb throughout most of Europe, and widely naturalized from* c. 60° *N. southwards; probably native in the east and south-east.* Al Au Be Bl Bu Co Cz Ga Ge Gr He Ho Hs Hu It Ju Po Rm Rs(B, C, W, K, E) Si Tu [Br No].

6. A. pistolochia L., *Sp. Pl.* 962 (1753). Stems 20–80 cm, simple or branched, pubescent; stock of numerous, fibrous, cylindrical, fasciculate tubers. Leaves 1–3(–5) cm, ovate-triangular, the margin and lower surface with cartilaginous teeth or papillae; petiole 1–5 mm. Flowers 2–5 cm, brownish, the limb dark purple. Capsule 2–3 cm, globose to pyriform. *S.W. Europe.* Ga Hs Lu.

7. A. rotunda L., *Sp. Pl.* 962 (1753). Stems 15–100 cm, simple or branched, glabrescent. Leaves 2–8(–13) cm, ovate-orbicular, amplexicaul; petiole not more than 0·5 cm. Flowers (2–)3–6 cm, yellow, the limb dark brown; pedicels longer than the petioles. Capsule 1–2 cm, globose. *S. Europe.* Al Bl Bu Co Ga Gr He Hs It Ju Sa Si Tu.

(a) Subsp. **rotunda**: Tuber globose or ovoid. Leaves up to 13 cm. Capsule 1–2 cm; seeds 4–6 mm. $2n = 12$. *Throughout the range of the species except Sardegna.*

(b) Subsp. **insularis** (Nardi & Arrigoni) Gamisans, *Cat. Pl. Vasc. Corse* 93 (1985): Tuber cylindrical. Leaves 2–8 cm. Capsule

1–1·5 cm; seeds 2–4 mm. $2n = 12$. ● *Sardegna, Corse, locally in Greece and Italy.*

Plants from calcareous gravels in N.W. Jugoslavia (island of Pag), like subsp. (**a**) but smaller in all their parts, have been called subsp. **reichsteinii** Nardi, *Bot. Helvet.* **97**: 162 (1987).

8. A. bianorii Sennen & Pau, *Butll. Inst. Catalana Hist. Nat.* **11**: 19 (1912). Stems 10–30 cm, branched, glabrous; tuber cylindrical or ovoid. Leaves 0·5–3·5 × 0·5–2 cm, ovate-oblong to ovate-triangular, cordate; petiole 0·1–0·4 cm. Flowers 1–3 cm, brownish-yellow, with brownish stripes, the limb reddish; pedicels equalling or longer than the petioles. Capsule 0·5–1·0 cm, globose. $2n = 12$. ● *Islas Baleares (Mallorca, Minorca).* Bl.

(9–11). A. paucinervis group (*A. longa* auct., non L.). Stems 20–90 cm, usually branched, glabrescent; tuber cylindrical. Leaves 2–6(–9) cm, ovate-triangular; petiole up to 1·5 cm. Flowers 2–5(–6) cm, brownish or yellowish-green, the limb brownish-purple; pedicels about equalling the petioles. Capsule 1·5–3 cm, globose to pyriform.

These species, together with **A. fontanesii** Boiss. & Reuter, *Pugillus* 108 (1852) (*A. longa* L.), from Algeria, form a complex group that requires further study, especially in N. Africa.

1 Leaves glabrous above, pubescent beneath; perianth not striped **11. navicularis**
1 Leaves pubescent on both surfaces; perianth striped
2 Perianth-tube straight, longer than limb **9. paucinervis**
2 Perianth-tube somewhat curved, equalling or shorter than limb **10. clusii**

9. A. paucinervis Pomel, *Bull. Soc. Sci. Phys. Algérie* **11**: 136 (1874). Leaves pubescent on both surfaces. Perianth striped; tube up to 3 cm, straight, slightly enlarged at apex; limb shorter than the tube. $2n = 36$. *S.W. France, Iberian peninsula, Islas Baleares.* Bl Ga Hs Lu.

10. A. clusii Lojac., *Fl. Sic.* **2(2)**: 314 (1907). Leaves pubescent on both surfaces. Perianth striped; tube up to 1·8 cm, somewhat curved, much enlarged at apex; limb longer than the tube. $2n = 12$. ● *S. Italy, Sicilia, Malta.* It Si.

11. A. navicularis Nardi, *Webbia* **38**: 261 (1984) (*A. pallida* auct., non Willd.). Leaves pubescent beneath, glabrous above. Perianth not striped; tube up to 2 cm, somewhat curved, slightly enlarged at apex; limb about equalling the tube. $2n = 24$. *Sardegna, Sicilia (Isole Egadi).* Sa Si.

12. A. parvifolia Sm. in Sibth. & Sm., *Fl. Graec. Prodr.* **2**: 222 (1816). Stems 10–40 cm, simple or branched, pubescent; tuber oblong or fusiform. Leaves 1–3 cm, ovate-oblong, cordate; petiole 0·1–1 cm. Flowers 1·5–8·5 cm, brownish to purplish; limb up to twice as long as the tube, greenish with reddish or violet-brown stripes. Capsule 1–2 cm, globose. $2n = 12$. *S. Aegean region; S. Greece.* Cr Gr.

(13–15). A. pallida group. Stems 15–60 cm, simple or branched, glabrous to puberulent. Leaves 2–7 cm, ovate-reniform, obtuse to retuse; petiole up to 1·5(–2·5) cm. Flowers (2–)2·5–7·5 cm, green, yellow, greyish or pale brown, with brownish or purplish stripes, the limb often darker. Capsule 2–3 cm, ovoid-oblong to pyriform.

1 Tuber cylindrical; perianth-tube dilated in middle part **15. elongata**
1 Tuber globose; perianth-tube dilated in upper part
2 Perianth-limb usually equalling or longer than, and broader than, the tube **13. pallida**
2 Perianth-limb shorter than, and narrower than, the tube
3 Perianth usually not more than 3·5 cm **13. pallida**
3 Perianth usually at least 3·5 cm **14. lutea**

13. A. pallida Willd., *Sp. Pl.* **4(1)**: 162 (1805) (*A. macedonica* Bornm., *A. melanoglossa* Bornm., non Speg.). Tuber globose. Flowers 2–6·5 cm. Perianth-tube straight; limb usually longer and broader than the tube. *S. Europe.* Al Bu Ga Gr Hs It Ju.

(**a**) Subsp. **pallida**: Flowers 2·5–6·5 cm. Perianth-limb 1–3 cm, longer than the tube. $2n = 10$. *Throughout the range of the species except Spain.*

(**b**) Subsp. **castellana** Nardi, *Webbia* **42**: 15 (1988): Flowers 2–3·5(–4) cm. Perianth-limb 0·5–1·5 cm, shorter than the tube. $2n = 10$. ● *C. Spain (Estremadura, Sierra de Gredos).*

Similar to **14** but distinguished by the smaller flowers.

14. A. lutea Desf., *Ann. Mus. Hist. Nat. (Paris)* **10**: 295 (1807) (*A. croatica* Horvatić). Tuber globose. Flowers (2·5–)3·5–7·5 cm. Perianth-tube straight; limb shorter and narrower than the tube. $2n = 8$. *S. Europe from Italy to Turkey-in-Europe.* Au Bu Gr Hu It Ju Rm Si Tu.

15. A. elongata (Duchartre) Nardi, *Webbia* **38**: 294 (1984) (*A. pallida* var. *elongata* Duchartre). Tuber cylindrical. Flowers 2·5–6·5 cm. Perianth-tube dilated in the middle; limb shorter and narrower than the tube. $2n = 10$. ● *S. part of Balkan peninsula.* Al Gr.

16. A. merxmuelleri Greuter & E. Mayer, *Bot. Jahrb.* **107**: 321 (1985). Stems 5–12 cm, usually simple, glabrous. Leaves 1·5–3 cm, broadly triangular, retuse, puberulent (especially beneath); petiole up to 0·5 cm. Flowers 2–3 cm, yellowish-green, often with faint violet stripes, the limb brownish-violet; pedicels very short. *On serpentine soil.* ● *C. Jugoslavia (N.E. of Djakova).* Ju.

17. A. tyrrhena Nardi & Arrigoni, *Boll. Soc. Sarda Sci. Nat.* **22**: 347 (1983). Stems up to 40 cm, simple or branched, puberulent; tuber narrowly cylindrical. Leaves 2–4 cm, ovate to ovate-reniform, cordate, obtuse to rounded, glabrous beneath; petiole 0·5–4 cm. Flowers 2–4 cm, greyish-green, the limb reddish-brown; pedicels shorter than the petioles. Capsule 0·8–1·2 cm, subglobose. $2n = 26$. *Rocky and stony places.* ● *Sardegna, Corse.* Co Sa.

18. A. microstoma Boiss. & Spruner in Boiss., *Diagn. Pl. Or. Nov.* **1(5)**: 50 (1844). Stems 10–40 cm, simple or branched, pubescent; tuber cylindrical. Leaves 1–4·5 cm, ovate-oblong, cordate; petiole 0·2–2 cm. Flowers 1·5–3 cm, clavate with apical pore, brownish; limb very small or absent. Capsule 1–2 cm, globose. ● *S. & C. Greece.* Gr.

19. A. sicula Tineo in Guss., *Fl. Sic. Syn.* **2**: 878 (1845). Stems up to 50 cm, usually simple; tuber globose. Leaves 2–12 cm, ovate-triangular to ovate, cordate, acute to acuminate; petiole 1–7 cm. Flowers 2–3 cm, yellow. Capsule 1–1·5 cm, subglobose. $2n = 16$. *Woods in the mountains.* ● *Sicilia.* Si.

45. RAFFLESIACEAE

(*Cytinaceae*)

EDIT. D. A. WEBB

Perennial herbs without chlorophyll, parasitic on the roots of other plants. Flowers unisexual. Perianth consisting of a single, more or less petaloid whorl. Ovary inferior; fruit a berry.

1. Cytinus L.

D. A. WEBB AND J. R. AKEROYD

Parasitic (in Europe) on members of the Cistaceae. Stem simple, stout, with fleshy scale-leaves. Flowers subsessile, in a dense spike, the lower female, the upper male. Perianth tubular, with 4 spreading lobes. Stamens 8, united into a column. Style single; stigma capitate. Seeds numerous, embedded in a sweet, viscid pulp.

Perianth and fruit yellow **1. hypocistis**
Perianth white or pale pink; fruit white **2. ruber**

1. C. hypocistis (L.) L., *Syst. Nat.* ed. 12, **2**: 602 (1767). Stem 3–7 cm, rarely longer. Scale-leaves densely imbricate, ovate-oblong, ciliate, yellow, orange or red. Flowers 5–10(–18), each subtended by 2 usually pubescent bracteoles of the same colour as the leaves; perianth usually pubescent, bright yellow. Fruit yellow. *S. & W. Europe, northwards to* c. 46° *N. in W. France.* Al Bl Co Cr Ga Gr Hs It Ju Lu Sa Si Tu.

1 Perianth *c.* 12 mm, scarcely exceeding the bracteoles (a) subsp. **hypocistis**
1 Perianth at least 16 mm, distinctly exceeding the bracteoles
2 Flowers usually more than 12 in number (d) subsp. **pityusensis**
2 Flowers less than 12 in number
3 Scale-leaves usually yellow or orange; flowers often more than 20 mm. (b) subsp. **macranthus**
3 Scale-leaves usually red; flowers *c.* 18 mm. (c) subsp. **orientalis**

(a) Subsp. **hypocistis**: Scale-leaves usually yellow or orange. Flowers 5–10. Perianth *c.* 12 mm, scarcely exceeding the bracteoles. *On white-flowered species of* Cistus; *perhaps also on* Halimium *species. Throughout the range of the species.*

(b) Subsp. **macranthus** Wettst., *Ber. Deutsch. Bot. Ges.* **35**: 95 (1910): Scale-leaves usually yellow or orange. Flowers 5–10. Perianth 20–25 mm, distinctly exceeding the bracteoles. *On* Halimium *species. Portugal; W. Spain.*

(c) Subsp. **orientalis** Wettst., *op. cit.* 97 (1910): Scale-leaves usually red. Flowers 5–10. Perianth *c.* 18 mm, distinctly exceeding the bracteoles. *On* Cistus parviflorus. *S. Greece & Kriti; rare.*

(d) Subsp. **pityusensis** Finschow, *Veröff. Überseemus. Bremen A*, **4**: 110 (1974): Scale-leaves red or yellow. Flowers 13–18. Perianth *c.* 18 mm, distinctly exceeding the bracteoles. *On* Cistus clusii. ● *Islas Baleares (Ibiza).*

2. C. ruber (Fourr.) Komarov, *Fl. URSS* **5**: 442 (1936) (*C. hypocistis* subsp. *kermesinus* (Guss.) Wettst.). Like **1** but with stem up to 12 cm; scale-leaves and bracteoles deep crimson or bright red; perianth *c.* 12–15 mm, slightly exceeding the bracteoles, ivory-white or pale pink. Fruit white. *On pink-flowered species of* Cistus. *Mediterranean region and Portugal.* Bl Co Cr Ga Gr Hs It Lu Sa Si Tu.

BALANOPHORALES

46. BALANOPHORACEAE

EDIT. D. A. WEBB

Herbs without chlorophyll, parasitic on the roots of other plants. Flowers usually unisexual, in dense, spicate or capitate inflorescences. Fruit a small, 1-seeded nut; seed with abundant endosperm and minute embryo.

1. Cynomorium L.

D. A. WEBB

Perennial. Male, female and hermaphrodite flowers present. Perianth of 1–5 linear to cuneate or oblanceolate segments, partly fused to the ovary in female and hermaphrodite flowers. Stamen 1, epigynous. Style and stigma 1.

Literature: J. Léonard, *Bull. Jard. Bot. Nat. Belg.* **56**: 301–304 (1986).

1. C. coccineum L., *Sp. Pl.* 970 (1753). Plant dark red to purplish-black. Stems 15–30 cm, arising from a branched, underground rhizome, erect, stout, fleshy, bearing numerous triangular-lanceolate scale-leaves. Inflorescence 6–12 × 2–4 cm, terminal, cylindrical-clavate, with a stout, fleshy axis on which are numerous dense, subcapitate cymes subtended by triangular-peltate, deciduous bracts. *Parasitic on various plants, usually in saline habitats. Mediterranean region, mainly in the west; S. Portugal; very local.* Bl Co Hs It Lu Sa Si.

POLYGONALES

47. POLYGONACEAE

EDIT. D. A. WEBB

Herbs, shrubs or climbers. Leaves nearly always alternate; stipules often united to form a membranous sheath (ochrea). Flowers hermaphrodite or unisexual; perianth 3- to 6-merous, herbaceous, often enlarging and becoming membranous in fruit. Stamens usually 6–9. Ovary superior, unilocular; styles 2–4; ovule solitary, basal. Fruit a trigonous or lenticular nut.

1 Erect, much-branched shrubs
 2 Stamens 12 or more **11. Calligonum**
 2 Stamens 6–8
  3 Leaves at least 2 cm; flowers in panicles **8. Rumex**
  3 Leaves less than 2 cm; flowers in short racemes **12. Atraphaxis**
1 Herbs, dwarf shrubs or woody climbers
 4 Perianth-segments 3 or 4
  5 Perianth-segments 3; stamens 3 **1. Koenigia**
  5 Perianth-segments 4; stamens 6 **6. Oxyria**
 4 Perianth-segments 5 or 6
  6 Woody climbers
   7 Leaves 3–6 cm wide, cordate **3. Fallopia**
   7 Leaves less than 3 cm wide, not cordate **10. Muehlenbeckia**
  6 Herbs or dwarf shrubs
   8 Outer fruiting perianth-segments with 3 stout spines **9. Emex**
   8 Outer fruiting perianth-segments not spiny
    9 Leaves palmately lobed; stamens 9 **7. Rheum**
    9 Leaves not palmately lobed; stamens 8 or fewer
     10 Perianth-segments 6, the inner much larger than the outer in fruit **8. Rumex**
     10 Perianth-segments 5, equal in fruit or the outer larger
      11 Outer perianth-segments winged or keeled in fruit
       12 Twining annuals; stigmas compact, subsessile **3. Fallopia**
       12 Erect, rhizomatous perennials; stigmas fimbriate; styles fairly long **4. Reynoutria**
      11 Outer perianth-segments not winged or keeled in fruit
       13 Leaves deltate-cordate, about as wide as long; fruit at least twice as long as perianth **5. Fagopyrum**
       13 Leaves distinctly longer than wide, never deltate and rarely cordate; fruit included in the perianth or protruding for less than ½ its length **2. Polygonum**

1. Koenigia L.

†T. G. TUTIN

Annual herbs. Leaves usually subopposite. Perianth-segments 3, sepaloid. Stamens 3, alternating with 3 gland-like staminodes. Styles 2.

1. K. islandica L., *Mantissa* 35 (1767). Stems 1–6 cm, erect, reddish. Leaves 3–5 mm, broadly elliptical; petiole *c.* 1 mm; stipules hyaline. Flowers in terminal and axillary clusters; perianth-segments *c.* 1 mm, pale green. $2n = 14, 28$. *Open habitats. Arctic Europe, extending southwards in the mountains to* 56°30′ N. *in Scotland.* Br Fa Fe Is No Rs(N) Sb Su.

2. Polygonum L.

D. A. WEBB & A. O. CHATER (EDITION 1) REVISED BY J. R. AKEROYD (EDITION 2)

Herbs or dwarf shrubs. Leaves variously shaped, always distinctly longer than wide. Perianth-segments usually more or less equal, free or united in lower part, petaloid at least in part, not winged or keeled. Stamens 8, rarely fewer. Stigmas 2 or 3. Nut lenticular or trigonous, not winged, enclosed in the persistent perianth or protruding from it for less than ½ its length.

The sections set out below are often elevated to generic rank, by reason mainly of constant differences in the structure of the pollen. Differences in habit, ochrea, inflorescence, floral parts and chromosome number show some correlation with the differences in pollen-structure, but not with sufficient constancy to provide satisfactory generic diagnoses.

A number of hybrids have been reported, especially in Sect. *Persicaria*; none appears to be widespread. Some, at least, of these supposed hybrids are referable either to one of the parents or to a third species, but it appears that **28** can hybridize with **25**, **26** and **29**, and **23** with **25** and **28**. These hybrids are intermediate in appearance between the parents.

In addition to the species described below, several robust, perennial species are cultivated in gardens. Among these **P. molle** D. Don, *Prodr. Fl. Nepal.* 72 (1825), from the Himalaya, and **P. weyrichii** Friedrich Schmidt Petrop. in Maxim., *Prim. Fl. Amur.* 234 (1859), from Sakhalin and Japan, are perhaps becoming naturalized.

1 Flowers in wide, diffuse panicles; perianth usually white or cream (Sect. *Aconogonon*)
 2 Leaves not more than 3 cm wide, cuneate at the base; ochreae hyaline, evanescent
  3 Branches few, ascending; nut 4–5 mm, exceeding the perianth **41. alpinum**
  3 Branches many, divaricate; nut 3–4 mm, enclosed within the perianth **42. laxmannii**
 2 Leaves 2–8 cm wide, ±truncate at the base; ochreae brown, persistent
  4 Flowers white, in lax panicles; leaves 3–8 cm wide, glabrous or with brownish tomentum beneath **43. polystachyum**
  4 Flowers usually pink, in dense panicles; leaves 2–4 cm wide, with dense pinkish-brown tomentum beneath **44. campanulatum**

1 Flowers in spikes or axillary clusters; perianth often pink or green
5 Ochreae entire or fimbriate but scarcely lacerate; flowers usually in dense spikes, rarely in lax, leafless spikes
6 Rhizomatous perennial
7 Rhizome extensively creeping; styles 2, rarely 3, united in lower part
8 Ochreae not ciliate; spikes stout, dense **31. amphibium**
8 Ochreae ciliate; spikes slender, rather lax **32. hydropiperoides**
7 Rhizome short, stout; styles 3, free (Sect. *Bistorta*)
9 Leaf-margin revolute; lower flowers of spike usually replaced by bulbils **40. viviparum**
9 Leaf-margin plane; bulbils not present
10 Petiole of lower leaves winged; upper leaves not amplexicaul **38. bistorta**
10 Petiole of lower leaves not winged; upper leaves amplexicaul **39. amplexicaule**
6 Annual, or perennial with rooting branches or epigeal stolons
11 Stem furnished, at least in lower part, with recurved prickles **37. sagittatum**
11 Stem without prickles
12 Ochreae and petioles densely hairy
13 Stems not more than 30 cm, decumbent; inflorescence capitate **36. capitatum**
13 Stems at least 50 cm, erect; inflorescence spicate
14 Leaves lanceolate to ovate-lanceolate, whitish; flowers pink **30. lanigerum**
14 Leaves ovate, green; flowers reddish **33. orientale**
12 Ochreae and petioles glabrous or sparsely hairy
15 Spikes dense and stout, with the flowers crowded and overlapping
16 Peduncles bearing numerous subsessile, yellow glands; perianth dull pink or greenish-white **29. lapathifolium**
16 Peduncles without glands; perianth bright or pale pink
17 Leaves lanceolate **28. persicaria**
17 Leaves ovate or triangular
18 All leaves broadly ovate, turning blue on drying; ochreae $\pm$hyaline; inflorescence not capitate **34. tinctorium**
18 Upper leaves triangular, not turning blue on drying; ochreae brown; inflorescence capitate **35. nepalense**
15 Spikes lax and slender, with each flower distinctly visible
19 Cleistogamous flowers present in axils of many of the leaves; perianth furnished with many glands **26. hydropiper**
19 No cleistogamous flowers present; perianth without or with few glands
20 Ochreae with a few short cilia or none **24. foliosum**
20 Ochreae conspicuously ciliate
21 Usually perennial, with epigeal stolons or rooting, decumbent branches; larger leaves often 10 cm or more **27. salicifolium**
21 Annual; leaves seldom more than 8 cm
22 Leaves linear-lanceolate to linear-oblong, seldom more than 8 mm wide; nut 1·8–2·5 mm **23. minus**
22 Leaves lanceolate, usually more than 10 mm wide; nut 2·8–4 mm **25. mite**
5 Ochreae usually becoming deeply lacerate; flowers in small, subsessile axillary clusters, or in lax, slender, often leafy spikes (Sect. *Polygonum*)
23 Annual; stems scarcely woody at the base
24 Perianth-segments shorter than the tube; leaves very narrowly linear, almost subulate **10. salsugineum**
24 Perianth-segments at least as long as the tube; leaves linear to elliptical
25 Bracts all leaf-like, longer than the flowers
26 Nut punctulate, dull, enclosed in the perianth or slightly exceeding it **(17–20). P. aviculare** group
26 Nut glossy, distinctly exceeding the perianth
27 Leaves very glaucous and minutely papillose; sap with acid taste **22. acetosum**
27 Leaves green or slightly glaucous, not papillose; sap without acid taste
28 Leaves more than 2 mm wide; nut usually more than 2·5 mm **15. oxyspermum**
28 Leaves less than 2 mm wide; nut not more than 2·5 mm **16. graminifolium**
25 At least the upper bracts inconspicuous and shorter than the flowers
29 Perianth pink or white
30 Leaves oblong to ovate, not more than 3 times as long as wide **9. floribundum**
30 Leaves linear-lanceolate, about 5 times as long as wide **13. arenarium**
29 Perianth green, sometimes with pink or white margins
31 Nut 3–5 mm, shorter than the perianth **11. bellardii**
31 Nut 1·5–3 mm, equalling or slightly exceeding the perianth
32 Flowers subsessile, crowded towards the ends of the branches; nut 2–3 mm **12. patulum**
32 Flowers distinctly pedicellate, not crowded towards the ends of the branches; nut 1·5–2 mm **14. albanicum**
23 Perennial; stems woody at the base
33 Ochreae in inflorescence longer than the internodes, those in middle part of the stem usually at least $\frac{1}{2}$ as long as the internodes
34 Stock forming a dense, branched, woody mat; stems 3–6 cm (mountains of Kriti) **7. idaeum**
34 Stock sparingly branched; stems usually more than 10 cm
35 Ochreae with 8–12 strong, branched veins; nut 3·5–5 mm **6. maritimum**
35 Ochreae with not more than 8 faint veins; nut 2–3 mm
36 Leaves usually more than 5 mm wide; perianth becoming hard in fruit **8. cognatum**
36 Leaves not more than 5 mm wide; perianth becoming scarious in fruit
37 Leaves linear to linear-oblong, acute; nut 2–3 mm **5. romanum**
37 Leaves oblong to lanceolate, obtuse or subacute; nut 1·5–2 mm **21. longipes**

33 Ochreae shorter than all internodes (except on unexpanded young stems) and much shorter than those in middle part of the stem
38 At least the upper bracts scarious and shorter than the flowers, or bracts caducous
39 Perianth less than 4 mm; ochreae ±truncate **1. scoparium**
39 Perianth 4–5 mm; ochreae lacerate **3. tenoreanum**
38 Bracts all leaf-like, at least as long as the flowers and usually longer, not caducous
40 Bracts much shorter than the lower leaves
41 Stem 30–100 cm; leaves 2–4 cm **2. equisetiforme**
41 Stem 10–25 cm; leaves less than 2 cm **4. icaricum**
40 Bracts scarcely shorter than the lower leaves
42 Leaves more than 6 mm wide **15. oxyspermum**
42 Leaves less than 6 mm wide
43 Leaves linear to linear-oblong, acute; nut 2–3 mm **5. romanum**
43 Leaves oblong to lanceolate, obtuse or subacute; nut 1·5–2 mm **21. longipes**

Sect. POLYGONUM. Herbs or dwarf shrubs. Leaves small; ochreae silvery- or whitish-hyaline, at least in upper part, usually eventually lacerate. Flowers axillary, solitary or in small clusters, forming very lax spikes. Perianth petaloid to more or less sepaloid; stamens 3–8; styles 3 (rarely 2), usually short; nut usually trigonous.

1. P. scoparium Req. ex Loisel., *Mém. Soc. Linn. Paris* **6**: 410 (1827). Perennial, with a branched, woody stock. Stems 50–120 cm, erect, sparingly branched. Leaves elliptical, caducous. Ochreae 3–5 mm, much shorter than the internodes, reddish-brown, more or less truncate. Flowers pink or white, solitary or in pairs, in very lax terminal spikes; bracts scarious, shorter than the flowers. Nut glossy, scarcely exceeding the perianth. $2n = 20$. *Damp, open habitats.* ● *Corse; Sardegna.* Co Sa.

2. P. equisetiforme Sm. in Sibth. & Sm., *Fl. Graec. Prodr.* **1**: 269 (1809). Perennial with a branched, woody stock. Stems 30–100 cm, procumbent or ascending, usually branched. Leaves 2–4 cm, oblong to linear, acute, with wrinkled margins, the lower often caducous. Ochreae hyaline, brownish towards the base, much shorter than the internodes. Flowers pink or white, in axillary clusters of 2–3, forming long, usually lax, terminal spikes; bracts herbaceous, equalling or slightly exceeding the flowers but much smaller than the lower leaves. Perianth 2·5–4 mm; styles 2–3. Nut 2–2·5 mm, glossy. $2n = 20$. *Chiefly ruderal. S. Europe; local.* Az Bu Cr Gr Hs Lu *Si Tu.

P. papillosum Hartvig, *Willdenowia* **19**: 75 (1989), from serpentine soils in C. Greece, like **2** but with stems erect, leaves 1–2 cm and conspicuously papillose-veined, and perianth-segments 2–3 mm, is related to **1–4**. This group of species requires further investigation.

3. P. tenoreanum Nardi & Raffaelli, *Webbia* **31**: 516 (1977) (*P. elegans* Ten., non Aiton fil., *P. equisetiforme* var. *elegans* Fiori). Like **2** but stems up to 150 cm, erect; leaves all caducous, margins revolute but not wrinkled; inflorescence usually more branched, the branches slender; bracts absent or inconspicuous; perianth 4–5 mm; nut 2·5–4 mm. $2n = 20$. *Open habitats, often on clay soils.* ● *S. Italy.* It.

4. P. icaricum Rech. fil., *Magyar Bot. Lapok* **33**: 8 (1934). Like **2** but smaller and more condensed; stock stouter, with very numerous crowded, erect, sparingly branched, slender stems 10–25 cm; leaves 7–15 mm, persistent; flowers solitary or in pairs in much shorter spikes; nut *c.* 2 mm. *Rock-crevices. N. Aegean region (Samothraki).* Gr.

Recorded elsewhere only from Ikaria.

5. P. romanum Jacq., *Obs. Bot.* **3**: 8 (1768). Perennial with woody stock. Stems (20–)40–100 cm, procumbent or ascending, densely leafy above, usually with leafy shoots in the leaf-axils. Leaves 10–30 × 1–5 mm, linear to linear-oblong, acute, somewhat glaucous; margins plane. Ochreae hyaline, with rather faint veins, much shorter than the internodes, at least in the middle of the stem. Flowers white or pink, in numerous axillary clusters; bracts leaf-like. Nut 2–3 mm, finely striate, dull. *Chiefly ruderal.* ● *C. & S. Italy; S. France; Islas Baleares.* Bl Ga It.

1 Leaves 1–2 mm wide **(b) subsp. balearicum**
1 Leaves 2–5 mm wide
2 Pedicels longer than the perianth; perianth-segments patent **(a) subsp. romanum**
2 Pedicels ±equalling the perianth; perianth-segments suberect **(c) subsp. gallicum**

(a) Subsp. **romanum**: Stems 40–100 cm, procumbent or weakly ascending, much branched. Leaves 10–20(–30) × 2–5 mm. Pedicels longer than the perianth. Perianth-segments patent; nut 2–3 mm. *C. & S. Italy.*

(b) Subsp. **balearicum** Raffaelli & L. Villar, *Collect. Bot. (Barcelona)* **17**: 50 (1988): Like subsp. (a) but stems 20–60 cm, more slender, little-branched; leaves 1–2 mm wide. *Islas Baleares (Mallorca).*

(c) Subsp. **gallicum** (Raffaelli) Raffaelli & L. Villar, *loc. cit.* (1988): Stems 20–60 cm, ascending, little branched. Leaves 10–15(–20) × 2–4 mm. Pedicels more or less equalling the perianth. Perianth-segments suberect; nut *c.* 2 mm. *Coast of S. France (Hérault).*

6. P. maritimum L., *Sp. Pl.* 361 (1753). Perennial, with stout, woody stock. Stems 10–50 cm, procumbent, branched, stout. Leaves 5–25 mm, narrowly elliptical, acute, glaucous, often blackening on drying; margins revolute. Ochreae reddish-brown at the base, with 8–12 conspicuous, branched veins, silvery-hyaline distally, longer than most of the internodes. Flowers pink or whitish, solitary or in axillary clusters of 2–4; bracts leaf-like. Nut 3·5–5 mm, equalling or slightly exceeding the perianth, glossy. $2n = 20$. *Maritime sand and shingle. Shores of Atlantic, Mediterranean and Black Sea, northwards to S.E. Ireland.* Al Az Bl Br Bu Co Cr Ga Gr Hb Ho Hs It Ju Lu Rm Rs(K) Sa Si Tu.

7. P. idaeum Hayek, *Prodr. Fl. Penins. Balcan.* **1**: 110 (1924). Like **6** but much condensed; stock forming a dense, woody mat; stems 3–6 cm; leaves 3–12 mm, elliptical, obtuse, with prominent veins on lower surface; ochreae silvery-hyaline throughout; flowers solitary. *Clayey depressions on limestone mountains.* ● *Kriti.* Cr.

8. P. cognatum Meissner, *Monogr. Gen. Polyg. Prodr.* 91 (1826) (*P. alpestre* C. A. Meyer). Perennial, with a branched, woody stock. Stems 5–30 cm, procumbent. Leaves 10–20 × 5–9 mm, oblong, subacute or mucronate. Ochreae hyaline, usually longer than the internodes. Flowers solitary or in axillary clusters of 2–5; bracts as large as the lower leaves. Perianth pinkish; tube at least as long as the segments, becoming hard in fruit. Nut 3 mm, glossy, included in the perianth. *Locally naturalized as a ruderal.* [Bu Ga.] (*S.W. Asia.*)

9. P. floribundum Schlecht. ex Sprengel, *Syst. Veg.* **2**: 257 (1825). Annual. Stems long, stout, freely branched. Leaves 7–25 mm, oblong or ovate, subacute. Flowers in axillary clusters of 3–10, most of them aggregated into numerous, fairly dense, stout, spike-like racemes at the ends of the branches; lower bracts leaf-like, upper ones shorter than the flowers. Pedicels slender, longer than the perianth; perianth 2 mm, pink, with patent segments, enclosing the lenticular nut. *Dry steppes and semi-deserts. S.E. Russia.* Rs(E). (*W. Asia.*)

10. P. salsugineum Bieb., *Beschr. Länd. Terek Casp.* 169 (1800). Erect, bushy annual with very slender, branched stems 5–20 cm. Leaves 3–12 mm, narrowly linear, almost subulate, very few. Ochreae reddish-brown. Flowers subsessile, solitary in the axils of minute scarious bracts. Perianth pink or yellowish; segments shorter than the tube. Nut long and narrow, exceeding the perianth, dull. *Gypsaceous and saline habitats. S.E. Russia and Krym.* Rs(C, K, E).

P. samarense H. Gross, *Bot. Jahrb.* **49**: 340 (1913), and **P. aschersonianum** H. Gross, *op. cit.* 341 (1913), described from S.E. Russia, both differ principally from **10** by the somewhat longer leaves and glossy nut. They may be variants of **10** or hybrids between **10** and **17** or **20**.

11. P. bellardii All., *Fl. Pedem.* **2**: 207 (1785) (*P. patulum* auct., non Bieb.; incl. *P. kitaibelianum* Sadler, *P. novoascanicum* Klokov). Annual. Stem 20–100 cm, erect, usually much-branched. Leaves 25–45 mm, linear-lanceolate to elliptic-oblong, caducous. Flowers solitary or in axillary clusters of 2–4, subsessile, forming lax spikes; lower bracts leaf-like, the upper scarious and shorter than the flowers. Perianth 3–5 mm, divided almost to the base; segments greenish, erect. Nut 3–5 mm, glossy, not exceeding the perianth. $2n = 20$. *Cultivated ground. C. & S. Europe, extending northwards to* c. 49° *N. in France and* c. 55° *N. in S. Ural.* Al Au Bl Bu Co Cz Ga Gr Hs Hu It Ju Rm Rs(C, W, K, E) Sa Si Tu.

12. P. patulum Bieb., *Fl. Taur.-Cauc.* **1**: 304 (1808) (*P. gracilius* (Ledeb.) Klokov, *P. bellardii* auct., non L., *P. bordzilowskii* Klokov, *P. cretaceum* Komarov, *P. kotovii* Klokov, *P. spectabile* Lehm.). Like **11** but more slender; leaves smaller and narrower, the uppermost linear; inflorescence more compact, with flowers crowded towards the ends of the branches; nut (1·5–)2–3 mm, equalling or slightly exceeding the perianth. *Dry and sandy habitats, and as a ruderal. E.C. & S.E. Europe.* Hu Rm Rs(C, W, K, E).

13. P. arenarium Waldst. & Kit., *Pl. Rar. Hung.* **1**: 69 (1801). Annual. Stems 20–50 cm, diffusely branched from the base, procumbent or ascending. Leaves linear-lanceolate, acute, usually caducous. Bracts small, inconspicuous. Perianth pink and white, conspicuous, with patent segments. Nut 2 mm. *Mainly in cultivated ground, especially on sandy soils. S. & E. Europe, northwards to* 53° *N. in E. Russia.* Al Bu Cz Ga Gr Hu It Ju Rm Rs(C, W, K, E) Tu.

(a) Subsp. **arenarium** (incl. *P. pseudoarenarium* Klokov, *P. janatae* Klokov, *P. venantianum* G. C. Clementi): Flowers rather distant, solitary or in pairs on the main branches, clustered and forming short, condensed terminal spikes on the lateral branches. Nut glossy. $2n = 20$. *S.E. & E.C. Europe.*

(b) Subsp. **pulchellum** (Loisel.) Thell., *Fl. Adv. Montpellier* 186 (1912) (*P. pulchellum* Loisel.): Flowers solitary or in pairs, forming lax spikes throughout the inflorescence. Nut dull or scarcely glossy. $2n = 20$. *Mediterranean region, mainly in the east.*

14. P. albanicum Jáv., *Bot. Közl.* **19**: 18 (1920). Annual. Stems 10–50 cm, diffusely branched from the base, erect. Leaves lanceolate to linear-lanceolate, acute, with a single vein, caducous. Flowers solitary or paired, rarely in clusters of 3; bracts inconspicuous, only the lowermost exceeding the flowers. Perianth 1·3–2 mm; segments longer than the tube, green with white or pink margins, connivent. Nut 1·5–2 mm, dark brown, rather glossy, distinctly exceeding the perianth. *Serpentine soils.* ● *W. part of Balkan peninsula; local.* Al Gr Ju.

15. P. oxyspermum C. A. Meyer & Bunge ex Ledeb., *Ind. Sem. Hort. Dorpat.*, suppl. 2, 5 (1824). Annual to perennial. Stems 10–100 cm, procumbent, rather stout and sometimes woody at the base. Leaves 10–30 mm, elliptical to linear-lanceolate; margins scarcely revolute. Ochreae much shorter than the internodes, mostly hyaline, with 4–6(–8) simple veins. Flowers in axillary clusters of 2–6, rarely solitary; bracts leaf-like. Nut glossy, exceeding the perianth. *Maritime sand and shingle. Coasts of most of Europe, but rare in the south-west.* Br Bu Da Fe Ga Ge Gr Hb Hs It No Po Rm Rs(N, B, W, K) Sa Si Su Tu.

1 Leaves linear-lanceolate, green; nut usually more than 5 mm, pale or greenish-brown (a) subsp. **oxyspermum**
1 Leaves elliptical, green or glaucous; nut usually not more than 5 mm, dark brown
 2 Nut up to 5·5 mm, distinctly exceeding the perianth (b) subsp. **raii**
 2 Nut not more than 3 mm, slightly exceeding the perianth (c) subsp. **robertii**

(a) Subsp. **oxyspermum**: Annual or biennial, green. Leaves linear-lanceolate. Perianth-segments with deep red margins; nut 5–6·5 mm, pale or greenish-brown, distinctly exceeding the perianth. $2n = 40$. *Baltic; Kattegat; S. Norway.*

(b) Subsp. **raii** (Bab.) D. A. Webb & Chater, *Feddes Repert.* **68**: 188 (1963) (*P. raii* Bab., *P. robertii* auct., non Loisel.; incl. *P. norvegicum* Sam. ex Lid, *P. mesembricum* Chrtek): Annual to perennial, green or somewhat glaucous. Leaves usually elliptical, very variable in width. Perianth-segments with white or pale pink margins; nut 2·5–5·5 mm, dark brown, exceeding the perianth. $2n = 40$. *N. Europe, from N.W. France to arctic Russia.*

(c) Subsp. **robertii** (Loisel.) Akeroyd & D. A. Webb, *Bot. Jour. Linn. Soc.* **97**: 344 (1988). Perennial, rather woody at the base, green or somewhat glaucous. Leaves elliptical. Perianth-segments with white margins; nut 2·5–3·5 mm, dark or blackish-brown, slightly exceeding the perianth. $2n = 40$. *W. & C. Mediterranean region, from N.E. Spain to N.E. Italy and Sicilia.*

Plants from the Black Sea and N. Aegean coasts, which have been called **P. mesembricum** Chrtek, *Preslia* **32**: 367 (1960), appear to be annual, with greyish-green or silvery leaves, ochreae with 6–8 veins, and white margins to the perianth. The nut is similar to that of subsp. (b), but in some cultivated specimens can approach that of subsp. (a). In spite of the geographical disjunction there are no constant differences between this and subsp. (b).

16. P. graminifolium Wierzb. ex Heuffel, *Verh. Zool.-Bot. Ges. Wien* **8**: 190 (1858). Annual. Stems 10–30 cm, branched, slender, decumbent. Leaves 10–20 × 1–1·5 mm, linear, subacute. Ochreae brownish, caducous, with *c.* 6 veins. Flowers solitary or in axillary clusters of 2 or 3 near the ends of the branches, subsessile. Perianth 1·5–2 mm, the segments longer than the tube. Nut 2–2·5 mm, exceeding the perianth, glossy. ● *Lower and middle Danube valley.* Cz Hu Ju Rm.

(17–20). P. aviculare group. Annuals. Stems erect, ascending or procumbent, 10–60 cm. Leaves variable in shape and size, entire, acute to obtuse, green or somewhat glaucous. Ochreae silvery-hyaline, often brownish near the base, with few, faint veins, lacerate. Flowers solitary or in axillary clusters of 2–6; bracts leaf-like. Perianth-segments greenish with white, pink or reddish margins. Nut 1·5–4 mm, punctulate, dull, included in or slightly exserted from the perianth.

A complex of variable, closely related species which are widespread in ruderal and other open habitats. Many taxa have been treated as species by different authors, but four principal variants can be distinguished. The treatment below follows B. T. Styles, *Watsonia* **5**: 177–214 (1962) and M. Raffaelli, *Webbia* **35**: 361–406 (1982).

1 Perianth-tube usually at least ½ as long as the segments; leaves uniform in size; plant usually procumbent **20. arenastrum**
1 Perianth-tube less than ½ as long as the segments; leaves on the main stem much longer than those on the branches
2 Larger leaves 1–5 mm wide; perianth-segments narrow, not overlapping **19. rurivagum**
2 Larger leaves usually 5–20 mm wide; perianth-segments broad, overlapping
3 Larger leaves lanceolate to ovate; petiole included in the ochrea; nut 2·5–3·5 mm **17. aviculare**
3 Larger leaves obovate; petiole distinct, partly exserted from the ochrea; nut 3·5–4·5 mm **18. boreale**

17. P. aviculare L., *Sp. Pl.* 362 (1753) (*P. monspeliense* Pers., *P. heterophyllum* Lindman, *P. littorale* auct.). Leaves 15–50 × 5–18 mm, lanceolate to ovate, acute to subacute, those on the main stem much larger than those on the branches. Ochreae silvery-hyaline, usually less than ½ as long as the internodes; petioles very short, included in the ochreae. Perianth-tube very short, the segments with pink or white margins, wide and overlapping. Nut 2·5–3·5 mm, included in the perianth. $2n = 60$. *Throughout Europe.* All territories, but only as a casual in Sb.

18. P. boreale (Lange) Small, *Bull. Torrey Bot. Club* **21**: 479 (1894). Like **17** but leaves obovate, those on the main stems 25–50 × 5–20 mm, those on the branches much smaller; petioles well exserted from the ochreae; perianth-segments with conspicuous, bright pink margins; nut 3·5–4·5 mm. $2n = 40$. *N.W. Europe, from N. Scotland northwards.* Br Fa Fe Is No Su.

19. P. rurivagum Jordan ex Boreau, *Fl. Centre Fr.* ed. 3, **2**: 560 (1857). Like **17** but usually erect; leaves linear to linear-lanceolate, acute or acuminate, those on the main stem 15–35 × 1–5 mm; ochreae longer (up to 12 mm), reddish-brown or brownish below; flowers few; perianth-segments narrow, scarcely or not overlapping, reddish; nut 2·5–3·5 mm, slightly exserted from the perianth. $2n = 60$. *Usually a weed in cultivated fields. S., W. & W.C. Europe, extending northwards to Sweden.* Au Be Bl Br Bu Co Cz Da Ga Ge Gr ?Ho Hs It Ju Lu Sa Si Su.

P. neglectum Besser, *Enum. Pl. Volhyn.* 45 (1821), originally described from Lithuania, is said to differ in its more or less uniformly sized leaves, its frequently procumbent habit, and its smaller nut. Plants from Russia and from C. Europe have been referred here, but the status of the taxon is not clear. **P. scythicum** Klokov in Kotov, *Fl. RSS Ucr.* **4**: 650 (1952) from the S. Ukraine is very similar.

20. P. arenastrum Boreau, *Fl. Centre Fr.* ed. 3, **2**: 559 (1857) (*P. aequale* Lindman, *P. littorale* auct., *P. aviculare* auct.; incl. *P. propinquum* Ledeb., *P. microspermum* Jordan ex Boreau, *P. calcatum* Lindman, *P. acetosellum* Klokov). Stems usually less than 30 cm, procumbent, freely branched, forming a mat. Leaves 5–20 × 2–5 mm, elliptical or lanceolate, more or less uniform in size, often obtuse. Ochreae silvery-hyaline with few, faint veins, less than ½ as long as the internodes. Flowers solitary or in few-flowered axillary clusters. Perianth-tube at least ½ as long as the segments. Nut 1·5–2·5 mm, included in the perianth, often with 1 side much narrower than the other 2. $2n = 40$. *Often in trampled habitats. Throughout most of Europe except the extreme north; precise distribution still unknown.*

21. P. longipes Halácsy & Charrel, *Österr. Bot. Zeitschr.* **90**: 164 (1890). Like **20** but perennial, somewhat woody at the base; stems up to 60 cm, with long internodes; leaves oblong to lanceolate, obtuse or subacute, greyish-green, crowded on the upper part of stems or on short, lateral branches; bracts usually smaller than the leaves; perianth 1·5–2 mm; nut 1·5–2 mm, finely punctulate-striate, rather glossy, usually slightly exceeding the perianth. *Dry, open ground.* ● *S. part of Balkan peninsula and Aegean region.* Al Cr Gr Ju Tu.

Similar plants from Sicilia and Calabria that have been called **P. gussonei** Tod., *Fl. Sicula Exsicc.* 102, 172 (1864), require further investigation.

22. P. acetosum Bieb., *Fl. Taur.-Cauc.* **1**: 304 (1808). Annual. Stems procumbent or ascending, branched; sap with acid taste. Ochreae silvery-hyaline, with very faint veins. Leaves 2–3 cm, linear-oblong, obtuse, rarely acute, somewhat fleshy, veinless, covered with minute, white tubercles and appearing glaucous. Flowers in clusters of 3–7. Perianth bluish-green; segments with white margins. Nut 1·7–3 mm, slightly glossy. *W. Kazakhstan; S.E. Russia.* Rs(E). (*S.W. Asia.*)

Sect. PERSICARIA (Miller) DC. Herbs. Ochreae usually brownish, entire or ciliate but not lacerate. Flowers in spikes; bracts few and inconspicuous. Perianth usually petaloid; stamens 5–8; styles 2, rarely 3, united below. Nut lenticular or trigonous. (Incl. Sect. *Cephalophilon* Meissner, *Echinocaulon* Meissner.)

23. P. minus Hudson, *Fl. Angl.* 148 (1762). Glabrous annual. Stems 10–40 cm, slender, decumbent or ascending, more rarely erect. Ochreae with long cilia. Leaves 25–75 × 3–9 mm, usually 6–9 times as long as wide, narrowly oblong-lanceolate, subsessile. Spikes slender, lax, often interrupted below, erect. Perianth 2–2·5 mm, deep pink (rarely white). Nut 1·8–2·5 mm, black, glossy. $2n = 40$. *Wet places. From N. Spain and Macedonia northwards to 67° N. in Russia.* Au Be Br Bu Cz Da Fe Ga Ge Gr Hb He Ho Hs Hu It Ju No Po Rm Rs(N, B, C, W, K, E) Su [Lu].

24. P. foliosum H. Lindb., *Meddel. Soc. Fauna Fl. Fenn.* **23**: 3 (1900). Like **23** but ochreae with short cilia or none; leaves linear, *c.* 10 times as long as wide; inflorescence sparser and more leafy; perianth and nut 1·5–2 mm. $2n = 20$. *Wet places. Fennoscandia; N.W. Russia.* Fe No Rs(?B, C, ?E) Su.

25. P. mite Schrank, *Baier. Fl.* **1**: 668 (1789). Glabrous annual. Stems 10–60 cm, slender, erect. Ochreae with long cilia. Leaves 3–10 × 0·5–2 cm, usually less than 6 times as long as wide, elliptic-lanceolate, subsessile. Inflorescence lax, slender, interrupted. Perianth 3–4·5 mm, pink (rarely white), often with a few glands. Nut

2·8–4 mm, black, glossy, lenticular or trigonous. $2n = 40$. *Damp places and lake shores. Most of Europe except the north.* Al Au Be Br Bu Co Cz Ga Ge Gr *Hb He Ho Hs Hu It Ju Po Rm Rs(B, C, W) Sa Si ?Tu.

Like **26** in general habit and leaf-shape but easily distinguished by the glossy nut and the absence of cleistogamous flowers, and usually fewer and less conspicuous glands on the perianth. It is usually distinguishable from **23** by the dimensions of the leaves and the size of the nut, but a few plants (possibly hybrids) are difficult to assign to either species and many records are ambiguous.

26. P. hydropiper L., *Sp. Pl.* 361 (1753). A glabrous, acrid-tasting annual. Stems 20–80 cm, more or less erect. Ochreae fringed with a few short cilia. Leaves 5–12 × 1–2·5 cm, lanceolate, acute or acuminate, subsessile. Flowers in numerous, usually very lax, slender, nodding spikes; a cleistogamous flower is also present in many of the leaf-axils. Perianth 3–5 mm, pink or greenish-white, covered with flat, transparent glands that become brownish and raised on drying. Nut *c.* 3 mm, punctulate, dull. $2n = 20$. *Damp places; calcifuge and somewhat nitrophilous. Europe, except the extreme north, Islas Baleares and Kriti.* All except Bl Cr Fa Is Sb.

27. P. salicifolium Brouss. ex Willd., *Enum. Pl. Horti Berol.* **1**: 428 (1809) (*P. serrulatum* Lag., *P. serrulatoides* H. Lindb.). Perennial, with rooting, procumbent or decumbent branches. Stems 30–70 cm, ascending, rather stout. Ochreae strongly ciliate. Leaves 7–15 cm, linear-lanceolate, glabrous except for stiff hairs on the margins and the veins beneath. Flowers pink, in long, lax, very slender spikes. Nut 2–2·5 mm, black, glossy, usually trigonous. *Wet places and river-banks. S. Europe.* Al Bl Bu Co Cr Ga Gr Hs It Lu Sa Si Tu.

28. P. persicaria L., *Sp. Pl.* 361 (1753). Annual. Stems 20–80 cm, erect or ascending. Ochreae shortly ciliate. Leaves up to 15 × 3·5 cm, lanceolate, often with a large, blackish spot, glabrous, or occasionally tomentose beneath. Spikes cylindrical, dense, usually stout. Perianth *c.* 3 mm, pink or white. Nut black, glossy, usually lenticular. $2n = 22, 44$. *Cultivated ground and waste places. Throughout Europe, but doubtfully native in some territories.* All except Cr Sb.

29. P. lapathifolium L., *Sp. Pl.* 360 (1753) (*P. andrzejowskianum* Klokov, *P. hypanicum* Klokov, *P. incanum* F. W. Schmidt, *P. paniculatum* Andrz., *P. zaporoviense* Klokov; incl. *P. nodosum* Pers., *P. scabrum* Moench and *P. linicola* Sutulov). Annual. Stems up to 80 cm, procumbent or erect, simple or branched, sometimes spotted with red. Ochreae entire or very shortly ciliate. Leaves ovate to linear-lanceolate, acute or obtuse, sometimes with a large, blackish spot, glabrous or densely tomentose beneath, with pellucid, often yellow glands visible from lower surface. Spikes stout and usually dense; peduncles, and sometimes also pedicels and perianth, bearing yellow, subsessile glands. Perianth dull pink or greenish-white. Nut black, glossy, usually lenticular. $2n = 22$. *Cultivated ground, waste places, river-gravels and beside lakes and ponds. Throughout Europe.* All except Cr Is Sb.

Extremely variable, especially in habit, colour of foliar glands and of perianth, and in indumentum. Its widespread dissemination as a weed has obscured any geographical pattern which may have existed, and there is scarcely any correlation of these characters that remains constant over a wide area. Furthermore, it is known that some of the characters, especially of habit and tomentum, which are partly determined genetically, are also very plastic phenotypically.

The only variant which is known both to retain its characteristics in cultivation and to show a well-defined geographical distribution is **P. brittingeri** Opiz, *Naturalientausch* **8**: 74 (1824) (*P. danubiale* A. Kerner), which grows on river-alluvia in the basins of the upper Danube and Rhine and perhaps elsewhere. It has procumbent, much-branched stems with very short internodes, and broadly elliptical or ovate leaves, densely tomentose beneath and with colourless glands. It may merit recognition as subsp. *brittingeri* (Opiz) Rech. fil.

P. nodosum Pers., *Syn. Pl.* **1**: 440 (1805) is a commonly recurring variant with glabrous leaves, yellow foliar glands, red-spotted stems, and pink flowers in a rather lax spike; it is characteristic of river-gravels and cultivated land in some regions. Similarly, **P. tomentosum** Schrank, *Baier. Fl.* **1**: 669 (1789) (*P. pallidum* With.), with a low habit, densely tomentose leaves and greenish-white flowers, is often seen on drying mud or as a weed in fields. In both cases, however, it is possible to find plants with some of these characteristics perfectly developed and others not at all, and it seems doubtful whether anything is gained by giving them taxonomic rank without further biosystematic study.

P. pensylvanicum L., *Sp. Pl.* 362 (1753), from North America, like **29** but somewhat more robust and erect, the peduncles usually with stipitate glands, and the perianth pink to purplish-pink, is a frequent casual in Britain and probably elsewhere.

30. P. lanigerum R. Br., *Prodr. Fl. Nov. Holl.* 419 (1810). Villous perennial; stems 80–200 cm, erect, stout, rooting at lower nodes. Leaves 10–25 × 2–6 cm, shortly petiolate, lanceolate to ovate-lanceolate, cuneate at base, acuminate, densely white-hairy especially beneath. Flowers pink, in slender, dense spikes up to 7 cm. Nut lenticular, glossy. *River-banks. S. Kriti (near Phaistos).* *Cr. (*Old World Tropics, extending to Egypt and the Jordan Valley.*)

31. P. amphibium L., *Sp. Pl.* 361 (1753). Perennial, with slender, far-creeping rhizome. Stems erect, ascending or floating, usually much-branched. Leaves of aquatic plants floating, long-stalked, oblong-ovate, truncate or subcordate at the base, glabrous; those of terrestrial plants oblong-lanceolate, rounded at the base, pubescent or appressed-hispid. Ochreae not ciliate. Spikes 2·5 × 1 cm, terminal, obtuse; flowers deep pink. Styles 2, long; nut lenticular, glossy. $2n = 66, 88, 94–96$. *Damp places or as an aquatic. Most of Europe, but rare in the extreme south.* All except Az Bl Cr Sb.

32. P. hydropiperoides Michx, *Fl. Bor.-Amer.* **1**: 239 (1803). Perennial, with slender, far-creeping rhizome. Stems 20–40 cm, decumbent or ascending; ochreae fringed with long cilia. Leaves 5–20 mm wide, linear-lanceolate, cuneate, glabrous. Flowers pink, in slender, lax, usually branched spikes. Nut trigonous, glossy. *Damp places. Açores.* Az. (*North and South America.*)

33. P. orientale L., *Sp. Pl.* 362 (1753). Densely pubescent annual. Stem 50–100 cm, erect, stout. Leaves up to 20 cm, ovate-acuminate, stalked, sometimes slightly cordate. Ochreae with green, leaf-like lobes on their distal margin. Flowers reddish-pink, in long, slender, moderately dense, branched spikes. Nut 3 mm, lenticular, glossy. $2n = 22$. *Cultivated for ornament, and locally naturalized, mainly in C. & S.E. Europe; a frequent casual elsewhere* [Al Ge ?Gr It Ju Rm Rs(C).] (*E. & S.E. Asia.*)

34. P. tinctorium Aiton, *Hort. Kew.* **2**: 31 (1789). Annual. Stems 30–80 cm, erect, usually simple, reddish. Ochreae more or less hyaline, ciliate. Leaves 4–10 × 3–4 cm, broadly ovate, glabrous, petiolate, turning dark bluish-green on drying. Flowers pink, in short, rather dense spikes, forming a terminal, leafy panicle. Nut trigonous, glossy. *Formerly cultivated in several countries as a dye-plant, and reported to be naturalized in Ukraine.* [Rs(W).] (*China.*)

35. P. nepalense Meissner, *Monogr. Gen. Polyg. Prodr.* 84 (1826). Sparsely hairy annual. Stems 5–40 cm, decumbent to weakly ascending, much branched, rooting at the lower nodes. Lower leaves ovate, auriculate-amplexicaul, acuminate; upper leaves triangular, cordate, acute; petiole winged. Ochreae with stiff, deflexed, white hairs at the base. Inflorescence capitate; bracts membranous. Perianth pink; nut 1–2 mm, lenticular or trigonous, yellowish to dark brown, dull, included within the perianth. *Naturalized in N. Italy; casual elsewhere.* [It.] (*Tropical Asia and Africa.*)

36. P. capitatum Buch.-Ham. ex D. Don, *Prodr. Fl. Nepal.* 73 (1825). Glandular-pubescent perennial. Stems up to 30 cm, decumbent, rooting at the lower nodes. Leaves 20–50 × 10–25 mm, ovate to elliptical, acute. Inflorescence 5–10 mm, capitate, dense; peduncles 10–30 mm. Perianth pink; nut *c.* 2 mm, trigonous. *Cultivated for ornament, and naturalized in Açores and N. Portugal.* [Az Lu.] (*Himalaya.*)

37. P. sagittatum L., *Sp. Pl.* 363 (1753). Annual. Stems 50–100 cm, weak, scrambling by numerous, hooked prickles. Leaves 3–6 cm, oblong-sagittate, glabrous. Flowers whitish, in small, long-stalked, subcapitate spikes. Nut glossy, trigonous. *Naturalized in S.W. Ireland (Kerry).* [Hb.] (*E. Asia, North America.*)

Sect. BISTORTA (Scop.) D. Don. Perennial herbs with short, stout rhizomes. Ochreae brown, not lacerate. Flowers in fairly dense, terminal spikes. Perianth petaloid; stamens 6–8; styles 3, long and slender. Nut trigonous.

38. P. bistorta L., *Sp. Pl.* 360 (1753). Rhizome stout, contorted. Stems 20–100 cm, erect, simple. Lower leaves 5–15 cm, ovate, obtuse, truncate at base; petioles long, broadly winged in upper part. Upper leaves triangular-lanceolate, acute or acuminate. Flowers bright pink, in a dense, cylindrical, terminal spike 20–70 × 10–15 mm. Nut 4–6 mm, glossy. $2n = 44$. *Damp, grassy places. Throughout a large part of Europe, but only on mountains in the south and absent as a native from most of Fennoscandia.* Al Au Be Br Bu Cz Ga Ge Gr He Ho Hs Hu It Ju Lu Po Rm Rs(N, B, C, W, K, E) [Da *Fe Hb No Su].

39. P. amplexicaule D. Don, *Prodr. Fl. Nepal.* 70 (1825). Stems 60–100 cm, erect. Leaves large, ovate-acuminate, cordate; the lower petiolate, the upper sessile-amplexicaul. Flowers deep red, in dense but rather slender, cylindrical spikes. Nut 4–5 mm, rarely produced in Europe. *Cultivated for ornament, and naturalized in Britain and Ireland.* [Br Hb.] (*Himalaya.*)

40. P. viviparum L., *Sp. Pl.* 360 (1753). Rhizome not contorted. Stems 5–40 cm, erect, simple. Leaves up to 10 cm but usually less, oblong- to linear-lanceolate, tapered at both ends; margins revolute. Flowers pale pink or white, in a slender, moderately dense, terminal spike 20–60 × 5–10 mm, in the lower part of which the flowers are usually replaced by brown or deep purple bulbils. Nut rarely produced. $2n = 66, 77, 88, 99$, *c.* 132. *N. Europe, and on mountains in C. and parts of S. Europe.* Al Au Br Bu Cz †Da Fa Fe Ga Ge Hb He Hs Is It Ju No Po Rm Rs(N, B, C, W) Sb Su.

Sect. ACONOGONON Meissner. Perennial herbs. Flowers in diffuse, terminal panicles. Perianth petaloid; stamens 8; styles 3. Nut trigonous.

41. P. alpinum All., *Mélang. Philos. Math. Soc. Roy. Turin (Misc. Taur.)* **5**: 94 (1774) (*P. undulatum* Murray). Stems 30–80 cm, erect, arising from a short, creeping underground rhizome. Leaves 10–30 mm wide, oblong-lanceolate, tapered at both ends. Ochreae mainly hyaline, soon disappearing. Flowers white or pale pink, in a diffuse panicle. Perianth-segments subequal. Styles very short, with large capitate stigmas. Nut 4–5 mm, exceeding the perianth, trigonous, pale brown, glossy. *Usually calcifuge. S., C. & E. Europe, mainly in the mountains.* Al Au Bu Co Ga Gr He Hs It Ju Rm Rs(*N, C, W, E) [Br Da].

42. P. laxmannii Lepechin, *Nova Acta Acad. Petrop.* **10**: 414 (1797). Like **41** but stems 10–30(–50) cm, divaricately branched from the base; leaves 3–8 mm wide, densely hairy (sometimes glabrous); nut 3–4 mm, included in the perianth. *River-gravels. N.E. Russia.* Rs(N). (*Siberia.*)

43. P. polystachyum Wall. ex Meissner in Wall., *Pl. Asiat. Rar.* **3**: 61 (1832) (*Reynoutria polystachya* (Meissner) Moldenke). Stems 60–120 cm, stout, erect, arising from a creeping underground rhizome. Leaves 10–20 × 3–8 cm, oblong-lanceolate, acuminate, truncate or shortly cuneate at the base, glabrous or hairy beneath, usually with red veins. Ochreae brown, thick, persistent, entire. Flowers white, in lax, somewhat leafy panicles with a red axis. Inner perianth-segments broadly obovate to orbicular; outer ones smaller, elliptical. Styles long and slender; stigmas small. Nut rarely produced in Europe. *Cultivated for ornament, and naturalized in C. & N.W. Europe.* [Au Br Da Ga Ge Hb He Ho No Su.] (*Himalaya.*)

44. P. campanulatum Hooker fil., *Fl. Brit. India* **5**: 51 (1886). Stoloniferous; stems up to 100 cm, erect, much branched, stout. Leaves 5–12 × 2–4 cm, lanceolate to ovate or ovate-elliptical, cuneate at the base, acute, with conspicuous whitish to pinkish-brown tomentum beneath. Inflorescence a compact, terminal panicle; flowers heterostylous. Perianth 4–5 mm, pink or white; nut *c.* 2 mm, brown, glossy, included in the perianth. *Cultivated for ornament, and naturalized in Britain and Ireland.* [Br Hb.] (*Himalaya and W. China.*)

3. Fallopia Adanson

D. A. WEBB

Stem twining or procumbent; leaves deltate or cordate-sagittate, petiolate. Ochreae truncate. Flowers in lax, spike-like or paniculate terminal and lateral inflorescences. Perianth-segments 5(6); the outer 3 larger, keeled or winged. Stamens 8. Stigmas capitate, subsessile. Not triquetrous, not exceeding the perianth. (*Bilderdykia* Dumort.)

1 Woody perennial; inflorescence paniculate **3. baldschuanica**
1 Annual; inflorescence narrow, scarcely branched
 2 Fruiting pedicels 1–3 mm; nut finely punctate, dull **1. convolvulus**
 2 Fruiting pedicels 5–8 mm; nut smooth, glossy **2. dumetorum**

1. F. convolvulus (L.) Á. Löve, *Taxon* **19**: 300 (1970). (*Polygonum convolvulus* L., *Bilderdykia convolvulus* (L.) Dumort., *Fagopyrum convolvulus* (L.) H. Gross). Puberulent or slightly mealy annual. Stem up to 1 m, angular, twining or spreading. Leaves

2–6 cm, acuminate. Flowers in narrow, spike-like inflorescences. Pedicels not more than 3 mm, even in fruit, articulated above the middle. Perianth-segments greenish-white, accrescent, the outer keeled or slightly winged in fruit, rarely with broad wings (var. *subalatum* (Lej. & Court.) Kent) as in **2**. Nut 4–5 mm, finely punctate, dull black. $2n = 40$. *Cultivated ground and waste places. Throughout Europe, but certainly introduced in the extreme north and perhaps elsewhere.* All except Is Sb.

2. F. dumetorum (L.) J. Holub, *Folia Geobot. Phytotax.* (*Praha*) **6**: 176 (1971) (*Polygonum dumetorum* L., *Bilderdykia dumetorum* (L.) Dumort., *Fagopyrum dumetorum* (L.) Schreber). Like **1** but often taller; pedicels 5–8 mm in fruit, articulated at or below the middle and usually deflexed; outer perianth-segments broadly winged in fruit, the wings decurrent on the pedicel; nut 2·5–3 mm, smooth and glossy. $2n = 20$. *Usually in hedges or among native vegetation. Most of Europe except the extreme north and many of the islands.* Al Au Be Br Bu Co Cz Da Fe Ga Ge Gr He Ho Hs Hu It Ju Lu No Po Rm Rs(N, B, C, W, K, E) Si Su.

3. F. baldschuanica (Regel) J. Holub, *loc. cit.* (1971) (*Polygonum baldschuanicum* Regel, *Bilderdykia baldschuanica* (Regel) D. A. Webb, *Fagopyrum baldschuanicum* auct.). A vigorous, woody climber, up to 6 m or more. Leaves 3–8 cm, cordate, obtuse to acuminate. Flowers in diffuse, drooping, terminal and axillary panicles. Perianth-segments white or pink in flower, pink in fruit, the outer with decurrent wings. Nut black, glossy. *Cultivated for ornament, and occasionally naturalized.* [Au Br Cz Hs Rm.] (*Tadzhikistan.*)

Some cultivated plants with white perianth-segments have been referred by many authors to **F. aubertii** (Louis Henry) J. Holub, *loc. cit.* (1971) (*Bilderdykia aubertii* (Louis Henry) Moldenke, *Polygonum baldschuanicum* auct., non Regel, *Reynoutria aubertii* (Louis Henry) Moldenke, *Polygonum aubertii* Louis Henry), from W. China. This has been distinguished by the shiny, somewhat undulate leaves, and erect panicles of flowers that show little or no pink coloration even in fruit. It seems very doubtful, however, that it is specifically distinct from **3**.

4. Reynoutria Houtt.

D. A. WEBB

Perennials with extensive subterranean rhizomes and stout, erect, annual stems. Plants functionally dioecious (rarely polygamous), but with conspicuous rudiments of stamens in the female and of gynoecium in the male flowers. Flowers in relatively small axillary panicles. Perianth accrescent; segments 5, the outer 3 keeled, eventually winged. Stamens 8. Styles 3, distinct; stigmas fimbriate. (*Fallopia* pro parte.)

Leaves seldom more than 12 cm, acuminate-cuspidate, truncate at the base; flowers white — **1. japonica**
Leaves usually more than 15 cm, acute or somewhat acuminate, somewhat cordate at the base; flowers greenish — **2. sachalinensis**

1. R. japonica Houtt., *Natuurl. Hist.* (*Handleid.*) **2(8)**: 640 (1777) (*Tiniaria japonica* (Houtt.) Hedberg, *Polygonum cuspidatum* Siebold & Zucc.). Stems 1–2·5 m, numerous, robust, glaucous, often reddish, branched above, forming a dense thicket. Leaves 5–15 cm, broadly ovate-triangular, acuminate-cuspidate, truncate at base, petiolate. Flowers white, in clusters of 2–4, arranged in axillary panicles 8–12 cm long, with slender, rather lax branches. Nut 4 mm, glossy. $2n = 44, 88$. *Cultivated for ornament, and extensively naturalized.* [Au Be Br Bu Cz Da Fe Ga Ge Hb He Ho Hs Hu It Ju Lu No Po Rm Rs(B, C) Su.] (*Japan.*)

2. R. sachalinensis (Friedrich Schmidt Petrop.) Nakai in T. Mori, *Enum. Pl. Corea* 135 (1922) (*Polygonum sachalinense* Friedrich Schmidt Petrop.). Like **1** but with stouter stems, often more than 3 m; leaves 15–40 cm, ovate-oblong, acute or slightly acuminate, somewhat cordate; panicles shorter with stouter, denser branches, the flowers greenish, usually in clusters of 4–7, sometimes polygamous. $2n = 44$. *Naturalized from gardens as* **1** *but much less frequently.* [Au Be Br Bu Cz Da Fe Ga Ge Hb Ho Po Rs(B, W) Su.] (*E. Asia.*)

The hybrid between **1** and **2** (R. × *bohemica* Chrtek & Chrtková) is established in England and Czechoslovakia, and perhaps elsewhere.

5. Fagopyrum Miller

D. A. WEBB

Erect annuals or perennials with hollow stems. Leaves triangular-sagittate, cordate. Ochreae short, truncate, entire. Flowers heterostylous, andromonoecious, in narrow, terminal and axillary, raceme-like panicles; pedicels equalling the perianth, articulated above the middle. Perianth campanulate, with 5 segments, not accrescent. Stamens 8, alternating with yellow nectaries on the disc. Styles 3, long and slender; stigmas capitate, small. Nut triquetrous, greatly exceeding the perianth.

Perianth 3–4 mm; nut with smooth faces and angles — **1. esculentum**
Perianth 2 mm; nut with rugose faces and sinuate-dentate angles — **2. tataricum**

1. F. esculentum Moench, *Meth.* 290 (1794) (*F. vulgare* T. Nees, *F. sagittatum* Gilib., *Polygonum fagopyrum* L.). Plant 15–60(–100 cm), glabrous or puberulent, tinged with red at maturity. Leaves up to 7 × 6 cm, usually slightly longer than wide, entire or sinuate, dark green, the lower stalked, the upper sessile. Axillary panicles short and compact, on long peduncles. Perianth-segments 3–4 mm, greenish-white tipped with pink, or sometimes all pink. Nut 5–6 mm, dark brown, dull at maturity, with smooth faces and acute, entire angles. $2n = 16$. *Formerly extensively cultivated (but now considerably less so) as a grain crop in most of Europe except Fennoscandia, and naturalized as a ruderal through much of Europe.* [Probably all except Cr Fa Gr Hb Is Rs(N) Sb Tu.] (*E.C. Asia.*)

2. F. tataricum (L.) Gaertner, *Fruct. Sem. Pl.* **2**: 182 (1790) (*Polygonum tataricum* L.). Like **1** but often taller, glabrous, seldom tinged with red; leaves usually wider than long, lighter green; panicles longer and laxer; perianth-segments 2 mm, usually entirely green; nut irregularly rugose with obtuse, sinuate-dentate angles. $2n = 16$. *Formerly cultivated in much of Europe, though less often than* **1**, *and widely distributed as a ruderal, and as a weed in cultures of* **1**. [Be Da Fe Ga He Ho Ju No Po Rm Rs(N, W, C, E) Su], casual elsewhere. (*C. Asia.*)

With the decrease in the cultivation of **1**, this species has probably disappeared from much of its former area of distribution.

F. dibotrys (D. Don) Hara, *Fl. East. Himalaya* 69 (1966), from the Himalaya, is perennial and up to 120 cm with larger leaves, and flowers in lax, axillary and terminal panicles. It is cultivated for ornament, and may be becoming naturalized at a few localities on the coasts of W. Europe.

6. Oxyria Hill

†T. G. TUTIN

Perennial herbs. Perianth-segments in 2 whorls of 2, the inner accrescent in fruit but not tuberculate. Stamens 6. Stigmas 2. Fruit lenticular, broadly winged.

1. O. digyna (L.) Hill, *Hort. Kew.* 158 (1768). Glabrous, up to 30 cm. Leaves 1–4 cm, almost all basal, reniform, petiolate. Inflorescence branched, leafless. Pedicels slender, articulated about the middle, thickened towards the top. Outer perianth-segments spreading or deflexed, the inner appressed. $2n = 14$. *Somewhat calcifuge. Arctic Europe, and on mountains southwards to Corse and C. Greece.* Al Au Br Bu Co Cz Fe Ga Ge Gr Hb He Hs Is It Ju No Po Rm Rs(N, W) Sb Su.

7. Rheum L.

D. A. WEBB

Robust perennial herbs with a woody rhizome. Leaves mostly basal, large, palmately lobed. Ochreae loose, persistent, not ciliate. Flowers hermaphrodite, in a panicle. Perianth-segments 6, free, equal, not accrescent. Stamens 9. Stigmas 3, subsessile. Fruit with 3 membranous wings.

Several species from temperate Asia are occasionally cultivated in gardens for their decorative foliage, and may be naturalized locally. **R. palmatum** L., *Syst. Nat.* ed. 10, 1010 (1759), from W. China, and **R. officinale** Baillon, *Adansonia* **10**: 246 (1872), **11**: 229 (1874), from the Himalaya, were formerly much cultivated as purgatives. **R. × hybridum** J. A. Murray, *Novi Comment. Gotting.* **2(5)**: 50 (1775), is cultivated extensively in Britain and occasionally elsewhere in N.W. Europe for its edible petioles (Rhubarb); it is a hybrid derived from **R. rhabarbarum** L., *Sp. Pl.* 372 (1753), from Mongolia and neighbouring territories.

Stem with a single leaf; inflorescence compact, subglobose **1. tataricum**
Stem with several leaves; inflorescence diffuse, pyramidal **2. rhaponticum**

1. R. tataricum L. fil., *Suppl.* 229 (1781). Basal leaves up to 50 cm wide, ovate to orbicular, cordate, entire, glabrous above, finely tuberculate on the lower surface and petiole. Stems stout, inclined, each bearing a single leaf. Inflorescence condensed, subglobose. Perianth-segments 3 mm, yellowish, with brown veins. Fruit *c.* 11 × 9 mm, dark brown, with rather narrow wings. *Dry, open places. S.E. Russia, Ukraine.* Rs(W, E). (*C. Asia.*)

2. R. rhaponticum L., *Sp. Pl.* 371 (1753). Like **1** in general habit, but leaves glabrous throughout and with undulate margin; stems with several leaves; panicle pyramidal, diffuse, somewhat leafy; fruit much smaller. $2n = 22$. *Wet mountain rocks.* ● *S.W. Bulgaria (Rila Planina); S. Norway (near Aurland).* *Bu No.

Closely related taxa, which have been considered to be conspecific, are found in Siberia and C. Asia; they are sometimes cultivated.

8. Rumex L.

K. H. RECHINGER (EDITION 1) REVISED BY J. R. AKEROYD (EDITION 2)

Herbs, rarely shrubs, usually with long, stout roots, sometimes rhizomatous. Leaves alternate; ochreae tubular. Flowers hermaphrodite or unisexual, arranged in whorls on simple or branched inflorescences, anemophilous. Perianth-segments in 2 whorls of 3, the outer remaining small and thin, the inner becoming enlarged and often hardened in fruit. Valves (fruiting inner perianth-segments) sometimes developing marginal teeth or dorsal tubercles as they mature. Stamens in 2 whorls of 3; anthers basifixed. Fruit a trigonous nut.

Literature: K. H. Rechinger, *Beih. Bot. Centr.* **49(2)**: 1–132 (1932). *Feddes Repert.* **31**: 225–283 (1933). *Candollea* **12**: 9–152 (1949).

1 Shrub up to 2 m; leaves broadly ovate to suborbicular **3. lunaria**
1 Herb (sometimes woody at base), or dwarf shrub up to 70 cm; leaves usually distinctly longer than wide
2 At least some leaves hastate or sagittate, with ±acute basal lobes; flowers usually unisexual
3 Valves ±equalling nut
4 Basal lobes of leaves conspicuous, sometimes numerous **1. acetosella**
4 Basal lobes of leaves minute **2. graminifolius**
3 Valves distinctly longer than nut
5 Basal leaves narrowly linear, without or with indistinct lobes **2. graminifolius**
5 Basal leaves various, if narrowly linear, then distinctly sagittate or hastate
6 Annual; valves usually more than 13 mm wide (S. Greece) **16. vesicarius**
6 Perennial; valves not more than 13 mm wide
7 Outer perianth-segments free, appressed to base of inner ones
8 Valves usually at least 9 × 10 mm; ochreae of inflorescence large, persistent; leaves often sinuate-crenate, sometimes pinnatisect **4. tingitanus**
8 Valves less than 9 × 10 mm; ochreae of inflorescence evanescent; leaves not sinuate-crenate or pinnatisect
9 Leaves narrowly linear **5. suffruticosus**
9 Leaves scutate or hastate, about as long as wide **6. scutatus**
7 Outer perianth-segments united at base, deflexed
10 Inflorescence usually lax, simple or with lateral branches sparingly branched
11 Ochreae ±entire
12 Stem not more than 20 cm, with at most 2 small cauline leaves; valves 3 mm wide **8. nivalis**
12 Stem usually more than 20 cm; cauline leaves numerous, large; valves (3–)3·5–4·5 mm wide **9. alpestris**
11 Ochreae fimbriate
13 Cauline leaves few, 6–12 times as long as wide, with long basal lobes **10. nebroides**
13 Cauline leaves fairly numerous, 2–6(–10) times as long as wide, with short basal lobes
14 Roots tuberous; valves 5–8 × 5–8 mm **7. tuberosus**

14 Roots not tuberous; valves 3–5 × 2–4 mm **11. acetosa**
10 Inflorescence usually dense, the branches repeatedly branched
15 Valves 6–9 mm wide
16 Valves rounded at apex, suborbicular **7. tuberosus**
16 Valves broadly emarginate, wider than long **15. thyrsoides**
15 Valves less than 6 mm wide
17 Upper cauline leaves 2–3 times as long as wide **12. rugosus**
17 Upper cauline leaves at least 4 times as long as wide
18 Lower leaves lanceolate to narrowly oblong, sagittate **13. thyrsiflorus**
18 Lower leaves narrowly linear, hastate **14. intermedius**
2 Leaves neither hastate nor sagittate; base of lamina cuneate or truncate, or cordate with the basal lobes short and rounded
19 Valves without corky tubercles
20 Leaves more than 4 times as long as wide
21 Stem erect; leaves undulate; flowers mostly hermaphrodite **23. pseudonatronatus**
21 Stem ±procumbent; leaves not undulate; flowers mostly unisexual **2. graminifolius**
20 Leaves not more than 4 times as long as wide
22 Basal leaves cuneate to truncate at base
23 Petiole of basal leaves shorter than lamina **22. balcanicus**
23 Petiole of basal leaves at least as long as lamina **20. aquaticus**
22 Basal leaves cordate at base
24 Basal leaves about as long as wide
25 Leaves and petioles densely papillose beneath; valves 6–8 × 8–10 mm **24. aquitanicus**
25 Leaves and petioles not papillose; valves 4·5–6 × 3·5–5 mm **19. alpinus**
24 Basal leaves distinctly longer than wide
26 Basal leaves 3–4 times as long as wide; valves almost as long as wide **25. longifolius**
26 Basal leaves $1\frac{1}{2}$–2 times as long as wide; valves distinctly longer than wide
27 Basal leaves ±acute; valves entire **20. aquaticus**
27 Basal leaves obtuse; valves minutely denticulate **21. azoricus**
19 At least 1 valve bearing a corky tubercle
28 Valves with conspicuous marginal teeth at least 1 mm long
29 Teeth of valves hooked or spirally curved at apex **35. nepalensis**
29 Teeth of valves straight or slightly curved
30 Lower leaves not more than $3\frac{1}{2}$ times as long as wide
31 Flowers in clusters of 2–4 **44. bucephalophorus**
31 Flowers in whorls of more than 4
32 Pedicels equalling or shorter than the valves, stout, articulated near the middle; leaves usually panduriform **36. pulcher**
32 Pedicels longer than the valves, slender, articulated near the base; leaves not constricted near the middle
33 Annual; teeth of valves 3–6 mm **38. dentatus**
33 Perennial; teeth of valves seldom more than 2 mm **37. obtusifolius**
30 Lower leaves at least 4 times as long as wide
34 Only 1 valve with a tubercle **42. marschallianus**
34 All valves with tubercles
35 Teeth of valves longer than width of valve
36 Stem branched from the base; infructescence reddish-brown **43. ucranicus**
36 Stem branched from a little distance above the base; infructescence golden-yellow **40. maritimus**
35 Teeth of valve ±equalling width of valve
37 Leaves obovate; valves up to 2·5 × 1 mm **43. ucranicus**
37 Leaves elliptical or lanceolate; valves more than 2·5 × 1 mm
38 Valves at least 3 mm, about $2\frac{1}{2}$ times as long as wide **39. palustris**
38 Valves not more than 3 mm, less than twice as long as wide **41. rossicus**
28 Valves entire, obscurely crenate, or with marginal teeth not more than 1 mm long
39 Lower leaves more than $3\frac{1}{2}$ times as long as wide
40 Leaves 50–100 cm; plant ±aquatic **27. hydrolapathum**
40 Leaves not more than 35 mm; plant of relatively dry places
41 Terminal inflorescence overtopped at fruiting stage by those from axillary branches; tubercles of valves rugose **17. salicifolius**
41 Terminal inflorescence not overtopped by those of lateral branches; tubercles of valves smooth
42 Valves oblong, distinctly longer than wide; tubercle occupying almost the entire width of valve **34. rupestris**
42 Valves almost as wide as long; tubercle occupying not more than $\frac{1}{2}$ the width of valve
43 Valves deltate, with short but distinct teeth **31. stenophyllus**
43 Valves suborbicular, entire or obscurely toothed
44 Leaf-margin ±plane; petiole flat above; tubercles of valves developing and hardening late **29. patientia**
44 Leaf-margin undulate; petiole canaliculate; tubercles of valves developing and hardening early **30. crispus**
39 Lower leaves not more than $3\frac{1}{2}$ times as long as wide
45 Branching strongly divaricate; leaves mostly less than 10 cm **36. pulcher**
45 Branching not strongly divaricate; larger leaves more than 10 cm
46 Valves at least 6 mm
47 Basal leaves only slightly longer than wide, usually deeply cordate; valves with 1 small tubercle **26. confertus**
47 Basal leaves at least twice as long as wide; valves with 1 large tubercle and sometimes 2 smaller
48 Valves not cordate **37. obtusifolius**
48 Valves ±cordate
49 Lateral veins joining midrib of leaf at an angle of 60–80°; valves irregularly crenate or toothed **28. cristatus**

49 Lateral veins joining midrib of leaf at an angle of 45–60°; valves entire or obscurely crenate **29. patientia**
46 Valves less than 6 mm
50 Stems less than 50 cm, arising from a creeping rhizome; leaves obovate, the margins undulate **18. frutescens**
50 Stems usually more than 50 cm; plant without creeping rhizome; leaves broadest at or below the middle, the margins not undulate
51 Basal leaves about as long as wide, deeply cordate **26. confertus**
51 Basal leaves at least twice as long as wide
52 Valves usually toothed
53 Basal leaves acute; petiole up to $\frac{1}{2}$ as long as lamina **28. cristatus**
53 Basal leaves obtuse or subacute; petiole slightly longer than lamina **37. obtusifolius**
52 Valves entire or obscurely crenulate
54 All the whorls in the lower $\frac{1}{2}$ of the inflorescence subtended by small, leaf-like bracts **32. conglomeratus**
54 Only the lowest 1 or 2 whorls in each branch of the inflorescence subtended by bracts
55 Valves ±cordate **29. patientia**
55 Valves not cordate
56 Basal leaves ±cordate **37. obtusifolius**
56 Basal leaves not cordate
57 Leaves glaucous; valves *c.* 4 mm, all with ellipsoidal tubercles **34. rupestris**
57 Leaves not glaucous; valves *c.* 3 mm, usually only 1 with a globose or subglobose tubercle **33. sanguineus**

Subgen. **Acetosella** (Meissner) Rech. fil. Dioecious (very rarely polygamous). Slender perennials with acid taste. Valves seldom much longer than the nut, without tubercles. Leaves hastate or sagittate, sometimes with several pairs of basal lobes.

1. R. acetosella L., *Sp. Pl.* 338 (1753) (R. *fascilobus* Klokov). Stems 5–75 cm, erect or ascending, branching at or above the middle. Leaves hastate; central lobe lanceolate or linear; basal lobes 2 (rarely 0) or 2–15 pairs. Inflorescence less leafy than the lower stem. Nut 0·8–1·5 mm, obtusely trigonous, adhering to (angiocarpous) or free from (gymnocarpous) the valves. *Calcifuge. Almost throughout Europe.* All except Bl, Rs(K); naturalized in Sb.

Very variable. Four subspecies can be recognized on the basis of the degree of adhesion of the perianth to the fruit, and the number of lobes at the base of the leaf. These subspecies occur in more or less distinct geographical areas, although they overlap in distribution and individual plants or populations may exhibit characters that are at variance with a particular regional facies. The following treatment is a summary of that given by J. C. M. den Nijs, *Feddes Repert.* **95**: 43–66 (1984).

1 Valves loosely enclosing and (when rubbed) easily separable from the nut
2 Leaves usually with a single pair of basal lobes (**a**) subsp. **acetosella**
2 Leaves with 2–15 pairs of basal lobes (**b**) subsp. **acetoselloides**
1 Valves adnate to and (when rubbed) not easily separable from the nut
3 Leaves usually with a single pair of basal lobes (**c**) subsp. **pyrenaicus**
3 Leaves with 2–12 pairs of basal lobes (**d**) subsp. **multifidus**

(**a**) Subsp. **acetosella** (incl. R. *tenuifolius* (Wallr.) Á. Löve): Leaves usually with a single pair of basal lobes. Valves loosely enclosing and easily separable from the nut. $2n = 28, 42$. *N.W., N. & C. Europe.*

(**b**) Subsp. **acetoselloides** (Balansa) den Nijs, *Feddes Repert.* **95**: 60 (1984): Like subsp. (**a**) but leaves with 2–15 pairs of basal lobes. $2n = 14, 28, 42$. *S.E. Europe; rarely elsewhere in C. & E. Europe.*

(**c**) Subsp. **pyrenaicus** (Pourret ex Lapeyr.) Akeroyd, *Bot. Jour. Linn. Soc.* **106**: 99 (1991) (R. *angiocarpus* auct., R. *salicifolius* auct.; incl. R. *australis* (Willk.) A. Fernandes): Leaves usually with a single pair of basal lobes. Valves adnate to and adhering tightly to the nut. *Often ruderal.* $2n = 14, 28, 42, 56$. *W. Europe.*

(**d**) Subsp. **multifidus** (L.) Arcangeli, *Comp. Fl. Ital.* 587 (1882) (R. *acetosella* subsp. *angiocarpus* (Murb.) Murb.): Like subsp. (**c**) but leaves with 2–12 pairs of basal lobes. $2n = 14, 28, 42$. *C. & S. Italy; Balkan peninsula, Romania.*

Plants with narrowly linear leaves up to 10 times as long as wide, and procumbent to ascending stems, usually branched from below the middle, have been called **R. tenuifolius** (Wallr.) Á. Löve, *Bot. Not.* **1941**: 99 (1941). The variation between these and typical plants of subsp. (**a**) is continuous, and they merit only varietal rank as var. *tenuifolius* Wallr.

Angiocarpous plants from Portugal and S. Spain, with slightly larger male flowers and pollen grains than subsp. (**c**), and a chromosome number of $2n = 56$, have been recognized as **R. australis** (Willk.) A. Fernandes, *Revista Biol.* **12**: 356 (1983), but probably merit only varietal rank.

2. R. graminifolius J. H. Rudolph ex Lamb., *Trans. Linn. Soc. London* **10**: 264 (1811) (R. *angustissimus* Ledeb.). Stem usually procumbent. Leaves small, narrowly linear, obtuse; basal lobes indistinct or absent. Valves from slightly longer than to nearly twice as long as the nut. Nut *c.* 1·8 mm, longer than wide. $2n = 56$. *Arctic Europe, extending southwards to 61° N. in N.W. Russia.* No Rs(N, C).

Subgen. **Acetosa** (Miller) Rech. fil. Dioecious or polygamous, rarely monoecious. Valves considerably longer than the nut, with or without tubercles. Leaves usually hastate or sagittate, often with acid taste.

3. R. lunaria L., *Sp. Pl.* 336 (1753). Shrub up to 2 m with flexuous branches. Leaves 2·5–5 × 2·5–6 cm, often wider than long, broadly ovate-spathulate, truncate to rounded at the base, rounded at the apex. Inflorescence a wide, compound panicle up to 15 cm. Outer perianth-segments 2 mm, lanceolate, deflexed. Valves 5 × 7 mm, reniform-orbicular, narrowly cordate at the base, rounded or truncate at the apex, with a small, flat, quadrangular tubercle near the base; margin entire. Nut 3 × 1·3 mm, widest a little below the middle. *Planted for hedges, and naturalized in S. Italy, Sicilia and Sardegna.* [Sa Si It.] (*Islas Canarias.*)

4. R. tingitanus L., *Syst. Nat.* ed. 10, **2**: 991 (1759). Rhizome emitting long stolons. Leaves ovate- to lanceolate-triangular, usually truncate at the base, gradually attenuate at the apex, sinuate-dentate or pinnatipartite (var. *lacerus* Boiss., non R. *lacerus* Balbis).

Ochreae of inflorescence large, persistent. Flowers usually polygamous. Outer perianth-segments *c.* 4 mm, free, obovate, deflexed. Valves (7–)9–11 × 10–13 mm, thin, pale, often purplish towards the margin, entire, with tubercles. Nut 3·5–4·2 × 1·5–2 mm, reddish, widest below the middle. *Sandy soils. S.W. Europe.* Ga Hs Lu.

5. R. suffruticosus Gay ex Meissner in DC., *Prodr.* **14**: 72 (1856). Stems densely branched, dark and woody at the base; flowering branches herbaceous, erect. Leaves narrowly linear, hastate, with long, narrow basal lobes diverging at right-angles. Outer perianth-segments free, appressed to the margins of the valves. Inflorescence small, narrow, usually lax; branches simple. Valves *c.* 3 × 3 mm, dark brown. ● *Mountains of N. & C. Spain; one station in E.C. Portugal.* Hs Lu.

6. R. scutatus L., *Sp. Pl.* 337 (1753) (incl. R. *aetnensis* C. Presl). Suffrutescent, with subterranean shoots, branching from the base; stems 20–60(–80) cm. Leaves scutate, panduriform or hastate, about as long as wide, green or glaucous. Inflorescence lax. Pedicels shorter than the valves. Flowers polygamous. Valves 4·5–6(–9) × 5–6(–10) mm, pale, cordate, without tubercles. Nut 3–3·5 mm, yellowish-grey. *Screes and other open, stony habitats. C. & S. Europe, extending northwards to the Netherlands; cultivated elsewhere as a vegetable, and locally naturalized.* Al Au Be Bu Co Cz Ga Ge Gr He Ho Hs It Ju Lu Po Rm Rs(W, K) Sa Si [Br Su].

(a) Subsp. **scutatus** (incl. subsp. *gallaecicus* Lago): Inflorescence more or less leafy, with few branches; flowering branches not becoming indurate in fruit, forming a loose mass. $2n = 20$. *Usually calcicole. Throughout most of the range of the species, except Portugal.*

(b) Subsp. **induratus** (Boiss. & Reuter) Nyman, *Consp.* 636 (1881) (R. *induratus* Boiss. & Reuter): Inflorescence almost leafless, much-branched; flowering branches becoming indurate in fruit, forming an intricate, globose mass. $2n = 20, 40$. *Calcifuge. Iberian peninsula, mainly in the south and west.*

7. R. tuberosus L., *Sp. Pl.* ed. 2, 481 (1762) (R. *euxinus* Klokov). At least some of the roots tuberous, fusiform. Stems 20–80 cm. Leaves hastate-sagittate. Valves 5–8 × 5–8 mm, more or less orbicular, cordate. *S. Europe, from Sicilia and Italy to S.E. Russia.* Bu Cr †Ga Gr It Ju Rm Rs(K, E) Si Tu.

(a) Subsp. **tuberosus**: Stems solitary or few, up to 80 cm, erect. Leaves acute. Inflorescence long, usually lax. Valves *c.* 5 × 5 mm. *Throughout most of the range of the species.*

(b) Subsp. **creticus** (Boiss.) Rech. fil., *Candollea* **12**: 30 (1949) (R. *creticus* Boiss.): Stems several, usually less than 30 cm, arcuate-ascending. Basal leaves more or less obtuse. Inflorescence short, dense. Valves up to 8 × 8 mm. $2n = 14$(♀), 15(♂). *Aegean region.*

8. R. nivalis Hegetschw., *Fl. Schweiz* 345 (1839). Stems 7–20 cm, several from a thick stock, leafless or with 1 or 2 cauline leaves. Leaves small; outermost (or sometimes all) basal leaves rounded at the apex, slightly cordate at the base, without basal lobes; cauline leaves (if present) and sometimes inner basal leaves more or less hastate. Inflorescence lax, usually simple, rarely with 1 or 2 short branches. Valves *c.* 3 mm. Nut pale brown. $2n = 14$(♀), 15(♂). *Stony places in the mountains above* 1500 *m; calcicole.* ● *Alps; S.W. Jugoslavia.* Al Au Ge He It Ju.

9. R. alpestris Jacq., *Enum. Stirp. Vindob.* 62 (1762) (R. *arifolius* All., R. *carpaticus* Rech. fil., R. *montanus* Desf.; incl. R. *amplexicaulis* Lapeyr.). Stems (10–)50–120 cm, with numerous cauline leaves. Basal leaves ovate, about twice as long as wide, cordate at the base, with short, truncate or rounded basal lobes; upper leaves sagittate. Ochreae more or less entire. Panicle lax, with flexuous branches, the lower sometimes branched. Valves 3·3–4·5 mm long and wide. Nut 2·5–3 mm, usually yellowish-grey (rarely brown), dull. $2n = 14$(♀), 15(♂). *Meadows; mainly on mountains in the south. N. Europe, southwards to S. Norway; C. & S. Europe, from C. Germany to N. Greece and N. Spain.* Au Bu Co Cz Fe Ga Ge Gr He Hs It Ju ?Lu No Po Rm Rs(N, W) ?Sa Su.

The Fennoscandian plants are generally smaller and more slender, with fewer cauline leaves. They have been called subsp. **lapponicus** (Hiitonen) Jalas, *Ann. Bot. Fenn.* **14**: 191 (1977) (R. *acetosa* subsp. *lapponicus* Hiitonen).

10. R. nebroides Campd., *Monogr. Rumex* 150 (1819) (R. *triangularis* sensu Guss., non DC., R. *gussonei* Arcangeli, R. *acetosa* var. *alpinus* Boiss.). Stems 10–50 cm, sometimes more or less caespitose. Basal leaves 2–3 times as long as wide, sagittate, acute, with acute, deflexed basal lobes; cauline leaves few, very narrow. Inflorescence lax, narrow, with few, short branches. Valves 3–5 mm, pale, membranous, with a small, deflexed tubercle near the base. Nut 2–2·5 × 1·5 mm, brown, widest below the middle. $2n = 14$(♀), 15(♂). *Rocks and stony pastures on limestone. Mountains of S. Europe, from E. Pyrenees to S. Greece.* Al Ga Gr It Ju Si.

Recorded elsewhere only from one station in N.W. Turkey.

11. R. acetosa L., *Sp. Pl.* 337 (1753). Stems 10–130 cm, erect or ascending. Basal leaves sagittate or slightly hastate, oblong-lanceolate to ovate-oblong or -lanceolate, 2–5 times as long as wide. Ochreae fimbriate. Branches of inflorescence simple, except the lowermost which sometimes have secondary branches. Valves 3–5 mm, suborbicular. Nut 1·8–2·5 mm, brown or blackish, glossy. $2n = 14$(♀), 15(♂). *Most of Europe, but rare in much of the south.* All except Az Bl Cr Gr Sb.

A very variable species that has been divided into a number of species and subspecies by different authors. The variation is generally of a continuous pattern, but the following subspecies can be recognized. Further study is required of the species across its whole range.

1 Plant purplish; leaves without acid taste; stems branched from below the middle (c) subsp. **vinealis**
1 Plant green (rarely somewhat purplish); leaves with acid taste; stems branched only in upper part
2 Plant usually glabrous; inflorescence lax; stems usually more than 30 cm (a) subsp. **acetosa**
2 Plant papillose-puberulent; inflorescence dense; stems 10–30 cm (b) subsp. **hibernicus**

(a) Subsp. **acetosa** (incl. subsp. *serpentinicola* (Rune) Nordh., R. *fontanopaludosus* Kalela): Plant glabrous, rarely papillose-puberulent (var. *hirtulus* Freyn): Stems solitary or few, 30–130 cm. Basal and lower cauline leaves sagittate, variable in the shape of the main lobe; upper cauline leaves triangular, up to 6 times as long as wide. Inflorescence lax, the lower branches up to 25 cm, often branched. *Grassland. Throughout the range of the species.*

(b) Subsp. **hibernicus** (Rech. fil.) Akeroyd, *Watsonia* **17**: 444 (1989) (R. *hibernicus* Rech. fil.): Plant papillose-puberulent. Stems usually several, 10–30 cm. Leaves small, somewhat fleshy; upper cauline leaves up to 10 times as long as wide. Inflorescence dense, simple or little-branched, with branches up to 5 cm. *Sand-dunes.* ● *Coasts of W. & S. Ireland, W. Britain.*

(c) Subsp. **vinealis** (Timb.-Lagr.) O. Bolòs & Vigo, *Butll. Inst. Catalana Hist. Nat.* **38** (Sec. Bot. 1): 85 (1974) (R. *vinealis* Timb.-Lagr.): Plant glabrous, purplish, without acid taste. Stems numerous from a thick stock, branched from below the middle. Basal leaves ovate, slightly sagittate. Inflorescence long, narrow. *Vineyards.* ● *S.W. France (near Toulouse).*

Plants with fleshy leaves, stout stems and a dense inflorescence, from coastal rocks and cliffs in W. Europe from N. Spain to W. Ireland, have been called subsp. **biformis** (Lange) Castroviejo & Valdés-Bermejo, *Anal. Inst. Bot. Cavanilles* **34**: 326 (1977), but cannot be separated satisfactorily from robust plants of subsp. (**a**).

12. R. rugosus Campd., *Monogr. Rumex* 113 (1819) (R. *ambiguus* Gren.). Stems 50–160 cm, erect, stout. Basal leaves up to 15 × 10 cm, pale green, somewhat fleshy when alive, very thin when dry, (1½–)2–4 times as long as wide, ovate-oblong, rounded at the apex, with deflexed basal lobes; cauline leaves decreasing slightly upwards, the uppermost sessile, triangular, with a broadly cordate base, tapering towards the apex. Panicle rather lax, the branches repeatedly branched. Valves 3–4 mm. Nut *c.* 2 mm, dark brown, shiny. $2n = 14$(♀), 15(♂). *Origin unknown; cultivated as a vegetable (Sorrel) and established here and there as an escape.*

Known only from cultivation. It has been treated by some authors as a variant or subspecies of **11**, to which vegetatively it bears a close resemblance, although the repeatedly branched inflorescence and the slightly smaller valves are more similar to those of **13**.

13. R. thyrsiflorus Fingerh., *Linnaea* **4**: 380 (1829) (R. *auriculatus* Wallr., R. *haematinus* Kihlman; incl. R. *nemorivagus* Timb.-Lagr.). Stems 30–120 cm. Basal leaves 3–6 times as long as wide, narrowly oblong; cauline leaves becoming progressively narrower upwards, up to 14 times as long as wide; basal lobes divergent, sometimes divided. Panicle dense, the branches repeatedly branched. Valves cordate-reniform to orbicular, sometimes wider than long, with a small, deflexed tubercle near the base. *E. & C. Europe, extending westwards to Norway and E. France; Spain and Portugal.* Al Au Be Bu Cz Da Fe Ga Ge Gr Ho Hs Hu It Ju Lu No Po Rm Rs(N, B, C, W, E) Su [He It].

(a) Subsp. **thyrsiflorus**: Stems and leaves usually glabrous. Valves (2·5–)3–4 mm. Nut 1·8–2·2 mm. $2n = 14$(♀), 15(♂). *Throughout the range of the species except Spain and Portugal.*

(b) Subsp. **papillaris** (Boiss. & Reuter) Sagredo & Malagarriga in Malagarriga, *Nuev. Comb. Subesp. Almeria* 9 (1974) (R. *papillaris* Boiss. & Reuter): Plant more slender. Stems and lower surface of leaves papillose-pubescent. Valves 2·2–3·5 mm. Nut 1·5–1·8 mm. $2n = 14$(♀), 15(♂). ● *Spain and Portugal.*

14. R. intermedius DC. in Lam. & DC., *Fl. Fr.* ed. 3, **5**: 369 (1815). Stems 60 cm, often several. Basal leaves 10 × 1–1·5 cm, rarely wider, linear-triangular, with narrow, divergent basal lobes; cauline leaves narrowly linear with sagittate base; all leaves often crispate at the margin. Valves 3–4 mm, somewhat wider than long, not or scarcely cordate at the base. Nut 2 × 1 mm, dark brown. $2n = 14$(♀), 15(♂). *S.W. Europe; S.E. Italy.* Bl Ga Hs It Lu.

15. R. thyrsoides Desf., *Fl. Atl.* **1**: 321 (1798). Roots tuberous. Stems 30–50 cm. Basal leaves thick, *c.* 2–3 times as long as wide, often constricted above the base, with short, acute, deflexed basal lobes; cauline leaves progressively smaller and narrower, the uppermost sometimes linear; all leaves often crispate at the margin. Inflorescence dense. Valves 4–6 × 7–9 mm, subcordate at the base. Nut 3 × 1·5 mm, dark purplish-brown. $2n = 14$(♀), 15(♂). *S. Europe, from Portugal to S.E. Italy.* Co Hs It Lu Sa Si.

16. R. vesicarius L., *Sp. Pl.* 336 (1753). Glaucous annual 5–30 cm. Leaves triangular, with long petioles. Each pedicel bearing paired flowers (primary and smaller secondary flower); valves of primary flowers 10–20 mm long, suborbicular, membranous, finely and densely reticulately veined, 2 valves of each flower with small tubercles, all with margins bent outwards so as to hide the tubercles and the secondary flower. Nut of primary flower 3·5–5 mm, greyish-brown; nut of secondary flower 2·8–4·5 mm, somewhat darker brown. *Rocks and stony ground. S. Greece (N.E. Peloponnisos).* Gr. (*N. Africa, S.W. Asia.*)

Subgen. **Rumex** (Subgen. *Lapathum* (Campd.) Rech. fil.). All or most flowers hermaphrodite. Valves several times as wide as the nut, with or without tubercles. Basal and lower cauline leaves never hastate or sagittate.

Many species of the subgenus hybridize freely. Most hybrids are highly sterile, the pollen grains being of irregular size or remaining clumped together in the anthers. Hybrids may be recognized by their lax and untidy habit, the stems often taller than those of their parents, the reddish and mostly sterile flowers shed prematurely from the primary inflorescence, the occurrence of secondary flowering branches below the primary inflorescence, the valves of different shapes and sizes on the same plant, and the nuts empty or failing to develop.

17. R. salicifolius (Danser) Hickman, *Madroño* **31**: 249 (1984) (R. *triangulivalvis* auct., non (Danser) Rech. fil., R. *mexicanus* auct., non Meissner). Perennial. Stem 30–50(–100) cm, rarely flexuous and decumbent at the base. Leaves *c.* 5 times as long as wide, linear-lanceolate, acute, pale green; petiole short. Valves 3–4 × 2·5–3 mm, triangular, entire, each with a tubercle. Nut 2 × 1·3 mm. *Naturalized in N. & C. Europe.* [Br Cz Fe Ge No.] (*North America.*)

This species is very variable in its native range. The plant naturalized in Europe is var. *triangulivalvis* (Danser) Hickman.

18. R. frutescens Thouars, *Mélang. Bot.* **1(5)**: 38 (1811) (R. *cuneifolius* Campd.). Rhizome long, creeping, jointed, branched, with erect, leafy flowering stems up to 25 cm. Leaves *c.* 1½ times as long as wide, obovate, thick; petiole short. Inflorescence small, crowded, leafless. Valves 4 × 3 mm, longer than the pedicels, ovate-triangular, entire, coriaceous, each with a fusiform tubercle. Nut 2·5 × 2 mm, broadest in the middle. *Naturalized on sand-dunes in S.W. Britain; casual elsewhere.* [Br ?Lu.] (*South America.*)

19. R. alpinus L., *Sp. Pl.* 334 (1753). Perennial with stout, creeping rhizome; stems 50–100 cm, erect. Basal leaves about as long as wide, deeply cordate, rounded at the apex; petiole longer than the lamina. Panicle crowded, leafless; branches fasciculate, repeatedly branched. Valves 4·5–5(–6) × 3·5–5 mm, entire, without tubercles. $2n = 20$. *Strongly nitrophilous. Mountains of C. & S. Europe.* Al Au Bu Cz Ga Ge Gr He Hs It Ju Po Rm Rs(W) [Br].

20. R. aquaticus L., *Sp. Pl.* 336 (1753). Perennial. Stem up to 200 cm, erect. Basal leaves triangular to elliptical, acute, 1½–2½ times as long as wide, widest at or below the middle; petiole at least as long as the lamina. Panicle large, dense. Valves (4–)6–8·5 mm, ovate-triangular, more or less acute, longer than wide, entire, without tubercles. *Wet places. N., C. & E. Europe, extending locally to W. Jugoslavia and C. France.* Au †Be Br Bu Cz Da Fe Ga Ge He †Ho Hu It Ju No Po Rm Rs(N, B, C, W, E) Su.

1 Panicle much-branched, dense (a) subsp. **aquaticus**
1 Panicle simple, lax
2 Stem more than 50 cm; leaves widest near the base (b) subsp. **protractus**
2 Stem up to 50 cm; leaves widest near the middle (c) subsp. **arcticus**

(a) Subsp. **aquaticus**: Stem usually more than 100 cm. Leaves widest near the base. Panicle much-branched, dense. Valves (5–)6–8·5 mm, ovate-triangular. $2n = 120$, *c*. 200. *Throughout the range of the species except arctic Russia.*

(b) Subsp. **protractus** (Rech. fil.) Rech. fil., *Candollea* **12**: 56 (1949): Stem 60–100 cm. Leaves widest near the base. Panicle lax, simple. Valves *c*. 5 mm, ovate-orbicular or -triangular. $2n = 120$. *Arctic Russia.*

(c) Subsp. **arcticus** (Trautv.) Hiitonen, *Suomen Kasvio* 297 (1933) (*R. arcticus* Trautv.): Stem 15–50 cm. Leaves widest near the middle. Panicle lax, usually simple. Valves 5–6 mm, ovate. $2n = 120$. *Arctic and subarctic Russia.*

Plants from the island of Kolgujev, intermediate between subspp. (**b**) and (**c**), have been described as subsp. **insularis** Tolm., *Fl. Arct. URSS* 158 (1966).

Many records from S. Europe are erroneous. The hybrid with **27** (described under that species) is frequent and fully fertile. The hybrid with **25** is frequent, especially in N. Sweden.

21. R. azoricus Rech. fil., *Candollea* **11**: 229 (1948). Like **20**(**a**) but stouter, with massive stock; basal leaves obtuse, the petiole shorter than the lamina; valves *c*. 5 mm, subacute, the margins minutely but distinctly denticulate. ● *Açores.* Az.

22. R. balcanicus Rech. fil., *Magyar Bot. Lapok* **33**: 5 (1934). Perennial with fleshy, creeping rhizome. Stems up to 100 cm, erect, with few cauline leaves. Basal leaves elliptical to rhomboidal, 1½–2 times as long as wide, equally tapered at both ends, somewhat fleshy when alive, very thin when dry; petiole shorter than the lamina. Inflorescence lax, usually leafless. Valves 5–6 × 4·5–6 mm, ovate-triangular, cordate at the base, obtuse, more or less entire, without tubercles. Nut *c*. 3 mm, brown. *Beside mountain streams.* ● *W. part of Balkan peninsula; local.* Al Gr Ju.

23. R. pseudonatronatus Borbás, *Értek. Term. Köréb. Magyar Tud. Akad.* **11(18)**: 21 (1880) (*R. fennicus* (Murb.) Murb.). Perennial; root fusiform. Stem 80–150 cm, stiffly erect, slender. Leaves 8–15 times as long as wide, narrowly lanceolate, undulate. Inflorescence long, narrow, dense, nearly leafless. Valves 3·5–5 × 3–5 mm, slightly cordate at the base, rounded at the apex, entire, without tubercles. Nut 2–3 × 1–1·5 mm. $2n = 40$. *Riversides and seashores. E. Europe, extending to E. Sweden and E. Austria.* Au Fe Hu Rm Rs(N, B, C, W, E) Su.

24. R. aquitanicus Rech. fil., *Feddes Repert.* **26**: 177 (1929) (*R. cantabricus* Rech. fil.). Perennial up to 100 cm. Petioles and leaves densely papillose-scabrid. Petiole of basal and lower cauline leaves about as long as the lamina; lamina distinctly cordate, widest below the middle, about as long as wide; lateral veins forming an angle of 50–60° with the midrib. Valves 6–8 × 8–10 mm, slightly cordate at the base, without tubercles; margin with many small teeth. Nut *c*. 3·5 × 2 mm, pale brown. ● *Cordillera Cantábrica and W. Pyrenees.* Ga Hs.

25. R. longifolius DC. in Lam. & DC., *Fl. Fr.* ed. 3, **5**: 368 (1815) (*R. domesticus* Hartman). Stout perennial 60–120 cm. Leaves broadly lanceolate, 3–4 times as long as wide. Valves 4·5–5·5 × 5·5–6·5 mm, slightly cordate at the base, rounded at the apex, entire, without tubercles. $2n = 60$. *Ruderal. N. Europe, extending southwards to 50° N. in Russia; S. France and N. Spain.* Br Da Fa Fe Ga Ge Hs Is No Rs(N, B, C, W, E) Su [Cz Po].

26. R. confertus Willd., *Enum. Pl. Horti Berol.* 397 (1809). Perennial up to 80 cm, papillose-pubescent. Basal leaves deeply cordate-triangular, scarcely longer than wide, the basal lobes and apex rounded; petiole often longer than the lamina. Axis of panicle flexuous; branches arcuate at the base. Valves *c*. 6 × 8 mm, broadly reniform-scutate, with small, irregular, indistinct teeth near the base, 1 with a small tubercle. $2n = 40$. *Mainly ruderal. E. & E.C. Europe; C. Appennini.* Cz Hu It ?Po Rm Rs(N, B, C, W, K, E) [Br Fe].

27. R. hydrolapathum Hudson, *Fl. Angl.* ed. 2, 154 (1778). Perennial up to 200 cm. Leaves flat, suberect, broadly lanceolate, tapering equally at both ends, lateral veins at a right-angle to the midrib; basal leaves 4–5 times as long as wide. Pedicels stiff, *c*. 1½ times as long as the valves. Valves triangular, each with a distinct, fusiform tubercle. Nut 4·5–5 mm. $2n = 200$. *Marshes and beside lakes and rivers. Most of Europe, from 62° N. in Finland to 40° N. in Italy, but very local in the south.* Al Au Be Br Bu Cz Da Fe Ga Ge Gr Hb He Ho Hs Hu It Ju No Po Rm Rs(N, B, C, W, E) Sa Su.

Outside Europe, known only from one station in the Caucasus.

The hybrid with **20** (*R. heterophyllus* C. F. Schultz, *R. maximus* Schreber, non C. C. Gmelin) is frequent in N.C. Europe and S. Fennoscandia. It is similar in habit to **27** but has wider and shorter leaves, rounded or slightly cordate at the base, and valves of somewhat irregular development, in part larger and wider than in **27**. The hybrid with **37** is rarer, but not infrequent in England and the Netherlands; it is similar in habit to **27** but easily distinguished by the distinct teeth near the base of the valves and by their rather long apex.

28. R. cristatus DC., *Cat. Pl. Horti Monsp.* 139 (1813) (*R. graecus* Boiss. & Heldr.). Perennial 40–200 cm. Basal leaves 20–30 × 10–15 cm, ovate-lanceolate, with a cordate base, acute at the apex; lateral veins joining the midrib of the leaf at an angle of 60–80°; petiole up to ½ as long as the lamina. Valves 5–8 × 5–8 mm, reddish-brown when ripe, cordate, with very small, irregular, acute teeth up to 1 mm; all with unequal tubercles. Nut 3–3·5 mm, dark brown. *Ruderal. Balkan peninsula and E.C. Europe; naturalized further west.* Al Bu Gr Hu Ju Rm Tu [Br Hs It Lu *Si].

(a) Subsp. **cristatus**: Plant up to 200 cm; branches often secondarily branched. Leaves minutely papillose beneath. Valves often with 3 tubercles, reddish-brown, with distinct, irregular teeth up to 1 mm. $2n = 80$. *Ruderal; mainly lowland. S. part of Balkan peninsula; locally naturalized elsewhere.*

(b) Subsp. **kerneri** (Borbás) Akeroyd & D. A. Webb, *Bot. Jour. Linn. Soc.* **106**: 104 (1991) (*R. confertoides* Bihari): Plant often less than 100 cm; branches short, usually without secondary branches. Leaves papillose-scabrid beneath. Valves usually with only 1 tubercle, usually dark brown, subentire or with teeth up to 0·5 mm. $2n = 80$. *Mainly in the mountains.* ● *From Hungary to C. Romania and C. Greece.*

29. R. patientia L., *Sp. Pl.* 333 (1753). Perennial 80–200 cm. Leaves pale green, 3–4 times as long as wide, ovate- or oblong-lanceolate, acute, truncate or broadly cuneate (rarely subcordate) at the base; lateral veins joining midrib of leaf at an angle of 45°–60°. Valves 4–10 mm long and wide, entire or subentire. Nut 3–4 mm, pale brown. *Mainly ruderal. From N.W. Jugoslavia to S.E. Russia; formerly cultivated elsewhere as a vegetable, and locally naturalized.* Al Au Bu Cz Gr Hu *It Ju Rm Rs(W, K, E) Tu [Br Ge He Po Rs(C) *Si].

1 Valves 6–10 × 8–10 mm (c) subsp. **orientalis**
1 Valves less than 8 mm wide
 2 Valves about as long as wide, obtuse, distinctly cordate at the base (a) subsp. **patientia**
 2 Valves longer than wide, acute, truncate or subcordate at the base (b) subsp. **recurvatus**

(a) Subsp. **patientia**: Valves 4–8 × 4–8 mm, obtuse, cordate at the base, with 1 subglobose to avoid tubercle 1·5–2 mm. 2*n* = 60. *From Czechoslovakia to N. Greece.*

(b) Subsp. **recurvatus** (Rech.) Rech. fil., *Feddes Repert.* **31**: 252 (1933): Valves 4–8 × 4–7 mm, acute, truncate to subcordate at the base, with 1 subglobose to ovoid tubercle 1·5–2 mm. 2*n* = 60. *Czechoslovakia; Romania; Bulgaria.*

(c) Subsp. **orientalis** (Bernh.) Danser, *Nederl. Kruidk. Arch.* **1923**: 281 (1924) (*R. lonaczewskii* Klokov, *R. orientalis* Bernh.): Valves 6–10 × 8–10 mm, obtuse to subacute, cordate at the base, with usually 1 ovoid tubercle 2–3 mm. 2*n* = 60. *Through most of the range of the species.*

30. R. crispus L., *Sp. Pl.* 335 (1753) (*R. odontocarpus* Sándor). Usually perennial; stems 30–150(–200) cm. Basal leaves 4–5 times as long as wide, narrowly lanceolate, acute, usually cuneate at the base; margins undulate; petiole usually shorter than the lamina. Branches of inflorescence erect or ascending; lower whorls usually remote. Pedicels 2–2½ times as long as the valves. Valves 3–6(–8) × 3–5(–7) mm, cordate, variable in outline and in development of tubercles, more or less entire. *In a wide range of open habitats; frequently as a ruderal. Almost throughout Europe.* All except Cr Sb, but only naturalized in Az Is.

Hybridizes commonly with other members of the section, especially **25**, **26**, **29**, **36** and **37**. The hybrid with **37** is the commonest in the genus, and was originally described as a species (*R. pratensis* Mert. & Koch). It is found wherever the parents grow together, and can be recognized by the different size and outline of the valves on the same plant, some of the flowers falling off before the valves have reached their normal size.

A very variable species. The following three subspecies can be distinguished, although intermediate populations occur, especially where habitats have been disturbed.

1 Nut 1·3–2·5 mm; tubercles of valves usually less than 2·5 mm, distinctly unequal in size, sometimes only 1 developed (a) subsp. **crispus**
1 Nut 2·5–3·5 mm; tubercles of valves up to 3·5 mm, often subequal
 2 Stems usually less than 100 cm; panicle dense in fruit (b) subsp. **littoreus**
 2 Stems 100–200 cm; panicle lax in fruit (c) subsp. **uliginosus**

(a) Subsp. **crispus**: Stems 30–120(–150) cm. Leaves with strongly undulate margins, not fleshy, green. Panicle usually lax in fruit. Tubercles of valves usually less than 2·5 mm, distinctly unequal, sometimes only 1 developed. Nut 1·3–2·5 mm. 2*n* = 60. *Ruderal, and as a weed of cultivation. Throughout the range of the species.*

(b) Subsp. **littoreus** (Hardy) Akeroyd, *Watsonia* **17**: 444 (1989) (*R. crispus* var. *littoreus* Hardy): Stems 30–80(–120) cm. Leaves with strongly undulate margins, fleshy, somewhat glaucous. Panicle dense in fruit, branches sometimes fastigiate. Tubercles of valves up to 3·5 mm, often subequal. Nut 2·5–3·5 mm. *Maritime sand and shingle. Coasts of W. & N. Europe.*

(c) Subsp. **uliginosus** (Le Gall) Akeroyd, *loc. cit.* (1989) (*R. crispus* var. *uliginosus* Le Gall): Stems 100–200 cm. Leaves flat or with slightly undulate margins, not fleshy, green. Panicle lax in fruit, with long branches. Tubercles of valves up to 3·5 mm, subequal to unequal. Nut 2·5–3·5 mm. *Muddy banks of tidal rivers.* ● *N.W. France, S. Ireland, S. Britain.*

31. R. stenophyllus Ledeb., *Fl. Altaica* **2**: 58 (1830). Like **30(a)** but leaves usually less undulate; pedicels 1½–2 times as long as the valves; valves 3·5–4(–5) mm long and wide, cordate-triangular, with numerous, small but distinct teeth 0·5–1 mm, the apex broadly triangular, entire. 2*n* = 60. *Damp, often saline habitats. C. & E. Europe.* Au Bu Cz Ge Hu Ju Rm Rs(C, W, K, E).

32. R. conglomeratus Murray, *Prodr. Stirp. Gotting.* 52 (1770) (*R. nemorosus* Schrader). Perennial 30–80(–120) cm. Leaves oblong to lanceolate, subacute. Panicle lax; whorls all remote and usually subtended by a leaf. Pedicels about as long as the valves, articulated near the middle. Valves 2·5–3·2 × 1–1·7 mm, oblong-ovate, entire, all tuberculate; tubercles ovoid, often covering nearly the whole valve. 2*n* = 20. *Damp, more or less open habitats. Europe, from Scotland, S. Sweden and C. Russia southwards.* All except Fa Is Rs(N, B, E) Sb, but only as an alien in Az Fe No.

33. R. sanguineus L., *Sp. Pl.* 334 (1753). Like **32** but more slender and erect; leaves oblong, acute or subacute; panicle leafy only at the base; pedicels longer than the valves, articulated below the middle; only 1 valve with a tubercle; tubercle globose or subglobose, much shorter than the valve. 2*n* = 20. *Damp and shady places. Most of Europe from* c. *60° N. southwards, except E. Russia; rare in the Mediterranean region.* Au Be Br Bu Co Cz Da Ga Ge Gr Hb He Ho Hs Hu It Ju No Po Rm Rs(B, C, W, E) Sa Si Su Tu.

The common wild plant is var. *viridis* (Sibth.) Koch; var. *sanguineus*, with purple-veined leaves and stems suffused with reddish-purple, is cultivated for ornament, and is locally naturalized.

Fertile intermediates between **32** and **33**, of presumed hybrid origin, occur commonly in Britain and perhaps elsewhere.

34. R. rupestris Le Gall, *Congr. Sci. Fr.* **16(1)**: 143 (1850). Like **32** but leaves oblong, narrowed at the base, obtuse, glaucous, lamina usually much longer than the petiole; usually only the lowest whorl on each branch subtended by a leaf; valves *c.* 4 mm, oblong, obtuse, with ellipsoidal tubercles 2·5–3 mm. 2*n* = 20. *Maritime rocks, sand and shingle.* ● *Coasts of W. Europe, from N.W. Spain to Wales.* Br Ga Hs.

35. R. nepalensis Sprengel, *Syst. Veg.* **2**: 159 (1825). Rhizome stout; stems 60–170 cm. Basal leaves 1½–2½ times as long as wide, oblong-ovate, pubescent beneath, cordate at the base, rounded or broadly acuminate at the apex. Panicle open; branches solitary, arcuate-divaricate, nearly leafless; flower-whorls remote. Valves 5–7 × 3·5–5 mm, ovate-triangular, all tuberculate, with numerous hooked teeth developing early. Nut 3–5 × 2–2·5 mm. 2*n* = 120. *Clearings in mountain woods. Balkan peninsula, C. & S. Italy; local.* Al Gr It Ju. (*Anatolia to S.W. China.*)

36. R. pulcher L., *Sp. Pl.* 336 (1753). Perennial 20–60 cm. Plant often papillose. Basal leaves small, fleshy, about twice as long as wide, usually panduriform, cordate at the base. Panicle open; branches usually divaricate, flexuous, often entangled or arcuate; whorls all remote and subtended by a leaf. Pedicels thick, shorter than to as long as the valves; fruits scarcely deciduous. Valves thick, coarsely foveolate-rugose, variable in outline; at least 1 valve with a large rugose or verrucose tubercle. *Mainly ruderal. S. & W. Europe, extending to Hungary and N. Romania.* Al Az Bl Br Bu Co Cr Ga Ge Gr Hb He Hs Hu It Ju Lu Rm Rs(K) Sa Si Tu [Au He].

1 Valves entire, or with a few teeth not more than 0·5 mm near the base (d) subsp. **anodontus**
1 Valves with several teeth on each margin; teeth more than 0·5 mm
2 Valves suborbicular or ovate-triangular, broadly acuminate at apex; teeth short, up to 8 on each margin (c) subsp. **woodsii**
2 Valves narrowly ovate or oblong with lingulate apex; teeth 4–6 on each margin
3 Valves 4–5 mm; teeth 1–2 mm, 4 on each margin (a) subsp. **pulcher**
3 Valves *c.* 6 mm; teeth 3–4 mm, 5 or 6 on each margin (b) subsp. **raulinii**

(a) Subsp. **pulcher**: Basal leaves usually panduriform. Branches angular-flexuous, often forming an intricate mass. Valves 4–5 mm, narrowly ovate or oblong, with lingulate apex; 1 or 2 meshes of venation on each side of the tubercle; teeth 1–2 mm, 4 on each margin. $2n = 20$. *Throughout the range of the species.*

(b) Subsp. **raulinii** (Boiss.) Rech. fil., *Beih. Bot. Centr.* **49(2)**: 39 (1932): Like subsp. (a) but valves *c.* 6 mm; teeth 3–4 mm, 5 or 6 on each margin. $2n = 20$. *Greece and Aegean region.*

(c) Subsp. **woodsii** (De Not.) Arcangeli, *Comp. Fl. Ital.* 585 (1882) (subsp. *divaricatus* (L.) Murb.): Basal leaves seldom panduriform. Branches arcuate-divaricate. Valves 4–5 mm, suborbicular or ovate-triangular, broadly acuminate at the apex; usually 3 meshes of venation on each side of the tubercle; teeth 0·5–1 mm, up to 8 on each margin. $2n = 20$. *S. Europe.*

(d) Subsp. **anodontus** (Hausskn.) Rech. fil., *Beih. Bot. Centr.* **49(2)**: 34 (1932): Basal leaves and branches similar to those of subsp. (c). Valves 4–5 mm, narrowly ovate-triangular, entire or with 1 or 2 teeth up to 0·5 mm near the base. *Kriti. (S.W. Asia, N. Africa.)*

37. R. obtusifolius L., *Sp. Pl.* 335 (1753). Perennial 60–120 cm. Basal leaves large, twice as long as wide, thin, obtuse or subacute, cordate at the base; petiole slightly longer than the lamina. Panicle open; branches arcuate; whorls mostly remote, the lower ones subtended by a leaf. Pedicels slender, up to $2\frac{1}{2}$ times as long as the valves; fruits readily deciduous. Valves 3–6 × 2–3 mm. *Europe, except the extreme north and south.* All except Cr Is Sb Si, but only as an alien in Az Fe and perhaps Rs(N).

1 Valves entire or with a few short or indistinct teeth near the base, all with tubercles (d) subsp. **silvestris**
1 Valves with several distinct, often long teeth
2 All valves with tubercles; tubercles often unequal (c) subsp. **transiens**
2 One valve with a tubercle, the other 2 without
3 Valves ovate or ovate-triangular, obtuse; leaves papillose beneath (a) subsp. **obtusifolius**
3 Valves narrowly triangular or lingulate, subacute; leaves usually not papillose beneath (b) subsp. **subalpinus**

(a) Subsp. **obtusifolius** (subsp. *agrestis* (Fries) Čelak.): Leaves papillose beneath. Valves ovate or ovate-triangular, with several distinct, often long teeth, obtuse; 1 valve with a tubercle. $2n = 40$. *Cultivated ground, pastures and waste places. Native in W. Europe, and widely naturalized elsewhere.*

(b) Subsp. **subalpinus** (Schur) Čelak., *Prodr. Fl. Böhm.* 159 (1873): Leaves smooth or minutely papillose beneath. Valves narrowly triangular, with several distinct teeth, subacute; 1 valve with a tubercle. *Streams and woodland margins. Carpathians and mountains of Balkan peninsula.*

(c) Subsp. **transiens** (Simonkai) Rech. fil., *Beih. Bot. Centr.* **49(2)**: 52 (1932): Leaves somewhat papillose beneath. Valves ovate to narrowly triangular, with a few teeth near the base, obtuse to subacute; all valves with tubercles. $2n = 40$. *Mainly ruderal. C. Europe, Balkan peninsula, S. Scandinavia; locally naturalized elsewhere.*

(d) Subsp. **silvestris** Čelak., *Prodr. Fl. Böhm.* 159 (1871) (subsp. *sylvestris* (Wallr.) Rech., *R. sylvestris* Wallr.): Leaves smooth beneath. Valves narrowly triangular or lingulate, entire or with a few short or indistinct teeth near the base, subacute; all valves with tubercles. $2n = 40$. *Riversides, woodland margins, and as a ruderal. Europe from Italy and Sweden eastwards; occasionally naturalized elsewhere.*

The species hybridizes easily with other members of the section, especially **25**, **26**, **28**, **29** and **30**.

38. R. dentatus L., *Mantissa Alt.* 226 (1771) (*R. limosus* Thuill.). Annual 20–70 cm. Basal leaves small, 2–3 times as long as wide, truncate or subcordate at the base; petiole equalling or shorter than the lamina. Pedicels somewhat longer than the valves. Valves 4–6 × 2–3 mm, all or only 1 with tubercle; teeth 3–6 mm. $2n = 40$. *S.E. Europe.* Bu Gr Rm Rs(W, E) [Hs].

(a) Subsp. **halacsyi** (Rech.) Rech. fil., *Beih. Bot. Centr.* **49(2)**: 16 (1932): Valves 4 × 2 mm, all with tubercles; teeth as long as width of valves. *Romania to Greece; naturalized in S.W. Spain and perhaps elsewhere.*

(b) Subsp. **reticulatus** (Besser) Rech. fil., *op. cit.* 18 (1932) (*R. reticulatus* Besser): Valves 5–6 × 3 mm, only 1 with a tubercle; teeth longer than width of valves. *Ukraine and S.E. Russia.*

Subsp. *dentatus*, described from Egypt, has not been recorded from Europe except as a casual.

R. obovatus Danser, *Nederl. Kruidk. Arch.* **1920**: 241 (1921), from temperate South America, like **38** but with obovate basal and lower cauline leaves and warty tubercles, has been widely recorded in N. Europe as a casual and may be becoming locally naturalized.

39. R. palustris Sm., *Fl. Brit.* **1**: 394 (1800). Annual or biennial 30–60 cm; whole plant brown when fruit is ripe. Basal leaves about 6 times as long as wide, lanceolate, acute, cuneate at the base; margin undulate. Branches of panicle arcuate; whorls remote, all subtended by leaves. Pedicels $1–1\frac{1}{2}$ times as long as the valves, slender. Valves 3–3·5(–4) × 1·2–1·5 mm (excluding teeth), narrowly lingulate; teeth 2 or 3 on each margin, about as long as the width of the valve. $2n = 40$. *Marshes and riversides. C. & S. Europe, extending northwards to Denmark.* Al Au Be Br Bu Cz Da Ga Ge Gr Ho Hu It Ju Po Rm Rs(B) Su Tu.

40. R. maritimus L., *Sp. Pl.* 335 (1753). Annual 10–50(–70) cm; whole plant golden-yellow when fruit is ripe. Basal leaves narrowly elliptical, tapering nearly equally at both ends. Branches of panicle arcuate; all whorls except the lower ones confluent when fruit is ripe. Pedicels filiform, somewhat longer than the valves. Valves 2·5–3 × 1·5–2 mm (excluding teeth), ovate-triangular, with a narrow, triangular apex; teeth 2–5 on each margin, setiform, longer than the width of the valve. Nut 1·3–1·5(–1·8) mm. $2n = 40$. *Wet places. Most of Europe except the south and the extreme north.* Au Be Br Cz Da Fe Ga Ge Hb He Ho Hu It Ju No Po Rm Rs(N, B, C, W, K, E) Su.

41. R. rossicus Murb., *Bot. Not.* **1913**: 221 (1913). Like **40** but pedicels mostly shorter than the valves; teeth 1 or 2 on each margin, about equalling the width of the valve; nut 1·1–1·3 mm. $2n = 40$. *Riversides. N. & C. Russia.* Rs(N, C).

42. R. marschallianus Reichenb., *Pl. Crit.* **4**: 58 (1826). Slender annual 5–25 cm, often with decumbent branches from the base. All leaves with comparatively long petioles, linear-lanceolate; lower ones with cordate or truncate base. Lower whorls remote; upper confluent. Valves 2·5–3 mm, excluding the long subulate apex, which is as long as the rest of the valve, unequal, the larger one with a narrow tubercle; teeth up to 5 mm, several times as long as the width of the valve. $2n = 40$. *Riversides. S.E. Russia, E. Ukraine.* Rs(C, W, E).

43. R. ucranicus Besser ex Sprengel, *Novi Provent.* 36 (1819). Slender, purplish annual 8–25 cm, branched from the base. Basal leaves small, obovate. Cauline leaves linear-lanceolate, 3–4 times as long as wide; margin often undulate; petiole longer than the width of the lamina. Petiole of leaves of inflorescence exceeding the whorl it subtends. Valves 2(–2·5) × 1 mm (excluding teeth), narrowly triangular, very acute, each with a large, orange tubercle; teeth 3 on each margin, at least as long as the width of the valve. $2n = 40$. *Riversides. C. & E. Russia southwards, extending to Poland and S.W. Ukraine.* Po Rs(C, W, E) [Rs(N)].

Subgen. **Platypodium** (Willk.) Rech. fil. Usually annual. Basal leaves small, lanceolate, ovate or spathulate. Flowers in clusters of 4 or fewer. Flowers, fruits and pedicels often heteromorphic; valves very small, usually with teeth and small tubercles.

44. R. bucephalophorus L., *Sp. Pl.* 336 (1753). Annual with 1 to several slender stems 5–50 cm. Flowers usually in clusters of 2 or 3. Pedicels usually heteromorphic, some cylindrical and short, others long, curved, convex and clavately thickened. Valves up to 5 × 3·5 mm, often heteromorphic; teeth straight to slightly curved, or uncinate. *Sandy and stony places. Mediterranean region and S.W. Europe.* Al Az Bl Co Cr Ga Gr Hs It Ju Lu Sa Si Tu.

Very variable. The following treatment follows J. R. Press, *Bot. Jour. Linn. Soc.* **97**: 344–355 (1988).

1 Pedicels mostly about as long as valves; valves with 4–8 teeth on each margin
2 All valves ovate-oblong, with up to 6 hooked teeth on each margin (c) subsp. **hispanicus**
2 Some valves narrowly triangular with up to 8 straight or slightly curved teeth on each margin, others lingulate, entire (d) subsp. **canariensis**
1 Pedicels longer or shorter than valves; valves with 2–4 teeth on each margin or some entire
3 All valves broadly triangular, with 2(3) stout teeth on each margin (a) subsp. **bucephalophorus**
3 Some valves narrowly triangular, with (2)3 or 4 slender teeth on each margin, others lingulate, entire (b) subsp. **gallicus**

(a) Subsp. **bucephalophorus** (incl. subsp. *graecus* (Steinh.) Rech. fil.): Stems robust, usually few, often erect, simple or sparsely branched. Thickened fruits absent at the base of the stems. Pedicels heteromorphic. Valves 2·5–5 × 1–3·5 mm, broadly triangular; teeth 2 or 3(4) on each margin, stout. $2n = 16$. *Mediterranean region; mainly in the centre and the east.*

(b) Subsp. **gallicus** (Steinh.) Rech. fil., *Bot. Not.* **1939**: 497 (1939) (subsp. *bucephalophorus* auct. eur.): Stems slender to robust, often many and branched near the base. Thickened fruits often present at the base of the stems. Pedicels heteromorphic. Valves 2–3 × 0·5–1·2 mm, lingulate, entire, or 1·5–2·5 × 0·5–1 mm, narrowly triangular, with (2)3 or 4 slender teeth on each margin. $2n = 16$. *Mediterranean region.*

(c) Subsp. **hispanicus** (Steinh.) Rech. fil., *op. cit.* 500 (1939): Stems procumbent to decumbent, somewhat robust. Thickened fruits absent at the base of the stems. Pedicels mostly short. Valves 2·5–3 × 1–1·5 mm, ovate-oblong, with 4–6 slender, uncinate teeth on each margin. $2n = 16$. ● *Coastal regions of Portugal, N. Spain and W. France.*

(d) Subsp. **canariensis** (Steinh.) Rech. fil., *op. cit.* 502 (1939): Stems decumbent to ascending, usually branched at the base, slender. Thickened fruits absent at the base of the stems. Pedicels mostly short. Valves 2–3(–3·5) mm, lingulate, entire, or 2–2·5 mm, narrowly triangular, with 4–8 slender teeth on each margin. *Açores. (Madeira, Islas Canarias.)*

Plants with thickened fruits at the base of the stems and both basal and cauline fruits toothed (var. *aegaeus* (Rech. fil.) Maire), occur in Greece and the Aegean region. Similar plants from the W. Mediterranean region and Sicilia, but with fruits both entire and toothed, have been called var. *subaegaeus* Maire.

9. **Emex** Campd.

†T. G. TUTIN

Monoecious annuals. Ochreae not ciliate, soon lacerate. Female flowers at the base of the inflorescence; perianth-segments 6, free in male flowers, connate in female flowers, the outer 3 spinescent and indurate in fruit. Stamens 4–6. Stigmas 3. Nut triquetrous, included in the perianth.

1. E. spinosa (L.) Campd., *Monogr. Rumex* 58 (1819). Plant glabrous; stems 30 cm or more, erect or ascending. Leaves ovate, truncate or subcordate at the base, petiolate. Male flowers in terminal and axillary, pedunculate clusters; female axillary, sessile. Fruit enclosed in the perianth; outer segments ending in patent spines; inner erect, tubercled. *Sandy shores and disturbed ground. Mediterranean region, extending to N.W. Spain.* Bl †Co Cr Gr Hs It Lu Sa Si [Az].

10. **Muehlenbeckia** Meissner

J. R. AKEROYD AND D. A. WEBB

Woody plants, usually climbers; dioecious or polygamous. Ochreae not ciliate, soon lacerate and evanescent. Perianth-segments 5, sepaloid, united at the base, accrescent and fleshy in fruit. Stamens 8, represented by staminodes in female flowers. Stigmas 3, subsessile. Nut triquetrous, partly fused with the accrescent perianth. (*Calacinum* Rafin.)

1. M. complexa (A. Cunn.) Meissner, *Pl. Vasc. Gen.* **1**: 227 (1839). Glabrous, straggling or climbing shrub 1–5 m, with slender, twining branches. Leaves 5–25 mm, broadly oblong to suborbicular, slightly apiculate. Flowers greenish, in short, axillary or terminal racemes. Fruiting perianth berry-like, white. Nut 2–2·5 mm, black, glossy. *Naturalized on coasts of W. Europe.* [Az Br Ga Lu.] (*New Zealand.*)

M. sagittifolia (Ortega) Meissner, *loc. cit.* (1839), from temperate South America, like **1** but with lanceolate leaves *c.* 5 mm, has been recorded as an escape from gardens in Açores and Portugal, and may be locally naturalized.

11. Calligonum L.

D. A. WEBB

Erect shrubs. Leaves small, caducous. Ochreae short, without veins. Flowers hermaphrodite, in axillary clusters. Perianth-segments 5 or 6, more or less petaloid, free, persistent but deflexed in fruit. Stamens *c.* 15. Stigmas 4; fruit furnished with wings or bristles.

1. C. aphyllum (Pallas) Gürke in K. Richter, *Pl. Eur.* **2**: 111 (1897). Stems up to 2 m; twigs numerous, slender, dark green, arising in bunches of 2–6 from each node. Leaves 2–3 mm, more or less subulate, caducous. Flowers usually in pairs, stalked. Outer perianth-segments 3 mm, broadly ovate, green with white margins; the inner smaller, white with pink mid-vein. Nut with 4 flat, membranous wings decurrent on the pedicel. *Sandy places. S.E. Russia.* Rs(C, E). (*W. & C. Asia.*)

12. Atraphaxis L.

D. A. WEBB

Erect, small-leaved shrubs. Ochreae brown proximally, hyaline distally, bifid, 2-veined. Flowers hermaphrodite, in short racemes. Perianth-segments 4 or 5, the inner 2 or 3, accrescent and surrounding the fruit. Stamens 6 or 8, united into a ring at the base. Stigmas 2 or 3; nut not winged.

1 Perianth-segments 4; stamens 6
 2 Branches mostly leafless and spiny towards the apex; fruiting perianth 5–6 mm wide **1. spinosa**
 2 Branches mostly leafy to the apex, not spiny; fruiting perianth 8–9 mm wide **2. replicata**
1 Perianth-segments 5; stamens 8
 3 Inner perianth-segments scarcely exceeding the nut **3. frutescens**
 3 Inner perianth-segments about twice as long as the nut **4. billardieri**

1. A. spinosa L., *Sp. Pl.* 333 (1753). Plant 30–80 cm, with numerous, slender, divaricate branches, many of them leafless towards the apex and more or less spiny. Leaves 7–11 mm, ovate or elliptical, mucronate, very shortly petiolate. Flowers in short-stalked, axillary racemes. Perianth-segments bright pink with white margins; segments 4, the 2 inner 4–5 × 5–6 mm in fruit. Nut broadly ovate, flattened, pale greenish-brown. *Steppes and semi-deserts. S.E. Russia, W. Kazakhstan.* Rs(E). (*W. & C. Asia.*)

2. A. replicata Lam., *Encycl. Méth. Bot.* **1**: 329 (1783) (*A. spinosa* var. *rotundifolia* Boiss.). Like **1** but with branches seldom spine-tipped and usually leafy to the apex; leaves 3–8 mm, sometimes almost orbicular; and fruiting perianth-segments 7–8 × 8–9 mm. *S.E. Russia, westwards to Krym.* Rs(K, E). (*W. & C. Asia.*)

3. A. frutescens (L.) C. Koch, *Dendrologie* **2(1)**: 360 (1872) (*A. lanceolata* Meissner). Plant 20–70 cm; branches usually leafy, rarely spine-tipped. Leaves 12–17 × 2–8 mm, fleshy, glaucous, linear-lanceolate to oblong-obovate, entire or bluntly toothed, with a short, white, apical mucro. Flowers in short, lax, terminal racemes. Perianth-segments 5, pinkish- or greenish-white, the 2 inner 4–5 × 5–6 mm, semi-circular, equalling or slightly exceeding the dark brown, glossy, acutely trigonous nut. *Steppes and semi-deserts. S.E. Russia.* Rs(C, E). (*W. & C. Asia.*)

4. A. billardieri Jaub. & Spach, *Ill. Pl. Or.* **2**: 14 (1844). Like **3** but branches spiny; leaves ovate or oblong, obtuse to acute, sometimes slightly acuminate but not mucronate; and perianth-segments pink, the 2 inner 5–9 × 4–8 mm in fruit and nearly twice as long as the nut. *Dry mountain rocks. C. Greece and Kriti; very local.* Cr Gr. (*S.W. Asia.*)

CENTROSPERMAE

48. CHENOPODIACEAE

EDIT. J. R. EDMONDSON

Herbs, shrubs or rarely small trees, often succulent; glabrous, pubescent or farinose with vesicular hairs. Leaves alternate or opposite, exstipulate. Flowers hermaphrodite or unisexual, often bracteolate, solitary or in dense cymose clusters in a spicate or paniculate inflorescence. Perianth absent or 1- to 5-merous, often accrescent in fruit; segments usually more or less connate. Stamens 1–5, opposite the perianth-segments. Ovary superior, rarely semi-inferior, unilocular; ovules solitary, basal; stigmas (1)2 or 3(–5). Fruit usually an achene. Seeds horizontal (compressed in the vertical plane) or vertical (compressed in the horizontal plane).

The majority of the species are halophytes or ruderals. The family is particularly predominant in maritime habitats, and in the steppe and semi-desert regions of S.E. Europe.

The interpretation of the structure of the segmented genera in this family (**18–22**, **29**, **30**) is a matter of considerable controversy. For accounts of the two interpretations see: A. Fahn & T. Arzee, *Amer. Jour. Bot.* **46**: 330–338 (1959); F. F. Leysle, *Bot. Žur.* **34**: 253–266 (1949). The latter interpretation is that accepted here.

1 Plant with spinose branches
 2 Annual herb; inflorescence of numerous dichasial cymes **3. Chenopodium**
 2 Small shrubs; flowers solitary
  3 Leaves not more than 8 mm, filiform or scale-like, mucronate **25. Salsola**
  3 Leaves up to 50 mm, semi-cylindrical, obtuse **26. Noaea**

1 Plant without spinose branches, sometimes with spine-tipped leaves and bracts
4 Leaves flat
5 Plant with yellow or brownish glands or glandular hairs **3. Chenopodium**
5 Plant eglandular
6 Leaves and bracts spinose-acuminate
7 Leaves 1–4 cm; flowers unisexual; bracteoles connate, enclosing fruit **10. Ceratocarpus**
7 Leaves to 8 cm; flowers hermaphrodite; bracteoles free, not enclosing fruit **15. Agriophyllum**
6 Leaves and bracts not spinose
8 Flowers all unisexual; female flowers mostly without perianth, but with 2 bracteoles which become enlarged and enclose the fruit
9 Plant pubescent **9. Krascheninnikovia**
9 Plant glabrous or farinose
10 Bracteoles free, at least in the upper $\frac{1}{2}$
11 Male flowers mostly 5-merous, bracteole 0; female bracteoles without subtending bract **6. Atriplex**
11 Male flowers mostly 4-merous, bracteole 1; female bracteoles much larger than subtending bract **8. Cremnophyton**
10 Bracteoles connate almost to the apex
12 Plant green; stigmas 4 or 5 **5. Spinacia**
12 Plant silvery-farinose; stigmas 2 or 3 **7. Halimione**
8 Flowers mostly hermaphrodite, rarely all unisexual and then the female with 3 or more perianth-segments
13 Perianth-segments in fruit with a tubercle, spine, horizontal wing or keel dorsally
14 Perianth-segments with a spine in fruit **13. Bassia**
14 Perianth-segments with a wing, keel or tubercle in fruit
15 Plant glabrous or farinose **23. Suaeda**
15 Plant pubescent at least above
16 Wings of the perianth connate and completely encircling the fruit; leaves dentate to sinuate-pinnatifid **4. Cycloloma**
16 Wings of the perianth free; leaves entire **13. Bassia**
13 Perianth-segments without appendages, or vertically keeled or winged
17 Ovary semi-inferior, connate with the swollen receptacle in fruit **2. Beta**
17 Ovary superior
18 Achenes encircled by a distinct wing or margin, sometimes emarginate at apex **14. Corispermum**
18 Achenes not winged or margined, rarely with a 2-lobed wing at apex
19 Plant pubescent
20 Perianth-segments mostly solitary, not enclosing the achene **14. Corispermum**
20 Perianth-segments 3 or more, enclosing the achene
21 Leaves up to 9 cm, broadly lanceolate; flowers unisexual **11. Axyris**
21 Leaves not more than 1·5 cm, linear or linear-lanceolate; flowers hermaphrodite, sometimes some female **12. Camphorosma**
19 Plant glabrous or farinose
22 Flowers unisexual, the male in terminal spicate inflorescences, the female axillary **5. Spinacia**
22 Flowers hermaphrodite, sometimes some female
23 Perianth-segments equal; lower leaves elliptical to ovate or triangular **3. Chenopodium**
23 Two lateral perianth-segments larger than the rest; lower leaves linear to linear-oblong **12. Camphorosma**
4 Leaves filiform, semi-cylindrical, scale-like or apparently absent
24 Stems segmented, with opposite branches; leaves rudimentary, opposite
25 Stamens (3–)5; perianth-segments in fruit usually with a transverse wing dorsally; embryo spiral
26 Staminodes present; stigmas 2(3); seeds vertical **29. Anabasis**
26 Staminodes absent; stigmas 4; seeds horizontal **30. Haloxylon**
25 Stamens 1 or 2; perianth-segments not winged; embryo annular or conduplicate
27 Annual; all branches terminated by an inflorescence
28 Flowers and perianth obvious, fleshy; seeds with soft, often minutely hairy, green or brown testa **21. Salicornia**
28 Flowers minute, ± hidden by the bracts; perianth thin, membranous; seeds with black, crustaceous, granular testa **22. Microcnemum**
27 Perennial, usually shrubs or small trees, with many non-flowering branches
29 Opposite pairs of bracts connate to form a segment **20. Arthrocnemum**
29 Opposite pairs of bracts free, reniform or orbicular
30 Shrub or small tree up to 350 cm, without subglobular axillary branches; bracts ± orbicular **18. Halostachys**
30 Small shrub not more than 50 cm, with numerous subglobular axillary branches; bracts reniform-orbicular **19. Halocnemum**
24 Stems not segmented, the branches opposite or alternate; leaves obvious, sometimes small but then alternate
31 Perianth-segments in fruit with a wing, keel or dorsal spine
32 Perianth-segments with 1 unbranched spine dorsally **13. Bassia**
32 Perianth-segments winged or keeled, rarely with spine dorsally
33 Leaves mostly opposite
34 Leaves semi-cylindrical or oblong; apex obtuse or mucronulate **25. Salsola**
34 Leaves broadly triangular or ovate; apex spinose **28. Girgensohnia**
33 Leaves all alternate
35 Perianth with 3 outer large segments and 2 inner very small segments (rarely 0); stamen 1 **27. Ofaiston**
35 Perianth-segments ± equal; stamens usually 3–5
36 Flowers (at least the outer ones in each cyme) with 2 conspicuous bracteoles
37 All flowers with bracteoles **25. Salsola**

37 Outer flowers of each cyme with bracteoles, the inner flowers ebracteolate **34. Halogeton**
36 Flowers ebracteolate or with minute bracteoles
38 Plant pubescent, sometimes sparsely so **13. Bassia**
38 Plant glabrous or papillose
39 Perianth-segments of the lower flowers keeled, the upper narrowly winged in fruit, the wings free **23. Suaeda**
39 Perianth-segments broadly winged in fruit, the wings connate **24. Bienertia**
31 Perianth without appendages
40 Bracteoles 2, connate in fruit, each with a long, terminal, patent, subulate appendage; flowers all unisexual **10. Ceratocarpus**
40 Bracteoles absent, or if present without a terminal subulate appendage; flowers hermaphrodite and female
41 Perianth-segments up to 10 mm in fruit, shining; pulvinate shrub **31. Nanophyton**
41 Perianth-segments smaller, not conspicuously shining; plant not pulvinate
42 Leaves with spinose apex
43 Flowers with 2 conspicuous bracteoles **1. Polycnemum**
43 Flowers with minute bracteoles **23. Suaeda**
42 Leaves not spinose
44 Cymes mostly 3-flowered, immersed in the axis of a spike-like inflorescence; lamina of leaves not more than 6 mm
45 Small shrubs; seeds 1–1·5 mm **16. Kalidium**
45 Annual; seeds *c.* 0·5 mm **17. Halopeplis**
44 Cymes with a variable number of flowers, not immersed in the axis of the inflorescence; leaves usually more than 6 mm
46 Achenes conspicuously margined or winged **14. Corispermum**
46 Achenes not winged or margined
47 Perianth-segments solitary **14. Corispermum**
47 Perianth-segments 2–5
48 Flowers with 2 conspicuous bracteoles; perianth-segments becoming hard in fruit (see also **27**)
49 Seeds vertical; hairs medifixed, irregular **32. Petrosimonia**
49 Seeds horizontal; hairs simple **33. Halimocnemis**
48 Flowers ebracteate or with minute bracteoles; perianth-segments fleshy or membranous
50 Perianth membranous, 2 segments larger than the others; embryo annular; plant pubescent, rarely glabrous **12. Camphorosma**
50 Perianth fleshy, the segments all ±equal; embryo spiral; plant glabrous or farinose **23. Suaeda**

1. Polycnemum L.

P. W. BALL

Glabrous or pubescent annuals. Leaves subulate, rigid, spine-tipped. Flowers hermaphrodite, solitary in the axils of leaf-like bracts; bracteoles 2; perianth-segments 5; stamens usually 3; stigmas 2. Seeds vertical.

1 Leaves not more than 0·3 mm in diameter, filiform, often glandular-pubescent **4. heuffelii**
1 Leaves at least 0·5 mm in diameter, triquetrous, usually glabrous
2 Lower leaves 10–20 mm; bracteoles up to twice as long as the perianth; seeds 1·5–2 mm **1. majus**
2 Lower leaves 3–10(–12) mm; bracteoles as long as the perianth; seeds 1–1·5 mm
3 Uppermost bracts at least 3 times as long as the perianth **2. arvense**
3 Uppermost bracts not more than 2–3 times as long as the perianth **3. verrucosum**

1. P. majus A. Braun, *Flora (Regensb.)* **24**: 151 (1841). Procumbent or erect, 5–30 cm, glabrous, not verrucose. Leaves (6–)10–20 mm. Bracts at least twice as long as the perianth; bracteoles 1–2 times as long as the perianth. Seeds 1·5–2 mm. *Dry places, often as a ruderal. C. & S. Europe, extending to N. France and to c. 42° E. in Russia.* Al Au Bu Cz Ga Ge Gr He Hs Hu It Ju Rm Rs(C, W, K, E) Tu [Be Po].

2. P. arvense L., *Sp. Pl.* 35 (1753). Procumbent or erect, with spirally twisted branches, up to 50 cm, usually verrucose. Leaves 3–10(–12) mm, softly spine-tipped. Bracts at least 3 times as long as the perianth; bracteoles about equalling the perianth. Seeds 1–1·5 mm. *Sandy soils, often as a weed. C. & S. Europe, extending to C. France and C. Russia.* Al Au Be Bu Co Cz Ga Ge Gr He Hs Hu It Ju Lu Po Rm Rs(B, C, W, E) Tu.

3. P. verrucosum A. F. Láng, *Syll. Pl. Nov. Ratisbon. (Königl. Baier. Bot. Ges.)* **1**: 179 (1824). Like **2** but branches stiffly erect at the apex; leaves glaucous; uppermost bracts not more than 2–3 times as long as the perianth. *Dry, sandy places. Danube basin, Ukraine, S.E. Russia.* Au Cz Ge Hu Ju Rm Rs(W, E) *Tu.

4. P. heuffelii A. F. Láng, *op. cit.* **2**: 219 (1828). Procumbent, 5–30 cm. Leaves 5–12 × 0·1–0·3 mm, filiform, recurved, densely glandular-pilose to subglabrous. Bracts at least twice as long as the perianth; bracteoles 1–2 times as long as the perianth. Seeds 1–1·3 mm. ● *S.E. & E.C. Europe from W. Poland and N.E. Ukraine to C. Greece.* Au Bu Cz Gr Hu Ju Po Rm Rs(C, W).

2. Beta L.

P. W. BALL AND J. R. AKEROYD

Glabrous herbs. Leaves flat, more or less entire. Flowers hermaphrodite, solitary or in few-flowered cymes arranged in a spicate inflorescence. Perianth-segments and stamens 5; ovary semi-inferior, connate with the receptacle in fruit; stigmas usually 2 or 3. Fruits often adhering together by the swollen perianth and receptacle. Seeds horizontal.

Literature: B. V. Ford-Lloyd & J. T. Williams, *Bot. Jour. Linn. Soc.* **71**: 89–102 (1975). K. P. Buttler, *Pl. Syst. Evol.* **128**: 123–136 (1977).

1 Stems not more than 10 cm, decumbent, leafless except for bracts subtending solitary flowers **6. nana**
1 Stems normally much more than 10 cm, leafy; cymes 1- to many-flowered
2 Perianth mostly white or yellow, petaloid **5. trigyna**

2 Perianth green or purplish, sometimes with a scarious margin
3 Inflorescence ebracteate or the bracts much smaller than the leaves
4 Inflorescence ebracteate or bracteate only in the lower ½; perianth segments not more than 3 mm in fruit **1. vulgaris**
4 Inflorescence bracteate to the apex; perianth segments up to 5 mm in fruit **2. macrocarpa**
3 Inflorescence bracteate with at least the lower bracts similar in size and shape to the cauline leaves
5 Stigmas 3; cymes 2- to many-flowered **3. adanensis**
5 Stigmas 2; cymes 1- to 3-flowered, usually only 1 flower forming a fruit **4. patellaris**

1. B. vulgaris L., *Sp. Pl.* 222 (1753). Annual to perennial up to 200 cm. Basal leaves ovate-cordate to rhombic-cuneate; cauline leaves rhombic to lanceolate. Inflorescence dense, becoming interrupted towards the base in fruit, ebracteate at least in the upper ½. Receptacle pelviform; segments not more than 3 mm in fruit, incurved. Stigmas usually 2. *S. & W. Europe northwards to* c. 59° *N. in W. Sweden; cultivated throughout most of Europe, and occasionally naturalized.* Al Az Be Bl Br Bu Co Cr Da Ga Ge Gr Hb Ho Hs It Ju Lu Sa Si Su Tu.

(a) Subsp. **vulgaris** (subsp. *cicla* (L.) Arcangeli): Usually biennial. Stem up to 200 cm, erect; root swollen; leaves up to 20 cm; cymes 2- to 8-flowered. $2n = 18$. *Widely cultivated, some cultivars for sugar (Sugar-beet), others for the edible leaf (Spinach-beet, Chard), hypocotyl and root (Beetroot), others for fodder; often persisting as a casual and occasionally naturalized.*

(b) Subsp. **maritima** (L.) Arcangeli, *Comp. Fl. Ital.* 593 (1882) (*B. maritima* L., *B. perennis* (L.) Freyn, ?*B. atriplicifolia* Rouy): Annual (especially in S. Europe) to perennial. Stem up to 80 cm, procumbent to erect, root usually not swollen; leaves up to 10 cm; cymes 1- to 3-flowered. $2n = 18$. *Coasts of S. & W. Europe, and as a weed inland in S. Europe.*

The many cultivars of subsp. (a) have sometimes been given subspecific or varietal rank.

2. B. macrocarpa Guss., *Fl. Sic. Prodr.* **1**: 302 (1827) (*B. bourgaei* Cosson). Annual 15–40 cm, procumbent or ascending. Basal leaves oblong-spathulate to triangular-ovate, sometimes subcordate; cauline leaves obovate to lanceolate or rhombic. Inflorescence lax, bracteate to the apex. Cymes 2- or 3(–5)-flowered. Receptacle pelviform and hard in fruit; perianth-segments up to 5 mm, erect, often incurved at the apex; stigmas 2 or 3. *S. part of Mediterranean region (very local) and S. Portugal.* Bl Cr Gr Hs It Lu Sa Si [Ga].

3. B. adanensis Pamukç. ex Aellen, *Notes Roy. Bot. Gard. Edinb.* **28**: 29 (1967). Annual or biennial; stems up to 100 cm, branched from base. Basal leaves deltoid; cauline leaves rhombic-spathulate, the lower ones petiolate. Inflorescence lax, composed of 2- to many-flowered axillary cymes; bracts oblong-ovate. Perianth-segments triangular-oblong, strongly keeled, spongy, incurved in fruit. Stigmas 3. *Dry places near the sea. Karpathos and E. Kriti; one station in N.E. Greece.* Cr Gr.

4. B. patellaris Moq. in DC., *Prodr.* **13(2)**: 57 (1849) (incl. *B. diffusa* Cosson). Annual up to 60 cm, procumbent. Leaves broadly triangular-ovate, usually cordate. Inflorescence very lax, with leaf-like bracts almost to the apex. Cymes 1- to 3-flowered, usually developing 1 fruit. Receptacle hemispherical in fruit; segments 1–1·5 mm, incurved or erect; stigmas 2. *Maritime rocks. E. Spain; Mallorca; Lampedusa.* Bl Hs Si.

5. B. trigyna Waldst. & Kit., *Pl. Rar. Hung.* **1**: 34 (1800). Perennial up to 100 cm, erect. Basal leaves up to 20 cm, broadly ovate; cauline leaves ovate to lanceolate, often cordate. Inflorescence usually dense in the upper part, the bracts small or absent. Cymes 1- to 3-flowered. Receptacle pelviform; perianth whitish-yellow; segments erect in fruit; stigmas 3. $2n = 54$. *Ruderal. S.E. Europe, from W. Jugoslavia to Krym; occasionally naturalized elsewhere.* Bu *Gr Ju Rm Rs(W, K) Tu [Br Ga].

6. B. nana Boiss. & Heldr. in Boiss., *Diagn. Pl. Or. Nov.* **1(7)**: 82 (1846–7). Decumbent perennial with stout, cylindrical root. Leaves all basal, 15–20 × 8–10 mm, ovate-oblong or rhombic. Flowering stems 5–10 cm, leafless but with a lax, bracteate inflorescence. Flowers solitary. Receptacle hemispherical, strongly ridged; segments incurved in fruit; stigmas 2. $2n = 18$. *Snow-patches.* ● *Mountains of S. & C. Greece.* Gr.

3. Chenopodium L.

†J. P. M. BRENAN (EDITION 1) REVISED BY J. R. AKEROYD (EDITION 2)

Annual, rarely perennial herbs, glabrous, pubescent, glandular or farinose. Leaves alternate, flat. Flowers hermaphrodite and female, usually in cymes, variously arranged. Bracteoles absent. Perianth-segments 2–5, more or less unaltered in fruit, rarely becoming fleshy; stamens 1–5; stigmas 2(–5). Seeds usually horizontal.

Most species grow in man-made habitats and it is often impossible to decide whether they fall into the category of casual, established alien or native in any particular country.

To see the markings on the testa, which are taxonomically important, the skin-like pericarp which closely invests the ripe seed must be removed; a magnification of × 40 is required.

Literature: P. Aellen in Hegi, *Ill. Fl. Mitteleur.* ed. 2, **3(2)**: 569–659 (1960). T. Kowal, *Polsk. Towarz. Bot. (Monogr. Bot.)* **1**: 87–163 (1954). A. Beaugé, Chenopodium album *et espèces affines.* Paris, 1974.

1 Ultimate branches of the axillary cymes ending in bare, subulate, spinescent points; leaves linear-oblanceolate, entire to slightly dentate **3. aristatum**
1 Ultimate branches of the inflorescence neither bare and subulate nor spinescent; leaves various
2 Plant pubescent, with yellow or amber-coloured glands or glandular hairs, aromatic, not farinose
3 Inflorescence usually composed of distinct though sometimes small, axillary, dichasial cymes
4 Glands on lower surface of leaves and outside of sepals very shortly stalked; sepals not keeled **1. botrys**
4 Glands on lower surface of leaves and outside of sepals sessile; sepals keeled **2. schraderianum**
3 Inflorescence composed of small, axillary, sessile clusters of flowers
5 Sepals in fruit connate to near apex, the calyx saccate and net-veined dorsally; seeds 0·9–1·5 mm **6. multifidum**
5 Sepals in fruit free in at least upper ½; the calyx neither saccate nor net-veined dorsally; seeds 0·5–0·8 mm
6 Inflorescence distinctly paniculate; leaves entire or dentate **4. ambrosioides**

6 Inflorescence not paniculate; leaves sinuate, with 3 or 4 lobes on each margin **5. pumilio**
2 Plant glabrous or farinose (at least on the young parts), eglandular, not aromatic (sometimes foetid)
7 Perennial with triangular-hastate or -sagittate leaves; stigmas 0·8–1·5 mm; seeds vertical except in terminal flowers **7. bonus-henricus**
7 Annual; leaves variable but not hastate or sagittate (except **8** and **9**); stigmas up to *c.* 0·5 mm; seeds vertical or horizontal
8 Seeds vertical except sometimes in the terminal flowers of the partial inflorescence; inflorescence glabrous
9 Leaves densely glaucous-farinose beneath **10. glaucum**
9 Leaves green (or reddish) and glabrous beneath
10 Cymes sessile, axillary along main stems, densely globose; perianth often fleshy and red in fruit
11 Leaves sharply dentate; perianth always red and fleshy in fruit **8. foliosum**
11 Leaves bluntly dentate; perianth usually green and dry in fruit **9. exsuccum**
10 Cymes usually in branched axillary and terminal panicles; perianth not markedly fleshy and red in fruit
12 Sepals of lateral flowers of partial inflorescence scarcely ridged dorsally, usually free in at least upper $\frac{1}{2}$ **11. rubrum**
12 Sepals of lateral flowers distinctly ridged or keeled dorsally (at least when young), connate almost to apex, forming a sac closely investing the fruit **12. chenopodioides**
8 Seeds usually all horizontal; inflorescence sometimes glabrous, but more often farinose
13 Larger cauline leaves ±cordate to subtruncate at base; seeds 1·75–2 mm in diameter, with coarsely pitted testa **13. hybridum**
13 Larger cauline leaves not cordate, ±cuneate; seeds usually less than 1·75 mm in diameter, the testa not coarsely pitted
14 Inflorescence-axes and outside of sepals glabrous (rarely sparsely farinose in **16**)
15 Leaves entire, or rarely with a single, inconspicuous tooth on each side, ovate-elliptical, cuneate; stems 4-angled **14. polyspermum**
15 Leaves ±dentate, the larger ones deltate; stems not 4-angled **16. urbicum**
14 Inflorescence-axes and outside of sepals ±conspicuously farinose, at least when young
16 Plant smelling strongly of decaying fish; leaves entire or with a single angle on 1 or both sides towards base **15. vulvaria**
16 Plant not smelling strongly of decaying fish; leaves often ±dentate or lobed
17 Seeds with acute margins; testa with minute, very close, rounded pits; leaves not at all 3-lobed **17. murale**
17 Seeds with obtuse margins; testa smooth, furrowed, or with much larger pits than **17**; leaves sometimes 3-lobed
18 Testa strongly pitted; leaves conspicuously 3-lobed
19 Middle lobe of leaves ±oblong and parallel-sided, slightly lobed to subentire; pits on testa close and radially elongate; plant not turning red **18. ficifolium**
19 Middle lobe of leaves ±triangular and acute, with a few, irregular teeth; pits less regular; plant soon turning red **19. acerifolium**
18 Testa not strongly pitted, usually ±smooth or furrowed; leaves 3-lobed or not **(20–25). album** group

1. C. botrys L., *Sp. Pl.* 219 (1753). Annual (4–)15–70 cm, aromatic, clothed all over with short, glandular hairs. Leaves pinnatifid, with up to 4(–6) lobes on each side. Inflorescence narrow, elongate, of many axillary cymes, the upper equalling or exceeding the bracts. Sepals not keeled. Seeds 0·5–0·8 mm in diameter. $2n = 18$. *S. Europe, extending northwards to N. France and C. Russia.* Al Bu Co Cz Ga Gr *He Hs Hu It Ju Lu Rm Rs(W, K, E) Sa Si Tu [Au Be Ge Po Rs(B, C)].

2. C. schraderianum Schultes in Roemer & Schultes, *Syst. Veg.* **6**: 260 (1820) (*C. foetidum* Schrader, non Lam.). Like **1** but up to 130 cm; smell unpleasant; glands on the lower side of the leaf and outside of sepals sessile; sepals each with a prominent, dentate keel outside from near apex to base; seeds 0·7–0·8 mm in diameter. $2n = 18$. *E.C.* & *E. Europe.* [Cz Hu Po Rm Rs(C, W, E).] (*E Africa.*)

3. C. aristatum L., *Sp. Pl.* 221 (1753). Annual 3–30 cm, bushy, much-branched, glabrous or sparsely glandular. Leaves up to 7 × 0·7 cm, linear-oblanceolate, entire or feebly dentate. Inflorescence of numerous axillary, dichasial cymes, whose ultimate branches are bare, subulate and spinescent. Sepals free to near the base, slightly keeled. Seeds 0·5–0·7 mm in diameter. *E. Russia and E. Ukraine; naturalized elsewhere.* Rs(*C, E) [Hu It]. (*N.* & *C. Asia, North America.*)

4. C. ambrosioides L., *Sp. Pl.* 219 (1753) (*C. anthelminticum* auct., non L., *C. integrifolium* Vorosch.). Annual, rarely short-lived perennial, up to 1·2 m, strongly aromatic, pubescent (rarely with longer hairs), with many sessile glands. Leaves usually lanceolate, entire, dentate or rarely laciniate. Inflorescence paniculate; cymes sessile along the ultimate branches, usually bracteate. Calyx not saccate or net-veined; sepals free in at least upper $\frac{1}{2}$, rounded dorsally. Seeds 0·5–0·8 mm in diameter. $2n = 32$. *Cultivated as a vermifuge, and naturalized throughout much of W., C.* & *S. Europe.* [Al Au Az Be Bl Co Cr Ga Ge Gr Hs Hu It Ju Lu Po Rm Rs(W) Sa Si.] (*Tropical America.*)

5. C. pumilio R. Br., *Prodr. Fl. Nov. Holl.* **1**: 407 (1810). Annual, 10–40 cm, procumbent to erect, aromatic, densely clothed with eglandular, white, multicellular hairs and sessile glands. Leaves 1–4 × 0·5–2 cm, sinuate with 3 or 4 lobes on each margin. Inflorescence of sessile, axillary clusters of flowers on the ultimate branches. Calyx not saccate or net-veined; sepals free almost to the base, rounded dorsally. Seeds 0·5–0·8 mm in diameter, red-brown, glossy. $2n = 18$. *Spreading and becoming naturalized in scattered localities in W.* & *C. Europe.* [Be Cz ?Ga Ge Hs Lu.] (*Australia*).

6. C. multifidum L., *Sp. Pl.* 220 (1753). Perennial up to 1 m, procumbent to erect, much-branched, pubescent and with sessile glands. Leaves mostly pinnatifid, with narrow lobes. Flowers sessile or almost so, in sessile, axillary clusters. Fruiting calyx enlarged and saccate; sepals connate nearly to the apex, net-veined dorsally. Seeds 0·9–1·5 mm in diameter. $2n = 32$. *Widely naturalized* *in S. Europe.* [Al Bl Bu Co Ga Gr Hs It Ju Lu Sa Si Tu.] (*South America.*)

7. C. bonus-henricus L., *Sp. Pl.* 218 (1753). Perennial 5–80 cm, erect or ascending, sparsely farinose. Leaves more or less triangular, hastate to sagittate, usually subentire except for basal lobes. Inflorescence mostly terminal, narrow, tapering, leafless above. Sepals not or scarcely keeled. Stigmas 0·8–1·5 mm. Seeds 1·5–2·2 mm in diameter, vertical except in the terminal flowers. $2n = 36$. ● *Much of Europe but rare in the east and doubtfully native in the north; mainly on mountains in the south.* Al Au *Be Bu Co Cz *Da Ga Ge Gr He *Ho Hs Hu It Ju Lu *No *Po Rm Rs(*B, C, W, E) Sa Si [Br Fe Hb].

8. C. foliosum (Moench) Ascherson, *Fl. Brandenb.* **1**: 572 (1864) (*Morocarpus foliosus* Moench, *C. virgatum* (L.) Ambrosi, non Thunb.). Annual (7–)25–100 cm, erect, glabrous or almost so. Lower leaves triangular-sagittate, coarsely and sharply dentate-serrate; bracts linear-lanceolate, with a projecting lobe on either side near the base. Inflorescence of many sessile, conspicuously bracteate cymes along the main stems. Calyx red and fleshy in fruit. Seeds 1–1·3 mm in diameter, mostly vertical, reddish-brown, not acutely keeled. $2n = 18$. *Mountains of S. & S.C. Europe; widely naturalized elsewhere.* Al Au Bu Ga *Ge Gr He Hs It Ju [Cz Ho Hu Po Rm Rs(B, C, W, K, E)].

C. capitatum (L.) Ambrosi, *Fl. Tirolo Mer.* **2**: 180 (1857), is like **8** but has less toothed leaves, and fewer and larger cymes in the inflorescence, which is ebracteate in the upper part. It has long been cultivated, and occurs as a casual in scattered localities in Europe.

9. C. exsuccum (C. Loscos) Uotila, *Ann. Bot. Fenn.* **16**: 237 (1979). Like **8** but smaller; stems up to 40 cm, ascending; lower leaves broadly triangular-hastate, bluntly dentate-serrate; upper leaves broader than those of **8**, hastate; calyx green and dry or sometimes red and fleshy in fruit; seeds 0·8–1·2 mm in diameter. *Spain and N.E. Portugal.* Hs Lu.

10. C. glaucum L., *Sp. Pl.* 220 (1753). Annual 2–40(–120) cm, procumbent or erect, much-branched, glabrous or almost so, except for the leaves beneath. Leaves elliptical, lanceolate, or rarely linear-elliptic, usually coarsely sinuate-serrate to sinuate, rarely subentire or entire, green above, densely glaucous-farinose beneath. Inflorescences axillary and terminal; cymes usually arranged in spikes or more or less paniculate. Sepals of lateral flowers 2–4, free to the middle or below, not or scarcely keeled. Seeds of lateral flowers 0·6–1·1 mm in diameter, vertical, reddish-brown. $2n = 18$. *Most of Europe except the extreme north and the islands.* Al Au Be *Br Bu Cz Da Fe He Ho Hs Hu It Ju Lu Po Rm Rs(N, B, C, W, K, E) Su Tu [No].

C. wolffii Simonkai, *Term. Füz.* **3**: 164 (1879), recorded from Romania (Transsylvania), is very like **10** but the leaves are mostly linear and entire or only slightly dentate; it is probably a variant of **10**.

11. C. rubrum L., *Sp. Pl.* 218 (1753). Annual 5–100 cm, procumbent to erect, usually much-branched, glabrous or almost so. Leaves ovate-rhombic to lanceolate, coarsely serrate-lobed to subentire. Cymes in variable, often pyramidal and leafy inflorescences. Flowers and seeds as in **10**. $2n = 36$. *Most of Europe except the extreme north, but rare in the Mediterranean region.* Au Be Br Bu Co Cz Da Fe Ga Ge Gr Hb He Ho Hs Hu It Ju Po Rm Rs(N, B, C, W, K, E) Su [No].

12. C. chenopodioides (L.) Aellen, *Ostenia* 98 (1933) (*C. botryodes* Sm., *C. concatenatum* Thuill., *C. crassifolium* Hornem., *C. rubrum* var. *crassifolium* (Hornem.) Moq.). Like **11** but stems more often spreading or decumbent; leaves deltate to rhombic, subentire or slightly dentate, except for basal angles; sepals of lateral flowers more or less distinctly ridged or keeled, connate almost to the apex and forming a sac closely investing the fruit. *Usually in saline habitats. C. & S. Europe, extending to England, Denmark and E. Ukraine.* Au Be Br Co Cz Da Ga Ge †Ho Hs Hu It Ju Lu Po Rm Rs(?B, W) Sa.

13. C. hybridum L., *Sp. Pl.* 219 (1753) (*C. angulosum* Lam.). Annual 10–100 cm, sparsely farinose to subglabrous. Leaves ovate, usually acuminate at the apex, more or less cordate to subtruncate at the base, with 1–5 coarse, angular teeth on each margin. Inflorescence cymose, often lax; ultimate branches divaricate. Sepals not keeled. Seeds 1·75–2 mm in diameter, horizontal; testa black, coarsely pitted. $2n = 18$. *Europe southwards from 60° N., but rare in the Mediterranean region.* Al Au Be Bu Cz Da Ga Ge Gr He Ho Hs Hu It Ju Po Rm Rs(?N, B, C, W, K, E) Si Tu [Br Fe No].

14. C. polyspermum L., *Sp. Pl.* 220 (1753). Annual (5–)15–100 cm, erect or procumbent, glabrous. Leaves up to 8 × 5 cm, ovate-elliptical, entire or rarely with a single, slight tooth just above the base on each side, cuneate. Inflorescence long, lax, tapering, of many axillary, mostly divaricately branched cymes. Sepals 5, not keeled. Seeds 0·8–1·25 mm in diameter, horizontal. $2n = 18$. *Most of Europe except the extreme north, but rare in the Mediterranean region.* All except Az Bl Cr Fa Hb Is Sb.

15. C. vulvaria L., *Sp. Pl.* 220 (1753) (*C. foetidum* Lam., *C. olidum* Curtis). Annual (4–)10–65 cm, procumbent, much-branched, grey-farinose, smelling of decaying fish. Leaves up to 2·5(–3) × 2·3(–2·7) cm, rhombic to ovate, entire or with an acute angle on each margin at the broadest part, more or less densely grey-farinose beneath. Inflorescences terminal and axillary, small, leafy. Sepals 5, not keeled. Seeds 1–1·5 mm in diameter, brownish-black, obtusely keeled. $2n = 18$. *Most of Europe northwards to c. 50° N.* Al Au Be Bl Br Bu Co Cr Cz Ga Ge Gr He †Ho Hs Hu It Ju Lu Po Rm Rs(B, C, W, K, E) Sa Si Su Tu.

16. C. urbicum L., *Sp. Pl.* 218 (1753). Annual (4–)15–100 cm, erect, glabrous or very sparingly farinose. Lower leaves up to 14 × 11 cm, deltate to ovate, attenuate into the petiole; margins shallowly dentate to coarsely dentate-lobate above the basal lobes. Panicles with crowded suberect branches. Sepals 5, not keeled. Seeds 0·9–1·3 mm in diameter, horizontal, black, obtusely keeled. $2n = 36$. *Most of Europe except the extreme north, but now decreasing.* Al Au Be Bu Co Cz †Da †Fe He Hs Hu It Ju Lu Po Rm Rs(B, C, W, K, E) Sa Si Tu [Br].

17. C. murale L., *Sp. Pl.* 219 (1753). Annual up to 90 cm, erect or spreading, usually much-branched, more or less farinose. Leaves usually rhombic-ovate and coarsely dentate, not 3-lobed. Inflorescences terminal and axillary, divaricately branched, leafy. Sepals 5, bluntly keeled above. Seeds 1·2–1·5 mm in diameter, black, horizontal, acutely margined; pericarp adherent; testa closely and minutely pitted. $2n = 18$. *S., W. & C. Europe, extending to C. Sweden and E. Ukraine, but doubtfully native in much of the north.* Al Au Az Be Bl Br Bu Co Cr Cz Da Ga Ge Gr He Ho Hs Hu It Ju Lu Po Rm Rs(B, C, W, K) Sa Si Su Tu.

18. C. ficifolium Sm., *Fl. Brit.* **1**: 276 (1800) (*C. serotinum* auct., non L.). Annual 30–90 cm, usually erect, more or less farinose, green. Leaves up to *c.* 7 × 4 cm, lanceolate to linear-oblong, usually with a single, prominent, divergent lobe on either side near the base; middle lobe oblong, more or less parallel-sided, slightly lobed to subentire. Inflorescence much-branched; branches usually slender. Sepals 5, keeled. Seeds 1–1·15 mm in diameter, obtusely keeled; testa with close, radially elongate pits. $2n = 18$. *S.* & *C. Europe, extending to England.* Au Be Br Bu Cz Ga Ge Gr He Ho Hs Hu It Ju Po Rm Rs(N, C, W, K, E) Tu [Da Rs(B) Su].

19. C. acerifolium Andrz., *Univ. Izv. (Kiev)* **7–8**: 132 (1862) (*C. album* subsp. *hastatum* (Klinggr.) J. Murr, *C. klinggraeffii* (Abromeit) Aellen). Like **18** but with more prominent lateral lobes to leaves, the middle lobe more or less triangular and acute, with few irregular teeth; pits on testa of seeds less deep and regular. Whole plant soon becoming red, and finally yellow. *From Poland to E. Russia, southwards to Ukraine.* Po Rs(B, C, W, E).

(20–25). C. album group. More or less farinose annuals. Stems 5–150(–300) cm, usually erect and much-branched, but variable in habit. Leaves variable in outline, often rhombic or deltate, entire or toothed, acute to obtuse. Inflorescence a spicate or cymosely branched panicle. Sepals 5, with or without keel. Seeds 1–1·85 mm in diameter, black; testa smooth or with radial furrows, rarely reticulate.

1 Leaves up to 14 cm; young shoots conspicuously tinged with reddish-purple **25. giganteum**
1 Leaves rarely more than 6 cm; young shoots not conspicuously tinged with reddish-purple
2 Stems 5–30 cm, ±procumbent; leaves 3-lobed (arctic Russia) **20. jenissejense**
2 Stems usually erect or ascending; leaves variable
3 Leaves (at least middle and lower cauline) ±as wide as long; inflorescence very grey-farinose **21. opulifolium**
3 Leaves (except the most juvenile) at least $1\frac{1}{2}$ times as long as wide; inflorescence usually not very grey-farinose
4 Plant usually rather bright glaucescent green; stems not red; larger cauline leaves ovate-rhombic, with acute, forwardly-directed teeth; testa with rather numerous close and deep furrows **24. suecicum**
4 Plant deep green to greyish; stems often red or red-striped; leaves usually ovate-lanceolate, dentate or entire; testa with shallow, widely-spaced, radial furrows
5 Leaves rhombic-ovate to lanceolate, usually acute; stems green or red **22. album**
5 Leaves oblong to elliptical, obtuse; stems red-striped **23. strictum**

20. C. jenissejense Aellen & Iljin in Komarov, *Fl. URSS* **6**: 873 (1936). Annual 5–30 cm, procumbent, rarely suberect. Leaves up to 4 × 2·5 cm, oblong- or elliptic-deltate, more or less 3-lobed, with tooth-like lateral lobes at the middle of the leaf, and with 1 or 2 teeth above, green. Inflorescence dense, spicate. Sepals slightly keeled and farinose. Seeds 1–1·25 mm in diameter, without clear pattern on testa. *N.E. Russia.* Rs(N). (*Siberia.*)

21. C. opulifolium Schrader ex Koch & Ziz, *Fl. Palat.* 6 (1814). Annual 60–150 cm, erect, much-branched, green to almost white, rarely red-tinged, more or less grey-farinose. Leaves (at least middle and lower) rhombic-ovate, almost as wide as long, usually with a short, prominent lobe on each side, otherwise entire or with several teeth. Inflorescence a very grey-farinose panicle. Sepals keeled. Seeds 1·1–1·5 mm in diameter, obtusely keeled; testa not pitted. $2n = 54$. *S., C. & S.E. Europe.* Al Au Bl Bu Co Cr Cz Ga Ge Gr *He Hs Hu It Ju Lu Po Rm Rs(C, W, K, E) Sa Si Tu [Be Su].

22. C. album L., *Sp. Pl.* 219 (1753). Annual 10–150 cm, usually erect, more or less grey-farinose; stems green or red. Leaves 1–8 × 0·3–5·5 cm, very variable, rhombic-ovate to lanceolate, mostly at least $1\frac{1}{2}$ times as long as wide, entire or shallowly dentate, sometimes more or less 3-lobed. Sepals weakly keeled. Seeds 1·2–1·6(–1·85) mm in diameter, obtusely margined; testa usually marked with faint, radial furrows or (var. *reticulatum* (Aellen) Uotila) a reticulate pattern, otherwise almost smooth. $2n$ = ?36, 54. *Most of Europe.* All except Sb; introduced in Fa and Is.

A very variable and complex species. The pattern of variation has not been fully resolved. There are a number of closely related taxa adventive in Europe, some of which form hybrids with **22.**

23. C. strictum Roth, *Nov. Pl. Sp.* 180 (1821) (*C. striatum* Krašan, *C. album* subsp. *striatum* (Krašan) J. Murr, *C. betaceum* Andrz.; incl. *C. striatiforme* J. Murr). Like **22** but stems red- or purple-striped; leaves oblong to elliptical, obtuse; leaf-margins entire or somewhat dentate; sepals not keeled; seeds 1·0–1·3 mm in diameter. $2n = 36$. *C. & S.E. Europe; a frequent casual elsewhere.* Au Cz Ge Hu Rm Rs(C, W, E).

C. zerovii Iljin in Kotov, *Fl. RSS Ucr.* **4**: 650 (1952), from the C. & S. Ukraine, is like **22** but has elongate, horizontally spreading and arcuate-decumbent lower branches, and rather thick leaves, very whitish-farinose beneath, ovate, subobtuse, and held upright. The status of this plant requires further investigation.

24. C. suecicum J. Murr, *Magyar Bot. Lapok* **1**: 341 (1902) (*C. pseudopulifolium* (J. B. Scholz) A. Nyárády, *C. viride* auct., non L.). Like **22** but usually a rather bright, glaucescent green, somewhat farinose when young; stems not tinged with red; cauline leaves larger, ovate-rhombic, with sharp ascending teeth; cymes rather lax; testa with more numerous, closer and deeper furrows. $2n = 18$. *N. Europe southwards to N. Italy and S.E. Russia.* Au Br Cz Da Fe Ga Ge He Ho ?Hu It No Po ?Rm Rs(N, B, C, W, E) Su.

25. C. giganteum D. Don, *Prodr. Fl. Nepal.* 75 (1825) (*C. album* subsp. *amaranticolor* Coste & Reyn., *C. amaranticolor* (Coste & Reyn.) Coste & Reyn.). Like **22** but 200–300 cm, with young parts of plant conspicuously tinged vivid reddish-purple, and with large rhombic-deltate leaves up to *c.* 14 cm long and wide. *Cultivated as a vegetable, and naturalized in S. Europe.* [Cr Ga Gr Si.] (*N. India.*)

4. Cycloloma Moq.

P. W. BALL

Like *Chenopodium* but pubescent and eglandular; perianth-segments in fruit with a transverse wing on the back, the wings connate and completely encircling the fruit.

1. C. atriplicifolia (Sprengel) J. M. Coulter, *Mem. Torrey Bot. Club* **5**: 143 (1894) (*C. platyphylla* (Michx) Moq.). Erect or ascending pubescent annual up to 80 cm. Leaves 3–6 cm, oblong-ovate or lanceolate, sinuate-pinnatifid or dentate. Perianth *c.* 2 mm in diameter in fruit; wing *c.* 0·5 mm wide. *Locally naturalized in S. & S.C. Europe.* [Al Cz Ga It.] (*C. & W. North America.*)

5. Spinacia L.

P. W. BALL

Annual or biennial glabrous herb. Leaves flat. Flowers unisexual. Male flowers 4- or 5-merous, in a dense, spicate inflorescence. Female flowers axillary, without perianth but with 2(–4) persistent bracteoles which become enlarged, connate and hardened in fruit; stigmas 4 or 5. Seeds vertical.

1. S. oleracea L., *Sp. Pl.* 1027 (1753) (*S. glabra* Miller, *S. inermis* Moench, *S. spinosa* Moench). Stems up to 1 m or more, erect. Leaves ovate to triangular-hastate, entire or dentate. Bracteoles in fruit orbicular-obovate, usually wider than long, often with a divergent spine at the apex. *Widely cultivated vegetable (Spinach); occurring throughout most of Europe as an escape from cultivation, but rarely persisting.* (*?W. Asia.*)

6. Atriplex L.

†P. AELLEN (EDITION 1) REVISED BY J. R. AKEROYD (EDITION 2)

Annuals or small shrubs; glabrous or farinose. Leaves flat. Flowers usually unisexual. Male flowers with 5 perianth-segments and 5 stamens. Female flowers without perianth but with 2 large persistent bracteoles, free or connate to half-way; rarely some female flowers with 4- or 5-lobed perianth; stigmas 2. Seeds vertical, often dimorphic, rarely some horizontal.

1 Shrubby perennial
2 Bracteoles reniform to orbicular, orbicular-ovate or rhombic-orbicular, not lobed
3 Plant erect, usually more than 50 cm; leaves entire, rarely dentate **1. halimus**
3 Plant prostrate, less than 50 cm; leaves dentate or lobed **13. recurva**
2 Bracteoles ovate-rhombic to orbicular-deltate, often 3-lobed
4 Bracteoles 4–5 mm, not or only slightly 3-lobed, with numerous filiform dorsal appendages **2. glauca**
4 Bracteoles 2–4 mm, 3-lobed, with 0–2 dorsal tubercles **3. cana**
1 Annual
5 Female flowers dimorphic, some with horizontal seeds and 4- or 5-lobed perianth, the rest with vertical seeds and 2 bracteoles
6 Leaves green or purplish-brown, glabrous, sometimes the uppermost slightly farinose **4. hortensis**
6 Leaves grey- to white-farinose, at least beneath
7 Stem cylindrical or obtusely angled; leaves thin, grey beneath, mostly broadly triangular-cordate **5. nitens**
7 Stem 4- or 5-angled; leaves thick, white beneath, narrowly triangular-cordate **6. aucheri**
5 Female flowers monomorphic, all with vertical seeds and 2 bracteoles
8 Bracteoles elliptical or orbicular-cordate, smooth and usually entire
9 Leaves mostly lanceolate to narrowly hastate **7. oblongifolia**
9 Leaves broadly triangular-hastate **8. micrantha**
8 Bracteoles rhombic or rhombic-orbicular to -ovate, usually with teeth, or appendages dorsally
10 Bracteoles becoming hard in fruit; stems terete or angled
11 Cymes in terminal, usually leafless, often long panicles **12. tatarica**
11 Cymes axillary or in leafy panicles
12 Bracteoles broadly rhombic, usually wider than long, 3-lobed with quadrangular lateral lobes and entire middle lobe, smooth or muricate dorsally **11. laciniata**
12 Bracteoles rhombic or angular-ovate, acute and irregularly toothed in the upper part, usually with large dorsal appendages
13 Stem and branches almost glabrous; flowers not more than 3 in each cyme; seeds with flat sides **9. sphaeromorpha**
13 Stem and branches densely farinose; flowers 5–10 in a cyme; seeds with convex sides **10. rosea**
10 Bracteoles not becoming hard in fruit; stems strongly ridged
14 Bracteoles serrate or laciniate **17. calotheca**
14 Bracteoles entire or dentate
15 At least some bracteoles stalked **(18–21). prostrata** group
15 Bracteoles sessile, sometimes cuneate
16 Lower leaves truncate at base **(18–21). prostrata** group
16 Lower leaves cuneate or attenuate at base
17 Bracteoles ovate-rhombic with long-acuminate apex; lower leaves linear or linear-lanceolate **14. littoralis**
17 Bracteoles rhombic or orbicular-rhombic; lower leaves rhombic-hastate to elliptic-lanceolate
18 Lower leaves at least 4 times as long as wide **15. patens**
18 Lower leaves 3–4 times as long as wide **16. patula**

1. A. halimus L., *Sp. Pl.* 1052 (1753). Erect, stout, shrubby perennial up to 250 cm, silvery-white. Leaves up to 4 cm, ovate-rhombic or deltate, almost coriaceous, entire or rarely dentate. Cymes remote, in a paniculate inflorescence. Bracteoles orbicular-ovate to orbicular or reniform, entire or dentate, usually without dorsal appendages. $2n = 18$. *Saline habitats, less often in dry waste places inland. S. Europe.* Bl Co Cr Ga Gr Hs It Lu Sa Si.

2. A. glauca L., *Sp. Pl.* ed. 2, 1493 (1763). Dwarf shrub up to 50 cm, grey-green to silvery. Leaves up to $1(-3) \times 0{\cdot}5(-1)$ cm, oblong-lanceolate to orbicular, entire or dentate, slightly fleshy, silvery, sometimes grey-green on the upper surface. Cymes in long, spicate inflorescences. Bracteoles 4–5 mm, ovate-rhombic to ovate-deltate, entire or dentate, sometimes almost 3-lobed, with numerous, large dorsal appendages. *C., E. & S. Spain, S. Portugal.* Hs Lu.

3. A. cana Ledeb., *Icon. Pl. Fl. Ross.* **1**: 11 (1829). Dwarf shrub up to 50 cm, much-branched from the base. Leaves $0{\cdot}5–3 \times 0{\cdot}2–0{\cdot}7$ cm, oblong-spathulate to linear-lanceolate, usually entire, thick, coriaceous, grey. Inflorescence paniculate, lax, shortly branched. Bracteoles up to 4×3 mm, fleshy, 3-lobed, campanulate to orbicular-deltate, entire, with or without dorsal appendages. *Steppes. S.E. Russia and Krym.* Rs(K, E).

4. A. hortensis L., *Sp. Pl.* 1053 (1753). Erect annual up to 250 cm. Leaves usually more than 10 cm, cordate- or hastate-triangular, slightly dentate or almost entire, the uppermost triangular or oblong-lanceolate, slightly farinose when young, more or less glabrous when mature, often purplish-brown. Inflorescence terminal, spicate. Female flowers dimorphic. Bracteoles 5–15 mm, orbicular-cordate. Seeds horizontal and vertical. *Cultivated throughout most of Europe as a vegetable* (*Orache*), *and naturalized mainly in the centre and south.* [Al Au Bl Bu Cr Ge Gr He Ho Hs Hu It Ju Lu Po Rm Rs(C, W, K, E) Tu.] (*C. Asia.*)

5. A. nitens Schkuhr, *Handb.* **3**: 541 (1803) (*A. acuminata* Waldst. & Kit.). Like **4** but the leaves white beneath, often irregularly and coarsely dentate; bracteoles oblong-cordate, conspicuously reticulate-veined. *Roadsides and waste places. C. & E. Europe.* Au Bu Cz Ge Hu *It Ju Po Rm Rs(B, C, W, K, E) [Ga].

6. A. aucheri Moq., *Chenop. Monogr. Enum.* 51 (1840) (*A. amblyostegia* Turcz., *A. desertorum* D. Sosn.). Like **4** but not more than 150 cm; leaves smaller, thick, distinctly bicolorous, the lower surface white, the upper green. *E. Ukraine; S.E. Russia.* Rs(W, E).

7. A. oblongifolia Waldst. & Kit., *Pl. Rar. Hung.* **3**: 278 (1812). Erect annual up to 120 cm. Lower leaves ovate-lanceolate, often hastate, usually sinuate-dentate, the upper lanceolate. Cymes few-flowered, axillary, or in long, terminal, interrupted spicate inflorescences. Bracteoles up to 13 mm, triangular- to rhombic-ovate, usually entire, smooth dorsally. $2n = 36$. *Ruderal. C., E. & S.E. Europe.* Au Bu Cz Ge Gr Hu It Ju Po Rm Rs(C, W, K, E) [Ga].

8. A. micrantha Ledeb., *Icon. Pl. Fl. Ross.* **1**: 11 (1829) (*A. micrantha* C. A. Meyer, *A. heterosperma* Bunge). Like **7** but the lower leaves up to 15 cm, sagittate, entire or irregularly dentate, the upper narrow, hastate; inflorescence lax, spicate; bracteoles up to 6 mm, orbicular or cordate, somewhat farinose. *Ruderal. Southern part of S.E. Russia and Ukraine; naturalized or casual elsewhere.* Rs(W, E) [?Au ?Ga Ge Hs Po ?Rm].

9. A. sphaeromorpha Iljin, *Bull. Jard. Bot. URSS* **26**: 414 (1927). Much-branched annual up to 80 cm. Leaves up to 3 × 2 cm, triangular-ovate or deltate, broadly cuneate, obtuse, crenate-serrate, grey-green. Female flowers few, axillary; male terminal, in short, leafless, spicate inflorescences. Bracteoles *c.* 6 mm, rhombic, irregularly dentate, ventricose, becoming hardened towards the base. *S.E. Russia and S. Ukraine.* Rs(W, E).

10 A. rosea L., *Sp. Pl.* ed. 2, 1493 (1763). Erect or ascending annual up to 100 cm, much-branched; stems smooth or angled. Leaves up to 6 × 3 cm, ovate-rhombic or rhombic-triangular, sinuate-dentate, white. Cymes axillary, except the uppermost. Bracteoles up to 12 mm, rhombic, dentate, becoming hard in the lower ½, usually with large dorsal appendages. *Ruderal. Europe northwards to 57° N. in C. Russia, but rare in the north-west.* Au Bl Bu Co Cr Cz Ga Ge Gr Hs Hu It Ju Lu Po Rm Rs(B, C, W, K, E) Sa Si Tu.

11. A. laciniata L., *Sp. Pl.* 1053 (1753) (*A. sabulosa* Rouy, *A. arenaria* J. Woods, *A. maritima* L.). Procumbent, much-branched, silvery annual up to 30 cm; stem smooth or angled. Leaves 1·5–2(–4) cm, rhombic to ovate, sinuate-dentate. Cymes axillary. Bracteoles 6–7 mm, broadly rhombic, rounded or toothed on the lateral angles, becoming hard in the lower ½. $2n = 18$. *Maritime sands and shingle. W. & N.W. Europe, extending to S. Sweden.* Be Br Da Ga Ge Hb Ho Hs No Su.

12. A. tatarica L., *Sp. Pl.* 1053 (1753) (*A. laciniata* auct., non L.; incl. *A. tornabenii* Tineo). Procumbent to erect, much-branched, whitish annual up to 150 cm; stem smooth or angled. Leaves up to 10 × 7 cm, silvery, triangular-rhombic to triangular-hastate, irregularly sinuate-lobed, the lowest lobe largest. Inflorescence terminal, paniculate or spicate, more or less leafless. Bracteoles up to 7 mm, orbicular to oblong-rhombic, reticulate-veined, becoming hard in the lower ½, with or without dorsal appendages. $2n = 18$. *S. & S.C. Europe; southern half of S. & S.C. Europe, extending to E. Russia.* Au Bl Bu Co Cz Ga Gr Hs Hu It Ju Rm Rs(C, W, K, E) Sa Si Tu [Ge Po Rs(N, B)].

13. A. recurva D'Urv., *Mém. Soc. Linn. Paris* **1**: 284 (1822) (*A. tatarica* subsp. *recurva* (D'Urv.) Rech. fil.). Prostrate, shrubby perennial 10–40 cm. Leaves up to 7 × 5 cm, similar in shape to those of **12**. Inflorescence terminal, leafless. Bracteoles up to 20 mm, rhombic to orbicular, without or with small dorsal appendages. $2n = 18$. • *Coasts of S. & C. Aegean region.* Cr Gr.

14. A. littoralis L., *Sp. Pl.* 1054 (1753) (*A. laciniata* auct., non L.). Erect, usually much-branched annual up to 150 cm; stems strongly ridged. Leaves linear to oblong-linear, entire or dentate, the lower shortly petiolate, the upper sessile. Inflorescence long, spicate, leafy only at the base. Bracteoles 3–6 mm, triangular-ovate, dentate or entire, dorsally muricate; apex acuminate or lingulate. $2n = 18$. *Seashores and saline places inland. Coasts of most of Europe; inland in C. & S.E. Europe.* Au Be Br Co Cz Da Fe Ga Ge Gr Hb Ho Hs Hu It Ju No Po Rm Rs(B, C, W, E) Su.

15. A. patens (Litv.) Iljin, *Bull. Jard. Bot. URSS* **26**: 415 (1927). Erect annual 15–70 cm. Lower and middle leaves up to 8 × 2 cm, narrowly oblong-hastate to elliptic-lanceolate, crenate-serrate, the upper linear-lanceolate, entire. Cymes small, in leafless, terminal, spicate inflorescences. Bracteoles orbicular-deltate, acuminate, entire or dentate, farinose, smooth or muricate dorsally. *S. Russia, Ukraine.* Rs(C, W, E).

16. A. patula L., *Sp. Pl.* 1053 (1753). Much-branched annual 10–150 cm, slightly farinose; stem strongly ridged. Lower leaves 3–14 × 1–6 cm, rhombic-hastate with cuneate base, the upper lanceolate to linear, entire. Cymes axillary or in long, spicate inflorescences. Bracteoles 3–7(–20) mm, broadly rhombic, entire or denticulate, smooth or tuberculate dorsally. $2n = 36$. *Cultivated ground and as a ruderal. Throughout Europe.* All except Sb.

17. A. calotheca (Rafn) Fries, *Nov. Fl. Suec.* ed. 2, *Mant.* **3**: 164 (1842). More or less glabrous annual 30–100 cm. Leaves triangular-hastate, coarsely and irregularly sinuate-dentate. Inflorescence spicate, acute. Bracteoles up to 30 mm, hastate or cordate-triangular, laciniate, without dorsal appendages. $2n = 18$. • *Coasts of the Baltic, extending to S. Norway and N.W. Germany.* Da Fe Ge No Po Rs(B, C) Su.

18–21. A. prostrata group. Erect or procumbent annuals up to 100 cm, farinose when young; stems strongly ridged. Lower leaves up to 10 × 7 cm. Inflorescence paniculate or spicate. Bracteoles smooth or tuberculate dorsally. Seeds dimorphic; the 'small' seeds 1–2 mm, black, smooth, the 'large' seeds 2–3 mm, brown, reticulate-rugose.

A difficult group not yet fully understood. The evidence available indicates that all the species within the group are interfertile, and many populations are made up of plants variously intermediate between two or more of the species. It is, therefore, often impossible to identify individual plants with certainty.

Literature: M. Gustafsson, *Op. Bot. (Lund)* **39**: 1–63 (1976). P. M. Taschereau, *Watsonia* **15**: 183–209 (1985); **17**: 247–264 (1989).

1 Bracteoles sessile, sometimes cuneate at base
2 Bracteoles thin, not inflated; plant usually erect **18. prostrata**
2 Bracteoles thick, inflated; plant usually procumbent **19. glabriuscula**
1 At least some bracteoles distinctly stalked
3 Bracteoles at least 5 mm, with stalks up to 30 mm **20. longipes**
3 Bracteoles 3–5 mm, with stalks up to 1·5 mm **21. praecox**

18. A. prostrata Boucher ex DC. in Lam. & DC., *Fl. Fr.*, ed. 3, 387 (1805) (*A. hastata* auct., non L., *A. triangularis* Willd., *A. deltoidea* Bab., *A. latifolia* Wahlenb.). Erect or procumbent; stems up to 100 cm. Lower leaves hastate, the base subcordate. Bracteoles 2–6 mm, triangular to triangular-ovate, dentate. Seeds nearly all small. $2n = 18$. *Ruderal, often near the coast; salt-marshes. Throughout Europe northwards to* c. 71° *N. in Norway*. All except ?Cr Fa Is Sb.

Very variable; one of the more distinct variants is subsp. **polonica** (Zapał.) Uotila, *Ann. Bot. Fenn.* **14**: 197 (1977), from S.E. Poland, Estonia and W. Ukraine. It has the lower leaves almost sagittate-hastate, deeply sinuate-dentate and bracteoles rhombic with acute dorsal appendages.

19. A. glabriuscula Edmondston, *Fl. Shetl.* 39 (1845). (*A. babingtonii* J. Woods). Usually procumbent; stems 20–90 cm. Leaves hastate; base truncate. Bracteoles 4–10 mm, rhombic, inflated, thick, entire or dentate. Large seeds numerous. $2n = 18$. *Maritime habitats. Coasts of W. & N.W. Europe*. Br Da Fa Ga Ge Hb Ho ?Hs Is No Po Rs(N, B) Su.

Hybrids between **19** and **20** are frequent in N. Britain.

20. A. longipes Drejer, *Fl. Excurs. Hafn.* 107 (1838). Erect; stems 20–90 cm. Lower leaves triangular-hastate and somewhat cuneate at base. Bracteoles 5–25 mm, at least some distinctly stalked up to 30 mm, triangular to rhombic, usually entire, smooth or muricate dorsally. Large seeds usually numerous. $2n = 18$. *Brackish salt-marshes. Coasts of the Baltic, arctic Russia and Britain; probably also occurring elsewhere in W. & N.W. Europe*. Br Da Fe Po Rs(N) Su [No].

A. × gustafssoniana Taschereau, *Watsonia* **17**: 256 (1989) (incl. *A. longipes* subsp. *kattegatensis* Turesson), the hybrid between **18** and **20**, is frequent on the coasts of Britain and W. Sweden.

21. A. praecox Hülphers in Lindman, *Svensk. Fanerogamfl.* 228 (1918) (incl. *A. nudicaulis* Boguslaw, *A. lapponica* Pojark.?, *A. kuzenerae* N. Semen.?). Erect or procumbent, 3–15 cm, often reddish. Leaves up to 3 cm, ovate to lanceolate, cuneate at the base. Bracteoles 3–6 mm, some with stalks up to 1·5 mm, rhombic- or triangular-ovate, entire, thin, smooth dorsally. $2n = 18$. *Shingle or sand on sheltered coasts. Fennoscandia; N. Britain; Iceland.* Br Fe Is No Rs(N, B) Su.

Plants intermediate between **20** and **21**, together with putative hybrids, occur in the Baltic and these two species are often treated at subspecific rank.

7. Halimione Aellen

P. W. BALL

Like *Atriplex* but the bracteoles connate almost to the apex in fruit.

1 Annual; fruit with pedicel up to 12 mm **3. pedunculata**
1 Small shrub; fruit ± sessile
2 Bracteoles 2·5–5 mm in fruit, obdeltate, usually 3-lobed distally **1. portulacoides**
2 Bracteoles *c.* 2 mm in fruit, suborbicular, obtuse with dentate margin **2. verrucifera**

1. H. portulacoides (L.) Aellen, *Verh. Naturf. Ges. Basel* **49**: 126 (1938) (*Atriplex portulacoides* L., *Obione portulacoides* (L.) Moq.). Decumbent small shrub 20–80(–150) cm, silvery-farinose; stems often rooting. Lower leaves opposite, oblong, elliptical or obovate, entire, thick and fleshy. Fruit sessile. Bracteoles 2·5–5 mm, obdeltate, usually 3-lobed at the apex. $2n = 36$. *Salt-marshes, particularly at the edges of channels and pools; occasionally on rocks and cliffs. Coasts of S. & W. Europe, from Bulgaria to Denmark.* Al Be Bl Br Bu Co Cr Da Ga Ge Gr Hb Ho Hs It Ju Lu Sa Si †Su Tu.

2. H. verrucifera (Bieb.) Aellen, *op. cit.* 129 (1938) (*Atriplex verrucifera* Bieb., *Obione verrucifera* (L.) Moq.). Like **1** but bracteoles *c.* 2 mm, suborbicular, obtuse, dentate, verrucose. *Saline habitats. S.E. Europe from Turkey to* c. 52° *N. in E. Russia.* Rm Rs(C, W, K, E) Tu.

3. H. pedunculata (L.) Aellen, *op. cit.* 123 (1938) (*Atriplex pedunculata* L., *Obione pedunculata* (L.) Moq.). Erect annual up to 50 cm, silvery-farinose. Leaves alternate, oblong to elliptical or obovate, entire. Fruiting pedicels up to 12 mm. Bracteoles 2–3 mm, obdeltate, 3-lobed, the middle lobe very small, the lateral lobes large and spreading. $2n = 18$. *Saline soils. Coasts of N.W. & N. Europe, from N. France to Estonia, and of the Black Sea, and in saline places inland.* Be Br Bu Da Ga Ge Ho Po Rm Rs(B, C, W, K, E) Su.

Rare and becoming extinct in many of its localities in N.W. Europe.

8. Cremnophyton Brullo & Pavone

J. R. AKEROYD

Small shrubs; silvery-lepidote. Leaves flat, entire. Flowers unisexual, sessile, in a lax panicle. Male flowers with 4 perianth-segments and 4 stamens; bracteole 1. Female flowers without perianth; bracteoles 2, free; bract 1. Seeds vertical.

Literature: S. Brullo & P. Pavone, *Candollea* **42**: 621–625 (1987).

1. C. lanfrancoi Brullo & Pavone, *Candollea* **42**: 622 (1987). Erect, branched shrub up to 100 cm. Leaves opposite or alternate, 10–30 × 1–4 mm, linear-lanceolate or linear, obtuse or subacute. Flowers in sessile clusters of usually 1 female and 4 male. Female bracteoles 5–9 mm in fruit, orbicular, rounded or retuse at the apex, with or without 2 small appendages dorsally. Seeds *c.* 2 mm, ovoid, brown. $2n = 20$. *Limestone sea-cliffs.* ● *Malta & Gozo.* Si.

9. Krascheninnikovia Gueldenst.

P. W. BALL

Perennial herbs or small shrubs with stellate hairs. Leaves alternate, flat, entire. Flowers unisexual. Male flowers in a dense, spicate inflorescence; perianth-segments 4; stamens 4. Female flowers axillary, without perianth but with 2 persistent bracteoles; stigmas 2. Seeds vertical. (*Ceratoides* Gagnebin, *Eurotia* auct.)

1. K. ceratoides (L.) Gueldenst., *Novi Comment. Acad. Sci. Petrop.* **16**: 555 (1792) (*Eurotia ceratoides* (L.) C. A. Meyer, *E. ferruginea* (Nees) Boiss., *Ceratoides latens* (J. F. Gmelin) Reveal & Holmgren). Plant 30–100 cm, erect, tomentose or lanate, with grey or reddish-brown hairs. Leaves 12–40 mm, linear-oblong or lanceolate, attenuate at the base into a short petiole. Flowers densely tomentose or lanate. Bracteoles *c.* 3 mm in fruit, obovate, pubescent. *Dry places. S., S.C. & S.E. Europe, from S. Spain to the middle Volga; very local.* Au †Cz Hs Hu Ju Rm Rs(C, W, K, E).

10. Ceratocarpus L.

P. W. BALL

Annual herbs with stellate hairs. Leaves entire, spine-tipped. Flowers unisexual. Male flowers solitary or in short, axillary cymes; perianth-segments 2; stamens 1(–3). Female flowers solitary, without perianth but with 2 persistent bracteoles; stigmas 2. Seeds vertical.

1. C. arenarius L., *Sp. Pl.* 969 (1753). Much-branched herb up to 30 cm with grey or reddish-brown tomentum. Leaves 1–4 cm, linear or linear-filiform, the lower opposite, the upper alternate. Bracteoles 5–7 mm in fruit, connate, forming a triangular-cuneate involucre; apex of bracteoles with a patent subulate appendage. *Sandy ground, and as a weed of cultivation. E. Europe from N.E. Bulgaria to E. Russia.* Bu Rm Rs(C, W, K, E).

11. Axyris L.

P. W. BALL

Stellate-pubescent annual herbs. Leaves alternate, flat. Flowers unisexual; the male in dense cymes forming a spicate inflorescence; the female 2-bracteolate, solitary or forming a cyme with the male flowers. Perianth-segments 3(–5); stamens 2–5; stigmas 2. Fruit usually with a short wing at the apex. Seeds vertical.

1. A. amaranthoides L., *Sp. Pl.* 979 (1753). Stems 15–80 cm. Leaves up to 9 × 3 cm, lanceolate, densely pubescent. Achenes 2–3 mm, often with 2-lobed wing at the apex. *Cultivated ground and waste places. C., S. & E. Russia; W. Kazakhstan.* Rs(C, E).

12. Camphorosma L.

P. W. BALL AND J. R. AKEROYD

Annual or perennial herbs or small shrubs, usually pubescent. Leaves alternate, linear or subulate. Flowers hermaphrodite, male or female, solitary or in dense ovoid cymes. Perianth-segments 4(5) with the 2 lateral segments larger. Stamens 4 or 5. Stigmas 2(3). Seeds vertical.

1 Dwarf shrub — **1. monspeliaca**
1 Annual herb
  2 Perianth ± glabrous, with divergent lateral segments — **2. annua**
  2 Perianth densely pubescent, with incurved lateral segments — **3. songorica**

1. C. monspeliaca L., *Sp. Pl.* 122 (1753) (*C. ruthenica* Bieb.; incl. *C. nestensis* Turrill). Caespitose perennial with short, woody, leafy branches at the base, and flowering stems 10–80 cm. Leaves 2–10 mm, linear, stiff, often fasciculate. *Saline soils and dry waste places. S. Europe, extending northwards to* c. 53° N. *in E. Russia.* Al Bu Co Ga Gr Hs It Ju Rm Rs(C, W, K, E) Sa Si.

(a) Subsp. **monspeliaca**: Plant smelling of camphor. Perianth 2–3·5 mm, glabrous or pubescent at the apex of the segments; 1 pair of segments larger, divergent at the apex. Seeds 1·5–2 mm. $2n = 12$. *Throughout the range of the species.*

(b) Subsp. **lessingii** (Litv.) Aellen, *Notes Roy. Bot. Gard. Edinb.* **28**: 31 (1967) (*C. lessingii* Litv., *C. ruthenica* auct., non Bieb.): Plant not smelling of camphor. Perianth 2–2·5 mm, villous-tomentose; segments subequal, not divergent at the apex. Seeds *c.* 1 mm. *W. Kazakhstan and adjacent parts of S.E. Russia. (C. & S.W. Asia.)*

2. C. annua Pallas, *Reise* **3**: 603 (1776) (*C. ovata* Waldst. & Kit.). Glabrous or sparsely pubescent annual or biennial 5–40 cm. Leaves 5–15 mm, linear. Perianth 2·5–4 mm, glabrous; lateral perianth-segments divergent at the apex. *Saline habitats.* ● *E.C. Europe, extending to Bulgaria and C. Ukraine.* Au Bu Cz Hu Ju Rm Rs(C, W).

3. C. songorica Bunge, *Acta Horti Petrop.* **6**: 415 (1880) (*C. annua* auct., non Pallas). Like **2** but lanate; perianth pubescent; apex of lateral segments incurved, usually villous. *Salt-marshes. S.E. Russia.* Rs(C, E). *(C. & W. Asia.)*

13. Bassia All.

P. W. BALL AND J. R. AKEROYD

Pubescent herbs or small shrubs. Leaves linear, flat or cylindrical, entire. Flowers hermaphrodite or female, solitary or in cymes arranged in a panicle. Perianth-segments 5, becoming enlarged in fruit and developing a spine or transverse wing (rarely a tubercle) dorsally. Stamens 5. Stigmas 2 or 3. Seeds horizontal. (*Echinopsilon* Moq.; incl. *Kochia* Roth.)

Literature: A. J. Scott, *Feddes Repert.* **89**: 101–109 (1978).

1 Fruiting perianth-segments with spine dorsally
  2 Leaves linear-lanceolate, flat, attenuate towards the base; spines of perianth 1–1·5 mm — **1. hyssopifolia**
  2 Leaves semi-cylindrical or filiform, rarely the lowest ± flat; spines of perianth usually not more than 1 mm
    3 Spines of perianth pubescent; stem usually procumbent or ascending — **2. hirsuta**
    3 Spines of perianth glabrous distally; stem erect — **3. sedoides**
1 Fruiting perianth-segments with transverse wing or tubercle dorsally
  4 Leaves flat, lanceolate or linear-lanceolate, 1- or 3-veined; fruiting perianth-segments with very short wing sometimes reduced to a tubercle — **7. scoparia**
  4 Leaves semi-cylindrical, linear-subulate, obscurely veined or 1-veined; fruiting perianth-segments with distinct, scarious wing
    5 Annual; fruiting perianth-segments with oblong or obovate wing not more than 1 mm wide — **6. laniflora**
    5 Small shrub; fruiting perianth-segments with suborbicular or triangular-orbicular wing usually at least 2 mm wide
      6 Leaves up to 12 mm, acute or mucronate, not fleshy; perianth-segments ovate, obtuse, densely pubescent — **4. prostrata**

6 Leaves up to 30 mm, obtuse, fleshy; perianth-segments orbicular, glabrous or sparsely pubescent **5. saxicola**

1. B. hyssopifolia (Pallas) O. Kuntze, *Revis. Gen.* **1**: 547 (1891) (*Kochia hyssopifolia* (Pallas) Schrader). Erect, white, hirsute or villous annual 10–100 cm; branches erect. Leaves up to 35 × 4 mm, linear-lanceolate, flat, attenuate at the base. Perianth villous-lanate; spines 1–1·5 mm, subulate, curved, glabrous. $2n = 18$. *Saline soils and cultivated ground. From Moldavia to W. Kazakhstan; E. Spain.* Hs ?It Rs(W, ?K, E) [Ga].

2. B. hirsuta (L.) Ascherson in Schweinf., *Beitr. Fl. Aethiop.* 187 (1867) (*Echinopsilon hirsutum* (L.) Moq., *Kochia hirsuta* (L.) Nolte). Procumbent or ascending, villous or glabrescent annual 10–60 cm; branches flexuous. Leaves up to 15 mm, linear, fleshy, semi-cylindrical, rarely the lower spathulate and more or less flat. Perianth hirsute; spines *c.* 1 mm, conical or hooked, pubescent. $2n = 18$. *Maritime sands and saline soils. S.E. Europe, from Turkey to S.E. Russia; W. Kazakhstan; C. Mediterranean region; N. Germany and Denmark.* ?Be Bu Co Da Ga Ge Gr †Ho It Rm Rs(W, K, E) Sa †Su Tu.

3. B. sedoides (Pallas) Ascherson in Schweinf., *loc. cit.* (1867) (*Kochia sedoides* (Pallas) Schrader). Erect, lanate annual 10–100 cm; branches erect. Leaves up to 12 mm, filiform, fleshy. Perianth villous; spines *c.* 1 mm, conical, usually straight, glabrous distally. *Saline habitats. E. Europe, from N.C. Russia to S.E. Romania; Hungary.* Hu Ju Rm Rs(C, W, K, E).

4. B. prostrata (L.) A. J. Scott, *Feddes Repert.* **89**: 108 (1978) (*Kochia prostrata* (L.) Schrader). Small shrub with procumbent stems up to 80 cm. Leaves up to 12 mm, linear-subulate, stiff, acute or mucronate. Fruiting perianth 3–4·5 mm, pubescent; segments ovate, obtuse; wing semi-circular, erose. *Dry places. S. Europe, extending northwards to Czechoslovakia and E.C. Russia.* Au Bu Cz Ga He Hs Hu It Ju Rm Rs(C, W, K, E).

5. B. saxicola (Guss.) A. J. Scott, *loc. cit.* (1978) (*Kochia saxicola* Guss.). Small shrub up to 30 cm. Leaves up to 30 mm, cylindrical, fleshy, obtuse. Fruiting perianth *c.* 4 mm, glabrous or sparsely pubescent; segments suborbicular, very obtuse; wing cuneate-obovate, apex obtuse or retuse. *Maritime rocks and cliffs.* ● *Islands of Ischia, Capri and Strombolichio (Isole Lipari).* It Si.

6. B. laniflora (S. G. Gmelin) A. J. Scott, *loc. cit.* (1978) (*Kochia laniflora* (S. G. Gmelin) Borbás, *K. arenaria* (P. Gaertner, B. Meyer & Scherb.) Roth). Erect annual up to 80 cm. Leaves up to 25 mm, filiform, soft. Fruiting perianth 4–6 mm, perianth-segments ovate; wing not more than 1 mm wide, oblong-obtuse or obovate. $2n = 18$. *Dry, sandy places. From S. France and W. Germany to E.C. Russia.* Al Au Bu Cz Ga Ge Gr Hs Hu It Ju Po Rm Rs(C, W, K, E).

7. B. scoparia (L.) A. J. Scott, *loc. cit.* (1978) (incl. *B. sicorica* (O. Bolòs & Masclans) Greuter & Burdet, *Kochia sicorica* O. Bolòs & Masclans, *K. scoparia* (L.) Schrader). Erect annual 20–150 cm. Leaves up to 50 mm, flat, lanceolate or linear-lanceolate, 1- or 3-veined. Fruiting perianth 3–4 mm, pubescent, with ovate, obtuse segments; wing very short or reduced to a tubercle. $2n = 18$. *Widely cultivated for its ornamental foliage, and naturalized in C., E. & S. Europe, but perhaps native in S. and E. Russia.* Rs(*C, *E) [Al Au Bu Cz Ga Ge Gr He Hs Hu It Ju Po Rm Rs(W, K) Si Tu]. (*Temperate Asia.*)

14. Corispermum L.

†P. AELLEN (EDITION 1) REVISED BY J. R. AKEROYD (EDITION 2)

Annual herbs, often pubescent with stellate hairs. Leaves alternate, linear or lanceolate, flat or semi-cylindrical. Flowers hermaphrodite, solitary in the axils of bracts, in spikes arranged in a branched inflorescence. Perianth-segments 0–3(–5). Stamens 1–5. Stigmas 2. Achene strongly compressed, usually with a marginal wing. Seeds vertical.

All species occur in open, gravelly or sandy habitats.

Literature: M. Iljin, *Bull. Jard. Bot. URSS* **28**: 637–654 (1929). M. Klokov, *Not. Syst.* (*Leningrad*) **20**: 90–136 (1960).

1 Base of style curved outwards **1. uralense**
1 Base of style straight
2 Achene not winged **11. orientale**
2 Achene winged, sometimes narrowly
3 Wing membranous
4 Flowers in long, lax spikes **4. canescens**
4 Flowers in short, dense spikes
5 Apex of wing distinctly emarginate **2. marschallii**
5 Apex of wing not emarginate **3. algidum**
3 Wing thick, or membranous only at margin, often very narrow
6 Perianth-segments absent from most flowers **10. intermedium**
6 Perianth-segments obvious in all flowers
7 Achene with triangular-acuminate apex **8. declinatum**
7 Achene with emarginate or obtuse apex
8 Achene with emarginate apex **5. aralocaspicum**
8 Achene with obtuse apex
9 Leaves filiform or semi-cylindrical
10 Achene 4–5 mm **7. filifolium**
10 Achene 2·5–3·5 mm **6. nitidum**
9 Leaves flat
11 Wing not more than $\frac{1}{10}$ as wide as achene **9. hyssopifolium**
11 Wing *c.* $\frac{1}{5}$ as wide as achene **10. intermedium**

1. C. uralense (Iljin) Aellen, *Feddes Repert.* **69**: 144 (1964). (*C. squarrosum* auct.) Plant 5–40 cm, often reddish, pubescent becoming glabrous; branches long. Leaves oblong-linear to linear. Spikes cylindrical, dense. Bracts ovate or oblong-ovate, pubescent. Perianth-segments 0 or 1. Achene 2·3–3·3 × 2–2·6 mm, ovate; base of the style curved outwards; wing very small, membranous, undulate. *S.E. Russia.* Rs(E). (*Asiatic Ural.*)

2. C. marschallii Steven, *Mém. Soc. Nat. Moscou* **5**: 336 (1814). Plant 30–40 cm, densely pubescent, becoming glabrous; branches long. Leaves narrowly lanceolate to linear. Spikes usually short, dense. Bracts broadly ovate, acute, longer but narrower than the achene, margin membranous. Perianth-segments 0(–2). Achene 4–5 × 3–4 mm, ovate to orbicular; wing broad ($\frac{1}{3}$–$\frac{1}{2}$ as wide as the achene), cordate at the base, membranous, erose, emarginate; style included in the notch. *E. & E.C. Europe.* *Bu *Cz *Ge Ju Po Rm Rs(C, W, E) [Ho It Rs(B)].

3. C. algidum Iljin, *Bull. Jard. Bot. URSS* **28**: 642 (1929). Like 2 but the plant 3–15 cm, pubescent; leaves oblong or linear-oblong; spikes usually ovoid, bracts oblong or ovate; achene 3–4 × 2·2–3 mm, ovate, or ovate-orbicular, obtuse; wing $\frac{1}{8}$–$\frac{1}{4}$ as wide as the achene, almost entire, or dentate towards the apex, scarcely emarginate. *N. Russia*. Rs(N).

4. C. canescens Kit. in Schultes, *Österreichs Fl.* ed. 2, **1**: 7 (1814). Plant 30–70 cm, simple or with long branches, usually pubescent. Leaves linear. Spikes long, lax. Bracts ovate, acute, broadly membranous-margined, covering the fruit. Perianth-segments 0–5. Achene 3·5–4 × 2·5–3·3 mm, ovate or orbicular; wing broad ($\frac{1}{8}$–$\frac{1}{4}$ as wide as the achene), translucent, erose. *C. & E. Europe*. Cz Hu Ju Rm Rs(C, W, E).

C. ucrainicum Iljin, *Not. Syst. (Leningrad)* **9**: 262 (1946), **C. stenopterum** Klokov, *Not. Syst. (Leningrad)* **20**: 93 (1960), **C. volgicum** Klokov, *op. cit.* 100 (1960) and **C. borysthenicum** Andrz., *Univ. Izv. (Kiev)* **7–8**: 134 (1862) should probably be included here.

5. C. aralocaspicum Iljin, *Bull. Jard. Bot. URSS* **28**: 637 (1929). Plant 10–60 cm, with long branches from the base, more or less glabrous. Leaves linear, flat. Spikes short, dense at the base. Bracts oblong-ovate, shorter than the fruit. Perianth-segments 1(–3). Achene 3·5–3·75 × 3·3–3·5 mm, almost orbicular, apex rounded, almost truncate; wing broad ($\frac{1}{4}$–$\frac{1}{2}$ as wide as the achene) towards the apex, undulate. *S.E. Russia; W. Kazakhstan*. Rs(E).

6. C. nitidum Kit. in Schultes, *Österreichs Fl.* ed. 2, **1**: 7 (1814). Plant 10–50 cm, branches procumbent, sparsely pubescent or glabrous, often reddish. Leaves linear, slightly fleshy. Spikes long, slender, lax. Upper bracts broadly ovate to orbicular, shortly acuminate, broadly membranous-margined. Perianth-segments 3. Achene 2·5–3·5 × 2–2·75 mm, ovate or almost orbicular, apex truncate and shortly bidentate; wing very narrow, thick, entire. *E.C. & E. Europe*. Bu Cz Gr Hu Ju Rm Rs(W, K, E).

C. nitidulum Klokov, *Not. Syst. (Leningrad)* **20**: 105 (1960) and **C. calvum** Klokov, *op. cit.* 112 (1960) are probably only variants of **6**.

7. C. filifolium C. A. Meyer, *Bull. Soc. Nat. Moscou* **2**: 455 (1854). Like **6** but leaves filiform; achene 4–5 × 3–4 mm, ovate, obtuse, the wing with a gap at the style-base. *S.E. Russia*. Rs(E).

8. C. declinatum Stephan ex Steven, *Mém. Soc. Nat. Moscou* **5**: 384 (1817). Plant 15–50 cm, with long branches from the base, glabrous or subglabrous. Leaves linear or linear-lanceolate. Spikes long, lax. Bracts oblong, long-acuminate, broadly membranous at the base. Perianth-segments rudimentary. Achene 3–4 × 1·3–2·5 mm, oblong-obovate, almost parallel-margined; wing very narrow, thick, slightly erose, with a triangular apex. *S.E. Russia*. Rs(*C, E).

9. C. hyssopifolium L., *Sp. Pl.* 4 (1753). Plant 15–60 cm, branched from the base. Leaves linear. Spikes long, narrow, more or less lax. Lower bracts longer than the achene, upper shorter and narrower. Perianth-segments 1(–3). Achene 4–5 × 3–4 mm, ovate; wing very narrow, thin, sometimes absent, apex rounded, entire. *S. & C. Russia, Ukraine*. *Po Rs(*B, C, W, E) [Cz].

C. hybridum Besser ex Andrz., *Univ. Izv. (Kiev)* **7–8**: 135 (1862), **C. glabratum** Klokov, *Not. Syst. (Leningrad)* **20**: 122 (1960), **C. czernajaevii** Klokov, *op. cit.* 120 (1960), and **C. insulare** Klokov, *op. cit.* 124 (1960) should probably be included in **9**.

10. C. intermedium Schweigger, *Königsb. Arch. Naturw.* **1**: 211 (1812) (incl. *C. leptopterum* (Ascherson) Iljin, *C. gallicum* Iljin, *C. hyssopifolium* auct. eur. centr., non L.). Plant 10–60 cm, with long branches, pubescent when young, or glabrous. Leaves lanceolate to linear. Spikes short or elongate, usually thick and dense. Bracts lanceolate to broadly ovate-lanceolate, narrower but slightly longer than the fruit. Perianth-segments 0 or 1(–3). Achene 3–4·5 × 3·5 mm, elliptical to suborbicular; wing thick or membranous, $\frac{1}{8}$–$\frac{1}{3}$ as wide as the achene, undulate or entire, apex usually rounded, projecting slightly at the base of the style. ● *S. & C. Europe*. Ga Ge It Po Rs(C, E) Si [Be Cz Ho Hs Rs(B)].

Plants from S. Europe, extending northwards to S. Germany, have been placed in a separate species, **C. leptopterum** (Ascherson) Iljin, *Bull. Jard. Bot. URSS* **28**: 653 (1929). They are generally taller and glabrous, with a more elongate, narrowly winged achene. The two taxa were perhaps formerly morphologically and geographically distinct, but introgression has probably obscured differences between them.

11. C. orientale Lam., *Encycl. Méth. Bot.* **2**: 111 (1786). Plant 5–60 cm, with patent branches, more or less pubescent. Leaves oblong to linear. Spikes short, dense. Bracts ovate, acuminate, membranous-margined, slightly longer than the fruit. Perianth-segment 1. Achene 2–3 × 1·5–2 mm, elliptic-ovate, bidentate at the apex; wing practically absent. *C. & S.E. Russia; W. Kazakhstan*. Rs(C, E).

15. Agriophyllum Bieb. ex C. A. Meyer

P. W. BALL

Stellate-pubescent annual herbs. Leaves alternate, flat, spine-tipped. Flowers hermaphrodite, in axillary cymes. Perianth-segments (0)1–3. Stamens 2 or 3. Stigmas 2. Achene compressed, narrowly winged, with the persistent stigmas forming a 2-lobed beak. Seeds vertical.

1. A. squarrosum (L.) Moq. in DC., *Prodr.* **13(2)**: 139 (1849) (*A. arenarium* Bieb. ex C. A. Meyer, *Corispermum squarrosum* L.). Leaves up to 8 cm, lanceolate or linear-lanceolate, acuminate, with spinose apex and with 5–15 more or less parallel veins. Bracts ovate, with spinose apex. Cymes *c.* 1 cm, ovoid. Achene 3–6 mm. *Mobile sands. S.E. Russia; W. Kazakhstan*. Rs(E). (*W. & C. Asia.*)

16. Kalidium Moq.

P. W. BALL

Small fleshy shrubs, not articulate. Leaves alternate, fleshy, small, decurrent. Flowers hermaphrodite or female, in usually 3-flowered cymes immersed in the fleshy axis of a spicate inflorescence. Perianth-segments 5. Stamens 2. Stigmas 2.

Lamina of leaves 2–6 mm, patent or recurved **1. foliatum**
Lamina of leaves reduced to a small tubercle at the apex of the decurrent-amplexicaul base **2. caspicum**

1. K. foliatum (Pallas) Moq. in DC., *Prodr.* **13(2)**: 147 (1849) (*Salicornia foliata* Pallas). Stems 10–75 cm, branches not fragile when dry; lamina of leaves 2–6 mm, linear to ovate, semi-terete, patent or recurved; base semi-amplexicaul. Terminal spikes 20–30 mm. Achene *c.* 1·25–1·5 mm. *Saline habitats. S.E. Russia, W. Kazakhstan; S. Spain*. Hs Rs(E). (*W. & C. Asia.*)

2. K. caspicum (L.) Ung.-Sternb., *Atti Congr. Int. Bot. Firenze* 317 (1876) (*Salicornia caspica* L.). Stems 15–25 cm, branches fragile when dry; lamina of leaves a small tubercle, convex below, flattened above; base amplexicaul-obovate. Terminal spikes 6–12 mm. Achene *c*. 1 mm. *Saline habitats. W. Kazakhstan.* Rs(E). (*S.W. & C. Asia.*)

17. **Halopeplis** Bunge ex Ung.-Sternb.

P. W. BALL

Annual herbs. Leaves alternate, subglobose and amplexicaul with rudimentary lamina. Inflorescence spicate, with a 3-flowered cyme in the axil of each bract. Flowers hermaphrodite or female, more or less connate with each other and with the surrounding bracts. Perianth-segments 3. Stamens 1 or 2. Stigmas 2.

1. **H. amplexicaulis** (Vahl) Ung.-Sternb. ex Cesati, Passer. & Gibelli, *Comp. Fl. Ital.* 271 (1874) (*Halostachys perfoliata* sensu Willk.). Stem erect, 5–30 cm, glaucous, with stout base. Leaves distant; lamina *c*. 2 mm. Spikes 5–15 mm; bracts ovate-orbicular. Seeds *c*. 0·5 mm, with cylindrical papillae. *Salt-marshes. S. Europe from S.E. Portugal to S. Italy.* Hs It Lu Sa Si.

H. pygmaea (Pallas) Bunge ex Ung.-Sternb., *Syst. Salicorn.* 105 (1866), a smaller, more slender plant with crowded leaves, has been doubtfully recorded from the European shore of the Caspian Sea.

18. **Halostachys** C. A. Meyer

P. W. BALL

Shrubs or small trees. Leaves opposite, connate, with a rudimentary lamina. Inflorescence spicate, with 3-flowered cymes in the axils of opposite, free bracts. Flowers free, hermaphrodite. Perianth-segments 3. Stamen 1. Stigmas 2 or 3.

1. **H. belangeriana** (Moq.) Botsch., *Not. Syst.* (*Leningrad*) **16**: 84 (1954) (*H. caspica* auct., *H. caspia* C. A. Meyer, *Halocnemum caspicum* Bieb. pro parte, *Arthrocnemum belangerianum* Moq., *Salicornia caspica* sensu Pallas, non L.). Shrub or small tree up to 3·5 m with erect branched trunk; ultimate branches articulated. Spikes 1–3 cm. Bracts more or less orbicular, slightly cordate. Seeds 0·5–1 mm. *Salt-marshes. W. Kazakhstan.* Rs(E). (*S.W. & C. Asia.*)

19. **Halocnemum** Bieb.

P. W. BALL

Small, fleshy shrubs with opposite and decussate leaves. Flowers hermaphrodite, in 2- or 3-flowered axillary cymes arranged in a spicate, articulate inflorescence. Perianth-segments 3. Stamen 1. Stigmas 2 or 3.

1. **H. strobilaceum** (Pallas) Bieb., *Fl. Taur.-Cauc.* **3**: 3 (1819) (*Halostachys perfoliata* (Forskål) Moq.). Stem up to 50 cm, becoming woody. Leaves connate at the base and enclosing the stem so that it is apparently articulate; free part of leaves *c*. 1 mm, broadly obovate, obtuse; leaves subtending short, globose branches. Bracts free, reniform-orbicular; cymes globose to cylindrical-globose. Seeds 0·5–0·8 mm. *Salt-marshes. Coasts of S. Europe, from S.W. Spain to W. Kazakhstan.* Al Cr Hs Rm Rs(W, K, E) Sa Si Tu.

20. **Arthrocnemum** Moq.

P. W. BALL

Articulate, glabrous, dwarf shrubs. Leaves opposite, scale-like, the bases of each pair amplexicaul, fused to form a segment. Inflorescence spicate, segmented; each fertile segment composed of two 3-flowered cymes immersed in the pair of bracts arising from the node above; the flowers of each cyme more or less equal in size, the central distinctly separating the lateral. Perianth-segments usually 4. Stamens 2. (Incl. *Sarcocornia* A. J. Scott.)

The delimitation of this genus from *Salicornia* is that proposed by C. E. Moss, *Jour. S. Afr. Bot.* **20**: 1–22 (1954), but there is much disagreement as to the limits of these genera.

The species occur mainly in salt-marshes.

1 Flowers of each cyme falling to leave an undivided hollow in the segment; seeds black; testa hard, tuberculate **3. macrostachyum**
1 Flowers of each cyme falling to leave a tripartite hollow in the segment; seeds greenish-brown to greyish; testa thin, membranous, covered with short conical or curved hairs
2 Plant usually with creeping, subterranean stems; green, often becoming red or orange-brown; seeds covered with curved or hooked hairs **1. perenne**
2 Plant always without creeping, subterranean stems; usually glaucous; seeds covered with ±conical hairs **2. fruticosum**

1. **A. perenne** (Miller) Moss, *Jour. S. Afr. Bot.* **14**: 40 (1948) (*Salicornia perennis* Miller, *S. radicans* Sm., *Sarcocornia perennis* (Miller) A. J. Scott). Small shrub with creeping, subterranean stems forming mats up to 1 m in diameter and short, erect, non-flowering and flowering stems, which are green, becoming red or orange-brown. Cymes completely immersed in and connate with the fertile segments, falling to leave a tripartite hollow in the segment; flowers almost reaching the upper edge of the segment. Seeds greenish-brown or greyish; testa thin, membranous, covered with curved or hooked hairs; endosperm almost absent. $2n = 18$. *Coasts of S. & W. Europe, northwards to* c. 55° *N. in Britain.* Al Bl Br Co Cr Ga Gr Hb Hs It Lu Sa Si.

Much confused with procumbent plants of **2**.

2. **A. fruticosum** (L.) Moq., *Chenop. Monogr. Enum.* 111 (1840) (*Salicornia fruticosa* (L.) L., *Salicornia arabica* auct.). Like **1** but the stems up to 1 m, erect or procumbent, usually stout and not rooting, glaucous; cymes clearly not reaching the upper edge of the fertile segment; testa covered with short conical hairs. $2n = 54$. *Coasts of S. & W. Europe, extending northwards to* 48° *N. in W. France.* Al Bl Co ?Cr Ga Gr Hs It Ju Lu Sa Si Tu.

3. **A. macrostachyum** (Moric.) C. Koch, *Hort. Dendrol.* 96 (1853) (*A. glaucum* (Delile) Ung.-Sternb., *Salicornia macrostachya* Moric., *Salicornia glauca* Delile). More or less erect shrub up to 1 m, glaucous, becoming yellowish-green or reddish. Cymes protruding, free, falling to leave 1 hollow in the segment; flowers extending to not more than $\frac{1}{3}$ of the length of the segment. Seeds black; testa hard, tuberculate; endosperm abundant. $2n = 36$. *Coasts of S. Europe.* Al Bl Co Cr Ga Gr Hs It Ju Lu Sa Si.

21. **Salicornia** L.

P. W. BALL AND J. R. AKEROYD

Annual, articulate herbs. Leaves and inflorescence like those of *Arthrocnemum*. Cymes (1–)3-flowered, connate with and completely immersed in the bract of the segment, the flowers arranged in a triangle, the lateral flowers usually meeting below the central.

Perianth 3- or 4-lobed. Stamens 1 or 2. Seeds with thin membranous testa; endosperm very sparse.

All species occur in saline habitats, and it is possible that immersion in salt water is essential for their growth.

It is usually impossible to identify dried specimens with any reasonable degree of accuracy. As a result most Floras recognize only one species (*S. europaea* L. or *S. herbacea* (L.) L.), although there can be little doubt that several others occur in Europe. In recent years accounts of the genus based on living material, but covering relatively small areas, have been published, but these cannot be satisfactorily correlated either taxonomically or nomenclaturally, even in N.W. Europe.

Literature: J. Nannfeldt, *Svensk Bot. Tidskr.* **49**: 97–109 (1955). P. W. Ball & T. G. Tutin, *Watsonia* **4**: 193–205 (1959). D. König, *Mitt. Fl.-Soziol. Arbeitsgem.* n.s. **8**: 5–58 (1960). K. R. Soó von Bere, *Acta Bot. Acad. Sci. Hung.* **6**: 397–403 (1960). P. W. Ball, *Feddes Repert.* **69**: 1–8 (1964). A. J. Scott, *Bot. Jour. Linn. Soc.* **75**: 367–374 (1977).

1 Cymes 1-flowered; infructescence disarticulating **5. pusilla**
1 Cymes 3-flowered; infructescence not disarticulating
2 Anthers 0·2–0·5 mm, often not exserted; fertile segments with convex sides; lateral flowers usually distinctly smaller than the central **(1–4). europaea** group
2 Anthers (0·5–)0·6–1 mm, always exserted; fertile segments ±cylindrical, sometimes with slightly concave sides; flowers subequal **(6–8). procumbens** group

Sect. SALICORNIA. Cymes usually 3-flowered. Lateral flowers usually distinctly smaller than the central. Stamens usually 1, rarely 0 or 2; anthers 0·2–0·5 mm. Seeds 1–1·7 mm. Diploid.

(1–4). S. europaea group. Plants usually red or purple in fruit; the uppermost primary branches straight, making an acute angle (usually less than 45°) with the main stem; terminal spike with 3–12(–22) fertile segments; fertile segments with convex sides, the lateral flowers of each cyme usually distinctly smaller than the central flower; flowers often cleistogamous, the anthers usually dehiscing before exsertion, or not exserted and persisting in the perianth.

The specific limits within this group are obscure. Inbreeding is predominant, the taxa consisting of one or a few homozygous lines.

Plants belonging to this group are found on almost all the coasts of Europe, and locally inland in C. Europe and the Ukraine and S. Russia. On the basis of the data available it is not possible to assign many records to a particular species within the group. Records are available for the following territories: Al Be Bl Br Bu Co Cr Cz Da Fe Ga Ge Gr Hb Ho Hs It Ju No Po Rm Rs(N, B, W, K, E) Sa Si Su Tu.

1 Lower fertile segments of the terminal spike 3–5 mm wide at the narrowest point; upper edge with an inconspicuous, scarious margin (not more than 0·1 mm wide)
2 Upper scarious margin of fertile segments up to 0·1 mm wide, cuspidate at apex; lateral flowers distinctly smaller than the central **1. europaea**
2 Upper scarious margin of fertile segments up to 0·05 mm wide, rounded at apex; lateral flowers smaller than the central, but not distinctly so **2. obscura**
1 Lower fertile segments of the terminal spike 2–3·5(–4) mm wide (or less) at the narrowest point; upper edge with a conspicuous, scarious margin (0·1–0·2 mm wide)
3 Lower fertile segments of the terminal spike 2·5–4 mm wide at the narrowest point and 3–5·3 mm wide at the widest **3. ramosissima**
3 Lower fertile segments of the terminal spike 2–2·5 mm wide at the narrowest point, and 2·5–3 mm wide at the widest **4. prostrata**

1. S. europaea L., *Sp. Pl.* 3 (1753) (*S. herbacea* (L.) L.). Plant 10–35 cm, erect or procumbent, simple to much-branched, primary branches often as long as the main stem; grass-green, sometimes becoming yellow-green, usually with red or pinkish-red colour appearing diffusely in the fertile segments. Terminal spike 10–50 mm; lower fertile segments 2·5–4·5 mm, 3–5 mm wide at the narrowest point and 3·5–6 mm wide at the widest; upper edge with an inconspicuous, narrow, scarious margin up to 0·1 mm wide, cuspidate at the apex. Flowers distinctly unequal in size. $2n = 18$. *Mostly in upper part of salt-marshes. N.W. Europe, probably elsewhere.*

Very variable. **S. stricta** Dumort., *Bull. Soc. Bot. Belg.* **7**: 334 (1868), **S. patula** Duval-Jouve, *Bull. Soc. Bot. Fr.* **15**: 175 (1868), **S. brachystachya** (G. F. W. Meyer) D. König, *Mitt. Fl.-Soziol. Arbeitsgem.* n.s. **8**: 11 (1960), pro parte, and **S. simonkaiana** (Soó) Soó, *Acta Bot. Acad. Sci. Hung.* **6**: 401 (1960) are probably variants of this species.

2. S. obscura P. W. Ball & Tutin, *Watsonia* **4**: 204 (1959). Like **1** but 10–45 cm, always erect, little-branched (without tertiary branches), primary branches not more than ½ as long as the main stem; rather glaucous-green becoming yellow, rarely becoming pink or red around the flowers; upper edge of fertile segments with very narrow, scarious margin up to 0·05 mm, rounded and not cuspidate at the apex; flowers unequal in size, but not distinctly so. $2n = 18$. *Bare sand and mud in salt-pans and channels of salt-marshes.* ● *W. Europe.*

3. S. ramosissima J. Woods, *Bot. Gaz.* (*London*) **3**: 29 (1851) (*S. appressa* Dumort., *S. prostrata* auct., non Pallas, *S. smithiana* Moss). Plant up to 40 cm, erect or procumbent, typically much-branched and bushy but sometimes forming pure stands of small, simple plants; dark green sometimes becoming yellowish-green, with dark purplish-red or purple first appearing round the flowers and along the upper edge of the fertile segments and sometimes eventually colouring the whole segment. Terminal spike 5–30(–40) mm; lower fertile segments 1·9–3·5 mm, 2·5–4 mm wide at the narrowest point and 3–5·3 mm wide at the widest; upper edge with a conspicuous, scarious margin 0·1–0·2 mm wide. $2n = 18$. *Less open mud in middle and upper parts of salt-marshes.* ● *W. Europe and W. Mediterranean region.*

Very variable in habit and colour.

4. S. prostrata Pallas, *Ill. Pl.* 3 (1803) (*S. ramosissima* auct. eur. orient., non J. Woods). Like **3** but more slender; lower fertile segments of the terminal spike 2–2·5 mm wide at the narrowest point and 2·5–3 mm wide at the widest. $2n = 18$. *E. & E.C. Europe.* Au Bu Hu Rm Rs(W, E).

5. S. pusilla J. Woods, *Bot. Gaz.* (*London*) **3**: 30 (1851) (*S. disarticulata* Moss). Plants up to 25 cm, erect or procumbent, usually much-branched, yellowish-green becoming brownish- or pinkish-yellow. Terminal spike up to 6(–15) mm, disarticulating in fruit; fertile segments 2–4(–12), with convex sides. Cymes 1-flowered.

Flowers almost orbicular. $2n = 18$. *Uppermost parts of salt-marshes and shores of estuaries.* ● *S. Britain, S. Ireland, N. France; probably elsewhere in W. Europe.* Br Ga Hb.

Intermediates with 1- to 3-flowered cymes, presumably of hybrid origin between **5** and **3**, occur sometimes when the two species grow together. Similar plants also occur in the W. Mediterranean region. The breeding system is similar to that of the other diploid species (**1–4**).

Sect. DOLICHOSTACHYAE A. J. Scott. Cymes 3-flowered. Infructescence not disarticulating. Flowers subequal. Stamens usually 2, exserted. Anthers at least (0·4–)0·5 mm. Seeds 1·4–2·2 mm. Tetraploid.

(**6–8**). **S. procumbens** group. Plants usually not pink or purple in fruit; uppermost primary branches curving upwards; terminal spike with (4–)6–30 fertile segments; fertile segments usually more or less cylindrical, the lateral flowers of each cyme almost as large as the central flower; flowers chasmogamous; stamens 1 or 2; anthers dehiscing after exsertion.

S. procumbens Sm. in Sowerby, *Engl. Bot.* t. 2475 (1813) must be identical with either **7** or **8** but information at present available does not permit of a decision.

1 Lower fertile segments of terminal spike with a minimum diameter 2–3·5 mm; plant usually becoming light brownish- or orange-purple **6. nitens**
1 Lower fertile segments of terminal spike with a minimum diameter usually exceeding 3·5 mm; plant rarely pale red or purple
2 Terminal spike with 6–16(–22) fertile segments, ±cylindrical; spikes of the primary lateral branches cylindrical; plant dull green to yellowish-green, often becoming bright yellow in fruit **7. fragilis**
2 Terminal spike with 12–30 fertile segments, distinctly tapering towards the apex; spikes of the primary lateral branches tapering; plant dark, dull green, becoming paler or dull yellow in fruit **8. dolichostachya**

6. S. nitens P. W. Ball & Tutin, *Watsonia* **4**: 204 (1959). Erect, 5–25 cm, usually with the primary branches less than ½ as long as the main stem; green to yellowish-green, usually soon becoming light brownish- or orange-purple. Terminal spike 1–4 cm, cylindrical, with 4–9 fertile segments. Lower fertile segments 2–3·5 × 2–3·5 mm at the narrowest point and 2·3–4·0 mm wide at the widest. $2n = 36$. *Mud in upper part of salt-marshes.* ● *W. Europe.* Br Hb Lu.

7. S. fragilis P. W. Ball & Tutin, *loc. cit.* (1959) (*S. stricta* auct., pro parte, non Dumort.). Erect, (10–)15–30(–40) cm, usually much-branched, primary branches usually not more than ½ as long as the main stem; green to yellowish-green, usually becoming bright yellow to brownish. Terminal spike (1·5–)2·5–8 cm, cylindrical, with 6–16(–22) fertile segments. Lower fertile segments 3–5 × 3·5–6 mm, cylindrical. $2n = 36$. *Often colonizing open sandy soils, but occurring in a wide range of salt-marshes.* ● *Known certainly only from England and Ireland but probably widespread in W. & S. Europe.* Br Hb.

S. emerici Duval-Jouve, *Bull. Soc. Bot. Fr.* **15**: 176 (1868), erect or prostrate and becoming pink or red, and **S. stricta** subsp. **decumbens** Aellen in Hegi, *Ill. Fl. Mitteleur.* ed. 2, **3(2)**: 730 (1961), pro parte (*S. stricta* subsp. *procumbens* sensu D. König, pro parte) are probably variants of this species. **S. veneta** Pignatti & Lausi, *Gior. Bot. Ital.* **103**: 185 (1969), from N.E. Italy (Laguna Veneta and adjacent coasts), is somewhat taller, up to 50 cm, with primary branches up to ⅔ as long as the main stem, terminal spike cylindrical, and plant green becoming yellowish. It is similar to **S. lutescens** P. W. Ball & Tutin, *Watsonia* **4**: 203 (1959), described from S. Britain. This group of taxa requires further investigation on a European basis.

8. S. dolichostachya Moss, *New Phytol.* **11**: 409 (1912) (*S. stricta* sensu D. König, pro max. parte, non Dumort., ?*S. leiosperma* K. Gram). Erect, 10–40(–45) cm, much-branched, primary branches almost as long as the main stem, straggling to fastigiate, dark green becoming yellowish and finally brown. Terminal spike (2·5–)5–12(–20) cm, tapering, with (8–)12–30 fertile segments. Lower fertile segments 3–6 × 3–6 mm, cylindrical. *Mud and sand at relatively low tidal levels.* ● *N. Europe.* Be Br Da Ga Ge Hb Ho No Rs(N) Su.

This species is said to differ from all others in the absence of spirally thickened or striated cells in the fertile segments. Some variants of **7** are, however, often almost without these cells, so the taxonomic value of this character is doubtful.

(**a**) Subsp. **dolichostachya** (*S. pojarkovae* N. Semen.): Plant straggling to fastigiate; uppermost branches arising at an angle of c. 60–90° with the axis and curving upwards. $2n = 36$. *Throughout the range of the species, except the Baltic region.*

(**b**) Subsp. **strictissima** (K. Gram) P. W. Ball, *Feddes Repert.* **69**: 7 (1964) (*S. strictissima* K. Gram). Plant erect, fastigiate; uppermost branches arising at an angle of 45° or less with the axis, straight. *Baltic region and possibly parts of the North Sea coasts.*

S. oliveri Moss, *Jour. Bot. (London)* **49**: 183 (1911), described from N.W. France, is possibly a dwarf variant of subsp. (**a**).

22. Microcnemum Ung.-Sternb.

P. W. BALL

Like *Salicornia* but flowers free, minute, with thin membranous perianth; central flower in each cyme hermaphrodite, laterals usually female. Stamen 1. Seeds with black, crustaceous, granular testa.

1. M. coralloides (Loscos & Pardo) Buen, *Anal. Soc. Esp. Hist. Nat.* **12**: 431 (1883) (*M. fastigiatum* Ung.-Sternb.). Erect, usually much-branched annual 5–10 cm, often becoming purplish. Terminal spike 1–4 cm, with up to 16 fertile segments. Fertile segments 2–3 mm, with broad membranous upper margin; flowers more or less concealed. Seeds obscurely tuberculate. *Saline soils. E. & C. Spain.* Hs. (*S.W. Asia.*)

In Europe represented only by the endemic subsp. **coralloides**, with tuberculate, not papillose, seeds.

23. Suaeda Forskål ex Scop.

P. W. BALL AND J. R. AKEROYD

Annual or perennial herbs or small shrubs, glabrous or farinose. Leaves usually alternate, semi-cylindrical or flat. Flowers hermaphrodite and female, solitary or in few-flowered cymes; bracteoles 2, minute. Perianth-segments 5, fleshy, sometimes with a small tubercle, horn, or narrow transverse wing dorsally in fruit. Stamens 5. Stigmas 2 or 3(–5). Seeds vertical or horizontal.

Several species produce dimorphic seeds. The first-formed are black, shining, smooth or with a fine pattern on the testa; the later-formed are larger, light brown or olive, dull, with a distinct, coarse reticulum on the testa. The seeds referred to in this account are always the first-formed seeds.

All the species occur in saline habitats.

1 Fruiting perianth-segments of the upper flowers transversely winged, those of the lower flowers usually keeled
2 Leaves linear-semicylindrical, not more than 1·5 mm in diameter; bracts ovate, acute, much shorter than the leaves **14. heterophylla**
2 Leaves and bracts similar, flat, ovate to oblong, up to 5 mm wide, obtuse **15. kossinskyi**
1 Fruiting perianth-segments not transversely winged
3 Fruiting perianth with 1 or 2 segments horned dorsally, the upper perianths often almost stellate **9. corniculata**
3 Fruiting perianth not horned (occasionally with tubercles)
4 Peduncle of the cyme connate with the bract
5 Small shrub; leaves 5–16 mm **5. dendroides**
5 Annual herb; leaves 12–40 mm
6 Leaves *c.* 0·5 mm in diameter, cylindrical **11. altissima**
6 Leaves 1–3 mm wide, flat
7 Fruiting perianth-segments with a slight tubercle dorsally; seeds granular **12. linifolia**
7 Fruiting perianth-segments with a transverse keel; seeds smooth **13. eltonica**
4 Peduncle of the cyme not connate with the bract
8 Seeds with a distinct fine pattern on the testa
9 Small shrub; perianth inflated in fruit **4. physophora**
9 Annual herb, sometimes ± woody at the base; perianth not inflated **8. maritima**
8 Seeds smooth or very indistinctly patterned
10 Leaves with a slender caducous apex (up to 1 mm); fruiting perianth rugose **10. splendens**
10 Leaves obtuse, acute or shortly acuminate, but without a caducous apex; fruiting perianth smooth
11 Perianth-segments very inflated, red in fruit **7. baccifera**
11 Perianth-segments not inflated in fruit, usually green
12 Annual, sometimes woody at base; longest leaves usually more than 20 mm; stigmas 2(3)
13 Leaves acute to mucronate; seeds completely smooth **6. confusa**
13 Leaves obtuse or subacute; seeds at least faintly reticulate **8. maritima**
12 Small shrub; longest leaves usually less than 18 mm; stigmas 3
14 Stigmas 0·2–0·4 mm, united at base, flat, or lobed, forming complex, peltate or capitate structure **1. vera**
14 Stigmas 0·5–1·0 mm, free from base, or almost so, undivided, subulate or filiform
15 Young axes with dense, appressed hairs; leaves obtuse **2. vermiculata**
15 All axes glabrous; leaves distinctly apiculate **3. pelagica**

1. S. vera J. F. Gmelin in L., *Syst. Nat.* ed. 13, **2(1)**: 503 (1791) (*S. fruticosa* auct.). Small shrub up to 120 cm, glabrous. Leaves 5–18 × (0·8–)1–2·5 mm, semi-cylindrical, usually shortly apiculate, sessile, glaucous. Cymes dense, shorter than the bracts. Stigmas 0·2–0·4 mm, united at the base, flat, lobed, forming a complex peltate or capitate structure. Fruiting perianth not inflated, usually green. Seeds smooth, usually vertical. $2n = 18, 36$. *S. & W. Europe northwards to* c. 53° *N. in England.* Al Bl Br Co Cr Ga Gr Hs It Ju Lu Sa Si.

2. S. vermiculata Forskål ex J. F. Gmelin in L., *Syst. Nat.* 503 (1791) (*S. pruinosa* Lange, *S. fruticosa* var. *brevifolia* Moq., pro parte, *S. vera* auct. lusit.). Like **1** but the young stems with dense, appressed hairs; leaves 2–5(–6) × 0·5–1·0 mm, obtuse, more or less cylindrical; middle and upper bracts not exceeding the cymes; stigmas 0·8–1·0 mm, filiform or subulate. *Spain; one station in Sicilia.* Hs Si.

3. S. pelagica Bartolo, Brullo & Pavone, *Willdenowia* **16**: 391 (1987). Compact shrub up to 100 cm, with divaricate branches forming an intricate mass. Leaves 3–6 × 1·5–2·5 mm, semi-cylindrical, arcuate, sessile, apiculate. Flowers *c.* 2 mm, sessile, shorter than the ovate bract, forming spikes. Fruiting perianth not inflated, green. Seeds smooth, black, mostly horizontal. *Marl cliffs.* ● *Sicilia (Lampedusa).* Si.

The species is very close to and probably not specifically distinct from *S. palaestina* Eig & Zohary, described from Palestine.

4. S. physophora Pallas, *Ill. Pl.* 51 (1803). Small shrub up to 120 cm. Leaves 20–40 × 1–2 mm, filiform, obtuse or subacute. Cymes very lax. Fruiting perianth spongy-inflated, reddish. Seeds punctate, mostly horizontal. *S.E. Russia, W. Kazakhstan.* Rs(E). (*C. Asia.*)

5. S. dendroides (C. A. Meyer) Moq., *Chenop. Monogr. Enum.* 126 (1840). Small shrub up to 60 cm. Leaves 5–16 × 1–2 mm, semicylindrical, obtuse, attenuate at the base. Cymes dense; peduncle connate with the base of the bract. Fruiting perianth inflated, green. Seeds more or less smooth, mostly horizontal. *S.E. Russia, W. Kazakhstan.* Rs(E). (*Caspian region.*)

6. S. confusa Iljin, *Acta Horti Petrop.* **26**: 284 (1930). Annual up to 50 cm, often woody at the base. Leaves 10–25 × *c.* 1 mm, semi-cylindrical, acute, acuminate or mucronate. Cymes dense. Fruiting perianth not inflated, green. Seeds smooth, mostly horizontal. *S.E. Russia, S. Ukraine.* Rs(W, K, E).

7. S. baccifera Pallas, *Ill. Pl.* 48 (1803). Annual up to 50 cm, often woody at the base. Leaves 12–25 × 0·6–1 mm, semi-cylindrical, mostly obtuse. Cymes dense. Fruiting perianth much enlarged, baccate, red. Seeds smooth, mostly horizontal. *S. Russia; one locality in E. Ukraine.* Rs(W, E).

8. S. maritima (L.) Dumort., *Fl. Belg.* 22 (1827) (*S. cavanillesiana* (Láz.-Ibiza) Coutinho). Annual up to 50(–100) cm, of very variable habit and colour. Leaves 10–50 × *c.* 1 mm, semi-cylindrical, acute or subobtuse. Cymes dense. Fruiting perianth not inflated, green or red. Seeds finely reticulate, usually horizontal. *Throughout most of Europe except the north-east and extreme north.* All except Az ?Cr Fa He Is Rs(N) Sb.

Very variable in habit, leaf-shape and colour. Very little of the variation is correlated with distribution. The following subspecies appear to be the most distinct.

1 Perianth-segments with a membranous margin, with a keel and small tubercle dorsally **(c)** subsp. **pannonica**

1 Perianth-segments without either a membranous margin, or keel or tubercle dorsally
2 Upper surface of the leaves flat; seeds distinctly reticulate (a) subsp. **maritima**
2 Upper surface of the leaves sulcate; seeds almost smooth (b) subsp. **salsa**

(a) Subsp. **maritima** (incl. *S. prostrata* Pallas): Plant up to 50 cm. Upper surface of leaves flat. Perianth-segments without a membranous margin and without a keel or dorsal tubercle. Seeds 1–1·5(–3) mm, with a distinctly reticulate surface. $2n = 36$. *Throughout the range of the species, but many records from* E.C. & E. *Europe may be referable to subsp.* (**b**).

(**b**) Subsp. **salsa** (L.) Soó in Soó & Jáv., *Magyar Növ. Kéz.* 785 (1951) (*S. salsa* (L.) Pallas): Plant 25–100 cm. Upper surface of leaves sulcate. Perianth-segments without a membranous margin and without a keel or dorsal tubercle. Seeds 1·3–1·5 mm, almost smooth. $2n = 36$. *E.C.* & *E. Europe.*

(**c**) Subsp. **pannonica** (G. Beck) Soó ex P. W. Ball, *Feddes Repert.* **69**: 44 (1964) (*S. pannonica* G. Beck): Plant up to 20 cm. Upper surface of leaves flat. Perianth-segments with a membranous margin and with a keel and small dorsal tubercle. Seeds *c.* 0·8 mm, almost smooth. ● *E.C. Europe.*

9. S. corniculata (C. A. Meyer) Bunge, *Acta Horti Petrop.* **6**: 429 (1880). Annual up to 50 cm. Leaves 12–20 × 1–1·5 mm, semi-cylindrical, more or less acute. Cymes dense, occurring in the axils of most leaves. Fruiting perianth with 1 or 2 segments horned dorsally, the upper perianths almost stellate. Seeds reticulate-punctate, usually horizontal. *S.E. Russia, E. Ukraine, W. Kazakhstan.* Rs(C, W, E).

10. S. splendens (Pourret) Gren. & Godron, *Fl. Fr.* **3**: 30 (1855) (*S. setigera* (DC.) Moq.). Diffuse annual up to 50 cm. Leaves 8–20 × 1–2 mm, semi-cylindrical, semi-translucent, acuminate or mucronate, usually with a fine caducous apex up to 1 mm. Cymes dense. Fruiting perianth very inflated, rugose. Seeds smooth, usually horizontal. $2n = 18, 36$. *Mediterranean region, extending to C. Portugal.* Al Bl Ga Gr Hs It Lu Sa Tu.

11. S. altissima (L.) Pallas, *Ill. Pl.* 49 (1803). Erect annual up to 150 cm. Leaves 12–30 × 0·5 mm, filiform, acute or subobtuse. Cymes lax or dense, usually shortly pedunculate, the peduncle connate with the base of the bracts. Fruiting perianth not inflated, rounded dorsally. Seeds almost smooth, vertical or horizontal. *S.E. Europe from E. Greece to W. Kazakhstan; Spain.* Bu Gr Hs Rs(W, K, E).

12. S. linifolia Pallas, *Ill. Pl.* 43 (1803). Erect annual up to 75 cm. Leaves 12–30 × 1–3 mm, linear, acute, flat. Cymes usually lax, very shortly pedunculate, the peduncle connate with the base of the bracts. Fruiting perianth-segments not inflated, with a slight tubercle dorsally. Seeds granular, vertical or horizontal. *S.E. Russia, W. Kazakhstan.* Rs(E). (*W. Asia.*)

13. S. eltonica Iljin, *Bull. Jard. Bot. URSS* **26**: 415 (1927). Like **12** but plant up to 150 cm; leaves 20–40 mm; cymes dense; fruiting perianth-segments transversely keeled; seeds smooth. *S.E. Russia.* Rs(E).

14. S. heterophylla (Kar. & Kir.) Bunge ex Boiss., *Fl. Or.* **4**: 943 (1879). Erect annual up to 80 cm. Leaves 10–16 × 1–1·5 mm, semi-cylindrical, acute. Cymes free, dense. Fruiting perianth-segments of the upper flowers with a narrow transverse wing, of the lower flowers with a transverse keel. Seeds finely punctate-reticulate, mostly horizontal. *S.E. Russia, W. Kazakhstan; E. Bulgaria.* Bu Rs(E).

15. S. kossinskyi Iljin, *Bull. Jard. Bot. URSS* **26**: 115 (1927). Like **14** but 5–15 cm, branched at the base; leaves up to 5 mm wide, oblong to obovate, obtuse; cymes in the axils of almost all the leaves. *S.E. Russia, W. Kazakhstan.* Rs(E).

24. **Bienertia** Bunge ex Boiss.

P. W. BALL

Like *Suaeda* but the fruiting perianth with transverse wings dorsally, the wings connate and encircling the perianth.

1. B. cycloptera Bunge ex Boiss., *Fl. Or.* **4**: 945 (1879). Annual 10–40 cm. Leaves 10–20 mm, linear to oblong, semi-terete, fleshy. Inflorescence a narrow panicle with short branches bearing 6- to 10-flowered cymes. Fruiting perianth 5–7 mm in diameter (including wing), lenticular; apices of the segments verrucose. *Gypsaceous and saline habitats. S.E. Russia (near Volgograd), W. Kazakhstan.* Rs(E). (*S.W. Asia.*)

25. **Salsola** L.

†P. AELLEN REVISED BY J. R. AKEROYD

Annuals or dwarf shrubs; glabrous or pubescent. Leaves filiform or semi-cylindrical, alternate or opposite. Flowers hermaphrodite, with 2 conspicuous bracteoles. Perianth-segments 5 (3 outer and 2 inner), usually developing a transverse wing dorsally in fruit. Stamens 5. Stigmas 2(3). Seeds usually horizontal. (*Darniella* Maire.)

Most species are restricted to more or less saline, sandy habitats, either maritime or in arid regions inland. A large proportion is essentially Asiatic, and occurs only on the extreme S.E. margin of Europe.

Measurements of the perianth in fruit include the wings.

1 Shrubs
2 Anther-appendage lanceolate or elliptic-oblong, acute
3 Leaves 2–3 mm; bracteoles reniform, shorter than the flower **15. nodulosa**
3 Leaves 5–13 mm; bracteoles elliptical, about equalling the flower **16. carpatha**
2 Anther-appendage short, obtuse
4 Leaves constricted just above the base
5 Bracteoles ovate **17. melitensis**
5 Bracteoles linear **18. arbuscula**
4 Leaves not constricted just above the base
6 First- and second-year stems with numerous longitudinal ridges; leaves fugacious **19. genistoides**
6 First- and second-year stems terete or with *c.* 4 angles or ridges; leaves usually ± persistent
7 Perianth-segments obtuse, fimbriate; leaves nearly all opposite **21. oppositifolia**
7 Perianth-segments acute, not fimbriate; leaves alternate or some opposite
8 Leaves 1–3 mm in diameter (S.E. Spain) **22. papillosa**
8 Leaves less than 1 mm in diameter (sometimes deciduous)
9 Leaves 10–25 mm; plant glabrous except for tufts of hairs in the axils of the leaves (S. Spain) **20. webbii**

9 Leaves 2–10(–15) mm; stems and often the leaves pubescent
10 Leaves 2–5 mm, glabrous **25. dendroides**
10 Leaves 5–10(–15) mm, usually pubescent, sometimes very sparsely so
11 Primary branches of the inflorescence up to 20–40 cm, with the secondary branches regularly arranged (S.W. Europe) **23. vermiculata**
11 Primary branches of the inflorescence not more than 10 cm, irregularly branched
12 Plant not more than 30 cm; branches divaricate **24. aegaea**
12 Plant up to 70 cm; branches erect **26. laricina**
1 Annual herbs
13 Upper leaves clavate; perianth in fruit spongy, inflated **1. foliosa**
13 Leaves not clavate; perianth in fruit not spongy and inflated
14 Leaves spine-tipped, mucronate or acuminate
15 Perianth-segments rigid, mucronulate; mid-vein distinct
16 Plant densely covered with brittle bristles **8. kali**
16 Plant glabrous or papillose
17 Leaves coriaceous, erect, yellowish, contrasting sharply with the purple or reddish stem **3. paulsenii**
17 Leaves soft, dark green, the lower usually deflexed; stem green **4. pellucida**
15 Perianth-segments flaccid; mid-vein weak or absent
18 Leaves opposite, pubescent **2. brachiata**
18 Leaves alternate, or sometimes opposite but then glabrous
19 Perianth tubular in fruit; lower leaves mostly opposite
20 Lower leaves 20–70 mm; uppermost bracts equalling the flowers **5. soda**
20 Lower leaves 10–20(–30) mm; uppermost bracts shorter than the flowers **6. acutifolia**
19 Perianth pelviform in fruit; lower leaves alternate or at most 1–3 pairs opposite
21 Bracteoles mucronulate **7. tamariscina**
21 Bracteoles cuspidate
22 Bracts ±appressed **9. collina**
22 Bracts ±patent **8. kali**
14 Leaves obtuse, not spine-tipped, mucronate or acuminate
23 Leaves not decurrent; perianth 5–10 mm in diameter in fruit
24 Lower leaves 5–8 mm; bracts shorter than or equalling the bracteoles **13. nitraria**
24 Lower leaves 8–16 mm; bracts exceeding the bracteoles **14. affinis**
23 Leaves shortly decurrent; perianth 10–20 mm in diameter in fruit
25 Plant lanate in the upper part; stigmas shorter than the style **10. lanata**
25 Plant pubescent or glabrous; stigmas equalling or longer than the style
26 Stigmas at least 3 times as long as the style; plant glaucous; hairs short, straight **11. turcomanica**
26 Stigmas usually less than 3 times as long as the style; plant grey or white; hairs long, crispate **12. crassa**

1. S. foliosa (L.) Schrader in Roemer & Schultes, *Syst. Veg.* **6**: 804 (1820) (*S. clavifolia* Pallas). Simple or virgate annual 15–80 cm, glabrous, blue-black when dry. Leaves 10–20 mm, semi-cylindrical, at least the upper clavate, obtuse, mucronulate. Perianth spongy, inflated in fruit, the segments ovate, the part above the wing short, triangular or rounded, brownish; wings reniform or orbicular, yellow, finely veined, the lowermost larger than the rest. *S.E. Russia, W. Kazakhstan.* Rs(E). (*C. Asia.*)

2. S. brachiata Pallas, *Ill. Pl.* 30 (1803). Virgate annual 10–30 cm, usually pubescent; branches mostly opposite. Leaves 15–20 mm, semi-cylindrical, semi-amplexicaul, acute, mucronate, opposite. Lower bracts acuminate. Perianth-segments small, with sparse appressed hairs; wings unequal, reniform or obovate, densely dark-striate. Stigma sessile. Seeds vertical. *S.E. Russia, W. Kazakhstan, Krym.* Rs(K, E).

3. S. paulsenii Litv., *Bull. Turkestan Sect. Russ. Geogr. Soc.* **4(5)**: 28 (1905). Annual 10–40 cm, glabrous or sparsely papillose; stem red. Leaves 15–30 mm, semi-cylindrical, mucronate, coriaceous, yellow, contrasting sharply with the stems. Bracts ovate at the base, with a linear, spinose apex; bracteoles partly connate with the solitary flowers. Perianth with a short tube and small, acute, incurved segments; wings membranous, with broad veins. Style short. *S.E. Russia.* Rs(E). (*C. Asia.*)

4. S. pellucida Litv., *Sched. Herb. Fl. Ross.* **8**: 16 (1922). Like **3** but up to 60 cm, sparsely but distinctly papillose; stem green; leaves dark green, the lower opposite; bracts and bracteoles linear, sometimes with a triangular base; perianth with short, broad tube, the segments long, narrow, acute, softly spine-tipped; wing flabellate. *S.E. Russia, W. Kazakhstan.* Rs(E). (*C. Asia.*)

5. S. soda L., *Sp. Pl.* 223 (1753). Erect glabrous annual up to 70 cm. Leaves 20–70 mm, semi-cylindrical, ovate at the base, with long, linear, mucronulate apex; the lower opposite. Upper bracts equalling the flowers; bracteoles ovate, with short acumen. Perianth-segments ovate, becoming hardened in fruit, the margin pectinate-ciliate, transversely keeled or with a small wing (*c.* 1 mm) dorsally. Stigma longer than the styles. Seeds 3–4 mm, vertical, oblique or horizontal. $2n = 18$. ***Europe, northwards to N.W. France, Hungary and E.C. Russia.*** Al Bl Bu Co Ga Gr Hs Hu It Ju Lu Rm Rs(W, K, E) Sa Si Tu.

6. S. acutifolia (Bunge) Botsch., *Not. Syst. (Leningrad)* **22**: 29 (1963) (***S. mutica*** C. A. Meyer). Like **5** but not more than 50 cm; leaves 10–20(–30) mm, semi-cylindrical becoming slightly wider at the base, acuminate; uppermost bracts shorter than the perianth; bracteoles orbicular; perianth-segments ovate-elliptical, with a transverse appendage dorsally; seeds always vertical. *S.E. Russia, S.E. Ukraine, W. Kazakhstan.* Rs(W, E).

7. S. tamariscina Pallas, *Reise* **3**: 604 (1776). Erect, virgate, papillose-scabrid annual 10–50 cm. Leaves 5–15 mm, semi-cylindrical, wider at the base and membranous-margined, mucronate. Bracts acuminate, not or only slightly longer than the flowers; bracteoles ovate-lanceolate, acuminate, mucronate, equalling the flowers. Perianth-segments ovate, with long, narrow, triangular apex; wing suborbicular. Stigmas longer than the style. *S.E. Russia, E. Ukraine.* Rs(W, K, E).

8. S. kali L., *Sp. Pl.* 222 (1753) (*S. aptera* Iljin, *S. pontica* Iljin, *S. praecox* Litv.). Erect or diffuse, glabrous or hispid annual up to 100 cm. Leaves 10–40 mm, linear-subulate, acuminate. Bracts patent; bracteoles ovate-triangular, with long, spine-like apex, longer than the flowers. Perianth-segments ovate, acuminate, usually with ovate to reniform wing. Stigmas longer than the style. *Often on non-saline sands, or as a ruderal. Throughout most of Europe northwards to 60° 30′ N., but almost exclusively coastal in the north.* All except Fa Is Sb Rs(N) [He].

A polymorphic species, comprising many subspecies and varieties. At least three subspecies can be recognized in Europe.

1 Perianth-segments stiff, ±spinose, with distinct mid-vein (a) subsp. **kali**
1 Perianth-segments soft, with obscure mid-vein
 2 Bracteoles swollen and ±connate at the base; perianth-segments not winged (b) subsp. **tragus**
 2 Bracteoles not swollen, free; perianth-segments usually winged (c) subsp. **ruthenica**

(a) Subsp. **kali**: Plant up to 60 cm, usually hispid, green. Bracteoles not swollen, free. Perianth-segments stiff, usually winged, more or less spinose, with a distinct mid-vein. *Coasts of Europe, northwards to 60° 30′ N.*

(b) Subsp. **tragus** (L.) Nyman, *Consp.* 631 (1881) (*S. tragus* L.): Plant up to 100 cm, glabrous or hispid, yellowish. Bracteoles swollen and more or less connate at the base. Perianth-segments soft, not winged, not spinose, with obscure mid-vein. *Coasts of S. Europe.*

(c) Subsp. **ruthenica** (Iljin) Soó in Soó & Jáv., *Magyar Növ. Kéz.* **2**: 786 (1951) (*S. ruthenica* Iljin; incl. *S. pestifer* A. Nelson): Plant up to 100 cm, glabrous or shortly hispid, grey-green. Bracteoles not swollen, free. Perianth-segments soft, usually winged, not spinose, with obscure mid-vein. $2n = 36$. *Throughout the range of the species, but only casual in much of the north.*

9. S. collina Pallas, *Ill. Pl.* 34 (1803). Like **8(c)** but leaves semi-amplexicaul; bracts appressed, imbricate, exceeding the bracteoles; perianth-segments often connate with the bracteoles; wing small, with erose margin, or absent. *C. & E. Russia.* Rs(C, E).

10. S. lanata Pallas, *Reise* **2**: 736 (1773). Erect annual 10–60 cm, lanate with long and short hairs. Leaves 10–25 × 1–2 mm, semi-cylindrical, obtuse, semi-amplexicaul, decurrent. Bracts oblong-triangular, somewhat deflexed; bracteoles triangular-ovate, broadly membranous-margined in the lower part. Perianth 10–20 mm in diameter in fruit, glabrous in the lower part, segments triangular, with lingulate apex; wing broadly reniform, brown, with numerous fine veins. Style *c.* 2 mm; stigmas much shorter than the style. *S.E. Russia, W. Kazakhstan.* Rs(E). (*C. & S.W. Asia.*)

11. S. turcomanica Litv., *Sched. Herb. Fl. Ross.* **2**: 10 (1900). Like **10** but glaucous, pubescent, with very short, patent hairs; bracts broadly ovate at the base, membranous-margined, narrowed to a semi-cylindrical, sulcate apex; perianth pubescent, the segments long, acute, with incurved or patent apex; stigmas at least 3 times as long as the style. *S.E. Russia, W. Kazakhstan.* Rs(E). (*C. & S.W. Asia.*)

12. S. crassa Bieb., *Mém. Soc. Nat. Moscou* **1**: 100 (1806). Like **10** but upper parts grey- or white-pubescent with long, crispate hairs, or glabrous; perianth-segments acute; style short; stigmas 3–4 mm, but usually less than 3 times as long as the style. *S.E. Russia, W. Kazakhstan.* Rs(E). (*S.W. Asia.*)

13. S. nitraria Pallas, *Ill. Pl.* 23 (1803). Erect, much-branched annual 5–40 cm; upper parts villous. Leaves 5–8 × 1–2 mm, linear, semi-cylindrical, somewhat enlarged at the base. Bracts broadly cordate, spurred, glabrous, shorter than or equalling the bracteoles; bracteoles orbicular-elliptical. Perianth 7–10 mm in diameter in fruit, glabrous, the segments triangular, acute, incurved; wing blackish-brown. Stigmas about as long as the style. *S.E. Russia, W. Kazakhstan.* Rs(E). (*C. & S.W. Asia.*)

14. S. affinis C. A. Meyer, *Bull. Phys.-Math. Acad.* (*Pétersb.*) **1**: 360 (1843). Erect annual up to 25(–40) cm, pubescent or subglabrous. Leaves 8–16 × 1·5–2 mm, semi-cylindrical, obtuse. Bracts elliptical or elliptic-orbicular, longer than the bracteoles; bracteoles oblong or lanceolate. Perianth 7–10 mm in diameter in fruit, the segments lanceolate, acute, glabrescent; wing reniform. Stigmas shorter than the style. *S.E. Russia (Astrakhan'), W. Kazakhstan.* Rs(E).

15. S. nodulosa (Moq.) Iljin, *Acta Horti Petrop.* **43(4)**: 222 (1930). Shrub 10–40 cm, grey-pubescent. Leaves 2–3 × 0·5 mm, narrowly triangular, obtuse, semi-amplexicaul. Bracts resembling the leaves but shorter and broader; bracteoles reniform, shortly acuminate, shorter than the flower. Perianth *c.* 10 mm in diameter in fruit, the segments oblong, with triangular apex, sparsely pubescent; wing brownish. Filaments with a lanceolate or elliptic-oblong, acute appendage. *Doubtfully recorded from S.E. Russia (near Astrakhan).* Rs(E). (*W. Caucasus.*)

16. S. carpatha P. H. Davis, *Notes Roy. Bot. Gard. Edinb.* **21**: 139 (1953). Erect grey-pubescent shrub 30–40 cm. Leaves 5–13 mm, oblong-linear, acute. Bracts narrowly oblong-elliptical; bracteoles similar, about equalling or shorter than the flower. Perianth-segments ovate-lanceolate, grey-pubescent. Filaments with lanceolate appendage *c.* 1·5 mm. Fruit not known. *Calcareous maritime rocks.* ● *Islands of S.E. Aegean region.* Cr Gr.

17. S. melitensis Botsch., *Nov. Syst. Pl. Vasc.* (*Leningrad*) **13**: 95 (1976). Glabrous shrub up to 100 cm, green or somewhat glaucous. Leaves up to 14 mm, linear, terete, acute, constricted at the base with hairy pulvinus. Bracts 2–9 mm, leaf-like, longer than the bracteoles and perianth; bracteoles ovate, acute. Flowers solitary. Perianth subglobose, 5–7 mm in diameter in fruit; segments ovate, obtuse, with 5 wings up to 2·5 mm. Stigmas shorter than the style. ● *Maltese Islands.* Si.

The only European representative of a group of closely related species from the southern part of the Mediterranean region from Morocco to Palestine, sometimes treated under *Darniella* Maire.

18. S. arbuscula Pallas, *Reise* **1**: 488 (1771). Glabrous shrub 20–100 cm. Leaves 5–35 mm, linear to oblong-linear, much enlarged at the base, with a distinct constriction above it. Bracts orbicular-ovate, becoming very hard in fruit; bracteoles linear, attenuate at the base, longer than the flower. Perianth 8–12 mm in diameter, becoming hard in fruit; segments broadly lanceolate-oblong, incurved, pale yellow or greenish-yellow; wing brown. Style shorter than the stigmas. *E. Russia.* Rs(C, E). (*C. Asia.*)

19. S. genistoides Juss. ex Poiret in Lam., *Encycl. Méth. Bot.* **7**: 294 (1806) (*S. tamariscifolia* Lag. pro parte). Glabrous, glaucous virgate shrub 30–50 cm; first- and second-year stems with numerous longitudinal ridges. Leaves up to 8 mm, fugacious, linear-filiform, mucronate, the uppermost scale-like. Bracts ovate-orbicular, acute, keeled, floccose on the upper surface. Perianth 10–15 mm in diameter in fruit, the segments elliptical, obtuse, the apex deflexed, 1-veined; wing reddish. ● *S.E. Spain.* Hs.

20. S. webbii Moq., *Chenop. Monogr. Enum.* 139 (1840). Erect shrub up to 200 cm, glabrous except for tufts of hairs in the axils of the leaves. Leaves 10–25 × *c.* 0·5 mm, filiform, mucronate, pale bluish-green, semi-amplexicaul, decurrent. Bracts ovate, subacute, shorter than the flowers; bracteoles orbicular-ovate, keeled, obtuse or shortly acuminate, about equalling the perianth. Perianth *c.* 13 mm in diameter in fruit, the segments lanceolate; wing brownish. *S. Spain.* Hs. (*N.W. Africa.*)

21. S. oppositifolia Desf., *Fl. Atl.* **1**: 219 (1798) (*S. longifolia* sensu Willk., *S. verticillata* auct., non Schousboe, *S. sieberi* auct.). Glabrous shrub 50–100 cm. Leaves up to 30 mm, nearly all opposite, linear or linear-oblong, attenuate towards the base and scarcely amplexicaul. Bracts linear-oblong, keeled, usually longer than the flowers; bracteoles oblong-elliptical. Perianth 12–20 mm in diameter in fruit, the segments obtuse, almost entirely membranous, fimbriate; wing orbicular-obovate, reniform. *S. & E. Spain, Islas Baleares; S. Italy, Sicilia, Lampedusa.* Bl Hs It Si.

22. S. papillosa (Cosson) Willk., *Strand-Steppengeb. Iber. Halbins.* 146 (1852). Divaricately branched, often procumbent shrub up to 60 cm, glabrous or pubescent. Leaves 4–12 × 1–3 mm, semi-cylindrical, obtuse, enlarged at the base, glabrous, pale bluish-green, whitish when dry. Bracts like the leaves; bracteoles oblong-lanceolate, obtuse, hyaline at the base. Perianth-segments oblong-lanceolate. *Rocky, calcareous hills.* ● *S.E. Spain.* Hs.

23. S. vermiculata L., *Sp. Pl.* 323 (1753) (*S. tamariscifolia* (L.) Lag.; incl. *S. hispanica* Botsch., *S. agrigentina* Guss., *S. flavescens* Cav.). Pubescent virgate shrub up to 100 cm. Leaves 5–10 × 0·5–1 mm, semi-cylindrical to filiform, expanding into an ovate base, semi-amplexicaul, obtuse, usually pubescent. Primary branches of the inflorescence up to 20–40 cm, with regularly arranged secondary branches. Bracts ovate with filiform apex; bracteoles ovate, keeled, fimbriate, shorter than the perianth. Perianth 6–12 mm in diameter in fruit, the segments oblong-ovate, acute, sparsely pubescent; wing obovate, finely veined. Stigmas shorter than the style, subulate, long-papillose. *W. Mediterranean region and S. Portugal.* Bl Hs Lu Sa Si.

24. S. aegaea Rech. fil., *Denkschr. Akad. Wiss. Math.-Nat. Kl.* (*Wien*) **105(2)**: 67 (1943) (*S. vermiculata* auct. balcan., non L.). Like **23** but not more than 30 cm, with much shorter, divaricate branches; lower leaves 5(–10) mm, oblong, glabrescent; upper leaves and bracts shortly triangular, usually pubescent; bracteoles orbicular to rhombic. *Calcareous maritime rocks. S. Aegean region.* Cr Gr. (*Rodhos.*)

25. S. dendroides Pallas, *Ill. Pl.* 22 (1803) (*S. brevifolia* D'Urv., *S. verrucosa* Bieb.). Virgate shrub 75–150 cm. Lower leaves 2–5 mm, semi-cylindrical, obtuse, broader at the base, glabrous, deciduous. Primary branches of the inflorescence 5–10 cm. Bracts like the leaves but shorter; bracteoles orbicular, glabrous. Perianth 7–9 mm in diameter in fruit, the segments narrowly triangular-ovate, acute, almost entirely membranous, glabrous, incurved. Stigmas lingulate. *S.E. Russia* (*N.W. shore of Caspian Sea*). Rs(E). (*C. & S.W. Asia.*)

26. S. laricina Pallas, *Ill. Pl.* 21 (1803). Erect shrub 20–70 cm. Leaves 5–12 mm, semi-cylindrical or filiform, slightly broader at the base, glaucous, sparsely pubescent, persistent. Primary branches of the inflorescence 5–10 cm, the secondary branches irregular. Bracts like the leaves; bracteoles ovate-orbicular, keeled, apex cucullate, glabrous. Perianth 5–7 mm in diameter in fruit, the segments orbicular to oblong, acute, almost entirely membranous. Stigmas lorate. *S.E. Russia, S. Ukraine, W. Kazakhstan.* Rs(W, K, E).

26. Noaea Moq.

P. W. BALL

Spiny, glabrous shrub. Leaves alternate, filiform. Flowers hermaphrodite, solitary, axillary. Perianth-segments 5 (3 outer and 2 inner), all developing a transverse wing dorsally in fruit. Stamens 5. Stigmas 2. Seeds vertical.

1. N. mucronata (Forskål) Ascherson & Schweinf., *Mém. Inst. Égypt.* **2**: 131 (1889) (*N. spinosissima* L. fil.). Stem 20–75 cm, much-branched, with the branches spine-tipped. Leaves up to 5 cm, filiform, obtuse. Perianth *c.* 4 mm, the wings 3–6 mm, obovate or obovate-orbicular with erose margin. *Maritime rocks. Aegean region, Turkey-in-Europe* (*Istanbul*); *local.* Cr Gr Tu.

27. Ofaiston Rafin.

P. W. BALL

Annual glabrous herbs; leaves alternate, nearly all basal. Flowers hermaphrodite, solitary, enclosed by 2 bracteoles. Perianth-segments 3 or 5, the 3 outer larger, developing a short transverse wing dorsally in fruit, the 2 inner very small or absent. Stamen 1. Stigmas 2. Seeds vertical.

1. O. monandrum (Pallas) Moq. in DC., *Prodr.* **13(2)**: 203 (1849). More or less erect, flexuous, up to 40 cm. Leaves 10–20 mm, fleshy, filiform. Bracts 2–3 mm, ovoid-triquetrous, acute, with broad, membranous margin; bracteoles ovate, becoming cartilaginous and keeled, enclosing the fruit. Perianth-segments membranous, with oblong or orbicular-ovate wing. *Salt-marshes. S.E. Russia, W. Kazakhstan.* Rs(?K, E). (*W. Asia.*)

28. Girgensohnia Bunge ex Fenzl

P. W. BALL

Annual glabrous herb. Leaves opposite. Flowers hermaphrodite. Perianth-segments 5 (3 outer and 2 inner), 2 or 3 segments developing a transverse wing dorsally in fruit. Stamens 5, alternating with 5 short staminodes. Stigmas 2. Seeds vertical.

1. G. oppositifolia (Pallas) Fenzl in Ledeb., *Fl. Ross.* **3**: 835 (1851). Stems 10–50 cm, more or less segmented, glaucous or reddish. Leaves up to 15 mm, broadly triangular or ovate, spine-tipped. Flowers solitary or in short, spicate inflorescences in the axils of broadly ovate bracts. Wing of perianth suborbicular, reddish. Seeds 1–2 mm. *Saline semi-deserts. On the borders of Europe in W. Kazakhstan.* Rs(E). (*C. & S.W. Asia.*)

29. Anabasis L.

P. W. BALL

Articulate, fleshy small shrubs. Leaves opposite, connate and amplexicaul, forming a segment. Flowers hermaphrodite and female, 1 or several in the axils of the upper leaves. Perianth-segments 5, usually developing a transverse wing dorsally in fruit. Stamens 5, alternating with 5 staminodes. Stigmas 2(3). Seeds vertical.

1 Perianth-segments not winged in fruit; anthers apiculate **1. salsa**
1 Perianth-segments developing a conspicuous wing in fruit; anthers not apiculate
2 Plant caespitose, with simple, deciduous flowering stems not more than 10 cm **3. cretacea**
2 Plant with branched, persistent stems usually more than 10 cm
3 Fruiting perianth segments unequally winged, 3 with larger wing than other 2 **2. aphylla**
3 Fruiting perianth segments equally winged **4. articulata**

1. A. salsa (C. A. Meyer) Bentham ex Volkens in Engler & Prantl, *Natürl. Pflanzenfam.* **3(1a)**: 87 (1893). Stems 5–25 cm, procumbent, woody at the base, with numerous erect, more or less simple, deciduous branches; upper segments up to 40 × 2–3 mm. Leaves 1–5 mm, subcylindrical, with deciduous apical seta. Perianth-segments not winged, becoming hard in the upper part in fruit; staminodes fimbriate. *Saline soils. S.E. Russia, W. Kazakhstan.* Rs(C, E).

2. A. aphylla L., *Sp. Pl.* 223 (1753). Stems 25–75 cm, persistent, branched, woody below; upper segments up to 20 × 1 mm. Leaves *c.* 1 mm, each pair forming a short, truncate cupule. Three perianth-segments winged; wing up to 4 mm, orbicular-ovate; staminodes ciliolate. *Saline soils. S.E. Russia, W. Kazakhstan.* Rs(E).

3. A. cretacea Pallas, *Reise* **1**: 442 (1771). Caespitose, with simple deciduous flowering stems up to 10 cm; upper segments of flowering stems *c.* 10 × 1 mm. Leaves *c.* 3 mm, triangular-ovate, often patent. Three perianth-segments with an orbicular-ovate wing and 2 without a wing or with a smaller narrower wing; staminodes fimbriate. *Calcareous soils. S.E. Russia, W. Kazakhstan.* Rs(E). (*W.C. Asia.*)

4. A. articulata (Forskål) Moq. in DC., *Prodr.* **13(2)**: 212 (1849) (*A. hispanica* Pau, *A. mucronata* (Lag.) C. Vicioso). Stems 10–30 cm, persistent, branched, woody below; upper segments up to 8 × 2 mm. Leaves 1–2 mm. Five perianth-segments winged; wing up to 4 × 7 mm, reniform-orbicular; staminodes villose. *Saline soils. S.E. Spain.* Hs. (*N. Africa, W. Asia.*)

30. Haloxylon Bunge

P. W. BALL

Small shrubs with articulate stems. Leaves opposite, connate and amplexicaul, forming segments; lamina rudimentary. Inflorescence spicate, usually interrupted; flowers usually solitary. Perianth-segments 5, developing a transverse wing dorsally in fruit. Stamens usually 5, connate. Stigmas 4. Seeds horizontal.

1. H. tamariscifolium (L.) Pau, *Bol. Soc. Aragon. Ci. Nat.* **2**: 71 (1903) (*Haloxylon articulatum* (Moq.) Bunge, *Hammada tamariscifolia* (L.) Iljin). Plant 10–60 cm, erect, caespitose, glaucous; segments of main branches *c.* 10 mm. Leaves with lamina up to 3 mm, triangular-subulate. Perianth-segments *c.* 1 mm, elliptical, obtuse; wing up to 3 × 4 mm, reniform-orbicular to orbicular-ovate, pinkish. *Dry saline soils. S.E. Spain.* Hs. (*N. Africa, S.W. Asia.*)

31. Nanophyton Less.

P. W. BALL

Glabrous small shrub. Leaves alternate, triangular-subulate. Flowers hermaphrodite, in groups of 1–7 at the apices of the branches; bracteoles 2. Perianth-segments 5, very enlarged and inflated in fruit. Stamens 5. Stigmas 2. Seeds vertical.

1. N. erinaceum (Pallas) Bunge, *Mém. Acad. Sci. Pétersb.* ser. 7, **4(11)**: 51 (1862). Plant 5–15 cm, pulvinate, with procumbent, tortuous woody stems. Leaves 3–7 mm, semi-amplexicaul, spine-tipped. Perianth-segments up to 10 mm in fruit, pale yellowish-green, shining. Achenes *c.* 3 mm. *Dry and saline places. S.E. Russia, W. Kazakhstan.* Rs(E). (*W. & C. Asia.*)

32. Petrosimonia Bunge

P. W. BALL

Annual, usually pubescent herbs. Leaves alternate or opposite, cylindrical or semi-cylindrical. Inflorescence a panicle, the flowers hermaphrodite, solitary, axillary; bracteoles 2, becoming hard and enclosing the fruit. Perianth-segments 2–5. Stamens 1–5. Stigmas 2. Seeds vertical.

All species occur in saline habitats.

1 Leaves all opposite; perianth-segments 5 **1. brachiata**
1 Middle and upper leaves alternate; perianth-segments almost always 2 or 3
2 Leaves *c.* 1 mm in diameter or less; stamens 1–3
3 Perianth-segments 3 or 4, all pubescent dorsally; stamens 1 or 2 **4. monandra**
3 Perianth-segments 2 or 3, glabrous (rarely one of them pubescent dorsally); stamens 2 or 3
4 Perianth-segments and stamens 3; lower leaves up to 50 mm **2. triandra**
4 Perianth-segments and stamens mostly 2; lower leaves usually not more than 10 mm **3. litwinowii**
2 Leaves (at least the lower) 2–3 mm in diameter; stamens 5
5 Apex of bracteoles deflexed, attenuate **7. glaucescens**
5 Apex of bracteoles erect, not attenuate
6 Perianth-segments glabrous or shortly pubescent, becoming cartilaginous in fruit **5. oppositifolia**
6 Perianth-segments long-ciliate, at least at the apex, not becoming cartilaginous in fruit **6. brachyphylla**

1. P. brachiata (Pallas) Bunge, *Mém. Acad. Sci. Pétersb.* ser. 7, **4(11)**: 59 (1862). Stems 5–40 cm, erect, branched from the base, pubescent. Leaves up to 30 × 2 mm, all opposite. Perianth-segments and stamens 5. *S.E. Europe, from S.E. Greece to W. Kazakhstan.* Bu Gr Rs(W, K, E).

2. P. triandra (Pallas) Simonkai, *Enum. Fl. Transs.* 466 (1887) (*P. volvox* (Pallas) Bunge). Stems 5–35 cm, procumbent or diffuse, pubescent or glabrous. Leaves up to 50 × 1 mm, the middle and upper alternate. Panicle lax in fruit. Perianth-segments 3, glabrous; stamens 3. *S.E. Europe.* Rm Rs(W, K, E).

3. P. litwinowii Korsh., *Mém. Acad. Sci. Pétersb.* ser. 8, **7(1)**: 358 (1898). Stems 5–30 cm, green, erect or procumbent, glabrous or pubescent. Leaves usually not more than 10 × 1 mm, alternate. Panicle lax in fruit. Perianth-segments 2, glabrous; stamens 2. *S.E. Russia, W. Kazakhstan.* Rs(E).

4. **P. monandra** (Pallas) Bunge, *Mém. Acad. Sci. Pétersb.* ser. 7, **4(11)**: 53 (1862). Stems 5–30 cm, erect or procumbent, often densely pubescent. Leaves up to 10 × 1 mm, crowded, alternate. Panicle dense in fruit. Perianth-segments 3(–5), pubescent dorsally; stamens 1 or 2. *S.E. Russia, W. Kazakhstan.* Rs(E). (*W.* & *C. Asia.*)

5. **P. oppositifolia** (Pallas) Litv., *Sched. Herb. Fl. Ross.* **7**: 13 (1911) (*P. crassifolia* auct.). Stems 5–40 cm, erect or procumbent, glabrous or pubescent. Leaves up to 40 × 3 mm, very fleshy, obtuse, the upper alternate. Bracteoles obtuse, not deflexed. Perianth-segments 2, glabrous or shortly pubescent, becoming cartilaginous in fruit; stamens 5. *S.E. Europe.* Al Rm Rs(W, K, E).

6. **P. brachyphylla** (Bunge) Iljin, *Acta Horti Petrop.* **43**: 234 (1930). Like **5** but the perianth-segments long-ciliate at least towards the apex, not becoming cartilaginous in fruit. *S.E. Russia, W. Kazakhstan.* Rs(E). (*W.* & *C. Asia.*)

7. **P. glaucescens** (Bunge) Iljin, *loc. cit.* 233 (1930) (*P. glauca* sensu Boiss.). Like **5** but plant densely pubescent; bracteoles acute, apex deflexed; perianth-segments hirsute dorsally, not becoming cartilaginous in fruit. *S.E. Russia, W. Kazakhstan.* Rs(E). (*W.* & *C. Asia.*)

33. Halimocnemis C. A. Meyer

P. W. BALL

Like *Petrosimonia* but the seeds horizontal.

1. **H. sclerosperma** (Pallas) C. A. Meyer in Ledeb., *Fl. Altaica* **1**: 384 (1829). Stems 5–30 cm, thick and flexuous. Leaves 15–25 mm, linear or oblong, mucronate. Bracts ovate, acuminate. Perianth-segments 3(–5), the 2 outer segments becoming 7–8 mm and hard in fruit. *Saline soils. S.E. Russia, W. Kazakhstan.* Rs(E).

34. Halogeton C. A. Meyer

P. W. BALL

Annual glabrous or papillose herbs. Leaves alternate, fleshy, more or less cylindrical. Flowers hermaphrodite or female, in dense cymes in the axils of leaf-like bracts; the outer in each cyme 2-bracteolate, the inner ebracteolate. Perianth-segments 5, developing a transverse wing in fruit. Stamens 3–5. Stigmas 2. Seeds vertical or horizontal.

1. **H. sativus** (L.) Moq., *Chenop. Monogr. Enum.* 158 (1840). Stems 15–80 cm. Leaves 5–16 × 1–1·5 mm, usually mucronate. Stamens 5. Perianth-wing up to 3 × 4 mm, reniform-orbicular to ovate, unequal, often pinkish. Seeds *c.* 1·5 mm. *Saline habitats, mainly maritime. S.E. Spain.* Hs. (*N.W Africa.*)

Formerly cultivated in the Mediterranean region, where it was grown as a vegetable and burned for the production of base-rich ash.

H. glomeratus (Bieb.) C. A. Meyer in Ledeb., *Fl. Altaica* **1**: 378 (1829), from S.E. Russia and W. Kazakhstan, is very doubtfully distinct from **1**. It is less robust (stems 3–40 cm), with the leaves up to 12 × 2 mm and 1–3(–5) stamens.

49. AMARANTHACEAE

EDIT. †T. G. TUTIN AND J. R. EDMONDSON

Herbs or rarely small shrubs. Leaves opposite or alternate, entire, exstipulate. Flowers usually hermaphrodite, often in spicate or capitate, usually bracteate inflorescences; bracteoles 2–5. Perianth almost always dry and scarious, usually 4- or 5-merous; segments free or connate at base. Stamens 1–5, opposite the perianth-segments and usually not exserted. Ovary superior, unilocular; ovules amphitropous. Fruit with a membranous, rarely fleshy wall, dehiscing irregularly, or circumscissile (sometimes indehiscent).

Literature: U. H. Eliasson, *Bot. Jour. Linn. Soc.* **96**: 235–283 (1988).

1 Leaves opposite
 2 Inflorescence long, lax; fruits deflexed — **3. Achyranthes**
 2 Inflorescence dense, capitate; fruits not deflexed — **4. Alternanthera**
1 Leaves alternate
 3 Filaments connate below into a tube; fruit with numerous seeds — **1. Celosia**
 3 Filaments free; fruit with 1 seed — **2. Amaranthus**

1. Celosia L.

†T. G. TUTIN (EDITION 1) REVISED BY J. R. AKEROYD (EDITION 2)

Leaves alternate. Bracteoles 3. Perianth-segments 5, oblong, acute or obtuse. Stamens 5, alternating with 5 teeth; filaments connate below into a membranous tube. Style long, filiform. Fruit circumscissile; seeds numerous.

1. **C. argentea** L., *Sp. Pl.* 205 (1753). Glabrous, branched annual 15–50 cm. Leaves broadly lanceolate to ovate; petioles short. Inflorescences subsessile, spicate, usually cylindrical, often fasciated. Flowers sessile; perianth much longer than the bracteoles, usually purple, sometimes red, orange, yellow or white. *Widely cultivated for ornament, and frequently occurring as a casual in S. Europe.* [It.] (*Tropics.*)

The plant cultivated in Europe is var. *cristata* (L.) O. Kuntze (*C. cristata* L.).

2. Amaranthus L.

†P. AELLEN (EDITION 1) REVISED BY J. R. AKEROYD (EDITION 2)

Annual, rarely perennial herbs; monoecious or dioecious. Leaves usually alternate. Bracteoles 3–5, small and herbaceous or membranous and spinescent. Perianth-segments 0–5, linear or lanceolate to spathulate. Styles and stigmas 2 or 3. Fruit dry, membranous, indehiscent or dehiscing transversely. Seeds vertically compressed.

In the following descriptions characters of the perianth refer to female flowers.

All the species are ruderals or weeds. Most of them have been introduced into Europe and are usually casual in the north and centre. This account includes only those species that are known to be naturalized.

A. caudatus L., *Sp. Pl.* 990 (1753), native in South America, with long, pendent, red or rarely green inflorescences, is commonly cultivated for ornament, and sometimes occurs as a casual.

Literature: A. Thellung in Ascherson & Graebner, *Syn. Mitteleur. Fl.* **5(1)**: 225–356 (1914). S. Priszter, *Ann. Sect. Horti-Viticult. Univ. Sci. Agr. (Budapest)* **20**: 121–262 (1953). P. Aellen in Hegi, *Ill. Fl. Mitteleur.* ed. 2, **3(2)**: 465–516 (1959). J. P. M. Brenan, *Watsonia* **4**: 261–280 (1961). J. D. Sauer, *Ann. Missouri Bot. Gard.* **54**: 103–137 (1967). J. L. Carretero, *Collect. Bot. (Barcelona)* **11**: 105–142 (1979).

1 Perianth-segments (4)5
  2 Inflorescence composed entirely of axillary cymose clusters, usually leafy to the apex
    3 Fruit dehiscing transversely; larger perianth-segments with a short acumen **6. blitoides**
    3 Fruit indehiscent; perianth-segments obtuse **7. crispus**
  2 Inflorescence a terminal panicle, the apical part leafless and often spicate
    4 Fruit dehiscing transversely
      5 Perianth-segments enlarged in the upper part, ± spathulate, obtuse or truncate **4. retroflexus**
      5 Perianth-segments lanceolate or narrowly ovate or elliptical, acute
        6 Longest bracteoles of the female flowers 1–1½ times as long as the perianth **2. cruentus**
        6 Longest bracteoles of the female flowers usually twice as long as the perianth
          7 Longest bracteoles not more than 4 mm; stamens 5 **1. hybridus**
          7 Longest bracteoles 4–6 mm; stamens usually 3 **3. powellii**
    4 Fruit indehiscent
      8 Bracteoles longer than the perianth, spinose; leaves ovate or ovate-oblong **3. powellii**
      8 Bracteoles shorter than the perianth, not spinose; leaves linear-lanceolate **5. muricatus**
1 Perianth-segments (2)3
  9 Fruit dehiscing transversely
    10 Bracteoles twice as long as the perianth, spinose **8. albus**
    10 Bracteoles not longer than the perianth, not spinose **9. graecizans**
  9 Fruit indehiscent or dehiscing irregularly
    11 Leaves obtuse; stems puberulent above; fruit inflated **10. deflexus**
    11 Leaves ± emarginate; stems usually glabrous; fruit not inflated
      12 Fruit feebly rugose, broadly ellipsoidal; stems glabrous **11. lividus**
      12 Fruit strongly muricate, subglobose; stems sometimes puberulent above **12. viridis**

1. A. hybridus L., *Sp. Pl.* 990 (1753) (*A. patulus* auct., non Bertol., *A. chlorostachys* Willd.). Stems 20–100 cm, erect, sparsely to densely pubescent above. Leaves 3–8 cm, rhombic-ovate. Inflorescence elongate-spicate, often compound, with long branches. Bracteoles (2–)3–4 mm, ovate, with a very long mucro, about twice as long as the perianth. Perianth-segments 5, 1·5–2(–2·4) mm, narrowly ovate, usually acute, about as long as the fruit. Stamens 5. Fruit dehiscing transversely. $2n = 32$. *Commonly naturalized in many parts of Europe.* [Al Au Az Bl Br Bu Co Cr Cz Ga Ge Gr He Hs Hu It Ju Lu Po Rm Rs(K) Sa Si.] (*Tropical and subtropical America.*)

There has been considerable taxonomic and nomenclatural confusion between species **1–3** which has arisen because of the great level of phenotypic plasticity in these species, hybridization with each other and other species (especially **4**), and the naming of taxa in the Old and New Worlds without reference to the total pattern of variation. The geographical distributions of these species reported here must therefore be regarded as provisional.

Their taxonomy is summarized in J. D. Sauer, *loc. cit.* (1967).

2. A. cruentus L., *Syst. Nat.* ed. 10, **2**: 1269 (1759) (incl. *A. paniculatus* L., *A. patulus* Bertol.). Like **1** but the terminal inflorescence usually dense and with short branches at the base; bracteoles 2–4 mm, with short mucro, 1–1½ times as long as the perianth. $2n = 34$. *Commonly naturalized, mainly in S. & C. Europe.* [Al Bl Bu Co Cr Cz Ga Ge He Hs Hu It Ju Lu Rm Rs(C, W, K) Sa Si.] (*Tropical and subtropical America.*)

Grown as a grain crop in Central America, but long cultivated for ornament in Europe, where red- or yellow-flowered plants have been known as **A. paniculatus** L., *Sp. Pl.* ed. 2, 1406 (1763).

3. A. powellii S. Watson, *Proc. Amer. Acad. Arts Sci.* **10**: 347 (1875) (*A. bouchonii* Thell., *A. chlorostachys* auct., non Willd.). Like **1** but the stem glabrous to sparsely pubescent; longest bracteoles 4–6 mm, linear-lanceolate, long-spinous; perianth-segments 3–5, (1·5–)2·5–3 mm, elliptic-lanceolate to linear; stamens 3(–5); fruit dehiscent or indehiscent. *Naturalized in S.W. & C. Europe; casual elsewhere.* [Ga Ge Hs It.] (*North and South America.*)

In North America this species usually has dehiscent fruits, and plants with indehiscent fruits occur sporadically within populations. Plants from S.W. Europe and elsewhere differ in the more branched inflorescence and the somewhat smaller, indehiscent fruits.

4. A. retroflexus L., *Sp. Pl.* 991 (1753) (incl. *A. bulgaricus* Kovacev). Stems 15–100 cm, erect, lanate in the upper part. Leaves 3–8 cm, rhombic-ovate. Inflorescence spicate, usually short and dense. Bracteoles 3–6 mm, stout and spinescent, 1⅓–2 times as long as the perianth. Perianth-segments 2–3 mm, linear-cuneate, truncate or obtuse, with short mucro. Fruit dehiscing transversely, feebly muricate. $2n = 34$. *Introduced and naturalized throughout most of Europe, but only casual in the north.* [Al Au Az Be Bl Br Bu Co Cz Da Ga Ge Gr He Ho Hs Hu It Ju Lu Po Rm Rs(B, C, W, K, E) Si Su Tu.] (*North America.*)

A. quitensis Kunth in Humb., Bonpl. & Kunth, *Nov. Gen. Sp.* **2**: 194 (1818), from South America, like **4** but glabrous or puberulent, the inflorescence with numerous distinct lateral branches, and the perianth-segments with green excurrent midrib, is a widespread casual that is perhaps becoming naturalized in Açores and Islas Baleares.

5. A. muricatus (Gillies ex Moq.) Hieron., *Bol. Acad. Nac. Ci.* **4**: 421 (1881) (*Euloxus muricatus* Gillies ex Moq.). Decumbent perennial up to 60 cm, usually glabrous. Leaves 2–5 cm, linear to lanceolate or narrowly ovate-lanceolate, long-petiolate. Inflorescence a long panicle, branched at the base. Bracteoles ½ as long as the perianth, ovate, acute. Perianth-segments 5, *c.* 2 mm, spathulate. Fruit indehiscent, strongly muricate. $2n = 34$. *Naturalized* *locally in S. Europe; casual elsewhere.* [Bl Gr Hs Lu Sa Si.] (*Temperate South America.*)

6. A. blitoides S. Watson, *Proc. Amer. Acad. Arts Sci.* **12**: 273 (1877). Stems 15–50 cm, procumbent, whitish, glabrous or pubescent in the upper part. Leaves 1·5–3 cm, oblong-lanceolate to obovate-spathulate, obtuse, with distinct membranous margin. Inflorescence of short, axillary cymose clusters, leafy to the apex. Bracteoles shorter than the perianth, lanceolate. Perianth-segments 4 or 5, unequal, resembling the bracteoles; the largest 2–2·5 mm, with short acumen. Fruit dehiscing transversely. $2n = 32$. *Naturalized in C. & S. Europe; casual throughout most of Europe.* [Al Au Az Bl Bu Co Cr Cz Ga Ge Gr Ho Hs Hu It Lu Po Rm Rs(B, C, W, K, E).] (*C. & W. North America.*)

7. A. crispus (Lesp. & Thév.) N. Terracc., *Rendic. Accad. Sci. Fis. Mat. (Napoli)* **4**: 188 (1890). Stems up to 40 cm, procumbent or ascending, much-branched, densely puberulent. Leaves mostly 0·5–1·5 cm, ovate- or lanceolate-rhombic, obtuse, mucronate; margin undulate-crenate. Inflorescence of ovoid or spherical, axillary cymose clusters, leafy to the apex. Bracteoles ½ as long as the perianth, ovate, acute. Perianth-segments 5, 0·6 mm wide at the apex, obovate to spathulate, obtuse. Fruit muricate, indehiscent. $2n = 34$. *Naturalized in S. and parts of C. Europe; casual elsewhere.* [Au Bu Cz Hu It Ju Rm.] (*Argentina.*)

A. standleyanus Parodi ex Covas, *Darwiniana* **5**: 339 (1941) (*A. vulgatissimus* auct.), from Argentina, also with indehiscent fruits, but with stems up to 70 cm, leaves 3–5 cm, entire, and distinctly clawed perianth-segments, has been reported from Spain and elsewhere, but is only casual.

8. A. albus L., *Syst. Nat.* ed. 10, **2**: 1268 (1759). Stems 10–50 cm, erect or procumbent, much-branched, glabrous or sparsely puberulent. Leaves 2–5 cm, oblong or spathulate, cuneate, obtuse and slightly emarginate, mucronate; margin undulate. Inflorescence of short, axillary cymose clusters. Bracteoles twice as long as the perianth, ensiform with long, spinose apex. Perianth-segments 3, narrowly elliptical. Fruit *c.* 1·5 mm, scarcely muricate, dehiscing transversely. $2n = 32$. *Naturalized or a casual in most of Europe.* [Al Au Be Bl Bu Co Cr Cz Ga Ge Gr He Ho Hs Hu It Ju Lu Po Rm Rs(B, C, W, K, E) Sa Si Tu.]

9. A. graecizans L., *Sp. Pl.* 990 (1753) (*A. angustifolius* Lam., *A. sylvestris* Vill., *A. blitum* auct., non L.). Stems up to 70 cm, usually erect, usually glabrous. Leaves 2–4 cm, ovate or elliptic-rhombic, usually acute. Inflorescence of axillary cymose clusters. Bracteoles ⅗–¾ as long as the perianth, ovate, mucronulate. Perianth-segments 3, 1·3–2 mm, ovate-lanceolate, acute. Fruit longer than the perianth, somewhat muricate, dehiscing transversely, with green, longitudinal veins when young. $2n = 32$. *S. Europe, perhaps native further north; introduced throughout most of Europe and sometimes naturalized.* Al Au Az Bl Bu *Co Cr *Cz *Ga Gr Hs Hu It Ju Lu *Rs(K) Si Tu [Au Be Ge He Hu Rm].

10. A. deflexus L., *Mantissa Alt.* 295 (1771). Procumbent perennial up to 40 cm, densely puberulent in the upper part. Leaves 3–5 cm, rhombic-ovate, obtuse; margin finely undulate. Inflorescence usually dense, terminal, spicate, becoming interrupted and leafy towards the base. Bracteoles ⅓–½ as long as the perianth, ovate, wide at the base, mucronate. Perianth-segments 2 or 3, 1·2–1·5 mm, linear to oblong-spathulate. Fruit 2·5–3 mm, oblong-ovate, inflated-membranous, smooth, with 3 dull green, longitudinal veins, not dehiscing transversely. Seeds much smaller than the fruit. $2n =$ 32, 34. *Naturalized in S. Europe; casual elsewhere.* [Al Az Bl Bu Co Cz Ga Gr He Hs Hu It Ju Lu Rm Rs(C, W, K) Sa Si Tu.]

11. A. lividus L., *Sp. Pl.* 990 (1753) (*A. blitum* L., *A. polygonoides* L., *A. ascendens* Loisel.). Stems 30–80 cm, procumbent to erect, glabrous. Leaves 2–6 cm, rhombic- to orbicular-ovate, with light or dark spots on the upper surface, subtruncate or emarginate, the margin often undulate. Inflorescence of axillary cymose clusters, forming a dense, more or less leafless spike towards the apex. Bracteoles ⅓–½ as long as the perianth, ovate with wide base, acute. Perianth-segments 3(–5), oblong-linear to spathulate. Fruit feebly rugose, without green veins, indehiscent or dehiscing irregularly. Seeds almost as large as the fruit. $2n = 34$. *S. Europe; introduced and sometimes naturalized elsewhere.* Al Au Az Bl Bu Cr *Cz Ga *Ge Gr *He Hs Hu It Ju Lu *Po Rm Sa Si Tu [Be Ho Rs(B, C, W, K, E) Su].

Many records, especially from S. Europe, may result from confusion with **9** and **12.**

12. A. viridis L., *Sp. Pl.* ed. 2, 1405 (1763) (*A. gracilis* Poiret). Stems up to 70 cm, procumbent to erect, glabrous or slightly hairy above. Leaves 3–8 cm, ovate-rhombic, somewhat emarginate, long-petiolate. Inflorescence of axillary cymose clusters, forming a long, slender, terminal spike. Bracteoles ⅓–⅔ as long as the perianth, ovate, acuminate. Perianth-segments 3, ovate-oblong. Fruit about as long as the perianth, subglobose, strongly muricate, dehiscing irregularly, or indehiscent. *Naturalized or casual in S. Europe.* [Az Bl Cr Gr Hs It Si.] (*South America*).

3. Achyranthes L.

†T. G. TUTIN

Leaves opposite. Inflorescence long, lax, spicate. Flowers erect, becoming deflexed. Perianth-segments 4 or 5, narrowly lanceolate, becoming indurate and ribbed. Stamens (2–)5 alternating with teeth; filaments connate below into a tube. Fruits deflexed.

Stem densely pubescent; leaves green beneath	**1. aspera**
Stem almost glabrous; leaves silvery beneath	**2. sicula**

1. A. aspera L., *Sp. Pl.* 204 (1753). Pubescent perennial, woody below. Leaves broadly ovate or broadly rhombic, shortly acuminate, green on both surfaces. Flowers pedicellate, somewhat shining, green. Bracteoles glabrous, awned. Perianth about twice as long as the bracteoles. Teeth alternating with the stamens, fimbriate at the apex. *Ruderal. Locally naturalized in S. Europe.* [?Az Hs Si.] (*Tropics.*)

2. A. sicula (L.) All., *Mélang. Philos. Math. Soc. Roy. Turin (Misc. Taur.)* **5**: 93 (1774) (*A. argentea* Lam.). Like **1** but the stem only slightly pubescent to nearly glabrous; leaves ovate to ovate-oblong, densely silvery-sericeous beneath; teeth alternating with stamens, entire at the apex. *Ruderal. S. part of W. Mediterranean region.* Hs It Sa Si.

4. Alternanthera Forskål

†T. G. TUTIN

Leaves opposite. Inflorescence capitate, terminal or axillary. Perianth-segments 5, free. Stamens 2–5; filaments connate below into a tube. Fruits not deflexed.

A number of species besides the following occur as casuals.

Literature: P. Aellen in Hegi, *Ill. Fl. Mitteleur.* ed. 2, **3(2)**: 523–528 (1959). S. J. van Ooststroom & Th. J. Reichgelt, *Gorteria* **1**: 2–6 (1961).

1 Annual; leaves linear; bracteoles and outer perianth-segments not pungent **3. nodiflora**
1 Perennial; leaves ovate or obovate; bracteoles and outer perianth-segments pungent
2 Two of the outer perianth-segments with points *c.* $\frac{1}{3}$ as long as the whole; filament about as long as the anther **1. pungens**
2 Two of the outer perianth-segments with points not more than $\frac{1}{10}$ as long as the whole; filament about twice as long as the anther **2. peploides**

1. A. pungens Kunth in Humb., Bonpl. & Kunth, *Nov. Gen. Sp.* **2**: 206 (1818) (*A. achyrantha* (L.) Swartz, *A. repens* auct.). Perennial up to 30 cm. Stems somewhat woody at base, pubescent. Leaves ovate or obovate, shortly petiolate. Inflorescences sessile, 2 or 3 together, ovoid. Flowers subtended by 2 boat-shaped, pungent bracteoles. Three outer perianth-segments longer than the 2 inner, 2 with points *c.* $\frac{1}{3}$ as long as the whole, the other and the inner ones with short, slender points; all with small tufts of short, retrorsely barbed hairs. Filaments and the teeth between them about as long as the anthers. *Ruderal. Naturalized in E. Spain.* [Hs.] (*Temperate South America.*)

2. A. caracasana Kunth in Humb., Bonpl. & Kunth, *Nov. Gen. Sp.* **2**: 165 (1818) (*A. peploides* (Humb. & Bonpl.) Urban, *A. achyrantha* (L.) Swartz, *A. repens* auct.). Like **1** but points of outer perianth-segments not more than $\frac{1}{10}$ as long as the whole; perianth-segments with more numerous, longer, retrorsely barbed hairs; filaments and the teeth between them about twice as long as the anthers. $2n = 96$. *Cultivated ground and as a ruderal. Naturalized in Açores and S. & E. Spain.* [Az Hs.] (*W. Indies, tropical America.*)

3. A. nodiflora R. Br., *Prodr. Fl. Nov. Holl.* **1**: 417 (1810). Annual. Stems glabrous or slightly hairy at the nodes, but with 2 lines of hairs when young. Leaves linear, sessile. Inflorescence sessile, axillary, globose. Bracteoles not spinose. Perianth-segments *c.* 3 mm, white. *Naturalized in S. Spain.* [Hs.] (*Australia, India, Africa.*)

50. NYCTAGINACEAE

EDIT. V. H. HEYWOOD

Herbs, shrubs or woody climbers, often with swollen nodes. Leaves simple, entire, exstipulate. Flowers solitary or in small, terminal umbels (rarely in whorls), subtended by bracts which often simulate a perianth. Perianth petaloid, tubular, with a short, 5-lobed limb; lower part of tube persistent and concrescent with the fruit to form an achene-like anthocarp. Stamens usually 3–5. Ovary superior, 1-celled, containing a single ovule.

Various cultivars and hybrids of *Bougainvillea*, woody climbers with alternate leaves and an involucre of 3 conspicuous, brightly coloured bracts, are extensively planted for ornament in parts of the Mediterranean region.

1 Bracts few and inconspicuous; flowers in umbels of 4–10
2 Leaves and stems viscid; anthocarp with distinct longitudinal ribs **1. Boerhavia**
2 Leaves and stems not viscid; anthocarp terete **2. Commicarpus**
1 Bracts conspicuous, united around the base of each flower or umbel to form a calyx-like involucre; flowers solitary or in umbels of 3–5
3 Perianth *c.* 25 mm in diameter; involucre tubular, not accrescent, or membranous, subtending only 1 flower **3. Mirabilis**
3 Perianth *c.* 8 mm in diameter; involucre broadly conical, accrescent, membranous in fruit, subtending several flowers **4. Oxybaphus**

1. Boerhavia L.

J. R. AKEROYD

Shrubs or somewhat woody herbs with opposite leaves. Flowers small, in whorls or umbels. Bracts few and inconspicuous. Perianth-tube contracted above the ovary and surmounted by a spreading, 5-lobed limb. Stamens 1–6. Anthocarp with distinct longitudinal ribs, glandular, viscid.

1. B. repens L., *Sp. Pl.* 3 (1753). Plant glandular-pubescent, viscid; rather woody. Stems up to 100 cm, decumbent to ascending, much-branched. Leaves 5–25(–50) mm, ovate to elliptical, with sinuate margins; petiole about as long as the lamina. Flowers in rather dense, 4- to 10-flowered umbels on peduncles 1–2 cm. Perianth 2·5–3 mm, pink or purple. Stamens 2. Anthocarp 3–4 mm, clavate. *Naturalized as a ruderal in Sicilia (Palermo).* [Si.] (*Tropical Africa and Asia.*)

The plant naturalized in Europe is subsp. **viscosa** (Choisy) Maire, *Sahara Central* 88 (1933).

2. Commicarpus Standley

D. A. WEBB

Like *Boerhavia*, but stamens usually 3 and anthocarp terete with hooked hairs.

1. C. plumbagineus (Cav.) Standley, *Contr. U.S. Nat. Herb.* **18**: 101 (1916) (*Boerhavia plumbaginea* Cav.). Stems up to 70 cm, woody, geniculate, straggling or procumbent with long internodes. Leaves *c.* 3 × 2 cm, deltate to ovate, stalked, puberulent beneath. Flowers in small, long-pedunculate, axillary, *c.* 10-flowered umbels, sometimes with a whorl of flowers on the peduncle. Bracts 1 or 2, very small. Perianth *c.* 13 × 6 mm, pubescent, sharply differentiated into a lower, blackish portion surrounding the ovary, and an upper portion, white or pale pink, narrowly tubular below and gradually expanded into a flat limb with 5 emarginate lobes; at the junction of the 2 portions are 5 black tubercles. Stamens and style exserted. Anthocarp 8 mm, obconical, striate, glandular-tuberculate, especially at the apex. *Rocky and stony places. S.E. Spain (hills near Orihuela).* Hs. (*S.W. Asia and tropical Africa.*)

3. Mirabilis L.

D. A. WEBB

Perennial herbs with opposite leaves. Flowers in axillary cymes, each flower surrounded by a tubular or narrowly campanulate, calyx-like involucre of 5 bracts. Perianth infundibuliform, with a long tube, contracted above the ovary, and a spreading, plicate, slightly 5-lobed limb. Stamens 5. Anthocarp coriaceous.

1. M. jalapa L., *Sp. Pl.* 177 (1753). Plant glabrous or slightly pubescent. Stems 50–100 cm, erect. Leaves ovate, acuminate. Flowers in crowded terminal cymes, opening in the afternoon. Perianth red, yellow, white, or parti-coloured; limb *c.* 25 mm in diameter; tube 25–35 mm. *Cultivated for ornament, and locally naturalized in S. Europe.* [Az Bl Bu Ga Gr Hs It.] (*Tropical America.*)

Plants with scented flowers and the anthocarp more strongly ridged and verrucose than usual are sometimes distinguished as **M. odorata** L., *Cent. Pl.* **1**: 7 (1755) (*M. dichotoma* L.), but are probably best regarded as a variety of **1. M. longiflora** L., *Kungl. Svenska Vet.-Akad. Handl.* **1755**: 176 (1755), from Mexico, like **1** but glandular-pubescent and viscid, with a very slender perianth-tube 60–100 mm and perianth-limb white to pink or lilac, is also cultivated for ornament and may occasionally be naturalized.

4. Oxybaphus L'Hér. ex Willd.

D. A. WEBB

Like *Mirabilis* but with involucre subtending several flowers and strongly accrescent, becoming membranous in fruit; stamens exserted, sometimes less than 5.

1. O. nyctagineus (Michx) Sweet, *Hort. Brit.* 334 (1826) (*Mirabilis nyctaginea* (Michx) MacMillan, *M. oxybaphus* Bordzil.). Stems erect, angular, dichotomously branched. Leaves 5–10 cm, broadly ovate, petiolate. Flowers in axillary and terminal cymes. Peduncles and pedicels hairy. Involucre up to 2 cm in diameter in fruit, very broadly conical, 5-lobed, membranous, with conspicuous veins, subtending 3–5 flowers. Perianth *c.* 12 × 8 mm, campanulate, red. Stamens 3–5, exserted. Anthocarp ribbed, hairy. $2n = 58$. *Established as a weed and ruderal in parts of C. & E. Europe.* [Cz Hu Rm Rs(C, W).] (*North America.*)

51. PHYTOLACCACEAE

EDIT. D. A. WEBB

Herbs, shrubs or trees with alternate, entire leaves. Flowers in racemes. Perianth-segments 5, free, persistent. Stamens hypogynous. Carpels united at least at the base (rarely free), each containing a single ovule.

1. Phytolacca L.

D. A. WEBB AND J. R. AKEROYD

Racemes usually opposite the leaves. Perianth-segments small, between sepaloid and petaloid in colour and texture. Stamens 10 or more; carpels 10 or fewer. Fruit a berry or cluster of drupelets.

Literature: H. Walter in Engler, *Pflanzenreich* **39 (IV. 83)**: 36–63 (1909).

1 Small tree; dioecious; stamens *c.* 25 — **3. dioica**
1 Herb, sometimes woody at the base; flowers hermaphrodite; stamens 8 or 10
 2 Stamens 10; carpels 10, united — **1. americana**
 2 Stamens 8; carpels 8, free — **2. esculenta**

1. P. americana L., *Sp. Pl.* 441 (1753) (*P. decandra* L.). Glabrous, perennial herb, somewhat woody at the base. Stems 1–3 m, subdichotomously branched, often red, with decurrent ridges from the leaf-bases. Leaves 12–25 × 5–10 cm, ovate-lanceolate, petiolate. Racemes *c.* 10 cm, more or less erect, drooping in fruit. Flowers hermaphrodite. Perianth-segments 2·5 mm, broadly ovate, greenish-white, turning reddish in fruit. Stamens 10. Carpels 10, united except for the styles. Fruit 10 mm in diameter, depressed-globose, purplish-black. $2n = 36$. *Waste places. Cultivated for ornament and for dye from the berries; widely naturalized in S. Europe, and locally in C. & W. Europe.* [Al Au Az Bl Bu Co Cr Cz Ga Ge Gr He Ho Hs Hu It Ju Lu Rm Rs(W, K) Sa Si Tu.] (*United States.*)

2. P. esculenta Van Houtte, *Fl. Serres Jard. Eur.* **4**: 398 (1848). Like **1** but not more than 2 m; leaves ovate-elliptical to suborbicular; racemes erect in fruit; stamens 8; carpels 8, free. *Cultivated for ornament and as a vegetable; naturalized in the Netherlands, and locally elsewhere.* [?Br Cz ?Da Ho Rm.] (*E. Asia.*)

3. P. dioica L., *Sp. Pl.* ed. 2, 632 (1762). Dioecious small tree with stout branches. Leaves 6–12 × 2·5–6 cm, evergreen, ovate to lanceolate, petiolate, glabrous. Racemes drooping. Perianth-segments 3·5 mm, oblong, green spotted with white. Stamens *c.* 25, exceeding the perianth. Carpels 7–10, united only in the lower part. Fruit subglobose, purplish-black. *Planted for ornament and shade in the Mediterranean region, and locally naturalized.* [Ga Gr Hs Si.] (*Temperate and subtropical South America.*)

52. AIZOACEAE

EDIT. D. A. WEBB

Herbs or small shrubs. Leaves usually opposite and fleshy. Calyx tubular or turbinate, often with fleshy lobes. Petals and stamens usually numerous. Ovary usually inferior or semi-inferior. Fruit woody and opening on moistening, or fleshy and indehiscent.

Most of the genera that occur in Europe originate from South Africa. They are cultivated for ornament, and many have become naturalized in milder, more or less frost-free, coastal regions of S. and W. Europe. The taxonomy and nomenclature of this family is confused, since many species and genera have been described from cultivated plants in European gardens without knowledge of the precise origin of the material. Furthermore, these succulent plants make poor herbarium specimens and type specimens have not always been preserved or have not survived. In addition to the genera described below, **Galenia secunda** Sonder in Harvey & Sonder, *Fl. Cap.* **2**: 474 (1862), from South Africa, a procumbent perennial with appressed-hairy stems, ovate-spathulate leaves and 5-merous, yellowish flowers, is becoming naturalized in S.W. Spain. **Ruschia caroli** (L. Bolus) Schwantes, *Zeitschr. Sukkulentenk.* **3**: 20 (1927), a perennial with linear, more or less triquetrous leaves, and flowers 1–5 together, 1–3 cm in diameter and with purple petals, is more or less established on walls in S.W. England (Isles of Scilly) and S.W. Portugal (Algarve).

Literature: G. Schwantes, *Flowering Stones and Mid-day Flowers*. London, 1957. H. Herre, *The Genera of the Mesembryanthemaceae*. Cape Town, 1971. C. D. Preston & P. D. Sell, *Watsonia* **17**: 217–245 (1989). J. R. Akeroyd & C. D. Preston, *Bot. Jour. Linn. Soc.* **103**: 197–200 (1990).

1 Petals absent; ovary superior — **1. Aizoon**
1 Petals present; ovary inferior or semi-inferior
 2 Stigmas 8–20; fruit fleshy; seeds embedded in mucilage — **2. Carpobrotus**
 2 Stigmas 4 or 5; fruit a woody capsule; seeds not embedded in mucilage
  3 At least the lower leaves petiolate
   4 Leaves cordate at base, finely papillose — **5. Aptenia**
   4 Leaves not cordate at base, covered with hyaline vesicles — **3. Mesembryanthemum**
  3 All leaves sessile
   5 Annual; flowers subsessile — **3. Mesembryanthemum**
   5 Perennial; flowers distinctly pedunculate
    6 Leaves in groups of 3 or more — **6. Disphyma**
    6 Leaves in pairs
     7 Stems hirsute or hispid; leaves papillose-tuberculate — **4. Drosanthemum**
     7 Stems and leaves glabrous
      8 Leaves obscurely 3-angled; petals longer than sepals — **7. Lampranthus**
      8 Leaves acutely 3-angled; petals shorter than sepals — **8. Erepsia**

1. Aizoon L.

†T. G. TUTIN

Annual. Leaves usually alternate. Flowers subsessile. Petals absent. Stamens *c.* 20, connate in 5 bundles. Ovary superior; stigmas 5. Capsule opening by 5 valves at the apex; seeds numerous.

1. A. hispanicum L., *Sp. Pl.* 488 (1753). Densely papillose annual, 5–15 cm. Leaves oblong-lanceolate, obtuse. Flowers solitary, usually in the dichotomies of the stem. Perianth-tube shortly obconical to hemispherical; lobes lanceolate, acute, yellowish above. *S. & E. Spain; Islas Baleares; S.W. Italy; Kriti.* Bl Cr Hs It.

Sesuvium portulacastrum L., *Syst. Nat.* ed. 10, **2**: 1058 (1759), from Tropical Africa and America, similar to **1** but with opposite, glabrous leaves and a circumscissile capsule, was formerly naturalized on maritime sands near Lisboa, but has not been seen recently.

2. Carpobrotus N. E. Br.

†T. G. TUTIN (EDITION 1) REVISED BY J. R. AKEROYD AND C. D. PRESTON (EDITION 2)

Procumbent, woody perennials forming large mats. Leaves opposite, 3-angled, fleshy. Flowers solitary, pedunculate. Petals and stamens numerous. Stigmas 8–20. Ovary inferior; fruit fleshy, indehiscent. Seeds embedded in mucilage.

Flowers usually more than 5 cm in diameter; petals purple or yellow at the base — **1. edulis**
Flowers usually 3–5 cm in diameter; petals white at the base — **2. glaucescens**

1. C. edulis (L.) N. E. Br. in E. P. Phillips, *Gen. S. Afr. Fl. Pl.* 249 (1926) (*Mesembryanthemum edule* L.). Stems up to several metres. Leaves not glaucous, not broadening above the middle, tapering gradually to the acute apex; dorsal angle serrulate. Flowers 4–9 cm in diameter; petals yellow, pinkish-purple (var. *rubescens* Druce, *C. acinaciformis* auct. eur., non (L.) L. Bolus, *Mesembryanthemum acinaciforme* auct. eur., non L.) or purple with a yellow base (var. *chrysophthalmus* C. D. Preston & P. D. Sell); anthers yellow, the filaments yellow or brownish. $2n = 18$. *Cultivated for ornament, and naturalized on coastal cliffs and sands in S. & W. Europe, northwards to N. Wales.* [Al Az Bl Br Co Ga Gr Hb Hs It Lu Si.] (*South Africa.*)

2. C. glaucescens (Haw.) Schwantes, *Gartenfl.* **77**: 69 (1928). Like **1** but smaller, somewhat glaucous; flowers 3–5(–6) cm in diameter; petals purple with white base; anthers yellow; filaments white. $2n = 18$. *Cultivated for ornament, and naturalized on coastal cliffs and rocks in the Channel Islands and locally in Britain.* [Br Ga.] (*Australia.*)

C. aequilaterus (Haw.) N. E. Br., *Jour. Bot.* (*London*) **66**: 324 (1928) (*C. chilensis* (Molina) N. E. Br., *Mesembryanthemum aequilaterus* Haw., *M. chilense* Molina), from South America, is like **2** but more slender with more acute leaves and petals without a white base. It is reported to be naturalized in Spain and Islas Baleares.

3. Mesembryanthemum L.

†T. G. TUTIN

Herbs. Leaves fleshy, flat or subterete, alternate or opposite. Flowers solitary or in cymes. Petals and stamens numerous. Stigmas 5. Ovary inferior; capsule with 5 valves, winged on the angles. (*Cryophytum* N. E. Br., *Gasoul* Adanson.)

Leaves narrowly oblong, subterete — **1. nodiflorum**
Leaves spathulate to broadly ovate, flat — **2. crystallinum**

1. M. nodiflorum L., *Sp. Pl.* 480 (1753) (*Gasoul nodiflorum* (L.) Rothm.). More or less procumbent annual, 3–20 cm. Leaves narrowly oblong, subterete, obtuse, glaucous and slightly crystalline-papillose. Flowers terminal and axillary, solitary, subsessile. Petals shorter than the sepals, yellowish or white. *Coastal sands and salt-marshes. Mediterranean region.* Bl Co Cr Ga Gr Hs It Ju Sa Si [Az Lu].

2. M. crystallinum L., *Sp. Pl.* 480 (1753) (*Cryophytum crystallinum* (L.) N. E. Br.). Like **1** but the whole plant densely crystalline-papillose; leaves spathulate to broadly ovate, flat; petals longer than the sepals. $2n = 18$. *Coastal sands and salt-marshes. W. & C. Mediterranean region.* Bl Co *Cr Ga *Gr Hs It Ju Sa Si [Az Lu].

4. Drosanthemum Schwantes

J. R. AKEROYD AND C. D. PRESTON

Like *Mesembryanthemum* but sprawling shrubs; leaves cylindrical or narrowly obovoid; flowers solitary or 2 or 3 together.

Flowers usually less than 2·5 cm in diameter; petals whitish or pink **1. floribundum**
Flowers 2·5–3 cm in diameter; petals purple **2. hispidum**

1. D. floribundum (Haw.) Schwantes, *Zeitschr. Sukkulentenk.* **3**: 29 (1927) (*D. candens* (Haw.) Schwantes). Stems procumbent or trailing, much-branched, hirsute, not rooting freely at the nodes. Leaves 0·5–2·5(–3) cm, almost cylindrical, densely tuberculate-papillose. Flowers solitary, 1–2·5 cm in diameter; petals whitish or pink. $2n = 36$. *Cultivated for ornament, and naturalized on coastal rocks and walls in W. Europe.* [Az Br Ga ?Hs Lu.] (*South Africa.*)

2. D. hispidum (L.) Schwantes, *loc. cit.* (1927) (*Mesembryanthemum hispidum* L.). Stems erect or trailing, hispid, the lower branches sometimes rooting at the nodes. Leaves 1·2–3 cm, tuberculate-papillose, often reddish. Flowers solitary or 2 or 3 together, 2·5–3 cm in diameter; petals purple. *Cultivated for ornament, and locally naturalized on coasts of W. Mediterranean region.* [Bl Hs It.] (*South Africa.*)

5. Aptenia N. E. Br.

†T. G. TUTIN

Like *Mesembryanthemum* but with 4 stigmas and a 4-locular capsule; valves without wings.

1. A. cordifolia (L. fil.) Schwantes, *Gartenfl.* **77**: 69 (1928). Procumbent perennial; stems 20–60 cm, much-branched. Leaves opposite, ovate-cordate, finely papillose, flat, not glaucous. Flowers solitary, axillary and terminal, pedunculate; petals purple. $2n = 18$. *Naturalized in open ground on coasts of S. & W. Europe.* [Az Bl Br Cr Ga Hs It Lu Si.] (*South Africa.*)

6. Disphyma N. E. Br.

D. A. WEBB (EDITION 1) REVISED BY J. R. AKEROYD (EDITION 2)

Like *Mesembryanthemum* but sprawling shrubs; glabrous, not papillose; wings on capsule widely separated at the base; apical tubercles of capsule bifid.

1. D. crassifolium (L.) L. Bolus, *Fl. Pl. S. Afr.* **7**: t. 276 (1927) (*Mesembryanthemum crassifolium* L.). Perennial; stems 20–100 cm, procumbent or trailing, rooting freely at the nodes. Leaves in groups of 3–10, 15–40 × 3–10 mm, oblanceolate, flat above, rounded or somewhat keeled beneath, with rounded apex, yellowish-green to reddish. Flowers solitary, axillary, 3–5 cm in diameter; petals purple. $2n = 36$. *Cultivated for ornament, and naturalized on coastal cliffs and sands in W. Europe, northwards to N. Wales.* [Az Br Ga Lu.] (*South Africa.*)

7. Lampranthus N. E. Br.

D. A. WEBB (EDITION 1) REVISED BY J. R. AKEROYD (EDITION 2)

Like *Mesembryanthemum* but shrubs; glabrous, not papillose; leaves opposite, slightly connate, more or less 3-angled; flowers solitary or 2 or 3 together, terminal, the petals longer than the sepals; stigmas large, conspicuous, surrounded by a raised nectary-ring; ovary semi-inferior. (*Oscularia* Schwantes.)

1 Leaves less than twice as long as wide; peduncles 2–3 mm **1. deltoides**
1 Leaves 4–8 times as long as wide; peduncles 5–20 mm
 2 Stems mostly erect; leaves dark bluish-green **2. multiradiatus**
 2 Stems mostly procumbent; leaves distinctly pruinose **3. falciformis**

1. L. deltoides (L.) Glen, *Bothalia* **16**: 55 (1986) (*Mesembryanthemum deltoides* L., *Oscularia deltoides* (L.) Schwantes). Stems up to 50 cm, decumbent or trailing. Leaves 5–18 × 3–10 mm, triangular-obovoid, broadest above the middle, denticulate on the angles, pruinose. Flowers 1–3 together, 1–2 cm in diameter; petals purple. $2n = 18$. *Cultivated for ornament, and naturalized on coastal rocks and walls in S.W. England (Isles of Scilly).* [Br.] (*South Africa.*)

2. L. multiradiatus (Jacq.) N. E. Br., *Gard. Chron.* ser. 3, **87**: 71 (1930) (*L. roseus* (Willd.) Schwantes, *L. glaucus* auct., non (L.) N. E. Br.). Stems up to 20 cm, decumbent, ascending or erect. Leaves 10–25 × 2–4 mm, linear-oblong, almost triangular in section, dark bluish-green. Flowers 1–3 together, 2–5 cm in diameter; petals pink to purplish-pink. $2n = 36$. *Cultivated for ornament, and naturalized on coastal cliffs in the Channel Islands and S.W. Britain.* [Br Ga.] (*South Africa.*)

3. L. falciformis (Haw.) N. E. Br., *Gard. Chron.* ser. 3, **87**: 212 (1930). Stems up to 80 cm, mostly procumbent but some erect. Leaves 6–8 × 1·5–4 mm, linear-oblong to falcate, generally more slender than those of **2**, flattened above, obscurely keeled beneath, pale bluish-green, pruinose. Flowers 1 or 2 together, 3·5–4·5 cm in diameter; petals pink or purple. *Cultivated for ornament, and naturalized on coastal cliffs and sands in W. Europe.* [Az Bl Br Ga Hb Lu.] (*South Africa.*)

2 and **3** are very similar and are perhaps not distinct.

8. Erepsia N. E. Br.

C. D. PRESTON

Like *Mesembryanthemum* but shrubs; glabrous, not papillose; leaves opposite, slightly connate, acutely 3-angled; flowers solitary, or 2 or 3 together; stigmas more or less globular, very small; nectary glands minute, inconspicuous.

1. E. heteropetala (Haw.) Schwantes, *Gartenfl.* **77**: 68 (1928). Stems up to 35 cm, erect or decumbent. Leaves 16–30 × 6–13 mm, narrowly obovoid, irregularly denticulate on the angles, pruinose. Flowers terminal, 1–1·5 cm in diameter; petals purple, shorter than the sepals; inner petals erect or inflexed; staminodes numerous. $2n = 18$. *Naturalized on coastal cliffs in S.W. England (Isles of Scilly).* [Br.] (*South Africa.*)

53. MOLLUGINACEAE

EDIT. †T. G. TUTIN AND S. M. WALTERS

Annual herbs. Leaves not fleshy. Perianth of 5 free sepals. Petals absent. Petaloid staminodes 0–20. Stamens 3–20. Ovary superior. Fruit a membranous, loculicidal capsule, opening to the base by 3–5 valves on drying.

Petaloid staminodes absent; plant ± glabrous **1. Mollugo**
Petaloid staminodes present; plant stellate-tomentose **2. Glinus**

1. Mollugo L.

†T. G. TUTIN

Flowers in axillary cymes or fascicles. Petaloid staminodes absent. Stamens 5. Stigmas 3. Capsule opening by 3 valves; seeds without a strophiole.

Flowers in axillary fascicles; leaves obovate-lanceolate to spathulate **1. verticillata**
Flowers in axillary, pedunculate, umbellate cymes; leaves linear **2. cerviana**

1. M. verticillata L., *Sp. Pl.* 89 (1753). Procumbent or ascending, branched, glabrous annual 5–10(–40) cm. Leaves in whorls of 3–5, obovate-lanceolate to spathulate. Flowers small, pedicellate, in axillary fascicles. Perianth-segments acute, with scarious margins. *Naturalized locally in S. Europe.* [Az Hs It Lu.] (*Tropical America.*)

2. M. cerviana (L.) Ser. in DC., *Prodr.* **1**: 392 (1824). Like **1** but rather taller; leaves in whorls of 3–10, linear; flowers in pedunculate, umbellate cymes; perianth-segments obtuse. *Sandy and gravelly places. S. Europe, extending northwards to* c. 52° *N. in C. Russia.* Al Bu Gr Hs It Lu Rm Rs(C, W, K, E) Tu.

2. Glinus L.

†T. G. TUTIN

Flowers in subsessile, axillary fascicles. Petaloid staminodes numerous. Stamens 12. Stigmas 5. Capsule opening by 5 valves; seeds with a large strophiole.

1. G. lotoides L., *Sp. Pl.* 463 (1753) (*Mollugo lotoides* (L.) Arcangeli). Softly stellate-tomentose, procumbent annual, 10–40 cm. Leaves obovate or oblong-spathulate, often fascicled. Sepals yellow above; staminodes white. *S. Europe.* Al Bu Cr Gr Hs It Ju Lu Rm Sa Si.

54. TETRAGONIACEAE

EDIT. †T. G. TUTIN AND S. M. WALTERS

Perianth of 4 or 5 sepaloid segments. Stamens 3–15, inserted in the perianth-tube. Ovary inferior; stigmas 3–8. Fruit an indehiscent drupe.

1. Tetragonia L.

†T. G. TUTIN

Leaves ovate-rhombic. Flowers subsessile, solitary. Stamens usually *c.* 15.

1. T. tetragonoides (Pallas) O. Kuntze, *Revis. Gen.* **1**: 264 (1891) (*T. expansa* Murray). Densely papillose, somewhat fleshy, procumbent or ascending annual. Leaves often more or less hastate. Sepals yellow above. Fruit with large tubercles near the top. $2n = 32$. *Cultivated as a vegetable (New Zealand spinach) in S.W. Europe; occurring as a casual, and sometimes becoming naturalized.* [Az Bl Hs It Lu.] (*Australia, New Zealand.*)

55. PORTULACACEAE

EDIT. S. M. WALTERS

Annual or perennial herbs, usually glabrous and fleshy. Leaves simple, entire. Flowers solitary or in cymes, hermaphrodite. Sepals 2; petals 4–6, free or joined below; stamens 3 to many. Ovary unilocular, with 1 to many campylotropous ovules on a basal placenta. Fruit a capsule.

Several species from America are grown in gardens. Species of *Calandrinia* Humb., Bonpl. & Kunth in particular are occasionally recorded as naturalized, but none seem to be at all widely established. **Talinum paniculatum** (Jacq.) Gaertner, *Fruct. Sem. Pl.* **2**: 219 (1791), may be becoming naturalized in Açores.

Literature: J. McNeill, *Canad. Jour. Bot.* **53**: 789–809 (1975).

1 Stamens numerous; ovary semi-inferior **1. Portulaca**
1 Stamens 3 or 5; ovary superior
2 Plant usually procumbent; leaves all cauline **2. Montia**
2 Plant erect; leaves mostly in basal rosette **3. Claytonia**

1. Portulaca L.

S. M. WALTERS

Glabrous, fleshy herbs with alternate or opposite leaves and small, setaceous stipules. Flowers often terminal. Petals fugacious; stamens numerous; ovary semi-inferior. Capsule unilocular, with thin wall, dehiscing with a transverse lid. Seeds numerous, reniform.

Literature: K. von Poellnitz, *Feddes Repert.* **37**: 240–320 (1934); A. Danin, I. Baker & H. G. Baker, *Israel Jour. Bot.* **27**: 177–211 (1978).

1. P. oleracea L., *Sp. Pl.* 445 (1753). Annual with branched stems 10–50 cm. Leaves mostly scattered and alternate, but subopposite and crowded below flowers, oblong-obovate, sessile with a cuneate base, shining. Flowers solitary or 2 or 3 together, often terminal. Sepals *c.* 4 mm, keeled, united into a short tube at the base; petals 5, 6–8 mm, yellow, obovate, slightly united; stamens 7–12. Capsule 3–9 mm, obovoid; seeds black. *Cultivated as a vegetable* (*Purslane*), *and a weed of cultivation. Europe, northwards to* c. 55° *N. Not native, except perhaps in parts of the south; casual only in N. Europe.* *Al *Au *Az *Bl *Bu *Co *Cr *Cz *Ga *Ge *Gr *He *Hs *Hu *It *Ju *Lu *Rm *Rs(C, W, K, E) *Sa *Si *Tu [Be Br Ho Po].

(**a**) Subsp. **oleracea** (*P. oleracea* subsp. *silvestris* (Montandon) Thell.): Procumbent or decumbent. Seeds *c.* 0·5 mm. $2n = 54$. *Widespread weed of cultivation in S. and C. Europe, rare and casual in N. Europe; native range unknown but probably Eurasia.*

(**b**) Subsp. **sativa** (Haw.) Čelak., *Prodr. Fl. Böhm.* 484 (1875) (*P. sativa* Haw.). Robust, erect, and larger in all its parts than subsp. (**a**). Seeds *c.* 1 mm. $2n = 54$. *Cultivated in S. and S.E. Europe, and occurring as an escape.*

A number of other subspecies have been recognized, mainly on the basis of the ornamentation of the testa, by A. Danin *et al.*, *loc. cit.*, but few data are available on their distribution.

P. grandiflora Hooker, *Bot. Mag.* **56**: t. 2885 (1829), with 4–5 cm reddish or orange flowers, usually *flore pleno*, is a native of South America commonly grown in gardens. It is occasionally established on roadsides and in waste places in S. and S.C. Europe.

2. Montia L.

S. M. WALTERS

Small, glabrous, somewhat fleshy herbs with opposite, exstipulate cauline leaves. Flowers in terminal cymes. Stamens 3. Ovary superior; ovules 3. Fruit a globose capsule dehiscing with 3 valves. Seeds broadly reniform, black.

1. M. fontana L., *Sp. Pl.* 87 (1753) (*M. minor* C. C. Gmelin, *M. rivularis* C. C. Gmelin, *M. verna* Necker). Annual or perennial, with weak, branching stems up to 50 cm or more in water, but short and erect, more or less caespitose on land. Leaves 3–30 mm, usually narrowly spathulate. Flowers *c.* 2 mm, very inconspicuous, in small, terminal cymes often overtopped by non-flowering branches. Capsule 1·5–2 mm; seeds usually 3. *In water, or on mud or seasonally wet open ground, usually calcifuge. Much of Europe, but rare in the east and parts of the south.* All except Al Az Bl Rs(W, K, E) Sb.

A very variable species, in which the seed characters show correlation with geographical distribution and, less satisfactorily, with habit and habitat differences. Four subspecies can be recognized:

1 Ripe seeds shining, smooth (**a**) subsp. **fontana**
1 Ripe seeds dull or somewhat shining, with at least some low tubercles on keel
2 Ripe seeds dull, entirely covered with broad, obtuse tubercles (**d**) subsp. **chondrosperma**
2 Ripe seeds somewhat shining, tubercles confined to keel
3 Ripe seeds with 3 or 4 rows of long, acute tubercles on keel (**c**) subsp. **amporitana**
3 Ripe seeds with variably developed, broad, low tubercles on keel (**b**) subsp. **variabilis**

(**a**) Subsp. **fontana** (*M. lamprosperma* Cham., *M. rivularis* auct., ?C. C. Gmelin): Annual or perennial, variable in habit. Ripe seeds 1·1–1·35 mm, shining, smooth. $2n = 18, 20$. *N. & C. Europe; occasional in the mountains of S. Europe.*

(**b**) Subsp. **variabilis** Walters, *Watsonia* **3**: 5 (1953) (*M. rivularis* auct., ?C. C. Gmelin): Annual or perennial, variable in habit. Ripe seeds 0·9–1·1 mm, with variously developed, broad, low tubercles on keel. *N., W. & C. Europe.*

(**c**) Subsp. **amporitana** Sennen, *Bull. Géogr. Bot.* (*Le Mans*) **21**: 110 (1911) (*M. lusitanica* Samp., *M. rivularis* auct., ?C. C. Gmelin, *M. limosa* Decker): Annual or perennial, variable in habit. Ripe seeds 0·85–1·1 mm, with 3 or 4 rows of long, acute tubercles on keel. *Europe eastwards to Romania and Greece, but mainly in the west.*

(**d**) Subsp. **chondrosperma** (Fenzl) Walters, *Watsonia* **3**: 4 (1953) (*M. verna* Necker, *M. minor* C. C. Gmelin): Usually annual, with erect, caespitose stems. Ripe seeds 1·0–1·2 mm, dull, entirely covered with broad, obtuse tubercles. $2n = 18, 20$. *In drier places than subspp.* (**a**), (**b**) *and* (**c**), *often on sandy ground or short, mossy turf. Europe northwards to Scotland and S. Sweden; rare in the south.*

3. Claytonia L.

S. M. WALTERS AND J. R. AKEROYD

Like *Montia* but leaves mostly basal; cauline leaves 2, subtending the inflorescence; stamens 5; ovules 3–6.

Cauline leaves connate; petals 2–3 mm **1. perfoliata**
Cauline leaves not connate; petals 8–10 mm **2. sibirica**

1. C. perfoliata Donn ex Willd., *Sp. Pl.* **1(2)**: 1186 (1798) (*Montia perfoliata* (Donn ex Willd.) Howell). Annual, 5–30 cm. Basal leaves in a rosette, elliptical to rhombic-ovate, fleshy; cauline leaves broadly connate. Inflorescence usually few-flowered. Petals 2–3 mm, white, entire to emarginate, slightly longer than the sepals. Capsule shorter than the persistent sepals. $2n = 36$. *On rather dry, sandy soils, sometimes locally abundant. Naturalized mainly in N.W. Europe.* [Be Br Cz Da Ga Ge Ho Lu Su.] (*W. North America.*)

2. C. sibirica L., *Sp. Pl.* 204 (1753) (*C. alsinoides* Sims, *Montia sibirica* (L.) Howell). Like **1** but basal leaves ovate, acuminate; cauline leaves not connate; petals 8–10 mm, pink or more rarely white, bifid. *Damp, shady places on acid, sandy soils. Extensively naturalized in N.W. Europe, and occasionally elsewhere.* [Br Bu Cz Fa Fe Ga Ho No.] (*W. North America.*)

56. BASELLACEAE

EDIT. S. M. WALTERS

Glabrous, perennial, twining herbs, usually fleshy. Leaves alternate, usually ovate with a more or less cordate base, entire. Flowers in spikes, racemes or racemose panicles, hermaphrodite. Sepals 2; petals 5; stamens 5, opposite petals. Ovary unilocular, with 3 styles and a single basal ovule. Fruit indehiscent, fleshy, surrounded by persistent sepals and petals.

A small family, mainly in America, and not native to Europe.

Basella alba L., *Sp. Pl.* 390 (1753), a very fleshy perennial with flowers in spikes, probably a native of tropical Asia, is widely cultivated as a vegetable in S. Europe and occasionally escapes, but does not seem to be naturalized.

1. **Boussingaultia** Humb., Bonpl. & Kunth

S. M. WALTERS

Filaments curved outwards in bud; sepals not winged in fruit.

Literature: E. Ulbrich in Engler & Prantl, *Natürl. Pflanzenfam.* ed. 2, **16c**: 263–271 (1934).

1. B. cordifolia Ten., *Ann. Sci. Nat.* ser. 3, **19**: 355 (1853) (*B. baselloides* auct., non Humb., Bonpl. & Kunth). Tall, slender, twining plant with numerous potato-like tubers on the rhizomes. Leaves 2–7 cm. Inflorescence 5–15 cm, more or less compound, axillary; flowers *c.* 2 mm, whitish. *Cultivated for ornament and as a vegetable, and naturalized on roadsides and in waste places in S. Europe.* [Az Ga Hs Ju Lu Si.] (*Subtropical South America.*)

57. CARYOPHYLLACEAE

EDIT. S. M. WALTERS

Herbs, more rarely small shrubs. Leaves usually opposite and decussate, more rarely alternate or verticillate, simple, entire, with or without scarious stipules. Flowers actinomorphic, usually hermaphrodite, often in bracteate dichasia. Sepals 4 or 5, free, or fused and often united by scarious strips of tissue (commissures) alternating with the calyx-teeth. Petals (0–)4 or 5, free. Stamens usually 8–10, obdiplostemonous. Ovary superior, unilocular at least above, with 1 to numerous campylotropous ovules on a basal or free-central placenta; stigmas (1)2–5. Fruit usually a capsule, dehiscing with teeth equalling the styles in number or twice as many; more rarely fruit a berry or achene.

The subdivision of the family, and also in most cases the generic circumscriptions, follow F. Pax & K. Hoffmann in Engler & Prantl, *Natürl. Pflanzenfam.* ed. 2, **16c**: 275–364 (1934).

1 Stipules present, though sometimes caducous (Subfam. Paronychioideae)
2 All leaves alternate
3 Sepals less than 3 mm; fruit an achene **14. Corrigiola**
3 Sepals 3–7 mm; fruit a capsule **24. Telephium**
2 Leaves opposite or verticillate (or some leaves apparently alternate)
4 Fruit with persistent inflated peduncle, and crowned by spiny processes **18. Pteranthus**
4 Fruit not as above
5 Stigmas 1 or 2
6 Bracts conspicuous, longer than flowers **15. Paronychia**
6 Bracts inconspicuous, shorter than flowers
7 Leaves aristate; plant erect **15. Paronychia**
7 Leaves not aristate; plant procumbent or erect
8 Sepals not white and spongy **16. Herniaria**
8 Sepals conspicuously white and spongy **17. Illecebrum**
5 Stigmas 3 or 5
9 Sepals with a single, large tooth on each side **21. Loeflingia**
9 Sepals entire
10 Leaves obovate or orbicular **19. Polycarpon**
10 Leaves subulate, linear or linear-lanceolate
11 Petals 0; style 1 **20. Ortegia**
11 Petals 5; styles 3 or 5
12 Styles 5; stipules not connate **22. Spergula**
12 Styles 3; stipules connate **23. Spergularia**
1 Stipules absent
13 Sepals free, or joined only at base (Subfam. Alsinoideae)
14 Styles 2
15 Petals absent; fruit indehiscent **13. Scleranthus**
15 Petals present; fruit a dehiscent capsule
16 Capsule dehiscing with 4 equal teeth; seeds with strophiole **2. Moehringia**
16 Capsule dehiscing with 2 entire or bifid teeth; seeds without strophiole
17 Basal leaves linear-spathulate; capsule with bifid teeth **1. Arenaria**
17 Basal leaves setaceous; capsule with entire teeth **5. Bufonia**
14 Styles 3–5
18 Capsule-teeth as many as styles
19 Fleshy maritime plant; seeds 3–4·5 mm **4. Honkenya**
19 Plant not fleshy; seeds less than 3 mm
20 Styles fewer than sepals **3. Minuartia**
20 Styles as many as sepals
21 Capsule-teeth shallowly bifid; leaves ovate **11. Myosoton**
21 Capsule-teeth entire; leaves subulate **12. Sagina**
18 Capsule-teeth twice as many as styles
22 Petals absent
23 Styles 3 **6. Stellaria**
23 Styles 5 **9. Cerastium**
22 Petals present
24 Petals bifid to at least half-way **6. Stellaria**
24 Petals bifid to less than half-way or entire
25 Plant with napiform tubers on rhizome **7. Pseudostellaria**
25 Plant without napiform tubers
26 Petals irregularly toothed **8. Holosteum**
26 Petals entire, emarginate or bifid
27 Styles 4 or 5
28 Annuals or perennials, usually hairy; petals (4)5, distinctly bifid **9. Cerastium**
28 Glabrous, glaucous annuals; petals 4(5), entire or emarginate **10. Moenchia**
27 Styles 3 on at least some flowers
29 Petals bifid to *c.* $\frac{1}{3}$ **9. Cerastium**
29 Petals emarginate (to less than $\frac{1}{4}$) or entire
30 Seeds without strophiole **1. Arenaria**
30 Seeds with persistent strophiole **2. Moehringia**
13 Sepals joined to form a distinct calyx-tube (Subfam. Silenoideae)
31 Calyx-tube with commissural veins alternating with the mid-veins of the sepals; styles 3–5
32 Fruit a berry **29. Cucubalus**
32 Fruit a capsule
33 Capsule dehiscing with a lid **30. Drypis**
33 Capsule dehiscing with teeth

34 Capsule-teeth twice as many as styles, or teeth bifid **28. Silene**
34 Capsule-teeth as many as styles, entire
35 Annual; calyx-teeth foliaceous **26. Agrostemma**
35 Perennial; calyx-teeth not foliaceous
36 Seeds glabrous; petals contorted in bud **25. Lychnis**
36 Seeds with a hair-tuft at hilum; petals imbricate in bud **27. Petrocoptis**
31 Calyx-tube without commissural veins; styles 2 or 3
37 Calyx-tube with 5 wings **34. Vaccaria**
37 Calyx-tube without wings
38 Calyx-tube with scarious commissures
39 Seeds scutate, with facial hilum **35. Petrorhagia**
39 Seeds reniform or comma-shaped, with lateral hilum
40 Calyx campanulate; petal-limb gradually narrowed into claw **31. Gypsophila**
40 Calyx tubular; petal-limb abruptly contracted into claw **32. Bolanthus**
38 Calyx-tube without scarious commissures
41 Epicalyx present **36. Dianthus**
41 Epicalyx absent
42 Seeds reniform, with lateral hilum **33. Saponaria**
42 Seeds scutate, with facial hilum **37. Velezia**

Subfam. **Alsinoideae**

Leaves opposite; stipules absent. Petals usually well-developed; sepals free or joined only at the base.

1. **Arenaria** L.

A. O. CHATER AND G. HALLIDAY

Annual, biennial or perennial herbs. Leaves opposite, entire, very variable in shape from orbicular to subulate. Flowers usually in few-flowered cymes, sometimes solitary, (4)5-merous. Sepals free; petals usually entire, white (rarely pink or purple); stamens (8–)10; styles (2)3(–5). Capsule conical to cylindrical, dehiscing with twice as many teeth as styles; teeth narrow, acute. Seeds several, reniform (rarely globose), usually dark brown, smooth or tuberculate. (Incl. *Gouffeia* Robill. & Cast. ex DC.)

Literature: J. McNeill, *Notes Roy. Bot. Gard. Edinb.* **24**: 102–129 (1962), 245–302 (1963).

1 Perennial, with leafy vegetative stems
2 Leaves usually more than 30 mm, linear, erect
3 Sepals acute or acuminate
4 Pedicels usually longer than sepals; petals 2–3 times as long as sepals **3. gypsophiloides**
4 Pedicels usually equalling or shorter than sepals; petals equalling or slightly exceeding sepals
5 Basal leaves 20–120 × 0·5–0·75 mm; inflorescence ± elongate; bracts 5–7 mm **1. rigida**
5 Basal leaves 100–400 × 1–1·3 mm; inflorescence a compact, hemispherical head; bracts *c.* 15 mm **2. cephalotes**
3 Sepals obtuse
6 Sepals 2·5–5 mm, not keeled **4. procera**
6 Sepals 2–3 mm, with a prominent keel **5. longifolia**
2 Leaves usually less than 30 mm, usually patent or recurved
7 Leaves 15–30 mm, linear-subulate, very stiff and spiny **13. pungens**
7 Leaves not stiff or spiny
8 Flowers sessile or subsessile
9 Leaves thick or subcoriaceous, with thickened margins
10 Leaves acute and mucronate
11 Inflorescence with not more than 10 flowers **11. aggregata**
11 Inflorescence usually with at least 15 flowers
12 Leaves of sterile and flowering stems ± reflexed **11. aggregata**
12 Leaves of sterile stems not reflexed **12. querioides**
10 Leaves obtuse
13 Flowers not more than half-exserted from the very hard, compact cushion
14 Sepals *c.* 2·5 mm **7. alfacarensis**
14 Sepals 4–6 mm **11. aggregata**
13 Flowers completely exserted from the foliage, or clustered on long peduncles
15 Plant ± densely covered with very short, white hairs, forming compact, silvery cushions 5–10 cm in diameter; sepals keeled for *c.* 1 mm at base, otherwise very obscurely veined **9. tomentosa**
15 Plant not silvery; sepals with strong, raised veins
16 Flowers solitary and terminal **8. tetraquetra**
16 Flowers (2–)4–12 in dense terminal clusters
17 Leaves distinctly obtuse; capsule usually longer than sepals **10. armerina**
17 Leaves subobtuse; capsule usually shorter than sepals **11. aggregata**
9 Leaves thin, without raised margins
18 Petals slightly exceeding sepals; leaves glabrous **21. humifusa**
18 Petals distinctly shorter than sepals; leaves scabrid-puberulent **36. serpyllifolia**
8 Flowers pedicellate; pedicels usually equalling or longer than sepals
19 Stems procumbent, rooting at nodes **20. biflora**
19 Stems not rooting at nodes
20 Sepals glabrous on surface, though often ciliate
21 Plant with slender, yellowish stolons; stems very slender **21. humifusa**
21 Plant without stolons
22 Capsule very long-exserted from calyx, cylindrical, shining **6. purpurascens**
22 Capsule slightly exserted from calyx, ovoid to globose
23 Leaves with revolute margins; stems 5–20 cm, straggling **14. valentina**
23 Leaves flat; stems not more than 8 cm, ± compact
24 Leaves finely denticulate **26. gracilis**
24 Leaves entire
25 Leaves not ciliate (Greece) **25. gionae**
25 Leaves ± ciliate at base
26 Leaves somewhat fleshy, obscurely 1-veined; not more than the lower $\frac{1}{3}$ of margin ciliate **22. norvegica**
26 Leaves not fleshy, distinctly 1-veined; the lower $\frac{1}{4}$–$\frac{2}{3}$ of margin ciliate **23. ciliata**
20 Sepals hairy on surface
27 Leaves linear-lanceolate, with prominent keel-like midrib beneath and a long arista **15. grandiflora**

27 Leaves without keel-like midrib or arista
28 Plant with ±filiform, procumbent stems forming dense mats; leaves broadly ovate or orbicular; sepals not more than 3 mm **19. balearica**
28 Plant with stout stems, not forming dense mats; leaves narrow; sepals usually more than 3 mm
29 Leaves linear-subulate (or rarely linear-lanceolate), not more than 1 mm wide (S. France) **34. hispida**
29 Leaves usually more than 1 mm wide
30 Sepals usually more than 5 mm
31 Petals less than $1\frac{1}{2}$ times as long as sepals **30. ligericina**
31 Petals more than $1\frac{1}{2}$ times as long as sepals
32 Leaves ovate **16. bertolonii**
32 Leaves oblanceolate or oblong to linear
33 Caespitose; leaves usually not more than 10 mm long, widest above middle **17. huteri**
33 Not caespitose; leaves usually more than 10 mm long, widest at or below middle (rarely a few in middle of stem widest above middle) **18. montana**
30 Sepals not more than 5 mm
34 Sepals eglandular
35 Pedicels not more than 3 times as long as sepals **23. ciliata**
35 Pedicels more than 3 times as long as sepals **33. cinerea**
34 Sepals glandular-pubescent
36 Petals about twice as long as sepals
37 Sepals acute to acuminate; capsule slightly shorter than sepals **27. filicaulis**
37 Sepals obtuse or subacute; capsule slightly exceeding sepals **31. cretica**
36 Petals not more than $1\frac{1}{2}$ times as long as sepals
38 Dwarf caespitose plant, less than 5 cm; pedicels 1–2 times as long as sepals; flowers solitary or paired **32. halacsyi**
38 Large plant, more than 5 cm; pedicels 2–5(–10) times as long as sepals; inflorescence with more than 2 flowers
39 Leaves broadly elliptical to obovate; petals very slightly exceeding sepals
40 Plant pubescent; stems, branches and pedicels slender and long **28. orbicularis**
40 Plant puberulent; stems, branches and pedicels stout and short **29. fragillima**
39 Leaves elliptical, lanceolate or oblanceolate; petals $1\frac{1}{3}$–$1\frac{1}{2}$ times as long as sepals
41 Seeds with acute tubercles; leaves with prominent midrib **27. filicaulis**
41 Seeds with low, obtuse tubercles or ridges; leaves without prominent midrib **30. ligericina**
1 Annual or biennial, lacking leafy vegetative stems at time of flowering (see also **33**)
42 Lower leaves lanceolate or wider, often with 3 or more veins
43 Petals equalling or exceeding sepals
44 Pedicels ±glabrous **37. conferta**
44 Pedicels pubescent or puberulent
45 Seeds globose and smooth; sepals often purplish-black at apex **43. hispanica**
45 Seeds reniform and tuberculate; sepals usually not purplish-black at apex
46 Seeds with acute tubercles
47 Sepals ovate-oblong, obtuse **47. conimbricensis**
47 Sepals lanceolate or ovate-lanceolate, acute or acuminate
48 Petals $1\frac{1}{2}$–2 times as long as sepals **52. retusa**
48 Petals scarcely exceeding or shorter than sepals
49 Sepals acute; seeds 0·4–0·6 mm **39. muralis**
49 Sepals acuminate; seeds 0·6–1 mm **40. graveolens**
46 Seeds with low, obtuse tubercles
50 Sepals dimorphic; petals equalling or exceeding inner sepals, shorter than the outer **44. pomelii**
50 Sepals not dimorphic; petals equalling or exceeding all sepals
51 Leaves ±densely pubescent **37. conferta**
51 Leaves not pubescent, though usually ciliate
52 Leaves more than 3 times as long as wide **22. norvegica**
52 Leaves less than 3 times as long as wide **24. gothica**
43 Petals shorter than sepals
53 Bracts forming an involucre around the inflorescence; stems less than 4 cm **41. saponarioides**
53 Bracts not forming an involucre; stems 3–20 cm
54 Sepals with patent or recurved apices **42. guicciardii**
54 Sepals straight
55 Sepals 4·5–7 mm
56 Lower leaves ovate, the upper linear-lanceolate; sepals lanceolate **38. nevadensis**
56 Leaves oblong to spathulate; sepals dimorphic, broadly ovate or oblong **44. pomelii**
55 Sepals 2–4·5 mm
57 Leaves obovate, with single vein; sepals obscurely veined **40. graveolens**
57 Leaves ovate to ovate-lanceolate, 3- to 5-veined; sepals with 3–5 veins
58 Petals not more than $\frac{2}{3}$ as long as sepals; seeds 0·3–0·7 mm **36. serpyllifolia**
58 Petals almost equalling sepals; seeds 0·7–1 mm **37. conferta**
42 All leaves linear-lanceolate or linear, usually 1-veined
59 Pedicels glabrous or subglabrous
60 Petals $1\frac{1}{2}$–2 times as long as sepals; sepals ovate, obtuse **48. obtusiflora**
60 Petals ±equalling sepals; sepals lanceolate, acute
61 Sepals 2·5 mm; capsule with 6 teeth **53. capillipes**
61 Sepals 3–4 mm; capsule with 2 bifid teeth **54. provincialis**
59 Pedicels pubescent or puberulent
62 Petals pink, truncate-emarginate, ±equalling sepals **45. emarginata**
62 Petals white

63 Petals emarginate, 2–3 times as long as sepals **46. algarbiensis**
63 Petals entire, not more than twice as long as sepals
64 Petals shorter than sepals
65 Inflorescence few- (often 3-)flowered; pedicels $1\frac{1}{2}$–2 times as long as sepals; sepals subacute **50. conica**
65 Inflorescence many-flowered; pedicels 2–5 times as long as sepals; sepals usually very finely acute **51. modesta**
64 Petals equalling or exceeding sepals
66 Sepals obtuse
67 Capsule included in, or scarcely exserted from sepals; pedicels with patent glandular hairs **47. conimbricensis**
67 Capsule $1\frac{1}{2}$–2 times as long as sepals; pedicels minutely scabrid-puberulent **48. obtusiflora**
66 Sepals acute
68 Plant eglandular; pedicels with minute deflexed hairs; leaves subtrigonous **49. controversa**
68 Plant glandular-pubescent; pedicels usually with patent glandular hairs; leaves thin, flat
69 Sepals usually very finely acute, very obscurely veined; petals ± equalling sepals **52. modesta**
69 Sepals acute, with 3(–5) raised veins; petals exceeding sepals
70 Petals obtuse; seeds rugose-granulate **35. fontqueri**
70 Petals truncate; seeds with acute tubercles **52. retusa**

Subgen. **Eremogone** (Fenzl) Fenzl. Perennials, woody at the base, with linear leaves more than 30 mm. Styles 3(–5).

1. A. rigida Bieb., *Fl. Taur.-Cauc.* **1**: 346 (1808). Glabrous perennial with woody, branched stock; stems 7–40 cm, erect, slender, rigid, simple. Basal leaves 30–120 × 0·5–0·75 mm, linear, acuminate; cauline shorter and wider, merging into lanceolate, acuminate bracts 5–7 mm. Inflorescence a several- to many-flowered, more or less strict, compact, oblong panicle; pedicels mostly equalling or shorter than sepals. Sepals 4–6 mm, lanceolate, acuminate, with very obtuse keel; petals equalling or slightly exceeding sepals. *Dry, stony or sandy ground.* ● *S. Ukraine and Black Sea coast.* Bu Rm Rs(W, E).

2. A. cephalotes Bieb., *loc. cit.* (1808). Like **1** but stems 20–50 cm, stout; basal leaves 100–400 × 1–1·3 mm; inflorescence a dense, terminal, hemispherical head; bracts *c.* 15 mm, coriaceous, conspicuous; pedicels shorter than sepals and flowers often subsessile; sepals 5·5–7 mm, lanceolate, acute. *Dry, open slopes.* ● *S.W. Ukraine and Moldavia.* Rs(W).

3. A. gypsophiloides L., *Mantissa* 71 (1767). Like **1** but stems stout; basal leaves *c.* 1 mm wide; inflorescence a lax, elongate panicle; pedicels 1–2 times as long as sepals; sepals 4–5 mm, ovate, acute or acuminate; petals 2–3 times as long as sepals. *S. & E. Bulgaria.* Bu. (*S.W. Asia.*)

The typical plant, with glandular-pubescent inflorescence, is recorded only from Aitos. Var. *rhodopaea* Velen., recorded from limestone rocks in two valleys in the Rodopi Planina and not collected since the turn of the century, has larger flowers and glabrous inflorescences.

4. A. procera Sprengel, *Hist. Rei Herb.* **2**: 153 (1808) (*A. biebersteinii* Schlecht.). Perennial with branched, woody stock; stems (10–)20–40 cm, erect, rigid, rather stout, simple, glabrous below, sometimes pubescent above. Basal leaves 30–120 × *c.* 1 mm, linear, acuminate; cauline leaves shorter and wider. Inflorescence usually a very lax, subcorymbose panicle, sometimes elongate or fasciculate; pedicels usually 2–4 times as long as sepals. Sepals 2·5–5 mm, ovate, very obtuse, not keeled; petals about twice as long as sepals. *C. & E. Europe.* Au Cz Hu Po Rm Rs(N, B, C, W, E).

(a) Subsp. **procera** (*A. procera* subsp. *glabra* (F. N. Williams) J. Holub, *A. graminifolia* Schrader, non Ard., *A. micradenia* Smirnov, *A. saxatilis* sensu Ikonn., non L.; incl. *A. koriniana* Fischer, *A. ucrainica* Sprengel ex Steudel, *A. stenophylla* Ledeb., *A. syreistschikowii* Smirnov, *A. polaris* Schischkin): Leaf-sheaths at base of stem *c.* 2·5–4 mm. Inflorescence-branches and pedicels glabrous, eglandular. $2n = 22, 44, 110$. *Throughout the range of the species.*

(b) Subsp. **pubescens** (Fenzl) Jalas, *Ann. Bot. Fenn.* **20**: 109 (1983) (incl. *A. pineticola* Klokov): Leaf-sheaths at base of stem *c.* 2 mm. Branches of inflorescence and pedicels glandular-pubescent. *Ukraine and S. Russia.*

Subsp. (a) is very variable, but it is not clear on the basis of the data available at present whether or not it should be divided into several subspecies. For further information, and full references, see Jalas & Suominen, *Atlas Florae Europaeae* **6**: 12 (1983).

5. A. longifolia Bieb., *Fl. Taur.-Cauc.* **1**: 345 (1808). Glabrous perennial with branched, slightly woody stock; stems 20–40 cm, erect, fairly stout, simple. Basal leaves 100–200 × *c.* 0·75 mm, linear, acuminate; cauline leaves shorter but scarcely wider. Inflorescence many-flowered, subcorymbose; pedicels usually 1–3 times as long as sepals. Sepals 2–3 mm, ovate, obtuse, with a prominent, blunt keel; petals about twice as long as sepals. *C., S. & E. Russia; E. Ukraine.* Rs(C, W, E).

Subgen. **Porphyrantha** (Fenzl) McNeill. Perennials with lanceolate leaves. Styles 3. Capsule cylindrical, long-exserted, shiny.

6. A. purpurascens Ramond ex DC. in Lam. & DC., *Fl. Fr.* ed. 3, **4**: 785 (1805). Stems 4–10 cm, diffuse, ascending, glabrous, with distinct scale-leaves towards the base; puberulent with deflexed hairs, branched and leafy in upper part. Leaves 5–10 mm, elliptic-lanceolate or lanceolate, acute, strongly 1-veined, glabrous, ciliate at base. Flowers 2–4 in rather dense clusters, rarely solitary; pedicels usually equalling sepals, pubescent, erect in fruit. Sepals 4·5–6·5 mm, lanceolate, acute, 3- to 5-veined, glabrous; petals $1\frac{1}{2}$–2 times as long as sepals, oblong, pale purplish or white. Capsule $1\frac{3}{4}$–2 times as long as sepals, cylindrical. $2n = 46$. *Damp, rocky places on mountains.* ● *Pyrenees; Cordillera Cantábrica; one station in S.E. France (Vercors).* Ga Hs.

Subgen. **Arenaria**. Annuals to perennials, of various habit. Leaves less than 30 mm. Styles 3(–5). Capsule more or less ovoid.

Species **7–12** form a variable complex that has had numerous different taxonomic treatments. The treatment below largely follows that of D. J. Goyder, *Bot. Jour. Linn. Soc.* **97**: 9–32 (1988).

7. A. alfacarensis Pamp., *Nuovo Gior. Bot. Ital.* nov. ser., **22**: 64 (1915) (*A. lithops* Heywood ex McNeill, *A. pulvinata* Huter, non Edgew.). Dioecious perennial, sparsely covered with very short, crisped hairs, forming very compact, hard cushions 5–100 cm in diameter which show growth-rings inside; stems usually completely covered by the 4-ranked, densely imbricate leaves. Leaves 1–2 mm, ovate-deltate, obtuse, recurved. Flowers 4-merous, solitary, terminal, sessile, not more than half-exserted from the cushion. Sepals *c.* 2·5 mm, ovate-lanceolate, obtuse; petals *c.* 1½ times as long as sepals. Seeds 0·7–0·9 mm. $2n = 40$. *Rocky mountain slopes.* ● *S.E. Spain.* Hs.

8. A. tetraquetra L., *Sp. Pl.* 423 (1753). Perennial, usually forming dense cushions 5–25 cm in diameter, not showing growth-rings, subglabrous; stems more or less densely covered with 4-ranked imbricate leaves. Leaves 1–4 mm, ovate, obtuse, recurved, glabrous, ciliate at base. Flowers solitary and terminal, completely exserted from foliage, on slightly (up to *c.* 1 cm) protruding stems; bracts ovate-lanceolate. Sepals 4–9 mm, ovate-lanceolate, gradually narrowed to obtuse apex, with prominent, raised veins; petals slightly exceeding or up to 1½ times as long as sepals. Seeds 0·8–1·2(–1·4) mm. *Dry places on mountains.* ● *C. & E. Pyrenees, N.C. & S.E. Spain.* Ga Hs.

(a) Subsp. **tetraquetra**: Leaves broadly ovate. Flowers 4-merous; petals 4·5–8 mm; stamens 8. $2n = 120$, *c.* 140. *C. & E. Pyrenees; N.C. Spain (Segovia and Guadalajara provs).*

(b) Subsp. **amabilis** (Bory) H. Lindb., *Acta Soc. Sci. Fenn.* nov. ser., B, **1**: 45 (1932) (var. *granatensis* Boiss.): Leaves ovate. Flowers 5-merous; petals 3·5–6·5 mm; stamens 10. $2n = 40$, 60, 80, 100. *S.E. Spain.*

9. A. tomentosa Willk., *Linnaea* **25**: 15 (1852). Perennial, forming compact, silvery cushions 5–10 cm in diameter, more or less densely covered with very short, crisped, white hairs; stems densely covered with imbricate leaves. Leaves 1–1·5 mm, ovate, obtuse, recurved. Flowers 5-merous, completely exserted from the cushion, solitary and terminal or 2 or 3 in a terminal cluster with 1 or 2 axillary flowers just below. Sepals 4–5 mm, lanceolate, gradually narrowed to obtuse apex, keeled for *c.* 1 mm at base, otherwise very obscurely veined; petals slightly exceeding sepals. ● *S.E. Spain (W. of Vélez Rubio).* Hs.

Possibly not specifically distinct from **10**.

10. A. armerina Bory, *Ann. Gén. Sci. Phys. (Bruxelles)* **3**: 5 (1820) (*A. armeriastrum* Boiss.; incl. *A. vitoriana* Uribe-Echebarría & Alejandre). Usually laxly caespitose perennial, with flowering stems protruding from the lax vegetative cushion; stems 5–20 cm, usually puberulent. Leaves 2–6 mm, linear-lanceolate to ovate, obtuse, recurved, usually folded, rarely imbricate, puberulent or subglabrous. Flowers 5-merous, 2–12 in dense terminal heads surrounded by linear-lanceolate bracts, sometimes with 1–3 axillary flowers below. Sepals 3–8(–9) mm, ovate-lanceolate, acute, with very prominent raised veins, subglabrous to sublanate; petals slightly exceeding or up to 1½ times as long as sepals. Seeds 0·7–1·3(–1·5) mm. $2n = 30$, 32, 36, 60. *S.E. & E.C. Spain.* Hs.

A variable species, in which several local populations have been described at specific or subspecific rank. However, these are linked to typical populations of **10** by intermediates and probably merit, at most, varietal rank. Densely puberulent plants with sublanate sepals are best treated as var. *caesia* (Boiss.) Pau.

11. A. aggregata (L.) Loisel. in Cuvier, *Dict. Sci. Nat.* **46**: 513 (1827). Caespitose perennial. Leaves 1–4 mm, usually acute and mucronate, recurved and folded, glabrous or sparsely puberulent. Flowers 5-merous. Sepals usually lanceolate or linear-lanceolate, with strong raised veins; petals equalling or slightly exceeding sepals. *S.W. Europe.* Ga Hs It ?Lu.

1 Leaves not imbricate; sepals ovate-lanceolate (c) subsp. **racemosa**
1 Leaves imbricate; sepals lanceolate or linear-lanceolate
2 Flowers 3–20 in head (a) subsp. **aggregata**
2 Flowers solitary, rarely 2 or 3 together (b) subsp. **erinacea**

(a) Subsp. **aggregata** (*A. capitata* Lam.; incl. *A. pseudarmeriastrum* Boiss.): Plant forming a lax cushion with stems 2–5 cm, covered with more or less imbricate, mostly triangular leaves. Flowering stems protruding usually 3 cm or more above the cushion, unbranched, with dense, terminal 3- to 20-flowered heads. Sepals 5–6·5 mm, acuminate. $2n = 26$, 30. *Iberian peninsula; S. France; N.W. Italy (Alpi Marittime).*

(b) Subsp. **erinacea** (Boiss.) Font Quer, *Anal. Jard. Bot. Madrid* **6(2)**: 487 (1946) (*A. erinacea* Boiss.): Plant forming a dense cushion 5–10 cm in diameter, with stems covered with densely imbricate, ovate leaves. Flowers solitary and terminal, rarely 2 or 3 together, on the surface of the cushion, not on protruding stems. Sepals 4–6 mm, acuminate. $2n = 20$, *c.* 28, 30. ● *E. Spain.*

(c) Subsp. **racemosa** (Willk.) Font Quer, *Arx. Secc. Ci. Inst. Est. Catalans* **15**: 30 (1948) (*A. racemosa* Willk.): Plant laxly caespitose, with acute, mucronate, more or less distant and not imbricate leaves. Flowers (1)2–5 in lax racemes. Sepals 4–6 mm, ovate-lanceolate, acute, strongly veined. $2n = 30$. ● *S. Spain.*

Intermediates between subspp. (a) and (b) have been reported from S.E. France (Mont Ventoux).

Other subspecies have been recognized in the Iberian peninsula; these are listed in W. Greuter, H. M. Burdet & G. Long, *Med-checklist* **1**: 168–169 (1984).

12. A. querioides Pourret ex Willk., *Bot. Zeit.* **5**: 239 (1847). Caespitose perennial. Leaves of flowering stems 2·5–7 mm, ovate, strongly keeled, acute to acuminate, reflexed, usually glabrous; leaves of sterile stems not reflexed. Flowers 5-merous, 10–20 in a terminal head; flowering stems often procumbent; bracts ovate. Sepals 4–8 mm, ovate-lanceolate; petals slightly exceeding sepals. Seeds 0·9–1·3 mm. $2n = 26$. ● *Mountains of N. & C. Spain and N. Portugal.* Hs Lu.

13. A. pungens Clemente ex Lag., *Gen. Sp. Nov.* 15 (1816). Perennial, forming dense, spiny cushions; stems 10–20 cm, branched, viscid-puberulent at least above. Leaves *c.* 15–30 mm, linear-subulate, subtriquetrous, very stiff and spiny, patent, at least the upper puberulent. Stems 1(–3)-flowered; pedicels *c.* 2–3 times as long as sepals, viscid-puberulent. Sepals 6–13 mm, linear-lanceolate, acuminate, stiff, spiny, viscid-puberulent; petals shorter than sepals, white, oblong-obovate. Capsule shorter than sepals, ovoid. $2n = 56$. *Dry screes,* c. 1900–3000 *m. S. Spain (Sierra Nevada).* Hs. (*Morocco.*)

14. A. valentina Boiss., *Diagn. Pl. Or. Nov.* **3(1)**: 90 (1854). Perennial, sometimes laxly caespitose; stems 5–20 cm, slender, straggling, branched at right-angles, bearing minute, deflexed hairs. Leaves 3–10 × 0·5–1·5 mm, linear or linear-lanceolate, rather rigid but not spiny, usually fasciculate, flat but often appearing subulate because the margins are revolute. Branches 1(–3)-flowered; pedicels 3–5 times as long as sepals. Sepals 2·5–5 mm, ovate, subacute to acuminate, with a distinct mid-vein, glabrous, often ciliate at base; petals about twice as long as sepals, oblong-spathulate, white. Capsule slightly exceeding sepals, ovoid-globose. *Calcareous rocks.* ● *S.E. Spain.* Hs.

15. A. grandiflora L., *Syst. Nat.* ed. 10, **2**: 1034 (1759). Laxly caespitose perennial; stems 5–15 cm, the vegetative stems with short, deflexed hairs; flowering stems with patent, glandular hairs. Leaves 5–10 × 0·75–1(–1·5) mm, linear-lanceolate, acuminate, green, with a long arista; midrib and margins prominent beneath and coriaceous. Stems 1- to 3(–6)-flowered; pedicels 2–6 times as long as sepals, glandular-pubescent. Sepals 3–5·5 mm, ovate, acute or acuminate, sparsely glandular-pubescent, often keeled, the mid-vein always prominent; petals $1\frac{1}{2}$–$2\frac{1}{2}$ times as long as sepals, oblong-obovate, white. Capsule $1\frac{1}{4}$(–2) times as long as sepals, ovoid. $2n = 44$. *Dry, rocky and stony places, mostly on mountains. S. & S.C. Europe, eastwards to* c. 17° *E.* Au Bl Cz Ga He Hs It Si.

Var. *incrassata* (Lange) Cosson (subsp. *incrassata* (Lange) C. Vicioso) differs from the typical plant in forming dense, more leafy tussocks, in the tetragonous, not terete, stems, and the broader, elliptical and densely fasciculate leaves. It replaces typical plants of **15** in N. Spain and may merit subspecific status.

16. A. bertolonii Fiori in Fiori & Paol., *Fl. Anal. Ital.* **1**: 346 (1898) (*A. saxifraga* (Bertol.) Fenzl). Laxly caespitose perennial; stems 5–12 cm, branched, pubescent or glabrous. Leaves 5–13 mm, ovate, acute or subobtuse, pubescent; the lower shortly petiolate, the upper sessile. Flowers solitary or in 2- or 3-flowered cymes; pedicels 2–4 times as long as sepals, slender. Sepals (4–)5–6·5 mm, ovate-lanceolate, obtuse or subacute, veinless, sparsely glandular-pubescent; petals 2–3 times as long as sepals, usually emarginate, white. Capsule equalling or slightly exceeding sepals, ovoid. $2n = 30$. *Mountain rocks and screes.* ● *C. Mediterranean region.* Co It Sa ?Si.

17. A. huteri A. Kerner, *Österr. Bot. Zeitschr.* **22**: 368 (1872). Like **16** but usually more caespitose; leaves oblanceolate, subobtuse; pedicels 3–5 times as long as sepals; sepals oblong-lanceolate; petals $1\frac{3}{4}$–2 times as long as sepals, entire; capsule shorter than sepals. *Crevices of dolomitic rocks.* ● *N.E. Italy (Alpi Carniche).* It.

18. A. montana L., *Cent. Pl.* **1**: 12 (1755). Robust, greyish-green, crisply and shortly pubescent perennial; stems 10–30 cm, spreading or erect. Leaves 10–20(–40) × 2–4 mm, oblong-lanceolate to linear, acute, 1-veined. Flowers solitary or in 2- to 11-flowered cymes; pedicels 2–6 times as long as sepals. Sepals (5–)6–9 mm, ovate-lanceolate to ovate, acute or subacute, 1-veined; petals twice as long as sepals, white. Capsule shorter than or equalling sepals. *S.W. Europe, extending northwards to C. & N.W. France.* Ga Hs Lu.

(a) Subsp. **montana**: Stems 10–30 cm, spreading or erect. Leaves 10–20(–40) × 2–4 mm, oblong- or linear-lanceolate. Pedicels usually eglandular. Sepals lanceolate to ovate, acute or subacute, usually with scarious margin. $2n = 28$. *Throughout the range of the species except S.E. Spain.*

(b) Subsp. **intricata** (Dufour) Pau, *Actas Soc. Esp. Hist. Nat.* **27**: 199 (1898): Stems 10–40 cm, straggling. Leaves 10–40 × 1–2 mm, linear. Pedicels usually glandular. Sepals ovate, acute or subobtuse, usually without scarious margin. ● *S.E. Spain.*

19. A. balearica L., *Syst. Nat.* ed. 12, **3**: 230 (1768) (*A. gayana* F. N. Williams). Perennial with slender, almost filiform, procumbent, branched stems forming dense mats. Leaves 2–4 mm, broadly ovate or orbicular, obtuse, shortly pubescent; petioles equalling lamina. Flowering stems up to 6 cm, ascending, glabrous or scabrid; flowers solitary; pedicels 5–10 times as long as sepals, filiform. Sepals 2·5–3 mm, ovate, subobtuse, distinctly 1-veined, shortly and sparsely pubescent; petals twice as long as sepals. Capsule slightly exserted. $2n = 18$. *Shaded rocky places, up to* 1450 *m.* ● *Islands of W. Mediterranean, from Mallorca to Montecristo; naturalized from gardens in the British Isles and N. France.* Bl Co It Sa [Br Ga Hb].

20. A. biflora L., *Mantissa* 71 (1767). Perennial; stems up to 20 cm, procumbent, slender, rooting at nodes, glabrous or rarely puberulent. Leaves 3–4 × 1·5–2·5 mm, obovate to orbicular, obtuse, distinctly 1-veined, abruptly contracted into a ciliate petiole. Flowering branches 2–3 cm; flowers solitary or 2(–5) together; pedicels 1–2 times as long as sepals. Sepals 3–4·5 mm, ovate, obtuse or subacute, rarely acute, usually glabrous, often ciliate near base, 1(–3)-veined; petals slightly exceeding sepals. Capsule more or less equalling sepals. $2n = 22$. *Damp places above* 1600 *m, often in snow-patches. Mountains of S. & C. Europe, from E. Carpathians to E. Pyrenees and C. Greece.* Al Au Bu Ga Gr He Hs It Ju Rm.

Plants from Romania and the northern part of the Balkan peninsula with lanceolate, acute sepals, often more-flowered branches, and petals slightly shorter than the sepals, have been called **A. rotundifolia** Bieb., *Fl. Taur.-Cauc.* **1**: 343 (1808), but there appears to be no satisfactory basis for specific separation. A chromosome number of $2n = 44$ has been recorded for this plant.

21. A. humifusa Wahlenb., *Fl. Lapp.* 129 (1812). Small, mat-forming perennial up to 3 cm, with procumbent, glabrous stems and slender, elongate stolons at or just above ground-level. Leaves 3–4 × 1–1·75 mm, elliptical to narrowly obovate, glabrous or very slightly ciliate at base. Flowers solitary or rarely 2 together; pedicels much shorter than or equalling sepals, slightly scabrid. Sepals 3–4 mm, ovate, usually glabrous, obscurely 3-veined, with narrow, scarious margin; petals slightly exceeding sepals; anthers pale purple. Capsule exserted, cylindrical. Seeds 0·7–0·75 mm, reddish-brown. $2n = 40, 44$. *Moist, open habitats on basic soil. Arctic Europe; subarctic Norway (very rare).* No Rs(N) Sb Su.

22. A. norvegica Gunnerus, *Fl. Norv.* **2**: 144 (1772). Annual to perennial forming lax tufts 3–7 cm high. Leaves oblanceolate, glabrous, often slightly ciliate below, obscurely veined. Cymes 1- or 2(–4)-flowered; pedicels 1–3 times as long as sepals, slightly scabrid. Sepals 3–4 mm, ovate, glabrous, 3-veined, the lateral veins often obscure; petals 4–5·5 mm, exceeding sepals; anthers white. Capsule equalling or slightly exceeding sepals, conical. Seeds 0·8–1 mm, black, dull, with low, broad tubercles. ● *N.W. Europe, extending to N. Finland.* Br Fe ?Hb Is No Su.

(a) Subsp. **norvegica**: Perennial. Leaves 3–4·5(–6) mm, somewhat fleshy, dark green. Sepals usually not ciliate; petals 4–4·5 mm. $2n = 80$. *Open habitats on basic soils in the mountains.* ● *Throughout the range of the species, except England.*

(b) Subsp. **anglica** Halliday, *Watsonia* **4**: 209 (1960) (*A. gothica* auct. angl., non Fries): Annual or biennial, with a laxer habit than subsp. (a). Leaves 4·5–5(–6) mm, lighter green. Sepals usually with a few basal cilia; petals 5–5·5 mm. $2n = 80$. *Damp, bare depressions on limestone pavement.* ● *N. England (W. Yorkshire).*

23. A. ciliata L., *Sp. Pl.* 425 (1753). Low-growing perennial with scabrid stems 2–15 cm. Leaves broadly elliptical, oblanceolate or spathulate, the midrib prominent below, usually ciliate at least in lower ½. Cymes 1- to 7(–9)-flowered; pedicels scabrid. Sepals ovate-lanceolate, usually ciliate at base; petals exceeding sepals; anthers white. Capsule swollen below. Seeds 0·85–1·05 mm, black, dull, with low, broad tubercles. *Open habitats on basic soil. Mountains of C. Europe, southwards to C. Spain and N. Italy; locally in N. Europe.* Au Cz Fe Ga Ge Hb He Hs It Ju No Po Rm Rs(N) Sb.

The following is a conservative treatment of the various geographical and chromosomal variants of this very polymorphic species.

1 Leaves 3–4 times as long as wide, usually ciliate throughout (a) subsp. **ciliata**
1 Leaves 2–3 times as long as wide, not ciliate throughout
 2 Flowers in 2- to 6(–9)-flowered cymes (b) subsp. **moehringioides**
 2 Flowers usually solitary (c) subsp. **pseudofrigida**

(a) Subsp. **ciliata** (incl. subsp. *tenella* (Kit.) Br.-Bl., subsp. *hibernica* Ostenf. & O. C. Dahl): Leaves 3–4 times as long as wide, scarcely to entirely ciliate. Flowers 1 or 2(3). Sepals 3·5–5 mm; petals 5–8 mm. $2n = 40, 80, 120, 160$. *C. & E. Alps, Carpathians; N.W. Ireland.*

(b) Subsp. **moehringioides** (J. Murr) Br.-Bl., *Sched. Fl. Raet. Exsicc.* 279 (1927) (incl. subsp. *polycarpoides* Br.-Bl., *A. multicaulis* L., nom. illeg.): Leaves 2–3 times as long as wide, ciliate for the basal ⅓–⅔. Cymes 2- to 6(–9)-flowered. Sepals 3·5–4 mm; petals 4–6 mm. $2n = 40$. • *N.W. Spain, Pyrenees, Jura, Alps eastwards to Vorarlberg, N. Appennini.*

(c) Subsp. **pseudofrigida** Ostenf. & O. C. Dahl, *Nyt. Mag. Naturvid. (Christiania)* **55**: 217 (1917) (*A. pseudofrigida* (Ostenf. & O. C. Dahl) Juz. ex Schischkin & Knorring): Like subsp. (a) but smaller and more compact; leaves weakly ciliate; flowers 1(2). $2n = 40$. *Finland, arctic Norway and Russia, Svalbard.*

Subsp. **bernensis** Favarger, *Ber. Schweiz. Bot. Ges.* **73**: 176 (1963), from the W. Alps, is similar to subsp. (a) but has larger parts and strongly ciliate leaves, and a chromosome number of $2n = c.\ 200, 240$. It is not, however, readily distinguished from the very variable subsp. (a).

Subsp. (b) is perhaps more closely related to **24**.

24. A. gothica Fries, *Nov. Fl. Suec.* ed. 2, *Mant.* 2: 33 (1839). Annual or biennial with ascending, rather robust, scabrid stems up to 12(–15) cm. Leaves 4–6 mm, broadly elliptical to obovate, less than 3 times as long as wide, the margin strongly ciliate at least in the lower ½. Cymes (1)2- to 6(–8)-flowered; pedicels strongly scabrid. Sepals 3–4 mm, ovate-lanceolate, ciliate below and often also on keel, 3-veined; petals 4–5·5 mm, slightly exceeding sepals. Seeds 0·8–0·9 mm, black, dull, with low, broad tubercles. $2n = 100$. *Dry limestone pavement and lake-shores.* • *S. Sweden (Gotland; one station in Västergötland); Jura (Lac de Joux).* He Su.

This description applies to var. *gothica* from Sweden. The plants from Switzerland (var. *fugax* (Gay ex Gren.) M. B. Wyse Jackson & J. Parnell) are taller, less ciliate, with more flowers in a cyme, and slightly larger seeds. Although their chromosome number is the same, it is quite possible that they are of different origin.

25. A. gionae L.-Å. Gustavsson, *Bot. Not.* **129**: 276 (1976). Glabrous, caespitose perennial; stems 1–2(–3) cm, slender, with a single pair of cauline leaves. Leaves 3–7 mm, crowded, linear-lanceolate to linear, acute, 1-veined. Flowers solitary; pedicels 1–2 times as long as sepals. Sepals 3·5–4·2 mm, obtuse or acute, keeled at base, the outer lanceolate with a narrow, scarious margin, the inner ovate with a scarious margin up to 0·7 mm wide; petals about twice as long as sepals, white. Capsule slightly exceeding sepals, ovoid. *Crevices of limestone rocks,* 1800–2200 *m.* • *S.C. Greece (Giona Oros).* Gr.

26. A. gracilis Waldst. & Kit., *Pl. Rar. Hung.* **3**: 305 (1812). Laxly (rarely densely) caespitose perennial; stem 2–8 cm, ascending or erect, slender, papillose-scabrid below, glabrous above. Leaves 3–8 mm, lanceolate to elliptic-lanceolate, rarely narrower, acute, 1-veined, usually glabrous, denticulate. Flowers 1 or 2(3); pedicels 2–5 times as long as sepals, glabrous. Sepals 3·5–5·5 mm, ovate, obtuse or acute, often strongly keeled below, obscurely 1- to 3-veined, glabrous; petals about twice as long as sepals, white. Capsule slightly exceeding sepals, ovoid. $2n = 24$. *Mountain rocks.* • *W. Jugoslavia.* Ju.

27. A. filicaulis Fenzl in Griseb., *Spicil. Fl. Rumel.* **1**: 203 (1843). Glandular-pubescent caespitose perennial; stems 5–15 cm, slender, fragile. Leaves acute, 1-veined, light green, the lower shortly petiolate, the upper sessile; midrib prominent beneath. Flowers 3–10 in a lax, leafy inflorescence with slender, patent branches; pedicels 2–5(–10) times as long as sepals. Sepals 3·5–4·5 mm; petals oblong-cuneate, white. Seeds 0·7–1 mm, with acute tubercles. *Limestone rocks. Mountains of Balkan peninsula southwards from S.C. Bulgaria; W. Kriti.* Bu Cr Gr Ju.

A variable species that can be divided into three rather poorly defined subspecies.

1 Lower bracts truncate or cordate at base (c) subsp. **teddii**
1 Lower bracts cuneate at base
 2 Sepals ovate-lanceolate, acuminate, slightly shorter than capsule (a) subsp. **filicaulis**
 2 Sepals ovate, acute, about as long as capsule (b) subsp. **graeca**

(a) Subsp. **filicaulis** (incl. subsp. *euboica* McNeill): Plant moderately hairy with rather short, slender hairs. Leaves 3–8 mm, elliptic-lanceolate or oblanceolate. Sepals ovate-lanceolate, acuminate. Lower bracts cuneate at the base. Capsule slightly exceeding sepals. *S.C. Bulgaria to E.C. Greece.*

(b) Subsp. **graeca** (Boiss.) McNeill, *Notes Roy. Bot. Gard. Edinb.* **24**: 270 (1963) (*A. graeca* (Boiss.) Halácsy, *A. graveolens* var. *graeca* Boiss.): Plant usually more densely hairy with longer, stouter hairs. Leaves 5–10 mm, ovate-elliptical. Lower bracts cuneate at base. Sepals ovate, acute. Capsule about as long as sepals. $2n = 40$. *Greece, W. Kriti; ?S. Bulgaria.*

(c) Subsp. **teddii** (Turrill) Strid, *Mount. Fl. Gr.* **1**: 86 (1986) (*A. teddii* Turrill): Plant more or less densely hairy with rather short, slender hairs. Leaves 3–7 mm, ovate. Lower bracts truncate or cordate at the base. Sepals ovate-lanceolate, acuminate. Capsule slightly shorter than sepals. • *N. Greece (Pangeon).*

A. phitosiana Greuter & Burdet, *Willdenowia* **12**: 37 (1982) (*A. litoralis* Phitos, non Salisb.) is finely but densely pubescent, with moderately thick, obovate-spathulate lower leaves and obovate median leaves. From limestone coastal cliffs in the Aegean (Voriai Sporadhes), it may be a variant of subsp. (a).

A. rhodopaea Delip., *Compt. Rend. Acad. Bulg. Sci.* **17**: 645 (1964), from Rodopi Planina, is like subsp. (b) but with suborbicular, fleshier and more hairy leaves, and somewhat larger flowers. It should perhaps be included within subsp. (b).

28. A. orbicularis Vis., *Fl. Dalm.* **3**: 180 (1850) (*A. deflexa* auct., non Decne, *A. pubescens* D'Urv., *A. graveolens* auct. balcan., non Schreber). Like **27(a)** but whole plant more densely glandular-pubescent; leaves thicker and wider, broadly elliptical or obovate; all leaves, including the upper, narrowed at base and more or less petiolate; sepals lanceolate, acuminate; petals only slightly exceeding sepals; seeds 0·4–0·6 mm. $2n = 44$. ● *N.W. Jugoslavia* (*Velebit*). Ju.

29. A. fragillima Rech. fil., *Feddes Repert.* **47**: 49 (1939) (*A. pamphylica* sensu Hayek). Like **27(a)** but plant less densely caespitose, greyish-green, and covered with very short, eglandular hairs, with or without glandular hairs; leaves thicker and shorter, all more or less petiolate; inflorescence dense, few-flowered; peduncles and pedicels shorter and stouter; sepals 4–5 mm, lanceolate, acuminate; petals only slightly exceeding sepals; seeds 0·5–0·6 mm. $2n = 22$. *Dry stony places.* ● *S. Aegean region.* Cr Gr.

30. A. ligericina Lecoq & Lamotte, *Cat. Pl. Centr. Fr.* 104 (1847) (*A. lesurina* Loret). Like **27(a)** but stems 7–40 cm, usually stouter; leaves with midrib obscure beneath; sepals 3·5–5·5 mm, more or less distinctly 1-veined, lacking or having obscure lateral veins; petals *c.* $1\frac{1}{3}$ times as long as sepals; seeds with low, obtuse tubercles or ridges. *Calcareous rocks and sands.* ● *S. France* (*Lozère and Aveyron*); *E. Pyrenees.* Ga Hs.

31. A. cretica Sprengel, *Syst. Veg.* **2**: 396 (1825) (incl. *A. pirinica* Stoj.). Densely caespitose perennial; stems 2–10 cm, slender, fragile, usually glandular-pubescent only in inflorescence. Leaves 3–10 mm, crowded, oblong-elliptical or oblanceolate, rather obtuse, glabrous or glandular-pubescent (var. *stygia* (Boiss. & Heldr.) Boiss.), 1-veined. Inflorescence 1- to 5-flowered, usually corymbose; pedicels 1–3 times as long as sepals. Sepals 2·5–5 mm, oblong or oblong-lanceolate, obtuse or subacute, veinless, keeled at the base, glandular-pubescent; petals 2 or more times as long as sepals, white. Capsule slightly exceeding sepals, ovoid. $2n = 24, 40$. *Limestone rocks.* ● *Mountains of S. part of Balkan peninsula; Kriti.* Al Bu Cr Gr.

32. A. halacsyi Bald., *Malpighia* **5**: 65 (1891). Dwarf caespitose perennial; stems 2–5 cm, slender, eglandular, with short, white, deflexed hairs especially above. Leaves 2–4 mm, ovate, lanceolate or obovate, shortly petiolate, 1-veined, subglabrous, strongly ciliate at base, crowded on vegetative stems. Flowers solitary or paired; pedicels 1–2 times as long as sepals. Sepals 3–4 mm, oblong-lanceolate, obtuse or subacute, glandular-hairy; petals scarcely exceeding sepals, broadly spathulate, white. Capsule slightly exceeding sepals, ovoid. *Rock-crevices.* ● *Crna Gora* (*Komovi, S.W. of Andrijevica*). Ju.

33. A. cinerea DC. in Lam. & DC., *Fl. Fr.* ed. 3, **5**: 611 (1815). Perennial, suffruticose at base, with many ascending flowering stems 10–20 cm, and with very few and inconspicuous vegetative leafy stems at time of flowering, so appearing annual; stems with short, deflexed hairs. Leaves 3–5 mm, 1-veined, the lower oblong-lanceolate, acute, the upper linear-lanceolate. Inflorescence lax, 3- to 10-flowered, eglandular; pedicels 4–6 times as long as sepals, rigid, erect, with short, deflexed hairs. Sepals 4–5 mm, ovate-lanceolate, acute, distinctly 1-veined, developing a strongly ciliate keel after flowering; petals about twice as long as sepals. Capsule slightly exceeding sepals, ovoid-globose. Seeds with low, flattened tubercles. *Dry and rocky places.* ● *S.E. France.* Ga.

34. A. hispida L., *Sp. Pl.* 608 (1753). Pubescent, caespitose perennial; stems 7–25 cm, many, ascending. Leaves 5–10 × 0·5(–1) mm, subulate, rarely linear-lanceolate and acuminate, not narrowed at base, 1-veined. Inflorescence lax, 3- to 10-flowered, glandular-pubescent. Sepals 3–4·5 mm, lanceolate, acuminate, 3- to 5-veined, with the mid-vein prominent; petals 5 mm, *c.* $1\frac{1}{3}$ times as long as sepals. Capsule slightly shorter than or equalling sepals, broadly ovoid. Seeds 0·6 mm, reniform, with acute tubercles. $2n = 40$. *Calcareous, often dolomitic rocks.* ● *S.E. France.* Ga.

35. A. fontqueri Cardona & J. M. Monts., *Biol. Écol. Médit.* **8**: 15 (1981). Like **34** but biennial; leaves linear-oblanceolate, acute; sepals ovate-lanceolate, acute, 3(–5)-veined; petals 4 mm; capsule ovoid-conical, more or less exceeding sepals; seeds 0·8–1 mm, rugose-granulate. $2n = 44, 66$. ● *N.E. Spain.* Hs.

Three subspecies have been recognized, based on minor differences, but the characters overlap to a large extent.

36. A. serpyllifolia L., *Sp. Pl.* 423 (1753) (*Minuartia olonensis* (Bonnier) P. Fourn.). A relatively robust, scabrid-puberulent annual, sometimes biennial or perennial; stems 2–30 cm, usually profusely branched at base, ascending or erect. Leaves 2·5–8 mm, broadly ovate to ovate-lanceolate, acute or acuminate, 3- to 5-veined, the lower petiolate, the upper sessile. Flowers in diffuse or dense dichasia; pedicels usually longer than sepals. Sepals ovate-lanceolate or ovate, acute, 3- to 5-veined; the inner with a scarious margin, the central green part usually being $\frac{1}{3}$–$\frac{1}{2}$ the width of the sepals; petals white, not more than $\frac{2}{3}$ as long as sepals. Capsule less than twice as long as wide, ovoid-conical or subovoid-cylindrical. Seeds 0·3–0·7 mm. *Dry, often sandy places. Almost throughout Europe.* All except Az Fa Is Sb.

1 Capsule ovoid-conical, ± swollen at base
  2 Inflorescence lax; pedicels longer than sepals; capsule distinctly swollen at base, 1·5–2 mm wide ... **(a)** subsp. **serpyllifolia**
  2 Inflorescence dense; pedicels shorter than sepals; capsule slightly swollen at base, at least 2 mm wide ... **(b)** subsp. **macrocarpa**
1 Capsule subovoid-cylindrical, not swollen at base
  3 Sepals 2–3 mm; plant lax ... **(c)** subsp. **leptoclados**
  3 Sepals 3·5–4·5 mm; plant compact
    4 Plant usually eglandular; leaves acute ... **(d)** subsp. **marschlinsii**
    4 Plant glandular, often densely; leaves subacute to obtuse ... **(e)** subsp. **aegaea**

(a) Subsp. **serpyllifolia**: Plant 3–30 cm, lax, usually eglandular-puberulent. Leaves ovate or elliptical, acute or acuminate. Pedicels longer than sepals. Sepals 3–4·5 mm, the inner with broad, scarious margins. Capsule ovoid-conical, distinctly swollen at base, slightly longer than sepals. Seeds 0·4–0·6(–0·7) mm. $2n = 40$. *Almost throughout the range of the species, but only on mountains in the south.*

(b) Subsp. **macrocarpa** (Lloyd) Perring & P. D. Sell, *Watsonia* **6**: 294 (1967): Like subsp. **(a)** but inflorescence dense, with pedicels shorter than sepals; capsule at least 2 mm wide; seeds 0·6(–0·7) mm. *Dunes on the Atlantic coasts of Europe.*

(c) Subsp. **leptoclados** (Reichenb.) Nyman, *Consp.* 115 (1878) (*A. leptoclados* (Reichenb.) Guss., *A. minutiflora* Loscos, *A. leptoclados* subsp. *minutiflora* (Loscos) H. Lindb., *A. brevifolia* Gilib., *A. uralensis* Pallas, *A. zozii* Kleopow): Stems and leaves like subsp. **(a)** but more slender. Sepals 2–3 mm. Capsule subovoid-cylindrical, not swollen at base, shorter than or equalling sepals. Seeds

0·3–0·6 mm. $2n = 20$. *S., W. & C. Europe, extending to S.E. Sweden and Ukraine.*

(d) Subsp. **marschlinsii** (Koch) Nyman, *loc. cit.* (1878) (*A. marschlinsii* Koch): Plant 2–10 cm, often perennial, compact, usually eglandular-puberulent. Leaves ovate or sometimes elliptical, acute. Inflorescences dense, few-flowered. Sepals 3·5–4·5 mm, the inner with narrow, scarious margin. Capsule like that of subsp. (c). Seeds 0·4–0·7 mm. $2n = 20$. *Open habitats above 1800 m.* ?● *Alps; Pyrenees.*

(e) Subsp. **aegaea** (Rech. fil.) Akeroyd, *Bot. Jour. Linn. Soc.* **97**: 337 (1988) (*A. aegaea* Rech. fil.): Plant 2–6 cm, compact, densely branched, glandular-puberulent, especially above. Leaves broadly elliptical or ovate to suborbicular, subacute to obtuse, somewhat fleshy. Sepals 3·5–4 mm, the inner with broad, scarious margin. Capsule like that of subsp. (c). Seeds 0·4–0·5 mm. *Maritime or submaritime rocky and stony habitats, usually on small islands. C. & S. Aegean region.*

37. A. conferta Boiss., *Diagn. Pl. Or. Nov.* **1(1)**: 51 (1843). Like **36(a)** but often biennial; plant with a short, greyish pubescence; inflorescence dense, the branches more or less strict, usually with pedicels shorter than sepals (rarely some pedicels up to 3 times as long as sepals); sepals more sharply acute; petals more or less equalling or longer than sepals; capsule slightly shorter than sepals; seeds 0·75–1 mm. *Stony places on mountains.* ● *Balkan peninsula, mainly in the west.* Al Gr Ju.

Very variable in habit, but probably always distinguishable from **36** by the longer petals and shorter capsules.

(a) Subsp. **conferta**: Stems procumbent to ascending. Leaves densely puberulent, greyish. Inflorescence usually dense, with both eglandular and glandular hairs. $2n = 20$. *Almost throughout the range of the species.*

(b) Subsp. **serpentini** (A. K. Jackson) Strid, *Ann. Bot. Fenn.* **20**: 113 (1983) (*A. serpentini* A. K. Jackson): Stems usually ascending to suberect. Leaves sparsely puberulent, green. Inflorescence lax to dense, eglandular-hairy or glabrous. *Mostly on serpentine.* ● *S. Albania to C. Greece.*

38. A. nevadensis Boiss. & Reuter in Boiss., *Diagn. Pl. Or. Nov.* **3(1)**: 90 (1853). Glandular-puberulent annual, usually purplish at base; stems 5–8 cm, branched. Leaves 4–7 mm, mostly 3-veined, the lower ovate and patent, the upper linear-lanceolate and erect. Flowers clustered in more or less dense corymbs; pedicels equalling or up to twice as long as sepals, always erect and strict. Sepals 4·5–6 mm, lanceolate, long-pointed, obtuse or subacute; petals *c.* $\frac{3}{4}$ as long as sepals, oblong, white. Capsule shorter than sepals, oblong-ovoid. ● *S. Spain (Sierra Nevada).* Hs.

39. A. muralis (Link) Sieber ex Sprengel, *Syst. Veg.* **2**: 397 (1825) (*A. oxypetala* auct., vix Sibth. & Sm.). Glandular-pubescent annual; stems 5–15 cm, ascending or erect, slender, diffusely branched from base. Leaves 5–15 mm, obovate, acute, 1-veined, narrowed into a distinct petiole. Inflorescence lax, 5- to 20-flowered; pedicels (2–)3–4 times as long as sepals, slender, rigid, patent or deflexed in fruit. Sepals 2·5–3·5 mm, lanceolate, acute, with obscure veins; petals scarcely exceeding sepals, linear-lanceolate. Capsule slightly exceeding sepals, narrowly ovoid. Seeds 0·4–0·6 mm with very acute tubercles. *Aegean region.* Cr Gr.

Recorded elsewhere only from the East Aegean Islands.

40. A. graveolens Schreber, *Nova Acta Acad. Leop.-Carol.* **3**: 478 (1767) (*A. oxypetala* Sibth. & Sm.). Like **39** but stems up to 30 cm; petals shorter than or equalling sepals; sepals narrowly lanceolate, acuminate; seeds (0·6–)0·8–1 mm. *E.C. Greece (Evvoia).* Gr. (*Anatolia.*)

41. A. saponarioides Boiss. & Balansa in Boiss., *Diagn. Pl. Or. Nov.* **3(6)**: 35 (1859) (*A. nana* Boiss. & Heldr., non Willd.). Glandular-pubescent dwarf annual; stems not more than 4 cm, solitary or 2 or 3 together. Leaves oblong-ovate, narrowed into the petiole, obtuse, 3- to 7-veined, the upper forming an involucre around the inflorescence. Flowers 1–5; pedicels shorter than sepals. Sepals 5–6 mm, lanceolate, acute, with 3 prominent veins, elongating in fruit; petals *c.* $\frac{1}{2}$ as long as sepals, ovate. Capsule shorter than sepals, ovoid. Seeds *c.* 1 mm, dark brown. *Stony places on mountains. Kriti.* Cr.

The plant from Kriti is referable to subsp. **boissieri** (Pax) McNeill, *Notes Roy. Bot. Gard. Edinb.* **24**: 289 (1963), which also occurs in Cyprus. Subsp. *saponarioides* is known only from Anatolia.

42. A. guicciardii Heldr. ex Boiss., *Diagn. Pl. Or. Nov.* **3(5)**: 60 (1856). Scabrid-puberulent, slightly glandular annual; stems 3–20 cm, usually many, ascending or erect, rather densely leafy. Leaves 3–5 mm, ovate or ovate-oblong, acute or acuminate, subsessile or petiolate, 3- to 5-veined. Flowers 5–20 in dense clusters; pedicels usually not more than $\frac{1}{2}$ as long as sepals, rather stout. Sepals 4–6 mm, lanceolate, long-acuminate with usually patent or recurved apex, 3- to 5-veined with thickened, prominent veins; petals $\frac{1}{3}$ as long as sepals. Capsule more or less equalling sepals, narrowly cylindrical. *Dry, stony places. S. Greece and Kriti.* Cr Gr.

Recorded elsewhere only from the East Aegean Islands (Rodhos and Samos).

43. A. hispanica Sprengel, *Syst. Veg.* **2**: 396 (1825) (*A. cerastioides* auct., non Poiret). Robust, pubescent and glandular annual; stems 10–40 cm, usually branched from base. Leaves 10–25 mm, broadly spathulate to obovate-oblanceolate, subacute to acuminate, 1-veined, the lower long-petiolate. Inflorescence 4- to many-flowered, lax; pedicels 3–5 times as long as sepals. Sepals 4–5·5 mm, ovate, obtuse or subacute, usually purplish-black at apex; petals twice as long as sepals, slightly emarginate, white; anthers blue or violet. Capsule $\frac{3}{4}$–$1\frac{1}{3}$ as long as sepals. Seeds 0·7–0·9(–1) mm, globose, black, almost smooth. *Disturbed ground. S.W. Spain.* Hs.

A. cerastioides Poiret, *Voy. Barb.* **2**: 166 (1789) (*A. spathulata* Desf.), from N.W. Africa, smaller than **43** in all its parts and with reniform, tuberculate seeds 0·4–0·7 mm, has been recorded in error from S.W. Spain.

44. A. pomelii Munby, *Bull. Soc. Bot. Fr.* **11**: 45 (1864). Puberulent annual; stems up to 16 cm, ascending, branched from the base. Leaves 7–14 mm, oblong to spathulate, obtuse. Inflorescence few-flowered. Pedicels 10–20 mm. Sepals dimorphic; 3 outer 5–7 mm, ovate; 2 inner 4·5–5·5 mm, oblong. Petals shorter than outer sepals, equalling or slightly exceeding the inner, oblong, acute, entire. Capsule shorter than outer sepals. Seeds 0·9–1 mm, reniform, tuberculate, grey or black. *Calcifuge. Mountains of S. Spain.* Hs. (*N.W. Africa.*)

45. A. emarginata Brot., *Fl. Lusit.* **2**: 202 (1804). Glandular-pubescent annual; stems 5–10 cm, erect, branched. Leaves 5–15 mm, linear or linear-subulate, obtuse, 1-veined, patent; the lower and middle equalling or exceeding internodes; the upper shorter. Flowers in a usually dense inflorescence; pedicels scarcely equalling or up to twice as long as sepals, patent in fruit. Sepals 3–6 mm, lanceolate, subobtuse, 1-veined; petals more or less equalling sepals, oblong, truncate-emarginate, pinkish. Capsule shorter than sepals, ovoid-oblong. *Dry, sandy places. S. Portugal; S.W. Spain.* Hs Lu. (*N.W. Africa.*)

46. A. algarbiensis Welw. ex Willk., *Icon. Descr. Pl. Nov.* **1(9)**: 93 (1855). Glandular, puberulent or shortly pubescent annual; stems 5–10 cm, ascending or erect, slender. Leaves 2–5 mm, linear-lanceolate, all shorter than internodes. Inflorescence lax, several- to many-flowered; pedicels 3–5 times as long as sepals. Sepals 3–4 mm, ovate or ovate-lanceolate, obtuse or subacute, with 3–5 prominent veins; petals 2–3 times as long as sepals, deeply emarginate, white. Capsule slightly shorter than sepals, ovoid-oblong. ● *S. Portugal; S.W. Spain.* Hs Lu.

47. A. conimbricensis Brot., *Fl. Lusit.* **2**: 200 (1804) (*A. tenuior* (Mert. & Koch) Gürke; incl. *A. loscosii* Texidor). Glandular, puberulent or pubescent annual; stems 5–15 cm, branched. Leaves 5–15 mm, linear to linear-oblanceolate, flat, obtuse and subapiculate, ciliate at base, the lower oblanceolate to spathulate. Inflorescence lax, several- to many-flowered; pedicels 3–5 times as long as sepals, very slender but scarcely capillary, patent in fruit, covered with patent, glandular hairs. Sepals 1·5–3·5 mm, ovate, obtuse, obscurely 1-veined, or veinless; petals 1½–2 times as long as sepals, obovate, entire, white; anthers vinous. Capsule about equalling sepals, ovoid-oblong. Seeds *c.* 0·6 mm, reniform, with obtuse or acute tubercles. $2n = 22$. *Sandy places or waste ground.* ● *C., E. & S. Spain; C. & S. Portugal.* Hs Lu.

48. A. obtusiflora G. Kunze, *Flora (Regensb.)* **29**: 632 (1846). Eglandular annual; stems 5–15 cm. Leaves 3–10 mm, linear-oblanceolate, acute, 1-veined, flat. Inflorescence glabrous or puberulent. Sepals 2–3·5 mm, ovate, obtuse, often apiculate; petals 1½–2 times as long as sepals; anthers vinous. Seeds 0·5–0·8 mm. *Screes.* ● *N. & E. Spain.* Hs.

(**a**) Subsp. **obtusiflora**: Stems usually simple and with short deflexed hairs below, branched and glabrous or subglabrous above. Inflorescence glabrous or subglabrous; pedicels 3–6 times as long as sepals, filiform, patent in fruit. Sepals 2–3 mm. Capsule more or less equalling sepals, ovoid. Seeds with acute tubercles. *S.E. Spain.*

(**b**) Subsp. **ciliaris** (Loscos) Font Quer, *Collect. Bot. (Barcelona)* **3**: 348 (1953) (*A. ciliaris* Loscos): Usually smaller and more bushy than subsp. (**a**); stems usually much-branched from the base, with short deflexed hairs throughout. Inflorescence-branches puberulent; pedicels 2–3 times as long as sepals, always erect, subglabrous or with minute deflexed hairs. Sepals 2·5–3·5 mm, subglabrous or puberulent. Capsule 1½–2 times as long as sepals, globose-conical. Seeds with obtuse tubercles. *N. & E.C. Spain.*

49. A. controversa Boiss., *Voy. Bot. Midi Esp.* **2**: 100 (1839). Like **48** but stems, branches and pedicels covered with minute deflexed hairs; leaves linear, trigonous; pedicels 2–3 times as long as sepals; sepals 2–4·5 mm, ovate-lanceolate, acute or subacute; capsule slightly exceeding sepals. ● *C. & W. France.* Ga.

50. A. conica Boiss., *op. cit.* 98 (1839). Annual; whole plant glandular-puberulent; stems 2·5–5 cm, branched. Leaves 5–15 mm, linear-subulate, obtuse. Inflorescence usually 3-flowered; pedicels 1½–2 times as long as sepals, erect or slightly patent. Sepals 3·5–4 mm, ovate-lanceolate, subacute, 3-veined; petals *c.* ⅔ as long as the sepals, oblong, white. Capsule slightly exceeding sepals, ovoid. Seeds 0·5–0·6 mm, with obtuse tubercles. *Cultivated ground.* ● *S. Spain.* Hs.

51. A. modesta Dufour, *Ann. Gén. Sci. Phys. (Bruxelles)* **7**: 291 (1821). Annual; stems 5–15 cm, often branched from base, glandular-pubescent. Leaves 3–10 mm, linear or linear-oblanceolate, acute, 1-veined. Inflorescence usually many-flowered, lax; pedicels 2–5 times as long as sepals, erect or patent in fruit, with patent glandular and minute deflexed eglandular hairs, rarely more or less eglandular. Sepals 3–4(–5) mm, ovate-lanceolate, acute with a very fine, sharp point, obscurely 3-veined, glandular-pubescent; petals more or less equalling sepals, oblong, obtuse, white. Capsule slightly exceeding sepals, ovoid-conical. Seeds *c.* 0·5 mm, reniform, with usually obtuse tubercles. $2n = 26$. *Dry places. E. & S. Spain; S. France.* Ga Hs.

Var. *purpurascens* Cuatrec., from S.E. Spain, has sepals less finely acute and seeds (0·6–)0·7–0·8 mm, with acute tubercles. In these respects it approaches **52**, but is quite distinct from it in the shape of its leaves. It may merit subspecific status under **51**.

52. A. retusa Boiss., *Voy. Bot. Midi Esp.* **2**: 99 (1839). Like **51** but leaves oblanceolate to obovate-oblanceolate, acuminate; sepals less finelv acute, with more prominent veins; petals truncate, twice as long as sepals; seeds *c.* 0·7 mm, with acute tubercles. *Stony and rocky places.* ● *S. Spain.* Hs.

53. A. capillipes (Boiss.) Boiss., *op. cit.* 98 (1839). Very slender annual; stems *c.* 5 cm, branched, capillary, glabrous above, scabrid with minute hairs below; internodes much longer than leaves. Leaves 2–4 mm, linear-lanceolate, acute, veinless. Inflorescence several- to many-flowered, very lax; pedicels 3–6 times as long as sepals, capillary, glabrous. Sepals 2·5 mm, lanceolate, acute, 3-veined; petals more or less equalling sepals. Capsule ovoid-globose. *Sandy places.* ● *S. Spain.* Hs.

Subgen. **Arenariastrum** F. N. Williams. Annuals. Styles 2.

54. A. provincialis Chater & Halliday, *Feddes Repert.* **69**: 50 (1964) (*A. gouffeia* Chaub. nom. illeg., *Gouffeia arenarioides* Robill. & Cast. ex DC.). Glabrous; stems 10–20 cm, many, slender, branched, ascending. Basal leaves linear-spathulate, obtuse; cauline leaves linear-lanceolate, acute, subconnate at base; all leaves ciliate at base, obscurely 3-veined. Flowers in a lax cyme, rarely solitary and terminal; bracts herbaceous; pedicels mostly longer than sepals, capillary. Sepals 3–4 mm, lanceolate, acute, 3- to 5-veined; petals equalling sepals, oblong, denticulate, white. Capsule 2–3 mm, oblong-ovoid, with 2 bifid teeth. $2n = 40$. *Rocky and sandy places.* ● *S.E. France (near Marseille and Toulon).* Ga.

2. Moehringia L.

G. HALLIDAY (EDITION 1) REVISED BY D. J. N. HIND (EDITION 2)

Annual (rarely biennial) or perennial herbs. Stems sometimes weak or fragile, often straggling. Leaves opposite, spathulate or ovate to linear, entire. Inflorescence a lax, usually few-flowered, simple dichasium, sometimes of solitary axillary flowers. Flowers 4- or 5-merous; petals white, usually entire, rarely rudimentary; stamens 8 or 10, rarely reduced to 5; styles 2 or 3. Capsule subglobose, rarely elongated or pyriform, dehiscing with 4 or 6 teeth. Seeds reniform to subreniform, dark reddish-brown to black, often shiny, smooth or tuberculate, with a persistent strophiole.

The genus differs from *Arenaria* in possessing strophiolate seeds, and in its basic chromosome number, $x = 12$. The shape and size of strophioles have been increasingly used in the discrimination of species, but the differences are not easily described and moreover are unreliably preserved in dried material. There are good illustrations in the account of the genus in S. Pignatti, *Flora d'Italia* **1**: 195–199 (1982).

Literature: H. Merxmüller & J. Grau, *Mitt. Bot. Staatssam. (München)* **6**: 257–273 (1967). S. M. Walters, *Bot. Jour. Linn. Soc.* **109**: 323–324 (1992).

1 Stems glabrous
2 Leaves ciliate at base
3 Decumbent perennial; stems often rooting **20. ciliata**
3 Annual or biennial; stems not rooting
4 Lower leaves spathulate; upper leaves linear-lanceolate **4. diversifolia**
4 All leaves ±ovate **5. minutiflora**
2 Leaves glabrous
5 Flowers solitary; pedicels without bracteoles **17. markgrafii**
5 Flowers usually in a simple cyme; pedicels bracteolate
6 Flowers mostly 4-merous
7 Plant green; leaves sometimes shiny when fresh, rarely glaucescent
8 Stems freely rooting, often procumbent; lower leaves never fleshy **15. muscosa**
8 Stems arising from apex of rootstock, often pendent, not rooting; lower leaves often fleshy **16. intermedia**
7 Plant grey (rarely green); leaves glaucous, never shiny when fresh
9 Leaves usually 5–10 mm, cylindrical; stems weak and slender **25. sedoides**
9 Leaves mostly more than 10 mm, usually flat; stems sturdy, not very slender
10 Petals not more than 4 mm; strophiole simply branched **18. papulosa**
10 Petals usually more than 5 mm; strophiole intricately branched **19. tommasinii**
6 Flowers mostly 5-merous
11 At least the upper leaves linear
12 Petals at least 4 mm
13 Upper leaves glaucous; stems slightly thickened at nodes; petals usually $1\frac{1}{2}$–2 times as long as sepals **22. bavarica**
13 Whole plant glaucous; stems strongly thickened at nodes; petals less than $1\frac{1}{2}$ times as long as sepals **23. insubrica**
12 Petals 2–4 mm
14 Sepals 1·8–2·2 mm; seeds tuberculate **10. pichleri**
14 Sepals 2·5–3 mm; seeds smooth **21. glaucovirens**
11 Leaves lanceolate, ovate or broadly spathulate
15 Pedicels erect in fruit
16 Leaves distinctly petiolate **6. intricata**
16 Leaves sessile or subsessile
17 Leaves less than 3 times as long as broad **7. tejedensis**
17 Leaves more than 3 times as long as broad **8. lebrunii**
15 Pedicels deflexed in fruit
18 Sepals obtuse **10. pichleri**
18 Sepals acute or acuminate
19 Leaves 2–3 mm wide; seeds 1·5–2·0 mm **24. dielsiana**
19 Leaves less than 2 mm wide; seeds not more than 1 mm
20 Leaves lanceolate; petals *c.* 7 mm **11. villosa**
20 Leaves oblanceolate-spathulate; petals *c.* 3 mm **14. hypanica**
1 Stems at least slightly pubescent
21 Stems with 2 distinct lines of hairs, at least at base
22 Perennial; strophiole large **13. pendula**
22 Annual; strophiole small
23 Lamina distinctly hairy (margin conspicuously ciliate); capsule ovoid; seeds smooth **2. trinervia**
23 Lamina with few scattered hairs; capsule ±globose; seeds tuberculate **3. pentandra**
21 Stems with scattered hairs, not in distinct lines even at base
24 Upper leaf-surface pubescent **9. grisebachii**
24 Upper leaf-surface glabrous or subglabrous
25 Leaves ovate to elliptic-lanceolate; strophiole small, ±entire **1. lateriflora**
25 Leaves linear to linear-lanceolate or lanceolate; strophiole large or medium-sized, papillate or branched
26 Stems procumbent, readily rooting **20. ciliata**
26 Stems ±erect, not rooting
27 Petals *c.* 7 mm; seeds smooth **11. villosa**
27 Petals 3–4 mm; seeds tuberculate **12. jankae**

1. M. lateriflora (L.) Fenzl, *Vers. Darstell. Alsin.* tab. ad p. 18 & p. 38 (1833) (*Arenaria lateriflora* L.). Rhizomatous perennial with erect, pubescent stems, 5–20 cm. Leaves 10–30 × 3·5–10 mm, much shorter than internodes, ovate to elliptic-lanceolate, obtuse, 1- to 3-veined, glabrous above, ciliate on midrib and on margins. Cymes 1- to 5-flowered; pedicels slender, puberulent. Flowers 5-merous; sepals 2–3 mm, ovate or obovate, obtuse or acute, glabrous with scarious margins; petals 4–6 mm; styles 3. Capsule 5 × 3 mm, ovoid, often exserted from calyx. Seeds 1–1·4 mm, dark brown, glossy, smooth; strophiole small, more or less entire. $2n = c.$ 52. *Woods, meadows and screes. N.E. Europe, extending southwards to 50° N. in C. Ukraine.* Fe No Rs(N, B, C, E) Su.

2. M. trinervia (L.) Clairv., *Man. Herb. Suisse* 150 (1811) (*Arenaria trinervia* L.). Annual, with stems 10–30(–45) cm, prostrate or ascending, diffusely branched, pubescent with hairs forming 2 distinct lines. Leaves 6–25(–45) × 8–17 mm, ovate, acute, 3(5)-veined; upper leaves more or less sessile, ovate to lanceolate; margins entire, ciliate. Inflorescence simple, of 1- to 7(–15)-flowered cymes; pedicels slender, pubescent, becoming recurved after anthesis. Flowers mostly 5-merous; sepals 3–5 mm, lanceolate, 3-veined, acuminate, with ciliate scarious margins; petals 3–5 mm; styles usually 3. Capsule 3·5 mm, ovoid, usually included in sepals. Seeds 0·8–1·2 × 0·6–1·0 mm, reniform, dark brown or blackish, shining, smooth, but with distinct cellular pattern; strophiole small, whitish. $2n = 24$. *Woods and shady places. Almost throughout Europe.* All except Az Bl Cr Fa Is Sb.

3. M. pentandra Gay, *Ann. Sci. Nat.* ser. 1, **26**: 230 (1832) (*M. trinervia* subsp. *pentandra* (Gay) Nyman, *Arenaria pentandra* (Gay) Ard.; incl. *M. thasia* Stoj. & Kitanov). Like **1** but more delicate; leaves glabrous (but petiole ciliate), rarely with scattered cilia; flowers solitary in axils of leaves and in forks of branching stems, never in simple cymes; sepals 2–3(–4) mm, with wider scarious margins and indistinct lateral veins; petals rudimentary, 0·5–1·2 mm; stamens usually 5; capsule 2–2·5 mm, subglobose, usually just exceeding calyx; seeds tuberculate. $2n = 48$. *In drier habitats than* 2, *in woods and open ground. S.W. Europe, extending locally eastwards to N. Greece.* Bl Co Ga Gr Hs It Lu Sa Si.

4. M. diversifolia Dolliner ex Koch, *Flora* (*Regensb.*) **22**: 2 (1839). More or less glabrous annual or biennial with prostrate to ascending stems 7–23 cm, branching profusely. Leaves bright green, dimorphic; lower leaves spathulate, with few, long cilia around the connate leaf-bases, the lamina cuneate, petiolate, usually glabrous, occasionally with scattered hairs, 5–10 × 2–6 mm, acute to obtuse; upper leaves 20–40(–50) × 1–3·5 mm, sessile, linear-lanceolate, acute. Inflorescence a terminal, glabrous, 3- to 7-flowered cyme; pedicels glabrous, bracteolate on upper flowers. Flowers 5-merous; sepals 1·8–2 mm, ovate, subobtuse to obtuse, with broad scarious margins; petals 1·5–2 mm; styles 3. Capsule 2 × 3 mm, subglobose. Seeds 0·8–1·1 × 0·6–0·8 mm, brown to black, shiny, smooth; strophiole large, white to yellowish. $2n = 24$. *Rock-crevices and screes; usually calcicole.* ● *S.E. Austria.* Au.

5. M. minutiflora Bornm., *Feddes Repert.* **16**: 183 (1919). Glabrous or subglabrous, very slender annual; stems 3–17 cm, procumbent to ascending, branching profusely below. Leaves 7–10(–15) × 4–8(–10) mm, ovate, 3- to 5-veined, with scattered cilia towards base, acute to subacute, more or less acuminate, often very pale green. Flowers solitary in axils of branches of the dichasium, or sometimes forming a simple 3- to 15-flowered cyme; pedicels ebracteolate, filiform. Flowers usually 5-merous; sepals 1–2 mm, glabrous, keeled, with broad hyaline margins; petals to 1·5 × 0·5 mm, rudimentary, white, entire or crenate; stamens usually 5; styles usually 3. Capsule 1·5 mm, subglobose, included within calyx. Seeds 0·5–0·8 mm, reniform, brown to blackish, shiny, smooth; strophiole minute, whitish to pale yellowish-brown. $2n = 24$. *Crevices under granite boulders and in fissures of granite rocks.* ● *Makedonija* (*Prilep*). Ju.

6. M. intricata R. de Roemer ex Willk., *Linnaea* **25**: 14 (1852). Caespitose, glabrous perennial, woody at base, with profusely branched stems, 6–15(–22) cm. Leaves spathulate to lanceolate, somewhat glaucous, petiolate; the lower 2–8 × 1·5–2·5 mm, distinctly fleshy, the upper 7–15 × 2–6 mm, less fleshy, narrowly elliptical to spathulate, acuminate. Inflorescence a simple, 3- to 15-flowered cyme; pedicels (7–)12–20(–35) mm, slender, bracteolate. Flowers 5-merous; sepals 2·5–3·5 mm, oblong-lanceolate, with narrow scarious margins, glabrous; petals 5–6 × 2–2·5 mm, ovate-lanceolate; styles 3. Capsule *c.* 3·5 mm, globose. Seeds *c.* 1·3 mm, brown, rather dull, tuberculate, especially on dorsal ridge; strophiole papillate, white becoming brown. $2n = ?26$. *Limestone crevices and caves.* ● *Mountains of C. & S.E. Spain.* Hs.

7. M. tejedensis Huter, Porta & Rigo ex Willk., *Suppl. Prodr. Fl. Hisp.* 275 (1893) (? incl. *M. fontqueri* Pau). Caespitose, usually glabrous perennial; stems 5–13(–15) cm, arising from a central rootstock, ascending, simple or basally branched, clothed in remains of dead leaves. Leaves 4–10 × 1·5–3·5 mm, glabrous, persistent, crowded below, bright green when young; lower leaves sessile to subsessile; upper leaves ovate, somewhat fleshy, apiculate. Inflorescence a terminal 1- to 7-flowered simple dichasium; pedicels 15–25 mm, bracteolate. Flowers 5-merous; sepals lanceolate, with broad scarious margins; petals obovate; styles 3. Capsule 3·5–4 × 2·5 mm, ovoid to subglobose. Seeds 1·2 × 0·9–1·0 mm, reniform, tuberculate especially on dorsal ridge; strophiole papillate, brownish. *Crevices of limestone rocks.* ● *S.E. Spain* (*Sierra Tejeda and Sierra Granada*). Hs.

8. M. lebrunii Merxm., *Monde Pl.* **50**: 5 (1965) (*M. dasyphylla* auct., pro parte, *M. papulosa* auct., non Bertol.). Caespitose, rather glaucous perennial; stems 5–30 cm, numerous, pendent or erect, profusely branched, glabrous, somewhat glaucous; old leaves persistent. Leaves 7–25(–35) × 1·5–4 mm, oblanceolate, cuneate, acute or mucronate; lower leaves rather fleshy, rigid, the upper leaves thinner. Inflorescence a 1- to 3-flowered cyme; pedicels 20–40(–80) mm, robust, bracteolate. Flowers (4)5-merous; sepals 3–4 mm, lanceolate, with scarious margins; petals 6–7 × 3–3·5 mm; styles (2)3. Capsule *c.* 3 mm wide, subglobose, included in the calyx. Seeds *c.* 1·5 mm, dark brown, shiny, smooth; strophiole branched, whitish. $2n = 24$. *Rock-fissures and ledges; calcicole.* ● *Maritime Alps.* Ga It.

9. M. grisebachii Janka, *Österr. Bot. Zeitschr.* **23**: 194 (1873). Caespitose perennial; stems 5–13 cm, glaucous, profusely branched, densely pubescent. Leaves (6–)9–33 × 0·5–0·8(–2·3) mm, linear-lanceolate or -oblong, sessile, slightly fleshy, pubescent, acuminate or acute. Inflorescence a simple (3–)7- to 10-flowered cyme. Flowers 5-merous; sepals 1·8–2·5 mm, glabrous, obtuse, with broad scarious margins; petals 2–2·5 mm, obtuse; styles 3. Capsule 1·5 × 2·5 mm, subglobose, generally included within sepals. Seeds 0·9 mm, black, dull, tuberculate; strophiole papillate, small. *Damp, rocky areas; calcifuge.* ● *N.E. part of Balkan peninsula; E. Romania.* Bu Rm Tu.

10. M. pichleri Huter, *Österr. Bot. Zeitschr.* **54(12)**: 449 (1904). Caespitose perennial with glabrous, profusely-branched stems 4–7 cm. Leaves 11–25 × 0·4–0·7 mm, linear to lanceolate, glabrous. Inflorescence a simple 3- to 7-flowered cyme; pedicels 4–10 mm, filiform, bracteolate, deflexed in fruit. Flowers 5-merous; sepals 1·8–2·2 mm, glabrous, with broad scarious margins, obtuse; petals *c.* 2 × 1 mm; styles 3. Capsule subglobose, 2 × 3 mm. Seeds 0·9 mm, tuberculate; strophiole large, papillate. *Shady rocks.* ● *S. Bulgaria* (*Rodopi Planina*). Bu.

11. M. villosa (Wulfen) Fenzl, *Vers. Darstell. Alsin.* tab. ad p. 46 (1833) (*Arenaria villosa* Wulfen). Caespitose perennial with glabrous or pubescent stems 4–12 cm, the nodes usually glabrous. Leaves 7–20 × 0·5–1·6 mm, lanceolate, subglabrous with scattered hairs on midrib and margin, the lowest somewhat fleshy. Inflorescence a simple (1–)3- to 5-flowered cyme; pedicels glabrous or pubescent, bracteolate, deflexed in fruit. Flowers 5-merous; sepals 2·7–4 × 1·2 mm, lanceolate, acute, ciliate on margin and midrib, with narrow scarious margins; petals *c.* 7 × 2 mm; styles 3. Capsule 2·7 × 4 mm. Seeds 1·0 × 0·8 mm, black, shiny, smooth, occasionally with cellular outlines; strophiole large, branched, whitish. *Limestone cliffs.* ● *Mountains of N.W. Jugoslavia* (*Slovenija*). Ju.

12. M. jankae Griseb. ex Janka, *Österr. Bot. Zeitschr.* **23**: 195 (1873). Caespitose perennial, with stems 5–11 cm, erect, generally sparsely branched, glabrous above, pubescent towards base; internodes swollen above ciliate nodes. Lower leaves 12–20(–30) × 1·2–3(–4) mm, acuminate, distinctly petiolate, glabrous, except for petiole-base and scattered hairs on upper surface; upper leaves 8–15 × 1–1·5 mm, linear-lanceolate, subglabrous, ciliate on margins. Inflorescence a (1–)3- to 7-flowered cyme; pedicels bracteolate, filiform, usually deflexed in fruit; bracteoles 0·5–1 mm, ciliate on margins in lower ½ only. Flowers 5-merous; sepals 2·2–3 mm, acuminate, the margins ciliate at base; petals 3–3·5 mm; styles 3. Capsule 2 × 2–3 mm. Seeds 1 × 0·8 mm, black, dull, tuberculate; strophiole medium-sized, papillate. *Shady calcareous rocks.* ● *E. Bulgaria, E. Romania.* Bu Rm.

13. M. pendula (Waldst. & Kit.) Fenzl, *Vers. Darstell. Alsin.* tab. ad p. 46 (1833) (*Arenaria pendula* Waldst. & Kit.). Caespitose, suffrutescent perennial with numerous creeping, often pendent stems 14–25(–50) cm, often purplish above, with abundant hairs in 2 distinct lines, becoming glabrous. Leaves 10–25(–40) × 1–3·5(–5) mm, linear, lanceolate, or narrowly elliptical, flat, patent, glabrous beneath, with scattered hairs on upper surface, and ciliate on margins only at base, acute to acuminate. Inflorescence an erect, 1- to 7-flowered simple cyme at ends of short side branches; pedicels 10–17 mm, often swollen towards apex, becoming deflexed in fruit, bracteolate. Flowers 5-merous; sepals 3–4 mm, glabrous, sometimes with scattered cilia at base, keeled, lanceolate to subulate, with broad scarious margins; petals 4·5–8 mm, oblong, obtuse, somewhat narrowed into a claw, patent; styles 3. Capsule 4 × 5 mm. Seeds 1·2–1·4 × 1 mm, reniform, dark brown to black, shiny, tuberculate; strophiole large, branched, white. *Rocks and screes.* ● *From C. Romania to N. Greece.* Bu Gr Ju Rm.

14. M. hypanica Grinj & Klokov, *Jour. Bot. Acad. Sci. Ukr.* **7(4)**: 56 (1951). Caespitose, glabrous perennial; stems 8–10 cm, prostrate to ascending, diffusely branched, glabrous. Leaves 8–12 × 0·7–1·7 mm, oblanceolate-spathulate, apex acute or acuminate. Inflorescence a terminal 1- to 7-flowered cyme; pedicels 5–12 mm, deflexed in fruit; bracteoles linear-lanceolate with scarious margins. Flowers 5-merous; sepals 2–2·5 mm, elliptic-lanceolate, acuminate; petals 2·5–3 mm, ovate-lanceolate; styles 3. Capsule *c.* 2·5 mm, subglobose. Seeds 0·8 mm, dark reddish, smooth; strophiole white. *Granite outcrops.* ● *S. Ukraine, N. of Odessa.* Rs(W).

15. M. muscosa L., *Sp. Pl.* 359 (1753). Laxly caespitose, glabrous perennial; stems 10–20(–35) cm, procumbent to ascending, rarely pendent or erect, branched, the internodes often swollen above node. Leaves 10–30(–70) × 0·5–2 mm, linear to filiform, bright green, sometimes shiny. Inflorescence a terminal, 3- to 15(or many)-flowered divaricate cyme; pedicels 12–25 mm, swollen at base, becoming patent in fruit, but not recurved. Flowers usually 4-merous; sepals 3 mm, with narrow scarious margins; petals 5–7 mm, patent, without a distinct claw, acute; styles usually 2. Capsule 2·5 × 2 mm, subglobose to ovoid, usually slightly exceeding calyx, teeth usually somewhat recurved. Seeds 1·2–1·5 mm, black, shiny, smooth; strophiole entire, white. $2n = 24$. *Damp rocks and woodland.* ● *Mountains of S. & C. Europe northwards to* 50° *N.* Al Au Bu Cz Ga Ge He Hs Hu It Ju Po Rm Rs(W) Si.

Very variable in habit and leaf size. Hybrids occur frequently with **22** (*M.* × *coronensis* Behrendsen) and with **20** (*M.* × *hybrida* A. Kerner ex Hand.-Mazz.). Hybrid plants are very variable, usually with a more or less equal ratio of 4- and 5-merous flowers on the same plant.

16. M. intermedia (Loisel.) Panizzi, *Nuovo Gior. Bot. Ital.* **21(3)**: 477 (1889) (*M. provincialis* Merxm. & Grau). Caespitose, glabrous perennial; stems up to 25 cm, usually arising from a central rootstock, often pendent. Leaves often dark green, rarely somewhat glaucescent, linear to narrowly oblanceolate, apex acute, somewhat fleshy, upper surface flat, lower keeled; upper leaves 15–25(–30) × 1·5–2·5 mm. Inflorescence a 1- to 7-flowered cyme; pedicels becoming recurved in fruit, bracteolate. Flowers 4(5)-merous; sepals 4 × 1·5 mm, lanceolate, acute, keeled, with rather wide scarious margins; petals 7 × 3 mm, broadly elliptical, truncate at base, obtuse; styles 2 or 3. Capsule 3–4 × 3 mm, subglobose. Seeds *c.* 1·3 mm, black, shiny, smooth; strophiole large, branched, white. $2n = 24$. *Limestone fissures.* ● *S.E. France* (*Alpes de Haute-Provence*). Ga.

17. M. markgrafii Merxm. & Guterm., *Phyton* (*Austria*) **7**: 1 (1957). Caespitose, glabrous perennial, becoming suffruticose; stems 10–15 cm, filiform, usually few, decumbent, arising from a single rootstock, becoming branched 2–4 cm above crown. Leaves on mature stems (3–)4–6 × 1 mm, semi-terete, but shrivelling at maturity; leaves on younger stems 15–30 × 1–1·5 mm, rather fleshy, distinctly keeled, linear or linear-lanceolate. Flowers solitary, axillary or subterminal, rarely terminal, ebracteolate; pedicel usually thinner than branches, but markedly swollen at apex below flower, becoming recurved in fruit. Flowers 4-merous; sepals 2·7–3 mm, with a distinctly keeled dorsal surface and broad scarious margins; petals *c.* 4 mm, somewhat longer than sepals; styles 2. Capsule 2–2·5 mm, globose, included within calyx. Seeds 1·4–1·7 mm, black, shiny, smooth; strophiole branched, white, becoming brownish. $2n = 24$. *Fissures of shady limestone cliffs.* ● *N. Italy* (*Alpi Bresciane*). It.

18. M. papulosa Bertol., *Fl. Ital.* **4**: 363 (1840). Suffrutescent, glabrous perennial; stems 5–15 cm, pendent, somewhat glaucous and fleshy. Leaves *c.* 20 × 2 mm, oblanceolate, usually fleshy. Inflorescence a 3- to 5-flowered cyme; pedicels swollen below calyx, bracteolate. Flowers 4-merous; sepals 3 mm, broadly lanceolate, acute; petals 4 × 2 mm, not or only indistinctly clawed; styles 2. Capsule 2·5–3 mm, subglobose. Seeds 1·2 × 1·0 mm, black, shiny, smooth; strophiole large, simply branched, white. $2n = 24$. *Limestone cliffs and overhanging rocks.* ● *C. Appennini.* It.

19. M. tommasinii Marchesetti, *Boll. Soc. Adr. Sci. Nat. Trieste* **5**: 327 (1880). Caespitose, suffrutescent perennial; stems 5–20 cm, pendent to procumbent, occasionally suberect, glabrous, glaucous, somewhat fleshy, occasionally swollen at nodes, becoming fragile when dry. Leaves 5–20(–25) × 1–1·5 mm, semi-terete, rather fleshy, the upper flat, linear or elongate-spathulate, keeled, acute. Inflorescence a terminal 3- to 7-flowered cyme; pedicels 8–20 mm, becoming markedly recurved in fruit, slightly swollen at apex, bracteolate. Flowers 4-merous; sepals 2–2·5 mm, lanceolate, with rather wide scarious margins; petals ovate-lanceolate, (4–)5–7 × 3 mm; styles 2(–4). Capsule 3 mm, subglobose to ovoid, more or less included within calyx. Seeds 1–1·2 mm, black, shiny; testa smooth; strophiole intricately branched, white. $2n = 24$. *Calcareous rocks and cliffs.* ● *E. part of Istrian peninsula.* It Ju.

Flowers with shorter petals are sometimes found and plants may also show some form of teratological aberration.

20. M. ciliata (Scop.) Dalla Torre in Hartinger, *Atlas Alpenfl.* (*Text*) 78 (1882) (*Stellaria ciliata* Scop., *M. polygonoides* (Wulfen) Mert. & Koch). Decumbent, caespitose, subglabrous or somewhat hairy perennial; stems 5–13 cm, often profusely branched, usually producing adventitious roots. Leaves 3–10 × 0·6–1·5 mm, usually linear to lanceolate, rather fleshy, glabrous except for margins usually ciliate towards base. Inflorescence a 1- to 5-flowered cyme; peduncle puberulent; pedicels 5–8 mm, deflexed in fruit, bracteolate. Flowers (4)5-merous; sepals 2·5–4 × 1 mm, ovate, glabrous, with broad scarious margins; petals 3·5–5 mm, elliptical, patent or recurved, acute; styles usually 3. Capsule 2–4 mm, subglobose, equalling or exceeding sepals. Seeds 1 × 1·2 mm, black, shiny, smooth; strophiole large, branched, white. $2n = 24$. *Limestone rocks and screes.* ● *Alps; W. part of Balkan peninsula.* Al Au Ga Ge He It Ju.

Very variable in habit and hairiness. Dwarf, very compact plants with subsessile flowers with petals almost twice as long as sepals have been called subsp. *nana* (St.-Lager) Schinz & R. Keller, but this variation does not seem to be worth taxonomic distinction.

21. M. glaucovirens Bertol., *Fl. Ital.* **6**: 626 (1847). Caespitose, glabrous and glaucous perennial, forming dense, fragile tufts up to 30 cm in diameter; stems up to 10 cm, slender, branched from base, somewhat fleshy, often swollen above and below nodes. Leaves 5–10(–15) mm, linear, semi-terete, somewhat fleshy, persistent. Inflorescence a terminal 1- to 7-flowered cyme; peduncles 10–15 mm, somewhat fleshy; pedicels filiform, reflexed in fruit, bracteolate. Flowers (4)5-merous; sepals 2·5–3 mm, ovate-lanceolate, with narrow scarious margins; petals usually 3–4 mm, sometimes shorter; styles usually 3. Capsule 2 × 2·5 mm, subglobose. Seeds *c.* 1 mm, reniform, black, shiny, smooth; strophiole branched, whitish. $2n = 24$. *Shady limestone cliffs.* ● *S. Alps from 10°30′ to 12°30′ E.* It.

22. M. bavarica (L.) Gren., *Mem. Soc. Émul. Doubs.*, sér. 1, **1(2)**: 37 (1841) (*Arenaria bavarica* L.). Caespitose, suffrutescent, glabrous and glaucous perennial; stems 10–20(–30) cm, clothed in remains of old leaves, fleshy, becoming fragile, especially when dry; internodes slightly swollen above and below node. Lower leaves 3–7 mm, linear to linear-lanceolate, cylindrical to keeled, the upper 7–35(–40) mm, somewhat flattened. Inflorescence a terminal, (1–)3- to 7-flowered cyme; pedicels 10–25 mm, robust, fleshy, swollen below calyx, becoming reflexed in fruit, bracteolate. Flowers 5-merous; sepals with broad, scarious margins; petals ovate to obovate, obtuse, usually $1\frac{1}{2}$–2 times as long as sepals; styles 3. Capsule *c.* 3 mm, globose to ellipsoidal. Seeds 1·2 mm, black, shiny, smooth; strophiole large, branched, white. $2n = 24$. *Limestone cliffs.* ● *E. Alps; mountains of W. part of Balkan peninsula.* Au Al It Ju.

23. M. insubrica Degen, *Magyar Bot. Lapok* **24**: 76 (1926) (*M. bavarica* subsp. *insubrica* (Degen) W. Sauer). Caespitose, suffrutescent, glabrous perennial, the young parts distinctly pruinose or glaucous; stems 10–20(–30) cm, pendulous, erect at apex; internodes distinctly swollen above and below nodes. Leaves 5–15(–25) × 1·5–2·5 mm, green or glaucous, often fleshy. Inflorescence a terminal, (1–)3- to 7-flowered cyme; pedicels 10–25 mm, usually swollen at base, less markedly swollen below calyx, erect, becoming reflexed in fruit, bracteolate. Flowers 5-merous; sepals ovate-lanceolate, with broad scarious margins; petals ovate-oblong, less than $1\frac{1}{2}$ times as long as sepals; styles 3. Capsule *c.* 2·5 mm, subglobose to pyriform. Seeds 1–1·5 mm, black, shiny, smooth; strophiole large, branched, white. $2n = 24$. ● *N. Italy (Alpi Bresciane on the east shore of Lago d'Iseo).* It.

24. M. dielsiana Mattf., *Ber. Deutsch. Bot. Ges.* **43**: 509 (1925). Caespitose, glabrous, somewhat fleshy perennial; stems 5–20 cm, glaucous, fragile, prostrate, ascending or pendent. Upper leaves 5–11 × 2–3 mm, oblanceolate (rarely linear-lanceolate), fleshy, acute or shortly mucronate, usually longer than internodes. Inflorescence a terminal 1- to 5-flowered cyme; pedicels 10–15 mm, more or less swollen at apex, becoming recurved in fruit, bracteolate. Flowers (4)5-merous; sepals *c.* 4 mm, oblanceolate, acute, glabrous, with broad scarious margins; petals *c.* 6 mm, obtuse; styles 3. Capsule 2·5 × 3 mm, globose to pyriform, usually included within calyx. Seeds 1·5–1·8 mm, black, shiny, smooth; strophiole branched, white, becoming pale brown. $2n = 24$. *Limestone cliffs.* ● *N. Italy (Alpi Bergamasche).* It.

25. M. sedoides (Pers.) Cumino ex Loisel., *Fl. Gall.* 725 (1807) (*M. sedifolia* Willd., *M. dasyphylla* Bruno ex Gren. & Godron). Densely to laxly caespitose, glabrous and glaucous perennial; stems arising from a rather woody stock, usually fleshy, becoming fragile when dry, either rather erect, up to 15 cm, producing hemispherical clumps, or pendent, producing larger plants. Leaves 5–15 mm, cylindrical, usually fleshy. Inflorescence a terminal, 1- to 7-flowered cyme; peduncle 8–15 mm; pedicels 6–15 mm, filiform, bracteolate. Flowers 4(5)-merous; sepals 3–4 mm, with scarious margins; petals 5–7 mm, patent; styles usually 2. Capsule 3–5 mm, subglobose to ellipsoidal. Seeds 1·2 × 1 mm, black, shiny, smooth; strophiole large, branched, whitish. $2n = 24$. *Shady limestone and conglomerate cliffs.* ● *Maritime Alps.* Ga It.

3. Minuartia L.

G. HALLIDAY

Like *Arenaria* but leaves usually narrowly lanceolate, setaceous or subulate; capsule dehiscing with as many teeth as styles; capsule-teeth wide and obtuse; seeds 1 to many, smooth, tuberculate or fimbriate. (Incl. *Alsine* auct., *Cherleria* L., *Queria* L.)

Literature: J. Mattfeld, *Feddes Repert. (Beih.)* **15**: 1–228 (1922). J. McNeill, *Notes Roy. Bot. Gard. Edinb.* **24**: 133–155 (1962); 311–401 (1963).

1 Annual or biennial, lacking vegetative stems at time of flowering
2 Sepals 1-veined
3 Stems and sepals glandular-pubescent; pedicels up to $\frac{1}{2}$ as long as sepals
4 Glandular hairs short and crisped; fascicles of flowers sessile **10. campestris**
4 Glandular hairs patent; fascicles of flowers pedunculate **12. glomerata**
3 Stems and sepals usually glabrous or sparsely pubescent; pedicels about equalling sepals
5 Stems erect; cymes in small, dense fascicles **13. rubra**
5 Stems ascending; cymes elongate, corymbose **14. funkii**
2 Sepals 3-veined
6 Stems slender; cymes usually lax; scarious margin of sepals usually narrow
7 Leaves ciliolate at base; sepals with a broad scarious margin **6. regeliana**
7 Leaves not ciliolate; sepals with a narrow scarious margin
8 Pedicels not longer than sepals; cymes crowded **5. mediterranea**
8 Pedicels usually longer than sepals; cymes lax
9 Sepals 2–3 times as long as broad; petals at least $\frac{3}{4}$ as long as sepals
10 Seeds obscurely tuberculate; stems decumbent or erect **1. mesogitana**
10 Seeds acutely tuberculate; stems erect **3. bilykiana**
9 Sepals 3–5 times as long as broad; petals not more than $\frac{3}{4}$ as long as sepals
11 Capsule usually shorter than sepals; seeds 0·3–0·4 mm; sepals 2–3(–3·5) mm **2. viscosa**
11 Capsule 1–$1\frac{1}{2}$ times as long as sepals; seeds (0·3–)0·4–0·6 mm; sepals (2·5–)3–5 mm **4. hybrida**
6 Stems robust; cymes condensed into dense fascicles;

scarious margin of sepals extending half-way to mid-vein
12 Bracts elongate, strongly curved; capsule with 1 seed **7. hamata**
12 Bracts neither elongate nor strongly curved; capsule with 5 or more seeds
13 Plant densely glandular-pubescent; pedicels $\frac{1}{2}$ as long as sepals **11. globulosa**
13 Plant with crisped, eglandular hairs; flowers subsessile
14 Fascicles of flowers forming a dense, wide, terminal capitulum; stamens 3 **8. dichotoma**
14 Fascicles of flowers axillary and terminal; stamens 10 **9. montana**
1 Perennial; vegetative shoots usually present at time of flowering
15 Sepals appearing veinless, densely glandular-pubescent; petals usually pale pink **53. geniculata**
15 Sepals usually distinctly veined; petals white (rarely pale lilac)
16 Sepals obtuse, oblong
17 Petals absent or rudimentary; flowers solitary **52. sedoides**
17 Petals equalling or exceeding sepals; cymes usually 3- to many-flowered
18 Leaves prominently ciliate; seeds 2–2·5 mm, fringed with long tubercles **49. macrocarpa**
18 Leaves not prominently ciliate; seeds not more than 2 mm, usually smooth or with low tubercles
19 Flowering stems 0·5–1 cm, mostly lateral, scarcely exserted from the cushion; flowers solitary **47. handelii**
19 Flowering stems at least 3 cm
20 Plant glabrous **43. wettsteinii**
20 Plant pubescent, at least above
21 Petals at least twice as long as the sepals
22 Plant densely caespitose **50. arctica**
22 Plant with conspicuous woody rhizomes **45. baldacii**
21 Petals usually less than twice as long as the sepals
23 Plant usually laxly caespitose; petals $1\frac{1}{2}$–2 times as long as the sepals
24 Leaves 10–20 mm; inflorescence densely glandular-pubescent; sepals 5–7 mm **44. capillacea**
24 Leaves 5–12 mm; inflorescence with dense crispate hairs, rarely glandular-pubescent; sepals 4–5·5 mm **48. laricifolia**
23 Plant densely caespitose; petals not more than $1\frac{1}{2}$ times as long as the sepals
25 Inflorescence with dense crispate hairs, rarely glandular-pubescent **48. laricifolia**
25 Inflorescence glandular-pubescent
26 Leaves acute; sepals obtuse **46. garckeana**
26 Leaves obtuse; sepals acuminate **51. biflora**
16 Sepals acute, ovate to linear-lanceolate
27 Sepals 1-veined; scarious margin extending nearly to vein
28 Pedicels and sepals glandular-pubescent
29 Sepals 2–4 mm; vegetative fascicles of leaves short, ovoid, compact **20. anatolica**
29 Sepals 4–6 mm; vegetative fascicles of leaves long
30 Petals less than $\frac{1}{2}$ as long as sepals **21. trichocalycina**
30 Petals more than $\frac{1}{2}$ as long as sepals
31 Petals $\frac{2}{3}$ as long as sepals **12. glomerata**
31 Petals equalling or slightly exceeding sepals **19. adenotricha**
28 Pedicels and sepals glabrous
32 Petals less than $\frac{1}{2}$ as long as sepals; stems procumbent **21. confusa**
32 Petals more than $\frac{1}{2}$ as long as sepals; stems erect
33 Cymes 1- to 4-flowered **18. krascheninnikovii**
33 Cymes usually many-flowered
34 Petals slightly exceeding sepals **16. setacea**
34 Petals not exceeding sepals
35 Sepals 3·5–5·5 mm, linear-lanceolate **15. mutabilis**
35 Sepals 3–4 mm, ovate-lanceolate **17. bosniaca**
27 Sepals 3- to 7(–9)-veined; scarious margin extending at most to lateral veins
36 Leaves linear-setaceous to -subulate
37 Sepals 3-veined
38 Leaves 1-veined; plant glabrous
39 Stems laxly caespitose; cymes 1- to 4-flowered; pedicels 15–50 mm **41. stricta**
39 Stems densely caespitose; flowers solitary, rarely present; pedicels 5–15 mm **42. rossii**
38 Leaves 3-veined; stems usually glandular-pubescent, at least above
40 Sepals erect at anthesis
41 Veins of leaf all equally prominent; pedicels patent to erecto-patent **38. pichleri**
41 Midrib of leaf more prominent than lateral veins; pedicels erect to erecto-patent
42 Leaves rigid, somewhat fleshy
43 Leaves 5–10 mm (Krym) **36. taurica**
43 Leaves 10–20 mm (Greece) **37. juniperina**
42 Leaves not rigid, not fleshy
44 Leaves glabrous
45 Flowers usually in pairs; capsule $1\frac{1}{2}$ times as long as sepals **33. austriaca**
45 Cymes 3- or more-flowered; capsule and sepals $\pm$equal
46 Sepals 2–3 mm **31. grignensis**
46 Sepals 3·5–5 mm **35. villarii**
44 Leaves glandular-pubescent
47 Flowers solitary or in pairs **34. helmii**
47 Cymes usually 3- or 4-flowered **35. villarii**
40 Sepals patent at anthesis
48 Leaves usually secund; stems black below **23. recurva**
48 Leaves somewhat recurved but never secund; stems not black below
49 Petals ovate, acute **39. verna**
49 Petals obovate, obtuse **40. rubella**
37 Sepals 5- to 7-veined at least near base
50 Stems laxly caespitose **22. hirsuta**
50 Stems densely caespitose; plant often pulvinate
51 Cymes 1- to 5(–8)-flowered, lax; pedicels 1–3 times as long as sepals **23. recurva**
51 Cymes many-flowered, dense; pedicels not longer than sepals **24. bulgarica**
36 Leaves linear, oblanceolate, lanceolate or oblong-elliptical

52 Leaves oblong-elliptical, cucullate above; flowers 4-merous **32. cherlerioides**
52 Leaves linear to lanceolate, oblanceolate or narrowly oblong, flat; flowers 5-merous
53 Leaves usually narrowly linear, 3-veined; plant glabrous **31. grignensis**
53 Leaves linear-lanceolate to lanceolate, oblanceolate or narrowly oblong, 5-veined; stems usually glandular-pubescent above
54 Sepals 1- to 3-veined in upper ½
55 Petals and capsule 1½ times as long as sepals; pedicels longer than sepals **29. cerastiifolia**
55 Petals and capsule equalling sepals **30. rupestris**
54 Sepals 5- to 7(–9)-veined in upper ½
56 Flowering stems 1–2 cm; flowers usually solitary **27. stellata**
56 Flowering stems 2–18 cm; cymes (1)2- to 12-flowered
57 Leaves oblanceolate to narrowly oblong; petals 1–2 times as long as sepals; flowering stems 2–5 cm **28. pseudosaxifraga**
57 Leaves ovate- to linear-lanceolate; petals ± equalling to slightly exceeding sepals; flowering stems often more than 5 cm
58 Leaves not rigid; outer veins of sepals curving outwards **25. saxifraga**
58 Leaves rigid; outer veins of sepals parallel **26. graminifolia**

Subgen. **Minuartia**. Annuals or perennials. Flowers usually white. Radicle incumbent.

Sect. SABULINA (Reichenb.) Graebner. Annuals. Leaves linear, rarely linear-spathulate, 3-veined at least at the base. Sepals 3-veined, patent at anthesis, not becoming indurated; scarious margin narrow. Petals seldom exceeding sepals; stamens 3–10.

1. M. mesogitana (Boiss.) Hand.-Mazz., *Ann. Naturh. Mus. (Wien)* **26**: 148 (1912) (*M. tenuifolia* subsp. *mesogitana* (Boiss.) Prodan). Annual; stems up to 12 cm. Leaves linear, somewhat fleshy. Cymes lax, sparsely glandular-pubescent; pedicels patent. Sepals 2·25–3·5 mm, ovate-lanceolate; petals and capsule up to 1–1½ times as long as sepals. Seeds obscurely tuberculate. *Dry places. S.E. Romania; Balkan peninsula; Aegean region.* Al Bu Cr Gr Ju Rm.

1 Stems usually simple; seeds 0·3–0·5 mm (b) subsp. **kotschyana**
1 Stems branched; seeds 0·5–0·7 mm
2 Plant decumbent; leaves 8–20 mm (a) subsp. **mesogitana**
2 Plant erect; leaves 4–8 mm (c) subsp. **velenovskyi**

(a) Subsp. **mesogitana**: Plant decumbent, branched at base. Leaves 8–20 mm. Sepals 3–3·5 mm, more or less equalling or shorter than petals. Seeds 0·5–0·65 mm. *S.E. Romania, N.E. Bulgaria.*

(b) Subsp. **kotschyana** (Boiss.) McNeill, *Notes Roy. Bot. Gard. Edinb.* **28**: 19 (1967): Plant erect, the stems simple or only slightly branched. Leaves 8–20 mm. Sepals 2–3 mm, more or less equalling or shorter than petals. Seeds 0·3–0·5 mm. *S. Greece, Aegean region; ?S.W. Bulgaria.*

(c) Subsp. **velenovskyi** (Rohlena) McNeill, *Notes Roy. Bot. Gard. Edinb.* **24**: 389 (1963) (*M. velenovskyi* Rohlena): Plant erect, much-branched. Leaves 4–8 mm. Sepals 2–3 mm, longer than petals. Seeds 0·55–0·7 mm. ● *Albania, S. Jugoslavia.*

M. thymifolia (Sm.) Bornm., *Beih. Bot. Centr.* **31(2)**: 193 (1914), like **1** but a smaller plant with linear-spathulate, fleshier leaves and ovate sepals, has been recorded from maritime sands in Kriti but probably in error.

2. M. viscosa (Schreber) Schinz & Thell., *Bull. Herb. Boiss.* ser. 2, **7**: 404 (1907) (*M. piskunovii* Klokov, *M. tenuifolia* subsp. *viscosa* (Schreber) Briq., *Alsine viscosa* Schreber). Annual with rather slender, erect stems branched usually from above the middle; branches usually erect, more rarely erecto-patent; plant glandular-pubescent, often densely so, rarely glabrous. Leaves 4–9(–12) mm, linear-subulate. Cymes lax; pedicels 3–8 mm, slender. Sepals 2–3(–3·5) mm, lanceolate; petals distinctly shorter than sepals. Capsule shorter than to more or less equalling sepals. Seeds 0·3–0·4 mm, almost smooth. $2n = 46$. *Dry, sandy places. Widely distributed, but local, from N. France to E. Ukraine, and from Denmark to N. Greece.* Al Au Bu Cz Da Ga Ge Gr He Hu It Ju Po Rm Rs(C, W, E) Su ?Tu.

3. M. bilykiana Klokov in Kotov, *Fl. RSS Ucr.* **4**: 654 (1952). Like **2** but stems somewhat fleshy; branches often patent; sepals 2·25–3 mm; petals and capsule slightly exceeding sepals; seeds 0·5–0·6 mm, acutely tuberculate. ● *S.W. Ukraine (near Izmail).* Rs(W).

4. M. hybrida (Vill.) Schischkin in Komarov, *Fl. URSS* **6**: 488 (1936) (*M. tenuifolia* (L.) Hiern, non Nees ex C. F. P. Mart., *M. birjuczensis* Klokov, *Alsine tenuifolia* (L.) Crantz, *Arenaria tenuifolia* L.). Annual, but more robust than **2**, with erect stems 3–12(–20) cm, branched at base and from above the middle; plant usually glandular-pubescent, at least above, or sometimes glabrous. Leaves up to 12 mm. Cymes lax, many-flowered; pedicels 5–20 mm. Sepals (2·5–)3–5 mm, linear to ovate-lanceolate, acuminate; petals slightly shorter than sepals. Capsule 1–1½ times as long as sepals. Seeds (0·3–)0·4–0·6 mm, minutely tuberculate, reddish-brown. $2n = 46, 70$. *Dry, open habitats. S. & W. Europe, northwards to N. England, and extending locally to S.E. Russia; occasionally naturalized elsewhere.* Be Bl Br Bu Co Cr Cz Ga Ge Gr He Ho Hs It Ju Lu Rm Rs(C, W, K, E) Sa Si Tu [Au Da Hb].

A very variable species, particularly in height, pubescence and size and shape of the sepals.

Plants with glabrous, ovate-lanceolate sepals occurring from Italy to the Netherlands (**M. tenuifolia** subsp. **vaillantiana** (DC.) Mattf.) do not justify subspecific recognition, since the sepal-shape is inconstant and glabrous plants occur throughout the range of the species.

Plants from Krym and C. Greece (Parnes) with lanceolate, narrowly acute sepals, the petals almost as long as the sepals, have been referred to subsp. **turcica** McNeill, *Notes Roy. Bot. Gard. Edinb.* **24**: 395 (1963), a principally S.W. Asian taxon.

Plants from E. Greece, with smaller sepals, have been called subsp. **lydia** (Boiss.) Rech. fil., but the difference is inadequate for subspecific distinction.

5. M. mediterranea (Ledeb.) K. Maly, *Glasn. Muz. Bosni Herceg.* **20**: 563 (1908) (*M. tenuifolia* subsp. *mediterranea* (Ledeb.) Briq.). Like **4** in habit; stems 4–12 cm, glabrous or subglabrous. Cymes dense; pedicels usually shorter than sepals. Sepals 3–5 × 0·5–0·7 mm, narrowly linear-lanceolate; petals ⅓–½ as long as sepals, sometimes absent. Capsule usually included. Seeds 0·4 mm, smooth. $2n = 24$. *Mediterranean region, Portugal; N.W. France*

(*W. coast of Normandie*). Al Bl Bu Co Cr Ga Gr Hs It Ju Lu ?Sa Si Tu.

6. M. regeliana (Trautv.) Mattf., *Bot. Jahrb.* **57** Beibl. 126: 29 (1921). Glabrous annual; stems 5–15 cm, branched from base. Leaves 2–10 mm, narrowly linear, often ciliolate at base. Pedicels 3–17 mm, filiform. Sepals 2·5–3·5 mm, ovate-lanceolate, 3-veined in lower ½, with broad scarious margins. Capsule usually exserted. Seeds 0·5 mm, finely tuberculate. *Salt steppe and desert soils. Borders of W. Kazakhstan and S.E. Russia.* Rs(E). (*Caucasus, C. Asia.*)

Sect. MINUARTIA. Annuals or perennials. Leaves linear to linear-lanceolate. Cymes lax (perennials) or in dense fascicles (annuals). Sepals 3-veined, with scarious margin extending to lateral veins, or 1-veined with margin extending almost to vein; sepals becoming indurated; stamens 3–10.

7. M. hamata (Hausskn.) Mattf., *Bot. Jahrb.* **57** Beibl. 126: 29 (1921) (*Queria hispanica* L., non *M. hispanica* sensu Mattf.). Annual, sparsely covered with crisped hairs; stems 3–7 cm, usually branched from base. Leaves 10–20 mm, linear-setaceous. Bracts 3- to 5-veined, shorter than leaves, abruptly narrowed from a very broad sheathing base, hamate, incurved above, with broad, scarious margin; fascicles of cymes in capitula which are deciduous at maturity; flowers sessile, the terminal rudimentary. Sepals *c.* 4 mm, lanceolate, becoming indurated, 3-veined; scarious margin broad; petals minute, subulate; stamens usually 10. Capsule with 1 seed. $2n = 30$. *Dry places. Spain; Balkan peninsula; Krym.* Al Bu Gr Hs Ju Rs(K) ?Tu.

8. M. dichotoma L., *Sp. Pl.* 879 (1753) (*M. hispanica* sensu Mattf., *Alsine dichotoma* (L.) Fenzl). Annual, covered with crispate hairs; stems 2–8 cm, erect, occasionally branched from the base; stems branched above to give a congested capitulum of fascicles of flowers. Leaves 8–15 mm, erecto-patent, much longer than internodes. Bracts scarious, exceeding the flowers, curved above. Sepals 3-veined, sparsely hairy; petals absent or rudimentary; stamens usually 3. Seeds 0·5–0·8 mm, 6–12, almost smooth. $2n = 60$. *Dry, sandy places. N. & C. Spain.* Hs.

9. M. montana L., *Sp. Pl.* 90 (1753) (*Alsine montana* (L.) Fenzl). Like **8** but more robust; stems often branched from base; leaves 15–20 mm, squarrose, sometimes recurved; fascicles of flowers terminal and axillary, shorter than bracts; sepals glabrous; stamens 10; seeds 0·6–0·8 mm, minutely tuberculate. *Dry, sandy and rocky places. C., E. & S. Spain; S.W. Bulgaria; Krym.* Bu Hs Rs(K).

(**a**) Subsp. **montana**: Inflorescence sparsely glandular-hairy. Petals 0·7–0·9 mm. $2n = 28$. *C., E. & S. Spain.*

(**b**) Subsp. **wiesneri** (Stapf) McNeill, *Notes Roy. Bot. Gard. Edinb.* **24**: 359 (1963) (*M. wiesneri* (Stapf) Schischkin): Inflorescence densely crispate-hairy. Petals absent. *S.W. Bulgaria; Krym.*

10. M. campestris L., *Sp. Pl.* **89** (1753) (*Alsine campestris* (L.) Fenzl). Glandular-pubescent annual; stems 2–10 cm. Leaves up to 15 mm, erect. Bracts and fascicles of flowers almost equal to and scarcely exceeding internodes; at maturity fascicles present at most nodes. Sepals 1-veined; petals rudimentary; stamens usually 5. Seeds 0·4–0·6 mm, minutely tuberculate. *Dry places. Spain.* Hs.

11. M. globulosa (Labill.) Schinz & Thell., *Bull. Herb. Boiss.* ser. 2, **7**: 403 (1907). Densely glandular-pubescent annual with erect stems up to 13 cm. Leaves linear-lanceolate. Bracts scarcely exceeding pedicels; cymes axillary and terminal; pedicels ½ as long as sepals, erecto-patent. Sepals lanceolate; petals ⅓ as long as sepals. Capsule included. Seeds 0·6–0·8 mm, tuberculate dorsally. *Dry, cultivated ground. W. Jugoslavia, E. Greece and Kriti.* Cr Gr Ju.

12. M. glomerata (Bieb.) Degen, *Mitt. Naturw. Ver. Steierm.* **46**: 319 (1910). Glandular-pubescent biennial or perennial (sometimes annual) with erect stems up to 20 cm. Leaves linear-setaceous, 3-veined at base. Fascicles of flowers terminal and axillary; pedicels up to 2(–3) mm. Sepals 4–6 mm, narrowly lanceolate, long-acuminate; petals ⅓–⅔ as long as sepals. *S.E. Europe, extending northwards to S. Czechoslovakia and N. Ukraine.* Al Bu Cz Gr Hu Ju Rm Rs(W, K) Tu.

(**a**) Subsp. **glomerata**: Annual or biennial. Leaves erect, appressed. Cymes dense, many-flowered. Petals ½ as long as sepals. $2n = 28$. *Throughout the range of the species, except the extreme south.*

(**b**) Subsp. **macedonica** (Degen & Dörfler) McNeill, *Notes Roy. Bot. Gard. Edinb.* **24**: 384 (1963) (subsp. *velutina* (Boiss. & Orph.) Mattf.; incl. *M. jordanovii* Panov, *M. rhodopaea* (Degen) Kožuharov & Kuzmanov): Perennial. Leaves often falcate. Cymes lax, few-flowered. Petals slightly shorter than sepals. $2n = 28$. *From Albania to Thrace.*

The relationship between subsp. (**b**) and *M. velutina* Boiss. & Orph. requires clarification.

13. M. rubra (Scop.) McNeill, *Feddes Repert.* **68**: 173 (1963) (*M. fastigiata* (Sm.) Reichenb., *M. fasciculata* auct., non (L.) Hiern, *M. cymifera* (Rouy & Fouc.) Graebner, *Alsine fasciculata* auct., non (L.) Wahlenb., *A. jacquinii* Koch). Biennial with erect stems up to 30 cm; branches erect, glabrous or sparsely hairy above, very rarely glandular-pubescent. Leaves 3-veined at base, linear-setaceous. Flowers in dense axillary and terminal fascicles; pedicels and sepals almost equal. Sepals 4–6 mm, linear, glabrous or sparsely hairy; petals ⅓ as long as sepals; stamens 5. Capsule included. Seeds 0·7–0·8 mm, distinctly but shortly tuberculate dorsally. $2n = 30$. *Dry, sandy places.* ● *S.C. Europe, northwards to* 50° N. *in W. Germany.* Au Cz Ga Ge He Hs Hu It Ju Rm.

14. M. funkii (Jordan) Graebner in Ascherson & Graebner, *Syn. Mitteleur. Fl.* **5**(1): 714 (1918) (*Alsine funkii* Jordan). Like **13** but with ascending stems and corymbose cymes. *E. Spain; S. France.* Ga Hs.

15. M. mutabilis (Lapeyr.) Schinz & Thell. ex Becherer, *Ber. Schweiz. Bot. Ges.* **48**: 296 (1938) (*M. rostrata* (Pers.) Reichenb., *M. lanuginosa* (Coste) P. Fourn., *Alsine mucronata* auct., non L., *A. rostrata* (Pers.) Fenzl, ?*M. trichocalycina* (Ten. & Guss.) Grande). Laxly caespitose, usually glabrous perennial with numerous, erect flowering stems 10–20 cm, branched above. Leaves erect to erecto-patent. Terminal cymes usually many-flowered; pedicels usually equalling sepals. Sepals 3·5–5·5 mm, narrowly lanceolate; petals slightly shorter than sepals. Capsule about equalling sepals. Seeds 0·7–0·8 mm, distinctly tuberculate dorsally. $2n = 28, 30$. *Dry places. Mountains of S.W. Europe, extending to C. France and to* 11° E. *in N. Italy.* Co Ga He Hs It.

16. M. setacea (Thuill.) Hayek, *Fl. Steierm.* **1**: 271 (1908) (*Alsine setacea* (Thuill.) Mert. & Koch). Caespitose perennial with numerous erect flowering stems up to 20 cm, with crispate hairs below, glabrous above. Leaves 8–15 × 0·2 mm, setaceous-filiform, 3-veined at base, usually erect, rarely patent or falcate, scabrid. Cymes 3- to many-flowered, lax (dense in var. *parviflora* (Velen.) Prodan); pedicels usually 2–4 times as long as sepals. Sepals 2–5 mm, ovate-lanceolate to lanceolate; petals and capsule usually slightly exceeding sepals. Seeds 0·5–0·7 mm, with low, rounded tubercles. $2n = 30$. *C. & S.E. Europe, extending to W. France and to 39° E. in S.C. Russia.* Al Au Bu Cz Ga Ge Gr Hu Ju Po Rm Rs(C, W, K, E).

Many variants of this very polymorphic species have been given taxonomic recognition but, pending a detailed study of the European material, only two subspecies are recognized here.

(a) Subsp. **setacea** (*M. aucta* Klokov, *M. leiosperma* Klokov, *M. thyraica* Klokov): Cymes 3- to many-flowered. Sepals 2–3·5 mm, ovate-lanceolate. $2n = 30$. *Throughout the range of the species.*

(b) Subsp. **bannatica** (Reichenb.) E. I. Nyárády, *Enum. Pl. Vasc. Cheia Turzii* 124 (1939) (incl. subsp. *stojanovii* (Kitanov) Strid): Cymes 3- to 5-flowered. Sepals 3·5–5 mm, linear-lanceolate. • *From C. Austria and C. Czechoslovakia to Bulgaria.*

The distribution of subsp. (**b**) is imperfectly known.

17. M. bosniaca (G. Beck) K. Malý, *Glasn. Muz. Bosni Herceg.* **20**: 563 (1908). Like **16** but sepals 3–4 mm; petals and sepals subequal; glands at base of the outer stamens elongate and bipartite; seeds 0·8 mm, acutely tuberculate. $2n = 28$. • *Balkan peninsula southwards to N. Greece.* Al Bu Gr Ju.

18. M. krascheninnikovii Schischkin, *Acta Inst. Bot. Acad. Sci. URSS (Ser. 1)* **3**: 170 (1937). Like **16** but leaves 0·2–0·5 mm wide; cymes 1- to 4-flowered; pedicels usually less than twice as long as sepals; sepals 3–4 mm; petals 1–1½ times as long as sepals; seeds 0·75–1 mm, acutely tuberculate. *E. Russia.* Rs(C).

19. M. adenotricha Schischkin, *loc. cit.* 169 (1937). Like **16** but leaves 0·3 mm, wide; cymes 1- to 6-flowered; pedicels 2–6 mm; pedicels and sepals sparsely pubescent; sepals 4–5 mm; petals scarcely exceeding sepals; seeds *c.* 0·75 mm, with low, rounded tubercles. *Mountain rocks.* • *Krym.* ?Bu Rs(K.)

20. M. anatolica (Boiss.) Woronow in Woronow & Schelkownikow, *Sched. Herb. Fl. Cauc.* **4**: 92 (1914). Densely pubescent, laxly caespitose perennial; flowering stems with compact, ovoid, axillary vegetative fascicles, often covered with arachnoid indumentum; fascicular leaves appressed. Cymes crowded, many-flowered; pedicels seldom longer than sepals. Sepals 3–4 mm, equalling or slightly exceeding petals. *E. part of Balkan peninsula; Samothraki.* Bu Gr Tu.

The closely related **M. erythrosepala** (Boiss.) Hand.-Mazz., *Ann. Naturh. Mus. (Wien)* **23**: 152 (1909), has been reported from Samothraki. Records from Turkey-in-Europe are probably referable to **20**.

21. M. confusa (Boiss.) Maire & Petitmengin, *Étude Pl. Vasc. Réc. Grèce* **4**: 49 (1908) (*M. trichocalycina* auct., non (Ten. & Guss.) Grande). Perennial with sparse, crispate hairs and procumbent stems. Leaves linear-subulate. Cymes dense; pedicels shorter than sepals. Sepals 3·5–6 mm, linear-lanceolate; petals less than ½ as long as sepals. Capsule included. $2n = 30$. • *Calcicole, 1500–2400 m. Greece.* Gr.

Sect. PLURINERVIAE McNeill (*Tryphane* sensu Mattf.). Perennials. Leaves linear-subulate to -setaceous, 3- to 5-veined towards base. Sepals usually 5- to 7-veined, patent at anthesis; petals 1–1½ times as long as sepals, usually oblong-obovate, narrowed towards base.

22. M. hirsuta (Bieb.) Hand.-Mazz., *Ann. Naturh. Mus. (Wien)* **23**: 152 (1909) (*M. recurva* subsp. *hirsuta* (Bieb.) Stoj. & Stefanov). Laxly caespitose, usually glandular-pubescent perennial; flowering stems erect, up to 20 cm. Cymes lax, 3- to many-flowered; pedicels longer than sepals. Sepals usually ovate-lanceolate, 5- to 7-veined; margin scarious. *Mountain rocks and stony places. E.C. & S.E. Europe.* Bu Cz Gr Hu Ju Rm Rs(K) Tu.

A variable species. Some authors regard it as a subspecies of **23**, whilst others give the following subspecies specific rank.

1 Plant glabrous (**d**) subsp. **eurytanica**
1 Plant glandular-pubescent, at least above
 2 Leaves 20–30 mm, slender, erect, glabrous; plant glandular-pubescent above (**c**) subsp. **frutescens**
 2 Leaves 5–15 mm, somewhat rigid, often falcate
  3 Plant densely glandular-pubescent throughout; leaves 0·5–1 mm wide; seeds acutely tuberculate dorsally (**a**) subsp. **hirsuta**
  3 Plant glandular-pubescent above; leaves not more than 0·3 mm wide; seeds obscurely tuberculate (**b**) subsp. **falcata**

(a) Subsp. **hirsuta** (subsp. *falcata* sensu Mattf. pro parte): Plant densely glandular-pubescent. Leaves 5–15 mm, somewhat rigid, often falcate. Sepals with narrow, scarious margins. Seeds acutely tuberculate dorsally. • *Krym.*

(b) Subsp. **falcata** (Griseb.) Mattf., *Bot. Jahrb.* **57** Beibl. 126: 13 (1921): Like subsp. (**a**) but leaves more slender; bracts lanceolate; seeds obscurely tuberculate. $2n = 30$. *Balkan peninsula.*

(c) Subsp. **frutescens** (Kit.) Hand.-Mazz., *Ann. Naturh. Mus. (Wien)* **23**: 152 (1909) (*M. frutescens* (Kit.) Tuzson; incl. *M. cataractarum* Janka): Plant glandular-pubescent above. Sepals with narrow, scarious margin. $2n = 30, 32$. • *From W. Carpathians to S. Jugoslavia and S. Bulgaria.*

(d) Subsp. **eurytanica** (Boiss. & Heldr.) Strid, *Ann. Bot. Fenn.* **20**: 113 (1983) (*M. eurytanica* (Boiss. & Heldr.) Hand.-Mazz.): Plant glabrous. Leaves 7–15 mm, somewhat rigid, falcate. Sepals lanceolate with wide, scarious margin. • *N. & C. Greece.*

23. M. recurva (All.) Schinz & Thell., *Bull. Herb. Boiss.* ser. 2, **7**: 404 (1907) (*Alsine recurva* (All.) Wahlenb.). Densely caespitose perennial; stems woody and often black below; flowering stems usually up to 12 cm; plant sparsely glandular-pubescent, at least above. Leaves 3–10 mm, 3-veined, falcate. Cymes 1- to 8-flowered; pedicels up to 3 times as long as sepals. Sepals 3–5·5(–6) mm, ovate-lanceolate, 5- to 7-veined; petals and capsule slightly exceeding sepals. Seeds 1·2 mm, weakly tuberculate. $2n = 30$. *Usually calcifuge. Mountains of S. & S.C. Europe; one station in S.W. Ireland.* Al Au Bu Ga Gr Hb He Hs It Ju Lu Rm Si Tu.

(a) Subsp. **recurva**: Cymes 1- to 3(–5)-flowered; bracts 3- to 5-veined. Outer sepals 3–4·25 mm, 5-veined. $2n = 30$. *Almost throughout the range of the species, but absent from Sicilia.*

(b) Subsp. **condensata** (C. Presl) Greuter & Burdet, *Willdenowia* **12**: 188 (1982) (subsp. *juressi* (Willd. ex Schlecht.) Mattf., *M. condensata* (J. & C. Presl) Hand.-Mazz.; incl. *M. engleri* Mattf.): More robust than subsp. (**a**); flowering stems up to 15 cm. Cymes 3- to 8-flowered; bracts 5- to 7-veined. Outer sepals 4–5·5(–6) mm, 7-veined. *Sicilia; S. & C. Italy; S. part of Balkan peninsula.* Al Gr It Ju Si.

Although subsp. (**b**) appears fairly distinct, and has often been accorded specific rank, plants intermediate between the two subspecies occur in the Iberian peninsula, S. France and the S.E. Alps.

24. M. bulgarica (Velen.) Graebner in Ascherson & Graebner, *Syn. Mitteleur. Fl.* **5(1)**: 727 (1918) (*M. recurva* subsp. *bulgarica*

(Velen.) Stoj. & Stefanov). Like **23** but cymes compact, many-flowered; pedicels not longer than sepals; sepals 3–4 mm, 5- to 9-veined, slightly shorter than petals. *Mountain rocks and screes.* ● *W.* & *C. Bulgaria.* Bu.

Sect. LANCEOLATAE (Fenzl) Graebner. Perennial. Leaves lanceolate to linear-lanceolate, rarely oblanceolate. Sepals 5- to 7(–9)-veined (often obscurely so), erect or suberect at anthesis.

25. M. saxifraga (Friv.) Graebner in Ascherson & Graebner, *Syn. Mitteleur. Fl.* **5(1)**: 756 (1918). Densely caespitose perennial with erect, densely glandular-pubescent flowering stems. Leaves lanceolate to ovate-lanceolate, not rigid. Flowers in compact, 3- to 12-flowered cymes. Sepals lanceolate, densely glandular-pubescent; veins 5–9, the outer curved towards the margin; petals obovate, slightly exceeding sepals. $2n = 32$. *Mountain rocks,* 1200–1800 *m. Bulgaria and N. Greece.* Bu Gr.

The European plants belong to subsp. *saxifraga.* Subsp. **tmolea** Mattf., *Feddes Repert. Beih.* **15**: 132 (1922), occurs in W. Anatolia (Boz Dağ).

26. M. graminifolia (Ard.) Jáv., *Sched. Fl. Hung. Exsicc.* **2**: 22 (1914). Densely pulvinate perennial with unbranched flowering stems 4–18 cm. Leaves 10–40 mm, linear-lanceolate, rigid. Cymes 2- to 7-flowered. Sepals 5–10 mm, lanceolate, 5- to 7-veined; petals exceeding sepals. Capsule included. $2n = 32$. *Mountain rocks.* ● *S. Europe, from N. Italy and S.W. Romania to Sicilia.* Al Gr It Ju Rm Si.

(a) Subsp. **graminifolia** (incl. subsp. *hungarica* Jáv.): Whole plant glandular-pubescent, often densely so. Leaves of non-flowering stems 20–40 mm, those of the flowering stems at least ½ as long as internodes. *Throughout the range of the species, except Jugoslavia and Albania.*

(b) Subsp. **clandestina** (Portenschl.) Mattf., *Bot. Jahrb.* **57** Beibl. 126: 31 (1921): Stems and leaves glabrous. Leaves of non-flowering stems 10–30 mm; those of the flowering stems rarely more than ⅓ as long as internodes. Pedicels and sepals glabrous or sparsely glandular-pubescent. *N. Albania; W.* & *S. Jugoslavia.*

27. M. stellata (E. D. Clarke) Maire & Petitmengin, *Mat. Étude Fl. Géogr. Bot. Or.* **4**: 48 (1908). Densely pulvinate perennial with columnar stems clothed with dead, imbricate leaves. Leaves up to 10 mm, in rosettes, deltate-lanceolate to lanceolate, rigid, glabrous. Flowering stems and sepals glandular-pubescent; flowers solitary, rarely in pairs. Sepals 4–6 mm, lanceolate, 5-veined; petals *c.* 1½ times as long as sepals. Capsule slightly shorter than sepals. $2n = 32$. *Limestone rocks above* 1700 *m.* ● *Mountains of Greece and S. Albania.* Al Gr.

28. M. pseudosaxifraga (Mattf.) Greuter & Burdet, *Willdenowia* **12**: 188 (1982) (*M. stellata* subsp. *pseudosaxifraga* Mattf.). Laxly caespitose, densely glandular-pubescent perennial, forming greyish-green cushions. Stems rather long, stout, woody, not clothed with persistent dead leaves. Leaves 8–18 mm, oblanceolate to narrowly oblong, rather soft. Flowering stems 2–5 cm; cymes (1- \3- to 5-flowered. Sepals 5·5–7·5 mm; petals 1–2 times as long as sepals. Capsule slightly shorter than sepals. $2n = 32$. *Limestone rocks,* 1800–2000 *m.* ● *N.W. Greece (Pindhos).* Gr.

Although described as a subspecies of **27**, this species is very distinct and perhaps more closely related to **25**.

29. M. cerastiifolia (Ramond ex DC.) Graebner in Ascherson & Graebner, *Syn. Mitteleur. Fl.* **5(1)**: 754 (1918) (*Alsine cerastiifolia* (Ramond ex DC.) Fenzl). Glandular-pubescent, caespitose perennial with stems up to 8 cm. Leaves 3–10 mm, lanceolate to lanceolate-elliptical; midrib very prominent beneath. Flowers 1(2); pedicels usually twice as long as sepals. Sepals 3–4 mm, ovate-lanceolate; petals 1½ times as long as sepals. Capsule exserted. Seeds 1–1·4 mm, reddish-brown. *Screes and rocks above* 1800 *m.* ● *W.C. Pyrenees.* Ga Hs.

30. M. rupestris (Scop.) Schinz & Thell., *Bull. Herb. Boiss.* ser. 2, **7**: 403 (1907) (*M. lanceolata* (All.) Mattf. subsp. *rupestris* (Scop.) Mattf., *Alsine rupestris* (Scop.) Fenzl). Laxly caespitose or creeping perennial; stems rooting intermittently at nodes, glandular-pubescent above. Leaves lanceolate, shortly ciliate. Sepals 3·5–5 mm, lanceolate; usually glandular-pubescent; petals and sepals about equal. Capsule included. Seeds 1–1·2 mm. *Screes and rocks above* 1600 *m.* ● *Alps.* Au Ga Ge He It Ju.

(a) Subsp. **rupestris**: Leaves 2–4 mm. Flowering stems 1–3 mm; flowers usually solitary; pedicels 1–4 mm. $2n = c.\ 72$. *Throughout much of the range of the species.*

(b) Subsp. **clementei** (Huter) Greuter & Burdet, *Willdenowia* **14**: 43 (1984) (*M. lanceolata* (All.) Mattf., *M. lanceolata* subsp. *clementei* (Huter) Mattf.): Plant more robust. Stem leaves up to 8 mm. Flowering stems 5–7 cm; flowers 1 or 2; pedicels 8–15 mm. $2n = 36$. *S.W. Alps (Alpes Cottiennes).*

31. M. grignensis (Reichenb.) Mattf., *Feddes Repert. (Beih.)* **15**: 141 (1922) (*Alsine villarii* var. *grignensis* (Reichenb.) Tanfani). Glabrous perennial with very crowded lower internodes and elongate flowering stems up to 10 cm. Leaves 5–15 mm, linear, erect or erecto-patent, 3-veined. Cymes 5- to 12-flowered; pedicels 8–15 mm, slender. Sepals 2–3 mm, ovate, acute, 3-veined; petals and capsule slightly exceeding sepals. Seeds 1–1·2 mm, fimbriate. $2n = 36$. *Limestone and dolomite cliffs,* 1300–1900 *m.* ● *N. Italy (Alpi Bergamasche).* It.

A very distinct species that falls between Sect. *Lanceolatae* and Sect. *Acutiflorae.*

Sect. ARETIOIDEAE Mattf. Perennials. Leaves oblong-elliptical. Flowers 4-merous. Sepals erect at anthesis. Seeds fimbriate.

32. M. cherlerioides (Hoppe) Becherer, *Denkschr. Schweiz. Naturf. Ges.* **81**: 167 (1956) (*M. aretioides* (Sommer.) Schinz & Thell., *Alsine octandra* (Sieber) A. Kerner). Densely caespitose perennial 2–5 cm; stems usually richly branched. Leaves 1·5–3 × 1 mm, oblong-elliptical, obtuse, somewhat fleshy; apex cucullate; veins 3, distinct and persistent. Flowers solitary; pedicels shorter than sepals. Sepals 2–4 mm, lanceolate, acute, 3-veined; petals slightly shorter than sepals, sometimes absent; styles usually 3. Capsule exserted. Seeds 0·9–1·2 mm. *Rocks and screes above* 2000 *m.* ● *C.* & *E. Alps.* Au Ge He It Ju.

(a) Subsp. **cherlerioides**: Leaves glabrous. *Calcicole. E. Alps, westwards to* c. 9°30′ *E.*

(b) Subsp. **rionii** (Gremli) Friedrich, *Feddes Repert.* **70**: 2 (1964) (*M. aretioides* subsp. *rionii* (Gremli) Schinz & Thell.): Leaves ciliolate. *Calcifuge. C. Alps, eastwards to* c. 10°30′ *E.*

Sect. ACUTIFLORAE (Fenzl) Hayek. Perennials. Leaves linear, 3-veined. Bracts herbaceous. Sepals erect at anthesis; petals exceeding sepals. Seeds rugose, or shortly tuberculate.

33. M. austriaca (Jacq.) Hayek, *Fl. Steierm.* **1**: 274 (1908) (*Alsine austriaca* (Jacq.) Wahlenb.). Laxly caespitose perennial with glabrous or sparsely glandular-pubescent stems 8–20 cm. Leaves 10–20 × 0·5–1 mm, glabrous. Flowers 1 or 2(3); pedicels more than 4 times as long as sepals. Sepals 4–6 mm, ovate-lanceolate; petals slightly emarginate; petals and capsule 1½–2 times as long as sepals. Seeds 1–1·5 mm, prominently tuberculate dorsally. $2n = 26$. *Calcareous rocks and screes.* ● *E. Alps, westwards to Alpi Bergamasche.* Au Ge It Ju.

34. M. helmii (Ser.) Schischkin, *Jour. Governm. Bot. Gard. Nikita* **10(2)**: 38 (1928). Like **33** but stems densely caespitose; plant densely glandular-pubescent; sepals obscurely veined; capsule only slightly exceeding sepals. $2n = 26$. *N. Ural.* Rs(N).

35. M. villarii (Balbis) Wilczek & Chenevard, *Annu. Cons. Jard. Bot. Genève* **15**: 256 (1912) (*M. flaccida* sensu Schinz & Thell., *Alsine villarii* (Balbis) Mert. & Koch). Like **33** in habit. Leaves 0·7–2 mm wide. Cymes 2- to 4(–7)-flowered; pedicels 1–5(–7) times as long as sepals. Sepals 3·5–5 mm; petals 1–1½ times as long as sepals. Capsule slightly exceeding sepals. $2n = 26$. *Rocks and screes.* ● *Mountains of S.W. Europe from N. Spain to S.E. Alps.* Ga Hs It.

Variable in degree of pubescence; completely glabrous plants are not uncommon, whilst plants with densely glandular-pubescent leaves occur in the S.W. Alps.

36. M. taurica (Steven) Graebner in Ascherson & Graebner, *Syn. Mitteleur. Fl.* **5(1)**: 758 (1918). Glandular-pubescent perennial. Leaves 5–10 × 1 mm. Cymes 1- to 5-flowered; pedicels 2–4 times as long as sepals. Sepals 4·5–5 mm, ovate; petals twice as long as sepals. Capsule usually included. Seeds *c.* 0·75 mm, slightly tuberculate dorsally. *Mountain rocks.* ● *Krym.* Rs(K).

37. M. juniperina (L.) Maire & Petitmengin, *Mat. Étude Fl. Géogr. Bot. Or.* **4**: 48 (1908). Sparsely glandular-pubescent, caespitose perennial; stems rigid, nodes thickened. Leaves 10–20 mm, rigid, almost spiny, usually patent, terete, with fascicles of small leaves in their axils. Cymes 3- to 7(–15)-flowered. Sepals 4–5 mm, lanceolate; petals and capsule *c.* 1½ times as long as sepals. $2n = 26$. *Mountain rocks above* 2000 *m. S. & W. Greece.* Gr. (*S.W. Asia.*)

38. M. pichleri (Boiss.) Maire & Petitmengin, *loc. cit.* (1908). Densely caespitose, glandular-pubescent perennial. Leaves *c.* 6 mm, 3-veined, somewhat falcate. Cymes 1- to 6-flowered; pedicels 3–5 times as long as sepals. Sepals 3–5 mm, ovate-lanceolate; petals 1½ times as long as sepals. Capsule slightly exserted. Seeds acutely tuberculate dorsally. $2n = 26$. *Mountain rocks above* 1200 *m.* ● *S. Greece; local.* Gr.

Although very similar in appearance to **39** it is readily distinguished by the dense pubescence, herbaceous bracts and longer petals.

M. favargeri Iatroú & Georgiadis, *Candollea* **40**: 129 (1985), from 500 to 600 m on Parnon Oros, differs from **38** in its somewhat sparser glandular pubescence, leaves 8–12 mm, smaller sepals and petals, and capsule distinctly exceeding sepals; the chromosome number is $2n = 26$. It is probably not specifically distinct.

Sect. TRYPHANE (Fenzl) Hayek. Perennials. Leaves linear-subulate to setaceous; veins 3, persistent. Sepals prominently 3-veined, patent at anthesis. Seeds rugose.

39. M. verna (L.) Hiern, *Jour. Bot. (London)* **37**: 320 (1899) (*M. caespitosa* (Ehrh.) Degen, *Alsine verna* (L.) Wahlenb., *Arenaria verna* L.). Laxly caespitose perennial; stems usually glandular-pubescent above. Leaves 5–20 mm, linear-lanceolate, setaceous or subulate, glabrous or glandular-pubescent, very rarely scabrid. Cymes lax, few- to many-flowered; pedicels longer than sepals. Sepals ovate-lanceolate, usually glandular-pubescent; petals ½–1½ times as long as sepals. Capsule equalling or slightly exceeding sepals. Seeds 0·5–1 mm, brown, varying from almost smooth to acutely tuberculate. Protandrous. *S., W. & C. Europe; N. Russia.* Al Au Be Br Bu Co Cr Cz Ga Ge Gr Hb He Hs Hu It Ju Po Rm Rs(N, W) Sa Si Tu.

A very variable species; the following treatment is provisional.

1 Petals obtuse, widest at or above middle
2 Stems woody below; leaves linear-lanceolate, the margins scabrid — (c) subsp. **paui**
2 Stems not woody below; leaves linear-subulate to setaceous, the margins entire
3 Axillary fascicles of leaves present on flowering stems; cymes 6- to many-flowered — (b) subsp. **collina**
3 Axillary fascicles of leaves absent; cymes 1- to 7-flowered — (a) subsp. **verna**
1 Petals acute, widest below middle
4 Plant not woody below, green; sepals 2·2–3 mm — (d) subsp. **oxypetala**
4 Plant woody below, glaucous; sepals (2–)3–6 mm
5 Sepals (2–)3–5 mm; leaves ± strict — (e) subsp. **attica**
5 Sepals 4–6 mm; leaves ± patent — (f) subsp. **grandiflora**

(a) Subsp. **verna** (incl. subsp. *gerardii* (Willd.) Fenzl, *M. zareczny*i (Zapał.) Klokov, *M. smejkalii* Dvořáková): Plant not woody below. Leaves up to 20 mm. Cymes 1- to 7-flowered. Petals obtuse, widest at or above middle, usually exceeding sepals. Sepals 3·5–4·5 mm. Anthers purplish. $2n = 24$. *Throughout the range of the species but only on mountains in the south.*

(b) Subsp. **collina** (Neilr.) Domin, *Věstn. Král. České Spolecn. Nauk. Tr. Mat.-Přír.* **1947(2)**: 24 (1947) (subsp. *montana* (Fenzl) Hayek): Like subsp. (a) but flowering stems with axillary fascicles of leaves; cymes 6- to many-flowered; petals equalling or slightly exceeding sepals; sepals 2·2–3·2 mm; anthers yellow. $2n = 48$. *S. & E.C. Europe, from N.E. Italy and Czechoslovakia eastwards.*

(c) Subsp. **paui** (Willk. ex Hervier) Rivas Goday & Borja, *Anal. Inst. Bot. Cavanilles* **19**: 331 (1961) (*Alsine paui* Willk. ex Hervier, *M. verna* subsp. *valentina* (Pau) Font Quer): Stems woody below. Leaves linear-lanceolate, stiffly patent, the margins scabrid. Cymes usually more than 7-flowered. Petals obtuse, widest at or above the middle. ● *E. Spain (Sierra Espadada, north of Valencia).*

(d) Subsp. **oxypetala** (Wołoszczak) Halliday, *Feddes Repert.* **69**: 13 (1964): Leaves up to 10 mm. Cymes 1- to 5-flowered. Petals acute, widest below middle, shorter than sepals. Sepals 2·2–3 mm. ● *E. Carpathians.*

(e) Subsp. **attica** (Boiss. & Spruner) Hayek, *Österr. Bot. Zeitschr.* **71**: 112 (1922) (*M. attica* (Boiss. & Spruner) Vierh.; incl. *M. idaea* (Halácsy) Pawł.): Plant woody below, glaucous. Leaves up to 10 mm. Cymes (1–)3- to 20-flowered. Sepals (2–)3–5 mm. Petals acute, widest below middle, more or less equalling or shorter than sepals. Anthers reddish. *S. Europe, from N.W. Jugoslavia and C. Italy to Bulgaria and Kriti.*

(f) Subsp. **grandiflora** (C. Presl) Hayek, *loc. cit.* (1922): Like subsp. (e) but leaves patent; cymes 3- to 5-flowered; sepals 4–6 mm, shorter than petals. ● *Sicilia, Sardegna.*

40. M. rubella (Wahlenb.) Hiern, *Jour. Bot. (London)* **37**: 320 (1899). Like small high montane variants of **39(a)** but usually smaller; leaves 4–8 mm, glabrous or glandular-pubescent; cymes 1 or 2(3)-flowered; petals usually $\frac{2}{3}$ as long as sepals, obtuse; capsule-teeth 3 or 4; seeds 0·55–0·7 mm. Homogamous. $2n = 24, 26$. *Open, stony habitats on base-rich soil.* ● *Arctic and subarctic Europe, extending southwards to* 56°30′ *N. in Scotland.* Br Fa Fe Is No Rs(N) Sb Su.

Sect. ALSINANTHE (Fenzl) Graebner. Perennial. Plant glabrous. Leaves 1-veined, linear to subulate. Sepals erect at anthesis; petals shorter than sepals.

41. M. stricta (Swartz) Hiern, *Jour. Bot. (London)* **37**: 320 (1899) (*Alsine stricta* (Swartz) Wahlenb.). Laxly caespitose perennial; flowering stems up to 10 cm, erect. Leaves 6–12 mm, obscurely 1-veined. Cymes 1- to 3(4)-flowered; pedicels 15–50 mm. Sepals 2·5–4 mm, ovate, acute, 3-veined when dry; petals slightly shorter than sepals. Capsule equalling or slightly exceeding sepals, dehiscing almost to base with patent teeth. Seeds 0·65–0·85 mm, reddish-brown, smooth. $2n = 22, 26, 30$. *Arctic and subarctic Europe southwards to* 60° *N. in Norway; isolated stations in N. England and N.W. Switzerland.* Br Fe †Ga †Ge He Is No Rs(N) Sb Su.

42. M. rossii (R. Br. ex Richardson) Graebner in Ascherson & Graebner, *Syn. Mitteleur. Fl.* **5(1)**: 772 (1918). Low, densely caespitose perennial with spreading flowering stems. Leaves 2–5 mm, subulate; vegetative propagation by short, detachable shoots in axils of upper leaves. Flowers solitary, rarely produced; pedicels 5–15 mm. Sepals 1·5–2·5 mm, ovate-deltate; petals and capsule slightly shorter than sepals. Seeds 0·6–0·7 mm. *Spitsbergen.* Sb. (*Arctic and montane North America, Greenland.*)

Sect. SPECTABILES (Fenzl) Hayek. Sepals linear-oblong, obtuse, erect at anthesis. Petals usually exceeding sepals. Capsule cylindrical.

43. M. wettsteinii Mattf., *Bot. Jahrb.* **57** Beibl. 127: 62 (1922). Caespitose, glabrous perennial. Leaves linear, obtuse, glaucous. Cymes few-flowered. Sepals 3·5–5 mm, 5-veined, somewhat fleshy; petals just exceeding sepals; glands at base of outer stamens swollen, hairy. *Rocks at* c. 1400 *m.* ● *E. Kriti* (*Afendi Kavousi, N.E. of Ierapetra*). Cr.

44. M. capillacea (All.) Graebner in Ascherson & Graebner, *Syn. Mitteleur. Fl.* **5(1)**: 757 (1918) (*Alsine liniflora* (L.) Hegetschw.). Laxly caespitose perennial with woody stock; stems 8–30 cm, densely glandular-pubescent above. Leaves 10–20 × 0·5 mm, linear-setaceous, rigid, 1- to 3-veined, obtuse, ciliolate. Cymes 1- to 6-flowered; pedicels and sepals densely glandular-pubescent. Sepals 5–7 mm, ovate-oblong; veins 3, the outer pair disappearing in upper $\frac{1}{3}$; petals and capsule $1\frac{1}{2}$–2 times as long as sepals. Seeds 1·5–2 mm, prominently tuberculate to almost fimbriate dorsally, the face rugose. $2n = 26$. *Calcareous rocks and screes.* ● *Mountains of S. & S.C. Europe from France to S. Jugoslavia.* Al Ga He It Ju.

45. M. baldaccii (Halácsy) Mattf. in Ascherson & Graebner, *Syn. Mitteleur. Fl.* **5(1)**: 940 (1919). Perennial with woody rhizomes up to 20 cm; stems (2–)5–20 cm, with crowded internodes; flowering stems glandular-pubescent, especially above. Leaves linear-subulate, acute, often secund, ciliolate, 1-veined. Cymes 1- to 7-flowered. Sepals 4–7 × 1·5 mm, oblong, 3-veined to apex; petals broadly obovate, 2–$2\frac{1}{2}$ times as long as sepals. Capsule equalling sepals. *Rocky and stony ground.* ● *Balkan peninsula, mainly in the west.* Al Gr Ju.

1 Leaves densely glandular-pubescent (c) subsp. **skutariensis**
1 Leaves glabrous
2 Leaves green; sepals 4–5 mm; petals 11–12 mm (b) subsp. **doerfleri**
2 Leaves glaucous; sepals 5–7 mm; petals 15–18 mm (a) subsp. **baldaccii**

(a) Subsp. **baldaccii**: Leaves 8–18 mm, glabrous, glaucous. Flowering stems 6–15(–22) cm; cymes 1- to 5-flowered. Sepals 5–7 mm; petals 15–18 mm. *Usually on serpentine.* *Albania; S. Jugoslavia; N.W. Greece.*

(b) Subsp. **doerfleri** (Hayek) Hayek, *Prodr. Fl. Penins. Balcan.* **1**: 193 (1924) (*M. doerfleri* Hayek): Leaves 3–8 mm, glabrous, green. Flowering stems 2–6 cm; cymes 1(–3)-flowered. Sepals 4–5 mm; petals 11–12 mm. *N.E. Albania; S.W. Jugoslavia; N.E. Greece* (*Falakron*).

(c) Subsp. **skutariensis** Hayek, *loc. cit.* (1924): Leaves 5–10 mm, densely glandular-pubescent. Flowering stems up to 12 cm; cymes 1- to 7-flowered. Sepals *c.* 5 mm; petals *c.* 10 mm. *N.W. Albania.*

46. M. garckeana (Ascherson & Sint. ex Boiss.) Mattf., *Feddes Repert. (Beih).* **15**: 192 (1922). Like **45** but stems densely caespitose, shortly rhizomatous; leaves somewhat fleshy; sepals 3–5 × 0·5–1 mm; petals 6 mm, slightly exceeding sepals. *Balkan peninsula, from E. Albania and N. Greece to E. Bulgaria.* Al Bu Gr Ju.

47. M. handelii Mattf., *Feddes Repert.* **19**: 193 (1923). Like **45** but flowering stems only 0·5–1 cm; flowers solitary; petals 5 mm, scarcely exceeding sepals. *Mountain rocks,* 2100 *m.* ● *Hercegovina* (*Čvrsnica*). Ju.

48. M. laricifolia (L.) Schinz & Thell., *Bull. Herb. Boiss.* ser. 2, **7**: 403 (1907) (*M. striata* (L.) Mattf., *Alsine laricifolia* (L.) Crantz, *A. striata* (L.) Gren.). Perennial with stems 8–30 cm, laxly or densely caespitose, woody below, but less robust than **44**. Leaves linear-setaceous, rigid, acute, somewhat falcate, obscurely 1-veined, glabrous, ciliolate at base. Cymes 1- to 6-flowered; pedicels and sepals with crispate hairs, rarely glandular-pubescent. Sepals 4–7 mm, linear-oblong, 3-veined to apex, the margin usually red; petals up to $1\frac{1}{2}$–2 times as long as sepals. *Mountain rocks and screes.* ● *Mountains of S. & C. Europe from N.E. Spain to the Carpathians.* Au Cz Ga He Hs It Po Rm.

1 Petals 6–7 mm, less than $1\frac{1}{2}$ times as long as sepals (c) subsp. **ophiolitica**
1 Petals 8–10 mm, $1\frac{1}{2}$–2 times as long as sepals
2 Capsule 1–$1\frac{1}{2}$ times as long as sepals; seeds 0·8–1 mm, rugose (a) subsp. **laricifolia**
2 Capsule $1\frac{1}{2}$–2 times as long as sepals; seeds 1·2–1·5 mm, tuberculate (b) subsp. **kitaibelii**

(a) Subsp. **laricifolia** (incl. subsp. *diomedis* (Br.-Bl.) Mattf., subsp. *striata* (L.) Mattf.): Laxly caespitose. Sepals 4–5·5 mm, sometimes glandular. Capsule 1–$1\frac{1}{2}$ times as long as sepals. Seeds 0·8–1 mm, rugose dorsally. $2n = 26$. *Calcifuge. From the C. Pyrenees to* 12° *E. in S.E. Alps.*

(b) Subsp. **kitaibelii** (Nyman) Mattf., *Feddes Repert. (Beih.)* **15**: 190 (1922). More robust than subsp. (a). Sepals 5–7 mm, eglandular. Capsule $1\frac{1}{2}$–2 times as long as sepals. Seeds 1·2–1·5 mm, tuberculate dorsally. $2n = 26$. *Calcicole. E. Austrian Alps; Carpathians.*

(c) Subsp. **ophiolitica** Pignatti, *Gior. Bot. Ital.* **107**: 208 (1973): Like subsp. (a) but densely caespitose; leaves greyish; petals 6–7 mm, less than $1\frac{1}{2}$ times as long as sepals; seeds smaller. *Serpentine rocks.* ● *N. Appennini.*

49. M. macrocarpa (Pursh) Ostenf., *Meddel. Grønl.* **37**: 226

(1920). Caespitose perennial; flowering stems 2–6 cm, glandular-pubescent. Leaves 5–12 mm, linear-oblong, obtuse, distinctly ciliate. Flowers solitary; pedicels 5–13 mm. Sepals 5–7 mm, linear-oblong, glandular-pubescent; petals usually 1½–2 times as long as sepals. Capsule 2–2½ times as long as sepals. Seeds 2–2·5 mm, fringed dorsally with very long, fine tubercles. *Arctic and subarctic Russia*. Rs(N). (*N. Asia, N.W. America.*)

50. M. arctica (Ser.) Graebner in Ascherson & Graebner, *Syn. Mitteleur. Fl.* **5(1)**: 772 (1918). Densely caespitose perennial; flowering stems 5–9 cm, glandular-pubescent. Leaves 6–16 mm, narrowly linear, with cilia usually absent; leaves of flowering stems shorter and wider. Flowers solitary, rarely in pairs. Sepals 5–7 mm, linear-oblong, glandular-pubescent; petals twice as long as sepals. Capsule 1½–2 times as long as sepals. Seeds 1–1·5 mm, bluntly tuberculate dorsally. *Arctic and N.E. Russia.* Rs(N). (*N. Asia, N.W. America.*)

51. M. biflora (L.) Schinz & Thell., *Bull. Herb. Boiss.* ser. 2, **7**: 404 (1907) (*Alsine biflora* (L.) Wahlenb.). Slender, caespitose perennial with flowering stems up to 10 cm. Leaves up to 10 mm, linear, obtuse, 1-veined. Flowers 1–3; pedicels and sepals sparsely pubescent. Sepals (3–)3·5–4·5(–5) mm, ovate-oblong, cucullate, 3-veined; petals 0·5 mm wide, 1–1½ times as long as sepals, occasionally pale lilac. Capsule 1½ times as long as sepals. Seeds 0·7 mm, smooth dorsally. *Damp, open habitats in the mountains, often in snow-patches. N. Europe, southwards to* 59° *N. in Norway; C. & E. Alps.* Au Fe He Is It No Rs(N) Sb Su.

M. olonensis (Bonnier) P. Fourn., described from W. France, has been shown by M. Laínz, *Candollea* **24**: 259 (1969), to be a variant of *Arenaria serpyllifolia.*

Sect. CHERLERIA (L.) Mattf. Perennial. Leaves somewhat fleshy. Sepals obtuse; petals absent or rudimentary; glands of outer stamens prominent, deeply divided.

52. M. sedoides (L.) Hiern, *Jour. Bot.* (*London*) **37**: 321 (1899) (*Alsine sedoides* (L.) Kittel, *A. cherleri* Fenzl). Glabrous, densely pulvinate perennial. Leaves 3·5–6 mm, densely imbricate, linear-lanceolate, 3-veined, furrowed above; margins smooth or ciliate-scabrid. Flowers solitary, scarcely exserted from cushion; pedicels up to 3(–5) mm. Sepals 2–5 mm, linear; either the stamens or the ovary may abort. Capsule 1½–2 times as long as sepals. Seeds 0·8–1 mm, smooth. $2n = 26$, 51–52. *Mountain ledges, exposed ridges, moraines.* • *Pyrenees; Alps; Carpathians; mountains of W. Jugoslavia; Scotland.* Au Br Cz Ga Ge He Hs It Ju Po Rm.

Subgen. **Rhodalsine** (Gay) Graebner. Perennial. Flowers usually pale pink. Radicle of embryo accumbent.

53. M. geniculata (Poiret) Thell., *Fl. Adv. Montpellier* 232 (1912) (*Rhodalsine geniculata* (Poiret) F. N. Williams; incl. *M. procumbens* (Vahl) Graebner). Glandular-pubescent perennial; stems up to 30 cm, woody at base, procumbent and ascending, geniculate, much-branched. Leaves (4–)6–10(–12) × (0·5–)1–2(–2·5) mm, elliptical to linear, with 1 prominent vein beneath. Flowers axillary and in terminal cymes; pedicels 1½–6 times as long as sepals. Sepals 3–4 mm, oblong-ovate, usually obtuse, appearing veinless, with a wide, scarious margin; petals and capsule about equalling sepals. Seeds 0·5–0·8 mm, the margin slightly rugose. *Dry, sandy places, often by the sea. W. Mediterranean region, S. Portugal.* Bl ?Gr Hs It Lu Sa Si.

Very variable, particularly in leaf-shape and pubescence.

4. Honkenya Ehrh.

G. HALLIDAY

Like *Minuartia* but succulent; capsule globose; seeds more than 3 mm, pyriform. (*Ammadenia* Rupr.)

1. H. peploides (L.) Ehrh., *Beitr. Naturk.* **2**: 181 (1788) (*Ammadenia peploides* (L.) Rupr., *Arenaria peploides* L., *Minuartia peploides* (L.) Hiern). Glabrous, hermaphrodite or dioecious perennial with trailing, succulent stems rooting at the nodes. Leaves 5–20 × 5–10 mm, ovate to oblong, 1-veined, acute. Flowers axillary and solitary, and in 1- to 6-flowered terminal cymes; pedicels usually 2–6 mm. Sepals 3–5 mm, ovate, 1-veined; petals greenish-white, equalling sepals in male flowers, shorter than sepals in female flowers. Capsule (4–)6–10 mm wide, globose, twice as long as sepals. Seeds 3–4(–4·5) mm, usually 3–8, dark reddish-brown, shiny, almost smooth. $2n = 66$, 68–70. *Maritime sands and shingle, very rarely inland. Coasts of N. & W. Europe, southwards to C. Portugal; Ural.* Be Br Da Fa Fe Ga Ge Hb Ho Hs Is Lu No Po Rs (N, B, C) Sb Su.

Plants from Iceland and arctic Europe usually have a more diffuse growth, pedicels 4–15 mm, and capsules with usually 4–12 seeds. They are best treated as var. *diffusa* (Hornem.) Ostenf.

5. Bufonia L.

G. HALLIDAY

Annual or perennial herbs, rarely biennial; stems usually branched from base; flowering stems erect. Leaves linear-setaceous, appressed, those of the flowering stems shorter than internodes. Inflorescence a cymose panicle; pedicels short; flowers self-pollinated. Sepals 4, the 2 outer being shorter, lanceolate, with a hyaline margin; petals 4, equalling or shorter than sepals, white; stamens 2–8; styles 2. Capsule included, dehiscing with 2 teeth. Seeds 1 or 2, oblong, compressed, tuberculate.

All species occur in dry, sandy or rocky habitats.

1 Annual, or rarely biennial, without vegetative stems at time of flowering
 2 Sepals with 5 veins, 3 of which continue into the upper ½; seeds usually more than 1·5 mm, with prominent, obtuse tubercles **1. paniculata**
 2 Sepals 3-veined; veins confluent in upper ½; seeds 1–1·5 mm, with small, acute tubercles **2. tenuifolia**
1 Perennial, with vegetative stems at time of flowering
 3 Flowering stems usually unbranched, arcuate **7. stricta**
 3 Flowering stems branched above, straight
  4 Petals exceeding outer sepals **4. macropetala**
  4 Petals not exceeding outer sepals
   5 Inflorescence dense; branches strict; flowers overlapping **3. tuberculata**
   5 Inflorescence lax; branches usually divergent; flowers not overlapping
    6 Outer sepals 2–2·5 mm; seeds 1–1·3 mm **5. willkommiana**
    6 Outer sepals 3–3·5 mm; seeds 1·5–2 mm **6. perennis**

1. B. paniculata F. Dubois in Delarbre, *Fl. Auvergne* ed. 2, 300 (1800) (*B. macrosperma* Gay ex Mutel). Annual up to 20 cm, usually richly and divaricately branched. Sepals lanceolate, acute; veins 5, distinct, 3 continuing almost to the apex; petals almost equalling outer sepals. Capsule 2–3 mm. Seeds 1·5–2·25 mm, oblong-elliptical; margin with prominent, obtuse tubercles; faces rugose. *S.W. Europe, extending to C. France and Switzerland; isolated stations in C. Italy and Greece.* Ga Gr He Hs It.

2. B. tenuifolia L., *Sp. Pl.* 123 (1753) (*B. parviflora* Griseb.). Like **1** but usually less robust and smaller in all its parts. Sepals strongly acuminate; veins 3, confluent in upper $\frac{1}{2}$; petals $\frac{1}{2}$ as long as outer sepals; stamens 2 or 3. Capsule 1·5 mm. Seeds 1–1·5 mm; margin with small, acute tubercles; faces almost smooth. *S. Europe, extending north-eastwards to E. Ukraine.* Bu Ga Gr Hs It Rm Rs(W, K).

The plants from Ukraine include specimens intermediate between **1** and **2**.

3. B. tuberculata Loscos, *Trat. Pl. Arag.* **3**, Suppl. 8: 104 (1886). Erect perennial. Leaves setaceous, serrulate, glaucous. Inflorescence dense; branches strict; flowers usually overlapping. Sepals lanceolate, the outer 3·5–4·5 mm; veins usually 5; petals and capsule slightly shorter than the outer sepals. Seeds 1·75–2 mm, 1 or 2 per capsule; margin with prominent, oblong-cylindrical tubercles. • *E. Spain.* Hs.

The available descriptions of this species are inadequate, and it may prove to be only a perennial variant of **1**.

4. B. macropetala Willk., *Flora (Regensb.)* **34**: 604 (1851). Perennial with glabrous, erect stems branched above. Leaves linear-subulate; margins ciliolate. Sepals lanceolate, the outer 3·5–4·5 mm; veins 5, the outer often obscure; petals oblong, about equalling inner sepals. Capsule often 1-seeded. Seeds tuberculate on margin. • *W. & C. Spain.* Hs.

5. B. willkommiana Boiss., *Diagn. Pl. Or. Nov.* **3(1)**: 83 (1853). Perennial with robust, erect stems, woody below, shortly branched above. Leaves linear-setaceous, margins ciliolate. Inflorescence lax; cymes 1- to 3(4)-flowered; branches patent. Sepals ovate-lanceolate, the outer 2–2·5 mm; veins 3–5; petals and capsule equalling outer sepals. Seeds 1–1·3 mm, 1 or 2 per capsule; margin with low tubercles. • *S. Spain, E. Portugal.* Hs Lu.

6. B. perennis Pourret, *Mém. Acad. Sci. Toulouse* **3**: 309 (1788). Perennial with robust stems up to 30 cm. Leaves rigid, erect, linear-setaceous, the basal 1–2 cm; margins serrulate. Inflorescence lax; cymes 1- or 2-flowered; branches patent. Sepals lanceolate-acuminate, the outer 3–3·5 mm, 5-veined below; petals about equalling outer sepals; stamens 8. Seeds 1·5–2 mm, 1 or 2 per capsule, with prominent tubercles on margin; faces rugose. • *S. France, N.E. Spain.* Ga Hs.

7. B. stricta (Sibth. & Sm.) Gürke in K. Richter, *Pl. Eur.* **2**: 247 (1899). Perennial with erect, arcuate, slender, usually unbranched flowering stems up to 20 cm, woody below. Leaves of flowering stems short, subulate, often closely appressed. Flowers few, often 1. Outer sepals 3·5–4 mm; veins 3–5; petals equalling outer sepals. Seeds small; margins tuberculate; faces smooth. *Limestone rocks and stony ground.* • *Mountains of S. Greece, Kriti.* Cr Gr.

The flowering stems bear a striking, superficial resemblance to those of *Parapholis strigosa* (Gramineae).

6. Stellaria L.

A. O. CHATER AND V. H. HEYWOOD

Annual or perennial herbs. Inflorescence usually a dichasium; flowers rarely solitary or 2 together; bracts scarious or herbaceous. Sepals 5; petals 5 or fewer, or absent, white, rarely greenish, usually deeply bifid; stamens 10(11) or fewer or absent; nectaries usually present; styles 3. Fruit a globose to cylindrical capsule, dehiscing with 6 teeth usually to about the middle.

1 At least the lower leaves distinctly petiolate; leaves never linear-lanceolate or linear
2 Stems hairy all round
3 Petals equalling or slightly exceeding sepals ... **(3–5). media** group
3 Petals about twice as long as sepals
4 Sepals broadly ovate, pubescent beneath all over, the margins narrowly scarious; ripe capsule equalling or only slightly exceeding sepals ... **2. bungeana**
4 Sepals lanceolate, pubescent only at base, the margins broadly scarious; ripe capsule $1\frac{1}{2}$–2 times as long as sepals ... **1. nemorum**
2 Stems with a single line of hairs down each internode, rarely with 2 lines or glabrous
5 Stems quadrangular; bracts scarious ... **7. uliginosa**
5 Stems terete; bracts herbaceous ... **(3–5). media** group
1 All leaves sessile
6 Bracts herbaceous
7 Petals bifid to about half-way or less
8 Leaves 30–80 mm ... **6. holostea**
8 Leaves not more than 10 mm ... **(12–14). longipes** group
7 Petals bifid almost to base
9 Bracts and leaves ciliate; largest leaves usually 15 mm or more ... **16. borealis**
9 Bracts and leaves not ciliate; leaves usually less than 15 mm
10 Leaves 2–8 mm; sepals 3·5–6 mm, equalling ripe capsule; seeds ±smooth ... **18. humifusa**
10 Leaves 6–15(–25) mm; sepals 2–3 mm, shorter than ripe capsule; seeds rugose ... **17. crassifolia**
6 Bracts scarious, with green midrib, or entirely scarious
11 Angles of upper part of stems or margins of leaves rough with papillae
12 Leaf-margins smooth; sepals 5–7 mm ... **8. palustris**
12 Leaf-margins rough with small papillae; sepals not more than 5 mm
13 Petals up to twice as long as sepals ... **9. fennica**
13 Petals equalling sepals ... **15. longifolia**
11 Stems and margins of leaves smooth
14 Leaves elliptical, ovate or ovate-lanceolate, usually less than 15 mm
15 Stems terete; leaves thick, rigid ... **(12–14). longipes** group
15 Stems quadrangular; leaves thin
16 Sepals not ciliate; ripe capsule pale brown ... **7. uliginosa**
16 Sepals ciliate; ripe capsule dark brown, or blackish ... **(12–14). longipes** group
14 Leaves attenuate-triangular to linear-lanceolate, usually more than 15 mm
17 Sepals pubescent beneath towards apex ... **10. hebecalyx**
17 Sepals glabrous (rarely puberulent) towards apex
18 Inflorescence usually 10- to 60-flowered; bracts usually ciliate ... **11. graminea**
18 Inflorescence usually 2- to 9-flowered; bracts not ciliate

19 Sepals lanceolate, acute; plant usually more than 25 cm **8. palustris**
19 Sepals ovate-lanceolate, obtuse; plant usually less than 25 cm **(12–14). longipes** group

1. S. nemorum L., *Sp. Pl.* 421 (1753). Stoloniferous perennial herb; stems up to 65 cm, terete, hairy all round, usually branched only in the lax, many-flowered inflorescence. Lower leaves ovate, acute, with long petioles; upper usually sessile, acuminate. Sepals 6–7 mm, lanceolate, glabrous, or pubescent only at the base, margins narrowly scarious; petals twice as long as sepals, bifid almost to base. Seeds 1–1·3 mm. *Most of Europe, but rare in much of the west and south.* Al Au Be Br Bu Co Cz Da Fe Ga Ge Gr He Ho Hs Hu It Ju No Po Rm Rs(N, B, C, W, E) Su.

(a) Subsp. **nemorum** (*S. nemorum* subsp. *montana* sensu Murb., *S. nemorum* subsp. *reichenbachiana* Wierzb.): Bracts decreasing gradually in size at each dichotomy, the second pair usually more than $\frac{1}{3}$ as long as the first. Ripe seeds with rows of hemispherical tubercles on margins. $2n = 26$. *Almost throughout the range of the species.*

(b) Subsp. **glochidisperma** Murb., *Lunds Univ. Årsskr.* **27(5)**: 156 (1892) (*S. glochidisperma* (Murb.) Freyn, *S. montana* Pierrat, *S. nemorum* subsp. *circaeoides* A. Schwarz): Bracts decreasing abruptly in size after first dichotomy, second pair usually less than $\frac{1}{3}$ as long as the first, third pair 1–2 mm, scale-like. Ripe seeds with cylindrical papillae on margins. *From Britain (Wales) and N.W. Spain eastwards to E. Sweden and C. Greece.*

2. S. bungeana Fenzl in Ledeb., *Fl. Ross.* **1**: 376 (1842). Like **1(a)** but upper leaves acute to subacuminate; sepals 5–6 mm, ovate, obtuse, pubescent beneath all over, margins narrowly scarious; seeds 1·3–1·5 mm. *Woods and shady ravines. N. & E. Russia.* Rs(N, C, E).

(3–5). S. media group. Annual; stems terete. At least the lower leaves petiolate. Petals bifid almost to base.

The species of this group are extremely variable and almost all their characters show some overlap, though they can usually be readily distinguished when all the relevant characters are considered together. They are mostly autogamous, and so far as is known separated by sterility barriers.

1 Stems hairy all round **3. media**
1 Stems with 1(2) lines of hairs, or rarely glabrous
 2 Stamens (2–)10(11); seeds usually 1·1–1·7 mm; petals equalling or exceeding sepals **4. neglecta**
 2 Stamens (0–)3–5(–10); seeds usually less than 1·3 mm; petals usually shorter than sepals, or absent
  3 Sepals usually less than 3 mm; seeds usually less than 0·8 mm, light yellowish-brown; petals absent (rarely minute) **5. pallida**
  3 Sepals usually more than 3 mm; seeds usually more than 0·8 mm, dark reddish-brown; petals present (rarely minute or absent) **3. media**

3. S. media (L.) Vill., *Hist. Pl. Dauph.* **3**: 615 (1789) (*S. media* subsp. *vulgaris* Raunk.). Stems up to 90 cm. Lower leaves ovate, acute or acuminate, with long petioles; upper more or less sessile. Inflorescence few- or many-flowered; pedicels patent or erect in fruit. Sepals 3–7 mm; petals shorter than or slightly longer than sepals; stamens (0–)3–10. Seeds usually 0·8–1·4 mm, usually dark reddish-brown. *Mainly ruderal and as a weed of cultivation. Throughout Europe.* All territories, but probably only as an alien in the extreme north.

(a) Subsp. **media** (*S. media* subsp. *glabra* Raunk.): Stems up to 40 cm, with 1(2) lines of hairs, or rarely glabrous. Leaves usually glabrous. Inflorescence glandular, usually lax. Sepals usually 3–5 mm; petals equalling or slightly shorter than sepals; stamens (0–)3–5(–10). Seeds usually 0·8–1·3 mm, with rounded, rarely conical tubercles. $2n = 40, 42, 44$. *Throughout the range of the species.*

(b) Subsp. **cupaniana** (Jordan & Fourr.) Nyman, *Consp.* 111 (1878) (*S. neglecta* var. *cupaniana* auct., *S. cupaniana* (Jordan & Fourr.) Béguinot; incl. *S. media* subsp. *postii* Holmboe, *S. media* var. *pubescens* Post): Stems up to 90 cm, pubescent all round and usually throughout, usually glandular. Leaves puberulent or pubescent, often glandular. Inflorescence densely glandular-pubescent, lax or dense. Sepals usually 4·5–7 mm; petals variable in length; stamens usually 5–10. Seeds usually 1·3–1·5 mm, with conical tubercles. *C. & E. Mediterranean region.*

Var. *apetala* Gaudin of subsp. **(a)**, with sepals 3–4 mm, petals minute or absent, and condensed inflorescences, has often been confused with **5**.

Subsp. **(b)** appears to be distinguishable from **4** by the pubescence of the stems and inflorescences, and intermediates between the species do not seem to occur, but their relationship requires further study.

4. S. neglecta Weihe in Bluff & Fingerh., *Comp. Fl. Germ.* **1**: 560 (1825) (*S. media* subsp. *major* (Koch) Arcangeli, *S. media* subsp. *neglecta* (Weihe) Gremli, *S. neglecta* subsp. *gracilipes* Raunk., *S. neglecta* subsp. *vernalis* Raunk.). Stems up to 80 cm, with 1 line of hairs down each internode, eglandular. Lower leaves ovate, subcordate at base, acute or acuminate, more or less glabrous, petiolate; upper leaves usually sessile. Inflorescence few- or many-flowered; pedicels long, slender, usually deflexed in fruit. Sepals usually 5–6·5 mm; petals equalling or slightly exceeding sepals; stamens (2–)10(–11). Seeds usually more than 1·1 mm, often 1·3–1·6 mm, usually dark reddish-brown, with conical tubercles. $2n = 22$. *W., S. & C. Europe, extending to S. Sweden and S. Ukraine.* Al Au Be Bl Br Bu Co Cr Cz Da Ga Ge Gr Hb He Ho Hs Hu It Ju Lu Po Rm Rs(B, W, K) Sa Si Su.

5. S. pallida (Dumort.) Piré, *Bull. Soc. Bot. Belg.* **2**: 49 (1863) (*S. apetala* auct., non Ucria, *S. media* subsp. *pallida* (Dumort.) Ascherson & Graebner). Stems up to 30 cm, very slender, with 1 line of hairs down each internode, rarely glabrous. Leaves ovate, subacute, glabrous, usually all petiolate. Inflorescence few- or many-flowered; pedicels short, filiform, patent or erect in fruit. Sepals usually less than 3 mm, often grey-tomentose; petals absent (or minute); stamens 1–3, very rarely 5. Seeds usually less than 0·8 mm, pale yellowish-brown, rarely dark brown, with rounded or conical tubercles. $2n = 22$. *Usually on sandy soils. W., S. & C. Europe, extending to S. Sweden, Latvia and Ukraine.* Al Au Be Bl Br Bu Co Cr Cz Da Ga Ge Gr Hb He Ho Hs Hu It Ju Po Rm Rs(B, W, K) Sa Si Su Tu.

6. S. holostea L., *Sp. Pl.* 422 (1753). Perennial; stems up to 60 cm, ascending, weak, diffuse, sharply quadrangular, usually rough. Leaves 30–80 mm, lanceolate-acuminate, rough on margins and on midrib beneath. Inflorescence lax; flowers 15–30 mm in diameter. Sepals 6–9 mm; petals up to twice as long as sepals, bifid to about half-way, rarely absent. $2n = 26$. *Much of Europe, but absent from N. & C. Fennoscandia and many of the islands, and rare in the*

Mediterranean region. Au Be Br Bu Cz Da Fe Ga Ge Gr Hb He Ho Hs Hu It Ju Lu No Po Rm Rs(N, B, C, W, E) Su Tu.

7. S. uliginosa Murray, *Prodr. Stirp. Gotting.* 55 (1770) (*S. alsine* Grimm, nom. inval.). Creeping perennial; stems up to 40 cm, procumbent to ascending, quadrangular, glabrous, smooth. Leaves up to 15(–20) mm, elliptical to ovate-lanceolate, acute, sessile, or long-petiolate on overwintering vegetative shoots. Bracts scarious, with green midrib, not ciliate. Calyx infundibular at base; sepals 2·5–3·5 mm, exceeding petals; stamens 10. $2n = 24$. *Most of Europe, northwards to 67° N. in Norway, but rare in the east and only on mountains in the south.* Au Be Br Bu Co Cz Da Fa Fe Ga Ge Gr Hb He Ho Hs Hu It Ju Lu No Po Rm Rs(N, B, C, ?W, E) Su.

Very variable in habit and in the size and shape of the leaves.

8. S. palustris Retz., *Fl. Scand. Prodr.* ed. 2, 106 (1795) (*S. barthiana* Schur, *S. dilleniana* Moench, *S. glauca* With.). Creeping perennial, entirely glabrous; stems up to 60 cm, erect, quadrangular, usually smooth. Leaves 15–20 mm, linear-lanceolate, smooth, glaucous or green. Inflorescence lax, usually 2- to 9-flowered; bracts scarious, with green midrib. Sepals 5–7 mm, lanceolate, acute; petals equalling or up to twice as long as sepals, bifid almost to base. *N. & C. Europe, extending southwards to Corse, Bulgaria and Ukraine.* Au Be Br Bu Co Cz Da Fe Ga Ge Hb †He Ho Hu It Ju No Po Rm Rs(N, B, C, W, E) Su.

9. S. fennica (Murb.) Perf., *Bot. Žur.* **2–3**: 152 (1941) (*S. palustris* var. *fennica* Murb.). Like **8** but the angles of the stems and the margins of the leaves rough with small papillae (as in **15**); sepals 3–5 mm. *N.E. Europe.* Fe Rs(N) Su.

In N. Fennoscandia intermediates are found between **8** and **9**.

10. S. hebecalyx Fenzl in Rupr., *Beitr. Pfl. Russ. Reich.* **2**: 26 (1845) (*S. ponojensis* A. Arrh.). Like **8** but leaves 8–25 mm, triangular-lanceolate, narrowed evenly to subacute or obtuse apex; bracts sometimes ciliate towards apex; sepals 4–5·5 mm, ciliate, ovate-lanceolate, pubescent beneath towards apex. *River-banks and ground subject to flooding. N. & E. Russia.* Rs(N, C) [Fe]. (*Siberia.*)

Very variable. See A. Kalela, *Arch. Soc. Zool.-Bot. Fenn. Vanamo* **9** (suppl.): 92–112 (1955).

11. S. graminea L., *Sp. Pl.* 422 (1753). Perennial; stems up to 90 cm, ascending, weak, diffuse, usually much-branched, quadrangular, smooth. Leaves up to 5 cm, linear- to elliptic-lanceolate, acute, smooth. Inflorescence lax, usually 10- to 60-flowered; bracts scarious, usually ciliate; flowers 5–12 mm in diameter, very variable. Sepals 3–7 mm, usually ciliate, sometimes pubescent beneath but only near base, glabrous (rarely puberulent) in upper $\frac{1}{2}$; petals shorter than or exceeding sepals. $2n = 26, 39, 52$. *Europe, except the extreme south and many of the islands.* All except Az Bl Cr Fa Sa Sb Si. Introduced in Is.

The flowers are large in hermaphrodite plants, and small in partly or completely male-sterile plants.

(12–14). S. longipes group. Short-creeping or caespitose perennials. Leaves ovate to linear-lanceolate, sessile.

This group has been reviewed by E. Hultén, *Bot. Not.* **1943**: 251 (1943), T. W. Böcher, *Bot. Tidsskr.* **48**: 401 (1951) and C. C. Chinnappa & J. K. Morton, *Rhodora* **78**: 488–502 (1976). The three species given below are morphologically quite distinct in Europe, but not in North America, and a satisfactory interpretation of the group cannot be made without further investigation throughout the range.

1 Sepals ciliate **14. ciliatisepala**
1 Sepals not ciliate
 2 Leaves ovate to ovate-lanceolate, thick, rigid; short, condensed non-flowering shoots usually present in axils of some upper leaves **13. crassipes**
 2 Leaves linear-lanceolate, thin; short, condensed non-flowering shoots not present in axils of upper leaves **12. longipes**

12. S. longipes Goldie, *Edinb. Philos. Jour.* **6**: 327 (1822) (?*S. peduncularis* Bunge). Creeping or caespitose perennial; stems up to 25 cm, usually ascending, laxly branched, quadrangular; sometimes with non-flowering branches in axils of upper leaves, but never with the condensed shoots typical of **13**. Leaves linear-lanceolate, thin, very acute. Inflorescence usually several-flowered. Sepals 3–7 mm, ovate-lanceolate, obtuse, not ciliate, glabrous; petals exceeding sepals. Ripe capsule $1\frac{1}{2}$–2 times as long as sepals, dark brown or blackish; teeth straight. *Spitsbergen and Kolguev.* Rs(N) Sb. (*Arctic Asia and America.*)

13. S. crassipes Hultén, *Bot. Not.* **1943**: 261 (1943) (*S. longipes* auct., *S. arctica* Schischkin). Shortly creeping or caespitose perennial; stems 2–10 cm, terete; short, condensed non-flowering shoots usually present in axils of some upper leaves. Leaves 3–15 × 2–4 mm, ovate to ovate-lanceolate, thick, rigid. Flowers usually solitary; pedicels short, stout. Sepals 3–4 mm, shorter than petals. Ripe capsule shorter than calyx, light brown; teeth recurved. $2n = 104$. *Arctic Europe; one station in C. Norway.* No Rs(N) Sb Su.

14. S. ciliatisepala Trautv. in Middendorff, *Reise Nord. Öst. Sibir.* **1(2)1**: 52 (1856) (*S. longipes* auct., *S. edwardsii* auct.). Very variable. Shortly creeping perennial; stems up to 20 cm, quadrangular, lacking condensed non-flowering shoots in axils of upper leaves. Leaves up to 10 mm, ovate to ovate-lanceolate. Inflorescence few-, rarely 1-flowered; bracts scarious, ciliate. Sepals ciliate, sometimes pubescent beneath but only near base. Ripe capsule exceeding calyx, dark brown or blackish. *E. part of arctic Russia.* Rs(N). (*Circumpolar.*)

In North America **13** and **14** cannot be separated satisfactorily and the two species should perhaps be treated at subspecific rank.

15. S. longifolia Muhl. ex Willd., *Enum. Pl. Horti Berol.* 479 (1809) (*S. friesiana* Ser.). Creeping perennial; stems 5–25 cm, ascending, branched, quadrangular; angles of upper part of stem, margins of leaves and underside of midrib rough with small papillae. Leaves 5–40 mm, linear-lanceolate to linear. Inflorescence many-flowered; bracts scarious, often with green midrib. Sepals 2–4 mm, glabrous; petals equalling sepals. Ripe capsule slightly exceeding sepals. $2n = 26$. *C. Europe; Fennoscandia; C. & N. Russia.* Au Cz Fe Ge He It No Po Rm Rs(N, B, C, W, E) Su.

In the opinion of some authors plants from Europe and Asia are distinct from plants from North America. They have been called **S. diffusa** Schlecht., *Ges. Naturf. Freunde Berlin Mag.* **7**: 195 (1815), but the characters used for separating them are unreliable.

16. S. borealis Bigelow, *Fl. Boston.* ed. 2, 182 (1824) (*S. calycantha* auct. eur., non (Ledeb.) Bong.). Shortly creeping or caespitose perennial; stems 10–45 cm, quadrangular, smooth or slightly rough. Leaves 10–30 × 2–8 mm, ovate- to linear-lanceolate, usually pale yellowish-green, the margins papillose and ciliate.

Flowers few or solitary; bracts herbaceous, ciliate. Sepals 2–3·5 mm; petals shorter than sepals, or absent. Ripe capsule nearly twice as long as sepals. $2n = 52$. *N. Europe, southwards to S. Norway.* Fe Is No Rs(N) Su.

17. S. crassifolia Ehrh., *Hannover. Mag.* **8**: 116 (1784). Creeping perennial; stems 3–45 cm, usually ascending, laxly branched, quadrangular, glabrous, smooth. Leaves 6–25 × 2–6 mm, ovate to linear-lanceolate, acute, sessile, usually somewhat fleshy, margins smooth, not ciliate. Flowers solitary or in few-flowered leafy cymes; bracts herbaceous, not ciliate; flowers 5–8 mm in diameter. Sepals 2–3 mm, linear-lanceolate, acute; petals exceeding sepals. Ripe capsule exceeding and up to twice as long as sepals. Seeds rugose. $2n = 26$. *N. Europe, extending locally southwards to S. Germany and C. Ukraine.* Da Fe Ge Is No Po Rs(N, B, C, W, E) Su.

18. S. humifusa Rottb., *Kong. Danske Vid. Selsk.* **10**: 447 (1770). Like **17** but stems usually procumbent, matted; leaves 2–8 × 1–4 mm, usually crowded; plant often reproducing by vegetative buds in axils of leaves; flowers 8–16 mm in diameter; sepals 3·5–6 mm, ovate-lanceolate, subacute; ripe capsule usually equalling sepals; seeds almost smooth. $2n = 26$. *Coasts of arctic and subarctic Europe, southwards to 64° N. in N.W. Russia.* ?Fe Is No Rs(N) Sb.

7. Pseudostellaria Pax

A. O. CHATER

Like *Stellaria*, but bearing small napiform tubers laterally on the rhizomes. (*Krascheninnikowia* Turcz., non Gueldenst.)

The other species in this genus are from Asia, and are further characterized by small, fertile cleistogamous flowers, without petals and with 4 sepals (the larger outer pair enclosing the smaller inner pair), which are borne on the lower parts of the plant. The European species rarely produces these cleistogamous flowers and, since its petaliferous flowers are probably always sterile, propagation is largely by the tubers.

1. P. europaea Schaeftlein, *Phyton (Austria)* **7**: 190 (1957) (*Stellaria bulbosa* Wulfen, non *Pseudostellaria bulbosa* (Nakai) Ohwi). Perennial herb; stems 5–20 cm, usually simple, terete, with 1 line of hairs. Leaves sessile, elliptical to lanceolate, acute. Flowers 1–3, 9–13 mm in diameter. Sepals 4–7 mm; petals equalling or exceeding sepals, bifid to less than half-way. $2n = 32$. *Damp and shady places.* ● *S.E. Austria and N. Jugoslavia; N.W. Italy (Vercelli prov.).* Au It Ju.

8. Holosteum L.

S. M. WALTERS AND J. R. AKEROYD

Annual. Inflorescence terminal, umbellate. Petals irregularly toothed; styles 3. Fruit a cylindrical capsule. Seeds asymmetrically reniform and laterally expanded.

1. H. umbellatum L., *Sp. Pl.* 88 (1753). Erect, simple or branched, 5–20(–35) cm, usually glaucous towards base, and variably glandular-viscid or with simple eglandular hairs. Basal leaves oblanceolate, narrowed into petiole; cauline elliptical, sessile; all leaves acute, entire. Flowers in simple umbels on slender pedicels of varying length; pedicels deflexed in young fruiting stage. Petals white or pale pink, about twice as long as sepals. Capsule cylindrical, somewhat narrowed above, twice as long as sepals. Seeds 0·5–1 mm, reddish-brown. *Light, often sandy, soils, usually in disturbed habitats. C., E. & S. Europe, extending northwards to S. Sweden.* Al Au Be †Br Bu *Co Cr Cz Da Ga Ge Gr He Ho Hs Hu It Ju Lu Po Rm Rs(B, C, W, K, E) Sa Si Su Tu.

Very variable in habit and indumentum.

1 Glandular hairs absent; seeds more than 1 mm — (c) subsp. **hirsutum**
1 Glandular hairs present; seeds up to 1 mm
 2 Lower part of stem glabrous; stamens usually 3–5 — (a) subsp. **umbellatum**
 2 Lower part of stem glandular-viscid; stamens usually 8–10 — (b) subsp. **glutinosum**

(a) Subsp. **umbellatum**: Glandular hairs more or less confined to middle part of stem and margin of leaves. Stamens usually 3–5. Seeds 0·5–1 mm. $2n = 20$. *Throughout the range of the species.*

(b) Subsp. **glutinosum** (Bieb.) Nyman, *Consp.* 112 (1878) (*H. glutinosum* (Bieb.) Fischer & C. A. Meyer, *H. umbellatum* subsp. *heuffelii* (Wierzb.) Dostál): Whole plant, except leaf-surfaces, more or less viscid with a dense glandular pubescence. Stamens usually 8–10. Seeds 0·5–1 mm. *Mainly in S.E. Europe.*

(c) Subsp. **hirsutum** (Mutel) Breistr., *Candollea* **25**: 95 (1970): Glandular hairs absent; margin and often upper surface of leaves with simple hairs. Stamens usually 10. Seeds 1·2–1·3 mm. ● *S.E. France.*

9. Cerastium L.

J. JALAS (PERENNIAL SPECIES); *C. fontanum*, M. B. WYSE JACKSON; P. D. SELL AND †F. H. WHITEHEAD (ANNUAL SPECIES)

Herbs, sometimes slightly woody at the base, usually hairy. Flowers usually in cymose inflorescences, sometimes solitary. Sepals free. Petals white, usually bifid or emarginate, sometimes absent; stamens 5–10, rarely fewer; nectaries present; styles usually 5, opposite the sepals, sometimes 3, 4 or 6. Fruit a cylindrical or oblong capsule, often more or less curved, dehiscing with twice as many teeth as styles. Seeds numerous, spherical or reniform.

Literature: A. Buschmann, *Feddes Repert.* **43**: 118–143 (1938). H. Gartner, *Feddes Repert. (Beih.)* **113**: 1–96 (1939). E. Hultén, *Svensk Bot. Tidskr.* **50**: 411–495 (1956). J. Jalas, *Arch. Soc. Zool.-Bot. Fenn. Vanamo* **18**: 57–65 (1963); *Ann. Bot. Fenn.* **3**: 129–139 (1966); **20**: 109–111 (1983). J. Jalas & P. D. Sell, *Watsonia* **6**: 292–294 (1967). A. Lonsing, *Feddes Repert.* **46**: 139–165 (1939). W. Möschl, *Feddes Repert.* **41**: 153–163 (1936); *Agron. Lusit.* **13**: 23–66 (1951).

1 Styles 3, 4, or 6; capsule-teeth 6, 8 or 12
 2 Petals shorter than sepals — **56. diffusum**
 2 Petals $1\frac{1}{3}$–2 times as long as sepals
  3 Stem decumbent or ascending, glabrous except for a line of eglandular hairs down each internode — **1. cerastoides**
  3 Stem erect, covered with minute viscid glands — **2. dubium**
1 Styles 5; capsule-teeth 10
 4 Perennial, often with short non-flowering branches in the axils of leaves
  5 Petals less than twice as long as sepals
   6 Stems 2–6 cm; leaves imbricate (C. Italy) — **25. thomasii**
   6 Stems more than 6 cm; if less, then leaves not imbricate
    7 Capsule 6–8 mm, straight; stem 20–40 cm
     8 Leaves 4–6 times as long as wide, lanceolate — **22. vagans**
     8 Leaves less than 4 times as long as wide, spathulate to ovate-lanceolate — **23. azoricum**
    7 Capsule more than 8 mm, straight or curved; if less, then capsule curved or stem not more than 15 cm

9 Capsule conspicuously curved; bracts with or without scarious margins
10 Seeds 1·5–2 mm, slightly rugose **32. uniflorum**
10 Seeds less than 1·5 mm, tuberculate
11 Lower cauline leaves petiolate **42. sylvaticum**
11 Leaves sessile **43. fontanum**
9 Capsule straight; bracts without scarious margins
12 Leaves usually 50–80 mm **4. dahuricum**
12 Leaves usually not more than 30 mm
13 Petals ciliate at base; seeds 1·7–2·5 mm **35. pyrenaicum**
13 Petals not ciliate; seeds 0·5–1·8 mm
14 Plant densely glandular-hairy; seeds 1·4–1·8 mm **33. theophrasti**
14 Plant with eglandular hairs or ±glabrous; seeds up to *c.* 1 mm
15 Stem with line of hairs down each internode; petals up to twice as long as sepals **1. cerastoides**
15 Stem ±uniformly pubescent or almost glabrous; petals only slightly longer than sepals
16 Sepals 7–8 mm, with wide scarious margins **36. runemarkii**
16 Sepals 5–7 mm, with narrow scarious margins **37. pedunculatum**
5 Petals at least twice as long as sepals
17 Plant with long, soft hairs (sometimes on stems and young leaves only)
18 Plant densely white- to yellow-tomentose to lanate
19 Ovary hairy; leaves 0·5–1·5 mm wide **9. grandiflorum**
19 Ovary glabrous; leaves usually more than 1·5 mm wide
20 Plant with branched hairs **10. candidissimum**
20 Plant without branched hairs
21 Capsule-teeth with flat margins **11. biebersteinii**
21 Capsule-teeth with revolute margins **12. tomentosum**
18 Plant not densely tomentose, greyish or greenish
22 Leaves less than 3 times as long as wide
23 Plant tall, laxly caespitose; flowers many; bracts and sepals obtuse **13. moesiacum**
23 Plant mat-forming; flowers 1–5; bracts and sepals acute **29. alpinum**
22 Leaves at least 3 times as long as wide
24 Seeds more than 1·6 mm
25 Leaves 10–20 × 1–5 mm, lanceolate **15. boissierianum**
25 Leaves 20–50 × 4–7 mm, elliptical to oblanceolate **40. transsilvanicum**
24 Seeds up to 1·6 mm
26 Leaves of flowering stems sparsely ciliate at margins, the surfaces glabrous or subglabrous; midrib thin **17. lineare**
26 Leaves tomentose and/or glandular, with soft hairs at least at margins; midrib strong and prominent beneath
27 Capsule twice as long as sepals **16. gibraltaricum**
27 Capsule as long as or slightly longer than sepals
28 Seeds 1–1·3 mm, strongly tuberculate **14. decalvans**
28 Seeds 1·5–2 mm, rugose **15. boissierianum**
17 Plant pubescent, glandular or glabrous, but never with long, soft hairs
29 Flowering stems with non-flowering axillary shoots or conspicuous buds
30 Fruiting pedicel curved just beneath calyx; capsule curved
31 Sepals (4–)5–8 mm; leaves usually more than 3 times as long as wide
32 Stems usually more than 6 cm; leaves not imbricate **24. arvense**
32 Stems 2–6 cm; leaves imbricate (C. Italy) **25. thomasii**
31 Sepals 3–5 mm; leaves not more than 3 times as long as wide
33 Petals up to $2\frac{1}{2}$ times as long as sepals **26. alsinifolium**
33 Petals not more than twice as long as sepals **27. gorodkovianum**
30 Fruiting pedicel and capsule straight
34 Bracts ciliate from base to apex
35 Leaves less than 4 mm wide **20. soleirolii**
35 Leaves usually at least 4 mm wide **21. scaranii**
34 Bracts not ciliate or only ciliate at base
36 Young stems glabrous except for a line of small hairs down each internode; leaves pale green, somewhat fleshy **1. cerastoides**
36 Young stems hairy all round, at least above; leaves dark to greyish-green, not fleshy
37 Leaves 6–10 mm wide, lanceolate to ovate-lanceolate (Açores)
38 Leaves 4–6 times as long as wide, lanceolate **22. vagans**
38 Leaves less than 4 times as long as wide, spathulate to ovate-lanceolate **23. azoricum**
37 Leaves linear to linear-lanceolate, up to 5 mm wide
39 Bracts pubescent, all with wide, scarious margins **18. banaticum**
39 Bracts glabrous or nearly so, the lowest herbaceous, the upper with narrow, scarious margins **19. julicum**
29 Non-flowering axillary shoots or buds not present on flowering stems
40 Bracts leaf-like, without scarious margins
41 Plant glaucous-pruinose; stems up to 100 cm; leaves usually 50–80 mm (N.E. Russia) **4. dahuricum**
41 Plant not pruinose; stems up to 20 cm; leaves up to 30 mm
42 Leaves of at least the non-flowering shoots ovate to obovate to spathulate, or suborbicular
43 Seeds 1–1·5 mm, the testa close **31. nigrescens**
43 Seeds 1·5–2 mm, the testa loose **32. uniflorum**
42 Leaves ovate to elliptical, acute
44 Seeds 1·7–2·5 mm **34. latifolium**
44 Seeds 1·3–1·7 mm **39. carinthiacum**
40 At least upper bracts sepal-like or minute, with scarious margins
45 Sepals 7–10 mm
46 Plant glabrous; leaves up to 20 mm **30. glabratum**
46 Plant pubescent; leaves 20–80 mm

47 Inflorescence umbel-like; pedicels mostly shorter than sepals **3. maximum**
47 Inflorescence a wide-spreading cyme; pedicels much longer than sepals **40. transsilvanicum**
45 Sepals 2–6 mm
48 Leaves usually 30–70 mm
49 Capsule-teeth revolute **44. pauciflorum**
49 Capsule-teeth erect
50 Leaves sessile **41. subtriflorum**
50 Lower cauline leaves petiolate **42. sylvaticum**
48 Leaves usually not more than 25 mm
51 Plant pulvinate, propagating by means of bulbil-like terminal buds (Arctic) **28. regelii**
51 Plant usually not pulvinate, without bulbil-like terminal buds
52 Leafy stolons absent
53 Seeds squamose **38. dinaricum**
53 Seeds with low tubercles **39. carinthiacum**
52 Leafy stolons present
54 All bracts with scarious margins **24. arvense**
54 Lowest bracts herbaceous
55 Leaves usually not more than 5 mm wide; seeds finely tuberculate **26. alsinifolium**
55 Leaves usually more than 5 mm wide; seeds strongly tuberculate **41. subtriflorum**
4 Annual, never with short non-flowering branches in axils of leaves
56 Plant glabrous
57 Cauline leaves connate **6. perfoliatum**
57 Cauline leaves not connate **54. semidecandrum**
56 Plant hairy
58 Pedicels with deflexed-appressed eglandular hairs, without glandular hairs
59 Sepals with eglandular hairs protruding well beyond apex
60 Petals equalling or longer than sepals; styles usually at least 2 mm **45. illyricum**
60 Petals shorter than sepals; styles 1–1·5 mm
61 Stem 1·5–12 cm; outer sepals without or with only narrow scarious margins **46. comatum**
61 Stem 15–25 cm; sepals with scarious margins **49. brachypetalum**
59 Sepals without eglandular hairs protruding beyond apex
62 Pedicels 8–15 mm; petals ±equalling sepals **45. illyricum**
62 Pedicels 10–70 mm; petals longer than sepals
63 Sepals 8–9·5 mm **47. pedunculare**
63 Sepals 4–5·5 mm **48. scaposum**
58 Pedicels without deflexed-appressed eglandular hairs, with or without glandular hairs
64 Sepals with eglandular hairs protruding well beyond apex
65 Pedicels shorter than sepals; flowers in dense clusters **50. glomeratum**
65 Pedicels usually longer than sepals; flowers in open dichasia
66 Sepals 10–11 mm **45. illyricum**
66 Sepals 3–6·5 mm **49. brachypetalum**
64 Sepals without eglandular hairs protruding beyond apex
67 Petals ciliate at base
68 Sepals with eglandular and glandular hairs **51. rectum**
68 Sepals with eglandular hairs only **5. nemorale**
67 Petals glabrous
69 Pedicels shorter than sepals; flowers clustered
70 Sepals with eglandular hairs **57. siculum**
70 Sepals without eglandular hairs
71 Sepals 2·5–7 mm; capsule not more than 15 mm **58. gracile**
71 Sepals 9–12 mm; capsule more than 15 mm **7. dichotomum**
69 Pedicels longer than sepals; flowers in open dichasia
72 Bracts with scarious margin
73 Upper bracts scarious at least in upper $\frac{1}{3}$ **54. semidecandrum**
73 Upper bracts scarious at most in upper $\frac{1}{5}$
74 Leaves 12–27 mm; plant up to 22 cm; petals at least $1\frac{1}{2}$ times as long as sepals **52. ligusticum**
74 Leaves 4–15 mm; plant not more than 14 cm; petals not more than $1\frac{1}{2}$ times as long as sepals **55. pumilum**
72 Bracts completely herbaceous
75 Sepals glandular-hairy, without eglandular hairs
76 Petals $1\frac{1}{2}$–2 times as long as sepals **52. ligusticum**
76 Petals shorter than to slightly longer than sepals **58. gracile**
75 Sepals with eglandular as well as glandular hairs
77 Petals shorter than sepals **56. diffusum**
77 Petals longer than sepals
78 Petals entire **8. vourinense**
78 Petals bifid
79 Sepals obtuse; seeds 0·5–0·9 mm **52. ligusticum**
79 Sepals acute; seeds 1–1·4 mm **53. smolikanum**

1. C. cerastoides (L.) Britton, *Mem. Torrey Bot. Club* **5**: 150 (1894) (*C. trigynum* Vill., *C. lapponicum* Crantz, *C. lagascanum* C. Vicioso). Loosely matted perennial, usually glabrous except for a line of small hairs down each internode. Vegetative stems creeping; flowering stems 5–15 cm, decumbent or ascending; all stems rooting. Leaves 6–20 × 1–2·5 mm, pale green, somewhat fleshy, linear to oblong, obtuse, usually curving to one side. Flowers 1–3; pedicels up to 8 cm, slender, glandular; bracts herbaceous. Sepals 4–6 mm; petals deeply bifid, up to twice as long as the sepals; styles usually 3. Capsule oblong, straight, up to twice as long as the sepals; teeth patent. Seeds 0·5 mm, tuberculate. $2n = 38$. *Snow-lies and other damp places. Arctic and subarctic Europe, southwards to N. England; principal mountain ranges of Europe.* Al Au Br Bu Cz Fa Fe Ga Ge Gr He Hs Is It Ju No Po Rm Rs(N, W) Sb Su.

2. C. dubium (Bast.) Guépin, *Fl. Maine Loire* ed. 2, **1**: 267 (1838) (*C. anomalum* Waldst. & Kit.). Annual with numerous erect, minutely viscid-glandular stems up to 40 cm. Leaves 10–30 × 1–2 mm, linear to linear-oblong, obtuse to subacute, glabrous or minutely viscid-glandular. Inflorescence of lax-flowered dichasia; pedicels up to 15 cm, slender, with numerous minute viscid glands; bracts herbaceous. Sepals 5–6 mm, ovate-lanceolate, the margin and apex scarious, minutely viscid-glandular; petals $1\frac{1}{3}$ times as long as the sepals, bifid; styles 3. Capsule twice as long as the sepals, oblong-ovoid; teeth patent. Seeds 0·6 mm, pale brown, tuberculate. $2n = 38$. *Mainly in C. & S.E. Europe, but extending to C. Spain, Sicilia, S. Poland and C. Russia.* Au Bu Cz Ga Ge Gr Hs Hu It Ju Po Rm Rs(C, W, K, E) Si Tu.

3. C. maximum L., *Sp. Pl.* 439 (1753). Robust, stoloniferous perennial 20–40 cm, puberulent and glandular above. Leaves 40–80 × 5–15 mm, lanceolate to linear-lanceolate. Flowers 3–13, in dense umbel-like cymes, with minute bracteoles. Petals up to 3 times as long as the sepals, obcordate, with crenate or dentate lobes. Capsule 17–22 × 6–8 mm, straight; teeth revolute. Seeds *c.* 1·8 mm, obtusely tuberculate. *E. part of arctic Russia.* Rs(N). (*N. Asia and North America.*)

4. C. dahuricum Fischer ex Sprengel, *Pugillus* **2**: 65 (1815). Glaucous perennial 50–100 cm. Leaves usually 50–80 × 15–30 mm, ovate, subobtuse. Flowers 11–17, in a large, spreading cyme; bracts herbaceous. Sepals 8–12 mm; petals up to twice as long as the sepals, ciliate. Capsule 10–15 mm, straight. Seeds *c.* 1·5 mm, acutely tuberculate. *Meadows and woodland-margins. E. Russia southwards to* 52° *N.* Rs(N, C).

5. C. nemorale Bieb., *Fl. Taur.-Cauc.* **3**: 317 (1819). Annual with weak, ascending stems (8–)15–60 cm, hairy. Leaves 20–70 × 3–20 mm, the basal obovate-spathulate, the cauline lanceolate, acute, weakly cordate at the base; all more or less hairy. Inflorescence furcate, many-flowered; pedicels hairy, 1–4 times as long as the sepals. Sepals 6–10 mm, lanceolate, acute, narrowly scarious at the apex, with dense eglandular hairs, without glandular hairs; petals bifid for $\frac{1}{3}$ their length, equalling or a little shorter than the sepals, with a ciliate claw; stamens hairy at the base; styles 5. Capsule twice as long as the sepals, more or less bent; teeth deflexed. Seeds 0·7 mm, reddish-brown, tuberculate. *S. & E. Ukraine; Moldavia.* Rs(W, K, E).

6. C. perfoliatum L., *Sp. Pl.* 437 (1753). Glaucous, glabrous annual up to 50 cm. Leaves up to 40 mm, very narrowly elliptical to ovate-lanceolate, subacute, connate at the base. Inflorescence furcate, with a dichasium of 2–7 flowers at the end of each branch; pedicels 1–2 times as long as the sepals; bracts herbaceous. Sepals 9–11 mm, lanceolate, acute, with a scarious margin at the apex; petals shorter than the sepals, bifid at the apex. Capsule twice as long as the sepals, bent near the apex; teeth deflexed. Seeds 1·2–1·5 mm, acutely tuberculate. $2n = c.\ 37$. *Black Sea region; E. & C. Spain.* Bu Hs Rs(W, K, E), Tu.

7. C. dichotomum L., *Sp. Pl.* 438 (1753). Annual up to 18 cm; stems densely viscid-glandular. Leaves up to 22 mm, more or less linear, the upper sometimes ovate-lanceolate, densely viscid-glandular. Pedicels shorter than the sepals; bracts herbaceous. Sepals 9–12 mm, more or less lanceolate, acute, densely viscid-glandular, without eglandular hairs; petals *c.* $\frac{1}{2}$ as long as the sepals, shortly bifid; stamens 5. Capsule 15–22 mm. Seeds 1 mm, chestnut-brown, tuberculate. *Spain; S. Greece.* Gr Hs.

8. C. vourinense Möschl & Rech. fil., *Bol. Soc. Brot.*, ser. 2, **35**: 129 (1961). Annual up to 18 cm, glandular- and eglandular-pubescent, pale green. Basal leaves spathulate, the cauline elliptical, sessile. Inflorescence very lax. Pedicels slender, 1–3 times as long as the calyx, deflexed after anthesis; bracts herbaceous, the lower ones leaf-like. Sepals 5–5·5 mm, the inner with scarious margins; petals *c.* 6 × 2 mm, narrowly oblong to narrowly ovate, entire. Capsule about as long as the sepals, subcylindrical, slightly curved; teeth circinnately revolute. Seeds 0·9–1 mm, tuberculate on the dorsal face. *Shady serpentine rocks and screes, above* 1500 *m.* ● *N.W. Greece.* Gr.

9. C. grandiflorum Waldst. & Kit., *Pl. Rar. Hung.* **2**: 183 (1804) (*C. nodosum* Buschm.). Perennial 15–30 cm, caespitose, white- to greyish-tomentose; uppermost internode up to 15 cm. Leaves 25–70 × 0·5–1·5 mm, narrowly linear, subacute, the margins strongly revolute. Inflorescence a fastigiate cyme of 7–15 flowers; peduncles 1–4 cm; bracts ovate to broadly lanceolate, with wide, scarious margins and apices. Sepals 6–8 mm, subobtuse, tomentose, with scarious margins; petals up to 18 mm, gradually tapering towards the base; ovary and capsule tomentose, at least at the base. Capsule straight, up to twice as long as the calyx; teeth flat, patent. *Mountain rocks.* ● *W. Jugoslavia, Albania.* Al Ju.

10. C. candidissimum Corr., *Österr. Bot. Zeitschr.* **59**: 171 (1909) (*C. tomentosum* sensu Boiss., non L., *C. tomentosum* sensu Halácsy, non L.). Perennial 15–30 cm, caespitose, white- to yellowish-lanate; uppermost internode not considerably longer than the others. Leaves *c.* 20 × 3 mm, lanceolate to linear-lanceolate, subobtuse, straight to curved, the margins revolute. Inflorescence a dense cyme with peduncles and pedicels not more than twice as long as the sepals; bracts ovate, acute to subobtuse, with wide scarious margins. Sepals 5–6 mm, obtuse, white-tomentose; petals up to 10 mm, truncate to auriculate; ovary glabrous. Capsule-teeth flat, recurved. *Mountain rocks and screes.* $2n = 36$. ● *C. & S. Greece.* Gr.

11. C. biebersteinii DC., *Mém. Soc. Phys. Hist. Nat. Genève* **1**: 436 (1822). Perennial 10–30 cm, mat-forming, white-lanate; uppermost internode up to 10 cm, considerably longer than the others. Leaves 20–50 × 3–8 mm, linear-lanceolate to lanceolate, margins not or slightly revolute. Inflorescence an elongate cyme of 3–15 flowers, with pedicels more than twice as long as the sepals; bracts lanceolate, tomentose, with scarious margins. Sepals 6–10 mm, lanceolate, tomentose, subacute to obtuse, with broad scarious margins; petals about twice as long as the sepals, gradually tapering towards the base. Capsule-teeth erect, with flat margins. ● *Krym; cultivated for ornament elsewhere, and becoming naturalized.* Rs(K) [Br].

12. C. tomentosum L., *Sp. Pl.* 440 (1753). Like **11** but up to 45 cm; leaves 10–30 × 2–5 mm, with slightly revolute margins; sepals 5–7 mm; capsule-teeth patent, with revolute margins. $2n = c.\ 108$. ● *C. & S. Appennini, Sicilia; widely cultivated for ornament elsewhere and often becoming naturalized.* It Si [Au Br Cz Da ?Ga Ge Hb He Ho Su].

Very variable in habit, hairiness and leaf-shape. Some of the naturalized plants recorded under this name may be **10** or **11** × **22**.

Var. *aetnaeum* Jan from Sicilia (Etna) is shorter, with narrower and shorter leaves which are sparingly tomentose to subglabrous beneath, and subglabrous sepals and a shorter capsule. It has a chromosome number of $2n = 72$.

13. C. moesiacum Friv., *Flora* (*Regensb.*) **19**: 435 (1836) (*C. tomentosum* sensu Boiss., non L.). Perennial up to 40 cm, laxly caespitose, with creeping, non-flowering basal shoots, tomentose; uppermost internode up to 10 cm, slightly longer than the others. Leaves up to 25 × 10 mm, elliptical to broadly lanceolate, those of the non-flowering shoots obovate, obtuse. Flowers many, in an elongate cyme; bracts obtuse, with wide, scarious margins, sparingly tomentose or sericeous. Sepals obtuse; petals more than twice as long as the sepals. Capsule 10–12 mm, straight; teeth almost erect, with revolute margins. Seeds 1–1·6 mm, yellowish-brown, rugose. $2n = c.$ 144. *Damp, montane grassland.* ● *C. & S. Jugoslavia, N. Albania, W. Bulgaria.* Al Bu Ju.

14. C. decalvans Schlosser & Vuk., *Fl. Croat.* 360 (1869) (*C. lanigerum* G. C. Clementi, non Desv., *C. grandiflorum* sensu Boiss., non Waldst. & Kit.). Perennial 7–40 cm. Leaves lanceolate to lanceolate-elliptical, attenuate, acute, sparsely tomentose to pubescent, with prominent, yellowish midrib. Bracts and sepals lanceolate, acute, with wide scarious margins. Sepals 6–8 mm, pubescent and often glandular; petals patent. Capsule 7–9 mm, broadly cylindrical; teeth erect, with revolute margins. Seeds 1–1·3 mm, strongly tuberculate. $2n = 36, 72, 126$. ● *Mountains of Balkan peninsula.* Al Bu Gr Ju.

(a) Subsp. **decalvans**: Stems up to 30 cm. Leaves 8–20 mm, sparsely tomentose, becoming more or less glabrous. Sepals often glandular. Capsule slightly curved. $2n = 36, 72$. *Almost throughout the range of the species.*

(b) Subsp. **orbelicum** (Velen.) Stoj. & Stefanov, *Fl. Bălg.* 416 (1948) (*C. orbelicum* Velen.). Stems up to 40 cm. Leaves 40–50 mm, persistently tomentose. Sepals eglandular. Capsule straight. $2n = 126$. *S.W. Bulgaria, N.E. Greece.*

15. C. boissierianum Greuter & Burdet, *Willdenowia* **14**: 41 (1984) (*C. boissieri* Gren.). Perennial 5–30 cm, laxly caespitose, mat-forming, with non-flowering stems arising from subterranean basal branches; basal internodes of flowering stems almost as long as the leaves, the uppermost much longer. Leaves usually 10–20 × 1–5 mm, lanceolate, stiff, patent, straight to recurved, narrowly acute or almost aristate, white- to greyish-tomentose; midrib strong. Flowers 1–7 in a lax cyme; upper internodes, peduncles, pedicels, bracts and sepals densely glandular. Sepals 6–8 mm, lanceolate, with wide scarious margins; petals up to 15 mm. Capsule straight, broadly cylindrical, the teeth erect, with slightly curved margins. Seeds 1·5–2 mm, yellowish-brown; testa loose. $2n = 72$. *Rocks and screes. S. & E. Spain; Corse, Sardegna.* Co Hs Sa.

16. C. gibraltaricum Boiss., *Elenchus* 24 (1838) (*C. boissieri* var. *gibraltaricum* (Boiss.) Gren.). Perennial 10–25 cm; non-flowering stems procumbent; flowering stems ascending, with leafy axillary shoots, glandular-pubescent. Leaves up to 45 × 8 mm, linear-lanceolate, subaristate, often somewhat curved, glandular, the margins ciliate with short and long hairs; midrib prominent beneath. Flowers 1–5, on long peduncles; bracts glandular, with scarious margins, especially distally. Sepals 6–8 mm, ovate-elliptical, glandular, with wide scarious margin at the apex. Capsule twice as long as the sepals; teeth short, straight, with somewhat revolute margins. Seeds 1–1·6 mm, rugose, dark brown. *S. Spain (Gibraltar).* Hs. (*N. Africa.*)

17. C. lineare All., *Fl. Pedem.* **2**: 365 (1785). Perennial 5–20 cm, stoloniferous, fragile; lower internodes 1–2 cm, the uppermost longer, all sparsely villous to slightly hairy, often on only 1 side. Leaves of procumbent non-flowering stems rosulate, lanceolate to elliptical, soon decaying and sparsely villous-tomentose; leaves of flowering stems 20–50 × 2–4 mm, linear, acute, with surfaces glabrous or nearly so, sparsely ciliate, with slender midrib. Flowers 1–3; peduncles and pedicels long, slender, villous especially distally; bracts lanceolate, connate at the base. Sepals 5–7 mm, lanceolate, acute, villous to slightly hairy, the outer with narrow, the inner with wide scarious margins. Capsule straight, cylindrical. Seeds 1·2–1·5 mm, pale yellowish-brown, strongly tuberculate. *Rocky grassland.* ● *S.W. Alps.* Ga It.

18. C. banaticum (Rochel) Heuffel, *Enum. Pl. Banat.* 41 (1858). Perennial up to 40 cm; stems, especially above, with deflexed hairs shorter than the diameter of the stem, sometimes glandular, with somewhat swollen nodes; internodes about as long as the leaves, the uppermost long. Leaves linear to linear-lanceolate, often pilose, especially along the prominent, rounded midrib, or glabrous, ciliate at the base. Flowers 3–7; bracts convex, subobtuse, pubescent beneath, with wide scarious margins. Sepals 5–9 mm, ovate to ovate-lanceolate, subobtuse, puberulent, the margins scarious. Capsule 8–12 mm; teeth erect, with revolute margins. Seeds 1·2–1·8 mm, longer than wide, acutely tuberculate. *Rocks. S. & E. part of Balkan peninsula; S.W. Romania.* Al Bu Gr Ju Rm.

(a) Subsp. **banaticum**: Hairs of uppermost internode short, appressed. Leaves up to 55 × 2·5 mm, sparsely hairy to glabrous. *Throughout most of the range of the species.*

(b) Subsp. **speciosum** (Boiss.) Jalas, *Ann. Bot. Fenn.* **20**: 109 (1983) (*C. speciosum* (Boiss.) Hausskn., *C. banaticum* subsp. *alpinum* (Boiss.) Buschm.). Hairs of uppermost internode longer and more patent than in subsp. (a). Leaves up to 25 × 5 mm, hairy to nearly glabrous. $2n = c.$ 120, $c.$ 144. *From C. Greece to N. Albania and N. Bulgaria.*

19. C. julicum Schellm., *Carinthia II* **48**: 69 (1938) (*C. rupestre* Krašan, non Fischer). Densely caespitose perennial; upper part of stems glandular with patent hairs about as long as the diameter of the stem, the basal part slightly hairy to glabrous; upper internode not or only slightly elongate. Leaves 10–30 × 1–2·5 mm, sparsely ciliate towards the base, otherwise glabrous, with keeled midrib and revolute margins. Flowers 1–3; bracts lanceolate, acute, glabrous or almost so, the lower herbaceous, the upper with narrow scarious margins. Sepals 5–8 mm, acute, pubescent. Capsule up to twice as long as the sepals, with thickened outer and inner cell-walls; teeth erect, with revolute margins. Seeds 1·3–1·8 mm, verrucose. $2n = 36$. *Calcareous rocks,* 1700–2400 *m.* ● *S.E. Alps.* Au Ju.

20. C. soleirolii Ser. ex Duby, *Bot. Gall.* **1**: 87 (1928) (incl. *C. stenopetalum* Fenzl, *C. thomasii* sensu Briq., non Ten.). Perennial 2·5–20 cm, more or less caespitose; stems pubescent and glandular, at least above. Leaves 10–20 × 1·5–4 mm, linear to linear-lanceolate, pubescent to glabrous. Flowers 1–7; bracts glandular-pubescent, ciliate, scarious at the apex. Sepals 7–9 mm, lanceolate, acute, glandular-pubescent, with scarious margins; petals obovate. Capsule with thin cell-walls; teeth erect, with flat margins. Seeds 1·5–1·8 mm; testa loose. $2n = 72$. *Rocks and screes above* 1900 *m.* ● *Corse.* Co.

21. C. scaranii Ten., *Prodr. Fl. Nap.* xxvii (1811) (*C. hirsutum* auct. ital., non Crantz, *C. kochii* (Wettst.) Ascherson & Graebner). Perennial 10–25 cm; stems strongly pubescent and sometimes glandular. Leaves 15–55 × 4–10 mm, ovate-lanceolate to elliptical, usually densely pubescent on both sides. Bracts ciliate, with wide scarious margins. Sepals 6–8 mm; petals obcordate. Capsule $c.$ 10 mm; teeth erect, with revolute margins. Seeds 0·9–1·3 mm, rugose; testa usually close. $2n = 36$. *Dry places; calcicole.* ● *Italy, from Alpi Apuane southwards; Sicilia.* It Si.

22. C. vagans Lowe, *Trans. Cambr. Philos. Soc.* **6**: 548 (1838). Perennial 20–40 cm, with robust, ascending flowering stems; uppermost internodes not conspicuously longer than the others. Leaves 6–10 mm wide, 4–6 times as long as wide, lanceolate, hirsute. Flowers 7–11 in divaricate, glandular-pubescent inflorescence; bracts ovate, acute, with scarious margins at the apex. Sepals

5–7 mm, hairy; outer ciliate in basal ½, with scarious margins at the apex; inner with scarious margins (0·2–0·3 mm wide) from base to apex; petals 9–10 mm. Capsule 6–8 mm; teeth erect to somewhat patent, with revolute margins. Seeds 0·9–1·4 mm, acutely tuberculate especially along the edges. *Açores* (*São Jorge*). Az. (*Madeira.*)

The Açores plants belong to var. *ciliatum* Tutin & E. F. Warburg. Var. *vagans* is confined to Madeira.

23. C. azoricum Hochst. ex Seub., *Fl. Azor.* 45 (1844). Like **22** but more hirsute; leaves less than 4 times as long as wide, spathulate- to ovate-lanceolate, with yellowish hairs; inflorescence narrower and more compact; seeds 0·7–1 mm. ● *Açores* (*Corvo, Flores*). Az.

Perhaps not specifically distinct from **22**.

24. C. arvense L., *Sp. Pl.* 438 (1753). (?*C. strictum* L.). Perennial 5–30 cm, loosely matted to caespitose, with procumbent leafy stems freely rooting at the lower internodes. Leaves linear-acicular to lanceolate-elliptical. Bracts ciliate, with scarious margins; petals 2(–3) times as long as the sepals. Seeds seldom more than 1 mm, tuberculate. *Most of Europe, but only as a naturalized alien in Fennoscandia.* Al Au Be Br Bu Cz *Da Ga Ge Hb He Ho Hs Hu It Ju Po Rm Rs(N, B, C, W, E) Su [Fe No].

1 Lower leaves attenuate, shortly petiolate; seeds 1·1–1·7 mm
 2 Sepals lanceolate, subacute (**e**) subsp. **lerchenfeldianum**
 2 Sepals ovate-lanceolate, subobtuse (**f**) subsp. **glandulosum**
1 Leaves sessile; seeds not more than 1 mm
 3 Plants mostly loosely matted; non-flowering shoots nearly as long as flowering stems (**a**) subsp. **arvense**
 3 Plants ±caespitose; non-flowering shoots much shorter than flowering stems
  4 Stems seldom more than 10 cm; capsule as long as or slightly longer than sepals (**b**) subsp. **strictum**
  4 Stems usually 10–30 cm; capsule up to twice as long as sepals
   5 Non-flowering axillary shoots long; leaves up to 40 mm, linear-filiform (**c**) subsp. **suffruticosum**
   5 Non-flowering axillary shoots short; leaves seldom more than 20 mm, linear to lanceolate (**d**) subsp. **molle**

(**a**) Subsp. **arvense**: Stems usually 10–20 cm. Leaves 10–30 × 1·5–3 mm, linear-lanceolate. Flowers 3–7. Sepals 5–7 mm. $2n = 72$. *Throughout the range of the species.*

(**b**) Subsp. **strictum** Gaudin, *Fl. Helv.* **3**: 245 (1828) (*C. strictum* Haenke, non L.): Plants 3–10 cm, caespitose. Leaves 6–15 × 1–4 mm, lanceolate. Flowers usually 1–3. Sepals 4–6 mm; petals up to 12 mm. $2n = 36$. *Pyrenees; Alps, Appennini, mountains of W. Jugoslavia.*

(**c**) Subsp. **suffruticosum** (L.) Hegi, *Ill. Fl. Mitteleur.* **3**: 375 (1911): Plant usually 10–30 cm, mat-forming, with weak stems. Leaves 10–40 mm, linear-filiform. Bracts with wide scarious margins. Sepals 5–7 mm. $2n = 36, 72, 108$. ● *Alps, Appennini.*

(**d**) Subsp. **molle** (Vill.) Arcangeli, *Comp. Fl. Ital.* 99 (1882) (incl. subsp. *matrense* (Kit.) Jáv., subsp. *calcicola* (Schur) Borza, subsp. *pallasii* (Vest) Walters, subsp. *rigidum* (Scop.) Hegi, subsp. *ciliatum* (Waldst. & Kit.) Hayek, *C. ciliatum* Waldst. & Kit., *C. pallasii* Vest): Plant densely matted and stiff. Bracts with wide, scarious margins. Sepals 5–8 mm. $2n = c. 72$. *E.C. Europe; Balkan peninsula.*

(**e**) Subsp. **lerchenfeldianum** (Schur) Ascherson & Graebner, *Syn. Mitteleur. Fl.* **5**(**1**): 611 (1917) (*C. lerchenfeldianum* Schur): Stems 10–30 mm. Lower leaves 10–20 × 2–3 mm, attenuate; upper leaves 10–20 × 6 mm, with rounded base, almost glabrous. Bracts scarious at the apex. Sepals 5–7 mm; petals twice as long as the sepals. Capsule 8 mm; teeth almost flat. ● *Carpathians, N.E. Jugoslavia.*

(**f**) Subsp. **glandulosum** (Kit.) Soó, *Acta Geobot. Hung.* **2**: 237 (1939) (*C. tatrae* Borbás): Like subsp. (**e**) but more densely caespitose and with stems 8–15 cm; bracts with narrow scarious margins; sepals wider and more obtuse; petals wider and up to 3 times as long as the calyx; capsule up to twice as long as the sepals. ● *W. Carpathians* (*Tatra*).

C. uralense Grubov, *Nov. Syst. Pl. Vasc.* (*Leningrad*) **1968**: 104 (1968), described from sandy banks of River Ufa in E.C. Russia, and said to differ by its longer capsules and larger leaves up to 8 mm wide, is probably a variant of subsp. (**a**).

Plants from Morocco have been referred to subsp. (**b**), but these require further investigation.

25. C. thomasii Ten., *Prodr. Fl. Nap., Suppl.* **4**: 21 (1823) (*C. arvense* subsp. *thomasii* (Ten.) Rouy & Fouc.; incl. *C. viscatum* (Montelucci) Jalas). Plant compact, stems 2–6 cm. Leaves more or less imbricate, 4–10 × 1–3 mm, linear to lanceolate, obtuse, hairy. Flowers 1–3. Sepals 5–6 mm; petals 8–10 mm. Capsule somewhat longer than the sepals. $2n = 36$. *Calcareous rocks*, 1700–2900 *m*. ● *C. Appennini.* It.

26. C. alsinifolium Tausch, *Syll. Pl. Nov. Ratisbon.* (*Königl. Baier. Bot. Ges.*) **2**: 243 (1828). Caespitose perennial 5–25 cm; non-flowering axillary shoots often lacking. Leaves of creeping basal stems 5–15 × 2–6 mm, ovate-elliptical, somewhat hairy to almost glabrous. Flowers few to many, in lax cymes; lowest bracts herbaceous. Sepals 3–4 mm; petals up to 2½ times as long as the sepals. Capsule up to 2½ times as long as the sepals, slightly curved or almost straight. *On serpentine.* ● *W. Czechoslovakia* (*near Mariánske Lazne*). Cz.

27. C. gorodkovianum Schischkin, *Fl. URSS* **6**: 883 (1936) (*C. jenisejense* Hultén, *C. beeringianum* sensu Schischkin, non Cham. & Schlecht.). Perennial 12–30 cm; flowering stems slender, with long internodes, sparsely hairy with short patent hairs, or subglabrous; basal branches stolon-like, leafy, with axillary bulbils usually present. Leaves 10 × 4 mm, elliptical, acute, pubescent at least on the margins and veins. Flowers 3–7, in lax cymes; pedicels glandular-pubescent; upper bracts small, acute, with narrow, scarious margins. Sepals 3–5 mm, acute, with scarious margins; petals about twice as long as the sepals. Capsule 8–12 mm, curved; teeth erect or somewhat patent, with revolute margins. Seeds *c.* 1 mm, strongly tuberculate. *Arctic and subarctic Russia.* Rs(N).

C. beeringianum Cham. & Schlecht., *Linnaea* **1**: 62 (1826) (*C. bialynickii* Tolm.) is like **27** but more compact and low-growing, with rigid, erect flowering stems, lacking long runners and axillary bulbils, and petals only slightly longer than the obtuse sepals. It has been recorded from Poluostrov Kanin but the identity of specimens available is not clear.

28. C. regelii Ostenf., *Skr. Vid.-Selsk. Christ.* **1908**(**8**): 10 (1910). Densely caespitose, subglabrous perennial 1–5 cm, often not flowering, and propagating by means of thin, more or less subterranean runners with deciduous, bulbil-like terminal buds. Leaves 3–7 × 1·5–3 mm, roundish to broadly spathulate or elliptical, often distinctly connate at the base, yellowish-green and somewhat fleshy, ciliate, otherwise glabrous and shining. Flowers mostly solitary on short, upright stems which are glandular in the upper part; bracts scarious at the apex. Sepals 4–6 mm, rounded at the apex, glandular, $\frac{1}{3}$ as long as the deeply bifid petals. Ripe capsules and seeds unknown. $2n = 72$. *N.E. Russia, Svalbard.* Rs(N) Sb. (*Circumpolar.*)

29. C. alpinum L., *Sp. Pl.* 438 (1753) (incl. *C. bekuravense* Jáv., *C. lanatum* Lam., *C. squalidum* Ramond). Perennial 5–20 cm. Leaves *c.* 10 × 5 mm, obovate to oblanceolate-elliptical, acute to obtuse at the apex, with long, soft hairs. Flowers 1–5; peduncles 1–4 cm; bracts acute, with scarious margins. Sepals 7–10 mm, acute, truncate at the base; petals about twice as long as the sepals. Capsule 8–14 mm, the upper $\frac{1}{2}$ narrower and slightly curved. Seeds 1–1·4 mm, acutely tuberculate. $2n = c.\ 54, 72\ (144)$. *Arctic and subarctic Europe, extending southwards to Britain and Estonia; most of the principal mountain ranges of Europe.* Al Au Br Bu Cz Fe Ga Ge He Hs Is It Ju No Po Rm Rs(N, B, W) Sb Su.

Very variable in habit, leaf-shape and hairiness. Plants with dense, lanate, eglandular indumentum occur throughout the range of the species and are best distinguished as var. *lanatum* (Lam.) Hegetschw. (*C. lanatum* Lam.). Pyrenean plants with densely glandular-hairy peduncles have been called subsp. *squalidum* (Ramond) Hultén, but similar variants have been recorded for the Alps, Carpathians and Balkans, and are best recognized, if at all, at varietal level.

30. C. glabratum Hartman, *Handb. Skand. Fl.* 180 (1820) (*C. alpinum* subsp. *glabratum* (Hartman) Á. & D. Löve). Like **29** but with slender subterranean runners and totally glabrous stems and leaves; habit suberect; upper internodes often somewhat elongated; leaves up to 20 mm, narrow; sepals somewhat saccate. *N.W. Russia, Fennoscandia, Iceland.* Fe Is No Rs(N) Su.

C. glaberrimum Lapeyr., *Hist. Abr. Pyr.* 265 (1813), from C. & E. Pyrenees, is like **29** but the petals do not exceed the sepals. It probably deserves subspecific recognition.

31. C. nigrescens (H. C. Watson) Edmondston ex H. C. Watson, *Part First Suppl. Cyb. Brit.* **1**: 81 (1860). Perennial 2–15 cm, with subterranean runners. Leaves usually 10–15 × 4–5 mm, mostly narrower than in **29(a)**. Flowers 1–3; bracts herbaceous. Sepals 6–9 mm, ovate to ovate-lanceolate with wide, scarious margins; petals at least twice as long as the sepals, shallowly bifid. Capsule broadly cylindrical, with a wide mouth, almost straight, up to twice as long as the sepals. Seeds 1–1·5 mm, rugose; testa close. *N.W. Europe, extending to Svalbard and N.W. Finland.* Br Fa Fe Is No Sb Su.

Very variable in leaf-shape, hairiness, etc., especially in the Arctic. To some extent this variation may be due to hybridization with **28** and **29**. At present it seems best to distinguish only the following two subspecies.

(a) Subsp. **nigrescens** (*C. edmondstonii* (Edmondston) Murb. & Ostenf., *C. arcticum* subsp. *edmondstonii* (Edmondston) Á. & D. Löve): Compact. Leaves short, ovate to suborbicular, somewhat fleshy, dark green, densely glandular. Pedicels short, densely glandular. Sepals obtuse. $2n = 108$. *Stony slopes on serpentine.* ● *Zetland (Unst).*

(b) Subsp. **arcticum** (Lange) Lusby, *Watsonia* **16**: 296 (1987). (*C. arcticum* Lange, *C. edmondstonii* auct., non (Edmondston) Murb. & Ostenf.): Leaves up to 15 mm, pale or yellowish-green; margins with short, few-celled cilia, otherwise leaves glabrous to slightly pubescent. Sepals acute. $2n = 108$. *Throughout the range of the species.*

32. C. uniflorum Clairv., *Man. Herb. Suisse* 147 (1811). Like **31(b)** but 3–10 cm with leaves 10–18 × 3–5 mm, those of non-flowering shoots obovate to spathulate, subobtuse, attenuate, soft, bright green; sepals 5–7 mm; petals up to twice as long as the sepals, bifid half-way to the base; capsule narrower, twice as long as the sepals; pedicel twice as long as the narrowly cylindrical, curved capsule; seeds 1·5–2 mm, smoother; testa loose. $2n = 36$. *Screes and stony slopes; calcifuge.* ● *Alps, W. Carpathians, mountains of C. Jugoslavia.* Au Cz Ga Ge He It Ju Po.

33. C. theophrasti Merxm. & Strid, *Bot. Not.* **130**: 469 (1977). Caespitose, densely glandular-hairy perennial; stems 2–9 cm. Leaves 7–11(–14) × 2·5–4 mm, narrowly elliptical to oblanceolate, obtuse. Flowers 1 or 2(3); pedicels deflexed at anthesis. Sepals 5–7 mm, subacute, with narrow scarious margins; petals 7–9·5 mm. Capsule 8–10 mm, tapering slightly above. Seeds 1·4–1·8 mm, with long acute tubercles. $2n = 36$. *Limestone screes above 2600 m.* ● *N. Greece (Olimbos).* Gr.

34. C. latifolium L., *Sp. Pl.* 439 (1753). Perennial 3–10 cm, laxly caespitose, glandular-pubescent. Leaves 12–30 × 5–10 mm, ovate to ovate-elliptical, acute, sessile, rigid, bluish-green. Flowers 1–3; bracts herbaceous. Sepals 5–7 mm, with wide scarious margins; petals more than twice as long as the sepals, shallowly bifid. Capsule often more than twice as long as the sepals, broadly cylindrical, almost straight; pedicel not longer than the capsule. Seeds 1·7–2·5 mm, shallowly rugose; testa loose. $2n = 36$. *Stony slopes; calcicole.* ● *Alps; isolated stations in Appennini and Carpathians.* Au Cz Ga Ge He It Po Rm.

35. C. pyrenaicum Gay, *Ann. Sci. Nat.* ser. 1, **26**: 231 (1832). Like **34** but petals ciliate at the base, and only slightly exceeding the sepals. $2n = 38$. ● *E. Pyrenees.* Ga Hs.

36. C. runemarkii Möschl & Rech. fil., *Anzeig. Akad. Wiss. (Wien)* **1962**: 231 (1962) (*C. coronense* Runemark, non Schur). Mat-forming or caespitose perennial 5–10 cm; stems, peduncles, leaves and sepals covered with long, straight, eglandular hairs. Leaves 5–10 × 2–4 mm, elliptical to oblanceolate. Flowers solitary, terminal or in leaf-axils, without sepal-like bracts; peduncles 1·5–3 cm. Sepals 7–8 mm, lanceolate, with wide scarious margins; petals 9–10 mm. Capsule not much longer than the sepals, scarcely curved. Seeds 0·7–0·8 mm, minutely muricate. $2n = 36$. *Limestone and schistose cliffs.* ● *Kikhlades (Naxos); Evvoia (Okhi).* Gr.

37. C. pedunculatum Gaudin, *Fl. Helv.* **3**: 251 (1828). Perennial 3–10 cm, laxly caespitose, sparsely pubescent or almost glabrous. Leaves 10–20 × 3 mm, lanceolate, stiff. Flowers campanulate, mostly solitary; pedicels up to 3 times as long as the capsule. Sepals 5–7 mm, acute, with narrow scarious margins; petals glabrous, only slightly exceeding the sepals, deeply bifid. Capsule up to twice as long as the sepals; teeth revolute. Seeds *c.* 1 mm; testa loose. $2n = 36$. *Stony slopes above 2000 m; calcifuge.* ● *Alps.* Au Ga He It.

38. C. dinaricum G. Beck & Szysz., *Rozpr. Akad. Um. (Mat.-Przyr.)* **19**: 62 (1889). Like **37** but up to 15 cm; upper bracts sepal-like, with scarious margins; sepals 3–5 mm, acute, with broad scarious margins; capsule more than twice as long as the sepals;

seeds *c.* 1·5 mm. ● *Mountains of N.W. part of Balkan peninsula.* Al Ju.

39. C. carinthiacum Vest, *Bot. Zeit.* (*Regensb.*) **6**: 120 (1807) (*C. latifolium* auct. carpat., non L.). Perennial up to 20 cm, loosely matted. Leaves 10–25 × 3–8 mm, ovate-elliptical to lanceolate, acute, often glabrous, shining. Flowers up to 7, in a spreading cyme. Sepals 5–6 mm, subobtuse. Capsule at least twice as long as the sepals, straight or slightly curved. Seeds 1·3–1·7 mm, with low tubercles; testa loose. ● *E. Alps.* Au He It Ju.

(**a**) Subsp. **carinthiacum**: Often nearly glabrous. All except the lowest bracts with wide scarious margins. $2n = 36$. *Mainly in the eastern part of the range of the species.*

(**b**) Subsp. **austroalpinum** (H. Kunz) H. Kunz in Janchen, *Cat. Fl. Austr.* **1**: 155 (1956) (*C. austroalpinum* H. Kunz): More densely glandular-hairy, especially above. Uppermost bracts herbaceous or with narrow scarious margins. $2n = 36$. *From S. Switzerland to E. Austria; local.*

40. C. transsilvanicum Schur in Griseb. & Schenk, *Arch. Naturgesch.* (*Berlin*) **1852**: 305 (1852). Glandular-pubescent, caespitose perennial with numerous flowering stems up to 40 cm, and short non-flowering shoots. Leaves of flowering stems 20–50 × 4–7 mm, elliptical to oblanceolate, acute, attenuate, sparsely pubescent to subglabrous, often with longer, somewhat villous hairs on the margins; leaves of non-flowering stems smaller, more densely hairy. Inflorescence a wide, many-flowered cyme; bracts scarious, obtuse. Sepals acute, the inner with wide scarious margins. Capsule up to twice as long as the sepals, straight; teeth erect, obtuse. Seeds *c.* 2 mm, acutely tuberculate. ● *E. & S. Carpathians.* Rm.

41. C. subtriflorum (Reichenb.) Pacher, *Jahrb. Naturh. Landesmus. Kärnten* **18**: 104 (1886) (incl. *C. sonticum* G. Beck). Biennial or perennial up to 45 cm, hirsute, glandular, with leafy stolons. Leaves elliptical to ovate-lanceolate, sessile. Inflorescence a 3- to many-flowered cyme; lowest bracts usually herbaceous, the upper small, with scarious margins. Sepals up to 5 mm, with scarious margins; petals up to 3 times as long as the sepals. Capsule twice as long as the sepals, straight; teeth erect. Seeds 0·7–1 mm, tuberculate; testa close. $2n = 36$. ● *N.W. Jugoslavia (Slovenija), just extending into Italy (Alpi Giulie).* It Ju.

42. C. sylvaticum Waldst. & Kit., *Pl. Rar. Hung.* **1**: 100 (1802) (*C. umbrosum* Kit.). Hirsute, glandular, biennial or perennial 15–70 cm, with runner-like rooting, leafy basal branches. Lower cauline leaves up to 75 × 18 mm, the lower oblanceolate, petiolate; upper sessile, elliptical to lanceolate-elliptical; leaves of non-flowering shoots oblanceolate to rhombic-elliptical. At least the upper bracts usually with scarious margins. Sepals usually 3–6(–8) mm; petals and stamens ciliate or glabrous. Capsule up to 10 mm, often conspicuously curved. Seeds 0·8–1·3 mm, densely tuberculate. $2n = 36$. ● *Mainly in E.C. Europe, extending from S. Italy, Albania and S.W. Ukraine northwards to the Baltic region.* Al Au Cz Hu It Ju Po Rm Rs(B, W).

43. C. fontanum Baumg., *Enum. Stirp. Transs.* **1**: 425 (1816). Short-lived perennial up to 90 cm, with short, basal non-flowering shoots. Leaves sessile. At least the upper bracts with scarious margins. Petals shorter than to somewhat longer than the sepals. Capsule usually curved. Seeds 0·4–1·3 mm, tuberculate. *Almost throughout Europe.* All territories except Cr, but only introduced in Sb.

An extremely variable species.

1 Plant glandular
 2 Lowest bracts without scarious margin; leaves usually 25–60 mm; petals usually ciliate (**b**) subsp. **lucorum**
 2 Lowest bracts with or without scarious margin; leaves usually less than 25 mm; petals usually without cilia (**c**) subsp. **vulgare**
1 Plant eglandular
 3 Lowest bracts with or without scarious margin; cymes usually with 8–40 flowers; petals shorter to slightly longer than sepals; seeds 0·4–0·8(–1·2) mm (**c**) subsp. **vulgare**
 3 Lowest bracts nearly always with scarious margin; cymes with 1–6(–11) flowers; petals usually conspicuously longer than sepals; seeds 0·8–1·2(–1·3) mm
  4 Flowering stems usually densely pubescent, with hairs at least 0·8 mm; sepals 6–9 mm, weakly keeled (**a**) subsp. **fontanum**
  4 Flowering stems usually sparsely pubescent, with hairs not more than 0·6 mm; sepals usually not more than 6 mm, usually strongly keeled at base (**d**) subsp. **scoticum**

(**a**) Subsp. **fontanum** (subsp. *scandicum* H. Gartner, subsp. *alpicum* H. Gartner, *C. macrocarpum* Schur nom. illeg., *C. vulgatum* subsp. *alpinum* (Mert. & Koch) Hartman, *C. longirostre* Wichura, *C. vulgatum* subsp. *fontanum* (Baumg.) Simonkai, *C. caespitosum* subsp. *fontanum* (Baumg.) Schinz & R. Keller, *C. caespitosum* subsp. *alpestre* (Lindblom ex Fries) Lindman, *C. vulgatum* subsp. *macrocarpum* Nyman, *C. fontanum* subsp. *alpestre* (Hegetschw.) Janchen, *C. fontanum* subsp. *alpinum* (Mert. & Koch) Janchen): Plant very densely to sparsely pubescent, without glandular hairs. Flowering stems 10–40(–50) cm. Cauline leaves (8–)10–24(–27) mm. Cymes with 1–6(–11) flowers; lowest bracts almost always with scarious margin; petals up to 1·4 times as long as sepals. Capsule 11–17(–18) mm. Seeds (0·8–)0·9–1·2(–1·3) mm, with tubercles up to 0·07 mm. $2n = 144$. *N. Europe; mountains of C. Europe.*

(**b**) Subsp. **lucorum** (Schur) Soó, *Acta Bot. Acad. Sci. Hung.* **15**: 340 (1969) (*C. fontanum* subsp. *macrocarpum* sensu Jalas, non *C. macrocarpum* Schur, *C. fontanum* subsp. *schurii* Borza, *C. vulgatum* subsp. *lucorum* (Schur) Soó, *C. lucorum* (Schur) Möschl): Plant often sparsely pubescent, the stems with glandular hairs. Flowering stems up to 70 cm, sometimes with long, non-flowering basal shoots. Cauline leaves (15–)25–60 × (6–)10–20(–25) mm, thin, somewhat translucent. Cymes with (9–)12–33 flowers; lowest bracts without scarious margin; sepals 6–8·5 mm. Capsule (12–)13–16(–18) mm. Seeds 0·8–1·1(–1·2) mm, with tubercles up to 0·08 mm. $2n = 144$. ● *Damp, shady habitats. Mainly in C. Europe, but extending westwards to N. Spain.*

(**c**) Subsp. **vulgare** (Hartman) Greuter & Burdet, *Willdenowia* **12**: 37 (1982) (subsp. *balcanum* H. Gartner, subsp. *hispanicum* H. Gartner, subsp. *pyrenaeum* H. Gartner, *C. caespitosum* subsp. *triviale* (Spenner) Hiitonen, *C. vulgatum* subsp. *caespitosum* (Ascherson) Dostál, *C. fontanum* subsp. *triviale* (Spenner) Jalas, *C. caespitosum* Gilib. ex Ascherson nom. illeg., *C. triviale* Link nom. illeg., *C. vulgare* Hartman, ?*C. viscosum* L. pro parte, ?*C. vulgatum* L. pro parte, *C. holosteoides* Fries, *C. vulgatum* subsp. *glabrescens* (G. F. W. Meyer) Janchen, *C. holosteoides* subsp. *pseudoholosteoides* Möschl, *C. fontanum* subsp. *glabrescens* (G. F. W. Meyer) Salman, van Omm. & de Voogd): Plant densely pubescent to subglabrous, the stems sometimes with some glandular hairs. Flowering stems 5–70(–90) cm, with leafy basal shoots. Cauline leaves

(4–)8–25(–36) mm × (1·5–)3–10(–13) mm. Cymes with few to many flowers; lowest bracts with or without scarious margin; sepals 3–7(–9) mm; petals shorter to slightly longer than the sepals. Capsule (7–)9–13(–15) mm. Seeds 0·4–0·85(–1·2) mm, with tubercles usually not more than 0·04 mm. 2*n* = 144, 162 (?72–180). *Throughout the range of the species.*

(d) Subsp. **scoticum** Jalas & P. D. Sell, *Watsonia* **6**: 293 (1967) (*C. triviale* var. *alpinum* auct. brit., non Mert. & Koch): Plant pubescent to subglabrous, without glandular hairs. Flowering stems 3–16 cm. Cauline leaves (5–)7–8(–10) × 1·5–3 mm. Cymes with (1–)3(–6) flowers; lowest bracts with or without scarious margin; petals 1·4–1·7 times as long as the sepals. Capsule 9–12·5 mm. Seeds 0·8–1·2 mm, with tubercles up to 0·05 mm. *Serpentine rocks.* ● *Scotland (Glen Clova).*

Subsp. (**b**) sometimes resembles **42** but has longer sepals and capsules and shorter petals.

Plants from N. Europe that have been called subsp. **scandicum** H. Gartner, *Feddes Repert. (Beih.)* **113**: 68 (1939), are often less hairy than those from C. Europe. Plants from Iceland and Færöer that have been assigned to subsp. *scandicum* are more or less intermediate between subspp. (**a**) and (**c**). Plants similar to subsp. (**a**) but with glandular hairs are hybrids with **29** or **31**.

Var. *holosteoides* (Fries) Jalas of subsp. (**c**) (*C. holosteoides* Fries, *C. fontanum* subsp. *glabrescens* (G. F. W. Meyer) Salman, van Omm. & de Voogd), from riversides in N.W. Europe, grades into var. *vulgare* but in its extreme form has subglabrous, eglandular flowering stems up to 90 cm, leaves up to 36 mm, sepals up to 9 mm, capsules up to 16 mm and seeds up to 1·2 mm.

44. C. pauciflorum Steven ex Ser. in DC., *Prodr.* **1**: 414 (1824). Perennial 20–60 cm, glandular-hairy, usually with a single flowering stem, without runner-like leafy shoots. Leaves lanceolate-elliptical, sessile or subsessile. Sepals 4–6 mm, obtuse; petals and capsule $2\frac{1}{2}$–3 times as long as the sepals. Capsule with revolute teeth. Seeds 0·7–0·8 mm, strongly tuberculate. *E. Russia.* Rs(C). (*N. & C. Asia.*)

45. C. illyricum Ard., *Animadv. Bot. Spec. Alt.* 26 (1763) (*C. illyricum* var. *macropetalum* Boiss., *C. pilosum* Sibth. & Sm.). Annual 1·5–20(–30) cm, very hairy throughout, the hairs of the stems and pedicels often deflexed-appressed. Leaves up to 14 mm, the lower spathulate or obovate, the upper oblong to ovate or elliptical, more or less obtuse. Bracts herbaceous. Sepals oblong-lanceolate to lanceolate, more or less acute; petals with a minute, acute auricle at the base, glabrous; stamens 5–10; styles 5. Seeds chestnut-brown, minutely tuberculate. ● *S. & W. Greece.* Gr.

1 Sepals 10–11 mm; stamens ciliate at base; capsule 11–12 mm (**d**) subsp. **crinitum**
1 Sepals 4–9 mm; stamens glabrous; capsule 6–7 mm
 2 Sepals 8–9 mm, with wide scarious margins; hairs not or scarcely extending beyond apex (**a**) subsp. **illyricum**
 2 Sepals 4–8 mm, with narrow or no scarious margins; hairs extending well beyond apex
  3 Plant 2·5–10 cm; sepals 5–5·5 mm; fruiting pedicels 5–13 mm (**b**) subsp. **decrescens**
  3 Plant 5–22 cm; sepals 5·5–8 mm; fruiting pedicels 10–50 mm (**c**) subsp. **brachiatum**

(**a**) Subsp. **illyricum** (*C. pelligerum* Bornm. & Hayek): Plant 6–20 cm. Fruiting pedicels 8–15(–18) mm, erect, with appressed hairs. Sepals 8–9 mm, with wide scarious margins; hairs not or scarcely exceeding the apex; petals about equalling the sepals; stamens glabrous. Capsule *c.* 6 mm. 2*n* = 34. ● *W. Greece (Ionioi Nisoi).*

(**b**) Subsp. **decrescens** (Lonsing) P. D. Sell & Whitehead, *Feddes Repert.* **69**: 15 (1964) (*C. pedunculare* sensu Hayek pro parte, non Bory & Chaub., *C. illyricum* subsp. *pilosum* sensu Rouy & Fouc.): Plant 2·5–10 cm. Fruiting pedicels 5–13 mm, more or less divaricate, the hairs mainly deflexed-appressed. Sepals 5–5·5 mm, the outer without or with only narrow, scarious margins; hairs much exceeding the apex; petals equalling or a little longer than the sepals; stamens glabrous; styles 2–3 mm. Capsule 6–7 mm. Seeds *c.* 0·6 mm. ● *S. Greece (Argolis, Arkadhia and Kefallinia).*

(**c**) Subsp. **brachiatum** (Lonsing) Jalas, *Ann. Bot. Fenn.* **20**: 110 (1983) (subsp. *prolixum* (Lonsing) P. D. Sell & Whitehead, *C. pedunculare* sensu Hayek, pro parte, non Bory & Chaub., *C. brachiatum* Lonsing): Plant 5–22 cm. Fruiting pedicels 10–50 mm, erect or divaricate, the hairs mostly deflexed-appressed. Sepals 5·5–8 mm, the outer without or with only narrow scarious margins; hairs much exceeding the apex; petals equalling or a little longer than the sepals; stamens glabrous; styles 3–4 mm. Capsule 6–7·5 mm. Seeds *c.* 0·6 mm. 2*n* = 34. ● *S. & W. Greece.*

(**d**) Subsp. **crinitum** (Lonsing) P. D. Sell & Whitehead, *Feddes Repert.* **69**: 15 (1964) (*C. crinitum* Lonsing): Plant up to 18 cm. Fruiting pedicels 8–15 mm, erect, with long, patent hairs. Sepals 10–11 mm, the outer without or with very narrow scarious margins; hairs much exceeding the apex; petals more or less equalling the sepals; stamens ciliate at the base; styles *c.* 3·5 mm. Capsule 11–12 mm. Seeds 0·6–1 mm. ● *W. Greece (Akarnania).*

46. C. comatum Desv., *Jour. Bot. Appl.* **3**: 228 (1816) (*C. illyricum* Ard. subsp. *comatum* (Desv.) P. D. Sell & Whitehead, *C. illyricum* sensu Hayek, non Ard.). Annual 1·5–12 cm; very hairy throughout. Leaves like those of **45**. Bracts herbaceous. Fruiting pedicels 2–20 mm, more or less erect, the hairs mostly deflexed. Sepals 4–7·5 mm, the outer without or with only narrow scarious margins; hairs much exceeding the apex; petals usually only $\frac{1}{2}$ but sometimes up to $\frac{3}{4}$ as long as the sepals; stamens glabrous; styles *c.* 1 mm. Capsule *c.* 7 mm. Seeds 0·5–0·8 mm. 2*n* = 34. *Dry places. Aegean region; Corse.* Co Cr Gr Tu.

47. C. pedunculare Bory & Chaub. in Bory, *Expéd. Sci. Morée* **3(2)**: 130 (1832) (*C. laxum* Boiss. & Heldr.). Annual up to 20 cm; stem with deflexed-appressed hairs and minute glands. Lower leaves spathulate, the cauline up to 15 mm, ovate or elliptical, hairy. Inflorescence lax; pedicels 10–60 mm, with deflexed-appressed hairs; bracts herbaceous. Sepals 8–9·5 mm, more or less lanceolate, subacute, the margin and apex scarious, with long hairs not exceeding the apex; petals $1\frac{1}{2}$ times as long again as the sepals, bifid for $\frac{1}{5}$ their length; stamens 10; styles 5. Capsule *c.* 8 mm. Seeds 0·8–1 mm, chestnut-brown, minutely tuberculate. ● *S.W. Greece (Messinia).* Gr.

48. C. scaposum Boiss. & Heldr. in Boiss., *Diagn. Pl. Or. Nov.* **2(8)**: 104 (1849). Annual up to 11 cm; stem very short, stiffly hairy. Leaves 3–12 mm, ovate, obovate or elliptical, more or less obtuse, hairy. Flowers usually solitary on slender, appressed-hairy pedicels 30–70 mm. Sepals 4–5·5 mm, ovate-lanceolate to lanceolate, more or less obtuse, with wide, scarious margins and apex, and long, appressed hairs not exceeding the apex; petals longer than the sepals, bifid for *c.* $\frac{1}{5}$ their length; stamens 10; styles 5. Capsule 6–8 mm. Seeds 0·6–1 mm, chestnut-brown, minutely tuberculate. 2*n* = 36. *Limestone rocks and screes.* ● *Mountains of Kriti.* Cr.

49. C. brachypetalum Pers., *Syn. Pl.* **1**: 520 (1805). Annual up to 40 cm; stem with long, deflexed, patent or ascending eglandular hairs, with or without glandular hairs. Leaves up to 20 mm, the lower spathulate or obovate, the upper ovate, elliptical or oblong, obtuse to acute, hairy. Inflorescence more or less lax; pedicels 3–27 mm, bent just below the flower, with patent or ascending-appressed eglandular hairs, with or without glandular hairs; bracts herbaceous. Sepals 3–6·5 mm, lanceolate to oblong-lanceolate, obtuse to acute, the margin scarious, with eglandular hairs exceeding the apex, with or without glandular hairs; petals shorter than or longer than the sepals, bifid for up to $\frac{1}{3}$ their length, with a small auricle at the base; stamens up to 10; styles 5. Capsule 6–9 mm. Seeds 0·4–1 mm, minutely tuberculate. *S., W. & C. Europe, extending northwards to S. Sweden and eastwards to Krym.* Al Au Be Bl *Br Bu Co Cz Da Ga Ge Gr He †Ho Hs Hu It Ju Lu No Po Rm Rs(C, ?W, K, E) Si Su Tu.

1 Pedicels and sepals with long eglandular hairs, without glandular hairs
2 Pedicels with ascending-appressed eglandular hairs (e) subsp. **tenoreanum**
2 Pedicels with patent, deflexed or slightly ascending hairs
3 Hairs of pedicels patent or deflexed; hairs on sepals up to 2·5 mm; capsule-teeth erect (f) subsp. **atheniense**
3 Hairs of pedicels usually slightly ascending; hairs on sepals up to 1·5 mm; capsule-teeth patent (a) subsp. **brachypetalum**
1 Pedicels and sepals with long glandular hairs mixed with eglandular hairs
4 Petals longer than sepals
5 Plant up to 7 cm, with numerous glandular and eglandular hairs; styles *c.* 1 mm; seeds *c.* 1 mm (d) subsp. **doerfleri**
5 Plant 7–25 cm, with numerous glandular and few eglandular hairs; styles 1·5–3 mm; seeds 0·5–0·7 mm
6 Stamens glabrous; styles 2–3 mm (b) subsp. **pindigenum**
6 Stamens ciliate; styles 1·5–2 mm (c) subsp. **corcyrense**
4 Petals shorter than or almost equalling sepals
7 Seeds 0·5–0·6 mm; glandular and eglandular hairs of pedicels and sepals usually more or less equally abundant and not dense (g) subsp. **tauricum**
7 Seeds 0·6–0·8 mm; glandular hairs usually denser and eglandular hairs fewer (h) subsp. **roeseri**

(a) Subsp. **brachypetalum**: Plant 5–40 cm; stems with slightly ascending eglandular hairs up to 1·5 mm, without glandular hairs. Pedicels 5–18 mm, clothed like the stem. Sepals 4–5·5 mm, clothed like the stem; petals *c.* $\frac{3}{4}$ as long as the sepals, glabrous or ciliate; stamens glabrous or ciliate; styles 0·75–1 mm. Capsule 6–7·5 mm; teeth patent. Seeds *c.* 0·5 mm. $2n = 90$. *W. & C. Europe, extending to Italy and Denmark.*

(b) Subsp. **pindigenum** (Lonsing) P. D. Sell & Whitehead, *Feddes Repert.* **69**: 18 (1964) (*C. pindigenum* Lonsing): Plant 7–25 cm; stems with dense glandular and few eglandular hairs. Pedicels 4–20 mm, clothed like the stem. Sepals 3–5 mm, with numerous glandular hairs and a few eglandular hairs mostly near the apex; petals longer than the sepals, cuneate at the base, glabrous or with a few hairs; stamens glabrous; styles 2–3 mm. Capsule 6–8 mm; teeth patent. Seeds 0·5–0·7 mm. ● *N. & C. Greece.*

(c) Subsp. **corcyrense** (Möschl) P. D. Sell & Whitehead, *loc. cit.* (1964) (*C. corcyrense* Möschl, *C. brachiatum* Lonsing): Plant 7–12 cm; stems with dense glandular and few eglandular hairs. Pedicels 3–13 mm, clothed like the stem. Sepals 4–5 mm, with numerous glandular hairs, and few eglandular hairs especially near the apex; petals longer than the sepals, obcordate, ciliate at the base; stamens ciliate at the base; styles 1·5–2 mm. Capsule 7–9 mm; teeth patent. Seeds 0·5–0·7 mm. ● *N.W. Greece (Kerkira).*

(d) Subsp. **doerfleri** (Halácsy ex Hayek) P. D. Sell & Whitehead, *loc. cit.* (1964) (*C. doerfleri* Halácsy ex Hayek, *C. brachiatum* Lonsing): Plant up to 7 cm; stems with eglandular and glandular hairs. Pedicels 3–11 mm, clothed like the stem. Sepals 3·5–6 mm, clothed like the stem; petals longer than the sepals, glabrous; stamens glabrous; styles *c.* 1 mm. Capsule *c.* 8 mm; teeth erect. Seeds *c.* 1 mm. ● *Mountains of Kriti.*

(e) Subsp. **tenoreanum** (Ser.) Soó in Soó & Jáv., *Magyar Növ. Kéz.* **2**: 761 (1951) (?*C. epiroticum* Möschl & Rech. fil., *C. tenoreanum* Ser.): Plant 5–18 cm; stems with ascending-appressed eglandular hairs, without glandular hairs. Pedicels 5–17 mm, clothed like the stem. Sepals 4–5 mm, with eglandular hairs up to 1·5 mm, without glandular hairs; petals *c.* $\frac{3}{4}$ as long as the sepals, ciliate at the base; stamens glabrous; styles 0·75–1 mm. Capsule 6·5–7·5 mm; teeth patent. Seeds 0·4–0·6 mm. $2n = 52$. ● *C. & S. Europe from Italy eastwards.*

(f) Subsp. **atheniense** (Lonsing) P. D. Sell & Whitehead, *Feddes Repert.* **69**: 19 (1964) (*C. atheniense* Lonsing): Plant 15–25 cm; stems with long, patent or deflexed eglandular hairs, without glandular hairs. Pedicels 5–25 mm, clothed like the stem. Sepals 4–6 mm, with eglandular hairs up to 2·5 mm, without glandular hairs; petals *c.* $\frac{1}{2}$ as long as the sepals, ciliate at the base; stamens glabrous or slightly ciliate; styles 1·2–1·5 mm. Capsule 6–9 mm; teeth erect. Seeds 0·6–0·8 mm. $2n = c.\ 90$. ● *E.C. Greece (Attiki).*

(g) Subsp. **tauricum** (Sprengel) Murb., *Lunds Univ. Årsskr.* **27(5)**: 159 (1892) (*C. luridum* Guss., *C. tauricum* Sprengel): Plant 5–37 cm; stems with long, patent or slightly ascending eglandular hairs mixed with long glandular hairs. Pedicels 5–27 mm, clothed like the stem. Sepals 4·5–5 mm, clothed like the stem; petals *c.* $\frac{3}{4}$ as long as the sepals, glabrous or ciliate; styles *c.* 0·75 mm. Capsule 6–8 mm; teeth patent. Seeds 0·5–0·6 mm. $2n = c.\ 90$. *W. & C. Europe, extending to S. Sweden, Balkan peninsula and Krym.*

(h) Subsp. **roeseri** (Boiss. & Heldr.) Nyman, *Consp.* 109 (1878) (*C. luridum* subsp. *mediterraneum* Lonsing): Plant 6–35 cm; stems with dense, long glandular and few eglandular hairs. Pedicels 3–12(–15) mm, clothed like the stem. Sepals 4–6·5 mm, clothed like the stem; petals nearly as long as the sepals, glabrous; stamens glabrous; styles 0·75–1·5 mm. Capsule 6–9 mm; teeth patent. Seeds 0·6–0·8 mm. $2n = 72$. *S. Europe.*

50. C. glomeratum Thuill., *Fl. Paris* ed. 2, 226 (1799) (*C. viscosum* auct. mult.). Annual up to 30(–45) cm; stem with eglandular and glandular hairs. Leaves 5–25 mm, the lower oblanceolate to obovate, the cauline ovate or ovate-elliptical, obtuse, hairy. Flowers in compact, cymose clusters; pedicels shorter than the sepals; bracts herbaceous. Sepals 4–5 mm, lanceolate, acute, with a narrow, scarious margin, with glandular hairs and eglandular hairs exceeding the apex; petals more or less equalling or shorter than the sepals (rarely absent), bifid for up to $\frac{1}{4}$ their length; stamens 10; styles 5. Capsule 6–10 mm. Seeds 0·4–0·5 mm, pale brown, finely tuberculate. $2n = 72$. *Dry, open habitats. Throughout Europe except the north-east.* All except Fe Rs(N) Sb.

51. C. rectum Friv., *Flora (Regensb.)* **19**: 435 (1836). Annual 8–80 cm; stem with glandular and short eglandular hairs. Leaves up to 60 mm, ovate, oblong-ovate or lanceolate, with more or less numerous eglandular and glandular hairs. Pedicels longer than the sepals; bracts herbaceous or with a narrow scarious margin. Sepals 3–10 mm, lanceolate, acute, with numerous glandular and eglandular hairs not exceeding the apex, the margin scarious; petals $\frac{2}{3}$–$1\frac{1}{3}$ as long as the sepals, strongly ciliate at the base; stamens 5–10; styles 5. Capsule up to twice as long as the sepals. Seeds 0·6–1 mm. ● *Balkan peninsula.* Al Bu Gr Ju.

(**a**) Subsp. **rectum**: Plant 25–80 cm. Leaves 25–60 × 10–25 mm, lanceolate. Inflorescence lax, many-flowered. Sepals 7–10 mm; stamens 10. Capsule 10–15 mm. Seeds *c.* 1 mm. *Almost throughout the range of the species.*

(**b**) Subsp. **petricola** (Pančić) H. Gartner, *Feddes Repert.* (*Beih.*) **113**: 38 (1939) (*C. petricola* Pančić): Plant 8–25 cm. Leaves 5–25 × 5–10 mm, ovate or oblong-ovate, rarely lanceolate. Inflorescence less lax, with fewer flowers than in subsp. (**a**). Sepals 3–7 mm; stamens 5–10. Capsule 7–10 mm. Seeds *c.* 0·6 mm. $2n$ = 36. ● *From N. Greece to W.C. Bulgaria.*

52. C. ligusticum Viv., *Elench. Pl. Horti Bot.* 15 (1802). Annual up to 30 cm; stem with fine glandular and occasional eglandular hairs. Lower leaves spathulate, the cauline elliptical or oblong, hairy. Pedicels much longer than the sepals; bracts herbaceous or scarious at most for $\frac{1}{5}$ their length. Sepals 3–8 mm, ovate-lanceolate, obtuse, the margin scarious, with numerous glandular hairs, with or without eglandular hairs usually not exceeding the apex; petals $1\frac{1}{2}$–2 times as long as the sepals, bifid for up to $\frac{1}{3}$ their length, glabrous; stamens 5–10; styles 5. Capsule 6–7 mm. Seeds 0·5–0·9 mm, brown, sharply and minutely tuberculate. ● *E. & C. Mediterranean region.* Al Co ?Ga It Ju Sa Si.

1 Bracts scarious for $\frac{1}{5}$ their length (**d**) subsp. **trichogynum**
1 Bracts herbaceous or very slightly scarious
 2 Sepals 7–8 mm; seeds 0·7–0·9 mm (**c**) subsp. **granulatum**
 2 Sepals 3–6 mm; seeds 0·5–0·65 mm
  3 Plant not more than 7 cm; leaves not more than 6 mm; sepals 3–4 mm (**b**) subsp. **palustre**
  3 Plant up to 15 cm; leaves up to 20 mm; sepals 4–6 mm (**a**) subsp. **ligusticum**

(**a**) Subsp. **ligusticum** (*C. campanulatum* Viv.): Rarely more than 15 cm. Leaves up to 20 mm. Bracts herbaceous. Sepals 4–6 mm; stamens 10. Seeds 0·5–0·65 mm. $2n$ = 34. *Italy, Sicilia, Corse.*

(**b**) Subsp. **palustre** (Moris) P. D. Sell & Whitehead, *Feddes Repert.* **69**: 20 (1964) (*C. palustre* Moris, *C. campanulatum* subsp. *palustre* (Moris) Möschl, *C. pumilum* subsp. *campanulatum* (Viv.) Briq.): Up to 7 cm. Leaves up to 6 mm. Bracts herbaceous. Sepals 3–4 mm; stamens 10. Seeds 0·5–0·6 mm. $2n$ = 34. ● *Sardegna.*

(**c**) Subsp. **granulatum** (Huter, Porta & Rigo ex Möschl) P. D. Sell & Whitehead, *Feddes Repert.* **69**: 21 (1964) (*C. campanulatum* subsp. *granulatum* Huter, Porta & Rigo ex Möschl): Up to 30 cm. Leaves up to 30 mm. Bracts herbaceous. Sepals 7–8 mm; stamens 10. Seeds 0·7–0·9 mm. ● *S. Italy.*

(**d**) Subsp. **trichogynum** (Möschl) P. D. Sell & Whitehead, *loc. cit.* (1964) (*C. trichogynum* Möschl): Up to 22 cm. Leaves 12–27 mm. Bracts scarious at most for $\frac{1}{5}$ their length. Sepals 4–5 mm; stamens 5–10. Seeds *c.* 0·7 mm. ● *Albania, W. Jugoslavia.*

53. C. smolikanum Hartvig, *Bot. Not.* **132**: 359 (1979). Like **52** but stems 5–15 cm; cauline leaves spathulate; bracts herbaceous, obovate to spathulate; sepals narrowly ovate, acute, only slightly exceeded by the petals; stamens 10; capsule up to 8 mm; seeds 1–1·4 mm, yellowish-brown, with longer tubercles. *Serpentine rocks and screes above* 2300 *m.* ● *N.W. Greece* (*Smolikas*). Gr.

54. C. semidecandrum L., *Sp. Pl.* 438 (1753) (*C. balearicum* F. Hermann, *C. dentatum* Möschl, *C. fallax* Guss., *C. heterotrichum* Klokov, *C. obscurum* Chaub., *C. pentandrum* L., *C. rotundatum* Schur). Procumbent to erect annual up to 20 cm; stem with eglandular and dense glandular hairs, rarely glabrous. Leaves up to 18 mm, the basal oblanceolate, the cauline ovate to broadly elliptical. Pedicels equalling or slightly longer than the sepals; pedicels and sepals usually with dense glandular and few eglandular hairs; bracts sometimes almost entirely scarious, always scarious in upper $\frac{1}{3}$. Sepals 3–5 mm, lanceolate, acute, with wide scarious margins; petals shorter than the sepals, slightly notched. Capsule 4·5–7 mm. Seeds 0·4–0·5 mm, yellowish-brown, finely tuberculate. $2n$ = 36. *Throughout Europe except Açores and parts of the north and north-east.* All except Az Fa Is Rs(N) Sb.

Subsp. **macilentum** (Aspegren) Möschl, *Feddes Repert.* **93**: 157 (1936), recorded from S. Sweden and differing from **54** only in being completely glabrous, is best treated as a variety of **54** and is probably extinct.

55. C. pumilum Curtis, *Fl. Lond.* **2(6)**: t. 30 (1777) (*C. ucrainicum* (Kleopow) Klokov, *C. varians* Cosson & Germ.). Annual up to 14 cm; stem with numerous glandular and some eglandular hairs. Leaves 4–15 mm, the lower oblanceolate, the upper ovate or ovate-oblong, obtuse, hairy. Pedicels longer than the sepals; bracts scarious for up to $\frac{1}{8}$ their length. Sepals 4–5 mm, lanceolate to oblong-lanceolate, acute, scarious for up to $\frac{1}{4}$ their length, with glandular hairs, and eglandular hairs reaching near to but not exceeding the apex; petals sometimes purple-tinged, equalling or slightly longer than the sepals, bifid for up to $\frac{1}{4}$ their length, with branched veins; stamens 5–10; styles 5. Capsule 6–8 mm. Seeds 0·5–0·6 mm, chestnut-brown, finely tuberculate. $2n$ = 90–100. *Most of Europe, but absent from large areas of the north and east.* Al Au Be Bl Br Bu Co Cz Da Fe He Ho Hs Hu It Ju Lu No Po Rm Rs(B, C, W) Sa Su Tu.

1 Only upper bracts with a scarious margin; petals 3 times as long as wide; stamens 5 (**a**) subsp. **pumilum**
1 All bracts with a scarious margin; petals not more than $2\frac{1}{2}$ times as long as wide; stamens (5)6–10
 2 Petals *c.* $2\frac{1}{2}$ times as long as wide; stamens (5)6–10; anthers 0·2–0·5 mm (**b**) subsp. **glutinosum**
 2 Petals about twice as long as wide; stamens 10; anthers 0·4–1 mm (**c**) subsp. **litigiosum**

(**a**) Subsp. **pumilum** (*C. glutinosum* auct., non Fries): Plant often suffused with red. Only upper bracts with a scarious margin. Petals narrow, *c.* 3 times as long as wide; stamens 5. $2n$ = 72, 90–100. *W. Europe; S. Sweden; possibly also in parts of C. & S. Europe.*

(**b**) Subsp. **glutinosum** (Fries) Jalas, *Ann. Bot. Fenn.* **20**: 110 (1983) (subsp. *pallens* (F. W. Schultz) Schinz & Thell., *C. atriusculum* Klokov, *C. kioviense* Klokov, *C. pallens* F. W. Schultz, *C. syvaschicum* Kleopow, *C. glutinosum* Fries): Plant often pale green. All bracts with a scarious margin. Petals *c.* $2\frac{1}{2}$ times as long as wide; stamens (5)6–10; anthers 0·2–0·5 mm. *Probably throughout the range of the species.*

(**c**) Subsp. **litigiosum** (De Lens) P. D. Sell & Whitehead, *Feddes Repert.* **69**: 22 (1964) (*C. litigiosum* De Lens): All bracts with a scarious margin. Petals about twice as long as wide and exceeding the sepals by at least 1 mm; stamens 10; anthers 0·4–1 mm. ● *C. Europe and N. Italy.*

Subsp. (**c**) has been included in **52**(**a**) by some authors.

56. C. diffusum Pers., *Syn. Pl.* **1**: 520 (1805) (*C. tetrandrum* Curtis). Annual up to 30 cm; stem with glandular and usually some eglandular hairs. Leaves 5–20 mm, the lower oblanceolate to spathulate, the upper ovate to elliptical, hairy. Pedicels much longer than the sepals; bracts usually herbaceous. Sepals 4–9 mm, ovate-lanceolate to lanceolate, acute or acuminate, with a scarious margin at the apex for up to $\frac{1}{10}$ their length, with glandular hairs, and some eglandular hairs not exceeding the apex. Petals shorter than the sepals, bifid for *c.* $\frac{1}{5}$ their length, with branched veins; stamens and styles 4 or 5. Capsule 5–10 mm. Seeds 0·4–0·7 mm, yellowish to chestnut-brown, bluntly tuberculate. *S., W. & C. Europe, extending northwards to S. Sweden and eastwards to Ukraine.* Au Be Br Co Cz Da Fa Ga Ge Gr Hb Ho Hs Hu It Ju Lu No Po Rs(W, K, ?E) Sa Si Su Tu.

1 Sepals 7–9 mm; capsule 7–10 mm (**c**) subsp. **subtetrandrum**
1 Sepals 4–7 mm; capsule 5–7 mm
 2 Flowers 5-merous; petals 4 times as long as wide (**b**) subsp. **gussonei**
 2 Flowers 4(5)-merous; petals 3 times as long as wide (**a**) subsp. **diffusum**

(**a**) Subsp. **diffusum** (*C. atrovirens* Bab., *C. tetrandrum* Curtis, nom. illeg.): Flowers 4(5)-merous. Sepals 4–7 mm; petals 3 times as long as wide. Capsule 5–7 mm. Seeds 0·5–0·7 mm. $2n = 36, 72$. *Mainly W. & C. Europe.*

(**b**) Subsp. **gussonei** (Tod. ex Lojac.) P. D. Sell & Whitehead, *Feddes Repert.* **69**: 23 (1964) (*C. gussonei* Tod. ex Lojac.): Flowers always 5-merous. Sepals *c.* 4 mm; petals 4 times as long as wide. Capsule *c.* 5 mm. Seeds *c.* 0·5 mm. *Sicilia.*

(**c**) Subsp. **subtetrandrum** (Lange) P. D. Sell & Whitehead, *loc. cit.* (1964) (*C. subtetrandrum* (Lange) Murb.): Flowers (4)5-merous. Sepals 7–9 mm; petals $2\frac{1}{2}$ times as long as wide. Capsule 7–10 mm. Seeds *c.* 0·5 mm. $2n = 72$. ● *E.C. Europe, extending to S. Sweden.*

Subsp. (**c**) has been included in **55**(**a**) by some authors.

57. C. siculum Guss., *Suppl. Fl. Sic. Prodr.* 137 (1832). Annual up to 12 cm; stems with minute glandular and a few eglandular hairs. Leaves 5–15 mm, the basal oblanceolate, the cauline ovate-lanceolate to broadly elliptical, obtuse or subacute, hairy. Inflorescence with a dense dichasium at the end of each branch; pedicels shorter than the sepals; bracts herbaceous. Sepals 5–6 mm, oblong or oblong-lanceolate, acute, scarious at the apex, with dense glandular hairs, and a few eglandular hairs not exceeding the apex; petals $\frac{2}{3}$ as long as the sepals, bifid for up to $\frac{1}{5}$ their length. Capsule 7–8 mm. Seeds 0·4–0·5 mm, pale brown, finely tuberculate. *W. Mediterranean region.* Bl Co Ga ?Hs It Sa Si.

58. C. gracile Dufour, *Ann. Gén. Sci. Phys. (Bruxelles)* **7**: 304 (1820) (*C. bulgaricum* Uechtr., *C. gayanum* Boiss., *C. schmalhausenii* Pacz., *C. velenovskyi* Hayek, *C. carpetanum* Lomax, *C. cavanillesianum* Font Quer & Rivas Goday, *C. durieui* Desmoulins, *C. lamottei* Legrand, *C. pseudobulgaricum* Klokov, *C. ramosissimum* Boiss., *C. riaei* Desmoulins). Annual up to 22 cm; stem, leaves, young capsule and petals sometimes suffused with purple; stems with numerous glandular and few eglandular hairs. Leaves 3–15 mm, the lower spathulate or obovate, the cauline linear-lanceolate to ovate-oblong, glandular and hairy. Inflorescence often much-branched; pedicels usually shorter, rarely longer, than the sepals; bracts herbaceous. Sepals 2·5–7 mm, ovate-lanceolate to ovate-oblong, obtuse to acute, the outer without or with a very narrow, scarious margin, the inner with a scarious margin, all glandular-hairy, the glands not exceeding the apex; petals shorter to slightly longer than the sepals, emarginate or shallowly bifid. Capsule 4–14 mm. Seeds 0·4–1 mm, pale to chestnut-brown, finely tuberculate. $2n = 36, 44–46, 54, 88–92$. *Spain and S. France; S.E. Europe.* Bu Ga Gr Hs Lu Rm Rs(W, K, E) Tu.

Very variable, especially in the size of the sepals, petals, capsules and seeds; perhaps divisible into several subspecies.

10. Moenchia Ehrh.

†A. R. CLAPHAM

Glabrous, usually glaucous annuals. Leaves linear to oblong-lanceolate. Flowers 4- or 5-merous, solitary or in few-flowered, spreading cymes; bracts white-margined. Sepals lanceolate, acute, with broad, white, membranous margins; petals entire or slightly emarginate; stamens 4(5) or 8(10); styles (3)4(5), opposite the sepals. Fruit a straight capsule, dehiscing with twice as many short, obtuse teeth as styles. Seeds numerous, reniform, papillose.

1 Petals exceeding the sepals; flowers 5-merous; capsule broadly ovoid **3. mantica**
1 Petals shorter than or equalling the sepals; flowers 4- or 5-merous; capsule cylindrical to narrowly ovoid
 2 Leaves linear to linear-lanceolate; pedicels not thickened distally; flowers 4-merous **1. erecta**
 2 Leaves oblong-lanceolate; pedicels thickened distally; flowers 5-merous **2. graeca**

1. M. erecta (L.) P. Gaertner, B. Meyer & Scherb., *Fl. Wetter.* **1**: 219 (1799) (*M. quaternella* Ehrh., *Cerastium erectum* (L.) Cosson & Germ.). Erect annual, usually with ascending basal branches. Basal leaves shortly petiolate, the upper sessile, ascending; all linear or linear-lanceolate, rigid, acute, glaucous. Flowers 1–3(–5), *c.* 8 mm in diameter. Petals 4 (rarely 5 or absent), narrow, entire; stamens 4 or 8; styles 4, short and recurved. Capsule equalling or somewhat exceeding the sepals, opening by 8 revolute teeth. *W., W.C. & S. Europe, northwards to England & eastwards to Turkey-in-Europe.* Be Br Bu Co †Cz Ga Ge Gr †Ho Hs It Ju Lu Sa Si Tu [Po].

(**a**) Subsp. **erecta**: Stem 2·5–10 cm. Flowers 1–3. Petals $\frac{2}{3}$ as long as the sepals; stamens 4. Capsule cylindrical, usually slightly exceeding the sepals. $2n = 36$. *Almost throughout the range of the species.*

(**b**) Subsp. **octandra** (Ziz ex Mert. & Koch) Coutinho, *Fl. Port.* 211 (1913) (*M. octandra* (Ziz ex Mert. & Koch) Gay): Stem 10–20 cm. Flowers 2–5. Petals almost equalling the sepals; stamens 8. Capsule narrowly ovoid, scarcely equalling the sepals. *W. Mediterranean region, Portugal; Turkey-in-Europe.*

2. M. graeca Boiss. & Heldr. in Boiss., *Diagn. Pl. Or. Nov.* **3**(**1**): 91 (1853). Like **1** but usually smaller and more branched; leaves oblong-lanceolate, scarcely acute; flowers 5-merous; pedicels flexuous, thickened distally; sepals exceeding the petals; styles about as long as the ovary in flower; capsule *c.* $\frac{2}{3}$ as long as the sepals. *E. & S. parts of Balkan peninsula; Aegean islands.* Al Bu Cr Gr Ju.

3. M. mantica (L.) Bartl., *Cat. Sem. Horti Gotting.* 5 (1839). Like **1** but usually 15–30 cm, and larger throughout; bracts very broadly white-margined; flowers 3–9 or more, 5-merous; petals equalling to much exceeding the sepals; stamens 10; styles 5 (rarely 3 or 4), at least as long as the ovary; capsule broadly ovoid, opening by 10 teeth. $2n = 36$. *S.E. Europe, extending to Austria and N.W. Italy.* Al Au Bu Gr He Hu It Ju Rm Tu.

(a) Subsp. **mantica**: Petals white, up to twice as long as the sepals. Capsule about equalling the sepals. *Throughout the range of the species.*

(b) Subsp. **caerulea** (Boiss.) Clapham, *Feddes Repert.* **69**: 49 (1964) (*M. caerulea* Boiss., *M. mantica* var. *violascens* Aznav.): Petals blue, 2–3 times as long as the sepals. Capsule exceeding the sepals. *Balkan peninsula; perhaps introduced elsewhere.*

11. **Myosoton** Moench

†A. R. CLAPHAM

Perennial. Leaves ovate. Inflorescence a leafy dichasium; flowers 5-merous. Petals deeply bifid, white; stamens 10; styles 5, alternating with the sepals. Fruit an ovoid capsule, dehiscing to almost half-way with 5 shortly bifid, blunt teeth. Seeds numerous, reniform, tuberculate. (*Malachium* Fries.)

1. M. aquaticum (L.) Moench, *Meth.* 225 (1794) (*Cerastium aquaticum* L., *Stellaria aquatica* (L.) Scop., *Malachium aquaticum* (L.) Fries). Flowering stems 20–120 cm, decumbent or ascending, weak, glandular-hairy above. Leaves 2–5(–8) cm, thin, ovate, acute to acuminate at the apex, truncate or cordate at the base, hairy or glabrous, shortly petiolate, or the upper sessile. Petals bifid to the base, white, exceeding the obtuse sepals. Ripe capsule exceeding the sepals, pendent from the patent pedicel. $2n = 28$. *Most of Europe, but absent from much of the north and parts of the Mediterranean region.* All except Az Bl Cr Fa Hb Is Sa Sb.

Variable, especially in habit, hairiness and leaf-shape. **Malachium calycinum** Willk., *Bot. Zeit.* **5**: 239 (1847), from S. Spain (Carratraca, N.W. of Málaga), has stems with a single row of hairs and petals only ½ as long as the sepals. It may merit subspecific rank within **1**.

12. **Sagina** L.

†A. R. CLAPHAM AND N. JARDINE

Small annual or perennial herbs, often caespitose, with slender, procumbent or ascending flowering stems and subulate to linear-lanceolate leaves in slightly connate pairs. Flowers 4- or 5-merous, almost globose in bud, solitary or in few-flowered cymes. Sepals free; petals usually white, entire, often minute, sometimes 0; stamens as many or twice as many as the sepals; styles 4 or 5, alternating with the sepals. Fruit a capsule splitting to the base into 4 or 5 valves. Seeds numerous.

1 Annual, without vegetative stems at time of flowering; flowers 4-merous
 2 Leaves with a well-marked arista; seeds usually less than 0·4 mm **11. apetala**
 2 Leaves muticous or very shortly mucronate; seeds usually more than 0·4 mm **12. maritima**
1 Usually perennial, with vegetative stems at time of flowering, rarely annual; flowers usually 5-merous, sometimes 4-merous (if annual, flowers 5-merous)
 3 Cauline leaves at the 2 uppermost nodes less than 2·5 mm, with short, dense axillary leaf-fascicles **1. nodosa**
 3 Cauline leaves at the 2 uppermost nodes not distinctly shorter than those at the lower nodes
  4 Leaves with a terminal arista ¾–1½ times as long as the maximum leaf-width
   5 Petals 1½–2 times as long as sepals (Corse and Sardegna) **5. pilifera**
   5 Petals less than 1½ times as long as sepals **6. subulata**
  4 Leaves muticous or with an arista less than ¾ as long as the maximum leaf-width
   6 Sepals and pedicels glandular (sometimes very sparsely so)
    7 Petals more than 1½ times as long as sepals **4. glabra**
    7 Petals less than 1½ times as long as sepals
     8 Plant more or less densely caespitose; flowers solitary **8. saginoides**
     8 Plant not densely caespitose; inflorescence 2- to 5-flowered **7. sabuletorum**
   6 Sepals and pedicels glabrous
    9 Plants forming small, dense tufts; sepals usually with purple margins; fruiting pedicels not recurved during ripening
     10 Flowers 4- or 5-merous, usually with fewer than 10 stamens; sepals longer than petals; fruiting pedicels curving outwards **2. nivalis**
     10 Flowers 5-merous; stamens 10; petals longer than sepals; fruiting pedicels straight **3. caespitosa**
    9 Plants forming lax tufts or mats; sepals with white margins; fruiting pedicels recurved at some stage during ripening
     11 Flowers usually 4-merous, occasionally 5-merous; petals often minute or absent, sometimes conspicuous; capsule less than 3 mm **10. procumbens**
     11 Flowers usually 5-merous, occasionally 4-merous; petals about equalling sepals; capsule 3 mm or more
      12 Capsule more than 3·5 mm; sepals more than 2·8 mm; plant fully fertile **8. saginoides**
      12 Capsule less than 3·5 mm; sepals less than 2·8 mm; plant with reduced fertility **9. × normaniana**

1. S. nodosa (L.) Fenzl, *Vers. Darstell. Alsin.* tab. ad 18 (1833) (incl. *S. merinoi* Pau). Perennial with short, non-flowering main stem and many procumbent or ascending flowering stems 5–15(–35) cm, each with a basal rosette and with cauline leaves markedly diminishing upwards. Upper leaves shorter than the internodes and mostly with short, dense axillary leaf-fascicles; all leaves narrowly linear, shortly mucronate. Flowers 1–3, 5-merous, 5–10 mm in diameter. Sepals 2–4 mm, ovate-oblong, obtuse; petals 2–3 times as long as the sepals; stamens 10. Ripe capsule 1⅓ times as long as the appressed sepals. $2n = 22–24, 44, 56$. *Damp places. Europe southwards to N. Portugal and Romania.* Au Be Br Cz Da Fa Fe Ga Ge Hb He Ho Hs Hu Is †It Lu No Po Rm Rs(N, B, C, W) Su.

Var. *moniliformis* (G. F. W. Meyer) Lange, procumbent and rarely flowering, but with numerous, deciduous, bulbil-like axillary fascicles, occurs in N. Europe.

2. S. nivalis (Lindblad) Fries, *Nov. Fl. Suec., Mant.* **3**: 31 (1842) (*S. intermedia* Fenzl). Perennial, forming small cushions 1·5–5 cm in diameter and 1–3 cm high. Basal leaf-rosettes only present during the first season; cauline leaves 3–6 mm, linear, usually shortly mucronate, usually glabrous, sometimes slightly ciliate. Flowers 3–6 mm in diameter, 4- or 5-merous, solitary on pedicels 2–5 mm. Sepals 1·5–2 mm, ovate, obtuse, with narrow, scarious, often violet margins; petals narrow, somewhat shorter than the sepals; stamens 8–10, sometimes fewer. Ripe capsule 2·5–3 mm, greenish- or whitish-yellow, 1½ times as long as the appressed sepals; fruiting pedicels usually curving outwards. $2n =$

88. *Arctic & N.W. Europe, southwards to Scotland and S. Norway.* Br Fa Fe Is No Rs(N) Sb Su.

3. S. caespitosa (J. Vahl) Lange in Rink, *Grønl. Geogr. Stat. Beskr.* **2(6)**: 133 (1857). Perennial, forming small cushions like those of **2** but with persistent dead leaves. Flowers usually 5-merous, 1 or 2 together on short peduncles scarcely exserted from the cushions. Sepals 1·8–3 mm, with violet margins and often with conspicuous veins; petals up to 4 mm, longer than the sepals; stamens 10. Ripe capsule 2–3·5 mm, less than $1\frac{1}{2}$ times as long as the appressed sepals; fruiting pedicels straight. $2n = 88$. *W. Fennoscandia; Iceland; Spitsbergen.* Is No Sb Su.

4. S. glabra (Willd.) Fenzl, *Vers. Darstell. Alsin.* tab. ad 57 (1833) (*S. repens* (Zumagl.) Burnat). Laxly caespitose perennial with procumbent or ascending rooting stems up to 2 cm. Upper leaves scarcely shorter than the lower, about equalling the internodes and with elongating axillary shoots; all leaves narrowly linear, muticous or shortly mucronate. Flowers solitary, 5-merous, 5–10 mm in diameter; pedicels 1–2 cm or more. Sepals ovate-oblong, obtuse; petals $1\frac{1}{2}$–2 times as long as the sepals; stamens 10. Ripe capsule $1\frac{1}{3}$ times as long as the appressed sepals. ● *Alps; Appennini; ?Pyrenees.* Ga He It.

5. S. pilifera (DC.) Fenzl, *loc. cit.* (1833) (*S. saginoides* (L.) Karsten var. *pilifera* (DC.) Fiori). Like **4** but more densely caespitose; leaves long-aristate; petals more than twice as long as the sepals. ● *Mountains of Corse and Sardegna.* Co Sa.

6. S. subulata (Swartz) C. Presl, *Fl. Sic.* 158 (1826). Perennial, mat-forming, with short, non-flowering main stem and numerous decumbent then erect or ascending flowering stems 2–7·5(–12·5) cm. All stems with basal rosettes of linear leaves 0·5–1·5 cm; cauline leaves narrowed to an aristate apex, the arista $\frac{3}{4}$–$1\frac{1}{2}$ times as long as the maximum leaf-width. Flowers usually solitary, 5-merous; pedicels 2–4 cm, filiform, glandular-hairy at least above, rarely glabrous. Sepals broadly ovate, obtuse, usually glandular; petals about equalling the sepals; stamens 10. Ripe capsule 3 mm, slightly exceeding the appressed sepals. $2n = 22$. *Dry, sandy, gravelly or rocky places. Europe eastwards to 23° E. in Czechoslovakia and Greece.* Al Au Br Co Cz Da Fa Ga Ge Gr Hb He †Ho Hs Hu Is It Ju Lu No Po ?Rm Sa Si Su.

Variable. Dwarf, short-lived variants from coastal localities in S.W. Europe (var. *pygmaea* Samp. and var. *gracilis* Fouc. & E. Simon primus) need further investigation, as do robust, long-lived Mediterranean variants with basal leaves up to 2·5 cm and petals much exceeding the sepals (**S. revelieri** Jordan & Fourr., *Brev. Pl. Nov.* **1**: 11 (1866)).

The hybrid between **6** and **10** is occasional.

7. S. sabuletorum (Gay) Lange, *Descr. Icon. Ill.* 3 (1864) (*S. loscosii* Boiss.). Annual or perennial, robust, with a single, non-flowering rosette from which many procumbent flowering stems radiate. Rosette-leaves 1·5–3 cm; cauline leaves 0·5–1·5 cm, all shortly mucronate. Flowers 5-merous, 2–4 per stem, borne singly in the axils of the uppermost leaves; pedicels and sepals densely glandular; pedicels not more than 1·5 cm. Fertile stamens 5. Capsule much longer than the sepals. *Sandy and gravelly places at low altitudes. Spain and N.E. Portugal.* Hs Lu [?Hu].

8. S. saginoides (L.) Karsten, *Deutsche Fl.* 539 (1882) (*S. linnaei* C. Presl, *S. macrocarpa* (Reichenb.) J. Maly, *S. rosoni* Merino, *S. saxatilis* (Wimmer & Grab.) Wimmer). Perennial. Rosette leaves up to 2 cm; cauline leaves 0·5–1 cm; leaves usually glabrous, sometimes slightly ciliate, muticous to shortly mucronate. Flowers usually solitary, sometimes 2, on slender pedicels. Sepals more than 2·8 mm, ovate-oblong, obtuse; petals about equalling the sepals; stamens (5–)10. Ripe capsule (3·5–)4–5 mm, pale straw-coloured, shining, about twice as long as the appressed sepals when dehisced. *Arctic and subarctic Europe, southwards to C. Scotland; principal mountain ranges southwards to S. Spain and C. Greece.* Al Au Br Bu Co Cz Fe Ga Ge Gr He Hs Is It Ju No Po Rm Rs(N) Sa Su.

Variable; the following treatment is tentative.

(**a**) Subsp. **saginoides**: Flowers sometimes 2. Pedicels glabrous, rarely flexuous. Sepals glabrous. Stamens (5–)10. $2n = 22$. *Throughout the range of the species, except Spain & Portugal.*

(**b**) Subsp. **nevadensis** (Boiss. & Reuter) Greuter & Burdet, *Willdenowia* **12**: 189 (1982) (*S. nevadensis* Boiss. & Reuter): More densely caespitose than subsp. (**a**). Flowers solitary. Pedicels glandular-puberulent, often flexuous. Sepals glandular-puberulent. Stamens 10. ● *N.C. & S. Spain; N.C. Portugal.*

9. S. × normaniana Lagerh., *Kong. Norske Vid. Selsk. Skr. (Trondhjem)* **1898(1)**: 1 (1898) (*S. procumbens × saginoides*, *S. saginoides* subsp. *scotica* (Druce) Clapham). Like **8** but with longer, more slender and rooting procumbent stems; rosette-leaves 0·5–3 cm; sepals 2–2·5 mm; fruiting pedicels 1·5–4 cm; capsules 3–3·5 mm, usually remaining undeveloped and setting no seed. *Mountains of Scandinavia, Scotland and Austria.* Au Br No Su.

Plants in cultivation often have some well-developed capsules with good seed.

Variants of uncertain status intermediate between **8** and **10** include **S. saginoides** subsp. **parviflora** Litard. & Maire, *Mém. Soc. Sci. Nat. Maroc.* **4**: 9 (1924), from S. Spain (Sierra Nevada), **S. muscosa** Jordan, *Mém. Acad. Roy. Sci. Lyon Sect. Sci.* ser. 2, **1**: 243 (1851), from S.E. France, and **S. pyrenaica** Rouy, *Ill. Pl. Eur. Rar.* **4**: 26 (1895), from the Pyrenees.

10. S. procumbens L., *Sp. Pl.* 128 (1753) (*S. fasciculata* Poiret). Mat-forming perennial with short, non-flowering main stem bearing a dense, central leaf-rosette, and numerous lateral stems up to 20 cm, ascending from procumbent rooting bases; usually glabrous, sometimes minutely ciliate. Leaves 5–12 mm, linear-subulate, shortly aristate, glabrous or rarely ciliate. Flowers solitary, 4(5)-merous; pedicels 5–20 mm, glabrous. Sepals 1–2·5 mm, broadly ovate, obtuse; petals usually minute or absent, sometimes conspicuous; stamens 4(5). Ripe capsule 2–3 mm, longer than the usually patent sepals. $2n = 22$. *Throughout Europe.* All except Bl ?Cr Sb.

Variable. Many poorly defined local taxa have been described.

S. boydii Buchanan-White, *Trans. Proc. Bot. Soc. Edinb.* **17**: 33 (1887), a densely caespitose perennial with short, erect, glabrous stems, crowded, imbricate, rigid, strongly recurved leaves, and 4- or 5-merous flowers, is presumed to have been collected near Braemar, Scotland, in 1878 but has not been seen since, though retained in cultivation. Ripe seeds are very rarely formed; the capsule remains enclosed in the tightly appressed sepals. $2n = 22$.

11. S. apetala Ard. *Animadv. Bot. Spec. Alt.* **2**: 22 (1763) (*S. patula* Jordan, *S. reuteri* Boiss.). Annual; stem (1–)3–10(–20) cm, erect or ascending, with non-persistent basal leaves in a lax cluster, simple or with decumbent (rarely procumbent) to ascending, non-rooting branches. Leaves linear, long-mucronate to aristate, usually more or less ciliate towards the base. Flowers usually 4-merous, solitary; pedicels filiform, often glandular-hairy at least above. Sepals ovate to ovate-oblong, rounded to acute at the apex, often hooded; petals minute, often falling early, rarely absent. Ripe capsule equalling or exceeding the patent or appressed sepals. Seeds usually less than 0·4 mm. $2n = 12$. *Dry, open habitats. W., C. & S. Europe, extending to S. Sweden; probably casual further east.* All except Fa Fe Is No Rs(N, ?B, K, E) Sb.

This largely autogamous species is extremely variable, and all combinations of characters involving the ciliation of the leaves and the glandular-hairiness of the sepals and pedicels are to be found in local variants. The following subspecies can be recognized.

(a) Subsp. **apetala** (*S. ciliata* Fries, *S. depressa* C. F. Schultz): Fruiting sepals appressed or slightly patent, subacute. Terminal capsule of well-grown plants *c.* $1\frac{1}{4}$ times as long as the sepals. *Throughout the range of the species.*

(b) Subsp. **erecta** (Hornem.) F. Hermann, *Fl. Deutschl. Fennosk.* 182 (1912) (*S. apetala* auct.). Fruiting sepals patent, subobtuse. Terminal capsule of well-grown plants more than $1\frac{1}{4}$ times as long as the sepals. *Almost throughout the range of the species, but rare in the Mediterranean region.*

Glabrous maritime variants of subsp. (a) with persistent basal leaves (*S. ambigua* auct., non Lloyd), occur in W. France and S. England.

Intermediates (*S. melitensis* Gulia ex Duthie, *S. filicaulis* auct., non Jordan, *S. ciliata* var. *minor* Rouy & Fouc.) are not uncommon, especially in the Mediterranean region where the subspecific distinction is difficult to maintain.

12. S. maritima G. Don, *Herb. Brit.* fasc. 7, 155 (1806) (*S. rodriguezii* Willk., *S. stricta* Fries). Annual, usually glabrous, with or without a central rosette of leaves; main stem flowering, it and the numerous lateral branches varying from procumbent to erect; sometimes densely caespitose. Leaves linear-lanceolate, somewhat fleshy, obtuse or mucronulate but not aristate, rarely ciliate. Flowers usually 4-merous, solitary; peduncles erect, glabrous, filiform. Sepals ovate, obtuse, not mucronate, often with purplish margin; petals minute, or absent. Ripe capsule equalling or slightly shorter than the obliquely erect but not appressed sepals. Seeds usually more than 0·4 mm. $2n = 22–24, 28$. *Coasts of Europe from Bulgaria and Estonia westwards, and northwards to 68° N. in Norway; rarely on mountains in Scotland.* Az Be Bl Br Bu Co Cr Da Fe Ga Ge Gr Hb Ho Hs It Ju Lu No Po Rs(B) Sa Si Su Tu.

Very variable, many local species and subspecies having been described. All combinations of characters may be found, and it is difficult to make any useful subdivision.

13. Scleranthus L.

P. D. SELL

Annual, biennial or perennial herbs; stems diffusely branched, hairy on 2 sides. Leaves connate at the base. Inflorescence of more or less dense terminal and axillary cymose clusters. Sepals usually 5, inserted on the rim of the urceolate perigynous zone (the whole being called the 'fruit' in the following account); petals absent; stamens up to 10; styles 2. Fruit an indehiscent 1(2)-seeded nutlet enclosed by the hardened wall of the perigynous zone and the persistent sepals, which are shed with it. Seeds lenticular, smooth.

Measurements of the 'fruit' are from the apex of the sepal to the base of the perigynous zone and vary little between flowering and fruiting.

Literature: W. Rössler, *Agron. Lusit.* **15**: 97–138 (1953); *Österr. Bot. Zeitschr.* **102**: 30–72 (1955).

1 Apex of sepals curved inwards and forming a hook — **3. uncinatus**
1 Apex of sepals patent, erect or slightly incurved, but not hooked
  2 Sepals obtuse, with a scarious margin 0·3–0·5 mm wide — **1. perennis**
  2 Sepals ±acute, with a scarious margin not more than 0·1 mm wide — **2. annuus**

1. S. perennis L., *Sp. Pl.* 406 (1753). Perennial herb often with woody stock and procumbent to erect stems up to 22 cm. Leaves linear to lanceolate, channelled, obtuse to acute, glabrous or with short cilia. Inflorescence of terminal clusters which are not usually exceeded by the bracts. Fruit 2–6·5 mm; sepals lanceolate, obtuse, with a scarious margin 0·3–0·5 mm wide. *Dry, open habitats. Most of Europe, but absent from much of the north and rare in the extreme west.* Al Au Be Br Bu Co Cz Da Ga Ge Gr He Ho Hs Hu It Ju No Po Rm Rs(B, C, W, K) ?Sa Si Su Tu [Fe].

1 Fruit 4·5–6·5 mm
  2 Stems procumbent, not dichotomously branched; fruit 4·5–5·5 mm — (f) subsp. **marginatus**
  2 Stems ascending, dichotomously branched; fruit 6–6·5 mm — (g) subsp. **dichotomus**
1 Fruit 2–4·5 mm
  3 Plant with procumbent or slightly ascending stems or forming a dense cushion; at least some of the inflorescences with more than 7 flowers in a cluster
    4 Plant with long procumbent or slightly ascending stems; fruit 2–3(–3·5) mm — (b) subsp. **prostratus**
    4 Plant forming a dense cushion with stems up to 5 cm
      5 Fruits 2–3 mm — (c) subsp. **burnatii**
      5 Fruits 3·5–4·5 mm — (d) subsp. **polycnemoides**
  3 Plant with ascending to erect stems 3–22 cm; all inflorescences with not more than 7 flowers in a cluster
    6 Stems up to 6 cm; leaves lanceolate, obtuse; internodes 2–4 mm — (e) subsp. **vulcanicus**
    6 Stems up to 22 cm; leaves linear, acute; internodes 6–10 mm — (a) subsp. **perennis**

(a) Subsp. **perennis** (?*S. dichotomus* sensu P. Fourn., non Schur): Stems up to 22 cm, ascending to erect; internodes 6–10 mm. Leaves 5–9 mm, linear, acute, shortly ciliate. Inflorescence with not more than 7 flowers in a cluster. Fruit (3–)3·5–4·5 mm; sepals erect or slightly incurved. $2n = 22$. *Throughout the range of the species.*

(b) Subsp. **prostratus** P. D. Sell, *Feddes Repert.* **68**: 168 (1963): Stems long, procumbent to slightly ascending; internodes 2–3(–10) mm. Leaves 3–5(–7) mm, linear, acute, glabrous or slightly ciliate. Inflorescence often with more than 7 flowers in a cluster. Fruit 2–3(–3·5) mm; sepals erect or slightly incurved. *Sandy heaths.* ● *E. England.*

(c) Subsp. **burnatii** (Briq.) P. D. Sell, *loc. cit.* (1963) (*S. burnatii* Briq.): Forming a more or less dense cushion, with stems up to 5 cm, ascending; internodes 1–5 mm. Leaves 4–6 mm, linear, more or less acute, ciliate. Inflorescence usually with more than 7 flowers in a cluster. Fruit 2–3 mm; sepals slightly incurved. ● *Mountains of Corse and S. Spain; ?Sardegna.*

(d) Subsp. **polycnemoides** (Willk. & Costa) Font Quer, *Butll. Inst. Catalana Hist. Nat.* **37**: 51 (1949) (*S. polycnemoides* Willk. & Costa): Forming a dense cushion, with stems up to 3 cm; internodes up to 5 mm. Leaves 4–6 mm, linear, acute, glabrous or slightly hairy. Inflorescence usually with more than 7 flowers in a cluster. Fruit 3·5–4·5 mm; sepals erect or incurved. ● *E. Pyrenees.*

(e) Subsp. **vulcanicus** (Strobl) Béguinot in Fiori & Béguinot, *Nuovo Gior. Bot. Ital.* nov. ser., **16**: 464 (1909): Compact, with stems up to 6 cm; internodes 2–4 mm. Leaves 3–6(–8) mm, lanceolate, obtuse, shortly ciliate. Inflorescence with not more than 5 flowers in a cluster. Fruit 3–4·5 mm; sepals more or less patent. ● *Sicilia (Etna).*

(f) Subsp. **marginatus** (Guss.) Nyman, *Consp.* 257 (1879) (*S. neglectus* Rochel ex Baumg., *S. perennis* subsp. *neglectus* (Rochel ex Baumg.) Stoj. & Stefanov): With numerous long, procumbent stems; internodes 5–10(–25) mm. Leaves 5–9 mm, linear, obtuse, shortly ciliate. Inflorescence with not more than 5 flowers in a cluster. Fruit 4·5–5·5 mm; sepals more or less patent. *Balkan peninsula, Romania; S. Italy, Sicilia.*

(g) Subsp. **dichotomus** (Schur) Nyman, *Consp., Suppl.* **2**: 126 (1889) (*S. dichotomus* Schur): Like subsp. (f) but stems fewer, ascending, dichotomously branched; fruit 5–6·5 mm. *Mountain rocks. From N. Romania to N. Greece.*

2. S. annuus L., *Sp. Pl.* 406 (1753). Annual or biennial herb with ascending stems up to 25 cm. Leaves linear, channelled, obtuse, ciliate at the base. Inflorescence of axillary and terminal clusters, which are often exceeded by the bracts. Fruit 1·2–5·5 mm; sepals lanceolate, more or less acute, with a scarious margin not more than 0·1 mm wide. $2n = 44$. *Almost throughout Europe except the extreme north.* All except Az Fa Is Sb.

1 Fruit 3·2–4·5(–5·5) mm; sepals patent (a) subsp. **annuus**
1 Fruit 1·2–3(–3·8) mm; sepals erect or connivent
2 Sepals unequal, connivent (c) subsp. **verticillatus**
2 Sepals equal, erect or connivent
3 Fruit 2·2–3(–3·8) mm; sepals erect or connivent (b) subsp. **polycarpos**
3 Fruit 1·2–1·6(–2·3) mm; sepals distinctly connivent (d) subsp. **delortii**

(a) Subsp. **annuus** (*S. glaucovirens* Halácsy, *S. velebiticus* Degen & Rossi): Stems 2–24 cm. Leaves 4–10(–20) mm. Fruit 3·2–4·5(–5·5) mm; sepals more or less equal, patent. *Usually calcifuge. Throughout the range of the species.*

(b) Subsp. **polycarpos** (L.) Thell. in Schinz & R. Keller, *Fl. Schweiz* ed. 3, **2**: 109 (1914) (*S. polycarpos* L.): Stems 2–17 cm. Leaves 4–8 mm. Fruit 2·2–3(–3·8) mm; sepals equal, erect or connivent. $2n = 44$. *Usually calcifuge. Most of Europe, but absent from much of the south and east.*

(c) Subsp. **verticillatus** (Tausch) Arcangeli, *Comp. Fl. Ital.* 110 (1882) (*S. polycarpos* subsp. *collinus* (Hornung ex Opiz) Pignatti, *S. collinus* Hornung ex Opiz, *S. verticillatus* Tausch, *S. syvashicus* Kleopow): Stems 3–14 cm. Leaves 4–6(–10) mm. Fruit 1·5–2·2(–3·0) mm; sepals unequal, connivent. *Somewhat calcicole. S. & S.C. Europe.*

(d) Subsp. **delortii** (Gren.) Meikle, *Fl. Cyprus* **1**: 286 (1977) (*C. delortii* Gren., *S. ruscinonensis* (Gillot & Coste) Rössler, *S. polycarpos* subsp. *ruscinonensis* (Gillot & Coste) Pignatti, *S. annuus* subsp. *ruscinonensis* (Gillot & Coste) P. D. Sell): Usually less than 10 cm. Leaves 4–6 mm. Fruit 1·2–1·6(–2·3) mm; sepals equal, connivent at the apex. ● *S.W. Europe.*

Subsp. **aetnensis** (Strobl) Pignatti, *Gior. Bot. Ital.* **107**: 207 (1973), from Sicilia (Etna), has fruits *c.* 3 mm and sepals with a somewhat narrower scarious margin, but otherwise resembles subsp. (a).

Hybrids between **1** and **2**, which are more or less sterile, sometimes occur where the two species grow together.

3. S. uncinatus Schur, *Verh. Mitt. Siebenb. Ver. Naturw.* **1**: 107 (1850). Annual with ascending stems up to 12 cm. Leaves 7–10 mm, linear, channelled, subacute, glabrous or slightly ciliate at the base. Inflorescence of axillary and terminal 2- to 6-flowered clusters; bracts often exceeding clusters. Fruit 5–6 mm; sepals linear-lanceolate, patent with an incurved hook at the apex, with a scarious margin not more than 0·1 mm wide. *S. Europe; Carpathians.* Al Bu Ga Hs It Ju Rm Rs(W, ?K) ?Tu.

Subfam. **Paronychioideae**

Leaves opposite, alternate or verticillate; stipules present. Petals often very small or absent; sepals free.

14. **Corrigiola** L.

S. M. WALTERS AND J. R. AKEROYD

Glabrous herbs with more or less decumbent stems and alternate leaves. Inflorescence compound, cymose; flowers numerous, small, more or less aggregated, slightly perigynous. Sepals, petals and stamens 5; stigmas 3. Fruit a trigonous achene enclosed in the persistent calyx.

Literature: M. N. Chaudhri, *Meded. Bot. Mus. Herb. Rijksuniv. Utrecht* **285**: 34–52 (1968).

1. C. litoralis L., *Sp. Pl.* 271 (1753). Glaucous, with decumbent or ascending stems up to 25 cm. Cauline leaves linear-oblanceolate, obtuse, entire; stipules small, scarious. Flowers in dense, terminal and axillary cymose clusters. Sepals *c.* 1 mm, slightly longer than the whitish petals. Fruit 1–2 mm. *On seasonally wet, sandy ground. W., C. & S. Europe; occasional as a casual elsewhere.* Al Be Bl Br Bu Co Cr Cz Ga Ge Gr Ho Hs It Ju Lu Po Sa Si Tu.

(a) Subsp. **litoralis** (incl. subsp. *foliosa* (Pérez-Lara ex Willk.) Chaudhri): Annual, with slender taproot; stems decumbent. Basal leaves similar to the cauline, not fleshy. Inflorescence-branches with leafy bracts at the base. Fruit 1–1·5 mm. $2n = 18$. *Throughout the range of the species.*

(b) Subsp. **telephiifolia** (Pourret) Briq., *Prodr. Fl. Corse* **1**: 481 (1910) (*C. telephiifolia* Pourret): Usually perennial, with thick, woody taproot; stems often ascending, stouter. Basal leaves narrowly obovate, more or less fleshy. Inflorescence-branches usually ebracteate. Fruit 1·5–2 mm. $2n = 18$. *S.W. Europe.*

C. imbricata Lapeyr., *Hist. Abr. Pyr.* 169 (1813) (*C. telephiifolia* subsp. *imbricata* (Lapeyr.) Greuter & Burdet), from S.W. Europe, is perennial with short stems and more or less densely imbricate leaves. It is, perhaps, best included within subsp. (b) as var. *imbricata* (Lapeyr.) DC.

15. **Paronychia** Miller

A. O. CHATER AND J. R. AKEROYD

Herbs, sometimes woody at the base, with erect to procumbent, usually much-branched stems. Leaves opposite (or sometimes apparently alternate), elliptical to linear; stipules usually conspicuous, scarious. Flowers small, slightly perigynous, in axillary, rarely terminal, spherical clusters; bracts scarious, usually silvery and often very conspicuous and concealing the flowers. Calyx very deeply 5-lobed, the lobes often cucullate and awned, and often with membranous margins; petals minute or absent; stamens 5; styles 2, or sometimes fused but bifid near the apex or with a bifid stigma (rarely conical and subentire). Fruit an achene with membranous pericarp.

Literature: M. N. Chaudhri, *Meded. Bot. Mus. Herb. Rijksuniv. Utrecht* **285**: 65–297 (1968).

1 Calyx-lobes with membranous margins, cucullate, awned on the back near the apex
2 Calyx-lobes very unequal, 3 outer enclosing 2 smaller, narrower, inner ones; leaves less than 1 mm wide **1. cymosa**
2 Calyx-lobes equal or subequal; leaves more than 1 mm wide
3 Bracts conspicuous, exceeding calyx; plant perennial
4 Flower-clusters usually more than 8 mm in diameter, well-defined; bracts 4–6 mm, ovate, concealing flowers **5. argentea**
4 Flower-clusters usually less than 6 mm in diameter, indefinite; bracts 2–4 mm, lanceolate, not concealing flowers **6. polygonifolia**
3 Bracts inconspicuous, shorter than calyx; plant perennial or annual
5 Perennial; calyx without hooked hairs at base, and with awn less than 0·5 mm **4. suffruticosa**
5 Annual; calyx with hooked hairs at base, and with awn *c.* 1 mm or more
6 Leaves ovate to oblong-ovate, reddish; awn on calyx-lobes usually straight **2. echinulata**
6 Leaves linear-oblong, green; awn on calyx-lobes curled or hooked at apex **3. rouyana**
1 Calyx-lobes entirely herbaceous, not cucullate or awned
7 Calyx-lobes very unequal in length
8 Bracts scarcely concealing flowers **17. macrosepala**
8 Bracts greatly exceeding and concealing flowers
9 Bracts 5–6 mm; calyx 1·6–2 mm
10 Stems ascending; leaves often recurved, ±obtuse **13. macedonica**
10 Stems procumbent; leaves not recurved, ±acute **14. albanica**
9 Bracts 6–10 mm; calyx 2·5–3·5 mm **16. capitata**
7 Calyx-lobes equal or subequal in length
11 Leaves *c.* 1·5 mm, imbricate, ±completely concealed by scarious stipules (E. & S. Spain) **15. aretioides**
11 Leaves usually more than 2 mm, not completely concealed by stipules
12 Calyx 2·5–4 mm, about twice as long as ripe fruit; lobes linear-lanceolate, acute, not incurved at apex in fruit **10. cephalotes**
12 Calyx 1·5–3 mm, not more than 1½ times as long as ripe fruit; lobes oblong to linear-oblong or linear-ellipsoidal, subacute, ±incurved at apex in fruit
13 Flower-clusters 17–21 mm in diameter, lax; bracts ovate-lanceolate, acute **9. taurica**
13 Flower-clusters 7–15 mm in diameter, dense; bracts orbicular to ovate, obtuse or subacute
14 Leaves linear to linear-oblong **12. pontica**
14 Leaves elliptical to ovate or suborbicular
15 Bracts 6–9 mm; leaves 4–8 mm **8. chionaea**
15 Bracts not more than 6·5 mm; leaves not more than 6 mm
16 Mat-forming; bracts up to 6·5 mm **11. rechingeri**
16 Usually not mat-forming; bracts not more than 5 mm **7. kapela**

1. P. cymosa (L.) DC. in Lam., *Encycl. Méth. Bot.* **5**: 26 (1804) (*Chaetonychia cymosa* (L.) Sweet). Annual; stems 3–7(–15) cm, usually erect. Leaves linear, usually in apparent whorls of 4. Flower-clusters 3–10 mm in diameter, globose, elongating up to 17 mm in fruit to form scorpioid cymes. Calyx 2–2·5 mm, exceeding the bracts; lobes unequal, the outer 3 broadly spathulate with alveolate hood, enclosing the 2 smaller inner lobes; awn stout, often hooked. *Dry, open habitats. S.W. Europe.* Co Ga Hs Lu Sa.

2. P. echinulata Chater, *Feddes Repert.* **69**: 52 (1964) (*P. echinata* auct., non Lam.). Annual; stems 2–20 cm, spreading or erect, usually much-branched. Leaves ovate to oblong-lanceolate, reddish, with pale, membranous margin, shortly aristate. Flower-clusters 3–8 mm in diameter, not elongating in fruit. Calyx 2–2·5 mm, equalling or exceeding the bracts; lobes equal or subequal, spathulate, with crispate or hooked hairs at the base; base and mid-vein reddish; margin and hood slightly alveolate; awn 1 mm or more, stout, patent, usually straight. *Dry, sandy or stony places; calcifuge. Mediterranean region, Portugal.* Bl Co Cr Ga Gr Hs It Lu Sa Si.

3. P. rouyana Coincy in Morot, *Jour. Bot. (Paris)* **8**: 65 (1894). Like **2** but plant light green, reddish only below; leaves linear-oblong, more strongly aristate; calyx-lobes pale greenish, sometimes reddish towards apex, with apex of awn strongly curved or hooked. ● *C. Spain.* Hs.

4. P. suffruticosa (L.) Lam., *Encycl. Méth. Bot.* **5**: 25 (1804) (*Herniaria polygonoides* Cav., *H. suffruticosa* (L.) Desf.). Perennial; stems 7–30 cm, erect, branched, subglabrous or scabrid with short, stiff, deflexed hairs. Leaves 5–10 mm, elliptical to ovate-elliptical, acute, aristate, coriaceous, reddish, with a thickened, pale margin. Flowers subsessile in small, dense clusters in a branched inflorescence; bracts ⅓–½ as long as the calyx, inconspicuous, ciliate, white. Calyx 1·2–1·6 mm; lobes equal, glabrous, reddish, with scarious margins and hood; awn *c.* 0·5 mm, patent, straight. ● *S.E. Spain.* Hs.

5. P. argentea Lam., *Fl. Fr.* **3**: 230 (1778). Perennial; stems 5–30 cm, usually procumbent, much-branched, mat-forming; internodes usually equalling or longer than the leaves. Leaves 4–8(–20) mm, ovate to lanceolate. Flower-clusters usually more than 8 mm in diameter, well-defined; bracts 4–6 mm, ovate, acute, silvery, concealing the flowers. Calyx 1·5–2·5 mm; lobes equal, oblong, cucullate; awn smooth or scarcely spinulose; membranous margins equalling or wider than the brownish, usually smooth mid-vein. *Mediterranean region, Portugal and N.W. Spain.* Bl Co Cr Ga Gr Hs It Lu Sa Si.

6. P. polygonifolia (Vill.) DC. in Lam. & DC., *Fl. Fr.* ed. 3, **3**: 403 (1805). Like **5** but leaves 2–3(–10) mm, often subspathulate, crowded, usually longer than the internodes; flower-clusters usually less than 6 mm in diameter, indefinite and inconspicuous; bracts (1·5–)2–4 mm, lanceolate, not concealing the flowers; calyx 1–1·75 mm; lobes with spinulose awn; membranous margins narrower than the spinulose mid-vein. *Calcifuge. Mountains of S. Europe.* Co Ga Gr Hs It Lu Sa Si.

P. arabica (L.) DC. in Lam., *Encycl. Méth. Bot.* **5**: 24 (1804) subsp. **cossoniana** (Gay) Batt., *Bull. Soc. Bot. Fr.* **46**: 267 (1900), from N. Africa, like **6** but leaves oblanceolate to linear-oblong and calyx-lobes with membranous margins somewhat wider than the mid-vein, has been recorded from S. Spain (Málaga) but perhaps only as a casual.

7. P. kapela (Hacq.) A. Kerner, *Österr. Bot. Zeitschr.* **19**: 367 (1869) (*P. capitata* auct., non (L.) Lam.). Perennial; stems 5–15 cm, much-branched. Leaves crowded, ciliate; stipules linear-lanceolate, usually equalling or shorter than the leaves. Flower-clusters 7–15 mm in diameter, very conspicuous; bracts 3–5 mm, elliptical to suborbicular, silvery. Calyx less than $1\frac{1}{2}$ times as long as the ripe fruit; lobes entirely herbaceous, equal or subequal, ovate to linear-oblong, usually obtuse, incurved at the apex in fruit. Fruit *c.* 1·5 mm, ovoid-subglobose. *S. Europe.* Al Bu Ga Hs It Ju Rm.

1 Leaves flattened in 1 plane — (b) subsp. **serpyllifolia**
1 Leaves in 4 ranks
 2 Leaves glabrous on both surfaces — (a) subsp. **kapela**
 2 Leaves pubescent on both surfaces
  3 Sepals 1·6–2·3 mm — (c) subsp. **galloprovincialis**
  3 Sepals 1·4–1·6 mm — (d) subsp. **baetica**

(a) Subsp. **kapela**: Stems procumbent to ascending, not mat-forming. Leaves 3–5 mm, in 4 ranks, narrowly elliptical to ovate-lanceolate, crowded, glabrous on both surfaces, ciliate. Calyx 1·5–2·5 mm; lobes ovate to linear-oblong. $2n = 18$. *Throughout the range of the species except C. & S. Spain.*

(b) Subsp. **serpyllifolia** (Chaix) Graebner in Ascherson & Graebner, *Syn. Mitteleur. Fl.* **5(1)**: 892 (1919) (*P. serpyllifolia* (Chaix) DC.): Stems procumbent, mat-forming. Leaves 1·5–3·5 mm, flattened in 1 plane, elliptical to ovate or suborbicular, densely crowded, pubescent, strongly ciliate. Calyx *c.* 1·5 mm; lobes ovate. $2n = 18, 36$. *Spain and Pyrenees; S.W. Alps.*

(c) Subsp. **galloprovincialis** Küpfer, *Boissiera* **23**: 148 (1974): Like subsp. (a) but leaves 2·5–3·5 mm, densely crowded, narrowly obovate, pubescent; stipules $1–1\frac{1}{2}$ times as long as the leaves; calyx 1·6–2·3 mm. $2n = 54$. ● *S.E. France (Mt. Ventoux).*

(d) Subsp. **baetica** Küpfer, *op. cit.* 150 (1974): Like subsp. (a) but leaves pubescent; stipules equalling to somewhat longer than the leaves; calyx 1·4–1·6 mm. $2n = 18$. ● *S.E. Spain.*

8. P. chionaea Boiss., *Diagn. Pl. Or. Nov.* **1(3)**: 9 (1843) (*P. kapela* subsp. *chionaea* (Boiss.) Borhidi). Like **7(a)** but stems 5–10(–15) cm, mat-forming; leaves 4–8 mm, densely pubescent; stipules lanceolate; bracts 6–9 mm, oblong to ovate or suborbicular; calyx 2·5–3 mm; fruit *c.* 2 mm, oblong-ellipsoidal. *W. part of Balkan peninsula.* Al ?Gr Ju.

P. sintenisii Chaudhri, *Acta Bot. Neerl.* **15**: 196 (1966), from N.W. Turkey, like **8** but with shorter stems, smaller leaves and bracts, and flower-clusters 7–9 mm, has been reported from N. Greece, apparently in error.

9. P. taurica Borhidi & Sikura, *Acta Bot. Acad. Sci. Hung.* **7**: 3 (1961). Like **7(a)** but stems ascending; flowering stems much longer than vegetative stems; flower-clusters 17–21 mm in diameter, laxer; bracts 7–11 mm, ovate-lanceolate, acute; calyx 2·5–3·5 mm. *Calcareous rocks. Mountains of Krym.* Rs(K).

10. P. cephalotes (Bieb.) Besser, *Enum. Horto Cremen.* 4 (1830). Like **7(a)** but leaves oblong to linear-lanceolate; bracts 5–7 mm; calyx 2·5–4 mm, twice as long as the ripe fruit; lobes linear-lanceolate, acute, not incurved at the apex in fruit. $2n = 36$. *E.C. & S.E. Europe.* Bu Gr Hu Ju Rm Rs(W, K) Tu.

P. bornmuelleri Chaudhri, *Meded. Bot. Mus. Herb. Rijksuniv. Utrecht* **285**: 259 (1968), from the N. Aegean region (Thasos), like **10** but with a more lax habit, leaves obovate, and sepal-lobes unequal, is known only from the type collection.

11. P. rechingeri Chaudhri, *op. cit.* 225 (1968). Perennial; stems 10–25 cm, slightly woody below, mat-forming. Leaves 3–6 mm, crowded, elliptical to obovate, subobtuse, glabrous on both surfaces, ciliate. Stipules lanceolate, up to as long as the leaves. Flower-clusters 10–14 mm in diameter, conspicuous; bracts up to 6·5 mm, silvery. Calyx 1·6–2 mm; lobes more or less equal, oblong, obtuse or subobtuse. Fruit ovoid, included within the sepals. *Limestone rocks.* ● *Mountains of N. & C. Greece and S.W. Bulgaria.* Bu Gr.

12. P. pontica (Borhidi) Chaudhri, *op. cit.* 241 (1968) (*P. cephalotes* subsp. *pontica* Borhidi). Perennial; stems 5–15 cm, ascending, much-branched; flowering stems much longer than vegetative. Leaves crowded, linear to linear-oblong, glabrous to pubescent, ciliate; stipules linear-lanceolate. Flower-clusters 10–15(–18) mm in diameter; bracts 6–8 mm, oblong to ovate or suborbicular. Calyx 2·5–3·5 mm; lobes equal, obtuse. *Black Sea region.* Bu Rs(W, K).

13. P. macedonica Chaudhri, *op. cit.* 248 (1968). Densely caespitose perennial; stems 5–12(–18) mm, ascending, woody at the base. Leaves 2–5 mm, imbricate, often recurved, oblong to oblanceolate, subobtuse, pubescent to subglabrous, ciliate. Stipules oblong to ovate, equalling or slightly shorter than the leaves. Flower-clusters 9–13 mm in diameter; bracts 5–6 mm, broadly ovate, obtuse, concealing the flowers. Calyx 1·6–2 mm; lobes very unequal, oblanceolate. ● *Greece, S. Jugoslavia.* Gr Ju.

14. P. albanica Chaudhri, *op. cit.* 250 (1968). Caespitose perennial; stems 3–10 cm, procumbent. Leaves 2·5–6 mm, not recurved, obovate-elliptical, subacute, subglabrous, ciliate. Stipules ovate-lanceolate, more or less equalling the leaves. Flower-clusters 10–18 mm in diameter; bracts 5–6 mm, broadly ovate, subacute, concealing the flowers. Sepals 1·6–2 mm; lobes somewhat unequal, oblanceolate. $2n = 36$. ● *S. & C. Greece, Albania, S. Jugoslavia.* Al Gr Ju.

The plants from C. & S. Greece are more densely caespitose and have smaller flowers, and have been distinguished as subsp. **graeca** Chaudhri, *op. cit.* 252 (1968).

15. P. aretioides DC., *Prodr.* **3**: 371 (1828). Caespitose perennial, woody at the base; stems up to 8 cm, much-branched, covered by densely imbricate leaves which are themselves more or less completely covered by the cucullate, scarious stipules. Leaves *c.* 1·5 mm, oblong or linear-oblong, more or less obtuse. Flower-clusters very conspicuous; bracts 2·5–3 mm, orbicular or often wider than long, very obtuse. Calyx 1–1·75 mm. *Calcicole.* ● *Mountains of E. & S. Spain.* Hs.

16. P. capitata (L.) Lam., *Fl. Fr.* **3**: 229 (1778) (*P. nivea* DC.). Laxly caespitose perennial; stems up to 15 cm, much-branched. Leaves 3–6 mm, oblong- to linear-lanceolate, acute, pubescent, ciliate, greyish-green. Flower-clusters *c.* 10 mm in diameter, conspicuous; bracts 6–10 mm, greatly exceeding and concealing the flowers. Calyx 2·5–3·5 mm; lobes very unequal, the shortest $\frac{2}{3}$–$\frac{3}{4}$ as long as the longest. *Mediterranean region.* Bl Co Ga Gr Hs It Si.

17. P. macrosepala Boiss., *Diagn. Pl. Or. Nov.* **1(3)**: 11 (1843) (*P. euboea* Beauverd & Top.). Like **16** but stems 2–10 cm; leaves elliptical to obovate-oblanceolate, subacute to obtuse; bracts 4·5–6 mm, scarcely concealing the flowers. $2n = 18$. *Aegean region.* Cr Gr Tu.

16. Herniaria L.

R. K. BRUMMITT AND V. H. HEYWOOD (EDITION 1) REVISED BY J. R. AKEROYD (EDITION 2)

Like *Paronychia*, but flowers sometimes 4-merous; bracts inconspicuous; stigma notched or bifid, subsessile or rarely on a distinct style.

Literature: F. N. Williams, *Bull. Herb. Boiss.* ser. 2, **4**: 556–570 (1896). F. Hermann, *Feddes Repert.* **42**: 203–224 (1937). M. N. Chaudhri, *Meded. Bot. Mus. Herb. Rijksuniv. Utrecht* **285**: 297–398 (1968).

1 Flowers 4-merous, with the 2 outer sepals ±enclosing 2 smaller inner; upper stipules and bracts with conspicuous purplish-black coloration
2 Leaves ±flat, elliptical, not in dense clusters; flower-clusters distributed along most upper branches **16. fontanesii**
2 Leaves ovoid to subglobose, in dense clusters; flower-clusters terminal or subterminal **17. fruticosa**
1 Flowers 4- or 5-merous, with sepals ±equal; stipules and bracts usually without purplish-black coloration
3 Annual, without a woody stock
4 Flowers 4-merous
5 Perigynous zone with patent, hooked hairs **14. polygama**
5 Perigynous zone without patent, hooked hairs **15. nigrimontium**
4 Flowers 5-merous
6 Leaves and sepals glabrous or ciliate **5. glabra**
6 Leaves and sepals conspicuously hairy
7 Leaves suborbicular or obovate-spathulate, with curved or curled hairs **13. algarvica**
7 Leaves elliptical or oblong to oblanceolate, with straight hairs
8 Leaves and sepals with ±stiff hairs; flowers 1–1·5 mm **11. hirsuta**
8 Leaves and sepals with soft, fine hairs; flowers usually more than 1·5 mm **12. lusitanica**
3 Perennial, with a ±woody stock
9 Sepals and leaves glabrous or ciliate
10 Sepals *c.* 0·5 mm; leaves usually entirely glabrous; flower-clusters usually contiguous on short lateral branches **5. glabra**
10 Sepals 0·7–1 mm; at least the younger leaves ciliate; flower-clusters usually not contiguous on short lateral branches
11 Plant usually compact; hairs on leaf-margins strongly curved forwards, appressed to leaf-margin **4. parnassica**
11 Plant not compact; hairs on leaf-margins not appressed **6. ciliolata**
9 Sepals densely hairy; leaves densely hairy or conspicuously ciliate
12 Plant compact, the internodes seldom more than 5 mm; flower-clusters terminal or subterminal
13 Leaves more than 2·5 × 1·2 mm, at least the older leaves glabrous on the surface **1. alpina**
13 Leaves up to 2·5 × 1·2 mm, densely covered with appressed hairs **2. boissieri**
12 Plant not compact, with most internodes more than 5 mm, or if not, then flower-clusters distributed along most of the younger branches
14 Leaves strongly ciliate, the surface ±glabrous except for a line of hairs on the prominent midrib **3. latifolia**
14 Leaves with hairs evenly distributed over the surface
15 Mature leaves mostly more than 6 mm, at least 3 times as long as wide; sepals with hairs about equalling the width of the sepals **10. incana**
15 Leaves rarely more than 6 mm, less than 3 times as long as wide; sepals with hairs much shorter than the width of the sepals
16 Flower-clusters lax, up to 6-flowered, mostly terminal or subterminal on the shorter branches; pubescence greyish-white **9. baetica**
16 Flower-clusters dense, more than 6-flowered, leaf-opposed or distributed along the shorter branches; pubescence usually yellowish
17 Older stems woody, the younger hairy on 1 side only **7. maritima**
17 Stems usually not woody, the younger hairy all round **8. scabrida**

Subgen. **Herniaria**. Annuals or perennials. Upper stipules and bracts usually without purplish-black coloration. Flowers 5-merous, or 4-merous in some annual species. Sepals more or less equal.

1. H. alpina Chaix in Vill., *Hist. Pl. Dauph.* **1**: 379 (1786). Caespitose perennial; younger stems densely leafy, with internodes not more than 5 mm. Leaves not more than 4·5 × 2 mm, elliptic-obovate, strongly ciliate, the surface glabrous, or rarely with few hairs. Flowers up to 2·5 mm, in few-flowered, lax, mostly terminal clusters, or solitary. Sepals with patent hairs. $2n = 18$. ● *Alps; E. Pyrenees.* Au Ga He ?Hs It.

H. olympica Gay in Duchartre, *Rev. Bot.* **2**: 370 (1847), from Anatolia, recorded from S.W. Bulgaria (Ali Botuš), is like **1** but has shorter, usually curved hairs on the sepals and a short, broad, bifid style.

2. H. boissieri Gay in Duchartre, *Rev. Bot.* **2**: 370 (1847) (*H. frigida* G. Kunze, nom. nud.). Like **1** but smaller in all parts; stem, leaves and flowers densely covered with minute, closely appressed, forwardly-directed, silky, white hairs; leaves up to 2·5 × 1·2 mm; flowers up to 1 mm, in fairly dense terminal or subterminal clusters of up to 12. *Mountains of S. Spain (Sierra Nevada).* Hs. (*N. Africa.*)

3. H. latifolia Lapeyr., *Hist. Abr. Pyr.* 127 (1813) (*H. pyrenaica* Gay). Perennial; internodes up to 15 mm, hairy on 1 side only. Leaves up to 9 × 4 mm, broadly elliptical, darkish green, conspicuously ciliate, the surface glabrous or with hairs mainly on the prominent midrib. Flowers 1·5–2 mm, in clusters which are leaf-opposed or contiguous on leafless lateral branches. Sepals with conspicuous patent hairs. ● *Pyrenees and mountains of N. & C. Spain; Corse, Sardegna.* Co Ga Hs Sa.

(a) Subsp. **latifolia**: Leaves subtruncate, with a few hairs on the surface. *Spain and Pyrenees.*

(b) Subsp. **litardieri** Gamisans, *Candollea* **36**: 6 (1981) (*H. litardieri* (Gamisans) Greuter & Burdet): Leaves cuneate, ciliate only; hairs more robust than in subsp. (a), and flowers smaller. *Corse, Sardegna.*

4. H. parnassica Heldr. & Sart. ex Boiss., *Diagn. Pl. Or. Nov.* **3(1)**: 95 (1853). Perennial, usually rather compact; stems 3–10(–15) cm. Leaves obovate to suborbicular, with minute, strongly curved hairs appressed to the margin, the surfaces glabrous. Flowers 1·5 mm, the clusters more or less distinct, not contiguous. Sepals 0·7–1 mm, ciliate, rarely with few hairs on the dorsal surface. Capsule slightly exceeding the sepals. ● *Mountains of S. Albania, Greece and Kriti.* Al Cr Gr.

(a) Subsp. **parnassica**: Adventitious roots conspicuous. Stems puberulent. Leaves densely ciliate. Sepals usually ciliate. Capsule 1–1·25 mm, narrowly ellipsoidal. *S. Albania; Greece.*

(b) Subsp. **cretica** Chaudhri, *Meded. Bot. Mus. Herb. Rijksuniv. Utrecht* **285**: 33 (1968). Adventitious roots absent or inconspicuous. Stems glabrous to slightly puberulent. Leaves smaller, sparsely ciliate to glabrous. Sepals usually glabrous. Capsule *c.* 0·75 mm, ovoid-globose. *Kriti.*

H. bornmuelleri Chaudhri, *op. cit.* 325 (1968), described from C. Italy (Abruzzi), is like **4(a)** but pulvinate, stems becoming glabrous and without adventitious roots, leaves obscurely ciliate and sepals glabrous. It may represent a third subspecies of **4**.

5. H. glabra L., *Sp. Pl.* 218 (1753) (*H. ceretana* Sennen, *H. ceretanica* (Sennen) Sennen, *H. corrigioloides* Lojac., *H. kotovii* Klokov, *H. rotundifolia* Vis., *H. suavis* Klokov, *H. vulgaris* Hill). Annual to perennial, sometimes woody at the base; stems subglabrous or with short hairs all round. Leaves up to 7 × 3·5 mm, elliptic-obovate, glabrous, or rarely more or less ciliate; upper stipules and bracts very rarely with purplish-black coloration. Flowers usually *c.* 1 mm, the clusters usually contiguous on short, leafless branches. Sepals *c.* 0·5 mm. Capsule usually distinctly exceeding the sepals. $2n = 18, 36, 72$. *Dry, sandy places; calcifuge. Most of Europe except the extreme north.* Al Au Be Br Bu Co Cz Da Ge Gr He Ho Hs Hu It Ju Lu Po Rm Rs(B, C, W, K, E) ?Sa Si Su Tu [Fe No Rs(N)].

Variable in duration, habit, presence of hairs on leaf-margins and sepals, size of flowers and length of fruit (*vide* H. W. Pugsley, *Jour. Bot.* (*London*) **68**: 214–218 (1930) and M. N. Chaudhri, *op. cit.* 315–320 (1968)). Perennial, suffruticose plants with ciliate leaves and larger flowers have been distinguished as subsp. **nebrodensis** Jan ex Nyman, *Consp.* **3**, in obs. post pag. 677 (1881) (*H. microcarpa* C. Presl). They occur on mountains of Sicilia, C. & S. Italy and perhaps elsewhere.

6. H. ciliolata Melderis, *Watsonia* **4**: 42 (1957) (*H. ciliata* Bab., non Clairv.). Often woody at the base; stems hairy on 1 side only. Leaves up to 7 × 4 mm, suborbicular to elliptical, at least the younger ones ciliate with patent hairs, the surfaces glabrous. Flowers *c.* 1·5 mm, the clusters mostly not contiguous. Sepals 0·7–1 mm, sometimes ciliate, usually with a single, minute, apical hair. Capsule equalling to slightly exceeding the sepals. $2n = 72$, 108, 126. ● *W. coast of Europe, from C. Portugal to S.W. England.* Br Ga Hs Lu.

7. H. maritima Link in Schrader, *Jour. für die Bot.* **1799 (1)**: 57 (1799). Stems up to 30 cm; older stems woody at the base, with prominent nodes; younger stems hairy on 1 side only. Leaves 2–4(–6) mm, broadly elliptical to rhombic, yellowish-green, with dense, appressed hairs on both surfaces, or the older sometimes only ciliate; upper stipules and bracts sometimes with purplish-black coloration. Flowers up to 2 mm; clusters not contiguous. Sepals covered with more or less appressed hairs. Capsule about equalling the sepals. ● *S.W. Portugal.* Lu.

H. regnieri Br.-Bl. & Maire, *Bull. Soc. Hist. Nat. Afr. Nord.* **16**: 28 (1925), like **7** but more compact with smaller, fleshy leaves and sepals with loosely spreading hairs, has been reported once from S. Spain (Málága prov.).

8. H. scabrida Boiss., *Elenchus* 42 (1838). Compact or elongate perennial; stems up to 45 cm. Leaves up to 6 × 3 mm, elliptical, yellowish-green, densely covered with appressed hairs; upper stipules and bracts often with some purplish-black coloration. Flowers *c.* 1·5 mm; clusters often contiguous on short, leafless branches. Sepals densely covered with usually very short hairs. ● *S.W. Europe.* Ga Hs Lu.

Very variable in habit, hairiness, and development of dark coloration on the stipules and bracts. More elongate and less hairy plants, intermediate between this species and **5**, have been referred to var. *glabrescens* Boiss.

9. H. baetica Boiss. & Reuter in Boiss., *Diagn. Pl. Or. Nov.* **3(1)**: 95 (1853). Woody at the base. Leaves broadly obovate to suborbicular, densely covered with whitish, appressed hairs. Flowers 2–2·5 mm (including short pedicel); clusters few-flowered, mostly more or less contiguous towards the ends of the branches. Sepals densely covered with white, patent hairs. ● *Mountains of S. Spain.* Hs.

10. H. incana Lam., *Encycl. Méth. Bot.* **3**: 124 (1789) (*H. besseri* (Fischer) DC., *H. macrocarpa* Sibth. & Sm.). Stock stout and woody. Leaves up to 12 × 3 mm, oblanceolate, at least 3 times as long as wide, densely covered with stiff, white hairs; stipules usually hairy on their outer surface. Flowers *c.* 2 mm (including short pedicel); clusters mostly leaf-opposed. Sepals 5, covered with white, patent hairs. $2n = 18$. *S. Europe, extending northwards to Czechoslovakia and to 52° N. in E. Russia.* Al †Au ?Bl Bu Cz Ga †Ge Gr Hs Hu It Ju ?Po Rm Rs(C, W, K, E) Tu.

11. H. hirsuta L., *Sp. Pl.* 218 (1753) (*H. diandra* Bunge; ?incl. *H. permixta* Guss.) Annual; stems slender, usually with regularly alternating branches, with patent hairs. Leaves elliptical to oblanceolate, hispid, or the older sometimes only ciliate; stipules ciliate. Flowers 1–1·5 mm, sessile, in dense clusters which are leaf-opposed or contiguous on short branches. Sepals 5, densely hairy. $2n = 18$, 36. *C. & S. Europe extending to N.W. France and C. Ukraine.* Al Au Be Bl Bu Co Cr Cz Ga Ge Gr He Hs Hu It Ju Lu Rm Rs(W) Sa Si Tu [Po].

(a) Subsp. **hirsuta**: Sepals with short, straight hairs; perigynous zone subglabrous. Stamens usually 3–5. Style 0·1–0·2 mm. *Throughout the range of the species except Portugal and most of Spain.*

(b) Subsp. **cinerea** (DC.) Coutinho, *Fl. Port.* 202 (1913) (*H. cinerea* DC.): Sepals with long, stout hairs; perigynous zone with hooked hairs. Stamens 2. Style 0·2–0·5 mm. *Mediterranean region; S. Portugal.*

H. micrantha A. K. Jackson & Turrill, *Kew Bull.* **1939**: 478 (1939), from Anatolia and Cyprus, like subsp. (**a**) but the leaves glabrous or ciliate and with a more compact habit, has been reported in error from Turkey-in-Europe.

12. H. lusitanica Chaudhri, *Meded. Bot. Mus. Herb. Rijksuniv. Utrecht* **285**: 341 (1968) (incl. *H. berlingiana* (Chaudhri) Franco). Annual to perennial; stems prostrate, branched, with patent or deflexed hairs, or subglabrous. Leaves up to 9 × 3 mm, oblong, oblanceolate or obovate, finely pubescent. Flowers (1–)1·75–2 mm, in often dense, leaf-opposed clusters. Sepals with fine, patent hairs. Fruit ovoid-globose. *Dry, often ruderal habitats.* ● *Portugal; W., C. & S. Spain.* Hs Lu.

13. H. algarvica Chaudhri, *op. cit.* 346 (1968). Annual; stems up to 10 cm, prostrate, much-branched, forming a dense mat. Leaves and flowers densely covered with curved or curled hairs. Leaves up to 4 × 2·5 mm, suborbicular or obovate-spathulate, obtuse. Stipules ciliate. Flowers 1·5–1·75 mm, in dense, leaf-opposed clusters, subsessile; perigynous zone usually subglabrous. *Sandy places.* ● *S.W. Portugal.* Lu.

14. H. polygama Gay in Duchartre, *Rev. Bot.* **2**: 371 (1847) (*H. euxina* Klokov, *H. odorata* Andrz.). Perennial; stems slender, with regularly alternating branches; hairs strongly recurved. Leaves glabrous or with few, minute, marginal hairs. Sepals 4, glabrous towards the apex but with patent, hooked hairs at the base and on the perigynous zone. Style slender, exceeding the sepals. *White Russia, Ukraine, and Russia northwards to* 57° *N.* ?Po Rs(C, W, E).

15. H. nigrimontium F. Hermann, *Feddes Repert.* **42**: 223 (1937). Like **14** but annual or biennial; leaves shortly ciliate; sepals with hairs over the outer surface and no patent, hooked hairs at the base. ● *C. part of Balkan peninsula.* Bu Gr Ju.

H. degenii (F. Hermann) Chaudhri, *Meded. Bot. Mus. Herb. Rijksuniv. Utrecht* **285**: 382 (1968), from the N. Aegean region (Samothraki), is like **15** but with young leaves strigose, stipules densely ciliate and flowers covered with long, stout, bristly hairs. It is probably only subspecifically distinct from **15**.

Subgen. **Heterochiton** (Graebner & Mattf.) F. Hermann. Perennials, woody at the base. Upper stipules and bracts with purplish-black coloration over most of the surface. Flowers 4-merous; 2 opposite sepals more or less enclosing the other 2.

16. H. fontanesii Gay in Duchartre, *Rev. Bot.* **2**: 371 (1847). Leaves mostly 2–4 mm, more or less flat, elliptical, mostly opposite and not clustered. Flowers 2–2·5 mm, in clusters distributed along all the younger branches. Outer sepals divergent above. *S.E. Spain; S. Sicilia.* Hs Si. (*N. Africa.*)

(a) Subsp. **fontanesii** (*H. empedocleana* Lojac.): Stems with strongly deflexed hairs. Leaves and sepals glabrous or rarely with short, crisped hairs. Outer sepals without purplish-black patches. *Maritime hills and sand-dunes. Sicilia (near Porto Empedocle).*

(b) Subsp. **almeriana** Brummitt & Heywood, *Feddes Repert.* **69**: 31 (1964): Stems with more or less patent hairs. Leaves and sepals with more or less silky hairs. Outer sepals with purplish-black patches on the lateral margins near the base. *Hills of S.E. Spain.*

17. H. fruticosa L., *Cent. Pl.* **1**: 8 (1755). Leaves up to 2 mm, ovoid to subglobose, in dense clusters at each node, glabrous or minutely puberulent. Flowers *c.* 1·5 mm, in terminal or subterminal clusters. Outer sepals more or less erect, not divergent above, glabrous or subglabrous. ● *C. & E. Spain.* Hs.

(a) Subsp. **fruticosa**: Procumbent. Flowers in dense, terminal clusters; purplish-black coloration, if present, restricted to small patches at the base of the sepals. *Probably throughout the range of the species.*

(b) Subsp. **erecta** (Willk.) Batt. in Batt. & Trabut, *Fl. Algér. (Dicot.)* 168 (1888) excl. descr.: Suberect. Flowers in lax, terminal and subterminal clusters. Outer sepals with purplish-black patches on margin near the base; inner sepals purplish-black on most of the margins. *S.E. Spain.*

17. Illecebrum L.

A. O. CHATER

Like *Paronychia*, but annuals; sepals white, spongy, persistent round the capsule; capsule 1-seeded, dehiscing with 5(–10) valves which remain more or less joined above.

1. I. verticillatum L., *Sp. Pl.* 206 (1753). Glabrous; stems usually 5–30 cm (sometimes 1 cm in dry, or up to 70 cm in flooded habitats), branched or simple, decumbent, slender, quadrangular, often rooting near the base. Leaves 2–5 mm, opposite, linear-spathulate to obovate, obtuse; stipules small. Flower-clusters 2 at each node, 4- to 6-flowered, forming small, very white, usually crowded pseudo-whorls; bracteoles *c.* 1 mm, silvery. Sepals 1·5–2·5 mm. $2n = 10$. *Seasonally damp, sandy or gravelly places. W. & N.C. Europe; C. & E. Mediterranean region, very local.* Au Az Be Br Co Cz Da Ga Ge Gr †He Ho Hs It Lu Po Sa [Su].

18. Pteranthus Forskål

P. D. SELL AND S. M. WALTERS

Small annuals. Flowers sessile, in groups of 3. 'Fruit' complex, 1-seeded, described below.

1. P. dichotomus Forskål, *Fl. Aegypt.* lxii & 36 (1775). Fleshy annual herb; stems 10–20 cm, procumbent to ascending, articulate at the nodes. Leaves linear, subverticillate; stipules minute, lanceolate. Flowers sessile, in groups of 3, situated at the apex of an oblanceolate, swollen, hollow, compressed peduncle, the 2 lateral flowers sterile; each flower situated between 2 minutely glandular bracts with spirally arranged curved spines. Petals absent; stamens 4; styles 2. Capsule indehiscent, 1-seeded, enclosed in the spiny bracts and borne on the persistent, compressed, inflated, wing-like peduncle, the whole 'fruit' 12–20 mm. *S.E. Spain; Malta.* Hs *Si. (*N. Africa, S.W. Asia.*)

19. Polycarpon Loefl. ex L.

A. O. CHATER AND J. R. AKEROYD

Small annuals to perennials; stems ascending or erect, usually branched, usually rough at the angles. Leaves more or less petiolate, opposite, often apparently verticillate. Flowers small, in cymose clusters with scarious bracts. Sepals 5; petals 5, hyaline, shorter than the sepals; stamens 1–5, the filaments more or less united at the base; style short, 3-lobed. Capsule dehiscing with 3 valves almost to the base; valves twisting spirally into tubes. Seeds several.

Annual to perennial, without a woody stock; stipules and bracts ±conspicuous and silvery; petals usually emarginate **1. tetraphyllum**

Perennial, with a woody stock; stipules and bracts inconspicuous, greyish; petals entire **2. polycarpoides**

1. P. tetraphyllum (L.) L., *Syst. Nat.* ed. 10, **2**: 881 (1759). Annual, biennial or perennial, without a woody stock; stems simple or much-branched. Leaves obovate. Inflorescence lax, spreading, much-branched. Sepals *c.* 2 mm; petals usually emarginate. *S. & W. Europe.* Al Az Bl Br Bu Co Cr Ga Ge Gr He Hs It Ju Lu Sa Si Tu [Be Cz].

1 Leaves not purplish-tinged; inflorescence lax, spreading, with many conspicuous branches (**a**) subsp. **tetraphyllum**
1 Leaves often purplish-tinged; inflorescence ±condensed, few-flowered, usually without conspicuous branches
 2 Stems usually simple; leaves usually paired, always purplish-tinged at least near base of plant; stamens 1–3 (**b**) subsp. **diphyllum**
 2 Stems often branched; usually at least some leaves near middle of stem in whorls of 4; leaves not purplish-tinged; stamens 5 (**c**) subsp. **alsinifolium**

(**a**) Subsp. **tetraphyllum**: Stems much-branched. Leaves green, mostly in whorls of 4. Inflorescence lax, much-branched. Stamens (1–)3–5. Seeds usually less than 0·5 mm, punctulate. *Sandy or rocky places, usually inland. Throughout the range of the species.*

(**b**) Subsp. **diphyllum** (Cav.) O. Bolòs & Font Quer, *Collect. Bot.* (*Barcelona*) **6**: 356 (1962) (*P. diphyllum* Cav., *P. alsinifolium* auct., non (Biv.) DC.): Annual; stems usually simple. Leaves purplish-tinged, at least near the base of the stems, usually paired. Inflorescence few-flowered, condensed. Stamens 1–3. Seeds like those of subsp. (**a**). *Sandy places, usually coastal. S.W. Europe; Aegean region.*

(**c**) Subsp. **alsinifolium** (Biv.) Ball, *Jour. Linn. Soc. London* (*Bot.*) **16**: 370 (1877) (*P. alsinifolium* (Biv.) DC.; incl. *P. rotundifolium* Rouy): Stems often branched. Leaves green or purplish-tinged, at least some near the middle of the stem in whorls of 4. Inflorescence few-flowered, condensed. Stamens 5. Seeds at least 0·5 mm, smooth. *Sandy places, usually coastal. W. & S. Europe, northwards to N.W. France.*

2. P. polycarpoides (Biv.) Zodda, *Nuovo Gior. Bot. Ital.* nov. ser., **15**: 347 (1908) (*P. alsinifolium* auct., non (Biv.) DC., *P. peploides* DC.; incl. *P. colomense* Porta). Perennial, with woody stock; stems branched. Leaves obovate to suborbicular, fleshy, usually paired, rarely verticillate; stipules small, greyish. Flowers few, in a lax inflorescence. Sepals 1·5–2 mm, obtuse or acute; petals entire, about as long as the sepals; stamens 5. Seeds 0·5 mm or more, tuberculate. *Rocks, usually coastal. W. Mediterranean region.* Bl Ga Hs It Si.

Subsp. **herniarioides** (Ball) Maire & Weiller in Maire, *Fl. Afr. Nord* **9**: 73 (1963), from N.W. Africa, has been reported from Sierra de Gádor, west of Almería. It differs from subsp. **polycarpoides** by the more compact habit with short, little-branched stems, petals ½ as long as the sepals and seeds 0·4–0·5 mm with smaller tubercles.

20. Ortegia L.

A. O. CHATER

Small plant with opposite leaves. Flowers small, shortly pedicellate in opposite, cymose clusters which are arranged in a panicle. Sepals 5, keeled, the margins scarious; petals absent; stamens 3; style 1; stigma 3-toothed or -lobed. Capsule dehiscing with 3 valves. Seeds numerous, minute, ovoid, acute.

1. O. hispanica L., *Sp. Pl.* 560 (1753) (*O. dichotoma* L.). Annual to perennial; stems up to 40 cm, with many strict branches, angular, often very rough above. Leaves 5–18 mm, opposite, linear or linear-lanceolate, obtuse or acute, much shorter than the internodes, erect; stipules setiform, caducous, with blackish glands. Sepals 2–3 mm, ovate-lanceolate, acute, keeled, green, with scarious margins. *Spain and Portugal.* Hs †It Lu.

21. Loeflingia L.

V. H. HEYWOOD

Annuals. Flowers small, sessile, in compound dichasial or monochasial cymes, often spicate. Sepals free, unequal, the outer 3 longer and wider than the inner 2, all or only the outer with a setiform appendage on each side; petals 3 or 5, very small; stamens 3 or 5; style 1; stigmas 3, or 1, more or less 3-lobed. Fruit a capsule dehiscing with 3 valves. Seeds numerous.

1 Inflorescence-branches narrow, often secund; style deeply 3-partite **3. tavaresiana**
1 Inflorescence-branches ±condensed, not secund; style usually almost undivided or with 3 short lobes
 2 Sepals (2–)2·5–3 mm, all with setiform appendages; style shorter than capsule **1. hispanica**
 2 Sepals 1·5–2 mm, only 3 with setiform appendages; style as long as capsule **2. baetica**

1. L. hispanica L., *Sp. Pl.* 35 (1753) (*L. hispanica* var. *pentandra* sensu Coutinho; incl. *L. pentandra* Cav.). Much-branched, glandular-pubescent annual 2–10 cm. Leaves linear, acute, connate at the base with the subulate stipules. Inflorescence-branches more or less condensed, not secund. Sepals (2–)2·5–3 mm, all with setiform appendages; stamens usually 3; style about ½ as long as the capsule; stigma usually only shallowly lobed. *Sandy soils. W. Mediterranean region.* Bl Ga Hs Lu Si.

2. L. baetica Lag., *Period. Soc. Méd. Cádiz* **4(1)**: 5 (1824) (*L. hispanica* var. *micrantha* (Boiss. & Reuter) Maire; incl. *L. gaditana* Boiss. & Reuter, *L. micrantha* Boiss. & Reuter). Like **1** but sepals 1·5–2 mm, only 3 with setiform appendages; stamens 5; style as long as the capsule. *Maritime sands. S. Spain; C. & S. Portugal.* Hs Lu.

3. L. tavaresiana Samp. in Nobre, *Anais Ci. Nat.* (*Porto*) **10**: 25 (1906). Like **1** but much-branched from the base; inflorescence-branches narrow, often secund and sometimes obviously recurved; stamens 5; style deeply 3-partite. *Maritime sands.* ● *S. Portugal.* Lu.

22. Spergula L.

J. A. RATTER AND J. R. AKEROYD

Annuals (rarely perennials); stems ascending, often decumbent and much-branched at the base. Leaves linear, obtuse, decussate; stipules scarious, not united to surround the node; leaf-fascicles (short, leafy, lateral branches) borne on both sides at each node. Perianth 5-merous; sepals free, green, with scarious margins; petals white, entire; stamens 5–10; styles 3 or 5. Capsule ovoid to subglobose, dehiscing with 3 or 5 valves. Seeds often winged.

1 Root thick and woody; plant densely glandular-hairy (N.W. Spain) **5. viscosa**
1 Root ±slender, not woody; plant glandular-hairy or not
2 Styles and capsule-valves 3; seeds compressed, with a broad, scarious wing **4. fallax**
2 Styles and capsule-valves 5; seeds subglobose or compressed, winged or not
3 Seeds subglobose or globose, keeled or with a narrow wing; leaves usually channelled beneath **1. arvensis**
3 Seeds compressed, broadly winged; leaves not channelled beneath
4 Seed wider than its brownish wing; petals ovate, obtuse, contiguous **2. morisonii**
4 Seed as wide as its transparent, shining wing; petals lanceolate, acute, not contiguous **3. pentandra**

1. S. arvensis L., *Sp. Pl.* 440 (1753) (*S. vulgaris* Boenn., *S. sativa* Boenn., *S. maxima* Weihe, *S. linicola* Boreau). Annual 5–70 cm, with ascending stems more or less branched at the base, moderately to densely glandular-hairy above. Leaves 1–3(–8) cm, linear, fleshy, channelled beneath. Sepals 3–5 mm, ovate; petals white, obovate, obtuse; stamens 5–10. Capsule 5 mm. Seeds 1–2 mm, subglobose, keeled or with a very narrow wing, grey-black, papillose or not. *Cultivated ground, especially on sandy soils; calcifuge. Throughout Europe.* All except Bl Rs(K) Sb.

Very variable in habit, pubescence, size of seed and presence of seed-papillae. Two subspecies can be recognized.

(a) Subsp. **arvensis**: Terminal inflorescence pedunculate. Petals more or less truncate at the base, about as long as or slightly longer than the sepals. Capsule distinctly exceeding the sepals. $2n = 18$. *Almost throughout the range of the species.*

(b) Subsp. **chieussiana** (Pomel) Briq., *Prodr. Fl. Corse* **1**: 494 (1910) (*S. chieussiana* Pomel): Terminal inflorescence sessile. Sepals and petals narrower; petals rounded at the base, usually longer than the sepals. Capsule about as long as or longer than the sepals. *W. Mediterranean region, C. & S. Portugal.*

Plants from maritime sands on Corse, that are like subsp. (**b**) but prostrate to weakly ascending, with dark, glandular hairs, have been called subsp. **gracilis** (E. Petit) Briq., *op. cit.* 495 (1910). They resemble coastal ecotypic variants of **1** from other parts of Europe.

2. S. morisonii Boreau in Duchartre, *Rev. Bot.* **2**: 424 (1847) (*S. vernalis* auct.). Annual with ascending stems 5–30 cm, glabrous or sparsely pubescent. Leaves 1–2 cm, linear, not channelled beneath; leaf-fascicles conspicuously shorter than the internodes. Sepals 4 mm, ovate, acuminate; petals about equalling the sepals, ovate, obtuse, overlapping at the margins; stamens 5–10. Capsule 5 mm, slightly exceeding the sepals. Seeds 1–1·5 mm (including wing), laterally compressed, brown, with a brownish, striate wing narrower than the seed. $2n = 18$. *W. & C. Europe, extending northwards to 64° N. in Sweden.* Be Cz Da Fe Ga Ge Ho Hs Lu No Po Rm Rs(B, C) Su.

3. S. pentandra L., *Sp. Pl.* 440 (1753). Like **2** but petals lanceolate, acute, not contiguous, somewhat exceeding the sepals; stamens usually 5; seeds with a white, shining wing about as wide as the seed. *W., S. & C. Europe.* Au †Be Bu Co Cz Ga Ge Gr Hs Hu It Ju Lu Po Rm Sa Si Tu [Rs(B)].

4. S. fallax (Lowe) E. H. L. Krause in Sturm, *Deutschl. Fl.* ed. 2, **5**: 19 (1901). Glabrous annual, with ascending stems 5–30 cm. Leaves 1–2·5 cm, linear, not channelled beneath. Sepals 3–4 mm, oblong; petals whitish, shorter than or about equalling the sepals; styles 3. Capsule 4–6 mm, distinctly exceeding the sepals, dehiscing by 3 valves. Seeds 1–1·5 mm (including wing), compressed, black or blackish-grey, glossy, with a broad, scarious wing. *Dry hills and fields. Sicilia (Linosa).* Si. (*N. Africa & S.W. Asia.*)

5. S. viscosa Lag., *Gen. Sp. Nov.* 15 (1816). Apparently perennial with woody taproot; stems many, ascending, densely glandular-hairy and viscid. Leaves densely fasciculate, channelled beneath. Petals much shorter than the sepals. Seeds 2·5–3 mm (including wing), laterally compressed, dark brown, with a brownish wing narrower than the seed. *Sandy places on mountains.* ● *N.W. Spain.* Hs.

23. Spergularia (Pers.) J. & C. Presl

P. MONNIER AND J. A. RATTER

Annual to perennial herbs, sometimes woody at the base; stems erect, decumbent or procumbent, dilated at the nodes, somewhat flattened. Leaves linear, decussate, with pale, scarious stipules united to surround the node, forming more or less triangular structures on either side of the stem; leaf-fascicles (short, leafy, lateral branches) when present borne on only 1 side at each node. Perianth 5-merous; sepals free, green, with scarious margins; petals entire; stamens 1–10; styles 3. Capsule dehiscing with 3 valves. Seeds often winged.

1 Plant robust, with thick, woody stock; sepals usually at least 4 mm
2 Stipules 6–10 mm, with setaceous apex; leaves awned; wing of seed (if present) deeply laciniate-fimbriate **2. fimbriata**
2 Stipules 2–6 mm, not setaceous; leaves mucronate or obtuse; wing of seed (if present) entire or somewhat toothed
3 Stipules wider than long; leaves flattened, widening above (Açores) **1. azorica**
3 Stipules longer than wide; leaves semi-cylindrical, linear
4 Seeds unwinged; plant often densely glandular-hairy throughout
5 Petals much shorter than sepals, white or pinkish; capsule subglobose **4. macrorhiza**
5 Petals about as long as sepals, pink; capsule ovoid **3. rupicola**
4 Seeds with ±developed wing (exceptionally absent); plant glandular-hairy only in the inflorescence
6 Seeds black; leaves narrowed towards base; stipules silvery **5. australis**
6 Seeds brown; leaves not narrowed towards base; stipules not silvery
7 Stipules on young stems connate for more than $\frac{1}{2}$ their length; capsule more than 6 mm **6. media**
7 Stipules on young stems connate for considerably less than $\frac{1}{2}$ their length; capsule less than 6 mm **13. nicaeensis**
1 Plant slender, with ±slender taproot; sepals usually less than 4 mm
8 Stipules on young shoots connate for about $\frac{1}{2}$ their length (forming a sheath); seeds light brown, often winged
9 Inflorescence little-branched; seeds smooth or tuberculate; petals at least 3 mm **7. marina**
9 Inflorescence much-branched; seeds sparsely papillose; petals less than 3 mm **8. tangerina**

8 Stipules on young shoots connate for considerably less than ½ their length; seeds ±dark brown, unwinged (or with vestigial wing)
10 Sepals mucronate, scarious except for a narrow, green vein; seeds less than 0·4 mm **9. segetalis**
10 Sepals not or scarcely mucronate, green, with scarious margins; seeds more than 0·4 mm
11 Inflorescence ebracteate above; petals narrowly elliptical; capsule globose **10. diandra**
11 Inflorescence bracteate (upper bracts often very short); petals ovate; capsule ovoid to subglobose
12 Petals much longer than sepals **11. purpurea**
12 Petals about equalling or shorter than sepals
13 Seeds black, smooth or densely spinulose
14 Stipules ±cordate; sepals acute to mucronulate; seeds ±smooth **16. heldreichii**
14 Stipules triangular; sepals obtuse; seeds densely spinulose **17. echinosperma**
13 Seeds brown, tuberculate
15 Petals white, or pink above and white beneath; seeds pale greyish-brown **15. bocconii**
15 Petals uniformly pink; seeds dark brown
16 Upper bracts almost as long as leaves **12. rubra**
16 Upper bracts much shorter than leaves
17 Plant rather robust, not rooting at nodes; stipules broadly triangular **13. nicaeensis**
17 Plant slender, often rooting at nodes; stipules lanceolate-acuminate **14. capillacea**

1. S. azorica (Kindb.) Lebel, *Mém. Soc. Nat. Sci. Cherbourg* **14**: 47 (1868) (*S. macrorhiza* sensu Trelease, non (Req.) Heynh.). Perennial; stems 8–15 cm, woody below, densely glandular-hairy. Leaves flat, widening towards the apex, mucronate; stipules wider than long, dull. Sepals and petals 3–5 mm; stamens 10. Capsule subglobose, equalling or exceeding the sepals. Seeds brown, pyriform, unwinged or with vestigial wing, the hilum subapical. ● *Açores*. Az.

2. S. fimbriata Boiss., *Diagn. Pl. Or. Nov.* **3(1)**: 94 (1854). Perennial; stems 5–35 cm, robust, often woody below, glandular-hairy above. Leaves awned; stipules very long (6–10 mm), with setaceous apices, silvery. Petals 4–6 mm, lilac, equalling or longer than the sepals; stamens 10. Capsule 3·5–5·5 mm. Seeds black, some unwinged, others usually with laciniate-fimbriate wings. *Coasts of S.W. Spain and S. Portugal; Ibiza*. Bl Hs Lu.

3. S. rupicola Lebel ex Le Jolis, *Mém. Soc. Nat. Sci. Cherbourg* **7**: 274 (1860) (*S. lebeliana* Rouy, *S. rupestris* Lebel). Perennial; stems 5–35 cm, robust, somewhat woody below, often glandular-hairy throughout. Leaves narrowly linear, mucronate; stipules ovate-triangular, acuminate, somewhat silvery. Sepals 4–4·5 mm; petals uniformly pink, about equalling the sepals; stamens 10. Capsule 4·5–7 mm, ovoid, equalling or somewhat exceeding the sepals. Seeds 0·5–0·7 mm, dark brown to black, unwinged, regularly tuberculate. $2n = 36$. *Rocky places by the sea. W. Europe, from S. Portugal to N.W. Scotland*. Br Ga Hb Hs Lu.

4. S. macrorhiza (Req.) Heynh., *Nomencl. Bot.* **2**: 689 (1846). Like **3** but leaves broadly linear and very crowded; petals white or pinkish, shorter than the sepals; capsule subglobose, much shorter than the sepals. $2n = 36$. *Rocky places by the sea. Corse; Sardegna; S.E. Italy*. Co It Sa.

5. S. australis (Samp.) Ratter, *Bot. Jour. Linn. Soc.* **109**: 322 (1992) (S. rupicola var. *australis* Samp.). Like **3** but leaves somewhat narrowed towards base; stipules very silvery; capsule more globose; seeds with vestigial or well-developed wings, smooth or slightly tuberculate, black. $2n = 40$. *Rocky places by the sea.* ● *Portugal, Spain*. Hs Lu.

6. S. media (L.) C. Presl, *Fl. Sic.* 161 (1826) (*S. maritima* (All.) Chiov., *S. marginata* Kittel, *S. marina* sensu Halácsy, non (L.) Griseb., *Spergula marginata* (Kittel) Murb.). Perennial; stems 5–40 cm, glabrous throughout or glandular-hairy in the inflorescence. Leaves mucronate; stipules broadly triangular, not acuminate. Sepals 4–6 mm; petals white or pink, equalling or somewhat exceeding the sepals; stamens (7–)10. Capsule 7–9 mm, usually much exceeding the sepals. Seeds 0·7–1 mm (excluding wing), dark brown, smooth or tuberculate, usually winged; margin of wing entire or only slightly denticulate. $2n = 18, 36$. *Seashores, salt-marshes, and saline habitats inland. Coasts of Europe (except the north-east) and inland saline areas.* Au Be Bl Br Bu Co Cz Da Ga Ge Gr Hb Ho Hs Hu It Ju Lu No ?Po Rm Rs(B, W, K, E) Sa Si Su Tu.

Populations of **6** with unwinged seeds occur in W. Europe and the Baltic region. They have been called subsp. **angustata** (Clavaud) Greuter & Burdet, *Willdenowia* **12**: 190 (1982) (*S. marginata* var. *angustata* Clavaud).

Subsp. **tunetana** (Maire) Greuter & Burdet, *op. cit.* 191 (1982) (*S. tunetana* (Maire) Jalas), from N. Africa, with winged and unwinged echinulate seeds, the wing entire or subentire, has been reported from W. Sicilia.

7. S. marina (L.) Griseb., *Spicil. Fl. Rumel.* **1**: 213 (1843) (*S. salina* J. & C. Presl, *S. dillenii* Lebel, *Spergula salina* (J. & C. Presl) D. Dietr.). Annual, biennial or rarely perennial, with slender or slightly fleshy stock. Leaves fleshy, mucronate; stipules short, obtuse, forming a sheath. Sepals 2·5–4 mm; petals pink above and white near the base, rarely entirely white, not exceeding the sepals; stamens 1–5(–8). Capsule (3–)4–6 mm, usually exceeding the sepals. Seeds 0·6–0·7(–0·8) mm, light brown, smooth or densely tuberculate, unwinged or mixed winged and unwinged; wing of seed when present erose to laciniate. $2n = 36$. *Coasts of Europe and inland saline areas.* All except Fa He Sb.

8. S. tangerina P. Monnier, *Feddes Repert.* **69**: 50 (1964). Like **7** but stems 5–10 cm, very slender; inflorescence very condensed; petals and sepals 1·5–3 mm; capsule 2–3 mm; seeds 0·5–0·65 mm, sparsely covered with long papillae. $2n = 18$. *Saline soils. W. Mediterranean region.* Co Ga Hs Lu.

9. S. segetalis (L.) G. Don fil., *Gen. Syst.* **1**: 425 (1831) (*Alsine segetalis* L., *Delia segetalis* (L.) Dumort.). Glabrous annual with slender taproot; stems 3–15 cm, ascending. Leaves awned; stipules long, silvery, much-divided at the apex. Inflorescence very slender, without upper bracts. Sepals 1·5–2 mm, acute, mucronate, with wide, scarious margin and narrow, green vein; petals white, shorter than the sepals. Capsule 1·5–3 mm, ovoid. Seeds 0·4 mm, black, ovoid, unwinged, tuberculate. $2n = 18$. *Cornfields. W. & W.C. Europe, extending to C. Italy.* †Be Ga Ge He †Ho Hs It Lu Po.

10. S. diandra (Guss.) Boiss., *Fl. Or.* **1**: 733 (1867) (*S. salsuginea* Fenzl). Annual or biennial with slender taproot; stems 3–30 cm, slender, ascending. Stipules short, triangular (rarely lanceolate). Inflorescence much-branched, very slender, without upper bracts. Sepals 2–3 mm; petals lilac (rarely white), narrowly elliptical, equalling the sepals; stamens 2 or 3. Capsule 1·5–3 mm, globose, about equalling the sepals; valves purple-black at maturity. Seeds 0·6–0·7 mm, unwinged, dark brown to black, rugulose or bristling with rigid papillae. 2*n* = 18. *Saline waste places. S. Europe; S.E. Russia.* Al *Az Bl Cr Ga Gr Hs It Sa Si Rs(E) Tu.

11. S. purpurea (Pers.) G. Don fil., *Gen. Syst.* **1**: 425 (1831) (*S. longipes* Rouy). Annual or biennial with slender taproot; stems 2–25 cm, slender. Stipules lanceolate, acuminate, silvery. Inflorescence often slender, bracteate. Sepals 2·5–4 mm; petals 3–4·5 mm, uniformly rose-purple, exceeding the sepals; stamens 10. Capsule 2·5–3·5 mm, shortly stipitate, equalling or somewhat exceeding the sepals. Seeds 0·45–0·6 mm, unwinged, dark brown or black, finely tuberculate. 2*n* = 18, 36. *Sandy waste places. Spain and Portugal.* Hs Lu.

12. S. rubra (L.) J. & C. Presl, *Fl. Čechica* 94 (1819) (*S. campestris* (L.) Ascherson, *Spergula rubra* (L.) D. Dietr.). Annual to perennial with slender to somewhat woody taproot; stems 5–25 cm, diffuse, decumbent or procumbent. Leaves very fasciculate; stipules lanceolate, acuminate, silvery. Bracts of the inflorescence almost as large as the leaves. Sepals and petals 3–4 mm; petals uniformly pink; stamens (5–)10. Capsule 4–5 mm, about equalling the sepals. Seeds 0·45–0·55 mm, unwinged, dark brown, subtrigonous. 2*n* = 36, 54. *Sandy soils. Throughout Europe.* All except Az ?Cr Fa Is Rs(K) Sb.

Plants from the mountains of W. Kriti (Lefka Ori) have been referred to the Anatolian species **S. lycia** P. Monnier & Quézel, *Candollea* **25**: 359 (1970), but the status of plants of **12** from high altitudes in the Mediterranean region, often treated as var. *alpina* (Boiss.) Willk. & Lange, is in need of clarification.

13. S. nicaeensis Sarato ex Burnat, *Fl. Alp. Marit.* **1**: 269 (1892). Like **12** but biennial to perennial with stout, woody stock; stems 10–35 cm, robust; stipules short, triangular; upper bracts very short; capsule slightly exceeding the sepals; seeds 0·5–0·65 mm, dark brown, ovoid, often with a vestigial wing near the micropyle. 2*n* = 36. *Saline soils and waste places; nitrophilous. W. & C. Mediterranean region.* Co Ga Hs It Sa ?Si.

14. S. capillacea (Kindb. & Lange) Willk. in Willk. & Lange, *Prodr. Fl. Hisp.* **3**: 163 (1874). Like **12** but stems 8–35 cm, procumbent, rooting at the nodes; inflorescence few-flowered, with very small bracts; seeds semicircular in outline. 2*n* = 18. *Granitic sands. N. & C. Spain; N. & C. Portugal.* Hs Lu.

15. S. bocconii (Scheele) Ascherson & Graebner, *Syn. Mitteleur. Fl.* **5(1)**: 849 (1919) (*S. atheniensis* (Heldr. & Sart.) Ascherson & Schweinf., *S. campestris* sensu Willk. & Lange, non (L.) Ascherson). Annual or biennial with slender taproot; stems 5–25 cm, slender, densely glandular-hairy in the inflorescence. Leaves not fasciculate; stipules triangular, not acuminate. Upper bracts much reduced. Sepals 2–3·5 mm; petals pink, with white base or entirely white, somewhat shorter than or equalling the sepals; stamens (0–)2–5(–8). Capsule 2–3·5 mm, equalling or shorter than the sepals. Seeds 0·35–0·45 mm, unwinged, light grey-brown, finely tuberculate. 2*n* = 36. *Ruderal. S. & W. Europe, mainly near the coast.* Az Bl Br Co Cr Ga Gr Hs It Ju Lu Sa Si Tu.

16. S. heldreichii Fouc. ex E. Simon secundus & P. Monnier, *Bull. Soc. Bot. Fr.* **105**: 263 (1958). Annual with slender taproot; stems 8–35 cm, glandular-hairy only in the inflorescence. Stipules more or less cordate, with short, often bifid, point. Sepals 2·5–3·5 mm, acute with nearly mucronate apex; petals pink to lilac, somewhat paler at the base, shorter than or equalling the sepals; stamens (2–)6–8(–10). Capsule 2·5–3·5 mm, about equalling the sepals. Seeds 0·5–0·6 mm, unwinged, metallic black, smooth or slightly rugulose. 2*n* = 36. *Maritime sands. S.W. Europe.* Bl Co Ga Hs It Lu.

17. S. echinosperma (Čelak.) Ascherson & Graebner, *Ber. Deutsch. Bot. Ges.* **11**: 517 (1893) (*S. rubra* subsp. *echinosperma* Čelak.). Annual (rarely perennial) with slender taproot; stems 4–10 cm, slender. Leaves slightly fasciculate; stipules triangular, often wider than long. Inflorescence few-flowered, with well-developed bracts. Sepals 2·5–3·5 mm, curved in fruit; petals pale pink, equalling or somewhat shorter than the sepals; stamens (0–)2–5. Capsule subglobose, exceeding the sepals. Seeds 0·45–0·55 mm, black, unwinged, densely spinulose. *Margins of lakes and streams. W.C. Europe and France; one station in S.W. Spain.* Au Cz Ga Ge Hs.

24. Telephium L.

A. O. CHATER

Procumbent perennials, woody at the base. Flowers in terminal cymes, 5-merous. Styles 3(4), free to the base; ovary septate below, with many ovules. Capsule usually trigonous, dehiscing with 3(4) valves.

1. T. imperati L., *Sp. Pl.* 271 (1753). Glabrous dwarf shrub; rhizome very stout, woody; stems 15–40 cm, rarely branched, procumbent to ascending, often rooting near the base, terete. Leaves up to 15 mm, rather fleshy and glaucous, alternate, usually secund. Flowers shortly pedicellate, in 5- to 20(–50)-flowered, terminal, capitate cymes; bracts small. Sepals oblong-lanceolate, obtuse, green with white, scarious margins; petals white, slightly exceeding the sepals. Capsule 15- to 20-seeded, exceeding the sepals. *S. Europe, extending northwards to the French Jura.* Cr Ga Gr He Hs It.

(**a**) Subsp. **imperati**: Leaves obovate to obovate-subspathulate, obtuse. Sepals 4–7 mm. Capsule rather abruptly contracted into a beak. 2*n* = 18. *Spain and S. France; very locally in N. Italy and S.W. Switzerland.*

(**b**) Subsp. **orientale** (Boiss.) Nyman, *Consp.* 254 (1879) (*T. orientale* Boiss.): Like subsp. (**a**) but smaller and more slender in all its parts; leaves narrower, the upper linear-lanceolate and rather acute; cymes few-flowered, denser; sepals 3–4·5 mm; capsule more gradually narrowed into a beak, and more exserted. 2*n* = 18. *Greece, Kriti.*

The plants from Kriti have been separated from subsp. (**b**) as subsp. **pauciflorum** (Greuter) Greuter & Burdet, *Willdenowia* **12**: 191 (1982) (*T. imperati* subsp. *orientale* var. *pauciflorum* Greuter), on the basis of their dwarf habit and fewer seeds, but are best included in subsp. (**b**).

Subfam. Silenoideae

Leaves opposite; stipules absent. Flowers 5-merous. Epicalyx-scales (bracteoles) sometimes present. Sepals joined in a tubular or campanulate calyx, sometimes with scarious commissures. Petals usually well-developed; stamens, petals and ovary often situated on a more or less elongated column which forms the carpophore in fruit.

25. Lychnis L.

A. O. CHATER

Erect perennials. Petals contorted in bud. Epicalyx absent. Calyx 10-veined, with 5 short teeth. Limb of petal more or less distinct from claw; coronal scales present. Stamens 10, rarely up to 20; styles usually 5, opposite calyx-teeth. Fruit a capsule, dehiscing with (usually) 5 teeth. Carpophore short or long. Seeds without a tuft of hairs at hilum. (Incl. *Coronaria* Schaeffer, *Polyschemone* Schott, *Viscaria* Bernh.)

1 Leaves and stems densely villous with white hairs
2 Inflorescence ±capitate; petals 2-lobed **3. flos-jovis**
2 Flowers long-stalked in a lax inflorescence; petals entire or emarginate **2. coronaria**
1 Stems and leaves not villous
3 Petals deeply 4-lobed with linear, patent lobes; rarely broadly 2-lobed with toothed lobes **4. flos-cuculi**
3 Petals entire, emarginate or 2-lobed with entire lobes
4 Leaves and lower part of stems sparsely hispid or pubescent
5 Calyx more than 10 mm; leaves sparsely hispid **1. chalcedonica**
5 Calyx less than 10 mm; leaves pubescent with crispate hairs **5. sibirica**
4 Leaves glabrous or ciliate; lower part of stems glabrous
6 Calyx less than 6 mm; inflorescence ±capitate **7. alpina**
6 Calyx more than 6 mm; inflorescence not capitate
7 Inflorescence many-flowered **6. viscaria**
7 Inflorescence 1- to 3-flowered **8. nivalis**

1. L. chalcedonica L., *Sp. Pl.* 436 (1753). Stems 30–45(–60) cm, simple, stout, hispid. Leaves sparsely hispid, ovate, acute; the cauline cordate-amplexicaul. Inflorescence 10- to 50-flowered, capitate. Calyx 14–17 mm. Petal-limb bright scarlet, bifid for ⅓ of its length. Carpophore 4–6 mm. *Woods and thickets. C. & E. Russia from* c. 49° *to* 56° *N.* Rs(C, E).

Cultivated widely for ornament.

2. L. coronaria (L.) Desr. in Lam., *Encycl. Méth. Bot.* **3**: 643 (1792) (*Coronaria coriacea* Schischkin ex Gorschk.). Whole plant densely villous with white hairs; stems 30–100 cm, often branched. Leaves ovate to ovate-lanceolate, acute. Flowers long-pedicellate in a few-flowered inflorescence. Calyx 15–20 mm. Petal-limb purplish, rarely very pale or white, entire or emarginate. Carpophore *c.* 2 mm. $2n = 24$. *Dry banks and scrub. E.C. & S.E. Europe, Italy; doubtfully native elsewhere in S. Europe; cultivated widely for ornament, and locally naturalized.* Al ?Au Bu Cz *Ga Gr *He *Hs Hu It Ju *Lu Rm Rs(W, K) Tu.

3. L. flos-jovis (L.) Desr. in Lam., *Encycl. Méth. Bot.* **3**: 644 (1792). Less densely villous than **2**; stems 20–90 cm, usually erect and little-branched. Basal leaves lanceolate-spathulate, acute; cauline lanceolate. Inflorescence 4- to 10-flowered, more or less capitate. Calyx 11–13 mm. Petal-limb purplish or scarlet, rarely white, bifid with wide, often lobed, lobes. Capsule *c.* 14 mm; carpophore *c.* 2 mm. *Dry, sunny slopes,* 1000–2000 *m.* ● *Alps, eastwards to* c. 11°15′ *E.; one station in C. Italy.* ?Au Ge He It [Cz].

4. L. flos-cuculi L., *Sp. Pl.* 436 (1753) (*L. cyrilli* K. Richter, *Agrostemma flos-cuculi* (L.) G. Don fil., *Coronaria flos-cuculi* (L.) A. Braun). Stems 20–90 cm, often branched, sparsely scabrid-puberulent. Basal leaves oblong-spathulate, petiolate, slightly scabrid, often ciliate at base; cauline linear-lanceolate, connate at base. Flowers in long-stalked dichasia. Calyx 6–10 mm. Capsule 6–10 mm; carpophore very short or absent. *Damp places. Most of Europe, but rare in parts of the south.* All except Az Bl Cr Rs(K) Sb ?Si.

(a) Subsp. **flos-cuculi**: Petal-limb deeply 4-lobed with linear, acute, patent lobes, usually purplish-pink. $2n = 24$. *Throughout the range of the species.*

(b) Subsp. **subintegra** Hayek, *Österr. Bot. Zeitschr.* **70**: 14 (1921): Petal-limb 2-lobed with oblong, obtuse lobes which are toothed near apex, usually white. ● *Balkan peninsula.*

5. L. sibirica L., *Sp. Pl.* 437 (1753). Whole plant pubescent with short, crispate hairs. Stems 8–30 cm, usually simple. Basal leaves crowded on woody stock, linear-lanceolate; cauline 3–8 pairs, remote. Inflorescence (1–)3- to 4(–8)-flowered, lax; pedicels mostly longer than calyx. Calyx 5–6(–8) mm. Petal-limb pale yellow or white, bifid to about half-way; coronal scales lanceolate, acute. Capsule 6–8 mm; carpophore very short. *Arctic and subarctic Russia.* Rs(N). (*N. Asia.*)

The European plant is subsp. **samojedorum** Sambuk, *Bull. Acad. Sci. URSS* ser. 7, **22**: 47 (1928) (*L. samojedorum* (Sambuk) Gorschk.). Subsp. **sibirica** occurs in Siberia and approaches close to the boundary of Europe north-east of Sverdlovsk.

6. L. viscaria L., *Sp. Pl.* 436 (1753) (*Viscaria viscosa* Ascherson, *V. vulgaris* Bernh.). Stems 15–90 cm, simple or branched above, glabrous or slightly hairy above, viscid below the upper nodes. Leaves mostly basal, linear- to ovate-lanceolate, petiolate, glabrous, usually ciliate at base. Flowers usually in 3- to 6-flowered, shortly stalked, opposite dichasia which are arranged in a lax, spike-like panicle. Calyx 6–15 mm. Petals usually dark- or pinkish-purple (rarely white), entire or emarginate. Capsule 7–10 mm. *Dry, sunny slopes; somewhat calcifuge. Much of Europe, except the south-west, but absent from the islands, except Britain.* Al Au Be Br Bu Cz Da Fe Ge Gr He Hu It Ju No Po Rm Rs(N, B, C, W, E) Su Tu [Ga].

Very variable, the centre of variation being in Makedonija. To the north and west of this region, the plants have a long carpophore, and in Greece only plants with a short carpophore occur. The shape of the calyx is more or less correlated with the length of the carpophore, but the other variable characters are not. Two subspecies can be recognized.

(a) Subsp. **viscaria**: Calyx clavate. Carpophore more than 3 mm, at least ½ as long as ripe capsule. $2n = 24$. *Throughout the range of the species, except the south and east parts of the Balkan peninsula.*

(b) Subsp. **atropurpurea** (Griseb.) Chater, *Feddes Repert.* **69**: 45 (1964) (*L. atropurpurea* (Griseb.) Nyman, *L. sartorii* (Boiss.) Hayek, *Viscaria atropurpurea* Griseb.; incl. *V. sartorii* Boiss.). Calyx tubular to campanulate, not clavate. Carpophore up to 2 mm, less than ⅓ as long as ripe capsule. $2n = 24$. ● *Balkan peninsula and Romania.*

7. L. alpina L., *Sp. Pl.* 436 (1753) (*Viscaria alpina* (L.) G. Don fil.). Caespitose, glabrous, with ciliate leaf-bases; stems 5–15 cm, simple, not viscid. Basal leaves crowded, linear-subspathulate to linear; cauline usually 2 or 3 pairs, often wider. Inflorescence (6–)10- to 20-flowered, more or less capitate, rarely elongate. Calyx (3–)4–5 mm. Petal-limb pink, rarely white, bifid to about half-way. Capsule *c.* 4 mm; carpophore 2 mm or less. Some flowers are large and hermaphrodite, some small and usually with abortive stamens. $2n = 24$. *Screes and stony grassland; somewhat calcifuge. N. Europe, southwards to* 55° *N.; mountains of S. & S.C. Europe, from W.C. Spain to W. Austria.* Au Br Fe Ga He Hs Is It No Rs(N, C) Su.

8. L. nivalis Kit. in Schultes, *Österreichs Fl.* ed. 2, **1**: 698 (1814). Caespitose, glabrous, with ciliate or serrulate leaves. Stems 5–20 cm, simple, not viscid. Basal leaves crowded, oblong-lanceolate to subspathulate; cauline 1 or 2 pairs, often linear. Flowers 1–3; pedicels equalling or shorter than calyx. Calyx 10–12 mm. Petal-limb purple or white, 2-lobed. Carpophore short. *Damp, stony places,* 1800–2300 *m.* ● *E. Carpathians (Munţii Rodnei).* Rm.

26. Agrostemma L.

A. O. CHATER

Like *Lychnis*, but annuals; calyx with 5 long, linear, foliaceous teeth; styles 5, alternating with calyx-teeth; carpophore absent.

Calyx-teeth longer than tube; petals shorter than calyx **1. githago**
Calyx-teeth equalling or shorter than tube; petals longer than calyx **2. gracile**

1. A. githago L., *Sp. Pl.* 435 (1753) (*Lychnis githago* (L.) Scop.; incl. *A. linicola* Terechov). Plant covered with long, appressed, greyish hairs; stems 30–100 cm. Leaves usually linear, slightly connate, acute, or the lower obtuse. Calyx 3–7 cm, the ovoid or oblong tube usually much shorter than the often caducous teeth. Petals shorter than calyx; limb not spotted, dull purple, rarely white. Seeds mostly 3 mm or more, with prominent, acute tubercles. $2n = 48$. *Probably native in Europe only in the E. Mediterranean region; introduced as a weed of cultivation almost throughout Europe, but much less abundant than formerly, and extinct in some parts.* All except Sb, but in Az Fa Is only as a rare casual.

Plants from flax-fields in Russia (var. *linicola* (Terechov) Hammer) have a narrower, less hairy calyx-tube and smaller and less tuberculate seeds.

2. A. gracile Boiss., *Diagn. Pl. Or. Nov.* **3(1)**: 80 (1854). Like **1** but smaller, more slender and less hairy; calyx-tube narrower, equalling or longer than teeth; petals longer than calyx; limb pale purple, with lines of black spots at base. *E.C. Greece (near Farsala).* Gr. (*Anatolia.*)

27. Petrocoptis A. Braun

†W. ROTHMALER (EDITION 1) REVISED BY S. M. WALTERS (EDITION 2)

Perennials, woody at the base, glabrous or nearly so, often glaucous and viscid above. Inflorescence cymose, 1- to 10-flowered. Epicalyx absent. Calyx conical-campanulate, 10-veined. Petals imbricate in bud, long-clawed; coronal scales present; stamens 10; styles 5 opposite the calyx-teeth; carpophore present. Seeds reniform, black, with a conspicuous strophiole consisting of a tuft of hairs at the hilum.

All species grow in crevices of calcareous rocks, mainly above 500 m, but occasionally in the lowlands.

Literature: W. Rothmaler, *Bot. Jahrb.* **72(1)**: 117–130 (1941). H. Merxmüller & J. Grau, *Collect. Bot.* (*Barcelona*), **7**: 787–797 (1968). P. Montserrat & J. Fernández Casas in S. Castroviejo et al. (eds.), *Flora Iberica* **2**: 304–312 (1990).

1 Seeds smooth and shining
2 Seeds *c.* 1·5 mm **3. hispanica**
2 Seeds *c.* 1 mm
3 Calyx 5–9 mm **1. pyrenaica**
3 Calyx 10–14 mm **2. grandiflora**
1 Seeds ± rugose, not or scarcely shining
4 Strophiole covering not more than ⅓ of seed; petals often white, slightly emarginate **4. crassifolia**
4 Strophiole covering more than ⅓ of seed; petals pink or red, entire **5. pardoi**

1. P. pyrenaica (J. P. Bergeret) A. Braun ex Walpers, *Repert. Bot. Syst.* **1**: 281 (1842) (*Lychnis pyrenaica* J. P. Bergeret). Plant more or less glaucous, with or without basal leaf rosettes. Stems up to 30 cm, more or less viscid above. Basal leaves up to 3 cm, ovate to ovate-lanceolate, petiolate, with more or less ciliate margins; cauline ovate to lanceolate, sessile. Bracts 1–4 mm, often greenish or purplish with scarious margins, sometimes entirely scarious. Calyx 5–9 mm. Petals 10–18 mm, usually emarginate, white, pink or purple. Seeds *c.* 1 mm, smooth and shining. ● *Pyrenees and mountains of N. Spain.* Ga Hs.

1 Plant without basal leaf-rosettes **(b)** subsp. **glaucifolia**
1 Plant with basal leaf-rosettes
2 Stems very viscid; bracts *c.* 1 mm **(c)** subsp. **viscosa**
2 Stems not very viscid; bracts up to 4 mm **(a)** subsp. **pyrenaica**

(a) Subsp. **pyrenaica**: Basal leaf-rosettes present; bracts 1–4 mm; calyx 5–8 mm. $2n = 24$. *W. & C. Pyrenees, extending westwards to* c. 2°30′ *W. in N. Spain.*

(b) Subsp. **glaucifolia** (Lag.) P. Monts. & Fernández Casas, *Anal. Jard. Bot. Madrid* **45**: 362 (1988) (*P. glaucifolia* (Lag.) Boiss., *P. lagascae* (Willk.) Willk., *P. wiedmannii* Merxm. & Grau, *Silene glaucifolia* Lag.): Basal leaf-rosettes absent; bracts 1–2 mm; calyx 7–9 mm. $2n = 24$. *Cordillera Cantabrica.*

(c) Subsp. **viscosa** (Rothm.) P. Monts. & Fernández Casas, *loc. cit.* (1988) (*P. viscosa* Rothm.): Basal leaf-rosettes present; leaves and stems very viscid; bracts *c.* 1 mm; calyx 7–9 mm. $2n = 24$. *N.W. Spain (near Ponferrada).*

2. P. grandiflora Rothm., *Cavanillesia* **7**: 111 (1935). Like **1(b)** but bracts *c.* 4 mm, green; calyx 10–14 mm, petals *c.* 2 cm, entire, deep purple. ● *N.W. Spain (Leon & Orense provs.).*

3. P. hispanica (Willk.) Pau, *Bol. Soc. Aragon. Ci. Nat.* **15**: 65 (1916) (*P. pyrenaica* var. *hispanica* Willk.). Plant with basal leaf-rosettes; stems up to 40(–60) cm. Basal leaves up to 8 cm, obovate, glaucous, thick in texture, petiolate, not ciliate; lower cauline elliptical to ovate; upper suborbicular. Bracts 1–2·5 mm, scarious with green midrib. Calyx 6–8(–10) mm. Petals emarginate, white or pale pink. Seeds *c.* 1·5 mm, rather shining. $2n = 24$. ● *W. & C. Pre-Pyrenees.* Hs.

P. pseudoviscosa Fernández Casas, *Cuad. Ci. Biol.* **2**: 44 (1973), described from the same region (gorge of R. Esera), is said to differ from **3** in the absence of leaf-rosettes and in the larger strophiole. It seems doubtfully distinct.

4. P. crassifolia Rouy, *Ill. Pl. Eur. Rar.* **4**: 26 (1895) (incl. *P. montserratii* Fernández Casas). Plant with basal leaf-rosettes; stems up to 40(–80) cm. Basal leaves ovate-spathulate, glaucous, thick in texture, petiolate, not ciliate; cauline ovate-lanceolate. Bracts 2–5 mm, acuminate, more or less scarious. Calyx 8–12 mm. Petals at least 2 cm, entire or slightly emarginate, white or pink. Seeds *c.* 2 mm, rugose, not shining, with small strophiole not covering ⅓ of seed. $2n = 24$. ● *C. Pyrenees.* Hs.

5. P. pardoi Pau, *Actas Soc. Esp. Hist. Nat.* **22**: 196 (1898) (incl.

P. guarensis Fernández Casas, *P. montsicciana* O. Bolòs & Rivas Martínez). Glaucous plant without basal leaf-rosettes; stems up to 40 cm, often branched. Leaves ovate to lanceolate, thick in texture. Bracts 2·5–5 mm, greenish or scarious. Calyx 9–13(–15) mm. Petals 15–22 mm, entire, pink or lilac. Seeds 1·2–1·7 mm, rugose, not or scarcely shining; strophiole covering at least $\frac{1}{3}$ of seed. $2n = 24$. • *Mountains of N.E. Spain.*

28. Silene L.

A. O. CHATER, S. M. WALTERS AND J. R. AKEROYD; SECT. *Otites*, F. WRIGLEY

Herbs or small shrubs. Epicalyx absent. Calyx with (5–)10–30(–60) veins and 5 short teeth. Limb of petal distinct from claw. Stamens 10; styles usually 3 (sometimes 4 on same plant), more rarely 5. Fruit with or without basal septa opening by 6 (more rarely 5, 8 or 10) teeth, twice the number of styles; carpophore present. Seeds glabrous. (Incl. *Eudianthe* (Reichenb.) Reichenb., *Heliosperma* (Reichenb.) Reichenb., *Melandrium* Röhling, *Otites* Adanson.)

A wide generic concept is adopted, following P. K. Chowdhuri (see below). Calyx-length normally changes little between flowering and fruiting, and provides a convenient measurement of flower-size. Measurements of the capsule and carpophore refer to the ripe fruiting stage. Seed-characters have been extensively used in the description of annual species, and ripe seed is necessary to confirm the identification in some groups. There is great variety in seed-form. Most seeds are reniform, and the two sides ('faces' in the descriptions) may be plane, concave, excavate (i.e. with an abrupt, relatively deep excavation), or concavo-convex (i.e. convex, but with a concavity near the hilum). The edge of the seed ('back' in the descriptions) may be plane, grooved, ridged, winged or variously sculpted.

The terminology of the inflorescence is difficult. In the perennial sections there are some paniculate inflorescences with more or less equal opposite branches bearing groups of 3-flowered dichasia, and it is possible to derive many of the other inflorescence types theoretically from this paniculate type. Chowdhuri's arrangement of the genus, largely adopted here, follows this logical pattern. Reduction of the typical panicle produces the narrow, spike-like inflorescence in which the opposite branches may be very short-stalked (the inflorescence is then pseudoverticillate) or may be reduced to single flowers.

Further reductions may suppress some or all of the branches, and this series of more or less irregular, alternately branched panicles finishes with those species with a solitary flower. In the annual species it is possible to describe most inflorescences as regular or irregular compound dichasia, simple or compound raceme-like monochasia, few-flowered, or solitary. Alar flowers are single in the apparent axil of a branch. As, however, environmental conditions affect the development of the inflorescence, no key which depends on inflorescence-form can work with more than moderate success, and small specimens, whether of annual or perennial species, may be wrongly identified. In the present state of knowledge, there appears to be no remedy to this.

Most annual species of *Silene* are weeds of ruderal or cultivated habitats and it is therefore difficult in many cases to decide on the limits of their range as natives. Most perennial species grow in open, stony or sandy ground; many are caespitose, mat-forming or, more rarely, pulvinate mountain-plants with relatively large flowers.

Literature: P. Rohrbach, *Monographie der Gattung Silene*. Leipzig, 1868. F. N. Williams, *Jour. Linn. Soc. London (Bot.)* **32**: 1–196 (1896). P. K. Chowdhuri, *Notes Roy. Bot. Gard. Edinb.* **22**: 221–278 (1957). V. Melzheimer, *Bot. Jahrb.* **98**: 1–92 (1977); **101**: 153–190 (1980). D. Jeanmonod, *Candollea* **39**: 195–259; 549–639 (1984). S. Talavera in S. Castroviejo et al. (eds.), *Flora Iberica* **2**: 313–406 (1990). F. Wrigley, *Ann. Bot. Fenn.* **23**: 69–81 (1986) (Sect. *Otites*).

1 Annual, without non-flowering shoots at time of flowering, and usually without a woody stock; inflorescence various, but never paniculate with regular and equal opposite branches
2 Styles 5, or flowers male only; capsule with 10 teeth or with 5 bifid teeth
3 Leaves ovate-lanceolate to obovate; petals usually white
4 Calyx (at least in female flowers) with 20 conspicuous veins; capsule usually with erect teeth **129. latifolia**
4 Calyx with 10 inconspicuous veins; capsule with recurved teeth **132. heuffelii**
3 Leaves linear-lanceolate; petals pink (Sect. *Eudianthe*)
5 Calyx less than 12 mm **161. laeta**
5 Calyx more than 12 mm **162. coeli-rosa**
2 Styles 3; capsule with 6 teeth
6 Calyx 15- to 60-veined
7 Calyx 15- to 20-veined
8 Calyx with 10 longer and 10 alternating shorter veins; petals white **71. csereii**
8 Calyx with 15–20 subequal veins; petals pink **190. ammophila**
7 Calyx 30- or 60-veined
9 Calyx 60-veined **191. macrodonta**
9 Calyx 30-veined
10 Capsule 12–18 mm; seeds more than 1 mm, dark brown **194. conoidea**
10 Capsule 7–12 mm; seeds 1 mm or less, pruinose
11 Calyx shortly pubescent **192. conica**
11 Calyx with short glandular hairs and ± dense, long eglandular hairs **193. lydia**
6 Calyx 10-veined
12 Seeds with back narrowly grooved between 2 undulate wings (Sect. *Dipterospermae*)
13 Carpophore more than 4 mm
14 Seeds 1–1·5 mm; calyx-teeth ovate, obtuse **186. colorata**
14 Seeds at least 2 mm; calyx-teeth lanceolate, acute **187. secundiflora**
13 Carpophore less than 4 mm
15 Carpophore 1–2 mm; petals often absent **188. apetala**
15 Carpophore 2·5–3·5 mm; petals always present **189. longicaulis**
12 Seeds without 2 undulate wings on back
16 Flowers solitary, terminal
17 Calyx not contracted at mouth
18 Seeds 0·6–0·8 mm **165. littorea**
18 Seeds 1–1·4 mm **178. sericea**
17 Calyx contracted at mouth (Iberian peninsula)
19 Calyx loose and inflated in fruit **164. psammitis**
19 Calyx not loose and inflated in fruit
20 Seeds ± smooth; carpophore (2·5–)4–9 mm **165. littorea**
20 Seeds tuberculate; carpophore 2–4 mm

21 Calyx not more than 15 mm; calyx-tube 2–3 times as long as teeth; petals white **168. germana**
21 Calyx usually more than 15 mm; calyx-tube 3–5 times as long as teeth; petals pale pink **169. almolae**
16 Flowers in monochasia or dichasia
22 Flowers in raceme-like, simple or branched monochasia, with or without alar flowers
23 Calyx not contracted at mouth
24 Seeds subglobose, umbilicate (Karpathos) **143. insularis**
24 Seeds reniform
25 Pedicels patent in fruit, slender, usually longer than calyx
26 Calyx 6–9 mm; seeds 0·3–0·5 mm **146. sedoides**
26 Calyx 10–25 mm; seeds at least 0·6 mm
27 Seeds 0·8–1·2 mm, deeply grooved and striate **166. oropediorum**
27 Seeds 0·6–1 mm, ±smooth
28 Calyx 10–19 mm **165. littorea**
28 Calyx (17–)19–25 mm **167. stockenii**
25 Pedicels erect in fruit, ±stout, usually shorter than calyx
29 Stems 1(–3)-flowered **178. sericea**
29 Stems more than 3-flowered
30 Seeds excavate on face; plant never villous
31 Carpophore less than 2 mm **179. nocturna**
31 Carpophore usually more than 2 mm
32 Calyx 10–14 mm; cauline leaves obtuse **176. obtusifolia**
32 Calyx 9–10·5 mm; cauline leaves acute **177. gaditana**
30 Seeds ±plane on face; plant usually villous, at least on buds and shoot-apices
33 Plant not villous (Karpathos) **175. discolor**
33 Plant villous, at least on buds and shoot-apices
34 Carpophore more than 5 mm; petal-limb more than 5 mm, reddish **173. scabriflora**
34 Carpophore less than 5 mm; petal-limb not more than 3 mm, white or pink **174. micropetala**
23 Calyx contracted at mouth
35 Fruiting pedicels patent or deflexed
36 Calyx not inflated **165. littorea**
36 Calyx inflated, widest at the middle
37 Cauline leaves ovate to ovate-lanceolate **163. pendula**
37 Cauline leaves linear-lanceolate to linear **164. psammitis**
35 Fruiting pedicels ±erect
38 Carpophore at least 5 mm; calyx with ascending, ±bulbous-based hairs on veins **160. gallinyi**
38 Carpophore not more than 5 mm; calyx without ascending, bulbous-based hairs
39 Seeds more than 1 mm; carpophore glabrous
40 Plant entirely glabrous **152. graeca**
40 Plant hairy, at least in part (Sect. *Dichotomae*)
41 Veins of calyx sparsely hispid **170. dichotoma**
41 Veins of calyx glabrous but serrate with ascending teeth **171. heldreichii**
39 Seeds less than 1 mm; carpophore pubescent (Sect. *Silene*)
42 Veins of calyx distinctly anastomosing above
43 Calyx usually less than 11 mm; calyx-tube about twice as long as the lanceolate, acute teeth; carpophore 2–4 mm **183. cerastoides**
43 Calyx at least 11 mm; calyx-tube about equalling the linear, acuminate teeth; carpophore 1–2 mm **184. tridentata**
42 Veins of calyx obscurely anastomosing or free
44 Calyx more than 13 mm **182. bellidifolia**
44 Calyx less than 13 mm
45 Inflorescence subcapitate **185. disticha**
45 Inflorescence ±elongate
46 Calyx 9–12 mm; carpophore 1·5–2 mm; seeds with ±flat face and narrow, acutely grooved back **180. ramosissima**
46 Calyx 7–10 mm; carpophore less than 1 mm; seeds with excavate face and wide, flat or concave back **181. gallica**
22 Flowers in regular or irregular dichasia, not in raceme-like monochasia
47 Calyx not contracted at mouth
48 Flowers usually 2–8, in sessile or shortly stalked, subcapitate dichasia arranged laterally on a simple or branched axis **172. nicaeensis**
48 Inflorescence more or less regularly dichasially branched
49 Calyx at least 11 mm
50 Carpophore not more than 4 mm
51 Capsule 10–11 mm; calyx-teeth acute **144. divaricata**
51 Capsule 7–8 mm; calyx-teeth obtuse **(140–142). diversifolia** group
50 Carpophore more than 4 mm
52 Calyx glabrous
53 Seeds with 1 or 2 rows of rounded tubercles on back; petals pink or whitish, without darker veins **134. portensis**
53 Seeds with 4 rows of acute tubercles on back; petals white, with purplish veins **135. echinosperma**
52 Calyx hairy
54 Carpophore 9–10 mm, pubescent **139. pseudatocion**
54 Carpophore 4·5–8 mm, glabrous
55 Carpophore 4·5–7 mm; inflorescence ±dense **138. fuscata**
55 Carpophore 7–8 mm; inflorescence ±lax **145. integripetala**
49 Calyx less than 11 mm
56 Carpophore more than 4 mm **149. laconica**
56 Carpophore less than 4 mm
57 Seeds with excavate face **(140–142). diversifolia** group
57 Seeds with plane or concave, but not excavate, face
58 Calyx glabrous
59 Petal-limb exserted; calyx 4–7·5 mm **136. pinetorum**

59 Petals absent, or with limb included; calyx 7–10 mm **137. inaperta**
58 Calyx hairy
60 Capsule exceeding calyx; carpophore *c.* 2 mm; seeds 0·5–0·75 mm **147. pentelica**
60 Capsule equalling calyx; carpophore *c.* 3 mm; seeds 0·75–1 mm **148. haussknechtii**
47 Calyx contracted at mouth
61 Veins of calyx with prominent, ascending, bulbous-based hairs
62 Hairs on veins of calyx *c.* 1·5 mm, with distal part longer than bulbous base **158. echinata**
62 Hairs on veins of calyx *c.* 1 mm, with distal part shorter than bulbous base **159. squamigera**
61 Veins of calyx without bulbous-based hairs
63 Plant glabrous (except sometimes for carpophore)
64 Calyx with 5 green wings **154. stricta**
64 Calyx without wings or with scarious wings
65 Carpophore not more than 2 mm
66 Inflorescence lax; flowers not overlapping **155. behen**
66 Inflorescence dense; flowers overlapping **156. holzmannii**
65 Carpophore more than 2 mm
67 Cauline leaves not amplexicaul; calyx-teeth acute **153. muscipula**
67 Cauline leaves amplexicaul; calyx-teeth obtuse
68 Upper cauline leaves not closely investing base of laxly or densely corymbose inflorescence **125. armeria**
68 Upper cauline leaves closely investing base of subcapitate inflorescence **126. compacta**
63 Plant hairy, at least at base
69 Flowers in a 3- to 8-flowered subcapitate inflorescence; calyx 8–9 mm **185. disticha**
69 Inflorescence lax; calyx at least 9 mm
70 Stem pubescent with patent hairs; calyx 20–30 mm **128. noctiflora**
70 Stem pubescent with crispate, deflexed hairs, or puberulent; calyx less than 20 mm
71 Carpophore pubescent
72 Seeds with ±plane face; capsule oblong **153. muscipula**
72 Seeds with excavate face; capsule ovoid **157. linicola**
71 Carpophore glabrous
73 Calyx smooth, glabrous, usually shorter than pedicel
74 Carpophore 1–5 mm **150. cretica**
74 Carpophore 6–8 mm **151. ungeri**
73 Calyx scabrid to pubescent, usually at least as long as pedicel
75 Calyx 11–15 mm; calyx-tube 2–3 times as long as teeth; petals white **168. germana**
75 Calyx (12–)15–18 mm; calyx-tube 3–5 times as long as teeth; petals very pale pink **169. almolae**

1 Perennial (more rarely biennial), usually with non-flowering shoots at time of flowering and a woody, branched stock; inflorescence, if compound, often paniculate with equal, opposite branches
76 Styles 5; dioecious; flowers large (calyx more than 10 mm)
77 Calyx inconspicuously 10-veined **132. heuffelii**
77 Calyx, at least in female flowers, 20-veined
78 Biennial or short-lived perennial; flowers white **129. latifolia**
78 Perennial; flowers usually pink or red
79 Stems densely glandular-hairy **130. marizii**
79 Stems ±eglandular
80 Leaves ovate; petals deeply bifid **131. dioica**
80 Leaves lanceolate; petals emarginate or shallowly bifid **133. diclinis**
76 Styles 3 (rarely 4 or 5); flowers usually hermaphrodite (or, if dioecious, calyx less than 8 mm)
81 Calyx 20-veined
82 Calyx pubescent; petals emarginate **77. procumbens**
82 Calyx glabrous and glaucous; petals bifid
83 Plant usually tall, ±erect; inflorescence many-flowered; bracts usually scarious **69. vulgaris**
83 Plant smaller, ±procumbent; inflorescence 1- to few-flowered; bracts usually herbaceous **70. uniflora**
81 Calyx (5–)10-veined
84 Flowers small, unisexual; calyx less than 8 mm; petals often entire, never deeply bifid (Sect. *Otites*) (see also **113**)
85 Petals spathulate, white
86 Petals ±entire **47. sendtneri**
86 Petals shallowly bifid
87 Calyx subglabrous, not inflated **46. roemeri**
87 Calyx pubescent, inflated **48. ventricosa**
85 Petals linear, greenish, yellow or white
88 Monoecious; carpophore *c.* 2 mm **45. sibirica**
88 Usually dioecious; carpophore less than 2 mm
89 Whole plant, including pedicels and calyx, ±densely pubescent
90 Calyx 2–3 mm; capsule *c.* 3 mm, subglobose **49. borysthenica**
90 Calyx 3·5–5 mm; capsule 7–9 mm, ovoid **51. hellmannii**
89 At least pedicels and calyx glabrous
91 Calyx 1·5–3 mm; capsule subglobose
92 Petal-claw ciliate; filaments hairy **50. media**
92 Petal-claw and filaments glabrous **53. velebitica**
91 Calyx more than 3 mm; capsule more than 3 mm, usually ovoid
93 Petals yellow or greenish; filaments glabrous **(56–61). otites** group
93 Petals white or greenish; filaments hairy
94 Hairs on lower part of stem both long and short; petal-claw glabrous **54. wolgensis**
94 Hairs on lower part of stem of same length; petal-claw ciliate
95 Petals white; seeds tuberculate **55. baschkirorum**
95 Petals greenish; seeds smooth **52. cyri**
84 Flowers larger, hermaphrodite; calyx often more than 8 mm; petals usually distinctly bifid

96 Inflorescence a ±spreading, racemose panicle with opposite branches, at least 1 branch with a 3- or more-flowered dichasium

97 Capsule less than $1\frac{1}{2}$ times as long as carpophore

98 Plant glabrous; flowers large; calyx more than 20 mm

99 Petals red; inflorescence short **20. fruticosa**

99 Petals white or pink; inflorescence long **30. bupleuroides**

98 Plant hairy, at least below; calyx often less than 20 mm

100 Inflorescence short, often subcorymbose **(4–11). mollissima** group

100 Inflorescence ±elongate

101 Calyx 8–13 mm **12. mellifera**

101 Calyx at least 14 mm

102 Calyx 25–30 mm; coronal scales lanceolate, acute **15. paradoxa**

102 Calyx less than 25 mm; coronal scales small and obtuse, or absent

103 Calyx usually glandular-pubescent, rarely ±glabrous **1. italica**

103 Calyx sparsely pubescent, eglandular **13. fernandezii**

97 Capsule at least $1\frac{1}{2}$ times as long as carpophore

104 Calyx glabrous or scabrid, sometimes with very sparse glandular or eglandular hairs

105 Terminal flower distinctly overtopped by long, slender lateral branches of inflorescence arising in axils of 2 uppermost bracts (Greece) **27. longipetala**

105 Terminal flower not distinctly overtopped by inflorescence-branches in axils of 2 uppermost bracts

106 Inflorescence a compound panicle of many small flowers; capsule 6–7 mm **25. catholica**

106 Inflorescence simpler, with fewer, larger flowers; capsule usually more than 7 mm

107 Capsule $1\frac{1}{2}$–2 times as long as carpophore

108 Calyx-teeth dimorphic, 2 acute, 3 obtuse **14. longicilia**

108 Calyx-teeth not dimorphic

109 Capsule 9–19 mm; calyx 17–22 mm **13. fernandezii**

109 Capsule 6·5–9 mm; calyx 8–13 mm

110 Stems viscid above; coronal scales small or absent **12. mellifera**

110 Stems not viscid; coronal scales lanceolate, acute **23. velutinoides**

107 Capsule at least $2\frac{1}{2}$ times as long as carpophore

111 Petals deep pinkish-purple; calyx 4–6 mm **26. viscariopsis**

111 Petals white or greenish; calyx more than 6 mm

112 Plant densely caespitose; leaves linear or linear-lanceolate **29. marschallii**

112 Plant usually not densely caespitose; basal leaves lanceolate or wider

113 Whole plant glabrous; coronal scales very small, obtuse **32. chlorantha**

113 Plant usually ±pubescent; coronal scales obvious, acute **22. nutans**

104 Calyx distinctly pubescent, with or without glandular hairs

114 Plant with a persistent, woody cushion of rigid, non-flowering branches (Greece) **19. spinescens**

114 Plant woody or not at base, without a cushion of rigid, non-flowering branches

115 Inflorescence short, often subcorymbose; plant ±densely tomentose **(4–11). mollissima** group

115 Inflorescence elongate; plant not tomentose

116 Capsule $1\frac{1}{2}$–2 times as long as carpophore

117 Petals white, with long, linear lobes; filaments ±hairy **28. niederi**

117 Petals rarely white, with shorter, broader lobes; filaments glabrous

118 Calyx eglandular, or at least with some eglandular hairs **(4–11). mollissima** group

118 Calyx with glandular hairs only

119 At least some calyx-teeth acute

120 Calyx-teeth dimorphic, 2 acute, 3 obtuse **14. longicilia**

120 Calyx-teeth all acute **23. velutinoides**

119 Calyx-teeth all obtuse

121 Calyx 8–13 mm **12. mellifera**

121 Calyx 17–22 mm **13. fernandezii**

116 Capsule at least $2\frac{1}{2}$ times as long as carpophore

122 Calyx cylindrical, ±attenuate at base; petal-claw distinctly exserted **24. viridiflora**

122 Calyx clavate, truncate or umbilicate at base; petal-claw only slightly exserted

123 Flowers erect; inflorescence not secund; capsule *c.* 3 times as long as carpophore **21. gigantea**

123 Flowers often inclined; inflorescence ±secund; capsule 3–5 times as long as carpophore **22. nutans**

96 Inflorescence not a diffuse panicle with opposite branches bearing dichasia

124 Inflorescence more than 5-flowered (rarely 4- or 5-flowered)

125 Inflorescence a ±regular dichasium, sometimes corymbose or subcapitate

126 Petals usually 4-toothed or 4-fid; seeds winged (Sect. *Heliosperma*) **(117–124). pusilla** group

126 Petals entire, emarginate or bifid; seeds not winged

127 Calyx 4–6 mm **115. rupestris**

127 Calyx more than 6 mm

128 Carpophore *c.* 2 mm; calyx ±inflated

129 Calyx strongly inflated; stem leafless in upper $\frac{1}{2}$ **74. fabarioides**

129 Calyx slightly inflated; stem leafy throughout

130 Capsule at least 10 mm; robust plant to 100 cm **72. fabaria**

130 Capsule 6–8·5 mm; low-growing plant **73. ionica**

128 Carpophore more than 3 mm; calyx not inflated

131 Flowering stems slender, obviously lateral;

calyx 9–12 mm; petal-limb deeply emarginate **116. lerchenfeldiana**
131 Flowering stem usually robust, terminal; calyx 12–20 mm; petal-limb entire or emarginate (Sect. *Compactae*)
132 Upper cauline leaves not closely investing base of laxly or densely corymbose inflorescence **125. armeria**
132 Upper cauline leaves (bracts) closely investing base of subcapitate inflorescence
133 Biennial; bracts herbaceous; calyx not purplish **126. compacta**
133 Perennial; bracts membranous; calyx purplish **127. asterias**
125 Inflorescence not a regular dichasium, and neither corymbose nor subcapitate
134 Calyx hairy (sometimes slightly so)
135 Calyx 7–9 mm
136 Capsule *c.* 8 mm
137 Calyx eglandular **78. flavescens**
137 Calyx glandular-pubescent **79. thessalonica**
136 Capsule 4–5 mm
138 Petals yellowish; calyx densely glandular-hairy **80. congesta**
138 Petals purple; calyx sparsely hairy **103. schmuckeri**
135 Calyx more than 9 mm
139 Leaves acicular, pungent **90. altaica**
139 Leaves not acicular
140 Plant with long, creeping stolons
141 Leaves linear **85. repens**
141 Leaves obovate to spathulate **86. sangaria**
140 Plant not stoloniferous
142 Coronal scales well-developed and conspicuous
143 Inflorescence a lax panicle **88. spergulifolia**
143 Inflorescence a raceme-like monochasial cyme
144 Leaves spathulate to obovate-spathulate; face of seeds excavate **176. obtusifolia**
144 Leaves linear to lanceolate-spathulate; face of seeds not excavate
145 Flowering stems arising terminally from centre of leaf-rosettes; carpophore pubescent **110. ciliata**
145 Flowering stems arising below leaf-rosettes; carpophore glabrous **111. legionensis**
142 Coronal scales small or absent
146 Petals yellow
147 Calyx eglandular **78. flavescens**
147 Calyx glandular-pubescent **79. thessalonica**
146 Petals not yellow
148 Inflorescence-branches alternate, with rather compact, long-peduncled, 3-flowered dichasia; basal leaves linear-lanceolate **81. cephallenia**
148 Inflorescence-branches not as above; basal leaves lanceolate, ovate or spathulate
149 Robust biennial (rarely perennial), covered with dense, viscid, glandular hairs **35. viscosa**
149 Perennial, not viscid except in inflorescence **34. multiflora**
134 Calyx glabrous (except for cilia on teeth)
150 Calyx 20–30 mm **30. bupleuroides**
150 Calyx less than 20 mm
151 Petals greenish (sometimes pinkish); carpophore 2–3 mm
152 Stem glabrous
153 Calyx teeth ovate, obtuse **32. chlorantha**
153 Calyx teeth ovate-lanceolate, acute **40. tatarica**
152 Stem scabrid with short, deflexed hairs
154 Calyx 7–8·5 mm **37. oligantha**
154 Calyx 10–12 mm **39. skorpilii**
151 Petals white, pink or reddish; carpophore at least 3 mm
155 Inflorescence very long and narrow, with short, opposite, 1-flowered branches **33. frivaldszkyana**
155 Inflorescence shorter, ±compound
156 Lower part of stem with patent hairs
157 Inflorescence narrow; lower part of stem usually with numerous short, leafy, non-flowering branches **40. tatarica**
157 Inflorescence a spreading, alternately-branched panicle; lower part of stem with few or no short leafy branches (*cf.* also **1**)
158 Calyx 14–20 mm **17. cythnia**
158 Calyx 11–13 mm **18. goulimyi**
156 Lower part of stem glabrous, or rough with minute, deflexed hairs
159 Coronal scales absent; capsule equalling or only slightly longer than carpophore **34. multiflora**
159 Coronal scales present, but small; capsule *c.* 3 times as long as carpophore **38. reichenbachii**
124 Inflorescence 1- to 5(–7)-flowered
160 Calyx hairy, sometimes only sparsely so
161 Calyx with glandular hairs
162 Petals usually 4-toothed or 4-fid; seeds winged (Sect. *Heliosperma*) **(117–124). pusilla** group
162 Petals ±bifid; seeds not winged
163 Calyx inflated, not closely investing ripe capsule
164 Basal leaves not thick and fleshy; capsule with basal septa **68. herminii**
164 Basal leaves thick and fleshy, in rosettes; capsule without basal septa (Sect. *Odontopetalae*)
165 Calyx 12–15 mm; basal leaves with long cilia 1–2 mm **62. auriculata**
165 Calyx more than 15 mm; basal leaves with shorter cilia *c.* 1 mm, or not ciliate
166 Petals dark reddish-purple **64. elisabethae**
166 Petals white or whitish

167 Calyx 15–17 mm, only slightly glandular **63. zawadzkii**
167 Calyx (20–)25–30 mm, densely glandular **65. requienii**
163 Calyx not or only slightly inflated, ±closely investing ripe capsule
168 Flowering stems ±procumbent
169 Leaves linear to linear-lanceolate **88. spergulifolia**
169 Leaves ovate to obovate or oblanceolate
170 Whole plant viscid with dense glandular pubescence **87. succulenta**
170 Plant pubescent and ±glandular, but not viscid **89. thymifolia**
168 Flowering stems erect or suberect
171 Calyx 8–11·5 mm
172 Basal leaves spathulate; carpophore up to 3 mm **79. thessalonica**
172 Basal leaves linear or linear-lanceolate; carpophore more than 3 mm **81. cephallenia**
171 Calyx at least 12 mm
173 Cauline leaves cordate-ovate **66. cordifolia**
173 Cauline leaves ovate-lanceolate to linear
174 Petal-limb emarginate **67. foetida**
174 Petal-limb bifid
175 Capsule scabrid or glandular-hairy (Sect. *Auriculatae*)
176 Capsule 8–10 mm, at least twice as long as carpophore **84. jailensis**
176 Capsule more than 10 mm, shorter than or equalling carpophore
177 Coronal scales small; capsule equalling carpophore **82. vallesia**
177 Coronal scales conspicuous; capsule much shorter than carpophore **83. boryi**
175 Capsule glabrous
178 Calyx less than 25 mm; coronal scales small and obtuse, or absent **2. sieberi**
178 Calyx 25–30 mm; coronal scales lanceolate, acute **15. paradoxa**
161 Calyx eglandular
179 Styles 5; calyx ±inflated (Sect. *Gastrolychnis*)
180 Calyx 14–18 mm; petals reddish-purple **43. uralensis**
180 Calyx 10–12 mm; petals white **44. furcata**
179 Styles 3; calyx usually not inflated
181 Calyx not more than 11 mm
182 Petals yellow; carpophore *c.* 1 mm **78. flavescens**
182 Petals pink or purple; carpophore at least 3 mm
183 Petals purple; basal leaves not papillose **103. schmuckeri**
183 Petals pink; basal leaves papillose and punctate **112. borderei**
181 Calyx more than 11 mm
184 Basal leaves obovate-spathulate or elliptical
185 Petals white **63. zawadzkii**
185 Petals greenish **95. schwarzenbergeri**
184 Basal leaves linear or lanceolate, more rarely lanceolate-spathulate
186 Calyx-teeth alternately acute and obtuse **104. multicaulis**
186 Calyx-teeth not alternately acute and obtuse
187 Calyx-teeth acute **94. linifolia**
187 Calyx-teeth obtuse
188 Inflorescence viscid (N. Greece and S. Albania) **3. damboldtiana**
188 Inflorescence not viscid
189 Flowering stems terminal to leaf-rosettes; veins of calyx usually obviously anastomosing **110. ciliata**
189 Flowering stems lateral to leaf-rosettes; veins of calyx scarcely anastomosing **111. legionensis**
160 Calyx glabrous
190 Calyx less than 10 mm
191 Plant densely pulvinate; petals pink or pale purple
192 Calyx 7–9 mm, campanulate; leaves stiffly ciliate **113. acaulis**
192 Calyx 5–7·5 mm, turbinate; leaves scabrid **109. barbeyana**
191 Plant not densely pulvinate; petals white, pink or greenish, rarely brownish-purple
193 Basal leaves spathulate, ±fleshy or not
194 Capsule 6–9(–12) mm; carpophore pubescent **36. radicosa**
194 Capsule 4·5–6 mm; carpophore glabrous **37. oligantha**
193 Basal leaves linear or linear-lanceolate to spathulate, not fleshy
195 Petals usually 4-toothed or 4-fid; seeds winged (Sect. *Heliosperma*) **(117–124). pusilla** group
195 Petals ±bifid; seeds not winged
196 Plant caespitose, with many slender flowering stems **(97–102). saxifraga** group
196 Plant not caespitose; stems solitary or few
197 Calyx not inflated (Maritime Alps) **96. campanula**
197 Calyx ±inflated (N.E. Russia)
198 Calyx 4–6 mm wide; petal-claw not ciliate **41. paucifolia**
198 Calyx 3–4 mm wide; petal-claw ciliate **42. graminifolia**
190 Calyx at least 10 mm
199 Cauline leaves lanceolate or wider
200 Calyx at least 18 mm
201 Plant glabrous and glaucous; cauline leaves cordate, amplexicaul **31. chlorifolia**
201 Plant not glaucous; stem somewhat pubescent; cauline leaves not cordate
202 Petals white; calyx-teeth obtuse; capsule *c.* 10 mm **16. nodulosa**
202 Petals pink; calyx-teeth acute; capsule 11–12 mm **17. cythnia**
200 Calyx not more than 15 mm
203 Stems shortly pubescent to tomentose

204 Capsule less than twice as long as carpophore **17. cythnia**
204 Capsule about twice as long as carpophore **93. taliewii**
203 Stems glabrous
205 Stems up to 40 cm; petals white or pink **75. caesia**
205 Stems up to 10 cm; arising from long, procumbent shoots; petals greyish-violet **76. variegata**
199 At least some cauline leaves linear
206 Calyx more than 18 mm
207 Leaves thick and rigid; coronal scales absent **91. cretacea**
207 Leaves not obviously thick and rigid; coronal scales present
208 Lower cauline leaves oblanceolate; coronal scales minute; carpophore 10–12 mm
209 Carpophore glabrous **16. nodulosa**
209 Carpophore pubescent **104. multicaulis**
208 Lower cauline leaves linear; coronal scales conspicuous; carpophore (11–)13–22 mm
210 Uppermost cauline leaves very close to base of calyx of single terminal flower **108. orphanidis**
210 Uppermost cauline leaves distinctly separated from flower(s)
211 Inflorescence 1- to 5-flowered, on stems up to 25 cm; petals white **105. waldsteinii**
211 Flowers solitary, on slender stems usually *c.* 10 cm; petals brownish **106. pindicola**
206 Calyx up to 18(–19) mm
212 Basal leaves spathulate, ±fleshy
213 Internodes rough with minute deflexed hairs, the upper glabrous **36. radicosa**
213 All internodes glabrous **76. variegata**
212 Basal leaves linear or narrowly spathulate, not fleshy
214 Capsule 6–8 mm; stems retrorsely aculeolate below **107. dionysii**
214 Capsule often more than 8 mm; stems not retrorsely aculeolate
215 Densely pulvinate or mat-forming
216 Calyx clavate **(97–102). saxifraga** group
216 Calyx oblong-campanulate **114. dinarica**
215 Plant not pulvinate or mat-forming
217 Not caespitose; stems few, robust (S.E. Russia) **92. suffrutescens**
217 Caespitose, with many slender flowering stems
218 Calyx 14–18 mm; carpophore pubescent
219 Petal-claw not or slightly exceeding calyx; calyx oblong-turbinate **(97–102). saxifraga** group
219 Petal-claw long-exserted from calyx; calyx narrowly clavate **104. multicaulis**
218 Calyx not more than 13 mm; carpophore usually glabrous
220 Lobes of petals long, linear; filaments ±hairy (Greece) **29. marschallii**
220 Lobes of petals broader, not linear; filaments glabrous **(97–102). saxifraga** group

Sect. SIPHONOMORPHA Otth (incl. Sect. *Paniculatae* (Boiss.) Chowdhuri, Sect. *Viridiflorae* Boiss.). Perennials with woody stock, more rarely biennials. Lower leaves lanceolate to spathulate. Inflorescence paniculate, with opposite, viscid branches bearing (1–)3- to 7-flowered dichasia; flowers usually large, erect or inclined at anthesis. Filaments glabrous. Seeds with plane face and usually shallowly grooved back.

1. S. italica (L.) Pers., *Syn. Pl.* **1**: 498 (1805). Stems 20–80(–100) cm, pubescent below. Basal leaves lanceolate- or oblanceolate-spathulate, obtuse. Inflorescence lax, with ascending branches usually bearing 3-flowered dichasia. Calyx 14–24 mm, glandular-pubescent, rarely more or less glabrous; teeth obtuse. Petals usually creamy-white above, reddish or greenish beneath, deeply bifid; coronal scales very small or absent. Capsule 8–13 mm, ovoid, usually more or less equalling pubescent carpophore. Seeds *c.* 1 mm, reniform. *S. & C. Europe; casual in N. Europe, rarely naturalized.* Al Au Bu ?Co Cz Ga Ge Gr Hs Hu It Ju Lu Po Rm Rs(W, K) Si Tu [Br Da].

1 Carpophore usually more than 7 mm
2 Plant perennial with several flowering stems; petal-claw ciliate **(a)** subsp. **italica**
2 Plant biennial with single flowering stem; petal-claw usually glabrous **(b)** subsp. **nemoralis**
1 Carpophore usually less than 7 mm
3 Plant robust with stems up to 100 cm; flowers white **(c)** subsp. **coutinhoi**
3 Plant slender with stems rarely more than 60 cm; flowers pinkish **(d)** subsp. **sicula**

(a) Subsp. **italica**: Perennial, with short stolons and several stems bearing lax panicles. Petal-claw usually ciliate. Carpophore 7–8(–10) mm. $2n = 24$. *S. Europe; casual in W., C. & N. Europe.*

(b) Subsp. **nemoralis** (Waldst. & Kit.) Nyman, *Consp.* 90 (1878) (*S. nemoralis* Waldst. & Kit., *S. jundzillii* Zapał.): Biennial, without stolons and with single stems bearing a rather dense panicle. Petal-claw usually glabrous. Carpophore 7–8(–10) mm. $2n = 24$. *C. Europe, and mountains in S. Europe.*

(c) Subsp. **coutinhoi** (Rothm. & P. Silva) Franco, *Nova Fl. Portugal* **1**: 143 (1971) (*S. coutinhoi* Rothm. & P. Silva): Usually robust perennial, with several stems bearing lax panicles. Calyx glabrous or nearly so. Petal-claw usually glabrous. Carpophore 5–7 mm. $2n = 24$. ● *Portugal, N.C. Spain.*

(d) Subsp. **sicula** (Ucria) Jeanmonod, *Willdenowia* **14**: 46 (1984) (*S. sicula* Ucria): Perennial, with several stems up to 60 cm bearing lax panicles. Petals usually pinkish; petal-claw ciliate or glabrous. Carpophore 5–7(–8) mm. $2n = 24$. ● *Sicilia, Calabria.*

Robust plants from Spain that have been called **S. crassicaulis** Willk. & Costa, *Linnaea* **30**: 91 (1859), are best treated as a variety of subsp. (**b**).

2. S. sieberi Fenzl, *Pugillus* 8 (1842). Like **1** but smaller, stems 10–30(–40) cm; stolons absent; panicle very lax, with a few long, divaricate branches, usually 1-flowered; small stems often 1- or 2-flowered. $2n = 24$. *Rocks and stony ground.* ● *Kriti.* Cr.

3. S. damboldtiana Greuter & Melzh., *Willdenowia* **8**: 614 (1979). Like **1** but smaller, up to 40 cm; stolons absent; inflorescence narrow with short, usually 1-flowered branches; calyx eglandular-pubescent. 2*n* = 24. ● *Mountains of N. Greece & S.E. Albania*. Al Gr.

(4–11). S. mollissima group. Robust, tomentose perennials with woody stock and flowering stems up to 120 cm. Basal leaves usually elliptical. Inflorescence usually relatively short and broad, often subcorymbose. Calyx 17–25 mm, more or less densely hairy with relatively long, multicellular glandular or eglandular hairs; teeth obtuse. Petals white, reddish, purplish or violet. Capsule 8–14 mm, ovoid, enclosed in calyx and as long as or somewhat longer than carpophore. Seeds usually 1·5–2 mm, reniform, dark brown.

The above description covers a group of closely related plants in S.W. Europe, mostly chasmophytes of sea-cliffs, which have been treated at specific rank by W. Rothmaler, *Feddes Repert.* **52**: 275 (1943), and D. Jeanmonod, *Candollea* **39**: 195–259 (1984).

S. rothmaleri P. Silva, *Agron. Lusit.* **9**: 18 (1956), from S.W. Portugal (Cabo de S. Vicente), with whitish flowers in rather dense panicles, and calyx 20–23 mm, may belong to this group, but its affinities are obscure. It is known only from the type collection.

1 Calyx glandular, at least in part
  2 Calyx with both glandular and eglandular hairs **5. velutina**
  2 Calyx with glandular hairs only
    3 Calyx 20–25 mm **7. andryalifolia**
    3 Calyx 16–20 mm
      4 Capsule 11–14 mm; inflorescence condensed **10. tyrrhenia**
      4 Capsule 9–11 mm; inflorescence lax **11. rosulata**
1 Calyx eglandular
  5 Basal leaves obtuse; petals whitish **4. mollissima**
  5 Basal leaves acute or acuminate; petals pink to purple
    6 Petals pale violet, with large coronal scales **9. tomentosa**
    6 Petals pink, red or purple, with small coronal scales
      7 Capsule not more than 11 mm **6. hicesiae**
      7 Capsule 11–13 mm **8. hifacensis**

4. S. mollissima (L.) Pers., *Syn. Pl.* **1**: 498 (1805). Basal leaves obtuse. Calyx 20–22 mm, densely eglandular-tomentose. Petals whitish, with ciliate claw and small coronal scales. Capsule 10–12 mm, as long as carpophore. 2*n* = 24. ● *Islas Baleares*. Bl.

5. S. velutina Pourret ex Loisel. in Desv., *Jour. Bot. Rédigé* **2**: 324 (1809) (*S. salzmannii* Otth). Basal leaves acute. Calyx (15–)18–25 mm, tomentose, with a rather variable proportion of glandular and eglandular hairs. Petals whitish or purplish-pink; coronal scales absent. Capsule 9–12 mm, usually somewhat longer than carpophore. 2*n* = 24. ● *Small islands between Sardegna and Corse; formerly on Corse*. Co Sa.

6. S. hicesiae Brullo & Signorello, *Willdenowia* **14**: 141 (1984). Basal leaves acuminate. Stems up to 120 cm; inflorescence elongate. Calyx (15–)18–25 mm, eglandular-pubescent. Petals pinkish-purple, with small coronal scales. Capsule 8–11 mm. 2*n* = 24. *Cliffs.* ● *Isole Eolie* (*Panarea*)*; one station in Sicilia* (*near Palermo*). Si.

7. S. andryalifolia Pomel, *Nouv. Mat. Fl. Atl.* 331 (1875) (*S. pseudovelutina* Rothm.). Basal leaves acuminate. Calyx 20–25 mm, densely tomentose with long glandular hairs only. Petals whitish, with glabrous claw; coronal scales absent. Capsule 8–12 mm, as long as carpophore. 2*n* = 24. *S. Spain*. Hs. (*N. Africa*.)

8. S. hifacensis Rouy ex Willk., *Ill. Fl. Hisp.* **1(10)**: 150 (1885). Basal leaves acute to acuminate. Calyx 17–22 mm, densely glandular-pubescent. Petals reddish or purplish, with small coronal scales. Capsule 11–13 mm, slightly longer than carpophore. 2*n* = 24. ● *Islas Baleares; E. Spain* (*Alicante prov.*). Bl Hs.

9. S. tomentosa Otth in DC., *Prodr.* **1**: 383 (1824) (*S. gibraltarica* Boiss.). Basal leaves spathulate, acuminate. Calyx 18–20 mm, densely eglandular-pubescent. Petals pale violet, with large coronal scales. Capsule *c.* 11 mm, slightly longer than carpophore. *Crevices of calcareous cliffs near the sea.* ● *S. Spain* (*Gibraltar*). Hs.

10. S. tyrrhenia Jeanmonod & Bocquet, *Candollea* **38**: 198 (1983) (*S. salzmannii* Badaro ex Moretti, non Otth). Whole plant densely puberulent. Basal leaves in dense rosettes, elliptical-subspathulate, acute or subacute. Stems up to 60 cm bearing condensed inflorescence. Calyx 16–20 mm, densely glandular-puberulent. Petals white or yellowish-white. Capsule 11–14 mm, oblong-ovoid. 2*n* = 24. *Coastal cliffs, rocks and screes.* ● *S.E. France and N.W. Italy*. Ga It.

11. S. rosulata Soyer-Willemet & Godron, *Monogr. Silene Algér.* 50 (1851). Basal leaves elliptical-spathulate, more or less acute. Stems 40–70(–90) cm, purple. Inflorescence lax, with ascending branches bearing usually 3-flowered dichasia. Calyx 16–19 mm, densely glandular-puberulent. Petals white. Capsule 9–11 mm, oblong-ovoid; carpophore 7–9 mm. Seeds *c.* 1·3 mm. *Maritime sands. N. Sardinia* (*near Segna Teresa Gallura*). Sa. (*N. Africa*.)

European plants have been referred to subsp. **sanctae-therasiae** (Jeanmonod) Jeanmonod, *Willdenowia* **14**: 47 (1984). Subsp. *rosulata* and other subspecies occur widely in N. Africa.

12. S. mellifera Boiss. & Reuter, *Diagn. Pl. Nov. Hisp.* 8 (1842) (incl. *S. nevadensis* Boiss.). Stems (20–)40–80 cm, eglandular-tomentose in lower $\frac{1}{2}$, viscid above. Basal leaves lanceolate-spathulate, upper cauline linear. Inflorescence lax, with branches bearing 1- to 3-flowered dichasia. Calyx 8–13 mm, glabrous or sparsely hairy, with anastomosing veins. Petal-limb 5–8 mm, bifid, greenish-white; coronal scales very small or absent. Capsule 6·5–9 mm; carpophore 3·5–6·5 mm, pubescent. 2*n* = 24. ● *C., E. & S. Spain; S. Portugal*. Hs Lu.

S. sennenii Pau, *Bol. Soc. Aragon. Ci. Nat.* **4**: 309 (1905), from N.E. Spain (Gerona), with linear-lanceolate to linear leaves, non-anastomosing calyx veins, and somewhat narrower capsule, seems doubtfully distinct from **12**.

13. S. fernandezii Jeanmonod, *Candollea* **39**: 619 (1984). Like **12** but densely caespitose; leaves narrowly oblong-elliptical; calyx 17–22 mm, sparsely eglandular-pubescent. *Stony and rocky places on serpentine.* ● *S. Spain* (*Sierra Bermeja*). Hs.

14. S. longicilia (Brot.) Otth in DC., *Prodr.* **1**: 377 (1824) (*S. patula* auct., non Desf.; incl. *S. cintrana* Rothm.). Like **12** but calyx 10–14 mm, subglabrous, with sparse glandular hairs; teeth usually dimorphic (2 triangular acute, 3 obtuse with broad hyaline margins); petals reddish-purple (more rarely whitish); capsule $1\frac{1}{2}$–2 times as long as carpophore. 2*n* = 24. ● *W. Portugal*. Lu.

15. S. paradoxa L., *Sp. Pl.* ed. 2, 1673 (1763). Stems up to 80 cm, pubescent below. Basal leaves lanceolate to lanceolate-obovate. Inflorescence lax, with ascending branches. Calyx

25–30 mm, cylindrical, with attenuate base, glandular; teeth lanceolate, acute. Petals pinkish above, yellowish beneath; coronal scales acute. Capsule 10–13(–18) mm, rather narrowly ovoid; carpophore 7–11 mm. Seeds *c.* 1·5 mm. $2n = 24$. ● *S. Europe, from S.E. France and Corse to Greece.* Al Co Ga Gr It Ju.

16. S. nodulosa Viv., *Fl. Cors.* 6 (1824) (*S. pauciflora* Salzm. ex DC., non Ucria). Perennial with rather slender, branched, woody stock and numerous slender flowering stems up to 30 cm; lower part of stem and leaves scabrid-puberulent. Lower leaves up to 30 × 5 mm, oblanceolate to narrowly spathulate; upper cauline leaves linear, distant. Inflorescence glabrous, (1)2- to 5-flowered; pedicels up to 7 cm, very slender. Calyx 18–20 mm, narrow, glabrous; teeth obtuse, more or less ciliate. Petals white, reddish beneath. Capsule *c.* 10 mm, ovoid; carpophore *c.* 12 mm, glabrous. Seeds *c.* 1 mm, black. $2n = 24$. ● *Corse; Sardegna.* Co Sa.

17. S. cythnia (Halácsy) Walters, *Feddes Repert.* **69**: 46 (1964) (*S. italica* var. *cythnia* Halácsy). Stems 20–80 cm. Basal leaves lanceolate-spathulate. Inflorescence a lax panicle, with alternate branches bearing 3-flowered dichasia. Calyx 14–20 mm, glabrous, with acute, ciliate teeth. Petals pink. Capsule 11–12 mm, ellipsoidal, up to twice as long as carpophore. $2n = 24$. *Aegean region (Kikladhes).* Gr.

18. S. goulimyi Turrill, *Kew Bull.* **1955**: 353 (1955). Like **17** but woodier at base, dwarf and few-flowered; calyx 11–13 mm; carpophore up to 4 mm. $2n = 24$. ● *S. Greece (Taÿetos).* Gr.

19. S. spinescens Sibth. & Sm., *Fl. Graec. Prodr.* **1**: 299 (1809). Stock very woody, much-branched; flowering stems 10–30 cm, erect, bearing numerous, non-flowering, rigid branches below, forming a spiny, persistent cushion. Basal leaves spathulate; cauline leaves linear, stiff, often spinose; leaves and stems greyish-tomentose. Inflorescence-branches mostly 1-flowered. Calyx *c.* 20 mm, pubescent; teeth obtuse. Petals white, deeply bifid; coronal scales ovate. Capsule *c.* $1\frac{1}{2}$ times as long as carpophore. $2n = 24$. ● *S.E. Greece.* Gr.

20. S. fruticosa L., *Sp. Pl.* 417 (1753). Robust perennial up to 50 cm, almost completely glabrous. Lower leaves obovate or narrowly spathulate, the upper lanceolate; all glabrous and shining above, ciliate. Inflorescence dense; flowers large, erect, on short peduncles. Calyx *c.* 25 mm, with patent glandular hairs; teeth acute. Petals pink or red, shallowly bifid; coronal scales obvious, acute, sometimes laciniate. Capsule *c.* 15 mm, ovoid, beaked, equalling or longer than carpophore. Seeds 1·5–2 mm. $2n = 24$. *Mediterranean region (Sicilia to Karpathos); very local.* Cr Gr Si.

21. S. gigantea L., *Sp. Pl.* 418 (1753). Robust biennial up to 100 cm, usually unbranched, with a well-developed basal rosette of thick, spathulate leaves; stem and leaves pubescent. Inflorescence-branches short and pseudo-verticillate (or long and rather diffuse in var. *incana* (Griseb.) Chowdhuri). Flowers rather small. Calyx 8–12 mm. Petals white, pink or greenish. Capsule *c.* 3 times as long as carpophore. $2n = 24$. *Balkan peninsula northwards to* 43°30′ *N.; Aegean region.* Al Bu Cr Gr Ju.

The typical plant is apparently less widespread than var. *incana* (*S. rhodopaea* Janka), which is less hairy, with narrow leaves and a laxly paniculate inflorescence.

22. S. nutans L., *Sp. Pl.* 417 (1753). Stems 20–60(–100) cm, usually unbranched, and obviously pubescent below, rarely glabrous. Lower leaves up to 10 cm, oblong-spathulate, the upper linear-lanceolate. Flowers usually inclined, in a lax, secund panicle. Calyx 9–12 mm, glandular-pubescent, truncate at base; teeth lanceolate, acute. Petals variable in colour, usually whitish above, greenish or reddish beneath, deeply bifid with narrow, inrolled lobes; coronal scales lanceolate, acute. Capsule 7–13(–18) mm, 3–5 times as long as pubescent carpophore. *Usually calcicole. Most of Europe except the extreme north and most of the islands.* Al Au Be Br Bu Cz Da Fe Ga Ge Gr He Ho Hs Hu It Ju Lu No Po Rm Rs(N, B, C, W, E) Su.

Extremely variable in habit, hairiness, leaf-shape, colour and size of flowers, and size of capsule.

(**a**) Subsp. **nutans** (incl. subsp. *smithiana* (Moss) Jeanmonod & Bocquet): Stems and leaves pubescent with spreading multicellular hairs (rarely glabrous, var. *infracta* (Kit.) Wahlenb.). Inflorescence-branches usually at least 3-flowered. Calyx-teeth usually subequal. Petals white, often veined yellow, green or red. Petal-claw 5–8 mm, usually without an auricle. Capsule 8–12(–18) mm, *c.* 3 times as long as carpophore. $2n = 24$. *Throughout the range of the species.*

(**b**) Subsp. **dubia** (Herbich) Zapał., *Rozpr. Wydz. Mat.-Przyr. Polsk. Akad. Um.* (*Biol.*) ser. 3, **11B**: 151 (1911) (*S. dubia* Herbich): Stems with short, retrorse hairs. Leaves with short marginal cilia and usually with some 1- or 2-cellular hairs on surface. Inflorescence-branches usually 1- to 3-flowered. Calyx-teeth usually distinctly unequal. Petals white to yellowish-green. Petal-claw 2·5–4·5 mm, with well-developed, acute auricle. Capsule 8–11 mm, 4–5 times as long as carpophore. $2n = 24$. *S. & E. Carpathians and mountains of Transylvania.*

Plants from S. France and N. Spain like subsp. (**a**) but with capsule 13–18 mm have been called **S. brachypoda** Rouy, *Ill. Pl. Eur. Rar.* **4**: 26 (1895), but capsule length is variable in this species and these plants are best treated as a variety of subsp. (**a**), var. *brachypoda* (Rouy) Molero.

Subsp. (**b**) has been treated as a separate species by several authors.

Plants from S. Europe, from S.E. France to C. Romania, that are robust, with stems up to 100 cm bearing a broad inflorescence and slightly larger flowers and capsules, the petals greyish-purple to greenish beneath, have been called subsp. **livida** (Willd.) Jeanmonod & Bocquet, *Candollea* **38**: 291 (1983) (*S. livida* Willd.). This pattern of variation may represent former introgressive hybridization with **24**.

23. S. velutinoides Pomel, *Nouv. Mat. Fl. Atl.* **1**: 208 (1874). Like **22(a)** but woody at the base, caespitose; calyx 8–10 mm; petals white with greenish veins; capsule 7–8 mm, not more than twice as long as carpophore. $2n = 24$. *Crevices of calcareous rocks. Sardegna.* Sa. (*Algeria.*)

24. S. viridiflora L., *Sp. Pl.* ed. 2, 597 (1762). Plant robust, 40–90 cm. Lower leaves oblong-spathulate, the upper ovate-lanceolate, acuminate. Inflorescence lax, with long peduncles; flowers inclined. Calyx 15–20 mm, gradually attenuate at base, pubescent; teeth lanceolate, acute. Petals greenish-white, with long-exserted claw and linear lobes; coronal scales oblong, acute. Capsule 12–14 mm, ovoid; carpophore 2–3·5 mm. $2n = 24$. *S. Europe, extending northwards to* c. 49° *N. in Czechoslovakia.* Al Bu Co Cz Ga Gr Hs Hu It Ju Rm Rs(K) Sa Si Tu.

25. S. catholica (L.) Aiton fil. in Aiton, *Hort. Kew.* ed. 2, **3**: 85 (1811). Slender perennial, 40–80 cm, very viscid above. Leaves ovate-lanceolate, scabrid and sparsely ciliate. Inflorescence-branches very long; flowers numerous and very small. Calyx 6–8 mm, glabrous. Petals small, white. Capsule 6–7 mm, ovoid, 1½–2 times as long as carpophore. ● *S. & C. Italy; one station in W. Jugoslavia*. It Ju.

26. S. viscariopsis Bornm., *Feddes Repert.* **17**: 38 (1921). Biennial or short-lived perennial with slender, erect stems 20–30(–40) cm, shortly pubescent below, glabrous and viscid above. Basal leaves linear-lanceolate; cauline linear. Panicle 3- to 20(–25)-flowered, with opposite, spreading, 1- to 3-flowered branches; pedicels 1–3 cm, slender. Calyx 4–6 mm, broadly campanulate, somewhat inflated and contracted at mouth in fruit, 10-veined, with no lateral veins, more or less pruinose and scabrid. Petals deep pinkish-purple, with long claw and deeply bifid lamina. Capsule 7–8 mm, broadly ovoid; carpophore very short. $2n = 24$. *Dry, open grassland.* ● *S. Makedonija (near Prilep)*. Ju.

J. Bornmüller described a new section to accommodate this species. It is very different from the other members of Sect. *Siphonomorpha*, though resembling them in inflorescence form. It could perhaps be classified with the annual species in Sect. *Behenantha* Otth (spp. **150–156**).

Sect. LASIOSTEMONES (Boiss.) Schischkin. Perennials with woody stock. Inflorescence paniculate; flowers nodding at anthesis. Filaments hairy.

27. S. longipetala Vent., *Descr. Pl. Jard. Cels* t. 83 (1802). Stock branched, procumbent; stems 30–85 cm. Basal leaves oblong-lanceolate, acute, scabrid, the cauline lanceolate or linear. Panicle large, subsecund, with long, spreading branches; terminal flower usually distinctly overtopped by lateral branches in axils of 2 subtending bracts. Calyx 10–12 mm, glabrous; teeth obtuse. Petals white, with long, linear lobes; coronal scales triangular. Capsule 7–10 mm, ovoid; carpophore 2–2·5 mm. *E. Greece*. Gr.

28. S. niederi Heldr. ex Boiss., *Diagn. Pl. Or. Nov.* **3(6)**: 32 (1859). Like **27** in habit, leaf-shape and flower-size, but calyx pubescent; coronal scales absent; capsule 1½ times as long as carpophore. $2n = 24$. ● *C. & N.W. Greece*. Gr.

29. S. marschallii C. A. Meyer, *Verz. Pfl. Cauc.* 214 (1831) (incl. *S. guicciardii* Boiss. & Heldr.). Plant densely caespitose; flowering-stems 15–30 cm, numerous, erect. Leaves linear-lanceolate to linear, glabrous. Inflorescence-branches with 1–3 flowers; pedicels very short. Calyx *c.* 12 mm, glabrous. Petals white, with linear lobes. Capsule *c.* 10 mm; carpophore 2–3·5 mm. *Mountain rocks. C. Greece (Parnassos)*. Gr. (*S.W. Asia*.)

Sect. SCLEROCALYCINAE (Boiss.) Schischkin (incl. Sect. *Chloranthae* (Rohrb.) Schischkin, Sect. *Tataricae* Chowdhuri). Perennials with woody stock, more rarely biennials. Inflorescence usually narrow, spike-like, with opposite branches bearing 1- to 3(–5)-flowered dichasia. Calyx often glabrous or subglabrous, usually rather sulcate, with conspicuous veins.

30. S. bupleuroides L., *Sp. Pl.* 421 (1753) (*S. longiflora* Ehrh.). Stems 25–80 cm, glabrous and glaucous. Basal leaves oblong-lanceolate to linear-lanceolate, long-attenuate into petiole. Inflorescence paniculate to pseudo-racemose; flowers large. Calyx 20–30 mm, glabrous and membranous-coriaceous. Petal-limb cuneate, deeply bifid, white, pinkish or greenish; coronal scales acute. Capsule 10–14 mm, included in calyx. Seeds 1·5–2 mm. *S.E. & E.C. Europe*. Al Bu Cz Gr Hu Ju Rm Rs(W, K).

(a) Subsp. **bupleuroides** (incl. *S. mariae* Klokov, *S. montifuga* Klokov, *S. odessana* Klokov, *S. ucrainica* Klokov): Plant laxly caespitose; flowering stems up to 80 cm. Basal leaves oblong-lanceolate. Inflorescence lax, the lower branches several-flowered. Capsule equalling carpophore. $2n = 24$. *Throughout the range of the species except for S. part of the Balkan peninsula.*

(b) Subsp. **staticifolia** (Sm.) Chowdhuri, *Notes Roy. Bot. Gard. Edinb.* **22**: 255 (1957) (*S. staticifolia* Sm.; incl. *S. regis-ferdinandi* Degen & Urum.). Plant densely caespitose; flowering stems up to 40 cm. Basal leaves linear. Inflorescence very narrow, pseudo-racemose. Capsule shorter than carpophore. $2n = 24$. *Balkan peninsula northwards to C. Jugoslavia.*

31. S. chlorifolia Sm., *Pl. Icon. Ined.* **1**: t. 13 (1789). Glabrous and glaucous; stems 25–50 cm, robust, erect. Basal leaves broadly elliptical to spathulate, attenuate at base, acute or acuminate, thick; cauline broadly cordate, amplexicaul, acute. Inflorescence 1- to 5-flowered; flowers very large, on slender pedicels which are sometimes opposite in upper leaf-axils. Calyx (15–)18–25(–30) mm, rather coriaceous; teeth alternately acute and obtuse. Petals whitish; claw exserted; coronal scales oblong, acute. Capsule oblong, longer than carpophore. *Turkey-in-Europe (near Tekirdağ); one station in Jugoslavia*. ?Gr Ju Tu. (*S.W. Asia*.)

32. S. chlorantha (Willd.) Ehrh., *Beitr. Naturk.* **7**: 144 (1792). Glabrous and usually glaucous; stems 30–80 cm. Basal leaves lanceolate-spathulate; cauline small, linear, much shorter than the slender internodes. Inflorescence typically narrow, subsecund; branches usually 1- to 3-flowered; flowers inclined or horizontal. Calyx 9–12 mm, glaucous, not sulcate; teeth ovate, obtuse, with wide, ciliate, hyaline margin. Petals pale yellowish-green; lobes linear; claw exserted; coronal scales very small. Capsule *c.* 8 mm, 3–4 times as long as carpophore. Seeds less than 1 mm. $2n = 24$. *E. & C. Europe, westwards to* c. 13° *E. in Germany and southwards to S. Bulgaria*. Bu Ge Ju Po Rm Rs(B, C, W, K, E).

33. S. frivaldszkyana Hampe, *Flora (Regensb.)* **20**: 226 (1837). Like **32** but inflorescence-branches short, erect, always 1-flowered; calyx 12–16 mm, teeth ovate-lanceolate, acute; petals whitish; coronal scales very small or absent; capsule *c.* 10 mm, about twice as long as carpophore. $2n = 24$. ● *Balkan peninsula from* 40°30′ *to* 43° *N*. Al Bu ?Gr Ju Tu.

Large specimens of **32** and **33** may have more profusely-branched inflorescences, in which the branches themselves have the structure here described for the whole inflorescence.

34. S. multiflora (Waldst. & Kit.) Pers., *Syn. Pl.* **1**: 496 (1805) (incl. *S. steppicola* Kleopow, *S. syvashica* Kleopow). Stems 30–60 cm, simple, rough with minute, crisped hairs or glabrous, not viscid. Basal leaves thick, spathulate, very obtuse; upper cauline linear. Inflorescence narrowly paniculate to pseudo-verticillate; branches short, erect, opposite, 1- to 3(–7)-flowered. Calyx 12–15 mm, glabrous or sparsely hairy; teeth ovate-lanceolate, acute, margins hyaline, ciliate. Petals white; coronal scales absent. Capsule (7–)8–10 mm, equalling or somewhat longer than pubescent carpophore. $2n = 24$. *E. & C. Europe*. Au Cz Hu Ju Rm Rs(C, W, E).

35. S. viscosa (L.) Pers., *Syn. Pl.* **1**: 497 (1805) (*Melandrium viscosum* (L.) Čelak.). Robust biennial, sometimes perennial, up to 60 cm. Whole plant glandular-tomentose, viscid. Basal and lower cauline leaves ovate-lanceolate, acute, undulate. Inflorescence tall, narrow, pseudo-verticillate; branches short, erect, opposite, 1- to

3(–7)-flowered; flowers large. Calyx 14–24 mm; teeth ovate, obtuse. Petals white; coronal scales absent. Capsule 12–14 mm, 3–4 times as long as carpophore. $2n = 24$. *C. & E. Europe; extending northwards to* 60°30′ *N. in Finland.* Au ?Bu Cz Da Fe Hu Ju Po Rm Rs(B, C, W, K, E) Su [Ge].

36. S. radicosa Boiss. & Heldr. in Boiss., *Diagn. Pl. Or. Nov.* **1(6)**: 24 (1846). Caespitose, with thick, woody stock and numerous erect flowering stems 15–40(–60) cm; internodes very long, the lower rough with minute deflexed hairs, the upper glabrous. Basal leaves somewhat fleshy, spathulate, mucronate, margin strongly ciliate; upper cauline very small, linear. Inflorescence small, (1)2- to 5-flowered. Calyx 7–11 mm, coriaceous, somewhat sulcate, with wide, simple veins; teeth acute, patent, ciliate. Petals reddish, greenish or brownish-purple; coronal scales small. Capsule 6–9(–12) mm, slightly (or sometimes distinctly) exceeding calyx; carpophore *c.* 3 mm, pubescent. ● *Mountains of S. part of Balkan peninsula.* Al Gr Ju.

1 Calyx 7–8 mm; capsule infundibular (c) subsp. **pseudoradicosa**
1 Calyx 9–11 mm; capsule ovoid to subglobose
2 Petiole of basal leaves ciliate; capsule narrowly ovoid to ovoid (a) subsp. **radicosa**
2 Petiole of basal leaves villous; capsule broadly ovoid to subglobose (b) subsp. **rechingeri**

(a) Subsp. **radicosa**: Petiole of basal leaves ciliate. Calyx 9–11 mm, not abruptly contracted at base of capsule. Petal-limb light greenish-yellow above, greenish-brown beneath. Capsule 6–8(–12) mm, narrowly ovoid to ovoid. $2n = 24$. *Throughout the range of the species, except Evvoia.*

(b) Subsp. **rechingeri** Melzh., *Bot. Jahrb.* **98**: 31 (1977): Petiole of basal leaves villous. Calyx 9–11 mm, abruptly contracted at base of capsule. Petal-limb brownish-purple. Capsule 6·5–8 mm, broadly ovoid to subglobose. $2n = 24$. *Usually on serpentine. N.W. Greece.*

(c) Subsp. **pseudoradicosa** Rech. fil., *Bot. Jahrb.* **80**: 320 (1961): Petiole of basal leaves shortly ciliate. Calyx 7–8 mm. Petal-limb greenish-white above, brownish-purple beneath. Capsule 7–9 mm, truncately infundibular, distinctly exceeding calyx. $2n = 24$. *E. Greece (Evvoia).*

37. S. oligantha Boiss. & Heldr. in Boiss., *Diagn. Pl. Or. Nov.* **3(1)**: 75 (1854). Like **36** but basal leaves broadly spathulate, densely and shortly pubescent; inflorescence 4- to 9-flowered, somewhat secund; flowers nodding, fragrant; calyx 7–8·5 mm; capsule 4·5–6 mm; carpophore 2–3 mm, glabrous. *Rocky places in openings of forest.* $2n = 24$. ● *N. Greece (Olimbos).* Gr.

38. S. reichenbachii Vis., *Fl. Dalm.* **3**: 169 (1850). Like **36** but basal leaves oblanceolate-spathulate, acute; upper cauline larger, linear-lanceolate; inflorescence 5- to 20-flowered, with 1- to 3-flowered branches; petals white. $2n = 24$. ● *W. Jugoslavia.* Ju.

39. S. skorpilii Velen., *Sitz.-Ber. Böhm. Ges. Wiss.* **1**: 39 (1890). Robust, 30–50 cm, with somewhat woody stock. Stem and leaves densely clothed with very short, stiff hairs. Basal leaves spathulate, withered at time of flowering; cauline up to 4 × 1·5 cm, ovate, more or less sessile, with cuneate base. Inflorescence usually narrow, (5–)7- to 13-flowered; upper flowers more or less aggregated; lower in more distant opposite pairs. Calyx 10–12 mm, clavate, glabrous, viscid; veins distinct, simple. Petals narrow, greenish. Capsule 7–9 mm, 2–3 times as long as carpophore. $2n = 24$. ● *Balkan peninsula from C. Bulgaria to C. Greece.* Bu Gr Tu.

40. S. tatarica (L.) Pers., *Syn. Pl.* **1**: 497 (1805). Stems 30–60 cm, with sparse patent hairs below, almost glabrous and not viscid above; often with short, leafy, non-flowering branches below. Leaves lanceolate or lanceolate-spathulate, acute, the basal smaller and withering early. Inflorescence narrow, with 1- to 3-flowered branches; flowers inclined; pedicels erect, slightly hairy. Calyx 10–13 mm, glabrous; teeth ovate-lanceolate, acute; veins distinct. Petals white, cream or greenish-white; lobes linear; coronal scales absent. Capsule 8–10 mm, 3–4 times as long as the slightly pubescent carpophore. *E. & E.C. Europe, northwards from* 47°30′ *N. in Ukraine, and extending to arctic Norway.* Fe Ge No Po Rs(N, B, C, W, E).

Sect. GRAMINIFOLIAE Chowdhuri. Small, narrow-leaved perennials with few, linear cauline leaves. Flowers few, hermaphrodite. Calyx glabrous, usually somewhat inflated. Styles 3. Capsule with basal septa.

41. S. paucifolia Ledeb., *Fl. Ross.* **1**: 306 (1842). Plant dwarf, subglabrous; stems 5–15 cm, few-flowered, scarcely viscid. Basal leaves few, sparsely hairy, oblong-linear; cauline linear. Calyx 6–9 × 4–6 mm, inflated, glabrous. Petal-limb ovate, white; claw glabrous; coronal scales very small or absent. $2n = 24$. *N.E. Russia.* Rs(N).

42. S. graminifolia Otth in DC., *Prodr.* **1**: 368 (1824). Like **41** but stems taller, more leafy, viscid above; flowers smaller; calyx 3–4 mm wide; petal-claw hairy. *N.E. Russia (near Pechora).* Rs(N). *(Siberia.)*

Sect. GASTROLYCHNIS (Fenzl) Chowdhuri. Small, narrow-leaved perennials. Flowers few, hermaphrodite. Styles 5. Calyx pubescent. Capsule without basal septa, dehiscing with 5 more or less bifid teeth.

43. S. uralensis (Rupr.) Bocquet, *Candollea* **22**: 26 (1967) (*S. wahlbergella* Chowdhuri, *Gastrolychnis uralensis* Rupr.). Stem simple, erect, pubescent above. Flowers usually solitary, nodding at anthesis. Calyx 14–18 mm, whitish, inflated. Petals not or only slightly exserted, usually dull reddish-purple. Seeds with prominent, swollen wings. *N. Europe southwards to S. Norway and C. Ural.* Fe No Rs(N, C) Sb Su.

(a) Subsp. **uralensis**: Petals distinctly exserted from calyx. $2n = 24$. *Spitsbergen and N.E. Russia. (Arctic Asia and North America.)*

(b) Subsp. **apetala** (L.) Bocquet, *Candollea* **22**: 26 (1967) (*Lychnis apetala* L., *Melandrium apetalum* (L.) Fenzl): Petals not exserted from calyx. $2n = 24$. ● *N. & W. Fennoscandia.*

44. S. furcata Rafin., *Autikon Bot.* **1**: 28 (1840) (*Lychnis affinis* J. Vahl ex Fries). Like **43** but stem often branched above, glandular, viscid; flowers smaller, erect; calyx 10–12 mm, scarcely inflated; petals whitish, exserted; seeds with narrow wings. *Arctic Europe.* Fe No Rs(N) Sb Su.

(a) Subsp. **furcata**: Stem with 1(2) pair of cauline leaves, much shorter than the basal. Calyx broadly campanulate, somewhat inflated in fruit. $2n = 48$. *Spitsbergen and N.E. Russia. (Arctic Asia.)*

(b) Subsp. **angustiflora** (Rupr.) Walters, *Feddes Repert.* **69**: 46 (1964) (*Wahlbergella angustiflora* Rupr.): Stem with 2 or 3 pairs of cauline leaves, the lower about equalling the basal. Calyx narrowly campanulate, not inflated in fruit. $2n = 48$. *Throughout the range of the species except Spitsbergen.*

Sect. OTITES Otth (incl. Subsect. *Sibiricae* Schischkin ex Chowdhuri). Biennials or perennials; often dioecious. Flowers small, usually unisexual, in dichasia aggregated in compound racemose panicles, often narrow with crowded verticillasters. Calyx less than 7 mm. Petals entire or shallowly bifid; coronal scales absent. Carpophore not more than 3 mm, sometimes almost absent. Seeds 0·7–1·4 mm, reniform.

Petal-colour and flower-scent provide useful characters in fresh material, and are particularly helpful in distinguishing **54** and **61**.

45. S. sibirica (L.) Pers., *Syn. Pl.* **1**: 497 (1805). Subglabrous perennial 40–60 cm; monoecious. Basal leaves oblong-lanceolate or oblong-linear; cauline leaves linear-lanceolate, in fascicles. Inflorescence narrow, interrupted, with long internodes separating many-flowered verticillasters; flowers mostly unisexual. Calyx 4–5 mm. Petals greenish, narrow, entire or emarginate. Capsule *c.* 6 mm, ovoid, 2–3 times as long as carpophore. Seeds reniform, with plane face and grooved back. *S.E. Russia, Ukraine.* Rs(C, W, E).

46. S. roemeri Friv., *Flora* (*Regensb.*) **19**: 439 (1836) (*Otites roemeri* (Friv.) J. Holub). Densely puberulent perennial 20–50 cm. Basal leaves oblong-spathulate, acute, attenuate into a long, ciliate petiole; cauline lanceolate to linear. Inflorescence subcapitate in small specimens but often, in larger plants, more or less interrupted with rather dense verticillasters. Calyx 3–6 mm, subglabrous; teeth ovate. Petal-limb white, broadly spathulate, shallowly bifid, with oblong lobes. Capsule 4·5–7 mm, broadly ovoid, 2–5 times as long as carpophore. Seeds *c.* 1 mm, reniform, with plane face and grooved back. $2n = 24$. ● *Mountains of Balkan peninsula and C. & S. Italy.* Al Bu Gr It Ju.

Apparently gynodioecious, with female and hermaphrodite plants, the latter with both male and hermaphrodite flowers.

47. S. sendtneri Boiss., *Fl. Or.* **1**: 608 (1867) (*Otites sendtneri* (Boiss.) J. Holub; incl. *S. velenovskyana* Jordanov & Panov). Like **46** but with shorter calyx, entire or only slightly emarginate petals, and more often capitate inflorescence. $2n = 24$. *Subalpine meadows.* ● *W. part of Balkan peninsula.* Al Bu Ju.

Apparently almost completely dioecious.

48. S. ventricosa Adamović, *Österr. Bot. Zeitschr.* **55**: 180 (1905) (*Otites ventricosa* (Adamović) J. Holub). Like **46** but dioecious, with more or less procumbent non-flowering shoots and short, erect flowering stems up to 30 cm; inflorescence subcapitate; calyx 4–5 × 3·5–4 mm, inflated, more or less puberulent, often purplish. ● *Mountains of S. Albania and S. Jugoslavia.* Al ?Gr Ju.

Species **46–48** are very similar and might be treated as subspecies of a single species, but there is insufficient information available.

49. S. borysthenica (Gruner) Walters, *Feddes Repert.* **69**: 47 (1964) (*S. otites* var. *borysthenica* Gruner, *Otites borysthenica* (Gruner) Klokov, *S. parviflora* (Ehrh.) Pers., non Moench, *S. ehrhartiana* Soó). Dioecious biennial (?rarely perennial); stems up to 80 cm, without viscid internodes; whole plant pubescent. Basal leaves narrowly spathulate, often withered at time of flowering; cauline linear-oblanceolate. Inflorescence usually a narrow panicle; flowers very numerous, small; hairs on pedicels short, dense, forwardly-directed and curled. Calyx 2–3 mm. Petals linear, greenish, entire; claw ciliate; filaments hairy. Capsule 2–3 mm, subglobose; carpophore very short. Seeds *c.* 0·75 mm, smooth. $2n = 24$. *Dry places, particularly in open coniferous woodland. C. & E. Europe.* Bu Cz Hu Ju Po Rm Rs(B, C, W, K, E) [Rs(N)].

50. S. media (Litv.) Kleopow, *Bull. Jard. Bot. Kieff* **9**: 64 (1929). Like **49** but stem puberulent below, glabrous above, without viscid internodes; leaves sparsely puberulent; cauline leaves linear, acute; pedicels glabrous; calyx glabrous, often only 5-veined; calyx of female flowers 1·5–2·5 mm. *S.E. Russia, S. Ukraine.* Rs(W, E). (*C. Asia.*)

51. S. hellmannii Claus, *Beitr. Pfl. Russ. Reich.* **8**: 289 (1851) (incl. *S. krymensis* Kleopow, *Otites graniticola* Klokov). Dioecious; stems up to 60 cm, with viscid internodes; whole plant, including pedicels and calyx, clothed with short, dense, more or less straight, deflexed hairs. Basal leaves broadly spathulate; cauline oblanceolate-spathulate. Inflorescence usually a narrow panicle of relatively few flowers. Calyx 3·5–5 mm. Petals linear, greenish, entire; claw ciliate; filaments hairy. Capsule 7–9 mm, ovoid; carpophore very short; seeds more than 1 mm, with undulate surface. ● *S.E. Russia, S. & E. Ukraine.* Rs(W, K, E).

52. S. cyri Schischkin in Grossh. *et al.*, *Trav. Mus. Géorgie* **3**: 202 (1925). Like **51** but densely puberulent below, glabrous above, including pedicels and calyx; seeds smooth. *S.E. Russia* (*N.W. shores of Caspian Sea*). Rs(E). (*W. Asia.*)

51 and **52** are very similar in appearance, but seem to be distinct species with different distributions and have been shown experimentally to exhibit intersterility.

53. S. velebitica (Degen) Wrigley, *Ann. Bot. Fenn.* **23**: 70 (1986) (*S. otites* subsp. *velebitica* Degen, *Otites velebitica* (Degen) J. Holub). Dioecious; stems 45–70(–85) cm, slender, shortly and densely puberulent below, without viscid internodes. Basal leaves narrowly oblanceolate; cauline leaves linear-lanceolate. Inflorescence subsimple, glabrous. Calyx 3 mm. Petals pale yellow; claw and filaments glabrous. Capsule 5–6 mm, subglobose; carpophore very short; seeds *c.* 1·2 mm, smooth with tuberculate margin. *Rocky grassland on limestone.* ● *N.W. Jugoslavia.* Ju.

Readily distinguished from the other members of Sect. *Otites* by its slender habit.

54. S. wolgensis (Hornem.) Otth in DC., *Prodr.* **1**: 370 (1824) (incl. *Otites orae-syvashicae* Klokov). Dioecious; stems often 100 cm or more, with very diffuse panicle of small, white flowers; lower part of stem with hairs of different lengths, 0·25–1·15 mm; glabrous and viscid above. Leaves pilose, oblanceolate-spathulate. Calyx 3–3·5 mm. Petal-claw glabrous; filaments hairy. Capsule 5–6 mm; carpophore very short. Seeds less than 1 mm, surface undulate. $2n = 24$. *S. Russia, S.E. Ukraine.* Rs(C, W, K, E).

The white petals, unpleasant flower-scent and wide-angled lower panicle-branches provide good characters in living material. The complete intersterility with **56** emphasizes the relationship of this species with species **49–53** rather than with **56–61**.

55. S. baschkirorum Janisch., *Ber. Saratow. Naturforscherges.* **3(1)**: 33 (1929) (incl. *S. polaris* Kleopow). Like **54** but stems usually 40–60 cm, puberulent below without long hairs, glabrous and viscid above; leaves puberulent; flowers with pleasant scent; petal-claw ciliate; seeds *c.* 1·25 mm, tuberculate. $2n = 24$. *E. Russia.* Rs(N, C, E).

(56–61). S. otites group. Dioecious biennials or short-lived perennials, possessing the general habit of **49–55** but with glabrous pedicels, petal-claws, filaments and calyx. Stems hairy below, glabrous and viscid above. Petals yellow. Capsule 4–7 mm; carpophore very short, up to 1 mm. Very variable in habit, inflorescence-form and pubescence. This group is characteristic of steppe communities and ruderal habitats in S.E. Europe.

1 Hairs on lower part of stem more than 1 mm; tall, leafy plants (up to 200 cm)
  2 Petals pale yellow; seeds more than 1 mm, ±smooth; inflorescence acuminate **60. exaltata**
  2 Petals dark yellow; seeds less than 1 mm, tuberculate; inflorescence rhombic **61. chersonensis**
1 Hairs on lower part of stem less than 1 mm; plants of variable height
  3 Rosette leaves conspicuously undulate, broadly spathulate, grey-green with soft pubescence
    4 Hairs on lower stem up to 0·9 mm; dichasia dense, ±sessile **59. densiflora**
    4 Hairs on lower stem up to 0·4 mm; dichasia not dense, stalked **58. colpophylla**
  3 Rosette leaves not undulate, oblanceolate-spathulate or narrowly spathulate, puberulent
    5 Lower ½ of stem with few leaves; 1–4 middle internodes slightly viscid; dichasia not dense, stalked **56. otites**
    5 Lower ½ of stem leafy in axils; 3–8 middle internodes very viscid; dichasia dense, ±sessile **57. donetzica**

56. S. otites (L.) Wibel, *Prim. Fl. Werthem.* 241 (1799). Stems 10–50(–140) cm. Basal leaves narrowly spathulate; cauline leaves few and small with little or no development of axillary leafy shoots. Calyx 3–4 mm, ovoid, glabrous; teeth obtuse. Petal-limb inconspicuous, greenish, linear. Capsule (3·5–)4–5(–5·5) mm, ovoid to subglobose, exceeding the calyx; carpophore usually *c.* 1 mm. *S.C. & W. Europe, from C. Spain eastwards to Lithuania and Bulgaria.* Al Au Br Bu Cz Da Ga Ge Gr He Ho Hs Hu It Ju Po ?Rm Rs(B) Tu.

A very variable species. At least two subspecies can be recognized.

(a) Subsp. **otites**: Stems usually less than 50 cm. Seeds smooth. $2n = 24$. *Almost throughout the range of the species.*

(b) Subsp. **hungarica** Wrigley, *Ann. Bot. Fenn.* **23**: 74 (1986): Stems usually at least 50 cm. Seeds tuberculate. • *E.C. Europe.*

These subspecies are linked by intermediate populations; subsp. (b) may have arisen through introgressive hybridization with **57, 61** and perhaps **60**.

S. × pseudotites Besser ex Reichenb., *Fl. Germ. Excurs.* 819 (1832), a tall plant with broadly spathulate basal leaves with undulate margins, spreading panicles and larger, subglobose capsules, occurs in N. Italy and S.E. France. This plant has sometimes been included in **56**, but has been shown to be a hybrid between **56** and **58** (*vide* F. Wrigley, *Ann. Bot. Fenn.* **23**: 69–81 (1986)).

57. S. donetzica Kleopow, *Jour. Inst. Bot. Acad. Sci. Ukr.* **9**: 116 (1936) (*S. sillingeri* Hendrych). Like **56** but stems robust, 90–150(–170) cm, with dense, short pubescence (up to 0·1 mm); internodes 4–8, very viscid; basal leaves oblanceolate-spathulate; cauline leaves numerous, with leafy shoots developed in the axils; inflorescence very dense with condensed verticillasters; ultimate dichasia with very short peduncles; flowers often subsessile; seeds larger, tuberculate. • *C. & E. Europe from E. Czechoslovakia to E. Ukraine.* Cz Hu Rm Rs(C, W, E).

58. S. colpophylla Wrigley, *Ann. Bot. Fenn.* **23**: 77 (1986). Stems usually 100–150 cm, pubescent below, glabrous above. Basal leaves broadly spathulate, with undulate margins, densely pubescent; cauline leaves few, with little development of axillary leafy shoots. Inflorescence somewhat lax, with distinctly stalked dichasia. Calyx 3·5–4·5 mm, cylindrical-campanulate, glabrous. Petals yellowish. Capsule 6–6·5 mm, ovoid; carpophore *c.* 0·7 mm. Seeds large, smooth, with tuberculate margin. • *S.E. France (Provence).* Ga.

59. S. densiflora D'Urv., *Mém. Soc. Linn. Paris* **1**: 303 (1822) (*Otites dolichocarpa* Klokov). Like **58** but smaller (up to 110 cm), the stem pilose below, with 2–7 very viscid internodes; inflorescence with more or less condensed verticillasters; ultimate dichasia with very short peduncles, and flowers often subsessile. *S. Ukraine.* Rs(W, K, E).

60. S. exaltata Friv., *Flora (Regensb.)* **18**: 333 (1835) (*Otites exaltata* (Friv.) J. Holub). Stems up to 200 cm, with hairs more than 1·5 mm in the lower part. Cauline leaves many, oblanceolate, becoming revolute, with leafy shoots developed in the axils. Inflorescence lax, acuminate, with subsessile dichasia. Flowers somewhat fragrant; petals pale yellow. Calyx 3–3·5 mm. Capsule *c.* 5·5 mm; carpophore *c.* 1 mm. Seeds *c.* 1·3 mm, more or less smooth. • *S.E. Europe.* Bu Gr Ju Rm.

61. S. chersonensis (Zapał.) Kleopow, *Bull. Jard. Bot. Kieff* **9**: 9 (1929) (incl. *Otites moldavica* Klokov). Like **60** but shorter (up to 150 cm); cauline leaves patent or somewhat deflexed; inflorescence not acuminate, the dichasia stalked; flowers fragrant; petals dark yellow; capsule *c.* 4·8 mm; seeds *c.* 0·8 mm, tuberculate. • *S.E. Europe.* Rm Rs(C, W, K, E).

The hybrid between **57** and **61** occurs in parts of Romania and along the River Dnester.

Sect. ODONTOPETALAE Schischkin ex Chowdhuri. Perennials with woody stock; flowering stems lateral in axils of basal leaves. Flowers large, solitary or in few-flowered dichasia. Calyx more or less inflated, hairy, 10-veined. Petals with bifid limb toothed near base, and auriculate claw; coronal scales present. Capsule unilocular, without basal septa.

62. S. auriculata Sm. in Sibth. & Sm., *Fl. Graec. Prodr.* **1**: 301 (1809) (incl. *S. lanuginosa* Bertol.). Caespitose, with short, branched, woody stock and robust, 1- to few-flowered, pubescent stems up to 30 cm. Basal leaves up to 12 cm, lanceolate, acute, thick, with roughly punctate surface, densely ciliate with undulate hairs 1–2 mm; cauline leaves 1–3 pairs, small, remote. Calyx 12–15 mm, densely glandular-hairy; teeth triangular, acute. Petals white or cream; limb 5–8 mm; claw exserted. Capsule *c.* 3 times as long as carpophore, ovoid, enclosed within calyx. $2n = 24$. *Crevices of limestone rocks.* • *Mountains of Italy (Alpi Apuane) and Greece.* Gr It.

63. S. zawadzkii Herbich in Zawadzki, *Enum. Pl. Galic. Bucow.* 191 (1835). Like **62** but basal leaves usually wider, elliptical; cilia shorter; flowers larger; calyx 15–17 mm, hairy, but not or slightly glandular; petal-limb 10 mm or more. $2n = 24$. • *E. Carpathians.* Rm Rs(W).

64. S. elisabethae Jan, *Flora (Regensb.)* **15**: 177 (1832) (*Meland-*

rium elisabethae (Jan) Rohrb.). Like **62** but basal leaves glabrous or only sparsely ciliate; cauline leaves 3–5 pairs; flowers very large, often solitary; calyx up to 20 mm, densely glandular-hairy; petal-limb *c.* 15 mm, dark red or reddish-purple; coronal scales laciniate; carpophore very short. *Calcareous rocks and screes,* 1500–2500 *m.* 2*n* = 24. ● *S. Alps, between* 9° *and* 11° *E.* It.

65. S. requienii Otth in DC., *Prodr.* **1**: 381 (1824). Like **62** but basal leaves wider, subspathulate, not roughly punctate; inflorescence several-flowered; calyx up to 30 mm, cylindrical-clavate; capsule oblong-acuminate, only slightly longer than carpophore. *Mountain rocks.* ● *Corse.* Co.

Sect. CORDIFOLIAE Chowdhuri. Caespitose, glandular-hairy perennials with woody stock and numerous ovate or cordate-ovate acuminate leaves. Inflorescence a few-flowered dichasium. Calyx glandular, somewhat inflated.

66. S. cordifolia All., *Fl. Pedem.* **2**: 82 (1785). Stems up to 20 cm. Basal leaves ovate, attenuate at base, withering early; cauline leaves larger, cordate-ovate, acuminate. Inflorescence of 1–4 erect flowers. Calyx 12–15 mm, glandular-pubescent, slightly inflated, with linear-lanceolate, acuminate teeth. Petal-limb bifid, white or pink; claw not auriculate; coronal scales small, obtuse. Capsule 8–10 mm, oblong, twice as long as glabrous carpophore. 2*n* = 24. ● *Maritime Alps.* Ga It.

67. S. foetida Link ex Sprengel, *Syst. Veg.* ed. 16, **2**: 406 (1825) (*S. acutifolia* Link ex Rohrb., *S. melandrioides* Lange). Like **66** but cauline leaves ovate-lanceolate; stems usually 3-flowered; petals pink with rather wide, emarginate limb, short auriculate claw and prominent coronal scales. 2*n* = 24. ● *Mountains of N. Portugal and N.W. Spain.* Hs Lu.

68. S. herminii (Welw. ex Rouy) Welw. ex Rouy, *Ill. Pl. Eur. Rar.* **4**: 17 (1895) (*S. foetida* auct., non Link, *S. macrorhiza* Gay & Durieu ex Lacaita). Like **66** but stock very woody; stems up to 30 cm, more or less procumbent; leaves more densely pubescent; calyx 22–30 mm, obviously inflated; petal-limb white or pale pink, deeply bifid; claw long, not auriculate; capsule *c.* 15 mm. ● *Mountains of Portugal and N.W. Spain.* Hs Lu.

Sect. INFLATAE (Boiss.) Chowdhuri. Usually perennial. Flowers solitary or in dichasia, sometimes many-flowered. Calyx 10- or 20-veined, with conspicuous reticulate venation, often markedly inflated. Petals with imbricate aestivation.

69. S. vulgaris (Moench) Garcke, *Fl. Nord- Mittel-Deutschl.* ed. 9, 64 (1869) (*S. cucubalus* Wibel, *S. inflata* Sm., *S. latifolia* (Miller) Britten & Rendle, non Poiret, *S. venosa* Ascherson). Perennial, sometimes caespitose and woody at base; stems up to 60 cm, erect, usually branched, glabrous or pubescent, often glaucous. Leaves up to 12 cm, variable in shape from broadly ovate to linear. Inflorescence with numerous flowers; bracts usually scarious. Flowers inclined, somewhat zygomorphic (especially the functionally female ones). Calyx 20-veined, inflated, persistent, loosely investing ripe capsule. Petals large, deeply bifid, usually whitish, with poorly-developed coronal scales. Ripe capsule up to 13 mm, with neck 2–3 mm wide and erect or patent teeth. Seeds 1–2 mm, usually tuberculate. *Throughout Europe except for some northern islands.* All except Fa Is Sb.

Extremely variable. The taxa here included in **69** and **70** have been variously treated by different authors. The treatment here takes into account the large body of experimental and other data summarized in E. M. Marsden-Jones & W. B. Turrill, *Bladder Campions* (1957) and in D. Aeschimann, *Candollea* **40**: 57–98 (1985).

1 Capsule usually more than 10 mm; petals usually pink or greenish
  2 Plant with long stolons (**d**) subsp. **macrocarpa**
  2 Plant suffruticose, without stolons (**e**) subsp. **suffrutescens**
1 Capsule up to *c.* 10 mm; petals usually white
  3 Lower cauline leaves broadly oblong-ovate to ovate-elliptical, often obtuse and apiculate (**c**) subsp. **commutata**
  3 Lower cauline leaves ovate-lanceolate to linear
    4 Lower cauline leaves ovate-lanceolate (**a**) subsp. **vulgaris**
    4 Lower cauline leaves linear-lanceolate to linear (**b**) subsp. **angustifolia**

(**a**) Subsp. **vulgaris**: Stems up to 60 cm, ascending or erect. Leaves 3–12 × 1–2·5 cm, glabrous or rarely pubescent, ovate-lanceolate. Capsule *c.* 10 mm; carpophore 2–3 mm. 2*n* = 24. *Throughout Europe except for some of the islands.*

(**b**) Subsp. **angustifolia** Hayek, *Prodr. Fl. Penins. Balcan.* **1**: 256 (1924) (*S. tenoreana* Colla): Like subsp. (**a**) but leaves 3–8 × 0·3–1 cm, linear-lanceolate or linear. *Usually on coastal sands and rocks. Mediterranean region, Portugal.*

(**c**) Subsp. **commutata** (Guss.) Hayek, *op. cit.* 258 (1924) (*S. commutata* Guss.; incl. subsp. *antelopum* (Vest) Hayek, subsp. *bosniaca* (G. Beck) Janchen): Like subsp. (**a**) but leaves 4–9 × 1·5–4 cm, glabrous or pubescent, broadly oblong-ovate to ovate-elliptical, often obtuse and apiculate. 2*n* = 48. *S. & S.C. Europe.*

(**d**) Subsp. **macrocarpa** Turrill, *Hook. Ic.* **36**: t. 3551 (1956) (*S. angustifolia* var. *carneiflora* sensu Clapham): Plant with long stolons; stems up to 50 cm. Leaves 4–7 × 0·7–1·6 cm, narrowly lanceolate. Flowers pink or greenish. Capsule 10–13 mm; carpophore 2–4 mm; seeds large, 1·6–2·1 mm. 2*n* = 48. *Usually a weed of cultivation. Mediterranean region. Introduced and long established in S.W. England (Plymouth).*

(**e**) Subsp. **suffrutescens** Greuter, Matthäs & Risse, *Willdenowia* **14**: 34 (1984): Like subsp. (**d**) but woody at base, without stolons; capsule up to 11 mm. 2*n* = 24. *Cliffs.* ● *Kriti; one station in S.E. Greece.*

Subsp. **aetnensis** (Strobl) Pignatti, *Gior. Bot. Ital.* **107**: 208 (1973), from Sicilia (W. slopes of Etna), is like subsp. (**c**) but has spathulate, acute, rather fleshy leaves and an indistinctly veined calyx *c.* 10 mm. It has not been recorded recently.

70. S. uniflora Roth, *Ann. Bot.* (*Usteri*) **10**: 46 (1794) (*S. maritima* With.). Like **69** but usually procumbent to decumbent-ascending; leaves often fleshy; inflorescence 1- to few-flowered; bracts usually herbaceous; coronal scales usually well developed; ripe capsule with neck 3–4 mm wide and patent or deflexed teeth, and seeds usually without tubercles. ● *Coasts of W. & N.W. Europe; mountains of C. & S. Europe.* ?Al Au Az Br Co Cz Da Fe Ga Ge Hb He Hs Is It Ju Lu No Po Rm Rs(N) Sa Si Su.

1 Flowers large, often solitary, ± erect and actinomorphic
  2 Leaves linear-lanceolate to subspathulate, glaucous, often not crowded (**a**) subsp. **uniflora**
  2 Leaves broadly elliptical to spathulate, not or only slightly glaucous, rather crowded (**b**) subsp. **thorei**
1 Flowers smaller, usually inclined and slightly zygomorphic, in (1)2–7-flowered inflorescences

3 Leaves lanceolate, glabrous; inflorescence often 5- to 7-flowered (c) subsp. **glareosa**
3 Leaves ovate to ovate-lanceolate, with short, dense pubescence; inflorescence 1- to 3(–5)-flowered (d) subsp. **prostrata**

(**a**) Subsp. **uniflora** (incl. subsp. *cratericola* (Franco) Franco): Glabrous, glaucous plant, often with rather diffuse habit. Leaves 0·5–2(–4) × 0·2–0·7 cm, linear-lanceolate to subspathulate, often not crowded. Flowers large, actinomorphic, often solitary. $2n$ = 24. *Usually on coastal rocks or shingle. N. & W. Europe, eastwards to the Kola peninsula.*

(**b**) Subsp. **thorei** (Dufour) Jalas, *Willdenowia* **14**: 48 (1984) (*S. thorei* Dufour): Like subsp. (**a**) but stems much-branched and usually embedded in sand below; leaves 0·5–2 × 0·3–1·2 cm, broadly elliptical to subspathulate, not or only slightly glaucous, rather crowded. $2n$ = 24. *Coastal sands. N. Spain and W. France, northwards to* 46°30′ *N.*

(**c**) Subsp. **glareosa** (Jordan) Chater & Walters, *Bot. Jour. Linn. Soc.* **103**: 216 (1990) (*S. glareosa* Jordan): Glabrous plant with slender, procumbent to ascending stems from a slightly woody stock. Leaves 1–3 × 0·2–0·6 cm, lanceolate. Flowers smaller than subsp. (**a**), rarely solitary, usually in 3- to 7-flowered inflorescences, inclined and slightly zygomorphic. $2n$ = 24. *Usually on calcareous screes. From the Pyrenees and E.C. France to the E. Carpathians.*

(**d**) Subsp. **prostrata** (Gaudin) Chater & Walters, *op. cit.* **103**: 215 (1990) (*S. alpina* auct., *S. angustifolia* subsp. *prostrata* (Gaudin) Briq., *S. inflata* subsp. *prostrata* Gaudin, *S. vulgaris* subsp. *marginata* (Kit.) Hayek and subsp. *megalosperma* (Sart.) Hayek, *S. willdenovii* Sweet): Like subsp. (**c**) but stems shorter; leaves 0·5–2 × 0·5–1·5 cm, ovate to ovate-lanceolate, covered in short, dense pubescence; flowers often solitary, sometimes in 2- or 3-flowered inflorescences. $2n$ = 24. *Mountain rocks. S. Europe, from N. Spain to E. Greece.*

Icelandic plants, often subcaespitose with small, imbricate, ovate-lanceolate to subspathulate leaves, have been distinguished as **S. maritima** With. subsp. **islandica** Á. & D. Löve, *Acta Horti Gothob.* **20**: 183 (1956). They seem best treated as local variants of subsp. (**a**).

On Öland and Gotland small-flowered, procumbent plants with few-flowered inflorescences have been distinguished as *S. maritima* With. var. *petraea* Fries. Such plants approach subsp. (**c**) and could be included in this subspecies, but the populations are apparently very variable and further study is required. See R. Sterner, *Acta Phytogeogr. Suec.* **9**: 97 (1938).

71. S. csereii Baumg., *Enum. Stirp. Transs.* **3**: 345 (1816). Glabrous annual to 65 cm, with few, spathulate basal leaves, and numerous pairs of ovate-lanceolate, sessile cauline leaves. Inflorescence variably composed, usually with several flowers. Calyx 9–13 mm, somewhat inflated in flower but closely investing ripe fruit, thin and scarious, with 20 green veins of which the alternate ones are shorter. Petal-limb not more than 5 mm, white, with 2 spathulate lobes; corona represented only by 2 small protuberances. Capsule 7·5–10 mm, ovate; carpophore *c.* 1·5 mm. Seeds with concave face. $2n$ = 24. *S.E. Europe; occasionally casual elsewhere.* Bu Rm Rs(W, K, E).

72. S. fabaria (L.) Sm. in Sibth. & Sm., *Fl. Graec. Prodr.* **1**: 293 (1809) (incl. *S. thebana* Orph. ex Boiss.). Robust, glabrous and glaucous perennial up to 100 cm. Leaves fleshy; basal more or less spathulate-ovate; cauline distant, smaller, elliptical or obovate-lanceolate, obtuse, often mucronate. Inflorescence usually branched. Calyx 8–10 mm, 10-veined, not markedly inflated; teeth triangular, acute. Petal-limb 2·5–5 mm, white; coronal scales present. Capsule 10–12 mm, ovoid, exceeding the closely appressed calyx; carpophore *c.* 2 mm. Seeds with concave face. $2n$ = 24. *Usually on maritime rocks. C. Greece and Aegean region.* Cr Gr.

V. Melzheimer, *Pl. Syst. Evol.* **155**: 251–256 (1987), has treated *S. thebana* as a subspecies of **72**, but the cited differences, mainly in coronal development, seem to be an inadequate basis for distinction.

73. S. ionica Halácsy, *Consp. Fl. Graec.* **1**: 158 (1900). Like **72** but a low-growing plant with ascending flowering stems, and often with overwintering stolons terminating in rosettes of green leaves; basal leaves spathulate-orbicular; petal-limb *c.* 2 mm, with no coronal scales; capsule 6–8·5 mm. $2n$ = 24. *Stony mountain slopes.* ● *W.C. Greece.*

74. S. fabarioides Hausskn., *Mitt. Thür. Bot. Ver.* nov. ser. **5**: 47 (1893). Like **72** but biennial, up to 60 cm; stems leafless in upper ½; calyx (9–)10–14 mm, markedly inflated; ripe capsule 6–9 mm, not exceeding the calyx. ● *Balkan peninsula.* Al Bu Gr.

Earlier described as perennial, but *fide* V. Melzheimer, *Bot. Jahrb.* **101**: 174–178 (1980), strictly biennial, and clearly separable from **72**.

75. S. caesia Sm. in Sibth. & Sm., *Fl. Graec. Prodr.* **1**: 294 (1909). Glabrous, rhizomatous perennial with ascending flowering stems up to 40 cm, and variably-developed sterile shoots often submerged in scree. Leaves fleshy, glaucous, the basal spathulate, the cauline elliptical to ovate. Inflorescence few-flowered; calyx 10–13 mm, not strongly inflated, with 10 veins very conspicuous in fruiting stage; petal-limb 2–6 mm, white or pink, often red beneath, with 2 linear lobes and small coronal scales. Capsule 6–9 mm, ovoid; carpophore *c.* 2 mm. *Mountain screes. W. Greece and S. Albania; N. Aegean region (Samothraki).* Al Gr.

(**a**) Subsp. **caesia**: Petal-limb 4–6 mm; capsule exceeding the calyx. $2n$ = 24. ● *Greece and S. Albania.*

(**b**) Subsp. **samothracica** (Rech. fil.) Melzh., *Bot. Jahrb.* **101**: 181 (1980): Petal-limb *c.* 2 mm; capsule equalling the calyx. *N. Aegean region (Samothraki). (E. Aegean (Chios).)*

76. S. variegata (Desf.) Boiss. & Heldr. in Boiss., *Diagn. Pl. Or. Nov.* **2(8)**: 82 (1849). Like **75** but smaller, with stems up to 10 cm; calyx 11–15 mm; petals greyish-violet; ripe capsule 8–11 mm, ovate-lanceolate; carpophore *c.* 3 mm. $2n$ = 24. *Mountain screes.* ● *Kriti.* Cr.

77. S. procumbens Murray, *Comment. Gotting.* ser. 2, 7: 83 (1786). Stems up to 30 cm, procumbent, branched, scabrid-puberulent. Leaves oblong-lanceolate, acute, subglabrous. Flowers solitary or 2 or 3 together; pedicels equalling calyx. Calyx 15–20 mm, hairy, inflated, oblong-campanulate, 20-veined, with anastomosing lateral veins. Petals white, cuneate, emarginate; coronal scales acute. Capsule 5–6 mm, ovoid-globose, *c.* 3 times as long as the glabrous carpophore. *S., C. & E. Russia, E. Ukraine.* Rs(C, W, E).

Sect. BRACHYPODAE (Boiss.) Chowdhuri. Perennials, sometimes suffruticose, with lanceolate to spathulate basal leaves and few-flowered stems. Capsule included in the hairy calyx, usually much longer than the short carpophore.

78. S. flavescens Waldst. & Kit., *Pl. Rar. Hung.* **2**: 191 (1804) (*S. subcorymbosa* Adamović). Plant with woody rootstock. Stems up to 30 cm, stiff, erect, shortly pubescent below, viscid above. Basal leaves rather densely hairy, spathulate; cauline few, linear or linear-lanceolate. Flowers usually solitary or paired; pedicels long. Calyx 8–11 mm, cylindrical, distinctly hairy. Petals yellow; coronal scales small. Capsule *c.* 8 mm; carpophore *c.* 1 mm. $2n = 24$. ● *Mountains of Balkan peninsula, extending northwards to Transylvania and Hungary.* Bu Gr Hu Ju Rm.

79. S. thessalonica Boiss. & Heldr. in Boiss., *Diagn. Pl. Or. Nov.* **3(1)**: 74 (1854). Like **78** but with a denser, more glandular indumentum; cauline leaves obovate or ovate to linear; inflorescence (1)2- to 7-flowered; calyx glandular-pubescent; carpophore 1–3 mm. ● *Albania and N. Greece; E. Kriti.* Al Cr Gr.

(a) Subsp. **thessalonica**: Stems several, up to 35 cm. Basal leaves with long petiole; cauline leaves in 5–8 pairs, obovate to linear. $2n = 24$. *Rock-crevices and rocky slopes. Albania, N. Greece.*

(b) Subsp. **dictaea** (Rech. fil.) Melzh., *Bot. Jahrb.* **98**: 45 (1977) (*S. dictaea* Rech. fil.): Stems usually 1–3, up to 17 cm. Basal leaves with short petiole; cauline leaves in 3–5 pairs, ovate to linear. *Crevices of limestone rocks. E. Kriti (Dhikti Ori).*

80. S. congesta Sm. in Sibth. & Sm., *Fl. Graec. Prodr.* **1**: 300 (1809). Stock woody, branched. Stems up to 65 cm, stiff, erect, woody, with dense pubescence on leaves and lower part of stem. Inflorescence-branches alternate, with terminal, subsessile groups of 3–5 rather small flowers. Calyx 7·5–9 mm, densely glandular-hairy. Petals yellowish; coronal scales absent. Capsule *c.* 5 mm, broadly ovoid; carpophore 1·5–2 mm. $2n = 24$. ● *Mountains of Greece.* Gr.

(a) Subsp. **congesta**: Stems up to 65 cm, with 8–11 pairs of cauline leaves. Pedicels and calyx densely glandular-hairy. Calyx-teeth acute, with narrow membranous margin. *N. & C. Greece.*

(b) Subsp. **moreana** Melzh., *Bot. Jahrb.* **98**: 48 (1977): Stems shorter, up to 40 cm, with 5–8 pairs of cauline leaves. Pedicels and calyx very densely glandular-hairy. Calyx-teeth obtuse, with broad membranous margin. *S. Greece.*

S. adelphiae Runemark, *Willdenowia* **14**: 45 (1984), from limestone cliffs in S. Kikladhes, like **80** but pulvinate, with flowering stems shorter, more woody and densely pubescent, and seeds tuberculate, may represent a third subspecies.

81. S. cephallenia Heldr., *Fl. Céphal.* 26 (1882). Woody at base; stems up to 50 cm, with linear or linear-lanceolate basal leaves. In robust specimens the flowers usually in rather crowded, 3-flowered cymes, the inflorescence often with long alternate branches spreading at a wide angle; in small specimens the inflorescence reduced to a single, few-flowered group. Petals white. Calyx 9·5–11·5 mm, glandular-hairy. Capsule 6–8 mm; carpophore 3·5–6 mm, hairy. $2n = 24$. *Rocks.* ● *S. Albania and W. Greece.* Al Gr.

(a) Subsp. **cephallenia**: Basal leaves 3–5 mm wide, lanceolate to spathulate, green. Inflorescence usually many-flowered. Carpophore densely hairy. $2n = 24$. *Kefallinia.*

(b) Subsp. **epirotica** Melzh., *Bot. Jahrb.* **98**: 51 (1877): Basal leaves 1–2 mm wide, linear to lanceolate, yellow-green; inflorescence usually few-flowered. Carpophore slightly hairy. *Throughout the range of the species except Kefallinia.*

S. paeoniensis Bornm., *Mitt. Thür. Bot. Ver.* nov. ser. **36**: 44 (1925), from S.W. Jugoslavia, has oblanceolate-spathulate basal leaves and an eglandular calyx, but otherwise resembles **81**. It may deserve subspecific distinction.

Sect. AURICULATAE (Boiss.) Schischkin. Caespitose, montane perennials with large flowers terminating short stems. Calyx cylindrical-clavate, glandular. Petal-claw auriculate.

82. S. vallesia L., *Syst. Nat.* ed. 10, **2**: 1032 (1759). Mat-forming perennial with ascending, glandular-pubescent stems up to 15 cm. Leaves oblong-lanceolate to linear, pubescent. Inflorescence 1- to 3-flowered. Petals usually pale pink above and red beneath, or white; claw exserted; coronal scales small, truncate. Capsule scabrid, equalling the glabrous carpophore. *Mountain rocks.* ● *S. & S.C. Europe from S.E. France to S.W. Jugoslavia.* Al Ga He It Ju.

(a) Subsp. **vallesia**: Leaves oblong-lanceolate or lanceolate. Calyx *c.* 25 mm. Petals usually pink. $2n = 24, 48$. *Alps.*

(b) Subsp. **graminea** (Vis. ex Reichenb.) Nyman, *Consp.* 92 (1878) (*S. graminea* Vis. ex Reichenb.): Leaves linear. Calyx 12–15 mm. Petals usually white. $2n = 24, 48$. *Jugoslavia and Albania; Appennini and Alpi Apuane, locally in S.W. Alps.*

83. S. boryi Boiss., *Elenchus* 19 (1838) (incl. *S. tejedensis* Boiss., *S. duriensis* Samp.). Like **82** but coronal scales conspicuous; carpophore up to 18 mm, much longer than the ripe capsule. $2n = 24, 48, 72$. *Spain, N. Portugal.* Hs Lu.

A variable species that has been divided into as many as five subspecies. Since some at least of these are very weakly differentiated, we have not given any subspecific treatment. J. Jalas & J. Suominen, *Atlas Florae Europaeae* **7**: 68 (1986), give full references.

84. S. jailensis Rubtsov, *Bjull. Gosud. Nikitsk. Bot. Sada* **2(24)**: 5 (1974). Densely-branched, glandular-pubescent small shrub; stems 10–20 cm. Leaves 3–4 cm, linear. Flowers solitary or sometimes paired. Calyx 20–25 mm. Petals white or yellowish. Capsule 8–10 mm, ovoid, more than twice as long as carpophore. *Mountain rocks,* c. 1400 *m.* ● *Krym (near Gurzuf).* Rs(K).

Sect. SPERGULIFOLIAE Boiss. (incl. Sect. *Suffruticosae* (Rohrb.) Schischkin and Sect. *Macranthae* Rohrb.). Perennials with more or less woody stock and ovate, lanceolate or linear leaves. Inflorescence few-flowered or flowers solitary; flowers usually large. Carpophore well-developed.

85. S. repens Patrin in Pers., *Syn. Pl.* **1**: 500 (1805). Pubescent perennial with long, creeping stolons and suberect stems up to 60 cm. Leaves up to 5 cm, linear, acute, the basal withering early. Inflorescence narrow, with short, few-flowered, opposite branches. Calyx 10–16 mm, densely pubescent, cylindrical-clavate, somewhat inflated in fruit. Petal-limb *c.* 10 mm, whitish, deeply lobed with broad, ovate lobes and glabrous, auriculate claw; coronal scales *c.* 1·5 mm. Capsule *c.* 6 mm, more or less equalling carpophore. $2n = 24$. *N., C. & E. Russia.* Rs(N, C, E).

86. S. sangaria Coode & Cullen, *Notes Roy. Bot. Gard. Edinb.* **28**: 2 (1967). Like **85** but leaves fleshy, the basal obovate to spathulate and the cauline narrowly obovate. *Maritime sands. Turkey-in-Europe.* Tu.

87. S. succulenta Forskål, *Fl. Aegypt.* lxvi, 89 (1775) (incl. *S. corsica* DC.). Fleshy perennial with woody stock and numerous procumbent or ascending stems, clothed throughout with dense, viscid, glandular pubescence. Leaves obovate or oblanceolate. Flowers large, solitary or paired (rarely 3 or more), in axils of leafy bracts. Calyx 15–20 mm, narrowly clavate, with conspicuous greenish or reddish veins. Petal-limb bifid, white; claw long-exserted; coronal scales conspicuously toothed. Capsule *c.* 10 mm, broadly oblong, equalling the deflexed-hairy carpophore. $2n = 24$. *Maritime sands. Corse, Sardegna; Kriti.* Co Cr Sa.

Plants from Corse and Sardegna have smaller leaves and rather shorter, more obtuse calyx-teeth than the typical plant, which occurs in Kriti; they have been distinguished as subsp. **corsica** (DC.) Nyman, *Consp.* 92 (1878).

88. S. spergulifolia (Willd.) Bieb., *Fl. Taur.-Cauc.* **3**: 305 (1820) (*Cucubalus spergulifolius* Willd., *S. armeniaca* Rohrb., *S. supina* Bieb.). Perennial with branched, woody stock. Stems numerous, 15–40 cm, procumbent or ascending to erect, scabrid-puberulent, somewhat glandular above. Leaves 15–40 × (0·5–)1–4 mm, linear to linear-lanceolate, usually with sterile shoots in the axils. Inflorescence a lax, (1)2- to 12-flowered panicle. Calyx (8–)11–20 mm, narrowly cylindrical, glandular-puberulent. Petals white or pale yellow; claw exserted, ciliate; coronal scales up to 1·5 mm. Capsule (3–)5–9 mm, ovoid, enclosed within the calyx, about as long as the pubescent carpophore. $2n = 24, 48$. *S.E. Europe, southwards to* 40° *N.* Bu Gr Ju Rm Rs(C, W, K, E) Tu.

The plants from Krym have been distinguished as **S. syreistschikowii** Smirnov, *Bjull. Mosk. Obšč. Isp. Prir., Otd. Biol.*, ser. 2, **49(2)**: 87 (1940), with somewhat broader leaves and almost eglandular calyx, but the differences do not merit specific distinction.

89. S. thymifolia Sm. in Sibth. & Sm., *Fl. Graec. Prodr.* **1**: 292 (1809) (*S. pontica* Brândză). Like **88** but leaves fleshy, ovate, conspicuously pubescent; inflorescence few-flowered, with alternate or opposite 2- or 3-flowered branches; calyx 12–15 mm, narrowly clavate, glandular-villous; capsule slightly longer than carpophore. $2n = 48$. *Maritime sands. N. & W. coasts of Black Sea.* Bu Rm Rs(W) Tu.

90. S. altaica Pers., *Syn. Pl.* **1**: 497 (1805). Stems up to 50 cm, much-branched, scabrid-puberulent below, glabrous and viscid above. Leaves up to 3 cm, acicular, pungent, scabrid. Inflorescence more or less compound, with alternate 1- to 3-flowered branches; pedicels equalling or longer than calyx. Calyx 10–14 mm, sparsely pubescent. Petals white; claw ciliate; coronal scales present. Capsule 8–10 mm, 1½–2 times as long as carpophore. *S.E. Russia.* Rs(C). (*Temperate Asia.*)

91. S. cretacea Fischer ex Sprengel, *Syst. Veg.* ed. 16, **2**: 405 (1825). Stems up to 30 cm, much-branched, with hoary pubescence below. Leaves up to 10 mm, linear, thick and rigid, acuminate. Inflorescence 1- to 3-flowered; pedicels usually shorter than calyx. Calyx 18–20 mm, glabrous or slightly scabrid on veins. Petals white; claw exserted; coronal scales absent. Capsule 10–12 mm; carpophore 5–7 mm, stout. Seeds with flat faces and grooved back. *Chalk hills.* ● *S. Russia, E. Ukraine.* Rs(C, W, E).

92. S. suffrutescens Bieb., *Beschr. Länd. Terek Casp.* 175 (1800). Stems up to 40 cm, suberect, shortly pubescent. Basal leaves up to 4 mm wide, linear-spathulate, mucronate; upper cauline linear. Inflorescence 1- or few-flowered; pedicels equalling or longer than calyx. Calyx 10–15 mm, glabrous or slightly scabrid. Petals white; coronal scales linear, acute. Capsule *c.* 10 mm; carpophore 6–7 mm. *S.E. Russia, W. Kazakhstan.* Rs(E).

93. S. taliewii Kleopow, *Jour. Bot. Acad. Sci. Ukr.* **9(17)**: 119 (1936). Like **92** but with stems less than 20 cm; leaves up to 5 mm wide, oblanceolate or lanceolate-spathulate, thick; calyx entirely glabrous; coronal scales absent; capsule about twice as long as carpophore, ovoid-globose. *S.E. Russia (west of Ural'sk).* Rs(E).

94. S. linifolia Sm. in Sibth. & Sm., *Fl. Graec. Prodr.* **1**: 301 (1809). Stems up to 65 cm, numerous, suberect, with short, hoary pubescence below, glabrous and viscid above. Basal leaves linear-lanceolate, puberulent. Inflorescence with few- or 1-flowered alternate branches; pedicels shorter than calyx. Calyx 15–20 mm, abruptly contracted below capsule, usually pubescent; teeth acute. Petals pinkish above, purplish beneath; claw ciliate; coronal scales truncate. Capsule 7–10 mm, equalling or somewhat longer than carpophore. $2n = 24$. ● *C. Greece.* Gr.

95. S. schwarzenbergeri Halácsy, *Denkschr. Akad. Wiss. Math.-Nat. Kl. (Wien)* **61**: 472 (1894) (incl. *S. horvatii* Micevski). Stems 10–45 cm, simple, erect from woody stock, slender, pubescent, viscid above. Basal leaves small, obovate-spathulate; upper cauline minute, linear. Inflorescence more or less racemose, 1- to 4-flowered; pedicels much shorter than calyx. Calyx 12–21 mm, puberulent; teeth ovate-lanceolate, more or less obtuse. Petals greenish; claw glabrous. Capsule equalling or somewhat shorter than carpophore. $2n = 24$. *Mountain rocks.* ● *N.W. Greece and adjacent parts of Albania and Jugoslavia.* Al Gr Ju.

96. S. campanula Pers., *Syn. Pl.* **1**: 500 (1805). Slender, glabrous perennial up to 20 cm, with the habit of *Campanula rotundifolia*. Leaves linear to linear-lanceolate, acute, ciliate at base. Flowers solitary or paired, terminal on slender pedicels. Calyx 7–8 mm, campanulate, vinous, glabrous. Petals white, reddish-purple beneath. Capsule *c.* 6 mm, slightly longer than or up to twice as long as the thick, pubescent carpophore. *Calcareous mountain rocks.* ● *Maritime Alps.* Ga It.

(97–102). S. saxifraga group. Caespitose perennials. Stems usually numerous, up to 20(–30) cm, suberect, slender, usually pubescent below, glabrous and viscid above. Leaves linear to linear-spathulate or linear-oblanceolate, often with scabrid or ciliate margins. Inflorescence 1- or 2(–4)-flowered. Calyx 6–16 mm, glabrous, pale, rarely vinous. Petals white above, reddish or greenish beneath; claw more or less exserted; coronal scales small. Capsule 5·5–12·5 mm; carpophore 2·5–12 mm, glabrous or slightly pubescent.

A difficult group of closely related, variable species that requires further revision, especially in the Balkan peninsula. The following treatment is provisional and it may be more appropriate for these taxa to be treated at subspecific rank.

1 Carpophore less than 5 mm
 2 Carpophore glabrous **99. conglomeratica**
 2 Carpophore slightly pubescent below
  3 Stems 10–25 cm; calyx oblong-turbinate **97. saxifraga**
  3 Stems usually less than 10 cm; calyx turbinate-clavate **100. dirphya**

1 Carpophore 5–12 mm
 4 Plant pulvinate; stems up to 10 cm **98. fruticulosa**
 4 Plant forming a ±lax mat; stems up to 30 cm

5 Petal-claw distinctly exserted (2–3 mm) from calyx; capsule considerably exceeding calyx **102. parnassica**
5 Petal-claw equalling calyx or slightly exserted (less than 2 mm); capsule ± equalling calyx
6 Stems up to 10 cm; carpophore glabrous **99. conglomeratica**
6 Stems usually more than 10 cm; carpophore slightly pubescent **101. balcanica**

97. S. saxifraga L., *Sp. Pl.* 421 (1753) (*S. petraea* Waldst. & Kit.; incl. *S. hayekiana* Hand.-Mazz. & Janchen, *S. stojanovii* Panov, *S. taygetea* Halácsy, *S. velcevii* Jordanov & Panov). More or less densely caespitose. Stems usually numerous, up to 25 cm, pubescent below, viscid above, 1- or 2(–4)-flowered. Basal leaves linear-oblanceolate, scabrid-ciliate on lower margins. Calyx 6–11 mm, oblong-turbinate. Petals white above, brownish-red beneath; claw not exceeding calyx. Capsule 6–10 mm, ovoid, not exceeding or slightly exceeding calyx; carpophore 2·5–4 mm, slightly pubescent below. $2n = 24$. *Rocks and screes.* ?● *S. Europe, mainly in the mountains, extending northwards to W. Austria.* Al Au Bu Ga Gr He Hs It Ju Rm Si.

98. S. fruticulosa Sieber in DC., *Prodr.* **1**: 376 (1824). Like **97** but pulvinate; stems up to 10 cm, 1(2)-flowered; basal leaves linear-spathulate; calyx (9–)10·5–13 mm, clavate; petal-limb greenish beneath; carpophore 5–9 mm, glabrous. $2n = 24$. *Rocks.* ● *E. Kriti.* Cr.

99. S. conglomeratica Melzh., *Willdenowia* **13**: 123 (1983). Like **97** but with woodier, extensively branched stock; branches procumbent or trailing; flowering stems few, up to 10 cm, 1(2)-flowered; calyx clavate; petals white or pink; capsule shorter than calyx; carpophore up to 5·5 mm, glabrous. *Conglomerate rocks.* ● *N. Peloponnisos (gorge of R. Vouraikos).* Gr.

100. S. dirphya Greuter & Burdet, *Willdenowia* **13**: 281 (1983) (*S. smithii* Boiss. & Heldr., non J. F. Gmelin). Like **97** but stems up to 10 cm, very slender, 1(2)-flowered; leaves subspathulate; calyx 7–8·5 mm, turbinate-clavate; capsule 5·5–6·5 mm, slightly exceeding calyx; carpophore stout, slightly pubescent. *Crevices of limestone rocks.* ● *E. Greece (N. Evvoia).* Gr.

101. S. balcanica (Urum.) Hayek, *Prodr. Fl. Penins. Balcan.* **1**: 270 (1924). Densely caespitose. Stems 5–30 cm, 1(2)-flowered. Leaves linear-spathulate. Calyx 8·5–16 mm, oblong-turbinate. Petal-claw slightly exceeding calyx. Capsule more or less equalling or somewhat longer than calyx; carpophore 5–7 mm, slightly pubescent. $2n = 24$. *Rocky slopes and rocky grassland.* ● *E. part of Balkan peninsula.* Bu Gr Ju.

102. S. parnassica Boiss. & Spruner in Boiss., *Diagn. Pl. Or. Nov.* **2(8)**: 91 (1849). Densely caespitose. Stems up to 20(–30) cm, 1-flowered. Leaves linear to linear-lanceolate or linear-spathulate, with serrulate or strongly scabrid margins; cauline leaves often with sterile shoots in the axils. Petal-claw exserted 2–3 mm from calyx. Calyx (8–)10–14(–16) mm, oblong to obconical. Capsule 7·5–10·5(–12) mm, distinctly and often completely exserted from calyx; carpophore 5–10(–12) mm. $2n = 24$. *Rocks.* ● *Mountains of W. part of Balkan peninsula and Italy.* Al Gr It Ju.

Plants from serpentine in N. Greece (Vourinos Oros), distinguished mainly by the pubescent calyx, have been described as subsp. **vourinensis** Greuter, *Willdenowia* **14**: 47 (1984).

103. S. schmuckeri Wettst., *Biblioth. Bot. (Stuttgart)* **26**: 30 (1892). Caespitose; stems up to 15 cm. Leaves up to 10 mm, linear-lanceolate, acute (or the lower obtuse), scabrid or puberulent, canescent. Inflorescence 1- to 6(–8)-flowered. Calyx 7–9 mm, pale greenish or vinous, slightly pubescent above. Petals purple; claw not exserted. Capsule 4–5 mm, equalling or slightly exceeding calyx; carpophore 3–4 mm, slender, slightly pubescent. ● *Mountains of N.W. Makedonija.* Ju.

104. S. multicaulis Guss., *Pl. Rar.* 172 (1826) (*S. saxifraga* var. *multicaulis* (Guss.) Fiori). Usually less densely caespitose than **97**; stems up to 45 cm, (1–)3- to 4(–8)-flowered. Leaves oblanceolate or linear-lanceolate to linear. Calyx 14–22 mm, narrowly clavate, pale greenish or vinous, rarely with anastomosing veins, sometimes slightly scabrid above; teeth alternately obtuse and acute. Petals whitish above, greenish to reddish beneath; claw long-exserted. Capsule 7–9·5 mm, equalling or slightly longer than calyx; carpophore 6–12 mm, pubescent, especially below. $2n = 24$. ● *Mountains of Balkan peninsula and Italy; E. Kriti.* Al ?Cr Gr It Ju.

1 Lower cauline leaves oblanceolate; calyx 17–22 mm **(c) subsp. sporadum**
1 Lower cauline leaves linear to lanceolate; calyx 12–18 mm
2 Calyx slightly pubescent; carpophore 6–8 mm **(a) subsp. multicaulis**
2 Calyx glabrous; carpophore 8–10 mm **(b) subsp. stenocalycina**

(a) Subsp. **multicaulis**: Lower cauline leaves 1–3 mm wide, linear to lanceolate. Calyx 13–18 mm, slightly pubescent. Capsule ovoid; carpophore 6–8 × 1–2 mm. *Throughout the range of the species except most of the Aegean region.*

(b) Subsp. **stenocalycina** (Rech. fil.) Melzh., *Bot. Jahrb.* **98**: 60 (1977) (*S. stenocalycina* Rech. fil.): Like subsp. (a) but calyx 13–16 mm, glabrous; carpophore 8–10 × 1 mm. *E. Greece (Evvoia).*

(c) Subsp. **sporadum** (Halácsy) Greuter & Burdet, *Willdenowia* **12**: 190 (1982) (*S. genistifolia* Halácsy): Lower cauline leaves 2·5–4 mm wide, oblanceolate. Calyx 17–22 mm, slightly pubescent. Capsule narrowly ovoid; carpophore 10–12 × 1–2 mm. *N. & C. Greece, N. Aegean region.*

A fourth subspecies, subsp. **cretica** Melzh., *Phyton (Austria)* **21**: 132 (1981), with stems 10–20 cm and 1- or 2-flowered, calyx 12–15 mm and carpophore 6–8 mm, has been described from E. Kriti (Dhikti Ori). It may represent a non-caespitose variant of **98**, which it otherwise closely resembles.

105. S. waldsteinii Griseb., *Spicil. Fl. Rumel.* **1**: 179 (1843) (*S. clavata* (Hampe) Rohrb., non Moench; incl. *S. macropoda* Velen.). Densely caespitose, with crowded, linear leaves, and 1- to 5-flowered, erect stems up to 25 cm. Calyx 20–25(–28) mm, narrowly clavate, with acute, distinctly patent teeth; veins anastomosing. Petals white; claw exserted. Capsule exceeding calyx; carpophore pubescent at least near base. $2n = 48$. ● *Mountains of Balkan peninsula.* Al Bu Gr Ju.

106. S. pindicola Hausskn., *Mitt. Thür. Bot. Ver.* **5**: 85 (1887). Dwarf, densely caespitose; flowering stems up to 20 cm, very slender, 1-flowered, pubescent below, glabrous and very viscid above. Leaves up to 15 mm, linear; cauline leaves often with sterile shoots in the axils. Calyx 18–24 mm, narrowly clavate, glabrous, vinous. Petals brownish; claw $\frac{1}{2}$- to $\frac{2}{3}$-exserted; coronal scales *c.* 1 mm. Capsule 7–12·5 mm, narrowly ovoid, $\frac{1}{2}$-exserted from calyx; carpophore 11–16 mm, pubescent. $2n = 24$. *Rocky and stony* *places, usually on serpentine.* ● *Mountains of N.W. Greece and S.E. Albania.* Al Gr.

107. S. dionysii Stoj. & Jordanov, *Annu. Univ. Sofia Phys.-Math. 3 (Sci. Nat.)* **34**: 175 (1938). Caespitose; stems woody below with numerous, persistent, short shoots. Flowering stems 20–35 cm, slender, retrorsely aculeolate below, 1- to 3-flowered. Leaves linear, margins scabrid. Calyx 16–19 mm. Petals white or pale greenish. Capsule 6–8 mm, more or less exserted from calyx; carpophore 13–18 mm. *Rocky places in ravines and clearings of* Pinus *woodland.* $2n = 24$. ● *E.C. Greece (Olimbos).* Gr.

108. S. orphanidis Boiss., *Fl. Or.* **1**: 651 (1867). Caespitose; stems up to 20 cm, very slender, viscid above, 1-flowered. Uppermost pair of leaves small, very close to the base of the calyx. Calyx 20–30 mm. Petals white or pale pink above, purplish beneath. Capsule 8–10 mm, almost completely exserted from calyx; carpophore 18–22 mm. *Crevices of limestone rocks.* ● *N. Greece (Athos).* Gr.

109. S. barbeyana Heldr. ex Boiss., *Fl. Or., Suppl.*, 107 (1888). Densely pulvinate, with short, branched stock densely clothed with whitish, persistent leaves. Leaves linear, glabrous, with scabrid margins. Flowering stems up to 9 cm, very slender, glabrous, 1-flowered, viscid above, with 1 or 2 pairs of linear leaves. Calyx 5–7·5 mm, turbinate, glabrous. Petal-limb obcordate, pale purple. Carpophore very short. $2n = 24$. *Crevices of limestone cliffs above 2000 m.* ● *Mountains of S.C. Greece.* Gr.

S. falcata Sm. in Sibth. & Sm., *Fl. Graec. Prodr.* **1**: 301 (1809), was erroneously reported from Greece in *Flora Europaea* ed. 1, **1**: 172 (1964). It is endemic to the mountains of W. Anatolia.

S. caryophylloides (Poiret) Otth in DC., *Prodr.* **1**: 369 (1824), a widespread and variable species in Anatolia, is recorded without detail from Krym in P. H. Davis, *Fl. Turkey* **2**: 221 (1967). It has not proved possible to substantiate this record.

Sect. FRUTICULOSAE (Rohrb.) Chowdhuri. Perennials with elongated, woody stock. Inflorescence a raceme-like monochasial cyme with the axis simple or forked below, without an alar flower. Calyx 10-veined, clavate and widest at mouth in fruit; teeth ovate, obtuse. Seeds reniform; faces somewhat concave, striate; back concave.

110. S. ciliata Pourret, *Mém. Acad. Sci. Toulouse* **3**: 329 (1788) (incl. *S. perinica* Hayek). Caespitose, with dense rosettes of linear- to lanceolate-spathulate leaves; flowering stems 5–30 cm, arising terminally from centre of rosettes, erect or ascending, pubescent. Inflorescence simple, (1)2- to 3(–7)-flowered, rarely branched. Calyx 11–20 mm, pubescent; veins usually anastomosing. Petals white or pink, usually greenish or reddish beneath; coronal scales well-developed. Capsule 5–10 mm, broadly ovoid; carpophore 5–8 mm, pubescent. Seeds *c.* 1 mm, black, almost smooth, with shallowly grooved back. $2n = 24, 36, 48, 72, 84$, *c.* 96, *c.* 120, 144, 168, 192, 216, 240. ● *Mountains of S. Europe; local.* Bu Ga Gr Hs It ?Ju Lu.

A very variable species, unique in the genus for its extraordinary range of polyploid chromosome numbers. The Italian plant is large-leaved with a 4- to 7-flowered inflorescence, and has been treated by some authors as a separate species, **S. graefferi** Guss., *Pl. Rar.* 170 (1826), as has also the small plant from N. Spain & N. Portugal, **S. elegans** Link ex Brot., *Fl. Lusit.* **2**: 185 (1805), with a very reduced 1- or 2-flowered inflorescence. Full details and references can be found in J. Jalas & J. Suominen, *Atlas Florae Europaeae* **7**: 80–81 (1986).

111. S. legionensis Lag., *Gen. Sp. Nov.* 14 (1816). Like **110** but flowering stems 15–40 cm, arising laterally from below leaf-rosettes; inflorescence usually simple, with 3–9 flowers; calyx 13–23 mm, veins scarcely anastomosing; petals yellowish or greenish; carpophore 7–12 mm, almost glabrous at maturity. $2n = 24, 48$. ● *Spain, N.E. Portugal.* Hs Lu.

112. S. borderei Jordan, *Ann. Soc. Linn. Lyon* nov. ser., **12**: 445 (1866). Dwarf, mat-forming perennial with thick, branching stock and erect, glabrous, 1- or 2(–5)-flowered stems up to 12 cm. Basal leaves with minute papillae on upper surface, ciliate, narrowly spathulate; cauline few, linear. Calyx 8–10 mm, pubescent, with obtuse teeth. Petal-limb pink, bifid; claw ciliate, not auriculate; coronal scales ovate. Capsule *c.* 6 mm, broadly ovoid, equalling or somewhat longer than the pubescent carpophore. Seeds reniform, with obtuse tubercles and shallowly grooved back. *Mountain rocks.* ● *E. & C. Pyrenees.* Ga Hs.

Variously associated with Sect. *Macranthae* and Sect. *Fruticulosae*; it resembles small plants of **110** (*S. elegans* Link ex Brot.), but differs in the papillose leaves and in the tuberculate seeds.

Sect. NANOSILENE Otth. Dwarf, mat-forming or pulvinate perennials with woody taproot, linear leaves, and usually 1-flowered stems bearing 1 or 2(–4) pairs of cauline leaves.

113. S. acaulis (L.) Jacq., *Enum. Stirp. Vindob.* **78**, 242 (1762). Glabrous, moss-like, mat-forming or pulvinate perennial with short, erect flowering stems 0·5–3(–10) cm. Leaves 6–12 mm, linear-subulate; margin cartilaginous, stiffly ciliate. Calyx 7–9 mm, glabrous, campanulate, faintly 10-veined, without lateral veins. Petal-limb emarginate to shallowly bifid, usually deep pink; claw not auriculate; coronal scales small. Capsule subcylindrical, up to twice as long as calyx and longer than the pubescent carpophore. *Arctic Europe and Ural; higher mountains of Europe.* Au Br Bu Cz Fa Fe Ga Ge Hb He Hs Is It Ju No Po Rm Rs(N, C) Sb Su.

Both unisexual and hermaphrodite plants occur. A variable species that can be divided into two subspecies in the Alps.

(a) Subsp. **acaulis** (subsp. *longiscapa* (A. Kerner ex Vierh.) Vierh., subsp. *pannonica* Vierh.): Flowering stems up to 5 cm. Calyx truncate at base. Capsule 6–13 mm, $1\frac{1}{2}$–2 times as long as calyx. $2n = 24$. *N. Europe, southwards to Wales; mountains of C. & S. Europe, southwards to C. Appennini and S.W. Bulgaria.*

(b) Subsp. **bryoides** (Jordan) Nyman, *Consp.* 93 (1878) (subsp. *exscapa* (All.) J. Braun, subsp. *norica* Vierh.): Flowering stems very short, rarely more than 0·5 cm. Calyx more or less cuneate at base. Capsule 3–5 mm, scarcely exceeding calyx. $2n = 24$. ● *Mountains of C. & S. Europe.*

114. S. dinarica Sprengel, *Syst. Veg.* ed. 16, **2**: 405 (1825). Like **113** but stems more or less pubescent, usually with 2–4 pairs of leaves; calyx 12–15 mm, oblong-campanulate, with reticulate lateral venation. ● *S. Carpathians.* Rm.

Sect. RUPIFRAGA Otth. Small, slender, glabrous perennials with numerous small flowers in regular dichasia. Seeds reniform with flat or shallowly grooved back.

115. S. rupestris L., *Sp. Pl.* 421 (1753). Stems up to 25 cm, erect, branched below. Basal leaves oblanceolate; cauline numerous, lanceolate, acute. Inflorescence a diffuse, compound dichasium with small flowers on long, slender pedicels. Calyx 4–6 mm, glabrous, obconical, 10-veined, without lateral veins. Petal-limb obovate, deeply emarginate, white or pink; coronal scales acute. Capsule *c.* 5 mm, oblong-ovoid; carpophore *c.* 1 mm, glabrous. $2n = 24$. ?● *Fennoscandia; mountains of C. & S. Europe, from the Vosges to S.E. Spain and S. Jugoslavia.* Au Co *Cz Fe Ga Ge He Hs It Ju No Rm Rs(N, C) ?Sa Su.

Doubtfully recorded outside Europe at one station in W. Siberia.

116. S. lerchenfeldiana Baumg., *Enum. Stirp. Transs.* **1**: 398 (1816). Like **115** but stems arcuate-ascending, obviously lateral, in axils of longer, narrower basal leaves; flowers fewer, larger, shortly pedicellate; calyx 9–12 mm, narrowly clavate, with some lateral venation; petal-limb oblong-linear, reddish or purplish, sometimes white (var. *macedonica* (Form.) Bornm.); carpophore 5–7·5 mm, almost equalling capsule. $2n = 24$. *Rock-crevices.* ● *S. Carpathians and mountains of Balkan peninsula southwards to* 41° N. Bu Gr Ju Rm.

Sect. HELIOSPERMA Reichenb. Like Sect. *Rupifraga*, but often hairy; seeds winged with a characteristic dorsal crest consisting of 2 or more rows of long papillae on the back. Endemic to mountains of S. Europe.

(117–124). **S. pusilla** group. Perennials with rather slender, sometimes woody stock, and slender, weak, branched stems up to 30 cm. Inflorescence a spreading dichasium of small flowers on long, slender pedicels. Calyx 3–10 mm, obconical, weakly 10-veined. Petal-limb obovate, usually (2–)4-toothed or -fid, white or pink. Capsule 5–8 mm, ovoid, equalling or exceeding calyx; carpophore 1–5 mm.

Plants of this group occur in damp, open habitats, often on calcareous rock, in mountain regions from the Pyrenees to the Carpathians and Balkan peninsula. The great variation in habit, leaf-shape and hairiness shows some geographical and ecological correlation, but has been treated differently by different authors. The following treatment attempts to reconcile the accounts of H. Neumayer in Hayek, *Prodr. Fl. Penins. Balcan.* **1**: 264–267 (1924), and of P. F. A. Ascherson & K. O. R. P. P. Graebner, *Syn. Mitteleur. Fl.* **5(2)**: 17–31 (1920), and takes into account more recent work incorporated in the treatment by J. Jalas & J. Suominen, *Atlas Florae Europaeae* **7**: 85–87 (1986).

1 Petal-claw ciliate **124. alpestris**
1 Petal-claw glabrous
2 Capsule at least 3 times as long as carpophore
3 Petals entire or shallowly emarginate; calyx *c.* 8 mm **119. macrantha**
3 Petals ± deeply 2- to 4-toothed or -fid; calyx 3–6(–7) mm
4 Subglabrous or somewhat glandular-hairy; dorsal crest of seed constituting more than $\frac{1}{3}$ of diameter of seed **117. pusilla**
4 Densely hairy, with long, multicellular glandular and eglandular hairs; dorsal crest of seed constituting not more than $\frac{1}{4}$ of diameter of seed **118. veselskyi**
2 Capsule not more than twice as long as carpophore
5 Plant caespitose; leaves less than 2 mm wide **122. intonsa**
5 Plant not caespitose; leaves usually at least 2 mm wide
6 Calyx less than 7 mm **123. chromodonta**
6 Calyx usually more than 7 mm
7 Sparsely hairy, with short glandular hairs; dorsal crest of seed constituting *c.* $\frac{1}{10}$ of diameter of seed **120. tommasinii**
7 Densely hairy, with long glandular and eglandular hairs; dorsal crest of seed constituting *c.* $\frac{1}{5}$ of diameter of seed **121. retzdorffiana**

117. S. pusilla Waldst. & Kit., *Pl. Rar. Hung.* **3**: 235 (1812) (*S. quadrifida* auct., non L., *S. quadridentata* subsp. *quadridentata* sensu Hayek, subsp. *pusilla* (Waldst. & Kit.) H. Neumayer, subsp. *albanica* (K. Malý) H. Neumayer; incl. *S. monachorum* Vis. & Pančić, *Heliosperma quadrifidum* sensu Hegi). Stems up to 15 cm, weak, ascending, subglabrous or rather sparsely hairy, viscid in upper part with sessile glands and occasionally some multicellular glandular hairs. Leaves 1–2(–4) mm wide, linear or linear-lanceolate. Calyx 3·5–7 mm, glabrous or somewhat glandular-hairy. Petals white, rarely pink or lilac; claw glabrous. Capsule broadly ovoid to subglobose, usually equalling or only slightly exceeding calyx, and 3–4 times as long as carpophore. Seeds *c.* 1 mm, papillate; dorsal crest constituting at least $\frac{1}{3}$ of diameter of seed. $2n = 24$. *Mountain rocks and streamsides, rarely below* 500 *m.* ● *S. & C. Europe, northwards to W. Carpathians.* Al Au Bu Cz Ga Ge Gr He Hs It Ju Po Rm Rs(W) Si.

Two plants described as subspecies of *S. quadridentata* auct., non (L.) Pers., subsp. **malyi** H. Neumayer, *Österr. Bot. Zeitschr.* **72**: 282 (1923), from Jugoslavia (Dinara Planina), and subsp. **candavica** H. Neumayer, *loc. cit.* (1923), from E. Albania (Jablanica), are characterized by unusually well-developed coronal scales, which reach or exceed the base of the petal-lobes. There is insufficient information to decide their status, and they are best treated as variants of **117**.

118. S. veselskyi (Janka) H. Neumayer, *Denkschr. Akad. Wiss. Math.-Nat. Kl.* (*Wien*) **94**: 143 (1917) (*S. quadridentata* subsp. *marchesettii* H. Neumayer, *Heliosperma eriophorum* Juratzka). Like **117** but densely hairy with long, multicellular glandular and eglandular hairs; sessile glands absent; capsule subglobose; dorsal crest of seed constituting $\frac{1}{8}$–$\frac{1}{4}$ of diameter of seed. $2n = 24$. *Cliffs and rocks, in drier habitats than* **117**, *mainly below* 500 *m.* ● *S.E. Alps.* Au It Ju.

119. S. macrantha (Pančić) H. Neumayer, *Denkschr. Akad. Wiss. Math.-Nat. Kl.* (*Wien*) **94**: 143 (1917). More robust than **117** and **118**, with glandular-hairy stems up to 15 cm, and obovate-lanceolate leaves. Flowers relatively large. Calyx 8 mm, slightly hairy. Petals usually pink, with obovate, truncate or slightly emarginate, long-exserted limb. Capsule subglobose, *c.* 4 times as long as carpophore. *Mountain rocks.* ● *N. Albania, S.W. Jugoslavia* (*Crna Gora*). Al Ju.

120. S. tommasinii Vis., *Flora* (*Regensb.*) **12**, *Erg.-Bl.* **1**: 12 (1829) (*S. quadridentata* subsp. *tommasinii* (Vis.) H. Neumayer). Stems up to 10 cm, slender, clothed with short glandular hairs. Leaves 2–3(–8) mm wide, linear-lanceolate to obovate-lanceolate. Calyx 7–9 mm. Petals white; claw glabrous. Capsule ovoid, only slightly longer than carpophore. Dorsal crest of seed very narrow, constituting *c.* $\frac{1}{10}$ of diameter of seed. *Mountain rocks.* ● *N. Albania, W. Jugoslavia.* Al Ju.

121. S. retzdorffiana (K. Malý) H. Neumayer, *Denkschr. Akad. Wiss. Math.-Nat. Kl.* (*Wien*) **94**: 143 (1917). Like **120** but more robust; stems up to 20 cm; glandular hairs abundant, long; calyx (6–)8–10 mm; dorsal crest of seed constituting *c.* $\frac{1}{5}$ of diameter of seed. *Wet rocks*, 200–300 *m.* ● *N. Albania, W. Jugoslavia.* Al ?Gr Ju.

122. S. intonsa Melzh. & Greuter, *Willdenowia* **12**: 29 (1982). Like **120** but caespitose with woody basal shoots, densely pilose, glandular-pubescent above; stems rarely more than 8 cm; leaves 0·7–1·5 mm wide, lanceolate to spathulate; calyx 5–8 mm, glandular-pubescent; capsule 3–4 mm; carpophore 1·5–3·5 mm. *Damp crevices of limestone rocks,* 450–550 *m.* ● *N.W. Greece (near Konitsa).* Gr.

123. S. chromodonta Boiss. & Reuter, *Diagn. Pl. Or. Nov.* **3(1)**: 71 (1854) (incl. *S. moehringifolia* Uechtr. ex Pančić, *S. trojanensis* (Velen.) Jordanov & Panov, *S. quadridentata* subsp. *phyllitica* H. Neumayer). Like **120** but calyx only 4–6 mm; calyx-teeth usually reddish; capsule up to twice as long as carpophore. *Shaded crevices of mountain rocks and screes.* ● *Balkan peninsula southwards to* 40° N. Al Bu Gr Ju.

124. S. alpestris Jacq., *Fl. Austr.* **1**: 60 (1773) (*S. quadrifida* L., nom. ambig., *Heliosperma alpestre* (Jacq.) Reichenb.; incl. *H. arcanum* Zapał.). More robust than **117–123**. Stock rather woody; stems up to 30 cm, subglabrous to somewhat hairy. Leaves up to 9 mm wide, obovate-lanceolate to linear-lanceolate. Inflorescence more or less viscid; flowers relatively large. Calyx 5–7 mm, scabrid or more or less glandular-puberulent. Petals usually white, 4- to 6-toothed; claw ciliate. Capsule ovoid, 3–4 times as long as carpophore, and much exceeding calyx. Seeds as in **117** but somewhat larger (1–1·3 mm). *Mountain rocks, rarely below* 1250 *m.* ● *E. Alps, N.W. Jugoslavia.* Au ?Cz It Ju ?Rs(W).

The record from Ukraine (*Heliosperma arcanum* Zapał.) was based on a single gathering from a gorge of the Dnestr river. It has not been confirmed.

Sect. COMPACTAE (Boiss.) Schischkin. Usually robust annuals, biennials or perennials with cymose, often densely corymbose, inflorescence. Pedicels often very short. Petals entire or emarginate, not bifid.

125. S. armeria L., *Sp. Pl.* 420 (1753) (*S. lituanica* Zapał.). Erect, glabrous and glaucous annual or biennial with usually simple stems up to 40 cm, viscid above. Basal leaves spathulate, withering early; cauline ovate-cordate to lanceolate, amplexicaul, decreasing gradually upwards. Inflorescence usually corymbose, often densely so. Calyx 12–15 mm, cylindrical-clavate; teeth obtuse. Petal-limb obovate, emarginate, usually pink; coronal scales lanceolate, acute. Capsule 7–10 mm, oblong, equalling or somewhat longer than the glabrous carpophore. $2n = 24$. *Certainly native in C., S. and parts of E. Europe; widely cultivated elsewhere and sometimes naturalized.* Al Bu Co Ga Gr He It Ju Po Rm Rs(B, C, W) Sa Tu [Az Be Cz Hs Hu].

Variable in habit, leaf-shape and inflorescence. Especially in the E. Mediterranean region, variants with a less compact inflorescence occur, which have been called var. *serpentini* G. Beck and var. *sparsiflora* Schur. Such plants may deserve subspecific recognition.

126. S. compacta Fischer, *Cat. Jard. Gorenki* ed. 2, 60 (1812). Like **125** but stems stouter, up to 80 cm, and leaves wider; upper cauline leaves closely investing base of subcapitate inflorescence; calyx 13–20 mm; petals entire, bright pink. $2n = 24$. *S.E. Europe, from S. Aegean region to S.W. Ukraine.* Bu Gr Rm Rs(W) Tu.

Plants from the Ukraine have been distinguished as **S. hypanica** Klokov, *Jour. Bot. Acad. Sci. Ukr.* **5**: 20 (1948); they are said to differ mainly in the upper cauline leaves being herbaceous, not membranous, and in the brown, not black, seeds. They seem best regarded as a variant of **126**.

127. S. asterias Griseb., *Spicil. Fl. Rumel.* **1**: 168 (1843). Like **125** but a rhizomatous perennial; stems up to 100 cm; inflorescence capitate, subtended by membranous bracts; cauline leaves few, lanceolate; calyx-teeth acute; petal-limb oblong, entire, deep purple; coronal scales short; capsule 5–7 mm, slightly shorter than the carpophore. *Marshy places.* ● *From N.W. Greece to N.W. Bulgaria.* Al Bu Gr Ju.

Sect. ELISANTHE (Fenzl) Fenzl (Sect. *Melandriformes* (Boiss.) Chowdhuri). Annuals, biennials or perennials with large flowers, solitary or in dichasia; dioecious (except **128**). Petal-limb bifid; coronal scales prominent; claw auriculate; styles 5 (3 in **128**). Capsule without basal septa, dehiscing with 6 or 10 teeth; carpophore very short.

128. S. noctiflora L., *Sp. Pl.* 419 (1753) (*Melandrium noctiflorum* (L.) Fries). Annual; stems up to 40 cm, erect, simple or with few basal branches, densely hairy below, viscid with abundant glandular hairs above. Leaves ovate or ovate-lanceolate; cauline sessile. Inflorescence few-flowered; flowers large, hermaphrodite, scented in evening. Calyx 20–30 mm, 10-veined; teeth long, slender. Petal-limb pink above, yellowish beneath, inrolled during day, opening in evening; styles 3. Capsule ovoid-conical, more or less enclosed by calyx, dehiscing with 6 teeth. $2n = 24$. *Cultivated ground. Widespread in Europe, but absent from or only casual in much of the south-west, north and many islands.* Au Be Br Bu Cz Ga Ge Gr He Ho Hu It Ju Po Rm Rs(B, C, W, K, E) Tu [Da Hb No Rs(N) Su].

129. S. latifolia Poiret, *Voy. Barb.* **2**: 165 (1789). Dioecious, short-lived perennial (sometimes annual) up to 80 cm, often much-branched, usually rather densely and softly hairy, and more or less glandular above. Leaves ovate or ovate-lanceolate; cauline sessile. Inflorescence a lax, compound dichasium of large flowers, opening in the evening and slightly scented. Calyx of male flowers 15–22 mm, 10-veined; of female 20–30 mm, 20-veined, more or less inflated and accrescent in fruit; calyx-teeth narrowly triangular. Petals usually white; styles 5. Capsule 10–25 mm, more or less ovoid, dehiscing with 10 teeth. *Disturbed or cultivated ground. Almost throughout Europe.* All except Bl Cr Fa Is Sb; probably not native in Hb; naturalized in Az.

1 Calyx eglandular; seeds with acute tubercles — (c) subsp. **eriocalycina**
1 Calyx glandular; seeds with obtuse tubercles
 2 Capsule 20–25 mm, subglobose — (d) subsp. **mariziana**
 2 Capsule 10–20 mm, ovoid
  3 Calyx-teeth obtuse; capsule-teeth erect — (b) subsp. **alba**
  3 Calyx-teeth acuminate; capsule-teeth ±patent or recurved — (a) subsp. **latifolia**

(a) Subsp. **latifolia** (*Lychnis divaricata* Reichenb., *L. macrocarpa* Boiss. & Reuter, *Melandrium boissieri* Schischkin, *M. latifolium* (Poiret) Maire): Fruiting calyx glandular, strongly accrescent and inflated; teeth very long, acuminate. Capsule-teeth more or less patent or recurved on dehiscence. Seeds with concave faces and obtuse tubercles. $2n = 24$. *S. Europe.*

(b) Subsp. **alba** (Miller) Greuter & Burdet, *Willdenowia* **12**: 189 (1982) (*S. alba* (Miller) E. H. L. Krause, *S. pratensis* (Rafn) Godron, *Melandrium album* (Miller) Garcke): Fruiting calyx glandular, slightly accrescent; teeth obtuse. Capsule-teeth erect on dehiscence. Seeds with plane faces and obtuse tubercles. $2n = 24$. *Almost throughout the range of the species.*

(c) Subsp. **eriocalycina** (Boiss.) Greuter & Burdet, *Willdenowia* **12**: 189 (1982) (*Melandrium eriocalycinum* Boiss.): Like subsp. (a) but calyx eglandular and seeds with acute tubercles. *S.E. Europe, from E. Jugoslavia to Turkey-in-Europe.*

(**d**) Subsp. **mariziana** (Gand.) Greuter & Burdet, *Willdenowia* **12**: 189 (1982) (*Melandrium marizianum* Gand.): Like subsp. (**a**) but plant densely glandular-tomentose; leaves thicker; capsule 20–25 mm, subglobose. *Coastal rocks.* ● *W. Portugal; N.W. Spain.*

S. astrachanica (Pacz.) Takht., *Krasnaya Kniga* 48 (1975) (*Melandrium astrachanicum* Pacz.), from S.E. Russia (Volga delta), is closely related to **129**. It is said to differ in being biennial, with rather narrow, ovate-lanceolate basal leaves and eglandular calyx 10–15 mm. Its relationship to **129** (**b**) and (**c**) requires further investigation.

130. S. marizii Samp., *Ann. Sci. Acad. Polyt. Porto* **4**: 126 (1909) (*Melandrium viscosum* Mariz, non (Pers.) Čelak., *M. glutinosum* Rouy). Like **129** but always perennial, foetid; stems densely glandular-hairy, with soft, flexuous, viscid hairs; all leaves sessile, glandular; basal leaves spathulate to oblanceolate; cauline leaves lanceolate to ovate-elliptical; flowers open in daytime; calyx densely glandular-hairy; calyx of male flowers 9–15 mm, that of female flowers 10–20 mm; corolla white or pale pink; capsule 10–20 mm, ovoid to conical. $2n = 24$. *Rocky ground.* ● *C. Spain, N. Portugal.* Hs Lu.

131. S. dioica (L.) Clairv., *Man. Herb. Suisse* 145 (1811) (*Melandrium dioicum* (L.) Cosson & Germ., *M. rubrum* (Weigel) Garcke, *M. silvestre* (Schkuhr) Röhling, *Lychnis rubra* (Weigel) Patze, E. H. F. Meyer & Elkan). Like **129** but always perennial; stems eglandular or almost so; leaves broadly ovate; flowers usually red, open in daytime; calyx 10–15 mm, teeth broadly triangular; capsule 10–15 mm, globose to broadly ovoid, dehiscing with recurved teeth. $2n = 24$. *Usually in woodland. Much of Europe, but rare in the south and east.* Au Be Br Bu Cz Da Fa Fe Ga Ge Hb He Ho Hs Hu Is It Ju No Po Rm Rs(N, B, C, W) Su [Is (Rs(K)].

Widespread and variable in habit; dwarf variants occur on mountains, exposed rocks and sand-dunes.

Hybrids with **129**, usually having pale pink flowers, occur frequently where the two species meet; they are fully fertile.

132. S. heuffelii Soó, *Feddes Repert.* **69**: 48 (1964) (*Melandrium nemorale* Heuffel ex Reichenb.). Like **129** but usually biennial; calyx 10–12(–15) mm, sparsely hairy, inconspicuously 10-veined, broadly campanulate in fruit, with triangular teeth; capsule *c.* 10 mm, subglobose, dehiscing with recurved teeth. $2n = 24$. *Woods in hilly regions.* ● *Romania and N. half of Balkan peninsula.* Au Bu Gr Ju Rm.

Resembles **129** in habit and flower colour, but **131** in capsule; the calyx, particularly in fruiting specimens, distinguishes it from both.

133. S. diclinis (Lag.) Laínz, *Bol. Inst. Estud. Astur.* (*Supl. Ci.*) **6**: 45 (1963) (*Lychnis diclinis* Lag., *Melandrium dicline* (Lag.) Willk.). Dioecious, rhizomatous perennial with decumbent, branched stems up to 20 cm, bearing lanceolate, 1-veined leaves and clothed with long, soft, white eglandular hairs. Flowers usually solitary in leaf-axils; pedicels long. Calyx of female flowers 11–14 mm, 20-veined; of male flowers 9–10 mm, 10-veined. Petal-limb deep pink, shallowly bifid (for up to $\frac{1}{4}$ of its length) in female flowers, emarginate and smaller in male; coronal scales prominent, lanceolate, acute; petal-claw long; auricles acute. Filaments glabrous; styles 5, hairy. Capsule *c.* 8 mm, ovoid; carpophore less than 1 mm. Seeds 1·5–2 mm, pale, reniform, with convex face and rounded back, tuberculate. $2n = 24$. *Calcareous rocky slopes.* ● *E. Spain (near Játiva).* Hs.

Sect. RIGIDULAE (Boiss.) Schischkin. Annuals; stems branched, rigid, often filiform, viscid above. Cauline leaves usually fasciculate. Inflorescence very lax, composed of divaricately-branched, equal dichasia; pedicels usually 1–4 times as long as calyx. Calyx not inflated, not contracted at mouth in fruit. Seeds broadly reniform, with flat or slightly convex faces.

134. S. portensis L., *Sp. Pl.* ed. 2, 600 (1762). Stems 15–40 cm, usually divaricately branched, puberulent or glabrous. Rosette-leaves oblong-spathulate, obtuse, rarely persistent; cauline linear, acute. Calyx 11–16 mm, clavate in fruit, glabrous; teeth short, ovate, obtuse or apiculate. Petal-limb *c.* 6 mm, bifid to *c.* half-way, pink, rarely whitish; claw slightly exserted. Capsule 5–10 mm; carpophore 6–9 mm, pubescent. Seeds 0·75–1 mm, blackish to pale brown; faces slightly concave, striate, with subconical tubercles at edge; back wide, plane or slightly convex, with 1 or 2 rows of subconical tubercles. *W. Europe, northwards to N.W. France; C. & S.E. Greece.* ?Co Ga Gr Hs Lu.

(**a**) Subsp. **portensis**: Capsule 5–6 mm, shorter than the carpophore. $2n = 24$. ● *W. Europe.*

(**b**) Subsp. **rigidula** Greuter & Burdet, *Willdenowia* **12**: 190 (1982) (*S. rigidula* Sm., non L., *S. vittata* Stapf): Capsule 6–10 mm, more or less equalling the carpophore. $2n = 24$. *C. & S.E. Greece.*

Plants from S. Greece (Korinthos), described as **S. corinthiaca** Boiss. & Heldr. in Boiss., *Fl. Or., Suppl.*, 96 (1888), differ especially in having the back of the seed distinctly grooved, and require further investigation.

135. S. echinosperma Boiss. & Heldr. in Boiss., *Diagn. Pl. Or. Nov.* **2**(**8**): 78 (1849). Like **134**(**a**) but usually taller; petal-limb white, with purplish veins; seeds 1–1·4 mm, the faces somewhat concavo-convex, the back wider, with 4 rows of acute, conical tubercles. *Shady rocks.* ● *S. Greece (Taÿetos).* Gr.

136. S. pinetorum Boiss. & Heldr. in Boiss., *Diagn. Pl. Or. Nov.* **2**(**8**): 75 (1849). Stems 5–20 cm, branched, puberulent below, glabrous above. Basal leaves often in a rosette, spathulate, obtuse; uppermost linear-lanceolate, acute. Calyx 4–7·5 mm, clavate in fruit, glabrous; teeth short, ovate, rounded. Petals pink; claw scarcely exserted; limb 2–3 mm, bifid. Capsule 3·5–4·5 mm; carpophore 2–3 mm, slender, glabrous. Seeds 0·6–0·8 mm, pale brown; faces slightly concave, very finely striate; back very wide, plane or slightly concave, smooth or finely striate. ● *Kriti.* Cr.

137. S. inaperta L., *Sp. Pl.* 419 (1753). Stems 15–40 cm, erect, often simple below, divaricately branched above, puberulent. Leaves all narrowly linear, acute. Calyx 7–10 mm, oblong-clavate in fruit, glabrous; teeth short, ovate, acute or apiculate. Petals absent or included in calyx. Capsule 5–8 mm; carpophore 2–3·5 mm, stout, slightly pubescent. Seeds 0·6–0·8 mm, dark chestnut or blackish-brown; faces more or less plane, striate; back narrow, more or less acutely grooved, striate. $2n = 24$. *S.W. Europe.* Co Ga Hs It Lu.

Sect. ATOCION Otth. Annuals; stems erect, usually slender, not rigid, viscid above. Inflorescence a dichasial or monochasial cyme. Calyx not contracted at mouth in fruit. Seeds not winged.

138. S. fuscata Link ex Brot., *Fl. Lusit.* **2**: 187 (1805). Stems 10–45 cm, simple or branched, not rigid, rather stout, ribbed or angled, pubescent with patent, stout, contorted hairs. Lower leaves oblong-lanceolate, undulate, sparsely hispid or subglabrous. Inflorescence several- to many-flowered, rather densely cymose-paniculate. Calyx 12–16 mm, clavate in fruit, glandular-pubescent, usually reddish; teeth ovate, obtuse. Petals deep pink. Capsule 6–8 mm; carpophore 4·5–7 mm, glabrous. Seeds *c.* 1 mm, reniform; faces concavo-convex or subexcavate, striate; back wide, with shallow median groove. $2n = 24$. *Mediterranean region; Portugal.* Cr ?Gr Hs It Lu Sa Si.

139. S. pseudatocion Desf., *Fl. Atl.* **1**: 353 (1798). Stems 30–60 cm, branched, pubescent with fine, patent hairs. Inflorescence a lax dichasium; pedicels $\frac{1}{4}$–$1\frac{1}{4}$ times as long as calyx. Calyx (13–)18–20 mm, clavate in fruit, glandular-pubescent, greenish; teeth lanceolate, acute. Petal-limb *c.* 10 mm, entire, pink. Capsule 7–9 mm; carpophore 9–10 mm, pubescent. Seeds 1·5–1·75 mm, reniform; faces very deeply excavate, striate; back very wide, plane or slightly concave, with 3 or 4 rows of low tubercles. $2n = 24$. *Naturalized in Mallorca and perhaps in S.W. Spain.* [Bl ?Hs]. (*N. Africa.*)

(140–142). S. diversifolia group. Stems 10–50 cm, with short, deflexed eglandular hairs above, subglabrous below. Lower leaves oblanceolate to spathulate, pubescent. Calyx 7–11 mm, the veins reddish or greenish; teeth very obtuse. Petal-limb up to 4 mm, emarginate or bifid, pink. Capsule 7–8 mm, 1–2 times as long as wide. Seeds 0·75–1 mm, reniform, with wide notch; faces excavate, sharply ridged or striate.

Plants of this group have been variously treated by different authors. The following account is based on that by B. Oxelman, *Bot. Jour. Linn. Soc.* **106**: 115–117 (1991).

1 Petal-limb emarginate; veins of calyx not anastomosing **140. diversifolia**
1 Petal-limb bifid; veins of calyx often anastomosing
 2 Middle cauline leaves acute **141. turbinata**
 2 Middle cauline leaves obtuse **142. bergiana**

140. S. diversifolia Otth in DC., *Prodr.* **1**: 378 (1824) (*S. rubella* auct., non L.; incl. *S. segetalis* Dufour). Lower leaves spathulate, the middle obovate, obtuse; leaves of third node from top 2–4 times as long as wide. Veins of calyx often reddish, never anastomosing; teeth without conspicuous colouring of veins. Petal-limb 3–4 mm, emarginate. Carpophore 2–3·5 mm. Capsule up to twice as long as wide, shiny. Face of seeds twice as wide as back. *W. part of Mediterranean region, extending to S.E. Italy.* Bl †Co Hs It Sa Si [Ju].

141. S. turbinata Guss., *Fl. Sic. Prodr.* **1**: 506 (1827) (*S. rubella* L. subsp. *turbinata* (Guss.) Chater & Walters). Lower leaves oblanceolate, the middle linear-lanceolate, acute; leaves of third node from top 5–7 times as long as wide. Veins of calyx reddish or greenish, sometimes anastomosing, often wide; teeth often appearing acute due to arrangement of conspicuously coloured veins. Petal-limb 1–3 mm, bifid. Carpophore 1·5–2·5 mm. Capsule $1\frac{1}{2}$ times as long as wide, dull. Face of seeds 1–$1\frac{1}{2}$ times as wide as back. *Sicilia.* Si.

142. S. bergiana Lindman, *Acta Horti Berg.* **1(6)**: 4 (1891) (*S. rubella* L. subsp. *bergiana* (Lindman) Graebner & Graebner fil.). Like **141** but lower leaves spathulate, the middle oblanceolate to lanceolate, obtuse; leaves of third node from top 4–6 times as long as wide; veins of calyx greenish, wide, anastomosing; teeth not appearing acute; petal-limb 1–2 mm; carpophore 2–3 mm; capsule slightly shiny; face of seeds $1\frac{1}{2}$–2 times as wide as back. ● *Portugal, S. Spain.* Hs Lu.

143. S. insularis W. Barbey, *Bull. Soc. Vaud. Sci. Nat.* ser. 3, **21**: 220 (1886). Stems 5–15 cm, procumbent-ascending, branched, pubescent with patent hairs. Flowers 2–5 in lax, raceme-like monochasial cymes; pedicels 1–3 times as long as calyx. Calyx 6–10 mm, oblong-ovoid in fruit; teeth ovate, obtuse. Petal-limb 2 mm, entire, pink. Capsule *c.* 8 mm; carpophore 1–2 mm, stout. Seeds *c.* 1·5 mm, subglobose, umbilicate at hilum. $2n = 24$. ● *S. Aegean region (Karpathos).* Cr.

144. S. divaricata Clemente, *Elench. Horti Matrit.* 103 (1806) (incl. *S. willkommiana* Gay, *S. ramosissima* sensu Willk., non Desf.). Stems 10–20 cm, branched, terete, densely hirsute with long, patent hairs. Inflorescence lax, irregularly dichotomously and divaricately branched; pedicels mostly 1–3 times as long as calyx. Calyx (11–)13–15 mm, clavate in fruit, hirsute, usually tinged with red; teeth ovate, acute. Petal-limb very short, emarginate, pink. Capsule 10–11 mm; carpophore 3–4 mm, very stout, glabrous. Seeds 0·75–1 mm, reniform; faces excavate, obtusely ridged; back wide, shallowly grooved. $2n = 24$. *S.E. Spain.* Hs. (*N.W. Africa.*)

145. S. integripetala Bory & Chaub. in Bory, *Expéd. Sci. Morée* **3(2)**: 123 (1832). Stems 10–40 cm, usually branched, glandular-pubescent with rather thick, patent or ascending hairs. Lower leaves often crowded. Inflorescence few- to many-flowered, dichotomously branched; pedicels $\frac{1}{2}$–3 times as long as calyx. Calyx 13–17 mm, narrowly clavate in fruit, reddish, shortly glandular-pubescent; teeth ovate, obtuse. Petal-limb pink. Capsule 7–8 mm; carpophore 7–8 mm, slender, glabrous. Seeds 0·6–0·75 mm, reniform; faces slightly concave, striate; back not wide, shallowly grooved. $2n = 24$. ● *S. Greece; Kriti.* Cr Gr.

(a) Subsp. **integripetala**: Plant lax; stems 10–30(–40) cm. Inflorescence many-flowered, spreading. Petal-limb 10–12 mm, entire. $2n = 24$. *S. Greece.*

(b) Subsp. **greuteri** (Phitos) Akeroyd, *Bot. Jour. Linn. Soc.* **97**: 341 (1988): Plant compact; stems 4–10 cm. Inflorescence few-flowered; flowers often solitary. Petal-limb 4–6 mm, emarginate. *W. Kriti.*

146. S. sedoides Poiret, *Voy. Barb.* **2**: 164 (1789). Stems 5–20(–30) cm, usually much-branched, erect, slender, densely pubescent with patent hairs. Inflorescence profusely and divaricately branched, the branches forming raceme-like monochasial cymes; pedicels mostly 1–2 times as long as calyx, patent in fruit. Calyx 6–8(–9) × 2–3(–3·5) mm, oblong-clavate in fruit, shortly and stiffly pubescent, greenish, rarely reddish; teeth ovate, obtuse or subacute. Petal-limb short, reddish or whitish. Capsule 5–6 mm; carpophore 1·5–2 mm. Seeds 0·3–0·5 mm, reniform; faces slightly concave, striate; back wide, obtusely grooved. $2n = 24$. *Mediterranean region, mainly in the east.* Al Bl Cr Ga Gr Hs It Ju Sa Si.

147. S. pentelica Boiss., *Diagn. Pl. Or. Nov.* **2(8)**: 74 (1849). Like **146** but inflorescence a more or less regular dichasium; pedicels erect in fruit; calyx usually reddish-tinged; capsule (5–)6–7·5 mm, often exceeding calyx by *c.* 2 mm; carpophore 2–2·5 mm; seeds 0·5–0·75 mm, more acutely striate on face and with slightly narrower back. *E. Greece (Attiki, Evvoia).* Gr.

148. S. haussknechtii Heldr. ex Hausskn., *Mitt. Thür. Bot. Ver.* nov. ser., **5**: 51 (1893). Like **146** but inflorescence a very lax, more or less regular dichasium; pedicels erect in fruit; calyx 7–10 mm, clavate in fruit; carpophore 2–3 mm; seeds 0·75–1 mm; back narrow, obtusely grooved. *Rocky and stony places on serpentine.* ● *Mountains of N.W. Greece.* Gr.

149. S. laconica Boiss. & Orph. in Boiss., *Diagn. Pl. Or. Nov.* **3(6)**: 34 (1859). Like **146** but inflorescence a more or less regular dichasium; pedicels erect in fruit; flowers larger; calyx usually reddish-tinged; petal-limb bilobed, pink; carpophore more than 4 mm, equalling capsule. ● *S. Greece (Parnon).* Gr.

Sect. BEHENANTHA Otth. Annuals; stems usually glabrous and viscid above. Calyx contracted at mouth in fruit.

150. S. cretica L., *Sp. Pl.* 420 (1753) (*S. clandestina* Jacq.; incl. *S. annulata* Thore, *S. tenuiflora* Guss.). Stems (15–)30–60 cm, usually branched, erect, puberulent at base with deflexed hairs, glabrous above, rarely entirely glabrous. Flowers in lax, usually narrow dichasia; pedicels 2–5 times as long as calyx. Calyx 9–16 mm, ovoid in fruit; teeth triangular, acute or acuminate. Petal-limb large, red, emarginate or bifid. Capsule 7–10 mm; carpophore 1–5 mm, glabrous. Seeds 0·75–2 mm, reniform; faces slightly concave, tuberculate-ridged; back wide, plane or slightly concave, with 3 or 4 rows of low, rounded tubercles. *Mediterranean region, westwards to N.W. Italy; locally naturalized in Spain and Portugal, and occasionally casual elsewhere.* Al Bu Cr Gr It Ju Si Tu [Hs Lu].

Plants from Greece and Italy with the calyx 14–16 mm, the carpophore 4–5 mm, the capsule *c.* 10 mm and the seeds *c.* 0·75 mm have been separated as **S. tenuiflora** Guss., *Pl. Rar.* 177 (1826). A predominantly western variant with the petals very deeply bifid, the carpophore less than 2 mm, the capsule 7–8 mm, and the seeds *c.* 2 mm, has been separated as **S. annulata** Thore, *Essai Chlor. Land.* 173 (1803) (*S. cretica* subsp. *annulata* (Thore) Hayek). These variations are of a clinal nature, and do not merit even subspecific status.

151. S. ungeri Fenzl in Unger, *Wiss. Ergeb. Reise Griech.* 136 (1862) (*S. aetolica* Heldr.). Stems 15–40 cm, often branched, erect, puberulent at base with deflexed hairs, glabrous and very viscid above, often purplish. Flowers in 3- to 12-flowered dichasial cymes; pedicels shorter than, or up to *c.* 4 times as long as calyx. Calyx 14–17 mm, clavate, glabrous; teeth ovate, acute or acuminate, often patent; veins distinctly anastomosing above, reddish. Petal-limb 7–8 mm, emarginate, crimson. Capsule 8–11 mm; carpophore 6–8 mm. Seeds *c.* 1 mm, reniform; faces concave, tuberculate-ridged; back not wide, more or less plane, with 3 rows of rounded tubercles. ● *W. Greece.* ?Al Gr.

152. S. graeca Boiss. & Spruner in Boiss., *Diagn. Pl. Or. Nov.* **1(1)**: 36 (1843). Whole plant glabrous; stems (10–)20–40(–60) cm, usually branched, erect. Leaves glaucous; cauline dense but not usually overlapping. Inflorescence dichotomously branched at base with 1 alar flower and 2 equal, raceme-like, monochasial cymes; pedicels very short; flowers not overlapping. Calyx 9–15 mm, oblong-subclavate in fruit; teeth ovate, obtuse, usually greenish. Petals white or flesh-coloured; limb 5–10 mm, bifid. Capsule 6–9 mm, ovoid-oblong; carpophore 2–4 mm, stout. Seeds 1–1·3 mm, reniform; faces slightly concave, tuberculate-striate; back very wide, more or less plane, with 4 or 5 rows of low, rounded tubercles. $2n = 24$. *S. part of Balkan peninsula.* Al Bu Gr Ju.

153. S. muscipula L., *Sp. Pl.* 420 (1753). (*S. arvensis* Loscos, non Salisb.; incl. *S. corymbifera* Bertol.). Stems 15–40 cm, usually simple below, erect, generally glabrous, rather rigid. Cauline leaves dense, always imbricate throughout most of stem. Inflorescence a more or less regularly branched dichasium; internodes rather long; branches strict; pedicels all very short. Calyx 13–17 mm, oblong-clavate; veins prominent, greenish or reddish, anastomosing; teeth ovate-lanceolate, acute. Petal-limb deeply emarginate, pink. Capsule 8–10 mm, oblong; carpophore 3–6(–7) mm, pubescent. Seeds 0·9–1·1 mm, reniform; faces plane, striate, tuberculate on edge; back narrow, obtusely grooved. $2n = 24$. *Cultivated ground. Mediterranean region, Portugal; doubtfully native in some regions.* Ga *Gr Hs It Lu *Si.

154. S. stricta L., *Cent. Pl.* **2**: 17 (1756). Like **153** but calyx with 5 herbaceous wings; calyx-teeth lanceolate-acuminate; capsule ovoid-conical, attenuate above; seeds *c.* 0·75 mm, with subacute tubercles on sides of faces and on back. $2n = 24$. *S. Spain, Portugal.* Hs Lu.

155. S. behen L., *Sp. Pl.* 418 (1753) (incl. *S. reinholdii* Heldr.). Whole plant glabrous; stems 15–20 cm, simple or branched, rather stout, erect or ascending. Cauline leaves glaucous, usually imbricate. Inflorescence a lax dichasium; lowest pair of internodes of fruiting inflorescence 5–10 cm; next pair also long; flowers not overlapping; pedicels stout, usually shorter than calyx. Calyx 11–17 mm, ovoid in fruit, whitish, with reddish, anastomosing veins above; teeth ovate, obtuse. Petal-limb 3–6 mm, pink. Capsule 9–10 mm, ovoid; carpophore 1–2 mm, very stout. Seeds *c.* 1·5 mm, reniform; faces concavo-convex, strongly tuberculate-ridged; back wide, plane or slightly convex, with 4 rows of acute, conical tubercles. $2n = 24$. *Mediterranean region.* Cr Gr Hs It *Sa Si Tu.

156. S. holzmannii Heldr. ex Boiss., *Fl. Or., Suppl.*, 91 (1888). Like **155** but lowest pair of internodes of fruiting inflorescence 2–4 cm; next pair very short; flowers (except the lowest one) overlapping; petals smaller; limb scarcely exserted; capsule indehiscent; seeds echinate with parallel-sided spines. $2n = 24$. *Islets in S. Aegean region.* Cr Gr.

Sect. LASIOCALYCINAE Chowdhuri. Annuals. Calyx contracted at mouth in fruit; veins papillose-scabrid or with ascending, bulbous-based hairs.

157. S. linicola C. C. Gmelin, *Fl. Bad.* **4**: 304 (1826). Whole plant scabrid-puberulent; stems erect, slender, with deflexed scales or hairs, simple below, strictly branched above. Inflorescence a lax, subcorymbose, more or less regularly-branched dichasium; pedicels $\frac{1}{2}$–3 times as long as calyx. Calyx 11–14 mm, broadly clavate in fruit, with ascending scales or hairs; veins broad, green, anastomosing; teeth triangular-ovate, obtuse. Petal-limb 2–4 mm, pink. Capsule 9–11 mm, ovoid; carpophore 3–5 mm. Seeds 1·75–2 mm, reniform; faces deeply excavate, ridged-striate; back very wide, slightly concave. *Flax-fields, becoming very rare. Italy; probably extinct elsewhere.* ?Ga It ?Ju.

A closely related species, **S. crassipes** Fenzl, *Pugillus* 8 (1842), from Syria and Palestine, has been reported from cultivated fields in Thrace, but there are no recent records.

158. S. echinata Otth in DC., *Prodr.* **1**: 380 (1824). Stems 20–60 cm, simple or branched, puberulent to pubescent-hispid with deflexed (or the longer patent) hairs. Inflorescence irregularly dichotomously branched; flowers subsessile, 4–6 in rather dense, long-stalked dichasia. Calyx 15–20 mm, clavate in fruit, greenish, sometimes purplish above; teeth ovate-lanceolate, acute; hairs on veins *c.* 1·5 mm, narrow, with bulbous base shorter than the narrow distal part. Petal-limb 6–8 mm, pink or whitish. Capsule (7–)8–11 mm, ovoid-oblong; carpophore (5–)7·5–9 mm, glabrous. Seeds 0·6–0·8 mm, reniform; faces plane or slightly concave, tuberculate; back concave or grooved, not wide. ● *S. Italy, casual further north.* It.

159. S. squamigera Boiss., *Diagn. Pl. Or. Nov.* **1(1)**: 38 (1843). Like **158** but branches more patent; flowers in lax, indistinct dichasia; calyx 14–15 mm, teeth ovate, obtuse; hairs on veins *c.* 1 mm, with bulbous base longer than the narrow distal part; carpophore pubescent; seeds 1–1·25 mm, with radially elongated tubercles on faces. *C. & S. Greece; E. Jugoslavia.* Gr Ju.

160. S. gallinyi Reichenb., *Fl. Germ. Excurs.* 815 (1832) (*S. trinervia* Sebastiani & Mauri). Stems 20–60 cm, usually divaricately branched, pubescent with appressed hairs. Inflorescence of 1 to several 3- to 8-flowered, long-stalked raceme-like monochasial cymes; flowers subsessile, usually slightly overlapping. Calyx 13–20 mm, clavate in fruit, greenish; hairs on veins less than 1 mm, the relative length of the bulbous base variable. Petal-limb *c.* 6–7 mm, pink. Capsule 8–10 mm; carpophore 5–8 mm, pubescent, slender. Seeds 0·75–0·9 mm, reniform; faces slightly concavo-convex, tuberculate-ridged; back wide, more or less plane, with 4 rows of low, rounded tubercles. $2n = 24$. *Balkan peninsula, extending to S.W. Romania and C. Italy.* Al Bu Gr It Ju Rm Tu.

Sect. EUDIANTHE (Reichenb.) A. Braun. Annuals. Inflorescence a very lax, irregular, several-flowered dichasium; lower pedicels 4–10, upper 2–6 times as long as calyx. Styles 5. Capsule dehiscing with 5 bifid teeth. Seeds 0·6–1 mm, subreniform, more or less umbilicate at hilum; faces convex, with obtuse, conical tubercles; back very broad and convex, with acute tubercles.

161. S. laeta (Aiton) Godron in Gren. & Godron, *Fl. Fr.* **1**: 220 (1847) (*Eudianthe laeta* (Aiton) Willk.; incl. *S. loiseleurii* Godron). Whole plant glabrous, or upper part of stems, pedicels and calyx-veins sparsely spinulose-papillose (var. *loiseleurii* (Godron) Rouy & Fouc.). Leaves linear-lanceolate. Calyx 6–10 mm, broadly campanulate and widest at mouth in fruit; bands of thickened tissue present between veins; tube 2–3 times as long as the triangular-acuminate, often patent teeth. Petal-limb 4–10 mm, pink. Capsule 5–10 mm, globose to ovoid; carpophore 1–2 mm, glabrous. $2n = 24$. *Damp places. S.W. Europe, extending eastwards to C. Italy.* Co Ga Hs It Lu Sa.

162. S. coeli-rosa (L.) Godron in Gren. & Godron, *op. cit.* 221 (1847) (*Eudianthe coeli-rosa* (L.) Reichenb.). Whole plant glabrous. Stems 20–50 cm. Leaves linear-lanceolate. Calyx 15–28 mm, subclavate and contracted at mouth in fruit, deeply sulcate between veins, with transverse undulations on each side of grooves; tube 2–3 times as long as the linear, acuminate or acute, usually patent teeth; petal-limb 1–2 cm, pink. Capsule (7–)10–17 mm; carpophore (5–)7–12 mm, glabrous. *W. & C. Mediterranean region.* Co Hs It Lu Sa Si [Hu].

Sect. ERECTOREFRACTAE Chowdhuri. Annuals; stems glandular-pubescent. Inflorescence of raceme-like monochasial cymes, or dichasial; angle of pedicels to axis changing after flowering. Calyx usually contracted at mouth.

163. S. pendula L., *Sp. Pl.* 418 (1753). Stems 15–40 cm, procumbent-ascending, branched, pubescent. Cauline leaves 2–5 cm, ovate to ovate-lanceolate, acute, pubescent. Inflorescence very lax, of raceme-like monochasial cymes. Pedicels usually erect in flower, patent or deflexed in fruit. Calyx 13–18 mm, obovoid in fruit, contracted at mouth, very inflated and loose, with wide hyaline bands between the prominent, narrow veins; teeth short, ovate or triangular, obtuse. Petal-limb 7–10 mm, pink, rarely white. Capsule 9–12 mm, ovoid-conical; carpophore 3–6 mm. Seeds 1·3–1·4 mm, subglobose, blackish, faces convex or plano-convex, with concentric rings of small tubercles; back convex or plano-convex, with 7 or 8 rows of tubercles. $2n = 24$. *Italy; cultivated for ornament, and a frequent casual elsewhere.* It [Hs.].

The plant commonly cultivated is a *flore pleno* variant.

164. S. psammitis Link ex Sprengel, *Novi Provent.* 39 (1818) (*S. agrostemma* Boiss. & Reuter; incl. *S. lasiostyla* Boiss.). Stems 10–20 cm, ascending, much-branched, glandular-pubescent. Basal and lower cauline leaves usually 1–3 cm, linear-lanceolate or linear, somewhat fleshy, often crowded. Inflorescence a few-flowered, usually condensed monochasial cyme; lower pedicels usually longer than calyx, erect in flower, patent or deflexed in fruit. Calyx 13–18 mm, contracted at mouth, straight, less distinctly inflated in fruit than in **163**, with wide hyaline bands between the prominent, narrow veins. Petals pink, rarely white. Capsule 9–12 mm, ovoid; carpophore 3–6 mm, glabrous. Seeds 0·8–1·2 mm, otherwise like those of **163**. $2n = 24$. *Spain and Portugal.* Hs Lu.

165. S. littorea Brot., *Fl. Lusit.* **2**: 186 (1805) (incl. *S. adscendens* Lag.). Stems 5–15 cm, ascending, much-branched, glandular-pubescent. Basal and lower cauline leaves oblong-obovate to linear-oblong. Inflorescence a few-flowered, usually condensed, leafy monochasial cyme; lower pedicels usually equalling or shorter than the calyx. Calyx 10–19 mm, subclavate but not loose and inflated in fruit, with narrow hyaline bands, not or scarcely contracted at mouth. Petal-limb purplish-pink or white. Capsule 6–9 mm, ovoid; carpophore (2·5–)4–9 mm. Seeds 0·6–0·8 mm, reniform; faces convex, reticulate-pitted, rather smooth; back slightly grooved or plane, smooth. $2n = 24$. *Maritime sands. Spain and Portugal.* Hs Lu.

Plants from maritime sands in Islas Baleares and one locality in E. Spain have been called **S. cambessedesii** Boiss. & Reuter, *Pugillus* 18 (1852). They differ from **165** principally in the longer calyx, capsule and carpophore, and are possibly subspecifically distinct.

166. S. oropediorum Cosson, *Ill. Fl. Atl.* **1**: 132 (1891). Like **165** but taller and more erect; seeds 0·8–1·2 mm, deeply grooved and striate. *Weed of cultivation. E. Spain (Albacete prov.).* Hs. *(N. Africa.)*

167. S. stockenii Chater, *Lagascalia* **3**: 219 (1973). Like **165** but basal leaves linear-spathulate; cauline leaves linear-oblanceolate or linear; inflorescence lax; calyx (17–)19–25 mm; petal-limb bright pink; claw yellowish-white; carpophore 5–9 mm; seeds 0·7–1 mm. $2n = 24$. ● *S. Spain (Prov. Cádiz).* Hs.

168. S. germana Gay in Cosson, *Not. Pl. Crit.* 31 (1849). (*S. boissieri* Gay). Stems 10–20 cm, erect, usually branched. Basal leaves ovate-spathulate; cauline linear-lanceolate to linear. Flowers 3–8 in more or less regular dichasia, sometimes solitary; pedicels of lower flowers often equalling, of the upper shorter than, calyx, usually nodding in flower, erect in fruit. Calyx 11–15 mm, oblong in flower, ovoid in fruit, toothed for $\frac{1}{4}$–$\frac{1}{3}$ of its length with lanceolate, acute or triangular, acuminate teeth; veins rather wide, not or obscurely anastomosing. Petal-limb 3–4 mm, white; claw exserted. Capsule 8–11 mm, ovoid; carpophore 2–4 mm, glabrous. Seeds *c.* 1 mm, reniform, chestnut or blackish-brown; faces slightly concavo-convex, tuberculate-ridged; back wide, more or less plane, with 3 or 4 rows of tubercles. $2n = 24$. ● *S. Spain.* Hs.

169. S. almolae Gay in Cosson, *loc. cit.* (1849). Like **168** but stems always simple, less viscid, sometimes subglabrous in part;

calyx (12–)15–18 mm, toothed for $\frac{1}{6}$–$\frac{1}{4}$ of its length with triangular, acute teeth; veins more distinctly anastomosing; petals very pale pink; claw included; seeds up to *c.* 1·4 mm, with 2 rows of tubercles on back. $2n = 24$. ● *C. & S.E. Spain.* Hs.

Sect. DICHOTOMAE (Rohrb.) Chowdhuri. Annuals; stems erect, pubescent. Inflorescence branched, composed of raceme-like monochasial cymes. Calyx contracted at mouth in fruit, with prominent simple veins. Seeds reniform, with plane or slightly concave, tuberculate faces, and bluntly tuberculate back.

170. S. dichotoma Ehrh., *Beitr. Naturk.* 7: 143 (1792). (*S. mathei* Pénzes). Stems 20–100 cm, branched above, puberulent to hispid-pubescent. Lower leaves spathulate to lanceolate; cauline lanceolate. Inflorescence usually with alar flowers; branches 5- to 10-flowered; pedicels all very short; upper bracts small, ovate, with scarious margins. Calyx 7–15 mm, ovoid-oblong in fruit, not inflated, sparsely and shortly pubescent between veins, sparsely hispid on veins; teeth ovate-lanceolate, acute. Petal-limb 7–8 mm, white, rarely pink; filaments and styles very long-exserted. Carpophore 1·5–4 mm, very stout, glabrous. Seeds 1·1–1·4 mm, blackish-brown. $2n = 24$. *Native in S. & E. Europe; widely naturalized elsewhere.* Bu Cr Gr Hu It Ju Rm Rs(B, C, W, K, E) Tu [Be Co Cz Ga Ge He Hs Po Rs(N) Su].

(a) Subsp. **dichotoma**: Basal leaves not in dense rosettes, green, pubescent. Flowers crowded, or sometimes separated by their own length; inflorescence 1 or 2(3) times dichotomously branched; upper bracts not ciliate. Petal-limb bifid for *c.* $\frac{2}{3}$ of its length. $2n = 24$. *Throughout the range of the species, except Greece and the Aegean region.*

(b) Subsp. **racemosa** (Otth) Graebner in Ascherson & Graebner, *Syn. Mitteleur. Fl.* **5(2)**: 93 (1920) (*S. racemosa* Otth, *S. sibthorpiana* Reichenb.): Basal leaves in dense rosettes, more or less densely grey-pubescent. Flowers less densely crowded; inflorescence often more branched; upper bracts ciliate. Petal-limb bifid almost to base. *S.E. Europe.*

Subsp. **euxina** (Rupr.) Coode & Cullen, *Notes Roy. Bot. Gard. Edinb.* **28**: 8 (1967) (*S. euxina* Rupr.), from maritime sands of the N. Aegean and Black Sea, is similar to subsp. (**b**) but has narrower leaves, up to 8 mm wide. It probably does not merit subspecific recognition. The European plants have white petals (*S. euxina* var. *bulgarica* Jordanov & Panov), whereas they are usually pink in Asiatic material.

171. S. heldreichii Boiss., *Diagn. Pl. Or. Nov.* **2(8)**: 81 (1849) (*S. remotiflora* Vis.). Like **170(a)** but lower leaves ovate-lanceolate; cauline linear-lanceolate; inflorescence 1–3 times dichotomously branched, without alar flowers; flowers very remote, 1–3 on each branch; pedicels up to $\frac{1}{2}$ as long as calyx; upper bracts larger, mostly equalling flowers; calyx scabrid between veins, serrate with ascending sharp teeth on veins. *W. part of Balkan peninsula; very local.* Al ?Gr Ju.

Sect. SCORPIOIDEAE (Rohrb.) Chowdhuri. Annuals, rarely perennials. Calyx not contracted at mouth in fruit; veins usually anastomosing. Seeds of various types, but never winged.

172. S. nicaeensis All., *Mélang. Philos. Math. Soc. Roy. Turin (Misc. Taur.)* **5**: 88 (1774). Often perennial; whole plant hirsute; stems 10–40 cm, erect, stout, usually branched, viscid. Lower leaves linear-spathulate to oblong-lanceolate; cauline linear-lanceolate to linear, often curved. Flowers on pedicels equalling or shorter than calyx, 2–8 together in sessile or shortly stalked, subcapitate dichasia which are arranged, the lower remote, the upper overlapping, on long axes in a simple or branched inflorescence. Calyx 10–13 mm, subclavate in fruit; teeth ovate, obtuse. Petal-limb very deeply bifid, with oblong-linear lobes, white or reddish-pink. Capsule 6–9 mm, ovoid-oblong; carpophore 3·5–5 mm, pubescent. Seeds 0·9–1·1 mm, reniform, dark brown; faces flat, finely striate; back narrow, more or less deeply grooved. $2n = 24$. *Maritime sands. Mediterranean region, mainly in the west; Portugal.* Bl Co Ga Gr Hs It Lu Sa Si.

173. S. scabriflora Brot., *Fl. Lusit.* **2**: 184 (1805) (*S. hirsuta* Lag., non Poiret; incl. *S. laxiflora* Brot.). Whole plant often densely villous, sometimes pubescent or subglabrous in part, the hairs always more or less ascending; buds and shoot-apices always villous; stems 5–35 cm, ascending or erect, branched. Flowers 4–9 in raceme-like monochasial cymes, all but the lower 1 or 2 usually overlapping and subsessile; inflorescence usually branched, never with alar flowers. Calyx (11–)13–25 mm, clavate in fruit, usually villous; veins not anastomosing; teeth obtuse. Petal-limb 8–10 mm, pale to deep red. Capsule 5–9 mm, ovoid-oblong; carpophore 7–16 mm. Seeds 0·6–0·8 mm, reniform, brown, slightly pruinose; faces plane, ridged; back wide, plane or concave, tuberculate. $2n = 24$. ● *Spain, Portugal.* Hs Lu.

S. mariana Pau, *Mem. Soc. Esp. Hist. Nat.* (tomo extr.) 292 (1921), from S. Spain, has a more robust habit and slightly larger capsules and seeds, but otherwise closely resembles **173**.

174. S. micropetala Lag., *Gen. Sp. Nov.* 15 (1816). Like **173** but calyx 10–15 mm, cuneate-subclavate in fruit; veins usually anastomosing above; petal-limb up to 3 mm, white or pink; capsule 9–10 mm, oblong; carpophore 3–4 mm; seeds with striate faces and less strongly tuberculate back. $2n = 24$. *C. & S. Spain, Portugal.* Hs Lu.

175. S. discolor Sm. in Sibth. & Sm., *Fl. Graec. Prodr.* **1**: 292 (1809) (*S. pompeiopolitana* Gay ex Boiss.). Like **173** but stems and calyx with both long, soft hairs and short, glandular hairs; plant never villous; calyx 10–12 mm; petal-limb 4–6 mm, pink above, greenish beneath; capsule 7–8 mm; carpophore 4 mm; seeds 0·4–0·6 mm, chestnut-brown; faces very finely striate; back narrow, concave or grooved with a median row of low tubercles. *Maritime sands. Karpathos.* Cr. (*E. Mediterranean region.*)

176. S. obtusifolia Willd., *Enum. Pl. Horti Berol.* 473 (1809). Plant sometimes perennial, never villous; stems 15–35 cm, erect, branched, densely pubescent with patent or ascending hairs. All leaves except uppermost spathulate to obovate-spathulate, obtuse, rather thick; uppermost oblanceolate, obtuse; all ciliate. Inflorescence as in **173**. Calyx 10–13(–14) mm, subclavate in fruit, pubescent with short, ascending hairs; veins somewhat raised, usually reddish, anastomosing; teeth oblong, obtuse. Petals pink or white; coronal scales well-developed. Capsule 8–9 mm, oblong; carpophore 3–5·5 mm, puberulent. Seeds 0·6–1 mm, elongate-reniform, dark brown; faces excavate; back very wide, with an obtuse median groove. $2n = 24$. *Maritime sands and rocks. S.W. Spain, S.W. Portugal.* Hs Lu. (*N.W. Africa; Islas Canarias.*)

177. S. gaditana Talavera & Bocquet, *Lagascalia* **5**: 50 (1975). Like **176** but always annual; leaves lanceolate-spathulate; cauline leaves acute to subacute; calyx 9–10·5 mm, usually glandular; capsule 7–8 mm. $2n = 24$. *Stony slopes.* ● *S.W. Spain.*

178. S. sericea All., *Fl. Pedem.* **2**: 81 (1785). Stems up to 50 cm, pubescent, ascending, branched. Leaves linear-lanceolate. Flowers solitary and terminal at ends of stems, rarely 2 or 3 in a reduced monochasial cyme. Calyx 12–20(–23) mm; teeth lanceolate, acute or subobtuse. Petals pink. Carpophore 4–10(–12) mm. Seeds 1–1·4 mm, reniform or elongate-reniform, dark brown; faces concave-excavate, striate; back grooved between rounded or angled (not winged and undulate) sides. $2n = 24$. *Maritime sands.* ?● *W. Mediterranean region.* Bl Co ?Ga It Sa.

Often confused with **186** and best distinguished by the seeds.

179. S. nocturna L., *Sp. Pl.* 416 (1753) (*S. micropetala* subsp. *boullui* (Jordan) Rouy & Fouc., *S. boullui* Jordan). Stems 10–60 cm, erect, usually branched. Lower leaves obovate- to lanceolate-spathulate, obtuse, pubescent. Flowers 5–15 in raceme-like monochasial cymes, very densely crowded above, remote below (or flowers 1–5 in the cymes, mostly remote, var. *brachypetala* (Robill. & Cast.) Vis.); inflorescence usually compound, sometimes with alar flowers. Calyx (6–)9–13 mm, oblong-obovoid in fruit, pubescent with ascending hairs; veins wide, green, anastomosing. Petal-limb usually pink, bifid, exserted (or emarginate and included, var. *brachypetala*). Capsule 8–11 mm, oblong or ovoid-oblong; carpophore 1–1·5 mm, puberulent. Seeds reniform; faces excavate-auriculate; back wide, with shallow, tuberculate groove. *Mediterranean region, extending to Portugal and N. Spain.* Al Bl Co Cr Ga Gr Hs It Ju Lu Sa Si Tu.

(**a**) Subsp. **nocturna**: Plant usually fairly sparsely pubescent below. Calyx-teeth usually triangular-acuminate or lanceolate and acute. Filaments glabrous. Seeds 0·5–0·7 mm, blackish. $2n = 24$. *Throughout the range of the species.*

(**b**) Subsp. **neglecta** (Ten.) Arcangeli, *Comp. Fl. Ital.* 88 (1882) (*S. neglecta* Ten., *S. reflexa* auct.): Plant usually densely hirsute below. Lower leaves often patent or deflexed. Calyx-teeth linear. Alternate filaments hairy at base. Seeds 0·8–0·9 mm, reddish-brown. *S. France, Italy, Sicilia.*

S. reflexa (L.) Aiton fil. in Aiton, *Hort. Kew.* ed. 2, **3**: 86 (1811), non Moench, is probably a cleistogamous variant of subsp. (**a**).

Sect. SILENE. Annuals. Inflorescence of monochasial cymes. Calyx contracted at mouth in fruit. Carpophore pubescent. Seeds less than 1 mm, reniform, not winged.

180. S. ramosissima Desf., *Fl. Atl.* **1**: 354 (1798). Whole plant viscid; stems 10–40 cm, stout, erect, branched, villous. Leaves oblong-spathulate, obtuse, pubescent. Flowers in monochasial cymes; lower pedicels 1–5 times as long as calyx. Calyx 9–12 mm, cylindrical, becoming ovoid in fruit, hairy, toothed for $\frac{1}{4}$ of its length with lanceolate-triangular, acute teeth. Petals pink or white. Capsule 8–10 mm; carpophore 1·5–2 mm, stout. Seeds pale brown; faces more or less plane and smooth; back very narrow, deeply and acutely grooved. $2n = 24$. *Maritime sands. S. Portugal, S. & E. Spain.* Hs Lu.

181. S. gallica L., *Sp. Pl.* 417 (1753) (*S. anglica* L.; incl. *S. giraldii* Guss., *S. linophila* Rothm., *S. transtagana* Coutinho). Stems 15–45 cm, erect, simple to much-branched, pubescent, viscid above, or (var. *giraldii* (Guss.) Walters) glabrous. Leaves pubescent. Flowers subsecund, in 1 to many raceme-like monochasial cymes; lower pedicels up to $1\frac{1}{2}$ times as long as calyx, the upper short. Calyx 7–10 mm, cylindrical-ovoid, becoming ovoid in fruit, hispid, toothed for $\frac{1}{4}$ of its length with triangular, acute teeth; veins scarcely anastomosing. Petals white or pink, often with crimson spot (var. *quinquevulnera* (L.) Koch); limb up to 6 mm, entire or emarginate. Capsule 6–9 mm; carpophore usually 1 mm or less. Seeds dark brown; faces deeply concave, striate; back wide, plane. $2n = 24$. *W., S. & S.C. Europe; naturalized further north.* Al Au Az Bl Br Bu Co Cr Ga Gr He Hs Hu It Ju Lu Rm Rs(W) Sa Si *Tu [*Be Cz Ge *Hb Ho Po Rs(C)].

182. S. bellidifolia Juss. ex Jacq., *Hort. Vindob.* **3**: 44 (1776) (*S. hispida* Desf., *S. vespertina* Retz.). Stems 30–60 cm, erect, branched, pubescent or puberulent above; hispid and simple, rarely branched, below. Leaves sparsely hispid. Flowers subsessile, usually secund, in dense, raceme-like, long-stalked monochasial cymes. Calyx 14–17 mm, cylindrical-clavate, becoming clavate in fruit, villous. Petal-limb 3–5 mm, bifid, pink. Capsule 9–11 mm; carpophore 4–5 mm. Seeds very small, dark brown; faces deeply concave, striate; back wide, shallowly and broadly grooved. $2n = 24$. *Mediterranean region; Portugal.* Al Bl Co Cr Gr Hs It Lu Sa Si Tu.

183. S. cerastoides L., *Sp. Pl.* 417 (1753). Whole plant somewhat hispid, glandular-viscid above; stems up to 25 cm, often branched at base. Flowers in lax, raceme-like monochasial cymes; pedicels not longer than calyx. Calyx 8–11 mm, cylindrical-clavate, glandular-pubescent, becoming very strongly contracted both above and below capsule in fruit, toothed for $\frac{1}{3}$ of its length with lanceolate, acute teeth; veins conspicuously anastomosing. Petal-limb small, bifid, pink. Capsule 6–8 mm, shortly attenuate above; carpophore 2–4 mm. Seeds dark brown; faces deeply concave, striate; back wide with narrow, shallow central groove. $2n = 24$. *S. & E. Spain, S. Portugal, Islas Baleares; a few localities in the E. Mediterranean.* Bl Cr Gr Hs Ju Lu.

184. S. tridentata Desf., *Fl. Atl.* **1**: 349 (1798). Whole plant hispidulous; stems 10–40 cm, simple or branched. Flowers subsessile, in lax, raceme-like monochasial cymes. Calyx 11–14 mm, ellipsoid-cylindrical, becoming ovoid in fruit, not so strongly contracted as in **183**, toothed to *c.* half-way, with acuminate or linear teeth; veins anastomosing. Petal-limb bifid, almost included in calyx, pink. Capsule 7–10 mm, attenuate above; carpophore 1–2 mm. Seeds dark brown; faces concave, striate; back wide, with wide, shallow central groove. $2n = 24$. *C., S. & E. Spain.* Hs.

185. S. disticha Willd., *Enum. Pl. Horti Berol.* 476 (1809). Whole plant hispid; stems 20–40 cm, erect, simple. Flowers subsessile or shortly pedicellate in a 3- to 8-flowered, subcapitate or fastigiate inflorescence. Calyx 8–9 mm, ellipsoidal, becoming ovoid and shortly clavate in fruit, toothed for $\frac{1}{4}$ of its length, with triangular, acute teeth; veins not anastomosing. Petals white. Capsule *c.* 7 mm; carpophore 1–2 mm, stout. Seeds dark brown; faces concave, striate; back very wide, with wide, shallow groove. $2n = 24$. *Islas Baleares; S. Spain, Portugal.* Bl Hs Lu.

Sect. DIPTEROSPERMAE (Rohrb.) Chowdhuri. Annuals. Inflorescence of various types. Calyx not or scarcely contracted at mouth in fruit. Seeds reniform, with flat faces, the back deeply and acutely grooved between 2 undulate wings.

186. S. colorata Poiret, *Voy. Barb.* **2**: 163 (1789). Whole plant pubescent or puberulent. Stems 10–50 cm, decumbent to erect, branched. Leaves linear to ovate-spathulate. Pedicels shorter than calyx; bracts of a pair usually unequal. Calyx 11–13(–17) mm, cylindrical, becoming broadly clavate in fruit; teeth ovate, obtuse, densely ciliate. Petal-limb 5–9 mm, pink or white. Capsule 7–9 mm, ovoid; carpophore 5–7 mm. Seeds 1–1·5 mm, dark chestnut-brown. $2n = 24$. *S. Europe.* Al *Co Cr Gr Hs It Lu Sa Si Tu.

Often confused with **178** and best distinguished by the seeds.

Plants from S. Italy, Sicilia and Greece with the calyx *c.* 10 mm, and carpophore and capsule 4–5 mm, have been called **S. canescens** Ten., *Prodr. Fl. Nap.* xxv (1811), and may be worth subspecific rank.

187. S. secundiflora Otth in DC., *Prodr.* **1**: 375 (1824) (*S. glauca* Pourret ex Lag., non Salisb.). Like **186** but leaves usually ovate-lanceolate and often subglabrous; lowest pedicels sometimes up to twice as long as calyx; bracts of a pair usually more or less equal; calyx 13–16 mm; teeth lanceolate, acute; veins more conspicuously anastomosing; carpophore 4·5–6 mm; seeds 2 mm or more. $2n = 24$. *S. & E. Spain; Islas Baleares.* Bl Hs ?Lu.

188. S. apetala Willd., *Sp. Pl.* **2(1)**: 703 (1799) (incl. *S. decipiens* Barc., non Ball). Whole plant pubescent; stems 10–35 cm, erect, branched, especially at base. Leaves lanceolate to linear-lanceolate; upper ovate-lanceolate. Lower pedicels as long as or up to 3 times as long as calyx. Calyx 7–10 mm, ellipsoid-cylindrical, becoming broadly campanulate in fruit; teeth triangular, acute. Petals absent, or included in calyx, or exserted, with bifid limb up to *c.* 3 mm. Capsule 6–7·5 mm; carpophore 1–2 mm. Seeds 1 mm or less, dull blackish-brown. $2n = 24$. *S.W. Europe; locally elsewhere in the Mediterranean region.* Al Bl Cr ?Gr Hs It Lu Sa Si.

189. S. longicaulis Pourret ex Lag., *Gen. Sp. Nov.* 15 (1816). Stems 10–40 cm, simple or little-branched, pubescent or hispid, glabrescent above; upper internodes very long. Leaves ovate-lanceolate to subspathulate, pubescent. Lower pedicels (1–)2–5 times as long as calyx. Calyx 9–13 mm, oblong-cylindrical, becoming campanulate-subclavate in fruit, pubescent on veins or glabrous; teeth triangular, acute. Petal-limb *c.* 5 mm, bifid, pink. Capsule 7·5–11 mm; carpophore 2·5–3·5 mm. Seeds 1·1–5 mm, chestnut-brown. $2n = 24$. *Sandy places, mainly maritime. C. & S. Portugal; S.W. Spain.* Hs Lu.

Sect. CONOMORPHA Otth. Annuals. Inflorescence of lax dichasia. Calyx 15- to 60-veined; veins usually prominent, parallel. Seeds reniform; faces plane or slightly concave, usually striate; back wide, with shallow groove.

190. S. ammophila Boiss. & Heldr. in Boiss., *Diagn. Pl. Or. Nov.* **2(8)**: 82 (1849). Whole plant grey-pubescent; stems up to 15 cm, branched at base, procumbent or ascending, viscid above. Lower leaves up to 1·5 cm, more or less obtuse. Flowers pink. Calyx cylindrical, becoming broadly campanulate in fruit, 15- to 20-veined; tube twice as long as the lanceolate, acute teeth. Capsule ovoid, subsessile. ● *S. Aegean region.* Cr.

(a) Subsp. **ammophila**: Calyx 5–7 mm. Petal-limb 1–2 mm, emarginate. Capsule 3–4 mm. *Maritime sands. E. Kriti and Gaidhouronisi.*

(b) Subsp. **carpathae** Chowdhuri, *Notes Roy. Bot. Gard. Edinb.* **22**: 278 (1957): Calyx 7–9·5 mm. Petal-limb 3–4 mm, entire. Capsule 5–7 mm. *Karpathos.*

191. S. macrodonta Boiss., *Diagn. Pl. Or. Nov.* **1(1)**: 37 (1843). Stems 10–30 cm, erect, densely and shortly pubescent. Leaves linear to lanceolate. Dichasia up to 15-flowered. Calyx 14–17 mm, 60-veined; veins slender, not prominent. Petal-limb white or pink, emarginate to bifid. Capsule 8–10 mm, shorter than calyx; carpophore 1–2·5 mm. *S. Aegean region (Karpathos).* Cr. (*S.W. Asia.*)

192. S. conica L., *Sp. Pl.* 418 (1753). Whole plant pubescent; dichasia 5- to 30-flowered. Calyx 30-veined, toothed for *c.* ⅓ of its length with acuminate teeth, shortly pubescent, glandular or eglandular. Petal-limb 3–5 mm. Capsule 7–12 mm, obovoid-conical. Seeds 0·75–1 mm, pruinose. *W., S. & S.C. Europe.* Al Au Be Br Bu Cr Cz Ga Ge Gr Ho Hs Hu It Ju Lu Rm Rs(W, K) Si Tu [Da He Po Su].

1 Carpophore 1–4 mm; calyx 13–18 mm (b) subsp. **subconica**
1 Carpophore less than 1 mm; calyx 8–15 mm
 2 Lower leaves obtuse; stems usually ascending (c) subsp. **sartorii**
 2 Lower leaves acute; stems erect (a) subsp. **conica**

(a) Subsp. **conica**: Stems 15–50 cm, erect. Lower leaves oblong- to linear-lanceolate, acute. Calyx 10–15 mm, obovoid-cylindrical in flower, broadly obovoid in fruit. Petal-limb comparatively narrow, pink, rarely white, bifid. Capsule 7–12 mm; carpophore less than 1 mm. $2n = 20$. *Almost throughout the range of the species, but absent from most of S.E. Europe and Ukraine.*

(b) Subsp. **subconica** (Friv.) Gavioli in Fiori & Béguinot, *Sched. Fl. Ital. Exsicc.*, ser. 3, **16**: 363 (1927) (*S. subconica* Friv.; incl. *S. juvenalis* Delile, *S. tempskyana* Freyn & Sint.): Plant glandular-pubescent. Stems 15–50 cm, erect. Lower leaves oblong- to linear-lanceolate, acute. Calyx 13–18 mm, cylindrical in flower, broadly obovoid in fruit. Petal-limb comparatively wide, pink, emarginate (claw sometimes much exceeding calyx, *S. subconica* var. *grisebachii* Davidov). Capsule 7–8 mm; carpophore 1–4 mm. $2n = 20$. *S.E. Europe; Italy.*

(c) Subsp. **sartorii** (Boiss. & Heldr.) Chater & Walters, *Feddes Repert.* **69**: 49 (1964) (*S. sartorii* Boiss. & Heldr.): Stems 5–20 cm, usually ascending. Lower leaves obovate-spathulate, obtuse. Calyx 8–13 mm, cylindrical in flower, obovoid in fruit. Petal-limb comparatively wide, pink, bifid. Capsule 8–10 mm, ovoid-conical; carpophore less than 1 mm. $2n = 20$. *S. part of Aegean region.*

193. S. lydia Boiss., *Diagn. Pl. Or. Nov.* **1(1)**: 37 (1843). Like **192(a)** but stems more strongly pubescent and viscid above; upper cauline leaves long-acuminate; calyx with short glandular hairs and long eglandular hairs; teeth longer, often up to ½ as long as calyx. *E. part of Balkan peninsula.* Bu Gr ?Ju Tu.

194. S. conoidea L., *Sp. Pl.* 418 (1753). Like **192(a)** but leaves wider and less densely pubescent, often subglabrous; calyx 15–28 mm, attenuate above in flower; petal-limb entire or slightly emarginate; capsule 12–18 mm, long-attenuate above; seeds 1·25–1·5 mm, dark chestnut-brown. $2n = 20$. *S.W. Europe; Turkey-in-Europe.* Ga Hs *It Tu.

29. Cucubalus L.

S. M. WALTERS

Like *Silene* but fruit indehiscent, somewhat fleshy and berry-like when ripe.

1. C. baccifer L., *Sp. Pl.* 414 (1753). Weak-stemmed, pubescent perennial herb up to 120 cm, with divaricate branches. Leaves ovate, acuminate, entire; petioles short. Inflorescence a lax, few-flowered dichasium with leaf-like bracts. Calyx 8–15 mm, broadly campanulate, indistinctly veined, with 5 long, obtuse teeth revolute in fruiting stage. Petals greenish-white, deeply bifid. Fruit 6–8 mm, black, globose; carpophore 2–3 mm. Seeds 1·5 mm, black. $2n = 24$. *Europe northwards to* c. 60° *N. in W. Russia, but rare in much of the Mediterranean region.* Al Au Be *Br Bu Cz Ga Ge ?Gr He Ho Hs Hu It Ju Lu Po Rm Rs(B, C, W, K, E) Si Tu.

30. Drypis L.

R. DOMAC

Perennial herbs. Leaves and bracts spinose. Inflorescence subcapitate. Petals with long claw and bifid limb; stamens 5; stigmas 3. Capsule 1(2)-seeded, transversely dehiscent.

1. D. spinosa L., *Sp. Pl.* 413 (1753). Stems 8–30 cm, numerous, stiff, much-branched, glabrous, quadrangular. Leaves spinose-subulate, canaliculate above. Inflorescence a subcapitate compound dichasium surrounded by spinose bracts. Calyx 5-toothed. Petals very small and narrow, white or pinkish, with small coronal scales; anthers bluish. Capsule ovoid-ellipsoidal. $2n = 60$. *Open stony, gravelly or sandy habitats.* ● *S. Europe, from N.W. Jugoslavia to S. Italy and S. Greece.* Al Gr It Ju.

(**a**) Subsp. **spinosa** (subsp. *linneana* Murb. & Wettst.): Outer bracts lanceolate, with long terminal spine considerably exceeding flowers. Calyx membranous below. Limb of petal bifid to base; claw exceeding calyx. *Mountain screes. Throughout the range of the species.*

(**b**) Subsp. **jacquiniana** Murb. & Wettst. ex Murb., *Lunds Univ. Årsskr.* **27**: 161 (1891): Outer bracts ovate, with short terminal spine scarcely exceeding flowers. Calyx coriaceous. Limb of petal bifid half-way to base; claw equalling calyx. *Rocks and sand-dunes. N.W. & W. Jugoslavia.*

31. Gypsophila L.

Y. I. BARKOUDAH & A. O. CHATER (EDITION 1) REVISED BY J. R. AKEROYD (EDITION 2)

Annual or perennial herbs. Inflorescence dichasial, paniculate to capitate; bracts scarious. Epicalyx absent. Calyx 5-veined, with scarious commissures. Petals exceeding calyx, usually gradually narrowed into claw; coronal scales absent; claw unwinged; stamens 10; styles 2; stigmatic surface terminal. Capsule dehiscing with 4 teeth; carpophore absent. Seeds subreniform, flattened on both sides, with lateral hilum; embryo peripheral, with prominent radicle.

Literature: F. N. Williams, *Jour. Bot.* (*London*) **27**: 321–329 (1889). G. Stroh, *Beih. Bot. Centr.* **59B**: 455–477 (1930). Y. I. Barkoudah, *Wentia* **9**: 1–157 (1962).

1 Annual, without woody stock and vegetative stems
2 Leaves not more than 3 mm wide, linear
3 Calyx 2 mm, glandular-puberulent; petals emarginate **23. linearifolia**
3 Calyx 2·5–4 mm, glabrous; petals entire **25. muralis**
2 Leaves more than 3 mm wide, lanceolate
4 Plant glabrous **24. elegans**
4 Stem villous or hispid **27. pilosa**
1 Perennial, with woody stock and vegetative stems
5 Flowers subsessile, in dense, globose clusters
6 Stem not more than 20 cm, simple, with solitary flower-cluster **14. petraea**
6 Stem usually more than 20 cm, always branched
7 Branches of inflorescence glabrous, rarely glandular-pubescent; leaves 10–50 mm, semi-terete **10. struthium**
7 Branches of inflorescence puberulent; leaves 20–100 mm, linear-lanceolate
8 Bracts triangular, acute, entire; seeds with long, acute, conical tubercles **15. pallasii**
8 Bracts ovate-orbicular, erose; seeds with obtuse tubercles **16. glomerata**
5 Flowers distinctly pedicellate
9 Whole inflorescence, including calyx and pedicels, glabrous
10 Leaves 10 mm or more wide, 3- to 7-veined
11 Petals emarginate; leaves ovate to oblong-lanceolate **20. perfoliata**
11 Petals entire; leaves lanceolate to ovate-lanceolate **22. tomentosa**
10 Leaves less than 10 mm wide, often 1-veined
12 Most leaves 3- to 7-veined
13 Calyx 3–3·5 mm, the teeth acuminate; seeds with long, acute, conical tubercles **19. acutifolia**
13 Calyx 2–2·5 mm, the teeth obtuse; seeds smooth **22. tomentosa**
12 Leaves usually 1-veined
14 Leaves *c.* 6 × 0·5 mm **26. macedonica**
14 Leaves at least 10 × 0·5 mm
15 Inflorescence lax, paniculate or corymbose; pedicels at least twice as long as calyx
16 Plant with procumbent rhizome; leaves usually subobtuse; calyx-teeth acute **3. repens**
16 Plant without procumbent rhizome; leaves acute; calyx-teeth obtuse
17 Calyx 1·5–2 mm **17. paniculata**
17 Calyx 2·5–4·5 mm
18 Petals less than twice as long as calyx; seeds with acute tubercles (Spain) **11. bermejoi**
18 Petals 2–3 times as long as calyx; seeds with flattened tubercles (E. Russia) **12. patrinii**
15 Inflorescence dense, corymbose; pedicels less than twice as long as calyx
19 Calyx usually not more than 2 mm; leaves obtuse or subacute (Spain) **10. struthium**
19 Calyx 2 mm or more; leaves acuminate
20 Bracts and calyx-teeth not ciliate; seeds with obtuse tubercles **8. papillosa**
20 Bracts and calyx-teeth ciliate; seeds with long, acute, conical tubercles **9. collina**
9 Inflorescence glandular-hairy, at least in part
21 Inflorescence dense; pedicels not more than twice as long as calyx
22 Leaves 1-veined, linear
23 Calyx 2–3 mm, glabrous **7. fastigiata**
23 Calyx 3–4·5 mm, glandular-pubescent **13. uralensis**
22 Most leaves 3(–5)-veined, not linear
24 Leaves oblanceolate to linear-oblanceolate **5. altissima**
24 Leaves lanceolate to linear-lanceolate **19. acutifolia**
21 Inflorescence lax; pedicels usually more than twice as long as calyx
25 Leaves not more than 20 mm; inflorescence rarely with more than 10 flowers
26 Leaves at least 2 mm wide **1. nana**
26 Leaves less than 2 mm wide
27 Seeds with flattened tubercles **2. spergulifolia**
27 Seeds with acute tubercles **4. montserratii**
25 Leaves at least 20 mm; inflorescence with more than 10 flowers
28 Leaves 10 mm or more wide
29 Seeds with blunt tubercles; plant glaucous **21. scorzonerifolia**

29 Seeds smooth; plant green **22. tomentosa**
28 Leaves less than 10 mm wide
30 Seeds smooth **22. tomentosa**
30 Seeds tuberculate
31 All pedicels glabrous; calyx not more than 2 mm **17. paniculata**
31 At least some pedicels pubescent; calyx at least 2 mm
32 Pedicels 2–3 times as long as calyx **6. litwinowii**
32 Pedicels 5–10 times as long as calyx **18. arrostii**

Sect. GYPSOPHILA. Perennials. Stems up to 25 cm, usually procumbent or ascending. Leaves linear. Inflorescence usually lax or few-flowered; pedicels mostly at least twice as long as calyx. Calyx campanulate. Capsule globose. Ovules 8–16.

1. G. nana Bory & Chaub. in Bory, *Expéd. Sci. Morée* 3(**2**): 116 (1832) (incl. *G. achaia* Bornm.). Whole plant viscid-pubescent (leaves rarely glabrous), more or less caespitose; stems 2–4(–15) cm, ascending. Leaves 5–20 × 2–3 mm, narrowly oblong, obtuse, rarely subacute. Flowers 3–7; pedicels mostly 2–3 times as long as calyx. Calyx 3–5 mm. Petals 2–3 times as long as calyx, lilac-pink, rarely white. Seeds with acute, conical tubercles. $2n = 34$. *Crevices of limestone rocks.* ● *Mountains of S. & C. Greece; Kriti.* Cr Gr.

2. G. spergulifolia Griseb., *Spicil. Fl. Rumel.* **1**: 183 (1843). Densely caespitose; stems 5–25 cm, much-branched, glabrous below, densely glandular-pubescent in the inflorescence. Leaves 5–15 × *c.* 0·5 mm. Pedicels glandular-pubescent. Calyx 2·5–4 mm, sparsely glandular-pubescent. Petals twice as long as calyx, white or pale purplish. Seeds with flattened tubercles. ● *C. Jugoslavia, Albania.* Al Ju.

3. G. repens L., *Sp. Pl.* 407 (1753). Plant glabrous; rhizome much-branched, with numerous vegetative and flowering stems up to 25 cm. Leaves 10–30 × 1·5–3 mm, often falcate. Inflorescence usually 5- to 30-flowered, subcorymbose; bracts scarious; pedicels mostly 2 or more times as long as calyx. Calyx 2·5–3·5 mm. Petals more than twice as long as calyx, white, lilac or pale purplish. Seeds with long, acute, conical tubercles. $2n = 34$. *Calcicole.* ● *Principal mountains of C. & S. Europe from the Jura, Alps and W. Carpathians to N.W. Spain and C. Italy.* Au Cz Ga Ge He Hs It Ju Po.

4. G. montserratii Fernández Casas, *Publ. Inst. Biol. Apl. (Barcelona)* **52**: 121 (1972). Stock woody, branched; stems 5–15(–20) cm, glandular-puberulent. Leaves 4–8 mm, somewhat fleshy, obtuse, with ciliate margins. Inflorescence 3- to 8-flowered. Pedicels and calyx glandular-puberulent. Petals about twice as long as calyx. Seeds with acute, conical tubercles. $2n = 26$. *Crevices of limestone rocks.* ● *S.E. Spain (Albacete prov.).* Hs.

Sect. CORYMBOSAE Barkoudah. Perennials. Stems usually more than 25 cm. Inflorescence usually more or less corymbose; pedicels usually short and rigid. Calyx campanulate. Capsule globose. Ovules 8–24.

5. G. altissima L., *Sp. Pl.* 407 (1753) (incl. *G. ucrainica* Kleopow). Stock stout; stems 25–80 cm, erect, branched above, glabrous below, glandular-pubescent above. Leaves 50–80 × 5–20 mm, oblanceolate to linear-oblanceolate, acute or subobtuse, 3(–5)-veined. Inflorescence dense; pedicels shorter than or equalling calyx, glandular-pubescent. Calyx 2–2·5 mm; teeth ovate, obtuse. Petals $1\frac{1}{2}$–2 times as long as calyx. *S. & E. Russia, Ukraine.* Rs(C, W, E).

6. G. litwinowii Kos.-Pol., *Sched. Herb. Fl. Ross.* **8**: 61 (1922). Like **5** but leaves 20–50 × 2–4·5 mm, linear to linear-lanceolate, obscurely veined; inflorescence a lax, diffuse panicle; pedicels 2–3 times as long as calyx. ● *S.C. Russia (Zemljansk, N.W. of Voronež).* Rs(C).

Possibly a hybrid between **5** and **17**.

7. G. fastigiata L., *Sp. Pl.* 407 (1753) (?*G. dichotoma* Besser). Rhizome woody; stems 5–100 cm, erect, often branched, glabrous below, glandular-pubescent above and in the inflorescence; internodes longer than leaves. Leaves 20–80 × 1–4 mm, linear, obtuse or abruptly acute, 1-veined. Flowers many, in a rather dense inflorescence; pedicels usually shorter than calyx, usually glabrous. Calyx 2–3 mm, glabrous; teeth ovate, obtuse, not apiculate. Petals *c.* $1\frac{1}{2}$ times as long as calyx, obovate, rounded at apex, white or pale purplish. Seeds with long, acute, conical tubercles. $2n = 34$. ● *Mainly in E.C. Europe, but extending locally to W. Germany, Sweden, arctic Russia and C. Ukraine.* Au Cz Fe Ge Hu ?Ju Po ?Rm Rs(N, B, C, W) Su.

Towards the south-east plants become larger, and have laxer, densely pubescent inflorescences with pubescent pedicels and rather larger flowers. These have been called **G. arenaria** Waldst. & Kit., *Pl. Rar. Hung.* **1**: 40 (1800–1), but the differences are mostly phenotypic, and the difference in size is the only character that remains constant in cultivation.

8. G. papillosa Porta, *Atti Accad. Agiati* ser. 3, **11** (**2**): 1 (1905) (*G. glandulosa* Porta, non Walpers, *G. fastigiata* auct. ital., non L.). Like **7** but entirely glabrous; internodes shorter than leaves; leaves linear-lanceolate, long-acuminate; calyx-teeth apiculate; petals emarginate; capsule purplish; seeds with obtuse tubercles. *Dry, stony ground.* ● *N. Italy (near Garda).* It.

9. G. collina Steven ex Ser. in DC., *Prodr.* **1**: 352 (1824) (*G. dichotoma* auct., non Besser, *G. arenaria* var. *leioclados* Borbás). Like **7** but entirely glabrous; internodes shorter than leaves; leaves linear, long-acuminate; inflorescence denser, often subcapitate; calyx-teeth obtuse and mucronate, ciliate. *C. Romania; S.W. Ukraine.* Rm Rs(W).

10. G. struthium Loefl., *Iter Hisp.* **79**: 303 (1758) (*G. iberica* Barkoudah). Plant more or less glabrous; rhizome woody, branched. Stems 20–100 cm, ascending to erect, branched. Leaves 10–50 × 1–2 mm. Inflorescence a loose panicle of dense clusters of flowers. Pedicels short, glabrous or glandular-pubescent. Calyx 1·5–3·5 mm. Petals *c.* 3 mm, cuneate, white. Seeds with long, acute, conical tubercles. *Gypsaceous soils.* ● *C., E. & N. Spain.* Hs.

(**a**) Subsp. **struthium**: Leaves semi-terete, acute. Flowers subsessile in capitate clusters *c.* 1 cm in diameter. Calyx 2–3·5 mm; teeth lanceolate. $2n = 24$. *C. & S.E. Spain.*

(**b**) Subsp. **hispanica** (Willk.) G. López, *Anal. Jard. Bot. Madrid* **41**: 36 (1984) (*G. hispanica* Willk., *G. struthium* sensu Barkoudah, non Loefl.): Leaves subtriquetrous or flat, obtuse or subacute. Flowers in distinctly pedicellate corymbs. Calyx 1·5–2 mm; teeth oblong-lanceolate. *C., E. & N. Spain.*

11. G. bermejoi G. López, *An. Jard. Bot. Madrid* **41**: 35 (1984). Plant glabrous; rhizome woody. Stems 40–80(–100) cm, ascending, branched. Leaves 10–50 × 2–8 mm, narrowly oblong to oblong-lanceolate, flat, acute. Inflorescence a many-flowered corymbose cyme. Calyx 2·5–3 mm; teeth ovate-lanceolate, sometimes ciliate. Petals 2·5–4·5 mm, ovate-oblong, entire or emarginate, pink. Seeds 1–1·5 mm, reniform, with acute tubercles, black. 2*n* = 68. *Gypsaceous soils.* ● *C. & N.C. Spain.* Hs.

Probably an allopolyploid derivative of the hybrid between **9** and **22** (*G.* × *castellana* Pau).

12. G. patrinii Ser. in DC., *Prodr.* **1**: 353 (1824). Plant entirely glabrous; stems up to 50 cm, ascending or erect, often branched. Leaves 10–50 × 1–4 mm, linear or linear-lanceolate, long-acuminate, flat or triquetrous. Inflorescence lax; pedicels 2–5 times as long as calyx. Calyx 2·5–3·5 mm; teeth obtuse, mucronate, ciliate. Petals 2–3 times as long as calyx, pale purplish. Seeds with flattened tubercles. *Rocky and stony ground. E. Russia.* Rs(C). (*Temperate Asia.*)

13. G. uralensis Less., *Linnaea* **9**: 172 (1834). Plant caespitose. Stems 5–20 cm, erect, simple, glabrous below, glandular-pubescent above. Leaves 10–50 × 0·5–2 mm, linear, sometimes triquetrous, falcate. Inflorescence dense, with 5–15 flowers; pedicels glandular-pubescent, not longer than calyx. Calyx 3–4·5 mm, glandular-pubescent; teeth oblong, obtuse. Petals 2–3 times as long as calyx, white. Seeds with very small tubercles. *Rocky and stony ground. N. Russia and Ural.* Rs(N, C).

Sect. CAPITULIFORMES F. N. Williams. Perennials. Leaves very narrow. Flowers in dense, capitate clusters. Calyx conic-campanulate. Capsule globose. Ovules 4–12.

14. G. petraea (Baumg.) Reichenb., *Fl. Germ. Excurs.* 801 (1832) (*G. transylvanica* Sprengel). Plant densely caespitose. Stems 4–20 cm, simple, glabrous or scabrid below, puberulent above. Leaves 25–50 × *c.* 1 mm, linear, glabrous. Inflorescence 1–2 cm in diameter, globose or subcapitate, surrounded at base by large, triangular-ovate, scarious bracts. Calyx 2–3·5 mm; teeth triangular-ovate, acute. Petals 1½ times as long as calyx, white or pale purplish. Seeds with spiny tubercles on back. 2*n* = 34. *Mountain rocks; calcicole.* ● *E. & S. Carpathians.* Rm.

15. G. pallasii Ikonn., *Nov. Syst. Pl. Vasc. (Leningrad)* **13**: 118 (1976) (*G. glomerata* auct., non Pallas ex Adams). Plant more or less caespitose; stems 20–80 cm, erect, branched above, glabrous, or sometimes puberulent above; upper branches and inflorescence puberulent, more or less viscid. Leaves 20–100 × 1–4 mm, linear to linear-lanceolate, glabrous, glaucous. Flower-heads long-pedunculate. Bracts at base of flower-heads large, triangular-acute, entire, ciliate, scarious. Calyx 2–3·5 mm; teeth oblong, obtuse, entire. Petals 1½ times as long as calyx, white. Seeds with long, acute, conical tubercles. *S.E. Europe, from E. Jugoslavia to Krym.* Bu Gr Rm Rs(W, K) Tu.

16. G. glomerata Pallas ex Adams in Weber fil. & Mohr, *Beitr. Naturk.* **1**: 54 (1805) (*G. globulosa* Steven ex Boiss.). Like **15** but bracts at base of flower-heads ovate-orbicular, obtuse, strongly erose; calyx-teeth erose; seeds with obtuse tubercles. *S.W. Ukraine (Reni); S.E. Russia (W. of Rostov).* Rs(W, E). (*Caucasus.*)

Sect. ROKEJEKA (Forskål) A. Braun. Perennials. Inflorescence a lax, diffuse panicle. Calyx campanulate. Capsule globose.

17. G. paniculata L., *Sp. Pl.* 407 (1753). Plant glaucous, glabrous (or pubescent below, var. *hungarica* Borbás), very rarely pubescent throughout except for pedicels and calyx; rhizome stout; stems 50–90 cm, diffusely branched. Leaves 20–70 × 2·5–10 mm, lanceolate, acute or acuminate. Pedicels 2–3 times as long as calyx. Calyx 1·5–2 mm; teeth ovate, obtuse. Petals 3–4 mm, linear-spathulate, white or pale reddish. Seeds with obtuse tubercles. *Dry, sandy and stony places. E. & C. Europe, westwards to C. Austria, southwards to Bulgaria, and northwards to* c. 57° *N. in Russia; cultivated for ornament and naturalized elsewhere.* Au Bu Cz Hu Ju Po Rm Rs(C, W, K, E) [Ge Rs(B)].

Plants from one station in White Russia, differing chiefly by their acute calyx-teeth, have been named **G. belorossica** Barkoudah, *Wentia* **9**: 99 (1962).

18. G. arrostii Guss., *Pl. Rar.* 160 (1826). Like **17** but some pedicels always pubescent and the rest of plant usually subglabrous; leaves linear to linear-lanceolate; pedicels 5–10 times as long as calyx; calyx 2–2·5 mm; teeth oblong, obtuse; petals elliptic-oblong, pale purplish; seeds with flattened tubercles. *S. Italy, Sicilia.* It Si.

European plants belong to subsp. **arrostii**, which is endemic. Subsp. **nebulosa** (Boiss. & Heldr.) Greuter & Burdet, *Willdenowia* **12**: 188 (1982), occurs in Anatolia.

19. G. acutifolia Steven ex Sprengel, *Novi Provent.* 21 (1818). Stems 20–170 cm, erect, branched, glabrous below, glandular-pubescent above. Leaves 20–80 × 2–10 mm, linear-lanceolate to lanceolate, long-acuminate. Inflorescence a rather dense panicle; pedicels 1–4 mm, glabrous or glandular-pubescent. Calyx 3–3·5 mm; teeth oblong, acuminate. Petals twice as long as calyx, white. Seeds with long, acute, conical tubercles. *W. Ukraine.* Rs(W) [Rm]. (*Caucasus.*)

G. scariosa Tausch, *Flora (Regensb.)* **14**: 213 (1831), should be excluded from the European flora; recorded in error from Switzerland.

20. G. perfoliata L., *Sp. Pl.* 408 (1753) (*G. trichotoma* Wenderoth, *G. scorzonerifolia* auct., non Ser.; incl. *G. paulii* Klokov and *G. tekirae* Stefanov). Plant yellow-green; stems 30–100 cm, ascending at base, glandular-pubescent below, glabrous above, rarely entirely glabrous. Leaves ovate or oblong-lanceolate, acute to obtuse, amplexicaul and shortly connate at base; the lower 20–80 × 10–35 mm, 3- to 7-veined, pubescent. Pedicels 4–15 mm, glabrous. Calyx 2–2·5 mm; teeth ovate, obtuse. Petals white to pale purple, emarginate. Seeds with very small tubercles. *S.E. Europe, from Bulgaria to E. Russia.* Bu Rm Rs(W, K, E) [Ge Rs(B)].

21. G. scorzonerifolia Ser. in DC., *Prodr.* **1**: 352 (1824). Like **20** but glaucous, glabrous below, glandular-pubescent above; leaves glabrous; pedicels mostly glandular-pubescent at base; petals darker purple; seeds with obtuse tubercles. *Saline and sandy soils. S.E. Russia, E. Ukraine.* Rs(E) [Cz Ge].

22. G. tomentosa L., *Cent. Pl.* **1**: 11 (1755) (*G. perfoliata* auct. hisp.). Like **20** but somewhat smaller; stems, inflorescence-branches and leaves usually glandular-pubescent; pedicels usually glabrous; leaves narrower, ovate-lanceolate to lanceolate; petals entire; seeds black, smooth, shining. *Saline places.* ● *C. & S.E. Spain.* Hs.

Sect. DICHOGLOTTIS (Fischer & C. A. Meyer) Fenzl. Annuals. Inflorescence a dichasial panicle; pedicels capillary; bracts more or less herbaceous. Calyx campanulate. Capsule globose. Ovules 12–24.

23. G. linearifolia (Fischer & C. A. Meyer) Boiss., *Fl. Or.* **1**: 550 (1867). Stems 3–25 cm, glandular-pubescent below, glabrous above. Leaves 10–20 × 1–2 mm, linear to linear-spathulate, pubescent. Inflorescence few-flowered; pedicels many times as long as calyx, filiform. Calyx 2·5 mm or less, glandular-puberulent. Petals 3–4 mm, white, bilobed. Seeds with long, acute, conical tubercles. *Gypsaceous semi-deserts. E. Russia, W. Kazakhstan.* Rs(E). (*W.C. Asia.*)

24. G. elegans Bieb., *Fl. Taur.-Cauc.* **1**: 319 (1808). Plant glabrous; stems 20–50 cm, branched above. Leaves 20–40 × 3–5(–15) mm, oblong- to linear-lanceolate, obscurely 1- to 3-veined. Inflorescence lax; bracts scarious, with dark midrib; pedicels many times as long as calyx. Calyx 3–5 mm. Petals 2–5 times as long as calyx, white with purple veins. Seeds with obtuse tubercles. *S. Ukraine; cultivated elsewhere for ornament, and naturalized in a few places.* Rs(W, K) [?Au Cz Ge Rs(B)]. (*Caucasus, Anatolia.*)

Sect. MACRORRHIZAEA Boiss. Calyx more or less tubular. Capsule oblong. Ovules 24–36.

25. G. muralis L., *Sp. Pl.* 408 (1753). Annual; stems 4–25(–40) cm, branched, glabrous above, puberulent (often very sparsely so) below. Leaves 5–25 × 0·5–3 mm, linear, acute, glaucous. Inflorescence diffusely corymbose-paniculate; pedicels filiform, several to many times as long as calyx. Calyx 3–4 mm. Petals about twice as long as calyx, entire, pink, with darker veins. Capsule longer than calyx. Seeds with very small tubercles. $2n = 34$. *C. & E. Europe, extending locally to W. France, S. Sweden and C. Greece.* Au Be Bu Cz Fe Ga Ge Gr He Ho Hs Hu It Ju Po Rm Rs(N, B, C, W, K, E) Su Tu.

G. muralis var. *stepposa* (Klokov) Schischkin (*G. stepposa* Klokov), from S. Russia, has a narrower calyx and shorter, white petals with pink spots.

26. G. macedonica Vandas, *Magyar Bot. Lapok* **4**: 111 (1905). Like **25** but entirely glabrous, caespitose perennial with stems up to 15 cm, little-branched; leaves *c.* 6 × 0·5 mm; calyx not more than 2 mm. ● *S. Jugoslavia (Bitola); N. Greece (Veroia).* Gr Ju.

Sect. HAGENIA A. Braun. Annuals. Inflorescence paniculate, dichasial. Calyx more or less tubular. Capsule long-ovoid. Ovules 4–20.

27. G. pilosa Hudson, *Philos. Trans. Roy. Soc. Lond.* **56**: 252 (1767) (*G. porrigens* (L.) Boiss.). Stems 20–65 cm, stout, usually glabrous at base and in the inflorescence, villous or hispid in the middle, branched above. Leaves 40–80 × 10–15 mm, lanceolate, obscurely 3-veined. Pedicels 3–4 times as long as calyx, filiform, becoming deflexed. Calyx 4–7 mm. Petals about twice as long as calyx, pink or pale purple. Seeds with obtuse tubercles. *Naturalized in Mallorca and S.E. Spain; an occasional casual elsewhere.* [Bl Hs *Tu.] (*W. Asia.*)

32. Bolanthus (Ser.) Reichenb.

Y. I. BARKOUDAH (EDITION 1) REVISED BY J. R. AKEROYD (EDITION 2)

Perennials; rather woody at base. Inflorescence dichasial, paniculate to subcapitate; bracts leafy. Epicalyx absent. Calyx with 5 projecting ribs and scarious commissures. Petals with long, linear claw and small, patent limb; coronal scales absent; claw delicately winged; stamens 10; styles 2, stigmatose all along their inner side. Capsule oblong-ovoid, dehiscing with 4 teeth; carpophore short. Seeds comma-shaped, with lateral hilum; embryo hook-shaped, with projecting radicle. (*Gypsophila* Sect. *Bolanthus* (Ser.) Boiss.)

All species occur in calcareous, rocky or stony habitats.

Literature: As for *Gypsophila*; also D. Phitos, *Bot. Chron.* **1**: 35–45 (1981).

1 Inflorescence usually lax, of solitary or paired flowers
  2 Petal-limb with transverse purple stripe; stem rigid, somewhat branched above; seeds with flattened tubercles **2. fruticulosus**
  2 Petal-limb entirely white; stem flexible, much-branched; seeds with acute tubercles **3. thessalus**
1 Inflorescence ± dense, of 2- to 20-flowered clusters
  3 Calyx 5–6·5 mm **4. graecus**
  3 Calyx 2·5–4 mm
    4 Stems not more than 6 cm; calyx 2·5–3·5 mm; petals purplish **6. creutzbergii**
    4 Stems usually more than 6 cm; calyx 3–4 mm; petals white
      5 Stems glabrous to puberulent; flowers up to 20 in cluster **1. laconicus**
      5 Stems densely pubescent; flowers up to 5 in cluster **5. thymifolius**

1. B. laconicus (Boiss.) Barkoudah, *Wentia* **9**: 163 (1962) (*Gypsophila laconica* (Boiss.) Boiss. & Heldr.). Stems 10–25(–30) cm, slender, mostly unbranched, glabrous to puberulent with glandular and eglandular hairs. Leaves 10–20 × 0·5–1 mm, linear to linear-subulate, subobtuse. Flowers 2–20 in dense, terminal, long-pedunculate clusters; pedicels shorter than calyx. Calyx 3–4 mm, with small triangular teeth, glandular-hairy. Petals white. Seeds with small tubercles. *Mountain rocks; calcicole.* ● *S. Greece (Peloponnisos).* Gr.

2. B. fruticulosus (Bory & Chaub.) Barkoudah, *op. cit.* 164 (1962) (*Gypsophila fruticulosa* (Bory & Chaub.) Boiss.). Stems 5–35 cm, branched in upper part, rigid, erect, hairy with a mixture of short hairs and long, patent hairs. Leaves 5–15 × 1·5–3 mm, oblanceolate, hirsute. Inflorescence usually lax; flowers solitary or in pairs; pedicels usually ½ as long as calyx, sometimes as long as calyx. Calyx 4·5–6 mm, glandular-hairy. Petals white, with a transverse purple stripe. Seeds with flattened tubercles. ● *C. & S. Greece.* Gr.

3. B. thessalus (Jaub. & Spach) Barkoudah, *op cit.* 164 (1962) (*Gypsophila thessala* (Jaub. & Spach) Halácsy, *G. hirsuta* (Labill.) Boiss.). Stems 10–30 cm, much-branched, flexible, with a mixture of short hairs and long, patent hairs. Leaves 5–12 × 0·5–1 mm, linear-oblanceolate, hirsute. Inflorescence lax, dichasial; flowers solitary or in pairs; pedicels as long as calyx. Calyx 4–5·5 mm, with a mixture of long and short hairs on the ribs. Seeds with acute tubercles. ● *C.E. Greece (Thessalia, Evvoia).* Gr.

B. chelmicus Phitos, *Bot. Chron.* **1**: 40 (1981), described from S. Greece (Chelmos Oros), is like **3** but has flowers mostly in 2- to 8-flowered clusters, and calyx 3·5–4·5 mm. It is known only from the type collection, and is perhaps not specifically distinct.

4. B. graecus (Schreber) Barkoudah, *Wentia* **9**: 166 (1962) (*Gypsophila polygonoides* (Willd.) Halácsy, *G. hirsuta* (Labill.) Boiss., *G. ocellata* Sm.). Stems 5–25 cm, usually unbranched, pubescent, sometimes glandular-hairy; hairs of more or less equal length. Leaves 5–10 × 1–2 mm, oblanceolate or oblong-spathulate, puberulent. Flowers usually in dense 3- to 10-flowered heads; pedicels usually shorter than calyx. Calyx 5–6·5 mm, puberulent, with long, patent hairs on the ribs. Petals white, with a transverse purple stripe. $2n = 30$. ● *C. Greece; Kikhlades (Naxos).* Gr.

B. intermedius Phitos, *Bot. Chron.* **1**: 39 (1981), described from serpentine rocks in N. Evvoia, differs from **4** mainly by the flowers being in terminal 2- to 7-flowered clusters or solitary or paired in the axils of cauline leaves, and having both long and short hairs on the stems. It is known only from the type collection, and is probably not specifically distinct.

5. B. thymifolius (Sm.) Phitos, *Bot. Chron.* **1**: 39 (1981) (*B. graecus* var. *thymifolia* (Sm.) Barkoudah). Plant densely pubescent, glandular-hairy above. Stems up to 25 cm, branched. Leaves up to 15 mm, linear-oblanceolate or sometimes obovate-spathulate. Flowers subsessile in 2- to 5-flowered dense heads, or with longer pedicels, solitary or paired in the axils of cauline leaves. Calyx 3–4 mm. Petals white. ● *N.C. & C. Greece.* ?Bu Gr.

6. B. creutzbergii Greuter, *Candollea* **20**: 210 (1965). Plant caespitose, densely hairy with glandular and longer eglandular hairs. Stems 1·5–6 cm, mostly unbranched. Leaves 2–5 × 0·5–1·5 mm, lanceolate-spathulate. Flowers in 2- to 10-flowered terminal clusters, subsessile or shortly pedicellate. Calyx 2·5–3·5 mm. Petals purplish. Seeds minutely papillose. $2n = 20$. *Calcareous rocks and screes*, 1700–2000 *m*. ● *Mountains of Kriti.* Cr.

Plants from coastal sands in W. Kriti (near Palaiochora) seem to belong within **6**, but require further investigation.

33. Saponaria L.

A. O. CHATER

Annual or perennial herbs. Inflorescence dichasial, paniculate to capitate; rarely flowers solitary. Epicalyx absent. Calyx cylindrical or oblong, 5-toothed, 15- to 25-veined, without scarious commissures. Petals 5, exceeding calyx, narrowed abruptly (or rarely gradually) into a long claw; coronal scales usually present; stamens 10; styles 2, rarely 3. Capsule cylindrical to ovoid, dehiscing with 4, rarely 6, teeth. Seeds reniform, with lateral hilum, more or less compressed.

Literature: G. Simmler, *Denkschr. Akad. Wiss. Math.-Nat. Kl. (Wien)* **85**: 433–509 (1910).

1 Petals yellow; inflorescence capitate
 2 Filaments yellow; leaves spathulate **1. bellidifolia**
 2 Filaments deep violet; leaves linear-lanceolate **2. lutea**
1 Petals pink to purple, rarely white, never yellow; inflorescence various
 3 Stems 1-flowered
  4 Styles 3 (rarely 2); calyx not more than 20 mm; teeth obtuse **4. pumilio**
  4 Styles 2; calyx 20–30 mm; teeth acute **6. sicula**
 3 Stems with 2 or more flowers
  5 Calyx not more than 15 mm
   6 Plant densely caespitose; basal leaves linear-lanceolate; calyx 12–15 mm **3. caespitosa**
   6 Plant not caespitose; basal leaves spathulate, ovate-lanceolate or ovate; calyx 7–13 mm
    7 Perennial, woody below, with conspicuous, non-flowering leafy stems **8. ocymoides**
    7 Annual, without conspicuous, non-flowering leafy stems
     8 Stems much-branched; pedicels and calyx glandular-hispid **7. calabrica**
     8 Stems little-branched; pedicels and calyx glabrous **10. chlorifolia**
  5 Calyx more than 15 mm
   9 Plant usually caespitose; leaves 1-veined; inflorescence-branches mostly alternate **6. sicula**
   9 Plant not caespitose; leaves 3-veined; inflorescence-branches mostly opposite
    10 Plant ±glabrous, or hairy only in the inflorescence; petal-limb *c.* 10 mm **9. officinalis**
    10 Plant glandular-hispid; petal-limb *c.* 5 mm **5. glutinosa**

1. S. bellidifolia Sm., *Spicil. Bot.* **1**: 5 (1791). Caespitose; flowering stems 20–40 cm, erect, unbranched, usually glabrous, usually with 1 pair of linear-lanceolate leaves. Basal leaves spathulate. Inflorescence capitate, with a pair of linear-lanceolate bracts; flowers sessile. Upper part of calyx hairy; teeth triangular, acute. Petal-limb *c.* 4 mm, yellow; filaments yellow, long-exserted. $2n = 28$. *Mountain rocks and pastures.* ● *Balkan peninsula; locally elsewhere in S. Europe from the Pyrenees to C. Romania.* Al Bu Ga Gr Hs It Ju Rm.

2. S. lutea L., *Sp. Pl.* ed. 2, 585 (1762). Caespitose; flowering stems 2–12 cm, erect, unbranched, shortly hairy, with 2 or more pairs of linear leaves. Basal leaves linear-lanceolate. Inflorescence more or less capitate; pedicels and calyx densely hairy. Calyx-teeth triangular, acute. Petal-limb *c.* 4 mm, yellow; filaments deep violet, long-exserted. ● *S.W. & S.C. Alps.* Ga He It.

3. S. caespitosa DC., *Rapp. Voy. Bot.* **2**: 78 (1808). Densely caespitose; flowering stems 5–15 cm, usually unbranched, glabrous, or slightly hairy above, with 2 or more pairs of small linear-lanceolate leaves. Basal leaves linear-lanceolate. Inflorescence condensed or capitate. Calyx densely hairy, purplish; teeth triangular-lanceolate, acute. Petal-limb 4–7 mm, purplish. *Mountain rocks and screes.* ● *C. Pyrenees.* Ga Hs.

4. S. pumilio (L.) Fenzl ex A. Braun, *Flora (Regensb.)* **26**: 801 (1843) (*Silene pumilio* (L.) Wulfen). Caespitose; stems very short, often shorter than the solitary, terminal flower. Leaves linear. Calyx 13–20 mm, rather inflated, hairy; teeth ovate, obtuse. Petals pale purplish, rarely white; limb 7–9 mm; styles 3, rarely 2. *Mountain pastures; calcifuge.* ● *E. Alps; S. Carpathians.* Au It Rm.

5. S. glutinosa Bieb., *Fl. Taur.-Cauc.* **1**: 322 (1808). Annual or biennial, usually glandular-hispid throughout; stem 25–50 cm, branched above in a pyramidal panicle. Basal leaves spathulate, rarely almost linear; cauline ovate-lanceolate, acute. Calyx 20–25 mm; teeth lanceolate, acuminate. Petal-limb *c.* 5 mm, purple, bifid. $2n = 56$. *S.E. Europe; Spain.* Al Bu Cr Gr Hs Ju Rm Rs(K) ?Tu.

6. S. sicula Rafin., *Specch. Sci.* **2**: 7 (1814). Caespitose perennial; stems procumbent to erect, glabrous below. Leaves 1–4 cm, narrowly spathulate to linear-oblanceolate, 1-veined. Inflorescence usually with alternate branches; branches, pedicels and calyx hairy. Calyx 20–30 mm, reddish-tinged, becoming somewhat inflated; teeth triangular, acute or acuminate. Petal-limb 5–8 mm, bifid, red. *Sicilia; Sardegna; C. part of Balkan peninsula.* Al Bu Gr Ju Sa Si.

1 Calyx eglandular-pubescent **(b)** subsp. **intermedia**
1 Calyx glandular-pubescent

2 Flowering stems 50–60 cm, mostly more than 8-flowered; inflorescence-branches glandular-hispid (c) subsp. **stranjensis**
2 Flowering stems 5–25 cm, 2- to 8-flowered; inflorescence-branches glandular-pubescent (a) subsp. **sicula**

(a) Subsp. **sicula** (*S. depressa* Biv.): Densely caespitose; flowering stems 5–15(–25) cm, with *c.* 2 pairs of linear-oblanceolate leaves, *c.* 2- to 8-flowered. Inflorescence-branches and pedicels densely glandular-pubescent. Calyx densely glandular-pubescent with rather stout hairs. *Sicilia; Sardegna.*

(b) Subsp. **intermedia** (Simmler) Chater, *Feddes Repert.* **69**: 52 (1964) (*S. intermedia* Simmler; incl. *S. haussknechtii* Simmler): Like subsp. (a) but usually less densely caespitose; flowering stems (5–)10–50 cm, more leafy, 1- to 30-flowered; inflorescence-branches and pedicels less densely pubescent, usually glandular; calyx less densely pubescent with more slender hairs, eglandular. ● *S. Jugoslavia, N. Greece, Albania.*

(c) Subsp. **stranjensis** (Jordanov) Chater, *Feddes Repert.* **69**: 52 (1964) (*S. stranjensis* Jordanov): Like subsp. (a) but less densely caespitose and very woody at base; flowering stems 50–60 cm, mostly more than 8-flowered; inflorescence-branches and pedicels glandular-hispid; petal-limb more deeply bifid. ● *S.E. Bulgaria.*

7. S. calabrica Guss., *Pl. Rar.* 164 (1826) (*S. aenesia* Heldr., *S. graeca* Boiss.). Annual; stem with many divaricate branches, glabrous below, glandular-hairy above. Lower leaves more or less spathulate, upper oblong-ovate. Inflorescence lax, spreading; pedicels glandular-hispid, often deflexed in fruit. Calyx 6–10 mm, glandular-hispid; teeth obtuse. Petal-limb 3–5 mm, pale purplish. *E. Mediterranean region.* Al Cr Gr It ?Si Tu.

S. orientalis L., *Sp. Pl.* 409 (1753), a native of S.W. Asia, which is like **7** in habit but without coronal scales and calyx-teeth acute, has been doubtfully recorded from Greece.

8. S. ocymoides L., *Sp. Pl.* 409 (1753). Much-branched perennial with procumbent or ascending hairy stems. Lower leaves subspathulate to ovate-lanceolate, obtuse; upper narrower, acute. Inflorescence lax, spreading; pedicels and calyx glandular-pubescent. Calyx 7–12 mm; teeth obtuse. Petal-limb 3–5 mm, pale purplish. $2n = 28$. ● *S.W. & S.C. Europe; N. & C. Italy.* Au Co Ga Ge He Hs It Ju Sa [Cz].

Plants from Corse and Sardegna that are less hairy and smaller in all their parts have been called subsp. **alsinoides** (Viv.) Arcangeli, *Comp. Fl. Ital.* ed. 2, 304 (1892).

9. S. officinalis L., *Sp. Pl.* 408 (1753). Perennial; stems 30–90 cm, erect, usually glabrous, simple or branched above. Leaves ovate to ovate-lanceolate, 3-veined, acute. Inflorescence condensed, with opposite branches bearing few-flowered dichasia; flowers large, shortly pedicellate, usually pale pink. Calyx *c.* 20 mm, glabrous or rarely hairy, green or reddish; teeth triangular, acute. Petal-limb *c.* 10 mm, more or less entire. $2n = 28$. *Europe, from N. Germany and C. Russia southwards; frequently cultivated, often as a variant* flore pleno, *and naturalized in many places in the north.* Al Au Be Bl Bu Co Cz Ga Ge Gr He Ho Hs Hu It Ju Lu Po Rm Rs(C, W, K, E) Sa Si Tu [Br Da Fe Hb No *Rs(B) Su].

10. S. chlorifolia (Poiret) G. Kunze, *Ind. Hort. Lips.* **1846**: 4 (1846) (*Cyathophylla chlorifolia* (Poiret) Bocquet & Strid). Glabrous, glaucous annual; stems 8–15 cm, procumbent or ascending. Basal leaves in a rosette, ovate or broadly spathulate, obtuse; cauline perfoliate, ovate to orbicular. Inflorescence capitate, of 2–6 shortly pedicellate flowers; involucre bowl-shaped, entire. Calyx 9–13 mm, with short, triangular teeth. Petals about equalling the calyx, pink. Seeds *c.* 1·2 mm, reddish-brown to blackish. $2n = 30$. *Limestone rocks and screes,* 1700–1850 *m. S. Greece (Chelmos Oros).* Gr. (*W. & C. Anatolia.*)

34. Vaccaria Medicus

A. O. CHATER

Annuals. Epicalyx absent. Calyx-tube inflated below, whitish, with 5 green wings at the angles, without scarious commissures. Petals 5, long-clawed; coronal scales absent; ovary unilocular (almost 2-locular at the base); styles 2. Capsule ovoid, with thick, papery exocarp dehiscing with 4 teeth, and with thinner endocarp dehiscing irregularly.

1. V. hispanica (Miller) Rauschert, *Wiss. Zeitschr.. Univ. Halle* **14**: 496 (1965) (*V. pyramidata* Medicus, *V. parviflora* Moench, *V. perfoliata* Gilib., *V. grandiflora* Jaub. & Spach, *V. vulgaris* Host, *V. segetalis* Garcke, *Gypsophila vaccaria* (L.) Sm., *Saponaria vaccaria* L., *S. hispanica* Miller). Stems 30–60 cm, erect, branched above, glabrous. Leaves *c.* 5 cm, ovate to lanceolate, glaucous. Flowers in much-branched dichasia; pedicels long; bracts scarious, with green midrib. Calyx 12–17 mm; teeth ovate, acute. Petal-limb 3–8 mm, sometimes exserted, pale or dark purplish, entire or bifid. Seeds 2–2·5 mm, numerous, subglobose. $2n = 30$. *Usually a weed of cultivated fields. Most of Europe, except Fennoscandia and the north-west, but only naturalized or casual in much of the northern part of its range.* Al †Au Be Bl Bu †Co Cr Ga Ge Gr He Hs Hu It Ju Lu Po Rm Rs(N, B, C, W, K, E) *Sa Si Tu [Cz Ho].

35. Petrorhagia (Ser. ex DC.) Link

P. W. BALL AND J. R. AKEROYD

Annual or perennial herbs. Inflorescence paniculate to capitate. Bracts present or absent. Calyx 5-toothed, 5- to 15-veined with scarious commissures. Petals 5, exceeding the calyx, with or without claw; coronal scales absent. Stamens 10; styles 2. Capsule dehiscing with 4 teeth. Seeds dorsiventrally compressed; embryo straight. (Incl. *Tunica* auct., *Kohlrauschia* Kunth.)

It is not practicable to make a clear distinction, as is done in *Dianthus*, between epicalyx-scales and bracts. The term 'bract' is therefore used to cover both.

All species occur in dry, calcareous or sandy habitats.

Literature: P. W. Ball & V. H. Heywood, *Bull. Brit. Mus. (Bot.)* **3**: 121–172 (1964).

1 Bracts enclosing the calyx; inflorescence often capitate or fasciculate
2 Flowers solitary or fasciculate; bracts usually distinctly shorter than the calyx; stem usually much-branched
3 Stems glandular-pubescent, at least at the base; calyx-teeth ± triangular, acute or subobtuse **8. fasciculata**
3 Stems eglandular-pubescent; calyx-teeth oblong, obtuse **9. saxifraga**
2 Flowers in a capitulum; bracts ± completely enclosing the calyx; stems simple or with a few branches
4 Largest bracts not more than 4 mm wide; petals linear-spathulate, white with pink or purple veins
5 Calyx-teeth glabrous **11. thessala**

5 Calyx-teeth pubescent or glandular
6 Petals 18–22 mm; calyx-teeth obtuse **12. grandiflora**
6 Petals 6·5–8 mm; calyx-teeth ±acute **13. dianthoides**
4 Largest bracts at least 4 mm wide; petals with a long claw and distinct limb, pink or purplish, sometimes with darker veins
7 Leaf-sheaths at least twice as long as wide; seeds 1–1·3 mm, covered with cylindrical papillae **16. velutina**
7 Leaf-sheaths less than twice as long as wide; seeds at least (1·1–)1·3 mm, reticulate, tuberculate or almost smooth
8 Largest bracts 15–22 mm; petal-limb laciniate to crenate; seeds with ±flat lateral margins **18. glumacea**
8 Largest bracts not more than 15(–16) mm; petal-limb obcordate, rarely crenate or laciniate; seeds with strongly incurved lateral margins
9 Petal-limb 3–6 mm wide, sometimes laciniate or crenate; largest bracts 10·5–16 × 6–13 mm, obtuse **17. obcordata**
9 Petal-limb 2–3·5 mm wide, entire; largest bracts 6–12 × 3–7(–8) mm, mucronate
10 Seeds reticulate **14. prolifera**
10 Seeds tuberculate **15. nanteuilii**
1 Bracts not enclosing the calyx; flowers solitary, rarely fasciculate
11 Petals emarginate
12 Calyx-teeth oblong, obtuse; petals usually pink or reddish **10. graminea**
12 Calyx-teeth triangular-oblong, ±acute or mucronate; petals white, sometimes with pink veins
13 Calyx-teeth 3-veined, the lateral veins sometimes obscure near the apex; seeds *c.* 1·5 mm, black, smooth **4. candica**
13 Calyx-teeth 1-veined; seeds *c.* 1 mm, blackish-brown, ±tuberculate **8. fasciculata**
11 Petals entire
14 Annual, without non-flowering rosettes
15 Glandular-pubescent; calyx-teeth 3-veined **5. cretica**
15 Glabrous; calyx-teeth 1-veined **6. alpina**
14 Perennial, ±caespitose and woody at the base, with non-flowering rosettes
16 Petals pink; seeds blackish-brown, reticulate-tuberculate **7. phthiotica**
16 Petals white or yellow, sometimes with pink veins or purplish spots at the base; seeds black, smooth
17 Stem glandular-pubescent at the base and apex, glabrous in the middle; calyx-teeth strongly 3-veined **3. armerioides**
17 Indumentum of stem variable, but never as above; calyx-teeth 1-veined or with obscure lateral veins
18 Stem densely glandular-pubescent throughout or glabrous below, rarely completely glabrous; petals white or pale yellow; anthers usually purple **1. illyrica**
18 Stem densely glandular-tomentose at the base, otherwise glabrous; petals yellow; anthers white **2. ochroleuca**

Sect. PSEUDOTUNICA (Fenzl) P. W. Ball & Heywood. Annuals or perennials. Leaves 3-veined. Bracts absent. Petals not clawed. Seeds black, smooth, with thin margins.

1. P. illyrica (Ard.) P. W. Ball & Heywood, *Bull. Brit. Mus. (Bot.)* **3**: 133 (1964) (*Tunica illyrica* (Ard.) Fischer & C. A. Meyer). Perennial up to 40 cm. Inflorescence lax or subfastigiate. Petals 5·5–10 mm, oblong-spathulate, entire, white or rarely pale yellow, spotted with purple at the base and sometimes pink-veined; anthers usually purple. Seeds 1·7–2·3 × 0·9–1·3 mm. *S.E. Europe, extending westwards to Sicilia.* Al Bu Cr Gr It Ju Rm Si.

1 Stem usually glabrous, at least at the base; calyx 5–8 mm (b) subsp. **haynaldiana**
1 Stem glandular-pubescent throughout; calyx 3·5–6 mm
2 Calyx-teeth 1-veined; inflorescence usually subfastigiate (a) subsp. **illyrica**
2 Calyx-teeth with weak lateral veins; inflorescence lax (c) subsp. **taygetea**

(a) Subsp. **illyrica**: Stems glandular-pubescent throughout. Inflorescence usually subfastigiate. Calyx 3·5–6 mm; calyx-teeth 1-veined. ● *From N.E. Jugoslavia to Greece.*

(b) Subsp. **haynaldiana** (Janka) P. W. Ball & Heywood, *op. cit.* 134 (1964) (*Tunica illyrica* subsp. *haynaldiana* (Janka) Prodan): Stems usually glabrous, at least at the base. Inflorescence lax or subfastigiate. Calyx 5–8 mm; calyx-teeth 1-veined. $2n = 26$. *Romania; Balkan peninsula southwards to S.C. Greece; Calabria; Sicilia.*

(c) Subsp. **taygetea** (Boiss.) P. W. Ball & Heywood, *op. cit.* 137 (1964) (*Tunica cretica* sensu Hayek, pro parte, *T. taygetea* (Boiss.) P. H. Davis): Stems glandular-pubescent throughout. Inflorescence lax. Calyx 3·5–5 mm; calyx-teeth 3-veined, lateral veins often weak. $2n = 26$. ● *S. Greece (Peloponnisos); Kriti.*

2. P. ochroleuca (Sibth. & Sm.) P. W. Ball & Heywood, *op. cit.* 138 (1964) (*Tunica ochroleuca* (Sibth. & Sm.) Fischer & C. A. Meyer). Very like **1** but the stems densely glandular-tomentose at the base, glabrous or rarely sparsely glandular in the upper part; pedicels 4–14 mm; petals 5–7 mm, linear-oblong, pale yellow, purplish at the base; anthers white. $2n = 30$. ● *E. Greece (Attiki and Evvoia).* Gr.

3. P. armerioides (Ser.) P. W. Ball & Heywood, *op. cit.* 139 (1964) (*Tunica armerioides* (Ser.) Halácsy, *T. sibthorpii* Boiss.). Perennial up to 30 cm, densely glandular-pubescent at the base and apex, glabrous in the middle. Inflorescence fastigiate; pedicels 1–5 mm. Calyx 4–6·5 mm, teeth strongly 3-veined. Petals 6–8(–9) mm, oblong-spathulate, white, often purple-spotted at the base; anthers usually purple. Seeds 1·8–2·1 × 1·0–1·3 mm. $2n = 26$. *Aegean islands and S.E. Greece.* Gr.

Some plants from the mainland of Greece and Evvoia are intermediate between **1** and **3** or **2** and **3**. They usually have the stems glandular-pubescent throughout, and sometimes lax inflorescences with pedicels up to 9 mm. These latter plants have been confused with **4**, from which they can be distinguished by the entire petals.

4. P. candica P. W. Ball & Heywood, *op. cit.* 141 (1964) (*Tunica cretica* Fischer & C. A. Meyer, pro parte, *T. taygetea* sensu P. H. Davis pro parte, non (Boiss.) P. H. Davis, *T. cretica* auct. balcan., non (L.) Fischer & C. A. Meyer). Perennial up to 20 cm, glabrous or sparsely glandular-pubescent. Inflorescence lax; pedicels up to 20 mm. Calyx 3–4·5 mm; teeth 3-veined, the lateral veins sometimes weak. Petals 4·5–7 mm, oblong-spathulate, emarginate, white, with pink veins. Seeds *c.* 1·5 mm. ● *Kriti.* Cr.

5. P. cretica (L.) P. W. Ball & Heywood, *op. cit.* 142 (1964) (*Tunica pachygona* Fischer & C. A. Meyer, *T. cretica* (L.) Fischer & C. A. Meyer). Annual up to 40 cm, densely glandular-pubescent. Inflorescence lax; pedicels 4·5–20(–30) mm, usually glabrous. Calyx 6–10·5 mm, with strongly 3-veined teeth. Petals linear-spathulate, white, sometimes reddish beneath, usually included in the calyx. Seeds 2–2·8 mm. *C. Greece (near Kalambaka).* Gr. (*S.W. Asia.*)

Sect. PSEUDOGYPSOPHILA (A. Braun) P.W. Ball & Heywood. Annuals. Leaves 3- to 5-veined. Bracts absent. Petals not clawed. Seeds blackish-brown, reticulate-tuberculate, with thickened margin.

6. P. alpina (Habl.) P. W. Ball & Heywood, *op. cit.* 145 (1964) (*Tunica stricta* (Ledeb.) Fischer & C. A. Meyer, *T. alpina* (Habl.) Bobrov, *T. olympica* Boiss., *Gypsophila alpina* Habl.). Up to 40 cm, glabrous. Inflorescence lax; pedicels up to 30 mm, patent. Calyx 2·5–4·5(–5·5) mm, with 1-veined teeth. Petals 3–6 mm, linear-oblong, white. Seeds 0·7–1·2 mm. *S.W. Bulgaria (Pirin Planina).* Bu.

In Europe represented by subsp. **olympica** (Boiss.) P. W. Ball & Heywood, *op. cit.* **3**: 146 (1964), which occurs mainly in W. & S. Anatolia. Subsp. **alpina** is widespread in the mountains of W. & C. Asia.

Sect. PETRORHAGIA. Perennials. Leaves 1(–5)-veined. Flowers with or without bracts. Petals not abruptly clawed. Seeds blackish-brown, tuberculate, with thickened margins.

7. P. phthiotica (Boiss. & Heldr.) P. W. Ball & Heywood, *op. cit.* 149 (1964) (*Tunica ochroleuca* var. *phthiotica* (Boiss. & Heldr.) Hayek, *T. phthiotica* Boiss. & Heldr.). Plant 3–20 cm, sparsely glandular-pubescent above. Lower leaves 1- to 3-veined. Inflorescence lax. Bracts absent. Calyx 3–5 mm, glabrous; teeth 1-veined, broadly triangular, acute to obtuse, mucronate. Petals 4–6·5 × 0·5–1 mm, entire, pink. Seeds 1·2–1·6 mm. $2n = 28$. *Mountain rocks, 1800–2150 m.* ● *S. & S.C. Greece.* Gr.

8. P. fasciculata (Margot & Reuter) P. W. Ball & Heywood, *op. cit.* 150 (1964) (*Tunica fasciculata* (Margot & Reuter) Boiss., *Gypsophila fasciculata* Margot & Reuter). Plant 5–30 cm, glandular-pubescent, sometimes glabrous above. Leaves rarely up to 5-veined. Flowers usually fasciculate and bracteate, rarely in a lax panicle and ebracteate. Calyx 2·5–5·5 mm, pubescent or sparsely hirsute, sometimes glandular; teeth 1-veined, acute or subobtuse. Petals 3·5–6 mm, linear-spathulate, emarginate, white. Seeds 0·8–1 mm. $2n = 30$. ● *W. Greece.* Gr.

9. P. saxifraga (L.) Link, *Handb.* **2**: 235 (1829) (*Tunica saxifraga* (L.) Scop., *Kohlrauschia saxifraga* (L.) Dandy; incl. *T. rigida* (L.) Boiss.). Up to 45 cm, glabrous or scabrid-pubescent. Inflorescence lax, occasionally fasciculate; flowers usually with 4 bracts. Calyx 3–6(–7) mm; teeth oblong, obtuse, 1-veined, with 2 weak lateral veins. Petals 4·5–10 × 1·2–3(–4) mm, white or pink; limb obcordate, narrowing gradually into the claw. Seeds 0·9–1·6 mm. $2n = 30, 60$. *C. & S. Europe, eastwards to W. Ukraine.* Al Au Bu Co Cz Ga Ge Gr He Hs Hu It Ju Lu Rm Rs(W) Sa Si ?Tu [Br ?Po Rs(B) Su].

10. P. graminea (Sm.) P. W. Ball & Heywood, *Bull. Brit. Mus. (Bot.)* **3**: 155 (1964) (*Tunica graminea* (Sm.) Boiss.). Up to 40 cm, shortly and densely pubescent, sometimes glabrous above. Inflorescence usually lax, sometimes with a few fasciculate flowers. Bracts absent. Calyx 3·5–5·5 mm, densely pubescent; teeth oblong, obtuse, 1-veined, with 2 weak lateral veins. Petals 5–10 × 1·8–2·5 mm, pink or reddish; limb obcordate, narrowing gradually into the claw. Seeds 1·1–1·5 mm. $2n = 60$. ● *S. Greece (Peloponnisos).* Gr.

Intermediates between this species and **9** occur in N. Peloponnisos, and require further investigation.

11. P. thessala (Boiss.) P. W. Ball & Heywood, *op. cit.* 156 (1964) (*Tunica thessala* Boiss.). Plant 10–35 cm, papillose-pubescent at least at the base. Inflorescence a capitulum, up to 10-flowered, with 6 to many bracts. Largest bracts 6–10 × 2·5–4 mm, brown with white, membranous margin, rarely 5–7 × 1·5–2 mm, almost completely white, membranous. Calyx 5·5–7 mm, glabrous or papillose, with oblong, obtuse teeth. Petals 6·5–8 mm, linear-spathulate, entire, emarginate or crenate, white, with pink or purple veins. Seeds 1·8–2·3 mm. $2n = 30$. *S. part of Balkan peninsula.* Bu Gr Ju.

12. P. grandiflora Iatroú, *Nordic Jour. Bot.* **5**: 441 (1985). Like **11** but stems up to 45 cm, glabrous; largest bracts 10–12 × 3–4 mm, white or pale brown, membranous, with a dark brown midrib; calyx *c.* 10 mm, glandular-pubescent; petals 18–22 mm, entire. $2n = 30$. ● *S. Greece (Parnon).* Gr.

13. P. dianthoides (Sm.) P. W. Ball & Heywood, *Bull. Brit. Mus. (Bot.)* **3**: 158 (1964) (*Tunica dianthoides* (Sm.) Boiss.). Like **11** but stems up to 40 cm, glabrous; heads with 1–6(–8) flowers; bracts 4 to many, the largest 4·5–10 × 1–2·5 mm, white-membranous except for the brown midrib; calyx pubescent, with triangular-lanceolate, more or less acute teeth; petals 6–10 mm, entire; seeds 1–1·3 mm. $2n = 30$. ● *W. Kriti.* Cr.

Sect. KOHLRAUSCHIA (Kunth) P. W. Ball & Heywood. Annuals. Leaves 3-veined. Inflorescence a capitulum, with very wide, brown, scarious bracts. Petals abruptly clawed. Seeds blackish-brown, reticulate to papillose, with thick margins.

14. P. prolifera (L.) P. W. Ball & Heywood, *op. cit.* 161 (1964) (*Dianthus prolifer* L., *Tunica prolifera* (L.) Scop., *Kohlrauschia prolifera* (L.) Kunth). Up to 50 cm, glabrous or scabrid-pubescent. Leaf-sheaths usually about as long as wide. Largest bracts 6–12 × 3–7(–8) mm, usually obtuse, mucronate. Petals 10–14 × 2–3·5 mm, pink with darker veins; limb obcordate. Seeds 1·3–1·9 × 0·8–1·1 mm, reticulate, with strongly incurved sides. $2n = 30$. *C. & S. Europe, extending northwards to S. Sweden, westwards to C. Spain and eastwards to Krym.* Al Au Be Bu Co Cz Da Ga Ge Gr He Ho Hs Hu It Ju Po Rm Rs(C, W, K) Sa Si Su Tu [*Br].

15. P. nanteuilii (Burnat) P. W. Ball & Heywood, *op. cit.* 164 (1964) (*Dianthus prolifer*, *Tunica prolifera*, *Kohlrauschia prolifera* auct. eur. occident., pro parte, *K. nanteuilii* (Burnat) P. W. Ball & Heywood). Like **14** but the middle part of the stem often tomentose; leaf-sheaths up to twice as long as wide; seeds tuberculate. $2n = 60$. *W. Europe, extending eastwards to C. Italy.* Bl Br Co Ga Hs It Lu Sa.

16. P. velutina (Guss.) P. W. Ball & Heywood, *op. cit.* 166 (1964) (*Dianthus velutinus* Guss., *Tunica velutina* (Guss.) Fischer & C. A. Meyer, *Kohlrauschia velutina* (Guss.) Reichenb.). Very like **14** but the middle part of the stem usually densely glandular-tomentose; leaf-sheaths at least twice as long as wide; all outer bracts acute or mucronate; petal-limb 1·2–2·5 mm wide; seeds 1–1·3 × 0·7–0·8 mm, with cylindrical papillae. $2n = 30$. *S. Europe.* Al Bu Co Cr Ga Gr Hs It Ju Lu Sa Si Tu.

This species sometimes has a glabrous stem, and such plants have been confused with **14**, particularly in Italy.

17. P. obcordata (Margot & Reuter) Greuter & Burdet, *Willdenowia* **12**: 188 (1982) (*Dianthus obcordatus* Margot & Reuter, *Kohlrauschia glumacea* (Chaub. & Bory) Hayek var. *obcordata* (Margot & Reuter) Hayek). Up to 50 cm, glabrous; stems more or less simple. Leaf-sheaths about as long as wide. Largest bracts 10·5–16 × 6–13 mm, obtuse, mucronate. Petals 12–18 × 3–6 mm, pink with darker veins, rarely bright pink and without conspicuous veins; limb obcordate, entire to laciniate. Seeds 1·1–2·2 × 0·7–1·7 mm, reticulate or tuberculate, with strongly incurved sides. $2n = 30$. ● *Balkan peninsula*. Al Gr Ju Tu.

Plants intermediate between **17** and **18** occur in Peloponnisos. Plants with reticulate seeds are said to be restricted to S. Greece, outside the range of **14**. Plants with entire petals occur mainly in the northern half of the range, whilst plants with crenate to laciniate petals occur mainly in Greece.

18. P. glumacea (Chaub. & Bory) P. W. Ball & Heywood, *Bull. Brit. Mus. (Bot.)* **3**: 169 (1964) (*Tunica glumacea* (Chaub. & Bory) Boiss.). Up to 50 cm, glabrous, simple or with few branches. Leaf-sheaths about as long as wide. Largest bracts 15–22 × 9–17 mm, obtuse. Petals 15–18 × 3–4·5 mm, bright pink, without conspicuous veins; limb crenate to laciniate. Seeds 1·5–3·1 × 1·3–2·1 mm, sparsely tuberculate to almost smooth, the sides not strongly incurved. $2n = 30$. ● *S. Greece (Peloponnisos)*. Gr.

36. Dianthus L.

†T. G. TUTIN AND S. M. WALTERS

Perennial herbs or small shrubs, more rarely annuals or biennials, with entire, usually linear, parallel-veined, often glaucous leaves. Flowering stems often swollen and brittle at nodes. Flowers usually conspicuous, solitary, in lax, few-flowered cymes, or in heads surrounded by bracts. Epicalyx-scales 2 to many, usually appressed to the calyx. Calyx tubular, 5-toothed, without scarious commissures. Petals 5, long-clawed, entire, dentate or laciniate, but not deeply bifid; coronal scales absent, but petals often with a tuft of hairs (bearded) at mouth of flower. Capsule dehiscing apically with 4 teeth; carpophore often present. Seeds numerous, concave on 1 side.

Male-sterile plants of a number of species occur sporadically and add to the difficulties of identification, as such plants are often dwarf, with flowers smaller in all their parts than normal, and sometimes with a reduced number of epicalyx-scales. Late flowers, particularly if borne on lateral branches produced by damaged main stems, may also be abnormal. Species which normally have capitate inflorescences frequently produce solitary flowers in these circumstances. It is therefore often difficult to identify plants flowering outside their normal season.

The shape and measurements of the calyx refer to the calyx at anthesis. The diameter of the stem is measured just below a node. Bracts occur in species with capitate inflorescences, and should be distinguished from epicalyx-scales which subtend a single flower.

Most of the species are more or less interfertile but, since they are usually geographically isolated, hybrids are rather local. They seem, however, to occur in most localities when two or more species grow together, and are particularly common in the Pyrenees.

In the absence of any recent monograph, the arrangement here adopted largely follows that of Pax & Hoffmann (see Literature) in dividing the species into two subgenera: *Dianthus* (*Caryophyllum*) and *Armeriastrum*. Whilst the position of several species in such an arrangement is quite arbitrary, the classification has the advantage that any *Dianthus* with several flowers in a head surrounded by involucral bracts is likely to be found in subgenus *Armeriastrum*, and most Mediterranean shrubby species also belong to this subgenus.

Literature: F. Pax & K. Hoffmann in Engler & Prantl, *Natürl. Pflanzenfam.*, ed. 2, **16c**: 356–361 (1934), which has a very extensive bibliography.

1 Shrubs with obvious woody branches
2 Leaves more than 3 mm wide, linear-oblong to oblanceolate or elliptical
3 Calyx less than 18 mm **39. uralensis**
3 Calyx at least 18 mm
4 Calyx 4–5 mm wide, almost cylindrical; flowers in distinct bracteate heads **18. rupicola**
4 Calyx not more than 3 mm wide, narrowed from about the middle; flowers not in bracteate heads **19. fruticosus**
2 Leaves not more than 3 mm wide, linear, often needle-like
5 Calyx at least 15 mm **17. aciphyllus**
5 Calyx less than 15 mm
6 Hummock-forming (E. Kriti) **16. pulviniformis**
6 Not hummock-forming
7 Petals white (S.E. Russia, Krym) **38. rigidus**
7 Petals pinkish, at least above
8 Epicalyx-scales obcordate, with subulate apex **15. juniperinus**
8 Epicalyx-scales ovate, acuminate **39. uralensis**
1 Herbs, sometimes woody at base, but without obvious woody branches
9 Sheaths of cauline leaves at least 3 times as long as diameter of stem; flowers in heads, usually with involucral bracts, more rarely solitary or few
10 Flowers yellow **5. knappii**
10 Flowers pink to purple
11 Uppermost leaves greatly widened in lower part **20. capitatus**
11 Uppermost leaves not greatly widened in lower part
12 Leaves with thin margins and without prominent submarginal veins beneath
13 Bracts entirely brown **31. cruentus**
13 Bracts green, at least in upper part
14 Epicalyx-scales acute or shortly aristate, brown or scarious, all much shorter than calyx **6. membranaceus**
14 Epicalyx-scales aristate, largely green, at least some about as long as calyx
15 Sheaths *c.* 3 times as long as diameter of stem; bracts 4 **3. trifasciculatus**
15 Sheaths 5–9 times as long as diameter of stem; bracts numerous **4. urumoffii**
12 Leaves with thick margins or prominent submarginal veins beneath
16 Calyx-teeth obtuse or subobtuse but sometimes mucronate
17 Calyx *c.* 8 mm **24. giganteiformis**
17 Calyx 10–22 mm

18 Leaves obtuse or subobtuse; calyx-teeth little longer than wide, not mucronate **25. diutinus**

18 Leaves acute or acuminate; calyx-teeth at least twice as long as wide, mucronate

19 Calyx not more than 14 mm; petal-limb 5–8 mm, pink **26. platyodon**

19 Calyx 18–22 mm; petal-limb 10–15 mm, dark purplish **27. bessarabicus**

16 Calyx-teeth acuminate

20 Plant glandular-viscid; calyx 6–7 × 1·5 mm **115. formanekii**

20 Plant sometimes puberulent but not glandular-viscid; calyx more than 7 mm or, if less, 2–3 mm wide

21 Flowers solitary or few together, usually distinctly pedicellate; bracts 0, or narrow and not closely subtending the flowers

22 Epicalyx-scales coriaceous, not broadly scarious; calyx ± coriaceous

23 Petals with short glandular hairs on upper surface **108. biflorus**

23 Petals with long eglandular hairs on upper surface **106. gracilis**

22 Epicalyx-scales membranous or broadly scarious; calyx membranous

24 Epicalyx-scales broadly scarious **34. brachyzonus**

24 Epicalyx-scales usually dark brown, not or narrowly scarious

25 Leaves 1–2 mm wide, usually flat; calyx usually dark purple **28. carthusianorum**

25 Leaves usually setaceous, up to *c.* 1 mm wide; calyx usually green with a green apex (E. & S. Carpathians)

26 Cauline leaves as long as or longer than internodes; some bracts at least as long as inflorescence **28. carthusianorum**

26 Cauline leaves shorter than internodes; bracts shorter than inflorescence **30. henteri**

21 Flowers usually numerous, subsessile, crowded in dense heads; bracts conspicuous, closely subtending the flowers

27 Calyx up to 10 mm

28 Epicalyx-scales gradually narrowed at apex, acute or shortly awned **23. giganteus**

28 Epicalyx-scales abruptly narrowed into an awn

29 At least the basal leaves setaceous

30 Petal-limb 5–8(–10) mm **22. pinifolius**

30 Petal-limb 3–4 mm

31 Leaves *c.* 0·5 mm wide; awn of epicalyx-scales usually patent **36. stenopetalus**

31 Leaves *c.* 1 mm wide; awn of epicalyx-scales erect **37. moesiacus**

29 Leaves flat

32 Bracts broadly ovate; awn of epicalyx-scales shorter than wide part of scale, usually patent or deflexed **35. pelviformis**

32 Bracts ovate-oblong; awn of epicalyx-scales as long as wide part of scale, erect or somewhat patent

33 Epicalyx-scales dark brown **32. tristis**

33 Epicalyx-scales pale purplish **33. stribrnyi**

27 Calyx more than 10 mm

34 Epicalyx-scales gradually narrowed at apex

35 Epicalyx-scales broadly scarious; calyx pink-tinged **29. borbasii**

35 Epicalyx-scales not or scarcely scarious; calyx dark purple

36 Calyx (15–)17–20 mm; petal-limb 5–8 mm **23. giganteus**

36 Calyx 10–14 mm; petal-limb 3–5 mm **24. giganteiformis**

34 Epicalyx-scales obcordate or almost truncate, abruptly aristate

37 Calyx 15–20 mm

38 Leaves less than 1 mm wide

39 Petal-limb 5–8(–10) mm **22. pinifolius**

39 Petal-limb 10–15 mm **28. carthusianorum**

38 Leaves at least 1 mm wide

40 Epicalyx-scales whitish with wide hyaline margin **34. brachyzonus**

40 Epicalyx-scales brown or blackish without or with a narrow hyaline margin

41 Wide part of epicalyx-scales distinctly longer than awn **28. carthusianorum**

41 Wide part of epicalyx-scales usually shorter than awn

42 Bracts with a green apex; epicalyx-scales coriaceous **21. ferrugineus**

42 Bracts without a green apex; epicalyx-scales not coriaceous **31. cruentus**

37 Calyx 10–15 mm

43 Leaves setaceous

44 Calyx usually more than 2 mm wide; epicalyx-scales at least $\frac{3}{4}$ as long as calyx **22. pinifolius**

44 Calyx usually less than 2 mm wide; epicalyx-scales *c.* $\frac{1}{2}$ as long as calyx **37. moesiacus**

43 At least the cauline leaves flat

45 Petal-limb subglabrous **32. tristis**

45 Petal-limb bearded

46 Awn of epicalyx-scales longer than wide part; calyx ± cylindrical **21. ferrugineus**

46 Awn of epicalyx-scales equalling or shorter than wide part; calyx ± tapering above middle

47 Epicalyx-scales whitish, ± purplish-tinged; awn ± equalling wide part of scale **31. cruentus**

47 Epicalyx-scales usually brown; awn shorter than wide part of scale **28. carthusianorum**

9 Sheaths of cauline leaves not more than twice (rarely 3 times) as long as diameter of stem; flowers mostly solitary or in small groups without involucral bracts (but see **8** and **22**)

48 Calyx pubescent or glandular-pubescent

49 Annual or biennial, with slender stock and root; non-flowering stems absent

50 Epicalyx-scales *c.* $\frac{1}{2}$ as long as calyx **(111–114). corymbosus** group

50 Epicalyx-scales about as long as calyx

51 Epicalyx-scales lanceolate; calyx 15–20 mm **1. armeria**
51 Epicalyx-scales ovate; calyx *c.* 10 mm **2. pseudarmeria**
49 Perennial with stout, woody stock or root; non-flowering stems present at anthesis
52 Lower leaves obtuse or subobtuse
53 Petal-limb bearded **93. deltoides**
53 Petal-limb glabrous (**111–114**). **corymbosus** group
52 Lower leaves acuminate
54 Cauline leaves shorter than internodes (S.W. Europe) **76. scaber**
54 Cauline leaves at least as long as internodes (Bulgaria, Greece, Turkey) **105. roseoluteus**
48 Calyx glabrous
55 Annual, with slender stock and root; non-flowering stems absent
56 Calyx verruculose **110. tripunctatus**
56 Calyx smooth
57 Cauline leaves narrowly oblong, mostly obtuse **47. viridescens**
57 Cauline leaves linear, acute or acuminate
58 Flowers usually in clusters of 2 or 3; calyx (10–)12–17 mm (**111–114**). **corymbosus** group
58 Flowers in many-flowered heads; calyx *c.* 10 mm **2. pseudarmeria**
55 Perennial, with stout stock or root; non-flowering stems usually present at anthesis
59 At least some of the leaves obtuse or subobtuse
60 Flowers numerous, in a dense head **8. barbatus**
60 Flowers solitary, or few together and distinctly pedicellate
61 Leaves of non-flowering stems *c.* 5 mm, elliptical
62 Leaves on flowering stems similar to those on non-flowering stems, longer than internodes **94. myrtinervius**
62 Leaves on flowering stems narrower than on non-flowering stems, shorter than internodes
63 Stock stout; calyx tapering upwards from below the middle **95. sphacioticus**
63 Stock slender; calyx cylindrical or widening upwards **93. deltoides**
61 Leaves of non-flowering stems more than 5 mm, or else not elliptical
64 Stems shortly pubescent
65 Petals glabrous
66 Calyx *c.* 18 mm; petal-limb *c.* 3 mm, cream above, cinnamon beneath **101. cinnamomeus**
66 Calyx 9–13 mm; petal-limb 7–10 mm, pink or pinkish-purple (**111–114**). **corymbosus** group
65 Petals bearded
67 Petals spotted; basal leaves usually narrowly oblong or lanceolate **93. deltoides**
67 Petals unspotted; basal leaves linear or spathulate (**111–114**). **corymbosus** group
64 Stems glabrous, smooth or rarely scabrid on the angles
68 Petal-limb laciniate
69 Leaves of non-flowering shoots 1·5–3 mm wide (Atlantic coast from N. Spain to N.W. France) **91. gallicus**
69 Leaves of non-flowering shoots usually less than 1 mm wide
70 Central, undivided part of petal-limb lanceolate, with a greenish or purplish spot at base **88. arenarius**
70 Central, undivided part of petal-limb narrowly obovate-cuneate, unspotted
71 Epicalyx-scales 4–6 (E.C. Russia) **89. krylovianus**
71 Epicalyx-scales 2–4
72 Calyx-teeth obtuse or apiculate, with broadly scarious margin **83. plumarius**
72 Calyx-teeth acute to acuminate, usually not broadly scarious (S. Czechoslovakia) **84. moravicus**
68 Petal-limb dentate to subentire
73 Epicalyx-scales not more than ⅓ as long as calyx
74 Calyx 6–10(–13) mm; petals glabrous **66. subacaulis**
74 Calyx 14–25 mm; petals usually bearded
75 Epicalyx-scales gradually narrowed at apex **78. cintranus**
75 Epicalyx-scales abruptly contracted at apex
76 Flowers usually 2–4 together; leaves with thin margins **40. seguieri**
76 Flowers solitary; leaves with thick margins
77 Epicalyx-scales green and purplish, herbaceous (W. & C. Europe) **62. gratianopolitanus**
77 Epicalyx-scales brown, coriaceous (Kriti and Karpathos) **63. xylorrhizus**
73 Epicalyx-scales at least ½ as long as calyx
78 Epicalyx-scales *c.* ½ as long as calyx
79 Cauline leaves at least 5 pairs; flowers usually 2 or 3 together **55. nitidus**
79 Cauline leaves 1–3(4) pairs; flowers nearly always solitary
80 Epicalyx-scales membranous, acute; cauline leaves scale-like **60. microlepis**
80 Epicalyx-scales herbaceous, apex subulate; cauline leaves not scale-like
81 Epicalyx-scales often 4, the inner ovate, abruptly narrowed at apex; calyx *c.* 5 mm wide **53. repens**
81 Epicalyx-scales 2, lanceolate, gradually narrowed at apex; calyx *c.* 3 mm wide **56. scardicus**
78 Epicalyx-scales more than ½ as long to as long as calyx
82 Leaf-margins thick and cartilaginous **107. haematocalyx**
82 Leaf-margins not thick and cartilaginous
83 Cauline leaves at least 5 pairs
84 Middle cauline leaves much smaller than basal **55. nitidus**
84 Middle cauline leaves much larger than basal **57. callizonus**
83 Cauline leaves 4 pairs or fewer
85 Basal leaves at least 3 mm wide, oblong-lanceolate **54. alpinus**
85 Basal leaves up to 2·5 mm wide, linear

86 Leaves soft, widest above the middle; calyx 12–16 mm **58. glacialis**

86 Leaves rigid, linear; calyx 9–10 mm **59. freynii**

59 All leaves acute or acuminate

87 Leaves thin and flat, without a thick margin or prominent submarginal veins

88 Sheaths of cauline leaves 2–4 times as long as diameter of stem

89 Flowers long-pedunculate **103. lanceolatus**

89 Flowers subsessile in heads

90 Calyx-teeth acuminate; epicalyx-scales gradually narrowed at apex **6. membranaceus**

90 Calyx-teeth obtuse; epicalyx-scales rounded and mucronate at apex **7. dobrogensis**

88 Sheaths of cauline leaves 1–2 times as long as diameter of stem

91 Petal-limb laciniate

92 Petal-limb divided more than $\frac{1}{4}$-way to middle, the undivided central portion narrowly oblong **92. superbus**

92 Petal-limb divided not more than $\frac{1}{4}$-way to middle, the undivided central portion suborbicular

93 Plant usually more than 20 cm; flowers in clusters **51. monspessulanus**

93 Plant usually less than 20 cm; flowers solitary **52. sternbergii**

91 Petal-limb dentate to subentire

94 Calyx verruculose, not longitudinally ribbed **109. strictus**

94 Calyx not verruculose, distinctly longitudinally ribbed

95 Petals glabrous

96 Epicalyx-scales 6 **78. cintranus**

96 Epicalyx-scales 4

97 Epicalyx-scales gradually narrowed into a green, subulate point **46. furcatus**

97 Epicalyx-scales abruptly acuminate or cuspidate, the point usually brownish **102. monadelphus**

95 Petals bearded

98 Outer epicalyx-scales at least as long as calyx, leaf-like

99 Leaves 7- to 9-veined, some lateral veins almost as prominent as midrib **3. trifasciculatus**

99 Leaves 3- to 5-veined, lateral veins all much weaker than midrib **50. guttatus**

98 Epicalyx-scales shorter than calyx

100 Leaves 0·5–1 mm wide; epicalyx-scales up to $\frac{1}{2}$ as long as calyx **78. cintranus**

100 Leaves 1–8 mm wide; epicalyx-scales usually at least $\frac{1}{2}$ as long as calyx

101 Stems solitary or few, stout, simple or branched in upper part; flowers often in small heads

102 Petal-limb narrowly obovate, purple with black spots at base **48. eugeniae**

102 Petal-limb broadly ovate, pinkish or purplish, unspotted or with whitish spots at base

103 Basal leaves present and ±rosette-forming at anthesis **40. seguieri**

103 Basal leaves withered at anthesis

104 Cauline leaves 3–8 mm wide, usually 10 or more pairs, tapering from about the middle **41. collinus**

104 Cauline leaves usually less than 3 mm wide, usually fewer than 10 pairs, tapering from near base **42. fischeri**

101 Stems numerous, slender, usually branched from near base; flowers usually solitary

105 Basal rosette well developed at anthesis (Pyrenees, N.W. Spain) **46. furcatus**

105 Basal leaves mostly or quite withered at anthesis

106 Stem and leaves glabrous; epicalyx-scales usually more than $\frac{1}{2}$ as long as calyx **43. pratensis**

106 Stem and leaves usually puberulent; epicalyx-scales usually less than $\frac{1}{2}$ as long as calyx

107 Epicalyx-scales shortly aristate, the margin not scarious **49. tesquicola**

107 Epicalyx-scales acuminate, the margin broadly scarious

108 Calyx tapering markedly from the middle **44. versicolor**

108 Calyx scarcely tapering upwards **45. pseudoversicolor**

87 At least the upper cauline leaves with a thick margin or strong submarginal veins, often ±convolute

109 Petal-limb laciniate, divided at least to half-way

110 Leaves not crowded at base of stems, rigid, mostly recurved **87. squarrosus**

110 Leaves crowded at base of stems; cauline few and distant, not recurved

111 Calyx 10–12 mm; outer epicalyx-scales at least $\frac{1}{2}$ as long as calyx, with a green, subulate point **46. furcatus**

111 Calyx at least 15 mm; epicalyx-scales up to $\frac{1}{2}$ as long as calyx, without a green, subulate point

112 Epicalyx-scales ovate, gradually long-acuminate

113 Epicalyx-scales $\frac{1}{3}$–$\frac{1}{2}$ as long as calyx, usually 6 or more; petal-limb *c.* 15 mm **75. serrulatus**

113 Epicalyx-scales $\frac{1}{5}$–$\frac{1}{3}$ as long as calyx, usually 4(–6); petal-limb 4–10 mm **79. petraeus**

112 Epicalyx-scales obovate, not long-acuminate

114 Calyx tapering from near base

115 Plant usually glaucous; cauline leaves (6–)8–14(–19) pairs; calyx-teeth lanceolate, apiculate **85. serotinus**

115 Plant usually green; cauline leaves rarely more than 6 pairs; calyx-teeth narrowly acuminate **79. petraeus**

114 Calyx ±cylindrical or tapering from about the middle

116 Petal-limb with a greenish or purplish spot near base **88. arenarius**

116 Petal-limb unspotted

117 Leaves usually not more than 0·5 mm wide (Ural) **86. acicularis**

117 Leaves usually at least 1 mm wide

118 Calyx 3–6 mm wide

119 Undivided central part of petal-limb cuneate; calyx 3–4 mm wide **82. spiculifolius**

119 Undivided central part of petal-limb suborbicular; calyx 4–6 mm wide

120 Calyx-teeth obtuse or apiculate, with broadly scarious margin **83. plumarius**

120 Calyx-teeth acute to acuminate, usually not broadly scarious (S. Czechoslovakia) **84. moravicus**

118 Calyx less than 3 mm wide

121 Calyx not more than 20 mm; epicalyx-scales $\frac{1}{6}$–$\frac{1}{4}$ as long as calyx; petal-limb $\frac{1}{2}$–$\frac{3}{4}$ as long as calyx **89. krylovianus**

121 Calyx *c.* 25 mm; epicalyx-scales $\frac{1}{4}$–$\frac{1}{3}$ as long as calyx; petal-limb less than $\frac{1}{2}$ as long as calyx **90. volgicus**

109 Petal-limb subentire to deeply dentate

122 Petal-limb bearded, sometimes only sparsely so

123 Densely pulvinate (Samothraki) **11. arpadianus**

123 Not pulvinate

124 Leaves without apparent veins, except at base **74. lusitanus**

124 Leaves with midrib apparent throughout their length

125 Flowers subsessile, in dense heads

126 Cauline leaves less than 1 mm wide, the sheaths (2–)3–8 times as long as diameter of stem; epicalyx-scales obovate, usually 6 **22. pinifolius**

126 Cauline leaves 2–2·5 mm wide, the sheaths as long as diameter of the stem; epicalyx-scales lanceolate, usually 4 **77. crassipes**

125 Flowers solitary or few, never in dense heads

127 Most flowers with 2 or 4 epicalyx-scales

128 Basal leaves and non-flowering stems absent at anthesis

129 Leaves almost setaceous, usually appressed to stem **99. carbonatus**

129 Leaves at least 1 mm wide, ±patent

130 Calyx less than 15 mm **96. pallidiflorus**

130 Calyx usually at least 15 mm

131 Epicalyx-scales *c.* $\frac{1}{2}$ as long as calyx **106. gracilis**

131 Epicalyx-scales up to $\frac{1}{3}$ as long as calyx

132 Stems 25–60 cm, without non-flowering branches **97. campestris**

132 Stems usually 10–20 cm, with short, non-flowering branches **98. hypanicus**

128 Basal leaves or non-flowering stems present at anthesis

133 Epicalyx-scales at least $\frac{3}{4}$ as long as calyx

134 Stems 4–10 cm; epicalyx-scales ±equalling calyx **61. pavonius**

134 Stems 15–40 cm; epicalyx-scales *c.* $\frac{3}{4}$ as long as calyx

135 Leaf-sheaths ±equalling diameter of stem **46. furcatus**

135 Leaf-sheaths *c.* 3 times as long as diameter of stem **106. gracilis**

133 Epicalyx-scales less than $\frac{3}{4}$ as long as calyx

136 Inner epicalyx-scales broadly ovate, or obovate, or almost truncate, apiculate or abruptly aristate

137 Stem usually puberulent **(111–114). corymbosus** group

137 Stem glabrous

138 Epicalyx-scales $\frac{1}{3}$–$\frac{1}{2}$ as long as calyx **78. cintranus**

138 Epicalyx-scales not more than $\frac{1}{3}$ as long as calyx

139 Calyx almost cylindrical; teeth acute to obtuse **62. gratianopolitanus**

139 Calyx tapering from below the middle; teeth acuminate **79. petraeus**

136 Inner epicalyx-scales ovate, attenuate, acuminate

140 Calyx 17–27 mm **78. cintranus**

140 Calyx 10–15 mm

141 Stem puberulent, at least below; basal leaves *c.* 1 mm wide **81. graniticus**

141 Stem glabrous; basal leaves mostly 0·5 mm wide **9. humilis**

127 Most flowers with 6 or more epicalyx-scales

142 At least some epicalyx-scales about as long as calyx-tube

143 Flowers in small heads **76. scaber**

143 Flowers solitary or few, not in heads

144 Cauline leaves 6 or more pairs **108. biflorus**

144 Cauline leaves 2 or 3 pairs **107. haematocalyx**

142 Epicalyx-scales distinctly shorter than calyx-tube

145 Epicalyx-scales obovate, almost truncate and apiculate **106. gracilis**

145 Epicalyx-scales ovate, attenuate to the acuminate or aristate apex

146 Leaves up to 1 mm wide

147 Calyx 17–28 mm **78. cintranus**

147 Calyx 10–18 mm

148 Petal-limb 2–3 mm **9. humilis**

148 Petal-limb at least 5 mm

149 Plant 15–40 cm, not densely caespitose; flowers solitary or in few-flowered inflorescences **106. gracilis**

149 Plant up to 10 cm, densely caespitose; flowers always solitary **12. nardiformis**

146 Leaves more than 1 mm wide

150 Flowers in small heads; petal-limb 5–8 mm **76. scaber**

150 Flowers solitary; petal-limb 9–12 mm **78. cintranus**
122 Petal-limb completely glabrous
151 Basal leaves withered at anthesis
152 Sheaths 2–3 times as long as diameter of stem **103. lanceolatus**
152 Sheaths about as long as diameter of stem
153 Petal-limb narrowly rhombic **104. leptopetalus**
153 Petal-limb obovate
154 Stems simple or with 1 or 2 branches; petal-limb cinnamon beneath **101. cinnamomeus**
154 Stems freely branched; petal-limb greenish or pink beneath
155 Calyx *c.* 15 × 3 mm **100. marschallii**
155 Calyx 17–22 × 4 mm **102. monadelphus**
151 Basal leaves or leafy non-flowering stems present at anthesis
156 Epicalyx-scales usually 8–12
157 Stem glabrous below; petal-limb 5–10 mm **10. ciliatus**
157 Stem puberulent below; petal-limb 2–4 mm **14. ingoldbyi**
156 Epicalyx-scales usually 4(–8)
158 At least the outer epicalyx-scales attenuate at apex
159 Leaves *c.* 0·5 mm wide
160 Calyx 4·5–6 mm wide **78. cintranus**
160 Calyx 2–4 mm wide
161 Stems 5–15(–20) cm, simple (E. Pyrenees) **67. pungens**
161 Stems 15–45 cm, simple or branched
162 Stems usually simple; calyx 9–15 mm **72. serratifolius**
162 Stems usually branched; calyx 15–30 mm **73. pyrenaicus**
159 Some leaves 1 mm or more wide
163 Calyx 17–32 mm
164 Calyx 4·5–6 mm wide; petals reddish-lilac **78. cintranus**
164 Calyx (2–)3–4 mm wide; petals white **79. petraeus**
163 Calyx 10–16 mm (see also **79(a)**)
165 Epicalyx-scales *c.* $\frac{1}{3}$ as long as calyx; calyx ± cylindrical, or narrowly ovoid **80. integer**
165 Epicalyx-scales *c.* $\frac{1}{2}$ as long as calyx; calyx tapering from middle or below
166 Cauline leaves 10–30 mm, usually patent; stock slender **46. furcatus**
166 Cauline leaves 5–10 mm, closely appressed to stem; stock stout and woody **69. costae**
158 At least the inner epicalyx-scales abruptly cuspidate to truncate
167 Calyx 15 mm or more
168 Calyx less than 3 mm wide **79. petraeus**
168 Calyx at least 4 mm wide
169 Calyx up to 20 mm, tapering markedly above the middle at anthesis
170 Petal-limb 3–5 mm **68. hispanicus**
170 Petal-limb 5–10 mm **71. laricifolius**
169 Calyx up to 30 mm, almost cylindrical, if tapering, more than 20 mm
171 Basal leaves usually less than 1 mm wide, convolute, wiry **64. sylvestris**
171 Basal leaves more than 1 mm wide, flat and rather soft
172 Petal-limb 4–6 mm, dirty white **63. xylorrhizus**
172 Petal-limb 10–15 mm, pink or purplish **65. caryophyllus**
167 Calyx less than 15 mm
173 Calyx-teeth up to 2(–2·5) times as long as wide, obtuse or subobtuse to acute, not acuminate
174 Calyx usually not more than 10 mm
175 Basal leaves *c.* 1 mm wide; calyx 3–4 mm wide, ventricose **66. subacaulis**
175 Basal leaves *c.* 0·5 mm wide; calyx *c.* 2 mm wide, almost cylindrical **70. langeanus**
174 Calyx at least 10 mm
176 Flowers pink **64. sylvestris**
176 Flowers white
177 Epicalyx-scales 2–4; petal-limb ± entire **80. integer**
177 Epicalyx-scales (4–)6; petal-limb subentire to dentate **13. anatolicus**
173 Calyx-teeth 3–4 times as long as wide, acuminate or apiculate
178 Inner epicalyx-scales less than twice as long as wide
179 Inner and outer epicalyx-scales similar; flowers solitary **68. hispanicus**
179 Outer epicalyx-scales narrow and gradually acuminate, inner obovate, cuspidate; flowers usually 2 or 3 together **71. laricifolius**
178 Inner epicalyx-scales at least twice as long as wide
180 Calyx 6–10(–13) mm; stem usually simple **66. subacaulis**
180 Calyx 14–20 mm; stem usually freely branched **71. laricifolius**

Subgenus **Armeriastrum** Ser. Flowers in groups or heads often surrounded by involucral bracts, rarely solitary. Plants often obviously woody at base.

1. D. armeria L., *Sp. Pl.* 410 (1753) (*D. epirotus* Halácsy). Usually pubescent and branched annual or biennial up to 40 cm. Basal leaves oblong, obtuse; cauline linear, acute; all flat and thin. Flowers clustered; bracts leaf-like, about as long as flowers. Epicalyx-scales 2, about as long as calyx, lanceolate, with green, subulate apex. Calyx 15–20 × 2–3 mm, narrowed above the middle. Petal-limb *c.* 5 mm, dentate, bearded, reddish with pale dots. *W., C. & S. Europe, extending to S. Sweden and E. Ukraine.* Al Au Be Br Bu Co Cz Da Ga Ge Gr He Ho Hs Hu It Ju Lu Po Rm Rs(B, C, W, ?E) Sa Si Su Tu.

(a) Subsp. **armeria**: Inner epicalyx-scales lanceolate. Calyx green. $2n = 30$. *Throughout the range of the species.*

(b) Subsp. **armeriastrum** (Wolfner) Velen., *Fl. Bulg., Suppl.* 42 (1898) (*D. armeriastrum* Wolfner): Inner epicalyx-scales ovate. Calyx purplish. $2n = 30$. *From Hungary and N. Romania southwards to N. Greece.*

2. D. pseudarmeria Bieb., *Fl. Taur.-Cauc.* **1**: 323 (1808). Like **1** but epicalyx-scales ovate, abruptly contracted to a subulate point; calyx *c.* 10 × 1·5 mm; petal-limb pink. *Lowlands of S.E. Europe, from S.E. Jugoslavia to S. Russia.* Bu Ju Rm Rs(W, K, E) Tu.

3. D. trifasciculatus Kit. in Schultes, *Österreichs Fl.* ed. 2, **1**: 654 (1814). Glabrous or puberulent perennial up to 80 cm. Leaves 4–10 mm wide, linear-lanceolate, thin; basal leaves few or none at anthesis; cauline usually 10–20 pairs; veins 7–9, at least some of the laterals almost as prominent as the midrib; sheaths 2–3 times as long as diameter of stem. Flowers usually numerous and usually in 3 pedunculate heads; bracts 4, green at least above, about as long as flowers. Epicalyx-scales lanceolate to ovate, with a green, subulate apex about as long as calyx. Calyx 10–16 mm. Petal-limb 4–10 mm, bearded. ● *S.E. Europe, from W.C. Bulgaria to C. Ukraine.* Bu Ju Rm Rs(W).

1 Petal-limb *c.* 4 mm, lilac (b) subsp. **parviflorus**
1 Petal-limb *c.* 10 mm, pink or purple
2 Heads shortly pedunculate; petals pink (a) subsp. **trifasciculatus**
2 Heads long-pedunculate; petals purple (c) subsp. **pseudobarbatus**

(a) Subsp. **trifasciculatus** (incl. subsp. *pseudoliburnicus* Stoj. & Acht.): Heads shortly pedunculate. Calyx *c.* 16 mm. Petal-limb *c.* 10 mm, pink. *Romania, Bulgaria, E. Jugoslavia.*

(b) Subsp. **parviflorus** Stoj. & Acht., *Sborn. Bălg. Akad. Nauk.* **29**: 40 (1935) (*D. desertii* Prodan, non Post): Heads subsessile. Calyx *c.* 11 mm. Petal-limb *c.* 4 mm, lilac. *Lower Danube valley.*

(c) Subsp. **pseudobarbatus** (Schmalh.) Jalas, *Ann. Bot. Fenn.* **22**: 219 (1985) (*D. euponticus* Zapał.): Heads long-pedunculate, often more than 3. Calyx 10–13 mm. Petal-limb *c.* 10 mm, purple. *Almost throughout the range of the species.*

4. D. urumoffii Stoj. & Acht., *Sborn. Bălg. Akad. Nauk.* **29**: 42 (1935). Like **3(a)** but sheaths 5–9 times as long as diameter of stem; bracts numerous; calyx 16–20 × 3–3·5 mm, narrowly cylindrical and narrowed towards apex; petal-limb red, with yellow spot at base. ● *W. Bulgaria.* Bu.

Not seen recently and possibly extinct.

5. D. knappii (Pant.) Ascherson & Kanitz ex Borbás, *Math. Term. Közl.* **13**: 196 (1877). Shortly pubescent perennial to 40 cm. Leaves linear-lanceolate, 5- to 7-veined; midrib more prominent than lateral veins; sheaths *c.* 3 times as long as diameter of stem. Heads usually 2, shortly pedunculate, many-flowered; outer bracts leaf-like, the inner more or less scarious. Epicalyx-scales ovate, with a green, subulate awn nearly as long as calyx. Calyx 15 mm; teeth ovate-acuminate. Petals sulphur-yellow. ● *W. Jugoslavia.* Ju.

6. D. membranaceus Borbás, *Österr. Bot. Zeitschr.* **26**: 125 (1876) (*D. euponticus* sensu Schischkin, non Zapał., *D. rehmannii* Błocki). Perennial 30–70 cm. Leaves linear-lanceolate, acute or acuminate; sheaths 2–4 times as long as diameter of stem. Heads usually 3- to 6-flowered, solitary or 2 or 3, shortly pedunculate; bracts lanceolate or ovate-lanceolate, acuminate, usually shorter than flowers. Epicalyx-scales ovate, broadly scarious, narrowed into a short awn, shorter than broad part of scale. Calyx 12–18 mm, narrowed upwards; teeth lanceolate, aristate, the margins pubescent, whitish. Petal-limb *c.* 5 mm, bearded, purple. ● *From N.E. Bulgaria to S.C. Russia.* Bu Rm Rs(C, W, E).

7. D. dobrogensis Prodan, *Bul. Acad. Stud. Agron. Cluj* **5(1)**: 97 (1934) (*D. membranaceus* sensu Stoj. & Stefanov, non Borbás). Like **6** but epicalyx-scales rounded at apex, mucronate; calyx *c.* 10 mm; teeth obtuse, sometimes shortly mucronate; petals glabrous, pink. ● *S.E. Romania and N.E. Bulgaria.* Bu Rm.

8. D. barbatus L., *Sp. Pl.* 409 (1753) (*D. girardini* Lamotte). Subglabrous perennial up to 60 cm. Leaves lanceolate, often narrowed into a short petiole, the lower obtuse, the upper acute or acuminate; midrib prominent; lateral veins obscure. Heads large, many-flowered; pedicels short; bracts herbaceous, about equalling flowers. Epicalyx-scales ovate, aristate. Petal-limb purple, shortly bearded. *From the Pyrenees to the E. Carpathians and Balkan peninsula; widely cultivated for ornament elsewhere, and locally naturalized.* Au Bu Cz Ga Hs Hu It Ju Po Rm Rs(W) Tu [Fe Ge Rs(B, C, E)].

(a) Subsp. **barbatus**: Leaves sessile. Bracts equalling flowers. Epicalyx-scales and calyx usually green. $2n = 30$. *Throughout the range of the species except E.C. Europe; mainly lowland and often cultivated and naturalized.*

(b) Subsp. **compactus** (Kit.) Heuffel, *Verh. Zool.-Bot. Ges. Wien* **8**: 68 (1858) (*D. compactus* Kit.): Lower leaves attenuate into a petiole. Bracts shorter than flowers. Epicalyx-scales and calyx usually purplish-brown. $2n = 30$. *Mountains of C. & S. Europe from the Carpathians to S. Italy.*

9. D. humilis Willd. ex Ledeb., *Fl. Ross.* **1**: 280 (1842). Densely caespitose perennial 5–35 cm; stems simple or little-branched. Leaves 0·5–1 mm wide, short, stiff, setaceous. Epicalyx-scales usually 6, *c.* $\frac{2}{3}$ as long as calyx; inner ovate, acuminate; outer much smaller and narrower. Calyx 10–15 × 2 mm, tapering above the middle. Petal-limb 2–3 mm, dentate, bearded. ● *S. Ukraine.* Rs(*W, K).

10. D. ciliatus Guss., *Ind. Sem. Horto Boccad.* **1825**: 5 (1825). Laxly caespitose perennial up to *c.* 60 cm. Leaves 1–2 mm wide, more or less flat, linear, acuminate, the basal often few at anthesis. Epicalyx-scales usually 8, *c.* $\frac{1}{2}$ as long as calyx, ovate, acuminate. Calyx 15–23 × 3 mm, tapering from below the middle. Petal-limb 5–10 mm, dentate to subentire, glabrous. ● *Italy, W. Jugoslavia, Albania.* Al It Ju.

(a) Subsp. **ciliatus**: Stock stout; stems little-branched. Cauline leaves usually 4–6 pairs. Petals shallowly dentate or subentire. *Italy, N.W. Jugoslavia.*

(b) Subsp. **dalmaticus** (Čelak.) Hayek, *Prodr. Fl. Penins. Balcan.* **1**: 246 (1924) (*D. dalmaticus* Čelak.): Stock slender; stems much-branched. Cauline leaves usually 7–13 pairs. Petals dentate. *S.W. Jugoslavia, W. Albania.*

11. D. arpadianus Ade & Bornm., *Feddes Repert.* **36**: 385 (1934). Pulvinate perennial 2–8 cm. Leaves usually 5–8 mm, linear, acuminate; cauline leaves 1–3 pairs. Flowers 1 or 2. Epicalyx-scales 4–6, $\frac{1}{2}$–$\frac{3}{4}$ as long as calyx, ovate, with a subulate apex. Calyx 8–10 × 2–3 mm, almost cylindrical. Petal-limb 2–3 mm, dentate, bearded. *N. Aegean region (Samothraki).* Gr. (*Anatolia.*)

12. D. nardiformis Janka, *Österr. Bot. Zeitschr.* **23**: 195 (1873). Caespitose perennial up to *c.* 10 cm, with slender, branched, woody stems. Basal leaves *c.* 10 mm, setaceous, semi-terete; cauline leaves 6–10 pairs, longer than internodes. Flowers solitary. Epicalyx-scales (4–)6, *c.* ½ as long as calyx, ovate, aristate. Calyx 15–18 mm, narrowed from about the middle. Petal-limb *c.* 5 mm, dentate, bearded, pink. 2*n* = 30. ● *E. Romania, N. Bulgaria.* Bu Rm.

13. D. anatolicus Boiss., *Diagn. Pl. Or. Nov.* **1**(**1**): 22 (1843). Caespitose perennial 8–35 cm. Leaves 1–1·8 mm wide, linear, acuminate. Flowers solitary or 2 or 3 together; pedicels 5 mm or more. Epicalyx-scales (4–)6, narrowly ovate, apiculate. Calyx 8·5–11 mm; petal-limb 2·5–3·5 mm, linear to linear-oblong, subentire to dentate, white. *Turkey-in-Europe* (*Gelibolu peninsula*). Tu. (*Anatolia*.)

14. D. ingoldbyi Turrill, *Kew Bull.* **1924**: 314 (1924). Caespitose perennial *c.* 30 cm, with a woody stock; stems puberulent below. Leaves *c.* 2 mm wide, linear, acuminate, coriaceous. Flowers 2–5 together, subsessile. Epicalyx-scales 10–12, ovate, gradually acuminate. Calyx 15–17 × 2 mm, tapering above the middle. Petal-limb 2–4 mm, denticulate, whitish- or greenish-yellow. ● *Turkey-in-Europe.* Tu.

15. D. juniperinus Sm., *Trans. Linn. Soc. London* ser. 1, **2**: 303 (1794). Small shrub with erect, woody, branched stems bearing herbaceous, usually unbranched flowering stems up to 20 cm. Leaves 0·5–1·5 mm wide, linear, abundant on non-flowering stems, 2 or 3 pairs on flowering stems. Epicalyx-scales 4–8, ⅓–½ as long as calyx, obcordate, with a subulate apex. Calyx 10–13 × 2–3·5 mm, nearly cylindrical; teeth acute. Petal-limb 4–8 mm, dentate, bearded. 2*n* = 30. *Limestone rocks.* ● *W. Kriti* (*Sfakia prov.*). Cr.

(a) Subsp. **juniperinus**: Leaves usually 2–3 cm. Flowering stems 10–20 cm, robust, 2- to 6-flowered. Calyx 10–11 mm. Petal-limb 3–4 mm wide, pale pink, unspotted, shortly dentate. *Mountains above* 1400 *m.*

(b) Subsp. **heldreichii** Greuter, *Candollea* **20**: 187 (1965): Leaves usually 1·5–2 cm. Flowering stems 6–10 cm, slender, 1- or 2-flowered. Calyx 12–13 mm. Petal-limb 5–6 mm wide, bright pink, with a few purple spots, shortly dentate or crenate. *Hills,* 600–650 *m.*

16. D. pulviniformis Greuter, *op. cit.* 189 (1965). Dwarf shrub forming rounded hummocks. Leaves of young branches 1–1·5 cm, needle-like, densely crowded. Flowering stems 3–5 cm, slender, 2- or 3-flowered. Epicalyx-scales *c.* ½ as long as calyx, with slender mucro 1–1·5 mm. Calyx 7–8 mm. Petal-limb 3–4 mm wide, obovate, cuneate, shortly and sharply dentate, white. *Limestone rocks,* c. 600 *m.* ● *S. Kriti* (*Kedros*). Cr.

Known only from the type collection.

17. D. aciphyllus Sieber ex Ser. in DC., *Prodr.* **1**: 358 (1824). (*D. arboreus* L., nom. ambig.). Small shrub up to 50 cm. Leaves 20–40 × 1·5–3 mm, needle-like, linear to linear-lanceolate, densely crowded. Flowering stems 15–30 cm, robust, many-flowered. Epicalyx-scales with strong mucro 2–4 mm. Calyx (15–)18–23 mm. Petal-limb *c.* 10 mm wide, suborbicular, pink, with purple spots, deeply and sharply dentate. *Limestone rocks.* ● *E. Kriti.* Cr.

18. D. rupicola Biv., *Sic. Pl. Cent.* **1**: 31 (1806). Small shrub 30–60 cm. Leaves 3–5 mm, wide, rather thick, linear-oblong, the lower obtuse, the upper acute. Flowers in bracteate heads; pedicels short. Epicalyx-scales 12–16, obovate, acuminate. Calyx 25–30 × 4–5 mm, almost cylindrical. Petal-limb 10–15 mm, dentate, bearded, pink. *S. Italy, Sicilia, Mallorca.* Bl It Si.

Plants from Islas Baleares (Mallorca) have been referred to the N. African subsp. **hermaeensis** (Cosson) O. Bolòs & Vigo, *Butll. Inst. Catalana Hist. Nat.* **38** (Sec. Bot. 1): 87 (1974).

19. D. fruticosus L., *Sp. Pl.* 413 (1753) (*D. arboreus* auct., non L.). Small shrub up to 50 cm, with tortuous branches. Leaves 3–8 mm wide, linear to oblanceolate or elliptical, obtuse, fleshy, glaucous. Flowers numerous; pedicels short. Epicalyx-scales (8–)10–20, obovate, shortly cuspidate. Calyx 18–25 × 2–3 mm, narrowed from about the middle. Petal-limb *c.* 10 mm, dentate, bearded, pink. 2*n* = 30. *S.* & *W. Greece, S. Aegean region.* Cr Gr.

In the European part of the distribution six subspecies have been recognized by H. Runemark, *Bot. Not.* **133**: 475–490 (1980). These represent well-marked local variants, but the characters show too great a degree of overlap for the subspecies to be maintained here.

20. D. capitatus Balbis ex DC., *Cat. Pl. Horti Monsp.* 103 (1813). Glaucous perennial up to 70 cm. Leaves 2–3 mm wide, linear, the upper 1 or 2 pairs greatly enlarged near base; sheaths more than 3 times as long as diameter of stem. Heads many-flowered, dense, solitary; bracts about as long as calyx, ovate, coriaceous, with a green or brown apex. Epicalyx-scales ½ as long as to nearly as long as calyx, ovate, shortly awned; margins scarious. Calyx 9–19 mm, narrowed upwards; teeth lanceolate to ovate, acute. Petal-limb *c.* 5 mm, bearded to almost glabrous, purple. *S.E. Europe, extending northwards to* c. 56° *N. in C. Russia.* Ju Rm Rs(C, W, K, E) Tu.

(a) Subsp. **capitatus**: Epicalyx-scales about as long as calyx. Calyx 16–19 mm. Petals distinctly bearded. *Scattered localities around the Black Sea.*

(b) Subsp. **andrzejowskianus** Zapał., *Rozpr. Wydz. Mat.-Przyr. Polsk. Akad. Um.* (*Biol.*) ser. 3, **11B**: 25 (1911) (*D. andrzejowskianus* (Zapał.) Kulcz.): Epicalyx-scales about ½ as long as calyx. Calyx 10–15 mm. Petals almost glabrous. *Almost throughout the range of the species.*

21. D. ferrugineus Miller, *Gard. Dict.* ed. 8, no. 9 (1768) (*D. balbisii* Ser.). Perennial with woody stock; stems 30–60 cm, numerous, simple. Leaves convolute to linear-lanceolate, acuminate; sheaths more than 3 times as long as diameter of stem. Heads few- to many-flowered, usually dense; bracts oblong to ovate, long-acuminate, herbaceous above, about equalling flowers. Epicalyx-scales 4, oblong to ovate, abruptly contracted into the awn, about as long as calyx. Calyx 14–20 × 3–5 mm, almost cylindrical; teeth lanceolate, acuminate or aristate. Petal-limb pink, purplish or sometimes yellowish, more or less bearded. *C. Mediterranean region, from S.E. France to Albania.* Al ?Co Ga It Ju.

1 Leaves 1·5–2·5 mm wide, usually convolute (c) subsp. **vulturius**
1 Leaves 2–5(–10) mm wide, flat
 2 Leaves 2–3 mm wide; petal-limb 8–10 mm (a) subsp. **ferrugineus**
 2 Leaves 3–5 mm wide; petal-limb 5–6 mm (b) subsp. **liburnicus**

(a) Subsp. **ferrugineus** (*D. liburnicus* sensu Hayek, non Bartl.): 2*n* = 30. *Throughout the range of the species.*

(b) Subsp. **liburnicus** (Bartl.) Tutin, *Feddes Repert.* **68**: 191 (1963) (*D. liburnicus* Bartl., *D. balbisii* subsp. *liburnicus* (Bartl.) Pignatti). ● *N.W. Italy, Jugoslavia.*

(c) Subsp. **vulturius** (Guss. & Ten.) Tutin, *loc. cit.* (1963) (*D. vulturius* Guss. & Ten.): *C.* & *S. Italy.*

Subsp. (**c**) approaches **22** in the leaves and the often rather small heads with more or less oblong bracts, but is larger in all its parts; intermediates between it and **21(a)** occur.

D. stamatiadae Rech. fil., *Bot. Not.* **124**: 7 (1971), from E.C. Greece (near Servia), is like **21(a)** but with a laxer habit and longer leaf-sheaths, and is probably not specifically distinct from **21**.

22. D. pinifolius Sm. in Sibth. & Sm., *Fl. Graec. Prodr.* **1**: 284 (1809) (*D. lilacinus* Boiss. & Heldr.; incl. *D. lydus* sensu Hayek, *D. rhodopeus* Davidov, non Velen., *D. serulis* Kulcz., ?*D. serresianus* Halácsy & Charrel). Densely caespitose perennial 15–45 cm. Leaves less than 1 mm wide, linear, often falcate, acuminate, usually rather rigid, with prominent midrib and strongly thickened margins; sheaths (2–)3–8 times as long as diameter of stem. Flowers in dense heads; bracts coriaceous, ovate- to obovate-oblong, aristate. Epicalyx-scales usually 6, obovate, aristate. Calyx 8–20 × 2–4 mm, tapering somewhat above the middle. Petal-limb 5–8(–10) mm, dentate, purple or lilac. *Balkan peninsula, S. Romania.* Al Bu Gr Ju Rm Tu.

A variable species. The following subspecies can be distinguished, but many intermediate plants occur.

1 Bracts distinctly shorter than calyx (**a**) subsp. **pinifolius**
1 Bracts (including awn) about as long as to slightly longer than calyx
2 Calyx 13–18 mm (**b**) subsp. **serbicus**
2 Calyx 9–14 mm (**c**) subsp. **lilacinus**

(**a**) Subsp. **pinifolius** (subsp. *smithii* Wettst.): Flowering stems 15–25 cm, rather slender. Leaf-sheaths 3–6 mm. Bracts small, distinctly shorter than calyx, contracted abruptly into an awn. Calyx 8–12(–15) mm. $2n = 30$. *Balkan peninsula and Aegean region southwards to* 40° *N.*

(**b**) Subsp. **serbicus** Wettst., *Biblioth. Bot.* (*Stuttgart*) **26**: 34 (1892): Flowering stems 20–45 cm. Leaf-sheaths 5–10 mm. Bracts about as long as calyx, gradually narrowed into an awn. Calyx 13–18 mm. ● *From N.W. Greece to S.C. Romania.*

(**c**) Subsp. **lilacinus** (Boiss. & Heldr.) Wettst., *op. cit.* 33 (1892): Flowering stems 15–40 cm, rather robust. Leaf-sheaths 5–10 mm. Bracts large and conspicuous, about as long as to somewhat longer than calyx, abruptly contracted into an awn. Calyx 9–14 mm. ● *S. Albania, Greece.*

D. androsaceus (Boiss. & Heldr.) Hayek, *Denkschr. Akad. Wiss. Math.-Nat. Kl.* (*Wien*) **94**: 141 (1917), from S. Greece (Killini), a densely pulvinate perennial, with smaller flowers 1–3 in a more or less acaulescent cluster, and epicalyx-scales elliptical, is probably not specifically distinct from **22**. It is known only from the type collection.

23. D. giganteus D' Urv., *Mém. Soc. Linn. Paris* **1**: 301 (1822). Robust, often pruinose perennial 20–100 cm. Leaves 0·5–8 mm wide, widest near the middle, linear, acuminate; sheaths several times as long as diameter of stem. Flowers usually numerous, in dense heads, occasionally with a pair of leaves close below the head. Epicalyx-scales usually ½ as long as calyx, brown or scarious, ovate, acute or acuminate. Calyx (8–)17–22 × (2–)3–4 mm, tapering from about the middle. Petal-limb 5–8 mm, dentate, purple. $2n = 30$. *Balkan peninsula, extending to N. Jugoslavia and N. Romania.* ?Al Bu Gr Ju Rm Tu.

1 Leaves *c.* 0·5 mm wide, often convolute (**f**) subsp. **vandasii**
1 Leaves 2–8 mm wide, flat
2 Epicalyx-scales acute to acuminate, glabrous (**a**) subsp. **giganteus**
2 Epicalyx-scales cuspidate, usually ±puberulent
3 Cauline leaves 5–7 mm wide (**d**) subsp. **haynaldianus**
3 Cauline leaves not more than 5 mm wide
4 Calyx 15–20 mm; cauline leaves 2–5 mm wide (**e**) subsp. **subgiganteus**
4 Calyx up to 17 mm; cauline leaves rarely more than 3 mm wide
5 Calyx 14–17 × 2·5–3 mm (**b**) subsp. **croaticus**
5 Calyx 12–15 × 3–4·5 mm (**c**) subsp. **banaticus**

(**a**) Subsp. **giganteus**: Plant very robust, sometimes pruinose. Cauline leaves 3–8 mm wide. Heads many-flowered. Epicalyx-scales acute or acuminate, glabrous, brownish. Calyx 17–20 mm. $2n = 30$. ● *From C. Romania to N. Greece.*

(**b**) Subsp. **croaticus** (Borbás) Tutin, *Feddes Repert.* **70**: 4 (1964) (*D. croaticus* Borbás): Plant usually slender, green. Cauline leaves rarely more than 3 mm wide. Heads rather few-flowered. Epicalyx-scales cuspidate, dark brown, awned; awn of epicalyx-scales usually 3–5 mm. Calyx 14–17 × 2·5–3 mm. ● *N. & C. Jugoslavia.*

(**c**) Subsp. **banaticus** (Heuffel) Tutin, *Feddes Repert.* **68**: 191 (1963) (*D. carthusianorum* var. *banaticus* Heuffel, *D. banaticus* (Heuffel) Borbás): Like subsp. (**b**) but awn of epicalyx-scales usually 1–2 mm; calyx 12–15 × 3–4·5 mm. ● *Romania, E. Jugoslavia.*

(**d**) Subsp. **haynaldianus** (Borbás) Tutin, *loc. cit.* (1963) (*D. haynaldianus* Borbás, *D. intermedius* Boiss.): Like subsp. (**b**) but cauline leaves 5–7 mm wide; calyx 20 mm. *S. Jugoslavia.*

(**e**) Subsp. **subgiganteus** (Borbás) Hayek, *Denkschr. Akad. Wiss. Math.-Nat. Kl.* (*Wien*) **94**: 138 (1917): Plant robust, green. Cauline leaves 2–5 mm wide. Heads usually many-flowered. Epicalyx-scales brown near apex, awned. Calyx 15–20 mm. ● *N. Greece; Bulgaria.*

(**f**) Subsp. **vandasii** (Velen.) Stoj. & Acht., *Sborn. Bălg. Akad. Nauk.* **29**: 43 (1935) (*D. vandasii* Velen.): Plant slender, sometimes pruinose. Leaves *c.* 0·5 mm wide, often convolute. Epicalyx-scales cuspidate. Calyx 8–14 × 2–3 mm. *Bulgaria, S. Romania.*

24. D. giganteiformis Borbás, *Math. Term. Közl.* **12**: 83 (1876). Like **23** but smaller and more slender; inner epicalyx-scales obovate, acute or shortly aristate; calyx 8–14 × 2–3 mm; petal-limb 3–5 mm. ● *E.C. Europe, extending to N. Italy and E. Bulgaria.* Au Bu Cz Hu ?It Ju Rm ?Rs(K).

1 Calyx *c.* 8 mm; teeth subobtuse and mucronate (**c**) subsp. **kladovanus**
1 Calyx 10 mm or more; teeth acute
2 Calyx *c.* 10 mm; outer epicalyx-scales brown, abruptly contracted into a short awn (**b**) subsp. **pontederae**
2 Calyx *c.* 14 mm; outer epicalyx-scales straw-coloured, acuminate (**a**) subsp. **giganteiformis**

(**a**) Subsp. **giganteiformis**: $2n = 30$. *Hungary, Romania, E. Jugoslavia.*

(**b**) Subsp. **pontederae** (A. Kerner) Soó, *Acta Bot. Acad. Sci. Hung.* **15**: 339 (1970) (*D. diutinus* Reichenb., non Kit., *D. urziceniensis* Prodan): $2n = 30$. *From C. Czechoslovakia to E. Romania.*

(**c**) Subsp. **kladovanus** (Degen) Soó, *Feddes Repert.* **83**: 161 (1972) (*D. kladovanus* Degen, *D. pontederae* subsp. *kladovanus* (Degen) Stoj. & Acht.): $2n = 30$. *Jugoslavia, Bulgaria, Romania.*

25. D. diutinus Kit. in Schultes, *Österreichs Fl.* ed. 2, **1**: 655 (1814). Somewhat glaucous perennial up to *c.* 50 cm. Basal leaves setaceous, the cauline up to 1 mm wide, linear, obtuse or subobtuse; sheaths several times as long as diameter of stem. Flowers 2 to several, in a dense head; bracts and epicalyx-scales ovate, coriaceous, apiculate or shortly aristate. Calyx (10–)12–15 × 2–3 mm, somewhat narrowed above the middle; teeth rounded, little longer than wide. Petal-limb 5–8 mm, dentate, glabrous, pinkish-lilac. ● *Hungary, E. Jugoslavia.* Hu Ju.

26. D. platyodon Klokov, *Jour. Bot. Acad. Sci. Ukr.* **5(1)**: 27 (1948). Perennial 10–30 cm. Leaves *c.* 1 mm wide, linear, acute; sheaths several times as long as diameter of stem. Flowers 1 to several in a head; bracts lanceolate, acuminate. Epicalyx-scales obovate, abruptly acuminate or mucronate, coriaceous, with scarious margins. Calyx 10–14 × 3–4 mm; teeth oblong, mucronate. Petal-limb 5–8 mm, bearded, deep pink. *Ukraine, S.E. Russia.* Rs(C, W, K, E).

27. D. bessarabicus (Kleopow) Klokov in Kotov, *Fl. RSS Ucr.* **4**: 659 (1952). Robust perennial 30–50 cm. Leaves linear, acute; lower cauline up to *c.* 2 mm wide; sheaths several times as long as diameter of stem. Flowers usually several in a head; bracts oblong, often with a green point. Epicalyx-scales obovate-oblong, mucronate, coriaceous, with hyaline margins. Calyx 18–22 × 5–6 mm; teeth ovate, obtuse. Petal-limb 10–15 mm, bearded, deep purplish. ● *E. Romania, Moldavia.* Rm Rs(W).

25–27 may be conspecific with **D. polymorphus** Bieb., *Fl. Taur.-Cauc.* **1**: 324 (1808).

28. D. carthusianorum L., *Sp. Pl.* 409 (1753) (*D. atrorubens* All., *D. bukovinensis* Klokov, *D. capillifrons* (Borbás) H. Neumayer, *D. carpaticus* Wołoszczak, *D. polonicus* Zapał., *D. vaginatus* Chaix, *D. velebiticus* Borbás; incl. *D. puberulus* (Simonkai) A. Kerner, *D. tenuifolius* Schur, *D. rogowiczii* Kleopow, *D. commutatus* (Zapał.) Klokov, *D. sanguineus* Vis., *D. giganteus* subsp. *italicus* Tutin, *D. leucophoeniceus* (Dörfler & Hayek) Tutin). Glabrous perennial up to *c.* 60 cm. Leaves 0·5–5 mm wide, linear, acuminate, flat; sheaths several times as long as diameter of stem. Flowers few to many in a usually dense head; bracts lanceolate to oblong, more or less herbaceous to coriaceous. Epicalyx-scales obovate to obcordate, abruptly aristate, coriaceous to membranous. Calyx 10–20 × 2–5 mm, narrowed above the middle. Petal-limb usually 10–15 mm, dentate, deep pink to purple, very rarely white, bearded. $2n = 30$. ?● *S., W. & C. Europe, extending eastwards to C. Ukraine.* Al Au Be Bu Cz Ga Ge He †Ho Hs Hu It Ju Po Rm Rs(C, W) *Si [Rs(*B) Su].

A very variable species which has been much subdivided by different authors. The variation in leaf-width, length of awn, colour, shape and texture of epicalyx-scales, length and colour of the calyx, and petal-size seems to be largely continuous and with little correlation between the different characters. The variation does not always fall into a geographical pattern, though distinct populations appear to exist, especially near the limits of the distribution of the species. A valuable summary of the taxonomy and distribution of the infraspecific taxa is given in J. Jalas & J. Suominen, *Atlas Florae Europaeae* **7**: 210–214 (1986).

29. D. borbasii Vandas, *Österr. Bot. Zeitschr.* **36**: 193 (1886). Perennial with an unbranched woody stock from which flowering stems and short leafy shoots arise. Leaves *c.* 1 mm wide, linear, acuminate, the basal mostly withered at anthesis. Stems 30–50 cm, simple, or little-branched near the top. Heads (1)2- to 8-flowered. Epicalyx-scales ovate, narrowed into a short awn, pale brown and broadly scarious. Calyx 14–17 × 3–4 mm, pink or purplish. Petal-limb *c.* 10 mm, dentate, pink, bearded. ● *E. Europe, from* c. 57° *N. in E.C. Russia southwards to Krym, and westwards to Lithuania.* Rs(B, C, W, K, E).

(a) Subsp. **borbasii**: Bracts and epicalyx-scales usually not more than ½ as long as calyx. Calyx-teeth 3·5–4·5 mm. *Throughout the range of the species.*

(b) Subsp. **capitellatus** (Klokov) Tutin, *Feddes Repert.* **68**: 192 (1963) (*D. capitellatus* Klokov): Bracts and epicalyx-scales usually not more than ½ as long as calyx. Calyx-teeth 3·5–4·5 mm. *S.E. Ukraine; perhaps elsewhere.*

30. D. henteri Heuffel ex Griseb. & Schenk, *Arch. Naturgesch. (Berlin)* **18(1)**: 303 (1852). Caespitose perennial 15–30 cm. Leaves *c.* 0·5 mm wide, linear, acuminate, rather rigid; cauline leaves shorter than internodes; sheaths 3–4 times as long as diameter of stem. Flowers usually 2 or 3 together; bracts narrow, often herbaceous, shorter than inflorescence. Epicalyx-scales broadly ovate to obovate, often somewhat retuse, aristate. Calyx 12–14 × 4–5 mm, nearly cylindrical, green or purplish with brown teeth. Petal-limb 8–12 mm, dentate. *Damp, grassy places.* ● *S. Carpathians.* Rm.

31. D. cruentus Griseb., *Spicil. Fl. Rumel.* **1**: 186 (1843) (incl. *D. lateritius* Halácsy, *D. calocephalus* Bald., non Boiss., *D. holzmannianus* Heldr. & Hausskn.). Usually more or less glaucous perennial up to 100 cm. Leaves 1–3 mm wide, linear, acuminate, flat; sheaths several times as long as diameter of stem. Flowers numerous, in a dense head; bracts puberulent, oblong to ovate-oblong, with an awn up to about as long as calyx. Epicalyx-scales similar but smaller, pale to reddish-brown. Calyx 13–20 × 3–4 mm, somewhat tapering from about the middle, puberulent, usually reddish-purple. Petal-limb 3–8 mm, dentate, more or less bearded. *Balkan peninsula.* Al Bu Gr Ju Tu.

(a) Subsp. **cruentus**: Awn considerably longer than the reddish-brown epicalyx-scales. Petals more or less glabrous, deep purple. $2n = 30$. ● *Bosna to Albania and W. Bulgaria.*

(b) Subsp. **turcicus** (Velen.) Stoj. & Acht. in Stoj. & Stefanov, *Fl. Bălg.* ed. 3, 405 (1948) (*D. turcicus* Velen.): Awn about as long as the pale epicalyx-scales. Petals bearded, pinkish-purple. $2n = 30$. *Albania and S.W. Greece to E. Bulgaria and Turkey.*

D. quadrangulus Velen., *Sitz.-Ber. Böhm. Ges. Wiss.* **1892**: 372 (1893), from the E. part of the Balkan peninsula, and **D. strymonis** Rech. fil., *Bot. Jahrb.* **69**: 450 (1939), from N.E. Greece (near Serrai), have been described as separate species but are difficult to separate from **31**. They are more or less intermediate between subspp. (**a**) and (**b**), which are themselves frequently difficult to distinguish.

32. D. tristis Velen., *Sitz.-Ber. Böhm. Ges. Wiss.* **1889(2)**: 41 (1890) (*D. pancicii* Velen., non F. N. Williams). Like **31** but up to 40 cm, less robust; cauline leaves 1–2 mm wide; bracts and epicalyx-scales glabrous, dark brown; awn of epicalyx-scales up to about as long as the scale, often patent; calyx 8–11 × 2–3 mm, dark purple; petal-limb pink, subglabrous. $2n = 30$. ● *C. part of Balkan peninsula, mainly in the mountains.* Al Bu Gr Ju.

33. D. stribrnyi Velen., *Sitz.-Ber. Böhm. Ges. Wiss.* **1893(37)**: 15 (1894) (*D. moesiacus* subsp. *stribrnyi* (Velen.) Stoj. & Acht.). Like **31** but smaller and less robust; cauline leaves 2–3 mm wide; bracts and epicalyx-scales purplish; awn of epicalyx-scales about as long as scale; calyx 8–10 × 2–3 mm, purplish; petal-limb pale pink, nearly or quite glabrous. ● *From N. Albania to C. Bulgaria.* Al Bu Ju.

34. D. brachyzonus Borbás & Form., *Verh. Naturf. Ver. Brünn* **35**: 194 (1897) (*D. hyalolepis* Acht. & Lindt.). Perennial up to 50 cm. Leaves 1–3 mm wide, linear, acuminate, flat; sheaths several times as long as diameter of stem. Flowers 1–7 together; bracts ovate-lanceolate, with a long, usually green, subulate point. Epicalyx-scales broadly obovate, almost truncate, emarginate, whitish-coriaceous and broadly scarious; awn 3–6 mm. Calyx 15–20 × 2–3 mm, almost cylindrical. Petal-limb 4–5 mm, dentate, dark purple, bearded. $2n = 30$. ● *S.W. Jugoslavia, S. Albania, N.W. Greece.* Al Gr Ju.

35. D. pelviformis Heuffel, *Flora (Regensb.)* **36**: 625 (1853) (incl. *D. zernyi* Hayek). Perennial up to 40 cm. Leaves *c.* 2 mm wide, linear, acute, flat; sheaths several times as long as diameter of stem. Flowers numerous, in a dense head; bracts broadly ovate, imbricate, usually emarginate with a patent or deflexed, sometimes green, subulate point. Epicalyx-scales ovate, broadly scarious, shortly awned. Calyx 7–8 × 2–3 mm, somewhat ventricose. Petal-limb 2–3 mm, dentate, glabrous, dark purple. $2n = 30$. ● *C. part of Balkan peninsula.* Al Bu Gr Ju.

36. D. stenopetalus Griseb., *Spicil. Fl. Rumel.* **1**: 189 (1843) (*D. geticus* Kulcz.). Caespitose perennial, almost stemless or up to *c.* 40 cm. Leaves *c.* 0·5 mm wide, convolute; sheaths several times as long as diameter of stem. Flowers few to many in a dense head; bracts broadly ovate, emarginate, awned, coriaceous. Epicalyx-scales ovate, awned, broadly scarious, pale. Calyx 5–7 × 2–3 mm, somewhat ventricose, purple. Petal-limb entire or obscurely dentate, reddish-purple. $2n = 30$. ● *Balkan peninsula, from c. 42° N. southwards.* Al Bu Gr Ju.

37. D. moesiacus Vis. & Pančić, *Mem. Ist. Veneto* **15**: 17 (1870) (incl. *D. burgasensis* Tutin). Caespitose perennial up to 60 cm. Leaves setaceous, up to 1 mm wide; sheaths several times as long as diameter of stem. Flowers numerous in rather narrow, dense heads; bracts oblong, emarginate, with a subulate point, coriaceous. Epicalyx-scales like the bracts but smaller, shortly awned, *c.* ½ as long as calyx to nearly equalling it. Calyx 10–12 × 1·5–2 mm, tapering from the middle. Petal-limb 3–4 mm, obovate, glabrous. ● *E. Jugoslavia, Bulgaria.* Bu Ju.

38. D. rigidus Bieb., *Fl. Taur.-Cauc* **1**: 325 (1808). Dwarf shrub 10–50 cm, with rigid, much-branched, erect woody stems and herbaceous flowering stems. Leaves not crowded at base of stem, linear, less than 1 mm wide, apparently 1-veined. Epicalyx-scales 4, ovate, acuminate, *c.* ¼ as long as calyx, the outer much smaller than the inner. Calyx 10–12 × 2 mm, cylindrical. Petal-limb *c.* 5 mm, dentate, bearded, white. *S.E. Russia; Krym.* Rs(K, E).

39. D. uralensis Korsh., *Mém. Acad. Sci. Pétersb.* ser. 8, **7**: 59 (1898). Like **38** but most leaves at least 1 mm wide, distinctly 3-veined; epicalyx-scales *c.* ⅓ as long as calyx; calyx *c.* 3 mm wide; petal-limb pink above, yellowish-green beneath. *S. Ural.* Rs(C).

Subgen. **Dianthus**. Flowers solitary or in groups, usually without involucral bracts. Plants usually not woody at base.

40. D. seguieri Vill., *Hist. Pl. Dauph.* **1**: 330 (1786). Laxly caespitose perennial (10–)30–60 cm, with a slender stock. Stem simple or branched. Leaves 1–4(–5) mm wide, more or less flat, linear-lanceolate; cauline leaves usually more than 4 pairs; sheaths up to as long as diameter of stem, somewhat inflated. Flowers 1 to several. Epicalyx-scales 2–6, ovate to ovate-lanceolate, gradually or abruptly narrowed to a subulate apex, ½–¾ as long as calyx. Calyx 14–20 mm, almost cylindrical. Petal-limb 7–12(–17) mm, dentate, reddish-pink, bearded and often whitish-spotted at base. ● *S.W. & W.C. Europe, from N.E. Spain to N. Italy and E. Germany.* Cz Ga Ge He Hs It.

1 Leaves 2–5 mm wide (c) subsp. **glaber**
1 Leaves usually 1–2 mm wide
 2 Leaves with midrib much more prominent than lateral veins beneath; epicalyx-scales often nearly as long as calyx (a) subsp. **seguieri**
 2 Leaves with all veins ±equally prominent beneath; epicalyx-scales *c.* ½ as long as calyx (b) subsp. **gautieri**

(a) Subsp. **seguieri** (incl. subsp. *italicus* Tutin): Leaves usually 1–2 mm wide, the midrib much more prominent than the lateral veins beneath. Flowers often in pairs. Epicalyx-scales almost as long as calyx. $2n = 90$. *S.E. France and N. Italy.*

(b) Subsp. **gautieri** (Sennen) Tutin, *Feddes Repert.* **68**: 189 (1963) (*D. gautieri* Sennen, *D. neglectus* auct. hisp., non Loisel.): Leaves usually 1–2 mm wide, all veins more or less equally prominent beneath. Flowers often solitary. Epicalyx-scales *c.* ½ as long as calyx. *N.E. Spain, E. Pyrenees.*

(c) Subsp. **glaber** Čelak., *Prodr. Fl. Böhm.* **3**: 507 (1875) (incl. *D. sylvaticus* Hoppe, *D. seguieri* subsp. *sylvaticus* (Hoppe) Hegi): Leaves 2–5 mm wide. Flowers 1 to several. Epicalyx-scales *c.* ½ as long as calyx. $2n = 60$. *S.C. France, Czechoslovakia, S. Germany.*

41. D. collinus Waldst. & Kit., *Pl. Rar. Hung.* **1**: 51 (1801). Shortly pubescent to glabrous perennial 20–80 cm. Leaves 3–8 mm wide, linear-lanceolate, thin; basal leaves few or none at anthesis; cauline usually 7–15 pairs; veins 3–5, the midrib usually distinctly more prominent than the laterals; sheaths 1–2 times as long as diameter of stem. Flowers usually 2–8, commonly in 2 rather lax heads; bracts green, shorter than flowers. Epicalyx-scales ovate, acuminate, ½ as long as calyx. Calyx 14–18 mm. Petal-limb 7–14 mm, pink or purplish, bearded. ● *E.C. Europe, extending to E. Jugoslavia and N.E. Romania.* Au Cz Hu Ju Po Rm Rs(W).

(a) Subsp. **collinus**: Plant shortly pubescent. $2n = 90$. *From Austria and E. Czechoslovakia to Jugoslavia.*

(b) Subsp. **glabriusculus** (Kit.) Thaisz, *Bot. Közl.* **8**: 252 (1910) (*D. piatr-neamtzui* Prodan): Plant glabrous. $2n = 90$. *From Poland to Hungary and Romania.*

42. D. fischeri Sprengel, *Pugillus* **2**: 62 (1815) (*D. mariae* (Kleopow) Klokov). Like **41** but cauline leaves *c.* 7 pairs, 2–3 mm wide, tapering from just above base. ● *W. & C. Russia, Ukraine.* Rs(C, W, K, E).

43. D. pratensis Bieb., *Fl. Taur.-Cauc.* **3**: 300 (1819). Glabrous perennial with a slender, creeping stock. Leaves 20–70 × 2–4 mm, linear-lanceolate; cauline leaves usually 3 or 4 pairs. Flowers 1 or 2; pedicels long. Epicalyx-scales 4(–6), ovate, gradually narrowed into a subulate, herbaceous apex reaching about to base of calyx-teeth. Calyx 10–17 × 4–5 mm, widest at about the middle. Petal-limb 5–7 mm, dentate, pale pink to pinkish-purple, bearded. ● *E. Europe, from C. Ural to E. Romania.* Rm Rs(C, W, E) [Rs(B)].

(a) Subsp. **pratensis**: Stems 15–40 cm, slender. Outer epicalyx-scales *c.* 5 mm. *Throughout the range of the species except Romania.*

(b) Subsp. **racovitzae** (Prodan) Tutin, *Feddes Repert.* **68**: 189

(1963) (*D. racovitzae* Prodan): Stems 40–60 cm, robust. Outer epicalyx-scales *c.* 10 mm. ● *E. Romania.*

44. D. versicolor Fischer ex Link, *Enum. Hort. Berol. Alt.* **10**: 420 (1821). Like **43** but stock very stout; stems and leaves more or less puberulent and rough; outer epicalyx-scales *c.* $\frac{1}{2}$ as long as calyx, abruptly contracted into a subulate apex; petal-limb pinkish-purple above, greenish beneath. *C. & E. Russia, N.E. Ukraine.* Rs(C, E).

45. D. pseudoversicolor Klokov in Kotov, *Fl. RSS Ucr.* **4**: 660 (1952). Like **43(a)** but pubescent; leaves 10–55 × 1–3 mm; epicalyx-scales mucronate; calyx 15–18 × 3–4·5 mm, almost cylindrical; petal-limb 18–25 mm, dentate, pinkish-purple. ● *S. Russia, N.E. Ukraine.* Rs(W, E).

46. D. furcatus Balbis, *Mém. Acad. Sci. (Turin)* **10–11**: 13 (1804) (*D. requienii* Gren. & Godron, *D. fallens* Timb.-Lagr., *D. alpester* Balbis, *D. benearensis* Loret). Laxly caespitose, not glaucous perennial (10–)15–30(–60) cm, with a slender stock. Stem simple or more commonly branched. Leaves 1–3 mm wide, more or less flat, linear-lanceolate, the cauline usually at least 3 pairs; sheaths up to as long as diameter of stem, somewhat inflated. Flowers 1 to several. Epicalyx-scales 2–6, ovate to ovate-lanceolate, narrowed to a subulate apex, usually *c.* $\frac{1}{2}$ as long as calyx. Calyx 10–17(–20) mm, somewhat attenuate above; teeth lanceolate, acuminate. Petal-limb 5–10 mm, shallowly laciniate to subentire, glabrous, rarely sparsely hairy, pink or whitish, unspotted, but dark at base. ● *Mountains of S.W. Europe.* Co Ga Hs It.

1 Stems up to *c.* 10 cm and less than 0·5 mm in diameter, simple; 2 pairs of cauline leaves; calyx 10–13 mm (b) subsp. **lereschii**
1 Stems usually 15–30 cm and more than 0·5 mm in diameter, commonly branched; 3 or more pairs of cauline leaves; calyx usually 14–19 mm
2 Leaves usually 2–3 mm wide, entire; epicalyx-scales as long as calyx-tube; petals whitish (d) subsp. **gyspergerae**
2 Leaves usually 1–2 mm wide, serrulate; epicalyx-scales shorter than calyx-tube; petals pink
3 Calyx 15–16 × 2·5 mm, tapering markedly from the middle, usually purplish; teeth narrowly triangular (a) subsp. **furcatus**
3 Calyx *c.* 17 × 4 mm, nearly cylindrical, usually green; teeth broadly triangular (c) subsp. **geminiflorus**

(a) Subsp. **furcatus**: Leaves usually 1–2 mm wide, serrulate. Epicalyx-scales shorter than calyx-tube. Calyx *c.* 15 × 2·5 mm, tapering markedly from the middle, usually purplish. Petals pink. $2n = 60$. *S.W. Alps.*

(b) Subsp. **lereschii** (Burnat) Pignatti, *Gior. Bot. Ital.* **107**: 209 (1973) (subsp. *tener* Tutin, non *D. tener* Balbis): Stems not more than 10 cm and less than 0·5 mm in diameter, simple. Cauline leaves 2 pairs. Leaves 1–2 mm wide, serrulate. Epicalyx-scales shorter than calyx-tube. Calyx 10–13 mm. Petals pink. *N.W. Italy (Piemonte).*

(c) Subsp. **geminiflorus** (Loisel.) Tutin, *Feddes Repert.* **68**: 189 (1963) (*D. geminiflorus* Loisel.): Leaves usually 1–2 mm wide, serrulate. Epicalyx-scales shorter than calyx-tube. Calyx *c.* 17 × 4 mm, nearly cylindrical, usually green, the teeth broadly triangular. Petals pink. $2n = 30$. *Pyrenees.*

(d) Subsp. **gyspergerae** (Rouy) Burnat ex Briq., *Prodr. Fl. Corse* **1**: 572 (1910) (*D. gyspergerae* Rouy): Leaves usually 2–3 mm wide, entire. Epicalyx-scales as long as calyx-tube. Calyx 11–14 mm. Petals whitish. *W. Corse (near Porto).*

Subsp. **dissimilis** (Burnat) Pignatti, *Gior. Bot. Ital.* **107**: 209 (1973), from Alpi Marittime, similar to subsp. (a) but with calyx 16–18 × 4–5 mm, and larger petals, may be a hybrid between **46** and **62**.

47. D. viridescens G. C. Clementi, *Atti 3 Riun. Sci. Ital.* 520 (1841). Annual up to 60 cm; stems *c.* 1·5 mm in diameter, freely branched above. Leaves mostly 2–3 mm wide, obtuse, thin, flat; basal leaves few or absent at flowering; cauline usually 5–7 pairs; sheaths short. Epicalyx-scales 2, obovate, with a green, subulate apex, about as long as calyx-tube. Calyx *c.* 17 mm, almost cylindrical; teeth long, acuminate. Petal-limb 7–10 mm, pink, dentate, bearded. ● *S. & W. Jugoslavia; N. Albania.* Al Ju.

48. D. eugeniae Kleopow, *Bull. Jard. Bot. Kieff* **14**: 103 (1932). Glabrous perennial 40–60 cm. Leaves 2–5 mm wide, linear-lanceolate or linear, the basal withered at anthesis; sheaths about as long as diameter of stem. Flowers usually solitary at ends of branches. Epicalyx-scales ovate, abruptly narrowed to the subulate apex, at least some reaching base of calyx-teeth; margins scarious. Calyx 15–17 mm. Petal-limb 8–10 mm, purple, with black spots at base. ● *N. & E. Ukraine, and adjacent regions of Russia.* Rs(W, C, E).

49. D. tesquicola Klokov, *Jour. Inst. Bot. Acad. Sci. Ukr.* **5(1)**: 26 (1948). Like **48** but usually smaller; stems and leaves more or less puberulent; leaves 1–2 mm wide; epicalyx-scales shortly aristate, up to $\frac{1}{2}$ as long as calyx; calyx 12–15 mm; petal-limb 3–5 mm. ● *C. & E. Ukraine.* Rs(W).

50. D. guttatus Bieb., *Fl. Taur.-Cauc.* **1**: 328 (1808) (*D. pseudogrisebachii* Grec.). Like **48** but 20–45 cm; leaves 1–3 mm wide; sheaths *c.* $\frac{1}{2}$ as long as leaf-width; outer epicalyx-scales leaf-like, at least as long as calyx; petal-limb 5–8 mm, pinkish with white spots above, yellow-green beneath. ● *S. Romania to E. Ukraine.* Rm Rs(W).

51. D. monspessulanus L., *Amoen. Acad.* **4**: 313 (1759) (incl. *D. hyssopifolius* L. pro parte). Laxly caespitose perennial 20–60 cm; stems slender, simple or little-branched. Leaves 1–3 mm wide, linear to linear-lanceolate, thin, acuminate; sheaths about as long as diameter of stem. Flowers (1)2–5(–7), fragrant; pedicels short. Epicalyx-scales 4, ovate, with a subulate, herbaceous apex, usually more than $\frac{1}{2}$ as long as calyx. Calyx 18–25 × 3–5 mm, nearly cylindrical; teeth usually ovate-lanceolate, mucronate. Petal-limb 10–25 mm, divided to *c.* $\frac{1}{4}$ into narrow lobes, the entire part suborbicular, white or pink, sometimes bearded. $2n = 30$. ● *S. Europe, extending northwards to the Jura and N.E. Alps; mainly in the mountains.* Al Ga He Hs It Ju Lu.

(a) Subsp. **monspessulanus**: Plant 20–60 cm. Epicalyx-scales usually more than $\frac{1}{2}$ as long as calyx. Petal-limb 10–15 mm. *Throughout the range of the species except C. & S. Italy.*

(b) Subsp. **marsicus** (Ten.) Novák, *Acta Fac. Rer. Nat. Carol.* **21**: 25 (1924) (*D. marsicus* Ten.): Plant usually *c.* 20 cm. Epicalyx-scales usually at least $\frac{1}{2}$ as long as calyx. Petal-limb 15–25 mm. *C. & S. Appennini.*

52. D. sternbergii Sieber ex Capelli, *Cat. Stirp.* 24 (1821) (*D. waldsteinii* Sternb., *D. monspessulanus* subsp. *sternbergii* Hegi): Like **51** but plant 10–22 cm; flowers 1(2); epicalyx-scales sometimes less than $\frac{1}{2}$ as long as calyx; calyx *c.* 18 mm; petal-limb 12–25 mm, pink. $2n = 90$. ● *E. Alps.* Au It Ju.

53. D. repens Willd., *Sp. Pl.* **2(1)**: 681 (1799). Glabrous perennial 5–22 cm. Leaves linear-oblong, at least the lower obtuse. Epicalyx-scales usually 4, $\frac{1}{2}$ as long as calyx, rarely longer; outer lanceolate with a subulate point, the inner ovate, abruptly narrowed at apex. Calyx 9–12 × 4–5 mm, widening upwards. Petal-limb 10–12 mm, bearded, pink or purplish. $2n = 60$. *N. & E. Russia.* Rs(N, C).

54. D. alpinus L., *Sp. Pl.* 412 (1753). Subcaespitose, glabrous perennial 2–20 cm. Leaves oblong to oblanceolate, obtuse, the basal 3–5 mm wide. Flowers borne above the leaves. Epicalyx-scales 2–4, herbaceous, ovate, with a subulate apex, more than $\frac{1}{2}$ as long as calyx. Calyx 15–18 × 6–7·5 mm, widening upwards; teeth ovate, acute, broadly scarious. Petal-limb 15–18 mm, bearded, purplish-red with white spots (rarely entirely white). $2n = 30$. ● *N.E. Alps.* Au.

55. D. nitidus Waldst. & Kit., *Pl. Rar. Hung.* **3**: 209 (1806). Like **54** but taller; stems often branched and flowers in pairs; calyx 10–12 mm; petal-limb 8–10 mm. $2n = 30$. *Calcicole.* ● *W. Carpathians.* Cz †Po.

56. D. scardicus Wettst., *Biblioth. Bot. (Stuttgart)* **26**: 31 (1892) (*D. nitidus* sensu Boiss. pro parte, non Waldst. & Kit.). Subcaespitose, with long non-flowering stems; stems 1·5–10 cm. Leaves *c.* 15 × 2–3 mm, 3-veined. Flowers usually solitary. Epicalyx-scales 2, about $\frac{1}{2}$ as long as calyx; apex 3-veined, patent or appressed. Calyx 10–12 mm, widening upwards. Petal-limb 7–8 mm, pink. *Mountain pastures.* ● *S.W. Jugoslavia; E. Albania.* Al Ju.

57. D. callizonus Schott & Kotschy, *Bot. Zeit.* **9**: 192 (1851). Stems 5–20 cm, 1-flowered, glabrous. Leaves linear-lanceolate, the upper acute or acuminate, the lower obtuse. Epicalyx-scales 2–4, ovate, with a green subulate apex, the outer about as long as calyx. Calyx *c.* 16 mm, widening upwards, dark red; teeth ovate-lanceolate, acuminate. Petal-limb 10–15 mm, bright pink, with dark spots near base. $2n = 30$. ● *S. Carpathians (Munţii Piatra Craiului, S.W. of Braşov).* Rm.

58. D. glacialis Haenke in Jacq., *Collect. Bot.* **2**: 84 (1789). Nearly glabrous, caespitose perennial usually *c.* 5 cm. Leaves 1–2 mm wide, linear, or widest above the middle, mostly obtuse; cauline 1–3 pairs. Flowers solitary, usually surrounded by the leaves. Epicalyx-scales 2–4, ovate, with a green, subulate point, about as long as calyx. Calyx 12–16 × 3–7 mm, widening upwards. ● *E. Alps and Carpathians.* Au Cz He It ?Ju Po Rm.

(a) Subsp. **glacialis**: Leaf-margin almost glabrous. Calyx 3–4 mm wide. Petal-limb usually 5–7 mm. $2n = 30$. *Almost throughout the range of the species.*

(b) Subsp. **gelidus** (Schott, Nyman & Kotschy) Tutin, *Feddes Repert.* **68**: 190 (1963) (*D. gelidus* Schott, Nyman & Kotschy): Leaf-margin ciliate towards base. Calyx 5–7 mm wide. Petal-limb 10–12 mm. *S. & E. Carpathians.*

Plants intermediate between subspp. (**a**) and (**b**) have been called *D. glacialis* var. *pawlowskianus* Soó.

59. D. freynii Vandas, *Sitz.-Ber. Böhm. Ges. Wiss.* **1890(1)**: 255 (1890). Like **58** but leaves usually less than 1 mm wide; flowers borne above leaves; calyx 8–10 × 3–4 mm; petal-limb 5–9 mm. ● *C. Jugoslavia (Bosna).* Ju.

60. D. microlepis Boiss., *Diagn. Pl. Or. Nov.* **1(1)**: 22 (1843). Caespitose perennial 1–10 cm; stems leafless or with 1 or 2 pairs of scale-like leaves. Basal leaves 10–20 × 1·2–1·5 mm, obtuse. Epicalyx-scales 2, more or less remote, membranous, ovate, the apex herbaceous or not. Calyx 9–10·5 mm, widening upwards. Petal-limb 6–7 mm, purple. $2n = 30$. ● *Mountains of Bulgaria and S. Jugoslavia.* Bu ?Gr Ju.

61. D. pavonius Tausch, *Flora (Regensb.)* **22**: 145 (1839) (*D. neglectus* Loisel. pro parte). Caespitose perennial 4–10 cm. Leaves linear, acuminate, 3-veined. Flowers 1(–3). Epicalyx-scales usually herbaceous, ovate, gradually narrowed to the subulate apex, about as long as calyx. Calyx 12–15 × 4–5 mm, narrowing somewhat upwards; teeth broadly scarious, acute or acuminate. Petal-limb 10–12 mm, bearded, purplish-red. $2n = 30$. ● *S.W. Alps; ?E. Alps.* Ga It.

62. D. gratianopolitanus Vill., *Hist. Pl. Dauph.* **3**: 598 (1789) (*D. caesius* Sm.). Subcaespitose, glabrous, more or less glaucous perennial 6–25 cm. Leaves 1–2 mm wide, linear-lanceolate, more or less flat, the cauline usually 2 or 3 pairs. Flowers usually solitary, fragrant. Epicalyx-scales herbaceous, ovate to obovate, cuspidate or mucronate, $\frac{1}{4}$–$\frac{1}{3}$ as long as calyx. Calyx 13–17(–20) mm, almost cylindrical; teeth ovate, little longer than wide, often obtuse. Petal-limb 7–15 mm, bearded, pink, unspotted. $2n = 60, 90$. ● *W. & C. Europe; one station in W. Ukraine.* Be Br Cz Ga Ge He Po Rs(W).

63. D. xylorrhizus Boiss. & Heldr. in Boiss., *Diagn. Pl. Or. Nov.* **2(8)**: 67 (1849). Perennial 5–15 cm, with a thick, woody stock. Basal leaves up to 4 mm wide, linear-oblong, obtuse or acute, rather thick and soft; midrib prominent beneath; lateral veins obscure; cauline leaves 3–6 pairs, shorter and narrower than the basal. Epicalyx-scales usually 4, ovate, acute or cuspidate, coriaceous, *c.* $\frac{1}{4}$ as long as calyx. Calyx 20–25 × 4–5 mm, widest below the middle. Petal-limb 4–6 mm, glabrous, dirty white. ● *W. Kriti.* Cr.

64. D. sylvestris Wulfen in Jacq., *Collect. Bot.* **1**: 237 (1786) (*D. boissieri* Willk., *D. virgineus* Gren. & Godron, *D. inodorus* (L.) Gaertner). Densely caespitose perennial with a short, stout, woody stock. Basal leaves usually numerous, 0·5–1·2 mm wide, green, wiry and often recurved. Flowers not or slightly fragrant. Epicalyx-scales 2–5(–8), broadly obovate, truncate or cuspidate, coriaceous, $\frac{1}{4}$–$\frac{1}{3}$ as long as calyx. Calyx 12–30 × 4–7 mm, nearly cylindrical. Petal-limb glabrous, usually pink, dentate to entire. $2n = 30$, *c.* 60. *S. & S.C. Europe, extending northwards to* 47°50′ *N. in E.C. France.* Al Au Co Ga Ge Gr He Hs It Ju Sa Si.

A polymorphic species closely related to **65**. The following subspecies may be recognized:

1 Plant somewhat glaucous; epicalyx-scales 2(–4); petals entire, usually not contiguous (**f**) subsp. **tergestinus**

1 Plant usually green; epicalyx-scales (2–)4–6(–8); petals dentate or denticulate, usually contiguous

2 Stems usually puberulent below; epicalyx-scales not green at apex; calyx-teeth 2–3 mm, obtuse (**d**) subsp. **nodosus**

2 Stems usually glabrous; epicalyx-scales with a green apex; calyx-teeth 3–5 mm, obtuse to acuminate

3 Calyx usually less than 15 mm; petals denticulate (**e**) subsp. **bertisceus**

3 Calyx usually more than 15 mm; petals dentate
4 Calyx usually 4–5 mm wide; epicalyx-scales usually 4
(a) subsp. **sylvestris**
4 Calyx 5–7 mm wide; epicalyx-scales 4–8
5 Petal-limb 10–15 mm; calyx-teeth acuminate
(b) subsp. **siculus**
5 Petal-limb 7–9 mm; calyx-teeth obtuse or subacute
(c) subsp. **longicaulis**

(a) Subsp. **sylvestris**: Variable; stems usually branched, but in alpine plants often simple and *c*. 10 cm high. Epicalyx-scales usually 4. Calyx (12–)15–22 × 4–5 mm; teeth ovate-lanceolate, obtuse. Petal-limb 8–12 mm, usually contiguous. *Throughout the range of the species except parts of the W. Mediterranean region.*

(b) Subsp. **siculus** (C. Presl) Tutin, *Feddes Repert.* **68**: 190 (1963) (*D. siculus* C. Presl, *D. longicaulis* Ten., *D. gasparrinii* Guss.): Robust, usually branched. Epicalyx-scales 4–8. Calyx 24–30 × 5–7 mm; teeth lanceolate, acuminate. Petal-limb 10–15 mm, usually contiguous, dentate or denticulate. *Sardegna and Sicilia.*

(c) Subsp. **longicaulis** (Ten.) Greuter & Burdet, *Willdenowia* **12**: 187 (1982) (*D. contractus* Jan ex Nyman, *D. godronianus* Jordan, *D. sylvestris* subsp. *garganicus* (Grande) Pignatti): Like subsp. (**b**) but often glaucous; calyx 18–28 mm, the teeth long, triangular, subacute or obtuse; petal-limb 7–9 mm. 2*n* = 30. *Mediterranean region, from S. Spain to W. Jugoslavia.*

(d) Subsp. **nodosus** (Tausch) Hayek, *Prodr. Fl. Penins. Balcan.* **1**: 247 (1924) (*D. nodosus* Tausch): Stems usually puberulent below. Epicalyx-scales (2–)4–6(–8). Calyx 15–20 × *c*. 4 mm. Petal-limb 8–10 mm, usually contiguous, dentate or denticulate. ● *W. part of Balkan peninsula.*

(e) Subsp. **bertisceus** Rech. fil., *Feddes Repert.* **38**: 150 (1935). Stems usually simple. Epicalyx-scales (2–)4–6(–8). Calyx (10–)12–16(–20) × 5 mm. Petal-limb 5–6 mm, dentate or denticulate. ● *W. & S. Jugoslavia, Albania.*

(f) Subsp. **tergestinus** (Reichenb.) Hayek, *loc. cit.* (1924) (*D. tergestinus* (Reichenb.) A. Kerner): Plant somewhat glaucous. Epicalyx-scales 2(–4). Calyx 15–20 × 4–5 mm. Petal-limb *c*. 6 mm, usually not contiguous, entire. 2*n* = 30. ● *W. Jugoslavia, Albania; E. Italy (Gargano).*

65. D. caryophyllus L., *Sp. Pl.* 410 (1753). Laxly caespitose, usually glaucous perennial up to 80 cm; woody basal part of stem rather long and not very thick. Leaves 2–4 mm wide, linear, nearly flat. Flowers strongly fragrant. Epicalyx-scales 4, *c*. ¼ as long as calyx, broadly obovate, cuspidate. Calyx usually 25–30 × 5–7 mm, nearly cylindrical; teeth *c*. 5 mm, acuminate to subobtuse. Petal-limb 10–15 mm, usually contiguous, glabrous, purple, rarely pink or white, dentate. 2*n* = 30. *Widely cultivated for ornament and occasionally naturalized, but apparently not known wild, except perhaps in some Mediterranean countries.* *Gr *It *Sa *Si [Ga Hs].

D. arrostii C. Presl in J. & C. Presl, *Del. Prag.* **1**: 60 (1822), from Sardegna and Sicilia, appears to differ only in the patent epicalyx-scales, somewhat shorter calyx, and uniformly pale pink petals.

D. multinervis Vis., *Fl. Dalm.* **3**: 164 (1852), described from the Adriatic island of Pomo, but now extinct, appears to have been close to, or possibly an abnormal variant of **65**, characterized by the 5- to 9-veined leaves which are described as broadly lanceolate, and by the shorter calyx.

66. D. subacaulis Vill., *Hist. Pl. Dauph.* **3**: 597 (1789) (incl. *D. brachyanthus* Boiss.). Usually densely caespitose perennial 3–20 cm, with a stout, woody stock; stems nearly always simple. Leaves linear-lanceolate, often obtuse, the basal usually *c*. 10 × 1 mm; cauline shorter, appressed. Epicalyx-scales 4, broadly ovate, apiculate or acuminate, *c*. ⅓ as long as calyx. Calyx 6–10(–13) × 3–4 mm, more or less ventricose. Petal-limb 3–5(–10) mm, glabrous, subentire, pale pink. *Mountains of S.W. Europe.* Ga Hs.

(a) Subsp. **subacaulis**: Plant often laxly caespitose. Epicalyx-scales usually acuminate, sometimes nearly ½ as long as calyx. Calyx-teeth acute or apiculate, *c*. 3 times as long as wide. *S.E. France, ?Spain.*

(b) Subsp. **brachyanthus** (Boiss.) P. Fourn., *Quatre Fl. Fr.* 331 (1936): Plant usually densely caespitose with thick, woody stock. Epicalyx-scales apiculate, *c*. ⅓ as long as calyx. Calyx-teeth obtuse, twice as long as wide. 2*n* = 30. *From C. France to S.E. Spain.*

Plants of subsp. (**a**) from S. France (near Narbonne) with long, slender, trailing stems, longer leaves, calyx 10–13 × 4 mm, and petal-limb *c*. 10 mm have been called var. *ruscinonensis* Boiss. They have a chromosome number of 2*n* = 60.

67. D. pungens L., *Mantissa Alt.* 240 (1771) (*D. acuminatus* Rouy, ?*D. serratus* Lapeyr.). Laxly caespitose perennial 5–20 cm, with slender, woody stock; stems simple. Leaves *c*. 20 × 0·5 mm, acuminate, rather rigid. Epicalyx-scales 4, ovate, gradually long-acuminate, *c*. ½ as long as calyx. Calyx *c*. 15 × 3–4 mm, nearly cylindrical. Petal-limb *c*. 5 mm, glabrous, pink. ● *E. Pyrenees.* Ga Hs.

68. D. hispanicus Asso, *Syn. Stirp. Arag.* 53 (1779) (*D. pungens* sensu Willk., non L.). Perennial 15–40 cm, with a stout, woody stock. Stems simple or little-branched. Leaves usually 2–3 cm, acute, flat or convolute, rigid, often recurved; cauline 2–4(–6) pairs. Epicalyx-scales 4, obovate, cuspidate, mainly coriaceous, *c*. ⅓ as long as calyx. Calyx 13–20 × 4–5 mm, tapering above the middle but becoming cylindrical or even widening upwards in fruit. Petal-limb 3–5 mm, glabrous. 2*n* = 30. ● *Spain.* Hs.

Plants intermediate between this species and **66** are not infrequent.

69. D. costae Willk. in Willk. & Lange, *Prodr. Fl. Hisp.* **3**: 683 (1878) (*D. pungens* sensu Willk., non L.; incl. *D. algetanus* Graells ex F. N. Williams, *D. turolensis* Pau). Perennial 10–40 cm, with stout, woody stock. Leaves linear, acuminate, flat; cauline usually 2–6 pairs, appressed to the stem. Flowers solitary or few together; pedicels short. Epicalyx-scales 4–6, sometimes with 1 or 2 pairs of scale-like leaves just below them, *c*. ½ as long as calyx, ovate, acuminate; margin broadly scarious; apex green. Calyx 10–16 × 3–4 mm, narrowed from the middle or below. Petal-limb 6–8 mm, dentate, glabrous. ● *C. & E. Spain.* Hs.

70. D. langeanus Willk. in Willk. & Lange, *Prodr. Fl. Hisp.* **3**: 690 (1878) (*D. gredensis* Pau ex Caball.). Laxly caespitose perennial, often with rather slender, trailing stems; stems 10–20 cm, slender, simple or little-branched. Leaves 1–2 cm, rather thick and rigid, acute, canaliculate when dry; cauline leaves 2–4 pairs, shorter than internodes, erect. Epicalyx-scales 4, *c*. ¼ as long as calyx, ovate, subacute, shortly apiculate or obtuse, the inner wider than the outer. Calyx 8–10 × 2 mm, almost cylindrical. Petal-limb 5–7 mm, dentate to entire, pink, glabrous. ● *N.W. & W.C. Spain.* Hs ?Lu.

71. D. laricifolius Boiss. & Reuter, *Biblioth. Univ. Genève* ser. 2, **38**: 199 (1842) (incl. *D. caespitosifolius* Planellas, *D. planellae* Willk.,

D. cintranus auct., non Boiss. & Reuter, *D. marizii* (Samp.) Samp.). Laxly caespitose perennial, woody at base; stems 20–60 cm. Basal leaves 10–30 mm, linear-subulate, fasciculate, with a distinct mid-vein and prominent marginal veins, erect and somewhat rigid; cauline leaves 4–6 pairs, erect. Flowers often 2 or 3 together. Epicalyx segments 4(–6), ovate, whitish, not veined, all more or less abruptly cuspidate with a green point, *c.* $\frac{1}{3}$ as long as calyx. Calyx 14–20 × 4–5 mm, narrowed above; teeth acute, mucronulate. Petal-limb 5–10 mm, bright reddish-purple to pale pink. ● *C. & W. Spain, N. & C. Portugal.* Hs Lu.

A very variable species variously divided into a number of subspecies and species by different authors.

72. D. serratifolius Sm. in Sibth. & Sm., *Fl. Graec. Prodr.* **1**: 287 (1809). Laxly to rather densely caespitose perennial 15–30 cm, with long, trailing, woody stems. Basal leaves usually 15–30 mm, linear, acuminate, with 3 prominent veins beneath; cauline smaller. Flowers usually solitary. Epicalyx-scales elliptical or ovate, acute or acuminate, usually entirely brown. Calyx 9–15 × *c.* 3 mm, tapering from near base, becoming cylindrical in fruit. Petal-limb 3–4 mm, dentate, pink. *S. & E.C. Greece.* Gr.

The plants from the mountains of S. Greece (Peloponnisos) have somewhat shorter basal leaves, fewer cauline leaves, (2–)4 somewhat longer epicalyx-scales, and a more deeply dentate petal-limb. They have been called subsp. **abbreviatus** (Heldr. & Halácsy) Strid, *Mount. Fl. Gr.* **1**: 179 (1986).

73. D. pyrenaicus Pourret, *Mém. Acad. Sci. Toulouse* **3**: 318 (1788) (*D. attenuatus* Sm., *D. requienii* sensu Willk. pro parte, non Gren. & Godron, *D. maritimus* Rouy). Laxly caespitose perennial 15–45 cm, usually with long, trailing, rather slender, woody stems. Leaves usually rigid, pungent, not densely crowded on flowering stems. Inflorescence usually branched. Epicalyx-scales (4–)6–8, ovate, gradually acuminate, *c.* $\frac{1}{3}$ as long as calyx. Calyx 15–30 × 2–3 mm, tapering markedly from about the middle. Petal-limb 3–8 mm, glabrous, dentate, pink. $2n = 30, 60$. ● *S. France, N.E. Spain.* Ga Hs.

(a) Subsp. **pyrenaicus**: Not glaucous. Calyx 15–20 mm. Petal-limb *c.* 3 mm, pale pink. *Pyrenees.*

(b) Subsp. **catalaunicus** (Willk. & Costa) Tutin, *Feddes Repert.* **68**: 190 (1963). Usually glaucous. Calyx 20–30 mm. Petal-limb 6–8 mm, deep pink. *Throughout the range of the species.*

74. D. lusitanus Brot., *Fl. Lusit.* **2**: 177 (1804) (*D. lusitanicus* auct.). Glaucous perennial 15–45 cm; stems erect or ascending, woody below. Leaves usually *c.* 10 mm, not crowded on flowering stems, somewhat fleshy and without obvious veins. Inflorescence usually branched. Epicalyx-scales 4, ovate, acuminate, *c.* $\frac{1}{3}$ as long as calyx. Calyx 20–23 × 2·5–3·5 mm, tapering markedly from about the middle. Petal-limb 7–10 mm, bearded, deeply dentate, pink. $2n = 30, 60$. *Spain and Portugal.* Hs Lu.

75. D. serrulatus Desf., *Fl. Atl.* **1**: 346 (1798) (*D. malacitanus* Haenseler ex Boiss., nom. inval., *D. broteri* Boiss. & Reuter, *D. valentinus* Willk., *D. fimbriatus* Brot., non Bieb.). Perennial up to 50 cm. Leaves 1–2 mm wide, linear, acuminate, glaucous. Epicalyx-scales usually 6–8, elliptical to ovate-lanceolate, acuminate, $\frac{1}{3}$–$\frac{1}{2}$ as long as calyx. Calyx *c.* 30 mm, tapering from near base. Petal-limb *c.* 15 mm, laciniate, bearded. $2n = 60$. *S. & E. Spain, S. & C. Portugal.* Hs Lu.

European plants are referable to subsp. **barbatus** (Boiss.) Greuter & Burdet, *Willdenowia* **13**: 281 (1984). Subsp. **serrulatus** occurs in N.W. Africa.

76. D. scaber Chaix in Vill., *Hist. Pl. Dauph.* **1**: 331 (1786) (*D. hirtus* Vill.). Laxly caespitose perennial 15–40 cm. Leaves 1–3 mm wide, flat, linear, acuminate; cauline leaves usually 3 or 4 pairs. Flowers in small heads; bracts usually absent. Epicalyx-scales usually 6, ovate, acuminate, from $\frac{1}{2}$ as long as to as long as calyx. Calyx 17–22(–25) mm, tapering above the middle. Petal-limb 4–8 mm, dentate, bearded. ● *S.W. Europe.* Ga Hs Lu.

1 Point of epicalyx-scales subulate, at least $\frac{1}{2}$ as long as wide part of scale, 5- to 7-veined at base (c) subsp. **toletanus**
1 Point of epicalyx-scales setaceous, $\frac{1}{4}$–$\frac{1}{3}$ as long as wide part of scale, 3-veined at base
2 Stems usually 10–20 cm; petal-limb 4–6 mm (a) subsp. **scaber**
2 Stems usually 20–30 cm; petal-limb 7–8 mm (b) subsp. **cutandae**

(a) Subsp. **scaber**: Stems usually 10–20 cm. Point of epicalyx-scales setaceous, $\frac{1}{4}$–$\frac{1}{3}$ as long as wide part of scale, 3-veined at base. Petal-limb 4–5 mm. *S.E. France; N.E. Spain.*

(b) Subsp. **cutandae** (Pau) Tutin, *Feddes Repert.* **68**: 190 (1963) (*D. cutandae* (Pau) Pau): Stems usually 20–30 mm. Epicalyx-scales as in subsp. (a). Petal-limb 7–8 mm. $2n = 30$. *C. & N.W. Spain.*

(c) Subsp. **toletanus** (Boiss. & Reuter) Tutin, *loc. cit.* (1963) (*D. toletanus* Boiss. & Reuter): Stems 15–40(–60) cm. Point of epicalyx-scales subulate, at least $\frac{1}{2}$ as long as wide part of scale, 5- to 7-veined at base. Petal-limb 5–8 mm. $2n = 30$. *C. & S.E. Spain; S. Portugal.*

77. D. crassipes R. de Roemer, *Linnaea* **25**: 11 (1852) (incl. *D. serenaeus* Coincy). Robust perennial with a stout, woody stock; stems 30–60 cm, puberulent, terete. Cauline leaves 2–2·5 mm wide, rigid, acute, puberulent; sheaths as long as diameter of stem. Flowers usually 20–30, subsessile in dense corymbose heads; bracts broadly scarious, with a green point, about as long as flowers. Epicalyx-scales 4, little shorter than calyx, lanceolate, similar to bracts but more abruptly and finely acuminate. Calyx 18–25 mm, narrowed above. Petal-limb 7–9 mm, dentate, bearded, purple. ● *S.C. Spain.* Hs.

78. D. cintranus Boiss. & Reuter, *Pugillus* 20 (1852). Glabrous perennial 10–50 cm, with trailing, woody stems. Leaves 0·5–2(–5) mm wide, flat, acute or acuminate, the lower sometimes obtuse. Inflorescence simple or branched. Epicalyx-scales 4–6, $\frac{1}{3}$–$\frac{1}{2}$ as long as calyx, ovate, cuspidate to acuminate. Calyx 17–28 × 4·5–6 mm, not purple-tinged. Petal-limb 4–12 mm, obovate, dentate, scarcely bearded to glabrous. *Spain and Portugal.* Hs Lu.

1 Stems 15–20 cm, usually simple; epicalyx-scales often 6 (d) subsp. **charidemii**
1 Stems usually more than 20 cm, branched; epicalyx-scales usually 4
2 Petal-limb usually less than 5 mm (b) subsp. **multiceps**
2 Petal-limb usually more than 5 mm
3 Petal-limb 5–8 mm wide, glabrous (a) subsp. **cintranus**
3 Petal-limb 3–4 mm wide, bearded (c) subsp. **barbatus**

(a) Subsp. **cintranus** (incl. *D. gaditanus* Boiss.): Stems usually more than 20 cm, branched. Epicalyx-scales usually 4, gradually contracted to an apex 2–3 mm wide at base. Petal-limb 7–12 × 5–8 mm; claw usually included. $2n = 60$. *C. Portugal, S. & N.W. Spain.*

(**b**) Subsp. **multiceps** (Costa ex Willk.) Tutin, *Feddes Repert.* **68**: 190 (1963) (*D. multiceps* Costa ex Willk.): Like subsp. (**a**) but petal-limb usually less than 5 mm, the claw often exserted. ● *N.E. Spain.*

(**c**) Subsp. **barbatus** R. Fernandes & Franco in Franco, *Nova Fl. Portugal* **1**: 159 (1971): Like subsp. (**a**) but leaves usually narrower; epicalyx-scales abruptly contracted to apex, 0·5–1 mm wide at base; petal-limb 5–10 × 3–4 mm, bearded. *Calcicole.* ● *N. & C. Portugal.*

(**d**) Subsp. **charidemii** (Pau) Tutin, *Feddes Repert.* **68**: 190 (1963) (*D. charidemii* Pau): Stems 15–20 cm, usually simple. Epicalyx-scales 4–6. Petal-limb usually more than 5 mm; claw usually exserted. ● *S.E. Spain (Cabo de Gata).*

D. anticarius Boiss. & Reuter, *Pugillus* 19 (1852), from S. Spain, usually has a simple inflorescence, but otherwise resembles **78(a)**, and is probably best included in this subspecies.

79. D. petraeus Waldst. & Kit., *Pl. Rar. Hung.* **3**: 246 (1808) (incl. *D. nicolai* G. Beck & Szysz., *D. prenjus* G. Beck, *D. strictus* auct., non Sibth. & Sm., nec Banks & Solander, *D. suendermannii* Bornm.). Green or somewhat glaucous, usually laxly caespitose perennial up to *c.* 30 cm. Leaves acuminate, tapering from near the middle; cauline leaves usually 2–5 pairs. Epicalyx-scales elliptical to ovate, acuminate or cuspidate, $\frac{1}{5}$–$\frac{1}{3}$ as long as calyx. Calyx (12–)20–32 × (2–)3–4 mm, tapering from below the middle; teeth narrowly triangular. Petal-limb 4–10 mm, glabrous or somewhat bearded, laciniate, dentate or subentire, white. Capsule more or less equalling or slightly longer than calyx. ● *Balkan peninsula, S.W. & C. Romania.* Al Bu Gr Ju Rm.

A species which shows great variability in size, division and pubescence of the petals, though the long, narrow, tapering calyx in particular is characteristic. At least three subspecies may be recognized:

1 Epicalyx-scales narrowed into an acute or acuminate herbaceous apex (a) subsp. **petraeus**
1 Epicalyx-scales shortly cuspidate
 2 Petal-limb *c.* 6 mm, obovate, dentate to subentire (b) subsp. **orbelicus**
 2 Petal-limb *c.* 8 mm, elliptical, laciniate (c) subsp. **noaeanus**

(**a**) Subsp. **petraeus** (*D. bebius* Vis., *D. kitaibelii* Janka, *D. noaeanus* auct. jugosl., non Boiss.): Epicalyx-scales 4–6(–8), narrowed into an acute or acuminate herbaceous apex. Petal-limb obovate, dentate to laciniate. 2*n* = 30. *Almost throughout the range of the species.*

(**b**) Subsp. **orbelicus** (Velen.) Greuter & Burdet, *Willdenowia* **12**: 187 (1982) (*D. strictus* subsp. *orbelicus* Velen.; incl. *D. stefanoffii* Eig, *D. strictus* Sibth. & Sm., non Banks & Solander, *D. petraeus* subsp. *simonkaianus* (Péterfi) Tutin, *D. simonkaianus* Péterfi): Epicalyx-scales 4–6(–8), shortly cuspidate. Petal-limb *c.* 6 mm, obovate, dentate or subentire. 2*n* = 30. *From N.C. Romania to N.E. Greece; S.W. & C. Bulgaria.*

(**c**) Subsp. **noaeanus** (Boiss.) Tutin, *Feddes Repert.* **68**: 190 (1963) (*D. noaeanus* Boiss.): Densely caespitose; leaves more pungent than those of subspp. (**a**) and (**b**). Epicalyx-scales 6–8, shortly cuspidate. Petal-limb *c.* 8 mm, elliptical, laciniate. 2*n* = 30. *Bulgaria and E. Jugoslavia; one station in N. Greece.*

80. D. integer Vis., *Flora (Regensb.)* **12**, *Erg.-Bl.* **1**: 11 (1829). Like **79** but epicalyx-scales 2–4, at least $\frac{1}{3}$ as long as calyx; calyx 8–20 × 3–4·5 mm, cylindrical to narrowly ovoid, usually ringed with purple; calyx-teeth ovate-triangular; petal-limb entire or subentire; capsule exceeding calyx. ● *Mountains of S. & W. parts of Balkan peninsula.*

(**a**) Subsp. **integer**: Epicalyx-scales acute to acuminate. Calyx 8–14 mm; teeth acute. Petal-limb pale pink. 2*n* = 60. *W. Jugoslavia, Albania.*

(**b**) Subsp. **minutiflorus** (Halácsy) Bornm., *Bot. Jahrb.* **49**: 388 (1935) (*D. minutiflorus* Halácsy, *D. petraeus* subsp. *minutiflorus* (Halácsy) Greuter & Burdet, *D. strictus* var. *brachyanthus* (Boiss.) Boiss.): Epicalyx-scales cuspidate. Calyx 12–20 mm; teeth obtuse. Petal-limb white. 2*n* = 60, 90. *From N. Albania to S. Greece.*

81. D. graniticus Jordan, *Obs. Pl. Crit.* **7**: 13 (1849). Like **80** but stems somewhat hairy below, with 6 or 7 cauline leaves; calyx 10–15 mm, cylindrical; petal-limb toothed, bearded, reddish-purple. 2*n* = 30. *Calcifuge.* ● *S.C. France.*

82. D. spiculifolius Schur, *Enum. Pl. Transs.* 98 (1866). Like **80** but petals pink, deeply laciniate, always bearded, with limb 10–15 mm. 2*n* = 90. *Calcicole.* ● *E. Carpathians.* Rm Rs(W).

83. D. plumarius L., *Sp. Pl.* 411 (1753) (*D. blandus* (Reichenb.) Hayek, *D. hoppei* Portenschl., *D. neilreichii* Hayek). A more or less glaucous perennial up to 40 cm; flowering stems with 1–4(–10) pairs of leaves. Leaves *c.* 1 mm wide, narrowed only in the upper part, acute or subacute. Flowers solitary (rarely more in wild plants), fragrant. Epicalyx-scales (2–)4(–6), obovate, often almost truncate, shortly apiculate, *c.* $\frac{1}{4}$ as long as calyx. Calyx 17–30 × 4–6 mm, almost cylindrical, green or slightly purple; teeth lanceolate to ovate, obtuse or apiculate; margin broadly scarious. Petal-limb 12–18 mm, divided to about the middle into narrow lobes, white or bright pink, usually bearded. ● *Calcareous mountains of E.C. Europe; widely cultivated for ornament, and locally naturalized.* Au Cz Hu Ju Po [Br Ge *It].

Many hybrids between this and **65** are grown in gardens. A number of isolated and slightly different populations have been given specific or subspecific rank by some authors. The most distinct of these are **D. hungaricus** Pers., *Syn. Pl.* **1**: 494 (1805), from the Tatra mountains, with broadly scarious epicalyx-scales, and violet filaments, and **D. lumnitzeri** Wiesb., *Bot. Centr.* **26**: 85 (1886), from E. Austria (Hainburger Berge) to N.E. Hungary (Bükk Hegyseg), with a narrow, rather attenuate calyx.

84. D. moravicus Kovanda, *Preslia* **54**: 223 (1982). Like **83** but leaves linear to narrowly triangular; calyx 15–20(–24) mm, the teeth acute to acuminate, usually not broadly scarious; petal-limb 12–15(–17) mm, pink to red. 2*n* = 60, 90. ● *S. Czechoslovakia (Moravia).*

85. D. serotinus Waldst. & Kit., *Pl. Rar. Hung.* **2**: 188 (1804). Like **83** but flowering stems usually more slender, with (6–)8–14(–19) pairs of leaves; leaves narrowed gradually from near base; flowers often 1–5; calyx 20–28 × 2·5–4 mm, tapering upwards from below the middle; teeth lanceolate, apiculate; petal-limb usually cream. 2*n* = 90. *Sandy ground.* ● *E.C. Europe.* Cz Hu Ju Rm.

86. D. acicularis Fischer ex Ledeb., *Fl. Ross.* **1**: 284 (1842). Caespitose perennial up to 30 cm; flowering stems quadrangular, with 2–4 pairs of leaves. Leaves *c.* 0·5 mm wide, narrowly linear, acuminate. Epicalyx-scales 4, obovate, obtusely mucronate, *c.* $\frac{1}{4}$ as long as calyx. Calyx 20–25 × 3–4 mm, tapering from the middle. Petal-limb 10–15 mm, laciniate; central part oblong, bearded. *C. & S. Ural, extending westwards to near Saratov.* Rs(C, E).

87. D. squarrosus Bieb., *Fl. Taur.-Cauc.* **1**: 331 (1808). Laxly caespitose perennial 15–30 cm. Leaves 10–20 mm, not crowded at base of flowering stems, linear, acuminate, recurved. Inflorescence branched. Epicalyx-scales 4, obovate, cuspidate, *c.* $\frac{1}{5}$ as long as calyx, the outer much smaller than the inner. Calyx *c.* 25 × 3 mm, tapering from below the middle. Petal-limb *c.* 8 mm, deeply laciniate, bearded, white. *E. Europe from C. Ukraine to W. Kazakhstan.* Rs(C, W, E).

88. D. arenarius L., *Sp. Pl.* 412 (1753). Caespitose perennial up to *c.* 45 cm; stems slender, subcylindrical, simple or branched. Leaves frequently less than 1 mm wide, linear, obtuse or acute. Flowers usually solitary, fragrant. Epicalyx-scales 2–4, broadly ovate to obovate, up to $\frac{1}{4}$ as long as calyx, often with a short, obtuse apiculus. Calyx 16–30 × 2–4·5 mm, nearly cylindrical. Petal-limb 10–15 mm, laciniate to beyond the middle, white, with a green spot, often with a purplish margin at base, bearded. • *E. & E.C. Europe, southwards to 46° N. in N. Jugoslavia, and extending westwards to Czechoslovakia, E. Germany and Sweden.* Au Cz Fe Ge Hu Ju Po Rm Rs(N, B, C, W, E) Su.

1 Stems usually 10–20 cm, simple or little-branched (a) subsp. **arenarius**
1 Stems usually more than 20 cm, freely branched
 2 Plant not glaucous; leaves usually 0·5 mm wide; petal-limb usually *c.* 10 mm (b) subsp. **borussicus**
 2 Plant glaucous; leaves usually *c.* 1 mm wide; petal-limb usually *c.* 15 mm
  3 Stem with 1–3 flowers; epicalyx-scales 4; calyx usually 4–4·5 mm wide (c) subsp. **pseudoserotinus**
  3 Stem with (1–)3–24 flowers; epicalyx-scales 4–6; calyx usually 3–4 mm wide (d) subsp. **pseudosquarrosus**

(a) Subsp. **arenarius**: Caespitose, usually not glaucous, the stems 10–20 cm, simple or little branched. Lower leaves usually obtuse. Calyx *c.* 20 mm. $2n = 60$. *S. Sweden.*

(b) Subsp. **borussicus** Vierh. in Fritsch, *Sched. Fl. Exsicc. Austr.-Hung.* **9**: 15 (1902) (*D. borussicus* (Vierh.) Juz.): Laxly caespitose, not glaucous, the stems usually more than 20 cm, flexuous. Lower leaves short, acute or subacute. Calyx 20–25 × 3 mm. *Almost throughout the range of the species.*

(c) Subsp. **pseudoserotinus** (Błocki) Tutin, *Feddes Repert.* **68**: 190 (1963) (*D. pseudoserotinus* Błocki): Laxly caespitose, glaucous; stems straight. Lower leaves long, acute or acuminate. Calyx *c.* 20 × 4–4·5 mm. *W. Ukraine.*

(d) Subsp. **pseudosquarrosus** (Novák) Kleopow, *Bull. Jard. Bot. Kieff* **12–13**: 35 (1931) (*D. pseudosquarrosus* Novák): Laxly caespitose, glaucous. Lower leaves long, acuminate, recurved. Calyx usually 22–30 × 3–4 mm. *White Russia and C. Ukraine.*

89. D. krylovianus Juz., *Not. Syst.* (*Leningrad*) **13**: 71 (1950). Caespitose perennial; stems 12–28 cm, branched, subcylindrical. Leaves up to 1·5 mm wide, often subobtuse. Epicalyx-scales 4–6, broadly ovate or obovate, $\frac{1}{6}$–$\frac{1}{4}$ as long as calyx, shortly mucronate. Calyx 17–20 × 2–3 mm, nearly cylindrical. Petal-limb 8–15 mm, laciniate usually to less than $\frac{1}{4}$, white, bearded. • *E.C. Russia (N. & W. of Kazan').* Rs(C).

90. D. volgicus Juz., *op. cit.* 73 (1950). Like **89** but laxly caespitose; epicalyx-scales 4, acute or acuminate, $\frac{1}{4}$–$\frac{1}{3}$ as long as calyx; calyx *c.* 25 mm; petal-limb less than $\frac{1}{2}$ as long as calyx. • *Middle Volga (N. & W. of Kujbyšev).* Rs(C, E).

91. D. gallicus Pers., *Syn. Pl.* **1**: 495 (1805). Laxly caespitose perennial up to 50 cm. Leaves *c.* 10–15 × 1·5–3 mm, obtuse or subacute, the cauline usually 6–10 pairs. Flowers 1–3 together, fragrant. Epicalyx-scales 4, broadly obovate, cuspidate, or the outer narrower, mostly green, *c.* $\frac{1}{3}$ as long as calyx. Calyx 20–25 × 3–4 mm, slightly tapering. Petal-limb 10–15 mm, laciniate, bearded, pink. $2n = 60$. *Maritime sands.* • *Coasts of W. France and N. Spain.* Ga Hs.

92. D. superbus L., *Fl. Suec.* ed. 2, 146 (1755). Perennial up to 90 cm; stems often decumbent below, branched above. Leaves linear-lanceolate. Flowers usually solitary or paired, fragrant. Epicalyx-scales 2–4, ovate, acuminate or shortly awned, $\frac{1}{4}$–$\frac{1}{3}$ as long as calyx. Calyx 15–30 × 3–6 mm, narrowed upwards. Petal-limb (6–)15–30 mm, laciniate to more than $\frac{1}{4}$, pink or purplish, rarely white, bearded. *Europe, except much of the west and south.* Au Bu Cz Da Fe Ga Ge Gr He †Ho Hu It Ju No Po Rm Rs(N, B, C, W, E) Su.

1 Calyx *c.* 24 × 3 mm, green or somewhat purplish (c) subsp. **stenocalyx**
1 Calyx 15–30 × 4–6 mm, purple or violet
 2 Not or slightly glaucous; petal-limb *c.* 20 mm (a) subsp. **superbus**
 2 Glaucous; petal-limb *c.* 30 mm (b) subsp. **alpestris**

(a) Subsp. **superbus**: Plant not or slightly glaucous. Calyx 15–30 × 4–6 mm, purple or violet. Petal-limb *c.* 20 mm. *Throughout the range of the species; lowland.*

(b) Subsp. **alpestris** Kablik ex Čelak., *Prodr. Fl. Böhm.* 508 (1881) (subsp. *speciosus* (Reichenb.) Pawł., *D. speciosus* Reichenb.): Like subsp. (a) but glaucous; petal-limb *c.* 30 mm. $2n = 30$. • *Higher mountains of Europe, from the S.W. Alps to the E. Carpathians.*

(c) Subsp. **stenocalyx** (Trautv. ex Juz.) Kleopow, *Bull. Jard. Bot. Kieff* **14**: 137 (1932) (*D. stenocalyx* Trautv. ex Juz.): Plant not glaucous. Calyx 20–25 × 3 mm, green or somewhat purplish. Petal-limb *c.* 30 mm. *S.C. Russia, N. Ukraine.*

93. D. deltoides L., *Sp. Pl.* 411 (1753). Subcaespitose green or glaucous perennial; stems puberulent. Flowers usually solitary. Epicalyx-scales 2(–4), ovate, *c.* $\frac{1}{2}$ as long as calyx, narrowed to a subulate apex, herbaceous, with a scarious margin. Epicalyx-scales and calyx puberulent. Calyx nearly cylindrical. Petal-limb 6–9 mm, bearded. *Most of Europe, but rare in parts of the south.* Al Au Be Br Bu Cz Da Fe Ga Ge Gr He Ho Hs Hu It Ju No Po Rm Rs(N, B, C, W, E) Si Su.

(a) Subsp. **deltoides**: Leaves of non-flowering stems 8–14 mm, broadly linear to narrowly elliptical, shortly pubescent on margins. Flowering stems 15–45 cm, with 5–10 pairs of linear cauline leaves. Epicalyx-scales puberulent. Calyx 14–18 mm. Petal-limb deep pink or rarely white, with a darker basal band and pale spots. $2n = 30$. *Throughout the range of the species.*

(b) Subsp. **degenii** (Bald.) Strid, *Willdenowia* **13**: 280 (1983) (*D. degenii* Bald.): Leaves of non-flowering stems 5–10 mm, elliptical, usually glabrous. Flowering stems 8–15 cm, usually with 4 pairs of narrowly oblong cauline leaves. Epicalyx-scales glabrous. Calyx 8–12 mm. Petal-limb with a purple spot near base. • *Mountains of W. part of Balkan peninsula.*

94. D. myrtinervius Griseb., *Spicil. Fl. Rumel.* **1**: 194 (1843) (incl. *D. oxylepis* (Boiss.) Kümmerle & Jáv., *D. kajmaktzalanicus* Micevski). Procumbent or caespitose, densely leafy perennial. Leaves 2–5 mm, ellliptical, longer than internodes. Flowers solitary, on very short pedicels. Epicalyx-scales 2–4, the outer usually leaf-like, at least $\frac{2}{3}$ as long as calyx. Calyx 5–9 mm, campanulate, herbaceous. Petal-limb 3–5 mm, bearded, pink. • *Mountains of W. Macedonia.* Gr Ju.

95. D. sphacioticus Boiss. & Heldr. in Boiss., *Diagn. Pl. Or. Nov.* **2(8)**: 70 (1849). Perennial 2–10(–15) cm with a stout, woody stock. Leaves 3–6 mm, oblong to oblong-lanceolate, those of the flowering stems shorter than internodes. Flowers solitary. Epicalyx-scales usually 6, *c.* $\frac{1}{3}$ as long as calyx; outer ovate and leaf-like; inner obovate, cuspidate. Calyx 14–18 mm, tapering from below the middle. Petal-limb 3–4 mm, obovate, subentire, bearded. • *W. Kriti (Levka Ori).* Cr.

96. D. pallidiflorus Ser. in DC., *Prodr.* **1**: 358 (1824) (*D. maeoticus* Klokov; incl. *D. aridus* Griseb. ex Janka). Freely-branched perennial up to 50 cm, with a stout stock. Basal leaves withered at anthesis; cauline 3·5–6 cm, usually 6 or 7 pairs, flat, rather soft. Flowers usually long-pedicellate. Epicalyx-scales usually 4, $\frac{1}{3}$–$\frac{3}{4}$ as long as calyx, ovate with scarious margins, shortly aristate. Calyx 10–14 × 2 mm, tapering above the middle. Petal-limb 4–6 mm, dentate, bearded, white or pale pink. *S.E. Europe, from S. Bulgaria to S.E. Russia.* Bu ?Gr Rs(?C, W, K, E).

97. D. campestris Bieb., *Fl. Taur.-Cauc.* **1**: 326 (1808) (*D. serbanii* Prodan). Usually puberulent perennial with a stout stock and numerous stems 20–60 cm. Leaves 1–5 mm wide, the lower withered at anthesis; sheaths 2–4 mm. Flowers solitary or in pairs, terminating all the branches. Epicalyx-scales 4–6, *c.* $\frac{1}{3}$ as long as calyx, ovate, acuminate, with scarious margins. Calyx 15–18 × 3–4 mm, nearly cylindrical. Petal-limb 6–7 mm, dentate, bearded, pink or purplish above, greenish-yellow beneath. *S.E. Europe from W. Ukraine to S. Ural.* Rm Rs(C, W, K, E).

1 Leaves not more than 2·5 mm wide; flowers *c.* 20 mm in diameter (a) subsp. **campestris**
1 Leaves 3–5 mm wide; flowers less than 20 mm in diameter
 2 Epicalyx-scales with a point up to 0·5 mm (b) subsp. **laevigatus**
 2 Epicalyx-scales with a point at least 1 mm (c) subsp. **steppaceus**

(a) Subsp. **campestris**: Plant up to 30 cm, puberulent. Branches of inflorescence long; flowers *c.* 20 mm in diameter. $2n = 30$. *Throughout the range of the species.*

(b) Subsp. **laevigatus** (Gruner) Klokov in Kotov, *Fl. RSS Ucr.* **4**: 625 (1952): Plant 40–60 cm, glabrous. Branches of inflorescence long; flowers less than 20 mm in diameter. • *Dnepr and Dnestr valleys.*

(c) Subsp. **steppaceus** Širj. in Širj. & Lavrenko, *Consp. Crit. Fl. Prov. Charkov.* **1** (1926): Plant up to 50 cm, glabrous or puberulent. Branches of inflorescence short; flowers less than 20 mm in diameter. *E. Ukraine.*

98. D. hypanicus Andrz., *Trudy Kommiss. Kiev. Učebn. Okr.* **4(1)**: 18 (1860). Like **97** but stems 10–20(–25) cm, slender, with numerous short, non-flowering branches; leaves 1–2 mm wide; sheaths up to 1·5 mm; flowers solitary; epicalyx-scales usually $\frac{1}{4}$ as long as calyx. • *S.C. Ukraine.* Rs(W).

99. D. carbonatus Klokov, *Sci. Mag. Biol. (Kharkov)* **1927**: 15 (1927). Like **97** but leaves rarely as much as 1 mm wide, almost setaceous; sheaths 1–3 mm; flowers often 2 or 3 together; epicalyx-scales usually $\frac{1}{2}$ as long as calyx which is usually 2·5 mm wide. • *Moldavia, S. Ukraine.* Rs(W, K).

100. D. marschallii Schischkin, *Jour. Governm. Bot. Gard. Nikita* **10(2)**: 39 (1928) (*D. bicolor* Bieb.). Perennial 20–30 cm, with a woody stock and no basal leaves at anthesis; stems usually with long branches. Cauline leaves usually 3–5 pairs, linear, acuminate. Epicalyx-scales ovate, cuspidate, *c.* $\frac{1}{3}$ as long as calyx. Calyx *c.* 15 × 3 mm, narrowed in upper $\frac{1}{2}$; teeth triangular. Petal-limb 7–10 mm, glabrous, subentire, yellowish-white above, pink beneath. • *Moldavia to Krym.* Rs(W, K).

101. D. cinnamomeus Sm. in Sibth. & Sm., *Fl. Graec. Prodr.* **1**: 287 (1809). Perennial 15–30 cm, with a stout stock; stem terete, more or less puberulent. Lower leaves usually obtuse, the basal often withered at anthesis. Epicalyx-scales obovate, apiculate to cuspidate, $\frac{1}{4}$–$\frac{1}{3}$ as long as calyx. Calyx *c.* 18 × 4 mm; teeth oblong-lanceolate. Petal-limb *c.* 3 mm, glabrous, cream above, cinnamon beneath. *Islands of S. Aegean region.* Cr Gr.

102. D. monadelphus Vent., *Descr. Pl. Jard. Cels* t. 39 (1801) (incl. *D. rhodopaeus* Velen.). Like **101** but up to 50 cm, usually pruinose; epicalyx-scales ovate, acute or acuminate; teeth lanceolate, acuminate; petal-limb greenish beneath. *S. & E. parts of Balkan peninsula; S.E. Romania.* Bu Gr Rm Ju Tu.

(a) Subsp. **monadelphus**: Epicalyx-scales $\frac{1}{2}$ as long as calyx, their apices patent, subulate. *Turkey-in-Europe (near Istanbul). (S.W. Asia.)*

(b) Subsp. **pallens** (Sm.) Greuter & Burdet, *Willdenowia* **12**: 187 (1982) (*D. pallens* Sm.): Epicalyx-scales $\frac{1}{4}$–$\frac{1}{3}$ as long as calyx, apiculate. $2n = 30$. *Almost throughout the range of the species.*

103. D. lanceolatus Steven ex Reichenb., *Pl. Crit.* **6**: 34 (1828) (*D. leptopetalus* auct., non Willd.). Robust perennial up to *c.* 50 cm, with a woody stock and no basal leaves at anthesis; stems usually branched above. Cauline leaves usually 5–7 pairs, linear, acuminate, thin; sheaths 2–3 times as long as diameter of stem. Epicalyx-scales 4, $\frac{1}{4}$–$\frac{1}{3}$ as long as calyx; outer ovate, abruptly acuminate; inner broadly obovate, cuspidate. Calyx 25–30 × 4–5 mm, narrowed in the upper $\frac{1}{2}$. Petal-limb *c.* 10 mm, dentate, whitish, glabrous. • *S.E. Russia, S. Ukraine.* Rs(C, W, K, E).

104. D. leptopetalus Willd., *Enum. Pl. Horti Berol.* 468 (1809) (*D. rhodopaeus* Velen.). Like **103** but calyx 20–28 × 3–4 mm, tapering markedly from below the middle; petal-limb narrowly rhombic, widest just above the middle, purplish beneath. $2n = 30$. *S.E. Europe, from Turkey-in-Europe to S. Ural.* ?Gr Rm Rs(C, W, E) Tu.

105. D. roseoluteus Velen., *Österr. Bot. Zeitschr.* **36**: 226 (1886) (*D. purpureoluteus* Velen.). Pubescent, branched perennial 30–50 cm. Basal leaves withered at anthesis; cauline 20–30 × 2–3 mm, usually 5–10 pairs, at least as long as internodes, flat, acuminate; sheaths about as long as diameter of stem. Epicalyx-scales usually 6; outer leaf-like and about as long as calyx; inner ovate, with a subulate point. Calyx 18–22 × *c.* 4 mm, tapering from about the middle; teeth usually 5–8 mm. Petal-limb *c.* 7 mm, dentate, bearded, pink above, yellowish beneath. *N.E. part of Balkan peninsula.* Bu Gr Tu.

106. D. gracilis Sm. in Sibth. & Sm., *Fl. Graec. Prodr.* **1**: 288 (1809). Glabrous perennial 15–40 cm; stems simple or branched. Basal leaves *c.* 1 mm wide, often absent at anthesis; cauline 4–6 pairs, shorter than internodes, flat, acuminate; sheaths *c.* 3 times as long as diameter of stem. Epicalyx-scales 4–6, $\frac{1}{4}$–$\frac{3}{4}$ as long as calyx, obovate, cuspidate or, rarely, gradually tapering from about the middle. Calyx (10–)13–18 mm, cylindrical or tapering slightly from below the middle. Petal-limb 5–10 mm, dentate, bearded with long hairs, deep pink above, yellow or purplish beneath. $2n = 60$. • *Balkan peninsula from S.C. Jugoslavia to S. Greece.* Al Bu Gr Ju.

1 Point of epicalyx-scales less than ½ as long as wide part of scale
2 Epicalyx-scales usually 4; flowers mostly solitary; pedicels long (a) subsp. **gracilis**
2 Epicalyx-scales usually 6; flowers mostly in groups of 2 or more; pedicels very short (b) subsp. **armerioides**
1 Point of epicalyx-scales more than ½ as long as wide part of scale
3 Basal leaves usually absent at flowering; cauline leaves soft; epicalyx-scales *c.* ½ as long as calyx (a) subsp. **gracilis**
3 Basal leaves usually present at flowering; cauline leaves rigid; epicalyx-scales *c.* ¾ as long as calyx (c) subsp. **drenowskianus**

(a) Subsp. **gracilis** (*D. athous* Rech. fil.; incl. subsp. *xanthianus* (Davidov) Tutin, *D. xanthianus* Davidov): *Almost throughout the range of the species.*

(b) Subsp. **armerioides** (Griseb.) Tutin, *Feddes Repert.* **68**: 191 (1963) (*D. gracilis* var. *armerioides* Griseb., *D. suskalovicii* Adamović, *D. albanicus* Wettst.; incl. subsp. *achtarovii* (Stoj. & Kitanov) Tutin, subsp. *friwaldskyanus* (Boiss.) Tutin): *N. Macedonia, Albania.*

(c) Subsp. **drenowskianus** (Rech. fil.) Strid, *Willdenowia* **13**: 281 (1983) (*D. drenowskianus* Rech. fil.): *N.E. Greece, S.W. Bulgaria.*

D. simulans Stoj. & Stefanov, *Magyar Bot. Lapok* **32**: 1 (1933), from E. Macedonia (Orvilos), densely caespitose with stems 2–6 cm, flowers solitary, and calyx 9–13 mm, may represent a variant of **106(c)**.

107. D. haematocalyx Boiss. & Heldr. in Boiss., *Diagn. Pl. Or. Nov.* **3(1)**: 65 (1853). Caespitose, glabrous perennial 1–30 cm. Leaves linear to linear-lanceolate, acute or obtuse. Epicalyx-scales 4–6, ovate-lanceolate, with a long, more or less patent apex. Calyx 16–26 × 4–7 mm, narrowed upwards; teeth acuminate. Petal-limb 6–12 mm, bearded, usually deep pinkish-purple above, yellow beneath. • *S. part of Balkan peninsula, mostly in the mountains.* Al Gr Ju.

1 Epicalyx-scales tapering gradually towards apex; stems up to 30 cm, usually branched
2 Plant not glaucous; leaves not markedly rigid; epicalyx-scales about as long as calyx (a) subsp. **haematocalyx**
2 Plant glaucous; leaves rigid; epicalyx-scales shorter than calyx (b) subsp. **pruinosus**
1 Epicalyx-scales abruptly contracted into a subulate apex; stems up to 10 cm, simple
3 Plant glaucescent; leaves linear (c) subsp. **sibthorpii**
3 Plant not glaucescent; leaves narrowly elliptical (d) subsp. **pindicola**

(a) Subsp. **haematocalyx**: Usually laxly caespitose; stems 5–25 cm. Leaves 15–40 mm, linear to linear-lanceolate, not glaucous. Flowers 1–5. Epicalyx-scales tapering gradually towards apex, about as long as calyx. Calyx 18–24 mm. $2n = 30$. *S. Jugoslavia, N. Greece.*

(b) Subsp. **pruinosus** (Boiss. & Orph.) Hayek, *Prodr. Fl. Penins. Balcan.* **1**: 240 (1924) (*D. pruinosus* Boiss. & Orph.).: Like subsp. (a) but plant glaucous; leaves rigid; epicalyx-scales shorter than calyx. *E. Greece (near Volos).*

(c) Subsp. **sibthorpii** (Vierh.) Hayek, *loc. cit.* (1924) (*D. ventricosus* Halácsy, nom. illeg., *D. sibthorpii* Vierh.): Densely caespitose; stems 1–4(–8) cm. Leaves up to 15 mm, linear, somewhat glaucous. Flowers solitary. Epicalyx-scales abruptly contracted into a subulate apex, shorter than calyx. Calyx 12–17 mm. Petals lilac-pink. $2n = 30$. *C. Greece.*

(d) Subsp. **pindicola** (Vierh.) Hayek, *loc. cit.* (1924) (*D. pindicola* Vierh.): More or less densely caespitose; stems 1–6(–10) cm. Leaves up to 15 mm, narrowly elliptical. Flowers solitary. Epicalyx-scales abruptly contracted into a subulate apex, shorter than calyx. Calyx 12–19 mm. *Often on serpentine. N.W. Greece; S. Albania.*

D. kapinaensis Markgraf & Lindt., *Glasn. Skopsk. Naučn. Društva* **18**: 125 (1938), from Jugoslavia, has been compared with **106(b)** and **107(b)**, but requires further study.

108. D. biflorus Sm. in Sibth. & Sm., *Fl. Graec. Prodr.* **1**: 285 (1809) (*D. cinnabarinus* Spruner ex Boiss.; incl. *D. samaritanii* Heldr. ex Halácsy, *D. mercurii* Heldr.). Laxly caespitose perennial 15–60 cm. Cauline leaves 3–6 pairs, acuminate; sheaths usually several times as long as diameter of stem. Flowers solitary or few in a head. Epicalyx-scales usually 6, *c.* ⅓–½ as long as, rarely as long as, calyx, obovate, aristate, coriaceous. Calyx 16–25 × 4–5 mm, tapering from about the middle. Petal-limb 6–15 mm, dentate, usually with numerous very short hairs and glands all over the upper surface. $2n = 30$. *Mountain rocks.* • *Greece, northwards to* 39°50′ *N.* Gr.

109. D. strictus Banks & Solander in A. Russell, *Nat. Hist. Aleppo* ed. 2, **2**: 252 (1794) (*D. multipunctatus* Ser.). Glabrous perennial 30–50 cm, with woody stock and simple or branched stems. Basal leaves withered at anthesis; cauline 1–3 mm wide, acuminate, 6–12 pairs, about as long as internodes. Epicalyx-scales ⅓ as long as calyx, ovate, cuspidate, broadly scarious. Calyx 15–18 mm, conical, verruculose, unribbed but with purple lines on the teeth continued downwards from sinuses. Petal-limb *c.* 10 mm, dentate, bearded. $2n = 30$. *W. Kriti.* Cr. (*S.W. Asia.*)

European plants belong to subsp. **multipunctatus** (Ser.) Greuter & Burdet, *Willdenowia* **12**: 187 (1982).

110. D. tripunctatus Sm. in Sibth. & Sm., *Fl. Graec. Prodr.* **1**: 286 (1809). Divaricately-branched, glabrous annual 15–40 cm. Basal leaves usually withered at anthesis, 3–8 mm wide, oblong-spathulate, obtuse; cauline leaves linear, acuminate. Epicalyx-scales *c.* ¾ as long as calyx, ovate with a long, usually green, subulate apex. Calyx *c.* 15 mm, conical, broadly ribbed and verruculose on sides of ribs, glabrous. Petal-limb *c.* 10 mm, dentate, bearded, with 3 red spots at base. *S. Greece and Aegean region; Italy (Calabria, Elba).* Cr Gr It [Lu].

(111–114). D. corymbosus group. Pubescent, often glandular annuals to perennials, usually ascending to erect. Basal leaves mostly withered at anthesis. Flowers solitary or in few-flowered clusters. Calyx 10–22 mm. Petal-limb usually bearded, pink or pinkish-purple.

A difficult group in the mountains of the Balkans and the Aegean Islands; for further details, see A. Strid, *Mount. Fl. Gr.*, **1**: 191–193 (1986).

1 Stems 4–10 cm, procumbent; epicalyx-scales almost as long as calyx **113. tymphresteus**
1 Stems usually more than 10 cm, ascending to erect; epicalyx-scales ½–¾ as long as calyx

2 Annual or biennial; calyx 2·5–3 mm wide **111. corymbosus**
2 Perennial; calyx 3–4 mm wide
3 Flowers usually in clusters of 2–7 **112. viscidus**
3 Flowers usually solitary **114. pubescens**

111. D. corymbosus Sm. in Sibth. & Sm., *Fl. Graec. Prodr.* **1**: 285 (1809) (*D. chalcidicus* Halácsy, *D. viscidus* sensu Hayek pro parte, non Bory & Chaub.). Annual or biennial 10–45 cm; upper cauline leaves linear, acuminate. Flowers usually in clusters of 2 or 3. Epicalyx-scales $\frac{1}{2}$–$\frac{3}{4}$ as long as calyx, ovate-lanceolate with a green apex usually about as long as broad part of scale. Calyx 10–17(–20) × 2·5–3 mm, usually glandular-pubescent, almost cylindrical. Petal-limb 7–10 mm, dentate, bearded, unspotted. $2n = 30$. ● *N. & N.C. Greece.* Gr.

112. D. viscidus Bory & Chaub., *Nouv. Fl. Pélop.* 26 (1838) (*D. corymbosus* auct., non Sibth. & Sm., *D. tenuiflorus* Griseb., *D. parnassicus* Boiss. & Heldr.). Like **111** but perennial; basal leaves acute; flowers usually in clusters of 2–7; epicalyx-scales more or less inflated, shiny, with a setaceous point; calyx 10–22 × 3–4 mm, glabrous or puberulent, tapering from about the middle; petal-limb 6–9 mm, usually with small, whitish spots. $2n = 30$. *Balkan peninsula.* Al Bu Gr Ju Tu.

113. D. tymphresteus (Boiss. & Spruner) Heldr. & Sart. ex Boiss., *Diagn. Pl. Nov. Or.* **2(6)**: 27 (1859). Like **111** but perennial; stems 4–10 cm, procumbent; basal leaves subobtuse; epicalyx-scales almost as long as calyx, broadly obovate, with a setaceous point; calyx 9–13 mm, glabrous or puberulent, tapering from about the middle; petal-limb glabrous, with purple spot at base. $2n = 30$. *Limestone.* ● *Mountains of C. Greece.* Gr.

114. D. pubescens Sm. in Sibth. & Sm., *Fl. Graec. Prodr.* **1**: 285 (1809) (incl. *D. diffusus* Sibth. & Sm., *D. glutinosus* Boiss. & Heldr.). Like **111** but perennial; flowers solitary, or 2 together but distinctly pedicellate; epicalyx-scales broadly ovate, with a wide scarious margin and setaceous point usually shorter than the wide part of the scale; calyx 18–22 × 3–4 mm, glabrous. $2n = 30$. *S. part of Balkan peninsula; Aegean region.* Gr Ju Tu.

115. D. formanekii Borbás ex Form., *Verh. Naturf. Ver. Brünn* **32**: 187 (1894). Glandular-viscid perennial up to *c.* 50 cm. Leaves 0·5–1 mm wide, linear, acute, flat; sheaths several times as long as diameter of stem. Flowers numerous, in small, dense heads; bracts ovate-oblong, shortly aristate. Epicalyx-scales like the bracts, $\frac{1}{2}$–$\frac{3}{4}$ as long as calyx. Calyx 6–7 × 1·5 mm, narrowed from below the middle. Petal-limb *c.* 2 mm, denticulate, purple. ● *Macedonia and Thessalia.* Gr Ju.

37. Velezia L.

A. O. CHATER

Annuals with rigid, dichotomously branched stems. Flowers shortly pedicellate, solitary or in clusters of 2 or 3 at the nodes. Epicalyx absent. Calyx narrowly tubular, 5-toothed, (5–)15-veined, without scarious commissures. Petals 5, with long claw and small limb; coronal scales small or absent; stamens 5 or 10; styles 2. Capsule cylindrical, very narrow, dehiscing with 4 teeth; carpophore absent. Seeds few, scutate, with facial hilum, usually with projecting radicle.

Limb of petal bifid; calyx linear, 5–6 times as long as pedicel; plant glandular-pubescent **1. rigida**
Limb of petal 4-toothed; calyx narrowly elliptical, 2–3 times as long as pedicel; plant sparsely glandular-pubescent or glabrous **2. quadridentata**

1. V. rigida L., *Sp. Pl.* 332 (1753). Glandular-pubescent; stems 5–15(–30) cm, procumbent or ascending, rigid, with many divaricate branches. Leaves 10–20 mm, linear-acuminate, the lower sometimes linear-subspathulate; all ciliate. Flowers solitary or paired, borne at most nodes. Calyx 10–14 mm, linear, 5–6 times as long as pedicel. Limb of petal bifid. $2n = 28$. *Dry, open habitats. S. Europe.* Al Bu †Co Cr Ga Gr Hs It Ju Lu Rs(K) Si Tu.

2. V. quadridentata Sm. in Sibth. & Sm., *Fl. Graec. Prodr.* **1**: 283 (1809). Like **1** but plant sparsely glandular-pubescent or glabrous; flowers solitary or 2 or 3 together; calyx 9–13 mm, slightly inflated near middle, narrowly elliptical, 2–3 times as long as pedicel; limb of petal 4-toothed. *Aegean region.* Gr Tu.

RANALES

58. NYMPHAEACEAE

EDIT. †T. G. TUTIN AND D. A. WEBB

Perennial, rhizomatous, aquatic herbs. Leaves broadly elliptic-ovate to orbicular, with a deep basal sinus, glabrous in European plants. Flowers solitary, hermaphrodite, actinomorphic. Perianth-segments free, numerous, the outer 4–6 more or less clearly differentiated as sepals, the remainder petaloid. Stamens numerous, the outer often petaloid. Carpels 8 or more; ovary superior to semi-inferior; ovules 1 to numerous, scattered all over the inner wall of the ovary. Fruit a rather dry berry.

All species grow in still or slowly flowing fresh water.

Flowers white, rarely pink or red; ovary semi-inferior **1. Nymphaea**
Flowers deep yellow; ovary superior **2. Nuphar**

1. Nymphaea L.

†T. G. TUTIN REVISED BY D. A. WEBB

Leaves all floating, often reddish beneath, stipulate; lateral veins anastomosing. Sepals green beneath. Petals white, rarely pink or red, the outermost longer than the sepals. Ovary semi-inferior.

Literature: H. S. Conard, *The Waterlilies.* Washington, D.C., 1905.

1 Sepals conspicuously white-veined; leaves spinous-dentate **4. lotus**
1 Sepals obscurely veined; leaves entire
2 Leaves 5–10 × 3–8 cm; flowers usually with not more than

12 petals; outermost stamens with fairly narrow filament, quite distinct from the petals **3. tetragona**

2 Leaves at least 10 × 7 cm; flowers with at least 12 petals; outermost stamens with very broad filament, showing a gradual transition to the petals

3 Filaments of innermost stamens narrower than the anthers; stigma ± flat **1. alba**

3 Filaments of innermost stamens wider than the anthers; stigma distinctly concave **2. candida**

1. N. alba L., *Sp. Pl.* 510 (1753) (*N. minoriflora* (Simonkai) Wissjul., *N. occidentalis* (Ostenf.) Moss, *Castalia alba* (L.) W. Wood). Rhizome horizontal, with a few stout branches. Leaves up to 30 × 25 cm, entire. Flowers 10–20 cm in diameter, open nearly all day, faintly scented. Receptacle cylindrical. Petals 20–25, passing gradually into the stamens; outermost stamens with a broad, petaloid filament; the innermost filiform filament narrower than the anther. Stigma more or less flat, with 14–20 rays. Seeds 2–3 mm. $2n = 84$, *c.* 105, 112. *Almost throughout Europe.* All except Az Cr Fa Is Rs(K) Sb; perhaps extinct in Sa Si.

Plants similar to **1** but with smaller flowers and leaves, recorded mainly from N.W. Europe, have been distinguished as *N. occidentalis* (Ostenf.) Moss, but the frequency of intermediate plants makes it difficult to give them taxonomic recognition at any rank.

2. N. candida C. Presl in J. & C. Presl, *Del. Prag.* 224 (1822) (*N. fennica* Mela, *Castalia candida* (C. Presl) Schinz & Thell.). Like **1** but usually slightly smaller in all its parts; flowers 6–11 cm in diameter; receptacle weakly 4-angled; filament of innermost stamens lanceolate, wider than the anther; stigma concave, with 6–14 rays; seeds 3–5 mm. $2n = 112$, *c.* 160. *N., N.E. & N.C. Europe, extending locally to E. France, S.W. Romania and S.E. Russia.* Au Be Cz Fe Ga Ge †He Ho No Po Rm Rs(N, B, C, W, E) Su.

Hybrids with **1** and **3** are fairly frequent in N. Europe.

3. N. tetragona Georgi, *Bemerk. Reise* **1**: 220 (1775) (*N. fennica* Mela). Rhizome vertical, unbranched. Leaves 5–10 × 3–8 cm, entire. Flowers 4–5 cm in diameter, open in the afternoon only, not scented. Receptacle strongly 4-angled. Petals 10–12(–15). Outer stamens with relatively broad filaments, but not transitional to the petals. Stigma more or less flat, with 7–10 rays. Seeds 2–3 mm. *Finland and N. Russia.* Fe Rs(N, C). (*N. & E. Asia, North America.*)

4. N. lotus L., *Sp. Pl.* 511 (1753). Rhizome stout, horizontal, with slender, stolon-like branches, from the ends of which new plants arise. Leaves 20–50 cm, acutely spinose-dentate and somewhat undulate, prominently veined beneath. Flowers 15–25 cm in diameter, open at night and early morning. Sepals conspicuously veined with white. Petals 12–25, white above, purplish beneath. Stamens with relatively wide, white filament, but quite distinct from the petals. Stigma concave, with 21–35 rays. *Hot springs in N.W. Romania (near Oradea).* Rm. (*Egypt, tropical Africa, tropical Asia.*)

The European plant, which is glabrous, belongs to var. *thermalis* (DC.) Tuzson (*N. thermalis* DC.), in which the leaves are glabrous. In var. *lotus* they are pubescent beneath.

N. rubra Roxb. ex Salisb., *Parad. Lond.* **1**: t. 14 (1806), from India, which is like **4** but with deep crimson flowers and young leaves, has been planted in hot springs in Hungary, and may perhaps be regarded as naturalized.

2. Nuphar Sm.

†T. G. TUTIN

Leaves partly submerged, partly floating, exstipulate; lateral veins forked, but not anastomosing. Sepals yellowish. Petals yellow, the outer shorter than the sepals. Ovary superior.

Literature: E. O. Beal, *Jour. Elisha Mitchell Soc.* **72**: 317–346 (1956).

Flowers 4–6 cm in diameter; stigma-rays 15–20 **1. lutea**

Flowers 1·5–3·5 cm in diameter; stigma-rays 8–10 **2. pumila**

1. N. lutea (L.) Sm. in Sibth. & Sm., *Fl. Graec. Prodr.* **1**: 361 (1809) (*Nymphaea lutea* L., *Nymphosanthus luteus* (L.) Fernald). Floating leaves 12–40 × 9–30 cm, ovate, with a deep, acute basal sinus; submerged leaves thin, cordate, with wider sinus. Flowers borne above the water, 4–6 cm in diameter, smelling of alcohol. Sepals 2–3 cm, obovate, persistent, 3 times as long as the broadly spathulate petals. Disc of stigma entire; rays 15–20, not quite reaching its margin. Seeds 5 mm. $2n = 34$. *Almost throughout Europe.* All except Az Bl Co Cr Fa Is Rs(K) Sb; extinct in Si.

2. N. pumila (Timm) DC., *Reg. Veg. Syst. Nat.* **2**: 61 (1821) (*N. tenella* Reichenb., *Nymphosanthus pumilus* (Timm) Fernald). Like **1** but smaller in all its parts; floating leaves not more than 15 × 13 cm; flowers 1·5–3·5 cm in diameter; sepals suborbicular; disc of stigma lobed; rays 8–10, reaching its margin; seeds 4 mm. $2n = 34$. *N., C. & E. Europe, southwards to* c. *43° N.* Au †Be Br Cz Da Fe Ga Ge He Hs Ju No Po Rs(N, B, C, E) Su.

The hybrid between **1 × 2** (*N. × spenneriana* Gaudin) occurs frequently in N. Europe, where local populations may consist exclusively of it.

59. NELUMBONACEAE

EDIT. D. A. WEBB

Perennial, rhizomatous, aquatic herbs. Flowers solitary, hermaphrodite, actinomorphic. Sepals 4 or 5; petals and stamens numerous. Carpels numerous, immersed in the obconical receptacle; ovules 1 or 2.

1. Nelumbo Adanson

†T. G. TUTIN

Leaves and flowers borne above the surface of the water. Foliage-leaves peltate.

The only genus.

1. N. nucifera Gaertner, *Fruct. Sem. Pl.* **1**: 73 (1788) (*Nelumbium nuciferum* auct.; incl. *N. caspicum* Fischer). Rhizome stout, branching, bearing numerous scale-leaves. Foliage leaves with petioles 1–2 m; lamina 30–100 cm in diameter, orbicular, entire, glaucous and unwettable. Flowers 16–23 cm in diameter, pink or almost white, scented, borne above the leaves. Seeds 1·7 × 1·3 cm, ovoid. *Still or slowly moving water. Lower Volga; cultivated in parts of S. Europe, and locally naturalized.* Rs(E) [It Rm Rs(W)]. (*S. & E. Asia, tropical Australia.*)

60. CERATOPHYLLACEAE

EDIT. D. A. WEBB

Monoecious herbs with whorled leaves. Perianth of 8–12 linear segments, united at the base. Stamens numerous; filaments short or absent; connective prolonged apically. Ovary superior, 1-celled; fruit a 1-seeded nut.

1. Ceratophyllum L.

D. A. WEBB

Submerged aquatics, perennating by dormant terminal buds. Leaves in whorls of 3–8, dichotomously divided. Flowers axillary, sessile, the male and female at different nodes. Perianth green in female flowers, whitish in male. Stamens up to 30. Nut usually armed with 1 or more spines.

All species inhabit still or slow-moving waters.

Literature: D. H. Les, *Syst. Bot.* **10**: 338–346 (1985); **11**: 549–558 (1986). M. Wilmot-Dear, *Kew Bull.* **40**: 243–271 (1985).

Most upper leaves forked once or twice, the segments conspicuously spinose-denticulate; terminal spine of nut (1·5–)3·5–12 mm — **1. demersum**
Upper leaves forked 3 or 4 times, the segments entire or minutely denticulate; terminal spine of nut usually 0·3–1(–2) mm — **2. submersum**

1. C. demersum L., *Sp. Pl.* 992 (1753). Stems 30–150 cm. Leaves 8–40 mm, rather stiff; upper leaves forked once or twice into linear, flattened, spinose-denticulate segments 0·2–0·7 mm thick; lower leaves similar or 3 or 4 times dichotomous. Male flowers 1–3 at a node; stamens often more than 20. Female flowers solitary. Nut (3–)4–5·5(–7) × 2–3·5 mm, with or without 2 basal spines; terminal spine (1·5–)3·5–12 mm. $2n = 24$. *Most of Europe.* All except Fa Is Sb.

(a) Subsp. **demersum**: Nut ovoid or ellipsoidal, rarely obovoid, not winged, usually not conspicuously flattened, the surface more or less smooth with scattered glands. Basal spines (1–)2–6(–10) mm, terete or flattened, sometimes absent; terminal spine terete or laterally flattened. *Throughout the range of the species.*

(b) Subsp. **platyacanthum** (Cham.) Nyman, *Consp.* 251 (1879) (*C. platyacanthum* Cham., *C. komarovii* Kuzen., *C. pentacanthum* Haynald): Nut more or less triangular, conspicuously flattened, surrounded by a broad wing between the bases of the usually flattened spines, the surface with numerous raised glands and usually a single spine or tubercle in the centre of each face; basal spines 7–10 mm; terminal spine 10–12 mm. *Widely scattered in Europe, from France to Russia, but very local.*

2. C. submersum L., *Sp. Pl.* ed. 2, 1409 (1763). Stems 20–80 cm. Leaves 20–40 mm, soft; all leaves forked 3 or 4(5) times, the segments 0·1–0·3 mm thick, entire or minutely denticulate. Stamens often fewer than 10. Nut 3–5 × (1·5–)2·5–3 mm, conspicuously flattened, surrounded by thickened or membranous wing, the surface papillose-tuberculate, rarely smooth; basal spines present or absent; terminal spine 0·3–2(–9) mm. $2n = 24$, 40. *Europe, northwards to 58° N. in Sweden, but rather rare in the Mediterranean region.* Al Au Be Br Bu Co Cz Da Ga Ge Gr He Ho Hs Hu It Ju Po Rm Rs(B, C, W, K, E) Si Su.

(a) Subsp. **submersum**: Nut ovoid-ellipsoidal to ellipsoidal; basal spines absent; terminal spine 0·3–1(–2) mm. *Throughout the range of the species.*

(b) Subsp. **muricatum** (Cham.) Wilmot-Dear, *Kew Bull.* **40**: 266 (1985) (*C. muricatum* Cham., *C. tanaiticum* Sapjegin): Nut obovoid, ellipsoidal or ovoid to almost triangular; basal spines (0·5–)2–6 mm; terminal spine (1–)4–9 mm. *Ukraine & S.E. Russia; naturalized in rice-fields in Bulgaria (near Plovdiv).*

61. RANUNCULACEAE

EDIT. †T. G. TUTIN AND A. O. CHATER

Herbs or rarely woody climbers. Leaves alternate, exstipulate, rarely opposite or stipulate. Flowers usually hermaphrodite and actinomorphic, hypogynous. Perianth petaloid or sepaloid, verticillate. Honey-leaves (petaloid structures bearing nectaries) often present, sometimes infundibuliform. Stamens numerous, usually spirally arranged, extrorse. Carpels 1 to many, usually free and spirally arranged. Fruit usually of 1 or more follicles or a head of achenes.

The term 'fruit' is used in the English sense of the product of a single flower, whether the ovary be apocarpous, syncarpous or intermediate between the two.

1 Flowers zygomorphic
2 Upper perianth-segment hooded — **11. Aconitum**
2 Upper perianth-segment not hooded
3 Follicles 2 or more — **12. Delphinium**
3 Follicle 1 — **13. Consolida**
1 Flowers actinomorphic

4 Leaves opposite; perianth-segments valvate in bud **17. Clematis**

4 Leaves alternate or verticillate; perianth-segments imbricate in bud

5 Flowers spurred

6 Leaves linear to spathulate, entire, all basal; achenes in a long head **21. Myosurus**

6 Leaves ternately divided, some cauline; follicles in a single whorl **22. Aquilegia**

5 Flowers not spurred

7 Perianth of 2 dissimilar whorls (or apparently so), the inner of which may consist of petaloid honey-leaves

8 Petaloid whorl of the perianth without nectaries

9 Leaves with 3–5 shallow, ovate lobes **15. Hepatica**

9 Leaves finely divided **18. Adonis**

8 Petaloid whorl of the perianth with nectaries at base

10 Fruit a group of follicles **9. Cimicifuga**

10 Fruit a group of achenes

11 Achenes with an empty cell on either side; beak 2–3 times as long as achene **20. Ceratocephala**

11 Achenes without empty cells; beak less than twice as long as achene

12 Leaves pinnately divided **3. Callianthemum**

12 Leaves simple or palmately divided **19. Ranunculus**

7 Perianth of 1 whorl of green or variously coloured segments; nectaries present or absent, but not petaloid

13 Perianth-segments and honey-leaves shorter than stamens

14 Fruit a berry **8. Actaea**

14 Fruit a group of achenes or follicles

15 Fruit a group of achenes **23. Thalictrum**

15 Fruit a group of follicles **9. Cimicifuga**

13 Perianth-segments or honey-leaves longer than stamens

16 Infundibuliform or spathulate honey-leaves present

17 Follicles united for at least $\frac{1}{3}$ of their length; annuals

18 Honey-leaves much shorter than the ±persistent perianth-segments **4. Nigella**

18 Honey-leaves longer than the caducous perianth-segments **5. Garidella**

17 Follicles free or united only at base; perennials

19 Perianth green, purplish or white

20 Leaves palmate or pedate **1. Helleborus**

20 Leaves 2-ternate **7. Isopyrum**

19 Perianth yellow

21 Cauline leaves in a whorl of 3 **2. Eranthis**

21 Cauline leaves not verticillate **6. Trollius**

16 Honey-leaves absent (nectar-secreting staminodes present in **16**)

22 Flowers subtended by a whorl of leaves or bracts

23 Flowers closely subtended by 3 sepaloid bracts **15. Hepatica**

23 Bracts ± leaf-like and some distance below flower

24 Styles not elongating in fruit; flowers without nectar-secreting staminodes **14. Anemone**

24 Styles elongating greatly and becoming feathery in fruit; nectar-secreting staminodes present **16. Pulsatilla**

22 Flowers not subtended by a whorl of leaves or bracts

25 Leaves simple **10. Caltha**

25 Leaves pinnate or ternate

26 Flowers in a raceme; fruit a head of follicles **9. Cimicifuga**

26 Flowers in a panicle; fruit a head of achenes **23. Thalictrum**

1. Helleborus L.

†T. G. TUTIN

Perennial herbs with rhizomes or erect, rather woody stems and digitate or pedate leaves. Inflorescence cymose. Flowers large; perianth-segments 5; honey-leaves 8–12, obliquely infundibuliform, smaller than the perianth-segments; involucre absent. Fruit of 3–8, several-seeded follicles.

Literature: H. Merxmüller & D. Podlech, *Feddes Repert.* **64**: 1–8 (1961). B. Mathew, *Hellebores*. Woking. 1989.

1 Rhizome absent; stems overwintering, ± woody; leaves all cauline

2 Leaf-segments 7–11; flowers campanulate **1. foetidus**

2 Leaf-segments 3; flowers with widely spreading perianth-segments **2. lividus**

1 Rhizome present; stems of short duration, herbaceous; basal leaves present

3 Bracts entire **11. niger**

3 Bracts divided and leaf-like

4 Follicles free to base, shortly stipitate

5 Flowers pale green, scented; leaves not overwintering **3. cyclophyllus**

5 Flowers cream at first, becoming greenish- to yellowish-brown, not scented; leaves overwintering **4. orientalis**

4 Follicles connate at base for about $\frac{1}{4}$ of their length

6 Flowers reddish or purplish, at least outside

7 Leaves entirely glabrous, even on the slender, inconspicuous veins beneath **9. dumetorum**

7 Leaves pubescent, at least on the stout, prominent veins beneath

8 Leaves digitate; flowers 4–7 cm in diameter **10. purpurascens**

8 Leaves pedate; flowers 3–4·5 cm in diameter **7. multifidus**

6 Flowers greenish with no reddish or purplish tinge

9 Flowers usually not more than 4 cm in diameter, nodding; perianth-segments not or scarcely imbricate

10 Leaves pubescent, at least on the prominent veins beneath; cauline leaves small **7. multifidus**

10 Leaves glabrous or nearly so, even on the slender veins beneath; cauline leaves large

11 Leaves very finely dentate; flowers not scented, over-topped by the leaves **9. dumetorum**

11 Leaves coarsely incised-dentate; flowers scented, over-topping the leaves **5. viridis**

9 Flowers at least 4 cm in diameter, not or slightly nodding at anthesis; perianth-segments usually imbricate

12 Flowers 4–5 cm in diameter, yellowish-green; veins slender, but prominent beneath **5. viridis**

12 Flowers 5–7 cm in diameter, yellowish; veins stout, very prominent beneath
13 Leaf-segments 7–11, undivided or with the marginal ones somewhat divided **6. odorus**
13 Leaf-segments usually 5–7, more deeply divided
14 Leaves entirely glabrous, the basal *c.* 10 cm wide, the cauline scarcely smaller **8. bocconei**
14 Leaves pubescent, at least on the veins beneath; basal 20–30 cm wide; cauline much smaller
15 Leaf-segments ovate-lanceolate, some undivided, the others 2- to 5-fid, finely dentate **6. odorus**
15 Leaf-segments linear-lanceolate, most or all 2- to 10-fid, coarsely dentate **8. bocconei**

1. H. foetidus L., *Sp. Pl.* 558 (1753). Plant foetid, glandular-puberulent above. Stems 20–80 cm, stout, leafy, overwintering. Basal leaves absent; lower cauline pedate, petiolate, with sheathing base; leaf-segments 7–11, narrowly lanceolate, serrate. Bracts broadly ovate, entire, sometimes leaf-like. Flowers 1–3 cm in diameter, numerous, nodding, campanulate. Perianth-segments green, usually with purplish margins. Follicles usually 3, connate below. $2n = 32$. ● *W. Europe, extending northwards to N. England and eastwards to S. Italy and C. Germany.* Be Bl Br Co Ga Ge He Hs It Lu ?Si.

2. H. lividus Aiton, *Hort. Kew.* **2**: 272 (1789) (*H. trifolius* Miller). Like **1** but leaf-segments 3, unequally ovate-lanceolate; flowers 2–5 cm in diameter; perianth-segments patent, pale green or purplish. *Dry scrub and rocky ground.* ● *Islands of W. Mediterranean.* Bl Co Sa.

(a) Subsp. **lividus**: Leaf-segments with small, distant teeth or entire. Inflorescence lax, up to 10-flowered; perianth pale green or purplish. $2n = 32$. *Islas Baleares (Mallorca).*

(b) Subsp. **corsicus** (Willd.) Tutin, *Feddes Repert.* **69**: 53 (1964) (*H. corsicus* Willd., *H. argutifolius* Viv.): Plant taller, up to 150 cm. Leaf-segments closely spinose-dentate. Inflorescence dense, 15- to 30-flowered; perianth pale green. $2n = 32$. *Corse, Sardegna.*

3. H. cyclophyllus Boiss., *Fl. Or.* **1**: 61 (1867). Stems 20–50 cm. Leaves not overwintering; basal usually very large, solitary, weakly pedate to almost digitate with 5–9, usually undivided, ovate-lanceolate, serrate segments; cauline smaller, 3- to 5-fid. Flowers 3 or 4, 4–7 cm in diameter, scented. Perianth-segments patent, pale green. Follicles free, narrowed to the shortly stipitate base. $2n = 32$. *Woods and thickets.* ● *Balkan peninsula.* Al Bu Gr Ju.

Plants intermediate between **3** and **6** occur occasionally. In N.W. Greece (Kerkira) plants occur which have some follicles united at the base, but which are otherwise indistinguishable from **3**.

4. H. orientalis Lam., *Encycl. Méth. Bot.* **3**: 96 (1789). Like **3** but leaves overwintering; leaf-segments 5–11, biserrate; flowers not scented; perianth-segments very pale greenish-cream, becoming brownish- to yellowish-green (rarely purplish). $2n = 32$. *Woods and scrub. Thrace.* ?Gr Tu.

5. H. viridis L., *Sp. Pl.* 558 (1753). Stems 20–40 cm. Leaves not overwintering; basal usually 2, digitate to weakly pedate, with 7–13 serrate segments and slender, prominent veins beneath; cauline smaller, more or less digitate. Flowers 2–4, 4–5 cm in diameter. Perianth-segments patent, yellowish-green. Follicles connate at the base. *Calcicole.* ● *W. & C. Europe, eastwards to N. Italy and E. Austria.* Au Be Br Ga Ge He Hs It [Cz *Ho Rs(W)].

(a) Subsp. **viridis**: Leaves pubescent beneath; segments finely serrate. Perianth-segments broadly ovate. $2n = 32$. *C. Europe and Maritime Alps.*

(b) Subsp. **occidentalis** (Reuter) Schiffner, *Bot. Jahrb.* **11**: 118 (1889): Leaves glabrous beneath; segments coarsely serrate. Perianth-segments narrower, ovate to elliptical. $2n = 32$. *W. Europe.*

6. H. odorus Waldst. & Kit. in Willd., *Enum. Pl. Horti Berol.* 592 (1809). Stems 20–60 cm. Basal leaves overwintering, usually solitary; segments (5–)7–11, ovate-lanceolate, undivided or rarely with a marginal segment divided, coriaceous, pubescent beneath. Flowers 2 or 3, 5–7 cm in diameter, scented. Perianth-segments patent, broadly ovate, clear green. Follicles connate at the base. ● *E., C. & S. Europe from Italy to S. Romania.* Al Hu It Ju Rm.

(a) Subsp. **odorus**: Leaf-segments 7–11, mostly undivided. *Throughout the range of the species except Italy.*

(b) Subsp. **laxus** (Host) Merxm. & Podl., *Feddes Repert.* **64**: 5 (1961): Leaf-segments usually 5–7, mostly 2- to 5-fid. $2n = 30 + 1–11B$. *N. & C. Italy, N.W. Jugoslavia.*

7. H. multifidus Vis., *Flora (Regensb.)* **12(1)** *Erg.-Bl*: 13 (1829) (*H. odorus* subsp. *multifidus* (Vis.) Hayek). Like **6** but leaf-segments 9–15, divided almost to the base into 3–12 linear lobes; flowers 3·5–5·5 cm in diameter; perianth-segments narrow, scarcely overlapping. *Scrub and grassland; calcicole.* ● *Jugoslavia, Albania.* Al Ju ?Rm.

1 Leaf-segments entire or divided to a little beyond the middle, usually finely dentate (c) subsp. **istriacus**
1 Leaf-segments all divided nearly to the base, rather coarsely and remotely dentate
2 Flowers green (a) subsp. **multifidus**
2 Flowers purplish-violet (b) subsp. **serbicus**

(a) Subsp. **multifidus**: Leaf-segments divided nearly to the base, rather coarsely and remotely dentate. Flowers 3·5–4·5 cm in diameter, green. $2n = 30 + 1–9B$. *Albania, W. Jugoslavia, ?Romania.*

(b) Subsp. **serbicus** (Adamović) Merxm. & Podl., *Feddes Repert.* **64**: 5 (1961) (*H. torquatus* Archer-Hind): Like subsp. (a) but flowers 4–5·5 cm in diameter, dark purplish-violet. $2n = 28$. *Srbija, Albania.*

(c) Subsp. **istriacus** (Schiffner) Merxm. & Podl., *loc. cit.* (1961): Leaf-segments entire or divided to a little beyond the middle, usually finely dentate. Flowers 4–5·5 cm in diameter, green. *Jugoslavia.*

8. H. bocconei Ten., *Corso Bot. Lez.* **4(1)**: 459 (1822). Stems 20–60 cm. Basal leaves overwintering, usually solitary; segments 5–7, lanceolate or linear-lanceolate, usually divided to the middle, coarsely and irregularly toothed, coriaceous, glabrous or pubescent beneath. Flowers 2 or 3, 5–7 cm in diameter, scented. Perianth-segments patent, broadly ovate, green. Follicles connate at the base. *Woods and shady places.* ● *C. & S. Italy, Sicilia.* It Si.

(a) Subsp. **bocconei**: Leaves pubescent, at least on the veins beneath; basal leaves 20–30 cm wide. *C. Italy.*

(b) Subsp. **intermedius** (Guss.) Greuter & Burdet, *Willdenowia* **19**: 44 (1989) (*H. viridis* subsp. *siculus* (Schiffner) Merxm. & Podl.): Leaves usually entirely glabrous; basal leaves 8–20 cm wide. *S. Italy, Sicilia.*

9. H. dumetorum Waldst. & Kit. in Willd., *Enum. Pl. Horti Berol.* 592 (1809) (*H. viridis* subsp. *dumetorum* (Waldst. & Kit.) Hayek). Stems 15–45 cm. Basal leaves 2 or 3, thin, not overwintering; leaf-segments 7–11(–13), distinctly pedate, undivided; veins not prominent beneath. Flowers 2 or 3, 3–6 cm in diameter. Perianth-segments patent, violet or green. Follicles connate at the base. $2n = 32$. ● *E.C. Europe.* Au Hu Ju Rm [Rs(W)].

(a) Subsp. **dumetorum**: Flowers 3–4 cm in diameter, green. *Throughout the range of the species.*

(b) Subsp. **atrorubens** (Waldst. & Kit.) Merxm. & Podl., *Feddes Repert.* **64**: 5 (1961) (*H. atrorubens* Waldst. & Kit.): Flowers 4–6 cm in diameter, violet. *N. Jugoslavia.*

10. H. purpurascens Waldst. & Kit., *Pl. Rar. Hung.* **2**: 105 (1802 or 1803). Stems 15–35 cm. Basal leaves usually 2, not overwintering; segments usually 5, not distinctly pedate, divided to the middle into 2–5 lobes, pubescent, at least on the prominent veins beneath. Flowers up to 3, 4–7 cm in diameter. Perianth-segments patent, purplish-violet. Follicles connate at the base. $2n = 32$. ● *E.C. Europe.* Cz Hu Ju Rm Rs(W) [It].

11. H. niger L., *Sp. Pl.* 558 (1753). Stems 10–30 cm. Basal leaves overwintering, pedate with 7–9 segments; cauline ovate, entire, pale green. Flowers 1–3, 5–11 cm in diameter. Perianth-segments white or pink-tinged, ovate. Follicles *c.* 7, connate at the base. ● *Woods and thickets, mainly in the mountains; calcicole. Alps, Appennini and adjacent regions; widely cultivated for ornament, and sometimes naturalized.* Au Ge He It Ju [Cz Ga].

(a) Subsp. **niger**: Leaf-segments oblong-cuneate, dark green, serrate near the apex. Flowers 5–8 cm in diameter. Perianth-segments often pinkish, broadly ovate. $2n = 32$. *Throughout the range of the species.*

(b) Subsp. **macranthus** (Freyn) Schiffner, *Bot. Jahrb.* **11**: 105 (1889): Leaf-segments broadly lanceolate, somewhat glaucous, spinulose-serrate. Flowers 8–11 cm in diameter. Perianth-segments usually white, ovate. $2n = 32$. *Italy, N. Jugoslavia.*

2. Eranthis Salisb.

†T. G. TUTIN

Perennials with tuberous rhizomes. Basal leaves petiolate, deeply palmately divided; cauline leaves 3, similar to basal but sessile and arranged in a whorl close to the solitary terminal flower. Perianth-segments usually 6, yellow; honey-leaves tubular, 2-lipped. Follicles usually 6.

1. E. hyemalis (L.) Salisb., *Trans. Linn. Soc. London* **8**: 304 (1807) (incl. *E. bulgaricus* (Stefanov) Stefanov, *E. cilicicus* Schott & Kotschy). Plant 5–15 cm, glabrous. Flowers 25–40 mm in diameter. Perianth-segments narrowly ovate; honey-leaves shorter than the stamens. Follicles *c.* 15 mm, shortly stalked. *S. Europe, from S.E. France to Bulgaria; cultivated for ornament, and naturalized in C. & W. Europe.* Bu *Ga It Ju [Be Br Cz Ge He Ho Hu Rm].

3. Callianthemum C. A. Meyer

†T. G. TUTIN

Small perennials of high mountains. Leaves imparipinnate, the divisions 2- or 3-pinnatifid. Perianth-segments sepaloid, shorter than the honey-leaves; honey-leaves 5–20, petaloid. Carpels numerous, very shortly stipitate; ovule 1, pendulous. Fruit of 1-seeded follicles.

Literature: J. Witasek, *Verh. Zool.-Bot. Ges. Wien* **49**: 316–356 (1899).

1 Honey-leaves broadly ovate; basal leaves developed at anthesis **3. coriandrifolium**
1 Honey-leaves linear-oblong; basal leaves developing after anthesis
 2 Plant 5–20 cm; fruit 4·5–5 × 2 mm, strongly reticulately veined **1. anemonoides**
 2 Plant 3–6 cm; fruit 3–4 × 2·5 mm, smooth **2. kernerianum**

1. C. anemonoides (Zahlbr.) Endl. ex Heynh., *Nomencl. Bot.* **2**: 106 (1846) (*C. rutifolium* auct., pro parte). Plant 5–20 cm. Basal leaves developing after anthesis, long-petiolate, more or less triangular in outline; segments linear-oblong; cauline leaves similar but sessile and less divided, the uppermost bract-like. Flowers 30–35 mm in diameter. Honey-leaves pink to white, with orange nectaries. Follicles 4·5–5 × 2 mm, strongly reticulately veined. $2n = 32$. *Open coniferous woods; somewhat calcicole.* ● *N.E. Alps.* Au.

2. C. kernerianum Freyn ex A. Kerner, *Sched. Fl. Exsicc. Austro-Hung.* **5**: 36 (1888). Like **1** but smaller; lower branches of the basal leaves shortly stalked; flowers *c.* 25 mm diameter; follicles 3–4 × 2·5 mm, smooth. $2n = 32$. *Stony mountain pastures; calcicole.* ● *S. Alps from* c. 10°30′ *to* c. 11°30′ *E.* It.

3. C. coriandrifolium Reichenb., *Fl. Germ. Excurs.* 727 (1832) (*C. rutifolium* auct., pro parte). Like **1** but basal leaves developed at anthesis; honey-leaves broadly ovate; follicles 3 × 2–2·5 mm. $2n = 16$. *Somewhat calcifuge.* ● *Mountains of S. & S.C. Europe, from N.W. Spain to the S. Carpathians.* Au Cz Ga He Hs It Ju Po Rm.

4. Nigella L.

†T. G. TUTIN (EDITION 1) REVISED BY J. R. AKEROYD (EDITION 2)

Annuals. Leaves usually 2- or 3-pinnatisect with linear segments. Perianth-segments 5, petaloid, persistent for a time after anthesis. Honey-leaves 5, opposite to and much smaller than the perianth-segments. Fruit of 5 (rarely 10) partly or completely united follicles; styles long, patent or rarely erect.

Literature: A. Strid, *Op. Bot. (Lund)* **28**: 1–169 (1970). M. Zohary, *Pl. Syst. Evol.* **142**: 71–107 (1983).

1 Flowers yellowish-white or greenish-yellow; seeds compressed
 2 Lower lip of honey-leaves very short, 2-fid; follicles usually slightly longer than styles **13. orientalis**
 2 Lower lip of honey-leaves with 2 filiform appendages; follicles up to twice as long as styles **14. oxypetala**
1 Flowers blue, greenish or reddish; seeds ovoid or triquetrous
 3 Anthers not mucronate
  4 Flowers without an involucre of leaves
   5 Stems procumbent; leaf-segments small, ovate; follicles united for ⅓ of their length **7. fumariifolia**
   5 Stems erect; leaf-segments linear; follicles united for nearly all of their length **10. sativa**
  4 Flowers with an involucre of leaves
   6 Perianth-segments not clawed; lobes of lower lip of honey-leaves *c.* 3 times as long as wide; fruit 5-celled **11. elata**
   6 Perianth-segments shortly clawed; lobes of lower lip of honey-leaves about as long as wide; fruit 10-celled **12. damascena**

3 Anthers mucronate, sometimes very shortly so
- 7 Cauline leaves with short, ovate or oblong segments, or sometimes undivided
  - 8 Follicles united to at least the middle
    - 9 Stems 15–30 cm, procumbent; anthers red **3. carpatha**
    - 9 Stems 5–15 cm, erect; anthers yellowish **8. doerfleri**
  - 8 Follicles united for *c.* $\frac{1}{3}$ of their length
    - 10 Stems usually more than 10 cm, erect; perianth-segments violet to greenish **6. degenii**
    - 10 Stems usually less than 10 cm, procumbent; perianth-segments reddish **9. stricta**
- 7 Cauline leaves deeply divided into linear segments
  - 11 Follicles 3-veined throughout, united for *c.* $\frac{1}{2}$ of their length **1. arvensis**
  - 11 Follicles 1-veined, united for usually more than $\frac{1}{2}$ of their length
    - 12 Claw of perianth-segments not more than $\frac{1}{4}$ as long as limb; follicles usually densely glandular **2. papillosa**
    - 12 Claw of perianth-segments $\frac{1}{3}$ as long to as long as limb; follicles eglandular or sparsely glandular
      - 13 Claw of perianth-segments usually more than $\frac{1}{2}$ as long as limb; lobes of lower lip of honey-leaves with a lingulate, obtuse apex **4. hispanica**
      - 13 Claw of perianth-segments $\frac{1}{3}$–$\frac{1}{2}$ as long as limb; lobes of lower lip of honey-leaves with an acute apex **5. segetalis**

1. N. arvensis L., *Sp. Pl.* 534 (1753). Stems 10–40 cm. Leaves with linear segments. Involucre absent (or present in some E. Mediterranean plants). Flowers 20–30 mm in diameter. Perianth-segments ovate-cordate, apiculate, bluish; claw usually at least as long as limb. Follicles 3-veined, united for *c.* $\frac{1}{2}$ of their length. Seeds granulate, black. *Cornfields and open ground. Most of Europe, except the north and much of the south-west, but rapidly becoming rarer, and probably extinct in several territories.* Al Au Be Bu †Cr Cz Ga Ge Gr †He Ho Hu It Ju Po Rm Rs(C, W, K, E) Sa Si Tu.

A very polymorphic species that has been divided into numerous taxa that are connected by intermediate variants. Taxa are treated as varieties or subspecies by different authors.

1 Cauline leaves forming an involucre below the flower **(b) subsp. aristata**
1 Cauline leaves not forming an involucre below the flower
- 2 Follicles smooth **(a) subsp. arvensis**
- 2 Follicles densely tuberculate **(c) subsp. glauca**

(a) Subsp. **arvensis** (*N. divaricata* Gaud.-Beaup.): Leaves green, the cauline not crowded to form an involucre below the flower. Follicles smooth. $2n = 12$. ● *Almost throughout the range of the species.*

(b) Subsp. **aristata** (Sibth. & Sm.) Nyman, *Consp.* 17 (1878) (*N. aristata* Sibth. & Sm., *N. arvensis* subsp. *rechingeri* (Tutin) Tutin): Leaves rather glaucous, the upper cauline crowded and forming an involucre below the flower. Follicles densely tuberculate. $2n = 12$. *C. & S. Greece, Kikladhes; Turkey-in-Europe.*

(c) Subsp. **glauca** (Boiss.) N. Terracc., *Boll. Orto Bot. Palermo* **2**: 34 (1898): Leaves green to glaucous, the cauline usually not crowded to form an involucre below the flower. Follicles densely tuberculate. *Turkey-in-Europe.* (*Anatolia.*)

Subsp. **brevifolia** Strid, *Op. Bot.* (*Lund*) **28**: 44 (1970) (*N. cretica* auct., non Miller), from Rodhos and adjacent small islands in the East Aegean, and characterized by short leaves with few, rather wide segments, was described from W. Kriti but has not been seen there since 1884.

2. N. papillosa G. López, *Anal. Jard. Bot. Madrid* **41**: 468 (1985) (*N. hispanica* auct., non L.). Like **1(a)** but flowers 35–70 mm in diameter, blue; claw of perianth-segments up to $\frac{1}{4}$ as long as limb; follicles 1-veined, usually densely glandular. $2n = 12$. *Spain, Portugal; ?Sicilia.* Hs Lu ?Si [Az].

(a) Subsp. **papillosa**: Flowers 5–7 mm in diameter. Claw of perianth-segments absent or very short. Follicles densely glandular. ● *Spain, Portugal.*

(b) Subsp. **atlantica** (Murb.) Amich ex G. López, *op. cit.* 468 (1985) (*N. hispanica* subsp. *atlantica* Murb.): Flowers 3·5–7 mm in diameter. Claw of perianth-segments up to $\frac{1}{3}$ as long as limb. Follicles sparsely glandular. *S. Spain, S. Portugal; ?Sicilia. Naturalized in Açores.*

3. N. carpatha Strid, *Op. Bot.* (*Lund*) **28**: 49 (1970). Stems 15–30 cm, procumbent, diffuse, slender. Flowers 12–17 mm in diameter. Perianth-segments reddish. Anthers very shortly mucronate, red. Follicles 6–10 mm, united for *c.* $\frac{1}{2}$ of their length, tuberculate. $2n = 12$. *Open, stony ground.* ● *Karpathos and Kasos.* Cr.

4. N. hispanica L., *Sp. Pl.* 534 (1753) (*N. gallica* Jordan). Stems 10–40 cm. Leaves with broadly linear segments. Involucre absent. Flowers 20–35 mm in diameter. Perianth-segments pale blue; claw usually more than $\frac{1}{2}$ as long as limb. Lobes of lower lip of honey-leaves ovate-lanceolate in lower $\frac{1}{2}$; upper $\frac{1}{2}$ lingulate, obtuse. Anthers distinctly mucronate. Follicles glabrous or very slightly glandular. *Cornfields.* ● *S.W. Europe; almost extinct in S.W. France.* Ga Hs Lu.

5. N. segetalis Bieb., *Fl. Taur.-Cauc.* **2**: 16 (1808). Like **4** but claw of perianth-segments $\frac{1}{3}$–$\frac{1}{2}$ as long as limb; lobes of lower lip of honey-leaves ovate, acute; anthers very shortly mucronate. *Cornfields. S. Ukraine.* Rs(W, K, E).

6. N. degenii Vierh., *Magyar Bot. Lapok* **25**: 148 (1926). Stems 10–40 cm, usually erect. Cauline leaves with oblong segments. Flowers 12–25 mm in diameter. Perianth-segments violet to greenish. Anthers mucronate. Follicles free for most of their length; beak slender, usually not longer than follicle. $2n = 12$. *Disturbed open habitats.* ● *Kikladhes.* Gr.

A very variable species that forms distinct local populations on different islands, some of which have been given taxonomic rank (see A. Strid, *Op. Bot.* (*Lund*) **28**: 1–169 (1970)).

7. N. fumariifolia Kotschy in Unger & Kotschy, *Ins. Cypern* 319 (1865). Stems numerous, procumbent. Leaf-segments ovate, very small. Perianth-segments ovate-oblong, cordate, clawed, greenish-white. Anthers not mucronate. Follicles united for $\frac{1}{3}$ of their length, tuberculate-rugose in lower $\frac{1}{2}$, indehiscent. $2n = 12$. *S. Aegean region.* Cr Gr.

8. N. doerfleri Vierh., *Magyar Bot. Lapok* **25**: 147 (1926). Stems 5–15 cm, erect, somewhat glaucous, with few, usually divaricate branches. Leaf-segments few, ovate. Flowers small, greenish-blue. Anthers shortly mucronate, yellowish. Follicles united for nearly all of their length, tuberculate-scabrid; beak *c.* $\frac{1}{2}$ as long as follicle, stout, patent. $2n = 12, 14$. ● *S. Aegean region.* Cr Gr.

9. N. stricta Strid, *Op. Bot. (Lund)* **28**: 63 (1970). Stems 5–12 cm, procumbent, with divaricate branches. Leaf-segments linear-lanceolate. Flowers 10–15 mm in diameter. Perianth-segments reddish. Anthers very shortly mucronate, red. Follicles 8–14 mm, united for *c.* $\frac{1}{3}$ of their length, tuberculate. $2n = 12$. ● *S.W. Kriti, Kithira.* Cr Gr.

10. N. sativa L., *Sp. Pl.* 534 (1753) (*N. cretica* Miller). Stems 20–60 cm, erect, branched. Leaf-segments linear-lanceolate. Perianth-segments ovate, shortly clawed, whitish. Anthers not mucronate. Follicles united for all of their length, tuberculate on the back. $2n = 12$. *Widely cultivated for its aromatic seeds, and frequently naturalized.* [Bu Ga Gr It Ju Rm Rs(C, W, K) Tu.] (*Origin unknown; possibly native in parts of S.E. Europe.*)

11. N. elata Boiss., *Diagn. Pl. Or. Nov.* **1(1)**: 66 (1843). Stems 20–100 cm, simple or little-branched. Leaf-segments linear. Involucre with short, rigid segments. Perianth-segments ovate-oblong, not clawed, blue. Lobes of lower lip of honey-leaves *c.* 3 times as long as wide. Anthers not mucronate. Follicles united for all of their length, forming an inflated 5-celled capsule. *S.E. Europe; local.* Bu ?Gr Rs(K) Tu. (*S.W. Asia.*)

12. N. damascena L., *Sp. Pl.* 534 (1753). Stems 10–60 cm. Segments of the lower leaves linear-lanceolate, those of the upper leaves linear; involucral leaves similar to cauline. Perianth-segments ovate-oblong, clawed, bluish. Lobes of lower lip of honey-leaves about as long as wide. Anthers not mucronate. Follicles united for all of their length, forming an inflated 10-celled capsule, the 5 outer loculi sterile. $2n = 12$. *S. Europe; cultivated for ornament, and a frequent casual elsewhere.* Al Bl Bu Co Cr Ga Gr Hs It Ju Lu Rs(K) Sa Si Tu.

13. N. orientalis L., *Sp. Pl.* 534 (1753). Stems up to 50 cm. Leaf-segments linear. Perianth-segments ovate, shortly clawed, yellowish-white with red dots, eventually deflexed. Flowers without involucre. Anthers not mucronate. Follicles united for $\frac{1}{2}$ of their length, slightly longer than the erect beaks. Seeds compressed. *S.E. part of Balkan peninsula.* Bu ?Gr Tu. (*S.W. Asia.*)

14. N. oxypetala Boiss., *Ann. Sci. Nat.* **16**: 357 (1841). Like **13** but with a lax involucre; perianth-segments oblong, scarcely clawed, greenish-yellow; lobes of lower lip of honey-leaves produced into long, filiform appendages; follicles up to twice as long as the beaks. *Krym.* Rs(K). (*S.W. Asia.*)

5. Garidella L.

†T. G. TUTIN (EDITION 1) REVISED BY J. R. AKEROYD (EDITION 2)

Like *Nigella* but perianth-segments caducous, shorter than the honey-leaves; follicles 2 or 3, united below, inflated; styles *c.* 1 mm.

Literature: A. Strid, *Bot. Not.* **122**: 330–332 (1969).

Honey-leaves 7–9 mm, oblong; anthers dark brown **1. nigellastrum**
Honey-leaves 10–14 mm, broadly ovate; anthers yellowish **2. unguicularis**

1. G. nigellastrum L., *Sp. Pl.* 425 (1753) (*Nigella nigellastrum* (L.) Willk., *N. garidella* Spenner). Stem 30–50 cm, slender, simple or with ascending branches. Flowers solitary on long pedicels. Perianth-segments 3–4 mm, greenish, tinged purplish or reddish. Honey-leaves 7–9 mm, the claw not exserted; lower lip of limb oblong, narrowed at base, divided into two linear-oblong lobes. Anthers dark brown. Seeds rugose with anastomosing ridges. *Mediterranean region and Krym; very local.* Cr Ga Gr Hs Rs(K).

2. G. unguicularis Poiret in Lam., *Encycl. Méth. Bot.* **10**: 709 (1812) (*Nigella unguicularis* (Poiret) Spenner). Like **1** but honey-leaves 10–14 mm, the claw long-exserted; lower lip of limb broadly ovate, cordate at base, 2-fid; anthers yellowish. $2n = 12$. *Kriti.* Cr. (*S.W. Asia.*)

6. Trollius L.

†T. G. TUTIN

Perennial herbs. Flowers with all parts spirally arranged. Perianth-segments 5–15, petaloid; honey-leaves 5–15, small, narrow, yellow. Fruit of many follicles.

Perianth-segments strongly incurved, the flower globose; honey-leaves as long as stamens **1. europaeus**
Perianth-segments not strongly incurved, the flower somewhat open; honey-leaves usually longer than stamens **2. apertus**

1. T. europaeus L., *Sp. Pl.* 556 (1753). Plant 10–80 cm, glabrous. Basal leaves long-petiolate, 3- to 5-lobed; lobes cuneate and more or less deeply lobed and serrate; cauline leaves smaller and more or less sessile. Flowers up to 5 cm in diameter, globose. Perianth-segments usually 10, incurved, pale yellow, rarely the outer green beneath. Honey-leaves about as long as the stamens. Follicles *c.* 12 mm; beak 0·5–5 mm. *Meadows and woodland-margins.* ● *Widespread in Europe, but only on mountains in the south.* Al Au Br Bu Cz Da Fe Ga Ge Gr Hb He Hs Hu It Ju No Po Rm Rs(N, B, C, W, E) Su [Be].

The length of the beak of the mature follicle and the shape of the basal leaves are rather variable, especially in E. Europe. The prevalence of self-pollination probably helps to emphasize this variation. Two subspecies can be recognized:

(a) Subsp. **europaeus**: Plant usually 30–80 cm. Stigma not more than 3 mm, straight. $2n = 16, 18$. *Almost throughout the range of the species.*

(b) Subsp. **transsilvanicus** (Schur) Domin in Domin & Podp., *Klíč Úplné Květeně Rep. Česk.* 25 (1928): Plant usually 10–20 cm; stigma 3–5 mm, more or less recurved. $2n = 16$. ● *From S. Alps to Carpathians.*

2. T. apertus Perf. ex Igoshina, *Bot. Žur.* **53**: 793 (1968) (*T. asiaticus* auct., non L., *T. uralensis* Gorodkov). Like **1** but plant 40–60 cm; flowers somewhat open; perianth-segments not incurved, often pale orange; honey-leaves usually longer than the stamens. *N.E. Russia.* Rs(N).

7. Isopyrum L.

†T. G. TUTIN

Perennials. Flowers solitary. Perianth-segments 5, petaloid; honey-leaves small or absent. Stamens shorter than the perianth-segments. Fruit of 2–20 sessile, free follicles.

1. I. thalictroides L., *Sp. Pl.* 557 (1753). Plant 10–30 cm, slender, glabrous, more or less glaucous. Rhizome creeping. Stem simple and leafless below, branched and leafy above. Basal leaves petiolate, ternate; leaflets 3-fid; segments ovate, more or less 3-lobed. Cauline leaves similar to the basal but sessile, with conspicuous stipules. Flowers 10–20 mm in diameter. Perianth-segments white. Honey-leaves 1–1·5 mm. Follicles usually 2, broadly ovate, compressed. $2n = 14$. ● *W. & C. Europe, extending to C. Italy, Bulgaria and C. Ukraine.* Au Bu Co Cz Ga †He Hs Hu It Ju Po Rm Rs (B, C, W).

8. Actaea L.

†T. G. TUTIN

Glabrous perennials. Flowers small, in short racemes. Perianth-segments 3–5, petaloid; honey-leaves 4–10, small. Carpel 1. Fruit a berry.

Fruit 12–13 mm, black when ripe **1. spicata**
Fruit 9–10 mm, red when ripe **2. erythrocarpa**

1. A. spicata L., *Sp. Pl.* 504 (1753). Plant 30–65 cm. Basal leaves large, 2-ternate or 2-pinnate; leaflets ovate, more or less dentate, more or less pubescent beneath; cauline leaves much smaller. Flowers white. Honey-leaves shorter than the white stamens. Fruit 12–13 mm, ovoid, black when ripe. $2n = 16$. *Shady places. Most of Europe, but only on mountains in the south.* Al Au Be Br Bu Cz Da Fe Ga Ge Gr He Ho Hs Hu It Ju No Po Rm Rs(N, B, C, W, K, E) Su.

2. A. erythrocarpa Fischer in Fischer & C. A. Meyer, *Ind. Sem. Horti Petrop.* **1**: 20 (1835). Like **1** but leaves usually 3-ternate; fruit 9–10 mm, red when ripe. $2n = 16$. *N.E. Europe.* Fe Rs(N, C) Su.

9. Cimicifuga L.

†T. G. TUTIN

Like *Actaea* but fruit of 2–8 dry, dehiscent follicles.

1. C. europaea Schipcz. in Komarov, *Fl. URSS* **7**: 85 (1937) (*C. foetida* auct., non L.). Foetid herb 40–100(–200) cm. Stem glabrous. Leaves very large, 2- or 3-pinnate; leaflets ovate, acuminate, irregularly 2-serrate; petioles sparsely pubescent. Raceme long and narrow. Flowers greenish. Bracts lanceolate, acuminate, shorter than pedicels. Follicles 10–17 mm, pubescent. $2n = 16$. *E. & C. Europe from Czechoslovakia to C. Ukraine.* †Au Cz Hu Po Rm Rs(C, W).

10. Caltha L.

†T. G. TUTIN (EDITION 1) REVISED BY J. R. AKEROYD (EDITION 2)

Perennial herbs with more or less cordate leaves. Inflorescence a few-flowered corymbose cyme. Perianth-segments 5 or more, yellow. Honey-leaves absent; carpels nectar-secreting. Follicles 5–15.

Literature: P. G. Smit, *Blumea* **21**: 119–150 (1973). S. R. J. Woodell & M. Kootin-Sanwu, *New Phytol.* **70**: 173–186 (1971).

1. C. palustris L., *Sp. Pl.* 558 (1753) (incl. *C. minor* Miller, *C. polypetala* Hochst. ex Lorent, *C. cornuta* Schott, Nyman & Kotschy, *C. laeta* Schott, Nyman & Kotschy, *C. longirostris* G. Beck). Stems 10–60 cm, hollow, glabrous, creeping and rooting to erect. Leaves crenate to dentate, the basal long-petiolate, the cauline smaller, shortly petiolate to subsessile. Flowers 1·5–5 cm in diameter; perianth-segments bright shining yellow above, often greenish beneath. Ripe follicles 9–18 mm, beaked, erect or recurved. $2n = 24, 30, 32–34, 36, 40, 42–44, 46, 48, 52–64, 72, 80$, with 0–6 B chromosomes. *Wet places. Most of Europe, but very rare in the Mediterranean region.* All except Az Bl Co Cr Sa Sb Si Tu.

This species varies greatly in many characters such as habit, size, shape of leaf and follicle and in chromosome number; many of these characters are phenotypically very plastic. Several taxa have been described from Europe and given specific or subspecific rank; these have been mostly based on follicle characters, which vary independently of other morphological characters and of chromosome number.

Plants from the Arctic and mountainous regions that are small, few-flowered and have stems that root at the nodes, have been called var. *radicans* (T. F. Forster) Hooker. The variation between this and var. *palustris* is continuous.

11. Aconitum L.

†T. G. TUTIN (EDITION 1) REVISED BY J. R. AKEROYD AND A. O. CHATER (EDITION 2)

Perennials, with stout leafy stems; stock tuberous, with brown, fragile roots. Leaves alternate, palmately or pedately divided, the segments dentate or lobed. Flowers zygomorphic, in a terminal raceme or racemose panicle. Perianth-segments 5, petaloid, the posterior forming a large, erect hood (helmet). Honey-leaves 2–10; posterior pair included in the helmet, the claw long and the limb prolonged into nectar-secreting spurs; others very small or absent. Stamens numerous. Follicles 2–5, free or shortly connate at base.

The measurements of the helmet given in the text are from the highest point of the straight or arched part of the base of the helmet to the apex (height), and across the most nearly parallel-sided part (width).

A great deal of variation occurs in all the species, perhaps as a result of hybridization. It does not in general fall into a recognizable morphological or geographical pattern, though many local populations can be recognized and have been given specific or subspecific rank. Much work remains to be carried out in this genus, and the following account must be regarded as a provisional outline of the grouping of the very many taxa which have been described.

Literature: G. Gáyer, *Magyar Bot. Lapok* **5**: 122–137 (1906); **6**: 286–303 (1907); **8**: 114–206, 310–327 (1909). H. G. L. Reichenbach, *Monographia Generis Aconiti*. Leipzig. 1820–1821. *A. vulparia* group: L. A. Lauener & M. Tamura, *Notes Roy. Bot. Gard. Edinb.* **37**: 113–124 (1978); M. Tamura & L. A. Lauener, *Notes Roy. Bot. Gard. Edinb.* **37**: 431–466 (1979). *A. variegatum* group: E. Götz, *Feddes Repert.* **76**: 1–62 (1967). *A. napellus* group: W. Seitz, *Feddes Repert.* **80**: 1–76 (1969).

1 Helmet *c.* 3 times as high as wide, conical-cylindrical; nectary-spurs ± spirally curved **1. lycoctonum**
1 Helmet not more than twice as high as wide, rounded or hemispherical; nectary-spurs ± straight
2 Flowers usually yellow; perianth persistent **2. anthora**
2 Flowers blue or purple, sometimes white, rarely variegated; perianth deciduous
3 Seeds winged on 3 angles, smooth or rugulose, and without lamellae on the sides
4 Inflorescence eglandular-hairy (rarely a few glandular hairs present) **5. napellus**
4 Inflorescence densely glandular-hairy **6. burnatii**
3 Seeds winged on 1 angle, and with prominent transverse lamellae on the sides
5 Bracteoles ovate **4. toxicum**
5 Bracteoles linear to linear-spathulate
6 Leaf-segments more than 3 mm wide **3. variegatum**

6 Leaf-segments 1–2(–3) mm wide **7. angustifolium**

1. A. lycoctonum L., *Sp. Pl.* 523 (1753). Leaf-segments divided into 3 or more incise-serrate to very deeply laciniate-dentate lobes. Inflorescence simple or branched. Perianth-segments deciduous; helmet 12–25 × 3–10 mm, *c.* 3 times as high as wide, conical-cylindrical; nectary-spurs spirally curved. Follicles (2)3(–5). Seeds brownish-black, 4-sided with convex sides and obtuse angles. *Most of Europe, but absent from most of the north-west and much of the Mediterranean region.* Al Au Be Bu Cz Fe Ga Ge Gr He Ho Hs Hu It Ju No Po Rm Rs(N, B, C, W, K, E) Su.

1 Helmet with a wide base, tapering abruptly into a narrow, elongate hood (N. & E. Europe)
 2 Flowers dark violet (a) subsp. **lycoctonum**
 2 Flowers yellowish (b) subsp. **lasiostomum**
1 Helmet usually saccate, not conspicuously wider at base (S. & C. Europe)
 3 Flowers blue (c) subsp. **moldavicum**
 3 Flowers yellowish
  4 Leaf-segments 3-fid to the middle; terminal inflorescence small and few-flowered (d) subsp. **vulparia**
  4 Leaf-segments divided beyond the middle into several lobes; terminal inflorescence large and many-flowered (e) subsp. **neapolitanum**

(a) Subsp. **lycoctonum** (*A. septentrionale* Koelle; incl. *A. excelsum* Reichenb.): Stems 60–150(–200) cm. Leaves dark green; segments 4–6, 3-fid. Inflorescence villous with eglandular and glandular hairs. Flowers dark violet; helmet 18–25 × 3–5 mm, tapering abruptly from the wide base, sparsely hairy. $2n = 16$. *N. & E. Europe from Norway to S.E. Russia.*

(b) Subsp. **lasiostomum** (Reichenb. ex Besser) Warncke, *Eur. Sippen* Aconitum lycoctonum-*Gruppe* 51 (1964) (*A. lasiostomum* Reichenb. ex Besser, *A. besserianum* Andrz. ex Trautv., *A. rogowiczii* Wissjul.): Like subsp. (a) but flowers yellowish. ● *C. Russia, extending to Estonia, Romania and Krym.*

(c) Subsp. **moldavicum** (Hacq.) Jalas, *Ann. Bot. Fenn.* **22**: 219 (1985) (*A. moldavicum* Hacq., *A. hosteanum* Schur): Stems 60–150(–200) cm. Leaves dark green; segments 4–6, 3-fid. Inflorescence covered with short, crispate eglandular hairs. Flowers blue; helmet 16–23 × 4–7 mm, not tapering abruptly from a wide base. *E.C. Europe, extending to Romania and W. Ukraine.*

(d) Subsp. **vulparia** (Reichenb.) Nyman, *Consp., Suppl.* **2**: 13 (1889) (*A. croaticum* Degen & Gáyer, *A. gracilescens* Gáyer, *A. lasianthum* (Reichenb.) Simonkai, *A. laxiflorum* (DC.) Reichenb., *A. lycoctonum* auct., non L., *A. pauciflorum* Host., *A. penninum* (Ser.) Gáyer, *A. puberulum* (Ser.) G. Grinţ., *A. thalianum* (Wallr.) Gáyer, *A. velebiticum* Degen, *A. velutinum* (Reichenb.) G. Grinţ., *A. vulparia* Reichenb.): Like subsp. (c) but stems 40–120 cm; leaf-segments 3-fid to the middle; inflorescence small, few-flowered; flowers yellow. $2n = 16$. ● *From France and the Netherlands eastwards to Romania.*

(e) Subsp. **neapolitanum** (Ten.) Nyman, *Consp.* 19 (1878) (*A. fallax* (Gren. & Godron) Gáyer, *A. lamarckii* Reichenb., *A. neapolitanum* Ten., *A. pantocsekianum* Degen & Bald., *A. pyrenaicum* auct., non L., *A. ranunculifolium* Reichenb., *A. stenotomum* Borbás, *A. vulparia* subsp. *neopolitanum* (Ten.) Muñoz Garmendia, *A. wagneri* Degen): Leaves light green; segments 5–8, several times divided to beyond the middle. Inflorescence large, many-flowered. Flowers yellow. $2n = 16$. *Mountains of S. & S.C. Europe.*

Populations more or less intermediate between subspp. (d) and (e) occur, particularly in the southern Alps, Jugoslavia and Romania; some of these have been called **A. dasytrichum** (Degen ex Gáyer) G. Grinţ. in Săvul., *Fl. Rep. Pop. Române* **2**: 506 (1953) and **A. planatifolium** Degen ex Gáyer, *Magyar Bot. Lapok* **6**: 118 (1907). Plants morphologically indistinguishable from subsp. (c) occur in the mountains of China.

2. A. anthora L., *Sp. Pl.* 532 (1753). Leaves divided 3 or 4 times completely to the base or midrib; segments not more than 3 mm wide, linear. Inflorescence simple or branched, with eglandular hairs. Flowers yellowish, rarely blue; perianth persistent; helmet 8–12 × 14–16 mm, more or less hemispherical; nectary-spurs not spirally curved. Follicles (3–)5. Seeds 3–5 mm, black, irregularly 4-sided, with acute angles. $2n = 32$. *S., C. & E. Europe, mainly in the mountains.* Au ?Co Cz Ga He Hs Hu It Ju Rm Rs(C, W, K, E).

3. A. variegatum L., *Sp. Pl.* 532 (1753). Leaves divided to the base, the wide segments often divided to beyond the middle, and deeply incise-dentate. Inflorescence nearly always branched, the terminal part not much larger than the branches. Flowers white, blue or variegated; helmet 8–15 × 9–14 mm, up to twice as high as wide, hemispherical to conical; nectary-spurs not spirally curved. Follicles 3–5. Seeds winged on 1 angle, with prominent transverse lamellae on the sides. *C. Europe, extending to C. Italy, Bulgaria and N.E. Greece; W. Pyrenees.* Au Bu Cz Ga Ge Gr He Hs Hu It Ju Po Rm Rs(W).

1 Inflorescence with glandular hairs
 2 Follicles 3–5, glabrous (d) subsp. **paniculatum**
 2 Follicles 5, hairy (e) subsp. **valesiacum**
1 Inflorescence glabrous or with eglandular hairs
 3 Helmet scarcely higher than wide (b) subsp. **pyrenaicum**
 3 Helmet about twice as high as wide
  4 Follicles 3–5, hairy at the suture (a) subsp. **variegatum**
  4 Follicles 3, glabrous (c) subsp. **nasutum**

(a) Subsp. **variegatum** (*A. cammarum* Jacq., *A. gracile* Reichenb., *A. judenbergense* (Reichenb.) Gáyer, *A. odontandrum* Wissjul., *A. rostratum* Bernh. ex DC.): Inflorescence glabrous or with short, crispate hairs only, eglandular; bracteoles linear-spathulate. Helmet about twice as high as wide. Follicles 3–5, hairy at the suture. $2n = 16$. ● *C. Europe, extending to N. Italy and C. Ukraine.*

(b) Subsp. **pyrenaicum** Vivant ex Delay, *Bull. Soc. Bot. Fr.* **127**, *Lettres Bot.*: 501 (1981): Like subsp. (a) but helmet scarcely higher than wide; follicles 3, glabrous. ● *W. Pyrenees.*

(c) Subsp. **nasutum** (Fischer ex Reichenb.) Götz, *Feddes Repert.* **76**: 36 (1967) (*A. balcanicum* Velen., *A. camptotrichum* Gáyer, *A. nasutum* Fischer ex Reichenb.): Like subsp. (a) but follicles 3, glabrous. *From the E. Alps to N. Greece.*

(d) Subsp. **paniculatum** (Arcangeli) Greuter & Burdet, *Willdenowia* **19**: 42 (1989) (*A. degenii* Gáyer, *A. paniculatum* Lam., nom. illeg.): Inflorescence more or less glandular-hairy; bracteoles linear. Helmet scarcely higher than wide. Follicles 3–5, glabrous. ● *Mountains of C. Europe.*

(e) Subsp. **valesiacum** (Gáyer) Greuter & Burdet, *Willdenowia* **19**: 42 (1989) (*A. hebegynum* DC., *A. paniculatum* subsp. *valesiacum* (Gáyer) Gáyer, *A. valesiacum* Gáyer): Like subsp. (d) but follicles 5, hairy. ● *Alps.*

Plants called **A. × oenipontanum** Gáyer, *Magyar Bot. Lapok* **10**: 201 (1911), are hybrids between subspp. (a) and (d). Those called **A. × zahlbruckneri** Gáyer, *op. cit.* **8**: 148 (1909), are hybrids between subsp. (a) and **5**(c), and those called **A. × intermedium**

DC., *Reg. Veg. Syst. Nat.* **1**: 374 (1817), and **A. × stoerckianum** Reichenb., *Uebersicht Acon.* 49 (1919), are hybrids between subsp. (**a**) and **5**(**a**).

4. A. toxicum Reichenb., *Uebersicht Acon.* 43 (1819) (*A. bosniacum* G. Beck). Like **3**(**a**) but inflorescence glandular; bracteoles ovate; helmet about twice as high as wide. ● *Romania; C. & W. Jugoslavia.* Ju Rm.

5. A. napellus L., *Sp. Pl.* 532 (1753) (*A. formosum* Reichenb., *A. lobelianum* Reichenb., *A. pyramidale* Miller, *A. strictum* Bernh. ex DC.). Stems 10–200(–300) cm. Leaves divided to the base, the usually narrow segments divided to half-way or more to the midrib into linear or linear-lanceolate lobes 1–8 mm wide. Inflorescence simple or branched, with simple, curved eglandular hairs (rarely a few glandular hairs present, or glabrous), the terminal part larger than the branches. Flowers violet or blue; helmet (12–)18–25(–30) × 11–18 mm, hemispherical; nectary-spurs straight. Follicles usually 3. Seeds 4–5 mm, winged at the angles, the sides smooth or rugulose. ● *W. & C. Europe, extending to the S. Carpathians; one station in C. Russia.* Au Be Br Co Cz †Da Ga Ge He Hs It Ju Lu Po Rm Rs(C, W) *Su [Rs(B)].

A very variable species in which numerous local variants have been recognized as species or subspecies. A number of distinct regional facies appear to merit subspecific rank; the treatment here is based on that of W. Seitz, *Feddes Repert.* **80**: 1–76 (1969). Measurements of bracteoles refer to those of the lower flowers in the terminal inflorescence.

Hybrids between **5** and **7** are cultivated for ornament and are locally naturalized from gardens.

1 At least some bracteoles 2- or 3- or multi-fid, or dentate (**i**) subsp. **firmum**
1 Bracteoles undivided, entire
 2 Leaf-lobes tapered to a long, acute apex (**a**) subsp. **napellus**
 2 Leaf-lobes usually acute, but not tapered to the apex
  3 Stalk of nectary straight; helmet ± triangular (**d**) subsp. **corsicum**
  3 Stalk of nectary curved; helmet not triangular
   4 Bracteoles 1–2(–3) mm; carpels usually 2 (**e**) subsp. **hians**
   4 Bracteoles more than 3 mm; carpels usually 3
    5 Inflorescence glabrous or very sparsely hairy; helmet ± glabrous
     6 Plant usually at least 80 cm; helmet 18–25(–30) mm (**g**) subsp. **fissurae**
     6 Plant 10–60(–80) cm; helmet 12–20 mm (**h**) subsp. **tauricum**
    5 Inflorescence hairy; helmet somewhat hairy
     7 Lower surface of cauline leaves with distinct reticulate venation; middle leaf-segment divided for not more than ½ of its length (**f**) subsp. **superbum**
     7 Lower surface of cauline leaves without reticulate venation; middle leaf-segment divided for more than ½ of its length
      8 Leaf-lobes at least (2–)3–6 mm wide (**b**) subsp. **lusitanicum**
      8 Leaf-lobes 1–2(–3) mm wide (**c**) subsp. **vulgare**

(**a**) Subsp. **napellus** (*A. anglicum* Stapf): Stems 50–150 cm. Middle leaf-segment divided for ⅗ of its length; leaf-lobes tapered to a long, acute apex. Inflorescence branched; bracteoles 2–2·5(–4) mm, entire. Flowering in spring and summer (earlier than the other subspecies). $2n = 32$. *S.W. Britain; ?S.W. France.*

(**b**) Subsp. **lusitanicum** Rouy, *Naturaliste* (*Paris*) **6**: 405 (1884) (*A. linnaeanum* Gáyer, *A. napellus* subsp. *neomontanum* (Wulfen) Gáyer, *A. neomontanum* Wulfen; incl. *A. napellus* subsp. *castellanum* Molero & C. Blanché): Stems up to 200(–300) cm. Middle leaf-segment divided for ⅔–¾ of its length; leaf-lobes (2–)3–6 mm wide, acute to subobtuse. Inflorescence branched, hairy or sometimes glabrous; bracteoles 3–5 mm, entire. $2n = 32$. *W. & C. Europe.*

(**c**) Subsp. **vulgare** (DC.) Rouy & Fouc., *Fl. Fr.* **1**: 142 (1893) (*A. compactum* (Reichenb.) Gáyer): Stems up to 170 cm. Middle leaf-segment divided for ⅔ of its length; leaf-lobes 1–2(–3) mm wide. Inflorescence usually simple, compact; bracteoles 3–8 mm, entire. Helmet (14–)18–32 mm. *Alps; Pyrenees, Cordillera Cantabrica.*

(**d**) Subsp. **corsicum** (Gáyer) W. Seitz, *Feddes Repert.* **80**: 44 (1969) (*A. corsicum* Gáyer): Like subsp. (**c**) but claw of honey-leaf and stalk of nectary straight; helmet more or less triangular. *Corse.*

(**e**) Subsp. **hians** (Reichenb.) Gáyer in Hegi, *Ill. Fl. Mitteleur.* **3**: 498 (1912) (*A. firmum* Reichenb. pro parte, *A. callibotryon* Reichenb.): Stems up to 200 cm. Middle leaf-segment divided for ½–⅔ of its length. Inflorescence little-branched; bracteoles 1–2(–3) mm, entire. Carpels usually 2. *C. Europe.*

(**f**) Subsp. **superbum** (Fritsch) W. Seitz, *Feddes Repert.* **80**: 41 (1969) (*A. superbum* Fritsch): Stems 50–200(–300) cm. Lower surface of cauline leaves with distinct reticulate venation; leaf-lobes wide; middle leaf-segment divided for *c.* ½ of its length. Inflorescence lax, much-branched; bracteoles 3–5 mm, entire. $2n = 32$. *N.W. & C. Jugoslavia.*

(**g**) Subsp. **fissurae** (E. I. Nyárády) W. Seitz, *op. cit.* 42 (1969) (incl. *A. flerovii* Steinb., *A. romanicum* Wołoszczak): Stems 60–150 cm. Inflorescence glabrous or very sparsely hairy; bracteoles 3–5 mm, entire. Helmet 18–25(–30) mm, glabrous. $2n = 32$. *N.W. Jugoslavia; Romania; C. Russia.*

(**h**) Subsp. **tauricum** (Wulfen) Gáyer in Hegi, *Ill. Fl. Mitteleur.* **3**: 496 (1912) (*A. eustachyum* Reichenb., *A. latemarense* Degen & Gáyer, *A. tauricum* Wulfen): Stems 10–60(–80) cm. Leaf-lobes 2–3 mm wide. Inflorescence glabrous or sometimes very sparsely hairy; bracteoles (2–)3–8 mm, entire. Helmet 12–20 mm, glabrous. $2n = 32$. *E. Alps and N.W. Jugoslavia; S. & E. Carpathians.*

(**i**) Subsp. **firmum** (Reichenb.) Gáyer, *op. cit.* 498 (1912) (*A. firmum* Reichenb., *A. napellus* subsp. *skerisorae* (Gáyer) W. Seitz, *A. skerisorae* Gáyer): Stems up to 150 cm. Middle leaf-segment divided for *c.* ⅗ of its length. Inflorescence little-branched, glabrous. Bracteoles 2·5–15 mm, mostly 2- or 3- or multi-fid or dentate. $2n = 32$. *E.C. Europe.*

6. A. burnatii Gáyer, *Magyar Bot. Lapok* **8**: 141 (1909). Like **5**(**c**) but somewhat smaller, the stems up to 140 cm; inflorescence densely glandular-pubescent. *Mountains of S. Europe.* Bu Ga Gr Hs It Ju.

(**a**) Subsp. **burnatii** (*A. delphinense* Gáyer, *A. nevadense* Uechtr. ex Gáyer): Inflorescence lax, usually branched below. *W. Mediterranean region.*

(**b**) Subsp. **pentheri** (Hayek) Jalas, *Ann. Bot. Fenn.* **22**: 220 (1985) (*A. divergens* Pančić, non Rafin., *A. pentheri* Hayek): Inflorescence dense, simple. *Balkan peninsula.*

7. A. angustifolium Bernh. ex Reichenb., *Monogr. Acon.* 95 (1820). Like **5**(**c**) but stems up to 130 cm; inflorescence glabrous; seeds with 7 or 8 transverse lamellae on each side. $2n = 48$. ● *N.E. Italy; N.W. Jugoslavia.* It Ju.

12. Delphinium L.

†B. PAWŁOWSKI

Leaves usually about as wide as long, and deeply palmate, the lobes pinnatisect. Flowers zygomorphic. Perianth-segments 5, the upper with a spur. Honey-leaves 4, free; the 2 upper (nectariferous) with spurs inserted into the spur of the uppermost perianth-segment, and with the limb exserted; the 2 lateral with a wide limb and a narrow claw. Stamens in 8 spirally arranged series. Follicles 3(–5), free.

1 Perennial; limb of lateral honey-leaves ciliate and bearded on upper surface
2 Tubers absent; seeds not covered with scales, winged at angles
3 Stems (especially above) and perianth-segments with mostly smooth, arcuate hairs; perianth-segments rather narrow
4 Perianth-segments usually 22–32 mm, acuminate; hairs not crowded near apex of perianth-segments **3. oxysepalum**
4 Perianth-segments rarely more than 23 mm, subobtuse or acute; hairs ±crowded near apex of perianth-segments
5 Perianth-segments pale blue; petiole, even of upper leaves, not or scarcely shorter than lamina; carpels usually densely pubescent **1. montanum**
5 Perianth-segments deep blue; petiole of upper and middle leaves usually many times shorter than lamina; carpels glabrous or sparsely (very rarely densely) puberulent **2. dubium**
3 Stems with straight hairs or minutely scabrid arcuate hairs, or stems glabrous; perianth-segments 1–2(–3) times as long as wide, suborbicular or obovate, rarely oblong
6 Honey-leaves blue, or the upper yellowish; leaves 3-lobed almost to base
7 Bracts and bracteoles linear; perianth-segments 10–16(–18) mm **7. dyctiocarpum**
7 Lower bracts usually 3-fid, bracteoles linear; perianth-segments 15–25 mm **8. middendorffii**
6 Honey-leaves blackish or dark brown (if blue then leaves not 3-lobed to base)
8 Leaves ±cordate at base; lobes rather wide, not free to base **4. elatum**
8 Leaves ±cuneate at base; lobes rather narrow, usually free to base
9 Inflorescence-axis and perianth-segments glabrous, more rarely patent-pubescent; leaf-lobes linear-lanceolate **5. simonkaianum**
9 Inflorescence-axis and perianth-segments covered with subappressed, minutely scabrid hairs; leaf-lobes oblong **6. cuneatum**
2 Tubers present; seeds covered with membranous scales, not winged
10 Base of petiole little dilated, surrounding less than $\frac{1}{2}$ circumference of stem; pedicels of lower flowers usually at least as long as flowers
11 Follicles usually 5 **9. pentagynum**
11 Follicles 3 **10. emarginatum**
10 Base of petiole strongly dilated, almost completely surrounding stem; pedicels of lower flowers usually shorter than flowers
12 Stem angled; petiole of middle and upper cauline leaves very short **15. schmalhausenii**
12 Stem ±terete; all cauline leaves ±long-petiolate
13 Flowers dirty blackish-violet or -purple, rather widely open; spur ±equalling perianth-segments **14. puniceum**
13 Flowers blue, violet-blue, lilac or whitish, rarely yellowish-violet, not widely open; spur usually distinctly longer than perianth-segments
14 Bracts and bracteoles ovate-lanceolate; flowers (incl. spur) 19–23 mm, pale blue or whitish **13. albiflorum**
14 Bracts and bracteoles linear; flowers (incl. spur) usually more than 23 mm, blue, lilac or yellowish-violet
15 Stems 3–6 mm thick; flowers (incl. spur) 22–27 mm **11. fissum**
15 Stems 7–10 mm thick; flowers (incl. spur) 30–36 mm **12. bolosii**
1 Annual or biennial; lateral honey-leaves glabrous and not ciliate (Mediterranean region)
16 Limb of upper honey-leaves unwinged; seeds few, rugose-areolate
17 Spur $\frac{1}{5}$–$\frac{1}{3}$ as long as perianth-segments; seeds 5·5–7·5 mm **23. staphisagria**
17 Spur at least $\frac{2}{5}$ as long as perianth-segments; seeds 3–4·5 mm
18 Inflorescence-axis, pedicels and outside of perianth-segments shortly pubescent; bracteoles inserted at base of pedicels **25. pictum**
18 Inflorescence-axis and outside of perianth-segments villous-hirsute; bracteoles inserted at some distance above base of pedicels **24. requienii**
16 Limb of upper honey-leaves with lateral wings; seeds numerous, covered with transverse membranous scales
19 Limb of lateral honey-leaves cuneate at base, gradually narrowed into claw
20 Limb of lateral honey-leaves ovate or elliptical; spur of perianth-segments up to twice as long as segments **16. peregrinum**
20 Limb of lateral honey-leaves subquadrate, with wide, truncate apex; spur of perianth-segments ±equalling segments **17. hirschfeldianum**
19 Limb of lateral honey-leaves abruptly contracted into claw
21 Limb of lateral honey-leaves not or little shorter than claw, usually distinctly exserted, not cordate at base **18. ambiguum**
21 Limb of lateral honey-leaves $\frac{1}{2}$–$\frac{3}{4}$ as long as claw, cordate or subcordate at base
22 Limb of lateral honey-leaves oblong, $1\frac{1}{4}$–2 times as long as wide, exserted
23 Limb of lateral honey-leaves usually $1\frac{1}{3}$–2 times as long as wide; perianth-segments usually 7–9 mm **21. gracile**
23 Limb of lateral honey-leaves up to $1\frac{1}{3}$ times as long as wide; perianth-segments usually 5–7 mm **22. hellenicum**
22 Limb of lateral honey-leaves suborbicular, included

24 Follicles and outside of perianth-segments minutely appressed-pubescent (rarely follicles glabrous) **19. halteratum**
24 Follicles and outside of perianth-segments densely covered with patent hairs **20. balcanicum**

Sect. DELPHINASTRUM DC. Perennials. Limb of lateral honey-leaves deflexed, ciliate, bearded on the upper surface. Upper honey-leaves not winged or clawed. Seeds numerous.

1. D. montanum DC. in Lam. & DC., *Fl. Fr.* ed. 3, **5**: 641 (1815). Tubers absent. Stems 15–50(–65) cm, densely covered, like the leaves, pedicels and perianth-segments, with erecto-patent hairs. Petioles not sheathing at the base, those of the upper leaves about as long as the lamina. At least the 3 lowest bracts divided; all bracts shorter, or the lower 1–4 bracts longer, than the flowers. Perianth-segments 12–20(–24) × 3·5–8 mm, pale blue. Honey-leaves about ½ as long as the perianth-segments. Carpels densely pubescent (rarely more or less glabrous). Seeds winged at angles. ● *E. & C. Pyrenees.* Ga Hs.

2. D. dubium (Rouy & Fouc.) Pawł., *Bull. Int. Acad. Sci. Cracovie* ser. B, **1933(1)**: 39 (1934). Like **1** but stems 30–100 cm; hairs usually appressed; petioles of upper leaves shorter than the lamina; lower bracts longer than the flowers; perianth-segments 16–23(–28) × 4–10(–14) mm, dark blue; carpels glabrous (rarely pubescent). ● *S.E. France, N. Italy.* Ga It.

3. D. oxysepalum Borbás & Pax in Borbás, *Term.-Tud. Közl.* **22**: 647 (1890). Tubers absent. Stems 10–50(–80) cm, subglabrous below, more or less densely pubescent above, with appressed hairs. Petioles scarcely sheathing at the base, those of the upper leaves about as long as the lamina. Usually not more than the 2 lowest bracts divided and all of them shorter than the flowers. Perianth-segments 20–32(–40) mm, deep blue or bluish-violet. Honey-leaves about ½ as long as the perianth-segments. Carpels glabrous (rarely sparsely pubescent). Seeds winged at angles. $2n = 32$. ● *W. Carpathians.* Cz Po.

4. D. elatum L., *Sp. Pl.* 531 (1753). Tubers absent. Stems 40–200 cm, with patent or slightly deflexed hairs below, glabrous or with patent hairs above, usually pruinose. Petioles not sheathing at the base, those of the upper leaves shorter than the lamina; lamina palmately lobed. All bracts usually shorter than the flowers, the lower 0–5 divided. Perianth-segments (9–)13–19(–22) mm, deep or dirty blue, or bluish-violet, glabrous or with patent hairs on the outside. Carpels glabrous or more or less patent-pilose. Seeds winged at angles. *Mountains of C. Europe extending to C. Jugoslavia; N. & C. Russia.* Au Cz Ga He It Ju Po Rm Rs(N, C, W, E) [Fe Rs(B) Su].

A polymorphic species that includes three subspecies and many varieties in Europe.

1 Honey-leaves the same colour as the perianth, or the upper paler or yellowish (c) subsp. **austriacum**
1 Honey-leaves blackish or dark brown
2 Lower perianth-segments 1½–2½ times as long as wide, the lateral distinctly wider (a) subsp. **elatum**
2 Lower perianth-segments 2½–3½ times as long as wide, the lateral little wider (b) subsp. **helveticum**

(a) Subsp. **elatum** (*D. intermedium* Aiton; incl. *D. alpinum* Waldst. & Kit., *D. tiroliense* A. Kerner, *D. cryophilum* Nevski, *D. elatum* subsp. *cryophilum* (Nevski) Jurtzev): Perianth-segments (excluding spur) (11–)14–19(–22) mm. Spur usually distinctly longer than the perianth-segments. $2n = 32$. *Throughout most of the range of the species.*

(b) Subsp. **helveticum** Pawł., *Bull. Int. Acad. Sci. Cracovie* ser. B, **1933(1)**: 99 (1934): Perianth-segments (excluding spur) (12–)15–19(–24) mm. Spur about equalling the perianth-segments. ● *S.W. & C. Alps.*

(c) Subsp. **austriacum** Pawł., *Fragm. Fl. Geobot.* **9**: 434 (1963): Flowers usually larger. Perianth-segments (excluding spur) (16–)17–21(–25) mm. Spur about equalling the perianth-segments. ● *Austria.*

5. D. simonkaianum Pawł., *Bull. Acad. Polon. Sci. Lett., Cl. Math. Nat. B.* (1) **1933**: 107 (1934) (*D. pyramidatum* auct., non Albov). Tubers absent. Stems very sparsely hairy below, otherwise glabrous (rarely puberulent in inflorescence), with few, short branches. Leaves shortly cuneate at the base, palmatisect, the segments elongate. Bracts all linear and entire. Perianth-segments with a tuft of hairs at the apex, otherwise glabrous (rarely pilose outside). Carpels usually glabrous. Seeds winged at angles. ● *C. Romania.* Rm.

6. D. cuneatum Steven ex DC., *Reg. Veg. Syst. Nat.* **1**: 359 (1817) (incl. *D. rossicum* Litv.). Tubers absent. Stems with deflexed hairs below; inflorescence-axis, pedicels and outside of perianth-segments puberulent with minutely scabrid hairs. Leaves distinctly cuneate at the base, deeply palmatisect, the segments elongate or not. Lowest 1–3 bracts divided; bracteoles linear to ovate-lanceolate. Carpels glabrous or puberulent. Seeds winged at angles. ● *C. & S.E. Russia, Ukraine.* Rs(C, W, E).

7. D. dyctiocarpum DC., *Reg. Veg. Syst. Nat.* **1**: 360 (1817) (*D. pubiflorum* Turcz.). Tubers absent. Stems 60–120 cm. Leaves rigid, with closed or narrow basal sinus, divided almost to the base into 3 lobes, the lateral lobes deeply divided into narrow segments. Bracts and bracteoles linear. Perianth-segments 10–16(–18) mm, pale blue; spur 10–12 mm. Honey-leaves blue, or the upper yellowish, the laterals usually with suborbicular limb, very densely long-ciliate and bearded. Seeds winged at angles. *E. Russia.* Rs(C, E).

(a) Subsp. **dyctiocarpum**: Stems glabrous, or patent-pilose at the base. Petioles glabrous or long-ciliate below. Perianth-segments glabrous outside. Follicles glabrous, or ciliate on suture. *S.E. Russia.* (*C. Asia, W. Siberia.*)

(b) Subsp. **uralense** (Nevski) Pawł., *Fragm. Fl. Geobot.* **9**: 435 (1963) (*D. uralense* Nevski): Stems, leaves, pedicels, perianth-segments and follicles covered with appressed hairs. *S. Ural.*

8. D. middendorffii Trautv. in Middendorff, *Reise Nord. Öst. Sibir.* **1(2)**: 63 (1847) (*D. cheilanthum* auct., non Fischer ex DC.). Tubers absent. Stems usually less than 50 cm, simple, more or less glabrous. Leaves cordate or reniform, divided almost to the base into 3 lobes, each 3-fid into linear-lanceolate segments, greyish, shortly puberulent beneath, usually glabrous above; petioles long, glabrous. Lower bracts usually 3-fid; bracteoles linear. Perianth-segments 15–25 mm, blue; spur 10–18 mm. Honey-leaves blue. Follicles usually appressed-pubescent, sometimes glabrous. $2n = 16$. *Arctic Russia (near Saratov).* Rs(N). (*Siberia.*)

9. D. pentagynum Lam., *Encycl. Méth. Bot.* **2**: 264 (1786) (incl. *D. gautieri* (Rouy) Pawł.). Tubers present. Stock surrounded by fibres. Stems 30–70(–100) cm, slender, slightly angled, with patent or somewhat deflexed hairs below, with subappressed hairs above, more rarely glabrous above. Base of petiole of cauline leaves surrounding not more than ½ the circumference of the stem. Upper leaves palmatisect, with more or less linear segments; lower leaves palmatipartite, with wider segments. Flowers pale or dirty violet-blue; lateral perianth-segments 13–16 mm, about equalling the subacute spur. Follicles 5, more or less divergent. Seeds 1·2–1·5 mm, obscurely angled, covered with long scales. $2n = 16$. *W. & S. Spain, C. & S. Portugal.* Hs Lu.

10. D. emarginatum C. Presl in J. & C. Presl, *Del. Prag.* **1**: 6 (1822). Tubers present. Stock surrounded by fibres. Stems robust, distinctly angled. Base of petiole of cauline leaves surrounding not more than ½ the circumference of the stem. Perianth-segments 10–14 mm, glabrous on the outside. Follicles 3, erect, glabrous. Seeds obscurely angled, covered with scales. *S.E. Spain; W. Sicilia.* Hs Si. (*N. Africa.*)

(**a**) Subsp. **emarginatum**: Stems up to 80 cm, glabrous or slightly pubescent above. Pedicels glabrous or subglabrous. Perianth-segments bluish-lilac, as long as or scarcely shorter than the spur. *W. Sicilia.*

(**b**) Subsp. **nevadense** (G. Kunze) C. Blanché & Molero, *Anal. Jard. Bot. Madrid* **41**: 469 (1985) (*D. nevadense* G. Kunze): Stems up to 160 cm, glabrous above. Pedicels glabrous. Perianth-segments bright blue, distinctly shorter than the spur. $2n = 16$. *Mountains of S.E. Spain.*

11. D. fissum Waldst. & Kit., *Pl. Rar. Hung.* **1**: 83 (1802) (*D. leiocarpum* Huth, *D. pallasii* Nevski, *D. velutinum* Bertol.). Tubers present. Stock without fibres. Stems 50–160 cm, terete, 3–6 mm thick, simple or little-branched, often pubescent. Leaves mostly 3-fid, acute. Base of petiole of lower cauline leaves almost surrounding the stem. Raceme dense. Bracts and bracteoles linear; margin not membranous. Flowers (including spur) 22–27 mm; perianth-segments directed forwards; spur more or less horizontal. Seeds 2–3·5 mm, densely covered with undulate scales. *S. Europe, extending northwards to N. Romania and C. Ukraine.* Al Bu Ga Gr Hs It Ju Rm Rs(W, K) Tu.

A polymorphic species which varies especially in the degree and nature of the pubescence on all parts; the bracts may be entire or the lower divided, and the leaf-segments may be linear to lanceolate.

(**a**) Subsp. **fissum**: Leaf-segments usually linear-lanceolate to linear. Flowers blue, blue-violet or lilac, the perianth-segments not connivent, variably but not densely pubescent. *Throughout the range of the species except Spain.*

(**b**) Subsp. **sordidum** (Cuatrec.) Amich, Rico & Sánchez, *Anal. Jard. Bot. Madrid* **38**: 153 (1981) (*D. sordidum* Cuatrec.): Leaf-segments broadly lanceolate. Flowers dirty yellowish-violet, the perianth-segments connivent, densely pubescent. $2n = 16$. ● *W. & S. Spain.*

12. D. bolosii C. Blanché & Molero, *Candollea* **38**: 710 (1983). Like **11** but stems 7–10 mm thick, more or less glabrous; leaves 5- to 7-fid; racemes lax; flowers (including spur) 30–36 mm, violet-blue, glabrous to scarcely pubescent. $2n = 18$. ● *Mountains of N.E. Spain (Tarragona prov.).*

13. D. albiflorum DC., *Reg. Veg. Syst. Nat.* **1**: 353 (1817) (*D. fissum* subsp. *albiflorum* (DC.) Greuter & Burdet). Like **11** but bracts and bracteoles ovate-lanceolate, usually membranous-margined; flowers (including spur) 19–23 mm; perianth-segments pale blue, whitish or yellowish-white; honey-leaves blue. *N. Greece (Athos).* ?Gr. (*Anatolia.*)

The presence of this species in Greece requires confirmation. Bulgarian records have been referred to *D. fissum* var. *alibotuchiensis* Rech. fil.

14. D. puniceum Pallas, *Reise* **3**: 736 (1776). Tubers present. Stock without fibres. Stems 30–80 cm, shortly appressed-pubescent. Leaf-segments linear. Base of petiole of lower cauline leaves almost surrounding the stem. Bracts shorter than the pedicels, linear. Perianth-segments dark purple-violet, puberulent on the outside. Follicles appressed-pubescent, more rarely glabrous. Seeds obscurely angled, covered with scales. *S.E. Russia, E. Ukraine.* Rs(E). (*W. & C. Asia.*)

15. D. schmalhausenii Albov, *Trudy Odessk. Obšč. Sad.* **1890**: 441 (1891) (*D. sergei* Wissjul.). Tubers present. Stock without fibres. Stems 40–130 cm, distinctly angled, more or less pubescent. Leaves with numerous linear segments; petioles of middle and upper leaves very short, widely membranous-margined, almost surrounding the stem. Inflorescence long and dense. Flowers deep bluish-violet; spur little longer than the perianth-segments, steeply ascending. Seeds obscurely angled, covered with scales. *S.E. Russia.* Rs(E). (*Caucasus, Anatolia.*)

Sect. DELPHINIUM. Annuals or biennials. Honey-leaves glabrous, the same colour as the perianth-segments, the upper winged, not clawed. Seeds numerous.

16. D. peregrinum L., *Sp. Pl.* 531 (1753). Stems (15–)30–80 cm, erect, pruinose, pubescent throughout or only at the base, with ascending branches. Leaves 1(2)-palmatisect, with linear-lanceolate or linear segments; upper leaves entire. Flowers dirty violet or bluish-violet. Perianth-segments puberulent on the outside, the lateral 7–10 mm, distinctly shorter than the spur. Honey-leaves longer than the perianth-segments; upper with the distal part (above the wings) 3–6 times as long as wide; lateral with elliptical, ovate or obovate limb, gradually narrowed into claw. Follicles usually pubescent. *C. & E. Mediterranean region.* Al Bu Cr Gr It Ju Tu.

17. D. hirschfeldianum Heldr. & Holzm. in Boiss., *Fl. Or., Suppl.* 19 (1888). Like **16** but perianth-segments deep blue, about as long as the spur; limb of lateral honey-leaves subquadrate, crenate-dentate, shortly cuneate at the base. ● *Aegean islands (Aiyina, Kithnos).* Gr.

18. D. ambiguum L., *Sp. Pl.* ed. 2, 749 (1762) (*D. nanum* DC., *D. obcordatum* DC.). Stems 10–30(–50) cm, slender, usually appressed-pubescent, with numerous, long arcuate-divaricate branches. Inflorescence 2- to 10(–15)-flowered, lax. Pedicels and outside of the perianth-segments appressed-pubescent. Flowers deep bluish-violet; spur about twice as long as the perianth-segments. Limb of lateral honey-leaves about as long as the claw, not cordate at the base. Follicles 6–11 mm, $3\frac{1}{2}$–5 times as long as wide, slender, densely appressed-puberulent, often more or less divaricate. $2n = 16$. *S. Spain, S. Portugal.* Hs Lu.

19. D. halteratum Sm. in Sibth. & Sm., *Fl. Graec. Prodr.* **1**: 37 (1809) (*D. longipes* Moris). Stems 10–80 cm, minutely appressed-pubescent (rarely glabrous), few- or many-branched; branches sometimes long and erecto-patent, but not virgate. Inflorescence up to 50-flowered, the flowers violet-blue. Perianth-segments 7–12 mm, minutely and more or less densely appressed-pubescent. Follicles 6–10(–11) mm, $2\frac{1}{2}$–$4\frac{1}{2}$ times as long as wide. *S.W. Europe, extending eastwards to S. Italy.* Ga Hs It Ju Lu Sa Si.

(a) Subsp. **halteratum**: Upper leaves often entire. Inflorescence up to 5-flowered, more or less lax. Limb of lateral honey-leaves weakly cordate at the base. Follicles usually appressed-pubescent. *From S.E. France to S. Italy and W. Jugoslavia.*

(b) Subsp. **verdunense** (Balbis) Graebner & Graebner fil. in Ascherson & Graebner, *Syn. Mitteleur. Fl.* **5(2)**; 703 (1929) (*D. verdunense* Balbis, *D. cardiopetalum* DC.): Upper leaves usually 3-fid. Inflorescence usually not more than 15-flowered, dense. Limb of lateral honey-leaves strongly cordate at the base. Follicles subglabrous. $2n = 16$. *Iberian peninsula; W. France.*

20. D. balcanicum Pawł., *Fragm. Fl. Geobot.* **9**: 439 (1963) (*D. peregrinum* sensu Willk. pro max. parte, non L.). Like **19(a)** but inflorescence usually dense; perianth-segments 6–9 mm, patent-pilose; follicles 5–7(–9) mm, 2–3 times as long as wide, densely covered with rather long patent hairs. $2n = 16$. *S. part of Balkan peninsula.* Bu Gr Ju Tu.

21. D. gracile DC., *Reg. Veg. Syst. Nat.* **1**: 347 (1817). Stems 20–90 cm, very slender, with numerous, long virgate branches, appressed-puberulent or more or less glabrous. Upper leaves mostly entire, shorter than the internodes. Inflorescence usually lax, (1–)3- to 15(–30)-flowered. Flowers often pale lilac-violet or whitish; perianth-segments (6–)7–9 mm, minutely appressed-puberulent outside; spur ($1\frac{1}{2}$–)2–$2\frac{1}{2}$ times as long as the perianth-segments. Lateral honey-leaves longer than the perianth-segments; limb $1\frac{1}{3}$–2 times as long as wide, oblong, exserted. Follicles 6–11 mm, $2\frac{1}{2}$–4 times as long as wide, glabrous or sparsely appressed-pubescent. *Spain, Portugal and Islas Baleares.* Bl Hs Lu.

22. D. hellenicum Pawł., *Fragm. Fl. Geobot.* **9**: 442 (1963) (*D. peregrinum* sensu Halácsy, non L.). Like **21** but stems up to 100 cm, sparsely branched; perianth-segments 5–7(–8) mm; spur $2\frac{1}{2}$–3 times as long as the perianth-segments; limb of lateral honey-leaves *c.* $1\frac{1}{4}$ times as long as wide, ovate or subquadrate; follicles glabrous. ● *S. Greece.* Gr.

Sect. STAPHISAGRIA DC. Annuals or biennials. Honey-leaves glabrous, the same colour as the perianth-segments, the upper paler or yellowish, shortly clawed, unwinged. Seeds few.

23. D. staphisagria L., *Sp. Pl.* 531 (1753). Stems 30–100 cm, stout, simple, patent-pilose. Leaves palmately 5- to 7-lobed, pubescent on both surfaces with mixed very short and longer hairs; segments entire or 3-lobed, with ovate-lanceolate, or oblong, acute lobes. Flowers deep blue. Perianth-segments 13–20 mm. Limb of lateral honey-leaves gradually narrowed into claw. Follicles 8–11 mm wide, inflated. *Mediterranean region; Portugal.* Al Bl Co Cr Ga Gr Hs It Ju Lu Sa Si.

24. D. requienii DC. in Lam. & DC., *Fl. Fr.* ed. 3, **5**: 642 (1815) (*D. pictum* subsp. *requienii* (DC.) C. Blanché & Molero). Like **23** but rhachis, pedicels and outside of perianth-segments villous-hirsute; inflorescence usually lax; lower and middle pedicels distinctly longer than the flowers; bracteoles inserted distinctly above the base of the pedicels; limb of lateral honey-leaves obovate, more or less gradually narrowed into claw. ● *S. France (Îles d'Hyères), Corse, Sardegna.* ?Co Ga Sa.

25. D. pictum Willd., *Enum. Pl. Horti Berol.* 574 (1809). Stems up to 80 cm, pubescent, sometimes with mixed long and short hairs. Rhachis, pedicels and outside of perianth-segments shortly pubescent. Inflorescence usually dense; pedicels shorter than the flowers (except sometimes the lower); bracteoles inserted at the base of the pedicels. Perianth-segments 9–14 mm, bluish-white, darkening to pale blue on drying; spur 6–8 mm. Limb of lateral honey-leaves suborbicular, abruptly contracted into claw. Follicles up to 6 mm wide, inflated. ● *Islands of W. Mediterranean region.* Bl Co Sa.

13. Consolida (DC.) S. F. Gray

A. O. CHATER

Like *Delphinium* but always annual; the 2 upper honey-leaves coalescent into a single structure (nectary), with a single spur; lateral honey-leaves absent; stamens in 5 spirally arranged series; fruit a single follicle.

Literature: E. Huth, *Bot. Jahrb.* **20**: 365–391 (1895). R. de Soó, *Österr. Bot. Zeitschr.* **71**: 233–246 (1922).

1 Nectary ovate, entire; follicle appressed-sericeous — **2. hellespontica**
1 Nectary lobed; follicle glabrous or pubescent but not sericeous
 2 Spur circinate-involute at apex — **1. aconiti**
 2 Spur straight or scarcely curved
  3 Follicle glabrous
   4 Perianth-segments 9 mm or more — **10. regalis**
   4 Perianth-segments less than 9 mm — **11. tenuissima**
  3 Follicle pubescent
   5 Spur less than 7 mm
    6 Lower pedicels more than 7 mm — **6. brevicornis**
    6 Lower pedicels not more than 5 mm
     7 Racemes with not more than 6 flowers — **8. tuntasiana**
     7 Racemes with more than 6 flowers — **9. raveyi**
   5 Spur 7 mm or more
    8 Lower bracts entire
     9 Inflorescence racemose, not or little-branched — **7. uechtritziana**
     9 Inflorescence paniculate, much-branched
      10 Upper lobe of nectary more than 2 mm wide; plant sparsely pubescent; flowers dark blue; seeds black — **9. regalis**
      10 Upper lobe of nectary not more than 2 mm wide; plant densely whitish-pubescent; flowers violet, pale blue or whitish; seeds grey or reddish
       11 Upper lobe of nectary 2 mm or more; spur 12–16 mm — **12. pubescens**
       11 Upper lobe of nectary less than 2 mm; spur 16–22 mm — **13. mauritanica**
    8 Lower bracts dissected
     12 Bracteoles reaching to or beyond base of flower; spur not more than 12 mm
      13 Lower pedicels 1–2 times as long as spur; follicle gibbous above — **3. orientalis**
      13 Lower pedicels shorter than spur; follicle not gibbous above — **4. phrygia**
     12 Bracteoles, at least of lower flowers, not reaching base of flower; spur 12 mm or more
      14 Inflorescence usually racemose; ripe follicle 15–20 mm — **5. ajacis**
      14 Inflorescence paniculate; ripe follicle 10–15 mm
       15 Upper lobe of nectary 2 mm or more; spur 12–16 mm — **12. pubescens**
       15 Upper lobe of nectary less than 2 mm; spur 16–22 mm — **13. mauritanica**

(A) Spur of perianth-segment circinate-involute at apex; nectary attached to pedicel by the extreme end, between the 2 lateral (lower) lobes; spur remote from lateral lobes and pedicel; intermediate lobes present above each lateral lobe; upper lobe small, bifid.

1. C. aconiti (L.) Lindley, *Jour. Hort. Soc. London* **6**: 55 (1851) (*Delphinium aconiti* L.). Stem up to 50 cm, much-branched, densely covered with short, deflexed hairs. Leaf-segments linear. Inflorescence paniculate. Perianth-segments 5–6 mm, blue; spur *c*. 12 mm, circinate-involute at apex. Nectary with triangular, acute intermediate lobes. Follicle 15–25 mm, linear, glabrous. *Turkey-in-Europe* (*Eceabat*). Tu. (*W. Anatolia.*)

C. thirkeana (Boiss.) Schrödinger, *Abh. Zool.-Bot. Ges. Wien* **4(5)**: 62 (1909) (*Delphinium thirkeanum* Boiss.), from Anatolia, like **1** but the nectary with obtuse intermediate lobes and follicle 7–8 mm, has been reported erroneously from Turkey-in-Europe.

(B) Spur of perianth-segment straight or slightly curved; nectary attached to pedicel by the middle, near base of spur which is adjacent to the 2 lateral (lower) lobes; intermediate lobes absent; upper lobe bifid or entire (in **2**, nectary, except for spur, often entire).

2. C. hellespontica (Boiss.) Chater, *Feddes Repert.* **69**: 55 (1964) (*C. olopetala* (Boiss.) Hayek var. *paphlagonica* sensu Hayek, non *Delphinium paphlagonica* Huth, *D. tomentosum* Boiss. 1867 pro parte, non Boiss. 1841, *D. hellesponticum* Boiss.). Stem 10–40 cm. Leaf-segments linear. Inflorescence racemose. Perianth-segments 12–15 mm, bluish-violet; spur 12–22 mm. Nectary ovate, entire, or minutely bifid at apex. Follicle 10–15 mm, appressed-sericeous. *S.E. Jugoslavia, S. Bulgaria, N.E. Greece*. Bu Gr Ju.

European plants have been distinguished as subsp. **macedonica** (Halácsy & Charrel) Chater, *Feddes Repert.* **69**: 55 (1964). The typical subspecies, from Asia Minor, is smaller in all its parts. However, the difference is inconstant.

3. C. orientalis (Gay) Schrödinger, *Abh. Zool.-Bot. Ges. Wien* **4(5)**: 25 (1909) (*Delphinium orientale* Gay). Stem up to 100 cm, simple or branched. Basal leaves with linear-oblong segments; cauline with linear segments. Inflorescence racemose; lower bracts dissected; bracteoles usually close to, and reaching beyond, base of flower. Lower pedicels usually more than 12 mm in fruit, up to twice as long as spur. Perianth-segments 10–14(–20) mm, purplish-violet; spur (8–)10–12 mm. Nectary 3-lobed, with bifid upper lobe. Follicle 15–20 × 5 mm, gibbous above, very abruptly contracted at apex, with beak *c*. 1 mm; seeds reddish-brown. $2n = 16$. *N, C. & E. Spain; S.E. Europe; locally naturalized elsewhere*. Al Bu Gr Hs *Hu Ju Rm Rs(W, K, E) Tu [Au Cz Ga Ge He *It Si].

4. C. phrygia (Boiss.) Soó, *Österr. Bot. Zeitschr.* **71**: 245 (1922). Like **3** but inflorescence laxer, with fewer, paler flowers; lower pedicels 5–10 mm in fruit, shorter than spur; perianth-segments and spur 10–17 mm; follicle not gibbous above; beak *c*. 2 mm. *N. Aegean region and Bulgaria, just extending to S.E. Jugoslavia*. Bu Gr Ju Tu.

(a) Subsp. **phrygia** (*C. orientalis* subsp. *phrygia* (Boiss.) Chater, *Delphinium phrygium* Boiss.): Lower bracts about equalling pedicels, with few divisions. Perianth-segments 7–11 mm, pale pink or mauve. Upper lobe of nectary shorter than lateral lobes. *Bulgaria and S.E. Jugoslavia.*

(b) Subsp. **thessalonica** (Soó) P. H. Davis, *Notes Roy. Bot. Gard. Edinb.* **26**: 174 (1965): Lower bracts exceeding pedicels, with many divisions. Perianth-segments 10–17 mm, pale violet. Upper lobe of nectary at least as long as lateral lobes. *N. Aegean region.*

Further work is required on the distinction between these subspecies in Europe, as well as confirmation of the identity of the plants from Bulgaria.

5. C. ajacis (L.) Schur, *Verh. Mitt. Siebenb. Ver. Naturw.* **4**: 47 (1853) (*C. ambigua* sensu P. W. Ball & Heywood, *Delphinium ajacis* L., *D. ambiguum* auct., non L.). Stem up to 100 cm, simple or branched. Basal leaves with oblong segments; cauline with linear segments. Inflorescence usually racemose, occasionally paniculate; lower bracts dissected; bracteoles small, remote, usually not reaching base of flower. Lower pedicels mostly more than 12 mm in fruit, but less than twice as long as ripe follicle. Perianth-segments 10–14(–20) mm, usually deep blue; spur 13–18 mm. Nectary 3-lobed, with bifid upper lobe. Follicle 15–20 × 5 mm, pubescent, gradually narrowed at apex; seeds black. *S. Europe, extending to N.W. France and N. Romania; locally naturalized elsewhere*. Al Bu Co Cr Ga Gr It Ju Rs(K) *Si Tu [Az Be Br Ge He Hs Rm Rs(W)].

Both **3** and **5** are frequently cultivated for ornament, and naturalized plants often belong to cultivars of these species.

6. C. brevicornis (Vis.) Soó, *Österr. Bot. Zeitschr.* **71**: 245 (1922). Like **5** but plant less robust; basal leaves often with linear segments; perianth-segments violet; spur 5–6 mm; follicle abruptly contracted at apex. ● *W. coast of Jugoslavia; N.W. Greece* (*Ionioi Nisoi*). Gr Ju.

7. C. uechtritziana (Huth) Soó, *Österr. Bot. Zeitschr.* **71**: 236 (1922) (*Delphinium uechtritzianum* Huth). Like **5** but lower bracts entire; spur almost twice as long as perianth-segments; seeds reddish-brown. ● *Jugoslavia, Greece; local*. Gr Ju.

According to Huth, this species becomes very similar to **5** in cultivation, but the available information is insufficient for any decision to be made as to its status.

8. C. tuntasiana (Halácsy) Soó, *Österr. Bot. Zeitschr.* **71**: 239 (1922). Stem less than 15 cm, simple or little-branched. Basal leaves with cuneate-obovate segments; cauline with linear segments. Flowers up to 6, in lax racemes; bracts dissected. Pedicels not more than 5 mm, very stout. Perianth-segments *c*. 10 mm, pale violet; spur *c*. 4 mm. Nectary with very broad, shallowly bifid upper lobe. Follicle 15–17 mm, pubescent. ● *S. Greece*. Gr.

9. C. raveyi (Boiss.) Schrödinger, *Abh. Zool.-Bot. Ges. Wien* **4(5)**: 62 (1909). Stems 10–40 cm, densely pubescent, simple or branched near base. Leaves with linear to linear-oblanceolate segments, appressed-pubescent. Flowers (7–)10–30 in lax racemes; bracts dissected. Pedicels less than 5 mm, stout. Perianth-segments 12–17 mm, violet; spur 2–4 mm. Nectary with narrow, deeply bifid upper lobe. Follicle 10–15 mm, appressed-pubescent. *Turkey-in-Europe* (*near Istanbul*). Tu. (*Anatolia.*)

10. C. regalis S. F. Gray, *Nat. Arr. Brit. Pl.* **2**: 711 (1821) (*C. segetum* (Lam.) S. F. Gray). Stem up to 50 cm, branched, covered with short deflexed hairs. Leaf-segments all linear. Inflorescence paniculate; all bracts linear, entire; lower pedicels longer than flower and ripe follicle. Spur 12–25 mm. Upper lobe of nectary more than 2 mm wide, entire, erose, or shallowly bifid with rounded lobes. Follicle 8–15 mm; seeds black. *Most of Europe,* *except the extreme north and south.* Al Au Be Bu Cz Da Ga Ge Gr He Hu It Ju Po Rm Rs(N, B, C, W, K, E) Su Tu [Fe Ho Hs].

(a) Subsp. **regalis** (*Delphinium consolida* L., *D. consolida* subsp. *arvense* (Opiz) Graebner, *C. regalis* subsp. *arvensis* (Opiz) Soó): Stem little-branched. Inflorescence rather dense. Perianth-segments 12–15 mm, usually light violet-blue. Follicle usually 3 times as long as wide, glabrous. $2n = 16$. *Mainly N. & C. Europe.*

(b) Subsp. **paniculata** (Host) Soó, *Österr. Bot. Zeitschr.* **71**: 243 (1922) (*C. segetum* subsp. *hostianum* sensu Dostál, *Delphinium paniculatum* Host; incl. *D. divaricatum* Ledeb., *C. divaricata* (Ledeb.) Schrödinger, *C. regalis* subsp. *divaricata* (Ledeb.) Munz): Stem much-branched. Inflorescence very lax. Perianth-segments 9–11 mm, dark blue. Follicle usually twice as long as wide, glabrous or appressed-pubescent. $2n = 16$. *S.E. & S.C. Europe, extending to N. Italy and S.E. France.*

11. C. tenuissima (Sibth. & Sm.) Soó, *Österr. Bot. Zeitschr.* **71**: 241 (1922) (*Delphinium tenuissimum* Sibth. & Sm.). Stem up to 50 cm, branched, covered with patent hairs. Segments of basal leaves oblong; segments of cauline leaves linear. Inflorescence lax, paniculate; lower bracts linear, entire; lower pedicels longer than flower and ripe follicle. Perianth-segments 5–7 mm, violet-blue; spur 7–10 mm. Upper lobe of nectary broad, bifid, with obtuse lobes. Follicle 5–8 mm, less than twice as long as wide, glabrous. ● *S.E. Greece.* Gr.

12. C. pubescens (DC.) Soó, *loc. cit.* (1922) (*Delphinium pubescens* DC.; incl. *D. loscosii* Costa). Whole plant densely appressed-pubescent; stem up to 50 cm, branched. Leaf-segments linear. Inflorescence lax, paniculate; lower bracts entire or dissected; bracteoles not reaching base of flower; pedicels longer than flower and ripe follicle. Perianth-segments 6–12 mm, violet, pale blue or whitish; spur 12–16 mm. Upper lobe of nectary 2–3(–4) mm, longer than lateral perianth-segments, not more than 2 mm wide, bifid, with acute lobes. Follicle 10–15 × 5 mm, sparsely pubescent, the suture not enlarged and tuberculate; seeds grey or reddish, with well-separated annular flanges 0·3–0·5 mm. *S.W. Europe, extending to S. Italy.* Ga Hs It †Si.

13. C. mauritanica (Cosson) Munz, *Jour. Arnold Arb.* **48**: 48 (1967) (*Delphinium mauritanicum* Cosson). Like **12** but upper lobe of nectary 0·5–1 mm, shorter than lateral perianth-segments; spur 16–22 mm; fruiting pedicel as long as or shorter than flower and ripe follicle; suture of follicle enlarged and tuberculate; seeds with more or less imbricate, annular flanges 0·2–0·3 mm. *C. & E. Spain.* Hs.

C. incana (E. D. Clarke) Munz, *Jour. Arnold Arb.* **48**: 181 (1967) (*C. rigida* (DC.) Hayek, *Delphinium rigidum* DC.), from W. Asia, has been reported as naturalized in Jugoslavia, but appears to have been never more than casual.

14. Anemone L.

†T. G. TUTIN (EDITION 1) REVISED BY A. O. CHATER (EDITION 2)

Perennial herbs. Flowering stems with a whorl of 3(4) often partially united leaves. Flowers conspicuous, usually solitary. Perianth-segments 5–19, petaloid, imbricate. Honey-leaves absent. Achenes numerous; style not elongated or feathery.

1 Cauline leaves sessile
2 Stems branched — **10. dichotoma**
2 Stems simple
3 Cauline leaves undivided or almost so
4 Perianth-segments 12–19, narrowly elliptical — **16. hortensis**
4 Perianth-segments 7–12, broader — **17. pavonina**
3 Cauline leaves much divided
5 Flowers 3–8 in an umbel; achenes glabrous, winged — **9. narcissifolia**
5 Flowers solitary; achenes lanate, not winged
6 Basal leaves deeply cut; flowers not yellow — **15. coronaria**
6 Basal leaves shallowly lobed, suborbicular; flowers yellow — **14. palmata**
1 Cauline leaves petiolate
7 Plant without a creeping or tuberous rhizome; achenes lanate
8 Primary divisions of leaves 5; perianth-segments usually 5 — **11. sylvestris**
8 Primary divisions of leaves 3; perianth-segments 7–10
9 Plant 5–12 cm (up to 20 cm in fruit); stock slender, with few or no remains of old petioles; head of achenes elongating in fruit — **12. baldensis**
9 Plant 25–30 cm; stock stout, clothed in remains of old petioles; head of achenes subglobose — **13. pavoniana**
7 Plant with a creeping or tuberous rhizome; achenes not lanate, but sometimes densely pubescent
10 Perianth-segments 1–1·5 mm wide, greenish-white, strongly deflexed — **8. reflexa**
10 Perianth-segments at least 2·5 mm wide, variously coloured or white but not greenish-white, patent
11 Leaves 3-partite and serrate but not further divided — **3. trifolia**
11 Leaves with primary divisions lobed
12 Anthers pale yellow, greenish or white; flowers usually blue
13 Perianth-segments pubescent beneath; head of achenes erect — **6. apennina**
13 Perianth-segments glabrous beneath; head of achenes nodding — **7. blanda**
12 Anthers deep yellow; flowers rarely blue
14 Perianth-segments with white hairs beneath
15 Cells of the upper epidermis of the perianth-segments not mamillate; flowers always yellow — **4. ranunculoides**
15 Cells of the upper epidermis of the perianth-segments mamillate; flowers reddish, purplish, white or yellow (C. Ural) — **5. uralensis**
14 Perianth-segments glabrous beneath
16 Rhizome brown; perianth-segments (5)6 or 7(–12) — **1. nemorosa**
16 Rhizome yellowish; perianth-segments 8–12 — **2. altaica**

Sect. SYLVIA Gaudin. Achenes not or slightly compressed, not lanate. Style shorter than achene, hooked.

1. A. nemorosa L., *Sp. Pl.* 541 (1753). Rhizome brown, creeping. Basal leaves 1 or 2, 3-partite, the divisions shortly stalked and again deeply lobed, pubescent beneath; petioles long. Flowering stems 6–30 cm; cauline leaves without buds in their axils, similar to basal leaves but smaller and petioles flattened; marginal hairs 0·5–0·75 mm. Perianth-segments (5)6 or 7(–12), white, flushed with pink or purple beneath (rarely entirely purple or blue), oblong-ovate, glabrous. Anthers yellow. Achenes downy. Head of achenes nodding. $2n$ = 28–32, 37, 42, 45, 46. *Woods. Most of Europe, but rare in the Mediterranean region.* Al Au Be Br Bu Co Cz Da Fe Ga Ge Gr Hb He Ho Hs Hu It Ju Rs(N, B, C, W, ?E) Su Tu.

An isolated population in arctic Russia (Poluostrov Kol'skij), with 9–12 perianth-segments, is somewhat intermediate between **1** and **2**. Sterile hybrids between **1** and **4** occur where the parents grow together.

2. A. altaica Fischer ex C. A. Meyer in Ledeb., *Fl. Altaica* **2**: 362 (1830). Like **1** but with a rather thick, yellow rhizome; marginal hairs on cauline leaves shorter; flowering stems up to 20 cm; flowers with 8–12 perianth-segments which are sometimes violet beneath. *Woodland margins. N., C. & E. Russia.* Rs(N, C, E). (*N. Asia.*)

Closely related to **1** and perhaps a subspecies of it.

3. A. trifolia L., *Sp. Pl.* 540 (1753). Like **1** but basal leaves usually absent; all leaves 3-partite and serrate, but not lobed; cauline leaves with buds in their axils; anthers white or blue; achenes shortly setose; head of achenes erect or nodding. $2n$ = 32. ● *S. & S.C. Europe, from Portugal to Hungary; one station in Finland.* Au Fe Ga Hs Hu It Ju Lu.

(**a**) Subsp. **trifolia**: Perianth-segments ovate; anthers blue; head of achenes erect. $2n$ = 16. *Usually calcicole. Throughout the range of the species, except the Iberian peninsula.*

(**b**) Subsp. **albida** (Mariz) Ulbr., *Bot. Jahrb.* **37**: 220 (1905). Perianth-segments elliptical; anthers white; head of achenes usually nodding. *Usually calcifuge. N.W. Spain and Portugal.*

4. A. ranunculoides L., *Sp. Pl.* 541 (1753). Rhizome brown. Basal leaves 0 or 1, deeply divided. Stem 7–30 cm; cauline leaves deeply divided; petioles short. Flowers 1·5–2 cm in diameter, solitary or 2(–5), yellow. Perianth-segments broadly ovate, slightly pubescent beneath; epidermal cells of upper surface not mamillate. Anthers yellow. $2n$ = 32, 48. *Woods. Most of Europe except the islands, but rare in the west and most of the Mediterranean region.* Au Be Bu Cz Da Fe Ga Ge He Ho Hs Hu It Ju No Po Rm Rs(N, B, C, W, K, E) Su Tu.

Small plants with short rhizomes, forming dense patches, have been called subsp. **wockeana** (Ascherson & Graebner) Ulbr., *Bot. Jahrb.* **37**: 215 (1905). They occur in Czechoslovakia, Germany and Poland.

5. A. uralensis Fischer ex DC., *Prodr.* **1**: 19 (1824). Like **4** but flowers reddish, purplish, yellow or white; epidermal cells of upper surface of perianth-segments mamillate. *Open coniferous woodland. C. & S. Ural* (55° *N. to* 58° *N.*). Rs(C).

Intermediate between **4** and *A. caerulea* DC. from Siberia, and perhaps of hybrid origin.

6. A. apennina L., *Sp. Pl.* 541 (1753). Rhizome tuberous, brown. Basal leaves 1 or 2, 3-partite, the divisions acute, stalked, and again deeply lobed, pubescent beneath; cauline leaves all petiolate, 2-ternate, subglabrous. Flowering stems 5–30 cm. Perianth-segments 8–14(–18), narrowly oblong-elliptical, blue (or white, var. *albiflora* Strobl), pubescent beneath near base. Anthers greenish-cream. Achenes downy. Head of achenes erect. ● *C. part of Mediterranean region, extending to S.E. Jugoslavia; cultivated elsewhere for ornament, and widely naturalized.* Al Co It Ju Si [Au Be Br Da Ga Ge Ho Su].

7. A. blanda Schott & Kotschy, *Österr. Bot. Wochenbl.* **4**: 129 (1854). Like **6** but leaves glabrous beneath, the primary divisions subsessile, more or less obtuse; perianth-segments usually more numerous, glabrous; head of achenes nodding. $2n$ = 14, 16. *Balkan peninsula.* Al Bu Gr Ju Tu.

Perhaps only subspecifically distinct from **6**.

8. A. reflexa Stephan in Willd., *Sp. Pl.* **2(2)**: 1282 (1799). Rhizome slender, yellowish-brown. Basal leaves with rhombic or ovate, long-petiolulate segments; cauline leaves petiolate, ternate, subglabrous. Perianth-segments 5(6), 6–7 × 1–1·5 mm linear-oblong, greenish-white, pubescent beneath, deflexed parallel to the peduncle. Achenes densely pubescent. *E. Russia* (*C. Ural*). Rs(C). (*N. Asia.*)

Sect. OMALOCARPUS DC. Achenes strongly compressed, often winged, not lanate. Style shorter than achene, hooked.

9. A. narcissifolia L., *Sp. Pl.* 542 (1753) (*A. narcissiflora* L. (1759), nom. illeg.). Plant 20–40 cm, pubescent. Stock stout. Basal leaves deeply palmately divided; petioles long. Cauline leaves sessile, deeply cut, the segments narrowly linear. Inflorescence umbellate, with 3–8 flowers. Flowers 2–3(–4) cm in diameter, white, sometimes pinkish outside; pedicels 3–4 cm. Perianth-segments 5 or 6, obovate. Anthers yellow. *Mountains of S. & C. Europe; Ural.* Al Au Bu Cz Ga Ge He Hs It Ju Po Rm Rs(N, C, W).

Variable in E. Europe and still more so in Asia, and probably divisible into a number of subspecies, but a thorough review of the whole complex is needed. Two subspecies can be recognized in Europe:

(**a**) Subsp. **narcissifolia** (incl. *A. laxa* (Ulbr.) Juz.): Primary divisions of leaves 5, sessile. $2n$ = 14. *Throughout the range of the species except Ural.*

(**b**) Subsp. **biarmiensis** (Juz.) Jalas, *Ann. Bot. Fenn.* **25**: 297 (1988) (*A. biarmiensis* Juz.): Primary divisions of leaves 3, distinctly petiolulate. $2n$ = 14. *Ural.*

Sect. ANEMONIDIUM Spach. Achenes strongly compressed and winged, not lanate. Style about as long as achene, straight.

10. A. dichotoma L., *Sp. Pl.* 540 (1753) (*A. pennsylvanica* sensu Ledeb., non L.). Rhizome slender. Stems 20–70 cm, repeatedly branched. Leaves deeply 5- to 7-lobed; petioles long. Lowest cauline leaves 3 in a whorl, deeply 2- or 3-lobed, the others 2 in a whorl; all sessile. Flowers *c.* 3 cm in diameter, white. Perianth-segments 5, unequal. Achenes glabrous. *E. Russia.* Rs(C). (*N. Asia.*)

Sect. ANEMONE. Achenes densely lanate with hairs longer than diameter of achene. Style straight, shorter than achene.

11. A. sylvestris L., *Sp. Pl.* 540 (1753). Plant 15–50 cm, pubescent, spreading by root-buds. Basal leaves deeply palmately divided, the petioles long; cauline leaves similar but smaller, the petioles short. Flowers solitary or (2), 4–7 cm in diameter, white. Perianth-segments usually 5, broadly ovate. Anthers yellow. $2n$ = 16. *C. & E. Europe, extending to S.E. Sweden and N.C. France.* Au †Be Bu Cz Ga Ge He Hu It Ju Po Rm Rs(N, B, C, W, ?K, E) Su.

12. A. baldensis L., *Mantissa* 78 (1767). Plant 5–12 cm, up to 20 cm in fruit, pubescent. Stock slender, with few or no remains of old petioles. Basal leaves 3 times 3-fid with 2- or 3-lobed segments;

cauline similar but smaller, with short, broad petioles. Flowers solitary or (2), 2·5–4 cm in diameter, white. Perianth-segments (6–)8–10, ovate to elliptical, acute. Anthers yellow. Head of achenes ovoid-oblong. $2n = 16$. *Alps, local; mountains of Jugoslavia, southwards to Crna Gora.* Au Ga He It Ju.

13. A. pavoniana Boiss., *Diagn. Pl. Or. Nov.* **3(1)**: 6 (1854). Like **12** but plant 25–50 cm, sparsely pubescent; stock stout, clothed in remains of old petioles; perianth-segments 6–8, elliptic-oblong, obtuse; head of achenes subglobose. $2n = 16$. ● *Mountains of N. Spain.* Hs.

14. A. palmata L., *Sp. Pl.* 538 (1753). Plant 10–60 cm. Rhizome tuberous. Stem pubescent. Basal leaves suborbicular, with 3–5 shallow, obtuse, dentate lobes. Cauline leaves united at base, with 3–5 linear-lanceolate divisions. Flowers 1 or 2, 2·5–3·5 cm in diameter, yellow. Perianth-segments 8–15, oblong. $2n = 16$. *S.W. Europe.* Ga ?Gr Hs Lu Sa Si.

15. A. coronaria L., *Sp. Pl.* 539 (1753). Plant 15–45 cm. Basal leaves biternate, the segments deeply lobed. Cauline leaves deeply cut into narrow divisions. Flowers 3·5–6·5 cm in diameter, solitary. Perianth-segments 5–8, elliptical, red, purple, pink, blue or white. Anthers blue. *Mediterranean region.* Bl Co Cr Ga Gr *Hs It Ju *Sa Si Tu.

16. A. hortensis L., *Sp. Pl.* 540 (1753) (*A. stellata* Lam.). Like **15** but basal leaves less divided (or only lobed), with broader segments; cauline leaves linear-lanceolate, usually undivided; perianth-segments 12–19, narrowly elliptical, patent, purplish-pink. ● *C. Mediterranean region, extending to the S. Aegean region.* Al Co Cr Ga Gr It Ju Sa Si.

(a) Subsp. **hortensis**: Flowers 3–4 cm in diameter. Perianth-segments often more than 14, usually purplish-pink. $2n = 16$. *C. Mediterranean region.*

(b) Subsp. **heldreichii** (Boiss.) Rech. fil., *Denkschr. Akad. Wiss. Math.-Nat. Kl. (Wien)* **105**: 74 (1943) (*A. hortensis* var. *heldreichii* Boiss.): Flowers 2–3·5 cm in diameter. Perianth-segments up to 14, narrower, white, bluish or pinkish beneath. ● *Kriti, Karpathos.*

17. A. pavonina Lam., *Encycl. Méth. Bot.* **1**: 166 (1783). Like **15** but lowest leaves less divided (or only lobed), with broader segments; cauline leaves linear-lanceolate, usually undivided; perianth-segments (7)8 or 9(–12), scarlet, pink or purple, often yellowish at base. $2n = 16$. *S. Europe, from S.W. France to Turkey, local.* Bu †Co Ga Gr *It Ju Tu.

A. fulgens Gay in DC., *Prodr.* **1**: 18 (1824), is probably a hybrid between **16** and **17**. It is more or less intermediate and does not breed true.

15. Hepatica Miller

†T. G. TUTIN

Like *Anemone* but with 3- to 5-lobed evergreen leaves with entire or crenate-dentate lobes, and a calyx-like involucre of 3 small, entire bracts usually close below the flower.

Rhizome short, thick; leaf-lobes entire (rarely with 1 or 2 teeth); flowers 2·5–3·5 cm diameter — **1. nobilis**
Rhizome long, slender; leaf-lobes with 3–5 large crenations; flowers *c.* 5 cm diameter — **2. transsilvanica**

1. H. nobilis Schreber, *Spicil. Fl. Lips.* 39 (1771) (*H. triloba* Chaix, *Anemone hepatica* L., *A. angulosa* auct., non Lam.). Rhizome short, thick, dark brown. Petioles and peduncles 5–15 cm. Leaves cordate at base, 3-lobed, often purplish beneath; lobes ovate, entire, or rarely with 1 or 2 teeth, sericeous when young and sometimes persistently villous beneath. Flowers 15–25 mm in diameter, bluish-violet, purple, white or pinkish. Bracts ovate, entire. Perianth of 6 or 7(–10) oblong-ovate segments. $2n = 14$. *Much of Europe, but absent from most of the islands.* Al Au Bu Co Cz Da Fe Ga Ge He Hs Hu It Ju No Po Rm Rs(N, B, C, W) Su [Be].

2. H. transsilvanica Fuss, *Verh. Mitt. Siebenb. Ver. Naturw.* **1**: 83 (1850) (*H. angulosa* auct., non (Lam.) DC.). Like **1** but with a long, slender rhizome; leaf-lobes crenate-dentate with 3–5 large crenations; flowers 25–40 mm in diameter, usually with 8 or 9 elliptical perianth-segments; and bracts with 2 or 3 small teeth near the top. $2n = 28$. *Mountain woods.* ● *C. Romania.* Rm.

The hybrid between **1** and **2** (*H.* × *media* Simonkai) occurs locally with the parents and has sterile pollen.

16. Pulsatilla Miller

†T. G. TUTIN (EDITION 1) REVISED BY J. R. AKEROYD (EDITION 2)

Caespitose perennial herbs with a stout stock. Leaves usually 2- to 4-pinnately or -palmately divided, often sericeous when young. Cauline leaves usually sessile and united at the base. Flowers solitary. Perianth-segments usually 6, sericeous beneath; nectar-secreting staminodes present. Styles elongating and feathery in fruit.

The distribution of most of the species is very fragmented, probably mainly owing to post-glacial climatic changes and the intolerance of the species to ploughing, shade and bad drainage; they display considerable variation in pubescence, dissection of leaves and the size and colour of the perianth-segments. These characters have been used as a basis for specific distinction, but correlations between characters that are reliable within a small area break down when larger areas are considered. The variation, in fact, appears to be of a 'dissected continuous' type, rather than the discontinuous kind met with in related genera, and does not often appear to fall into the geographical pattern characteristic of subspecies.

Literature: A. Zāmels, *Acta Horti Bot. Univ. Latv.* **1**: 81–108 (1926). A. Zāmels & B. Paegle, *op. cit.* **2**: 133–161 (1927). D. Aichele & H. W. Schwegler, *Feddes Repert.* **60**: 1–230 (1957). K. Krause, *Bot. Jahrb.* **78**: 1–68 (1958).

1 Cauline leaves shortly petiolate, resembling the basal but smaller
2 Terminal segments of mature leaves not divided to midrib; lamina distinctly pubescent — **1. alpina**
2 Terminal segment of mature leaves divided to midrib; lamina glabrous or subglabrous — **2. alba**
1 Cauline leaves sessile, divided into linear segments and not closely resembling the basal
3 Basal leaves palmately divided — **9. patens**
3 Basal leaves pinnately divided
4 Basal leaves evergreen, 1-pinnatifid, the segments lobed; flowers usually white — **3. vernalis**
4 Basal leaves withering in autumn, 1- to 4-pinnate; flowers usually purple
5 Flowers erect
6 Basal leaves 3- or 4-pinnatisect, with 7–9 primary

segments; plant at first sericeous, becoming glabrous **7. vulgaris**
6 Basal leaves pinnate, with 3–5 primary segments; plant persistently lanate **8. halleri**
5 Flowers nodding
7 Perianth-segments less than $1\frac{1}{2}$ times as long as stamens, recurved at apex **4. pratensis**
7 Perianth-segments at least twice as long as stamens, not recurved at apex
8 Cauline leaves with *c.* 25 lobes; flowers bluish to dark violet **5. montana**
8 Cauline leaves with *c.* 20 lobes; flowers reddish or sometimes dark violet **6. rubra**

1. P. alpina (L.) Delarbre, *Fl. Auvergne* ed. 2, 552 (1800) (*Anemone alpina* L.). Stem 20–45 cm. Basal leaves long-petiolate, distinctly pubescent, 2-pinnate; terminal segment not divided quite to the midrib, the lobes often recurved; cauline leaves similar, but with short, broad petioles. Flowers 4–6 cm in diameter, more or less erect; perianth-segments ovate, the outer purplish, the inner white, or all pale yellow, sericeous. *Mountains of C. & S. Europe, from S. Germany to C. Spain.* Au Co Ga Ge He Hs It Ju.

(**a**) Subsp. **alpina** (incl. subsp. *cantabrica* Laínz, subsp. *font-queri* Laínz & P. Monts.): Outer perianth-segments white or purplish, the inner white. Achenes *c.* 5 mm. $2n = 16$. *Usually calcicole.*

(**b**) Subsp. **apiifolia** (Scop.) Nyman, *Consp.* 2 (1878) (subsp. *sulphurea* (DC.) Ascherson & Graebner): Perianth-segments pale yellow. Achenes *c.* 4 mm. $2n = 16$. *Usually calcifuge.*

The plants from Corse have been distinguished as subsp. **cyrnea** Gamisans, *Candollea* **32**: 58 (1977), but differ from subsp. (**a**) only in having broader, less hairy leaf-lobes and in being calcifuge.

2. P. alba Reichenb., *Fl. Germ. Excurs.* 732 (1832) (*P. alpina* subsp. *alpicola* H. Neumayer, nom. inval.; incl. *P. alpina* subsp. *austriaca* Schwegler). Like **1** but usually smaller; terminal segments of mature leaves divided quite to the midrib, the lobes not recurved; lamina glabrous or subglabrous; flowers 2·5–4·5 cm in diameter, always white. $2n = 16$. *Usually calcifuge. Mountains of C. Europe, extending to N. Spain and C. Jugoslavia.* Au Cz Ga Ge Hs Ju Po Rm Rs(W).

3. P. vernalis (L.) Miller, *Gard. Dict.* ed. 8, no. 3 (1768) (*Anemone vernalis* L.). Stem 5–15 cm (up to 35 cm in fruit). Basal leaves 1-pinnatifid with 3–5 dentate segments, evergreen, shortly petiolate, subglabrous; cauline leaves sericeous, the segments linear. Flowers 4–6 cm in diameter, nodding at first, then erect; perianth-segments narrowly ovate, the outer flushed pink, violet or blue, the inner white, sericeous. $2n = 16$. ● *From* c. 63° *N. in Scandinavia to N. Spain, N. Italy and Bulgaria.* Au Bu Cz Da Fe Ga Ge He Hs It Ju No Po Su.

Hybrids with **9**(**a**) occur in Finland.

4. P. pratensis (L.) Miller, *op. cit.*, no. 2 (1768) (*P. nigricans* Störck, *Anemone pratensis* L.). Stem *c.* 10 cm (up to 45 cm in fruit). Basal leaves pubescent, petiolate, usually 3-pinnate, the segments deeply cut into *c.* 150 narrow lobes; cauline leaves united below, pubescent, with *c.* 30 lobes. Flowers 3–4 cm in diameter, more or less cylindrical, nodding, dark purple, reddish, pale violet, greenish-yellow or rarely white; perianth-segments recurved at the apex, less than $1\frac{1}{2}$ times as long as the stamens. $2n = 16, 32$. ● *C. & E. Europe, extending westwards to S.E. Norway, W. Denmark and N.W. Jugoslavia.* Au Bu Cz Da Ge Hu Ju No Po Rm Rs(B, C, W, E) Su.

The variation in flower-colour is correlated with distribution but is probably not of subspecific significance. Plants with dark purple flowers occur in the north, and have been called subsp. *nigricans* (Störck) Zāmels, those with the flowers dirty yellow or pale greyish-violet inside occur in the south-east (subsp. *hungarica* Soó), and those with pale violet flowers (subsp. *pratensis*) occur in the intervening area.

5. P. montana (Hoppe) Reichenb., *Fl. Germ. Excurs.* 733 (1832) (*Anemone montana* Hoppe, *A. balkana* Gürke). Like **4** but the cauline leaves with *c.* 25 lobes; flowers bluish to dark violet; perianth-segments not recurved at the apex but soon spreading from the base, about twice as long as the stamens. $2n = 16$. ● *From S.W. Alps to E. Romania and N. Greece.* Bu Ga ?Gr He It Ju Rm.

This species has been divided into a number of subspecies and varieties by different authors; these are described in J. Rummelspacher, *Feddes Repert.* **71**: 1–49 (1965).

6. P. rubra (Lam.) Delarbre, *Fl. Auvergne* ed. 2, 553 (1800). Like **4** but the cauline leaves with *c.* 20 lobes; flowers dark reddish-brown to purplish- or blackish-red, or reddish-violet; perianth-segments not recurved at the apex but soon spreading from the base, $2–2\frac{1}{2}$ times as long as the stamens. ● *C. & S. France; N. & E. Spain.* Ga Hs.

The plants from Spain have been described as subsp. **hispanica** Zimm. ex Aichele & Schwegler, *Feddes Repert.* **60**: 158 (1957). They differ from subsp. *rubra* principally in having reddish-violet flowers, and merit at most varietal rank.

7. P. vulgaris Miller, *Gard. Dict.* ed. 8, no. 1 (1768) (*P. ucrainica* (Ugr.) Wissjul., *Anemone pulsatilla* L.). Stem 3–12 cm (up to 45 cm in fruit). Basal leaves more or less sericeous when young, becoming glabrous or subglabrous, pinnately divided into 7–9, 2- or 3-pinnatisect segments; lobes linear to linear-lanceolate; cauline leaves united below, sericeous. Flowers 5·5–8·5 cm in diameter, campanulate, erect or suberect, dark to pale purple; perianth-segments usually straight, acute, 2–3 times as long as the stamens. ● *W. & C. Europe, extending northwards to* 60° *N. in Sweden, and eastwards to E. Ukraine.* Au Be Br Cz Da Ga Ge He †Ho Hu Ju Po Rm Rs(C, W, E) Su.

A very variable complex which has been divided into numerous species and subspecies which are, however, ill-defined and appear to represent the more distinct of the numerous isolated populations.

The smaller and more isolated of these, particularly towards the edge of the range of the species, are fairly homogeneous, but in other areas there is considerable variation within populations. While it is likely that most populations are statistically separable from one another, there is a large amount of overlap between them and intermediates occur frequently. Two subspecies can be recognized:

(**a**) Subsp. **vulgaris**: Leaves appearing with the flowers; basal leaves usually with more than 100 lobes; perianth-segments narrowly elliptical. $2n = 32$. *W. & C. Europe; S. Scandinavia.*

(**b**) Subsp. **grandis** (Wenderoth) Zāmels, *Acta Horti Bot. Univ. Latv.* **1**(**2**): 104 (1926) (*P. grandis* Wenderoth): Leaves appearing after the flowers; basal leaves with *c.* 40 lobes; perianth-segments broadly elliptical. *C. Europe; Ukraine.*

An isolated population on Gotland, which seems to belong to subsp. (**b**), has been called subsp. **gotlandica** (K. Joh.) Zāmels & Paegle, *Acta Horti Bot. Univ. Latv.* **2**: 159 (1927).

8. P. halleri (All.) Willd., *Enum. Pl. Horti Berol.* 580 (1809) (*Anemone halleri* All.). Stems 3–12 cm (up to 45 cm in fruit). Basal leaves persistently and often densely lanate, 1-pinnate with 3–5 segments, the terminal long-stalked; segments pinnatifid; lobes oblong-lanceolate; cauline leaves united below, sericeous. Flowers 5·5–8·5 cm in diameter, campanulate, erect or suberect, dark violet; perianth-segments usually straight, acute, 2–3 times as long as the stamens. ● *Alps; W. Carpathians; Balkan peninsula; Krym.* ?Al Au Bu Cz Ga He It Ju Po Rs(K).

1 Plant rarely more than 5 cm at anthesis; basal leaves with 50–100 lobes
2 Primary divisions of basal leaves often petiolate (**b**) subsp. **rhodopaea**
2 Primary divisions of basal leaves usually sessile (**c**) subsp. **taurica**
1 Plant usually more than 5 cm at anthesis; basal leaves usually with fewer than 50 lobes
3 Primary divisions of basal leaves usually 3 (**d**) subsp. **slavica**
3 Primary divisions of basal leaves usually 5
4 Lamina 3–7 cm (**a**) subsp. **halleri**
4 Lamina 5–11 cm (**e**) subsp. **styriaca**

(**a**) Subsp. **halleri**: Plant usually more than 5 cm at anthesis. Primary divisions of the basal leaves usually 5, sessile; lamina 3–7 cm, usually with fewer than 50 lobes, more or less lanate. $2n = 32$. *S.W. & S.C. Alps.*

(**b**) Subsp. **rhodopaea** (Stoj. & Stefanov) K. Krause, *Bot. Jahrb.* **78**: 44 (1958) (*P. latifolia* Rupr., *P. velezensis* (G. Beck) Aichele & Schwegler, *Anemone rhodopaea* (Stoj. & Stefanov) Stoj. & Stefanov): Plant rarely more than 5 cm at anthesis. Primary divisions of the basal leaves usually 5, often petiolate; lamina with 50–100 lobes, densely lanate. $2n = 32$. *S. Jugoslavia, Bulgaria.*

(**c**) Subsp. **taurica** (Juz.) K. Krause, *op. cit.* 45 (1958) (*P. taurica* Juz.): Like subsp. (**a**) but primary divisions of the basal leaves usually sessile. $2n = 32$. *Krym.*

(**d**) Subsp. **slavica** (G. Reuss) Zāmels, *Acta Horti Bot. Univ. Latv.* **1(2)**: 104 (1926) (*P. slavica* G. Reuss): Plant usually more than 5 cm at anthesis. Primary divisions of the basal leaves usually 3, sessile; lamina with usually fewer than 50 lobes, more or less lanate. $2n = 32$. *W. Carpathians.*

(**e**) Subsp. **styriaca** (G. A. Pritzel) Zāmels, *loc. cit.* (1926) (*P. styriaca* (G. A. Pritzel) Simonkai): Like subsp. (**a**) but the lamina of the basal leaves 5–11 cm. $2n = 32$. *E. Austria (Steiermark).*

9. P. patens (L.) Miller, *Gard. Dict.* ed. 8, no. 4 (1768) (*Anemone patens* L.). Stem 3–12 cm (up to 45 cm in fruit). Basal leaves more or less sericeous when young, becoming glabrous or subglabrous, with 3–7 palmate primary divisions, the middle segment sometimes shortly stalked. Flowers 5·5–8·5 cm in diameter, campanulate, erect or suberect, usually bluish-violet; perianth-segments spreading widely from the base, 2–3 times as long as the stamens. *E. & E.C. Europe, extending westwards to Sweden and S. Germany.* Cz Fe Ge Hu Po Rm Rs(N, B, C, W, E) Su.

1 Lobes of basal leaves usually 17–30, 5–12 mm wide (**a**) subsp. **patens**
1 Lobes of basal leaves usually 30–80, 1–4 mm wide
2 Flowers bluish-violet (**b**) subsp. **multifida**
2 Flowers yellow to yellowish-white (**c**) subsp. **flavescens**

(**a**) Subsp. **patens** (*P. latifolia* sensu Wissjul., non Rupr., *P. patens* subsp. *teklae* (Zāmels) Zāmels, *P. teklae* Zāmels, *P. wolfgangiana* Besser): Middle segment of the basal leaves sessile; lobes usually 17–30, 5–12 mm wide. Flowers bluish-violet. $2n = 16$. *Throughout the range of the species.*

(**b**) Subsp. **multifida** (G. A. Pritzel) Zāmels, *Acta Horti Bot. Univ. Latv.* **1(2)**: 98 (1926) (*P. multifida* (G. A. Pritzel) Juz.): Middle segment of the lower basal leaves with a long stalk (rarely sessile); lobes usually 30–80, 1–4 mm wide. Flowers bluish-violet. *E. Russia. (Siberia.)*

(**c**) Subsp. **flavescens** (Zucc.) Zāmels, *op. cit.* 95 (1926) (*P. flavescens* (Zucc.) Juz.): Like subsp. (**b**) but the flowers yellow to yellowish-white. *E. Russia. (N. Asia.)*

17. Clematis L.

†T. G. TUTIN (EDITION 1) REVISED BY J. R. AKEROYD (EDITION 2)

Woody climbers or perennial herbs. Leaves opposite, simple, ternate or 1- or 2-pinnate. Perianth-segments 4 (rarely 5, 6 or 8), petaloid, valvate. Honey-leaves absent. Petaloid staminodes sometimes present. Achenes numerous; style persistent, often long and plumose.

1 Plant erect, not climbing; stems herbaceous, or woody at base only
2 Leaves simple; flowers deep violet or blue, very rarely white **9. integrifolia**
2 Leaves pinnate; flowers white
3 Leaves 1-pinnate; leaflets up to 9 cm **4. recta**
3 Leaves usually 2-pinnate; leaflets 3–5 cm **5. pseudoflammula**
1 Plant climbing or scrambling by means of tendrils; stems usually woody
4 Spathulate petaloid staminodes present **10. alpina**
4 Petaloid staminodes absent
5 Flowers blue, purple or violet-tinged; style not plumose in fruit
6 Style glabrous; flowers rotate **7. viticella**
6 Style pubescent in the lower $\frac{2}{3}$; flowers ± campanulate **8. campaniflora**
5 Flowers white or yellowish, rarely reddish; style plumose in fruit
7 Plant evergreen; bracteoles connate, forming a 2-lipped involucre beneath the flower **6. cirrhosa**
7 Plant deciduous; bracteoles free, not forming an involucre beneath the flower
8 Perianth-segments yellowish, acuminate; leaves glaucous **3. orientalis**
8 Perianth-segments white or greenish-white, obtuse; leaves not glaucous
9 Leaves 2-pinnate; perianth-segments glabrous on upper surface **1. flammula**
9 Leaves 1-pinnate; perianth-segments tomentose on both surfaces **2. vitalba**

Sect. CLEMATIS. Bracteoles free. Flowers white or yellowish, usually numerous, in large panicles. Staminodes absent. Style plumose in fruit.

1. C. flammula L., *Sp. Pl.* 544 (1753). Deciduous, more or less woody climber 3–5 m. Leaves 2-pinnate; leaflets narrowly oblong to suborbicular, entire or 3-lobed. Flowers *c.* 2 cm in diameter, fragrant. Perianth-segments obtuse, pubescent beneath and on the margins, white. Anthers 3–4 mm. Receptacle sparsely puberulent to glabrous. Achenes strongly compressed. $2n = 16$. *S. Europe.* Al Bl Bu Co Ga Gr Hs It Ju Lu Sa Si Tu [Az Cz Hu Rs(K)].

2. C. vitalba L., *loc. cit.* (1753). Deciduous, woody climber up to 30 m. Leaves 1-pinnate; leaflets ovate, rarely linear-lanceolate, dentate or subentire. Flowers *c.* 2 cm in diameter, fragrant. Perianth-segments obtuse, greenish-white, pubescent on both surfaces. Anthers 1–2 mm. Receptacle pubescent. Achenes scarcely compressed. $2n = 16$. *S., W. & C. Europe.* Al Au Be Br Bu Co Cz Ga Ge Gr He Ho Hs Hu It Ju Lu Rm Rs(W, K) Sa Si Tu [Da Hb Po Su].

3. C. orientalis L., *Sp. Pl.* 543 (1753). Like **2** but leaflets oblong or linear, 3-lobed, often dentate, glaucous; perianth-segments acuminate, yellowish. Stamens ciliate. *S.E. Russia and Ukraine; naturalized locally elsewhere.* Rs(*W, E) [Cz Hs It]. (*Temperate Asia.*)

4. C. recta L., *Sp. Pl.* 544 (1753) (*C. lathyrifolia* Besser). Erect herb with fistular stems 1–1·5 m. Leaves up to 25 cm, 1-pinnate; leaflets 5–9 cm, ovate, acute, usually entire, stalked. Flowers *c.* 2 cm in diameter, white, erect, fragrant. Perianth-segments 0·5–1·8 cm, with tomentose margins but otherwise glabrous. *C., S. & E. Europe, northwards to* 57°30′ *N. in C. Russia and westwards to N. Spain.* Au Bu Co Cz Ga Ge He Hs Hu It Ju Po Rm Rs(*B, C, W, K, E) [Be].

C. elisabethae-caroli Greuter, *Candollea* **20**: 213 (1965), from Kriti (Sfakia prov.), is like **4** but is a shrub up to 2 m with wiry stems, simple lower leaves, leaflets 3–6 cm, and flowers *c.* 3 cm in diameter. It is known only from the type locality and its status is uncertain.

5. C. pseudoflammula Schmalh. ex Lipsky, *Mém. Soc. Nat. Kieff* **12**: 230 (1894). Like **4** but the stems not more than 70 cm; leaves almost always 2-pinnate with leaflets 3–5 cm. *S. Russia, S. & E. Ukraine.* Rs(W, E).

Sect. CHEIROPSIS DC. Bracteoles united into a 2-lipped involucre. Flowers white, yellowish or sometimes red-spotted, solitary in the leaf-axils; staminodes absent. Style plumose in fruit.

6. C. cirrhosa L., *Sp. Pl.* 544 (1753) (*C. calycina* Aiton). Evergreen climber up to 4 m. Leaves 2·5–5 cm, simple, dentate or 3-lobed, or 1- or 2-ternate with dentate or lobed leaflets. Flowers 4–7 cm in diameter, nodding; pedicels 2–5 cm. Perianth-segments yellowish-white, sometimes with red spots, ovate, pubescent beneath. Flowering in winter. *Mediterranean region, S. Portugal.* Bl Co Cr Gr Hs It Lu Sa Si Tu.

Sect. VITICELLA Link. Bracteoles free. Flowers blue, purple, violet or violet-tinged, 1–3 in the leaf-axils; staminodes absent. Style not plumose in fruit.

7. C. viticella L., *Sp. Pl.* 543 (1753). A deciduous, more or less woody climber 3–4 m. Leaves 10–13 cm, 1-pinnate, the primary divisions with 3 lanceolate to ovate, often lobed leaflets 2–7 cm. Flowers 4 cm in diameter, rotate, blue or purple, fragrant. Style glabrous. *S. Europe, eastwards from Italy; often cultivated for ornament, and locally naturalized elsewhere.* Al Bu Gr It Ju Tu [Be Cz Ga Ge Ho].

In S. Italy and elsewhere plants transitional to **8** have been recorded. They require further investigation.

8. C. campaniflora Brot., *Fl. Lusit.* **2**: 359 (1804) (*C. viticella* subsp. *campaniflora* (Brot.) Font Quer ex O. Bolòs & Vigo). Like **7** but the stems up to 7 m, very slender; flowers broadly campanulate, pale violet; styles pubescent in the lower $\frac{2}{3}$. ● *Portugal, W. & S. Spain.* Hs Lu.

Sect. VIORNA (Reichenb.) Prantl. Bracteoles absent. Flowers dark violet or blue, rarely white, solitary, terminal. Staminodes absent. Style plumose in fruit.

9. C. integrifolia L., *Sp. Pl.* 544 (1753). Erect herb 30–70 cm. Stems usually simple. Leaves up to 9 × 5 cm, simple, ovate, acute, entire. Flowers solitary, rarely 2 or 3, terminal, nodding, stellate-campanulate. Perianth-segments 3–5 cm, glabrous except near the margins. *S.C. & S.E. Europe, northwards to* c. 53° *N. in C. Russia.* Al Au Bu Cz Hu It Ju Rm Rs(C, W, K, E).

Sect. ATRAGENE (L.) DC. Bracteoles free, basal. Flowers violet or yellowish-white, solitary, terminal on short axillary shoots; staminodes present, petaloid. Style plumose in fruit.

10. C. alpina (L.) Miller, *Gard. Dict.* ed. 8, no. 9 (1968). Deciduous. Stems 1–4 m, scrambling, woody. Leaves 2-ternate; leaflets 2·5–5 cm, ovate-lanceolate, coarsely serrate. Flowers solitary, nodding, stellate-campanulate. Perianth-segments 2·5–4 cm, silky beneath. Staminodes up to $\frac{1}{2}$ as long as the perianth-segments. *N. Europe and mountains of C. & S. Europe.* Au Bu Cz Fe Ga Ge He Hu It Ju No Po Rm Rs(N, C, W).

(a) Subsp. **alpina** (*Atragene alpina* L.): Flowers violet (rarely white). $2n = 16$. ● *C. & S. Europe.*

(b) Subsp. **sibirica** (Miller) O. Kuntze, *Verh. Bot. Ver. Brandenb.* **26**: 162 (1885) (*C. sibirica* Miller, *Atragene sibirica* L., nom. ambig.): Flowers yellowish-white (rarely pale violet). $2n = 16$. *N.E. Europe, extending to S. Norway.*

18. Adonis L.

†T. G. TUTIN (EDITION 1) REVISED BY J. R. AKEROYD (EDITION 2)

Herbs with 1- to 3-pinnate leaves with more or less linear segments. Sepals 5(–8), often somewhat petaloid. Petals 3–20, glossy; nectaries absent. Achenes numerous, rugose, forming an elongated head at maturity.

There is much diversity of opinion amongst recent authors about the taxonomic treatment of the annual species (Sect. *Adonis*), and the following treatment should be regarded as provisional.

Literature: H. Riedl, *Ann. Naturh. Mus.* (*Wien*) **66**: 51–90 (1963). C. Steinberg, *Webbia* **25**: 299–351 (1971).

1 Annual; petals not more than 8; anthers blackish-purple
  2 Inner margin of achene almost straight, without a projection **7. annua**
  2 Inner margin of achene with a rounded or angular projection
    3 Sepals hirsute; beak of achene black at apex **8. flammea**
    3 Sepals glabrous or slightly hairy; beak of achene green throughout
      4 Upper projection on inner margin of achene at $\frac{1}{3}$ of the distance from apex to base **9. aestivalis**
      4 Upper projection on inner margin of achene close to beak **10. microcarpa**
1 Perennial; petals more than 8; anthers yellow
  5 Basal leaves reduced to scales

6 Leaves at least twice as long as wide; petals obovate **3. sibirica**
6 Leaves little longer than wide; petals elliptical or narrowly lanceolate
7 Leaf-lobes narrowly linear, entire **1. vernalis**
7 Leaf-lobes linear-lanceolate, dentate **2. volgensis**
5 Basal foliage-leaves present
8 Sepals pubescent; beak of achene *c.* 1 mm; stems less than 20 cm **5. distorta**
8 Sepals usually glabrous; beak of achene at least 2 mm; stems usually 25–60 cm
9 Beak *c.* 2 mm, distinctly shorter than achene, stout and flattened **4. pyrenaica**
9 Beak 3–4 mm, almost as long as achene, slender, terete **6. cyllenea**

Sect. CONSILIGO DC. Perennials. Petals 8 or more, yellow or white. Anthers yellow.

All species occur on calcareous, rocky ground.

1. A. vernalis L., *Sp. Pl.* 547 (1753). Stems 10–45 cm, scaly at the base. Cauline leaves sessile, not or scarcely longer than wide, 2-pinnatisect with narrowly linear, entire lobes. Flowers 4–8 cm in diameter. Sepals ½ as long as the petals, broadly obovate, pubescent. Petals 10–20, elliptical, yellow. Achenes 3–4·5 mm, almost globose, reticulately rugose and rather densely pubescent; beak short, curved and appressed to the achene. $2n = 16$. *E., C. & S. Europe, northwards to C. Ural and S.E. Sweden.* Au Bu Cz Ga Ge He Hs Hu It Ju Po Rm Rs(C, W, K, E) Su.

2. A. volgensis Steven ex DC., *Reg. Veg. Syst. Nat.* **1**: 545 (1817). Like **1** but the stems more branched; leaf-lobes linear-lanceolate, dentate; sepals ⅓ as long as the petals; petals more numerous, narrowly lanceolate; achenes pubescent only at the base. *S.E. Europe, extending northwards to C. Ural, and locally westwards to S.E. Hungary.* Bu Hu Rm Rs(C, W, K, E).

The hybrid between **1** and **2** (*A.* × *hybrida* H. Wolff) occurs with the parents.

3. A. sibirica Patrin ex Ledeb., *Ind. Sem. Horti Dorpat.* suppl., 2 (1824). Plant glabrous; stems 20–60 cm. Leaves at least twice as long as wide, the segments 2- or 3-pinnatifid with linear lobes. Flowers 4–6 cm in diameter. Petals obovate, yellow. Achenes *c.* 4·5 mm, more or less pubescent. *Forest-margins. N.W. & E. Russia.* Rs (N, C). (*N. Asia.*)

4. A. pyrenaica DC. in Lam. & DC., *Fl. Fr.* ed. 3, **5**: 635 (1815). Stems 20–45 cm, without scales at the base. Leaves 2- to 4-pinnatisect, the basal long-petiolate. Sepals glabrous or sparsely pubescent, *c.* ½ as long as the petals. Petals 10–15, yellow. Achenes *c.* 6 mm, sparsely pubescent or almost glabrous; beak 2–3 mm, flattened, curved, not appressed to the achene. $2n = 16$. ● *Pyrenees and Cordillera Cantábrica; one station in Alpes Maritimes.* Ga Hs.

5. A. distorta Ten., *Fl. Nap.* **4**: 337 (1830). Like **4** but plant seldom more than 15 cm, with curved leaf-segments; petals 8–18, sometimes white; sepals pubescent; achenes glabrous, the beak *c.* 1 mm, strongly curved. *Calcareous rocks and screes, 2000–2500 m.* ● *C. Appennini.* It.

6. A. cyllenea Boiss., Heldr. & Orph. in Boiss., *Diagn. Pl. Or. Nov.* **3(5)**: 5 (1856). Like **4** but the stems often up to 60 cm and freely branched; sepals almost as long as the petals; beak 3–4 mm, almost as long as the achene, slender, terete, hooked at the apex. $2n = 16$. ?● *S. Greece (Killini Oligirtos).* Gr.

Elsewhere known only from one collection from Pontus in N.E. Turkey which has been referred to var. *paryadrica* Boiss. Its status requires investigation.

Sect. ADONIS. Annuals. Petals 8 or fewer, usually red, though often yellow when dried. Anthers blackish-purple.

All species occur as weeds of cultivation or in similar open habitats; although not strictly calcicole, they are found mainly on base-rich soils.

7. A. annua L., *Sp. Pl.* 547 (1753) (*A. autumnalis* L.). Stems 10–70 cm. Leaves 3-pinnate, the segments linear, mucronate. Flowers 15–25 mm in diameter. Sepals ovate, patent. Petals 5–8, suberect, concave, oblong, somewhat longer than the sepals, bright scarlet, with a dark basal spot. Inner margin of achene nearly straight, the outer curved; beak short, straight, green throughout. *Cultivated land. S. Europe, extending northwards to N. France; sometimes naturalized and often a casual elsewhere.* Al †Be Bl Bu Co Cr Ga Gr He Hs It Ju Lu Sa Si Tu [Br].

(a) Subsp. **annua** (incl. *A. baetica* Cosson): Sepals glabrous. Achenes 4·5–5·5(–6) mm. *Almost throughout the range of the species.*
(b) Subsp. **cupaniana** (Guss.) C. Steinb., *Webbia* **25**: 324 (1971): Sepals hairy. Achenes 3–4·5 mm. *S. Europe.*

Plants from Spain and S. France, similar to subsp. (**b**) but hairy throughout, have been distinguished as subsp. **castellana** (Pau) C. Steinb., *op. cit.* 332 (1971).

8. A. flammea Jacq., *Fl. Austr.* **4**: 29 (1776). Like **7** but the flowers 20–30 mm in diameter; sepals more or less hirsute, appressed to the deep scarlet (rarely yellow), linear-oblong petals; achenes 2·5–4 mm, the inner margin with a rounded projection just below the beak; beak ascending, black at the apex. *S. & C. Europe, extending to N. France; sometimes casual elsewhere.* Al Au †Be Bu Cz Ga Ge Gr He Hs Hu It Ju Po Rm Rs(W, K) Tu.

9. A. aestivalis L., *Sp. Pl.* ed. 2, 771 (1762). Stems 20–50(–70) cm. Leaves 3- or 4-pinnate, the segments linear, mucronate. Flowers 10–35 mm in diameter. Sepals glabrous, ovate, patent. Petals 5–8, suberect, concave, oblong, somewhat longer than the sepals, red or yellow. Achenes 3·5–6 mm, with a transverse ridge about the middle; inner margin with 2 projections, the lower acute, the upper obtuse and ⅓ of the distance from apex to base. *Most of Europe, southwards from 55° N.* Au Be Bl Bu Co Cz Ga Ge Gr He †Ho Hs Hu It Ju Po Rm Rs(C, W, K, E) Sa Si Tu.

(a) Subsp. **aestivalis**: Plant glabrous above. Sepals usually glabrous. Achenes with an indistinct transverse ridge. *Almost throughout the range of the species.*
(b) Subsp. **squarrosa** (Steven) Nyman, *Consp.* 4 (1878) (incl. subsp. *provincialis* (DC.) C. Steinb.): Plant somewhat smaller, sparsely hairy above. Sepals hairy, sometimes with a few glandular hairs. Achenes with a distinct transverse ridge. *S. Europe.*

Plants with smaller flowers and smaller achenes with inconspicuous projections have been recorded from Kazakhstan and elsewhere in W.C. Asia, and perhaps from S.E. Russia. They have been distinguished as subsp. **parviflora** (Fischer ex DC.) N. Busch in Kusn., N. Busch & Fomin, *Fl. Cauc. Crit.* **3**: 201 (1903), but the distinctions from subsp. (**a**) appear to be inconstant.

10. A. microcarpa DC., *Reg. Veg. Syst. Nat.* **1**: 223 (1817) (*A. dentata* auct. eur., non Delile, *A. annua* subsp. *carinata* Vierh.). Like **9** but the stems 10–25(–50) cm; petals usually yellow; achenes 2·5–4 mm, with or without a transverse ridge; upper projection from inner edge of achene close to the beak. *S. Europe*. Al Bl †Bu Co Cr Ga Gr Hs It Ju Lu Rs(W) Sa Si Tu.

19. Ranunculus L.

†T. G. TUTIN (EDITION 1) REVISED BY †T. G. TUTIN AND J. R. AKEROYD (EDITION 2); SUBGEN. *Batrachium*, C. D. K. COOK

Annual or perennial herbs, sometimes aquatic. Flowers solitary or in cymose panicles. Perianth-segments (3–)5(–7). Honey-leaves (0–)5(–12), usually deciduous, petaloid, yellow or white, rarely red or purple. Achenes numerous, usually with a persistent glabrous style.

1 Flowers white, pink, purple or red
2 Terrestrial; leaves mostly basal; achene not transversely rugose
3 Basal leaves hastate (S. Spain) **98. acetosellifolius**
3 Basal leaves not hastate
4 Most basal leaves entire to dentate, but not more deeply divided
5 Cauline leaves lanceolate to ovate, amplexicaul
6 Sepals ±glabrous
7 Achene smooth **119. parnassiifolius**
7 Achene strongly veined
8 Cauline leaves few or 0; pedicels hairy **115. kuepferi**
8 Cauline leaves present; pedicels glabrous **121. amplexicaulis**
6 Sepals pubescent
9 Petiole not widening into lamina; main veins usually 7; achene strongly veined **119. parnassiifolius**
9 Petiole widening into lamina; main veins usually 5; achene smooth **120. wettsteinii**
5 Cauline leaves linear to linear-oblong or narrowly lanceolate, not amplexicaul
10 Basal leaves never crenate or dentate **114. pyrenaeus**
10 Basal leaves crenate or sometimes dentate at apex
11 Honey-leaves rounded at apex **93. crenatus**
11 Honey-leaves emarginate to deeply notched at apex **94. bilobus**
4 All basal leaves ±deeply lobed
12 Sepals 3 **84. pallasii**
12 Sepals 5 or more
13 Sepals densely ferruginous-villous **99. glacialis**
13 Sepals glabrous or nearly so
14 Flowers at least 30 mm in diameter; honey-leaves 7 or more; roots tuberous **69. asiaticus**
14 Flowers 10–20(–25) mm in diameter; honey-leaves 5; roots not tuberous
15 Cauline leaves differing markedly from basal, simple or rarely 3-fid; receptacle glabrous **92. alpestris**
15 Cauline leaves like basal, lobed and dentate; receptacle pubescent
16 Plant rarely as much as 20 cm; segments of basal leaves narrowly oblong; flowers 1 to few **97. seguieri**
16 Plant usually more than 20 cm; segments of basal leaves ovate-cuneate; flowers numerous
17 Middle segment of basal leaves free to base **95. aconitifolius**
17 Middle segment of basal leaves not free to base **96. platanifolius**
2 Aquatic or marsh plants; leaves all cauline; achene transversely rugose
18 Laminate leaves present; capillary leaves present or absent
19 Capillary leaves absent
20 Achene pubescent (sometimes sparsely and minutely so) **126. peltatus**
20 Achene entirely glabrous
21 Receptacle glabrous
22 Leaf-segments widest at the sinus; honey-leaves not or scarcely longer than sepals **122. hederaceus**
22 Leaf-segments narrowest at the sinus; honey-leaves 2–3 times as long as sepals **123. omiophyllus**
21 Receptacle pubescent
23 Honey-leaves less than 6 mm, not contiguous; receptacle globose **124. tripartitus**
23 Honey-leaves more than 6 mm contiguous; receptacle ovoid **126. peltatus**
19 Capillary leaves present
24 Honey-leaves less than 6 mm, not contiguous **124. tripartitus**
24 Honey-leaves more than 6 mm, contiguous
25 Immature achene completely glabrous; laminate leaves 3-lobed to $\frac{2}{3}$ or more
26 Capillary leaves collapsing when removed from water; mature achene unwinged; receptacle globose (fresh water) **125. ololeucos**
26 Capillary leaves not collapsing when removed from water; mature achene winged; receptacle ovoid (brackish water) **126. peltatus**
25 Immature achene pubescent (sometimes minutely so); laminate leaves 3- to 5-lobed, usually to less than $\frac{2}{3}$
27 Pedicel in fruit usually less than 50 mm, rarely exceeding petiole of opposing laminate leaf; honey-leaves rarely more than 10 mm
28 Segments of laminate leaves crenate; nectaries ±pyriform **126. peltatus**
28 Segments of laminate leaves dentate; nectaries circular **129. aquatilis**
27 Pedicel in fruit usually more than 50 mm, often exceeding petiole of opposing laminate leaf; honey-leaves usually more than 10 mm
29 Capillary leaves shorter than internodes **126. peltatus**
29 Capillary leaves longer than internodes **127. penicillatus**
18 Laminate leaves absent
30 Leaves usually exceeding 8 cm (up to 30 cm), longer than internodes, their segments usually parallel to the main axis
31 Receptacle distinctly hairy **127. penicillatus**
31 Receptacle nearly or quite glabrous **133. fluitans**
30 Leaves usually much less than 8 cm, shorter than

internodes, their segments rarely parallel to the main axis

32 Leaf-segments lying in 1 plane; leaves circular in outline **132. circinatus**

32 Leaf-segments not lying in 1 plane

33 Honey-leaves rarely exceeding 5 mm, not contiguous; nectaries lunate

34 Achene more than 1·5 mm **130. trichophyllus**

34 Achene less than 1 mm **131. rionii**

33 Honey-leaves usually exceeding 5 mm, contiguous; nectaries not lunate

35 Immature achene glabrous; mature achene winged **126. peltatus**

35 Immature achene pubescent (sometimes minutely so); mature achene unwinged

36 Pedicel in fruit usually less than 50 mm; honey-leaves rarely more than 10 mm; nectaries circular **129. aquatilis**

36 Pedicel in fruit usually more than 50 mm; honey-leaves usually more than 10 mm; nectaries elongate, ± pyriform

37 Achene more than 1·5 mm **126. peltatus**

37 Achene less than 1 mm **128. sphaerospermus**

1 Flowers yellow

38 Aerial leaves (including the cauline) entire, serrate, crenate or dentate, but not lobed

39 Achene puberulent **86. ficaria**

39 Achene glabrous or rarely hispid

40 Sepals 3 **87. ficarioides**

40 Sepals 5

41 Cauline leaves at least as wide as long, entire below, increasingly dentate towards the sometimes almost 3-lobed apex

42 Basal leaves not developed at anthesis **89. thora**

42 Basal leaves present at anthesis **90. hybridus**

41 Cauline leaves distinctly longer than wide, not becoming conspicuously more strongly dentate towards apex, rarely absent

43 Leaves mostly crowded at base of stem; cauline small and few, or absent

44 Lower leaves linear to narrowly lanceolate, sessile

45 Roots fibrous; leaves not cucullate at apex; honey-leaves 5 **117. gramineus**

45 Roots tuberous; leaves cucullate at apex; honey-leaves 8–14 **118. abnormis**

44 Lower leaves ovate to suborbicular, petiolate

46 Lower leaves cuneate at base **116. bupleuroides**

46 Lower leaves rounded to subcordate at base

47 Roots fibrous; stolons present; petiole *c.* twice as long as lamina **85. cymbalaria**

47 Roots tuberous; stolons absent; petiole shorter than lamina **88. bullatus**

43 All or most of leaves cauline

48 Flowers sessile or subsessile

49 Leaves abruptly narrowed at base; beak about as long as achene **111. lateriflorus**

49 Leaves gradually narrowed at base; beak *c.* $\frac{1}{2}$ as long as achene **112. nodiflorus**

48 Flowers all distinctly pedicellate

50 Achene less than 1 mm; flowers 3–8 mm in diameter

51 Submerged leaves divided into filiform segments; lowest aerial leaves often shallowly 3-lobed **113. polyphyllus**

51 Submerged leaves absent; lowest leaves spathulate, entire **104. batrachioides**

50 Achene more than 1 mm; flowers 5–50 mm in diameter

52 Stems filiform, procumbent; flowers solitary; achene up to 1·5 mm **103. reptans**

52 Stems stouter, usually erect; flowers rarely solitary; achene more than 1·5 mm

53 Flowers 30–50 mm in diameter (rarely less); achene 2·5 mm; plant at least 50 cm **105. lingua**

53 Flowers up to 15(–20) mm in diameter; achene 1·5–2 mm; plant rarely more than 50 cm

54 Achene smooth

55 Perennial; pedicels sparsely appressed-pubescent **102. flammula**

55 Annual; pedicels glabrous **108. fontanus**

54 Achene granulate or tuberculate

56 Sepals longer than honey-leaves

57 Pedicels sulcate, not thickened in fruit, the lower shorter than subtending leaves **107. thracicus**

57 Pedicels terete, ± thickened in fruit, the lower about as long as subtending leaves **110. revelieri**

56 Sepals equalling or shorter than honey-leaves

58 Sepals shorter than honey-leaves; lower pedicels at least as long as subtending leaves **106. ophioglossifolius**

58 Sepals equalling honey-leaves; lower pedicels shorter than subtending leaves **109. longipes**

38 At least some aerial leaves distinctly lobed or even more deeply divided

59 Stems creeping, rooting at nodes; leaves not crowded at base of stem (small arctic plants, except **6**)

60 Honey-leaves 3 **80. hyperboreus**

60 Honey-leaves 5 or more

61 Sepals 3 **83. lapponicus**

61 Sepals 5

62 Flowers 20–30 mm in diameter; achene *c.* 3 mm **6. repens**

62 Flowers *c.* 10 mm in diameter; achene *c.* 1·5 mm **82. gmelinii**

59 Stems not creeping and rooting at nodes; leaves mostly crowded at base; plant sometimes with stolons from a well-marked rosette

63 Roots of 2 kinds, some fibrous, some fleshy, forming fusiform to ovoid tubers

64 Sepals 3 **87. ficarioides**

64 Sepals 5

65 Sepals deflexed in flower

66 Leaves or leaf-segments linear-lanceolate, entire **50. illyricus**

66 Leaf-segments cuneate, variously cut or lobed, or leaves suborbicular and shallowly lobed

67 Achene tuberculate, with a compressed appendage at base **49. isthmicus**

67 Achene without an appendage

68 Pedicels sulcate
69 All but the lowest cauline leaves small and bract-like **32. neapolitanus**
69 Several of the cauline leaves similar to basal, but smaller
70 Middle segment of leaf sessile; beak of achene *c.* 1 mm **33. pratensis**
70 Middle segment of leaf usually stipitate; beak of achene *c.* 0·5 mm **34. bulbosus**
68 Pedicels terete
71 Leaves appressed-pubescent or sericeous
72 Tubers ovoid or shortly oblong **46. rumelicus**
72 Tubers fusiform or cylindrical
73 Achene glabrous, verrucose **45. psilostachys**
73 Achene slightly pubescent, not verrucose **47. monspeliacus**
71 Leaves not appressed-pubescent
74 Leaf-segments deeply divided into numerous linear-lanceolate lobes **48. miliarakesii**
74 Leaf segments obtusely lobed or incise-serrate
75 Leaves longer than wide; beak of achene straight **44. oxyspermus**
75 Leaves usually wider than long; beak of achene curved
76 Leaves sericeous, deeply divided; middle segment stipitate **47. monspeliacus**
76 Leaves not sericeous, usually not divided to base **43. gracilis**
65 Sepals not deflexed in flower
77 Basal leaves glabrous, pedately 3- to 5-partite; segments linear-lanceolate, entire or deeply 2-lobed **51. pedatus**
77 Basal leaves not as above
78 Achene circular, flat, winged; beak very short, hooked **52. platyspermus**
78 Achene not as above
79 Anthers purplish-black **69. asiaticus**
79 Anthers yellow
80 Stock swollen, usually densely and persistently fibrous
81 Lobes of basal leaves *c.* 1 mm wide, linear **55. pseudomillefoliatus**
81 Lobes of basal leaves more than 1 mm wide
82 Beak of achene curved or distinctly hooked **54. barceloi**
82 Beak of achene straight (often somewhat hooked at tip)
83 Flowers 1–8; receptacle glabrous; beak about as long as achene **53. paludosus**
83 Flowers usually solitary; receptacle pubescent to subglabrous; beak shorter than achene **56. gregarius**
80 Stock not swollen and densely and persistently fibrous
84 Basal leaves ±lobed for up to *c.* $\frac{3}{4}$ of their length
85 Basal leaves shallowly lobed; segments contiguous or overlapping
86 Basal leaves coriaceous, rugose
87 Leaves *c.* 2 cm across, ±glabrous; flowers few, *c.* 25 mm in diameter **61. nigrescens**
87 Leaves *c.* 15 cm across, densely appressed-villous; flowers numerous, *c.* 40 mm in diameter **62. cortusifolius**
86 Basal leaves thin, not rugose
88 Leaves *c.* 8 cm across; flowers several **64. creticus**
88 Leaves *c.* 5 cm across; flowers 1(2) **63. spicatus**
85 Basal leaves lobed to *c.* $\frac{3}{4}$; segments distant
89 Plant stout; basal leaves ±patent-pubescent **68. sprunerianus**
89 Plant slender; basal leaves appressed-pubescent or nearly or quite glabrous
90 Basal leaves appressed-pubescent
91 Honey-leaves at least 15 mm or more **63. spicatus**
91 Honey-leaves *c.* 10 mm **57. ollissiponensis**
90 Basal leaves nearly or quite glabrous
92 Stems much-branched; beak about as long as achene, stout, curved **65. thasius**
92 Stems simple or subsimple; beak shorter than achene, slender, hooked
93 Honey-leaves 5–7 mm; leaves divided to base **66. subhomophyllus**
93 Honey-leaves 7–9 mm; leaves divided up to $\frac{4}{5}$ of their length **67. incomparabilis**
84 Basal leaves lobed nearly or quite to base
94 Segments of basal leaves dentate or lobed but not 2- or 3-pinnatisect **57. ollissiponensis**
94 Segments of basal leaves 2- or 3-pinnatisect
95 Honey-leaves copper-coloured beneath (Kriti) **58. cupreus**
95 Honey-leaves yellow on both surfaces
96 Honey-leaves 7–12 **60. millii**
96 Honey-leaves 5
97 Receptacle pubescent at insertion of stamens **57. ollissiponensis**
97 Receptacle glabrous at insertion of stamens **59. millefoliatus**
63 Roots all fibrous; fibres sometimes thick but not obviously tuberous
98 Achene not or slightly compressed, with strongly convex sides
99 Annual; head of achenes cylindrical; achenes 70–100, *c.* 1 mm **81. sceleratus**
99 Perennial; head of achenes globose; achenes fewer and larger
100 Achene puberulent
101 Basal and cauline leaves markedly dissimilar **(70–75). auricomus** group
101 Basal and cauline leaves ±similar **76. polyrhizos**
100 Achene glabrous
102 Most leaves entire at base and then increasingly strongly dentate to the wide apex, wider than long
103 Basal leaves usually 2; cauline 2 or 3; achene 2–3 mm; beak short **90. hybridus**

103 Basal leaves several; cauline 1(2); achene 3–4 mm; beak long **91. brevifolius**

102 Most leaves ±regularly lobed all round, not wider than long

104 Cauline leaves divided to base into linear or linear-lanceolate, entire segments

105 Sepals pubescent **(70–75). auricomus** group

105 Sepals villous **78. nivalis**

104 Cauline leaves less divided, or segments wider

106 Stems slender, flexuous; upper cauline leaves simple, bract-like, or cauline leaves absent

107 Hairs all appressed; sepals deflexed; achene smooth **100. cymbalariifolius**

107 Hairs at base of stems and petioles patent; sepals patent; achene strongly veined **101. weyleri**

106 Stems stout, erect; upper cauline leaves at least 3-lobed

108 Flowers not more than 10 mm in diameter; stem rarely exceeding 4 cm, about as long as basal leaves at anthesis **77. pygmaeus**

108 Flowers usually at least 15 mm in diameter; stem 8–60 cm, much longer than basal leaves at anthesis

109 Basal leaves reniform, deeply lobed; head of achenes cylindrical; beak about as long as achene **78. nivalis**

109 Basal leaves cuneate, shallowly lobed; head of achenes broadly ovoid; beak about $\frac{1}{2}$ as long as achene **79. sulphureus**

98 Achene ±strongly compressed

110 Achene distinctly spiny or muricate with either numerous or long projections (annual)

111 Honey-leaves 2–3 times as long as sepals **37. marginatus**

111 Honey-leaves equalling or slightly longer than sepals

112 Achene up to *c.* 3 mm

113 Receptacle pubescent **38. trilobus**

113 Receptacle glabrous

114 Pedicel not thickened in fruit; achene more than twice as long as beak **41. parviflorus**

114 Pedicel greatly thickened in fruit; achene at most twice as long as beak **42. chius**

112 Achene *c.* 7 mm; receptacle pubescent

115 Sepals deflexed; leaves shallowly lobed **39. muricatus**

115 Sepals patent; leaves, except the lowest, deeply lobed **40. arvensis**

110 Achene smooth, punctulate or with few, small, obtuse tubercles (usually perennial)

116 Receptacle glabrous

117 Sepals deflexed at anthesis

118 Leaf-segments cuneate-obovate **1. velutinus**

118 Leaf-segments broadly ovate **2. constantinopolitanus**

117 Sepals not deflexed at anthesis

119 Honey-leaves orange-yellow; basal leaves usually 3-lobed to about $\frac{3}{4}$; stock truncate **7. lanuginosus**

119 Honey-leaves golden-yellow; basal leaves usually more deeply divided, or rhizome stout, creeping

120 Achene at most twice as long as beak; leaf-segments ±stipitate

121 Leaves ±sericeous beneath; achene twice as long as beak **11. serbicus**

121 Leaves pubescent to nearly glabrous beneath; achene as long as beak **12. brutius**

120 Achene at least 4 times as long as beak; leaf-segments sessile

122 Filaments glabrous (rarely with a few hairs near base); plant often with short rhizome **8. acris**

122 Filaments pubescent; plant with a stout, long, ±horizontal rhizome

123 Beak of achene usually at least 0·8 mm **9. granatensis**

123 Beak of achene 0·5–0·7 mm **10. strigulosus**

116 Receptacle pubescent

124 Sepals deflexed at anthesis

125 Annual

126 Honey-leaves pale yellow; achene 2·5–3 mm; beak 0·5 mm **35. sardous**

126 Honey-leaves deep yellow; achene 3–4 mm; beak 1 mm **37. marginatus**

125 Perennial

127 Pedicels terete; sepals at first patent, later ±strongly deflexed **26. macrophyllus**

127 Pedicels sulcate; sepals strongly deflexed soon after anthesis

128 Basal leaves suborbicular-cordate, incised crenate **36. cordiger**

128 Basal leaves 3-partite

129 Cauline leaves (except the lowest) small and bract-like **32. neapolitanus**

129 Several of the lower cauline leaves similar to the basal, but smaller

130 Basal leaves lobed to $\frac{3}{4}$; segments united at base; beak of achene *c.* 1 mm **33. pratensis**

130 Basal leaves lobed to base; middle segment often stipitate; beak of achene *c.* 0·5 mm **34. bulbosus**

124 Sepals not deflexed at anthesis

131 Achene strongly keeled or sulcate

132 Pedicel sulcate

133 Stolons present **6. repens**

133 Stolons absent

134 Leaves divided into linear- or oblong-lanceolate lobes **3. polyanthemos**

134 Leaves divided into cuneate-obovate segments with ovate, dentate lobes

135 Middle segment of basal leaves sessile **4. serpens**

135 Middle segment of basal leaves long-stipitate **5. caucasicus**

132 Pedicel terete

136 Plant rarely more than 15 cm; achene with keeled but not sulcate margin **(13–24). montanus** group

136 Plant often at least 30 cm, achene with keeled and sulcate margin

137 Cauline leaves conspicuously smaller than basal; uppermost 3-fid with entire segments, or simple **26. macrophyllus**
137 Cauline leaves about as large as basal; uppermost 3-fid with lobed and dentate segments **25. carpaticus**
131 Achene neither keeled nor sulcate
138 Sepals glabrous **29. marschlinsii**
138 Sepals pubescent
139 Leaves hairy beneath
140 Basal leaves hairy beneath only; beak of achene 0·8–1 mm, hooked **30. dissectus**
140 Basal leaves hairy on both surfaces; beak of achene 1·5–2 mm, circinate **31. radinotrichus**
139 Leaves glabrous or subglabrous beneath
141 Basal leaves with shortly stipitate segments; flowers usually solitary **27. demissus**
141 Basal leaves not divided to base; flowers usually several **28. hayekii**

Subgen. **Ranunculus.** Usually terrestrial plants. Leaves various but very rarely divided into capillary segments. Flowers yellow or white, rarely red or purple. Achenes very rarely transversely rugose.

Sect. RANUNCULUS. Honey-leaves yellow; nectary covered by a flap which is more or less free laterally. Achenes glabrous, distinctly beaked, compressed. Receptacle in fruit not more than 3 times its length in flower.

(A) Pedicel terete. Sepals deflexed. Receptacle glabrous.

1. R. velutinus Ten., *Ind. Sem. Horti Neap.* 12 (1825). Perennial 40–80 cm. Petioles and lower part of stems patent-pubescent; otherwise appressed-sericeous. Basal leaves broadly ovate, deeply 3-fid; segments cuneate-obovate, 3-lobed and dentate; cauline leaves small, linear. Pedicels slender. Flowers 15–25 mm in diameter. Head of achenes depressed-globose; achenes compressed, strongly bordered; beak very short, straight. $2n = 14$. *Damp places. C. & E. Mediterranean region.* Al Bu Co Ga Gr It Ju Sa Si Tu.

2. R. constantinopolitanus (DC.) D'Urv., *Mém. Soc. Linn. Paris* **1**: 317 (1822) (R. *villosus* subsp. *constantinopolitanus* (DC.) Jelen.). Perennial *c.* 45 cm, densely patent-pubescent below, appressed-pubescent above. Basal leaves triangular-ovate, 3-fid to $\frac{3}{4}$; segments broadly ovate, again cut into dentate lobes; cauline leaves small, linear. Flowers 15–25 mm in diameter. Head of achenes globose; achenes 4–5 mm, ovate, rounded; beak 1 mm, falcate. *S.E. Europe.* Bu Gr Ju Rm Rs(K) Tu.

(B) Pedicel sulcate. Sepals patent. Receptacle pubescent.

3. R. polyanthemos L., *Sp. Pl.* 554 (1753) (R. *meyerianus* Rupr.). A usually much-branched perennial 10–130 cm. Petioles and lower part of stems patent-pubescent or glabrescent, appressed-pubescent above. Basal leaves (3–)5-fid; segments cut into linear-lanceolate or linear, dentate lobes; cauline leaves like the basal, but smaller. Flowers 18–25 mm in diameter; honey-leaves 7–14 mm, golden-yellow. Head of achenes globose; achenes 3–5 mm, compressed, bordered; beak *c.* 0·5 mm, curved. *Europe, except the west and parts of the south.* Al Au Be Bu Co Cz Da Fe Ga Ge Gr Ho Hu It Ju No Po Rm Rs(N, B, C, W, K, E) Su Tu.

Species **3** and **4** are very closely related and their precise delimitation and that of their subspecies is still uncertain.

1 Basal leaves deeply 3-fid; stems up to 130 cm (c) subsp. **polyanthemoides**
1 Basal leaves 5-fid; stems rarely more than 60 cm
2 Stems usually much-branched; plant 30–60 cm (a) subsp. **polyanthemos**
2 Stems simple or little-branched; plant 10–30 cm (b) subsp. **thomasii**

(a) Subsp. **polyanthemos**: Rarely more than 60 cm. Basal leaves 5-fid; segments all more or less equal, more or less pubescent. $2n = 16$. *From Norway, C. Germany and Albania eastwards.*

(b) Subsp. **thomasii** (Ten.) Tutin, *Feddes Repert.* **69**: 53 (1964) (R. *thomasii* Ten.): Stems 10–30 cm, simple or little-branched. Basal leaves 5-fid; middle segments usually longest. $2n = 16$. ● *S. Italy.*

(c) Subsp. **polyanthemoides** (Boreau) Ahlfvengren in Neuman, *Sver. Fl.* 502 (1901) (R. *polyanthemoides* Boreau): Up to 130 cm. Basal leaves very deeply 3-fid; middle segment very long, narrow, cuneate, 3-lobed, densely pubescent. $2n = 16$. ● *Scattered through Europe from C. Fennoscandia to N. Greece.*

Subsp. (c) has often been placed under **4** on vegetative characters, but can be distinguished with certainty by the achene, which has a short beak and closely resembles that of **3**.

4. R. serpens Schrank, *Baier. Fl.* **2**: 101 (1789) (R. *amansii* Jordan, R. *mixtus* Jordan, R. *tuberosus* Lapeyr., R. *nemorosus* DC., R. *breyninus* auct., non Crantz). Like **3** but basal leaves 3-fid, with broadly ovate, lobed and dentate segments; honey-leaves 15–20 mm; achenes with beak *c.* 1·5 mm, strongly curved. *Woods and scrub.* ● *Much of Europe, but absent from most of the north and the islands.* Al Au Be Bu Co Cz Da Ga Ge Gr He Ho Hs It Ju Po Rm Rs(B, C, W, E) Su.

1 Stems becoming procumbent and rooting at nodes; lower cauline leaves resembling the basal (a) subsp. **serpens**
1 Stems always erect, never rooting at nodes; lower cauline leaves with narrow, ±entire segments
2 All basal leaves pentagonal in outline; segments lobed and dentate (b) subsp. **nemorosus**
2 Outer basal leaves orbicular in outline, more divided and with narrow lobes (c) subsp. **polyanthemophyllus**

(a) Subsp. **serpens** (R. *nemorosus* proles *radicescens* (Jordan) Rouy & Fouc., R. *radicescens* Jordan): Biennial 10–20 cm; stems at first erect, later procumbent and rooting at the nodes, patent-pubescent below. Basal leaves pentagonal in outline, deeply 3-fid; segments obovate-cuneate, lobed and dentate; lower cauline leaves resembling the basal. $2n = 16$. *From C. Germany to the S. Alps.*

(b) Subsp. **nemorosus** (DC.) G. López, *Anal. Jard. Bot. Madrid* **41**: 470 (1985): Erect perennial 20–80 cm. Stem sparsely appressed- or slightly patent-pubescent. Basal leaves pentagonal in outline, deeply 3-fid; segments obovate-cuneate, lobed and dentate; cauline leaves with narrow, more or less entire segments. $2n = 16$. *Throughout the range of the species.*

(c) Subsp. **polyanthemophyllus** (Walo Koch & H. Hess) M. Kerguélen, *Lejeunia* **120**: 149 (1987) (R. *polyanthemophyllus* Walo Koch & H. Hess): Like subsp. (b) but often 50–100 cm; leaves orbicular in outline, divided more or less to base and with narrower lobes; middle segment often stipitate; later leaves similar to those of subsp. (a). $2n = 16$. *Alps and adjacent regions.*

5. R. caucasicus Bieb., *Fl. Taur.-Cauc.* **2**: 27 (1808). A slender, pubescent perennial *c.* 30 cm. Stock erect, with fibres. Basal leaves with 3 segments; middle segments long-stipitate; lateral segments usually 3-lobed, the lobe on the lower side downward-directed and more or less covering the petiole; all lobes cuneate-obovate, dentate; cauline leaves less divided, with narrower lobes. Flowers 10–20 mm in diameter. Achenes *c.* 4 mm, strongly compressed, 2-veined on upper margin; beak *c.* 1 mm, stout, curved, 2-veined at base. $2n = 16$. *Woods and mountain meadows. Krym.* Rs(K).

6. R. repens L., *Sp. Pl.* 554 (1753) (R. *oenanthifolius* Ten. & Guss., R. *pubescens* Lag.). A nearly glabrous to somewhat pubescent perennial 15–60 cm. Stolons long, rooting at the nodes. Basal leaves triangular-ovate, 3-partite; middle lobe long-stipitate; segments usually cut into 3-dentate lobes; cauline leaves smaller and less divided. Flowers 20–30 mm in diameter. Head of achenes globose; achenes *c.* 3 mm, compressed, bordered; beak *c.* 1·5 mm, falcate. $2n = 32$. *Damp places, especially on disturbed ground. Almost throughout Europe.* All except Bl Cr Sb.

A very variable species in need of detailed investigation.

(C) Pedicel terete. Sepals patent. Receptacle glabrous.

7. R. lanuginosus L., *Sp. Pl.* 554 (1753) (R. *umbrosus* Ten.). A pubescent perennial 30–50 cm; stock truncate. Stem with long patent hairs. Basal leaves usually about as wide as long, 3(–5)-lobed to *c.* $\frac{3}{4}$; segments broadly ovate, irregularly dentate; cauline leaves similar but smaller. Flowers (15–)20–30(–40) mm in diameter. Honey-leaves deep orange-yellow. Head of achenes globose; achenes 4–5 mm, compressed, bordered; beak *c.* 1·5 mm, broad, strongly recurved. $2n = 28$. *Woods.* • *C. & S. Europe from S.E. France to Romania, and extending to Denmark and Estonia.* Al Au Bu Co Cz Da Ga Ge Gr He Hu It Ju Po Rm Rs(B, ?C, W) Sa Si Tu.

8. R. acris L., *Sp. Pl.* 554 (1753) (R. *acer* auct., R. *stevenii* auct., non Andrz. ex Besser). Perennial 10–100 cm, more or less glabrous, appressed-pubescent or with stiff patent or deflexed hairs on stem. Stock truncate, oblique, or forming a stout, creeping rhizome. Basal leaves more or less deeply divided into 3–7 sessile segments, which are ovate-cuneate in outline, dentate and simple or again divided; cauline leaves resembling the basal but smaller. Flowers 15–25 mm in diameter. Honey-leaves golden-yellow. Head of achenes globose; achenes 2–3·5 mm, broadly elliptical or suborbicular; beak short, hooked. *Grassland. Most of Europe.* All except Az Bl Cr Lu Rs(K) Si ?Tu; introduced in Sa Sb.

This very variable species is divisible into a number of subspecies that differ in the presence or absence of a rhizome, the shape and dissection of the leaves, pubescence, and the number of flowers in the inflorescence. Intermediates occur between the various taxa where their distributions overlap.

1 Stock a stout ± horizontal rhizome; plant often more than 50 cm ... (d) subsp. **friesianus**
1 Stock short, truncate; plant usually less than 50 cm
 2 Plant usually more than 20 cm; basal leaves much-divided, lobes narrow; flowers numerous ... (a) subsp. **acris**
 2 Plant less than 20 cm; basal leaves with 3(–5) wide segments with a few coarse teeth; flowers 1–3
  3 At least spring leaves glabrous; flowers usually 1 or 2 ... (b) subsp. **pumilus**
  3 Leaves pubescent, petioles with deflexed hairs; flowers 1–3(–5) ... (c) subsp. **borealis**

(a) Subsp. **acris** (R. *stevenii* auct. angl., hung. et roman., non Andrz. ex Besser): Plant 20–50 cm, appressed-pilose (sometimes villous below), with short truncate rootstock. Basal leaves much-divided; leaf-lobes linear-lanceolate. Inflorescence with numerous flowers. $2n = 14$. *Mainly in meadows. Throughout most of the range of the species, but mainly on mountains in the south.*

(b) Subsp. **pumilus** (Wahlenb.) Á. & D. Löve, *Taxon* **34**: 164 (1985): Plant 10–20 cm, somewhat glaucous. Basal leaves with 3 segments with a few coarse teeth; spring leaves glabrous. Flowers 1 or 2(3). $2n = 14$. *Fennoscandia, Iceland, Scotland.*

(c) Subsp. **borealis** (Regel) Nyman, *Consp.* 12 (1878) (R. *borealis* Trautv., nom. illeg., R. *glabriusculus* Rupr., R. *stevenii* sensu Hiitonen, non Andrz. ex Besser): Like subsp. (**b**) but all leaves pubescent; flowers 1–3(–5). $2n = 14$. *N.E. Europe, southwards to 54° N. in C. Ural.*

(d) Subsp. **friesianus** (Jordan) Rouy & Fouc., *Fl. Fr.* **1**: 103 (1893) (subsp. *despectus* Laínz, R. *stevenii* auct. gall. non Andrz. ex Besser): Plant 50–100 cm, with a stout, more or less horizontal rhizome. Basal leaves much-divided; leaf-lobes ovate to linear-lanceolate. Inflorescence with numerous flowers. *C. Europe, and locally in S. Europe; introduced elsewhere.*

9. R. granatensis Boiss., *Diagn. Pl. Or. Nov.* **3(1)**: 8 (1853) (R. *acris* subsp. *granatensis* (Boiss.) Nyman, R. *stevenii* sensu Freyn). Perennial 40–100(–120) cm, with a stout more or less horizontal rhizome. Leaves coriaceous, thick, strigose; lobes ovate or ovate-oblong, with few teeth. Flowers numerous, 15–25 mm in diameter. Honey-leaves golden-yellow. Sepals with more or less appressed, whitish hairs all of about the same length. Filaments pubescent. Achenes with more or less straight beak 0·8–1·5 mm. $2n = 28$. *C. & S.E. Spain.* Hs.

10. R. strigulosus Schur, *Enum. Pl. Transs.* 17 (1866) (R. *stevenii* auct. ross., non Andrz. ex Besser, R. *acris* subsp. *strigulosus* (Schur) Hyl.). Like **9** but leaf-lobes narrower, with broader sinuses; sepals with more or less patent reddish hairs, short at base and long at apex of sepal; beak of achenes 0·5–0·7 mm. $2n = 28$. *E. & E.C. Europe, northwards to White Russia.* Bu Cz Hu Ju ?Po Rm Rs (?C, W, E).

11. R. serbicus Vis., *Mem. Ist. Veneto* **9**: 168 (1858) (?R. *kladnii* Schur). Perennial 40–120 cm; rhizome long, stout. Leaves sericeous, at least when young; basal leaves with 3 cuneate, sometimes stipitate, segments which are variously cut and dentate; cauline leaves resembling the basal but smaller. Filaments pubescent. Beak rather shorter than the achene, nearly straight. $2n = 28$. *Damp and shady places.* • *Balkan peninsula, S.W. Italy.* Al Bu Gr It Ju ?Rm [Su].

12. R. brutius Ten., *Prodr. Fl. Nap.* **1**: lxi (1811) (R. *calabrus* Grande). Patent-pubescent perennial 30–50 cm, with short-creeping stock. Leaves subglabrous; basal leaves large, pentagonal-reniform in outline, 3-partite or pedately 5-partite; segments often stipitate, 2- or 3-fid and strongly serrate; cauline leaves resembling the basal but smaller and sessile. Filaments glabrous. Achenes compressed, keeled; beak about as long as the achene, slender, circinate. *Woods. S. part of Balkan peninsula; S.W. Italy.* Gr It Tu.

R. crimaeus Juz., *Not. Syst.* (*Leningrad*) **22**: 5 (1950), from Krym, appears to be closed related to **12**, from which it is said to differ in its more dissected leaves and the shorter beak of the achene.

(D) Pedicel terete. Sepals patent. Receptacle pubescent.

13–24. R. montanus group. Glabrous or pubescent perennials 5–50 cm. Rhizome often with fibres and sometimes with hairs. Basal leaves 3- to 5(–7)-lobed; segments obovate, dentate; cauline leaves deeply 3- to 5-lobed with narrower, usually entire segments. Flowers 1–3(–5), 20–40 mm in diameter. Achenes 2·5–3·5 mm, compressed, sharply keeled; beak 0·5–2·5 mm, hooked.

There are a number of closely related species in the mountains of C., E. & S. Europe. The following key may help with their identification.

Literature: E. Landolt, *Ber. Schweiz. Bot. Ges.* **64**: 9–83 (1954); **66**: 92–117 (1956).

1 All leaves glabrous or nearly so
 2 Stems glabrescent at base; beak $\frac{1}{3}$–$\frac{1}{2}$ as long as achene (Corse) **17. clethraphilus**
 2 Stems pubescent at base; beak not more than $\frac{1}{3}$ as long as achene
  3 Segments of cauline leaves linear **16. carinthiacus**
  3 Segments of cauline leaves linear-lanceolate or wider
   4 Beak $\frac{1}{4}$–$\frac{1}{3}$ as long as achene **18. montanus**
   4 Beak of achene very short **19. pseudomontanus**
1 Leaves pubescent, except occasionally the first basal ones
 5 Beak very short, appressed to achene; receptacle pubescent at insertion of stamens; segments of cauline leaves linear **23. oreophilus**
 5 Beak $\frac{1}{4}$–$\frac{2}{3}$ as long as achene; receptacle usually glabrous at insertion of stamens or, if pubescent, segments of cauline leaves not linear
  6 Apex of rhizome with numerous hairs 3–4 mm; cauline leaves often semi-amplexicaul (Pyrenees) **13. gouanii**
  6 Rhizome glabrous or rarely with few short hairs, sometimes with fibres
   7 Hairs of sepals at least 2 mm; beak at least $\frac{1}{2}$ as long as achene (S.W. Alps, Spain) **24. aduncus**
   7 Hairs of sepals less than 2 mm; beak usually less than $\frac{1}{2}$ as long as achene
    8 Segments of cauline leaves linear (or, if lanceolate, widest near base)
     9 Beak of achene slender; lateral segments of basal leaves divided for *c.* $\frac{1}{2}$ of their length **22. grenierianus**
     9 Beak of achene rather stout and rigid, lateral segments of basal leaves usually divided for more than $\frac{1}{2}$ of their length
      10 Leaves distinctly pubescent **14. ruscinonensis**
      10 Leaves sparsely pubescent to subglabrous **15. sartorianus**
    8 Segments of cauline leaves not linear, widest near the middle
     11 Leaves sparsely pubescent; beak of achene slender **18. montanus**
     11 Leaves rather densely pubescent; beak of achene stout, rigid
      12 Receptacle usually pubescent at insertion of stamens (Balkan peninsula) **20. concinnatus**
      12 Receptacle glabrous at insertion of stamens (N.E. Italy) **21. venetus**

13. R. gouanii Willd., *Sp. Pl.* **2(2)**: 1322 (1800). Pubescent. Rhizome apex with hairs 3–4 mm, usually with fibres. Cauline leaves usually semi-amplexicaul. Sepals rather densely pubescent. Honey-leaves 10–25 mm. Beak at least $\frac{1}{3}$ as long as the achene, rigid. $2n = 16$. ● *Pyrenees and Cordillera Cantábrica.* Ga Hs.

14. R. ruscinonensis Landolt, *Ber. Schweiz. Bot. Ges.* **66**: 101 (1956). Pubescent. Rhizome without hairs, usually without fibres at apex. Cauline leaves not amplexicaul. Sepals rather densely pubescent. Honey-leaves 8–15 mm. Beak *c.* $\frac{1}{3}$ as long as the achene, rigid. $2n = 16$. ● *C. & E. Pyrenees.* Ga Hs.

15. R. sartorianus Boiss. & Heldr. in Boiss., *Diagn. Pl. Or. Nov.* **3(1)**: 8 (1853). Like **14** but sparsely pubescent to subglabrous; rhizome with fibres at apex; honey-leaves up to 20 mm. $2n = 16$, 32. *Italy and Balkan peninsula.* Al Bu Gr Ju.

Plants from Appennini and the mountains of C. & S. Italy, like **15** but with appressed-pubescent basal leaves, the segments of the single cauline leaf lanceolate, and the beak *c.* $\frac{1}{5}$ as long as the achene have been called **R. apenninus** (Chiov.) Pignatti, *Gior. Bot. Ital.* **116**: 94 (1983), and plants from the same area like **15** but with the basal leaves more strongly pubescent have been called **R. pollinensis** (N. Terracc.) Chiov., *Bull. Soc. Bot. Ital.* **1892**: 279 (1892); their status is uncertain.

16. R. carinthiacus Hoppe in Sturm, *Deutschl. Fl.* Abth. **1** **(12/46)**, no. 10 (1826). Rhizome usually without fibres. Stems pubescent at base. Leaves glabrous but ciliate, the basal deeply lobed. Sepals slightly pubescent, sometimes only at base. Honey-leaves 5–15 mm. Beak of achenes very short, appressed. $2n = 16$. ● *Mountains of S. & S.C. Europe.* ?Al Au Bu Ga Ge He Hs It Ju.

17. R. clethraphilus Litard., *Bull. Acad. Int. Géogr. Bot.* (*Le Mans*) **19**: 94 (1909). Like **16** but stems more or less glabrous at base; basal leaves obtusely dentate; beak $\frac{1}{3}$–$\frac{1}{2}$ as long as the achene. ● *Corse.* Co.

18. R. montanus Willd., *Sp. Pl.* **2(2)**: 1321 (1800) (*R. geraniifolius* sensu Schinz & Thell.). Leaves glabrous or sparsely pubescent; segments of cauline leaves linear-lanceolate to cuneiform or elliptical, dentate, often semi-amplexicaul. Sepals finely pubescent. Honey-leaves 7–20 mm. Receptacle glabrous at insertion of the stamens. Beak $\frac{1}{4}$–$\frac{1}{3}$ as long as the achene. $2n = 32$. *Somewhat calcicole.* ● *Alps, Jura, Schwarzwald.* Au Ga Ge He It Ju.

19. R. pseudomontanus Schur, *Verh. Naturf. Ver. Brünn* **15(2)**: 42 (1877). Like **18** but achene with a very short beak. $2n =$ 16, 32. ● *Carpathians, Balkan peninsula.* Bu Cz Gr Ju Po Rm Rs(W).

20. R. concinnatus Schott, *Österr. Bot. Wochenbl.* **7**: 182 (1857) (*R. croaticus* Schott). Like **18** but leaves pubescent; beak of achene stouter and more rigid; receptacle usually pubescent at insertion of the stamens. ● *W. part of Balkan peninsula.* Al Ju.

21. R. venetus Huter ex Landolt, *Feddes Repert.* **70**: 3 (1964). Like **18** but leaves pubescent; beak of achene stouter and more rigid. $2n = 32$. ● *S.E. Alps.* It.

22. R. grenierianus Jordan in F. W. Schultz, *Arch. Fl. Fr. Allem.* 304 (1854). Rhizome often with abundant fibres. Leaves densely pubescent; segments of cauline leaves linear to lanceolate, widest near base. Sepals finely pubescent. Honey-leaves 5–15 mm. Beak up to $\frac{1}{2}$ as long as the achene. $2n = 16$. *Calcifuge.* ● *Alps.* Au Ga Ge He It.

23. R. oreophilus Bieb., *Fl. Taur.-Cauc.* **3**: 383 (1819) (*R. hornschuchii* Hoppe). Rhizome with fibres and hairs in upper part. Basal leaves pubescent; young folded lamina deflexed. Sepals usually finely patent-pubescent. Honey-leaves 7–15 mm. Receptacle pubescent at insertion of the stamens. Beak of achene very short, appressed. $2n = 16$. *Mountains of C. & S. Europe, eastwards from 6° E.* Au Co Cz Ga Ge He It Ju Po Rm Rs(W, K).

24. R. aduncus Gren. in Gren. & Godron, *Fl. Fr.* **1**: 32 (1847) (*R. villarsii* auct., ?an DC.). Rhizome glabrous and usually without fibres. Leaves pubescent, lower cauline petiolate or with sheathing base. Sepals with long hairs more than 2 mm. Honey-leaves 8–15 mm. Beak at least ½ as long as the achene, stout. $2n = 16$. • *S.W. Alps; E. Spain.* Ga Hs It.

25. R. carpaticus Herbich, *Sel. Pl. Rar. Galic.* 15 (1836) (*R. dentatus* (Baumg.) Freyn). Glabrescent or appressed-pubescent perennial 15–40 cm. Rhizome creeping. Basal leaves 1–3, up to 15 cm wide, 3- to 5-fid; segments broad, rhombic-ovate, dentate; cauline leaves similar, shortly petiolate. Sepals pubescent. Honey-leaves *c.* 20 mm. Achenes *c.* 3 mm, broadly obovate, sharply keeled; beak short, curved. $2n = 16$. • *Carpathians.* Cz Rm Rs(W).

26. R. macrophyllus Desf., *Fl. Atl.* **1**: 437 (1798) (*R. byzantinus* P. H. Davis, *R. palustris* auct., non L. ex Sm.). Robust, pubescent perennial 30–60 cm. Stock stout, erect, surrounded by fibres. Basal leaves *c.* 4–7 × 5–10 cm, pentagonal-orbicular, deeply cordate, 3- to 5-partite; segments wide, irregularly incised and crenate-dentate; upper cauline leaves 3-fid with entire segments, or undivided. Flowers 25–30 mm in diameter. Achenes 4 mm, compressed, sharply keeled, bordered, smooth, minutely pitted, or with scattered bulbous-based hairs; beak *c.* 1 mm, stout, more or less curved. $2n = 16$. *W. Mediterranean region, S. Portugal.* Bl Co Hs Lu Sa [Ga].

27. R. demissus DC., *Reg. Veg. Syst. Nat.* **1**: 275 (1817). Glabrous or sparsely pubescent perennial up to *c.* 20 cm. Rhizome oblique, crowned with fibres. Basal leaves orbicular, 3- to 5-partite; segments shortly stipitate, divided into linear-oblong, obtuse lobes; cauline leaves sessile; segments linear, entire. Stems slender, decumbent. Flowers usually solitary. Sepals pubescent. Achenes 2·5 mm, broadly obovate, not keeled; beak very short, hooked. $2n = 16$. *Damp and shady places in the mountains. S.E. Spain; W. part of Balkan peninsula.* Al ?Gr Hs Ju.

28. R. hayekii Dörfler, *Denkschr. Akad. Wiss. Math.-Nat. Kl.* (*Wien*) **94**: 146 (1918). Like **27** but basal leaves less deeply divided; cauline leaves entire or 3- to 5-fid, with lanceolate segments; flowers usually several; achenes *c.* 3 mm. *Mountain screes.* • *Balkan peninsula; local.* Al Bu ?Ju.

29. R. marschlinsii Steudel, *Nomencl. Bot.* ed. 2, **2**: 434 (1841). Like **27** but basal leaves reniform; segments united at base and irregularly dentate at apex; sepals glabrous; achenes 1–2·5 mm. $2n = 16$. • *Mountains of Corse.* Co.

30. R. dissectus Bieb., *Fl. Taur.-Cauc.* **2**: 25 (1808). More or less erect perennial 10–20 cm, usually with long, white hairs below. Basal leaves ovate-orbicular, cut into oblong-lanceolate lobes, hairy beneath; petioles short. Flowers solitary or few, 20–30 mm in diameter. Sepals pubescent, obtuse. Achenes 3–3·5 mm, elliptical, smooth, compressed, not keeled; beak 0·8–1 mm, hooked. $2n = 16$, $16+4B$. *Mountains of Krym.* Rs(K). (*S.W. Asia.*)

Represented in Europe by subsp. *dissectus* only; several other subspecies occur in Anatolia.

31. R. radinotrichus Greuter & Strid, *Willdenowia* **11**: 267 (1981). Rhizomatous perennial with thick, fibrous roots. Basal leaves divided into 3 obovate segments with a few teeth at the distal end, densely white-sericeous. Cauline leaves 1 or 2, inconspicuous. Stems up to 12 cm, erect, unbranched. Flowers solitary. Sepals pilose; honey-leaves 7–10 mm, broadly obovate, yellow. Achenes orbicular, compressed, smooth, glabrous; beak 1·5–2 mm, circinate. *Limestone cliffs and screes,* 2000–2300 *m.* • *W. Kriti* (*Levka Ori*).

(E) Pedicel sulcate. Sepals deflexed. Receptacle pubescent.

32. R. neapolitanus Ten., *Ind. Sem. Horti Neap.* 11 (1825) (*R. eriophyllus* C. Koch, *R. heucherifolius* sensu Arcangeli, non C. Presl, *R. panormitanus* Tod., *R. bulbosus* subsp. *neapolitanus* (Ten.) H. Lindb.). Patent-pubescent perennial 20–50 cm, without a corm-like stock but often with thick, fusiform roots. Basal leaves 3-partite; segments cuneate-obovate, the lateral 2-lobed, acutely dentate; cauline leaves mostly bract-like. Flowers 20–30 mm in diameter. Achenes orbicular, smooth; beak 0·5–0·75 mm, triangular to somewhat uncinate. $2n = 16$. *Damp grassland. Italy and Sicilia; S.E. Europe.* Al Bu Cr Gr It Ju Rs(K) Si Tu.

33. R. pratensis C. Presl in J. & C. Presl, *Del. Prag.* 9 (1822). Like **32** but more robust, with a more or less corm-like stock; stems appressed-pubescent, often sparsely so; cauline leaves not all small and bract-like; achenes often tuberculate and sparsely hispid; beak 0·8–1·2 mm, uncinate. • *Sardegna, Sicilia.* Sa Si.

34. R. bulbosus L., *Sp. Pl.* 554 (1753) (*R. heucherifolius* C. Presl). Pubescent perennial up to 80 cm. Basal leaves 3-partite, variously dentate and lobed; middle segment abruptly contracted into a stalk, or cuneate and sessile; lower cauline leaves resembling the basal but smaller; upper sessile with linear-lanceolate lobes. Flowers 20–30 mm in diameter. Achenes 2–4 mm; margin keeled and grooved; beak short, curved. *Much of Europe but absent from much of the north and east.* Au Be Bl Br Bu Co Cz Da Fe Ga Ge Hb He Ho Hs Hu It Ju Lu No Po Rm Rs(B, C, W) Sa ?Si Tu.

1 Stock not or scarcely swollen and corm-like; roots thick, fleshy — (c) subsp. **aleae**
1 Stock conspicuously swollen and corm-like; roots slender
 2 Leaf-bases persistent, sheathing; leaves sparsely pubescent or glabrescent above — (b) subsp. **castellanus**
 2 Leaf-bases not persistent or sheathing; leaves usually pubescent above — (a) subsp. **bulbosus**

(a) Subsp. **bulbosus** (incl. subsp. *bulbifer* (Jordan) P. Fourn., *R. bulbifer* Jordan): Plant up to 80 cm, appressed- or patent-pubescent below. Stock conspicuously swollen and corm-like, usually without fibres; roots not swollen. Leaves more or less densely hairy above; middle segment of basal leaves usually stipitate. Pedicel strongly sulcate. Receptacle subglobose. Achenes 2–3 mm, finely punctate. $2n = 16$. *Dry grassland. Widespread in N., W. & C. Europe.*

(b) Subsp. **castellanus** (Freyn) P. W. Ball & Heywood, *Feddes Repert.* **66**: 151 (1962) (*R. castellanus* Freyn): Like subsp. (a) but leaf-bases persistent, very widely sheathing; basal leaves small, cordate-ovate or widely pentagonal in outline, sparsely hairy or glabrescent above; branches short, divaricate. *Mountain grassland,* 1400–2000 *m.* • *N.W. Spain.*

(c) Subsp. **aleae** (Willk.) Rouy & Fouc., *Fl. Fr.* **1**: 106 (1893) (*R. aleae* Willk.; incl. subsp. *adscendens* (Brot.) J. Neves, *R. adscendens* Brot., subsp. *gallecicus* (Willk.) P. W. Ball & Heywood, *R. gallecicus*

Willk., subsp. *broteri* (Freyn) Vasc., *R. broteri* Freyn, *R. occidentalis* Freyn): Plant 20–80 cm, robust; stock not or scarcely swollen and corm-like, usually with abundant fibres; roots thick, fleshy, fusiform or cylindrical. Middle segments of basal leaves stipitate. Pedicel sulcate near the top, striate below. Receptacle elongate-conical. Achenes 2·5–4 mm, smooth. $2n = 16$. *Damp places. S. Europe, extending to Hungary.*

35. R. sardous Crantz, *Stirp. Austr.* **2**: 84 (1763) (*R. philonotis* Ehrh., *R. pseudobulbosus* Schur; incl. subsp. *xatardii* (Lapeyr.) Rouy & Fouc.). More or less pubescent annual 10–45 cm, with the habit of **34** but without fleshy roots and with a corm-like stock not or feebly developed. Flowers 12–25 mm in diameter. Honey-leaves pale yellow. Achenes 2·5–3 mm, smooth or minutely punctate, with small, obtuse tubercles near the margin, or sometimes smooth; beak *c.* 0·5 mm, upward curving. $2n = 16$. *From Scotland and N. Spain eastwards to S.C. Russia.* Al Au Be Bl Br Bu Co Cz Da Ga Ge Gr He Ho Hs Hu It Ju Po Rm Rs(B, C, W, K) Sa Si Su [No].

36. R. cordiger Viv., *Fl. Cors.* 8 (1824) (incl. *R. cordiger* subsp. *diffusus* (Moris) Arrigoni). Like **35** but perennial; basal leaves suborbicular-cordate, incised-crenate, appressed-pubescent and long-petiolate; petioles broadly winged at base. *Streamsides and other wet places, 700–1600 m.* ● *Corse, Sardegna.* Co Sa.

37. R. marginatus D'Urv., *Mém. Soc. Linn. Paris* **1**: 318 (1822) (incl. *R. angulatus* C. Presl). Like **35** but honey-leaves 3–4 mm, golden-yellow; margin of achenes wider; beak *c.* 1 mm, lanceolate and 2-veined. $2n = 16$. *S.E. Europe; Sicilia.* Al Cr Gr Ju Rs(K) Si Tu [Br].

The achenes vary from almost smooth, to strongly tuberculate in var. *trachycarpus* (Fischer & C. A. Meyer) Aznav. (*R. trachycarpus* Fischer & C. A. Meyer).

38. R. trilobus Desf., *Fl. Atl.* **1**: 437 (1798). Nearly glabrous annual 10–50 cm. Lowest leaves simple; the next 3-partite, the segments simple, cuneate-obovate, dentate; upper with more numerous, linear-oblong lobes. Flowers 10–15 mm in diameter. Honey-leaves about as long as the sepals. Achenes *c.* 2 mm, with numerous small tubercles; beak short, triangular. $2n = 32$. *Damp, often disturbed ground. S. Europe.* Al Az Bl Co Ga Gr Hs It Lu Sa Si.

39. R. muricatus L., *Sp. Pl.* 555 (1753). More or less glabrous annual up to *c.* 50 cm. Leaves all similar, coarsely crenate-dentate and often shallowly lobed. Honey-leaves little longer than the sepals. Achenes 7–8 mm, ovate, spiny, with a broad, smooth margin; beak 2–3 mm, nearly straight. $2n = 32, 48, 64$. *Damp, often disturbed ground. S. Europe.* Al Bl Bu Co Cr Ga Gr Hs It Ju Lu Rm Rs(K) Sa Si Tu [Az Br].

R. cornutus DC., *Reg. Veg. Syst. Nat.* **1**: 300 (1817) (*R. lomatocarpus* Fischer & C. A. Meyer), from S.W. Asia, like **39** but with honey-leaves 2–3 times as long as the sepals, and achenes tuberculate, has been recorded from Turkey-in-Europe, but requires confirmation.

Sect. ECHINELLA DC. Nectary covered by an entire, truncate scale, attached laterally or free to base. Achenes beaked, strongly bordered and with spines, hooked hairs or tubercles.

40. R. arvensis L., *Sp. Pl.* 555 (1753). More or less pubescent annual 15–60 cm. Lowest leaves simple, the others 3-lobed and often again divided into narrow, entire or dentate segments. Pedicel terete. Flowers 4–12 mm in diameter, pale greenish-yellow. Sepals patent. Receptacle pubescent. Achenes 6–8 mm, few (4–8), in 1 whorl, spiny or tuberculate, rarely only ribbed-reticulate, with a broad sulcate border; beak 3–4 mm, straight. $2n = 32$. *Cultivated or disturbed ground. S., W. & C. Europe, extending to S.E. Sweden and E. Ukraine.* Al Au Be Bl Br Bu Co Cr Cz Da Ga Ge Gr Hb He Ho Hs Hu It Ju Lu Po Rm Rs(B, C, W, K) Sa Si Su Tu [No].

41. R. parviflorus L. in Loefl., *Iter Hisp.* 291 (1758). Pubescent, spreading to decumbent annual 5–40 cm. Lower leaves 3- to 5-lobed; segments obovate-cuneate, dentate; upper simple or with entire, oblong segments. Flowers 3–6 mm in diameter. Sepals deflexed. Receptacle glabrous. Achenes 2–3 mm, bordered, with hooked spines; beak short, hooked at apex. $2n = 32$. *Mediterranean region and W. Europe, northwards to N. England and eastwards to the middle Danube.* Az Bl Br Co Ga Gr Hb Hs It Ju Lu Sa Si [Be].

42. R. chius DC., *Reg. Veg. Syst. Nat.* **1**: 299 (1817). Like **41** but leaves 3-lobed; middle segment entire to 3(–5)-dentate; fruiting pedicel greatly thickened; beak at least $\frac{1}{2}$ as long as the achene. *Lowland grassland and scrub. S. Europe.* Bu Co Cr Gr It Ju Rs(K) Sa Si Tu.

Sect. RANUNCULASTRUM DC. Roots of 2 sorts, some tuberous and some fibrous. Achenes compressed, keeled. Receptacle elongating in fruit and becoming more or less cylindrical.

(A) Sepals deflexed at flowering.

43. R. gracilis E. D. Clarke, *Travels* **2(2)**: 336 (1814) (*R. chaerophyllos* sensu Hayek, non L., *R. agerii* Bertol.). Simple or little-branched perennial with ovoid tubers. Stock usually without fibres. Basal leaves glabrous or slightly patent-pilose, the outer suborbicular, shallowly 3-lobed, the inner more deeply cut into obovate-cuneate, dentate segments. Achenes ovoid; beak somewhat curved, nearly as long as the achene. $2n = 16$. *S. Europe, eastwards from Sicilia.* Bu Cr Gr It Ju Si Tu.

44. R. oxyspermus Willd., *Sp. Pl.* **2(2)**: 1328 (1800). Like **43** but tubers fusiform; all the leaves patent-hirsute, mostly conspicuously longer than wide, deeply lobed with deeply divided segments; achenes tapering into a straight, sharp beak longer than the achene. $2n = 16$. *S.E. Europe.* Bu Ju Rm Rs(W, K, E).

45. R. psilostachys Griseb., *Spicil. Fl. Rumel.* **1**: 304 (1843). Like **43** but tubers fusiform; leaves sericeous beneath, mostly conspicuously longer than wide; achenes verrucose. $2n = 16, 32$. ● *Balkan peninsula, extending northwards to Hungary.* Al Bu Gr Hu Ju Tu.

46. R. rumelicus Griseb., *op. cit.* 305 (1843). Like **43** but tubers oblong; leaves sericeous beneath, distinctly longer than wide; achenes tuberculate; beak straight. $2n = 16, 24$. *Balkan peninsula, Kriti.* Bu Cr Gr Ju Tu.

47. R. monspeliacus L., *Sp. Pl.* 553 (1753) (*R. cylindricus* Jordan). Whitish-lanate or sericeous perennial 20–50 cm, with cylindrical or fusiform tubers. Basal leaves variable, ovate-cordate, more or less deeply 3-lobed and dentate to 3-fid with oblong-cuneate, laciniate, stipitate segments. Flowers 25–30 mm in diameter. Receptacle glabrous. Head of achenes cylindrical. Achenes ovoid, sparsely pubescent; beak *c.* $\frac{1}{2}$ as long as the achene, curved. $2n = 16$. ● *W. Mediterranean region.* ?Bl Co Ga Hs It ?Sa Si [Ho Su].

48. R. miliarakesii Halácsy, *Magyar Bot. Lapok* **11**: 116 (1912).

Like **47** but very slender; tubers ovoid; basal leaves 3-partite; segments often stipitate, cuneate, again 3-partite; lobes linear-lanceolate; flowers *c.* 15 mm in diameter; head of achenes globose; beak of achene very short, hooked. ● *Thessalia.* Gr.

49. R. isthmicus Boiss., *Diagn. Pl. Or. Nov.* **1(6)**: 4 (1846) (incl. R. *orientalis* auct., non L.). Small, spreading, appressed-pubescent perennial with fusiform tubers. Basal leaves 2- or 3-fid, or the outer 3-sect or 3-fid; lobes linear-lanceolate. Achenes tubercled, with an appressed appendage at base; beak as long as achene, wide, falcate. *Sicilia; Greece.* Gr Si.

Represented in Europe by subsp. *isthmicus* only.

50. R. illyricus L., *Sp. Pl.* 552 (1753) (R. *scythicus* Klokov). Appressed-sericeous perennial up to *c.* 50 cm, with ovoid tubers. Basal leaves 3-fid; segments linear-lanceolate, entire or 2- or 3-fid into linear-lanceolate, entire lobes. Receptacle glabrous. Achenes *c.* 3 mm, more or less triangular, strongly compressed and almost winged, finely punctate; beak about as long as achene, nearly straight. $2n = 32$. *Dry grassland; calcicole. C. & S.E. Europe extending to Italy; Öland.* Au Bu Cz Ge Gr Hu It Ju Po Rm Rs(C, W, K, E) Su Tu.

R. stojanovii Delip., *Feddes Repert.* **81**: 715 (1971), known only from the type collection from Bulgaria (Rodopi Planina) is like **50** but has lanceolate leaf-segments and verrucose, sparsely hairy (not glabrous) achenes; its status is uncertain.

(B) Sepals not deflexed at anthesis.

51. R. pedatus Waldst. & Kit., *Pl. Rar. Hung.* **2**: 112 (1802 or 1803). Almost glabrous perennial up to *c.* 30 cm, with oblong tubers. Basal leaves glabrous, pedately 3- to 5-partite; segments linear-lanceolate, entire or deeply 2-lobed. Achenes membranous, strongly compressed and almost winged; beak straight, except for the hooked tip. $2n = 16$. *E. & E.C. Europe.* Bu Cz Gr Hu Ju Rm Rs(W).

Plants from S. Ukraine, like **51** but with basal leaves palmatisect and with narrowly linear segments have been called **R. odessanus** Klokov fil., *Nov. Syst. Pl. Vasc. (Leningrad)* **1968**: 120 (1968), and plants from E. Ukraine and S. Russia, like **51** but with stems more pubescent, basal leaves with fewer, wider segments, larger flowers and subglabrous (not densely hairy) sepals have been called **R. silvisteppaceus** Dubovik, *op. cit.* 116 (1968); their status is uncertain.

52. R. platyspermus Fischer ex DC., *Prodr.* **1**: 37 (1824). Like **51** but villous; basal leaves 1- or 2-ternate, with pinnatifid segments. Achenes circular, strongly compressed and winged; beak very short, hooked. *S.E. Russia.* Rs(E). (*C. Asia.*)

53. R. paludosus Poiret, *Voy. Barb.* **2**: 184 (1789) (R. *flabellatus* Desf., R. *chaerophyllos* sensu Coste, non L., R. *chaerophyllos* sensu Arcangeli, non L.; incl. R. *heldreichianus* Jordan, R. *winkleri* Freyn). Perennial 10–50 cm, with fusiform tubers. Stock usually short, with persistent fibres. Outer basal leaves shallowly 3-lobed; inner 3-fid; middle segment long-stipitate; all divided into narrow, dentate segments. Flowers 1–8. Receptacle glabrous. Achenes *c.* 2 mm, minutely punctate, sparsely pubescent, strongly keeled on back, tapering into a nearly straight beak about as long as the achene and hooked at tip. $2n = 16, 32$. *Mediterranean region and W. Europe, northwards to the Channel Islands.* Co Cr Ga Gr Hs It Ju Lu Sa Si Tu.

54. R. barceloi Grau, *Mitt. Bot. Staatssamm. (München)* **20**: 54 (1984) (R. *chaerophyllos* var. *balearicus* Barc.). Like **53** but with more slender tuberous roots; stem 15–45 cm; outer basal leaves sub-orbicular, coarsely crenate; flowers 1–4(5); achenes with curved, hooked beak. *Streamsides.* ● *Islas Baleares (Mallorca).* Bl.

55. R. pseudomillefoliatus Grau, *loc. cit.* (1984). Like **53** but stems not more than 30 cm; basal leaves withered at anthesis; leaves multifid, the segments *c.* 1 mm wide, linear; cauline leaves 1 or 2, the upper subsimple; achenes pubescent, with straight beak. ● *C. & S. Spain.* Hs.

56. R. gregarius Brot., *Fl. Lusit.* **2**: 369 (1804) (R. *hollianus* Reichenb.). Slender, pubescent perennial 10–25 cm, with cylindrical tubers. Basal leaves reniform or pentagonal, entire to 3-fid; segments more or less lobed and dentate, sometimes divided into narrow sessile or stipitate lobes, appressed-hirsute to more or less glabrous; cauline leaves 0 or few with lanceolate segments. Flowers 1(–3), *c.* 20 mm in diameter. Head of achenes *c.* 10 mm. Receptacle pubescent to nearly glabrous. Achenes broadly orbicular, sparsely setose or rarely glabrous; beak shorter than achene, hooked. *S. Spain, Portugal.* Hs Lu.

R. henriquesii Freyn, *Bot. Centr.* **6(3)**: 21 (1881), from C. & N. Portugal, up to 50 cm, with narrow, acute, villous, distant leaf-lobes and densely pubescent receptacle, and **R. malessanus** Degen & Hervier, *Bull. Acad. Int. Géogr. Bot. (Le Mans)* **16**: 222 (1906), from S. Spain (Sierra de Cazorla, Sierra de Segura), with basal leaves 3- to 5-fid, the leaf-lobes oblong, obtuse, more or less glabrous and overlapping, stems scapose, and receptacle nearly glabrous, are probably best included in **56**.

57. R. ollissiponensis Pers., *Syn. Pl.* **2**: 106 (1806) (R. *gregarius* sensu Tutin, non Brot.). Slender perennial 5–40 cm, with cylindrical or fusiform tubers. Basal leaves ovate, 3- to 5-fid, the segments 3-fid; cauline leaves few, simple to subsimple. Achenes pubescent, with straight beak. ● *Spain and Portugal.* Hs Lu.

(a) Subsp. **ollissiponensis** (R. *nevadensis* Willk., R. *escurialensis* Boiss. & Reuter, R. *carpetanus* Boiss. & Reuter, R. *suborbiculatus* Freyn): Tubers fusiform. Middle segment of basal leaves sessile or subsessile. Stem usually with several flowers. Achenes *c.* 2·5 mm, sparsely pubescent. *Throughout the range of the species.*

(b) Subsp. **alpinus** (Boiss. & Reuter) Grau, *Mitt. Bot. Staatssamm. (München)* **20**: 53 (1984) (R. *blepharicarpos* auct., non Boiss., R. *carpetanus* var. *alpinus* Boiss. & Reuter): Tubers cylindrical. Middle segment of basal leaves stipitate. Stem usually with a single flower. Achenes *c.* 3 mm, glabrous or subglabrous. *Mountains of C. & N. Spain.*

58. R. cupreus Boiss. & Heldr. in Boiss., *Diagn. Pl. Or. Nov.* **2(8)**: 3 (1849). A somewhat pubescent perennial 5–20 cm, with ovoid tubers. Basal leaves 2-pinnatisect; segments obovate. Flowers *c.* 15 mm in diameter. Honey-leaves copper-coloured beneath. Achenes 2·5 mm, with a broad, hooked beak *c.* 1 mm. *Shady, limestone rocks.* ● *Kriti.* Cr.

59. R. millefoliatus Vahl, *Symb. Bot.* **2**: 63 (1791) (incl. R. *garganicus* Ten.). More or less pubescent perennial 10–30 cm, with ovoid or oblong tubers. Basal leaves 2- or 3-pinnatisect, with linear-lanceolate acute or obtuse lobes. Sepals usually glabrous. Honey-leaves 5, yellow on both surfaces. Receptacle glabrous, less frequently pubescent. Achenes *c.* 3·5 mm, rounded, broadly keeled on back; beak $\frac{1}{2}$ as long as the achene, wide, hooked to nearly straight. $2n = 16, 32$. *S. & E.C. Europe.* Al Bu Cz Ga Gr It Ju Rm Si Tu.

60. R. millii Boiss. & Heldr. in Boiss., *Fl. Or.* **1**: 35 (1867). Like **59** but tubers ovoid; honey-leaves 7–12, oblong-ovate; achenes *c.* 2 mm; beak strongly hooked, *c.* ¼ as long as the achene. *Dry hills.* ● *S. Greece.* Gr.

61. R. nigrescens Freyn in Willk. & Lange, *Prodr. Fl. Hisp.* **3**: 921 (1880). Somewhat pubescent, erect, little-branched perennial up to 10–40 cm, with cylindrical tubers. Basal leaves *c.* 2 cm across, reniform, 3-fid to the middle, crenate, rather thick and rugose, nearly or quite glabrous; cauline leaves few, all bract-like. Flowers 1 or 2, *c.* 25 mm in diameter. Head of achenes *c.* 12 mm. Achenes glabrous, orbicular; beak shorter than achene, curved. ● *N. Spain, N. & C. Portugal.* Hs Lu.

62. R. cortusifolius Willd., *Enum. Pl. Horti Berol.* 558 (1809) (*R. megaphyllus* Steudel). Robust, densely villous perennial up to *c.* 100 cm, with fusiform tubers. Basal leaves up to *c.* 30 cm wide, coriaceous, orbicular-cordate, shallowly lobed; segments nearly parallel-sided, shallowly lobed and dentate at apex; lower cauline leaves resembling basal but smaller; upper sessile, bract-like. Inflorescence corymbose. Flowers up to 50 mm in diameter. Receptacle nearly glabrous. Achenes 3 mm, glabrous, smooth; beak 1 mm, curved. *Açores.* Az. (*Madeira, Islas Canarias.*)

63. R. spicatus Desf., *Fl. Atl.* **1**: 438 (1798) (incl. *R. aspromontanus* Huter, *R. blepharicarpos* Boiss., *R. rupestris* Guss., *R. spicatus* subsp. *aspromontanus* (Huter) Greuter & Burdet, *R. spicatus* subsp. *blepharicarpos* (Boiss.) Grau, *R. spicatus* subsp. *rupestris* (Guss.) Maire). Pubescent perennial up to *c.* 40 cm, with fusiform tubers. Basal leaves reniform or orbicular, shallowly 3-lobed; middle segment smaller than the lateral; all more or less shallowly lobed and crenate, thin, densely appressed-hirsute on both surfaces. Flowers 30–40 mm in diameter, few. Receptacle pubescent. Achenes obovate, glabrous or with short setae; beak *c.* ½ as long as the achene, broad-based, recurved and hooked. *S. Spain; S.W. Italy, Sicilia.* Hs It Si.

Three subspecies have been recognized in Europe, with the typical subspecies confined to N. Africa, but these are very weakly characterized and are not worth maintaining.

64. R. creticus L., *Sp. Pl.* 550 (1753). Branched, patent-pubescent perennial up to 50 cm. Basal leaves reniform, crenate and shallowly lobed; cauline leaves 3-fid; segments lanceolate, entire or dentate at apex. Flowers 20–30 mm in diameter, numerous. Receptacle nearly glabrous. Achenes 4–5 mm, pubescent, compressed; beak *c.* 0·5 mm, hooked. $2n = 16$. *Rocky places. S. Aegean region.* Cr Gr.

65. R. thasius Halácsy, *Österr. Bot. Zeitschr.* **42**: 412 (1892). Slender, flexuous, much-branched, more or less glabrous perennial, with oblong tubers. Basal leaves orbicular-cordate, 3-fid; segments divided into ovate, obtuse lobes. Flowers numerous. Achenes about as long as the stout, slightly curved beak. *E. Greece; Aegean region.* Gr.

Outside Europe known only from the E. Aegean island of Ikaria.

66. R. subhomophyllus (Halácsy) Vierh., *Österr. Bot. Zeitschr.* **84**: 131 (1935) (*R. cadmicus* auct., non Boiss.). Slender, not or little-branched, more or less glabrous perennial, with oblong tubers. Basal leaves orbicular or wider than long, 3-sect to base; segments with obtuse lobes. Flowers 1 or 2; honey-leaves 5–7 mm. Achene longer than the slender, hooked beak. *Rocky places on limestone. S. Aegean region.* Cr Gr.

67. R. incomparabilis Janka, *Österr. Bot. Zeitschr.* **22**: 174 (1872). Like **66** but tubers cylindrical to fusiform; stems unbranched; leaves 3-sect to *c.* ⅘; honey-leaves 7–9 mm, broader; achene with longer hooked beak. $2n = 16$. *Damp rock-ledges and crevices on schist and granite.* ● *C. part of Balkan peninsula.* Bu Gr Ju.

68. R. sprunerianus Boiss., *Diagn. Pl. Or. Nov.* **1(1)**: 64 (1843). Much-branched, patent-pubescent to subglabrous perennial, with cylindrical tubers. Basal leaves orbicular-cordate, 3- to 5-fid. Flowers numerous. Achenes with tubercle-based hairs; beak about as long as the achene, nearly straight but hooked at apex. $2n = 16$, 32. *Balkan peninsula and Aegean region.* Bu Cr Gr Ju Tu.

69. R. asiaticus L., *Sp. Pl.* 552 (1753) (*Cyprianthe asiatica* (L.) Freyn). Pubescent perennial with tuberous roots. Outer basal leaves 3-lobed and dentate; inner 3-sect; segments stipitate, pinnatisect. Flowers 30–50 mm in diameter; honey-leaves white, less often yellow, red or purple. Head of achenes *c.* 10 mm. Achenes ovate, attenuate into a wide, hooked beak. $2n = 16$. *S. Aegean region.* Cr Gr.

Sect. AURICOMUS Spach. Roots all fibrous. Achenes slightly compressed, often puberulent; receptacle not elongating in fruit.

(70–75). R. auricomus group. Perennial herbs with simple to deeply palmately divided, reniform to suborbicular-cordate basal leaves; cauline leaves smaller, sessile, deeply dissected or rarely simple. Pedicels terete. Sepals patent, pubescent. Honey-leaves yellow, often unequal or some or all lacking. Achenes slightly compressed, puberulent (rarely glabrous), borne on peg-like projections from the glabrous or pubescent receptacle. *Most of Europe but absent from much of the Mediterranean region.*

Species **70–75** include a small number of sexual diploids and a large number of polyploids which are facultative or obligate apomicts and which have perhaps arisen from hybrids between the sexual species. Most of the apomicts are very local but a few have a fairly wide distribution. Detailed investigations have been made in certain geographical areas, but much work remains to be done before a complete understanding of the group can be attained.

Literature: E. Borchers-Kolb, *Mitt. Bot. Staatssamm. (München)* **19**: 363–429 (1983). P. A. Haas, *Ber. Bayer. Bot. Ges.* **29**: 5–12 (1952); **30**: 27–32 (1954). A. Jasiewicz, *Fragm. Fl. Geobot.* **2**: 62–110 (1956). E. Julin & J. A. Nannfeldt, *Ark. Bot.* **6**: 163–241 (1967). G. Marklund, *Mem. Soc. Fauna Fl. Fenn.* **16**: 45–53 (1939). G. Marklund & A. Rousi, *Evolution* **15**: 510–522 (1961). E. I. Nyárády, *Bul. Grăd. Bot. Cluj* **13**, no. 1–4 (1934). A. Rousi, *Ann. Bot. Soc. Zool.-Bot. Fenn. Vanamo* **29(2)** (1956). M. Rozanowa, *Trudy Petergof. Est.-Nauč. Inst.* no. 8 (1932). Z. Schiller, *Math. Term. Értesitő* **35**: 361–447 (1917). O. Schwarz, *Mitt. Thür. Bot. Ges.* **1(1)**: 120–143 (1949). An extensive bibliography is given in J. Jalas & J. Suominen, *Atlas Florae Europaeae* **8**: 161–176 (1989).

The treatment here is very conservative and includes only the principal species that have been recognized in Europe. A detailed synonymy and a valuable synthesis of the distribution of some of the segregate taxa can be found in J. Jalas & J. Suominen, *loc. cit.* (1989).

1 Achenes glabrous
 2 Outer basal leaves deeply lobed **74. affinis**

2 Outer basal leaves crenate **75. degenii**
1 Achenes puberulent
3 Scale-like sheaths present at base outside foliage leaves
4 Basal leaves usually 2 or 3, at least some lobed or coarsely dentate **71. fallax**
4 Basal leaves usually 1 or 2, serrate-crenate but not lobed
5 Basal leaves pubescent beneath; segments of cauline leaves dentate; beak of achene long **70. cassubicus**
5 Basal leaves glabrous beneath; segments of cauline leaves entire or slightly dentate; beak of achene short **72. monophyllus**
3 Scale-like sheaths absent
6 Basal leaves usually at least 5; segments of cauline leaves usually entire **73. auricomus**
6 Basal leaves usually 2 or 3; segments of cauline leaves dentate and often lobed **71. fallax**

70. R. cassubicus L., *Sp. Pl.* 551 (1753) (incl. R. *flabellifolius* Heuffel & Reichenb., ?R. *allemannii* Br.-Bl.). Tall (*c.* 40 cm), stout. Scale-like sheaths present. Basal leaves 1 or 2, up to 10 cm wide, orbicular-cordate, serrate-crenate, pubescent beneath; segments of cauline leaves dentate. Receptacle densely pubescent. Beak of achene long. $2n = 16, 24, 32, 40, 44, 48, 64$. *C. & E. Europe*. Au Bu Cz Fe Ge He Hu It Ju Po Rm Rs(N, B, C, W, E) Su.

71. R. fallax (Wimmer & Grab.) Sloboda, *Rostlinnictví* 679 (1852) (R. *megacarpus* Walo Koch, pro parte). Stems *c.* 40 cm, stout. Scale-like sheaths small and soon decaying. Basal leaves 2 or 3, at least some lobed or coarsely dentate, pubescent beneath; segments of cauline leaves dentate or lobed. Receptacle more or less pubescent. Achenes densely pubescent; beak medium. $2n = 16, 32(-34)$. ● *C. & E. Europe, extending to Sweden*. Au Bu Cz Fe Ge He Hu ?It Ju Po Rm Rs(N, C, W) Su.

Intermediate between **70** and **73**.

72. R. monophyllus Ovcz., *Not. Syst.* (*Leningrad*) **3**: 54 (1922). Like **71** but stems *c.* 20 cm, slender; scale-like sheaths larger and persistent; basal leaves glabrous beneath; segments of cauline leaves entire or slightly dentate; achenes sparsely and shortly pubescent. $2n = 32$. *Fennoscandia and N.E. Russia*. Fe No Rs(N, C, E) Su.

73. R. auricomus L., *Sp. Pl.* 551 (1753) (R. *binatus* Kit.). Variable in size. Scale-like sheaths absent. Basal leaves commonly 5 or more, some or all deeply 3- to 5-lobed, usually glabrous beneath; cauline leaves with narrow, usually entire segments. Receptacle pubescent or glabrous. Achenes densely pubescent; beak short. $2n = 32(-34), 40, 48$. *Widespread in Europe*. All except Al Az Bl Cr Lu Sa Si Tu.

74. R. affinis R. Br. in Parry, *Jour. Voy. N.W. Pass.* (*Suppl. App.*) 265 (1824) (R. *auricomus* var. *glabratus* Lynge, R. *pedatifidus* auct., non Sm.). Scale-like sheaths absent. Basal leaves deeply 3- to 5-lobed; cauline leaves with linear segments. Achenes glabrous. *Arctic Europe*. Rs(N) Sb.

75. R. degenii Kümmerle & Jáv., *Bot. Közl.* **19**: 19 (1921). Scale-like sheaths absent. Outer basal leaves crenate, the inner divided to $\frac{2}{3}$; cauline leaves with narrowly oblanceolate, entire segments. Achenes glabrous. ● *N.E. Albania, S.W. Jugoslavia*. Al Ju.

A similar plant with densely pubescent achenes occurs in S. Jugoslavia.

76. R. polyrhizos Stephan ex Willd., *Sp. Pl.* **2(2)**: 1324 (1800). Almost glabrous perennial 8–15 cm. Basal leaves 3-lobed or 3-sect; segments narrow, not markedly different from the smaller, sessile, 3- to 5-sect cauline leaves. Sepals patent. Receptacle pubescent. Achenes puberulent. *Moldavia; E. Ukraine, S.E. Russia*. Rs(C, W, E).

77. R. pygmaeus Wahlenb., *Fl. Lapp.* 175 (1812). Nearly glabrous perennial 1·5–4(–7) cm with a stout stock. Stems simple. Basal leaves reniform in outline, 3-lobed; segments wide, obtuse, the lateral usually shallowly lobed; cauline leaves usually sessile, 3(–5)-lobed, the segments ovate-lanceolate, entire, obtuse. Flowers 5–10 mm in diameter. Sepals villous at base. Achenes *c.* 1 mm, ovoid, smooth; beak short, hooked. $2n = 16, 32$. *N. Europe, southwards to 59° N. in Norway; E. Alps; W. Carpathians*. Au Cz Fe He Is It No Po Rs(N) Sb Su.

78. R. nivalis L., *Sp. Pl.* 553 (1753). Usually unbranched perennial 8–15(–25) cm. Basal leaves orbicular-reniform in outline, deeply pedately lobed; cauline leaves sessile, with linear-lanceolate, entire lobes. Receptacle nearly glabrous, cylindrical. Sepals patent, villous. Achenes ovoid, smooth; beak about as long as the achene, slender. $2n = 40, 48, 56$. *Arctic Europe, extending southwards to 62° N. in Norway*. Fe No Rs(N) Sb Su.

79. R. sulphureus C. J. Phipps, *Voy. N. Pole* 202 (1774) (R. *altaicus* sensu Resvoll-Holmsen, non Laxm.). Like **78** but stouter; basal leaves cuneate, shallowly lobed; receptacle densely pubescent, ovoid; sepals with abundant, brown hairs; beak $\frac{1}{2}$ as long as the achene. $2n = 96$. *Arctic Europe, extending southwards to 59° N. in Ural*. Fe No Rs(N) Sb Su.

Sect. HECATONIA (Lour.) DC. Marsh- or water-plants. Nectary-scale often forked or completely surrounding the nectary. Achenes small, scarcely compressed; beak usually very short.

80. R. hyperboreus Rottb., *Skr. Kiøbenhavnske Selsk. Lærd. Vid.* **10**: 458 (1770). Small, slender, creeping or floating, nearly glabrous perennial. Leaves 5-lobed, ovate; petioles short. Flowers *c.* 5 mm in diameter, solitary, axillary. *Arctic and subarctic Europe, southwards to 61° N. in Norway*. Fe Is No Rs(N) Sb Su.

(a) Subsp. **hyperboreus**: Scarcely caespitose; leaf-lobes ovate, divaricate; head of achenes 4–5 mm in diameter. $2n = 32$. *Throughout the range of the species*.

(b) Subsp. **arnellii** Scheutz, *Kungl. Svenska Vet.-Akad. Handl.* nov. ser., **22(10)**: 75 (1888) (R. *samojedorum* Rupr.): Somewhat caespitose; leaf-lobes linear-oblong, not divaricate; head of achenes *c.* 2 mm in diameter. *Arctic Russia*.

81. R. sceleratus L., *Sp. Pl.* 551 (1753). Stout, erect, more or less glabrous annual up to 60 cm. Basal leaves reniform, 3-lobed, the lateral lobes often again lobed, all crenate; upper cauline leaves sessile, up to 3-lobed; segments entire. Flowers 5–10 mm in diameter, numerous. Sepals deflexed, pubescent beneath. Receptacle pubescent, much elongated in fruit. Achenes *c.* 1 mm, ovoid, glabrous but faintly rugose; beak very short, obtuse. *Almost throughout Europe*. All except Az Cr Fa Is Sa Sb.

(a) Subsp. **sceleratus**. Stems more or less glabrous. Sepals and honey-leaves 5. $2n = 32$. *Throughout the range of the species*.

(b) Subsp. **reptabundus** (Rupr.) Hultén, *Bot. Not.* **1947**: 352 (1947) (R. *reptabundus* Rupr.): Stems densely pubescent. Sepals 3 or 4; honey-leaves 5 or absent. *N. Russia*.

82. R. gmelinii DC., *Reg. Veg. Syst. Nat.* **1**: 303 (1817). Creeping, floating or decumbent, rooting at the nodes. Leaves orbicular or reniform, with 3–5 cuneate segments divided into linear lobes. Flowers few, *c.* 10 mm in diameter. Achenes *c.* 1·5 mm, ovoid; beak slender. $2n=16$, 32. *N. Russia.* Rs(N, C).

Sect. COPTIDIUM (Prantl) Tutin. Perennial. Roots fibrous. Sepals 3. Honey-leaves 6–8, yellow. Achenes scarcely compressed, the upper part filled with spongy tissue; beak long, slender. Receptacle scarcely enlarging in fruit.

83. R. lapponicus L., *Sp. Pl.* 553 (1753) (R. *altaicus* Laxm.). Slender, creeping, rooting at the nodes. Leaves reniform in outline, 3-partite; lateral segments spreading; all obovate-cuneate, shallowly lobed and crenate-dentate. Flowers axillary, solitary, long-pedicellate. Sepals deflexed. Honey-leaves yellow. Achenes 3–4·5 mm, ovoid-fusiform, constricted about the middle, keeled; beak hooked, scarcely shorter than the achene. $2n = 16$. *N. Europe southwards to C. Sweden and C. Ural.* Fe No Rs (N, C) Sb Su.

Sect. PALLASIANTHA (L. Benson) Tutin. Like Sect. *Coptidium* but with reddish or white honey-leaves, the upper part of the achenes not filled with spongy tissue, and the receptacle distinctly enlarged in fruit.

84. R. pallasii Schlecht., *Animadv. Ranunc.* **1**: 15 (1819). Like **83** but leaves deeply 3-lobed, obovate-cuneate in outline; lobes narrow, entire; flowers larger, with 6–12, reddish-violet or white honey-leaves; achenes 5–7 mm. $2n = 32$. *Arctic Russia, Spitsbergen.* Rs(N) Sb.

R. × spitzbergensis Hadač, *Norges Svalb.-Ishavs-Undersøk. Skr.* **87**: 36 (1944) (R. *pallasii* var. *minimus* Rupr.), from Spitsbergen and arctic Russia, with narrow, spreading, usually yellow leaf-lobes, is a hybrid between **83** and **84**.

Sect. HALODES (A. Gray) L. Benson. Perennial. Roots fibrous. Sepals 5. Honey-leaves 5, yellow; margins of nectary-scale free from honey-leaf. Achenes compressed, veined or striate. Receptacle elongating in fruit.

85. R. cymbalaria Pursh, *Fl. Amer. Sept.* **2**: 392 (1814). Slender, with ascending stems and creeping stolons. Leaves obtuse, reniform or truncate at base, crenate or dentate. Scapes up to 25 cm, 1- to 10-flowered. Flowers 6–9 mm in diameter, bright yellow. Achenes thin-walled, obovate, longitudinally ribbed; beak very asymmetrical, short, slender. *Naturalized on coasts of Fennoscandia.* [Fe No Su.] (*North America, N. Asia.*)

Sect. FICARIA (Schaeffer) Boiss. Some roots tuberous, some fibrous. Leaves broadly ovate, cordate. Flowers yellow. Sepals 3. Honey-leaves 8–12. Achenes ovoid, scarcely compressed, keeled; beak minute.

86. R. ficaria L., *Sp. Pl.* 550 (1753) (*Ficaria verna* Hudson, *F. ranunculoides* Roth, *F. degenii* Hervier). Perennial 5–40 cm. Some roots fibrous, some forming fusiform tubers. Leaves broadly ovate, cordate, obtusely angled, shallowly crenate or rarely dentate. Flowers 15–50 mm in diameter. Honey-leaves ovate. Receptacle pubescent. Achenes *c.* 2·5 mm, ovoid, keeled, pubescent. *Mostly in shady places. Almost throughout Europe, but rare in the north-east.* All except Az Is Sb; only naturalized in Fa.

1 Stems at anthesis very short, the leaves crowded at base; achenes hirsute (b) subsp. **calthifolius**
1 Stems elongated at anthesis, leaves not crowded in a rosette; achenes puberulent or pubescent
  2 Bulbils present in the leaf-axils after anthesis
   3 Flowers up to *c.* 20 mm in diameter, with obovate-oblong, not overlapping honey-leaves (c) subsp. **bulbilifer**
   3 Flowers more than 20 mm in diameter, with broadly obovate, overlapping honey-leaves (d) subsp. **ficariiformis**
  2 Bulbils not present in the leaf-axils
   4 Sepals green, with narrow scarious margin; flowers 20–30 mm in diameter; achenes puberulent (a) subsp. **ficaria**
   4 Sepals yellowish-white; flowers 30–50 mm in diameter; achenes pubescent (e) subsp. **chrysocephalus**

(a) Subsp. **ficaria**: Plant up to 30 cm, rather stout. Leaves not in a rosette. Basal sinus of leaves usually wide. Bulbils absent. Flowers 20–30 mm in diameter. Sepals herbaceous. Achenes well developed, puberulent. $2n = 16$. *W. Europe, eastwards to S. Italy.*

(b) Subsp. **calthifolius** (Reichenb.) Arcangeli, *Comp. Fl. Ital.* 11 (1882) (*Ficaria calthifolia* Reichenb.): Plant small. Stems at anthesis very short. Leaves crowded in a rosette. Basal sinus of leaves narrow. Flowers up to 20 mm in diameter. Sepals green, with a narrow scarious margin. Achenes well developed, hirsute. *S.C. & E. Europe.*

(c) Subsp. **bulbilifer** Lambinon, *Bull. Jard. Bot. Nat. Belg.* **51**: 462 (1981) (R. *ficaria* subsp. *bulbifer* Lawalrée, nom. illeg.): Plant up to 30 cm, rather slender. Leaves not in a rosette. Basal sinus of leaves usually wide. Bulbils present in leaf-axils. Flowers up to 20 mm in diameter. Sepals herbaceous. Achenes mostly sterile, puberulent. $2n = 32$. *N. & C. Europe, extending locally to Spain, Albania and E.C. Russia.*

(d) Subsp. **ficariiformis** (F. W. Schultz) Rouy & Fouc., *Fl. Fr.* **1**: 73 (1893) (R. *calthifolius* (Guss.) Jordan, R. *ficariiformis* F. W. Schultz, *Ficaria grandiflora* Robert): Plant very robust. Bulbils present in leaf-axils. Flowers 30–50 mm in diameter. Sepals yellowish-white. Achenes pubescent. *S. Europe.*

(e) Subsp. **chrysocephalus** P. D. Sell, *Bot. Jour. Linn. Soc.* **106**: 117 (1991): Like subsp. (d) but bulbils absent. *Greece, Kriti.*

87. R. ficarioides Bory & Chaub., *Nouv. Fl. Pélop.* 34 (1838). Like **86** but smaller (up to *c.* 7 cm); pedicels shorter than the basal leaves; leaves coarsely crenate-dentate or shallowly lobed; sepals yellowish; achenes glabrous. $2n = 16$. *Damp places in the mountains. Greece; Karpathos.* Cr Gr.

Plants from Karpathos have small and slender leaves and smaller flowers (var. *gracilis* W. Barbey).

Sect. PHYSOPHYLLUM Freyn. Roots tuberous. Leaves all basal. Receptacle ovoid. Achenes inflated.

88. R. bullatus L., *Sp. Pl.* 550 (1753). Perennial. Leaves ovate, crenate, hispid beneath, more or less bullate. Pedicels 5–25 cm, pubescent, 1- or 2-flowered. Flowers *c.* 25 mm in diameter, scented. Sepals greenish, pubescent. Honey-leaves 5–12, oblong, yellow. Receptacle glabrous. Achenes narrowly bordered; beak short, curved. $2n = 16$. *Mediterranean region, Portugal.* Bl Co Cr Gr Hs It Lu Sa Si.

Unusual in the genus in Europe in flowering in autumn.

Sect. THORA DC. Roots tuberous. Leaves broader than long. Receptacle ovoid. Achenes scarcely compressed, strongly veined.

89. R. thora L., *Sp. Pl.* 550 (1753) (R. *tatrae* Borbás; incl. R. *scutatus* Waldst. & Kit.). Perennial 10–30(–50) cm. Leaves glaucous, glabrous; the lower reniform, entire at base and then increasingly coarsely serrate to apex; basal petiolate, appearing after flowering; the lower cauline sessile; the upper cauline small, lanceolate, mostly 3-lobed. Flowers 1 to several, 10–20 mm in diameter, yellow. Sepals glabrous. Honey-leaves ovate. Receptacle sparsely hairy. Achenes *c.* 4 mm, few, glabrous, subglobose; beak short, hooked. $2n = 16$. *Calcicole.* ● *Mountains of C. & S. Europe, from N.W. Spain to E. Carpathians.* Al Au Cz Ga He Hs It Ju Po Rm Rs(W).

90. R. hybridus Biria, *Hist. Renonc.* 38 (1811). Like **89** but smaller; basal leaves usually 2, present at flowering, truncate or broadly cuneate at base, coarsely dentate at apex, otherwise entire; flowers 12–25 mm in diameter; achenes 2–3 mm. ● *E. Alps, extending to W. Jugoslavia.* Au Ge It Ju ?Rm.

91. R. brevifolius Ten., *Prodr. Fl. Nap.* **1**: lxviii (1815). Like **89** but smaller; basal leaves usually several; cauline leaves 1(2), 3-sect, smaller; flowers 15–25 mm in diameter; achenes 3–4 mm; beak long, curved. $2n = 16$. *Damp, stony ground on limestone. C. Italy; Balkan peninsula and Kriti.* Al Cr Gr It Ju.

Plants from Greece and Kriti have been called subsp. **pindicus** (Hausskn.) E. Mayer, *Razpr. Mat.-Prir. Akad. Ljubljani* **5**: 26 (1959), on the basis of more numerous, more deeply lobed basal leaves and smaller achenes, but these differences are not consistent.

Sect. LEUCORANUNCULUS Boiss. Roots fibrous. Basal leaves lobed. Honey-leaves white (rarely pink), caducous. Receptacle glabrous. Achenes subglobose, smooth; beak long, nearly straight.

92. R. alpestris L., *Sp. Pl.* 553 (1753) (R. *caballeroi* Losa & P. Monts.). Glabrous, caespitose perennial 3–15 cm. Leaves shiny; basal leaves several, 3- to 5-lobed; cauline leaves 1 or 2. Flowers 15–25 mm in diameter, solitary or 2 or 3. Sepals glabrous. Honey-leaves obcordate. Achenes *c.* 1·5 mm, glabrous, obovoid; beak slender. *Calcicole.* ● *Mountains of C. & S. Europe.* Au Cz Ga Ge He Hs It Ju Po Rm.

(a) Subsp. **alpestris**: Plant caespitose. Leaves shiny; basal leaves usually 5-lobed, deeply crenate with obtuse teeth; cauline leaves 1 or 2, the lower 3-fid. Flowers *c.* 20 mm in diameter, often 2 or 3. Honey-leaves emarginate. $2n = 16$. *Almost throughout the range of the species.*

(b) Subsp. **traunfellneri** (Hoppe) P. Fourn., *Quatre Fl. Fr.* 354 (1936). Plant not caespitose, usually smaller. Leaves dull; basal leaves usually 3-lobed, lobes deeply divided with obtuse to subacute teeth; cauline leaf 1, usually linear, simple. Flowers *c.* 15 mm in diameter, solitary. Honey-leaves emarginate to truncate. $2n = 16$. *S.E. Alps.*

R. cacuminis Strid & Papanicolaou, *Ann. Mus. Goulandris* **4**: 221 (1978), from moist rocks above 2500 m on Kaimakçalan on the border between Greece and Jugoslavia, differs from **92(b)** by the somewhat swollen roots, acute teeth of leaf-margin, and slightly smaller achenes, and may represent a third subspecies.

93. R. crenatus Waldst. & Kit., *Pl. Rar. Hung.* **1**: 9 (1799) (incl. R. *magellensis* Ten.). Perennial 5–15 cm, glabrous. Basal leaves petiolate, suborbicular, weakly cordate, crenate, sometimes 3-lobed at apex, obscurely veined; cauline leaves 1 or 2, lanceolate to linear, bract-like. Flowers 1 or 2, 20–25 mm in diameter. Sepals narrowly ovate. Honey-leaves broadly ovate or oblong, obcordate. Achenes 1·5–2 mm, glaucous, ovoid; beak about as long as the achene, slender, hooked at apex. $2n = 16$. ● *E. Alps; C. Appennini; S. & E. Carpathians, Balkan peninsula.* Al Au Bu It Ju Rm Rs(W).

94. R. bilobus Bertol., *Misc. Bot.* **19**: 5 (1858). Like **93** but basal leaves with prominent veins; sepals wider; honey-leaves emarginate or more deeply notched. $2n = 16$. *Rocks and stony grassland; calcicole.* ● *S.E. Alps.* It.

Sect. ACONITIFOLII Tutin. Roots fibrous. Basal leaves lobed. Honey-leaves white, caducous. Receptacle pubescent. Achenes subglobose, veined; beak short, curved.

95. R. aconitifolius L., *Sp. Pl.* 551 (1753). Perennial up to *c.* 50 cm. Leaves palmately 3- to 5-lobed; lobes serrate-dentate, the middle one free to base; cauline leaves sessile. Pedicels 1–3 times as long as subtending leaf, pubescent above. Flowers 10–20 mm in diameter. Sepals reddish or purple beneath, glabrous, caducous. Honey-leaves 5, ovate, weakly obcordate. Achenes 5 mm, slightly compressed; beak slender. $2n = 16$. ● *C. Europe, extending to C. Spain, N. Italy and W. Jugoslavia.* Au Cz Ga Ge He Hs It Ju.

96. R. platanifolius L., *Mantissa* 79 (1767). Like **95** but often larger (up to 130 cm); leaves 5- to 7-lobed, the middle lobe not free to base; pedicels 4–5 times as long as the subtending leaves, nearly or quite glabrous above. $2n = 16$. ● *C. & S. Europe, Belgium, W. Fennoscandia.* Al Au Be Bu Co Cz Ga Ge Gr He Hs It Ju No Po Rm Rs(W) Sa Su.

97. R. seguieri Vill., *Prosp. Pl. Dauph.* 50 (1779). Perennial 8–20 cm, at first pubescent, later glabrescent. Basal leaves palmately 3- to 5-lobed, the segments again divided; cauline leaves similar but smaller. Flowers up to 25 mm in diameter. Sepals glabrous. Honey-leaves emarginate. Achenes *c.* 4 mm, broadly ovoid; beak slender. *Calcicole.* ● *Alps, Cordillera Cantábrica, C. Appennini; S.W. Jugoslavia.* Au Ga He Hs It Ju.

(a) Subsp. **seguieri**: Slightly pubescent. Flowers 20–25 mm in diameter. Achenes inflated. $2n = 16$. *Throughout the range of the species except Jugoslavia.*

(b) Subsp. **montenegrinus** (Halácsy) Tutin, *Feddes Repert.* **69**: 55 (1964). Tomentose. Flowers smaller. Achenes ovate, sulcate. *S.W. Jugoslavia.*

Sect. ACETOSELLIFOLII Tutin. Roots tuberous, fusiform. Leaves all basal, palmately 5- to 7-veined, hastate. Honey-leaves white, caducous. Receptacle glabrous. Achenes keeled on inner margin, veined; beak very short.

98. R. acetosellifolius Boiss., *Biblioth. Univ. Genève* nov. ser., **13**: 406 (1838). Glabrous perennial 3–20 cm. Stock stout, fibrous. Stems decumbent or ascending, simple or divaricately branched. Leaves hastate and irregularly laciniate at base; middle lobe the longest. Flowers 15–25 mm in diameter. Honey-leaves obovate. Achenes subglobose; beak curved. $2n = 16$. *Damp grassland, 2400–3000 m.* ● *S. Spain (Sierra Nevada).* Hs.

Sect. CRYMODES (A. Gray) Tutin. Roots fibrous. Leaves deeply lobed. Honey-leaves white to purple, persistent. Receptacle glabrous. Achenes strongly compressed and winged on 2 sides.

99. R. glacialis L., *Sp. Pl.* 553 (1753) (*Oxygraphis vulgaris* Freyn). Perennial 4–25 cm, usually glabrous except for the sepals. Stem stout, erect or ascending. Basal leaves thick, 3-sect; segments usually stipitate, deeply divided into elliptical or oblong lobes;

cauline leaves similar but smaller, sessile or shortly petiolate. Sepals with abundant, reddish-brown hairs beneath. Achenes 2·5 mm, glabrous; beak *c.* 1·5 mm, nearly straight. $2n = 16$. *Snow-lies; calcifuge. Arctic and subarctic Europe, W. Fennoscandia; higher mountains of C. Europe; Pyrenees; Sierra Nevada.* Au Cz Fa Fe Ga Ge He Hs Is It No Po Rm Rs(N) Sb Su.

R. kamchaticus DC., *Reg. Veg. Syst. Nat.* **1**: 302 (1817) (*Oxygraphis glacialis* (Fischer) Bunge) occurs in N. Ural just within Europe. It is a glabrous perennial up to 6 cm, with broadly ovate, entire or weakly and remotely dentate leaves in a basal rosette. The solitary flowers have 5 broadly ovate sepals which are appressed to the 11–15, narrowly oblong, yellow honey-leaves.

Sect. INSULARES Tutin. Roots fleshy, cylindrical. Honey-leaves yellow, caducous. Receptacle pubescent. Achenes ovoid, with a sulcate border, smooth; beak short, curved.

100. R. cymbalariifolius Balbis ex Moris, *Stirp. Sard.* **1**: 2 (1827) (*R. balbisii* Moris). Caespitose, appressed-pilose perennial up to *c.* 5 cm. Basal leaves up to *c.* 1 cm, orbicular, 3-lobed usually to half-way or less; segments ovate, strongly crenate; cauline leaves small, few, near the base of the slender, strigose pedicels. Petioles enlarged and sheathing at base. Flowers *c.* 10 mm in diameter, solitary. Sepals deflexed, shortly pubescent. Honey-leaves narrowly elliptical, not contiguous, pale glossy yellow. Achenes smooth, strongly keeled; beak short, curved. $2n = 16$. ● *Sardegna.* Sa.

101. R. weyleri Marès, *Bull. Soc. Bot. Fr.* **12**: 232 (1865). Like **100** but plant up to 20 cm, with patent hairs at base of stems and petioles; leaves up to 2 cm, ovate-elliptical, deeply 3-lobed; segments often stipitate; petioles not enlarged and sheathing at base; sepals patent; achenes rough, with prominent veins. $2n = 16$. ● *Islas Baleares (Mallorca).* Bl.

Sect. FLAMMULA Webb. Roots fibrous. Leaves all simple. Honey-leaves usually yellow, caducous. Receptacle glabrous. Achenes slightly compressed; beak very short.

102. R. flammula L., *Sp. Pl.* 548 (1753). Usually glabrous perennial 5–80 cm, erect, or creeping and rooting at the nodes. Leaves variable, orbicular to subulate, entire or serrate; the lower petiolate; the upper smaller and sessile. Flowers 1 to many, 7–20 mm in diameter. Pedicel sparsely appressed-pubescent, sulcate. Achenes glabrous, minutely pitted, weakly bordered; beak very short. *Wet places. Most of Europe, but rare in the Mediterranean region.* All except Bl Cr Is Rs(K) Sb Tu.

1 Lamina of basal leaves suborbicular-cordate, fleshy; stems not rooting at nodes (c) subsp. **minimus**
1 Lamina of basal leaves at least twice as long as wide, thin; stems often rooting at nodes
2 Basal leaves persistent; lamina ovate to lanceolate (a) subsp. **flammula**
2 Basal leaves caducous; lamina subulate or very small (b) subsp. **scoticus**

(a) Subsp. **flammula**: Stem erect to procumbent, often rooting at the nodes; internodes 4–7 cm. Basal leaves about twice as long as wide, usually cuneate or rounded at base, thin. Achenes *c.* $1\frac{1}{2}$ times as long as wide. $2n = 32$. *Throughout the range of the species.*

(b) Subsp. **scoticus** (E. S. Marshall) Clapham in Clapham, Tutin & E. F. Warburg, *Fl. Brit. Is.* 91 (1952) (*R. scoticus* E. S. Marshall): Like subsp. (a) but stems always erect; basal leaves caducous; petioles long; lamina subulate or very narrow, obtuse to rounded; flowers few, usually 1; honey-leaves more or less truncate. ● *N. Scotland, W. Ireland.*

(c) Subsp. **minimus** (Ar. Benn.) Padmore, *Watsonia* **4**: 21 (1957): Stem more or less procumbent, not rooting at the nodes; internodes 0·5–2 cm. Basal leaves suborbicular-cordate, thick and fleshy. Achenes scarcely longer than wide. $2n = 32$. ● *N. Scotland, W. Ireland.*

103. R. reptans L., *Sp. Pl.* 549 (1753) (*R. flammula* subsp. *reptans* (L.) Syme). Like small forms of **102** but with very slender, always procumbent stems rooting at every node; leaves all petiolate, spathulate to narrowly elliptical; flowers 5(–10) mm in diameter, solitary. $2n = 32, 48$. *N. Europe, extending southwards locally to C. Italy, Bulgaria and N. Ukraine.* Au †Br Bu ?Cz Da Fa Fe Ga Ge He Is It No ?Po Rs(N, B, C, W, E) Su.

Hybrids between **102** and **103**, **R. × levenensis** Druce ex Gornall, *Watsonia* **14**: 383 (1987), occur in N. Europe and W. Alps, and also in N. Britain outside the present range of **103**.

104. R. batrachioides Pomel, *Nouv. Mat. Fl. Atl.* 249 (1874) (*R. xantholeucos* Cosson & Durieu). Slender, branched, glabrous annual 3–6 cm. Leaves spathulate, entire. Flowers 5–8 mm in diameter. Sepals patent. Achenes 0·5 mm, ovoid, keeled; beak very short, slender. *Wet places. C. Spain, Sardegna; local.* Hs Sa. (*N.W. Africa.*)

The plants from Spain, with smaller flowers and shorter peduncles, have been described as subsp. **brachypodus** G. López, *Anal. Jard. Bot. Madrid* **41**: 470 (1985).

105. R. lingua L., *Sp. Pl.* 549 (1753). Robust, stoloniferous perennial 50–120 cm. Basal leaves ovate, cordate, long-petiolate, soon withering; cauline leaves oblong-lanceolate, shortly petiolate to sessile; all more or less serrate. Flowers (20–)30–50 mm in diameter, few; pedicels terete. Achenes *c.* 2·5 mm, glabrous, minutely pitted, bordered; beak short and wide. $2n = 128$. *Marshes and shallow water. Most of Europe but rare in the Mediterranean region and the far north.* Al Au Be Br Bu Cz Da Fe Ga Ge Hb He Ho Hs Hu It Ju No Po Rm Rs(N, B, C, W, E) Si Su Tu.

106. R. ophioglossifolius Vill., *Hist. Pl. Dauph.* **3**: 731 (1789). Annual 10–40 cm. Basal leaves ovate or suborbicular, cordate, long-petiolate; upper leaves smaller, narrower, shortly petiolate or sessile; all entire or obscurely serrate; lower pedicels as long as subtending leaves. Flowers 5–9 mm in diameter, numerous; pedicel somewhat sulcate, appressed-pubescent. Sepals glabrous. Honey-leaves obovate, nearly twice as long as the sepals. Achenes *c.* 1·5 mm, compressed, minutely tuberculate; beak *c.* $\frac{1}{10}$ as long as the achene. $2n = 16$. *Wet places. S. & W. Europe, northwards to England; Gotland.* Al Bl Br Bu Co Cr Ga Gr Hs Hu It Ju Lu Rm Rs (W, K) Sa Si Su Tu.

107. R. thracicus Aznav., *Bull. Soc. Bot. Fr.* **46**: 136 (1899). Like **106** but lower leaves ovate-elliptical, not cordate; lower pedicels shorter than the subtending leaves; honey-leaves shorter than the sepals. *Wet places. C. Greece; Turkey-in-Europe.* Gr Tu.

108. R. fontanus C. Presl in J. & C. Presl, *Del. Prag.* 6 (1822). Like **106** but smaller and more or less procumbent, rooting at the lower nodes; glabrous; flowers smaller; achenes smooth, with longer beak. $2n = 48$. *C. Mediterranean region; Balkan peninsula.* Al Bu Co Gr It Ju Si.

109. R. longipes Lange ex Cutanda, *Fl. Comp. Madrid* 103

(1861) (*R. dichotomiflorus* Freyn). Like **106** but the basal leaves mostly elliptical or lanceolate, rarely ovate; lower pedicels $\frac{1}{3}$–$\frac{1}{2}$ as long as the leaves; honey-leaves oblong, about as long as the sepals. Beak *c.* $\frac{1}{4}$ as long as the achene. ● *Spain, N. Portugal.* Hs Lu.

110. R. revelieri Boreau, *Mém. Soc. Acad.* (*Angers*) **1**: 85 (1857). Annual 10–20 cm. Leaves lanceolate, petiolate. Pedicels terete, thickened in fruit; lower about as long as the leaves. Sepals usually pubescent beneath, longer than the honey-leaves. Achenes 1·5 mm, minutely tuberculate, scarcely compressed; beak very short. $2n =$ 32. ● *Corse, Sardegna, S.E. France* (*Var*). Co Ga Sa.

(a) Subsp. **revelieri**: Pedicels stout, fistular. Sepals densely pubescent beneath. *Corse, Sardegna.*

(b) Subsp. **rodiei** (Litard.) Tutin, *Feddes Repert.* **69**: 55 (1964). Pedicels less stout, scarcely fistular. Sepals glabrous or with few scattered hairs beneath. *S.E. France* (*Var*).

Sect. MICRANTHUS (Ovcz.) A. Nyárády. Like Sect. *Flammula* but flowers sessile or nearly so, not more than 3 mm in diameter, yellow. Achenes tuberculate; beak $\frac{1}{2}$ as long to as long as the achene.

111. R. lateriflorus DC., *Reg. Veg. Syst. Nat.* **1**: 251 (1817). Much-branched annual 5–25 cm. Lower leaves ovate, broadly cuneate at base; petioles very long; upper leaves narrower, with shorter petioles; all remotely denticulate. Flowers 2·5–3 mm in diameter, sessile at the nodes or in the dichotomies of the stem, pale yellow. Sepals glabrous, yellow. Honey-leaves about as long as the sepals. Achenes 25–30, *c.* 2·5 mm, narrowed into a beak about as long as the achene, curved at tip. $2n = 16$. *Wet or seasonally flooded ground. C., S. & S.E. Europe.* †Au Bu Cr Cz Ga Gr Hs Hu It Ju Rm Rs(W, K, E) Si Tu.

112. R. nodiflorus L., *Sp. Pl.* 549 (1753). Like **111** but leaves narrower, tapering gradually into the petiole, entire or slightly serrulate; achenes 12–28; beak *c.* $\frac{1}{2}$ as long as the achene. ● *S.W. Europe.* Co Ga Hs Lu.

Sect. XANTHOBATRACHIUM (Prantl) Ovcz. Like Sect. *Flammula* but submerged leaves with filiform, verticillate segments; flowers 3–6 mm in diameter, yellow; achenes less than 1 mm; beak almost absent.

113. R. polyphyllus Waldst. & Kit. ex Willd., *Sp. Pl.* **2(2)**: 1331 (1800). Plant 10–15 cm, creeping then ascending. Stems stout, divaricately branched. Aerial leaves entire or 3-lobed, more or less elliptical in outline; lamina 5–10 mm. Flowers 3–6 mm in diameter. Achenes numerous, minutely papillose. *E.C. & E. Europe.* Cz Hu Rm Rs(C, W, E).

Sect. RANUNCELLA (Spach) Freyn. Roots fibrous or rarely tuberous. Leaves entire, more or less parallel-veined. Flowers yellow, white or pinkish. Achenes more or less compressed, veined, keeled.

114. R. pyrenaeus L., *Mantissa Alt.* 248 (1771) (*R. plantagineus* All., nom. illeg.). Perennial 5–15 cm. Stock fibrous. Leaves linear to lanceolate; cauline leaves few, sessile. Flowers 1–10; pedicels glabrous below, pubescent above. Flowers 10–20 mm in diameter, white. Sepals glabrous, whitish. Honey-leaves obovate, often imperfect or some lacking. Receptacle ovoid, pubescent. Achenes inflated, nearly smooth; beak short, nearly straight. ● *Pyrenees, mountains of Spain.* Ga Hs.

(a) Subsp. **pyrenaeus**: Stems often branched. Leaves linear; cauline 1–3. Flowers 10–20 mm in diameter. $2n = 16$. *Pyrenees and N. & E. Spain.*

(b) Subsp. **angustifolius** (DC.) Rouy & Fouc., *Fl. Fr.* **1**: 81 (1893) (*R. alismoides* Bory): Stems usually simple. Leaves lanceolate; cauline 0 or 1. Flowers 10–13 mm in diameter. $2n = 16$. *S. Spain* (*Sierra Nevada*).

115. R. keupferi Greuter & Burdet, *Willdenowia* **16**: 452 (1987) (*R. pyrenaeus* subsp. *plantagineus* (All.) Rouy & Fouc.). Perennial 5–30(–40) cm, more robust than **114**. Leaves lanceolate; cauline leaves few, sessile. Flowers 1–10; pedicels hirsute or tomentose. Flowers 10–30 mm in diameter, white. Sepals glabrescent, yellowish-green. Honey-leaves obovate, often imperfect or lacking. Receptacle ovoid to spherical. Achenes inflated, strongly veined; beak recurved. $2n = 16, 24, 32, 40$. *Alps, Corse.* Au Co Ga He It.

116. R. bupleuroides Brot., *Fl. Lusit.* **2**: 365 (1804). Like **114** but up to 50 cm, more branched; basal leaves ovate or lanceolate-ovate with slender petioles; plant glabrous except for petioles and base of stem; flowers pale yellow; achenes transversely rugose. $2n = 16$. *Dry places; somewhat calcifuge.* ● *N.W. Spain, Portugal.* Hs Lu.

117. R. gramineus L., *Sp. Pl.* 549 (1753). Glaucous, glabrous or pubescent perennial 20–50 cm. Stock stout, fibrous. Basal leaves linear to lanceolate, flat, sessile; cauline leaves few, small, sessile. Pedicels 1- to few-flowered. Flowers *c.* 20 mm in diameter, deep yellow. Sepals glabrous, yellowish. Honey-leaves broadly obovate. Receptacle ovoid, glabrous. Achenes 3 mm, slightly compressed, keeled, veined; beak 0·5 mm, stout, nearly straight. $2n = 16$. *S.W. Europe extending to C. France and C. Italy.* Ga He Hs It Lu Sa.

118. R. abnormis Cutanda & Willk., *Linnaea* **30**: 83 (1859). Perennial. Roots tuberous. Stem 50–20 cm, glabrous or pubescent. Leaves linear-lanceolate, cucullate at apex; cauline leaves small, sessile. Pedicels 1- to 3-flowered, more or less pubescent. Flowers 20–25 mm in diameter, yellow. Sepals yellowish, more or less pubescent. Honey-leaves 8–14, oblong-obovate. Receptacle oblong, glabrous. Achenes ovate, slightly compressed, smooth; beak very short, curved. $2n = 16$. ● *W.C. & N.W. Spain, N. Portugal.* Hs Lu.

119. R. parnassiifolius L., *Sp. Pl.* 549 (1753). Perennial (4–)10–20(–30) cm. Basal leaves entire, ovate-cordate or broadly lanceolate, at first pubescent at base, on margins and beneath; main veins usually 7; petioles not widened into the lamina; cauline leaves amplexicaul. Pedicels pubescent above, 1- to several-flowered. Flowers 20–30 mm in diameter, white or reddish. Sepals glabrous or pubescent. Honey-leaves broadly ovate. Achenes smooth or with raised veins, inflated; beak very short, hooked. $2n = 16, 32,$ 40. ● *Alps, Pyrenees, N. Spain.* Au Ga Ge He Hs It.

1 Basal leaves hairy beneath; sepals densely villous; achenes strongly veined (b) subsp. **cabrerensis**
1 Basal leaves glabrous beneath, except sometimes at base; sepals glabrous or sparsely hairy; achenes smooth or with inconspicuous veins
2 Petals unequal or absent; stamens fewer than carpels (c) subsp. **heterocarpus**
2 Petals ±equal; stamens more than carpels
3 Basal leaves uniformly hairy above; peduncles mostly less than 15 mm in fruit (a) subsp. **parnassiifolius**
3 Basal leaves more densely hairy on the veins than

elsewhere above; peduncles mostly more than 15 mm in fruit (d) subsp. **favargeri**

(a) Subsp. **parnassiifolius**: Basal leaves rounded to subcordate at base, glabrous beneath, more or less uniformly hairy above. Sepals glabrous or sparsely hairy. Honey-leaves 5, subequal, white or pinkish. Peduncles mostly less than 15 mm in fruit. Achenes more or less smooth. *E. Pyrenees.*

(b) Subsp. **cabrerensis** Rothm., *Bol. Soc. Esp. Hist. Nat.* **34**: 148 (1934): Basal leaves cordate at base, hairy on both surfaces. Sepals densely villous. Honey-leaves 5, subequal, white. Peduncles mostly less than 15 mm in fruit. Achenes with prominent venation. *Mountains of N.W. Spain.*

(c) Subsp. **heterocarpus** Küpfer, *Boissiera* **23**: 192 (1975): Basal leaves cuneate to subcordate at base, hairy on both surfaces, more densely so on the veins than elsewhere above. Sepals glabrous or sparsely hairy. Honey-leaves 0–5, unequal in size, white. Peduncles mostly more than 15 mm in fruit. Achenes smooth. *Throughout most of the range of the species.*

(d) Subsp. **favargeri** Küpfer, *op. cit.* 191 (1975): Basal leaves rounded to cordate at base, glabrous beneath, more densely so on the veins than elsewhere above. Sepals glabrous or sparsely hairy. Honey-leaves 5, white. Peduncles mostly more than 15 mm in fruit. Achenes smooth. *W. Pyrenees; Picos de Europa.*

120. R. wettsteinii Dörfler, *Anzeig. Akad. Wiss. (Wien)* **55**: 282 (1918). Like **119** but petiole widened gradually into the lamina; main veins usually 5; leaves often coarsely 2- to 4-dentate; sepals pubescent; achenes with prominent venation. ● *W. Jugoslavia (N.E. of Korab).* Ju.

This plant, known only from the type collection, should perhaps be treated as a subspecies of **119**.

121. R. amplexicaulis L., *Sp. Pl.* 549 (1753). Perennial 8–30 cm. Basal leaves ovate-lanceolate; cauline leaves amplexicaul. Pedicels glabrous, 1- to several-flowered. Flowers *c.* 20 mm in diameter, white. Sepals glabrous, greenish, caducous. Honey-leaves obovate-orbicular. Achenes inflated, strongly veined; beak curved. $2n = 16$. ● *Pyrenees, mountains of N. & C. Spain.* Ga Hs.

Subgen. **Batrachium** (DC.) A. Gray. Aquatic plants. Leaves all with a broad lamina (laminate) or all divided into capillary segments (capillary), or of both kinds. Honey-leaves white, sometimes with yellow claw. Achenes transversely rugose. Hybrids occur between most species in the subgenus.

Literature: C. D. K. Cook, *Mitt. Bot. Staatssamm. (München)* **6**: 47–237 (1966).

122. R. hederaceus L., *Sp. Pl.* 556 (1753). Procumbent annual to perennial growing on mud or in shallow water. Leaves all laminate, 1–3·5 cm wide, reniform or suborbicular-cordate, with 3(–5), shallow, semi-orbicular or triangular, obtuse, entire segments which are widest at their base. Honey-leaves up to 4·5 mm, scarcely longer than the sepals, not contiguous. Receptacle glabrous. Achenes glabrous; beak lateral to subterminal. $2n = 16$. ● *W. Europe, extending eastwards to C. Germany and S. Sweden.* Be Br Da Ga Ge Hb Ho Hs Lu †No †Rs(B) Su.

123. R. omiophyllus Ten., *Fl. Nap.* **4**: 338 (1830) (*R. lenormandii* F. W. Schultz, *R. homoiophyllus* auct.). Like **122** but leaf-segments narrowest at their base, with wide, shallow crenations; honey-leaves up to 7 mm, about twice as long as the sepals; beak of achene subterminal. $2n = 16, 32$. *W. Europe, northwards to Scotland; S.W. Italy and Sicilia.* Br Ga Hb †Ho Hs It Lu Si.

124. R. tripartitus DC., *Icon. Pl. Gall. Rar.* **1**: 15 (1808) (*R. obtusiflorus* (DC.) Moss, *R. petiveri* Koch; incl. *R. lutarius* (Revel) Bouvet). Annual or perennial growing on mud or in water. Laminate leaves up to 4 cm wide, reniform or suborbicular, with 3(–5) cuneate, distant, entire or crenate segments. Capillary leaves with very slender, sometimes compressed segments, absent in terrestrial plants. Honey-leaves up to 6 mm, not more than twice as long as the sepals. Receptacle globose, pubescent. Achenes glabrous; beak lateral or subterminal. $2n = 40, 42, 44, 46, 48$. *W. Europe, northwards to Britain and the Netherlands; S. Aegean (Mykonos).* †Be Br Ga Ge Gr Hb Ho Hs Lu.

125. R. ololeucos Lloyd, *Fl. Loire-Inf.* 3 (1844) (incl. *R. lusitanicus* Freyn). Like **124** but honey-leaves usually more than 6 mm, more than twice as long as the sepals, contiguous. $2n = 16$. ● *W. Europe, from S. Spain to the Netherlands.* Be Ga Ge Ho Hs Lu.

126. R. peltatus Schrank, *Baier. Fl.* **2**: 103 (1789). Annual or perennial with either laminate or capillary leaves, or both. Laminate leaves semi-orbicular with a truncate base, to orbicular, with 3–7 usually crenate segments. Capillary leaves shorter than the internodes. Honey-leaves (3–)6–15 mm, contiguous; nectaries more or less pyriform. Receptacle pubescent. Achenes usually more than 2 mm, ovate. *Most of Europe, but rare in the east.* All except Al Az Fa Is Rs(W, K, E) Sb.

1 Laminate leaves with 3(–5) segments; achene usually with dorsal wing (b) subsp. **baudotii**
1 Laminate leaves with 3–7 segments; achene without wing
2 Achenes usually hairy, with lateral or subterminal beak; pedicels usually more than 50 mm at anthesis (a) subsp. **peltatus**
2 Achenes glabrous, with subterminal to terminal beak; pedicels usually less than 50 mm at anthesis (c) subsp. **fucoides**

(a) Subsp. **peltatus** (*R. petiveri* auct., non Koch, *R. floribundus* Bab., *Batrachium dichotomum* Schmalh., *B. langei* F. W. Schultz, *B. triphyllos* (Wallr.) Dumort., *R. capillaceus* Thuill., *R. carinatus* Schur, *R. diversifolius* sensu Willk. & Lange, non Gilib., *R. trichophyllus* Wallr.): Laminate leaves with (3–)5(–7) shallow segments; segments of capillary leaves rigid or flaccid. Pedicels usually more than 50 mm at anthesis, longer than the petiole of the subtending leaf. Sepals 3–6 mm, patent. Honey-leaves usually more than 10 mm. Achenes pubescent or hispid, rarely glabrous (var. *microcarpus* Meikle), not winged. $2n = 16, 32, 48$. *Throughout most of the range of the species.*

(b) Subsp. **baudotii** (Godron) Meikle ex C. D. K. Cook, *Anal. Jard. Bot. Madrid* **40**: 473 (1984) (*R. baudotii* Godron, *R. confusus* Godron, *R. hololeucos* auct., *R. leontinensis* Freyn, *R. dubius* Freyn): Laminate leaves with 3(–5) deep, cuneate segments; segments of capillary leaves rigid. Pedicels usually more than 50 mm at anthesis, longer than the petiole of the subtending leaf. Sepals 2·5–4·5 mm, patent or reflexed. Honey-leaves 5·5–10 mm. Receptacle elongating in fruit. Achenes glabrous, with dorsal wing. $2n = 32$. *Usually in brackish water. Coasts of W. & S. Europe and the Baltic region; locally inland in W. & C. Europe.*

(c) Subsp. **fucoides** (Freyn) Muñoz Garmendía, *Anal. Jard. Bot. Madrid* **41**: 477 (1985) (*R. peltatus* subsp. *saniculifolius* (Viv.)

C. D. K. Cook, R. *saniculifolius* Viv.): Laminate leaves with (3–)5(–7) deep, cuneate segments; segments of capillary leaves somewhat rigid. Pedicels usually less than 50 mm at anthesis, shorter than the petiole of the subtending leaf. Sepals 2·5–4·5 mm, usually reflexed. Honey-leaves less than 10 mm. Achenes glabrous or somewhat pubescent, not winged. $2n = 16, 32, 40$. *In eutrophic and brackish water. S. Europe.*

127. R. penicillatus (Dumort.) Bab., *Man. Brit. Bot.* ed. 7, 7 (1874) (*Batrachium carinatum* Schur, *B. kaufmannii* (Clerc) V. Krecz.). Like **126** but always perennial, larger, more robust; capillary leaves often longer than the internodes. *Usually in flowing water. Throughout much of Europe but distribution imperfectly known.* Al Be Br Bu Co Cz Da Ga Ge Gr Hb ?He Ho Hs Hu It Lu Po Rs (N, B, C, W) ?Sa Su.

This species is both genetically and phenotypically variable. On the basis of cultivation experiments, S. D. Webster, *Watsonia* **17**: 1–22 (1988), has recognized the following two subspecies in N.W. Europe, but more information is required on their distribution elsewhere.

(**a**) Subsp. **penicillatus**: Plant with both laminate and capillary leaves (laminate leaves present in summer). Capillary leaves on mature stems longer than the internodes; segments subparallel, flaccid. Pedicel in fruit exceeding the petiole of the opposed leaf. $2n = 32, 48$. *Almost throughout the range of the species.*

(**b**) Subsp. **pseudofluitans** (Syme) S. Webster, *Watsonia* **17**: 20 (1988) (R. *aquatilis* subsp. *peltatus* var. *pseudofluitans* Syme, R. *pseudofluitans* (Syme) Newbould ex Baker & Foggitt, R. *kauffmannii* Clerc, R. *sphaerospermus* auct. angl., non Boiss. & Blanche, R. *vaginatus* Freyn, R. *calcareus* Butcher): Plant with capillary leaves only. Leaves on mature stems shorter to longer than the internodes; segments divergent or subparallel, rigid or flaccid. $2n = 16, 24, 32, 40, 48$. *Britain, W. France, Iberian peninsula; Greece; probably more widespread in W. Europe.*

128. R. sphaerospermus Boiss. & Blanche in Boiss., *Diagn. Pl. Or. Nov.* 3(**5**): 6 (1856) (R. *peltatus* subsp. *sphaerospermus* (Boiss. & Blanche) Meikle). Like **126** but without laminate leaves; achenes not more than 1 mm, subglobose. *S. part of Balkan peninsula.* Bu Gr Tu.

129. R. aquatilis L., *Sp. Pl.* 556 (1753) (R. *capillaceus* Thuill., R. *heterophyllus* Weber, R. *diversifolius* Gilib., R. *radians* Revel, *Batrachium gilibertii* V. Krecz.). Like **126** but segments of laminate leaves with dentate margins; pedicel in fruit rarely more than 50 cm and usually shorter than the petiole of the opposed leaf; honey-leaves up to *c.* 10 mm; nectaries circular. $2n = 48$. *Most of Europe, but rare in the north and east.* All except Az Bl Fa Is Lu Rs(N, K, E) Sb Tu.

130. R. trichophyllus Chaix in Vill., *Hist. Pl. Dauph.* **1**: 335 (1786) (R. *paucistamineus* Tausch, R. *brattius* G. Beck, R. *divaricatus* Schrank, R. *flaccidus* Pers., *Batrachium divaricatum* (Schrank) Wimmer, R. *drouetii* F. W. Schultz ex Godron, *B. trichophyllum* (Chaix) van den Bosch). Annual or perennial without laminate leaves. Capillary leaves rarely more than 4 cm. Pedicel in fruit usually less than 40 mm. Honey-leaves rarely more than 5 mm, not contiguous. Achenes more than 2 mm, ovate. *Most of Europe.* All except Az Fa Sb.

(**a**) Subsp. **trichophyllus**: Robust, erect, rooting only at the lower internodes. Flowers chasmogamous. $2n = 32, 40, 48$. *Throughout the range of the species.*

(**b**) Subsp. **eradicatus** (Laest.) C. D. K. Cook, *Mitt. Bot. Staatssamm. (München)* **6**: 622 (1967) (R. *eradicatus* (Laest.) Nevski, R. *confervoides* Fries, *Batrachium eradicatum* (Laest.) Fries, R. *lutulentus* Perr. & Song.): Delicate, procumbent, rooting at most internodes. Flowers usually cleistogamous. $2n = 16, 32$. *N. Europe; Alps; mountains of Spain and Portugal.*

131. R. rionii Lagger, *Flora (Regensb.)* **31**: 49 (1848) (*Batrachium rionii* (Lagger) Nyman). Like **130(a)** but short-lived annual; achenes not more than 1 mm, subglobose. $2n = 16$. *E. & C. Europe, extending southwards to N.W. Greece.* Au Cz Ga Ge Gr He Hu †It Ju Rm Rs(W, K, E).

132. R. circinatus Sibth., *Fl. Oxon.* 175 (1794) (R. *divaricatus* sensu Coste, non Schrank, R. *capillaceus* Thuill., *Batrachium foeniculaceum* auct., non (Gilib.) V. Krecz.). Perennial without laminate leaves. Capillary leaves *c.* 3 cm, circular or reniform in outline; segments rigid, divergent, lying in 1 plane. Honey-leaves rarely more than 10 mm, contiguous. Receptacle pubescent. $2n = 16$. *Much of Europe, but rare in the north and the Mediterranean region.* Au Be Br Cz Da Fe Ga Ge Hb He Ho Hu It Ju Po Rm Rs(N, B, C, W, E) ?Si Su.

133. R. fluitans Lam., *Fl. Fr.* **3**: 184 (1779). Perennial without laminate leaves. Capillary leaves usually more than 8 cm and frequently more than 25 cm, rarely more than 3-fid; segments not diverging, collapsing when removed from water. Receptacle nearly or quite glabrous. $2n = 16, 24, 32$. ● *From Scotland and Sweden southwards to Sardegna and Romania.* Au Be Br Co Cz Da Ga Ge Hb He Ho Hu It Po Rm Rs(B, W) Sa Su.

20. Ceratocephala Moench

†T. G. TUTIN

Like *Ranunculus* but achenes with an empty cell on either side of the seed, with an acuminate, more or less up-curved beak 2–3 times as long as the achene.

Beak of achene falcate, broad; empty cells distant **1. falcata**
Beak of achene nearly straight, narrow; empty cells nearly touching **2. testiculata**

1. C. falcata (L.) Pers., *Syn. Pl.* **1**: 341 (1805) (*Ranunculus falcatus* L.). Pubescent annual 2–10 cm. Leaves 3-fid, the lobes forked into linear-oblong segments. Flowers 10–15 mm in diameter, solitary. Honey-leaves yellow; nectaries *c.* $\frac{1}{3}$ as long as the honey-leaf. Receptacle elongate in fruit. Achenes 9–10 mm, numerous, with broad, falcate beak and empty cells distant. *Cultivated fields and waste places. S. Europe.* †Au Bu Ga Gr Hs ?Hu It Ju Rs(K, E) Tu.

2. C. testiculata (Crantz) Roth, *Enum.* **1**: 1014 (1827) (*C. orthoceras* DC., *Ranunculus testiculatus* Crantz). Like **1** but less hairy and rather glaucous; flowers 5–10 mm in diameter; achenes 5–6 mm, with narrow, nearly straight beak and empty cells nearly touching. *Cultivated fields and waste places. E.C. & E. Europe.* Au Bu Cz Hu Ju ?Po Rm Rs(C, W, K, E).

21. Myosurus L.

†T. G. TUTIN (EDITION 1) REVISED BY J. R. AKEROYD (EDITION 2)

Small annuals. Flowers solitary, small. Perianth-segments 5 or more. Honey-leaves 5–7, tubular, sometimes absent. Stamens few. Achenes numerous; receptacle greatly elongated in fruit.

Pedicels slender in fruit, longer than fruiting receptacle **1. minimus**

Pedicels becoming much thickened in fruit, shorter than fruiting receptacle **2. sessilis**

1. M. minimus L., *Sp. Pl.* 284 (1753). Glabrous. Leaves linear, entire, in a basal rosette. Flowers pale greenish-yellow. Perianth-segments 3–4 mm, linear-oblong, spurred at the base. Honey-leaves with a short, oblong limb. Pedicels 5–12 cm, slender in fruit, longer than the fruiting receptacle. Achenes 1–1·5 mm, keeled, very shortly beaked. $2n = 16, 28$. *Open habitats, especially when seasonally flooded. Most of Europe, but rare in the extreme north and south and in the islands.* Au Be Bl Br Bu Cz Da Fe Ga Ge Gr He Ho Hs Hu It Ju No Po Rm Rs(N, B, C, W, K, E) Si Su Tu.

2. M. sessilis S. Watson, *Proc. Amer. Acad. Arts Sci.* **17**: 362 (1882) (*M. breviscapus* Huth, *M. heldreichii* Léveillé). Like **1** but the pedicels much thickened in fruit, shorter than the fruiting receptacle; beak of achene longer, straight. *Seasonally flooded ground. Mediterranean region; very local.* Ga †Gr Hs Tu.

22. Aquilegia L.

J. CULLEN AND V. H. HEYWOOD (EDITION 1) REVISED BY J. R. AKEROYD (EDITION 2)

Perennial herbs with erect, woody stock. Leaves 1- to 3-ternate. Flowers hermaphrodite. Perianth-segments 5, petaloid. Honey-leaves 5, more or less tubular, each with a flat limb and a backwardly directed nectar-secreting spur. Stamens numerous, the innermost represented by scarious staminodes. Carpels free. Follicles several, free.

Many species and hybrids are cultivated for ornament.

The European species are in need of revision, and the following account must be regarded as provisional.

Literature: P. A. Munz, *Gentes Herb.* **7**: 1–150 (1946).

1 Flowers yellow **11. aurea**
1 Flowers violet, purple, blue or white
2 Spur shorter than limb of honey-leaf
3 Perianth-segments 22–40 × 16–22 mm **2. transsilvanica**
3 Perianth-segments less than 20 × 11 mm
4 Spur straight or slightly curved; basal leaves 2-ternate **12. kitaibelii**
4 Spur strongly hooked; basal leaves ternate (rarely 2-ternate)
5 Flowers violet-blue; perianth-segments elliptic-ovate **1. litardierei**
5 Flowers pale blue to white; perianth-segments elliptic-lanceolate **3. vulgaris**
2 Spur at least as long as limb of honey-leaf
6 Flowers bicolorous
7 Limb of honey-leaf at least 9 mm
8 Basal leaves 2-ternate; perianth-segments 18–24 mm **7. ottonis**
8 Basal leaves 1-ternate; perianth-segments 25–30 mm **8. dinarica**
7 Limb of honey-leaf usually not more than 9 mm
9 Leaves 3-ternate, with numerous glandular hairs on both surfaces **4. pancicii**
9 Leaves 1- or 2-ternate, with few glandular hairs
10 Spur strongly hooked; stems usually at least 30 cm **3. vulgaris**
10 Spur straight or slightly curved; stems not more than 30 cm **16. pyrenaica**
6 Flowers concolorous
11 Stamens exserted at least 1 mm beyond limb of honey-leaf
12 Spur strongly hooked
13 Perianth-segments 15–24 mm; leaves glabrous beneath **9. atrata**
13 Perianth-segments 25–35 mm; leaves pubescent beneath **10. nigricans**
12 Spur straight or somewhat curved
14 Limb of honey-leaf 5–9 mm; spur 6–12 mm **16. pyrenaica**
14 Limb of honey-leaf usually 8–20 mm; spur more than 8 mm
15 Perianth-segments greenish at apex; honey-leaves pale blue **3. vulgaris**
15 Perianth-segments not greenish at apex; honey-leaves purple to violet **18. grata**
11 Stamens not exserted beyond limb of honey-leaf, or exserted for less than 1 mm
16 Plant subscapose; all cauline leaves ±undivided, linear, sessile
17 Perianth-segments 9–14 mm wide; staminodes 6–7 mm **15. bertolonii**
17 Perianth-segments 7–8 mm wide; staminodes *c.* 5 mm
18 Leaves glabrous beneath; limb of honey-leaf 8–10 mm **13. einseleana**
18 Leaves pubescent beneath; limb of honey-leaf 11–13 mm **14. thalictrifolia**
16 Plant not subscapose; cauline leaves divided like the basal leaves
19 Spur strongly hooked
20 Flowers scarcely nodding; spur less than 15 mm **5. nuragica**
20 Flowers nodding; spur at least 15 mm
21 Staminodes obtuse; stamens exserted **3. vulgaris**
21 Staminodes acute; stamens not or somewhat exserted **6. viscosa**
19 Spur straight or somewhat curved, but never strongly hooked
22 Follicles 10–18 mm **16. pyrenaica**
22 Follicles 20–30 mm
23 Spur 15–17 mm; styles 10–15 mm **17. bernardii**
23 Spur 18–25 mm; styles 6–7 mm **19. alpina**

1. A. litardierei Briq., *Prodr. Fl. Corse* **1**: 589 (1910). Stems up to 12 cm, ascending, simple, glabrous below, with sparse, long eglandular hairs above. Basal leaves ternate; leaflets 3-fid, crenate, subglabrous above, with short eglandular hairs beneath. Flowers violet-blue, nodding. Perianth-segments 14 × 7 mm, elliptic-ovate; limb of honey-leaf *c.* 12 × 7 mm, the spur up to 8 mm, hooked. Stamens slightly exserted. Follicles puberulent. ● *Corse.* Co.

Not seen since its original discovery in 1908.

2. A. transsilvanica Schur, *Verh. Mitt. Siebenb. Ver. Naturw.* **4**: 31 (1853). Stems 15–45 cm, subglabrous below, with long eglandular hairs above. Basal leaves 2-ternate; leaflets 2- or 3-fid or more, glabrous above, with sparse, short eglandular hairs beneath. Flowers blue-violet, nodding. Perianth-segments 22–40 ×

16–22 mm; limb of honey-leaf 20–24 mm, the spur strongly hooked. Follicles pubescent. ● *Carpathians*. Rm Rs(W).

3. A. vulgaris L., *Sp. Pl.* 533 (1753) (*A. collina* Jordan, *A. longisepala* Zimmeter, *A. mollis* Timb.-Lagr., *A. nemoralis* Jordan). Stems 30–70(–100) cm, subglabrous to eglandular- or glandular-hairy. Basal leaves 2-ternate; leaflets 2- or 3-fid, usually glabrous above, hairy beneath. Flowers nodding, usually violet or blue, occasionally red or white. Perianth-segments 18–25 × 10–12 mm; limb of honey-leaf 6–13 × 9–12 mm, the spur 5–22 × 7–9 mm, strongly hooked, rarely curved. Stamens occasionally somewhat exserted; staminodes obtuse. Follicles 15–25 mm, glandular-pubescent. *W., C. & S. Europe, extending to W. Ukraine*. Al Au Be Br Co Cz Ga Ge Gr Hb He Ho Hs Hu It Ju Lu Po *Rm Rs(B, C) Si [Az *Da *Fe No Rs(N, W, K, E) *Su].

Several subspecies can be recognized within this polymorphic species; for further information on the variation see P. A. Munz, *op. cit.* (1946).

1 Leaves densely glandular-pubescent; follicles viscid (b) subsp. **nevadensis**
1 Leaves glabrous, eglandular- or slightly glandular-pubescent; follicles glandular-pubescent, but not viscid
 2 Perianth-segments usually at least 20 × 8 mm (a) subsp. **vulgaris**
 2 Perianth-segments usually less than 20 × 8 mm
  3 Limb of honey-leaf about as long as spur; leaves 2-ternate (c) subsp. **dichroa**
  3 Limb of honey-leaf distinctly longer than spur; leaves 1(2)-ternate (d) subsp. **paui**

(a) Subsp. **vulgaris** (incl. subsp. *hispanica* (Willk.) Heywood): Stems subglabrous to slightly glandular-pubescent. Leaves usually glabrous on the upper surface. Perianth-segments 18–30(–35) × 8–12(–16) mm, ovate to ovate-lanceolate. Limb of honey-leaf 5–8(–12) × 5–8 mm; spur 7–22 mm. Follicles glandular-pubescent. $2n = 14$. *Throughout the range of the species.*

(b) Subsp. **nevadensis** (Boiss. & Reuter) Díaz, *Anal. Jard. Bot. Madrid* **41**: 211 (1984) (*A. nevadensis* Boiss. & Reuter): Stems and leaves densely glandular-pubescent. Flowers blue. Perianth-segments 14–25(–35) × 5–9(–12) mm, lanceolate, with a greenish apex. Limb of honey-leaves 7–15(–20) × 6–9(–13) mm; spur 10–18(–21) mm, somewhat curved. Follicles viscid. *S. Spain.*

(c) Subsp. **dichroa** (Freyn) Díaz, *loc. cit.* (1984) (*A. dichroa* Freyn): Stems glandular-pubescent, especially above. Leaves subglabrous to somewhat glandular-pubescent. Perianth-segments 11–20 × 4–8 mm, lanceolate, blue, usually with a white apex; limb of honey-leaf 5–10 × 5–9 mm, white; spur 9–15(–17) mm. Follicles glandular-pubescent. $2n = 14$. ● *W. Spain, Portugal; naturalized in Açores.*

(d) Subsp. **paui** (Font Quer) O. Bolòs & Vigo, *Butll. Inst. Catalana Hist. Nat.* **38**: 65 (1974) (*A. paui* Font Quer): Stems and leaves glabrous to glandular-pubescent; leaves 1(2)-ternate. Perianth-segments 10–22 × 5–7 mm, elliptic-lanceolate, usually with a white apex; limb of honey-leaf 6–11(–13) × 4–7 mm, blue or white; spur 5–13 mm. Follicles glandular-pubescent. ● *Mountains of N.E. Spain (Tarragona).*

A. barbaricina Arrigoni & Nardi, *Boll. Soc. Sarda Sci. Nat.* **16**: 265 (1977), from a single locality on Sardegna, is similar to **3(a)** but has white flowers and a curved spur. Its status is uncertain.

4. A. pancicii Degen, *Magyar Bot. Lapok* **4**: 118 (1905). Like **3(c)** but leaves 3-ternate, densely glandular-pubescent; staminodes acute. ● *E. Jugoslavia.* Ju.

5. A. nuragica Arrigoni & Nardi, *Boll. Soc. Sarda Sci. Nat.* **17**: 215 (1978). Like **3(a)** but more or less glabrous, slightly glandular-puberulent above; stems 20–35 cm; flowers scarcely nodding, blue, the spur 11–13 mm, more slender; staminodes mucronate. $2n = 14$. *Shady rocks.* ● *Sardegna (S.E. of Oliena).* Sa.

6. A. viscosa Gouan, *Fl. Monsp.* 267 (1764). Stems glandular-hairy, viscid. Basal leaves 1- or 2-ternate; leaflets 2- or 3-fid, glandular-hairy; cauline leaves few, small. Flowers nodding, blue to white. Perianth-segments 18–27 × 8–12 mm, ovate-lanceolate; limb of honey-leaf (7–)10–14 × 6–9 mm, the spur (10–)15–23 mm, strongly hooked. Staminodes acute. Follicles glandular-pubescent. ● *N.E. Spain, S. France.* Ga Hs.

(a) Subsp. **viscosa**: Stems 30–120 cm. Stamens exserted. Follicles 22–28 mm. *Woods. S. France.*

(b) Subsp. **hirsutissima** (Lapeyr.) Breistr., *Bull. Soc. Bot. Fr.* **128**, *Lettres Bot.*: 66 (1981) (*A. hirsutissima* (Lapeyr.) Timb.-Lagr. ex Gariod; incl. *A. montisicciana* Font Quer): Stems 10–35 cm. Stamens not exserted. Follicles 12–20 mm. *S. France, N.E. Spain.*

7. A. ottonis Orph. ex Boiss., *Diagn. Pl. Or. Nov.* **3(1)**: 11 (1854). Stems (10–)15–45 cm, glandular-hairy, often branched. Basal leaves 1- or 2-ternate; leaflets 2- or 3-fid, almost glabrous to rather densely glandular-pubescent. Flowers nodding, bicolorous. Perianth-segments 18 × 8–9 mm, pale blue-violet; limb of honey-leaf 13–20 × 8 mm, the spur 13–14 × 3–4 mm, pale violet, strongly hooked. Follicles 11–19 mm, glandular-pubescent. *Damp, shady rock-ledges and screes.* ● *S. part of Balkan peninsula; C. Appennini.* Al Gr It Ju.

1 Limb of honey-leaf at least 15 mm, truncate at apex; follicles 15–19 mm (a) subsp. **ottonis**
1 Limb of honey-leaf less than 15 mm, rounded or subtruncate at apex; follicles 11–15 mm
 2 Flowering-stems usually at least 20 cm, branched; leaves ± glabrous (b) subsp. **amaliae**
 2 Flowering-stems not more than 20 cm, unbranched; leaves densely pubescent (c) subsp. **taygetea**

(a) Subsp. **ottonis** (incl. var. *unguisepala* Borbás): Flowering stems 15–30 cm. Leaves sparsely glandular-pubescent. Limb of honey-leaf 16–20 mm, truncate at apex, whitish or pale blue. Stamens exserted. Follicles 15–19 mm. $2n = 14$. *C. & S. Greece; C. Italy.*

(b) Subsp. **amaliae** (Heldr. ex Boiss.) Strid, *Mount. Fl. Gr.* **1**: 227 (1986) (*A. amaliae* Heldr. ex Boiss.): Flowering stems 20–45 cm. Leaves more or less glabrous. Limb of honey-leaf 13–14 mm, rounded or subtruncate at apex, white. Stamens not exserted. Follicles 12–15 mm. *S. part of Balkan peninsula.*

(c) Subsp. **taygetea** (Orph.) Strid, *op. cit.* 228 (1986) (*A. taygetea* Orph.): Like subsp. (b) but flowering stems 10–20 cm, unbranched; leaves rather densely glandular- and eglandular-pubescent; limb of honey-leaf cream; follicles 11–14 mm, densely glandular-pubescent. *S. Greece (Taÿetos).*

Plants similar to subsp. (a) from S.W. Italy (Monte Terminio), with glabrous leaves and somewhat larger flowers with spurs curved but not hooked, have been described as **A. champagnatii** Moraldo, Nardi & La Valva, *Webbia* **35**: 84 (1981).

8. A. dinarica G. Beck, *Ann. Naturh. Mus. (Wien)* **6**: 341 (1891).

Stems up to 20 cm, with patent, villous hairs in the lower ½. Basal leaves ternate; leaflets more or less deeply 3-partite, greyish, covered with soft, patent hairs. Flowers nodding, bicolorous. Perianth-segments 25–30 mm, intense blue; limb of honey-leaf 11–20 × 11 mm, white or bluish inside, the spur 13–15 mm, blue, hooked. Follicles glandular-pubescent. *Calcareous rocks and screes, 1200–2000 m.* ● *W. Jugoslavia; N. Albania.* Al Ju.

9. A. atrata Koch, *Flora (Regensb.)* **13**: 119 (1830) (*A. atroviolacea* Avé-Lall.). Stems 40–80 cm, densely pilose, at least above. Basal leaves 2-ternate; leaflets 2- or 3-fid, more or less glabrous. Flowers nodding, dark purple-violet. Perianth-segments 15–24 × 8–9 mm; limb of honey-leaf 8–12 × 7–9 mm, the spur 10–15 × 4 mm, hooked. Stamens long-exserted. Follicles glandular-pubescent. 2*n* = 14. ● *Alps, Appennini.* Au Ga Ge He It Ju.

A number of variants of this species have been described, but they seem of doubtful status.

10. A. nigricans Baumg., *Enum. Stirp. Transs.* **2**: 104 (1816) (?*A. ullepitschii* Pax). Like **9** but stems glandular-hairy; leaves and leaflets smaller; perianth-segments 25–35 mm; limb of honey-leaf 11–14 × 8–10 mm, the spur 13–15 × 5–6 mm. ● *C. & S.E. Europe.* Au Bu Gr Hu It Ju Po Rm Rs(W).

(a) Subsp. **nigricans**: Stems 40–80 cm. Flowers numerous, purplish. 2*n* = 14. *Throughout the range of the species.*

(b) Subsp. **subscaposa** (Borbás) Soó, *Acta Geobot. Hung.* **5**: 206 (1943) (*A. subscaposa* Borbás): Stems up to 30 cm. Flowers 1–3, bright blue. *Romania.*

A. blecicii Podobnik, *Biosistematika* **12**: 16 (1986), from Crna Gora, like subsp. (a) but with larger, bicolorous flowers, may be another subspecies of **10**.

11. A. aurea Janka, *Österr. Bot. Zeitschr.* **22**: 174 (1872). Stems 10–40 cm, subglabrous to densely hairy. Basal leaves 2-ternate; leaflets 2- or 3-fid, glabrous above, pubescent beneath. Flowers suberect, yellow. Perianth-segments 20–30 × 10–15 mm; limb of honey-leaf 15–20 × 10–12 mm, the spur 13–15 × 4 mm, hooked. Follicles pubescent. *S.E. Jugoslavia, S.W. Bulgaria.* ● Bu ?Gr Ju.

12. A. kitaibelii Schott, *Verh. Zool.-Bot. Ges. Wien* **3**: 129 (1853). Stems 15–30 cm, glandular-hairy, densely so above. Basal leaves 2-ternate; leaflets hairy above and beneath. Flowers suberect, red- to blue-violet. Perianth-segments 16–20 × 8–11 mm; limb of honey-leaf 10–13 × 7–10 mm, the spur 5–10 × 3–4 mm, straight or slightly curved, not hooked. Follicles glandular-pubescent. ● *N.W. Jugoslavia.* Ju.

13. A. einseleana F. W. Schultz, *Arch. Fl. Fr. Allem.* 135 (1848) (*A. aquilegioides* auct.). Stems 10–45 cm, subscapose, subglabrous below, sparsely glandular above. Basal leaves 2-ternate; leaflets shallowly 2- or 3-fid, subglabrous above, glabrous beneath; cauline leaves more or less entire, linear. Flowers nodding, blue-violet. Perianth-segments 15–19 × 7–8 mm; limb of honey-leaf 8–10 × 6–9 mm, the spur 7–10 × 2–3 mm, straight. Staminodes *c.* 5 mm. Follicles glandular-pubescent. ● *C. & E. Alps.* Au Ge He It Ju.

14. A. thalictrifolia Schott & Kotschy, *Verh. Zool.-Bot. Ges. Wien* **3**: 130 (1853). Like **13** but the leaves, petioles and stems glandular-pubescent; perianth-segments *c.* 20 mm; limb of honey-leaf 11–13 × 7–9 mm, the spur 8–11 × 3–4 mm. 2*n* = 14. *Shady rocks.* ● *N. Italy.* It.

15. A. bertolonii Schott, *op. cit.* 127 (1853) (*A. reuteri* Boiss.). Stems 10–30 cm, subscapose, glandular-pubescent above. Basal leaves 2-ternate, the leaflets 2- to 30-fid; cauline leaves more or less entire, linear. Flowers nodding, blue-violet. Perianth-segments 18–33 × 9–14 mm; limb of honey-leaf 10–14 × 6–8 mm, the spur 10–14 × 3–4 mm, straight or somewhat curved. Staminodes 6–7 mm. Follicles pubescent. 2*n* = 14. ● *S.E. France, N.W. Italy.* Ga It.

16. A. pyrenaica DC. in Lam. & DC., *Fl. Fr.* ed. 3, **5**: 640 (1815). Stems 10–30 cm, subglabrous to glandular-pubescent, sometimes subscapose. Basal leaves usually 2-ternate; leaflets 2- or 3-fid, glabrous to glandular-pubescent; cauline leaves simple or 3-fid, the segments linear. Flowers 1–6, usually bright blue, nodding. Spur at least as long as the limb of the honey-leaf, straight or slightly curved. Follicles 10–18 mm, glandular-pubescent. ● *Pyrenees, Spain.* Ga Hs.

A variable species that has been divided into a number of separate species by many authors.

1 Limb of honey-leaf usually at least 9 × 8 mm (a) subsp. **pyrenaica**
1 Limb of honey-leaf usually not more than 9 × 8 mm
 2 Stamens exserted; leaves usually 2-ternate (d) subsp. **cazorlensis**
 2 Stamens not exserted; leaves 1-ternate
  3 Stems usually not more than 15 cm, simple, subglabrous below (b) subsp. **discolor**
  3 Stems usually at least 15 cm, branched, glandular-pubescent throughout (c) subsp. **guarensis**

(a) Subsp. **pyrenaica** (incl. *A. aragonensis* Willk.): Stems 10–30(–35) cm, simple, glandular-pubescent above, glabrous or subglabrous below. Leaves (1)2-ternate, subglabrous. Flowers bright blue. Perianth-segments ovate to ovate-lanceolate; limb of honey-leaf 9–15(–17) × 8–11(–12) mm; spur 14–20(–25) mm, straight or slightly curved. Stamens not exserted. 2*n* = 14. *Pyrenees and E. part of Cordillera Cantábrica.*

(b) Subsp. **discolor** (Levier & Leresche) Pereda & Laínz, *Bol. Inst. Estud. Astur.* (*Supl. Ci.*) **5**: 12 (1962) (*A. discolor* Levier & Leresche): Stems 5–15(–20) cm, simple, glandular-pubescent above, subglabrous below. Leaves 1-ternate, glabrous or subglabrous. Perianth-segments ovate-lanceolate, blue; limb of honey-leaf 5–9(–10) × 4–5 mm, blue or white; spur 6–12 mm, straight. Stamens not exserted. 2*n* = 14. *W. part of Cordillera Cantábrica.*

(c) Subsp. **guarensis** (Losa) Rivas Martínez, *Bol. Soc. Esp. Hist. Nat., Secc. Biol.* **65**: 108 (1967) (*A. guarensis* Losa): Stems 15–30 cm, branched, glandular-pubescent throughout. Leaves 1-ternate, puberulent. Perianth-segments lanceolate, blue or bluish-white; limb of honey-leaf 6–9 × 3–5 mm; spur 9–12 mm, straight or curved. Stamens not exserted. 2*n* = 14. *C. Pyrenees.*

(d) Subsp. **cazorlensis** (Heywood) Galiano & Rivas Martínez, *loc. cit.* (1967) (*A. cazorlensis* Heywood): Stems 15–30 cm, usually branched, subglabrous or glandular-pubescent. Leaves (1)2-ternate, subglabrous to glandular-pubescent. Flowers bright blue. Perianth-segments lanceolate; limb of honey-leaf 5–8(–9) × 5–8 mm; spur 6–12 mm, slightly curved. Stamens exserted. 2*n* = 14. *S.E. Spain (Sierra de Cazorla, Sierra de Segura).*

17. A. bernardii Gren. & Godron, *Fl. Fr.* **1**: 45 (1847). Stems 50–80 cm, glabrous below, glandular-pubescent above. Basal leaves 2-ternate; leaflets 2- or 3-fid or more, glabrous. Flowers nodding, pale blue. Perianth-segments 25–35 × 15 mm; limb of honey-leaf

15–20 × 15 mm, the spur 15–17 mm, straight or somewhat curved. Follicles 20–25 mm, glandular-pubescent; styles 10–15 mm. $2n = 14$. ● *Corse*. Co.

A. nugorensis Arrigoni & Nardi, *Boll. Soc. Sarda Sci. Nat.* **17**: 220 (1978), described from the mountains of E.C. Sardegna, is like **17** but the whole plant is pubescent and the flowers are somewhat smaller.

18. A. grata Zimmeter, *Jahresb. Staats-ober-Realschule Steyr* **5**: 46 (1875). Stems 15–45 cm, glandular-hairy. Basal leaves 2-ternate; leaflets 3-fid, glandular-pubescent above and beneath. Flowers nodding, purplish-violet. Perianth-segments *c.* 30 × 9–11 mm; limb of honey-leaf 6–10 mm, the spur 14–20 mm, straight. Stamens exserted. Follicles glandular-pubescent. $2n = 14$. ● *Jugoslavia*. Ju.

19. A. alpina L., *Sp. Pl.* 533 (1753). Stems 15–80 cm, with long, sparse hairs below, densely pubescent above. Basal leaves 2-ternate; leaflets 2- or 3-fid, subglabrous. Flowers nodding, bright blue. Perianth-segments 30–45 × 14–22 mm; limb of honey-leaf 14–17 × 8–11 mm, the spur 18–25 × 6–7 mm, straight to curved. Follicles 20–28 mm; styles 6–7 mm. ● *Alps, N. Appennini*. Au Ga He It.

23. Thalictrum L.

†T. G. TUTIN (EDITION 1) REVISED BY J. R. AKEROYD (EDITION 2)

Perennial herbs. Leaves 2- or 3-pinnate or -ternate, stipulate and sometimes with stipels. Flowers small, in panicles or racemes. Perianth-segments 4 or 5, usually caducous. Honey-leaves absent. Stamens numerous, conspicuous. Achenes few, stipitate or sessile, ribbed, angled or winged.

Literature: J. C. Lecoyer, *Bull. Soc. Bot. Belg.* **24**: 78–325 (1885).

1 Filaments enlarged above, at least as wide as the anthers
 2 Filaments at least twice as wide as the anthers **3. uncinatum**
 2 Filaments about as wide as the anthers
  3 Achenes pendent, 3-angled, winged; most leaflets about as wide as long **1. aquilegiifolium**
  3 Achenes erect, ribbed but not winged; most leaflets nearly twice as wide as long **2. calabricum**
1 Filaments filiform, much narrower than the anthers
 4 Inflorescence a simple raceme **7. alpinum**
 4 Inflorescence a panicle
  5 Perianth-segments at least 8 mm, white
   6 Leaves evenly spaced along stem; achenes 2–6; roots not tuberous **6. orientale**
   6 Leaves crowded at base of stem; achenes (6–)10–30; roots tuberous **5. tuberosum**
  5 Perianth-segments *c.* 3 mm, greenish, purplish or yellowish
   7 Achenes *c.* 10 mm, with anastomosing ribs **4. macrocarpum**
   7 Achenes less than 5 mm, with longitudinal ribs only
    8 Leaves 3- or 4-ternate, about as wide as long
     9 Plant not or scarcely glandular **9. minus**
     9 Plant densely glandular
      10 Stigma fimbriate; achenes compressed, strongly ribbed **8. foetidum**
      10 Stigma not fimbriate; achenes not or scarcely compressed, ± weakly ribbed **9. minus**
    8 Leaves 2- or 3-pinnate, distinctly longer than wide
     11 Stamens pendent; anthers apiculate **10. simplex**
     11 Stamens erect; anthers not or very shortly apiculate
      12 Rhizome not far-creeping; stipels absent on young leaves **11. lucidum**
      12 Rhizome far-creeping; stipels present on young leaves
       13 Leaflets of upper leaves dentate or lobed; inflorescence oblong-ovoid; stem not shining
        14 Plant not glaucous; leaflets without prominent veins beneath **14. flavum**
        14 Plant glaucous; leaflets with prominent veins beneath **15. speciosissimum**
       13 Leaflets of upper leaves entire; inflorescence wide, with patent branches; stem shining
        15 Plant green or slightly glaucous; achenes 1·8–2·3 mm **12. morisonii**
        15 Plant glaucous, bluish; achenes 4–4·5 mm **13. maritimum**

(A) Stamens longer than the perianth; filaments widened towards the top.

1. T. aquilegiifolium L., *Sp. Pl.* 547 (1753). Glabrous perennial 40–150 cm, with a short stock. Leaves 2- or 3-ternate, with stipels; leaflets obovate-cuneate, dentate. Panicle much-branched, rather dense, corymbose. Flowers usually numerous, erect. Filaments as wide as the anthers, lilac or whitish. Achenes *c.* 7 mm, long-pedicellate, pendulous, with 3 winged angles; beak short, appressed. $2n = 14$. *C., E. & S. Europe, extending to S. Sweden*. Al Au Bu Cz Fe Ga Ge Gr He Hs Hu It Ju Po Rm Rs(N, B, C, W) Su Tu.

2. T. calabricum Sprengel, *Pugillus* **1**: 37 (1813). Like **1** but leaves without stipels; flowers few; achenes shortly pedicellate, 7- or 8-ribbed, not winged. $2n = 42$. *Deciduous woods*. ● *S.W. Italy; Sicilia*. It Si.

3. T. uncinatum Rehmann, *Spraw. Kom. Fizyogr. Krakow.* **7**: 90 (1873) (*T. petaloideum* auct., non L.). Glabrous perennial 15–40 cm. Leaves 2- or 3-pinnate, with stipels; leaflets suborbicular, entire or 3-lobed, sessile, glaucous. Panicle compact and umbel-like. Flowers erect. Filaments white, much wider at top than the anthers. Achenes 3–4 mm, sessile, ovoid-ellipsoidal, ribbed; beak *c.* 1 mm, hooked. ● *W. Ukraine (Upper Dnestr valley)*. Rs(W).

T. podolicum Lecoyer, *Bull. Soc. Bot. Belg.* **24**: 173 (1885), was described from a single collection from the same region and has not been rediscovered. It is said to differ from **3** in having the leaves pubescent on the veins beneath, a broad pyramidal inflorescence, yellowish filaments, and achenes with a curved, not hooked beak.

(B) Stamens longer or shorter than the perianth; filaments filiform.

(a) Carpels longer than the stamens and perianth-segments.

4. T. macrocarpum Gren., *Séances Publiq. Acad. Sci. Besançon* **1838**: 117 (1838). Glabrous perennial 20–80 cm, with a short stock. Basal leaves 3- or 4-ternate, without stipels; leaflets broadly ovate, coarsely dentate. Panicle with divaricate branches. Flowers yellowish, few, long-pedicellate, erect. Achenes 1–5, sessile, erect, compressed; beak long, curved. *Damp limestone rocks*. ● *W. & C. Pyrenees*. Ga Hs.

(b) Carpels shorter than the stamens and perianth-segments.

(i) Perianth-segments longer than the stamens.

5. T. tuberosum L., *Sp. Pl.* 545 (1753). Glabrous perennial 20–50 cm, with a short stock and ovoid tuberous roots. Leaves mostly basal, 2- or 3-pinnate, without stipels; leaflets suborbicular-cuneate, dentate. Flowers yellowish-white, few, erect. Perianth-segments usually 8–15 mm. Achenes 7–12, sessile, fusiform, sulcate; beak short. *Dry rocky places.* ● *N. & E. Spain, Pyrenees, Corbières.* Ga Hs.

6. T. orientale Boiss., *Ann. Sci. Nat.* ser. 2, Bot. **16**: 349 (1841). Glabrous perennial 10–30 cm, shortly rhizomatous, with fibrous roots. Leaves like those of **5** but mostly cauline. Flowers few, erect, white. Perianth-segments not more than 10 mm. Achenes 2–6, subsessile, narrowly oblong, sulcate; beak short. *S. Greece (foothills of Taÿetos).* Gr. (*S. Anatolia, N. Syria.*)

(ii) Perianth-segments shorter than the stamens; anthers not or scarcely apiculate.

7. T. alpinum L., *Sp. Pl.* 545 (1753). Glabrous perennial 5–20 cm, with a short stock. Leaves almost all basal, 2-ternate, without stipels; leaflets suborbicular, dentate. Flowers few, purplish-green, at first pendent, later erect, in a simple raceme. Achenes 2 or 3, pendent, narrowly oblong, ribbed, shortly stalked; beak short, curved. $2n = 14$. *N. & N.W. Europe, southwards to Italy and C. Ural; locally on mountains of C. & S. Europe from Sierra Nevada to E. Carpathians.* Au Br Fa Fe Ga Hb He Hs Is It Ju No Rm Rs(N, C) Su.

(iii) Perianth-segments shorter than the stamens; stamens pendent; anthers apiculate.

8. T. foetidum L., *loc. cit.* (1753) (*T. alpestre* Gaudin). Glandular, foetid, shortly rhizomatous perennial 20–40 cm, with some long eglandular hairs. Basal leaves 3- or 4(5)-ternate, without stipels; leaflets 2–4 mm, suborbicular or broadly ovate, irregularly dentate in the upper $\frac{1}{2}$. Inflorescence with long branches. Flowers yellow, pendent. Filaments slightly thickened. Stigma minutely fimbriate. Achenes *c.* 10, sessile, erect, compressed, ovoid, strongly ribbed; beak nearly as long as the achene. $2n = 14$. *Mountains of E., C. & S. Europe, from E. Spain to C. Ural.* Au Bu Cz Ga Gr He Hs Hu It Ju Rm Rs(C, W).

The plants from the Pyrenees and Spain lack eglandular hairs and have been called subsp. **valentinum** O. Bolòs & Vigo, *Butll. Inst. Catalana Hist. Nat.* **38**: 65 (1974), but are probably worth only varietal rank.

9. T. minus L., *Sp. Pl.* 546 (1753) (*T. silvaticum* Koch). Glabrous or glandular perennial 15–150 cm, subcaespitose or rhizomatous. Basal leaves 3- or 4-ternate, without stipels; leaflets suborbicular or broadly ovate, irregularly lobed or dentate in the upper $\frac{1}{2}$. Inflorescence with long branches. Flowers yellowish, pendent, becoming erect. Filaments filiform. Achenes 3–15, erect, sessile, not or slightly compressed, broadly ovoid to narrowly oblong-ovoid, the ribs not very strong; beak much shorter than the achene. *Most of Europe.* All except Az Bl Cr Fa Is Sb Tu; only as an alien in Fe.

Species **9–15** form a variable complex and it is sometimes difficult to identify a given specimen with certainty. Many more or less uniform populations have been described as species but they appear generally to be connected with one another by numerous intermediates. The variation-pattern is complicated by the common occurrence of the plants in small isolated populations which may, in rhizomatous species, often consist of single clones. All the populations which have been examined cytologically are polyploid, and inbreeding or possible apomixis may occur.

The following grouping of **9** into subspecies should be regarded as provisional, in the absence of thorough experimental and cytological investigation.

1 Leaves crowded at or below middle of stem; leaflets with prominent veins beneath (b) subsp. **olympicum**
1 Leaves ±evenly spaced; leaflets without prominent veins beneath
  2 Plant densely glandular, at least on lower surface of leaflets (c) subsp. **pubescens**
  2 Plant not or scarcely glandular
    3 Branches of leaf-rhachis terete or flattened (d) subsp. **pseudominus**
    3 Branches of leaf-rhachis strongly ribbed
      4 Stigma minutely papillose; inflorescence many-flowered (a) subsp. **minus**
      4 Stigma strongly fimbriate; inflorescence few-flowered (e) subsp. **kemense**

(a) Subsp. **minus** (*T. elatum* Jacq., *T. flexuosum* Bernh., *T. majus* Crantz, *T. minus* subsp. *majus* (Crantz) Rouy & Fouc.): Stems 25–120 cm. Leaves evenly spaced; leaflets 4–30 mm wide. $2n = 42$. *Widespread except in the north.*

(b) Subsp. **olympicum** (Boiss. & Heldr.) Strid, *Mount. Fl. Gr.* **1**: 229 (1986) (subsp. *saxatile* Schinz & R. Keller, non Hooker fil., *T. minus* sensu Hayek, non L., *T. saxatile* DC.): Stems 15–50(–80) cm. Leaves crowded at or below the middle of the stem; leaflets coriaceous, with veins prominent beneath. $2n = 42$. *C., E. & S.E. Europe.*

(c) Subsp. **pubescens** (Schleicher ex DC.) Rouy & Fouc., *Fl. Fr.* **1**: 14 (1893) (*T. minus* sensu Willk. & Lange, non L.; incl. subsp. *madritense* (Pau) P. Monts.): Plant densely glandular, at least on the lower surface of the leaflets; leaflets 4–15 mm, about as long as wide. *S. France and Spain.*

(d) Subsp. **pseudominus** (Borbás) Soó in Soó & Jáv., *Magyar Növ. Kéz.* **1**: 226 (1951) (*T. pseudominus* (Borbás) Jáv.): Plant glaucous; stems 10–40 cm. Branches of the leaf-rhachis terete or flattened, not ribbed; leaflets 10–20 mm wide. *E.C. Europe.*

(e) Subsp. **kemense** (Fries) Cajander in Mela, *Suomen Kasvio* ed. 5, 276 (1906) (*T. kemense* Fries): Stems up to 60 cm. Inflorescence narrow. Flowers few. Stigma strongly fimbriate. $2n = 70$. *N.E. Europe.*

Plants from coastal sands in N.W. Europe, with glaucous leaves that are densely glandular especially beneath, are similar to subsp. (d) but have somewhat smaller leaflets. They have been called subsp. **arenarium** (Butcher) Clapham in Clapham, Tutin & E. F. Warburg, *Fl. Brit. Is.* 107 (1952).

10. T. simplex L., *Mantissa* 78 (1767). Glabrous perennial 20–120 cm, with a long rhizome. Basal leaves 2- or 3-pinnate, without stipels; leaflets ovate-cuneate to linear, lobed or dentate to entire. Inflorescence usually narrowly oblong, with short branches. Flowers yellowish, pendent, becoming erect. Fruit ellipsoidal, ribbed; stigma sagittate. *Much of continental Europe, but absent from the islands and most of the west.* Au Bu Cz Da Fe Ga Ge He Hu It Ju No Po Rm Rs(N, B, C, W, ?K, E) Su.

1 Leaflets of upper leaves linear-lanceolate to linear, entire (d) subsp. **galioides**
1 Leaflets of upper leaves oblong- to ovate-cuneate, lobed or dentate
2 Inflorescence a lax, ovoid-pyramidal panicle (c) subsp. **gallicum**
2 Inflorescence narrowly ovoid or oblong
3 Leaflets of upper leaves oblong-cuneate (a) subsp. **simplex**
3 Leaflets of upper leaves obovate-cuneate (b) subsp. **boreale**

(a) Subsp. **simplex** (*T. strictum* Ledeb.): Leaflets oblong-cuneate, lobed or dentate. Inflorescence usually leafy to apex. Flowers numerous; pedicels short. $2n = 56$. *Throughout the range of the species except the extreme north.*

(b) Subsp. **boreale** (F. Nyl.) A. & D. Löve, *Bot. Not.* **114**: 52 (1961) (*T. boreale* F. Nyl.): Leaflets ovate-cuneate, lobed or minutely dentate. Inflorescence very leafy except at apex. Flowers few; pedicels long. Stigma papillose. $2n = 56$. ● *N. Fennoscandia.*

(c) Subsp. **gallicum** (Rouy & Fouc.) Tutin, *Feddes Repert.* **69**: 55 (1964) (*T. gallicum* Rouy & Fouc.): Leaflets obovate-cuneate, usually 3-lobed, the margins recurved. Inflorescence with rather long, patent or ascending branches. Flowers numerous; pedicels long. ● *N. & E. France.*

(d) Subsp. **galioides** (Nestler) Korsh., *Tent. Fl. Russ. Or.* **1**: 41 (1898) (*T. galioides* Nestler; incl. subsp. *bauhinii* (Crantz) Tutin, *T. bauhinii* Crantz): Leaflets linear, entire. Inflorescence leafless at apex. Flowers numerous; pedicels short. *From Estonia to S.E. France and S. Romania.*

Plants from S. Bulgaria, like subsp. (a) but hairy on the upper part of the stems, petioles and lower surfaces of the leaves, and with more erect flowers, have been called subsp. **rhodopaeum** (Rech. fil.) Panov ex Kožuharov & Petrova, *Ann. Bot. Fenn.* **25**: 390 (1988); their status is uncertain.

(iv) Perianth-segments shorter than the stamens; stamens erect; anthers not or shortly apiculate.

11. T. lucidum L., *Sp. Pl.* 546 (1753) (*T. angustifolium* auct.; incl. *T. bulgaricum* Velen.). Caespitose perennial 60–120 cm. Leaves mostly sessile, 2- or 3-pinnate, without stipels; leaflets of lower leaves ovate-cuneate to linear-oblong, those of upper leaves lanceolate to linear, all entire, or 2- or 3-lobed at apex. Inflorescence ovoid, the branches long, ascending. Flowers yellowish, in rather dense clusters. Achenes oblong-ovoid, with 8–10 strong ribs; beak short. $2n = 28$. *C. & E. Europe.* Al Au Bu Cz Ge Gr Hu ?It Ju Po Rm Rs(N, B, C, W, E) Tu [Fe Rs(K)].

12. T. morisonii C. C. Gmelin, *Fl. Bad.* **4**: 422 (1826) (*T. exaltatum* Gaudin, *T. angustifolium* auct., *T. lucidum* auct. hisp., non L.). Perennial 50–190 cm, with far-creeping rhizome. Stem shiny. Leaves 2- or 3-pinnate, with stipels when young; leaflets of lower leaves obovate-cuneate, lobed, those of upper leaves linear to linear-lanceolate, entire. Inflorescence ovoid, the branches long, patent. Flowers yellow, not densely clustered. Anthers shortly apiculate. Achenes 1·8–2·3 mm, broadly ovoid, with 8–10 strong ribs; beak short. ● *C. Mediterranean region; Alps and N. Italy.* Co Ga He Hs It.

(a) Subsp. **morisonii**: Leaves glabrous and eglandular beneath. *Alps and N. Italy.*

(b) Subsp. **mediterraneum** (Jordan) P. W. Ball, *Feddes Repert.* **66**: 153 (1962) (*T. mediterraneum* Jordan): Leaves pubescent or sparsely glandular beneath. *C. Mediterranean region.*

13. T. maritimum Dufour, *Bull. Bot. Soc. Fr.* **7**: 221 (1860). Like **12** but plant glabrous, glaucous, bluish; stems 60–80 cm, more slender; leaflets of lower leaves mostly narrowly elliptical; anthers not apiculate; achenes 4–4·5 mm, fusiform. *By saline and brackish lagoons.* ● *E. Spain (Valencia and Castellón provs.).* Hs.

14. T. flavum L., *Sp. Pl.* 546 (1753). Perennial 50–130 cm, not glaucous, with far-creeping rhizomes and stolons. Stem angled and somewhat winged, not shiny, often with adventitious roots at base. Leaves 2- or 3-pinnate, with stipels when young; leaflets obovate-cuneate to oblong, 3- or 4-lobed at apex, without prominent veins beneath. Inflorescence narrowly oblong-ovoid. Flowers yellow, in dense clusters. Achenes ovoid, with 6 ribs. $2n = 84$. *Much of Europe, but local in the south.* Al Au Be Br Bu Cz Da Fe Ga Ge Hb He Ho Hs Hu It Ju No Po Rm Rs(N, B, C, W, E) Su.

15. T. speciosissimum L. in Loefl., *Iter Hisp.* 303: 57 (1758) (incl. *T. glaucum* Desf., *T. flavum* subsp. *glaucum* (Desf.) Batt., *T. costae* Timb.-Lagr. ex Debeaux). Like **14** but up to 180 cm, glaucous; leaflets with prominent veins beneath; inflorescence sometimes broadly corymbose. *Spain and Portugal.* Hs Lu.

62. PAEONIACEAE

EDIT. D. A. WEBB

Perennial herbs or shrubs. Leaves alternate, exstipulate. Flowers usually solitary, hermaphrodite, actinomorphic, hypogynous. Sepals 5, free. Petals 5–10(–13), free. Stamens numerous. Carpels 2–8, free, borne on a fleshy disc. Fruit a group of 2–8 follicles, each with several seeds.

1. Paeonia L.

J. CULLEN & V. H. HEYWOOD (EDITION 1) REVISED BY J. R. AKEROYD (EDITION 2)

Robust perennials, with erect, tuberous stock and fleshy roots. Leaves large, simply biternate or further divided. Flowers 6–14 cm in diameter; petals white, pink or red in the European species. Stamens numerous. Follicles spreading horizontally; seeds in 2 rows.

A number of species and hybrids are widely cultivated for ornament.

Literature: F. C. Stern, *A Study of the Genus* Paeonia. London. 1946. W. T. Stearn & P. H. Davis, *Peonies of Greece*. Kifissia. 1984.

1 Segments of lower leaves more than 40, each segment less than 5 mm wide **1. tenuifolia**
1 Segments of lower leaves fewer than 40, each segment more than 5 mm wide
2 Petals white, occasionally flushed with pink **6. clusii**
2 Petals red or deep pink
3 Most leaflets divided; lower leaves with narrowly elliptical to lanceolate segments

4 Leaves without minute bristles along the veins of upper surface
5 Segments of lower leaves fewer than 15; petals blackish-red **4. parnassica**
5 Segments of lower leaves usually more than 15; petals pink to red
6 Leaves green or brownish and pubescent beneath; filaments red **5. officinalis**
6 Leaves glaucous and glabrous beneath; filaments greenish-white to yellow **7. broteroi**
4 Leaves with minute bristles along the veins of upper surface
7 Ultimate leaf-segments narrowly triangular, deltate or subulate; filaments yellow **2. anomala**
7 Ultimate leaf-segments broadly triangular; filaments pink or red **3. peregrina**
3 Most leaflets undivided; lower leaves with lanceolate, elliptical, ovate or orbicular segments
8 Leaflets ovate to orbicular; margins undulate **8. mascula**
8 Leaflets lanceolate to ovate; margins not undulate
9 Follicles 5–8, *c.* 6 cm, glabrous, purple **10. cambessedesii**
9 Follicles fewer than 5, less than 6 cm, glabrous or pubescent, sometimes reddish but not purple
10 Follicles (3·5–)4·5 cm, glabrous, with apex attenuate into the stigma **9. coriacea**
10 Follicles 2·5–4 cm, usually pubescent, with obtuse or rounded apex and sessile stigma
11 Leaves glabrous beneath; leaflets 2–4 cm wide **7. broteroi**
11 Leaflets usually pubescent beneath; leaflets 5–10 cm wide **8. mascula**

1. P. tenuifolia L., *Syst. Nat.* ed. 10, **2**: 1079 (1759). Stems 20–40 cm. Leaves divided into numerous linear segments, each less than 5 mm wide, glabrous above, pubescent beneath. Flowers 6–8 cm in diameter, red, appearing to rest on the leaves. Filaments yellow. Follicles 2 or 3, *c.* 2 cm, tomentose. $2n = 10$. *S.E. Europe.* Bu Ju Rm Rs(C, W, K, E).

2. P. anomala L., *Mantissa Alt.* 247 (1771). Stems 50–100 cm. Leaf-segments numerous, the ultimate segments deltate or subulate, more than 5 mm wide, glabrous beneath, with minute bristles along the main veins on the upper surface. Flowers 7–13 cm in diameter, red. Filaments yellow. Follicles 3–5 , *c.* 2 cm. $2n = 10$. *N.E. Russia.* Rs(N, C). (*N. Asia.*)

Most European plants of this species belong to var. *intermedia* (C. A. Meyer) B. Fedtsch., with villous follicles, and are regarded by some Russian authors as var. *intermedia* (C. A. Meyer) Krylov of *P. hybrida* Pallas. Var. *anomala*, with glabrous follicles, is restricted in Europe to parts of Ural.

3. P. peregrina Miller, *Gard. Dict.* ed. 8, no. 3 (1768) (*P. decora* G. Anderson). Stems 30–50 cm, glabrous. Lower leaves divided into 17–30 narrowly elliptical segments, the ultimate segments short, broadly triangular, spreading, giving the apex of the leaflet a serrate appearance, sparsely villous to glabrous beneath, with minute bristles along the main veins above. Flowers 7–13 cm in diameter, dark red, less widely open than in other species. Filaments pink or red. Follicles 2 or 3, 2–3·5 cm, tomentose. $2n = 10, 20$. *Balkan peninsula, E. Romania, Moldavia; one station in C. Italy.* Al Bu Gr It Ju Rm Rs(W) Tu.

4. P. parnassica Tzanoudakis, *Kytt. Meleti Gen.* Paeonia Ell. 43 (1977) (*P. arietina* auct., non G. Anderson). Stems 30–65 cm, pubescent. Lower leaves divided into 9–13 obovate to narrowly elliptical or lanceolate segments, densely pilose beneath, giving a greyish-green appearance; ultimate segments acute or acuminate, entire. Flowers 8–12 cm in diameter, blackish-red. Filaments purplish. Follicles 2 or 3, tomentose. $2n = 20$. ● *Mountains of S.C. Greece.* Gr.

5. P. officinalis L., *Sp. Pl.* 530 (1753). Lower leaves divided into 17–30 segments, pubescent beneath, glabrous above; petiole deeply channelled on the upper surface. Flowers 7–13 cm in diameter, red. Filaments red. Follicles 2 or 3, 2–3·5 cm, usually tomentose, sometimes glabrous. *S. & S.C. Europe.* Al Ga He Hs Hu It Ju Lu Rm [Ge].

1 At least the central leaflet divided almost to base
2 Most of the leaflets divided almost to base (a) subsp. **officinalis**
2 Only the central leaflet divided almost to base (b) subsp. **banatica**
1 Leaflets divided not more than $\frac{1}{3}$ of the distance to base
3 Follicles pubescent; stems and petioles floccose (c) subsp. **villosa**
3 Follicles glabrous; stems and petioles pubescent (d) subsp. **microcarpa**

(a) Subsp. **officinalis**: Stems and petioles floccose. Most leaflets divided almost to base. Follicles tomentose. $2n = 20$. *From S.E. France to Albania.*

(b) Subsp. **banatica** (Rochel) Soó, *Növényföldrajz* 146 (1945) (*P. banatica* Rochel). Like subsp. (a) but only the central leaflet deeply divided. $2n = 20$. *Hungary, Jugoslavia, Romania.*

(c) Subsp. **villosa** (Huth) Cullen & Heywood, *Feddes Repert.* **69**: 34 (1964): Stems and petioles floccose. Leaflets divided to not more than $\frac{1}{3}$ of the distance to base. Follicles tomentose. *From S. France to C. Italy.*

(d) Subsp. **microcarpa** (Boiss. & Reuter) Nyman, *Consp.* 22 (1878) (*P. microcarpa* Boiss. & Reuter, *P. humilis* Retz., *P. officinalis* subsp. *humilis* (Retz.) Cullen & Heywood, *P. paradoxa* G. Anderson): Stems and petioles pubescent. Leaflets divided to not more than $\frac{1}{3}$ of the distance to base. Follicles glabrous. *S.W. Europe.*

6. P. clusii F. C. Stern, *Bot. Mag.* **162**: t. 9594 (1940) (*P. officinalis* var. *glabra* (Boiss.) Hayek). Stems 20–30 cm. Lower leaves with 30 or more narrowly oblong to narrowly elliptical, acute segments, glabrous above and beneath, or slightly pubescent beneath. Flowers 7–10 cm in diameter, white, rarely flushed with pink. Filaments pink. Follicles 2–5, *c.* 3 cm, densely tomentose. $2n = 10, 20$. *Kriti and Karpathos.* Cr.

Outside Europe, known only from the E. Aegean islands, where the plants have been distinguished as **P. rhodia** Stearn, *Gard. Chron.*, ser. 3, **110**: 159 (1941) (subsp. *rhodia* (Stearn) Tzan.).

7. P. broteroi Boiss. & Reuter, *Diagn. Pl. Nov. Hisp.* 4 (1842) (*P. lusitanica* auct., non Miller). Stems up to 50 cm. Lower leaves divided into (9–)17–20 narrowly elliptical, glabrous segments. Flowers 8–13 cm in diameter, purplish-pink. Filaments greenish-white to yellow. Follicles 2–4, 3–4 cm, densely tomentose. $2n = 10$. ● *S., C. & W. Spain; Portugal.* Hs Lu.

8. P. mascula (L.) Miller, *Gard. Dict.* ed. 8, no. 1 (1768) (*P. corallina* Retz., *P. caucasica* (Schipcz.) Schipcz.; incl. *P. banatica* auct., non Rochel). Stems 20–80 cm. Leaves simply biternate, or with a few leaflets divided; segments 9–16(–21), narrowly to broadly elliptical to ovate or orbicular, glabrous or pubescent beneath. Flowers 8–14 cm in diameter, red or rarely white. Follicles (2)3–5(6), 2–4 cm, usually villous, rounded at apex, with a sessile stigma. *S. Europe, extending northwards to N.C. France and Austria.* Al Bu Co Ga Gr Hs It Ju Rm Rs(K) Sa Si [Au Br].

1 Flowers white or pinkish-white (**d**) subsp. **hellenica**
1 Flowers pink to red
 2 Lower leaves with 9 or 10(11) broadly elliptical to orbicular leaflets
  3 Leaflets broadly ovate to orbicular, with undulate margins, glabrous beneath (**b**) subsp. **triternata**
  3 Leaflets elliptical to broadly elliptical with plane margins, pubescent beneath (**c**) subsp. **russoi**
 2 Lower leaves usually with more than 12 narrowly elliptical to ovate leaflets
  4 At least stem and petioles without hairs; leaflets elliptical to ovate (**a**) subsp. **mascula**
  4 Stem and petioles and lower surface of leaves pubescent; leaflets narrowly elliptical (**e**) subsp. **arietina**

(**a**) Subsp. **mascula**: Leaves glabrous or pubescent beneath; lower leaves with 9–16(–21) elliptical to ovate leaflets. Flowers red. Filaments purple. $2n = 20$. *From C. France to E. Greece.*

(**b**) Subsp. **triternata** (Pallas ex DC.) Stearn & P. H. Davis, *Peonies of Greece* 107 (1984) (*P. triternata* Pallas ex DC., *P. daurica* Andrews, *P. taurica* auct.): Leaves glabrous beneath; leaflets usually 9, broadly ovate to suborbicular, with undulate margins. Flowers pale purplish-red. Filaments yellow. Follicles 2 or 3. *Romania; Krym.*

(**c**) Subsp. **russoi** (Biv.) Cullen & Heywood, *Feddes Repert.* **69**: 35 (1964) (*P. russi* Biv., *P. ovatifolia* Rouy & Fouc., *P. corallina* var. *pubescens* Moris, *P. corsica* Sieber): Leaves sparsely pubescent beneath; lower leaves with 9 or 10 broadly elliptical leaflets. Flowers purplish-pink. Filaments white or pink. $2n = 10$. ● *Mediterranean region, from C. Spain to W. Greece, mainly in the islands.*

(**d**) Subsp. **hellenica** Tzanoudakis, *Kytt. Meleti Gen.* Paeonia *Ell.* 36 (1977) (*P. flavescens* C. Presl): Leaves glabrous or pubescent beneath; lower leaves with 9–13 broadly ovate or elliptical leaflets. Flowers white. Filaments purplish. $2n = 20$. *S. & S.E. Greece.*

(**e**) Subsp. **arietina** (G. Anderson) Cullen & Heywood, *Feddes Repert.* **69**: 35 (1964) (*P. arietina* G. Anderson): Leaves pubescent beneath; lower leaves with 12–15 narrowly elliptical leaflets. Flowers red. Filaments purple. *S.E. Europe.*

9. P. coriacea Boiss., *Elenchus* 7 (1838). Like **8(a)** but stems 50–100 cm; leaves coriaceous; follicles 2(3), somewhat longer (3·5–4·5 cm), glabrous, with the apex attenuate into the stigma. *S. Spain; Corse, Sardegna.* Co Hs Sa.

10. P. cambessedesii (Willk.) Willk. in Willk. & Lange, *Prodr. Fl. Hisp.* **3**: 976 (1880), *in obs.*, et Willk., *Ill. Fl. Hisp.* **1**: 104, t. 65_A (1883). Stems 20–50 cm. Lower leaves biternate; leaflets lanceolate to ovate, occasionally elliptical, glabrous above and beneath, flushed with purple beneath. Flowers 6–10 cm in diameter, red. Follicles 5–8, *c.* 6 cm, glabrous, purplish. $2n = 10$. ● *Islas Baleares.* Bl.

63. BERBERIDACEAE

EDIT. D. A. WEBB

Shrubs or herbs. Leaves alternate or basal. Flowers hermaphrodite, (2)3-merous. Perianth-segments 6–9, in 3 or 4 whorls, at least the inner ones petaloid; honey-leaves 4–6, petaloid or nectariform. Stamens 4–6, opposite the honey-leaves; anthers dehiscing by apically-hinged valves. Carpel solitary, superior.

1 Shrubs; fruit a berry
 2 Leaves simple **5. Berberis**
 2 Leaves pinnate **6. Mahonia**
1 Herbs; fruit dry
 3 Stamens 4; ovules numerous, on 2 lateral placentae **4. Epimedium**
 3 Stamens 6; ovules 2–8, on a basal placenta
  4 Leaves all basal; honey-leaves flat, petaloid **3. Bongardia**
  4 Stem ± leafy; honey-leaves small, convolute
   5 Several cauline leaves on each stem, most of them with an axillary raceme; fruit 20–40 mm **1. Leontice**
   5 Each flowering stem with a single cauline leaf and raceme; fruit *c.* 5 mm **2. Gymnospermium**

1. Leontice L.

W. T. STEARN AND D. A. WEBB

Glabrous herbs with tuberous rhizome. Leaves 2- or 3-ternate, not all basal. Flowers in terminal and axillary, bracteate racemes. Perianth-segments 6(–8), conspicuous, petaloid, yellow; honey-leaves 6, much smaller, convolute. Stamens 6. Fruit 1- to 4-seeded, with inflated, membranous pericarp, dehiscing irregularly by decay. Seeds without aril.

1. L. leontopetalum L., *Sp. Pl.* 312 (1753). Stem 20–50(–80) cm, erect, branched above. Leaves up to 20 cm wide, the lower with long petioles, the upper sessile, all 2- or 3-ternate; ultimate segments broadly obovate, entire. Racemes usually numerous, in axils of upper leaves, pedunculate, with conspicuous bracts, of which the lower are often compound or lobed. Flowers 15–40, on long pedicels, crowded at first, later distant. Perianth-segments *c.* 8 mm, ovate-oblong; honey-leaves 1·5 mm, less than ½ as long as the stamens. Fruit 20–40 mm, ovoid. $2n = 16$. *Ploughed fields and waste places. S.E. part of Balkan peninsula, Aegean region.* Bu Cr Gr Tu.

2. Gymnospermium Spach

W. T. STEARN AND D. A. WEBB

Like *Leontice*, but each stem with 1 basal and 1 cauline leaf and a single raceme; fruit dehiscing apically by rounded lobes before the seeds are ripe; seeds with a membranous aril.

1. G. altaicum (Pallas) Spach, *Hist. Vég. (Phan.)* **8**: 67 (1838)

(*Leontice altaica* Pallas; incl. *L. odessana* (DC.) Fischer ex G. Don fil.). Stems 5–20 cm, slender, each with a long-petiolate basal leaf and a subsessile cauline leaf immediately below the inflorescence. Leaves ternate, the primary divisions divided palmately into 4–7 entire, oblong, obtuse segments 15–35 mm long. Raceme short, with 6–12 flowers. Perianth-segments 8–10 mm, oblong; honey-leaves shorter than the stamens. Fruit 5–6 mm, subglobose, pendent; seeds 1–4, remaining attached for some time after dehiscence of the fruit. *Black Sea region, from S. Ukraine to E. Romania; one station in S. Greece.* Gr Rm Rs(W).

In spite of the wide geographical separation, typical material from the Altai Mountains agrees well with plants from Europe, and the separation of the latter at more than varietal level does not seem to be justified.

3. **Bongardia** C. A. Meyer

W. T. STEARN AND D. A. WEBB

Like *Leontice*, but leaves all basal, pinnatisect; perianth-segments small; honey-leaves large, petaloid.

1. **B. chrysogonum** (L.) Griseb., *Spicil. Fl. Rumel.* **1**: 294 (1843) (*Leontice chrysogonum* L.). Tuber large, subglobose. Leaves 10–25 cm, all basal, spreading horizontally, deeply pinnatisect into 7–17 sessile, oblong-cuneate segments, which are usually 3- to 5-toothed at the apex but sometimes bifid to the base, glaucous-green, often reddish near the rhachis. Stem up to 60 cm, the upper $\frac{1}{2}$ consisting of a large panicle with ascending branches. Perianth-segments 6, small, suborbicular, sepaloid, caducous; honey-leaves 8–12 × 3·5–5 mm, irregularly crenate distally, petaloid, golden yellow. Fruit *c.* 15 mm, ovoid. *Ploughed fields. S. Greece (N. Peloponnisos); ?S. Aegean region (Astipalaia).* Gr. (*S.W. Asia.*)

4. **Epimedium** L.

W. T. STEARN AND D. A. WEBB

Rhizomatous herbs. Leaves compound. Flowers 2-merous, in panicles. Perianth-segments 8, the outer 4 sepaloid, the inner 4 petaloid. Honey-leaves 4, flat proximally and tubular distally (slipper-shaped). Stamens 4. Fruit many-seeded, dry, dehiscing into 2 unequal valves.

Literature: W. T. Stearn, *Jour. Linn. Soc. London (Bot.)* **51**: 409 (1938).

Inflorescence shorter than cauline leaf; inner perianth-segments dark red **1. alpinum**
Inflorescence over-topping cauline leaf; inner perianth-segments pink **2. pubigerum**

1. **E. alpinum** L., *Sp. Pl.* 117 (1753). Stems 15–30 cm, numerous, arising from a slender, extensively creeping rhizome, each bearing a single leaf which over-tops the lax, terminal, nodding panicle; basal leaves also present. Leaves 2- or 3-ternate; leaflets 5–10, ovate, cordate, acute, spinose-ciliate, pubescent beneath at first but glabrous at maturity. Outer 4 perianth-segments pinkish-grey, caducous; inner 4 dull, dark red, longer; honey-leaves bright yellow, lying in the concavity of the inner perianth-segments. $2n = 12$. *Shady places. Foothills and lower slopes of mountains, from N. & C. Italy to Austria and Albania; cultivated for ornament in W. & C. Europe, and locally naturalized.* Al Au It Ju [Be Br Da Ga Ge Ho].

2. **E. pubigerum** (DC.) Morren & Decne, *Ann. Sci. Nat.* sér. 2, **2**: 355 (1834). Like **1** but with shorter, stouter rhizome; mature leaflets usually white-pubescent beneath; inflorescence more erect, raising the flowers above the leaves; inner perianth-segments pale pink. $2n = 12$. *Thrace.* Bu Tu. (*Anatolia and Caucasus.*)

5. **Berberis** L.

J. R. AKEROYD AND D. A. WEBB

Shrubs; wood bright yellow. Leaves on long shoots (in European species) transformed into spines, simple or palmately divided to the base into 3 or 5. Foliage-leaves simple (deciduous in European species), clustered on short shoots in the axils of the spines. Flowers in racemes, sometimes condensed to small, umbel-like clusters. Perianth-segments usually 9; the 3 outermost (sometimes interpreted as bracteoles) small and sepaloid, the 6 inner bright yellow, petaloid; honey-leaves petaloid, similar to the inner perianth-segments. Stamens 6, usually sensitive to touch. Fruit a 2-seeded berry.

Numerous species, mostly from E. Asia or temperate South America, are cultivated for ornament. **B. darwinii** Hooker, *Hook. Ic.* **7**: t. 672 (1844), from southern Chile, with evergreen, spiny leaves, bright orange flowers and pruinose black berries, has been reported as locally naturalized in W. Ireland and W. Scotland.

Literature: C. K. Schneider, *Bull. Herb. Boiss.* ser. 2, **5**: 33–48, 133–148, 391–403, 449–464, 655–670, 800–831 (1905); **8**: 192–204, 258–266 (1908). L. W. S. Ahrendt, *Jour. Linn. Soc. London (Bot.)* **57**: 1–410 (1961).

Plant not suckering; racemes usually more than 15 mm **1. vulgaris**
Plant suckering; racemes not more than 15 mm **2. cretica**

1. **B. vulgaris** L., *Sp. Pl.* 330 (1753). Plant 30–300 cm. Spines usually 3-fid; segments 5–25 cm. Leaves 8–40(–60) mm, elliptical-obovate to elliptical, usually spinulose-serrate. Racemes 10–50 mm, pendent, with 5–30 flowers. Honey-leaves 3–6 mm. Berry 5–10 × 3–5 mm, oblong; stigma sessile or very shortly stipitate. *W., C. & S.E. Europe, extending to Estonia and E.C. Russia; extensively naturalized elsewhere.* Al Au Be Bu Co Cz Ga Ge Gr He *Ho Hs Hu It Ju *Po Rm Rs(B, C, W, K, E) Sa Si [Br Da Fe Hb No Rs(N) Su Tu].

The limits of the native range of this species cannot be precisely determined. On the one hand, it has been extensively planted, originally for its edible fruit and more recently for ornament, and has become naturalized; on the other hand, in some regions attempts to extirpate it (as the intermediate host of the Wheat Rust, *Puccinia graminis*) have been more or less successful.

A variable species that can be divided into a number of subspecies, which have been treated as distinct species by some authors.

1 Honey-leaves 4·5–6 mm; berry orange-red; stigma sessile **(a) subsp. vulgaris**
1 Honey-leaves 3–4 mm; berry dark red to black; stigma very shortly stipitate
2 Plant not more than 60 cm; leaves spinulose-serrate **(b) subsp. aetnensis**
2 Plant up to 150 cm; leaves entire or with up to 6 marginal teeth **(c) subsp. australis**

(a) Subsp. **vulgaris**: Plant 150–300 cm; twigs yellowish. Spines

6–18 mm. Leaves 25–60 mm, spinulose-serrate. Racemes 30–50 mm, with (12–)15–30 flowers. Honey-leaves 4·5–6 mm. Berry bright orange-red; stigma sessile. $2n = 28$. *Throughout the range of the species, but rare in the Mediterranean region.*

(**b**) Subsp. **aetnensis** (C. Presl) Rouy & Fouc., *Fl. Fr.* **1**: 148 (1893) (incl. *B. boissieri* C. K. Schneider): Plant 30–60 cm; twigs reddish. Spines 12–25 mm. Leaves 15–45 mm, spinulose-serrate. Racemes 20–30 mm, with 3–15 flowers. Honey-leaves 3–4 mm. Berry blackish-red; stigma very shortly stipitate. $2n = 28$. ● *Mountains of Corse, Sardegna, Sicilia and S. Italy.*

(**c**) Subsp. **australis** (Boiss.) Heywood, *Feddes Repert.* **64**: 49 (1961) (*B. hispanica* Boiss. & Reuter): Plant 60–150 cm; twigs dark purple. Spines 12–20 mm. Leaves 8–30 mm, entire or with 1–6 marginal teeth. Racemes 10–25 mm, with 3–10 flowers. Honey-leaves 3–4 mm. Berry bluish-black; stigma very shortly stipitate. $2n = 28$. *Mountains of S. Spain, above* 1000 *m.*

Plants with characters intermediate between subspp. (**a**) and (**b**) in the Alps and S. France have been described as *B. vulgaris* var. *alpestris* Rikli.

Plants with characters intermediate between subspp. (**a**) and (**c**) in E. & C. Spain have been described as subsp. **seroi** O. Bolòs & Vigo, *Butll. Inst. Catalana Hist. Nat.* (*Ser. Bot.*) **38**: 65 (1974) (*B. garciae* Pau).

2. B. cretica L., *Sp. Pl.* 331 (1753). Like **1** but plant suckering; leaves nearly always entire; racemes 7–15 mm, scarcely exceeding the leaves, with 3–8(–12) flowers; honey-leaves *c.* 4·5 mm, longer than the inner perianth-segments; fruit 6–8 mm, bluish-black. *Mountains of C. & S. Greece and the Aegean region.* Cr Gr.

6. Mahonia Nutt.

D. A. WEBB

Like *Berberis* but with unarmed stems and pinnate, evergreen leaves.

1. M. aquifolium (Pursh) Nutt., *Gen. N. Amer. Pl.* **1**: 212 (1818) (*Berberis aquifolium* Pursh). Plant 50–100(–200) cm, suckering. Stems stout, sparingly branched. Leaves evergreen, composed of 5–9(–13) leaflets; leaflets 4–8 × 2–4 cm, ovate, distantly spinose-dentate, coriaceous, dark green, shining. Racemes 5–8 cm, suberect, in groups of 3–5; flowers yellow. Honey-leaves *c.* 8 mm, slightly exceeding the inner perianth-segments. Berries globose, black, strongly bluish-pruinose. *Frequently cultivated, especially in W. & C. Europe, and locally naturalized.* [Au Be Br Cz Da Ga Ge Ho Hs Hu No Rs(W, K) Su.] (*W. North America.*)

64. MAGNOLIACEAE

EDIT. †T. G. TUTIN

Trees or shrubs with alternate, simple leaves. Stipules large, deciduous, leaving annular scars at the nodes. Flowers actinomorphic, hermaphrodite, large, solitary. Perianth-segments free, arranged in whorls of 3–6, usually not differentiated into sepals and petals. Stamens numerous, spirally arranged. Carpels unilocular, numerous, spirally arranged. Fruiting carpels dry or fleshy, with 1 or several seeds, usually dehiscent. Embryo minute; endosperm copious.

Apart from *Liriodendron*, several species of *Magnolia*, with entire leaves which are not truncate or emarginate at the apex, and with dehiscent fruiting carpels from which the seeds hang on long, silky threads, are cultivated for ornament.

1. Liriodendron L.

†T. G. TUTIN

Deciduous trees. Leaves with a broad, truncate or emarginate apex and 1 or 2 large lobes on each side; stipules enclosing the bud and each of the young leaves. Perianth petaloid. Fruiting carpels brown, densely imbricate, each with a terminal wing, indehiscent, ultimately separating from the axis.

1. L. tulipifera L., *Sp. Pl.* 535 (1753). Glabrous tree up to 60 m. Leaves 7–12 cm; petiole 5–10 cm. Flowers campanulate. Perianth-segments 4–5 cm, greenish-white, the inner with a broad, orange band near the base. Fruit 5–8·5 cm; fruiting carpels up to 5 cm. *Planted for timber and as an ornamental tree.* [Au Da Ge It Rs(W, K).] (*E. North America.*)

65. LAURACEAE

EDIT. †T. G. TUTIN

Trees or shrubs. Leaves evergreen, entire, gland-dotted, alternate. Dioecious, or flowers hermaphrodite or polygamous; flowers actinomorphic, small, greenish. Perianth deeply 4- to 6-lobed. Stamens usually in 4 whorls; anthers dehiscing by apically hinged valves. Ovary superior, 1-celled; style simple. Fruit a berry.

Inflorescence subsessile; perianth 4-lobed — **1. Laurus**
Inflorescence long-pedunculate; perianth 6-lobed — **2. Persea**

1. Laurus L.

†T. G. TUTIN

Dioecious. Inflorescence subsessile. Perianth 4-lobed. Stamens 8–12; anthers all introrse, opening by 2 valves.

Leaves oblong-lanceolate, glabrous; young twigs glabrous — **1. nobilis**
Leaves suborbicular to lanceolate, tomentose-hirsute beneath, at least on the midrib, when young; young twigs densely tomentose-hirsute — **2. azorica**

1. L. nobilis L., *Sp. Pl.* 369 (1753). Shrub or small tree 2–20 m, with slender, glabrous twigs. Leaves 5–10 × 2–4(–7·5) cm, narrowly oblong-lanceolate, acute or acuminate, glabrous. Male flowers with 8–12 stamens, all or most with 2 glands at base. Female flowers with 2–4 staminodes. Fruit 10–15 mm, ovoid, black when ripe. $2n = 42$. *Mediterranean region; cultivated elsewhere, and locally naturalized.* Al *Bl Co Cr Ga Gr *Hs It Ju *Lu Sa Si Tu [Az Rs(K)].

2. L. azorica (Seub.) Franco, *Anais Inst. Sup. Agron.* (*Lisboa*) **23**: 96 (1960) (*L. canariensis* Webb & Berth., non Willd., *Persea azorica* Seub.). Like **1** but twigs usually stout, densely tomentose-hirsute when young; leaves 5–10(–17) × 3–8 cm, suborbicular to lanceolate, tomentose-hirsute beneath, at least on the midrib, when young. *Açores.* Az. (*Madeira, Islas Canarias.*)

2. Persea Miller

†T. G. TUTIN

Inflorescence with a long peduncle. Perianth 6-lobed; flowers hermaphrodite or polygamous. Stamens 9; staminodes 3; anthers opening by 4 valves, those of the third whorl extrorse.

1. P. indica (L.) Sprengel, *Syst. Veg.* **2**: 268 (1825) (*Laurus indica* L.). Tree up to 20 m, with a broad, rounded crown and stout twigs, finely sericeous when young. Leaves 8–25 × 3–8 cm, lanceolate, obtuse or acute, glabrous. Perianth finely sericeous. Fruit *c.* 20 mm, ellipsoidal, bluish-black when ripe. *Naturalized in Açores.* [Az.] (*Madeira, Islas Canarias.*)

RHOEADALES

66. PAPAVERACEAE

(Incl. *Fumariaceae*)

EDIT. J. R. EDMONDSON

Herbs with or without latex. Leaves usually spirally arranged, variously dissected. Flowers hermaphrodite, actinomorphic or zygomorphic, hypogynous. Sepals 2(3), caducous. Petals 4–6. Stamens 2, 4 or numerous, whorled. Ovary superior, 1-celled, 2-celled or imperfectly many-celled; carpels 2 to many.

Argemone mexicana L., *Sp. Pl.* 508 (1753), a glaucous, spiny annual up to 80 cm with large, solitary, pale yellow to orange flowers, is cultivated for ornament. It has been reported to be naturalized in S. Europe, but rarely persists.

1 Corolla actinomorphic
  2 Sap watery; sepals connate, forming a hood — **6. Eschscholzia**
  2 Latex present; sepals free
    3 Capsule less than 10 times as long as wide, narrowed at base
      4 Style absent; stigmas on a sessile disc at top of ovary — **1. Papaver**
      4 Style short; stigmas distinct — **2. Meconopsis**
    3 Capsule more than 10 times as long as wide, ±parallel-sided
      5 Flowers violet; capsule opening by 2–4 valves — **3. Roemeria**
      5 Flowers yellow or red; capsule opening by 2 valves
        6 Flowers solitary; petals at least 2 cm; capsule 2-celled — **4. Glaucium**
        6 Flowers in a simple umbel; petals up to 1 cm; capsule 1-celled — **5. Chelidonium**
1 Corolla zygomorphic or bisymmetrical
  7 Corolla weakly zygomorphic; petals not spurred or saccate — **7. Hypecoum**
  7 Corolla strongly zygomorphic or bisymmetrical; upper petal spurred or saccate at base
    8 Corolla bisymmetrical — **8. Dicentra**
    8 Corolla zygomorphic
      9 At least the upper fruits 2- to many-seeded, dehiscent
        10 Plant annual; tendrils present — **16. Ceratocapnos**
        10 Plant usually perennial; tendrils absent
          11 Style deciduous, translucent
            12 Plant much-branched but not forming dense cushion; fruit 3- to 14-seeded — **11. Pseudofumaria**
            12 Plant forming dense cushion; fruit usually 2-seeded — **12. Sarcocapnos**
          11 Style persistent, green
            13 Inflorescence racemose; bracts conspicuous — **9. Corydalis**
            13 Inflorescence cymose; bracts inconspicuous — **10. Capnoides**
      9 All fruits 1-seeded, indehiscent
        14 Upper petal saccate; stigma 3-fid, the middle lobe deeply notched, the lateral patent or deflexed — **14. Platycapnos**
        14 Upper petal spurred; stigma 2-lobed, with a small tooth between lobes
          15 Cauline leaves numerous; flowers in racemes — **13. Fumaria**
          15 Cauline leaves few; flowers in corymbs — **15. Rupicapnos**

Subfam. Papaveroideae

Latex usually present. Flowers solitary or umbellate, actinomorphic. Sepals entire; petals entire, not spurred. Stamens numerous. Capsule dehiscing by pores or longitudinal valves.

1. Papaver L.

A. B. MOWAT & S. M. WALTERS (EDITION 1) REVISED BY J. W. KADEREIT (EDITION 2)

Annual, biennial or perennial herbs with latex. Sepals 2 or 3, usually caducous. Petals 4 or 6, usually entire, often bright red, crumpled in bud, usually caducous. Stigmas 4–16, sessile over the placentae. Capsule clavate to globose, opening by valves or pores below the stigmatic disc, or disc deciduous.! Seeds reniform, without aril.

Literature: Sect. *Meconella*: F. Markgraf, *Phyton* (*Austria*) **7**: 302–314 (1958); G. Knaben, *Op. Bot.* (*Lund*) **2**: 1–74 (1959); **3**: 1–96 (1959); U. Rändel, *Feddes Repert.* **84**: 655–732 (1974); J. W.

Kadereit, *Bot. Jahrb.* **112**: 79–97 (1990). Sect. *Pseudopilosa*: W. Vent *et. al.*, *Feddes Repert.* **77**: 47–56 (1968). Sect. *Macrantha*: P. Goldblatt, *Ann. Missouri Bot. Gard.* **61**: 264–296 (1974). Sect. *Papaver*: J. W. Kadereit, *Bot. Jahrb.* **106**: 221–244 (1986); **108**: 1–16 (1986). Sect. *Argemonidium*: J. W. Kadereit, *Notes Roy. Bot. Gard. Edinb.* **44**: 25–43 (1986). Sect. *Rhoeadium*: J. W. Kadereit, *Notes Roy. Bot. Gard. Edinb.* **45**: 225–286 (1989).

1 Scapose perennial (arctic-alpine)
2 Capsule narrowly obovoid to cylindrical (arctic or sub-arctic Europe)
3 Flower-buds narrowly ovoid, usually less than 5 mm broad; stamens 30–40 **13. lapponicum**
3 Flower-buds broadly ellipsoidal to subglobose, usually more than 5 mm broad; stamens 20–30 **14. radicatum**
2 Capsule ellipsoidal to obovoid (C. or S. Europe)
4 Stamens not longer than ovary; petals not overlapping in flower **15. lapeyrousianum**
4 Stamens at least as long as ovary; petals overlapping in flower **16. alpinum**
1 Annual, biennial or perennial with at least some cauline leaves
5 Perennial; non-flowering stems present
6 Stamens pale yellow; petals 2–3 cm, light orange, without basal black spot; capsule clavate **11. rupifragum**
6 Stamens dark violet; petals 5–9 cm, dark orange, usually with basal black spot; capsule broadly obovoid to subglobose **12. pseudoorientale**
5 Annual; all stems bearing flowers
7 Filaments clavate; capsule usually setose; stigmatic disc convex
8 Mature capsule at least twice as long as wide, often without setae in lower part; sepals with short subapical processes **10. argemone**
8 Mature capsule at most twice as long as wide, usually with setae throughout; sepals without subapical processes
9 Setae patent in all parts of capsule; stigmatic disc not vaulted between stigmatic rays; petals usually with a small, basal black spot **8. hybridum**
9 Lower part of capsule with ±appressed setae; mature stigmatic disc strongly vaulted between stigmatic rays; petals without a basal black spot **9. apulum**
7 Filaments filiform or clavate; capsules not setose; stigmatic disc flat or rarely umbonate
10 Upper cauline leaves amplexicaul; petals white to violet **7. somniferum**
10 Upper cauline leaves not amplexicaul; petals red
11 Sepals with ±conspicuous subapical processes; ultimate segments of upper leaves usually less than 1 mm wide **4. arenarium**
11 Sepals very rarely with subapical processes and then inconspicuous; if present, ultimate segments of upper leaves more than 1 mm wide
12 Anthers light yellow; upper leaves with cuneate to rounded base and basal pair of large, acute, triangular, antrorse lobes, or leaves entire or with usually 1 pair of recurved lobes
13 Sepal-margin never dark violet; upper leaves very rarely entire, usually with many strongly antrorse lobes or teeth **1. pinnatifidum**
13 Sepal-margin usually dark violet; upper leaves often entire or with a single pair of recurved lobes **2. purpureomarginatum**
12 Anthers dull yellowish-brown or greenish; upper leaves not as above
14 Terminal segment of basal leaves much larger than lateral segments, usually less than twice as long as wide; plants usually with distinct red tinge on axis and leaves, at least on some cauline setae; capsules gradually narrowing into clavate base **5. guerlekense**
14 Terminal segment of basal leaves variable, if much larger than lateral segments, then usually more than twice as long as wide; plants usually without red tinge, if with tinge, then capsules ±abruptly contracted at base
15 Pedicel setae always densely appressed; capsules usually much more than twice as long as wide; stigmatic disc not wider than capsule **3. dubium**
15 Pedicel setae usually patent, sometimes loosely appressed; capsules usually less than twice as long as wide; stigmatic disc usually wider than capsule **6. rhoeas**

Sect. RHOEADIUM Spach. Annuals with few to many usually setose hairs. Leaves pinnatisect to pinnatipartite; upper cauline leaves not amplexicaul; lower leaves petiolate. Filaments filiform, dark violet. Capsule glabrous.

1. P. pinnatifidum Moris, *Fl. Sard.* 1: 74 (1837). Plant 10–85 cm, erect. Leaves incised to pinnatipartite, the lower sometimes pinnatisect. Upper leaves sessile with cuneate or rounded base, with a large basal pair of acute, triangular lobes. Pedicels appressed-setose above. Flower-buds densely appressed-setose. Petals 14–30 mm, without distinct basal spots. Anthers pale yellow. Capsule 1–2·7 × 0·4–0·8 cm, $2\frac{1}{2}$–4 times as long as wide, clavate, distinctly contracted below pores. Disc narrower than capsule, with 4–9 rays. $2n = 28$. *W. Mediterranean region; Açores.* Az Bl Co Ga Hs It Sa Si.

2. P. purpureomarginatum Kadereit, *Notes Roy. Bot. Gard. Edinb.* **45**: 235 (1989). Plant 8–45 cm, erect. Leaves entire to bipinnatipartite. Upper leaves sessile, often entire or with 1 pair of recurved basal lobes, less often with more than 1 pair of lobes. Pedicels sparsely appressed-setose above. Flower-buds subglabrous; sepals with dark violet margin. Petals 10–20 mm, without distinct basal spots. Anthers pale yellow. Capsule 1–1·8 × 0·5–0·7 cm, 2–$2\frac{1}{2}$ times as long as wide, ellipsoidal to obovoid, distinctly contracted below pores. Disc narrower than capsule, mostly slightly concave, with marginal dark violet marks and 4–7 rays. $2n = 28$. *S. Greece, S. Aegean region.* Cr Gr.

3. P. dubium L., *Sp. Pl.* 1196 (1753) (incl. *P. obtusifolium* Desf., *P. modestum* Jordan, *P. hirtodubium* Fedde). Plant 10–90 cm, erect. Leaves pinnatifid to pinnatipartite. Upper leaves shortly petiolate to sessile, tripartite with pinnatifid to pinnatisect lobes. Pedicels appressed-setose above. Flower-buds glabrous to densely appressed-setose. Petals 10–40 mm, red or white without or with basal black spots. Anthers dull yellow, brown or green. Capsule 0·8–2·6 × 0·4–1·1 cm, 2–$4\frac{1}{2}$ times as long as wide, narrowly obovoid to clavate. Disc not wider than capsule, with 4–11 rays. *Throughout Europe, except parts of the north.* All except Cr Fa Fe Is No Rs(N, C) Sb.

1 Plants with very sparse indumentum, strongly glaucous (c) subsp. **laevigatum**

1 Plants with more or less dense indumentum, green or only slightly glaucous

2 Latex white or cream, brown to black when dry; lobes of upper leaves usually wider than 1·5 mm (a) subsp. **dubium**

2 Latex yellow or turning yellow, red when dry; if latex colourless, petals white; if latex white, lobes of upper leaves usually narrower than 1·5 mm (b) subsp. **lecoqii**

(a) Subsp. **dubium**: Upper leaves sessile with cuneate base, rarely shortly petiolate. Petals red, mostly without basal spot. Latex white or cream, black to brown when dry. $2n = 42$. *Throughout the range of the species.*

(b) Subsp. **lecoqii** (Lamotte) Syme, *Engl. Bot.* ed. 3, **1**: 30 (1863) (*P. lecoqii* Lamotte; incl. *P. albiflorum* (Besser) Pacz.): Upper leaves usually shortly petiolate. Petals red or (var. *albiflorum* Besser) white, often with basal black spot. Latex yellow or turning yellow, red when dry; if colourless, petals white; if white, lobes of upper leaves narrower than 1·5 mm. $2n = 28$. *W., C. & S. Europe.*

(c) Subsp. **laevigatum** (Bieb.) Kadereit, *Notes Roy. Bot. Gard. Edinb.* **45**: 244 (1989) (*P. laevigatum* Bieb., *P. maeoticum* Klokov, *P. nothum* Steven): Like subsp. (a) but with very sparse indumentum, strongly glaucous. Petals red, with often very large, basal black spot. *S.E. Europe.*

4. P. arenarium Bieb., *Fl. Taur.-Cauc.* **3**: 364 (1819). Plant 10–70 cm, erect. Leaves 1- to 3-pinnatipartite, finely divided; ultimate segments often less than 1 mm wide. Upper leaves usually shortly petiolate. Pedicels appressed- or sometimes patent-setose above. Flower-buds patent- or retrorse-setose; sepals with usually conspicuous apical process. Petals 18–45 mm, red, with usually large, basal black spot. Anthers brown or green. Capsule 1–1·8 × 0·5–0·8 cm, 2–2·8 times as long as wide, obovoid. Disc at least as wide as capsule, with 7–9 rays. $2n = 14$. *S.E. Russia, W. Kazakhstan.* Rs(E). (*S.W. Asia.*)

5. P. guerlekense Stapf, *Denkschr. Akad. Wiss. Math.-Nat. Kl. (Wien)* **51**: 359 (1886) (*P. stipitatum* Fedde). Plant 20–40 cm, erect, usually with distinct red tinge. Leaves pinnatipartite to pinnatisect. Terminal lobe of lower leaves much bigger than lateral, usually less than twice as long as wide; upper leaves sessile, lateral lobes usually becoming smaller towards leaf apex. Pedicels appressed- or patent-setose. Flower-buds patent-setose. Petals 30–38 mm, red, with or without basal or sub-basal black spots. Anthers brown. Capsule 0·8–1·7 × 0·3–0·7 cm, 1·8–4 times as long as wide, narrowly obovoid to obpyriform, narrowing gradually into clavate base. Disc narrower to wider than capsule, with 5–9 rays. $2n = 14$. *S. Aegean region.* Cr Gr.

6. P. rhoeas L., *Sp. Pl.* 507 (1753) (incl. *P. insignitum* Jordan, *P. intermedium* G. Beck, *P. roubiaei* Vig., *P. trilobum* Wallr., *P. commutatum* Fischer & C. A. Meyer, *P. strigosum* (Boenn.) Schur, *P. tenuissimum* Fedde, *P. tumidulum* Klokov). Plant 10–60 cm, erect or ascending. Leaves pinnatifid to pinnatipartite. Terminal lobe of lateral leaves usually more than twice as long as wide; upper leaves sessile, 3-fid with larger middle than basal lobes. Pedicels patent- or (var. *strigosum* Boenn.) appressed-setose. Flower-buds patent- or appressed-setose. Petals 13–50 mm, red, with or without basal black or black and distally white spot. Anthers brown or rarely yellow. Capsule 0·5–1·6 × 0·4–1·4 cm, 1–2 times as long as wide, usually broadly obovoid, cylindrical or almost globose, sometimes shortly stipitate, usually more or less abruptly contracted at base. Disc usually wider than capsule, sometimes slightly umbonate, with 6–16 rays. $2n = 14$. *Throughout most of Europe as a weed of cultivation and probably native in the south, but only as a casual in the north.* All except Fa Is Rs(N) Sb.

Sect. PAPAVER. Annuals with few to many setose hairs. Leaves incised to pinnatipartite; upper cauline leaves amplexicaul. Filaments filiform or clavate, pale to dark violet. Capsules glabrous.

7. P. somniferum L., *Sp. Pl.* 508 (1753). Plant 15–150 cm, erect, often strongly glaucous. Leaves serrate, dentate, incised or pinnatipartite. Lower leaves usually petiolate; upper leaves sessile, with amplexicaul base. Pedicels glabrous or patent- or appressed-setose. Flower-buds glabrous or patent- or appressed-setose. Petals 20–80 mm, white to violet, darker at base but without distinct basal spot. Filaments filiform or clavate. Anthers yellow to green or brown. Capsule 1·4–9·5 × 0·8–6·5 cm, 1–2 times as long as wide, ovoid, ellipsoidal, cylindrical or globose, distinctly stipitate, dehiscent or not. Disc not wider than capsule, with 5–18(–22) rays. *Cultivated since ancient times in most of Europe except the extreme north for its medicinal and narcotic latex (Opium) and its seeds, which are used as a condiment and a source of oil, and more recently grown for ornament; widespread as a naturalized alien or as a casual, but probably native in parts of the Mediterranean region.* Bl Co ?Cr Ga Hs It Lu Sa Si [Az Br Ge Hb Hu Ju Rm Rs(B, C, W, K, E) Si Tu]; casual elsewhere.

(a) Subsp. **somniferum** (incl. subsp. *hortense* (Hussenot) Corb., subsp. *songaricum* Basil.): Subglabrous, strongly glaucous, usually unbranched. Upper leaves usually little divided. Sepals glabrous. Filaments clavate, pale violet. Capsule often indehiscent. $2n = 22$. *Cultivated and naturalized throughout most of the range of the species.*

(b) Subsp. **setigerum** (DC.) Arcangeli, *Comp. Fl. Ital.* 25 (1882) (*P. setigerum* DC.): With more or less densely setose indumentum, weakly glaucous. Upper leaves usually pinnatifid. Sepals with at least some setae. Filaments filiform or clavate, dark violet. Capsules always dehiscent. $2n = (22), 44$. *Mainly W. Mediterranean region, where it is probably native.*

Subsp. (a) is very variable in shape and colour of petals, shape and size of capsules, colour of seeds, etc.

Sect. ARGEMONIDIUM Spach. Annuals with few to many setose or arachnoid hairs. Leaves strongly dissected, the lower petiolate and the upper sessile. Filaments usually clavate, dark violet. Capsules usually setose.

8. P. hybridum L., *Sp. Pl.* 506 (1753). (*P. hispidum* Lam., *P. siculum* Guss.). Plant 5–60 cm, erect or ascending, usually with arachnoid hairs below, setose hairs above. Leaves 1- to 3-pinnatipartite. Pedicels usually appressed-setose. Flower-buds more or less patent-setose. Petals 18–25 mm, pink with distinct basal black spot. Anthers blue. Capsule 0·7–1·7 × 0·5–1·1 cm, obovoid to ellipsoidal to subglobose, with patent setae. Disc vaulted, narrower than capsule, with 5–10 rays. $2n = 14$. *Native in S. Europe, and widespread as a weed of cultivation further north.* Al Bl Bu Co Cr Ga Gr Hs *Hu It Ju Lu No *Rm Rs(K) Sa Si Tu [Br ?Ge Hb Rs(W)].

9. P. apulum Ten., *Fl. Neap. Prodr. App. Quinta* 16 (1826). Plant 10–55 cm, erect or ascending, setose throughout. Leaves 1- to 3-pinnatipartite. Pedicels appressed-setose. Flower-buds more or less patent-setose. Petals 16–35 mm, pale reddish-orange to bluish-red, darker at base but without distinct basal black spot. Anthers violet. Capsule 0·6–1·3 × 0·4–0·7 cm, ellipsoidal to subglobose, with more or less appressed setae below and patent setae above. Disc conspicuously vaulted between rays, narrower than capsule, with 4–6 rays. $2n = 12$. ● *Mediterranean region, from N.W. Italy to E. Bulgaria and Kriti.* Al Br Cr Gr It Ju Si.

10. P. argemone L., *Sp. Pl.* 506 (1753). Plant 10–75 cm, erect or ascending, arachnoid below and setose above. Leaves 1- to 3-pinnatipartite. Pedicels usually appressed-setose. Flower-buds almost glabrous or more or less patent-setose; sepals usually with inconspicuous apical processes. Petals 10–30 mm, reddish-orange with distinct basal black spot. Anthers blue, green or yellow. Capsule 0·9–2·5 × 0·3–0·6 cm, narrowly obovoid to ellipsoidal or cylindrical, with patent to half-appressed setae mainly above. Disc vaulted to distinctly pointed, with 4–6 rays, not wider than capsule. *Native in S. Europe, and widespread as a weed of cultivation further north.* Al Au Be Br Co Cr Cz Da Ga Ge Gr Hb He Ho Hs Hu It Ju Lu Po Rm Rs(B, C, W, K) Sa Su Tu.

(a) Subsp. **argemone**: Plants mostly tall and erect. Capsules usually longer than 1·5 cm. Stigmatic disc vaulted, margin without teeth or teeth not subdividing pores. $2n = 40, 42$. *Throughout the range of the species, except Aegean region.*

(b) Subsp. **nigrotinctum** (Fedde) Kadereit, *Notes Roy. Bot. Gard. Edinb.* **44**: 37 (1986) (*P. nigrotinctum* Fedde): Plants usually smaller, up to 25 cm; stems often ascending. Capsules usually shorter than 1·5 cm, ellipsoidal. Stigmatic disc pointed, teeth on margin subdividing pores. $2n = 14$. *S. Aegean region.*

Sect. PSEUDOPILOSA M. Popov ex K. Günther. Subscapose perennials, more or less densely setose-hairy. Leaf vernation revolute; leaves pinnatifid to pinnatisect. Filaments pale yellow, filiform. Petals orange. Capsule usually clavate, glabrous.

11. P. rupifragum Boiss. & Reuter, *Pugillus* 6 (1852). Plant 10–60 cm, erect or ascending. Leaves pinnatifid to pinnatisect, usually petiolate, usually setose on veins of lower surface and teeth. Pedicel glabrous or half-appressed-setose. Flower-buds glabrous. Petals 20–28 mm, orange, without distinct basal spot. Anthers light yellow. Capsule 1·3–2·5 × 0·4–0·7 cm, clavate. Disc as wide as capsule, with 5–8 rays. $2n = 14, 28$. *Crevices of mountain rocks. S. Spain (near Grazalema).* Hs.

P. atlanticum (Ball) Cosson, *Ill. Fl. Atl.* **1**: 11 (1882), from Morocco, with more or less densely appressed-setose leaves, pedicels and sepals, is reported to be locally naturalized in Austria, Britain and Denmark.

Sect. MACRANTHA Elkan. Robust perennials, densely setose-hairy. Leaves pinnatisect to pinnatipartite. Filaments dark violet, clavate. Petals orange or orange-red. Capsule usually subglobose, glabrous.

12. P. pseudoorientale (Fedde) Medw., *Izv. Kavkaz. Muz.* **11**: 204 (1918) (*P. orientale* auct., non L.). Plant 40–100 cm, erect. Leaves pinnatisect to pinnatipartite. Lower leaves petiolate, upper leaves sessile. Pedicels usually appressed-setose. Flower-buds with or without dentate to entire subtending bracts, usually erect during development, with subpatent, slender setae. Petals 4 or 6, 50–90 mm, usually with distinct basal spots. Anthers violet. Capsule 2·0–2·5 × 1·5–2·5 cm, broadly obovoid to subglobose. Disc at least as wide as capsule, with 9–19 rays. $2n = 42$. *Widely cultivated for ornament, and occasionally more or less naturalized.* [Cz Ga Hu.]

Most cultivated material in this section must be referred to **12**. Two other species, **P. bracteatum** Lindley, *Coll. Bot.* t. 23 (1821), with dark red petals, 3–8 bracts subtending the flowers, and calyx setae with broadly triangular bases, and **P. orientale** L., *Sp. Pl.* 508 (1753), with ebracteate flowers, and petals usually unmarked, or petals with pale violet to white marks, are sometimes cultivated.

Sect. MECONELLA Spach (Sect. *Scapiflora* Reichenb.). Scapose, erect, often caespitose perennials with more or less densely setose indumentum. Leaves pinnatifid to pinnatipartite. Filaments filiform, yellow, sometimes with dark tinge. Petals white, yellow or sometimes reddish-orange. Capsule cylindrical to obovoid or ellipsoidal, setose, dehiscing by valves.

Cultivars referable to the Asiatic *P. nudicaule* L., sensu lato (*P. croceum* Ledeb.), are widely grown for ornament, and sometimes become locally naturalized. They are robust plants with flowering stems up to 50 cm, large flowers (up to 6 cm), and broadly lobed leaves. They all have a chromosome number of $2n = 14$.

13. P. lapponicum (Tolm.) Nordh., *Bergens Mus. Aarb., Naturv. Rekke* **2**: 45 (1931). Plants 15–30 cm. Leaves 1- or 2-pinnatipartite. Pedicel half-appressed-setose throughout, or appressed-setose below and patent-setose above. Flower-buds densely patent-setose. Petals 15–17 mm, yellow, caducous. Stamens 30–40. Capsule 1·1–1·8 × 0·5–0·6 cm, narrowly obovoid to cylindrical. Disc with 5 or 6 rays. $2n = 56$. *Arctic Europe and Ural.* No Rs(NC.).

14. P. radicatum Rottb., *Skr. Kiobenhavnske Selsk. Lærd. Vid.* **10**: 455 (1767) (*P. steindorssonianum* Á. Löve, *P. chibinense* N. Semen., *P. radicatum* subsp. *relictum* (E. Lundström) Tolm., *P. relictum* (E. Lundström) Nordh., *P. radicatum* subsp. *intermedium* (Nordh.) Knaben, subsp. *oeksendalense* Knaben, subsp. *subglobosum* Nordh., subsp. *hyperboreum* Nordh., subsp. *macrostigma* (Nordh.) Nordh., subsp. *ovatilobum* Tolm., subsp. *dahlianum* (Nordh.) Rändel, *P. dahlianum* Nordh., *P. jugoricum* (Tolm.) Stankov, *P. radicatum* subsp. *brachyphyllum* Tolm., subsp. *polare* Tolm. pro max. parte, *P. polare* (Tolm.) Perf.). Plants 6–20 cm. Leaves pinnatifid to pinnatipartite. Pedicel half-appressed to patent-setose. Flower-buds densely patent-setose. Petals 13–18 mm, white or yellow, persistent or sometimes caducous. Stamens 20–30. Capsule 1–1·8 × 0·6–0·7 cm, obovoid to cylindrical. Disc with 5 or 6 rays. $2n = 70$. *Arctic and subarctic Europe, southwards to 61°30′ N. in Norway.* Fa Is No Rs(N) Sb Su.

This is a highly variable species in which ten subspecies have been recognized in Europe. Subsp. *dahlianum* (Nordh.) Rändel, from Svalbard and Arctic Europe, is often treated at specific rank.

P. laestadianum (Nordh.) Nordh., *Bot. Not.* **1939**: 693 (1939), from Arctic Norway and Sweden, originally described as a subspecies of **14**, is distinguished as an independent species by its relatively tall size and persistent petals. This taxon seems to have $2n = 56$ chromosomes.

15. P. lapeyrousianum Guterm., *Österr. Bot. Zeitschr.* **122**: 268 (1974) (*P. suaveolens* Lapeyr., nom. illeg.; incl. *P. suaveolens* subsp. *endressii* Ascherson). Plants 5–12 cm. Leaves 1- or 2-pinnatipartite. Pedicel appressed-setose. Flower-buds patent-setose. Petals 6–14 mm, yellow, not overlapping in flower. Stamens shorter to as long as ovary. Capsules 0·8–1·2 × 0·4–0·7 cm, narrowly to broadly ellipsoidal; disc acute, with 5 or 6 rays. $2n = 14$. *Pyrenees; Sierra Nevada.* Ga Hs.

16. **P. alpinum** L., *Sp. Pl.* 507 (1753) (incl. subsp. *kerneri* (Hayek) Fedde, *P. kerneri* Hayek, *P. alpinum* subsp. *rhaeticum* (Leresche) Markgraf, *P. rhaeticum* Leresche, *P. pyrenaicum* subsp. *rhaeticum* (Leresche) Fedde, *P. alpinum* subsp. *corona-sancti-stephani* (Zapał.) Borza, *P. corona-sancti-stephani* Zapał., *P. alpina* subsp. *degenii* (Urum. & Jáv.) Markgraf, *P. pyrenaicum* subsp. *degenii* Urum. & Jáv., *P. alpinum* subsp. *sendtneri* (A. Kerner ex Hayek) Schinz & R. Keller, *P. sendtneri* A. Kerner ex Hayek, *P. pyrenaicum* subsp. *sendtneri* (A. Kerner ex Hayek) Fedde, *P. alpinum* subsp. *ernesti-mayeri* Markgraf, subsp. *tatricum* E. I. Nyárády, *P. occidentale* (Markgraf) H. Hess, Landolt & Hirzel). Plants 4–23 cm. Leaves 1- to 3-pinnatipartite, glabrous to densely setose. Pedicel and flower-buds appressed- to half-appressed- or patent-setose. Petals 10–30 mm, white, yellow, sulphur-yellow or orange-red, overlapping in flower. Stamens as long as to longer than ovary. Capsules 0·7–1·6 × 0·4–0·7 cm, narrowly to broadly ellipsoidal to obovoid; disc acute or obtuse, with 4–6 rays. $2n = 14$. *Mountains of C. & S. Europe.* Au Bu Cz Ga He ?Hs It Ju Po Rm.

In view of the limited degree of morphological differentiation in the *P. alpinum* complex, the recognition of its constituent taxa at specific rank is scarcely justified. Often difficult to identify material without a knowledge of its geographical provenance, its treatment is discussed fully by J. W. Kadereit, *Bot. Jahrb.* **112**: 79–97 (1990).

2. **Meconopsis** Vig.

A. B. MOWAT (EDITION 1) REVISED BY J. W. KADEREIT (EDITION 2)

Perennials with yellow latex. Sepals usually 2, free, caducous. Petals usually 4, entire, crumpled in bud, caducous. Stigmas on usually distinct style over placentae. Capsule opening by short valves. Seeds ovoid to reniform, without aril.

1. **M. cambrica** (L.) Vig., *Hist. Pavots Argém.* 48 (1814). Plant 30–60 cm, erect, glabrous or sparsely pubescent. Leaves pinnatifid to pinnatisect, all usually petiolate. Flowers solitary, long-pedicellate, in axils of upper leaves. Pedicel and flower-buds glabrous or pubescent. Petals 2–4 cm, yellow. Anthers and filaments yellow. Capsule 2·0–4·0 × 0·7–1·1 cm, ovoid to ellipsoid-oblong, ribbed, glabrous. $2n = 14, 22, 28, 56$. *Shady places. W. Europe; cultivated for ornament there and elsewhere, and sometimes naturalized.* Br Ga Hb Hs [Da No].

3. **Roemeria** Medicus

A. B. MOWAT (EDITION 1) REVISED BY J. W. KADEREIT (EDITION 2)

Annuals with yellow latex. Sepals 2, free, caducous. Petals 4, entire, crumpled in bud, caducous. Stigmas 3 or 4, sessile over the placentae. Capsule linear, opening by 2–4 long valves from top to bottom. Seeds reniform, without aril.

Literature: J. W. Kadereit, *Flora* **179**: 135–153 (1987).

1. **R. hybrida** (L.) DC., *Reg. Veg. Syst. Nat.* **2**: 92 (1821) (*R. violacea* Medicus). Erect or ascending annual 20–45 cm, usually with arachnoid indumentum, sometimes with some setae. Leaves 1- to 3-pinnatipartite. Lower leaves petiolate, upper sessile. Pedicels and flower-buds usually with few arachnoid hairs. Petals 9–30 mm, violet, darker at base but without distinct basal spot. Anthers violet or yellow. Capsule 4·5–9·5 × 0·2–0·4 cm, rarely shorter, linear, usually with setae only in upper ½ but sometimes throughout. $2n = 22, 24$. *S. Europe; casual elsewhere.* ?Al Bl Co Cr Ga Gr Hs It Tu.

Only subsp. **hybrida** occurs in Europe. Subsp. **dodecandra** (Forskål) Maire in Jahandiez & Maire, *Cat. Pl. Maroc* **2**: 257 (1932), with shorter capsules setose throughout, is widespread in S.W. Asia.

4. **Glaucium** Miller

A. B. MOWAT (EDITION 1) REVISED BY J. R. AKEROYD (EDITION 2)

Glaucous; latex yellow. Sepals free. Petals 4, entire. Stigma 2-lobed, sessile over placentae. Capsule linear, 2-celled, opening from above almost to base by 2 valves. Seeds embedded in septum, without aril.

Petals yellow; capsule glabrous, but often tuberculate **1. flavum**
Petals orange or reddish; capsule hispid **2. corniculatum**

1. **G. flavum** Crantz, *Stirp. Austr.* **2**: 133 (1763). Sparsely pubescent biennial or perennial. Stems 30–90 cm. Basal leaves 15–35 cm, lyrate-pinnatifid; lower segments smaller and more or less entire, the upper coarsely dentate; cauline leaves smaller, ovate, lobed, amplexicaul. Flowers solitary, terminal or axillary. Sepals more or less pubescent. Petals 3–4 cm, yellow, broadly obovate. Stamens yellow. Ovary tuberculate, at least towards apex. Capsule 10–30 cm, often curved, glabrous. $2n = 12$. *Sandy and gravelly seashores; also as a ruderal. Coasts of S. & W. Europe, northwards to S. Norway locally naturalized in C. Europe.* Al Be Bl Br Bu Co Cr Da Ga Ge Gr Hb Ho Hs It Ju Lu No Rm Rs(K) Sa Si Tu [Cz ?Ge He].

Plants from S.E. Europe, with deeper yellow petals and capsules smooth in lower part and slightly constricted between the seeds, have been distinguished as **G. leiocarpum** Boiss., *Fl. Or.* **1**: 122 (1867). Experimental studies by W. B. Turrill, *Kew Bull.* **1933**: 174–184 (1933), suggest that they are best treated as *G. flavum* var. *leiocarpum* (Boiss.) Stoj. & Stefanov.

2. **G. corniculatum** (L.) J. H. Rudolph, *Fl. Jen. Pl.* 13 (1781) (*G. grandiflorum* sensu Hayek, non Boiss. & Huet). Pubescent annual, rarely biennial. Stems 30–40 cm. Leaves lyrate-pinnatifid; segments unequal, dentate; basal leaves petiolate, upper sessile. Flowers solitary, terminal or axillary. Petals up to 3(–4) cm, obovate, orange or reddish, often with dark spot at base. Stamens with dark anthers; filaments yellow. Capsule up to 20 cm, usually straight, pubescent. *Waste places and cultivated ground. S. Europe, extending northwards to Czechoslovakia and N. Ukraine; sometimes naturalized elsewhere.* Bl Bu Co Cr Ga Gr Hs Hu It Ju Lu Rm Rs(C, W, K, E) Sa Si Tu [Au Cz Ge He].

5. **Chelidonium** L.

A. B. MOWAT

Latex orange. Sepals free. Petals 4, entire. Stigma 2-lobed; style very short. Capsule linear, 1-celled, opening from below by 2 valves. Aril crested.

Literature: A. Krahulcová, *Folia Geobot. Phytotax.* (*Praha*) **17**: 237–268 (1982).

1. **C. majus** L., *Sp. Pl.* 505 (1753). Glaucous, sparsely pubescent perennial. Stems 30–90 cm. Leaves pinnate; leaflets 5–7, ovate to oblong, the terminal usually 3-lobed, the lateral often with a stipule-like lobe at base on lower side, all crenate or (var. *laciniatum* (Miller) Syme) deeply incised. Inflorescence an umbel of 2–6 flowers. Petals up to 1 cm, obovate, bright yellow. Stamens yellow; filaments expanded above. Capsule 3–5 cm, more or less straight, glabrous; aril white. $2n = 12$. *Shady banks and walls. Europe, except*

the extreme north, but in some districts only as an escape from gardens. All except Cr Fa Is Sb; only as a naturalized alien in Hb and perhaps elsewhere.

6. **Eschscholzia** Cham.

A. B. MOWAT

Latex absent. Sepals connate, forming a hood which is shed when the flower opens, leaving 2 rims at base of ovary. Petals 4, entire. Capsule ribbed, opening from below by 2 valves, which separate from the placentae. Aril absent.

1. **E. californica** Cham. in Nees, *Horae Phys. Berol.* 74 (1820). Glaucous, glabrous annual or perennial. Stems 20–60 cm, erect or spreading. Leaves ternately divided; ultimate segments linear. Flowers solitary, long-pedicellate, terminal. Petals 1–6 cm, orange to yellow, darker at base, obovate. Stamens yellow; filaments not expanded. Capsule 7–10 cm, straight, glabrous. *Widely cultivated for ornament and locally naturalized in S.W. Europe; a frequent casual elsewhere.* [Bl Co Ga]. (*S.W. United States.*)

Subfam. **Hypecooideae**

Latex absent. Flowers solitary or in cymes. Sepals free. Corolla slightly zygomorphic; petals not spurred. Capsule linear, usually straight.

7. **Hypecoum** L.

A. B. MOWAT AND †T. G. TUTIN (EDITION 1) REVISED BY Å. E. DAHL (EDITION 2)

Glabrous, often glaucous annuals. Basal leaves lanceolate, pinnate; ultimate leaflets narrow. Sepals 2, free. Petals 4, at least the inner pair 3-lobed. Stamens 4. Stigmas 2; style short. Capsule a striate lomentum, usually breaking up into 1-seeded portions. Seeds rugose, without aril.

Literature: Å. E. Dahl, *Pl. Syst. Evol.* **163**: 227–280 (1989).

1 Fruit pendent; plant smelling strongly of curry on drying **1. pendulum**
1 Fruit erect; plant not smelling strongly of curry on drying
 2 Central lobe of inner petals unguiculate; inner filaments linear to narrowly elliptical; fruit 1–2 mm wide **2. littorale**
 2 Central lobe of inner petals not unguiculate; inner filaments narrowly triangular or ovate; fruit 1·5–3·5 mm wide
  3 All or most ultimate leaflets of upper primary leaf-segments wider towards the 3-fid apex, often with 1 to several lateral teeth; pollen orange-yellow.
   4 Outer petals not more than 6 mm wide, always longer than wide; lateral lobes often involute **3. torulosum**
   4 Outer petals usually at least 6 mm wide, as wide as or wider than long; lateral lobes flat **4. imberbe**
  3 All or most ultimate leaflets of upper primary leaf-segments not wider towards the entire apex, with 0(1) lateral teeth; pollen yellowish-white
   5 Filaments of median stamens narrowly triangular; outer petals shortly unguiculate or not **5. pseudograndiflorum**
   5 Filaments of median stamens narrowly ovate; outer petals distinctly unguiculate **6. procumbens**

1. **H. pendulum** L., *Sp. Pl.* 124 (1753) (*H. tetragonum* Bertol.). Plant 5–40 cm, greyish-green to glaucous. Ultimate leaflets linear. Inflorescence erect, with 1–15 flowers. Sepals (1–)2–3(–5) × 1–2·5 mm, ovate, acute, entire to dentate, caducous. Petals pale yellow, the outer (3–)5–8(–9) × (1·5–)2·5–5(–6) mm, rhombic, with more or less involute margins and keeled apex; inner petal with black spots, the lateral lobes not exceeding ½ length of lamina of central lobe; central lobe 1–2·5 × 1–2·5 mm, circular-elliptical, unguiculate. Filaments linear-elliptical, sometimes with black spots. Pollen yellow. Fruit 2–3 mm wide, pendent, straight, not thickened at septa, rectangular in section; seeds pale brown. $2n = 16, 32$. *Cultivated and waste ground, sometimes in saline habitats. Mediterranean region; rare in C. Europe.* Bl Ga Ge Gr Ju Hs.

2. **H. littorale** Wulfen in Jacq., *Collect. Bot.* **2**: 205 (1789) (*H. geslinii* Cosson & Kralik, *H. deuteroparviflorum* Fedde). Plant 5–20(–25) cm, glaucous. Ultimate leaflets usually linear, cuneate at base, the apex acute, with 0 to several lateral teeth. Inflorescence decumbent, with (2–)5–10(–25) flowers. Sepals (1·5–)3–4 × 1–1·6 mm, persistent, entire, crenulate or dentate. Petals yellow; outer petals 6–8·5(–9·5) × 2–4·5 mm, rhombic to shallowly 3-lobed, concave, the apical lobe greenish dorsally; inner petals without black spots, the lateral lobes *c.* ½ as long as lamina of central lobe; central lobe 1·2 × 1·5 mm, ovate-elliptical, unguiculate. Filaments linear, without black spots. Pollen yellow. Fruit 1–2 mm wide, compressed, erect, more or less arcuate, not thickened at septa; seeds dark brown. $2n = 32$. *Sandy ground near the sea. S.W. Spain, S. Portugal.* Hs Lu.

3. **H. torulosum** Å. E. Dahl, *Pl. Syst. Evol.* **163**: 268 (1989) (*H. littorale* G. C. Clementi, non Wulfen, *H. glaucescens* auct., non Guss.). Plant 5–12(–16) cm, glaucous. Ultimate leaflets of upper primary leaf-segments dilated towards the 3-fid apex, often with 1 to several lateral teeth. Inflorescence decumbent to erect, with 1–7 flowers. Sepals 2–5 × 1–2 mm, persistent, entire, crenulate or dentate, sometimes with a 2-fid or 3-fid apex. Petals lemon-yellow; outer petals (4–)5–8 × 3–6 mm, unguiculate, distinctly 3-lobed, always longer than wide, the lateral lobes involute giving petal a rhombic appearance; inner petals with black spots, the lateral lobes linear, usually not exceeding central lobe; central lobe 1·7–3·3 × 1–2·8 mm, obovate, not unguiculate. Filaments of median stamens narrowly triangular, without black spots. Pollen orange-yellow. Fruit strongly torulose, 2–3·5 mm wide at widest septa; seeds dark brown. $2n = 16$. *Sand-dunes. Coasts of Black Sea and Mediterranean region.* Bu Co Cr Ga Gr It Rm Rs(W, K, E) Si Tu.

4. **H. imberbe** Sm. in Sibth. & Sm., *Fl. Graec. Prodr.* **1**: 107 (1806) (*H. grandiflorum* Bentham, *H. glaucescens* Guss., *H. aequilobum* auct., non Viv.). Plant 1–40 cm, green or glaucous. Ultimate leaflets linear to flabellate, with 1 to several lateral teeth. Inflorescence erect, with 1–18 flowers. Sepals 2·5–6 × 1·3–3 mm, entire, erose or dentate, seldom 2-fid. Petals orange-yellow; outer petals 6·5–13 × 5–13·5 mm, distinctly 3-lobed, usually somewhat longer than wide, the lateral lobes flat, obtuse. Inner petals with black spots, the lateral lobes obovate, obtuse, sometimes exceeding central lobe; central lobe 4–6·5(–9) × 1·7–3·5(–5·5) mm, obovate, fimbriate with truncate base. Filaments of median stamens narrowly triangular, without black spots. Pollen orange-yellow. Fruit erect, only slightly torulose, more or less arcuate, 2–3·5 mm at widest septa; seeds dark brown. $2n = 16$. *Cultivated ground, pastures and waste places. S. Europe.* Al Bl Co Ga Gr Hs It Ju Lu Sa Si.

5. H. pseudograndiflorum Petrović, *Addit. Fl. Agri Nyss.* 25 (1886). Plant 5–40 cm, green or glaucous. Most or all ultimate leaflets lanceolate. Inflorescence decumbent to erect with 1–18(–28) flowers. Sepals 2·5–5 × 1–2·5 mm, erose, entire or dentate. Petals lemon-yellow; outer petals (6·5–)8–11 × (4–)8–10(–11) mm, not or shortly unguiculate, distinctly 3-lobed, not or scarcely longer than wide, the lateral lobes flat; inner petals with or without black spots, the lateral lobes obovate, obtuse, not exceeding central lobe; central lobe 4–6(–7·5) × 1·5–3 mm, obovate, fimbriate with truncate base. Filaments of median stamens narrowly triangular. Pollen yellowish-white. Fruit erect to patent, not torulose, more or less arcuate, 1·5–2·5 mm at widest septa. $2n = 16$. *Usually ruderal. Balkan peninsula.* Bu Gr Ju Tu.

6. H. procumbens L., *Sp. Pl.* 124 (1753) (incl. *H. ponticum* Velen.). Plant up to 40 cm, green or glaucous. Ultimate leaflets linear to lanceolate, with 1 lateral tooth. Inflorescence decumbent to erect, with 1–7(–14) flowers. Sepals 2–7 × 1–2·2 mm, erose, entire or dentate. Petals lemon-yellow or yellowish-orange; outer petals 4–12 × 1·5–11·5 mm, unguiculate, rhombic to distinctly 3-lobed, usually as long as or longer than wide, the lateral lobes flat; inner petals without black spots, the lateral lobes obovate, obtuse, varying in length relative to central lobe; central lobe 2–5 × 1–3 mm, obovate, with truncate base. Filaments of median stamens narrowly ovate. Pollen yellowish-white. Fruit scarcely torulose, more or less arcuate, 1–3·5 mm at widest septa; seeds dark brown. $2n = 16$. *Seashores and sandy habitats. S. Europe.* Bl Bu Co Cr Ga Gr Hs It Ju Lu Sa Si Tu.

A very variable species. Plants from the Kikladhes (Mikonos and Rhinia), with up to 14 strongly scented flowers per inflorescence and slightly larger anthers, have been described as subsp. **fragrantissimum** A. E. Dahl, *Pl. Syst. Evol.* **163**: 254 (1989). They are self-incompatible, and hybrids with subsp. *procumbens* (which is usually self-compatible) show reduced fertility.

Subfam. **Fumarioideae**

Usually glabrous herbs. Sepals 2, free or absent. Inflorescence racemose or cymose, bracteate. Corolla transversely zygomorphic or bisymmetrical. Upper exterior petal spurred or saccate. Stamens 6, united into two groups of 3.

8. **Dicentra** Bernh.

D. A. WEBB AND J. R. AKEROYD

Perennial herbs with ternate leaves and cymose inflorescence. Outer petals alike, saccate at the base. Fruit a 2-valved capsule with several seeds; style persistent. Seeds with an aril in Europe.

Literature: K. R. Stern, *Brittonia* **13**: 1–57 (1961).

Stems leafy; outer petals more than 20 mm, with spathulate, deflexed apex **1. spectabilis**
Stems leafless, or with a single leaf subtending inflorescence; outer petals less than 20 mm, with ovate, spreading apex **2. formosa**

1. D. spectabilis (L.) Lemaire, *Fl. Serres Jard. Eur.* ser. 1, **3**: t. 258 (1847). Stems 30–40 cm, leafy, reddish. Leaves biternate; leaflets often ternatisect; ultimate segments obovate-cuneate, irregularly dentate. Cymes 18–25 cm, raceme-like, nearly horizontal, with 7–12 pendent flowers. Outer petals *c.* 25 × 9 mm, bright reddish-pink, cucullate, with broad, gibbous base and narrow, spathulate, deflexed apex. Inner petals about as long as outer, but not deflexed at apex and therefore projecting beyond them, white, keeled, sharply constricted near the middle; apices coherent. Filaments geniculate at junction of free and fused portions. $2n = 36$. *Cultivated for ornament, naturalized in Czechoslovakia and perhaps elsewhere.* [Cz.] (*E. Asia.*)

2. D. formosa (Haw.) Walpers, *Repert. Bot. Syst.* **1**: 118 (1841). Stems 20–40 cm, usually leafless, arising from slender, fleshy rhizome. Leaves mostly basal, biternate, glaucous beneath; leaflets ternatisect; ultimate segments oblong, deeply and irregularly dentate. Cymes 5–10 cm, compound, 5- to many-flowered, nodding. Outer petals *c.* 16 × 4 mm, pink to pinkish-purple, with ovate, obtuse, spreading apex. Inner petals slightly shorter than outer, the same colour and not projecting beyond them. *Cultivated for ornament, naturalized in shady places in N.W. Europe.* [Br Da Ho.] (*W. North America.*)

D. eximia (Ker-Gawler) Torrey, *Fl. N. York* **1**: 46 (1843), a native of North America, and differing from **2** in its narrower outer petals and more conspicuous inner petals, is cultivated for ornament and may be becoming naturalized in Scotland.

9. **Corydalis** Vent.

A. B. MOWAT AND A. O. CHATER (EDITION 1) REVISED BY A. O. CHATER (EDITION 2)

Glabrous annuals to perennials with pinnate or ternate leaves and racemose inflorescences. Bracts usually conspicuous. Flowers zygomorphic. Sepals absent. Petals 4, the 2 inner similar, oblong, broadened and coherent at apex; the 2 outer dissimilar, the lower usually somewhat saccate at base, expanded into broad limb at apex, the upper with distinct spur at base and a more or less expanded limb at apex. Style persistent; stigma flattened, with marginal papillae. Fruit an oblong, 2-valved capsule with many seeds. Seeds with conspicuous aril.

Literature: M. Ryberg, *Acta Horti Berg.* **16(7)**: 233–240 (1953); **17(5)**: 115–175 (1955). K. v. Poellnitz, *Feddes Repert.* **44**: 154–157 (1938); **45**: 96–112 (1938). F. Fedde in Engler & Prantl, *Natürl. Pflanzenfam.* ed. 2, **17B**: 123 (1936). M. Lidén, *Op. Bot.* **88**: 21–29 (1986); *Notes Roy. Bot. Gard. Edinb.* **45**: 349–363 (1988).

1 Annual or biennial, or perennial with rhizomatous stock; stems leafy
2 Racemes 5- to 8-flowered, lax **1. capnoides**
2 Racemes 20- to 30-flowered, capitate **2. nobilis**
1 Perennial with tuberous stock; stems, or each branch of stem, with 1 or 2(3) leaves
3 Stem without a conspicuous scale below lowest leaf
4 Racemes 1- to 3-flowered; leaves opposite **3. uniflora**
4 Racemes with more than 5 flowers; leaves alternate
5 Stems at least 10 cm; corolla 20–30 mm **4. cava**
5 Stems usually less than 10 cm; corolla 17–21 mm **5. parnassica**
3 Stem with a conspicuous scale below lowest leaf
6 All bracts entire **8. intermedia**
6 At least the lower bracts dissected or 3-fid
7 Lower pedicels less than 5 mm **9. pumila**
7 Lower pedicels more than 5 mm

8 Ripe fruit 2–3 times as long as pedicel
9 Flowers purple; ultimate leaf-segments narrowly oblong **6. paczoskii**
9 Flowers white or cream, rarely suffused with pale purple; ultimate leaf-segments linear-lanceolate **7. angustifolia**
8 Ripe fruit about as long as pedicel
10 Fruit elliptical **10. solida**
10 Fruit linear **11. integra**

(A) Annual, biennial, or rhizomatous perennial herbs. Stems usually branched, with many leaves.

1. C. capnoides (L.) Pers., *Syn. Pl.* **2**: 270 (1806) (*C. gebleri* Ledeb.; incl. *C. alba* Mansfeld). Robust, ascending or erect annual or biennial up to 40 cm. Leaves 2-ternate, all petiolate; ultimate segments 3-fid or variously dissected. Racemes 5- to 8-flowered, lax. Lower bracts compound, like the upper leaves; upper bracts entire. Corolla 11–16 mm, cream or white, yellow at apex; spur 5–7 mm, slender; style persistent, straight. Fruit 20–30 mm, slender. $2n = 16$. *E. Alps (local); Carpathians; N.E. Russia.* Au Cz *It Po Rm Rs(N, ?C) [Fe].

2. C. nobilis (L.) Pers., *Syn. Pl.* **2**: 269 (1806). Robust, rhizomatous perennial 30–50 cm. Leaves 2- or 3-pinnate, glaucous. Racemes 20- to 30-flowered, capitate. Bracts cuneate, dissected, much longer than pedicels. Lower pedicels 5–15 mm. Corolla *c.* 20 mm, yellow or white; spur *c.* 7 mm. Fruit 10–20 mm. Seeds black, shiny. $2n = 16$. *Cultivated for ornament, and naturalized in Finland and Sweden.* [Fe Su.] (*C. Asia.*)

(B) Perennials with tuberous stock. Stems simple, or with 1 or 2 branches, with 1–3 leaves on each stem or branch. Racemes terminal.

3. C. uniflora (Sieber) Nyman, *Syll.* 185 (1855) (*C. rutifolia* (Sm.) DC. subsp. *uniflora* (Sieber) Cullen & P. H. Davis). Stem 5–15 cm, erect, without a scale below leaves; tuber solid. Leaves opposite, usually 2-ternate, glaucous. Raceme 1- to 3-flowered. Bracts broadly ovate, entire. Corolla 20–25 mm, whitish to pale pink, veined with purple; spur 10–13 mm. *Stony places above* 1500 *m.* ● *Kriti.* Cr.

4. C. cava (L.) Schweigger & Koerte, *Fl. Erlang.* **2**: 44 (1811) (*C. bulbosa* auct., non DC.). Stem 10–35 cm, erect, without a scale below lowest leaf. Leaves 2-ternate, usually alternate. Raceme 10- to 20-flowered, more or less dense. Bracts ovate, entire. Corolla 20–30 mm; spur curved at apex; ovary at anthesis gradually narrowed into straight style. Fruit 20–25 mm, 3–4 times as long as pedicels, pendent when ripe. *Much of Europe, but absent from the islands and rare in the west and north.* Al Au Be Bu Cz Da Ga Ge Gr He Hs Hu It Ju Lu Po Rm Rs(B, C, W, K, E) Su [Ho].

(a) Subsp. **cava** (incl. subsp. *blanda* (Schott) Chater): Tuber hollow. Leaves 2- or 3-ternate, glaucous or green; median leaflets long-cuneate at base, usually more or less sessile; segments dentate. Corolla purplish or sometimes white. $2n = 16$. *Almost throughout the range of the species.*

(b) Subsp. **marschalliana** (Pallas) Chater, *Feddes Repert.* **69**: 56 (1964) (*C. marschalliana* (Pallas) Pers.): Tuber often solid. Leaves 2-ternate, glaucous, rarely green; median leaflets shortly stalked; segments more or less entire. Corolla whitish, cream or yellow, rarely purplish. $2n = 32$. *S.E. Europe.*

5. C. parnassica Orph. & Heldr. ex Boiss., *Diagn. Pl. Or. Nov.* **2(6)**: 9 (1859). Like **4** but stems 3–7 cm; leaves 3–4 cm, fleshy, glaucous, the segments lanceolate; racemes 5- to 12-flowered; corolla 17–21 mm; fruit 15–20 mm. $2n = 16$. *Calcareous screes above* 1800 *m.* ● *S. & C. Greece.* Gr.

6. C. paczoskii N. Busch, *Trudy Tifliss. Bot. Sada* **9** suppl.: 55 (1903). Tuber solid. Stem 10–20 cm, erect, simple or branched, with an ovate scale below lowest leaf. Leaves usually 3-ternate; segments narrowly oblong. Racemes 4- to 10-flowered, lax. Lower bracts 3- to 8-fid, the upper mostly 3-fid. Corolla 15–25 mm, purple; spur straight; style straight. Fruit 17–25 mm, 2–3 times as long as pedicels. $2n = 16$. ● *S. Ukraine.* Rs(W, K).

7. C. angustifolia (Bieb.) DC., *Reg. Veg. Syst. Nat.* **2**: 120 (1821). Like **6** but leaf-segments linear-lanceolate; lower bracts always 3-fid, the upper 3-fid or entire; corolla 20–30 mm, yellow or cream. *Naturalized near St. Petersburg.* [Rs(C).] (*Caucasus.*)

C. bracteata (Stephan) Pers., *Syn. Pl.* **2**: 269 (1807), native in Siberia, is also reported as an alien and perhaps naturalized near St. Petersburg. It is like **6** but is more robust with broader, yellow petals and less deeply divided bracts.

8. C. intermedia (L.) Mérat, *Nouv. Fl. Env. Paris* 272 (1812) (*C. fabacea* Pers.). Tuber solid. Stem 7–20 cm, erect, with an ovate scale below lowest leaf. Leaves 2-ternate with variously dissected segments. Racemes 1- to 8-flowered, dense. Bracts ovate, entire. Pedicels *c.* 5 mm. Corolla 10–17 mm, purple, rarely white; spur straight or curved; ovary at anthesis gradually narrowed into straight, persistent style. Fruit 15–20 mm, ovate-lanceolate. $2n = 16, 20, 40$. *Woods.* ● *N. & C. Europe, extending to C. Spain, S. Italy and the middle Volga.* Au Cz Da Fe Ga Ge He Hs Hu It Ju No Po Rm Rs(N, B, C, W) ?Si Su.

Robust plants, similar to **8** but with 5- to 15-flowered racemes, broader outer petals and long-attenuated fruit, have been described from Sweden as **C. gotlandica** Lidén, *Nordic Jour. Bot.* **11**: 132 (1991).

9. C. pumila (Host) Reichenb., *Fl. Germ. Excurs.* 698 (1832) (*C. fabacea* var. *digitata* Gren. & Godron). Tuber solid. Stem 7–20 cm, erect, with an ovate scale near base. Leaves 2-ternate. Racemes (3)4- to 12-flowered, dense. Bracts cuneate, digitate. Pedicels usually less than 5 mm. Corolla 12–17 mm, purplish; spur almost straight; ovary at anthesis gradually narrowed into straight, persistent style. Fruit 15–20 mm. $2n = 16$. *Woods.* ● *From S. Fennoscandia to N. Greece.* Au Co Cz Da Ge Gr Hu It Ju No Po Rm Sa Su.

10. C. solida (L.) Clairv., *Man. Herb. Suisse* 371 (1811) (*C. bulbosa* (L.) DC., *C. halleri* Willd., *C. tenuis* Schott, Nyman & Kotschy; incl. *C. tenella* Ledeb. ex Nordm.). Tuber solid. Stem 10–30 cm, erect, with ovate scale near base. Leaves 2(3)-ternate. Racemes (4–)10- to 20-flowered, usually dense. Lower bracts always lobed. Pedicels 5–15 mm. Corolla 15–25(–30) mm; spur usually slightly curved; style often geniculate. Fruit 15–20 mm, broadly elliptical, pendent when ripe. *Most of Europe except the north-west and most of the islands.* Al Au Be Bu Co Cz Fe Ga Ge Gr He Ho Hs Hu It Ju Po Rm Rs(N, B, C, W, E) Si Su Tu [Br Da No].

Very variable. Four subspecies are given below, but the treatment is provisional, especially because of the occasional close similarity between specimens of subspp. **(a)** and **(b)**, and because of the very great variability of subsp. **(a)**.

1 Bracts entire or divided into entire segments
2 Leaves 2-ternate; upper bracts often entire; corolla yellow or yellowish-purple (b) subsp. **slivenensis**
2 Leaves 3-ternate; upper bracts ±divided; corolla usually purple or pink (a) subsp. **solida**
1 Bracts divided into lobate or dentate segments
3 Outer petals obtuse; style very short, straight; pedicels 4–8 mm (d) subsp. **densiflora**
3 Outer petals emarginate; style long, usually geniculate at base; pedicels 5–15 mm (c) subsp. **incisa**

(a) Subsp. **solida** (incl. subsp. *laxa* (Fries) Nordstedt, *C. laxa* Fries): Stem (7–)10–25(–30) cm. Leaves (2)3-ternate; segments obovate to oblanceolate, obtuse. Racemes short, 5–20(–25)-flowered, rarely lax. Bracts shallowly to deeply palmatisect or distally dentate. Corolla 18–23 mm, purple, rarely white, red, pink or purplish-blue; lower petal usually prominently gibbous at base, broadly emarginate. Fruit 11–20 mm. $2n = 16, 24, 32$. *Deciduous forests. Almost throughout the range of the species.*

(b) Subsp. **slivenensis** (Velen.) Hayek, *Prodr. Fl. Penins. Balcan.* **1**: 364 (1925): Stem 6–15(–20) cm. Leaves 2-ternate with broadly obovate segments. Racemes (3–)5–10(–20)-flowered. Bracts about as long as or longer than pedicels, shallowly toothed or lobed, or entire. Corolla 20–25 mm, pale yellow to pale purple, the lower petal broadly emarginate. Fruit 10–15 mm. ● *Bulgaria, Jugoslavia.*

(c) Subsp. **incisa** Lidén, *Notes Roy. Bot. Gard. Edinb.* **45**: 351 (1989) (*C. solida* subsp. *densiflora* (C. Presl) Hayek pro parte): Stem 6–15(–20) cm. Leaves 3-ternate with lobed leaflets; ultimate segments narrowly oblong to linear-lanceolate. Racemes (3–)8- to 22-flowered, longer and narrower than in the other subspecies. Bracts large, deeply divided into narrow lobes, these again lobed or dentate into acute segments. Corolla 21–27 mm, usually pale pink; lower petal emarginate, often with a small apiculus in the sinus. $2n = 16$. ● *Mountains of Balkan peninsula.*

(d) Subsp. **densiflora** (C. Presl) Hayek, *Prodr. Fl. Penins. Balcan.* **1**: 364 (1925) (*C. densiflora* C. Presl): Leaves often 2-ternate with leaflets very broad and apically incised or deeply divided into lanceolate segments. Racemes very dense. Bracts short and broad; lobes apically divided into short lobules. Pedicels 4–8 mm. Corolla 16–21 mm; lower petal 10–13 mm, broadly obtuse, cuneate. Fruit short and broad with very short, straight style. *S. Italy, Sicilia.*

11. C. integra W. Barbey & Major in Stefani, Major & W. Barbey, *Samos* 30 (1892) (*C. wettsteinii* Adamović, *C. solida* subsp. *wettsteinii* (Adamović) Hayek): Like **10** but stems not more than 20 cm; fruit 13–30 mm, linear. $2n = 16$. *Aegean region; S. Bulgaria.* Bu Gr.

C. zetterlundii Lidén, *Willdenowia* **21**: 178 (1991), has been described from Makedonija. It is said by its author to differ from **10** by its bracts, fruits and seeds, but to be most closely related to **11**, differing in its bracts, pedicels and outer petals.

10. Capnoides Miller

M. LIDÉN

Like *Corydalis* but always annual or biennial; inflorescence a cyme, with inconspicuous bracts; fruit narrowly linear; seeds with elaiosome.

1. C. sempervirens (L.) Borkh., *Arch. Bot. (Roemer)* **1**: 44 (1797) (*Corydalis sempervirens* (L.) Pers.). Stems 10–100 cm, erect, branched above, glaucous. Leaves 1- to 3-pinnate, glaucous, petiolate, the upper subsessile. Cymes *c.* 10-flowered, lax, or inflorescence sometimes almost paniculate. Bracts 3–5 mm, much shorter than pedicels, lanceolate, sometimes toothed. Corolla 11–18 mm, white to pinkish-purple, yellow at apex; spur 3–4 mm, saccate; style persistent. Fruit 30–50 mm, erect, purple. $2n = 16$. *Naturalized in Norway.* [No.] (*North America.*)

11. Pseudofumaria Medicus

M. LIDÉN

Glabrous perennials, densely branched with leafy angular stems and 2- or 3-pinnate leaves. Inflorescence racemose with short, linear scarious bracts. Flowers zygomorphic; outer petals broadly winged apically; upper petal with short, blunt spur; style deciduous. Fruit 3- to 13-seeded; seeds with elaiosome.

Flowers yellow; fruit pendent **1. lutea**
Flowers white or cream tipped yellow; fruit erect **2. alba**

1. P. lutea (L.) Borkh., *Arch. Bot. (Roemer)* **1**: 45 (1797) (*Corydalis lutea* (L.) DC.). Stems 10–40 cm, erect. Leaves green above, glaucous beneath. Racemes (4–)6- to 20-flowered, dense at first, later elongating. Bracts oblong-lanceolate, entire, much shorter than pedicels. Corolla 14–20 mm, yellow; spur 2–4 mm. Fruit *c.* 10 mm, pendent; seeds 1·5 mm, smooth. $2n = 64$. *Rocks and walls; usually on limestone.* ● *Southern foothills of S.W. & C. Alps; cultivated for ornament and widely naturalized elsewhere.* He It [Au Be Br Cz Da Ga Ge Hb Ho No Po Rs(B) Su].

2. P. alba (Miller) Lidén, *Op. Bot.* **88**: 32 (1986) (*Corydalis ochroleuca* Koch). Like **1** but leaves 3-pinnate, usually glaucous on both surfaces; petiole narrowly but distinctly winged; corolla cream or white, the inner petals tipped yellow, fruit erect. *Italy and W. part of Balkan peninsula; naturalized in W. & C. Europe.* Al ?Gr It Ju [Be Ga Ge Ho].

1 Bracts at least $\frac{1}{3}$ as long as pedicel; seeds smooth, shiny (c) subsp. **leiosperma**
1 Bracts less than $\frac{1}{3}$ as long as pedicel; seeds tuberculate, dull
2 Plant usually more than 10 cm; leaves thin (a) subsp. **alba**
2 Plant not more than 10 cm; leaves fleshy (b) subsp. **acaulis**

(a) Subsp. **alba**: Stems up to 40 cm. Leaflets thin, acute. Bracts less than $\frac{1}{3}$ as long as pedicels. Corolla 14–17 mm. Fruit 3- to 11-seeded; seeds 1·6 mm, tuberculate, dull. $2n = 32$. *N.W. part of Balkan peninsula, N. & C. Italy; widely naturalized elsewhere.*

(b) Subsp. **acaulis** (Wulfen) Lidén, *loc. cit.* (1986) (*Corydalis acaulis* (Wulfen) Pers.). Like subsp. (a) but stems 5–10 cm, glaucous. Leaves fleshy, fragile, with obtuse leaflets. Corolla 13–16 mm. Fruit 3- to 7-seeded. *W. Jugoslavia.*

(c) Subsp. **leiosperma** (Conrath) Lidén, *loc. cit.* (1986) (*Corydalis ochroleuca* subsp. *leiosperma* (Conrath) Hayek, *C. leiosperma* Conrath). Like subsp. (a) but stems 10–20 cm; bracts at least $\frac{1}{3}$ as long as pedicels; corolla 13–16 mm; fruit 8- to 13-seeded; seeds 1·2 mm, smooth, shiny. $2n = 32$. *W. part of Balkan peninsula.*

12. Sarcocapnos DC.

V. H. HEYWOOD

Caespitose, perennial herbs with entire or 1- to 3-ternatisect or pinnatisect leaves. Flowers in racemes, zygomorphic, white, yellow or pale pink, in short corymb-like racemes with long, slender pedicels. Upper petal spurred or gibbous; style caducous. Fruit ovate-elliptical, compressed, indehiscent, usually 2-seeded; valves 3-veined.

Literature: V. H. Heywood, *Bull. Brit. Mus. (Bot.)* **1(4)**: 90–92 (1954).

1 Flowers 5–6 mm; upper petal gibbous but not spurred **3. baetica**
1 Flowers 12–20 mm; upper petal spurred
2 Flowers 12–17 mm; sepals ovate; leaves 2- or 3-ternatisect or pinnatisect **1. enneaphylla**
2 Flowers 15–20 mm; sepals ovate-lanceolate; leaves 1- or 2-ternatisect or ± simple **2. crassifolia**

1. S. enneaphylla (L.) DC., *Reg. Veg. Syst. Nat.* **2**: 129 (1821). Stems 5–15 cm, much-branched, woody at base. Leaves 2- or 3-ternatisect or pinnatisect, long-petiolate; the segments ovate-rounded, obtuse, often apiculate, the terminal often reniform-cordate. Pedicels slender. Flowers 12–17 mm, white to yellowish, purple-tipped; upper petal spurred, the spur much shorter than rest of corolla. Sepals ovate. $2n = c.$ 32. *Shady, calcareous rock-crevices. C. & E. Spain; E. Pyrenees.* Ga Hs.

2. S. crassifolia (Desf.) DC., *op. cit.* **2**: 130 (1821) (incl. *S. speciosa* Boiss.). Like **1** in habit but more robust; leaves usually more fleshy, 1- or 2-ternatisect or more or less simple; flowers 15–20 mm, usually pinkish; sepals ovate-lanceolate. $2n = c.$ 32. *Rock-crevices in the mountains. S. & S.E. Spain.* Hs.

The European plants have been distinguished as subsp. **speciosa** (Boiss.) Rouy, *Bull. Soc. Bot. Fr.* **31**: 53 (1884). They differ from plants from N. Africa in their laxer habit and larger flowers and fruits.

S. saetabensis Mateo & Figuerola, *Fl. Analit. Prov. Valencia* 371 (1987), from E. Spain (Valencia prov.), is more or less intermediate in morphology between **1** and **2** and may represent stable hybrid plants.

3. S. baetica (Boiss. & Reuter) Nyman, *Consp.* 26 (1878). Like **1** but leaves simple or ternatisect, glaucous; flowers 5–6 mm, white to yellowish; upper petal not spurred. *Rock-crevices; usually calcicole.* ● *S. Spain.* Hs.

(a) Subsp. **baetica**: Leaves ternatisect, thin; segments 2- to 5(–7)-partite, cordate. $2n = 32$. *Almost throughout the range of the species.*

(b) Subsp. **integrifolia** (Boiss.) Nyman, *Consp. Suppl.* **2**: 17 (1889) (*S. integrifolia* (Boiss.) Cuatrec., *S. baetica* var. *integrifolia* (Boiss.) Lange): Leaves simple, more or less ovate, somewhat fleshy, cuneate to subcordate. *La Sagra; Sierra de Mágina.*

13. Fumaria L.

P. D. SELL

Annuals with erect, diffuse or climbing stems. Leaves cauline, 2- to 4-pinnatisect. Inflorescence racemose, bracteate, with short pedicels. Sepals 2, lateral. Corolla zygomorphic, consisting of a spurred upper petal, 2 inner petals, and a lower one. Fruit 1-seeded, indehiscent, showing 2 apical pits in the mesocarp when dry.

In W. Europe mostly weeds of arable land, but in S. & E. Europe found also in rocky, sandy and grassy places.

In this account the measurement of the corolla is that of the largest flowers; cleistogamous or more or less depauperate corollas produced by plants grown in an unfavourable environment are not considered.

Orientation of the wings of the upper petals and the margins of the lower petals can only be determined with certainty when the plant is fresh.

Literature: H. W. Pugsley, *Jour. Bot. (London)* **50**, Suppl. 1: 1–76 (1912); *Jour. Linn. Soc. London (Bot.)* **44**: 233–354 (1919); **47**: 427–469 (1927); **49**: 93–113 (1932); **49**: 517–529 (1934); **50**: 541–559 (1937). A. Soler, *Lagascalia* **11**: 141–228 (1983). M. Lidén, *Op. Bot.* **88**: 5–133 (1986).

1 Corolla not more than 9 mm
2 Sepals 0·2–1·5 × 0·2–0·7 mm, up to ⅕ as long as corolla
3 Bracts equalling or longer than pedicels; corolla usually white **39. parviflora**
3 Bracts shorter than pedicels; corolla pink
4 Pedicels 4 mm; bracts not more than ⅓ as long as pedicels **34. schleicheri**
4 Pedicels up to 3 mm; bracts more than ⅓ as long as pedicels
5 Fruit 2–2·2 × 2–2·2 mm; bracts ¾ as long as pedicels
6 Spur of corolla long and straight **36. vaillantii**
6 Spur of corolla short and ascending **38. pugsleyana**
5 Fruit 1·7–2 × 1·7–2 mm; bracts ½ as long as pedicels
7 Raceme 20- to 35-flowered; upper petals not dorsally compressed, with wings turned upwards **35. microcarpa**
7 Raceme 6- to 15-flowered; upper petals dorsally compressed, with wings patent **37. schrammii**
2 Sepals 1·5–3·5 × 0·7–3 mm, more than ¼ and often more than ⅓ as long as corolla
8 Sepals 1·5–2 × 0·7–1 mm
9 Corolla 7–9 mm **29. officinalis**
9 Corolla 4·5–7 mm
10 Fruit subglobose, rounded-obtuse or emarginate at apex
11 Corolla *c.* 4·5 mm **27. bracteosa**
11 Corolla 6–7 mm **32. caroliana**
10 Fruit ovoid, subacute at apex
12 Corolla 4–5 mm; sepals *c.* 1·5 × 0·7 mm **31. jankae**
12 Corolla 6–7 mm; sepals 1·5–2 × 1–1·2 mm **33. segetalis**
8 Sepals 2–3·5 × 1–3 mm
13 Bracts shorter than fruiting pedicels
14 Fruit 2–2·5 × 1·7–2 mm, subglobose or ovoid **28. rostellata**
14 Fruit 2 × 2·5–3 mm, wider than long
15 Raceme longer than peduncle; sepals 2·5–3·5 × 1–1·5 mm; upper petal obtuse; apical pits of fruit without black spots **29. officinalis**
15 Raceme scarcely longer than peduncle; sepals 2 × 1·5 mm; upper petal subacute; apical pits of fruit black-spotted **30. ragusina**
13 Bracts as long as or longer than fruiting pedicels
16 Fruiting pedicels arcuate-recurved
17 Fruit nearly smooth when dry **23. kralikii**
17 Fruit rugose when dry **21. petteri**
16 Fruiting pedicels erect or patent
18 Sepals 2–2·2 mm **25. faurei**
18 Sepals 2·5–3·5 mm
19 Corolla 7·5–9 mm **24. mirabilis**
19 Corolla 6–7 mm **26. densiflora**
1 Corolla at least 9 mm
20 Sepals more than 3 mm wide, wider than corolla
21 Fruiting pedicels arcuate-recurved **9. capreolata**

21 Fruiting pedicels ±erect **14. macrosepala**
20 Sepals not more than 3 mm wide, usually narrower than corolla
22 Fruit more than 3 mm long and wide
23 Sepals 3·5–5·5 × 1–2 mm **1. agraria**
23 Sepals 2·5–3 × 0·5–1 mm **8. macrocarpa**
22 Fruit not more than 3 mm long and wide
24 Sepals 1·5–2 mm
25 Corolla white or pale pink; bracts $\frac{1}{3}$ to $\frac{1}{2}$ as long as fruiting pedicels **7. amarysia**
25 Corolla purplish-pink; bracts $\frac{1}{2}$ to as long as fruiting pedicels **30. ragusina**
24 Sepals more than 2 mm
26 Fruit ±smooth when dry
27 Raceme shorter than or equalling peduncle
28 Sepals obtuse to subacute; pedicels often arcuate-recurved
29 Corolla usually white or cream (rarely flushed with pink, or deep red); upper petal laterally compressed; wings not exceeding keel **9. capreolata**
29 Corolla purplish; upper petal not laterally compressed; wings exceeding keel **13. purpurea**
28 Sepals acute to acuminate; pedicels erect or erecto-patent
30 Corolla 12–14 mm, pinkish-white **18. sepium**
30 Corolla 9–12 mm, pink **19. muralis**
27 Raceme longer than peduncle
31 Sepals 2·5–3 mm
32 Fruit 2–2·5 × 2–2·5 mm **16. bastardii**
32 Fruit *c.* 2 × 1·7–2 mm **21. petteri**
31 Sepals 3–4·5 mm
33 Sepals subentire
34 Corolla more than 11 mm **17. martinii**
34 Corolla less than 11 mm
35 Sepals *c.* 3 × 1·5–2 mm **19. muralis**
35 Sepals 3·5–4·5 × 2–2·5 mm **20. reuteri**
33 Sepals dentate
36 Raceme at least as long as peduncle **19. muralis**
36 Raceme sessile or with very short peduncle **21. petteri**
26 Fruit rugose when dry
37 Pedicels arcuate-recurved
38 Flowers creamy-white, sometimes flushed red
39 Fruit 2–2·7 × 2–2·7 mm, tuberculate-rugose, keeled, emarginate **10. flabellata**
39 Fruit 2–2·2 × 2–2·2 mm, finely rugose, scarcely keeled, obtuse or subacute at apex **11. munbyi**
38 Flowers reddish, pink or purple
40 Sepals 2·5–3 mm
41 Fruit more than 1·7 mm wide **21. petteri**
41 Fruit less than 1·7 mm wide **22. calcarata**
40 Sepals 3–6·5 mm
42 Sepals 4·5–6·5 mm, entire; corolla purple or whitish-purple **13. purpurea**
42 Sepals 2–4 mm, dentate; corolla white or reddish
43 Corolla 12–14 mm, white or reddish **11. munbyi**
43 Corolla 10–12 mm, pink **12. melillaica**
37 Pedicels erect or patent
44 Sepals 2–3 × 1–2 mm
45 Corolla white, sometimes becoming tinged with pink
46 Fruit 2·5–3 × 2·5–3 mm **6. judaica**
46 Fruit 2–2·2 × 1·5–1·7 mm **15. bicolor**
45 Corolla pink
47 Corolla 9–13 mm
48 Raceme usually shorter than peduncle; apex of fruit subacute **15. bicolor**
48 Raceme longer than peduncle; apex of fruit obtuse **16. bastardii**
47 Corolla 7·5–9 mm **24. mirabilis**
44 Sepals 3–5 × 1–3 mm
49 Corolla white or pinkish-white
50 Fruit slightly rugose when dry, with obtuse apex **18. sepium**
50 Fruit distinctly rugose when dry, with emarginate beak
51 Sepals 1–2 mm wide; wings of upper petals not blotched with dark purple **1. agraria**
51 Sepals 2–3 mm wide; wings of upper petal blotched with dark purple, with a white margin **2. occidentalis**
49 Corolla pink
52 Corolla 9–10·5 mm **21. petteri**
52 Corolla 10·5–14 mm
53 Wings of upper petal pale pink; sepals subentire or denticulate **3. rupestris**
53 Wings of upper petal dark purple or red; sepals dentate
54 Sepals 4–5 × 2·5–3 mm **19. muralis**
54 Sepals 3–4 × 1·5–2 mm
55 Corolla 12–14 mm; upper petal obtuse; fruit subglobose to obovoid **4. barnolae**
55 Corolla 10–12 mm; upper petal subacute; fruit subglobose-quadrate **5. gaillardotii**

Sect. CAPREOLATAE Hammar. Leaf-segments flat and relatively broad, from broadly ovate to oblong or lanceolate. Corolla at least 9 mm (except **21(b)**), the wings of the upper petal turned upwards, the lower petal not or obscurely spathulate.

1. F. agraria Lag., *Gen. Sp. Nov.* 21 (1816). Raceme 14- to 25(–30)-flowered, longer than peduncle. Bracts about $\frac{2}{3}$ as long to as long as the patent fruiting pedicels. Sepals 3·5–5·5 × 1–2 mm, dentate to subentire. Corolla 12–16 mm, white or pinkish-white, only the inner petals with dark purple apices; lower petal with patent margin. Fruit 2·5–3·5 × 2·5–3·5 mm, ovoid, strongly keeled, rugose when dry; apex with emarginate beak. $2n = 80$. *Mediterranean region, S. Portugal.* Bl ?Co *Ga *Gr Hs It ?Ju Lu Sa Si.

2. F. occidentalis Pugsley, *Jour. Bot.* (*London*) **42**: 218 (1904). Raceme 12- to 20-flowered, as long as peduncle. Bracts shorter than or nearly as long as the suberect fruiting pedicels. Sepals 4–5·5 × 2–3 mm, incise-dentate. Corolla 12–14 mm, pinkish-white; apex of inner petals dark purplish; wings of upper petal dark purplish with white margin; lower petal with patent or slightly deflexed margin. Fruit *c.* 3 × 3 mm, subglobose, distinctly keeled, tuberculate-rugose when dry; apex with short, emarginate beak. $2n = 112$. ● *S. England.* Br.

3. F. rupestris Boiss. & Reuter, *Pugillus* 4 (1852) (*F. arundana* Boiss. ex Willk. & Lange). Raceme 6- to 15(–20)-flowered, longer than the rather short peduncle. Bracts equalling the suberect fruiting pedicels. Sepals 3–5·5 × 1–2 mm, narrowly lanceolate, subentire or slightly dentate. Corolla 10–14 mm, pale pink; apex of inner petals dark purple; margin of lower petal commonly almost absent, rarely narrow and patent. Fruit 2·5–2·75 × 2–2·5 mm, subglobose-ovoid, slightly keeled, tuberculate-rugose when dry; apex more or less acute, obscurely apiculate. $2n = 64$. *S. Spain*. Hs.

4. F. barnolae Sennen & Pau, *Treb. Inst. Catalana Hist. Nat.* **3**: 63 (1917) (*F. bella* P. D. Sell, *F. major* Badaro, non Roth, *F. agraria* sensu Coste, non Lag.). Racemes 10- to 25-flowered, the upper ones very often shorter than peduncle. Bracts more or less equalling the suberect fruiting pedicels. Sepals 3–3·5 × 1·5–2 mm, irregularly incise-dentate. Corolla 12–14 mm, pink; apex of inner petals and wings of upper one dark purple; lower petal with broad, patent margin. Fruit 2·5–3 × 2·5–3 mm, subglobose to obovoid, slightly keeled, densely tuberculate-rugose when dry; apex very obtuse or subtruncate. $2n = 80$. *Mediterranean region*. Bl Co Ga Hs It *Ju Sa.

5. F. gaillardotii Boiss., *Fl. Or.* **1**: 139 (1867). Like **4** but sepals 3–4 × *c.* 2 mm; corolla 10–12(–13) mm, paler, the upper petal with narrower wings and larger spur; fruiting pedicels short and thick; fruit subglobose-quadrate and more markedly keeled. $2n = 112$. *Mediterranean region*. Al Bl Cr Gr Hs It Ju Sa Si.

6. F. judaica Boiss., *Diagn. Pl. Or. Nov.* **2(8)**: 15 (1849). Raceme 8- to 29-flowered, often slightly longer than peduncle. Bracts $\frac{1}{3}$–$\frac{1}{2}$ as long as the suberect fruiting pedicels. Sepals 2–3 × 1–1·5 mm, more or less dentate. Corolla 9–13 mm, white, rarely becoming tinged with pink; apex of inner petals dark purple; lower petal with rather broad, patent margin. Fruit 2·5–3 × 2·5–3 mm, subglobose, obscurely keeled, tuberculate-rugose when dry, often with black-spotted apical pits; apex very obtuse or slightly emarginate. *Mediterranean region, westwards to Sicilia*. Al Cr Gr It Ju ?Si.

(a) Subsp. **judaica**: Raceme 8- to 16(–29)-flowered. Sepals 2–3 × 1–1·5 mm. Corolla 9–12 mm, white, not rostellate and often straight. Fruit 2·5–3 × 2·5–3 mm. $2n = 48$. *Almost throughout the range of the species*.

(b) Subsp. **insignis** (Pugsley) Lidén, *Op. Bot.* **88**: 48 (1986). Raceme up to 12-flowered. Sepals 2–3 × 1–1·5 mm. Corolla 12–13 mm, white, curved upwards and rostellate at apex. Fruit 2·7–3 × 2·5–2·7 mm. ● *W. Jugoslavia*.

7. F. amarysia Boiss. & Heldr. in Boiss., *Fl. Or.* **1**: 138 (1867) (*F. judaica* subsp. *amarysia* (Boiss. & Heldr.) Lidén). Like **6** but raceme up to 20-flowered; sepals 1·5–2 × *c.* 1 mm; corolla 9–11 mm, white or pale pink, not rostellate and often straight; fruit 2–2·5 × 2–2·5 mm. $2n = 48$. ● *S. Greece*. Gr.

8. F. macrocarpa Parl., *Pl. Nov.* 5 (1842). Raceme 4- to 10(–15)-flowered, about as long as or shorter than peduncle. Bracts $\frac{1}{2}$ as long as to longer than the patent fruiting pedicels. Sepals (1–)2·5–3 × 0·5–1 mm, strongly dentate. Corolla 9–11 mm, white or pale pink without any dark purple colouring or on adaxial side only; lower petal with patent margin. Fruit 3–4 × 3–4 mm, subglobose, obscurely keeled, strongly rugose when dry; apex very obtuse. $2n = 16$. *Mediterranean region, westwards to Sicilia*. Cr Gr Ju Si.

9. F. capreolata L., *Sp. Pl.* 701 (1753). Raceme up to 25(–35)-flowered, shorter than peduncle. Fruiting pedicels usually rigidly arcuate-recurved. Sepals more or less dentate. Corolla 10–14 mm, creamy-white or pinkish, rarely deep red; wings of upper petal and apex of inner petals blackish-red; lower petal with very narrow, erect margin. Fruit usually smooth, obscurely keeled. *S., W. & S.C. Europe*. Al Be Bl Br Co Cr Ga Gr Hb He Hs It Ju Lu Sa Si Tu [Az Cz Ge Ho Po].

(a) Subsp. **capreolata**: Bracts usually shorter than fruiting pedicels. Sepals 4–6 × 2–4 mm. Fruit 2 × 2 mm or smaller, subglobose; apex very obtuse but not truncate. $2n = 64$. *Throughout the range of the species, except Britain and Ireland*.

(b) Subsp. **babingtonii** (Pugsley) P. D. Sell, *Feddes Repert.* **68**: 176 (1963) (*F. capreolata* var. *babingtonii* Pugsley). Bracts more or less equalling fruiting pedicels. Sepals frequently narrower. Upper petals more acute. Fruit 2·5 × 2·5 mm, often truncate at apex and more or less rectangular in outline, smooth or somewhat rugulose. $2n = 64$. ● *Britain and Ireland*.

10. F. flabellata Gaspar., *Rendic. Accad. Sci. Fis. Mat. (Napoli)* **1**: 51 (1842). Like **9** but raceme 10- to 30-flowered, at first as long as, then longer than, peduncle; sepals 3–5 × 1·5–2·5(–3) mm; lower petal with wide, patent margin; fruit 2–2·7 × 2–2·7 mm, keeled, densely tuberculate-rugose when dry; apex emarginate. $2n = 64$. *C. & E. Mediterranean region*. Al Bl Co Gr It Ju Sa Si.

11. F. munbyi Boiss. & Reuter, *Pugillus* 5 (1852). Like **9** but raceme up to 25-flowered; sepals 2–4 × 1·5–2·5 mm; corolla 12–14 mm, white, often flushed purplish-red; lower petal with narrow, suberect margin; fruit 2–2·2 × 2–2·2 mm, scarcely keeled, finely rugose; apex obtuse or subacute. $2n = c.\ 72(–75)$. *E. Spain (Islas Columbretes)*. Hs. (*N.W. Africa*.)

12. F. melillaica Pugsley, *Jour. Linn. Soc. London (Bot.)* **50**: 547 (1937). Like **9** but sepals 3·5–4 × 2·5–2·7 mm, deeply incise-dentate; corolla 10–12 mm, pink; fruit 2–2·5 × 2–2·5 mm, rugose. $2n =$ 72–75. *S.E. Spain*. Hs. (*Morocco*.)

13. F. purpurea Pugsley, *Jour. Bot. (London)* **40**: 135 (1902). Raceme up to 25-flowered, about as long as peduncle. Bracts as long as or longer than the sometimes rigidly patent-recurved fruiting pedicels. Sepals 4·5–6·5 × 2–3 mm, entire. Corolla 10–13 mm, purple or whitish-purple; wings of upper petal and apex of inner petals darker purple; lower petal with narrow, erect margin. Fruit 2·5 × 2·5 mm, more or less quadrate, faintly rugose when dry; apex truncate. $2n = 80$. ● *Britain and Ireland; Guernsey*. Br Ga Hb.

14. F. macrosepala Boiss., *Elenchus* 8 (1838) (incl. *F. malacitana* Hausskn. & Fritze). Raceme 4- to 11-flowered, dense, shorter than peduncle. Bracts as long as or longer than the more or less erect fruiting pedicels. Sepals 4–7 × 3–5 mm, wider than corolla, subentire. Corolla 10–14 mm, white, often tinged with pinkish-red; wings of upper petal and apex of inner petals dark purple; lower petal with narrow, erect margin. Fruit 2·5–3 × 2–2·75 mm, elliptic-ovate or subglobose, rugulose when dry; apex subacute and mucronulate or obtuse. $2n = 32$. *Mostly limestone cliffs and screes*, 300–2000 *m*. *S. Spain*. Hs.

The European plants are subsp. *macrosepala*; it and other subspecies occur in N.W. Africa.

15. F. bicolor Sommier ex Nicotra, *Fum. Ital.* 55 (1897). Raceme 8- to 15(–20)-flowered, shorter than or equalling peduncle. Bracts $\frac{1}{4}$–$\frac{1}{2}$ as long as the more or less suberect fruiting pedicels. Sepals 2–2·5 × *c.* 1 mm, more or less dentate. Corolla 10–13 mm, white or pinkish-white, finally all pink; apex of inner petals dark purple; lower petal with narrow, patent margin. Fruit 2–2·2 × 1·5–1·7(–2) mm, subglobose-ovoid, obscurely keeled, rugose when dry; apex subacute. $2n = 32$. *Islands of W. Mediterranean region; rare on the mainland.* Bl Co Ga It Sa Si.

16. F. bastardii Boreau in Duchartre, *Rev. Bot.* **2**: 359 (1847). Raceme usually 15- to 25-flowered, longer than peduncle. Bracts $\frac{1}{3}$–$\frac{1}{2}$ as long as the suberect or erecto-patent, thickened fruiting pedicels. Sepals 2–3 × 1–2 mm, more or less serrate. Corolla 9–12 mm, pale pink with apex of inner petals and sometimes wings of the upper dark purple; upper petal narrow and laterally compressed; lower petal with narrow, patent margin. Fruit 2–2·5 × 2–2·5 mm, ovoid, slightly keeled and usually rugose when dry; with wide, flattish base and more or less obtuse apex. $2n = 48$. *S. & W. Europe.* Bl Br Co Ga Gr Hb Hs It Ju Lu Sa Si.

17. F. martinii Clavaud, *Actes Soc. Linn. Bordeaux* **42**: lxix (1889) (*F. reuteri* subsp. *martinii* (Clavaud) Soler). Raceme up to 20-flowered, much longer than peduncle. Bracts $\frac{1}{2}$–$\frac{2}{3}$ as long as the arcuate-deflexed, becoming erecto-patent, fruiting pedicels. Sepals 3–5 × 1·5–2·5 mm, subentire. Corolla 11–13 mm, pink; wings of upper petal and apex of inner blackish-red; lower petal with very narrow, patent margin. Fruit 2·5–2·7 × 2–2·5 mm, subglobose, obscurely keeled, nearly smooth when dry; apex more or less acute. $2n = 48$. ● *W. Europe, from S. England to N. Portugal.* Br Ga Hs Lu.

18. F. sepium Boiss. & Reuter in Boiss., *Diagn. Pl. Or. Nov.* **2(1)**: 16 (1853) (*F. gaditana* Hausskn.). Raceme 6- to 16(–19)-flowered, shorter than to about as long as peduncle. Bracts subequalling or shorter than the erecto-patent fruiting pedicels. Sepals 3–5 × 1·5–3 mm, subentire. Corolla 12–14 mm, pinkish-white, apex often tinged red; wings of upper petal and apex of inner dark purple; lower petal with very narrow erect or subpatent margin. Fruit 2·2 × 2 mm, subglobose, keeled, smooth or slightly rugose when dry; apex very obtuse. $2n = 32$. *S.W. Spain, Portugal.* Hs Lu.

19. F. muralis Sonder ex Koch, *Syn. Fl. Germ.* ed. 2, 1017 (1845) (*F. media* sensu Merino). Raceme about as long as or longer than peduncle. Bracts $\frac{1}{2}$ as long as to about as long as the more or less erect fruiting pedicels. Sepals 3–5 × 1·5–3 mm. Corolla pink; wings of upper petal and tips of inner ones blackish-red; lower petal with narrow, erect margin. Fruit obscurely keeled. $2n = 48$. *W. Europe.* Az Be Bl Br Da Ga Ge Hb Hs Lu No Si [Da Ho].

1 Corolla 10–12 mm; fruit 2·2–2·5 × *c.* 2 mm, obscurely rugose when dry — (b) subsp. **boraei**
1 Corolla 9–10 mm; fruit *c.* 2 × 2 mm, smooth when dry
2 Racemes fewer than 15-flowered; sepals dentate; fruit subacute at apex — (a) subsp. **muralis**
2 Racemes up to 20-flowered; sepals subentire; fruit truncate at apex — (c) subsp. **neglecta**

(a) Subsp. **muralis**: Plant slender. Raceme usually with fewer than 15 flowers. Sepals dentate. Corolla 9–10(–11) mm. Fruit *c.* 2 × 2 mm, subglobose-ovoid, smooth when dry; apex subacute. $2n = 48$. *Almost throughout the range of the species.*

(b) Subsp. **boraei** (Jordan) Pugsley, *Jour. Bot.* (*London*) **40**: 180 (1902) (*F. boraei* Jordan): Plant slender or robust. Raceme with fewer than 15 flowers. Sepals dentate. Corolla 10–12 mm. Fruit 2·2–2·5 × *c.* 2 mm, more or less obovoid, often obscurely rugulose when dry; apex obtuse. $2n = 48$. *Almost throughout the range of the species.*

(c) Subsp. **neglecta** Pugsley, *Jour. Bot.* (*London*) **50**, Suppl. **1**: 24 (1912): Plant robust. Raceme with up to 20 flowers. Sepals subentire. Corolla *c.* 10 mm. Fruit *c.* 2 × 2 mm, shortly obovoid, nearly smooth when dry; apex almost truncate. ● *S.W. England* (*W. Cornwall*).

20. F. reuteri Boiss., *Diagn. Pl. Or. Nov.* **2(8)**: 13 (1849) (*F. apiculata* Lange). Like **19** but racemes with very short peduncles; pedicels erect; sepals 3·5–4·5 × 2–2·5(–3) mm, subentire; corolla 9–11 mm, the upper petal narrow and acute; fruit smooth when dry; apex with persistent apiculus. $2n = 48$. *Spain and Portugal.* Hs Lu.

21. F. petteri Reichenb., *Icon. Fl. Germ.* **3**: 1 (1838). Raceme shortly pedunculate or subsessile. Bracts about equalling the fruiting pedicels. Sepals usually more or less dentate. Corolla pink; wings of upper petal and apex of inner dark purple; lower petal with very narrow, erect margin. Fruit ovoid, subglobose-ovoid or turbinate, obscurely keeled, more or less rugose when dry; apex obtuse, but sometimes apiculate. *S. Europe.* Al Bu Cr Ga Gr Hs It Ju Lu Rm.

(a) Subsp. **petteri**: Raceme usually 12- to 18-flowered; fruiting pedicels patent or erecto-patent. Sepals 3·5–4·5 × 2–2·5 mm. Corolla 9–10·5 mm. Fruit 2·5–2·7 × *c.* 2 mm. ● *Jugoslavia.*

(b) Subsp. **thuretii** (Boiss.) Pugsley, *Jour. Linn. Soc. London* (*Bot.*) **50**: 550 (1937) (*F. thuretii* Boiss., *F. pikermiana* Boiss. & Heldr.): Raceme 15- to 35-flowered. Fruiting pedicels arcuate-recurved. Sepals 2·5–3 × 1·5–2 mm. Corolla 7–10 mm. Fruit *c.* 2 × 1·7–2 mm. *Throughout the range of the species.*

22. F. calcarata Cadevall, *Mem. Real Acad. Ci. Artes Barcelone Ser. 3*, **5(12)**: 12 (1905) (*F. transiens* P. D. Sell, *F. reuteri* auct., non Boiss., *F. petteri* subsp. *calcarata* (Cadevall) Lidén & Soler). Like **21(b)** but sepals subentire; corolla 9–10 mm, often paler; fruit 2–2·2 × 1·5–1·7 mm, more or less acute, apical pits narrower. $2n = 32$. *S.W. Europe.* Ga Hs Lu.

Sect. FUMARIA. Leaf-segments flat or channelled, relatively narrow, from oblong or lanceolate to linear or setaceous. Peduncles usually short. Corolla not more than 9 mm and generally much smaller; wings of upper petal sometimes patent; lower petal more or less spathulate.

23. F. kralikii Jordan, *Cat. Jard. Dijon* 19 (1848) (*F. anatolica* Boiss.). Raceme 10- to 20-flowered; peduncle very short. Bracts more or less equalling the recurved fruiting pedicels. Sepals 2–3 × 1·5–2 mm, dentate. Corolla 5–6(–8) mm, pale pink with wings of upper petal and apex of inner petals dark purple; lower petal with patent margin. Fruit *c.* 1·7 × 1·7 mm, subglobose, nearly smooth or slightly rugose when dry; apex obtuse. $2n = 32$. *S.E. Europe.* Al Bu Gr Rm Rs(K) ?Sa Tu [Ga].

24. F. mirabilis Pugsley, *Jour. Linn. Soc. London* (*Bot.*) **47**: 432 (1927). Raceme 8- to 15(–25)-flowered, sessile. Bracts longer than the erect fruiting pedicels. Sepals 2·5–3 × *c.* 1·5 mm, shallowly serrate, long-persistent. Corolla 7·5–9 mm, pink; inner petals tipped dark purple. Fruit 2–2·2 × 2·5–2·7 mm, subrotund, keeled, tuberculate-rugose when dry; apex obtuse or slightly retuse. $2n = 64$. *S.E. Spain* (*Cabo de Gata*). Hs. (*N. Africa.*)

25. F. faurei (Pugsley) Lidén, *Lagascalia* **9**: 133 (1980) (*F. mirabilis* var. *faurei* Pugsley, *F. mirabilis* auct., non Pugsley). Raceme 15- to 30-flowered, subsessile or shortly pedunculate. Bracts about equalling or just exceeding the erecto-patent fruiting pedicels. Sepals 2–2·2 × 1–1·2 mm, irregularly dentate. Corolla 7–8·5 mm, pale pink; apex of inner petals dark purple; lower petal with rather broad, patent margin. Fruit 2–2·5 × 2·2–2·5 mm, subglobose, obscurely keeled, rugose when dry; apex very obtuse. 2*n* = 80. *Often on gypsaceous soils. S., C. & E. Spain, S. Portugal.* Hs Lu.

26. F. densiflora DC., *Cat. Pl. Horti Monsp.* 113 (1813) (*F. micrantha* Lag.). Raceme 15- to 30(–35)-flowered, much exceeding the very short or obsolete peduncle. Bracts normally exceeding the erecto-patent fruiting pedicels. Sepals (2–)2·5–3·5 × (1·5–)2–3 mm, subentire or laciniate. Corolla 6–7 mm, pink; wings of upper petal and apex of inner blackish-red; lower petal with patent margin, subspathulate. Fruit 2–2·5 × 2–2·5 mm, subglobose, keeled, rugose when dry; apex rounded-obtuse. 2*n* = 32. *W. & S. Europe, extending to N.E. Romania.* Al Be Bl Br Bu Co Ga Gr Hb Hs It Ju Lu Rm Sa Si Tu.

27. F. bracteosa Pomel, *Nouv. Mat. Fl. Atl.* 239 (1874). Raceme 15- to 36-flowered, longer than the short peduncle. Bracts as long as or longer than the erecto-patent fruiting pedicels. Sepals 1·5–2·5 × 1–1·2 mm, dentate or subentire. Corolla *c.* 4·5 mm, deep pink. Fruit 2–2·2 × 2–2·2 mm, subrotund, strongly keeled, rugose when dry; apex emarginate. 2*n* = 16. *S.E. Spain* (*Almeria, Málaga*); *Islas Baleares.* Bl Hs. (*N. Africa, S.W. Asia.*)

28. F. rostellata Knaf, *Flora* (*Regensb.*) **29**: 290 (1846). Raceme 15- to 30(–40)-flowered, longer than the short peduncle. Bracts $\frac{1}{2}$ to $\frac{2}{3}$ as long as the patent fruiting pedicels. Sepals 2–3 × 1–1·5 mm, dentate. Corolla 5–7(–9) mm, purplish-pink; wings of upper petal and apex of inner ones blackish-purple; lower petal with patent margin, spathulate. Fruit 2–2·5 × 1·7–2 mm, subglobose to ovoid, keeled, nearly smooth or faintly rugulose when dry; apex subtruncate. *C. & S.E. Europe.* Al Au Bu Cz Ge Gr Hu Ju Po Rm Rs(C, W) Tu.

29. F. officinalis L., *Sp. Pl.* 700 (1753). Raceme longer than peduncle. Bracts $\frac{1}{2}$ as long to nearly as long as the erecto-patent fruiting pedicels. Sepals 1·5–3·5 × 0·7–1·5 mm, irregularly dentate. Corolla 7–9 mm, purplish-pink; wings of upper petal and apex of inner ones blackish-red; lower petal with patent margin, spathulate. Fruit *c.* 2 × 2·5–3 mm, more or less obreniform, obscurely keeled, rugose when dry; apex truncate or slightly emarginate. *Almost throughout Europe.* All except Az Sb; only as a casual in Fa Is.

(a) Subsp. **officinalis**: Raceme normally more than 20-flowered. Sepals 2·5–3·5 × 1–1·5 mm. 2*n* = 32. *Throughout the range of the species.*

(b) Subsp. **wirtgenii** (Koch) Arcangeli, *Comp. Fl. Ital.* 27 (1882) (*F. wirtgenii* Koch): Raceme 10- to 20(–24)-flowered. Sepals 1·5–2 × 0·7–1 mm. 2*n* = 48. *W., C. & S. Europe.*

30. F. ragusina (Pugsley) Pugsley, *Jour. Linn. Soc. London* (*Bot.*) **49**: 524 (1934). Like **29** but habit more slender; foliage less dissected; peduncle not much shorter than raceme; sepals 2 × 1·5 mm, subglobose-ovate; corolla *c.* 9 mm; upper petal usually subacute or shortly rostellate; fruit rounded-truncate or truncate and mucronulate at apex; apical pits black-spotted. ● *Jugoslavia and Albania.* Al Ju.

31. F. jankae Hausskn., *Flora* (*Regensb.*) **56**: 491 (1873). Like **29** but sepals *c.* 1·5 × 0·7 mm; corolla 4–5 mm; fruit *c.* 2 × 2·5 mm, ovoid, acuminate. ● *N.W. Romania* (*Săcueni*). Rm.

32. F. caroliana Pugsley, *Jour. Linn. Soc. London* (*Bot.*) **47**: 448 (1927). Raceme 10- to 15(–20)-flowered, much longer than the short peduncle. Bracts about $\frac{1}{2}$ as long as the suberect fruiting pedicels. Sepals 1·5–2 × 0·7–1 mm, irregularly incise-dentate. Corolla 6–7 mm, pink; inner petals with dark purple apex; lower petal obovate-spathulate, with patent margin. Fruit *c.* 2 × 2·5 mm, subglobose, obscurely keeled, rugose when dry; apex rounded-obtuse with a short apiculus when young. ● *N. France* (*near Arras*). Ga.

33. F. segetalis (Hammar) Coutinho, *Fl. Port.* 246 (1913). Raceme 12- to 20(–36)-flowered, sessile. Bracts slightly shorter than or equalling the erecto-patent fruiting pedicels. Sepals 1·5–2 × 1–1·2 mm, dentate. Corolla 6–7 mm, pink or pale pink, rarely almost white, the inner petals tipped dark purple. Fruit 2–2·4 × 2–2·4 mm, triangular-ovoid with a broad base, finely rugose when dry; apex subacute. 2*n* = 48. *S. Spain.* Hs. (*N. Africa.*)

34. F. schleicheri Soyer-Willemet, *Observ. Pl. Fr.* 17 (1828). Raceme usually 12- to 25-flowered, at first equalling but later exceeding peduncle. Bracts about $\frac{1}{3}$ as long as the suberect fruiting pedicels. Sepals 0·5–1 × 0·2–0·7 mm, irregularly incise-dentate. Corolla 5–6 mm, deep pink; wings of upper petal and apex of inner dark purple; lower petal with patent margin. Fruit *c.* 2 × 2 mm, subglobose, keeled, obscurely rugose when dry; apex rounded-obtuse and apiculate. *C. Europe, extending to S. France, Bulgaria and E. Russia.* Au Bu Cz Ga Ge ?Gr He Hu It Ju Po Rm Rs(C, W, K, E).

35. F. microcarpa (Hausskn.) Pugsley, *Jour. Linn. Soc. London* (*Bot.*) **44**: 312 (1919). Like **34** but with dwarfer and more erect habit; raceme 20- to 35-flowered; bracts $\frac{1}{2}$ as long as pedicels; flowers smaller and less deeply coloured; fruit *c.* 1·75 × 1·75 mm. *S.E. Russia* (*near Volgograd*). Rs(E).

36. F. vaillantii Loisel. in Desv., *Jour. Bot. Rédigé* **2**: 358 (1809). Lobes of leaves not channelled. Raceme usually 6- to 20(–25)-flowered, longer than the short peduncle. Bracts about $\frac{3}{4}$ as long as the more or less erect fruiting pedicels. Sepals 0·7–1 × 0·3–0·5 mm, more or less laciniate-dentate. Corolla 5–6 mm, pale to deep pink; apex of inner petals and wings of upper petal often tinged with blackish-red; lower petal with patent margin, spathulate; spur long and straight. Fruit 2–2·2 × 2–2·2 mm, subglobose, obscurely keeled, granular-rugose when dry; apex rounded and mucronulate when young. 2*n* = 32. *Most of Europe, but rare N. of 55° N.* Al Au Be Br Co Cz Da Fe Ga Ge Gr He Ho Hs Hu It Ju Po Rm Rs(B, C, W, K, E) Si Su Tu [Lu].

37. F. schrammii (Ascherson) Velen., *Fl. Bulg.* 22 (1891) (*F. vaillantii* subsp. *schrammii* (Ascherson) Hausskn.). Like **36** but leaf-segments narrower; raceme 6- to 15-flowered, more or less sessile; flower paler; fruit 1·7–2 × 1·7–2 mm; apex obtuse and apiculate. *Scattered through Europe from France and Spain to Greece and Ukraine.* Al Bu Ga Ge Gr He Hs Hu Rs(W).

Perhaps not specifically distinct from **36**, but in some respects intermediate between **34** and **36**.

38. F. pugsleyana (Maire ex Pugsley) Lidén, *Anal. Jard. Bot. Madrid* **41**: 222 (1984) (*F. schrammii* var. *pugsleyana* Maire ex Pugsley). Raceme 5- to 15-flowered, sessile. Bracts about $\frac{3}{4}$ as long as the thickened, short fruiting pedicels. Sepals 0·2–0·7 × 0·1–0·5 mm, linear-lanceolate and entire or broader and of 2 or 3 lobes. Corolla 5–6 mm, white, often suffused pinkish-red; lower petal spathulate; spur very short and ascending. Fruit 2–2·2 × 2–2·5 mm, ovoid, densely and finely rugose when dry; apex strongly apiculate. 2*n* = 32. *S. Spain* (*Almeria and Málaga*). Hs. (*Morocco.*)

39. F. parviflora Lam., *Encycl. Méth. Bot.* **2**: 567 (1788) (*F. caespitosa* Loscos). Lobes of leaves very narrowly linear, channelled. Raceme 7(–10)- to 15(–22)-flowered, subsessile. Bracts as long as or longer than the fruiting pedicels. Sepals 0·5–0·7 × 0·5–0·7 mm, laciniate-dentate. Corolla white, occasionally flushed pink; lower petal spathulate with patent margin. Fruit 1·6–2·3 × 1·7–2·5 mm, subglobose, distinctly keeled and often subacute. $2n = 32$. *W., S. & S.C. Europe, extending locally to C. Ukraine.* Be Bl Br Bu Co Cr Ga Ge Gr Hs Hu It Ju Lu Rm Rs(W) Sa Si Tu.

14. Platycapnos (DC.) Bernh.

†T. G. TUTIN (EDITION 1) REVISED BY M. LIDÉN (EDITION 2)

Annual or short-lived perennials. Leaves 2-pinnatisect, the segments linear. Inflorescence a subglobose, oblong or ovoid raceme. Flowers zygomorphic, very shortly pedicellate; upper petal basally saccate; style deciduous. Fruit a 1-seeded nut, elliptical, strongly compressed.

1 Perennial; flowering stems leafless above — **3. saxicola**
1 Annual; flowering stems leafy above
2 Flowers 5–6 mm; racemes rounded to ovoid — **1. spicatus**
2 Flowers 7–8 mm; racemes oblong — **2. tenuilobus**

1. P. spicatus (L.) Bernh., *Linnaea* **8**: 471 (1833) (*Fumaria spicata* L.). More or less glaucous annual. Stems up to 30 cm, stout, branched, leafy. Ultimate leaf-segments 4–5 × 0·4–0·5 mm, mucronulate. Racemes 2·5–5 cm, subglobose to ovoid; bracts lanceolate to ovate, about twice as long as pedicels; pedicels recurved in fruit. Flowers 5–6 mm, whitish or yellowish, dark purple at apex of petals; sepals $\frac{1}{6}$–$\frac{1}{3}$ as long as petals, entire or denticulate, caducous. Fruit ovate, flat or convex, obtuse, mucronate. $2n = 32$. *Cultivated ground. S.W. Europe, extending eastwards to S. Italy.* Bl Ga Hs It Lu Si.

Subsp. **echeandiae** (Pau) Heywood, *Feddes Repert.* **64**: 51 (1961) (*P. echeandiae* Pau), poorly defined by subglobose racemes, short sepals, and fruit with convex sides, has been recognized from E. and S.E. Spain.

2. P. tenuilobus Pomel, *Bull. Soc. Sci. Phys. Algérie* **11**: 240 (1874). Like **1** but racemes somewhat laxer, oblong; flowers 7–8 mm; stigma appendages curved. *E., C. & S. Spain, just extending into S. France.* Ga Hs.

(a) Subsp. **tenuilobus**: Leaflets divaricate. Racemes long, up to 80-flowered. Fruit convex and smooth. $2n = 32$. *Dry places. Throughout the range of the species.*
(b) Subsp. **parallelus** Lidén, *Anal. Jard. Bot. Madrid* **41**: 222 (1984): Leaflets parallel. Racemes shorter. Fruit alveolate. *Screes. S. Spain (Málaga* prov.). Hs.

3. P. saxicola Willk., *Bot. Zeit.* **1848**: 367 (1848). Glaucous perennial, with stems up to 20 cm arising from a slender, vertical stock. Cauline leaves few. Ultimate leaf-segments 8–10 × 0·6–0·9 mm, with acute, somewhat hooded apex. Racemes 1–1·5 cm, subglobose; bracts ovate, about as long as pedicels; pedicels arcuate in fruit. Petals 5 mm, pale pink or lilac, dark purple at apex; sepals $\frac{1}{2}$–$\frac{2}{3}$ as long as petals, entire; stigma without appendage. Fruit obovate, slightly rugulose, slightly emarginate. $2n = 28$. *Mountain screes above* 2000 *m. S.E. Spain (Granada and Jaén provs.).* Hs.

15. Rupicapnos Pomel

V. H. HEYWOOD

Flowers in corymbose racemes. Pedicels elongate and deflexed in fruit. Upper petal spurred. Fruit a more or less globose, 1-seeded, thickened, rugose, tuberculate nut, lacking apical germination pores.

1. R. africana (Lam.) Pomel, *Nouv. Mat. Fl. Atl.* 240 (1874) (*Fumaria africana* Lam.). Glaucous annual to short-lived perennial with elongate root and thick, branched, decumbent stock. Leaves mainly basal, more or less fleshy, very long-petiolate, 2-pinnatisect into oblong-elliptical or cuneate lobes. Flowers 13–16 mm, white to pinkish; spur *c.* 4 mm. Fruit 3–3·75 × *c.* 3 mm, shortly mucronate. $2n = 32$. *Crevices in limestone cliffs. S.W. Spain.* Hs. (*N.W. Africa.*)

The European plants belong to subsp. **decipiens** (Pugsley) Maire in Jahandiez & Maire, *Cat. Pl. Maroc* **2**: 261 (1932) (*Rupicapnos decipiens* Pugsley, *Fumaria africana* auct. hisp., non Lam.). Subsp. *africana* and other subspecies occur in N.W. Africa.

16. Ceratocapnos Durieu

P. W. BALL (EDITION 1) REVISED BY M. LIDÉN (EDITION 2)

Slender, climbing or scrambling annuals. Leaves 1- or 2-ternate or -pinnate, the upper with tendrils. Flowers small, zygomorphic, in leaf-opposed racemes. Style deciduous.

All fruits 2- to 4-seeded, smooth — **1. claviculata**
At least some fruits 1-seeded, with muricate ribs — **2. heterocarpa**

1. C. claviculata (L.) Lidén, *Anal. Jard. Bot. Madrid* **41**: 221 (1984) (*Corydalis claviculata* (L.) DC.). Delicate, climbing, much-branched annual 20–100 cm. Racemes 6- to 10-flowered, dense. Bracts 1–3 mm, oblong-lanceolate. Lower pedicels 1–2 mm. Corolla 5–6 mm; spur short, saccate. Fruit 6–9 mm, 2- to 4-seeded, smooth, the apex triangular. *Shady places; calcifuge.* ● *W. Europe, extending eastwards to E. Denmark.* Be Br Da Ga Ge Hb Ho Hs Lu No [Su].

(a) Subsp. **claviculata**: Plant glaucous. Petals white. Fruit glabrous. $2n = 32$. *Throughout the range of the species.*
(b) Subsp. **picta** (Samp.) Lidén, *loc. cit.* (1984) (*Corydalis claviculata* subsp. *picta* P. Silva & Franco): Plant not glaucous. Outer petals pinkish-violet. Fruit puberulent. *N. Portugal (Vila-Nova-de-Paiva).*

2. C. heterocarpa Durieu in Parl., *Gior. Bot. Ital.* **1(1)**: 336 (1844) (*Corydalis heterocarpa* (Durieu) Ball). Like **1** but racemes 5- to 12-flowered; corolla purple; fruit dimorphic, the lower 3·5–4 mm, 1-seeded, with muricate ribs, the upper 10–12 mm, 2- to 5-seeded, smooth, with curved beak. $2n = 32$. *Stony ground and cliffs. Portugal, S. Spain.* Hs Lu. (*N.W. Africa.*)

67. CAPPARACEAE

EDIT. V. H. HEYWOOD

Herbs, shrubs or trees. Leaves simple, or ternate to palmate, alternate, stipulate or not; stipules sometimes spinose. Flowers solitary or in racemes, actinomorphic or more usually zygomorphic, hermaphrodite. Sepals 4. Petals 4. Stamens 6 to many. Ovary 1, superior, on a gynophore; carpels 2. Fruit a 2-valved, unilocular capsule, or a berry.

1 Leaves all simple; flowers solitary; fruit a berry **1. Capparis**
1 At least some leaves compound; flowers in racemes; fruit a capsule
 2 Petals entire; stamens 6 **2. Cleome**
 2 Petals bifid; stamens 12 or more **3. Polanisia**

1. Capparis L.

V. H. HEYWOOD (EDITION 2) REVISED BY R. N. HIGTON AND J. R. AKEROYD (EDITION 2)

Shrubs with simple leaves, and usually with spinose stipules. Flowers solitary, showy, zygomorphic; petals white or pink; stamens numerous. Fruit a berry borne on a long gynophore.

Literature: M. Zohary, *Bull. Res. Counc. Israel* **8D**: 49–64 (1960). M. Jacobs, *Blumea* **12**: 385–541 (1965).

1. C. spinosa L., *Sp. Pl.* 503 (1753) (incl. *C. ovata* Desf., *C. sicula* Duh.). Stems trailing or ascending, branched. Leaves petiolate, ovate to suborbicular, obtuse or emarginate, mucronate, sometimes obscurely. Stipules setaceous to spinose, often recurved, but sometimes weakly developed. Flowers 5–7 cm in diameter, slightly zygomorphic. Fruit 1–4 cm, oblong to somewhat pyriform. *Rocks, walls and other dry places. Mediterranean region and Krym.* Al *Bl Co Cr *Ga Gr Hs It Ju Rs(K) Sa Si Tu.

A very variable species with variants treated as species or varieties by different authors. The treatment here follows R. N. Higton & J. R. Akeroyd, *Bot. Jour. Linn. Soc.* **106**: 104–112 (1991).

(**a**) Subsp. **spinosa**: Leaves obovate, ovate or elliptical, rarely cordate at the base, obtuse to acuminate with mucro of varying length, glabrous to densely pubescent, chartaceous. Stipules at least 2 mm, often spinose, conspicuous. *Throughout the range of the species.*

(**b**) Subsp. **rupestris** (Sm.) Nyman, *Consp.* 68 (1878) (*C. rupestris* Sm., *C. inermis* Turra): Branches more pendent. Leaves ovate to suborbicular, rounded to cordate at the base, obtuse or emarginate with mucro vestigial (less than 0·25 mm), glabrous to sparsely pubescent, coriaceous or fleshy. Stipules less than 2 mm, setaceous, inconspicuous. Ovary longer and narrower. *Coastal rocks and cliffs, usually on limestone. Almost throughout the range of the species.*

The common plant in Europe is subsp. (**a**) var. *canescens* Cosson (*C. ovata* var. *canescens* (Cosson) Heywood), with leaves obovate or elliptical, petiole sulcate and midrib immersed on the adaxial surface. The flowers are somewhat smaller and more strongly zygomorphic than those of var. *spinosa*.

There has been selection for large flower-buds in subsp. (**a**), which is probably introduced in the western part of its range, for their use as a condiment (Capers).

2. Cleome L.

A. O. CHATER AND J. R. AKEROYD

Erect, usually glandular-pubescent, viscid annuals with exstipulate, simple or ternate leaves. Flowers in long, bracteate racemes; sepals 4, deciduous; petals 4, contracted into a distinct claw, more or less dimorphic; stamens 6. Receptacle short. Fruit a linear, 2-valved, many-seeded capsule on a long gynophore.

Literature: A. Carlström, *Willdenowia* **14**: 119–130 (1984).

1 Ripe fruit more than 50 mm; upper bracts at least as long as pedicels **3. violacea**
1 Ripe fruit less than 50 mm; upper bracts *c.* $\frac{1}{3}$ as long as pedicels
 2 Sepals and pedicels glandular-pubescent; sepals lanceolate, acuminate **1. ornithopodioides**
 2 Sepals and pedicels glabrous or with a few ±sessile glands; sepals oblong, obtuse or cuspidate **2. iberica**

1. C. ornithopodioides L., *Sp. Pl.* 672 (1753) (incl. *C. aurea* Čelak., non Torrey & A. Gray). Plant glandular-pubescent, at least on pedicels and sepals. Stems 15–50 cm, erect, usually branched. Leaves petiolate, ternate, except for the uppermost which are simple; leaflets oblong-lanceolate, entire. Bracts of upper flowers *c.* $\frac{1}{3}$ as long as pedicels. Sepals lanceolate, acuminate, unequal. Petals *c.* 3 mm, whitish or yellow, sometimes red-striped, slightly dimorphic. Ripe fruit 15–25 × 1·5–2·5 mm, torulose, deflexed. *Open, sandy and stony habitats. S. part of Balkan peninsula.* Al Bu Gr Ju Tu.

2. C. iberica DC., *Prodr.* **1**: 240 (1824) (*C. ornithopodioides* L. subsp. *canescens* (Steven ex DC.) Tzvelev). Like **1** but glabrous or sparsely glandular-pubescent; sepals oblong, obtuse or cuspidate, subequal. *Open, often disturbed habitats. S.E. Russia, Krym; Turkey-in-Europe.* Rs(W, K) Tu.

3. C. violacea L., *Sp. Pl.* 672 (1753). Like **1** but the middle and upper cauline leaves simple; leaflets linear or linear-lanceolate; upper bracts at least as long as pedicels; petals violet, spotted with yellow, rarely entirely yellow, strongly dimorphic (the longer petals 4–6 mm); ripe fruit 50–100 × 2 mm, not torulose. *S. & W. Spain, Portugal.* Hs Lu.

3. Polanisia Rafin.

J. R. AKEROYD

Like *Cleome* but stamens 12–16; capsule linear to oblong, sessile or on a very short gynophore.

1. P. trachysperma Torrey & A. Gray, *Fl. N. Amer.* **1**: 669 (1840) (*P. dodecandra* (L.) DC. var. *trachysperma* (Torrey & Gray) Iltis). Stems 20–50 cm, erect. Leaves petiolate, ternate; leaflets 3–7 cm, elliptical or oblong, obtuse, entire. Petals 8–12 mm, yellowish-white, bifid, the blade usually shorter than the claw. Stamens 12–16; filaments unequal, up to 20 mm, long-exserted, purple. Fruit 3–7 cm, erect, linear to oblong. Seeds *c.* 2 mm, areolate. *Naturalized in N.W. & N.C. Italy.* [It.] (*North America.*)

68. CRUCIFERAE

EDIT. P. W. BALL, V. H. HEYWOOD AND J. R. AKEROYD

Annual to perennial herbs, rarely small shrubs. Leaves alternate, exstipulate. Flowers usually hermaphrodite, actinomorphic, hypogynous. Sepals 4, free, in 2 decussate pairs. Petals 4, rarely absent, free, clawed, imbricate or contorted, alternating with the sepals. Stamens usually 6, rarely 4, 2 or 0, tetradynamous (an outer pair with short filaments, and 2 inner pairs, 1 posterior and 1 anterior, with long filaments); filaments sometimes winged or with a tooth-like appendage. Nectarial glands of various sizes, shapes, colours and dispositions around the base of the stamens and ovary. Ovary of 2 carpels, syncarpous, with 2 parietal placentas, usually bilocular through the formation of a membranous false septum by the union of outgrowths of the placentas, sometimes transversely plurilocular. Stigma capitate to bilobed. Fruit usually a dehiscent capsule opening by 2 valves from below, called a *siliqua* when at least 3 times as long as wide or a *silicula* if less than 3 times as long as wide; sometimes indehiscent, breaking into 1-seeded portions or not; rarely transversely articulate with dehiscent and indehiscent segments; sometimes dividing at maturity into 1-seeded portions (*lomentum*).

There is great diversity in the form and structure of the fruit in this family, often affording an easy means of identification, especially in genera which possess a distinctive siliqua or silicula. The seeds are always inserted in 2 rows in each loculus, but where the diameter of the seed is approximately the same as that of the fruit they appear to be in 1 row, and are so described below. Unless otherwise stated, descriptions of pedicels refer to those of the infructescence.

The following notes of diagnostic features of particular genera may assist in identification.

Flowers zygomorphic: *Iberis*, *Teesdalia*, *Teesdaliopsis*, *Jonopsidium* (1 sp.), *Calepina*.

Spiny small shrubs: *Alyssum* (2 spp.), *Vella*.

Filaments of the inner stamens connate in pairs: *Leptaleum*, *Sterigmostemum*, *Euzomodendron*, *Vella*, *Boleum*.

Many species in this family, especially among the annuals, are found in Europe principally or exclusively as weeds of cultivation or as ruderals. In many cases it is difficult to determine the native range and to discriminate between naturalized and casual status, and the information on the status of such plants given below must in many cases, therefore, be regarded as tentative.

1 Fruit with 2 segments, the upper flat, foliaceous or lingulate, the lower with 1 or 2 seeds
2 Annual herbs; filaments free
3 Fruit erect or erecto-patent, 12–35 mm **90. Eruca**
3 Fruit pendent or patent, not more than 8 mm **94. Carrichtera**
2 Small shrubs; filaments of the inner stamens united in pairs
4 Lower segment of fruit glabrous or sparsely setose; ovary shortly stipitate **95. Vella**
4 Lower segment of fruit densely setose; ovary sessile **97. Boleum**
1 Fruit without a terminal, flat, foliaceous or lingulate segment, often not segmented or with more than 2 segments
5 Fruit covered with conical spines 1–3 mm long **96. Succowia**
5 Fruit not covered with conical spines
6 Fruit pendent, flat or ±hemispherical, not more than 12 times as long as wide
7 Fruit strongly compressed, convex beneath, concave with incurved margins above so as to be ±hemispherical **15. Tauscheria**
7 Fruit not convex beneath and not concave with incurved margins above
8 Hairs branched or stellate; petals yellow **55. Clypeola**
8 Plant glabrous or with a few unbranched or medifixed hairs
9 Petals yellow; fruit with a wide, often inflated wing surrounding the loculus; loculus usually with a distinct longitudinal rib **14. Isatis**
9 Petals white, pink or violet; fruit not winged or with a very narrow wing, without a longitudinal rib
10 Petals less than 5 mm, white; fruit up to 10(–13) mm, reticulately veined or emarginate **46. Peltaria**
10 Petals at least 5 mm, pink or violet; fruit at least 10 mm, neither reticulately veined nor emarginate **44. Ricotia**
6 Fruit erect or patent, rarely pendent and ±flattened, but then more than 12 times as long as wide (see also **87**)
11 Pedicels pendent, with the fruit erect or erecto-patent on the end of the pedicel **17. Goldbachia**
11 Pedicels various, but never pendent with the fruit erect or erecto-patent
12 Fruit a siliqua, at least 3 times as long as wide
13 Glabrous or with unbranched hairs only
14 Fruit ±compressed laterally, with 2 triangular wings at the apex, indehiscent **61. Andrzeiowskia**
14 Fruit compressed dorsally, or with convex valves
15 Stigma deeply 2-lobed, the lobes often erect or connivent
16 Petals not more than 10 mm; fruits of 2 kinds, the lower ones articulate, indehiscent, the upper ones opening by 2 valves **31. Diptychocarpus**
16 Petals at least 10 mm; fruits all similar
17 Plant scapose **27. Parrya**
17 Flowering stem leafy
18 Plant glabrous; leaves entire **85. Moricandia**
18 Plant pubescent or glandular, rarely glabrous and then with toothed leaves
19 Siliqua with a beak 5–30 mm
20 Siliqua with a beak 12–30 mm, almost as long as the rest of the siliqua; seeds in 1 row in each loculus **32. Chorispora**
20 Siliqua with a beak 5–12 mm, shorter than the rest of the siliqua; seeds in 2 rows in each loculus **90. Eruca**
19 Siliqua not beaked, but with a style less than 3 mm

21 Siliqua 1·5–3 mm wide; radicle incumbent **20. Hesperis**
21 Siliqua 1·2–1·5 mm wide; radicle accumbent **26. Clausia**
15 Stigma entire, capitate or shallowly 2-lobed
22 Petals yellow, sometimes with purple or brown veins
23 Valves of the siliqua with 3–7 distinct longitudinal veins
24 Cauline leaves cordate-amplexicaul, entire **84. Conringia**
24 Cauline leaves not cordate-amplexicaul, usually toothed or lobed
25 Siliqua with a short stipitate segment at the base, separated from the rest of the siliqua by an articulation
26 Racemes bracteate, at least in the lower $\frac{1}{2}$ **106. Enarthrocarpus**
26 Racemes ebracteate **107. Raphanus**
25 Siliqua without a short stipitate segment at the base
27 Siliqua clavate at the apex, constricted in the middle; stigma sessile **2. Lycocarpus**
27 Siliqua not clavate at the apex, usually with a distinct beak or style
28 Sepals erect (calyx closed)
29 Siliqua (including beak) more than 20 mm **92. Coincya**
29 Siliqua (including beak) less than 20 mm
30 Siliqua patent to deflexed **92. Coincya**
30 Siliqua erect and appressed **93. Hirschfeldia**
28 Sepals patent or erecto-patent
31 Siliqua not beaked, with a style not exceeding 2(–4) mm **1. Sisymbrium**
31 Siliqua with a beak at least 7 mm **89. Sinapis**
23 Valves of the siliqua 1-veined, sometimes with reticulate lateral veins, or without veins
32 Siliqua not more than 7 times as long as wide
33 Siliqua beaked **88. Brassica**
33 Siliqua not conspicuously beaked
34 Leaves pinnate **37. Rorippa**
34 Leaves simple **57. Draba**
32 Siliqua more than 7 times as long as wide
35 Siliqua strongly compressed, the valves flat
36 Valves coiling spirally from the base on dehiscence **40. Cardamine**
36 Valves not coiling spirally on dehiscence
37 At least the basal leaves sinuate-dentate **42. Arabis**
37 All leaves entire **84. Conringia**
35 Siliqua terete or 4-angled, the valves convex
38 Valves with the median vein weak or absent
39 Siliqua attenuate; seeds mucilaginous when moistened **36. Sisymbrella**
39 Siliqua not attenuate; seeds not mucilaginous
40 Seeds in 1 row in each loculus **35. Barbarea**
40 Seeds in 2 rows in each loculus **37. Rorippa**
38 Valves with distinct median vein
41 Seeds in 2 or more rows in each loculus **87. Diplotaxis**
41 Seeds in 1 row in each loculus, rarely somewhat 2-rowed towards the apex
42 Cauline leaves simple, entire, amplexicaul
43 Sepals erect; all leaves simple, entire **84. Conringia**
43 Sepals erecto-patent or patent; lower leaves lobed, sinuate-dentate or sinuate-crenate **88. Brassica**
42 Cauline leaves toothed or lobed, rarely entire, then not amplexicaul
44 Leaves all or nearly all basal **88. Brassica**
44 Flowering stem leafy
45 Valves of the siliqua rounded on the back; seeds globose **88. Brassica**
45 Valves of the siliqua keeled; seeds ovoid, ellipsoidal or oblong
46 Valves of the siliqua not or only slightly torulose **35. Barbarea**
46 Valves of the siliqua torulose **91. Erucastrum**
22 Petals white, pink or purple
47 Small, much-branched shrub with crowded, pinnatisect leaves divided into ± fleshy segments; filaments of the inner stamens united in pairs **86. Euzomodendron**
47 Habit and leaves not as above; filaments free
48 Valves of the siliqua with 3–7 distinct longitudinal veins
49 Leaves reniform to triangular-ovate, dentate, cordate; plant smelling of garlic when crushed **6. Alliaria**
49 Leaves longer than wide, entire to pinnate, not cordate; plant not smelling of garlic
50 Lower segment of fruit indehiscent, either sterile and stipitate or very small with 0–2 seeds **107. Raphanus**
50 Lower segment of most fruits dehiscent, with 2 or more seeds
51 Petals white with violet veins; siliqua at least 2 cm, erecto-patent **92. Coincya**
51 Petals purple; siliqua not more than 2 cm, erect and closely appressed to the stem **98. Erucaria**
48 Valves of the siliqua 1-veined, sometimes with reticulate lateral veins, or without veins
52 Siliqua with a distinct transverse articulation, the lower segment sometimes sterile and stipitiform **99. Cakile**
52 Siliqua not transversely articulated
53 Siliqua not more than 7 times as long as wide
54 Siliqua indehiscent, oblong-pyriform; leaves suborbicular to ovate, cordate **12. Sobolewskia**
54 Siliqua dehiscent, not oblong-pyriform;

at least the cauline leaves distinctly longer than wide
55 Seeds in 1 row in each loculus
56 Basal leaves ovate to orbicular; petals 2–4 mm **7. Eutrema**
56 Basal leaves ovate or obovate; petals 4–11 mm **42. Arabis**
55 Seeds in 2 rows in each loculus
57 Stem hairy, at least at the base; radicle incumbent **10. Braya**
57 Stem glabrous; radicle accumbent
58 Leaves pinnate **38. Nasturtium**
58 Leaves simple **57. Draba**
53 Siliqua more than 7 times as long as wide
59 Valves flat, the siliqua strongly compressed
60 Valves coiling spirally from the base on dehiscence **40. Cardamine**
60 Valves not coiling spirally on dehiscence
61 Petals pale yellow or yellowish
62 Leaves sinuate-dentate **42. Arabis**
62 Leaves entire **84. Conringia**
61 Petals white, pink or purple
63 Cauline leaves sessile **42. Arabis**
63 At least the middle and lower cauline leaves distinctly petiolate
64 Valves of the siliqua smooth, not torulose **42. Arabis**
64 Valves of the siliqua ± torulose **41. Cardaminopsis**
59 Valves convex, the siliqua not strongly compressed
65 Valves with the median vein weak or absent
66 Siliqua 30–60 mm; pedicels not more than 3 mm in fruit, almost as thick as the siliqua **36. Sisymbrella**
66 Siliqua not more than 30 mm; pedicels up to 10 mm or more in fruit, thinner than the siliqua **38. Nasturtium**
65 Valves with a distinct median vein
67 Seeds in 2 rows in each loculus
68 Siliqua without a beak, the style very short; cauline leaves with rounded auricles at the base **9. Thellungiella**
68 Siliqua with a distinct beak or style up to 7 mm; cauline leaves cuneate or hastate at the base **87. Diplotaxis**
67 Seeds in 1 row in each loculus
69 Siliqua with a beak at least 5 mm **88. Brassica**
69 Siliqua not beaked, the style up to 3·5 mm
70 Cauline leaves sessile, cordate-amplexicaul **84. Conringia**
70 Cauline leaves not amplexicaul
71 Most flowers subtended by a bract **1. Sisymbrium**
71 Most flowers ebracteate
72 Petals emarginate, sometimes only shallowly; inner sepals saccate at the base **3. Murbeckiella**
72 Petals entire or absent; inner sepals not saccate at the base
73 Basal leaves truncate or cordate at the base; seeds *c.* 3 mm, winged **7. Eutrema**
73 Basal leaves attenuate at the base; seeds less than 1 mm, not winged **8. Arabidopsis**
13 Hairs stellate, branched, or a mixture of branched and unbranched
74 Stigma deeply 2-lobed, the lobes sometimes erect and connate to form a beak on the siliqua
75 Lobes of the stigma with a dorsal swelling or horn **28. Matthiola**
75 Lobes of the stigma without a swelling or horn
76 Style at least ½ as long as the rest of the siliqua **32. Chorispora**
76 Style much less than ½ as long as the rest of the siliqua
77 Petals yellow
78 Sepals not saccate; filaments of the inner stamens connate in pairs; siliqua articulated **25. Sterigmostemum**
78 Inner sepals strongly saccate at the base; filaments all free; siliqua not articulated
79 Hairs all medifixed; style at least 2 mm in fruit **18. Erysimum**
79 Hairs various, but not all medifixed; style not exceeding 1 mm in fruit **20. Hesperis**
77 Petals white, pink or violet, rarely reddish
80 Filaments of the inner stamens connate in pairs; seeds in 2 rows in each loculus **24. Leptaleum**
80 All filaments free; seeds in 1 row in each loculus
81 Lobes of the stigma free **20. Hesperis**
81 Lobes of the stigma erect, connate **21. Malcolmia**
74 Stigma capitate, retuse or slightly 2-lobed
82 Siliqua with 2 or 4 horns at the apex
83 Siliqua with 2 horns at the apex; petals white **29. Notoceras**
83 Siliqua with 4 horns at the apex; petals yellow **30. Tetracme**
82 Siliqua without horns at the apex
84 Leaves 2- to 4-pinnatisect
85 Petals shorter than the sepals; siliqua 0·5–1 mm wide; seeds 0·8–1 mm **4. Descurainia**
85 Petals exceeding the sepals; siliqua 1·2–1·5 mm wide; seeds *c.* 2 mm **5. Hugueninia**
84 Leaves entire to pinnatisect
86 Petals yellow
87 Siliqua with mostly medifixed hairs lying transversely across it **19. Syrenia**
87 Hairs on siliqua not as above
88 Cauline leaves cuneate to truncate at the base, not amplexicaul; plant sometimes scapose (see also **42**)
89 Seeds in 2 rows in each loculus; siliqua 3–7 times as long as wide **57. Draba**

89 Seeds in 1 row in each loculus; siliqua at least 10 times as long as wide
90 Scapose annual; siliqua not more than 0·8 mm wide **58. Drabopsis**
90 Flowering stem leafy; siliqua usually more than 0·8 mm wide **18. Erysimum**
88 Cauline leaves sagittate or cordate, amplexicaul; plant never scapose
91 Siliqua not more than 6 times as long as wide **11. Chrysochamela**
91 Siliqua at least 10 times as long as wide
92 Siliqua pubescent **8. Arabidopsis**
92 Siliqua glabrous **42. Arabis**
86 Petals white, pink or purple
93 Plant ±densely covered with medifixed hairs **18. Erysimum**
93 Plant with unbranched, branched or stellate hairs, sometimes mixed with a few medifixed
94 Style at least 2·5 mm **43. Aubrieta**
94 Style not more than 2 mm
95 Siliqua less than 10 times as long as wide
96 Seeds in 1 row in each loculus
97 Siliqua torulose **10. Braya**
97 Siliqua not or scarcely torulose **42. Arabis**
96 Seeds in 2 rows in each loculus
98 Radicle incumbent; hairs unbranched and bifid **10. Braya**
98 Radicle accumbent; usually with at least some stellate or branched hairs **57. Draba**
95 Siliqua at least 10 times as long as wide
99 Valves of the siliqua flat, rarely keeled
100 Valves of the siliqua ±torulose, with a distinct median vein **41. Cardaminopsis**
100 Valves of the siliqua not torulose, without a median vein **42. Arabis**
99 Valves of the siliqua rounded or angled
101 Siliqua pubescent, at least when immature
102 At least some hairs unbranched or bifid, patent, hispid **22. Neotorularia**
102 Hairs all stellate and ±appressed **23. Maresia**
101 Siliqua glabrous
103 Petals ±emarginate; inner sepals ±saccate at the base **3. Murbeckiella**
103 Petals entire or truncate; inner sepals not saccate at the base
104 Siliqua up to 1 mm wide **8. Arabidopsis**
104 Siliqua at least 1 mm wide **10. Braya**
12 Fruit a silicula, less than 3 times as long as wide
105 Silicula with a transverse articulation, the lower segment sometimes sterile and stipitate
106 Leaves and flowers all basal; pedicels elongating and curving downwards to bury the fruit **104. Morisia**
106 Plant with obvious aerial stem; pedicels not curving downwards to bury the fruit
107 Upper segment of fruit oblong-ovoid, mostly at least twice as long as wide **99. Cakile**
107 Upper segment of fruit globose or ovoid-globose
108 Upper segment of fruit not beaked; style absent **102. Crambe**
108 Upper segment of fruit beaked, or with a distinct persistent style
109 Cauline leaves amplexicaul, with acute auricles **103. Calepina**
109 Cauline leaves not amplexicaul
110 Petals white; upper segment of fruit with a pungent beak **101. Didesmus**
110 Petals yellow; upper segment of fruit not pungent
111 Upper segment of fruit variously ribbed, but not winged, usually glabrous or with appressed hairs **100. Rapistrum**
111 Upper segment of fruit with 8 longitudinal winged ribs, sparsely hispid **105. Guiraoa**
105 Silicula not articulated
112 Silicula didymous, flat
113 Petals yellow; style long **78. Biscutella**
113 Petals white or pink; stigma sessile **79. Megacarpaea**
112 Silicula not didymous (rarely didymous, then not flat)
114 Silicula with 3 loculi, the upper 2 side by side, sterile, the lower 1-seeded **13. Myagrum**
114 Silicula without 2 sterile loculi side by side
115 Silicula with 4 longitudinal wings or ridges or covered with irregular protuberances **16. Bunias**
115 Silicula without 4 wings or ridges and not covered with irregular protuberances
116 Silicula latiseptate (compressed parallel to the septum which is therefore as wide as the widest diameter of the silicula), rarely indehiscent
117 Small, scapose aquatic; ovary surrounded at the base by a glandular ring **83. Subularia**
117 Terrestrial or, if aquatic, not scapose; ovary without a glandular ring
118 Glabrous or with unbranched hairs
119 Sepals erect or nearly so (calyx closed)
120 Petals at least 10 mm, red, purple or rarely white; silicula at least 20 mm **45. Lunaria**
120 Petals less than 10 mm, yellow or white; silicula not more than 12 mm **65. Camelina**
119 Sepals patent or erecto-patent (calyx ±open)
121 Petals yellow
122 Plant with at least some leaves pinnate, pinnatifid or coarsely toothed **37. Rorippa**
122 Leaves usually entire
123 Leaves linear to narrowly spathulate,

rarely obovate; plant scapose, rarely with few cauline leaves **57. Draba**
123 Leaves ovate, cordate or truncate at the base; cauline leaves present **62. Cochlearia**
121 Petals white, lilac, pink or reddish
124 Silicula longitudinally 6-veined, the veins in 2 groups of 3 which diverge from the base **34. Litwinowia**
124 Silicula not longitudinally 6-veined
125 Plant grey-pubescent **60. Petrocallis**
125 Plant green, glabrous or sparsely hairy
126 Silicula with a short, broad, obtuse beak; petals unequal, the inner distinctly longer than the outer **103. Calepina**
126 Silicula not beaked, but sometimes with a persistent style; petals ±equal
127 Robust plants up to 100 cm; roots fusiform **39. Armoracia**
127 Slender plants up to 40 cm; roots not fusiform
128 Filaments with a tooth-like appendage at the base (Balkan peninsula) **53. Bornmuellera**
128 Filaments without appendages
129 Petals deeply bifid **59. Erophila**
129 Petals entire or emarginate
130 Valves of the silicula ±flat **57. Draba**
130 Valves of the silicula convex
131 Sepals persistent in fruit; valves of the silicula not veined **64. Rhizobotrya**
131 Sepals deciduous; valves of the silicula usually with distinct veins, at least below
132 Filaments straight **62. Cochlearia**
132 Filaments strongly curved **63. Kernera**
118 Hairy, at least some of the hairs branched or stellate
133 Sepals erect (calyx closed)
134 Silicula indehiscent, subglobose or sometimes compressed, reticulate-rugose **66. Neslia**
134 Silicula dehiscent, variously shaped, not reticulate-rugose
135 Silicula stipitate **47. Alyssoides**
135 Silicula sessile
136 Petals pink, purple or violet **43. Aubrieta**
136 Petals yellow or white
137 Silicula compressed (petals short-clawed) **50. Fibigia**
137 Silicula inflated
138 Seeds usually numerous; cauline leaves usually with amplexicaul auricles; branched hairs dendritic or with only 2 or 3 rays **65. Camelina**
138 Seeds not more than 8 in each loculus; cauline leaves attenuate at the base; hairs stellate, with more than 4 rays, or lepidote
139 Petals 10–12 mm **48. Degenia**
139 Petals less than 9 mm **49. Alyssum**
133 Sepals patent or erecto-patent (calyx open)
140 Petals deeply 2-lobed
141 Scapigerous; leaves mainly in a basal rosette **59. Erophila**
141 Not scapigerous; cauline leaves present
142 Flower-buds globose; basal leaves with the petioles enlarged at the base and persistent **52. Aurinia**
142 Flower-buds oblong or ellipsoidal; basal leaves, if present, with the petioles not enlarged at the base and persistent **51. Berteroa**
140 Petals entire or emarginate
143 Petals yellow
144 Valves of the silicula reticulately veined; cauline leaves sagittate-amplexicaul **11. Chrysochamela**
144 Valves of the silicula not reticulately veined; cauline leaves not sagittate-amplexicaul
145 Petals at least 12 mm; seeds and ovules 4–8 in each loculus **47. Alyssoides**
145 Petals not more than 8 mm
146 Seeds and ovules numerous in each loculus; plant dwarf, ±scapose, usually perennial **57. Draba**
146 Seeds and ovules 1 or 2(–8) in each loculus; plant with a leafy stem **52. Aurinia**
143 Petals white, pink or purple
147 Silicula indehiscent, longitudinally 4-veined; pedicels stout, about as thick as the silicula **33. Euclidium**
147 Silicula dehiscent, not longitudinally 4-veined; pedicels much thinner than the silicula
148 Hairs nearly all medifixed
149 Silicula glabrous; filaments with a short tooth at the base **53. Bornmuellera**
149 Silicula pubescent, sometimes becoming ±glabrous; filaments without appendages **54. Lobularia**
148 Hairs stellate or dendritic
150 Ovules and seeds 1 or 2 in each loculus (seeds usually winged) **52. Aurinia**
150 Ovules and seeds at least 4 in each loculus
151 Filaments of the inner stamens winged **56. Schivereckia**

151 Filaments of the inner stamens not winged **57. Draba**
116 Silicula angustiseptate (compressed at right-angles to the septum)
152 Fruit didymous, reniform or cordate
153 Cauline leaves sessile, amplexicaul; petals 2·5–4 mm **81. Cardaria**
153 Cauline leaves shortly petiolate; petals 0·5–2 mm **82. Coronopus**
152 Fruit not didymous, reniform or cordate, but sometimes obcordate
154 Outer petals conspicuously larger than the inner
155 Style distinct, at least 0·5 mm
156 Seeds 2 or 3 in each loculus **71. Jonopsidium**
156 Seeds solitary in each loculus **77. Iberis**
155 Style inconspicuous, up to 0.3 mm
157 Annual; stigma 2-lobed **73. Teesdalia**
157 Caespitose perennial; stigma capitate **76. Teesdaliopsis**
154 Petals (when present) equal in size
158 Filaments with a wing or tooth-like appendage
159 Leaves usually pinnatifid, mostly confined to a basal rosette; sepals erecto-patent **73. Teesdalia**
159 Leaves entire, not ±confined to a basal rosette; sepals erect **75. Aethionema**
158 Filaments not appendaged
160 Valves of fruit winged or strongly keeled
161 Inflorescence bracteate, at least at the base
162 Flowers white, pink or purple **71. Jonopsidium**
162 Flowers yellow **72. Bivonaea**
161 Inflorescence ebracteate
163 Seeds 2–8 in each loculus **74. Thlaspi**
163 Seeds solitary in each loculus
164 Style 0·7–3 mm **74. Thlaspi**
164 Style short or absent **80. Lepidium**
160 Valves of fruit not winged or keeled, or with a weak keel
165 Fruits triangular-obcordate **67. Capsella**
165 Fruits not triangular-obcordate
166 Leaves simple, entire or with a few teeth or shallow lobes
167 Valves of fruit convex, not strongly compressed **62. Cochlearia**
167 Valves of fruit ±strongly compressed **69. Hymenolobus**
166 Leaves pinnate or pinnatisect
168 Perennial; seeds 1 or 2 in each loculus **68. Pritzelago**
168 Annual or biennial
169 Seeds 3–10 in each loculus; hairs unbranched when present **69. Hymenolobus**
169 Seeds 1 or 2 in each loculus; hairs stellate when present **70. Hornungia**

1. Sisymbrium L.

P. W. BALL

Annuals to perennials; glabrous or with unbranched hairs. Leaves entire to pinnate. Sepals not saccate at the base; petals yellow, rarely white, entire. Fruit a siliqua; valves usually 3-veined; style distinct or indistinct; stigma more or less 2-lobed. Seeds small (usually less than 2·5 mm).

Many of the widespread lowland species occur as weeds, and are often of uncertain status in many parts of Europe.

Literature: O. E. Schulz in Engler, *Pflanzenreich* **86 (IV.105)**: 46–157 (1924).

1 Inflorescence bracteate, at least in the lower part
2 Petals white; valves of the siliqua with a distinct median vein and reticulate lateral veins **1. supinum**
2 Petals pale yellow; valves of the siliqua 3-veined
3 Petals 2·5–3·5 mm; style almost as thick as the siliqua **16. runcinatum**
3 Petals 1·5–2·5 mm; style distinctly thinner than the siliqua
4 Pedicels 0·5–1 mm in fruit; siliqua straight to recurved; style 0·5–1 mm **14. polyceratium**
4 Pedicels 2–3 mm in fruit; siliqua ±straight; style 1–2 mm **15. confertum**
1 Inflorescence ebracteate (rarely lowest 1–3 flowers bracteate)
5 Siliqua not more than 20 mm, closely appressed to the stem
6 Siliqua (7–)10–20 mm, straight, conical-cylindrical; petals 2–4 mm **18. officinale**
6 Siliqua 7–10 mm, curved or twisted, strongly compressed at the apex; petals *c.* 1·5 mm **19. cavanillesianum**
5 Siliqua usually more than 20 mm, not closely appressed to the stem
7 Lower leaves ovate, entire or dentate
8 Annual; petals less than 3 mm, not exceeding the sepals **17. erysimoides**
8 Biennial or perennial; petals more than 3 mm, exceeding the sepals
9 Lower pedicels 5–15 mm in fruit; petals 4·5–10 mm **2. strictissimum**
9 Lower pedicels 3–6 mm in fruit; petals 3–5 mm **8. austriacum**
7 At least the lower leaves deeply lobed or divided
10 Petals less than 3 mm, shorter than or only slightly exceeding the sepals; anthers *c.* 0·5 mm
11 Pedicels 3–6 mm in flower, up to 20 mm in fruit, and much thinner than the siliqua **3. irio**
11 Pedicels 1–2 mm in flower, up to 5 mm in fruit, and almost as thick as the siliqua
12 Lower leaves lyrate-pinnatifid, the lateral lobes obovate to lanceolate; stamens *c.* 3 mm, longer than the petals **17. erysimoides**
12 Lower leaves sinuate-pinnatifid, the lateral lobes ±triangular; stamens *c.* 1·5 mm, shorter than the petals **15. confertum**
10 Petals 3 mm or more, distinctly exceeding the sepals; anthers (0·7–)1–3 mm
13 Pedicel about as thick as the siliqua, usually at least 0·7 mm in diameter at the base

14 Ovules not more than 30 in each loculus; siliqua rarely more than 60 mm **8. austriacum**
14 Ovules 40–60 in each loculus; siliqua 40–100(–180) mm
15 Uppermost leaves pinnatisect, with the terminal lobe linear **12. altissimum**
15 Uppermost leaves entire or hastate, with the terminal lobe oblong or lanceolate **13. orientale**
13 Pedicel thinner than the siliqua, rarely more than 0·5 mm in diameter at the base
16 Inflorescence very contracted, the young siliquae distinctly overtopping the flowers and buds; anthers *c.* 0·7 mm **3. irio**
16 Inflorescence elongate, the young siliquae not or scarcely overtopping the flowers; anthers 1–3 mm
17 Siliqua less than 15 mm, usually strongly contorted **8. austriacum**
17 Siliqua more than 15 mm, straight or only slightly contorted
18 Septum of the siliqua white or yellowish, opaque; stem glabrous or with short, upwardly curving hairs; seeds 1–1·5 mm
19 Valves of the siliqua with slender veins; septum not or only slightly foveolate; petals usually less than 6·5 mm **8. austriacum**
19 Valves of the siliqua with prominent, usually very thick veins; septum foveolate; petals more than 6·5 mm **(9–11). crassifolium** group
18 Septum of the siliqua ± hyaline; stem glabrous or hispid
20 Seeds 0·7–1 mm; petals 4–7 mm
21 Stem and leaves pubescent, at least in the lower part; basal leaves not distinctly larger than the cauline, often dead at anthesis **4. loeselii**
21 Stem and leaves glabrous or subglabrous; basal leaves 15–35 cm, persistent, much larger than the cauline **5. assoanum**
20 Seeds 1–1·5 mm; petals 6–10 mm
22 Lower leaves deeply pinnatisect with *c.* 4 pairs of lobes; siliqua not more than 1 mm in diameter, the values ± distinctly 3-veined **6. polymorphum**
22 Lower leaves triangular-hastate with prominent basal lobes; siliqua up to 1·2 mm in diameter, the valves with weak lateral veins **7. volgense**

Sect. KIBERA (Adanson) DC. Inflorescence bracteate. Petals white. Valves of the siliqua with a median vein and reticulate lateral veins.

1. S. supinum L., *Sp. Pl.* 657 (1753) (*Braya supina* (L.) Koch). Annual 5–35 cm, shortly pubescent with patent hairs. Leaves sinuate-pinnatisect, the terminal lobe oblong to obovate, the laterals linear to oblong; lobes often shallowly lobed or coarsely toothed. Pedicels 1–2 mm in flower, 2–4 mm in fruit. Petals 3–4 mm; anthers 0·5–0·7 mm. Siliqua 10–30 × 1·5–2 mm, obtuse, pubescent; style 1–2·5 mm. $2n = 42$. ● *From Spain to N.W. Russia; local and in many regions transitory.* †Be Ga †Ge He †Ho Hs Rs(B, C) Su.

Sect. NORTA (Adanson) DC. Leaves simple. Inflorescence ebracteate. Petals yellow. Valves of the siliqua 3-veined.

2. S. strictissimum L., *Sp. Pl.* 660 (1753). Perennial 50–100(–150) cm, glabrescent or shortly pubescent. Leaves ovate or lanceolate, acute to acuminate, dentate or entire. Pedicels (4–)8–15 mm; Petals 4·5–10 mm; stamens 4–7 mm; anthers 1–2 mm. Siliqua (15–)30–80 × 0·7–2 mm. Seeds 1·5–3 mm. $2n = 28$. ● *From France and Italy eastwards to E.C. Russia and Bulgaria; introduced further west.* Al Au Bu Cz Ga Ge He Hu It Ju Po Rm Rs(C, W, E) [Br Hb ?Hs].

Sect. IRIO DC. Inflorescence ebracteate. Petals yellow. Valves of the siliqua 3-veined.

3. S. irio L., *Sp. Pl.* 659 (1753). Annual up to 60 cm, glabrescent or with long, curved hairs. Lower leaves sinuate-pinnatifid, the terminal lobe larger than the laterals; cauline leaves lobed or entire and hastate, shortly petiolate. Inflorescence condensed, the young siliquae overtopping the flowers and buds; pedicels 3–6 mm in flower, up to 20 mm in fruit. Petals 2·5–3·5(–6) mm; anthers *c.* 0·7 mm. Siliqua 25–65 × 0·7–1·2 mm, torulose; style 0·3–0·7 mm; septum more or less hyaline. Seeds 0·7–1·1 mm. $2n = 14$, 42. *Native in S. Europe; widely naturalized elsewhere.* *Au Bl Bu Co Cr Ga Gr He Hs It Ju Lu *Rm Sa Si Tu [Br Cz Hb Ho Hu Rs(B, C, W)].

4. S. loeselii L., *Cent. Pl.* **1**: 18 (1755) (*S. laeselii* auct.). Annual 20–100(–180) cm, usually more or less hispid. Leaves lyrate- or sinuate-pinnatisect, the terminal lobe triangular-ovate to oblong; upper shortly petiolate. Pedicels 5–14 mm. Petals 4–7 mm; anthers *c.* 1·5 mm. Siliqua (10–)15–45 × 0·6–1 mm; style 0·5–1·2 mm; septum more or less hyaline. Seeds 0·7–1 mm. $2n = 14$. *C. & E. Europe, westwards to Germany and Italy; often introduced in the west and north.* Au Bu Cz Ge Gr Hu It Ju Po Rm Rs(B, C, W, K, E) Tu [Be Br Da Ga He Ho Hs Su].

5. S. assoanum Loscos & Pardo, *Ser. Pl. Arag.* 6 (1863). Like **4** but the stem glabrous or shortly pubescent at the base; leaves glabrous or subglabrous; basal leaves 15–35 cm, much larger than the cauline, the terminal lobe oblong-elliptical; cauline leaves often simple; petals 4–6 mm; anthers *c.* 1 mm. ● *E. & C. Spain.* Hs.

6. S. polymorphum (Murray) Roth, *Man. Bot.* **2**: 946 (1830) (*S. junceum* Bieb.). Annual or perennial 20–100 cm, glabrescent, or hispid at the base. Lower leaves pinnatisect, the terminal lobe large, ovate, the laterals linear; upper leaves entire, linear. Petals 6–9 mm; anthers 1.5–3 mm. Siliqua (12–)20–40 × 0.7–1 mm, the style rarely more than 0.7 mm; septum more or less hyaline. Seeds 1–1.5 mm. *E. & E.C. Europe.* Bu Hu Po Rm Rs(C, W, K, E) [Rs(N)].

7. S. volgense Bieb. ex E. Fourn., *Rech. Fam. Crucif.* 97 (1865). Like **6** but rhizomatous perennial, more or less glabrous or shortly pubescent at the base; lower leaves ovate or elliptic-triangular, hastate, with prominent basal lobes; petals 6–10 mm; siliqua 25–60 × 0·7–1·2 mm, the style 0·4–1 mm; valves with weak lateral veins. $2n = 14$. ● *S.E. Russia; locally naturalized elsewhere.* Rs(E) [Br Cz Ge Po Rs(B, C, W)].

The leaves of this species are sometimes entire.

8. S. austriacum Jacq., *Fl. Austr.* **3**: 35 (1775) (*S. pyrenaicum* (L.) Vill., non L., *S. multisiliquosum* Willk.). Biennial or perennial 10–100 cm, almost glabrous or with upward-curving, setose hairs. Leaves entire to sinuate-pinnatisect. Petals 3–7(–8) mm; anthers

1–2 mm. Siliqua up to 60 × 0·5–1·5 mm; septum yellowish, opaque. Seeds 1–1·5 mm. ● *W.C. & S.W. Europe, from* c. 15° *E. to the Iberian peninsula.* Au Be Cz Ga Ge He Hs It *Lu [Ho Rs(W)].

1 Middle and upper cauline leaves entire or dentate
  2 Basal leaves pinnatifid or pinnatisect; sepals 2–3 mm (c) subsp. **contortum**
  2 Basal leaves entire or sinuate-pinnatifid; sepals 1·5–2·5 mm (d) subsp. **hispanicum**
1 Middle and upper cauline leaves pinnatifid
  3 Siliqua 7–15 mm, strongly contorted (b) subsp. **chrysanthum**
  3 Siliqua (10–)15–50 mm, not or slightly contorted
    4 Stem glabrous or sparsely setose; siliqua 0·7–1·7 mm in diameter (a) subsp. **austriacum**
    4 Stem densely setose; petals 3–5·5 mm; siliqua 0·7–1 mm in diameter (c) subsp. **contortum**

(**a**) Subsp. **austriacum** (*S. acutangulum* DC.): Stem glabrous or sparsely setose. Basal leaves pinnatifid or pinnatisect; middle and upper cauline leaves pinnatifid. Sepals (2–)2·5–4 mm; petals (3·5–)4–7 mm. Siliqua (10–)15–50 × 0·7–1·7 mm, not or slightly contorted. *Throughout the range of the species except for parts of the Iberian peninsula; frequently casual elsewhere.*

(**b**) Subsp. **chrysanthum** (Jordan) Rouy & Fouc., *Fl. Fr.* **2**: 17 (1895) (*S. pyrenaicum* auct., *S. contortum* auct. gall., non Cav.): Stem and leaves as in subsp. (**a**). Sepals 1·5–3 mm; petals 3·5–4·5 mm. Siliqua 7–15 × 0·8–1·4 mm, strongly contorted. *C. & W. Pyrenees, mountains of N. Spain.*

(**c**) Subsp. **contortum** (Cav.) Rouy & Fouc., *op. cit.* 19 (1895) (*S. contortum* Cav.): Stem densely setose. Basal leaves pinnatifid or pinnatisect; middle and upper cauline leaves entire or dentate, rarely pinnatifid. Sepals 2–3 mm; petals 3–5·5 mm. Siliqua 15–50 × 0·5–1(–1·2) mm, slightly contorted. *N. & C. Spain.*

(**d**) Subsp. **hispanicum** (Jacq.) P. W. Ball & Heywood, *Feddes Repert.* **64**: 17 (1961) (*S. hispanicum* Jacq.): Stems usually more or less glabrous. Basal leaves entire or sinuate-pinnatifid; middle and upper cauline leaves entire. Sepals 1·5–2·5 mm; petals 3–4·5 mm. Siliqua 15–35 × 0·5–1 mm, not or slightly contorted. *C. & S.E. Spain.*

(**9–11**). **S. crassifolium** group. Stem glabrescent or with upward-curving hairs. Lower leaves sinuate-pinnatifid to pinnatisect. Petals 6·5–11 mm, pale yellow. Siliqua usually 1–1·5 mm in diameter, the valves with very prominent veins; style up to 2 mm; septum thick, opaque, foveolate. Seeds 1–1·5 mm.

1 Inflorescence condensed, at least the upper flowers overtopping the buds; valves of the siliqua with very thick veins **9. crassifolium**
1 Inflorescence lax; valves of the siliqua with thin but prominent veins
  2 Seeds ovate-oblong; lower leaves usually sinuate-pinnatifid, the lateral lobes triangular; uppermost leaves dentate or entire **10. laxiflorum**
  2 Seeds cylindrical; lower leaves sinuate-pinnatisect, the lateral lobes oblong- to ovate-triangular; uppermost leaves usually with linear lobes at the base **11. arundanum**

9. S. crassifolium Cav., *Descr. Pl.* 437 (1803) (*S. granatense* Boiss.). Biennial or perennial 30–70(–100) cm. Lower leaves sinuate-pinnatifid, the lateral lobes usually triangular; uppermost leaves dentate to entire. Inflorescence condensed; pedicels 2–6 mm, up to 12 mm in fruit. Siliqua 35–90 mm, the valves with very thick, broad veins. Seeds ovoid-oblong to oblong. *Spain & Portugal.* Hs Lu.

10. S. laxiflorum Boiss., *Elenchus* 9 (1838). Like **9** but inflorescence lax; pedicels 1·5–4 mm in flower, up to 10 mm in fruit; valves of the siliqua with slender, very prominent veins. ● *Mountains of S. Spain.* Hs.

11. S. arundanum Boiss., *Voy. Bot. Midi Esp.* **2**: 30 (1839). Like **9** but lower leaves sinuate-pinnatisect, the lateral lobes ovate- to oblong-triangular; upper leaves with linear lobes at their base; racemes lax; seeds cylindrical. ● *Mountains of S. Spain.* Hs.

Sect. SISYMBRIUM. Inflorescence ebracteate. Petals yellow. Siliqua patent, with 80–120 seeds.

12. S. altissimum L., *Sp. Pl.* 659 (1753) (*S. sinapistrum* Crantz, *S. pannonicum* Jacq.). Annual up to 100 cm, usually hispid at the base. First-formed basal leaves minute, oblong-obovate, subentire; later-formed basal and lower cauline leaves sinuate-pinnatifid; upper pinnate, with linear lobes. Pedicels 4–10 mm. Outer sepals with a short horn at the apex; petals 5–11 mm, pale yellow; anthers 1–2 mm. Siliqua (35–)50–100 × 1–1·5 mm; style 0·5–2·5 mm, cylindrical, with shortly 2-lobed stigma. Seeds 0·7–1·2 mm. $2n = 14$. *C. & E.C. Europe; widely naturalized elsewhere.* Al Bu Gr Hu Ju Po Rm Rs(*B, C, W, K, E) Tu [Au Be Br Cz Da Fe Ga Ge He Ho Hs Is *It Lu No Rs(N) *Si Su].

13. S. orientale L., *Cent. Pl.* **2**: 24 (1756) (*S. columnae* Jacq., *S. longesiliquosum* Willk.; incl. *S. costei* Fouc. & Rouy). Like **12** but the stem softly pubescent; upper leaves petiolate, simple, 3-lobed or hastate, the terminal lobe linear or lanceolate; sepals usually without a short horn at the apex; petals (6–)7–10 mm; style and stigma 0·2–3·5 mm, clavate, the apex about as thick as the siliqua. $2n = 14$. *S. Europe; widely naturalized elsewhere.* Al Bl Bu Co Cr Ga Gr *Hu Hs *It Ju *Rm *Sa *Si Tu [Au Be Br Cz ?Ge Hb He Ho No Po Rs(W, *K) Su].

Sect. CHAMAEPLIUM (Wallr.) Thell. Inflorescence usually bracteate. Petals pale yellow. Valves of the siliqua 3-veined.

14. S. polyceratium L., *Sp. Pl.* 658 (1753) (*Chamaeplium polyceratium* (L.) Wallr.). Usually glabrous, foetid annual up to 75 cm. Lower leaves sinuate-pinnatifid, the lobes triangular. Inflorescence bracteate to the apex; flowers usually fasciculate; pedicels 0·5–1 mm in fruit. Petals 1·5–2 mm, equalling or slightly exceeding the sepals; anthers 0·3–0·5 mm. Siliqua 10–25 × 0·7–1·2 mm, straight to recurved, torulose; style 0·5–1 mm, thinner than the siliqua. Seeds 0·6–1 mm. *S. Europe.* Al Bl Bu Co Cr Ga Gr It Ju Sa Si Tu [Lu].

15. S. confertum Steven ex Turcz., *Bull. Soc. Nat. Moscou* **27(2)**: 304 (1854) (*S. anomalum* Aznav., *S. austriacum* subsp. *thracicum* Aznav., *S. lagascae* Amo). Like **14** but the inflorescence often ebracteate in the upper part; pedicels 2–3 mm in fruit; petals 2–2·5 mm; siliqua up to 45 mm, usually more or less straight; style 1–2 mm. *Krym; Turkey-in-Europe.* Rs(K) Tu. (*S.W. Asia.*)

16. S. runcinatum Lag. ex DC., *Reg. Veg. Syst. Nat.* **2**: 478 (1821) (*Chamaeplium runcinatum* (Lag. ex DC.) Hayek). Like **14** but the lower cauline leaves subentire to sinuate-pinnatifid, the lobes oblong; flowers usually solitary; petals 2·5–3·5 mm; siliqua 10–35 × 1–2 mm, straight or slightly recurved; style 1–2 mm, almost as thick as the siliqua. $2n = 56$. *S.W. Europe; sometimes casual elsewhere.* Ga Hs Lu.

Sect. OXYCARPUS Paol. Inflorescence ebracteate. Petals yellow, drying to white; stamens longer than the petals. Siliqua patent.

17. S. erysimoides Desf., *Fl. Atl.* **2**: 84 (1798). Annual 10–80 cm, glabrous or shortly pubescent. Leaves lyrate-pinnatifid, rarely entire, ovate-lanceolate, serrate. Pedicels 1–2 mm in flower, up to 5 mm in fruit. Petals 1–3 mm; anthers *c.* 0·5 mm. Siliqua 20–50 × 0·7–1·3 mm, attenuate into the style; style not more than 1 mm. Seeds 0·8–1·2 mm. *W. Mediterranean region.* Bl Hs *Lu Sa [Az].

Sect. VELARUM DC. Inflorescence ebracteate. Petals yellow. Siliqua closely appressed to the stem.

18. S. officinale (L.) Scop., *Fl. Carn.* ed. 2, **2**: 26 (1772) (*Chamaeplium officinale* (L.) Wallr.). Annual or biennial 5–100(–130) cm. Lower leaves pinnatisect, more or less ovate in outline, with a large terminal lobe. Pedicels 1–2 mm. Petals 2–4 mm; anthers *c.* 0·7 mm; ovary with 10–20 ovules. Siliqua (7–)10–20 × 1–1·6 mm, conical-cylindrical, straight, attenuate into the style; style 0·5–2 mm. Seeds 1–1·7 mm. $2n = 14$. *Europe northwards to* c. 68° *N.* All except Fa Sb; introduced in Is.

19. S. cavanillesianum Castroviejo & Valdés-Bermejo, *Anal. Inst. Bot. Cavanilles* **34**: 327 (1977) (*S. matritense* P. W. Ball & Heywood, non Pau, *S. corniculatum* Cav., non Lam.). Like **18** but the lower leaves oblong in outline, the terminal lobe only slightly larger than the laterals; petals *c.* 1.5 mm; ovary with 6 or 7 ovules; siliqua 7–10 mm, curved or contorted, the apex compressed. $2n = 14$. ● *C. Spain.* Hs.

2. Lycocarpus O. E. Schulz

P. W. BALL

Like *Sisymbrium* but the siliqua clavate at the apex, constricted in the middle; stigma sessile.

1. L. fugax (Lag.) O. E. Schulz in Engler, *Pflanzenreich* **86 (IV.105)**: 164 (1924) (*Sisymbrium fugax* Lag.). Subglabrous or pubescent annual 10–40 cm. Leaves sinuate-pinnatifid. Racemes ebracteate; pedicels 2–5 mm in fruit, only slightly thinner than the siliqua. Petals 2·5–4 mm, yellow. Siliqua 8–20 mm, 1 mm in diameter at the base, the apex up to 1·2 mm in diameter. Seeds 0·5–0·8 mm. ● *S.E. Spain.* Hs.

3. Murbeckiella Rothm.

A. O. CHATER

Perennials; glabrous, or with stellate or long, unbranched hairs. Leaves entire to pinnatisect. Sepals unequal, the inner more or less saccate at the base; petals white, emarginate. Fruit a siliqua; valves with a distinct median vein; style very short; stigma slightly 2-lobed. Seeds small (not more than 1·5 mm), often winged at the apex. (*Phryne* sensu O. E. Schulz.)

Literature: W. Rothmaler, *Bot. Not.* **1939**: 467–476 (1939). R. Fernandes, *Mem. Soc. Brot.* **6**: 79–91 (1950). O. E. Schulz in Engler, *Pflanzenreich* **86 (IV.105)**: 169–175 (1924).

1 Basal leaves entire to crenate; pedicels 3–8 mm in fruit
2 Cauline leaves 6–9; stems leafy to the apex **1. pinnatifida**
2 Cauline leaves 3–5; stems leafy only in lower ½ **2. boryi**
1 Basal leaves pinnatifid; pedicels 5–15 mm in fruit
3 Cauline leaves 4–8, pinnatifid, with 4–6 lobes on each side **3. zanonii**
3 Cauline leaves 2 or 3, subentire to dentate **4. sousae**

1. M. pinnatifida (Lam.) Rothm., *Bot. Not.* **1939**: 469 (1939) (*Braya pinnatifida* (Lam.) Koch, *Sisymbrium pinnatifidum* (Lam.) DC., *S. dentatum* All., *Arabis pinnatifida* Lam.). Plant sparsely to densely covered with stellate hairs; stems more or less densely leafy throughout. Basal leaves 1–2 cm, entire to repand-dentate; cauline leaves 6–9, pinnatifid, with 4–6 lobes on each side. Pedicels 3–4 mm in fruit. Sepals 1·5–2·5 mm; petals 3·5–4 mm, slightly emarginate. Siliqua 10–30 × 1 mm; style 0·25 mm, conical. Seeds 1 × 0·5 mm, wingless or winged only at the apex. $2n = 16$. *Mountain rocks.* ● *W. Europe, from C. Pyrenees to* c. 8° *E. in W.C. Alps.* Ga He Hs It.

2. M. boryi (Boiss.) Rothm., *loc. cit.* (1939) (*Braya pinnatifida* sensu Willk. pro parte; incl. *M. glaberrima* (Rothm.) Rothm.). Plant sparsely to densely covered with stellate hairs, or with only unbranched hairs, sometimes subglabrous; stems sparsely leafy up to the middle. Basal leaves 1·5–3 cm, entire or sinuate-dentate; cauline leaves 3–5, pinnatifid, with 1–6 lobes on each side. Pedicels 3–8 mm in fruit. Sepals 2–3·5 mm; petals 3·5–8 mm, deeply emarginate. Siliqua 10–35 × 0·5–1 mm; style 0·5 mm. Seeds 1–1·5 × 0·5–0·75 mm, winged all round or only at the apex. *C. & S. Spain, C. Portugal.* Hs Lu.

3. M. zanonii (Ball) Rothm., *op. cit.* 471 (1939) (*Sisymbrium zanonii* (Ball) Gay). Plant usually densely covered with stellate hairs; stems leafy to the apex. Basal leaves 2–5 cm, pinnatifid; cauline leaves 4–8, pinnatifid, with 4–6 lobes on each side. Pedicels 5–14 mm in fruit. Sepals 2·5 mm; petals 5–7 mm, emarginate. Siliqua 20–50 × 0·75 mm, somewhat patent; style 0.5–1 mm, cylindrical. Seeds 0·8 × 0·5 mm, wingless or winged only at the apex. ● *N. Appennini.* It.

4. M. sousae Rothm., *op. cit.* 474 (1939). Plant glabrous or with sparse stellate hairs; stems leafless or with 2 or 3 leaves. Basal leaves 4–7 cm, pinnatifid, with 2–4 lobes on each side; cauline leaves 2 or 3, subentire or sinuate-dentate. Pedicels 10–15 mm in fruit. Sepals 3–3·5 mm; petals 6–7 mm, slightly emarginate. Siliqua 40–55 × 1 mm; style 0·5 mm, conical. Seeds 1 × 0·6 mm, winged all round. ● *C. & N. Portugal (Serra de Lousã, Serra de Marão).* Lu.

4. Descurainia Webb & Berth.

P. W. BALL

Annual or biennial; hairs of 2 kinds, dendritic and unbranched. Leaves 2- to 4-pinnatisect. Sepals not saccate; petals pale yellow, smaller than the sepals. Fruit a siliqua; valves with a distinct median vein; style very short; stigma subcapitate. Seeds small (not more than 1 mm), not mucilaginous when moistened.

1. D. sophia (L.) Webb ex Prantl in Engler & Prantl, *Natürl. Pflanzenfam.* **3(2)**: 192 (1891) (*Sisymbrium sophia* L.). Plant 20–100 cm. Cauline leaves with linear or oblong lobes. Pedicels 5–50 mm in fruit. Sepals 2–3 mm; stamens usually exceeding the sepals. Siliqua 8–45 × 0·5–1 mm, usually erecto-patent, torulose; valves reticulately veined. Seeds 0·8–1·1 mm. $2n = 28$. *Usually ruderal. Throughout Europe northwards to* c. 70° *N., but doubtfully native in much of the north.* All except Az Fa Sa Sb.

5. Huguenінia Reichenb.

P. W. BALL

Perennial; hairs more or less stellate. Leaves 2-pinnatisect. Sepals not saccate at the base; petals yellow, exceeding the sepals. Fruit a siliqua; valves 1-veined; style very short; stigma slightly 2-lobed. Seeds *c.* 2 mm, not winged, mucilaginous when moistened.

1. **H. tanacetifolia** (L.) Reichenb., *Fl. Germ. Excurs.* 691 (1832) (*Sisymbrium tanacetifolium* L.). Stout, tomentose, pubescent or subglabrous perennial 30–70 cm. Lower leaves up to 30 cm, long-petiolate; segments 8–10 pairs, broadly linear to lanceolate, serrate to pinnatifid, the ultimate lobes 4–10 mm wide. Petals *c.* 4 mm. Siliqua 6–15 × 1·2–1·5 mm, oblong-oblanceolate, erecto-patent; valves with a strong mid-vein. Seeds *c.* 2 mm. ● *S.W. Alps, eastwards to* c. 7° 45′ *E.; C. Pyrenees, mountains of N. Spain.* Ga He Hs It.

(a) Subsp. **tanacetifolia**: Stem and lower surface of leaves grey-pubescent or subglabrous; segments of the lower leaves with 4–8(–10) pairs of teeth or lobes. Pedicels 5–8 mm in fruit. *S.W. Alps.*

(b) Subsp. **suffruticosa** (Coste & Soulié) P. W. Ball, *Feddes Repert.* **68**: 194 (1963): Stem and lower surface of leaves grey-tomentose; segments of the lower leaves with 1–4(5) pairs of teeth or lobes. Pedicels 7–11 mm in fruit. *Pyrenees and mountains of N. Spain.*

6. Alliaria Scop.

P. W. BALL

Biennial, more rarely perennial; hairs unbranched. Basal leaves undivided. Sepals not saccate at the base; petals white. Fruit a siliqua; valves 3-veined; style distinct; stigma more or less entire. Seeds large (2·5–4·5 mm).

1. **A. petiolata** (Bieb.) Cavara & Grande, *Boll. Orto Bot. Napoli* **3**: 418 (1913) (*A. officinalis* Andrz. ex Bieb., *Sisymbrium alliaria* (L.) Scop.). Plant up to 120 cm, smelling of garlic when crushed. Leaves reniform to triangular-ovate, cordate, crenate to coarsely sinuate-dentate, the basal long-petiolate. Pedicels 2·5–13 mm in fruit, about as thick as the siliqua. Petals 4–8(–9) mm. Siliqua 20–70 × 1·5–3·5 mm, patent or erecto-patent. $2n = 36, 42$. *Europe from about* 68° *N. southwards, but rarer in the extreme south.* All except Az Bl Cr Fa Is Rs(N) Sa Sb.

7. Eutrema R. Br.

P. W. BALL

Perennial; glabrous or with very sparse unbranched hairs. Basal leaves simple. Sepals not saccate at the base; petals white. Fruit a siliqua; valves 1-veined; style distinct; stigma slightly 2-lobed. Seeds large (*c.* 3 mm), winged.

1. **E. edwardsii** R. Br., *Chloris Melv.* 9 (1823). Plant 5–40 cm, rhizomatous. Basal leaves ovate to orbicular, broadly cuneate to cordate at the base, usually entire, long-petiolate; cauline leaves oblong-ovate to linear-lanceolate, cuneate, more or less sessile; all fleshy. Petals 2–4 mm. Siliqua 6–20 × 1.5–3 mm, linear or oblong, erecto-patent; style 0.25–0.9 mm. $2n = 42$. *Arctic Europe.* Rs(N) Sb.

8. Arabidopsis (DC.) Heynh.

P. W. BALL

Annual to perennial; glabrous or with unbranched and branched hairs. Leaves entire to pinnatifid. Sepals not saccate at the base; petals white, pale purple or yellow, sometimes absent. Fruit a siliqua; valves 1-veined; style short; stigma subcapitate. Seeds small (less than 1 mm). (*Stenophragma* Čelak, *Hylandra* Á. Löve.)

Literature: O. E. Schulz in Engler, *Pflanzenreich* **86** (**IV.105**): 268–285 (1924).

The limits of this genus are uncertain. For a full discussion of the problem see N. Hylander, *Bull. Jard. Bot. Bruxelles* **27**: 591–604 (1957), F. Laibach, *Planta* **51**: 148–166 (1958), and Á. Löve, *Svensk Bot. Tidskr.* **55**: 211–217 (1961). It seems probable that *Cardaminopsis* should be combined with *Arabidopsis*, and some species of *Arabidopsis* may have to be removed from the enlarged genus. Schulz's delimitation is maintained here.

1 Cauline leaves sagittate-amplexicaul at the base; most hairs bifid or stellate
  2 Petals 4–8 mm, white; siliqua glabrous **4. toxophylla**
  2 Petals 2–3 mm, yellow; siliqua pubescent with stellate hairs **5. pumila**
1 Cauline leaves cuneate at the base; most hairs unbranched and bifid or absent
  3 Glabrous; petals absent; siliqua abruptly contracted into a very short style **3. parvula**
  3 Pubescent or hispid, at least at the base; petals 2 mm or more; siliqua ± attenuate into style
    4 Petals 2–4(–4·5) mm; anthers 0·3–0·5 mm; siliqua 5–20(–30) × 0·5–0·8 mm, laterally compressed **1. thaliana**
    4 Petals 4 mm or more; anthers 0·5–0·8 mm; siliqua 20–30 × 0·7–1 mm, not strongly compressed **2. suecica**

1. **A. thaliana** (L.) Heynh. in Holl & Heynh., *Fl. Sachs.* **1**: 538 (1842) (*Sisymbrium thalianum* (L.) Gay). Annual or biennial up to 40(–70) cm, sparsely pubescent in the lower parts with mostly unbranched hairs on the stem and 2(3)-fid hairs on the leaves. Basal leaves entire to dentate; cauline usually entire, sessile, cuneate at the base. Petals 2–4(–4·5) mm, white; stamens 4–6; anthers 0·3–0·5 mm. Siliqua 5–20(–30) × 0·5–0·8 mm, glabrous, attenuate into style 0·3–0·5 mm. $2n = 10$. *Open habitats. Europe northwards to* c. 68° *N.* All except Az Fa Is Sb.

2. **A. suecica** (Fries) Norrlin, *Meddel. Soc. Fauna Fl. Fenn* **2**: 12 (1878) (*Arabis suecica* Fries, *Cardaminopsis suecica* (Fries) Hiitonen, *Hylandra suecica* (Fries) Á. Löve). Like **1** but usually not more than 20 cm; basal leaves dentate to pinnatifid; the lower cauline usually dentate; petals 4–8 mm; anthers 0·5–0·8 mm; siliqua 20–30 × 0·7–1 mm; style 0·3–0·7 mm. $2n = 26$. *Gravelly places. Fennoscandia and Baltic region.* Fe *Ge *No Rs(C) Su [Rs(N, B)].

This species, intermediate between **1** and *Cardaminopsis arenosa* (L.) Hayek, is very probably an allopolyploid derived from these two species. It is often confused with *C. arenosa* but may be distinguished by the dentate to pinnatifid (not pinnatipartite) basal leaves, the always white flowers, the petals without teeth on the claw, and the siliqua not strongly compressed.

3. **A. parvula** (Schrenk) O. E. Schulz in Engler, *Pflanzenreich* **86** (**IV.105**): 269 (1924). Like **1** but stem 3–15 cm, glabrous; leaves fleshy; sepals 2 mm; petals absent; siliqua 6–15 mm, torulose,

abruptly contracted into a very short style. *Saline habitats. S.E. Russia (Ozero El'ton).* Rs(E). (*C. & S.W. Asia.*)

4. A. toxophylla (Bieb.) N. Busch in Kusn., N. Busch & Fomin, *Fl. Cauc. Crit.* **3(4)**: 457 (1909) (*Sisymbrium toxophyllum* (Bieb.) C. A. Meyer). Biennial or perennial 10–45 cm, subglabrous to densely tomentose with stellate hairs. Basal leaves usually sinuate-dentate; cauline denticulate or entire, sagittate-amplexicaul. Petals 4–8 mm, white; anthers *c.* 1 mm. Siliqua 10–25 mm, glabrous; style 0·5–0·8 mm. $2n = 24$. *Saline habitats. From S.W. Ukraine to E. Russia and W. Kazakhstan.* Rs(W, ?K, E).

5. A. pumila (Stephan) N. Busch in Kusn., N. Busch & Fomin, *loc. cit.* (1909) (*Sisymbrium pumilum* (Stephan) Boiss.). Annual up to 50 cm, with bi- to multifid hairs. Basal leaves dentate to dentate-pinnatifid; cauline denticulate, sagittate-amplexicaul at the base. Petals 2–3 mm, yellow; anthers *c.* 0·25 mm. Siliqua 10–25(–35) mm, more or less pubescent with stellate hairs; style 0·2–0·8 mm. $2n = 32$. *Saline habitats. S.E. Russia, W. Kazakhstan.* Rs(E). (*S.W. Asia.*)

9. Thellungiella O. E. Schulz

P. W. BALL

Glabrous annual or biennial. Leaves usually entire. Sepals not saccate at the base. Petals white or pink. Fruit a siliqua; valves 1-veined; style very short; stigma slightly 2-lobed. Seeds in 2 rows in each loculus, small (*c.* 0.5 mm).

1. T. salsuginea (Pallas) O. E. Schulz in Engler, *Pflanzenreich* **86(IV.105)**: 252 (1924) (*Sisymbrium salsugineum* Pallas). Plant erect, divaricately branched, up to 35 cm. Cauline leaves oblong-ovate, obtuse with rounded auricles. Pedicels 4–8 mm in fruit, patent or erecto-patent. Petals 3–4 mm. Siliqua 12–16 × 0·7–1 mm, linear. *Saline habitats. S.E. Russia; perhaps only casual.* *Rs(E). (*N. & C. Asia, W. North America.*)

10. Braya Sternb. & Hoppe

P. W. BALL

Perennials; hairs of 2 kinds, branched and unbranched. Leaves undivided. Sepals not saccate at the base; petals white or purplish, truncate. Fruit a siliqua or silicula; valves 1-veined; style short; stigma slightly 2-lobed. Seeds in 1 or 2 rows in each loculus, small (*c.* 1 mm). Radicle incumbent.

Literature: O. E. Schulz in Engler, *Pflanzenreich* **86 (IV.105)**: 226–238 (1924).

1 Cauline leaves 0 or 1; siliqua 4–10 × 1–3 mm, elliptical or oblong-ovate **3. purpurascens**
1 Cauline leaves 1 to several; siliqua 8–15 × 1–1·7 mm, linear or broadly linear
2 Siliqua 5–11 × 1–1·7 mm, 4–7 times as long as wide **1. alpina**
2 Siliqua 8–15 × 1–1·2 mm, 7–12 times as long as wide **2. linearis**

1. B. alpina Sternb. & Hoppe, *Denkschr. Bayer. Bot. Ges. Regensb.* **1**: 66 (1815). Laxly caespitose; flowering stems up to 10 cm. Cauline leaves 1 to several; lower leaves lanceolate, entire or denticulate. Petals 3–4 mm. Siliqua 5–11 × 1–1·7 mm, 5–7 times as long as wide, broadly linear; style 0·2–0·3 mm. Seeds and ovules 12–18. *Usually calcicole; 2000–3000 m.* ● *E. Alps.* Au It.

2. B. linearis Rouy, *Ill. Pl. Eur. Rar.* **11**: 84 (1899) (*B. glabella* auct. scand., non Richardson). Like **1** but flowering stems up to 20 cm; lower leaves linear; siliqua 8–15 × 1–1·2 mm, 7–12 times as long as wide, linear; style 0·2–0·8 mm; seeds and ovules up to 24. $2n = 42$. *Calcareous screes and gravel. Mountains of Norway and Sweden from 67° to 70° N.* No Su.

3. B. purpurascens (R. Br.) Bunge in Ledeb., *Fl. Ross.* **1**: 195 (1841). Laxly caespitose; flowering stems up to 10 cm. Cauline leaves 0 or 1; lower leaves oblong-linear, entire or remotely dentate. Petals 3–4(–5) mm. Siliqua 4–10 × 1–3 mm, elliptical or oblong-ovate; style 0·7–1 mm. $2n = 56$. *Calcicole. Arctic and subarctic Europe.* Is No Rs(N) Sb.

11. Chrysochamela (Fenzl) Boiss.

P. W. BALL

Annual; hairs branched. Leaves undivided. Sepals not saccate at the base; petals pale yellow. Fruit a siliqua or silicula; valves with a distinct median vein; style short; stigma subcapitate. Seeds in 2 rows in each loculus, small (*c.* 0·25 mm), mucilaginous when moistened.

1. C. draboides Woronow, *Acta Horti Petrop.* **43**: 397 (1931). 10–25 cm, pubescent below, glabrous above. Leaves linear to oblong, the cauline sagittate, amplexicaul. Petals *c.* 1 mm. Silicula 5–7 × *c.* 2 mm, oblong to elliptic-oblong, obtuse; valves with a median vein and reticulate lateral veins. ● *S. Ural (near Sterlitamak).* Rs(E).

12. Sobolewskia Bieb.

P. W. BALL

Perennial; glabrous or with sparse unbranched hairs. Leaves simple. Sepals patent, not saccate at the base; petals white. Fruit an indehiscent siliqua; valves with a median vein; style very short; stigma subcapitate. Seeds large (up to 5 mm).

1. S. sibirica (Willd.) P. W. Ball, *Feddes Repert.* **68**: 194 (1963) (*S. lithophila* Bieb., *Cochlearia sibirica* Willd.). Plant 20–40 cm, glabrous or papillose, with woody rhizome. Leaves suborbicular to ovate-cordate, crenate or serrate. Petals 3·5–5 mm. Siliqua 5–10 mm, 1·5–2·5 mm in diameter at the apex, oblong-pyriform. Seeds 3·5–5 mm. *Rocky and stony places.* ● *Krym.* Rs(K).

13. Myagrum L.

P. W. BALL

Glabrous annual. Cauline leaves sessile, amplexicaul. Sepals erect, the inner slightly saccate at the base; petals yellow, not clawed. Ovules 2, one aborting. Fruit an indehiscent, 1-seeded silicula with 3 loculi; the upper 2 loculi side by side, sterile, the lower containing the seed.

1. M. perfoliatum L., *Sp. Pl.* 640 (1753). Plant 15–100 cm, glaucous. Basal leaves petiolate, oblanceolate, sinuate-dentate to pinnatifid; cauline leaves sessile, oblong-lanceolate, entire or denticulate, sagittate- to cordate-amplexicaul. Petals 3–5 mm. Silicula 5–8 mm, more or less compressed, broadly clavate; seeds *c.* 3 mm. *Probably native in S. & S.C. Europe; frequently naturalized or casual elsewhere.* Al Bu Ga Ge Gr Hs Hu It Ju Rm Rs(K) ?Sa Tu [Au Cz He It No Si Su].

14. Isatis L.

P. W. BALL AND J. R. AKEROYD

Annuals to perennials; glabrous or with unbranched hairs, glaucous. Cauline leaves simple, entire, usually amplexicaul. Sepals not saccate; petals yellow, shortly clawed. Fruit an indehiscent 1(2)-seeded silicula with a wide, often inflated, wing surrounding the loculus; loculus usually with a distinct longitudinal rib; stigma sessile.

A difficult genus requiring revision. Species **4–8** are all very similar and some are probably better regarded as subspecies of **7**.

1 Cauline leaves attenuate at the base, not amplexicaul **3. sabulosa**
1 Cauline leaves amplexicaul
 2 Silicula orbicular-elliptical, not more than twice as long as wide, rounded at the base and apex or emarginate at the apex
  3 Perennial; cauline leaves cordate at the base; silicula 15–25 mm **1. allionii**
  3 Annual; cauline leaves sagittate at the base; silicula 8–10 mm **2. platyloba**
 2 Silicula at least twice as long as wide, attenuate at the base or apex, or at both
  4 Annual; silicula 6–8 times as long as wide **9. lusitanica**
  4 Biennial or perennial; silicula 2–5 times as long as wide
   5 Central rib of silicula conspicuously 3-ridged **5. costata**
   5 Central rib of silicula simple
    6 Silicula 2–3 times as long as wide
     7 Central rib of silicula very wide **4. laevigata**
     7 Central rib of silicula slender (see also 7) **6. praecox**
    6 Silicula mostly at least 3 times as long as wide
     8 Silicula narrowly elliptical, obovate or oblanceolate; cauline leaves with acute auricles **7. tinctoria**
     8 Silicula oblong; cauline leaves with obtuse auricles **8. arenaria**

1. I. allionii P. W. Ball, *Feddes Repert.* **69**: 57 (1964) (*I. alpina* All. et auct. mult., non Vill.). Perennial with branched rhizome and vegetative shoots; flowering stems 10–30 cm. Cauline leaves with obtuse auricles. Petals (3–)4–5 mm. Silicula 15–25 × 8–13 mm, orbicular-elliptical, sometimes subcordate at the base. *Calcareous rocks and screes.* ● *S.W. Alps; C. Appennini.* Ga It.

2. I. platyloba Link ex Steudel, *Nomencl. Bot.* **1**: 440 (1821). Annual 30–100 cm. Basal leaves obovate-spathulate, obtuse; cauline leaves with acute auricles. Petals 2–2·5 mm. Silicula 8–10 mm, orbicular-elliptical, emarginate or obcordate. *Granite cliffs.* ● *W. C. Spain (near Pereña); N.E. Portugal (Miranda do Douro).* Hs Lu.

3. I. sabulosa Steven ex Ledeb., *Fl. Ross.* **1**: 212 (1841). Biennial 40–80 cm. Cauline leaves attenuate at the base, without auricles. Petals 4·5–6 mm. Silicula 17–30 × 4–9 mm, linear-oblong to oblanceolate. *Sandy soils. N.W. shore of Caspian Sea.* Rs(E).

4. I. laevigata Trautv., *Del. Sem. Horti Kiov.* **1840**: 6 (1840). Biennial 50–100 cm. Cauline leaves with acute auricles. Petals 2–2·5 mm. Silicula 6–11 × 2·5–5 mm, elliptical, 2–3 times as long as wide, glabrous; central rib wide and thick. *Steppes. S.E. Russia.* Rs(E). (*C. Asia.*)

5. I. costata C. A. Meyer in Ledeb., *Fl. Altaica* **3**: 204 (1831). Biennial up to 120 cm. Cauline leaves with acute auricles. Petals 2·5–3 mm. Silicula 9–14 × 3–5 mm, oblong-obovate, *c.* 3 times as long as wide, glabrous or pubescent; central rib wide, distinctly 3-ridged. *Dry, stony places. S.E. Russia; Krym.* Rs(K, E).

6. I. praecox Kit. ex Tratt., *Arch. Gewächsk.* **1**: 40 (1812) (incl. *I. lasiocarpa* Ledeb.). Biennial 50–100 cm. Cauline leaves with acute auricles. Petals 2·5–3 mm. Silicula 7–14 × 3–6 mm, obovate-elliptical, 2–3 times as long as wide, glabrous or pubescent; central rib slender. *E., C. & S.E. Europe.* Bu Cz Hu It Ju Rm Rs(W, E).

Plants intermediate between **6** and **7** are not infrequent.

7. I. tinctoria L., *Sp. Pl.* 670 (1753) (*I. alpina* Vill.; incl. *I. canescens* DC., *I. taurica* Bieb., *I. vermia* Papanicolaou). Biennial 50–120 cm. Basal leaves oblong-lanceolate, subacute; cauline leaves with acute auricles. Petals 2·5–4 mm. Silicula 11–27 × 3–7(–10) mm, usually oblong-obovate or elliptic-obovate, (2½–)3–5 times as long as wide, glabrous to tomentose, or hirsute; central rib usually slender. $2n = 28$. *Often ruderal. Native to S.W. Asia and possibly to parts of S.E. Europe; cultivated since ancient times as a source of dye (Woad), and widely naturalized elsewhere in Europe.* All except Az Bl Cr Fa Hb Is Sb.

This species is very variable, particularly in the size, shape and pubescence of the silicula, and has been divided up into a number of taxa at different rank. There is no general consensus as to the specific limits of **7**, and the characters used to separate segregate species overlap considerably. The following provisional treatment is based on P. H. Davis, *Notes Roy. Bot. Gard. Edinb.* **26**: 11–25 (1964).

1 Perennial; silicula with prominent central rib, pubescent **(d) subsp. athoa**
1 Usually biennial; silicula usually with slender central rib, glabrous to puberulent
 2 Silicula 15–22(–27) × 4–7(–10) mm **(a) subsp. tinctoria**
 2 Silicula 9–15 × 2·5–4 mm
  3 Silicula 12–15 mm, usually puberulent **(b) subsp. tomentella**
  3 Silicula *c.* 10 mm, glabrous or puberulent **(c) subsp. corymbosa**

(a) Subsp. tinctoria: Plant usually biennial, up to 120 cm. Silicula 15–22(–27) × 4–7(–10) mm, glabrous to puberulent; central rib usually slender. $2n = 28$. *Throughout the range of the species.*

(b) Subsp. tomentella (Boiss.) P. H. Davis, *Notes Roy. Bot. Gard. Edinb.* **26**: 22 (1964) (*I. tomentella* Boiss.): Like subsp. **(a)** but silicula 12–15 mm, usually puberulent; central rib slender. *Greece; Krym.*

(c) Subsp. corymbosa (Boiss.) P. H. Davis, *loc. cit.* (1964) (*I. corymbosa* Boiss.): Like subsp. **(a)** but silicula *c.* 10 mm, usually glabrous; central rib slender. *Greece.* (*Anatolia.*)

(d) Subsp. athoa (Boiss.) Papanicolaou in Strid, *Mount. Fl. Gr.* **1**: 238 (1986) (*I. athoa* Boiss.): Perennial with vegetative shoots; stems 20–50(–80) cm. Silicula 10–16 × 3.5 mm, puberulent; central rib prominent. $2n = 28$. *Calcareous, rocky grassland.* ● *N. Greece (Athos).*

I. littoralis Steven ex DC., *Reg. Veg. Syst. Nat.* **2**: 568 (1821), from S. Ukraine, has the silicula 15–27 × 5–10 mm, truncate or emarginate, with a prominent central rib, but cannot be separated satisfactorily from subsp. **(a)**.

I. vermia Papanicolaou, *Nordic Jour. Bot.* **2**: 555 (1982), from

N. Greece (Vermion), is autumn-flowering and has angled stems and silicula 9–14 × 3–4 mm, glabrous, with a slender central rib. The chromosome number is $2n = 28$. It is more or less intermediate between subspp. (**b**) and (**c**) and requires further study.

8. I. arenaria Aznav., *Bull. Soc. Bot. Fr.* **46**: 138 (1899). Perennial with vegetative rosettes and flowering stems 30–70 cm. Cauline leaves with obtuse auricles. Petals 3–4 mm. Silicula 15–25 × 5–8 mm, oblong, glabrous or pubescent; central rib slender. *Maritime sands. Turkey (near Kilyos, N. of Istanbul).* Tu.

Endemic to the coasts of the Bosporus region.

9. I. lusitanica L., *Sp. Pl.* 670 (1753) (*I. aleppica* Scop.). Annual 20–70(–90) cm. Basal leaves pinnately lobed, glabrous or puberulent; cauline leaves with acute auricles. Petals 3–5 mm. Silicula (12–)15–28 × (1–)2·5–5 mm, linear-oblanceolate, 6–8 times as long as wide, truncate or retuse, usually somewhat pubescent. *Karpathos.* Cr. (*S.W. Asia; N.W. Africa.*)

15. Tauscheria Fischer ex DC.

P. W. BALL

Annual; hairs unbranched. Leaves simple, entire, the cauline amplexicaul. Sepals erect, not saccate; petals yellow, not clawed. Fruit an indehiscent, more or less hemispherical, 2-seeded silicula.

1. T. lasiocarpa Fischer ex DC., *Reg. Veg. Syst. Nat.* **2**: 563 (1821). Glabrous except for the inflorescence, glaucous, 15–40 cm. Basal leaves cuneate, the cauline cordate-amplexicaul. Pedicels curved downwards in fruit. Silicula 3–5 × 2·5–4 mm, pubescent, strongly compressed, convex beneath, concave with incurved margins above so as to be more or less hemispherical; apex with upwardly curving beak. *Dry places. S.E. Russia.* Rs(E). (*N. & C. Asia.*)

16. Bunias L.

P. W. BALL

Annuals to perennials; glabrous or with glandular, unbranched or branched hairs. Leaves entire to pinnatisect. Sepals erecto-patent, the inner not or scarcely saccate at the base; petals white or yellow. Fruit an indehiscent silicula with irregular longitudinal wings, ridges or protuberances, and with 1–4 1-seeded loculi; style distinct; stigma capitate.

1 Petals white; silicula 3–4 mm — **3. cochlearioides**
1 Petals yellow; silicula 5–12 mm
2 Silicula 10–12 mm, with 4 irregularly dentate or lobed longitudinal wings; petals usually 8–13 mm — **1. erucago**
2 Silicula 5–10 mm, unwinged, but with small, irregular protuberances; petals 4–8 mm — **2. orientalis**

1. B. erucago L., *Sp. Pl.* 670 (1753) (*B. tricornis* Lange). Hispid, glandular annual or biennial 30–60(–100) cm. Lower leaves entire to sinuate-pinnatifid or pinnatisect; upper leaves entire or dentate. Petals (6–)8–13 mm, yellow, obcordate. Silicula 10–12 mm, quadrangular, with irregularly dentate or lobed wings on the angles; loculi 4; style 3–5 mm. *Ruderal. S. Europe; often introduced elsewhere.* Al Bu Co Cr Ga Gr He Hs It Ju Lu Sa Si Tu [Au Cz Ge *Rm].

2. B. orientalis L., *Sp. Pl.* 670 (1753). Glabrous or sparsely glandular biennial or perennial 25–120 cm. Lower leaves pinnate with 1 or 2 pairs of lateral leaflets and a large, pinnatifid terminal leaflet, sometimes entire; upper leaves entire to pinnatifid. Petals 4–8 mm, yellow, entire or truncate. Silicula 5–10 mm, asymmetrically ovoid and covered with small, irregular protuberances; loculi 1 or 2; style 0·5–2 mm. $2n = 14$. *E. Europe, extending westwards to Hungary; often naturalized as a weed elsewhere.* Bu Hu Ju Po Rm Rs(N, C, W, K, E) [Au Be Br Cz Da Fe Ga Ge He Ho It No Rs(B) Su].

3. B. cochlearioides Murray, *Novi Comment. Gotting.* **8**: 42 (1778). Glabrous or sparsely pubescent biennial 15–30 cm. Leaves dentate to lyrate-pinnatifid. Petals 3–4 mm, white, entire or truncate. Silicula 3–4 mm, orbicular-ovate with 4 longitudinal ridges; loculi 2; style *c.* 1 mm. *S.E. Russia.* Rs(E). (*W. Asia.*)

17. Goldbachia DC.

P. W. BALL

Glabrous annuals. Leaves entire, sinuate, the cauline amplexicaul. Sepals erect, not saccate; petals white to purple, not clawed. Fruit an articulated silicula or siliqua, sometimes with only 1 segment; segments 1-seeded; style distinct; stigma shortly 2-lobed.

1. G. laevigata (Bieb.) DC., *Reg. Veg. Syst. Nat.* **2**: 577 (1821). Plant 5–40 cm. Leaves oblong to obovate, the basal petiolate, the cauline with acute auricles. Pedicels curved downwards in fruit. Petals 4–6 mm. Silicula 10–12 × 2–4 mm, erect, 4-angled, attenuate at the base and apex, with 1–3 segments. *Semi-deserts. S.E. Russia; rarely as a casual elsewhere.* Rs(E). (*C. & S.W. Asia.*)

18. Erysimum L.

P. W. BALL

Annuals to perennials, with branched hairs. Sepals erect, the inner usually saccate at the base; petals yellow, rarely purple or brownish, long-clawed; median nectaries usually present. Fruit a siliqua; valves usually 1-veined, the hairs not lying transversely across the valves; style distinct, not more than $\frac{1}{3}$ as long as the rest of the siliqua, rarely absent; stigma usually weakly 2-lobed. Seeds in 1(2) rows in each loculus. (Incl. *Cheiranthus* L.)

Descriptions of the hairs apply primarily to those found on the leaves. Usually the stem has a higher proportion of hairs with fewer branches than is found on the leaves, but other parts of the plant have hairs similar to those found on the leaves unless otherwise stated. In virtually all species a few stellate hairs can be found on the style of the fruit, even those in which the hairs are consistently only medifixed (bifid) elsewhere on the plant.

Descriptions of fruit attitude refer only to the angle at which the siliqua diverges from the inflorescence axis, not the pedicel. The pedicel may arise at the same or at a somewhat larger angle.

A difficult genus which, despite a number of recent publications, is still not understood. The following account is based primarily on the published work of A. Polatschek (*loc. cit.* below).

Species groups have generally not been utilized because in many instances groups cannot be clearly defined or keyed out. However the following species form difficult complexes and particular care is needed when determining members of these: **6–13, 26–29, 35–42, 46–50, 51–54**. In addition **14–25** are not always clearly distinct from **6–13**.

Literature: C. Favarger & M. Goodhue, *Bull. Soc. Neuchâtel. Sci. Nat.* **100**: 93–105 (1977). C. Favarger, *Anal. Inst. Bot. Cavanilles* **35**: 361–393 (1978). P. Correvon & C. Favarger, *Pl. Syst. Evol.* **131**:

53–69 (1979). C. Favarger, *Bull. Soc. Neuchâtel. Sci. Nat.* **103**: 85–90 (1980). A. Polatschek, *Ann. Naturh. Mus.* (*Wien*) **78**: 171–182 (1974) (Italy); **82**: 352–362 (1979) (Spain); *Ann. Mus. Goulandris* **1**: 113–126 (1973) (Crete). A. Polatschek in F. Ehrendorfer, *Liste der Gefasspflanzen Mitteleuropas* ed. 2, 106. Stuttgart. 1973 (C. Europe). A. Polatschek in A. Strid, *Mount. Fl. Gr.* **1**: 239–247. Cambridge. 1986 (Greece). S. Snogerup, *Op. Bot.* (*Lund*) **13**: 1–70 (1967); **14**: 1–86 (1967) (Greece and Aegean region).

1 Annual; anthers not more than 1·2(–1·6) mm (see also **55**)
2 Petals 6–10 mm; style 2–5 mm; siliqua patent **56. repandum**
2 Petals (2–)3–6 mm; style 0–1·5 mm; siliqua erect to erecto-patent
3 Pedicels not more than 4 mm in flower, almost as thick as the siliqua in fruit **57. incanum**
3 Pedicels 4–8 mm in flower, distinctly thinner than the siliqua in fruit **58. cheiranthoides**
1 Perennial or biennials; anthers 1–4 mm
4 Hairs on leaves mostly bifid
5 Petals purple
6 Petals with bifid hairs on the back; stem with very dense sessile leaves up to the inflorescence **20. cazorlense**
6 Petals glabrous; stem either not densely leafy, or densely leafy only at the base; lower leaves petiolate
7 Style 2–3(–4) mm
8 Pedicel and siliqua erecto-patent, not appressed to the stem **26. linifolium**
8 Pedicel and siliqua erect, ±appressed to the stem **27. baeticum**
7 Style 4–7·5 mm
9 Perennial with branched stock and persistent basal leaf-rosettes; pedicels 2–2·5 mm in flower, up to 4 mm in fruit **28. popovii**
9 Biennial without persistent basal leaf-rosettes; pedicels *c.* 3·5 mm in flower, 4–5 mm in fruit **29. favargeri**
5 Petals yellow, rarely reddish or purple on the back, or variegated
10 Petals glabrous on the back
11 Petal-limb 5–10 mm wide
12 Style less than 2 mm
13 Plant with long, leafy procumbent shoots at the base **14. linariifolium**
13 Plant with simple or shortly branched stock
14 Plant usually without leaf-rosettes; upper cauline leaves often with axillary fascicles **13. nevadense**
14 Plant with basal leaf-rosettes; upper cauline leaves without axillary fascicles
15 Sepals 6·5–7 mm; basal and lower leaves long-petiolate **10. jugicola**
15 Sepals 7–16 mm; basal and lower leaves mostly short-petiolate
16 Style 0·5–1·5 mm **6. sylvestre**
16 Style (1·2–)1·5–2 mm **25. duriaei**
12 Style at least 2 mm
17 Siliqua 1·5–2·3 mm wide; seeds 3–4 mm; style (2·5–)3–8 mm **24. humile**
17 Siliqua 0·7–1·5(–2) mm wide; seeds 1·5–3 mm; style 2–3(–5) mm
18 Biennial or short-lived perennial, without or with few leaf-rosettes at the base; seeds 1·5–2 mm **13. nevadense**
18 Rhizomatous perennial, usually with much-branched stock bearing numerous leaf-rosettes; seeds 2–3 mm
19 Basal and lower leaves with petiole about $\frac{1}{2}$ as long as the lamina **7. montosicola**
19 Basal and lower leaves shortly petiolate **24. duriaei**
11 Petal-limb 2·5–5 mm wide
20 Style 0·5–2 mm
21 Plant with stout, little-branched, woody stock, often 5–10 mm in diameter; leaves 6–9 mm wide, very crowded **17. metlesicsii**
21 Plant with slender, often branched stock, usually less than 5 mm in diameter; cauline leaves not more than 5 mm wide, not crowded
22 Plant with long, procumbent leafy stems at the base **14. linariifolium**
22 Plant with leaf-rosettes at the base, sometimes dead at anthesis
23 Plant with much-branched, woody rhizome, the branches long, procumbent and bearing persistent leaf-rosettes **24. duriaei**
23 Plant with short-branched rhizome with ±erect branches, often without persistent leaf-rosettes
24 Siliqua 0·7–1 mm wide, erecto-patent (*c.* 50°) **12. pseudorhaeticum**
24 Siliqua 1–1·3 mm wide, erecto-patent (15–40°) **13. nevadense**
20 Style at least 2 mm
25 Plant with woody rhizome with long procumbent branches bearing numerous leaf-rosettes **24. duriaei**
25 Plant biennial or perennial with short-branched, ±erect rhizome
26 Cauline leaves obtuse, green, about 5–8 times as long as wide; siliqua *c.* 2 mm wide **25. penyalarensis**
26 Cauline leaves acute, grey or greyish-green, at least 10 times as long as wide; siliqua 0·7–1·3 mm wide
27 Angles of the siliqua glabrescent, otherwise the siliqua densely hairy **11. rhaeticum**
27 Siliqua uniformly pubescent **13. nevadense**
10 Petals with hairs on the back, mostly at the centre of the limb and at the top of the claw
28 Petals yellow with buff or purple on the back; calyx 4–5(–6) mm; siliqua 6–24 mm **22. mutabile**
28 Petals yellow, rarely somewhat variegated with buff or purple; calyx (5–)6–16 mm; siliqua usually (20–)30–100 mm
29 Siliqua 2–6 mm wide; style very distinctly 2-lobed, the lobes usually at least 1 mm; median nectaries absent **(1–5). cheiri** group
29 Siliqua 0·5–2 mm wide; style not or only shortly 2-lobed; median nectaries usually present
30 Siliqua ±patent, forming an angle of at least 50° with the inflorescence axis
31 Stock covered with dense petiole remains **18. comatum**

31 Stock not densely covered with petiole remains
32 Calyx 8–13 mm; petals 5–9 mm wide **8. majellense**
32 Calyx up to 9 mm; petals 2–4(–5) mm wide
33 Biennial, usually without persistent leaf-rosettes at the base
34 Siliqua with mainly stellate hairs; at least the lower leaves coarsely sinuate-dentate **35. creticum**
34 Siliqua with mainly bifid hairs; leaves entire or dentate
35 Pedicels as thick as the siliqua, both at an angle of *c.* 90° to the axis **33. graecum**
35 Pedicels somewhat thinner than the siliqua, both at an angle of 60–90° to the axis **34. crassistylum**
33 Perennial, usually with persistent leaf-rosettes at the base or with the remains of old flowering stem
36 Plant with numerous persistent leaf-rosettes; siliqua dorsally compressed **15. pusillum**
36 Plant without or with few basal leaf-rosettes; siliqua square in transverse section
37 Plant with long, procumbent leafy shoots; siliqua erecto-patent (30–50°) **14. linariifolium**
37 Plant without sterile leafy shoots or with few basal rosettes; siliqua patent (60–70°) **16. drenowskii**
30 Mature siliqua erect or erecto-patent
38 Siliqua and pedicel erect, closely appressed to the axis
39 Leaves filiform, canaliculate, grey **19. myriophyllum**
39 Leaves flat, green, the basal oblanceolate **21. raulinii**
38 Siliqua and pedicel not closely appressed to the axis
40 Pedicels 6–10 mm in flower, up to 16 mm in fruit **30. ucrainicum**
40 Pedicels not more than 6(–8) mm in flower, up to 8(–11) mm in fruit
41 Petals (4·5–)5–9 mm wide
42 Style 0·8–1·5(–1·8) mm
43 Petals glabrous or sparsely pubescent; siliqua erecto-patent (20–30°) **6. sylvestre**
43 Petals pubescent; siliqua erecto-patent to patent (30–60°) **18. comatum**
42 Style 1·5–2·5 mm
44 Sepals not more than 9 mm; petals not more than 16 mm; siliqua dorsally compressed **15. pusillum**
44 Sepals 8–13 mm; petals 15–21 mm; siliqua ± square in transverse section
45 Cauline leaves without small axillary shoots; siliqua 1·3–2 mm wide **7. montosicola**
45 Cauline leaves usually with short axillary shoots; siliqua 1–1·3 mm wide **8. majellense**
41 Petals 2–5 mm wide
46 Biennial with a simple taproot and a basal leaf-rosette, usually dead at anthesis
47 Sepals (7·5–)9–11 mm, conspicuously saccate; stem usually with numerous, very dense leaf-remains at the base
48 Cauline leaves 6–9 mm wide, oblong, the upper with short axillary shoots **17. metlesicsii**
48 Cauline leaves 1–6(–8) mm wide, linear or linear-lanceolate, the upper without short axillary shoots **18. comatum**
47 Sepals 5–9(–9·5) mm, not or only weakly saccate at the base
49 Pedicels 1–2 mm in flower, up to 5 mm in fruit; style 1·5–4·5 mm **31. krynkense**
49 Pedicels (2–)3–6(–8) mm in flower, up to 11 mm in fruit; style 0·5–2·5 mm
50 Siliqua 0·5–1 mm wide; plant usually densely grey-pubescent **32. diffusum**
50 Siliqua 1–1·5 mm wide; plant usually green and not densely pubescent **36. calycinum**
46 Perennial, usually with branched stock bearing conspicuous leaf-rosettes, or with leafy shoots
51 Pedicels (2–)3–5 mm in flower, up to 11 mm in fruit; siliqua usually glabrescent on the angles
52 Plant with long, procumbent leafy shoots; pedicels not more than 7 mm in fruit **14. linariifolium**
52 Plant without long, procumbent leafy shoots; pedicels up to 11 mm in fruit
53 Siliqua 30–55 mm, with numerous trifid as well as bifid hairs; sepals *c.* 2·5 mm wide, with frequent trifid hairs **36. calycinum**
53 Siliqua (50–)60–100 mm, without or with few trifid hairs; sepals *c.* 1 mm wide, with few trifid hairs **37. olympicum**
51 Pedicels 0·5–2(–2·5) mm in flower, up to 4·5 mm in fruit; siliqua glabrescent or pubescent on the angles
54 Petals 9–14 mm **15. pusillum**
54 Petals 14–18 mm
55 Siliqua ± glabrescent on the angles; style 2–4 mm **9. bonannianum**
55 Siliqua pubescent on the angles; style 1·8–2·5 mm **15. pusillum**
4 Hairs on the leaves stellate or with numerous stellate hairs mixed with bifid hairs
56 Anthers 1–2 mm
57 Caespitose perennial with branched, woody stock bearing numerous persistent leaf-rosettes; leaves with numerous bifid as well as stellate hairs **45. pulchellum**
57 Biennial or short-lived perennial, without or with few persistent leaf-rosettes at the base; leaves with almost all hairs stellate
58 Siliqua patent or deflexed **43. aureum**
58 Siliqua erect or erecto-patent
59 Siliqua 0·4–0·8 mm in diameter, the valves rounded

on the back and without a prominent mid-vein **44. leucanthemum**
59 Siliqua 0·8–2 mm wide, the valves with a prominent mid-vein
60 Sepals 2–6 mm
61 Siliqua erecto-patent; leaves 4–15 mm wide, sinuate-dentate; seeds 1·7–2·1 mm **52. hieracifolium**
61 Siliqua erect; leaves 2–5 mm wide, entire or denticulate; seeds 1·2–1·8 mm **53. marschallianum**
60 Sepals 6–8·5(–10) mm
62 Leaves 5–30 mm wide, the middle cauline not more than 6 times as long as wide; siliqua erect to erecto-patent, diverging at an angle of 10–30° **51. hungaricum**
62 Leaves 1·5–6 mm wide, the middle cauline at least 6 times as long as wide; siliqua erect **54. virgatum**
56 Anthers at least 2 mm
63 Petals 5–10 mm wide
64 Siliqua with mainly 3- or 4-fid hairs; cauline leaves sinuate-dentate to pinnatisect **48. carniolicum**
64 Siliqua with 2(3)-fid hairs; cauline leaves denticulate to sinuate-dentate
65 Lower cauline leaves sessile or shortly petiolate; siliqua 0·8–1·2 mm wide; style 1–2 mm **46. odoratum**
65 Lower cauline leaves long-petiolate; siliqua 1·2–1·5 mm wide; style *c.* 1 mm **47. witmannii**
63 Petals not more than 5 mm wide
66 Style 3–6(–10) mm; siliqua strongly laterally compressed, with keeled or winged valves **55. cuspidatum**
66 Style not more than 2·5 mm; siliqua-valves not keeled or winged
67 Middle cauline leaves sinuate-pinnatifid to -pinnatisect, most divided to at least half-way to the mid-vein
68 Usually biennial; petals 3–5 mm wide; siliqua with mainly stellate hairs **49. kuemmerlei**
68 Perennial with branched, woody stock; petals 2–3 mm wide; siliqua with mainly bifid hairs **50. pectinatum**
67 Middle cauline leaves entire to coarsely sinuate-dentate, not divided to more than ⅓ of the way to the mid-vein
69 Hairs on the leaves mostly 4- or 5-fid **42. horizontale**
69 Hairs on the leaves (2)3(4)-fid
70 Siliqua with numerous bifid as well as trifid hairs
71 Leaves with numerous bifid as well as trifid hairs **41. exaltatum**
71 Leaves with mainly stellate hairs **46. odoratum**
70 Siliqua with mainly stellate hairs
72 Leaves with numerous bifid as well as stellate hairs
73 Leaves 50–90 mm; siliqua patent (70–90°) **35. creticum**
73 Leaves 15–60 mm; siliqua erect to erecto-patent (10–40°)
74 Hairs on leaves mainly trifid **38. asperulum**
74 Hairs on leaves mainly bifid **39. crepidifolium**
72 Leaves with mainly stellate hairs
75 Calyx with stellate hairs; anthers not exceeding 2·2 mm, stellate-hairy **51. hungaricum**
75 Calyx with bifid as well as stellate hairs; anthers 2–3 mm, glabrous **40. leptostylum**

(1–5). E. cheiri group. Perennials, with a stout, branched, woody stock, or dwarf shrubs, usually with persistent leaf-rosettes; hairs almost all bifid. Leaves lanceolate to narrowly obovate. Pedicels 4–20 mm in fruit. Petals yellow, sometimes variegated with red or purple in cultivars of **3**, with bifid hairs on the back; median nectaries absent. Siliqua 20–100 mm; style 0.5–4 mm, with distinctly 2-lobed stigma, the lobes usually at least 1 mm.

A closely related group of taxa native to the Aegean region.

1 Seeds narrower than the width of the siliqua, so clearly in 2 rows in each loculus; style stout; sepals with dark glands **3. cheiri**
1 Seeds almost as wide as the siliqua, so ± in 1 row in each loculus; style thin; sepals eglandular
2 Leaves sparsely to moderately pubescent, distinctly petiolate, the longest petioles *c.* ½ as long as the lamina
3 Lower part of stems without broad, persistent leaf-scars; inflorescence lateral, without or with few branches; flowers 2–30 on each branch **4. senoneri**
3 Lower part of stems with broad, persistent leaf-scars; inflorescence terminal, usually with several branches; flowers 25–70 on each branch **5. naxense**
2 Leaves densely pubescent, sessile or with a short petiole
4 Petals 20–27 × 8–12 mm; siliqua ± square in transverse section, with a thick, prominent mid-vein on each valve **2. corinthium**
4 Petals 13–22 × 5–8.5 mm; siliqua ± terete, with a weak mid-vein on each valve **1. candicum**

1. E. candicum Snogerup, *Op. Bot.* (*Lund*) **13**: 34 (1967). Small shrub 25–50 cm. Leaves 20–120 × 3–13 mm, entire or with a few serrations, densely pubescent; petiole much shorter than the lamina. Inflorescence usually simple, with not more than 25 flowers. Petals 5–8·5 mm wide. Siliqua 20–60 mm, somewhat compressed, with a thin mid-vein on each valve. Style 1·5–4 mm. Seeds in 1 row in each loculus. ● *S. Aegean region.* Cr.

(a) Subsp. **candicum**: Inflorescence simple. Sepals 8–11·5 mm; petals 19–22 mm. Siliqua 4–4·5 mm wide. Seeds 5–6 mm. *Kriti.*
(b) Subsp. **carpathum** Snogerup, *op. cit.* 38 (1967): Inflorescence sometimes branched. Sepals 7–8·5 mm; petals 13–20 mm. Siliqua 2·5–4 mm wide. Seeds 3·5–5·5 mm. *Karpathos and Saria.*

2. E. corinthium (Boiss.) Wettst., *Österr. Bot. Zeitschr.* **39**: 283 (1889) (*Cheiranthus corinthius* Boiss.). Small shrub or subshrub 25–75 cm. Leaves 25–90 × 2·5–14 mm, entire, densely pubescent; petiole much shorter than the lamina. Inflorescence usually simple, with 5–25(–50) flowers; pedicels 4–8 mm in fruit. Sepals 7–10·5 mm; petals 20–27 × 8–12 mm. Siliqua 30–80 × 3–4·5 mm, more or less square in transverse section, the valves with a broad, conspicuous mid-vein; style 1–4 mm. Seeds 4–6 mm, in 1 row in each loculus. ● *S. Greece.* Gr.

3. E. cheiri (L.) Crantz, *Class. Crucif.* 116 (1769) (*Cheiranthus cheiri* L.). Perennial or small shrub 25–80 cm (sometimes biennial in cultivation); lower part of stems often with conspicuous leaf-scars. Leaves 40–220 × 6–20 mm, entire or with a few short teeth, sparsely to moderately pubescent; petiole much shorter than the lamina. Inflorescence usually branched, with up to 60 flowers on each branch; pedicels 5–20 mm in fruit. Sepals 9.5–14 mm, with dark glands; petals 22–35 × 10–20 mm, yellow or orange-yellow, sometimes variegated with red or purple. Siliqua 40–100 × (4–)5–6(–7) mm, with a conspicuous mid-vein on the valves; style 0·5–4 mm, stout. Seeds 2·5–4 mm, in 2 rows in each loculus. $2n$ = 14. *Cultivated for ornament, and widely naturalized on walls, cliffs and rocks in C., W. & S. Europe; not known in the wild.* [Al Au Be Bl Br Cz Ga Ge Gr Hb He Ho Hs Hu It Ju Lu Rm Sa Si Tu.]

Probably derived by selection and hybridization from two or more of the native species of **1–5**.

4. E. senoneri (Heldr. & Sart.) Wettst., *Österr. Bot. Zeitschr.* **39**: 283 (1889) (*Cheiranthus senoneri* Heldr. & Sart.). Small shrub 15–50 cm, without conspicuous leaf-scars on the lower parts of the stems. Leaves 15–50 × 2·5–16 mm, entire or serrate, sparsely to moderately pubescent; petiole distinct, up to ½ as long as the lamina. Inflorescence simple or with few branches, with 2–30 flowers per branch; pedicels 4–11 mm in fruit. Sepals 6·5–11 mm; petals 12·5–27 × 3–12·5 mm. Siliqua 30–85 × 2–4 mm, compressed, with a narrow mid-vein on the valves; style 1–3·5 mm. Seeds 2·5–5·5 mm, in 1 row in each loculus. *Kikladhes*. Gr.

Plants from Amorgos have been recognized as a distinct subspecies, subsp. **amorginum** Snogerup, *Op. Bot.* (*Lund*) **13**: 55 (1967), but they cannot be satisfactorily distinguished on the basis of morphology.

5. E. naxense Snogerup, *Op. Bot.* (*Lund*) **13**: 45 (1967). Small shrub 30–70 cm, with conspicuous, broad, persistent leaf-scars on the lower parts of the stems. Leaves 50–140 × 5–23 mm, serrate (sometimes minutely), sparsely to moderately pubescent; petiole distinct, up to ½ as long as the lamina. Inflorescence terminal, branched, with 25–70 flowers per branch; pedicels 5–13 mm in fruit. Sepals 8–11 mm; petals 15–24 × 5–9 mm. Siliqua 40–90 × 2·5–4 mm, compressed, with a narrow mid-vein on the valves. Seeds 4·5–5·5 mm, in 1 row in each loculus. ● *Kikladhes* (*Naxos*). Gr.

6. E. sylvestre (Crantz) Scop., *Fl. Carn.* ed. 2, **2**: 28 (1772) (*E. cheiranthus* Pers.). Perennial, with branched stock bearing leaf-rosettes; hairs 2(3)-fid. Leaves 20–120 × 1·5–7 mm, linear or linear-lanceolate, entire or with few teeth. Pedicels 2·5–5 mm in flower, up to 8 mm in fruit. Petals 12–23 × 4·5–7 mm, yellow, glabrous or sparsely pubescent on the back. Siliqua 33–80 × 1–1·5 mm; style 0·5–1·5 mm. Seeds *c.* 2 mm. ● *E. Alps and N.W. Jugoslavia*. Au It Ju.

(a) Subsp. **sylvestre**: Flowering stems 10–40 cm, green or brown at the base. Leaves not more than 3·5 mm wide, linear, sessile or very shortly petiolate. Sepals 9–16 mm, with most hairs bifid; petals 16–23 mm. Siliqua erecto-patent (20–30°); style 0·8–1·5 mm. $2n$ = 14. *Almost throughout the range of the species.*

(b) Subsp. **aurantiacum** (Leybold) P. W. Ball, *Bot. Jour. Linn. Soc.* **103**: 203 (1990) (*E. aurantiacum* (Leybold) Leybold): Flowering stems 25–65 cm, often violet at the base. Leaves linear-lanceolate, the lower distinctly petiolate. Sepals 7–10 mm, with some trifid hairs; petals 12–19 mm, always glabrous. Siliqua erecto-patent to more or less patent (45–75°); style 0·5–1 mm. $2n$ = 14. *Foothills of Alps, N. of Lago di Garda.*

7. E. montosicola Jordan in Billot, *Annot.* 123 (1858). Like **6** but basal leaves with petiole about ½ as long as the lamina, sometimes sinuate-dentate or -lobed; lower cauline leaves shortly petiolate; sepals 8–10 mm; petals 17–20 × 5–7 mm, usually sparsely pubescent on the back; siliqua 60–90 × 1·3–2 mm, erecto-patent (20–30°); style (1–)2–2·5 mm. $2n$ = 28. ● *S.W. Alps* (*Hautes-Alpes, Isère*). Ga.

8. E. majellense Polatschek, *Ann. Naturh. Mus.* (*Wien*) **78**: 177 (1974). Perennial, with branched stock bearing leaf-rosettes; stems 8–30 cm in flower, up to 40 cm in fruit; hairs 2(3)-fid. Leaves 10–60(–80) × 1–6(–8) mm, linear to oblanceolate, the basal petiolate, the rest sessile, entire or the lower sinuate-dentate. Pedicels 1·5–4 mm in flower, up to 4(–6) mm in fruit, more or less patent. Sepals 8–13 mm; petals 15–21 × 5–9 mm, yellow, pubescent on the back. Siliqua 40–90 × 1–1·3 mm, erecto-patent (*c.* 50°), grey with glabrescent angles; style (1–)1·5–3 mm. $2n$ = 22, 28. ● *C. & S. Appennini*. It.

9. E. bonannianum C. Presl, *Fl. Sic.* **1**: 78 (1826). Like **8** but pedicels 1–2 mm in flower, up to 4·5 mm in fruit; sepals 6–8·5 mm; petals 14–18 × 3–4(–4·5) mm; siliqua 25–60 mm, erecto-patent (20–50°); style 2–4 mm. $2n$ = 22, 24. ● *Mountains of N. & E. Sicilia*. Si.

10. E. jugicola Jordan, *Diagn.* **1**: 173 (1864) (*E. pumilum* (Murith) Gaudin, non (Hornem.) DC.). Caespitose perennial with numerous basal leaf-rosettes; stems 6–25 cm in flower, up to 30 cm in fruit; hairs 2(3)-fid. Leaves 10–50 × 2–5 mm, linear-spathulate to lanceolate, usually denticulate, the basal with petiole at least ½ as long as the lamina. Pedicels 1·5–2 mm in flower, up to 5(–7) mm in fruit, erecto-patent. Sepals 6·5–7 mm; petals 14–20 × 5–10 mm, yellow, glabrous. Siliqua 25–55 × 1–1·2 mm, erect (10–20°); style 1–1·8 mm. $2n$ = 18. ● *S.W. Alps*. Ga ?He It.

11. E. rhaeticum (Schleicher ex Hornem.) DC., *Reg. Veg. Syst. Nat.* **2**: 503 (1821) (*E. helveticum* auct., non (Jacq.) DC., *E. lanceolatum* auct. ital.). Perennial with shortly branched stock, with or without basal leaf-rosettes; stems up to 70 cm in flower, up to 110 cm in fruit; hairs 2(3)-fid. Leaves 30–75 × 2–6 mm, linear-lanceolate, entire or denticulate, with clusters of small leaves in the upper axils. Pedicels 2.5–3 mm in flower, up to 6 mm in fruit. Sepals 7.5–9 mm; petals 16–20 × 4–5 mm, yellow, glabrous. Siliqua 45–60 × *c.* 1 mm, erecto-patent (25–35°), grey with glabrescent angles; style 2·6–3·2 mm. $2n$ = 56, *c.* 59. ● *Alps*. Au Ga He It.

12. E. pseudorhaeticum Polatschek, *Ann. Naturh. Mus.* (*Wien*) **78**: 179 (1974). Like **11** but sepals 7–12 mm; petals 13–18·5 × 3·5–5 mm; siliqua 35–100 × 0·7–1 mm, erecto-patent (*c.* 50°); style 0·5–1·5(–2) mm. $2n$ = 14. ● *Mountains of C. Italy*. It.

13. E. nevadense Reuter, *Cat. Graines Jard. Bot. Genève* **1855**: 4 (1855) (*E. grandiflorum* auct. eur., non Desf., *E. lanceolatum* R. Br., nom. illeg., *E. longifolium* DC., nom. illeg., *E. australe* Gay, nom. illeg., *E. bocconei* (All.) Pers., nom. illeg.). Biennial or perennial, usually without basal leaf-rosettes; stems 10–60 cm in flower, up to 70 cm or more in fruit; hairs 2(3)-fid. Leaves 20–80(–110) mm, linear to lanceolate, entire to denticulate, often with fascicles of small leaves in the upper axils. Pedicels 1–4 mm in flower, up to 6 mm in fruit. Sepals 7–11 mm; petals 12–17(–20) × 3–5(–6) mm, yellow, glabrous. Siliqua 20–70 × 0·7–1·2(–1·4) mm; style 1–3·5(–9) mm. Seeds *c.* 2 mm. *S.W. Europe*. ?Bl ?Co Ga Hs It Lu ?Sa.

A difficult complex considered by A. Polatschek, *loc. cit.* (1979), to consist of at least 7 species in the Iberian peninsula. Favarger, *loc. cit.* (1978), considered the group to consist of a single species with several subspecies, but he did not propose any formal classification. The following treatment is conservative.

The name **E. grandiflorum** Desf., *Fl. Atl.* **2**: 85 (1798), has been applied to this species in recent years, but this North African species appears distinct from this complex, being more similar morphologically to **8** and **9**.

1 Style 1–1·8 mm, with mainly trifid hairs; cauline leaves 0·8–2 mm wide (c) subsp. **gomez-campoi**
1 Style (1·5–)2·6 mm or more, with numerous bifid as well as some trifid hairs; cauline leaves variable, often much wider
2 Petals 9–12 mm; style 4–6(–9) mm (d) subsp. **fitzii**
2 Petals 12–20 mm; style 1·5–4(–5) mm
3 Petals (14–)17–20 mm; seeds 2–2·5 mm; cauline leaves with excurrent tip (S.W. Spain) (b) subsp. **rondae**
3 Petals 12–17(–18) mm; seeds 1·5–2 mm; cauline leaves without an excurrent tip
4 Larger cauline leaves at least 5 mm wide, green (e) subsp. **merxmuelleri**
4 Larger cauline leaves not more than 4(–5) mm wide, green to grey
5 Siliqua 0·8–1 mm wide, erect (0–20°); sepals 7–8 mm (f) subsp. **mediohispanicum**
5 Siliqua 1–1·4 mm wide, usually erecto-patent (15–30°); sepals (7–)8–11 mm
6 Perennial, rarely biennial; stem and leaves green or greyish-green; sepals 8–11 mm (g) subsp. **collisparsum**
6 Biennial or short-lived perennial; stem and leaves grey or greyish-green; sepals 8–9 mm (a) subsp. **nevadense**

(a) Subsp. **nevadense**: Biennial or short-lived perennial. Leaves 1–5 mm wide, linear or lanceolate, usually entire, grey or greyish-green. Sepals 8–9 mm; petals 13–16(–18) × 3·5–6 mm. Siliqua 30–60 × 1–1·2 mm, erecto-patent (15–30°), the valves with a weak mid-vein; style (1·6–)2–3·5 mm. $2n = 14$. ● *S. Spain.*

(b) Subsp. **rondae** (Polatschek) P. W. Ball, *Bot. Jour. Linn. Soc.* **103**: 206 (1990) (*E. rondae* Polatschek): Like subsp. (a) but leaves 2·5–8 mm wide, lanceolate or oblanceolate, with excurrent tip, sinuate-denticulate; sepals 8–11 mm; petals (14–)17–20 × 4–5 mm; siliqua (35–)50–75 mm, erect (0–10°), with distinct mid-vein on the valves; style 3–5 mm. $2n = 28$. ● *S.W. Spain.*

(c) Subsp. **gomez-campoi** (Polatschek) P. W. Ball, *op. cit.* 205 (1990) (*E. gomez-campoi* Polatschek): Usually perennial. Leaves 0·8–2 mm wide, linear or linear-lanceolate, usually entire, grey. Sepals 8–9 mm; petals 12–16 × 3·5–5 mm. Siliqua 33–75 × 1–1·3 mm, erecto-patent (25–40°); style 1–1·8 mm, with mainly trifid hairs. $2n = 14$. ● *E. Spain.*

(d) Subsp. **fitzii** (Polatschek) P. W. Ball, *op. cit.* 205 (1990) (*E. fitzii* Polatschek): Like subsp. (c) but sepals 6–7·5 mm; petals 9–12 × 2·5–3·5 mm; siliqua 22–26 × 0·7–1 mm, erect (0–10°); style 4–6·5(–9) mm, with mainly bifid hairs. $2n = 14$. ● *S.E. Spain (Sierra de Pandera).*

(e) Subsp. **merxmuelleri** (Polatschek) P. W. Ball, *op. cit.* 206 (1990) (*E. merxmuelleri* Polatschek, *E. helveticum* sensu Samp., non (Jacq.) DC.): Biennial or short-lived perennial. Leaves (1·5–)4–10 mm wide, lanceolate, entire or shallowly sinuate-denticulate, green. Sepals *c.* 9 mm; petals 12–14 × 3–5 mm. Siliqua 20–60 × *c.* 1 mm, erecto-patent to erect (10–30°), the valves weakly veined; style 2–3 mm. $2n = 14$. ● *W. Spain, E. Portugal.*

(f) Subsp. **mediohispanicum** (Polatschek) P. W. Ball, *op. cit.* 205 (1990) (*E. mediohispanicum* Polatschek): Usually biennial. Leaves 1–4 mm wide, linear-lanceolate or lanceolate, entire or sinuate-denticulate, grey. Sepals 7–8 mm; petals 13–17 × 3–5 mm. Siliqua 30–50 × 0·8–1 mm, erect (0–20°), the valves distinctly veined; style 2·5–4 mm. $2n = 28$. ● *C. & N.C. Spain from N. Granada and Murcia provinces to the W. Pyrenees.*

(g) Subsp. **collisparsum** (Jordan) P. W. Ball, *op. cit.* 205 (1990) (*E. collisparsum* Jordan): Biennial or perennial. Leaves 2–5 mm wide, linear-lanceolate, green or greyish-green, entire or sinuate-denticulate. Sepals 8–11 mm; petals 14–16 × 3·5–5 mm. Siliqua 25–70 × *c.* 1 mm, usually erecto-patent (15–30°); style (1·5–)2–3 mm. $2n = 14$. *From E. Pyrenees to S.E. France and N.W. Italy.*

In S. France there is considerable variation in density of indumentum, flower size and siliqua posture. It may be possible to recognize other subspecies in this area.

14. E. linariifolium Tausch, *Flora (Regensb.)* **14**: 212 (1831) (*E. linearifolium* auct., non Moench). Perennial with branched, woody stock bearing long, procumbent leafy shoots; stems 10–40 cm in flower, up to 45 cm in fruit; hairs 2(3)-fid. Leaves 20–60 × (0·5–)2·5–5 mm, linear to oblanceolate, entire to sinuate-dentate. Pedicels 2–4 mm in flower, up to 7 mm in fruit. Sepals 6–9 mm; petals 12–18 × 4–5 mm, yellow, glabrous or sparsely pubescent on the back. Siliqua 40–80 × 1–1·5 mm, erecto-patent (30–50°), grey, sometimes with glabrescent angles; style 0·5–2 mm. Seeds 1·5–2 mm. $2n = 14$. *Limestone slopes, cliffs and rocks.* ● *Mountains of W. part of Balkan peninsula.* Al Ju.

Not always clearly distinct from **15** in the southern part of its range.

15. E. pusillum Bory & Chaub. in Bory, *Expéd. Sci. Morée* **3**: 190 (1832) (*E. boryanum* Boiss. & Spruner, *E. trichophyllum* Halácsy). Perennial with branched, woody stock bearing leaf-rosettes or long, procumbent leafy shoots; stems up to 30 cm in flower, up to 40 cm in fruit; hairs 2(3)-fid. Leaves 10–110 mm, linear to linear-spathulate, entire to sinuate-denticulate, rarely sinuate-dentate. Pedicels 0·5–2 mm in flower, up to 4 mm in fruit. Sepals 5–9 mm; petals 9–15(–16) × 2·5–5(–6) mm, yellow, pubescent on the back. Siliqua 20–60(–90) mm, compressed; style 0·5–4 mm. Seeds 1·5–3 mm. ● *S. part of Balkan peninsula; Aegean region.* Al Bu Gr Ju.

1 Siliqua spreading at an angle of 45–90° to the axis
2 Plant with thick, woody stock with long branches; siliqua 20–90 × 1–1·6 mm, with glabrescent angles (d) subsp. **rechingeri**
2 Plant usually caespitose with short, slender, branched stock or with short leafy shoots; siliqua 20–45 × 0·8–1·3 mm, the angles persistently pubescent
3 Siliqua spreading at an angle of 50–80° to the axis; stock with short leafy branches; siliqua tapering to the style (c) subsp. **hayekii**
3 Siliqua spreading at an angle of 40–50° to the axis; stock with leaf-rosettes; siliqua abruptly contracted to the style (a) subsp. **pusillum**
1 Siliqua arising at an angle of 0–45° to the axis
4 Style 0.5–1.5 mm, the siliqua tapering to the style; stock usually without persistent leaves (f) subsp. **microstylum**

4 Style 1–4 mm, the siliqua abruptly contracted to the style; stock usually with persistent leaf-rosettes or with long leafy shoots
5 Stock with long leafy shoots, sometimes also with leaf-rosettes (e) subsp. **cephalonicum**
5 Stock with leaf-rosettes only
6 Stamens sparsely hairy; stigma about as thick as the style; petals 3–4 mm wide (g) subsp. **parnassi**
6 Stamens glabrous; stigma capitate, wider than the style; petals usually 2–3 mm wide
7 Style (1·5–)2·5–4 mm; basal leaves runcinate to coarsely sinuate-dentate (often dead at anthesis); siliqua 1·5–2 mm wide (b) subsp. **atticum**
7 Style 1–2(–3) mm; all basal leaves entire or denticulate; siliqua 0·8–1·5 mm wide (a) subsp. **pusillum**

(**a**) Subsp. **pusillum**: Stock with numerous leaf-rosettes; leaves 0·5–4 mm wide, entire or sinuate-denticulate. Sepals 5–7 mm; petals 9–14 × 2–3·5 mm. Siliqua 25–50 × 0·5–1(–1·5) mm, erecto-patent to patent (10–50°), abruptly contracted to the style; style 1–2(–3) mm. 2*n* = 14. *S. Greece.*

(**b**) Subsp. **atticum** (Heldr. & Sart. ex Boiss.) P. W. Ball, *Bot. Jour. Linn. Soc.* **103**: 210 (1990) (*E. atticum* Heldr. & Sart. ex Boiss.): Like subsp. (**a**) but leaves 1–4(–6) mm wide, the basal coarsely sinuate-dentate or runcinate, the rest mostly entire; sepals 6–9 mm; siliqua 1·5–2 mm wide, usually erect (10–20°); style (1·5–)2·5–3·5 mm. 2*n* = 14. *S.E. Greece.*

(**c**) Subsp. **hayekii** Jáv. & Rech. fil., *Ann. Naturh. Mus.* (*Wien*) **43**: 296 (1929): Stock much-branched with leaf-rosettes and long leafy shoots. Leaves 1–4 mm wide, entire. Sepals 5·5–7 mm; petals 10–13 × 3–4 mm. Siliqua 20–45 × 0·8–1·3 mm, patent (50–70°), with pubescent angles, tapering to the style; style 1–2·5 mm. *Kikladhes.*

(**d**) Subsp. **rechingeri** (Jáv.) P. W. Ball, *loc. cit.* (1990) (*E. rechingeri* Jáv.): Like subsp. (**c**) but the stock with long, woody branches; siliqua 20–90 × 1·2–1·6 mm, pubescent with glabrescent angles; style 1·5–3 mm. *Limnos.*

(**e**) Subsp. **cephalonicum** (Polatschek) P. W. Ball, *loc. cit.* (1990) (*E. cephalonicum* Polatschek): Stock branched, with leaf-rosettes and long leafy shoots. Leaves 1·5–4 mm wide, denticulate or sinuate-denticulate. Sepals 5–9 mm; petals (11–)12–15(–16) × 3·5–5(–6) mm. Siliqua (20–)30–60(–80) × 1·2–1·4 mm, erecto-patent (15–30°), abruptly contracted to the style; style 1·8–2·5 mm. 2*n* = 26, 28, 30. *Mountains of N.W. & C. Greece and S. Albania.*

(**f**) Subsp. **microstylum** (Hausskn.) Hayek, *Prodr. Fl. Penins. Balcan.* **1**: 379 (1925): Stock branched, but usually without leaf-rosettes; leaves 1–3 mm wide, usually entire. Sepals 6–7 mm; petals 9–12 × 2·5–4 mm. Siliqua 25–75 × 1–1·3 mm, erecto-patent (15–25°), tapering to the style; style 0·5–1·5 mm. 2*n* = 14. *Mountains of C. & N. Greece, S. Jugoslavia, S.W. Bulgaria and S. Albania.*

(**g**) Subsp. **parnassi** (Boiss. & Heldr.) Hayek, *op. cit.* 380 (1925) (*E. parnassi* (Boiss. & Heldr.) Hausskn.): Laxly caespitose with some basal leaf-rosettes; leaves 0·7–2 mm wide, entire or subentire. Sepals 5–7 mm; petals 9–12 × 3–4 mm; anthers with a few hairs. Siliqua 25–50 × 1–2 mm, more or less erect (10–20°), abruptly contracted to the style; style 2–4 mm, with a minute stigma. 2*n* = 18. *S.C. Greece (Parnassos).*

Subsp. (**e**) appears to be the source of records of **14** from Greece.

Subsp. (**g**) is distinguished from the other subspecies by the hairs on the anthers and by the chromosome number.

16. E. drenowskii Degen, *Magyar Bot. Lapok* **33**: 73 (1934) (*E. helveticum* auct. bulg., non (Jacq.) DC.). Perennial with shortly branched stock, with or without basal leaf-rosettes; stems 10–25 cm; hairs 2(3)-fid. Leaves 20–40 × 1–5 mm, linear to linear-oblanceolate, entire or minutely denticulate, greyish-green. Pedicels 2–3·5 mm in flower, up to 6 mm in fruit. Sepals (5–)6–9 mm; petals (10–)11–14 × 3–4(–5) mm, yellow, pubescent on the back. Siliqua 33–70 × *c.* 1 mm, patent (60–70°), grey with glabrescent angles; style 1·5–2·5 mm. 2*n* = 14. ● *S. Bulgaria, N.E. Greece.* Bu Gr.

17. E. metlesicsii Polatschek, *Ann. Naturh. Mus.* (*Wien*) **78**: 178 (1974) (*E. suffruticosum* auct., non Sprengel, *E. elatum* auct., non Pomel, nec Nutt., *E. murale* auct., non Desf., *E. lanceolatum* auct. ital. pro parte). Biennial with long, stout, woody stock, often with dense petiole remains; stems 15–45 cm in flower, up to 60 cm in fruit, with numerous crowded leaves; hairs 2(3)-fid. Leaves 60–90 × 6–9 mm, oblong, entire or sinuate-denticulate, greyish-green. Pedicels 2–3 mm in flower, up to 6 mm in fruit. Sepals 10–11 mm; petals 14–18 × 3·5–5 mm, pale yellow, glabrous or pubescent on the back; anthers sparsely hairy. Siliqua 60–80 × *c.* 1 mm, erecto-patent (35–50°); style 1·3–1·7 mm. 2*n* = 14. ● *Sicilia* (*E. of Palermo*). Si.

18. E. comatum Pančić, *Fl. Princ. Serb.* 131 (1874) (*E. banaticum* auct., *E. saxosum* E. I. Nyárády). Biennial, perhaps sometimes perennial, with stout, woody stock, usually covered with dense petiole remains; stems 5–40 cm in flower, up to 60 cm in fruit, usually with dense crowded leaves; hairs 2(3)-fid. Leaves (10–)30–90 × 1–6(–8) mm, linear to linear-lanceolate, entire, green to grey. Pedicels 2·5–6 mm in flower, up to 8 mm in fruit. Sepals 7·5–11 mm; petals 13–19 × 3·5–5·5 mm, yellow, pubescent on the back. Siliqua 40–100 × 1·1–1·7 mm, erecto-patent to patent (30–60°); style 0·8–2 mm. 2*n* = 14. ● *From S.W. Romania to S. Jugoslavia.* Bu ?Gr Ju Rm.

19. E. myriophyllum Lange, *Vid. Meddel. Dansk Naturh. Foren. Kjøbenhavn* **1881**: 102 (1882). Biennial, with simple, woody stock with dense persistent leaf remains; stems 10–30 cm in flower, up to 60 cm in fruit, with numerous crowded leaves; hairs bifid. Leaves 18–50 × 1–2·5 mm, linear-lanceolate to filiform, entire, grey or white, erect and appressed to the stem. Pedicels 1·5–2 mm in flower, up to 6(–9) mm in fruit. Sepals 7–11 mm; petals 11–17 × 2–4·5 mm, yellow, pubescent on the back. Siliqua 20–45 × 0·8–1·5 mm, erect (0–10°), appressed to the stem; style 2·5–4 mm. 2*n* = 28. ● *S. Spain.* Hs.

20. E. cazorlense (Heywood) J. Holub, *Folia Geobot. Phytotax.* (*Praha*) **9**: 273 (1974) (*E. linifolium* subsp. *cazorlense* Heywood). Like **19** but petals purple; style (3–)4–7 mm. 2*n* = 28. ● *S.E. Spain.*

21. E. raulinii Boiss., *Fl. Or.* **1**: 192 (1867). Biennial; stems up to 30 cm in flower, up to 60 cm in fruit; hairs 2(–4)-fid. Leaves 20–80 × 1–10 mm, oblanceolate to linear, the lower sinuate or denticulate, the rest entire, green. Pedicels 1–2·5 mm in flower, up to 6 mm in fruit. Sepals 6–8 mm; petals 11–17 × 3–5·5 mm, yellow, pubescent on the back. Siliqua 20–50 × 1–1·3 mm, erect (0–10°), appressed to the stem; style 2–4 mm. 2*n* = 12. ● *W. Kriti.* Cr.

22. E. mutabile Boiss. & Heldr. in Boiss., *Diagn. Pl. Or. Nov.* **2(8)**: 24 (1849). Laxly caespitose perennial with thin, procumbent stems, without leaf-rosettes; stems up to 10 cm in flower, up to 20 cm in fruit; hairs 2(–4)-fid. Lower leaves up to 20 × 4 mm, oblanceolate, petiolate; upper leaves 4–10 × *c.* 1 mm, linear, entire, green. Pedicels 0·5–1 mm in flower, up to 4 mm in fruit. Sepals 4–5 mm; petals 7–11 × 1·5–3·5 mm, yellow on the upper surface, buff or purple and pubescent on the back; anthers pubescent. Siliqua 6–24 × 1–4·5 mm, patent (60–80°); style 1–4 mm. $2n = 14$. ● *Kriti*. Cr.

23. E. humile Pers., *Syn. Pl.* **2**: 200 (1806) (*E. decumbens* (Schleicher ex Willd.) Dennst., *E. ochroleucum* DC., nom. illeg., *E. dubium* (Suter) Thell., non DC.). Perennial with branched, woody stock forming mats with leaf-rosettes; stems 10–30 cm in flower, up to 40 cm in fruit; hairs 2(3)-fid. Leaves 20–80 × 2–8 mm, linear to linear-oblanceolate, entire or denticulate, green, the basal distinctly petiolate. Pedicels 3–7 mm in flower, up to 10 mm in fruit. Sepals 9–12 mm; petals 16–27 × 5–10 mm, yellow, glabrous. Siliqua 45–90 × 1·5–2·3 mm, erecto-patent (20–30°); style 3–8 mm. Seeds 3–4·5 mm. ● *S.W. Alps*. Ga He.

24. E. duriaei Boiss., *Diagn. Pl. Or. Nov.* **3(1)**: 26 (1854) (incl. *E. procumbens* auct., pro parte, *E. ochroleucum* auct., pro parte). Perennial with branched, woody stock, caespitose or forming mats, with leaf-rosettes; stems up to 40 cm in flower, up to 60 cm in fruit; hairs 2(3)-fid. Leaves 10–60 × 1–10 mm, linear to lanceolate, entire or denticulate, green, the lower shortly petiolate. Pedicels 1·5–4 mm in flower, up to 6 mm in fruit. Sepals 7–10(–12) mm; petals 12–20 mm, yellow, glabrous. Siliqua 25–110 × 1–1·5 mm, erect to erecto-patent (10–45°). Seeds 2–3 mm. ● *Pyrenees and mountains of N. Spain*. Ga Hs.

Often included in **23**, but clearly distinguished by the smaller flowers, narrower fruit and smaller seeds.

Cytologically very variable and divided by Polatschek, *op. cit.* (1979), into four species, each with a different chromosome number. However Correvan & Favarger, *op. cit.* (1979), demonstrated that plants with different chromosome numbers could hybridize to give offspring with various levels of fertility. They also reported plants with anomalous chromosome numbers, especially in the Pyrenees. Morphologically the four taxa are difficult to distinguish, so they are here treated at subspecific rank.

1 Style 3–3·5(–5) mm — (d) subsp. **gorbeanum**
1 Style 1–3 mm
 2 Petals 3–4(–5·5) mm wide; middle and upper leaves with clusters of small leaves in their axils; hairs on leaves all bifid — (a) subsp. **duriaei**
 2 Petals (4–)5–8 mm wide; middle and upper leaves usually without axillary clusters of leaves; hairs on leaves 2(3)-fid
  3 Stock with short, erect branches; style 1·5–3 mm; siliqua distinctly 4-angled in transverse section — (b) subsp. **pyrenaicum**
  3 Stock with long, often procumbent branches; style 1·2–2(–3) mm; siliqua rounded in transverse section — (c) subsp. **neumannii**

(a) Subsp. **duriaei**: Stock with erect or procumbent branches. Leaves 1–4 mm wide, linear to linear-lanceolate, the middle and upper with short axillary shoots; hairs all bifid. Sepals 7–8 mm; petals 12–13 × 3–4 mm. Siliqua 25–65 × *c.* 1 mm, rounded in transverse section; style 1·5–2 mm. $2n = 14$. *N. Spain (Oviedo and León)*.

(b) Subsp. **pyrenaicum** (O. Bolòs & Vigo) P. W. Ball, *Bot. Jour. Linn. Soc.* **103**: 207 (1990) (*E. pyrenaicum* Jordan, non (L.) Vill., *E. seipkae* Polatschek): Stock shortly branched. Leaves 1·5–7 mm wide, lanceolate, the hairs 2(3)-fid. Sepals *c.* 9 mm; petals 15–21 × 5–7 mm. Siliqua 40–110 × *c.* 1·2 mm, square in transverse section; style 1·5–3 mm. $2n = 26, 28, 34, 51$. *Pyrenees*.

(c) Subsp. **neumannii** (Polatschek) P. W. Ball, *op. cit.* 207 (1990) (*E. neumannii* Polatschek): Like subsp. (**b**) but stock with some long, procumbent branches; sepals 8–10 mm; petals 14–18 × (4–)5–8 mm; siliqua 1–1·5 mm wide, rounded in transverse section; style 1·2–2(–2·5) mm. $2n = 26, 28$. *N. Spain (Santander, Oviedo, León)*.

(d) Subsp. **gorbeanum** (Polatschek) P. W. Ball, *op. cit.* 207 (1990) (*E. gorbeanum* Polatschek): Stock with long, procumbent branches. Leaves 2·5–10 mm wide, lanceolate. Sepals 8–10(–12) mm; petals 15–18 × 4·5–7 mm. Siliqua 35–50 × *c.* 1·2 mm, square in transverse section; style 3–3·5(–4) mm. $2n = 42$. *N. Spain (Peña Gorveya, Vizcaya prov.)*.

Subsp. (**c**) is reported also from Logroño, but the plants are relatively robust and have the style *c.* 3 mm long. A few collections from Picos de Europa are distinctive in having short, axillary shoots on the stem.

25. E. penyalarensis (Pau) Polatschek, *Ann. Naturh. Mus. (Wien)* **82**: 349 (1979). Perennial with branched, woody stock with leaf-rosettes; stems up to 20 cm in flower, up to 25 cm in fruit; hairs 2(–4)-fid. Leaves 25–45 × 3–7 mm, spathulate to obovate, obtuse, entire or denticulate, the lower petiolate, green. Pedicels 2·5–3·5 mm in flower, up to 4 mm in fruit. Sepals 8–11 mm; petals 15–16 × 3–5 mm, yellow, glabrous. Siliqua 35–70 × *c.* 2 mm, erecto-patent (20–40°), square in transverse section; style 3–4·5 mm. $2n = 48$. ● *C. Spain*. Hs.

Perhaps only a subspecies of **24**, but somewhat disjunct in distribution from that species.

26. E. linifolium (Pers.) Gay, *Erysim. Nov.* 3 (1842). More or less caespitose perennial with branched, woody stock; stems up to 60 cm in flower, up to 70 cm in fruit; hairs 2(3)-fid. Leaves 20–90 × 1–6 mm, linear to lanceolate, usually sinuate-dentate, green. Pedicels 1·5–3 mm in flower, up to 7 mm in fruit. Sepals 6–9 mm; petals 12–18 × 5–5·5 mm, purple, glabrous. Siliqua 35–75 × 1–1·7 mm, erecto-patent (*c.* 20°); style 2–3 mm. $2n = 14$. ● *N. Portugal, C. Spain*. Hs Lu.

27. E. baeticum (Heywood) Polatschek, *Ann. Naturh. Mus. (Wien)* **82**: 329 (1979) (*E. linifolium* subsp. *baeticum* Heywood). Like **26** but usually biennial, with unbranched stock; petals 14–20 mm; siliqua erect (0–10°); style 2–4(–5) mm. $2n = 28$. ● *S. Spain*. Hs.

Some specimens are clearly perennial with the remains of old flowering stems, a branched, woody stock and persistent basal leaf-rosettes. These sometimes occur with plants which are clearly biennial.

28. E. popovii Rothm., *Feddes Repert.* **49**: 180 (1940). Perennial with branched, woody stock and leaf-rosettes; stems up to 20 cm in flower, up to 60 cm in fruit; hairs 2(3)-fid. Leaves 25–60 × 1·5–5 mm, oblanceolate, entire or sinuate-denticulate. Pedicels 1–2.5 mm in flower, up to 4 mm in fruit. Sepals *c.* 9 mm; petals 14–16 × 4–4·5 mm, purple, glabrous. Siliqua 20–40 × 1–1·4 mm, erect (*c.* 10°); style 4–7·5 mm. $2n = 42$. ● *S.E. Spain (Sierra de Alfacar, Cerro Jabalcuz)*. Hs.

29. E. favargeri Polatschek, *Ann. Naturh. Mus. (Wien)* **82**: 334 (1979). Like **28** but biennial, up to 90 cm, sometimes branched at the base; pedicels *c.* 3·5 mm in flower, up to 5 mm in fruit; petals

3·5–5 mm wide; siliqua up to 40 mm; style 4·5–6 mm. $2n = 56$. ● *S.E. Spain (Sierra de Cazorla, Sierra de Pozo)*. Hs.

30. E. ucrainicum Gay, *Erysim. Nov.* 3 (1842) (*E. cretaceum* (Rupr.) Schmalh.). Perennial with shortly branched stock; stems 25–50 cm, with dense persistent petiole-remains at the base; hairs 2(3)-fid. Leaves 20–30 × 1–3 mm, linear or linear-lanceolate, entire, grey. Pedicels 5–10 mm in flower, up to 16 mm in fruit. Sepals 6–10 mm; petals 12–18 × 3·5–6 mm, yellow, pubescent on the back. Siliqua 20–60 × 1–1·3 mm, erect (*c.* 10°), always with some trifid hairs; style 0·5–2·5 mm. *Don Basin*. Rs(C, E).

31. E. krynkense Lavrenko, *Ind. Sem. Horti Charkov* **1925**: 7 (1926). Like **30** but biennial; stems up to 80 cm, with sparse petiole-remains; pedicels 1–2 mm in flower, up to 5 mm in fruit; siliqua 15–40 mm, with style 1·5–4·5 mm. ● *S.E. Ukraine (near Donetsk)*. Rs(E).

Known from only a single locality.

32. E. diffusum Ehrh., *Beitr. Naturk.* **7**: 157 (1792) (*E. canescens* Roth, *E. australe* auct. balcan.). Biennial or short-lived perennial; stems up to 120 cm; hairs 2(3)-fid. Leaves 15–70 × 2–8 mm, rarely wider, linear to linear-lanceolate, entire or sinuate-denticulate, grey. Pedicels 3–6 mm in flower, up to 9 mm in fruit. Sepals 6–8(–9) mm; petals (9–)10–15 × 2·5–4 mm, yellow, pubescent on the back. Siliqua (20–)35–80 × 0·5–1 mm, erect or erecto-patent (10–30°), square in transverse section, grey-green or grey with glabrescent angles; style 0·5–1·8 mm. $2n = 28, 70, 72$. *C. & S.E. Europe*. Al Au Bu Cz Gr Hu Ju ?Po Rm Rs(C, W, K, E) ?Tu.

Plants with $2n = 28$ have narrow, usually entire leaves, the petals (9–)10–12(–13) mm and the style *c.* 0.5 mm. Those with $2n = 70$, 72 have broader, usually denticulate leaves, the petals (13–)14–15 mm and the style 0·8–1·8 mm. These two entities are sympatric in C. Europe, but nothing is known of their occurrence in other areas.

In the Balkan peninsula plants with patent siliquae sometimes occur, these being determined as either **32** or **33**. They appear to resemble **34** more closely, a species which is supposedly endemic to S. Italy.

33. E. graecum Boiss. & Heldr. in Boiss., *Diagn. Pl. Or. Nov.* **3(1)**: 27 (1853). Biennial; stems up to 50 cm in flower, up to 100 cm in fruit; hairs 2(3)-fid. Leaves 15–90 × 1·5–7 mm, linear to oblanceolate, entire or sinuate-denticulate, greyish-green. Pedicels 2–4 mm in flower, 4–5(–8) mm in fruit and as thick as the siliqua. Sepals 6–8 mm; petals 9–13 × 2·5–3·5 mm, yellow, pubescent on the back; anthers sometimes hairy. Siliqua 30–90 × 0·8–1·1 mm, patent (70–90°), square in transverse section, grey with glabrescent angles; style 0·5–2 mm. $2n = 24$. ● *S.E. Greece, W. Kriti*. Cr Gr.

Records of this species from elsewhere in the Balkan peninsula are erroneous.

34. E. crassistylum C. Presl, *Fl. Sic.* **1**: 77 (1826) (*E. canescens* auct. ital., *E. diffusum* auct. ital., *E. graecum* auct. ital.). Biennial or short-lived perennial; stems 20–35 cm in flower, up to 70 cm or more in fruit; hairs 2(3)-fid. Leaves 20–90 × 0·5–4 mm, linear to linear-spathulate, entire, greyish-green. Pedicels 3–4 mm in flower, up to 5·5 mm in fruit, sometimes almost as thick as the siliqua. Sepals 4·5–8 mm; petals 11–13 × 2·8–4 mm, yellow, pubescent on the back. Siliqua 30–80 × *c.* 1 mm, erecto-patent to patent (60–70°), greyish-green with glabrescent angles; style *c.* 1 mm. $2n = 42$. ● *S. Italy*. It.

Similar plants, usually determined as **32** or **33**, occur in the central part of the Balkan peninsula.

35. E. creticum Boiss. & Heldr. in Boiss., *Diagn. Pl. Or. Nov.* **3(1)**: 26 (1853). Biennial; stems 10–45 cm in flower, up to 60 cm in fruit; hairs 2- or 3(–5)-fid. Leaves 50–90 × 3–10 mm, entire to sinuate-dentate, grey. Pedicels 1·5–4 mm in flower, up to 7·5 mm in fruit. Sepals 6–8·5 mm; petals 9–14 × 3–4·5 mm, yellow, pubescent on the back. Siliqua 30–65 × 1–1·4 mm, patent (70–90°), grey with glabrescent angles, most hairs tri- or quadrifid; style 0·5–1·2 mm. $2n = 14$. ● *E. Kriti*. Cr.

36. E. calycinum Griseb., *Spicil. Fl. Rumel.* **1**: 260 (1843). Like **35** but sometimes branched at the base and perhaps then perennial; hairs 2(3)-fid; leaves entire to denticulate; pedicels up to 10 mm in fruit; petals 2·5–3 mm wide, with (2)3-fid hairs on the back; siliqua erecto-patent (20–35°), with bi- or trifid hairs; style 1·3–2·5 mm. ● *N. Greece (Athos)*. Gr.

37. E. olympicum Boiss., *Fl. Or.* **1**: 191 (1867). Perennial with branched woody stock, without leaf-rosettes; stems 15–35 cm; hairs 2(3)-fid. Leaves 15–120 × 1–4 mm, linear, entire or the lower sinuate-denticulate. Pedicels (2–)3–5 mm in flower, up to 11 mm in fruit. Sepals 6–8 mm; petals 11–14 × 3–5 mm, yellow, pubescent on the back, with mainly bifid hairs. Siliqua (50–)60–100 × 1–1·5 mm, erect (10–20°), rarely erecto-patent (*c.* 35°), grey with glabrescent angles; hairs mostly bifid; style 1(–2) mm. $2n = 56$. ● *E.C. Greece (Olimbos)*. Gr.

38. E. asperulum Boiss. & Heldr. in Boiss., *Diagn. Pl. Or. Nov.* **3(6)**: 11 (1859). Biennial; stems 10–50 cm in flower, up to 70 cm in fruit; hairs bi- or trifid. Leaves 15–50 × 2–6 mm, linear-lanceolate, sinuate-dentate to entire, grey. Pedicels 2–6 mm in flower, up to 8 mm in fruit. Sepals 6–8 mm; petals 12–16 × 2·5–4 mm, yellow, pubescent on the back, with mainly trifid hairs. Siliqua 30–80 × 1–1·5 mm, erect or erecto-patent (10–40°), grey with glabrescent angles, hairs tri- or quadrifid; style 0·5–2 mm. $2n = 24$. ● *Albania and C. & S. Greece*. Al Gr.

Plants from W.C. Greece (Evritania) are intermediate between **37** and **38**. They appear to be biennial, with mainly bifid hairs on the leaves and petals, and with mixed bifid and trifid hairs on the siliqua. Their status is not clear.

39. E. crepidifolium Reichenb., *Pl. Crit.* **1**: 8 (1823) (*E. banaticum* sensu E. I. Nyárády). Biennial or short-lived perennial; stems 15–60 cm, or more in fruit; hairs 2- to 4-fid. Leaves 15–60 × 2–10 mm, linear to linear-lanceolate, sinuate-dentate or the upper entire, green or greyish-green. Pedicels 1–4 mm in flower, up to 6 mm in fruit. Sepals 5–8(–9) mm; petals 10–18 × 2–4 mm, yellow, pubescent on the back with mainly tri- or quadrifid hairs. Siliqua 25–60 × 0·8–2(–2·5) mm, erect or erecto-patent (10–30°), with tri- or quadrifid hairs; style 0·8–2(–2.5) mm. ● *C. & S.E. Europe*. Cz Ge Hu Rm.

40. E. leptostylum DC., *Reg. Veg. Syst. Nat.* **2**: 494 (1821). Like **39** but leaves oblong-lanceolate with tri- or quadrifid hairs, green; pedicels up to 7 mm in fruit; sepals 7–9 mm; petals 15–18 mm, with orbicular-obovate limb; siliqua erect; style *c.* 2 mm. *From Moldavia to the Caspian Sea*. Rs(W, K, E).

Possibly not distinct from **39**.

41. E. exaltatum Andrz. ex Besser, *Enum. Pl. Volhyn.* 71 (1822) (*E. hieracifolium* var. *exaltatum* (Andrz. ex Besser) Hayek). Biennial or perhaps sometimes perennial; stems 40–120 cm, with dense petiole-remains at the base; hairs bi- or trifid. Leaves 40–90 × 3–10 mm, oblong-lanceolate, subentire to sinuate-dentate. Pedicels 4–10 mm. Sepals 7·5–10 mm; petals 12–19 × 3·5–4·5 mm, yellow, pubescent on the back, with mainly trifid hairs. Siliqua 60–100 × 1–1·2 mm, erect to erecto-patent (10–20°), grey with glabrescent angles; style 1·5–2·5 mm. ● *From Moldavia to N. Bulgaria.* Bu Rm Rs(W).

Clearly distinct from **52** with anthers 2·2–3·5 mm. This species is part of a difficult complex consisting of species **35–42**. The distinction between it and **39** and **40** needs to be more thoroughly investigated. Little is known of the range of variability found in **40** and **41**, and it is possible that they will be found to be conspecific.

42. E. horizontale P. Candargy, *Bull. Soc. Bot. Fr.* **44**: 154 (1897) (*E. smyrnaeum* auct.). Biennial; stems up to 60 cm; hairs (3)4- to 5(6)-fid. Basal leaves pectinate-pinnatifid, usually dead at anthesis; cauline leaves 30–60 × 3–7(–10) mm, oblong, sinuate-dentate to entire, green. Pedicels 2–4 mm in flower, up to 8 mm in fruit. Sepals 6–8 mm; petals 12–14 × 3–3·5 mm, yellow, pubescent on the back. Siliqua 40–70 × 1–1·5 mm, patent (70–90°), grey with glabrescent angles; style 0·5–1·5 mm. *Turkey-in-Europe.* Tu. (*E. Aegean Islands.*)

Some specimens from Turkey-in-Europe have erect (*c.* 10°) siliquae, but are otherwise very similar to **42**. Not certainly distinct from **E. smyrnaeum** Boiss. & Balansa in Boiss., *Diagn. Pl. Or. Nov.* 3(5): 23 (1856), from W. Anatolia, which has bi- or trifid hairs, petals 13–17 × 4–4·5 mm, and the siliqua 45–80 mm, erecto-patent (30–60°), grey, with tri- or quadrifid hairs.

43. E. aureum Bieb., *Fl. Taur.-Cauc.* **2**: 117 (1808) (incl. *E. sylvaticum* Bieb.). Biennial; stems 40–100 cm; hairs (2)3- or 4-fid. Leaves oblong to oblong-lanceolate, sinuate-dentate or serrate to denticulate, green. Pedicels 6–15 mm in fruit. Sepals 4·5–6 mm; petals 8–12 × *c.* 3 mm, yellow, pubescent on the back. Siliqua 20–50 × 1–2 mm, patent or deflexed (80° or greater), more or less square in transverse section; style (0·5–)1·5–3 mm. *C. & S. Russia, Ukraine.* Rs(C, W, E).

44. E. leucanthemum (Stephan) B. Fedtsch., *Acta Horti Petrop.* **23**: 413 (1905) (*E. versicolor* Andrz. ex DC.). Biennial; stems 10–65 cm; hairs (2)3- or 4-fid. Leaves 20–30 × *c.* 2 mm, linear or linear-oblong, the basal runcinate, usually dead at anthesis, the upper runcinate, serrate or subentire. Pedicels 1·5–3 mm in flower, up to 5 mm in fruit. Sepals 4·5–6 mm; petals 8–12 × *c.* 2 mm, pale yellow or whitish, sparsely pubescent on the back or glabrous. Siliqua 18–45 × 0·4–0·8 mm, erect (0–10°), terete; style *c.* 1 mm. *E. & S. Russia, Ukraine.* Rs(C, W, K, E).

45. E. pulchellum (Willd.) Gay, *Erysim. Nov.* 10 (1842) (incl. *E. korabense* Kümmerle & Jáv.). Caespitose perennial with branched stock and numerous leaf-rosettes; stems 5–15 cm; hairs bi- to quadrifid. Leaves 3–6 mm wide, oblong-spathulate or -lanceolate, entire to sinuate-dentate, green. Pedicels *c.* 2 mm in flower. Sepals 5–7 mm; petals 11–16 × 3·5–5 mm, yellow, pubescent on the back. Siliqua 20–30 × *c.* 1 mm, erect (10–20°), sparsely pubescent, green; style 1·5–4 mm. *Grassland. Mountains of E. Albania, S. Jugoslavia, Bulgaria.* Al Bu Ju.

46. E. odoratum Ehrh., *Beitr. Naturk.* **7**: 157 (1792) (*E. pannonicum* Crantz, *E. erysimoides* (L.) Fritsch ex Janchen, non (Kar. & Kir.) O. Kuntze, *E. pallescens* Herbich, *E. aureum* auct. roman., non Bieb.). Biennial; stems 20–80 cm in flower, up to 130 cm in fruit; hairs (2)3(4)-fid. Leaves 20–75 × 4–15 mm, elliptical to obovate, sinuate-dentate, the lower sessile or shortly petiolate, the basal usually dead at anthesis. Pedicels 2·5–8 mm in flower, up to 11 mm in fruit. Sepals 6–10 mm; petals 12–18 × 4–8 mm, yellow, pubescent on the back. Siliqua (20–)40–70 × 0·8–1·2 mm, usually erect (10–20°), glabrescent to pubescent with glabrescent angles, the hairs mostly bifid; style 1–2 mm. $2n = 24$. ● *C. Europe, extending to E. France, N. Italy and S. Bulgaria.* ?Al Au Bu Cz Ga Ge Hu It Ju Po Rm Rs(W) [He].

Records from the central part of the Balkan peninsula may be erroneous due to confusion with **49**.

47. E. witmannii Zawadzki, *Enum. Pl. Galic. Bucow.* 81, 194 (1835) (incl. subsp. *pallidiflorum* (Jáv.) Jáv., subsp. *transsilvanicum* (Schur) P. W. Ball, *E. transsilvanicum* Schur, *E. baumgartenianum* Schur). Biennial; stems 10–50 cm; hairs bi- or trifid. Leaves 20–70 × 3–15 mm, linear to oblong-lanceolate, sinuate-dentate, rarely almost entire, the lower and often the middle leaves long-petiolate. Pedicels 3–6 mm in flower, up to 10 mm in fruit. Sepals 8–12 mm; petals 15–24 × 5–9 mm, yellow or pale greenish-yellow, pubescent on the back. Siliqua (25–)40–120 × 1–1·5 mm, erecto-patent to patent, the hairs on the angles mostly bifid; style *c.* 1 mm. $2n = 14$. ● *Carpathians, Hungary and N.W. Bulgaria.* Bu Cz Hu Po Rm Rs(W).

48. E. carniolicum Dolliner, *Flora (Regensb.)* **10**: 254 (1827) (*E. pannonicum* var. *sinuatum* (Neilr.) Janchen). Biennial; stems 10–60 cm; hairs bi- or trifid. Leaves 20–50 × 4–12 mm, linear-oblong to oblong, pectinate-pinnatifid to sinuate-dentate, the basal usually persistent, the lower and middle leaves sessile or very shortly petiolate. Pedicels 3–8 mm in flower, up to 11 mm in fruit. Sepals 8·5–12 mm; petals 16–24 × 5–10 mm, yellow, pubescent on the back. Siliqua 60–110 × *c.* 1·5 mm, erecto-patent (20–30°), sparsely pubescent; style 0·5–1 mm. $2n = 32$. ● *W. Jugoslavia, S.E. Austria.* Au Ju.

49. E. kuemmerlei Jáv., *Bot. Közl.* **19**: 20 (1921). Like **48** but hairs mostly tri- or quadrifid; sepals 5–8 mm; petals (9–)10–14 × 3–5 mm; siliqua 30–80 × 1–1·5 mm; style 0·8–2 mm. $2n = 14$. ● *W. part of the Balkan peninsula, from C. Jugoslavia to N. Greece.* Al Gr Ju.

50. E. pectinatum Bory & Chaub. in Bory, *Expéd. Sci. Morée* **3**: 189 (1832). Perennial with branched, woody stock, usually without basal leaf-rosettes; stems 10–50(–80) cm; hairs (2)3(–5)-fid. Lower and middle leaves 15–50 × 3·5–6 mm, pinnatifid to pinnatisect, sessile. Pedicels 1–2(–4) mm in flower, up to 5 mm in fruit. Sepals 4·5–6·5 mm; petals 10–13 × 2–3 mm, yellow, pubescent on the back. Siliqua 20–45 × 0·8–1·2 mm, erect (0–10°), grey with glabrescent angles, the hairs 2(3)-fid; style 0·5–2 mm. $2n = 12$. ● *S. Greece (Peloponnisos).* Gr.

Records from elsewhere in the Balkan peninsula are referable to **49**.

51. E. hungaricum Zapał., *Bull. Int. Acad. Sci. Cracovie, ser. B* **1913**: 49 (1913) (incl. *E. wahlenbergii* (Ascherson & Engler) Simonkai, *E. pieninicum* (Zapał.) Pawł.). Usually biennial; stems 30–100 cm; hairs mostly tri- or quadrifid. Leaves 20–90 × 5–30 mm, linear- to oblong-lanceolate, entire to sinuate-dentate, green. Pedicels 3–5 mm in flower, up to 15 mm in fruit.

Sepals 6–8·5(–10) mm; petals (8–)10–15(–20) × 3–4 mm, yellow, pubescent on the back. Siliqua 45–90 × 0·9–1·9 mm, erect or erecto-patent (10–30°); style 0.8–2 mm. ● *W. & N.E. Carpathians.* Au Cz Po Rm ?Rs(W).

52. E. hieracifolium L., *Cent. Pl.* **1**: 18 (1755) (*E. strictum* P. Gaertner, B. Meyer & Scherb.). Biennial or short-lived perennial; stems 30–70 cm in flower, up to 120 cm in fruit; hairs 3(4)-fid. Leaves 25–90 × 4–15 mm, linear- to oblong-lanceolate, sinuate-dentate, green. Pedicels 2·5–5 mm in flower, up to 10 mm in fruit. Sepals 4–6 mm; petals (7–)8–10(–12) × 2–3·5 mm, yellow, pubescent on the back; anthers 1–2·2 mm. Siliqua (20–)30–60 × 0·8–1·5 mm, erecto-patent (10–30°), usually grey; style 1–2 mm. $2n = 32$. *N., C. & E. Europe, extending westwards to Belgium and E. France.* Al Au Be Bu Cz Da Fe Ga Ge Ho ?Hu Is Ju No Po Rm Rs(N, B, C, W, E) Su.

The southern and eastern limits of the distribution of this species are uncertain owing to confusion with **53** and **54**.

53. E. marschallianum Andrz. ex Bieb., *Fl. Taur.-Cauc.* **3**: 441 (1819) (*E. durum* J. & C. Presl). Like **52** but biennial; hairs tri- or quadrifid; leaves 15–40 × 2–5 mm, linear or linear-oblong, entire or denticulate, grey; sepals 2–5 mm; petals 5–8·5(–9) × 2–5 mm; siliqua 20–75 × 1–1·8 mm, erect (0–10°), usually grey with glabrescent angles; style 0·5–1·5 mm. *E.C. & E. Europe.* Au Cz Ge Po Rs(N, B, C, W, E).

54. E. virgatum Roth, *Catalecta Bot.* **1**: 75 (1797). Like **52** but leaves 15–50 × 1·5–6 mm, linear-elliptical or -lanceolate, entire or denticulate; pedicels 2–3·5 mm in flower, up to 8 mm in fruit; sepals 6–8(–9) mm; petals 9–14 × 2–4 mm, glabrous; siliqua erect (0–10°). $2n = 48$. ● *Alps.* Au Ge He It Ju.

55. E. cuspidatum (Bieb.) DC., *Reg. Veg. Syst. Nat.* **2**: 493 (1821) (*Syrenia cuspidata* (Bieb.) Reichenb.; incl. *E. goniocaulon* auct., *E. pulchellum* auct. turc., ?*E. tetovense* Rohlena). Annual or biennial; stems 20–70(–100) cm; hairs tri- or quadrifid. Lower leaves oblong- or ovate-lanceolate, sinuate-dentate or dentate, green. Pedicels 1–3 mm in flower, up to 5 mm in fruit. Sepals 5–7 mm; petals 9–13 × 2·5–4 mm, yellow, pubescent on the back. Siliqua 10–30 × 2·5–3·5 mm, strongly laterally compressed, erect, grey with glabrescent angles; valves keeled or narrowly winged on the back; style 3–6(–10) mm. *S.E. Europe.* Al Bu Gr Ju Rm Rs(W, K, E) Tu.

56. E. repandum L., *Demonstr. Pl.* 17 (1753) (*E. patens* Loscos). Annual; stems up to 35(–60) cm; hairs bi- or trifid. Lower leaves 10–70 × 1–1·3 mm, linear to lanceolate, entire to sinuate-dentate, green. Pedicels 1–3 mm in flower, up to 5(–7) mm in fruit. Sepals 3–6 mm; petals 6–10 × 1–2 mm, yellow, pubescent on the back. Siliqua 30–100 × 1–1·5 mm, patent (*c.* 90°), green; style 2–5 mm, about as wide as the siliqua. $2n = 16$. *C., E. & S. Europe.* Al Au Bu Cz Ge Gr Hu Ju Po Rm Rs(C, W, K) Tu [He].

A sometimes frequent casual in most other countries.

57. E. incanum G. Kunze, *Flora (Regensb.)* **29**: 753 (1846) (incl. *E. aurigeranum* Jeanb. & Timb.-Lagr., *E. kunzeanum* Boiss. & Reuter). Annual; stems up to 60 cm in flower, up to 75 cm in fruit; hairs bi- or trifid. Leaves 10–60 × 1·5–9 mm, linear-lanceolate or lanceolate, the basal pinnatifid, usually dead at anthesis, the cauline subentire to sinuate-dentate, green. Pedicels 0·5–4 mm in flower, 1·5–6 mm in fruit. Sepals 2·5–4(–5) mm; petals 3–6(–9) × 0·5–1(–2) mm, yellow, slightly pubescent on the back. Siliqua 15–60 × 0·7–1·1 mm, erect to erecto-patent (10–60°), grey; style *c.* 0·5 mm. $2n = 16$. *C., S. & E. Spain, extending northwards to the Pyrenees.* Ga Hs.

58. E. cheiranthoides L., *Sp. Pl.* 661 (1753). Annual or rarely biennial; stems up to 100 cm; hairs (2)3- to 5-fid. Lower leaves oblong-lanceolate, entire to sinuate-dentate, green. Pedicels 4–8 mm in flower, up to 16 mm in fruit, slender, patent (75–90°). Sepals 2–4 mm; petals (2–)3–6 × 1–2 mm, yellow, pubescent on the back. Siliqua 10–50 × 1–1·5 mm, erecto-patent (45–60°), green; style up to 1·5 mm. *Throughout much of Europe, but absent from much of the south.* Au Be Br Cz Da Fe Ga Ge Hb He Ho Hu Is Ju No Po Rm Rs(N, B, C, W, E) Su.

(a) Subsp. **cheiranthoides**: Annual, usually 20–50 cm, sparsely pubescent; base of the stem 2–6 mm in diameter, with 5–10 internodes between the basal rosette and the inflorescence; cauline leaves acute or subacute, widest at the middle, erecto-patent. $2n = 16$. *Throughout the range of the species except the extreme north.*

(b) Subsp. **altum** Ahti, *Arch. Soc. Zool.-Bot. Fenn. Vanamo* **16**: 24 (1962): Biennial, usually 40–100 cm, densely pubescent; base of the stem up to 15 mm in diameter, with 20–40 internodes between the basal rosette and the inflorescence; cauline leaves acuminate, widest near the base, appressed to the stem. $2n = 16$. *Fennoscandia and N. Russia, southwards to* c. 55° *N.*

19. Syrenia Andrz.

P. W. BALL

Like *Erysimum* but median nectaries absent; hairs on the siliqua usually medifixed, lying transversely across the valves; seeds in 2 rows in each loculus.

1 Pedicels 3–5 mm in flower, up to 6 mm in fruit, about ½ as long as the calyx and siliqua **3. siliculosa**
1 Pedicels less than 3 mm in flower, up to 3(–5) mm in fruit, not more than ⅓ as long as the calyx and siliqua
2 Siliqua compressed laterally; valves narrowly winged on the back; style 6–12 mm, about as long as the siliqua **4. montana**
2 Siliqua square in cross-section; valves keeled but not winged; style 3–7 mm, usually shorter than the siliqua
3 Siliqua 5–20(–25) mm; pedicels not more than 2 mm; hairs medifixed **1. cana**
3 Siliqua 15–40 mm; pedicels 2–5 mm; hairs medifixed and trifid **2. talijevii**

1. S. cana (Piller & Mitterp.) Neilr., *Aufz. Ung. Slav. Gefäss., Nachtr.* 73 (1870) (*S. angustifolia* (Ehrh.) Reichenb., *S. ucrainica* Klokov, *Erysimum canum* (Piller & Mitterp.) Polatschek). Biennial 25–80 cm; hairs medifixed. Leaves linear, entire, grey or grey-green. Pedicels 0·5–2 mm. Sepals 6–10 mm, saccate at the base; petals (10–)12–22 × 3–6 mm, pale yellow, glabrous on the back; anthers 2–3 mm. Siliqua 5–25(–30) × 1·5–4 mm, square in cross-section, white, with green angles; valves keeled; style 3·5–7 mm. Seeds 1–1·8 mm. *Sandy soils. E. & E.C. Europe, from Czechoslovakia and Jugoslavia to S. Russia.* Bu Cz Hu Ju Rm Rs(C, W, K, E).

2. S. talijevii Klokov, *Trav. Inst. Bot. (Charkov)* **1**: 107 (1936). Like **1** but (35–)50–90 cm; hairs medifixed and trifid; pedicels 1·5–2 mm in flower, up to 5 mm in fruit; siliqua 15–40 mm; style 3–7 mm. ● *E. Ukraine; ?S. Russia.* Rs(E).

3. S. siliculosa (Bieb.) Andrz. ex C. A. Meyer in Ledeb., *Fl.*

Altaica **3**: 162 (1831). Biennial 40–80 cm; hairs medifixed. Leaves linear, entire, grey-green. Pedicels 3–5 mm in flower, up to 6 mm in fruit. Sepals 7–10(–12) mm, saccate at the base; petals (13–)15–18(–20) mm, pale yellow. Siliqua 5–10(–12) × 2–3 mm, square in cross-section; valves keeled; style 5–10(–12) mm, about as long as the siliqua. *S. & S.E. Russia; S. & E. Ukraine.* Rs(C, K, E).

4. S. montana (Pallas) Klokov in Klokov & Wissjul., *Fl. RSS Ucr.* **5**: 505 (1953) (*S. sessiliflora* Ledeb., *S. aucta* Klokov, *S. dolichostylos* Klokov, *Erysimum sessiliflora* R. Br.). Biennial 30–90 cm; hairs medifixed. Leaves linear, entire, grey-green. Pedicels 0.5–1(–2) mm in flower, up to 2(–2·5) mm in fruit. Sepals (6–)8–13 mm, saccate at the base; petals 15–21 × 3–7 mm, pale yellow. Siliqua 6–15 × 2–4 mm, laterally compressed, white with green angles; valves narrowly winged; style 6–12(–14) mm. Seeds *c.* 1 mm. *S. Russia; S. Ukraine.* Rs(C, K, E).

20. Hesperis L.

P. W. BALL

Biennials to perennials; glabrous or with unbranched, branched or glandular hairs. Sepals erect, the inner saccate at the base; petals yellow, purple or white, long-clawed. Fruit a siliqua; valves with a distinct median and lateral veins; style short; stigma deeply 2-lobed, the lobes erect, free. Seeds in 1 row in each loculus. Radicle incumbent.

Literature: V. Borbás, *Magyar Bot. Lapok* **1**: 161 *et seq.* (1902); **2**: 12–23 (1903). N. Tzvelev, *Not. Syst. (Leningrad)* **19**: 114–155 (1959).

1 Flowers yellow, sometimes purple or suffused with purple
 2 Lower pedicels usually 40–90 mm in fruit; petal-limb 2–4 mm wide, linear-lanceolate **1. tristis**
 2 Lower pedicels usually less than 20 mm in fruit; petal-limb 3–9 mm wide, usually ovate-oblong **2. laciniata**
1 Flowers purple, pink or white
 3 Stems hispid; pedicels usually much shorter than the calyx **2. laciniata**
 3 Stems not hispid; pedicels ± equalling the calyx **(3–13). matronalis** group

1. H. tristis L., *Sp. Pl.* 663 (1753). Stout, hispid biennial or perennial 25–50(–70) cm; indumentum of branched, unbranched and glandular hairs. Leaves ovate-lanceolate to lanceolate, entire to denticulate; cauline leaves subcordate. Lower pedicels 40–90 mm in fruit. Sepals 9–15 mm; petals 20–25(–30) mm, yellow, rarely purplish; limb 2–4 mm wide, narrowly lanceolate. Siliqua 40–180 × 2·5–4·5 mm; valves flat; style *c.* 0.5 mm. $2n = 14$. *Dry, sunny places and scrub. C. & E. Europe.* Au Bu Cz Hu Ju Rm Rs(C, W, K, E) ?Tu.

2. H. laciniata All., *Fl. Pedem.* **1**: 271 (1785) (*H. dalmatica* E. Fourn., ?*H. dauriensis* Amo; incl. *H. glutinosa* Vis.). Biennial or perennial 10–80 cm; indumentum of branched, unbranched and often glandular hairs. Leaves pinnatifid to sinuate-dentate, the cauline sessile. Pedicels shorter than the sepals in flower, up to 15 mm in fruit. Sepals 5–12 mm; petals 15–30(–40) mm, yellow variably suffused with purple, or entirely purple; limb 3–9 mm wide, oblong to obovate. Siliqua 50–150 × 1·5–3(–4) mm; valves flat; style *c.* 1 mm. *Cliffs and rocks. S. Europe.* Al Bu Ga Gr Hs It Ju Lu Sa Si.

(**a**) Subsp. **laciniata** (incl. subsp. *spectabilis* (Jordan) Rouy & Fouc.): Plant usually glandular-pubescent. Siliqua pubescent. *Almost throughout the range of the species.*

(**b**) Subsp. **secundiflora** (Boiss. & Spruner) Breistr., *Mém. Soc. Bot. Fr.* **1952**: 84 (1952) (incl. subsp. *scabricarpa* (Boiss.) Dvořák): Plant eglandular-pubescent. Siliqua glabrous. *S. part of Balkan peninsula.*

Subsp. (**a**) is variable and often divided into a number of subsidiary taxa. A variant that has been called subsp. *spectabilis* (Jordan) Rouy & Fouc., with the petals 25–40 × 5–9 mm and purple, is often confused with **3**. It generally occurs only in the southern half of the range of subsp. (**a**).

(3–13). H. matronalis group. Biennials or perennials 40–120 cm; hairs branched, unbranched and sometimes glandular. Pedicels 5–20 mm in fruit, about equalling the sepals in flower. Sepals 5–8(–10) mm; petals 14–25 mm, white or purple; limb 3–7 mm wide, oblong to obovate. Siliqua 25–100 × 1·5–3 mm; valves rounded; style 1·5–2·5 mm.

A group of closely related taxa of which the status is in many cases doubtful.

1 Middle and upper cauline leaves ± shortly petiolate, never amplexicaul **3. matronalis**
1 Middle and upper cauline leaves sessile, ± amplexicaul
 2 Flowers white
  3 Upper leaves and upper part of stem glandular
   4 Leaves with numerous branched hairs, usually densely glandular **10. dinarica**
   4 Leaves with mainly unbranched hairs, sparsely glandular **11. vrabelyiana**
  3 Plant entirely eglandular
   5 Stem and leaves green, glabrous or pubescent; leaves dentate to entire **12. nivea**
   5 Stem and leaves white-tomentose; leaves coarsely sinuate-dentate or serrate **13. inodora**
 2 Flowers pink or purple
  6 Plant eglandular
   7 Calyx and pedicels glabrous **7. oblongifolia**
   7 Calyx and pedicels pubescent
    8 Stem and leaves with long unbranched hairs **8. steveniana**
    8 Stem and leaves without conspicuous, long unbranched hairs **9. pycnotricha**
  6 Plant glandular-pubescent
   9 Plant without long unbranched hairs **5. sylvestris**
   9 Plant with long unbranched hairs
    10 Plant sparsely pubescent, green; siliqua eglandular **4. elata**
    10 Plant grey-tomentose; siliqua usually glandular **6. theophrasti**

3. H. matronalis L., *Sp. Pl.* 663 (1753). Indumentum variable, glandular or eglandular, rarely glabrous. Lower leaves serrate to lyrate; upper leaves entire or dentate, shortly petiolate to subsessile, not amplexicaul. Petals purple or white. $2n = 24$. *Damp or shaded habitats. Scattered through C. & S. Europe but absent from the extreme south and much of the west; widely cultivated for ornament, and locally naturalized throughout a large part of Europe.* Al Au Bu Cz Ga *Hs Hu It Ju Po Rm Rs(W, K) Tu [Be Br Co Da Fe Ge Hb He Ho Is No Rs(B) Sa Si Su].

1 Flowers white
2 Lower leaves with only unbranched hairs, or glabrous (c) subsp. **candida**
2 Lower leaves with branched and unbranched hairs (d) subsp. **voronovii**
1 Flowers purple
3 Lower leaves with mainly unbranched hairs (a) subsp. **matronalis**
3 Lower leaves with numerous branched hairs (b) subsp. **cladotricha**

(**a**) Subsp. **matronalis** (*H. oblongipetala* Borbás; incl. *H. sibirica* L.): *Italy, S. France, ?N. Spain; widespread as an alien.*

(**b**) Subsp. **cladotricha** (Borbás) Hayek, *Prodr. Fl. Penins. Balcan.* **1**: 416 (1925) (*H. cladotricha* Borbás; incl. *H. obtusa* Moench): *Balkan peninsula and Romania.*

(**c**) Subsp. **candida** (Kit.) Hegi & E. Schmid in Hegi, *Ill. Fl. Mitteleur.* **4(1)**: 467 (1919) (*H. candida* Kit., *H. nivea* auct. gall., non Baumg.; incl. *H. moniliformis* Schur): *Mountains, from the Carpathians to Pyrenees.*

(**d**) Subsp. **voronovii** (N. Busch) P. W. Ball, *Feddes Repert.* **68**: 194 (1963) (*H. voronovii* N. Busch): *Krym.*

Plants of subsp (**a**) from Italy are usually glandular-pubescent while those from France are eglandular. The latter is the plant usually cultivated and naturalized.

4. H. elata Hornem., *Hort. Hafn.* suppl., 74 (1819) (*H. sibirica* auct. ross., non L.). Indumentum of numerous long and short unbranched hairs, and glandular hairs; branched hairs more or less absent from the lower part of the plant. Lower leaves serrate to lyrate; upper leaves sessile, semi-amplexicaul, entire to dentate. Pedicels glandular-pubescent. Petals purple. Siliqua glabrous or hairy. *E. & S. Russia, E. Ukraine.* Rs(C, W, E).

5. H. sylvestris Crantz, *Stirp. Austr.* **1**: 34 (1762) (*H. runcinata* Waldst. & Kit.). Indumentum dense, of short unbranched, branched and glandular hairs. Lower leaves lyrate; upper leaves sessile, semi-amplexicaul, more or less serrate. Petals purple or pinkish. Siliqua glabrous. *Woods and scrub.* • *C. & S.E. Europe.* Al Au Bu Cz Gr Hu Ju Po Rm Rs(W, K).

(**a**) Subsp. **sylvestris** (incl. *H. siliqua-glandulosa* (Rohlena) Dvořák): Pedicels glandular-pubescent. *E.C. Europe, extending southwards to Macedonia.*

(**b**) Subsp. **velenovskyi** (Fritsch) Borza, *Consp. Fl. Roman.* 124 (1947) (*H. suaveolens* (Andrz.) Steudel): Pedicels and often the whole inflorescence glabrous. *S.E. Europe.*

6. H. theophrasti Borbás, *Magyar Bot. Lapok* **1**: 267 (1902) (*H. matronalis* sensu Halácsy, non L.; incl. *H. macedonica* Adamović, *H. graeca* Dvořák, *H. rechingeri* Dvořák, *H. verroiana* Dvořák). Like **5(a)** but eglandular, with long unbranched hairs; lower leaves dentate; pedicels and usually the siliqua densely glandular. *Woods, grassy and rocky places.* • *Balkan peninsula.* Al Bu Gr Ju Tu.

A variable species divided by F. Dvořák, *Preslia* **38**: 57–64 (1966), into several closely related species. The characters used to separate these do not seem to be sufficiently consistent to recognize the taxa at specific rank.

7. H. oblongifolia Schur, *Enum. Pl. Transs.* 52 (1866). Indumentum of unbranched and branched hairs, eglandular. Leaves dentate, the cauline semi-amplexicaul, the lower often glabrous. Pedicels glabrous or hairy at the base. Petals purple. Siliqua glabrous. • *E. Carpathians.* Rm.

8. H. steveniana DC., *Reg. Veg. Syst. Nat.* **2**: 452 (1821). Indumentum of long and short unbranched and branched hairs, eglandular. Lower leaves usually lyrate; upper leaves more or less sessile, semi-amplexicaul, dentate to entire. Pedicels densely pubescent. Petals pale purple, sometimes almost white. Siliqua more or less pubescent. • *Krym.* Rs(K).

9. H. pyconotricha Borbás & Degen, *Magyar Bot. Lapok* **1**: 269 (1902). Indumentum of short branched hairs, with a few unbranched hairs, eglandular. Leaves dentate to entire, the cauline sessile, slightly amplexicaul. Pedicels pubescent. Petals purple. Siliqua sparsely pubescent at the base, glabrous at the apex. *C. & S. Russia, E. & S. Ukraine; Thrace; E. Czechoslovakia.* Bu Cz Rs(C, W, K, E) Tu.

10. H. dinarica G. Beck in Dörfler, *Jahres-Kat. Wien. Bot. Tauschver.* **1894**: 6 (1894) (incl. *H. degeniana* Borbás). Indumentum of long and short unbranched and branched hairs and glandular hairs. Leaves dentate, the cauline sessile, semi-amplexicaul. Pedicels glandular-pubescent, rarely glabrous. Petals white. Siliqua glabrous. • *Stony places and woods. E.C. & S.E. Europe.* Al Bu Cz Gr Ju Rm.

11. H. vrabelyiana (Schur) Borbás, *Magyar Bot. Lapok* **2**: 21 (1903). Indumentum of long and short unbranched hairs and glandular hairs, with few branched hairs. Leaves dentate, the cauline sessile, semi-amplexicaul. Pedicels pubescent. Petals white. Siliqua pubescent. $2n = 24$. • *Mountains of N. Hungary (Bükk hegység).* Hu.

12. H. nivea Baumg., *Enum. Stirp. Transs.* **2**: 278 (1816). Indumentum of long and short unbranched and branched hairs, the latter absent from the lower part of the plant. Leaves dentate to entire, the cauline sessile, semi-amplexicaul. Pedicels pubescent. Petals white. Siliqua glabrous or pubescent. • *Carpathians.* Cz Rm Rs(W).

13. H. inodora L., *Sp. Pl.* ed. 2, 927 (1763) (*H. subsinuata* Borbás). Indumentum dense, of long and short unbranched and branched hairs, eglandular. Leaves coarsely sinuate-dentate or serrate, the cauline sessile, semi-amplexicaul. Pedicels densely pubescent or glabrous. Petals white. • *Alpes Maritimes.* Ga.

21. Malcolmia R. Br.

P. W. BALL AND J. R. AKEROYD

Annuals to perennials; hairs variable or plant subglabrous. Sepals erect, the inner usually saccate at the base; petals pink to violet, rarely white, long-clawed; median nectaries absent. Fruit a siliqua; valves 3-veined; style absent or indistinct; stigma deeply 2-lobed or retuse, the lobes erect and connate. Seeds in 1 row in each loculus. (*Wilckia* Scop.)

Literature: A. L. Stork, *Op. Bot. (Lund)* **33**: 1–118 (1972); *Svensk Bot. Tidskr.* **66**: 239–256 (1972).

1 Sepals not or only slightly saccate at the base
2 Lower pedicels 2–7 mm in fruit; siliqua 12–35 × 1 mm, torulose, terete, with appressed hairs **3. ramosissima**
2 Lower pedicels 0–2 mm in fruit; siliqua usually larger, not torulose, ±4-angled, glabrous or with patent hairs

3 Siliqua densely hispid; stigma acute; lower pedicels usually 1–2 mm in fruit **4. africana**
3 Siliqua with short, sparse, patent hairs or glabrous; stigma retuse; lower pedicels 0–1 mm in fruit **5. taraxacifolia**
1 At least 2 of the sepals strongly saccate at the base
4 Cauline hairs with numerous rays, usually very crowded
5 Basal leaves ±sessile; siliqua 30–65 mm, not torulose **1. littorea**
5 Basal leaves shortly petiolate; siliqua 20–40 mm, torulose **2. lacera**
4 Cauline hairs with 2–4 rays, not crowded
6 Calyx (8–)10–17 mm; style 5–10 mm in fruit **10. macrocalyx**
6 Calyx not more than 10 mm; style not more than 5 mm in fruit
7 Racemes bracteate, at least in the lower part **9. orsiniana**
7 Racemes ebracteate
8 Indumentum entirely of tri- or quadrifid hairs (each ray not more than 0·15 mm) **8. graeca**
8 Indumentum including numerous bifid hairs (each ray at least 0·25 mm)
9 Style more than 1·5 mm in fruit; petals 10–25 mm; sepals 5–16 mm
10 Pedicels in fruit 0·5–1 mm in diameter at their base, distinctly narrower than the siliqua **6. maritima**
10 Pedicels in fruit 1–2·5 mm in diameter at their base, about as wide as the siliqua **7. flexuosa**
9 Style not more than 1·5 mm in fruit; petals 4·5–12 mm; sepals 3–6 mm
11 Petals pink with yellow base; siliqua 10–45 mm **8. graeca**
11 Petals entirely pink or violet; siliqua (25–)30–70 mm, patent or erecto-patent **11. chia**

1. M. littorea (L.) R. Br. in Aiton, *Hort. Kew.* ed. 2, **4**: 121 (1812). Perennial 10–40 cm, woody, with numerous non-flowering stems, densely white-pubescent or -tomentose; hairs stellate. Leaves entire to sinuate-dentate, more or less sessile. Petals 14–22 mm, purple. Siliqua 30–65 × 1–1·5 mm, not torulose, the apex attenuate; stigma 2–6 mm. $2n = 20$. *S.W. Europe.* Ga Hs It Lu.

2. M. lacera (L.) DC., *Reg. Veg. Syst. Nat.* **2**: 445 (1821) (incl. *M. patula* DC.). Annual or perennial up to 40 cm, grey-green, pubescent or tomentose; hairs stellate. Basal leaves entire to sinuate-pinnatifid, shortly petiolate. Petals 8–16 mm, purple. Siliqua 20–40 × 0·7–1·2 mm, torulose; stigma 2·5–6 mm. *C. & S. Spain and Portugal.* Hs Lu.

3. M. ramosissima (Desf.) Thell., *Fl. Adv. Montpellier* 285 (1912) (*Hesperis ramosissima* Desf., *M. parviflora* (DC.) DC., *M. arenaria* auct. hisp.). Annual 5–20 cm; hairs stellate. Leaves oblong, entire or sinuate-dentate. Pedicels 2–7 mm in fruit. Sepals 2·5–5 mm, not saccate; petals 4–8 mm, violet or pink. Siliqua 15–35 × *c.* 1 mm, terete, torulose, pubescent; septum almost completely opaque; stigma 1–2 mm. $2n = 14$. *Mediterranean region, Portugal.* Bl Co Ga Gr Hs It Lu Sa.

Often confused with *Maresia nana* (DC.) Batt., from which it can most readily be distinguished by the stigma and the almost completely opaque septum of the siliqua.

4. M. africana (L.) R. Br. in Aiton, *Hort. Kew.* ed. 2, **4**: 121 (1812). Annual up to 40 cm, with branched or stellate hairs; branches usually shorter than the main stem. Leaves lanceolate, entire or sinuate-dentate. Pedicels 1–2 mm in fruit. Sepals 3–5 mm, not saccate; petals (5–)8–10(–12) mm, violet. Siliqua 25–65 × 1–1·5 mm, patent, rigid, more or less 4-angled, densely hispid; stigma 0·5–1·5 mm, acute. *S. Europe; naturalized or casual elsewhere.* Cr Gr Hs Rs(W, K) Si Tu [Cz Ga Hu].

5. M. taraxacifolia Balbis, *Cat. Stirp. Hort. Bot. Taur.* App. 4, 10 (1814). Like **4** but not more than 15 cm, with unbranched and branched hairs or subglabrous; branches often longer than the main stem; leaves serrate to pinnatisect; pedicels 0–1 mm; petals 6–8 mm, white or pale bluish-pink; siliqua (15–)30–60 × (1–)1·5–2 mm, glabrous or with short patent hairs; stigma retuse. *S.E. Russia.* Rs(E).

6. M. maritima (L.) R. Br. in Aiton, *Hort. Kew.* ed. 2, **4**: 121 (1812). Annual 10–35 cm, with bi- to quadrifid hairs. Leaves obovate to oblong, cuneate at the base, entire or dentate. Inflorescence ebracteate; pedicels 4–15 × 0·5–0·8(–1) mm in diameter in fruit, thinner than the siliqua. Sepals 6–10 mm, with medifixed hairs; petals 12–25 mm, pink to violet. Siliqua 35–80 × 1–2 mm, with mostly medifixed hairs; stigma 2–5 mm. ● *S. & W. Greece, S. Albania; naturalized on maritime sands elsewhere in S. Europe.* Al Gr [Bl Co Ga Hs *It Ju Si].

Widely cultivated for ornament, and often occurring as a casual.

7. M. flexuosa (Sm.) Sm. in Sibth. & Sm., *Fl. Graeca* **7**: 33, t. 634 (1830) (*M. maritima* auct. eur. merid.; incl. *M. naxensis* Rech. fil.). Like **6** but the pedicels 1–2·5 mm in diameter at the base, about as wide as the siliqua; siliqua up to 3 mm in diameter; stigma 1·5–4 mm. $2n = 16$. *Sea-shores and maritime cliffs. Greece and the Aegean region; naturalized elsewhere in S. Europe, where it has been confused with* **6**. Cr Gr *It Tu [Ga].

8. M. graeca Boiss. & Spruner in Boiss., *Diagn. Pl. Or. Nov.* **1(1)**: 71 (1843). Annual 5–30 cm; hairs bi- to quadrifid. Leaves ovate-oblong to oblong-lanceolate, entire to lyrate-pinnatifid, cuneate at the base. Inflorescence ebracteate; pedicels 3–10 mm in fruit. Petals 4–17 mm, purple, violet or pink. Siliqua 10–50 × 1–1·5 mm; stigma 0·5–2·5(–3) mm. ● *C., S. & N.W. Greece; one station in S. Albania.* Al Gr.

A very variable species. Three subspecies can be recognized, although they intergrade in C. Greece.

1 Basal leaves densely covered with tri- or quadrifid hairs; hairs on calyx tri- or quadrifid **(a) subsp. graeca**
1 Basal leaves with sparse indumentum of bi- to quadrifid hairs; hairs on calyx usually bifid
2 Leaves usually lyrate-pinnatifid; siliqua patent, curved **(b) subsp. hydraea**
2 Leaves entire to serrulate; siliqua erecto-patent, straight **(c) subsp. bicolor**

(a) Subsp. graeca: Basal leaves lyrate-pinnatifid, densely covered with tri- or quadrifid hairs. Hairs on calyx tri- or quadrifid. Flowers (6–)9–16 mm in diameter; petals purple, white at the base of limb. Siliqua 20–50 mm, patent, curved. $2n = 16$. *E.C. Greece (Attiki).*

(b) Subsp. hydraea (Heldr. & Halácsy) A. L. Stork, *Op. Bot. (Lund)* **33**: 40 (1972) (*M. hydraea* Heldr. & Halácsy): Basal leaves lyrate-pinnatifid (sometimes dentate), with sparse indumentum of bi- to quadrifid hairs. Hairs on calyx usually bifid. Flowers 4–12(–14) mm in diameter; petals white, pink, violet or purplish, yellowish at the base of limb. Siliqua 20–45 mm, patent, curved. $2n = 16$. *E. Greece*.

(c) Subsp. **bicolor** (Boiss. & Heldr.) A. L. Stork, *op. cit.* 39 (1972) (*M. bicolor* Boiss. & Heldr., *M. graeca* sensu Hayek pro parte): Basal leaves entire to serrulate, with a sparse indumentum of bi- to quadrifid hairs. Hairs on calyx usually bifid. Flowers 3–9 mm in diameter; petals white to pink or violet, yellowish at the base of limb. Siliqua 10–30(–40) mm, erecto-patent, straight. $2n = 16$. *S., W. & W.C. Greece; S. Albania.*

9. M. orsiniana (Ten.) Ten., *Fl. Nap.* **5**: 67 (1835) (*M. chia* auct. balcan.). Annual to biennial 10–60 cm, with bi- to quadrifid hairs. Leaves oblong to obovate or ovate, dentate to shallowly lobed, the basal cordate (but soon dead). Inflorescence bracteate in the lower part; pedicels 5–10 mm in fruit. Sepals with bi- to quadrifid hairs. Siliqua 25–75 mm; stigma 1–2·5 mm. *Shady, usually calcareous rocks. Mountains of Balkan peninsula; C. Appennini.* Al Bu Gr It Ju.

(a) Subsp. **orsiniana**: Sepals 3–5 mm; petals 4–10 mm, usually entire. Siliqua 25–35 mm. • *C. Appennini.*

(b) Subsp. **angulifolia** (Boiss. & Orph.) A. L. Stork, *Svensk Bot. Tidskr.* **66**: 245 (1972) (*M. angulifolia* Boiss. & Orph., *M. illyrica* Hayek, *M. cymbalaria* Heldr. & Sart., *M. maritima* var. *serbica* sensu Hayek, *M. orsiniana* auct. balcan., *M. serbica* Pančić): Sepals 6–10 mm; petals 11–25 mm, emarginate. Siliqua 35–75 mm. *Throughout the range of the species except Italy.*

Many plants from Albania and N.W. Greece are intermediate between subspp. (**a**) and (**b**).

10. M. macrocalyx (Halácsy) Rech. fil., *Ann. Naturh. Mus. (Wien)* **43**: 29 (1929). Annual up to 40 cm, with bi- to quadrifid hairs. Basal leaves ovate to ovate-orbicular, cuneate at the base, entire or dentate. Sepals (8–)10–17 mm, with bifid hairs; petals 15–30 × 5–10 mm, pink or purple. Siliqua 50–80 mm; stigma 5–10 mm. • *W. Aegean region.* Gr.

(a) Subsp. **macrocalyx**: Inflorescence ebracteate (or bracteate in the lower part only). *N. Sporadhes.*

(b) Subsp. **scyria** (Rech. fil.) P. W. Ball, *Feddes Repert.* **68**: 181 (1963) (*M. scyria* Rech. fil.): Inflorescence bracteate, usually almost to the apex. $2n = 16$. *Skiros, Evvoia, Andros.*

11. M. chia (L.) DC., *Reg. Veg. Syst. Nat.* **2**: 440 (1821). Annual up to 20 cm, with bi- to quadrifid hairs. Leaves ovate-oblong, the basal cuneate, entire or dentate. Inflorescence ebracteate; pedicels 4–10 mm in fruit. Sepals with bifid hairs; petals 6–10 × 1·5–2(–2·5) mm, pale pink to violet. Siliqua 25–70 × *c.* 1 mm, erecto-patent; stigma 0·5–1·5 mm. *Greece and Aegean region, generally near the coast.* Cr Gr Tu.

22. **Neotorularia** Hedge & J. Léonard

P. W. BALL (EDITION 1) REVISED BY R. R. MILL (EDITION 2)

Like *Malcolmia* but always annuals; hairs unbranched and bifid; inner sepals not saccate at the base; style distinct; stigma capitate or retuse. (*Torularia* (Cosson) O. E. Schulz, non Bonnemaison.)

Literature: O. E. Schulz in Engler, *Pflanzenreich* **86** (**IV.105**): 213–226 (1924). J. Léonard, *Bull. Jard. Bot. Nat. Belg.* **56**: 389–395 (1986).

Petals white; siliqua 10–25 mm; pedicels 0·5–1 mm in fruit **1. torulosa**
Petals pink; siliqua 20–35 mm; pedicels (1–)2–8 mm in fruit **2. contortuplicata**

1. N. torulosa (Desf.) Hedge & J. Léonard, *Bull. Jard. Bot. Nat. Belg.* **56**: 395 (1986) (*Malcolmia torulosa* (Desf.) Boiss., *Torularia torulosa* (Cosson) O. E. Schulz). Stem up to 30 cm, hispid with unbranched and bifid hairs. Basal leaves dentate or pinnatifid, the upper cauline sometimes pectinate-pinnatifid. Pedicels not more than 1 mm. Petals 2·5–4 mm, white. Siliqua 10–25 mm, usually erect, straight or contorted, torulose; style up to 0·5 mm. *Krym (Sudak).* Rs(K). (*C. & S.W. Asia; N. Africa.*)

N. rossica (O. E. Schulz) Hedge & J. Léonard, *op. cit.* 394 (1986) (*T. rossica* O. E. Schulz), described from E.C. Russia (near Uljanovsk), is said to differ from **1** in having the hairs all bifid, leaves entire or denticulate, petals absent and siliqua *c.* 10 mm.

2. N. contortuplicata (Stephan ex Willd.) Hedge & J. Léonard, *op. cit.* 393 (1986) (*T. contortuplicata* (Stephan ex Willd.) O. E. Schulz, *Malcolmia contortuplicata* (Stephan) Boiss.). Stem 5–30 cm, hispid. Basal leaves sinuate-pinnatifid, the upper cauline usually more or less entire. Pedicels (1–)2–8 mm in fruit. Petals 3–8 mm, pink. Siliqua 20–35 mm, erecto-patent, straight or spirally coiled; style 1–1·5 mm. *S.E. Russia (near Astrakhan').* Rs(E). (*Caucasus; C. Asia.*)

23. **Maresia** Pomel

P. W. BALL

Like *Malcolmia* but always annuals; hairs stellate; inner sepals slightly saccate at the base; style distinct; stigma capitate or emarginate.

1. M. nana (DC.) Batt. in Batt. & Trabut, *Fl. Algér.* (*Dicot.*) 68 (1888) (*Malcolmia nana* (DC.) Boiss., *M. binervis* (C. A. Meyer) Boiss., *M. confusa* Boiss., *M. parviflora* auct. pro parte). Shortly pubescent, 5–20 cm. Leaves oblong or linear-oblong, entire or sinuate-dentate. Petals 4–5·5 mm, violet to pink. Siliqua 10–28 × *c.* 0·7 mm, terete, torulose; style 0·5–1 mm. $2n = 26$. *Maritime sands. Mediterranean and shores of Black Sea northwards to S.E. Bulgaria.* Al Bl Bu Co Cr Ga Gr Hs It Si Tu.

Often confused with *Malcolmia ramosissima* (Desf.) Thell., from which it can be distinguished by the style and stigma, and by the septum of the siliqua being 2-veined with broad hyaline margins.

24. **Leptaleum** DC.

P. W. BALL

Annuals; hairs branched and unbranched. Sepals erect, not saccate; petals white, becoming reddish, not clawed; filaments of the inner stamens connate in pairs. Fruit a tardily dehiscent siliqua; valves with a distinct median vein; style very short; stigma 2-lobed, the lobes erect, connate at the apex. Seeds in 2 rows in each loculus.

1. L. filifolium (Willd.) DC., *Reg. Veg. Syst. Nat.* **2**: 511 (1821). Sparsely pubescent or glabrescent annual 5–15 cm. Leaves linear or filiform, entire or pinnatisect with 3–5 filiform lobes. Pedicels 2–3 mm in fruit. Petals 6–10 mm, linear. Siliqua 15–30 × 1·5–3 mm, compressed; valves with reticulate-striate lateral veins. Seeds *c.* 0·75 mm. *Steppes. N. of Astrakhan', on borders of S.E. Russia and Kazakhstan.* Rs(E). (*C. & S.W. Asia.*)

25. **Sterigmostemum** Bieb.

P. W. BALL

Annuals; hairs branched. Sepals erect, not saccate; petals yellow, with a short, broad claw; filaments of the inner stamens connate in pairs. Fruit an articulated siliqua, breaking irregularly into 2-seeded segments; style short; stigma 2-lobed, the lobes erect. Seeds in 1 row in each loculus.

1. S. tomentosum (Willd.) Bieb., *Fl. Taur.-Cauc.* **3**: 444 (1819). Plant 20–40 cm, grey-tomentose. Leaves entire to pinnatisect with lanceolate lobes. Pedicels 5–10 mm. Petals 6–10 × 3–5 mm, truncate. Siliqua 20–60 × 1·5–2 mm, cylindrical, torulose. Seeds *c.* 2 mm, olive-green. *Saline or sandy soils. S.E. Russia, W. Kazakhstan.* Rs(E). (*C. & S.W. Asia.*)

26. **Clausia** Trotzky

P. W. BALL

Perennials; hairs unbranched, glandular or not. Sepals erect, the inner saccate at the base; petals long-clawed. Fruit a tardily dehiscent siliqua; valves with a distinct median and lateral veins; style short; stigma 2-lobed, the lobes erect, free. Seeds in 1 row in each loculus. Radicle accumbent.

1. C. aprica (Stephan ex Willd.) Trotzky, *Ind. Sem. Horti Casan.* (1834) (*Hesperis aprica* (Stephan) Poiret). Rhizomatous perennial up to 40 cm. Leaves oblong to oblanceolate, entire to dentate. Sepals 5–10 mm; petals 10–20 × 4–9 mm. Siliqua 30–80 × 1·2–1·5 mm; style 1·5–1·8 mm. *Calcareous hills. From N. Ukraine to C. Ural.* Rs(C, W, E).

27. **Parrya** R. Br.

P. W. BALL

Perennials. Sepals erect, not saccate; petals white or purplish, clawed. Fruit a siliqua; valves with distinct median vein; style very short; stigma 2-lobed, the lobes erect, connate. Seeds in 2 rows in each loculus.

1. P. nudicaulis (L.) Boiss., *Fl. Or.* **1**: 159 (1867). Glabrous or glandular-tomentose, with long, stout rhizome; flowering stems scapose, up to 40 cm. Leaves broadly spathulate to linear-oblong, entire or serrate. Pedicels 1·5 cm. Petals 12–20 mm, obovate, emarginate. Siliqua 20–65 × 3·5–5·5 mm, compressed, shortly stipitate, attenuate at base and apex. Seeds 4–5·5 × 3–4·5 mm (including wing), elliptical, broadly winged. *Arctic Russia.* Rs(N) ?Sb. (*Arctic Asia and America.*)

Eremoblastus caspicus Botsch., *Bot. Žur.* **65**: 425 (1980), has recently been described from W. Kazakhstan.

28. **Matthiola** R. Br.

P. W. BALL

Annuals to perennials; hairs branched. Sepals erect, the inner saccate at the base; petals purple, white, or yellowish, long-clawed. Fruit a siliqua; valves 1-veined; style absent; stigma deeply 2-lobed, the lobes erect, each with a dorsal swelling or horn. Seeds in 1 row in each loculus.

A number of species are cultivated for ornament.

Literature: P. Conti, *Mém. Herb. Boiss.* **1(18)**: 6–86 (1900).

1 Siliqua with 3 ±equal triangular horns at the apex, all at least 2 mm long **10. tricuspidata**
1 Siliqua without conspicuous horns or with only 2 longer than 2 mm
2 Siliqua compressed, at least 2 mm wide, without horns or with horns not more than 1·5 mm; lower pedicels up to 25 mm in fruit
3 Siliqua with large, conspicuous blackish or yellowish glands, easily visible even when immature **2. sinuata**
3 Siliqua without conspicuous glands, but usually with minute glandular hairs
4 Stigma conical in fruit; petals 2·5–3 mm wide **5. fragrans**
4 Stigma 2-lobed in fruit; petals usually 3–12 mm wide
5 Petals white, pink or purple; lower pedicels at least 7 mm in fruit; stigma narrower than the siliqua **1. incana**
5 Petals yellowish- or brownish-purple; lower pedicels usually less than 7 mm
6 Leaves ±sinuate to sinuate-pinnatisect; stigma at least as wide as or wider than the siliqua; lower pedicels 3–5(–8) mm in fruit **3. odoratissima**
6 Leaves irregularly serrate or divided; stigma narrower than the siliqua; lower pedicels 2–3(–4) mm in fruit **4. tatarica**
2 Siliqua ±cylindrical, usually not more than 2 mm in diameter; stigmatic horns usually more than 1·5 mm; pedicels not more than 3(–8) mm
7 Perennial, with vegetative leaf-rosettes and often a long, woody stock **6. fruticulosa**
7 Annual or rarely biennial, without vegetative leaf-rosettes, but sometimes woody at the base
8 Petals 6–10(–12) × 1–2 mm; sepals 4–6 mm **9. parviflora**
8 Petals (12–)15–25 × 2–7 mm; sepals at least 6 mm
9 Petals linear, yellowish; horns of siliqua recurved or patent **7. longipetala**
9 Petals obovate or oblong, pink or purple; horns of siliqua curved upwards
10 Nectaries filiform; siliqua not or only slightly torulose **7. longipetala**
10 Nectaries small, inconspicuous; siliqua ±torulose **8. lunata**

1. M. incana (L.) R. Br. in Aiton, *Hort. Kew.* ed. 2, **4**: 119 (1812). Stout perennial 10–80 cm, woody at the base. Leaves entire, rarely sinuate-pinnatifid. Lower pedicels 7–25 mm in fruit. Sepals 9–15 mm; petals 20–30 × 4–12 mm, purple, pink or white. Siliqua 45–160 × 3–5 mm, erecto-patent, compressed; stigma narrower than the siliqua, without conspicuous horns. *Coasts of S. & W. Europe; often naturalized or casual elsewhere.* Al Az Br Co Ga *Gr *Hs It Ju Sa Si *Tu [Bl Co Ho Lu].

(**a**) Subsp. **incana** (*M. annua* (L.) Sweet, *M. fenestralis* (L.) R. Br.): Plant densely white-tomentose to subglabrous. Lower leaves 5–22 mm wide, linear-lanceolate, rarely oblong-lanceolate, obtuse or subacute. Sepals 9–13 mm. $2n = 14$. *Throughout the range of the species.*

(**b**) Subsp. **rupestris** (Rafin.) Nyman, *Consp.* 30 (1878) (*M. rupestris* (Rafin.) DC., *M. undulata* Tineo; incl. *M. incana* subsp. *pulchella* (P. Conti) Greuter & Burdet): Plant glabrescent or sparsely pubescent. Lower leaves up to 25–40 mm wide, lanceolate, acute or acuminate. Sepals 11–15 mm. ● *S. Italy, Sicilia.*

2. M. sinuata (L.) R. Br. in Aiton, *Hort. Kew.* ed. 2, **4**: 120

(1812) (*M. glandulosa* Vis.). Densely white-tomentose, stout biennial (rarely annual or perennial) 8–60 cm, woody at the base. Leaves sinuate-dentate to pinnatifid, the lobes oblong, obtuse; uppermost leaves entire. Lower pedicels 4–15(–18) mm in fruit. Sepals 8–12 mm; petals 17–25 × 3–8 mm, pale purple. Siliqua 50–150 × 3–5 mm, erecto-patent, compressed, the valves with large, conspicuous, stipitate yellow or black glands; stigma narrower than the siliqua, without conspicuous horns. *Coasts of S. & W. Europe*. $2n = 28$. Al Bl Br Co Cr Ga Gr †Hb Hs It Ju Lu Sa Si Tu.

3. M. odoratissima (Bieb.) R. Br. in Aiton, *loc. cit.* (1812). Densely white-tomentose perennial 15–75 cm, woody at the base. Leaves usually sinuate to sinuate-pinnatisect. Pedicels 3–5(–8) mm in fruit. Sepals 6–15 mm; petals 20–30 × 3–5 mm, brown or yellow tinged with purple. Siliqua 85–180 × 2–4 mm, erecto-patent, compressed; stigma capitate, as wide as or wider than the siliqua, without conspicuous horns. $2n = 12$. *N.E. Bulgaria; Krym*. Bu Rs(K). (*Caucasus, Iran*.)

4. M. tatarica (Pallas) DC., *Reg. Veg. Syst. Nat.* **2**: 170 (1821). Like **3** but the stem subglabrous to grey-tomentose; leaves irregularly serrate or incised; pedicels 2–3(–5) mm in fruit; siliqua 50–150 × 2–3 mm, glabrous or sparsely pubescent; stigma narrower than the siliqua. *S.E. Russia, W. Kazakhstan*. Rs(E).

5. M. fragrans (Fischer) Bunge, *Del. Sem. Horti Dorpat.* 8 (1839). Subglabrous to grey-tomentose perennial 20–50 cm, woody at the base. Leaves subentire to sinuate-pinnatisect. Pedicels 3–8(–18) mm in fruit. Sepals 10–12 mm; petals 18–25 × 2.5–3(–4) mm, dull yellow tinged with purple. Siliqua 75–130 × 2.5–3 mm, erecto-patent, compressed; stigma conical, narrower than the siliqua, without horns. *S. Ukraine to W. Kazakhstan*. Rs(W, E).

6. M. fruticulosa (L.) Maire in Jahandiez & Maire, *Cat. Pl. Maroc.* **2**: 311 (1932) (*M. tristis* R. Br., *M. provincialis* (L.) Markgraf, *M. thessala* Boiss. & Orph., *Cheiranthus fruticulosus* L.). Sparsely pubescent to densely white-tomentose perennial up to 60 cm, woody at the base. Leaves linear or oblong, entire to sinuate-pinnatifid. Pedicels 0–3(–8) mm in fruit. Sepals 6–14 mm; petals 12–28 mm, yellow to purplish-violet. Siliqua 25–120 × 1–2(–3) mm, more or less cylindrical; stigma without horns or with horns up to 3 mm. *S. Europe*. Al Bl Bu Ga Gr He Hs It Ju Lu Si Tu.

1 Siliqua 2–3 mm wide; valves 3- to 5-veined — (c) subsp. **perennis**
1 Siliqua 1–2 mm wide; valves 1- to 3-veined
 2 Siliqua patent or deflexed, eglandular; plant with branched stock, lax — (a) subsp. **fruticulosa**
 2 Siliqua ±erect with yellow glandular hairs; plant caespitose — (b) subsp. **valesiaca**

(a) Subsp. **fruticulosa** (incl. *M. varia* (Sm.) DC. pro parte): Plant with branched stock, lax. Siliqua 1–2 mm wide, patent or deflexed, eglandular; valves 1- to 3-veined. $2n = 12, 24$. *Throughout the range of the species except the Alps*.

(b) Subsp. **valesiaca** (Gay ex Gaudin) P. W. Ball, *Feddes Repert.* **66**: 157 (1962) (*M. varia* (Sm.) DC. pro parte): Plant caespitose. Siliqua 1–2 mm wide, more or less erect, with yellow glandular hairs; valves 1- to 3-veined. $2n = 12, 24$. • *Pyrenees, N. & E. Spain; S. Alps; Balkan peninsula*.

(c) Subsp. **perennis** (P. Conti) P. W. Ball, *loc. cit.* (1962) (*M. perennis* P. Conti): Plant caespitose. Siliqua 2–3 mm wide, eglandular, more or less erect; valves 3- to 5-veined. $2n = c. 12$. *N. Spain (Picos de Europa)*.

7. M. longipetala (Vent.) DC., *Reg. Veg. Syst. Nat.* **2**: 174 (1821). Annual, rarely overwintering, up to 50 cm. Lower leaves sinuate-dentate to pinnatifid. Lower pedicels not more than 3 mm in fruit. Sepals 8–12 mm; petals 15–25 × 2–7 mm; nectaries filiform. Siliqua 50–150 × 1–2(–2·5) mm, cylindrical; horns 2–10 mm. *C. & S. Greece, S. Aegean region*. Gr [Rs(W, K)].

(a) Subsp. **longipetala** (*M. oxyceras* DC., *Cheiranthus longipetalus* Vent.): Petal-limb 2–4 mm wide, yellowish. Horns of the siliqua usually patent or recurved. *Naturalized in S. Ukraine*.

(b) Subsp. **bicornis** (Sm.) P. W. Ball, *Feddes Repert.* **68**: 194 (1963) (*M. bicornis* (Sm.) DC.): Petal-limb 3–7 mm wide, pink or purple. Horns of the siliqua curved upwards. *C. & S. Greece; S. Aegean region*.

Subsp. **pumilio** (Sm.) P. W. Ball, *loc. cit.* (1963) (*M. pumilio* (Sm.) DC.), with shorter, broader siliqua with obtuse horns, was once recorded from the S. Aegean region (Sikinos), probably in error.

8. M. lunata DC., *Reg. Veg. Syst. Nat.* **2**: 176 (1821). Annual 10–50 cm. Leaves sinuate-dentate to entire. Pedicels up to 4 mm in fruit. Sepals 6–12 mm; petals (12–)15–25 × 3–6 mm, oblong-obovate, purple; nectaries very small. Siliqua 30–70 × 1–2 mm, cylindrical, patent or deflexed, torulose; horns 2–4 mm, subobtuse, curved upwards. *S. Spain*. Hs.

9. M. parviflora (Schousboe) R. Br. in Aiton, *Hort. Kew.* ed. 2, **4**: 121 (1812). Annual up to 20 cm. Lower leaves sinuate-dentate to pinnatisect. Sepals 4–6 mm; petals 6–10(–12) × 1–2 mm, purple or brownish-purple; nectaries very small. Siliqua 25–70 × 1·5–2 mm, slightly torulose; horns 1·5 mm, straight, acute. $2n = 12$. *S. Spain, S. Portugal*. ?Bl Hs Lu.

10. M. tricuspidata (L.) R. Br. in Aiton, *Hort. Kew.* ed. 2, **4**: 120 (1812). Annual 7–40 cm. Leaves sinuate-crenate to pinnatisect, the lobes ovate, rounded at the apex. Lower pedicels 2–5 mm in fruit. Sepals 7–11 mm; petals 15–22 mm, purple. Siliqua 25–100 × 2–3 mm, patent or deflexed, cylindrical; stigma-lobes connate to form a third horn, the 3 horns equal, 2–6 mm, triangular, acute. $2n = 14$. *Maritime sands, rarely inland. Mediterranean region*. Al Bl Co Cr Ga Gr Hs It Sa Si Tu.

29. Notoceras R. Br.

P. W. BALL

Annuals; hairs bifid. Sepals erecto-patent, not saccate; petals white, not clawed. Fruit a short, tardily dehiscent siliqua with 2 horns at the apex; valves keeled or narrowly winged; style distinct; stigma capitate. Seeds in 1 row in each loculus.

1. N. bicorne (Aiton) Amo, *Fl. Iber.* **6**: 536 (1873) (*N. canariense* R. Br.). Procumbent or erect, up to 30 cm. Leaves lanceolate to linear-lanceolate, entire. Racemes short, dense, leaf-opposed; pedicels up to 2 mm in fruit, stout. Siliqua 4–6 × *c.* 1·5 mm, erect; valves torulose; horns 0·5–1 mm. *Dry places. S.E. Spain*. Hs (*N. Africa; S.W. Asia*.)

30. Tetracme Bunge

P. W. BALL

Annuals; hairs branched. Sepals patent, not saccate; petals yellow, clawed. Fruit a siliqua with 4 horns at the apex; valves 3-veined; style very short; stigma more or less capitate. Seeds in 1 row in each loculus.

1. **T. quadricornis** (Willd.) Bunge, *Del. Sem. Horti Dorpat.* 8 (1836). Plant 5–20 cm, grey-tomentose. Leaves linear-oblong, entire or remotely dentate. Pedicels 1–1·5 mm in fruit, clavate. Siliqua 6–10 × 1–1·5 mm, patent and curved, torulose; horns 0·5–2·5 mm, divergent. *S.E. Russia; rarely as a casual elsewhere.* Rs(E). (*C. Asia.*)

31. Diptychocarpus Trautv.

P. W. BALL

Annuals; hairs unbranched. Sepals erect, the inner saccate at the base; petals white or purplish, not clawed. Fruits dimorphic; the upper siliquae dehiscent, compressed, with winged seeds, the lower indehiscent, articulated, with the seeds not winged; style distinct; stigma 2-lobed.

1. **D. strictus** (Fischer) Trautv., *Bull. Soc. Nat. Moscou* **33(1)**: 108 (1860). Plant 10–50 cm, sparsely pubescent. Leaves linear to oblong, the basal dentate to pinnatisect. Petals 7–10 mm, linear. Upper siliquae 35–80 × 2·5–4 mm, obtuse, with style 3–5 mm, the lower 25–60 × 2·5–4 mm, fleshy, gradually attenuate into the style. *Saline soils. W. Kazakhstan.* Rs(E). (*C. & S.W. Asia.*)

32. Chorispora R. Br. ex DC.

P. W. BALL

Annuals; hairs glandular, and sometimes also eglandular, branched or unbranched. Sepals erect, the inner saccate at the base; petals purple, clawed. Fruit an articulated siliqua, breaking into 2-seeded segments; style at least ½ as long as the rest of the siliqua; stigma 2-lobed. Seeds in 1 row in each loculus.

1. **C. tenella** (Pallas) DC., *Reg. Veg. Syst. Nat.* **2**: 435 (1821). Plant 10–60 cm, with sparse glandular and sometimes longer unbranched and branched, eglandular hairs. Leaves lanceolate, entire or serrate, the lower often pinnatifid. Petals 10–12 mm, entire. Siliqua 15–30 × 1·5–3 mm (excluding beak), more or less cylindrical, torulose; beak 12–30 mm. *Waste places. S.E. Europe; casual elsewhere.* Bu Ju Rm Rs(C, W, K, E).

33. Euclidium R. Br.

P. W. BALL

Annuals; hairs of 2 kinds, branched and unbranched. Sepals erecto-patent, not saccate; petals white, not clawed. Fruit an indehiscent, 2-seeded silicula, longitudinally 4-veined; style long; stigma 2-lobed.

1. **E. syriacum** (L.) R. Br. in Aiton, *Hort. Kew.* ed. 2, **4**: 74 (1812). Plant 10–40 cm, densely grey-pubescent throughout. Leaves lanceolate to ovate-oblong, sinuate-dentate, the basal often pinnatifid. Petals *c.* 1–1·5 mm, emarginate. Silicula 2·5–4 mm, obliquely ellipsoidal or ovoid; style 1–2 mm. *Waste places. Ruderal. E. & E.C. Europe; rarely casual elsewhere.* Au Bu Cz Hu Ju Rm Rs(C, W, K, E).

34. Litwinowia Woronow

P. W. BALL

Like *Euclidium* but all the hairs unbranched; inner sepals slightly saccate at the base; silicula compressed, longitudinally 6-veined; stigma subentire.

1. **L. tenuissima** (Pallas) Woronow ex Pavlov, *Fl. Centr. Kazakhstana* **2**: 302 (1935) (*Euclidium tataricum* (Willd.) DC., *Vella tenuissima* Pallas). Slender, hispid annual 10–30 cm. Leaves linear- or oblong-lanceolate, the basal dentate to lyrate-pinnatisect. Petals 2.5–5 mm, entire. Silicula 2.5–3 × 2.5–3 mm, broadly obovate or almost orbicular, glabrous, with the 6 veins in two lateral groups of 3 which diverge from the base; style 2–3 mm, often persistent. *Steppes. S.E. Russia.* Rs(E). (*W. & C. Asia.*)

35. Barbarea R. Br.

P. W. BALL

Biennials to perennials (rarely annuals); glabrous or with unbranched hairs. Leaves pinnatifid or pinnatisect, the cauline auriculate. Inner sepals slightly saccate at the base; petals yellow, indistinctly clawed. Fruit a siliqua, terete to 4-angled; valves usually with a strong median vein and reticulate lateral veins; style distinct; stigma entire or slightly 2-lobed. Seeds in 1 row in each loculus.

All species occur usually in wet or damp habitats, and some are weeds or ruderals.

1 Inflorescence bracteate, at least in the lower ½ — **7. bracteosa**
1 Inflorescence ± completely ebracteate
 2 Infructescence corymbose (Greece) — **9. conferta**
 2 Infructescence elongate, racemose
  3 Siliqua hairy; style (2–)3–5 mm — **10. longirostris**
  3 Siliqua glabrous; style usually less than 3 mm
   4 Uppermost leaves simple, dentate
    5 Style 2–4 mm in fruit; buds glabrous — **1. vulgaris**
    5 Style usually not more than 2 mm in fruit; buds pubescent — **2. stricta**
   4 Uppermost leaves pinnatifid or pinnatisect
    6 Basal leaves with 6–10 pairs of lateral lobes — **3. verna**
    6 Basal leaves simple or with 1–5(6) pairs of lateral lobes
     7 Style 0·5–2 mm in fruit
      8 Petals 7–10 mm; siliqua 30–60(–90) mm — **8. rupicola**
      8 Petals less than 7 mm; siliqua 10–30 mm
       9 Pedicels stout, not more than ¼ of the length of the siliqua; siliqua *c.* 2 mm in diameter; style usually less than 1.5 mm — **4. intermedia**
       9 Pedicels slender, ¼–⅓ of the length of the siliqua; siliqua 1–1·5 mm in diameter; style at least 1·5 mm — **5. sicula**
     7 Style at least 2 mm in fruit
      10 Valves of the siliqua without a median vein — **6. bosniaca**
      10 Valves of the siliqua with a distinct median vein
       11 Pedicels in fruit not more than ¼ of the length of the siliqua; siliqua 15–30(–40) × 1–2 mm — **1. vulgaris**
       11 Pedicels in fruit ¼–⅓ of the length of the siliqua; siliqua 10–20 × 1–1·5 mm — **5. sicula**

1. **B. vulgaris** R. Br. in Aiton, *Hort. Kew.* ed. 2, **4**: 109 (1812)

(*B. iberica* (Willd.) DC., *B. lyrata* Ascherson; incl. *B. arcuata* (Opiz ex J. & C. Presl) Reichenb., *B. macrophylla* Halácsy). Usually glabrous biennial or perennial up to 100 cm. Basal leaves with 1–5(6) pairs of lateral lobes, rarely simple; uppermost leaves simple, dentate, rarely pinnatifid. Pedicels 3–6 mm in fruit, slender. Buds glabrous; petals 5–8 mm. Siliqua 15–30(–40) × 1–2 mm; valves with distinct median vein; style 2–3·5(–4) mm. $2n = 16, 18$. *Throughout Europe except for some islands.* All except Az Bl Cr Fa Sb; introduced in Is.

A variable species with many infraspecific taxa described. **B. lepuznica** E. I. Nyárády, *Bul. Grăd. Bot. Univ. Cluj* **14**: 97 (1934), from Romania (Munţii Retezatului), with upper leaves pinnatifid, pedicels 6–8 mm in fruit, siliqua 15–20 mm, and style *c.* 3 mm, is probably a variant of **1**.

B. balcana Pančić, *Srpska Kralj. Bot. Bašta Beograd* **1888**: 6 (1888), from the central part of the Balkan peninsula, differs from **1** only by the pinnatifid upper leaves, pedicels 6–8 mm, and slightly longer style.

2. B. stricta Andrz. in Besser, *Enum. Pl. Volhyn.* 72 (1822). More or less glabrous biennial up to 100 cm. Basal leaves usually with 1 or 2 pairs of lateral lobes; uppermost leaves simple, sinuate-dentate. Pedicels 3–7 mm in fruit, stout. Buds pubescent; petals 3·5–6 mm. Siliqua 18–30(–35) × 1·5–2 mm; valves with distinct median vein; style 0·5–1·6(–2·3) mm. $2n = 16$. *C., E. & N. Europe; introduced elsewhere and of uncertain status in a number of countries.* Au *Br Bu Cz Da Fe Ge Ho Hu It Ju No Po Rm Rs(N, B, C, W, E) Su [Be Ga Hb].

3. B. verna (Miller) Ascherson, *Fl. Brandenb.* **1**: 36 (1860) (*B. praecox* (Sm.) R. Br., *Campe verna* (Miller) A. A. Heller). Glabrous or sparsely pubescent biennial up to 75 cm. Basal leaves usually with 6–10 pairs of lateral lobes; uppermost leaves pinnatifid. Pedicels 2–8 mm in fruit, stout. Petals 5–8 mm. Siliqua 30–70 × 1·5–2 mm; valves with distinct median vein; style 0·6–2 mm. *S.W. Europe; cultivated as a salad, and for the oil from its seeds, and naturalized elsewhere.* Az Co Ga Hs It Lu Sa [Au Br Da Hb He Ho Rs(W)].

4. B. intermedia Boreau, *Fl. Centre Fr.* **2**: 48 (1840) (*B. pinnata* Lebel, *B. sicula* auct., non C. Presl, *B. sicula* var. *prostrata* (Gay) Gren. & Godron). Glabrous or sparsely pubescent biennial up to 60 cm. Basal leaves with 2–5(6) pairs of lateral lobes; uppermost leaves pinnatifid. Pedicels 3–6 mm in fruit, stout. Petals 4–6 mm. Siliqua 10–30(–35) × 2 mm; valves with distinct median vein; style 0·6–1·7 mm. *S. & C. Europe from S. Germany to N. Portugal and S. Jugoslavia; frequently introduced elsewhere.* Au Ga Ge Hs It Ju Lu [Be Br Da Hb He Ho Po].

5. B. sicula C. Presl in J. & C. Presl, *Del. Prag.* 17(1822). Like **4** but smaller (up to 40 cm); basal leaves with (1)2–4 pairs of lateral lobes; pedicels (3–)4–8 mm in fruit, slender; siliqua 10–20 × 1–1·5 mm; style 1·5–2·8(–3·5) mm. ● *S. Italy, Sicilia; Greece.* Gr It Si.

6. B. bosniaca Murb., *Lunds Univ. Årsskr.* **27**: 169 (1891). Like **4** but the pedicels 4–6 mm in fruit, slender; valves of the siliqua without a median vein; style 2–4 mm. ● *C. Jugoslavia (Bosna).* Ju.

7. B. bracteosa Guss., *Fl. Sic. Prodr.* **2**: 257 (1828) (?*B. intermedia* var. *gautieri* Rouy & Fouc.). Glabrous biennial or perennial up to 50 cm. Lower leaves with 2–4 pairs of lateral lobes; upper leaves pinnatifid. Racemes bracteate at least in the lower ½; pedicels 2–4(–6) mm in fruit. Petals 4–5 mm. Siliqua 15–30 × 1–2 mm; valves with distinct median vein; style 0·5–2 mm. $2n = 16$. *C. Mediterranean region, extending to Bulgaria.* Al Bu Ga Gr It Ju Si.

8. B. rupicola Moris, *Stirp. Sard.* **1**: 55 (1827). Glabrous perennial up to 45 cm. Lower leaves simple or with up to 2 pairs of lateral lobes; upper leaves pinnatifid. Pedicels 6–12 mm in fruit. Petals 7–10 mm. Siliqua 30–60(–90) × 1·5–2·5 mm; style 1–2 mm. $2n = 16$. *Wet, siliceous rocks.* ● *Corse, Sardegna.* Co Sa.

9. B. conferta Boiss. & Heldr. in Boiss., *Fl. Or., Suppl.*, 36 (1888). Glabrous biennial up to 30 cm. Upper leaves pinnatifid. Pedicels 2–4 mm. Petals *c.* 5 mm. Infructescence corymbose; siliqua 24–30 mm; valves with distinct median vein; style 2–3 mm. ● *Mountains of C. & S. Greece (Peristeri, Killini).* Gr.

10. B. longirostris Velen., *Sitz.-Ber. Böhm. Ges. Wiss.* **29**: 1 (1898). Perennial 20–50 cm, hirsute. Basal leaves with 2 or 3 pairs of lateral lobes; upper pinnatifid. Pedicels 4–6 mm in fruit. Petals 6–9 mm. Siliqua 10–20 × 1·5–2 mm, hirsute; style (2–)3–5 mm. ● *C. part of Balkan peninsula.* Al Bu Gr Ju.

36. **Sisymbrella** Spach

V. H. HEYWOOD

Annuals to short-lived perennials; glabrous or with unbranched hairs. Leaves lyrate to pinnatisect. Inner sepals saccate at the base; petals yellow or white. Fruit a siliqua, attenuate at the apex; valves without distinct median vein; style distinct; stigma capitate. Seeds in 1 or 2 rows in each loculus.

Differs from *Nasturtium* and *Rorippa* in the attenuate siliqua and in the seeds becoming mucilaginous when moistened.

Literature: V. H. Heywood, *Bull. Brit. Mus. (Bot.)* **1**: 106–110 (1954). R. Virot, *Cahiers Nat. Paris.* nov. ser. **15**: 89–96 (1959).

Petals white; stems with unbranched hairs, smooth — **1. dentata**
Petals yellow; stems ±glabrous or scabrid — **2. aspera**

1. S. dentata (L.) O. E. Schulz in Engler, *Pflanzenreich* **86(IV. 105)**: 144 (1924) (*Sisymbrium bursifolium* L., *Barbarea dentata* (L.) Paol.). Stems 10–40 cm, with unbranched hairs. Leaves lyrate to deeply pinnatifid, the terminal lobe longer than the laterals. Pedicels 2–3 mm, thick. Petals *c.* 5 mm, white. Siliqua 30–60 mm. Seeds in 1 row in each loculus. ● *S.W. Italy, Sicilia.* It Si.

2. S. aspera (L.) Spach, *Hist. Vég. (Phan.)* **6**: 426 (1838) (*Nasturtium asperum* (L.) Boiss.). Stems 10–40 cm, smooth or scabrid with white tubercles. Leaves pectinate-pinnatisect. Pedicels long or short. Petals yellow. Siliqua 15–50 mm. Seeds in 1 or 2 rows in each loculus. *S.W. Europe.* Ga Hs Lu.

1 Seeds in 1 row in each loculus — (d) subsp. **pseudoboissieri**
1 Seeds in 2 rows in each loculus
  2 Petals 7–12 mm; pedicels (8–)10–15 mm; siliqua 25–50 mm — (c) subsp. **boissieri**
  2 Petals 3–7 mm; pedicels 3–8(–10) mm; siliqua 15–25(–30) mm
    3 Siliqua 15–20(–26) × 2–2·5 mm, ±scabrid — (a) subsp. **aspera**
    3 Siliqua 25–30 × 1·7 mm, ±linear, almost entirely smooth — (b) subsp. **praeterita**

(a) Subsp. **aspera** (*Rorippa aspera* (L.) Maire, *Nasturtium asperum* (L.) Boiss.): Pedicels 3–8(–10) mm. Petals 3–7 mm. Siliqua 15–20(–26) × 2–2·5 mm, relatively thick and short, narrowing from below the middle to the apex, more or less scabrid. Seeds in 2 rows in each loculus. *Moist, sandy soil, often near streams.* ● *Most of Spain except the south, Portugal; S. & W. France.*

(b) Subsp. **praeterita** Heywood, *Feddes Repert.* **69**: 143 (1964) (*Sisymbrium laevigatum* Willd.): Pedicels 3–8(–10) mm. Petals 3–7 mm. Siliqua 25–30 × 1·7 mm, more or less linear, slightly attenuate at the apex, almost smooth. Seeds in 2 rows in each loculus. *Alluvial gravels and damp soils.* ● *Pyrenees and S.W. France.*

(c) Subsp. **boissieri** (Cosson) Heywood, *Bull. Brit. Mus. (Bot.)* **1**: 107 (1954) (*Nasturtium boissieri* Cosson, *Rorippa aspera* prol. *boissieri* Samp., *Nasturtium asperum* subsp. *boissieri* (Cosson) Coutinho): Pedicels (8–)10–15 mm. Petals 7–12 mm. Siliqua 25–50 × 2–2·5 mm, narrowing from below the middle to the apex, more or less scabrid. Seeds in 2 rows in each loculus. *Damp, grassy or rocky places, usually in the mountains. S. Spain, Portugal (around Lisboa).*

(d) Subsp. **pseudoboissieri** (Degen) Heywood, *op. cit.* 108 (1954) (*Sisymbrium pseudoboissieri* Degen): Like subsp. (**c**) but seeds in 1 row in each loculus. *Damp turf and near streams, on calcareous soil in mountains.* ● *S.E. Spain (Sierra de Cazorla).*

Species **2** is highly polymorphic in Spain, where populations are largely intermediate between subspp. (**a**) and (**b**).

37. Rorippa Scop.

†D. H. VALENTINE (EDITION 1) REVISED BY B. JONSELL (EDITION 2)

Annuals to perennials; glabrous or with unbranched hairs. Leaves simple to pinnate. Sepals non-saccate; petals yellow. Fruit a siliqua or silicula; valves veinless or with a weak median vein; style short but distinct; stigma flat to slightly 2-lobed. Seeds in 1 or 2 rows in each loculus.

Literature: B. Jonsell, *Symb. Bot. Upsal.* **19**: 1–222 (1968).

1 Petals shorter than or ±equalling sepals
2 Inflorescence with numerous branches forming a corymb **7. prolifera**
2 Inflorescence not corymbose
3 Sepals less than 1·6 mm; fruit-valves thin **9. islandica**
3 Sepals at least 1·6 mm; fruit-valves rather thick **10. palustris**
1 Petals at least $1\frac{1}{2}$ times as long as the sepals
4 All leaves entire or irregularly toothed, with conspicuous auricles **1. austriaca**
4 At least some leaves pinnatifid to pinnate, with or without conspicuous auricles
5 Basal leaf-rosette present at anthesis; seeds finely reticulate
6 Fruit not more than 6 mm, broadly ellipsoidal **11. pyrenaica**
6 Fruit at least 6 mm, lanceolate to linear
7 Fruit 10–20 × *c.* 1 mm; petals (3–)4–5 mm **12. lippizensis**
7 Fruit 6–10 × *c.* 2 mm; petals less than 3 mm **13. thracica**
5 Basal leaf-rosette withered at anthesis
8 Cauline leaves with conspicuous auricles **2. × armoracioides**
8 Cauline leaves not or only slightly auriculate
9 Uppermost cauline leaves undivided **3. amphibia**
9 Uppermost cauline leaves more or less pinnatifid
10 Fruit less than 5 mm **8. brachycarpa**
10 Fruit at least 5 mm
11 Cauline leaves ±pinnatifid with long and broad terminal lobe; style more than 1·2 mm in fruit **4. × anceps**
11 Cauline leaves pinnate with short and narrow terminal lobe; style not more than 1·2 mm in fruit
12 Fruit at least 8 mm, linear **5. sylvestris**
12 Fruit not more than 8 mm, narrowly ellipsoidal **6. kerneri**

1. R. austriaca (Crantz) Besser, *Enum. Pl. Volhyn.* 103 (1822) (*Nasturtium austriacum* Crantz). Perennial 30–100 cm, stoloniferous, nearly glabrous. Leaves elliptical, simple, entire or irregularly toothed, sessile with clasping auricles. Pedicels 7–15 mm, ascending, sometimes recurved in outer part. Petals 3–4·5 mm, about $1\frac{1}{2}$ times as long as the sepals. Fruit globose, *c.* 3 mm (often not setting seed and much smaller); style 1–2 mm. $2n = 16$. *C. & E. Europe, frequently naturalized or casual further west and north.* Al Au Bu Cz Ge Hu Ju Po Rm Rs(C, W, K, E) Tu [Br Da Fe Ga He Ho It No Su].

2. R. × armoracioides (Tausch) Fuss, *Fl. Transsilv. Exc.* 47 (1866) (*R. austriaca × sylvestris*, *R. cracoviensis* Zapal). Perennial 40–80 cm, stoloniferous, finely hairy below. Leaves pinnatifid-pinnatisect with large terminal lobe, sometimes simple, irregularly toothed, petiolate, with clasping auricles. Pedicels straight, horizontal-ascending. Petals 3–4·5 mm, about twice as long as the sepals. Fruit 3.5–9 × 1·5–2 mm, oblong-ellipsoidal, often setting good seed; style 1–1·5 mm. $2n = 32, 40$. *C., E. & N. Europe.* Au Br Cz Da Fe Ga Ge Hu Ju No Po Rm Rs(B, C, W, E) Su; probably elsewhere.

This plant spreads independently of the parent species along rivers and on rather dry, cultivated ground. It is very variable and forms a series of intermediates between **1** and **5**.

3. R. amphibia (L.) Besser, *op. cit.* 27 (1822) (*Nasturtium amphibium* (L.) R. Br.). Perennial 40–120 cm with prostrate, rooting basal stem parts, glabrous. Lowest leaves pectinate, caducous; upper leaves pinnatifid to simple, elliptical, irregularly toothed, usually without auricles. Pedicels 6–17 mm, deflexed from the base. Petals 3·5–5·5 mm, about twice as long as the sepals. Fruit 2·5–6 × 1·7–3 mm (often not setting seed and much smaller), oblong-ovoid; style 1–2·5 mm. $2n = 16, 32$. *Wet places. Much of Europe, but rare in the north and the Mediterranean region.* Al Au Be Br Bu Co Cz Da Fe Ga Ge Gr Hb He Ho Hs Hu It Ju Lu Po Rm Rs(N, B, C, W, K, E) Sa ?Si Su Tu.

4. R. × anceps (Wahlenb.) Reichenb., *Icon. Fl. Germ.* **2**: 15 (1837–8) (*R. amphibia × sylvestris*, *R. prostrata* (J. P. Bergeret) Schinz & Thell., *R. hybrida* Klokov, *Nasturtium anceps* (Wahlenb.) Reichenb., *N. stenocarpum* Godron). Perennial 35–100 cm, glabrous or finely hairy below. Leaves pinnatisect-pinnatifid, with large terminal lobe, auricles small or absent. Pedicels horizontal-deflexed. Petals 3·5–5·5 mm, about twice as long as the sepals. Fruit 5–10 × 1·2–2·5 mm, oblong-ellipsoidal, setting good seed; style 1·2–3 mm. $2n = 32, 40$. *Much of Europe, but absent from many islands and most of the Mediterranean region.* Au Be Br Cz Da Fe Ga Ge Hb He Ho Hs Hu It Ju Lu Po Rm Rs(C, W, K, E) Su [Be Br ?Fe Rs(N, B)].

This plant grows largely independently of the parent species, spreading along river systems. It forms a series of intermediates between **3** and **5**.

5. R. sylvestris (L.) Besser, *Enum. Pl. Volhyn.* 27 (1822) (*Nasturtium sylvestre* (L.) R. Br.). Ascending perennial 15–90 cm, with extensive stolons, often mat-forming, glabrous. Leaves petiolate, pinnate-pinnatifid with short terminal lobe; lobes lanceolate-rhombic, entire, toothed or lobulate. Pedicels 4–12 mm, erecto-patent. Petals 2·5–5·5 mm, about twice as long as the sepals. Fruit 9–22 × 1–1·2 mm (often not setting seed and much shorter and narrower), linear, sometimes curved; style 0·5–1·2 mm. $2n$ = 32, 40, 48. *Almost throughout Europe.* All except Az Bl Fa ?Sa Sb; introduced in Fe Is No.

6. R. kerneri Menyh., *Kal. Vid. Növén.* 39 (1877). Like **5** but 30–50 cm, erect, with short stolons; leaves pinnate with lanceolate lobes; pedicels 8–10 mm, patent; petals *c.* 4 mm; fruit 4–9 mm, narrowly ellipsoidal; style *c.* 1 mm. $2n$ = 16. *Saline habitats.* • *E.C. Europe and N. part of Balkan peninsula.* Bu Cz Hu Ju.

7. R. prolifera (Heuffel) Neilr., *Aufz. Ung. Slav. Gefäss.* 263 (1866). Annual to perennial 40–80 cm. Leaves auriculate, pinnate with 2–5 pairs of oblong-lanceolate, coarsely toothed or entire lobes. Racemes clustered in a corymb with densely set siliquae on very thin pedicels 4·5–8 mm. Petals 2–3 mm, equalling or slightly longer than the sepals. Fruit 4·5–9 × *c.* 1 mm, linear; style 0·5–0·8 mm. *S.E. Europe.* Al Bu Gr Ju Rm Rs(W, E) Tu.

8. R. brachycarpa (C. A. Meyer) Hayek, *Prodr. Fl. Penins. Balcan.* **1**: 390 (1925), quoad basion. (*Nasturtium brachycarpum* C. A. Meyer). Annual to short-lived perennial 15–40 cm, erect and somewhat flexuous, glabrous. Leaves lyrate-pinnatifid, the upper pinnate with 2–7 pairs of rhombic-linear lobes. Pedicels 5–9 mm, patent-ascending. Petals 2·5–4 mm, about 1½ times as long as the sepals. Fruit 3–4·5 × 0·9–1·3 mm, narrowly ellipsoidal; style 0·6–1 mm. *C. & S. Russia, Ukraine.* Rs(C, W, K, E).

In addition to **2** and **4**, a number of other hybrids involving most of species **1–8** have been reported, and some have been treated as species in Floras. They are partly fertile and back-cross products appear to be frequent.

9. R. islandica (Oeder) Borbás, *Balaton Növényföldr.* **2**: 392 (1900) (*R. terrestris* (Curtis) Fuss). Annual 5–30 cm, decumbent to ascending, glabrous. Leaves petiolate, not auriculate, pinnatisect with 3–5 pairs of lobes. Petals 1–1·5(–1·7) mm, about equalling the sepals. Fruit 6–12 × 2–3 mm, shortly cylindrical, 2–3 times as long as the patent-deflexed pedicels; valves of fruit thin; style 0·3–0·7 mm. Seeds finely colliculate. $2n$ = 16. *W.C. & S. Europe extending to S.E. Russia.* Au Br Ga Gr Hb He Hs Is It Ju No Rs(?C, W, E).

(a) Subsp. **islandica**: Leaf-segments finely serrate or entire. Epidermal cells of seed about as high as broad. *W., C. & S. Europe.*

(b) Subsp. **dogadovae** (Tzvelev) Jonsell, *Symb. Bot. Upsal.* **19**: 156 (1968): Leaf-segments toothed or incised. Epidermal cells of seed much higher than broad. *S. Russia, Ukraine.*

10. R. palustris (L.) Besser, *Enum. Pl. Volhyn.* 27 (1822) (*Nasturtium palustre* (L.) DC., *R. islandica* auct.). Annual to short-lived perennial 10–110 cm, ascending to erect, usually glabrous. Leaves more or less petiolate, auriculate, pinnatisect to lyrate-pinnatifid with 2–6 pairs of serrate lobes. Petals (1·4–)1·5–2·8 mm, about equalling the sepals. Fruit 5–12 × 1·7–3 mm, shortly cylindrical to oblong-ellipsoidal, often curved, slightly shorter to twice as long as the horizontal to deflexed pedicels; valves of fruit rather thick; style 0·4–1 mm. Seeds coarsely colliculate. $2n$ = 32. *Europe, except S. part of Balkan peninsula and many of the islands.* Au Be Br Bu Cz Da Fa Fe Ga Ge Hb He Ho Hs Hu It Ju Lu No Po Rm Rs(N, B, C, W, K, E) Su.

The sterile hybrid between **3** and **10** is fairly frequent, while that between **5** and **10**, in spite of numerous alleged reports, is very rare.

11. R. pyrenaica (Lam.) Reichenb., *Icon. Fl. Germ.* **2**: 15 (1837) (*Nasturtium pyrenaicum* (Lam.) R. Br., *R. stylosa* (Pers.) Mansfeld & Rothm.). Perennial 5–40 cm, shortly pubescent at the base. Lower leaves in a rosette, with long petiole, ovate, entire, or lyrate-pinnatifid with roundish segments; cauline leaves more or less sessile, with narrow, amplexicaul auricles, pinnate with 2–8 pairs of linear segments. Petals 2–4 mm. Fruit 2–6 × 1·5–2 mm, shorter than or equalling the pedicels; style 0·7–2 mm. $2n$ = 16. *C. & S. Europe, northwards to C. France.* Al *Be Bu Cz Ga *Ge Gr He Hs It Ju Lu Rm Rs(W).

Very variable in the shape of the fruit, which may be subglobose, ellipsoidal or subcylindrical. Plants with subcylindrical fruits from Spain and the Pyrenees have been called **R. hispanica** (Boiss. & Reuter) Willk. in Willk. & Lange, *Prodr. Fl. Hisp.* **3**: 845 (1880).

12. R. lippizensis (Wulfen) Reichenb., *Icon. Fl. Germ.* **2**: 15 (1837) (*Nasturtium lippizense* (Wulfen) DC.). Perennial 10–20 cm, shortly pubescent at the base. Lower leaves in a rosette, with long petioles, orbicular-ovate, entire, or pinnatifid with 1–4 pairs of narrow lateral lobes and a large terminal lobe; cauline leaves sessile, with narrow, amplexicaul auricles, pinnatifid with 1–4 pairs of linear to lanceolate lobes. Petals (3–)4–5 mm. Fruit 10–20 × *c.* 1 mm, longer than the erect or ascending pedicels; style 0·5–1·5 mm. $2n$ = 32. • *Balkan peninsula and N.W. Jugoslavia.* Al Bu Gr Ju Tu [It].

13. R. thracica (Griseb.) Fritsch, *Verh. Zool.-Bot. Ges. Wien* **44**: 316 (1894). Perennial 15–40 cm, shortly pubescent at the base; stem erect, somewhat branched above. Lower leaves in a rosette, petiolate, all lyrate-pinnatifid with an orbicular terminal lobe and few or no lateral lobes; upper cauline leaves auriculate, with 4–8 pairs of narrowly linear lobes. Petals 2–2.5 mm. Fruit 6–10 × *c.* 2 mm, narrowly lanceolate, shorter than or more or less equalling the patent pedicels; style 2 mm. $2n$ = 32. • *Balkan peninsula.* Bu Gr Ju ?Rm Tu.

Species **11–13** form a group clearly deviating from the rest of the genus, as evident from their distinctly reticulate seeds. Their relationship and status are not fully clear. While **11** seems distinct though variable, **12** and **13** may perhaps constitute one variable species.

38. Nasturtium R. Br.

†D. H. VALENTINE

Perennials; glabrous or with few unbranched hairs. Leaves pinnate to pinnatisect. Petals white, rarely pale purplish. Fruit a siliqua; valves with weak median vein; style short; stigma capitate, slightly 2-lobed. Seeds in 1 or 2 rows in each loculus. Radicle accumbent.

Seeds in 2 rows in each loculus, most with fewer than 12 depressions across the width **1. officinale**

Seeds usually in 1 row in each loculus, most with more than 12 depressions across the width **2. microphyllum**

1. N. officinale R. Br. in Aiton, *Hort. Kew.* ed. 2, **4**: 111 (1812) (*Rorippa nasturtium-aquaticum* (L.) Hayek). Perennial 10–100 cm with fleshy, hollow stems, procumbent and rooting at the lower nodes, upwards ascending, sometimes floating; more or less glabrous. Leaves petiolate, often auriculate, pinnate, with 2–9 pairs of lobes, which are rounded to broadly elliptical and more or less entire. Pedicels 8–12 mm, patent to somewhat reflexed. Petals 3·5–5(–6) mm, white to pale purplish, about twice as long as the sepals. Fruit 10–18(–24) × 2–2·7(–3) mm, straight or curved; style 0·5–1·8(–2) mm. Seeds in 2 distinct rows in each loculus, with *c*. 25 polygonal depressions on each face. $2n = 32$. *Wet places. Europe, northwards to Scotland, S. Sweden and to* c. 55° *N. in Russia.* All except Fa Fe Is No Rs(N) Sb; *doubtfully native in Az.*

2. N. microphyllum (Boenn.) Reichenb., *Fl. Germ. Excurs.* 683 (1832) (*N. uniseriatum* Howard & Manton, *Rorippa microphylla* (Boenn.) Hyl.). Like **1** but racemes often longer in fruit; flowers somewhat larger (petals 4–5·5 mm); fruit 15–30 × (1–)1·3–2 mm; seeds usually in 1 row in each loculus, with *c*. 100 depressions on each face. $2n = 64$. *Wet places. Mainly in W. Europe but extending eastwards to Sweden and Poland.* Be Br Cz Da Ga Ge Gr Hb He Ho Hs It Ju Po Su.

The distribution of **2** is still imperfectly known; at least in Denmark and the Netherlands it is much commoner than **1**. The hybrid **1 × 2** (R. × *sterilis* Airy Shaw, with $2n = 48$) is fairly common. It propagates vegetatively and also spreads outside the range of the parent species. Both **1** and **1 × 2** are cultivated as a salad (Water-cress), and this may obscure their native distribution.

39. Armoracia P. Gaertner, B. Meyer & Scherb.

P. W. BALL

Glabrous perennials. Leaves simple to pinnatifid. Sepals not saccate; petals white, shortly clawed. Fruit a subglobose to ellipsoidal silicula; valves weakly reticulate-veined; style short; stigma capitate to slightly 2-lobed. Seeds in 2 rows in each loculus.

Silicula 4–6 mm; ovules usually 4–6 in each loculus **1. rusticana**
Silicula 10–15 mm; ovules *c*. 10 in each loculus **2. macrocarpa**

1. A. rusticana P. Gaertner, B. Meyer & Scherb., *Fl. Wetter.* **2**: 426 (1800) (*A. lapathifolia* auct., *Cochlearia armoracia* L., *Nasturtium armoracia* (L.) Fries). Plant with stout, branched stock and fleshy, fusiform roots. Stems up to 1 m or more. Basal leaves with lamina 30–50 cm, ovate or ovate-oblong, crenate-serrate, petiole up to 50 cm; cauline leaves sessile or shortly petiolate, the lower often pinnatifid, the upper crenulate or entire. Petals 5–8 mm. Silicula rarely developed, 4–6 mm, globose or ovoid; ovules 4–6 in each loculus. $2n = 32$. *Cultivated as a condiment* (*Horse-radish*), *and naturalized throughout much of Europe; origin unknown.* [Al Au Be Br Bu Cz Da Fe Ga Ge Hb He Ho Hs Hu It Ju No Po Rm Rs(N, B, C, W, E) Si Su Tu].

2. A. macrocarpa (Waldst. & Kit.) Kit. ex Baumg., *Enum. Stirp. Transs.* **2**: 240 (1816). Like **1** but the basal leaves usually entire; cauline leaves lanceolate, the lower serrate, the upper crenate; silicula 10–15 mm, ovoid or ellipsoidal; ovules *c*. 10 in each loculus. $2n = 32$. *Marshes.* ● *C. Danube basin.* Cz Hu Ju Rm.

40. Cardamine L.

B. M. G. JONES (EDITION 1) REVISED BY J. R. AKEROYD; *C. amara* & *C. pratensis*, K. MARHOLD (EDITION 2)

Annuals to perennials; glabrous or with simple hairs. Leaves simple to pinnate. Inner sepals saccate or not. Petals white, pink or purple (rarely pale yellow). Fruit a strongly compressed siliqua; valves coiling spirally from the base at dehiscence, veinless or with an indistinct median vein; style short or distinct; stigma slightly 2-lobed. Seeds in 1 row in each loculus. (Incl. *Dentaria* L.)

Literature: O. E. Schulz, *Bot. Jahrb.* **32**: 280–623 (1903).

1 Petioles with small, amplexicaul auricles
2 Petals pink or violet, caducous; leaves ternate or pinnate, with large, lanceolate leaflets **26. chelidonia**
2 Petals white, not caducous (rarely absent); leaves various, with small, ovate or obovate leaflets
3 Lower leaves entire or ternate **20. resedifolia**
3 Lower leaves pinnate
4 Petals 2–4·5 mm, scarcely exceeding the sepals, sometimes absent; siliqua with a subsessile stigma **28. impatiens**
4 Petals 4–11 mm, at least twice as long as the sepals; siliqua with a flattened beak
5 Sepals apiculate; funicle of seeds slender **22. maritima**
5 Sepals obtuse; funicle of seeds broad, flattened **25. graeca**
1 Petioles without auricles
6 Petals not more than 5 mm, sometimes absent
7 Plant entirely glabrous
8 All leaves simple and entire, oblong-spathulate **21. bellidifolia**
8 At least some leaves pinnate
9 Basal leaves entire to 3-lobed; petals at least 4 mm; siliqua usually 1·5–1·8 mm wide **19. glauca**
9 Basal leaves pinnate with 3–5 pairs of lateral leaflets; petals not more than 2·5 mm; siliqua 0·8 mm wide **27. parviflora**
7 Plant at least sparsely hairy
10 Lowest leaves not in a rosette; rhizome with numerous stolons **13. amara**
10 Lowest leaves in distinct rosette; rhizome not giving rise to stolons, or absent
11 Pedicels 4–8 mm at anthesis; siliqua 20–40 mm **29. caldeirarum**
11 Pedicels not more than 3 mm at anthesis; siliqua not more than 25 mm
12 Most flowers with 6 stamens; stem usually hairy, flexuous; cauline leaves (3)4–10 **30. flexuosa**
12 Most flowers with 4 or 5 stamens; stem glabrous, not flexuous; cauline leaves 0–4(5) **31. hirsuta**
6 Petals more than 5 mm
13 Leaves in a whorl below the inflorescence
14 Leaves pinnate
15 Rhizome scale-leaves 2–3 mm, appressed; petals purple **4. quinquefolia**
15 Rhizome scale-leaves 5–8 mm, concave; petals yellow **5. kitaibelii**
14 Leaves ternate or digitate
16 Petals yellowish or white; stamens ± exserted, longer than the petal-claw **6. enneaphyllos**

16 Petals purple; stamens included, about as long as the petal-claw **7. glanduligera**
13 Leaves not in a whorl below the inflorescence
17 All leaves simple and entire
18 Leaves reniform, more than 10 mm wide **12. asarifolia**
18 Leaves oblong-spathulate, less than 10 mm wide **21. bellidifolia**
17 At least some leaves compound
19 Rhizome present, long, or giving rise to slender stolons
20 Basal leaves simple, reniform
21 Plant up to 15 cm; siliqua 12–16 mm **15. pratensis**
21 Plant at least 20 cm; siliqua 20–40 mm
22 Middle cauline leaves ternate, with rounded segments **12. asarifolia**
22 Middle cauline leaves, if present, with 1–3 pairs of narrowly ovate or oblong leaflets **17. tenera**
20 Basal leaves compound
23 Axils of upper leaves with brownish-purple bulbils **1. bulbifera**
23 Axils of leaves without bulbils
24 Leaves all ternate or digitate
25 Leaflets serrate; petals more than 13 mm; siliqua more than 2·5 mm wide **3. pentaphyllos**
25 Leaflets obscurely dentate; petals less than 13 mm; siliqua less than 2·5 mm wide
26 Rhizome with scale-leaves only; cauline leaves 3 or more; anthers violet **8. waldsteinii**
26 Rhizome with foliage leaves (or withered petioles) as well as scale-leaves; cauline leaves 0–3; anthers yellow **11. trifolia**
24 At least some leaves pinnate
27 Rhizome 4–10 mm in diameter; petals 14–20 mm; siliqua more than 3 mm wide **2. heptaphylla**
27 Rhizome usually less than 4 mm in diameter or absent; petals 5–13 mm; siliqua less than 3 mm wide
28 Leaflets crenate-apiculate or serrate, with appressed hairs; ovary and siliqua sparsely hairy **9. macrophylla**
28 Leaflets crenate to entire, pubescent (especially on the margin) or glabrous; ovary and siliqua glabrous
29 Rhizome without stolons; stems usually 30–80 cm
30 Cauline leaves with 1–6 pairs of leaflets, the terminal leaflet distinctly larger; siliqua 1·4–2 mm wide **14. raphanifolia**
30 Cauline leaves with 4–9 pairs of leaflets, the terminal leaflet slightly larger; siliqua 1–1·5 mm wide **16. uliginosa**
29 Rhizome with stolons; stems rarely more than 30 cm
31 Anthers yellow; stigma conspicuous **15. pratensis**
31 Anthers blackish-violet or rarely yellowish-white; stigma indistinct **13. amara**
19 Rhizome absent, or short, not giving rise to stolons, or plant annual to biennial
32 Ovary and siliqua with appressed hairs; seeds *c.* 3 mm **23. carnosa**
32 Ovary and siliqua glabrous; seeds 1–2 mm
33 Lowest leaves simple, sometimes 3- to 5-lobed
34 Lowest leaves in a rosette; stems not flexuous; petals 8–12 mm **15. pratensis**
34 Lowest leaves not in a rosette; stems flexuous; petals 4–8 mm
35 Plant not glaucous; leaflets of upper leaves lobed; infructescence always symmetrical **18. plumieri**
35 Plant glaucous; leaflets of upper leaves entire; infructescence usually secund **19. glauca**
33 Lowest leaves ternate, pinnate or pinnatipartite
36 Cauline leaves mostly ternate; some roots forming apical tubers **10. trifida**
36 Cauline leaves pinnate to pinnatifid; roots not forming apical tubers
37 Siliqua *c.* 4 mm wide **24. monteluccii**
37 Siliqua 1–2·5 mm wide
38 Leaflets with a few prominent apiculate teeth; pedicels hairy **29. caldeirarum**
38 Leaflets subentire; pedicels glabrous
39 Leaves all similar, lyrate-pinnate, with the terminal leaflet at least twice as large as the lateral leaflets; all leaflets orbicular to reniform; siliqua without a thickened border **14. raphanifolia**
39 Leaves dissimilar; the lower pinnate, with the terminal leaflet only a little larger than the lateral leaflets; at least the upper cauline leaves with oblong to linear-lanceolate leaflets; siliqua with a thickened border **15. pratensis**

Subgen. **Dentaria** (L.) Hooker fil. Perennials; rhizome subterranean, with scale-leaves. Cotyledons petiolate.

1. C. bulbifera (L.) Crantz, *Class. Crucif.* 127 (1769) (*Dentaria bulbifera* L.). Perennial 25–70 cm; rhizome 2–5 mm in diameter, with fleshy, deltate scale-leaves. Leaves alternate, more than 6; lower leaves pinnate, with 1–3 pairs of lanceolate, serrate leaflets; uppermost leaves simple. Small brownish-purple bulbils present in the upper leaf-axils. Petals 12–20 mm, pale purple. Siliqua 15–35 × 2–2·5 mm (formed occasionally, mainly in the southern part of its range). $2n = 96$. *Woods. Much of Europe, but local in the west and south.* Al Au Be Br Bu Cz Da Fe Ga Ge Gr He It Ju No Po Rm Rs(B, C, W, K) Su Tu.

2. C. heptaphylla (Vill.) O. E. Schulz, *Feddes Repert.* **46**: 116 (1939) (*C. pinnata* (Lam.) R. Br., *C. baldensis* Fritsch). Perennial 30–60 cm; rhizome 4–10 mm in diameter, with lunate scale-leaves 1–2 mm. Leaves usually 3 or more, remote, pinnate; lower leaves with 3–5 pairs of ovate-lanceolate, serrate, sparsely hairy or glabrescent leaflets; uppermost leaves with 2 or 3 pairs of similar leaflets. Petals 14–20 mm, white, pink or purplish. Siliqua 40–80 × 3·5–5 mm. $2n = 48$. *Woods. W. Europe, extending to S.W. Germany and S.W. Italy.* Ga Ge He Hs It.

The hybrid between **2** and **3** (*C.* × *digenea* Gremli) occurs where the parent species grow together. It is sterile and morphologically intermediate between the parents, and spreads vegetatively.

3. C. pentaphyllos (L.) Crantz, *Class. Crucif.* 127 (1769) (*Dentaria pentaphyllos* L., *D. digitata* Lam.). Like **2** but rhizome 1·5–2·5 mm in diameter, with triangular, trifid, concave scale-leaves 6–10 mm; leaves all digitate or ternate, glabrous except for the margins; petals (14–)18–22 mm, white or pale purple; siliqua 2·5–4 mm wide. $2n = 48$. *Mountain woods.* ● *W. & C. Europe, from the Pyrenees to S. Germany and N. Jugoslavia.* Au Ga Ge He Hs It Ju.

4. C. quinquefolia (Bieb.) Schmalh., *Fl. Sred. Juž. Ross.* **1**: 51 (1895). Perennial 15–40 cm; rhizome slender at the apex but in older parts 5–7 mm in diameter, with deltate, membranous, appressed scale-leaves 2–3 mm. Leaves usually 3, in a lax whorl, pinnate, with 2 or 3 pairs of ciliate (rarely entirely glabrous), irregularly biserrate, lanceolate leaflets. Petals 14–18 mm, deep purple, twice as long as the stamens. Siliqua 35–50 × 2·5 mm. *E. Europe.* Bu Rm Rs(C, W, K, E) Tu.

5. C. kitaibelii Becherer, *Ber. Schweiz. Bot. Ges.* **43**: 57 (1934) (*C. polyphylla* (Waldst. & Kit.) O. E. Schulz, non D. Don). Perennial 20–30(–50) cm; rhizome 3–6 mm in diameter, with concave scale-leaves 5–8 mm. Leaves 3(–5), more or less whorled, pinnate, with 2–6 pairs of serrate, lanceolate leaflets. Petals 15–22 mm, pale yellow, twice as long as the stamens. Siliqua 40–65 × 2·5–3 mm. $2n = 48$. *Woods.* ● *Alps and adjacent regions; Appennini; N.W. Jugoslavia.* He It Ju.

The hybrid between **3** and **5** (*C.* × *killiasii* Brügger) occurs where the parent species grow together.

6. C. enneaphyllos (L.) Crantz, *Class. Crucif.* 127 (1769). Perennial 20–30 cm; rhizome nodular, up to 6 mm in diameter, with inconspicuous scale-leaves 1–2 mm. Leaves 2–4 in a lax whorl, ternate or sometimes digitate, with ovate-lanceolate, irregularly biserrate leaflets. Flowers somewhat pendent; petals 12–16 mm, pale yellow or white, scarcely exceeding the stamens. Siliqua 40–75 × 3·5–4 mm. $2n = 52–54, 80$. *Mountain woods.* ● *From the W. Carpathians and E. Alps to S. Italy and S. Jugoslavia.* Al Au Cz Ge Hu It Ju Po Rm.

7. C. glanduligera O. Schwarz, *Feddes Repert.* **46**: 188 (1939) (*Dentaria glandulosa* Waldst. & Kit.). Perennial 12–25 cm; rhizome 4–6 mm in diameter, with deltate scale-leaves 2–3 mm, but slender (1–2 mm) and non-scaly in parts. Leaves usually 3 in a single whorl, ternate, with lanceolate, irregularly biserrate leaflets with ciliate margins. Petals 12–22 mm, purple, twice as long as the stamens. Siliqua 35–60 × 2–3 mm. $2n = 48$. *Mainly in* Fagus *woods. Carpathians and adjacent parts of E.C. Europe.* Cz Hu Ju Po Rm Rs(W) [Au].

8. C. waldsteinii Dyer, *Kew Handlist Herb. Pl.* 97 (1891) (*Dentaria trifolia* Waldst. & Kit., *C. savensis* O. E. Schulz). Perennial 20–40 cm; rhizome 3–5 mm in diameter, brown, with distant, deltate scale-leaves 4–5 mm. Leaves 3 or more, remote, ternate, with subsessile, obtuse, narrowly rhombic to ovate or almost lanceolate-acuminate leaflets, obscurely apiculate-dentate, glabrous or somewhat hairy. Petals 10–12 mm, white; anthers violet. Siliqua 20–35 × 2 mm. *Woods.* ● *N. Jugoslavia, extending to S. Austria and S. Hungary.* Au Hu Ju.

Subgen. **Cardamine.** Annuals or biennials, or perennials with rhizome on the surface of the ground, with scale-leaves few or absent. Cotyledons sessile or subsessile.

9. C. macrophylla Willd., *Sp. Pl.* **3(1)**: 484 (1800). Perennial 30–100 cm, with a long horizontal rhizome *c.* 2 mm in diameter. Basal leaves 10–25 cm, with 2–4 pairs of subsessile, ovate lateral leaflets and a somewhat larger terminal leaflet; cauline leaves 5–8, with narrower leaflets; leaflets apiculate-crenate to serrate, appressed-hairy. Petals (6–)8–11 mm, pale pink to purple. Siliqua 25–40 × 1·5–2·5 mm, sparsely hairy; beak 1·5–3·5 mm. $2n = 80$. *N.E. Russia.* Rs(N, C).

10. C. trifida (Lam. ex Poiret) B. M. G. Jones, *Feddes Repert.* **69**: 57 (1964) (*C. tenuifolia* (Ledeb.) Turcz.). Perennial 7–30 cm; rhizome short, *c.* 5 mm in diameter, not scaly, bearing adventitious roots some of which are swollen at the apex to form flattened tubers. Basal leaves pinnate with 1 or 2 pairs of ovate to orbicular, stalked, crenate-lobed or trifid leaflets, withered at anthesis; cauline leaves 1–4, mostly ternate with linear leaflets. Petals (6–)8–10(–13) mm, lilac or white. Siliqua 22–35 × 1·5–2 mm. $2n = 32$. *C. & E. Russia; very local.* Rs(C). (*Altai and N.E. Asia.*)

11. C. trifolia L., *Sp. Pl.* 654 (1753). Perennial 20–30 cm; rhizome 2–4 mm in diameter, nodular, creeping, branched, with remote scale-leaves; stem simple. Rhizomal leaves ternate; leaflets broadly rhombic to suborbicular, obscurely dentate-apiculate, sparsely hairy above and purplish beneath, long-petiolate; cauline leaves 0–3, small, sessile, simple or ternate. Petals 9–11 mm, white or pink; anthers yellow. Siliqua 20–25 × 2 mm. $2n = 16$. *Mountain woods.* ● *C. Europe from the French Jura to the W. Carpathians, extending southwards to N.C. Italy and C. Jugoslavia.* Au Cz Ga Ge He Hu It Ju Po.

12. C. asarifolia L., *loc. cit.* (1753). Perennial 20–40 cm; rhizome horizontal, short, stout, stoloniferous. Lower leaves reniform, sinuate or somewhat apiculate-crenate, long-petiolate; upper leaves similar, shortly petiolate; rarely some cauline leaves ternate with rounded segments. Petals 6–10 mm, white; anthers violet. Siliqua 20–30 × 1·5–2 mm. *Streamsides and other damp places; calcifuge.* ● *Pyrenees; Alps; N. Appennini.* Ga He Hs It.

13. C. amara L., *Sp. Pl.* 656 (1753). Perennial 10–60 cm; rhizome short, horizontal, with numerous stolons; stem angular, simple or branched. Basal leaves not in a rosette; lower cauline leaves pinnate with 2–11 pairs of ovate to orbicular, shortly stalked leaflets and a somewhat larger terminal leaflet; upper cauline leaves shortly petiolate with 2–5 pairs of lanceolate to ovate, angular leaflets. Petals 5–11 mm, white, rarely purplish; anthers blackish-violet or rarely yellowish-white. Siliqua 20–40 × 1–2 mm; style 0·5–2·5 mm; stigma narrower than the style. *Damp places and streamsides. Most of Europe, but rare in the north and extreme south.* Au Be Br Bu Co Cz Da Fe Ga Ge Hb He Ho Hs Hu It Ju Lu No Po Rm Rs(N, B, C, W, E) Su.

(a) Subsp. **amara**: Plant usually sparsely hairy at the base of the stem; rhizome slender; stem simple or branched with (2)3–14(–24) leaves, the leaves not congested under the inflorescence; lower cauline leaves with 2–6(7) pairs of leaflets. $2n = 16, 32$. *Throughout the range of the species.*

(b) Subsp. **opicii** (J. & C. Presl) Čelak., *Prodr. Fl. Böhm.* 449 (1875) (*C. opicii* J. & C. Presl): Plant glabrous or densely hairy; rhizome thick; stem simple with (10–)13–46(–53) leaves, the leaves congested under the inflorescence; lower cauline leaves with (4)5–9(–11) pairs of leaflets. $2n = 16$. *Mountains of C. Europe.*

Tetraploid plants require further study.

14. C. raphanifolia Pourret, *Mém. Acad. Sci. Toulouse* **3**: 310 (1788) (*C. latifolia* Vahl, non Lej.). Perennial 30–80 cm; rhizome long, horizontal; stem robust, glabrous. Leaves large, thick, clear green, lyrate-pinnate, long-petiolate; lower cauline leaves with (0)1–6 pairs of ovate to orbicular-reniform, sinuate, lateral leaflets, decreasing in size from the very large, reniform, terminal leaflet 3–7 cm wide. Petals 8–13 mm; anthers yellow. Siliqua 15–30(–40) × 1·4–2 mm; style (1·5–)2–5 mm, gradually attenuate. *Streamsides and wet meadows. Mountains of S. Europe.* Al Bu Ga Gr Hs It Ju [Br].

A variable species that requires critical evaluation across its range. The subspecies below must be regarded as provisional.

1 Lower leaves usually pubescent, with 2–4(5) pairs of lateral leaflets (c) subsp. **acris**
1 Lower leaves glabrous, with 1–3 pairs of lateral leaflets
2 Petals reddish-violet; siliqua about as long as the pedicel (a) subsp. **raphanifolia**
2 Petals white; siliqua about twice as long as the pedicel (b) subsp. **barbareoides**

(a) Subsp. **raphanifolia** (incl. subsp. *merinoi* (Laínz) Laínz, subsp. *gallaecica* Laínz): Leaves glabrous, the lower with 1–3 pairs of lateral leaflets. Petals reddish-violet. Infructescence short and dense; siliqua strict, about as long as the pedicel; stigma much wider than the style. $2n = 44, 46$. • *N. Spain to S. France.*

(b) Subsp. **barbareoides** (Halácsy) Strid, *Mount. Fl. Gr.* **1**: 257 (1986) (*C. barbareoides* Halácsy, *C. amara* subsp. *barbareoides* (Halácsy) Stoj. & Stefanov): Like subsp. (a) but petals white; siliqua about twice as long as the pedicel. • *C. & N. Greece; ?Albania.*

(c) Subsp. **acris** (Griseb.) O. E. Schulz, *Bot. Jahrb.* **33**: 497 (1903) (*C. acris* Griseb.): Leaves often pubescent, the lower with 2–4(5) pairs of lateral leaflets. Petals lilac. Infructescence long and lax; siliqua patent, twice as long as the pedicel; stigma scarcely wider than the style. *Balkan peninsula.*

Plants from Italy are in some respects intermediate between subspp. (b) and (c). The more widespread variant, from the Appennini, has dense compound racemes, white petals, and patent siliquae 35–40 × 1 mm; the other, from Calabria, has a simple raceme, violet petals, and erecto-patent siliquae 14–25 × 1·5 mm.

15. C. pratensis L., *Sp. Pl.* 656 (1753). Perennial; rhizome usually short and non-stoloniferous; stem usually erect, simple, glabrous. Basal leaves usually forming a rosette, with (0–)2–8(–15) pairs of ovate to orbicular or reniform, sessile or shortly stalked lateral leaflets and a larger or equal terminal leaflet, sparsely hairy, rough-surfaced, often bearing adventitious shoots; cauline leaves 2–18, pinnatisect or pinnate, with roundish, ovate, lanceolate to linear leaflets or segments, usually glabrous; leaves usually dull or grey-green. Sepals 2·2–6·0 mm; petals 5–16(–19) × 2·5–10(–12) mm, emarginate to rounded, white, lilac or purple. Siliqua 10–50 × 0·5–2 mm; style short, 0·25–0·8 mm wide, flattened, not attenuate; stigma conspicuous, enlarged. *Almost throughout Europe.* Al Au Az Be Br Bu Co Cz Da Fe Ga Ge Gr Hb He Ho Hs Hu It Ju Lu No Po Rm Rs(N, B, C, W, E) Su ?Tu.

A variable species that can be divided into a number of subspecies, many of them treated at specific rank by some authors. The following treatment is provisional.

1 Rhizome stoloniferous; stem ascending; basal leaflets not in a rosette (c) subsp. **crassifolia**
1 Rhizome not stoloniferous; stem erect; basal leaves in a rosette
2 Leaves thick, with impressed veins (d) subsp. **polemonioides**
2 Leaves thin, with raised veins
3 Ends of some roots tuberous; basal leaves simple or with 3–5 leaflets (b) subsp. **granulosa**
3 Ends of roots not tuberous; basal leaves mostly with more than 5 leaflets
4 All cauline leaves pinnate, leaflets distinctly stalked (g) subsp. **dentata**
4 At least upper cauline leaves pinnatisect
5 At least some basal leaves hairy
6 Hairs on rhachis of basal leaves patent (90°) (a) subsp. **pratensis**
6 Hairs on rhachis of basal leaves appressed
7 Flowers white or pale lilac; anthers yellow before dehiscence (e) subsp. **matthioli**
7 Flowers purple; anthers purplish before dehiscence (f) subsp. **rivularis**
5 All basal leaves glabrous (sometimes shrivelled at anthesis)
8 Lower segments of mid-cauline leaves somewhat deflexed (e) subsp. **matthioli**
8 Lower segments of mid-cauline leaves slightly ascending
9 Terminal leaflet of basal leaves approximately the same size as lateral ones; anthers purplish before dehiscence (f) subsp. **rivularis**
9 Terminal leaflet of basal leaves distinctly larger than lateral ones; anthers yellow before dehiscence (a) subsp. **pratensis**

(a) Subsp. **pratensis**: Stem 15–50 cm, simple or branched. Basal leaves with 1–10(–15) pairs of lateral leaflets, the terminal leaflet usually larger than the lateral ones; cauline leaves 2–10(–13), the lower with 2–8(–11) pairs of lateral segments or leaflets. Petals 6–17 mm, white, or pale lilac. Siliqua 15–40 × 0·9–1·6(–1·9) mm. $2n = 16, 30, 32, 38, 44, 48, 56$ (other aneuploid chromosome numbers also occur). *Throughout the range of the species except the extreme north and parts of the south-east.*

(b) Subsp. **granulosa** (All.) Arcangeli, *Comp. Fl. Ital.* ed. 2, 260 (1894) (*C. granulosa* All.): Ends of some roots tuberous; stem up to 40 cm, simple. Basal leaves simple or with 1 or 2 pairs of leaflets; cauline leaves 1–3, with 1 or 2 pairs of lateral segments. Petals white, 8–12 mm. $2n = 16$. *Wet meadows.* • *N.W. & N. Italy.*

(c) Subsp. **crassifolia** (Pourret) P. Fourn., *Quatre Fl. Fr.* 413 (1936) (*C. crassifolia* Pourret): Rhizome stoloniferous; stem 10–15 cm, ascending. Basal leaves not forming a rosette, simple or ternate; cauline leaves 1–3, with 1–3 pairs of lateral segments. Petals 6–8 mm, purplish. Siliqua 12–16 × 1·2 mm. • *Pyrenees.*

(d) Subsp. **polemonioides** Rouy in Rouy & Fouc., *Fl. Fr.* 234 (1893) (*C. nymanii* Gand.): Plant usually caespitose. Stem 5–20(–35) cm. Basal leaves with 4–6(–10) pairs of lateral leaflets; cauline leaves with 2–4(–7), the lower with 5–7(–10) pairs of lateral segments or leaflets; leaflets thick with impressed veins and often bearing adventitious shoots. Petals 9–13 mm, pale lilac. Siliqua (10–)15–25 × (1·2–)1·4–2·0 mm. $2n = 56, 60, 62, 64, 68, 72, 80, 90$. *N. Europe.*

(e) Subsp. **matthioli** (Moretti) Nyman, *Consp.* 36 (1878) (*C. matthioli* Moretti, *C. pratensis* L. var. *hayneana* Welw.): Plant simple or caespitose. Stem 11–50 cm, usually much-branched. Basal leaves with (2–)5–8 pairs of lateral leaflets, terminal leaflet larger than the lateral ones; cauline leaves (2)3–14(–18), the lower with 2–14 pairs of lateral segments or leaflets. Petals 5–9(–12) mm, usually less than 5·5 mm wide, white, rarely pale lilac. Siliqua (12–)25–34 × 0·5–1·3 mm. 2*n* = 16. *C. & S. Europe, northwards to the Alps and Carpathians.*

(f) Subsp. **rivularis** (Schur) Nyman, *loc. cit.* (1878) (*C. rivularis* Schur): Stem 12–30(–45) cm, usually simple. Basal leaves with 5–11 pairs of lateral leaflets, terminal leaflet usually the same size as the lateral ones; cauline leaves (3)4–12, the lower with (6)7–10 pairs of lateral segments or leaflets. Petals 6·5–10·5 mm, purple. Siliqua 15–25 × 1·5–2(–2·5) mm. 2*n* = 16, 24. ● *Wet mountain meadows in S. Romania and Bulgaria.*

(g) Subsp. **dentata** (Schultes) Čelak., *Prodr. Fl. Böhm.* 450 (1875) (*C. palustris* (Wimmer & Grab.) Peterm., *C. dentata* Schultes): Stem (20–)30–50 cm. Basal leaves with 3–12 pairs of lateral leaflets; cauline leaves 5–12, the lower with 3–10 pairs of lateral leaflets; leaflets of cauline leaves often deciduous. Petals (9–)12–16(–19) mm, whitish. Siliqua (20–)30–50 × 1·1–1·5(–2) mm. 2*n* = 56, 64, 72, 76, 80, 84, *c*. 96 (other aneuploid numbers occur through hybridization). *Wet places. N. Europe, extending locally southwards to N. Bulgaria.*

Diploid and tetraploid (2*n* = 16, 32) plants of subsp. (**a**), occurring in the E. Carpathians, Alps, Appennines, Massif Central, Vosges and perhaps Pyrenees, resemble subsp. (**f**) but are clearly distinct in the shape of hairs on the basal leaves and the colour of anthers. Previously treated as *C. rivularis* Schur (*C. crassifolia* auct., non Pourret), they require further study.

Plants from E. Czechoslovakia, E. Hungary, W. Ukraine and Romania that are like subsp. (**e**) but have larger flowers, the petals (8–)9–15(–16·5) mm, usually more than 5·5 mm wide, larger pollen grains, and a chromosome number of 2*n* = 32, have been called **C. majovskii** Marhold & Záborský, *Preslia* **58**: 194 (1986).

Plants from Belgium and N. & E. France, like subsp. (**a**) but usually taller, with petals 14–17 mm, have been described as **C. pratensis** L. subsp. **picra** De Langhe & D'hose, *Gorteria* **8**: 48 (1976). These plants are diploid (2*n* = 16) and, in common with the apparently distinct diploid central European **C. nemorosa** Lej., *Fl. Spa* 62 (1813), require further study.

16. C. uliginosa Bieb., *Fl. Taur.-Cauc.* **3**: 438 (1819). Perennial with erect, leafy stems up to 50 cm; rhizome long, without stolons. Basal leaves not forming a distinct rosette; leaves pinnate with 4–9 pairs of oblong to narrowly elliptical leaflets, the terminal leaf slightly larger (especially that of the basal leaves), glabrous to pubescent, rather fleshy and glaucous. Petals 8–11 mm, white or rarely pale yellowish, sometimes with purplish claw. Siliqua 20–35 × 1–1·5 mm, rather strict. *Damp, marshy ground. Turkey-in-Europe (N. of Istanbul).* *Tu. (*S.W. Asia.*)

17. C. tenera J. G. Gmelin ex C. A. Meyer, *Verz. Pfl. Cauc.* 179 (1831). Almost glabrous perennial 25–40 cm, with short, slender rhizome; long stolons from the axils of the lowest leaves. Basal leaves 2–4 cm wide, simple, reniform; lower cauline leaves few, with 1–3(4) remote pairs of narrowly ovate or oblong, sessile lateral leaflets and a distinctly larger terminal leaflet; uppermost cauline leaf simple. Pedicels 12–15 mm, patent to recurved at anthesis, patent in fruit. Petals 8–14 mm, narrowly cuneate, white, dark pink or violet. Siliqua 30–40 × 1·2 mm; style 0·5–1 mm. 2*n* = 16. *Woods. Krym.* Rs(K).

18. C. plumieri Vill., *Prosp. Pl. Dauph.* 38 (1779). Biennial or perennial 5–12(–20) cm; almost glabrous but not glaucous. Lowest leaves hederiform with 3–5 obtuse lobes; intermediate cauline leaves ternate with stalked, obovate lateral leaflets and a long terminal leaflet; upper cauline leaves pinnate with ternate, obtuse leaflets. Petals 6–8 mm, rounded or slightly emarginate at the apex, white. Siliqua 18–25 × (0·5–)1–1·5 mm, erect; style 1–2 mm. *Shady rocks and screes.* ● *S. Europe, from S.E. France to N. Greece.* Al Co Ga Gr It.

19. C. glauca Sprengel in DC., *Reg. Veg. Syst. Nat.* **2**: 266 (1821). Annual to perennial (5–)12–25 cm; entirely glabrous and glaucous. Lowest leaves entire to 3-lobed; intermediate leaves pinnate with 1–3 pairs of ovate-elliptical, obtuse, equal, sessile lateral leaflets, and an obovate, entire or 3-lobed, subsessile terminal leaflet; upper leaves smaller, imparipinnate with 2–5 pairs of oblong to linear, entire and acute leaflets. Petals 4–8 mm, deeply emarginate, white. Siliqua (10–)20–35 × (0·8–)1·5–1·8 mm, secund; style 1–2 mm. 2*n* = 16. *S. Europe, from the S. Carpathians to N. Greece and Sicilia.* Al Bu Gr It Ju Rm Si.

Variable; some alpine plants are dwarf and smaller in all their parts, with leaves in perennial rosettes and strict infructescences.

C. pancicii Hayek, *Denkschr. Akad. Wiss. Math.-Nat. Kl. (Wien)* **94**: 149 (1918), from S. Italy, Albania and Jugoslavia, with 4–8 pairs of leaf-segments and petals 4–5 mm, may represent a subspecies of **19**.

20. C. resedifolia L., *Sp. Pl.* 656 (1753) (*C. gelida* Schott). Perennial (2–)5–15(–23) cm; glabrous. Leaves thick, usually trifoliate, auriculate at the base of the petiole, the lowest spathulate; intermediate leaves trifid or ternate with ovate-lanceolate lateral lobes and an ovate, obtuse terminal lobe; upper 3- to 7-lobed. Petals 5–6 mm, entire, white. Siliqua 12–22 × 1–1·4 mm, strictly erect; style absent. Seeds broadly winged. 2*n* = 16. *Screes; usually calcifuge. Mountains of C. & S. Europe.* Au Bu Co Cz Ga Ge He Hs It Ju Po Rm.

21. C. bellidifolia L., *Sp. Pl.* 654 (1753). Glabrous perennial 1–8(–11) cm. Leaves all similar in shape, thick, entire (occasionally 2-lobed); rosette-leaves long-spathulate, the cauline sometimes absent. Petals 3·5–5 mm, white. Siliqua 10–25 × 1–1·5 mm, strictly erect; style 0·5–1·5 mm. Seeds wingless or narrowly winged. *Arctic and subarctic Europe; Alps; Pyrenees.* Au Fe Ga Ge He Hs Is It No Rs(N) Sb Su.

(**a**) Subsp. **bellidifolia**: Cauline leaves 0 or 1(–3), on petioles longer than the lamina. Flowers 2–8. Siliqua 10–25 × 1–1·5 mm, purplish-brown. Seeds 1·5 × 1 mm, narrowly winged. 2*n* = 16. *Arctic and subarctic Europe.*

(**b**) Subsp. **alpina** (Willd.) B. M. G. Jones, *Feddes Repert.* **69**: 59 (1964) (*C. alpina* Willd.): Cauline leaves (1)2 or 3(4), subsessile or on petioles less than ½ as long as the lamina. Flowers 3–8. Siliqua 10–15 × *c*. 1 mm, brown. Seeds 1·2 × 0·7 mm, unwinged. 2*n* = 16. *Alps and Pyrenees.*

22. C. maritima Portenschl. ex DC., *Reg. Veg. Syst. Nat.* **2**: 266 (1821). Annual or biennial 12–30 cm, subglabrous to pubescent. Leaves glabrous, pinnate with cuneate-obovate, subsessile leaflets, frequently lobed or incised, auriculate at the base of the petiole. Sepals glabrous, apiculate; petals (4–)7–11 mm, white. Siliqua 20–40 × 2–3 mm, glabrous, patent; style (3–)5–7(–10) mm. Seeds on a narrow funicle. ● *N. & W. Jugoslavia.* Ju.

C. fialae Fritsch, *Österr. Bot. Zeitschr.* **47**: 44 (1897), from Jugoslavia (Hercegovina), with hairy stems, leaves, pedicels and sepals, leaflets deeply lobed with obtuse segments, petals 10–12 mm, siliqua 40–55 × 1·4–1·8 mm, and beak 8–15 mm, is perhaps a subspecies of **22**.

23. C. carnosa Waldst. & Kit., *Pl. Rar. Hung.* **2**: 137 (1803). Perennial 20–30 cm, puberulent above. Leaves 4–6, thick, with appressed hairs, pinnate with 3–7 pairs of entire, obovate-cuneate leaflets. Sepals apiculate; petals 6–9 mm, white with indistinct yellow claw. Siliqua 20–30(–40) × 2–3 mm, strigose; style 1–3 mm. Seeds 2·5–3 mm, on a broad funicle. *Mountain screes.* ● *Balkan peninsula, mainly in the west.* Al Gr Ju.

24. C. monteluccii Br.-Catt. & Gubell., *Webbia* **39**: 398 (1986). Erect annual or sometimes biennial 10–20(–30) cm, glabrous or subglabrous. Leaves 2- or 3-pinnatipartite, with 2–4 pairs of lateral leaflets; petioles not auriculate at the base. Petals 10–12 mm, obcordate, emarginate. Siliqua (25–)35–50 × *c.* 4 mm, somewhat secund; style 5–8 mm. Seeds on a broad funicle. $2n = 16$. ● *C. & S. Italy; Sicilia (Madonie).* It Si.

25. C. graeca L., *Sp. Pl.* 655 (1753). Annual or biennial 10–30 cm, glabrous or hairy. Leaves pinnate; lower leaves with 4 or 5 pairs of cuneate-obovate leaflets, each with 3 or more obtuse lobes; upper cauline leaves with 2 or 3 pairs of 3-lobed or entire, lanceolate leaflets; petioles auriculate at the base. Sepals obtuse; petals (3–)4–6 mm, white. Siliqua 35–50 × (2·5–)3–4 mm, glabrous, papillose or setose, somewhat secund; style 4–8 mm, winged. Seeds on a broad funicle. $2n = 16, 18$. *S. Europe from Corse eastwards.* Al Bu Co Cr Gr It Ju Rm Rs(K) Si Tu.

26. C. chelidonia L., *loc. cit.* (1753). Annual to perennial 20–50 cm, almost glabrous. Leaves pinnate with 1–3 pairs of lanceolate to ovate-lanceolate, incised or lobed leaflets 2·5–3 cm wide; petioles auriculate at the base. Petals 5–9 mm, pink or violet, caducous. Siliqua 30–40 × 1·5 mm, setose; style 2–3 mm. Seeds on a broad funicle. *Woods.* ● *C. part of Mediterranean region.* Co It Ju Si.

27. C. parviflora L., *Syst. Nat.* ed. 10, **2**: 1131 (1759). Annual 7–30(–40) cm, glabrous. Leaves pinnate, the lower with 3–5 pairs, the upper with 5–8 pairs of linear to linear-oblong, cuneate, entire leaflets. Petals 1·8–2·5 mm, obovate, white. Siliqua (8–)12–20 × 0·8 mm, erect on patent pedicels. Seeds 0·7–0·8 × 0·5 mm, narrowly winged. $2n = 16$. *Usually in seasonally flooded places. Widespread in Europe, but local; absent from the islands and the Balkan peninsula.* Au Bu Cz Fe Ga Ge Hs Hu It Lu Po Rm Rs(N, B, C, W, ?K, E) Su.

28. C. impatiens L., *Sp. Pl.* 655 (1753). Biennial up to 80 cm, more or less glabrous. Rosette-leaves not persistent; lower cauline leaves petiolate, with 3–5 pairs of ovate, 2- to 5-lobed leaflets; upper leaves sessile with (3–)5–11 pairs of dentate, lateral leaflets and a somewhat larger terminal leaflet; leaves auriculate at the base, ciliate, otherwise glabrous. Petals narrow, white, usually scarcely exceeding the sepals, sometimes absent; anthers greenish. Siliqua 18–30 mm. Seeds 1·1–1·4 × 0·8–0·9 mm, unwinged. $2n = 16$. *Woods. Widespread in Europe, but absent from some areas, especially many of the islands.* Al Au Be Br Bu Co Cz Da Fe Ga Ge Gr Hb He Ho Hs Hu It Ju No Po Rm Rs(N, B, C, W, K, E) Su Tu.

(a) Subsp. **impatiens**: Stem 25–60 cm. Upper cauline leaves with 5–9 pairs of lanceolate lateral leaflets. Petals 2–3 mm. Siliqua 1–1·2 mm wide, erecto-patent. $2n = 16$. *Throughout the range of the species except parts of the south-east.*

(b) Subsp. **pectinata** (Pallas ex DC.) Stoj. & Stefanov, *Fl. Bălg.* ed. 3, 509 (1948) (*C. pectinata* Pallas ex DC.): Stem 15–30 cm. Upper cauline leaves with 3 or 4 pairs of ovate-lanceolate lateral leaflets. Petals 3·5–4·5 mm. Siliqua 0·8–1·2 mm wide, almost horizontal. *Bulgaria, N. and C. Greece.*

29. C. caldeirarum Guthnick in Seub., *Fl. Azor.* 43 (1844). Biennial or perennial 15–50 cm, sparsely hairy. Lowest leaves in a rosette, pinnate with 5 or 6 pairs of obliquely ovate lateral leaflets, and an ovate, truncate, distinctly stalked terminal leaflet; cauline leaves 3–6, similar to the lowest leaves but with 2–4 pairs of leaflets; leaflets coarsely crenate-dentate with apiculate teeth. Pedicels 4–8 mm at anthesis. Petals 4–7 mm, white, greatly exceeding the sepals. Siliqua 20–40 × 1–1·2 mm, on patent pedicels; style 1–2 mm, cylindrical. Seeds 1·2 × 0·8 mm. *Damp, shady places.* ● *Açores.* Az.

30. C. flexuosa With. in Stokes, *Arr. Br. Pl.* ed. 3, **3**: 578 (1796) (*C. sylvatica* Link). Biennial to perennial, occasionally annual, 10–50 cm; stem flexuous, hairy, especially at the base. Lowest leaves pinnate, with 2–7(8) pairs of ovate to reniform lateral leaflets and a somewhat larger terminal leaflet; cauline leaves 4–10, larger than the basal, with 2–6(7) pairs of ovate-lanceolate leaflets; leaflets dentate to entire. Pedicels 2–3 mm at anthesis. Petals 2·5–4(–5) mm, about twice as long as the sepals; stamens 6 (sometimes fewer). Siliqua (8–)12–25 × 1–1·5 mm, erecto-patent, scarcely overtopping the flowers; style 1–1·5 mm, conical. Seeds 1–1·4 × 0·8–0·9 mm. $2n = 32$, *c.* 50. *Damp, usually shady places. Europe, from about* 25° *E. westwards.* All except Az Bl Cr Gr Is Rs(N, C, K, E) Sb Tu.

31. C. hirsuta L., *Sp. Pl.* 655 (1753) (*C. umbrosa* DC., non Lej., *C. multicaulis* Hoppe). Annual 3–30 cm; stem usually glabrous. Lowest leaves in a distinct rosette, pinnate, with 1–6(7) pairs of obovate to orbicular lateral leaflets and a larger, reniform terminal leaflet; cauline leaves 0–4(5), smaller than the basal leaves, with 2–5 pairs of smaller leaflets; leaflets angular, hairy on the upper surface. Pedicels 1–2 mm at anthesis. Petals 2·5–4(–5) mm or absent, about twice as long as the sepals; stamens 4 or 5(6). Siliqua 10–25 × 0·8–1·4 mm, erect, overtopping the flowers; style 0·2–1 mm, conical. Seeds 0·7–1·2 × 0·7–0·8 mm. $2n = 16$. *Throughout Europe except the Arctic and* N. *Russia.* All except Rs(N) Sb.

41. Cardaminopsis (C. A. Meyer) Hayek

B. M. G. JONES (EDITION 1) REVISED BY J. R. AKEROYD (EDITION 2)

Annuals to perennials; glabrous or with branched and unbranched hairs. Leaves simple to pinnatisect. Inner sepals slightly saccate; petals white, pinkish or purplish. Fruit a strongly compressed siliqua; valves more or less torulose, with a distinct median vein; style short; stigma capitate. Seeds in 1 row in each loculus.

Several of the species in this genus are known to be interfertile.

1 Basal leaves orbicular, or pinnate with an orbicular terminal leaflet **5. halleri**
1 Basal leaves lanceolate to obovate, subentire to pinnatisect
 2 Cauline and basal leaves similar in shape and size (or the cauline somewhat larger) **4. croatica**
 2 Cauline and basal leaves dissimilar in shape and size
  3 Basal leaves shallowly pinnatifid to serrate, or almost entire; cauline leaves entire **2. petraea**
  3 Basal leaves pinnatisect; cauline leaves pinnatifid or distinctly serrate-dentate

4 Stem hairy; pedicels 3–5 mm at anthesis **1. arenosa**
4 Stem glabrous or subglabrous; pedicels 5–8 mm at anthesis **3. neglecta**

1. C. arenosa (L.) Hayek, *Fl. Steierm.* **1**: 478 (1908) (*C. suecica* sensu Lawalrée, *Arabidopsis arenosa* (L.) Lawalrée, *Arabis arenosa* (L.) Scop.). Annual to perennial, without stolons. Stem 5–80 cm, erect, robust, usually branched above and below, hairy. Basal leaves pinnatisect; cauline leaves lanceolate, pinnatifid to dentate. Flowers numerous; pedicels 3–5 mm at anthesis. Petals 6–8 mm, with a pair of small teeth on the claw. Siliqua 10–45 × 1 mm; pedicels 5–13 mm, patent. $2n = 16+0\text{–}4B$, 18–19, 28, 30–32, $32+0\text{–}1B$, 34, 39–40, 42. *Most of Europe except the south-west, the islands and part of the Mediterranean region.* Au Be Bu Cz Da Fe Ga Ge He Ho Hu It Ju No Po Rm Rs(N, B, C, W, E) Su.

Sometimes confused with *Arabidopsis suecica* (Fries) Norrlin, which has the basal leaves dentate to pinnatifid, petals without teeth on the claw, and the siliqua not strongly dorsally compressed.

(a) Subsp. **arenosa**: Often annual or biennial. Basal leaves with terminal segment distinctly larger than the lateral segments; lateral segments 1–6 pairs. Corolla white or pale lilac. Seeds very narrowly winged or wingless. *N. Europe; naturalized as a ruderal further south.*

(b) Subsp. **borbasii** (Zapał.) Pawł., *Fl. Tatr.* **1**: 339 (1956): Usually perennial. Basal leaves with terminal segment scarcely larger than the lateral segments; lateral segments 4–9 pairs. Corolla bright lilac. Seeds conspicuously winged. ● *C. Europe, extending northwards to Denmark.*

Further subspecies can perhaps be recognized in this polymorphic species. For an account of the cytological variation see J. Měsíček, *Preslia* **42**: 225–248 (1970).

2. C. petraea (L.) Hiitonen in Hyl., *Fört. Skand. Växt.* ed. 3, 62 (1941) (*Arabis septentrionalis* N. Busch, *Arabidopsis petraea* (L.) Lam.; incl. *C. hispida* (L.) Hayek). Perennial, stoloniferous; stem 5–30(–45) cm, sparsely branched, glabrous, or hairy below with simple and forked hairs. Basal leaves long-petiolate, pinnatifid to serrate, sparsely hairy or glabrous; cauline leaves few, lanceolate, entire or toothed. Flowers few; pedicels 2–6 mm at anthesis. Petals 3–7(–9) mm, white or purplish. Siliqua 10–45 × 1–1·5(–2) mm; pedicels 4–15 mm, patent or ascending. $2n = 16$, 32. *Rocks and gravelly ground in the mountains, but sometimes at sea-level. N. & C. Europe.* Au Br Cz Fa Ge Hb Hu Is It No Po ?Rm Rs(N, W) Su.

Plants from C. Europe are sometimes regarded as a separate species, **C. hispida** (L.) Hayek, *Fl. Steierm.* **1**: 478 (1908), and those from N.E. Russia are distinguished as **C. septentrionalis** (N. Busch) O. E. Schulz in Engler & Prantl, *Natürl. Pflanzenfam.* ed. 2, **17b**: 541 (1936), but the characters separating them from **2** are by no means clear.

3. C. neglecta (Schultes) Hayek, *Fl. Steierm.* **1**: 480 (1908). Perennial; stem 5–20 cm, sparsely branched, glabrous. Basal leaves petiolate, lyrate-pinnatifid, with few lobes or almost entire; cauline leaves ovate, entire or with 1 or 2 basal lobes; all leaves thick, coriaceous, glabrous or sparsely hairy. Flowers few; pedicels 5–8 mm at anthesis. Petals 5–6 mm, purple, somewhat cuneate. Siliqua 15–25 × *c.* 1·5 mm; pedicels 8–12 mm, ultimately deflexed. $2n = 16$, 32. ● *Carpathians.* Cz Po Rm Rs(W).

4. C. croatica (Schott, Nyman & Kotschy) Jáv., *Magyar Fl.* 435 (1924). Perennial; stem 8–18 cm, erect, flexuous, divaricately branched, glabrous. Leaves all long-petiolate, obovate, denticulate to entire (lowest sometimes lyrate-pinnatifid), glabrous or sparsely hairy, glaucous and fleshy. Flowers few; pedicels 4–8 mm at anthesis. Petals 4–6 mm, pinkish or purple. Siliqua 18–50 × 1 mm, widely patent; pedicels 7–10 mm in fruit, deflexed or horizontal. *Rocks.* ● *N.W. Jugoslavia (Velebit).* Ju.

5. C. halleri (L.) Hayek, *Fl. Steierm.* **1**: 479 (1908) (*Arabidopsis halleri* L.). Perennial, stoloniferous. Stem 10–50 cm, ascending, hairy or glabrous. Basal leaves long-petiolate, simple or pinnate, with 1–7 orbicular lateral leaflets, the terminal larger; cauline leaves shortly petiolate, oblong to ovate. Flowers rather numerous. Petals 4–6 mm, white or lilac. Siliqua 10–25 × 1 mm, patent; pedicels slender. $2n = 16$. *Mountains of C. Europe, extending to the S.W. Alps and Crna Gora.* Au Cz Ga Ge He It Ju Po Rm Rs(W) [Be].

(a) Subsp. **halleri**: Stolons few, short; flowering stems 20–40 cm, numerous, branched at the base and frequently also above. Basal leaves usually pinnate, the lower cauline oblong, dentate. Petals white or lilac. Pedicels 5–10 mm in fruit. *Throughout the range of the species.*

(b) Subsp. **ovirensis** (Wulfen) Hegi & E. Schmid in Hegi, *Ill. Fl. Mitteleur.* **4(1)**: 424 (1919) (*C. ovirensis* (Wulfen) O. Schwarz): Stolons numerous, slender and elongate; flowering stems 10–20 cm, few, only sparsely branched above. Basal leaves simple or pinnate with weakly developed lateral leaflets; lower cauline leaflets orbicular to ovate, entire or crenate. Petals lilac or purplish. Pedicels 10–14 mm in fruit. *S.E. Alps; Carpathians; N. part of Balkan peninsula.*

Intermediates between subspp. (a) and (b) occur in the Tatra.

42. Arabis L.

B. M. G. JONES (EDITION 1) REVISED BY J. R. AKEROYD (EDITION 2)

Annuals or perennials; hairs simple, dendritic or stellate; rarely the plant glabrous. Leaves simple. Inner sepals often slightly saccate at the base; petals usually white, pink or purple. Fruit a siliqua; valves flat, sometimes with a median vein; style distinct; stigma capitate or emarginate. Seeds usually in 1 row in each loculus, usually winged. (Incl. *Turritis* L.)

1 Cauline leaves absent
2 Plant conspicuously pubescent; petals 6–7 mm; style not more than 1 mm **25. bryoides**
2 Plant subglabrous or glabrous; petals 3–4 mm; style 1·5–2 mm **26. longistyla**
1 At least 1 cauline leaf present
3 Basal leaves pubescent; all the cauline leaves glaucous and usually glabrous
4 Cauline leaves sagittate; petals cream; siliqua usually not more than 70 mm **1. glabra**
4 Cauline leaves cordate at the base; petals white or lilac; siliqua at least 70 mm **2. laxa**
3 Basal and at least the lower cauline leaves with a similar indumentum, or glabrous
5 Siliqua usually more than 9 cm, distinctly arcuate at maturity **15. turrita**
5 Siliqua up to 9 cm, straight or scarcely arcuate at maturity
6 All leaves (except the uppermost cauline leaves) long-petiolate, cuneate or truncate (mountains of S. France and N.W. Italy)
7 Plant usually pubescent; basal leaves longer than wide, acute; petals 7–10 mm, violet or white **33. cebennensis**

7 Plant usually glabrous; basal leaves as wide as long, obtuse; petals 6–7 mm, white **34. pedemontana**
6 At least the upper cauline leaves sessile or shortly attenuate-petiolate
8 Infructescence conspicuously unilateral; siliqua pendent or deflexed **16. pendula**
8 Infructescence not conspicuously unilateral; siliqua erect to patent
9 Annual, the lower leaves withered by anthesis or shortly after flowering
10 Cauline leaves attenuate at the base; siliqua always puberulent **19. parvula**
10 Cauline leaves cordate or auriculate-amplexicaul at the base; siliqua often glabrous
11 Petals pale violet; siliqua 1·5–2 mm wide; pedicels short, thick **20. verna**
11 Petals white; siliqua 1·5 mm wide or less; pedicels slender
12 Petals 2–4 mm; siliqua less than 1 mm wide; pedicels up to 5 mm **17. recta**
12 Petals 4–6 mm; siliqua more than 1 mm wide; pedicels more than 5 mm **18. nova**
9 Biennial, the lower leaves present at anthesis but withering during the ripening of the fruit; or perennial with vegetative leaf-rosettes
13 Leaves entire, with medifixed hairs
14 Hairs covering both sides of the leaves; cauline leaves lanceolate **23. ferdinandi-coburgi**
14 Hairs restricted to the margin and veins of the lower surface; cauline leaves ovate to oblong
15 Basal leaves acuminate; cauline leaves ovate, subcordate at the base **21. procurrens**
15 Basal leaves obtuse; cauline leaves oblong, attenuate-petiolate **22. vochinensis**
13 Leaves dentate to entire, pubescent to glabrous, but never with medifixed hairs
16 Leaves apiculate; plant glabrous, except for unbranched, setiform hairs on the leaf-margins; siliqua 3–4 times as long as wide **24. scopoliana**
16 Leaves acute to obtuse; plant pubescent, or if glabrous, the leaf-margin with some branched hairs; siliqua many times as long as wide
17 Cauline leaves cordate, auriculate or sagittate
18 Flowers yellowish; siliquae patent on erect pedicels **28. subflava**
18 Flowers white or pink; siliquae not patent on erect pedicels
19 Pedicels erect in fruit; siliquae usually closely appressed to the axis of the infructescence **(4–10). hirsuta** group
19 Pedicels patent in fruit; siliquae not appressed
20 Plant entirely glabrous and glaucous **3. brassica**
20 Plant pubescent
21 Biennial without vegetative leaf-rosettes; petals 4–6 mm **18. nova**
21 Perennial with vegetative leaf-rosettes; petals 6–18 mm **32. alpina**
17 Cauline leaves rounded at the base to attenuate-petiolate
22 Cauline leaves 1–4; flowers 3–10(–12)
23 Stem pilose; petals pink **13. cretica**
23 Stem glabrous or pubescent, not pilose; petals white, pale blue or yellowish, rarely pink
24 Leaves sinuate-dentate, coriaceous; hairs mostly unbranched **27. scabra**
24 Leaves dentate towards the apex or entire, soft or brittle; hairs mostly branched
25 Basal leaves with 1 or 2 obscure apical teeth, or ±entire; petals white or pink; siliqua up to 2·2 mm wide, green **29. caerulea**
25 Basal leaves with 1 or 2 obscure apical teeth, or ±entire; petals white or pink; siliqua up to 2·2 mm wide, green
26 Lowermost cauline leaf rounded at the base; infructescence compact; siliqua at least 1·5 mm wide **30. pumila**
26 Lowermost cauline leaf attenuate at the base; infructescence lax; siliqua up to 1·2 mm wide **12. serpillifolia**
22 Cauline leaves 4 or more; flowers usually more than 10
27 Plant glabrous, dark green and shining, rarely leaves somewhat hairy on the surface or margin; siliqua 25–50 × 1·8–2·2 mm; seeds broadly winged all round **31. soyeri**
27 Plant pubescent or, if glabrous, dull green; siliqua more than 50 mm long or less than 1·8 mm wide; seed unwinged or narrowly winged on edges or at apex
28 Pedicels erecto-patent to patent in fruit; seeds unwinged
29 Hairs all or mostly unbranched or bifid
30 Siliqua not more than 30 mm; flowers *c.* 30 **11. ciliata**
30 Siliqua usually more than 30 mm; flowers 3–12 **13. cretica**
29 Hairs mostly stellate **12. serpillifolia**
28 Pedicels erect in fruit; seed winged at least at the apex
31 Apex of leaves acute or obtuse; flowers usually more than 20; petals 4–7(–8) mm **(4–10). hirsuta** group
31 Apex of leaves rounded; flowers not more than 20; petals 6–10 mm **14. collina**

1. A. glabra (L.) Bernh., *Syst. Verz. Erfurt* **1**: 195 (1800) (*A. perfoliata* Lam., *Turritis glabra* L.; incl. *A. pseudoturritis* Boiss. & Heldr.). Biennial to perennial 60–120 cm, unbranched and stiffly erect, pubescent below. Basal leaves oblong to lanceolate, sinuate-dentate to lyrate, with stellate hairs; cauline leaves ovate-lanceolate, sagittate, glabrous (or the lower sparsely pubescent) and glaucous. Petals 4–8 mm, cream or whitish. Infructescence strict; siliqua 40–80 × 1–1·5(–1·7) mm, somewhat compressed, with prominent median veins; pedicels 6–25 mm, erect. Seeds usually in 2 rows in each loculus. $2n = 12, 16, 32$. *Most of Europe except the extreme north and south.* Al Au Br Bu ?Co Cz Da Fe Ga Ge Gr He Ho Hs Hu It Ju Lu No Po Rm Rs(N, B, C, W, K, E) Si Su Tu.

2. A. laxa Sm. in Sibth. & Sm., *Fl. Graec. Prodr.* **2**: 28 (1813) (*A. doefleri* Halácsy, *Turritis laxa* (Sm.) Hayek). Like **1** but 50–80 cm, often branched above; cauline leaves oblong, cordate, often sparsely ciliate; petals *c.* 4 mm, white or lilac; siliqua 70–120 × 1·5–2 mm, compressed-tetragonous, patent or unilaterally deflexed; seeds in 1 or 2 rows in each loculus. *Greece and S. Jugoslavia.* Gr Ju. (*S.W. Asia.*)

3. A. brassica (Leers) Rauschert, *Feddes Repert.* **83**: 648 (1973) (*A. brassicaeformis* Wallr., *A. pauciflora* (Grimm) Garcke, *Turritis brassica* Leers, *Fourraea alpina* (L.) Greuter & Burdet). Perennial 30–100 cm, glabrous and glaucous. Basal leaves ovate, entire, long-petiolate; cauline leaves sessile, panduriform; upper lanceolate, auriculate and amplexicaul. Petals 4–7 mm, white or pink. Siliqua 30–80 × 1·5–2 mm, somewhat compressed, erect; pedicels patent. *S.W., S. & S.C. Europe, eastwards to Jugoslavia.* Au Be Co Cz Ga Ge He Hs It Ju.

(**4–10**). **A. hirsuta** group. Plant stiffly erect with stem tapering upwards. Basal leaves attenuate-petiolate, entire to dentate; cauline leaves ovate to linear-lanceolate, subentire to serrate-dentate. Flowers (15–)25–60; petals 4–6·5(–8) mm, white. Infructescence compact; siliquae 18–60(–70) × 0·6–1·9 mm, erect, not exceeding the open flowers; pedicels erect to patent; valves flat to somewhat torulose; style obconical or stigma sessile. Seeds winged at least at the apex.

1 Biennial; cauline leaves sagittate at the base
2 Auricles spreading; base of stem with numerous unbranched hairs and few bi- or trifid hairs **5. sagittata**
2 Auricles amplexicaul; base of stem with numerous bi- or trifid hairs and few or no unbranched hairs
3 Pedicels not more than 7 mm in fruit; seeds winged at the apex **4. planisiliqua**
3 Pedicels up to 20 mm in fruit; seeds winged all round **6. borealis**
1 Usually perennial; cauline leaves cordate or rounded at the base
4 Siliqua at least 40 mm
5 Stem appressed-hairy below; siliqua 40–55 mm **8. sadina**
5 Stem glabrous; siliqua 60–70 mm **10. juressi**
4 Siliqua 15–40 mm
6 Petals not more than 6 mm; pedicels 3–8 mm in fruit **7. hirsuta**
6 Petals at least 6 mm; pedicels 6–12 mm in fruit **9. allionii**

4. A. planisiliqua (Pers.) Reichenb., *Icon. Fl. Germ.* **2**: 13 (1847) (*A. gerardii* Besser ex Koch, *A. lusitanica* Boiss.). Biennial (30–)50–80 cm, branched above, frequently reddish; lower part of the stem with small, almost sessile bi- or trifid hairs whose long branches are appressed; upper part glabrous. Cauline leaves 20–55, sagittate with appressed auricles; at least the lower leaves with bi- or trifid hairs whose branches are shorter than their stalks. Petals 4–5 mm. Siliqua 30–50 × 0·6–0·9 mm, torulose; pedicels 3–7 mm, erect; valves with weak median vein or almost veinless; style *c.* 1 mm, cylindrical. Seeds winged at the apex. $2n = 16$. *Calcareous fens. Most of Europe except Scandinavia and the islands.* Au Co Cz Ga Ge He †Ho Hs Hu It Ju Po Rm Rs(N, B, C, W, K, E) Sa [Su].

5. A. sagittata (Bertol.) DC. in Lam. & DC., *Fl. Fr.* ed. 3, **5**: 592 (1815). Biennial 35–80 cm, branched above; lower part of the stem with mostly patent unbranched hairs, the upper part glabrous. Cauline leaves 15–30, cordate to sagittate with spreading auricles, pubescent with simple or bifid hairs. Petals 5–6·5 mm. Siliqua 25–50 × 0·8–1·1 mm, erect; pedicels 4–6 mm; valves with a distinct median vein; style *c.* 0·5 mm, obconical, or stigma sessile. Seeds narrowly winged, at least at the apex. $2n = 12, 16$. *Dry, calcareous slopes. C., S. & E. Europe.* Al Au ?Be Co Cz Ga Ge Gr He Ho Hu It Ju Rm Rs(N, B, C, K, E) Sa Tu.

The hybrid between this species and **7** is sterile.

6. A. borealis Andrz. ex Ledeb., *Fl. Altaica* **3**: 25 (1831). Like **5** but the lower part of the stem with numerous bi- or trifid and a few unbranched hairs; leaves and upper part of stem pubescent with bi- to 5-fid hairs; auricles amplexicaul; siliqua 40–65 × 0·5–1·5 mm; pedicels 5–20 mm; style 0·5–0·8 mm; seeds distinctly winged all round. *Dry, calcareous slopes. N. Russia.* Rs(N).

7. A. hirsuta (L.) Scop., *Fl. Carn.* ed. 2, **2**: 30 (1772) (*A. ciliata* R. Br.). Short-lived perennial, rarely biennial, (4–)10–60(–110) cm, often with several flowering stems and branched above; lower part of stem with mostly patent, unbranched or bifid hairs, sometimes glabrous, the upper part usually with bi- to 5-fid hairs. Cauline leaves 10–22(–30), subcordate, rounded or truncate at the base. Petals 4–5·5 mm. Siliqua 15–40 × 1–1·9 mm; pedicels 3–8 mm, erect; valves with distinct median vein; style *c.* 0·5 mm, conical, or stigma sessile. Seeds 0·8–1·3(–1·5) mm, usually winged all round. $2n = 32$. *Calcicole. Most of Europe.* All except Az Cr Fa Gr Is Sb.

A very variable species with many local taxa described. Plants from the Carpathians and the N. part of the Balkan peninsula have been distinguished as **A. hornungiana** Schur, *Enum. Pl. Transs.* 43 (1866). They have a shorter stem, narrower cauline leaves and a broader siliqua.

Glabrous plants from sand-dunes in W. Ireland have been distinguished as var. *brownii* (Jordan) Titz (*A. brownii* Jordan), but similar plants occur in other parts of Europe, especially Scandinavia.

8. A. sadina (Samp.) Coutinho, *Fl. Port.* 253 (1913). Like **7** but hairs of lower part of stem bi- to quadrifid, appressed; petals 5–8 mm; siliqua 40–55 mm, erecto-patent. ● *C. Portugal (Estremadura).*

9. A. allionii DC. in Lam. & DC., *Fl. Fr.* ed. 3, **4**: 676 (1805) (*A. glabrata* Wahlenb.; incl. *A. sudetica* Tausch, *A. constricta* Griseb.). Perennial 10–45 cm, usually unbranched; stem glabrous. Basal leaves glabrous or sparsely pubescent with simple and bifid hairs on the margin and veins beneath; cauline leaves *c.* 10, subcordate, glabrous except for the ciliate margin. Petals 6–7 mm. Siliqua 25–35 × 1·2–1·8 mm, erect; pedicels 6–12 mm; valves with distinct median vein. Seeds almost unwinged. $2n = 16$. *Wet alpine rocks. S.W. Alps, Carpathians and mountains of the Balkan peninsula.* Al Au Bu Cz Ga Ge Gr It Ju Po Rs(W).

10. A. juressi Rothm., *Agron. Lusit.* **2**: 79 (1940). Like **9** but siliqua 60–70 × 1·5 mm, rigid, torulose, obtuse; pedicels up to 15 mm. $2n = 32$. ● *N. Portugal (Serra de Gerês).* Lu.

11. A. ciliata Clairv., *Man. Herb. Suisse* 222 (1811) (*A. arcuata* R. J. Shuttlew., *A. corymbiflora* Vest). Biennial or perennial 6–30 cm, stiffly erect, pubescent with mostly unbranched and bifid hairs, or glabrous. Basal leaves shortly attenuate-petiolate, obovate, denticulate, with unbranched or bi- to quadrifid hairs, or glabrous with branched hairs along the margin; cauline leaves 3–9, rounded at the base, ovate to elliptical, entire or denticulate. Flowers *c.* 30; petals 3·5–5 mm, white. Infructescence somewhat compact; siliqua 12–22(–30) × 1–1·3 mm, more or less erect, exceeding the flowers when mature, often arcuate; pedicels 4–7 mm, erecto-patent; valves somewhat rounded on the back, with a distinct median vein; style conical. Seeds unwinged. $2n = 16$. *Alpine rocks, slopes and gravels.* ● *Pyrenees, Appennini and mountains of C. Europe.* Al Au Cz Ga Ge He Hs It Ju Po ?Rm [Da].

12. A. serpillifolia Vill., *Prosp. Pl. Dauph.* 39 (1779) (incl. *A. nivalis* Guss.). Biennial or perennial 5–25 cm, laxly caespitose; stem slender, usually flexuous, glabrous or minutely stellate-pubescent. Basal leaves long-petiolate, oblong, stellate-pubescent or glabrous, with unbranched and bifid hairs on the margin; cauline leaves 3–5, oblong-ovate, entire, the lower attenuate, the upper rounded at the base. Flowers 7–16; petals 4–6 mm, white or rarely pink. Infructescence more or less lax; siliqua 15–30 × 1–1·2 mm, equalling the flowers when immature; valves flat, with indistinct vein. Seeds unwinged. ● *Mountains of S. Europe, extending northwards through the Alps to the French Jura.* Al Ga He Hs It Ju.

(a) Subsp. **serpillifolia**: Stem pubescent at the base. Leaves pubescent. Petals 5–6 mm, white. Siliqua 20–30 mm; pedicels 3–4·5 mm. *Pyrenees, Jura, Alps.*

(b) Subsp. **nivalis** (Guss.) B. M. G. Jones, *Feddes Repert.* **69**: 60 (1964) (*A. nivalis* Guss., *A. surculosa* N. Terracc.). Leaves more or less glabrous with a ciliate margin. Petals 4–5 mm, white or pink. Siliqua 15–25 mm; pedicels 2–3 mm. *C. & S. Appennini and W. part of Balkan peninsula.*

13. A. cretica Boiss. & Heldr. in Boiss., *Diagn. Pl. Or. Nov.* **1(8)**: 20 (1849) (*A. serpillifolia* subsp. *cretica* (Boiss. & Heldr.) B. M. G. Jones). Like **12(a)** but stem pilose; leaves with mostly unbranched hairs; flowers 3–12; petals 5–7 mm, pink; siliqua 35–40 × *c.* 1·2 mm. $2n = 16$. ● *Kriti.* Cr.

14. A. collina Ten., *Prodr. Fl. Nap.* xxxix (1811) (*A. muralis* Bertol., *A. rosea* DC.). Perennial 10–30 cm, pale green, often branched; stem pubescent below, glabrous above or with tri- to 6-fid, stalked hairs. Basal leaves attenuate-petiolate, obovate, obtusely and deeply dentate, rounded at the apex, with branched hairs (sometimes setose on the margin); cauline leaves 6–14, rounded at the base, oblong to ovate, the uppermost lanceolate, scarcely broader than the stem. Flowers 8–18; petals 6–10 × 2–4 mm, erect, cuneate to spathulate, white to pink or purplish. Infructescence strict; siliqua 30–90 × 1·2–2·2 mm; pedicels 4–12 mm; valves with an indistinct median vein; style 0·5–2 mm. $2n = 16$. ● *S. & S.C. Europe.* Al Bl Bu Co Ga Gr He Hs It Ju Sa Si [Be].

Very variable and divided by some authors into three distinct species. Intermediates, however, are common. Further study may permit the recognition of subspecies.

15. A. turrita L., *Sp. Pl.* 665 (1753). Biennial or perennial 20–80 cm, pubescent. Basal leaves long-petiolate, obovate, regularly sinuate or denticulate; cauline leaves sessile, with a cordate, amplexicaul base, panduriform below to oblong above, the apex obtuse. Petals 6–9 mm, pale yellow. Infructescence elongate, unilateral; siliqua 80–140 × 2–2·7 mm, with thickened margins, at first erect, arcuate when ripe; pedicels erect. $2n = 16$. *C. & S. Europe.* Al Au Be Bu Co Cz Ga Ge Gr He Hs Hu It Ju Rm Rs(C, W, K) Si [Br].

16. A. pendula L., *Sp. Pl.* 665 (1753). Hispid biennial 50–90 cm. Lower leaves long-petiolate, rhombic, the upper sessile, auriculate and amplexicaul, oblong-lanceolate. Petals 3–4 mm, white. Siliqua 50–80 × 2–3 mm, unilaterally deflexed; pedicels 15–25 mm, slender, deflexed; valves with a prominent median vein. Seeds in 1 or 2 rows in each loculus. *S.E. Russia.* Rs(C, W, E).

A. mollis Steven, *Mém. Soc. Nat. Moscou* **3**: 270 (1812) (*A. christiani* N. Busch), from the Caucasus, like **16** but perennial, with larger flowers, and siliqua erect to patent and not more than 55 mm, has been reported in error from Bulgaria.

17. A. recta Vill., *Hist. Pl. Dauph.* **3**: 319 (1788) (*A. auriculata* sensu DC., non Lam.). Annual 10–30 cm; stem simple or branched above; pubescent, the hairs 2- to 5-fid. Basal leaves petiolate, ovate or obovate, entire; cauline leaves 5–14, 6–20 mm, sessile, auriculate, ovate to oblong, rounded at the apex. Flowers 7–40; pedicels up to 5 mm at anthesis; petals 2–3·5 mm, white. Infructescence lax, flexuous; siliqua 10–35 × 0·6–1 mm, glabrous to puberulent, patent; pedicels 2–5 mm; style up to 0·5 mm. Seeds up to 1 mm, unwinged, brown. *Europe, except the north and many of the islands.* Al Au Bu Cr Cz Ga Ge Gr He Hs Hu It Ju Po Rm Rs(B, C, W, K, E) Sa Si [Be].

18. A. nova Vill., *Prosp. Pl. Dauph.* 39 (1779) (*A. saxatilis* All., *A. auriculata* Lam.). Like **17** but often biennial, 20–50 cm, unbranched; cauline leaves 10–40 mm, ovate to lanceolate, acute; pedicels more than 3 mm at anthesis; petals 4–6 mm; infructescence straight; siliqua 25–70 × 1·5 mm; pedicels slender, 5–16 mm; style 0·5–1 mm; seeds 1·2–1·5 mm, narrowly winged. ● *Pyrenees, Alps, Jura and Balkan peninsula.* Au Bu Ga He Hs It Ju.

A. reverchonii Freyn in Willk., *Suppl. Prodr. Fl. Hisp.* 302 (1893), from E.C. Spain, is like **18** but up to 70 cm, branched above, the leaves dentate, and seeds *c.* 0·5 mm, unwinged. Its status is doubtful.

19. A. parvula Dufour in DC., *Reg. Veg. Syst. Nat.* **2**: 228 (1821). Like **17** but 10–20 cm; cauline leaves 10–25 mm, attenuate at the base, ovate; flowers subsessile at anthesis; infructescence rigid, slightly flexuous, more elongate; siliqua 20–30 × 1·5–2 mm, erect, stellate-puberulent; pedicels 2–4 mm, as thick as the siliqua; seeds dark brown. *C. & S. Spain.* Hs. (*N.W. Africa.*)

20. A. verna (L.) R. Br. in Aiton, *Hort. Kew.* ed. 2, **4**: 105 (1812). Annual 5–40 cm, frequently with several flowering stems from the rosette. Basal leaves petiolate, ovate or obovate; cauline leaves few (often only 1 or 2), 6–20 mm, cordate at the base, ovate, serrate. Flowers up to 10; pedicels less than 2 mm at anthesis; petals 5–8 mm, pale violet with a yellowish claw, or white. Infructescence lax, flexuous; siliqua (25–)45–60 × (1–)1·5–2 mm erecto-patent, glabrous to puberulent; pedicels thick; stigma sessile. Seeds up to 1 mm, narrowly winged, brown. *Mediterranean region.* Al Bl Co Cr Ga Gr Hs It Ju Sa Si Tu.

21. A. procurrens Waldst. & Kit., *Pl. Rar. Hung.* **2**: 154 (1803). Perennial 8–30 cm, with long stolons; stems glabrous or with sparse, medifixed hairs. Basal leaves 20–30 mm, obovate to oblanceolate, acuminate, entire, glabrous except for the margin and the veins of the lower surface which have medifixed hairs; cauline leaves rounded at the base, ovate. Petals 8–10 mm, white. Infructescence lax; siliqua 12–35 × 1–1·5 mm, patent; pedicels more than 10 mm. ● *Carpathians and mountains of the Balkan peninsula.* Bu Cz Ju Rm.

22. A. vochinensis Sprengel, *Pugillus* **1**: 46 (1813). Like **21** but without long stolons; stem pubescent, with appressed, medifixed hairs; basal leaves not more than 15 mm, obtuse, in compact rosettes, the cauline attenuate at the base, oblong; petals 5–7 mm; siliqua 8–15 × 1·3–2 mm; pedicels not more than 10 mm. • *S.E. Alps.* Au It Ju.

23. A. ferdinandi-coburgi J. Kellerer & Sünd. in Sünd., *Allgem. Bot. Zeitschr.* **1903**: 62 (1904). Like **21** but the basal leaves long-petiolate, narrowly oblong to lanceolate; cauline leaves lanceolate; leaves pubescent on both surfaces; siliqua 15–20 × 0·8 mm. 2*n* = 18. *Calcareous rocks.* • *S.W. Bulgaria* (*Pirin*). Bu.

24. A. scopoliana Boiss., *Ann. Sci. Nat.* sér. 2, **17**: 56 (1842). Perennial 3–10 cm, without stolons, glabrous except for the leaf-margins. Basal leaves up to 15 mm, attenuate-petiolate, obovate, acuminate, entire, setose-ciliate; cauline leaves attenuate at the base, oblong. Petals 7–11 mm, white. Infructescence compact; siliqua 6–10 × 2–2·5 mm, subterete, erect; valves keeled. *Calcareous rocks.* • *S.E. Alps and mountains of W. part of Balkan peninsula.* Al ?It Ju.

25. A. bryoides Boiss., *loc. cit.* 55 (1842). Perennial 2–6 cm, tomentose. Basal leaves attenuate at the base, ovate, acuminate, the lamina usually densely pubescent, the hairs unbranched, ciliate at the apex; cauline leaves absent. Flowers 3–6; petals 6–7 mm, white. Infructescence lax; siliqua 10–20 × 1·5 mm; style up to 1 mm. 2*n* = 16. *Mountain rocks.* • *S. part of Balkan peninsula.* Al Gr Ju.

26. A. longistyla Rech. fil., *Feddes Repert.* **43**: 150 (1938). Scapose perennial 6–15 cm; almost glabrous; stems several from the rosette, glabrous except at the base. Leaves attenuate-petiolate, obovate-lanceolate, fleshy, denticulate, glabrous. Petals 3–4 mm, cuneate, white. Siliqua 12–18 × *c.* 1·5 mm; pedicels 3–6 mm, patent; style 1·5–2 mm. • *Karpathos.* Cr.

27. A. scabra All., *Auct. Syn. Stirp. Horti Taur.* 22 (1773) (*A. stricta* Hudson). Perennial 5–25 cm, scabrid, with simple and a few branched hairs. Flowering stems often several from the rosette, unbranched. Basal leaves dark green, glossy, attenuate-petiolate, obovate, sinuate-dentate with 2–8 teeth; cauline leaves 1–4(5), rounded to cuneate at the base. Flowers 3–9(–12); petals 5–8 mm, yellowish or white. Siliqua 35–50 × 1·5–1·8 mm, erect; pedicels patent. 2*n* = 16. • *Mountains of N. Spain, Pyrenees, S.W. Alps, Jura; one station in S.W. England.* Br Ga He Hs.

28. A. subflava B. M. G. Jones, *Feddes Repert.* **69**: 60 (1964) (*A. ochroleuca* Boiss. & Heldr., non (Lam.) Lam.). Like **27** but 10–30 cm; pubescent on the upper surface of leaves and on veins beneath, the hairs branched; basal leaves runcinate-pinnatifid; cauline leaves 2–5, sagittate; siliqua 20–30 × *c.* 1 mm, patent; pedicels erect. • *Mountains of S. Greece.* Gr.

29. A. caerulea (All.) Haenke in Jacq., *Collect. Bot.* **2**: 56 (1789). Perennial 5–15 cm, pubescent or glabrous. Basal leaves attenuate-petiolate, obovate, with 2–5 distinct, obtuse, apical teeth; cauline leaves 1–3, cuneate at the base. Flowers 4–10; petals 4–5 mm, pale blue. Infructescence compact; siliqua 10–30 × 2·4–3·2 mm, strict, at first exceeding the flowers, bluish when immature. 2*n* = 16. *Alpine rocks and moraines.* • *Alps.* Au Ga Ge He It Ju.

30. A. pumila Jacq., *Fl. Austr.* **3**: 44 (1775). Perennial 5–18 cm, usually pubescent below. Leaves with branched hairs, at least on the margin; basal leaves attenuate-petiolate, obovate, with 1 or 2 faint teeth, or entire; cauline leaves 1–4, rounded at the base, oblong-ovate. Flowers 3–10; petals 6–7 mm, white. Siliqua 20–40 × 2 mm, at first exceeding the flowers; pedicels 4–9 mm, erect. • *Alps, Appennini.* Au Ga Ge He It Ju.

31. A. soyeri Reuter & Huet, *Ann. Sci. Nat.* sér. 3, **19**: 251 (1853) (*A. bellidifolia* Jacq., non Crantz, *A. jacquinii* G. Beck). Perennial 15–50 cm, almost glabrous; leaves occasionally pubescent. Basal leaves dark green and glossy, attenuate-petiolate, obovate, denticulate; cauline leaves 4–10, rounded at the base, ovate to oblong-lanceolate, entire. Flowers 10–20; petals 5·5–7 mm, white. Infructescence compact; siliqua 25–50 × 1·8–2·2 mm; pedicels 8–15 mm; valves with a distinct median vein. • *Pyrenees, Alps, W. Carpathians.* Au Cz Ga Ge He Hs It Ju Po Rm.

(a) Subsp. **soyeri**: Stem sparsely hairy. Leaves thin, ciliate on the margin with simple hairs; cauline leaves more or less amplexicaul. *Pyrenees.*

(b) Subsp. **subcoriacea** (Gren.) Breistr., *Bull. Soc. Sci. Dauph.* **61**: 615 (1947) (subsp. *jacquinii* (G. Beck) B. M. G. Jones): Stem glabrous; leaves fleshy, glabrous. Cauline leaves not amplexicaul. 2*n* = 16. *Alps, W. Carpathians.*

32. A. alpina L., *Sp. Pl.* 664 (1753) (*A. merinoi* Pau, *A. pieninica* Wołoszczak). Perennial 5–40 cm; branched below, with vegetative rosettes and unbranched flowering stems. Basal leaves attenuate-petiolate, oblong or obovate, dentate; cauline leaves ovate to lanceolate. Outer sepals conspicuously saccate at the base; petals 6–18 mm, white (rarely pink). Infructescence lax; siliqua 20–70 × 1·5–2·5 mm, patent. *Much of Europe.* Al Au Br Bu Co Cr Cz Fe Ga Ge Gr He Hs Hu Is It Ju No Po Rm Rs(N, W, K) Sb Si Su [Be].

(a) Subsp. **alpina**: Plant erect or ascending, with few rosettes, green, usually sparsely and coarsely stellate-hairy. Cauline leaves cordate to auriculate at the base. Sepals 3–5 mm; petals 6–10 × 2–3·5 mm. Siliqua 20–40 mm; valves with an indistinct median vein. Seeds conspicuously winged. 2*n* = 16, 32. *Moist, montane rocks and gravels; tundra. Almost throughout the range of the species.*

(b) Subsp. **caucasica** (Willd.) Briq., *Prodr. Fl. Corse* **2(1)**: 48 (1913) (*A. caucasica* Willd., *A. albida* Steven, *A. flavescens* Griseb.): Plant procumbent, with many rosettes, greyish-green, usually softly and densely pubescent. Cauline leaves auriculate to sagittate. Sepals 5–8 mm; petals 9–18 × 5–8 mm. Siliqua 40–70 mm; valves with a distinct median vein. Seeds usually wingless. 2*n* = 16. *Dry rocks. S. Europe; widely cultivated for ornament, and naturalized elsewhere.*

33. A. cebennensis DC., *Reg. Veg. Syst. Nat.* **1**: 234 (1817). Perennial 40–80 cm, sparsely stellate-pubescent, branched above. Basal leaves petiolate-ovate, longer than wide, coarsely dentate, with an acute apex and cuneate or truncate base; uppermost cauline leaves attenuate-petiolate. Petals 7–10 mm, pale or deep violet or white, somewhat patent. Siliqua 30–45 × 1–1·5 mm; pedicels patent. • *Mountains of S. France.* Ga.

34. A. pedemontana Boiss., *Diagn. Pl. Or. Nov.* **1(1)**: 69 (1843). Like **33** but 15–30 cm, smaller in all its parts, glabrous or almost so; basal leaves lyrate-pinnatifid to hederiform or orbicular, as wide as long, with an obtuse apex; petals 6–7 mm, white. • *N.W. Italy* (*Alpi Cozie*). It.

Species **33** and **34** may be better placed in *Cardaminopsis*.

43. Aubrieta Adanson

J. R. AKEROYD AND P. W. BALL

Perennial herbs; hairs stellate or both stellate and simple or forked, rarely glabrous. Leaves simple. Inner sepals saccate; petals pink, purple or violet, long-clawed. Filaments of the outer stamens with a dentate appendage. Fruit a siliqua, rarely a silicula; valves with a median vein; style distinct; stigma capitate. Seeds in 2 rows in each loculus.

All the species are montane or alpine, occurring on rocks and screes and in open coniferous woods.

Species limits within the genus are critical, and the characters used to define the species require experimental investigation. Some species are known to hybridize readily in cultivation, but the majority are allopatric, so that hybrids do not often occur in the wild.

Several species and hybrids are cultivated widely for ornament.

Literature: J. Mattfeld, *Blätt. Staudenk.* **1**: fols. 1–7 (1937); *Quart. Bull. Alp. Gard. Soc.* **7**: 157–181, 217–227 (1939). D. Phitos, *Candollea* **25**: 69–87 (1970).

1 Siliqua with both stellate and distinctly longer simple and forked hairs **1. deltoidea**
1 Siliqua with stellate hairs only, sometimes with a few forked and simple hairs that are not distinctly longer
 2 Siliqua not more than 12 mm
  3 Sepals 4–5·5 mm; petals white or pinkish **4. erubescens**
  3 Sepals 6–8 mm; petals purple or violet
   4 Stem and leaves hairy, green or greyish-green; siliqua usually reticulately veined **2. columnae**
   4 Stem and leaves densely hairy, whitish-grey; siliqua not reticulately veined **3. scyria**
 2 Siliqua more than 12 mm
  5 Siliqua $2\frac{1}{2}$–4 times as long as wide, not more than 16 mm long **2. columnae**
  5 Siliqua 5–15 times as long as wide, up to 35(–45) mm long
   6 Siliqua-valves usually reticulately veined; sepals 5–7·5 mm **5. gracilis**
   6 Siliqua-valves not or scarcely reticulately veined; sepals 7–14 mm **6. thessala**

1. A. deltoidea (L.) DC., *Reg. Veg. Syst. Nat.* **2**: 294 (1821) (incl. *A. intermedia* Heldr. & Orph. ex Boiss.). Caespitose to straggling. Leaves linear-spathulate to obovate-cuneate or rhombic, entire or with 1–3 pairs of teeth. Sepals 6–10 mm; petals 12–28 mm, reddish-purple to violet, rarely white. Siliqua 6–16(–22) × 2·5–4·5 mm, 2–5(–7) times as long as wide, not or only slightly reticulately veined, with long unbranched and forked as well as stellate hairs; style 4–8 mm. $2n = 16$. *S. part of Balkan peninsula and Aegean region; Sicilia (Madonie); widely cultivated for ornament and naturalized in S. & W. Europe.* Bu Cr Gr Ju Si [Br Ga Hs].

A variable species with a number of varieties, some of which may merit subspecific rank. A full account of these is given in Phitos, *Candollea* **25**: 69–87 (1970).

2. A. columnae Guss., *Pl. Rar.* 266 (1826). More or less caespitose to straggling. Leaves variable. Sepals 5·5–8 mm; petals 11–18 mm, purple or violet. Siliqua 5–16 × 2–4·5 mm, $2\frac{1}{2}$–4 times as long as wide, with stellate hairs only, the valves usually reticulately veined; style 3–10 mm. ● *S. Europe, from C. Italy to Bulgaria and Romania.* Al Bu It Ju Rm.

1 Leaves oblong-spathulate, ±entire; style 7–10 mm in fruit (a) subsp. **columnae**
1 Leaves obovate-cuneate or rhombic, usually with 1(2) pairs of teeth; style 3–7 mm in fruit
 2 Plant straggling; leaves with 1(2) pairs of wide teeth near the apex, sometimes entire (b) subsp. **italica**
 2 Plant ±caespitose; leaves with a pair of small teeth near the apex (c) subsp. **croatica**

(a) Subsp. **columnae**: More or less caespitose, with slender stems. Leaves oblong-spathulate, usually entire. Petals 11–17 mm. Siliqua 5–12 × 2–4·5 mm; style 7–10 mm. ● *C. & S. Appennini.*

(b) Subsp. **italica** (Boiss.) Mattf., *Blätt. Staudenk.* **1**: fols. 1–7 (1937): Straggling, with stout, long stems. Leaves broadly obovate-cuneate, with 1(2) pairs of wide teeth near the apex or entire. Petals 15–18 mm. Siliqua 8–11 × 3–4 mm; style 4–6 mm. ● *S. Italy (Monte Gargano).*

(c) Subsp. **croatica** (Schott, Nyman & Kotschy) Mattf., *loc. cit.* (1937) (*A. croatica* Schott, Nyman & Kotschy): More or less caespitose. Leaves broadly obovate-cuneate or rhombic, with a pair of short teeth just below the apex, the central tooth wider than long. Petals 12–18 mm. Siliqua 7–16 × 2·5–4·5 mm; style 3–7 mm. ● *Albania, W. Jugoslavia; S.W. Romania.*

Plants from S. Bulgaria (Pirin Planina) have been described as subsp. **pirinica** Assenov in Jordanov, *Fl. Rep. Pop. Bulg.* **4**: 707 (1970). They differ from **2** by the leaves with 1–3 pairs of teeth and petals 14–20 mm, and would thus seem to be closer to **1**; they have a chromosome number of $2n = 16$. Their status is uncertain.

3. A. scyria Halácsy, *Österr. Bot. Zeitschr.* **60**: 115 (1910). Caespitose, the whole plant whitish-grey, with dense, more or less appressed hairs. Leaves obovate-cuneate, entire or denticulate at the apex. Sepals 6–8 mm; petals 11–14 mm, purple. Siliqua 6–10(–12) × 2·5–3·5 mm, with stellate hairs, not reticulately veined; style 2·5–3·5 mm. ● *W. Aegean region (Skiros).* Gr.

4. A. erubescens Griseb., *Spicil. Fl. Rumel.* **1**: 268 (1843). Laxly caespitose. Leaves oblong-spathulate, entire or with a pair of small teeth. Sepals 4–5·5 mm; petals 8–11 mm, white or pinkish. Siliqua 7–12 × 2–3·5 mm, 2–4 times as long as wide, with stellate hairs, strongly compressed; style 4–7 mm. $2n = 16$. *Calcareous rocks.* ● *N. Greece (Athos).* Gr.

5. A. gracilis Spruner ex Boiss., *Diagn. Pl. Or. Nov.* **1(1)**: 74 (1843). More or less caespitose. Leaves linear-lanceolate to oblong-obovate, entire or with a few teeth. Sepals 5–7·5 mm; petals 12–18 mm. Siliqua 13–35(–45) mm, ($3\frac{1}{2}$–)6–16 times as long as wide, the valves usually reticulately veined; style 4–7 mm. ● *S. half of Balkan peninsula.* Al ?Bu Gr.

1 Plant glabrous or subglabrous; petals violet (c) subsp. **glabrescens**
1 Plant hairy; petals purple
 2 Leaves linear to lanceolate, 3–$7\frac{1}{2}$ times as long as wide (a) subsp. **gracilis**
 2 Leaves broadly lanceolate to oblong-obovate, 2–3(–4) times as long as wide (b) subsp. **scardica**

(a) Subsp. **gracilis**: Leaves linear to lanceolate, 3–$7\frac{1}{2}$ times as long as wide, hairy; hairs stellate and forked, with simple hairs on the margins. Petals purple. $2n = 16$. *C. Greece.*

(b) Subsp. **scardica** (Wettst.) Phitos, *Candollea* **25**: 84 (1970) (*A. scardica* (Wettst.) L.-Å. Gustavsson): Leaves broadly lanceolate to oblong-obovate, 2–3(–4) times as long as wide, hairy; hairs stellate or both stellate and forked, sometimes with simple hairs on the margins. Petals purple. $2n = 32$. *Throughout most of the range of the species.*

(c) Subsp. **glabrescens** (Turrill) Akeroyd, *Bot. Jour. Linn. Soc.* **106**: 100 (1991) (*A. glabrescens* Turrill): Leaves lanceolate to elliptical, 2–5½ times as long as wide, glabrous or with a few short, simple, forked or stellate hairs on the margins. Petals violet. *Serpentine rocks and screes. N.W. Greece (Smolikas).*

6. A. thessala Boissieu, *Bull. Soc. Bot. Fr.* **43**: 288 (1896). Laxly caespitose. Leaves obovate to broadly spathulate, 2–3½ times as long as wide, with 1–3 pairs of teeth. Sepals 7–14 mm; petals 12–20 mm, purple. Siliqua 14–26 × 2–4 mm, 5–10 times as long as wide, scarcely compressed, with stellate hairs, sometimes with a few short, forked hairs, not or only slightly reticulately veined; style 5–9 mm. *Limestone rocks.* ● *N. Greece (Olimbos).* Gr.

44. Ricotia L.

P. W. BALL

Annuals or perennials; glabrous or with unbranched hairs. Leaves entire to pinnatisect. Sepals erect, the inner saccate at the base; petals pink or violet, clawed. Fruit a pendent siliqua or latiseptate silicula; style short; stigma capitate.

Literature: B. L. Burtt, *Kew Bull.* **1951**: 123–132 (1951).

Petals 10–12 mm; fruit 5–8 times as long as wide, not winged **1. cretica**

Petals *c.* 5 mm; fruit about twice as long as wide, with a narrow marginal wing **2. isatoides**

1. R. cretica Boiss. & Heldr. in Boiss., *Diagn. Pl. Or. Nov.* **2(8)**: 29 (1849). Annual 10–25 cm. Leaves 2-pinnatifid or -pinnatisect, the upper with ovate or elliptical segments. Petals 10–12 mm, pink or violet. Siliqua 30–50 × 8–9 mm, 5–8 times as long as wide, not winged. Seeds up to 10. *Calcareous screes and stony ground.* ● *Kriti.* Cr.

2. R. isatoides (W. Barbey) B. L. Burtt, *Kew Bull.* **1951**: 131 (1951) (*Peltaria isatoides* W. Barbey). Fleshy perennial 15–20 cm. Leaves entire, crenate or pinnatifid with 3–5 lobes. Petals *c.* 5 mm, pale violet or with violet veins. Silicula 10–12 × 6–8 mm, about twice as long as wide, with very narrow marginal wing. Seeds usually solitary. *Calcareous screes.* ● *Karpathos.* Cr.

45. Lunaria L.

P. W. BALL

Biennials or perennials; hairs simple. Leaves simple, toothed. Sepals erect, the inner saccate at the base; petals long-clawed. Fruit a strongly compressed, latiseptate silicula; style long; stigma slightly lobed.

1 Upper leaves sessile or subsessile **3. annua**
1 Upper leaves distinctly petiolate
2 Saccate base of sepal less than 1 mm; carpophore of silicula 25–40 mm; leaves ±spinulose-dentate **1. rediviva**
2 Saccate base of sepal *c.* 2·5 mm; carpophore of silicula *c.* 1 mm; leaves crenate-dentate **2. telekiana**

1. L. rediviva L., *Sp. Pl.* 653 (1753). Perennial up to 140 cm. Leaves ovate, acuminate, spinulose-dentate, the uppermost distinctly petiolate. Saccate base of sepal less than 1 mm; petals (10–)12–20 mm, lilac to violet. Silicula 35–90 × 15–35 mm, elliptical, rarely ovate-elliptical, subacute at the base and apex; valves glabrous on the margin; carpophore (20–)25–40 mm; style 1–7 mm. $2n = 30$. ● *Most of Europe except the extreme north and south.* Al Au Be Bu Cz Da Ga Ge He Hs Hu It Ju Lu Po Rm Rs(B, C, W) Sa Su.

2. L. telekiana Jáv., *Magyar Bot. Lapok* **19**: 1 (1922). Like **1** but the leaves dentate-crenate; saccate base of sepal *c.* 2·5 mm; petals *c.* 12 mm; silicula 30–50 mm; valves densely ciliate on the margin; carpophore *c.* 1 mm. ● *N.E. Albania (Hekurave, Shkëlzen).* Al.

3. L. annua L., *Sp. Pl.* 653 (1753) (*L. biennis* Moench). Annual to perennial up to 100 cm. Leaves ovate to lanceolate, acuminate, coarsely and irregularly dentate, the upper sessile or subsessile. Petals 15–25 mm, purple, rarely white. Silicula 20–70 × (10–)15–35 mm, oblong-elliptical to suborbicular, rounded at the base and apex; carpophore (3–)5–20 mm; style 4–10(–12) mm. ● *S.E. Europe and Italy; cultivated and naturalized or casual elsewhere.* Al Bu Cr Gr It Ju Rm [Au Be Br Co Ga Ge He Ho Hs Hu Lu No Po Rs(W) Su].

Frequently cultivated for ornament, particularly for the white, shining, persistent septum of the silicula.

(a) Subsp. **annua**: Annual to biennial, without fusiform tubers. *Origin unknown; cultivated and naturalized in many parts of Europe.*

(b) Subsp. **pachyrhiza** (Borbás) Hayek, *Prodr. Fl. Penins. Balcan.* 1: 425 (1925) (*L. annua* subsp. *corcyrea* (DC.) Vierh.): Perennial, with fusiform tubers. *S.E. Europe and Italy.*

46. Peltaria Jacq.

P. W. BALL

Perennials; glabrous or with a few bifid hairs. Leaves simple, entire. Sepals patent, not saccate; petals white, shortly clawed. Fruit a pendent, indehiscent, strongly compressed, latiseptate silicula; style short; stigma capitate. (Incl. *Leptoplax* O. E. Schulz.)

Upper leaves amplexicaul, sessile; silicula rounded at the apex **1. alliacea**

Upper leaves cuneate at the base, shortly petiolate; silicula deeply emarginate **2. emarginata**

1. P. alliacea Jacq., *Enum. Stirp. Vindob.* 117 (1762) (*P. perennis* (Ard.) Markgraf). Stem 20–60 cm, glabrous. Cauline leaves ovate or lanceolate, cordate-sagittate, sessile, amplexicaul. Petals 3·5–4·5 mm. Silicula 6–10 × 5–9 mm, orbicular to ovate-elliptical, distinctly reticulately veined when mature. ● *From E. Austria to S. Romania and Albania.* Al Au Hu Ju Rm.

2. P. emarginata (Boiss.) Hausskn., *Mitt. Thür. Bot. Ver.* nov. ser. **3–4**: 111 (1893) (*Leptoplax emarginata* (Boiss.) O. E. Schulz, *Ptilotrichum emarginatum* Boiss.). Like **1** but often sparsely hairy; cauline leaves oblong-spathulate, cuneate, shortly petiolate; petals *c.* 3 mm; silicula deeply emarginate, obscurely reticulately veined when mature. *Rocky and stony ground on serpentine.* ● *C. & E. Greece.* Gr.

This species frequently forms hybrids with *Bornmuellera baldaccii* (Degen) Heywood and *B. tymphaea* (Hausskn.) Hausskn.

47. Alyssoides Miller

V. H. HEYWOOD AND P. W. BALL

Perennials; hairs branched or stellate. Sepals erect or erecto-patent, the inner saccate at the base; petals yellow, long-clawed. Fruit a latiseptate silicula; valves inflated, without a conspicuous median vein; style long; stigma capitate or emarginate. Seeds 4–8 in each loculus, usually winged. (*Vesicaria* Lam.)

Silicula stipitate; sepals erect **1. utriculata**
Silicula sessile; sepals erecto-patent **2. cretica**

1. A. utriculata (L.) Medicus, *Philos. Bot.* **1**: 189 (1789) (incl. *A. graeca* (Reuter) Jáv.). Woody and much-branched at the base; stems 40 cm, simple. Leaves green, those of non-flowering branches petiolate, densely crowded, rosulate, oblong-spathulate, with stellate hairs, those of flowering stems sessile, lanceolate, glabrous, sometimes ciliate. Sepals 8–12 mm; petals 16–20 mm; limb suborbicular, entire. Silicula 10–15 mm, ovoid-globose; valves strongly inflated; style 7–10 mm, filiform. $2n = 16$. *Rocks and crevices. S.W. & W.C. Alps, Appennini, Balkan peninsula, S. Romania.* Al Bu Ga Gr It Ju Rm [Ge].

Plants from the central part of the Balkan peninsula with bifid hairs on the cauline leaves and long patent hairs on the pedicels are regarded as var. *bulgarica* (Sagorski) Hayek (*A. bulgarica* (Sagorski) Assenov). Other variable characters in this complex are the shape of the petals and the length of the pedicels but they do not correlate with the indumentum characters, and no satisfactory division can be made.

2. A. cretica (L.) Medicus, *Philos. Bot.* **1**: 189 (1789) (*Alyssum creticum* L., *Lutzia cretica* (L.) Greuter & Burdet). Diffuse, woody and much-branched at the base with many leaf-rosettes, grey or white with appressed stellate hairs; flowering stems up to 20 cm. Leaves oblanceolate to obovate. Sepals 7–11 mm; petals 12–20 mm, entire. Silicula 10–15 mm, globose or ovoid-globose, densely stellate-pubescent; valves strongly inflated; style *c.* 2 mm. *Rocks, cliffs and walls.* ● *Kriti and Karpathos; Astipalea.* Cr Gr.

Recorded from Rodhos in the E. Aegean islands but perhaps only introduced there.

48. Degenia Hayek

R. DOMAC

Perennials; hairs stellate. Sepals erect, the inner slightly saccate at the base; petals yellow, long-clawed. Fruit a latiseptate, ellipsoidal silicula; valves inflated; style long; stigma slightly 2-lobed. Seeds 2 in each loculus, broadly winged.

1. D. velebitica (Degen) Hayek, *Österr. Bot. Zeitschr.* **60**: 93 (1910). Caespitose, silver-grey perennial with non-flowering rosettes. Stems up to 10 cm. Leaves few, linear-lanceolate. Petals 10–12 mm. Silicula covered with dense, stellate hairs. $2n = 16$. *Screes and among boulders.* ● *N.W. Jugoslavia (Velebit).* Ju.

49. Alyssum L.

P. W. BALL AND T. R. DUDLEY

Annuals to perennials, rarely small shrubs; hairs branched or stellate, sometimes mixed with unbranched hairs or the indumentum lepidote. Flowering stems terminal. Basal leaves usually similar to the cauline; petioles of basal leaves not grooved, not persistent or swollen at the base. Flower-buds ellipsoidal or oblong-ellipsoidal; sepals erect, not saccate at the base; petals yellow (white or purplish in Sect. *Ptilotrichum*), entire to shallowly bifid; filaments of the long stamens usually winged or toothed, those of the short stamens usually with an appendage. Fruit a latiseptate silicula; valves without a conspicuous median vein; style distinct but often short; stigma capitate or emarginate. Seeds 1 or 2(–8) in each loculus, often winged or margined. (Incl. *Ptilotrichum* C. A. Meyer.)

The indumentum is described as dimorphic when it is composed of appressed stellate hairs mixed with long, patent, unbranched or branched hairs, or with stellate hairs with some long, patent rays. The measurements given for the diameter of the stellate hairs are for the usually peltate tops of these hairs. Basal leaves are those at the base of non-flowering rosettes or stems.

Most species occur in dry, rocky, stony, or sandy places or on cliffs, although species **10–20** often occur as weeds.

A taxonomically difficult genus in which the status and circumscription of many of the taxa differ widely in published treatments, particularly in the perennial species of Sect. *Alyssum* and in Sect. *Odontarrhena*.

Literature: Sect. *Alyssum*: J. Baumgartner, *Jahresb. Landes-Lehrersem. Wiener-Neustadt* **34**: 1–35 (1907); **35**: 1–58 (1908); **36**: 1–38 (1909); *Jahresb. Landes-Lehrersem. Baden bei Wien* **48**: 1–18 (1911). Sect. *Odontarrhena*: E. I. Nyárády, *Bul. Grăd. Bot. Univ. Cluj* **7**: 1–51, 65–160 (1927); **8**: 152–156 (1928); **9**: 1–68 (1929); *Anal. Acad. Rep. Pop. Române (Sect. Geol., Geogr., Biol.)* ser. A, **1(3)**: 1–133 (1949). Sect. *Ptilotrichum*: T. R. Dudley, *Jour. Arnold Arb.* **45**: 358–373 (1964).

1 Petals white, rarely pink or purple (Sect. *Ptilotrichum*)
2 Petals purple; plant 2–5 cm **10. purpureum**
2 Petals white or rarely pink; plant usually at least 12 cm
3 Branches becoming spiny at the apex
4 Spines simple; petals abruptly contracted into claw **8. macrocarpum**
4 Spines branched; petals gradually narrowed into claw **9. spinosum**
3 Branches not spiny at the apex
5 Silicula pubescent or lepidote **1. pyrenaicum**
5 Silicula glabrous
6 Style less than $\frac{1}{10}$ as long as the silicula **6. longicaule**
6 Style more than $\frac{1}{10}$ as long as the silicula
7 Silicula cochleariform **5. baeticum**
7 Silicula not cochleariform
8 Leaves broadly spathulate
9 Style 1–1·5 mm, less than $\frac{1}{4}$ as long as the silicula **7. cadevallianum**
9 Style 2–4 mm, $\frac{1}{3}$–$\frac{1}{2}$ as long as the silicula **2. reverchonii**
8 Leaves oblong or linear
10 Seeds narrowly winged; infructescence lax, elongate **3. lapeyrousianum**
10 Seeds broadly winged; infructescence short, dense, corymbose **4. ligusticum**
1 Petals yellow
11 Ovules and seeds 4–8 in each loculus **11. linifolium**
11 Ovules and seeds 1 or 2 in each loculus
12 Ovules and seeds usually solitary in each loculus (Sect. *Odontarrhena*)
13 Leaves bicolorous, green or greyish-green on the upper surface, grey or white beneath
14 Silicula very densely pubescent, the hairs completely covering the valves

15 Silicula attenuate at the apex; style 1·5–3 mm in fruit, glabrous (Krym) **66. longistylum**
15 Silicula subacute or obtuse to emarginate; style not more than 1·5(–2) mm in fruit, usually pubescent
16 Petals 3·5–4 mm; hairs on the silicula *c.* 0·5 mm in diameter **50. argenteum**
16 Petals 2–3 mm; hairs on the silicula *c.* 0·3 mm in diameter
17 Leaves of non-flowering stems flat **65. tortuosum**
17 Leaves of the non-flowering stems plicate **68. serpyllifolium**
14 Silicula glabrous or pubescent, the valves visible beneath the hairs
18 Silicula acute or subacute
19 Basal leaves 4–7 mm wide, broadly obovate; plant with numerous non-flowering rosettes **55. smolikanum**
19 Basal leaves up to 4 mm wide, spathulate or obovate-spathulate; plant with long non-flowering stems
20 Seeds not winged; style of silicula 1·5–3 mm **66. longistylum**
20 Seeds winged; style of silicula not more than 2 mm
21 Basal and cauline leaves similar, obovate-spathulate; seeds 2·5–3 mm **54. robertianum**
21 Basal and cauline leaves dissimilar, the basal spathulate, the cauline oblanceolate, larger than the basal; seeds 1·8–2 mm **56. bertolonii**
18 Silicula obtuse, truncate or emarginate
22 Petals 3–3·5 mm; sepals 2–2·5 mm
23 Petals emarginate; silicula pubescent **48. murale**
23 Petals entire; silicula glabrous or sparsely pubescent
24 Basal leaves linear-oblanceolate; upper cauline leaves similar, up to 3·5 mm wide **53. heldreichii**
24 Basal leaves spathulate; upper cauline leaves oblanceolate, up to 5 mm wide **56. bertolonii**
22 Petals 2–3 mm; sepals 1·5–2 mm
25 Filaments with a multi-dentate appendage; septum of the silicula frequently asymmetrical in outline **51. fallacinum**
25 Filaments with an entire or 1- or 2-dentate appendage; septum of the silicula symmetrical in outline
26 Valves of the silicula asymmetrically inflated **58. corymbosoides**
26 Valves of the silicula flat or symmetrically inflated
27 Valves of the silicula not undulate
28 Basal leaves linear-oblanceolate; cauline leaves up to 3·5 mm wide **53. heldreichii**
28 Basal leaves oblanceolate to obovate-spathulate or spathulate; cauline leaves up to 6 mm wide
29 Silicula broadly obovate, pubescent; valves not inflated; seeds *c.* 3 mm **49. tenium**
29 Silicula suborbicular, glabrous; valves inflated; seeds *c.* 1·7 mm **57. markgrafii**
27 Valves of the silicula usually undulate
30 Branches of the inflorescence distant, very unequal; pedicels ±straight, rigid **48. murale**
30 Branches of the inflorescence subumbellate; pedicels strongly curved, flexuous
31 Silicula 4–6 × 3·5–4·5 mm, sparsely pubescent or glabrescent; leaves of non-flowering stems linear-oblanceolate **53. heldreichii**
31 Silicula smaller, pubescent; leaves of non-flowering stems obovate-spathulate to oblanceolate-spathulate **48. murale**
13 Leaves concolorous, but the cauline sometimes differing from the basal in colour
32 Silicula glabrous or glabrescent
33 Petals 2·5–3 mm; basal leaves orbicular-spathulate **52. corsicum**
33 Petals 3–3·5 mm; basal leaves spathulate or oblanceolate **56. bertolonii**
32 Silicula pubescent
34 Hairs on the leaves at least 0·5 mm in diameter
35 Seeds with or without wing; leaves all oblanceolate, the cauline ±decreasing in size towards the apex of the stem **61. caliacrae**
35 Seeds not winged; cauline and basal leaves usually conspicuously different in shape, at least the basal obovate- to orbicular-spathulate, the cauline ±increasing in size towards the apex of the stem
36 Sepals *c.* 2 mm; silicula (3·5–)4–6 mm; seeds *c.* 1·75 mm **69. nebrodense**
36 Sepals up to 1·5 mm; silicula 2–4 mm; seeds *c.* 1·5 mm
37 Stellate hairs on silicula 0·4–0·7 mm in diameter; stellate hairs on leaves of non-flowering stems with unequal rays up to 1·5 mm in diameter **60. sibiricum**
37 Stellate hairs on silicula 0·2–0·4 mm in diameter; stellate hairs on leaves of non-flowering stems with equal rays not more than 0·5 mm long **62. borzaeanum**
34 Hairs on leaves not more than 0·4 mm in diameter
38 Silicula 2–2·5 mm wide
39 Plant 1–3 cm, procumbent; largest cauline leaves 3–4 × 1·5 mm (Kriti) **70. fragillimum**
39 Plant at least 6 cm, procumbent to erect; cauline leaves larger than 3–4 × 1·5 mm
40 Silicula glabrous or sparsely pubescent **56. bertolonii**
40 Silicula densely grey- or white-pubescent
41 Silicula subacute **68. serpyllifolium**
41 Silicula truncate to emarginate
42 Basal leaves obtuse; silicula 2–3 mm **62. borzaeanum**
42 Basal leaves ±acute; silicula 3–4 mm **65. tortuosum**
38 Silicula at least 2·5 mm wide
43 Silicula sparsely pubescent, the valves easily visible beneath the hairs
44 Cauline leaves shorter than the basal; racemes simple (Evvoia) **59. euboeum**
44 Cauline leaves longer than the basal; racemes compound

45 Seeds winged; silicula 4–6 mm **56. bertolonii**
45 Seeds not winged; silicula 2·3–3·6 mm **63. obtusifolium**
43 Silicula very densely pubescent, the valves completely covered by the hairs
46 Petals 3–3·5 mm; sepals 2–2·5 mm
47 Flowering stems ±erect or ascending; basal leaves obovate-spathulate to suborbicular; petiole of cauline leaves 3–5 mm (Ukraine, Russia) **64. obovatum**
47 Flowering stems usually procumbent; basal leaves oblanceolate to obovate-spathulate; petiole of cauline leaves not more than 2 mm (Alps) **67. alpestre**
46 Petals 2–3 mm; sepals usually not more than 2 mm
48 Largest cauline leaves near the apex of the stem and distinctly larger than the basal **69. nebrodense**
48 Largest cauline leaves near the base of the stem, smaller than or about equalling the basal
49 Leaves of non-flowering stems flat; hairs on the silicula 20- to 25-rayed (Ukraine, Russia) **64. obovatum**
49 Leaves on non-flowering stems plicate; hairs on the silicula mostly 12- to 16-rayed **68. serpyllifolium**
12 Ovules and seeds 2 in each loculus
50 Sepals with a tuft of long, divergent-rayed, stellate hairs at the apex
51 Raceme ±elongate in fruit; short stamens with a long (1·5–2 mm) appendage; style 2·5–5 mm **45. doerfleri**
51 Raceme corymbose in fruit; short stamens with a small appendage; style not more than 3 mm **46. taygeteum**
50 Sepals without a tuft of hairs at the apex
52 Silicula glabrous or glabrescent
53 Indumentum of leaves lepidote **43. idaeum**
53 Indumentum of leaves consisting of branched or stellate hairs
54 Style 1·5–3·5 mm
55 Petals 3·5–4 mm; leaves oblong-spathulate; annual **17. smyrnaeum**
55 Petals 6–7 mm; leaves linear-oblong or linear-lanceolate, acute; perennial **25. lenense**
54 Style not more than 1(–1·5) mm
56 Upper leaves forming an involucre around the inflorescence; silicula ovate, the valves inflated; petals 3–3·5 mm **16. foliosum**
56 Upper leaves not forming an involucre; silicula ±orbicular, the valves inflated in the middle, with strongly flattened margin; petals 2–3 mm
57 Stellate hairs 0·2–0·3 mm in diameter; sepals deciduous **15. desertorum**
57 Stellate hairs more than 0.3 mm in diameter; sepals persistent **18. minutum**
52 Silicula pubescent
58 Indumentum of the inflorescence or the silicula consisting of long patent hairs, often mixed with appressed stellate hairs
59 Anthers less than 0·5 mm; annual
60 Filaments without appendage; indumentum dimorphic only on the silicula
61 Petals 2·5–3 mm; seeds 1·25–1·5 mm, without wing **12. dasycarpum**
61 Petals 3–4 mm; seeds 1·5–1·9 mm, with wing *c.* 0·2 mm wide **14. granatense**
60 Filaments winged or with appendage; indumentum ±dimorphic on all parts of the plant
62 Petals attenuate towards the base; style 0·7–1·3 mm **20. minus**
62 Petals constricted at the middle; style 1·2–2·5 mm **22. hirsutum**
59 Anthers more than 0·5 mm; biennial or perennial
63 Erect biennial up to 80 cm; middle cauline leaves usually at least 10 mm wide; hairs 5- to 7-rayed, sparse **29. wierzbickii**
63 Perennial up to 40(–60) cm; middle cauline leaves usually less than 10 mm wide, the hairs 8- to many-rayed
64 Raceme short and dense in fruit
65 Petals entire; silicula without flattened margin; seeds 2·5 mm, narrowly winged **42. lassiticum**
65 Petals deeply emarginate; silicula with flattened margin; seeds 1·5–2 mm, not winged **25. lenense**
64 Raceme elongate in fruit
66 Petals entire; filaments of the long stamens entire; plants densely caespitose or pulvinate (Bulgaria and Romania) **31. pulvinare**
66 Petals ±emarginate; filaments of the long stamens usually with 1 or 2 teeth; plants diffuse or laxly caespitose
67 Seeds 3–3·5 mm; sepals ±persistent (Krym) **30. calycocarpum**
67 Seeds 1·5–2 mm; sepals deciduous **24. repens**
58 Indumentum of the inflorescence and silicula consisting of appressed hairs
68 Petals up to 3·5 mm
69 Perennial; style usually 2–3·5 mm
70 Silicula 3–5 mm **32. montanum**
70 Silicula 2·5–3 mm **33. fastigiatum**
69 Annual; style 0·3–1·8(–2) mm
71 Raceme subumbellate in fruit; silicula ovate, the valves inflated **19. umbellatum**
71 Raceme not umbellate in fruit; silicula orbicular to elliptical, the valves with a flattened margin
72 Sepals persistent; style 0·3–0·6 mm **13. alyssoides**
72 Sepals deciduous; style 0·5–1·6 mm
73 Plant up to 40 cm; style 0·7–1·6 mm **20. minus**
73 Plant not more than 12 cm; style 0·5–0·8 mm **21. siculum**
68 Petals more than 3·5 mm
74 Annual 20–60 cm; pedicels 7–20 mm in fruit; raceme very long and lax in fruit **23. rostratum**
74 Perennial 5–25 cm; pedicels up to 8 mm in fruit, if longer, then raceme short and dense in fruit
75 Seeds at least 4 mm; wing 0·5–0·8 mm wide **44. handelii**
75 Seeds up to 3 mm; wing up to 0·5 mm wide

76 Hairs on the leaves not more than 5- to 10-rayed
77 Silicula elliptical or elliptic-orbicular, distinctly longer than wide **26. scardicum**
77 Silicula ±orbicular
78 Basal leaves linear-lanceolate to oblanceolate-spathulate, the cauline linear or linear-spathulate; hairs on leaves usually not more than 0·5(–0·7) mm in diameter **32. montanum**
78 Basal leaves obovate to elliptical, the cauline elliptical, cuneate; hairs on leaves 0·7–1 mm in diameter **35. diffusum**
76 At least some hairs on leaves 10- to many-rayed
79 Lower pedicels shorter than the silicula; valves of silicula dissimilar **47. densistellatum**
79 Lower pedicels equalling or longer than the silicula; valves of silicula similar
80 Lower pedicels 10–15 mm in fruit, often deflexed **41. sphacioticum**
80 Lower pedicels less than 10 mm in fruit, patent or erecto-patent
81 Silicula at least 6 mm, elliptical or elliptic-orbicular, rarely suborbicular
82 Silicula densely grey-pubescent
83 Basal leaves oblong-obovate; silicula elliptic-orbicular, truncate **36. cuneifolium**
83 Basal leaves suborbicular; silicula suborbicular, emarginate **37. arenarium**
82 Silicula green, sparsely pubescent
84 Petals glabrous; silicula 6–6·5 mm; basal leaves oblong-obovate, gradually attenuate into petiole **27. wulfenianum**
84 Petals pubescent on the back; silicula 6·5–8 mm; basal leaves orbicular-obovate, abruptly contracted into petiole **28. ovirense**
81 Silicula not more than 6 mm, orbicular, rarely elliptic-orbicular
85 Basal leaves ±abruptly contracted into petiole
86 Hairs on the leaves *c.* 0.4 mm in diameter **39. moellendorfianum**
86 Hairs on the leaves 0·5–0·7 mm in diameter
87 Basal leaves suborbicular, the cauline ovate-spathulate **37. arenarium**
87 Basal leaves broadly spathulate to obovate, the cauline oblanceolate or spathulate
88 Lower pedicels $1\frac{1}{2}$–2 times as long as the orbicular silicula; non-flowering stems long **40. stribrnyi**
88 Lower pedicels about equalling the elliptic-orbicular silicula; non-flowering stems very short, terminated by a rosette **36. cuneifolium**
85 Basal leaves gradually attenuate towards base, not obviously petiolate
89 Petals 6–8 mm; leaves usually white or grey **38. atlanticum**
89 Petals 3–6 mm; leaves usually green or grey-green
90 Silicula 2·5–3 mm **33. fastigiatum**
90 Silicula 3–5·5 mm
91 Silicula with dense, 12- to 24-rayed stellate hairs **32. montanum**
91 Silicula with sparse, 8- to 12-rayed stellate hairs **34. gustavssonii**

Sect. PTILOTRICHUM (C. A. Meyer) Hooker fil. Perennial herbs or small shrubs. Filaments usually not toothed or winged. Petals white, pink or purple. Ovules 1 or 2 in each loculus.

1. A. pyrenaicum Lapeyr., *Hist. Abr. Pyr.* 371 (1813) (*Ptilotrichum pyrenaicum* (Lapeyr.) Boiss.). Small, caespitose shrub up to 50 cm, with a woody stock branched above, and bearing clusters of leaves or erect flowering stems up to 20 cm. Leaves obovate-lanceolate, attenuate at the base, silvery with stellate hairs. Flowers in a dense corymb; petals *c.* 6 mm, white, obovate-orbicular, abruptly contracted into a claw. Silicula 6–8 mm, rhomboid-obovoid, compressed, pubescent; style about as long as the silicula. Seeds winged. *Limestone cliffs.* ● *E. Pyrenees* (*Font de Comps*). Ga.

2. A. reverchonii (Degen & Hervier) Greuter & Burdet, *Willdenowia* **13**: 86 (1983) (*Ptilotrichum reverchonii* Degen & Hervier). Like **1** but the leaves broadly spathulate; petals very shortly clawed; silicula glabrous; style $\frac{1}{3}$–$\frac{1}{2}$ as long as the silicula. *Limestone cliffs.* ● *S.E. Spain* (*Sierra de Cazorla and adjacent mountains*). Hs.

3. A. lapeyrousianum Jordan, *Obs. Pl. Crit.* **1**: 5 (1846) (*Ptilotrichum lapeyrousianum* (Jordan) Jordan, *P. peyrousianum* Willk.). Small, caespitose shrub up to 30 cm, with branched, woody stock. Flowering stems 7·5–15 cm, erect. Leaves oblong to oblong-lanceolate, obtuse, attenuate at the base, silvery with appressed stellate hairs. Inflorescence racemose; petals 3–4 mm, white, obovate. Infructescence lax, elongate; silicula 4–6 mm, obovoid, glabrous; style up to $\frac{1}{4}$ as long as the silicula. Seeds narrowly winged. *Limestone cliffs and screes.* ● *E. Pyrenees, E. Spain.* Ga Hs.

4. A. ligusticum Breistr., *Bull. Soc. Sci. Dauph.* **61**: 616 (1947) (*Ptilotrichum halimifolium* Boiss., *Alyssum halimifolium* auct., non L.). Like **3** but with the siliculae in short, dense, corymbose clusters; style $\frac{1}{3}$–$\frac{1}{2}$ as long as the silicula; seeds broadly winged. *Rocks.* ● *S.E. France, N.W. Italy.* Ga It.

5. A. baeticum (Küpfer) Greuter & Burdet, *Willdenowia* **13**: 85 (1983) (*Hormathophylla baetica* Küpfer). Small shrub 15–30(–40) cm, silvery-grey with appressed stellate hairs. Flowering stems 5–10 cm, not or slightly branched. Leaves 4–12(–16) × 1·5–4(–6) mm, oblong-spathulate, gradually tapering to a short petiole. Inflorescence dense; petals 4·5–5 × 2–2·5(–3) mm, white or creamy-white. Infructescence 5–10 cm, lax; silicula elliptical, cochleariform, glabrous; style 0·7–1·2 mm. Seeds 2·2–2·6 × 1·7–2 mm, with narrow wing, mucilaginous. $2n = 22$. ● *S.E. Spain.* Hs.

6. A. longicaule Boiss., *Biblioth. Univ. Genève* sér. 2, **13**: 407 (1838) (*Ptilotrichum longicaule* (Boiss.) Boiss.). Laxly caespitose perennial, woody at the base. Stems (12–)20–60 cm, fragile. Basal leaves obovate-spathulate, acute, attenuate into the petiole; cauline leaves few, linear-lanceolate; all grey with dense, appressed stellate hairs. Petals 2–4 × 0·5–1 mm, white. Silicula 4–6 mm, sessile, obovoid, glabrous; style up to 0·5 mm. Seeds broadly winged. *Limestone rocks and cliffs.* ● *S. Spain.* Hs.

7. A. cadevallianum Pau, *Mem. Mus. Ci. Nat. Barcelona,* (Bot.) **1**: 9 (1925) (*Ptilotrichum cadevallianum* (Pau) Heywood, *Hormathophylla cadevalliana* (Pau) T. R. Dudley). Like **6** but leaves suborbicular or obovate-spathulate; petals 4–6 × 2–3 mm; silicula up to 9 mm, shortly stipitate; style 1–1.5 mm. ● *S. Spain (El Maimón, near Vélez Rubio).* Hs.

8. A. macrocarpum DC., *Reg. Veg. Syst. Nat.* **2**: 321 (1821) (*Ptilotrichum macrocarpum* (DC.) Boiss.). Small, much-branched shrub, the branches intertwined and becoming more or less spiny. Stems up to 20 cm. Leaves oblong to oblong-ovate, obtuse, crowded at the base of the flowering branches or forming non-flowering rosettes, silvery-green to white, with stellate hairs. Petals 5–6 mm, white, abruptly contracted into a claw. Silicula 8–10 mm, suborbicular, inflated, glabrous; style $\frac{1}{3}$–$\frac{1}{2}$ as long as the silicula. Seeds 2–4 in each loculus, broadly winged. *Limestone rocks.* ● *S. France (from 1° 30′ to 4° 45′ E.)* Ga.

9. A. spinosum L., *Sp. Pl.* 650 (1753) (*Ptilotrichum spinosum* (L.) Boiss.). Small, convex, much-branched shrub up to 60 cm, the branches becoming spiny with branched spines. Leaves of non-flowering rosettes obovate-spathulate; those of the flowering branches linear-lanceolate; all silvery, with appressed stellate hairs. Petals *c.* 3 mm, white or purplish, gradually narrowing into a claw. Silicula 4–6 mm, obovoid, glabrous; style $\frac{1}{6}$–$\frac{1}{4}$ as long as the silicula. Seeds 2 in each loculus, broadly winged. *Rocks and screes. E. & S. Spain, S. France.* Ga Hs.

10. A. purpureum Lag. & Rodr., *Anal. Ci. Nat.* **5**: 275 (1802) (*Ptilotrichum purpureum* (Lag. & Rodr.) Boiss.). Small, inconspicuous, densely caespitose, spineless perennial 2–5 cm, half-hidden in screes. Leaves spathulate to linear, whitish or greyish, with stellate hairs. Petals 3–4 mm purple. Silicula 4–5 mm, ellipsoid, stellate-hairy; style $\frac{1}{3}$–$\frac{1}{2}$ as long as the silicula. *Screes.* ● *Mountains of S. & S.E. Spain.* Hs.

Sect. MENIOCUS (Desv.) Hooker fil. Annual. Inflorescence a simple raceme. Filaments with appendages. Ovules 4–8 in each loculus.

11. A. linifolium Stephan ex Willd., *Sp. Pl.* **3(1)**: 467 (1800) (*Meniocus linifolius* (Stephan ex Willd.) DC.). Erect or ascending, densely grey-pubescent annual up to 35 cm. Leaves linear. Raceme elongate in fruit; pedicels 2·5–7 mm, patent or erecto-patent. Sepals 1·5–2 mm; petals 2–3 mm, emarginate. Silicula 4–7 × 3–4·5 mm, elliptical or obovate-elliptical, glabrous; valves slightly convex; style (0·1–)0·3–0·6 mm. Seeds 1·2–1·3 mm, 4–6(–8) in each loculus, not winged. *S.E. Europe; E. & S. Spain.* Hs Ju Rm Rs(C, W, K, E) Tu.

Sect. PSILONEMA (C. A. Meyer) Hooker fil. Annuals or rarely biennials. Inflorescence a simple raceme. Filaments without appendages, sometimes the filaments of the longer stamens with a filiform nectary at the base. Ovules 2 in each loculus.

12. A. dasycarpum Stephan ex Willd., *Sp. Pl.* **3(1)**: 469 (1800). Erect grey-pubescent annual 10–25 cm. Leaves oblong-lanceolate to obovate. Raceme short and dense in fruit. Sepals *c.* 2 mm, more or less persistent; petals 2·5–3 mm. Silicula 3–3·5 × 2·5–3 mm, elliptical or orbicular-elliptical, pubescent, the indumentum dimorphic; valves inflated, with flattened margin; style (1–)1·5–1·8(–2) mm. Seeds 1·25–1·5 mm, not winged. *S.E. Russia.* Rs(E).

13. A. alyssoides (L.) L., *Syst. Nat.* ed. 10, **2**: 1130 (1759) (*A. calycinum* L., *A. conglobatum* Fil. & Jáv., *Clypeola alyssoides* (L.) L.). Erect or ascending, grey-pubescent annual or biennial 5–30(–40) cm. Lower leaves obovate to oblanceolate, the upper narrower. Fruiting raceme long, or short and dense; pedicels 2–4 mm, patent. Sepals 2–2·5 mm, persistent; petals 3–4 mm, emarginate. Silicula 3–4(–4·5) mm, orbicular, pubescent; valves inflated, with flattened margin; style 0.3–0.6 mm. Seeds 1·2–2 mm, narrowly winged. $2n = 32$. *Most of Europe, but only as an introduction in the north.* Al Au Be Bu Co Cz Ga Ge Gr He Ho Hs Hu It Ju Lu Po Rm Rs(B, C, W, K, E) Sa ?Si Tu [Br Da Fe No Rs(N) Su].

14. A. granatense Boiss. & Reuter, *Pugillus* 9 (1852) (*A. hispidum* Loscos & Pardo, *A. marizii* Coutinho). Annual up to 20 cm; indumentum monomorphic, dimorphic on the silicula, grey. Leaves elliptic-lanceolate or oblong. Raceme long and dense in fruit; pedicels 2·5–4 mm, erecto-patent. Sepals 2–3 mm, persistent; petals 3–4 mm, emarginate. Silicula 3–5 mm, orbicular or orbicular-elliptical, obtuse or emarginate; valves inflated, with flattened margin; style 0·7–1·5 mm. Seeds 1·5–1·9 mm; wing *c.* 0·2 mm wide. *E. & S. Spain; Portugal.* Hs Lu.

Sect. ALYSSUM. Annuals to perennials. Inflorescence a simple raceme. Most filaments winged or with appendages. Ovules 2 in each loculus.

15. A. desertorum Stapf, *Denkschr. Akad. Wiss. Math.-Nat. Kl. (Wien)* **51**: 302 (1886) (*A. minimum* auct.). Grey-green annual up to 20 cm. Cauline leaves linear-lanceolate. Raceme long and dense in fruit; pedicels 2–5 mm, erecto-patent. Sepals 1·5–2 mm, deciduous; petals 2–3 mm, entire or emarginate. Silicula 3–4·5(–5) mm, ovate-orbicular or orbicular, emarginate, glabrous; valves inflated, with flattened margin; style 0·5–1 mm. Seeds 1–1·4 mm, narrowly winged. $2n = 32$. *E., E.C. & S.E. Europe.* Al Au Bu Cz Gr Hu Ju Rm Rs(B, C, W, K, E) Tu.

16. A. foliosum Bory & Chaub. in Bory, *Expéd. Sci. Morée* **3**: 185 (1832). Grey-green annual up to 10 cm. Lower leaves ovate-orbicular, long-petiolate, the upper oblong. Raceme subumbellate in fruit, upper leaves forming an involucre. Sepals 2–2·5 mm, persistent; petals 3–3·5 mm, linear, emarginate. Silicula 4–5(–6·5) mm, ovate, glabrous; valves inflated; style 0·7–1 mm. Seeds 2–2·5 mm, narrowly winged. *Greece; Aegean region.* Cr Gr Tu.

17. A. smyrnaeum C. A. Meyer, *Bull. Sci. Acad. Imp. Sci. Pétersb.* **7**: 132 (1840). Grey-pubescent annual up to 10 cm. Leaves oblong-spathulate. Raceme short in fruit. Sepals 2·5–3·5 mm, persistent; petals 3·5–4 mm, emarginate. Silicula 3·5–4·5 mm, ovate or orbicular, glabrous, turgid; style *c.* 2 mm. Seeds without or with a narrow wing. *S.E. Europe.* Cr Gr Rs(K) Tu.

18. A. minutum Schlecht. ex DC., *Reg. Veg. Syst. Nat.* **2**: 316 (1821) (*A. compactum* De Not., ?*A. marginatum* Steudel; incl. *A. psilocarpum* Boiss., *A. ponticum* Velen.). Annual up to 12 cm; indumentum dimorphic, grey-green. Leaves obovate to oblong. Raceme more or less elongate in fruit, dense; pedicels 4–6 mm, erecto-patent or erect. Sepals *c.* 2 mm, persistent; petals 2·3–3 mm, emarginate, densely pubescent. Silicula 3–4 mm, orbicular, emarginate, glabrous; valves inflated, with flattened margin; style

0·5–1 mm. Seeds 1·5–1·7 mm, narrowly winged. $2n = 16$. *S. & E. Europe*. Bu Cr Gr Hs It Ju Lu Rm Rs(C, W, K, E) Si Tu.

19. A. umbellatum Desv., *Jour. Bot. Appl.* **3**: 173 (1814). Grey annual up to 10 cm. Leaves linear or linear-oblong. Raceme subumbellate in fruit, dense; pedicels 3–5 mm, erecto-patent. Sepals 2–2·2 mm, deciduous; petals *c*. 3 mm, emarginate. Silicula 4–5·5 mm, ovate, pubescent; valves inflated; style 0·7–1·2 mm. Seeds 1·7–2·1 mm, narrowly winged. $2n = 16$. *S.E. Europe*. Bu Cr Gr Ju Rs(K) Tu.

20. A. minus (L.) Rothm., *Feddes Repert.* **50**: 77 (1941) (*A. campestre* auct., *A. micranthum* C. A. Meyer, *A. micropetalum* Fischer & C. A. Meyer; incl. *A. parviflorum* Bieb.). Grey-green annual up to 40 cm. Leaves oblong-obovate to oblong-lanceolate. Raceme elongate in fruit; pedicels 3–6 mm, erecto-patent. Sepals 1·5–2·5 mm, caducous; petals 2–3·5 mm, gradually attenuate towards the base, entire to emarginate; filaments of the short stamens 1·5–2·5 mm, with connate appendages; nectaries 0·1 mm. Silicula 3·5–6 mm, suborbicular, pubescent; valves inflated, with flattened margin; style 0·7–1·6 mm. Seeds 1·7–2 mm; wing up to 0·4 mm wide. *S. & E. Europe*. Al Bu Cr Ga Gr Hs It Ju Lu Rm Rs(C, W, K, E) Sa Si Tu.

(a) Subsp. **minus**: Short stamens with appendages ½ as long as the filament. Pedicels and silicula with only stellate hairs; style pubescent. $2n = 16$. *Almost throughout the range of the species.*

(b) Subsp. **strigosum** (Banks & Solander) Stoj. in Jordanov, *Fl. Rep. Pop. Bulg.* **4**: 503 (1970) (*A. strigosum* Banks & Solander, *A. cephalotes* auct. balcan.): Short stamens up to ¾ as long as the filament. Pedicels and silicula with bifid as well as stellate hairs; style glabrous or sparsely pubescent. $2n = 16$. *Mainly in S.E. Europe.*

21. A. siculum Jordan, *Diagn.* **1**: 202 (1864). Like **20(a)** but not more than 12 cm; sepals deciduous; style 0·5–0·8 mm; nectaries 0·3–0·6 mm. $2n = 48$. *Mountains of C. & S. Greece; Sicilia*. Gr Si.

Perhaps an allopolyploid derived from **13** and **20**.

22. A. hirsutum Bieb., *Fl. Taur.-Cauc.* **2**: 106 (1808). Annual up to 40 cm. Leaves ovate- to lanceolate-oblong. Raceme long and dense in fruit; pedicels 4–5 mm, erecto-patent. Sepals 2–3·5 mm, deciduous, petals 3–5 mm, constricted at the middle, bifid; filaments of the short stamens 2·5–3 mm, with usually free appendages as long as the filament. Silicula 5–7 mm, orbicular, with appressed stellate hairs and patent, tubercle-based, unbranched or bifid hairs with unequal branches; valves inflated, with flattened margin; style 1·2–2·5 mm. Seeds 1·7–2·5 mm; wing 0·2–0·5 mm wide. $2n = 46$. *S.E. Europe*. Bu Ju Rm Rs(W, K, E).

23. A. rostratum Steven, *Mém. Acad. Sci. Pétersb.* **3**: 295 (1809–10). Grey or grey-green annual or biennial 20–60 cm. Lower leaves obovate-spathulate, the upper oblong-lanceolate. Raceme very long and lax in fruit; pedicels 7–20 mm, erecto-patent. Sepals 2·5–3·5 mm, deciduous; petals 4·5–6 mm, emarginate, densely pubescent on the back. Silicula 2·5–5·5 mm, orbicular-ovate, pubescent; valves inflated, with flattened margin; style 2–3 mm. Seeds *c*. 2 mm; wing 0·3–0·5 mm wide. *S.E. Europe*. Ju Rm Rs(W, K, E).

24. A. repens Baumg., *Enum. Stirp. Transs.* **2**: 237 (1816) (*A. virescens* Halácsy). Diffuse or erect perennial up to 60 cm, the non-flowering stems terminated by rosettes; indumentum dimorphic, grey-green. Basal leaves obovate-spathulate; cauline leaves lanceolate or linear-lanceolate, acute. Raceme elongate in fruit; pedicels 4·5–10 mm, patent. Sepals 2·5–4 mm; petals 4·5–7 mm, emarginate, usually glabrous on the back. Silicula 3–6 mm, suborbicular to orbicular-obovate, pubescent, truncate or emarginate; valves inflated, with a narrow, flattened margin; style 1·5–3·5 mm. Seeds 1·5–2 mm; wing usually 0·1–0·2 mm wide. *S.E.& E.C. Europe*. Al Au Bu Gr Ju Rm Rs(K) Tu.

(a) Subsp. **repens** (*A. transsilvanicum* Schur): Stems usually not more than 15 cm, procumbent or ascending. Petals 5–7 × 2–3 mm. Silicula up to 6 mm, suborbicular. $2n = 16$. *E.C. Europe.*

(b) Subsp. **trichostachyum** (Rupr.) Hayek, *Prodr. Fl. Penins. Balcan.* **1**: 436 (1925) (*A. trichostachyum* Rupr.): Stems up to 50 cm, erect or ascending. Petals 4–6 × 1·5–2·5 mm. Silicula 3·5–4·5 mm, orbicular-obovate. $2n = 48$. *Balkan peninsula; Krym.*

25. A. lenense Adams, *Mém. Soc. Nat. Moscou* **5**: 110 (1817) (?*A. fischerianum* DC.). Ascending perennial 10–30 cm, grey; hairs long, patent. Leaves linear-oblong or linear-lanceolate, acute. Raceme short and dense in fruit; pedicels 5–9 mm, erecto-patent. Sepals 3–4 mm; petals 6–7 mm, deeply emarginate; long stamens with a long tooth on the filament. Silicula (3–)4–7 mm, oblong-obovate or obovate, emarginate, glabrescent; valves inflated, flattened on the margin; style 1·5–3(–3·5) mm. Seeds 1·5–2 mm, not winged. *C. & E. Russia*. Rs(?N, C, W, E).

26. A. scardicum Wettst., *Biblioth. Bot. (Stuttgart)* **26**: 24 (1892). Green or grey-green, diffuse perennial 5–20 cm, with short or long non-flowering stems. Lower leaves elliptic-oblanceolate, obtuse; cauline leaves linear; hairs 0·4–0·6 mm in diameter, 5- to 9-rayed. Raceme short or long in fruit; pedicels up to 8 mm, patent. Sepals 2·5–3·5 mm; petals 4·5–6 mm, entire. Silicula 3–6(–7) × 2–4 mm, elliptical to elliptic-orbicular, densely pubescent; valves inflated, with a flattened margin; style 3–4 mm. Seeds *c*. 1·7 mm; wing *c*. 0·2 mm wide. $2n = 32$. *Stony places and cliffs*, 1700–2800 *m*. ● *Balkan peninsula*. Al Bu Gr Ju.

27. A. wulfenianum Bernh. in Willd., *Enum. Pl. Horti Berol. Suppl.*, 44 (1814). Erect or procumbent, diffuse, grey-green to almost white perennial up to 20 cm; non-flowering stems with rosettes. Basal leaves oblong-obovate, obtuse, gradually attenuate into the petiole; cauline leaves oblanceolate, larger than the basal; hairs *c*. 0·5 mm in diameter, 10- to 20-rayed. Raceme elongate in fruit; pedicels up to 8 mm, patent. Sepals 2·7–3·2 mm; petals 5·5–6·5 mm, obcordate, glabrous. Silicula 6–6·5 × 3–4 mm, elliptical, obtuse, sparsely pubescent; valves inflated; style 2–3 mm. Seeds *c*. 2·7 mm; wing 0·2 mm wide. ● *S.E. Alps*. Au It Ju.

28. A. ovirense A. Kerner, *Sched. Fl. Exsicc. Austro-Hung.* **2**: 99 (1882). Procumbent, diffuse, green or grey-green perennial up to 12 cm; non-flowering stems with rosettes. Basal leaves orbicular-obovate, abruptly contracted into the petiole; cauline leaves oblong-lanceolate, obtuse, about the same size as the basal; hairs *c*. 0·5 mm in diameter, 10- to 16-rayed. Raceme elongate in fruit; pedicels 6–7 mm, erecto-patent. Sepals 3·5–4 mm; petals 6–8 mm, entire or emarginate, stellate-pubescent on the back. Silicula 6·5–8 × 3·5–6 mm, elliptical or obovate, obtuse, sparsely pubescent; valves inflated, with flattened margin; style 2·5–3 mm. Seeds *c*. 2·5 mm. *Screes*. ● *S.E. Alps and W. Jugoslavia*. Au It Ju.

29. A. wierzbickii Heuffel, *Flora (Regensb.)* **18**: 242 (1835). Robust, erect, green or grey-green biennial or short-lived perennial 40–80 cm; non-flowering stems usually absent. Leaves ovate-elliptical to lanceolate, acute, green; hairs *c.* 0·5 mm in diameter, 5- to 7-rayed. Raceme dense in fruit; pedicels up to 10–13 mm, erecto-patent, with long patent hairs. Sepals 2·5–4 mm; petals 5–7 mm, truncate. Silicula 4–6 mm, orbicular or orbicular-ovate, pubescent, emarginate; valves inflated, flattened on the margin; style 3–4·5 mm. Seeds *c.* 3 mm; wing *c.* 0·1 mm wide. ● *C. Danube basin.* Bu Ju Rm.

30. A. calycocarpum Rupr., *Mém. Acad. Sci. Pétersb.* **15(2)**: 103 (1869). Erect perennial 15–35 cm; indumentum dimorphic on the inflorescence, grey-white; non-flowering stems more or less long. Lower leaves obovate or obovate-orbicular, obtuse, the upper oblanceolate. Raceme elongate in fruit; pedicels up to 9 mm, patent or erecto-patent. Sepals 2·5–4 mm; petals 4–7 mm. Silicula 4·5–6·5 mm, orbicular or orbicular-ovate, densely pubescent; valves inflated, with flattened margin; style 2–3·5 mm. Seeds *c.* 3 mm, winged. *Dry, rocky slopes.* ● *Krym.* Rs(K).

31. A. pulvinare Velen., *Sitz.-Ber. Böhm. Ges. Wiss.* **1889**: 30 (1889). Caespitose perennial 6–15 cm, with numerous rosettes and short non-flowering stems; indumentum dimorphic, grey. Basal leaves oblanceolate or spathulate, the upper lanceolate-spathulate, long-petiolate; hairs *c.* 1 mm in diameter. Racemes elongate in fruit; pedicels up to 8 mm. Sepals 3–4 mm; petals 5–6 mm, obovate, abruptly attenuate into claw. Silicula 4–5·5 mm, elliptical to suborbicular, pubescent; valves inflated, with a flattened margin; style 3–5 mm. Seeds *c.* 1·5 mm, broadly winged. $2n = 32$. ● *C. part of Balkan peninsula southwards to C. Greece; Romania.* Bu Gr Ju Rm.

32. A. montanum L., *Sp. Pl.* 650 (1753) (incl. *A. thessalum* Halácsy, *A. vourinonense* T. R. Dudley & Rech. fil.). Procumbent to erect, green to almost white perennial 5–25 cm, with non-flowering rosettes or short stems. Basal leaves oblong or oblanceolate-spathulate, the upper linear or linear-spathulate; hairs 0·4–0·7 mm in diameter, 6- to 24-rayed. Raceme elongate in fruit; pedicels 4–6 mm, patent or erecto-patent. Sepals 2·5–3·5 mm; petals emarginate. Silicula 3–5·5 mm; valves inflated, with flattened margin. Seeds 1·5–2 mm; wing 0·2–0·4 mm wide. *Most of Europe except the north, much of the west and the islands.* Al Au Bu Cz Ga Ge Gr He Hs Hu It Ju Po Rm Rs(B, C, W, K, E) Tu.

(**a**) Subsp. **montanum** (*A. thessalum* Halácsy): Procumbent or ascending with numerous non-flowering rosettes. Petals 4.5–6 mm; long stamens with a unilateral appendage. Silicula 3.5–5.5 mm, orbicular; style 1–3 mm. $2n = 16$. *Stony or rocky places, usually in the mountains. C. & S. Europe.*

(**b**) Subsp. **gmelinii** (Jordan) E. Schmid in Hegi, *Ill. Fl. Mitteleur.* **4(1)**: 451 (1919) (*A. gmelinii* Jordan): Ascending or erect with few non-flowering rosettes. Petals usually 3·5–4 mm; long stamens with a bilateral appendage. Silicula 3–5 mm, obovate-orbicular; style 2–3·5 mm. $2n = 32$. *Sandy places, usually lowland. C. & E. Europe.*

Subsp. (**a**) is very variable and possibly divisible into a number of other subspecies.

Dwarf plants from S. Spain (Sierra Nevada) which are like **32(a)** but with the basal leaves lanceolate or linear-lanceolate, the cauline linear, and hairs (0·5–)0·7–1 mm in diameter have been named **A. nevadense** Wilmott ex P. W. Ball & T. R. Dudley, *Jour. Arnold Arb.* **45**: 364 (1964).

33. A. fastigiatum Heywood, *Bull. Brit. Mus.* (*Bot.*) **1**: 92 (1954). Like **32(a)** but the stems 15–25 cm, numerous, erect, fastigiate; sepals 2–3 mm; petals 3–4·5 mm, deeply emarginate; silicula 2·5–3 mm; style 1·5–2·5 mm. ● *S.E. Spain* (*Sierra de Cazorla*). Hs.

34. A. gustavssonii Hartvig in Strid, *Mount. Fl. Gr.* **1**: 293 (1986). Like **32(a)** but caespitose, the flowering stems 2–5(–8) cm, ascending to erect; raceme condensed in fruit; pedicels 2·5–4 mm; sepals *c.* 2 mm; petals 3·5–4 mm, entire; silicula with sparse, 8- to 12-rayed stellate hairs; seeds wingless or very narrowly winged. *Limestone rocks and screes, 1850–2500 m.* ● *C. Greece.* Gr.

35. A. diffusum Ten., *Cat. Pl. Horti Neap.* app., ed. 1: 58 (1815). Diffuse, grey-green perennial up to 10 cm, with long, procumbent non-flowering stems. Basal leaves obovate or elliptical; upper leaves elliptical, cuneate or lanceolate; hairs 0·7–1 mm in diameter, sparse. Raceme elongate in fruit; pedicels 4–5 mm. Sepals 3–4 mm; petals 5–7 mm, emarginate. Silicula 4–6 mm, orbicular to ovate-orbicular; valves inflated, with a flattened margin; style 2·5–3·5 mm. Seeds 1·5–2 mm; wing *c.* 0·3 mm wide, or absent. ● *Mountains of S.W. Europe; C. & S. Italy.* Ga ?Gr Hs It.

36. A. cuneifolium Ten., *Prodr. Fl. Nap.* xxxvii (1811) (*A. brigantiacum* Jordan & Fourr.). Grey or almost white, caespitose perennial 5–15 cm with numerous non-flowering rosettes, and diffuse, flexuous flowering stems. Basal leaves oblong-obovate, attenuate into the petiole; cauline leaves usually narrower, obtuse or subacute; hairs *c.* 0·6 mm in diameter, multi-radiate, very crowded. Raceme very short and dense in fruit; pedicels up to 6 mm, patent. Sepals 3–4 mm; petals 5–8 mm, emarginate. Silicula (3–)5–7 × 3·5–4·5 mm, orbicular-elliptical, truncate, densely grey-pubescent; valves inflated, with a flattened margin; style 2–4 mm. Seeds 1·5–3 mm; wing *c.* 0·3 mm. ● *Mountains of S. Europe from the E. Pyrenees to the C. Appennini; very local.* Ga Hs It.

Similar plants have been reported from Macedonia but their identity requires confirmation.

37. A. arenarium Loisel., *Fl. Gall.* 401 (1807) (*A. loiseleurii* P. Fourn.). Like **36** but the basal leaves suborbicular; cauline leaves ovate-spathulate, obtuse; silicula 5–8 mm, suborbicular, emarginate. *Maritime sands.* ● *S.W. France and N. Spain.* Ga Hs.

38. A. atlanticum Desf., *Fl. Atl.* **2**: 71 (1798). Greyish-white, caespitose perennial 5–15 cm, with numerous non-flowering rosettes. Basal leaves oblong-obovate or oblong-oblanceolate, obtuse or subacute, not obviously petiolate; cauline leaves linear-oblanceolate; hairs *c.* 0.6 mm in diameter, multi-radiate, very crowded. Raceme short and crowded in fruit; pedicels 4–7 mm, patent or erecto-patent. Sepals 3·5–5·5 mm; petals 6–8 mm. Silicula 3–5·5 mm, orbicular-truncate or emarginate, pubescent; valves inflated, with a flattened margin; style 2·5–3 mm. Seeds *c.* 1·7 mm; wing 0–0·1 mm wide. *S. & E. Spain.* Hs.

39. A. moellendorfianum Ascherson ex G. Beck, *Ann. Naturh. Mus.* (*Wien*) **2**: 73 (1887) (?*A. galicicae* (Form.) Hayek). Silvery- or grey-lepidote, caespitose perennial 5–15 cm, with non-flowering rosettes or short stems. Leaves oblong-spathulate to orbicular-obovate, petiolate; hairs *c.* 0·4 mm in diameter. Raceme elongate in fruit; pedicels 5–7 mm, patent or erecto-patent. Sepals 2–2·5 mm; petals 5–6 mm, emarginate. Silicula 4·5–5 mm, orbicular, truncate or emarginate, densely lepidote; valves inflated, with a flattened margin; style 2–2·5 mm. Seeds *c.* 1·7 mm; wing *c.* 0·2 mm wide. ● *W. Jugoslavia.* Ju.

40. A. stribrnyi Velen., *Fl. Bulg.* 640 (1891) (*A. mildeanum* Podp.). Grey or silvery, diffuse perennial 6–20 cm, with few, long non-flowering stems. Basal leaves spathulate or obovate, subobtuse; cauline leaves oblanceolate, subobtuse or acute; hairs 0·6–0·7 mm in diameter, multiradiate. Raceme elongate in fruit; pedicels up to 8 mm, patent. Sepals 3·5–4·5 mm, persistent; petals 5·5–6·5 mm, emarginate. Silicula 4–5 mm, orbicular, truncate, densely pubescent; valves inflated, with a flattened margin; style 2–3·5 mm. Seeds *c.* 1·7 mm; wing absent or narrow. *C. part of Balkan peninsula.* Bu Ju Rm Tu.

41. A. sphacioticum Boiss. & Heldr. in Boiss., *Diagn. Pl. Or. Nov.* **2(8)**: 35 (1849). Silvery or grey perennial; flowering stems 5–10 cm, with procumbent non-flowering stems. Basal leaves obovate-orbicular to obovate; upper leaves linear or oblong, imbricate and appressed to the stem; hairs *c.* 0·5 mm in diameter, multi-radiate. Raceme short and dense in fruit; pedicels 10–15 mm, patent or deflexed. Sepals 3–4 mm; petals 5·5–6 mm, entire. Silicula 5–7 mm, obcordate to obovate or orbicular, emarginate, pubescent; valves inflated, with a flattened margin; style 3–4 mm. Seeds 2·5–3 mm, wider than long; wing 0·2–0·5 mm wide. *Screes.* ● *Kriti (Levka Ori).* Cr.

42. A. lassiticum Halácsy, *Consp. Fl. Graec. Suppl.* **1**, 10 (1908). White, dwarf shrub; flowering stems 10–20 cm, stout, with procumbent non-flowering stems. Basal leaves obovate, acute; cauline leaves linear-oblanceolate to obovate, obtuse; hairs *c.* 0·5 mm in diameter, very dense. Raceme short and dense in fruit; pedicels 4·5–6·5 mm, erecto-patent, with dimorphic indumentum. Sepals 4·5 mm; petals 6 mm, entire. Silicula 6–7 mm, orbicular, obtuse or truncate, densely pubescent; valves inflated; style 3·5–4 mm. Seeds 2·5 mm, orbicular, narrowly winged. ● *E. Kriti (Dhikti).* Cr.

43. A. idaeum Boiss. & Heldr. in Boiss., *Diagn. Pl. Or. Nov.* **2(8)**: 35 (1849). Diffuse, procumbent, grey-green to white perennial with simple flowering stems up to *c.* 5 cm. Leaves obovate-orbicular to ovate-oblong; hairs *c.* 0·3 mm in diameter, lepidote. Raceme short and dense in fruit; longest pedicels 4–5 mm, erecto-patent. Sepals 2·5–3·5 mm; petals 4–6 mm, entire. Silicula 5–7 mm, orbicular or ovate-orbicular, retuse, glabrous or minutely lepidote; valves inflated, with a scarcely flattened margin; style *c.* 2 mm. Seeds 2–2·5 mm; wing 0·1–0·2 mm wide. ● *E. Kriti.* Cr.

Incorrectly reported from the Kikladhes (Andros).

44. A. handelii Hayek, *Beih. Bot. Centr.* **45**: 279 (1928). Laxly caespitose, silvery-grey or grey-green perennial 5–15 cm, with elongate non-flowering stems. Leaves obovate, long-petiolate; hairs *c.* 0·3 mm in diameter, multi-radiate. Raceme short and dense in fruit; pedicels 6–9 mm, patent. Sepals 3·5–4 mm; petals 7–8 mm. Silicula 8–9 mm, obovate or broadly elliptical, subacute to truncate, pubescent; valves inflated; style 2–3 mm. Seeds 4–5 mm; wing 0·5–0·8 mm wide. *Limestone screes and rocks above* 2000 *m.* ● *N. Greece (Olimbos).* Gr.

45. A. doerfleri Degen, *Denkschr. Akad. Wiss. Math.-Nat. Kl. (Wien)* **64**: 708 (1897). Densely caespitose, white or silvery perennial 5–12 cm. Leaves linear-lanceolate to linear, acute, lepidote. Raceme slightly elongate in fruit; pedicels 3·5–5 mm, erecto-patent. Sepals 4–6 mm, with a tuft of long, divergent-rayed hairs at the apex; petals 8–9 mm; short stamens with free appendage 1·5–2 mm. Silicula 4–7·5 mm, elliptical, emarginate, white; valves inflated, with flat margin; style 2·5–5 mm. Seeds *c.* 2 mm, not winged. *Limestone screes and rocks.* ● *Macedonia and C. Greece; local.* Gr Ju.

46. A. taygeteum Heldr., *Sched. Herb. Graec. Norm.* no. 1405 (1897). Like **45** but the raceme not elongating in fruit; short stamens with a small appendage at the base; silicula elliptical to elliptic-orbicular, entire or shallowly emarginate; style up to 3 mm. ● *S. & C. Greece (Taÿetos, Giona).* Gr.

47. A. densistellatum T. R. Dudley, *Notes Roy. Bot. Gard. Edinb.* **24**: 160 (1962) (*A. praecox* auct. graec., non Boiss.). Diffuse to subcaespitose, grey or grey-green perennial up to 10 cm, the non-flowering stems with leaf-rosettes. Basal leaves obovate-spathulate, acute, with petiole up to 4 mm; cauline oblanceolate; hairs *c.* 0·7 mm in diameter, multi-radiate. Raceme elongate in fruit; pedicels 3–5 mm, patent. Sepals 3–4 mm; petals 5–6 mm, deeply emarginate or almost bifid. Silicula 4·5–6 mm, orbicular-elliptical, emarginate, grey-green; valves dimorphic, one strongly convex, with scarcely flattened margin, the other flat or slightly convex, with a wide flattened or deflexed margin; style 2·5–3·5 mm. Seeds *c.* 2 mm; wing 0–0·1 mm wide. *Serpentine rocks and stones.* ● *Greece (C. & N. Evvoia).* Gr.

Sect. ODONTARRHENA (C. A. Meyer) Koch. Perennials. Inflorescence a simple or compound raceme. Filaments with wings or appendages. Ovules solitary in each loculus.

48. A. murale Waldst. & Kit., *Pl. Rar. Hung.* **1**: 5 (1799) (*A. argenteum* auct. balcan., *A. argenteum* sensu E. I. Nyárády, non All., *A. decipiens* E. I. Nyárády, *A. gracile* Form., *A. orphanides* Janka ex E. I. Nyárády, *A. punctatum* E. I. Nyárády). Caespitose perennial 25–70 cm, with long non-flowering stems or dense leaf-rosettes. Basal leaves obovate- or oblanceolate-spathulate; cauline leaves usually 10–20 × 3–6 mm, lanceolate or oblanceolate, grey-green above, white or grey beneath, larger then the basal leaves. Petals 2–3·5 mm, entire, rarely emarginate. Silicula 2–5·5 × 1·5–4 mm, obtuse to emarginate, pubescent, the hairs with 6–10(–13) rays; valves flat, but often undulate; style 0·5–2 mm. Seeds *c.* 3 mm; wing 0·2–0·8 mm wide. $2n = 30$. *S.E. Europe.* Al Bu Gr Ju Rm Rs(W, K) [Au Ge].

A very variable species. The following is a key to some of the more distinct variants that have been described, but many intermediates occur and their distributions do not permit a satisfactory subspecific treatment.

1 Plant densely hirsute; hairs with long, spreading rays — **A. pichleri**
1 Plant appressed-pubescent; hairs with appressed rays
2 Style of silicula 0·5–1 mm; inflorescence subumbellate; pedicels filiform, flexuous — **A. chalcidicum**
2 Style of silicula (0·8–)1–2 mm; inflorescence corymbose, the branches not congested; pedicels stout, rigid
3 Hairs on the silicula at least 0·35 mm in diameter, ± dense; leaves acute — **A. murale**
3 Hairs on the silicula 0·2–0·3 mm in diameter, sparse; cauline leaves ± obtuse
4 Decumbent; silicula elliptic- to orbicular-obovate — **A. degenianum**
4 Usually erect; silicula elliptical or orbicular
5 Petals 2–2·5 mm, obtuse, sparsely pubescent; style glabrous — **A. chlorocarpum**
5 Petals 3–3·5 mm, emarginate; style sparsely pubescent — **A. subvirescens**

A. chalcidicum Janka, *Österr. Bot. Zeitschr.* **22**: 175 (1872). *W. & C. part of Balkan peninsula.*

A. chlorocarpum Hausskn., *Mitt. Thür. Bot. Ver.* nov. ser., **3–4**: 113 (1893). $2n = 16$. *W. & C. part of Balkan peninsula.*

A. degenianum E. I. Nyárády, *Bul. Grăd. Bot. Univ. Cluj* **7**: 74 (1927). *S. Bulgaria (Rodopi); Samothraki.*

A. murale *sensu stricto. Throughout the range of the species.*

A. pichleri Velen., *Fl. Bulg.* 38 (1891). *E. part of Balkan peninsula, S. Romania.*

A. subvirescens Form., *Verh. Naturf. Ver. Brünn* **37**: 195 (1898). *C. part of Balkan peninsula.*

49. A. tenium Halácsy, *Consp. Fl. Graec.* **1**: 93 (1900). Like **48** but leaves obovate-spathulate or broadly spathulate, very obtuse; silicula usually broadly obovate; valves not undulate. ● *S. Greece and Kikladhes.* Gr.

50. A. argenteum All., *Mélang. Philos. Math. Soc. Roy. Turin (Misc. Taur.)* **5**: 73 (1774). Plant 15–50 cm, erect, with long non-flowering stems. Cauline leaves 10–15(–20) × 3–4 mm, oblanceolate, greenish above, grey beneath, much larger than the basal leaves. Inflorescence lax. Petals 3·5–4 mm, entire. Silicula 3·5–6 × 2–4 mm, variable in shape, the apex usually obtuse, densely pubescent; hairs *c.* 0·5 mm in diameter, the rays 15–22; style 1–1·5(–2) mm. Seeds *c.* 2·5 mm; wing 0·5–0·8 mm wide. ● *S.W. Alps.* It.

51. A. fallacinum Hausskn., *Mitt. Thür. Bot. Ver.* nov. ser., **3–4**: 114 (1893) (*A. baldaccii* Vierh. ex E. I. Nyárády). Plant 20–50 cm, with long non-flowering stems. Cauline leaves 15–25 × 2–2·5 mm, oblanceolate, obtuse, greenish above, white beneath; basal leaves slightly smaller. Petals 2·5–3 mm; filaments with a multi-dentate appendage. Silicula 4–4·5 × 2·5–3·3 mm, broadly elliptical to obovate, obtuse or emarginate, septum frequently asymmetrical in outline; valves asymmetrically inflated, sparsely pubescent; hairs with *c.* 15 rays; style 1–1·5 mm. Seeds 1·5–2 mm; wing *c.* 0·2 mm. ● *Greece and Kriti.* Cr Gr.

52. A. corsicum Duby, *Bot. Gall.* **1**: 34 (1828). Plant 30–60 cm, grey, with numerous non-flowering rosettes on branched, woody stems. Basal leaves up to 15 × 7 mm, orbicular-spathulate, obtuse or emarginate; cauline leaves up to 35 × 7 mm, obovate to oblong-spathulate. Sepals *c.* 2 mm, subglabrous; petals 2·5–3 mm. Silicula 3–4 mm, obovate, obtuse, glabrous; style *c.* 1 mm. Seeds 1·8–2 mm, narrowly winged. $2n = 16$. *Naturalized in Corse.* [Co]. (*W. Anatolia.*)

53. A. heldreichii Hausskn., *Mitt. Thür. Bot. Ver.* nov. ser., **3–4**: 113 (1893). Plant 15–40 cm, with numerous, long non-flowering stems. Basal leaves linear-oblanceolate, obtuse, greyish-green above, silvery beneath; cauline leaves up to 30 × 3·5 mm, similar but greenish. Sepals *c.* 2 mm; petals *c.* 3 mm. Silicula 4–6 × 3·5–4·5 mm, obovate or orbicular, green, sparsely pubescent or glabrescent; hairs *c.* 9-rayed; valves flat or symmetrically inflated; style 0·7–1·5 mm. Seeds *c.* 2 mm; wing 0·2–0·3 mm wide. ● *C. & N. Greece.* Gr.

54. A. robertianum Bernard ex Gren. & Godron, *Fl. Fr.* **1**: 117 (1847). Plant 20–40 cm, procumbent, much-branched at the base, with long non-flowering stems. Leaves up to 8 × 4 mm, obovate-spathulate, obtuse, greenish above, grey beneath. Sepals *c.* 2 mm, usually subglabrous; petals 3–4 mm. Silicula 5–6 × 3–4 mm, broadly obovate to rhombic, acute, sparsely pubescent, green; style *c.* 2 mm. Seeds 2·5–3 mm; wing *c.* 0·25 mm wide. $2n = 16$. ● *Corse, Sardegna.* Co Sa.

Plants from Sardegna have been described as **A. tavolarae** Briq., *Prodr. Fl. Corse* **2**: 58 (1913). They have slightly larger sepals and petals, the sepals and silicula pubescent, and seeds *c.* 1·5 mm.

55. A. smolikanum E. I. Nyárády, *Bul. Grăd. Bot. Univ. Cluj* **9**: 43 (1928). Plant up to 15 cm, pulvinate; lower stem stout, woody, branched; non-flowering rosettes numerous. Basal leaves 7–11(–15) × 4–7 mm, broadly obovate, obtuse or truncate, grey-green above, white beneath; cauline leaves smaller. Sepals 2–2·5 mm; petals 3–4 mm, with orbicular limb. Silicula 5–6 × 2·5–3 mm, narrowly elliptical, acute or subacute, green, glabrous or sparsely pubescent; style 1·7 mm. Seeds winged. ● *N.W. Greece and S. Albania.* Al Gr.

56. A. bertolonii Desv., *Jour. Bot. Appl.* **3**: 172 (1814) (*A. rigidum* (E. I. Nyárády) E. I. Nyárády). Plant (10–)20–35 cm, erect, woody at the base; non-flowering stems long. Basal leaves up to 12 × 3 mm, spathulate, green above, almost white beneath, rarely green on both surfaces; upper cauline leaves up to 25 × 5 mm, oblanceolate. Sepals *c.* 2 mm; petals 3–3·5 mm, entire. Silicula 4–6 mm, variable, usually elliptical or orbicular-obovate, subacute or truncate, glabrous or sparsely pubescent; hairs 0·2–0·3 mm in diameter; valves asymmetrically or symmetrically inflated; style 0·8–2 mm. Seeds 1·8–2 mm, winged. ● *N. & C. Italy and N.W. part of Balkan peninsula.* Al Gr It Ju.

(a) Subsp. **bertolonii**: Silicula 2·3–3 mm wide. Seed-wing *c.* 0·2 mm wide. *Italy.*

(b) Subsp. **scutarinum** E. I. Nyárády, *Bul. Grăd. Bot. Univ. Cluj* **7**: 87 (1927) (*A. balkanicum* E. I. Nyárády, *A. janchenii* E. I. Nyárády, *A. kosaninum* E. I. Nyárády, *A. nebrodense* auct. balcan., non Tineo): Silicula 2·8–5·5 mm wide. Seed-wing up to 0·5 mm wide. $2n = 16$. *N.W. part of Balkan peninsula.*

57. A. markgrafii O. E. Schulz, *Ber. Deutsch. Bot. Ges.* **49**: 422 (1926). Plant 20–50 cm, erect, with non-flowering rosettes or short stems; cauline hairs 1–3 mm in diameter, 3- to 5(–9)-rayed. Cauline leaves up to 22 × 2–6 mm, oblanceolate, greenish above, grey beneath, distinctly larger than the basal. Sepals *c.* 1·5 mm; petals 2–2·5 mm, entire. Silicula 2–3·4 × 1·7–3·4 mm, suborbicular, glabrous; valves symmetrically inflated; style 1–1·2 mm. Seeds *c.* 1·7 mm; wing 0·2–0·3 mm wide. ● *Albania, and adjacent mountains of Jugoslavia.* Al Ju.

58. A. corymbosoides Form., *Verh. Naturf. Ver. Brünn* **34**: 329 (1895) (*A. rhodopense* Form., *A. rechingeri* E. I. Nyárády, *A. vranjanum* E. I. Nyárády). Erect, short-lived perennial (5–)15–35 cm, with long non-flowering stems. Cauline leaves 12–16 × 1·8–2·8 mm, usually oblanceolate, acute or subobtuse, greenish above, grey-green beneath; basal leaves 4·5–6 × 1·6–2·3 mm, obtuse. Sepals 1·5–2 mm, pubescent; petals 2–3 mm. Silicula 2·5–4 × 2–2·8 mm. elliptical to obovate, obtuse or emarginate, pubescent, rarely glabrescent; hairs 0·3 mm in diameter, 5- to 7-rayed; valves asymmetrically inflated; style 0·5–1 mm. Seeds *c.* 1·5 mm, not or very narrowly winged. $2n = 32$. ● *S. & E. part of Balkan peninsula; Romania.* Bu Gr Ju Rm.

59. A. euboeum Halácsy, *Consp. Fl. Graec.* **1**: 93 (1900). Plant erect, 8–25 cm, woody and leafless at the base, with long non-flowering stems. Basal leaves up to 18 × 5 mm, oblanceolate or obovate, greyish-white; cauline leaves smaller, grey-green. Sepals *c.* 2 mm; petals *c.* 3·5 mm. Silicula 3·5–4·5 × 2·5–3 mm, elliptical, obovate or rarely suborbicular, obtuse, greyish-green; valves asymmetrically inflated; style 1·2–2 mm. Seeds 1·5–1·7 mm; wing absent or up to 0·3 mm wide. ● *E. Greece* (*Evvoia*). Gr.

60. A. sibiricum Willd., *Sp. Pl.* 3(1): 465 (1800) (*A. suffrutescens* (Boiss.) Halácsy, *A. montanum* subsp. *epirotium* (Baumg.) Hayek, *A. epirotium* (Baumg.) E. I. Nyárády, *A. halacsyi* E. I. Nyárády, *A. lepidulum* E. I. Nyárády, *A. minutiflorum* Boiss., *A. murale* subsp. *dramense* E. I. Nyárády, *A. novakii* E. I. Nyárády). Plant 5–20 cm, erect or procumbent, stout and woody at the base, with non-flowering rosettes or short stems. Basal leaves up to 7(–15) × 3·5 mm, spathulate, grey-green to white; cauline leaves oblanceolate to suborbicular, grey-green, the largest near the apex of the stem; hairs with unequal, often patent rays up to 1·5 mm in diameter. Sepals 1–1·5 mm; petals 2–2·5 mm. Silicula 3–4 × 2–3·5 mm, broadly obovate, obtuse or emarginate, grey; hairs 0·4–0·7 mm in diameter, with 10–14 unequal rays; valves asymmetrically inflated; style 0·8–1·1 mm. Seeds *c.* 1·5 mm, not winged. *Balkan peninsula.* Al Gr Ju Tu.

61. A. caliacrae E. I. Nyárády, *Bul. Grăd. Bot. Univ. Cluj* 6: 92 (1926) (*A. obtusifolium* subsp. *cordatocarpum* E. I. Nyárády, *A. eximeum* (E. I. Nyárády) E. I. Nyárády, *A. racemosum* (E. I. Nyárády) E. I. Nyárády). Plant 10–30 cm, procumbent or ascending, woody at the base, with short or elongate non-flowering stems. Leaves up to 7 × 3 mm, the smallest near the apex of the stem, oblanceolate, white; hairs 0·5–1 mm in diameter. Sepals 1·9–2 mm; petals 2–2·8 mm. Silicula 3–5 × 2·5–4 mm, obovate to obovate-orbicular, emarginate, grey-green, punctate; hairs 0·4–0·5 mm in diameter, persistent; valves asymmetrically inflated; style 0·6–1·5 mm. Seeds *c.* 1·5 mm; wing absent or 0·2–0·5 mm wide. *S.E. Europe.* Bu Gr Ju Rm Rs(K).

62. A. borzaeanum E. I. Nyárády, *op. cit.* 90 (1926). Plant 10–30 cm, ascending or erect, with short or long non-flowering stems. Basal leaves 6–7 × 3–5 mm, obovate-spathulate to suborbicular, obtuse; cauline leaves oblanceolate, the largest near the apex of the stem, whitish; hairs 0·4–1·0 mm in diameter. Sepals *c.* 1·5 mm; petals 2–2·3 mm. Silicula 2–3 × 2–3 mm, obcordate, grey; hairs 0·2–0·4 mm in diameter, easily displaced; valves asymmetrically inflated; style *c.* 1 mm. Seeds *c.* 1·5 mm, not winged. *Black Sea coasts.* Bu Rm Rs(W, K).

63. A. obtusifolium Steven ex DC., *Reg. Veg. Syst. Nat.* 2: 305 (1821) (*A. tortuosum* subsp. *savranicum* E. I. Nyárády pro parte). Plant up to 30(–35) cm, ascending, with long non-flowering stems. Basal leaves (5–)7–10 × 4·5–5(–6) mm, spathulate to suborbicular, obtuse; cauline leaves 10–15 × 3–4 mm, oblanceolate-spathulate, white. Sepals 2–3 mm; petals 2·5–3·5 mm. Silicula 2·3–3·6 × 2·5–3·3 mm, elliptical to suborbicular, rarely obovate, obtuse or truncate, sparsely pubescent; valves asymmetrically inflated; style 1–1·5 mm. Seeds not winged. *S.E. Europe.* Bu Gr Rm Rs(K).

(a) Subsp. **obtusifolium**: Stem 10–30(–35) cm. Basal leaves 7–10 mm, spathulate. $2n = 32$. *Throughout the range of the species except Thasos.*

(b) Subsp. **helioscopioides** E. I. Nyárády, *Bul. Grăd. Bot. Univ. Cluj* 7: 167 (1927): Stem 5–10 cm. Basal leaves 5–7 mm, suborbicular, very crowded. ● *Thasos.*

64. A. obovatum (C. A. Meyer) Turcz., *Bull. Soc. Nat. Moscou* 1: 57 (1837) (*A. fallax* E. I. Nyárády, *A. odessanum* E. I. Nyárády). Like subsp. (a) but plant not more than 10(–15) cm, with short non-flowering stems; basal leaves 6–8 × *c.* 3 mm, obovate-spathulate to suborbicular; cauline leaves smaller; silicula 3·5–5 × 2·5–3·5 mm, elliptical, densely white-pubescent; hairs *c.* 0·3 mm in diameter, 20- to 25-rayed. *S. Ukraine, S.E. Russia.* Rs(W, E).

65. A. tortuosum Willd., *Sp. Pl.* 3(1): 466 (1800) (*A. grintescui* E. I. Nyárády). Plant 6–35(–60) cm, procumbent or ascending, with few, long non-flowering stems. Largest leaves 9–25 × 2–5 mm, spathulate or oblanceolate, more or less acute, grey or white; cauline leaves sometimes larger than the basal. Sepals *c.* 2 mm; petals *c.* 2·5 mm. Silicula 3–4 × 2–2·5 mm, elliptical or elliptic-obovate, truncate or emarginate, densely grey-pubescent; valves usually asymmetrically inflated; style 0·5–1 mm, pubescent. Seeds *c.* 1·5 mm, not winged. $2n = 32$. *E. & E.C. Europe.* Bu Cz Gr Hu Ju Rm Rs(C, W, K, E) Tu.

66. A. longistylum (Sommier & Levier) Grossh. in Grossh. & Schischkin, *Sched. Herb. Pl. Or. Exsicc.* 1: 18 (1924) (*A. tortuosum* var. *longistylum* (Sommier & Levier) N. Busch, *A. cuneipetalum* E. I. Nyárády). Like **65** but the petals 3–3·5 mm; silicula 4–6 × 2·5–3 mm, attenuate towards the apex; style 1·5–3 mm, glabrous. *Krym.* Rs(K). (*Caucasus.*)

67. A. alpestre L., *Mantissa* 92 (1767). Plant 5–15(–20) cm, usually procumbent, with numerous non-flowering rosettes. Leaves 4–7 × 1·5–3·5 mm, oblanceolate or obovate-spathulate, white or rarely grey-green. Sepals 2–2·5 mm; petals 3–3·5 mm, entire. Silicula 2·5–4·5 × 2·5–3·5 mm, elliptical, subacute to emarginate, white-pubescent; hairs 12- to 16-rayed; valves asymmetrically inflated; style 0·7–1·7 mm. Seeds 1·5–2 mm, not or narrowly winged. ● *S.W. & C. Alps, eastwards to* 7°50′ *E.* Ga He It.

68. A. serpyllifolium Desf., *Fl. Atl.* 2: 70 (1798) (*A. argenteum* sensu E. I. Nyárády, non All., *A. mijasense* E. I. Nyárády, *A. pyrenaicum* (Jordan & Fourr.) E. I. Nyárády). Plant up to 30 cm, procumbent to erect, with numerous non-flowering rosettes or short stems. Leaves up to 18 × 4 mm, oblanceolate or obovate-spathulate, grey or white beneath, grey or grey-green above, plicate on the non-flowering stems. Sepals 1·5–2 mm; petals 2–2·5(–3) mm, entire. Silicula 2·5–4·5 × (1·5–)2–3·5 mm, broadly elliptical or elliptic-rhombic to obovate, usually subacute, densely white-pubescent; hairs 12- to 16-rayed; valves asymmetrically inflated; style 0·8–1·5 mm. Seeds 1·3–1·8 mm, not or only narrowly winged. $2n = 32$. *S.W. Europe.* Ga Hs Lu.

Very variable and possibly divisible into a number of subspecies.

69. A. nebrodense Tineo, *Pl. Rar. Sic. Pug.* 12 (1817). Plant 3–12(–16) cm, procumbent, caespitose, with numerous, short non-flowering stems. Basal leaves orbicular-spathulate, obtuse, white; cauline leaves up to 8–9 × 3 mm, usually spathulate, grey-green. Sepals 2 mm; petals 2·5–3 mm. Silicula (3·5–)4–6 × 2·5–4 mm, elliptical, obtuse or emarginate, white-pubescent; hairs 12- to 16-rayed; valves asymmetrically inflated; style 0·7–2 mm. Seeds *c.* 1·75 mm, not winged. *Calcareous rocks and screes.* ● *Sicilia; C. Greece.* Gr Si.

(a) Subsp. **nebrodense**: Stems somewhat woody at the base. Stellate hairs on leaves 0·5–0·75 mm in diameter, those on the silicula 0·25–0·5 mm in diameter. *Sicilia* (*Madonie*).

(b) Subsp. **tenuicaule** Hartvig in Strid, *Mount. Fl. Gr.* 1: 300 (1986): stems not woody at the base, more slender than those of subsp. (a). Stellate hairs on leaves 0·4–0·5 mm in diameter, those on the silicula 0·18–0·25 mm in diameter. Pedicels somewhat shorter in fruit than in subsp. (**a**). *S.C. Greece* (*Vardhousia*).

70. A. fragillimum (Bald.) Rech. fil., *Denkschr. Akad. Wiss. Math.-Nat. Kl.* (*Wien*) **105(2)**: 77 (1943) (*A. nebrodense* var. *fragillimum* (Bald.) Halácsy). Like **69** but plant 1–3 cm; basal leaves lanceolate- to orbicular-spathulate, subacute; cauline leaves 3–4 × 0·5–1·5 mm, orbicular- to linear-spathulate; silicula 3·5–5 × 2–2·5 mm, sparsely pubescent; hairs (16–)18- to 23-rayed. ● *Kriti.* Cr.

50. Fibigia Medicus

P. W. BALL

Perennials; hairs mostly stellate, rarely a few simple. Sepals erect; petals yellow, shortly clawed. Fruit a strongly compressed latiseptate silicula; valves flat, not veined; style long; stigma more or less capitate. Seeds 2–8 in each loculus, winged.

1 Plant usually more than 30 cm, without numerous vegetative shoots at the base; sepals 4·5–7 mm; raceme usually more than 10 cm in fruit **1. clypeata**
1 Plant usually less than 30 cm, with numerous vegetative shoots at the base; sepals 7–10 mm; raceme less than 10 cm in fruit
2 Leaves with lanate indumentum; silicula at least 9 mm wide, not more than twice as long as wide **2. lunarioides**
2 Leaves with tomentose indumentum; silicula not more than 8 mm wide, usually at least twice as long as wide **3. triquetra**

1. F. clypeata (L.) Medicus, *Pflanzengatt.* **1**: 91 (1792) (*Farsetia clypeata* (L.) R. Br.). Stems 30–75 cm. Lower leaves oblong or oblanceolate, usually green or grey-green, with lanate indumentum. Raceme 10–20 cm in fruit; pedicels 2–5 mm. Sepals 4·5–7 mm; petals 8–13 mm. Silicula 14–28 × 9–13 mm, elliptical. $2n = 16$. *N. & C. Italy; Balkan peninsula southwards to N.E. Greece; Krym.* Al Bu Gr It Ju Rs(K) [Au Ga].

F. eriocarpa (DC.) Boiss., *Fl. Or.* **1**: 258 (1867), from N.W., C. & S. Greece, differs from **1** only in the long unbranched hairs on the silicula, and is probably not specifically distinct.

2. F. lunarioides (Willd.) Sm. in Sibth. & Sm., *Fl. Graeca* **7**: 22 (1830). Stems 5–30 cm, woody at the base, subcaespitose, with numerous leafy shoots. Lower leaves obovate-lanceolate to linear-spathulate, with ash-white, lanate indumentum. Raceme not more than 5 cm in fruit; pedicels 5–15 mm. Sepals 7–9 mm; petals 12–16 mm. Silicula 12–22 × 9–18 mm, elliptical to orbicular. *Cliffs.* ● *Kikladhes, Kriti.* Cr Gr.

3. F. triquetra (DC.) Boiss. ex Prantl in Engler & Prantl, *Natürl. Pflanzenfam.* **3(2)**: 196 (1891). Stems 5–20 cm, woody at the base, subcaespitose, with numerous leafy shoots. Lower leaves obovate to oblanceolate, with very closely appressed, silver-grey indumentum. Raceme up to 8 cm in fruit; pedicels 3–10 mm. Sepals 8–10 mm; petals 13–19 mm. Silicula 12–17 × 5–8 mm, elliptical or elliptic-oblong. *Cliffs.* ● *W. Jugoslavia.* Ju.

51. Berteroa DC.

P. W. BALL

Annuals to perennials; hairs stellate and sometimes also simple. Sepals erect to erecto-patent, not saccate at the base; petals white or pale yellow, sometimes becoming reddish, deeply bifid; outer filaments with a tooth at the base, the inner broad towards the base or toothed. Fruit a latiseptate silicula; valves flat or somewhat inflated, without a conspicuous median vein; style distinct; stigma capitate. Seeds 2–6 in each loculus, sometimes winged.

All species occur in sandy, stony or rocky places, or as ruderals.

1 Petals pale yellow; silicula about as long as wide, orbicular or obovate-orbicular **3. orbiculata**
1 Petals white or very pale cream, sometimes becoming reddish; silicula 1·5–3 times as long as wide, elliptical or elliptic-orbicular
2 Style 0·5–1 mm (Crna Gora) **5. gintlii**
2 Style 1–4 mm
3 Silicula inflated; seeds margined, but not conspicuously winged **4. incana**
3 Silicula flat; seeds winged
4 Silicula pubescent with stellate hairs; style usually at least 2 mm **1. obliqua**
4 Silicula ±glabrous; style 1–2 mm **2. mutabilis**

1. B. obliqua (Sm.) DC., *Reg. Veg. Syst. Nat.* **2**: 292 (1821) (*B. mutabilis* auct. ital., pro parte, *Alyssum mutabile* auct. ital., pro parte; incl. *B. stricta* f. *pindicola* (Halácsy) Hayek). Biennial or perennial 10–50 cm. Leaves lanceolate. Petals 5–8 mm, white becoming reddish. Silicula 8–12 × 4–5 mm, elliptical, flat, pubescent with stellate and sometimes with unbranched hairs; style 1·5–4 mm. Seeds broadly winged. *Balkan peninsula southwards from c. 43° N.; C. & S. Italy.* Al Bu Gr It Ju Tu.

2. B. mutabilis (Vent.) DC., *loc. cit.* (1821). Like **1** but leaves lanceolate to obovate-elliptical; petals 4–6 mm; silicula 6–12 × 3·5–5 mm, more or less glabrous; style 1–2 mm. $2n = 16$. *Balkan peninsula; S. Italy.* Al Bu Gr It Ju Tu.

Possibly only a subspecies of **1**.

3. B. orbiculata DC., *op. cit.* 293 (1821) (*B. samolifolia* sensu Hayek). Biennial or perennial 10–100 cm. Leaves oblong to ovate, entire or sinuate-dentate. Petals 4–7 mm, pale yellow. Silicula 7–11 × 6–9 mm, orbicular to obovate-orbicular, flat, pubescent with stellate hairs; style 2–4 mm. Seeds winged. $2n = 16$. *Macedonia, E. Albania.* ?● Al Gr Ju.

4. B. incana (L.) DC., *op. cit.* 291 (1821) (*Alyssum incanum* L., *Farsetia incana* (L.) R. Br.; incl. *B. stricta* Boiss. & Heldr.). Annual to perennial up to 70 cm. Basal leaves obovate; cauline leaves lanceolate or oblong, usually entire. Petals 4·5–7·5 mm, white. Silicula 4–10 mm, elliptical or ovate, inflated, with convex valves, pubescent; style 1–4 mm. Seeds not winged, sometimes margined. $2n = 16$. *C. & E. Europe, extending to Denmark and Italy, but precise native distribution uncertain owing to widespread introduction.* Al Au Bu Cz Da Ge Gr He Hu It Ju Po Rm Rs(N, B, C, W, K, E) Tu [Be Br Fe Ga Ho Hs No Su].

5. B. gintlii Rohlena, *Sitz.-Ber. Böhm. Ges. Wiss.* **1904**: 24 (1905). Like **4** but the silicula ovate-elliptical, pubescent with stellate hairs when young, more or less glabrous when mature; style 0·5–1 mm; seeds winged. ● *Crna Gora.* Ju.

52. Aurinia (L.) Desv.

J. R. AKEROYD

Perennials, more or less woody at the base; hairs stellate, branched or lepidote. Basal leaves considerably larger than the cauline; petioles of basal leaves grooved on the upper surface, with persistent, swollen bases. Flowering stems usually axillary. Flower-buds globose. Sepals erecto-patent, not saccate at the base; petals yellow or white, emarginate to deeply bifid, sometimes entire; filaments not winged or toothed, with very small, suborbicular appendage at the base. Fruit a latiseptate, usually globose silicula; valves without a conspicuous median vein; style short; stigma capitate. Seeds (1)2–6(–8) in each loculus, usually winged. (*Alyssum* Sect. *Aurinia* (Desv.) Koch; incl. *Lepidotrichum* Velen. & Bornm.)

Literature: T. R. Dudley, *Jour. Arnold Arb.* **45**: 390–400 (1964).

1 Petals white; seeds 1 or 2 in each loculus
 2 Petals entire; stems 5–20 cm, unbranched **8. rupestris**
 2 Petals deeply bifid; stems usually more than 20 cm, divaricately branched **9. uechtritziana**
1 Petals yellow; seeds 2–6 in each loculus
 3 Silicula at least 6 mm
  4 Seeds (excluding wing) less than 3 mm; valves of silicula flat **5. saxatilis**
  4 Seeds (excluding wing) 3–4·5 mm; valves of silicula inflated
   5 Sepals 1·5–2·5 mm **2. gionae**
   5 Sepals 2·5–3·5 mm **3. leucadea**
 3 Silicula less than 6 mm
  6 Seeds 2 in each loculus
   7 Valves of silicula inflated; seeds less than 2 mm, the wing not more than 0·3 mm **4. petraea**
   7 Valves of silicula flat; seeds 2–3 mm, the wing at least 0·3 mm **5. saxatilis**
  6 Seeds (2–)4–8 in each loculus
   8 Stems less than 15 cm; valves of silicula almost flat **6. moreana**
   8 Stems usually more than 15 cm; valves of silicula inflated
    9 Petals *c.* 4 mm, bifid; seeds 2–4 in each loculus **1. corymbosa**
    9 Petals 5–8 mm, emarginate; seeds 4–8 in each loculus **7. sinuata**

1. A. corymbosa Griseb., *Spicil. Fl. Rumel.* **1**: 271 (1843) (*Alyssum corymbosum* (Griseb.) Boiss.). Stems 20–50 cm; hairs branched or more or less stellate. Basal leaves oblanceolate to lanceolate or obovate, entire or sinuate-dentate. Inflorescence corymbose. Sepals (1·5–)2–2·5 mm; petals *c.* 4 mm, bifid, yellow. Silicula (3·5–)4·5–5·5 mm, orbicular to elliptic-orbicular, glabrous; valves strongly inflated; style 1–2 mm. Seeds 1·7–2·5 mm, 2–4 in each loculus; wing 0·2–0·5 mm wide. $2n = 16$. *Rocks.* ● *W. & S. part of Balkan peninsula.* Al Gr Ju.

2. A. gionae (Quézel & Contandriopoulos) Greuter & Burdet, *Willdenowia* **13**: 86 (1983) (*Alyssum gionae* Quézel & Contandriopoulos). Like **1** but basal leaves longer, oblanceolate to linear-lanceolate; inflorescence more or less paniculate; petals 4–5·5 mm; silicula 6–11(–13) mm; seeds 3–4·5 mm, (2–)4(–6) in each loculus, the wing (0·2–)0·5–1 mm wide. $2n = 16$. *Limestone rocks above 1800 m.* ● *N.W. & C. Greece.* Gr.

Perhaps only subspecifically distinct from **1**.

3. A. leucadea (Guss.) C. Koch, *Hort. Dendrol.* 23 (1853) (*Alyssum leucadeum* Guss.; incl. *A. medium* Host). Plant 10–40 cm, usually woody at the base; hairs stellate. Basal leaves oblong-lanceolate, entire or sinuate-dentate. Inflorescence racemose. Sepals 2·5–3·5 mm; petals 5–6 mm, deeply emarginate, yellow. Silicula 7–10 mm, globose or ovoid-globose, glabrous; valves inflated; style 1–2 mm. Seeds 3–4 mm, 4(–6) in each loculus; wing 0·5–0·7 mm wide. *Cliffs and rocks.* ● *Coasts of the Adriatic.* It Ju.

4. A. petraea (Ard.) Schur, *Enum. Pl. Transs.* 61 (1866) (*Alyssum petraeum* Ard., *A. gemonense* L., *A. edentulum* Waldst. & Kit.; incl. *A. microcarpum* Vis.). Stems 15–60 cm; hairs branched or more or less stellate. Basal leaves obovate-oblong, sinuate or pinnatifid. Inflorescence racemose. Sepals *c.* 2 mm; petals 4–4·5 mm, bifid, yellow. Silicula 3–5 mm, elliptical to obovate, glabrous; valves inflated but with narrow flattened margin; style 1–1·5 mm. Seeds 1·5–1·8 mm, 2 in each loculus; wing 0·1(–0·3) mm wide. ● *From N. Italy to Romania and N. Greece.* Al Gr It Ju Rm [Au Ga Ge].

5. A. saxatilis (L.) Desv., *Jour. Bot. Appl.* **3**: 162 (1815) (*Alyssum saxatile* L.). Plant 10–40(–50) cm, often woody at the base; hairs stellate. Basal leaves obovate to oblanceolate, sinuate-pinnatifid to entire. Inflorescence corymbose. Sepals 2–3(–4) mm; petals 3–6(–8) mm, yellow, emarginate or bifid. Silicula glabrous; valves almost flat. Seeds 2–2·7 mm, 2 in each loculus; wing 0·3–1·1 mm wide. *Limestone rocks. C. & S.E. Europe, extending to S.W. Italy.* Al Au Bu Cr Cz Ge Gr Hu It Ju Po Rm Rs(W, E) Tu [Ga].

1 Silicula usually longer than wide, with rounded apex; basal leaves entire to somewhat dentate (a) subsp. **saxatilis**
1 Silicula wider than long or as wide as long, with emarginate or truncate apex; basal leaves usually dentate to sinuate-pinnatifid
 2 Silicula not more than 6 × 6 mm (b) subsp. **orientalis**
 2 Silicula usually more than 6 × 6 mm (c) subsp. **megalocarpa**

(a) Subsp. **saxatilis**: Basal leaves entire to somewhat dentate. Silicula 3·5–5(–6) × 2·5–4(–5) mm, usually longer than wide, rounded at the apex; style 0·3–0·8 mm. $2n = 16, 48$. ● *C. Europe and N. part of Balkan peninsula; widely cultivated for ornament, and occasionally becoming naturalized.*

(b) Subsp. **orientalis** (Ard.) T. R. Dudley, *Jour. Arnold Arb.* **45**: 394 (1964) (*Alyssum orientale* Ard.): Basal leaves dentate or sinuate-pinnatifid, sometimes entire. Silicula 3·5–5·5 × 3·5–6 mm, wider than or as wide as long, emarginate or truncate at the apex; style 0·5–1·5 mm. $2n = 16$. *Balkan peninsula, mainly in the S. part.*

(c) Subsp. **megalocarpa** (Hausskn.) T. R. Dudley, *op. cit.* 397 (1964) (*Alyssum orientale* var. *megalocarpum* Hausskn.): Basal leaves dentate or sinuate-pinnatifid, sometimes entire. Silicula 6–9 × 6·5–10 mm, wider than or as wide as long, emarginate or truncate at the apex; style 1–2·5 mm. $2n = 16$. *S. Italy; Aegean region.*

Intermediates between subspp. (b) and (c) occur in the Ionioi Nisoi and on the coasts of S. Greece.

6. A. moreana Tzanoudakis & Iatroú, *Bot. Chron.* **1**: 22 (1981) (*Alyssum orientale* Ard. var. *alpinum* Halácsy). Like **5(a)** but with stouter rootstock; flowering stems up to 12 cm, unbranched; basal leaves oblanceolate to spathulate, entire or obscurely sinuate-dentate, greyish green; inflorescence a simple raceme; seeds 3–6 in each loculus. $2n = 16$. *Limestone rocks, 1500–2000 m.* ● *S. Greece (N. Peloponnisos).* Gr.

7. A. sinuata (L.) Griseb., *Spicil. Fl. Rumel.* **1**: 271 (1843) (*Alyssum sinuatum* L., *Alyssoides sinuata* (L.) Medicus, *Vesicaria* *sinuata* (L.) Cav.). Stems 15–50 cm, little-branched, woody only at the extreme base, sparsely to densely grey-pubescent. Leaves oblanceolate to lanceolate, the lower often sinuate-dentate. Sepals 3–4 mm; petals 5–8 mm, emarginate, pale yellow. Silicula 7–12 mm, globose or ellipsoidal; valves strongly inflated; style 2–4 mm. Seeds 4–8 in each loculus, usually winged. *Rocks and cliffs.* ● *W. part of Balkan peninsula; S.E. Italy.* Al Gr It Ju.

8. A. rupestris (Heynh.) Cullen & T. R. Dudley, *Jour. Arnold Arb.* **45**: 399 (1964) (*Alyssum rupestre* Ten., non Willd., *Ptilotrichum rupestre* (Heynh.) Boiss., *P. cyclocarpum* auct., non Boiss., *P. cyclocarpum* subsp. *pindicum* Hartvig). Caespitose; stems 5–20 cm, unbranched, woody at the base. Basal leaves lanceolate or linear; cauline leaves linear; all silvery-tomentose with stellate hairs. Petals 5 mm, entire, white. Silicula 5 mm, obovoid-globose, lepidote or glabrous; style 0·5 mm. Seeds 1 or 2 in each loculus, very narrowly winged. *Calcareous screes and cliffs. Mountains of C. Italy and W. part of Balkan peninsula.* Al Gr It Ju.

9. A. uechtritziana (Bornm.) Cullen & T. R. Dudley, *op. cit.* 398 (1964) (*Lepidotrichum uechtritzianum* (Bornm.) Velen.). Stems 20–60 cm, divaricately branched, grey-pubescent; hairs lepidote. Basal leaves oblong-lanceolate, entire; cauline leaves linear. Inflorescence very lax in fruit; pedicels 4–6 mm. Petals 4–5 mm, deeply bifid, white. Silicula 2–4 × 2–2·5 mm, elliptic-orbicular, appressed to the axis of the inflorescence; style *c.* 1 mm. Seeds 1 or 2 in each loculus, not winged. *Maritime sands. Black Sea coasts of Bulgaria and Turkey-in-Europe.* Bu Tu.

53. Bornmuellera Hausskn.

V. H. HEYWOOD (EDITION 1) REVISED BY J. R. AKEROYD (EDITION 2)

Perennial herbs or small shrubs; hairs bi- to 6-fid, rarely the plant glabrous. Sepals erecto-patent, not saccate at the base; petals entire, white; filaments with a tooth-like appendage at the base. Fruit a latiseptate silicula; valves flat or inflated, without a conspicuous median vein; style short; stigma capitate or emarginate. Seeds 1 or 2 in each loculus, winged or not.

1 Leaves not more than 3 mm wide, linear to linear-lanceolate or oblanceolate **2. baldaccii**
1 Leaves (3–)4–8 mm wide, oblong-spathulate
 2 Leaves glabrous above, sparsely hairy beneath; silicula 4 mm, ± compressed **1. dieckii**
 2 Leaves densely hairy; silicula (3·5–)5–7 mm, inflated **3. tymphaea**

1. B. dieckii Degen, *Österr. Bot. Zeitschr.* **50**: 313 (1900) (*Ptilotrichum dieckii* (Degen) Hayek). Leaves oblong-spathulate, glabrous above, with sparse bifid hairs beneath. Infructescence branched, dense; silicula 4 mm, ovoid, somewhat compressed. *Rocks.* ● *S. Jugoslavia (near Prizren).* Ju.

2. B. baldaccii (Degen) Heywood, *Feddes Repert.* **69**: 61 (1964) (*Ptilotrichum baldaccii* Degen). Flowering stems up to 10 cm, leafless. Basal leaves 5–12(–15) × 1–3 mm, linear to linear-lanceolate, green, glabrous or sparsely hairy with bi- to 6-fid hairs, densely crowded. Petals 3–4(–4·5) mm. Infructescence unbranched; pedicels patent or deflexed; silicula 4–6(–7) mm, ovoid, inflated. Seeds unwinged. *Serpentine rocks*, 1600–2000 *m*. ● *N. Greece, S. & C. Albania.* Al Gr.

Plants from N. Greece (Lingos mountains) which are sparsely hairy, the hairs mostly bifid, have been distinguished as subsp. **rechingeri** Greuter, *Candollea* **30**: 19 (1975).

3. B. tymphaea (Hausskn.) Hausskn., *Mitt. Thür. Bot. Ver.* nov. ser. **11**: 72 (1897) (*Ptilotrichum tymphaeum* (Hausskn.) Halácsy). Flowering stems up to 25 cm. Basal leaves 20–40 × 3–8 mm, oblong-spathulate, densely grey-hairy with bifid hairs. Petals 3·5–4·5 mm. Infructescence branched; pedicels patent; silicula (3·5–)5–7 mm, inflated. Seeds winged. *Serpentine rocks.* ● *N. Greece.* Gr.

54. Lobularia Desv.

P. W. BALL

Annuals or perennials; hairs bifid. Sepals patent, not saccate at the base; petals usually white, entire; filaments not winged and without appendages. Fruit a latiseptate silicula; valves slightly inflated, with a more or less distinct median vein; style distinct; stigma capitate. Seeds 1–5 in each loculus.

Silicula obovate or suborbicular; valves convex; seeds 1 in each loculus **1. maritima**
Silicula ovate; valves flat; seeds 4 or 5 in each loculus **2. libyca**

1. L. maritima (L.) Desv., *Jour. Bot. Appl.* **3**: 162 (1815) (*L. strigulosa* (G. Kunze) Willk., *Koniga maritima* (L.) R. Br., *Alyssum maritimum* (L.) Lam.). Greyish-white, more or less pubescent perennial (5–)10–40 cm, branched at the base. Leaves linear-lanceolate, acute, rarely obtuse. Petals 2·5–4 mm. Silicula 2–4 mm, obovate or suborbicular; valves convex, pubescent or glabrescent. Seeds 1 in each loculus. $2n = 24$. *Dry, open habitats. S. Europe; widely cultivated for ornament elsewhere, and often naturalized or casual.* Al Bl Co Cr Ga Hs It Lu Sa Si [Au Be Br Cz Da Gr Ho Hu Ju No Po Rm Rs(K)].

Plants in cultivation often do not perennate, and may have purple or pink petals.

2. L. libyca (Viv.) Webb & Berth., *Phyt. Canar.* **1**: 90 (1837) (*Koniga libyca* (Viv.) R. Br., *Alyssum libyca* (Viv.) Cosson). Like **1** but annual; leaves obtuse; silicula 3–7 mm, ovate; valves flat; seeds 4 or 5 in each loculus. *Maritime sands. S. Spain; Sicilia; Kikladhes (Thira).* Gr Hs *Si. (*From Islas Canarias to Iran.*)

55. Clypeola L.

P. W. BALL

Annuals; hairs branched or stellate. Sepals erecto-patent, not saccate; petals yellow, shortly clawed. Fruit an indehiscent, compressed, 1-seeded, pendent silicula; style short; stigma obtuse.

Literature: M. Breistroffer, *Candollea* **7**: 140–166 (1936); **10**: 241–280 (1946). D. A. Chaytor & W. B. Turrill, *Kew Bull.* **1935**: 1–24 (1935).

Silicula 2–5 mm, glabrous or pubescent; petals 1–2 mm **1. jonthlaspi**
Silicula 5·5–6 mm, with a very dense, long, white indumentum; petals *c.* 3 mm **2. eriocarpa**

1. C. jonthlaspi L., *Sp. Pl.* 652 (1753) (*C. pyrenaica* Bord.; incl. *C. microcarpa* Moris). Plant 2–20(–30) cm, erect, grey-pubescent. Leaves linear-oblanceolate to obovate. Raceme elongate in fruit; pedicels 1·5–3 mm. Petals 1–2 mm, glabrous. Silicula 2–5 mm, obovate- or elliptic-orbicular, with a distinct wing, entire or notched, pubescent or glabrous; style up to 0·25 mm. *S. Europe.* Al Bl Bu Co Cr Ga Gr He Hs It Ju Rm Rs(K) Sa Si Tu.

Variable in the size and pubescence of the silicula, but most of the variants occur sporadically throughout the range of the species. Some of these variants are recognized as subspecies by Breistroffer, *loc. cit.* (1936, 1946).

2. C. eriocarpa Cav., *Descr. Pl.* 401 (1802). Like **1** but petals *c.* 3 mm, stellate-pubescent on the back; silicula 5·5–6 mm, orbicular, scarcely winged, crenulate, with a very dense, long, white indumentum; style *c.* 0·5 mm. ● *C. & S. Spain.* Hs.

56. Schivereckia Andrz.

A. O. CHATER

Perennials; hairs stellate, minute. Sepals patent, the inner slightly saccate at the base; petals white; filaments of the inner stamens winged, the wing with a short tooth at the apex. Fruit a latiseptate silicula; valves convex; style distinct; stigma capitate or slightly 2-lobed. Seeds 4–7 in each loculus.

Literature: M. Alexeenko, *Not. Syst.* (*Leningrad*) **9(4–12)**: 215–231 (1946); *Trudy Nauč.-Issled. Inst. Biol.* (*Har'kov*) **13**: 95 (1950). M. V. Kazakova, *Biol. Nauki* **4**: 57–62 (1984).

Cauline leaves semi-amplexicaul, with 1–4 teeth on each side; plant densely caespitose **1. podolica**
Cauline leaves narrowed at the base, not or scarcely amplexicaul, entire (rarely with 1 tooth on each side); plant usually laxly caespitose **2. doerfleri**

1. S. podolica (Besser) Andrz. in DC., *Reg. Veg. Syst. Nat.* **2**: 300 (1821) (*S. berteroides* Fischer ex Alexeenko, *S. kuznezovii* Alexeenko, *S. mutabilis* (Alexeenko) Alexeenko). Densely caespitose; flowering stems up to 25 cm, simple or branched. Basal leaves rosulate, oblanceolate to oblong-spathulate, usually with 2–5 teeth on each side; cauline leaves oblong-ovate, semi-amplexicaul, with 1–4 teeth on each side. Raceme up to 30-flowered. Sepals 2–2·5 mm; petals 3–5 × 1·5–2·25 mm. Silicula 2–6 × 2–2·75 mm; style 0·75–2 mm; stigma capitate, entire. Seeds 0·75–1 × 0·5–1 mm, dark brown, rugulose. ● *E. Russia and N. Ukraine; N.E. Romania.* Rm Rs(N, C, W, E).

A number of species have been described by Alexeenko, *loc. cit.* (1946, 1950), which are in many ways intermediate between **1** and **2**. Studies by M. V. Kazakova, *loc. cit.* (1984), suggest that these are best included under **1**.

2. S. doerfleri (Wettst.) Bornm., *Feddes Repert.* **17**: 36 (1921) (*S. podolica* sensu Boiss., non (Besser) Andrz., *Draba doerfleri* Wettst.). Laxly caespitose; flowering stems up to 15 cm, simple or branched. Basal leaves more or less rosulate, oblanceolate, entire (or rarely with 1 or 2 teeth on each side); cauline leaves linear-lanceolate, attenuate at the base, obtuse, entire (rarely with 1 tooth on each side). Raceme up to 15-flowered. Sepals 2–2·25 mm; petals 4–6 × 2–3 mm. Silicula 4–5·5 × 2–4 mm; style 0·5–1 mm; stigma shortly 2-lobed. Seeds *c.* 1 mm. *Rocky and grassy places. S.W. Jugoslavia; S. Albania.* Al Ju. (*Anatolia.*)

57. Draba L.

S. M. WALTERS (EDITION 1) REVISED BY J. R. AKEROYD (EDITION 2)

Annuals to perennials. Leaves simple, entire or dentate. Sepals erecto-patent, the inner not or only slightly saccate at the base; petals white or yellow, entire or emarginate; filaments not or only slightly dilated at the base. Fruit a latiseptate silicula or siliqua; valves more or less flat, with a median vein in the lower $\frac{1}{2}$; stigma capitate. Seeds in 2 rows in each loculus, not winged. Radicle accumbent.

Most of the species are arctic or montane, growing in open, rocky or gravelly places. Several are widely grown for ornament as rock-plants in gardens.

The nature and density of the indumentum is used a great deal in the taxonomy of *Draba*. The hairs may be unbranched, branched, stellate or medifixed. The leaves are nearly always ciliate even when they are otherwise glabrous. The length of the stem given in the description refers to the fruiting stage; there is often considerable elongation after flowering.

The limits and particularly the rank of many described taxa are disputable, and further experimental studies are needed in the genus.

Literature: O. E. Schulz in Engler, *Pflanzenreich* **89(IV.105)**: 1–396 (1927). K.-P. Buttler, *Mitt. Bot. Staatssamm. München* **6**: 275–362 (1967); **8**: 539–566 (1969). G. A. Mulligan, *Canad. Jour. Bot.* **54**: 1386–1393 (1976).

1 Perennial, usually scapose; flowers yellow, rarely white
2 Plant with numerous long, procumbent leafy stems; hairs on leaves and base of stem medifixed **25. sibirica**
2 Plant without long, procumbent leafy stems, often densely caespitose; hairs not medifixed
3 Leaves lanceolate to obovate, not keeled, soft in texture, without stiff cilia
4 Leaves with unbranched hairs only **44. crassifolia**
4 Leaves with some branched and stellate hairs
5 Style not more than 0·7(–0·9) mm (Arctic) **(18–23). alpina** group
5 Style 0·7–1·2 mm (Alps) **24. ladina**
3 Leaves linear or more rarely lanceolate or oblanceolate, keeled, ±rigid, with stiff cilia
6 Scape hairy
7 Petals about equalling the stamens
8 Scape hispid with hairs not more than 0·5 mm **3. aspera**
8 Scape villous with some long hairs up to 1 mm **(5–8). lasiocarpa** group
7 Petals distinctly longer than the stamens
9 Silicula densely stellate-hairy **14. cretica**
9 Silicula with unbranched or branched hairs only
10 Scape villous with hairs *c.* 1 mm
11 Style 1–1·5 mm **12. parnassica**
11 Style 0·3–0·8 mm **15. dedeana**
10 Scape hispid with hairs mostly not more than 0·75 mm
12 Style 2–4 mm; silicula ±inflated at the base **11. hispanica**
12 Style *c.* 1 mm; silicula flat **13. loiseleurii**
6 Scape glabrous
13 Style not more than 1 mm in fruit
14 Plant laxly caespitose, with ±elongated stems clothed with scale-like leaf-remains below the rosette **16. sauteri**
14 Plant densely caespitose or pulvinate, without obvious persistent scale-like leaf-remains
15 Infructescence ±elongate

16 Leaves 1–2·5 mm wide, straight; silicula 6–9 mm, flat **9. lacaitae**
16 Leaves less than 1 mm wide, incurved; silicula 3·5–5 mm, inflated **17. heterocoma**
15 Infructescence subcapitate
17 Silicula distinctly hairy (**5–8**). **lasiocarpa** group
17 Silicula subglabrous or glabrous
18 Leaves 0·5–1·5 mm wide; style (0·5–)0·7–1·3 mm **2. hoppeana**
18 Leaves 1·5–4 mm wide; style (0·3–)0·4–0·8 mm **4. dolomitica**
13 Style more than 1 mm (usually more than 1·5 mm) in fruit
19 Style less than 3 mm in fruit
20 Leaves 1·5–4 mm wide (**5–8**). **lasiocarpa** group
20 Leaves less than 1·5 mm wide
21 Leaves usually at least 1 mm wide; silicula usually glabrous, flat
22 Scape 5–10 cm; petals 4–6 mm **1. aizoides**
22 Scape 1–5 cm; petals 2·5–4 mm **2. hoppeana**
21 Leaves less than 1 mm wide; silicula densely hairy, ±inflated at the base **10. haynaldii**
19 Style at least 3 mm in fruit
23 Silicula flat
24 Leaves ciliate, otherwise glabrous; silicula usually glabrous **1. aizoides**
24 Leaves ±hispid-setose all over; silicula (or siliqua) usually densely setose (**5–8**). **lasiocarpa** group
23 Silicula ±inflated (**5–8**). **lasiocarpa** group
1 Annual, biennial or perennial, usually with ±leafy stems; flowers usually white or cream
25 Annual (sometimes overwintering), rarely perennial; style less than 0·2 mm, or absent
26 Pedicels in fruit shorter than the silicula
27 Stem and silicula hairy **43. lutescens**
27 Stem and silicula glabrous **44. crassifolia**
26 Pedicels in fruit as long as or longer than the silicula
28 Silicula glabrous; cauline leaves cordate at the base, amplexicaul **41. muralis**
28 Silicula usually hairy; cauline leaves ±cuneate at the base, not or scarcely amplexicaul **42. nemorosa**
25 Perennial, more rarely biennial; style usually more than 0·2 mm
29 Stem glabrous
30 Infructescence compact, corymbose
31 Petals at least $1\frac{1}{2}$ times as long as the sepals **4. dolomitica**
31 Petals scarcely exceeding the sepals **44. crassifolia**
30 Infructescence ±elongated
32 Silicula lanceolate-elliptical; style usually less than 0.4 mm **35. fladnizensis**
32 Silicula elliptical; style 0·4–0·8 mm **36. dorneri**
29 Stem hairy (sometimes only sparsely so)
33 Style 1–2 mm in fruit **26. stellata**
33 Style less than 1 mm in fruit
34 Leaves densely clothed with very small stellate hairs only; a few cilia present
35 Silicula (or siliqua) usually glabrous, flat **27. nivalis**
35 Silicula with stellate hairs, strongly inflated **39. cinerea**
34 Unbranched and branched hairs present on leaves; stellate hairs present or not
36 Silicula hairy
37 Leaves linear, keeled, rigid **15. dedeana**
37 Leaves lanceolate or ovate-lanceolate to obovate, not keeled, soft in texture
38 Whole plant canescent with a dense stellate tomentum
39 Plant laxly caespitose; cauline leaves 0–3(4) **33. tomentosa**
39 Plant forming a dense mat; cauline leaves 1–8 **34. korabensis**
38 Plant not completely canescent; stellate hairs relatively sparse
40 Often scapose; silicula 3–6(–7) mm **28. norvegica**
40 Rarely scapose; silicula (5–)7–15 mm
41 Silicula straight; petals cream; filaments dilated at the base **37. glabella**
41 Silicula often somewhat twisted; petals white; filaments not or slightly dilated at the base **40. incana**
36 Silicula glabrous (or with sparse cilia)
42 Stem with minute stellate hairs only (rarely with a few unbranched or branched hairs)
43 Basal leaves hairy; petals 4–6 mm, cream **37. glabella**
43 Basal leaves sometimes glabrous; petals 3–4 mm, white **38. pacheri**
42 Stem with hairs of various types, always including some branched ones
44 Infructescence corymbose or subcapitate **29. subcapitata**
44 Infructescence ±elongate
45 Pedicels suberect in fruit; infructescence therefore very narrow
46 Stem 1–5(–10) cm; silicula 4–7 mm, straight **28. norvegica**
46 Stem up to 35 cm; silicula 5–15 mm, often somewhat twisted **40. incana**
45 Pedicels erecto-patent in fruit; infructescence therefore broader
47 Pedicels and upper part of stem glabrous **30. siliquosa**
47 Pedicels and stem at least sparsely hairy
48 Rosette-leaves entire **31. dubia**
48 Rosette-leaves coarsely toothed **32. kotschyi**

Sect. AIZOPSIS DC. Scapose, caespitose perennials, often densely pulvinate. Leaves rigid, entire, linear or narrowly spathulate, rosulate, ciliate. Flowers yellow (except in **15**).

1. D. aizoides L., *Mantissa* 91 (1767) (*D. affinis* Host). Caespitose; scape usually 5–10 cm, glabrous. Leaves 0·5–1·5 mm wide, linear to narrowly oblong, ciliate but otherwise glabrous. Inflorescence initially more or less condensed, later rather lax, 4- to 20-flowered. Petals 4–6 mm, yellow, obovate-cuneate, equalling the stamens. Silicula 5–12 mm, ellipsoidal, flat, usually glabrous; style 1–4(–6) mm. $2n = 16$. ● *Mountains of C. & S. Europe; Britain* (*coast of S.W. Wales*). Au Be *Br Cz Ga Ge He Hs It Ju Po Rm Rs(W).

Variable in habit, size of flower and silicula, hairiness of silicula, style-length, and other characters, but the many named taxa are hardly sufficiently geographically differentiated to merit subspecific rank. A possible exception is subsp. **zmudae** Zapał., *Rozpr. Wydz. Mat.-Przyr. Polsk. Akad. Um.* (*Biol.*) ser. 3, **12**$_{B}$: 230 (1912), from the Carpathians, with a silicula up to 12 mm and style 6 mm.

D. brachystemon DC., *Reg. Veg. Syst. Nat.* **2**: 334 (1821) (*D. bertolonii* sensu Thell. quoad pl. pyren., non Nyman), with stamens only 2–2·5 mm, and small, usually hairy siliculae, is recorded from the E. Pyrenees and the Alps (Monte Rosa). It is very rare and its relationship to **1** or to other species is obscure.

2. D. hoppeana Reichenb. in Moessler, *Handb.* ed. 2, **2**: 1132 (1828) (*D. zahlbrueckneri* Host). Like **1** but 1–5 cm, densely caespitose; petals 2·5–4 mm; infructescence subcapitate; style (0·5–)0·7–1·3 mm. $2n = 16$. *Mountain rocks above* 2200 *m*. • *Alps*. Au Ga Ge He It.

3. D. aspera Bertol., *Amoen.* 384 (1819) (*D. armata* Schott, Nyman & Kotschy; incl. *D. longirostra* Schott, Nyman & Kotschy, *D. bertolonii* Nyman). Like **1** but scape hispid; leaves *c.* 1 mm wide; inflorescence usually 4- or 5-flowered; silicula inflated; style 3–7 mm. $2n = 16$. *Mountain rocks.* • *Mountains of N. part of Balkan peninsula, C. & S. Italy and Sicilia; E.C. Pyrenees.* Al Ga It Ju Si.

4. D. dolomitica Buttler, *Mitt. Bot. Staatssamm. München* **8**: 541 (1969). Caespitose; scape 1–4·5 cm, glabrous. Leaves 1·5–4 mm wide, oblanceolate, ciliate but otherwise glabrous. Inflorescence 2- to 6(–12)-flowered, dense, corymbose. Petals 3–4 mm, yellow to whitish. Silicula 3·5–7 mm, narrowly ovate, glabrous; style 0·3–0·8 mm. $2n = 32$. • *S.E. Alps (Dolomiti); C. Alps (Brenner).* Au It.

A tetraploid derived from **2** × **37**.

(5–8). D. lasiocarpa group. Caespitose; scape up to 15 cm. Leaves linear to linear-oblong, setose-ciliate. Petals 3–9 mm, yellow. Silicula 3–10(–15) mm, elliptical; style 2–7 mm.

A group of similar taxa which some authors do not accept as distinct species.

1 Silicula 3–4(–6) mm — **5. scardica**
1 Silicula 6–13 mm
  2 Scape villous
    3 Petals 6–9 mm; style 4–7 mm — **6. cuspidata**
    3 Petals 4–6(–8) mm; style 2–3 mm — **8. lasiocarpa**
  2 Scape glabrous
    4 Petals 6–7 mm; style at least 3 mm — **7. athoa**
    4 Petals 4–6(–8) mm; style 0·2–2·5 mm — **8. lasiocarpa**

5. D. scardica (Griseb.) Degen & Dörfler, *Denkschr. Akad. Wiss. Math.-Nat. Kl.* (*Wien*) **64**: 70 (1897). Somewhat laxly caespitose; scape 2–5(–10) cm, glabrous. Leaves 1–2 mm wide, linear. Petals *c.* 4 mm, yellow. Silicula 3–4(–6) mm, inflated, usually glabrous; style 3·5–5 mm. • *Mountains of Balkan peninsula.* Al Bu Gr Ju.

6. D. cuspidata Bieb., *Fl. Taur.-Cauc.* **3**: 424 (1819). Pulvinate; scape 2–7 cm, densely villous with some hairs up to 1 mm long. Leaves 6–15 mm, narrowly linear. Inflorescence compact but elongating considerably in fruit. Petals 6–9 mm. Silicula 6–10 mm, more or less inflated, densely hairy with mostly unbranched hairs; style 4–7 mm. *Mountain rocks. Krym.* Rs(K).

7. D. athoa (Griseb.) Boiss., *Diagn. Pl. Or. Nov.* **3(1)**: 33 (1853) (*D. affinis* sensu Halácsy, non Host). Rather robust; scape up to 12 cm, glabrous. Leaves 7–9 × 1·5–3 mm wide, broadly linear, the surface glabrous to hispid. Inflorescence compact, elongating somewhat in fruit. Petals *c.* 7 mm. Silicula 6–10 × 2·5–3 mm, glabrous to setose; style 3 mm. $2n = 16$. • *Greece and W. part of Balkan peninsula.* Al Gr Ju.

8. D. lasiocarpa Rochel, *Sched. Pl. Hung. Exsicc.* (1810) (*D. aizoon* Wahlenb.; incl. *D. compacta* Schott, Nyman & Kotschy, *D. elongata* Host). Robust, usually densely caespitose; scape up to 15 cm. Leaves 5–20 × 2–3 mm wide, broadly linear, the surface glabrous. Inflorescence fairly compact even in fruit. Petals 3–8 mm. Silicula 4–10(–13) × 2–3·5 mm, flat, glabrous to hispid-setose especially on margins; style (0·5–)2–2·5 mm. $2n = 16$. *Carpathians and mountains of Balkan peninsula; E. Austria.* Al Au Bu Cz Gr Hu Ju Rm.

Plants from S. Albania and C. Greece with a villous scape 1–7 cm and a densely hispid-setose silicula have been distinguished as subsp. **dolichostyla** (O. E. Schulz) Buttler in Strid, *Mount. Fl. Gr.* **1**: 311 (1986).

D. boueana Zahlbr. ex O. E. Schulz in Engler, *Pflanzenreich* **89(IV.105)**: 47 (1927), described from Crna Gora and N. Albania, is a rather slender, dwarf plant with shorter leaves and smaller flowers and fruit than **8**, which it otherwise closely resembles. It needs further study.

9. D. lacaitae Boiss., *Fl. Or., Suppl.* 53 (1888). Densely caespitose, with filiform, ascending scape up to 15 cm. Leaves 1·5–2·5 mm wide, broadly linear, obtuse, sparsely hispid or ciliate. Inflorescence at first dense, later elongating; pedicels (4–)7–9 mm, slender. Petals *c.* 4 mm, obovate, rather pale yellow; stamens *c.* 2·5 mm. Silicula 6–9 × 2·5–3 mm, oblong-elliptical, setose; style 0·25–0·7 mm. $2n = 16$. *Mountain rocks.* • *S. & C. Greece.* Gr.

10. D. haynaldii Stur, *Österr. Bot. Zeitschr.* **11**: 186 (1861). Pulvinate; scape up to 6 cm, 3- to 8-flowered. Leaves 5–7 × 1 mm, linear, acute. Petals 4–4·5 mm; stamens 3–3·5 mm. Silicula 5–7 × 2–3 mm, broadly ovoid, inflated towards the base, more or less densely hispid; style 1–2 mm. $2n = 16$. • *S. Carpathians.* Rm.

11. D. hispanica Boiss., *Elenchus* 13 (1838) (*D. atlantica* Pomel). Dwarf, pulvinate, with short, distinctly hairy scape. Petals 5–9 mm, yellow, distinctly longer than the stamens. Silicula slightly inflated at the base, densely hispid with unbranched and branched hairs; style usually 2–4 mm. $2n = 16$. *Mountain rocks above* 1400 *m. E. & S. Spain.* Hs.

Somewhat variable in the size of the flower and fruit, and length of the style.

12. D. parnassica Boiss. & Heldr. in Boiss., *Diagn. Pl. Or. Nov.* **3(1)**: 34 (1853). Like **11** but scape with long villous hairs; style less than 2 mm. $2n = 16$. • *Mountains of C. Greece.* Gr.

13. D. loiseleurii Boiss., *op. cit.* **3(1)**: 34 (1854) (*D. bertolonii* sensu Thell. quoad pl. cors., non Nyman). Like **11** but silicula flat, scabrid, with hairs less than 0·25 mm; style *c.* 1 mm. $2n = 16$. *Siliceous rocks above* 2300 *m.* • *Corse.* Co.

14. D. cretica Boiss. & Heldr. in Boiss., *op. cit.* **2(8)**: 27 (1849). Like **11** but petals 3·5–4·5 mm; silicula densely stellate-hairy; style 0·75 mm. $2n = 16$. • *Mountains of Kriti.* Cr.

15. D. dedeana Boiss. & Reuter in Boiss., *Voy. Bot. Midi Esp.* **2**: 718 (1845) (incl. *D. cantabrica* Willk., *D. zapateri* Willk.). Densely pulvinate, with very woody stock; scape villous, elongating after flowering, rather lax in fruit. Petals 4–6 mm, white, broadly obovate, emarginate, much longer than the stamens. Silicula hispid; style *c.* 0·5 mm. $2n = 32$. • *Mountains of N. & E. Spain.* Hs.

A variant with yellow petals and $2n = 16$ has been described as **D. cantabriae** (Laínz) Laínz, *Candollea* **24**: 259 (1969).

16. D. sauteri Hoppe, *Flora (Regensb.)* **6**: 425 (1823). Laxly caespitose, with decumbent, rather elongated, branched stems, clothed with yellowish, appressed, scale-like, persistent leaf-bases, and terminating in lax rosettes; scape glabrous, up to 3 cm. Petals 4–6 mm, yellow. Silicula 3–7 mm, flat, usually glabrous and often somewhat asymmetrical; style 0·3–0·6 mm. $2n = 32$. • *N.E. Alps.* Au Ge.

17. D. heterocoma Fenzl, *Pugillus* 13 (1842). Densely pulvinate, with narrow, hairy, incurved leaves; scape 3–10 cm, glabrous. Inflorescence becoming lax in fruit. Petals 5–6 mm, yellow, obovate-cuneate, much longer than the stamens. Silicula 3·5–5 × 3 mm, ovoid, inflated, densely hairy; style 0·5 mm. *Karpathos.* Cr. (*Anatolia.*)

Sect. CHRYSODRABA DC. Scapose perennials. Leaves lanceolate to obovate, not rigid, often more or less densely hairy. Flowers yellow.

(18–23). D. alpina group. Densely rosulate, scapose perennials. Leaves elliptic-oblong, variably clothed with unbranched, branched and stellate hairs. Inflorescence dense. Petals yellow, rarely cream. Silicula 5–11 × 2·5–4 mm.

A taxonomically complex group confined to the Arctic, and particularly variable in Spitsbergen. The following treatment attempts to utilize the work of E. Ekman, *Svensk Bot. Tidskr.* **35**: 135 (1941), and A. Tolmatchev in Komarov, *Fl. URSS* **8**: 390–401 (1939).

1 Inflorescence elongating considerably after flowering; style not more than 0·2 mm
- 2 Petals 2–3 mm, pale yellow or cream — **22. adamsii**
- 2 Petals *c.* 5 mm, bright yellow — **23. glacialis**

1 Inflorescence not or only slightly elongating after flowering; style at least 0·2 mm
- 3 Silicula glabrous or subglabrous
  - 4 Petals bright yellow; leaves entire — **18. alpina**
  - 4 Petals pale yellow; leaves often denticulate — **19. gredinii**
- 3 Silicula hairy
  - 5 Branched and stellate hairs on underside of leaf obviously stalked; filaments slightly dilated at the base — **20. corymbosa**
  - 5 Branched and stellate hairs on underside of leaf nearly sessile; filaments broadly dilated at the base — **21. kjellmanii**

18. D. alpina L., *Sp. Pl.* 642 (1753). Scape up to 15 cm, hairy, more or less dense in fruit. Leaves usually rather densely hairy, with some simple hairs beneath, entire. Petals 3·5–5 mm, bright yellow. Silicula lanceolate- or ovate-elliptical, glabrous; style 0·2–0·75 mm. $2n = 62–64, 80$. *Arctic and subarctic Europe.* Fe Is No Rs(N) Sb Su.

19. D. gredinii Elis. Ekman, *Svensk Bot. Tidskr.* **27**: 102 (1933). Like **18** but leaves usually sparsely hairy and often denticulate; petals pale yellow. *Spitsbergen.* Sb. (*Greenland.*)

20. D. corymbosa R. Br. ex DC., *Reg. Veg. Syst. Nat.* **2**: 343 (1821) (*D. bellii* Holm, *D. macrocarpa* Adams). Scape up to 15 cm. Leaves with stalked, stellate and branched hairs on lower surface. Petals 3·5–5 mm, bright or pale yellow; filaments slightly dilated at the base. Silicula hairy. Style 0·2–0·75 mm. *Spitsbergen.* Sb. (*Circumpolar.*)

21. D. kjellmanii Lid ex Elis. Ekman, *Svensk Bot. Tidskr.* **25**: 478 (1931). Like **20** but the lower surface of leaves with nearly sessile stellate hairs; filaments broadly dilated at the base. *Spitsbergen, Vajgač.* Rs(N) Sb.

22. D. adamsii Ledeb., *Fl. Ross.* **1**: 147 (1842) (*D. oblongata* auct., non R. Br. ex DC.). Scape up to 20 cm in fruit, with more or less elongated infructescence. Leaves with stellate hairs on the lower surface. Petals 2–3 mm, scarcely longer than the sepals, pale yellow or cream. Silicula hairy; style up to 0·2 mm. *Spitsbergen.* Sb. (*Circumpolar.*)

23. D. glacialis Adams, *Mém. Soc. Nat. Moscou* **5**: 106 (1817). Like **22** but leaves more densely stellate-hairy; petals *c.* 5 mm, bright yellow; silicula subglabrous. *Arctic Russia; ?arctic Norway.* ?No Rs(N).

24. D. ladina Br.-Bl., *Verh. Schweiz. Naturf. Ges.* **1919(2)**: 117 (1920). Scape up to 5 cm, subglabrous. Leaves linear-lanceolate, entire, rather sparsely hairy with branched and stellate hairs, particularly beneath, and with sparse cilia. Inflorescence not elongating in fruit. Petals 3·5–5 mm, pale yellow, slightly pubescent. Silicula 6–10 mm, ovate-lanceolate, subglabrous, ciliate; style 0·7–1·2 mm. $2n = 32$. *Calcareous mountain rocks*, 2600–3050 *m*. • *E. Switzerland.* He.

A remarkable endemic, which is a tetraploid possibly derived from **1** × **31** or **33**, but has been referred by some authors to the arctic complex of species **18–23**.

25. D. sibirica (Pallas) Thell., *Mitt. Bot. Mus. Zürich* **28**: 318 (1906). Stem long, slender, creeping, producing erect or ascending scape with 8–20 yellow flowers. Leaves oblong-lanceolate, acute, variably clothed with appressed, medifixed hairs (also present towards the base of the scape). Petals 4–6 mm, obovate-cuneate, yellow with brownish veins; stamens 2–2·5 mm. Silicula 4–7·5 × 1·5–2 mm, oblong-ellipsoidal, often somewhat curved, glabrous; style 0·5–0·75 mm. $2n = 16$. *E. & C. Russia.* Rs(N, C, W) [Rs(B)].

Very different in habit from all other species in Europe.

Sect. DRABA. Caespitose biennials or perennials with erect flowering stems and variably hairy leaves, often canescent. Cauline leaves usually present. Flowers usually rather small; petals white or cream, longer than the stamens; filaments sometimes slightly dilated at the base.

26. D. stellata Jacq., *Enum. Stirp. Vindob.* 113 (1762) (*D. austriaca* Crantz). Stem up to 10 cm, stellate-hairy below. Basal leaves 4–8 mm, oblanceolate, mostly entire, stellate-canescent and ciliate; cauline leaves 0–3, often toothed. Inflorescence 3- to 12-flowered. Petals 4·5–8 mm; filaments slightly dilated at the base. Silicula 4–10 mm, lanceolate-elliptical, glabrous; style 1–2 mm; stigma capitate. $2n = 16$. *Calcareous mountain rocks.* • *N.E. Alps.* Au.

D. simonkaiana Jáv., *Bot. Közl.* 9: 281 (1910), from one station in the S. Carpathians in Romania, differs from **26** principally in its petals 3–4 mm, 2-lobed stigma 0·7–1 mm, and a chromosome number of $2n = 32$. It is probably not specifically distinct from **26**.

27. D. nivalis Liljeblad, *Kungl. Svenska Vet.-Akad. Handl.* **1793**: 208 (1793). Densely pulvinate, with flowering stems up to 5 cm. Basal leaves obovate-cuneate, usually entire, with a dense covering of uniform, minute stellate hairs, and variably developed cilia; cauline leaves often absent. Inflorescence dense, 2- to 5(–9)-flowered; pedicels very short. Petals 2·5–3 mm. Silicula 4–9 × 2 mm, usually glabrous; style 0·2–0·3 mm. $2n = 16$. *Arctic and subarctic Europe southwards to* c. 61° *N. in Norway.* Fe Is No Rs(N) Sb Su.

28. D. norvegica Gunnerus, *Fl. Norv.* **2**: 106 (1772) (incl. *D. rupestris* R. Br.). Very variable in habit, with somewhat hairy, slender, often flexuous stems 1–5(–10) cm. Basal leaves oblong-lanceolate, entire or 1- to 3-toothed, ciliate and variably covered with unbranched and branched hairs; cauline leaves 0–3. Inflorescence at first dense, but typically with remote basal flower, rather lax in fruit. Petals 2·5–3·8 mm. Silicula 4–7 mm, lanceolate-elliptical, glabrous or hairy, suberect on a short pedicel; style up to 0·5 mm. $2n = 48$. *Arctic and subarctic Europe southwards to Scotland; N.E. Alps (local).* Au Br Fa Fe Is No Rs(N) Sb Su.

29. D. subcapitata Simmons, *Vasc. Pl. Ellesmereland* 87 (1906) (*D. pauciflora* sensu O. E. Schulz, vix R. Br.). Dwarf, with a scapose, subcapitate or corymbose infructescence; stem pubescent. Leaves with numerous branched hairs and long silky hairs, sometimes subglabrous on upper side. Petals 2–3 mm. Silicula 4–8 mm, elliptical, usually glabrous; style up to 0·3 mm. *Arctic Islands (Spitsbergen, Björnöya, Vajgač).* Rs(N) Sb. (*Circumpolar.*)

30. D. siliquosa Bieb., *Fl. Taur.-Cauc.* **2**: 94 (1808) (*D. carinthiaca* Hoppe, *D. johannis* Host). Caespitose, often rather laxly so, with stems up to 5–15(–20) cm. Stem glabrous above, sometimes glabrous throughout (var. *glabrata* (Koch) Sauter). Basal leaves lanceolate, usually entire, covered with branched and stellate hairs, ciliate towards the base; cauline leaves 0–3. Inflorescence 4- to 8-flowered. Petals 2–4 mm. Silicula 3–8 mm, rather narrowly oblong-elliptical, glabrous; style not more than 0·4 mm; stigma capitate. $2n = 16$. *Mountains of C. & S. Europe.* Au Bu Cz Ga Ge He Hs It Rm Rs(W).

A variable species.

31. D. dubia Suter, *Fl. Helv.* **2**: 46 (1802) (*D. tomentosa* sensu Hegi, non Clairv., *D. laevipes* DC.). Laxly caespitose with stems up to 16 cm, stellate-hairy throughout. Basal leaves narrowly obovate; cauline leaves 0–3(–4); leaves entire. Petals 3–5 mm. Silicula 5–14 mm, oblong-elliptical, sometimes slightly hairy; style 0·1–0·4(–0·8) mm; stigma emarginate. $2n = 16$. ● *Mountains of C. & S. Europe from the Sierra Nevada and Pyrenees to the E. Alps and W. Carpathians (Tatra).* Au Co Cz Ga Ge He Hs It Ju Po.

A variable species. Plants from the Pyrenees have been referred to subsp. **laevipes** (DC.) Br.-Bl., *Commun. Stat. Int. Géobot. Médit. Alp.* **87**: 226 (1945), which is of dwarfer and more slender habit, with rather large flowers and pedicels elongate in fruit.

32. D. kotschyi Stur, *Österr. Bot. Zeitschr.* **9**: 33 (1859). Like **31** but stem up to 10 cm, lax, ascending, flexuous; basal leaves lanceolate to elliptical; cauline leaves ovate, toothed. $2n = 32$. ● *S. & E. Carpathians.* Rm.

Records from Austria refer to **28**.

33. D. tomentosa Clairv., *Man. Herb. Suisse* 217 (1811). Like **31** but stem up to 10 cm; dense indumentum of stellate hairs extending to the pedicels; silicula ovate, hairy, often slightly inflated when ripe. $2n = 16$. ● *Mountains of C. & S. Europe.* Au Bu Cz Ga Ge He It Ju Po.

Plants from the E. Pyrenees have been distinguished as **D. subnivalis** Br.-Bl., *Commun. Stat. Int. Géobot. Médit. Alp.* **87**: 226 (1945). They differ from **33** in the slender habit, narrower leaves, lanceolate silicula 4·5–10·5 mm, and somewhat longer style. They have a chromosome number of $2n = 32$.

34. D. korabensis Kümmerle & Degen ex Jáv., *Bot. Közl.* **19**: 22 (1921) (*D. tomentosa* auct. bulg., non Clairv.). Plant forming a dense mat; stem up to 16 cm. Basal leaves ovate-lanceolate, densely stellate-hairy; cauline leaves 1–8. Silicula ovate-lanceolate to elliptical, hairy; style 0·4–0·6(–0·9) mm. $2n = 32$. ● *Mountains of N. part of Balkan peninsula.* Al Bu Ju.

35. D. fladnizensis Wulfen in Jacq., *Misc. Austr. Bot.* **1**: 147 (1779) (*D. wahlenbergii* Hartman). Flowering stem up to 8 cm, usually subglabrous. Cauline leaves usually 1(2); basal leaves linear-spathulate, obtuse, ciliate, otherwise glabrous, or variably covered with usually simple hairs. Inflorescence 2- to 12-flowered. Petals 2–3·5 mm, white. Silicula 2·5–7·5 mm, lanceolate-elliptical, glabrous; style *c.* 0·2 mm. $2n = 16$. *Arctic Europe and mountains of Scandinavia and S. & C. Europe, from the Pyrenees to the E. Carpathians.* Au Cz Fe Ga Ge He Is It No Rm Rs(N, C) Sb Su.

Very variable and divided by some authors into several species. In northern Europe, the North American species **D. lactea** Adams, *Mém. Soc. Nat. Moscou* **5**: 104 (1817), is recognized as distinct on the basis of having basal leaves with some branched hairs and usually no cauline leaves, petals 3–5 mm, and style up to 0·5 mm. In the Alps, however, this distinction does not seem to be practicable.

36. D. dorneri Heuffel, *Österr. Bot. Zeitschr.* **8**: 25 (1858). Like **35** but basal leaves acute, glabrous on the surfaces, with the margins ciliate; cilia unbranched near the base of the leaf, branched towards the leaf-apex; cauline leaves 0–3; silicula elliptical; style 0·4–0·8 mm. $2n = 32$. ● *S. Carpathians (Munţii Retezatului).* Rm.

37. D. glabella Pursh, *Fl. Amer. Sept.* **2**: 434 (1814) (*D. hirta* auct., *D. daurica* DC., *D. magellanica* auct., non Lam.; incl. *D. cacuminum* Elis. Ekman). Robust, laxly caespitose perennial; stem up to 30 cm, often flexuous, usually simple, stellate-canescent at least below, with up to 10 somewhat toothed cauline leaves. Basal leaves up to 2 cm, oblanceolate, acute, entire or with 1 or 2 small teeth, ciliate and rather densely stellate-canescent. Inflorescence 8- to 30-flowered. Petals 4–6 mm, white or cream; filaments dilated at the base. Silicula 6–14 mm, ovate-lanceolate, flat, glabrous or with simple hairs; style 0·2–0·5 mm. $2n = 64$. *Arctic and subarctic Europe, southwards to* 61° *N. in Norway.* Fe Is No Rs(N) Sb Su.

38. D. pacheri Stur, *Österr. Bot. Wochenbl.* **5**: (49–50), 156 (1855) (*D. norica* Widder). Like **37** but stems not more than 20 cm; basal leaves sometimes glabrous; cauline leaves 0–7; petals 3–4 mm, white; silicula glabrous. $2n = 64$. ● *Mountains of E. Austria; Tatra; very local.* Au Cz.

39. D. cinerea Adams, *Mém. Soc. Nat. Moscou* **5**: 103 (1817) (*D. magellanica* subsp. *cinerea* (Adams) Elis. Ekman; incl. *D. arctica* J. Vahl). Like **37** but cauline leaves 0–5, densely stellate-canescent in all parts; silicula (3–)6–9 mm, elliptical, inflated, with stellate hairs; style 0·2–1 mm. $2n = 48, 80$. *Arctic Europe.* Fe No Rs(N) Sb.

A variable species.

D. dovrensis Fries, *Summa Veg. Scand.* 32 (1846) from S. Norway (Dovrefjell), with some forked as well as stellate hairs on the silicula, the style 0·4–0·5 mm, and a chromosome number of $2n = 64$, is sometimes placed here.

40. D. incana L., *Sp. Pl.* 643 (1753) (*D. bernensis* Moritzi, *D. stylaris* Gay ex Koch, *D. thomasii* Koch). Robust biennial, sometimes perennial, with erect, simple or branched, usually leafy, hairy stems up to 35 cm. Basal leaves up to 3 cm, lanceolate, subobtuse, canescent with unbranched, branched, and stellate hairs on the lower surface. Inflorescence 10- to 40-flowered; pedicels 2–3 mm in fruit, suberect. Petals 2·5–5 mm, white; filaments not or little dilated at the base. Silicula 5–15 mm, oblong, elliptical or lanceolate, often somewhat twisted, glabrous or hairy; style up to 0·5 mm. $2n = 32$. *N. Europe and in the mountains of W. & C. Europe from the Pyrenees to the E. Alps.* Au Br Da Fa Fe Ga Hb He Hs Is No Rs(N, B).

A variable species.

The Asian species **D. lanceolata** Royle, *Ill. Bot. Himal. Mount.* **1**: 72 (1834), a perennial with a narrow, stellate-hairy silicula, has been recorded from N.W. Russia (near Murmansk). This may, however, represent the North American species **D. cana** Rydb., *Bull. Torrey Bot. Club* **29**: 241 (1902) (*D. lanceolata* auct.). Another plant of this complex, a perennial with a lanceolate, acute, glabrous silicula, has also been described from N.W. Russia (shores of the White Sea) as **D. valida** Pissjauk. in Pojark., *Fl. Murmansk.* **3**: 370 (1956), non Goodding (1940).

Sect. DRABELLA DC. Annuals, biennials or perennials, with leafy stems and small, pale yellow to white flowers. Style very short.

41. D. muralis L., *Sp. Pl.* 642 (1753). Erect annual or biennial 10–50 cm, with slender taproot; stem simple or branched, more or less stellate-hairy. Cauline leaves broadly ovate, semi-amplexicaul; basal leaves obovate-cuneate. Inflorescence many-flowered. Petals 1·5–3 mm, white. Silicula 3–6 mm, oblong-elliptical, glabrous; pedicels patent, as long as or slightly longer than the silicula. $2n = 32$. *Dry, open ground and walls.* ● *Most of Europe from C. Fennoscandia southwards, but probably as an alien in parts of the north.* All except Az Bl Fa Is Rs(N) Sa Sb Si; only as a casual in Po Rs(C).

42. D. nemorosa L., *Sp. Pl.* 643 (1753). Like **41** but always annual; stem with stellate and simple hairs; cauline leaves oblong-ovate to broadly ovate, often remote; petals *c.* 2 mm, pale yellow, fading to whitish; silicula with short, simple hairs, rarely glabrous; pedicels much longer than the silicula; stigma subsessile. *Much of Europe, but absent from much of the north-west, and local in the Mediterranean region.* Au Cz Fe Ga Ge He Hs It Ju No Po Rm Rs(N, B, C, W, K, E) Su.

43. D. lutescens Cosson, *Not. Pl. Crit.* 147 (1852). Slender annual up to 12 cm; stem simple or branched from the base, hairy. Cauline leaves sessile, ovate, cuneate at the base. Silicula longer than the pedicels, rough with unbranched hairs. *S. Spain (Granada prov.).* ?Ga Hs. (*Morocco.*)

44. D. crassifolia R. C. Graham, *Edinb. New Philos. Jour.* **7**: 182 (1829). Perennial, with glabrous, usually leafless, flowering stem up to 6 cm. Basal leaves oblong-spathulate, rather thick, glabrous or with a few unbranched or branched hairs. Inflorescence 3- to 9-flowered. Petals pale yellow, narrow, scarcely longer than the sepals. Silicula 4–7 mm, glabrous; stigma subsessile. *Arctic Scandinavia.* No Su. (*Greenland, North America.*)

Perhaps better included in Sect. *Draba.*

58. Drabopsis C. Koch

J. R. AKEROYD

Like *Draba* but scapose annuals; petals yellowish; seeds in 1 row in each loculus.

Literature: J. Léonard, *Publ. Cairo Univ. Herb.* **7–8**: 295–297 (1977).

1. D. nuda (Bél.) Stapf, *Denkschr. Akad. Wiss. Math.-Nat. Kl. (Wien)* **51**: 298 (1886) (*D. verna* C. Koch). Stem 1–2 cm, pubescent. Leaves up to 6 mm, spathulate to oblong-spathulate, entire, with branched hairs especially on margins. Sepals *c.* 1·5 mm; petals *c.* 2 mm. Siliqua up to 8 mm, pubescent. *Open, stony ground,* 1800 *m. W. Kriti (Levka Ori).* Cr. (*S.W. Asia.*)

The plants from Kriti are smaller in their parts than those from Asia, and the siliqua in the latter is frequently glabrous.

59. Erophila DC.

S. M. WALTERS

Like *Draba* but annual or overwintering scapose herbs with basal leaf-rosette; petals deeply bifid.

A taxonomically difficult genus in which there is no general agreement as to specific limits. Local populations often consist of self-pollinating pure lines exhibiting some morphological distinctness, and many such variants have been distinguished at specific or infraspecific rank. The cytotaxonomic investigations of Ö. Winge, *Compt. Rend. Trav. Lab. Carlsb.* (*Sér. Physiol.*) **23**: 41–73 (1940), were mainly conducted on material from N.W. Europe. These, and recent work by S. A. Filfilan and T. T. Elkington on British material, have been used in the account given in A. R. Clapham et al., *Fl. Brit. Is.* ed. 3, 90–91 (1987), which distinguished three species. Unfortunately it does not seem possible to apply this taxonomy to the variation in Europe as a whole, and the treatment given here, whilst taking this recent work into account, remains essentially based on that of F. Markgraf in Hegi, *Ill. Fl. Mitteleur.* ed. 2, **4(1)**: 320–327 (1960–1962), where full references to the earlier literature are given.

Leaves spathulate to lanceolate; seeds 0·3–0·8 mm, numerous **1. verna**
Leaves linear; seeds 0·75–1 mm, few **2. minima**

1. E. verna (L.) Chevall., *Fl. Gén. Env. Paris* **2**: 898 (1827) (*Draba verna* L.; incl. *Erophila stenocarpa* Jordan, *E. vulgaris* DC.). Stems solitary or several, up to 20 cm. Leaves in a basal rosette, spathulate to lanceolate, obtuse or acute, entire or distally toothed, variably clothed with branched and stalked-stellate hairs, sometimes mixed with unbranched hairs. Inflorescence racemose, initially dense, later elongating; pedicels 1·5–6 mm in flower, slender, elongating in fruit. Sepals 1·5–2·5 mm; petals 1·5–6 mm, white or reddish. Silicula 3–12 mm, usually flat, linear or narrowly elliptical to obovate or suborbicular, glabrous. Seeds 0·3–0·8 mm, numerous. $2n = 14–64$. *Rocks, walls, sandy soils and disturbed ground. Throughout Europe except the Arctic.* All except Az Fa Sb.

1 Silicula not more than 5 mm, broadly obovate or suborbicular (d) subsp. **spathulata**
1 Silicula often more than 5 mm, oblanceolate, elliptical or linear
2 Leaves with many unbranched and a few branched hairs (c) subsp. **praecox**
2 Leaves with many stellate and branched hairs, but few or no unbranched hairs
3 Silicula oblanceolate or elliptical; pedicels usually more than 12 mm in fruit (a) subsp. **verna**
3 Silicula linear; pedicels often less than 12 mm in fruit (b) subsp. **macrocarpa**

(a) Subsp. **verna** (incl. *E. cuneifolia* Jordan, *E. krockeri* Andrz., *E. majuscula* Jordan, *E. stenocarpa* Jordan, *Draba obconica* (De Bary) Hayek): Stems solitary or several, up to 20 cm in fruit. Leaves broadly lanceolate or elliptical, more or less densely stellate-hairy on upper surface, with few or no unbranched hairs. Sepals 1–2 mm; petals *c.* 2·5 mm. Silicula 5–10 mm, elliptical or oblanceolate; pedicels (5–)10–25 mm in fruit. *Widespread in W., C. & N. Europe, and probably throughout the range of the species, but less common in S. & S.E. Europe.*

(b) Subsp. **macrocarpa** (Boiss.) Walters, *Feddes Repert.* **69**: 57 (1964) (*E. macrocarpa* Boiss., *Draba macrocarpa* Boiss. & Heldr.): Like subsp. (a) but silicula 7–12 mm, linear; pedicels 4–15 mm in fruit. *E. Mediterranean region.*

(c) Subsp. **praecox** (Steven) Walters, *loc. cit.* (1964) (*Draba praecox* Steven; incl. *E. adriatica* Degen, *E. verna* subsp. *oblongata* (Jordan) Janchen): Stems often solitary, up to 9 cm in fruit. Leaves obovate-lanceolate, more or less densely covered on upper surface with unbranched hairs, often with a few branched or stellate hairs. Sepals 1–1·5 mm; petals 2–2·5 mm. Silicula 4–6 mm, elliptical or oblanceolate; pedicels 2–12 mm in fruit. *Mainly in the Mediterranean region; rather rare in N. & C. Europe.*

(d) Subsp. **spathulata** (A. F. Láng) Walters, *Feddes Repert.* **69**: 57 (1964) (*E. spathulata* A. F. Láng; incl. *E. boerhaavii* (Van Hall) Dumort.): Stems usually several, up to 10 cm in fruit. Leaves obovate-spathulate, more or less densely clothed with mixed stalked-stellate and dendritic hairs. Sepals 1–1·5 mm; petals *c.* 2 mm. Silicula 4–5 mm, broadly obovate or suborbicular; pedicels 3–18 mm in fruit. *Distribution like that of subsp.* (a).

Diploid plants, with $2n = 14$ (*E. simplex* Winge), may be referable to subsp. (a).

Plants with intermediate chromosome numbers (*E. duplex* Winge) seem to be mainly referable to subspp. (a) or (d).

The high polyploid taxon described by Winge, **E. quadruplex** Winge, *Compt. Rend. Trav. Lab. Carlsb.* (*Sér. Physiol.*) **23**: 71 (1940), with $2n = 52–64$, a more robust plant with a single stem, and narrowly cuneate, usually toothed leaves, has been equated with **E. glabrescens** Jordan, *Pug. Pl. Nov.* 10 (1852), but the validity of this distinction on a European scale seems doubtful.

E. setulosa Boiss. & Blanche in Boiss., *Diagn. Pl. Or. Nov.* **3**(**5**): 31 (1856), described from Syria and recorded from Thrace (near Alexandroupolis and Edirne), is like subsp. (c) but is more sparsely hairy and has a narrower silicula.

2. E. minima C. A. Meyer, *Verz. Pfl. Cauc.* 184 (1831). Like **1** but stems not more than 10 cm; leaves linear, clothed with unbranched hairs only; sepals *c.* 1 mm; silicula 3–5 × 2–3 mm, obovoid, more or less inflated; seeds 0·75–1 mm, few. *Greece.* Gr. (*Anatolia.*)

60. Petrocallis R. Br.

S. M. WALTERS

Like *Draba* but leaves digitately lobed; hairs all unbranched; ovary with 2 ovules in each loculus.

1. P. pyrenaica (L.) R. Br. in Aiton, *Hort. Kew.* ed. 2, **4**: 93 (1812) (*Draba pyrenaica* L.). Caespitose, often pulvinate perennial, resembling many species of *Draba* or *Saxifraga* in habit; scape usually 2–3 cm, hairy. Leaves 4–6 mm, in compact rosettes, digitately 3(–5)-lobed, ciliate, stiff and greyish. Corymbs few-flowered. Petals 4–5 mm, pale lilac or pink, rarely white; anthers yellow. Silicula obovate to elliptical, glabrous, with 1(2) seeds in each loculus. $2n = 14$. *Mountain rocks and screes; calcicole.* ● *Pyrenees, Alps, Carpathians.* Au Cz Ga Ge He Hs It Ju ?Rm.

61. Andrzeiowskia Reichenb.

P. W. BALL

Glabrous annuals. Leaves pinnate. Sepals erect, not saccate; petals white; stamens without appendages. Fruit an indehiscent siliqua with 2 triangular wings at the apex; style long; stigma capitate. Seeds *c.* 5 in each loculus.

1. A. cardamine Reichenb., *Pl. Crit.* **1**: 15 (1823) (*A. cardaminifolia* Prantl, pro parte). Stems 25–60 cm. Leaves with up to 5 pairs of ovate-oblong, crenate segments, and a larger terminal segment; petiole with amplexicaul, suborbicular auricles. Petals 2–2·5 mm. Siliqua 12–20 mm, linear, keeled; style longer than the wings. *Wet places. Thrace.* Gr Tu.

62. Cochlearia L.

A. O. CHATER AND V. H. HEYWOOD (EDITION 1) REVISED BY P. S. WYSE JACKSON AND J. R. AKEROYD (EDITION 2)

Annual to perennial herbs, glabrous or with unbranched hairs. Leaves simple. Raceme bracteate or not. Sepals erecto-patent; petals short-clawed; filaments straight. Fruit a laterally compressed, swollen silicula with convex valves; valves membranous when ripe with dorsal mid-vein distinct.

Literature: E. Pobedimova, *Nov. Syst. Pl. Vasc.* (*Leningrad*) **6**: 67–106 (1970); **7**: 167–195 (1971). R. Vogt, *Mitt. Bot. Staatssamm. München* **23**: 393–421 (1987).

1 Stems fistular, usually more than 50 cm tall and more than 5 mm in diameter
2 Pedicels in fruit $1\frac{1}{2}$–2 times as long as the silicula; seeds less than 1 mm, 15–30 in a silicula **9. glastifolia**
2 Pedicels in fruit 3–4 times as long as the silicula; seeds 1–1·5 mm, 6–8 in a silicula **10. megalosperma**
1 Stems not fistular, not more than 50 cm tall and less than 5 mm in diameter
3 Uppermost cauline leaves linear **7. aragonensis**
3 Uppermost cauline leaves oblong to ovate, or absent
4 Silicula ± truncate at the apex (**2–4**). **officinalis** group
4 Silicula not truncate at the apex
5 Upper cauline leaves mostly petiolate, never auriculate; annual **1. danica**

5 Upper cauline leaves sessile or auriculate; usually biennial or perennial
6 Petals pale yellow **5. tatrae**
6 Petals white or lilac
7 Silicula 2–4 times as long as wide, oblong-ellipsoidal or ellipsoidal **7. fenestrata**
7 Silicula less than twice as long as wide, ovoid-ellipsoidal
8 Leaves mostly truncate or cordate at the base; silicula not more than 7 mm **(2–4). officinalis** group
8 Leaves cuneate at the base; silicula 8–15 mm **6. anglica**

1. C. danica L., *Sp. Pl.* 647 (1753). Slender, somewhat fleshy annual to biennial; stems 1·5–20(–30) cm, ascending. Basal leaves long-petiolate, lamina *c.* 1 cm wide, orbicular to triangular-cordate; cauline leaves mostly petiolate, the lower ones palmately 3- to 7-lobed. Petals 2–3·5(–4) mm, lilac or white. Silicula 3–6 × 2·5–4 mm, ovoid-globose to ellipsoidal, often attenuate at both ends, finely reticulately veined when mature. $2n = 42$. *Coastal habitats, more rarely on disturbed ground inland.* ● *W. & N. Europe.* Be Br Da Fe Ga Ge Hb Ho Hs Lu No Rs(B) Su.

(2–4). C. officinalis group. Biennial or perennial, rarely annual, usually somewhat fleshy. Basal leaves long-petiolate, reniform or ovate-cordate, more rarely truncate to cuneate; cauline leaves sessile or only the lowest petiolate, the upper often auriculate. Petals (2–)4–9 mm, white or lilac. Silicula globose to ovoid-ellipsoidal or obovoid.

1 Basal leaves ovate, truncate to cuneate at the base, usually longer than wide; silicula ± truncate at the apex **4. aestuaria**
1 Basal leaves usually reniform, cordate at the base (sometimes truncate), about as long as wide; silicula not truncate at the apex
2 Plant with leafy stems **2. officinalis**
2 Plant scapose or subscapose **3. groenlandica**

2. C. officinalis L., *loc. cit.* (1753). Stems 5–50 cm. Basal leaves usually reniform. Raceme usually lax in fruit; pedicels usually longer than the silicula. Petals 2–8 mm; lateral veins free or anastomosing to form only 1 mesh on each side of the mid-vein. Silicula 2·5–7 mm, ovoid to globose, rounded or attenuate at both ends. *Coasts of N.W. Europe; also in W. & C. Europe mainly in the mountains.* Au Be Br Cz Da Fa Ga Ge Hb He Ho Hs Is No Po Rm Rs(W) Su [It].

1 Plant not more than 10 cm; leaves usually less than 1·5 cm wide; petals often lilac **(c) subsp. scotica**
1 Plant usually more than 10 cm; leaves usually at least 1·5 cm wide; petals usually white
2 Basal leaves distinctly fleshy, entire; silicula usually subglobose **(a) subsp. officinalis**
2 Basal leaves not markedly fleshy, sometimes obscurely lobed; silicula usually ovoid or ellipsoidal, rarely subglobose **(b) subsp. pyrenaica**

(a) Subsp. **officinalis** (?incl. *C. atlantica* Pobed.): Stems often more than 10 cm. Lamina of basal leaves 1·5–9(–12) cm wide, orbicular to reniform, entire, cordate at the base. Sepals green; petals 4–8 mm, white. Silicula 3–6 × 2·5–5 mm, usually subglobose, sometimes ovoid or ellipsoidal. $2n = 24$. *N.W. Europe, mainly coastal.*

(b) Subsp. **pyrenaica** (DC.) Rouy & Fouc., *Fl. Fr.* **2**: 200 (1895) (*C. pyrenaica* DC.; incl. *C. alpina* (Bab.) H. C. Watson, *C. excelsa* Zahlbr., *C. macrorrhiza* (Schur) Pobed., *C. borzaeana* (Coman & E. I. Nyárády) Pobed.): Stems 5–40 cm. Lamina of basal leaves 1–4(–6) cm wide, reniform, sometimes obscurely lobed, less fleshy than those of subspp. **(a)** and **(c)**, cordate at the base. Sepals green; petals 4·5–8 mm, white. Silicula 3–7 × 2–5 mm, ovoid or ellipsoidal, attenuate at both ends, less commonly globose. $2n = 12, 24, 48$. *Wet flushes and beside streams, mainly in the mountains.* ● *C. & W. Europe.*

(c) Subsp. **scotica** (Druce) P. S. Wyse Jackson, *Bot. Jour. Linn. Soc.* **106**: 119 (1991) (*C. scotica* Druce, *C. groenlandica* auct., non L.): Stems up to 10 cm. Lamina of basal leaves 0·6–1·6 mm wide, reniform to triangular-ovate, entire, often truncate at the base, more fleshy and darker green than in subspp. **(a)** and **(b)**. Sepals green to purplish; petals 2–3(–4) mm, often lilac. Silicula 2·5–3·5 × 1·5–2·5 mm, ovoid to broadly ellipsoidal or globose. $2n = 24$. *Maritime rocks and sands.* ● *Scotland; N. & W. Ireland; S.W. England (Isles of Scilly).*

Small plants from above 800 m in N. Britain with a low growth habit, somewhat woody at the base, and dark green, suborbicular basal leaves, have been called **C. micacea** E. S. Marshall, *Jour. Bot. (London)* **32**: 289 (1894). They are morphologically very difficult to separate satisfactorily from subsp. **(b)**, although chromosome counts of $2n = 26$ support the recognition of a distinct taxon.

C. polonica A. Fröhlich, *Pl. Polon. Exsicc.* ser. 2, cent. **3**: 11 (1936), from damp sand by streams in S. Poland (near Olkusz), a perennial with more or less erect stems 15–50 cm, is similar to **2(b)** but differs principally by the petals 5·5–9·5 mm, and a chromosome number of $2n = 36$. It may merit recognition as another subspecies of **2**.

3. C. groenlandica L., *Sp. Pl.* 647 (1753) (incl. *C. islandica* Pobed.). Stems many, 3–30(–50) cm, procumbent or ascending, stout, usually leafless. Basal leaves reniform to ovate-cordate. Petals 3 mm, oblong, contracted at the base into a distinct claw, white, rarely reddish. Silicula 3·5–5 mm, ovoid, ellipsoid-ovoid or obovoid. $2n = 14$. *Maritime shingle and cliffs. Arctic and subarctic Europe.* Is No Rs(N) Sb.

4. C. aestuaria (Lloyd) Heywood, *Feddes Repert.* **70**: 5 (1964) (*C. officinalis* var. *aestuaria* Lloyd). Perennial up to 40 cm. Basal leaves broadly ovate, truncate to somewhat cuneate at the base, usually slightly longer than broad. Pedicels in fruit longer than the silicula. Petals 4–9 mm, white. Silicula 4–6 mm, obovoid or subglobose, narrowed at the base, truncate and sometimes slightly emarginate at the apex. $2n = 12$. ● *Atlantic coasts of France and N. Spain.* Ga Hs.

5. C. tatrae Borbás, *Pallas Nagy Lexikona* **10**: 28 (1875). Perennial. Basal leaves ovate, cordate or truncate at the base. Raceme lax in fruit; lower pedicels usually longer than but not more than twice as long as the silicula; pedicels somewhat thickened. Petals pale yellowish; lateral veins not anastomosing, or rarely forming a single mesh on each side of the mid-vein. Silicula obovoid-ellipsoidal, narrowed at the base, rounded at the apex. $2n = 42$. *Granite rocks.* ● *W. Carpathians (Tatra).* Cz Po.

6. C. anglica L., *Syst. Nat.* ed. 10, **2**: 1128 (1759). Robust biennial to perennial up to 40 cm. Basal leaves long-petiolate, ovate, obovate or oblong, cuneate at the base, almost entire. Pedicels about equalling the silicula. Petals 5–10 mm, white. Silicula 8–15 mm, ellipsoidal or ovoid-ellipsoidal, often strongly compressed laterally, with long, narrow septum. $2n = 48, 54$. *Muddy shores, saltmarshes and saline meadows.* ● *Coasts of W. Europe, extending to S.E. Sweden.* Br Da Ga Ge Hb Ho Hs No Su.

Hybrids and backcrosses between **2(a)** and **6**, with basal leaves generally truncate at the base and petals 4–6 mm, frequently occur throughout the range of **6**, often in the absence of one or both parents.

7. C. fenestrata R. Br. in J. Ross, *Voy. Disc. Baffin* app. 3, 163 (1819) (incl. *C. arctica* Schlecht.). Biennial or perennial up to 15 cm. Basal leaves usually ovate, cordate or truncate at the base. Pedicels usually equalling the silicula. Petals 2·5–6 mm, white, rarely purplish. Silicula 5–8 mm, oblong-ellipsoidal or narrowly ellipsoidal, 2–4 times as long as wide; style 0·2–0·5 mm. *Arctic and subarctic Europe.* Fa No Rs(N) Sb.

There has been much confusion between **7** and small plants of **3**, and the assignment of some records is uncertain.

8. C. aragonensis Coste & Soulié, *Bull. Géogr. Bot.* (*Le Mans*) **21**: 7 (1911). Biennial with slender, branched stems 10–40 cm. Basal leaves 4–8 mm, as broad as long, ovate-cordate, entire or with a small callose tooth on either side and at the apex; cauline leaves lanceolate to linear. Petals 3–5 mm, white or violet. Silicula 3–6(–8) mm, oblong-ellipsoidal or ellipsoidal, 2–4 times as long as wide, attenuate at both ends, with narrow septum; valves caducous; style *c.* 1 mm. *Calcareous screes.* ● *Mountains of N.E. Spain.* Hs.

9. C. glastifolia L., *Sp. Pl.* 648 (1753). Robust annual or perennial, with single stem 50–100(–150) cm, fistular, leafy. Basal leaves oblong to elliptical; cauline leaves oblong-lanceolate, auriculate-amplexicaul. Pedicels in fruit $1\frac{1}{2}$–2 times as long as the silicula. Petals 3–4·5 mm, obovate, white. Silicula 2·5–4 mm, globose or broadly ellipsoidal; style 0·3–0·4 mm. Seeds 0·7–0·9 mm, ellipsoidal, echinate, 15–30 in a silicula. ● *C. & E. Spain; naturalized elsewhere in S. Europe.* Hs ?Lu [Ga It Ju].

10. C. megalosperma (Maire) Vogt, *Mitt. Bot. Staatssamm. München* **23**: 199 (1987) (*C. glastifolia* var. *megalosperma* Maire). Like **9** but pedicels in fruit 3–4 times as long as the silicula; silicula 2–3·2 mm; style 0·2–0·3 mm; seeds 1–1·5 mm, verrucose, 6–8 in a silicula. *Mountains of S. Spain.* Hs.

63. Kernera Medicus

A. O. CHATER AND V. H. HEYWOOD

Like *Cochlearia* but filaments curved; valves of fruit rigid when ripe, with mid-vein absent or present only in the lower part.

1. K. saxatilis (L.) Reichenb. in Moessler, *Handb.* ed. 2, **2**: 1142 (1828) (*Cochlearia saxatilis* L.). Perennial; stems 10–30(–40) cm, usually branched. Basal leaves petiolate, entire to deeply toothed, obovate-lanceolate to spathulate, obtuse or narrowed to the apex; cauline leaves lanceolate or ovate, narrowed to the base or sagittate and amplexicaul. Raceme many-flowered. Petals 2–4 mm, white. Silicula 2–4·5 mm, ovoid and stipitate, or ovoid-globose to ellipsoid-ovoid and sessile; valves smooth, with a more or less prominent mid-vein. Septum hyaline and slender, or spongy. *Mountains of S. & C. Europe.* Al Au Bu Co Cz Ga Ge Gr He Hs It Ju Rm Po.

(**a**) Subsp. **saxatilis** (incl. *K. auriculata* (Lam.) Reichenb., *K. decipiens* (Willk.) Nyman): Stems usually more than 15 cm. Leaves variable in shape, entire to lobed, usually pubescent; cauline leaves often auriculate. Silicula usually more than 2 mm, often stipitate; mid-vein usually present. $2n = 16, 32$. *Throughout the range of the species, except S.E. Spain.*

(**b**) Subsp. **boissieri** (Reuter) Nyman, *Consp.* 51 (1878) (*K. boissieri* Reuter): Stems up to 15 cm. Basal leaves broadly spathulate, usually entire, glabrous; cauline leaves rarely auriculate. Silicula usually not more than 2·2 mm, not stipitate; mid-vein rarely present. ● *S.E. Spain.*

64. Rhizobotrya Tausch

V. H. HEYWOOD

Like *Cochlearia* but raceme bracteate; sepals persistent, surrounding the fruit; filaments of stamens curved; valves of the siliqua without mid-vein.

1. R. alpina Tausch, *Flora* (*Regensb.*) **19**: 34 (1836) (*Cochlearia brevicaulis* Facch., *Kernera alpina* (Tausch) Prantl). Caespitose perennial; stems 2–4 cm. Leaves oblong-spathulate, obtuse, long-petiolate. Inflorescence condensed. Petals 2 mm, white. Siliqua *c.* 2·5 mm ovate, obtuse. $2n = 14$. *Dolomitic rocks, 1900–2800 m.* ● *S.E. Alps* (*W. Dolomiti*). It.

65. Camelina Crantz

R. D. MEIKLE (EDITION 1) REVISED BY J. R. AKEROYD (EDITION 2)

Annuals or biennials; hairs unbranched or branched. Cauline leaves sessile, often auriculate. Inflorescence ebracteate. Sepals erect; petals yellow or white. Fruit a more or less inflated silicula; style distinct. Seeds numerous, in 2 rows in each loculus.

Literature: N. Zinger, *Trav. Mus. Bot. Acad. Pétersb.* **6**: 1–303 (1909). Z. Mirek, *Fragm. Fl. Geobot.* **27**: 445–507 (1981).

1 Seeds 0·8–1·5 mm; stem and leaves with dense, unbranched or branched and unbranched hairs
2 Petals less than 5 mm; basal leaves withered by anthesis **2. microcarpa**
2 Petals 6–9 mm; basal leaves usually persistent at anthesis **3. rumelica**
1 Seeds (1·5–)1·6–3 mm; stem and leaves glabrous or with branched hairs (rarely also with unbranched hairs)
3 Silicula 10–12 mm **4. alyssum**
3 Silicula 6–10 mm
4 Leaves entire or denticulate **1. sativa**
4 Leaves deeply toothed or lobed **4. alyssum**

1. C. sativa (L.) Crantz, *Stirp. Austr.* **1**: 17 (1762) (*C. glabrata* (DC.) Fritsch, *C. pilosa* (DC.) Vassilcz., *C. sativa* auct. ital., *Myagrum sativum* L.). Erect, unbranched or sparingly branched annual 30–80(–100) cm; stems and leaves subglabrous (or hairy, var. *pilosa* DC.). Leaves 3–9 cm, lanceolate or narrowly oblong, entire or remotely denticulate, the cauline sessile, with acute auricles. Petals 4–5·5 mm, yellow. Raceme elongate in fruit, rather dense; pedicels 1–2 cm in fruit, ascending. Silicula usually 7–9 mm; valves strongly convex, not very rigid or woody and often flattened in dried specimens. Seeds 1–2·5 mm. $2n = 40$. *Cultivated land, especially flax-*

fields. Throughout Europe, but doubtfully native in most territories, and often only casual. Al Au Be Br Bu Co Cr Cz Da Fe Ga Ge Gr He ?Hs Ho Hu It No Po Rm Rs(N, B, C, W, K, E) Sa Si Su.

Variable in habit, hairiness and size and shape of the silicula.

2. C. microcarpa Andrz. ex DC., *Reg. Veg. Syst. Nat.* **2**: 517 (1821) (*C. sylvestris* Wallr., *C. sativa* auct. ital.). Annual or biennial, like **1** in general appearance but often less robust, with rather densely hairy stems and leaves; indumentum of long unbranched hairs mixed with short branched hairs. Petals 2·5–4 mm, pale yellow. Raceme very elongate in fruit and rigid with numerous siliculae; pedicels 1–1·5(–2) cm in fruit, ascending. Silicula *c.* 5–7 mm; valves less strongly convex than in **1**, usually hard and woody and not flattened in dried specimens. Seeds 0·8–1·4 mm. $2n = 40$. *Cultivated land and waste places. Throughout Europe, but in many regions only as a casual.* Au Be Br Bu Cr Cz Da Fe Ga Ge Gr He Ho Hu Is It Ju No Po Rm Rs(N, B, C, W, K, E) Su.

A variable species, considered by some authors to be a subspecies of **1**.

3. C. rumelica Velen., *Sitz.-Ber. Böhm. Ges. Wiss.* **1887**: 448 (1887) (*C. albiflora* Kotschy, *C. sativa* auct. ital., *C. sativa* var. *hirsuta* Boiss., *C. sylvestris* var. *mediterranea* Pau). Erect annual or biennial 15–40(–60) cm, usually with several patent or ascending branches; lower part of plant more or less densely hispidulous, the upper part glabrous or subglabrous. Basal leaves forming a distinct rosette, usually persistent until anthesis or later. Petals 5–9 mm, whitish or very pale yellow. Raceme elongate in fruit, often rather lax; pedicels 0·7–1(–1·4) cm in fruit, ascending or sometimes almost patent. Silicula 5–8 mm; valves rather compressed, rigid and woody. Seeds 1·2–1·5 mm. $2n = 12, 26$. *Cultivated land and waste places. S. & E.C. Europe, but probably native only in parts of the south-east.* Al Bu Gr Ju Tu [Au Ga Hs Hu It Rm Rs(W, K)].

4. C. alyssum (Miller) Thell., *Verz. Tausch Säm. Früchte Zürich* **1906**: 10 (1906) (*C. foetida* (Schkuhr) Fries, *C. linicola* C. Schimper & Spenner, *C. macrocarpa* Wierzb. ex Reichenb., *C. dentata* (Willd.) Pers., *C. pinnatifida* Hornem, *C. sativa* auct. ital., *C. sativa* var. *sublinicola* N. Zinger, *Myagrum alyssum* Miller). Slender, erect, sparingly branched annual 15–60(–100) cm; stems and leaves subglabrous. Leaves deeply toothed or lobed (sometimes subentire, var. *integrifolia* (Fries) Hayek). Petals 4–6 mm, pale yellow. Raceme rather short and lax in fruit; pedicels 1·5–3(–4) cm in fruit, patent, erecto-patent or sometimes flexuous. Silicula 6–12 mm, depressed-globose, with a flattened or rounded apex; valves strongly convex, not hard or woody. Seeds 1·5–2·8(–3) mm. $2n = 40$. *A weed of flax-fields. Widespread in Europe.* Au Be Bu Co Cz Da Fe Ga Ge He Ho Hu It Ju No Po Rm Rs(N, B, C, W, K, E) Su.

66. **Neslia** Desv.

P. W. BALL

Annuals; hairs branched. Leaves simple. Inflorescence an ebracteate raceme. Sepals erect, not saccate; petals yellow; stamens without appendages. Fruit an indehiscent, latiseptate silicula; style distinct; stigma minute, slightly 2-lobed. Seeds 1–3. (*Vogelia* Medicus, non J. F. Gmelin.)

Literature: P. W. Ball, *Feddes Repert.* **64**: 11–13 (1961).

1. N. paniculata (L.) Desv., *Jour. Bot. Appl.* **3**: 162 (1814) (*Vogelia paniculata* (L.) Hornem.). Pubescent annual 15–80 cm. Leaves oblong or lanceolate, entire or remotely dentate, the basal petiolate, the cauline sessile, amplexicaul, with acute auricles. Petals 2–3 mm. Pedicels 5–13 mm in fruit. Silicula 1·5–3 mm in diameter, subglobose or compressed and almost lenticular, reticulate-rugose; style 0·7–1·1 mm. *Most of Europe except the extreme north, but of doubtful status in many territories.* Al Au Bl Bu Co Cr Cz *Da Ga Ge Gr He Hs Hu It Ju Lu Po Rm Rs(N, B, C, W, K, E) Sa Si *Su Tu.

(**a**) Subsp. **paniculata**: Silicula usually broader than long, not apiculate (excluding the often persistent style), with truncate base and 2 longitudinal ribs on the margin. $2n = 14$. *Throughout the range of the species except parts of the Mediterranean region.*

(**b**) Subsp. **thracica** (Velen.) Bornm., *Österr. Bot. Zeitschr.* **44**: 125 (1894) (*N. apiculata* Fischer, C. A. Meyer & Avé-Lall., *N. paniculata* auct. eur. merid.): Silicula about as long as broad, apiculate, with carpophore and 4 longitudinal ribs. *S. Europe; a rare casual elsewhere.*

These two subspecies have been very much confused so that their precise distribution is uncertain. Plants intermediate between subspp. (**a**) and (**b**) occur where the two subspecies meet (approximately 43–48° N.).

67. **Capsella** Medicus

A. O. CHATER

Annuals or biennials; glabrous or with branched and unbranched hairs. Basal leaves entire to pinnatifid; cauline leaves sagittate-amplexicaul. Inflorescence racemose, ebracteate. Sepals erect, not saccate; petals white, pink or yellowish; stamens without appendages. Fruit an angustiseptate silicula, usually triangular-obcordate; valves keeled, reticulately veined; style distinct; stigma minute, capitate. Seeds up to 12 in each loculus.

This account treats as species the four taxa most readily recognized in Europe. Within them there is extreme polymorphism; autogamy is frequent, and all are ruderals. Species **1** is especially polymorphic, and its variants incorporate many of the characters of the other three species, particularly as regards the shape and size of the silicula.

The species are usually plants of cultivated or waste ground.

Literature: E. Almquist, *Acta Horti Berg.* **7**: 41–95 (1921). G. H. Schull, *Amer. Jour. Bot.* **10**: 221–228 (1923).

1 Petals not or scarcely exceeding the sepals, usually reddish-tinged **2. rubella**
1 Petals usually distinctly exceeding the sepals, white or yellowish
  2 Petals pale yellowish; lower part of plant densely grey-hairy **4. orientalis**
  2 Petals white; whole plant sparsely hairy or glabrous
    3 Petals 4–5 mm **3. grandiflora**
    3 Petals 2–3 mm (rarely absent) **1. bursa-pastoris**

1. C. bursa-pastoris (L.) Medicus, *Pflanzengatt.* 85 (1792) (*Thlaspi bursa-pastoris* L.). Plant sparsely hairy, especially below, or glabrous. Flowers scentless. Sepals green, sometimes reddish or purplish, often pubescent; petals 2–3 mm, $1\frac{1}{2}$–2 times as long as the sepals, white (rarely absent). Silicula 4–10 × 4–9 mm, usually longer than wide, scarcely attenuate at the base; lateral margins usually straight or convex; apical lobes usually subacute. $2n = 16, 32$. *Throughout Europe as a ruderal; native range unknown.* All territories.

(**a**) Subsp. **bursa-pastoris**: Silicula usually only slightly emarginate; style *c.* 0·25 mm. $2n = 32$. *Throughout the range of the species.*

(**b**) Subsp. **thracica** (Velen.) Stoj. & Stefanov, *Fl. Bălg.* ed. 3, 513 (1948) (*C. thracica* Velen.): Silicula deeply emarginate; style *c.* 0·75 mm. ● *S. Bulgaria.*

Numerous variants of this species have been described by Almquist, *loc. cit.* (1921).

C. heegeri Solms-Laub., *Bot. Zeit.* **58**: 167 (1900) (*Solmsiella heegeri* (Solms-Laub.) Borbás), a profusely branched plant with an ellipsoidal, not compressed silicula, is an atavistic mutant of **1** first noted in 1897 in W. Germany. It became extinct in the original locality, but was cultivated in botanic gardens, and was later found naturalized near Berlin.

2. C. rubella Reuter, *Compt. Rend. Soc. Hallér.* **2**: 18 (1854). Plant sparsely hairy or glabrous. Flowers scentless. Sepals usually reddish at least at the apex, glabrous; petals 1·5–2 mm, shorter than or scarcely exceeding the sepals, usually reddish at least on margins. Silicula *c.* 6 × 6 mm, distinctly attenuate at the base, usually fairly deeply emarginate at the apex; lateral margins concave; apical lobes obtuse; style *c.* 0·25 mm. *S. Europe; naturalized in C. Europe, and occasionally casual elsewhere.* Al Bl *Br Co Ga Hs It Ju Lu Sa Si ?Tu [Au Ge He].

Sterile plants intermediate between **1** and **2**, presumably of hybrid origin, have been called **C. gracilis** Gren., *Mém. Soc. Émul. Doubs* ser. 3, **2**: 403 (1858) (*C. gelmii* J. Murr). Fertile intermediates also sometimes occur, and there is much difference of opinion as to whether **2** should be recognized even at subspecific rank. Specimens of **1** with reddish sepals have frequently been identified as **2**, but this character is variable even within a single plant.

3. C. grandiflora (Fauché & Chaub.) Boiss., *Diagn. Pl. Or. Nov.* **1(1)**: 76 (1843). Plant up to 75 cm, sparsely hairy. Flowers fragrant. Sepals green, glabrous; petals 4–5 mm, *c.* $2\frac{1}{2}$ times as long as the sepals, white. Silicula *c.* 6 × 6 mm, scarcely attenuate at the base, deeply emarginate at the apex; lateral margins straight; apical lobes very obtuse; style 0·25–0·7 mm. *Rocky limestone slopes and as a ruderal.* ● *Greece and Albania; naturalized in N. Italy.* Al Gr [It].

4. C. orientalis Klokov, *Bull. Soc. Nat. Voronèje* **1**: 122 (1926). Plant densely hairy in lower part, greyish-green. Sepals green; petals 1·5–2 mm, less than twice as long as the sepals, pale yellowish. Silicula 5–6 × 4–4·5 mm, deeply emarginate at the apex; style 0·25–0·5 mm. *S. Russia, W. Ukraine.* Rs(W, E).

68. Pritzelago O. Kuntze

V. H. HEYWOOD AND B. M. G. JONES

Small perennials; glabrous or with branched and unbranched hairs. Inflorescence ebracteate. Petals white, clawed. Fruit an angustiseptate silicula, elliptical to lanceolate. Seeds 1 or 2 in each loculus. (*Hutchinsia* auct., non R. Br., *Noccaea* auct., non Moench.)

Literature: G. Melchers, *Österr. Bot. Zeitschr.* **81**: 81 (1932). F. K. Meyer, *Wiss. Zeitschr. Friedrich-Schiller Univ., Math. Nat. Reihe* **31**: 267–276 (1982).

1. P. alpina (L.) O. Kuntze, *Revis. Gen. Pl.* **1**: 35 (1891) (*Hutchinsia alpina* (L.) R. Br., *Noccaea alpina* (L.) Reichenb.). Stock branched; stems up to 15 cm. Basal leaves pinnatisect. Petals 3–5 mm. *Rocks and stony ground. Mountains of C. & S. Europe, southwards to N. Spain, C. Italy and Macedonia.* Al Au Cz Ga Ge He Hs It Ju Lu Po Rm.

1 Flowering stems flexuous, leafy — (**b**) subsp. **auerswaldii**
1 Flowering stems ± leafless, straight
 2 Petals 3 mm wide, abruptly contracted into the claw; style 1 mm — (**a**) subsp. **alpina**
 2 Petals 1–2 mm wide, gradually attenuate into the claw; stigma sessile — (**c**) subsp. **brevicaulis**

(**a**) Subsp. **alpina**: Flowering stems 5–12 cm, sparsely hairy. Leaves all basal, glabrous, with 5–9 ovate-lanceolate segments. Petals 4–5 × 3 mm. Silicula 4–6 × 1·5–2 mm, ovate, acute. $2n = 12$. *Calcicole. Alps and Jura; Pyrenees; N. & C. Italy.*

(**b**) Subsp. **auerswaldii** (Willk.) Greuter & Burdet, *Willdenowia* **15**: 68 (1985): Like subsp. (**a**) but flowering stems up to 15 cm, flexuous, leafy. Cauline leaves with lanceolate to linear segments. ● *N. Spain (Cordillera Cantábrica).*

(**c**) Subsp. **brevicaulis** (Hoppe) Greuter & Burdet, *op. cit.* 69 (1985): Flowering stems 2–5 cm, glabrous. Leaves with 3–7 ovate-lanceolate segments. Petals 3–4 × 1–2 mm. Silicula 3·5–4 × 1–2 mm, obtuse. $2n = 12$. *Alps and Balkan peninsula.*

Plants from the Pyrenees and S.W. Alps with leaves like subsp. (**c**) but fruits like subsp. (**a**) have been distinguished as **Hutchinsia affinis** Gren. ex F. W. Schultz, *Arch. Fl. Fr. Allem.* 274 (1853).

69. Hymenolobus Nutt.

V. H. HEYWOOD

Like *Pritzelago* but all hairs unbranched and seeds 3–10 in each loculus.

Literature: R. Pampanini, *Nuovo Gior. Bot. Ital.* **16**: 36 (1909).

Stems up to 15(–30) cm; seeds 0·4–0·6 mm — **1. procumbens**
Stems not more than 6 cm; seeds 0·6–0·8 mm — **2. pauciflorus**

1. H. procumbens (L.) Nutt. in Torrey & A. Gray, *Fl. N. Amer.* **1**: 117 (1838) (*Hornungia procumbens* (L.) Hayek, *Hutchinsia procumbens* (L.) Desv., *Capsella procumbens* (L.) Fries). Annual or biennial with sparse unbranched hairs; stems 2–15(–30) cm, procumbent or erect, branched. Lower leaves deeply lyrate-pinnatifid to entire; upper leaves entire. Petals 1–3 mm, spathulate, equalling or slightly exceeding the sepals. Silicula 2–5 mm, elliptical to obovate or orbicular; valves translucent, with reticulate veins. Seeds 0·4–0·6 mm. $2n = 12, 24$. *Saline habitats. Mainly S. Europe, extending northwards to C. Germany.* Au Bl Co Ga Ge Gr He Hs It Ju Lu Rm Rs(W, K, E) Sa Si.

An extremely variable species. A distinctive variant, var. *revelieri* (Jordan) Heywood (*H. procumbens* subsp. *revelieri* (Jordan) Greuter & Burdet), with leaves all ovate-elliptical, entire, petals *c.* 1·5 mm, and silicula 3 mm, orbicular, occurs in Italy, Sicily, Corse and Malta. Other variants, about which there is wide disagreement in their characters, make the distinction between **1** and **2** difficult, and some authors have treated these two species as subspecies.

2. H. pauciflorus (Koch) Schinz & Thell., *Viert. Naturf. Ges. Zürich* **66**: 285 (1921) (*Capsella pauciflora* Koch). Like **1** but stems 2–6 cm, erect; leaves 3-lobed or entire, spathulate; inflorescence few-flowered; silicula orbicular to elliptical; seeds 0·6–0·8 mm. ● *Cevennes; Alps; Calabria and Sicilia; isolated localities in E. Spain.* Au Ga He Hs It Si.

70. **Hornungia** Reichenb.

V. H. HEYWOOD

Small annuals; hairs stellate or absent. Inflorescence ebracteate. Petals white, short-clawed. Fruit an angustiseptate silicula, elliptical to oblong-ovate. Seeds 1 or 2 in each loculus, mucilaginous.

Petals scarcely exceeding the sepals; silicula 2–2·5(–3) mm **1. petraea**

Petals twice as long as the sepals; silicula 3 mm **2. aragonensis**

1. H. petraea (L.) Reichenb., *Deutschl. Fl.* **1**: 33 (1837) (*Hutchinsia petraea* (L.) R. Br.). Stems 2–15 cm, slender. Leaves all pinnate; basal leaves rosulate, petiolate, with 3–15 or more ovate, lanceolate or obovate, acute segments; cauline leaves few, sessile. Petals 0·5–1 mm, as long as or slightly longer than the sepals. Silicula 2–2·5(–3) mm, narrowly elliptical to ovate or obovate; valves compressed, with distinct median vein. $2n = 12$. *Dry, open habitats. S., W. & C. Europe, extending to Sweden, Estonia and Ukraine.* Al Au Be Bl Br Bu Co Cr Cz Ga Ge Gr He Hs Hu It Ju Lu No Rs(B, W, K) Sa Si Su Tu.

2. H. aragonensis (Loscos & Pardo) Heywood, *Feddes Repert.* **66**: 155 (1962) (*Hutchinsia aragonensis* (Loscos & Pardo) Loscos & Pardo, *H. petraea* var. *aragonensis* Loscos & Pardo). Like **1** but petals twice as long as the sepals, broadly obovate, long-clawed; silicula 3 mm, ovate-lanceolate. ● *Mountains of N.E. Spain (Aragón, W. Cataluña).* Hs.

71. **Jonopsidium** Reichenb.

V. H. HEYWOOD

Slender, glabrous annuals. Inflorescence leafy, or bracteate at the base. Sepals patent; petals white, purple or pink; stamens free, without appendages. Fruit an angustiseptate silicula, the valves keeled. Seeds 2–6 in each loculus, covered with transparent papilliform glands. (Incl. *Pastorea* Tod., *Minaea* Lojac.)

Literature: A. Chiarugi, *Nuovo Gior. Bot. Ital.* nov. ser., **34**: 1452–1459 (1928).

1 Petals equal
2 Petals usually lilac or purple, 2–3 times as long as the calyx; petiole of basal leaves several times as long as the lamina **1. acaule**
2 Petals white, $1\frac{1}{2}$ times as long as the calyx; petiole of basal leaves not longer than the lamina **2. albiflorum**
1 Petals unequal
3 Sepals with a translucent margin; style 0·5 mm **5. savianum**
3 Sepals with a white margin; style 1 mm
4 Silicula 5–7 × 5–7 mm; 10–12 siliculae developed on each side of the raceme **3. prolongoi**
4 Silicula 3–4 × 2–2·5 mm; 16–20 siliculae developed on each side of the raceme **4. abulense**

Sect. JONOPSIDIUM (Sect. *Ionopsis* (DC.) Cosson). Inflorescence leafy. Petals equal. Septum of the silicula with a membranous margin.

1. J. acaule (Desf.) Reichenb., *Pl. Crit.* **7**: 26, t. 649 (1829). Plant caespitose, stemless or sometimes with short stem. Basal leaves rounded-ovate, entire or 3-lobed. Flowers usually solitary, borne on long pedicels in the axils of the basal leaves. Petals 4–5 mm, lilac or purple, sometimes white, 2–3 times as long as the calyx. Silicula obovoid-orbicular. Seeds 2–5 in each loculus. $2n = 24$. ● *Portugal. Cultivated in gardens, and naturalized in parts of S. Europe.* Lu [Ga Hs].

Sect. PASTORAEA (Tod.) Cosson. Inflorescence with leaf-like bracts. Petals equal. Septum of silicula without a membranous margin.

2. J. albiflorum Durieu in Duchartre, *Rev. Bot.* **2**: 433 (1847) (*Bivonea albiflora* (Durieu) Prantl). Stem 5–15 cm, simple or branched from the base. Basal leaves entire or sinuate-dentate. Petals *c.* 2 mm, white, $1\frac{1}{2}$ times as long as the calyx. Silicula oblong. Seeds 5 or 6 in each loculus. *Calcifuge. S.E. Italy (Puglia), Sicilia.* It Si. (*N.W. Africa.*)

Sect. MINAEA (Lojac.) Batt. Inflorescence bracteate at the base only. Petals unequal. Septum of silicula with a membranous margin.

3. J. prolongoi (Boiss.) Batt., *Bull. Soc. Bot. Fr.* **43**: 259 (1896) (*Bivonia prolongoi* (Boiss.) Samp., *Thlaspi prolongoi* Boiss.). Plant with many stems, 5–15 cm. Basal leaves in lax rosette, oblong-spathulate, shortly petiolate, irregularly toothed. Inflorescence racemose, the ultimate racemes subcorymbose, becoming very lax; 10–12 siliculae developed on each side of the raceme; pedicels horizontal, equalling or shorter than the fruits. Sepals narrowly white-margined; petals 4–5 mm, white. Silicula 4–7 × 4–7 mm, more or less orbicular; style 1 mm. Seeds 2 or 3 in each loculus. $2n = 22$. *Calcicole. S. Spain.* Hs. (*N.W. Africa.*)

4. J. abulense (Pau) Rothm., *Cavanillesia* **7**: 112 (1935) (*J. heterospermum* sensu Chiarugi, pro parte, *J. prolongoi* subsp. *abulense* (Pau) Laínz, *Bivonea abulensis* (Pau) Samp., *Thlaspi abulense* Pau, *Thlaspi montanum* sensu Coutinho, *Thlaspi prolongoi* prol. *abulense* (Pau) Samp.). Like **3** but petals 2–3 mm; 16–20 siliculae developed on each side of the raceme; pedicels $1\frac{1}{2}$–3 times as long as the silicula; silicula 3–4 × 2–2·5 mm. *Calcifuge.* ● *N. & C. Spain, N.E. & C. Portugal.* Hs Lu.

5. J. savianum (Caruel) Ball ex Arcangeli, *Comp. Fl. Ital.* 58 (1882) (*Bivonea saviana* Caruel). Like **3** but stems not more than 10 cm; ultimate racemes of the inflorescence not becoming very lax; sepals with translucent margin; petals 3–4 mm; pedicels slightly deflexed in fruit; silicula retuse at the apex, attenuate at the base; style 0·5 mm. $2n = 32$. ● *C. Italy; local.* It.

72. **Bivonaea** DC.

P. W. BALL

Like *Jonopsidium* but petals yellow; valves of the silicula winged; seeds smooth, without papilliform glands.

1. B. lutea (Biv.) DC., *Reg. Veg. Syst. Nat.* **2**: 555 (1821). Stem 4–20 cm, simple or branched from the base. Basal leaves oblong-spathulate, shortly petiolate, entire or toothed. Petals *c.* 3 mm, equal, $1\frac{1}{2}$ times as long as the sepals. Silicula 5–7 mm, obovoid-oblong, emarginate; wing 1–2 mm wide. *Sicilia; Sardegna.* Sa Si. (*N.W. Africa.*)

73. **Teesdalia** R. Br.

†J. DE CARVALHO E VASCONCELLOS

Annuals; glabrous or with unbranched hairs. Leaves mostly basal, usually pinnatifid. Sepals erecto-patent; petals white; stamens 4 or

6, the filaments with a white basal scale. Fruit an angustiseptate silicula, obtuse or obcordate, with thin-walled valves, narrowly winged in the upper part; style very short or absent. Seeds (1)2 in each loculus.

Basal leaves with obtuse lobes; petals unequal, the outer 1½–2 times as long as the sepals; style short **1. nudicaulis**
Basal leaves with acute lobes, or entire; petals ± equal, as long as the sepals; style absent **2. coronopifolia**

1. T. nudicaulis (L.) R. Br. in Aiton, *Hort. Kew.* ed. 2, **4**: 83 (1812). Stem 5–45 cm, often with ascending basal branches. Basal leaves 1–5 cm, petiolate, narrowly lyrate-pinnatifid, with few short, acute or obtuse lateral lobes and a broader, often 3-lobed, terminal segment; cauline leaves (if present) less lobed or more or less entire. Petals unequal, the inner slightly longer than the sepals, the outer 1½–2 times as long. Silicula 3–4(–4·5) mm; style absent or up to 0·3 mm. $2n = 36$. *W. & C. Europe, extending to Sweden, White Russia and Jugoslavia.* Au Be Br Cz Da Ga Ge Hb He Ho Hs Hu ?It Ju Lu No Po ?Rm Rs(B, C, W) ?Si Su.

2. T. coronopifolia (J. P. Bergeret) Thell., *Feddes Repert.* **10**: 289 (1912) (*T. lepidium* DC., *T. nudicaulis* var. *regularis* (Sm.) Fiori). Like **1** but more slender; rosette-leaves narrowly oblanceolate, usually pinnatifid with acute lobes; petals subequal, as long as the sepals; silicula not more than 3 mm, style absent. $2n = 36$. *S. Europe.* Bu Co Cr Ga Gr Hs It Ju Lu Sa Si Tu.

74. Thlaspi L.

†A. R. CLAPHAM (EDITION 1) REVISED BY J. R. AKEROYD (EDITION 2)

Annuals or perennials, with sessile, more or less amplexicaul cauline leaves; hairs unbranched or absent. Inflorescence racemose, ebracteate. Sepals erect, not saccate; petals usually white or purplish, shortly clawed; stamens without appendages. Fruit an angustiseptate silicula, with or without an apical notch, the valves keeled and usually winged; stigma capitate, somewhat 2-lobed. Seeds 1–8 in each loculus. (Incl. *Noccaea* Moench.)

In species described as having violet anthers, the violet coloration may not appear until a late stage of dehiscence.

A difficult genus requiring revision. It has been treated as a number of smaller genera by F. K. Meyer, *Feddes Repert.* **84**: 449–470 (1973).

T. macrophyllum Hoffm., *Comment. Soc. Phys.-Med. Mosq.* **1**: 7 (1805) (*Pachyphragma macrophyllum* (Hoffm.) N. Busch), from the Caucasus and N.E. Turkey, a more or less glabrous, rhizomatous perennial smelling of garlic, with large, long-petiolate, ovate to reniform or suborbicular leaves, flowering stems 15–40 cm, and white petals 8–9 mm, is cultivated for ornament and naturalized at one locality in S.W. England and perhaps elsewhere.

1 Annual, without non-flowering leaf-rosettes
2 Style long, exceeding the notch of the ripe fruit **24. macranthum**
2 Style very short, included within the notch of the ripe fruit
3 Upper cauline leaves ovate, cordate-amplexicaul at the base; seeds smooth
4 Silicula 5–7 mm, with wing up to 1·5 mm wide; seeds 3 or 4 in each loculus **3. perfoliatum**
4 Silicula usually more than 7 mm, with wing 2·5–3·5 mm wide; seeds 6–8 in each loculus **4. kotschyanum**
3 Upper cauline leaves lanceolate to oblong, hastate- or sagittate-amplexicaul at the base; seeds ridged or alveolate
5 Fruit usually more than 10 mm, broadly elliptical to suborbicular, broadly winged all round **1. arvense**
5 Fruit not more than 10 mm, obovate, narrowly winged below, broadly winged above **2. alliaceum**
1 Biennial to perennial, usually with non-flowering leaf-rosettes
6 Style included within the notch of the ripe fruit, or stigma sessile
7 Petals 5–7 mm **(15–19). praecox** group
7 Petals 1–4·5 mm
8 Plant more than 10 cm
9 Petals up to 2 mm; anthers white or whitish
10 Raceme elongating in fruit **(5–8). caerulescens** group
10 Raceme remaining short and compact in fruit **9. brevistylum**
9 Petals more than 2 mm; anthers yellow, sometimes becoming reddish to dark violet after dehiscence
11 Main stem with long decumbent or ascending basal branches **10. rivale**
11 Main stem without long basal branches **(5–8). caerulescens** group
8 Plant not more than 10 cm
12 Apical notch of fruit acute at the base **10. rivale**
12 Apical notch of fruit rounded at the base
13 Petals 3–4·5 mm **(5–8). caerulescens** group
13 Petals 2 mm **9. brevistylum**
6 Style equalling or exceeding the notch of the ripe fruit, or notch absent
14 Ripe fruit strongly keeled but not winged; notch very narrow or absent; inflorescence scarcely elongating in fruit
15 Petals white (Spain) **27. nevadense**
15 Petals purple
16 Plant caespitose; rosette-leaves spathulate, dentate; all cauline leaves alternate **26. bellidifolium**
16 Plant with stolons; rosette-leaves elliptical to suborbicular; lower cauline leaves opposite **25. cepaeifolium**
14 Ripe fruit distinctly, though sometimes very narrowly, winged; notch broad or absent; inflorescence usually elongating in fruit
17 Sepals 1–2(–2·2) mm; petals about equalling or shorter than the stamens **(5–8). caerulescens** group
17 Sepals (1·5–)2–3 mm; petals exceeding the stamens
18 Petals 2–4 mm
19 Plant 10–50 cm; raceme elongating in fruit **(5–8). caerulescens** group
19 Plant 1–2(–4) cm; raceme not or scarcely elongating in fruit **12. microphyllum**
18 Petals 4–8·5 mm
20 Roots tuberous; flowers violet (Greece) **14. bulbosum**
20 Roots not tuberous; flowers white or purplish
21 Wing of fruit 0·5(–1) mm wide, so forming a very shallow notch which is much exceeded by the style **(21–23). alpinum** group
21 Wing of fruit-valve more than 1 mm wide; notch very variable in depth and width

22 Stock with stolon-like branches; plant mat-forming **20. montanum**
22 Stock with short branches; plant ±caespitose
23 Anthers yellow **(15–19). praecox** group
23 At least some anthers violet
24 Inflorescence almost sessile; style 0·7–1·2 mm **12. microphyllum**
24 Inflorescence distinctly stalked; style 1–5 mm
25 Flowers white; style 1–2·5 mm **11. graecum**
25 Flowers purplish; style 3–5 mm **13. stylosum**

Sect. NOMISMA DC. Silicula suborbicular, strongly compressed, broadly winged, with a deep and narrow notch; stigma subsessile. Seeds concentrically ridged.

1. T. arvense L., *Sp. Pl.* 646 (1753). Annual 10–60 cm, erect, glabrous, more or less foetid. Basal leaves oblanceolate to obovate, petiolate; upper cauline leaves lanceolate to oblong with sagittate-amplexicaul base; all entire or sinuate-dentate. Sepals 1·5–3 mm, narrow; petals 3–5 mm, white; stamens shorter than the petals; anthers yellow. Silicula (6–)9–18 mm in diameter, broadly elliptical to almost orbicular, broadly winged all round; pedicels (5–)8–13 mm, ascending; style *c.* 0·3 mm, included within the deep, narrow notch. Seeds 3–8 in each loculus, concentrically ridged. $2n = 14$. *Weed of arable land and waste places. Most of Europe, except the extreme north and much of the Mediterranean region, but doubtfully native in a large part of this area.* All except Az Co Cr Fa Sa Sb Si; perhaps only casual in Is.

Sect. THLASPI. Silicula obovate or orbicular, plano-convex, narrowly to broadly winged, with or without a notch; style included to exserted. Seeds alveolar to smooth.

2. T. alliaceum L., *Sp. Pl.* 646 (1753). Annual 20–60 cm, erect, with a few long hairs at the base of the stem, smelling of garlic. Basal leaves not forming a rosette, lanceolate to oblong-obovate, petiolate, sinuate-dentate to almost lyrate; upper cauline leaves oblong with sagittate-amplexicaul base, entire to sinuate-dentate; all glabrous and glaucous. Sepals *c.* 1·5 mm; petals 2·5–3 mm, white; stamens shorter than the petals; anthers yellow. Silicula 5–10 mm, narrowly obcordate-obovate, convex beneath, narrowly winged below, more broadly winged above; style *c.* 0·3 mm, included within the shallow notch. Seeds 3–5 in each loculus, alveolate. *Weed of arable land and waste places. C. & S. Europe.* Al Au ?Co Ga Ge Hs Hu It Ju Rm Rs(W) Si Tu [Br Po].

3. T. perfoliatum L., *Sp. Pl.* 646 (1753) (*T. rotundifolium* Tineo; incl. *T. tinei* Nyman). Annual 5–20(–30) cm, glabrous, glaucous. Basal leaves forming a loose rosette, obovate, petiolate; upper cauline leaves ovate-cordate, sessile, with rounded, amplexicaul auricles; all entire or sinuate-denticulate. Sepals 1–1·7 mm, with broad, white margins; petals 1·5–3 mm, white; stamens shorter than the petals; anthers yellow. Silicula 3–7 mm, broadly obcordate, convex beneath; wing wider above, up to 1·5 mm wide; style *c.* 0·3 mm, included within the wide and fairly deep notch. Seeds 3 or 4 in each loculus, almost smooth. $2n = 14, 42, 70$. *In open vegetation on limestone, loess and base-rich loams, and a weed of arable land and waste places. Europe northwards to* c. 60° N. *in Sweden.* All except Az Fa Fe Hb Is No Sb.

4. T. kotschyanum Boiss. & Heldr. in Boiss., *Diagn. Pl. Or. Nov.* **1(8)**: 39 (1849). Glabrous annual, 5–20 cm. Basal leaves soon withering. Cauline leaves ovate, obtuse, entire, with acute to obtuse, cordate-amplexicaul auricles. Sepals 1·2–1·6 mm; petals 1·4–2·5 mm, white; anthers yellow. Silicula 8–11(–14) mm, orbicular, broadly winged (2·5–3·5 mm wide) above; stigma sessile; notch narrow. Seeds 6–8 in each loculus, smooth. *Calcareous rocky grassland,* c. 2100 *m. E.C. Greece (Oiti).* Gr. (*S.W. Asia.*)

(5–8). T. caerulescens group. Biennials or perennials 10–50 cm, with shortly branched stock and crowded leaf-rosettes. Basal leaves rosulate, petiolate; cauline leaves sessile, auriculate-amplexicaul. Inflorescence a compact, almost corymbose raceme, usually much elongating in fruit. Sepals 1–2(–2·2) mm; petals 1–5 mm, white or purplish; stamens about equalling or exceeding the petals. Silicula 5–10(–12) × 3–6 mm; wing narrow below, but usually broadening upwards and forming a notch variable in width and depth; style variable in length. Seeds (2–)4–6 in each loculus.

1 Petals ±equalling or up to 1⅓ times as long as the sepals **5. brachypetalum**
1 Petals at least 1½ times as long as the sepals
2 Anthers reddish to deep violet, at least after dehiscence **6. caerulescens**
2 Anthers remaining yellowish even after dehiscence
3 Fruit very narrowly winged all round; style much exceeding the notch **7. stenopterum**
3 Fruit broadly winged above; style included within or about equalling the notch
4 Raceme elongating considerably in fruit **6. caerulescens**
4 Raceme scarcely elongating in fruit **8. dacicum**

5. T. brachypetalum Jordan, *Obs. Pl. Crit.* **3**: 5 (1846). Biennial 20–50 cm, glabrous, glaucous. Leaves entire or denticulate, the basal elliptical, the cauline oblong with obtuse to subacute auricles. Sepals 1–1·5 mm; petals white, equalling or slightly exceeding the sepals; stamens somewhat exceeding the petals; anthers whitish. Raceme much elongating in fruit. Silicula 6–9 mm, narrowly obcordate, broadly winged above, with rounded apical lobes and a deep notch usually exceeding the short style. Seeds 4–6 in each loculus. *Mountain woods and pastures; usually calcifuge.* ● *Pyrenees to S.W. Alps; C. Appennini.* Ga He Hs It [Su].

6. T. caerulescens J. & C. Presl, *Fl. Čechica* 133 (1819) (*T. alpestre* L., non Jacq., *T. guadinianum* Jordan, *T. huteri* A. Kerner, *T. mureti* Gremli, *T. occitanicum* Jordan, *T. pratulorum* Gand., *T. rhaeticum* Jordan, *T. virgatum* Gren. & Godron, *T. villarsianum* Jordan, *T. vogesiacum* Jordan, *T. vulcanorum* Lamotte, *T. suecicum* Jordan). Biennial or perennial 10–50 cm, glabrous, often glaucous. Leaves entire or denticulate, the basal elliptical to obovate-spathulate, the cauline oblong-cordate. Sepals (1·1–)1·4–2·2 mm; petals white or purplish, 1½–3 times as long as the sepals and equalling or somewhat exceeding the stamens; anthers usually reddish to dark violet, at least after dehiscence, but sometimes remaining yellowish. Raceme much elongating in fruit. Silicula 4–8·5(–10) mm, narrowly to broadly obcordate with wing broadening upwards, with or without a notch; style included or exserted. Seeds (2)3–6 in each loculus. *Often on soils rich in heavy metals. S., W. & C. Europe, eastwards to Poland and Jugoslavia; naturalized in parts of N. Europe.* Au Be Br Cz Ga Ge He Ho Hs Hu It Ju Po [Da Fe Is No Su].

(a) Subsp. **caerulescens** (incl. subsp. *occitanicum* (Jordan) Laínz): Usually biennial and monocarpic. Petals 2–3(–4) mm, white or purplish. Silicula 4–9 mm, broadly winged above, with rounded or subacute apical lobes and a distinct notch, or truncate at the apex; style 0·4–1·5 mm, equalling or exceeding the notch. $2n = 14$, 21, 28. *Throughout the range of the species.*

(**b**) Subsp. **virens** (Jordan) Hooker fil., *Stud. Fl. Brit. Is.* 38 (1870) (*T. virens* Jordan): Perennial, laxly caespitose, not glaucous. Petals 3·5–4 mm, white. Silicula 5–6 × 3–4 mm, rather narrowly winged all round; apex truncate or with a very shallow notch; style (1·5–)1·75–2·5(–4) mm, long-exserted. *Grassland.* ● *S.W. Alps.*

Subsp. (**a**) is very variable; many local variants have been given specific or subspecific status, among them **T. lereschii** Reuter, *Compt. Rend. Soc. Hallér.* **2**: 17 (1854) (*T. alpestre* subsp. *lereschii* (Reuter) Thell.), **T. salisii** Brügger, *Zeitschr. Ferdinand. Tirol (Innsbruck)* ser. 3, **9**: 45 (1860), **T. sylvestre** Jordan, *Obs. Pl. Crit.* **3**: 9 (1846) (*T. alpestre* subsp. *sylvestre* (Jordan) Hooker), **T. occitanicum** Jordan, *op. cit.* 12 (1846), **T. oligospermum** (Merino) Greuter & Burdet, *Willdenowia* **13**: 96 (1983), and **T. calaminare** (Lej.) Lej. & Court., *Comp. Fl. Belg.* **2**: 307 (1831) (*T. alpestre* subsp. *calaminare* (Lej.) O. Schwarz).

7. T. stenopterum Boiss. & Reuter in Boiss., *Diagn. Pl. Or. Nov.* **2(8)**: 40 (1849). Perennial 30–60 cm, caespitose. Leaves entire, the basal obovate, the cauline oblong, cordate. Petals white, twice as long as the sepals; anthers yellow. Silicula narrowly obcordate, cuneate below; wing narrow above, very narrow below; style more than twice as long as the shallow notch. Seeds 2 in each loculus. $2n = 28$. *Grassland and mountain woods.* ● *N. & C. Spain.* Hs.

8. T. dacicum Heuffel, *Österr. Bot. Zeitschr.* **8**: 26 (1858) (*T. korongianum* Czetz.). Perennial 5–25(–35) cm, glabrous. Leaves entire, the basal elliptic or obovate, the cauline more or less broadly elliptical, sagittate-amplexicaul. Petals white, at least 1½ times as long as the sepals; anthers yellow. Raceme scarcely elongating in fruit. Silicula broadly winged, with a wide, shallow notch; style at most equalling the notch. Seeds 3 or 4 in each loculus. ● *E. & S. Carpathians.* Rm Rs(W).

(**a**) Subsp. **dacicum**: Petals 3 mm. Raceme 1–2(–5) cm in fruit. Silicula 6–9 mm; style more than ½ as long as the notch. *Throughout the range of the species.*

(**b**) Subsp. **banaticum** (Uechtr.) Jáv., *Magyar Fl.* 406 (1924) (*T. banaticum* Uechtr.): Petals *c.* 4·5 mm. Raceme 2·5–10 cm in fruit. Silicula 9–12 mm; style not more than ½ as long as the notch. *S. Carpathians.*

9. T. brevistylum (DC.) Mutel, *Fl. Fr.* **1**: 99 (1834). Glabrous biennial 2–10(–20) cm. Leaves small, thick, glaucous; rosette-leaves broadly elliptical to obovate, long-petiolate; cauline leaves oblong-cordate, obtuse, the lower petiolate, the upper sessile. Sepals *c.* 1 mm; petals 2 mm; stamens shorter than the petals; anthers white, becoming greyish. Raceme short and compact in fruit. Silicula narrowly obovate, the wing very narrow below but broadening upwards to become ½ the width of the valve; stigma subsessile at the base of the wide, rounded, apical notch. Seeds 3 or 4 in each loculus. *Mountain rocks and pastures*, 1000–2400 *m*. ● *Corse and Sardegna.* Co Sa.

10. T. rivale J. & C. Presl, *Del. Prag.* 12 (1822) (*T. pygmaeum* Jordan). Like **9** but erect main stem often with long decumbent or ascending basal branches; petals 2·5–3·5(–4) mm, 1½–2 times as long as the sepals and about equalling the stamens; anthers yellow; apical notch of ripe fruit acute (45–90°); stigma sessile or style up to 0·5 mm, included within the notch. $2n = 26$. *Mountain rocks and pastures. S. Albania, N. & C. Greece; S. Italy, Sicilia.* Al Gr It ?Sa Si.

11. T. graecum Jordan, *Obs. Pl. Crit.* **3**: 30 (1846) (*T. taygeteum* Boiss.). Caespitose perennial 2–14(–20) cm, with shortly branched stock, sometimes with short stolons. Rosette-leaves obovate to elliptical, long-petiolate; cauline leaves oblong-amplexicaul; all sinuate-denticulate, rarely entire. Sepals 2–3 mm; petals 4·5–6·5 mm, white, much exceeding the stamens; anthers violet. Raceme elongating in fruit. Silicula broadly winged above; style 1–2·5 mm, much exceeding the shallow apical notch. Seeds 2–4 in each loculus. $2n = 14$. *Stony, calcareous ground.* ● *Mountains of S. Greece.* Gr.

12. T. microphyllum Boiss. & Orph. in Boiss., *Diagn. Pl. Or. Nov.* **3(6)**: 19 (1859). Dwarf, caespitose perennial 1–2(–4) cm. Rosette-leaves 3–9(–13) mm, obovate, entire or denticulate, petiolate. Inflorescence almost sessile. Sepals 1·5–2·3 mm, violet; petals 3–5 mm, white, *c.* 3 times as long as the sepals; anthers violet. Racemes not or scarcely elongating in fruit. Silicula broadly winged; style 0·7–1·2 mm, exceeding the shallow apical notch. Seeds 1–4 in each loculus. $2n = 14$. *Rocky grassland and cliffs often near snow-patches, above* 1900 *m*. ● *Albania, Jugoslavia, N. & C. Greece.* Al Gr Ju.

T. creticum (Degen & Jáv.) Greuter & Burdet, *Willdenowia* **13**: 95 (1983) (*T. microphyllum* subsp *creticum* Degen & Jáv.), from the high mountains of Kriti, with petals 4–5 mm and style 0·3–0·5 mm, is a rarely collected plant that is perhaps more closely related to **11**. Further study is required to establish whether it should be treated as a distinct species.

13. T. stylosum (Ten.) Mutel, *Fl. Fr.* **1**: 99 (1834). Dwarf, caespitose perennial 1–2·5(–6) cm. Rosette-leaves 5–10 mm, elliptic-spathulate, petiolate. Petals 5 mm, purplish; anthers violet. Silicula rather broadly winged; style 3–5 mm, much exceeding the small apical notch. *Mountain pastures; calcicole.* ● *C. & S. Appennini.* It.

14. T. bulbosum Spruner ex Boiss., *Diagn. Pl. Or. Nov.* **1(1)**: 74 (1843). Caespitose perennial 5–15 cm, with stout, woody stock and several tuberous, long-tapering roots. Stems many, ascending. Rosette-leaves broadly ovate, abruptly petiolate; cauline leaves ovate-oblong with obtuse auricles; all glabrous, glaucous. Petals 6–8 mm, twice as long as the sepals, dark violet or purplish; anthers bright violet. Silicula obcordate, broadly winged above; style 1–2 mm, exceeding the wide apical notch. Seeds (2–)4 or 5, in each loculus. *Mountain woods. C. Greece.* Gr.

(15–19). T. praecox group. More or less densely caespitose perennials. Stock much-branched, with crowded leaf-rosettes and usually 2 or more flowering stems. Petals 5–7(–8·5) mm, white, 2 or more times as long as the sepals; stamens much shorter than the petals; anthers yellowish even after dehiscence. Silicula obcordate to triangular, broadly winged above; apex with a wide notch, or truncate. Seeds 2–10 in each loculus.

1 Style 2–5 mm, more than ½ as long as the ripe fruit; petals white, becoming distinctly yellow when dry — **19. ochroleucum**

1 Style 1–4 mm; petals white, not becoming distinctly yellow when dry

2 Inflorescence often branched; petals 6–8 mm — **17. goesingense**

2 Inflorescence simple; petals 5–7 mm

3 Plant 3–6 cm — **18. epirotum**

3 Plant usually 10–20 cm
4 Ripe fruit usually 7–9 mm, obcordate; angle at base of notch not more than 90° **15. praecox**
4 Ripe fruit usually 5–8 mm, triangular-obcordate; angle at base of notch more than 90° **16. jankae**

15. T. praecox Wulfen in Jacq., *Collect. Bot.* **2**: 124 (1789) (*T. affine* sensu Boiss., non Schott & Kotschy). Stems (5–)10–20(–35) cm, erect, glabrous, more or less glaucous. Rosette-leaves oblong to broadly ovate, petiolate, often violet beneath; cauline leaves ovate-oblong, amplexicaul with obtuse auricles; all entire or sinuate-denticulate, coriaceous. Inflorescence much elongating in fruit. Sepals 2–3 mm, violet-tipped; petals 5–7 mm, narrow, white. Silicula usually 7–9 mm, narrowly obcordate; angle at base of notch not more than 90°; style (1–)2–3·5 mm, always exceeding the notch at least slightly. Seeds 2–4 in each loculus. *Stony, shady slopes and dry grassland. S. Europe, from S.E. France to Krym.* Al Au Bu Ga Gr It Ju Rs(K) Tu.

(a) Subsp. **praecox**: Rosette-leaves oblong to broadly ovate; cauline leaves ovate. Style (1–)2–3·5 mm. $2n = 14$. *Throughout the range of the species.*

(b) Subsp. **cuneifolium** (Griseb. ex Pant.) Clapham, *Feddes Repert.* **70**: 4 (1964) (*T. cuneifolium* Griseb. ex Pant.): Rosette-leaves ovate-cuneate to spathulate; cauline leaves oblong. Style 2–4 mm. *S. Jugoslavia, Albania and S.W. Bulgaria.*

T. albanicum (F. K. Meyer) Greuter & Burdet, *Willdenowia* **13**: 95 (1983) and **T. cikaeum** (F. K. Meyer) Greuter & Burdet, *loc. cit.* (1983), from the mountains of Albania, may represent variants of **15**.

16. T. jankae A. Kerner, *Österr. Bot. Zeitschr.* **17**: 35 (1867) (incl. *T. hungaricum* Dvořáková). Like **15** but somewhat taller; silicula usually 5–8 mm, triangular-obcordate; angle at base of notch more than 90°; style 1–1·5(–2·5) mm, about equalling or only slightly exceeding the notch; seeds 3–10 in each loculus. $2n = 14$, 28. • *Czechoslovakia, Hungary.* Cz Hu.

17. T. goesingense Halácsy, *Österr. Bot. Zeitschr.* **30**: 173 (1880) (*T. tymphaeum* Hausskn.; incl. *T. umbrosum* Waisb.). Stock with many very short stolons (longer in shade), the numerous leaf-rosettes densely crowded; stems up to 40 cm, erect. Rosette-leaves 4–10 cm, elliptical to obovate-spathulate, petiolate; cauline leaves ovate-oblong, sagittate-amplexicaul; all glaucous, entire or almost so. Inflorescence commonly branched. Petals 6–8 mm, white. Silicula narrowly obovate to triangular, very narrowly winged below, the wing broadening upwards into apical lobes of varying form, with a notch between them, or into a broad truncate apex; style (1·5–)3 mm, much-exserted. Seeds 3–6 in each loculus. *Stony and shady slopes or mountain grassland.* $2n = 56$. • *Balkan peninsula and E.C. Europe.* Al Au Bu Hu Ju.

18. T. epirotum Halácsy, *Consp. Fl. Graec.* **1**: 109 (1900). Stock with many short stolons up to 10 mm; stems 3–6 cm, erect. Rosette-leaves obovate, shortly petiolate; cauline leaves oblong with obtuse auricles. Inflorescence remaining compact in fruit. Petals white, twice as long as the yellowish sepals. Silicula obovate, with narrow wing; style 1–2 mm, exceeding the shallow notch. Seeds 1 or 2 in each loculus. *Serpentine screes above 2000 m.* • *N.W. Greece.* Gr.

19. T. ochroleucum Boiss. & Heldr. in Boiss., *Diagn. Pl. Or. Nov.* **2(8)**: 39 (1849) (*T. balcanicum* Janka). Stock with many short stolons up to 4(–6) cm bearing crown of leaves 5–15 mm; stems 10–20(–35) cm, ascending, slender. Rosette-leaves ovate to oblong, gradually and shortly petiolate; cauline leaves oblong, amplexicaul with obtuse auricles; all entire or somewhat denticulate. Inflorescence elongating in fruit. Petals 5–8·5 mm, white, becoming pale yellow when dry, 2–2½ times as long as the green or reddish sepals; anthers yellow. Silicula 6–10 mm, obcordate, broadly winged; style 2–5 mm, long-exserted from the shallow notch. Seeds 1–4 in each loculus. $2n = 14$. *Grassy mountain slopes. Balkan peninsula.* Al Bu Gr Ju.

A variable species that is probably divisible into several subspecies.

T. tymphaeum Hausskn., *Mitt. Thür. Bot. Ver.* **3–4**: 115 (1893) (*T. pindicum* Hausskn.), from N. & C. Greece and S. Albania, usually on serpentine, is similar to **19** but has rather stout stolons 1·5–4·5 cm, bearing crown of leaves, more robust stems up to 40 cm, and a longer fruiting raceme. It is probably not specifically distinct from **19**.

T. lutescens Velen., *Sitz.-Ber. Böhm. Ges. Wiss.* **1903 (28)**: 2 (1904), a non-stoloniferous, somewhat glaucous, glabrous perennial, with erect stems 12–15 cm, and obovate, abruptly petiolate rosette-leaves, from N. Greece (Macedonia), is possibly not distinct from **19**.

20. T. montanum L., *Sp. Pl.* 647 (1753) (*T. lotharingum* Jordan). Mat-forming perennial, the branches of the stock usually elongated, but sometimes shorter and covered with dead leaf-bases; stems (7–)10–30 cm, erect, glabrous. Rosette-leaves with lamina 1–2·5 cm, ovate to orbicular, rather abruptly narrowed into the long petiole; cauline leaves ovate-oblong, amplexicaul with rounded or subacute auricles; all entire or sinuate-denticulate, more or less coriaceous, glabrous, somewhat glaucous. Inflorescence elongating in fruit. Sepals 2–3 mm; petals 5–7 mm, with limb 3 mm wide and narrow claw, much exceeding the stamens, white; anthers pale yellowish. Silicula (4–)7–8 mm, obcordate, broadly winged above, with rounded apical lobes and a wide notch; style 1·5–2 mm, exceeding the notch. Seeds 1(2) in each loculus, smooth, dull. $2n = 28$. *Cliff-ledges, screes, shaded rocky slopes and open grassland, mainly on limestone. C. Europe, extending to E. Pyrenees and S. Jugoslavia.* Au Be Cz Ga Ge He ?It Ju.

(21–23). T. alpinum group. Mat-forming to caespitose perennials, the stock with elongated and stolon-like or quite short branches; stems erect, glabrous. Rosette-leaves 1–2·5 cm, long-petiolate; cauline leaves amplexicaul; all glabrous, more or less coriaceous, entire or nearly so. Sepals 2–3 mm; petals 4–8 mm, much exceeding the stamens, white; anthers yellowish. Silicula about twice as long as wide; wing up to 0·5 mm wide, not or shallowly notched; style long-exserted. Seeds 2–8 in each loculus.

All plants of this group occur in mountain screes, grassland and rocky slopes.

1 Petals 3·5–5 mm; seeds 3–8 in each loculus **23. kovatsii**
1 Petals 5–8 mm; seeds 1–3 in each loculus
2 Petals 5 mm; style 1–1·5 mm; inflorescence usually not more than 3 cm in fruit **22. alpestre**
2 Petals 6–7 mm; style 2–3 mm; inflorescence usually more than 3 cm in fruit **21. alpinum**

21. T. alpinum Crantz, *Stirp. Austr.* **1**: 25 (1762). Stem usually 10–15 cm. Inflorescence more than 3 cm in fruit. Petals 6–7 mm.

Silicula narrowly triangular-obcordate, with or without a very shallow notch; style 2–3 mm. Seeds (1)2 or 3 in each loculus. $2n = c.\ 54$. ● *Alps*. Au Ga He It.

(a) Subsp. **alpinum**: Branches of stock long, stolon-like. Rosette-leaves ovate to suborbicular, often violet beneath; cauline leaves oblong, obtuse. Style 2 mm. *E. & C. Alps.*

(b) Subsp. **sylvium** (Gaudin) P. Fourn., *Quatre Fl. Fr.* 394 (1936) (*T. sylvium* Gaudin): Branches of stock usually short, not stolon-like; plant caespitose. Rosette-leaves oblong-spathulate; cauline leaves broadly ovate. Style 2–3 mm. $2n = 14$. *C. & S.W. Alps.*

22. T. alpestre Jacq., *Enum. Stirp. Vindob.* 116, 260 (1762) (*T. kerneri* Huter, *T. minimum* Ard.). Stock with elongated branches, the plant forming lax mats; stems usually 5–10 cm, erect. Rosette-leaves orbicular, the lamina abruptly contracted into the petiole; cauline leaves ovate, acute; all glaucous. Inflorescence rarely more than 3 cm in fruit. Petals 5 mm. Silicula 6 mm, narrowly obovate; apex truncate or with shallow notch; style 1–1·5 mm. Seeds 1–3 in each loculus. $2n = 14$. ● *S.E. Alps, N. & C. Jugoslavia.* Au It Ju.

23. T. kovatsii Heuffel, *Flora (Regensb.)* **36**: 624 (1853) (*T. avalanum* Pančić, *T. affine* Schott & Kotschy, *T. trojagense* Zapał.). Stock with stoloniferous branches; stems 8–25(–60) cm, erect. Rosette-leaves broadly elliptical to suborbicular, long-petiolate; cauline leaves ovate-oblong, obtuse, with obtuse auricles. Inflorescence much elongating in fruit. Sepals 2–2·5 mm; petals (3·5–)4–5 mm, white. Silicula 5–7 mm, narrowly obovate to triangular-obcordate, emarginate above; style 1–2 mm, exceeding the notch. Seeds 3–8 in each loculus. $2n = 14$. *Carpathians; Balkan peninsula.* Al Bu Ju Rm Rs(W).

T. cochleariforme DC., *Reg. Veg. Syst. Nat.* **2**: 381 (1821), from near the boundary of Europe in C. Ural (Kyštym region) and possibly not occurring within European Russia, is like **23** but has larger, unequal petals (the outer 5–6·5 mm, the inner 6–7·5 mm), the silicula 5–9·5 mm, and the style 1–2·25 mm.

24. T. macranthum N. Busch, *Acta Horti Bot. Univ. Jurjev.* **6**: 142 (1906). Annual 10–40 cm, glabrous, glaucous; stems branched at the base. Basal leaves obovate-elliptical, usually entire, petiolate; cauline leaves ovate-lanceolate, with amplexicaul auricles. Petals 6–7 mm; anthers yellow. Silicula 7–10 mm, obovate-oblong or -cuneate, emarginate at the apex; style long, exceeding the notch. Seeds 2–6 in each loculus, smooth. *Krym.* Rs(K). (*N.W. Caucasus.*)

Sect. APTERYGIUM Ledeb. Silicula narrowly obovate, strongly keeled but not winged, with or without a shallow notch; style exserted from the notch. Seeds smooth.

25. T. cepaeifolium (Wulfen) Koch in Röhling, *Deutschl. Fl.* ed. 3, **4**: 534 (1833) (*T. rotundifolium* subsp. *cepaeifolium* (Wulfen) Rouy & Fouc., *T. cepaeifolium* (Wulfen) Koch). Stock usually with long stolons. Petals purple. Style usually 1–2 mm. *Mountains from E. France to Italy and N. Jugoslavia.* Au Ga Ge He It Ju.

(a) Subsp. **cepaeifolium**: Basal leaves *c.* 1 cm, not distinctly rosulate; upper cauline leaves numerous, crowded, not or very slightly auriculate; all smaller than in subsp. (b), obovate to suborbicular, sinuate-dentate. Inflorescence elongating up to *c.* 3 cm in fruit. Style 1–2 mm. Seeds 4–6 in each loculus. *On calcareous and metalliferous scree.* ● *S.E. Alps.*

(b) Subsp. **rotundifolium** (L.) Greuter & Burdet, *Willdenowia* **15**: 71 (1985) (*T. rotundifolium* (L.) Gaudin, non Tineo, *T. lereschianum* Rouy & Fouc., *T. limosellifolium* Reuter; incl. subsp. *grignense* (F. K. Meyer) Greuter & Burdet): Basal leaves in a more or less distinct rosette; cauline leaves distant, auricled, the lower opposite; all more or less entire. Inflorescence remaining compact in fruit. Seeds 1–4 in each loculus. $2n = 14$. *Throughout the range of the species.*

In addition to the typical widespread calcicolous plant, several distinctive endemic variants of subsp. (b) occur on igneous rocks and siliceous soils in the W. Alps, among them var. *limosellifolium* Burnat, var. *lereschianum* Burnat, var. *corymbosum* (Gay) Gaudin (subsp. *corymbosum* (Gay) Gremli, non *T. corymbosum* Molina), and subsp. *cenisium* Rouy & Fouc.

26. T. bellidifolium Griseb., *Spicil. Fl. Rumel.* **2**: 505 (1845). Plant densely caespitose; stock with very short branches. Basal leaves rosulate, oblong-spathulate, dentate; cauline leaves alternate. Petals dark purple; anthers yellow. Silicula obovate-oblong, truncate. Seeds 2 in each loculus. *Alpine pastures.* ● *N. part of Balkan peninsula.* Al Bu ?Gr Ju.

27. T. nevadense Boiss. & Reuter, *Pugillus* 11 (1852). Caespitose perennial with much-branched stock and crowded leaf-rosettes; stems 5–10 cm. Rosette-leaves 0·5–1·5 cm, petiolate, elliptical to obovate; cauline leaves linear-oblong, acute, amplexicaul with short, rounded auricles, more or less erect; all more or less entire. Inflorescence up to 4 cm in fruit. Petals 6–7 mm, twice as long as the sepals, white. Silicula 7–9 mm, narrowly obovate-oblong with distinct but narrow wing, truncate or very slightly emarginate above; style 1·5–2·5 mm, exserted. Seeds 2–4 in each loculus. *Screes and rock-crevices.* ● *Spain (Sierra Nevada; Sierra de Guadarrama).* Hs.

75. Aethionema R. Br.

A. O. CHATER (EDITION 1) REVISED BY J. R. AKEROYD (EDITION 2)

Glabrous, annuals or perennials, with entire, sessile leaves. Inflorescence racemose. Sepals erect, the lateral saccate at the base; petals entire; the 4 inner stamens with winged and bent filaments, the wing sometimes ending in a tooth above. Silicula flattened and winged, bilocular and angustiseptate, with 1–4 seeds in each loculus, opening by 2 valves, or unilocular, 1-seeded and indehiscent.

1 Petals at least 6 mm, yellow; style 2·5–3·5 mm **3. cordatum**
1 Petals not more than 6 mm, not yellow; style less than 2·5 mm
 2 Silicula entire, subacute at the apex **1. orbiculatum**
 2 Silicula emarginate at the apex
  3 Annual; upper leaves cordate-amplexicaul **5. arabicum**
  3 Perennial; upper leaves not cordate-amplexicaul
   4 At least the lower leaves obtuse **(6–9). saxatile** group
   4 All leaves acute or subacute
    5 Silicula 4–5 mm, orbicular, 1-seeded **2. polygaloides**
    5 Silicula more than 5 mm, ovate-obcordate, 4-seeded **4. iberideum**

1. A. orbiculatum (Boiss.) Hayek, *Prodr. Fl. Penins. Balcan.* **1**: 472 (1925) (*Crenularia orbiculata* Boiss.). Suffruticose perennial 2–10 cm, arising from woody rootstock. Leaves opposite, crowded, 4–8(–13) mm, ovate-orbicular, obtuse, thick. Sepals 2–2·8 mm;

petals white with pink stripes. Silicula 3·5–5 mm, ovate, cordate at the base, narrowed to the subacute, entire apex, 1-seeded; style (0·1–)0·3–0·5 mm. $2n = 24$. *Limestone mountain rocks.* ● *N. Greece* (*Athos*). Gr.

2. A. polygaloides DC., *Reg. Veg. Syst. Nat.* **2**: 562 (1821) (*Crenularia polygaloides* (DC.) Boiss.). Suffruticose perennial with short stems. Leaves alternate (the lower sometimes opposite), 5–8 mm, subacute; lower leaves ovate, the upper oblong-linear. Petals pale purplish. Silicula 4–5 mm, suborbicular, emarginate at the apex, 1-seeded; style less than 2 mm. *E. Greece* (*Poros, Evvoia*). Gr.

3. A. cordatum (Desf.) Boiss., *Fl. Or.* **1**: 350 (1867). Suffruticose perennial with erect, simple stems 5–20 cm. Lower leaves opposite, ovate, cordate-amplexicaul, obtuse; upper leaves alternate, acute. Sepals 4 mm; petals 6·5–8 mm, yellow. Silicula *c.* 7 mm, obovate, truncate at the apex, 4-seeded; style 2·5–3·5 mm. *Limestone rocks. C. & S. Greece* (*Oiti, Chelmos Oros*). Gr. (*S.W. Asia.*)

4. A. iberideum (Boiss.) Boiss., *op. cit.* **1**: 351 (1867). Suffruticose perennial with erect or ascending, branched stems 5–7 cm; lower part of stems densely clothed with leaves and their remains. Leaves elliptical, acute, the lower 3–5 mm, the upper 6–8 mm. Sepals 2 mm, petals 5 mm, white. Silicula 7 mm, ovate-obcordate and emarginate, 4-seeded; style 0·5–0·7 mm. *Mountain rocks. E. Greece* (*C. Evvoia*). Gr. (*Anatolia.*)

5. A. arabicum (L.) Andrz. ex O. E. Schulz in Engler & Prantl, *Natürl. Pflanzenfam.* ed. 2, **17b**: 442 (1936) (*A. buxbaumii* (Fischer ex Hornem.) DC., *A. cappadocicum* Sprengel). Annual with slender, simple or branched stems 10–15 cm. Lower leaves *c.* 15 mm, ovate, acute, the upper ovate, acute, cordate-amplexicaul at the base. Petals 2–3 mm, purplish. Silicula 6–12 mm, suborbicular, emarginate at the apex, with up to 6 seeds, very densely crowded, imbricate; style 0·5–1 mm, shorter than notch. *Dry, rocky places. Thrace.* Bu Tu. (*S.W. Asia.*)

(6–9). A. saxatile group. Perennials, often woody at the base; stems 5–35 cm, ascending to erect, usually simple. Lower leaves often opposite, obtuse (rarely retuse), the upper acute or obtuse. Petals white, pink, purplish or lilac. Silicula 1- to 8-seeded, emarginate at the apex. *S. & S.C. Europe.* Al Bu Cr Cz Ga Ge Gr He Hs Hu It Ju Rm Sa Si.

1 All siliculae unilocular, usually 1-seeded
 2 Plant slightly woody at the base; silicula orbicular; style less than 0·5 mm — **6. carlsbergii**
 2 Plant strongly woody at the base; silicula obovate to obcordate; style more than 0·5 mm — **7. retsina**
1 At least a few siliculae bilocular
 3 Raceme somewhat lax; siliculae variably uni- and bilocular — **8. saxatile**
 3 Raceme dense; siliculae almost all unilocular — **9. thomasianum**

6. A. carlsbergii Strid & Papanicolaou, *Bot. Not.* **133**: 521 (1980). Short-lived perennial with ascending to suberect, usually simple stems 6–10 cm. Leaves 6–10 mm, obtuse, rather thick, glaucous; lower leaves opposite, ovate-orbicular to broadly elliptical, obtuse, the upper alternate, ovate to broadly oblong. Sepals 2–2·6 mm; petals 3·5–5·5 mm, pale pink. Silicula 5–6 mm, orbicular, emarginate at the apex, 1-seeded; style 0·3–0·4 mm. *Limestone screes,* 2100–2400 *m.* ● *S. Greece* (*Taïyetos*). Gr.

7. A. retsina Phitos & Snogerup, *Bot. Not.* **126**: 142 (1973). Suffruticose perennial up to 20 cm, arising from woody rootstock. Leaves opposite, crowded, 5–15 mm, spathulate to obovate or elliptical, obtuse to slightly retuse, thick, glaucous. Sepals 2·7–3·2 mm; petals 4–6 mm, white or purplish. Silicula 4–5 mm, obovate to obcordate, emarginate at the apex, usually 1-seeded; style 1–1·3 mm. $2n = 24$. *Crevices of limestone cliffs.* ● *C. Aegean region* (*Skiros*). Gr.

8. A. saxatile (L.) R. Br. in Aiton, *Hort. Kew.* ed. 2, **4**: 80 (1812) (*A. creticum* Boiss. & Heldr., *A. subcapitatum* Bornm.; incl. *A. gracile* DC., *A. pyrenaicum* Bout.). Perennial up to 35 cm, ascending or erect, simple or branched. Lower leaves more than 5 mm, ovate to suborbicular or oblong and obtuse, the upper narrower, often acute. Sepals 1–3(–3·8) mm; petals 2–6·5(–8·5) mm, white, purplish or lilac. Silicula 5–10 mm, obovate to suborbicular, emarginate, sometimes a little broader than long, up to 8-seeded, in lax or rather dense racemes; unilocular siliculae 3–6·5 mm, the wing 1–3 mm wide at the apex; bilocular fruits 4·5–10 mm, the wing 1·5–4 mm wide at the apex; style 0·2–2 mm, usually shorter than but sometimes equalling or exceeding notch. *S. & S.C. Europe, mainly in the mountains.* Al Bu Cr Cz Ga Ge Gr He Hs Hu It Ju Rm Sa Si.

A very variable species; some variants have been treated as species in regional floras. The account below is based on the study by I. A. Andersson *et al.*, *Willdenowia* **13**: 3–42 (1983).

1 Lower leaves distinctly petiolate; leaves up to 20 mm — **(b) subsp. scopulorum**
1 Lower leaves usually sessile or subsessile; leaves usually not more than 15 mm
 2 Plant strongly woody at the base; cauline leaves broadly elliptical to suborbicular, fleshy — **(e) subsp. creticum**
 2 Plant slightly woody at the base; cauline leaves oblong, obovate or elliptical, rarely fleshy
  3 Sepals 1–1·6 mm; cauline leaves at least 3½ times as long as wide — **(a) subsp. saxatile**
  3 Sepals more than 1·6 mm; cauline leaves 2–4 times as long as wide
   4 Cauline leaves obtuse — **(c) subsp. ovalifolium**
   4 Cauline leaves acute — **(d) subsp. graecum**

(a) Subsp. **saxatile**: Slightly woody at the base. Leaves 5–15 mm, at least 3½ times as long as wide, coriaceous, rather thin, the lower obtuse, the upper acute. Sepals 1–1·6 mm; petals 1·8–3·2(–4) mm. Most siliculae bilocular; style 0·2–0·4(–0·5) mm. $2n = 24, 48$. *Stony ground, usually in the mountains. Throughout the range of the species, except S. part of Balkan peninsula and Aegean region.*

(b) Subsp. **scopulorum** (Ronniger) I. A. Andersson *et al.*, *Willdenowia* **13**: 9 (1983) (*A. saxatile* var. *scopulorum* Ronniger): Strongly woody at the base. Leaves 10–20 mm, 3–5½ times as long as wide, fleshy, the lower distinctly petiolate, obtuse. Sepals 1·7–2·2 mm; petals 4–5 mm. At least 40% of siliculae unilocular; style 0·4–0·6 mm. *Limestone cliffs by the sea.* ● *W. Jugoslavia* (*near Dubrovnik*).

(c) Subsp. **ovalifolium** (DC.) Nyman, *Consp.* 63 (1878) (*A. ovalifolium* (DC.) Boiss.): Variable in habit, but usually somewhat woody at the base. Leaves 6–14 mm, 1½–3 times as long as wide, obtuse, coriaceous. Sepals 1·7–2·6 mm; petals (2·3–)3–5 mm. Usually most siliculae bilocular; style 0·3–0·7(–0·9) mm. *Limestone mountains cliffs. Spain, S.W. France; Sardegna.*

(d) Subsp. **graecum** (Boiss. & Spruner) Hayek, *Prodr. Fl.*

Penins. Balcan. **1**: 472 (1925) (*A. graecum* Boiss. & Spruner, *A. glaucescens* Halácsy, *A. saxatile* subsp. *oreophilum* I. A. Andersson *et al.*): Plant 5–13 cm, somewhat woody at the base. Leaves 5–12 mm, 2–4 times as long as wide, acute to subacute, coriaceous to somewhat fleshy. Sepals 1·8–2·5(–3·3) mm; petals 3–6·5(–8·5) mm. Most siliculae bilocular; style 0·3–0·9(–2) mm. $2n$ = 24, *c.* 36. *Rocky and stony ground. S. part of Balkan peninsula, N. Aegean region.*

(**e**) Subsp. **creticum** (Boiss. & Heldr.) I. A. Andersson *et al.*, *Willdenowia* **13**: 18 (1983) (*A. creticum* Boiss. & Heldr.): Long-lived perennial 3–10 cm, strongly woody at the base. Leaves 4–10 mm, 1½–2½ times as long as wide, broadly elliptical to suborbicular, obtuse to subacute, fleshy. Sepals 1·5–2·5 mm; petals 2·5–4·8 mm. Usually at least 40% of siliculae unilocular; style 0·3–0·9 mm. $2n$ = 24. *Limestone cliffs and stony ground. S. Aegean region; S.E. Peloponnisos.*

Var. *ovalifolium* of subsp. (**c**), a compact plant up to *c.* 20 cm with mainly bilocular siliculae, is widespread. Var. *monospermum* (R. Br.) Thell. (*A. monospermum* R. Br.), a taller plant with mainly unilocular siliculae, occurs on the northern slopes of the Pyrenees.

Plants from E.C. Greece have larger floral parts and longer styles in fruit than plants from the rest of the range of subsp. (**d**), and have been distinguished as subsp. **oreophilum** I. A. Andersson *et al.*, *Willdenowia* **13**: 14 (1983). The existence of intermediate populations makes it very difficult to separate these two variants and they are best subsumed within subsp. (**d**).

9. A. thomasianum Gay, *Ann. Sci. Nat.* ser. 3, **4**: 81 (1845). Like **8** but usually smaller and more compact; stems up to 13 cm; raceme very dense with almost all siliculae unilocular; unilocular siliculae 7–10 mm, the wing 3–5 mm wide at the apex; bilocular siliculae 10–12 mm, the wing 4·5–5 mm wide at the apex; style 0·6–0·8 mm. $2n$ = 24. *Open habitats on schistose soil.* ● *S.W. Alps (Val d' Aosta).* It.

Similar plants are known from the mountains of Algeria (Djuradjura), but their status is uncertain.

76. Teesdaliopsis (Willk.) Gand.

V. H. HEYWOOD

Inflorescence corymbose-racemose; flowers zygomorphic. Petals white; outer petals large, radiating; inner petals small. Fruit an angustiseptate silicula, scarcely winged at the apex, emarginate. Seed single, pendent.

1. T. conferta (Lag.) Rothm., *Feddes Repert.* **49**: 178 (1940) (*Iberis conferta* Lag.). Caespitose, suffruticose, glabrous perennial. Leaves in crowded rosettes, lanceolate, acute, entire. Flowering branches 5–15 cm, leafless, lateral; non-flowering branches terminal. Silicula 5 × 4 mm, ellipsoid-obovoid; valves with a thick median vein; style very short. ● *Mountains of N.W. Spain and N. Portugal.* Hs Lu.

77. Iberis L.

A. R. PINTO DA SILVA AND J. DO AMARAL FRANCO

Annuals or perennials, rarely dwarf shrubs; glabrous or with unbranched hairs. Inflorescence corymbose or racemose, often elongating in fruit. Sepals not saccate; petals white, pink or purple, the 2 outer much larger than the 2 inner; median nectaries absent. Fruit an angustiseptate silicula; valves keeled and usually winged at the apex; style long; stigma capitate. Seeds solitary in each loculus, often winged.

All species are usually found in calcareous habitats.

The following key refers to well-developed plants only. The perennial species sometimes flower in the first year and may then appear to be annual.

1 Silicula much wider than long; flowering in winter **1. semperflorens**
1 Silicula longer than wide or suborbicular; flowering in spring and summer
 2 Perennial with woody or herbaceous stems, usually with non-flowering rosettes
  3 Siliculae in racemes
   4 Leaves 2·5–5 mm wide, oblong-spathulate, flat, obtuse; flowering stems lateral **2. sempervirens**
   4 Leaves 1–2 mm wide, linear, semi-cylindrical on the non-flowering shoots, ± flat on the flowering stems, acute, mucronulate; flowering stems terminal **3. saxatilis**
  3 Siliculae in corymbs
   5 Plant 10–40 cm, usually with erect stems (low altitudes, up to 800 m)
    6 Cauline leaves linear to narrowly linear-spathulate, entire **4. contracta**
    6 Cauline leaves oblong-lanceolate or ± broadly spathulate, often denticulate at least at the apex
     7 Outer petals 8–10 mm; cauline leaves up to 6 mm wide **8. procumbens**
     7 Outer petals 10–18 mm; cauline leaves up to 12 mm wide
      8 Stems glabrous; seeds *c.* 5 mm **9. gibraltarica**
      8 Stems pubescent; seeds *c.* 2·5 mm **10. nazarita**
   5 Dwarf plants up to 15 cm, usually diffuse with ascending stems (high mountains, 800–2800 m)
    9 Silicula 6–8 mm, rectangular-elliptical, with wide, erect, acute lobes **5. pruitii**
    9 Silicula 3–5 mm, broadly ovate
     10 Silicula with short, convergent, ± obtuse lobes **6. spathulata**
     10 Silicula with acute, very divergent lobes **7. aurosica**
 2 Annual, without non-flowering rosettes
  11 Siliculae in racemes
   12 Leaves all pinnatifid to dentate (rarely entire), ± ciliate; silicula 3–5 mm **11. amara**
   12 Upper leaves entire, linear-lanceolate to linear, glabrous; silicula (5–)6–9 mm **12. linifolia**
  11 Siliculae in corymbs or umbels
   13 Silicula with irregularly serrulate lobes **16. fontqueri**
   13 Silicula with entire lobes
    14 Leaves entire or shallowly toothed
     15 Leaves linear- to obovate-spathulate
      16 Lower cauline leaves at least 3 mm wide, obovate-oblong to oblong-spathulate, often toothed **5. pruitii**
      16 Lower cauline leaves not more than 3 mm wide, linear-spathulate **15. ciliata**
     15 Leaves narrowly linear to linear-lanceolate
      17 Silicula 3–5 mm, with small, acute or acuminate, divergent lobes forming a shallow notch; leaves narrowly linear **13. stricta**

17 Silicula 7–10 mm, with erect, acuminate lobes forming an acute, deep notch; leaves linear-lanceolate **14. umbellata**

14 Leaves all, or at least the lower ones, pectinate-toothed, pinnatifid or pinnatisect with linear segments

18 Stems erect, branched above; flowering-branches long, straight, leafless; corymbs convex

19 Leaves pinnatifid or pinnatisect with linear segments; silicula glabrous; plants rugose-papillose or scarcely pubescent **17. pinnata**

19 Leaves pectinate-toothed; silicula papillose-verrucose; plant hispid **18. crenata**

18 Stems diffuse and branched from the base; flowering branches flexuous and leafy, or leafless for a short distance only; corymbs flat

20 Lower leaves pinnatifid with 1 or 2 pairs of segments; outer petals scarcely radiating **19. odorata**

20 Lower leaves regularly pectinate-pinnatifid with 3 or 4 pairs of segments; outer petals distinctly radiating **20. sampaioana**

1. I. semperflorens L., *Sp. Pl.* 648 (1753). Small, evergreen, procumbent, diffuse, glabrous shrub up to 80 cm, branched above, with herbaceous flowering stems. Leaves thick, entire, flat; lower leaves 30–70 × 7–18 mm, broadly spathulate; upper leaves oblong-spathulate and much smaller. Inflorescence corymbose, but elongating in fruit. Petals white. Silicula 5–8 × 10–14 mm, ovate-rhombic, very narrowly winged; notch almost absent. $2n = 22$ (44). *Rock-crevices on maritime cliffs.* ● *Sicilia and W. coast of Italy; cultivated elsewhere for ornament, and sometimes naturalized.* It Si [Rm].

The only winter-flowering species.

2. I. sempervirens L., *loc. cit.* (1753) (*I. garrexiana* All., *I. serrulata* Vis.). Small, evergreen, procumbent, diffuse, glabrous shrub with herbaceous, flexuous, lateral shoots 10–25 cm. Leaves 2·5–5 mm wide, oblong-spathulate, obtuse, entire, thick, flat. Inflorescence racemose, elongating in fruit. Petals white, sometimes flushed with lilac. Silicula 6–7 mm, orbicular-ovate, broadly winged from the base; lobes subobtuse; notch acute. $2n = 22$. *Rock-crevices on high mountains. Mediterranean region; cultivated elsewhere for ornament, and occasionally naturalized.* Al Cr Ga Gr Hs It Ju [Br Rm].

3. I. saxatilis L., *Cent. Pl.* **2**: 23 (1756). Small, evergreen, procumbent, diffuse shrub with straight, terminal flowering branches. Leaves up to 20 mm, semi-cylindrical on the non-flowering shoots but flat on the flowering stems, entire, linear, rather acute, mucronulate, at first ciliate but soon glabrous. Inflorescence corymbose in flower but elongating in fruit. Petals white. Silicula 5–8(–9) × 4·5–6(–7) mm, obovate, broadly winged from the base; lobes rounded. ● *S. Europe.* Ga Gr He Hs It Ju Rm Rs(K).

(a) Subsp. **saxatilis** (*I. vermiculata* Willd., *I. zanardinii* Vis.): Glabrous, up to 15 cm, with reddish twigs. Leaves up to 1·5 mm wide. Silicula with a shallow notch. $2n = 22$. *Rock-crevices. Throughout the range of the species.*

(b) Subsp. **cinerea** (Poiret) P. W. Ball & Heywood, *Feddes Repert.* **64**: 62 (1961) (*I. cinerea* Poiret, *I. subvelutina* DC., *I. latealata* Porta & Rigo): Greyish-green, velutinous, up to 40 cm. Leaves up to 2 mm wide. Notch of silicula acute. *Gypsaceous hills. C. & S. Spain.*

Subglabrous plants that otherwise resemble subsp. (**b**), from E. Spain (Valencia prov.), have been called subsp. **valentina** Mateo & Figuerola, *Fl. Analit. Prov. Valencia* 370 (1987).

4. I. contracta Pers., *Syn. Pl.* **2**: 186 (1806). Laxly caespitose perennial with a woody stock bearing numerous, erect, herbaceous, puberulent stems 15–30 cm. Leaves up to 30 × 2 mm, rather fleshy, usually ciliolate at the base, lower ones oblong-cuneate, with a few teeth near the apex, the cauline linear to narrowly linear-spathulate, entire. Inflorescence corymbose. Silicula 4–6 mm, orbicular-obovate, broadly winged from the base; lobes triangular, subacute; notch acute or obtuse. *W. & C. Iberian peninsula.* Hs Lu. (*N. Africa.*)

(a) Subsp. **contracta** (*I. linifolia* Loefl., non L., *I. reynevalii* Boiss. & Reuter): Petals deep purple. $2n = 14$. *Dry, calcareous places.* ● *C. Spain and E. Portugal.*

(b) Subsp. **welwitschii** (Boiss.) Moreno, *Collect. Bot.* (*Barcelona*) **15**: 348 (1984) (*I. welwitschii* Boiss., *I. linifolia* subsp. *welwitschii* (Boiss.) Franco & P. Silva): Petals white or pink. $2n = 14$. *Sandy, acid soils near the sea.* ● *S.W. Spain, W. & S. Portugal.*

5. I. pruitii Tineo, *Pl. Rar. Sic. Pug.* **1**: 11 (1817) (incl. *I. carnosa* Waldst. & Kit., non Willd., *I. lagascana* DC., *I. tenoreana* DC., *I. integerrima* Moris, *I. jordanii* Boiss., *I. granatensis* Boiss. & Reuter, *I. candolleana* Jordan, *I. spruneri* Jordan, *I. petraea* Jordan, *I. epirota* Halácsy, *I. grossii* Pau, *I. thracica* Stefanov, *I. contracta* sensu Willk., *I. glabrescens* Porta, *I. hegelmairei* Willk., *I. paularensis* Pau). Low, caespitose, procumbent-ascending perennial or annual 3–15 cm. Leaves rather fleshy, entire or with a few teeth near the apex, obtuse, the lower obovate-spathulate, the upper narrower. Inflorescence corymbose, rather dense in fruit. Petals white to lilac. Silicula 6–8 mm, rectangular-elliptical, broadly winged; lobes triangular-acute, erect; notch deep, acute. $2n = 22$. *Rock-crevices on high mountains. Mediterranean region.* Al Bu Ga Gr Hs It Ju Sa Si.

Extremely variable in habit and duration. Many variants have received specific names and some appear to be of restricted distribution.

6. I. spathulata J. P. Bergeret, *Phytonomat.* **3**: 3 (1784) (*I. nana* All., *I. carnosa* Willd., ?*I. pilosa* Desv.). Small, glabrous or more or less pubescent perennial with 1 or more erect or ascending, simple stems 3–10 cm, leafy to the apex. Leaves rather fleshy, entire to sublobulate-dentate, obtuse, the lower broadly spathulate, the upper obovate-spathulate. Inflorescence a short, dense corymb. Petals purplish to white. Silicula 4–5 × 5–5·5 mm, broadly ovate, narrowly winged; lobes short, subobtuse, convergent; notch very acute. *Rock-crevices and gravelly places on high mountains* (1500–2800 *m*). ● *Pyrenees; Alpi Marittime; Appennini Ligure.* Hs It.

(a) Subsp. **spathulata**: Stems, leaves and pedicels more or less pubescent. Leaves entire or with 1 or 2 teeth near the apex. *Pyrenees.*

(b) Subsp. **nana** (All.) Heywood, *Feddes Repert.* **69**: 61 (1964) (*I. nana* All.): Stems, leaves and pedicels glabrous. Much of leaf-margin distinctly and obtusely denticulate to sublobulate-dentate. *Alpi Marittime; Appennini Ligure.*

I. bernardiana Gren. & Godron, *Fl. Fr.* **1**: 138 (1847), from the Pyrenees, is of uncertain affinity and may be another subspecies of **6**. It is annual or biennial, with linear-oblong, crenate or entire leaves, violet petals and a broadly ovate, hispid silicula contracted at the apex and with narrow, erect, acute lobes.

I. bubanii Deville, *Bull. Soc. Bot. Fr.* **6**: 69 (1859), from rock-crevices in the French Pyrenees, is like **6** but up to 18 cm, more robust, erect, with leaves all oblong-spathulate, petals twice as long, pink, and silicula 3–4 × 4 mm.

7. I. aurosica Chaix in Vill., *Hist. Pl. Dauph.* **1**: 349 (1786). Small perennial with stems 4–15 cm, leafless above. Leaves rather fleshy, the lower oblong-spathulate, entire or with 1 or 2 teeth on either side near the apex, the upper entire. Inflorescence a short corymb, dense even in fruit. Silicula 4–5 mm, broadly ovate, broadly winged above; lobes acute, horned, very divergent, much shorter than the style; notch shallow. *Gravelly places on mountains* (800–2600 *m*). ● *S.W. Alps; N. Spain.* Ga Hs.

(a) Subsp. **aurosica**: Subglabrous. Upper leaves linear-spathulate, acute, glabrous. Petals purple-lilac. ● *S.W. Alps.*

(b) Subsp. **cantabrica** Franco & P. Silva, *Feddes Repert.* **68**: 195 (1963): Sparsely hirsute. Upper leaves oblong-spathulate, obtuse, strongly ciliate. Petals white. ● *N. Spain (Vizcaya).*

8. I. procumbens Lange, *Ind. Sem. Horto Haun.* **1861**: 29 (1861) (*I. contracta* auct. lusit., *I. sempervirens* auct. lusit., non L.). Bushy, puberulent perennial, with a woody stock from which arise both vegetative and flowering flexuous stems 10–30 cm. Leaves up to 25 × 6 mm, broadly spathulate, fleshy, entire or with 1 or 2 pairs of teeth near the apex, subobtuse, hispid or glabrous above. Inflorescence corymbose, flat in flower, contracted in fruit. Petals lilac or rarely white, the outer 8–10 mm. Silicula ovate, broadly winged from the base; lobes triangular, acute or obtuse; notch acute. ● *Portugal and N.W. Spain.* Hs Lu.

(a) Subsp. **procumbens**: Usually pulvinate, flowering profusely. Corymbs with 35–40 flowers. Silicula 6–7·5 mm. *Sea-shores. Throughout the range of the species.*

(b) Subsp. **microcarpa** Franco & P. Silva, *Feddes Repert.* **68**: 195 (1963): Lax, with fewer stems. Corymbs with 20–25 flowers. Silicula 4–6 mm. *Hills near the sea. W. Portugal (from Cabo Mondego to Serra de Arrábida).*

9. I. gibraltarica L., *Sp. Pl.* 649 (1753). Evergreen, densely caespitose, glabrous perennial with a woody stock from which arise leaf-rosettes and numerous ascending flowering stems 15–40 cm. Leaves fleshy, entire or with 1–4 teeth on either side near the apex, subobtuse, the lower up to 25 × 12 mm, broadly spathulate, the upper smaller. Inflorescence 40–50 mm in diameter, corymbose, flat in flower but contracted in fruit. Petals lilac to white, the outer 15–18 mm. Silicula 10–12 × 8–10 mm, orbicular, broadly winged from the base; lobes triangular, acute; notch deep, acute. Seeds *c.* 5 mm, winged. $2n = 14$. *Shady rock-crevices. Gibraltar.* Hs. (*Morocco.*)

10. I. nazarita Moreno, *Trab. Dep. Bot. Fisiol. Veg.* **12**: 95 (1983). Like **9** but smaller, the flowering stems up to 30 cm, pubescent; leaves oblong-lanceolate, denticulate; silicula 6 × 5 mm, broadly elliptical; seeds *c.* 2·5 mm, unwinged. $2n = 14$. ● *S. Spain (Andalucía).* Hs.

11. I. amara L., *Sp. Pl.* 649 (1753) (*I. pinetorum* Pau). Erect, leafy annual 10–40 cm, corymbosely branched above, more or less hairy below. Leaves spathulate, distantly pinnatifid or toothed, sometimes entire, more or less ciliate. Inflorescence corymbose, elongating in fruit. Petals white or purplish. Silicula 3–5 mm, suborbicular; lobes triangular, erect or more or less divergent; notch acute or shallow. *Cornfields and dry hillsides, mainly on calcareous or dolomitic soils. Europe eastwards to Germany and Italy.* Be Br Ga Ge He Ho Hs It [Au ?Cz Hu It Lu Po Rm Rs(C, W, K, E)].

A variable species of which a number of cultivars are used for ornament (e.g. *I. coronaria* hort.). Some of them probably arose through hybridization with **14**. At least two subspecies can be recognized.

(a) Subsp. **amara**: Lobes of the silicula erect, forming a deep, narrow notch. *Throughout the range of the species.*

(b) Subsp. **forestieri** (Jordan) Heywood, *Feddes Repert.* **69**: 61 (1964) (*I. forestieri* Jordan): Lobes of the silicula more or less divergent, forming a shallow, broad notch. *Pyrenees; E.C. Spain.*

12. I. linifolia L., *Syst. Nat.* ed. 10, 1129 (1759). Erect, glabrous annual or biennial up to 80 cm, branched above. Leaves 15–25(–50) × 3–4 mm, linear-lanceolate to linear, acuminate, the basal somewhat toothed, the cauline entire. Inflorescence shortly racemose, more or less dense, elongating in fruit. Petals white to purplish, the outer usually strongly radiating. Silicula ovate, more or less winged from the base; lobes acuminate, divergent; notch shallow. *Bushy and rocky places and calcareous slopes.* ● *From S. Germany to N.E. Spain and W. Jugoslavia.* Ga He Hs It Ju [Be].

1 Silicula *c.* 5 mm
  2 Silicula scarcely winged at the base; lobes short — (c) subsp. **prostii**
  2 Silicula broadly winged from the base; lobes long — (d) subsp. **violletii**
1 Silicula 6 mm or more
  3 Silicula winged from the base; lobes long — (a) subsp. **linifolia**
  3 Silicula scarcely winged at the base; lobes short — (b) subsp. **timeroyi**

(a) Subsp. **linifolia** (*I. intermedia* Guersent, *I. ciliata* sensu Willk., non All., *I. dunalii* (Bubani) Cadevall): Silicula 6–9 mm, winged from the base; lobes long. *Throughout the range of the species.*

(b) Subsp. **timeroyi** (Jordan) Moreno, *Collect. Bot.* **15**: 349 (1984): Silicula 6–9 mm, scarcely winged at the base; lobes short. *S. France.*

(c) Subsp. **prostii** (Soyer-Willemet ex Godron) Moreno, *op. cit.* 348 (1984): Silicula *c.* 5 mm, scarcely winged at the base; lobes short. *France, Spain.*

(d) Subsp. **violletii** (Soyer-Willemet ex Godron) Valdés, *Willdenowia* **15**: 67 (1985) (*I. violletii* Soyer-Willemet ex Godron): Silicula *c.* 5 mm, broadly winged from the base; lobes long. *E. France.*

13. I. stricta Jordan, *Diagn.* **1**: 278 (1864) (*I. leptophylla* Jordan). Erect, glabrous annual 30–60 cm; stems slender, branched above. Leaves linear, the cauline narrower, mucronate. Inflorescence corymbose even in fruit, small, dense. Petals rather small, pink to lilac. Silicula 3–5 mm, suborbicular; notch shallow. *Dry, calcareous soils.* ● *S.E. France and Liguria.* Ga It.

(a) Subsp. **stricta**: Basal leaves shallowly toothed. Silicula ovate, winged from about the middle, the lobes acuminate, $\frac{1}{2}$ as long as the style. *France (Hautes-Alps).*

(b) Subsp. **leptophylla** (Jordan) Franco & P. Silva, *Feddes Repert.* **68**: 195 (1963) (*I. linifolia* L. pro parte, non Loefl.): Basal leaves almost entire. Silicula suborbicular, winged only above the middle; lobes acute, much shorter than the style. *Throughout the range of the species.*

14. I. umbellata L., *Sp. Pl.* 649 (1753) (*I. roseopurpurea* Sagorski). Erect, glabrous annual 20–70 cm, corymbosely branched above. Leaves linear-lanceolate, acuminate, entire or almost so. Inflorescence dense, umbellate even in fruit. Petals pink to purplish. Silicula up to 10 mm, ovate, broadly winged from the base; lobes triangular, acuminate, erect; notch deep, acute. *Bushy and rocky places, on calcareous and serpentine soils.* ● *Mediterranean region; widely cultivated for ornament, and naturalized elsewhere.* Al Ga Gr It Ju [Au Az Be Br Cz Ge He Ho Hs Hu Lu Po Rm].

15. I. ciliata All., *Auct. Fl. Pedem.* 15 (1789). Erect annual or short-lived perennial 20–30 cm, often branched above. Leaves linear-spathulate, entire. Inflorescence dense, umbellate even in fruit. Petals white to pale purple. Silicula up to 6 mm, rectangular-elliptical, broadly winged from the base; lobes triangular, acute or obtuse, erect; notch deep, acute. ● *S.E. France; C. Spain.* Ga Hs.

I. simplex DC. in Lam. & DC., *Fl. Fr.* ed. 3, **5**: 597 (1815) (*I. taurica* DC.), from Krym and S. Russia (near Taganrog), appears to differ only in its more spathulate leaves.

16. I. fontqueri Pau, *Mem. Mus. Ci. Nat. Barcelona (Bot.)* **1(1)**: 22 (1922) (*I. pinnata* sensu Boiss., non L.). Annual, simple or branched from the base; stems up to 15 cm, erecto-patent, more or less branched, slender, very shortly puberulent. Leaves spathulate to linear-spathulate, with 1 or 2 pairs of teeth near the apex, or subentire, the lower obtuse, attenuate into a rather long petiole, caducous. Inflorescence a short, dense corymb. Petals white or lilac. Silicula 5 mm, broadly orbicular, truncate at the base, winged from the base; lobes subacute, erect, irregularly serrulate (at least in the upper part); notch shallow; style much exserted. ● *Mountains of S. Spain (Málaga prov.).* Hs.

17. I. pinnata L., *Cent. Pl.* **1**: 18 (1755). Erect annual 10–30 cm, rugose-papillose, scarcely pubescent, corymbose above; branches long, straight, leafless. Leaves obovate-oblong, pinnatifid or pinnatisect, with 1–3 pairs of linear segments. Inflorescence a short, usually dense, convex corymb; flowers fragrant. Petals white to lilac. Silicula 5–6 mm, almost square, glabrous; lobes entire, obtuse (or acute, var. *rollii* (Ten.) Fiori), erect; notch shallow. $2n = 14$. *Cereal fields. S. Europe.* Bl Ga Gr He Hs It Ju Rs(K) ?Si [Au Be Cz Ge Rm].

18. I. crenata Lam., *Encycl. Méth. Bot.* **3**: 223 (1789) (*I. pectinata* Boiss., *I. bourgaei* Boiss.). Erect, hispid annual; stems 15–30 cm, 1 or several, corymbosely branched above; branches long, straight, leafless. Leaves all, or at least the lower, linear-spathulate, pectinate-toothed, the upper narrower, entire. Inflorescence a many-flowered, dense, convex corymb. Petals white. Silicula *c.* 5 mm, almost square, winged only in the upper part, papillose-verrucose; lobes broad, triangular, acute; notch deep, acute. $2n = 14$. *Dry, calcareous or gypsaceous soils.* ● *C. & S. Spain.* Hs.

19. I. odorata L., *Sp. Pl.* 649 (1753) (*I. acutiloba* Bertol.). Diffuse, very shortly hispid annual; stems 15–30 cm, numerous, arising from the base, flexuous, leafy. Leaves linear-spathulate, pinnatifid with 1 or 2 pairs of segments near the apex. Inflorescence a flat, dense corymb, shortly stalked and surrounded by the upper leaves. Petals white, the outer ones scarcely radiating. Silicula ovate, glabrous; lobes erect, acute, entire; notch acute. *Limestone mountain slopes. Greece, Turkey-in-Europe, Kriti.* Cr Gr Tu [Be]. (*N. Africa, S.W. Asia.*)

20. I. sampaioana Franco & P. Silva, *Feddes Repert.* **68**: 195 (1963) (*I. pectinata* auct. lusit., non Boiss.). Procumbent hispid annual; stems up to 20 cm, numerous. Leaves spathulate (except sometimes the upper), regularly pectinate-pinnatifid with 3 or 4 pairs of oblong, subobtuse segments. Inflorescence a flat corymb, not very dense in fruit. Petals white, the outer ones strongly radiating. Silicula 5 × 4 mm, ovate-rectangular, glabrous or very rarely papillose-verrucose; lobes large, acute, entire, erect; notch obtuse. *Calcareous soils.* ● *Coast of S.W. Portugal.* Lu.

78. Biscutella L.

†E. GUINEA AND V. H. HEYWOOD

Annuals, perennials or small shrubs with entire to pinnatifid leaves. Petals usually clawed, yellow. Fruit a strongly compressed, didymous silicula, indehiscent but with the 1-seeded loculi eventually breaking away from the axis; valves glabrous or with simple or clavate hairs; style long. Seeds unwinged.

The taxonomy of this genus is made difficult by the relative uniformity of most of the floral and fruiting characters and by the unreliability of those that do vary. Reliance has to be placed on vegetative features, but there is inadequate knowledge about their range of plasticity. The two available monographs differ widely in their treatment, the one taking a broad, the other a narrow view of the species. The narrow species concept is adopted here, largely to force attention on the variation in the genus, so that a future synthetic treatment may be possible. Most of the species recognized here have a characteristic facies and a well-circumscribed distribution. Admittedly intermediates between some of them occur, but to treat them as subspecies would make the classification unwieldy and such treatment has been deferred until further information may be available.

Literature: E. Malinowski, *Bull. Int. Acad. Sci. Cracovie* **1910**: 111–139 (1910). B. Machatschki-Laurich, *Bot. Arch. (Königsberg)* **13**: 1–115 (1926). J. D. Olowokudejo & V. H. Heywood, *Pl. Syst. Evol.* **145**: 291–309 (1984).

1 Petals up to 15 mm, long-clawed, patent; silicula with a translucent margin (Sect. *Iondraba*)
  2 Silicula not emarginate at the apex, the wings excurrent on the style — **39. auriculata**
  2 Silicula emarginate at the apex, the wings not excurrent on the style — **40. cichoriifolia**
1 Petals not more than 8 mm, short-clawed, erect; silicula with a narrow membranous margin
  3 Petals gradually attenuate at the base, not auriculate
    4 Filaments with a wide membranous wing — **34. lyrata**
    4 Filaments unwinged
      5 Leaves lyrate
        6 Annual; silicula 3·5–5 × 7–10 mm — **36. maritima**
        6 Perennial; silicula 6–7 × 11–13 mm — **38. raphanifolia**
      5 Leaves toothed or entire, not lyrate
        7 Infructescence lax; pedicels patent — **35. baetica**
        7 Infructescence dense; pedicels erecto-patent — **37. didyma**
  3 Petals abruptly contracted into a claw, auriculate-dilated above the base
    8 Leaves tomentose
      9 Silicula 9–11 × 16–19 mm — **10. vincentina**
      9 Silicula 3–7 × 7–12 mm
        10 Basal leaves 1·5–2 cm wide; petals 3 mm; sepals 1·5 mm — **18. sclerocarpa**

10 Basal leaves usually more than 2 cm wide; petals at least 4 mm; sepals 2–3·5 mm
11 Inflorescence 10–15 cm; sepals *c.* 2 mm **9. frutescens**
11 Inflorescence less than 8 cm; sepals 3–3·5 mm **11. sempervirens**
8 Leaves glabrous to hirsute, not tomentose
12 Leaves more than 4 times as long as wide
13 Leaves deeply toothed, lobed or pinnatifid
14 Stems not more than 30 cm
15 Silicula 3–4 × 6–9 mm
16 Inflorescence dense, corymbose; petals 6 mm **17. arvernensis**
16 Inflorescence lax; petals 4–5 mm **21. brevicaulis**
15 Silicula 6–7 × 9–12 mm
17 Petals 6 mm
18 Basal leaves 5–7 cm, with 1–3 opposite teeth **19. divionensis**
18 Basal leaves 9–13 cm, with 3–5 subopposite teeth **20. controversa**
17 Petals 4–5 mm
19 Basal leaves polymorphic, spathulate, triangular or lanceolate **7. gredensis**
19 Basal leaves all lanceolate or linear to oblong **1. laevigata**
14 Stems 30–70 cm
20 Basal leaves 0·2–0·5 cm wide, excluding teeth **33. valentina**
20 Basal leaves more than 0·5 cm wide
21 Cauline leaves absent or all minute **2. scaposa**
21 Cauline leaves well developed, but decreasing in size towards apex of stem
22 Silicula 4–6 × 7–9(–10) mm
23 Basal leaves with 2–4 remote teeth on each side; petals 4 mm **16. lamottii**
23 Basal leaves with 3–5 remote teeth each side interspersed with smaller teeth; petals 5–6 mm **(22–27). coronopifolia** group
22 Silicula 6–10 × 9–14 mm
24 Basal leaves ±spathulate; sinuate-dentate **3. flexuosa**
24 Basal leaves linear-oblong or lanceolate; pinnatifid or deeply toothed
25 Basal leaves deeply toothed but not pinnatifid
26 Stem branches erect; basal leaves dentate, hispid **28. lusitanica**
26 Stem branches patent; basal leaves sinuate-dentate, sparsely tomentose **32. guillonii**
25 Basal leaves ±pinnatifid
27 Stems up to 70 cm, with patent branches; petals 4–5 mm **29. mediterranea**
27 Stems up to 50 cm, with erecto-patent branches; petals 6 mm **30. nicaeensis**
13 Leaves entire or slightly sinuate-dentate
28 Basal leaves up to 0·5 cm wide (excluding teeth)
29 Plant 7–15 cm; silicula 3 × 6 mm **12. glacialis**
29 Plant usually more than 15 cm; silicula 3·5–8 × 6–14 mm
30 Basal leaves 3–5 cm **1. laevigata**
30 Basal leaves 6–8 cm **33. valentina**
28 Basal leaves more than 0·5 cm wide
31 Basal leaves spathulate, without 2 glands at the base, the indumentum often villous **8. neustriaca**
31 Basal leaves various, if spathulate, then polymorphic or with 2 glands at the base, the indumentum not villous
32 Stems not more than 30 cm
33 Raceme dense, corymbose **17. arvernensis**
33 Raceme dense or lax, not corymbose
34 Leaves with 2 glands at the base **1. laevigata**
34 Leaves without 2 glands at the base **7. gredensis**
32 Stems usually more than 30 cm
35 Leaves with 2 glands at the base; stems usually ±leafy **1. laevigata**
35 Leaves without glands at the base; stems sparingly leafy
36 Basal leaves pubescent to hirsute (on volcanic debris) **16. lamottii**
36 Basal leaves setose
37 Cauline leaves absent or minute **2. scaposa**
37 Cauline leaves present, the lower similar to basal leaves **3. flexuosa**
12 Leaves not more than 4 times as long as wide
38 Stems at least 30 cm
39 Basal leaves ±regularly toothed **(4–6). variegata** group
39 Basal leaves pinnatifid or pinnatilobed
40 Leaf-lobes intermixed with small lobes
41 Petals 5–6 mm; silicula *c.* 5 mm **(22–27). coronopifolia** group
41 Petals 4–5 mm; silicula 6–7 mm **29. mediterranea**
40 Leaf-lobes without smaller lobes intermixed
42 Basal leaves 4–6 cm **(22–27). coronopifolia** group
42 Basal leaves 6–8 cm
43 Silicula not more than 5 mm **(22–27). coronopifolia** group
43 Silicula 6–7 mm
44 Stems intricately branched from the base; petals 5–6 mm **(22–27). coronopifolia** group
44 Stems branched from about the middle; petals 4–5 mm **29. mediterranea**
38 Stems not more than 30 cm
45 Basal leaves 1–3 cm, less than 1 cm wide
46 Silicula not more than 6 mm wide; basal leaves narrowly linear to oblong **12. glacialis**
46 Silicula at least 7 mm wide; basal leaves obovate-lanceolate **13. brevifolia**
45 Basal leaves 2–8 cm, at least 1 cm wide
47 Petals 3 mm; silicula 3·5–4 × 7 mm **18. sclerocarpa**
47 Petals at least 4 mm; silicula 5–7 × 8–13 mm
48 Basal leaves pinnatifid or pinnatilobed **(22–27). coronopifolia** group
48 Basal leaves sinuate-dentate to ±lobed, but not pinnatifid
49 Basal leaves 2–3·5 mm **31. intermedia**
49 Basal leaves at least 4 mm
50 Silicula 6 × 10 mm; basal leaves polymorphic **7. gredensis**
50 Silicula 5 × 8–9 mm; basal leaves cuneate or obovate-oblong, not polymorphic
51 Basal leaves with 1 or 2 teeth on each side, and 3 lobes at the apex **14. fontqueri**
51 Basal leaves with 3 or 4 teeth on each side **15. rotgesii**

Sect. BISCUTELLA (Sect. *Thlaspidium* DC.). Lateral sepals not saccate or spurred; petals small or medium, short-clawed, or gradually attenuate at the base, erect. Silicula with a narrow membranous margin.

Ser. *Laevigatae* Malinovski. Perennial. Petals abruptly contracted into a claw, auriculate-dilated above the claw. All nectaries extrastaminal.

1. B. laevigata L., *Mantissa Alt.* 255 (1771). Stems 10–50(–70) cm, simple or branched. Basal leaves 1·5–13 × 0·3–2 cm, forming an obvious rosette or not, linear- to ovate-lanceolate, ovate or rarely spathulate, entire, sinuate-dentate or denticulate, glabrous or hairy; cauline leaves 2–10, resembling the basal leaves or small, linear, entire. Flowers in lax or dense racemes. Petals 4–8 mm, clawed. Silicula 4–8(–15) × 8–14(–30) mm, glabrous or hairy; style 2–6 mm. *Usually on rocks or stony ground. C. & S. Europe, extending northwards to Belgium.* Au Be Cz Ga Ge He Hs Hu It Ju Po Rm Rs(W).

One of the most polymorphic species in the European flora. Studies by I. Manton, *Zeitschr. Indukt. Abstamm. Vererbungslehre* **47**: 41–57 (1933); *Ann. Bot.* nov. ser. **1**: 439–462 (1937), have shown that the group contains a series of both diploid and tetraploid populations. Diploids, with $2n = 18$, are known from a few localities at low altitudes from France (? and Portugal and Spain) to the Balkan peninsula; tetraploids, with $2n = 36$, are found mainly in the Alps and the mountains of S. & S.E. Europe. Morphological differentiation of the populations is weak, and they are best regarded as subspecies of a single species. Union of all the diploids into one species and all the tetraploids into another as suggested by Manton has not proved feasible.

The more distinct taxa are keyed below. This treatment is a modification of that given by Machatschki-Laurich, *loc. cit.* (1926), and is similar to the one adopted by Markgraf in Hegi, *Ill. Fl. Mitteleur.* ed 2, **4(1)**: 395–401 (1963).

1 Plant glabrous, shining; petals 7 mm (n) subsp. **lucida**
1 At least the leaves hairy; petals usually less than 7 mm
2 Leaves hispid
3 Silicula up to 15 × 30 mm; basal leaves rigid (m) subsp. **montenegrina**
3 Silicula less than 15 mm; basal leaves not rigid
4 Basal leaves ovate to cuneate-lanceolate, narrowed towards the base but not petiolate, forming a rosette which is ±appressed to the ground
5 Basal leaves ovate or spathulate, rounded; silicula 6–7 × 12–13 mm (l) subsp. **illyrica**
5 Basal leaves cuneate-lanceolate or obovate, apiculate; silicula 5(–8) × 10(–12) mm (b) subsp. **austriaca**
4 Basal leaves lanceolate or linear-lanceolate, petiolate, erect
6 Stems pubescent or glabrous at the base (j) subsp. **tirolensis**
6 Stems hispid at the base
7 Basal leaves up to 13 × 2 cm, lanceolate, sinuate-dentate; stems branched (a) subsp. **laevigata**
7 Basal leaves 3–5 × *c.* 0·3 cm, linear-lanceolate, entire; stems unbranched (k) subsp. **angustifolia**
2 Leaves finely pubescent
8 Leaves deeply sinuate-dentate
9 Basal leaves up to 8 × 1·2 cm, lanceolate (h) subsp. **varia**
9 Basal leaves 3–4 × 0·5–0·7(–1) cm, linear to oblong (i) subsp. **subaphylla**
8 Leaves dentate, subentire or entire
10 Basal leaves subentire or with acute teeth; cauline leaves few, minute
11 Basal leaves with acute teeth, pubescent (c) subsp. **kerneri**
11 Basal leaves subentire, ±glabrous (d) subsp. **hungarica**
10 Basal leaves entire or somewhat denticulate; cauline leaves numerous
12 Basal leaves up to 8 × 1–1·5 cm, entire (g) subsp. **guestphalica**
12 Basal leaves not more than 5 × 0·8(–1) cm, dentate
13 Basal leaves lanceolate, with acute teeth, ±long-petiolate; stems branched (e) subsp. **gracilis**
13 Basal leaves oblong, with obtuse teeth, shortly petiolate; stems unbranched (f) subsp. **tenuifolia**

(a) Subsp. **laevigata** (subsp. *longifolia* (Vill.) Rouy & Fouc., *B. longifolia* Vill.): $2n = 36$. *Almost throughout the range of the species.*
(b) Subsp. **austriaca** (Jordan) Mach.-Laur., *Bot. Arch. (Königsberg)* **13**: 67 (1926) (*B. austriaca* Jordan): $2n = 18$. *N.E. Alps, S. & E. Carpathians, N. Hungary.*
(c) Subsp. **kerneri** Mach.-Laur., *op. cit.* 68 (1926): $2n = 18$. *E.C. Europe.*
(d) Subsp. **hungarica** Sóo, *Acta Bot. Acad. Sci. Hung.* **10**: 373 (1964): *Hungary.*
(e) Subsp. **gracilis** Mach.-Laur., *op. cit.* 69 (1926): $2n = 18$. *C. Europe, from S.W. Poland to N.W. Jugoslavia.*
(f) Subsp. **tenuifolia** (Bluff & Fingerh.) Mach.-Laur., *loc. cit.* (1926): $2n = 18$. *C. Germany.*
(g) Subsp. **guestphalica** Mach.-Laur., *op. cit.* 70 (1926): $2n = 18$. *N.C. Germany (Wesergebirge).*
(h) Subsp. **varia** (Dumort.) Rouy & Fouc., *Fl. Fr.* **2**: 110 (1895) (*B. varia* Dumort., *B. alsatica* Jordan): $2n = 18$. *W. Germany, N.E. France, S.E. Belgium.*
(i) Subsp. **subaphylla** Mach.-Laur., *Bot. Arch. (Königsberg)* **13**: 70 (1926): $2n = 18$. *E. France, S. & W. Germany, Switzerland.*
(j) Subsp. **tirolensis** (Mach.-Laur.) Heywood, *Feddes Repert.* **69**: 147 (1964) (*B. laevigata* var. *tirolensis* Mach.-Laur.): *E. Alps.*
(k) Subsp. **angustifolia** (Mach.-Laur.) Heywood, *loc. cit.* (1964): *E. Alps, ?Hungary, Jugoslavia.*
(l) Subsp. **illyrica** Mach.-Laur., *Bot. Arch. (Königsberg)* **13**: 67 (1926): *Jugoslavia; isolated localities in Czechoslovakia and Hungary.*
(m) Subsp. **montenegrina** Rohlena, *Sitz.-Ber. Böhm. Ges. Wiss.* **1903**: 17 (1903): *Crna Gora.*
(n) Subsp. **lucida** (Balbis ex DC.) Mach.-Laur., *Bot. Arch. (Königsberg)* **13**: 66 (1926) (*B. lucida* Balbis ex DC.): $2n = 36$. *E. Alps, C. Appennini, N.W. Jugoslavia.*

2. B. scaposa Sennen ex Mach.-Laur., *Bot. Arch. (Königsberg)* **13**: 93 (1926) (*B. strictifolia* Pau, *B. subscaposa* Sennen). Stems 40–60 cm, simple or branched. Basal leaves 5–10 × 1–2 cm, lanceolate, obovate-lanceolate or subspathulate, sinuate-dentate with 3–5 acute or obtuse teeth on each side, more or less densely setose; cauline leaves few, minute. Petals 4 mm. Silicula 4–8 × 8–11 mm. $2n = 18$. ● *C. & N.W. Spain; E. Pyrenees.* Ga Hs.

3. B. flexuosa Jordan, *Diagn.* **1(1)**: 300 (1864). Stems 40–65 cm, simple or branched, robust. Basal leaves 6–11 × 1·5–2 cm, subspathulate or oblong-obovate, remotely sinuate-dentate, petiolate, setose, the apex somewhat obtuse; cauline leaves few, semi-amplexicaul, the lower like the basal leaves, the upper linear, entire. Silicula 6–7 × 9–12 mm. *Schistose rocks.* • *E. & C. Pyrenees.* Ga Hs.

(4–6). **B. variegata** group. Glabrous or pubescent, rhizomatous perennials 40–100 cm, with broadly obovate, more or less toothed basal leaves.

1 Silicula 5–7 × 7·5–10 mm; sepals 2–2·5 mm **4. variegata**
1 Silicula 8–13 × 10–20 mm; sepals 3–4 mm
 2 Cauline leaves few; silicula 11 × 20 mm **5. megacarpaea**
 2 Cauline leaves numerous; silicula 8 × 12 mm **6. foliosa**

4. B. variegata Boiss. & Reuter in Boiss., *Diagn. Pl. Or. Nov.* **3(1)**: 44 (1853) (*B. laevigata* auct. hisp. pro parte, non L.). Rhizome woody, covered with remains of old leaves; stems 40–70 cm. Basal leaves laxly rosulate, obovate to obovate-oblong, more or less abruptly contracted into petiole, regularly and deeply toothed or irregularly sinuate-dentate or crenate, pubescent or subglabrous; cauline leaves semi-amplexicaul. Sepals 2–2·5 mm; petals 4–5 mm. Silicula 5–7 × 7·5–10 mm. $2n = 54$. *Dry places and mountain rocks.* • *S. Spain.* Hs.

5. B. megacarpaea Boiss. & Reuter, *loc. cit.* (1853). Stems up to 1 m. Basal leaves 7–10(–20) × 2–2·5(–6) cm, densely rosulate, sinuate-dentate, attenuate into petiole, with long and short hairs on upper surface; cauline leaves few, the lower resembling the basal in shape and size, abruptly decreasing in size at about the middle of the stem, the upper linear, entire. Sepals 3–4 mm; petals 6–8 mm. Silicula 8–13 × 13–20 mm. $2n = 54$. *Limestone rocks.* • *S. Spain.* Hs.

6. B. foliosa Mach.-Laur., *Bot. Arch. (Königsberg)* **13**: 97 (1926). Like **5** but basal leaves not rosulate, up to 6 × 1–2 cm, remotely and regularly toothed; cauline leaves numerous; silicula 5–8 × 10–12 mm. $2n = 54$. *Crevices of limestone rocks.* • *S.W. Spain (Serranía de Ronda).* Hs.

7. B. gredensis Guinea, *Anal. Inst. Bot. Cavanilles* **21**: 398 (1963). Stems up to 30 cm, erect, pubescent. Basal leaves polymorphic, spathulate, triangular or lanceolate, entire, 2-dentate or 2- to 4-lobulate, the apex obtuse; cauline leaves few, oblong to oblanceolate, entire or toothed. Raceme very long. Sepals 2 mm; petals 4 mm. Silicula 6 × 10 mm. $2n = 18$. *Granite rocks.* • *W.C. Spain (Sierra de Gredos).* Hs.

8. B. neustriaca Bonnet, *Bull. Soc. Dauph. Éch. Pl.* **6**: 222 (1879). Stems 20–40 cm, simple or branched. Basal leaves 4–8 × 1–1·5 cm, erect, spathulate, slightly sinuate-dentate with 3 more or less equal teeth on each side, pubescent or villous, attenuate into a long petiole; cauline leaves broad, entire or dentate. Raceme dense, compact in flower, elongating in fruit. Sepals 2·5 mm; petals 4 mm. Silicula 6–7 × 10–11 mm. • *N.E. France (Seine valley).* Ga.

9. B. frutescens Cosson, *Not. Pl. Crit.* 27 (1849) (*B. 'suffrutescens'* Willk., sphalm.). Rhizome black, woody; whole plant covered with dense, white tomentum; stems simple or branched, up to 50 cm. Basal leaves 20 × 6 cm, ovate and sinuate-dentate or lyrate; cauline leaves sessile or amplexicaul, the lower resembling the basal leaves. Inflorescence much-branched. Sepals 2 mm; petals 4 mm. Silicula 3·5–4 × 7 mm, with oblique valves, the margin continuous, with small swellings. $2n = 18$. *S.W. Spain; Islas Baleares.* Bl Hs.

B. incana Ten., *Fl. Neap. Prodr. App. Quinta* 19 (1826), from S. Italy (Puglia and Calabria), like **9** but with leaves oblong, regularly sinuate-dentate, with obtuse teeth, densely white-strigose, and silicula glabrous, the margin subundulate, has been seldom collected; its status is uncertain.

10. B. vincentina (Samp.) Rothm. ex Guinea, *Feddes Repert.* **69**: 148 (1964) (*B. laevigata* prol. *vincentina* Samp.). Rhizome thick, woody; stems 20–30 cm, erect. Basal leaves 5–6 × 1·5–2·5 cm, numerous, densely rosulate, obovate to obovate-oblong, sinuate-undulate, tomentose; cauline leaves minute. Raceme dense, elongating in fruit. Sepals 4 mm; petals 6 mm. Silicula very large, 9–11 × 16–19 mm, with a membranous margin. *Sandy soil.* • *S.W. Portugal.* Lu.

11. B. sempervirens L., *Mantissa Alt.* 255 (1771) (incl. *B. montana* Cav., *B. tomentosa* Lag. ex DC., *B. rosularis* Boiss. & Reuter, *B. gibraltarica* Wilmott ex Guinea, *B. laevigata* auct. hisp. pro parte). Rhizome thick, woody, simple or branched, producing 1–several rosettes of leaves; stems up to 50 cm, simple or branched. Basal leaves 3–12 × 1–5 cm, oblong-lanceolate to broadly obovate or elliptic-spathulate, subentire to deeply sinuate-dentate, usually densely villous or tomentose; cauline leaves few, the lower resembling the basal in size and shape, the upper minute, linear. Raceme more or less dense-flowered. Sepals 3–3·5 mm; petals 5–7 mm. Silicula 4–7 × 7·5–12 mm. $2n = 18$. *Limestone rocks. S. & S.E. Spain; Islas Baleares.* Bl Hs.

12. B. glacialis (Boiss. & Reuter) Jordan, *Diagn.* **1(1)**: 310 (1864) (*B. sempervirens* auct., non L., *B. secunda* Jordan). Rhizome long, thick and woody; stems 7–15 cm. Basal leaves 1–3 × 0·5 cm, densely rosulate, usually narrowly linear, rarely oblong, densely villous, the apex obtuse; cauline leaves 1–3, minute. Silicula 3 × 6 mm. $2n = 18$. *Mountains of S. Spain.* Hs.

13. B. brevifolia (Rouy & Fouc.) Guinea, *Anales Inst. Bot. Cavanilles* **21**: 394 (1963). Rhizomatous perennial 10–20 cm. Basal leaves 2 × 0·7 cm, densely rosulate, obovate-lanceolate, sinuate-dentate (rarely subentire), with 2 or 3 obtuse teeth on each side, sparsely hairy, the apex obtuse; cauline leaves few or absent. Silicula 3–4 × 7–8 mm. *C. Pyrenees.* Ga ?Hs.

14. B. fontqueri Guinea & Heywood, *loc. cit.* (1963) (*B. cuneata* (Font Quer) Font Quer ex Mach.-Laur., non Lag.). Rhizomatous perennial up to 20 cm. Basal leaves 4 × 1·5–2 cm, sparsely pubescent, laxly rosulate, cuneate, with 1 or 2 teeth on either side and 3-lobed at the apex, narrowed to the base. Petals 4 mm. Silicula 5 × 8–9 mm. *Limestone rocks and screes*, c. 1300 *m.* • *N.E. Spain (near Tortosa).* Hs.

15. B. rotgesii Fouc., *Bull. Soc. Bot. Fr.* **47**: 85 (1900). Stems 18–26 cm, slender, branched. Basal leaves 4–5 × 1·5–1·7 cm, obovate-oblong, deeply sinuate-dentate, with 3 or 4 obtuse teeth on each side, sparsely hispid. Petals 4–4·5 mm. Silicula 5 × 8·5 mm. $2n = 18$. *Schistose rocks.* • *Corse.* Co.

16. B. lamottei Jordan, *Diagn.* **1(1)**: 302 (1864). Rhizome long, woody, producing rosette-bearing branches; stems 25–40 cm, slender, simple or branched, hairy below, glabrous above. Basal leaves (4–)8–9 × 1–1·5 cm, pubescent to hirsute, more or less densely rosulate, linear-lanceolate to oblanceolate-spathulate, sinuate-dentate, with 2–4 remote teeth on each side, obtuse or subacute at the apex, attenuate into a long petiole at the base; cauline leaves few, semi-amplexicaul. Raceme long. Sepals 2 mm; petals 4 mm. Silicula 4–6 × 7–10 mm; style 2·5 mm. $2n = 18$. *Volcanic debris.* • *S.C. France.* Ga.

17. B. arvernensis Jordan, *op. cit.* 298 (1864). Like **16** but stems 10–30 cm, leafy; raceme dense, corymbose; petals 6 mm; style up to 4 mm. *Volcanic debris and rocks.* 2*n* = 18. ● *S.C. France (Puy-de-Dôme).* Ga.

18. B. sclerocarpa Revel, *Congr. Sci. Fr. (Rodez)* **40(1)**: 262 (1874). Rhizome long, woody, terminating in a rosette or divided and producing rosette-bearing branches; stems up to 30 cm. Basal leaves 5–7 × 1·5–2 cm, more or less tomentose, laxly rosulate, subpinnatifid, with 3–6 large obtuse teeth on each side; cauline leaves several, semi-amplexicaul. Inflorescence branched, lax. Sepals 1·5 mm; petals 3 mm. Silicula 3·5–4 × 7 mm. 2*n* = 18. ● *S. & S.C. France.* Ga.

19. B. divionensis Jordan, *Diagn.* **1**: 305 (1864). Rhizome long, woody, terminating in a rosette or divided and producing rosette-bearing branches; stems up to 25 cm. Basal leaves 5–7 × 1–1·5 cm, densely puberulent or canescent, densely rosulate, cuneate or spathulate, with 1–3 large, opposite, acute teeth on each side; apex subacute; cauline leaves few. Inflorescence branched, somewhat dense. Petals 6 mm. Silicula 6–7 × 9–11 mm. 2*n* = 18. *Limestone rocks.* ● *E.C. France (near Dijon).* Ga.

20. B. controversa Boreau, *Fl. Centre Fr.* ed. 3, **2**: 56 (1857). Like **19** but basal leaves 9–13 × 1·5 cm, obovate-oblong, with 3–5 subopposite teeth; silicula 6–7 × 10–12 mm. 2*n* = 18. ● *C. France.* Ga.

21. B. brevicaulis Jordan, *Diagn.* **1**: 303 (1864). Like **19** but stems 10–20 cm; basal leaves oblong or lanceolate-cuneate, subpinnatifid with 1–3 acute lobes on each side, the apex subacute; inflorescence lax; petals 4–5 mm; silicula 3–4 × 6–9 mm. 2*n* = 18. ● *S.W. Alps.* Ga It.

(22–27). B. coronopifolia group. Stems (20–)30–60(–80) cm. Basal leaves pinnatifid or pinnatilobed, glabrous to hirsute but not tomentose.

1 Stems intricately branched from the base; cauline leaves numerous
2 Leaves pinnatilobed, the lobes ovate; petals 4–5 mm; silicula 4–5 × 8·5–9 mm **23. intricata**
2 Leaves pinnatifid, the lobes ovate-oblong; petals 5–6 mm; silicula 6–7 × 9–11 mm **27. polyclada**
1 Stems not intricately branched from the base; cauline leaves usually few to several
3 Leaf-lobes large, 5–7 on each side, unequal **26. pinnatifida**
3 Leaf-lobes small, 2–4(5) on each side
4 Leaf-lobes ±equal, 2 or 3 on each side **22. coronopifolia**
4 Leaf-lobes unequal, 3–5 on each side
5 Basal leaves 4–6 cm; silicula 7 × 10–11 mm **24. apricorum**
5 Basal leaves 8–9 cm; silicula 5–6 × 8–9 mm **25. granitica**

22. B. coronopifolia L., *Mantissa Alt.* 255 (1771) (*B. lima* Reichenb., *B. tarraconensis* Sennen). Stems up to 30 cm, hispid to hirsute. Basal leaves 3–4 × 1·5 cm, hispid or hirsute, laxly rosulate, pinnatifid, with 2–4 long, remote, somewhat acute lobes on each side; lower cauline leaves about as large as the basal, with large teeth, the upper linear-setaceous. Inflorescence several times branched. Sepals 2·5 mm; petals 5 mm. Silicula 5–7 × 9–13 mm. 2*n* = 18. ● *S. & S.C. France and E. Spain; N. Italy.* Ga Hs It.

23. B. intricata Jordan, *Diagn.* **1**: 308 (1864). Stems up to 50 cm, intricately branched from the base, the branches open and patent in a corymbose manner, hirsute below. Basal leaves 6–8 × 1·5–2 cm, pubescent, oblong, pinnatilobed, with 3–5 ovate, slightly acute lobes on each side; cauline leaves lobed or dentate. Raceme quite dense, elongating during anthesis. Petals 4·5 mm. Silicula 5 × 9 mm. ● *S.E. France (S. & W. of Lyon).* Ga.

24. B. apricorum Jordan, *op. cit.* 307 (1864). Stems 35–40 cm, sparsely hispid or hirsute. Basal leaves 4–6 × 1–1·5 cm, hispid-setose, oblong-spathulate, with 3 or 4 large, obtuse teeth on each margin; cauline leaves semi-amplexicaul, with large teeth. Sepals 2–3 mm; petals 5–6 mm. Silicula 7 × 10–11 mm. *Dry places.* ● *S.E. France.* Ga.

25. B. granitica Boreau ex Pérard, *Bull. Soc. Bot. Fr.* **16**: 353 (1869). Like **24** but basal leaves 6–9 × 1·5–2 cm, lanceolate, with 3–5 large teeth on each margin interspersed with smaller teeth; silicula 5 × 9 mm. *Siliceous rocks.* ● *S.C. France.* Ga ?Hs.

26. B. pinnatifida Jordan, *Diagn.* **1**: 312 (1864). Stems 30–50 cm, with erecto-patent branches. Basal leaves up to 6 cm, hispid, oblong, pinnatifid, with 4 or 5(–7) very unequal, oblong lobes, interspersed with a few smaller lobes; cauline leaves similar. Raceme somewhat dense, elongating in fruit. Petals 5 mm. Silicula 5–8 mm. *Dry places.* ● *S. France.* Ga.

27. B. polyclada Jordan, *loc. cit.* (1864). Like **23** but basal leaves obovate-oblong to oblong-subpinnatifid, with 3–5 ovate-oblong or oblong lobes on each side, hispid; petals 5–6 mm; silicula 6–7 × 9–11 mm. *Rocks.* ● *S.E. France, westwards to Narbonne.* Ga.

28. B. lusitanica Jordan, *op. cit.* 315 (1864). Stems 40–60 cm, stout, erect, branched, hispid. Basal leaves up to 8 cm, regularly and deeply toothed, with 3–6 teeth on each side; cauline leaves numerous or few. Sepals 2·5 mm; petals 5 mm. Silicula 6–10 × 11–14 mm. 2*n* = 54. ● *Portugal; W. Spain.* Hs Lu.

A large-fruited variant, with silicula 8–12 × 14–18 mm, has been described from Portugal (Estremadura).

29. B. mediterranea Jordan, *op. cit.* 313 (1864) (*B. stricta* Jordan). Stems 40–70 cm, robust, hispid below, branched from about the middle, with patent branches. Basal leaves rosulate, 6–8 × 1–2 cm, linear-oblong, pinnatifid, with 3–6 remote teeth on each side, sometimes interspersed with smaller teeth; lower cauline leaves numerous, upper cauline leaves few. Raceme rather dense. Sepals 2·5–3 mm; petals 4–5 mm. Silicula 6–7 × 9–12 mm. *Dry places.* 2*n* = 18. ● *S. France, E. Spain.* Ga Hs.

30. B. nicaeensis Jordan, *op. cit.* 314 (1864). Stems 40–50 cm, strict, hirsute and scabrid below, with erecto-patent branches above. Basal leaves up to 13 × 2 cm, hispid, scabrid, lanceolate, acute, with 4–6 short lobes on each side; cauline leaves numerous. Petals 6 mm. Silicula 6–7 × 11–12 mm. *Dry places.* ● *S.E. France; N.W. & C. Italy.* Ga It.

31. B. intermedia Gouan, *Obs. Bot.* 42 (1773) (*B. pyrenaica* Huet). Stems 10–20 cm. Basal leaves 2–3·5 × 1–1·5 cm, hispid or rarely subglabrous, obovate, or cuneate, very obtuse and subentire or sinuate-lobed, with 1–3 lobes on each side; cauline leaves few, small. Petals 4·5 mm. Silicula 5–6 × 9–10 mm. 2*n* = 18. ● *Pyrenees; N. & C. Spain.* Ga Hs.

32. B. guillonii Jordan, *Diagn.* **1**: 302 (1864). Stems 40–60 cm, slightly pubescent, with patent branches. Basal leaves deeply sinuate-dentate, with triangular-acute, sinuate lobes; cauline leaves numerous, more or less pinnatifid or deeply toothed. Sepals 2·5–3 mm; petals 4–6 mm. Silicula 6–7 × 10–12 mm. *Dry places.* ● *W. France.* Ga.

33. B. valentina (L.) Heywood, *Feddes Repert.* **66**: 155 (1962) (*B. stenophylla* Dufour, *B. tenuicaulis* Jordan, *Sisymbrium valentinum* L.). Rhizome woody; stems 30–50 cm, simple or slightly branched, hispid below. Basal leaves 6–8 × 0·2–0·5 cm (excluding the teeth), linear, deeply toothed, with 1 or 2 acute teeth on each side (or entire), usually with setiform hairs or subglabrous; cauline leaves few, linear-setaceous. Sepals 2–3 mm; petals 5–6 mm. Silicula 3·5–6 × 6–11 mm. $2n = 18$. *Dry places. C., E. & S. Spain.* Hs.

Variable in the size and toothing of the leaves, and in the indumentum.

Smaller, less woody plants with subentire or shallowly toothed basal leaves and glabrous stems have been called **B. atropurpurea** Mateo & Figuerola, *Fl. Analit. Prov. Valencia* 370 (1987). They are probably not specifically distinct from **33**.

Ser. *Lyratae* Malinovski. Annuals or rarely perennials. Petals gradually attenuate at the base, not auriculate. Lateral nectaries intrastaminal.

34. B. lyrata L., *Mantissa Alt.* 254 (1771) (*B. microcarpa* DC., *B. scutulata* Boiss. & Reuter). Annual; stems 15–40 cm, slender, with divaricate branches as long as the main stem arising from the base. Raceme rather dense, elongating in fruit. Basal leaves few, up to 10 × 1·5 cm, lyrate or sinuate-dentate, the apex obtuse. Sepals 2 mm; petals 4 mm; filaments with a wide membranous wing below or above the middle. Silicula 2·5–4 mm. $2n = 12$. *Schistose soils. S.W. Spain (Cádiz prov.).* Hs. (*N.W. Africa.*)

35. B. baetica Boiss. & Reuter in Boiss., *Diagn. Pl. Or. Nov.* **3(1)**: 42 (1853) (*B. apula* auct. non L.). Annual; stems 15–50 cm, simple or branched. Basal leaves 6(–10) × 1·8–3 cm, usually rosulate, obovate-cuneate, sinuate-dentate or denticulate with acute teeth, roughly hairy, mainly on the veins. Raceme dense, elongating greatly in fruit; pedicels patent. Sepals 2 mm; petals 4–4·5 mm; filaments filiform. Silicula 3–5 × 6–9·5 mm. $2n = 16$. *Dry places. S. Spain.* Hs.

36. B. maritima Ten., *Prodr. Fl. Nap.* 38 (1811) (*B. lyrata* auct., non L.). Annual; stems up to 60 cm, simple or branched. Basal leaves up to 18 × 5–6 cm, numerous, densely rosulate, lyrate, attenuate into petiole; terminal lobe broadly ovate; all lobes entire or toothed, hairy. Raceme rather dense, elongate in fruit. Sepals 3 mm; petals 6 mm; filaments filiform. Silicula 3·5–5 × 7–10 mm. *Dry places.* $2n = 16$. *W. Italy, Sardegna, Sicilia.* It Sa Si.

37. B. didyma L., *Sp. Pl.* 653 (1753) (*B. apula* L., *B. ciliata* DC., *B. columnae* Ten.). Annual; stems up to 40 cm, simple or branched. Basal leaves up to 8 × 2·5 cm, rosulate or not, obovate-cuneate, dentate or denticulate. Raceme dense even in fruit; pedicels erect. Sepals 2 mm; petals 4 mm; filaments filiform. Silicula 4·5–7 × 9–12·5 cm. $2n = 16$. *Dry places. C. & E. Mediterranean region.* Al Co Cr Gr It Ju Sa Si.

B. eriocarpa DC., *Ann. Mus. Hist. Nat. (Paris)* **18**: 298 (1811), probably native in N.W. Africa, and doubtfully distinct from **37**, has been recorded in error from Italy and Corse.

38. B. raphanifolia Poiret, *Voy. Barb.* **2**: 198 (1789) (*B. radicata* Cosson & Durieu). Perennial with woody rhizome; stems up to 1 m, branched, hirsute below. Basal leaves up to 24 × 6 cm, laxly rosulate, lyrate-pinnatipartite, with a large terminal, ovate, toothed segment and 2–6 small, entire or sinuate-dentate lobes on each margin. Raceme lax; pedicels erecto-patent. Sepals 2 mm; petals 5 mm; filaments more or less filiform. Silicula 6–7 × 11–13 mm. $2n = 16$. *S.W. Italy; Sicilia.* It Si. (*N.W. Africa.*)

Sect. JONDRABA Reichenb. (*Jondraba* Medicus). Lateral sepals saccate or spurred at the base; petals large, long-clawed, the limb patent. Silicula with a translucent margin.

39. B. auriculata L., *Sp. Pl.* 652 (1753). Stems 25–50 cm, hispid below, glabrous or subglabrous above. Basal leaves oblong, sinuate-dentate (or subentire, var. *erigerifolia* DC.), long-petiolate; cauline leaves auriculate-amplexicaul, sessile. Raceme many-flowered, remaining compact in fruit. Sepals with a short spur; petals up to 15 mm. Silicula 7–10 × 12–18 mm, broadly cordate, not emarginate at the apex, the wings excurrent on the style; style up to 10 mm. $2n = 16$. *Cultivated fields and dry places.* ● *W. Mediterranean region; Portugal.* Bl Ga Hs ?It Lu.

40. B. cichoriifolia Loisel., *Fl. Gall.* 167 (1810) (*B. hispida* DC.). Like **39** but stems villous or hispid above; raceme elongating in fruit; spur of sepals long and slender; silicula emarginate at both the base and the apex, the wings not excurrent on the style. $2n = 16$. *Rocks and dry places.* ● *S. Europe, from the Pyrenees to Crna Gora, extending northwards to the Jura.* Ga He Hs It Ju.

79. Megacarpaea DC.

D. A. WEBB

Perennial; hairs unbranched. Sepals not saccate. Fruit a large, angustiseptate, deeply 2-lobed silicula; lobes flat, broadly winged, 1-seeded.

1. M. megalocarpa (Fischer ex DC.) Schischkin ex B. Fedtsch. in Komarov, *Fl. URSS* **8**: 543 (1939) (*M. laciniata* DC.). Stem 20–40 cm, branched above; hairs stiff or crisped. Basal leaves up to 15 cm, petiolate, oblong-elliptical in outline, pinnatisect with many irregularly and acutely toothed lobes; cauline leaves similar but smaller and sessile. Inflorescence a panicle. Flowers irregularly monoecious, those in the apical part of each branch male, in the basal part female. Female flowers without perianth; male flowers with sepals and with petals *c.* 10 mm, linear-oblong. Intermediate flowers sometimes present, with smaller, white petals and variously developed stamens and gynoecium. Lobes of silicula 15–20 mm in diameter, suborbicular; wing 4–6 mm wide. *Dry steppes and semi-deserts. S.E. Russia, W. Kazakhstan.* Rs(E). (*C. Asia.*)

80. Lepidium L.

†J. DE CARVALHO E VASCONCELLOS (EDITION 1) REVISED BY J. R. AKEROYD AND T. C. G. RICH (EDITION 2)

Annuals to perennials, sometimes small shrubs; papillose or with usually unbranched hairs. Flowers small, in dense, terminal, ebracteate racemes. Sepals not saccate; petals white (rarely yellow or absent); stamens 2, 4 or 6. Fruit an angustiseptate silicula; valves strongly keeled, winged; style short or absent. Seeds usually 2, one pendent from the apex of each loculus.

In addition to the species described below, several others occur in Europe as casuals.

Literature: A. Thellung, *Neue Denkschr. Schweiz. Naturf. Ges.* **41**: 1–340 (1907). B. Jonsell, *Bot. Not.* **128**: 20–46 (1975).

1 Wing of the silicula connate with the lower part of the style; silicula at least 4 mm
2 Middle and upper cauline leaves cuneate at the base
3 Pedicels 1–2 mm in fruit **5. spinosum**
3 Pedicels (2–)3–6 mm in fruit **6. sativum**
2 Middle and upper cauline leaves amplexicaul
4 Annual or biennial; silicula densely covered with small, scale-like vesicles; style usually not projecting beyond the apical notch of the wing **1. campestre**
4 Perennial; silicula with few or no vesicles; style projecting beyond the apical notch of the wing
5 Plant grey-pubescent; silicula hairy, at least when young **4. hirtum**
5 Plant sparsely or shortly hairy; silicula usually glabrous even when young
6 Pedicels glabrous in fruit **2. villarsii**
6 Pedicels hairy in fruit **3. heterophyllum**
1 Wing of the silicula not connate with the style; silicula usually less than 4 mm
7 Silicula ± conspicuously winged and notched; style usually shorter than or equalling notch
8 Small shrub; all leaves linear, entire **7. subulatum**
8 Annual or biennial; at least the lower and middle cauline leaves toothed or pinnately divided
9 Petals longer than the sepals
10 Upper cauline leaves pinnatifid; silicula *c.* 1·5 mm **8. cardamines**
10 Upper cauline leaves entire to dentate; silicula 2–4 mm **9. virginicum**
9 Petals shorter than the sepals or absent
11 Upper cauline leaves pinnatifid; stem with long, unbranched hairs **11. bonariense**
11 Upper cauline leaves entire or dentate; stem glabrous, papillose or with scale-like trichomes
12 Pedicels erect, appressed to stem, glabrous (*vide* **13. africanum**)
12 Pedicels erecto-patent, not appressed to stem, papillose at least on upper side
13 Most cauline leaves entire or subentire; silicula with entire or slightly notched apex
14 Stem papillose; silicula usually more than 3 mm; sepals *c.* 1 mm **10. densiflorum**
14 Stem glabrous; silicula less than 3 mm; sepals 0·5–0·75 mm **14. pinnatifidum**
13 Middle cauline leaves often pinnatifid or dentate; silicula distinctly notched at the apex
15 Silicula not more than 2 mm wide
16 Petals 0 or very minute; pedicels papillose all round **12. ruderale**
16 Petals $\frac{1}{2}$–$\frac{2}{3}$ as long as the sepals; pedicels papillose on the upper side **13. africanum**
15 Silicula at least 2 mm wide
17 Apex of infructescence with curved, somewhat appressed papillae **9. virginicum**
17 Apex of infructescence with minute, straight papillae **10. densiflorum**
7 Silicula usually not winged, entire or slightly notched; if notched, style longer than notch
18 Petals yellow; leaves strongly dimorphic, the lower 2- or 3-pinnatifid, the upper suborbicular, entire, with broad, rounded lobes completely enclosing the stem and overlapping **15. perfoliatum**
18 Petals white; leaves not strongly dimorphic, the upper without broad, rounded lobes enclosing the stem
19 Annual or biennial; petals usually shorter than the sepals **14. pinnatifidum**
19 Perennial; petals equalling or longer than the sepals
20 Cauline leaves linear to linear-spathulate or pinnatisect, with narrow lobes
21 Basal leaves pinnatisect, linear to linear-lanceolate, with narrow lobes; petals about twice as long as the sepals; silicula *c.* 2 × 1–1·25 mm **18. lyratum**
21 Basal leaves oblanceolate to oblong, dentate or pinnately lobed with broad lobes; petals equalling or up to $1\frac{1}{2}$ times as long as the sepals; silicula 2·5–4 × 1·5–3 mm **19. graminifolium**
20 Cauline leaves ovate, elliptical or lanceolate, entire
22 Stem papillose; ripe silicula distinctly reticulate-alveolate **16. cartilagineum**
22 Stem glabrous; ripe silicula smooth or faintly reticulate **17. latifolium**

Sect. LEPIA (Desv.) DC. Middle and upper cauline leaves amplexicaul. Silicula broadly winged above, the wing connate with the lower part of the style. Stamens 6.

1. L. campestre (L.) R. Br. in Aiton, *Hort. Kew.* ed. 2, **4**: 88 (1812). Annual or biennial 20–60 cm, densely and shortly hairy. Basal leaves petiolate, ovate to obovate, obtuse, entire to slightly lobed; upper cauline leaves numerous, ovate to oblong, more or less dentate, with narrow, acute auricles. Sepals *c.* 1·5 mm; petals 1·5–2·6 mm; anthers yellow. Pedicels 5–6 mm in fruit, shortly hairy. Silicula 4–6·5 × 3·5–5·5 mm, sparsely to densely covered with small, scale-like vesicles, otherwise glabrous; style 0·1–0·7 mm, not or only slightly projecting beyond the apical notch. $2n = 16$. *Throughout Europe except for a few islands and much of the north-east.* All except Az Bl Cr Fa Rs(N) Sa Sb; only as a casual in Is.

2. L. villarsii Gren. & Godron, *Fl. Fr.* **1**: 150 (1847). Perennial 15–45 cm, grey-green, sparsely hairy. Basal leaves long-petiolate, obovate or broadly elliptical, entire; middle and upper cauline leaves numerous, narrowly triangular. Sepals 2–2·5 mm; petals almost twice as long as the sepals; anthers violet. Pedicels 5–7 mm in fruit, glabrous. Silicula 4–7 × 2·5–5 mm, winged for about $\frac{1}{4}$ of its total length, usually glabrous; style at least 1 mm, projecting beyond the apical notch. ● *French Alps and mountains of N.E. Spain.* Ga Hs.

(a) Subsp. **villarsii** (*L. pratense* (Serres ex Gren. & Godron) Rouy & Fouc.): Apex of silicula broad, distinctly notched even when young. *Mountain grassland. French Alps.*

(b) Subsp. **reverchonii** (Debeaux) Breistr., *Bull. Soc. Sci. Dauph.* **61**: 640 (1947) (*L. reverchonii* Debeaux): Apex of silicula gradually narrowed into the style when young, rarely with a slight notch when mature. *Mountains of* E. & N.E. *Spain.*

Plants with leaves glabrous or subglabrous, pedicels 5–6 mm in fruit, and silicula *c.* 7 × 6 mm, from S.E. Spain, have been called **L. ramburei** Boiss., *Voy. Bot. Midi Esp.* **2**: 52 (1939), and may represent a third subspecies.

3. L. heterophyllum Bentham, *Cat. Pl. Pyr. Bas-Languedoc* 95

(1826) (*L. smithii* Hooker). Perennial herb up to 50 cm, shortly hairy. Basal leaves petiolate, oblanceolate or elliptical, entire; upper cauline leaves sessile, narrowly triangular-ovate, with acute auricles; all shortly hairy to glabrous. Sepals *c.* 2 mm; petals $1\frac{1}{2}$–$3\frac{1}{2}$ times as long as the sepals; anthers violet. Pedicels 2–6 mm in fruit, hairy. Silicula 4–7 × 3–6 mm, ovate, usually glabrous, with few or no vesicles, winged for about $\frac{1}{3}$ of the total length, usually with a small apical notch, rarely entire; style 0·4–1 mm, projecting beyond the notch. $2n = 16$. *Dry, more or less open habitats; somewhat calcifuge.* ● *W. Europe, extending eastwards to Czechoslovakia.* Br Cz Da Ga Hb Hs Lu [Be Ge Ho No Su].

4. L. hirtum (L.) Sm., *Comp. Fl. Brit.* ed. 3, 98 (1818). Like **3** but usually strongly grey-pubescent; anthers yellow (blackish-red in subsp. (**d**)); silicula hairy, at least when young; wings up to $\frac{1}{2}$ of the total length of the silicula. *Dry places. Mediterranean region.* Co Cr Ga Gr Hs It Si.

1 Anthers blackish-red; pedicels not more than 3 mm in fruit (**d**) subsp. **stylatum**
1 Anthers yellow; pedicels more than 3 mm in fruit
2 Silicula less than 5 mm, very narrowly winged (**e**) subsp. **oxyotum**
2 Silicula 5–8 mm, ± distinctly winged
3 Style 1·5–2 mm, much exceeding the notched apex of the silicula; sepals 2–2·5 mm
4 Silicula abruptly contracted towards the apex; raceme short (**c**) subsp. **petrophilum**
4 Silicula gradually narrowed towards the apex; raceme long (**f**) subsp. **calycotrichum**
3 Style not more than 1·5 mm and only just exceeding the notched apex of the silicula; sepals 1·5–2 mm
5 Stems usually ascending; raceme short (**a**) subsp. **hirtum**
5 Stems procumbent; racemes long (**b**) subsp. **nebrodense**

(**a**) Subsp. **hirtum** (*L. scapiflorum* (Viv.) P. Fourn.): Stems usually ascending, very hairy. Raceme short. Sepals 1·5–2 mm; anthers yellow. Pedicels 5–6 mm in fruit. Silicula 6–7 × 4–4·6 mm, ovate, winged for about $\frac{1}{2}$ of its total length; style projecting just beyond the notched or truncate apex. *Almost throughout the range of the species.*

(**b**) Subsp. **nebrodense** (Rafin.) Thell., *Viert. Naturf. Ges. Zürich* **51**: 154 (1906) (*L. microstylum* Boiss. & Heldr., *L. nebrodense* (Rafin.) Guss.): Stems procumbent, glabrous to grey-pubescent. Raceme long. Sepals *c.* 1·5 mm; anthers yellow. Pedicels 4–6 mm in fruit. Silicula 6–8 × 4–5 mm, elliptical, winged for about $\frac{1}{3}$ of its total length, the wings rounded or obtuse; style short, just exceeding the notched apex. ● *C. & E. Mediterranean region.*

(**c**) Subsp. **petrophilum** (Cosson) Thell., *loc. cit.* (1906) (*L. petrophilum* Cosson): Stems diffuse or ascending, hairy. Raceme short. Sepals 2–2·5 mm; anthers yellow. Pedicels 4–5 mm in fruit. Silicula 5–6 × 3–4 mm, ovate, acuminate, sharply keeled and winged above the middle; style *c.* 2 mm, exceeding the notched apex. ● *S. Spain (Granada prov.)*

(**d**) Subsp. **stylatum** (Lag. & Rodr.) Thell., *op. cit.* 155 (1906) (*L. stylatum* Lag. & Rodr.): Stems ascending, grey-pubescent (or rarely green and glabrous). Raceme very short, often subcorymbose. Sepals *c.* 1·5 mm; anthers blackish-red. Pedicels 2–3 mm in fruit. Silicula 4–6 × 2–2·5 mm, narrowly elliptical, narrowly winged; style *c.* 2 mm, much exceeding the slightly notched apex. ● *S. Spain (Sierra Nevada).*

(**e**) Subsp. **oxyotum** (DC.) Thell., *op. cit.* 156 (1906) (*L. humifusum* (Loisel.) Req.): Stems ascending, hairy. Raceme short. Sepals 1·5–2 mm; anthers yellow. Pedicels 3–4 mm in fruit. Silicula *c.* 4·5 × 3·5 mm, orbicular-elliptical or obovate, very narrowly winged; style short, just exceeding the notched apex. ● *Corse; Kriti.*

(**f**) Subsp. **calycotrichum** (G. Kunze) Thell., *loc. cit.* (1906) (*L. calycotrichum* G. Kunze): Stems procumbent or ascending, hairy. Raceme very long. Sepals *c.* 2·5 mm; anthers yellow. Pedicels 4–6 mm in fruit. Silicula *c.* 7 × 4 mm, narrowly elliptical or ovate, winged for $\frac{1}{5}$–$\frac{1}{2}$ of its total length; style 1·5–2 mm, much exceeding the slightly notched apex. *S. & E. Spain.*

Sect. LEPIOCARDAMON Thell. Cauline leaves not amplexicaul. Silicula obovate, very deeply notched, broadly winged above. Stamens 6.

5. L. spinosum Ard., *Animadv. Bot. Spec. Alt.* 34 (1763) (*L. carrerassii* Rodr., *L. cornutum* Sm.). Glabrous annual or biennial. Basal and lower cauline leaves pinnately divided, the lobes linear and lobed again at the base; middle and upper cauline leaves linear or linear-oblanceolate. Sepals 1–1·5 mm, white-margined; petals $1\frac{1}{2}$ times as long as the sepals. Branches of inflorescence spiny in fruit. Pedicels 1–2 mm in fruit. Silicula 5–6 × *c.* 3 mm, obovate, very deeply notched at the apex, glabrous, the wings very wide above; style not exceeding notch. *Dry grassland. E. Mediterranean region; E. Bulgaria.* Bu Cr Gr Tu [Bl Hs].

Sect. CARDAMON DC. Cauline leaves not amplexicaul. Silicula broadly emarginate, winged above. Stamens (4–)6.

6. L. sativum L., *Sp. Pl.* 644 (1753). Usually glabrous, often glaucous annual with a single, erect stem 20–100 cm. Basal leaves long-petiolate, lyrate, with toothed obovate to linear lobes; upper cauline leaves linear, entire. Sepals 1–1·8 mm; petals up to twice as long as the sepals, sometimes pink or reddish. Pedicels (2–)3–6 mm in fruit. Silicula 5–6·5 × 3–6 mm, ovate or broadly elliptical to orbicular, glabrous; style 0·2–0·8 mm, included in the deep apical notch. *Cultivated throughout Europe as a salad (Cress); frequent as a casual and locally naturalized. (Egypt and W. Asia.)*

It is impossible from the available information to distinguish between naturalized and casual distribution.

Sect. DILEPTIUM DC. Cauline leaves not amplexicaul. Silicula not winged, or, if narrowly winged above, then with the style quite free and not exceeding the apical notch. Stamens 2–4.

7. L. subulatum L., *Sp. Pl.* 644 (1753) (*L. lineare* DC.). Dwarf shrub 10–20 cm, more or less glaucous and much-branched. All leaves linear, rigid, channelled and very acute, with varying numbers of short, stiff hairs. Sepals *c.* 1 mm; petals *c.* $1\frac{1}{2}$ times as long as the sepals; anthers yellow. Pedicels 2–4 mm in fruit, erecto-patent, glabrous or hairy. Silicula 2–2·5 × 1·5–2 mm; style not exceeding the apical notch. *Gypsaceous and saline soils. Spain.* Hs.

Plants from S., C. & E. Spain tend to be more hairy and glaucous and have smaller fruits and longer pedicels than those from elsewhere.

8. L. cardamines L., *Cent. Pl.* **1**: 17 (1755). Glaucous biennial; stem erect, branched, densely puberulent with short, crispate hairs. Basal leaves lyrate-pinnatifid; cauline leaves numerous, pinnate with large terminal and 1 or 2 lateral lobes. Sepals *c.* 1 mm; petals twice as long as the sepals. Pedicels *c.* 4 mm in fruit, minutely hairy.

Silicula *c.* 2–2·3 × 1·5–2 mm, ovate, glabrous; style very short, scarcely exceeding the small apical notch. *Gypsaceous and saline soils.* ● *C. Spain.* Hs.

9. L. virginicum L., *Sp. Pl.* 645 (1753). Annual or biennial, usually branched above; stem 30–50 cm, erect; papillae curved or appressed. Basal leaves up to 8 cm, rough with short bristles, lyrate; middle and upper cauline leaves sharply toothed to entire, ciliate. Sepals 0·6–1 mm; petals longer than the sepals, or absent. Pedicels 2·5–5 mm in fruit, papillose on the upper side, glabrous beneath. Silicula 2–4 × 2–4 mm, more or less orbicular, narrowly winged above, the apical notch broad but shallow; style 0·1–0·2 mm, not exceeding the notch. *Widely naturalized in Europe.* [Al Az Be Cz Da Fe Ga Ge He Hs Hu It Ju Lu No Po Rm Su.] (*North America.*)

10. L. densiflorum Schrader, *Ind. Sem. Horti Gotting.* 4 (1832) (*L. apetalum* sensu N. Busch in Komarov, non Willd.; incl. *L. neglectum* Thell., *L. virginicum* subsp. *neglectum* (Thell.) P. Fourn.). Annual or biennial; stem single, erect, branched above; papillae straight. Basal leaves long-petiolate, elliptical, usually deeply toothed; upper cauline leaves linear to linear-lanceolate, entire or remotely toothed, ciliate. Sepals *c.* 1 mm; petals filiform, shorter than the sepals or absent. Pedicels 3–5·5 mm in fruit, papillose on the upper side, glabrous beneath. Silicula 2·5–4 × 2–3 mm, ovate to obovate, narrowly winged in the upper $\frac{1}{3}$, narrowly notched; style very short. *Widely naturalized in Europe.* [Au Be Cz Da Fe Ga Ge He Hu It No Po Rs(C, W, E) Si Su.] (*North America.*)

11. L. bonariense L., *Sp. Pl.* 645 (1753). Unbranched annual or biennial; stem with long, unbranched hairs. All leaves 1- or 2-pinnate and more or less hairy. Sepals 0·5–0·9 mm; petals shorter than the sepals. Pedicels (2–)3–4·5 mm in fruit, with long papillae on the upper side. Silicula 2–3·5(–4) × 2–3 mm, ovate or orbicular, narrowly winged above, with narrow, shallow apical notch. *Naturalized in W. & C. Europe.* [Be Ge Ho Hs Hu.] (*S.E. South America.*)

12. L. ruderale L., *Sp. Pl.* 645 (1753) (*L. ambiguum* Lange). Annual or biennial, more or less foetid, with a single, erect or ascending stem 10–30 cm, minutely papillose. Basal leaves 5–8 cm, long-petiolate, 2- or 3-pinnatisect; upper cauline leaves linear, obtuse, entire. Sepals 0·7–0·9 mm; petals 0 or very inconspicuous. Pedicels 2–4 mm in fruit, papillose all round. Silicula (1·5–)2–2·5 × 1·5–2 mm, ovate or elliptical, almost wingless, deeply notched; style *c.* 0·1 mm and not exceeding notch. $2n = 16, 28, 32$. *Throughout most of Europe but perhaps native only in the south.* All except Az Co Hb ?Sa Sb Si.

13. L. africanum (Burm. fil.) DC., *Reg. Veg. Syst. Nat.* **2**: 552 (1821). Annual; stem papillose above. Basal leaves oblanceolate, subpinnate; upper cauline leaves lanceolate to oblanceolate, acute, dentate. Sepals 0·5–0·75 mm; petals $\frac{1}{2}$–$\frac{2}{3}$ as long as the sepals. Pedicels 2–4 mm in fruit, erecto-patent, papillose above; petals absent or up to 0·5 mm. Silicula 2–3 × 1·5–2 mm, ovate to elliptical, the apical notch shallow and narrow; style very short. *Naturalized in W. Europe.* [Be Ga Ge He Ho.] (*E.C. & S. Africa.*)

In Europe only subsp. *africanum* occurs. Subsp. **divaricatum** (Solander) Jonsell, *Bot. Not.* **128**: 41 (1975) (*L. divaricatum* Solander), from S. Africa, is perhaps present as a casual.

L. schinzii Thell., *Viert. Naturf. Ges. Zürich* **51**: 167 (1906), like **13** but the stem with scale-like trichomes, lower cauline leaves pinnatifid, with entire or rarely toothed segments, and pedicels erect, glabrous and appressed to the stem, has been reported to be naturalized in Europe but is probably only casual.

14. L. pinnatifidum Ledeb., *Fl. Ross.* **1**: 206 (1841). Like **13** but sometimes biennial; stem glabrous; cauline leaves almost entire, oblanceolate-spathulate; silicula 2–2·75 × 1·75–2·3 mm, broadly elliptical to suborbicular, with an entire to slightly notched apex. *Saline places. S.E. Russia.* Rs(E).

Sect. LEPIDIUM. Silicula not or very slightly winged, not or slightly notched. Stamens 6.

15. L. perfoliatum L., *Sp. Pl.* 643 (1753). Annual or biennial with single, erect, sparsely hairy stem 20–40 cm. Leaves strongly dimorphic, the basal long-petiolate, 2- or 3-pinnatifid, with linear segments; upper cauline leaves ovate to subrotund, more or less acute, entire, the broad, rounded lobes completely enclosing the stem and overlapping. Sepals *c.* 1 mm; petals pale yellow, narrowly spathulate, only a little longer than the sepals. Pedicels 4–5 mm in fruit, ascending, glabrous. Silicula usually *c.* 4 × 4 mm, very narrowly winged above; style short, equalling or exserted from the small apical notch. $2n = 16$. *Native in C., E. & S.E. Europe; naturalized or casual elsewhere.* Al Au Bu Cz Gr Hu Ju Po Rm Rs(C, W, K, E) Tu [Be Da Ga Ge He Hs].

16. L. cartilagineum (J. Mayer) Thell., *Viert. Naturf. Ges. Zürich* **51**: 173 (1906). Perennial; stems 5–30 cm, very flexuous, branched above, papillose. Leaves leathery-succulent, broadly elliptical, ovate, lanceolate or linear-lanceolate, more or less obtuse, entire, cuneate at the base; cauline leaves sometimes with a sagittate base. Sepals *c.* 1·25 mm; petals about twice as long as the sepals. Pedicels *c.* 4 mm, thickened in fruit. Silicula 2·5–3 × 2·5–3 mm, ovate, distinctly reticulate-alveolate when ripe; style short, exserted. *Saline areas of C. & E. Europe.* Au ?Cz Ge Hu Ju Po Rm Rs(C, W, K, E).

(a) Subsp. **cartilagineum** (incl. *L. crassifolium* Waldst. & Kit., *L. pumilum* Boiss. & Balansa): Plant usually 20–30 cm, branched. Leaves broadly elliptical or ovate, the cauline with distinct auricles at the base. $2n = 16$. *Throughout the range of the species.*

(b) Subsp. **pumilum** (Boiss. & Balansa) Hedge in Rech. fil., *Fl. Iran.* **57**: 67 (1968): Plant 5–15 cm, little branched. Leaves lanceolate or linear-lanceolate, the cauline without or with minute auricles at the base. *Jugoslavia; Ukraine.*

L. borysthenicum Kleopow, *Jour. Inst. Bot. Acad. Sci. Ukr.* **21–22**: 251 (1939) (from C. & E. Ukraine), with stems 30–40 cm, basal leaves 10–16 × 0·3–0·6 cm, and **L. syvaschicum** Kleopow, *loc. cit.* (1939) (from S. Ukraine), with stems up to 25 cm, basal leaves 4–8 × 0·5–1·3 cm, oblong-elliptical, may represent further subspecies of **16**.

17. L. latifolium L., *Sp. Pl.* 644 (1753). Perennial with creeping rhizome; stems 50–130 cm, erect, glabrous, much-branched above. Leaves coriaceous, the basal and lower cauline up to 30 cm, long-petiolate, ovate, toothed or sometimes pinnatilobed, glabrous or sparsely hairy; upper leaves sessile, ovate to lanceolate, entire. Flowers in a large, dense, pyramidal panicle. Sepals 0·9–1·5 mm, broadly white-margined; petals 1·8–2·5 mm, up to twice as long as the sepals, with rounded-obovate limb. Pedicels 4–5 mm in fruit. Silicula 1·5–2·5 × *c.* 2 mm, elliptical to orbicular, not winged or notched; style *c.* 0·2 mm, with a capitate stigma. $2n = 24$. *Most of Europe but rare in the north.* All except Bl Cr Fa Fe Is No Rs(N) Sb.

Plants from near the Baltic in Sweden seem to differ in having dark purple sepals and narrower, more toothed upper cauline leaves with a longer, acute apex. Specimens from Greece and Sicilia have the underside of the leaves densely covered with stellate hairs and are apparently referable to var. *velutinum* Hayek ex Thell. Further study may indicate that both these variants should be regarded as subspecies.

18. L. lyratum L., *Sp. Pl.* 644 (1753). Perennial up to 30 cm, sometimes woody at the base; stems very flexuous, glabrous. Leaves linear to linear-lanceolate, pinnate or pinnatisect, the upper cauline sometimes entire and bract-like. Sepals *c.* 1 mm; petals about twice as long as the sepals. Silicula *c.* 2 × 1–1·5 mm, ovate, acute. *S.E. Russia, Krym.* Rs(K, E).

(a) Subsp. **lacerum** (C. A. Meyer) Thell., *Neue Denkschr. Schweiz. Ges. Naturw.* **41**: 166 (1906) (incl. *L. turczaninovi* Lipsky, *L. meyeri* Claus): Plant woody at the base. Basal leaves pinnatipartite, the lobes remote, narrow; cauline leaves small, often bract-like, linear and entire, or the lower with a few linear, obtuse pinnae. Style $\frac{1}{4}$–$\frac{2}{3}$ as long as the ovary when in flower, $\frac{1}{8}$–$\frac{1}{4}$ as long as the silicula. ● *Throughout the range of the species.*

(b) Subsp. **coronopifolium** (Fischer) Thell., *op. cit.* 167 (1906) (*L. coronopifolium* Fischer): Plant not woody at the base. Basal and cauline leaves linear to narrowly linear-lanceolate, 1-pinnatisect, with a long, narrow terminal lobe and 2 to many narrow lateral lobes, long-petiolate; numerous linear bract-like leaves on the branches. Style $\frac{1}{2}$–$\frac{2}{3}$ as long as the ovary when in flower, *c.* $\frac{1}{2}$ as long as the silicula. *S.E. Russia.*

Subsp. *lyratum* occurs in the Caucasus and S.W. Asia.

19. L. graminifolium L., *Syst. Nat.* ed. 10, **2**: 1127 (1759). Perennial with erect glabrous or sparsely hairy stems up to 50 cm, branched above. Basal leaves up to 10 cm, long-petiolate, oblanceolate to oblong, toothed or pinnately lobed; upper cauline leaves linear or linear-spathulate. Sepals 0·5–1·4 mm, narrowly white-margined above the middle; petals up to $1\frac{1}{2}$ times as long as the sepals, with obovate-spathulate limb. Pedicels 2·5–4 mm in fruit. Silicula 2·5–4 × 1·5–3 mm, acute to acuminate, not notched, not or scarcely winged; style *c.* 0·2 mm, exceeding the notch. *S. & C. Europe, extending northwards to the Netherlands.* Al Au Bl Bu Co Cz Ga Ge Gr He Ho Hs Hu It Ju Rm Rs(K) Sa Si Tu [Be Br].

(a) Subsp. **graminifolium** (*L. iberis* L.): Perennial, sometimes woody; almost or quite glabrous. Basal leaves incise-dentate to pinnately lobed, long-petiolate. Sepals greenish. $2n = 48$. *Throughout the range of the species.*

(b) Subsp. **suffruticosum** (L.) P. Monts., *Feddes Repert.* **69**: 6 (1964) (*L. suffruticosum* L.): Perennial with woody rhizome protruding above the surface of the ground; stem, branches and pedicels minutely hairy. Basal leaves very long-petiolate, with a few teeth near the apex. Sepals usually purplish. *W. Mediterranean region.*

81. Cardaria Desv.

P. W. BALL

Like *Lepidium* but the inflorescence a dense corymbose panicle; silicula cordate, indehiscent.

Literature: G. A. Mulligan & C. Frankton, *Canad. J. Bot.* **40**: 1411–1426 (1962).

1. C. draba (L.) Desv., *Jour. Bot. Appl.* **3**: 163 (1814) (*Lepidium draba* L.) Glabrous to pubescent perennial 15–60(–90) cm, producing adventitious buds on the roots. Leaves obovate to ovate-oblong, sinuate-dentate, the basal cuneate, petiolate, the cauline sessile, amplexicaul. Petals 2·5–4 mm, white. Silicula 3–4·5 × 3·5–5 mm, emarginate, inflated. $2n = 62, 64$. *Cultivated and waste ground. Probably native in S. Europe, but established throughout most of Europe as a weed.* Al *Au Bl Bu Co Cr Ga Gr Hs *Hu It Ju *Lu Rm Rs(W, K, E) Sa Si Tu [Be Br Cz Da Fe Ge Hb He Ho No Po Rs(B, C) Su].

The common plant in Europe is subsp. *draba*. Subsp. **chalepensis** (L.) O. E. Schulz in Engler & Prantl, *Natürl. Pflanzenfam.* ed. 2, **17b**: 417 (1936) (*C. chalepensis* (L.) Hand.-Mazz., *Lepidium chalepense* L., *L. draba* subsp. *chalepense* (L.) Thell.), from S.W. Asia, with the fruit more or less cuneate at the base, occurs as a casual.

82. Coronopus Haller

P. W. BALL

Glabrous or with unbranched hairs. Raceme leaf-opposed. Sepals patent; petals white, small or absent; stamens 2–6. Fruit an angustiseptate silicula, indehiscent or breaking in 2 halves; valves subglobose, verrucose or reticulate. Seeds 2. (*Senebiera* DC.)

1 Silicula *c.* 1·5 × 2–3 mm, emarginate **3. didymus**
1 Silicula 2–3 × 3–4·5 mm, rounded or apiculate
 2 Perennial; pedicels at least 2·5 mm, longer than the silicula **1. navasii**
 2 Annual or biennial; pedicels not more than 2 mm, shorter than the silicula **2. squamatus**

1. C. navasii Pau, *Butll. Inst. Catalana Hist. Nat.* **22**: 31 (1922). Diffuse perennial up to 30 cm. Leaves ovate or lanceolate, pinnatifid. Inflorescence elongate in fruit. Pedicels at least 2·5 mm, longer than the silicula. Petals *c.* 2 mm. Silicula *c.* 2·5 × 3 mm, ovoid. ● *S. Spain (Sierra de Gádor).* Hs.

Possibly not distinct from **C. violaceus** (Munby) O. Kuntze, *Revis. Gen.* **1**: 27 (1891), from N.W. Africa.

2. C. squamatus (Forskål) Ascherson, *Fl. Brandenb.* **1**: 62 (1860) (*C. procumbens* Gilib., *C. ruellii* All., *Senebiera coronopus* (L.) Poiret). Procumbent annual or biennial up to 40 cm. Lower leaves pinnatipartite, the segments usually pinnatifid. Raceme usually crowded in fruit. Pedicels up to 2 mm. Petals 1–2 mm, longer than the sepals; fertile stamens 6. Silicula 2–3 × 3–4·5 mm, almost reniform, apiculate, strongly reticulate or ridged, or verrucose; style 0·3–0·5 mm. *Ruderal. W., C. & S. Europe, northwards to c. 66° N. in Fennoscandia; status uncertain in the northern part of its range.* All except Bl Fa Is Rs(N, C, E) Sb.

3. C. didymus (L.) Sm., *Fl. Brit.* **2**: 691 (1800) (*C. pinnatifida* DC., *Senebiera didyma* (L.) Pers.). Procumbent or ascending annual or biennial up to 40 cm. Lower leaves pinnatipartite, the segments usually pinnatifid. Raceme somewhat elongate in fruit. Pedicels 1·5–3·5 mm, usually longer than the silicula. Petals *c.* 0·5 mm, shorter than the sepals, sometimes absent; fertile stamens usually 2. Silicula 1·3–1·7 × 2–3 mm, emarginate, reticulate; style absent, or up to 0·2 mm and included in notch. $2n = 32$. *Ruderal. Naturalized in W., C. & S. Europe; a common casual elsewhere.* [Al Au Az Be Bl Br Co Cz Ga Ge Gr Hb He Ho Hs Hu It Lu Po Rm Sa Si.] (*?South America.*)

83. Subularia L.

†T. G. TUTIN

Glabrous, scapigerous annual. Sepals erect; petals white, sometimes absent; stamens 6, without appendages; ovary surrounded by a fleshy ring. Fruit a silicula; valves strongly convex. Seeds in 2 rows, 2–6 in each loculus.

Literature: G. A. Mulligan & J. A. Calder, *Rhodora* **66**: 127–135 (1964).

1. S. aquatica L., *Sp. Pl.* 642 (1753). Leaves 2–7 cm, numerous, rosulate, subulate, terete, entire. Flowers 2–12. Sepals caducous. Pedicels eventually 2–5 mm. Silicula 2–5 × 1·5–2·5 mm, elliptic-oblong. $2n = 28 + 0\text{–}4B$, *c.* 36. *In shallow water in base-poor pools and lakes. N. Europe, extending locally southwards in the mountains to the Pyrenees and Bulgaria.* Be Br Bu Da Fa Fe Ga Ge Hb †Ho Hs Is No Rs(N, B, C, E) Su.

84. Conringia Adanson

P. W. BALL

Glabrous annuals. Leaves simple, glaucous. Inflorescence a raceme. Inner sepals saccate; petals pale yellow to greenish-white. Fruit a siliqua; valves 1- or 3-veined; style short; stigma slightly 2-lobed. Seeds in 1 row in each loculus.

Literature: O. E. Schulz in Engler, *Pflanzenreich* **84(IV.105)**: 84–94 (1923).

1 Siliqua strongly compressed, torulose; petals with violet veins **3. planisiliqua**
1 Siliqua 4- or 8-angled, not torulose; petals without violet veins
2 Valves of the siliqua 1-veined; petals (7–)8–14 mm **1. orientalis**
2 Valves of the siliqua 3-veined; petals 6–8(–10) mm **2. austriaca**

1. C. orientalis (L.) Dumort., *Fl. Belg.* 123 (1827) (*Erysimum orientale* (L.) R. Br.). Plant 10–60(–80) cm. Basal leaves obovate, shortly petiolate; cauline leaves obovate to elliptical, sessile, cordate-amplexicaul. Pedicels 4–9 mm in flower, 6–20 mm in fruit. Petals (7–)8–14 mm, yellowish- or greenish-white. Siliqua (4·5–)6–14 cm, strongly 4-angled, the valves 1-veined; style 0·8–3·5 mm, cylindrical. Seeds 2–3 mm. $2n = 14$. *C. & E. Europe; frequently naturalized or casual elsewhere.* Al Au Bu Cz Ge Gr Hu It Ju Po Rm Rs(C, W, K, E) ?Si Tu [Bl Cz Da Fe Ga He Ho Hs Is No Rs(N) Sb].

2. C. austriaca (Jacq.) Sweet, *Hort. Brit.* 25 (1826) (*Goniolobium austriacum* (Jacq.) G. Beck). Like **1** but up to 100 cm; cauline leaves obovate-oblong; pedicels 2–5 mm in flower, 2–9 mm in fruit; petals 6–8(–10) mm, lemon-yellow; siliqua 5–8(–10) cm, 8-angled, the valves 3-veined. $2n = 28$. *S.E. & E.C. Europe, extending westwards to Austria and Sardegna.* Au Bu Gr Hu It Ju Rm Rs(K) Sa.

3. C. planisiliqua Fischer & C. A. Meyer, *Ind. Sem. Hort. Petrop.* **3**: 32 (1837). Plant 15–50 cm. Basal leaves oblong, shortly petiolate; cauline leaves oblong-ovate to -elliptical, cordate-amplexicaul, obtuse, mucronate. Pedicels 5–15 mm in fruit. Petals 6–8 mm, pale yellow with violet veins. Siliqua 4–9 cm, torulose, strongly compressed, the valves without veins; style 1–2 mm. Seeds 1–2 mm. *N.W. Greece (lower slopes of Timfi Oros).* *Gr. (*S.W. Asia.*)

85. Moricandia DC.

V. H. HEYWOOD

Glabrous annuals or perennials. Leaves simple, fleshy. Inflorescence corymbose. Inner sepals saccate at base; petals purple, rarely whitish. Fruit a siliqua; valves with a distinct mid-vein; style short; stigma 2-lobed. Seeds in 1 or 2 rows in each loculus.

Literature: O. E. Schulz in Engler, *Pflanzenreich* **84(IV.105)**: 64–72 (1923). A. de Bolòs, *Anal. Inst. Bot. Cavanilles* **6(2)**: 451–461 (1946).

1 Upper cauline leaves minute, triangular-ovate, very acute to acuminate; petals 12 mm, whitish **3. foetida**
1 Upper cauline leaves cordate, ±acute; petals 18–22 mm, usually purple-violet
2 Outer sepals markedly horned at the apex; raceme 20- to 40-flowered; seeds up to 2·5 mm, uniseriate **2. moricandioides**
2 Outer sepals scarcely horned at the apex; raceme 10- to 20(–25)-flowered; seeds *c.* 1 mm, biseriate **1. arvensis**

1. M. arvensis (L.) DC., *Reg. Veg. Syst. Nat.* **2**: 626 (1821) (*Brassica arvensis* L.). Short-lived perennial up to 65 cm, with branched stems. Lower leaves obovate, repand-crenate, obtuse at the apex, narrowed at the base; upper cauline leaves cordate, entire, widened and amplexicaul at the base, more or less acute. Raceme with 10–20 large, showy flowers, becoming lax. Petals *c.* 20 mm, violet-purple. Siliqua 30–80 × 2–3 mm, compressed, 4-angled. Seeds *c.* 1 mm, biseriate, brown. $2n = 28$. *Calcicole. Mediterranean region.* Bl Co Gr Hs It Si.

M. longirostris Pomel, *Nouv. Mat. Fl. Atl.* 367 (1875), recorded from S. Italy and Sicilia, which differs in its longer siliquae (up to 12 cm), with longer, often recurved beaks and very numerous seeds, may deserve recognition as a subspecies.

2. M. moricandioides (Boiss.) Heywood, *Feddes Repert.* **66**: 154 (1962) (*Moricandia ramburei* Webb, *Brassica moricandioides* Boiss.). Like **1** but raceme with more numerous flowers; outer sepals with a pronounced cucullate prolongation up to 1·2 mm; siliqua 80–130 mm, more or less terete; seeds 2–2·5 mm (1–1·5 mm in var. *microsperma*), uniseriate, distinctly winged, the wing narrow or broad. ● *S.C. & E. Spain.* Hs.

Two fairly well-marked variants sometimes cause difficulty in identifying this species: var. *microsperma* (Willk.) Heywood has more shortly cucullate sepals and seeds less than 2 mm, and var. *cavanillesiana* (Font Quer & A. Bolòs) Heywood has seeds broadly crispate-winged.

3. M. foetida Bourgeau ex Cosson, *Not. Pl. Crit.* 143 (1852). Stems much branched from the base. Lower leaves obovate; upper cauline leaves minute, triangular-ovate, very acute to acuminate. Raceme 5- to 12-flowered; sepals all obtuse; petals 12 mm, whitish. Siliqua 40–60 × 1·5–2 mm, compressed, 4-angled, long-pedicellate. Seeds *c.* 1·5 mm, brownish, uniseriate, distinctly white-margined. *Base-rich soils.* ● *S. & S.E. Spain.* Hs.

Plants combining the characters of **2** and **3** are found in the region of Murcia.

86. Euzomodendron Cosson

V. H. HEYWOOD

Sepals erect, hispid; petals very long-clawed; filaments of the inner stamens united in pairs. Fruit a linear-lanceolate siliqua, dorsally compressed; valves more or less convex, with 3–5 equidistant, parallel veins, attenuate into a triangular to lanceolate, seedless beak.

1. **E. bourgaeanum** Cosson, *Not. Pl. Crit.* 145 (1852). Small, erect shrub 20–50 cm, much branched at the base. Leaves crowded, pinnatisect into 2 or 3 pairs of linear, obtuse, somewhat fleshy segments, rarely undivided, with scattered, patent, setiform hairs. Flowers in ebracteate racemes on short pedicels, showy. Petals 12–16 mm, whitish, with brown veins. Siliqua 20–40 × 3–4 mm, ascending, glabrous; beak up to 10 mm. *Dry, calcareous soils.* ● *S. Spain* (*Almería prov.*). Hs.

87. Diplotaxis DC.

V. H. HEYWOOD (EDITION 1) REVISED BY J. R. AKEROYD AND J. B. MARTINEZ-LABORDE (EDITION 2)

Annuals to perennials. Leaves entire to pinnatipartite. Sepals erecto-patent; petals clawed, usually yellow. Gynophore usually short. Fruit an elongate, linear siliqua with a short beak; valves compressed, with a prominent median vein. Seeds usually in 2 rows in each loculus, ovoid or ellipsoidal. (Incl. *Pendulina* Willk.)

Literature: O. E. Schulz in Engler, *Pflanzenreich* **70(IV.105)**: 149–180 (1919). R. Nègre, *Mém. Soc. Sci. Nat. Phys. Maroc Bot.* nov. ser., no. 1 (1960). R. B. Fernandes, *Bol. Soc. Brot.* ser. 2, **58**: 235–248 (1985).

1 Mature siliqua pendent; gynophore usually at least 2 mm, conspicuous **5. harra**
1 Mature siliqua erecto-patent or rarely patent; gynophore usually less than 2 mm
2 Petals white, violet-veined or violet after anthesis **8. erucoides**
2 Petals yellow, sometimes violet-veined
3 Petals 3–4 mm; outer stamens sterile **10. viminea**
3 Petals 4–15 mm; outer stamens fertile
4 Leaves foetid when crushed; petal venation pinnate and looped at margin; beak of siliqua seedless
5 Plant scapose or subscapose **11. muralis**
5 Stems and branches leafy
6 Perennial; leaves entire to pinnatisect **6. tenuifolia**
6 Annual or biennial; leaves pinnatisect to bipinnatisect **7. cretacea**
4 Leaves not foetid; petal venation not pinnate and looped at margin; beak of siliqua with 1 or 2 seeds or dried ovules, more rarely seedless
7 Petals pale yellow; seeds ± spherical **3. siifolia**
7 Petals sulphur-yellow; seeds ellipsoidal
8 Seeds in (2)3 or 4 rows in each loculus; leaves with short, thin, patent hairs on adaxial surface, glabrous on the abaxial **2. siettiana**
8 Seeds in 1 or 2 rows in each loculus; leaf pubescence not as above or none
9 Short-lived perennial; beak of siliqua usually without seeds **4. ibicensis**
9 Annual; leaves more or less pubescent; beak of siliqua usually with 1 or 2 seeds
10 Stems and leaves conspicuously hirsute; leaves pinnatilobed to pinnatisect **9. virgata**
10 Stems and leaves with sparse, short, conical hairs; leaves pinnatisect to bipinnatisect, the upper ones with linear segments **1. catholica**

1. **D. catholica** (L.) DC., *Reg. Veg. Syst. Nat.* **2**: 632 (1821) (*Hugueninia balearica* (Porta) O. E. Schulz). Annual, sometimes overwintering, 5–90 cm, green or subglaucous. Stems glabrous or with deflexed hairs at the base. Leaves glabrous or sparsely hispid at the margin; basal leaves (including juvenile) usually 1- or 2-pinnatisect, the segments with acute lobes. Petals 7–8(–12) mm, sulphur-yellow. Siliqua (14–)20–25(–45) × 1·5–2 mm, torulose or not, seeds in 2 rows; beak 2–5 mm with (0)1 or 2 seeds. $2n = 18$. *S.W. Europe.* Bl Hs Lu.

2. **D. siettiana** Maire, *Bull. Soc. Hist. Nat. Afr. Nord* **24**: 198 (1933). Like **1** but leaf-lobes obtuse; petals 9–10 mm; siliqua 3–4 mm wide; seeds in 3 or 4 rows; beak seedless. $2n = 16$. *Sandy soil.* ● *S.E. Spain* (*Isla de Alborán*); *perhaps extinct.* Hs.

3. **D. siifolia** G. Kunze, *Flora* (*Regensb.*) **29**: 685 (1846). Like **1** but leaves with short conical hairs on both surfaces, lyrate-pinnatisect; segments unequal, more or less entire or dentate, the terminal much larger than the lateral. $2n = 20$. *S. Portugal, S. Spain.* Hs Lu.

Often treated as a subspecies of **1.**

D. vicentina (Coutinho) Rothm., *Agron. Lusit.* **2**: 84 (1940), from S. Portugal (Cabo de S. Vicente), with more or less leafless stems, petals up to 6 mm and globose seeds, appears to be a coastal ecotypic variant of **3.**

4. **D. ibicensis** (Pau) Gómez-Campo, *Anal. Jard. Bot. Madrid* **38**: 32 (1981) (*D. catholica* var. *ibicensis* Pau). Like **1** but short-lived perennial; juvenile leaves lobate to pinnatifid, the segments with obtuse lobes; petals 5–6 mm; seeds in 1 or 2 rows; beak with 0 or 1 seeds. $2n = 16$. *Limestone rocks by the sea; mainly on small islands.* ● *Islas Baleares; one station off Cabo de S. Antonio.* Bl Hs.

5. **D. harra** (Forskål) Boiss., *Fl. Or.* **1**: 388 (1867) (incl. *Pendulina intricata* Willk., *P. webbiana* Willk., *P. hispida* auct. hisp.). Subglaucous, more or less fleshy, suffruticose perennial, glabrous or hispid. Lower leaves toothed, lobed or pinnatipartite, the lobes entire or toothed. Petals 6–10 mm, bright yellow. Siliqua 12–60 × 1·8–3 mm; gynophore 2·5–7 mm, conspicuous. $2n = 26$. *W. Mediterranean region.* Hs Si.

(a) Subsp. **crassifolia** (Rafin.) Maire, *Bull. Soc. Hist. Nat. Afr. Nord* **24**: 198 (1933) (*D. crassifolia* (Rafin.) DC.): Lower leaves elliptical, 2–4 times as long as wide, acute to obtuse, cuneate at the base, somewhat lobed to dentate; teeth irregular, distinct, acute to subacute. *Sicilia.*

(b) Subsp. **lagascana** (DC.) O. Bolòs & Vigo, *Fl. Països Catal.* **2**: 57 (1990) (*D. lagascana* DC., *Pendulina lagascana* (DC.) Amo): Lower leaves oblong, (3–)4–6 times as long as wide, lobed or pinnatipartite, entire to obscurely toothed. *S.E. Spain.*

Subsp. *harra*, from N.W. Africa and S.W. Asia, does not occur in Europe.

6. **D. tenuifolia** (L.) DC., *Reg. Veg. Syst. Nat.* **2**: 629 (1821). Perennial, but sometimes flowering in the first year, 20–80 cm, woody at the base. Lower leaves not forming a rosette, petiolate, pinnatipartite with 4–8 segments, glabrous or with some hairs at the margin, foetid when crushed. Petals 7·5–15 mm, sulphur-yellow. Siliqua 20–60 × 1–2·5 mm, borne erect on patent pedicels 1–5 cm; gynophore 0·5–6·5 mm. $2n = 22$. *W., S. & C. Europe; naturalized or casual elsewhere.* Al Au Be Bl Bu Co Cz Ga Ge He Ho Hs Hu It Ju Lu Po Rm Rs(K) Sa Si Tu [*Br Da Fe No Rs(B, W) Su].

7. D. cretacea Kotov, *Ukr. Bot. Žur.* **3**: 17 (1926). Annual or biennial 40–60(–80) cm; stems much-branched, with deflexed hairs in the lower part, glabrous above. Leaves mostly basal, pinnatipartite or pinnatisect; segments obtuse. Petals 7–9 mm, yellow. Siliqua 30–40 × 2–2·5 mm, with short (0·5–1 mm), fleshy style; gynophore 1 mm. ● *Bare, chalky slopes. Donets basin (N.E. Ukraine and adjacent parts of Russia).* Rs(C, E).

8. D. erucoides (L.) DC., *Reg. Veg. Syst. Nat.* **2**: 631 (1821) (incl. *D. valentina* Pau, *D. virgata* var. *platystylos* (Willk.) Willk.). Annual or overwintering 5–50 cm; stems 1 or many, erect or ascending, leafy. Leaves of two types, the lower 5–15 cm, in a lax basal rosette, oblong-lyrate or pinnatisect, with 6–10 lobes, or the lowermost more or less entire or irregularly toothed; upper leaves sessile, more or less amplexicaul; all with sparse, slender hairs. Petals (5–)7–13 mm, white, violet-veined or becoming violet after anthesis. Siliqua (10–)18–50 × 1·5–3 mm, ascending, with a usually 1-seeded, conical beak up to 6 mm and almost as wide as the valves; gynophore more or less absent. $2n = 14$. *S.W. Europe, extending eastwards to S.E. Italy; one station in the Danube delta. A frequent casual in C. Europe.* Bl Co Ga Hs It Lu Rm Sa Si.

9. D. virgata (Cav.) DC., *loc. cit.* (1821). Annual 5–90 cm, yellowish- or reddish-green; stems erect or ascending, densely setose-hispid below. Leaves almost all basal in a rosette, or stems and branches leafy; petiolate or sessile, pinnatipartite, pinnatilobed or sinuate-dentate. Petals 5–8 mm, sulphur-yellow, often violet-veined. Siliqua 15–40 × 1–2 mm, erect or erecto-patent; beak 3–7 mm, long-attenuate, 1-seeded or seedless. $2n = 18$. *C. & S. Spain; Portugal.* Hs Lu.

Plants from E. & S.E. Spain have been called **D. gomez-campoi** Martínez-Laborde, *Willdenowia* 21: 66 (1991) (*D. platystylos* auct., non Willk., *D. virgata* var. *platystylos* (Willk.) Willk.); they are generally smaller, with the beak of the siliqua 1–4 mm, compressed and as broad as the valves, and $2n = 16$, and may represent at least a subspecies of **9**.

10. D. viminea (L.) DC., *op. cit.* 635 (1821). Slender annual 5–30 cm, glabrous or slightly hairy. Leaves confined to a basal rosette, petiolate, lyrate-pinnatilobed with entire lobes (or spathulate and toothed, var. *integrifolia* Guss.). Petals 3–4 mm, sulphur- or lemon-yellow. Siliqua 10–35(–40) × 1·2–1·8 mm, erecto-patent, borne on spreading pedicels; beak short (1–2 mm), narrow. $2n = 20$. *S. Europe; often casual and locally naturalized in C. Europe.* Az Bl Bu Ga Gr Hs It Ju Lu Sa Si Rs(K) Tu [Ge Rm].

11. D. muralis (L.) DC., *op. cit.* 634 (1821) (incl. *D. scaposa* DC.). Foetid annual, biennial or perennial 10–50 cm, usually many-stemmed; stems glabrous or sparsely hispid below. Leaves usually more or less confined to a basal rosette, with petioles up to 3 cm, lyrate-pinnatipartite, pinnatifid or sinuate-dentate, sometimes spathulate, entire. Cauline leaves 0–4(–6); when present, subsessile, coarsely toothed. Petals 4–7·5(–8·5) mm, bright sulphur-yellow or sometimes becoming violet. Siliqua (12–)18–45 × 1·5–2·5(–3) mm, erecto-patent; gynophore absent; beak more or less conical, seedless. $2n = 42$, ?44. *W.C. & S. Europe.* Al Bl Co Ga Ge Gr He Ho Hs Hu It Ju Lu Po Rm Rs(K) Sa Si Tu [Br Da Hb Su].

Extremely variable in habit and leaf-shape. This species is probably an allotetraploid derived from **6** and **10**.

88. Brassica L.

V. H. HEYWOOD (EDITION 1) REVISED BY J. R. AKEROYD (EDITION 2)

Annuals to perennials or small shrubs. Leaves entire to pinnatipartite. Sepals erect or patent, the inner larger than the outer; petals yellow or white, clawed. Gynophore short or absent. Fruit a siliqua with a long or short beak; valves convex, with a prominent median vein. Seeds usually in 1 row in each loculus, at least in the basal ½ of the siliqua, globose or rarely ellipsoidal.

Literature: O. E. Schulz in Engler, *Pflanzenreich* **70(IV.105)**: 21–84 (1919). M. Onno, *Österr. Bot. Zeitschr.* **82**: 309–334 (1933).

1 Cauline leaves very few or absent
 2 Small shrub with long, thick, woody stock **3. balearica**
 2 Perennial herb, usually caespitose
  3 Leaves all basal; seeds ellipsoidal
   4 Siliqua ± straight, broadest in the middle **21. repanda**
   4 Siliqua usually somewhat curved, broadest towards the apex **22. glabrescens**
  3 A few cauline leaves often present; seeds globose or subglobose
   5 Plant glabrous or subglabrous; siliqua up to 2 mm wide **2. nivalis**
   5 Plant hairy; siliqua more than 2 mm wide **20. gravinae**
1 Stems ± leafy
 6 Upper cauline leaves amplexicaul at the base
  7 Siliqua 5–12 mm; beak filiform **18. souliei**
  7 Siliqua more than 15 mm; beak conical
   8 Siliqua 5–12 mm wide, with thick, woody valves; seeds ± biseriate **4. macrocarpa**
   8 Siliqua 2–5 mm wide, with herbaceous valves; seeds uniseriate
    9 Sepals erecto-patent; annual (sometimes biennial) with herbaceous stems
     10 Open flowers usually not overtopping the buds of the inflorescence; petals 11–18 mm **11. napus**
     10 Open flowers usually overtopping the buds of the inflorescence; petals 5–12 mm **12. rapa**
    9 Sepals connivent; biennial or perennial with woody stems
     11 Lower leaves pubescent to tomentose
      12 Siliqua 25–40(–55) mm, tetragonal, abruptly contracted into beak **7. villosa**
      12 Siliqua 40–80 mm, terete, gradually attenuate into beak **8. incana**
     11 Leaves glabrous or with sparse hairs
      13 Basal leaves ± undivided, broadly ovate, spathulate or suborbicular, sinuate-crenate or -dentate to pinnatifid
       14 Petiole at least ½ as long as the lamina; cauline leaves usually truncate at the base **9. insularis**
       14 Petiole not more than ⅓ as long as the lamina; cauline leaves usually auriculate at the base **10. cretica**
      13 Basal leaves lyrate-pinnatipartite or -pinnatisect
       15 Basal leaves lyrate-pinnatipartite, the terminal lobe entire or pinnatifid, obtuse; siliqua terete **5. oleracea**
       15 Basal leaves lyrate-pinnatisect, the terminal lobe pinnatipartite, ± acute; siliqua tetragonal **6. rupestris**

6 Upper cauline leaves sessile or petiolate, not amplexicaul at the base
16 Siliqua appressed to stem **19. nigra**
16 Siliqua erecto-patent to recurved
17 Siliqua distinctly stipitate, the gynophore 1·5–4·5 mm
18 Lower leaves not lyrate; beak of siliqua 0·5–2·5 mm; plant usually not woody at the base **1. elongata**
18 Lower leaves ±lyrate; beak of siliqua 2–7 mm; plant usually woody at the base **14. fruticulosa**
17 Siliqua sessile or very shortly stipitate, the gynophore less than 1·5 mm
19 Beak of the siliqua nearly as long as the valves; flowers nodding after anthesis **16. barrelieri**
19 Beak of siliqua shorter than the valves; flowers remaining erect after anthesis
20 Petals 5–7 × 1·5 mm, small and narrow; beak of siliqua 10–20 mm **17. tournefortii**
20 Petals 6–13 × 2–7·5 mm, conspicuous; beak of siliqua 2–10 mm
21 Lower leaves runcinate-pinnatifid, with 7–10 pairs of lobes **16. barrelieri**
21 Lower leaves lyrate-pinnatifid to pinnatisect, with 1–8 pairs of lobes
22 Sepals glabrous or with few scattered hairs at the apex
23 Sepals erecto-patent; siliqua sessile, the beak 5–10 mm **13. juncea**
23 Sepals suberect; siliqua stipitate, the beak 2–7 mm **14. fruticulosa**
22 Sepals shortly villous
24 Lower leaves with 5–8 pairs of lobes; terminal lobe sinuate-dentate, not callose **14. fruticulosa**
24 Lower leaves with 3–5 pairs of lobes; terminal lobe unequally crenate-callose **15. cadmea**

1. B. elongata Ehrh., *Beitr. Naturk.* **7**: 159 (1792). Biennial or short-lived perennial. Lower leaves petiolate, obovate to elliptical, sinuate-pinnatifid or entire, covered with curved bristles on both surfaces. Petals 7–10 mm, yellow. Siliqua 15–22 mm, attenuate into a seedless beak 0·5–2·5 mm, with gynophore 1·5–4·5 mm. *S.E. & E.C. Europe; naturalized elsewhere as a weed or ruderal.* Au Bu Cz Hu Ju *Po Rm Rs(W, K, E) [Da Ga Ge Ho It Rs(C)].

(a) Subsp. **elongata** (*B. elongata* subsp. *armoracioides* (Czern.) Ascherson & Graebner): Leaves sinuate-pinnatifid. $2n = 22$. *Western part of the range of the species.*

(b) Subsp. **integrifolia** (Boiss.) Breistr., *Bull. Soc. Sci. Dauph.* **60**: 139 (1944) (as preprint p. 13 (1942)) (*B. elongata* subsp. *persica* (Boiss. & Hohen.) Thell., *B. persica* Boiss. & Hohen., *B. armoracioides* Czern.): Leaves entire. *S. Russia, Ukraine; one station in S.E. Poland.*

2. B. nivalis Boiss. & Heldr. in Boiss., *Diagn. Pl. Or. Nov.* **3(1)**: 32 (1856) (*Brassicella nivalis* (Boiss. & Heldr.) O. E. Schulz, *Rhynchosinapis nivalis* (Boiss. & Heldr.) Heywood, *Hutera nivalis* (Boiss. & Heldr.) Gómez-Campo, *Coincya nivalis* (Boiss. & Heldr.) Greuter & Burdet). Glabrous or subglabrous, caespitose perennial, somewhat woody at the base; stems 10–50 cm, usually unbranched. Cauline leaves few, small, linear, often absent. Sepals 4–5 mm, obtuse; petals 8–11 mm, yellow. Pedicels erecto-patent. Siliqua 25–60 × 1·5–2 mm, somewhat curved, constricted between the seeds; beak 1–3 mm, without seeds. Seeds 10–18 mm. *Limestone screes, usually above* 1800 *m.* ● *N. Greece; S.W. Bulgaria; local.* Bu Gr.

(a) Subsp. **nivalis**: Basal leaves deeply pinnatifid to lyrate-pinnatifid, the lamina abruptly attenuate into the petiole; beak of siliqua 1·5–3 mm. $2n = 20$. *N. Greece (Olimbos).*

(b) Subsp. **jordanoffii** (O. E. Schulz) Akeroyd & Leadlay, *Bot. Jour. Linn. Soc.* **106**: 102 (1991) (*B. jordanoffii* O. E. Schulz): Basal leaves coarsely dentate to shallowly pinnatifid, the lamina gradually attenuate into the petiole; beak of siliqua 1–1·5 mm. $2n = 22$. *S.W. Bulgaria (Pirin Planina).*

3. B. balearica Pers., *Syn. Pl.* **2**: 206 (1806). Glabrous small shrub with long, thick, woody stock; flowering stems up to 30 cm. Leaves mostly basal, long-petiolate, lyrate or pinnatilobed, resembling those of *Quercus robur*. Petals 10–15 mm, yellow. Siliqua 20–60 × 2–2·5 mm, linear, terete, constricted at intervals, terminating in a 0- or 1-seeded beak 1·5–3 mm. $2n = 32$. ● *Islas Baleares (Mallorca).* Bl.

4. B. macrocarpa Guss., *Ind. Sem. Horto Boccad.* **1824/5**: 3 (1825). Glabrous small shrub; stems 30–60 cm. Basal leaves 10–18 × 4–10 cm, lyrate-pinnatifid or pinnatilobed, the lobes crenate-dentate, petioles up to 8 cm; cauline leaves oblong to lanceolate, entire or shallowly toothed. Petals 10–15 mm, yellow. Siliqua 20–40 × 5–12 mm, very thick and fleshy; valves thick and woody, 1-veined, boat-shaped, attenuate into a conical, 0- to 2-seeded beak 8–13 mm. *Maritime rocks.* ● *W. Sicilia (Isole Egadi).* Si.

B. drepanensis (Caruel) Damanti, *Nat. Sicil.* **10(4)**: 91 (1891), from the adjacent mainland of Sicilia near Trapani, is like **4** but with densely tomentose leaves and more slender fruits with a shorter beak. It may not be specifically distinct from **4**.

5. B. oleracea L., *Sp. Pl.* 667 (1753) (*B. sylvestris* (L.) Miller). Glabrous biennial to perennial up to 300 cm; lower part of stems somewhat woody. Basal leaves up to 40 cm, usually petiolate, lyrate-pinnatipartite, crenate; cauline leaves ovate-lanceolate or oblong, entire, sessile (petiolate in some cultivars). Petals 15–30 mm, pale yellow. Siliqua 35–85(–100) × 2–4(–5) mm, linear-terete, with a short conical beak 4–8(–10) mm, as wide as the valves at the base, (0)1(2)-seeded. $2n = 18$. *Maritime cliffs.* ● *Europe, extending to E. Italy. Widely cultivated as a vegetable.* *Br Ga Hs It [Bu Cz Ge Ju].

Many cultivars have been derived from **5** and related cross-fertile species. There is great diversity in their taxonomic treatment, some authors regarding the wild plants as a separate species, **B. sylvestris** (L.) Miller, *Gard. Dict.* ed. 8, no. 4 (1768), while others regard the cultivars as subspecies and varieties of **5**; Lizgunova (*Bull. Appl. Bot. Pl.-Breed. (Leningrad)* **32**: 37–70 (1959)) recognizes five separate cultivated species with numerous subspecies. Some of the cultivars occur as escapes.

(a) Subsp. **oleracea**: Terminal lobe of basal leaves usually entire. $2n = 18$. *Atlantic coasts; naturalized elsewhere.*

(b) Subsp. **robertiana** (Gay) Bonnier & Layens, *Tabl. Syn. Pl. Vasc.* 21 (1894) (*B. oleracea* subsp. *pourretii* Fouc. & Rouy; incl. *B. montana* Pourret): Terminal lobe of basal leaves usually pinnatifid. $2n = 18$. *Mediterranean coasts.*

6. B. rupestris Rafin., *Caratt.* 77 (1810). Glabrous, branched small shrub; stems up to 150 cm. Basal leaves 20–40 cm, lyrate-pinnatisect, with terminal lobe pinnatipartite, more or less acute;

petiole about as long as the lamina. Siliqua 35–50(–70) × 3–5 mm, 4-angled, attenuate into a slender beak 3–6(–12) mm. $2n = 18$. *Calcareous rocks.* ● *W. Italy; Sicilia.* It Si.

7. **B. villosa** Biv., *Stirp. Rar. Sic. Descr.* **4**: 20 (1816) (incl. *B. tinei* Lojac.). Like **6** but terminal lobe of basal leaves more or less undivided, obtuse, pubescent, villous or white-tomentose; petiole shorter than the lamina; siliqua 25–40(–55) mm, abruptly contracted into beak 5 mm. *Calcareous rocks.* ● *Sicilia.* Si.

Usually very distinct from **6** but transitional variants are known.

8. **B. incana** Ten., *Prodr. Fl. Nap.* xxxix (1811). Perennial, woody at the base, branched; stems up to 100 cm, glabrous except at the base. Basal leaves up to 40 cm, pubescent to tomentose, shortly petiolate, lyrate, with the terminal lobe undivided, usually obtuse. Petals 16–18 mm, yellow. Siliqua 40–80 × 2–3 mm, linear, constricted at intervals, terete, gradually attenuate into the beak. *Calcareous rocks, usually coastal.* ● *W. & S. Italy, Sicilia; W. Jugoslavia.* Al It Ju Si.

In the islands of the Adriatic a number of endemic variants, distinguished principally by fruit characters, have been recognized as species: **B. botteri** Vis., *Fl. Dalm.* **3**: 135 (1850), with siliqua 30–40 × 4–5 mm; **B. cazzae** Ginzberger & Teyber, *Österr. Bot. Zeitschr.* **19**: 238 (1921), with siliqua 35–50(–60) × 3–3·5 mm; and **B. mollis** Vis., *Fl. Dalm.* **3**: 359 (1852), with siliqua 40–60 × 3–4 mm. Their status is uncertain.

9. **B. insularis** Moris, *Fl. Sard.* **1**: 168 (1837) (*B. oleracea* subsp. *insularis* (Moris) Rouy & Fouc.). Perennial up to 150 cm. Basal leaves 10–15 cm, lyrate or spathulate; terminal lobe broad and rounded or narrowly elongate, crenate or dentate, glabrous or with a few hairs on margins; petiole usually at least ½ as long as the lamina. Cauline leaves usually truncate at the base. Petals white or pale yellow. Siliqua 30–70(–90) × 3–5 mm, linear, attenuate into beak 5–10 mm, which is narrower than the siliqua. $2n = 18$. ● *Corse, Sardegna; Sicilia (Pantellaria).* Co Sa Si.

10. **B. cretica** Lam., *Encycl. Méth. Bot.* **1**: 747 (1785). Like **9** but petioles of the basal leaves not more than ⅓ as long as the lamina; cauline leaves usually auriculate at the base; beak usually as broad as the siliqua at the base. *S.E. Greece and Aegean region.* Cr Gr.

Subsp. *botrytis* (L.) O. Schwarz (Cauliflower) is regarded as having been derived from this species in cultivation.

1 Petals usually white; stem branched mainly in upper part — (b) subsp. **nivea**
1 Petals yellow, whitish- or brownish-yellow; stem branched from near base
2 Seeds faintly reticulate — (a) subsp. **cretica**
2 Seeds distinctly reticulate — (c) subsp. **laconica**

(a) Subsp. **cretica**: Plant short, the stem branching from the base. Basal leaves with short petiole (petiole rarely absent), pinnatifid or almost entire. Petals yellow, rarely whitish. Seeds faintly reticulate. *S.E. Greece.*

(b) Subsp. **nivea** (Boiss. & Spruner) M. Gustafsson & Snogerup, *Bot. Chron.* **3**: 8 (1983) (*B. nivea* Boiss. & Spruner): Plant up to 150 cm, the stem branched mainly in the upper part. Basal leaves with long petiole, lyrate or entire. Petals white, rarely yellowish-white. Seeds distinctly reticulate. *N.E. Peloponnisos; Kriti.*

(c) Subsp. **laconica** M. Gustafsson & Snogerup, *loc. cit.* (1983): Plant 50–100 cm, the stem branched from near the base. Basal leaves petiolate, crenate-serrate, deeply pinnatifid, or lyrate with prominent, rounded lobes. Petals whitish- to pale brownish-yellow. Seeds distinctly reticulate. $2n = 18$. *Limestone cliffs.* ● *S.E. Greece.*

11. **B. napus** L., *Sp. Pl.* 666 (1753). Annual or biennial, with slender or stout, often fusiform or tuberous taproot; stems up to 150 cm. Basal leaves lyrate, sometimes ciliate, petiolate, glaucous and glabrous, or with a few bristly hairs especially along the veins, upper cauline leaves sessile, more or less entire, amplexicaul. Open flowers usually not overtopping buds of inflorescence. Sepals erecto-patent; petals 11–18 mm, yellow, sometimes tinged with pink. Siliqua 35–95(–110) × (2·5–)3–5 mm, suberect to spreading, attenuate into a slender beak 5–25(–30) mm. $2n = 38$. *Known only from cultivation; cultivated in most European countries, and widely naturalized.*

Several cultivars are extensively cultivated. There are numerous different classifications of these, and interpretations of their relationship to the wild subspecies. A summary is given in E. N. Sinskaja, *Bull. Appl. Bot. Pl.-Breed. (Leningrad)* **33**: 233–250 (1960). These include subsp. *pabularia* (DC.) Janchen (Leaf-rape), an annual with crispate, dissected leaves; and subsp. *rapifera* Metzger (incl. var. *napobrassica* (L.) Reichenb.) (Swede), a biennial with a thickened, more or less globose, edible stem-base and taproot, with pinkish-yellow flowers.

Subsp. *napus* (subsp. *oleifera* DC.), an annual or biennial with a non-tuberous root and lyrate-pinnatifid leaves, is grown as a fodder crop (Rape) and for the oil extracted from its seeds.

It is thought that crosses of **5(a)** ($2n = 18$) with **11** ($2n = 20$) gave rise to subsp. *pabularia* ($2n = 38$), from which subsp. *napus* ($2n = 38$) and subsp. *rapifera* ($2n = 38$) and other cultivars were derived.

12. **B. rapa** L., *Sp. Pl.* 666 (1753) (*B. campestris* L., *B. asperifolia* Lam.). Like **11** but basal leaves bright green, with setiform hairs; open flowers usually overtopping buds of the inflorescence; petals 5–12 mm, bright yellow. *Weed or ruderal in much of Europe; native distribution not known.* *[Al Bl Br Bu Co Cz Da Fe Ge Gr Hb Ho Hs Hu Is It Ju No Rm Rs(K, E) Sa Si Su.]

(a) Subsp. **rapa**: Taproot tuberous. $2n = 20$. *Cultivated for the edible taproot (Turnip), and sometimes escaping.*

(b) Subsp. **sylvestris** (L.) Janchen in Janchen & Wendelberger, *Kleine Fl. Wien* 55 (1953): Taproot non-tuberous. $2n = 20$. *Throughout the range of the species.*

Subsp. **oleifera** DC., *Prodr.* **1**: 214 (1824), is grown as a fodder crop (Turnip rape) and for the oil extracted from its seeds. It differs from subsp. (a) by its larger, reddish-brown (rather than grey or blackish) seeds.

13. **B. juncea** (L.) Czern., *Consp. Pl. Charc.* 8 (1859). Annual up to 150 cm; branches long, erecto-patent. Lower leaves petiolate, lyrate-pinnatisect, with 1–3 pairs of lobes on each side and a larger, sparsely setose, terminal lobe; upper leaves subentire, shortly petiolate, glabrous. Sepals erecto-patent; petals 6–13 mm. Siliqua (20–)30–60 × 2–5 mm, constricted at intervals, sessile, attenuate into a tapering, seedless beak 5–10 mm. *Cultivated for its seeds in S. Russia; naturalized in S. Europe and a casual elsewhere.* [Bu Hs Rm Rs(E).] (*S. & E. Asia.*)

14. **B. fruticulosa** Cyr., *Pl. Rar. Neap.* **2**: 7 (1792). Annual to perennial, usually becoming woody at the base; stems up to 50 cm,

erect. Lower leaves long-petiolate, lyrate-pinnatifid, with 2–8 pairs of rounded or obtuse lobes, hispid; upper leaves smaller, pinnatilobed to entire. Sepals suberect; petals 6–10 mm, yellow. Siliqua 15–40 × 1·5–2 mm, constricted at intervals, stipitate, the gynophore 1–3 mm; beak 2–7 mm, 0- or 1-seeded. *Ruderal. Mediterranean region.* Ga Hs It Sa Si.

(a) Subsp. **fruticulosa**: Usually biennial to perennial, woody at the base. Sepals glabrous or sparsely hairy. $2n = 16$. *Throughout the range of the species.*

(b) Subsp. **cossoniana** (Boiss. & Reuter) Maire in Jahandiez & Maire, *Cat. Pl. Maroc* 2: 287 (1932) (*B. cossoniana* Boiss. & Reuter): Annual, scarcely woody at the base. Sepals densely villous. $2n = 32$. *S. Spain. (N.W. Africa.)*

15. B. cadmea Heldr. ex O. E. Schulz in Engler, *Pflanzenreich* **70(IV.105)**: 63 (1919). Like **14(b)** but basal and lower cauline leaves shortly petiolate, pinnatipartite, with 3–5 pairs of irregularly crenate, callose-tipped lobes; siliqua with short, broad, 1-seeded beak. *Clayey hills.* ● *C. Greece (near Thivai).* Gr.

B. procumbens (Poiret) O. E. Schulz, *Bot. Jahrb.* **54** Beibl. **119**: 55 (1916) (*Sinapis procumbens* Poiret), like **15** but with spreading or ascending stems, the lower part of the inflorescence bracteate, and the siliqua distinctly hairy, was recorded from C. Italy (Arcipelago Toscano) but is apparently extinct there; it has recently been reported as an alien in Corse.

16. B. barrelieri (L.) Janka, *Term. Füz.* **6**: 179 (1882) (*B. laevigata* Lag., *B. sabularia* Brot.). Annual, rarely perennial; stems up to 50 cm. Lower leaves numerous, rosulate, very shortly petiolate, runcinate-pinnatifid, with 7–10 pairs of lanceolate, acuminate lobes, hispid, especially along the thick, white mid-vein, and ciliate. Cauline leaves few, sessile, more or less entire, glabrous. Sepals erect, sparsely hispid or glabrous; petals 7–12 mm, obovate, yellow with livid veins, or whitish. Siliqua 25–60 × 1·5–2·5 mm, constricted at intervals, very shortly stipitate, attenuate into a 0- to 2-seeded beak 5–30 mm long. *Sandy soils. S.W. Europe.* Bl Hs Lu.

(a) Subsp. **barrelieri**: Petals 9–12 mm. Beak much shorter than the valvar portion of the siliqua, 0- or 1-seeded. *Throughout the range of the species.*

(b) Subsp. **oxyrrhina** (Cosson) Regel in Regel *et al.*, *Ind. Sem. Horti Petrop.* **1856**: 34 (1857) (*B. oxyrrhina* Cosson): Petals 7–8 mm. Beak nearly as long as the valvar portion of the siliqua or longer, usually 2-seeded. $2n = 18$. *S. Spain, S. Portugal.*

17. B. tournefortii Gouan, *Obs. Bot.* 44 (1773). Like **16** but leaf-lobes obtuse; sepals suberect; petals 5–7 mm, narrowly oblong-obovate, pale yellow, usually violaceous at the base, becoming whitish; siliqua 35–65 × 2·5–3 mm; beak 10–20 mm, $\frac{1}{3}$–$\frac{1}{2}$ as long as the valvar portion. $2n = 20$. *Mainly on maritime sands. Mediterranean region.* Cr Gr Hs It ?Lu Sa Si.

18. B. souliei (Batt.) Batt., *Bull. Soc. Bot. Fr.* **40**: 262 (1893). Annual 10–40 cm, many-stemmed and branched from the base. Lower leaves petiolate, obovate or oblong, sinuate-dentate, obtuse at the apex, glabrous or sparsely hairy, ciliate; upper leaves amplexicaul at the base. Petals 5–7 mm, yellow. Siliqua 5–12 × 1–2 mm, attenuate into a filiform, seedless beak, and borne on patent or recurved pedicels 5–15(–40) mm. *Dry, waste places. Sicilia.* Si.

The European plant is subsp. **amplexicaulis** (Desf.) Greuter & Burdet, *Willdenowia* **13**: 86 (1983) (*B. amplexicaulis* (Desf.) Pomel, non Hochst. ex A. Richard). Subsp. *souliei* occurs in N.W. Africa.

19. B. nigra (L.) Koch in Röhling, *Deutschl. Fl.* ed. 3, **4**: 713 (1833). Annual; stems up to 200 cm, branched from the middle or from near the base. Lower leaves lyrate-pinnatisect, with 1–3 pairs of lateral lobes and a much larger terminal lobe, hispid on both surfaces; upper leaves linear-oblong, entire or sinuate, glabrous; all leaves petiolate. Petals 7–13 mm, yellow. Siliqua 8–30 × 1·5–4·5 mm, attenuate into a slender, seedless beak. Pedicels 2·5–6 mm in fruit, appressed to the stem. $2n = 16$. *Most of Europe but commonest in the centre and south; perhaps native in parts of S. & W. Europe. Cultivated for its seeds which are used as a condiment (Black mustard).* All except Bl Is Sb.

20. B. gravinae Ten., *Prodr. Fl. Nap.* xxxix (1811). Caespitose perennial; stems 10–30–(50) cm, usually simple, hispid or pubescent. Basal leaves densely rosulate, shortly petiolate, oblong or obovate, lyrate, sinuate-dentate or pinnatifid; terminal lobe obovate, obtuse, crenate-dentate, somewhat fleshy; cauline leaves few, subsessile, slightly dentate. Raceme 25- to 40-flowered. Petals 9 mm, yellow. Siliqua 20–50 × 2·5–3 mm, subterete, attenuate into a seedless beak 2·5–4 mm. Seeds globose. *Calcareous rocks. C. & S. Appennini.* It.

21. B. repanda (Willd.) DC., *Reg. Veg. Syst. Nat.* **2**: 598 (1821) (*Diplotaxis saxatilis* DC.). Laxly to densely caespitose perennial; stems up to 50 cm, usually leafless, glabrous or hairy. Basal leaves 1–15 cm, entire to pinnatipartite. Raceme 2- to 45-flowered. Petals 7–30 mm, yellow. Siliqua 10–80 × 1–4·5 mm, variable in shape and proportions, gradually or abruptly attenuate into a seedless beak. Seeds ellipsoidal. $2n = 20$. *Usually in rocky or stony places. S.W. Europe.* Ga Hs It.

Very variable. A key is given below to the more important geographical variants, but anomalous populations occur in the Pyrenees and neighbouring regions.

1 Petals 15–25(–30) mm
  2 Petals cuneate, the apex truncate; leaves glaucous, fleshy — (d) subsp. **maritima**
  2 Petals obovate, the apex rounded; leaves green, membranous — (c) subsp. **galissieri**
1 Petals not more than 14 mm
  3 Siliqua 1–2 mm wide
    4 Plants densely caespitose with woody, branched stock; siliqua usually more than 30 mm — (e) subsp. **confusa**
    4 Plants laxly caespitose, with the stock ± unbranched; siliqua 10–30 mm
      5 Stems 30–60 cm; raceme 10- to 30-flowered — (j) subsp. **nudicaulis**
      5 Stems 10–15 cm; raceme 2- to 8-flowered — (k) subsp. **almeriensis**
  3 Siliqua 2·5–4·5 mm wide
    6 Raceme 2- to 8(–12)-flowered; siliqua 20–30(–60) × 3–3·5 mm, abruptly contracted into the beak; leaves sinuate-dentate — (a) subsp. **repanda**
    6 Racemes with 12–35 flowers, or, if fewer-flowered, siliqua less than 3 mm wide; siliqua (20–)25–80 × 2–4·5 mm, gradually attenuate into the beak
      7 Siliqua not more than 2·5 mm wide — (b) subsp. **saxatilis**
      7 Siliqua at least 2·5 mm wide
        8 Leaves 2–4(–6) cm, entire or slightly toothed, or, if pinnatipartite, 1·5–2 cm — (i) subsp. **latisiliqua**

8 Leaves 4·5–10 cm, sinuately lobed, pinnatifid to pinnatipartite
9 Leaves 4·5–8 cm, deeply pinnatipartite, the lobes rounded, denticulate (f) subsp. **cantabrica**
9 Leaves 6–10 cm, sinuate-lobed to pinnatifid, with the lobes entire
10 Siliqua 40–60 × 3·5–4 mm, pendent (g) subsp. **cadevallii**
10 Siliqua 50–80 × 2·5–3 mm, patent or ascending (h) subsp. **blancoana**

(a) Subsp. **repanda**: Stems 2–10(–15) cm. Leaves 5–40 mm, spathulate or obovate, sinuate-dentate, rarely pinnatifid. Raceme 2- to 12-flowered. Siliqua 20–30(–60) × 3–3·5 mm. $2n = 20$. ● *S.W. Alps.*

(b) Subsp. **saxatilis** (DC.) Heywood, *Feddes Repert.* **69**: 151 (1964) (*Diplotaxis saxatilis* DC., *D. humilis* Gren. & Godron): Stems 2–16 cm. Leaves 2–6 cm, usually pinnatipartite. Raceme 2- to 10-flowered. Siliqua 20–50 × 2–2·5 mm. $2n = 20$. ● *S. France (near Aix-en-Provence and Montpellier).*

(c) Subsp. **galissieri** (Giraud.) Heywood, *Feddes Repert.* **69**: 150 (1964) (*Diplotaxis galissieri* Giraud.): Stems 30–40 cm. Leaves 6–10 cm, pinnatipartite; petiole usually longer than the lamina. Raceme up to 35-flowered, dense. Siliqua 40–55 × 2·5–3 mm. ● *S. France (Ariège).*

(d) Subsp. **maritima** (Rouy) Heywood, *Feddes Repert.* **66**: 153 (1962) (*Diplotaxis maritima* Rouy): Distinguishable from the other subspecies by its fleshy, glaucous leaves, and compact, many-flowered corymb of conspicuous flowers, with broad, cuneate petals, truncate at the apex. *Maritime rocks.* ● *E. Spain (Alicante prov.).*

(e) Subsp. **confusa** (Emberger & Maire) Heywood, *loc. cit.* (1962) (*B. saxatilis* subsp. *confusa* Emberger & Maire, *B. humilis* sensu Willk. pro parte, *Diplotaxis brassicoides* Rouy pro parte): Densely caespitose with woody, branched stock; stems 10–35 cm. Leaves 3–8 cm, sinuate-dentate to pinnatifid. Raceme 10- to 30-flowered. Siliqua 30–50 × 1–2 mm. *Mountains of S. and S.E. Spain.*

(f) Subsp. **cantabrica** (Font Quer) Heywood, *Feddes Repert.* **69**: 151 (1964): Stems 10–20 cm. Leaves 4·5–8 cm, pinnatipartite or pinnatisect. Raceme 15- to 25-flowered. Siliqua 25–35 × 3–4 mm. $2n = 20$. *Limestone rocks.* ● *N.W. Spain (Peña Major).*

(g) Subsp. **cadevallii** (Font Quer) Heywood, *Feddes Repert.* **66**: 154 (1962): Stems 30–35 cm. Leaves 8–10 cm, pinnatifid or pinnatipartite. Raceme 10- to 30-flowered. Siliqua 40–60 × 3·5–4 mm, pendent. *Limestone rocks.* ● *N.E. Spain.*

(h) Subsp. **blancoana** (Boiss.) Heywood, *Feddes Repert.* **66**: 153 (1962) (*Brassica blancoana* Boiss., *Diplotaxis saxatilis* var. *longifolia* (Rouy) Willk., *D. brassicoides* Rouy pro parte): Stems 30–45 cm. Leaves 6–10 cm, sinuate to pinnatifid. Raceme 10- to 20-flowered. Siliqua 50–80 × 2·5–3 mm. *Rocks and screes.* ● *Mountains of S.E. Spain.*

(i) Subsp. **latisiliqua** (Boiss. & Reuter) Heywood, *loc. cit.* (1962) (*Brassica latisiliqua* Boiss. & Reuter, *Diplotaxis brassicoides* Rouy pro parte): Stems 10–30 cm. Leaves 2–4(–6) cm, entire, slightly toothed or pinnatipartite. Raceme 8- to 20-flowered. Siliqua 25–50 × 2·5–3·5 mm. *Mountains of S. Spain (Sierra Nevada, Sierra Tejeda).*

(j) Subsp. **nudicaulis** (Lag.) Heywood, *Feddes Repert.* **69**: 151 (1964) (*Brassica barrelieri* auct. hisp. mult., non (L.) Janka): Slightly caespitose habit with slender stock. Stems 30–60 cm. Leaves pinnatipartite. Raceme 10- to 30-flowered. Siliqua 20–30 × 1–2(–2·5) mm. $2n = 20$. *C. Spain.*

(k) Subsp. **almeriensis** Gómez-Campo, *Anal. Inst. Bot. Cavanilles* **33**: 154 (1976): Similar to subsp. (j) in habit, but sometimes biennial. Stems 10–15 cm. Leaves subentire to shallowly pinnatifid. Raceme 2- to 8-flowered. Siliqua 20–25 × *c.* 1·5 mm. *Stony, calcareous ground.* ● *S.E. Spain (N.W. of Vélez Blanco).*

Similar plants to subsp. (c) occur in the Aragonese Pre-Pyrenees.

Plants similar to subsp. (e) but with wider fruits occur in the Spanish Pyrenees.

22. B. glabrescens Poldini, *Gior. Bot. Ital.* **107**: 181 (1973). Caespitose perennial, arising from branched, woody stock; stems 10–25 cm, erect, usually leafless, glabrous, reddish. Leaves 2–9 cm, pinnatipartite, glabrous except for a marginal setose hair on each of the lobes. Raceme 2- to 12-flowered. Petals 7–10 mm, yellow. Siliqua 25–60 × 1·5–3 mm, subclavate, somewhat curved, attenuate into the seedless beak 3–5 mm. Seeds ellipsoidal. *Calcareous river-gravels.* ● *N.E. Italy (Rivers Meduno and Cellina, north of Pordenone).* It.

89. Sinapis L.

A. O. CHATER

Annuals or perennials. Leaves pinnatifid or pinnatisect. Sepals patent, equal or subequal; petals yellow, clawed. Fruit a siliqua with a long beak; valves distinctly 3- to 7-veined. Seeds in 1 row in each loculus, globose.

1 Siliqua with (7)8–17 seeds; beak cylindrical or conical, not or scarcely compressed, shorter than to about equalling valvar portion of siliqua
2 Siliqua at least 25 mm, glabrous or with short, stiff, deflexed hairs **1. arvensis**
2 Siliqua not more than 25 mm, densely covered with long, fine, ascending hairs **2. pubescens**
1 Siliqua with 4–8 seeds; beak strongly compressed, usually longer than valvar portion of siliqua
3 Sepals more than 4·5 mm; leaves variably covered with short, stiff hairs, but scarcely scabrid **3. alba**
3 Sepals less than 4·5 mm; leaves scabrid with short, stiff hairs **4. flexuosa**

1. S. arvensis L., *Sp. Pl.* 668 (1753) (incl. *S. orientalis* L., *S. schkuhriana* Reichenb., *Brassica arvensis* (L.) Rabenh., non L., *B. sinapistrum* Boiss.). Annual; stems up to 80 cm, usually hispid at least below, sometimes glabrous. Leaves up to 20 cm, usually hispid; lower leaves stalked, lyrate, with large, coarsely toothed terminal lobe, usually with several smaller lateral lobes; upper leaves more or less sessile, usually simple, ovate to lanceolate. Petals 10–16 × 5–8 mm. Siliqua 25–45(–55) × (1·5–)2·5–4 mm, patent to erect; valves glabrous or with short, stiff, deflexed hairs; beak 7–16 mm, straight. Seeds (7)8–13, reddish-brown or blackish. $2n = 18$. *Usually a weed of arable land. Probably native in the Mediterranean region, but introduced throughout Europe, rarer in the north.* All except Az Sb but only casual in Fa Is.

A very variable species. Var. *orientalis* (L.) Koch & Ziz, with hairy siliqua-valves, is as widespread as the typical form.

2. S. pubescens L., *Mantissa* 95 (1767) (*Brassica pubescens* (L.) Ard.). Perennial; stems up to 80 cm, densely pubescent or villous. Leaves appressed-pubescent; lower stalked, lyrate-pinnatisect; upper sessile, oblong, less divided or simple. Siliqua 15–25 × 2–2·5 mm, erect, covered to the apex with long, fine, ascending hairs; beak 9–12 mm, curved. Seeds usually 12–17, blackish. $2n = 18$. *C. & S. Italy, Sardegna and Sicilia; formerly in S.E. France.* ?Al †Ga It Sa Si.

3. S. alba L., *Sp. Pl.* 668 (1753). Annual; stems up to 80 cm, usually with stiff, deflexed hairs (rarely glabrous). Leaves usually hispid but not scabrid, all petiolate. Petals 8–15 mm. Siliqua 20–40 × 3–6·5 mm, patent; beak 10–30 mm, flat, attenuate. Seeds 4–8. *Usually a weed of cultivated land. Perhaps native in much of Europe but widespread as an alien in other parts.* All except Sb but only casual in Az Fa Fe Is Rs(N).

(a) Subsp. **alba** (*Brassica foliosa* (Willd.) Samp.): Leaves lyrate-pinnatifid or -pinnate. Siliqua 20–40 × 3–4 mm; valves usually hispid; beak 10–30 mm. $2n = 24$. *Throughout the range of the species.*

(b) Subsp. **dissecta** (Lag.) Bonnier, *Fl. Compl. Fr.* **1**: 58 (1912) (*S. dissecta* Lag.): Leaves 2-pinnatifid, not lyrate; terminal lobe oblong-ovate, not or scarcely larger than the oblong-linear lateral lobes. Siliqua 25–30 × 3·5–6·5 mm; valves slightly hairy or glabrous; beak 10–20 mm. *Mostly in flax-fields. S. Europe; sometimes casual further north.*

Subsp. (a) is widely cultivated as a forage crop and for the condiment obtained from the seeds (White mustard).

4. S. flexuosa Poiret in Lam., *Encycl. Méth. Bot.* **4**: 341 (1797) (*S. hispida* Schousboe). Like **3** but the stems more strongly hispid; leaf-surfaces scabrid with very rough, short hairs; pedicels shorter in fruit; sepals and petals smaller; siliqua 25–50 × 2·5–4 mm; beak 15–30 mm, often curved, scarcely attenuate; seeds usually 7 or 8, greyish-brown. *S. Spain.* Hs.

90. Eruca Miller

†T. G. TUTIN

Like *Sinapis* but the sepals erect, unequal, the inner somewhat saccate at the base; valves of the siliqua 1-veined; seeds in 2 rows in each loculus.

1. E. vesicaria (L.) Cav., *Descr. Pl.* 426 (1802) (*E. sativa* Miller). Foetid, usually hispid annual 20–100 cm. Leaves lyrate-pinnatifid (rarely pinnate), with a large terminal lobe and 3–5 narrow lateral lobes on each side. Petals 15–24 mm, whitish or yellowish with violet veins. Siliqua 12–35(–40) × 3–6 mm, erect; beak ensiform, seedless; pedicels short. *Mediterranean region; cultivated as a salad plant (Rocket) and often naturalized, so that the native distribution is obscured.* Bl Bu Co Cr Ga *Gr He Hs Hu It *Ju Lu Rm Rs(C, W, K, E) Sa Si Tu [Au Cz Ge No Po].

(a) Subsp. **vesicaria**: Sepals all cucullate, persistent until the fruit is nearly ripe; anthers subacute. $2n = 22$. *Spain, Islas Baleares.*

(b) Subsp. **sativa** (Miller) Thell. in Hegi, *Ill. Fl. Mitteleur.* **4(1)**: 201 (1918) (*E. longirostris* Uechtr., *E. orthosepala* (Lange) Lange): Sepals caducous, the inner not cucullate; anthers obtuse. $2n = 22$. *Throughout the range of the species.*

91. Erucastrum C. Presl

†T. G. TUTIN (EDITION 1) REVISED BY J. R. AKEROYD (EDITION 2)

Annuals or perennials. Flowers in terminal racemes. Sepals erect to patent, the inner somewhat saccate; petals yellow, clawed. Lateral nectaries semilunar or bilobed. Fruit a linear, torulose siliqua; valves keeled, strongly 1-veined, with a lateral network of smaller veins; beak more or less conical, with 0–3 seeds. Seeds ovoid or oblong, in 1 row in each loculus.

Literature: C. Gómez-Campo, *Anal. Jard. Bot. Madrid* **40**: 63–72 (1983).

1 Leaves somewhat fleshy, the upper cauline conspicuously different from the basal — **1. virgatum**
1 Leaves not fleshy, the upper cauline similar to the basal
  2 Raceme bracteate below; siliqua not stipitate — **4. gallicum**
  2 Raceme ebracteate; siliqua stipitate
    3 Plant somewhat hairy; lower cauline leaves lyrate-pinnatisect with 6–8 lobes on each side — **2. nasturtiifolium**
    3 Plant glabrous; lower cauline leaves pectinate-pinnatisect, with 10–18 lobes on each side — **3. palustre**

1. E. virgatum (J. & C. Presl) C. Presl, *Fl. Sic.* **1**: 94 (1826) (*E. pubescens* (L.) Willk.; incl. *E. laevigatum* (L.) O. E. Schulz, *E. pseudosinapis* Lange, *Brassica baetica* Boiss., *Sinapis laevigata* L.). Biennial or perennial, often somewhat glaucous; stem 30–150 cm, much-branched. Basal and lower cauline leaves lyrate to pinnatisect, hispid to subglabrous; upper cauline leaves smaller, more or less linear, entire. Sepals patent; petals 6–11 mm. Siliqua 10–35 × 1–2 mm, erect to erecto-patent; beak 3–8 mm, with 0–2 seeds. ● *S.E. Spain; S.W. Italy, Sicilia.* Hs It Si.

1 Siliqua not more than 2·5 cm, ± erect
  2 Siliqua contracted at base of beak; beak about as thick as valvar portion — (a) subsp. **virgatum**
  2 Siliqua not contracted at base of beak; beak usually thicker than valvar portion — (b) subsp. **brachycarpum**
1 Siliqua usually at least 2·5 cm, erecto-patent
  3 Annual or biennial; lower leaves deeply pinnatisect; siliqua usually more than 3 cm — (c) subsp. **pseudosinapis**
  3 Usually perennial; lower leaves lyrate; siliqua usually less than 3 cm — (d) subsp. **baeticum**

(a) Subsp. **virgatum**: Biennial or perennial. Lower leaves lyrate. Siliqua 1–2 cm, erecto-patent, contracted at the base of the beak; beak 3 mm. $2n = 14$. *S.W. Italy, Sicilia.*

(b) Subsp. **brachycarpum** (Rouy) Gómez-Campo, *Anal. Jard. Bot. Madrid* **40**: 68 (1983): Usually perennial. Lower leaves lyrate. Siliqua 1·5–2·5 cm, more or less erect, not contracted at the base of the beak; beak often nearly as long as, and usually thicker than, the valvar portion. $2n = 14$. *E. Spain (Alicante to Castellón).*

(c) Subsp. **pseudosinapis** (Lange) Gómez-Campo, *loc. cit.* (1983) (*Sinapis pseudosinapis* Lange): Annual or biennial. Lower leaves deeply pinnatisect, the terminal segment narrowed. Siliqua 2·5–3·5 cm, erecto-patent, not contracted at the base of the beak; beak up to $\frac{1}{2}$ as long as the valvar portion. $2n = 14, 28$. *S.E. Spain (Granada to Murcia).*

(d) Subsp. **baeticum** (Boiss.) Gómez-Campo, *loc. cit.* (1983) (*Sinapis baeticum* Boiss., *E. baeticum* (Boiss.) Nyman): Usually perennial. Lower leaves lyrate. Siliqua 1·5–3 cm, erecto-patent, not contracted at the base of the beak; beak up to $\frac{1}{3}$ as long as the valvar portion. $2n = 14$. *S. Spain (Málaga prov. to Gibraltar); one station in Alicante prov.*

2. E. nasturtiifolium (Poiret) O. E. Schulz, *Bot. Jahrb.* **54** Beibl. **119**: 56 (1916) (*E. obtusangulum* (Schleicher) Reichenb. fil., *Brassica erucastrum* L., *B. nasturtiifolium* Poiret, *B. obtusangula* Bertol., *Hirschfeldia obtusangula* (Reichenb.) Fritsch). Annual to perennial 20–80 cm, densely hispid with deflexed hairs at least below. Basal leaves lyrate-pinnatisect with *c.* 3 lobes on each side; cauline leaves similar to the basal but smaller, the upper with 6–8(–12) lobes on each side, the two basal lobes deflexed and clasping the stem. Raceme ebracteate or rarely the lowest flower with a bract. Sepals 5 mm, patent; petals *c.* 9 mm, yellow; limb broadly obovate. Siliqua 23–45 × 1·2–1·5 mm, patent, with a short gynophore; beak 3–6 mm, with 1(2) seeds. $2n = 16$. *S.W. Europe, extending to N. France and S. Germany; doubtfully native farther east, but widely naturalized.* *Al Ga Ge He Hs *Hu It *Ju Lu [Au Cz Po Rm].

3. E. palustre (Pirona) Vis., *Linnaea* **28**: 365 (1857). Like **2** but glabrous; lower cauline leaves pectinate-pinnatisect, with 10–18 lobes on each side; beak without seeds. *Marshes and rice-fields.* ● *N.E. Italy.* It.

Perhaps not specifically distinct from **2**.

4. E. gallicum (Willd.) O. E. Schulz, *Bot. Jahrb.* **54** Beibl. **119**: 56 (1916) (*E. pollichii* Schimper & Spenner). Like **2** but basal lobes of cauline leaves not deflexed or clasping the stem; raceme bracteate below; sepals erect to patent; petals 6–10 mm, pale yellow or whitish; siliqua 1·2–1·8 mm wide, not stipitate; beak cylindrical, seedless. $2n = 30$. ● *C. & S.W. Europe.* Au Cz Ga Ge He Ho Hs Hu It [Be Br Da No Po Rm Rs(N, B) Su].

92. Coincya Rouy

V. H. HEYWOOD (EDITION 1) REVISED BY E. A. LEADLAY AND J. R. AKEROYD (EDITION 2)

Annual to perennial, rhizomatous or with fusiform taproot; glabrous or with simple hairs. Inflorescence racemose, ebracteate. Sepals erect, the inner saccate at the base; petals clawed, yellow or whitish, sometimes with violet or reddish veins. Fruit a linear siliqua terminating in an ensiform or attenuate beak with 1–6(–11) seeds; valves convex, 3-veined; stigma capitate. Seeds in 1 row in each loculus, numerous. (*Hutera* Porta, *Rhynchosinapis* Hayek, *Brassicella* Fourr. ex O. E. Schulz)

Literature: O. E. Schulz in Engler, *Pflanzenreich* **70(IV.105)**: 106–116 (1919) (sub *Brassicella*). E. A. Leadlay & V. H. Heywood, *Bot. Jour. Linn. Soc.* **102**: 313–398 (1990).

1 Leaves simple, oblong-elliptical or spathulate, entire or repand-dentate; inflorescence a dense corymb **1. richeri**
1 Leaves pinnately divided; inflorescence lax
 2 Beak of siliqua shorter than valvar portion; petal with yellow or sometimes violet or brownish veins
  3 Annual to short-lived perennial; basal leaves pinnatisect (rarely lyrate); siliqua 1·5–2·5(–3) mm wide **5. monensis**
  3 Biennial to perennial; basal leaves lyrate; siliqua 2·5–4 mm wide **6. wrightii**
 2 Beak of siliqua as long as or longer than valvar portion; petals with violet or brownish veins
  4 Annual, branched at the base; beak of siliqua curved **2. transtagana**
  4 Biennial to perennial, branched above; beak of siliqua straight
   5 Stems hispid below; beak of siliqua 20–55 mm, not inflated **3. longirostra**
   5 Stems tomentose-lanate below; beak of siliqua 13–27 mm, inflated **4. rupestris**

1. C. richeri (Vill.) Greuter & Burdet, *Willdenowia* **13**: 87 (1983) (*Brassicella richeri* (Vill.) O. E. Schulz, *Rhynchosinapis richeri* (Vill.) Heywood, *Brassica richeri* Vill.). Glabrous, sometimes scapose perennial, with branched, stoloniferous stock; flowering stems 15–40 cm (up to 70 cm in fruit), unbranched. Leaves lanceolate to elliptical or spathulate, entire or repand-dentate. Raceme densely corymbose. Sepals 8 mm, obtuse; petals 14–18(–20) mm, yellow. Siliqua 20–60(–80) × 2·5–5 mm, erecto-patent or slightly pendent, straight; veins of valves often anastomosing; beak 7–20 mm, with 1–4 seeds. $2n = 24$. *Alpine meadows and rocks*, 1750–2500 *m*. ● *S.W. Alps.* Ga It.

2. C. transtagana (Coutinho) M. Clemente & J. E. Hernandez-Bermejo, *Lagascalia* **14**: 138 (1986) (*Brassicella valentina* O. E. Schulz pro parte, *Brassica longirostra* var. *transtagana* (Coutinho) Samp., *Sisymbrium valentinum* auct., non L., *Rhynchosinapis hispida* (Cav.) Heywood subsp. *transtagana* (Coutinho) Heywood, *Coincya hispida* subsp. *transtagana* (Coutinho) Greuter & Burdet). Annual, divaricately branched from the base, hispid-arachnoid; stems 15–40 cm. Leaves pinnatipartite or pinnatisect, with 5–9 pairs of very narrow pinnatipartite to entire lobes, the terminal lobe equalling the laterals. Petals 14–16·5 mm, pale yellow or whitish, with violet veins. Siliqua 20–50 × 1·5–2 mm; apex of valves retuse; beak 15–30 mm, recurved, up to as long as the valvar portion, with 3–5 seeds. *Rocky and disturbed ground.* ● *S.W. Spain, S. Portugal.* Hs Lu.

3. C. longirostra (Boiss.) Greuter & Burdet, *Willdenowia* **13**: 87 (1983) (*Brassica longirostra* Boiss., *Brassicella valentina* O. E. Schulz pro parte, *Rhynchosinapis longirostra* (Boiss.) Heywood, *Sinapis longirostra* (Boiss.) Boiss ex Amo). Biennial to short-lived perennial; stems 20–100 cm, branched, hispid below. Leaves pinnatipartite or pinnatisect, velutinous; lobes pinnatifid to dentate. Petals 15–20 mm, pale yellow, with violet-brown veins. Siliqua (10–)30–65 mm, patent to deflexed, puberulent; beak 20–55 mm, with 2–7(–11) seeds, considerably longer than the valvar portion. $2n = 24$. *Rocky slopes.* ● *S. Spain (Sierra Morena).* Hs.

4. C. rupestris Porta & Rigo ex Rouy in Deyrolle, *Naturaliste (Paris)* **2**: 248 (1891) (*Hutera rupestris* Porta & Rigo ex Porta). Biennial to short-lived perennial; stems 20–100 cm, erect, divaricately branched, tomentose-lanate below. Leaves pinnatipartite to pinnatisect, with 4–9 pairs of pinnatifid to dentate-crenate lobes, the terminal lobe larger than the laterals, tomentose to velutinous. Inflorescence corymbose. Petals 12–20 mm, yellow, with brownish veins. Siliqua 18–35 mm, straight or recurved, glabrous; beak 13–27 mm, inflated, usually longer than the valvar portion. *Rocks and screes*, 800–1200 *m*. ● *S. Spain; very local.* Hs.

(a) Subsp. **rupestris**: Valvar portion of the siliqua 5–8(–10) mm; beak 4–8 mm wide, pyriform, with 4–6(–8) seeds. $2n = 24$. *Calcicole. S. Spain (Sierra de Alcaraz, ?Sierra de Cazorla).*

(b) Subsp. **leptocarpa** (González-Albo) Leadlay, *Bot. Jour. Linn. Soc.* **102**: 364 (1990) (*Hutera leptocarpa* González-Albo, *Coincya leptocarpa* (González-Albo) Greuter & Burdet): Valvar portion of the siliqua (4–)7–17 mm; beak 2–4·5 mm wide, moniliform, with 3–6 seeds. $2n = 24$. *Calcifuge. S. Spain (Sierra Morena).*

5. C. monensis (L.) Greuter & Burdet, *Willdenowia* **13**: 87 (1983) (*Rhynchosinapis monensis* (L.) Dandy ex Clapham, *Brassica monensis* (L.) Hudson). Annual to perennial, prostrate to erect; stems up to 100 cm, usually branched and hispid below. Leaves pinnatisect to pinnatipartite or rarely lyrate-pinnatisect, with 3–9(–11) pairs of pinnatipartite to dentate lobes. Petals (7–)10–22(–26) mm, yellow, sometimes with brown or purple veins. Siliqua (10–)30–90 × 1·5–2·5(–3) mm, straight, rarely curved; beak

5–25(–35) mm, with (0)1–5 seeds. *Open habitats, often on rocks or sand; somewhat calcifuge. W. Europe, extending eastwards to W.C. Italy; mainly in the mountains.* Be Br Co Ga Ge Hs It Lu Sa [Ho].

A very variable species that includes a number of ecological and geographical variants. For an account of the taxa that have been recognized, see E. A. Leadlay & V. H. Heywood, *Bot. Jour. Linn. Soc.* **102**: 313–398 (1990). Several of these have been treated at specific or subspecific rank in regional Floras but the characters that have been used to separate them show considerable overlap. The following subspecies can be distinguished.

1 Plant densely hispid or puberulent; leaves pinnatisect
 2 Plant puberulent (sometimes slightly hispid below); basal leaves withered by anthesis (**d**) subsp. **puberula**
 2 Plant hispid; basal leaves usually present at anthesis
  3 Veins of petals usually yellow (**b**) subsp. **recurvata**
  3 Veins of petals often violet or brownish (**c**) subsp. **hispida**
1 Plant glabrous to sparsely hispid; leaves pinnatisect or lyrate
 4 Stems up to 100 cm, usually erect; petals (12–)15–22 mm (**b**) subsp. **recurvata**
 4 Stems not more than 60 cm, procumbent to ascending; petals 10–15(–20) mm
  5 Leaves pinnatisect; beak of siliqua with (0–)1–5 seeds; seeds 1·3–2 mm (**a**) subsp. **monensis**
  5 Leaves often lyrate; beak of siliqua with 1 or 2 seeds; seeds 0·8–1·2 mm (**e**) subsp. **nevadensis**

(**a**) Subsp. **monensis**: Biennial to short-lived perennial; stock sometimes rhizomatous or stoloniferous; stems 10–60 cm, procumbent to ascending, glabrous to sparsely hispid below. Leaves mostly basal, with 4–9 pairs of dentate to somewhat pinnatifid lobes, slightly hispid. Petals 10–20 mm, yellow. Siliqua 25–75(–80) mm; beak (5–)8–26 mm, with (0)1–5 seeds. Seeds 1·3–2 mm. $2n = 24$. *Maritime sands.* ● *W. Britain and Isle of Man.*

(**b**) Subsp. **recurvata** (All.) Leadlay, *Bot. Jour. Linn. Soc.* **102**: 370 (1990) (*Sinapis recurvata* All., *S. cheiranthus* subsp. *recurvata* (All.) Arcangeli, *Rhynchosinapis cheiranthos* (Vill.) Dandy subsp. *cheiranthos*, *R. erucastrum* (L.) Dandy ex Clapham & subsp. *cintrana* (Coutinho) Franco & P. Silva, *R. johnstonii* (Samp.) Heywood, *R. granatensis* (O. E. Schulz) Heywood, *R. pseuderucastrum* (Brot.) Franco subsp. *setigera* (Gay ex Lange) Heywood, *Brassicella erucastrum* O. E. Schulz pro parte, *B. valentina* O. E. Schulz var. *granatensis* O. E. Schulz, *Brassica cheiranthos* Vill., *B. erucastrum* O. E. Schulz pro parte, *B. monensis* auct., non (L.) Hudson, *B. johnstonii* Samp., *B. setigera* (Gay ex Lange) Willk., *B. pulverula* Pau, *Coincya cheiranthos* (Vill.) Greuter & Burdet, *C. johnstonii* (Samp.) Greuter & Burdet, *C. erucastrum* subsp. *setigera* (Gay ex Lange) Greuter & Burdet, *C. granatensis* (O. E. Schulz) Greuter & Burdet, *Sinapis setigera* Gay ex Lange): Annual to short-lived perennial; stems 10–100 cm, usually erect, sparsely to densely hispid below (rarely glabrous). Leaves pinnatisect or pinnatipartite to lyrate, with 3–9 pairs of lobes, sparsely to densely hispid, sometimes coriaceous and glaucous. Petals 12–26 mm, yellow. Siliqua 30–80(–85) mm; beak 8–22(–35) mm, with 1–3(4) seeds. Seeds 0·8–1·4(–1·6) mm. $2n = 24, 48$. *Sandy and rocky habitats. Throughout most of the range of the species; naturalized in Britain.*

(**c**) Subsp. **hispida** (Cav.) Leadlay, *op. cit.* 381 (1990) (*Rhynchosinapis hispida* (Cav.) Heywood, *Coincya hispida* (Cav.) Greuter & Burdet, *Brassicella valentina* O. E. Schulz pro parte, *Brassica valentina* DC., *Sisymbrium valentinum* auct., non L.): Annual to short-lived perennial. Stems 10–100 cm, erect, hispid, especially below. Leaves pinnatisect or pinnatipartite, with 5–11 pairs of pinnatipartite to pinnatifid lobes, hispid. Petals whitish to yellow, often with brown or violet veins. Siliqua 30–70(–80) mm, straight or rarely curved; beak 5–20 mm, with 1–4 seeds. Seeds 0·8–1·4 mm. $2n = 24$. *Ruderal.* ● *N. & C. Spain; C. Portugal.*

(**d**) Subsp. **puberula** (Pau) Leadlay, *op. cit.* 383 (1990) (*Brassica puberula* Pau, *Coincya puberula* (Pau) Greuter & Burdet): Annual to short-lived perennial; stems 20–60 cm, little-branched, densely puberulent, sometimes sparsely hispid below. Basal leaves withered by anthesis; leaves pinnatisect to pinnatipartite, puberulent or sparsely hispid. Petals 7–18 mm, yellow. Siliqua 20–50(–65) mm; beak 7–16 mm, with 1 or 2 seeds. Seeds 0·8–1·2 mm. $2n = 24$. *Usually ruderal.* ● *N. Spain; N.W. Portugal.*

(**e**) Subsp. **nevadensis** (Willk.) Leadlay, *op. cit.* 384 (1990) (*Rhynchosinapis cheiranthos* subsp. *nevadensis* (Willk.) Heywood, *Coincya cheiranthos* subsp. *nevadensis* (Willk.) Greuter & Burdet, *Brassica cheiranthos* var. *nevadensis* Willk.): Subscapose biennial to perennial; stems 10–25(–40) cm, prostrate, glabrous or sparsely hispid below. Leaves forming persistent rosette, lyrate to pinnatisect, with entire to dentate lobes, sparsely setose. Petals 10–15 mm, yellow. Siliqua 15–50 mm; beak 5–12 mm, with 1(2) seeds. Seeds 0·8–1·2 mm. $2n = 24$. *Rocks, screes and stony slopes; 2300–3200 m. S. Spain (Sierra Nevada, Sierra de los Filabres).*

6. C. wrightii (O. E. Schulz) Stace, *Watsonia* **17**: 443 (1989) (*Rhynchosinapis wrightii* (O. E. Schulz) Dandy ex Clapham). Biennial to perennial, somewhat woody at the base; stems 20–100 cm, densely pubescent at least below. Leaves lyrate, with 1–5(6) pairs of pinnatifid to dentate lobes, appressed-pubescent, especially beneath. Petals 14–20 mm, yellow. Ovary sparsely pubescent. Siliqua (20–)40–80 × 2·5–4 mm, straight to somewhat curved; beak 7–16 mm, with 1–3 seeds. $2n = 24$. *Cliffs and stony slopes.* ● *S.W. England (Lundy).* Br.

93. Hirschfeldia Moench

†T. G. TUTIN (EDITION 1) REVISED BY J. R. AKEROYD (EDITION 2)

Like *Sinapis* but the sepals almost erect, the inner pair slightly saccate; siliqua with a short, swollen beak; valves 3-veined when mature but obscurely veined when dry; seeds subglobose-ovoid.

1. H. incana (L.) Lagrèze-Fossat, *Fl. Tarn Gar.* 19 (1847) (*Sinapis incana* L.). Annual to biennial; stem up to 1·5 m, usually much-branched, leafy, densely white-hairy to somewhat glabrous below. Lower leaves runcinate-pinnatifid, with an ovate, obtuse, shallowly dentate terminal lobe, the upper simple, more or less sessile. Petals 5–10 mm, pale yellow, sometimes with dark veins. Siliqua 7–17 × 1–1·5(–1·8) mm, erect and appressed to the stem, 2-valved; beak *c.* $\frac{1}{2}$ as long as the valves, swollen at the base, usually with 1 seed. Seeds 2–6 in each loculus. $2n = 14$. *S. Europe; becoming widely naturalized further north.* Al Bl Co Cr Ga Gr Hs It Ju Lu Sa Si Tu [Au Be Br Da Ge Hb He Ho].

Often confused with *Brassica nigra*, which never has a seed in the beak.

94. Carrichtera DC.

V. H. HEYWOOD

Annual; hairs unbranched. Inflorescence leaf-opposed. Sepals erect; filaments free. Fruit a transversely articulate, deflexed silicula; lower segment ellipsoidal, with 2 naviculiform, 3-veined valves; upper segment sterile, strongly compressed, foliaceous, cochleariform.

1. C. annua (L.) DC., *Mém. Mus. Hist. Nat.* (*Paris*) **7**: 250 (1821) (*C. vellae* DC., *Vella annua* L.). Stems 5–40(–60) cm, branched from the base, with setiform, deflexed hairs; hairs and branches pungent in fruit. Leaves 2- or 3-pinnatisect, with obtuse, linear segments. Raceme with 10–15(–20) flowers. Petals 6–8 mm, pale yellow, violet-veined. Silicula 5–8 mm, deflexed or patent, on pedicels 2–3 mm; lower segment bilocular. Seeds 3 or 4 in each loculus. *Mediterranean region.* Bl Cr Gr Hs Sa Si.

95. Vella L.

V. H. HEYWOOD

Small, much-branched shrubs with sessile, entire leaves; hairs unbranched. Flowers in lax, spicate racemes. Sepals erect; petals long-clawed, yellow, sometimes violet-veined; filaments of inner stamens connate in pairs. Fruit a transversely articulate silicula; the lower segment 2-valved, ellipsoidal, subdidymous; valves convex, 3-veined; the upper segment sterile, strongly compressed, in the form of a foliaceous, lingulate beak, 5-veined. Seeds 1(2) in each loculus.

Literature: O. E. Schulz in Engler, *Pflanzenreich* **84(IV.105)**: 44–47 (1923).

Plant spineless, with obovate or obovate-lanceolate leaves; raceme with 10–40 flowers **1. pseudocytisus**
Plant very spiny, with linear to linear-lanceolate leaves; raceme with 3–5 flowers **2. spinosa**

1. V. pseudocytisus L., *Sp. Pl.* 641 (1753). Shrub 30–100 cm; setose-hispid or glabrous (var. *badalii* (Pau) Heywood). Leaves obovate or obovate-lanceolate, somewhat coriaceous. Raceme with 10–40 flowers. Petal-limb 3–4 mm wide, orbicular or shortly obovate, with indistinct veins. Lower segment of silicula 3·5–4·5 × 3 mm, sparsely setose, hispid or glabrous, bilocular; upper segment obtuse. *C., E. & S.E. Spain.* Hs.

(a) Subsp. **pseudocytisus** (*V. monosperma* Menéndez Amor): Leaves densely hispid; silicula with short dense hairs on valves. $2n = 68$. ● *Madrid and Almería provs.*

(b) Subsp. **paui** Gómez-Campo, *Bot. Jour. Linn. Soc.* **82**: 174 (1981) (*V. badallii* Pau ex Pau): Leaves glabrous, except on margin; silicula glabrous. $2n = 34$. ● *Teruel and Zaragoza provs.*

2. V. spinosa Boiss., *Biblioth. Univ. Genève* ser. 2, **13**: 407 (1838). Very spiny, caespitose, intricately branched shrub, 10–100 cm; stems with scattered long, patent hairs, the upper branches dichotomously branched and transformed into rigid, patent spines. Lower leaves linear-lanceolate; upper leaves linear; all setose-ciliate at the margin. Raceme with 3–5 flowers. Petal-limb 4–6 mm, obovate, strongly violet-veined. Lower segment of silicula 4·5–5·5 × 2·5–3 mm, glabrous, bilocular; upper segment acuminate. *Limestone mountains.* ● *S. & S.E. Spain.* Hs.

96. Succowia Medicus

P. W. BALL

Hairs unbranched. Sepals erect, not saccate; petals shortly clawed, yellow. Fruit a 2-seeded silicula, covered with long, conical spines; style long, conical.

1. S. balearica (L.) Medicus, *Pflanzengatt.* **1**: 65 (1792). Glabrous or scabrid annual 20–70 cm. Leaves ovate, pinnatisect, the segments dentate to 2-pinnatifid. Petals 7–10 mm. Silicula 3–6 mm in diameter, spines 1–3 mm; style 4–8 mm. Seeds *c.* 2 mm, pitted. $2n = 36$. *W. Mediterranean region.* Bl Co Hs It Sa Si.

97. Boleum Desv.

V. H. HEYWOOD

Hairs unbranched. Flowers in short, dense racemes, later elongating. Sepals erect; petals long-clawed, pale yellow, with reddish-brown veins and claw; stamens without appendages, the inner united in pairs. Fruit transversely articulate, the lower segment ovoid-globose, 2-valved, septate, indehiscent, the upper in the form of a sterile, lingulate beak.

1. B. asperum (Pers.) Desv., *Jour. Bot. Appl.* **3**: 163 (1814). Densely setose shrub up to 50 cm with a woody stock and erect, fasciculate, leafy stems and branches. Leaves 10–25 × 2–3·5 mm, linear-lanceolate, entire, or sometimes deeply pinnatifid with 1 or 2 pairs of narrow lobes. Lower segment of fruit deeply sculpted and covered with long, white, setiform, unicellular hairs, 2-locular, with 1 seed in each loculus. *Saline soils.* ● *E. Spain.* Hs.

98. Erucaria Gaertner

V. H. HEYWOOD

Annual or biennial; hairs unbranched. Sepals erect; petals with a long claw, lilac. Fruit a transversely articulate siliqua with both segments fertile, the lower 2-valved and dehiscent, the upper abruptly contracted into the style.

1. E. hispanica (L.) Druce, *Rep. Bot. Exch. Club Brit. Is.* **3**: 418 (1914) (*E. myagroides* (L.) Halácsy, *E. teñuifolia* DC., *E. aleppica* Gaertner). Annual or biennial; stem erect, glabrous or slightly hairy at the base. Leaves petiolate, pinnatisect with linear to oblong lobes, pinnatifid, or entire. Petals 10–15 mm. Siliqua 1–2 cm, longitudinally veined, appressed to the axis; lower segment elongate, cylindrical, with 2–4 seeds; upper segment compressed, ensiform, obtuse or truncate at the apex, with 1–3 seeds; style 2–5 mm, filiform. *Cultivated land and waste places. S. Greece and Aegean region.* Cr Gr [Hs ?It].

99. Cakile Miller

P. W. BALL

Glabrous, glaucous annuals, with succulent leaves. Petals clawed, violet, pink or white. Fruit a transversely articulate siliqua; upper segment larger, ovoid, more or less 4-angled, not attenuate into style, with 1(2) seeds; lower obconical, not evidently 2-valved, usually with a single seed, indehiscent.

Literature: E. G. Pobedimova, *Nov. Syst. Pl. Vasc.* (*Leningrad*) **1964**: 90–128 (1964). P. W. Ball, *Feddes Repert.* **69**: 35–40 (1964). J. E. Rodman, *Contr. Gray Herb.* **205**: 3–146 (1974). R. Elven & T. Gjelsås, *Blyttia* **39**: 87–106 (1981).

There is considerable diversity of opinion as to the number of taxa that can be recognized in Europe and as to the rank to which these should be assigned.

1 Lower segment of silicula with 2 lateral projections near the apex **1. maritima**
1 Lower segment of silicula without lateral projections
2 Pedicels 5–10 mm in fruit; upper segment of silicula 14–20 mm, terete **1. maritima**

2 Pedicels not more than 5(–8) mm in fruit; upper segment of silicula 7–15 mm, 4-angled **2. edentula**

1. C. maritima Scop., *Fl. Carn.* ed. 2, **2**: 35 (1772) (*C. monosperma* Lange). Plant 15–60 cm, usually decumbent. Leaves subentire to pinnatisect. Petals 5–14 mm. Silicula 12–25(–30) mm; lower segment often with 2 lateral, more or less deflexed projections at the apex; upper segment 4-angled to terete, often expanding into a broad membranous margin at the base. *Coasts of Europe.* All except Au Az He Hu; introduced in Cz.

A variable species that is divisible into at least 4 subspecies in Europe.

1 Leaves 1- or 2-pinnatisect, the primary segments mostly 8–20 times as long as wide
2 Lower segment of silicula ±concave at the apex (when viewed in the plane of the projections); projections usually less than 0.5 mm **(b) subsp. euxina**
2 Lower segment of silicula not concave at the apex (when viewed in the plane of the projections); projections often exceeding 1 mm, conspicuous, deflexed **(c) subsp. baltica**
1 Leaves subentire to pinnatisect, the primary divisions rarely more than 6 times as long as wide
3 Lower segment of silicula flat at the apex (when viewed in the plane of the projections); projections less than 0.5 mm, sometimes obscure **(a) subsp. maritima**
3 Lower segment of silicula ±concave at the apex (when viewed in the plane of the projections); projections often exceeding 1 mm, conspicuous **(d) subsp. aegyptiaca**

(a) Subsp. **maritima** (*C. maritima* subsp. *integrifolia* (Hornem.) Hyl.): Leaves subentire to pinnatisect, the primary segments not more than 5 times as long as wide. Pedicels 1·5–5(–7) mm in fruit. Lower segment of the silicula flat (when viewed in the plane of the projections) and convex (when viewed in the plane at right-angles to the projections); projections usually less than 0·5 mm, sometimes obscure. Upper segment of the silicula expanding into a membranous margin at the base. $2n = 18$. *W. & N.W. Europe from C. Portugal to* c. 68° *N. in Norway; rarely in the Baltic and Mediterranean regions in the vicinity of ports, where it is probably introduced.*

(b) Subsp. **euxina** (Pobed.) E. I. Nyárády in Săvul., *Fl. Rep. Pop. Române* **3**: 480 (1955) (*C. euxina* Pobed.): Leaves 1- or 2-pinnatisect, the primary segments up to 15 times as long as wide. Pedicels 2–4·5 mm in fruit. Lower segment of the silicula more or less concave at the apex (when viewed in the plane of the projections), with projections usually less than 0·5 mm; articulating surface with 2 erect triangular teeth. Upper segment of the silicula expanding into a membranous margin at the base. $2n = 18$. *Black Sea region.*

(c) Subsp. **baltica** (Jordan ex Rouy & Fouc.) Hyl. ex P. W. Ball, *Feddes Repert.* **69**: 37 (1964) (*C. baltica* Jordan ex Pobed.): Like **(b)** but primary leaf segments 10–20 times as long as wide; pedicels 3–5 mm in fruit; lower segment of the silicula not concave at the apex, with deflexed projections usually exceeding 1 mm; articulating surface without erect teeth. $2n = 18$. *Baltic region and S.E. Norway.*

(d) Subsp. **aegyptiaca** (Willd.) Nyman, *Consp.* 29 (1878): Leaves subentire to pinnatisect, the primary divisions rarely more than 6 times as long as wide. Pedicels 1·5–6 mm in fruit. Lower segment of the silicula more or less concave at the apex (when viewed in the plane of the projections), with deflexed projections usually exceeding 0·5 mm; articulating surface sometimes with 2 erect triangular teeth. Upper segment of the silicula expanding into a membranous margin at the base. $2n = 18$. *Mediterranean region and S. Portugal.*

Populations in the Aegean region are mostly intermediate between subspp. **(b)** and **(d)**.

Subsp. **(d)** is not always clearly distinct from subsp. **(a)** on the basis of the fruit characters given above. However, these subspecies are apparently distinct phytochemically (J. E. Rodman, *Syst. Bot.* **1**: 137–148 (1976)), and in nectary structure (M. Clemente & J. E. Hernández-Bermejo, *Anal. Inst. Bot. Cavanilles* **35**: 279–296 (1978)). The situation is further complicated by the occurrence of plants apparently identical with **(a)** in the vicinity of ports in the Mediterranean region.

In subsp. **(a)** the lower segment of the silicula is often sterile, and resembles an extension of the pedicel.

Plants in which the apex of the lower segment of the silicula is flat, without projections, and in which the upper segment is contracted to a flat base without a membranous margin, have been referred to **C. edentula** subsp. **islandica** (Gand.) A. & D. Löve, *Bot. Not.* **114**: 52 (1961) (*C. arctica* Pobed., *C. lapponica* Pobed.), but may justify recognition as a fifth subsp. of *C. maritima*. Reported from N. Russia, N. Norway, Svalbard, Faeröer and Iceland, intermediates between it and subsp. **(a)** are frequent in N. Norway between 66° N. and 68° N.

2. C. edentula (Bigelow) Hooker, *Fl. Bor.-Amer.* **1**: 59 (1830) (*C. americana* Nutt.). Plant up to 60 cm. Leaves subentire to pinnatisect. Pedicels 1·5–5(–8) mm in fruit. Petals 4·5–8 mm. Silicula 12–25 mm; lower segment flat at the apex, without lateral projections; upper segment 7–15 mm, 4(–8)-angled, tapered towards a flat base, without a membranous margin. *Açores.* *Az. (*E. North America.*)

100. Rapistrum Crantz

†J. DE CARVALHO E VASCONCELLOS (EDITION 1) REVISED BY J. R. AKEROYD (EDITION 2)

Annuals to perennials, with unbranched hairs. Sepals erecto-patent; petals yellow, with a short claw. Fruit a transversely articulate silicula; lower segment cylindrical, with (0)1–3 seeds; upper segment ovoid to globose, wider than the lower, indehiscent, caducous at maturity, with a single erect seed; beak conical or filiform.

Biennial or perennial; upper segment of fruit gradually attenuate into a beak 0·5–1 mm **1. perenne**
Annual; upper segment of fruit abruptly contracted into a beak 1–3(–5) mm **2. rugosum**

1. R. perenne (L.) All., *Fl. Pedem.* **1**: 258 (1785) (*R. diffusum* All.). Biennial or perennial 20–80 cm, densely hispid below, glabrous above. Lower leaves pinnate or pinnatipartite, coarsely serrate; upper pinnatifid or dentate, sessile or shortly petiolate. Petals 5–7 mm, bright yellow. Silicula 5–10 mm; upper segment 3–4·5 mm (including beak), ovoid, strongly longitudinally ribbed, gradually attenuate into a conical beak 0·5–1 mm; lower segment narrower, usually more or less cylindrical, about ½ as long as the pedicel. *Probably native in parts of C. & E. Europe; casual and often naturalized elsewhere in C. & W. Europe.* Au Bu Cz Ge Hu It Ju Rm Rs(C, W, E) [Br Ga He Po].

2. R. rugosum (L.) All., *Fl. Pedem.* **1**: 257 (1785). Annual 15–60(–100) cm, hispid below, often glabrous above. Lower leaves pinnate; upper usually dentate, sessile to petiolate. Petals 6–10 mm, pale yellow. Silicula 3–10 mm; upper segment 3–6 mm (including beak), ovoid to globose, abruptly contracted into a filiform beak 1–3(–5) mm; lower segment (0·7–)1·2–3 mm, cylindrical to ellipsoidal. *S. Europe; naturalized or casual in much of C. & N. Europe.* Al Az Bl Bu Co Cr Ga Gr Hs It Ju Lu Rs(K, E) Sa Si Tu [Au Be Br Cz Ge Hb He Ho Hu Po Rm Rs(C, W)].

A variable species, with several variants treated at subspecific rank or as species. The following subspecies are more or less distinct, although intermediates occur.

(**a**) Subsp. **rugosum**: Pedicels 1·5–3·2 mm, thick (up to 1·3 mm in diameter). Lower segment of the fruit 1–$1\frac{1}{2}$ times as wide as the pedicel, obconical, not constricted above; upper segment obovoid, usually deeply ribbed and rugose. $2n = 16$. *S. Europe; widely naturalized elsewhere.*

(**b**) Subsp. **orientale** (L.) Arcangeli, *Comp. Fl. Ital.* 49 (1882) (*R. orientale* (L.) Crantz; incl. *R. rugosum* subsp. *linnaeanum* Rouy & Fouc., *R. linnaeanum* Boiss. & Reuter, nom. illeg., *R. hispanicum* (L.) Crantz): Pedicels 2·3–5 mm, slender (up to 0·7 mm in diameter). Lower segment of the fruit $1\frac{1}{2}$–2 times as wide as the pedicel, ellipsoidal, constricted above; upper segment broadly ellipsoidal, usually shallowly ribbed and slightly rugose. *Mediterranean region; widely naturalized or casual elsewhere.*

Plants of subsp. (**b**) with the lower segment of the fruit sterile, 0·7–1·2 mm, and less than $1\frac{1}{2}$ times as wide as the pedicel, have been called subsp. **linnaeanum** Rouy & Fouc., *Fl. Fr.* **2**: 73 (1895), but the variation between them and subsp. (**b**) is continuous, and they probably do not merit formal taxonomic recognition. The lower segment of the fruit of subsp. (**a**) is usually fertile.

101. Didesmus Desv.

P. W. BALL

Like *Rapistrum* but petals white; upper segment of silicula as wide as the lower segment, tetragonal; beak long, pungent.

1. D. aegyptius (L.) Desv., *Jour. Bot. Appl.* **3**: 160 (1814). Erect annual 10–40 cm, sparsely hispid. Leaves oblong-elliptical, dentate to lyrate-pinnatisect with 2–5 pairs of lobes, or rarely 2-pinnatisect. Pedicels 3–5 mm in fruit. Sepals 3·5–4·5 mm; petals 6–10 mm, slightly emarginate. Silicula 6–10 mm; lower segment with 4–6 swellings at the base; upper segment ovoid, often tetragonal; beak 2·5–4 mm. *S. Aegean region.* Cr Gr.

102. Crambe L.

P. W. BALL

Annuals to perennials; glabrous or with unbranched hairs. Sepals erecto-patent; petals white, with a short claw or cuneate at the base; filaments of inner stamens usually with a tooth-like appendage. Fruit a transversely articulate silicula; lower segment short, sterile, forming a stalk together with the gynophore; upper segment ovoid to globose, indehiscent, caducous, with a single pendent seed; stigma sessile.

Literature: O. E. Schulz in Engler, *Pflanzenreich* **79(IV.105)**: 228–249 (1919).

1 Lower segment of silicula longer than the upper; filaments of the inner stamens without appendage **8. filiformis**
1 Lower segment of silicula shorter than the upper; filaments of the inner stamens usually with a tooth-like appendage
2 Annual; lower leaves with a large, $\pm$reniform terminal lobe and 0–2 pairs of small lateral lobes **7. hispanica**
2 Stout perennial; lower leaves entire to 2-pinnatisect, the terminal lobe neither reniform nor much larger than the lateral lobes
3 Plant densely hispid **(2–5). tataria** group
3 Plant glabrous or sparsely hispid
4 Petals 6–10 mm; upper segment of silicula 7–12 mm
5 Upper segment of silicula not 4-angled; lower cauline leaves pinnatifid or irregularly toothed **1. maritima**
5 Upper segment of silicula 4-angled; lower cauline leaves 1- or 2- pinnatisect **(2–5). tataria** group
4 Petals not more than 6 mm; upper segment of silicula less than 7 mm
6 Lower segment of silicula slender (less than 0·5 mm in diameter); pedicels slender (*c.* 0·3 mm in diameter) **6. koktebelica**
6 Lower segment of silicula relatively stout (more than 0·5 mm in diameter); pedicels relatively stout (*c.* 0·5 mm or more in diameter) **(2–5). tataria** group

1. C. maritima L., *Sp. Pl.* 671 (1753) (*C. pontica* Steven ex Rupr.). Stout, glabrous, glaucous perennial 30–75 cm, with branched, fleshy stock. Lower leaves sinuate-dentate to irregularly pinnatifid. Petals 6–10 mm. Lower segment of silicula 1–4 mm, stout; upper segment 7–12 mm, globose to ovoid; pedicels 8–26 mm; seeds 4–5 mm. $2n = 30, 60$. *Maritime sand and shingle. Atlantic, Baltic and Black Sea coasts of Europe; sometimes naturalized elsewhere.* Be Br Bu Da Fe Ga Ge Hb Ho Hs No Rm Rs(B, W, K, E) Su Tu [Au Cz Hu].

(**2–5**). **C. tataria** group. Stout perennials up to 150 cm, with a fleshy, fusiform stock branched at the apex. Petals (3–)3·5–7 mm. Lower segment of silicula *c.* 1 mm, stout; upper segment (3–)3·5–8 mm; pedicels relatively stout (*c.* 0·5 mm or more in diameter). Seeds 2–3·5 mm.

1 Upper segment of silicula 7–9 mm; petals 5–7 mm
2 Plant glabrous or sparsely hispid; upper segment of silicula 4-angled **4. grandiflora**
2 Plant densely hispid; upper segment of silicula not 4-angled **5. aspera**
1 Upper segment of silicula 3–7 mm; petals 3–6 mm
3 Leaves pinnatifid or pinnatisect, the ultimate divisions obtuse **2. tataria**
3 Leaves 2-pinnatisect, the ultimate divisions acute **3. steveniana**

2. C. tataria Sebeók, *Medico-Bot. Tatar.* 7 (1779) (incl. *C. pinnatifida* R. Br., *C. litwinowii* Grossh., *C. tatarica* Pallas). Plant subglabrous to densely hispid. Lower leaves pinnatifid to pinnatisect, the ultimate divisions obtuse; primary segments ovate to triangular in outline. Petals (3–)3·5–6 mm; inner stamens with appendage not more than 0·8 mm. Upper segment of silicula 3–6(–7) mm, 4-angled or not; pedicels usually less than 12 mm. $2n = 30$. *E. Europe, extending westwards to N.E. Italy and Czechoslovakia.* Au Bu Cz Hu It Ju Rm Rs(W, K, E).

3. C. steveniana Rupr., *Mém. Acad. Sci. Pétersb.* ser. 7, **15(2)**: 136 (1869). Like **2** but the leaves 2-pinnatisect, the ultimate divisions acute, the primary segments oblong or linear in outline;

inner stamens with appendage *c.* 1 mm long; upper segment of silicula 6–7 mm, apiculate. *Krym.* Rs(?W, K). (*Caucasus.*)

4. C. grandiflora DC., *Reg. Veg. Syst. Nat.* **2**: 652 (1821). Like **2** but the ultimate divisions of the leaves acute; petals 5·5–7 mm; upper segment of the silicula 7–8 mm; pedicels up to 18 mm. *Krym.* Rs(K). (*Caucasus.*)

5. C. aspera Bieb., *Fl. Taur.-Cauc.* **2**: 90 (1808). Plant densely hispid. Leaves deeply pinnatifid to 2-pinnatisect. Petals 5–7 mm. Upper segment of silicula 7–9 mm, not 4-angled; pedicels up to 13 mm. ● *S.E. Russia, S. & E. Ukraine.* Rs(K, E).

This species has been erroneously recorded from a number of other territories. Such plants usually have the upper segment of the silicula smaller and 4-angled and are included here in **2**.

C. gibberosa Rupr., *Mém. Acad. Sci. Pétersb.* ser. 7, **15(2)**: 136 (1869), has been recorded from S.E. Russia (near the northern margin of the Caucasus). It is like **5** but 30–50 cm (not 40–100 cm); upper segment of silicula smooth, obtuse (not rugose-tuberculate and slightly apiculate). It is doubtful if it is specifically distinct from **5**.

6. C. koktebelica (Junge) N. Busch in Kusn., N. Busch & Fomin, *Fl. Cauc. Crit.* **3(4)**: 294 (1908) (*C. mitridatis* Juz., *C. orientalis* var. *koktebelica* (Junge) O. E. Schulz). Stout perennial 150–250 cm, hispid at the base; stock fusiform. Lower leaves subentire to sinuate, with dentate lobes, sparsely setose to glabrous beneath. Petals 4–6·5 mm. Lower segment of silicula 0·5–1 mm, slender; upper segment 4–4·5 mm, more or less globose, very obscurely angled; pedicels slender (*c.* 0·3 mm in diameter). *S.E. Krym.* Rs(K). (*N.W. Caucasus.*)

Possibly a subspecies of **C. orientalis** L., *Sp. Pl.* 671 (1753), which is widespread in Anatolia.

7. C. hispanica L., *Sp. Pl.* 671 (1753). Slender, usually densely hispid annual 25–100 cm. Lower leaves lyrate-pinnatisect, with large reniform-orbicular terminal lobe, and 0–2 pairs of small lateral lobes. Petals 3–4 mm. Lower segment of silicula *c.* 1 mm; upper segment 3–4·5 mm, globose; pedicels 4–12 mm. *Mediterranean region, S. Portugal.* Al Co Gr Hs It Ju Lu Sa Si.

8. C. filiformis Jacq., *Icon. Pl. Rar.* **3**: 8 (1795) (*C. reniformis* Desf.). Slender, densely hispid perennial 30–100 cm. Lower leaves lyrate-pinnatifid, with more or less orbicular terminal lobe and 3–6 or more pairs of lateral lobes. Petals 3–6 mm. Lower segment of silicula 2–3 mm; upper segment 1–2 mm, globose; pedicels 2–3 mm. *S. Spain.* Hs. (*N. Africa.*)

103. Calepina Adanson

P. W. BALL

Like *Crambe* but glabrous annual or biennial; flowers zygomorphic; filaments without appendages; silicula ovoid-globose ending in a short, thick beak; lower segment absent.

1. C. irregularis (Asso) Thell. in Schinz & R. Keller, *Fl. Schweiz* ed. 2, **1**: 218 (1905) (*C. corvini* (All.) Desv.). Plant 15–80 cm. Basal leaves obovate, entire to lyrate-pinnatifid, petiolate; cauline leaves amplexicaul with acute, patent auricles. Shorter petals *c.* 2 mm, the longer 2·5–3 mm. Silicula 2·5–4 × 2–3 mm, ovoid-globose to ellipsoidal, shortly beaked, 4-veined, and reticulate-rugose when dry. $2n = 28$. *S. Europe and parts of C. & W. Europe; naturalized elsewhere.* Al Be Bu Co *Cr Ga Gr He *Ho Hs Hu It Ju Lu Rm Rs(K) Sa Si Tu [Au Ge].

104. Morisia Gay

†T. G. TUTIN

Sepals suberect; petals yellow; stamens without appendages. Fruit a transversely articulate silicula; the lower segment bilocular, larger than the upper.

1. M. monanthos (Viv.) Ascherson in W. Barbey, *Fl. Sard. Comp.* 173 (1885) (*M. hypogaea* Gay). Pubescent perennial; stock stout. Leaves rosulate, oblong-lanceolate in outline, pinnatisect. Flowers solitary. Petals 9–12 mm. Pedicels 5–25 mm, erect at flowering, elongating up to 60 mm, curving downwards and burying the fruit. Lower segment of fruit subglobose, bilocular, eventually dehiscing by 2 valves, with 3–5(–12) seeds; the upper segment 3 × 3 mm, ovoid-conical, indehiscent, with 1 or 2 seeds. $2n = 14$. *Sandy places.* ● *Corse, Sardegna.* Co Sa.

105. Guiraoa Cosson

V. H. HEYWOOD

Raceme narrow, ebracteate. Sepals erecto-patent; petals yellow, with long claw; filaments without appendages. Fruit an erect, transversely articulate silicula; upper segment bilocular, larger than the lower.

1. G. arvensis Cosson, *Not. Pl. Crit.* 98 (1851). Annual, with erect, simple or branched stems up to 60 cm, with deflexed white hairs below. Lower leaves oblong, irregularly toothed or pinnatifid; upper leaves linear. Raceme elongating during flowering. Petals 6–9 mm. Pedicels 6–8 mm in fruit. Lower segment of fruit *c.* 2 mm, cylindrical, as thick as the pedicel, bilocular, with 1 seed in each loculus; upper segment *c.* 6 × 4–5 mm, eventually deciduous, subglobose, with 8 prominent, winged ribs, bilocular, indehiscent, with 1 or 2 seeds, attenuate into the conical beak. $2n = 18$. *Dry places.* ● *S.E. Spain.* Hs.

106. Enarthrocarpus Labill.

V. H. HEYWOOD

Annuals, with unbranched hairs. Raceme bracteate. Sepals erecto-patent; petals clawed, yellow, with violet veins. Fruit a transversely articulate, non-septate siliqua; the lower segment 2-valved, indehiscent, seedless or with 1 or 2 seeds; the upper segment linear, elongate, caducous, with 3–15 seeds, constricted between the seeds and separating into 1-seeded portions.

Literature: O. E. Schulz in Engler, *Pflanzenreich* **70(IV.105)**: 210–218 (1919). A. Béguinot, *Nuovo Gior. Bot. Ital.* nov. ser., **21**: 361 (1914).

Petals 10–13 mm; raceme bracteate only at the base; lower segment of fruit 1–6 mm, seedless or with 1 seed **1. arcuatus**
Petals 6–7 mm; raceme bracteate almost to the apex; lower segment of fruit 7–14 mm, with 2 or 3 seeds **2. lyratus**

1. E. arcuatus Labill., *Icon. Pl. Syr.* **5**: 4 (1812). Stems up to 50 cm, hispid; branches procumbent or ascending. Lower leaves usually rosulate, petiolate, lyrate- or runcinate-pinnatipartite; lateral lobes alternate, oblong, with few teeth. Raceme bracteate at the base only. Petals 10–13 mm. Siliqua 3–9 cm, recurved, shortly setose; lower segment 1–6 mm, usually 1-seeded; upper segment much longer, with 3–15 seeds; beak 5–10 mm, conical. *Maritime sands and rocks. Bulgaria, Greece and Aegean region.* Bu Cr Gr.

2. E. lyratus (Forskål) DC., *Reg. Veg. Syst. Nat.* **2**: 661 (1821). Like **1** but raceme bracteate almost to the apex; petals 6–7 mm; lower segment of siliqua 7–14 mm, with 2 or 3 seeds; upper segment twice as long, with 3–6 seeds; beak 4–5 mm. *Cultivated ground. S. Greece (Methoni, S. Peloponnisos).* *Gr. (*E. Mediterranean.*)

E. pterocarpus (Pers.) DC., *Reg. Veg. Syst. Nat.* **2**: 661 (1821), from N. Africa, may be becoming naturalized in Malta. It has racemes bracteate to the apex, the lower segment of the fruit with 1 or 2 seeds, the upper segment broadly winged, with up to 10 seeds, and the beak 7–13 mm.

107. Raphanus L.

A. O. CHATER

Annuals to perennials. Raceme ebracteate. Sepals erect; petals abruptly contracted into claw; filaments without appendages. Fruit a transversely articulate siliqua; lower segment very short, slender, seedless, indehiscent; upper circular in section, indehiscent, straight-sided, or lomentaceous (at least in part), or constricted between the seeds but not breaking up into 1-seeded portions; beak narrow, seedless.

Literature: A. Thellung in Hegi, *Ill. Fl. Mitteleur.* **4(1)**: 272–286 (1918). O. E. Schulz in Engler, *Pflanzenreich* **70(IV.105)**: 194–210 (1919).

R. sativus L., *Sp. Pl.* 669 (1753), cultivated for its edible root (Radish), is of unknown origin; many cultivars exist, and the plant is a frequent casual over much of Europe. The root is thick, napiform or cylindrical, and the siliqua is 20–90 × 8–15 mm, not lomentaceous, and not or scarcely constricted between the (1–)5–12 seeds. Some of its variants are transitional to the wild plants in Europe and hybrids probably occur. Similar hybrids between R. *sativus* and **1(a)** have been demonstrated in North America by C. A. Panatsos & H. G. Baker, *Genetica* **38**: 243–274 (1967).

1. R. raphanistrum L., *Sp. Pl.* 669 (1753) (R. *sylvestris* Lam.). Plant usually more or less hispid; stem 15–150 cm, erect, branched. Basal and lower cauline leaves lyrate; upper usually entire. Sepals 5–10 mm, elliptic-lanceolate, obtuse; petals about twice as long as the sepals, white, yellow, lilac or violet. Pedicels 1–5 cm in fruit. Siliqua erecto-patent. Seeds 1·5–4 mm, ovoid to subglobose, usually reticulate. *Throughout Europe, but only as an alien in the extreme north.* All territories except Sb.

The following treatment is largely based on that of Thellung, *loc. cit.* (1918); the distribution-pattern, especially in the cases of subspp. (**a**) and (**e**), is confused by widespread introductions.

1 Siliqua less than 5 mm in diameter
 2 Siliqua 4 mm in diameter; seeds 2–5 (**c**) subsp. **rostratus**
 2 Siliqua 1·5–4 mm in diameter; seeds 4–11
  3 Siliqua 3–4 mm in diameter (**a**) subsp. **raphanistrum**
  3 Siliqua 1·5–2 mm in diameter (**b**) subsp. **microcarpus**
1 Siliqua at least 5 mm in diameter
 4 Leaves with contiguous lateral lobes (**d**) subsp. **maritimus**
 4 Leaves with distant lateral lobes
  5 Petals 15–25 mm, pale lilac with darker veins (**c**) subsp. **rostratus**
  5 Petals 10–15 mm, white or yellowish (**e**) subsp. **landra**

(**a**) Subsp. **raphanistrum** (R. *raphanistrum* subsp. *segetum* Clavaud): Annual or biennial. Lateral lobes of basal and lower cauline leaves distant. Petals 12–20 mm, usually white or yellow, usually dark-veined. Siliqua 30–90 × 3–4 mm, lomentaceous at least above, with 3–8 segments separated by irregular, long, rather shallow constrictions; segments usually longer than wide, strongly veined; beak 6–30 mm. $2n = 18$. *A weed of cultivated fields. Throughout Europe, but probably not native in N. Europe and rare or absent in the extreme north.*

(**b**) Subsp. **microcarpus** (Lange) Thell. in Hegi, *Ill. Fl. Mitteleur.* **4(1)**: 275 (1918) (R. *microcarpus* Lange): Like subsp. (**a**) but leaves and petals smaller; petals white or reddish; siliqua 25–35 × 1·5–2 mm, with deeper constrictions and beak 5–10 mm. *Weed of cultivated fields. C. & W. Spain; Portugal; Açores.*

(**c**) Subsp. **rostratus** (DC.) Thell. in Hegi, *op. cit.* 279 (1918) (R. *rostratus* DC.): Annual. Lateral lobes of basal and lower cauline leaves very small and distant. Petals 15–25 mm, pale lilac, with darker veins. Siliqua 30–130 × 4–6 mm, lomentaceous at least above, with 2–7 segments separated by irregular, very long, rather shallow constrictions; segments usually longer than wide, strongly veined; beak 15–75 mm. *Sea-shores. Aegean region.*

(**d**) Subsp. **maritimus** (Sm.) Thell. in Hegi, *op. cit.* 278 (1918) (R. *maritimus* Sm.): Usually perennial; root napiform. Basal leaves usually in a distinct rosette; lateral lobes often irregular in size, contiguous. Petals 15–25 mm, yellow, scarcely veined. Siliqua 15–45 × 5–8 mm, lomentaceous at least above, with 1–6 segments separated by deep constrictions of irregular length; segments as wide or wider than long; beak 6–20 mm. $2n = 18$. *Sandy and rocky sea-shores. Coasts of W. Europe northwards to N. Britain; W. Mediterranean region; Black Sea.*

(**e**) Subsp. **landra** (Moretti ex DC.) Bonnier & Layens, *Tabl. Syn. Pl. Vasc.* 21 (1894) (R. *landra* Moretti ex DC.): Like subsp. (**d**) but sometimes annual; root slender; lateral lobes of basal leaves distant, not interspersed with smaller lobes; petals 10–15 mm, white or yellowish; siliqua 25–60 × 5–8 mm; beak 15–40 mm. $2n = 18$. *Weed of fields and waste places, rarely on the coast. Mediterranean region and Açores, but absent from most of the Aegean region. Probably only casual in several localities further north and on the W. Black Sea coast.*

The petals of subsp. (**a**) are often yellow in the northern and white in the southern part of the range.

Plants from Britain and France with the siliqua *c.* 3 mm wide are intermediate between subspp. (**a**) and (**b**).

Populations intermediate between subspp. (**a**) and (**d**), of hybrid origin, occur frequently on the coasts of N.W. Europe.

69. RESEDACEAE

EDIT. J. R. EDMONDSON

Annual to perennial herbs, rarely woody, with alternate, simple or pinnatifid leaves. Flowers in terminal, bracteate racemes or spikes. Sepals 4–8; petals 4–8, free, entire or laciniate; stamens 7–25, inserted on a hypogynous or perigynous, often eccentric disc.

Carpels 3–7, superior, free and uniovulate or united into a unilocular ovary which is open above, with numerous ovules on parietal placentae. Fruit a capsule open at the top or consisting of 4–7 1-seeded, radiating carpels. Seeds suborbicular or reniform, without endosperm.

Literature: M. S. Abdallah & H. C. D. de Wit, *Belmontia* **8**: 1–416 (1978).

Fruit a unilocular capsule with numerous seeds **1. Reseda**
Fruit of 4–7, ±free, stellate-patent, 1-seeded carpels **2. Sesamoides**

1. Reseda L.

P. F. YEO

Annuals to perennials. Leaves entire, toothed or pinnatifid. Petals usually with dilated base (claw) and lobed distal portion (limb); bases of stamens forming a disc which is often produced dorsally; carpels and stigmas 3 or 4; ovary unilocular; capsule opening more widely at maturity.

Leaf-shape is very plastic in most species. It is necessary to observe carefully the shape and dissection of the petals when using the key and care should be taken to distinguish between the upper and the lateral petals.

1 All leaves entire or with a few minute, hyaline teeth near the base
2 Pedicels not more than 2·5 mm
3 Sepals and petals 4; leaves entire, at least some of them more than 4 mm wide **1. luteola**
3 Sepals and petals 6; leaves less than 3 mm wide, mostly with some minute teeth near the base
4 Leaves with 2–5 pairs of teeth near the base (always some leaves with more than 2 pairs) **5. virgata**
4 Leaves with 1 or 2 pairs of teeth near the base
5 Sepals 1–1·5 mm; petals 2–3 mm; capsule 1·5–3 × 2–3·5 mm **4. complicata**
5 Sepals 1·5–2 mm; petals 3·5–5 mm; capsule 3–4 × 4–5 mm
6 Leaves up to 60 mm, linear; lateral branches not reduced to axillary fascicles of leaves on the main stems **2. glauca**
6 Leaves not more than 12 mm, lanceolate; stems with axillary fascicles of leaves **3. gredensis**
2 Pedicels at least 5 mm
7 Leaves linear, not more than 1 mm wide **5. virgata**
7 Leaves oblanceolate to spathulate, at least 4 mm wide
8 Upper petals 3-lobed; capsule 3–5 times as long as wide **19. lanceolata**
8 Upper petals with numerous divisions; capsule less than 3 times as long as wide
9 Capsule 13–14 × 5·5–9 mm, obovoid-cylindrical; leaves spathulate **11. phyteuma**
9 Capsule 9–11 × 7–11 mm, subglobose; leaves oblanceolate to obovate **15. odorata**
1 Some leaves ternate or pinnatifid
10 Filaments persistent until fruit is ripe; carpels 4
11 Sepals and petals 6; petal-limb rectangular or triangular, not more than twice as long as the claw **10. suffruticosa**
11 Sepals and petals 5, rarely 6; petal-limb cuneate, at least 3 times as long as the claw
12 Petals without a distinct claw **9. decursiva**
12 Petals (at least the 2 upper) with a distinct claw
13 Capsule 8–16 mm; inflorescence rarely longer than the leafy part of the stem **6. alba**
13 Capsule 4–6 mm; inflorescence usually much longer than the leafy part of the stem
14 Stamens 10–12, shorter than the petals **7. undata**
14 Stamens 13–20, as long as or exceeding the petals **8. paui**
10 Filaments caducous long before fruit is ripe; carpels 3
15 Limb of the 2 upper petals 3-lobed, the lateral lobes entire or shallowly cleft
16 Lateral lobes of upper petal shorter than the claw; capsule 3–5 times as long as wide **19. lanceolata**
16 Lateral lobes of upper petals longer than the claw; capsule not more than $2\frac{1}{2}$ times as long as wide
17 Most of the leaf-segments widest in middle; bracts persistent; capsule 12–15 mm, nodding **20. jacquinii**
17 Many of the leaf-segments linear; bracts usually caducous; capsule 6–12 mm, rarely nodding **18. lutea**
15 Limb of the 2 upper petals (5–)9- to 19-lobed
18 Bracts caducous; pedicels capillary; flowers pendent **17. stricta**
18 Bracts persistent; pedicels not capillary; flowers not pendent
19 Petal-lobes spathulate
20 Capsule 9–11 × 7–11 mm, subglobose **15. odorata**
20 Capsule 7–13 × 5–6 mm, ovoid or obovoid **16. orientalis**
19 Petal-lobes linear or linear-oblanceolate
21 Capsule almost square in outline
22 Leaf-segments oblanceolate to narrow-lanceolate; pedicels 5–8 mm; capsule 8–10 mm **13. inodora**
22 Leaf-segments broadly lanceolate; pedicels 3–4·5 mm; capsule 10–12 mm **14. tymphaea**
21 Capsule obovoid or broadly cylindrical
23 Sepals strongly accrescent in fruit, finally 5–13 × 1·25–1·5(–3·5) mm; lateral lobes of upper petals unilaterally pinnatifid **11. phyteuma**
23 Sepals not strongly accrescent in fruit, finally 4–6 × 0·3–1·25 mm; limb of upper petals digitate
24 Leaves or leaf-segments usually not more than 8 mm wide; few leaves entire and petiolate; terminal lobe of pinnatifid leaves usually less than $\frac{1}{2}$ as long as the leaf **12. media**
24 Leaves or leaf-segments up to 7–12 mm wide; many leaves entire and petiolate; terminal lobe of ternate or pinnatifid leaves at least $\frac{1}{2}$ as long as the leaf
25 Pedicels 5–8 mm **13. inodora**
25 Pedicels less than 5 mm **14. tymphaea**

Sect. LUTEOLA Dumort. Biennials. Leaves entire. Sepals and petals 4. Filaments persistent. Carpels 3.

1. R. luteola L., *Sp. Pl.* 448 (1753). Erect, 50–130 cm. Leaves mostly 25–120 × 4–15 mm. Pedicels *c.* 1 mm. Petals yellow, the upper one clawed, with 4- to 8-lobed limb; the two laterals and the lower one clawed or clawless with entire or 4-lobed limb. Capsule

3–4 × 5–6 mm. $2n$ = 24, 28. *Stony and sandy places. S., W. and parts of C. Europe; formerly cultivated as a source of dye, and naturalized northwards to S. Sweden.* Al Az Be Bl Bu Co Cr Cz Da Ga Gr Ho Hs It Ju Lu Rm Rs(K) Sa Si Tu *[Au Br Ge Hb He Hu Po Rs(B) Su].

Sect. GLAUCORESEDA DC. Perennials. Leaves entire, with small whitish teeth near the base. Sepals 6; petals white or whitish, the limb cuneate; the two upper 3- or 5-lobed; the two laterals narrower, entire to 3-lobed; the two lower entire or 2-lobed; claw distinct in upper petals. Filaments persistent. Carpels 4. Capsule wider than long.

2. R. glauca L., *Sp. Pl.* 449 (1753). Stems 10–40 cm, numerous, with short branches. Leaves 15–60 × 1–2 mm, linear, with 0–2 pairs of teeth, glaucous. Bracts 1–3 mm; pedicels 1·5–2·5 mm. Sepals 2 mm; petals 3·5–5 mm, with the limbs of the upper ones wide, 3-lobed to ⅓ of their length, the lobes toothed. Capsule 3–4 × 4–5 mm. $2n$ = 28. *Alpine and subalpine pastures, rocks and screes.* ● *E. & C. Pyrenees, Cordillera Cantábrica.* Ga Hs.

3. R. gredensis (Cutanda & Willk.) Müller Arg. in DC., *Prodr.* **16(2)**: 582 (1868). Like **2** but stems 10–25 cm, with axillary fascicles of leaves; leaves 3–12 × 0·5–1(–1·5) mm, lanceolate or narrow-lanceolate. $2n$ = 28. *Montane pastures and gravelly places.* ● *W.C. Spain (Sierra de Gredos).* Hs.

4. R. complicata Bory, *Ann. Gén. Sci. Phys. (Bruxelles)* **3**: 13 (1820). Like **2** but stems 40–70 cm; stems and branches arcuate-ascending; leaves 15–25 × 1–2·5 mm; bracts 0·5 mm; sepals 1–1·5 mm; petals 2–3 mm, the upper with (3–)5 deeper and narrower lobes; capsule 1·5–3 × 2–3·5 mm, subsessile. $2n$ = 28. *Montane pastures and damp, gravelly places.* ● *S. Spain (Sierra Nevada).* Hs.

5. R. virgata Boiss. & Reuter, *Diagn. Pl. Nov. Hisp.* 6 (1842). Stems 20–60 cm, erect, branched above. Leaves 15–40 × 0·5–1 mm, with 2–5 pairs of teeth. Bracts 2 mm; pedicels 1·5–2(–5·5) mm. Sepals 1·5 mm; petals 3 mm, the upper as in **2** but with proportionately narrower limb and smaller claw. Capsule 3 × 4 mm. $2n$ = 28. *Waste sandy fields and roadsides.* ● *C. Spain, N. Portugal.* Hs Lu.

Sect. LEUCORESEDA DC. Leaves pinnatifid; lobes usually varying irregularly in length. Sepals 5 or 6; petals 5 or 6 (two upper, two lateral, one or two lower), white, more or less clawed. Filaments persistent. Carpels 4.

6. R. alba L., *Sp. Pl.* 449 (1753) (*R. fruticulosa* L., *R. suffruticulosa* L.). Annual to perennial (10–)30–80 cm, erect, branching above. Leaf-lobes 5–15 on each side, entire. Pedicels 1–8 mm. Sepals 5 or 6, (1·5–)3–4 mm; petals 5 or 6, 3·5–6 mm; limb triangular, 3–5 times as long as claw, lobed to ⅓–⅔ of its length; lobes 3, the lateral often again lobed. Stamens 10–12, shorter than the petals. Capsule 8–16 mm, narrowly obovate or elliptical, constricted at the apex. $2n$ = 40. *Disturbed ground. S. Europe, eastwards to Kriti; occasionally cultivated for ornament and naturalized in parts of C. & N.W. Europe.* Bl Co Cr Ga Gr Hs It Ju Lu Sa Si *Tu [Br Cz Ge Hb Rm].

7. R. undata L., *Syst. Nat.* ed. 10, **2**: 1046 (1759) (*R. gayana* Boiss., *R. leucantha* Hegelm. ex Lange). Like **6** but stems less leafy above; pedicels 1·5–3 mm; sepals 5, 1·5–2·5 mm; petals 5, 3–4·5 mm, the lower bilobed; capsule 4–6 × 3·5 mm, turbinate or subglobose. $2n$ = 20. *Dry places.* ● *S., E. & C. Spain.* Hs.

8. R. paui Valdés-Bermejo & Kaercher, *Anal. Jard. Bot. Madrid* **41**: 198 (1984). Like **6** but stamens 13–20, as long as or exceeding the petals; petal-limb obovate or subtriangular, petal-claw strongly dilated, concave. $2n$ = 20. ● *S.E. Spain.* Hs.

9. R. decursiva Forskål, *Fl. Aegypt.* lxvi (1775) (*R. propinqua* R. Br.). Annual or biennial up to 20 cm. Leaf-lobes 5–15 on each side, entire. Pedicels 1–2·5 mm. Sepals 5, 2–2·5 mm; petals 5, 3–4·5 mm; dilation of claw indistinct or absent; lower petal retuse or shallowly bilobed. Capsule 4·5–7 × 4·5 mm, turbinate or elliptical, usually constricted below the prominent stigmatic lobes. *Maritime sands. S. Spain (near Gibraltar).* Hs. (*N. Africa, S.W. Asia.*)

10. R. suffruticosa Loefl., *Reise Span. Länd.* 113 (1766) (*R. baetica* (Müller Arg.) Gay ex Lange, *R. macrostachya* Lange). Biennial or perennial 25–220 cm, robust, papillose. Leaf-lobes *c.* 25 on each side, usually pinnatifid or toothed. Pedicels 0·5–1·5 mm. Sepals 6; petals 6; claw and limb rectangular or triangular; limb subentire or 1- or 2-cleft, 1–2 times as long as the claw. Capsule 11–14 × 5–8 mm, obovoid. $2n$ = 20. *Dry places.* ● *S., C. & E. Spain, N.E. Portugal.* Hs Lu.

Sect. RESEDA (Sect. *Resedastrum* Duby). Leaves entire or pinnatifid. Sepals 6; petals 6, clawed, the two upper with limb 3-lobed; lateral lobes multifid or subentire; mid-lobe narrow, often indistinct; the two lateral petals similar, but with the lower lateral lobe absent or reduced; the two lower petals with the limb usually entire. Filaments usually caducous. Carpels 3.

11. R. phyteuma L., *Sp. Pl.* 449 (1753) (*R. aragonensis* Loscos & Pardo, *R. litigiosa* Sennen & Pau). Annual to perennial 10–50 cm, with ascending branches near the base. Leaves 50–100 × 5–15 mm, spathulate, long-cuneate, sometimes some with 1(2) lobes on each side. Bracts 2·5–3 mm. Sepals accrescent, 3–4·5 × *c.* 0·75 mm at anthesis, 5–13 × 1·25–1·5(–3·5) mm in fruit; petals 3–5 mm, white, the upper with claw investing disc; limb inserted dorsally on the claw below the middle, stipitate, the lateral lobes pinnatifid, each with 5–9 linear-oblanceolate segments. Capsule 13–14 × 5·5–9 mm, nodding, obovoid-cylindrical. $2n$ = 12, 24. *Cultivated and disturbed ground. S. Europe, extending northwards to 47° N. in France.* Al Bl Co Ga Gr He Hs Hu It Ju Lu Po Rm [Au Cz Ge].

12. R. media Lag., *Gen. Sp. Nov.* 17 (1816) (*R. macrosperma* Reichenb.). Like **11** but many leaves with 1–4(–8) pairs of lobes and small terminal lobe; the entire leaves 30–50 × 4–8 mm; the intermediate and upper leaves more shortly cuneate or rounded at the base; bracts 1·5–2·5 mm; sepals finally 4–6 mm; upper petals with claw less incurved and limb sessile, digitate; capsule 13–16 × 6–8·5 mm, nodding. $2n$ = 12. *Dry places. S.W. Europe.* *Az Bl Hs Lu.

13. R. inodora Reichenb., *Icon. Fl. Germ.* **2**: 22 (1838). Biennial or perennial 20–60 cm; stems erect, branched above. Leaves mostly 50–80 × 7–12 mm, lanceolate, erect, some of the upper with 1 or 2 pairs of lateral lobes and a large terminal lobe. Bracts 3 mm; pedicels 5–8 mm, patent. Sepals finally 5 mm; petals 3 mm, the upper with limb inserted dorsally on the claw above the middle; limb sessile, digitate, with *c.* 11–17 linear-oblanceolate segments. Orientation of capsule variable; capsule 8–10(–13) mm. *Cultivated ground and waste places.* ● *S.E. Europe, westwards to N.E. Italy.* Bu Hu It Ju Rm Rs(C, W, K, E) [Ho].

Similar plants from S. Jugoslavia and N. Greece (*R. inodora* var. *macrocarpa* Fischer & C. A. Meyer), which perhaps deserve subspecific rank, differ in often being papillose and in having stems 20–30 cm, sometimes ascending. They also resemble **14** and R. *alopecuros* Boiss. from S.W. Asia.

14. R. tymphaea Hausskn., *Mitt. Thür. Bot. Ver.* **1887**: 10 (1887). Perennial, like **13** but stems ascending or erect; inflorescence longer; pedicels 3–4·5 mm; capsule 10–12(–17) × 8–9 mm, slightly nodding, obovoid or almost square in outline above the abruptly tapered base; stigmatic lobes smaller. *Gravelly places and rocks.* ● *Greece (local and mainly in the west).* Gr.

15. R. odorata L., *Syst. Nat.* ed. 10, **2**: 1046 (1759). Annual to perennial 10–50 cm, with ascending branches near the base. Leaves oblanceolate to obovate, sometimes the upper with 2 or 3 lobes. Sepals attaining 4–5 mm; petals 4–4·5 mm, the upper with claws flat; limb inserted near apex of claw, digitate; segments *c.* 9–15, spathulate. Capsule 9–11 × 7–11 mm, nodding, subglobose. *Commonly cultivated for its fragrant flowers, and locally naturalized in S. & C. Europe; sometimes found as a casual further north.* [Au Bl Cz Ga Hs It Rm Rs.] (?*Libya*.)

16. R. orientalis (Müller Arg.) Boiss., *Fl. Or.* **1**: 427 (1867). Annual 20–70 cm. Lower leaves up to 100 × 11 mm, oblanceolate, obtuse, petiolate; the upper pinnatifid with 1 or 2 pairs of lobes, the lobes mostly 1·5–3·5 mm wide, ovate-lanceolate to linear. Sepals 3–4·5 mm; petals like those of **15** but 3–3·5 mm. Capsule 7–13 × 5–6 mm, nodding, ovoid or obovoid. *S.E. Greece (Salamis); W. Kriti (Gavdhos).* Cr Gr.

17. R. stricta Pers., *Syn. Pl.* **2**: 10 (1806). Biennial; stems 30–50 cm, erect, with long, erect branches. Leaves with 1 or 2 pairs of often bifid pinnae; segments up to 6 mm wide. Bracts caducous; pedicels filiform; flowers pendent. Sepals 1–1·5 mm, ovate; petals 2–3 mm, yellow; limb of the upper shorter than the claw, 5-partite, of the lower bilobed, with lobes 1- or 2-lobulate at the base. Capsule 7–16 mm, erect, obovoid or cylindrical. $2n = 24$. *Gypsaceous soils. E. & C. Spain.* Hs.

18. R. lutea L., *Sp. Pl.* 449 (1753) (R. *ramosissima* Pourret ex Willd., R. *gracilis* Ten., R. *reyeri* Porta & Rigo; incl. R. *truncata* Fischer & C. A. Meyer). Annual to perennial, bushy. Leaves mostly pinnatifid, with 1 or 2(–4) pairs of pinnae (or leaves sometimes bi- or ternate); segments elongate. Bracts usually caducous. Petals yellow; lateral lobes of the upper falcate, subentire; mid-lobe shorter; limb of the lower entire or with 2 or 3 linear-spathulate lobes. Capsule 6–12 × 4·5–5·5 mm, rarely nodding, oblong, oblong-obovoid or ellipsoidal, rarely subglobose (var. *vivantii* (P. Monts.) Fernández Casas). $2n = 24, 48$. *Cultivated and disturbed ground. S. & W. Europe; probably a naturalized alien further north and east.* Al Au Bl Bu Co Cr Ga Gr Ho Hs It Ju Lu Po Rm ?Sa ?Si Tu *[Be Br Cz Da Ge Hb He Hu Rs(N, C, W, K, E)].

19. R. lanceolata Lag., *Gen. Sp. Nov.* 17 (1816) (R. *constricta* Lange). Erect annual or biennial 40–120 cm. Leaves mostly 20–90 × 4–18 mm, oblanceolate or spathulate, sometimes some with 1 or 2 lateral lobes. Bracts caducous; pedicels 3–7(–18) mm. Sepals 6–8; petals white, similar to those of **18**, but limb of the upper with lateral lobes wide, shorter than the claw; limb of the lower entire. Capsule 16–28 × 5–6 mm. $2n = 24$. *Dry places. S. & E. Spain.* Hs.

20. R. jacquinii Reichenb., *Icon. Fl. Germ.* **2**: 22 (1838). Annual or biennial, branched at the base. Leaves 20–50 × 5–20 mm, subsessile, spathulate, obtuse, some with 1(2) pairs of oblong to elliptical lobes. Bracts persistent. Petals 4 mm, the upper with lateral lobes shallowly to deeply crenate. Capsule 12–15 × 6·5 mm, nodding, obovoid. *Dry places.* ● *E. Pyrenees; S. France (Cevennes).* Ga Hs.

R. arabica Boiss., *Diagn. Pl. Or. Nov.* **1(1)**: 6 (1843), differing from **20** by its annual habit, persistent filaments and capsule 6–12 mm, subglobose to broadly ovoid, has been recorded once from E. Kriti (Sitia).

2. Sesamoides Ortega

V. H. HEYWOOD (EDITION 1) REVISED BY J. R. AKEROYD (EDITION 2)

Usually perennials, often woody at the base, with simple leaves. Flowers in terminal, bracteate racemes. Calyx 5- or 6-lobed. Petals 5 or 6, the upper two multilaciniate, the others less divided. Stamens 7–15 on an eccentric, urceolate, hypogynous disc. Carpels 4–7 on a short gynophore, more or less free, stellate-patent in fruit, more or less dorsally gibbous making the style lateral or subterminal. (*Astrocarpa* Necker ex Dumort.)

Calyx-lobes narrow, obtuse, at least as long as the tube, with narrow sinuses **1. clusii**
Calyx-lobes broadly triangular, acute, shorter than the tube, with broad sinuses **2. purpurascens**

1. S. clusii (Sprengel) Greuter & Burdet, *Willdenowia* **19**: 47 (1989) (*Reseda sesamoides* L., *Astrocarpa sesamoides* (L.) DC., *A. suffruticosa* Lange, *A. minor* Lange, *S. pygmaea* (Scheele) O. Kuntze). Many-stemmed perennial with a more or less woody base; stems 5–15 cm. Leaves lanceolate to linear-lanceolate, the basal ones in dense rosettes. Calyx divided to at least ½-way into narrow, obtuse lobes separated by narrow sinuses. Stamens 7–12; filaments usually glabrous, sometimes papillose-hairy. Carpels 4–6, obovoid, the style subterminal, overtopping the dorsal gibbosity of the carpel. $2n = 20$. *S.W. Europe, from N.W. Spain to Sardegna and N.W. Italy.* Co Ga Hs It Sa.

2. S. purpurascens (L.) G. López, *Anal. Jard. Bot. Madrid* **42**: 321 (1986) (*Reseda purpurascens* L., *Astrocarpa purpurascens* (L.) Dumort., *S. canescens* auct., non (L.) O. Kuntze, *A. cochleariformis* Nyman). Like **1** but rosettes lax; calyx-lobes broadly triangular, acute, with broad sinuses, shorter than the tube; stamens 10–15; carpels 4–5; style lateral, not overtopping the dorsal gibbosity of the carpel. $2n = 20, 40, 60$. *S.W. Europe.* Co Ga Hs It Lu Sa.

A very variable species divided into several subspecies or varieties by different authors (*vide* M. S. Abdallah & H. C. D. de Wit, *op. cit.* 354–362 (1978)).

SARRACENIALES

70. SARRACENIACEAE

EDIT. D. A. WEBB

Perennial, insectivorous herbs. Flowers regular, hermaphrodite. Sepals and petals free; stamens numerous, hypogynous; ovary superior. Fruit a capsule.

1. Sarracenia L.

D. A. WEBB

Leaves all basal, each in the form of a hollow, curved, inverted cone, with the orifice partly covered by a broad flap or hood. (In the fluid contained in the leaf, insects are trapped, and the organic matter released by their decay is absorbed.) Flowers solitary on long peduncles, closely subtended by 3 bracts. Sepals and petals 5; ovary 5-celled; style expanded at the apex into a large, peltate disc which fills the centre of the flower. Seeds numerous.

Literature: J. M. Macfarlane in Engler, *Pflanzenreich* **34(IV.110)**: 27–38 (1908).

1. S. purpurea L., *Sp. Pl.* 510 (1753). Leaves up to 20 cm, numerous, green, suffused or marbled with dark red, the outer ones spreading almost horizontally, all strongly winged on the adaxial side; terminal flap reniform, nearly erect, covered on the inner side with downwardly pointing hairs. Peduncles 20–60 cm; flowers nodding. Sepals 3 × 2 cm, ovate to rhombic, dark purplish-red outside, pale green inside; petals 3 × 1·5 cm, obovate, tapered to a short claw, purplish-red on both surfaces; apical disc of style *c.* 3 cm in diameter, pentagonal, greenish. *Wet peat-bogs. Planted and thoroughly naturalized in a few places in W. Switzerland and C. Ireland.* [Hb He.] (*North America.*)

71. DROSERACEAE

EDIT. D. A. WEBB

Perennial, insectivorous herbs. Flowers regular, hermaphrodite, usually 5-merous. Sepals united at the base; petals free; stamens hypogynous; ovary superior, 1-celled; styles 2–5. Fruit a capsule.

Literature: L. Diels in Engler, *Pflanzenreich* **26(IV.112)**: 1–136 (1906).

1 Rootless aquatic with whorled leaves — **1. Aldrovanda**
1 Roots present; leaves alternate or basal
  2 Flowers white; stamens equal in number to the petals — **2. Drosera**
  2 Flowers yellow; stamens twice as many as the petals — **3. Drosophyllum**

1. Aldrovanda L.

D. A. WEBB

Stems submerged, rootless. Leaves whorled. Flowers axillary; stamens 5; styles 5.

1. A. vesiculosa L., *Sp. Pl.* 281 (1753). Leaves 10–15 mm, in crowded whorls of 6–9; each with a cuneate basal part, and terminating in 4–6 setaceous segments and an orbicular lobe, hinged along the midrib, which can close rapidly to entrap and digest small animals. Flowers shortly pedicellate, rarely produced in Europe and usually cleistogamous. Petals 4–5 mm, greenish-white. Perennation by persistent terminal buds. *Usually in shallow, still water. Dispersed irregularly and sparsely over a large part of Europe, mainly in the centre and east; declining over much of its European range.* †Au †Bu Cz Ga Ge He Hu It Ju Po Rm Rs(C, W, E).

2. Drosera L.

D. A. WEBB

Small plants of wet places. Leaves all basal; upper surface of lamina covered with long, red, gland-tipped, motile hairs, which entrap and digest insects. Scape slender, bearing a slender, spike-like, ebracteate cyme. Flowers usually 5-merous, sometimes 4- or 6–8-merous, open for a short time only, sometimes cleistogamous. Petals white; stamens equal in number to the petals. Carpels usually 3; styles free, deeply bifid.

1 Scape clearly lateral, scarcely longer than the leaves — **3. intermedia**
1 Scape apparently terminal, considerably longer than the leaves
  2 Lamina orbicular; petiole hairy — **1. rotundifolia**
  2 Lamina linear-oblanceolate; petiole glabrous — **2. anglica**

1. D. rotundifolia L., *Sp. Pl.* 281 (1753). Plants solitary. Leaves usually spreading horizontally, sometimes semi-erect; lamina 5–8 mm, orbicular or slightly wider than long; petiole 15–30 mm, hairy. Scape 4–8 cm, apparently terminal, arising from the axil of the highest leaf, bearing 6–10(–15) flowers. Petals 3·5–4·5 mm. Capsule smooth. Testa reticulate, very loose-fitting. $2n = 20$. *Usually on peat or* Sphagnum. *Europe, except for some of the islands and the extreme south.* Au Be Br Bu Co Cz Da Fa Fe Ga Ge Hb He Ho Hs Hu Is It Ju Lu No Po Rm Rs(N, B, C, W, K, E) Su.

2. D. anglica Hudson, *Fl. Angl.* ed. 2, 135 (1778) (*D. longifolia* L.). Plants solitary, larger than **1** and **3**. Leaves erect or inclined; lamina *c.* 30 × 7 mm, linear-oblanceolate; petiole 5–10 cm, glabrous. Scape apparently terminal, as in **1**, but 10–18 cm, bearing 3–6 flowers. Petals 4·5–6 mm. Capsule smooth. Testa reticulate, loose-fitting. $2n = 40$. *Wet places; more tolerant of basic conditions than the* *other species. N. & C. Europe, and very rarely in the south.* Au Be Br Cz Da Fe Ga Ge Gr Hb He Ho Hs †Hu It Ju No Po Rm Rs(N, B, C, W, K, E) Su.

D. × obovata Mert. & Koch in Röhling, *Deutschl. Fl.* ed. 3, **2**: 502 (1826) (*D. anglica × rotundifolia*), is fairly frequent, but sterile. Its leaves are rather like those of **3**, but the scape is apparently terminal, the capsule very small, and the seeds empty.

3. D. intermedia Hayne in Dreves, *Bot. Bilderb.* **3**: t. 3, fig. B (1798) (*D. longifolia* auct., non L.). Plants gregarious, from pseudo-dichotomous branching of the stem with subsequent decay; sometimes forming floating mats. Leaves inclined or almost erect; lamina *c.* 7 × 4 mm, obovate; petiole 2·5–4 cm, glabrous. Scape 2–3 cm, scarcely as long as the leaves, ascending, obviously lateral, from the axil of one of the lower leaves of the current year, bearing 3–7 flowers. Petals 5 mm. Capsule longitudinally grooved. Testa granular, close-fitting. $2n = 20$. *Usually on peat or* Sphagnum, *or in shallow bog-pools. N., W. & C. Europe.* Au Be Br Cz Da Fe Ga Ge Hb He Ho Hs It Ju Lu No Po Rm Rs(N, B, C) Su.

3. Drosophyllum Link

D. A. WEBB

Leaves with glandular, digestive hairs, as in *Drosera*. Flowers 5-merous, in a corymbose cyme. Petals yellow. Stamens 10, styles 5.

1. D. lusitanicum (L.) Link in Schrader, *Neues Jour. Bot.* **1(2)**: 53 (1805). Stem slender, ascending, sometimes branched. Leaves 10–20 × 0·2–0·3 cm, numerous, alternate, crowded, narrowly linear, tapering from the sheathing base to the filiform apex, covered throughout with red-tipped glandular hairs. Scape 15–30 cm, erect, bearing a corymb of 5–10 flowers. Petals 18–25 mm, broadly obovate, bright yellow. *Dry places; calcifuge. S. Spain and Portugal.* Hs Lu.

ROSALES

72. CRASSULACEAE

EDIT. D. A. WEBB

Annual, biennial or perennial herbs, rarely small shrubs. Leaves undivided, exstipulate, more or less succulent. Flowers regular, usually in cymes, less often in spikes or racemes or solitary in the leaf-axils. Sepals 3 to *c.* 20, united or free; petals as many, united or free; stamens hypogynous or epipetalous, equal in number to the petals or, more frequently, twice as many. Carpels superior, equal in number to the petals, free or slightly connate at the base, developing into follicles. Scale-like nectaries usually present between the stamens and carpels.

No really satisfactory basis for the division of the family into genera has yet been proposed; Berger's treatment in Engler & Prantl, *Natürl. Pflanzenfam.* ed. 2, **18a**: 352–483 (1930), has been followed with slight modifications.

1 Stamens equal in number to the petals
 2 Leaves opposite, connate; petals 3 or 4 (rarely 5) — **1. Crassula**
 2 Leaves not connate; petals 5 (rarely 4) — **10. Sedum**
1 Stamens twice as many as the petals
 3 Corolla-tube longer than, or only slightly shorter than the lobes
  4 Annual
   5 Stamens and style included; corolla-tube nearly as wide as long — **5. Mucizonia**
   5 Stamens and style exserted; corolla-tube very long and narrow — **4. Pistorinia**
  4 Perennial
   6 Leaves mostly cauline; inflorescence cymose — **2. Kalanchoe**
   6 Leaves mostly basal; cauline leaves much smaller; inflorescence racemose
    7 Basal leaves suborbicular, with long petioles — **3. Umbilicus**
    7 Basal leaves oblong-spathulate, sessile — **13. Rosularia**
 3 Petals free, or united for considerably less than ½ their length
  8 Petals 4 or 5
   9 Leaves mostly basal; cauline leaves much smaller; inflorescence racemose
    10 Basal leaves petiolate, not spiny; corolla tubular, with erect lobes — **3. Umbilicus**
    10 Basal leaves sessile, with a terminal spine; corolla-lobes spreading — **12. Orostachys**
   9 Leaves mostly cauline; inflorescence cymose
    11 Flowers usually 5-merous, hermaphrodite; rhizome slender and leafless, or absent — **10. Sedum**
    11 Flowers usually 4-merous, unisexual; rhizome stout, with persistent scale-leaves — **11. Rhodiola**
  8 Petals more than 5
   12 Annual
    13 Flowers yellow; leaves flat — **9. Aichryson**
    13 Flowers pink, white or blue; leaves ±terete — **10. Sedum**
   12 Perennial
    14 Leaves subulate — **10. Sedum**
    14 Leaves broad, flat on upper surface
     15 Soft-wooded shrub with erect, perennial stems — **8. Aeonium**
     15 Acaulescent in vegetative phase; flowering stems annual
      16 Petals entire, patent — **6. Sempervivum**
      16 Petals ±fimbriate, erect — **7. Jovibarba**

1. Crassula L.

D. A. WEBB AND J. R. AKEROYD

Small, glabrous annuals (outside Europe including perennials and shrubs), with opposite, connate, often reddish leaves. Flowers 3- to

5-merous. Petals free. Stamens equal in number to the petals. (Incl. *Tillaea* L.)

Literature: H. R. Tölken, *Contr. Bolus Herb.* **8**: 1–595 (1977).

The characteristic habitat of the native European species is ground flooded in winter and dry in summer.

Several shrubby species of *Crassula* from South Africa are widely cultivated, and of these **C. lactea** Aiton, *Hort. Kew.* ed. 1, **1**: 306 (1789), and **C. multicava** Lemaire, *Revue Hort.* (*Paris*) **11**: 97 (1862) (*C. quadrifida* Baker), are reported as escaped in Açores and are perhaps becoming naturalized. **C. decumbens** Thunb., *Prodr. Pl. Cap.* 54 (1794), a small annual species from South Africa, occurs locally as a weed in bulb-fields in S.W. England (Isles of Scilly), and may be naturalized.

1 Flowers 5-merous; petals red (N.E. Spain) **6. campestris**
1 Flowers 3- or 4-merous; petals white or pink
 2 Petals shorter than the sepals; follicles 2-seeded
  3 Leaves lanceolate, mucronate; flowers pedicellate **1. alata**
  3 Leaves ovate, subacute; flowers sessile **2. tillaea**
 2 Petals longer than the sepals; follicles usually with more than 2 seeds
  4 Leaves obtuse; flowers in cymes **5. vaillantii**
  4 Leaves acute; flowers solitary
   5 Stems less than 8 cm; flowers subsessile **3. aquatica**
   5 Stems usually more than 8 cm; flowers distinctly pedicellate **4. helmsii**

1. C. alata (Viv.) A. Berger in Engler & Prantl, *Natürl. Pflanzenfam.* ed. 2, **18a**: 389 (1930) (*Tillaea alata* Viv.). Stems 3–10 cm, procumbent, ascending or erect, winged. Leaves 3–7 mm, lanceolate, mucronate. Flowers 3- or 4-merous, on slender pedicels up to 2·5 mm, forming dense cymes in the leaf-axils. Petals 0·8–1·5 mm, lanceolate, acuminate, shorter than the sepals, white or pink. Follicles 2-seeded. *Kriti and Kikladhes.* Cr Gr.

2. C. tillaea Lester-Garland, *Fl. Jersey* 87 (1903) (*C. muscosa* Roth, non L., *Tillaea muscosa* L.). A minute, moss-like plant; stems 1–5 cm, procumbent or ascending. Leaves 1–2 mm, ovate, concave, crowded, almost imbricate. Flowers 3-merous (rarely 4-merous), sessile in small groups in the leaf-axils. Petals 1 mm, shorter than the sepals, narrowly lanceolate, white or pale pink. Follicles usually 2-seeded. *S. & W. Europe, extending locally to N.E. Germany.* Az Be Bl Br Bu Co Cr Ga Ge Gr Ho Hs It Lu Sa Si Tu.

3. C. aquatica (L.) Schönl. in Engler & Prantl, *Natürl. Pflanzenfam.* **3** (**2a**): 37 (1890) (*Tillaea aquatica* L.). Stems 2–6 cm, decumbent, rooting. Leaves 4–6 mm, linear, acute. Flowers usually 4-merous, subsessile, solitary in the leaf-axils. Petals rather longer than the sepals, ovate-oblong, obtuse, white. Follicles with numerous seeds. $2n = 42$. *N. & C. Europe.* Au Br Cz Da Fe Ge Is No Po Rs(N, B, C) Su [Lu].

4. C. helmsii (Kirk) Cockayne, *Trans. Proc. N.Z. Inst.* **39**: 349 (1907) (*Tillaea helmsii* Kirk). Stems 10–30 cm, aquatic or creeping, rooting at the nodes. Leaves 4–20 mm, linear-lanceolate to ovate-lanceolate, acute. Flowers 4-merous, solitary in the leaf-axils; pedicels 2–8 mm, recurved. Petals slightly longer than the sepals, ovate, obtuse, white or pale pink. Follicles 2- to 5-seeded. *Margins of ponds. Widely naturalized in Britain.* [Br.] (*Australia and New Zealand.*)

5. C. vaillantii (Willd.) Roth, *Enum.* **1**: 992 (1827) (*Tillaea vaillantii* Willd.). Stems 2–6 cm, erect or ascending. Leaves linear-oblong, obtuse. Flowers 4-merous, on slender pedicels longer than the leaves, forming small, irregular cymes. Petals longer than the sepals, ovate-lanceolate, acute, pink or white. Follicles with numerous seeds. *S. Europe, extending northwards to N. France, S. Ukraine and the lower Volga; rare and local over most of its range.* Bl Co Cr Ga Gr Hs It Lu Rs(W, E) Sa Si.

C. peduncularis (Sm.) Meigen, *Bot. Jahrb.* **17**: 239 (1893) (*C. bonariensis* (DC.) Camb.), introduced from South America, has been reported as naturalized in rice-fields in C. Portugal. It is like **5** but with pale, hyaline petals and much longer pedicels.

6. C. campestris (Ecklon & Zeyher) Endl. in Walpers, *Repert. Bot. Syst.* **2**: 253 (1843). Stems 2–7 cm, ascending to erect. Leaves 3–5 mm, lanceolate-subulate, acute. Flowers 5-merous, in axillary groups, subsessile. Petals 1 mm, shorter than the sepals, red. Follicles 2-seeded. $2n = 16$. *Dry, siliceous rocks. Naturalized at one station in N.E. Spain (near Barcelona).* [Hs.] (*Southern Africa.*)

2. Kalanchoe Adanson

D. A. WEBB

Perennials with opposite leaves. Flowers 4-merous, in cymes; corolla tubular, with short lobes; stamens 8, inserted near the base of the corolla. (Incl. *Bryophyllum* Salisb.)

1. K. pinnata (Lam.) Pers., *Syn. Pl.* **1**: 446 (1805) (*Bryophyllum calycinum* Salisb., *B. pinnatum* (Lam.) Oken). Stems 1 m or more, erect, woody at the base. Leaves petiolate, simple, ternate or pinnate; leaflets up to 10 cm, ovate, crenate, often with young plants arising on the margins between the crenations. Flowers pendent, in compound cymes. Calyx 3 cm, tubular, with short, triangular lobes; corolla similar but 4–5 cm, and with longer, more acute lobes; both pale green mottled with red. Follicles erect. *Cultivated as a house-plant, and naturalized in Açores.* [Az.] (*Madagascar.*)

3. Umbilicus DC.

D. A. WEBB

Perennials, with tuberous or rhizomatous rootstock. Basal leaves petiolate, usually suborbicular, at least 2 cm in diameter, glabrous. Cauline leaves much smaller. Flowers 5-merous, numerous, in a terminal, bracteate raceme or panicle. Calyx small; corolla tubular or campanulate, sympetalous; lobes more or less erect. Stamens (5–)10, epipetalous; filaments short. Follicles slender.

1 Lobes of corolla at least as long as the tube
 2 Flowers 9–13 mm; lobes of corolla about equalling the tube **3. erectus**
 2 Flowers 3–6 mm; lobes of corolla considerably longer than the tube
  3 Flowers 4–6 mm, obconical; carpels gradually tapered to a straight, fairly long style **1. parviflorus**
  3 Flowers 3–4 mm, campanulate or subglobose; carpels abruptly narrowed to a very short, curved style **2. chloranthus**
1 Lobes of corolla considerably shorter than the tube
 4 Corolla bright yellow, conspicuously constricted at the mouth of the tube; stamens usually 5 **6. heylandianus**
 4 Corolla pale yellow, greenish or reddish, scarcely constricted at the mouth of the tube; stamens 10

5 Flowers usually pendent; raceme occupying more than ½ the stem; corolla lobes ovate, mucronate **4. rupestris**
5 Flowers horizontal; raceme occupying not more than ½ the stem; corolla-lobes lanceolate, acuminate **5. horizontalis**

1. **U. parviflorus** (Desf.) DC., *Prodr.* **3**: 400 (1828) (*Cotyledon parviflora* Desf.). Stem 10–35 cm, erect or ascending, often flexuous. Basal leaves 2–5 cm in diameter, orbicular, cordate or subpeltate, sinuate or subentire. Flowers 4–6 mm, erect or horizontal, obconical, in a dense, usually narrow panicle. Bracts minute, about equalling the pedicels. Sepals ½ as long as the corolla. Corolla yellow, with lanceolate, acute lobes 1½–2 times as long as the tube. Carpels gradually tapered to a nearly straight style. *Rocks. S. Greece; Kriti.* Cr Gr.

2. **U. chloranthus** Heldr. & Sart. ex Boiss., *Fl. Or.* **2**: 768 (1872) (*Cotyledon chlorantha* (Heldr. & Sart. ex Boiss.) Halácsy). Like **1** but leaves more often peltate; flowers 3–4 mm, campanulate or subglobose, horizontal or somewhat drooping, in a laxer and more diffuse panicle; carpels abruptly narrowed to a very short, deflexed style. *Rocks and walls. S. & W. parts of Balkan peninsula, Aegean region.* Gr Ju..

3. **U. erectus** DC. in Lam. & DC., *Fl. Fr.* ed. 3, **4**: 384 (1805) (*U. umbilicus-veneris* sensu Stoj. & Stefanov, *Cotyledon umbilicus-veneris* L.; incl. *U. lassithiensis* Gand., *Cotyledon lassithiensis* (Gand.) Hayek). Stem 20–60 cm, stout, erect, simple. Basal leaves up to 7 cm in diameter, deltate-orbicular, cordate, sinuate-crenate; cauline leaves progressively smaller, with shorter petioles, dentate. Flowers very numerous, more or less erect, in a dense raceme 8–25 cm, sometimes branched at the base; pedicels 1–3 mm. Bracts usually narrow-lanceolate, small, entire or with a tooth on each side, but sometimes (var. *lassithiensis* (Gand.) Stoj.) broad, leafy, dentate. Sepals linear-lanceolate. Corolla 9–13 mm, tubular, bright or greenish-yellow, drying red-brown; lobes narrow-lanceolate, acuminate, about equalling the tube. *Damp or shady rocks. S. part of Balkan peninsula; S. Italy.* Al Bu Cr Gr It Ju Tu.

4. **U. rupestris** (Salisb.) Dandy in Riddelsd., Hedley & Price, *Fl. Gloucestershire* 611 (1948) (*U. pendulinus* DC., *U. vulgaris* Knoche pro parte, *Cotyledon umbilicus-veneris* auct., non L., *C. pendulina* (DC.) Batt.). Stem 20–50 cm, erect. Basal leaves orbicular, peltate, concave above, sinuate-crenate; cauline leaves progressively smaller, mostly reniform, dentate, the uppermost sometimes linear. Bracts usually linear, about equalling the pedicels, but occasionally large and leaf-like. Pedicels 3–9 mm. Flowers 7–10 mm, tubular, usually pendent; raceme fairly dense, sometimes branched at the base, occupying more than ½ the stem. Sepals ovate, acuminate. Corolla whitish-green or straw-coloured, sometimes tinged with pink; tube about 4 times as long as the ovate, mucronate lobes. Carpels tapered to a fairly long style. *Rocks and walls, rarely on trees; somewhat calcifuge. S. & W. Europe, northwards to Scotland.* Al Az Bl Br Bu Co Cr Ga Gr Hb Hs It Ju Lu Sa Si Tu.

A very variable species; some of the variants resemble **3**, **5** and **6** in single characters and have caused much confusion. The most distinct is **U. neglectus** (Coutinho) Rothm. & P. Silva, *Agron. Lusit.* **2**: 88 (1940) (*Cotyledon neglecta* Coutinho), from Portugal, S. Spain and Islas Baleares, which possibly deserves subspecific status. It is a robust plant, with basal leaves often cordate rather than peltate, and a narrow corolla approaching that of **5**.

U. intermedius Boiss., *Fl. Or.* **2**: 769 (1872) (*Cotyledon intermedia* (Boiss.) Stefanov), from S.W. Asia, like **4** but with a shorter raceme, short pedicels and slightly more pointed corolla-lobes, has been recorded in error from the Balkan peninsula (Bulgaria and Thrace).

5. **U. horizontalis** (Guss.) DC., *Prodr.* **3**: 400 (1828) (*Cotyledon horizontalis* Guss., *C. umbilicus-veneris* auct., non L., *Umbilicus vulgaris* Knoche pro parte). Like **4** but the raceme occupying not more than ½ the stem; cauline leaves more numerous and crowded, many of them usually linear; flowers narrower and somewhat shorter (*c.* 7 × 3 mm), subsessile, horizontal; corolla-lobes triangular, lanceolate, acuminate, dull crimson. $2n = 24$. *Rocks and walls. Mediterranean region, Bulgaria, Açores.* Al Az Bl Bu Co Cr Gr Hs It Ju Sa Si.

The plants from Spain are sometimes separated as **U. gaditanus** Boiss., *Diagn. Pl. Or. Nov.* **1(6)**: 58 (1846), and are in some features transitional to **4**, but neither their taxonomic status nor their correct name can be ascertained with certainty from the material available.

6. **U. heylandianus** Webb & Berth., *Phyt. Canar.* **1**: 176 (1840) (*U. praealtus* (Mariz) Samp., *U. erectus* sensu Willk., non DC., *Cotyledon coutinhoi* (Mariz) Coutinho, *C. strangulata* Font Quer, *C. praealta* (Mariz) Samp.). Stem 60–100 cm, stout, erect. Leaves as in **4**. Raceme 12–35 cm, dense, simple, somewhat secund. Bracts 6–10 mm, linear. Flowers horizontal or drooping; pedicels 2–4 mm at anthesis, often longer in fruit. Sepals small, ovate-lanceolate. Corolla 10–12 mm, bright yellow; tube distinctly 5-angled and constricted at the mouth; lobes ovate-lanceolate, acuminate, about ⅓ as long as the tube. Stamens usually reduced to 5. Carpels obtuse, with a short style. *Shady places. Spain; Portugal; very local.* Hs Lu.

4. Pistorinia DC.

D. A. WEBB

Erect annuals with alternate leaves. Flowers 5-merous, shortly stalked, in a dense, subcorymbose cyme. Corolla infundibuliform, with a long, narrow tube. Stamens 10, exserted.

Corolla-lobes pink; tube cylindrical, expanded abruptly into the limb **1. hispanica**
Corolla-lobes yellow; tube somewhat conical, expanded gradually into the limb **2. breviflora**

1. **P. hispanica** (L.) DC., *Prodr.* **3**: 399 (1828) (*Cotyledon hispanica* L., *Sedum hispanicum* (L.) Hamet, non L.). Plant 5–15 cm, glabrous below, glandular-pubescent above. Leaves 6–10 mm, oblong, terete, very obtuse, suberect. Sepals 2·5 mm, linear, acute. Corolla (12–)20–25 mm, reddish-brown outside, pink inside, pubescent; tube of uniform diameter (*c.* 1·5 mm), expanded abruptly into the deeply 5-lobed limb. Styles long and slender, equalling the stamens. *Sandy and gravelly places. Spain and Portugal.* Hs Lu.

2. **P. breviflora** Boiss., *Elenchus* 74 (1838) (*P. salzmannii* Boiss.). Like **1** but more robust and more densely hairy; leaves up to 20 mm, subacute; corolla yellow, tinged with brownish-red on the tips of the lobes and the outside of the tube, with a broader, conical tube 3 mm wide at its upper end. *S. Spain.* Hs. (*N.W. Africa.*)

5. Mucizonia (DC.) A. Berger

D. A. WEBB

Diffuse or erect annuals with alternate leaves. Flowers 5-merous. Corolla campanulate, with the lobes about equalling the tube. Stamens 10, included.

Leaves 2–4 mm, imbricate; stem seldom branched **1. sedoides**
Leaves 12–18 mm, patent, distant; bushy, with freely branched stem **2. hispida**

1. M. sedoides (DC.) D. A. Webb, *Feddes Repert.* **64**: 22 (1961) (*Cotyledon sedoides* DC., *C. sediformis* Lapeyr., *Sedum candollei* Hamet, *S. sedoides* (DC.) Rothm., non Hamet, *Umbilicus sedoides* (DC.) DC.). Glabrous; usually gregarious, forming dense tufts. Stem 2–6 cm, erect. Leaves 2–4 mm, numerous, oblong, obtuse, concavo-convex, imbricate. Flowers erect, subsessile, in a crowded terminal cyme, their lower part concealed by leaves. Corolla 6–7 mm, purplish-pink; lobes erect, longer than the tube. *Rocks and stony ground.* ● *Pyrenees, and higher mountains of Spain and Portugal.* Ga Hs Lu.

2. M. hispida (Lam.) A. Berger in Engler & Prantl, *Natürl. Pflanzenfam.* ed. 2, **18a**: 420 (1930) (*Cotyledon hispida* Lam., *C. mucizonia* Ortega, *Sedum hispidum* (Lam.) Hamet, *S. mucizonia* (Ortega) Hamet, *Umbilicus hispidus* (Lam.) DC.). Usually glandular-pubescent, at least in the inflorescence, but sometimes glabrous. Stem 8–15 cm, diffuse, branched from the base. Leaves 12–18 mm, oblong, terete, streaked with red. Flowers long-stalked, in lax cymes. Corolla 8–13 mm, yellowish-green tinged with pink; lobes rather shorter than the tube. *Rocks and walls. C. & S. Spain and Portugal.* Hs Lu.

6. Sempervivum L.

C. FAVARGER AND F. ZÉSIGER (EDITION 1) REVISED BY J. PARNELL AND C. FAVARGER (EDITION 2)

Perennials with monocarpic rosettes; vegetative reproduction by axillary stolons. Leaves alternate, entire, sessile, ciliate, usually glandular-pubescent, very fleshy. Flowers 8- to 16-merous, in a terminal cyme. Sepals lanceolate, united at the base, pubescent. Petals free, patent, pink, purple or yellowish, narrowly lanceolate, pubescent. Stamens twice as many as petals; filaments usually pubescent at least at the base. Ovary pubescent; style distinct, usually curved. Seeds pyriform, finely striate.

The habitat of all species, unless otherwise indicated, is in rocky places, mainly in the mountains.

Hybrids in this genus are very common both in natural habitats and in cultivation. The most widespread are *S.* × *barbulatum* Schott (**10 × 11**), *S.* × *fauconnettii* Reuter (**10 × 18**), and *S.* × *schottii* C. B. Lehm. & Schnittspahn (**11 × 18**). The hybrids usually occur only in the presence of both parents, but in the French Jura, the *locus classicus* of *S.* × *fauconnettii*, one of the parents (**10**) is not now present. *S. funckii* Braun ex Koch, a triple hybrid between **10, 11** and **18**, probably of garden origin, is naturalized in a few places in Austria, Germany and Belgium.

Literature: R. L. Praeger, *An Account of the* Sempervivum *Group*. London. 1932.

1 Upper surface of petals yellowish, at least in the apical ½
2 Leaves usually glabrous except for marginal cilia, glaucous, purplish-red at the base **1. wulfenii**
2 Leaves glandular-pubescent, not glaucous, not purplish-red at the base
3 Cilia of rosette-leaves 2–4 mm, stiff, interwoven with those of neighbouring leaves **3. ciliosum**
3 Cilia of rosette-leaves less than 2 mm, not interwoven
4 All leaf-hairs (including cilia) subequal
5 Rosette-leaves 8–15 mm wide **2. grandiflorum**
5 Rosette-leaves 3–4(–6) mm wide **11. montanum**
4 Some leaf-hairs (including cilia) distinctly longer than others
6 Petals not tinged with purple or pink
7 Stolons usually 5–8 cm; some cilia more than 1 mm **8. leucanthum**
7 Stolons usually 2–3 cm; all cilia less than 1 mm **9. pittonii**
6 Petals tinged with pink or purple towards the base
8 Flowering stem not more than 10 cm; flowers 9- or 10-merous **4. octopodes**
8 Flowering stem 10–30 cm; flowers 12- to 14-merous
9 Petals pale yellow or cream-coloured **7. kindingeri**
9 Petals bright greenish-yellow
10 Flowering stem *c.* 10–15 cm; nectarial scales less than 0·2 mm, truncate to emarginate **5. zeleborii**
10 Flowering stem *c.* 15–30 cm, nectarial scales *c.* 0·5 mm, rounded at the apex **6. ruthenicum**
1 Upper surface of petals predominantly red, pink or purple
11 Mature leaves glabrous, apart from marginal cilia and sometimes a few fairly long hairs on the surface
12 Rosettes *c.* 3 cm in diameter; leaves not more than 7 mm wide (S. Spain) **17. nevadense**
12 Rosettes usually more than 3 cm in diameter; leaves more than 7 mm wide
13 Young leaves puberulent on both surfaces **16. marmoreum**
13 Young leaves glabrous except for marginal cilia
14 Leaves with a well-defined reddish-brown spot near the apex; petals 7–8 mm **19. calcareum**
14 Leaves variably suffused with red towards the apex, but without a well-defined spot; petals 9–10 mm **18. tectorum**
11 Mature leaves pubescent on both surfaces
15 Apical cilia of leaves extremely long and flexuous, interwoven to form an arachnoid tomentum over the rosette **10. arachnoideum**
15 Apical cilia short, not arachnoid
16 Cilia of rosette-leaves scarcely longer than the other hairs; petals 12–20 mm **11. montanum**
16 Cilia of rosette-leaves at least twice as long as the other hairs; petals 7–11 mm
17 Rosette-leaves sparsely pubescent
18 Stem usually more than 20 cm, bearing at least 40 flowers; leaves without a conspicuous tuft of cilia at the apex **18. tectorum**
18 Stem 8–16 cm, bearing 12–30 flowers; leaves usually with a conspicuous tuft of cilia at the apex
19 Rosette-leaves 10–15 mm, acuminate, bright green **12. dolomiticum**
19 Rosette-leaves *c.* 25 mm, mucronate, dark green **15. cantabricum**
17 Rosette-leaves densely pubescent
20 Stolons stout; flowers 11- to 14-merous

21 Rosette-leaves oblanceolate, acuminate; flowers 13- or 14-merous **13. kosaninii**
21 Rosette-leaves obovate-spathulate to oblong, abruptly mucronate; flowers 11- to 13-merous **16. marmoreum**
20 Stolons slender, flowers 9- to 12-merous
22 Rosettes *c.* 3–5 cm in diameter; flowers 11- or 12-merous; stolons long **14. macedonicum**
22 Rosettes *c.* 3 cm in diameter; flowers 9- to 11-merous; stolons short **17. nevadense**

1. S. wulfenii Hoppe ex Mert. & Koch in Röhling, *Deutschl. Fl.* ed. 3, **3**: 386 (1831). Stolons long, stout, woody. Rosettes 4–5(–9) cm in diameter; inner leaves connivent. Leaves *c.* 30 × 12 mm, oblong-spathulate, cuspidate, glabrous except for cilia, glaucous; margins somewhat involute towards the apex. Flowering stem 15–25 cm. Flowers 11- to 15-merous; petals 10 × 1–2 mm, lemon-yellow, with a basal purple spot; filaments purple. 1500–2700 *m; usually calcifuge.* ● *E. Alps; N.W. Jugoslavia.* Au He It Ju.

(a) Subsp. **wulfenii**: Rosette-leaves glabrous on both surfaces. $2n = 36$. *E. Alps.*

(b) Subsp. **juvanii** (Strgar) Favarger & J. Parnell, *Bot. Jour. Linn. Soc.* **103**: 217 (1990) (*S. juvanii* Strgar): Rosette-leaves glandular-pubescent on both surfaces. $2n = 36$. *E. Slovenija.*

2. S. grandiflorum Haw., *Revis. Pl. Succ.* 66 (1821) (*S. gaudinii* Christ). Stolons 10–20 cm, stout. Rosettes 2–5 cm in diameter, rather lax and flat. Leaves oblong-cuneate, cuspidate, densely pubescent, dark green, with a reddish-brown apex, with strong resinous or foetid odour. Flowering stem 10–20 cm. Flowers 12- to 14-merous; petals 10–18 × 2–3 mm, yellow with a basal purple spot; filaments purple. $2n = 80$. *Calcifuge.* ● *W.C. part of S. Alps, from Susa to Simplon.* He It.

3. S. ciliosum Craib, *Kew Bull.* **1914**: 379 (1914) (*S. borisii* Degen & Urum.). Stolons slender. Rosettes 2–3·5(–5) cm in diameter, depressed-globose. Leaves *c.* 10 × 4 mm, oblong-oblanceolate, acute, strongly incurved, pubescent, bearing towards the apex stiff cilia 2–4 mm long, which are interwoven with those of adjoining leaves. Flowering stem 4–10 cm. Flowers 12- to 14-merous; petals 10–12 × 1·5 mm, lemon-yellow; filaments white (rarely purple). $2n = 34$. ● *Macedonia.* Bu Gr Ju.

The plants from the western end of the range (mountains S.E. of Ohrid) show some approach to **4**.

4. S. octopodes Turrill, *Gard. Chron.* ser. 3, **102**: 303 (1937). Stolons usually long, slender. Rosettes 2–2·5 cm in diameter. Leaves 7 × 3 mm, more or less erect, oblanceolate to obovate, pubescent, with cilia becoming longer towards the reddish-brown apex. Flowering stem up to 9 cm, slender. Flowers few, 9-merous (rarely 10-merous); sepals purplish; petals 8 × 1·5 mm, pale yellow, tinged with lilac at the base; filaments purple. $2n = 34$. ● *S.W. Makedonija (Baba Planina).* Ju.

S. thompsonianum Wale, *Quart. Bull. Alp. Gard. Soc.* **8**: 210 (1940), is probably a hybrid between **4** and **14**.

5. S. zeleborii Schott, *Österr. Bot. Wochenbl.* **7**: 245 (1857) (*S. ruthenicum* auct., non Schnittspahn & C. B. Lehm.). Stolons short. Rosettes *c.* 3–5 cm in diameter, more or less globose. Leaves *c.* 20 × 8 mm, oblong-obovate, shortly apiculate, ciliate, grey-green. Flowering stem *c.* 10–15 cm. Flowers 12- to 14-merous; calyx teeth ovate to ovate-lanceolate; petals 9 × 1·5 mm, bright greenish-yellow, lilac at the base; filaments purple; nectarial scales less than 0·2 mm, truncate or emarginate. $2n = 64$. *Dry places in the mountains.* ● *S.E. Europe, from E. Jugoslavia to S. Bulgaria and N.W. Ukraine.* Bu Ju Rm Rs(W).

6. S. ruthenicum Schnittspahn & C. B. Lehm., *Flora (Regensb.)* **38**: 5 (1855). Stolons 3–5 cm. Rosettes *c.* 8 cm in diameter, closed or open. Leaves *c.* 50 × 15 mm, oblong-obovate, shortly apiculate, ciliate, green. Flowering stem *c.* 15–30 cm. Flowers 12- to 14-merous; calyx-teeth narrowly lanceolate; petals *c.* 10 × 1·5 mm, bright greenish-yellow; filaments white to pink; nectarial scales *c.* 0·5 mm, rounded. *Dry places; mainly lowland.* ● *S.E. Europe, from Romania to C. Russia.* ?Bu Rm Rs(C, W).

For a discussion of the taxonomy and nomenclature of **5** and **6** see C. W. Muirhead, *Notes Roy. Bot. Gard. Edinb.* **26**: 279–285 (1965).

7. S. kindingeri Adamović, *Denkschr. Akad. Wiss. Math.-Nat. Kl. (Wien)* **74**: 125 (1904). Stolons short, few. Rosettes 4–6·5 cm in diameter, open. Leaves 20–25(–40) × 5–8(–15) mm, spathulate, shortly acuminate, pale green, glandular-hairy, unequally ciliate. Flowering stem 20–25 cm, stout, softly hairy. Flowers numerous, 12- to 14-merous; petals *c.* 10 × 2 mm, pale yellow or cream, tinged with pink or purple towards the base; filaments pale, with purple streaks. ● *Macedonia.* Gr Ju.

8. S. leucanthum Pančić, *Elem. Fl. Bulg.* 30 (1883). Stolons 5–8 cm, stout. Rosettes 2·5–5 cm in diameter, subglobose, forming a loose mat. Leaves *c.* 20 × 8 mm, oblong-spathulate, apiculate, finely pubescent, unequally ciliate (the shorter cilia *c.* 0·3 mm, the longer *c.* 1·5 mm), dark green with a dark red apex. Flowering stem 15–20 cm. Flowers 11- to 13-merous; petals *c.* 10 × 1 mm, pale greenish-yellow; filaments pale (rarely pink), almost glabrous. $2n = 64$. ● *Mountains of Bulgaria.* Bu.

9. S. pittonii Schott, Nyman & Kotschy, *Analecta Bot.* 19 (1854). Like **8** but with stolons 2–3 cm, and rosettes crowded in dense tufts; leaves somewhat smaller, more densely pubescent and often grey-green; cilia *c.* 0·1 and 0·5 mm, mostly bearing a red gland; flowering stem 12–15 cm; flowers 9- to 12-merous. $2n = 64$. *Serpentine rocks.* ● *E. Alps (near Kraubath in C. Steiermark).* Au.

10. S. arachnoideum L., *Sp. Pl.* 465 (1753). Rosettes small, compact, crowded, more or less covered above by an arachnoid veil of long, flexuous, interwoven hairs, which are the greatly elongate apical cilia of the leaves. Leaves 7–12 × 3–5 mm, oblanceolate to oblong-obovate, apiculate, pubescent. Flowering stem 4–12 cm; cauline leaves red-tipped, with an apical tuft of arachnoid hairs. Flowers usually 8- to 10-merous; petals 7–10 × 3 mm, broadly lanceolate or rhombic, bright reddish-pink with a purple mid-vein; filaments purple. *Usually calcifuge.* ● *Alps, Appennini, Pyrenees, Cordillera Cantábrica; Corse.* Au Co Ga Ge He Hs It.

Very variable. Two rather ill-defined subspecies may be recognized; they have some geographical basis but show considerable overlap, and intermediates are common.

(a) Subsp. **arachnoideum** (subsp. *doellianum* (C. B. Lehm.) Schinz & R. Keller): Rosettes not more than 1·5 cm in diameter, ovoid or globose; arachnoid hairs variable in quantity, sometimes rather scanty. $2n = 32, 64$. *Mainly in the eastern part of the range.*

(b) Subsp. **tomentosum** (C. B. Lehm. & Schnittspahn) Schinz

& Thell. in Schinz & R. Keller, *Fl. Schweiz* ed. 4, 325 (1923) (*S. tomentosum* C. B. Lehm. & Schnittspahn): Rosettes 1·5–2·5 cm in diameter, somewhat flattened above; arachnoid hairs always abundant and persistent. Leaves broader towards the apex than those of subsp. (**a**). $2n = 32, 64$. *Mainly in the south-western part of the range.*

11. S. montanum L., *Sp. Pl.* 465 (1753) (*S. candollei* Rouy & Camus). Stolons slender, with persistent leaves. Rosette-leaves usually *c.* 10 × 3 mm (rarely larger), oblanceolate, acute, pubescent above and beneath and ciliate with very short, subequal hairs, dull green, viscid. Flowering stem 5–10(–20) cm, bearing 2–8(–13) flowers, which are usually 11- to 13-merous. Petals 12–20 × 2 mm, vinous red (rarely yellowish); filaments nearly glabrous, pale. *Calcifuge.* ● *Pyrenees, Alps, Appennini; Carpathians; Corse.* Au Co Cz Ga He Hs It Po Rm Rs(W) [No].

1 Rosettes not more than 2 cm in diameter (**a**) subsp. **montanum**
1 Rosettes usually more than 2 cm in diameter
2 Leaves entirely green, obovate-cuneate, up to 7 mm wide (**b**) subsp. **burnatii**
2 Leaves tipped with red, oblanceolate, *c.* 3 mm wide (**c**) subsp. **stiriacum**

(**a**) Subsp. **montanum**: Rosettes not more than 2 cm in diameter. Leaves oblanceolate, not acuminate, entirely green; cilia scarcely longer than hairs on leaf-surface. Petals 12–15 mm. $2n = 42$. *Throughout most of the range of the species.*

(**b**) Subsp. **burnatii** Wettst. ex Hayek in Hegi, *Ill. Fl. Mitteleur.* **4(2)**: 554 (1922): Rosettes up to 8 cm in diameter. Leaves obovate-cuneate, up to 7 mm wide, entirely green. $2n = 42$. *S.W. Alps.*

(**c**) Subsp. **stiriacum** Wettst. ex Hayek in Hegi, *loc. cit.* (1922) (*S. braunii* Funck): Rosettes 2–4·5 cm in diameter. Leaves oblanceolate, distinctly acuminate, reddish at the apex; cilia distinctly longer than hairs on leaf-surface. Petals 16–20 mm. $2n = 84$. *E. Austria.*

Plants from the Carpathians, somewhat intermediate in appearance between subspp. (**b**) and (**c**) and with $2n = 42$, have been named subsp. **carpaticum** Wettst. ex Hayek in Hegi, *loc. cit.* (1922). Their status is uncertain.

12. S. dolomiticum Facch., *Zeitschr. Ferdinand. Tirol (Innsbruck)* ser. 3, **5**: 56 (1855). Stolons 2 cm, slender. Rosettes 2–4 cm in diameter, subglobose. Leaves 10–15 × 3–5 mm, oblong-lanceolate, acuminate, sparsely pubescent, bright green with a brownish apex; apical cilia coarser than the lateral. Flowering stem *c.* 10 cm, bearing 12–20 flowers; cauline leaves 10–20 × 3–5 mm. Flowers usually 10- to 12-merous; petals 9–10 × 2 mm, deep reddish-pink, with a central stripe of reddish-brown above and of green beneath; filaments more or less glabrous. $2n = 72$. *Dolomite and basalt rocks, 1600–2500 m.* ● *S.E. Alps.* It.

13. S. kosaninii Praeger, *Bull. Inst. Jard. Bot. Univ. Beograd* **1**: 210 (1930). Stolons up to 12 cm, stout, leafy. Rosettes 4–8 cm in diameter, open. Leaves 15–40 × 7–15 mm, oblanceolate, shortly acuminate, densely pubescent, dark green with a red apex; cilia about twice as long as the other hairs. Flowering stem *c.* 15 cm; cauline leaves 15–35 × 5–8 mm. Flowers 13- or 14-merous. Sepals very hairy; petals 10 × 1 mm, purple, paler beneath and with a narrow white margin; filaments slightly hairy at the base. *Calcicole.* ● *Mountains of S. Jugoslavia.* Ju.

14. S. macedonicum Praeger, *op. cit.* 212 (1930). Stolons 4–7 cm, giving a loose mat of flattish rosettes 3–5 cm in diameter. Leaves 15–20 × 5 mm, broadly oblanceolate, shortly acuminate, densely but minutely pubescent, reddish towards the apex; cilia as in **13**. Flowering stem 7–10 cm. Inflorescence compact, corymbose. Flowers 11- or 12-merous; petals 8–10 mm, dull pinkish-lilac. Filaments lilac, almost glabrous. $2n = 34$. ● *S.W. Jugoslavia (N.W. Makedonija).* ?Al Ju.

15. S. cantabricum J. A. Huber, *Feddes Repert.* **33**: 364 (1934). Stolons up to 6 cm, stout. Rosettes 2–5 cm in diameter, open. Leaves 20–35 × 10 mm, oblong-obovate, mucronate, more or less pubescent, dark green with a red apex; cilia about twice as long as the other hairs. Flowering stem up to 16 cm. Flowers 15–30, 9- to 12-merous; sepals obtuse; petals 9–10 × 2 mm, red, pink or white; filaments glabrous or pubescent. ● *N. & C. Spain.* Hs.

Three subspecies may be recognized; although geographically separated they show some morphological overlap.

1 Red apex of leaves less than 2 mm, not conspicuous (**b**) subsp. **urbionense**
1 Red apex of leaves more than 2 mm, conspicuous
2 Filaments pubescent; leaves densely pubescent (**a**) subsp. **cantabricum**
2 Filaments glabrous; leaves sparsely pubescent (**c**) subsp. **guadarramense**

(**a**) Subsp. **cantabricum**: Red apex of leaves more than 2 mm; leaf densely pubescent and with numerous apical cilia. Filaments pubescent. $2n = 68, 72$. *Cordillera Cantábrica.*

(**b**) Subsp. **urbionense** M. C. Sm., *Lagascalia* **10**: 21 (1981): Red apex of leaves less than 2 mm; leaf densely pubescent and with numerous apical cilia. Filaments glabrous or pubescent. $2n = 82$. *Sierra de Urbión.*

(**c**) Subsp. **guadarramense** M. C. Sm., *op. cit.* 20 (1981). Red apex of leaves more than 2 mm; leaf sparsely pubescent and with few apical cilia. Filaments glabrous. *Sierra de Guadarrama.*

S. giuseppii Wale, *Quart. Bull. Alp. Gard. Soc.* **9**: 115 (1941), from N. Spain (Cordillera Cantábrica), appears to be a hybrid between **10** and **15(a)**.

16. S. marmoreum Griseb., *Spicil. Fl. Rumel.* **1**: 329 (1843) (*S. tectorum* auct. eur. orient., non L.; incl. *S. ballsii* Wale, *S. banaticum* Domokos, *S. erythraeum* Velen., ?*S. italicum* Ricci, *S. reginae-amaliae* Heldr. & Guicc. ex Halácsy, *S. schlehanii* Schott). Stolons short or long. Rosettes 3–6 cm in diameter, open. Leaves 15–25 × 9–15 mm, obovate, mucronate, densely and finely pubescent (sometimes glabrous when mature), usually with stout, deflexed marginal cilia, dull green to purplish-grey, often tinged with red. Flowering stem 10–20 cm; cauline leaves 20–30 mm, oblong-lanceolate, acuminate. Inflorescence up to 7 cm wide. Flowers 11- to 13-merous; petals 10 mm, red with white margin; filaments usually purple, hairy at the base. $2n = 34$. ● *S.E. & E.C. Europe; S. & C. Italy.* Al Bu Cz Gr Hu ?It Ju Rm Rs(W).

Very variable, especially in leaf-colour and persistence of pubescence on the leaves. The variation is, however, largely continuous and reticulate, and it does not seem possible to recognize infraspecific taxa, except perhaps at varietal level.

S. balcanicum Stoj., *Izv. Bot. Inst. (Sofia)* **2**: 263 (1951), from C. Bulgaria, is said to differ principally in its pale lilac petals and more or less glabrous leaves. It is not clear from the information available whether it should be considered a distinct species or a variant of **16**.

17. S. nevadense Wale, *Quart. Bull. Alp. Gard. Soc.* **9**: 109 (1941). Stolons short, slender. Rosettes *c.* 3 cm in diameter. Leaves 12–18 × 5–7 mm, obovate, mucronate, glandular-puberulent when young, sometimes glabrous at maturity; cilia stout, often bent. Flowering stem 3–7(–12) cm, stout; cauline leaves imbricate. Inflorescence compact, but sometimes with additional flowers in axils of upper leaves. Flowers 9- to 11-merous; calyx puberulent, with some longer hairs; petals reddish-pink, occasionally with a darker central band; filaments hairy in lower $\frac{1}{2}$; anthers reddish-pink. $2n = 108$. 1650–2150 *m*. ● *S. Spain (Sierra Nevada)*. Hs.

18. S. tectorum L., *Sp. Pl.* 464 (1753) (incl. *S. arvernense* Lecoq & Lamotte, *S. glaucum* Ten., *S. andreanum* Wale, *S. alpinum* Griseb. & Schenk, *S. schottii* Baker, non C. B. Lehm. & Schnittspahn). Stolons up to 4 cm, stout. Rosettes usually large (3–8 cm in diameter), open. Leaves 20–40 × 10–15 mm, oblong-lanceolate to obovate, with a stout, pungent mucro, glabrous or with a few scattered hairs on the surface, dark or somewhat glaucous green, variably tinged with red; cilia conspicuous, white. Flowering stem 20–50 cm, stout, hairy; lower cauline leaves glabrous apart from the cilia, the upper ones pubescent. Inflorescence large, with from 40 to over 100 usually 13-merous flowers. Sepals acute; petals 9–10 × 2 mm, ciliate and pubescent beneath, dull pink or purple, glabrous or slightly pubescent at the base. $2n = 36, 40, 72$. ● *Mountains of W., C. & S. Europe, from C. Pyrenees to S.E. Alps and S. Appennini*. Au Ga Ge He Hs It Ju [Su].

Extremely variable, and divided variously by different authors into species, subspecies and varieties. Much of the variation is phenotypic, and a large part of the genetically determined variation, both in wild and cultivated plants, may be presumed to arise from hybridization with **19** and perhaps **11**.

A cultivar to which the Linnean type is referable, distinguished by its large size and frequent partial sterility, has been extensively cultivated in most of Europe, especially on the roofs of cottages as a charm against fire, and is perhaps locally naturalized.

19. S. calcareum Jordan, *Obs. Pl. Crit.* **7**: 26 (1849) (*S. arvernense* sensu Coste, pro parte). Like **18**, in which it is sometimes included, but distinct from all variants of the latter in its usually broad, glabrous (rarely pubescent when young), glaucous or pale green rosette-leaves with a well-defined purple-brown apical spot on both surfaces, its broad-based, subamplexicaul cauline leaves, all glabrous, and its pale pink or greenish-white, 10- to 12-merous flowers with obtuse sepals and petals 7–8 mm. $2n = 38$. *Limestone rocks*. ● *S.W. Alps*. Ga It.

7. Jovibarba Opiz

C. FAVARGER AND F. ZÉSIGER (EDITION 1) REVISED BY J. PARNELL AND C. FAVARGER (EDITION 2)

Perennials, like *Sempervivum* in habit, but with 6-merous (rarely 5- or 7-merous), campanulate flowers, and erect, pale yellow petals, keeled dorsally and fringed with glandular hairs. (*Diopogon* Jordan & Fourr.)

Stolons absent; rosette-leaves spinose-mucronate; petals ciliate, dorsally keeled but not winged — **1. heuffelii**
Stolons present; rosette-leaves not spinose-mucronate; petals fimbriate, with the dorsal keel produced into a wing — **2. globifera**

1. J. heuffelii (Schott) Á. & D. Löve, *Bot. Not.* **114**: 39 (1961) (*J. velenovskyi* (Česchm.) J. Holub, *Sempervivum heuffelii* Schott, *S. patens* Griseb. & Schenk, *S. hirtum* auct., non L., *Diopogon heuffelii* (Schott) H. Huber). Rosettes 5–7(–12) cm in diameter, open, with patent leaves 25–60 × 10–15 mm, oblong-obovate, spinose-mucronate, dark or glaucous green with white margins, ciliate with stiff, often deflexed hairs. Stolons absent; vegetative reproduction by division of the rosette. Flowering stem 10–20 cm. Sepals glandular-ciliate; petals 10–12 mm, oblong-obovate, usually tricuspidate, ciliate but scarcely fimbriate, keeled, but not dorsally winged. Styles 2 mm, relatively stout. $2n = 38$. *Mountain rocks; calcicole*. ● *E. Carpathians and mountains of Balkan peninsula*. Al Bu Gr Ju Rm.

2. J. globifera (L.) J. Parnell, *Bot. Jour. Linn. Soc.* **103**: 219 (1990) (*Sempervivum globiferum* L.). Rosettes 1–7 cm in diameter, open or closed. Leaves 8–20 × 3–7(–10) mm, patent, erect or incurved, lanceolate to obovate-oblanceolate, often acuminate but not spinose-mucronate, ciliate, usually glandular-pubescent on the surface. Stolons present. Sepals usually hairy, but sometimes glandular-puberulent or glabrous; petals 12–17 mm, fimbriate, with dorsal keel produced into a wing. Styles usually *c.* 4 mm. $2n = 38$. *Mountain rocks; or on dry, sandy or stony ground in the lowlands*. ● *C. & S.E. Europe, southwards to N. Albania, and extending to the S.W. Alps and to N.W. & C. Russia*. Al Au Cz Ga Ge Hu It Ju ?No Po Rm Rs(?N, B, C, W).

1 Sepals minutely glandular-puberulent on surface
 2 Rosette-leaves glandular-pubescent on surface — **(c)** subsp. **allionii**
 2 Rosette-leaves glabrous on surface — **(d)** subsp. **arenaria**
1 Sepals not glandular-puberulent on surface, although sometimes with some longer hairs
 3 Surface of sepals glabrous; rosette leaves widest above the middle — **(a)** subsp. **globifera**
 3 Surface of sepals usually ± hairy; rosette-leaves widest at or below the middle — **(b)** subsp. **hirta**

(a) Subsp. **globifera** (*Jovibarba sobolifera* (Sims) Opiz, *Diopogon hirtus* subsp. *borealis* H. Huber, *Sempervivum soboliferum* Sims): Rosettes 1–3 cm in diameter, their leaves widest above the middle, often with red apex, glandular-ciliate but otherwise glabrous; cauline leaves similar, or somewhat broader. Surface of sepals glabrous. *Calcifuge; usually at low altitudes. Throughout most of the range of the species.*

(b) Subsp. **hirta** (L.) J. Parnell, *loc. cit.* (1990) (*J. hirta* (L.) Opiz, *J. hirta* subsp. *glabrescens* (Sabr.) Soó & Jáv., *Diopogon hirtus* (L.) H. P. Fuchs ex H. Huber subsp. *hirtus*, *Sempervivum hirtum* L., *S. hirtum* subsp. *preissianum* Dostál & subsp. *tatrense* Dostál): Rosettes 2·5–5 cm in diameter, their leaves widest at or below the middle, without red apex, glandular-ciliate but otherwise glabrous; cauline leaves broader. Surface of sepals usually hairy. *Somewhat calcicole; mainly in the mountains. Throughout most of the southern part of the range of the species.*

(c) Subsp. **allionii** (Jordan & Fourr.) J. Parnell, *loc. cit.* (1990) (*J. allionii* (Jordan & Fourr.) D. A. Webb, *J. hirta* subsp. *allionii* (Jordan & Fourr.) Soó, *Diopogon allionii* Jordan & Fourr., *D. hirtus* subsp. *allionii* (Jordan & Fourr.) H. Huber, *Sempervivum allionii* (Jordan & Fourr.) Nyman, *S. hirtum* sensu Coste et auct. ital., non L.): Rosettes 1·5–2·5 cm in diameter, their leaves often with a red apex, glandular-puberulent; cauline leaves similar. Surface of sepals glandular-puberulent. *Calcifuge. S.W. Alps; S.E. Alps (Kärnten).*

(**d**) Subsp. **arenaria** (Koch) J. Parnell, *loc. cit.* (1990) (*J. arenaria* (Koch) Opiz, *Diopogon hirtus* subsp. *arenarius* (Koch) H. Huber, *Sempervivum arenarium* Koch): Rosettes 0·5–2 cm in diameter, their leaves usually widest at or below the middle, sometimes with red apex, glandular-ciliate but otherwise glabrous; cauline leaves similar. Sepals glandular-puberulent on surface. *Calcifuge. E. Alps, eastwards from* c. 11°45′ *E.*

8. Aeonium Webb & Berth.

D. A. WEBB

Perennials, with erect, somewhat woody stems and alternate leaves crowded into terminal rosettes. Flowers 9- to 11-merous, in a panicle. Petals free, yellow. Stamens twice as many as the sepals.

1. A. arboreum (L.) Webb & Berth., *Phyt. Canar.* **1**: 185 (1840) (*Sempervivum arboreum* L.). Stem 50–80 cm, stout, with suberect branches, marked by conspicuous leaf-scars. Leaves *c.* 6 × 2 cm, oblanceolate-cuneate, ciliate-denticulate but otherwise glabrous, bright shining green. Flowers numerous, in a compact, ovoid panicle. Petals 6–7 mm, bright yellow. *Cultivated for ornament, and widely naturalized on rocks and walls near the coast in S. Europe.* [Bl Ga Gr Hs Lu Sa Si.] (*Morocco.*)

9. Aichryson Webb & Berth.

D. A. WEBB

Annuals; leaves alternate, entire, petiolate, caducous. Flowers 6- to 10-merous, in a dichotomous cyme. Petals free, yellow. Stamens twice as many as the petals.

Petals usually 8; branches ± horizontal **1. villosum**
Petals usually 10; branches suberect, pseudodichotomous **2. dichotomum**

1. A. villosum (Aiton) Webb & Berth., *Phyt. Canar.* **1**: 181 (1840) (*Sempervivum villosum* Aiton). Plant up to 15 cm, bushy, with horizontal branches. Stem and leaves softly and densely glandular-hairy. Leaves spathulate-rhombic; lamina 15–20 mm. Flowers usually 8-merous, pedicellate, in a lax cyme. Petals 5–6 mm, ovate, bright golden-yellow. *Rocks. Açores.* Az. (*Madeira.*)

2. A. dichotomum (DC.) Webb & Berth., *loc. cit.* (1840) (*Sempervivum dichotomum* DC., *S. annuum* Coutinho). Like **1** but often up to 30 cm; stems usually reddish, with suberect, pseudo-dichotomous branches; and petals usually 10, paler yellow. *Naturalized in Portugal* (*Serra de Sintra*). [Lu.] (*Islas Canarias.*)

10. Sedum L.

D. A. WEBB, J. R. AKEROYD AND H. 'T HART

Leaves usually alternate, seldom crowded into rosettes, and rarely (in European species) into the dense, globose rosettes characteristic of *Sempervivum*. Inflorescence usually cymose. Flowers hermaphrodite, usually 5-merous, but sometimes 4- or 6- to 9-merous. Petals free or slightly connate at the base, usually patent. Stamens usually twice as many as the petals, sometimes equal in number. Carpels equal in number to the petals.

The habit of most species is characteristic, but it is difficult to frame a definition of the genus which delimits it clearly from *Crassula*, *Sempervivum*, and other genera, and the assignment of some species is still disputed.

In addition to those described below several cultivated species are locally naturalized near gardens.

Most European species occur on dry, rocky or stony ground.

Literature: R. Ll. Praeger, *Jour. Roy. Hort. Soc.* **46**: 1–134 (1921). H. Fröderström, *Acta Horti Gothob.* **5**: Bihang 1–75 (1930), **6**: Bihang 1–111 (1931), **7**: Bihang 1–126, **10**: Bihang 1–262 (1936). H. 't Hart, *Biosystematic Studies in the* acre-*group and the Series* Rupestria *Berger of the Genus* Sedum (*Crassulaceae*). Utrecht, 1978; *Fl. Medit.* **1**: 31–61 (1991).

1 Leaves flat, at least 8 mm wide
2 Petals bright yellow
3 Erect shrub with very stout stems **1. praealtum**
3 Stems usually herbaceous except at the extreme base, fairly slender
4 Stems erect, usually all flowering **2. aizoon**
4 Stems procumbent at the base, bearing non-flowering and flowering shoots **3. hybridum**
2 Petals red, purple or whitish
5 Petals 10–12 mm, erect **7. spurium**
5 Petals 3–5 mm, patent
6 Follicles stellate-patent **45. stellatum**
6 Follicles ± erect
7 Flowers whitish **4. telephium**
7 Flowers pink or purple
8 Leaves dentate **4. telephium**
8 Leaves entire
9 Leaves opposite **5. ewersii**
9 Leaves alternate **6. anacampseros**
1 Leaves less than 8 mm wide, often ± terete
10 Petals yellow, cream or greenish-white
11 Follicles erect
12 Flowers 5-merous
13 Petals up to 12 mm, bright yellow **10. montanum**
13 Petals 5 mm, greenish yellow **49. rubens**
12 Flowers 6- to 9-merous
14 Flowers few, in lax, unilateral cymes; non-flowering shoots in midsummer leafless except for a terminal summer-dormant bud
15 Leaves of the non-flowering shoots amplexicaul, with a broad, 3-lobed, hyaline spur; petals 6–8 mm **13. amplexicaule**
15 Leaves of the non-flowering shoots not amplexicaul, with a short, truncate spur; petals 8–12 mm **14. pruinatum**
14 Flowers numerous, in crowded cymes; plant without summer-dormant buds
16 Living leaves on non-flowering shoots confined to a terminal, tassel-like cluster at anthesis; dead leaves persistent on lower part of shoot **12. forsterianum**
16 Living leaves on non-flowering shoots not confined to a terminal cluster; dead leaves mostly not persistent
17 Petals greenish-white to very pale yellow
18 Sepals 2·5 mm, glabrous; petals patent **8. sediforme**
18 Sepals 5–7 mm, glandular-puberulent; petals erect throughout anthesis **9. ochroleuchum**
17 Petals bright yellow

19 Sepals 5–7 mm; inflorescence erect in bud **10. montanum**
19 Sepals 3–4 mm; inflorescence usually drooping in bud **11. rupestre**
11 Follicles ±patent
20 Perennial **(15–22). acre** group
20 Annual or biennial
21 Petals 2–3 times as long as the sepals; stamens 10 **46. annuum**
21 Petals scarcely exceeding the sepals; stamens usually 5 **47. litoreum**
10 Petals blue, pink or pure white
22 Stamens equal in number to the petals
23 Flowers sessile or subsessile
24 Leaves ciliate-denticulate **48. aetnense**
24 Leaves glabrous, entire
25 Leaves linear to elliptical, patent **49. rubens**
25 Leaves ovoid, imbricate **50. caespitosum**
23 Flowers distinctly pedicellate
26 Biennial; leaves glandular-pubescent **37. villosum**
26 Annual; leaves glabrous
27 Leaves ovoid or subglobose **51. andegavense**
27 Leaves linear-oblong **52. nevadense**
22 Stamens twice as many as the petals
28 Perennial (except **32**), with leafy non-flowering shoots
29 Plant glabrous throughout
30 Flowers 4-merous **27. stefco**
30 Flowers 5- to 9-merous
31 Leaves distinctly flattened, at least on upper side
32 Flowers 5-merous; styles *c.* 0·25 mm; follicles erect **31. magellense**
32 Flowers 6- to 9-merous; styles *c.* 0·75 mm; follicles stellate-patent **53. hispanicum**
31 Leaves ±terete
33 Leaves of non-flowering shoots opposite, 4-ranked **29. brevifolium**
33 Leaves of non-flowering shoots alternate
34 Leaves subulate, acuminate **25. subulatum**
34 Leaves linear to ovoid-globose, obtuse
35 Leaves spurred **26. anglicum**
35 Leaves scarcely spurred **29. brevifolium**
29 Plant pubescent or puberulent-papillose in part
36 Leaves mostly opposite or whorled
37 Leaves ovoid, never whorled **28. dasyphyllum**
37 Leaves oblong-lanceolate, the upper whorled **30. monregalense**
36 Leaves all alternate
38 Leaves glabrous **23. album**
38 Leaves pubescent or puberulent-papillose
39 Leaves not much longer than wide
40 Leaves semiterete, puberulent-papillose, sessile **24. gypsicola**
40 Leaves flat, glandular-pubescent, the lower petiolate
41 Plant annual; petals 5–6 mm **32. alsinifolium**
41 Plant perennial; petals 3–5 mm **33. fragrans**
39 Leaves at least twice as long as wide
42 Flowers 6- to 9-merous; follicles stellate-patent **53. hispanicum**
42 Flowers 5(6)-merous; follicles erect
43 Leaves flat **32. alsinifolium**
43 At least the cauline leaves terete or semiterete
44 Peduncles arising from axils of rosette-leaves
45 Sepals ovate-triangular; follicles erect **34. tristriatum**
45 Sepals oblong; follicles stellate-patent **35. tymphaeum**
44 Peduncles terminal
46 Buds drooping; carpels white; on dry rocks **36. hirsutum**
46 Buds erect; carpels green or purple; in wet places **37. villosum**
28 Annual or biennial; no non-flowering shoots present at flowering season
47 Petals blue **55. caeruleum**
47 Petals white, pink or purple
48 Leaves dentate or crenate **45. stellatum**
48 Leaves entire
49 Leaves mostly opposite or whorled **39. cepaea**
49 Leaves alternate
50 Leaves pubescent
51 Leaves oblong-spathulate, flat **38. creticum**
51 Leaves linear to linear-oblong, terete or semi-terete
52 Flowers 6- to 9-merous; follicles stellate-patent **53. hispanicum**
52 Flowers 5-merous; follicles erect
53 Biennial; inflorescence rather compact and narrow; flowers stellate **37. villosum**
53 Annual; inflorescence lax and broad; flowers campanulate **40. lagascae**
50 Leaves glabrous
54 Leaves terete, at least in apical $\frac{1}{2}$
55 Pedicels much longer than the calyx **41. pedicellatum**
55 Pedicels to as long as the calyx
56 Leaves ovoid or ellipsoidal; branches diffuse, ascending **42. arenarium**
56 Leaves oblong or obovoid-clavate, flattened towards the base; branches erect **43. atratum**
54 Leaves flat, at least on upper surface
57 Leaves oblong-spathulate **38. creticum**
57 Leaves linear
58 Flowers 6- to 9-merous; follicles stellate-patent **53. hispanicum**
58 Flowers 5-merous; follicles ±erect
59 Sepals united for $\frac{1}{2}$ their length; petals subacute **44. confertiflorum**
59 Sepals free almost to base; petals acuminate
60 Plant glandular-pubescent above **49. rubens**
60 Plant usually glabrous **54. pallidum**

1. S. praealtum A. DC. in DC. & A. DC., *Not. Pl. Rar. Jard. Bot. Genève* **10**: 21 (1847). Bushy, evergreen, glabrous shrub up to 75 cm; stems thick, terete, green above, grey and woody below. Leaves *c.* 6 × 1·5 cm, oblanceolate, obtuse, thick but flattened. Flowers 5-merous, subsessile, in large, terminal panicles. Petals 6–9 mm, lanceolate, acute, bright yellow. Stamens 10. Follicles suberect, yellowish. *Cultivated for ornament in parts of the Mediterranean region, and naturalized on coastal rocks in S.E. France and N.W. Italy.* [Ga It.] (*Mexico.*)

2. S. aizoon L., *Sp. Pl.* 430 (1753). Perennial; stems 30–40 cm, few, erect. Leaves 5–8 cm, alternate, lanceolate-cuneate, subacute, unequally serrate or dentate. Flowers 5-merous, in dense, compound, terminal cymes, subtended by leaves. Sepals tapered to a subulate apex: Petals 7–10 mm, acuminate, golden-yellow. Stamens 10. *Locally naturalized from gardens in N. & C. Europe.* [Cz Fe Ga Ge No Su.] (*N. Asia.*)

3. S. hybridum L., *Sp. Pl.* 431 (1753). Perennial; stems woody, branched, creeping and rooting, with short non-flowering shoots and ascending flowering stems 15–20 cm. Leaves 2–3 cm, oblong-cuneate, obtuse, obtusely dentate; teeth reddish. Flowers 5-merous, numerous, in lax, terminal corymbs. Sepals obtuse. Petals 6–9 mm, acute, golden-yellow. Stamens 10. *C. & S. Ural; naturalized from gardens elsewhere in N. & C. Europe.* Rs(C) [Au Cz Da Fe Ga Ge Hu No Su]. (*N. Asia.*)

4. S. telephium L., *Sp. Pl.* 430 (1753) (*S. complanatum* sensu Rouy & Camus). Perennial, with tuberous roots; stems 15–80 cm, usually simple. Leaves 2–8 cm, suborbicular to narrowly oblong. Flowers 5-merous, in large, dense, terminal corymbs. Petals 3–5 mm, ovate-lanceolate, acute. Stamens 10, equalling or slightly exceeding the petals. Follicles erect; styles rather short. $2n = 24, 30, 36, 48$ (other numbers reported from garden plants). *Almost throughout Europe.* All except Az Bl Cr Fa Is Sb Si.

A polymorphic complex which may be divided in Europe into four subspecies. Intermediates, some of them garden hybrids subsequently naturalized, are not uncommon, and the considerable confusion in taxonomy and nomenclature does not permit precise distributional data for the subspecies to be given. Typical members of subspp. (**b**), (**c**) and (**d**) appear to be characterized by chromosome numbers of $2n = 24$, 48, and 48 respectively, but no counts are available for plants with intermediate characters; counts of $2n = 36$ and 48 have been reported for subsp. (**a**).

1 Leaves ±ovate, sessile, often amplexicaul, usually subentire; flowers usually greenish- or yellowish-white
2 Stems 30–80 cm, erect; leaves 5–10 cm, oblong-ovate, green or glaucous — (**c**) subsp. **maximum**
2 Stems decumbent or procumbent, 15–40 cm; leaves 2–4 cm, suborbicular, pruinose-glaucous — (**d**) subsp. **ruprechtii**
1 Leaves ±oblong, usually strongly dentate, not amplexicaul, at least the lower ones cuneate at the base; flowers usually purplish-red or lilac
3 Upper leaves truncate at the base, sessile; follicles dorsally grooved — (**a**) subsp. **telephium**
3 Upper leaves cuneate at the base, sometimes stalked; follicles not grooved — (**b**) subsp. **fabaria**

(**a**) Subsp. **telephium** (*S. purpurascens* Koch, *S. complanatum* sensu Coutinho, *S. purpureum* Schultes): Stems 25–60 cm, erect. Leaves 5–8 cm, narrowly ovate-oblong, irregularly dentate, the upper truncate, the lower cuneate; not glaucous, often red-spotted. Flowers purplish-red or lilac, rarely white. Carpels grooved on outer side. *Chiefly in C. & E. Europe.*

(**b**) Subsp. **fabaria** (Koch) Kirschleger, *Fl. Alsace* **1**: 284 (1852) (*S. fabaria* Koch, *S. carpaticum* G. Reuss): Stems 20–40 cm, erect. Leaves 4–7 cm, oblong-lanceolate, dentate, cuneate, the lower ones stalked; not glaucous. Flowers purplish-red or lilac. Carpels not grooved. *Chiefly in W. & C. Europe.*

(**c**) Subsp. **maximum** (L.) Krocker, *Fl. Siles.* **2**: 64 (1790) (*S. maximum* (L.) Suter, *S. stepposum* Boiss., *S. haematodes* Miller): Stems 30–80 cm, erect. Leaves 4–10 cm, ovate or broadly oblong, bluntly dentate or subentire, truncate or cordate-amplexicaul, alternate, opposite or whorled, sometimes glaucous. Flowers greenish- or yellowish-white, rarely purple. *Much of Europe, but local in the west.*

(**d**) Subsp. **ruprechtii** Jalas, *Ann. Bot. Soc. Zool.-Bot. Fenn. Vanamo* **26**: 33 (1954) (*S. telephium* sensu Boriss.): Stems 15–40 cm, decumbent or procumbent. Leaves 2–5 cm, suborbicular, dentate or subentire, cordate-amplexicaul, opposite, glaucous-pruinose. Flowers whitish. *N.E. Europe.*

5. S. ewersii Ledeb., *Icon. Pl. Fl. Ross.* **1**: 14 (1829). Perennial, with procumbent woody stems from which arise ascending non-flowering and flowering shoots 10–20 cm. Leaves 15–20 mm, broadly ovate to orbicular, subentire, cordate, opposite, glaucous, spotted. Flowers 5-merous, in a dense, convex corymb. Petals 4–5 mm, acute, pink or mauve. Stamens 10. Follicles erect, with short, out-turned styles. *Naturalized from gardens in Fennoscandia.* [Fe No Su.] (*C. Asia, Himalaya.*)

6. S. anacampseros L., *Sp. Pl.* 430 (1753). Perennial; non-flowering shoots procumbent, with terminal leaf-rosettes; flowering stems 15–25 cm, ascending. Leaves 12–25 mm, alternate, ovate-elliptical, obtuse, entire, glaucous. Inflorescence and flowers as in **5**, but flowers sometimes 4-merous; petals glaucous-lilac outside and dull, deep red inside; styles straight. $2n = 36$. *Calcifuge.* ● *Pyrenees, S.W. Alps, Appennini.* Ga He Hs It [No].

7. S. spurium Bieb., *Fl. Taur.-Cauc.* **1**: 352 (1808) (*S. oppositifolium* Sims). Perennial; stems procumbent, rooting; non-flowering shoots short; flowering stems longer, procumbent or ascending. Leaves *c.* 25 × 12 mm, obovate or rhombic-cuneate, shortly petiolate, crenate or bluntly dentate towards the apex. Flowers 5-merous, subsessile, in rather dense corymbs. Sepals subulate, obtuse. Petals 10–12 mm, erect, linear-oblong, acuminate, reddish-purple, pink or white. Stamens 10. $2n = 28$. *Naturalized from gardens in many parts of Europe.* [Au Be Br Cz Da Fe Ga Ge Hb He Ho Hs Hu It No Po Rs(K) Su.] (*Caucasus.*)

8. S. sediforme (Jacq.) Pau, *Actas Mem. Prim. Congr. Nat. Esp. Zaragoza* 246 (1909), non Hamet (*S. altissimum* Poiret, *S. nicaeense* All., *S. ochroleucum* auct.). Robust, glabrous, somewhat glaucous perennial with ascending flowering stems 25–60 cm and shorter non-flowering shoots, both woody at the base. Leaves oblong or narrowly ellipsoidal, thick but somewhat flattened on the upper surface, usually apiculate or mucronate, shortly spurred, suberect, closely imbricate in spiral rows on the non-flowering shoots. Inflorescence without bracts, erect and subglobose in bud, with strongly recurved branches, concave in fruit. Flowers 5- to 8-merous, on very short pedicels. Sepals *c.* 2·5 mm, ovate, obtuse or mucronate, glabrous. Petals 5–8 mm, patent, greenish-white or straw-coloured. Stamens 10–16. Follicles erect, pale yellow or greenish-white. $2n = 32, 64, 96$. *Mediterranean region, extending to Portugal, N. Spain and C. France.* Al Bl Co Cr Ga Gr Hs It Ju Lu Sa Si Tu.

9. S. ochroleucum Chaix in Vill., *Hist. Pl. Dauph.* **1**: 325 (1786) (*S. anopetalum* DC.). Fairly robust perennial, with short, more or less procumbent non-flowering shoots and erect or ascending flowering stems 15–40 cm, both woody at the base. Leaves 10–15 mm, terete, linear-cylindrical, acuminate, spurred, suberect, loosely imbricate on the non-flowering shoots. Inflorescence bracteate, erect in bud, corymbose, with branches scarcely recurved. Flowers 5- to 8-merous, subsessile. Sepals 5–7 mm, triangular-lanceolate, acuminate, glandular-puberulent. Petals 8–10 mm, cream or greenish-white, erect throughout anthesis. Follicles greenish, erect. $2n = 34, 68, 102$. *S. Europe*. Al Bu Ga Gr He Hs It Ju Rm.

10. S. montanum Perr. & Song., *Billotia* **1**: 77 (1864) (*S. ochroleucum* subsp. *montanum* (Perr. & Song.) D. A. Webb, *S. rupestre* auct., non L.). Like **9** but flowers usually 5-merous; petals up to 12 mm, bright yellow, patent throughout anthesis. $2n = 34, 51$. ● *Pyrenees, Appennini, Alps and adjacent regions; Hungary*. Au Ga He Hs Hu It Ju.

11. S. rupestre L., *Sp. Pl.* 431 (1753) (*S. reflexum* L., *S. albescens* Haw.). Perennial; stems procumbent, intricate, somewhat woody; flowering stems 15–35 cm, ascending; non-flowering shoots much shorter. Leaves linear-terete, suberect (the upper ones sometimes recurved), apiculate, spurred, evenly distributed on the non-flowering shoots; dead leaves not persistent. Inflorescence bracteate, subglobose and usually drooping, in bud, concave in fruit. Flowers (5–)7(–9)-merous. Sepals 3–4 mm, lanceolate, subacute, glabrous or sparsely glandular-puberulent. Petals 6–7 mm, linear-lanceolate, patent, bright (rarely pale) yellow. Stamens 10–18. Follicles yellow, erect. $2n = 88, 112$. ● *From S. Fennoscandia southwards to the Pyrenees, Sicilia and N. Greece*. Al Be Co Cz Da Fe Ga Ge Gr He Ho Hs Hu It Ju No Po Rm Rs(B, C) Sa Si Su [Br Hb].

There has been much confusion in the literature between **9**, **10** and **11**, and many records, especially from the Balkan peninsula, require confirmation.

12. S. forsterianum Sm. in Sowerby, *Engl. Bot.* **26**: t. 1802 (1808) (*S. elegans* Lej., *S. pruinatum* auct., non Link ex Brot., *S. rupestre* auct.). Like **11** but leaves flat on the upper surface and aggregated on the non-flowering shoots into dense, cone-like terminal rosettes, the rest of the shoots bearing numerous, persistent dead leaves; inflorescence ebracteate, sometimes flat-topped; sepals 2–3 mm, oblong-lanceolate, obtuse. $2n = 48, 60, 72, 96$. *Often in damper places than* **11**. *W. Europe*. *Az Be Br Ga Ge †Ho Hs Lu [Hb].

13. S. amplexicaule DC., *Rapp. Voy. Bot.* 80 (1808) (*S. tenuifolium* (Sm.) Strobl). Slender perennial with ascending non-flowering shoots covered with erect, imbricate, terete, glaucous leaves, tapered to a long, filiform apex, and expanded at the base into a broad, amplexicaul, 3-lobed spur. In midsummer those on the lower part of the shoot withered; those near the apex persistent as a compact, summer-dormant bud which unfolds and roots in autumn. Flowering stems 7–20 cm, erect; leaves less crowded and with a smaller spur. Flowers 5- to 8(–12)-merous, rather few, subsessile, in lax, unilateral bracteate cymes drooping in bud. Petals 6–8 mm, yellow with red mid-vein. Stamens 10–16(–24). Follicles yellow, erect. $2n = 24, 48, 60, 72$. *Mediterranean region, extending to Bulgaria, Portugal and N. Spain*. Bu Cr Ga Gr Hs It Ju Lu Sa Si Tu.

14. S. pruinatum Link ex Brot., *Fl. Lusit.* **2**: 209 (1805). Like **13** but strongly pruinose throughout; non-flowering shoots long, slender and procumbent, the leaves with a truncate (not 3-lobed) spur and an acuminate but scarcely filiform apex; the leaves of summer-dormant buds in 5 rows; inflorescence erect in bud; petals 8–12 mm, straw-coloured; follicles whitish. $2n = 26–28, 30, 36$. ● *Portugal*. Lu.

(15–22). S. acre group. Perennials with short non-flowering shoots and flowering stems up to 20 cm, but often less. Leaves 3–10(–15) mm, obtuse to subacute, usually thick, often more or less terete, usually spurred at the base. Inflorescence a terminal cyme, the bracts often leaf-like. Flowers 5-merous. Sepals unequal. Petals yellow, patent. Stamens 10. Follicles gibbous, stellate-patent.

This group represents a complex that is still imperfectly understood. The treatment given here is provisional, and much more observation of the very varied populations of the Balkan peninsula, both in cultivation and in the field, is needed before the status and relationships of the various taxa can be fully resolved.

1 Leaves ovate to triangular-lanceolate, widest below the middle
  2 Persistent dead leaves (if present) entirely white, soft and papery **15. acre**
  2 Persistent dead leaves white at the base, grey or black towards the apex, rather coriaceous **16. urvillei**
1 Leaves linear, oblong, narrowly elliptical or oblanceolate, widest at or above the middle
  3 One or more small, axillary cymes of single flowers present on flowering stem below the terminal cyme **17. laconicum**
  3 Flowers confined to the terminal cyme
    4 Leaves with a few fairly large, hyaline papillae at the apex **18. grisebachii**
    4 Leaves smooth or finely papillose, but without large papillae at the apex
      5 Leaves widest above the middle **19. alpestre**
      5 Leaves widest at the middle, or of uniform width throughout their length
        6 Non-flowering shoots *c.* 5 mm in diameter, thickened below ground or arising from cylindrical tubers, covered with whitish scale-leaves; leaves on flowering stem with a 3-lobed spur **20. tuberiferum**
        6 Underground stems slender, not forming tubers; leaves with an entire spur or none
          7 Leaves light green, smooth, *c.* 4 times as long as wide **21. sexangulare**
          7 Leaves dull bluish-green, minutely papillose, *c.* $2\frac{1}{2}$ times as long as wide **22. borissovae**

15. S. acre L., *Sp. Pl.* 432 (1753) (*S. neglectum* Ten.). Glabrous, laxly caespitose, with short non-flowering shoots and flowering stems 5–12 cm. Leaves 3–6 mm, thick, elliptical in section, broadest below the middle (usually triangular-ovoid, rarely sub-conical), obtuse, with short, obtuse spur, rather densely imbricate on the non-flowering shoots; usually deciduous (if persistent after death, then pure white, soft and papery). Flowers 5-merous, pedicellate or subsessile, in small cymes with usually spreading branches. Sepals free, spurred. Petals 5–9 mm, acute to acuminate, bright yellow, patent. Stamens 10. Follicles gibbous, stellate-patent, yellowish-white, with long slender styles. $2n = 40, 60, 80, 100, 120$. *Almost throughout Europe*. All except Az Bl Sa Sb Tu; only as an alien in Co.

Very variable in size, leaf-shape, and size of inflorescence and flowers. Many local populations have been given specific or subspecific rank, but their supposedly diagnostic features can usually be matched in regions very remote from those from which they were described. Most of the plants from the Mediterranean region have longer, more acute leaves, more numerous flowers and narrower petals than are found in the plants of northern Europe, and have been distinguished as subsp. **neglectum** (Ten.) Rouy & Camus, *Fl. Fr.* **7**: 112 (1901), but the two subspecies are connected by a wide band of intermediates in S.C. Europe.

16. S. urvillei DC., *Prodr.* **3**: 408 (1828) (*S. sartorianum* Boiss., *S. sartorianum* subsp. *stribrnyi* (Velen.) D. A. Webb, *S. sartorianum* subsp. *hillebrandtii* (Fenzl) D. A. Webb, *S. sartorianum* subsp. *ponticum* (Velen.) D. A. Webb, *S. laconicum* sensu Stoj. & Stefanov). Glabrous; stems 5–20 cm, erect or ascending. Leaves 4–8(–15) mm, conical to oblong-linear, spurred, rather densely imbricate on the usually clavate non-flowering shoots; lower part of the stems clothed with rather coriaceous dead leaves which are white in their basal ½ but black or grey towards the apex. Cymes with 3–8(–15) flowers. Follicles brown. Seeds reddish-brown. $2n = 32, 48, 64, 80, 96, 112, 128$. *Danube basin and Balkan peninsula.* Al Au Bu Cz Gr Hu Ju Rm Rs(K) Tu.

Very variable, but representing an almost continuous range of variation. Plants with more numerous flowers in each cyme and pale yellowish-brown seeds have been described as *S. hillebrandtii* Fenzl. They occur in the Danube basin and may deserve recognition as a subspecies of **16.**

17. S. laconicum Boiss. & Heldr. in Boiss., *Diagn. Pl. Or. Nov.* **1(6)**: 55 (1846). Glabrous. Leaves 5–8 mm, more or less terete, broadly linear or narrowly ellipsoidal, acute with hyaline papillae at the apex. Flowering stems 4–15 cm, simple, bearing below the usually small and crowded terminal cyme one or more small, subsidiary, axillary cymes of 1–3 flowers. Petals 3–5 mm, apiculate-aristate, bright yellow with a red mid-vein or red streaks. Follicles erecto-patent, with white margins. Seeds pale brown. $2n = 16$. *Mountains of Greece and Kriti.* Cr Gr.

Var. *insulare* Rech. fil. (*S. idaeum* D. A. Webb), from the mountains of Kriti (Levka Ori), is perennial with flowering stems 1–4(–6) cm and an inflorescence with fewer flowers, sometimes having red anthers.

18. S. grisebachii Boiss. & Heldr. in Boiss., *Diagn. Pl. Or. Nov.* **3(2)**: 61 (1856) (*S. racemiferum* (Griseb.) Halácsy). Glabrous, with erect or ascending, sometimes numerous, slender flowering stems usually 2–5 cm, and a few short non-flowering shoots. Leaves 5–6 mm, linear-oblong to oblanceolate, semiterete, bearing at the apex a few large, hyaline papillae. Petals 4–6 mm, about 1½ times as long as the sepals, pale yellow. Styles slender. Seeds reddish-brown. ● *From E. Albania to S.W. Bulgaria and E.C. Greece; S. Carpathians.* Al Bu Gr Ju Rm.

A variable species in which at least two subspecies can be recognized.

(a) Subsp. **grisebachii** (incl. *S. kostovii* Stefanov): Annual or perennial. Flowering stems up to 15 cm. Inflorescence lax with up to 35 flowers. $2n = 16, 32$. *Mostly in the lowlands. Bulgaria and N. Greece.*

(b) Subsp. **flexuosum** (Wettst.) Greuter & Burdet in Greuter, Burdet & Long, *Med-Checklist* **3**: 20 (1986) (*S. flexuosum* Wettst., *S. horakii* (Rohlena) Rohlena): Perennial. Stems not more than 6 cm, bearing about 8 flowers. $2n = 16$. *Mountains, usually above 1500 m. Throughout most of the range of the species.*

19. S. alpestre Vill., *Prosp. Pl. Dauph.* 49 (1779) (*S. repens* Schleicher, *S. erythraeum* Griseb.). Non-flowering shoots numerous, very short; flowering stems 5–8 cm, ascending. Leaves 4–6 mm, somewhat flattened, oblong- to elliptic-oblanceolate, very shortly spurred, often streaked with red. Flowers in short, dense, terminal cymes. Petals 3·5–4 mm (only slightly exceeding the sepals), oblong, subacute, dull yellow. Follicles dark red, gibbous, stellate-patent, with a double wing on the adaxial side; styles very short, deflexed. Seeds pale brown. $2n = 16$. *Somewhat calcifuge. Mountains of C. & S. Europe, from the Vosges and the Carpathians to the Pyrenees, Sardegna and Macedonia.* Au Bu Co Cz Ga Ge Gr He Hs It Ju Po Rm Rs(W) Sa.

20. S. tuberiferum Stoj. & Stefanov, *Notizbl. Bot. Gart. Berlin* **11**: 1013 (1934). Stems up to 12 cm, sparingly branched, arising from subterranean, tuberous rhizome. Leaves minutely hyaline-papillose; on the non-flowering shoots 3–4 mm, lanceolate-spathulate; on the flowering stems 6–8 mm, oblong-linear, with a 3-lobed, whitish spur. Flowers shortly pedicellate, in a rather lax cyme. Petals oblong-lanceolate, acuminate, at least twice as long as the sepals, pale yellow. Follicles brown, distinctly gibbous, stellate-patent; styles fairly long, upturned. Seeds brown. $2n = 32$. ● *Mountains of W. Bulgaria and N. Greece; local.* Bu Gr.

21. S. sexangulare L., *Sp. Pl.* 430 (1753) (*S. boloniense* Loisel., *S. mite* auct.). Glabrous, laxly caespitose; flowering stems 6–15 cm. Leaves 3–6 mm, bright green, cylindrical-linear, spurred, closely imbricate on the non-flowering shoots, usually in 5 or 6 regular rows. Flowers subsessile in moderately lax cymes with spreading branches. Petals 4–5 mm, acute or acuminate, bright yellow. Follicles dark brown, nearly symmetrical, erecto-patent; styles long and slender. Seeds 0·3 mm, subglobose, pale brown. $2n = 74, 111, 148, 185$. ● *Mainly in C. Europe, extending locally to C. France, C. Italy, S.W. Jugoslavia and W. Ukraine; doubtfully native in S. Fennoscandia.* ?Al Au Be ?Bu Cz *Da *Fe Ga Ge ?Gr He Ho Hu It Ju Po Rm Rs(*B, C, W) *Su [Br].

An outlying population in N.E. Bulgaria has been distinguished as **S. tschernokolevii** Stefanov ex Vălev in Jordanov, *Fl. Rep. Pop. Bulg.* **4**: 629 (1970), but the differences scarcely justify specific separation.

Plants from the mountains of C. Greece that have somewhat compressed, oblong-ellipsoidal glaucous leaves, slightly larger sepals and narrower petals, and a chromosome number of $2n = 44$, have been distinguished as **S. apoleipon** 't Hart, *Willdenowia* **13**: 310 (1983).

22. S. borissovae Balk., *Not. Syst.* (*Leningrad*) **15**: 85 (1953). Like **21** but leaves grey-green and minutely papillose, ellipsoid-oblong and not or shortly spurred; petals pale yellow; follicles larger; seeds 0·7 mm, 2–3 times as long as wide. *Granite rocks.* ● *S. Ukraine* (*near Dolinsky*). Rs(W).

23. S. album L., *Sp. Pl.* 432 (1753) (*S. vermiculifolium* P. Fourn.; incl. *S. athoum* DC., *S. serpentini* Janchen). Laxly caespitose perennial, densely glandular-pubescent at base, with creeping, woody stems bearing short non-flowering shoots and erect flowering stems 5–18 cm. Leaves 4–20 mm, subterete but somewhat flattened on upper surface, obtuse, scarcely spurred, linear-cylindrical to ovoid-globose, usually reddish, patent or suberect, alternate. Flowers 5-merous, shortly stalked, in a freely branched, rather dense, bracteate subcorymbose cyme. Sepals united at the base. Petals 2–4 mm, subacute, white (rarely pink). Stamens 10. Follicles whitish-pink, erect. $2n = 34, 51, 68, 102, 136$. *Europe, except for parts of the north and east.* All except Az Fa Is Rs(N, C, E) Sb; only as a naturalized alien in Hb.

A very variable plant, especially in the size and shape of the leaves. Some segregates, especially **S. micranthum** DC. in Lam. & DC., *Fl. Fr.* ed. 3, **5**: 523 (1815) (*S. album* subsp. *micranthum* (DC.) Syme), from S.W. & S.C. Europe, with short stems, small flowers and rather short leaves, are sufficiently striking to have been given specific or subspecific rank, but intermediate plants are so numerous as to make their diagnosis impossible.

24. S. gypsicola Boiss. & Reuter, *Diagn. Pl. Nov. Hisp.* 14 (1842). Like **23** in habit, inflorescence and flowers, but with flatter, ovate-rhombic, greyish, densely puberulent leaves, closely imbricate in 5 rows on the sterile shoots. $2n = 68, 102$. *C. & S. Spain.* Hs.

25. S. subulatum (C. A. Meyer) Boiss., *Fl. Or.* **2**: 783 (1872). Like larger variants of **23** but erect flowering shoots glabrous throughout; often taller, with linear-subulate, acuminate leaves 10 × 1·5 mm; inflorescence more compact, with shorter pedicels; petals 5–6 mm; styles almost equalling the ovaries. $2n = 18$. *S.E. Russia.* Rs(E). (*S.W. Asia.*)

26. S. anglicum Hudson, *Fl. Angl.* ed. 2, 196 (1778). Glabrous, laxly caespitose perennial with creeping stems bearing short non-flowering shoots and flowering stems up to 15 cm, but usually much less. Leaves 3–5 mm, cylindrical, ovoid or subglobose, gibbous, spurred, alternate, often pink. Flowers 5-merous, usually rather crowded in a small cyme with 2 or 3 branches and leaf-like bracts. Sepals free, spurred. Petals 2·5–4·5 mm, acute to acuminate, pink or white. Stamens 10, with blackish anthers. Follicles red, stellate-patent. Seeds orange-brown to red. *Calcifuge. W. Europe, extending to S.W. Sweden.* Br Ga Hb Hs Lu No Su.

Variable in the southern part of its range; three subspecies, differing chiefly in size, can be recognized.

1 Flowering stems 10–15 cm; leaves *c.* 5 mm; cymes lax (**b**) subsp. **pyrenaicum**
1 Flowering stems seldom more than 7 cm; leaves 3–4(–5) mm; cymes short, dense
 2 Flowering stems 2–3·5 cm, very slender; petals *c.* 3 mm (**c**) subsp. **melanantherum**
 2 Flowering stems 3–7 cm, stouter; petals 4–4·5 mm (**a**) subsp. **anglicum**

(**a**) Subsp. **anglicum**: Flowering stems 3–7 cm, slender. Leaves 3–4(–5) mm. Cymes short, dense. Petals 4–4·5 mm. $2n = 120$, *c.* 144. ● *N. Spain to Sweden.*

(**b**) Subsp. **pyrenaicum** Lange, *Ind. Sem. Horto Haun.* **1857**: 27 (1857) (*S. pyrenaicum* Lange): Flowering stems 10–15 cm. Leaves *c.* 5 mm. Cymes lax. Petals 2·5–4·5 mm. $2n = 24$–36. ● *S. France, N. Spain, Portugal.*

(**c**) Subsp. **melanantherum** (DC.) Maire in Jahandiez & Maire, *Cat. Pl. Maroc* **2**: 324 (1932) (*S. melanantherum* DC.): Flowering stems 2–3·5 cm, very slender (more slender than those of subsp. (**a**)). Cymes short, dense. Petals *c.* 3 mm. $2n = 26$. *Mountains of S. Spain.*

27. S. stefco Stefanov, *God. Sof. Univ.* (*Agro.-Les. Fak.*) **24(2)**: 105 (1946). Small, glabrous perennial; flowering stems 7–10 cm. Leaves 4–5 × 1·5 mm, cylindrical, alternate, bright pink. Flowers 4-merous, shortly stalked, in a rather dense cyme. Petals 4–5 mm, acute, suberect, pale pink. Stamens 8. Follicles stellate-patent; styles very short. Seeds brown. $2n = 14$. ● *Mountains of S.W. Bulgaria and N. Greece.* Bu Gr.

28. S. dasyphyllum L., *Sp. Pl.* 431 (1753) (*S. nebrodense* Gaspar.). Perennial 3–12 cm, glandular-pubescent at least on the inflorescence and sometimes all over, usually pruinose and ringed with greyish-pink. Leaves ovoid or suborbicular, flattened on upper surface, mostly opposite, loosely imbricate on the non-flowering shoots. Flowers 5- or 6-merous, stalked, in small bracteate cymes. Petals 3 mm, white streaked with pink. Stamens 10 or 12. Follicles nearly erect; style short, turned sharply outwards. Seeds ovoid, pale brown. $2n = 28, 42, 56$. *S. Europe and parts of C. Europe; naturalized in the north-west.* Al Au Bl Bu Co Ga Ge Gr He Hs It Ju Rm Sa Si [Be Br Da Hb Ho].

29. S. brevifolium DC., *Rapp. Voy. Bot.* **2**: 79 (1808) (*S. cineritium* Pau). Like **28** but quite glabrous, with wiry, somewhat woody stems; leaves globose, usually opposite, closely imbricate in 4 rows on the non-flowering shoots, rarely alternate; styles longer; seeds brown, pyriform. *S.W. Europe.* Co Ga Hs Lu Sa.

30. S. monregalense Balbis, *Mém. Acad. Sci.* (*Turin*) **7**: 339 (1804) (*S. cruciatum* Desf.). Perennial, with suberect non-flowering shoots and erect flowering stems 7–15 cm high, glandular-pubescent above. Leaves 6 mm, opposite below, in whorls of 4 above, oblong-lanceolate, flat on upper surface, obtuse, glabrous. Flowers 5-merous, on long pedicels in a lax panicle. Petals white, lanceolate, acuminate, with pubescent mid-vein. Stamens 10. Follicles suberect, with long, straight styles. *Shady mountain rocks.* ● *S.W. Alps; Appennini; Corse.* Co Ga It.

31. S. magellense Ten., *Prodr. Fl. Nap.* xxvi (1811). Glabrous perennial with ascending flowering stems 6–15 cm. Leaves 6–10 mm, alternate or opposite, obovate-oblong, flat, obtuse. Flowers 5-merous, pedicellate; inflorescence simple, racemose. Petals lanceolate, acute, whitish. Stamens 10. Follicles erect, obtuse, shortly mucronate. $2n = 28, 30$. *Mountains of S.E. Europe, extending to C. Italy.* Al Bu Cr Gr It Ju.

Plants from S. Italy and the Balkan peninsula differ from the typical plants of C. Italy in their smaller, broader and constantly opposite leaves, and have been distinguished as subsp. **olympicum** (Boiss.) Greuter & Burdet in Greuter, Burdet & Long, *Med-Checklist* **3**: 24 (1986).

32. S. alsinifolium All., *Fl. Pedem.* **2**: 119 (1785). Delicate, glandular-pubescent annual; flowering stems 10–15 cm, weak and spreading. Leaves 10–15(–20) mm, alternate, oblong to elliptical, relatively thin, the lower petiolate. Flowers 5-merous, on long, slender pedicels, in a leafy, very lax panicle. Petals 5–6 mm, ovate-rhombic, acuminate, white. Stamens 10; filaments 3·5–5 mm. Follicles whitish, erect, pubescent. $2n = 26$. *Shady, schistose rocks and caves.* ● *N.W. Italy.* It.

33. S. fragrans 't Hart, *Bot. Helvet.* **93**: 277 (1983). Like **32** but perennial, with short, rosette-like non-flowering shoots; leaves elliptical to suborbicular; petals 3–5 mm; filaments 2–3 mm. $2n = 20$. *Shady, calcareous rocks and caves.* ● *Maritime Alps.* Ga It.

34. S. tristriatum Boiss., *Diagn. Pl. Or. Nov.* **2(10)**: 16 (1849). Glandular-pubescent perennial, with caespitose, rosette-like non-flowering shoots; flowering stems 2–10 cm, ascending, arising in the axils of lower leaves of the rosette. Rosette-leaves oblong-spathulate, almost flat; cauline leaves obovate, semiterete, alternate. Flowers 5(6)-merous, borne in groups of 4–7(–15) in a lax cyme on each flowering shoot. Sepals 1–2 mm, ovate-triangular. Petals narrowly ovate-elliptical, aristate, pale pink with darker stripes, or white. Stamens 10. Follicles whitish, erect. $2n = 22, 44$. ● *Mountains of S. Greece and Kriti*. Cr Gr.

35. S. tymphaeum Quézel & Contandriopoulos, *Taxon* **16**: 240 (1967). Like **34** but rosettes solitary, globose, densely and minutely glandular-puberulent; pedicels up to 8 mm; sepals up to 3 mm, lanceolate; petals acute, with only one dark stripe; follicles yellowish-brown, stellate-patent. $2n = 14$. *Limestone rocks*. ● *Mountains of N. Greece*. Gr.

36. S. hirsutum All., *Fl. Pedem.* **2**: 122 (1785) (*Rosularia hirsuta* (All.) Eggli). Densely glandular-hairy perennial with caespitose non-flowering shoots and terminal, erect flowering stems 5–12 cm. Leaves 5–15 mm, alternate, terete, oblanceolate. Flowers 5(6)-merous, pedicellate; buds nodding. Petals 5–7 mm, ovate, acuminate or apiculate, white or pinkish. Follicles erect, pubescent on inner side, white. *Usually calcifuge. S.W. Europe, extending to N. France and N. Italy*. Ga Hs It Lu.

(**a**) Subsp. **hirsutum**: Leaves not more than 10 mm. Cyme with not more than 15 flowers. Petals not more than 6 × 2·5 mm, usually with a red mid-vein, free or scarcely united at the base. Styles straight. *Throughout the range of the species*.

(**b**) Subsp. **baeticum** Rouy, *Bull. Soc. Bot. Fr.* **34**: 441 (1887) (*S. winkleri* (Willk.) Wolley-Dod, *Umbilicus winkleri* Willk.): Leaves up to 15 mm, forming large, densely caespitose rosettes, almost with the habit of *Sempervivum*. Cyme with up to 24 flowers. Petals *c.* 9 × 4·5 mm, pure white or with green veins, united for ¼ of their length. Styles curved outwards. $2n = 58$. *S. Spain*.

Subsp. (**b**) is very distinct in its extreme form, but apparently connected to subsp. (**a**) in S. & C. Spain by a full range of intermediates.

37. S. villosum L., *Sp. Pl.* 432 (1753). Usually perennial with small leafy offsets, but sometimes biennial or annual; usually glandular-pubescent, but occasionally glabrous in subarctic Europe (var. *glabratum* Rostrup); stem 5–15 cm, erect. Leaves 4–7 mm, alternate, erect, linear-oblong, semiterete. Flowers 5-merous, stalked, in a small, lax panicle; lower bracts leaf-like; buds erect. Petals 4–6 mm, acute, lilac or pale pink. Stamens 10, rarely 5. Follicles erect, whitish or dark purplish-brown. $2n = 30$. *Wet places. W. & C. Europe, extending to W. Finland, White Russia and N. Italy*. Au Br Co Cz Fa Fe Ga Ge He Hs Is It Ju No Po ?Rm Rs(B, C) Sa Su.

S. campanulatum (Willk.) F. Fernández & Cantó, *Lazaroa* **6**: 187 (1985) differs from **37** mainly by the somewhat larger, obtuse perianth-segments. It is probably best retained as *S. villosum* var. *campanulatum* Willk.

38. S. creticum C. Presl, *Isis (Oken)* **21**: 273 (1828) (*S. hierapetrae* Rech. fil.). Perennial or biennial; stem 3–15(–20) cm, erect, usually simple, or branched from near the base, glandular-pubescent. Leaves up to 4 mm, alternate, mostly in a basal rosette, oblong-spathulate, flat, glandular-pubescent or glabrous. Flowers 5(6)-merous, on short pedicels, in a narrow, densely glandular-pubescent panicle which occupies most of the stem. Petals 5–7 mm, lanceolate, acute, greenish-white, tinged with pink. Stamens 10. Follicles erect, pubescent. $2n = 22$. ● *Kriti, Karpathos*. Cr.

39. S. cepaea L., *Sp. Pl.* 431 (1753) (*S. spathulatum* Waldst. & Kit.). Annual or biennial, perhaps sometimes perennial; stem 15–40 cm, weak, erect or ascending, pubescent. Leaves mostly opposite or whorled, obovate or oblanceolate-spathulate, flat, glabrous, the lower ones petiolate. Flowers 5-merous, pedicellate, in a long, diffuse panicle. Petals 5 mm, obovate-ovate, aristate, pale pink or white with a red, pubescent mid-vein. Follicles erect, longitudinally grooved, whitish. $2n = 20, 22$. *Shady places. S. & S.C. Europe, mainly in the mountains*. Al Bu Co Ga Gr He Hs It Ju Rm Sa Si [Ge Ho].

40. S. lagascae Pau, *Not. Bot. Fl. Esp.* **6**: 53 (1895) (*S. villosum* auct. lusit.). Bushy annual 6–15 cm high and about as wide, glandular-pubescent all over. Leaves 6–9 mm, alternate, broadly linear, terete. Flowers 5-merous, numerous, on rather long pedicels. Sepals 2 mm, ovate-lanceolate. Petals *c.* 7 mm, ovate-oblong, pale pink, yellowish at the connate base. Stamens 10. Follicles erect, whitish; styles fairly long. *C. & S. Spain, Portugal*. Hs Lu.

41. S. pedicellatum Boiss. & Reuter, *Diagn. Pl. Nov. Hisp.* 13 (1842). Glabrous annual or sometimes perennial; stem 3–12 cm, erect, much branched above. Leaves 2–5 mm, alternate, patent, ovoid-cylindrical, glaucous. Flowers 5-merous, rather few, pedicellate, in a lax cyme. Petals 3 mm, elliptical, white with a pink dorsal line. Stamens 10. Follicles erect, brown; styles straight, rather short. Seeds pyriform, brown. *Gravelly places in the mountains*. ● *N. Portugal & C. Spain*. Hs Lu.

Plants from Portugal differ from those from C. Spain in their smaller, more strongly ridged seeds, and their usually more compact habit, and have been separated as subsp. **lusitanicum** (Mariz) Laínz, *Aport. Fl. Gallega* **6**: 16 (1968) (*S. willkommianum* R. Fernandes).

42. S. arenarium Brot., *Fl. Lusit.* **2**: 212 (1805). Glabrous annual, branched from the base, with ascending branches. Leaves 3–5 mm, ovoid, spurred. Flowers 5-merous, subsessile, in short, spreading cymes, the bracts leaf-like. Sepals often unequal. Petals *c.* 3 mm, white tinged with pink. Stamens 10. Styles straight, rather short. Seeds orange-brown to red. ● *Portugal, W. & C. Spain*. Hs Lu.

Often included under **26** but apparently quite distinct in its annual habit.

43. S. atratum L., *Sp. Pl.* ed. 2, 1673 (1763). Glabrous annual; stem 3–12 cm, with erect branches. Leaves 3–9 mm, alternate, oblong or obovoid-clavate, tapered and flattened towards the base. Flowers 5- or 6-merous, shortly pedicellate, in a crowded corymb. Stamens 10–12. Follicles stellate-patent; styles short, deflexed. *Somewhat calcicole*. ● *Mountains of C. & S. Europe, from the French Jura and Carpathians to N. Spain, Calabria and Greece*. Al Au Bu Cz Ga Ge Gr He Hs It Ju Po Rm Rs(W).

(**a**) Subsp. **atratum**: Dwarf, with stem unbranched below. Petals cream-coloured, lined or suffused with red, acute, only slightly exceeding the sepals. Plant usually dark red. $2n = 16$. *Throughout the range of the species, but rare in the Balkan peninsula*.

(**b**) Subsp. **carinthiacum** (Hoppe ex Pacher) D. A. Webb, *Feddes Repert.* **69**: 62 (1964): Taller, with stem usually branched from the base. Petals pale greenish-yellow, scarcely tinged with red, obtuse, twice as long as the sepals. Plant greenish, only slightly flushed with red. $2n = 32$. *E. Alps and Balkan peninsula*.

44. S. confertiflorum Boiss., *Diagn. Pl. Or. Nov.* **1 (3)**: 15 (1843). Glabrous annual; stem 2–8(–10) cm, erect or ascending. Leaves 4–6(–10) mm, alternate, oblong-linear, semiterete. Flowers 5-merous, subsessile, in a crowded corymb. Sepals glandular-papillose, connate in their lower ½. Petals 3–4 mm, subacute, white. Stamens 10. Follicles suberect, glandular-papillose; styles fairly long, straight. *N.E. Greece*. Gr. (*Turkey*.)

45. S. stellatum L., *Sp. Pl.* 431 (1753). Glabrous annual; stem 3–15 cm, stout, erect or ascending. Leaves 10–15 mm, flat, with a short, broad petiole and suborbicular lamina, crenate or bluntly dentate, the upper alternate, the lower often opposite. Flowers 5-merous (rarely 4-merous), sessile. Sepals broadly linear, accrescent, papillose at the apex. Petals 4–5 mm, acute, white or pink. Stamens 8 or 10. Follicles stellate-patent; styles very short. $2n = 10$. *Mediterranean region*. Al Bl Co Ga Gr It Ju Sa Si.

46. S. annuum L., *Sp. Pl.* 432 (1753). Glabrous annual or biennial, often spotted or streaked with red; stem simple or branched from the base with ascending, eventually flexuous branches 4–20 cm. Leaves *c.* 6 mm, alternate, distant, linear-oblong, thick but flattened on both surfaces, with a whitish, truncate spur. Flowers 5-merous, subsessile or on short pedicels, in a lax, compound cyme. Petals oblanceolate, acute to acuminate, twice as long as the sepals, yellow. Stamens 10. Follicles green or red, stellate-patent. Seeds orange-red. $2n = 22, 24$. *N. Europe, and in most of the mountain ranges of C. & S. Europe*. Al Au Bu Co Cz Fe Ga Ge Gr He Hs Is It Ju No Rm Rs(N, B, W) Su.

Normally biennial in Fennoscandia; apparently annual in C. Europe. Var. *perdurans* Murb. (*S. zollikoferii* F. Hermann & Stefanov), a perennial variant with decumbent branches that root in the soil, occurs sporadically in populations in the Balkan peninsula and the Carpathians.

47. S. litoreum Guss., *Pl. Rar.* 185 (1826) (incl. *S. praesidis* Runemark & Greuter). Glabrous annual; stem 4–15 cm, simple or with erect branches. Leaves 10–20 mm, alternate, obovate-spathulate, flattish, shortly spurred. Flowers 5-merous, subsessile, in long, lax cymes. Petals 2·5–4 mm, equalling or slightly exceeding the sepals, lanceolate, acute to acuminate, pale yellow. Stamens 5, rarely 10. Follicles erecto-patent; styles short. $2n = 20, 40, 60$. *E. & C. Mediterranean region*. Co Cr Gr It Ju Sa Si [Ga].

48. S. aetnense Tineo in Guss., *Fl. Sic. Syn.* **2**: 826 (1845). Somewhat glaucous annual; stem 2–6 cm, erect, with short, ascending branches. Leaves 3–4 mm, alternate, erect-appressed, terete, conical-oblong, ciliate-denticulate, with a scarious spur at the base. Flowers 4- or 5-merous, sessile, axillary. Sepals ciliate. Petals 2–3 mm, white or pink. Stamens 4 or 5. Follicles erect, dark red. $2n = 26$. • *S. Europe; very local*. Al Bu Hs Ju Rs(W, K) Si.

49. S. rubens L., *Sp. Pl.* 432 (1753) (*Crassula rubens* (L.) L.). Erect annual 2–15 cm, glandular-pubescent above, somewhat glaucous and usually reddish. Leaves 10–20 mm, alternate, or in whorls of 4 below, patent, linear, semiterete. Flowers 5-merous, sessile, in a leafy, corymbose cyme. Petals 5 mm, sharply acuminate. Follicles usually divergent at the base but suberect in upper part, glandular-tuberculate; styles long, straight. *S. & W. Europe, extending to Switzerland and S.W. Germany*. Al Be Bl Bu Co Cr Ga Ge Gr He Hs It Ju Lu Rm Rs(K) Sa Si.

(a) Subsp. **rubens**: Petals white or pink. Stamens 5(–10). Follicles somewhat spreading. $2n = 20, 40–42, 60, 80, 94, 100$. *Throughout the range of the species*.

(b) Subsp. **delicum** Vierh., *Verh. Zool.-Bot. Ges. Wien* **69**: 224 (1919) (*S. delicum* (Vierh.) A. Carlström): Petals greenish-yellow. Stamens 10. Follicles erect. • *S. Greece, Kikladhes*.

50. S. caespitosum (Cav.) DC., *Prodr.* **3**: 406 (1828) (*S. rubrum* (L.) Thell., non Royle ex Edgew.). Glabrous, usually reddish annual; stem 2–5 cm, erect. Leaves 3–6 mm, alternate, imbricate, subterete, broadly ovoid. Flowers 4- or 5-merous, sessile in short cymes. Petals 3 mm, mucronate, white tinged with pink. Stamens 4 or 5. Follicles patent, glabrous. $2n = 12$. *S. & S.C. Europe*. Bu Co Cr Ga Gr Hs Hu It Ju Rm Rs(K) Sa Si.

51. S. andegavense (DC.) Desv., *Obs. Pl. Angers* 150 (1818). Glabrous annual; stem 3–7 cm, usually erect, branched above. Leaves alternate (or lower opposite), ovoid or subglobose, imbricate, shortly spurred. Flowers 4- or 5-merous, on short pedicels in a corymbose cyme. Petals broadly ovate, mucronate, whitish or pink. Stamens 4 or 5. Follicles erect, rugose, dark brown; styles short. Seeds pyriform, brown. *S.W. Europe, extending to N.W. France and C. Italy* (*Capraia, Isole Ponziane*). Co Ga Hs It Lu Sa.

52. S. nevadense Cosson, *Not. Pl. Crit.* **2**: 163 (1849). Glabrous annual; stem 4–10 cm, erect. Leaves erect, semiterete, linear-oblong, not spurred. Flowers 5-merous, pedicellate, in a raceme-like cyme. Petals 3–4 mm, acute, connate at the base, white or pale pink. Stamens 5. Follicles erect, with short, straight styles. *Mountains of S., E. & N. Spain*. Hs.

53. S. hispanicum L., *Amoen. Acad.* **4**: 273 (1759) (*S. glaucum* Waldst & Kit., *S. sexfidum* Bieb.). Usually annual, but sometimes biennial or perennial; glabrous or wholly or partly covered with glandular hairs; stem 7–15 cm, branched, ascending. Leaves 7–18 mm, alternate, linear, semiterete, subacute, glaucous. Flowers 6- or 7(–9)-merous, subsessile, numerous, in unilateral bracteate cymes, the bracts leaf-like. Petals 5–7 mm, lanceolate, acuminate, white with pink mid-vein. Follicles stellate-patent, smooth or with glandular hairs, pale pink or whitish. $2n = 14, 28, 42$. *S.E. Europe, extending to Sicilia, Switzerland and E. Carpathians; occasionally naturalized further north*. Al Au Bu Cr Gr He Hu It Ju Rm Rs(W) Si Tu [Ge Su].

54. S. pallidum Bieb., *Fl. Taur.-Cauc.* **1**: 353 (1808). Like **52** but usually glabrous; flowers 5-merous; petals rather smaller, pink, suberect; follicles usually suberect, dark red, glandular-tuberculate. *S. part of Balkan peninsula; ?Aegean region; Krym*. Al Bu ?Cr ?Gr Ju Rs(K) Tu.

55. S. caeruleum L., *Mantissa Alt.* 241 (1771) (*S. heptapetalum* Poiret). Bushy, erect annual 5–20 cm, often pubescent above, usually strongly tinged with red. Leaves *c.* 10 mm, alternate, terete, linear-oblong. Flowers many, usually 7-merous, on slender pedicels in a broad, lax panicle. Petals lanceolate, sky-blue, white at the base. Stamens twice as many as the petals. Follicles erecto-patent; styles long, straight. *Islands of W. Mediterranean*. Co Sa Si.

11. Rhodiola L.

D. A. WEBB

Dioecious perennials with a thick, fleshy rhizome, bearing persistent, broad-based scale-leaves and flowering stems with alternate leaves and flowers in terminal cymes. Flowers usually 4-merous. Petals free. Stamens 8. Carpels 4. Follicles erect.

Leaves at least 5 mm wide, flat	**1. rosea**
Leaves less than 2 mm wide, terete	**2. quadrifida**

1. R. rosea L., *Sp. Pl.* 1035 (1753) (*Sedum rosea* (L.) Scop.,

S. rhodiola DC., *S. scopolii* Simonkai; incl. R. *arctica* Boriss., R. *iremelica* Boriss, R. *scopolii* Simonkai.). Rhizome fragrant when cut. Flowering stems 5–35 cm × 2–6 mm, erect, not persistent when dead. Leaves orbicular-ovate to linear-oblong, usually dentate, broad-based, sometimes amplexicaul, glabrous, rather glaucous. Cyme somewhat corymbose. Petals 3–4 mm, usually dull yellow, sometimes absent. Male flowers with conspicuous but abortive carpels. Follicles reddish. $2n = 22$. *N. Europe, and in most of the mountains of C. Europe, southwards to the Pyrenees, N. Italy and Bulgaria.* Au Br Bu Cz Fa Fe Ga Hb He Hs Is It Ju No Po Rm Rs(N, C, W) Sb Su.

Very variable. Variation in leaf-shape shows some geographical consistency (long and narrow in the south, usually short and broad in the north), but variation in other features is poorly correlated, and a satisfactory basis for division into subspecies is not yet apparent.

2. R. quadrifida (Pallas) Fischer & C. A. Meyer, *Enum. Pl. Nov.* **1**: 69 (1841) (*Sedum quadrifidum* Pallas). Flowering stems 3–12 cm × 0·5–1 mm, erect, wiry, numerous, persistent when dead. Leaves 5–10 cm × 1–1·5 mm, narrowly linear, terete, acute. Cyme small, few-flowered. Petals 4 mm, yellow, often tipped with red. Follicles dark red. *Mossy and stony tundra. N.E. Russia.* Rs(N, C). (*N. Asia.*)

The plants from Tibet and Himalaya often cited under this name appear to be very different.

12. Orostachys (DC.) Fischer ex Sweet

D. A. WEBB

Glabrous biennials with leaves in a compact, subglobose, basal rosette, from the centre of which an unbranched flowering stem arises in the second year. Leaves with a whitish cartilaginous border towards the apex, and a terminal spine. Flowers 5-merous, in a long, dense thyrsoid raceme. Petals slightly connate at the base, patent. Stamens 10. Follicles free, erect; styles slender, straight.

Flowers subsessile, petals greenish-yellow; anthers yellow **1. spinosa**
Pedicels of lower flowers 3–8 mm; petals white or pink; anthers dark purple **2. thyrsiflora**

1. O. spinosa (L.) Sweet, *Hort. Brit.* ed. 2, 225 (1830) (*Sedum spinosum* L.). Rosette-leaves 15–25 × 3–5 mm, oblong, tapered suddenly to a terminal spine 2–4 mm. Stem 10–30 cm; cauline leaves alternate, similar to the basal leaves but smaller. Raceme 5–20 cm, very dense; pedicels 0–1 mm. Petals 4 mm, lanceolate, acute, greenish-yellow. Anthers yellow. Follicles 5–6 mm. *Dry rocks and stony ground. E. Russia (S. Ural).* Rs(E). (*N. & C. Asia.*)

2. O. thyrsiflora (DC.) Fischer ex Sweet, *loc. cit.* (1830). Like **1** but rosette-leaves 6–8 mm, broadly elliptical or suborbicular, with a shorter spine; cauline leaves strongly keeled, almost triquetrous; raceme broader and laxer, the lower flowers clearly pedicellate; petals white or pinkish; anthers dark purple. *Dry rocks and stony ground. E. Russia.* Rs(C, E). (*C. Asia.*)

13. Rosularia (DC.) Stapf

D. A. WEBB

Perennials, with a basal rosette of sessile leaves and several peduncles arising from their axils. Flowers as in *Umbilicus*.

Literature: U. Eggli, *Bradleya* **6**, suppl.: 1–119 (1988).

1. R. serrata (L.) A. Berger in Engler & Prantl, *Natürl. Pflanzenfam.* ed. 2, **18a**: 465 (1930) (*Cotyledon serrata* L., *Umbilicus serratus* (L.) DC.). Leaves 10–30 × 5–8 mm, oblong-spathulate, obtuse or subacute, glabrous; margin cartilaginous, verrucose-denticulate or almost entire. Peduncles 10–18 cm, ascending, hairy above. Flowers 6–8 mm, pedicellate, in narrow panicles. Corolla pale purplish-red to whitish, tubular-campanulate, with erect, lanceolate, acuminate lobes somewhat longer than the tube. Styles slender, straight. $2n = 18$. *Limestone rocks. Kriti and Karpathos.* Cr. (*E. Aegean islands, Anatolia.*)

73. SAXIFRAGACEAE

EDIT. D. A. WEBB

Herbs, mostly perennial. Flowers 4- or 5-merous, usually in cymes (rarely solitary or in racemes). Petals usually 4 or 5; sometimes absent. Stamens twice as many as the sepals, or rarely equal in number or fewer. Carpels 2; united below but usually divergent above; styles free. Ovary superior, semi-inferior or almost inferior. Fruit a capsule. Seeds numerous.

Darmera peltata (Torrey ex Bentham) Voss, *Gärtn. Zentral-Bl.* **1**: 645 (1899) (*Saxifraga peltata* Torrey ex Bentham, *Peltiphyllum peltatum* (Torrey ex Bentham) Engler, nom. illeg.), from California, a robust, rhizomatous perennial with peltate, orbicular leaves and corymbs of white or pink flowers, is cultivated for ornament, and is becoming locally naturalized in damp, shady places in Britain.

1 Petals absent **3. Chrysosplenium**
1 Petals present
 2 Stamens 3; petals filiform **5. Tolmeia**
 2 Stamens 10; petals not filiform
  3 Petals laciniate-pinnatifid **4. Tellima**
  3 Petals entire or emarginate
   4 Leaves not more than 6 cm wide; hypanthium united with lower part of ovary, or absent **1. Saxifraga**
   4 Leaves 10–15 cm wide; ovary superior, free from the basin-shaped hypanthium **2. Bergenia**

1. Saxifraga L.

D. A. WEBB

Herbs, sometimes rather woody at the base; usually perennial, rarely annual or biennial. Leaves simple, but often deeply dissected and sometimes apparently ternate; usually alternate or basal, rarely opposite, exstipulate; foliar glands (other than hydathodes), if present, superficial or on hairs, not immersed as in *Bergenia*. Flowers 5-merous, usually in cymes or panicles, rarely solitary in leaf-axils. Petals present; stamens 10; ovary superior or, more often, semi-inferior or inferior; placentation axile.

Plants of this genus are frequently cultivated, especially in C. & N.W. Europe. The majority of garden plants are complex hybrids, often of uncertain origin; among the species which have certainly contributed to such hybrids are **24, 29, 31, 32, 63, 78, 79, 80, 85** and **98**. The species or primary hybrids most commonly cultivated are **3, 18, 20, 21, 29, 32, 36, 41, 44, 50** and **84**; also **24 × 32, 24 × 33, 29 × 32, 29 × 34, 32 × 33** and **44 × 45**.

The non-European species most widely cultivated is **S. stolonifera** Meerb., *Afbeeld. Zeldz. Gewass.* t. 23 (1777) (*S. sarmentosa* L.), from China and Japan, with orbicular leaves, long, filiform stolons and strongly zygomorphic flowers. It has been reported as occasionally naturalized on walls in W., C. & S. Europe.

Literature: A. Engler & E. Irmscher, *Pflanzenreich* **67** & **69** (**IV.117**): 1–709 (1916–19). H. W. Pugsley, *Jour. Linn. Soc. London* (*Bot.*) **50**: 267–289 (1936). D. A. Webb, *Proc. Roy. Irish Acad.* **53B**: 207–240 (1950). I. K. Ferguson & D. A. Webb, *Bot. Jour. Linn. Soc.* **63**: 295–311 (1970). K. Kaplan, *Biblioth. Bot.* (*Stuttgart*) **134** (1981). D. A. Webb & R. J. Gornall, *Saxifrages of Europe*. London. 1989.

1 Flowers entirely replaced by bulbils or leafy buds
2 Leaves reniform; petiole slender **62. cernua**
2 Leaves oblanceolate; petiole broad, scarcely distinct **12. foliolosa**
1 At least 1 flower present in inflorescence
3 Ovary superior, or nearly so (united to the hypanthium for less than ⅓ of its height)
4 Sepals deflexed, at least in fruit
5 Petals yellow or orange
6 Leaves linear-lanceolate, entire; petals 9–16 mm **1. hirculus**
6 Leaves reniform to orbicular, crenate or lobed; petals 5–7 mm **4. sibthorpii**
5 Petals white, often with red or yellow spots
7 Leaves with a narrow hyaline margin
8 Petals without red spots; ovary almost white; leaves usually entire in basal ⅓ of lamina **15. cuneifolia**
8 Petals usually with red spots; ovary pink (rarely green); leaves toothed or crenate almost to the base of lamina (**16–21**). **umbrosa group**
7 Leaves without a hyaline margin
9 Basal leaves with a long, slender petiole **9. nelsoniana**
9 Basal leaves with a short, broad petiole or none
10 Flowers few; inflorescence largely composed of leafy buds crowded towards the ends of the branches **12. foliolosa**
10 Flowers numerous; leafy buds, if present in the inflorescence, distributed throughout a lax panicle
11 Lower bracts large and leaf-like; most of the basal leaves with 5–10 teeth on each side **10. clusii**
11 All bracts small; most of the basal leaves with 3–5 teeth on each side **11. stellaris**
4 Sepals erect or patent, even in fruit
12 Leaves linear to oblong-lanceolate
13 Plant with long, leafless, filiform stolons **2. flagellaris**
13 Plant without stolons; procumbent shoots leafy
14 Axillary buds on leafy shoots not conspicuous at flowering time **56. bronchialis**
14 Axillary buds on leafy shoots conspicuous at flowering time
15 Leaves of non-flowering shoots straight, much longer than their axillary buds; leaves of flowering stems *c.* 10 mm, patent **54. aspera**
15 Leaves of non-flowering shoots incurved, scarcely longer than their axillary buds; leaves of flowering stems not more than 5 mm, nearly erect **55. bryoides**
12 Leaves reniform to suborbicular
16 Flowers solitary, in the axils of foliage-leaves on diffuse stems
17 Petals 2–3 mm, white or pale yellow **5. hederacea**
17 Petals 4·5–6 mm, bright yellow **3. cymbalaria**
16 Flowers in terminal cymes or panicles, or solitary at the apex of an erect stem
18 Flower solitary, with bulbils below it **62. cernua**
18 Flowers in a cyme or panicle; inflorescence without bulbils
19 Petals with red spots
20 Largest leaves not more than 25 mm wide, with 5–11 crenations or teeth **14. taygetea**
20 Largest leaves at least 25 mm wide, with at least 15 crenations or teeth **13. rotundifolia**
19 Petals without red spots
21 Basal leaves not palmately lobed, or very slightly lobed, with dentate lobes; flowers numerous, in a lax panicle **13. rotundifolia**
21 Basal leaves clearly palmately lobed, with entire lobes; flowers not more than 7, in a fairly compact cyme
22 Petals 8–14 mm **58. sibirica**
22 Petals 6–7 mm **59. carpatica**
3 Ovary inferior or semi-inferior (united to the hypanthium for at least ⅓ of its height)
23 Petals bright yellow or orange
24 Larger leaves at least 7 mm wide **52. mutata**
24 Leaves not more than 6 mm wide
25 Leaves fleshy and fairly soft, usually with a single hydathode **57. aizoides**
25 Leaves stiff and hard, with 3–7 hydathodes
26 Stamens at least as long as the petals; leaves apiculate
27 Flowering stem, pedicels and hypanthium villous **31. juniperifolia**
27 Flowering stem, pedicels and hypanthium glabrous **32. sancta**
26 Stamens shorter than the petals; leaves obtuse or shortly mucronate
28 Inflorescence usually with not more than 6 flowers; most of the leaves with a short, erect point **34. aretioides**
28 Inflorescence usually with more than 6 flowers; leaves obtuse, or with a short, incurved point **33. ferdinandi-coburgi**
23 Petals white, greenish, pink, purple, red or pale yellow
29 Bulbils present, at or below ground-level, in the axils of the basal leaves
30 Petals glandular-hairy on upper surface, at least towards the base
31 Bulbils present in the axils of cauline leaves and bracts **69. bulbifera**
31 Bulbils confined to the axils of the lower leaves
32 Lower leaves deeply lobed, with oblong or oblanceolate lobes

33 Inflorescence diffuse, with divaricate branches **65. haenseleri**
33 Inflorescence relatively compact, with suberect branches **66. dichotoma**
32 Lower leaves not lobed or with very short, broad lobes
34 Cauline leaves fairly numerous, sessile, at least the lower somewhat pinnatifid **67. carpetana**
34 Cauline leaves (if present) shortly petiolate, entire or shortly 3-toothed at the apex **68. cintrana**
30 Petals glabrous
35 Basal leaves deeply dissected, with 3 primary lobes, which are stalked, or at least narrowed at the base
36 Stems erect, forming a fairly compact tuft; nearly all the leaves basal **70. gemmulosa**
36 Stems diffuse, ascending, leafy **71. bourgaeana**
35 Basal leaves divided for not more than $\frac{2}{3}$ of the distance to the base; lobes broad-based
37 Petals 7–16 mm
38 Leaves crenate, but scarcely lobed; flowering stem usually branched only in upper $\frac{1}{2}$ **63. granulata**
38 Leaves distinctly 3-lobed; flowering stem usually branched from near the base **64. corsica**
37 Petals 2–6 mm
39 Lower leaves obovate **65. haenseleri**
39 Lower leaves reniform to semicircular
40 Plant up to 15 cm; leaves with truncate or cordate base, mostly 5-lobed; petals white or pink, rarely red **60. rivularis**
40 Plant not more than 4 cm; leaves somewhat tapered to the petiole, mostly 3-lobed; petals red **61. hyperborea**
29 Basal leaves without bulbils in their axils
41 Leaves sessile, not lobed, usually coriaceous, often lime-encrusted
42 Leaves opposite
43 Sepals not ciliate **42. retusa**
43 Sepals ciliate
44 Leaves hard and rigid, usually lime-encrusted, keeled beneath; petals usually contiguous **41. oppositifolia**
44 Leaves fairly soft, rather fleshy, rarely lime-encrusted, not keeled beneath; petals narrow, widely separated **43. biflora**
42 Leaves alternate or basal
45 Calyx campanulate or urceolate, enclosing and largely concealing the petals
46 Petals yellow **40. corymbosa**
46 Petals pink or purple
47 Most of the flowers sessile; inflorescence not branched **(35–37). porophylla** group
47 Most of the flowers distinctly pedicellate; inflorescence often branched
48 Inflorescence sparingly branched, with 1 or 2 flowers on each primary branch **38. media**
48 Inflorescence freely branched, with 3 or 4 flowers on each primary branch **39. stribrnyi**
45 Calyx cup-shaped or saucer-shaped, not concealing the petals
49 Larger leaves at least 15 mm long
50 Rosette solitary, without offsets; plant monocarpic
51 Leaves glaucous, lime-encrusted, subacute; flowers white **44. longifolia**
51 Leaves dull green, not lime-encrusted, apiculate; flowers dull pink **53. florulenta**
50 Flowering rosettes producing offsets, which persist after the flowering rosette has died
52 Leaves spathulate, obtuse **46. cochlearis**
52 Leaves linear to oblong; if somewhat expanded near the apex, then subacute
53 Leaves ± entire; hyaline margin very narrow or absent; hydathodes opening on the margin, not on the upper surface
54 Lower branches of panicle bearing 5 or more flowers **45. callosa**
54 Lower branches of panicle bearing 1–4 flowers **47. crustata**
53 Leaves with a distinct, toothed or crenate hyaline margin; hydathodes opening near the margin, but clearly on the upper surface
55 Flowering stem branched from below the middle, and usually from near the base; primary branches of inflorescence with more than 12 flowers, which are not crowded together near the apex **50. cotyledon**
55 Flowering stem branched from the middle or above it; primary branches of inflorescence with not more than 12 flowers, which are crowded together near the apex
56 Basal leaves tending to curve downwards near the apex, giving a flat or slightly convex rosette; most of the lower branches of the panicle with at least 5 flowers **48. hostii**
56 Basal leaves tending to curve upwards, giving a concave, sometimes almost hemispherical rosette; lower branches of the panicle with 1–3 (rarely 4) flowers **49. paniculata**
49 All leaves less than 15 mm long
57 Petals purple or deep pink
58 Flowers solitary; leaves with 1–5 hydathodes **41. oppositifolia**
58 Flowers 5–14, in a terminal cyme; leaves with 9–13 hydathodes **27. scardica**
57 Petals white, sometimes with red spots
59 Leaves mucronate to apiculate
60 Leaves not more than 5 mm long
61 Mucro of leaves sharply incurved **26. tombeanensis**
61 Mucro of leaves straight
62 Central part of leaf-margin finely toothed **27. scardica**
62 Central part of leaf-margin entire
63 Leaves distinctly keeled beneath, curved outwards near apex **24. marginata**
63 Leaves convex beneath but scarcely keeled, suberect even at apex **25. diapensioides**
60 Leaves at least 6 mm long

64 Stem with 1 or 2 flowers **29. burseriana**
64 Stem with 3–14 flowers
65 Leaves deep green, without calcareous incrustation, triangular-lanceolate, apiculate **30. vandellii**
65 Leaves glaucous, usually with calcareous incrustation, oblong to obovate-elliptical, mucronate
66 Central part of leaf-margin finely toothed **27. scardica**
66 Central part of leaf-margin entire **24. marginata**
59 Leaves obtuse to subacute
67 Petals 7–12 mm
68 Pedicels glabrous, or with a few very shortly stalked glands **46. cochlearis**
68 Pedicels densely glandular-pubescent
69 Leaves usually more than 1·5 mm wide, broadest above the middle **24. marginata**
69 Leaves usually less than 1·5 mm wide, broadest at or below the middle **25. diapensioides**
67 Petals 3–6 mm
70 Leaves finely toothed in distal ½ **49. paniculata**
70 Leaves entire in distal ½
71 Leaves glandular-hairy beneath **28. spruneri**
71 Leaves glabrous except for a few marginal hairs near the base
72 Upper surface of leaf irregularly pitted with hydathodes **51. valdensis**
72 Upper surface of leaf smooth, except for regular rows of hydathodes near the margin
73 Leaves curved outwards from near the base; upper part of flowering stems more densely pubescent than lower part **22. caesia**
73 Leaves curved outwards only near the apex; lower part of flowering stems more densely pubescent than upper part **23. squarrosa**
41 Leaves often lobed or petiolate; if sessile and undivided, then soft or fleshy, not coriaceous; very seldom lime-encrusted
74 Leaves opposite **44. biflora**
74 Leaves alternate or basal
75 Leaves linear-oblong, fleshy; flowers dark red **54. aizoides**
75 Leaves not fleshy
76 Leaves all entire
77 Petals inconspicuous, dull in colour and scarcely longer than the sepals
78 Leaves 10–30 mm wide; flowering stem leafless **6. hieracifolia**
78 Leaves not more than 5 mm wide; flowering stem usually with some leaves
79 Some of the leaves apiculate **116. sedoides**
79 All leaves obtuse or subacute
80 Flowering stems not more than 2·5 cm, scarcely exceeding the leafy shoots **114. facchinii**
80 Flowering stems at least 4 cm; flowers standing well above the level of the leaves
81 Petals tapered gradually to a narrow base, emarginate at the apex **115. presolanensis**
81 Petals not tapered at the base, rounded at the apex
82 Leafy shoots fairly long, forming a rounded cushion; basal leaves withered **85. exarata**
82 Leafy shoots very short, forming a thick mat rather than a cushion, with all leaves appearing ± basal **(109–112). androsacea** group
77 Petals conspicuous, larger than the sepals and usually pure white
83 Dormant axillary buds on leafy shoots conspicuous at flowering time
84 Leaves obovate, obtuse; petals turning pink after pollination **75. erioblasta**
84 Leaves linear-oblong, apiculate; petals remaining white **77. conifera**
83 Leafy shoots without conspicuous summer-dormant axillary buds
85 Leaves sparingly ciliate or completely glabrous
86 Leaves 3–6 mm wide; flowering stem with 1 or 2 cauline leaves and 1–3 flowers **(109–112). androsacea** group
86 Leaves 1–2 mm wide; flowering stems with 4–7 cauline leaves and 3–9 flowers
87 Leaves and sepals apiculate **119. tenella**
87 Leaves and sepals obtuse **118. glabella**
85 Leaves ± hairy on the surface
88 Annual; petals 2–3 mm **(120–123). tridactylites** group
88 Perennial; petals 4–7 mm
89 Leaves firm, usually with calcareous incrustation and with hyaline margin; inflorescence with 6–12 flowers **28. spruneri**
89 Leaves soft, without calcareous incrustation or hyaline margin; inflorescence with 1–4 flowers
90 Leafy shoots up to 10 cm, numerous, crowded, forming a compact cushion; leaves mostly less than 2·5 mm wide; petals contiguous **113. muscoides**
90 Leafy shoots short, relatively few, forming a dense but low mat; leaves mostly more than 2·5 mm wide; petals not contiguous **(109–112). androsacea** group
76 Some of the leaves toothed, crenate or lobed
91 Petals greenish, cream-coloured, dull or pale yellow or dark red
92 All leaves in a basal rosette **6. hieracifolia**
92 Plant with leafy stems and without basal rosette
93 Leaves divided to base into 3 lobes, each further subdivided **95. pentadactylis**
93 Leaves not divided to base
94 Larger leaves more than 10 mm wide, usually wider than long

95 Leaves covered with long, arachnoid hairs **107. arachnoidea**
95 Leaves ±glabrous **108. paradoxa**
94 Leaves not more than 10 mm wide, longer than wide
96 Leaf-segments furrowed on upper surface
97 Leaf-segments obtuse or acute, glandular-pubescent **85. exarata**
97 Leaf-segments mucronate, ±glabrous **86. hariotii**
96 Leaf-segments flat on upper surface
98 Petals linear, much narrower than the sepals
99 Some of the leaf-segments or undivided leaves acute to apiculate **116. sedoides**
99 All leaf-segments and undivided leaves obtuse **117. aphylla**
98 Petals ovate to oblong, as wide as the sepals or only slightly narrower
100 Leafy shoots ±procumbent; some of the leaf-segments mucronate; petals mucronate or emarginate **116. sedoides**
100 Leafy shoots ±erect; leaf-segments and petals obtuse **85. exarata**
91 Petals white, pink or bright red
101 Mature leaves glabrous, though often with sessile glands
102 Leaf-segments broadly triangular, 4–8 mm wide at the base **100. cuneata**
102 Leaf-segments linear-oblong to lanceolate, 1–3 mm wide at the base
103 All leaf-segments obtuse or subacute
104 Leaf-segments usually not more than 2 mm wide, parallel-sided, furrowed on upper surface; petals 3·5–5 mm **95. pentadactylis**
104 Leaf-segments usually more than 2 mm wide, often with curved margins, flat on upper surface; petals 7–14 mm **96. fragilis**
103 Some of the leaf-segments mucronate or apiculate
105 Larger leaves with (9–)11–17 segments; segments often recurved, crowded and overlapping **98. trifurcata**
105 Leaves with not more than 11 segments, which are not recurved, crowded or overlapping
106 Leaf-segments oblong to elliptical, $1\frac{1}{2}$–$2\frac{1}{2}$ times as long as wide **97. camposii**
106 Leaf-segments linear, 3–6 times as long as wide **99. canaliculata**
101 Mature leaves hairy, at least on margin or on petiole (hairs sometimes very short and visible only with a lens)
107 Leafy shoots bearing conspicuous, usually more or less dormant, axillary buds at flowering time
108 Petals 11–20 mm
109 Leaves deeply divided, with 3 primary lobes which are narrowed to a stalk-like base **72. biternata**
109 Leaves palmately lobed, the lobes not narrowed at the base **76. rigoi**
108 Petals 4–10(–12) mm
110 Leaf-segments mucronate or apiculate
111 Dormant axillary buds stalked, their outer leaves entirely scarious **78. continentalis**
111 Dormant axillary buds ±sessile, their outer leaves partly herbaceous **79. hypnoides**
110 Leaf-segments obtuse to acute
112 Leaves deeply divided, with 3 primary lobes which are stalked or narrowed at the base **71. bourgaeana**
112 Leaves palmately lobed, the lobes not narrowed at the base
113 Dormant buds with dense white, woolly hairs; petals turning pink after pollination **75. erioblasta**
113 Dormant buds hairy, but without dense white, woolly hairs; petals remaining white
114 Sepals 2 × 1·5 mm; petals plane; inflorescence usually with at least 4 flowers **73. globulifera**
114 Sepals 4 × 3 mm; margins of petals deflexed; inflorescence with 1–3 flowers **74. reuteriana**
107 Leafy shoots without conspicuous, dormant axillary buds at flowering time
115 Petals conspicuously emarginate, often unequal
116 Biennial; basal leaves divided almost to the base; petals 8–10 mm, contiguous **105. petraea**
116 Perennial; basal leaves divided not more than half-way to the base; petals 4–8 mm, not contiguous **106. berica**
115 Petals equal, not or only slightly emarginate
117 Annual or biennial
118 Basal leaves reniform, with long, clearly defined petiole **104. latepetiolata**
118 Basal leaves oblanceolate, with short, scarcely distinct petiole **(120–123). tridactylites** group
117 Perennial
119 Petals at least 7 mm
120 Leaves with more than 11 ultimate segments
121 Plant aquatic; flowering stems 2–5 mm in diameter **101. aquatica**
121 Plant terrestrial; flowering stems less than 2 mm in diameter
122 Stems woody at the base, spreading, forming a lax cushion **91. geranioides**
122 Stems not woody at the base, forming a compact tuft **103. irrigua**
120 Leaves with not more than 11 ultimate segments
123 Some of the leaves entire

124 Leafy shoots mostly procumbent; leaf-segments apiculate **79. hypnoides**
124 Leafy shoots ±erect, forming a cushion; leaf-segments obtuse **84. cebennensis**
123 All leaves lobed
125 Petals oblanceolate, usually at least 10 mm, and at least 2·5 times as long as wide **89. pedemontana**
125 Petals obovate, not more than 10 mm, and not more than twice as long as wide
126 Flowering stems terminal **80. rosacea**
126 Flowering stems axillary
127 Leafy shoots ±erect, forming a compact tuft; leaf-segments apiculate **90. babiana**
127 Leafy shoots procumbent or ascending, forming a wide mat; leaf-segments subacute to shortly mucronate **102. ×capitata**
119 Petals not more than 7 mm
128 Leaves crenate to shortly toothed on lateral margins, but not lobed
129 Flowering stem up to 2 mm in diameter, the lower part ±villous **7. nivalis**
129 Flowering stem not more than 1 mm in diameter, the lower part shortly glandular-pubescent or almost glabrous **8. tenuis**
128 Leaves distinctly lobed or shortly 3-toothed at the apex
130 Leaf-segments furrowed on upper surface
131 Hairs on leaves and calyx not more than 0·15 mm, the stalk scarcely longer than the diameter of the gland, giving a finely verrucose rather than a pubescent appearance
132 Plant forming cushions up to 30 cm in diameter; leaves fresh green; petals distinctly longer than wide, not contiguous **92. moncayensis**
132 Plant forming cushions seldom more than 10 cm in diameter; leaves rather dark green; petals almost as wide as long, contiguous **94. intricata**
131 Hairs on leaves and calyx mostly at least 0·3 mm, with the stalk considerably longer than the gland, giving a pubescent appearance
133 Leaves with rather short, sometimes few glandular hairs; petals about twice as long as wide, not contiguous **85. exarata**
133 Leaves densely covered with rather long glandular hairs; petals about $1\frac{1}{2}$ times as long as wide, usually contiguous
134 Leaves rather dark green; rosettes up to 20–25 mm in diameter; leaves on flowering stems usually lobed **82. pubescens**
134 Leaves fresh green; rosettes not more than 15 mm in diameter; leaves on flowering stems usually entire **84. cebennensis**
130 Leaf-segments not furrowed on upper surface
135 Leafy shoots very short or absent, or if well developed then diffuse or procumbent; habit of plant not cushion-like
136 Leaves cordate (N. Europe) **60. rivularis**
136 Leaves cuneate or truncate
137 Young stems and petioles covered with long, viscid, arachnoid hairs **107. arachnoidea**
137 Plant without long, viscid, arachnoid hairs
138 Leafy shoots very short, rosette-like **(109–112). androsacea** group
138 Leafy shoots long, ±procumbent
139 Inflorescence usually elongated, with more than 6 flowers **102. ×capitata**
139 Inflorescence fairly compact, with 1–6 flowers
140 Flowering stems arising from a rosette **80. rosacea**
140 Flowering stems arising from the axils or the apex of a procumbent leafy shoot
141 Most of the leaf-segments acute or mucronate **87. praetermissa**
141 All leaf-segments obtuse **88. wahlenbergii**
135 Leafy shoots at least 2 cm long, ±erect, crowded together to form a fairly compact cushion
142 Leaf-segments apiculate or mucronate
143 Flowering stems axillary **90. babiana**
143 Flowering stems terminal **80. rosacea**
142 Leaf-segments subacute to obtuse
144 Glandular hairs on leaves mostly not more than 0·15 mm; foliage with a strong spicy scent; plant forming cushions up to 20 cm in diameter **93. vayredana**
144 Glandular hairs on leaves mostly more than 0·15 mm; foliage not

strongly scented; cushions seldom more than 10 cm in diameter
145 Calyx and hypanthium dark red; petals often red-veined **83. nevadensis**
145 Calyx and hypanthium green; petals not red-veined
146 Petals not more than 5 mm; leaves very sparsely hairy **85. exarata**
146 Petals more than 5 mm; leaves densely hairy
147 Rosettes ±flat, with patent leaves; some leaves usually with 5 lobes, and sometimes with 7; petals up to 10 mm, pure white **80. rosacea**
147 Rosettes ±concave, with ascending or semi-erect leaves; leaves mostly with 3 and never more than 5 lobes; petals dull white, slightly tinged with green or yellow, not more than 6·5 mm **81. cespitosa**

Sect. CILIATAE Haw. (*Hirculus* Tausch). Evergreen perennials. Leaves alternate, entire, without calcareous incrustation. Petals yellow. Ovary superior to $\frac{1}{4}$-inferior.

1. S. hirculus L., *Sp. Pl.* 402 (1753) (*S. autumnalis* L.). Plant laxly caespitose or shortly stoloniferous; stem 20–35 cm, erect, leafy below. Leaves 10–25 mm, lanceolate, obtuse, tapered to a long, sheathing petiole, hairy, especially towards the base, with long, red-brown hairs. Flowers solitary or 2–4 in a lax corymb. Sepals deflexed in fruit; petals 9–16 mm, bright yellow, sometimes with red spots, and with 2 prominent callosities near the base. $2n = 32$. *Bogs and other wet places. N., E. & C. Europe, rapidly becoming rarer in the western and southern parts of its range.* Au Br †Cz Da Fe Ga Ge Hb He †Ho Is ?It No Po Rm Rs(N, B, C, W, E) Sb Su.

Possibly extinct in Au, Ga and He.

2. S. flagellaris Willd. in Sternb., *Revis. Saxifr.* 25 (1810) (*S. platysepala* Trautv.). Stem 5–20 cm, erect, emitting from its base filiform, epigeal stolons up to 15 cm long, each terminating in a leaf-rosette. Leaves mainly in a basal rosette, oblong-lanceolate, sessile, the upper densely, the lower sparsely glandular-hairy. Flowers usually solitary, terminal. Sepals suberect; petals 7–9 mm, obovate-oblong, bright yellow, without basal callosities; ovary about $\frac{1}{4}$-inferior. $2n = 32$. *Spitsbergen; Vajgač.* Rs(N) Sb.

The European plant, which has a circumpolar-arctic distribution, belongs to subsp. **platysepala** (Trautv.) Porsild. Other subspecies occur in W. North America, and in Asia from the Caucasus and Himalaya to the Bering Strait.

Sect. CYMBALARIA Griseb. Usually annuals. Stems diffuse, ascending, leafy. Leaves mostly alternate, palmately lobed to entire, long-petiolate. Petals white, yellow or orange. Ovary superior or very nearly so.

3. S. cymbalaria L., *Sp. Pl.* 405 (1753). Stems weak, sometimes decumbent, slightly hairy below. Leaves glabrous, somewhat fleshy, shining, reniform or orbicular, petiolate, with 7–9 triangular, subacute teeth. Flowers axillary, pedicellate. Sepals patent; petals 5 mm, yellow. *Romania (E. Carpathians, near Bacău).* Rm. (*S.W. Asia, N. Africa.*)

The plants from Romania belong to var. *cymbalaria*. Var. *huetiana* (Boiss.) Engler & Irmscher (*S. huetiana* Boiss.), from S.W. Asia, with smaller flowers and bluntly crenate or subentire leaves, is often cultivated and is locally naturalized in parts of Europe. It has a chromosome number of $2n = 18$.

4. S. sibthorpii Boiss., *Diagn. Pl. Or. Nov.* **1(3)**: 22 (1843). Stems decumbent, glabrous or with a few glandular hairs. Leaves fleshy, shining, reniform, broadly crenate, stalked. Flowers axillary, pedicellate, often deflexed in fruit. Sepals deflexed; petals 5–7 mm, orange-yellow. $2n = 18$. *Damp or shady rocks above* 1500 *m. Greece.* Gr.

5. S. hederacea L., *Sp. Pl.* 405 (1753). Stems ascending or decumbent. Leaves mainly basal, reniform to ovate, petiolate, crenate or slightly lobed, shining, sparingly glandular-hairy. Flowers solitary on filiform pedicels. Sepals erect or patent; petals 3 mm, white or very pale yellow. *Damp or shady rocks. E. Mediterranean region, westwards to Sicilia.* Cr Gr Ju Si.

Sect. MICRANTHES (Haw.) D. Don (*Boraphila* Engler, pro parte). Evergreen perennials. Leaves mostly basal, entire, crenate or toothed, somewhat fleshy or coriaceous. Petals usually white. Ovary superior to semi-inferior.

6. S. hieracifolia Waldst. & Kit. ex Willd., *Sp. Pl.* **2(1)**: 641 (1799). Stem underground, usually simple. Leaves all basal, 3–7 × 1–3 cm, ovate-elliptical, entire or obtusely dentate, ciliate, tapered to a short, winged petiole. Scape 10–40 cm, densely glandular-pubescent, purplish-red, bearing a narrow, congested panicle of numerous subsessile flowers; bracts rather large and leaf-like. Sepals deflexed in fruit; petals 1·5–3 mm, ovate-oblong, about equalling the sepals, greenish tinged with purplish-red. $2n = 112$, 120. *Damp rocks or tundra, or by mountain streams. Arctic Europe; Carpathians; very locally in mountains elsewhere (Norway, Steiermark, Auvergne).* Au Cz Ga No Po Rm Rs(N) Sb.

7. S. nivalis L., *Sp. Pl.* 401 (1753). Rhizome often branched, with several leaf-rosettes. Leaves 15–35 mm, ovate-oblong to rhombic, crenate or obtusely dentate, hairy on margins only, often dark red beneath; petiole broad, usually short. Scape 5–20 cm, densely glandular-pubescent, with some longer, white, crisped hairs below; inflorescence a small, often capitate, congested panicle of almost sessile flowers; bracts small. Sepals erect; petals *c.* 3 mm, suborbicular, somewhat exceeding the sepals, white or pink. $2n = 60$. *Arctic and subarctic Europe, extending locally southwards to N. Wales; one station in S.W. Poland.* Br Fa Fe Hb Is No Po Rs(N) Sb Su.

8. S. tenuis (Wahlenb.) K. A. H. Sm. ex Lindman, *Svensk Fanerogamfl.* 300 (1918). Like **7** but leaves 10–15 mm; scape seldom more than 12 cm and not more than 1 mm in diameter, pubescent, and without the longer hairs below; bracts somewhat larger; flowers usually more distinctly pedicellate in a laxer cyme. $2n = 20$. *Damp tundra, flushes and late snow-patches. Arctic and subarctic Europe.* Fa Fe Is No Rs(N) Sb Su.

9. S. nelsoniana D. Don, *Trans. Linn. Soc. London (Bot.)* **13**: 355 (1821) (*S. punctata* auct., non L.). Rhizome underground; leaves all basal, reniform, coarsely crenate, slightly fleshy, subglabrous; petiole long. Flowering stem 20–40 cm, glandular-hairy above, bearing a usually lax panicle of small flowers. Sepals deflexed; petals 3 mm, oblong, white or pale pink. Stamens often partly replaced by petaloid staminodes. Ovary superior. *N.E. Russia*. Rs(N). (*N. Asia and N.W. America*.)

The European plant belongs to subsp. **aestivalis** (Fischer & C. A. Meyer) D. A. Webb, *Feddes Repert.* **69**: 154 (1964).

Very similar to **18** in most characters; it is best distinguished by the absence of a hyaline margin to the leaf, and by the leaf-hairs, which are all filamentous and uniseriate, whereas in **18** many of them are multiseriate and broad-based.

10. S. clusii Gouan, *Obs. Bot.* 28 (1773). Stem very short. Leaves 4–12 cm, suberect or spreading, in a lax rosette, obovate-cuneate, coarsely and irregularly dentate, with rather numerous long, eglandular hairs; petiole broad but distinct. Flowering stem 12–30 cm, fragile, divaricately branched from the middle or below, to give a broad, diffuse panicle of numerous flowers; lower bracts usually large and leaf-like. Sepals deflexed; petals white, unequal (the 3 upper longer, with a conspicuous claw, and with 2 yellow spots). Ovary superior. *Damp, shady rocks and by mountain streams.* • *S.W. Europe*. Ga Hs Lu.

(**a**) Subsp. **clusii**: Plant not viviparous. *Cévennes, Pyrenees.*
(**b**) Subsp. **lepismigena** (Planellas) D. A. Webb, *Feddes Repert.* **68**: 199 (1963) (*S. lepismigena* Planellas, *S. paui* Merino): Plant viviparous, with many of the flowers replaced by leafy buds. Usually more hairy and more robust than subsp. (**a**). *N. & C. Portugal, N.W. Spain.*

Plants intermediate between subspp. (**a**) and (**b**), with the leafy buds few and inconstant, are found in N. Spain (Asturias).

11. S. stellaris L., *Sp. Pl.* 400 (1753). Densely or laxly caespitose; stem horizontal, usually short. Leaves usually in rosettes, obovate or somewhat spathulate, tapered to the base but scarcely with distinct petiole, hispid, slightly fleshy but not coriaceous. Flowering stem 4–20 cm, leafless, not fragile, branched in upper part, with usually suberect branches, giving a lax but fairly narrow panicle; bracts small. Sepals deflexed; petals 3–7 mm, lanceolate, occasionally unequal as in **10**, white with 2 yellow spots; anthers pink. Ovary superior. *By mountain streams and other damp places. In all the major mountain-ranges of Europe, and at low altitudes in the extreme north.* Au Br Bu Co Fa Fe Ga Ge Gr Hb He Hs Is It Ju Lu No Rm Rs(N, W) Su.

Very variable in habit. Two rather ill-defined subspecies may be recognized, but transitional forms are found in the Carpathians and elsewhere.

(**a**) Subsp. **stellaris**: Plant densely caespitose, with few rosettes; leaves rather hairy; petals 5–7 mm; pedicels not more than twice as long as capsule. $2n = 28$. *N. Europe.*
(**b**) Subsp. **alpigena** Temesy, *Phyton (Austria)* **7**: 40 (1957): Plant laxly caespitose, with more numerous rosettes and with creeping, sometimes leafy stems; leaves almost glabrous, usually smaller; petals 3–5 mm; pedicels more than twice as long as capsule. $2n = 28, 56$. *Mountains of C. & S. Europe, from the Vosges and the Carpathians to S. Spain, N. Appennini and N. Greece.*

In both subspecies viviparous plants are known, with many of the flowers replaced by leafy buds as in **10(b)**; these are less fleshy and bulbil-like than in **12**, and are dispersed throughout the lax panicle. Such plants are known from S.E. Austria & Slovenija, W. Ireland, and possibly elsewhere.

12. S. foliolosa R. Br., *Chloris Melv.* 17 (1823) (*S. stellaris* var. *comosa* Retz., *S. foliosa* auct.). Like smaller plants of **11** but rosettes usually solitary; leaves rather narrower, glabrous except at the margins; inflorescence narrow, with few, rather short branches bearing usually a single, terminal flower (sometimes several or none), the remaining flowers being replaced by small, reddish bulbils with fleshy, incurved leaves, densely crowded towards the ends of the panicle branches. $2n = 56$. *Arctic and subarctic Europe.* Fe Is No Rs(N) Sb Su.

Sect. COTYLEA Tausch (*Miscopetalum* (Haw.) Sternb.). Evergreen perennials with underground rhizome or stock. Leaves alternate, mostly basal, long-petiolate, entire to crenate, dentate or slightly lobed. Flowering stem leafy; flowers in a panicle. Sepals erect or patent; petals white, usually with red or yellow spots. Ovary superior.

13. S. rotundifolia L., *Sp. Pl.* 403 (1753) (incl. *S. olympica* Boiss., *S. heucherifolia* Griseb. & Schenk). Rhizome short, bearing lax rosettes of reniform-orbicular, crenate or coarsely dentate leaves, 30–85 mm wide, with at least 13 teeth or crenations, usually somewhat hairy. Petiole long, slender. Flowering stem 15–60(–100) cm, rather stout, usually with a few dentate or slightly lobed cauline leaves and a rather narrow panicle of numerous flowers. Petals 6–11 mm, narrowly oblong to broadly elliptical, with a distinct claw, white, usually spotted with yellow at the base and with red higher up. Ovary green or pink. Styles very short, divergent. *Damp or shady places in or near the mountains of C. & S. Europe.* Al Au Bu Co Cr Cz Ga Ge Gr He Hs It Ju Rm Sa Si [Be Br].

(**a**) Subsp. **rotundifolia** (incl. *S. heucherifolia* Griseb. & Schenk, *S. lasiophylla* Schott, *S. repanda* auct., vix Willd.): Petiole narrow right up to junction with lamina. Lamina with a narrow cartilaginous border. $2n = 22$. *Almost throughout the range of the species, but rare in the extreme south.*
(**b**) Subsp. **chrysosplenifolia** (Boiss.) D. A. Webb in Jordanov, *Fl. Rep. Pop. Bulg.* **4**: 658 (1970) (*S. chrysosplenifolia* Boiss.): Lamina of basal leaves decurrent along the petiole, which is broadened at the top; margin of lamina not cartilaginous, usually ciliate. $2n = 22$. • *S. part of Balkan peninsula and Aegean region.*

14. S. taygetea Boiss. & Heldr. in Boiss., *Diagn. Pl. Or. Nov.* **2(10)**: 19 (1849). Like **13(a)** but smaller (stem seldom 20 cm); basal leaves 15–25 mm wide, with 5–9 broad, shallow crenations, or subacute teeth; cauline leaves few and small; petals 7–9 mm, white with numerous red spots; ovary crimson. $2n = 22$. *Exposed mountain rocks.* • *S. & W. Greece, Albania, Crna Gora.* Al Gr Ju.

Sect. GYMNOPERA D. Don (*Robertsonia* (Haw.) Engler). Evergreen perennials. Leaves alternate, mostly in basal rosettes connected by short stolons or rhizomes, petiolate, crenate or dentate, somewhat succulent and coriaceous, without calcareous incrustation. Stem scapose; flowers numerous, in panicles. Sepals deflexed; petals white, usually spotted with yellow at the base and red higher up. Ovary superior.

15. S. cuneifolia L., *Sp. Pl.*, ed. 2, 574 (1762). Laxly caespitose, with procumbent stems bearing small rosettes of glabrous, orbicular-spathulate leaves, entire or bluntly toothed in upper ½, tapered gradually to a broad, sparsely ciliate petiole. Peduncles 15 cm, glandular-pubescent, slender. Panicle small; petals 2·5–4 mm, white, usually without red spots. Ovary green or pale pink; styles very short, erect. *Woods and shady rocks, mostly 800–1700 m.* ● *Pyrenees and N. Spain; Cévennes; Alps; N. Appennini; E. Carpathians; N.W. Jugoslavia.* Au Ga He Hs It Ju Rm.

(a) Subsp. **cuneifolia**: Plant diffuse, with fairly long stolons (rosettes usually 3–6 cm apart). Leaves not more than 20(–25) mm, including petiole, nearly entire. Panicle with not more than 10 flowers. $2n = 22$. *Maritime Alps, N. Appennini.*

(b) Subsp. **robusta** D. A. Webb, *Bot. Jour. Linn. Soc.* **97**: 355 (1988): Plant fairly compact, with short stolons (rosettes usually less than 2 cm apart). Larger leaves at least 25 mm, including petiole, distinctly crenate-dentate. Panicle with usually more than 10 flowers. $2n = 22$. *Almost throughout the range of the species, but rare in the area of subsp.* (a).

16–21. S. umbrosa group. Laxly or somewhat densely caespitose, with short, procumbent stems. Leaves mainly in basal rosettes, reniform to obovate-cuneate, dentate or crenate, with narrow hyaline margin. Petals *c.* 4 mm, white, with 2 yellow and usually several crimson spots. Anthers and ovary pink.

The three species included in this group are not difficult to distinguish, but the situation is confused by their hybrids, which are common in nature and widely diffused from garden culture. Secondary hybrids certainly occur, and introgressive hybridization has taken place in several areas.

1 Petiole subcylindrical, not much wider than thick; leaves usually hairy on both surfaces **18. hirsuta**
1 Petiole distinctly flattened; leaves usually glabrous at least on lower surface
 2 Petiole densely ciliate but glabrous on upper surface, usually shorter than the oblong, slightly crenate lamina **17. umbrosa**
 2 Petiole rather sparsely hairy, usually at least as long as the lamina
  3 Lamina sharply dentate; petiole very sparsely ciliate, glabrous on upper surface **16. spathularis**
  3 Lamina crenate or obtusely dentate; petiole rather densely ciliate, or with some hairs on upper surface
   4 Leaves rather deeply dentate or crenate, without conspicuous cartilaginous margin **19. × polita**
   4 Teeth or crenations broad and low; cartilaginous margin conspicuous
    5 Petiole narrow, considerably longer than the lamina; lamina sometimes slightly hairy **20. × geum**
    5 Petiole very broad, only slightly longer than the glabrous lamina **21. × urbium**

16. S. spathularis Brot., *Fl. Lusit.* **2**: 172 (1805) (*S. umbrosa* auct. lusit., non L.). Rather laxly caespitose. Leaves erecto-patent, orbicular, spathulate or obovate-cuneate, coarsely dentate; petiole broad and flat, longer than the lamina, very sparsely ciliate. Petals with numerous red spots. Ovary large, bright pink; styles divergent. $2n = 28$. *Somewhat calcifuge.* ● *N. Portugal and N.W. Spain; Ireland.* Hb Hs Lu.

17. S. umbrosa L., *Sp. Pl.* ed. 2, 574 (1762). Densely caespitose, with flat rosettes. Leaves patent, obovate-oblong, with broad, rather low crenations and a conspicuous cartilaginous border; petiole broad and flat, scarcely as long as the lamina, densely ciliate. Petals with numerous red spots. Ovary rather small, bright pink; styles suberect. $2n = 28$. *Shady banks, streamsides and mountain grassland.* ● *W. & C. Pyrenees.* Ga Hs [?Au Br Da].

18. S. hirsuta L., *Syst. Nat.* ed. 10, **2**: 1026 (1759) (*S. geum* auct., non L.). Laxly caespitose. Leaves somewhat fleshy, scarcely coriaceous, reniform, orbicular or ovate-oblong, crenate or rarely serrate-dentate, usually hispid on both surfaces, often red beneath; petiole narrow, hairy. Petals usually with few, rather weak red spots. Ovary pale pink. ● *Pyrenees and N. Spain; S.W. Ireland; occasionally naturalized elsewhere.* Ga Hb Hs [Br].

(a) Subsp. **hirsuta**: Lamina reniform or orbicular, cordate, much shorter than the petiole, with 17–25 crenations or teeth. $2n = 28$. *Shady places, or by mountain streams. Throughout the range of the species.*

(b) Subsp. **paucicrenata** (Leresche ex Gillot) D. A. Webb, *Feddes Repert.* **68**: 201 (1963): Leaves 10–30 mm, oblong-elliptical, with only 7–11 crenations; petiole equalling or shorter than the lamina. $2n = 28$. *Exposed limestone rocks and screes. W. Pyrenees and N. Spain.*

Subsp. (b) is a dwarf, very distinct ecotypic variant, but connected to subsp. (a) by intermediates.

19. S. × polita (Haw.) Link, *Enum. Hort. Berol. Alt.* **1**: 414 (1821) (*S. hirsuta × spathularis*). Leaves usually suborbicular to broadly obovate, slightly hairy; petiole moderately narrow, rather hairy. $2n = 28$. *Common wherever the parent species grow together, and grading imperceptibly into both.* ● *N. Spain and W. Ireland (including districts in Ireland where* S. hirsuta *does not now exist).* Hb Hs.

20. S. × geum L., *Sp. Pl.* 401 (1753) (*S. hirsuta × umbrosa; S. hirsuta* auct., non L., *S. umbrosa* auct. eur. med. non L.). Intermediate between the parents and variable, but usually with the lamina nearer that of **17** (broadly oblong, with low crenations and a conspicuous cartilaginous border) and the petiole nearer that of **18** (long, hairy, and fairly narrow). $2n = 28$. ● *Occasional, with the parents, in the Pyrenees; widely cultivated and naturalized in many places in W. & C. Europe.* Ga Hs [Au Be ?Br Ge It].

21. S. × urbium D. A. Webb, *Feddes Repert.* **68**: 199 (1963) (*S. spathularis × umbrosa, S. umbrosa* auct., non L.). Exactly intermediate between the parents; usually sterile and therefore constant, but fertile variants are known. $2n = 28$. *Unknown in nature, as the parent species are allopatric; of obscure garden origin, widely cultivated in Britain and Ireland, less often elsewhere. Naturalized in Britain, Ireland and France.* [Br Ga Hb.]

Sect. PORPHYRION Tausch (incl. *Kabschia* Engler). Evergreen perennials, forming cushions or mats. Leaves small, opposite or alternate, usually hard and coriaceous, entire or finely toothed, usually with calcareous incrustation. Flowers solitary or in small cymes, arising terminally from a leaf-rosette which does not die after flowering. Ovary semi-inferior to almost inferior.

Subsect. *Kabschia* (Engler) Rouy & Camus. Leaves alternate. Petals conspicuous, usually white or yellow, exceeding the usually patent sepals.

22. S. caesia L., *Sp. Pl.* 399 (1753). Leafy shoots densely caespitose, forming dense cushions, each shoot terminating in a small leaf-rosette. Leaves 3–6 mm, oblong-spathulate to elliptical, obtuse, arcuate-recurved from near the base, very glaucous and lime-encrusted, without hyaline border, ciliate near the base. Flowering stem 4–10 cm, very slender, with 2–5 flowers. Calyx and pedicels glandular-hairy; lower part of flowering stem often nearly glabrous. Petals 4–6 mm, broadly obovate, white. *Rocks, screes and stony ground; calcicole.* $2n = 26$. ● *Mountains of C. & S. Europe, from the Alps and W. Carpathians to the Pyrenees, C. Appennini and Crna Gora.* Au Cz Ga Ge He Hs It Ju Po.

23. S. squarrosa Sieber, *Flora (Regensb.)* **4**: 99 (1821). Like **22** but leafy shoots more densely packed, giving deeper, harder cushions; leaves slightly smaller, linear-oblong, subacute, suberect in lower part and arcuate-recurved only near the apex; flowering stem usually densely glandular-hairy towards the base and nearly glabrous in the inflorescence. $2n = 26$. *Limestone rocks.* ● *S.E. Alps.* Au It Ju.

24. S. marginata Sternb., *Revis. Saxifr. Suppl. I*, 1 (1822) (incl. *S. boryi* Boiss & Heldr., *S. rocheliana* Sternb.). Densely caespitose, with rosette-like or columnar leafy shoots. Leaves 3–13 × 1–5 mm, very variable, linear-oblong to obovate-spathulate, more or less obtuse (rarely shortly mucronate), glandular-ciliate towards the base, with a distinct hyaline border throughout, plane or recurved distally. Flowering stem 3–12 cm, leafy, pubescent, bearing a compact corymbose cyme of 2–8 flowers. Petals 5–13 mm, obovate, white, or rarely pale pink. *Mountain rocks; calcicole.* ● *Balkan peninsula, S. Carpathians, C. & S. Italy.* Al Bu Gr It Ju Rm.

The contrast between the extreme varieties of this species (var. *rocheliana* (Sternb.) Engler & Irmscher, with flat rosettes of leaves *c.* 10 × 3·5 mm, and var. *coriophylla* (Griseb.) Engler, with columnar shoots of imbricate, recurved leaves *c.* 5 × 1 mm) has led many authors to treat them as species or subspecies, but they are connected to each other and to the type by a full range of intermediates, and are not clearly separated geographically. Extreme examples of var. *coriophylla* can scarcely be distinguished from **25**.

25. S. diapensioides Bellardi, *App. Fl. Pedem.* 21 (1792). Leafy shoots numerous, columnar, caespitose, forming a deep, rigid cushion. Leaves 4–6 × 1–1·5 mm, suberect, densely imbricate, oblong, obtuse, thick, glaucous, glabrous except for a few cilia near the base; margin hyaline. Flowering stem 3–7 cm, glandular-hairy, bearing 2–6 flowers. Petals *c.* 8 × 5 mm, white. $2n = 26$. *Mountain rocks; calcicole.* ● *S.W. Alps, eastwards to Monte Rosa.* Ga He It.

26. S. tombeanensis Boiss. ex Engler, *Monogr. Gatt. Saxifr.* 268 (1872). Like **25** but differing chiefly in the leaves, which are slightly shorter, lanceolate, elliptical or rhombic, cucullate, with a distinct incurved mucro, thinner, scarcely glaucous, seldom lime-encrusted and more conspicuously ciliate. Petals often up to 12 mm. *Mountain rocks; calcicole.* $2n = 26$. ● *Italian Alps, from* 10°30′ *to* 11°15′ *E.* It.

27. S. scardica Griseb., *Spicil. Fl. Rumel.* **1**: 332 (1843). Leafy shoots up to 5 cm, forming a rather dense cushion. Leaves 5–15 × 2–4 mm, oblong, mucronate, convex below and keeled near the apex; margin hyaline, serrulate or ciliate except near the apex, somewhat glaucous. Flowering stem 4–12 cm, with a compact cyme of 5–14 flowers. Petals 7–12 mm, white or pale pink, rarely crimson. $2n = 26$. *Limestone rocks and screes.* ● *Mountains of Balkan peninsula, from Crna Gora to S. Greece.* Al Gr Ju.

28. S. spruneri Boiss., *Diagn. Pl. Or. Nov.* **1(3)**: 18 (1843). Leafy shoots numerous, caespitose, columnar, forming large cushions. Leaves 4–6 mm, obovate-oblong, obtuse, ciliate in basal ½, with a narrow, hyaline border in apical ½, glandular-hairy on lower surface, erect, appressed. Flowering stem 4–8 cm, with 6–12 flowers in a somewhat corymbose cyme. Petals *c.* 5 × 2·5 mm, white. $2n = 28$. *Crevices of limestone rocks, mostly above* 2000 *m.* ● *Mountains of C. Greece and S. Bulgaria.* ?Al Bu Gr ?Ju.

29. S. burseriana L., *Sp. Pl.* 400 (1753). Leafy shoots short, forming a low cushion or thick mat. Leaves 6–12 × 1·5–2 mm, linear-subulate, tapering from the base to an acute, pungent apex, glaucous, with a narrow, hyaline margin. Flowers solitary; peduncles 2–5 cm, reddish. Petals 7–15 × 5–10 mm, white. $2n = 26$. *Rocks and screes; calcicole.* ● *E. Alps, from* 10°20′ *E. eastwards.* Au Ge It Ju.

30. S. vandellii Sternb., *Revis. Saxifr.* 34 (1810). Leafy shoots numerous, columnar, caespitose, forming a dense, hard, deep cushion. Leaves 8–11 × 1·5–2 mm, suberect, lanceolate-subulate, strongly apiculate, pungent, keeled below, not glaucous, ciliate-denticulate towards the base; margin hyaline. Flowering stem 4–8 cm, bearing 3–8 flowers in a compact corymb. Petals *c.* 9 × 4 mm, oblanceolate-cuneate, white. $2n = 26$. *Limestone cliffs.* ● *Italian Alps, from* 9°15′ *to* 10°45′ *E.* It.

31. S. juniperifolia Adams in Weber fil. & Mohr, *Beitr. Naturk.* **1**: 53 (1806) (*S. juniperina* Bieb.; incl. *S. pseudosancta* Janka). Leafy shoots numerous, crowded, forming a compact cushion. Leaves 10–14 × 1·5–2·5 mm, strongly mucronate or apiculate, not glaucous or lime-encrusted, with a narrow hyaline border, which is denticulate below but entire in the apical ½. Flowering stem 3–6 cm, leafy, bearing 4–8 flowers in an oblong cyme. Pedicels and hypanthium villous. Petals 5–6 × 2–3 mm, obovate-cuneate, acute, suberect, bright yellow, shorter than the stamens. *Damp or shady mountain rocks. Bulgaria.* Bu ?Ju. (*Caucasus.*)

32. S. sancta Griseb., *Spicil. Fl. Rumel.* **1**: 333 (1843) (*S. juniperifolia* subsp. *sancta* (Griseb.) D. A. Webb). Like **31** but leaves seldom more than 10 mm, with longer and stouter marginal hairs, which extend at least ¾ of the way to the apex; flowering stem and inflorescence glabrous; cyme corymbose or more or less spherical; petals about equalling the stamens. *Crevices in marble rocks,* 1450–2050 *m. N. Greece (Athos, Pangaion, ?Olimbos).* Gr.

33. S. ferdinandi-coburgi J. Kellerer & Sünd., *Allgem. Bot. Zeitschr.* **1901**: 116 (1901). Shoots numerous, columnar, caespitose, forming a dense, hard cushion. Leaves 5–7 × 1–1·5 mm, suberect, linear-oblong, with a short, incurved mucro and a narrow hyaline border, scarcely glaucous but with abundant calcareous incrustation, glabrous except for a few cilia at the base. Flowering stem 3–7 cm, with usually 7–13 flowers. Sepals red; petals 5–7 mm, longer than the stamens, bright yellow. $2n = 26$. ● *Mountains of Bulgaria and N. Greece.* Bu Gr.

34. S. aretioides Lapeyr., *Fig. Fl. Pyr.* 28 (1801). Like **33** but leaves slightly shorter and wider, obtuse or with a straight mucro; inflorescence with only 3–5(–7) flowers; petals somewhat narrower. ● *Pyrenees, Cordillera Cantábrica.* Ga Hs.

Subsect. *Engleria* (Sünd.) Gornall. Leafy shoots short, crowded. Leaves alternate, entire, glaucous, lime-encrusted. Petals pink, purple or yellow, largely enclosed and concealed by erect, densely glandular-pubescent sepals.

(**35–37**). **S. porophylla** group. Leaves linear to spathulate. Flowers subsessile in a spike-like cyme. Calyx and hypanthium dull pink to deep crimson-purple. Petals purplish-pink, about equalling the sepals.

A troublesome complex of closely related taxa, very variously treated by different authors.

1 Leaves scarcely expanded towards the apex, acute or apiculate, not more than 2·5 mm wide **37. sempervivum**
1 Leaves oblanceolate to spathulate, obtuse or mucronate, usually more than 2·5 mm wide
2 Leaves 4–15 × 2–3 mm; flowers 4–7(–12); inflorescence dull pink, often tinged with green **35. porophylla**
2 Leaves 12–35 × 3–8 mm; flowers 10–20; inflorescence bright red or dark purplish-red **36. federici-augusti**

35. S. porophylla Bertol. in Desv., *Jour. Bot. Appl.* **4**: 76 (1814) (*S. media* auct. ital. non Gouan). Rosettes fairly flat. Leaves 4–15 × 2–3 mm, oblanceolate, mucronate, with 5–11 hydathodes. Flowering stem 3–8 cm, with 4–7(–12) flowers. Calyx and hypanthium dull, rather pale pink, sometimes tinged with green, rarely red. *Limestone rocks and screes above* 1000 *m*. ● *C. & S. Italy*. It.

36. S. federici-augusti Biasol., *Relaz. Viaggio* 109 (1841) (*S. frederici-augusti* auct., *S. media* subsp. *porophylla* sensu Hayek, pro parte; incl. *S. porophylla* var. *montenegrina* (Halácsy & Bald.) Engler & Irmscher). Rosettes regular, flat. Leaves 12–35 × 3–8 mm, oblanceolate to spathulate, usually mucronate. Flowering stem 7–20 cm, with 10–20 flowers. Inflorescence bright or dark red. *Limestone rocks*. ● *Balkan peninsula, from central Crna Gora to E.C. Greece*. Al Gr Ju.

(**a**) Subsp. **federici-augusti**: Leaves not more than 18 mm, obovate-oblanceolate, rather gradually widened at the apex. Inflorescence dark purplish-red, with usually not more than 15 flowers. Flowering stem fairly slender. *Mainly in the north-western part of the range of the species, and mostly at high altitudes*.

(**b**) Subsp. **grisebachii** (Degen & Dörfler) D. A. Webb, *Bot. Jour. Linn. Soc.* **95**: 235 (1987) (*S. grisebachii* Degen & Dörfler): Leaves up to 35 mm, spathulate, widened rather abruptly at the apex. Inflorescence crimson to bright red, with up to 20 flowers. $2n = 26$. *Mainly in the south-eastern part of the range of the species, and descending to lower altitudes*.

37. S. sempervivum C. Koch, *Linnaea* **19**: 40 (1846) (*S. porophylla* var. *sibthorpiana* (Griseb.) Engler & Irmscher; incl. *S. media* subsp. *porophylla* sensu Hayek, pro parte). Rosettes rather squarrose. Leaves 5–20 × 1–3 mm, linear to narrowly oblong, acute or apiculate. Flowering stem 6–20 cm, with 8–20 flowers. Inflorescence dark purplish-crimson. $2n = 12$. *Limestone rocks, usually above* 1800 *m*. *Balkan peninsula, from Crna Gora and C. Bulgaria southwards*. Al Bu Gr Ju.

38. S. media Gouan, *Obs. Bot.* 27 (1773). Rosettes somewhat squarrose. Leaves 6–17 × 2–4 mm, linear-oblong to oblanceolate, acute. Flowering stem 3–12 cm, with a raceme-like inflorescence of 2–12 flowers, all pedicellate, but with pedicels varying from 3 to 25 mm. Inflorescence reddish-pink, often tinged with green. Petals 2·5–3·5 mm, bright pinkish-purple, about equalling the sepals. $2n = 26$. *Rocks; somewhat calcicole*. ● *E. Pyrenees, westwards to* 0°20′ *E*. Ga Hs.

39. S. stribrnyi (Velen.) Podp., *Verh. Zool.-Bot. Ges. Wien* **52**: 652 (1902) (*S. media* subsp. *stribrnyi* (Velen.) Hayek). Rosettes more or less flat. Leaves 12–25 × 3–6 mm, oblanceolate to somewhat spathulate, obtuse or shortly mucronate, glaucous, usually with abundant calcareous incrustation. Flowering stem 3–9 cm, usually branched from near the middle to give a lax panicle of 8–25 clearly pedicellate flowers. Inflorescence deep crimson. Petals 3–4 mm, slightly exceeding the sepals, purple. $2n = 26$. *Mountain rocks; calcicole*. ● *Bulgaria and N. Greece*. Bu Gr.

40. S. corymbosa Boiss., *Diagn. Pl. Or. Nov.* **1(3)**: 17 (1843) (*S. luteo-viridis* Schott & Kotschy). Rosettes flat. Leaves 7–22 × 2·5–4 mm, linear-oblong to subspathulate, acute to mucronate, often purple beneath. Flowering stem 3–10 cm, bearing a rather crowded panicle of 2–12 shortly pedicellate flowers. Inflorescence usually pale green, but in strongly illuminated plants deep crimson. Petals 3 mm, shorter than the sepals, pale yellow. *Limestone rocks above* 1500 *m*. *E. & S. Carpathians; Bulgaria and N. Greece*. Bu Gr.

Subsect. *Oppositifoliae* Hayek. Leaves usually opposite, with or without calcareous incrustation. Flowering stem short, with 1–3 flowers. Sepals patent, not concealing the purple, pink or white petals.

41. S. oppositifolia L., *Sp. Pl.* 402 (1753). Stems woody, branched, more or less procumbent, forming a mat or low, lax cushion. Leaves 1·5–5(–8) × 1–2 mm, oblong to obovate, obtuse or subacute, narrowed at the base but scarcely petiolate, keeled beneath, hard and rigid, dull, dark green, usually crowded and imbricate; hydathodes 1–3(–5); calcareous incrustation moderate. Flowering stem 1–2(–5) cm; flower solitary. Sepals 2–4 mm, ciliate; petals 5–12(–20) × 2–7 mm, pale pink to deep purple, fading to violet. *Arctic and subarctic Europe, and on most of the mountain ranges southwards to the Sierra Nevada, C. Italy and N.W. Greece*. Al Au Br Bu Cz Fa Fe Ga Ge Gr Hb He Hs Is It Ju No Po Rm Rs(N) Sb Su.

1 Leaves mostly alternate (**b**) subsp. **paradoxa**
1 Leaves mostly opposite
2 Leaves 1·5–2 × 0·7–1·3 mm; flowers subsessile (**c**) subsp. **rudolphiana**
2 Leaves at least 2·5 × 1·5 mm; flowering stem developed
3 Leaf-margin ciliate to the apex, the apical hairs the longest (**d**) subsp. **blepharophylla**
3 Cilia absent from apex of leaf, or, if present, shorter than those near the base
4 Leaves ciliate almost to the apex; cartilaginous margin not widened at the apex (**a**) subsp. **oppositifolia**
4 Leaf-apex not ciliate, but with a wide cartilaginous margin (**e**) subsp. **speciosa**

(**a**) Subsp. **oppositifolia** (incl. *S. latina* (N. Terracc.) Hayek, *S. murithiana* Tiss., *S. pulvinata* Small): Rather lax in habit. Leaves 2·5–8 mm, opposite, except sometimes on upper part of flowering stem, ciliate almost to the apex, with longest hairs at the base. Flowering stem 1–2(–5) cm. Hairs on sepals usually eglandular. *Almost throughout the range of the species, but rare in the Pyrenees and C. Appennini*.

(**b**) Subsp. **paradoxa** D. A. Webb, *Bot. Jour. Linn. Soc.* **95**: 237 (1987): Like subsp. (**a**) but leaves alternate, except rarely on a few of the weaker shoots. Flowers usually fairly large. ● *Pyrenees*.

(**c**) Subsp. **rudolphiana** (Hornsch.) Engler & Irmscher, *Pflanzenreich* **69(IV.117)**: 638 (1919) (*S. rudolphiana* Hornsch.): Leafy shoots densely crowded to give a dense, flat cushion. Leaves 1·5–2 × 0·7–1·3 mm, opposite, subacute, with slender marginal hairs, often in proximal ½ only. Flowers subsessile. Petals 5–7 mm; sepals glandular-ciliate. ● *E. Alps; E. & S. Carpathians.*

(d) Subsp. **blepharophylla** (A. Kerner ex Hayek) Engler & Irmscher, *loc. cit.* (1919) (*S. blepharophylla* A. Kerner ex Hayek): Compact, often with columnar leafy shoots. Leaves 3–5 mm, opposite, broadly obovate, with marginal hairs extending to the rounded apex, those at the apex being the longest. Flowering stem short but distinct. Marginal hairs on sepals very long, non-glandular; petals 5–8 mm. ● *E. Alps, mainly in Austria.*

(e) Subsp. **speciosa** (Dörfler & Hayek) Engler & Irmscher, *op. cit.* 639 (1919) (*S. speciosa* (Dörfler & Hayek) Dörfler & Hayek): Fairly compact. Leaves 4–5 mm, broadly obovate to suborbicular, with short marginal hairs mainly in proximal part; apex broad, rounded, with a wide cartilaginous margin. Petals 8–12 mm. ● *C. Appennini.*

Plants generally similar to subsp. (a) but with glandular cilia on the sepals are found in the Alps, Pyrenees and Sierra Nevada, and have been distinguished as subsp. *glandulifera* Vacc., but they are united to subsp. (a) by too many intermediates to make recognition as a subspecies profitable. Plants from unusually low altitude on the shores of the Bodensee, usually submerged in spring and early summer, have been named subsp. *amphibia* (Sünd.) Br.-Bl., but their rather slight morphological distinction was probably due to their peculiar environment. They appear to be now extinct.

42. S. retusa Gouan, *Obs. Bot.* 28 (1773) (*S. purpurea* All.). Leafy shoots numerous, short, crowded, forming a low but very compact cushion. Leaves 2–4 × 1·5–2 mm, opposite, oblong, obtuse, entire, the basal ½ erect, somewhat membranous, often purple, the upper ½ sharply bent outwards, keeled beneath, flat and dark, shining green. Calcareous incrustation variable. Flowering stem 5–50 mm, bearing a small, umbel-like cyme of 1–3(–5) flowers. Petals 4–5 × 2–2·5 mm, widely separated, pinkish-purple. Stamens longer than the petals; anthers orange. $2n = 26$. *Usually above* 1800 *m, often on exposed, snow-free rocks or screes.* ● *Pyrenees, Alps, Carpathians, Bulgaria; rather local.* Au Bu Cz Ga He Hs It Po Rm.

(a) Subsp. **retusa** (*S. baumgartenii* Schott, *S. wulfeniana* Schott): Sepals and hypanthium glabrous; peduncles 1·5 cm or less, with 1–3 flowers; petals *c.* 4 mm. *Mainly calcifuge. Throughout the range of the species, except for parts of S.W. Alps.*

(b) Subsp. **augustana** (Vacc.) P. Fourn., *Quatre Fl. Fr.* 474 (1936): Sepals and hypanthium densely glandular-pubescent; peduncles 2–5 cm, with 2–5 flowers; petals *c.* 5 mm. *Mainly calcicole. S.W. Alps, eastwards to Monte Rosa.*

43. S. biflora All., *Auct. Syn. Stirp. Horti Taur.* 34 (1773) (incl. *S. macropetala* A. Kerner). Stems procumbent to ascending, forming a mat or loose cushion. Leaves 5–9 × 3–6 mm, opposite, broadly obovate to suborbicular, obtuse, sessile, scarcely coriaceous and seldom lime-encrusted, usually with a single hydathode, often reddish beneath. Flowers solitary or 2–5(–8) in terminal corymbose cymes. Sepals ciliate, glandular-pubescent; petals reddish-purple or dull white; disc yellow, conspicuous; filaments whitish, shorter than the petals; anthers pink to orange, eventually black. *Damp screes, moraines and river-gravels, usually above* 2000 *m.* ● *Alps; one station in N.W. Greece.* Au Ga Ge Gr He It.

The typical plant has oblanceolate petals, 5–6 × 2 mm. In var. *kochii* Kittel (subsp. *macropetala* (A. Kerner) Rouy & Camus) the petals are broadly elliptical, 6–9 × 4–5 mm, although variation in petal-size seems to be continuous, and other supposedly correlated characters are very variable. Some of these plants may be hybrids between **43** and **41**.

(a) Subsp. **biflora**: Leaves with a single hydathode; calcareous incrustation usually absent. Sepals and upper cauline leaves hairy. Flowers usually 2 or more; pedicels very short, so that the flowers are partly overlapped by the leaves. Ovary without hooked bristles. $2n = 26$. *Alps.*

(b) Subsp. **epirotica** D. A. Webb, *Bot. Jour. Linn. Soc.* **95**: 237 (1987): Leaves often with 3–5 hydathodes; calcareous secretion plentiful. Sepals and upper cauline leaves glabrous beneath. Flowers solitary, on pedicels long enough to carry them clear of the leaves. Upper part of ovary with several stout, hooked, bristles. *N.W. Greece (near Timfi Oros).*

Sect. LIGULATAE Haw. (*Aizoonia* Tausch). Leaves mostly basal in compact rosettes, coriaceous, usually glaucous, entire or finely toothed, usually lime-encrusted. Flowers in terminal panicles; after flowering the rosette bearing the panicle dies. Petals white, rarely orange or pink. Ovary semi-inferior to almost inferior.

44. S. longifolia Lapeyr., *Fig. Fl. Pyr.* 26 (1801). Plant long-lived, without offsets, and therefore monocarpic. Leaves 30–100 × 3–8 mm, linear, somewhat expanded below the subacute apex, entire or weakly erose-crenate, glabrous, glaucous, lime-encrusted, very numerous, forming a large, flat, regular rosette. Flowering stem 25–50 cm, stout, branched from near the base to form a large, pyramidal panicle with very many flowers. Pedicels and hypanthium glandular-hairy. Petals 5–6 mm, white, rarely with fine red spots. $2n = 28$. *Mountain cliffs; calcicole. Pyrenees and E. Spain.* Ga Hs.

45. S. callosa Sm. in Dickson, *Coll. Dried Pl.* **3**: no. 63 (1791). Leaves entire or feebly erose-crenate, glaucous, lime-encrusted, often dark red near the base; rosettes proliferating by short stolons which produce new rosettes. Flowering stem 15–40 cm. Flowers numerous, in a long, fairly narrow panicle, which usually occupies about ½ the flowering stem. Pedicels slender, often reddish. Petals 7–9 mm, white, often with crimson spots. *Limestone rocks.* ● *W. Mediterranean region.* Ga Hs It Sa Si.

(a) Subsp. **callosa** (*S. lingulata* Bellardi): Leaves 25–90 × 2·5–6 mm, linear or oblanceolate, usually subacute. Peduncle and inflorescence eglandular, or with a few scattered glandular hairs. $2n = 28$. *Maritime Alps, Appennini, Sardegna, Sicilia.*

(b) Subsp. **catalaunica** (Boiss.) D. A. Webb, *Feddes Repert.* **68**: 208 (1963) (*S. catalaunica* Boiss.): Leaves 15–45 × 8–9 mm, oblanceolate, often acute. Inflorescence rather densely covered by subsessile glands. Peduncle glabrous or with glandular hairs. *N.E. Spain; S.E. France (hills near Marseille).*

Typical plants of subsp. (a), with long, narrow, linear leaves, are restricted to the eastern part of the Maritime Alps and the N. & C. Appennini. Plants from the western Maritime Alps, and from the C. & S. Appennini and the islands, with broader, usually shorter, oblanceolate leaves may be distinguished as var. *australis* (Moric.) D. A. Webb (*S. australis* Moric.; incl. *S. lantoscana* Boiss.).

The plant from near Marseille is transitional to subsp. (a), but seems best placed in subsp. (b).

46. S. cochlearis Reichenb., *Fl. Germ. Excurs.* 559 (1832). Like **45(a)** but smaller and more delicate; leaves 8–40 × 2·5–6 mm, spathulate, with an expanded, suborbicular apical part; flowering stem usually 10–25 cm, with a small, open panicle of 15–25(–60) flowers. $2n = 28$. *Limestone rocks.* ● *Maritime Alps; Liguria.* Ga It.

47. S. crustata Vest, *Bot. Zeit.* (*Regensburg*) **3**: 314 (1804) (*S. incrustata* auct.). Leaves 15–60 × 2·4 mm, linear, entire, glaucous and lime-encrusted, in flat, densely caespitose rosettes. Flowering stem 12–30 cm; panicle smaller and with fewer flowers than in **45**, occupying about ½ the stem or less, each branch bearing 1–3 flowers. Pedicels glandular-hairy; hypanthium and calyx usually glabrous. Petals 4–6 mm, white, sometimes with red spots. $2n = 28$. *Limestone rocks.* ● *E. Alps, extending southwards to C. Jugoslavia.* Au It Ju.

48. S. hostii Tausch, *Syll. Pl. Nov. Ratisbon.* (*Königl. Baier. Bot. Ges.*) **2**: 240 (1828). Rosettes rather distant. Leaves 30–100 × 4–10 mm, linear-oblong, tending to curve downwards at the apex, finely serrate, somewhat glaucous, lime-encrusted. Flowering stem 25–60 cm; panicle rather shorter than the peduncle, its primary branches with (3–)5–12 flowers. Petals 5–8 mm, white, usually with numerous red spots. $2n = 28$. *Calcicole.* ● *E. Alps, mainly on S. side.* Au ?Cz ?Hu It Ju [No].

(a) Subsp. **hostii** (subsp. *dolomitica* Br.-Bl.; incl. *S. altissima* A. Kerner): Basal leaves up to 10 cm, and usually at least 6 mm wide, with obtuse, rounded apex; marginal teeth distinct. Primary branches of panicle with at least 5 flowers. *From* 10°30′ *E. in Austria to c.* 15° *E. in Slovenija.*

(b) Subsp. **rhaetica** (A. Kerner) Br.-Bl. in Hegi, *Ill. Fl. Mitt.-eleur.* **4(2)**: 590 (1922): Leaves not more than 50 × 3–7 mm, with acute apex; marginal teeth often obscure. Primary branches of panicle with 3–5 flowers. *Italian Alps from* 9°15′ *to* 10°45′ *E.*

49. S. paniculata Miller, *Gard. Dict.* ed. 8, no. 3 (1768) (*S. aizoon* Jacq.). Rosettes rather crowded. Leaves (5–)12–40(–60) × 2–8 mm, more or less upwardly curved at the apex, forming hemispherical rosettes, obovate to oblong-lingulate, obtuse or acute, finely serrate, glaucous and usually lime-encrusted. Flowering stem (4–)12–30 cm, branched only in its upper ⅓ to form a small panicle; branches patent or suberect, 1- to 3(4)-flowered, glandular-hairy. Petals 4–6 mm, white or pale cream, sometimes with small red spots. $2n = 28$. *Throughout the mountains of C. & S. Europe, from the Vosges and C. Poland to N. Spain, S. Italy and C. Greece; also very locally in Norway.* Al Au Bu Co Cz Ga Ge Gr He Hs Hu Is It Ju No Po Rm Rs(W).

Very variable in size and in leaf-shape. Subsp. **cartilaginea** (Willd.) D. A. Webb, *Feddes Repert.* **68**: 208 (1963), with apiculate or acuminate leaves, occurs in the Caucasus and Anatolia; certain plants from the Balkan peninsula approach this subspecies.

50. S. cotyledon L., *Sp. Pl.* 398 (1753) (incl. *S. montavoniensis* A. Kerner). Leaves 20–80 × 6–20 mm, oblong to oblanceolate, finely and regularly serrate, only slightly glaucous, sparsely lime-encrusted. Flowering rosette 7–12 cm, usually accompanied by small, rather distant daughter-rosettes. Flowering stem 15–70 cm, branched from near the base to form a large, pyramidal panicle of very numerous flowers. Pedicels and calyx glandular-hairy. Petals 6–10 mm, white. $2n = 28$. *Rocks; calcifuge.* ● *S. Alps; C. Pyrenees; W. Fennoscandia and Iceland.* Au Ge He ?Hs Is It No Su.

The populations from N. Europe, the Alps and the Pyrenees differ slightly, but scarcely enough to justify the recognition of subspecies.

51. S. valdensis DC. in Lam. & DC., *Fl. Fr.* ed. 3, **5**: 517 (1815). Leafy shoots erect, densely caespitose, terminating in small leaf-rosettes, forming a deep, hard, rather irregular cushion. Leaves 4–12 mm, linear to oblanceolate-spathulate, obtuse, entire, very glaucous and lime-encrusted; upper surface irregularly pitted. Flowering stem 5–12 cm; cauline leaves longer than those of the rosettes. Flowers 5–12, in a corymbose panicle. Petals 5–6 mm, white. $2n = 28$. *Rock-crevices; generally calcicole.* ● *S.W. Alps, from* 44°30′ *to* 45°30′ *N.* Ga It.

52. S. mutata L., *Sp. Pl.*, ed. 2, 570 (1762). Flowering rosettes 5–13 cm in diameter, sometimes solitary, so that the plant is monocarpic, but more often producing a few daughter rosettes on offsets. Leaves 20–70 × 7–15 mm, oblong-oblanceolate, obtuse, dark green, not glaucous, shining, not lime-encrusted, with a conspicuous hyaline margin, entire distally but fimbriate-ciliate towards the base. Petals 5–8 mm, linear, acute, orange; floral disc and top of ovary red. ● *Alps and adjacent regions; W. & S. Carpathians.* Au Cz Ga He It Rm.

(a) Subsp. **mutata**: Flowering stem 20–50 cm, stout, bearing a lax, densely glandular-hairy panicle; peduncle short but distinct. $2n = 28$. *Alps and adjacent regions; Tatra.*

(b) Subsp. **demissa** (Schott & Kotschy) D. A. Webb, *Feddes Repert.* **68**: 209 (1963) (*S. demissa* Schott & Kotschy): Peduncle very short or absent, so that each rosette bears numerous racemes or slightly branched panicles 6–20 cm. *S. Carpathians* (25° *to* 26° *E.*).

53. S. florulenta Moretti, *Gior. Fis.* (*Brugnat*) ser. 2, **6**: 468 (1823). Monocarpic; rosettes solitary, very slow-growing and with persistent dead leaves, so that the rosette eventually forms a low cylinder. Leaves 30–60 × 4–7 mm in mature plants, but considerably shorter in young ones, oblanceolate, apiculate or mucronate, coriaceous, dull, dark green, not lime-encrusted; hydathodes inconspicuous; margin cartilaginous, ciliate except near the apex. Inflorescence a long, narrow, glandular-hairy, thyrsoid panicle, shortly pedunculate. Petals 5–7 mm, oblanceolate, acute, flesh-coloured. Carpels often 3; sometimes 5 in the terminal flower. $2n = 28$. *Shady, vertical rocks above* 1900 *m; calcifuge.* ● *Central part of Maritime Alps; rare and apparently decreasing.* Ga It.

Sect. TRACHYPHYLLUM (Gaudin) Koch. Evergreen, mat-forming perennials with procumbent to decumbent leafy shoots. Leaves narrow, stiff, entire, not lime-encrusted. Flowering stem terminal, leafy. Sepals patent, even in fruit; petals white to pale yellow, sometimes with red or orange spots. Ovary superior.

54. S. aspera L., *Sp. Pl.* 402 (1753) (incl. *S. etrusca* Pignatti). Leafy shoots long, procumbent, forming a loose mat. Leaves patent, oblong-lanceolate, aristate, glabrous except for stiff marginal cilia; axillary buds conspicuous but considerably shorter than the leaves. Flowering stem 8–20 cm, erect, nearly glabrous, with 2–5 (rarely more) flowers in a lax cyme; lower cauline leaves *c.* 10 mm, patent. Petals 4–7 mm, with a very short but distinct claw, white or cream-coloured, usually yellow at the base and often with red or orange spots higher up. $2n = 26$. *Rocks and stony places, mostly* 1200–2200 *m; somewhat calcifuge.* ● *Alps, N. Appennini and Alpi Apuane; E. Pyrenees.* Au Ga He Hs It.

55. S. bryoides L., *Sp. Pl.* 400 (1753) (*S. aspera* subsp. *bryoides* (L.) Engler & Irmscher). Like **54** but with shorter and much more densely matted leafy shoots, bearing smaller, incurved leaves which scarcely exceed the axillary buds; flowers solitary, on stems 3–8 cm, with lower leaves *c.* 5 mm, suberect. 2*n* = 26. *More strictly calcifuge than* **54**, *and usually at a greater altitude.* ● *Mountains of Europe, from Auvergne and the Carpathians to the Pyrenees, S. Alps and Bulgaria.* Au Bu Cz Ga Ge He Hs Hu It Po Rm Rs(W).

56. S. bronchialis L., *Sp. Pl.* 400 (1753) (incl. *S. spinulosa* Adams). Leafy shoots procumbent or decumbent, forming a dense mat or low cushion. Leaves 8–15 mm, loosely imbricate, linear-subulate to oblong-lanceolate, mucronate, shortly ciliate, without conspicuous axillary buds. Flowering stem 6–15 cm, nearly glabrous, bearing several flowers in a corymbose cyme. Petals 3–6 mm, elliptical, without a claw, pale yellow with red spots in upper ½. *N. Ural.* Rs(N). (*N. Asia; North America.*)

Sect. XANTHIZOON Griseb. Evergreen perennials. Leaves fleshy, narrow, more or less entire, sometimes lime-encrusted. Flowering stem terminal, leafy; flowers in a fairly small, lax cyme. Petals yellow to red. Ovary semi-inferior.

57. S. aizoides L., *Sp. Pl.* 403 (1753) (*S. autumnalis* auct., vix L.; incl. *S. aizoidoides* Miégeville). Laxly caespitose, with leafy, ascending flowering stem up to 25 cm, but often much less; non-flowering shoots much shorter. Leaves 10–25 × 2–4 mm, linear-oblong, sessile, entire or rarely very shortly 3-lobed at the apex, acute, glabrous or spinose-ciliate in proximal ½, convex below, flat or convex above. Petals 3–7 mm, narrow, bright yellow or orange, often spotted with red, rarely dark red all over. 2*n* = 26. *Damp, stony places and by mountain streams; more rarely in crevices of limestone rocks. Arctic and subarctic Europe, and southwards in the mountains to the Pyrenees, C. Italy and N. Greece.* Al Au Br Bu Cz Fe Ga Ge Gr Hb He Hs Is It Ju No Po Rm Rs(N, W) Sb Su.

Sect. MESOGYNE Sternb. Perennials, usually winter-dormant, with bulbils in the axils of the basal leaves and sometimes in the inflorescence. Leaves herbaceous, the basal with slender petiole and palmately lobed lamina. Flowers solitary or in a small, terminal cyme. Petals white or pink. Ovary superior to ⅓-inferior.

58. S. sibirica L., *Sp. Pl.* ed. 2, 577 (1762) (incl. *S. mollis* Sm.). Stems 7–20 cm, erect or ascending, flexuous. Leaves reniform, petiolate, palmately 5- to 9-lobed, with triangular-ovate lobes, glabrous or slightly hairy, the basal ones with axillary bulbils. Flowers 2–7. Petals 8–14 × 4–6 mm, white. Ovary superior or very slightly sunk in hypanthium. *Rocky slopes and open woods. E. Russia; S.E. Bulgaria and N.E. Greece.* Bu Gr Rs(C, E). (*N. & C. Asia.*)

59. S. carpatica Sternb., *Revis. Saxifr. Suppl. II*, 32 (1831) (*S. carpathica* Reichenb.). Like **58** but slightly smaller in its vegetative parts; flowers 1–4; petals 6–7 mm, impure white, sometimes tinged with pink; styles stout, very short. 2*n* = *c.* 48. *Damp and shady rocks and stabilized screes; calcifuge. Carpathians; mountains of S.W. Bulgaria.* Bu Cz Po Rm Rs(W).

60. S. rivularis L., *Sp. Pl.* 404 (1753). Delicate, slender plant forming small tufts. Basal leaves 5–12(–20) × 9–17(–30) mm, semicircular to reniform with 3–7 obtuse lobes, subglabrous; axillary bulbils often germinating early to give rise to slender stolons. Flowering stem 3–15 cm, glandular-hairy, with 2–5 flowers. Petals 3·5–5 mm, white, sometimes tinged with pink. Ovary ⅓-inferior; styles very short, stout. 2*n* = 52. *Wet, mossy tundra, streamsides and under overhanging rocks. Arctic and subarctic Europe, southwards to C. Scotland and S. Norway.* Br Fa Fe Is No Rs(N) Sb Su.

61. S. hyperborea R. Br., *Chloris Melv.* 16 (1823). Like **60** but smaller and often suffused with red; stolons absent; basal leaves mostly 3-lobed, with truncate or shortly cuneate base; flowering stem seldom more than 4 cm; petals *c.* 25 mm, deep pink or red. 2*n* = 26. *Arctic Europe, southwards to Iceland.* Is No ?Rs(N) Sb.

Not always easy to distinguish from **60**; many records are ambiguous, and the distribution is imperfectly known.

62. S. cernua L., *Sp. Pl.* 403 (1753) (*S. bulbifera* auct. ross., non L.). Plants solitary or in small tufts. Basal leaves 5–18 × 9–25 mm, semicircular or reniform, with 3–7 short, subacute lobes, glabrous except at the base of the petiole. Flowering stem 3–30 mm; cauline leaves numerous, the lower lobed and the upper entire, each with a red or purplish-black axillary bulbil. Flowers solitary, terminal, sometimes abortive. Petals 7–12 mm, white; ovary superior to ¼-inferior. 2*n* = 36, *c.* 56, 60, 64. *Damp or shady places. Arctic and subarctic Europe, southwards to c.* 57° *N. in Scotland and Ural; locally in the Alps and Carpathians.* Au Br Cz Fe He Is It No Po Rm Rs(N, C) Sb Su.

S. × opdalensis Blytt, *Christ. Vidensk.-Sels. Forh.* **1892**: 52 (1892), a hybrid between **62** and **60**, is locally frequent in parts of Norway and Sweden. It is more like **62** in general habit, but has smaller petals and green cauline bulbils.

Sect. SAXIFRAGA. Usually perennial; habit varied, but usually with numerous leafy shoots, forming a mat or cushion. Leaves alternate, fairly soft, often lobed or crenate, without calcareous incrustation. Petals usually white, more rarely pale or dull yellow, pink or red. Ovary semi-inferior to inferior.

Subsect. *Saxifraga*. Always perennial. Most of the leaves lobed, or boldly crenate.

Series *Saxifraga*. Evergreen or summer-dormant, with bulbils in the axils of the basal leaves, rarely also of the upper leaves. Petals white, rarely tinged with pink.

63. S. granulata L., *Sp. Pl.* 403 (1753). Leaves mostly basal, petiolate, reniform, crenate, rarely dentate, often rather fleshy, hairy at least on the petiole, subtending numerous small, subterranean, axillary bulbils. Cauline leaves similar to the basal but with shorter petioles and without bulbils; sometimes absent. Stem up to 50 cm, erect, usually simple in lower ½, branched above; lower bracts usually large and leaf-like. Flowers in a lax cyme. Sepals triangular-ovate or oblong; petals 9–16 × 3–8 mm, white, glabrous. *Grassland and rocky places. N., C. & W. Europe, extending eastwards to N.W. Russia, N.W. Ukraine, W. Hungary and Sicilia.* Au Be Br Cz Da Fe Ga Ge Hb He Ho Hs Hu It Ju Lu No Po Rs(N, B, C) ?Sa Si Su.

Very variable, especially in size, degree of hairiness, extent of branching of stem, and width of petals; variable also in habitat, growing often in dry, rocky situations in S. Europe and often in damp grassland in N. & C. Europe. Although many local populations have a characteristic facies, there is little constant correlation of characters over wide areas, nor is habitat consistently correlated with form. One regional population of very distinct facies is here recognized as a subspecies, but intermediate variants, linking it to the variable subsp. *granulata*, are not uncommon.

(a) Subsp. **granulata** (incl. *S. glaucescens* Boiss. & Reuter): Basal leaves 18–40 mm wide, rarely less, crenate or obtusely dentate (rarely slightly 3-lobed), glabrous or hairy. Stem usually stout; branching variable, but most often from about the middle;

branches suberect, giving a rather narrow inflorescence. Sepals triangular-ovate; petals 11–16 × 4–8 mm. $2n$ = 32–60, perhaps usually 52. *Throughout the range of the species.*

(**b**) Subsp. **graniticola** D. A. Webb, *Feddes Repert.* **68**: 207 (1963) (*S. granulata* var. *gracilis* Lange, pro parte): Leaves 8–12 mm wide, regularly and rather deeply crenate, hairy. Stem 7–25 cm, slender; branching very variable; branches usually flexuous, somewhat patent. Sepals oblong; petals 7–13 × 3–5 mm. *Calcifuge, usually on granite.* ● *C. Spain, N. Portugal.*

64. S. corsica (Ser.) Gren. & Godron, *Fl. Fr.* **1**: 642 (1849) (*S. russii* auct., non J. & C. Presl, *S. granulata* subsp. *russii* Engler & Irmscher). Like **63**(**a**), but basal leaves more or less 3-lobed, with deeply crenate lobes, and stems branched at least from the middle and often from near the base, with rather patent branches, giving a much more diffuse inflorescence. *Shady rocks, walls and screes.* ● *W. Mediterranean region, from Corse and Sardegna to E. Spain.* Bl Co Hs Sa.

(**a**) Subsp. **corsica** (*S. granulata* subsp. *russii* Engler & Irmscher, non *S. russii* J. & C. Presl): Plant rather delicate in habit; basal leaves seldom divided more than half-way to base; stems branched almost from base. $2n$ = 52, 62–66. *Corse, Sardegna.*

(**b**) Subsp. **cossoniana** (Boiss.) D. A. Webb, *Feddes Repert.* **68**: 203 (1963) (*S. cossoniana* Boiss.): Plant rather robust; basal leaves divided about $\frac{2}{3}$ the distance to the base with overlapping lobes; lower $\frac{1}{3}$ of stem often unbranched. $2n$ = 64–66. *E. Spain.*

Plants from Islas Baleares are intermediate between the two subspecies. Some of those from Sardegna are intermediate between **63**(**a**) and **64**(**a**).

65. S. haenseleri Boiss. & Reuter, *Diagn. Pl. Nov. Hisp.* 13 (1842). A delicate plant forming small tufts; rather like members of Subsect. *Tridactylites* in habit, but with bulbils in the axils of the basal leaves, which are cuneate-obovate, rather deeply divided into 3–7 oblong lobes, shortly petiolate, glandular-hairy. Flowering stem 6–15(–30) cm, slender, branched usually from below the middle to form a lax cyme with long pedicels. Petals 5–6 mm, white, often glandular-hairy on the upper side near the base, but sometimes glabrous. *Shady but dry limestone rocks and debris.* ● *Mountains of S. Spain.* Hs.

66. S. dichotoma Willd. in Sternb., *Revis. Saxifr.* 51 (1810) (incl. *S. arundana* Boiss., *S. kunzeana* Willk.). Basal leaves glandular-hairy, petiolate; lamina up to 20 × 30 mm, but often much less, deeply divided into 5–15 oblong-oblanceolate to obovate-cuneate, obtuse segments. Cauline leaves similar, but smaller and sessile. Flowering stem 6–25 cm, branched usually above the middle, to give a narrow but fairly lax cyme of 2–7 flowers. Petals 6–10 mm, white, veined or tinged with pink, glandular-hairy. *Spain and Portugal.* Hs Lu.

(**a**) Subsp. **dichotoma**: Basal leaves 10–30 mm wide. Flowering stem 12–25 mm, fairly stout. $2n$ = 32+2B. *Mountains of S. & E. Spain.*

(**b**) Subsp. **albarracinensis** (Pau) D. A. Webb, *Feddes Repert.* **68**: 207 (1963) (*S. albarracinensis* Pau): Basal leaves 4–10 mm wide, rather less deeply divided than in subsp. (**a**). Flowering stem 6–10 cm, slender. ● *E. & C. Spain, N.E. Portugal.*

67. S. carpetana Boiss. & Reuter, *Diagn. Pl. Nov. Hisp.* 12 (1842). Basal leaves reniform-cordate to ovate; axillary bulbils present but usually few. Lower cauline leaves ovate, sessile, pinnately lobed, the upper smaller and entire. Flowering stem 12–25 mm, branched above the middle to give a compact cyme of 4–13 flowers. Petals 8–12 mm, white, glandular-hairy. *Grassland, mostly below* 1500 *m. S. Europe, from N.E. Portugal to the Aegean region.* Al Bu Cr Gr Hs It Ju Lu Si.

(**a**) Subsp. **carpetana** (incl. *S. blanca* Willk.): Basal leaves mostly short-stalked, ovate, with cuneate to truncate base, deeply crenate or lobed, generally similar to the lower cauline leaves; a few of the first-formed are, however, sometimes like the basal leaves of subsp. (**b**). Inflorescence few-flowered, very compact. *Spain and Portugal.*

(**b**) Subsp. **graeca** (Boiss. & Heldr.) D. A. Webb, *Bot. Jour. Linn. Soc.* **95**: 243 (1987) (*S. graeca* Boiss. & Heldr., *S. granulata* subsp. *graeca* (Boiss. & Heldr.) Engler, *S. bulbifera* var. *pseudogranulata* Lacaita, *S. atlantica* Boiss. & Reuter): Basal leaves nearly all orbicular to reniform, cordate, with long, slender petiole. Inflorescence larger and less compact than in subsp. (**a**). *S. Italy, Balkan peninsula, Aegean region; rarely in Spain.*

Plants intermediate between subspp. (**a**) and (**b**) are found in Sicilia.

68. S. cintrana Kuzinský ex Willk., *Österr. Bot. Zeitschr.* **39**: 318 (1889) (*S. hochstetteri* (Engler) Coutinho, *S. granulata* L. subsp. *hochstetteri* (Engler) Engler & Irmscher). Stem 10–17 cm, erect, branched, rather stout. Basal leaves suborbicular or rhombic, rounded or cuneate at base, obscurely crenate, hairy, with rather large, subterranean, axillary bulbils. Cauline leaves similar in shape but shortly petiolate, few, sometimes absent. Petals *c.* 10 mm, white, glandular-hairy on the upper surface. *Limestone rocks and walls.* ● *Portugal* (*N. & W. of Lisboa, from Sintra to Bombarral*). Lu.

69. S. bulbifera L., *Sp. Pl.* 403 (1753) (*S. russii* J. & C. Presl). Stem 20–40 cm, erect, simple, arising from a rather large bulb. Basal leaves reniform, crenate, cordate, petiolate, glandular-hairy. Cauline leaves 10–20, ovate, incised-dentate, cuneate, the lower shortly stalked, the upper sessile. Small bulbils present in the axils of cauline leaves and bracts. Flowers rather few, in a small, compact cyme. Petals 7–10 mm, white, glandular-hairy on the upper surface. $2n$ = 28. ● *C. & S. Europe, from Czechoslovakia, Switzerland and Sardegna eastwards.* Al Au Bu Co Cz Gr He Hu It Ju Rm Rs(W) Sa Si Tu.

70. S. gemmulosa Boiss., *Biblioth. Univ. Genève*, ser. 2, **13**: 409 (1838). Rather densely caespitose; stem and leaves often reddish. Leaves mostly basal, with their axillary bulbils at ground-level, glabrous or somewhat hairy. Petiole long, slender; lamina up to 15 × 18 mm, but usually much less, ternately lobed or divided, with the ultimate segments oblanceolate-cuneate and tapered to a narrow, stalk-like base. Flowering stem 5–15 cm, erect. Petals 5–8 mm, broadly obovate, white, glabrous. $2n$ = *c.* 64. *Crevices of ultrabasic igneous rocks*, 450–1200 *m.* ● *S. Spain* (*Málaga and Sevilla provs.*). Hs.

71. S. bourgaeana Boiss. & Reuter in Boiss., *Diagn. Pl. Or. Nov.* **3(2)**: 71 (1856) (*S. boissieri* Engler). Stems woody at the base, ascending or decumbent, branched, forming a loose cushion, with reddish, winter-dormant, bulbil-like buds at the base and in the lower leaf-axils. Leaves glandular-hairy, ovate or semicircular in outline, very deeply ternatisect (usually appearing ternate), the principal divisions crenate, dentate or ternately lobed; ultimate lobes fairly broad-based; petiole long, with a broad, sheathing base. Flowers rather numerous, cup-shaped, in lax, leafy cymes. Petals 4–7(–10) mm, white, glabrous. $2n$ = 64. *Damp, shady, limestone rocks.* ● *S.W. Spain* (*mountains west of Ronda*). Hs.

72. S. biternata Boiss., *Voy. Bot. Midi Esp.* **2**: 231 (1840). Stems woody, decumbent, bearing leafy shoots up to 15 cm, forming a very lax cushion. Leaves very deeply 3-lobed, the lobes narrow-based and those of larger leaves themselves deeply 3-lobed, velvety with rather long glandular hairs; many of them bear in their axils summer-dormant, bulbil-like buds up to 8 × 6 mm. Flowering stem up to 10 cm, terminal and axillary, with usually 2–6 narrowly campanulate flowers in a lax cyme. Petals 12–20 mm, white, glabrous. 2*n* = 66. *Dry limestone cliffs*, 1000–1100 *m*. • *S. Spain* (*S. & E. of Antequera*). Hs.

Series *Gemmiferae* (Willk.) S. Pawł. Evergreen perennials without bulbils, usually with summer-dormant buds in the axils of the leafy shoots. Leaves lobed or entire. Petals white, rarely tinged with pink, glabrous.

73. S. globulifera Desf., *Fl. Atl.* **1**: 342 (1798) (incl. *S. granatensis* Boiss. & Reuter, *S. gibraltarica* Boiss. & Reuter). Leafy shoots erect or spreading, forming a lax cushion. Leaves up to 8 × 17 mm but usually much less, semicircular in outline, rather deeply lobed with 3–7 acute lobes, glandular-hairy; petiole longer than the lamina. Summer-dormant axillary buds numerous, *c.* 3 mm wide, obconical, usually stalked, hairy but not white-villous. Flowering stem 7–12 cm, slender, bearing 2–7(–11) flowers. Sepals 2 mm; petals 5–7 mm, white. *Limestone rocks and cliffs. S. Spain, from* 4°35′ *to* 5°30′ *W*. Hs.

74. S. reuteriana Boiss., *Voy. Bot. Midi Esp.* **2**: 730 (1845). Like **73** but leaves rather larger (up to 12 × 20 mm) and with the primary lobes divided, giving 5–11 segments in all; summer-dormant buds 4–6 mm wide; flowering stem 2–5 cm, with 1–3(4) flowers; sepals 4 mm; petals 8–10 mm, with the margins deflexed near the apex, somewhat greenish-white. *Limestone cliffs*. • *S. Spain* (*mountains S. & E. of Antequera*). Hs.

75. S. erioblasta Boiss. & Reuter in Boiss., *Diagn. Pl. Or. Nov.* **3(2)**: 67 (1856). Leafy shoots short, densely caespitose, forming a compact cushion. Leaves 3–6 mm, entire or shortly 3-lobed with obtuse lobes, glandular-hairy. Summer-dormant buds very numerous, subglobular, sessile, usually densely white-villous at the apex. Flowering stem *c.* 5 cm, slender, with 3–5 flowers. Petals 3·5–5 mm, white at first, becoming bright pink after pollination. 2*n* = 32–34. *Limestone rocks and screes*, 1300–2250 *m*. • *S. Spain, from* 2°50′ *to* 4°05′ *W*. Hs.

76. S. rigoi Porta, *Atti Accad. Agiati* ser. 2, **9**: 26 (1891). Leafy shoots short, ascending, rather densely caespitose. Leaves with a long petiole, ciliate with subarachnoid hairs; lamina with 3–5 obtuse or subacute lobes furnished with shorter glandular hairs. Summer-dormant buds numerous, obovoid, stalked, their outer leaves ciliate with long white hairs. Flowering stem *c.* 6 cm, bearing 2 or 3 large, campanulate flowers. Petals 12–15 × 5 mm, white. • *Mountains of S.E. Spain* (*Granada and Jaén provs.*). Hs.

77. S. conifera Cosson & Durieu, *Bull. Soc. Bot. Fr.* **11**: 332 (1864). Shoots 1–3 cm, procumbent, forming a close mat. Leaves 3–7 mm, linear-lanceolate, strongly apiculate, pale, silvery green; those of the dormant buds fimbriate and ciliate with long arachnoid hairs, the remainder ciliate with short glandular hairs. Summer-dormant buds numerous, 9–12 mm, oblong-conical. Flowering stem 4–9 cm, slender, bearing 3–7 flowers. Petals 3–4 mm, white. *Limestone rocks and gravel*, 1000–2400 *m*. • *N. Spain* (*Cordillera Cantábrica*). Hs.

78. S. continentalis (Engler & Irmscher) D. A. Webb, *Proc. Roy. Irish Acad.* **53B**: 222 (1950) (*S. hypnoides* auct. gall. et iber., non L., *S. hypnoides* subsp. *continentalis* Engler & Irmscher). Leafy shoots long, slender, procumbent, bearing entire to 5-lobed leaves, with terminal and numerous axillary summer-dormant buds, which expand in autumn into rosettes of leaves with 5–13 lobes. Entire leaves narrow-linear, almost filiform; lobed leaves semicircular, with linear to elliptic-lanceolate, strongly apiculate, often recurved segments. Dormant buds shortly stalked, narrowly ellipsoidal, their outer leaves consisting of silvery, completely scarious scales. Flowering stem 8–25 cm, slender, with a small panicle of 4–11 flowers. Petals 4–8 mm, white. 2*n* = 26, 52. *Rocks, walls and screes*, 700–1700 *m*. • *S.W. Europe, from S.E. France to N. Portugal.* Ga Hs Lu.

79. S. hypnoides L., *Sp. Pl.* 405 (1753) (*S. hypnoides* subsp. *boreali-atlantica* Engler & Irmscher). Like **78** but leaf-segments linear to narrowly oblong, not recurved; summer-dormant buds sessile, with the outer leaves partly herbaceous, often absent; petals 7–10(–12) mm. 2*n* = 26, 52. *Grassland, screes, streamsides and mountain-ledges.* • N.*W. Europe, from Iceland to* N.*E. France*. Be Br Fa Ga Hb Is No.

Subsect. *Triplinervium* (Gaudin) Gornall. Evergreen perennials, without bulbils or summer-dormant buds. Most of the leaves lobed.

80. S. rosacea Moench, *Meth.* 106 (1794) (*S. decipiens* Ehrh., *S. cespitosa* auct. eur. med., non L., *S. cespitosa* subsp. *decipiens* (Ehrh.) Engler & Irmscher, *S. groenlandica* sensu P. Fourn., non L.). Very variable in foliage and habit; leafy shoots varying from short and suberect, forming a compact cushion, to procumbent and rather long, forming a loose mat. Leaves 8–25 mm, petiolate; lamina rhombic-cuneate, usually 5-lobed, densely hairy to nearly glabrous, but always with some hairs, at least on the petiole. Flowering stem 4–25 cm, fairly stout, with 1–4 cauline leaves and 2–6 flowers erect in bud. Petals 6–10 × 3–7 mm, pure white. *Rocks, screes and grassy ledges.* • *N.W. & C. Europe; local.* Be †Br Cz Fa Ga Ge Hb Is Po.

1 Most of the hairs on lamina glandular (c) subsp. **hartii**
1 Hairs on lamina eglandular or absent
  2 Leaf-segments obtuse, acute or slightly mucronate (a) subsp. **rosacea**
  2 Leaf-segments strongly mucronate to apiculate (b) subsp. **sponhemica**

(a) Subsp. **rosacea**: Leafy shoots procumbent to suberect. Leaves subglabrous to densely hairy with non-glandular hairs; segments obtuse, acute or shortly mucronate, fairly broad. Seeds variable. 2*n* = 52, 56, 64. *S. & C. Germany and E. France; Ireland; Iceland and Faeröer; extinct in Britain.*

(b) Subsp. **sponhemica** (C. C. Gmelin) D. A. Webb, *Feddes Repert.* **68**: 210 (1963) (*S. sponhemica* C. C. Gmelin): Leafy shoots ascending to erect. Leaves rather sparsely furnished with non-glandular hairs; segments strongly mucronate to apiculate, narrow. Seeds coarsely papillose. 2*n* = 50, 52. *Belgium*, *E. France and W. Germany; Czechoslovakia and S.W. Poland.*

(c) Subsp. **hartii** (D. A. Webb) D. A. Webb, *Bot. Jour. Linn. Soc.* **95**: 246 (1987) (*S. hartii* D. A. Webb): Leafy shoots suberect. Leaves patent, forming flat rosettes, usually rather densely glandular-hairy; segments 5–7, subacute, fairly broad. Seeds finely tuberculate. 2*n* = *c.* 50. *N.W. Ireland* (*Arranmore Island*).

Subsp. (c) is intermediate between subsp. (a) and **81**, but closer to the former.

81. S. cespitosa L., *Sp. Pl.* 404 (1753) (*S. groenlandica* L.). Leafy shoots short, more or less erect, forming lax or rather dense cushions. Leaves mostly 3-lobed (rarely simple or 5-lobed), cuneate, with a broad but distinct petiole and oblong, obtuse or subacute lobes, densely covered with rather short glandular hairs; young leaves somewhat incurved. Flowering stem 2–10 cm, bearing 1–3(–5) flowers. Petals *c.* 6 mm, dull white, often slightly greenish or cream-coloured. Seeds very finely tuberculate. $2n = 78, 80$. *Arctic and subarctic Europe, extending locally southwards in the mountains to* 53° *N. in Britain and* 50° *N. in C. Ural.* Br Fa Fe Is No Rs(N, C) Sb Su.

82. S. pubescens Pourret, *Mém. Acad. Sci. Toulouse* **3**: 327 (1788) (*S. mixta* Lapeyr., pro parte). Very variable in habit; usually densely caespitose in compact cushions, but sometimes lax and spreading. Rosette-leaves rather deeply 5-fid (less often 3-fid) with parallel-sided, obtuse, sulcate segments, rather dark green, densely covered with long glandular hairs. Cauline leaves usually divided into narrow segments, rarely entire. Petals 4–6 mm, broadly obovate, usually contiguous or overlapping. *Siliceous rocks and screes, mostly above* 2400 *m.* ● *E. & C. Pyrenees.* Ga Hs.

(a) Subsp. **pubescens** (subsp. *pourretiana* Engler & Irmscher, *S. obscura* Gren. & Godron): Habit usually relatively lax, with short leafy shoots on which the dead leaves do not persist for long. Leaves 10–18 mm, with long, narrow petiole and deeply sulcate segments 3–4 times as long as wide, often somewhat divergent. Petals pure white; anthers yellow; top of ovary green. $2n = 28$. *Screes, often in late snow patches. E. Pyrenees.*

(b) Subsp. **iratiana** (F. W. Schultz) Engler & Irmscher in Engler, *Pflanzenreich* **67(IV.117)**: 401 (1916) (*S. iratiana* F. W. Schultz, *S. groenlandica* auct. pyren., non L.): Habit very compact, with long, slender, columnar leafy shoots, on which the dead leaves persist for many years. Leaves 4–10 mm, with short, broad petiole and nearly parallel segments, twice as long as wide, less deeply sulcate than in subsp. (a). Petals sometimes veined with red; anthers and top of ovary reddish. *Exposed rocks. C. Pyrenees.*

83. S. nevadensis Boiss., *Diagn. Pl. Or. Nov.* **3(2)**: 67 (1856) (*S. pubescens* subsp. *nevadensis* (Boiss.) Engler & Irmscher). Like **82**, and intermediate in habit between the two subspecies, but differing by its leaves with usually 3 non-sulcate segments; calyx deep purplish-red; petals often red-veined; anthers red or orange. $2n = 58$. *Shady rocks and screes above* 2600 *m.* ● *S. Spain (Sierra Nevada).* Hs.

84. S. cebennensis Rouy & Camus, *Fl. Fr.* **7**: 55 (1901) (*S. prostiana* (Ser.) Luizet). Leafy shoots numerous, slender, forming a fairly large, domed, compact but soft cushion. Leaves (including petiole) *c.* 12 mm, light green, mostly 3-lobed, but sometimes 5-lobed or entire, rather densely covered with glandular hairs; segments sulcate, obtuse. Flowering stem 5–8 cm, with 3 or 4 usually entire cauline leaves and 2 or 3 flowers. Petals 6–8 mm, broadly obovate, contiguous, pure white. $2n = 26, 32$. *Shady limestone rocks.* ● *S. France (Cevennes).* Ga.

85. S. exarata Vill., *Prosp. Pl. Dauph.* 47 (1779) (*S. muscoides* Wulfen, non All.; incl. *S. varians* Sieber, *S. moschata* Wulfen, *S. tenuifolia* Rouy & Camus, *S. adenophora* C. Koch). Leafy shoots numerous, more or less erect, forming a fairly dense but sometimes rather flat cushion. Leaves 4–20 mm, usually with fairly distinct petiole, subglabrous to rather densely glandular-hairy. Lamina usually 3-lobed; more rarely 5- to 7-lobed or entire; lobes oblong, obtuse. Flowering stem 3–10 cm, branched in upper $\frac{1}{3}$ to give a small, often corymbose cyme of up to 8 flowers. Petals variable in size, shape and colour. $2n = 20, 22, 24, 26, 28, 32, 34, 36, 44, 48, 52$, *c.* 68. *Rocks, screes and mountain slopes with semi-open vegetation. Mountains of C. & S. Europe, from the Riesengebirge (Krkonoše) southwards to N. Spain, C. Italy and S. Greece.* Al Au Bu Cz Ga Ge Gr He Hs It Ju Po Rm.

Very variable. The taxa treated below as subspp. (a) and (c) are treated by most authors as separate species, but the supposed correlation of distinguishing characters is so imperfect that many specimens cannot be assigned to either. The arrangement here proposed must be considered provisional, and it may be that some other subspecies deserve recognition, but it deals with the variation on a European scale in a manner not attempted by most national Floras.

No clear correlation between chromosome number and morphology or geographical distribution is as yet apparent.

1 Leaf-segments sulcate
  2 Petals 4–6 mm, broadly obovate, nearly contiguous, white — (a) subsp. **exarata**
  2 Petals 3·5–4·5 mm, oblong to narrowly elliptical, widely separated, pale, dull yellow — (b) subsp. **pseudoexarata**
1 Leaf-segments not sulcate
  3 Petals 3–4 mm, oblong to narrowly elliptical, widely separated, yellowish, sometimes tinged with red — (c) subsp. **moschata**
  3 Petals 4–5 mm, obovate, nearly contiguous, white
    4 Inflorescence usually with 3–5 flowers; leaves not fleshy, entire to 5-lobed, usually with distinct petiole — (d) subsp. **lamottei**
    4 Inflorescence usually with 1–3 flowers; leaves fleshy, undivided or with very short, parallel lobes; petiole not distinct — (e) subsp. **ampullacea**

(a) Subsp. **exarata** (incl. subsp. *leucantha* (Thomas) Br.-Bl.): Leaves rather densely glandular-hairy, with 3–5(–7) sulcate, often widely divergent segments. Petals 4–6 mm, broadly obovate, nearly contiguous, white or pale cream. *Somewhat calcifuge. S.W. & C. Alps; Balkan peninsula from Crna Gora to Greece; perhaps locally elsewhere.*

(b) Subsp. **pseudoexarata** (Br.-Bl.) D. A. Webb, *Bot. Jour. Linn. Soc.* **95**: 247 (1987) (*S. moschata* subsp. *pseudoexarata* Br.-Bl.). Leaves moderately densely glandular-hairy, with usually 3 sulcate, slightly divergent segments. Petals 3·5–4·5 mm, oblong to narrowly elliptical, widely separated, pale, dull yellow. ● *E. & S.C. Alps; N. Appennini; S. part of Balkan peninsula.*

(c) Subsp. **moschata** (Wulfen) Cavillier in Burnat, *Fl. Alp. Marit.* **5**: 81 (1913) (*S. moschata* Wulfen, *S. planifolia* Lapeyr.). Leaves subglabrous to moderately densely glandular-hairy, entire or with 3 subparallel segments, which are not sulcate. Petals 3–4 mm, oblong to narrowly elliptical, widely separated, yellowish, sometimes tinged with red. *Somewhat calcicole. Throughout most of the range of the species but apparently absent from the Appennini and the S. part of the Balkan peninsula.*

(d) Subsp. **lamottei** (Luizet) D. A. Webb, *Bot. Jour. Linn. Soc.* **95**: 248 (1987) (*S. lamottei* Luizet). Leaves fairly thin, entire to 5-lobed, sparsely glandular-hairy. Inflorescence usually with 3–5 flowers. Petals 4–5 mm, obovate, nearly contiguous, white. ● *S.C. France (Auvergne).*

(e) Subsp. **ampullacea** (Ten.) D. A. Webb, *loc. cit.* (1987) (*S. ampullacea* Ten.): Leaves fleshy, subglabrous, undivided or with 3 very short, subparallel lobes, not sulcate; petiole not distinct from undivided part of lamina. Inflorescence with 1–3 flowers. Petals 4–5 mm, obovate, nearly contiguous, white. ● *Higher peaks of C. Appennini.*

86. S. hariotii Luizet & Soulié, *Bull. Soc. Bot. Fr.* **58**: 638 (1912). Leafy shoots up to 6 cm, but often less, more or less erect, forming a dense to rather lax cushion. Leaves 5–9 mm, undivided or with 3 conspicuously sulcate, mucronate segments, subglabrous to sparsely glandular-hairy. Flowering stem 3–7 cm, with or without a single cauline leaf, and with a narrow panicle of 3–12 flowers. Petals *c.* 4 mm, oblong, not contiguous, creamy-white, usually with reddish-brown veins. *Limestone rocks and screes*, 1600–2500 *m*. ● *W. Pyrenees.* Ga Hs.

87. S. praetermissa D. A. Webb, *Feddes Repert.* **68**: 204 (1963) (*S. ajugifolia* auct., non L.). Leafy shoots decumbent, forming a lax mat or low cushion. Leaves up to 15 × 10 mm (including petiole), sparsely hairy; lamina with 3–5 oblong-oblanceolate, usually acute or mucronate lobes. Flowering stem 6–15 cm, slender, arising from the axils of the leafy shoots, usually 4–7 cm below the apex; cauline leaves few and small. Flowers 1–3. Petals 4–5 mm, oblong-elliptical, white. Capsule long and narrow. *Fine screes and rock-debris, usually in late snow-lies, above* 1500 *m; somewhat calcifuge.* ● *Pyrenees, Picos de Europa.* Ga Hs.

88. S. wahlenbergii Ball, *Bot. Zeit.* **4**: 401 (1846) (*S. perdurans* Kit.). Glabrous except for a few inconspicuous, fleshy, appressed hairs on lower side of leaves and in inflorescence. Stems decumbent, forming a mat or low cushion. Leafy shoots short; leaves 7–12 × 3–5 mm; petiole broad, confluent with the cuneate lamina, which is divided distally into 3–5 short, oblong-triangular, obtuse lobes. Flowering stem 4–7 cm, slender, terminal but appearing axillary because of the growth of an axillary shoot from near the base. Petals 4–5 mm, white. Capsule long and narrow. $2n = 66$. *Damp, grassy slopes*, 1000–2500 *m*. ● *W. Carpathians.* Cz Po.

89. S. pedemontana All., *Fl. Pedem.* **2**: 73 (1785). Stems woody at the base; leafy shoots ascending to suberect, forming rather lax cushions. Leaves somewhat fleshy or coriaceous; lamina shortly glandular-pubescent; petiole with much longer, partly non-glandular hairs. Petiole 6–18 mm; lamina 8–15 × 9–20 mm, divided into 3–9 segments. Flowering stem 5–18 cm, terminal, sparingly branched above to give a narrow panicle of 2–12 flowers; cauline leaves few, often absent. Flowers infundibuliform. Petals 10–21 × 4–8 mm, oblanceolate, white, sometimes veined or tinged with red. Styles 5–7 mm, suberect. *Shady rocks. S. Europe, mainly in the mountains, extending northwards to the Ukrainian Carpathians.* Bu Co Ga Gr It Ju Rm Rs(W) Sa.

Very variable, and composed of four geographically isolated populations in Europe; a fifth is found in N. Africa. Each of these has a generally distinctive facies, but variation is sufficient to bring about overlap of all characters, and they are therefore given subspecific rather than specific rank.

1 Leaf-segments obtuse to subacute, usually not more than twice as long as wide, seldom divaricate
2 Leaves somewhat fleshy or coriaceous; segments subacute; flowering stem up to 18 cm; petals 15–21 mm (a) subsp. **pedemontana**
2 Leaves soft and fairly thin; segments obtuse; flowering stem seldom more than 8 cm; petals 9–15 mm (b) subsp. **cymosa**
1 Leaf-segments acute to mucronate, rarely obtuse, usually more than twice as long as wide, often divaricate
3 Rosette-leaves incurved in bud, usually sparsely hairy; petiole usually shorter than the lamina (c) subsp. **cervicornis**
3 Rosette-leaves not incurved in bud, usually fairly densely hairy; petiole usually longer than the lamina (d) subsp. **prostii**

(a) Subsp. **pedemontana**: Robust, and larger in all its parts than the other subspecies. Lamina flabelliform, tapered gradually into a fairly narrow petiole; segments short and fairly wide, subacute. Flowering stem up to 18 cm. Petals 15–21 × 6–8 mm, sometimes tinged with pink at the base. $2n = 42$. ● *S.W. Alps.*

(b) Subsp. **cymosa** Engler in Engler & Prantl, *Natürl. Pflanzenfam.* **3(2a)**: 55 (1891) (*S. cymosa* auct., nom. invalid.): Leaves soft and fairly thin; lamina tapered gradually into a wide petiole; segments short and wide, forwardly directed, obtuse. Flowering stem seldom more than 8 cm. Petals 9–15 × 3–5 mm. ● *E. & S. Carpathians; mountains of the Balkan peninsula from E.C. Jugoslavia to N. Greece.*

(c) Subsp. **cervicornis** (Viv.) Engler in Engler & Prantl, *loc. cit.* (1891) (*S. cervicornis* Viv.): Leaves somewhat coriaceous; lamina semicircular, narrowed suddenly into a slender petiole; segments 2½–4 times as long as wide, divaricate, obtuse to mucronate; rosette-leaves incurved when young. Flowering stem up to 15 cm. Petals 10–13 × 4–5 mm. $2n = 26, 42, 44$. ● *Corse, Sardegna.*

(d) Subsp. **prostii** (Sternb.) D. A. Webb, *Feddes Repert.* **68**: 205 (1963) (*S. ajugifolia* L., nom. ambiguum, *S. prostii* Sternb., *S. pedatifida* auct., vix Sm.): Lamina cuneate; segments 2–3 times as long as wide, not divaricate, acute to mucronate; rosette-leaves not incurved. Flowering stem up to 18 cm. Petals 9–12 × 2·5–4 mm. $2n = 32$. ● *S. France (Cevennes).*

90. S. babiana Díaz & Fernández Prieto, *Anal. Jard. Bot. Madrid* **39**: 249 (1983). Leafy shoots numerous, erect, crowded, forming a dense cushion; young leaves erect, forming a brush-like tuft. Leaves up to 37 mm (including petiole), furnished with rather long glandular hairs, at least on the petiole and sometimes all over. Lamina deeply divided into 3 primary lobes, each consisting of 3 short, strongly apiculate segments. Lower side of petiole crimson, at least towards the base. Flowering stem 10–20 cm, slender, ascending, axillary; cauline leaves *c.* 5. Flowers 5–9. Petals 6–8 mm, white. *Limestone rocks*, 1050–1750 *m*. ● *N.W. Spain (W. part of Cordillera Cantábrica).* Hs.

The typical plant has leaves glandular-hairy all over, and with the petiole crimson up to the lamina. Var. *septentrionalis* Díaz & Fernández Prieto, which is at least as common, has the hairs and the crimson pigment confined to the base of the petiole. It may possibly have arisen by introgression from **98**.

91. S. geranioides L., *Cent. Pl.* **1**: 10 (1755). Stems woody, branched, bearing lax cushions of light green foliage. Leaves herbaceous; petiole up to 40 mm, slender; lamina 15 × 25 mm, semicircular to suborbicular, divided rather deeply into 3 primary lobes which are further divided, giving 13–27 triangular-lanceolate, acute segments. Lamina glandular-puberulent with very short hairs (*c.* 0·15 mm). Flowering stem 15–25 cm, terminal, bearing a corymb of up to 20 slightly scented, campanulate flowers. Sepals up to 7 mm in fruit; petals 12 × 4 mm, oblanceolate, white. $2n = c.\ 52$. *Rocks and screes; calcifuge.* ● *E. Pyrenees and N.E. Spain.* Ga Hs.

Dwarf plants, mainly from high altitudes, with greyish leaves bearing much longer hairs, have been described as var. *palmata* (Lapeyr.) Engler & Irmscher, but it is possible that they are hybrids.

92. S. moncayensis D. A. Webb, *Feddes Repert.* **68**: 201 (1963). Leafy shoots long and very numerous, forming soft, fairly dense, deep cushions up to 150 cm across of light green, slightly scented foliage. Plant covered with very short glandular hairs (0·15–0·2 mm) giving a verruculose rather than pubescent appearance to the naked eye. Petiole 8–13 mm; lamina 8–11 × 9–15 mm, deeply divided into 3 primary lobes, which are narrowly oblong to oblanceolate, obtuse and deeply sulcate; the lateral lobes occasionally bear short secondary lobes. Flowering stem 5–10 cm, terminal; inflorescence with 3–7 primary branches, each bearing 3–5 flowers. Petals 6–7 mm, broadly oblong. Ovary almost inferior, surmounted by a bright yellow nectariferous disc. $2n = c.\ 60$. *Shady, siliceous rocks at* 1000–1600 *m*. ● *Mountains of* N.E. *Spain*. Hs.

93. S. vayredana Luizet, *Bull. Soc. Bot. Fr.* **60**: 413 (1913). Like **92** but foliage with a much stronger spicy or balsamic scent; leaves somewhat smaller, and with 3–9 acute segments, which are not sulcate; panicle smaller and more compact, with not more than 9 flowers; petals obovate. $2n = c.\ 64$. *Shady, siliceous rocks and screes,* 700–1600 *m*. ● N.E. *Spain* (*Sierra de Montseny*). Hs.

94. S. intricata Lapeyr., *Fig. Fl. Pyr.* 58 (1801) (*S. nervosa* Lapeyr., *S. mixta* Lapeyr., pro parte). Stems woody; leafy shoots short, forming small tufts or cushions. Leaves rather rigid, darkish green, very shortly glandular-pubescent, with hairs as in **92** and **93**. Lamina 5–7 × 5–10 mm, with 3–5(–9) linear-oblong, obtuse, sulcate segments. Flowering stem 5–10 cm, bearing a panicle of 3–12 flowers. Petals 4–6 mm, broadly obovate to suborbicular, contiguous, patent, white. $2n = 34$. *Exposed, often dry, siliceous rocks. E. & C. Pyrenees; one station in the Cordillera Cantábrica*. Ga Hs.

95. S. pentadactylis Lapeyr., *op. cit.* 64 (1801). Glabrous; leaves, shoots, flowering stem and hypanthia covered with sessile glands and often viscid. Stems woody at the base; leafy shoots rather few, forming small cushions. Petiole usually longer than the lamina, which is divided to the base into 3 lobes, each further divided into 2 or 3 segments; segments 3–10 × 0·5–1·5 mm, linear to oblong, obtuse, sulcate. Flowering stem 7–17 cm, terminal, bearing 5–50 flowers. Petals 3·5–5 mm, obovate to oblong. Ovary inferior. *Rocks, usually shady, but often dry, usually above* 1800 *m*. ● *E. Pyrenees and mountains of* N. & C. *Spain*. Ga Hs.

A variable species, of which the circumscription is still rather uncertain. The following treatment is provisional; more information is needed, especially about subspp. (**c**) and (**d**).

1 Petals *c.* 3·5 mm, greenish-yellow (**d**) subsp. **almanzorii**
1 Petals usually 4·5–5 mm, white
 2 Petiole longer than the lamina (**a**) subsp. **pentadactylis**
 2 Petiole about as long as the lamina
  3 Leaves fairly soft; petiole not wider than the widest leaf-segment; calcifuge (**b**) subsp. **willkommiana**
  3 Leaves hard and rigid; petiole much wider than leaf-segments, sometimes nearly as wide as the lamina; calcicole (**c**) subsp. **losae**

(**a**) Subsp. **pentadactylis**: Cushions small, rather lax; leaves dark green, hard and rigid; petiole *c.* 1 mm wide, longer than the lamina; leaf-segments 1–1·5 mm wide, linear-oblong, divaricate. Petals *c.* 4·5 mm, usually contiguous, white. $2n = 16 + 5B$. *Siliceous rocks. Pyrenees, and rarely in the mountains further south.*

(**b**) Subsp. **willkommiana** (Boiss. ex Engler) Rivas Martínez, *Anal. Inst. Bot. Cavanilles* **21**: 229 (1963): Cushions larger, fairly compact; leaves medium green, fairly soft; petiole *c.* 1·5 mm wide, about equalling, or shorter than, the lamina; leaf-segments 1·5–2 mm wide, oblong, divaricate. Petals *c.* 5 mm, white. $2n = 32$. *Siliceous rocks. Mountains of* N. & C. *Spain, from the Cordillera Cantábrica and the Cordillera Ibérica to the Sierra de Gredos.*

(**c**) Subsp. **losae** (Sennen) D. A. Webb, *Bot. Jour. Linn. Soc.* **95**: 249 (1987) (*S. losae* Sennen, *S. pentadactylis* subsp. *losana* (Sennen) Malagarriga): Cushions small, very compact, with more or less columnar leafy shoots. Leaves hard and rigid; petiole about equalling the lamina and sometimes nearly as wide, much wider than any of its segments, which are short and forwardly directed. Petals probably *c.* 4 mm, white. *Limestone rocks, mostly at* c. 1000 *m. Mountains of* N.E. *Spain, from Alava prov. to Teruel prov.*

(**d**) Subsp. **almanzorii** Vargas, *Anal. Jard. Bot. Madrid* **43**: 457 (1987): Petiole much longer than the lamina, and much wider than its segments, which are 0·5–1 mm wide, short and slightly divaricate. Petals 3·5 × 1·5 mm, not contiguous, greenish-yellow. *Siliceous rocks. Sierra de Gredos.*

96. S. fragilis Schrank, *Pl. Rar. Horti Monac.*, t. 92 (1822) (*S. corbariensis* Timb.-Lagr.). Glabrous and with sessile glands, as in **95**, but only slightly viscid. Stems woody at the base, sparingly branched, forming rather lax cushions. Leaves shining, rigid, brittle and rather coriaceous, 12–35 mm wide, divided almost to the base into 3 primary lobes; these are further divided, giving 5–9(–13) narrowly oblong, flat (not sulcate), obtuse to subacute segments. Flowering stem 10–22 cm, with 5–20 flowers. Petals 7–14 × 4–7 mm, white. *Limestone rocks and screes, usually shaded, up to* 2000 *m, but mostly below* 1000 *m*. ● *E. Spain and S. France*. Ga Hs.

(**a**) Subsp. **fragilis** (*S. geranioides* subsp. *corbariensis* (Timb.-Lagr.) Rouy & Camus): Leaves 18–30 mm, wide, usually with 7–11 segments. Flowering stem 10–25 cm. Petals 9–13 mm, oblanceolate; stamens at least 3 mm longer than the sepals. $2n = 60–66$. *S. France (Corbières) and* N.E. *Spain.*

(**b**) Subsp. **valentina** (Willk.) D. A. Webb, *Bot. Mag.* **180**: 186 (1975) (*S. valentina* Willk., *S. corbariensis* subsp. *valentina* (Willk.) D. A. Webb): Leaves 12–25 mm wide, usually with 3–7 segments. Flowering stem 6–15 cm. Petals 7–11 mm, obovate-elliptical; stamens exceeding the sepals by 1 mm or less. $2n = 64$. *E. Spain, from C. Teruel to E. Jaén.*

Between c. 40°30′ and 42° N. plants intermediate between the two subspecies are common. It is to such plants that the name *S. paniculata* Cav. (non Miller) was originally given.

97. S. camposii Boiss. & Reuter, *Pugillus* 47 (1852). Like **96**(**b**) but leaves more rigid, seldom more than 10 mm wide, with 3–5(–9) acute to mucronate or shortly apiculate segments. $2n = 64$. *Shady limestone rocks and screes,* 900–2100 *m*. ● *S. & S.E. Spain, from the eastern edge of Málaga prov. to the Sierra de Alcaraz*. Hs.

The typical plant is found mainly on the south-western part of the range; it has a cuneate lamina, usually with 5 segments, tapered gradually to a broad petiole. Plants referable to var. *leptophylla* Willk. (subsp. *leptophylla* (Willk.) D. A. Webb) occur in the central and north-eastern parts of the range; they have a narrow petiole and a nearly semicircular lamina, often wider than long. In plants of this variety many of the leaf-segments are feebly or obscurely mucronate, making it hard to distinguish them from **96(b)**. It might, perhaps, be better to reduce **97** to a third subspecies of **96**.

98. S. trifurcata Schrader, *Hort. Gotting.* 13 (1809). Glabrous and with sessile glands, as in **95–97**; occasionally very viscid, but more usually not at all. Leafy shoots numerous, forming a compact, rather hard cushion. Leaves 10–20(–30) mm wide, reniform to semicircular, with (9–)11–17 flat, triangular, apiculate segments, which are often crowded and sometimes overlapping, the lateral ones usually falcate-recurved. Flowering stem 8–30 cm, axillary. Flowers 5–15, in a lax cyme. Petals 8–11 mm, elliptic-oblong. $2n = 28$. *Limestone rocks, walls and roofs.* ● *N. Spain, from* 1°30′ *to* 7° *W.* Hs.

99. S. canaliculata Boiss. & Reuter ex Engler, *Monogr. Gatt. Saxifr.* 169 (1872) (*S. paui* Merino). Like **98** but always very viscid; leaves rhombic to semicircular, with 5–11 linear-oblong, straight, apiculate segments, broadly and deeply channelled above, not overlapping or falcate-recurved; flowering stem terminal, not more than 15 cm; petals broadly obovate, contiguous, reflexed at the apex. $2n = 52$. *Limestone rocks.* ● *W. part of Cordillera Cantábrica.* Hs.

100. S. cuneata Willd., *Sp. Pl.* **2(1)**: 658 (1799). Glabrous and with sessile glands, as in **95–99**; usually viscid. Leafy shoots forming a fairly lax cushion. Leaves up to 22–26 mm, rigid and rather coriaceous, rhombic to flabelliform, divided about half-way into 3 broadly triangular to ovate, obtuse or mucronate lobes, which sometimes bear short secondary lobes. Flowering stem 7–30 cm, axillary, bearing a narrow panicle of 7–15 flowers. Petals 5–7 mm, obovate, white. $2n = 28, 36$. *Limestone rocks, screes and walls,* 750–1900 *m.* ● *W. Pyrenees and N. Spain, westwards to* 4°45′ *W.* Ga Hs.

101. S. aquatica Lapeyr., *Fig. Fl. Pyr.* 53 (1801) (*S. adscendens* auct., non L.). Larger and coarser in habit than most other species. Leafy shoots decumbent, freely branched, forming deep mats of foliage up to 2 m across. Leaves furnished with fairly long hairs; petiole up to 35 mm; lamina up to 25 × 35 mm, more or less semicircular, deeply 3-lobed, the lobes repeatedly but not very deeply divided, giving 15–27 triangular, acute segments. Flowering stem 25–60 cm, 2–3(–5) mm in diameter, axillary, with numerous cauline leaves and narrow panicle with numerous, rather crowded flowers. Petals 7–9 mm, narrowly obovate, white (rarely pale yellow). $2n = 28, 66$. *Margins of fast-flowing streams on siliceous rocks,* 1500–2400 *m.* ● *E. & C. Pyrenees.* Ga Hs.

102. S. × capitata Lapeyr., *op. cit.* 55 (1801) (*S. aquatica* × *praetermissa*). Intermediate between the parent species, but in most characters closer to **101**. Leaves 12–20 mm wide, rhombic, mostly 5- to 7-lobed. Flowering stem 15–25 cm, usually few, not more than 2 mm in diameter; panicle lax, often nodding, with 7–25 flowers. Petals 6–7 mm. *Margins of mountain streams; fairly frequent where both parents grow together, and sometimes seen in the vicinity of one parent only.* ● *Pyrenees.* Ga Hs.

Although this hybrid is fertile it flowers sparingly, and may be distinguished at a distance from **101** by this character, as well as by its less robust growth and foliage.

103. S. irrigua Bieb., *Fl. Taur.-Cauc.* **2**: 460 (1808). Leafy shoots few, forming a dense tuft. Leaves up to 40 mm wide, with usually numerous, rather long hairs, 3-lobed almost to the base, the lobes again divided, giving 9–25 elliptical, oblong or triangular, subacute segments. Petiole long and broad. Flowering stem 10–30 cm, usually branched in the upper ½ only, bearing a panicle of numerous flowers. Sepals 4–5 mm, linear-oblong; petals 12–15 mm, oblanceolate, white. $2n = 44$. *Damp or shady places in the mountains.* ● *Krym.* Rs(K).

104. S. latepetiolata Willk. in Willk. & Lange, *Prodr. Fl. Hisp.* **3**: 120 (1874). A glandular-hairy, somewhat viscid biennial, consisting in the first year of a hemispherical dome of regularly arranged leaves; from the centre of this arises in the second year a stout, reddish flowering stem 15–30 cm, branched from near the base to form a narrowly pyramidal, many-flowered panicle. Petiole of basal leaves 15–50 × 3–4 mm, stiff and rather brittle; lamina 8–15 × 10–27 mm, reniform, divided half-way to the base into 5–7 cuneate, truncate to mucronate lobes. Petals 7–10 mm, narrowly obovate, white. Ovary almost completely inferior. $2n = c. 66$. *Vertical limestone rocks, usually shaded but often dry.* ● *E. Spain (Valencia region); local.* Hs.

105. S. petraea L., *Sp. Pl.* ed. 2, 578 (1762) (*S. alpina* Degen). Biennial, perhaps rarely perennial, covered with soft, glandular hairs. Leaves of basal rosette petiolate, semicircular or rhombic in outline, divided almost to the base into numerous toothed lobes; cauline leaves much less deeply divided. Stems 10–30 cm, weak, often decumbent, with numerous, long, divaricate, interwoven branches, forming a very lax, wide, leafy panicle. Petals 8–10 mm, contiguous, sometimes slightly unequal, emarginate, white. Seeds papillose. *Shady rocks; calcicole.* ● *Southern foothills of the E. Alps.* It Ju.

106. S. berica (Béguinot) D. A. Webb, *Feddes Repert.* **68**: 202 (1963) (*S. petraea* var. *berica* Béguinot). Like **105** but usually perennial; hairs shorter; basal leaves regularly crenate or dentate, or divided less than half-way to the base into 5–11 entire lobes; leaves often tinged with brown; petals 4–8 mm, not contiguous, usually very unequal; seeds smooth. *Shady recesses in limestone rocks.* ● *N. Italy (Colli Berici, near Vicenza).* It.

107. S. arachnoidea Sternb., *Revis. Saxifr.* 23 (1810). Stems weak, decumbent; whole plant covered with an intricate tomentum of long, viscid, arachnoid hairs. Leaves rhombic or ovate-cuneate, very variable, mostly palmately 3- to 7-lobed; lobes obtuse; petioles short. Flowers on long pedicels in small terminal cymes. Petals 3 mm, translucent, dirty white or straw-coloured. Ovary inferior. $2n = 56$. *Caverns in limestone rocks.* ● *S. Alps (a small region N.W. of Lago di Garda).* It.

108. S. paradoxa Sternb., *Revis. Saxifr.* 22 (1810) (*Zahlbrucknera paradoxa* (Sternb.) Reichenb.). Nearly glabrous, with fragile, ascending or decumbent stems. Leaves reniform, palmately lobed or crenate, petiolate, shining, thin. Flowers in small, lax, leafy, axillary cymes. Petals 1·5 mm, linear-oblong, pale green, similar to the sepals but smaller, not narrowed at the base. Ovary inferior, surmounted by a large disc. $2n = 64$. *Shady recesses in non-calcareous rocks.* ● *Foothills of S.E. Alps (Kärnten, Steiermark, Slovenija).* Au Ju.

Subsect. *Holophyllae* Engler & Irmscher. Evergreen perennials, usually of small size; leaves entire or shortly 3-lobed at the apex. Leafy shoots forming cushions or mats. Bulbils and summer-dormant buds absent. Petals white or pale yellow.

109–112. S. androsacea group. A group of four closely related species of distinctive habit. The leafy shoots are short, with terminal flowering stems, and the growth of the following year has its origin in axillary buds towards the base of the previous year's growth. The foliage, therefore, forms a flat turf rather than a mat or cushion, and all the leaves appear to be more or less basal.

1 Petals dull yellow **112. seguieri**
1 Petals white
 2 Larger leaves 6–9 mm wide; inflorescence with 3–7 flowers **110. depressa**
 2 Leaves not more than 6 mm wide; inflorescence with 1–3 flowers
  3 Leaves mostly 3-lobed, densely covered with short glandular hairs; flowering stem 2–4 cm **111. italica**
  3 Many of the leaves usually entire, with fairly long, mainly non-glandular hairs on the margin alone; flowering stem up to 10 cm **109. androsacea**

109. S. androsacea L., *Sp. Pl.* 399 (1753). Leaves 8–30 × 3–6 mm, including the scarcely distinct petiole, oblanceolate-spathulate, mostly entire, but occasionally some or all are very shortly 3-lobed, furnished with fairly long, mostly eglandular hairs on the margins; the upper surface of mature leaves more or less glabrous. Flowering stem 2–10 cm, bearing 1 or 2 cauline leaves and 1–3 flowers. Petals 4–7 mm, oblong to narrowly obovate, white. $2n = 16, 56, 66, 88$, *c.* 112, *c.* 120, *c.* 128, 154, *c.* 192, 206–220. *Snow-patches and damp, stabilized screes, mainly* 2000–2800 *m. Mountains of C. & S. Europe, from Auvergne and the W. Carpathians to the Pyrenees and S.W. Bulgaria.* Al Au Bu Cz Ga Ge He Hs It Ju Po Rm Rs(W).

110. S. depressa Sternb., *Revis. Saxifr.* 42 (1810). Leafy shoots rather longer than in other species of the group. Leaves 10–30 × 6–9 mm, including the usually ill-defined petiole, broadest near the apex, which is divided into 3 short, obtuse lobes, densely glandular-pubescent. Flowering stem up to 10 cm, with or without a cauline leaf and with 3–7 flowers. Petals 4–5 mm, oblong, white. *Shady ledges and damp screes,* 2000–2850 *m, mostly on porphyritic volcanic rock.* ● *S. Alps (Alpi Dolomitiche).* It.

111. S. italica D. A. Webb, *Feddes Repert.* **68**: 209 (1963) (*S. tridens* (Jan ex Engler) Engler & Irmscher, non Haw.). Like **110** but smaller in all its parts; leafy shoots *c.* 1 cm; leaves 6–15 × 2–4 mm, without a distinct petiole, oblanceolate, mostly shortly 3-lobed; flowering stem 2–4 cm with 1 or 2 cauline leaves and 1–3 flowers. $2n = 64$, 66. *Limestone rocks and screes,* 2000–2500 *m. C. Appennini.* It.

112. S. seguieri Sprengel, *Fl. Hal.* 40 (1807). Leaves 10–25 × 1–3 mm, oblanceolate, entire, obtuse, densely glandular-pubescent. Flowering stem 2–7 mm, usually with 1 cauline leaf and 1–3 flowers. Petals 2–3 mm, dull, pale yellow, narrower than the sepals but usually slightly longer. $2n = 66$. *Damp screes and stony slopes, often in late snow-lies,* 2000–3000 *m; somewhat calcifuge.* ● *Alps.* Au Ga He It.

113. S. muscoides All., *Auct. Syn. Stirp. Horti Taur.* 35 (1773) (*S. planifolia* auct., non Lapeyr.). Shoots short, erect, crowded, forming a dense, soft cushion. Leaves 6 × 1·5 mm, linear-lanceolate, entire, obtuse, glandular-pubescent but not very viscid; dead leaves persistent, conspicuously silvery-grey in apical ½ when dry. Flowering stem 5 cm or less, with several leaves and 1–3 flowers. Petals 4 mm, broadly obovate, contiguous, obtuse or truncate, white or pale lemon-yellow. $2n = 38$. *Rocks and screes, usually above* 2200 *m; somewhat calcicole.* ● *Alps, from Monte Viso to Hohe Tauern.* Au Ga He It.

114. S. facchinii Koch, *Flora (Regensb.)* **25**: 624 (1842). Like **113** but forming smaller cushions; flowering stem very short, scarcely projecting above leafy shoots, and often 1-flowered; petals oblong-cuneate, 2 × 1 mm or less, scarcely exceeding the sepals, dull yellow tinged to a variable extent with dull purplish-red. *Rocks and screes above* 1800 *m; calcicole.* ● *S. Alps (Alpi Dolomitiche).* It.

115. S. presolanensis Engler, *Pflanzenreich* **67(IV.117)**: 302 (1916). Leafy shoots numerous, erect, columnar, with numerous persistent dead leaves, forming a deep, soft, dense cushion. Leaves *c.* 15 mm including petiole, elongate-spathulate or oblanceolate, entire, obtuse to acute, pale green, rather densely covered with long, viscid, glandular hairs; dead leaves greyish-white in apical ½ when dry. Flowering stem 6–12 cm, slender, weak, viscid, bearing 2–8 flowers in a lax cyme. Petals 3–4 mm, widely separated, oblong-cuneate, emarginate, often with excurrent mid-vein, dull greenish-yellow. $2n = 16$. *Vertical, shaded limestone rocks,* 1800–2000 *m.* ● *N. Italy (Alpi Bergamasche).* It.

116. S. sedoides L., *Sp. Pl.* 405 (1753). Leafy shoots procumbent, forming a loose mat. Leaves 6–12 × 2–4 mm, oblanceolate to narrowly oblong, tapered to a broad, scarcely distinct petiole, acute to apiculate at the apex, or shortly 3-lobed, with fairly long glandular hairs, mainly on the margins. Flowering stem 1–7 cm, with 1–6 flowers in a lax cyme. Petals 1·5–3 mm, narrower than the sepals and about as long, dull yellow. Ovary inferior. $2n = 64$. *Snow-patches and shady screes; calcicole.* ● *E. Alps; C. Appennini; mountains of W. Jugoslavia and N. Albania.* Al Au It Ju.

Three subspecies may be recognized, though intermediate plants are frequent in some areas.

1 Flowering stems with 3–5 cauline leaves and 3–6 flowers **(b) subsp. hohenwartii**
1 Flowering stems without cauline leaves (rarely with 1 or 2) and with 1–3 flowers
 2 Petals lanceolate to narrowly ovate, acute, shorter than the sepals **(a) subsp. sedoides**
 2 Petals linear, truncate or emarginate, longer than the sepals **(c) subsp. prenja**

(a) Subsp. **sedoides**: Leaves nearly all entire. Cauline leaves usually absent. Flowers 1–3; petals lanceolate to narrowly ovate, acute, shorter than the sepals. *Northern and western parts of E. Alps; Appennini.*

(b) Subsp. **hohenwartii** (Vest ex Sternb.) O. Schwarz, *Mitt. Thür. Bot. Ges.* **1**: 104 (1949) (*S. hohenwartii* Vest ex Sternb.): Some leaves usually 3-lobed. Flowering stem with 3–5 leaves and 3–6 flowers. Petals linear, acute, equalling or exceeding the sepals. *Eastern and southern parts of E. Alps.*

(c) Subsp. **prenja** (G. Beck) G. Beck, *Fl. Bosn. Herceg.* 474 (1923) (*S. prenja* G. Beck): Some leaves 3-lobed. Cauline leaves absent; flowers 1–3. Petals linear, truncate or emarginate, longer than the sepals. *Balkan peninsula.*

117. S. aphylla Sternb., *Revis. Saxifr.* 40 (1810). Like **116** but rather more compact in habit, with fairly short leafy shoots; leaves mostly with 3 short, obtuse lobes at the apex, sparsely glandular-pubescent; flowering stem terminal and axillary, with no cauline leaves and a single flower; petals narrow-linear, slightly exceeding the sepals, greenish-yellow. $2n = c.$ 62, 64. *Screes, snow-lies and stony ground, usually on limestone, mostly* 2100–2800 *m.* ● *C. & E. Alps, mostly in the northern part.* Au Ge He It.

118. S. glabella Bertol., *Gior. Arcad. Sci.* (*Roma*) **21**: 192 *bis* (1824). Shoots short, decumbent or ascending, laxly caespitose. Leaves *c.* 8 mm, glabrous, narrowly spathulate, entire, obtuse. Flowering stem 3–10 cm, sparingly glandular-pubescent, with several flowers in a corymbose cyme. Sepals glabrous; petals 2·5 mm, obovate-orbicular, white. *Shady slopes and gullies on limestone, mostly* 2000–2500 *m.* ● *Balkan peninsula; C. Appennini.* Al Gr It Ju.

119. S. tenella Wulfen in Jacq., *Collect. Bot.* **3**: 144 (1791). Leafy shoots 5–10 cm, procumbent, forming a rather dense mat. Leaves 10 × 1–2 mm, linear-subulate, aristate, with a narrow hyaline margin, usually ciliate, especially near the base. Flowering stem 5–15 cm, erect, slender, glabrous, with 2–8 flowers in a lax cyme. Petals 3 mm, creamy-white, obovate. $2n = 66$. *Shady rocks and screes; calcicole.* ● *S.E. Alps.* Au It Ju.

Subsect. *Tridactylites* (Haw.) Gornall. Erect annuals or biennials, more or less glandular-hairy. Leaves entire or shortly 3- to 5-lobed. Flowers small, on long pedicels, forming a lax, leafy panicle. Petals white (rarely yellowish or red), obtuse, truncate or slightly emarginate. Ovary almost completely inferior.

120–123. S. tridactylites group. The characters of the group are those of the subsection.

1 Spring-flowering annual; basal leaves entire, not forming a distinct rosette **120. tridactylites**
1 Biennial (though sometimes flowering in the first autumn); basal leaves mostly lobed and forming a distinct rosette, at least up to the beginning of anthesis
 2 Capsule subglobose to broadly ellipsoidal, rounded at the base
  3 Basal rosette withering just before anthesis; pedicels 10–20 mm, very slender **122. blavii**
  3 Basal rosette persisting until near the end of anthesis; pedicels 2–8 mm **121. adscendens**
 2 Capsule obovoid, tapered at the base
  4 Lobes of basal leaves forwardly-directed; most pedicels not more than 5 mm; seeds finely tuberculate **121. adscendens**
  4 Lobes of basal leaves somewhat divergent; pedicels up to 8 mm; seeds coarsely papillose **123. osloensis**

120. S. tridactylites L., *Sp. Pl.* 404 (1753). Annual, varying greatly in size according to situation. Basal leaves spathulate, entire, not forming a rosette, and withering before anthesis; cauline leaves cuneate, entire or divided distally into 3–5 triangular-oblong lobes. Flowers in a very diffuse, leafy cyme; lower pedicels 10–20 mm, the upper shorter. Petals 2·5–3 mm, entire or slightly emarginate. Seeds coarsely papillose. $2n = 22$. *Walls, talus, railway-tracks and other open habitats; somewhat calcicole. Most of Europe except the north-east and extreme north.* All except Az Fa Is Rs(N, E) Sb.

121. S. adscendens L., *Sp. Pl.* 405 (1753) (*S. controversa* Sternb.). Biennial. Basal leaves 6–15 mm, mostly oblanceolate-cuneate with 3 forwardly-directed lobes, but some may be entire or 5-lobed; they form a compact rosette which persists until anthesis. Flowering stem 4–25 cm, rather stout, with several cauline leaves similar to the basal and a fairly compact panicle. Pedicels not more than 3 mm in flower and 7 mm in fruit. Petals 3–5 mm, white, rarely tinged with yellow or red. Seeds finely tuberculate. *Fennoscandia, N. Russia, Estonia; mountains of C. & S. Europe, from the W. Carpathians to the Pyrenees, Sicilia and S. Greece.* Al Au Bu Cz Fe Ga Gr He Hs It Ju No Po Rm Rs(N, B, W) Si Su.

(**a**) Subsp. **adscendens** (*S. tridactylites* subsp. *adscendens* (L.) Blytt; incl. *S. adscendens* subsp. *discolor* (Velen.) Kuzmanov): Plant robust, up to 25 cm, with few, suberect branches. Sepals at least 1½ times as long as wide. Capsule of most of the flowers considerably longer than wide, tapered at the base rather gradually into the pedicel. $2n = 22$. *Throughout the range of the species except the extreme south.*

(**b**) Subsp. **parnassica** (Boiss. & Heldr.) Hayek, *Prodr. Fl. Penins. Balcan.* **1**: 638 (1925) (*S. tridactylites* subsp. *parnassica* (Boiss. & Heldr.) Engler & Irmscher): Plant 10–15 cm, and rather less robust. Sepals scarcely longer than broad. Capsule of all flowers subglobose and rounded or truncate at the base. *Greece and S. Albania; C. & S. Italy and Sicilia.*

122. S. blavii (Engler) G. Beck, *Ann. Naturh. Mus.* (*Wien*) 2: 93 (1887) (*S. tridactylites* subsp. *blavii* (Engler) Engler & Irmscher, *S. adscendens* subsp. *blavii* (Engler) Hayek). Biennial. Basal leaves not more than 12 mm, oblanceolate, entire or shortly 3-lobed, forming a rather ill-defined rosette and withering shortly before anthesis; cauline leaves similar but larger. Stem 5–20 cm, slender, with ascending, somewhat flexuous branches. Pedicels 10–20 mm, mostly very slender. Petals 5 × 3·5 mm, broadly obovate, contiguous or overlapping. Capsule rounded or truncate at the base. Seeds coarsely papillose. *Limestone rocks and screes,* 1000–2000 *m.* ● *W. Jugoslavia and N. Albania.* Al Ju.

123. S. osloensis Knaben, *Nytt Mag. Bot.* **3**: 118 (1954). A natural amphidiploid hybrid between **120** and **121**(**a**), resembling **120** in its long pedicels, rather small petals and coarsely papillose seeds, but closer to **121** in the shape of leaves and capsule. *Moist or shady rocks.* ● *Sweden and E. Norway.* No Su.

This plant usually behaves as a biennial, but sometimes flowers in its first autumn as well as in the following summer.

2. Bergenia Moench

D. A. WEBB

Perennials, with stout, fleshy rhizome. Leaves large, somewhat coriaceous, exstipulate, punctulate with totally immersed glands. Flowers 5-merous. Petals entire. Stamens 10. Ovary superior, surrounded by but free from a basin-shaped hypanthium; carpels 2, united only at the base.

1. B. crassifolia (L.) Fritsch, *Verh. Zool.-Bot. Ges. Wien* **39**: 587 (1889). Glabrous, with creeping rhizome *c.* 2 cm in diameter. Leaves 15–30 cm, all basal, thick, shining, broadly obovate to oblong, obscurely sinuate-crenate or remotely denticulate, punctulate on both surfaces; petiole sheathing at the base. Flowering stem up to 30 cm, leafless, with numerous drooping flowers in a rather dense panicle. Petals 10–12 mm, obovate, tapered to a broad claw, erect, bright purplish-pink. Styles exceeding stamens. *Cultivated for ornament, and locally naturalized.* [Au Ga Ge.] (*N. & C. Asia.*)

3. **Chrysosplenium** L.

S. PAWŁOWSKA

Perennials. Leaves petiolate, entire or crenate, exstipulate. Flowers small, 4-merous, in terminal, leafy, corymbose cymes. Petals absent. Stamens 8 (rarely 4). Ovary partly or almost wholly inferior, 1-locular, surmounted by a conspicuous disc.

Literature: H. Hara, *Jour. Fac. Sci. Tokyo Univ.* (*Bot.*) **7**: 1–90 (1957).

1 Leaves alternate
 2 Stamens 8; plant somewhat hairy **1. alternifolium**
 2 Stamens 4; plant almost glabrous **2. tetrandrum**
1 Leaves opposite
 3 Capsule asymmetrical, much exceeding the sepals; seeds suborbicular, furnished with conspicuous, bristle-like papillae **5. dubium**
 3 Capsule symmetrical, scarcely exceeding the sepals; seeds ovoid, very minutely papillose
  4 Plant hairy, at least in lower part; leaves crenulate **3. oppositifolium**
  4 Plant glabrous; leaves ±entire **4. alpinum**

1. C. alternifolium L., *Sp. Pl.* 398 (1753). Somewhat hairy, at least towards the base. Stems up to 20 cm, 3-angled, arising from a mainly underground rhizome. Leaves mostly basal, orbicular-reniform, deeply cordate, with numerous crenations or shallow lobes; petiole long. Cauline leaves alternate; bracts yellowish. Sepals *c.* 2 mm, patent, ovate, obtuse, yellowish. Stamens 8. Filaments and styles much shorter than the sepals. Disc slightly 8-lobed. Seeds smooth, shining, brown or almost black. $2n = 48$. *Damp and shady places. Much of Europe, but absent from the extreme north and west and most of the Mediterranean region.* Au Be Br Bu Cz Da Fe Ga Ge Gr He Ho ?Hs Hu It Ju No Po Rm Rs(N, B, C, W, E) Su.

2. C. tetrandrum (N. Lund) Th. Fries, *Bot. Not.* **1858**: 193 (1858). Like **1** but very nearly or quite glabrous; creeping stems epigeal, leafy; leaves with fewer crenations; inflorescence smaller, with green bracts; sepals green, erect, suborbicular- or triangular-ovate; stamens 4 (opposite the sepals); disc 4-lobed. $2n = 24$. *Wet places. Arctic Europe, extending southwards to* 66° N. *in Finland.* Fe No Rs(N) Sb Su.

3. C. oppositifolium L., *Sp. Pl.* 398 (1753). Laxly caespitose, with decumbent, rooting, leafy stems but without underground rhizome; hairy, at least towards the base. Leaves opposite, orbicular-ovate, crenulate or sinuate, truncate or shortly cuneate at the base, somewhat hairy on upper surface, thin, dark glaucous-green; petiole rather long. Flowering stem up to 20 cm, 4-angled. Flowers *c.* 4 mm in diameter. Sepals ovate or triangular-ovate, yellow. Stamens 8; filaments slightly shorter than the sepals. Capsule free from hypanthium in its upper $\frac{1}{3}$ only. Seeds blackish, covered with glandular hairs about twice as long as thick. $2n = 42$. *Damp and shady places.* ● *W. Europe and parts of C. Europe, extending eastwards to W. Poland and C. Czechoslovakia.* Be Br Cz Da Ga Ge Hb He Ho Hs It ?Ju Lu No Po.

4. C. alpinum Schur, *Verh. Mitt. Siebenb. Ver. Naturw.* **10**: 133 (1859) (*C. oppositifolium* auct. roman., non L.). Like **3** but usually more densely caespitose and quite glabrous; leaves thicker, often wider than long, bright yellowish-green; petioles shorter; flowering stem up to 10(–15) cm; sepals suborbicular, often wider than long, with cucullate apex; seeds verrucose with very short hairs, which are scarcely longer than their diameter. *Mountain springs and flushes.* ● *E. Carpathians.* Rm Rs(W).

5. C. dubium Gay ex Ser. in DC., *Prodr.* **4**: 48 (1830) (*C. macrocarpum* auct. eur. medit., non Cham.). Glabrous, stoloniferous. Leaves opposite, ovate or suborbicular, crenate, cuneate or rounded at the base; petiole equalling or shorter than the lamina. Flowering stem up to 15 cm. Sepals 1·5–2 mm, yellowish-green. Stamens 8. Disc distinctly 8-lobed. Filaments and styles much shorter than the sepals. Capsule deeply bilobed, with unequal lobes, free from hypanthium in its upper $\frac{2}{3}$, considerably longer than the sepals. Seeds suborbicular, furnished with conspicuous, bristle-like papillae in longitudinal rows. *Mountains of S. Italy.* It. (*N. Africa and S.W. Asia.*)

4. **Tellima** R. Br.

D. A. WEBB

Perennial herbs. Leaves simple, palmately lobed, stipulate. Flowers 5-merous, regular, in lax racemes. Petals laciniate-pinnatifid. Stamens 10. Ovary semi-inferior, 1-locular.

1. T. grandiflora (Pursh) Douglas ex Lindley, *Bot. Reg.* t. 1178 (1828). Basal leaves long-petiolate; lamina up to 10 cm, suborbicular, cordate, somewhat obscurely 5- to 9-lobed, the lobes irregularly dentate or crenate. Petiole densely and upper surface of lamina sparsely hirsute. Cauline leaves few, smaller, shortly petiolate. Raceme 15–25 cm, secund, glandular-pubescent. Hypanthium 5 mm, campanulate, united below to lower part of the ovary. Petals cream, often reddish later. *Woods and shady walls. Naturalized in Britain and Ireland, and locally elsewhere.* [Br Da Ge Hb.] (*W. North America.*)

5. **Tolmiea** Torrey & A. Gray

D. A. WEBB

Perennial herbs. Leaves simple, palmately lobed, stipulate. Flowers zygomorphic, in lax racemes. Petals 4, filiform, entire. Stamens 3. Ovary superior, 1-locular.

1. T. menziesii (Pursh) Torrey & A. Gray, *Fl. N. Amer.* **1**: 582 (1840). Basal leaves long-petiolate; lamina up to 10 cm, broadly ovate to suborbicular, cordate, somewhat obscurely 5- to 7-lobed, the lobes irregularly dentate or crenate. Petiole densely and upper surface of lamina sparsely hirsute. Cauline leaves smaller, shortly petiolate. Raceme 10–30 cm, secund, glandular-pubescent. Hypanthium 5–9 mm, tubular, split down upper side almost to the base, free from the ovary. Petals dark reddish-brown. Capsule 9–14 mm, protruding through the slit in the hypanthium. *Shady habitats. Naturalized in Britain.* [Br.] (*W. North America.*)

Cultivated as a house-plant. Leafy buds can arise from the leaf-sinus, especially in cultivated plants, and serve as a means of vegetative reproduction.

74. PARNASSIACEAE

EDIT. D. A. WEBB

Perennial herbs with undivided leaves. Flowers solitary, 5-merous. Stamens 5, alternating with 5 staminodes. Ovary superior; carpels 4; placentae parietal. Fruit a capsule.

1. Parnassia L.

D. A. WEBB

Leaves mostly basal. Stigmas 4, sessile. Seeds numerous.

1. P. palustris L., *Sp. Pl.* 273 (1753). Glabrous and somewhat glaucous. Stem 5–40 cm, erect, usually bearing a single sessile leaf. Remaining leaves basal, petiolate, ovate to deltate-orbicular, cordate, often spotted with red beneath. Flowers 15–30 mm in diameter. Petals white, with darker, semi-transparent veins; staminodes spathulate, terminating in 9–13 linear processes tipped with greenish-yellow glands. *Wet places. Most of Europe, but rare in the south.* Al Au Be Br Bu Cz Da Fe Ga Ge Gr Hb He Ho Hs Hu Is It Ju No Po Rm Rs(N, B, C, W, E) Su.

A variable species, in which the relation between phenotypic, ecotypic and geographical variation, and their relation to chromosome number, is not yet clear. Two European subspecies may be recognized, but intermediates are found in the Alps and elsewhere.

(a) Subsp. **palustris**: Cauline leaf ovate-orbicular, cordate-amplexicaul, not far below the middle of the stem. Sepals considerably shorter than the petals or capsule. $2n = 18, 36$. *Throughout the range of the species, except the extreme north.*

(b) Subsp. **obtusiflora** (Rupr.) D. A. Webb, *Feddes Repert.* **64**: 25 (1961) (*P. obtusiflora* Rupr.): Cauline leaf often absent; if present, near the base of the stem and usually somewhat deltate, with truncate base, not amplexicaul. Sepals nearly as long as the petals and capsule. $2n = 36$. *Arctic and subarctic Europe.*

75. HYDRANGEACEAE

D. A. WEBB

Deciduous shrubs with opposite, simple, exstipulate leaves. Flowers regular. Petals 4 or 5 , free. Ovary inferior, 2- to 4-locular. Fruit a capsule opening at the top. Seeds numerous.

1 Flowers in corymbs, the outermost sterile with much-enlarged sepals **3. Hydrangea**
1 Flowers in racemes or panicles, all fertile and similar
2 Sepals and petals 4; stamens *c.* 25; styles 4, united except at the apex **1. Philadelphus**
2 Sepals and petals 5; stamens 10; styles 3, free **2. Deutzia**

1. Philadelphus L.

D. A. WEBB

Sepals and petals 4. Stamens numerous. Ovary 4-locular; styles united except at the apex.

Literature: S. Y. Hu, *Jour. Arnold Arb.* **36**: 52–109 (1955).

1. P. coronarius L., *Sp. Pl.* 470 (1753) (*P. pallidus* Hayek ex C. K. Schneider). Plant 1–3 m, with numerous slender, dark brown twigs. Leaves 5–8 cm, ovate to oblong-elliptical, acuminate, remotely and finely toothed, shortly stalked, glabrous or sparsely hairy beneath. Flowers in short terminal racemes, strongly scented. Sepals triangular, acute; petals 12–18 mm, oblong-elliptical, creamy white. Stamens *c.* 25, shorter than the petals. *Scrub or woodland on warm slopes. N. & C. Italy; Austria.* Au It [Cz Ga Rm].

The plant has been a favourite in European gardens for centuries, and its origin is subject to some doubt. Stations where it is undoubtedly native are very few, and the cultivated plant, on which the Linnaean species is based, differs somewhat from wild specimens.

2. Deutzia Thunb.

D. A. WEBB

Sepals and petals 5. Stamens 10. Ovary 3-locular; styles free.

1. D. scabra Thunb., *Nova Gen. Pl.* **1**: 20 (1781) (*D. crenata* Siebold & Zucc.). Plant 2–3 m; principal branches erect, with peeling bark. Leaves 4–8 cm, ovate, acute, finely serrate, shortly stalked, scabrid on both surfaces with white, stellate hairs. Flowers in small, erect panicles. Calyx and hypanthium tomentose. Petals 6–10 mm, suberect, oblong or narrow-elliptical, white, often tinged with pink outside. Filaments broad, usually with conspicuous shoulders or lobes near the top. *Cultivated for ornament, and locally naturalized in C. Europe.* [Au.] (*China and Japan.*)

3. Hydrangea L.

D. A. WEBB

Sepals and petals 4 or 5. Stamens 10. Ovary 3-locular; styles free.

1. H. macrophylla (Thunb.) Ser. in DC., *Prodr.* **4**: 15 (1830). Plant 1–3 m, with stout, pale green branches. Leaves 10–20 cm, ovate to obovate, shortly petiolate, acute to acuminate, serrate, almost glabrous. Flowers in large, terminal corymbs; some of the outer flowers sterile, up to 4 cm in diameter, consisting mainly of a greatly enlarged calyx; remaining flowers perfect, with 5 short, triangular sepals; petals 5, 3–4 mm, narrowly ovate, blue or pink. Stamens longer than the petals. *Widely cultivated for ornament and as a hedge-plant; locally naturalized in Açores.* [Az.] (*Japan.*)

The plant naturalized in Europe belongs to var. *normalis* Wilson. The species was originally described from a variant in which all the flowers are enlarged and sterile.

76. ESCALLONIACEAE

EDIT. D. A. WEBB

Shrubs with alternate, simple, exstipulate leaves. Flowers hermaphrodite, 5-merous, regular. Petals free; stamens 5. Ovary inferior, of 2 carpels; placentae parietal, but large, so that the central cavity of the ovary is occluded. Fruit a capsule.

1. **Escallonia** Mutis ex L. fil.

D. A. WEBB

Evergreen shrubs. Flowers in racemes or panicles. Petals with their claws juxtaposed to form a tube; limb patent or revolute. Stigma capitate or somewhat 2-lobed. Seeds numerous.

1. **E. rubra** (Ruiz & Pavón) Pers., *Syn. Pl.* **1**: 235 (1805). Up to 3 m, branches glandular-pubescent. Leaves 2–6 cm, obovate, subsessile, biserrate, glabrous. Flowers 8–18 mm in diameter, in small panicles. Corolla bright pinkish-red. Stamens very slightly exserted. *Cultivated for ornament, and as a hedge-plant in coastal districts, in W. & S. Europe, and locally naturalized.* [Br Ga Hb.] (*Chile and Argentina.*)

The variety most widely cultivated is var. *macrantha* (Hooker & Arnott) Reiche, which differs from var. *rubra* in its larger, more viscid and aromatic leaves, larger flowers and glandular-pubescent calyx.

77. GROSSULARIACEAE

EDIT. D. A. WEBB

Deciduous shrubs with alternate, exstipulate leaves. Flowers regular, 5-merous, in racemes or small axillary clusters. Petals free. Stamens 5. Ovary inferior; styles 2, united below. Fruit a berry.

1. **Ribes** L.

D. A. WEBB

Small or medium-sized shrubs. Leaves palmately 3- or 5-lobed. Petals small, often greenish.

R. aureum Pursh, *Fl. Amer. Sept.* **1**: 164 (1814), from W. North America, with bright yellow flowers and glabrous leaves, is cultivated for ornament, and is becoming locally naturalized in W. & C. Europe. **R. sanguineum** Pursh, *loc. cit.* (1814), also from W. North America, with conspicuous red or pink flowers, is frequently planted in hedges, and may perhaps be locally naturalized.

Literature: E. de Janczewski, *Mém. Soc. Phys. Hist. Nat. Genève* **35**: 199 (1907).

1 Flowers in axillary clusters of 1–3
  2 Leaves usually more than 20 mm wide; spines usually present; bracteoles present — **6. uva-crispa**
  2 Leaves rarely 15 mm wide; spines and bracteoles absent — **7. sardoum**
1 Flowers in racemes
  3 Flowers functionally dioecious; bracts 4–10 mm; leaves not more than 6 cm wide; axis of inflorescence glandular-hairy
    4 Buds ovoid, obtuse; fruit glandular-hairy — **9. orientale**
    4 Buds elongate, acute; fruit glabrous — **8. alpinum**
  3 Flowers hermaphrodite; bracts 1–2 mm; larger leaves 7–15 cm wide; axis of inflorescence without glandular hairs
    5 Leaves covered beneath with sessile, aromatic glands — **5. nigrum**
    5 Leaves without sessile glands
      6 Sepals ciliate; leaves up to 15 cm wide — **4. petraeum**
      6 Sepals not (or very sparsely) ciliate; leaves 6–10 cm wide
        7 Sepals ligulate, deflexed — **1. multiflorum**
        7 Sepals obovate or spathulate, erect or patent
          8 Receptacle nearly flat, with a raised ring; anther-lobes widely separated — **2. rubrum**
          8 Receptacle basin-shaped, without a raised ring; anther-lobes contiguous on inner side — **3. spicatum**

1. **R. multiflorum** Kit. ex Roemer & Schultes, *Syst. Veg.* **5**: 493 (1819). Plant 1·5–2 m. Leaves *c.* 10 × 10 cm, usually glabrous; petiole pubescent. Racemes drooping, up to 12 cm, with 30–50 yellowish-green flowers. Bracteoles 1 mm. Receptacle pelviform with, between the stamens and styles, a raised ring bearing 5 large protuberances. Sepals ligulate, deflexed; petals much shorter, also deflexed. Fruit red, acid, glabrous. *Damp woods or shady places in the mountains.* ● *Balkan peninsula, C. Italy, Sardegna.* Bu Gr It Ju Sa.

2. **R. rubrum** L., *Sp. Pl.* 200 (1753) (*R. vulgare* Lam., *R. sylvestre* (Lam.) Mert. & Koch, *R. sativum* Syme). Plant 1–1·5 m. Leaves *c.* 6 × 7 cm, cordate, nearly glabrous. Racemes inclined or drooping, rather lax, with 10–20 pale green flowers, slightly tinged with purple. Receptacle nearly flat, with a raised ring between stamens and styles. Sepals orbicular-spathulate, patent; petals very small. Anther-lobes separated by a connective as wide as themselves. Fruit red, acid, glabrous. $2n = 16$. ● *Cultivated throughout Europe for its fruits (Red Currants), and naturalized in many countries; native only in the west.* Be *Br Ga Ge Ho It.

3. **R. spicatum** Robson in With. (ed. Stokes), *Arr. Br. Pl.* ed. 3, **2**: 265 (1796) (*R. rubrum* sensu Jancz. et auct. recent. nonnull., non L.; incl. *R. pubescens* (Hartman) T. Hedl., *R. hispidulum* (Jancz.) Pojark., *R. heteromorphum* Ţopa, *R. scandicum* T. Hedl., *R. schlechtendalii* Lange). Like **2** and often confused with it, but leaves usually larger and less cordate, varying from glabrous to tomentose on lower surface; receptacle pelviform, without a raised ring; anthers about as long as wide, with the lobes almost contiguous on their inner face. $2n = 16$. *Damp woods and streamsides. N. & E. Europe; sometimes cultivated in the east, but rarely naturalized outside its native territory.* *Au Br Da Fe ?Ge No Po Rm Rs(N, B, C, W, E) Su [Cz].

Variable in pubescence, leaf-shape and inflorescence. Several local populations have been described as species or subspecies, but the diagnoses are usually unsatisfactory in view of the variability of the species elsewhere. It seems probable that the arctic and subarctic populations (**R. glabellum** (Trautv. & C. A. Meyer) T. Hedl., *Bot. Not.* **1901**: 98 (1901)), with nearly glabrous leaves, and those of E. European Russia (**R. hispidulum** (Jancz.) Pojark., *Bull. Appl. Bot. Pl.-Breed.* (*Leningrad*) **22**: 339 (1929)), with small, usually glandular-hairy leaves, may be entitled to subspecific rank.

4. R. petraeum Wulfen in Jacq., *Misc. Austr. Bot.* **2**: 36 (1781) (*R. carpaticum* Schultes). Plant 1–3 m. Leaves up to 15 × 15 cm, glabrous or pubescent, sometimes bullate above and glandular beneath. Racemes *c.* 10 cm, horizontal or drooping, with 20–35 pinkish, campanulate flowers. Sepals ciliate, orbicular-spathulate, the lower part erect, the upper patent; petals ½ as long as the sepals. Upper part of ovary protruding above the disc. Fruit dark purplish-red, acid. *Woods and scrub. Mountains of C. Europe, extending southwards to the Pyrenees, C. Italy and Bulgaria.* Au Bu Cz Ga Ge He Hs Hu It Ju Po Rm Rs(W).

5. R. nigrum L., *Sp. Pl.* 201 (1753). Plant 1–2 m. Leaves up to 10 cm wide, glabrous above, slightly pubescent and with numerous sessile, aromatic glands beneath. Racemes drooping. Flowers campanulate, reddish- or brownish-green. Sepals oblong, recurved, pubescent; petals smaller, erect, whitish. Fruit up to 12 mm in diameter, black, sweetish and aromatic. $2n = 16$. *Damp woods. Most of Europe, except the Mediterranean region; native certainly in C. & E. Europe; cultivated for its fruits (Black Currants), and widely naturalized in the west, where its native limits are hard to ascertain.* *Au *Be *Br Bu *Cz *Da Fe Ga Ge *Ho *Hu It Ju *No Po Rm Rs(N, B, C, W, E) Su [Co Hb He].

6. R. uva-crispa L., *Sp. Pl.* 201 (1753) (*R. grossularia* L., *Grossularia reclinata* (L.) Miller). Plant 1–1·5 m, freely and intricately branched, and armed at the nodes with stout spines, usually in groups of 3 (very rarely absent). Leaves 2–5 cm wide, rarely more, rather deeply lobed, glabrous or pubescent. Flowers in axillary clusters of 1–3; pedicels with 2 bracteoles near the middle. Sepals 5–7 mm, ligulate, pale or pinkish-green; petals white, smaller. Fruit *c.* 10 mm in diameter, green, yellow or purplish-red, usually hispid. $2n = 16$. *Hedges and wood-margins. Native in S., C. & W. Europe; extensively cultivated for its fruits (Gooseberries), and frequently naturalized by bird-dispersal in other areas.* Au Be Br Bu Cr Cz Ga Ge Gr He Ho Hs Hu It Ju Po Rm Rs(W) [Co Da Fe Hb Lu No Rs(N, B, C) Su].

7. R. sardoum U. Martelli, *Malpighia* **8**: 384 (1894). Plant *c.* 1 m, without spines. Leaves 10–15 mm wide, ovate to semi-circular, shortly 3-lobed, with numerous subsessile glands; petiole wide, sheathing, glandular-ciliate. Flowers solitary, axillary; pedicels short, bracteoles absent. Sepals greenish-yellow, ovate, deflexed; petals minute. Fruit bright red, glabrous. *Calcareous rocks, 1000–1200 m.* ● *Sardegna (near Oliena).* Sa.

8. R. alpinum L., *Sp. Pl.* 200 (1753) (*R. lucidum* Kit.). Plant 1–2 m. Leaves 2–6 cm, usually longer than wide, rather deeply 3-lobed, glabrous or sparsely hairy. Buds up to 10 mm, acute. Dioecious, but with rudimentary organs of the other sex. Flowers in glandular-hispid racemes (longer in male plants), each flower subtended by a conspicuous bracteole 4–8 × 1–2 mm. Flowers small, rotate, greenish; sepals elliptical. Fruit scarlet, glabrous, insipid. $2n = 16$. *Wood-margins and by mountain streams. N. & C. Europe, extending southwards in the mountains to N. Spain, C. Italy and Bulgaria.* Au Br Bu Cz Da Fe Ga Ge He Hs Hu It Ju No Po Rm Rs(N, B, C) Su.

9. R. orientale Desf., *Hist. Arb.* **2**: 88 (1809). Like **8** but with shorter, ovoid, obtuse buds; leaves usually as wide as long, and more thickly covered with glandular hairs; fruit paler and glandular-hispid. *C. & S. Greece.* Gr. (*W. & C. Asia.*)

78. PITTOSPORACEAE

EDIT. †T. G. TUTIN AND J. R. EDMONDSON

Trees or shrubs. Leaves simple, usually entire, alternate, exstipulate. Flowers regular, usually hermaphrodite, solitary or in cymes. Sepals 5, imbricate; petals 5, imbricate, often connate by the claws; limb recurved or patent. Stamens 5, hypogynous, free. Ovary superior, with 2, rarely 3–5 carpels. Fruit a capsule or berry.

1. Pittosporum Gaertner

J. DO AMARAL FRANCO (EDITION 1) REVISED BY D. A. WEBB AND J. R. AKEROYD (EDITION 2)

Leaves often clustered at the ends of the shoots, usually evergreen. Flowers solitary or in terminal, corymbose or umbellate cymes. Fruit a capsule, with 2–4 leathery or woody valves. Seeds not winged, immersed in a viscid substance.

1 Petals blackish-purple; leaves white-tomentose beneath — **1. crassifolium**
1 Petals white or yellowish; leaves glabrous
 2 Shrub; leaves obovate, obtuse — **2. tobira**
 2 Tree; leaves elliptic-oblong, acute — **3. undulatum**

1. P. crassifolium Banks & Solander ex A. Cunn., *Ann. Nat. Hist.* **4**: 106 (1839). An erect shrub or small tree up to 10 m, with black bark and a subfastigiate crown. Leaves 5–8 × 2–3 cm, obovate-oblong to oblong, obtuse to subacute, with revolute margins, dark green above and white-tomentose beneath, very coriaceous. Flowers in few-flowered umbellate cymes. Petals blackish-purple, oblong. Capsule 12–20 mm, subglobose, white-tomentose, with 3 or 4 woody valves. *Cultivated for ornament and shelter in parts of W. Europe; naturalized in S.W. England (Isles of Scilly).* [Br.] (*New Zealand.*)

2. P. tobira (Thunb.) Aiton fil., *Hort. Kew.* ed. 2, **2**: 27 (1811). Shrub up to 5 m. Leaves 4–10 × 2–5 cm, obovate, coriaceous, glabrous. Flowers in small, subterminal corymbs, *c.* 2·5 cm in diameter, very fragrant. Petals broadly oblong, creamy white, later yellow. Capsule 10–12 mm, subglobose, with 3 valves. *Widely cultivated for ornament in S. Europe; naturalized in Açores and parts of the Mediterranean region.* [Az Ga.] (*E. Asia.*)

3. P. undulatum Vent., *Descr. Pl. Jard. Cels* t. 76 (1802). Tree up to 20 m, with grey bark and a pyramidal crown, glabrous except for puberulent inflorescence. Leaves 7–13 × 3–6 cm, elliptic-oblong, acute, cuneate at the base, shining green, thin, usually undulate. Flowers fragrant, in few-flowered umbellate cymes. Petals white, oblong. Capsule 10–12 mm, obovoid, glabrous, orange when ripe, with 2 valves. *Widely cultivated for ornament in S. & W. Europe; extensively naturalized in Açores and locally elsewhere.* [Az ?Ga Lu.] (*S.E. Australia.*)

79. PLATANACEAE

EDIT. †T. G. TUTIN AND J. R. EDMONDSON

Trees with scaling bark. Monoecious; flowers arranged in globose, unisexual capitula. Perianth 4- or 6-merous, arranged in 2 whorls. Stamens 4–6, opposite the outer perianth-segments. Ovary superior; carpels 3–6, free, each with 1(2) pendulous, orthotropous ovules.

1. Platanus L.

†T. G. TUTIN

Leaves alternate; lamina palmately lobed; petiole with dilated base, enclosing the bud. Filaments of stamens very short. Carpels in fruit obpyramidal, indehiscent, surrounded at the base by long hairs.

Leaves lobed to beyond the middle, cuneate at the base; female capitula usually 3–6 **1. orientalis**
Leaves lobed at most to the middle, truncate or cordate at the base; female capitula usually 2 **2. acerifolia**

1. P. orientalis L., *Sp. Pl.* 999 (1753). Up to 30 m. Leaves 5- to 7-lobed; central lobe much longer than its width at the base; all lobes coarsely dentate, rarely entire. Capitula (2)3–6(7) on a long pendulous axis. *Damp woods and streamsides. Balkan peninsula southwards from* c. 42° N.*; Kriti; S. Italy and Sicilia; often planted elsewhere.* Al Bu Cr Gr It Ju Si *Tu.

2. P. acerifolia (Aiton) Willd., *Sp. Pl.* **1(1)**: 474 (1797) (*P. hybrida* auct., an Muenchh.?, *P. hybrida* auct., an Brot.?, *P. cuneata* Willd.). Like **1** but leaves lobed at most to the middle; central lobe little longer than its width at the base; capitula nearly always 2. $2n = 42$. *Commonly planted in much of Europe, especially as a roadside tree and in cities; widely naturalized in S. Europe.*

The origin and taxonomic status of this plant have been much discussed but are still uncertain. Some authors consider it to be a hybrid, which arose in S.W. Europe in the seventeenth century, between **1** and *P. occidentalis* L. from North America. Others regard it as a cultivar of **1**. It is fully fertile.

APPENDICES

NOTE TO APPENDICES I–III

Considerable variation is found in the orthography of the names of many authors, especially of the earlier ones and of those whose names are transliterated from Cyrillic script. Variant spellings are normally only given here if they are likely to give rise to doubts about their identity.

The initials used by some authors vary according to whether the vernacular or latinized form of a Christian name is used (e.g. *Karl* or *Carolus*); the form most frequently used by the author is adopted in these lists with variants being given in parentheses.

The dates given for books and periodicals indicate, as far as can be ascertained, the date of effective publication; where this differs from dates on the title-page or elsewhere in the work itself, there is usually a reference to explain the dates given.

Certain publications are of a character intermediate between books and periodicals (e.g. seed-lists, *schedae*). The assignment of these to Appendix II or Appendix III is inevitably somewhat arbitrary; in general, those with a definite authorship will be found in Appendix II, while those which are anonymous, or compiled by numerous workers, will be found in Appendix III.

In order to maintain continuity with *Flora Europaea* ed. 1, abbreviations used in these Appendices correspond to those used in the most recent volume of ed. 1 where the author or work is cited. No attempt has been made to adopt abbreviations recommended by R. D. Meikle, *Draft Index of Author Abbreviations compiled at The Herbarium, Royal Botanic Gardens, Kew* (H.M.S.O., 1980), R. K. Brummitt & C. E. Powell, Authors of Plant Names (Kew, 1992 — published just before the final text of these *Appendices* was sent to press), F. A. Stafleu & R. S. Cowan, *Taxonomic Literature* ed. 2 (Utrecht, 1976–1988) or *Botanico-Periodicum-Huntianum* (Pittsburgh, 1968 with *Supplement* in 1992), although most of these works have been heavily utilised in the compilation of these lists and due acknowledgement is made here to their authors or editors. Additional information on periodicals has been obtained from Burdet et al., *Catalogue des Périodiques de la Bibliothèque des Conservatoire et Jardin botaniques de la Ville de Genève* (Genève, 1980) and its *Supplément* 1980–1987 (Maiullari & Burdet, 1988).

APPENDIX I

KEY TO THE ABBREVIATIONS OF AUTHORS' NAMES

Abromeit J. Abromeit (1857–1946)
Acht. B. Achtarov (1885–1959)
Adamović L. Adamović (1864–1935)
Adams M. F. Adams (J. F. Adam) (1780–1829/32)
Adanson M. Adanson (1727–1806)
Ade A. Ade (1876–1968)
Aellen P. Aellen (1896–1973)
Ahlfvengren F. E. Ahlfvengren (1862–1921)
Ahti T. T. Ahti (b. 1934)
Aichele D. E. Aichele (b. 1928)
Airy Shaw H. K. Airy Shaw (1902–1985)
Aiton W. Aiton (1731–1793)
Aiton fil. W. T. Aiton (1766–1849)
Akeroyd J. R. Akeroyd (b. 1952)
Albov N. M. Albov (Alboff) (1866–1897)
Aldén B. G. Aldén (b. 1948)
Alejandre J. A. Alejandre Sáenz (b. 1947)
Alexeenko M. I. Alexeenko (Alexejenko) (b. 1905)
All. C. L. Allioni (1728–1804)
Alston A. H. G. Alston (1902–1958)
Ambrosi F. Ambrosi (1821–1897)
Amich F. Amich García (b. 1953)
Amo Mariano del Amo y Mora (1820–1896)
Anderson, G. G. W. Anderson (d. 1817)
Andersson, I. A. I. A. Andersson (b. 1959)
Andersson, N. J. N. J. Andersson (1821–1880)
Andrásovszky J. Andrásovszky (1889–1943)
Andrews H. C. Andrews (1794–1830)
Andrz. A. L. Andrzejowski (1785–1868)
Ångström J. Ångström (1813–1879)
Antoine F. Antoine (1815–1886)
Arcangeli G. Arcangeli (1840–1921)
Archer-Hind T. H. Archer-Hind (1814–1911)
Ard. P. Arduino (1728–1805)
Arnold J. F. X. Arnold (possibly a pseudonym; fl. 1780–1785)
Arnott G. A. W. Arnott (1799–1868)
Arrh., A. J. I. A. Arrhenius (1858–1950)
Arrigoni P. V. Arrigoni (b. 1932)
Ascherson P. F. A. Ascherson (1834–1913)
Aspegren G. C. Aspegren (1791–1828)
Assenov I. Assenov (fl. 1966)
Asso I. J. de Asso y del Rio (1742–1814)
Aublet J. B. C. F. Aublet (1720–1778)
Avé-Lall. J. L. E. Avé-Lallemant (1803–1867)
Avr. N. A. Avrorin (b. 1906)
Aznav. G. V. Aznavour (1861–1920)
Bab. C. C. Babington (1808–1895)
Badaro G. B. Badaro (1793–1831)
Bailey L. H. Bailey (1858–1954)
Baillon H. E. Baillon (1827–1895)
Bailly E. Bailly (1829–1894)
Baker J. G. Baker (1834–1920)
Balansa B. Balansa (1825–1891)
Balbis G.-B. Balbis (1765–1831)
Bald. A. Baldacci (1867–1950)
Balf. J. H. Balfour (1808–1884)
Balk. B. E. Balkovsky (1899–1985)
Ball J. Ball (1818–1889)
Ball, P. W. P. W. Ball (b. 1932)
Bange A. J. Bange (1896–1950)
Banks J. Banks (1743–1820)
Barbey, W. W. Barbey (later W. Barbey-Boissier) (1842–1914)
Barc. F. Barceló y Combis (1820–1889)
Barkoudah Y. I. Barkoudah (b. 1933)
Bartl. F. G. Bartling (1798–1875)
Bartolo G. Bartolo (b. 1948)
Basil. N. A. Basilevskaja (Bazilevskaja) (b. 1902)
Bast. T. Bastard (1784–1846)
Batt. J. A. Battandier (1848–1922)
Baumg. J. C. G. Baumgarten (1765–1843)
Beauv. A. M. F. J. Palisot de Beauvois (1752–1820)
Beauverd G. Beauverd (1867–1942)
Becherer A. Becherer (1897–1977)
Bechst. J. M. Bechstein (1757–1822)
Beck, G. G. Beck von Mannagetta (1856–1931)
Béguinot A. Béguinot (1875–1940)
Behrendsen W. Behrendsen (d. 1923)
Bél. C. P. Bélanger (1805–1881)
Bellardi C. A. L. Bellardi (1741–1826)
Benl G. Benl (b. 1910)
Benn., Ar. Arthur Bennett (1843–1929)
Bennert H. W. Bennert (b. 1945)
Benson, L. L. D. Benson (b. 1909)
Bentham G. Bentham (1800–1884)
Berger, A. A. Berger (1871–1931)
Bergeret, J. P. J. P. Bergeret (1751–1813)
Bernard P. F. Bernard (1749–1825)
Bernh. J. J. Bernhardi (1774–1850)
Berth. S. Berthelot (1794–1880)
Bertol. A. Bertoloni (1775–1869)
Besser W. S. J. G. von Besser (1784–1842)
Biasol. B. Biasoletto (1793–1858)
Bieb. F. A. Marschall von Bieberstein (1768–1826)
Bigelow J. Bigelow (1787–1879)
Bihari G. Bihari (b. 1889)
Billot P. C. Billot (1796–1863)
Bir S. S. Bir (b. 1929)
Biria J. A. J. Biria (b. 1789)
Biv. A. de Bivona-Bernardi (1774–1837)
Blanche E. Blanche (1824–1908)
Blanché, C. C. Blanché i Vergés (b. 1958)
Blasdell R. F. Blasdell (b. 1929)
Błocki B. Błocki (1857–1917)
Bluff M. J. Bluff (1805–1837)

Blume C. L. von Blume (1796–1862)
Blytt M. N. Blytt (1789–1862)
Bobrov E. G. Bobrov (1902–1983)
Böcher T. W. Böcher (1909–1979)
Bocquet G. Bocquet (1927–1986)
Boenn. C. M. F. von Boenninghausen (1785–1864)
Boguslaw I. A. Boguslaw (fl. 1846)
Boiss. P. E. Boissier (1810–1885)
Boissieu C. V. Boissieu de la Martinière (1784–1868)
Bolle, C. C. A. Bolle (1821–1909)
Bolòs, A. A. de Bolòs (Bolós) y Vayreda (1889–1975)
Bolòs, O. O. de Bolòs (Bolós) i Capdevila (b. 1924)
Bolton J. Bolton (fl. 1758–1799)
Bolus, L. H. M. L. Bolus (L. H. M. Bolus, Mrs F. Bolus) (1877–1970)
Bong. A. G. H. von Bongard (H. G. Bongard) (1786–1839)
Bonjean J. L. Bonjean (1780–1846)
Bonnemaison T. Bonnemaison (1774–1829)
Bonnet E. Bonnet (1848–1922)
Bonnier G. E. M. Bonnier (1851–1922)
Bonpl. A. J. A. Bonpland (1773–1858)
Borbás V. [von] Borbás (1844–1905)
Bord. H. Bordère (1825–1889)
Bordzil. E. I. Bordzilowski (1875–1949)
Boreau A. Boreau (1803–1875)
Borhidi A. Borhidi (b. 1932)
Boriss. A. G. Borissova-Bekrjaševa (1903–1970)
Borja J. Borja Carbonell (b. 1903)
Borkh. M. B. Borkhausen (Borckhausen) (1760–1806)
Börner C. J. B. Börner (1880–1953)
Bornm. J. F. N. Bornmüller (1862–1948)
Bory J. B. G. M. Bory de Saint-Vincent (1778–1846)
Borza A. Borza (1887–1971)
Borzi A. Borzi (1852–1921)
Bosc L. A. G. Bosc (1759–1828)
Botsch. V. P. Botschantzev (1910–1990)
Boucher J. A. G. Boucher de Crèvecoeur (1757–1844)
Bourgeau E. Bourgeau (1813–1877)
Bout. J. F. D. Boutigny (1820–1884)
Boutelou, E. E. Boutelou y Soldevilla (1823–1883)
Bouvet G. Bouvet (1874–1929)
Bowdich T. E. Bowdich (1791–1824)
Br., N. E. N. E. Brown (1849–1934)
Br., R. R. Brown (1773–1858)
Brackenr. W. D. Brackenridge (1810–1893)
Brândză D. Brândză (1846–1895)
Braun C. F. W. ('Friedrich') Braun (1800–1864)
Braun, A. A. C. H. Braun (1805–1877)
Braun, J. J. Braun (later J. Braun-Blanquet) (1884–1980)
Br.-Bl. J. Braun-Blanquet (1884–1980)
Breistr. M. A. F. Breistroffer (b. 1910)
Br.-Catt. A. J. B. [Markgraf De Planta-Salis] Brilli-Cattarini (b. 1923)
Briq. J. I. Briquet (1870–1931)
Britten J. Britten (1846–1924)
Britton N. L. Britton (1859–1934)
Brongn. A. T. Brongniart (1801–1876)
Brot. F. da S. de Avellar Brotero (1744–1828)
Brouss. P. M. A. Broussonet (1761–1807)
Browicz K. Browicz (b. 1925)
Brownsey P. J. Brownsey (b. 1948)
Brügger C. G. Brügger (1833–1899)
Bruhin Pater T. von Aquin Bruhin (1835–1896/1899)
Brullo S. Brullo (b. 1947)
Brummitt R. K. Brummitt (b. 1937)
Bruno —— Bruno (fl. 1760)
Bubani P. Bubani (1806–1888)
Buch.-Ham. F. Buchanan-Hamilton (1762–1829)
Buchanan-White F. Buchanan-White (1842–1894)
Buchholz F. Buchholz (1872–1924)
Buen O. de Buen y del Cos (1863–1945)
Buffon G. L. L. de Buffon (1707–1788)
Bug. W. Bugała (b. 1924)
Bunge A. A. von Bunge (1803–1890)
Burdet H. M. Burdet (b. 1939)
Burgsd. F. A. L. von Burgsdorff (1747–1802)
Burm. fil. N. L. Burman (N. L. Burmannus) (1733/1734–1793)
Burnat E. Burnat (1828–1920)
Burtt, B. L. B. L. Burtt (b. 1913)
Busch, N. N. A. Busch (1869–1941)
Buschm. A. Buschmann (b. 1908)
Butcher R. W. Butcher (1897–1971)
Buttler K. P. Buttler (b. 1942)
Caball. A. Caballero Segares (1877–1850)
Cadevall J. Cadevall y Diars (Cadeval i Diars) (1846–1921)
Cajander A. K. Cajander (1879–1943)
Camb. J. Cambessèdes (1799–1863)
Campd. F. Campderá (Campderà) i Camins (1793–1862)
Camus E. G. Camus (1852–1915)
Camus, A. A. A. Camus (1879–1965)
Candargy, P. P. C. Candargy (b. 1870)
Cantó P. Cantó Ramos (b. 1956)
Capelli C. M. Capelli (1763–1831)
Cardona M. de los Angeles Cardona y Florit (b. 1940)
Cariot A. Cariot (1820–1883)
Carlström, A. A. L. Carlström (b. 1957)
Carrière E. A. Carrière (1818–1896)
Caruel T. Caruel (1830–1898)
Cast. J. L. M. Castagne (1785–1858)
Castroviejo S. Castroviejo Bolibar (b. 1946)
Cav. A. J. Cavanilles (1745–1804)
Cavara F. Cavara (1857–1929)
Cavillier F. G. Cavillier (1868–1953)
Čelak. L. J. Čelakovský (1834–1902)
Cesati V. de Cesati (1806–1883)
Česchm. I. V. Češmedžiev (Cheshmedjiev, Cheshmejiyév) (b. 1930)
Chaix D. Chaix (1730/1–1799/1800)
Cham. L. K. A. von Chamisso (L. C. A. Chamisseau de Boncourt) (1781–1838)
Charrel L. Charrel ('Abd-ur-Rahmān-Nadji) (fl. 1888–1890)
Chater A. O. Chater (b. 1933)
Chaub. L. A. Chaubard (1781–1854)
Chaudhri M. N. Chaudhri (b. 1932)
Chenevard P. Chenevard (1839–1919)
Chevall. F. F. Chevallier (1796–1840)
Chiarugi A. Chiarugi (1901–1960)

Ching, R.-C. Ching Ren-Chang (Ch'in Jên-ch'ang, Ren-Chang Ching) (1898–1986)
Chiov. E. Chiovenda (1871–1940)
Choisy J. D. Choisy (1799–1859)
Chowdhuri P. K. Chowdhuri (b. 1923)
Chr., C. C. F. A. Christensen (1872–1942)
Christ K. H. H. Christ (1833–1933)
Chrtek J. Chrtek (b. 1930)
Chrtková A. Chrtková (formerly A. Žertková) (b. 1930)
Clairv. J. P. de Clairville (1742–1830)
Clapham A. R. Clapham (b. 1904)
Clarke, E. D. E. D. Clarke (1779–1822)
Claus K. E. Claus (1796–1864)
Clavaud A. Clavaud (1828–1890)
Clemente S. de Rojas Clemente y Rubio (1777–1827)
Clemente, M. M. Clemente Muñoz (b. 1949)
Clementi, G. C. G. C. Clementi (1812–1873)
Clerc O. E. Clerc (1845–1920)
Cockayne L. Cockayne (1855–1934)
Coincy A. H. C. de la Fontaine de Coincy (1837–1903)
Colla L. A. Colla (1766–1848)
Colmeiro M. Colmeiro y Penido (1816–1901)
Coman A. Coman (1881–1972)
Conrath P. Conrath (1861–1931)
Contandriopoulos J. Contandriopoulos (b. 1922)
Conti, P. P. Conti (1874–1898)
Coode M. J. E. Coode (b. 1937)
Cook, C. D. K. C. D. K. Cook (b. 1933)
Copel. E. B. Copeland (1873–1964)
Corb. F. M. L. Corbière (1850–1941)
Corley H. V. Corley (b. 1914)
Corr. C. F. J. E. Correns (1864–1933)
Cosent. F. Cosentini (1769–1840)
Cosson E. S. C. Cosson (1819–1889)
Costa A. C. Costa y Cuxart (1817–1886)
Coste H. J. Coste (1858–1924)
Coulter, J. M. J. M. Coulter (1851–1928)
Court. R. J. Courtois (1806–1835)
Coutinho A. X. Pereira Coutinho (1851–1939)
Covas G. Covas (b. 1915)
Craib W. G. Craib (1882–1933)
Crantz H. J. N. von Crantz (1722–1799)
Cuatrec. J. Cuatrecasas y Arumí (b. 1903)
Cullen J. Cullen (b. 1936)
Cumino P. Cumino (fl. 1807)
Cunn., A. A. Cunningham (1791–1839)
Curtis W. Curtis (1746–1799)
Cutanda V. Cutanda y Jarauta (1804–1866)
Cuvier G. F. Cuvier (1773–1838)
Cyr. D. M. L. Cirillo (Cyrillus, 'Cyrillo') (1739–1799)
Czecz. H. Czeczott (1888–1982)
Czern. V. M. Czernajew (Czernjaew) (1796–1871)
Czetz A. Czetz (1801–1865)
Dahl, Å. E. Å. E. Dahl (b. 1916)
Dahl, O. C. O. C. Dahl (1862–1940)
Dalla Torre K. W. von Dalla Torre von Thurnberg-Sternhoff (1850–1928)
Damanti P. Damanti (b. 1858)
Dandy J. E. Dandy (1903–1976)
Danser B. H. Danser (1891–1943)
Davidov B. Davidov (1870–1927)
Davis, P. H. P. H. Davis (1918–1992)
DC. A. P. de Candolle (1778–1841)
DC., A. A. L. P. P. de Candolle (1806–1893)
De Bary H. A. de Bary (1831–1888)
Debeaux J. O. Debeaux (1826–1910)
Decker P. Decker (b. 1867)
Decne J. Decaisne (1807–1882)
Degen A. von Degen (1866–1934)
De Langhe J. E. de Langhe (b. 1907)
Delarbre A. Delarbre (1724–1813)
Delay J. Delay (fl. 1960–1981)
De Lens A. J. De Lens (fl. 1828)
Delile A. R. Delile (1778–1850)
Delip. D. D. Delipavlov (b. 1919)
Dennst. A. W. Dennstedt (1776–1826)
De Not. G. de Notaris (1805–1877)
den Nijs J. C. M. den Nijs (b. 1946)
Desf. R. L. Desfontaines (1750–1833)
Desmoulins C. R. A. Desmoulins (Des Moulins) (1798–1875)
Desr. L. A. J. Desrousseaux (1753–1838)
Desv. N. A. Desvaux (formerly A. N. Desvaux) (1784–1856)
Deville L. Deville (fl. 1859)
De Voogd W. B. de Voogd (fl. 1976)
Deyrolle E. Deyrolle (fl. 1891)
D'hose R. D'hose (fl. 1976)
Díaz T. E. Díaz González (b. 1949)
Dickson J. Dickson (1738–1822)
Diels F. L. E. Diels (1874–1945)
Dietr., A. A. G. Dietrich (1795–1856)
Dietr., D. D. N. F. Dietrich (1799–1888)
Dingler H. Dingler (1846–1935)
Dippel L. Dippel (1827–1914)
Dode L. A. Dode (1875–1943)
Döll J. C. Döll (1808–1885)
Dolliner G. Dolliner (1794–1872)
Domin K. Domin (1882–1953)
Domokos J. Domokos (b. 1904)
Don, D. D. Don (1799–1841)
Don, G. G. Don (1764–1814)
Don fil., G. G. Don (1798–1856)
Donn J. Donn (1758–1813)
Dörfler I. Dörfler (1866–1950)
Dostál J. Dostál (b. 1903)
Douglas D. Douglas (1799–1834)
Drejer S. T. N. Drejer (1813–1842)
Dreves J. F. P. Dreves (1772–1816)
Druce G. C. Druce (1850–1932)
Dubois, F. F. N. A. Dubois (1752–1824)
Dubovik O. N. Dubovik (b. 1935)
Duby J. É. Duby (1798–1885)
Duchartre P.-E.-S. Duchartre (1811–1894)
Dudley, T. R. T. R. Dudley (b. 1936)
Dufour J.-M. L. Dufour (1780–1865)
Duh. H. L. Duhamel du Monceau (1700–1782)
Dumort. B. C. J. Dumortier (1797–1878)
Dunal M. F. Dunal (1789–1856)
Durieu M. C. Durieu de Maisonneuve (1796–1878)

Duroi J. P. Du Roi (1741–1785)
D'Urv. J. S. C. D. d'Urville (1790–1842)
Duthie J. F. Duthie (1845–1922)
Du Tour — Du Tour de Salvert (fl. 1803–1815)
Duval-Jouve J. Duval-Jouve (1810–1883)
Dvořák F. Dvořák (b. 1921)
Dvořáková M. Dvořáková (b. 1940)
Dyer W. T. Thiselton-Dyer (1843–1928)
Ecklon C. F. Ecklon (1795–1868)
Edgew. M. P. Edgeworth (1812–1881)
Edmondston T. B. Edmondston (1825–1846)
Eggli U. Eggli (b. 1959)
Ehrh. J. F. Ehrhart (1742–1795)
Eig A. Eig (1894–1938)
Ekman, Elis. H. M. E. A. E. Ekman (1862–1936)
Elkan L. Elkan (1815–1851)
Emberger M. L. Emberger (1897–1969)
Enander S. J. Enander (1847–1928)
Endl. S. F. L. Endlicher (1804–1849)
Engelm. G. Engelmann (1809–1884)
Engler H. G. A. Engler (1844–1930)
Facch. F. Facchini (1788–1852)
Farwell O. A. Farwell (1867–1944)
Fauché M. Fauché (fl. 1832)
Favarger C. P. E. Favarger (b. 1913)
Fedde F. K. G. Fedde (1873–1942)
Fedtsch., B. B. A. Fedtschenko (1872–1947)
Fée A. L. A. Fée (1789–1874)
Fenzl E. Fenzl (1808–1879)
Fernald M. L. Fernald (1873–1950)
Fernandes, A. A. Fernandes (b. 1906)
Fernandes, R. R. M. S. B. Fernandes (b. 1916)
Fernández, F. F. Fernández González (b. 1956)
Fernández Casas F. J. Fernández Casas (b. 1945)
Fernández Prieto J. A. Fernández Prieto (b. 1950)
Fieschi V. Fieschi (b. *c.* 1910)
Figuerola R. Figuerola Lamata (b. 1953)
Fil. N. Filarszky (1858–1941)
Fingerh. C. A. Fingerhuth (1798–1876)
Finschow G. Finschow (b. 1926)
Fiori A. Fiori (1865–1950)
Fischer F. E. L. von Fischer (1782–1854)
Fitschen J. Fitschen (1869–1947)
Flod., B. B. G. O. Floderus (1867–1941)
Flüggé J. Flüggé (1775–1816)
Foggitt W. Foggitt (1835–1917)
Fomin A. V. Fomin (1869–1935)
Font Quer P. J. M. Font y Quer (Font i Quer) (1888–1964)
Form. E. Formánek (1845–1900)
Forskål P. Forskål (Forsskål, Forskahl) (1732–1763)
Forster, G. J. G. A. Forster (1754–1794)
Forster, T. F. T. F. Forster (1761–1825)
Fouc. J. Foucaud (1847–1904)
Foug. A. D. Fougeroux de Bondaroy (1732–1789)
Fourn., E. E. P. N. Fournier (1834–1884)
Fourn., P. P. V. Fournier (1877–1964)
Fourr. J.-P. Fourreau (1844–1871)
Franco J. M. A. P. do Amaral Franco (b. 1921)
Franklin J. Franklin (1786–1847)
Fraser-Jenkins C. R. F. Jenkins (later C. R. Fraser-Jenkins) (b. 1948)
Freyc. H. L. C. de Saulces de Freycinet (1779–1842)
Freyn J. F. Freyn (1845–1903)
Friedrich H. C. Friedrich (b. 1925)
Fries E. M. Fries (1794–1878)
Fries, Th. T. M. Fries (1832–1913)
Fritsch K. Fritsch, jr. (1864–1934)
Fritze R. Fritze (1841–1903)
Friv. E. Frivaldsky von Frivald (I. Frivaldsky) (1799–1870)
Fröhlich, A. A. Fröhlich (1882–1969)
Fuchs, H. P. H. P. Fuchs Eckert (b. 1928)
Funck H. C. Funck (1771–1839)
Fuss J. M. Fuss (1814–1883)
Gaertner J. Gaertner (1732–1791)
Gaertner, P. P. G. Gaertner (1754–1825)
Gagnebin A. Gagnebin de la Ferrière (1707–1800)
Galiano E. Fernández-Galiano Fernández (b. 1921)
Gamisans J. Gamisans (b. 1944)
Gand. M. Gandoger (1850–1926)
Garcke C. A. F. Garcke (1819–1904)
Gariod C. H. Gariod (1836–1892)
Gars. F. A. P. de Garsault (1691–1778)
Gartner, H. H. Gartner (fl. 1939)
Gaspar. G. Gasparrini (1804–1866)
Gaud.-Beaup. C. Gaudichaud-Beaupré (1789–1854)
Gaudin J. F. A. T. G. P. Gaudin (1766–1833)
Gaussen H. M. Gaussen (1891–1981)
Gavioli O. Gavioli (1871–1944)
Gay J. É. Gay (1786–1864)
Gay, H. Hippolyte Gay (fl. 1885–1894)
Gáyer G. Gáyer (1883–1932)
Geltman D. V. Geltman (b. 1957)
Genn. P. Gennari (1820–1897)
Georgescu C. C. Georgescu (1898–1968)
Georgi J. G. Georgi (1729–1802)
Georgiadis T. Georgiadis (fl. 1977–1985)
Germ. J. N. E. Germain de Saint Pierre (Saint-Pierre) (1815–1882)
Gibby M. Gibby (M. Ambrose) (b. 1949)
Gibelli G. Gibelli (1831–1898)
Gilib. J. E. Gilibert (1741–1814)
Gillies J. Gillies (1792–1834)
Gillot F. X. Gillot (1842–1910)
Ginzberger A. Ginzberger (1873–1940)
Giraud. L. Giraudias (1848–1922)
Gled. J. G. Gleditsch (1714–1786)
Glen H. F. Glen (b. 1950)
Gmelin, C. C. C. C. Gmelin (1762–1837)
Gmelin, J. F. J. F. Gmelin (1748–1804)
Gmelin, J. G. J. G. Gmelin (1709–1755)
Gmelin, S. G. S. G. Gmelin (1744/1745–1774)
Godet C. H. Godet (1797–1879)
Godron D. A. Godron (1807–1880)
Goldie J. Goldie (1793–1886)
Gómez-Campo C. Gómez-Campo (b. 1933)
González-Albo J. González-Albo Campillo (1913–1990)
Goodding L. N. Goodding (1880–1967)
Gornall R. J. Gornall (b. 1951)

Gorodkov B. N. Gorodkov (1890–1953)
Gorschk. S. G. Gorschkova (1889–1972)
Görz, R. R. Görz (1879–1935)
Götz E. Götz (b. 1940)
Gouan A. Gouan (1733–1821)
Grab. H. E. Grabowski (1792–1842)
Graebner K. O. R. P. P. Graebner (1871–1933)
Graebner fil. P. Graebner (1900–1978)
Graells M. de la P. Graells y de la Agüera (1809–1898)
Graham, R. C. R. C. Graham (1786–1845)
Gram, K. K. J. A. Gram (1897–1961)
Grande L. Grande (1878–1965)
Grau H. R. J. Grau (b. 1937)
Gray, A. A. Gray (1810–1888)
Gray, S. F. S. F. Gray (1766–1828)
Grec. D. Grecescu (1841–1910)
Greene, E. L. E. L. Greene (1843–1915)
Gremli A. Gremli (1833–1899)
Gren. J. C. M. Grenier (1808–1875)
Greuter W. R. Greuter (b. 1938)
Grev. R. K. Greville (1794–1866)
Grimm J. F. K. Grimm (1737–1821)
Grinj F. A. Grinj (F. O. Grynj) (b. 1902)
Grinţ., G. G. P. Grinţescu (1870–1947)
Griseb. A. H. R. Grisebach (1814–1879)
Gröntved J. Gröntved (1882–1956)
Gross, H. H. Gross (b. 1888)
Grossh. A. A. Grossheim (1888–1948)
Grubov V. I. Grubov (b. 1917)
Gruner L. F. Gruner (b. 1838)
Gubell. L. Gubellini (b. 1954)
Gueldenst. J. A. von Gueldenstaedt (1745–1781)
Guépin J. P. Guépin (1779–1858)
Guérin J. X. B. Guérin (1775–1850)
Guersent L. B. Guersent (1776–1848)
Guicc. G. Guicciardi (fl. 1855)
Guinea E. Guinea López (1907–1985)
Guinier P. Guinier (1876–1962)
Gulia G. Gulia (1835–1889)
Gunnarsson J. G. Gunnarsson (1866–1944)
Gunnerus J. E. Gunnerus (1718–1773)
Günther, K. K.-F. Günther (b. 1941)
Gürke R. L. A. M. Gürke (1854–1911)
Guss. G. Gussone (1787–1866)
Gustaffson, M. M. A. Gustaffson (b. 1941)
Gustavsson, L.-Å. L.-Å. Gustavsson (b. 1946)
Guterm. W. Gutermann (b. 1935)
Guthnick H. J. Guthnick (1800–1870)
Habl. C. von Hablitz (1752–1821)
Hacq. B. A. Hacquet (1739–1815)
Hadač E. F. L. Hadač (1914–1987)
Haenke T. P. X. Haenke (1761–1816)
Haenseler F. Haenseler (1766–1841)
Hahne A. H. Hahne (1873–1942)
Halácsy E. von Halácsy (1842–1913)
Haller A. von Haller (1708–1777)
Halliday G. Halliday (b. 1933)
Hamet R. Hamet (Raymond-Hamet) (1890–1972)
Hammar O. N. Hammar (1821–1875)
Hammer K. Hammer (b. 1944)
Hampe G. E. L. Hampe (1795–1880)
Hand.-Mazz. H. F. von Handel-Mazzetti (1882–1940)
Hara H. Hara (1911–1986)
Hardy D. S. Hardy (b. 1931)
Hartig H. J. A. R. Hartig (1839–1901)
Hartinger A. Hartinger (1806–1890)
Hartman C. J. Hartman (1790–1849)
Hartvig P. Hartvig (b. 1941)
Hartweg C. T. Hartweg (1812–1871)
Harvey W. H. Harvey (1811–1866)
Hausskn. H. C. Haussknecht (1838–1903)
Haw. A. H. Haworth (1768–1833)
Hawksworth, F. G. F. G. Hawksworth (b. 1926)
Hayek A. von Hayek (1871–1928)
Haynald S. F. L. Haynald (1816–1891)
Hayne F. G. Hayne (1763–1832)
Häyrén E. F. Häyrén (1878–1957)
Hedberg K. O. Hedberg (b. 1923)
Hedge I. C. Hedge (b. 1928)
Hedl., T. J. T. Hedlund (1861–1953)
Hedley G. W. Hedley (1871–1941)
Heer O. von Heer (1809–1883)
Hegelm. C. F. Hegelmaier (1834–1906)
Hegetschw. J. J. Hegetschweiler-Bodmer (1789–1839)
Hegi G. Hegi (1876–1932)
Heldr. T. H. H. von Heldreich (1822–1902)
Heller, A. A. A. A. Heller (1867–1944)
Hendrych R. Hendrych (b. 1926)
Henry, A. A. Henry (1857–1930)
Henry, Louis Louis Henry (1853–1913)
Herbich F. Herbich (1791–1865)
Hermann, F. F. Hermann (1873–1967)
Hernández-Bermejo, J. E. J. E. Hernández-Bermejo (b. 1949)
Herter W. G. F. Herter (1884–1958)
Hervier G. M. J. Hervier-Basson (1846–1900)
Hess, H. H. E. Hess (b. 1920)
Heuffel J. A. Heuffel (1800–1857)
Heynh. G. Heynhold (1800–1860)
Heywood V. H. Heywood (b. 1927)
Hickman J. C. Hickman (b. 1941)
Hiern W. P. Hiern (1839–1925)
Hieron. G. H. E. W. Hieronymus (1846–1921)
Hiitonen H. I. A. Hiitonen (Hidén) (1898–1986)
Hill J. Hill (1714–1778)
Hirzel R. Hirzel (fl. 1967–1980)
Hitchc., E. E. Hitchcock (1793–1864)
Hochst. C. F. Hochstetter (1787–1860)
Hoffm. G. F. Hoffmann (1760–1826)
Hohen. R. F. Hohenacker (1798–1874)
Holl F. Holl (fl. 1820–1850)
Holm T. Holm (1880–1943)
Holmberg O. R. Holmberg (1874–1930)
Holmboe J. Holmboe (1880–1943)
Holmen K. A. Holmen (1921–1974)
Holmgren N. H. Holmgren (b. 1937)
Holub, J. J. Holub (b. 1930)
Holzm. T. Holzmann (b. 1843)
Hooker W. J. Hooker (1785–1865)

Hooker fil. J. D. Hooker (1817–1911)
Hoppe D. H. Hoppe (1760–1846)
Hornem. J. W. Hornemann (1770–1841)
Hornsch. C. F. Hornschuch (1793–1850)
Hornung E. G. Hornung (1795–1862)
Horvatić S. Horvatić (1899–1975)
Horvátovszky S. Horvátovszky (fl. 1770–1774)
Host N. T. Host (1761–1834)
Houtt. M. Houttuyn (1720–1798)
Houtzagers G. Houtzagers (1888–1957)
Howard H. W. Howard (b. 1913)
Howell T. J. Howell (1842–1912)
Huber, H. Hans Huber (b. 1919)
Huber, J. A. J. A. Huber (b. 1899)
Hudson W. Hudson (1730–1793)
Huet A. Huet du Pavillon (1829–1907)
Hülphers K. A. Hülphers (1882–1948)
Hultén O. E. G. Hultén (1894–1981)
Humb. F. W. H. A. von Humboldt (1769–1859)
Hussenot L. C. S. L. Hussenot (1809–1845)
Huter R. Huter (1834–1909)
Huth E. Huth (1845–1897)
Hy F. C. Hy (1853–1918)
Hyl. N. Hylander (1904–1970)
Iatroú G. A. Iatroú (b. 1949)
Igoshina K. N. Igoshina (b. 1894)
Ikonn. S. S. Ikonnikov (b. 1931)
Iljin M. M. Iljin (Il'in, Ilyin) (1889–1967)
Iltis H. H. Iltis (b. 1925)
Ionescu M. A. Ionescu (b. 1900)
Iranzo J. Iranzo Reig (b. 1942)
Irmscher E. Irmscher (1887–1968)
Iwatsuki K. Iwatsuki (b. 1934)
Jackson, A. B. A. B. Jackson (1876–1947)
Jackson, A. K. A. K. Jackson (b. 1914)
Jacq. N. J. von Jacquin (1727–1817)
Jahandiez É. Jahandiez (1876–1938)
Jakobsen K. Jakobsen (b. 1924)
Jalas A. J. J. Jalas (b. 1920)
Jan G. Jan (1791–1866)
Janchen E. E. A. Janchen (1882–1970)
Jancz. E. Janczewski von Glinka (1846–1918)
Janisch. D. E. Janischewsky (1875–1944)
Janka V. Janka von Bulcs (1837–1890)
Jaub. H. F. Jaubert (1798–1874)
Jáv. S. A. Jávorka (1883–1961)
Jeanb. E.-M.-J. Jeanbernat (1835–1888)
Jeanmonod D. Jeanmonod (b. 1953)
Jelen. A. G. Jelenevsky (Elenevsky) (b. 1928)
Jermy A. C. Jermy (b. 1932)
Joh., K. K. Johansson (1856–1928)
Jones, B. M. G. B. M. G. Jones (b. 1933)
Jonsell B. Jonsell (b. 1936)
Jordan C. T. A. Jordan (1814–1897)
Jordanov D. Jordanov (1893–1978)
Junge P. Junge (1881–1919)
Juratzka J. Juratzka (1821–1878)
Jurtzev B. A. Jurtzev (b. 1932)
Juss. A. L. de Jussieu (1748–1836)
Juz. S. V. Juzepczuk (1893–1959)
Kablík J. Kablík (J. Kablíková) (1787–1863)
Kadereit J. W. Kadereit (b. 1956)
Kaercher W. Kaercher (b. 1933)
Kalela A. A. A. Kalela (1908–1977)
Kalenicz. J. Kaleniczenko (1805–1876)
Kanitz A. Kanitz (1843–1896)
Kar. G. S. S. Karelin (1801–1872)
Karsten G. K. W. H. Karsten (1817–1908)
Kato M. Kato (b. 1946)
Kaulfuss G. F. Kaulfuss (1786–1830)
Keller, R. R. Keller (1854–1939)
Kellerer, J. J. Kellerer (b. 1859)
Kent D. H. Kent (b. 1920)
Ker-Gawler J. B. Ker (J. Gawler) (1764–1842)
Kerguélen, M. M. F.-J. Kerguélen (b. 1928)
Kerner, A. A. J. R. Kerner von Marilaun (1831–1898)
Kersten O. Kersten (1839–1900)
Kihlman A. O. Kihlman (Kairamo) (1858–1938)
Kindb. N. C. Kindberg (1832–1910)
Kir. I. P. Kirilow (1821–1842)
Kirk T. Kirk (1828–1898)
Kirschleger F. R. Kirschleger (1804–1869)
Kit. P. Kitaibel (1757–1817)
Kitanov B. P. Kitanov (b. 1912)
Kittel B. M. Kittel (1798–1885)
Kleopow J. D. Kleopow (1902–1942)
Klika J. Klika (1888–1957)
Klinggr. C. J. M. von Klinggräff (1809–1879)
Klokov M. V. Klokov (1896–1981)
Klokov fil. V. M. Klokov (fl. 1968–1972)
Klotzsch J. F. Klotzsch (1805–1860)
Knaben G. Knaben (b. 1911)
Knaf J. Knaf (1801–1865)
Knight J. Knight (1777–1855)
Knoche E. L. H. Knoche (1870–1945)
Knorring O. E. Knorring (later O. E. Knorring-Neustrujeva) (1887–1977)
Koch W. D. J. Koch (G. D. I. Koch) (1771–1849)
Koch, C. K. (C.) H. E. Koch (1809–1879)
Koch, Walo Walo Koch (1896–1956)
Koehne B. A. E. Koehne (1848–1918)
Koelle J. L. C. Koelle (1763–1797)
Koerte F. Koerte (1782–1845)
Koidz. Gen-Iti (Gen'ichi) Koidzumi (1883–1953)
Komarov V. L. Komarov (1869–1945)
Kondrat. E. N. Kondratjuk (b. 1914)
König, D. D. König (b. 1909)
Korsh. S. I. Korshinsky (1861–1900)
Kos.-Pol. B. M. Koso-Poliansky (1890–1957)
Kotov M. I. Kotov (1896–1978)
Kotschy C. G. T. Kotschy (1813–1866)
Kotula, A. A. Kotula (1822–1891)
Kovacev I. G. Kovacev (Kovačev, Kovatschev) (b. 1927)
Kovanda M. Kovanda (b. 1936)
Kožuharov S. I. Kožuharov (b. 1933)
Kralik J. L. Kralik (1813–1892)
Krašan F. Krašan (1840–1907)
Krause, E. H. L. E. H. L. Krause (1859–1942)

Krause, K. K. Krause (1883–1963)
Krecz., V. V. I. Kreczetowicz (1901–1942)
Krocker A. J. Krocker (1744–1823)
Krylov P. N. Krylov (1850–1931)
Kühlew. P. E. Kühlewein (1798–1870)
Kuhn F. A. M. Kuhn (1842–1894)
Kukkonen I. T. K. Kukkonen (b. 1926)
Kulcz. S. Kulczyński (1895–1975)
Kümmerle J. B. Kümmerle (1876–1931)
Kunth C. S. Kunth (1788–1850)
Kuntze, O. K. (C.) E. O. Kuntze (1843–1907)
Kunz, H. H. Kunz (b. 1904)
Kunze, G. G. Kunze (1793–1851)
Küpfer P. Küpfer (b. 1942)
Kurata S. Kurata (b. 1922)
Kusn. N. I. Kusnezow (Kuznetzov) (1864–1932)
Kuzen. O. I. Kuzeneva (1887–1978)
Kuzinský P. A. von Kuzinský (fl. 1889)
Kuzmanov B. A. Kuzmanov (1934–1991)
L. C. von Linné (C. Linnaeus) (1707–1778)
L. fil. C. von Linné (1741–1783)
Labill. J. J. H. de Labillardière (1755–1834)
Lacaita C. C. Lacaita (1853–1933)
Laest. L. L. Laestadius (1800–1861)
Lag. M. Lagasca (La Gasca) y Segura (1776–1839)
Lagerh. N. G. von Lagerheim (1860–1926)
Lagger F.-J. Lagger (1799–1870)
Lago M. E. Lago Canzobre (b. 1952)
Lagrèze-Fossat A. R. A. Lagrèze-Fossat (1814–1874)
Laínz P. M. Laínz Gallo (b. 1923)
Lam. J. B. A. P. Monnet de Lamarck (la Marck) (1744–1829)
Lamb. A. B. Lambert (1761–1842)
Lambinon J. Lambinon (b. 1936)
Lamotte M. Lamotte (1820–1883)
Landolt E. Landolt (b. 1926)
Láng, A. F. A. F. Láng (1795–1863)
Lange J. M. C. Lange (1818–1898)
Langsd. G. H. von Langsdorff (1774–1852)
Lapeyr. P. Picot de Lapeyrouse (1744–1818)
La Pylaie A. J. M. B. de la Pylaie (1786–1856)
Lauche F. W. G. Lauche (1827–1883)
Launert G. O. E. Launert (b. 1926)
La Valva V. La Valva (b. 1947)
Lausi D. Lausi (b. 1923)
Lavrenko E. M. Lavrenko (1900–1987)
Lawalrée A. G. C. Lawalrée (b. 1921)
Lawson, C. C. Lawson (1794–1873)
Lawson, P. P. Lawson (d. 1820)
Laxm. E. Laxman (Laxmann) (1737–1796)
Layens G. de Layens (1834–1897)
Láz.-Ibiza Blas Lázaro é Ibiza (1858–1921)
Leadlay E. A. Leadlay (fl. 1990)
Lebel J. E. Lebel (1801–1878)
Lecoq H. Lecoq (1802–1871)
Lecoyer J.-C. Lecoyer (1835–1899)
Ledeb. C. F. von Ledebour (1785–1851)
Leers J. G. D. Leers (1727–1774)
Le Gall N. J. M. le Gall de Kerlinov (1787–1860)
Legrand A. Legrand (le Grand) (1839–1905)
Lehm. J. G. C. Lehmann (1792–1860)
Lehm., C. B. C. B. Lehmann (1811–1875)
Lej. A. L. S. Lejeune (1779–1858)
Le Jolis A. F. Le Jolis (1823–1904)
Lemaire A. C. Lemaire (1800–1871)
Léman D. S. Léman (1781–1829)
Léonard, J. J. J. G. Léonard (b. 1920)
Lepechin I. I. Lepechin (1737–1802)
Leresche L. F. J. R. Leresche (1808–1885)
Lesp. J. M. G. Lespinasse (1807–1876)
Less. C. F. Lessing (1809–1862)
Lester-Garland L. V. Lester-Garland (1860–1944)
Léveillé A. A. H. Léveillé (1863–1918)
Levier E. Levier (1838–1911)
Leybold F. Leybold (1827–1879)
L'Hér. C.-L. L'Héritier de Brutelle (1746–1800)
Lid J. Lid (1886–1971)
Lidén M. Lidén (b. 1951)
Liebl. F. K. Lieblein (1744–1810)
Liebm. F. M. Liebmann (1813–1856)
Liljeblad S. Liljeblad (1761–1815)
Lindb., H. H. Lindberg (1871–1963)
Lindblad M. A. Lindblad (1821–1899)
Lindblom A. E. Lindblom (1807–1853)
Lindley J. Lindley (1799–1865)
Lindman C. A. M. Lindman (1856–1928)
Lindquist S. B. G. Lindquist (1904–1963)
Lindt. V. Lindtner (1904–1965)
Link J. H. F. Link (1767–1851)
Lipsky V. I. Lipsky (1863–1937)
Litard. R. V. de Litardière (1888–1957)
Litv. D. I. Litvinov (Litwinow) (1854–1929)
Lloyd J. Lloyd (1810–1896)
Loefl. P. Loefling (1729–1756)
Loisel. J. L. A. Loiseleur-Deslongchamps (1774–1849)
Lojac. M. Lojacono-Pojero (1853–1919)
Lomax A. E. Lomax (1861–1894)
Long G. Long (b. 1928)
Lonsing A. Lonsing (fl. 1939)
López, G. G. A. López González (b. 1950)
Lorent J. A. Lorent (1812–1884)
Loret H. Loret (1811–1888)
Losa T. M. Losa España (1893–1966)
Loscos F. Loscos y Bernál (1823–1886)
Loscos, C. C. Loscos (fl. 1886)
Loudon J. C. Loudon (1783–1843)
Lour. J. de Loureiro (1717–1791)
Löve, Á. Á. Löve (b. 1916)
Löve, D. D. B. M. Löve (b. 1918)
Lovis J. D. Lovis (b. 1930)
Lowe R. T. Lowe (1802–1874)
Luerssen C. Luerssen (1843–1916)
Luizet D. Luizet (1851–1930)
Lund, N. N. Lund (1814–1847)
Lundström, E. C. E. Lundström (b. 1882)
Lusby P. S. Lusby (b. 1953)
Lynge B. A. Lynge (1884–1942)
MacMillan C. MacMillan (1867–1929)
Mach.-Laur. B. Machatschki-Laurich (fl. 1926)

Maillard P. N. Maillard (1813–1883)
Maire R. C. J. E. Maire (1878–1949)
Major C. I. F. Major (C. I. Forsyth-Major) (1843–1923)
Malagarriga T. L. R. M. de P. Malagarriga [i] Heras (1904–1990)
Malinovski E. Malinovski (Malinowski) (1885–1979)
Maly, J. Joseph Karl Maly (1797–1866)
Malý, K. Karl Malý (1874–1951)
Mansfeld R. Mansfeld (1901–1960)
Manton I. Manton (1904–1988)
Marchesetti C. de Marchesetti (1850–1926)
Marès P. Marès (1826–1900)
Margot H. Margot (1807–1894)
Marhold K. Marhold (b. 1959)
Mariz J. de Mariz (1847–1916)
Markgraf F. Markgraf (1897–1987)
Marshall H. Marshall (1722–1801)
Marshall, E. S. E. S. Marshall (1858–1919)
Mart., C. F. P. C. F. P. von Martius (1794–1868)
Martelli, U. U. Martelli (1860–1934)
Martínez-Laborde J. B. Martínez-Laborde (b. 1955)
Masclans F. Masclans [i] Girvès (b. 1905)
Masters M. T. Masters (1833–1907)
Mateo G. Mateo Sanz (later G. Mateo-Sanz) (b. 1953)
Mattei G. E. Mattei (1865–1943)
Mattf. J. Mattfeld (1895–1951)
Matthäs U. Matthäs (b. 1949)
Mattuschka H. G. von Mattuschka (1734–1779)
Mauri E. Mauri (1791–1836)
Maxim. K. (C.) J. Maximowicz (1827–1891)
Maxon W. R. Maxon (1877–1948)
Mayer, E. E. Mayer (b. 1920)
Mayer, J. J. C. A. Mayer (1747–1801)
McClell. J. McClelland (1805–1883)
McNeill J. McNeill (b. 1933)
Medicus F. C. Medicus (F. K. Medikus) (1736–1808)
Medw. J. S. Medwedew (1847–1923)
Meerb. N. Meerburgh (1734–1814)
Meigen F. C. Meigen (b. 1864)
Meikle R. D. Meikle (b. 1923)
Meissner C. F. Meissner (before 1861, Meisner) (1800–1874)
Mela A. J. Mela (A. J. Malmberg) (1846–1904)
Melderis A. Melderis (1909–1986)
Melville R. Melville (1903–1985)
Melzh. V. Melzheimer (b. 1939)
Mendes E. J. S. M. Mendes (b. 1924)
Menéndez Amor J. Menéndez Amor (b. 1916)
Menyh. L. Menyhárth (1849–1897)
Mérat F. V. Mérat de Vaumartoise (1780–1851)
Merino B. Merino y Román (1845–1917)
Merr. E. D. Merrill (1876–1956)
Mert. F. K. Mertens (1764–1831)
Merxm. H. Merxmüller (1920–1988)
Mett. G. H. Mettenius (1823–1866)
Metzger J. Metzger (1789–1852)
Meyer, B. B. Meyer (1767–1836)
Meyer, C. A. C. A. von Meyer (1795–1855)
Meyer, D. E. D. E. Meyer (1926–1982)
Meyer, E. H. F. E. H. F. Meyer (1791–1858)
Meyer, F. K. F. K. Meyer (b. 1926)
Meyer, G. F. W. G. F. W. Meyer (1782–1856)
Micevski K. Micevski (Mitsevski) (b. 1926)
Michx A. Michaux (1746–1802)
Michx fil. F. A. Michaux (1770–1855)
Middendorff A. T. von Middendorff (1815–1894)
Miégeville J. Miégeville (1819–1901)
Milde C. A. J. Milde (1824–1871)
Miller P. Miller (1691–1771)
Mirbel C. F. B. de Mirbel (1776–1854)
Mitch., D. D. S. Mitchell (b. 1935)
Mitterp. L. Mitterpacher von Mitterburg (1734–1814)
Moench C. Moench (1744–1805)
Moessler J. C. Moessler (Mössler) (fl. 1805–1835)
Mohr D. M. H. Mohr (1780–1808)
Mokry F. Mokry (fl. 1986)
Moldenke H. N. Moldenke (b. 1909)
Molero J. Molero Briones (b. 1946)
Molina J. I. Molina (G. I. Molina) (1737–1829)
Monnier, P. P. C. J. Monnier (b. 1922)
Montandon F. J. Montandon (fl. 1856)
Montelucci G. Montelucci (1899–1983)
Monts., J. M. J. M. Montserrat Martí (b. 1955)
Monts., P. P. Montserrat Recoder (b. 1918)
Moore, T. T. Moore (1821–1887)
Moq. C. H. B. A. Moquin-Tandon (1804–1863)
Morais, T. A. A. Taborda de Morais (1900–1959)
Moraldo B. Moraldo (b. 1938)
Moreno M. Moreno Sanz (b. 1948)
Moretti G. L. Moretti (1782–1853)
Mori, T. T. Mori (1884–1962)
Moric. M. E. ('Stefano') Moricand (1779–1854)
Moris G. G. Moris (1796–1859)
Moritzi A. Moritzi (1806–1850)
Morot L. R. M. F. Morot (1854–1915)
Morren C. J. E. Morren (1833–1886)
Morton, C. V. C. V. Morton (1905–1972)
Möschl W. Möschl (1906–1981)
Moss C. E. Moss (1870–1930)
Motelay L. Motelay (1830–1917)
Mueller, F. F. J. H. von Mueller (1825–1896)
Mueller, O. F. O. F. Mueller (Müller) (1730–1784)
Muenchh. O. von Muenchhausen (1716–1774)
Muhl. G. H. E. Muhlenberg (1753–1815)
Müller Arg. J. Müller of Aargau (Argoviensis) (1828–1896)
Munby G. Munby (1813–1876)
Münch E. Münch (1876–1946)
Muñoz Garmendia J. F. Muñoz Garmendia (b. 1949)
Munz P. A. Munz (1892–1974)
Murb. S. S. Murbeck (1859–1946)
Murith L. J. Murith (1742–1816)
Murr, J. J. Murr (1864–1932)
Murray J. A. Murray (1740–1791)
Murray, A. A. Murray (1812–1878)
Mutel P. A. V. Mutel (1795–1847)
Mutis J. C. B. Mutis y Bosio (1732–1808)
Nakai T. Nakai (1882–1952)
Namegata T. Namegata (fl. 1952–1961)
Nardi E. Nardi (b. 1942)
Nasarow M. I. Nasarow (1882–1942)

Necker N. J. de Necker (1730–1793)
Nees C. G. D. Nees von Esenbeck (1776–1858)
Nees, T. T. F. L. Nees von Esenbeck (1787–1837)
Neilr. A. Neilreich (1803–1871)
Nelson, A. A. Nelson (1859–1952)
Nestler C. G. Nestler (1778–1832)
Neuman L. M. Neuman (1852–1922)
Neumann, A. A. Neumann (1916–1973)
Neumayer, H. H. Neumayer (1887–1945)
Neves, J. J. de Barros Neves (1914–1982)
Nevski S. A. Nevski (1908–1938)
Newbould W. W. Newbould (1819–1886)
Newman E. Newman (1801–1876)
Nicotra L. Nicotra (1846–1940)
Nitz. T. Nitzelius (b. 1914)
Nobre A. Nobre (b. 1865)
Nolte E. F. Nolte (1791–1875)
Nordh. R. Nordhagen (1894–1979)
Nordm. A. D. von Nordmann (1803–1866)
Nordstedt C. F. O. Nordstedt (1838–1924)
Norrlin J. P. Norrlin (1842–1917)
Notø A. Notø (1865–1948)
Novák F. A. Novák (1892–1964)
Nutt. T. Nuttall (1786–1859)
Nyárády, A. A. Nyárády (b. 1920)
Nyárády, E. I. E. I. Nyárády (1881–1966)
Nyl., F. F. Nylander (1820–1880)
Nyman C. F. Nyman (1820–1893)
Oberholzer E. Oberholzer (fl. 1886–1950)
Oeder G. C. Oeder (1728–1791)
Ohwi J. Ohwi (1905–1977)
Oken L. Oken (1779–1851)
Olivier G. A. Olivier (1756–1814)
Opiz P. M. Opiz (1787–1858)
Orlova N. I. Orlova (b. 1921)
Orph. T. G. Orphanides (1817–1886)
Örsted A. S. Örsted (Ørsted, Oersted) (1816–1872)
Ortega C. Gómez de Ortega (1740–1818)
Ostenf. C. E. H. Ostenfeld (1873–1931)
Otth K. A. Otth (1803–1839)
Ovcz. P. N. Ovczinnikov (1903–1979)
Pacher D. Pacher (1816–1902)
Pacz. I. K. Paczoski (1864–1942)
Padmore P. A. Padmore (b. 1929)
Paegle B. Paegle (fl. 1927)
Page, C. N. C. N. Page (b. 1942)
Paiva J. A. Rodrigues de Paiva (b. 1933)
Palassou P. B. Palassou (1745–1830)
Pallas P. S. Pallas (1741–1811)
Pamp. R. Pampanini (1875–1949)
Pamukç. A. Pamukçuoğlu (b. 1929)
Pančić J. Pančić (1814–1888)
Panizzi F. Panizzi-Savio (1817–1893)
Panov P. P. Panov (b. 1932)
Pant. J. Pantocsek (1846–1916)
Paol. G. Paoletti (1865–1941)
Papanicolaou K. Papanicolaou (b. 1947)
Pardo J. Pardo y Sastrón (1822–1909)
Parfenov V. I. Parfenov (b. 1934)
Parl. F. Parlatore (1816–1877)
Parnell, J. J. A. N. Parnell (b. 1954)
Parodi L. R. Parodi (1895–1966)
Parry W. E. Parry (1790–1855)
Passer. G. Passerini (1816–1893)
Patrin E. L. M. Patrin (1742–1815)
Patze C. A. Patze (1808–1892)
Pau C. Pau y Español (1857–1937)
Pavlov N. V. Pavlov (1893–1971)
Pavón J. A. Pavón y Jiménez (1754–1844)
Pavone P. Pavone (b. 1948)
Pawł. B. Pawłowski (1898–1971)
Pawł., S. S. Pawłowska (b. 1905)
Pax F. A. Pax (1858–1942)
Pénzes A. Pénzes (b. 1896)
Pérard A. J. C. Pérard (1834–1887)
Pereda J. M. de Pereda Sáez (1909–1972)
Pérez-Lara J. M. Pérez-Lara (1841–1918)
Perf. I. A. Perfiljew (1882–1942)
Perr. E. P. Perrier de la Bâthie (1825–1916)
Perring F. H. Perring (b. 1927)
Pers. C. H. Persoon (1761–1836)
Péterfi M. Péterfi (1875–1922)
Peterm. W. L. Petermann (1806–1855)
Petit, E. E. C. N. Petit (1817–1893)
Petitmengin M. G. C. Petitmengin (1881–1908)
Petrov V. A. Petrov (1896–1955)
Petrova A. V. Petrova (fl. 1973→)
Petrovič S. Petrovič (1839–1889)
Phillips, E. P. E. P. Phillips (1884–1967)
Phipps, C. J. C. J. Phipps (1744–1792)
Phitos D. Phitos (b. 1930)
Pichi Serm. R. E. G. Pichi Sermolli (b. 1912)
Pierrat D. Pierrat (1835–1895)
Pignatti S. Pignatti (b. 1930)
Pilger R. K. F. Pilger (1876–1953)
Piller M. Piller (1733–1788)
Piré L. A. H. J. Piré (1827–1887)
Pirona G. A. Pirona (1882–1895)
Pissjauk. V. V. Pissjaukowa (b. 1906)
Planchon J. E. Planchon (1823–1888)
Planellas J. Planellas Giralt (1821–1888)
Pobed. E. G. Pobedimova (1898–1973)
Podl. D. Podlech (b. 1931)
Podobnik A. Podobnik (fl. 1987)
Podp. J. Podpěra (1878–1954)
Poeppig E. F. Poeppig (1798–1868)
Poggenb. J. F. Poggenburg (1840–1893)
Poiret J. L. M. Poiret (1755–1834)
Pojark. A. I. Pojarkova (1897–1980)
Polatschek A. Polatschek (b. 1932)
Poldini L. Poldini (b. 1930)
Pomel A. N. Pomel (1821–1898)
Popl. G. I. Poplavskaja (Poplawska) (1885–1956)
Popov, M. M. G. Popov (1893–1955)
Porsild M. P. Porsild (1872–1956)
Porta P. Porta (1832–1893)
Portenschl. F. E. von Portenschlag-Ledermayer (1772–1822)
Post G. E. Post (1838–1909)

Pourret P. A. Pourret de Figeac (1754–1818)
Pouzar Z. Pouzar (b. 1932)
Prada M. del Carmen I. Prada Moral (b. 1953)
Praeger R. L. Praeger (1865–1953)
Prantl K. A. E. Prantl (1849–1893)
Presl, C. C. (K.) B. Presl (1794–1852)
Presl, J. J. S. Presl (1791–1849)
Preston, C. D. C. D. Preston (b. 1955)
Price W. R. Price (1886–1975)
Pritzel, G. A. G. A. Pritzel (1815–1874)
Prodan J. Prodan (1875–1959)
Pugsley H. W. Pugsley (1868–1947)
Purkyně E. Purkyně (1831–1882)
Pursh F. T. Pursh (1774–1820)
Quézel P. Quézel (b. 1926)
Rabenh. G. L. Rabenhorst (1806–1881)
Racib. M. Raciborski (1863–1917)
Raffaelli M. Raffaelli (b. 1944)
Rafin. C. S. Rafinesque-Schmaltz (1783–1840)
Rafn C. G. Rafn (1769–1808)
Ramond L. F. E. de Carbonnières Ramond (1753–1829)
Rändel U. Rändel (b. 1941)
Rasbach, H. H. Rasbach (b. 1924)
Rasbach, K. K. Rasbach (b. 1923)
Ratter J. A. Ratter (b. 1934)
Raunk. C. C. Raunkiær (1860–1938)
Rauschert S. Rauschert (1931–1986)
Rech. K. Rechinger (1867–1952)
Rech. fil. K. H. Rechinger (b. 1906)
Regel E. A. von Regel (1815–1892)
Rehder A. Rehder (1863–1949)
Rehmann A. Rehmann (Rehman) (1840–1917)
Reichard J. J. Reichard (1743–1782)
Reiche K. F. Reiche (1860–1929)
Reichenb. H. G. L. Reichenbach (1793–1879)
Reichenb. fil. H. G. Reichenbach (1824–1889)
Reichstein T. Reichstein (b. 1897)
Rendle A. B. Rendle (1865–1938)
Req. E. Requien (1788–1851)
Resvoll-Holmsen H. Resvoll-Holmsen (1873–1943)
Retz. A. J. Retzius (1742–1821)
Reuss, G. G. Reuss (1818–1861)
Reuter G. F. Reuter (1805–1872)
Reveal J. L. Reveal (b. 1941)
Revel J. Revel (1811–1883)
Reyn. A. Reynier (1845–1932)
Ricci A. M. Ricci (1777–1850)
Richard, A. A. Richard (1794–1852)
Richard, L. C. M. L. C. M. Richard (1754–1821)
Richardson J. Richardson (1787–1865)
Richter H. E. F. Richter (1808–1876)
Richter, K. K. Richter (1855–1891)
Rico E. Rico Hernández (b. 1953)
Riddelsd. H. J. Riddelsdell (1866–1941)
Riedl, H. H. Riedl (b. 1936)
Rigo G. Rigo (1841–1922)
Rikli M. A. Rikli (1868–1951)
Rink H. J. Rink (1819–1893)
Risse H. Risse (1948–1989)
Risso J. A. Risso (1777–1845)
Ritter H. Ritter-Studnička (b. 1911)
Rivas Goday S. Rivas Goday (1905–1981)
Rivas Martínez S. Rivas Martínez (b. 1935)
Robert G. N. Robert (1776–1857)
Robill. L. M. A. Robillard d'Argentelle (1777–1828)
Robson E. Robson (1763–1813)
Rochel A. Rochel (1770–1847)
Rodr. J. D. Rodriguez (1780–1846)
Roemer J. J. Roemer (1763–1819)
Roemer, R. de R. B. de [von] Roemer (fl. 1828–1852)
Rogow. A. S. Rogowicz (1812–1878)
Rohlena J. Rohlena (1874–1944)
Röhling J. C. Röhling (1757–1813)
Rohrb. P. Rohrbach (1847–1871)
Ronniger K. Ronniger (1871–1954)
Ross, J. J. Ross (1777–1856)
Rossi L. Rossi (1850–1932)
Rössler W. Rössler (b. 1909)
Rostrup F. G. E. Rostrup (1831–1907)
Roth A. W. Roth (1757–1834)
Rothm. W. H. P. Rothmaler (1908–1962)
Rottb. C. F. Rottboell (Rottbøll) (1727–1797)
Rouleau J. A. E. Rouleau (1916–1991)
Rouy G. C. C. Rouy (1851–1924)
Roxb. W. Roxburgh (1751–1815)
Royle J. F. Royle (1798–1858)
Rozier F. (J.-F.) Rozier (1734–1793)
Rubtsov N. I. Rubtsov (Rubtzov) (1907–1988)
Rudolph, J. H. J. H. Rudolph (1744–1809)
Ruiz H. Ruiz López (1754–1816)
Rune N. O. Rune (b. 1919)
Runemark H. Runemark (b. 1927)
Rupr. F. J. Ruprecht (1814–1870)
Russell, A. A. Russell (*c.* 1715–1768)
Rydb. P. A. Rydberg (1860–1931)
Rylands T. G. Rylands (1818–1900)
Sabine J. Sabine (1770–1837)
Sabr. H. Sabransky (1864–1915)
Sadler J. Sadler (1791–1849)
Sagorski E. A. Sagorski (1847–1929)
Sagredo R. Sagredo (b. 1899)
Salisb. R. A. Salisbury (1761–1829)
Salman A. H. P. M. Salman (fl. 1976–1977)
Salter S. J. A. Salter (1825–1897)
Salvo A. E. Salvo Tierra (b. 1957)
Salzm. P. Salzmann (1781–1851)
Sam. G. Samuelsson (1885–1944)
Sambuk F. V. Sambuk (1900–1942)
Samp. G. A. da Silva Ferreira Sampaio (1865–1937)
Sánchez J. Sánchez Sánchez (b. 1942)
Sándor I. Sándor (b. 1853)
Santi, G. G. Santi (1746–1822)
Sapjegin A. A. Sapjegin (1883–1946)
Sarato C. Sarato (1830–1893)
Sarg. C. S. Sargent (1841–1927)
Sart. G. B. Sartorelli (1780–1853)
Sarvela J. Sarvela (b. 1914)
Sauer, W. W. Sauer (b. 1935)

Sauter A. E. Sauter (1800–1881)
Sauzé J. C. Sauzé (1815–1889)
Săvul. T. Săvulescu (1889–1963)
Scaling W. Scaling (fl. 1863–1882)
Schaeffer J. C. Schaeffer (1718–1790)
Schaeftlein H. Schaeftlein (1886–1973)
Scheele G. H. A. Scheele (1808–1864)
Schelkownikow A. B. Schelkownikow (Schelkovnikov) (1870–1933)
Schellm. C. Schellmann (fl. 1938)
Schenk J. A. Schenk (1815–1891)
Scherb. J. Scherbius (1769–1813)
Scheutz N. J. W. Scheutz (1836–1889)
Schiffner V. F. Schiffner (1862–1944)
Schimper G. H. W. Schimper (1804–1878)
Schimper, C. C. F. Schimper (1803–1867)
Schinz H. Schinz (1858–1941)
Schipcz. N. V. Schipczinski (1886–1955)
Schischkin B. K. Schischkin (1886–1963)
Schkuhr C. Schkuhr (1741–1811)
Schlecht. D. F. L. von Schlechtendal (1794–1866)
Schleicher J. C. Schleicher (1768–1834)
Schlosser J. C. Schlosser von Klekovski (1808–1882)
Schmalh. J. T. Schmalhausen (I. F. Schmal'hausen) (1849–1894)
Schmid, E. Emil Schmid (b. 1891)
Schmidel C. C. Schmidel (1718–1792)
Schmidt, F. W. Franz Wilibald Schmidt (1764–1796)
Schmidt Petrop., Friedrich Friedrich [Karl] Schmidt of St Petersburg (1832–1908)
Schneider, C. K. C. K. Schneider (1876–1951)
Schnittspahn G. F. Schnittspahn (1810–1865)
Scholz, J. B. J. B. Scholz (1858–1915)
Schönl. S. Schönland (Schonland) (1860–1940)
Schott H. W. Schott (1794–1865)
Schousboe P. K. A. Schousboe (1766–1832)
Schrader H. A. Schrader (1767–1836)
Schrank F. von Paula von Schrank (1747–1835)
Schreber J. C. D. von Schreber (1739–1810)
Schrenk A. G. von Schrenk (1816–1876)
Schrödinger R. Schrödinger (1857–1919)
Schultes J. A. Schultes (1773–1831)
Schultz, C. F. C. F. Schultz (1765/6–1837)
Schultz, F. W. F. W. Schultz (1804–1876)
Schulz, O. E. O. E. Schulz (1874–1936)
Schumacher A. Schumacher (1893–1975)
Schur P. J. F. Schur (1799–1878)
Schwantes M. H. G. Schwantes (1881–1960)
Schwarz, A. A. F. Schwarz (1852–1915)
Schwarz, O. O. Schwarz (1900–1983)
Schwegler H. W. Schwegler (b. 1929)
Schweigger A. F. Schweigger (1783–1821)
Schweinf. G. A. Schweinfurth (1836–1925)
Scop. J. (G.) A. Scopoli (1723–1788)
Scott, A. J. A. J. Scott (b. 1950)
Sebastiani F.-A. Sebastiani (1782–1821)
Sebeók A. Sebeók de Szent-Miklós (fl. 1779–1780)
Seem. B. C. Seeman (1825–1871)
Séguier J. F. Séguier (1703–1784)
Seitz, W. W. Seitz (b. 1940)
Selander N. S. E. Selander (1891–1957)
Sell, P. D. P. D. Sell (b. 1929)
Semen., N. N. Z. Semenova-Tjan-Schanskaja (1906–1960)
Sennen Frère Sennen (E. M. Grenier-Blanc) (1861–1937)
Ser. N. C. Seringe (1776–1858)
Serres J. J. Serres (1790–1858)
Seub. M. A. Seubert (1818–1878)
Shivas M. G. Shivas (b. 1926)
Shuttlew., R. J. R. J. Shuttleworth (1810–1874)
Sibth. J. Sibthorp (1758–1796)
Sieber F. W. Sieber (1789–1844)
Siebold P. F. von Siebold (1796–1866)
Signorello P. Signorello (b. 1939)
Sikura J. J. Sikura (b. 1932)
Silva, P. A. R. Pinto da Silva (b. 1912)
Sim, R. R. Sim (1791–1878)
Simmler G. Simmler (b. 1884)
Simmons H. G. Simmons (1866–1943)
Simon primus, E. E. Simon (1848–1924)
Simon secundus, E. E. E. Simon (1871–1967)
Simonkai L. von Simonkai (L. P. Simkovics) (1851–1910)
Simon-Louis L. L. Simon-Louis (1834–1913)
Sims J. Sims (1749–1831)
Sint. P. E. E. Sintenis (1847–1907)
Sipliv. V. N. Siplivinskij (b. 1937)
Širj. G. I. Širjaev (Schirjaev) (1882–1954)
Skvortsov, A. A. K. Skvortsov (b. 1920)
Sloboda D. Sloboda (1809–1888)
Slosson M. Slosson (b. 1872)
Sm. J. E. Smith (1759–1828)
Sm., J. John Smith (1798–1888)
Sm., K. A. H. K. A. H. ('Harry') Smith (1889–1971)
Sm., M. C. M. C. C. Smith (1933–1984)
Small J. K. Small (1869–1938)
Smirnov P. A. Smirnov (1896–1980)
Snogerup S. E. Snogerup (b. 1939)
Soczava V. B. Soczava (b. 1905)
Solander D. C. Solander (1733–1782)
Solemacher J. V. L. A. G. Solemacher-Antweiler (b. 1889)
Soler A. Soler Hernando (b. 1940)
Solms-Laub. H. M. C. L. F. zu Solms-Laubach (1842–1915)
Sommer. I. Sommerauer (d. 1854)
Sommier C. P. S. Sommier (1848–1922)
Sommerf. S. C. Sommerfelt (1794–1838)
Sonder O. W. Sonder (1812–1881)
Song. A. Songeon (1826–1905)
Soó K. R. Soó von Bere (1903–1980)
Sosn., D. D. I. Sosnowsky (1886–1952)
Soulié J. A. L. Soulié (1868–1930)
Sowerby J. Sowerby (1757–1822)
Soyer-Willemet H. F. Soyer-Willemet (1791–1867)
Spach E. Spach (1801–1879)
Spampinato G. Spampinato (b. 1958)
Speg. C. L. Spegazzini (1858–1926)
Spenner F. C. (K.) L. Spenner (1798–1841)
Sprengel K. P. J. Sprengel (1766–1833)
Spring A. F. Spring (1814–1872)
Spruner W. von Spruner (1805–1874)
Stace C. A. Stace (b. 1938)

Standley P. C. Standley (1884–1963)
Stankov S. S. Stankov (1892–1962)
Stapf O. Stapf (1857–1933)
Stearn W. T. Stearn (b. 1911)
Stefani C. de Stefani (1851–1924)
Stefanov B. Stefanov (Stefanoff) (1894–1979)
Steinb. E. I. Steinberg (1884–1963)
Steinb., C. C. H. Steinberg (1923–1981)
Steinh. A. Steinheil (1810–1839)
Stephan C. F. Stephan (1757–1814)
Stern, F. C. F. C. Stern (1884–1967)
Sternb. K. (C.) M. von Sternberg (1761–1838)
Sterns, E. E. E. E. Sterns (1846–1926)
Steudel E. G. von Steudel (1783–1856)
Steven C. C. von Steven (1781–1863)
St-Lager J. B. Saint-Lager (1825–1912)
Stoj. N. A. Stojanov (1883–1968)
Stokes J. Stokes (1755–1831)
Störck A. von Störck (1731–1803)
Stork, A. L. A. L. Stork (b. 1937)
Strempel J. C. (K.) F. Strempel (1800–1872)
Strgar V. Strgar (fl. 1971)
Strid P. A. K. Strid (b. 1943)
Strobl P. G. Strobl (1846–1925)
Stur D. R. J. Stur (1827–1893)
Sturm J. Sturm (1771–1848)
Suckow, G. G. A. Suckow (1751–1813)
Sudworth G. B. Sudworth (1864–1927)
Suk. V. N. Sukaczev (Sukatschew) (1880–1967)
Sünd. F. Sündermann (1864–1946)
Suominen J. Suominen (b. 1936)
Suter J. R. Suter (1766–1827)
Sutulov A. N. Sutulov (fl. 1914)
Svob. P. Svoboda (1908–1978)
Swartz O. P. Swartz (1760–1818)
Sweet R. Sweet (1783–1835)
Syme J. T. I. Boswell Syme (formerly Boswell) (1822–1888)
Szov. A. J. Szovits (d. 1830)
Szysz. I. von Szyszylowicz (1857–1910)
Takht. A. L. Takhtajan (Takhtadzhan) (b. 1910)
Talavera S. Talavera Lozano (b. 1945)
Tanfani E. Tanfani (1848–1892)
Tardieu-Blot M.-L. Tardieu-Blot (M. L. Tardieu) (b. 1902)
Taschereau P. M. Taschereau (b. 1939)
Tausch I. F. Tausch (1793–1848)
Tavel R. F. von Tavel (1863–1941)
Temesy E. Temesy (E. Schönbeck-Temesy) (b. 1930)
Ten. M. Tenore (1780–1861)
Tepl. F. A. Teplouchow (1845–1905)
Terechov A. F. Terechov (1890–1974)
Terracc., N. N. Terracciano (Terraciano) (1837–1921)
Texidor J. Texidor y Cos (1836–1885)
Teyber A. Teyber (1846–1913)
't Hart H. 't Hart (b. 1944)
Thaisz L. de Thaisz (L. von Thaisz) (1867–1937)
Thell. A. Thellung (1881–1928)
Thév. A. V. Thévenau (1815–1876)
Thomas A. L. E. Thomas (1788–1859)
Thore J. Thore (1762–1823)
Thouars L. M. A. Aubert du Petit-Thouars (1758–1831)
Thuill. J. L. Thuillier (1757–1822)
Thunb. C. P. Thunberg (1743–1828)
Timb.-Lagr. P. M. E. Timbal-Lagrave (1819–1888)
Timm J. C. Timm (1734–1805)
Tineo V. Tineo (1791–1856)
Tiss. P. G. Tissière (1828–1868)
Titz W. Titz (b. 1941)
Tod. A. Todaro (1818–1892)
Tolm. A. I. Tolmatchev (Tolmachev) (1903–1979)
Top. S. Topali (1900–1944)
Ţopa E. Ţopa (1900–1987)
Torrey J. Torrey (1796–1873)
Trabut L. C. Trabut (1853–1929)
Tratt. L. Trattinnick (1764–1849)
Trautv. E. R. von Trautvetter (1809–1889)
Trelease W. Trelease (1857–1945)
Trew C. J. Trew (1695–1769)
Trotzky P. Kornuch Trotzky (1803–1877)
Tryon jun., R. M. R. M. Tryon jun. (b. 1916)
Turcz. N. S. Turczaninow (1796–1864)
Turesson G. W. Turesson (1892–1970)
Turra A. Turra (1730–1796)
Turrill W. B. Turrill (1890–1961)
Tutin T. G. Tutin (1908–1987)
Tuzson J. Tuzson (1870–1943)
Tzanoudakis D. B. Tzanoudakis (b. 1950)
Tzvelev N. N. Tzvelev (b. 1925)
Ucria B. de Ucria (Michelangelo Aurifici) (1739–1796)
Uechtr. R. K. F. von Uechtritz (1838–1886)
Ugr. K. A. Ugrinsky (fl. 1920)
Ulbr. O. E. Ulbrich (1879–1952)
Underw. L. M. Underwood (1853–1907)
Unger F. J. A. N. Unger (1800–1870)
Ung.-Sternb. F. Ungern-Sternberg (1808–1885)
Uotila P. J. Uotila (b. 1943)
Urban I. Urban (1848–1931)
Uribe-Echebarría P. M. Uribe-Echebarría Díaz (b. 1953)
Urum. I. K. Urumoff (1857–1937)
Vacc. L. Vaccari (1873–1951)
Vahl M. H. Vahl (1749–1804)
Vahl, J. J. L. M. Vahl (1796–1854)
Valck.-Suringar J. Valckenier-Suringar (1864–1932)
Valdés B. Valdés Castrillón (b. 1942)
Valdés-Bermejo E. Valdés-Bermejo (b. 1945)
Valentine D. H. Valentine (1912–1987)
Vălev S. T. Vălev (1910–1974)
Vandas K. Vandas (1861–1923)
Van den Bosch R. B. van den Bosch (1810–1862)
Van Hall H. C. van Hall (1801–1874)
Van Houtte L. B. Van Houtte (1810–1876)
Van Omm. G. van Ommering (fl. 1976–1977)
Vargas P. Vargas (fl. 1987)
Vasc. J. de Carvalho e Vasconcellos (1897–1972)
Vassilcz. I. T. Vassilczenko (b. 1903)
Vavilov N. I. Vavilov (1887–1943)
Velen. J. Velenovský (1858–1949)
Vent. E. P. Ventenat (1757–1808)
Vest L. C. von Vest (1776–1840)

Viane R. L. L. Viane (b. 1951)
Vicioso, C. M. C. Vicioso Martínez (1897–1968)
Vida G. Vida (b. 1935)
Vierh. F. K. M. Vierhapper (1876–1932)
Vig. L. G. A. Viguier (1790–1867)
Vigo J. Vigo Bonada (b. 1937)
Vill. D. Villars (Villar) (1745–1814)
Villar, H. del E. Huguet del Villar y Serratacó (1871–1951)
Villar, L. L. Villar Peréz (b. 1946)
Vis. R. de Visiani (1800–1878)
Vivant J. Vivant (fl. 1948–1981)
Viv. D. Viviani (1772–1840)
Vogler J. A. Vogler (fl. 1781)
Vogt R. M. Vogt (b. 1957)
Volkens G. L. A. Volkens (1855–1917)
Vollmann F. Vollmann (1858–1917)
Vorosch. V. N. Voroschilov (b. 1908)
Voss A. Voss (1857–1924)
Vuk. L. von F. Vukotinović (1813–1893)
Wahlenb. G. Wahlenberg (1780–1851)
Waisb. A. Waisbecker (1835–1916)
Waldst. F. de P. A. von Waldstein-Wartemberg (1759–1823)
Wale R. S. Wale (d. 1952)
Walker, S. S. Walker (1924–1985)
Wall. N. Wallich (1786–1854)
Wallr. C. (K.) F. W. Wallroth (1792–1857)
Walpers W. G. Walpers (1816–1853)
Walters S. M. Walters (b. 1920)
Wangenh. F. A. J. von Wangenheim (1749–1800)
Warburg, E. F. E. F. Warburg (1908–1966)
Warncke K. Warncke (b. 1937)
Watson, H. C. H. C. Watson (1804–1881)
Watson, S. S. Watson (1826–1892)
Watt D. A. P. Watt (1830–1917)
Webb P. B. Webb (1793–1854)
Webb, D. A. D. A. Webb (b. 1912)
Weber G. H. Weber (1752–1828)
Weber fil. F. Weber (1781–1823)
Webster, S. S. D. Webster (b. 1959)
Weigel C. E. von Weigel (1748–1831)
Weihe C. (K.) E. A. Weihe (1779–1834)
Weiller M. Weiller (1880–1945)
Welw. F. M. J. Welwitsch (1806–1872)
Wendelberger G. Wendelberger (b. 1915)
Wenderoth G. W. F. Wenderoth (1774–1861)
Wesmael, A. A. Wesmael (1832–1905)
Weston R. Weston (1733–1806)
Wettst. R. Wettstein von Westersheim (1863–1931)
Wettst., W. W. Wettstein (fl. 1952)
Whitehead F. H. Whitehead (b. 1913)
Wibel A. W. E. C. Wibel (1775–1813)
Wichura M. E. Wichura (1817–1866)
Widder F. J. Widder (1892–1974)
Wiens D. Wiens (b. 1932)
Wierzb. P. P. Wierzbicki (1794–1847)
Wiesb. J. B. Wiesbaur (1836–1906)
Wilce J. H. Wilce (b. 1931)
Wilczek E. Wilczek (1867–1948)
Willd. C. L. Willdenow (1765–1812)
Williams, F. N. F. N. Williams (1862–1923)
Willk. H. M. Willkomm (1821–1895)
Wilmot-Dear C. M. Wilmot-Dear (b. 1952)
Wilmott A. J. Wilmott (1888–1950)
Wilson E. H. Wilson (1876–1930)
Wimmer C. F. H. Wimmer (1803–1868)
Winge Ø. Winge (1886–1964)
Wissjul. E. D. Wissjulina (1898–1972)
With. W. Withering (1741–1799)
Wolff, H. K. F. A. H. Wolff (1866–1929)
Wolfner W. Wolfner (fl. 1858)
Wolley-Dod A. H. Wolley-Dod (1861–1948)
Wołoszczak E. Wołoszczak (1835–1918)
Wood, W. W. Wood (1745–1808)
Woods, J. J. Woods (1776–1864)
Woronow J. N. Woronow (Voronov) (1874–1931)
Woynar H. K. Woynar (1865–1917)
Wrigley F. A. Wrigley (b. 1936)
Wulf E. V. Wulf (E. W. Wulff, E. V. Vul'f) (1885–1941)
Wulfen F. X. von Wulfen (1728–1805)
Wünsche F. O. Wünsche (1839–1905)
Wyse Jackson, M. B. M. B. Wyse Jackson (b. 1962)
Wyse Jackson, P. S. P. S. Wyse Jackson (b. 1955)
Záborský J. Záborský (b. 1928)
Zaffran J. Zaffran (b. 1935)
Zahlbr. J. Zahlbruckner (1782–1851)
Zāmels A. Zāmels (from 1928, Zāmelis) (1897–1943)
Zapał. H. Zapałowicz (1852–1917)
Zawadzki A. Zawadzki (1798–1868)
Zenari S. Zenari (1895–1956)
Zeyher C. L. P. Zeyher (1799–1858)
Zieliński J. Zieliński (b. 1943)
Zimm. W. M. Zimmermann (1892–1980)
Zimmeter A. Zimmeter (1848–1897)
Zinger, N. N. W. Zinger (1866–1923)
Ziz J. B. Ziz (1779–1829)
Zodda G. Zodda (1877–1968)
Zoega J. Zoega (1742–1788)
Zohary M. Zohary (1898–1983)
Zucc. J. G. Zuccarini (1797–1848)
Zumagl. A. M. Zumaglini (1804–1865)

APPENDIX II

KEY TO THE ABBREVIATIONS OF TITLES OF BOOKS CITED IN VOLUME 1

Aiton, *Hort. Kew.*
W. Aiton, *Hortus kewensis, or a Catalogue of the Plants cultivated in the Royal Botanic Garden at Kew.* Ed. 1. London. 1789. (**1–3** in 1789.) Ed. 2, by W. T. Aiton. London. 1810–1813. (**1** in 1810; **2** & **3** in 1811; **4** in 1812; **5** in 1813. Cf. F. A. Stafleu & R. S. Cowan, *Taxonomic Literature* ed. 2, **1**: 25–26 (1976).)

All., *Auct. Fl. Pedem.*
C. Allioni, *Auctuarium ad Floram pedemontanam cum Notis et Emendationibus.* Augustae Taurinorum. 1789.

All., *Auct. Syn. Stirp. Horti Taur.*
C. Allioni, *Auctuarium ad Synopsim methodicam Stirpium Horti regii taurinensis.* [Torino.] 1773. (Preprint from *Mélanges Philos.-Mat. Soc. Roy. Turin* **5** (1770–1773): 53–96 (1774). Allioni in *Fl. Pedem.* always refers to the page numbers of the *journal*; the preprint was independently paginated. Cf. J. E. Dandy. *Taxon* **19**: 617–626 (1970) and F. A. Stafleu & R. S. Cowan, *Taxonomic Literature* ed. 2, **1**: 35 (1976).)

All., *Fl. Pedem.*
C. Allioni, *Flora pedemontana.* Augustae Taurinorum. 1785. (**1–3** in 1785.)

Ambrosi, *Fl. Tirolo Mer.*
F. Ambrosi, *Flora del Tirolo meridionale* (*Flora Tiroliae australis*). Padova. 1854–1857. (**1** in 1854–?; **2** in 1857–?. Published in parts; precise dates unknown. Cf. F. A. Stafleu & R. S. Cowan, *Taxonomic Literature* ed. 2, **1**: 40–41 (1976).)

Amo, *Fl. Iber.*
Mariano del Amo y Mora, *Flora fanerogámica de la Península ibérica.* Granada. 1871–1873. (**1** & **2** in 1871; **3** & **4** in 1872; **5** & **6** in 1873.)

Antoine, *Kupress.-Gatt.*
F. Antoine, *Die Kupressineengattungen* Arceuthos, Juniperus, *und* Sabina. Wien. 1857–1860.

Arcangeli, *Comp. Fl. Ital.*
G. Arcangeli, *Compendio della Flora italiana.* Ed. 1. Torino. 1882. Ed. 2. Torino & Roma. 1894.

Ard., *Animadv. Bot. Spec. Alt.*
P. Arduino, *Animadversionum botanicarum Specimen alterum.* Venetiis. 1763.

Arnold, *Reise Mariazell*
J. F. X. Arnold, *Reise nach Mariazell in Steyermark.* Wien. 1785. (For author's initials cf. J. H. Barnhart, *Bibliographic Notes upon Botanists* **1**: 77 (1965) and Castroviejo et al., *Flora Iber.* **2**: 727 (1990), but possibly a pseudonym.)

Ascherson, *Fl. Brandenb.*
P. F. A. Ascherson, *Flora der Provinz Brandenburg, der Altmark und des Herzogthums Magdeburg.* Berlin. 1859–1864. (**1**: pp. 1–320 in 1860; pp. i–xxii, 321–1034, 1–146 in 1864; **2** & **3** in 1859. Cf. F. A. Stafleu & R. S. Cowan, *Taxonomic Literature* ed. 2, **1**: 73–74 (1976).) For Ed. 2, cf. Ascherson & Graebner, *Fl. Nordostd. Flachl.*

Ascherson & Graebner, *Fl. Nordostd. Flachl.*
P. F. A. Ascherson & K. O. P. P. Graebner, *Flora des nordostdeutschen Flachlandes* (*ausser Ostpreussen*). Berlin. 1898–1899. (Pp. 1–480 in 1898; pp. 481–875 in 1899.) This is Ed. 2 of Ascherson, *Fl. Brandenb.*

Ascherson & Graebner, *Syn. Mitteleur. Fl.*
P. F. A. Ascherson & K. O. R. P. P. Graebner, *Synopsis der mitteleuropäischen Flora.* Ed. 1. Leipzig. 1896–1938. (For dates cf. F. A. Stafleu & R. S. Cowan, *Taxonomic Literature* ed. 2, **1**: 75–77 (1976).) Ed. 2. Leipzig. 1912–1920. (**1**: pp. 1–480 in 1912; pp. 481–630 in 1913; **2**(**1**): pp. 1–80 in 1919; pp. 81–160 in 1920.)

Asso, *Syn. Stirp. Arag.*
I. J. de Asso y del Rio, *Synopsis Stirpium indigenarum Aragoniae.* Massiliae. 1779.

Bab., *Man. Brit. Bot.*
C. C. Babington, *Manual of British Botany.* Ed. 1. London. 1843. Ed. 2. London. 1847. Ed. 3. London. 1851. Ed. 4. London. 1856. Ed. 5. London. 1862. Ed. 6. London. 1867. Ed. 7. London. 1874. Ed. 8. London. 1881–1882. (Pp. i–xlviii, 1–485, [2, err.] in 1881; pp. [2], Addenda in 1883.) Ed. 9. London. 1904. Ed. 10, edit. A. J. Wilmott. London. 1922.

Balbis, *Cat. Stirp. Hort. Bot. Taur.*
G. B. Balbis, *Catalogus Stirpium Horti botanici taurinensis.* Taurini. 1807. Ed. 1810, titled *Catalogus Plantarum ... ad Annum 1810.* Taurini. 1810. Ed. 1812, titled *Catalogus Stirpium ... ad Annum mdcccxii.* Taurini. 1812. Ed. 1813, titled *Catalogus Stirpium ... ad Annum mdcccxiii.* Taurini. 1813. ***App.***, *Ad Catalogum Stirpium Horti academici taurinensis editum Anno mdcccxiii Appendix prima.* Taurini. 1814. (Cf. F. A. Stafleu & R. S. Cowan, *Taxonomic Literature* ed. 2, **1**: 108 (1976).)

Barbey, W., *Fl. Sard. Comp.*
W. Barbey-Boissier, *Florae Sardoae Compendium. Catalogue raisonné des Végétaux observés dans l'Île de Sardaigne.* Lausanne. 1885. (Possibly in parts, 1884–1885; cf. internal evidence and F. A. Stafleu & R. S. Cowan, *Taxonomic Literature* ed. 2, **1**: 119 (1976).)

Bartl., *Cat. Sem. Horti Gotting.*
F. G. Bartling, *Catalogus Seminum Horti gottingensis.* Göttingen. 1838–1841, 1843, 1848, 1850. (Descriptions of new species reprinted as follows: **1838** as *Index Seminum Horti academici gottingensis* in *Linnaea* **13**: *Litt.-Ber.* 95–96 (1839); **1839** in *op. cit.* **14**: *Litt.-Ber.* 123–126 (1840); **1840** in *op. cit.* **15**: *Litt.-Ber.* 93 (1841); **1841** in *op. cit.* **16**: *Litt.-Ber.* 102–103 (1842); **1843** in *op. cit.* **17**: 158 (1844); **1848** in *Ann. Sci. Nat.* ser. 3, **11**: 254 (1849); **1850** in *Linnaea* **24**: 201–208 (1851).)

Batt. & Trabut, *Fl. Algér.*
J. A. Battandier & L. Trabut, *Flore de l'Algérie* [**1**(**1**): *Ancienne Flore d'Alger transformée*] *contenant la Description de toutes les Plantes signalées jusqu'à ce Jour comme spontanées en Algérie* [**1**(**2**): *et Catalogue des Plantes du Maroc.*] **Dicot.**, **1**(**1**), *Dicotylédones* by J. A. Battandier. Alger & Paris. 1888–1890. (Pp. 1–184 in 1888; pp. 185–576 in 1889; pp. 577–825 in 1890.) ***Monocot.***, **1**(**2**), *Monocotylédones* by J. A. Battandier & L. Trabut. Alger & Paris. 1895. (Cf. W. T. Stearn, *Jour. Soc. Bibl. Nat. Hist.* **1**: 145 (1938) and F. A. Stafleu & R. S. Cowan, *Taxonomic Literature* ed. 2, **1**: 142 (1976).)

Batt. & Trabut, *Fl. Algér. Tunisie*
J. A. Battandier & L. Trabut, *Flore analytique et synoptique de l'Algérie et de la Tunisie*. Alger. 1905.

Baumg., *Enum. Stirp. Transs.*
J. C. G. Baumgarten, *Enumeratio Stirpium magno Transsilvaniae Principatui praeprimis indigenarum*. Vindobonae & Cibinii. 1816–1846. (**1** & **2**, Vindobonae in 1816; **3**, Vindobonae in 1817; **4**, edit. M. Fuss, Cibinii in 1846). ***Mant.*, Mantissa,** by M. Fuss. Cibinii. 1846. *Indices*, by M. Fuss. Cibinii. 1846. (Cf. F. A. Stafleu & R. S. Cowan, *Taxonomic Literature* ed. 2, **1**: 151 (1976).)

Beauv., *Prodr. Aethéog.*
A. M. F. J. Palisot de Beauvois, *Prodrome des cinquième et sixième Familles de l'Aethéogamie. Les Mousses. Les Lycopodes*. Paris. 1805. (Part of text previously published in *Mag. Encycl.* 9e Année **5**: 289–339, 471–483 (1804).)

Beck, G., *Fl. Bosn. Herceg.*
G. Beck von Mannagetta, *Flora Bosne, Hercegovine i Novipazarskog Sandžaka*. Beograd & Sarajevo. 1903–1950. (*Pteridophyta* in *Glasn. Muz. Bosni Herceg.* **28**: 311–336 (1916). *Embryophyta siphonogama*: **1** as *op. cit.* **15**: 1–48, 185–230 (1903); **2(1)** as *op. cit.* **18**: 69–81 (1906); **2(2)** as *op. cit.* **18**: 137–150 (1906); **2(3)** as *op. cit.* **18**: 469–495 (1906); **2(4)** as *op. cit.* **19**: 15–29 (1907); **2(5)** as *op. cit.* **21**: 135–165 (1909); **2(6)** as *op. cit.* **26**: 451–475 (1914); **2(7)** as *op. cit.* **28**: 41–168 (1916); **2(8)** as *op. cit.* **30**: 177–217 (1918); **2(9)** as *op. cit.* **32**: 83–127 (1920); **2(10)** as *op. cit.* **33**: 1–17 (1921); **2(11)** as *op. cit.* **35**: 49–74 (1923); **3**, titled *Flora Bosnae Hercegovinae et Regionis Novipazar.* / Флора Босне, Херцеговине и Области Новога Пазара [*Flora Bosne, Hercegovine i Oblasti Novoga Pazara*], Beograd & Sarajevo in 1927; **4(1)**, titled *Flora Bosnae et Hercegovinae*, by G. Beck von Mannagetta & K. Malý, Sarajevo in 1950; **4(2)**, by Z. Bjelčič, Z. Slavnič & P. Fukarek, Sarajevo in 1967.) *Pteridophyta* (pp. 1–26), and **1** & **2** (with continuous pagination, pp. 1–484) also reprinted separately with double pagination. Cf. F. A. Stafleu & R. S. Cowan, *Taxonomic Literature* ed. 2, **1**: 158–159 (1976).)

Bellardi, *App. Fl. Pedem.*
C. A. L. Bellardi, *Appendix Ludovici Bellardi ad Floram pedemontanam*. Augustae Taurinorum. 1792. (Reprinted as *Mém. Acad. Sci.* (*Turin*) **10**: 209–286 (1793).)

Bentham, *Cat. Pl. Pyr. Bas-Languedoc*
G. Bentham, *Catalogue des Plantes indigènes des Pyrénées et du Bas-Languedoc, avec des Notes et Observations sur les Espèces nouvelles ou peu connues; précédés d'une Notice sur un Voyage botanique fait dans les Pyrénées pendant l'Été de 1825*. Paris. 1826.

Bergeret, J. P., *Phytonomat.*
J. P. Bergeret, *Phytonomatotechnie universelle*. Paris. 1783–1786. (**1** in 1783; **2** in 1784–1786; **3** in 1786?.)

Bernh., *Syst. Verz. Erfurt*
J. J. Bernhardi, *Systematisches Verzeichniss der Pflanzen, welche in der Gegend um Erfurt gefunden werden*. Erfurt. 1800.

Bertol., *Amoen.*
A. Bertoloni, *Amoenitates italicae, sistentes Opuscula ad Rem herbariam et Zoologiam Italiae spectantia*. Bononiae. 1819.

Bertol., *Fl. Ital.*
A. Bertoloni, *Flora italica, sistens Plantas in Italia et Insulis circumstantibus sponte nascentes*. Bononiae. 1833–1854. (**1–10**: for dates cf. F. A. Stafleu & R. S. Cowan, *Taxonomic Literature* ed. 2, **1**: 204 (1976).)

Bertol., *Misc. Bot.*
A. Bertoloni, *Miscellanea botanica*. Bononiae. 1842–1863. (**1–24** in 1842–1863; for dates cf. F. A. Stafleu & R. S. Cowan, *Taxonomic Literature* ed. 2, **1**: 205–206 (1976).)

Besser, *Enum. Horto Cremen.*
W. S. J. G. von Besser, *Enumeratio Plantarum in Horto botanico cremeneci*. Krzemieniec. 1816, 1819–1821, 1823 & 1830. (Preceded by W. S. J. G. von Besser, *Catalogus des Plantes du Jardin botanique de Krzemieniec en Volhynie*, 1810; *Catalogus ... du Gymnase de Volhynie à Krzemieniec* in 1811, with *Suppl.* **1** in 1812, *Suppl.* **2** in 1814, *Suppl.* **3** in 1814, *Suppl.* **4** in 1815; *Catalogus Plantarum in Horto botanico Gymnasii volhyniensis Cremeneci cultarum* in 1814; *Catalogus Plantarum in Horto botanico volhyniensis Cremeneci cultarum* in 1815; and *Catalogus Plantarum in Horto botanico Gymnasii volhyniensis Cremeneci cultarum* in 1816.)

Besser, *Enum. Pl. Volhyn.*
W. S. J. G. von Besser, *Enumeratio Plantarum hucusque in Volhynia, Podolia, Gub. kiioviensi, Bessarabia cis-tyraica et circa Odessam collectarum, simul cum Observationibus in Primitias Florae Galiciae austriacae*. Vilnae. 1822. (Definitive ed.; a preliminary issue, comprising pp. 1–79 only, was published in 1821. The text of this is identical to pp. 1–79 of the 1822 edition. Cf. F. A. Stafleu & R. S. Cowan, *Taxonomic Literature* ed. 2, **1**: 209 (1976).)

Besser, *Prim. Fl. Galic.*
W. S. J. G. von Besser, *Primitiae Florae Galiciae austriacae utriusque. Encheiridion ad Excursiones botanicas concinnatum*. Viennae. 1809. (**1** & **2** in 1809.)

Biasol., *Relaz. Viaggio*
B. Biasoletto, *Relazione del Viaggio fatto nella Primavera dell'Anno 1838 dalla Maestia del Re Federico Augusto di Sassonia, nell'Istria, Dalmazia e Montenegro*. Trieste. 1841.

Bieb., *Beschr. Länd. Terek Casp.*
F. A. Marschall von Bieberstein, *Beschreibung der Länder zwischen den Flüssen Terek und Kur am caspichen Meere. Mit einem botanischen Anhang*. Frankfurt. 1800.

Bieb., *Fl. Taur.-Cauc.*
F. A. Marschall von Bieberstein, *Flora taurico-caucasica, exhibens Stirpes phaenogamas in Chersonenso taurica et Regionibus caucasicis sponte crescentes*. Charkoviae. 1808–1819. (**1** & **2** in 1808; **3**, *Supplementum*, in 1819.)

Bigelow, *Fl. Boston.*
J. Bigelow, *Florula bostoniensis. A Collection of Plants of Boston and its Environs*. Boston. 1814. Ed. 2 (titled *Florula ... and its Vicinity*). Boston. 1824. Ed. 3 (titled *Florula ... and its Vicinity*). Boston. 1840.

Billot, *Annot.*
P. C. Billot, *Annotations à la Flore de France et d'Allemagne*. Haguenau. 1855–1862. (Pp. 1–38 in 1855; pp. 39-100 in 1856; pp. 101–116 in 1857; pp. 117–140 in 1858; pp. 141–210 in 1859; pp. 211–297 in 1862. Cf. F. A. Stafleu & R. S. Cowan, *Taxonomic Literature* ed. 2, **1**: 215–216 (1976).)

Biria, *Hist. Renonc.*
J. A. J. Biria, *Histoire naturelle et médicale des Renoncules, précédée de quelques Observations sur la Famille des Renonculacées*. Montpellier. 1811.

Biv., *Sic. Pl. Cent.*
A. de Bivona-Bernardi, *Sicularum Plantarum Centuria*. Panormi. 1806–1807/8. (**1** in 1806; **2** in 1807 or early 1808. Cf. F. A. Stafleu & R. S. Cowan, *Taxonomic Literature* ed. 2, **1**: 223 (1976).)

Biv., *Stirp. Rar. Sic. Descr.*
A. de Bivona-Bernardi, *Stirpium rariorum minusque cognitarum in Sicilia sponte provenientium Descriptiones nonnullis Iconibus auctae*. Panormi. 1813–1816. (**1** in 1813; **2** in 1814; **3** in 1815; **4** in 1818. Cf. F. A. Stafleu & R. S. Cowan, *Taxonomic Literature* ed. 2, **1**: 223 (1976).)

Bluff & Fingerh., *Comp. Fl. Germ.*
M. J. Bluff & K. A. Fingerhuth, *Compendium Florae Germaniae*. Norimbergae. 1825–1833. (**1** in 1825; **2** in 1825 or more probably 1826; **3** by K. F. W. Wallroth in 1831; **4** by K. F. W. Wallroth in 1833.) Ed. 2, by M. J. Bluff, C. G. D. Nees von Esenbeck & J. K. Schauer. Norimbergae. 1836–1839. (**1(1)** in 1836; **1(2)** in 1837; **2** in 1838; index in 1839.) (Cf. F. A. Stafleu & R. S. Cowan, *Taxonomic Literature* ed. 2, **1**: 233–234 (1976).)

Böcher, Holmen & Jakobsen, *Grønlands Fl.*
T. W. Böcher, K. Holmen & K. Jakobsen, *Grønlands Flora*. København. 1957.

Boiss., *Diagn. Pl. Or. Nov.*
P. E. Boissier, *Diagnoses Plantarum orientalium novarum*. Lipsiae & Parisiis. 1843–1859. (Ser. 1, **1(1–3)** in 1843; **1(4–5)** in 1844; **1(6–7)** in 1846; **2(8–11)** in 1849; **2(12)** in 1853; **2(13)** in 1854; ser. 2, **3(1)** in 1854; **3(2–3)** in 1856; **3(4)** in 1859; **3(5)** in 1856; **3(6)** in 1859. Cf. F. A. Stafleu & R. S. Cowan, *Taxonomic Literature* ed. 2, **1**: 258 (1976).)

Boiss., *Elenchus*
P. E. Boissier, *Elenchus Plantarum novarum minusque cognitarum, quas in Itinere hispanico legit*. Genevae. 1838 (before Webb, *Iter. Hisp.*).

Boiss., *Fl. Or.*
P. E. Boissier, *Flora orientalis*. Basileae, Genevae & Lugduni. 1867–1884. (**1**, Basileae & Genevae; **2–5**, Basileae, Genevae & Lugduni. **1** in 1867; **2** in 1872; **3** in 1875; **4**: pp. 1–280 in 1875; pp. 281–1276 in 1879; **5**: pp. 1–428 in 1882; pp. 429–868 in 1884.) ***Suppl.***, *Supplementum*, by R. Buser. Basileae, Genevae & Lugduni. 1888.

Boiss., *Voy. Bot. Midi Esp.*
P. E. Boissier, *Voyage botanique dans le Midi de l'Espagne pendant l'Année 1837*. Paris. 1839–1845. (**1**: pp. 1–40, tt. 1–135 in 1839; pp. 41–96, tt. 136–181 in 1840; pp. 97–248, i–x, t. 4a in 1845; **2**: pp. 1–96 in 1839; pp. 97–352 in 1840; pp. 352–544 in 1841; pp. 545–640 in 1842; pp. 641–710 in 1844; pp. 711–757 in 1845. Cf. F. A. Stafleu & R. S. Cowan, *Taxonomic Literature* ed. 2, **1**: 256–257 (1976).) All new species are described in **2**.

Boiss. & Reuter, *Diagn. Pl. Nov. Hisp.*
P. E. Boissier & G. F. Reuter, *Diagnoses Plantarum novarum hispanicarum, praesertim in Castella nova lectarum*. Genevae. 1842.

Boiss. & Reuter, *Pugillus*
P. E. Boissier & G. F. Reuter, *Pugillus Plantarum novarum Africae borealis Hispaniaeque australis*. Genevae. 1852.

Bolòs, O. & Vigo, *Fl. Països Catal.*
O. de Bolòs & J. Vigo, *Flora dels Països Catalans*. Barcelona. 1984→. (**1** in 1984; **2** in 1990.)

Bong. & C. A. Meyer, *Verz. Saisang-Nor*
A. G. H. von Bongard & C. A. von Meyer, *Verzeichniss der im Jahre 1838 am Saisang-nor und am Irtysch gesammelten Pflanzen. Ein zweites Supplement zur Flora altaica*. St. Petersburg. 1841.

Bonnier, *Fl. Compl. Fr.*
G. E. M. Bonnier, *Flore complète illustrée en Couleurs de France, Suisse et Belgique comprenant la Plupart des Plantes d'Europe*. Paris, Neuchâtel & Bruxelles. 1911–1935. (**1–8** Paris, Neuchâtel & Bruxelles; **9** Paris & Bruxelles; **10–13** Paris. For dates cf. F. A. Stafleu & R. S. Cowan, *Taxonomic Literature* ed. 2, **1**: 273 (1976).) **7–13** partly by R. C. V. Douin.

Bonnier & Layens, *Tabl. Syn. Pl. Vasc.*
G. E. M. Bonnier & G. de Layens, *Tableaux synoptiques des Plantes vasculaires de la Flore de la France*. (*La Végétation de la France*, **1**.) Paris. 1894.

Borbás, *Balaton Növényföldr.*
V. von Borbás, *A Balaton Tavának és Partmellékének Növényföldrajza és Edényes Növényzete*. Budapest. 1900. (Comprising *A Balaton tudományos Tanulmányozásának Eredményei* **2**; *A Balaton Flórája* **2**.)

Bordzil., or **Bordzil. & Lavrenko, *Fl. RSS Ucr.***
Cf. Fomin, *Fl. RSS Ucr.*

Boreau, *Fl. Centre Fr.*
A. Boreau, *Flore du Centre de la France*. Ed. 1. Paris. 1840. (**1** & **2** in 1840.) Ed. 2. Angers. 1849. (**1** & **2** in 1849.) Ed. 3. Angers. 1857. (**1** & **2** in 1857.) Cf. F. A. Stafleu & R. S. Cowan, *Taxonomic Literature* ed. 2, **1**: 279 (1976).

Bory, *Dict. Class. Hist. Nat.*
J. B. G. M. Bory de Saint-Vincent (edit.), *Dictionnaire classique d'Histoire Naturelle*. Paris. 1822–1831. (**1** & **2** in 1822; **3** & **4** in 1823; **5** & **6** in 1824; **7**–**9** in 1825; **10** in 1826; **11** & **12** in 1827; **13** & **14** in 1828; **15** in 1829; **16** in 1830; **17** in 1831. Cf. F. A. Stafleu & R. S. Cowan, *Taxonomic Literature* ed. 2, **1**: 285–286 (1976).)

Bory, *Expéd. Sci. Morée*
J. B. G. M. Bory de Saint-Vincent, *Expédition scientifique de Morée*. **3(2)**, *Botanique*. Paris. 1832–1836. (Pp. 1–336 in 1832; pp. 337–367 in 1833; *Atlas* in 1835–1836.)

Bory & Chaub., *Nouv. Fl. Pélop.*
J. B. G. M. Bory de Saint-Vincent & L. A. Chaubard, *Nouvelle Flore du Péloponnèse et des Cyclades*. Paris & Strasbourg. 1838.

Bory & Durieu, *Expl. Sci. Algér. (Bot.)*
J. B. G. M. Bory de Saint-Vincent & M. C. Durieu de Maisonneuve, *Exploration scientifique de l'Algérie pendant les Années 1840, 1841, 1842. Botanique*. **1**. *Cryptogamie*, *Atlas*. Paris. 1846–1869. (For dates cf. F. A. Stafleu & R. S. Cowan, *Taxonomic Literature* ed. 2, **1**: 711 (1976).)

Borza, *Consp. Fl. Roman.*
A. Borza, *Conspectus Florae Romaniae Regionumque affinium*. Cluj. 1947–1949. (Pp. i–vii, 1–160 in 1947; pp. 161–360 in 1949.)

Br., R., *Chloris Melv.*
R. Brown, *Chloris melvilliana. A List of Plants collected in Melville Island (latitude 74°–75° N, longitude 110°–112° W) in the Year 1820; by the Officers of the Voyage of Discovery and the Orders of Captain Parry. With Characters and Descriptions of the new Genera and Species by Robert Brown*. London. 1823. (Preprint of the *Botanical Appendix* to Parry, *J. Voy. N.W. Pass.* (1824).)

Br., R., *Prodr. Fl. Nov. Holl.*
R. Brown, *Prodromus Florae Novae Hollandiae et Insulae Van Diemen*. Ed. 1. Londini. 1810. Ed. 2. Londini. 1819. Ed. 3, titled *Editio secunda* ... by C. G. D. Nees von Esenbeck. Norimbergae. 1827.

Braun, A., *Ind. Sem. Horti Berol.*
A. Braun, *Index Seminum Horti botanici berolinensis*. Berolini. 1859.

Br.-Bl., *Sched. Fl. Raet. Exsicc.*
J. Braun-Blanquet, *Schedae ad Floram raeticam exsiccatam*. Chur. **1–13**, 1918–1938. (Published in *Jahres-Bericht der Naturforschenden Gesellschaft Graubündens*, various issues.)

Briq., *Prodr. Fl. Corse*
J. I. Briquet, *Prodrome de la Flore corse*. Genève, Bâle, Lyon & Paris. 1910–1955. (**1**, Genève, Bâle & Lyon in 1910; **2(1)**, Genève, Bâle & Lyon in 1913; **2(2)**, Paris in 1936; **3(1)**, Paris in 1938; **3(2)**, Paris in 1955.) **2(2)** & **3** (**1** & **2**) by R. V. de Litardière.

Britton, E. E. Sterns & Poggenb., *Prelim. Cat.*
Preliminary Catalogue of Anthophyta and Pteridophyta reported as

growing spontaneously within one hundred Miles of New York City. Compiled by the following Committee of the Torrey Botanical Club ... The Nomenclature revised and corrected by N. L. Britton, E. E. Sterns and Justus F. Poggenburg. New York. 1888. (The committee comprised N. L. Britton, A. Brown, C. A. Hollick, J. F. Poggenburg, T. C. Porter & E. E. Sterns. Cf. F. A. Stafleu & R. S. Cowan, *Taxonomic Literature* ed. 2, **1**: 333 (1976).)

Brot., *Fl. Lusit.*
F. de Avellar Brotero, *Flora lusitanica.* Olissipone (**2**, Olisipone). 1804–1805. (**1** in 1804; **2** in 1805. Cf. F. A. Stafleu & R. S. Cowan, *Taxonomic Literature* ed. 2, **1**: 359 (1976).)

Buffon, *Hist. Nat. Pl.*
G. L. L. de Buffon, *Histoire naturelle ... nouvelle Édition ... Ouvrage ... rédigé par C. S. Sonnini. Histoire naturelle, générale et particulière des Plantes; par C. F. Brisseau-Mirbel, ... continué par N. Jolyclerc.* Paris. 1806.

Bunge, *Del. Sem. Horti Dorpat.*
A. A. von Bunge, *Delectus Seminum e Collectione Anni 1836 quae Hortus botanicus dorpatensis ... offert.* Dorpati Livonorum. 1836; titled *Delectus Seminum e Collectione Anno ... quae Hortus botanicus dorpatensis pro mutua Commutatione offert*, 1839 & 1840; titled *Delectus Seminum quae Anno ... in Horto botanico Universitatis Caesareae dorpatensis collecta pro mutua Commutatione offeruntur*, 1841 & 1842; titled *Delectus Seminum e Collectione Anni 1845 quae Hortus botanicus dorpatensis pro mutua Commutatione offert*, 1845; titled *Delectus Seminum Horti botanici dorpatensis*, 1846 & 1850. (Cf. Lipschitz, *Botanicorum russicorum Lexicon biographo-bibliographicum* **1**: 302 (1947) and *Author Catalogue of the Royal Botanic Gardens Library, Kew* **1**: 420 (1974).) Descriptions of new species from some issues reprinted by Bunge in *Linnaea* **14**: *Litt.-Ber.* 116–122 (1840); *op. cit.* **15**: *Litt.-Ber.* 85–90 (1841); *op. cit.* **16**: *Litt.-Ber.* 98–102 (1842); *Bot. Zeit.* **1**: 165–168 (1843); *Ann. Sci. Nat.* ser. 3 (*Bot.*) **5**: 367–370 (1846); *op. cit.* **7**: 190–191 (1847); *Linnaea* **19**: 394–397 (1847); *Ann. Sci. Nat.* ser. 3 (*Bot.*), **12**: 363–364 (1849); *op. cit.* **14**: 350 (1850).

Burnat, *Fl. Alp. Marit.*
E. Burnat, *Flore des Alpes maritimes.* Genève, Bâle & Lyon. 1892–1931. (**1–6**, Genève, Bâle & Lyon. **7**, Genève. **1** in 1892; **2** in 1896; **3**: pp. 1–172 in 1899; pp. 173–332 in 1902; **4** in 1906; **5**: pp. 1–96, *Supplément*, in 1913; pp. 97–376 in 1915; **6**: pp. 1–170 in 1916; pp. 171–344 in 1917; **7** in 1931.) **3**: pp. i–xxxvi by J. Briquet; **5**: pp. 1–96 by F. G. Cavillier; **5**: pp. 97–376, **6** & **7** by J. Briquet & F. G. Cavillier.

Cadevall, or Cadevall & Font Quer, *Fl. Catalunya*
J. Cadevall y Diars (with collaboration of P. Font i Quer, W. Rothmaler, and A. Sallent y Gotés), *Flora de Catalunya.* Barcelona. 1913–1937. (**1** in 1913–1915; **2** in 1915–1919; **3**: pp. 1–96 in 1919; pp. 97–192 in 1922; pp. 193–522 in 1923; **4** in 1932; **5** in 1935 (as '1933'); **6** in 1937.)

Cajander, *Suomen Kasvio*
A. K. Cajander, *A. J. Melan Suomen Kasvio.* Helsingissä. 1906. Comprises ed. 5 of A. J. Mela, *Suomen Kasvio.*

Campd., *Monogr. Rumex*
F. Campderá, *Monographie des* Rumex, *précédée de quelques Vues générales sur la Famille des Polygonées.* Paris. 1819.

Capelli, *Cat. Stirp.*
C. M. Capelli, *Catalogus Stirpium quae aluntur in Regio Horto botanico taurinensi.* Torino. 1821.

Carrière, *Traité Gén. Conif.*
É. A. Carrière, *Traité général des Conifères ou Description de toutes les Espèces et Variétés aujourd'hui connues.* Paris. 1855. Ed. 2. Paris. 1867.

Castroviejo et al., *Fl. Iberica*
S. Castroviejo, M. Laínz, G. López González, P. Montserrat, F. Muñoz Garmendia, J. Paiva & L. Villar (eds.), *Flora iberica. Plantas vasculares de la Península Ibérica e Islas Baleares.* Madrid. **1**→, 1986→. (**1** in 1986; **2** in 1990.)

Cav., *Descr. Pl.*
A. J. Cavanilles, *Descripción de las Plantas, que D. Antonio Josef Cavanilles demonstró en las Lecciones públicas del Año 1801, precedida de los Principios elementales de la Botánica.* Ed. 1. Madrid. 1802–1803. (Pp. i–cxxxvi in 1802; pp. 1–284, titled *Géneros y Especies de Plantas demonstradas en las Lecciones públicas del Año 1801*, in 1802; pp. 285–625, titled ... *del Año 1802*, in 1802.) Ed. 2. Madrid. 1827.

Čelak., *Prodr. Fl. Böhm.*
L. J. Čelakovsky, *Prodromus der Flora von Böhmen.* Prague. 1867–1881. (Published as *Arch. Naturw. Landesf. Böhm.* **1**, *Bot. Abth.*: 1–112 (1867); **2(2)**, *Bot. Abth.*: 113–388 (1871); **3**, *Bot. Abth.*: 389–691 (1875); **4**, *Bot. Abth.*: 693–955 (1881).) A Czech version was published 1868–1883. (Cf. F. A. Stafleu & R. S. Cowan, *Taxonomic Literature* ed. 2, **1**: 476–477 (1976).)

Cesati & De Not., *Ind. Sem. Horti Genuensis*
V. de Cesati & G. de Notaris, *Index Seminum Horti regii botanici genuensis, 1858.* Genuae. 1859. (For date cf. *Author Catalogue of the Royal Botanic Gardens Library, Kew* **1**: 504 (1974).)

Cesati, Passer. & Gibelli, *Comp. Fl. Ital.*
V. de Cesati, G. Passerini & G. Gibelli, *Compendio della Flora italiana.* Milano. 1868–1889. (Pp. 1–48 in 1868; pp. 49–120 in 1869; pp. 121–168 in 1870; pp. 169–208 & 5 pp. index in 1871; pp. 209–256 in 1872; pp. 257–320 in 1874; pp. 321–376 in 1875; pp. 377–393 in 1876; pp. 394–416 in 1881?; pp. 417–440 in 1877; pp. 441–520 in 1878; pp. 521–560 in 1879; pp. 561–616 in 1880; pp. 617–640 in 1881; pp. 641–720 in 1882; pp. 721–784 in 1883; pp. 785–816 in 1884; pp. 817–888 in 1885; pp. 889–906 in 1886; tt. 100–105 in 1888; tt. 106–111 in 1889. *Index and Corrigenda*, by G. Gibelli & O. Mattirolo in 1901. Cf. F. A. Stafleu & R. S. Cowan, *Taxonomic Literature* ed. 2, **1**: 479–480 (1976).)

Chevall., *Fl. Gén. Env. Paris*
F. F. Chevallier, *Flore générale des Environs de Paris.* Ed. 1. Paris. 1826–1828. (**1** in 1826; **2(1)**: pp. 1–512 (or 416?) in 1827; **2(2)**: pp. 513 (or 417?)–980 in 1828. The division of **2** into parts is uncertain; it may have been published in more than 2 parts. Cf. F. A. Stafleu & R. S. Cowan, *Taxonomic Literature* ed. 2, **1**: 494 (1976).) Ed. 2. Paris. (**1** & **2** in 1836.)

Clairv., *Man. Herb. Suisse*
J. P. de Clairville, *Manuel d'Herborisation en Suisse et en Valais.* Winterthur. 1811. Ed. 2. Genève & Paris. 1819.

Clapham, Tutin & E. F. Warburg, *Fl. Brit. Is.*
A. R. Clapham, T. G. Tutin & E. F. Warburg, *Flora of the British Isles.* Ed. 1. Cambridge. 1952. Reprinted 1957 & 1958 with some corrections. Ed. 2. Cambridge. 1962. Ed. 3, by A. R. Clapham, T. G. Tutin & D. M. Moore. Cambridge. 1987. Reprinted 1989 with major corrections.

Clarke, E. D., *Travels*
E. D. Clarke, *Travels in various Countries of Europe, Asia and Africa.* Ed. 1. London. 1810–1823. (**1** in 1810; **2(1)** in 1812; **2(2)** in 1814; **2(3)** in 1816; **3(1)** in 1819; **3(2)** in 1823.) Several later editions.

Clemente, *Elench. Horti Matrit.*
S. de Rojas Clemente y Rubio, *Elenchus Plantarum Horti matritensis.* Matriti. 1806.

Cosson, *Ill. Fl. Atl.*
E. S. C. Cosson, *Illustrationes Florae atlanticae.* Parisiis. 1882–1897. (**1**: pp. 1–36, tt. 1–25 in 1882; pp. 37–72, tt. 26–50 in 1884; pp.

73–120, tt. 51–73 in 1889; pp. 121–159, tt. 74–98 in 1891; **2**: pp. 7–42, tt. 99–123 in 1892; pp. 43–82, tt. 124–148 in 1893; pp. 1–6, 83–125, tt. 149–175 in 1897.)

Cosson, *Not. Pl. Crit.*
E. S. C. Cosson, *Notes sur quelques Plantes critiques, rares et nouvelles, et Additions à la Flore des Environs de Paris.* Paris. 1849–1852. (5 fascicles with continuous pagination: pp. 1–24 in 1849; pp. 25–48 in 1849; pp. 49–92 in 1850; pp. 93–140 in 1851; pp. 141–184 in 1852.) Author of Corsican plants, pp. 49–72: J. L. Kralik. Co-author of notes on Algerian plants, pp. 133–139: M. C. Durieu de Maisonneuve.

Coutinho, *Fl. Port.*
A. X. Pereira Coutinho, *A Flora de Portugal* (*Plantas vasculares*). Ed. 1. Paris, Lisboa, Rio de Janeiro, S. Paulo & Bello Horizonte. 1913. Ed. 2, titled *Flora de Portugal*, by R. T. Palhinha. Lisboa. 1939.

Crantz, *Class. Crucif.*
H. J. N. von Crantz, *Classis Cruciformium emendata cum Figuris aeneis in necessarium Institutionum Rei Herbariae Supplementum.* Lipsiae. 1769.

Crantz, *Stirp. Austr.*
H. J. N. von Crantz, *Stirpium Austriacarum.* Ed. 1. Viennae & Lipsiae. 1762–1767. (**1**, Viennae in 1762; **2**, Viennae in 1763; **3**, Lipsiae in 1767.) Ed. 2. Viennae. 1769. (**1** & **2**, with continuous pagination, in 1769.)

Curtis, *Fl. Lond.*
W. Curtis, *Flora londinensis: or, Plates and Descriptions of such Plants as grow wild in the Environs of London.* London. 1775–1798(–1779?). For detailed accounts and dates of plates cf. A. Stevenson, *Catalogue of botanical Books in the Collection of Rachel McMasters Miller Hunt,* **2(2)**: 389–412 (1961) and F. A. Stafleu & R. S. Cowan, *Taxonomic Literature* ed. 2, **1**: 575–577 (1976).)

Cutanda, *Fl. Comp. Madrid*
V. Cutanda, *Flora compendiada de Madrid y su Provincia.* Madrid. 1861.

Cuvier, *Dict. Sci. Nat.*
G.-F. Cuvier (ed.), *Dictionnaire des Sciences naturelles dans lequel on traite méthodiquement des differentes Êtres de la Nature.* Ed. 1. Paris. 1804–1806. (**1** & **2** in 1804; **3** in 1805; **4–6** in 1806.) Ed. 2. Paris. 1816-1845. (**1** & **2** in 1816; **3–9** in 1817; **10** in 1818; **11–15** in 1819; **16–17** in 1820; **18–22** in 1821; **23–25** in 1822; **26–29** in 1823; **30–33** in 1824; **34–38** in 1825; **39–44** in 1826; **45–51** in 1827; **52–53** in 1828; **54** in 1829 [sic]; **55–57** in 1828; **58–59** in 1829; **60** in 1830; **61** in 1845.)

Cyr., *Pl. Rar. Neap.*
D. M. L. Cirillo (Cyrillus), *Plantarum rariorum Regni neapolitani Fasciculus primus* [*secundus*]. Neapoli. 1788–1792. (**1** in 1788; **2** in 1792.)

Czern., *Consp. Pl. Charc.*
V. M. Czernajew, *Conspectus Plantarum circa Charcoviam et in Ucrania sponte crescentium et vulgo cultarum.* 1859.

Davis, P. H., *Fl. Turkey*
P. H. Davis (edit.), *Flora of Turkey and the East Aegean Islands.* Edinburgh. **1–10**, 1965–1988. (**1** in 1965; **2** in 1967; **3** in 1970; **4** in 1972; **5** in 1975; **6** in 1979; **7** in 1982; **8** in 1984; **9** in 1985; **10**, *Supplement*, edit. P. H. Davis, R. R. Mill & Kit Tan in 1988. For date of **6** cf. P. H. Davis & J. R. Edmondson, *Notes Roy. Bot. Gard. Edinb.* **37**: 282 (1979).)

DC., *Cat. Pl. Horti Monsp.*
A. P. de Candolle, *Catalogus Plantarum Horti botanici monspeliensis, addito Observationum circa Species novas aut non satis cognitas Fasciculo.* Monspelii. 1813.

DC., *Icon. Pl. Gall. Rar.*
A. P. de Candolle, *Icones Plantarum Galliae rariorum.* Parisiis. 1808.

DC., *Prodr.*
A. P. de Candolle, *Prodromus Systematis naturalis Regni vegetabilis.* Parisiis. [Also Strasbourg & London.] 1824–1874. (**1** in 1824; **2** in 1825; **3** in 1828; **4** in 1830; **5** in 1836; **6** in 1838; **7(1)** in 1838; **7(2)** in 1839; **8** in 1844; **9** in 1845; **10** in 1846; **11** in 1847; **12** in 1848; **13(1)** in 1852; **13(2)** in 1849; **14**: pp. 1–492 in 1856; pp. 493–706 in 1857; **15(1)** in 1864; **15(2)**: pp. 1–188 in 1862; pp. 189–1286 in 1866; **16(1)** in 1869; **16(2)**: pp. 1–160 in 1864; pp. 161–691 in 1868; **17** in 1873. *Index* **1–4** in 1843; **5–7(1)** in 1840; **7(2)–13** in 1858–1859; **14–17** in 1874.)

DC., *Rapp. Voy. Bot.*
A. P. de Candolle, *Rapports sur les Voyages botaniques et agronomiques faits dans les Départements de l'Empire d'après les Ordres de S. E. le Ministère de l'Intérieur.* Paris. 1813. (First published under the above title in 1813; the individual reports were published earlier, with different titles, in *Mém. Agric. Soc. Agric. Dép. Seine* **10–15** (= *Mem. Soc. Agric. Paris* **39–44**) in 1808–1813. For details cf. *Mem. Agric. Soc. Agric. Dép. Seine.*)

DC., *Reg. Veg. Syst. Nat.*
A. P. de Candolle, *Regni vegetabilis Systema naturale.* Parisiis. 1817–1821. (**1** in 1817; **2** in 1821.)

DC. & A. DC., *Not. Pl. Rar. Jard. Bot. Genève*
A. P. de Candolle & A. L. P. P. de Candolle, *Rapport*[*s*] (*Notice*[*s*]) *sur les Plantes rares cultivées dans le Jardin botanique de Genève par August Pyramus et Alphonse de Candolle.* Genève & Paris. 1823–1847. (**1–4** by A. P. de Candolle; **5–8** by A. P. & A. L. P. P. de Candolle; **9** & **10** by A. L. P. P. de Candolle.) The 10 parts, each with its own pagination, were published separately under various titles, some of them in *Mém. Soc. Phys. Hist. Nat. Genève.* All 10 were later (?1847) published with a new title page (*Notices* ...), the abbreviation for which is used with the original dates unless it is certain that the particular part was first published in the journal. For dates of publication, both in *Mém. Soc. Phys. Hist. Nat. Genève* and of the combined work, cf. F. A. Stafleu & R. S. Cowan, *Taxonomic Literature* ed. 2, **1**: 445 (1976).

Delarbre, *Fl. Auvergne*
A. Delarbre, *Flore d'Auvergne, ou Recueil des Plantes de cette ci-devant Province.* Ed. 1. Clermont-Ferrand. 1795; re-issued, Paris & Clermont-Ferrand, 1797. Ed. 2. Riom & Clermont-Ferrand. 1800. (**1** & **2** in 1800; **2**, pp. 509–891, titled *Flore de la ci-devant Auvergne.*) Ed. 2 [*sic*]. Clermont-Ferrand. 1836 (**1** & **2** in 1836; **2** titled *Flore de la haute et basse Auvergne.*) Cf. F. A. Stafleu & R. S. Cowan, *Taxonomic Literature* ed. 2, **1**: 612–613 (1976).

Desf., *Fl. Atl.*
R. L. Desfontaines, *Flora atlantica, sive Historia Plantarum, quae in Atlante, Agro tunetano et algeriensi crescunt.* Parisiis. 1798–1799. (**1** in 1798; **2**: pp. 1–160 in 1798; pp. 161–458 in 1799. Cf. F. A. Stafleu & R. S. Cowan, *Taxonomic Literature* ed. 2, **1**: 628–629 (1976).

Desf., *Hist. Arb.*
R. L. Desfontaines, *Histoire des Arbres et Arbrisseaux qui peuvent être cultivés en pleine Terre sur le Sol de la France.* Paris. 1809. (**1** & **2** in 1809.)

Desv., *Jour. Bot. Appl.*
N. A. Desvaux (edit.), *Journal de Botanique, appliquée à l'Agriculture, à la Pharmacie, à la Médecine et aux Arts.* Paris. 1813–1816. (**1** & **2** in 1813; **3(1–3)** in 1814; **3(4)** in 1815; **3(5)** in 1816; **4(1–3)** in 1814; **4(4–6)** in 1815. Cf. F. A. Stafleu & R. S. Cowan, *Taxonomic Literature* ed. 2, **1**: 634 (1976).)

Desv., *Jour. Bot. Rédigé*
N. A. Desvaux *et al.* (edit.), *Journal de Botanique, rédigé par une Société de Botanistes.* Paris. 1808–1809. (**1**: pp. 1–192 in 1808; pp. 193–384 in 1809; **2** in 1809. Cf. F. A. Stafleu & R. S. Cowan, *Taxonomic Literature* ed. 2, **1**: 633–634 (1976).)

Desv., *Obs. Pl. Angers*
N. A. Desvaux, *Observations sur les Plantes des Environs d'Angers, pour servir de Supplément à la Flore de Maine et Loire, et de Suite à l'Histoire naturelle et critique des Plantes de France.* Angers & Paris. 1818.

Dickson, *Coll. Dried Pl.*
J. Dickson, *A Collection of dried Plants named on the Authority of the Linnean Herbarium and other original Collections.* London. 1789–1801. (Fasc. 1 in 1789; fasc. 2 in 1790; fasc. 3 in 1791; fasc. 4 in 1801.)

Domin, *Pl. Čechosl. Enum.*
K. Domin, *Plantarum Čechoslovakiae Enumeratio Species vasculares indigenas et introductas exhibens.* Praha. 1935. (Also published in *Preslia* **13–15** (1936).)

Domin, *Pterid.*
K. Domin, *Pteridophyta: soustavný přehled žijících i vyhynulých kaprodorostů.* Praze. 1929.

Domin & Podp., *Klíč Úplné Květeně Rep. Česk.*
K. Domin & J. Podpěra, *Klíč k úplné Květeně Republiky Československé.* Olomouci. 1928.

Don, D., *Prodr. Fl. Nepal.*
D. Don, *Prodromus Florae nepalensis, sive Enumeratio Vegetabilium, quae in Itinere per Nepaliam proprie dictam et Regiones conterminas, Ann. 1802–1803. Detexit atque legit D. D. Franciscus Hamilton (olim Buchanan) ... Accedunt Plantae a D. Wallich nuperius missae.* Londini. 1825. (For date cf. W. T. Stearn, *Jour. Arnold Arb.* **26**: 168 (1945).)

Don, G., *Herb. Brit.*
G. Don, *Herbarium britannicum, consisting of Fasciculi of dried British Plants, with their appropriate Names and particular Habitats annexed.* Edinburgh. Fasc. **1–9**, nos. 1–225. 1804–1812/13. (**1** & **2**, nos. 1–50, in 1804; **3** & **4**, nos. 51-100, in 1805; **5–8**, nos. 101–200, in 1806; **9** ('1806'), nos. 201–225, in 1812/1813. Cf. G. C. Druce, *Notes Roy. Bot. Gard. Edinb.* **3**: 144-182 (1904).)

Don fil., G., *Gen. Syst.*
G. Don, *A general System of Gardening and Botany.* Also titled *A general History of the dichlamydeous Plants.* London. 1831–1838. (**1** in 1831; **2** in 1832; **3** in 1834; **4** in 1837–1838. Cf. F. A. Stafleu & R. S. Cowan, *Taxonomic Literature* ed. 2, **1**: 669–670 (1976).)

Drejer, *Fl. Excurs. Hafn.*
S. T. N. Drejer, *Flora excursoria hafniensis.* Hafniae. 1838.

Dreves, *Bot. Bilderb.*
J. F. P. Dreves, *Botanisches Bilderbuch für die Jugend und Freunde der Pflanzenkunde.* Leipzig. 1794–1801.

Druce, *List Br. Pl.*
G. C. Druce, *List of British Plants.* Oxford. 1908. Ed. 2, titled ***Brit. Pl. List***, *British Plant List.* Arbroath. 1928.

Duby, *Bot. Gall.*
J. É. Duby, *Aug. Pyrami de Candolle Botanicon gallicum, seu Synopsis Plantarum in Flora gallica descriptarum. Editio secunda. Ex Herbariis et Schedis candollianis propriisque digestum a J. É. Duby.* Paris. 1828–1830. (**1** in 1828; **2** in 1830.)

Duchartre, *Rev. Bot.*
P.-E.-S. Duchartre (edit.), *Revue botanique. Recueil mensuel renfermant l'Analyse des Travaux publiés en France et à l'Étranger sur la Botanique et sur ses Applications à l'Horticulture, l'Agriculture, la Médecine etc.* Paris. 1845–1847. (**1** in 1845–1846; **2** in 1846–1847.)

Dumort., *Fl. Belg.*
B. C. J. Dumortier, *Florula belgica, Operis majoris Prodromus. Staminacia.* Tornaci Nerviorum. 1827. (Not reviewed until 1830 and perhaps not published until then. Cf. F. A. Stafleu & R. S. Cowan, *Taxonomic Literature* ed. 2, **1**: 699 (1976).)

Edmondston, *Fl. Shetl.*
T. B. Edmondston, *A Flora of Shetland.* Aberdeen, London, Edinburgh, Glasgow & Belfast. 1845. Ed. 2, edited by C. F. Argyll Saxby. Edinburgh & London. 1903.

Ehrend., *Liste Gefässpfl. Mitteleur.*
F. Ehrendorfer, *Liste der Gefässpflanzen Mitteleuropas.* Graz. 1967. Ed. 2, revised by W. Gutermann. Stuttgart. 1973.

Ehrh., *Beitr. Naturk.*
J. F. Ehrhart, *Beiträge zur Naturkunde.* Hannover & Osnabrück. 1787–1792. (**1** in 1787; **2** & **3** in 1788; **4** in 1789; **5** in 1790; **6** in 1791; **7** in 1792. Cf. F. A. Stafleu & R. S. Cowan, *Taxonomic Literature* ed. 2, **1**: 732–733 (1976).)

Enander, *Salic. Scand. Exsicc.*
S. J. Enander, *Salices Scandinaviae exsiccatae.* Uppsala. 1905–1910. (**1** & **2** in 1905; **3** in 1910.)

Endl., *Syn. Conif.*
S. L. Endlicher, *Synopsis Coniferarum.* Sangalli. 1847.

Engler, *Monogr. Gatt. Saxifr.*
H. G. A. Engler, *Monographie der Gattung* Saxifraga L. *mit besonderer Berücksichtigung der geographischen Verhältnisse.* Breslau. 1872.

Engler, *Pflanzenreich*
H. G. A. Engler (edit.), *Das Pflanzenreich. Regni vegetabilis Conspectus.* Berlin. 1900–1953. (For authors and dates of individual volumes cf. M. T. Davis, *Taxon* **6**: 161–182 (1957) and F. A. Stafleu & R. S. Cowan, *Taxonomic Literature* ed. 2, **1**: 785–797 (1976).) In citation the number of the *Heft* is placed first, followed by the systematic numbers (both Roman and Arabic) in brackets.

Engler & Prantl, *Natürl. Pflanzenfam.*
H. G. A. Engler & K. A. E. Prantl, *Die natürlichen Pflanzenfamilien nebst ihren Gattungen und wichtigeren Arten, insbesondere den Nutzpflanzen.* Ed. 1. Leipzig. 1887–1915. Ed. 2. Leipzig & Berlin. 1925→. (For dates of both editions cf. F. A. Stafleu & R. S. Cowan, *Taxonomic Literature* ed. 2, **1**: 764–783 (1976).) Photographic reprint. Berlin. 1960.

Fée, *Mém. Fam. Foug.*
A. L. A. Fée, *Mémoires sur les Familles des Fougères.* Strasbourg. 1844–1866. (**1** in 1844 (t.p.) or 1845; **2** in 1845; **3** & **4** in 1852; **5**, titled *Genera Filicum. Exposition des Genres de la Famille des Polypodiacées (Classe des Fougères)*, in 1850–1852 [pp. 3–30, 35–39, originally published in *Mém. Soc. Hist. Nat. Strasbourg* **4(1)** in 1850, pp. 40–388 in *op. cit.* **5** in 1852]; **6** in 1854; **7–9** in 1857 or 1858; **10** in 1865 or 1866; **11** in 1866. Cf. F. A. Stafleu & R. S. Cowan, *Taxonomic Literature* ed. 2, **1**: 820 (1976).)

Fenzl, *Pugillus*
E. Fenzl, *Pugillus Plantarum novarum Syriae et Tauri occidentalis primus.* Vindobonae. 1842.

Fenzl, *Vers. Darstell. Alsin.*
E. Fenzl, *Versuch einer Darstellung der geographischen Verbreitungs- und Vertheilungs-Verhältnisse der natürlichen Familie der Alsineen in der Polarregion und eines Theile der gemässigten Zone der alten Welt.* Wien. 1883.

Fernandes, A. & R., *Icon. Select. Fl. Azor.*
A. Fernandes & R. [M. S.] B. Fernandes, *Iconographia selecta Flora azoricae a Societate broteriana elaborata.* Conimbriga. 1980→. (**1(1)** in 1980; **1(2)** in 1983; **2(1)** in 1987.)

Fiori & Béguinot, or **Fiori, Béguinot & Pampanini, *Sched. Fl. Ital. Exsicc.***

A. Fiori, A. Béguinot & R. Pampanini, *Schedae ad Floram Italiam exsiccatam*. Ser. 1 & 2, Fasc. **1-11**. Florentiae. 1914. Also published in *Nuovo Gior. Bot. Ital.* as Ser. 1 Vols. **1** & **2**, Ser. 2 Vols. **1** & **2**, Centuriae I–XX, as follows: Ser. 1, **1**, Cent. I-II as *op. cit.* **12(2)**: 144–216 (1905); Cent. III as *op. cit.* **13(1)**: 5–50 (1905); Cent. IV as *op. cit.* **13(2)**: 163–205 (1906); Cent. V and Index to Cent. I–V as *op. cit.* **13(4)**: 289–346 (1906); **2**, Cent. VI as *op. cit.* **14(2)**: 69–116 (1907); Cent. VII as *op. cit.* **14(3)**: 247–292 (1907); Cent. VIII as *op. cit.* **15(3)**: 307–354 (1908); Cent. IX–X and Index to Cent. VI–X as *op. cit.* **15(4)**: 445–543 (1908). Ser. 2, **1**: Cent. XI as *op. cit.* **16(4)**: 443–495 (1909); Cent. XII as *op. cit.* **17(1)**: 62–122 (1910); Cent. XIII–XIV as *op. cit.* **17(4)**: 563–668 (1910); Cent. XV as *op. cit.* **18(3)**: 279–319 (1911); Cent. XVI as *op. cit.* **18(4)**: 459–513 (1911); Cent. XVII–XVIII as *op. cit.* **19**: 517–607 (1912); Cent. XIX–XX as *op. cit.* **20(1)**: 15–109 (1914). Ser. 3. Fasc. **12** (Cent. XXI– XXII, pp. 1–94). Padova. 1914. Fasc. **13** (Cent. XXIII–XXIV; pp. 95–175), place and date not traced but between 1915 and 1920. Fasc. **14** (Cent. XXV–XXVI, pp. 176–244). Sancasciano, Val di Pisa. 1921. Fasc. **15** (Cent. XVII–XVIII, pp. 245–326), place not traced, 1923 or 1924. Fasc. **16** (Cent. XXIX–XXX, pp. 327–436). Forli. 1927. (For dates of **12**, **14**, and **16** cf. *Just's Bot. Jahrb.* **44**: 1269 (1927), **51(2)**: 200 (1933) and **55(1)**: 443 (1935). *Österr. Bot. Zeitschr.* **78**: 72 (1924) gives date of issue of the herbarium sheets (*Flora Italiae Exsiccatae*) corresponding to Cent. XVII–XVIII as 1923, while *Just's Bot. Jahrb.* **54(1)**: 793 (1932) gives 1924.) Ser. 1 by A. Fiori, A. Béguinot & R. Pampanini; Ser. 2 & 3 by A. Fiori & A. Béguinot.

Fiori & Paol., *Fl. Anal. Ital.*

A. Fiori & G. Paoletti, *Flora analitica d'Italia*. Padova. 1896–1909. (**1**: pp. i–c, [i–vii] in 1908; pp. 1–256 in 1896; pp. 257–607 in 1898; **2**: pp. 1–224 in 1900; pp. 225–304 in 1901; pp. 305–492 in 1902; **3**: pp. 1–272 in 1903; pp. 273–524 in 1904; **4**: pp. 1–217, index pp. 1–16 in 1907; index pp. 17–192 in 1907; index pp. 193-330 in 1908; **5** in 1909. Cf. F. A. Stafleu & R. S. Cowan, *Taxonomic Literature* ed. 2, **1**: 833 (1976).)

Fischer, *Cat. Jard. Gorenki*

F. E. L. von Fischer, *Catalogue du Jardin des Plantes du Comte Alexis de Razoumoffsky à Gorenki près de Moscou*. Ed. 1. Moscou. 1808. Ed. 2. Moscou. 1812.

Fischer & C. A. Meyer, *Enum. Pl. Nov.*

F. E. L. von Fischer & C. A. von Meyer, *Enumeratio Plantarum novarum a clarissimo Schrenk lectarum*. Petropoli & Lipsiae. 1841–1842. (**1** in 1841; **2** (*Enumeratio altera* ...) in 1842.)

Fischer & C. A. Meyer, or **Fischer, C. A. Meyer & Avé-Lall., *Ind. Sem. Horti Petrop.***

F. E. L. von Fischer & C. A. von Meyer, or F. E. L. von Fischer & E. R. von Trautvetter, or F. E. L. von Fischer, C. A. von Meyer & J. L. E. Avé-Lallemant, or J. L. E. Avé-Lallemant, *Index [secundus ... undecimus] Seminum, quae Hortus botanicus imperialis petropolitanus pro mutua Commutatione offert. Accedunt Animadversiones botanicae nonnullae*. Petropoli. 1835–1850. (**1** in 1835; **2** in 1836; **3** in 1837; **4** in 1838; **5** in 1839; **6** in 1840; **7** in 1841; **8** in 1842; **9** in 1843; **9** (*Suppl.*) in 1844; **10** in 1845; **11** in 1846; **11** (*Suppl.*) in 1846 or later.) Co-authors: **1–8**, C. A. von Meyer; **3** & **4**, E. R. von Trautvetter; **6–11**, J. L. É. Avé-Lallemant. For continuation see Regel et al., *Ind. Sem. Horti Petrop.*

Fomin, *Fl. RSS Ucr.*

A. V. Fomin, *Flora RSS Ucr./Florae Reipublicae sovieticae socialisticae ucrainicae*. / Флора УРСР [*Flora URSR*]. Ed. 1 Kioviae. 1936–1965. (**1**, edit. A. V. Fomin in 1936; **2**, edit. E. I Bordzilowski & E. M. Lavrenko in 1940.) Titled Флора УРСР [*Flora URSR*]: **3**, edit. M. L. Kotov & A. I. Barbarich in 1950 **4**, edit. M. I. Kotov in 1952; **5**, edit. M. V. Klokov & E. D Wissjulina in 1953; **6**, edit. D. K. Zerov in 1954; **7**, edit. M. V Klokov & E. D. Wissjulina in 1955; **8**, edit. M. I. Kotov & A. I. Barbarich in 1957; **9**, edit. M. I. Kotov in 1960; **10**, edit M. I. Kotov in 1961; **11**, edit. E. D. Wissjulina in 1962; **12**, edit E. D. Wissjulina in 1965.) Ed. 2, titled *Flora RSS Ucr.* (*Florae Reipublicae sovieticae socialisticae ucrainicae*). / Флора УРСР [*Flora URSR*]. Kioviae. 1938. (**1**, edit. E. I. Bordzilowski in 1938.)

Forskål, *Fl. Aegypt.*

P. Forskål, *Flora aegyptiaco-arabica*. Hauniae. 1775. (Posthumous edit. C. Niebuhr.)

Forster, G., *Fl. Ins. Austral. Prodr.*

J. G. A. Forster, *Florulae Insularum australium Prodromus*. Gottingae. 1786.

Fourn., E., *Rech. Fam. Crucif.*

E. P. N. Fournier, *Recherches anatomiques taxonomiques sur la Famille des Crucifères et sur le Genre* Sisymbrium *en particulier* Paris. 1865.

Fourn., P., *Quatre Fl. Fr.*

P. V. Fournier, *Les quatre Flores de la France, Corse comprise* Poinson-les-Grancey. 1934–1940. (Pp. 1–64 in 1934; pp. 65–256 in 1935; pp. 257–576 in 1936; pp. 577–832 in 1937; pp. 833–896 in 1938; pp. 897–992 in 1939; pp. 993–1092 in 1940. [Pp 993–1062 presumably in 1940 although neither the 1946 edition nor F. A. Stafleu & R. S. Cowan, *Taxonomic Literature* ed. 2, **1**: 867 (1976) give the date of their publication; the date is inferred from the fascicle numbers given in the 1946 ed.].) Photographic reprint with additions and corrections at end. Paris. 1946. Reprint with more additions and corrections. Paris. 1961.

Franco, *Abetos*

J. M. A. P. do Amaral Franco, *Abetos*. Lisboa. 1950. (Reprinted from *Anais Inst. Sup. Agron. (Lisboa)* **17**: 1–260 (1950).)

Franco, *Conif. Duar. Nom.*

J. M. A. P. do Amaral Franco, *De Coniferarum duarum Nominibus*. Lisboa. 1950.

Franco, *Dendrologia Florestal*

J. M. A. P. do Amaral Franco, *Dendrologia florestal*. Lisboa. 1943.

Franco, *Nova Fl. Portugal*

J. M. A. P. do Amaral Franco, *Nova Flora de Portugal (continente e Açores)*. Lisboa. 1971 →. (**1** in 1971; **2** in 1984.)

Franklin, *Narr. Journey*

J. Franklin, *Narrative of a Journey to the Shores of the Polar Sea in the Years 1819, 20, 21 and 22*. London. 1823. *Botanical Appendix* by J. Richardson.

Freyc., *Voy. Bot.*

H. L. C. De Saulces de Freycinet, *Voyage autour du Monde, entrepris par Ordre du Roi ... exécuté sur les Corvettes de S. M. l'Uranie et la Physicienne, pendant les Années 1817, 1818, 1819 et 1820*. Paris. 1826–1830. *Botanique* by C. Gaudichaud-Beaupré. (Pp. 1–80, tt. 1–20 in 1826; pp. 81–208, tt. 21–50 in 1827; pp. 209–352, tt. 51–80 in 1828; pp. 353–464, tt. 81–110 in 1829 (t. 85 possibly in 1826); pp. 465–522, tt. 111–120 in 1830. *Atlas*: pp. 1–22 by A. Poiret fils in 1830. Cf. F. A. Stafleu & R. S. Cowan, *Taxonomic Literature* ed. 2, **1**: 922 (1976).)

Fries, *Nov. Fl. Suec.*

E. M. Fries, *Novitiae Florae suecicae*. Ed. 1. Lundae. 1814–1824. (**1** & **2** in 1814; **3** & **4** in 1817; **5** in 1819; **6** in 1823; **7** in 1824.) Ed.

2. Londini Gothorum. 1828. ***Mant.***, *Mantissa* **1**. Lundae. 1832–1835. (Pp. 1–56 in 1832; pp. 57–64 in 1834; pp. 65–84 in 1835.) *Mantissa* **2**. Upsaliae. 1839. *Mantissa* **3**. Upsaliae. 1842–1845. (Pp. 1–48 in 1842; pp. 49–96 in 1843; pp. 97–204 in 1845.) Cf. F. A. Stafleu & R. S. Cowan, *Taxonomic Literature* ed. 2, **1**: 883 (1976).

Fries, *Summa Veg. Scand.*
E. M. Fries, *Summa Vegetabilium Scandinaviae. Sectio prior.* (Pp. 1–258.) Upsaliae. 1845. (Reissued with different title, *Summa ... Plantarum quum cotyledonearum, tum nemearum ... lectarum, indicata simul Distributione geographica*, Upsaliae, Holmiae & Lipsiae in 1846.) *Sectio posterior.* (Pp. 259–572; same title as 1846 reissue of *Sectio prior.*) Upsaliae, Holmiae & Lipsiae. 1849.

Fritsch, *Sched. Fl. Exsicc. Austr.-Hung.*
Cf. A. Kerner, *Sched. Fl. Exsicc. Austr.-Hung.*

Fuss, *Fl. Transsilv. Exc.*
J. M. Fuss, *Flora transsilvaniae Excursoria ....* Hermannstadt. 1866.

Gaertner, *Fruct. Sem. Pl.*
J. Gaertner, *De Fructibus et Seminibus Plantarum.* Stuttgardiae, Tuebingae & Lipsiae. 1788–1807. (**1**, Stuttgardiae; **2**, Tuebingae; **3**, Lipsiae. **1** in 1788; **2**: pp. 1–184 in 1790; pp. 185–504 in 1791; pp. 505–520 in 1792; **3**: pp. 1–56 in 1805; pp. 57–128 in 1806; pp. 129–256 in 1807. Cf. F. A. Stafleu & R. S. Cowan, *Taxonomic Literature* ed. 2, **1**: 900–901 (1976).) **3** by C. F. von Gaertner.

Gaertner, P., B. Meyer & Scherb., *Fl. Wetter.*
P. G. Gaertner, B. Meyer & J. Scherbius, *Oekonomisch-technische Flora der Wetterau.* Frankfurt. 1799–1803. (**1** in 1799; **2** in 1800; **3(1)** in 1801; **3(2)** in 1803.)

Gamisans, *Cat. Pl. Vasc. Corse*
J. Gamisans, *Catalogue des Plantes vasculaires de la Corse.* Ajaccio. 1985.

Garcke, *Fl. Nord- Mittel-Deutschl.*
C. A. F. Garcke, *Flora von Nord- und Mittel-Deutschland.* Ed. 1. Berlin. 1849. Ed. 2. Berlin. 1851. Ed. 3. Berlin. 1854. Ed. 4. Berlin. 1858. Ed. 5. Berlin. 1860. Ed. 6. Berlin. 1863. Ed. 7. Berlin. 1865. Ed. 8. Berlin. 1867. Ed. 9. Berlin. 1869. Ed. 10. Berlin. 1871. Ed. 11. Berlin. 1873. Ed. 12. Berlin. 1875. Ed. 13, titled *Flora von Deutschland.* Berlin. 1878. Ed. 14. Berlin. 1882. Ed. 15. Berlin. 1885. Ed. 16. Berlin. 1890. Ed. 17, titled *Illustrierte Flora von Deutschland.* Berlin. 1895. Ed. 18. Berlin. 1898. Ed. 19. Berlin. 1903. Ed. 20, titled *August Garcke's Illustrierte Flora von Deutschland*, by F. Niedenzu. Berlin. 1908. Ed. 21. Berlin. 1912. Ed. 22. Berlin. 1922. Ed. 23, titled *August Garcke Illustrierte Flora von Deutschland und angrenzenden Gebiete*, by K. von Weihe. Berlin & Hamburg. 1972.

Gaudin, *Fl. Helv.*
J. F. A. T. G. P. Gaudin, *Flora helvetica.* Turici. 1828–1833. (**1–3** in 1828; **4** & **5** in 1829; **6** in 1830; **7** in 1833.)

Gay, *Erysim. Nov.*
J. É. Gay, *Erysimorum quorundum novorum Diagnoses simulque* Erysimi muralis *Descriptionem praemittit, Monographiam Generis editurus J. Gay.* Parisiis. 1842.

Georgi, *Bemerk.* *Reise*
J. G. Georgi, *Bemerkungen einer Reise im russischen Reich in Jahre 1772.* St Petersburg. 1775. (**1** & **2** in 1775, with continuous pagination.)

Gmelin, C. C., *Fl. Bad.*
C. C. Gmelin, *Flora badensis, alsatica et confinium Regionum cis et transrhenana, Plantas a Lacu bodamico usque ad Confluentem Mosellae et Rheni sponte nascentes, exhibens.* Carlsruhae. 1805–1826. (**1** in 1805; **2** in 1806; **3** in 1808; **4** in 1826.)

Gmelin, S. G., *Reise Russl.*
S. G. Gmelin, *Reise durch Russland zur Untersuchung der drey Natur-Reiche.* St. Petersburg. 1770–1784. (**1** in [possibly 1770–]1774; **2** & **3** in 1774; **4** in 1784. **4** posthumously published by P. S. Pallas. Cf. F. A. Stafleu & R. S. Cowan, *Taxonomic Literature* ed. 2, **1**: 958 (1976).)

Gorodkov, *Fl. Murmansk.*
B. N. Gorodkov (edit.), **Флора Мурманской Области** [*Flora Murmanskoj Oblasti*]. Moskva & Leningrad. 1953–1966. (**1** in 1953; **2** in 1954; **3** in 1956; **4** in 1959; **5** in 1966.) **2–5** edit. A. I. Pojarkova.

Gouan, *Fl. Monsp.*
A. Gouan, *Flora monspeliaca.* Lugduni. 1764. (For date cf. F. A. Stafleu & R. S. Cowan, *Taxonomic Literature* ed. 2, **1**: 976 (1976).)

Gouan, *Obs. Bot.*
A. Gouan, *Illustrationes et Observationes botanicae ad Specierum Historiam facientes, seu rariorum Plantarum indigenarum, pyrenaicarum, exoticarum Adumbrationes.* Tiguri. 1773.

Gray, A., *Man. Bot.*
A. Gray, *A Manual of the Botany of the northern United States, from New England to Wisconsin and south to Ohio and Pennsylvania inclusive.* Ed. 1. Boston & Cambridge. 1848. (Salicaceae, pp. 535–567, by J. Carey.) Ed. 2. New York. 1856. (Also 3 'School & College eds.' in 1857, 1858 & 1859 and a reissue of Ed. 2 in 1859.) Ed. 3 (as 'School & College ed.'). New York & Chicago. 1862. Ed. 4. New York & Chicago. 1863. ('School & College eds.' in 1864, 1865, 1867, 1869; reissue of 'Manual' Ed. 4 in 1865). Ed. 5. New York & Chicago. 1867. (*Pinus*, *Isoetes*, *Callitriche* etc. by G. Engelmann; Ferns by D. C. Eaton.) (Eight reissues, 1868–1880.) Ed. 6. New York & Chicago. 1890. (*Salix* by M. S. Bebb; Pteridophytes by D. C. Eaton. Co-authors: J. M. Coulter, S. Watson, L. H. Bailey, D. C. Eaton & L. M. Underwood.) Ed. 7. New York, Cincinnati & Chicago. 1908. Co-authors: B. L. Robinson & M. L. Fernald. Ed. 8, titled *Gray's Manual of Botany, eighth (Centennial) Edition – illustrated. A Handbook of the flowering Plants and Ferns of the central and northeastern United States and adjacent Canada*, edit. M. L. Fernald. New York, Cincinnati, Chicago, Boston, Atlanta, Dallas, San Francisco. 1950 (reprinted 1970, with corrections; reprinted, 1988).

Gray, S. F., *Nat. Arr. Brit. Pl.*
S. F. Gray, *A natural Arrangement of British Plants.* London. 1821. (**1** & **2** in 1821. Cf. F. A. Stafleu & R. S. Cowan, *Taxonomic Literature* ed. 2, **1**: 994 (1976).)

Gren. & Godron, *Fl. Fr.*
J. C. M. Grenier & D. A. Godron, *Flore de France, ou Description des Plantes qui croissent naturellement en France et en Corse.* Paris, London, Besançon, New York & Madrid. 1847–1856. (**1** & **2**, Paris, London & Besançon; **3**, Paris, London, New York, Madrid & Besançon. **1(1)**, pp. 1–330 in 1847; **1(2)**, pp. 331–766 in Dec. 1848–Jan. 1849; **2**: pp. 1–392 in 1850; pp. 393–760 in 1853; **3**: pp. 1–384 in 1855; pp. 385–779 in 1856. Cf. F. A. Stafleu & R. S. Cowan, *Taxonomic Literature* ed. 2, **1**: 999–1000 (1976).)

Greuter, Burdet & Long, *Med-Checklist*
W. Greuter, H. M. Burdet & G. Long, *Med-Checklist. A critical Inventory of vascular Plants of the circum-mediterranean Countries. / Inventaire critique des Plantes vasculaires des Pays circumméditerranéens.* Genève. **1**→, 1984→. (**1** in 1984; **3** in 1986; **4** in 1989. Text of Pteridophyta first published as a preliminary version titled *Med-Checklist ... I. Pteridophyta*, Genève & Berlin, 1981.) Taxonomy and novelties to be attributed to Greuter & Burdet only.

Griseb., *Spicil. Fl. Rumel.*
A. H. R. Grisebach, *Spicilegium Florae rumelicae et bithynicae*. Brunsvigae. 1843–1846. (**1** in 1843; **2**: pp. 1–160 in 1844; pp. 161–548 in 1846. Cf. F. A. Stafleu & R. S. Cowan, *Taxonomic Literature* ed. 2, **1**: 1008–1009 (1976).)

Grossh. & Schischkin, *Sched. Herb. Pl. Or. Exsicc.*
A. A. Grossheim & B. K. Schischkin, *Schedae ad Herbarium* "Plantae orientales exsiccatae". Tiflis. 1924–1928. (**1**: pp. 1–52, fasc. 1–8, nos. 1–200 in 1924 (nos. 1-25 possibly in 1923: cf. *Österr. Bot. Zeitschr.* **73**: 72 (1924)); **2**: pp. 1–50, fasc. 9–16, nos. 201–400 in 1928. (Cf. *Just's Bot. Jahrb.* **53(2)**: 686 (1934) and *op. cit.* **56(1)**: 697 (1937).)

Guépin, *Fl. Maine Loire*
J. P. Guépin, *Flore de Maine et Loire,... Tome premier*. Angers. 1830. (**2**, Cryptogams, never published.) Ed. 2. Angers. 1838. ***Suppl.***, *Supplément à la Flore de Maine et Loire*. Angers. 1842. Ed. 3. Angers & Paris. 1845. ***Suppl.***, *Supplément à la troisième Édition de la Flore de Maine et Loire*. Angers. 1850. ***Suppl. 2***, titled *Notice sur une Flore Angevine manuscrite, suivie d'un second Supplément à la Flore de Maine et Loire*. Angers. 1854.

Gunnerus, *Fl. Norv.*
J. E. Gunnerus, *Flora norvegica*. Nidrosiae & Hafniae. 1766–1772. (**1** in 1766; **2** in 1772.)

Guss., *Cat. Pl. Boccad.*
G. Gussone, *Catalogus Plantarum, quae asservantur in Regio Horto ser. Fr. Borbonii Principis Juventutis in Boccadifalco prope Panormum*. Neapoli. 1821.

Guss., *Fl. Sic. Prodr.*
G. Gussone, *Flora siculae Prodromus*. Neapoli. 1827–1828. (**1** in 1827; **2** in 1828.) ***Suppl. Fl. Sic. Prodr.***, *Supplementum ad Florae siculae Prodromus*. Neapoli. 1832–1834. (Pp. 1–166 in 1832; pp. 167–242 in 1834.)

Guss., *Fl. Sic. Syn.*
G. Gussone, *Florae siculae Synopsis*. Neapoli. 1843–1845. (**1** in 1843; **2(1)**: pp. 1–529 in 1844; **2(2)**: pp. 527[*sic*]–668 in 1844; pp. 669–920 in 1845. Cf. F. A. Stafleu & R. S. Cowan, *Taxonomic Literature* ed. 2, **1**: 1026 (1976).)

Guss., *Ind. Sem. Horto Boccad.*
Index Seminum in Horto Boccadifalci. Neapoli. By G. Gussone in 1825 & 1826.

Guss., *Pl. Rar.*
G. Gussone, *Plantae rariores quas in Itinere per Oras jonii ac adriatici Maris et per Regiones Samnii ac Aprutii collegit*... Neapoli. 1826.

Guss., *Suppl. Fl. Sic. Prodr.*
Cf. Guss., *Fl. Sic. Prodr.*

Halácsy, *Consp. Fl. Graec.*
E. von Halácsy, *Conspectus Florae graecae*. Lipsiae. 1900–1904. (**1**: pp. 1–576 in 1900; pp. 577–825 in 1901; **2** in 1902; **3** in 1904. Cf. F. A. Stafleu, *Taxonomic Literature* ed. 1, 189 (1967).) ***Suppl.***, *Supplementum* [**1**]. Lipsiae. 1908. *Supplementum secundum* [**2**]. Budapest. 1912. (Published as *Magyar Bot. Lapok* **11**: 114–202 (1912) and also issued separately with independent pagination.)

Hara, *Fl. East. Himalaya*
H. Hara, *The Flora of eastern Himalaya: Results of the botanical Expedition to eastern Himalaya organized by the University of Tokyo 1960 and 1963*. Tokyo. 1966. *Second Report*. Tokyo. 1971. *Third Report*, compiled by H. Ohashi. Tokyo. 1975.

Hartinger, *Atlas Alpenfl. (Text)*
A. Hartinger, *Atlas der Alpenflora. Text* by K. W. von Dalla Torre. Wien. Ed. 1. 1881–1884. (Lfg. 1–4 in 1881; Lfg. 5–16 in 1882; Lfg. 17–28 in 1883; Lfg. 29–36 in 1884. Text in 1882.) Ed. 2. Graz. 1896–1897. (**1** in 1896; **2** in 1897.) Cf. F. A. Stafleu & R. S. Cowan, *Taxonomic Literature* ed. 2, **2**: 64 (1979).

Hartman, *Handb. Skand. Fl.*
C. J. Hartman, *Handbok i Skandinaviens Flora*. Ed. 1. Stockholm. 1820. Ed. 2. Stockholm. 1832. Ed. 3. Stockholm. 1838. (**1**, *Botanologien*, & **2**, *Floran*, in 1838.) Ed. 4. Stockholm. 1843. Ed. 5. Stockholm. 1849 (Dec.) or 1850. Ed. 6 by C. Hartman. Stockholm. 1854. Ed. 7 by C. Hartman. Stockholm. 1858. Ed. 8 by C. Hartman. 1861. Ed. 9 by C. Hartman. Stockholm. 1864. (**1** & **2** in 1864.) Ed. 10 by C. Hartman. Stockholm. 1870–1871. (**1** in 1870; **2** in 1871.) Ed. 11, titled *C. J. Hartmans Handbok i Skandinaviens Flora*, by C. Hartman. Stockholm. 1879. Ed. 12, titled *C. J. och C. Hartmans Handbok i Skandinaviens Flora*, by T. O. B. N. Krok. Stockholm. 1889. For a continuation, see Holmberg, *Skand. Fl.*

Harvey & Sonder, *Fl. Cap.*
W. H. Harvey & O. W. Sonder, *Flora capensis, being a systematic Description of the Plants of the Cape Colony, Caffraria and Port Natal*. Ashford, Dublin, London and Cape Town. 1860–1933. (**1–3** by W. H. Harvey & O. W. Sonder, **4–5(2)** & **5(3)–7** by W. T. Thiselton-Dyer, **5(2)** *Suppl.* by A. J. Hill. **1**, Dublin & Cape Town in 1860; **2**, Dublin & Cape Town in 1862; **3**, Dublin & Cape Town in 1865; **4(1)**, London: pp. 1–336 in 1905; pp. 337–480 in 1906; pp. 481–672 in 1907; pp. 673–864 in 1908; pp. 865–1168 in 1909; **4(2)**, London in 1904; **5(1)**, London: pp. 1–224 in 1901; pp. 225–448 in 1900; pp. 449–747 in 1912; **5(2)**, London: pp. 1–384 in 1915; pp. 385–528 in 1920; pp. 529–600 in 1925. **5(2)** (*Supplement*), Ashford in 1933. **5(3)**, London: pp. 1–192 in 1912; pp. 193-332 in 1913. **6**, London: pp. 1–384 in 1896; pp. 385–563 in 1897. **7**, Ashford: pp. 1-192 in 1897; pp. 193–384 in 1898; pp. 385–576 in 1899; pp. 577–791 in 1900.)

Haw., *Revis. Pl. Succ.*
A. H. Haworth, *Saxifragearum Enumeratio. Accedunt Revisiones Plantarum succulentarum*. Londini. 1821.

Hayek, *Fl. Steierm.*
A. von Hayek, *Flora von Steiermark*. Berlin & Graz. 1908–1956. (**1**: pp. 1–480 in 1908; pp. 481–960 in 1909; pp. 961–1200 in 1910; pp. 1201–1271 in 1911; **2(1)**: pp. 1–160 in 1911; pp. 161–480 in 1912; pp. 481–640 in 1913; pp. 641–870 in 1914; **2(2)** in 1956. Cf. F. A. Stafleu & R. S. Cowan, *Taxonomic Literature* ed. 2, **2**: 111 (1979).)

Hayek, *Prodr. Fl. Penins. Balcan.*
A. von Hayek, *Prodromus Florae Peninsulae balcanicae*. (Published as *Feddes Repert.* (*Beih.*) **30**.) (**1**: pp. 1–352 in 1924; pp. 353–672 in 1925; pp. 673–960 in 1926; pp. 961–1193 in 1927; **2**: pp. 1–96 in 1928; pp. 97–336 in 1929; pp. 337–576 in 1930; pp. 577–1152 in 1931; **3**: pp. 1–368 in 1932; pp. 369–472 in 1933. Cf. F. A. Stafleu & R. S. Cowan, *Taxonomic Literature* ed. 2, **2**: 111 (1979).) **2** & **3** edit. F. Markgraf.

Hegetschw., *Fl. Schweiz*
J. J. Hegetschweiler-Bodmer, *Die Flora der Schweiz*. First issue. Zürich. 1838–1840. (Pp. 1–144 in 1838; pp. 145–456, *Extrabeilage* in Dec. 1838 or Jan. 1839; pp. 457–684 in 1839; pp. 685–1008 in 1840.) Second issue, with ***Supplementum*** by O. Heer with continuous pagination (pp. 1009–1135). Zürich. 1840. (Pp. 1–1135 in 1840).

Hegi, *Ill. Fl. Mitteleur.*
G. Hegi, *Illustrierte Flora von Mitteleuropa*. Ed. 1. München. 1906–1931. (**1**: pp. 1–72 in 1906; pp. 73–312 in 1907; pp. 313–412 in 1908; **2**: pp. 1–128 in 1908; pp. 129–405 in 1909; **3**: pp. 5–36 in 1909; pp. 37–328 in 1910; pp. 329–472 in 1911; pp. 473–607 in 1912; **4(1)**: pp. 5–96 in 1913; pp. 97–144 in 1914; pp. 145–192

in 1916; pp. 193–320 in 1918; pp. 321–491 in 1919; **4(2)**: pp. 497–540 in 1921; pp. 541–908 in 1922; pp. 909–1112b in 1923; **4(3)**: pp. 1113–1436 in 1923; pp. 1437–1748 in 1924; **5(1)**: pp. 1–316 in 1924; pp. 317–674 in 1925; **5(2)**: pp. 679–994 in 1925; pp. 995–1562 in 1926; **5(3)**: pp. 1567–1722 in 1926; pp. 1723–2250 in 1927; **5(4)**: pp. 2255–2630 in 1927; **6(1)**: pp. 5–112 in 1913; pp. 113–304 in 1914; pp. 305–352 in 1915; pp. 353–400 in 1916; pp. 401–496 in 1917; pp. 497–544 in 1918; **6(2)**: pp. 549–1152 in 1928; pp. 1153–1386 in 1929; **7** in 1931.) Ed. 2. München, Berlin & Hamburg. 1936→. (**1**, München in 1936; **2**, München in 1939; **3(1)**, München: pp. 1–240 in 1957; pp. 241–452 in 1958; **3(2)**: pp. 453–532, München in 1959; pp. 533–692, München in 1960; pp. 693–772, München in 1961; pp. 773–852, München in 1962; pp. 853–932, München in 1969; pp. 933–1012, München in 1971; pp. 1013–1092, Berlin & Hamburg in 1978; pp. 1093–1265, Berlin & Hamburg in 1979; **3(3)**: pp. 1–80, München in 1965; pp. 81–356, Berlin & Hamburg in 1974; **4(1)**, München: pp. 1–80 in 1958; pp. 81–160 in 1959; pp. 161–320 in 1960; pp. 321–480 in 1962; pp. 481–548 in 1963; **4(2A)**, München: pp. 1–80 in 1961; pp. 81–224 in 1963; pp. 225–304 in 1964; pp. 305–384 in 1965; pp. 385–448 in 1966; **4(2B)**, München: pp. 1–248 in 1990; **4(3)**, München in 1964 (unchanged reprint of ed. 1), reprinted with adaptations, Berlin & Hamburg in 1974; **5(1)**, München in 1965 (unchanged reprint of ed. 1), reprinted with adaptations, Berlin & Hamburg, 1975; **5(2)**, München in 1965 (unchanged reprint of ed. 1), reprinted with adaptations, Berlin & Hamburg in 1975; **5(3)**, München in 1966 (unchanged reprint of ed. 1), reprinted with adaptations, Berlin & Hamburg in 1975; **5(4)**, München in 1964 (unchanged reprint of ed. 1), reprinted with adaptations, Berlin & Hamburg in 1975; **5** (*Nachträge*), München in 1968; **6(1)**, München: pp. 1–80 in 1965; pp. 81–161 in 1966; pp. 161–240 in 1968; pp. 241–320 in 1969; Lfg. 5 in 1972; Lfg. 6 in 1974; Lfg. 7/8 in 1974; **6(2)**: pp. 1–96, München in 1966; pp. 97–176, München in 1970; pp. A1–A36, Berlin & Hamburg in 1979; **6(3)**: pp. 1–80, München in 1964; pp. 81–160, München in 1965; pp. 161–240, München in 1966; pp. 241–320, München in 1968; pp. 321–366, Berlin & Hamburg in 1979; **6(4)**, edit. G. Wagenitz: pp. 580–1483 in 1987.) Ed. 3. München, Berlin & Hamburg. 1966→. **1(1)**, Berlin & Hamburg in 1984; **1(2)**, Berlin & Hamburg in 1981; **1(3)**, Berlin & Hamburg: pp. 1–80 in 1979; pp. 81–160 in 1983; pp. 161–240 in 1985; pp. 241–320 in 1987; pp. 321–400 in 1989; **2(1)**: pp. 1–80, München in 1966; pp. 81–160, München in 1968; pp. 161–240, München in 1969; pp. 241–320, Berlin & Hamburg in 1977; pp. 321–400, Berlin & Hamburg in 1979; pp. 401–440, Berlin & Hamburg in 1980; **3(1)** in 1981; **4(1)**, Berlin & Hamburg in 1986. *Bibliographie zur Flora von Mitteleuropa*, by V. Hamann & G. Wagenitz: Berlin & Hamburg in 1977.

Heldr., *Fl. Céphal.*
T. H. H. von Heldreich, *Flore de l'Île de Céphalonie*. Lausanne. 1882; reissued in 1883. (Cf. F. A. Stafleu & R. S. Cowan, *Taxonomic Literature* ed. 2, **1**: 143 (1979).)

Heldr., *Sched. Herb. Graec. Norm.*
T. H. H. von Heldreich, *Schedae Plantarum ad Herbaria graeca normalia*. Ser. 1 (nos. 1–812) in 1854–1861; ser. 2 (nos. 813–1599) in 1885–1900; cf. *Bull. Soc. Bot. Suisse* **50a**: 259–266 (1940).)

Herbich, *Sel. Pl. Rar. Galic.*
F. Herbich, *Selectus Plantarum rariorum Galiciae et Bucovinae*. Czernovicii. 1836.

Hermann, F., *Fl. Deutschl. Fennosk.*
F. Hermann, *Flora von Deutschland und Fennoskandinavien sowie von Island und Spitzbergen*. Ed. 1. Leipzig. 1912. Ed. 2, titled *Flora von Nord- und Mitteleuropa*. Stuttgart. 1956.

Heuffel, *Enum. Pl. Banat.*
J. A. Heuffel, *Enumeratio Plantarum in Banatu temesiensi sponte crescentium et frequentius cultarum*. Viennae. 1858.

Heynh., *Nomencl. Bot.*
G. Heynhold, *Nomenclator botanicus hortensis, oder alphabetische und synonymische Aufzählung der in den Gärten Europa's cultivirten Gewächse*. Dresden & Leipzig. 1840–1841. (**1**: pp. 1–408 in 1840; **2**: pp. 409–888 in 1841.) *Zweiter Band*, titled ***Alph. Aufz. Gew.***, *Alphabetische und synonymische Aufzählung der in der Jahren 1840 bis 1846 in den europäischen Gärten eingeführten Gewächse* .... Dresden & Leipzig. 1846–1847. (Pp. 1–384[?] in 1846; pp. 385[?]–774 in 1847.)

Hiitonen, *Suomen Kasvio*
H. I. A. Hiitonen, *Suomen Kasvio*. Helsingissä. 1933.

Hill, *Hort. Kew.*
J. Hill, *Hortus kewensis, sistens Herbas exoticas indigenasque rariores in Area botanica Hortorum augustissimae Principissae Cambriae dotissae, apud Kew in Comitatu surreiano cultas*. Londini. 1768.

Hoffm., *Deutschl. Fl.*
G. F. Hoffmann, *Deutschlands Flora, oder botanisches Taschenbuch für das Jahr 1791*. Ed. 1. Erlangen. 1791. Ed. 2. ***Crypt.*** *...für das Jahr 1795. Cryptogamie*. Erlangen. 1796. Ed. 3. *...für das Jahr 1800... Dritter Jahrgang. I. Abtheilung*... Erlangen. 1800. Ed. 4. *...für das Jahr 1804 ... Vierter Jahrgang oder der III Jahrgangs II Abtheilung*. Erlangen. 1804. (Cf. F. A. Stafleu & R. S. Cowan, *Taxonomic Literature* ed. 2, **2**: 240 (1979).)

Holl & Heynh., *Fl. Sachs.*
F. Holl & G. Heynhold, *Flora von Sachsen*. **1**, *Phanerogamen* by G. Heynhold. Dresden. 1842. (Heynhold was sole author of this, the only volume which was published.)

Holmberg, *Skand. Fl.*
O. R. Holmberg, *Hartmans Handbok i Skandinaviens Flora*. Stockholm. 1922–1931. (**1**: pp. 1–160 in 1922; **1(b, 1)**: pp. 1–160 in 1931; **2**: pp. 161–320 in 1926; **2(a)** in 1928.)

Hooker, *Brit. Fl.*
W. J. Hooker, *The British Flora, comprising the phaenogamous Plants and the Ferns*. Ed. 1. London. 1830. Ed. 2. London. 1832. Ed. 3. London. 1835. Ed. 4. London. 1838. Ed. 5. London. 1842.

Hooker, *Fl. Bor.-Amer.*
W. J. Hooker, *Flora boreali-americana; or, the Botany of the northern Parts of British America*. London. 1829–1840. (**1**: pp. 1–48 in 1829; pp. 49–96 in 1830; pp. 97–160 in 1831; pp. 161–272 in 1832; pp. 273–328 in 1833; pp. 329–351 in 1834; **2**: pp. 1–48 in 1834; pp. 49–96 in 1837; pp. 97–192 in 1838; pp. 193-240 in 1839; pp. 241–328 in 1840. Cf. F. A. Stafleu & R. S. Cowan, *Taxonomic Literature* ed. 2, **2**: 292 (1979).)

Hooker, *Gen. Fil.*
W. J. Hooker, *Genera Filicum; or Illustrations of the Ferns, and other allied Genera*. London. 1842.

Hooker & Grev., *Icon. Fil.*
W. J. Hooker & R. K. Greville, *Icones Filicum. Figures and Descriptions of Ferns*. Londini. [1827-]1831[–1832]. (**1(1–3)**: tt. 1–60 in 1827; **1(4 & 5)**: tt. 61–100 in 1828; **1(6)**: tt. 101-120 in 1829; **2(7)**: tt. 121–140 in 1829; **2(8)**: tt. 141–160 in 1829 or 1830; **2(9)**: tt. 161-180 in 1830; **2(10)**: tt. 181–200 in 1830 or 1831; **2(11)**: tt. 201-220 in 1831; **2(12)**: tt. 221-240 in 1832.)

Hooker fil., *Fl. Brit. India*
J. D. Hooker, *The Flora of British India*. London. 1872–1897. (**1–7**; for dates cf. F. A. Stafleu & R. S. Cowan, *Taxonomic Literature* ed. 2, **2**: 274–282 (1979).)

Hooker fil., *Stud. Fl. Brit. Is.*
J. D. Hooker, *The Student's Flora of the British Islands*. Ed. 1. London. 1870. Ed. 2. London. 1878. Ed. 3. London. 1884. Reprints in 1897, 1930 & 1937.

Hornem., *Hort. Hafn.*
J. W. Hornemann. *Hortus regius botanicus hafniensis*. Hauniae. 1813–1815. (**1** in 1813; **2** in 1815.) ***Suppl.***, *Supplementum*. Hafniae. 1819.

Horvátovsky, *Fl. Tyrnav.*
S. Horvátovsky, *Florae tyrnaviensis indigenae Pars prima*. Tyrnavii. 1774.

Host, *Fl. Austr.*
N. T. Host, *Flora austriaca*. Viennae. 1827–1831. (**1** in 1827; **2** in 1831.) (Cf. F. A. Stafleu & R. S. Cowan, *Taxonomic Literature* ed. 2, **2**: 334 (1979).)

Houtt., *Natuurl. Hist. (Handleid.)*
M. Houttyn, *Natuurlijke Historie of Uitvoerige Beschrijving der Dieren, Planten en Mineraalen. Tweede Deel*. Amsterdam. 1773–1783. Botany in **2**. (**2(1)** in 1773; **2(2** & **3)** in 1774; **2(4** & **5)** in 1775; **2(6)** in 1776; **2(7** & **8)** in 1777; **2(9)** in 1778; **2(10** & **11)** in 1779; **2(12)** in 1780; **2(13)** in 1782; **2(14)** in 1783; **2(18)** of *Eerste Deels* (1st series) in 1773.) (Cf. F. A. Stafleu & R. S. Cowan, *Taxonomic Literature* ed. 2, **2**: 344 (1979).) Also published, apparently simultaneously, under the title *Handleiding tot de Plant en Kruidkunde*.

Hudson, *Fl. Angl.*
W. Hudson, *Flora anglica*. Ed. 1. London. 1762. Ed. 2. London. 1778. Ed. 3. London. 1798.

Humb., Bonpl. & Kunth, *Nov. Gen. Sp.*
F. W. H. A. von Humboldt, A. J. A. Bonpland & K. (C.) S. Kunth, *Nova Genera et Species Plantarum quas in Peregrinatione Orbis novi collegerunt, descripserunt, partim adumbraverunt Amatus Bonpland et Alexander de Humboldt. Ex Schedis autographis Amati Bonpland in Ordinem digesset Carolus Sigismund Kunth*. Lutetiae Parisiorum. 1816–1825. (**1–7**; for dates cf. F. A. Stafleu & R. S. Cowan, *Taxonomic Literature* ed. 2, **2**: 369 (1979).)

Hyl., *Fört. Skand. Växt.*
N. Hylander, *Förteckning över Skandinaviens Växter*. **1**. *Kärlväxter*. Ed. 3. Lund. 1941. (*Ytterligare Tillägg och Rättelser* published as *Bot. Not.* **1945**: 445–460 (1945).)

Jacq., *Collect. Bot.*
N. J. von Jacquin, *Collectanea ad Botanicam, Chemiam et Historiam naturalem spectantia*. Vindobonae. 1787–1797. (**1** in 1787; **2** in 1789; **3** in 1791; **4** in 1791; **5** (*Supplementum*) in 1797. (Cf. F. A. Stafleu & R. S. Cowan, *Taxonomic Literature* ed. 2, **2**: 412 (1979).)

Jacq., *Enum. Stirp. Vindob.*
N. J. von Jacquin, *Enumeratio Stirpium plerarumque, quae sponte crescunt in Agro vindobonensi, Montibusque confinibus*. Vindobonae. 1762.

Jacq., *Fl. Austr.*
N. J. von Jacquin, *Florae austriacae, sive Plantarum selectarum in Austriae Archiducatu sponte crescentium Icones*. Viennae. 1773–1778. (**1** in 1773; **2** in 1774; **3** in 1775; **4** in 1776; **5** in 1778.)

Jacq., *Hort. Vindob.*
N. J. von Jacquin, *Hortus botanicus vindobonensis*. Vindobonae. 1770–1777. (**1**: pp. 1–44, tt. 1–30 in 1770, tt. 31–100 in 1771; **2**: pp. 45–95, tt. 101–200 in 1772–1773; **3** in 1776–1777 (probably pp. 1–28. tt. 1–50 in 1776; pp. 29–52, tt. 51–100 in 1777). Cf. F. A. Stafleu & R. S. Cowan, *Taxonomic Literature* ed. 2, **2**: 410 (1979).)

Jacq., *Icon. Pl. Rar.*
N. J. von Jacquin, *Icones Plantarum rariorum*. Vindobonae. 1781–1795. (**1(1)** in 1781; **1(2)** in 1782; **1(3)** in 1783; **1(4)** in 1784; **1(5–8)** in 1787; *Text-List* of **1** in ?1787; **2–3(1)** in 1788; **2–3(2** & **3)** in 1789; **2–3(4** & **5)** in 1790; **2–3(6** & **7)** in 1791; **2–3(8)** in 1792; **2–3(9–11)** in 1792; **2–3(12)** in 1793; **2–3(13–15)** in 1794; **2–3(16)**, *Text-List* of **2–3** in 1795. (Cf. F. A. Stafleu & R. S. Cowan, *Taxonomic Literature* ed. 2, **2**: 411 (1979).)

Jacq., *Misc. Austr. Bot.*
N. J. von Jacquin, *Miscellanea austriaca ad Botanicam, Chemiam et Historiam naturalem spectantia*. Vindobonae. 1779–1781(–1782?). (**1** in 1779; **2** in 1781 or 1782.)

Jacq., *Obs. Bot.*
N. J. von Jacquin, *Observationum botanicorum Iconibus ab Auctore delineatis illustratarum Pars* 1[–4]. Vindobonae. 1764–1771. (**1** in 1764; **2** in 1767; **3** in 1768; **4** in 1771.)

Jahandiez & Maire, *Cat. Pl. Maroc.*
É. Jahandiez & R. C. J. E. Maire, *Catalogue des Plantes du Maroc (Spermatophytes et Ptéridophytes)*. Alger. 1931–1941. (**1** in 1931; **2** in 1932; **3** in 1934; **4** (*Supplément*) by L. Emberger & R. C. J. E. Maire, in 1941.)

Jalas & Suominen, *Atlas Florae Europaeae*
A. J. J. Jalas & J. K. K. Suominen, *Atlas Florae Europaeae. Distribution of vascular Plants in Europe*. Helsinki. 1972→. (**1** in 1972; **2** in 1973; **3** in 1976; **4** in 1979; **5** in 1980; **6** in 1983; **7** in 1986; **8** in 1989. (**1–7** republished, Cambridge, in 1988.)

Janchen, *Cat. Fl. Austr.*
K. Höfler & F. Knoll (edit.), *Catalogus Florae Austriae*. **1** *Pteridophyten und Anthophyten* by E. Janchen. Wien. 1956–1960. (Pp. i–viii, 1–176 in 1956; pp. 177–440 in 1958; pp. 441–710 in 1959; pp. 711–999 in 1960.)

Janchen & Wendelberger, *Kleine Fl. Wien*
E. Janchen & G. Wendelberger, *Kleine Flora von Wien, Niederösterreich und Burgenland*. Wien. 1953.

Jaub. & Spach, *Ill. Pl. Or.*
H. F. Jaubert & E. Spach, *Illustrationes Plantarum orientalium, ou Choix de Plantes nouvelles ou peu connues de l'Asie occidentale*. Paris. 1842–1857. (**1–5**; for dates cf. F. A. Stafleu & R. S. Cowan, *Taxonomic Literature* ed. 2, **2**: 438 (1979).)

Jáv., *Magyar Fl.*
S. Jávorka, *Magyar Flóra (Flora hungarica)*. Budapest. 1924–1925. (Pp. 1–800 in 1924; pp. 801–1307, i–cii in 1925.) No vol. no. is used, as pagination is continuous.

Jordan, *Cat. Jard. Dijon*
C. T. A. Jordan, *Catalogue des Graines récoltées au Jardin botanique de la Ville de Dijon en 1848, offertes en Échange*. Dijon. 1848.

Jordan, *Diagn.*
C. T. A. Jordan, *Diagnoses d'Espèces nouvelles ou méconnues, pour servir de Matériaux à une Flore réformée de la France et des Contrées voisines*. Paris & Leipzig. 1864. (Pp. 5–150 first published in *Ann. Soc. Linn. Lyon* ser. 2, **7**: 373–518 (1861).)

Jordan, *Obs. Pl. Crit.*
C. T. A. Jordan, *Observations sur plusieurs Plantes nouvelles, rares ou critiques de la France*. Paris & Leipzig. 1846–1849. (**1–4**, Paris & Leipzig in 1846; **5**, Paris in 1847; **6**, Paris in 1847; **7**, Paris in 1849.)

Jordan, *Pug. Pl. Nov.*
C. T. A. Jordan, *Pugillus Plantarum novarum praesertim gallicarum*. Paris. 1852.

Jordan & Fourr., *Brev. Pl. Nov.*
C. T. A. Jordan & J. P. Fourreau, *Breviarum Plantarum novarum, sive Specierum in Horto plerumque Cultura recognitarum Descriptio contracta ulterius amplianda*. Parisiis. 1866–1868. (**1** in 1866; **2** in 1868.)

Jordanov, *Fl. Rep. Pop. Bulg.*
D. Jordanov (edit.), **Флора на народна Република България** [*Flora na narodna Republika Bălgarija*]. / *Flora Reipublicae popularis Bulgariae*. Sofia/ Serdicae. **1**→, 1963→. (**1** in 1963; **2** in 1964; **3** in 1966; **4** in 1970; **5** in 1973; **6** in 1976; **7** in 1979; **8** in 1982; **9** in 1989.) **1** & **2** edit. D. Jordanov, B. Kitanov & S. Vălev; **3** edit. D. Jordanov & B. Kuzmanov; **4** edit. D. Jordanov & S. Kožuharov; **5** edit. S. Vălev & I. Assenov; **6** edit. D. Jordanov; **7** edit. B. Kuzmanov; **8** edit. S. Kožuharov; **9** edit. B. Kuzmanov.

Karsten, *Deutsche Fl.*
G. K. W. H. Karsten, *Deutsche Flora. Pharmaceutisch-medicinische Botanik*. Ed. 1. Berlin. 1880–1883. (Pp. 1–128 in 1880; pp. 129-528 in 1881; pp. 529–1008 in 1882; pp. 1009–1284 in 1883). Ed. 2, titled *Flora von Deutschland, Oesterreich und der Schweiz. Mit ein Einschluss der fremdländischen medicinisch und technisch wichtigen Pflanzen, Droguen und deren chemisch-physiologischen Eisenschaften*. Gera-Untermhaus. 1894–1895. (Pp. 1–672 in 1894; pp. 673–790 in 1895. Cf. F. A. Stafleu & R. S. Cowan, *Taxonomic Literature* ed. 2, **2**: 562 (1979).)

Kerner, A., *Sched. Fl. Exsicc. Austro-Hung.*
A. J. Kerner von Marilaun, *Schedae ad Floram exsiccatam austro-hungaricam*. Vindobonae. 1881–1913. (**1**: pp. 1–62 in 1881; pp. 63–136 in 1882; **2** in 1882; **3** in 1884; **4** in 1886; **5** in 1888 and 1889; **6** in 1893; **7** in 1896; **8** in 1899; **9** in 1902; **10** in 1913.) **1-7** by A. J. Kerner von Marilaun; **8** & **9** by K. (C.) Fritsch; **10** by H. von Handel-Mazzetti & I. Dörfler, edit. R. von Wettstein. Cf. F. A. Stafleu & R. S. Cowan, *Taxonomic Literature* ed. 2, **2**: 528 (1979).

Kersten, *Reise Ost. Afr. Bot.*
O. Kersten, *Baron Carl Claus von der Decken's Reisen in Ost-Afrika in den Jahren 1859 bis 1861*. Leipzig & Heidelberg. 1869–1879. Botany in **3**(**3**), 1879, separately paged and titled *Botanik von Ost-Afrika. Bearbeitet von P. Ascherson, O. Böckeler, F. W. Klatt, M. Kuhn, P. G. Lorentz, W. Sonder*. (Cf. F. A. Stafleu & R. S. Cowan, *Taxonomic Literature* ed. 2, **2**: 531 (1979).)

Kirschleger, *Fl. Alsace*
F. R. Kirschleger, *Flore d'Alsace et des Contrées limitrophes*. Strasbourg & Paris. 1850–1862. (**1**: pp. 1–288 in 1850; pp. 289–432 in 1851; pp. 433–624 in 1851 or 1852; pp. 625–662, i–xvii in 1852; **2** in parts between 1853 and 1857; **3**: pp. 1–188 in 1857–1858; pp. 189–396 in 1859–1860; pp. 397–456 in 1861–1862. Cf. F. A. Stafleu & R. S. Cowan, *Taxonomic Literature* ed. 2, **2**: 553–554 (1979).)

Klika et al., *Jehličnaté*
J. Klika, K. Šiman, F. Novák & B. Kavka, *Jehličnaté*. Praha. 1953.

Klokov & Wissjul., *Fl. RSS Ucr.*
Cf. Fomin, *Fl. RSS Ucr.*

Koch, *Syn. Fl. Germ.*
W. D. J. Koch, *Synopsis Florae germanicae et helveticae*. Ed. 1. Francofurti a. M. 1835–1837. (Pp. 1–352 in 1835; pp. 353–844 in 1837, index pp. [1–]102 in 1838.) Ed. 2. Francofurti a. M. & Lipsiae. 1843–1845. (Pp. 1–452, Francofurti a. M. in 1843; pp. 451 *bis*–964, Lipsiae in 1844; pp. 963 *bis*–1164, Lipsiae in 1845.) Ed. 3. Lipsiae. 1856–1857. (Pp. 1–400 in 1856; pp. 401–875 in 1857.) Cf. F. A. Stafleu & R. S. Cowan, *Taxonomic Literature* ed. 2, **2**: 592–593 (1979).)

Koch & Ziz, *Fl. Palat.*
W. D. J. Koch & J. B. Ziz, *Catalogus Plantarum quas in Ditione Florae Palatinatus legerunt*. Moguntiae. 1814.

Koch, C., *Dendrologie*
C. (K.) H. E. Koch, *Dendrologie. Bäume, Sträucher und Halbsträucher, welche in Mittel- und Nordeuropa im Freien kultivirt werden*. Erlangen. 1869–1873. (**1** in 1869; **2**(**1**) in 1872; **2**(**2**) in 1873. Cf. W. T. Stearn, *Jour. Soc. Bibl. Nat. Hist.* **3**: 175 (1957).)

Koch, C., *Hort. Dendrol.*
C. (K.) H. E. Koch, *Hortus dendrologicus*. Berlin. 1853.

Komarov, *Fl. URSS*
V. L. Komarov *et al.* (edit.), **Флора СССР** [*Flora SSSR*]. / *Flora URSS*. Leningrad & Mosqua. 1934–1964. (**1–3**, Leningrad; **4–30**, Mosqua & Leningrad. For full details of dates and editors of volumes cf. I. A. Linczevski, *Novit. Syst. Pl. Vasc. (Leningrad)* **1966**: 316–330 (1966) and F. A. Stafleu & R. S. Cowan, *Taxonomic Literature* ed. 2, **2**: 612–644 (1979).) **Свод дополнений и изменений к "Флоре СССР"** (тт. I–XXX) [*Svod dopolnenij i izmenenij k 'Flore SSSR' (TT. I–XXX)*]. / *Addimenta et Corrigenda ad 'Floram URSS' (Tomi I–XXX)*, by S. K. Czerepanov. Leningrad. 1973.

Korsh., *Tent. Fl. Russ. Or.*
S. I. Korshinsky, *Tentamen Florae Rossiae orientalis, id est Provinciarum Kazan, Wiatka, Perm, Ufa, Orenburg, Samara Partis borealis atque Simbirsk*. St.-Pétersbourg. 1898.

Kotov or **Kotov & Barbarich, *Fl. RSS Ucr.***
Cf. Fomin, *Fl. RSS Ucr.*

Kotschy, *Eichen*
T. Kotschy, *Die Eichen Europa's und des Orients*. Wien & Ollmütz. 1858–1862. (**1** in 1858; **2–4** in 1859; **5** (with text for tt. 16–25 of 1859–1860) in 1860; **6** in 1861; **7/8** in 1862.)

Krocker, *Fl. Siles.*
A. J. Krocker, *Flora silesiaca*. Vratislaviae. 1787–1823. (**1** in 1787; **2**(**1** & **2**) in 1790; **3** in 1814; **4** (*Supplementum*) in 1823.) Editio altera. Vratislaviae. 1796. (**1** & **2** in 1796.)

Kuntze, O., *Revis. Gen.*
K. E. O. Kuntze, *Revisio Generum Plantarum vascularium omnium*. Leipzig, London, Milan, New York, Paris. (**3**(**2**) also Melbourne.) 1891–1898. **1** & **2** in 1891; **3**(**1**), pp. [1*], [clvii]–ccccxx, [1–2, index] in 1893; **3**(**2**) & **3**(**3**), pp. [i–]vi, [1–]201, [202], [1–]576 in 1898. Cf. F. A. Stafleu & R. S. Cowan, *Taxonomic Literature* ed. 2, **2**: 700 (1979).)

Kunze, G., *Ind. Hort. Lips.*
G. Kunze, *Hortus Universitarum Literarum lipsiensis Seminum Anno 1839 (...) perceptorum offert Delectum*. Lipsiae. 1839, 1841, 1843–1850. (Descriptions of new taxa also published in *Linnaea* as follows: **1839** as **16**: *Litt.-Ber.* 106–108 (1842); **1841** as **16**: *Litt.-Ber.* 108–110 (1842); **1843** as **18**: 163 (1844); **1844** as **18**: 510 (1844); **1845** as **19**: 404 (1847); **1846** as **24**: 161–163 (1851); **1847** as **24**: 176–179 (1851); **1848** as **24**: 191–193 (1851); **1849** as **24**: 201–202 (1851); **1850** as **24**: 208–209 (1851). Commonly referred to as *Index Horti Lipsiae* or *Index Seminum Horti lipsiensis* but the title as cited in *Linnaea* is as given above until at least 1845, after which year the exact titles of the seed lists ceased to be quoted.)

Kusn., N. Busch & Fomin, *Fl. Cauc. Crit.*
N. I. Kusnezow, N. A. Busch & A. V. Fomin, *Flora caucasica critica*. / **Матеріалы для Флоры Кавказа** [*Materialy dlja Flory Kavkaza*]. Jurjev. 1901–1916. (**1**(**1**): pp. 1–96 in 1911; pp. 97–224 in 1912; pp. 225–248, i–xlvi in 1913; **2**(**1**): pp. 1–43 in 1911; **2**(**4**): pp. 1–64 in 1906; pp. 65–144 in 1912; pp. 145–176 in 1913; **2**(**5**): pp. 1–32 in 1916; **3**(**3**): pp. 1–32 in 1901; pp. 33–112 in 1902; pp. 113–256, i–xix in 1903; **3**(**4**): pp. 1–16 in 1904; pp. 17–64 in 1905; pp. 65–144 in 1907; pp. 145–304 in 1908; pp. 305–544 in 1909; pp. 545–820, i–lxxiv in 1910; **3**(**5**): pp. 1–32 in 1913; pp. 33–48 in 1915; **3**(**7**): pp. 1–80 in 1908; pp. 81–96 in 1910; pp. 97–112 in 1912; **3**(**8**): pp. 1–16 in 1911; pp. 17–48 in 1912; **3**(**9**): pp. 1–64 in 1906; pp. 65–224 in 1909; pp. 225–288 in

1910; pp. 289–320 in 1912; pp. 321–352 in 1913; **4(1)**: pp. 1–128 in 1901; pp. 129–208 in 1902; pp. 209–352 in 1903; pp. 353–384 in 1904; pp. 385–464 in 1905; pp. 465–512 in 1906; pp. 513–560 in 1907; pp. 561–590, i–lxii in 1908; **4(2)**: pp. 1–32 in 1912; pp. 33–208 in 1913; pp. 209–256 in 1914; pp. 257–320 in 1915; pp. 321–400 in 1916; **4(3)**: pp. 1–48 in 1916; **4(6)**: pp. 1–16 in 1903; pp. 17–48 in 1904; pp. 49–80 in 1905; pp. 81–144 in 1906; pp. 145–157, i–xviii in 1907.)

L., *Amoen. Acad.*
C. von Linné, *Amoenitates academicae*. Holmiae. 1749–1769. (**1** in 1749; **2** in 1751, with an Ed. 2 in 1762; **3** in 1756; **4** in 1759; **5** in 1760; **6** in 1763; **7** in 1769.) Amsterdam edition. Amstelaedami. 1752-1762. (**2** in 1752, with an Ed. 2 in 1764; **3** in 1756; **1** not issued.) Leiden edition. Lugduni Batavorum. 1760–1769. (**4** & **5** in 1760; **6** in 1764; **7** in 1769.) Erlangen edition. Erlangae. 1785–1790. (10 vols., being reprints of **1–7** (**1–3** in 1787; **4** & **5** in 1788; **6** & **7** in 1789) and three additional vols.: **8** in 1785; **9** in 1785; **10** in 1790.) Cf. F. A. Stafleu & R. S. Cowan, *Taxonomic Literature* ed. 2, **3**: 87–88 (1981).

L., *Cent. Pl.*
C. von Linné, *Centuria Plantarum*. Upsaliae. 1755–1756. (**1** in 1755; **2** in 1756.)

L., *Demonstr. Pl.*
C. von Linné, *Demonstrationes Plantarum in Horto upsaliensi, 1753* ... Upsaliae. 1753.

L., *Fl. Palaest.*
C. von Linné, *Flora palaestina*. Upsaliae. 1756. (Reprinted in L., *Amoen. Acad.* **4(69)**: 443–467 (1759).

L., *Fl. Suec.*
C. von Linné, *Flora suecica*. Ed. 2. Stockholmiae. 1755.

L., *Mantissa*
C. von Linné, *Mantissa Plantarum*. (Pp. 1–142.) Holmiae. 1767.

L., *Mantissa Alt.*
C. von Linné, *Mantissa Plantarum altera*. (Pp. 143–587.) Holmiae. 1771.

L., *Opobalsam. Decl.*
C. von Linné, *Opobalsamum declaratum in Dissertatione medica*. Upsaliae. 1764.

L., *Sp. Pl.*
C. von Linné, *Species Plantarum*. Ed. 1. Holmiae. 1753. Ed. 2. Holmiae. 1762–1763. (Pp. 1–784 in 1762; pp. 785–1684 in 1763.)

L., *Syst. Nat.*
C. von Linné, *Systema Naturae*. Ed. 10. Botany in **2**. Holmiae. 1759. Ed. 11. Botany in **1** & **2**. Lipsiae. 1762. Ed. 12. Botany in **2** & **3**. Holmiae. 1767–1768. (**2** in 1767; **3** in 1768.) Ed. 13, by J. F. Gmelin. Botany in **2**. Lipsiae. 1791–1792. (Pp. 1–884 in 1791; pp. 885–1661 in 1792. Cf. F. A. Stafleu & R. S. Cowan, *Taxonomic Literature* ed. 2, **3**: 100 [ed. 10], 106–107 [ed. 12] (1981).) Titled ***Syst. Veg.***, *Systema Vegetabilium*. Ed. 13, edit. J. A. Murray. Gottingae & Gothae. 1774. Ed. 14, by J. A. Murray (new taxa attributable to C. von Linné). Gottingae. 1784. Ed. 15, by C. H. Persoon. Gottingae. 1797. Ed. nov. (15), by J. J. Roemer & J. A. Schultes. Studtgartiae. 1817–1830. (**1** & **2** in 1817; **3** in 1818; **4** in 1819; **5** in 1819; **6** by J. A. Schultes & C. P. J. Sprengel in 1820; **7** by J. A. & J. H. Schultes: pp. i–xliv, 1–754 in 1829; pp. xlv–cviii, 755–1816 in 1830. *Mantissa* **1** by J. A. Schultes in 1822; *Mantissa* **2** by J. A. Schultes in 1824; *Mantissa* **3** by J. A. Schultes in 1827. Cf. F. A. Stafleu & R. S. Cowan, *Taxonomic Literature* ed. 2, **4**: 847–848 (1983).) Ed. 16, by C. P. J. Sprengel. Gottingae. 1824–1828. (**1** in 1824; **2** in 1825; **3** in 1826; **4** (**1** & **2**) in 1827; **5** & *Suppl.* in 1828.)

L. fil., *Suppl.*
C. von Linné fil., *Supplementum Plantarum Systematis Vegetabilium Editionis decimus tertiae, Generum Plantarum Editionis sextae, et Specierum Plantarum Editionis secundae*. Brunsvigae. 1781.

Labill., *Icon. Pl. Syr.*
J. J. H. de Labillardière, *Icones Plantarum Syriae rariorum*. Lutetiae Parisiorum (**1**) / Parisiis (**2–5**). 1791–1812. (**1** & **2** in 1791; **3** in 1809; **4** & **5** in 1812. Cf. F. A. Stafleu & R. S. Cowan, *Taxonomic Literature* ed. 2, **2**: 711 (1979).)

Labill., *Nov. Holl. Pl.*
J. J. H. de Labillardière, *Novae Hollandiae Plantarum Specimen* ... Parisiis. 1804–1807. (**1(1)**: pp. 1–8, tt. 1–10 in 1804; **1(2–14)** pp. 9–112, tt. 11–140[–142?] in 1805; **2(15–23)**: pp. 1–?, tt. 141–230 in 1806; **2(24–27)**: pp. ?–130, tt. 231–265 in 1807. Cf. F. A. Stafleu & R. S. Cowan, *Taxonomic Literature* ed. 2, **2**: 712–713 (1979).)

Lag., *Gen. Sp. Nov.*
M. Lagasca y Segura, *Genera et Species Plantarum quae aut novae sunt aut nondum recte cognoscuntur*. Matriti. 1816.

Lagrèze-Fossat, *Fl. Tarn Gar.*
A. R. A. Lagrèze-Fossat, *Flore de Tarn et Garonne*. Montauban. 1847.

Laínz, *Aport. Fl. Gallega*
M. Laínz, *Aportaciones al Conocimiento de la Flora gallega*. (**1** in *Broteria* **24** [**51**]: 108–143 (1955); **2** in *Anal. Inst. Bot. Cav.* **14**: 529–554 (1956); **3** in *Broteria* **26** [**53**]: 90–97 (1957); **4** in *Anal. Inst. For. Invest. Exp.* **1965**: 299–332 (1965); **5** in *Anal. Inst. For. Invest. Exp.* **1967**: 1–52 (1967); **6**, Madrid. 1968. (Reprinted from an unpublished vol. of *Anal. Inst. For. Invest. Exp.*; cf. Castroviejo et al., *Fl. Iberica* **2**: 778 (1990).) **7**, pp. 1–39. Madrid. 1971. **8** in *Comm. INIA Sér. Recurs. Nat.* **2**: 1–26 (1974).

Lam., *Encycl. Méth. Bot.*
J. B. A. P. Monnet de la Marck, *Encyclopédie méthodique. Botanique*. Paris. 1783–1817. (For dates cf. F. A. Stafleu & R. S. Cowan, *Taxonomic Literature* ed. 2, **2**: 732–733 (1979).) **1–2, 3** (up to P) by J. B. A. P. Monnet de la Marck; **3** (P–end) by J. B. A. P. Monnet de la Marck & L. A. J. Desrousseaux; **4** by L. A. J. Desrousseaux, J. L. M. Poiret & M. J. C. L. de Savigny; **5** by J. L. M. Poiret and (occasionally) A. P. de Candolle; **6–13** by J. L. M. Poiret.

Lam., *Fl. Fr.*
J. B. A. P. Monnet de la Marck, *Flore française*. Ed. 1. Paris. 1779. (**1–3** in 1779. Cf. F. A. Stafleu & R. S. Cowan, *Taxonomic Literature* ed. 2, **2**: 731 (1979).) Ed. 2. Paris. 1795. (**1–3** in 1795.) Ed. 3 by J. B. A. P. Monnet de la Marck & A. P. de Candolle. Paris. 1805–1815. (**1–4** in 1805; **5** in 1815. **4** issued in two parts with continuous pagination, the second called '*Tome quatrième. Seconde partie*' and '*Vol. V*' ; **5** called '*Tome cinquième, ou sixième Volume*' and '*Vol. VI*' ; **1–4** also reprinted in 1815.)

Lam. & DC., *Fl. Fr.*
Cf. Lam., *Fl. Fr.*

Lamb., *Descr. Gen. Pinus*
A. B. Lambert, *A Description of the Genus* Pinus. Ed. 1. London. 1803–1824. (**1**: pp. 1–86 in 1803; pp. 87–94 in 1806; pp. 95–98 in 1807; **2** in 1824.) Ed. 2. London. 1828–1837. (**1** in 1828; **2** by D. Don in 1828; **3**, completing both editions, by D. Don in 1837.) Ed. 3. London. 1832. (**1** & **2** in 1832.) Ed. 4. London. 1837–1842. (Cf. F. A. Stafleu & R. S. Cowan, *Taxonomic Literature* ed. 2, **2**: 736–737 (1979).)

Lange, *Descr. Icon. Ill.*
J. M. C. Lange, *Descriptio Iconibus illustrata Plantarum novarum vel minus cognitarum praecipue e Flora hispanica*. Hauniae. 1864–1866.

(**1**, pp. 1–8, tt. 1–12 in 1864; **2**, pp. 9–12, tt. 13-24 in 1865; **3**, pp. 13–20, tt. 25–35 in 1866.)

Lange, *Ind. Sem. Horto Haun.*
J. M. C. Lange, *Index Seminum in Horto academico hauniensis collectorum*. Hauniae. 1854, 1855, 1856, 1857, 1861, 1865–1868, 1870–1872.

Lapeyr., *Fig. Fl. Pyr.*
P. Picot de Lapeyrouse, *Figures de la Flore des Pyrénées, avec des Descriptions, des Notes critiques et des Observations*. Paris. 1795–1801. (Tt. 1–10 in 1795; tt. 11–43 in 1801.)

Lapeyr., *Hist. Abr. Pyr.*
P. Picot de Lapeyrouse, *Histoire abrégée des Plantes des Pyrénées et Itinéraire des Botanistes dans ces Montagnes*. Toulouse. 1813. *Supplément*. 1818.

Lavrenko, *Ind. Sem. Horti Charkov.*
E. M. Lavrenko, *Index Seminum Horti botanici charkoviensis*. 1926.

Lawson, P. & C., *Agric. Man.*
P. & C. Lawson, *The Agriculturist's Manual, being a familiar Description of the agricultural Plants cultivated in Europe*. Edinburgh. 1836.

Lecoq & Lamotte, *Cat. Pl. Centr. Fr.*
H. Lecoq & M. Lamotte, *Catalogue raisonné des Plantes vasculaires du Plateau central de la France. Comprenant l'Auvergne, le Velay, la Lozère, les Cévennes, une Partie du Bourbonnais et du Vivarais*. Paris. 1848. (Title page dated 1847, but cover dated 1848; cf. F. A. Stafleu & R. S. Cowan, *Taxonomic Literature* ed. 2, **2**: 805 (1979).)

Ledeb., *Fl. Altaica*
C. F. von Ledebour, *Flora altaica*. Berolini. 1829–1834. (**1** in 1829; **2** in 1830; **3** in 1831; **4** in 1833; Index in 1833 or 1834.)

Ledeb., *Fl. Ross.*
C. F. von Ledebour, *Flora rossica*. Stuttgartiae. 1841–1853. (**1**: pp. 1–240 in 1841; pp. 241–480 in 1842; pp. 481–790 in 1843; **2**: pp. 1–204 in 1843; pp. 205–462 in 1844; pp. 463–718 in 1845; pp. 719–937 in 1846; **3**: pp. 1–256 in 1847; pp. 257–492 in 1849; pp. 493–684 in 1850; pp. 685–866 in 1851; **4**: pp. 1–464 in 1852; pp. 465–741 in 1853. Cf. F. A. Stafleu & R. S. Cowan, *Taxonomic Literature* ed. 2, **2**: 807–808 (1979.)

Ledeb., *Icon. Pl. Fl. Ross.*
C. F. von Ledebour, *Icones Plantarum novarum vel imperfecte cognitarum Floram rossicam, imprimis altaicam, illustrantes*. Rigae. 1829–1834. (**1** in 1829; **2** in 1830; **3**: tt. 201–250 in 1831; tt. 251–300 in 1832; **4** in 1833 or 1834; **5** in 1834. Cf. F. A. Stafleu & R. S. Cowan, *Taxonomic Literature* ed. 2, **2**: 807 (1979).)

Ledeb., *Ind. Sem. Horti Dorpat.*
C. F. von Ledebour, *Index Seminum Horti botanici dorpatensis*. Dorpat. 1818–1835; *Appendix* in 1821, *Suppl.* **1–3** in 1823–1825. For continuation cf. Bunge, *Del. Sem. Horti Dorpat.*

Lej., *Fl. Spa*
A. L. S. Lejeune, *Flore des Environs de Spa*. Liège. 1811–1813. (**1** in 1811; **2** in 1813.)

Lej. & Court., *Comp. Fl. Belg.*
A. L. S. Lejeune & R. J. Courtois, *Compendium Florae belgicae*. Leodii & Verviae. 1828–1836. (**1**, Leodii in 1828; **2**, Leodii in 1831; **3**, Verviae in 1836.) **3** edit. A. L. S. Lejeune. (Cf. F. A. Stafleu & R. S. Cowan, *Taxonomic Literature* ed. 2, **2**: 831 (1979).)

Lester-Garland, *Fl. Jersey*
L. V. Lester-Garland, *A Flora of the Island of Jersey with a List of Plants of the Channel Islands in general, and Remarks upon their Distribution and geographical Affinities*. London. 1903.

Liebl., *Fl. Fuld.*
F. K. Lieblein, *Flora fuldensis, oder Verzeichniss der in dem Fürstenthume Fuld wildwachsenden Bäume, Sträuche und Pflanzen*. Frankfurt a. M. 1784.

Lindley, *Coll. Bot.*
J. Lindley, *Collectanea botanica; or, Figures and botanical Illustrations of rare and curious exotic Plants*. London. 1821(–?1826). (**1–7**, tt. 1–36, in 1821; **8**, tt. 37–41, A–B in ?1826. Cf. F. A. Stafleu & R. S. Cowan, *Taxonomic Literature* ed. 2, **3**: 52 (1981).)

Lindman, *Svensk Fanerogamfl.*
C. A. M. Lindman, *Svensk Fanerogamflora*. Ed. 1. Stockholm. 1918. Ed. 2. Stockholm. 1926.

Link, *Enum. Hort. Berol. Alt.*
J. H. F. Link, *Enumeratio Plantarum: Horti regii botanici berolinensis altera*. Berolini. 1821–1822. (**1** in 1821; **2** in 1822.)

Link, *Fil. Sp.*
J. H. F. Link, *Filicum Species in Horto regio botanico berolinensi cultae*. Berolini. 1841.

Link, *Handb.*
J. H. F. Link, *Handbuch zur Erkennung der nutzbarsten und am häufigsten vorkommenden Gewächse*. Berlin. 1829–1833. (**1** in 1829; **2** in 1829; **3** in 1833. Cf. F. A. Stafleu & R. S. Cowan, *Taxonomic Literature* ed. 2, **3**: 69 (1981).)

Lloyd, *Fl. Loire-Inf.*
J. Lloyd, *Flore de la Loire-inférieure*. Nantes. 1844.

Loefl., *Iter Hisp.*
P. Loefling, *Iter hispanicum, eller Resa til Spanska Länderna*. Stockholm. 1758.

Loefl., *Reise Span. Länd.*
P. Loefling, *Reise, nach den spanischen Ländern in Europa und Amerika in den Jahren 1751 bis 1756. Aus dem schwedischen übersetzet durch A. B. Kölpin*. Berlin & Stralsund. 1766.

Loisel., *Fl. Gall.*
J. L. A. Loiseleur-Deslongchamps, *Flora gallica*. Ed. 1. Lutetiae. 1806–1807. (Pp. 1–336 in 1806; pp. 337–742 in 1807.) Ed. 2. Paris. 1828. For *Suppl.*, cf. *Not. Pl. Fr.*

Loisel., *Not. Pl. Fr.*
J. L. A. Loiseleur-Deslongchamps, *Notice sur les Plantes à ajouter à la Flore de France* (Flora gallica); *avec quelques Corrections et Observations*. Paris. 1810.

Lojac., *Fl. Sic.*
M. Lojacono-Pojero, *Flora sicula o Descrizione delle Piante vascolari spontanee o indigenate in Sicilia*. Palermo. (1886–)1888–1909. (**1(1)** in [1886–]1889; **1(2)** in 1891; **2(1)** in 1903; **2(2)** apparently in parts, 1904–1907; **3** apparently in parts, 1908–1909. Parts of **1(1)** published previously, from 1886, in *Giornale del Comizio Agrario di Palermo*. Cf. F. A. Stafleu & R. S. Cowan, *Taxonomic Literature* ed. 2, **3**: 151 (1981).) *Addenda* in *Malpighia* **20**: 37–48, 95–119, 180–218 & 290–300 (1906).

Loscos, *Trat. Pl. Arag.*
F. Loscos y Bernál, *Tratado de Plantas de Aragón*. Madrid. 1876–1886. (**1** in 1876–1877; **2** & *Suppl.* **1–4** in 1880; **3** & *Suppl.* **5–8** in 1883–1886. Cf. F. A. Stafleu & R. S. Cowan, *Taxonomic Literature* ed. 2, **3**: 165 (1981).) The last pages of **8** were in part written after F. Loscos's death by his son, C. Loscos.

Loscos & Pardo, *Ser. Pl. Arag.*
F. Loscos y Bernál & J. Pardo y Sastrón, *Series inconfecta Plantarum indigenarum Aragoniae praecipue meridionalis, e Lingua castellana in latinam vertit, recensuit, emendavit, Observationibus suis auxit atque edendam curavit Mauritius Willkomm*. Dresdae. 1863.

Loudon, *Arbor. Fruticet. Brit.*
J. C. Loudon, *Arboretum et Fruticetum Britannicum: or the Trees and Shrubs of Britain, native and foreign*. London. 1835–1838. (**1–8**, in 63 parts, in 1835–1838.) Ed. 2. London. 1844.

Loudon, *Hort. Brit.*
J. C. Loudon, *Hortus britannicus. A Catalogue of all the Plants indigenous, cultivated in, or introduced into Britain.* Ed. 1. London. 1830. (**1** & **2** in 1830.) Ed. 2 & *Suppl.* **1** (pp. 579–602). London. 1832. Ed. 3 & *Suppl.* London. 1839. (*Suppl.* **1** reprinted, pp. 579–604; *Suppl.* **2** by W. H. Baxter revised by G. Don, pp. 603 [*sic*] –742). Ed. 4, with *Suppl.* by W. H. Baxter & D. Wooster under the direction of J. W. Loudon. London. 1850.

Maire, *Fl. Afr. Nord*
R. C. J. E. Maire, *Flore de l'Afrique du Nord (Maroc, Algérie, Tunisie, Tripolitaine, Cyrénaique et Sahara).* Paris. 1952→. (An entirely posthumous work; **1–3** edit. M. Guinochet & L. Faurel; **4** edit. M. Guinochet; **5** edit. M. Guinochet & P. Quézel; **6→** edit. P. Quézel. **1** in 1952; **2** in 1953; **3** in 1955; **4** in 1957; **5** in 1958; **6** in 1959; **7** in 1961; **8** in 1962; **9** in 1963; **10** in 1963; **11** in 1964; **12** in 1965; **13** in 1967; **14** in 1977; **15** in 1980; **16** in 1987.)

Maire, *Sahara Central*
R. C. J. E. Maire, *Études sur la Flore et la Végétation du Sahara central.* **1** & **2** (*Mém. Soc. Hist. Nat. Afrique Nord* **1933**(**3**): 1–272). Alger. 1933. **3.** (*ibid.* **1940**(**3**): 273–433.) Alger. 1940.

Maire & Petitmengin, *Étude Pl. Vasc. Réc. Grèce*
R. C. J. E. Maire & M. G. C. Petitmengin, *Étude des Plantes vasculaires récoltées en Grèce (1904).* In R. C. J. E. Maire, *Matériaux pour servir à l'Étude de la Flore et de la Géographie botanique de l'Orient* **2.** 46 pp. Nancy. 1907. *Étude ... (1906).* In *op. cit.* **4.** 239 pp. Nancy. 1908.

Malagarriga, *Nuev. Comb. Subesp. Almeria*
R. M. de P. Malagarriga Heras, *Nuevo Combinaciones de Subespecies de la Provincia de Almeria.* Barcelona. 1974.

Marshall, *Arbust. Amer.*
H. Marshall, *Arbustum americanum: the American Grove, or an alphabetical Catalogue of Forest Trees and Shrubs, Natives of the American United States.* Philadelphia. 1785.

Mart., C. F. P., *Fl. Brasil.*
C. F. P. Martius, *Flora brasiliensis. Enumeratio Plantarum in Brasilia hactenus detectarum quas suis aliorumque botanicis Studiis descriptas.* Leipzig. 1840–1906. (**1–15** in 1840–1906. For details of the very complex volumation and pagination cf. F. A. Stafleu & R. S. Cowan, *Taxonomic Literature* ed. 2, **3**: 333–337 (1981).)

Mateo & Figuerola, *Fl. Analit. Prov. Valencia*
G. Mateo Sanz & R. Figuerola Lamata, *Flora analitica de la Provincia de Valencia.* Valencia. 1987.

Maxim., *Prim. Fl. Amur.*
C. J. Maximowicz, *Primitiae Florae amurensis. Versuch einer Flora des Amur-Landes.* St. Petersburg. 1859.

Medicus, *Pflanzengatt.*
F. C. Medicus, *Pflanzen-Gattungen nach dem Inbegriffe sämtlicher Fruktifikations-Theile gebildet, und nach dem Sexual-Pflanzen-Register geordnet.* Mannheim. 1792. (For date cf. F. A. Stafleu, *Taxon* **12**: 74 (1963).)

Medicus, *Philos. Bot.*
F. C. Medicus, *Philosophische Botanik.* Mannheim. 1789–1791. (**1** in 1789 [part of this previously published in *Vorl. Churpf. Phys.-Ökon. Ges.* **4**(**1**): 167–282 (1788)]; **2** in 1791. Cf. F. A. Stafleu, *Taxon* **12**: 74 (1963).)

Meerb., *Afbeeld. Zeldz. Gewass.*
N. Meerburgh, *Afbeeldingen van zeldzaame Gewassen.* (By Johannes le Mair.) Leyden. 1775–1780. (**1** in 1775; **2** in 1776; **3** & **4** in 1777; **5** in 1780.) Reissue in 1789, titled *Plantae rariores vivis Coloribus depictae.*

Meikle, *Fl. Cyprus*
R. D. Meikle, *Flora of Cyprus.* Kew. 1977–1985. (**1** in 1977; **2** in 1985.)

Meissner, *Monogr. Gen. Polyg. Prodr.*
C. F. Meissner, *Monographiae Generis* Polygoni *Prodromus.* Genevae. 1826.

Meissner, *Pl. Vasc. Gen.*
C. F. Meissner, *Plantarum vascularium Genera.* Lipsiae. 1837–1843. (**1** (= *Tab. Diagn.*): pp. i–iv, 1–104 in 1837; pp. 105–176 in 1838; pp. 177–256 in 1839; pp. 257–312 in 1840; pp. 313–344 in 1841; pp. 345–408 in 1842; pp. 409–442 in 1843; **2** (= *Comm.*): pp. 1–72 in 1837; pp. 73–120 in 1838; pp. 121–160 in 1839; pp. 161–224 in 1840; pp. 225–252 in 1841; pp. 253–308 in 1842; pp. 309–402 in 1843. Cf. H. W. Rickett & F. A. Stafleu, *Taxon* **10**: 118 (1961).)

Mela, *Suomen Kasvio*
Cf. Cajander, *Suomen Kasvio.*

Menyh., *Kal. Vid. Növén.*
L. Menyhárth, *Kalocsa vidékének növénytenyészete.* Budapest. 1877.

Mérat, *Nouv. Fl. Env. Paris*
F. V. Mérat de Vaumartoise, *Nouvelle Flore des Environs de Paris.* Ed. 1. Paris. 1812. Ed. 2. Paris. 1821. (**1** & **2** in 1821.) Ed. 3. Paris. 1831–1834. (**1** in 1834; **2** in 1831.) Ed. 4. Paris. 1836. (**1** & **2** in 1836.) Ed. 5. Bruxelles. 1837–1838. (**1** in 1837; **2** in 1838.) Ed. 6. Bruxelles. 1841. (**1** & **2** in 1841.) (In each edition, **1** comprised cryptogams, **2** phanerogams.)

Meyer, C. A., *Verz. Pfl. Cauc.*
C. A. von Meyer, *Verzeichniss der Pflanzen, welche während der, auf allerhöchsten Befehl, in den Jahren 1829 und 1830 unternommenen Reise im Caucasus und in den Provinzen am westlichen Ufer des caspischen Meeres gefunden und eingesammelt worden sind.* St Petersburg. 1831.

Michx, *Fl. Bor.-Amer.*
A. Michaux, *Flora boreali-americana.* Parisiis & Argentorati. 1803. (**1** & **2** in 1803.)

Middendorff, *Reise Nord. Öst. Sibir.*
A. T. von Middendorff, *Reise in den äussersten Norden und Osten Sibiriens während der Jahre 1843 und 1844.* St Petersburg. 1847–1875. **1** (**2**) 1, titled *Die Phänogame Pflanzen aus dem Hochnorden* by E. R. von Trautvetter. 1847. **1** (**2**) **3**, titled *Florula ochotnensis phaenogama* by E. R. von Trautvetter & C. A. von Meyer. 1856.

Milde, *Fil. Eur.*
C. A. J. Milde, *Filices Europae et Atlantidis, Asiae minoris et Sibiriae.* Leipzig. 1867.

Milde, *Höheren Sporenpfl. Deutschl. Schweiz*
C. A. J. Milde, *Die höheren Sporenpflanzen Deutschlands' und der Schweiz.* Leipzig. 1865.

Miller, *Gard. Dict.*
P. Miller, *The Gardeners Dictionary.* Ed. 8. London. 1768. Ed. 9, by T. Martyn. London. 1797–1804. (**1** in 1797; **2** in 1804.)

Moench, *Meth.*
C. Moench, *Methodus Plantas Horti botanici et Agri marburgensis a Staminum Situ describendi.* Marburgi Cattorum. 1794. ***Suppl.***, *Supplementum.* Marburgi Cattorum. 1802.

Moench, *Verz. Ausl. Bäume Weissenst.*
C. Moench, *Verzeichniss ausländischer Bäume und Stauden des Lustschlosses Weissenstein bey Cassel.* Frankfurt a. M. & Leipzig. 1785.

Moessler, *Handb.*
J. C. Moessler, *Gemeinnütziges Handbuch der Gewächskunde.* Altona. 1815. Ed. 2, by H. G. L. Reichenbach. Ed. 1. Altona. 1827–1830. (**1** in 1827; **2**(**1**) in 1828; **2**(**2**) in 1829; **3** in 1830.) Ed. 3, by

H. G. L. Reichenbach. Altona. 1833–1835. (**1** in 1833; **2(1)** in 1833; **2(2 & 3)** in 1834; **3** in 1835.) Cf. F. A. Stafleu & R. S. Cowan, *Taxonomic Literature* ed. 2, **3**: 539 (1981).

Moq., *Chenop. Monogr. Enum.*
C. H. B. A. Moquin-Tandon, *Chenopodearum monographica Enumeratio*. Parisiis. 1840.

Mori, T., *Enum. Pl. Corea*
T. Mori, *An Enumeration of Plants hitherto known from Corea*. Seoul. 1922.

Moris, *Fl. Sard.*
G. G. Moris, *Flora sardoa*. Taurini. 1837–1859. (**1** in 1837; **2** in 1840–1843; **3** in 1858–1859.)

Moris, *Stirp. Sard.*
G. G. Moris, *Stirpium sardoarum Elenchus*. Carali & Taurini. 1827–1829. (**1**, Carali in 1827; **2**, Carali in 1828; **3**, Taurini in 1829.) ***App.***, *Appendix*. Taurini. 1828. (Cf. F. A. Stafleu & R. S. Cowan, *Taxonomic Literature* ed. 2, **3**: 586 (1981).)

Muenchh., *Hausv.*
O. von Muenchhausen, *Der Hausvater. Eine ökonomische Schrift*. Hannover. 1765–1774. (**1(1-3)** in 1765; **2(1)**: pp. 1–367 in 1765; **2(3)**: pp. 369–935 in 1766; **3(1 & 2)** in 1767; **3(3)** in 1768; **4(1)** in 1769; **4(2)** in 1772; **5(1 & 2)** in 1770; **6(1)** in 1773; **6(2)** in 1774.)

Murray, A., *Bot. Exped. Oreg.*
A. Murray, *Botanical Expedition to Oregon*. Edinburgh. 1849–1859. (**1** in 1849; **2** in 1850; **3** in 1851; **4** & **5** in 1852; **6–8** in 1853; **9** in 1854; **10** in 1857; **11** in 1859.) **1** & **3** by J. H. Balfour; **6** by J. Jeffrey. Cf. F. A. Stafleu & R. S. Cowan, *Taxonomic Literature* ed. 2, **3**: 668 (1981).

Murray, *Prodr. Stirp. Gotting.*
J. A. Murray, *Prodromus Designationis Stirpium gottingensium*. Goettingae. 1770.

Mutel, *Fl. Fr.*
A. P. V. Mutel, *Flore française destinée aux Herborisations*. Paris. 1834–1838. (**1** in 1834; **2** in 1835; **3** in 1836; **4** in 1837; **5** in 1838.) *Atlas*. Paris. 1835–1838. (**1** in 1835; **2** & **3** in 1836; **4** in 1837; *Suppl.* in 1838.)

Namegata & Kurata, *Coll. Cult. Ferns Fern Allies*
T. Namegata & S. Kurata, *Collection and Cultivation of our Ferns and Fern Allies*; and *An Enumeration of the Japanese Pteridophytes*. Tokyo. 1961. (Part 2, *An Enumeration* ..., also available as a separate.)

Nees, *Horae Phys. Berol.*
C. G. D. Nees von Esenbeck (edit.), *Horae physicae berolinenses*. Bonnae. 1820.

Neilr., *Aufz. Ung. Slav. Gefäss.*
A. Neilreich, *Aufzählung der in Ungarn und Slavonien bisher beobachteten Gefässpflanzen nebst einer pflanzengeographischen Uebersicht*. Wien. 1865 ('1866'). *Nachtr.*, *Aufzählung ... Nachträge und Verbesserungen*. Wien. 1870.

Neuman, *Sver. Fl.*
L. M. Neuman, *Sveriges Flora (Fanerogamerna)*. Lund. 1901.

Nicotra, *Fum. Ital.*
L. Nicotra, *Le Fumariacee italiane, Saggio di una Continuazione della* Flora Italiana *di Filippo Parlatore*. Firenze. 1897.

Nutt., *Gen. N. Amer. Pl.*
T. Nuttall, *The Genera of North-American Plants, and a Catalogue of the Species, to the Year 1817*. Philadelphia. 1818. (**1** & **2** in 1818.) Facsimile ed. New York. 1971.

Nutt., *N. Amer. Sylva*
T. Nuttall, *The North American Sylva: or, a Description of the Forest Trees of the United States, Canada, and Nova Scotia*. Three-volume edition: Philadelphia. 1842–1849. (**1(1 & 2)** in 1842; **2** in 1846; **3** in 1849. Reissues in 1849, 1852, 1853 & 1855.) Two-volume edition: Philadelphia. 1857. (**1** & **2** in 1857. Reissues in 1859 & two in 1865.) Cf. F. A. Stafleu & R. S. Cowan, *Taxonomic Literature* ed. 2, **3**: 786–787 (1981).

Nyárády, E. I., *Enum. Pl. Vasc. Cheia Turzii*
E. I. Nyárády, *Enumerarea Plantelor vasculare din Cheia Turzii*. (Published as *Mem. Comis. Monum. Nat. România*, **1**.) Bucureşti. 1939.

Nyman, *Consp.*
C. F. Nyman, *Conspectus Florae europaeae*. Örebro. 1878–1882. (Pp. 1–240 in 1878; pp. 241–493 (part) in 1879; pp. 493–677 in 1881; pp. 677–858 in 1882; p. 256 *errata* in 1881; p. 493 (part) and p. 677 published twice. Cf. W. T. Stearn, *Jour. Bot. (London)* **76**: 113 (1938).) ***Suppl.***, *Supplementum* **1** in 1883–1884 (pp. 1–21, 1–166 titled *Acotyledoneae vasculares et Characeae Europae* in 1883; both parts combined as pp. 859–1046 in 1884); **2(1)**: pp. 1–224 in 1889; **2(2)**: pp. 225–404 in 1890. ***Addit.***, *Additamenta ad Conspectus Florae europaeae. / Beiträge zu C. F. Nyman's* Conspectus Florae europaeae, by E. Roth. Berlin. 1885. (Cf. F. A. Stafleu & R. S. Cowan, *Taxonomic Literature* ed. 2, **3**: 798–799 (1981).)

Nyman, *Syll.*
C. F. Nyman, *Sylloge Florae europaeae*. Oerobroae. 1854–1855. ***Suppl.***, *Supplementum*. Oerebroae. 1865.

Olivier, *Voy. Emp. Othoman*
G. A. Olivier, *Voyage dans l'Empire Othoman, l'Égypte et la Perse, fait par Ordre du Gouvernement, pendant les six premières Années de la République*. Quarto edition. Paris. 1800–1807. (**1** in 1800–1801; **2** in 1804; **3** in 1807.) Octavo edition (= ed. min.). Paris. 1800–1807. (**1** & **2** in 1800–1801; **3** & **4** in 1803–1804; **5** & **6** in 1807.) *Atlas* (both editions) in 1800–1801.

Opiz, *Naturalientausch*
P. M. Opiz, *Naturalientausch*. Prag. 1823–1830. (**1–5** in 1823 [**1** & **2** published twice]; **6–8** in 1824; **9–10** in 1825; **11** in 1826(–1827); **12**, titled *Beiträge zur Naturgeschichte*, in 1828(–1830). **1** & **2** originally with separate pagination, reissued with **3**, with continuous pagination to **11** and continuing through **12**. Cf. F. A. Stafleu & R. S. Cowan, *Taxonomic Literature* ed. 2, **3**: 841 (1981).)

Opiz, *Tent. Fl. Crypt. Boem.*
P. M. Opiz, *Tentamen Florae cryptogamicae Boemiae*. Prag. 1819–1820.

Pallas, *Fl. Ross.*
P. S. Pallas, *Flora rossica*. Petropoli. 1784–1815(–1831?). (**1(1)** in 1784; **1(2)** in 1788 or 1789; **2(1)** in 1815 (title page) but possibly not published until 1831. Cf. F. A. Stafleu & R. S. Cowan, *Taxonomic Literature* ed. 2, **4**: 25–26 (1983).)

Pallas, *Ill. Pl.*
P. S. Pallas, *Illustrationes Plantarum imperfecte vel nondum cognitarum*. Lipsiae. 1803[–1806].

Pallas, *Reise*
P. S. Pallas, *Reise durch verschiedene Provinzen des russischen Reichs*. Ed. 1. St Petersburg. 1771–1776. (**1** in 1771; **2** in 1773; **3** in 1776.) Ed. 2. St Petersburg. 1801.

Pančić, *Elem. Fl. Bulg.*
J. Pančić, Граħа за Флору Кнежевине Бугарске [*Graća za Floru Kneževine Bugarske*]. / *Elementa ad Floram Principatus Bulgariae*. Beograd. 1883.

Pančić, *Fl. Princ. Serb.*
J. Pančić, Флора Кнежевине Србиjе [*Flora Kneževine Srbije*]. / *Flora Principatus Serbiae*. Beograd. 1874. ***Addit.***, Додатак [*Dodatak*]. / *Additamenta*. Beograd. 1884.

Pančić, *Srpska Kralj. Bot. Bašta Beograd*
J. Pančić, Српска кралевска ботаницка Башта у Београду ... [*Srpska kraljevska botanička Bašta u Beogradu* ...]. / *Samenkatalog der belgrader botanischer Gartens*. Beograd. 1888.

Parl., *Pl. Nov.*
F. Parlatore, *Plantae novae vel minus notae*. Parisiis. 1842.

Parry, *Jour. Voy. N. W. Pass.*
W. E. Parry, *Journal of a Voyage for the Discovery of a North West Passage from the Atlantic to the Pacific; performed in 1819–20 in H. M. S. Hecla and Griper*. London. 1821. ***Suppl. App.***, *Botanical Supplement* to the *Appendix* by R. Brown. London. 1824. (The *Botanical Supplement* was preprinted as ***Chloris Melv.***, *Chloris Melvilliana*, in 1823; cf. R. Br., *Chloris Melv.* Cf. F. A. Stafleu & R. S. Cowan, *Taxonomic Literature* ed. 2, **4**: 82 (1983).)

Pau, *Not. Bot. Fl. Esp.*
C. Pau y Español, *Notas botánicas à la Flora española*. Madrid & Segorbe. 1887–1895. (**1**, **2**, **4–6** Madrid; **3** Segorbe. **1** in 1887; **2** in 1888; **3** in 1889; **4** in 1891; **5** in 1892; **6** in 1895. Cf. F. A. Stafleu & R. S. Cowan, *Taxonomic Literature* ed. 2, **4**: 108 (1981).)

Pavlov, *Fl. Centr. Kazakhstana*
N. V. Pavlov, Флора централъного Казахстана [*Flora central'nogo Kazakhstana*]. Alma-Ata. 1928–1938. (**1** in 1928; **2** in 1935; **3** in 1938.)

Pawł., *Fl. Tatr.*
B. Pawłowski, *Flora Tatr. Rósliny naczyniowe*. / *Flora Tatrorum, Plantae vasculares*. Warszawa. 1956→. (**1** in 1956.)

Pers., *Syn. Pl.*
C. H. Persoon, *Synopsis Plantarum*. Parisiis Lutetiorum & Tuebingae. 1805–1807. (**1** in 1805; **2**: pp. 1–272 in 1806; pp. 273–657 in 1807.) Ed. 2, titled *Species Plantarum*. Petropoli. 1817–1822. (**1** in 1817; **2** & **3** in 1819; **4** & **5** in 1821; **6** in 1822.)

Petrovič, *Fl. Agri Nyss.*
S. Petrovič, Флора Околине Ниша [*Flora Okoline Niša*]. / *Flora Agri nyssani*. Beograd. 1882. ***Addit.***, Додатак Флори Околине Ниша [*Dodatak Flori Okoline Niša*]. / *Additamenta ad Floram Agri nyssani*. Beogradu. 1885.

Phillips, E. P., *Gen. S. Afr. Fl. Pl.*
E. P. Phillips, *The Genera of South African flowering Plants*. Ed. 1. Cape Town. 1926. Ed. 2. Cape Town. 1951.

Phipps, C. J., *Voy. N. Pole*
C. J. Phipps, *A Voyage towards the North Pole undertaken by His Majesty's Command 1773*. London. 1774. *Appendix* with lists of plants.

Poiret, *Voy. Barb.*
J. L. M. Poiret, *Voyage en Barbarie, ou Lettres écrites de l'ancienne Numidie, pendant les Années 1785 & 1786*. Paris. 1789.

Pojark., *Fl. Murmansk.*
Cf. Gorodkov, *Fl. Murmansk.*

Pomel, *Nouv. Mat. Fl. Atl.*
A. Pomel, *Nouveaux Matériaux pour la Flore atlantique*. Paris & Alger. 1874–1875. (Pp. i–iii, 1–260 in 1874; pp. [257–]261–399 in 1875.)

Presl, C., *Fl. Sic.*
C. B. Presl, *Flora sicula*. Pragae. 1826.

Presl, C., *Tent. Pteridogr.*
C. B. Presl, *Tentamen Pteridographiae, seu Genera Filicacearum praesertim juxta Venarum Decursum et Distributionem exposita*. Pragae. 1836. ***Suppl.***, *Supplementum*. Pragae. 1845.

Presl, J. & C., *Del. Prag.*
J. S. & C. B. Presl, *Deliciae pragenses, Historiam naturalem spectantes*. Pragae. 1822.

Presl, J. & C., *Fl. Čechica*
J. S. & C. B. Presl, *Flora čechica*. / *Kwětena česká*. Pragae. 1819.

Pursh, *Fl. Amer. Sept.*
F. T. Pursh, *Flora Americae septentrionalis; or a systematic Arrangement and Description of the Plants of North America*. Ed. 1. London. 1814. (**1** & **2**, with continuous pagination, in 1814.) Ed. 2. London. 1816. (**1** & **2**, with continuous pagination, in 1816.)

Rafin., *Autikon Bot.*
C. S. Rafinesque-Schmaltz, *Autikon Botanikon, or botanical Illustrations of 2500 new, rare or beautiful Trees, Shrubs, Plants, Vines, Lilies, Grasses, Ferns etc., of all Regions, but chiefly North America, with Descriptions and 2500 self Figures or Specimens*. (Also with an alternative title page partly in Latin.) Philadelphia. 1840. (Only one vol., describing 1500 species, appeared. No illustrations actually published; the work was to have been accompanied by botanical specimens.)

Rafin., *Caratt.*
C. S. Rafinesque-Schmaltz, *Caratteri de alcuni nuovi Generi e nuove Specie di Animali e Piante della Sicilia*. Palermo. 1810.

Rech. fil., *Fl. Iran.*
K. H. Rechinger, *Flora iranica. Flora des iranischen Hochlandes und der umrahmenden Gebirge*. Graz. **1**→, 1963→. (**1–3** in 1963; **4–8** in 1964; **9–15** in 1965; **17–39** in 1966; **40–48** in 1967; **49–57** in 1968; **58–66** in 1969; **67–74** in 1970; **75–89** in 1971; **90–100** in 1972; **101–102**, **104** in 1973; **103**, **105–110** in 1974; **111**, **113–115** in 1975; **112**, **116–121** in 1976; **122–125** in 1977; **126–138** in 1978; **139a**, **140–143** in 1979; **139b**, **144–146** in 1980; **147–148** in 1981; **149–156** in 1982; **157** in 1984; **158–161** in 1986; **162** in 1987; **163** in 1988; **164** in 1989; **165–167** in 1990; **168** in 1991.) By K. H. Rechinger except as follows: **1**, **3**, **12**, **14**, **48**, **50**, **59**, **60** & **71** by H. Riedl; **4**, **5**, **41** & **42** by H. Schiman-Czeika; **6**, **13**, **56** & **108** by K. H. Rechinger & H. Schiman-Czeika; **7** by P. H. Raven; **8** by T. G. Yuncker & K. H. Rechinger; **9**, **10**, **11**, **53**, **67**, **76**, **110** & **151** by P. Wendelbo, **15** & **43** by A. Patzak & K. H. Rechinger; **19**, **20**, **23**, **47**, **69** & **100** by E. Schönbeck-Temesy; **34** by J. Cullen; **36** by C. C. Townsend; **49** by N. K. B. Robson; **51** & **54** by A. Polatschek & K. H. Rechinger; **52** & **55** by E. Murray; **57** by I. C. Hedge; **58** by S. J. Casper; **61** by E. Murray & K. H. Rechinger; **65** by A. Neumann & A. K. Skvortsov; **66** by K. Browicz, A. Fröhner, A. Gilli, G. Nordborg, H. Riedl, H. Schiman-Czeika, E. Schönbeck-Temesy & L. T. Vassilczenko; **68**, **113** & **114** by I. C. Hedge & J. M. Lamond; **70** by N. L. Bor; **72** by C. A. Jansson & K. H. Rechinger; **74** by L. T. Vassilczenko; **75** by S. Snogerup; **77** by K. Browicz & G. L. Menitsky; **78–86** by J. E. Dandy; **87**, **92**, **96–97**, **101–102**, **121**, **148**, **153** & **159** by K. Browicz; **88** & **89** by D. Podlech; **91** by P. Aellen; **93–94** by B. Křísa; **95** by B. Peterson; **98** by M. N. El Hadidi; **99** & **105** by J. Chrtek; **111** & **125** by K. Browicz & J. Zieliński; **112** by P. Wendelbo & B. Mathew; **115** & **124** by J. Chrtek & B. Křísa; **118** by H. D. Schotsman; **120** by I. Riedl; **122** by K. H. Rechinger, H. W. Lack & J. L. van Soest; **123** by J. S. Andersen; **126** by J. Renz; **135** by M. Iranshahr; **139a** & **b** by M. Dittrich, F. Petrak, K. H. Rechinger & G. Wagenitz; **140** by A. Chrtková-Žerková, J. G. van der Maesen & K. H. Rechinger; **142** & **152** by J. Zieliński; **143** by C. Grey-Wilson; **144** by M. N. Chaudhri, J. A. Ratter & K. H. Rechinger; **145** by E. Georgiadou, H. W. Lack, H. Merxmuller, K. H. Rechinger & G. Wagenitz; **146** by H. J. Moore; **147** by M. A. Fischer, J. Grau, A. Huber-Morath, K. H. Rechinger, P. Wendelbo & P. F. Yeo; **149** by M. S. Abdallah, H. C. D. de Wit & K. H. Rechinger; **150** by K. H. Rechinger, I. C. Hedge, J. H. Ietswaart, J. Jalas, J. Mennema & S. Seybold; **154** by A. J. C. Grierson & K. H.

Rechinger; **155** by A. Polatschek; **157** by K. H. Rechinger, S. I. Ali, K. Browicz, A. Chrtková-Žertková, D. Heller, C. C. Heyn, M. Thulin & I. T. Vassilczenko; **158** by D. Podlech, A. Huber-Morath, M. Iranshahr & K. H. Rechinger; **162** by I. C. Hedge, J. M. Lamond, K. H. Rechinger, R. Alava, D. F. Chamberlain, L. Engstrand, I. Herrnstadt, C. C. Heyn, G. H. Leute, I. Mandenova, D. Peev, M. G. Pimenov, S. Snogerup & S. G. Tamamschian; **163** by K. H. Rechinger, V. Melzheimer, W. Möschl & H. Schiman-Czeika; **164** by M. Dittrich, B. Nordenstam & K. H. Rechinger; **165** by K. H. Rechinger, K. Browicz, K. Persson & P. Wendelbo; **168** by K. H. Rechinger & H. W. Lack.

Regel et al., *Ind. Sem. Horti Petrop.*
E. A. von Regel et al., *Index Seminum quae Hortus botanicus imperialis petropolitanus pro mutua Commutatione offert. Accedunt Animadversiones botanicae nonnullae.* St. Petersburg. 1856–1869. (**1855–1868** in 1856–1869. Continuation of Fischer & C. A. Meyer, *Ind. Sem. Horti Petrop.*)

Reichenb., *Deutschl. Fl.*
H. G. L. Reichenbach, *Deutschlands Flora.* Leipzig. 1837–1870. (German ed. of Reichenb. & Reichenb. fil., *Icon. Fl. Germ.*; at least early vols. published concurrently with Latin ed.)

Reichenb., *Fl. Germ. Excurs.*
H. G. L. Reichenbach, *Flora germanica excursoria.* Lipsiae. 1830–1833. (Pp. i–viii, 1–136 in 1830; pp. 137–140, 141^{1}–140^{20} in 1831 or 1832; pp. 141–184 in 1831; pp. 185–434 in 1831 or 1832; pp. 435–438 in 1832; **2**: pp. [i*–iv*], ix–xlviii, 435 [*bis*]–878 in 1832; **3**, titled *Reichenbachianae Florae germanicae Clavis synonymica*, in 1833.)

Reichenb., or **Reichenb. fil., *Icon. Fl. Germ.***
H. G. L. Reichenbach, *Icones Florae germanicae et helveticae.* Lipsiae & Gerae. 1834–1914. (**1–19(1)**, **20** & **21**, Lipsiae; **19(2)**, **22**, **24** & **25** (**1** & **2**), Lipsiae & Gerae; **23**, Gerae. **1**, titled *Agrostographia germanica, sistens Icones Graminearum et Cyperoidearum.* (= *Pl. Crit.* **11**) in 1834–1835 (Decas. 1–5 in 1834; Decas. 6–10 in 1835); **2** in 1837–1838 (Decas. 1–5 in 1837; Decas. 6–10 in 1838); **3** in 1838–1839; **4** in 1840; **5** in 1841–1842; **6** in 1842–1844; **7** in 1845; **8** in 1846; **9** in 1847; **10** in 1848; **11** in 1849; **12** in 1850. By H. G. Reichenbach: **13/14**: pp. 1–32, tt. 1–60 in 1850; pp. 33–194 in 1851; **15**: tt. 1–80 in 1852; tt. 81–160 in 1853; **16**: tt. 1–100 in 1853; tt. 101–150 in 1854; **17** in 1854–1855; **18** in 1856–1858; **19(1)**: tt. 1–60 in 1858; tt. 61–150 in 1859; tt. 151–260 in 1860. By G. Beck von Mannagetta, **19(2)**: pp. 1–10 in 1904; pp. 11–48 in 1905; pp. 49–104 in 1906; pp. 105–152 in 1907; pp. 153–184 in 1908; pp. 185–240 in 1909; pp. 241–288 in 1910; pp. 289–341 in 1911. By H. G. Reichenbach, **20**: pp. 1–48, tt. 1–120 in 1861; pp. 49–125, tt. 121–220 in 1862; **21**: tt. 1–70 in 1863; tt. 71–110 in 1864; tt. 111–190 in 1866; tt. 191–220 in 1867; **22**: pp. 1–16, tt. 1–40, 46, 47, 57 & 79 in 1867; pp. 17–24, tt. 41–45, 48–56, 58–60 in 1869; pp. 25–48 in 1869–1872; pp. 49–56 in 1872; pp. 57–88 in 1872–1885; pp. 89–96 in 1885; pp. 97–104 in 1886. By G. Beck von Mannagetta, **22**: pp. 105–112 in 1900; pp. 113–136 in 1901; pp. 137–176 in 1902; pp. 177–230 in 1903. By F. G. Kohl, **23**: pp. 1–32 in 1896; pp. 33–44 in 1897; pp. 45–68 in 1898; pp. 69–83 in 1899. By G. Beck von Mannagetta, **24**: pp. 1–16 in 1903; pp. 17–48 in 1904; pp. 49–64 in 1905; pp. 65–80 in 1906; pp. 81–112 in 1907; pp. 113–152 in 1908; pp. 153–160 in 1909; pp. 161–216 in 1909; **25(1)**: pp. 1–12 in 1909; pp. 13–32 in 1910; pp. 33–48 in 1911; pp. 49–72 in 1912; **25(2)**: pp. 1–24 in 1913; pp. 25–40 in 1914. Cf. F. A. Stafleu & R. S. Cowan, *Taxonomic Literature* ed. 2, **4**: 674–689 (1983).)

Reichenb., *Monogr. Acon.*
H. G. L. Reichenbach, *Monographia Generis Aconiti.* Lipsiae. 1820–1821. (Pp. i–iv, 1–72, tt. A, 1–6 in 1820; pp. 73–100, tt. 7–18 in 1821. Cf. F. A. Stafleu & R. S. Cowan, *Taxonomic Literature* ed. 2, **4**: 669 (1983).)

Reichenb., *Pl. Crit.*
H. G. L. Reichenbach, *Iconographia botanica seu Plantae criticae. Icones Plantarum rariorum et minus rite cognitarum.* Lipsiae. 1823–1832. (**1** in 1823; **2** in 1824; **3** in 1825; **4** in 1826; **5** in 1827; **6** in 1828; **7** in 1829; **8** in 1830; **9** in 1831; **10** in 1832. For **11** cf. Reichenb., *Icon. Fl. Germ.* **1**.)

Reichenb., *Uebersicht Acon.*
H. G. L. Reichenbach, *Uebersicht der Gattung* Aconitum. *Grundzüge einer Monographie derselben ...* Regensburg. 1819.

Reichenb. & Reichenb. fil., *Icon. Fl. Germ.*
Cf. Reichenb., *Icon. Fl. Germ.*

Retz., *Fl. Scand. Prodr.*
A. J. Retzius, *Florae Scandinaviae Prodromus.* Ed. 1. Holmiae. 1779. Ed. 2. Lipsiae. 1795. *Suppl.* 1. Lund. 1805. *Suppl.* 2. Lund. 1809.

Reuter, *Cat. Graines Jard. Bot. Genève*
G. F. Reuter, *Catalogues des Graines recueillies en 1852 [1853, 1854, 1855, 1856, 1857, 1861, 1863, 1865, 1867, 1868] et offertes en Échange par le Jardin botanique de Genève.* Genève. (**1852** in 1853; **1853** in 1854; **1854** in 1854; **1855** in 1855; **1856** in 1856; **1857** in 1858; **1861** in 1861; **1863** in 1864; **1865** in 1866; **1867** in 1868; **1868** in 1869. The descriptions of new and critical species were reprinted as *Notulae in Species novas vel criticas Plantarum Horti botanici genevensis publici Juris Annis 1852–1868 factae, collectae et iterum editae anno 1916*, edit. J. Briquet, in *Annu. Cons. Jard. Bot. Genève* **18/19**: 239–254 (1916); this gives exact publication dates of the original *Catalogues.*)

Richter, *Pl. Eur.*
K. Richter, *Plantae Europeae. Enumeratio systematica et synonymica Plantarum phanerogamicarum in Europa sponte crescentium vel mere inquilinarum.* Leipzig. 1890–1903. (**1** in 1890; **2**, titled *Plantae Europaeae ...*, by R. L. A. M. Gürke: pp. i–vi, 1–160 in 1897; pp. 161–320 in 1899; pp. 321–480 in 1903. Cf. W. T. Stearn, *Jour. Bot. (London)* **77**: 89–91 (1939).)

Riddelsd., Hedley & Price, *Fl. Gloucestershire*
H. J. Riddelsdell, G. W. Hedley & W. R. Price, *Flora of Gloucestershire.* Arbroath. 1948.

Rink, *Grønl. Geogr. Stat. Beskr.*
H. J. Rink, *Grønland geografisk og statistisk beskrevet.* Kjøbenhavn. Botany in **1(1)** by H. J. Rink, 1852; and in **2(6)** by J. M. C. Lange, 1857.

Rochel, *Sched. Pl. Hung. Exsicc.*
A. Rochel, *Schedae Plantarum ad Floram hungaricam exsiccatam.* [Pesth?] 1810.

Roemer & Schultes, *Syst. Veg.*
Cf. L., *Syst. Nat.*

Röhling, *Deutschl. Fl.*
J. C. Röhling, *Deutschlands Flora. Ein Taschenbuch.* Ed. 1. Bremen. 1796. Ed. 2. Frankfurt a. M. 1812–1813. (**1** & **2** in 1812; **3** in 1813.) Ed. 3. by F. K. Mertens & W. D. J. Koch. Frankfurt a. M. 1823–1839. (**1** in 1823; **2** in 1826; **3** in 1831; **4** in 1833; **5** in 1839.)

Ross, J., *Voy. Disc. Baffin*
J. Ross, *A Voyage of Discovery made under the Orders of the Admiralty, in H. M. S. Isabella and Alexander, for the Purpose of exploring Baffin's Bay, and inquiring into the Probability of a North-West Passage.* London. 1819. Appendix 3 contains *List of Plants*

collected on the Coasts of Baffin's Bay and at Possession Bay by R. Brown.

Roth, *Catalecta Bot.*
A. W. Roth, *Catalecta botanica, quibus Plantae novae et minus cognitae describuntur atque illustrantur*. Lipsiae. 1797–1806. (**1** in 1797; **2** in 1800; **3** in 1806.)

Roth, *Enum.*
A. W. Roth, *Enumeratio Plantarum phaenogamarum in Germania sponte nascentium*. Lipsiae. 1827.

Roth, *Man. Bot.*
A. W. Roth, *Manuale botanicum, Peregrinationibus botanicis accommodatum. Sive Prodromus Enumerationis Plantarum phaenerogamarum in Germania sponte nascentium*. Lipsiae. 1830. (**1–3** in 1830.)

Roth, *Nov. Pl. Sp.*
A. W. Roth, *Novae Plantarum Species praesertim Indiae orientalis*. Halberstadii. 1821. (Most names published earlier by Roth in Roemer & Schultes, *Syst. Veg.* **3–5** (1818–1820).)

Roth, *Tent. Fl. Germ.*
A. W. Roth, *Tentamen Florae germanicae*. Lipsiae. 1788–1800. (**1** in 1788; **2(1)** in 1789; **2(2)** in 1793; **3(1)(1)**, pp. 1–102 in 1799; **3(1)(2)**, pp. 103–578 in 1800. Cf. F. A. Stafleu & R. S. Cowan, *Taxonomic Literature* ed. 2, **4**: 918 (1983).)

Rouy, or **Rouy & Camus**, or **Rouy & Fouc., *Fl. Fr.***
G. C. C. Rouy, *Flore de France*. Asnières, Paris & Rochefort. 1893–1913. (**1** in 1893; **2** in 1895; **3** in 1896; **4** in 1897; **5** in 1899; **6** in 1900; **7** in 1901; **8** in 1903; **9** in 1905; **10** in 1908; **11** in 1909; **12** in 1910; **13** in 1912; **14** in 1913.) **1–3** in collaboration with J. Foucaud, **6** & **7** with E. G. Camus.

Rouy, *Ill. Pl. Eur. Rar.*
G. C. C. Rouy, *Illustrationes Plantarum Europae rariorum*. Paris. 1895–1905. (**1–4** in 1895; **5–7** in 1896; **8** in 1897; **9** & **10** in 1898; **11** & **12** in 1899; **13** & **14** in 1900; **15** & **16** in 1901; **17** in 1902; **18** in 1903; **19** in 1904; **20** in 1905. Cf. F. A. Stafleu & R. S. Cowan, *Taxonomic Literature* ed. 2, **4**: 950–951 (1983).)

Royle, *Ill. Bot. Himal. Mount.*
J. F. Royle, *Illustrations of the Botany and other Branches of the natural History of the Himalayan Mountains and of the Flora of Cashmere*. London. 1833–1840. (**1–2**; for dates cf. F. A. Stafleu & R. S. Cowan, *Taxonomic Literature* ed. 2, **4**: 963 (1983).)

Rudolph, J. H., *Fl. Jen. Pl.*
J. H. Rudolph, *Florae jenensis Plantas ad Polyandriam Monogyniam Linnaei pertinentes*. Jenae. 1781.

Russell, A., *Nat. Hist. Aleppo*
A. Russell, *The natural History of Aleppo, and Parts adjacent*. Ed. 1. London. 1756. Ed. 2. London. 1794. (**1** & **2** in 1794.)

Salisb., *Parad. Lond.*
R. A. Salisbury, *The* Paradisus londinensis: *or coloured Figures of Plants cultivated in the Vicinity of the Metropolis*. London. 1805–1808. (**1(1)** in 1805 and 1806; **1(2)** in 1806 and 1807; **2(1)** in 1807 and 1808; **2(2)** in 1808. Exact dates of publication on individual plates. Descriptions by R. A. Salisbury, plates by W. Hooker.)

Salisb., *Prodr.*
R. A. Salisbury, *Prodromus Stirpium in Horto ad Chapel Allerton vigentium*. Londini. 1796.

Sarg., *Silva N. Amer.*
C. S. Sargent, *The Silva of North America: a Description of the Trees which grow naturally in North America exclusive of Mexico*. Boston & New York. 1890–1902. (**1** in 1890; **2** in 1891; **3** & **4** in 1892; **5** in 1893; **6** in 1894; **7** & **8** in 1895; **9** & **10** in 1896; **11** in 1898; **12** in 1899; **13** & **14** in 1902.)

Săvul., *Fl. Rep. Pop. Române*
T. Săvulescu, *Flora Republicii populare Române. / Flora Reipublicae popularis romanicae*. Bucureşti. 1952–1972. (**1** in 1952; **2** in 1953; **3** in 1955; **4** in 1956; **5** in 1957; **6** in 1958; **7** in 1960; **8** in 1961; **9** in 1964; **10** in 1965; **11** in 1966; **12** in 1972. **3–10** titled *Flora Republicii Populare Romîne. / Flora Reipublicae popularis romanicae.* **11–12** titled *Flora Republicii Socialiste România. / Flora Reipublicae socialisticae România*.) **9–11** edit. E. I. Nyárády; **12** edit. E. I. Nyárády, A. Beldie, I. Molariu & A. Nyárády.

Schinz & R. Keller, *Fl. Schweiz*
H. Schinz & R. Keller, *Flora der Schweiz*. Ed. 1. Zürich. 1899–1900. Ed. 2. Zürich. 1905. (**1** *Exkursionsflora* in 1905; **2** *Kritische Flora* in 1905.) Ed. 3. Zürich. 1909–1914. (**1** *Exkursionsflora* in 1909; **2** *Kritische Flora*, edit. H. Schinz & A. Thellung, in 1914.) Ed. 4. Zürich. 1923. (**1** *Exkursionsflora*, edit. H. Schinz & A. Thellung, in 1923.)

Schischkin, or **Schischkin & Bobrov**, or **Schischkin & Juz., *Fl. URSS***
Cf. Komarov, *Fl. URSS*.

Schkuhr, *Handb.*
C. Schkuhr, *Botanisches Handbuch der mehresten theils in Deutschland wildwachsenden, theils ausländischen in Deutschland unter freyem Himmel ausdauernden Gewächse*. Ed. 1. Wittenbergae. 1791–1803. (**1** in 1791; **2** in 1796; **3** in 1803.) Ed. 2. Leipzig. 1806–1814. (**1–4** in 1806–1814; title pages all dated 1808. Cf. F. A. Stafleu & R. S. Cowan, *Taxonomic Literature* ed. 2, **4**: 184 (1983).)

Schlecht., *Animadv. Ranunc.*
D. F. L. von Schlechtendal, *Animadversiones botanicae in Ranunculaceas Candollii*. Berolini. 1819–1820. (**1** in 1819; **2** (= *Sect. post.*) in 1820.)

Schlosser & Vuk., *Fl. Croat.*
J. C. Schlosser von Klekovski & L. von F. Vukotinović, *Flora croatica*. Zagrabiae. 1869.

Schmalh., *Fl. Sred. Juž. Ross.*
J. T. (I. F.) Schmal'hausen, Флора средней и южной Россіи, Крыма и сѣвернаго Кавказа. [*Flora srednej i južnoj Rossii, Kryma i sěvernago Kavkaza*]. Kiev. 1895–1897. (**1** in 1895; **2** in 1897.)

Schneider, C. K., *Ill. Handb. Laubholzk.*
C. K. Schneider, *Illustriertes Handbuch der Laubholzkunde*. Jena. 1904-1912. (**1**: pp. 1–304 in 1904; pp. 305–592 in 1905; pp. 593–810 in 1906; **2**: pp. 1–240 in 1907; pp. 241–496 in 1909; pp. 497–816 in 1911; pp. 817–1070 in 1912. Cf. F. A. Stafleu & R. S. Cowan, *Taxonomic Literature* ed. 2, **4**: 271 (1983).)

Schott, *Gen. Fil.*
H. W. Schott, *Genera Filicum*. Vindobonae. 1834.

Schott, Nyman & Kotschy, *Analecta Bot.*
H. W. Schott, C. F. Nyman & T. Kotschy, *Analecta botanica*. Vindobonae. 1854.

Schrader, *Hort. Gotting.*
H. A. Schrader, *Hortus gottingensis, seu Plantae novae et rariores Horti regii botanici gottingensis descriptae et Iconibus illustratae Opera*. Goettingae. 1809–1811. (**1**: pp. 1–14, tt. 1–8 in 1809; **2**: pp. 15–22, tt. 9–16 in 1811. Cf. F. A. Stafleu & R. S. Cowan, *Taxonomic Literature* ed. 2, **5**: 319 (1985).)

Schrader, *Ind. Sem. Horti Gotting.*
H. A. Schrader, *Index Seminum Horti academici gottingensis*. Goettingae. 1821–1835 (*academici* not in title of all issues.) New taxa in 1830, 1831, 1832, 1835 issues also published in *Linnaea, Litt.-Ber.*, but not simultaneously. From 1837 the *Index Seminum* was prepared by F. G. Bartling, with Schrader responsible for some descriptions.

Schrader, *Jour. für die Bot.*
H. A. Schrader, *Journal für die Botanik.* Göttingen. 1799–1804. (**1799(1)** in 1799; **1799(2)**: pp. 1–200 in 1799; pp. 201–502 in 1800; **1800(1)**: pp. 1–220 in 1800; pp. 221–446 in 1801; **1800(2)** in 1801; **1801(1)**: pp. 1–272 in 1803; pp. 273–504 in 1803–1804. Cf. F. A. Stafleu & R. S. Cowan, *Taxonomic Literature* ed. 2, **5**: 318 (1985).)

Schrader, *Neues Jour. Bot.*
H. A. Schrader, *Neues Journal für die Botanik.* Erfurt. 1805–1810. (**1**(**1** & **2**) in 1805; **1**(**3**) in 1806; **2**(**1**) in 1807; **2**(**2** & **3**) in 1808; **3**(**1–4**) in 1809; **4**(**1** & **2**) in 1810. Cf. F. A. Stafleu & R. S. Cowan, *Taxonomic Literature* ed. 2, **5**: 319 (1985).)

Schrader, *Sert. Hannov.*
H. A. Schrader, *Sertum hannoveranum seu Plantae rariores quae in Hortis regiis Vicinis coluntur.* Goettingae. 1795–1798. (**1** in 1795; **2** in 1796; **3** in 1797; **4** by J. C. Wendland in 1798. J. C. Wendland also co-author of **2** & **3**. Cf. F. A. Stafleu & R. S. Cowan, *Taxonomic Literature* ed. 2, **5**: 317 (1985).)

Schrank, *Baier. Fl.*
F. von Paula von Schrank, *Baiersche Flora.* München. 1789. (**1** & **2** in 1789.)

Schrank, *Baier. Reise*
F. von Paula von Schrank, *Baiersche Reise.* München. 1786.

Schrank, *Pl. Rar. Horti Monac.*
F. von Paula von Schrank, *Plantae rariores Horti academici monacensis.* Monachii. 1817–1822. (**1**(**1–2**): tt. 1–20 in 1817; **1**(**3–5**): tt. 21–50 in 1819; **2**(**6**): tt. 51–60 in 1820; **2**(**7–8**): tt. 61–80 in 1821; **2**(**9–10**): tt. 81–100 in 1822. Cf. F. A. Stafleu, *Taxonomic Literature* ed. 1, 433 (1967); the year for **2**(**9–10**) is misprinted as 1821 in *op. cit.* ed. 2, **5**: (1985).)

Schrank & C. F. P. Mart., *Hort. Monac.*
F. von Paula von Schrank & C. F. P. von Martius, *Hortus regius monacensis. Verzeichniss der im königlichen botanischen Garten zu München wachsenden Pflanzen.* München & Leipzig. 1829.

Schreber, *Spicil. Fl. Lips.*
J. C. D. von Schreber, *Spicilegium Florae lipsicae.* Lipsiae. 1771.

Schultes, *Österreichs Fl.*
J. A. Schultes, *Östreichs Flora.* Ed. 1. Wien. 1794. (**1** & **2** in 1794.) Ed. 2, titled *Österreichs Flora.* Wien. 1814. (**1** & **2** in 1814.)

Schultz, F. W., *Arch. Fl. Fr. Allem.*
F. W. Schultz, *Archives de la Flore de France et d'Allemagne.* Bitche, Hagenau & Deux-Ponts. 1842–1855. (Pp. 1–48 in 1842; pp. 49–76 in 1844; pp. 77–98 in 1946; pp. 99–154 in 1848; pp. 155–166 in 1850; pp. 167–194 in 1851; pp. 195–258 in 1852; pp. 259–282 in 1853; pp. 283–326 in 1854; pp. 327–350 in 1855.)

Schur, *Enum. Pl. Transs.*
P. J. F. Schur, *Enumeratio Plantarum Transsilvaniae.* Vindobonae. 1866.

Schweigger & Koerte, *Fl. Erlang.*
A. F. Schweigger & F. Koerte, *Flora erlangensis.* Erlangae. 1811. (**1** & **2** in 1811. **1**: pp. 1–136, by A. F. Schweigger, were previously published in 1804 as *Specimen Florae erlangensis*; **1**: pp. 137–160, *Addenda*, by F. Koerte, and **2**: pp. 1–143, by A. F. Schweigger & F. Koerte, first appeared in 1811. Cf. F. A. Stafleu & R. S. Cowan, *Taxonomic Literature* ed. 2, **5**: 429 (1985).)

Schweinf., *Beitr. Fl. Aethiop.*
G. A. Schweinfurth, *Beitrag zur Flora Aethiopiens.* Berlin. 1867.

Scop., *Fl. Carn.*
J. A. Scopoli, *Flora carniolica.* Ed. 1. Viennae. 1760. (Generic names provided with a description or reference to earlier description validly published; no binary names for species.) Ed. 2. Viennae. 1771–1772. (**1** in 1771, reissued 1772; **2** in 1772. Cf. F. A. Stafleu & R. S. Cowan, *Taxonomic Literature* ed. 2, **5**: 455 (1985).)

Sebeók, *Medico-Bot. Tatar.*
A. Sebeók de Szent-Miklós, *Syn Theo Dissertatio inauguralis medico botanica de Tataria hungarica.* Viennae. 1779.

Seub., *Fl. Azor.*
M. A. Seubert, *Flora azorica, quam ex Collectionibus Schedisque Hochstetteri Patris et Filii elaboravit* ... Bonnae. 1844.

Sibth., *Fl. Oxon.*
J. Sibthorp, *Flora oxoniensis.* Oxonii. 1794.

Sibth. & Sm., *Fl. Graeca*
J. Sibthorp & J. E. Smith, *Flora graeca: sive Plantarum rariorum Historia, quas in Provinciis aut Insulis Graeciae legit, investigavit, et depingi curavit Johannes Sibthorp ... Characteres omnium, Descriptiones et Synonyma, elaboravit Jacobus Edvardus Smith* [**1–7**] ... *elaboravit Johannes Lindley* [**8–10**]. Londini. 1806–1840. (For dates see F. A. Stafleu & R. S. Cowan, *Taxonomic Literature* ed. 2, **5**: 579–580 (1985).)

Sibth. & Sm., *Fl. Graec. Prodr.*
J. Sibthorp & J. E. Smith, *Florae graecae Prodromus: sive Plantarum omnium Enumeratio, quas in Provinciis aut Insulis Graeciae invenit Johannes Sibthorp ... Characteres et Synonyma omnium cum Annotationibus elaboravit Jacobus Edvardus Smith.* Londini. 1806–1816. (**1**(**1**): pp. 1–218 in 1806; **1**(**2**): pp. 219–442 in 1809; **2**(**1**): pp. 1–210 in 1813; **2**(**2**): pp. 211–422 in 1816. Cf. F. A. Stafleu & R. S. Cowan, *Taxonomic Literature* ed. 2, **5**: 579 (1985).)

Siebold & Zucc., *Fl. Jap.*
P. F. von Siebold & J. G. Zuccarini, *Flora japonica.* Lugduni Batavorum. 1835–1870. (**1**: pp. 5–28 in 1835; pp. 29–40 in 1836; pp. 49–64 in 1837 or ?1838; pp. 65–72 in 1838; pp. 73–120 in 1839; pp. 121–140 in 1840; pp. 141–193 in 1841. **2**: pp. 1–28 in 1842; pp. 29–44 in 1844; pp. 45–89 in 1870. Cf. F. A. Stafleu & R. S. Cowan, *Taxonomic Literature* ed. 2, **5**: 589 (1985).)

Simmons, *Vasc. Pl. Ellesmereland*
H. G. Simmons, *The vascular Plants in the Flora of Ellesmereland.* Kristiania. 1906.

Simonkai, *Enum. Fl. Transs.*
L. von Simonkai, *Enumeratio Florae transsilvanicae vesculosae critica. / Erdély edényes Flórájának helyesbitett Foglalata.* Budapest. 1887. (Cf. F. A. Stafleu & R. S. Cowan, *Taxonomic Literature* ed. 2, **5**: 610 (1985).)

Širj. & Lavrenko, *Consp. Crit. Fl. Prov. Charkov.*
G. I. Širjaev & E. M. Lavrenko, *Conspectus criticus Florae Provinciae charkoviensis.* 1926→. (**1** in 1926.)

Sloboda, *Rostlinnictví*
D. Sloboda, *Rostlinnictví čili Návod k snadnéma určení a Pojmenování Rostlin v Čechach, Moravě a jiných zemích rakouského Mocnářství domácích* ... Praze (Praha). 1852.

Sm., *Comp. Fl. Brit.*
J. E. Smith, *Compendium Florae britannicae.* Ed. 1. Londini. 1800. German ed. Erlangae. 1801. Ed. 2. Londini. 1816. Ed. 3. Londini. 1818. Ed. 4. Londini. 1825. Ed. 5. Londini. 1828. Ed. 6. Londini. 1829.

Sm., *Fl. Brit.*
J. E. Smith, *Flora britannica.* Londini. 1800–1804. (**1**: pp. 1–436 in 1800; **2**: pp. 437–914 in 1800; **3** in 1804. Cf. S. W. Greene, *Jour. Soc. Bibl. Nat. Hist.* **3**: 281 (1957).)

Sm., *Pl. Icon. Ined.*
J. E. Smith, *Plantarum Icones hactenus ineditae, plerumque ad Plantas in Herbario linnaeano conservatas delineatae.* Londini. 1790–1791. (Fasc. **1**: tt. 1–25 in 1789; fasc. **2**: tt. 26–50 in 1790; fasc. **3**: tt. 51–75 in 1791. Cf. F. A. Stafleu, *Taxon* **12**: 78 (1963).)

Sm., *Spicil. Bot.*
J. E. Smith, *Spicilegium botanicum*. Londini. 1791–1792. (**1** in 1791; **2** in 1792. Some copies with English title *Gleanings of Botany*; text issued in Latin, English, or in both languages). Cf. F. A. Stafleu, *Taxon* **12**: 79 (1963).)

Sm., J., *Ferns Brit. Foreign*
John Smith, *Ferns: British & foreign*. London. 1866. Ed. 2. London. 1877, reissued 1879 and 1895 or 1896.

Soó, *Növényföldrajz*
R. de Soó, *Növényföldrajz*. Ed. 1. Budapest. 1945. Ed. 2. Budapest. 1953. Ed. 3. Budapest. 1956. Ed. 4. Budapest. 1961.

Soó & Jáv., *Magyar Növ. Kéz.*
R. de Soó & S. Jávorka, *A Magyar Növényvilág Kézikönyve*. Budapest. 1951. (**1** & **2** in 1951.)

Sowerby, *Engl. Bot.*
J. Sowerby, *English Botany*. Ed. 1. *Text* by J. E. Smith. London. 1790–1814. (**1–36** in 1790–1814.) *Supplement* by various authors. 1829–1866. (**1–5** in 1829–1866.) Ed. 2. London. 1829–1846. (**1–12** in 1829–1846.) Ed. 3[A]. London. 1847–1854. (**1–7** in 1847–1854, being reissues of ed. 2, **1–7**.) Ed. 3[B], with new descriptions by J. T. I. Boswell-Syme. London. 1863–1886. (**1–12** in 1863–1886.) (For dates of all eds. cf. F. A. Stafleu & R. S. Cowan, *Taxonomic Literature* ed. 2, **5**: 682–684 (1985).)

Soyer-Willemet, *Obs. Pl. Fr.*
H. F. Soyer-Willemet, *Observations sur quelques Plantes de France, suivies du Catalogue des Plantes vasculaires des Environs de Nancy*. Nancy. 1828.

Soyer-Willemet & Godron, *Monogr. Silene Algér.*
H. F. Soyer-Willemet & D. A. Godron, *Monographie des* Silene *de l'Algérie*. Nancy. 1851. (Reprinted or preprinted from *Mém. Soc. Sci. Lettr. Arts Nancy* **1850**: 139–184 (1850).)

Spach, *Hist. Vég. (Phan.)*
E. Spach, *Histoire naturelle des Végétaux. Phanérogames*. Paris. 1834–1848. (**1–3** in 1834; **4** in 1835; **5** in 1836; **6–9** in 1838; **10** & **11** in 1841; **12** & **13** in 1846; **14** in 1848; *Atlas* in parts, 1834–1847. For dates cf. F. A. Stafleu & R. S. Cowan, *Taxonomic Literature* ed. 2, **5**: 767 (1985).)

Sprengel, *Fl. Hal.*
C. (K.) P. J. Sprengel, *Florae halensis Tentamen novum*. Ed. 1. Halae Saxonum. 1806. Ed. 2. Halae. 1832. ***Mant.***, *Mantissa prima Florae halensis, addita novarum Plantarum Centuria*. Halae. 1807. Ed. 2. Halae. 1832. ***Mant. Alt.***, *Mantissa altera*, also titled *Observationes botanicae ad Floram halensem. Mantissa secunda*. Halae. 1811. *Supplementum tertium*, or *Annus botanicus*, by F. W. Wallroth. Halae. 1815. (For dates Cf. F. A. Stafleu & R. S. Cowan, *Taxonomic Literature* ed. 2, **5**: 810, 811 & 814 (1985).)

Sprengel, *Hist. Rei Herb.*
C. (K.) P. J. Sprengel, *Historia Rei Herbariae*. Amsteldami. 1807–1808. (**1** in 1807 or 1808; **2** in 1808.) Another issue in 1808. (**1** & **2** in 1808).) Cf. F. A. Stafleu & R. S. Cowan, *Taxonomic Literature* ed. 2, **5**: 810 (1985).

Sprengel, *Novi Provent.*
C. (K.) P. J. Sprengel, *Novi Proventus Hortorum academicorum halensis et berolinensis. Centuria Specierum minus cognitarum, quae vel per Annum 1818 in Horto halensi et berolinensi floruerunt, vel siccae missae fuerunt*. Halae. 1818. (Cf. F. A. Stafleu & R. S. Cowan, *Taxonomic Literature* ed. 2, **5**: 812 (1985).)

Sprengel, *Pugillus*
C. (K.) P. J. Sprengel, *Plantarum minus cognitarum Pugillus primus [secundus]*. Halae. 1813–1815. (**1** in 1813; **2** in 1815.)

Sprengel, *Syst. Veg.*
Cf. L., *Syst. Nat.*

Stearn & P. H. Davis, *Peonies of Greece*
W. T. Stearn & P. H. Davis, *Peonies of Greece: a taxonomic and historical Survey of the Genus* Paeonia *in Greece*. Kifissia. 1984.

Stefani, Major & W. Barbey, *Samos*
C. de Stefani, C. I. F. Major & W. Barbey-Boissier, *Samos. Étude géologique, paléontologique et botanique*. Lausanne. 1891 (title page) or 1892. Cf. F. A. Stafleu & R. S. Cowan, *Taxonomic Literature* ed. 2, **5**: 867 (1985).

Sternb., *Revis. Saxifr.*
K. (C.) M. von Sternberg, *Revisio Saxifragarum Iconibus illustrata*. Ratisbonae. 1810. *Suppl.* **1**. Ratisbonae. 1822. *Suppl.* **2**. Pragae. 1831.

Steudel, *Nomencl. Bot.*
E. G. von Steudel, *Nomenclator botanicus*. Ed. 1. Stuttgardtiae & Tuebingae. 1821–1824. (**1** & **2** in 1821; **3** in 1824.) Ed. 2, titled *Nomenclator botanicus, seu Synonymia Plantarum universalis enumerans* ... Stuttgartiae & Tuebingae. 1840–1841. (**1** in 1840; **2**: pp. 1–48 in 1840; p. 49–810 in 1841. (Cf. F. A. Stafleu & R. S. Cowan, *Taxonomic Literature* ed. 2, **5**: 910 (1985).)

Stoj. & Stefanov, *Fl. Bălg.*
N. Stojanov & B. Stefanov, *Flore de la Bulgarie*. Ed. 1. Sofija. 1924–1925. (**1**: pp. 1–608 in 1924; **2**, pp. 609–1367, i–x in 1925.) Ed. 2, titled **Флора на България** [*Flora na Bălgarija*]. Sofija. 1933. Ed. 3. Sofija. 1948. Ed. 4, with alternative title *Flora bulgarica*, by N. Stojanov, B. Stefanov & B. Kitanov. Sofija. 1966–1967. (**1** in 1966; **2** in 1967.)

Stokes, *Arr. Brit. Pl.*
Cf. With., *Arr. Brit. Pl.*

Strid, *Mount. Fl. Gr.*
P. A. K. Strid, *Mountain Flora of Greece*. Cambridge & Edinburgh. 1986–1991. (**1**, Cambridge in 1986; **2**, Edinburgh in 1991.)

Sturm, *Deutschl. Fl.*
J. Sturm, *Deutschlands Flora*. Nürnberg. 1796–1862. (**1–21** in 1796–1855; *Caricologica germanica*: **47** in 1827; **50** in 1827; **53** in 1829; **55** in 1830; **57** in 1831; **61** in 1833; **69** in 1835. No Heften dealing with Pteridophytes or Spermatophyta were published after 1855. For dates cf. F. A. Stafleu & R. S. Cowan, *Taxonomic Literature* ed. 2, **6**: 66–72 (1986).) Ed. 2. Stuttgart. 1900–1907. (For dates cf. F. A. Stafleu & R. S. Cowan, *Taxonomic Literature* ed. 2, **6**: 71 (1986).) **1**, **4–15** by E. H. L. Krause; **2** by E. R. Missbach & E. H. L. Krause; **3** by K. G. Lutz.

Suter, *Fl. Helv.*
J. R. Suter, *Flora helvetica*. Ed. 1. Turici. 1802. (**1** & **2** in 1802.) Ed. 2. Turici. 1822. (**1** & **2** in 1822.)

Sweet, *Hort. Brit.*
R. Sweet, *Hortus britannicus*. Ed. 1. London. 1826. Reissued 1827. Ed. 2. London. 1830. Ed. 3, edit. G. Don. London. 1839. Cf. F. A. Stafleu & R. S. Cowan, *Taxonomic Literature* ed. 2, **6**: 124 (1986).

Syme, *Engl. Bot.*
Cf. Sowerby, *Engl. Bot.*

Takht., *Krasnaja Kniga*
A. L. Takhtadzhan (ed.), **Красная Книга: дикорастущие Виды Флоры СССР нуждающиеся в охране** [*Krasnaja Kniga: dikorastuščie Vidy Flory SSSR nuždajuščiesja v okhrane*]. / *Red Book: native Plant Species to be protected in the USSR*. Leningrad. 1975.

Ten., *Cat. Pl. Hort. Neap.*
M. Tenore, *Catalogus Plantarum Horti regii neapolitani ad Annum 1813*. Neapoli. 1813. ***App.***, *Appendix*. Ed. 1. Neapoli. 1815. Ed. 2. Neapoli. 1819.

Ten., *Corso Bot. Lez.*
M. Tenore, *Corso delle botaniche Lezione*. Ed. 1. Napoli. 1806–1822. (**1** in 1806; **2** in 1810; **3** in 1821; **4** in 1822. Cf. F. A. Stafleu & R. S. Cowan, *Taxonomic Literature* ed. 2, **6**: 213–214 (1986). Also later editions.

Ten., *Fl. Nap.*
M. Tenore, *Flora napolitana*. Napoli. 1811–1838. (**1** in 1811–1815, including ***Prodr.***, *Prodromo della Flora napolitana* followed by ***Suppl.* 1,** *Supplimento primo* and ***Suppl.* 2,** *Supplimento secondo* with continuous pagination; **2** in 1820, including ***Prodr. Suppl.* 3,** *Prodromo della Flora napolitana, Supplimento terzo*; **3** in 1824–1829, including reprint of ***Prodr. Suppl.* 4,** *Prodromo della Flora napolitana, Supplimento quarto*; **4** in 1830, including ***Syll.***, *Florae neapolitanae Sylloge* (reprinted separately in 1831 as *Sylloge Plantarum vascularium Florae neapolitanae*) followed by ***Syll. App.* 1,** *Appendix 1* (*Addenda et Emendanda*) and ***Syll. App.* 2,** *Appendix 2* (*Addenda et Emendanda altera*) with continuous pagination, and ***Syll. App.* 3,** *Ad Florae neapolitanae Plantarum vascularium Syllogem Appendix tertia*; **5** in 1835–1838, including ***Syll. App.* 4,** *Ad Florae neapolitanae Syllogem Appendix quarta*.) The included works were in most cases reprinted separately later, with different pagination. The ***Fl. Neap. Syll.*** of 1831 is cited in **4**, 1830. Cf. F. A. Stafleu & R. S. Cowan, *Taxonomic Literature* ed. 2, **6**: 213–214 (1986).

Ten., *Fl. Neap. Prodr. App. Quinta*
M. Tenore, *Ad Florae neapolitanae Prodromum Appendix quinta*. Neapoli. 1826. This work was never included in the volumes of *Flora napolitana*.

Ten., *Fl. Neap. Syll.*
Cf. Ten., *Fl. Nap.*

Ten., *Ind. Sem. Horti Neap.*
M. Tenore, or M. Tenore & G. Gussone, *Index Seminum Horti botanici neapolitani*. Neapoli. 1825–1840.

Ten., *Prodr. Fl. Nap.*
Cf. Ten., *Fl. Nap.*; ***Prodr. Suppl.* 4** was published in 1823, and later reprinted in *Fl. Nap.* For *Suppl.* **5** cf. Ten., *Fl. Neap. Prodr. App. Quinta*.

Thell., *Fl. Adv. Montpellier*
A. Thellung, *La Flore adventice de Montpellier*. Cherbourg. 1912.

Thore, *Essai Chlor. Land.*
J. Thore, *Essai d'une Chloris du Département des Landes*. Dax. 1803.

Thouars, *Mélang. Bot.*
L. M. A. Aubert du Petit-Thouars, *Mélanges de Botanique et des Voyages*. Paris. 1811.

Thuill., *Fl. Paris*
J. L. Thuillier, *La Flore des Environs de Paris*. Ed. 1. Paris. 1790. Ed. 2. Paris. 1799. Reissued 1824.

Thunb., *Nova Gen. Pl.*
C. P. Thunberg, *Nova Genera Plantarum*. Upsaliae. 1781–1801. (**1** in 1781; **2** in 1782; **3** in 1783; **4** & **5** in 1784; **6** & **7** in 1792; **8** & **9** in 1798; **10–12** in 1800; **13–16** in 1801.)

Thunb., *Prodr. Pl. Cap.*
C. P. Thunberg, *Prodromus Plantarum capensium*. Upsaliae. 1794–1800. (Pars prior, pp. i–xii, 1–84 in 1794; pars posterior, pp. i–viii, 85–192 in 1800.)

Tineo, *Pl. Rar. Sic. Pug.*
V. Tineo, *Plantarum rariorum Siciliae minus cognitarum Pugillus primus*. Panormi. 1817.

Tod., *Fl. Sicula Exsicc.*
A. Todaro, *Flora sicula Exsiccata*. Palermo. 1864. (The work contains a list of names for numbers 1–200, but 1600 exsiccatae numbers were issued from 1864 onwards.)

Tolm., *Fl. Arct. URSS*
A. I. Tolmatchev, *Flora arctica URSS*. / Арктическая Флора СССР [*Arktičeskaja Flora SSSR*]. Moskva & Leningrad. 1960–1987. (**1** in 1960; **2** in 1964; **3** in 1966; **4** in 1963; **5** in 1966; **6** in 1971; **8**(**1**) in 1980; **8**(**2**) in 1983; **9**(**1**) in 1984; **9**(**2**) in 1986; **10** in 1987.)

Torrey, *Fl. N. York*
J. Torrey, *A Flora of the State of New York*. Albany. 1843. (**1** & **2** in 1843 but **1** dated Dec. 1842. Cf. F. A. Stafleu & R. S. Cowan, *Taxonomic Literature* ed. 2, **6**: 406 (1986).)

Torrey & A. Gray, *Fl. N. Amer.*
J. Torrey & A. Gray, *A Flora of North America*. New York & London. 1838–1843. (**1**: pp. 1-360 in 1838; pp. 361–711 in 1840; **2**: pp. 1–184 in 1841; pp. 185–392 in 1842; pp. 293–504 in 1843. Cf. F. A. Stafleu & R. S. Cowan, *Taxonomic Literature* ed. 2, **6**: 405 (1986).)

Tratt., *Arch. Gewächsk.*
L. Trattinick, *Archiv der Gewächskunde*. Wien. 1811–1818. (Uncoloured plates ed.: **1** in 1811; **2** in 1812 (preface in 1814); **3** in 1813 or 1814; **4** in 1814; **5** in 1818. Coloured plates ed.: **1** in 1812; **2** in 1813; **3** & **4** in 1814. Cf. F. A. Stafleu & R. S. Cowan, *Taxonomic Literature* ed. 2, **6**: 438 (1986).)

Trautv., *Del. Sem. Horti Kiov.*
E. R. Trautvetter, *Delectus Seminum in Horto kiovensi ... Anno 1840 [1841] collectorum*. Kioviae. **1840** in 1840; **1841** in 1842.

Trotzky, *Ind. Sem. Horti Casan.*
P. K. Trotzky, *Index Seminum Horti casaniensis*. [Kazan'.] 1834 and 1839.

Tzanoudakis, *Kytt. Meleti Gen.* Paeonia *Ell.*
D. B. Tzanoudakis, Κυτταροταξινομικί Μελέτι τοῦ Γένουσ *Paeonia* L. ἐν Ἑλλάδι [*Kyttarotaksinomiki Meleti tou Genous* Paeonia L. *en Elladi*]. / *Cytotaxonomic Study of the Genus* Paeonia L. *in Greece*. Patras. 1977.

Unger, *Wiss. Ergeb. Reise Griech.*
F. J. A. N. Unger, *Wissenschaftliche Ergebnisse einer Reise in Griechenland und in den jonischen Inseln*. Wien. 1862.

Unger & Kotschy, *Ins. Cypern*
F. J. A. N. Unger & T. Kotschy, *Die Insel Cypern, ihrer physischen und organischen Natur nach, mit Rücksicht auf ihre frühere Geschichte*. Wien. 1865.

Ung.-Sternb., *Syst. Salicorn.*
F. Ungern-Sternberg, *Versuch einer Systematik der Salicornieen*. Dorpat. 1866.

Vahl, *Symb. Bot.*
M. H. Vahl, *Symbolae botanicae, sive Plantarum, tam earum, quas in Itinere, imprimis orientali, collegit Petrus Forskål*. Hauniae. 1790–1794. (**1** in 1790; **2** in 1791; **3** in 1794. Cf. F. A. Stafleu & R. S. Cowan, *Taxonomic Literature* ed. 2, **6**: 630 (1986).)

Velen., *Fl. Bulg.*
J. Velenovský, *Flora bulgarica*. Pragae. 1891. ***Suppl.***, *Supplementum*. 1898.

Vent., *Descr. Pl. Jard. Cels*
E. P. Ventenat, *Description des Plantes nouvelles ou peu connues cultivées dans le Jardin de J. M. Cels*. Paris. 1800–1803 (Tt. 1–10 in 1800; tt. 11–20 in 1800 or 1801; tt. 21–60 in 1801; tt. 61–90 in 1802; tt. 91–100 in 1803. Cf. F. A. Stafleu & R. S. Cowan, *Taxonomic Literature* ed. 2, **6**: 701 (1986).)

Vent., *Tabl. Règne Végét.*
E. P. Ventenat, *Tableau du Règne végétal*. Paris. 1799. (**1–4** in 1799.)

Vig., *Hist. Pavots Argém.*
L. G. A. Viguier, *Histoire naturelle, médicale et économique, des Pavots et des Argémones*. Montpellier. 1814.

Vill., *Hist. Pl. Dauph.*
D. Villars, *Histoire des Plantes de Dauphiné*. Grenoble. 1786–1789. (**1** in 1786; **2** in 1787; **3**(**1**): pp. 1–580 in 1788; **3**(**2**): pp. 581-1091 in 1789. Cf. F. A. Stafleu & R. S. Cowan, *Taxonomic Literature* ed. 2, **6**: 740 (1986).)

Vill., *Prosp. Pl. Dauph.*
D. Villars, *Prospectus de l'Histoire des Plantes de Dauphiné*. Grenoble. 1779.

Vis., *Fl. Dalm.*
R. de Visiani, *Flora dalmatica*. Lipsiae. 1842–1851. (**1** in 1842; **2** in 1847; **3**(**1**): pp. 1–185 in 1850; **3**(**2**): pp. 185–390 in 1851. ***Suppl.* 1**, *Supplementum*. Lipsiae. 1872. ***Suppl.* 2**, *Supplementum alterum*. Lipsiae. 1877–1881. (**2**(**1**) in 1877; **2**(**2**) in 1881.) Cf. F. A. Stafleu & R. S. Cowan, *Taxonomic Literature* ed. 2, **6**: 757 (1986).

Viv., *Elench. Pl. Horti Bot.*
D. Viviani, *Elenchus Plantarum Horti botanici J. Car. Dinegro Observationibus quod novas, vel rariores Species passim interjectis*. Genuae. 1802.

Viv., *Fl. Cors.*
D. Viviani, *Florae Corsicae Specierum novarum, vel minus cognitarum Diagnosis*. Genuae. 1824. ***App.* 1**, *Appendix ad Florae Corsicae Prodromum*. Genuae. 1825. ***App.* 2**, *Appendix altera ad Florae Corsicae Prodromum*. Genuae. 1830. Cf. F. A. Stafleu & R. S. Cowan, *Taxonomic Literature* ed. 2, **6**: 762 (1986).

Viv., *Fl. Ital. Fragm.*
D. Viviani, *Florae italicae Fragmenta*. Genuae. 1808.

Viv., *Fl. Lib.*
D. Viviani, *Florae libycae Specimen*. Genuae. 1824.

Vollmann, *Fl. Bayern*
F. Vollmann, *Flora von Bayern*. Stuttgart. 1914.

Wahlenb., *Fl. Lapp.*
G. Wahlenberg, *Flora lapponica*. Berolini. 1812.

Waldst. & Kit., *Pl. Rar. Hung.*
F. A. von Waldstein-Wartemberg & P. Kitaibel, *Descriptiones et Icones Plantarum rariorum Hungariae*. Viennae. 1799–1812. (For dates cf. F. A. Stafleu & R. S. Cowan, *Taxonomic Literature* ed. 2, **7**: 31 (1988).

Wall., *Pl. Asiat. Rar.*
N. Wallich, *Plantae asiaticae rariores; or, Descriptions and Figures of a select Number of unpublished East Indian Plants*. London. 1829–1832. (**1**: pp. 1–22 [= part 1] in 1829; pp. 23–84 [= parts 2–4] in 1830; **2**: pp. 1–20 [= part 5] in 1830; pp. 21-86 [= parts 6–8] in 1831; **3**: pp. 1–? [= part 9] in 1831; pp. ?–117 [= parts 10–12] in 1832. Cf. F. A. Stafleu & R. S. Cowan, *Taxonomic Literature* ed. 2, **7**: 40 (1988).)

Walpers, *Repert. Bot. Syst.*
W. G. Walpers, *Repertorium Botanices systematicae*. Lipsiae. 1842–1848. (**1**: pp. 1–768 in 1842; pp. 769–947 in 1843; **2** in 1843; **3**: pp. 1–768 in 1844; pp. 769–1002 in 1845; **4**: pp. 1–192 in 1845; pp. 193–576 in 1847; pp. 577–821 in 1848; **5**: pp. 1–192 in 1845; pp. 193–982 in 1846; **6**: pp. 1–384 in 1846; pp. 385–834 in 1847. Cf. F. A. Stafleu & R. S. Cowan, *Taxonomic Literature* ed. 2, **7**: 46–47 (1988).)

Walters, S. M. et al., *Eur. Garden Fl.*
S. M. Walters, A. Brady, C. D. Brickell, J. Cullen, P. S. Green, J. Lewis, V. A. Matthews, D. A. Webb, P. F. Yeo & J. C. M. Alexander (eds.), *The European Garden Flora*. Cambridge etc., 1984→. (**1** in 1986; **2** in 1984; **3** in 1989.) **3** ed. S. M. Walters, J. C. M. Alexander, A. Brady, C. D. Brickell, J. Cullen, P. S. Green, V. H. Heywood, V. A. Matthews, N. K. B. Robson, P. F. Yeo & S. G. Knees.

Warncke, *Eur. Sippen* Aconitum lycoctonum-*Gruppe*
K. Warncke, *Die europäischen Sippen der* Aconitum lycoctonum-*Gruppe*. München. 1964. (Inaugural-Dissertation; in typescript, with printed title page.)

Watson, H. C., *Part First Suppl. Cyb. Brit.*
H. C. Watson, *Part first of a Supplement to the Cybele britannica*. London. 1860.

Webb & Berth., *Phyt. Canar.*
P. B. Webb & S. Berthelot, *Phytographia canariensis*. (Vol. **3**(**2**) of *Histoire naturelle des Îles Canaries*.) Paris. 1836–1850. (For dates cf. F. A. Stafleu & R. S. Cowan, *Taxonomic Literature* ed. 2, **7**: 119–120 (1988).)

Weber fil. & Mohr, *Beitr. Naturk.*
F. Weber & D. M. H. Mohr, *Beiträge zur Naturkunde. In Verbinding mit ihren Freunden verfasst und herausgegeben*. Kiel. 1805–1810. (**1** in parts in 1805–1806; **2** in 1810. Cf. F. A. Stafleu & R. S. Cowan, *Taxonomic Literature* ed. 2, **7**: 128 (1988).)

Wibel, *Prim. Fl. Werthem.*
A. W. E. C. Wibel, *Primitiae Florae werthemensis*. Jenae. 1799.

Willd., *Berlin. Baumz.*
C. L. Willdenow, *Berlinische Baumzucht, oder Beschreibung der in den Gärten um Berlin, im Freien ausdauernden Bäume und Sträucher*. Ed. 1. Berlin. 1796. Ed. 2. Berlin. 1811.

Willd., *Enum. Pl. Horti Berol.*
C. L. Willdenow, *Enumeratio Plantarum Horti regii botanici berolinensis*. Berolini. 1809. ***Suppl.***, *Supplementum* by D. F. L. von Schlechtendal. Berolini. 1814. For dates cf. F. A. Stafleu & R. S. Cowan, *Taxonomic Literature* ed. 2, **7**: 305 (1988).

Willd., *Sp. Pl.*
C. L. Willdenow, ed. 4 of C. von Linné, *Species Plantarum*. Berolini. 1797–1810(–1830). (**1**(**1**): pp. 1–495 in 1797; **1**(**2**): pp. 496–1068 in 1798; **2**(**1** & **2**) in 1799; **3**(**1**): pp. 1–847 in 1800; **3**(**2**): pp. 849–1474 in 1802; **3**(**3**): pp. 1475–2409 in 1803; **4**(**1**): pp. 1–629 in 1805; **4**(**2**): pp. 631-1157 in 1806; **5**(**1**): pp. 1–544 in 1810. By C. F. Schwägrichen: **5**(**2**): pp. 1–122 in 1830, also titled *Species Muscorum frondosorum*. By H. F. Link: **6**(**1**): pp. 1–162 in 1824; **6**(**2**): pp. 1-128 in 1825. Cf. F. A. Stafleu & R. S. Cowan, *Taxonomic Literature* ed. 2, **7**: 303 (1988).)

Willk., *Icon. Descr. Pl. Nov.*
H. M. Willkomm, *Icones et Descriptiones Plantarum novarum criticarum et rariorum Europae austro-occidentalis praecipue Hispaniae*. Lipsiae. 1852–1862. (For dates cf. F. A. Stafleu & R. S. Cowan, *Taxonomic Literature* ed. 2, **7**: 339 (1988).)

Willk., *Ill. Fl. Hisp.*
H. M. Willkomm, *Illustrationes Florae Hispaniae Insularumque Balearium*. Stuttgart. 1881–1892. (For dates cf. F. A. Stafleu & R. S. Cowan, *Taxonomic Literature* ed. 2, **7**: 342–343 (1988).)

Willk., *Strand-Steppengeb. Iber. Halbins.*
H. M. Willkomm, *Die Strand- und Steppengebiete der iberischen Halbinsel und deren Vegetation*. Leipzig. 1852.

Willk., *Suppl. Prodr. Fl. Hisp.*
H. M. Willkomm, *Supplementum Prodromi Florae hispanicae*. Stuttgartiae. 1893.

Willk. & Lange, *Prodr. Fl. Hisp.*
H. M. Willkomm & J. M. C. Lange, *Prodromus Florae hispanicae*. Stuttgartiae. 1861–1880. (**1**: pp. i–vii, 1–192 in 1861; pp. ix–xxx, 193–316 in 1862; **2**: pp. 1–272 in 1865; pp. 273–480 in 1868; pp. 481–680 in 1870; **3**: pp. 1–240 in 1874; pp. 241–512 in 1877; pp. 513–736 in 1878; pp. 737–1144 in 1880. Cf. F. A. Stafleu & R. S. Cowan, *Taxonomic Literature* ed. 2, **7**: 340 (1988).)

Wimmer, *Salices Eur.*
C. F. H. Wimmer, *Salices europaeae*. Vratislavae. 1866.

With., *Arr. Brit. Pl.*
W. Withering, *A botanical Arrangement of all the Vegetables naturally growing in Great Britain, with Descriptions of the Genera and Species.* Ed. 1. Birmingham & London. 1776. Ed. 2, titled *A botanical Arrangement of British Plants* by J. Stokes. Birmingham & London. 1787–1792. (**1** & **2** in 1787; **3(1)**: pp. i–clvii & chart in 1789; **3(2)**: pp. 9–503 & tt. 13–19 in 1792.) Ed. 3, titled *An Arrangement of British Plants* by J. Stokes. London. 1796. Ed. 4, titled *A systematic Arrangement of British Plants* by W. Withering fil. London. 1801. Ed. 5. Birmingham. 1812. Ed. 6, titled *An Arrangement of British Plants.* London. 1818. Ed. 7. London. 1830.

Woronow & Schelkownikow, *Sched. Herb. Fl. Cauc.*
G. N. Woronow & A. B. Schelkownikow, *Schedae Herbarium Florae caucasicae.* 1914–1916. (Published in *Trudy Tifliss. Bot. Sada* **12** & **19**.)

Wulf, *Fl. Kryma*
E. V. Wulf, **Флора Крыма** [*Flora Kryma*]. / *Flora taurica.* Jalta. 1927–1969. (**1(1)** in 1927; **1(2)** in 1929; **1(3)** in 1930; **1(4)** in 1951; **2(1)** in 1947; **2(2)** in 1960; **2(3)** in 1953; **3(1)** in 1957; **3(2)** in 1966; **3(3)** in 1969.) **1(2–3)**, Leningrad; **1(4)**, **2(2–3)** & **3(1–2)**, Moskva; **2(1)**, Moskva & Leningrad. **Дополнения к I Тому "Флоры Крыма"** [*Dopolnenija k I Tomu 'Flory Kryma'*]. / *Addenda et Corrigenda ad Vol. I 'Flora tauricae'.* Jalta. 1959.

Wünsche, *Pfl. Deutschl.*
F. O. Wünsche, *Schulflora von Deutschland.* Leipzig. 1871. Ed. 2. Leipzig. 1877. Ed. 3. Leipzig. 1881. Ed. 4. Leipzig. 1884. Ed. 5, titled *Schulflora von Deutschland ... Die höheren Pflanzen.* Leipzig. 1888. Ed. 6. Leipzig. 1892. Ed. 7, titled *Die Pflanzen Deutschlands.* Leipzig. 1897. Ed. 8. Leipzig & Berlin. 1902. By J. Abromeit: Ed. 9. Leipzig & Berlin. 1909. Ed. 10. Leipzig & Berlin. 1916. Ed. 11. Leipzig & Berlin. 1924. Ed. 12. Leipzig & Berlin. 1928. Ed. 13. Leipzig & Berlin. 1932.

Zawadzki, *Enum. Pl. Galic. Bucow.*
A. Zawadzki, *Enumeratio Plantarum Galiciae & Bucowinae, oder die in Galizien und in der Bukowina wildwachsenden Pflanzen mit genauer Angabe ihrer Standorte.* Breslau. 1835.

APPENDIX III

KEY TO THE ABBREVIATIONS OF TITLES OF PERIODICALS AND ANONYMOUS WORKS CITED IN VOLUME 1

Abh. Akad. Wiss. (München)
Abhandlungen der mathematisch-physikalischen Classe der königlich bayerischen Akademie der Wissenschaften. München. **1–29**, 1829–1922. (Minor variations of title occur.) Titled *Abhandlungen der königlich-bayerischen Akademie der Wissenschaften Math.-Phys. Kl.*, **30–32**, 1924–1928. Nov. ser., **1**→, 1929→, titled ***Abh. Bayer. Akad. Wiss.***, *Abhandlungen der bayerischen Akademie der Wissenschaften, Mathematisch-naturwissenschaftliche Abteilung.*

Abh. Bayer. Akad. Wiss.
Cf. *Abh. Akad. Wiss.* (*München*).

Abh. Böhm. Ges. Wiss.
Abhandlungen der böhmischen Gesellschaft der Wissenschaften in Prag. Prague. Ser. 1, **1785–1786**, 1785–1786; **3–4**, 1788–1789. Ser. 2, titled *Neues* [*Neuere*] *Abhandlungen der k.* (**3**, 1798, *königlichen*) *böhmischen Gesellschaft der Wissenschaften*, **1–3**, 1790–1798. Ser. 3, titled *Abhandlungen der königlichen böhmischen Gesellschaft der Wissenschaften*, **1–8**, 1804–1824. Nov. ser. [= Ser. 4], **1–5**, 1827–1837. Ser. 5, **1–14**, 1838–1875. Ser. 6, titled *Abhandlungen der königlichen-Böhmischen Gesellschaft der Wissenschaften* (*Math.-Nat. Kl.*) / *Pojednáni královské české Společnosti Nauk*, **1–12** (= **1867–1884**), 1868–1885. Ser. 7, titled *Abhandlungen der königlichen-Böhmischen Gesellschaft der Wissenschaften* (*Math.-Nat. Kl.* / *Rozpravy královské české Společnosti Nauk*, **1–4** (= **1885–1891**), 1886–1892. Continued as *Rozpravy královské české Společnosti Nauk. Třídy mathematicko-přírodovědecké.* / *Travaux scientifiques de la Société royale des Sciences de Bohême.* **1–3**, 1928–1929.

Abh. Senckenb. Naturf. Ges.
Abhandlungen herausgegeben von der Senckenbergischen Naturforschenden Gesellschaft. Frankfurt a. M. 1854→.

Abh. Zool.-Bot. Ges. Wien
Abhandlungen der kaiserlich-königlichen zoologisch-botanischen Gesellschaft in Wien. Wien. **1**→, 1901→.

Acta Bot. Acad. Sci. Hung.
Acta botanica Academiae Scientiarum hungaricae. Budapest. **1–28**, 1954–1982; titled *Acta Botanica hungarica*, **29**→, 1983→.

Acta Bot. Fenn.
Acta botanica fennica. Helsingforsiae. **1**→, 1925→.

Acta Bot. Malac.
Acta botanica malacitana. Málaga. **1**→, 1975→. (**1**, 1975, possibly not published until 1976.)

Acta Bot. Neerl.
Acta botanica neerlandica. Amsterdam. **1**→, 1952→.

Acta Fac. Rer. Nat. Carol.
Acta Facultatis Rerum naturalium Universitatis Carolinae. / *Spisy Vydávané Přírodovědeckou Fakultou Karlovy University.* / *Publications de la Faculté des Sciences de l'Université Charles.* Praha. **1–195**, 1923–1949.

Acta Geobot. Hung.
Acta geobotanica hungarica. / *Debreceni Tisza István tudományos Társaság III. Matematikai-természettudományi Osztályának Munkái.* Debrecen. **1–6**, 1936–1949.

Acta Horti Berg.
Acta Horti bergiani. / *Meddelanden från Kungl.* [*Kongl.*] *Svenska Vetenskaps-Akademiens Trädgård, Bergielund.* Stockholm. **1**→, 1890→.

Acta Horti Bot. Univ. Jurjev.
Труды ботаническаго Сада императорскаго Юрьевскаго Университета [*Trudy botaničeskago Sada imperatorskago Jur'evskago Universiteta*]. / *Acta Horti botanici Universitatis imperialis jurjevensis.* Jurjev. **1–14**, 1900–1914.

Acta Horti Bot. Univ. Latv.
Acta Horti botanici Universitatis latviensis. / *Latvijas Universitātes botaniskā darzā raksti.* Riga. **1–13**, 1926–1939; titled *Acta Horti botanici Universitatis.* / *Schriften des Botanischen Gartens der Universität.* / *Universitātes botaniskā darzā raksti*, **14**, 1944.

Acta Horti Gothob.
Acta Horti gothoburgensis [*gotoburgensis*]. / *Meddelanden från Göteborgs botaniska Trädgård.* Göteborg. **1**→, 1924→.

Acta Horti Petrop.
Acta Horti petropolitani. Peterburgi. **1–43**, 1871–1931. (With alternative titles Труды императорскаго С.-Петербургскаго ботаническаго Сада [*Trudy imperatorskago S.-Peterburgskago botaničeskago Sada*], **1–30**(**1**), 1871–1909, and **31**(**1–2**), 1912–1913, **32**, 1912; Труды императорскаго ботаническаго Сада Петра Великаго [*Trudy Imperatorskago botaničeskago Sada Petra Velikago*], **30**(**2**), 1913 & **31**(**3**), 1915, and Труды главнаго ботаноическаго Сада [*Trudy glavnago botaničeskago Sada*], Petrograd, **33–42**(**1**), 1915–1929, **43**(**1**), 1930. Titled Труды ботанического Сада Академии Наук СССР [*Trudy botaničeskogo Sada Akademii Nauk SSSR*]. / *Acta Horti botanici Academiae Scientiarum* ⟨*ante petropolitani*⟩, **42**(**2**), 1931, **43**(**2**)–**44**, 1931.)

Acta Inst. Bot. Acad. Sci. URSS (Ser. 1)
Труды ботанического Института Академии Наук СССР. Серия 1. Флора и Систематика высщих Растений [*Trudy botaničeskogo Instituta Akademii Nauk SSSR. Serija 1. Flora i Sistematika vysšikh Rastenij*]. / *Acta Instituti botanici Academiae Scientiarum URSS. Ser. 1. Flora et Systematica Plantae vasculares.* Leningrad & Mosqua. **1–13**, 1933–1964. (Minor variations of title in later volumes.)

Acta Phytogeogr. Suec.
Acta phytogeographica suecica. Uppsala. **1**→, 1929→.

Acta Phytotax. Geobot. (Kyoto)
Acta phytotaxonomica et geobotanica. 植物分類地理 [*Shokubutsu bunrul chiri*]. Kyōtō. **1**→, 1932→.

Acta Soc. Sci. Fenn.
Acta Societatis Scientiarum fennicae. Helsingforsiae. Ser. 1, **1–50**, 1842–1926. Nov. ser., B, **1**→, 1931→.

Acta Univ. Carol. (Biol.)
Universitas carolina biologica. Pragae. **1–3**, 1954–1957; titled *Acta Universitatis carolinae. 3. Biologica.* Pragae/Praha. **4**→, 1958→. ***Suppl.***, Supplementum. 1961–1967. (Published at irregular intervals as follows: **1960** in 1961; **1962** in 1962; **1964/1** & **1964/2** in 1964; **1965** in 1965; **1966 1/2** (single issue) in 1967.)

Actas Mem. Prim. Congr. Nat. Esp. Zaragoza
Actas y Memorias del primer Congreso de Naturalistas españoles

celebrado en Zaragoza los Dias 7–10 de Octobre de 1908. Zaragoza. 1909.

Actas Soc. Esp. Hist. Nat.
Actas de la Sociedad española de Historia natural. Madrid. 1897–1900. (Earlier issues, from 1872, were incorporated in *Anales de la Sociedad española de Historia natural*.)

Actes Soc. Linn. Bordeaux
Actes de la Société linnéenne de Bordeaux. Bordeaux. **4–106**, 1830–1969. (**1-3**, 1826–1829, titled *Bulletin de l'Histoire naturelle de la Société linnéenne de Bordeaux*; of this, an ed. 2, **1–2** was issued 1830–1845.)

Adansonia
Adansonia; *Recueil périodique d'Observations botaniques*. Paris. **1–12**, 1860–1879. Nov. ser., titled *Adansonia*, **1–20**, 1961–1981; titled *Bulletin du Muséum national d'Histoire naturelle. 4ᵉ Série. Section B. Adansonia. Botanique Phytochimie*, **3→**, 1981→.

Agron. Lusit.
Agronomia lusitana. Sacavém. **1→**, 1939→. (**24(3)→**, 1962→, at Oeiras.)

Allgem. Bot. Zeitschr.
Allgemeine botanische Zeitschrift für Systematik, Floristik, Pflanzengeographie, etc. Karlsruhe. **1–26/33**, 1895–1927.

Amer. Fern Jour.
American Fern Journal. A Quarterly devoted to Ferns. (From **62**, 1972, subtitled *Quarterly Journal of the American Fern Society*.) Port Richmond, New York. **1→**, 1910→. (**3–13**, 1913–1926, at Auburndale, Massachusetts; **14–24**, 1914–1934, at Lancaster, Pennsylvania; **25–28(2)**, 1935–1938, at Lancaster, Pennsylvania and Brattleboro, Vermont; **28(3)–37(1)**, 1938–1947, at Lancaster, Pennsylvania and Brooklyn, New York; **37(2)–49**, 1947–1959, at Lancaster, Pennsylvania; **50–62**, 1960–1972, at Baltimore, Maryland; **63–74**, 1973–1984, at Washington, D.C.; **75→**, 1985→, at Burlington, Vermont.)

Amer. Jour. Bot.
American Journal of Botany. Lancaster, Pennsylvania. 1914→. (**26–36**, 1939–1949, at Burlington, Vermont; **37–58**, 1950–1971, at Baltimore, Maryland; **59**, 1972, at Lawrence, Kansas; **60→**, 1973→, at Columbus, Ohio.)

Amer. Jour. Sci. Arts
American Journal of Science and Arts. New Haven, Connecticut. **2–118**. 1820–1879. (**1**, 1818, and **119→**, 1880→, titled *American Journal of Science*.)

Anais Ci. Nat. (Porto)
Anais de Sciências naturaes. / *Anais de Ciências naturais*. Porto. **1–10**, 1894–1906.

Anais Inst. Sup. Agron. (Lisboa)
Anais do Instituto superior de Agronomia da Universidade technica de Lisboa. Lisboa. **1→**, 1920→.

Anal. Acad. Rep. Pop. Rom.
Cf. *Anal. Acad. Române*

Anal. Acad. Române
Analile Societâţei Academice române. Bucureşti. Ser. 1, **1–11**, 1867–1878. Ser. 2, titled *Analele Academiei române*, **1–41**, 1879–1922. Ser. 3, titled ***Mem. Secţ. Şţi. (Acad. Româna)***, *Academia româna. Memoriile Secţiunei ştiinţifice*, **1-23**, 1922–1948. Titled ***Anal. Acad. Rep. Pop. Rom.***, *Analele Academiei Republicii populare române* [*romîne*], *Sect. Biol., Geogr., Geol.*, **1–3**, 1948–1950. Nov. ser., **1→**, 1950→.

Anal. Ci. Nat.
Cf. *Anal. Hist. Nat.*

Anal. Hist. Nat.
Anales de Historia natural. Madrid. **1–2**, 1799–1800; titled ***Anal. Ci. Nat.***, *Anales de Ciencias naturales*, **3–7**, 1801–1804.

Anal. Inst. Bot. Cavanilles
Anales del Instituto botánico A. J. Cavanilles. Madrid. **10–35**, 1951–1978. (For **1–9** & **36→**, cf. *Anal. Jard. Bot. Madrid*).

Anal. Jard. Bot. Madrid
Anales del Jardin botánico de Madrid. Madrid. **1–9**, 1941–1950, and **36→**, 1979→. (For **10–35**, 1951–1978, cf. *Anal. Inst. Bot. A. J. Cavanilles*).

Anal. Soc. Esp. Hist. Nat.
Anales de la Sociedad española de Historia natural. Madrid. **1–30**, 1872/1873–1901/1902. (Volumes **21–30** also numbered **1–10**. From 1872–1896, this journal also incorporated *Actas Soc. Esp. Hist. Nat.*)

Ann. Bot.
Annals of Botany. London. **1–50**, 1887–1936. Nov. ser., **1→**, 1937→. (Nov. ser. **1-38**, 1937–1974, at Oxford; **39→**, 1975→, at London etc.)

Ann. Bot. Fenn.
Annales botanici fennici. Helsinki. **1→**, 1964→.

Ann. Bot. Soc. Zool.-Bot. Fenn. Vanamo
Suomalaisen eläin- ja kasvitieteellisen Seuran Vanamon Julkaisuja. / Annales Societatis zoologicae-botanicae fennicae 'Vanamo'. Helsinki. **1–15**, 1923–1931. (Botany combined with zoology.) Titled *Suomalaisen eläin- ja kasvitieteellisen Seuran Vanamon kasvitieteellisiä Julkaisuja. / Annales botanici Societatis zoologicae-botanicae fennicae 'Vanamo'*, **1–35**, 1931–1964. (For continuation cf. *Ann. Bot. Fenn.*)

Ann. Bot. (Usteri)
Annalen der Botanick. Herausgegeben von P. Usteri. **1–24**. Zürich. 1791–1800. (**8–24**, 1794–1800, titled *Annalen der Botanik*; **14-24**, 1795–1800, at Leipzig.)

Ann. Gén. Sci. Phys. (Bruxelles)
Annales générales des Sciences physiques. Bruxelles. **1–8**, 1819–1821.

Ann. Missouri Bot. Gard.
Annals of the Missouri Botanical Garden. St Louis, Missouri. **1→**, 1914→.

Ann. Mus. Firenze
Annali del Musèo imperiale di Fisica e Storia naturale di Firenze. Firenze. Ser. 1, **1–2**, 1808–1810. Nov. ser., titled *Annali del r. Museo di Fisica e Storia naturale di Firenze*, **1**, 1866.

Ann. Mus. Goulandris
'Επετηρισ Μουσείου Γουλανδρῆ. Συμβολὴ εἰζ τὴν φυσικὴν ἱστορίαν τῆς 'Ελλάδος καὶ τῆς περιοχῆς τῆς Μεσογείου. Ἔκδοσις τοῦ Μουσείου Γουλανδρῆ φυσικῆς 'Ιστορίας [*Epetēris Mouseion Goulandrē. Sumbolē eis tēn phusikēn istorian tēs Ellados kai tēs periokhēs tēs Mesogeion. Ekdosis tou Mouseion Goulandrē phusikēs Istorias*]. / *Annales Musei Goulandris. Contributiones ad Historiam naturalem Graeciae et Regiones mediterranae a Museo Goulandris Historiae naturalis editae*. Kifissia. **1→**, 1973→.

Ann. Mus. Hist. Nat. (Paris)
Annales du Muséum d'Histoire naturelle. Paris. **1–20**, 1802–1813. **21** (*Index* to **1–20**), 1827.

Ann. Nat. Hist.
Annals of natural History; *or, Magazine of Zoology, Botany and Geology*. London. Ser. 1, **1–5**, 1838–1840; titled *The Annals and Magazine of natural History, including Zoology, Botany and Geology*, **6–20**, 1841–1847. Ser. 2, **1–20**, 1848–1857. Ser. 3, **1–20**, 1858–1867. Ser. 4, **1-20**, 1868–1877. Ser. 5, **1-20**, 1878–1887. Ser. 6, **1–20**, 1888–1897. Ser. 7, **1–20**, 1898–1907. Ser. 8, **1-20**, 1908–1917. Ser. 9, **1-20**, 1918–1927. Ser. 10, **1–20**, 1928–1937. Ser. 11, **1→**, 1938→.

Ann. Naturh. Mus. (Wien)
Annalen des k.k. naturhistorischen Hofmuseums. Wien. **1–31**,

1886–1917; titled *Annalen des naturhistorischen Hofmuseums*, **32**, 1918; titled *Annalen des naturhistorischen Museums in Wien*, **33**→, 1919→. (From **84**→, 1983→, Botany in *Serie B für Botanik und Zoologie*.)

Ann. Sci. Acad. Polyt. Porto
Annaes scientificos [*Anais cientificos*] *da Academia polytechnica do Porto*. Coimbra. **1–14**, 1905–1922; titled ***Anais Fac. Ci. Porto***, *Anais da Faculdade de Sciências do Porto*, **15–46**, 1927–1963; titled *Anais da Faculdade de Ciências. Universidade do Porto*, **47**→, 1964→. (**15**→, 1927→, at Porto.)

Ann. Sci. Nat.
Annales des Sciences naturelles. Paris. Ser. 1, **1–30**, 1824–1833. Ser. 2, **1–20**, 1834–1843. Ser. 3, **1–20**, 1844–1853. Ser. 4, **1–20**, 1854–1863. Ser. 5, **1–20**, 1864–1874. Ser. 6, **1–20**, 1875–1884. Ser. 7, **1–20**, 1885–1895. Ser. 8, **1–20**, 1895–1904. Ser. 9, **1–20**, 1905–1917. Ser. 10, **1–20**, 1919–1938. Ser. 11, **1–20**, 1939–1959. Ser. 12, **1–19**, 1960–1978. Ser. 13, **1**→, 1979→. (Ser. 2–10, *Botanique*. Ser. 11→, *Botanique et Biologie végétale*.)

Ann. Sect. Horti-Viticult. Univ. Sci. Agr. (Budapest)
Cf. *Mitt. Kgl. Ungar. Gartenb.-Lehranst.*

Ann. Soc. Linn. Lyon
Annales de la Société linnéenne de Lyon. Lyon. Ser. 1, **1–4** [**1836–1850/1852**], 1836–1852. Nov. ser., **1–80**, 1853–1937.

Annu. Cons. Jard. Bot. Genève
Annuaire du Conservatoire et du Jardin botaniques de Genève. Genève. **1–21**, [1896–]1897–1922. Continued as *Candollea*.

Annu. Soc. Dendrol. Pologne
Annuaire de la Société dendrologique de la Société botanique de Pologne. / *Rocznik Polskiego Towarzystwa dendrologicznego*. Lwów. 1926–1935; titled *Rocznik Sekcji dendrologicznej Polskiego Towarzystwa Botanicznego*, Warszawa, **1**→, 1946→.

Annu. Univ. Sofia Phys.-Math. 3. (Sci. Nat.)
Годишник на Софийския Университетъ [*Godišnik na Sofijskija Universitet*]. / *Jahrbuch der Universität Sofia, phys.-mathematische Fakultät*. / *Annuaire de l'Université de Sofia, Faculté physico-mathématique. Livre 3*. (*Sciences Nat.*) Sofia. **1**→, 1904→. (From **5**, divided into sections.)

Anzeig. Akad. Wiss. (Wien)
Anzeiger der kaiserliche Akademie der Wissenschaften. Mathematisch-naturwissenschaftliche Classe. Wien. **1–51**, 1864–1914; titled *Anzeiger. Kaiserliche Akademie der Wissenschaften in Wien. Mathematisch-naturwissenschaftliche Klasse*, **52–54**, 1915–1917; titled *Anzeiger. Akademie der Wissenschaften in Wien. Mathematisch-naturwissenschaftliche Klasse*, **55–83**, 1918–1946; titled *Anzeiger. Österreichische Akademie der Wissenschaften. Mathematisch-naturwissenschaftliche Klasse*, **84**→, 1947→.

Arboret. Kórnickie
Arboretum Kórnickie. Poznán. **1**→, 1955→.

Arch. Bot. (Roemer)
Archiv für die Botanik. (*Herausgegeben von D. Johann Jacob Römer*.) Leipzig. **1–3**, 1796–1805. (**1**(**1**): pp. 1–134 in 1796; **1**(**2**): pp. 1–122 in 1797; **1**(**3**): pp. 1–212 in 1798; **2**(**1**): pp. 1–131 in 1800; **2**(**2 & 3**): pp. 133–514 in 1801; **3**(**1/2**): pp. 1–310 in 1803; **3**(**3**): pp. 311–438 in 1805.)

Arch. Naturgesch. (Berlin)
Archiv für Naturgeschichte. Berlin. **1–92**, 1835–1928. Nov. ser., as *Abteilung B* of *Zeitschrift für wissenschaftliche Zoologie*, Leipzig, 1932–1944.

Arch. Soc. Zool.-Bot. Fenn. Vanamo
Suomalaisen eläin- ja kasvitieteellisen Seuran Vanamon Tiedonannot ja Pöytäkirjat (from **6**, 1951, titled *Suomalaisen eläin- ja kasvitieteellisen Seuran Vanamon Tiedonannot*). / *Archivum Societatis zoologicae-botanicae fennicae 'Vanamo'*. Helsinki. **1**→, 1946→.

Ark. Bot.
Arkiv för Botanik. Stockholm. Ser. 1, **1–33**, 1903–1949. Ser. 2, **1–7**, 1949–1974. (Ser. 1, **4–13**, 1905–1914, at Uppsala & Stockholm; **14–33**, 1915–1948, at Stockholm. Ser. 2, **1–3**, 1949–1956, & **6–7**, 1964–1974, at Stockholm; Ser. 2, **4–5**, 1957–1963, at Stockholm, Göteborg & Uppsala.)

Arx. Secc. Ci. Inst. Est. Catalans
Arxivs de la Secció de Ciéncies Institut d'Estudis Catalans. Barcelona. **1–12**, 1911–1924; **13**→, 1947→ (suspended 1924–1947). (Cf. *Cat. Périod. Biblioth. Cons. Jard. Bot. Genève Suppl. 1980–1987*: 14 (1988).)

Atti Accad. Agiati
Atti dell'imperiali regia Accademia di Scienze, Lettere ed Arti degli Agiati di Rovereto. Rovereto. **1**–?, 1826–?. Ser. 2, **1–12**, 1883–1894. Ser. 3, **1–18**, 1895–1912. Ser. 4, titled *Atti dell'i.r. Accademia roveretena di Scienze, Lettere ed Arti degli Agiati*, **1**→, 1913→.

Atti Congr. Int. Bot. Firenze
Atti del Congresso internazionale botanico tenuto in Firenze nel Mese di Maggio 1874. Pubblicati per Cura della R. *Società toscana di Orticultura*. Firenze. 1876.

Atti Riun. Sci. Ital.
Atti della Riunione degli Scienziati italiani. Pisa, etc. Ed. 1. **1–11**, 1840–1875. With minor variations of title and different places of publication for each volume. Ed. 2, **1**. 1840.

Beih. Bot. Centr.
Beihefte zum botanischen Centralblatt. Cassel. **1–62**, 1891–1944. (**1–11**, 1891–1902, at Cassel; **12–17**, 1903–1904, at Jena; **18–19**, 1905–1906, at Leipzig; **20–62**, 1906–1944, at Dresden.)

Beitr. Pfl. Russ. Reich.
Матеріалы къ ближайшему Познанію Прозябаемости россійской Имперіи [*Materialy k bližajšemu Poznaniju Prozjabaemosti rossijskoj Imperii*]. / *Beiträge zur Pflanzenkunde des russischen Reiches*. St. Petersburg. **1–11**, 1844–1859.

Belmontia
Belmontia. Miscellaneous Publications in Botany. I. Taxonomy. Wageningen. **1–14** (1–95), 1957–1973. (Other sections: *II. Ecology*, **1–17** (1–125), 1957–1972; *III. Horticulture*, **1–9** (1–54), 1957–1970; *IV. Incidental*, **1-13** (1–25), 1957–1970.) New Series, incorporating all four sections, **1**→ (1→), 1973→.

Ber. Bayer. Bot. Ges.
Berichte der bayerischen botanischen Gesellschaft zur Erforschung der heimischen Flora. München. **1**→, 1891→.

Ber. Deutsch. Bot. Ges.
Berichte der deutschen botanischen Gesellschaft. Berlin. **1–100**, 1883–1987. (**59–61**, 1941–1944, at Jena; **62–100**, 1949–1987, at Stuttgart.) Titled *Botanica Acta. Berichte der deutschen botanischen Gesellschaft*. / *Journal of the German botanical Society*, Stuttgart & New York, **101**(**1**)→, 1988→. (The holding at Genève is said to commence from **101**(**0**) [*sic*], 1987; this has not been seen. Cf. *Cat. Périod. Bibl. Cons. Jard. Bot. Genève Suppl. 1980–1987*: 24 (1988).)

Bergens Mus. Aarb.
Bergens Museums Aarbog [*Årbog*, *Aarbok*, *Årbok*]. *Afhandlinger og Aarsberetning*. Bergen. 1883–1948. From **1917–1918**→, 1918→, ***Naturv. Rekke***, *Naturvidenskabelig* [*Naturvitenskapelik*] *Række* [*Rekke*] is added to the title.) Titled *Universitet i Bergen. Årbok. Naturvitenskapelig Rekke*, 1948–1959; titled *Årbok for Universitet i Bergen, Matematisk-naturvitenskapelig Serie*, 1960→.

Bergens Mus. Skr.
Bergens Museums Skrifter. Bergen. **1–22**, 1878–1943.

Ber. Saratow. Naturforscherges.
Известия Саратовского Общества естествоиспытателей

[*Izvěstija Saratovskogo Obščestva estestvoispytatelej*]. / *Bericht der saratower Naturforschergesellschaft*. Saratow. **1–3**, 1924–1929.

Ber. Schweiz. Bot. Ges.
Berichte der schweizerischen botanischen Gesellschaft. / Bulletin de la Société botanique suisse. Basel & Genf. 1891–1980. (**1** & **2**, 1891–1892, at Basel & Genf; **3–18**, 1893–1909, at Bern; **19–26/29**, 1910–1920, at Zürich; **30/31**, 1922, at Bern; **32–47**, 1923–1947, at Bern & Zürich; **48–50**, 1938–1940, at Bern; **50A**, 1940, at Genève; **51–53**, **53A**, **54–69**, 1941–1959, at Bern; **70–81**, 1960–1972, at Wabern; **82–90**, 1973–1980, at Wetzikon.) Titled ***Bot. Helvet.***, Botanica helvetica, **91**→, 1981→. (**91–93**, 1981–1983, at Teufen; **94**→, 1984→, at Basel.)

Biblioth. Bot. (Stuttgart)
Bibliotheca botanica. Abhandlungen aus dem Gesammtgebiete der Botanik. Cassel. **1**→, 1886→. (**11**→, 1889→, at Stuttgart.)

Biblioth. Univ. Genève
Bibliothèque universelle des Sciences, Belles-lettres, et Arts, faisant suite à la Bibliothèque britannique. Partie des Sciences. Genève. Ser. 1, **1–60**, 1816–1835. Ser. 2, titled *Bibliothèque universelle de Genève*, **1–60**, 1836–1845.

Billotia
Billotia; *ou, Notes de Botanique*. Besançon. **1**, 1864–1869.

Biol. Écol. Médit.
(*Revue de*) *Biologie et Écologie méditerranéenne. Annales de l'Université de Provence*. ('*Revue de*' only in running heads and title pages, not on wrappers.) Marseille. **1–10**, 1974–1983.

Biol. Glasn. (Zagreb)
Glasnik biološke Sekcije. Hrvatsko prirodoslovno Društvo. / Periodicum biologorum. Societas Scientiarum naturalium croatica. Zagreb. **1–7**, 1947–1955; titled *Biološki Glasnik. Hrvatsko prirodoslovno Društvo. / Periodicorum biologorum. Societas Scientiarum naturalium croatica*, **8–21**, 1955–1969; titled *Periodicum biologorum*, **72**→, 1970→.

Biol. Nauki
Биологические Науки. Научные Доклады Высшеи Школы [*Biologičeskie Nauki. Naučnye Doklady Vysšei Školy*]. Moskva. 1958→.

Biosistematika
Biosistematika. (*Acta biologica iugoslavica Seriya G*.) Belgrade. **1**→, 1975→.

Bjull. Gosud. Nikitsk. Bot. Sada
Бюллетень государственного Никитского Опытного ботанического Сада [*Bjulleten' gosudarstvennogo Nikitskogo opytnogo Botaničeskogo Sada*]. / Бюуллетен' государственнуй Никитский Опытный ботанический Сад [*Bjulleten' gosudarstvennyj Nikitskij opytnyj botaničeskij Sad*]. / *Bulletin. Government botanical Garden Nikita*. Yalta. **1–19**, 1929–1938; titled Бюуллетен' государственного Никитского ботанического Сада [*Bjulleten' gosudarstvennogo Nikitskogo botaničeskogo Sada*], **1**(**7**)→, 1968→.

Bjull. Mosk. Obšč. Isp. Prir., Otd. Biol.
Cf. *Bull. Soc. Nat. Moscou*.

Bjull. Obšč. Estestv. Voronež. Univ.
Бюллетень Общества естествоиспытателей при Воронежскомь государственномь Универсйтеть [*Bjulleten' Obščestva Estestvoispytatelej pri Voronežskom gosudarstvennom' Universjtet'*]. / *Bulletin de la Société des Naturalistes de Voronèje*. Voronež. **1–2**(**2**), 1925–1928 & **3–9**, 1939–1955. (**2**(**3/4**), 1929, titled ***Bull. Soc. Nat. Voronèje***, Бюллетень Воронежского Общества Естествоиспытателей при государственномь Университеть [*Bjulleten' Voronežskogo Obščestva Estestvoispytatelej pri gosudarstvennom' Universjtet'*]. / *Bulletin de la Société des Naturalistes de Voronèje*.)

Blätt. Staudenk.
Blätter für Staudenkunde. Berlin. **1–2**, 1937–1940.

Blumea
Blumea. Tijdschrift voor de Systematiek en de Geografie der Planten. (*A Journal of Plant Taxonomy and Plant Geography*.) Leiden. **1**→, 1934→.

Blyttia
Blyttia. Oslo. **1**→, 1943→.

Boissiera
Boissiera. Mémoires du Conservatoire de Botanique et de l'Institut de Botanique systématique de l'Université de Genève. Genève. **1–7**, 1936–1948; titled *Boissiera* (without subtitle), **8–13**, 1949–1967; titled *Boissiera. Mémoires des Conservatoire et Jardin botaniques de la Ville de Genève*, **14–30**, 1969–1979; titled *Boissiera. Mémoires de Botanique systématique*, **31**→, 1980→. (Supplement of *Candollea*.)

Bol. Acad. Nac. Ci.
Boletín de la Academia nacional de Ciencias. Córdoba (Argentina). **1**→, 1875→.

Bol. Inst. Estud. Astur. (Supl. Ci.)
Boletín del Instituto de Estudios asturianos. Suplemento de Ciencias. Oviedo. **1–24**, 1960–1978 (**19–24**, 1974–1978, titled *Suplemento de Ciéncias del Boletín del Instituto de Estudios asturianos*); titled *Boletín de Ciéncias de la Naturaleza. Instituto de Estudios asturianos*, 1979→. (Cf. Castroviejo et al., *Fl. Iberica* **2**: 752 (1990).)

Bol. Soc. Aragon. Ci. Nat.
Boletín de la Sociedad aragonesa de Ciencias naturales. Zaragoza. **1–17**, 1902–1918; titled ***Bol. Soc. Ibér. Ci. Nat.***, *Boletín de la Sociedad ibérica de Ciencias naturales*, **18–33**, 1919–1934.

Bol. Soc. Brot.
Boletim da Sociedade broteriana. Coimbra. Ser. 1, **1–28**, 1880–1920. Ser. 2, **1**→, 1922→.

Bol. Soc. Esp. Hist. Nat.
Boletín de la Sociedad española de Historia natural. Madrid. **1-47**, 1901–1949; **48**→, 1950→ (divided into *Actas*, ***Secc. Biol.***, *Seccion biologica* and *Seccion geologica*). From **4–31**, 1904–1931, and **38**→, 1940→, titled *Boletín de la Real Sociedad española de Historia natural*.

Boll. Orto Bot. Napoli
Bollettino dell'Orto botanico della Regia Università di Napoli. Napoli. **1–17**, 1899–1947; titled *Delpinoa*, **1–11**, 1948–1958; nov. ser. **1**→, 1959→.

Boll. Orto Bot. Palermo
Bollettino delle Reale Orto botanico e Giardino coloniale di Palermo. Palermo. **1–11**, 1897–1912; nov. ser. **1–2**, 1914–1921.

Boll. Soc. Adr. Sci. Nat. Trieste
Bollettino della Società adriatica di Scienze naturali di Trieste. Trieste. **1**→, 1875→. (**32–45**, 1934–1950, at Udine; **46–49**, 1951–1958, at Rocca San Casciano.)

Boll. Soc. Sarda Sci. Nat.
Bollettino della Societá sarda di Scienze naturali. Sassari. **1**→, 1967→.

Bonplandia
Bonplandia. Zeitschrift für die gesammte Botanik. Hannover, London, New York & Paris. **1–10**, 1853–1862.

Bot. Arch. (Berlin)
Botanisches Archiv. Zeitschrift für die gesammte Botanik. Berlin & Königsberg. **1–45**, 1922–1944. (**21–45**, 1928–1944, at Leipzig as ***Bot. Arch. (Leipzig)***, *Botanisches Archiv. Zeitschrift für die gesammte Botanik*.)

Bot. Arch. (Königsberg)
Cf. *Bot. Arch.* (*Berlin*).

Bot. Arch. (Leipzig)
Cf. *Bot. Arch.* (*Berlin*).

Bot. Centr.
Botanisches Centralblatt. Cassel. **1–173**, 1880–1938; titled *Botanisches Zentralblatt*, **174–179**, 1938–1945. (**101–179**, 1906–1945, at Jena.)

Bot. Chron.
Βοτανικὰ Χρονικά [*Botanika Chronika*]. Patras. **1**→, 1981→.

Bot. Gaz. (London)
The botanical Gazette, a Journal of the Progress of British Botany and the contemporary Literature of the Science, by A. Henfrey. London. **1–3**, 1849–1851. (**1** in 1849; **2** in 1850; **3** in 1851.)

Bot. Helvet.
Cf. *Ber. Schweiz. Bot. Ges.*

Bot. Jahrb.
Botanische Jahrbücher für Systematik, Pflanzengeschichte und Pflanzengeographie. Leipzig. **1**→, 1880→. (**69**→, 1938→, at Stuttgart.)

Bot. Jour. Linn. Soc.
Cf. *Jour. Linn. Soc. London (Bot.)*

Bot. Közl.
Cf. *Növ. Közl.*

Bot. Macar.
Botánica Macaronésica. Las Palmas, Gran Canaria. **1**→, 1976→. (**1** (Jan. 1976) in 1976; **2** (Oct. 1976) in 1976; **3** (Jul. 1977) in 1978; **4** (Oct. 1977) in 1979; **5** (Nov. 1978) in 1980; **6** (Dec. 1978) in 1980; **7** (Nov. 1980) in 1982; **8/9** (1981) in Sept. 1982; **10** (1982) in 1984; **11** (1983) in 1984; **12/13** (1984) in 1986; **14** (1985) in 1986; **15** (1986) in 1986; nov. ser. **16** (no year in title) in 1988. Publication was often apparently not effective until considerably later than the year stated on the cover. Dates given in parentheses are those printed as part of the title on the outside front cover; others are (**1–7**) dates of receipt at E; (**8/9**→) date of publication as stated on inside front cover. **8/9** and **12/13** were single issues given double volume numbers.)

Bot. Mag.
The botanical Magazine (from **15**, 1801, *Curtis's botanical Magazine*); *or, Flower-Garden displayed: in which the most ornamental foreign Plants, cultivated in the open Ground, the Greenhouse, and the Stove, are accurately represented in their natural Colours*... London. **1–184**, 1787–1983. (For publication dates up to **180**, 1974, cf. F. A. Stafleu & R. S. Cowan, *Taxonomic Literature* ed. 2, **1**: 578–584 (1976); **181(1–2)** in 1976; **181(3–4)** in 1977; **182(1–2)** in 1978; **182(3–4)** in 1979; **183(1)** in 1980; **183(2–4)** in 1981; **184(1–2)** in 1982; **183(3–4)** in 1983. Continuous volume numbers are used in citations and series are ignored.) Continued as *The Kew Magazine, incorporating Curtis's Botanical Magazine.* Kew. **1**→, 1984→. (**1** in 1984; **2** in 1985; **3** in 1986; **4** in 1987; **5** in 1988; **6** in 1989; **7** in 1990; **8** in 1991.)

Bot. Mag. (Tokyo)
The botanical Magazine. / 植物學雜誌 [*Shokubutsu-gaku zasshi*]. Tokyo. **1–19**, 1887–1905; titled *The botanical Magazine published by the Tokyo botanical Society*, **20–45**, 1906–1931; titled *The botanical Magazine published by the botanical Society of Japan*, **46**→, 1932→. (**68**→, 1955→, also titled *The botanical Magazine, Tokyo.*) The Japanese title has remained unchanged throughout.

Bot. Not.
Botaniska Notiser. Lund. **1–133**, 1839–1980. (Year used as volume number until **108**→, 1955→.) Continued as ***Nordic Jour. Bot.***, *Nordic Journal of Botany.* Copenhagen. **1**→, 1981→.

Bot. Reg.
The botanical Register; consisting of coloured Figures of exotic Plants cultivated in British Gardens; with their History and Mode of Treatment. London. **1–14**, 1815–1829; titled *Edwards' Botanical Register; or, Flower Garden and Shrubbery*, **15–33**, 1829–1847. (For dates of publication cf. F. A. Stafleu & R. S. Cowan, *Taxonomic Literature* ed. 2, **1**: 724–725 (1976).)

Bot. Tidsskr.
Botanisk Tidsskrift. Kjøbenhavn. **1–75**, 1868–1981. Continued as ***Nordic Jour. Bot.***, *Nordic Journal of Botany.* Copenhagen. **1**→, 1981→.

Bot. Zeit.
Botanische Zeitung. Berlin. **1–68**, 1843–1910. (**14–68**, 1856–1910, at Leipzig. From **51**, 1893, 2 *Abteilung* is added to the title.)

Bot. Zeit. (Regensburg)
Botanische Zeitung welche Recensionen, Abhandlungen, Aufsätze, Neuigkeiten und Nachrichten, die Botanik betreffend, enthält. Regensburg. **1–6**, 1802–1807. (**1–5**, 1802–1805, also issued under the (part) title *Allgemeine botanische Bibliothek des neunzehnten Jahrhunderts*, Erlangen.)

Bot. Žur.
Журнал русскаго ботаническаго Общества при императорской Академии Наук [*Žurnal russkago botaničeskago Obščestva pri imperatorskoj Akademii Nauk*]. / *Journal de la Société botanique de Russie.* Petrograd. **1**, 1916–1917; titled Журнал русскаго [русского] ботаническаго [ботанического] Общества при Академии Наук [*Žurnal russkago [russkogo] botaničeskago [botaničeskogo] Obščestva pri Akademii Nauk*]. / *Journal de la Société botanique de Russie*, **2–9**, 1917–1924 ('*russkogo*' and '*botaničeskogo*' in **3–9**); titled Журнал русского ботанического Общества при Академии Наук СССР [*Žurnal russkogo botaničeskogo Obščestva pri Akademii Nauk SSSR*], Moskva & Leningrad, **10–13**, 1925–1928; titled Журнал русского ботанического Общества [*Žurnal russkogo botaničeskogo Obščestva*], **14–16**, 1930–1931; titled Ботанический Журнал СССР [*Botaničeskij Žurnal SSSR*]. / *Journal botanique de l'URSS*, **17-32**, 1932–1947; titled Ботанический Журнал [*Botaničeskij Žurnal*], **33**→, 1948→. (**33–51**, 1948–1966, at Moscow & Leningrad; **52**→, 1967→, at Leningrad.)

Bothalia
Bothalia. A Record of Contributions from the National Herbarium, Union of South Africa. Pretoria. **1**→, 1921→. (**7** & **8**, 1958–1965, subtitled *A Record ... Republic of South Africa*; subtitle omitted from **9**, 1966–1968, onwards.)

Bradleya
Bradleya. Yearbook of the British Cactus and Succulent Society. Botley. **1**→, 1983→. (**6–7**, 1988–1989, at Summerseat Bury; **8**→, 1990→, at Kew.)

Brit. Fern Gaz.
British Fern Gazette. Kendal. **1–10**, 1959–1973; titled ***Fern Gaz.***, *The Fern Gazette. The Journal of the British pteridological Society*, **11**→, 1974→. (**1–4**, 1909–1921, at Kendal; **5–8**, 1923–1958, at Reading; **9**→, 1959→, at London.)

Brittonia
Brittonia. New York. **1**→, 1931→.

Bul. Acad. Stud. Agron. Cluj
Buletinul Academiei de înalte Studii agronomice din Cluj. Cluj. **1–6**, 1930–1937; titled ***Bul. Fac. Agron. Cluj***, *Buletinul Facultății de Agronomie din Cluj*, 1938–1943; titled *Analele Facultății de Agronomie din Cluj*, 1944–1947; titled *Anuar lucrărilor științifice. Institutul agronomic 'Dr Petru Groza'*, 1957→.

Bul. Grăd. Bot. Univ. Cluj
Buletinul de Informații al Grădinii botanice și al Muzeului botanic dela Universitatea din Cluj. / *Bulletin d'Informations du Jardin et du Musée botanique de l'Université de Cluj*, Roumanie. Cluj. **1–5**, 1921–1925; titled *Buletinul Grădinii botanice și al Muzeului botanic dela Uni-*

versitatea din Cluj. / Bulletin du Jardin et du Musée botaniques de l'Université de Cluj, Roumanie, **6–28**, 1926–1948. With minor changes of title.

Bull. Acad. Int. Géogr. Bot. (Le Mans)
Cf. *Monde Pl.*

Bull. Acad. Polon. Sci. Lett., Cl. Math. Nat. B
Cf. *Bull. Int. Acad. Sci. Cracovie.*

Bull. Acad. Sci. URSS
Известия Академии Наук СССР [*Izvestija Akademii Nauk SSSR*]. / *Bulletin de l'Académie des Sciences de l'URSS.* Leningrad. 1928→. (In several series with different titles.)

Bull. Alp. Gard. Soc.
Bulletin of the Alpine Garden Society. London. **1–2(1)**, 1930–1933; titled ***Quart. Bull. Alp. Gard. Soc.***, *Quarterly Bulletin of the Alpine Garden Society*, **2(2)→**, 1933→. (**2(2)–15**, 1933–1947, at London; **16–27**, 1948–1959, at Farnborough; **28**, 1960, at Orpington; **29–39**, 1961–1971, at London; **40→**, 1972→, at Woking.)

Bull. Appl. Bot. Pl.-Breed. (Leningrad)
Труды по прикладной Ботанике и Селекции [*Trudy po prikladnoj Botanike i Selekcii*]. / *Bulletin of applied Botany and Selection* (later, ... *and Plant Breeding.*) Petrograd. **11(5/6)–17(2)**, 1918–1927; titled Труды по прикладной Ботанике, Генетике и Селекции [*Trudy po prikladnoj Botanike, Genetike i Selekcii*]. / *Bulletin of applied Botany, of Genetics, and Plant-breeding*, Moskva & Leningrad, **17(3)–27**, 1927–1931; **28(3)→**, 1950→. (Several concurrent series.)

Bull. Brit. Mus. (Bot.)
Bulletin of the British Museum (*Natural History*). *Botany.* London. **1→**, 1951→.

Bull. Fan Mem. Inst. Biol. Bot. (Peking)
Bulletin of the Fan Memorial Institute of Biology. Botany. Peking. **1-11**, 1929–1941. Nov. ser. **1**, 1943–1948.

Bull. Géogr. Bot. (Le Mans)
Cf. *Monde Pl.*

Bull. Herb. Boiss.
Bulletin de l'Herbier Boissier. Genève. Ser. 1, **1–8**, 1893–1900. Ser. 2, **1-8**, 1900–1909. (Ser. 1, **8**, 1900, only, at Genève & Bâle.)

Bull. Inst. Jard. Bot. Univ. Beograd
Гласник ботаничког Завода и Баште Универзитета у Београду [*Glasnik botaničkog Zavoda i Bašte Univerziteta u Beogradu*]. / *Bulletin de l'Institut et du Jardin botaniques de l'Université de Belgrade.* Beograd. **1–4**, 1928–1937.

Bull. Int. Acad. Sci. Cracovie
Bulletin international de l'Académie des Sciences de Cracovie. Classe des Sciences mathématiques et naturelles. Série B: Sciences naturelles. / Anzeiger der Akademie der Wissenschaften in Krakau. Mathematisch-naturwissenschaftliche Klasse. Reihe B: Biologische Wissenschaften. Cracovie. 1889–1919. Dates are used as vol. nos. (*Classe ... naturelles* and *Mathematisch-naturwissenschaftliche Klasse* only from 1902–1919; *Série/Reihe B* ... only from 1911– 1919.) Continued as ***Bull. Acad. Polon. Sci. Lett., Cl. Math. Nat. B***, *Bulletin international de l'Académie polonaise des Sciences et des Lettres*, Cracow, 1920–1953. (To the above title is added *Classe des Sciences mathématiques et naturelles. Série B. Sciences naturelles*, 1920–1928; *Classe ... Série B 1. Botanique*, 1928–1953.)

Bull. Jard. Bot. Bruxelles
Bulletin du Jardin botanique de l'État à Bruxelles. Bruxelles. **1–36**, 1920–1966. (**15–36**, 1938–1966, with alternative title *Bulletin van den* [*de*] *Rijksplantentuin.* Brussel); titled ***Bull. Jard. Bot. Nat. Belg.***, *Bulletin du Jardin botanique national de Belgique. / Bulletin van de nationale Plantentuin van Belgie*, **37→**, 1967→.

Bull. Jard. Bot. Kieff
Вісник Київського ботанічного Саду [*Visnyk Kyjivs'kogo botaničnogo Sadu*]. / Известия Киевъского ботанического Сада [*Izvestija Kievs'kogo botaničeskogo Sada*]. / *Bulletin du Jardin botanique de Kieff.* Kiev. **1–17**, 1924–1934; titled Вісник Київського ботанічного Саду імену Акад. 'О. В. Фоміна' [*Visnyk Kyjivs'kogo botaničnogo Sadu imeny Akad. 'O. V. Fomina'*]. / Известия Киевського ботанического Сада имени Акад. 'А. В. Фомин' [*Izvestija Kievs'kogo botaničeskogo Sada imeni Akad. 'A. V. Fomin'*], **18**, 1949; titled Труди ботанічного Саду ім. Акад. 'О. В. Фомин' [*Trudy botaničnogo Sadu im. Akad. 'O. V. Fomin'*], **19** & **20**, 1948–1949; titled Труді ботанічного Саду Академії Наук Юкрайінськой РСР [*Trudy botaničnogo Sadu Akademii Nauk Ukrajins'koji RSR*], **1**, 1949; titled Труды ботанического Сада [*Trudy botaničeskogo Sada*], **2–7**, 1953–1960; titled Праці Центрального республіканського ботанічного Саду [*Praci Central'nogo respublikans'kogo botaničnogo Sadu*], **8→**, 1962→.

Bull. Jard. Bot. Nat. Belg.
Cf. *Bull. Jard. Bot. Bruxelles.*

Bull. Jard. Bot. Pétersb.
Bulletin du Jardin impérial botanique de St.-Petersburg. / Извѣстія императорскаго С.-Петербургскаго ботаническагѡ Сада [*Izvestija imperatorskago S.-Peterburgskago botaničeskago Sada*]. S.-Peterburg. **1–12**, 1901–1912; titled *Bulletin du Jardin impérial botanique de Pierre le Grand.* / Извѣстия императорскаго ботаническаго Сада Петра Великаго [*Izvestija imperatorskago botaničeskago Sada Petra Velikago*], **13–17**, 1912–1917; titled ***Bull. Jard. Bot. URSS***, *Bulletin du principal Jardin botanique de la République Russe.* / Известия главного ботанического Сада РСФСР [*Izvestija glavnogo botaničeskogo Sada RSFSR*], **18–22**, 1918–1923; titled *Bulletin du Jardin botanique principal de la République Russe.* / [Russian title as before], **23–24**, 1924–1925; titled *Bulletin du Jardin botanique principal de l'URSS.* / Известия главного ботанического Сада СССР [*Izvestija glavnogo botaničeskogo Sada SSSR*], **25–29**, 1926–1930; titled *Bulletin du Jardin botanique de l'Académie des Sciences de l'URSS.* / Известия ботанического Сада Академии Наук СССР [*Izvestija botaničeskogo Sada Akademii Nauk SSSR*], **30**, 1932; titled Советскаи[й]а Ботаника [*Sovietskai[j]a Botanika*], Moscow & Leningrad, 1933–1947.

Bull. Jard. Bot. URSS
Cf. *Bull. Jard. Bot. Pétersb.*

Bull. Phys.-Math. Acad. (Pétersb.)
Bulletin de la Classe physico-mathématique de l'Académie impériale des Sciences de Saint-Pétersbourg. St Petersburg & Leipzig. Ser. 2, **1–17**, 1843–1859; titled *Bulletin de l'Académie impériale des Sciences de Saint Pétersbourg*, Ser. 3, **1–32**, 1860–1888; Ser. 4, **1–4** (also numbered **33–36**), 1890–1894; titled Известия императорской Академии Наук [*Izvestija imperatorskoj Akademii Nauk*]. / *Bulletin de l'Académie impériale des Sciences* (*de Saint Pétersbourg*), Ser. 5, **1–15**, 1894–1906/1907; Ser. 6, **1–11(4)**, 1907–1917; titled Известия Аладемии Наук [*Izvestija Akademii Nauk*]. / *Bulletin de l'Académie des Sciences*, Petrograd, **11(5)–11(11)**, 1917; titled Известия Российск. Академии Наук [*Izvestija Rossijsk. Akademii Nauk*]. / *Bulletin de l'Académie des Sciences de Russie*, **11(12)–19(9/11)**, 1917–1925; titled Известия Академии Наук СССР [*Izvestija Akademii Nauk SSSR*]. / *Bulletin de l'Académie des Sciences de l'URSS*, **19(12)-21**, 1925–1927; titled Известия Академии Наук СССР. Сер. 7. Отделение физико-математических Наук [*Izvestija Akademii Nauk SSSR. Ser.* 7. *Otdelenie fiziko-matematičeskikh*

Nauk]. / *Bulletin de l'Académie des Sciences de l'URSS. Classe des Sciences physico-mathématiques*, Moscow & Leningrad, 1928–1930; titled Известия Академии Наук СССР. Сер. 7. Отделение математических и естественных Наук [*Izvestija Akademii Nauk SSSR. Ser. 7. Otdelenie matematičeskikh i estestvennykh Nauk*]. / *Bulletin de l'Académie des Sciences de l'URSS. Classe des Sciences mathématiques et naturelles*, 1931–1935; titled Известия Академии Наук СССР. Отделение математических и естественных Наук. Серия биологическая [*Izvestija Akademii Nauk SSSR. Otdelenie matematičeskikh i estestvennykh Nauk. Serija biologičeskaja*]. / *Bulletin de l'Académie des Sciences de l'URSS. Classe des Sciences mathématiques et naturelles. Série biologique*, 1936–1938. (French title not in all vols.); titled Известия Академии Наук СССР. Серия биолгическая [*Izvestija Akademii Nauk SSSR. Serija biologičeskaja*]. / *Bulletin de l'Académie des Sciences de l'URSS. Série biologique*, 1939→.

Bull. Res. Counc. Israel
Bulletin of the Research Council of Israel. Jerusalem. **1–11** (from **5**, 1955, onwards *Botany* in *Section D*), 1951–1963. (Sect. D, *Botany*, being continuation of *Palestine Journal of Botany and horticultural Science*.) Titled ***Israel Jour. Bot.***, *Israel Journal of Botany*, **12→**, 1963→.

Bull. Sci. Acad. Imp. Sci. Pétersb.
Bulletin scientifique publié par l'Académie impériale des Sciences de Saint-Pétersbourg. Saint-Pétersbourg & Leipzig. **1–10**, 1836–1842.

Bull. Sci. Soc. Philom. Paris
Bulletin des Sciences, par la Société Philomatique. Paris. **1–3**, 1791–1805; titled *Nouveau Bulletin des Sciences, publié par la Société Philomatique*, **1–3** (= années **1–6**), 1807–1813. Continued under original title, 1814–1824; as *Nouv. Bull.* ..., 1825–1826; suspended 1827–1831; titled *Nouv. Bull.* ..., 1832–1833.

Bull. Soc. Bot. Belg.
Bulletin de la Société royale de Botanique de Belgique. Bruxelles. **1→**, 1862→. (**18–42**, 1879–1905, in two parts: 1. *Mémoires* and 2. *Comptes rendus des Séances*. **51–79**, 1912–1947, also numbered Ser. 2, **1-29**.)

Bull. Soc. Bot. Fr.
Bulletin de la Société botanique de France. Paris. **1→**, 1854→. From 1978/1979 divided into 2 parts: ***Actu. Bot.***, *Actualités botaniques* (from **125→**, 1978→) and ***Lettres Bot.***, *Lettres botaniques* (from **126→**, 1979→.)

Bull. Soc. Bot. Ital.
Bullettino della Società botanica italiana. Firenze. **1892–1926**, 1892–1926. (From 1927, published in *Nuovo Gior. Bot. Ital.*)

Bull. Soc. Dauph. Éch. Pl.
Bulletin de la Société dauphinoise pour l'Échange des Plantes. Grenoble. **1–16**, 1874–1889. Ser. 2, **1–3**, 1890–1892.

Bull. Soc. Hist. Nat. Afr. Nord
Bulletin de la Société d'Histoire naturelle de l'Afrique du Nord. Alger. **1→**, 1909→.

Bull. Soc. Hist. Nat. Toulouse
Bulletin de la Société d'Histoire naturelle de Toulouse. Toulouse. **1→**, 1867→.

Bull. Soc. Linn. Lyon
Bulletin mensuel de la Société linnéenne de Lyon. Lyon. Nos. 1–18, 1883–1884?; titled *Bulletin bi-mensuel de la Société linnéenne de Lyon*, **1–10**, 1922–1931; titled *Bulletin mensuel de la Société linnéenne et des Sociétés botaniques de Lyon, d'Anthropologie et de Biologie de Lyon réunies*, **1→**, 1932→.

Bull. Soc. Nat. Moscou
Bulletin de la Société impériale des Naturalistes de Moscou. Moscow. **1–62**, 1829–1886. Nov. ser., **1–30**, 1887–1917; titled ***Bjull. Mosk. Obšč. Isp. Prir., Otd. Biol.***, Бюллетень [Императорскаго] Московского [Москповскаго] Общества испытателей Природы, Отдел биологический [*Bjulleten' (Imperatorskago) Moskovskogo (Moskovskago) Obščestva ispytatelej Prirody. Otdel biologičeskij*]. / *Bulletin de la Société Impériale des Naturalistes de Moscou. Section biologique*. Nov. ser. **31→**, 1922→. (Nov. ser. **52(5)→**, 1947→, with Russian title only; from **46(3)→**, 1991→ with English subtitle *Bulletin of Moscow Society of Naturalists. Biological Series*.)

Bull. Soc. Nat. Voronèje
Cf. *Bjull. Obšč. Estestv. Voronež. Univ.*

Bull. Soc. Neuchâtel Sci. Nat.
Cf. *Bull. Soc. Sci. Nat. Neuchâtel*.

Bull. Soc. Sci. Dauph.
Bulletin de la Société de Statistique, des Sciences naturelles, et des Arts industriels du Département de l'Isère. Grenoble. Ser. 1, **1–4**, 1838 or 1840–1846 or 1848. **5–11** (ser. 2, **1–7**), 1851–1864. **12–26** (ser. 3, **1–15**), 1867–1890. **27–41** (ser. 4, **1–15**), 1892–1920. Titled ***Bull. Soc. Sci. Isère***, *Bulletin de la Société scientifique de l'Isère, ancienne Société de Statistique des Sciences naturelles et des Arts industriels*. **42** (ser. 5, **1**)–**45** (ser. 5, **4**), 1921–1924. Titled ***Bull. Soc. Sci. Dauph.***, *Bulletin de la Société scientifique du Dauphiné, ancienne Société de Statistique des Sciences naturelles et des Arts industriels du Département de l'Isère*, **46** (ser. 5, **5**)–**60** (ser. 5, **18**), 1925–1944; **61→** (ser. 6, **1→**), 1945→.

Bull. Soc. Sci. Isère
Cf. *Bull. Soc. Sci. Dauph.*

Bull. Soc. Sci. Nat. Neuchâtel
Bulletin de la Société des Sciences naturelles de Neuchâtel. Neuchâtel. **1–25**, 1847–1897; titled ***Bull. Soc. Neuchâtel Sci. Nat.***, *Bulletin. Société neuchâteloise des Sciences naturelles*, **26→**, 1898→.

Bull. Soc. Sci. Phys. Algérie
Bulletin de la Société des Sciences physiques, naturelles et climatologiques de l'Algérie. Alger. 1864–1889; titled *Bulletin de la Société climatologique algérienne*, 1890→.

Bull. Soc. Vaud. Sci. Nat.
Bulletin de la Société vaudoise des Sciences naturelles. Lausanne. **1→**, 1842→.

Bull. Torrey Bot. Club
Bulletin of the Torrey botanical Club. New York. **1→**, 1870→. (**73–78**, 1946–1951, titled *Bulletin of the Torrey botanical Club and Torreya*. **20–54**, 1893–1927, at Lancaster, Pennsylvania; **55–66**, 1928–1939, at Menasha, Wisconsin; **67**, 1940, at Burlington, Vermont.)

Bull. Turkestan Sect. Russ. Geogr. Soc.
Извѣстія Туркестанскаго Отдѣла императорскаго географическаго Общества [*Izvěstija Turkestanskago Otděla imperatorskago geografičeskago Obščestva*]. Taškent'. **1–12**, 1898–1916; titled Извѣстия Туркестанского Отдѣла Русского географического Общества [*Izvestija Turkestanskogo Otděla Russkogo geografičeskogo Obščestva*]. / *Journal of the Turkestan Branch of the Russian Geographical Society*. / *Bulletin de la Section du Turkestan de la Société russe de Géographie*, **13–17**, 1917–1924; titled Известия Средне-Азиатского Отдела Государственного русского географического Общества [*Izvestija Sredne-Aziatskogo Otdela Gosudarstvennogo russkogo geografičeskogo Obščestva*], **18**, 1929; titled Известия Средне-Азиатского Географического Общества [*Izvestija Sredne-Aziatskogo Geografičeskogo Obščestva*], **19–20**, 1930?–1932; titled Труды Узбекистанского географического Общества [*Trudy Uzbekistanskogo geografičeskogo Obščestva*], **21–22** (= **1–2**), 1937–1948.

Butll. Inst. Catalana Hist. Nat.
Butlleti de la Institució catalana d'Historia natural. Barcelona. **1→**, 1901→.

Cahiers Nat. Paris.
La Feuille des jeunes Naturalistes. Rennes & Paris. **1–44**, 1870–1915; titled *La Feuille des Naturalistes. Revue mensuelle d'Histoire naturelle. Bulletin de la Société des Naturalistes parisiens*, **45–47**, 1924–1926. Nov. ser., **1–7**, 1946–1952. Titled *Cahiers des Naturalistes. Bulletin des Naturalistes parisiens.* Nov. ser. **8→**, 1953→.

Canad. Jour. Bot.
Canadian Journal of Research. Ottawa. **1–12**, 1929–1934; titled *Canadian Journal of Research. Section C. (Botanical Sciences)*, **13-28**, 1935–1950; titled *Canadian Journal of Botany*, **29–50**, 1951–1972; titled *Canadian Journal of Botany. / Journal canadien de Botanique*, **51→**, 1973→.

Canad. Nat. Quart. Jour. Sci.
Canadian Naturalist and quarterly Journal of Science, with Proceedings of the Natural History Society of Montreal. (Subtitle varies.) Montreal. **1–8**, 1856–1863. Nov. ser. **1–10**, 1864–1883.

Candollea
Candollea. Organe du Conservatoire et du Jardin botaniques de la Ville de Genève. Genève. **1→**, 1922→.

Carinthia II
Carinthia II. / Mitteilungen des naturhistorischen Landesmuseums für Kärnten. / Naturwissenschaftliche Beiträge zur Heimatkunde Kärntens. / Mitteilungen des Vereines naturkundlicher Landesmuseum für Kärnten. Klagenfurt. **81→**. 1891→. (Continuation of *Carinthia. Ein Wochenblatt zum Nutzen und Vergnügen.* Klagenfurt. **1-80**, 1811–1890.)

Cavanillesia
Cavanillesia. Rerum botanicarum Acta. Barcinone. **1–8**, 1928–1938.

Christ. Vidensk.-Sels. Forh.
Kristiana Videnskabs Selskabs Forhandlingar. Oslo. 1886→. ('*Christiania*' in some vols.)

Collect. Bot. (Barcelona)
Collectanea botanica a barcinonensi botanico Instituta edita. Barcinone. **1→**, 1946→.

Comment. Gotting.
Commentarii Societatis regiae Scientiarum gottingensis. Gottingae. **1–4**, 1751–1754. Continued as **Novi Comment. Gotting.**, *Novi Commentarii Societatis regiae Scientiarum gottingensis.* Gottingae & Gotha, **1–8**, 1771–1778. Continued as **Comment. Gotting.**, Ser. 2. *Commentationes Societatis regiae Scientiarum gottingensis.* Gottingae. **1–16**, 1779–1808. Continued as **Comment. Gotting. Recent.**, *Commentationes Societatis regiae Scientiarum gottingensis recentiores.* Gottingae. **1–8**, 1811–1841.

Comment. Soc. Phys.-Med. Mosq.
Commentationes Societatis physico-medicae mosquensis. Mosquae. **1** in 1805. Titled **Comment. Soc. Phys.-Med. Univ. Mosq.**, *Commentationes Societatis physico-medicae apud Universitatem Literarum caesaream mosquensem Institutae.* Mosquae. **1–3**, 1808–1823.

Comment. Soc. Phys.-Med. Univ. Mosq.
Cf. *Comment. Soc. Phys.-Med. Mosq.*

Commun. Stat. Int. Géobot. Médit. Alp.
Communications de la Station internationale de Géobotanique méditerranéenne et alpine. Montpellier. **1→**, 1931→.

Compt. Rend. Acad. Bulg. Sci.
Comptes rendus de l'Académie bulgare des Sciences. Sofia. **1→**, 1948→. (**3→**, 1951→, also titled **Доклады болгарской Академии Наук** [*Dokladi bolgarskoj Akademii Nauk*].)

Compt. Rend. Acad. (Paris)
Compte[s] rendu[s] hebdomadaire[s] des Séances de l'Académie des Sciences. Paris. **1→**, 1835→.

Compt. Rend. Soc. Hallér.
Compte-rendu des Travaux de la Société hallérienne. Genève. **1–4**, 1852–1856. (**1**: pp. 1–12 in 1852–1853; **2**: pp. 13–76 in 1853–1854; **3**: pp. 77–90 in 1854–1855; **4**: pp. 93–184 in 1854–1856.)

Compt. Rend. Trav. Lab. Carlsb. (Sér. Physiol.)
Comptes-rendus des Travaux du Carlsberg Laboratoriet. Copenhagen. **1–20**, 1878–1935; titled *Comptes-rendus des Travaux du Carlsberg Laboratoriet; Série physiologique*, **21→**. 1935→. (The earlier volumes, not in series, are sometimes titled *Meddelelser frå Carlsberg Laboratoriet.*)

Congr. Sci. Fr.
Sessions des Congrès scientifiques de France. Rouen, Paris etc. (Place of publication changed each year.) **1–42, 44**. 1833–1879. (**43** not published.)

Contr. Bolus Herb.
Contributions from the Bolus Herbarium. Rondebosch C.P. **1→**, 1969→.

Contr. Gray Herb.
Contributions from the Gray Herbarium of Harvard University. Cambridge, Massachusetts. Nov. ser., **1–200**, 1891–1970 ('nov. ser.' omitted from **76–200**, 1926–1970); titled *Contributions from the Gray Herbarium*, **201–214**, 1971–1984. Superseded by *Harvard Papers in Botany*, **1→**, 1989→.

Contr. U.S. Nat. Herb.
Contributions from the United States National Herbarium. Washington. **1→**, 1890→.

Cuad. Ci. Biol.
Cuadernos de Ciencias biológicas. Universidad de Granada. Granada. **1–7**, [1971] 1972–1980.

Danske Vid. Selsk.
Cf. *Skr. Kiøbenhavnske Selsk. Laerd. Vid.*

Darwiniana
Darwiniana. Carpeta del "Darwinion". (From **3**, 1937/1939, subtitle changes to *Revista del Instituto do Botánica Darwinion.*) Buenos Aires. **1→**, 1922→.

Denkschr. Akad. Wiss. Math.-Nat. Kl. (Wien)
Denkschriften der kaiserlichen Akademie der Wissenschaften, mathematisch-naturwissenschaftliche Klasse. Wien. **1–95**, 1850–1918. (**93–95**, 1917–1918, titled **Kaiserl. Akad. Wiss. Wien Math.-Nat. Kl. Denkschr.**, *Kaiserliche Akademie der Wissenschaften in Wien. Mathematisch-naturwissenschaftliche Klasse. Denkschriften.*) Titled *Akademie der Wissenschaften in Wien. Mathematisch-naturwissenschaftliche Klasse. Denkschriften*, **96–107**, 1919–1951 (**105** in 1945–1951; **106** in 1942–1946); titled *Österreichische Akademie der Wissenschaften. Mathematisch-naturwissenschaftliche Klasse. Denkschriften*, **108→**, 1948→.

Denkschr. Bayer. Bot. Ges. Regensb.
Denkschriften der königlich-baierischen [königlich bayerischen] botanischen Gesellschaft in [zu] Regensburg. Regensburg. **1-22**, 1815–1947, with minor variations of title as indicated in parentheses. (**7–22**, 1898–1947, also called nov. ser., **1-15**.)

Denkschr. Schweiz. Naturf. Ges.
Cf. *Neue Denkschr. Schweiz. Ges. Naturw.*

Edinb. New Philos. Jour.
The Edinburgh new philosophical Journal, exhibiting a View of the progressive Improvements and Discoveries in the Sciences and the Arts. Edinburgh. Ser. 1, **1–57**, 1826–1854. Nov. ser., **1–19**, 1855–1864.

Edinb. Philos. Jour.
The Edinburgh philosophical Journal, exhibiting a View of the Progress of Discovery in natural Philosophy, Chemistry, natural History, practical Mechanics, Geography, Statistics, and the fine and useful Arts. Edinburgh. **1**, 1819; titled ... *Geography, Navigation, Statistics*, ... , **2-10**, 1820–1824 (**2** & **3** in 1820; **4** & **5** in 1821; **6** & **7** in 1822; **8** & **9** in 1823; **10** in 1824); titled ... *natural History, comparative Anatomy, practical Mechanics, Geography, Navigation, Statistics* ... , **11–13**, 1824–1825. (**11** in 1824; **12** & **13** in 1825.)

Értek. Term. Köréb. Magyar Tud. Akad.
Értekezések a Természettudományi[ok] Köréböl Magyar Tudományos Akadémia. Budapest. **1–23**, 1867–1893.

Evolution
Evolution, International Journal of organic Evolution. Lancaster, Pennsylvania. **1**→, 1947→. (**15**→, 1961→, at Lawrence, Kansas.)

Feddes Repert.
Repertorium novarum Specierum Regni vegetabilis. Berlin. **1–8**: p. 144, 1905–1910; titled *Repertorium Specierum novarum Regni vegetabilis*, **8**: p. 145–**51**, 1910–1942; titled *Feddes Repertorium Specierum novarum Regni vegetabilis*, **52–69**, 1943–1964; titled *Feddes Repertorium. Zeitschrift für botanische Taxonomie und Geobotanik*, **70**→, 1965→. (This work includes *Repertorium Europaeum et Mediterraneum*, indicated by dual pagination.)

Feddes Repert. (Beih.)
Repertorium Specierum novarum Regni vegetabilis. Beiheft. Berlin. **1–127**, 1914–1942; titled *Feddes Repertorium Specierum novarum Regni vegetabilis. Beiheft*, **128–141**, 1952–1964; titled *Feddes Repertorium. Zeitschrift für botanische Taxonomie und Geobotanik. Beiheft*, **142**→, 1965→. ***Sonderbeih.***, *Sonderbeiheft*. **A**–**E**, 1914–1939. (**A** in parts, 1928–1939; **B**, 1914–1929; **C** in parts, 1930–1936; **D** in parts, 1936–1937; **E** in 1938.)

Fern Gaz.
Cf. *Brit. Fern Gaz.*

Fl. Medit.
Flora mediterranea. Acta Herbarii mediterranei panormitani sub Auspiciis Societatis Botanicorum mediterraneorum "OPTIMA" nuncupanda edita. Palermo. **1**→, 1991→.

Fl. Pl. S. Afr.
The flowering Plants of South Africa. A Magazine containing hand-coloured Figures with Descriptions of the flowering Plants indigenous to South Africa. London, Johannesburg & Cape Town. **1–24**, 1920–1944; titled *The flowering Plants of Africa. A Magazine containing coloured Figures with Descriptions of the flowering Plants indigenous in Africa.* Pretoria & Ashford. **25**→, 1945→. This work has no page numbers, only plates.

Fl. Serres Jard. Eur.
Flore des Serres et des Jardins de l'Europe. Gand. **1–23**, 1845–1880. (**1** in 1845; **2** in 1846; **3** in 1847; **4** in 1848; **5** in 1849; **6** in 1851; **7** in 1851–1852; **8** in 1853; **9** in 1854; **10** in 1855; **11** in 1856; **12** in 1857; **13** in 1858; **14** in 1859; **15** in 1865; **16** in 1865–1867; **17** in 1868–1869; **18** in 1869–1870; **19** in 1873; **20** in 1874; **21** in 1875; **22** in 1877; **23** in 1880.) The full title of this work is very variable, but the first part (given above) is more or less constant. Earlier volumes have no page numbers, but nearly every leaf is numbered, and these are treated as page numbers, sometimes with a & b added.

Flora
Cf. *Flora* (*Regensb.*)

Flora (Regensb.)
Flora, oder allgemeine botanische Zeitung. Regensburg. **1**→, 1888→. (**72–95**, 1889–1905, at Marburg; **96**→, 1906→, at Jena. ***Erg.-Bl.***, *Ergänzungsblätter*. **1** in 1829.) Divided into 2 Abteilungen, *Abt. A, Physiologie und Biochemie*, **156–160**, 1966–1969, and *Abt B, Morphologie und Geobotanik*, **156–158**, 1966–1969. Abt. A continued as *Biochemie und Physiologie der Pflanzen*, Jena, **159**→, 1970→. Abt. B continued as ***Flora***, *Flora. Morphologie, Geobotanik, Oekophysiologie*, Jena, **159–179**, 1970–1987; titled *Flora. Morphologie, Geobotanik, Ökologie*, **180**→, 1988→.

Folia Geobot. Phytotax. (Praha)
Folia geobotanica & phytotaxonomica bohemoslovaca. Průhonice, Praha. **1**→, 1966→. (*bohemoslovaca* only in title of **1**.)

Fragm. Fl. Geobot.
Fragmenta floristica et geobotanica. / Materiały floristiczne i geoboticzne. Kraków. **1**→, 1954→. (**18(2)**→, 1972→, at Warszawa & Kraków).

Gard. Chron.
Gardeners' Chronicle and agricultural Gazette. London. Ser. 1, **1841–1873**, 1841–1873. Ser. 2, titled *The Gardeners' Chronicle. A weekly illustrated Journal of Horticulture and allied Subjects*, **1–26**, 1874–1886. Ser. 3, **1**→, 1887→. (**140–154**, 1956–1963, titled *Gardeners' Chronicle and Gardening illustrated*; **155–159** (**15**), 1964–1966, titled *Gardeners' Chronicle, Gardening illustrated and the Greenhouse*; **159**(**16**)–**199**(**1**), 1966–1986, titled *Gardeners' Chronicle*, with various subtitles.) Titled *Horticulture Week*, with various subtitles, **199**(2), 1986→.

Gard. Farm. Jour.
Gardeners' and Farmers' Journal. London. 1847–1880.

Gard. Mag. (Loudon)
The Gardeners' Magazine and Register of rural and domestic Improvement (conducted by J. C. Loudon). London. **1–19**, 1826–1844.

Gartenfl.
Gartenflora. Monatsschrift für deutsche und schweizerische Garten- und Blumenkunde. Erlangen. **1–87**, 1852–1938; Nov. ser., 1938–1940. With minor changes of title. (**23–34**, 1874–1885, at Stuttgart; **35–87** and Nov. ser., 1886–1940, at Berlin.)

Gärtn. Zentral-Bl.
Gärtnerisches Zentral-Blatt [*Centralblatt*]. *Referierendes und forschendes Organ für den gesammten Gartenbau, für Nomenklatur und Pflanzenkunde, Pflanzenernährung und Pflanzenschutz, Gewerbliches und Hygienisches, Unterrichtswesen und Litteratur.* Berlin. **1**, 1899. (Cf. R. Schmid & M. D. Turner, *Madroño* **24**: 68–74 (1977) and *Just's Botanischer Jahresbericht* **27** (Abt. 2): 194 (1899).)

Genetica
Genetica. Nederlandsch Tijdschrift voor Erfelijkheids- en Afstammingsleer. The Hague. **1**→, 1919→.

Gentes Herb.
Gentes Herbarum. Occasional Papers on the Kinds of Plants. Ithaca, N.Y. **1**→, 1920→.

Ges. Naturf. Freunde Berlin Mag.
Der Gesellschaft naturforschender Freunde zu Berlin, Magazin für die neuesten Entdeckungen in der gesammten Naturkunde. Berlin. **1–8**, 1807–1818. Other titles at other periods.

Gior. Arcad. Sci. (Roma)
Giornale arcadico di Scienze, Lettere ed Arti. Roma. **1–212**, 1819–1868.

Gior. Bot. Ital.
Giornale botanico italiano. Firenze. **1–2**. 1844–1847; titled ***Nuovo Gior. Bot. Ital.***, *Nuovo Giornale botanico italiano*. Ser. 1, **1–25**, 1869–1893. Nov. ser. **1-68**, 1894–1961; titled ***Gior. Bot. Ital.***, *Giornale botanico italiano*, **69–73**, 1962–1966. Volumes then revert to original numbering as **101**→, 1967→.

Gior. Fis. (Brugnat.)
Giornale di Fisica, Chimica e Storia naturale, ossia Raccolta di Memorie sulle Scienze, Arti e Manifatture ad esse relative di L. Brugnatelli.

Pavia. Ser. 1, **1–5**, 1808–1812; titled *Giornale di Fisica, Chimica, Storia naturale, Medicina ed Arti del* Regno *italico*, **6–10**, 1813–1817. Ser. 2, titled *Giornale di Fisica, Chimica e Storia naturale, Medicina ed Arti*, **1–10**, 1818–1827.

Gior. Ital. Sci. Nat. Agric. Arti Commerc.
Giornale d'Italia, spettante alla Scienze naturale, e principalmente all'Agricoltura, alle Arti ed al Commercio. Edit. F. Griselini. Venezia. **1–11**, 1764–1775; titled *Nuovo Giornale ...*, 1777–1780.

Gior. Sci. Nat. Econ. Palermo
Giornale di Scienze naturali ed economiche di Palermo. Palermo. **1→**, 1866→.

Glasn. Muz. Bosni Herceg.
Гласник Земальског Муз[с]еја у Босни и Херцеговини [*Glasnik Zemaljskog Muz(s)eja u Bosni i Hercegovini*]. Sarajevo. **1–49**, 1889–1937; titled Гласник Земалског Музеја краљевине Југолавије [*Glasnik Zemaljskog Muzea kraljevine Jugoslavije*], **50–51**, 1937–1942; titled Гласник Хрватских Земалског Музеа у Сарајебу [*Glasnik Hrvatskih zemaljskog Muzea u Sarajevu*], **53–54**, 1942–1944; titled Гласник Др. Жавног Музеа Сарајево. Придрне Науке [*Glasnik Dr. Žavnog Muzeja Sarajevo. Prirodne Nauke*], Nov. ser. **1**, 1945; titled Гласник Земалског Музеја а Сарајеву [*Glasnik Zemaljskog Muzeja a Sarajevu*]. / *Bulletin de Musée de la République populaire de Bosnie et Hercégovine à Sarajevo*, **2→**, 1947→.

Glasn. Skopsk. Naučn. Društva
Гласник Скопског научног Друштва [*Glasnik skopskog naunčog Društva*]. / *Bulletin de la Société scientifique de Skopje.* Skoplje. **1–22**, 1925–1941.

God. Sof. Univ. (Agro.-Les. Fak.)
Годишник на Софийския Университет. Агрономически-Лесоводство Факультет. [*Godišnik na Sofijskija Universitet. Agronomičeski-Lesovodstvo Fakul'tet*]. / *Annuaire de l'Université de Sofia.* Sofija. **1→**, 1904→. From **5**, 1909 in sections; titled Годишник ... Университет. Агрономически Факультет [*Godišnik ... Universitet. Agronomičeski Fakul'tet*]. / *Annuaire ... Sofia. Faculté agronomique*, 1923→.

Gorteria
Gorteria. Mededelingenblad ten dienste van de Floristiek en het Vegetatie-Onderzoek van Nederland. Leiden. **1**, 1961; titled *Gorteria. Tijdschrift ten dienste van de Floristiek, de Oecologie en het Vegetatie-Onderzoek van Nederland*, **2–3**, 1964–1967; titled *Gorteria. Tijdschrift voor de Floristiek, de Plantenoecologie en het Vegetatie-Onderzoek van Nederland*, **4–12**, 1968–1985; titled *Gorteria. Tijdschrift voor de Floristiek*, **13→**, 1986→.

Hannover. Mag.
Hannoverisches Magazin. Hannover. **1–28**, 1764–1791, and 1815–1850. Several series; often with many volumes published in the same year. Titled ***Neues Hannover. Mag.***, *Neues Hannoverisches Magazin, worin kleine Abhandlungen ... gesamlet und aufbewahrt sind*, **1–23**, 1791–1813.

Hedwigia
Hedwigia. Ein Notizblatt für kryptogamisch Studien. Dresden. **1–82**, 1852–1944. (**24–26**, 1885–1887, subtitled *Organ für specielle Kryptogamenkunde nebst Repertorium für kryptogamische Literatur*; **27–36**, 1888–1897, subtitled *Organ für Kryptogamenkunde nebst Repertorium für kryptogamische Literatur*; **37–82**, 1893–1944, subtitled *Organ für Kryptogamenkunde und Phytopathologie nebst Repertorium für Literatur.*)

Hilgardia
Hilgardia. A Journal of agricultural Science published by the California agricultural Experimental Station. Berkeley, California. **1→**, 1925→.

Hook. Ic.
Icones Plantarum; or Figures, with brief descriptive Characters and Remarks of new and rare Plants. London. **1–10**, 1836–1864; titled *Hooker's Icones Plantarum; ...*, **11→**, 1867→.

Isis (Oken)
Isis, oder encyclopädische Zeitung von Oken. Edit. L. Oken. Jena. **1–[5]**, 1817–1819; titled *Isis von Oken*, **[6–19]**, 1820–1827; titled *Isis*, **20–[41]**, 1828–1848. (**21–41**, 1828–1848, at Leipzig. **2–19**, 1818–1826, and **23–41**, 1830–1848, originally without volumation, the volume nos. being assigned later.)

Israel Jour. Bot.
Cf. *Bull. Res. Counc. Israel.*

Izv. Bot. Inst. (Sofia)
Известия на ботаническия Институт [*Izvestiya na botaničeskiya Institut.*] / *Bulletin de l'Institut botanique.* Sofia. **1–25**, 1950–1974. Continued as Фитология [*Fitologiya*] / *Phytology*, **1→**, 1975→.

Izv. Kavkaz. Muz.
Извѣстія Кавказскаго Музея [*Izvěstija Kavkazskago Muzeja*]. / *Bulletin. Musée du Caucase.* / *Mitteilungen des kaukasischen Museum.* St. Petersburg. **1–12**, 1897–1919. Continued as *Bull. Mus. Géorgie* and *Trav. Mus. Géorgie.*

Jahrb. Naturh. Landesmus. Kärnten
Jahrbuch des naturhistorischen Landesmuseums [*Landes-Museums*] *von Kärnten.* Klagenfurt. **1–29**, 1852–1918.

Jahresb. Landes-Lehrersem. Baden bei Wien
Jahresbericht des nieder-österreichischen Landes-Lehrerseminars Baden bei Wien. Wien. (**48** in 1911; full volumation and publication dates unavailable.)

Jahresb. Landes-Lehrersem. Wiener-Neustadt
Jahresbericht des nieder-österreichischen Landes-Lehrerseminars in Wiener-Neustadt. Wien. (**34–36** in 1907–1909; full volumation and publication dates unavailable.)

Jahresb. Schles. Ges. Vaterl. Kult.
Uebersicht der Arbeiten und Veränderungen der schlesischen Gesellschaft für vaterländische Cultur [*Kultur*]. Breslau. **1–28**, 1825–1851; titled *Jahresbericht* [*Jahres-Bericht*] *der schlesischen Gesellschaft für vaterländische Kultur* [*Cultur*], **28–114**, 1851–1941; *Sammelheften* in 1940 and 1943.

Jahresb. Staats-ober-Realschule Steyr
Jahresbericht der k.k. Staats-ober-Realschule zu Steyr. Steyr. [**5** in 1875; full volumation and publication dates unavailable.)

Jahres-Kat. Wien. Bot. Tauschver.
Jahres-Katalog pro ... des wiener botanischen Tauschvereins. Wien. 1894–1895; titled *Jahres-Katalog pro ... des wiener botanischen Tauschanstalt*, 1896–1914.

Jour. Arnold Arb.
Journal of the Arnold Arboretum. Cambridge, Massachusetts. **1–71**, 1919–1990. (**2(3)–13**, 1921–1932, at Lancaster, Pennsylvania; **14–36(2–3)**, 1933–1955, at Jamaica Plain, Massachusetts. Publication was suspended in 1990 but a *Supplementary Series* is being issued, **1→**, 1991→.)

Jour. Bot. Acad. Sci. Ukr.
Ботаничніу Журнал [*Botaničniy Žurnal*]. / *Journal botanique de l'Académie des Sciences de la RSS d'Ukraine.* Kieff. **1–12**, 1940–1955.

Jour. Bot. (London)
The Journal of Botany, British and foreign. London. **1–80**, 1863–1942.

Jour. Bot. (Paris)
Journal de Botanique. [Edit. by L. Morot.] Paris. Ser. 1, **1–20**, 1887–1906. Ser. 2, **1–3**, 1907–1925 ('1913'). (Ser. 2, **3**, for 1913, was unfinished, only pp. 1–104 being printed, in 1914; it was not

in fact issued until 1925. Cf. M. L. Green in *Kew Bull.* **1928**: 155–156 (1928).)

Jour. Elisha Mitchell Soc.
Journal of the Elisha Mitchell Scientific Society. Chapel Hill, North Carolina. **1**→, 1883→.

Jour. Fac. Sci. Tokyo Univ. (Bot.)
Journal of the Faculty of Science, Tokyo Imperial University. Sect. 3. *Botany*. Tokyo. **1**→, 1925→.

Jour. Governm. Bot. Gard. Nikita
Записки государственного Никитского опытного ботанического·Сада [*Zapiski gosudarstvennogo Nikitskogo opytnogo botaničeskogo Sada*]. / *Journal of the Government Botanic Garden, Yalta, Crimea*. Yalta. **8** ('*opytnogo*' omitted), **9–14(1)**, **16**, 1925–1930 and 1931. (**14(2)** & **15**, 1930, and **18–25**, 1934–1953, titled Труды государственного Никитского ботанического Сада [*Trudy gosudarstvennogo Nikitskogo botaničeskogo Sada*]. / *Arbeiten aus dem botanischen Garten Nikita, Jalta, Krim*.) Continuation of Записки императорскаго Никитскаго Сада [*Zapiski imperatorskago Nikitskago Sada*]. Yalta, **1–7**, 1908–1916; titled Труды государственный Никитский ботанический Сад [*Trudy gosudarstvenn'ij Nikitskij botaničeskij Sad*], **26**→, 1956→.

Jour. Hort. Soc. London
The Journal of the Horticultural Society of London. London. **1–9**, 1846–1855.

Jour. Inst. Bot. Acad. Sci. Ukr.
Журнал Інституту ботаніки Акад. Наук УРСР [*Žurnal Instytutu Botaniky Akad. Nauk URSR*]. / *Journal de l'Institut botanique de l'Académie des Sciences de la RSS d'Ukraine*. Kieff. **9–31**, 1933–1939.

Jour. Linn. Soc. London (Bot.)
The Journal of the Proceedings of the Linnean Society. Botany. London. **1–7**, 1856–1864; titled *The Journal of the Linnean Society. Botany*, **8–46**, 1865–1924; titled *The Journal of the Linnean Society of London. Botany*, **47–61**, 1925–1968; titled ***Bot. Jour. Linn. Soc.***, *Botanical Journal of the Linnean Society*, **62**→, 1969→.

Jour. Roy. Hort. Soc.
The Journal of the Royal Horticultural Society of London. London. **1–4**, 1865–1877; titled *The Journal of the Royal Horticultural Society*, **5–100(5)**, 1869–1975; titled *The Garden. Journal of the Royal Horticultural Society*, **100(6)**→, 1975→.

Jour. S. Afr. Bot.
Journal of South African Botany. Kirstenbosch. **1–50**, 1935–1984. Superseded by *South African Journal of Botany*. / *Suid-Afrikaanse Tydskrif vir Plantkunde*. Pretoria. **1–3**, 1982–1984 (as an independent journal from *Jour. S. Afr. Bot.*); **51**→, 1985→ (*Jour. S. Afr. Bot.* amalgamated with *S. Afr. Jour. Bot.* and title of the latter but volumation of the former used.)

Kaiserl. Akad. Wiss. Wien Math.-Nat. Kl. Denkschr.
Cf. *Denkschr. Akad. Wiss. Math.-Nat. Kl.* (*Wien*).

Kew Bull.
Bulletin of miscellaneous Information. Royal Gardens, Kew. London. **1887–1941**, 1887–1942; titled *Kew Bulletin*, **1**→, 1946→.

Kew Handlist Herb. Pl.
Kew Handlist of herbaceous Plants. Kew. Various dates.

Kong. Danske Vid. Selsk.
Cf. *Skr. Kiøbenhavnsk. Selsk. Lærd. Vid.*

Kong. Norske Vid. Selsk. Skr. (Trondhjem)
Det Kongelige norske Videnskabers Selskabs Skrifter. Trondhjem. 1878→. Year and part number are used as volume number.

Königsb. Arch. Naturw.
Königsberger Archiv für Naturwissenschaft und Mathematik. Königsberg. **1**, 1811–1812.

Kungl. Svenska Vet.-Akad. Handl.
Swenska Wetenskaps Academiens Handlingar. Stockholm. **1**, 1739; titled *Kongl. swenska Wetenskaps Academiens Handlingar*, **2–7**, 1743–1746; titled *Kongl. svenska Vetenskaps Academiens Handlingar*, **8–17**, 1747–1756; titled *Kongl. Vetenskaps Academiens Handlingar*, **18–40**, 1757–1779. Nov. ser., titled *Kongl. Vetenskaps Academiens nya Handlingar*, **1–33**, 1780–1812; titled *Kongl. Vetenskaps Academiens* [*Vetenskaps-academiens*] *Handlingar* **1813–1844**, 1813–1846; titled *Kongl. Vetenskaps-Akademiens Handlingar*, **1845–1855**, 1846–1855. Nov. ser., titled *Kongliga svenska Vetenskaps-Akademiens Handlingar. Ny Följd*, **1–35**, 1855–1902; titled *Kungliga svenska Vetenskaps-Akademiens Handlingar*, **36–63**, 1902–1923. Ser. 3, **1–25**, 1924–1948. Ser. 4, **1**→, 1951→.

Lagascalia
Lagascalia. Sevilla. **1**→, 1971→.

Lazaroa
Lazaroa. Madrid. **1**→, 1979→.

Lejeunia
Lejeunia. Revue de Botanique. (Subtitle varies.) Liège. **1–23**, 1937–Mar. 1961. Nov. ser., **1**→, Oct. 1961→.

Linnaea
Linnaea. Ein Journal für die Botanik in ihrem ganzen Umfange. Berlin. **1–43**, 1826–1882. (**9–34**, 1834–1866, at Halle. From **17–34**, 1843–1866, *Oder Beiträge zur Pflanzenkunde* was added to the subtitle; *Neue Folge Erster Band* [... *Band IX*] was additionally added from **35–43**, 1867–1882. From **3–16**, 1828–1842, with a separately paginated section, ***Litt.-Ber.***, *Litteratur-Bericht zu Linnaea für das Jahr 1828* (... *1842*). (For dates see F. A. Stafleu & R. S. Cowan, *Taxonomic Literature* ed. 2, **5**: 193–200 (1985).)

Lunds Univ. Årsskr.
Lunds Universitets Årsskrift. / *Acta Universitatis lundensis*. Lund. Ser. 1, **1–40**, 1864–1902. Nov. ser., **1–59**, 1905–1963. *Sectio 2 Medica, Mathematica, Scientiae Rerum naturalium*, **1**→, 1964→.

Madroño
Madroño. Journal of the California botanical Society. Berkeley, California. **1–2**, 1916–1934; titled *Madroño. A West American Journal of Botany*, **3**→, 1935→.

Magyar Bot. Lapok
Magyar Botanikai Lapok. / *Ungarische botanischer Blätter*. Budapest. **1–33**, 1902–1934.

Malpighia
Malpighia. Rassegna mensuale [*mensile*] *di Botanica*. Messina. **1–34**, 1886–1937. (**3–23**, 1889–1909, at Genova; **24–29**, 1911–1923, at Catania; **30-31**, 1927–1928, at Palermo; **32–34**, 1932–1937, at Bologna.)

Mat. Étude Fl. Géogr. Bot. Or.
Matériaux pour servir à l'Étude de la Flore et de la Géographie botanique de l'Orient (*Missions du Ministère de l'Instruction publique en 1904 et en 1906*.) Nancy. **1–7**, 1906–1922.

Math. Term. Értesitő
Mathematikai és természettudományi Értesitő. Budapest. **1–63**, 1882–1944 (from **40**, 1923, titled *Matematikai* ...; **43–63**, 1926–1944, with German subtitle *Mathematischer und naturwissenschaftlicher Anzeiger der ungarischen Akademie der Wissenschaften*).

Math. Term. Közl.
Mathematikai és természettudományi Közlemények, vonatkozólag a hazai Viszonyokra. Budapest. **1–40**, 1861–1944 (from **35(5)**, 1926, titled *Matematikai* ...).

Meddel. Grønl.
Meddelelser om Grønland, af Kommissionen for Ledelsen af de geologiske og geographiske Undersøgelser i Grønland. København. **1–206(5)**,

1879–1976. From 1979 divided into sections, *Bioscience*, *Geoscience* and *Man and Society*; Botany in *Meddelelser om Grønland, Bioscience*. Copenhagen. **1→**, 1979→.

Meddel. Soc. Fauna Fl. Fenn.
Meddelanden af Societas pro Fauna et Flora fennica. Helsingfors. **1–50**, 1876–1925.

Meded. Bot. Mus. Herb. Rijksuniv. Utrecht
Mededeelingen van het botanisch Museum en Herbarium van de rijks Universiteit te Utrecht. **1–90**, 1932–1942 (occasional parts, from **14**, 1934, titled ... *rijksuniversiteit*); titled *Mededelingen ... rijksuniversiteit ...*, **91→**, 1943→.

Mélang. Philos. Math. Soc. Roy. Turin (Misc. Taur.)
Miscellanea philosophica-mathematica Societatis privatae taurinensis. Augustae Taurinorum. **1**, 1759; titled *Mélanges de Philosophie et de Mathématique de la Société royale de Turin. Miscellanea taurinensia*, Turin, **2–5**, 1760–1774. Continued as *Mem. Acad. Sci.* (*Turin*).

Mem. Acad. Barcelona
Memorias de la Real Academia de Ciencias y Artes de Barcelona. Barcelona. **1–2**, 1876–1885. Nov. ser., **1→**, 1892→, titled ***Mem. Real Acad. Ci. Arts Barceloné***, *Memorias de la real Academia de Ciencias y Artes de Barcelona*. (Some vols. titled *Memòries de la reial Acadèmia de Ciències i Arts de Barceloné. Real/reial* omitted from **22(18)–25(16)**, 1931–1936. There are also alternative volume numbers in several series. Cf. Castroviejo et al., *Fl. Iberica* **2**: 759 (1990).)

Mem. Acad. Roy. Sci. Lyon Sect. Sci.
Mémoires de l'Académie royale des Sciences, Belles-Lettres et Arts de Lyon. Section de Sciences. Lyon. **1–2**, 1845–1847. Ser. 2, **1–31**, 1851–1892. (**1**, 1851, titled *Mémoires de l'Académie nationale des ...*; **3**, 1853, titled *Mémoires de l'Académie impériale ...*; **18**, 1870–1871, titled *Mémoires de l'Académie des Sciences ...*) Ser. 3, with section title *Sciences et Lettres*, **1→**, 1893→.

Mém. Acad. Sci. Pétersb.
Записки имп. Академіи Наукъ (по фисико-математическому Отдѣленію) [*Zapiski imp. Akademii Nauk (po fiziko-matematičeskomu Otdéleniju)*]. / *Mémoires de l'Académie impériale des Sciences de St.-Pétersbourg* (*Classe* [*des Sciences*] *physico-mathématiques*). St. Pétersbourg. Ser. 5, **1–11**, 1809–1830. Ser. 6, **1–10**, 1831–1859 (**3–10**, 1835–1859, also numbered **1–8** and with second title *Mémoires de l'Académie impériale des Sciences de Saint Pétersbourg. Sixième série. Sciences naturelles*.) Ser. 7, **1–42**, 1859–1897. Ser. 8, **1–21(6)**, 1894–1914; **22–34**, 1907–1916.

Mém. Acad. Sci. Toulouse
Histoire et Mémoires de l'Académie royale des Sciences, Inscriptions et Belles Lettres de Toulouse. Toulouse. Ser. 1, **1–4**, 1782–1792. Ser. 2, **1–6**, 1827–1843. Ser. 3, titled *Mémoires. Académie des Sciences, Inscriptions et Belles Lettres de Toulouse*, **1–6**, 1842–1850. Ser. 4, **1–6**, 1851–1856. Ser. 5, **1–6**, 1857–1862. Ser. 6, **1–6**, 1863–1868. Ser. 7, **1–10**, 1869–1878. Ser. 8, **1–10**, 1879–1888. Ser. 9, **1–9**, 1888–1897. Ser. 10, **1–12**, 1901–1912. Ser. 11, **1–10**, 1913–1922. Ser. 12, **1–?**, 1923–1937. (With numerous minor changes of title.) The *Mémoires* were suspended 1898–1900, *Bulletin de l'Académie des Sciences, Inscriptions et Belles-lettres de Toulouse* **1-3**, 1897–1900, being issued instead.

Mém. Acad. Sci. (Turin)
Mémoires de l'Académie royale des Sciences de Turin. Turin. Ser. 1, **1–5**, 1786–1793; titled *Mémoires de l'Académie des Sciences de Turin*, **6**, 1801; titled *Mémoires de l'Académie des Sciences, Littérature et Beaux-Arts de Turin. Sciences, Physiques et Mathématiques*, **1802–1803** (also numbered **7**), 1804; titled *Mémoires de l'Académie Impériale de Sciences, Littérature et Beaux-Arts de Turin. Sciences, Physiques et Mathématiques*, 1805–1813; titled *Mémoires de l'Académie royale des Sciences de Turin*, **1813-1814** (also numbered **22**), 1816; titled ***Mem. Accad. Sci. Torino***, *Memorie della reale Accademia delle Scienze di Torino*, **23–40**, 1818–1838 (each volume contains different classes, etc.). Ser. 2, **1–6**, **23→** (volume nos. **7–22** not used), 1839→.

Mém. Agric. Soc. Agric. Dép. Seine
Mémoires d'Agriculture, d'Économie rurale et domestique, pub. par la Société d'Agriculture du Département de la Seine. Paris. **1–15** (also numbered as *Mém. Soc. Agric. Paris* **29–44**). 1801–1813. Many other titles at other periods, from 1761–1800 and 1814–1916. (These vols. contain the original versions of **DC.**, ***Rapp. Voy. Bot.***, as follows: *Rapports sur deux Voyages botaniques et agronomiques dans les Départemens de l'Ouest et du Sud-Ouest, op. cit.* **10**: 228–? & **11**: 1–? (1808). *Rapports ... du Sud-Est et de l'Est, op. cit.* **12**: 210–260 & **13**: 203–253 (1810). *Rapports ... de Nord-Est et du Centre, op. cit.* **14**: 213–? & **15**: 200–? (1813). The six reports were reprinted under a different title in 1813 (cf. DC., *Rapp. Voy. Bot.* in Appendix II).)

Mém. Herb. Boiss.
Mémoires de l'Herbier Boissier. Genève, **1–22**, 1900.

Mém. Inst. Égypt.
Mémoires d'Institut Égyptien. Paris & Le Caïre. **1–9**, 1862–1916; titled *Mémoires de l'Institut d'Égypte*. Cairo. Nov. ser., **1→**. 1918→.

Mem. Ist. Veneto
Memorie dell'i. r. Istituto Veneto di Scienze, Lettere ed Arti. Venezia. **1–13**, 1843–1866; titled *Memorie del reale Istituto Veneto di Scienze, Lettere ed Arti*, **14→**, 1868→.

Mem. Mus. Ci. Nat. Barcelona (Bot.)
Memòries del Museu de Ciències naturals de Barcelona. Sèrie botanica. Barcelona. **1(1)**, 1922; titled *Memorias del Museo de Ciencias naturales de Barcelona. Serie botanica*, **1(2–3)**, 1924–1925. (**1(2)** in 1924; **1(3)** in 1925. Cf. *Cat. Périod. Bibl. Cons. Jard. Bot. Genève Suppl. 1980–1987*: 63 (1988).)

Mém. Mus. Hist. Nat. (Paris)
Mémoires du Muséum d'Histoire naturelle. Paris. **1–20**, 1815–1832.

Mem. Real Acad. Ci. Artes Barceloné
Cf. *Mem. Acad. Barcelona*.

Mém. Sav. Étr. Pétersb.
Mémoires des Savants étrangers. Mémoires présentés à l'Académie impériale des Sciences de St. Pétersbourg par divers Savants et lus dans ses Assemblées. St. Pétersbourg. **1–9**, 1831–1859.

Mém. Soc. Acad. (Angers)
Mémoires de la Société académique de Maine et Loire. Angers. **1–38**, 1857–1883.

Mém. Soc. Bot. Fr.
Mémoires de la Société botanique de France. Paris. 1905–1921, as part of *Bull. Soc. Bot. Fr.*; 1949→. Year is used as volume number.

Mem. Soc. Brot.
Memórias da Sociedade broteriana. Coimbra. **1→**, 1930→.

Mém. Soc. Émul. Doubs
Mémoires et Comptes rendus de la Société d'Émulation du Doubs (with minor variations of title). Besançon. Ser. 1, **1–3**, 1841–1849. Sér. 2, titled *Mémoires de la Société libre d'Émulation du Doubs*, **1–5**, 1850–1854; titled *Mémoires de la Société d'Émulation du Département du Doubs*, **6–8**, 1855–1857. Ser. 3, titled *Mémoires de la Société d'Émulation du Département du Doubs*, **1–7**, 1856–1864; titled *Mémoires de la Société d'Émulation du Doubs*, **8–10**, 1864–1869. Ser. 4, with same title, **1–10**, 1866–1876. Ser. 5, **1–10**, 1877–1886. Ser. 6, **1–10**, 1887–1896. Ser. 7, **1–10**, 1897–1906. Ser. 8, **1–10**, 1907–1920. Ser. 9, **1–10**, 1922–1931. Ser. 10, **1–8**, 1931–1939.

Mem. Soc. Esp. Hist. Nat.
Memorias de la real Sociedad española de Historia natural. Madrid. **1-17**, 1903–1935. ***Tomo extr.***, *Tomo extraordinario*.

Mem. Soc. Fauna Fl. Fenn.
Memoranda Societatis pro Fauna et Flora fennica. Helsingforsiae. **1**→, 1927→.

Mém. Soc. Linn. Paris
Mémoires de la Société linnéenne de Paris, précédé de son Histoire. Paris. **1799**, 1799; titled *Mémoires de la Société linnéenne de Paris*, **1–6**, 1822–1828.

Mém. Soc. Nat. Kieff
Записки Кіевскаго Общества естествоіспытателей [*Zapiski Kievskago Obščestva estestvoispytatelej*]. / *Mémoires de la Société des Naturalistes de Kieff* [*Kiew*]. Kiev. **1–26(1)**, 1870–1917; titled Запуску Куйівського товаруства прurодознавчів [*Zapysky Kyjivs'kogo tovarystva pryrodoznavčiv*], **26(2)–27**, 1919–1929.

Mém. Soc. Nat. Moscou
Записки общества испытателей природы, основаннаго при императорском Московском Юниверситетє [*Zapiski obščestva ispytatelej prirody, osnovannago pri imperatorskom Moskovskom Universitetě*]. / *Mémoires de la Société des Naturalistes de l'Université impériale de Moscou*. Moscou. **1**, 1806; titled *Mémoires de la Société impériale des Naturalistes de Moscou*, **2–6**, 1809–1823. Ed. 2. **1** in 1811; **2–4** in 1830. Ed. 3. **1** in 1830.

Mém. Soc. Nat. Sci. Cherbourg
Mémoires de la Société des Sciences naturelles de Cherbourg. Cherbourg. **1–2**, 1852–1854; titled *Mémoires de la Société impériale des Sciences naturelles de Cherbourg*, **3–15**, 1855–1870; titled *Mémoires de la Société nationale des Sciences naturelles de Cherbourg*, **16–21**, 1871–1878; titled *Mémoires de la Société nationale des Sciences naturelles et mathématiques de Cherbourg*, **22**→, 1879→.

Mém. Soc. Phys. Hist. Nat. Genève
Mémoires de la Société de Physique et d'Histoire naturelle de Genève. Genève & Paris. **1**→, 1821→.

Mém. Soc. Sci. Nat. Maroc
Mémoires de la Société des Sciences naturelles et physiques du Maroc. Rabat. **1– 50**, 1921–1952. Nov. ser., titled ***Mém. Soc. Sci. Nat. Phys. Maroc Bot.***, *Mémoires de la Société des Sciences naturelles et physiques du Maroc. Botanique*, **1**→, 1960→.

Mém. Soc. Sci. Nat. Phys. Maroc Bot.
Cf. *Mém. Soc. Sci. Nat. Maroc.*

Mem. Torrey Bot. Club
Memoirs of the Torrey botanical Club. New York. **1**→, 1889–1977, 1990→. (**18–19**, 1931–1941, at Menasha, Wisconsin; **20**, 1943–1954, at Lancaster, Pennsylvania; **21–23**, 1958–1977, at Durham, North Carolina; **24**→, 1990→ at New York.) Publication was suspended from **23**(**3**), January 1977 until **24**(**1**), June 1990.

Mitt. Bot. Mus. Zürich
Mitteilungen aus dem botanischen Museum der Universität Zürich. Zürich. **1**→, 1893→. (**223**, 1963, and **225**, 1964, titled *Mitteilungen aus dem botanischen Garten und Museum der Universität Zürich*.)

Mitt. Bot. Staatssamm. (München)
Mitteilungen der botanischen Staatssammlung München. München. **1**→, 1950→.

Mitt. Deutsch. Dendrol. Ges.
Mitteilungen der deutschen dendrologischen Gesellschaft. Berlin. **1**→, 1893→. (**2–21**, 1894–1912, at Poppelsdorf-Bonn; **22–44**, 1913–1932, at Thyrow; **45**, 1933, at Berlin; **46**, 1934, at Berlin & Dortmund; **47–55**, 1935–1942, at Dortmund; **56**→, 1950→, at Darmstadt.) Ed. 2, Berlin. **1–14**, 1909–1913. ***Jahrb.***, *Jahrbuch*, is added to title from 1919–*c*. 1968. A separate *Jahrbuch für Staudenkunde* was also issued, with 2 parts, in 1913.

Mitt. Fl.-Soziol. Arbeitsgem.
Mitteilungen der Floristisch-soziologischen Arbeitsgemeinschaft. Stolzenau. **1–5**, 1928–1939. Nov. ser., **1**→, 1949→.

Mitt. Kl. Ungar. Gartenb.-Lehranst.
A m.[*agyar*] *kir.*[*ályi*] *Kertészeti Tanintézet Közleményei.* / *Mitteilungen der kgl. ungarischen Gartenbau-Lehranstalt.* / *Bulletin de l'École royale hungroise d'Horticulture.* / *Bulletin of the royal Hungarian horticultural College*. Budapest. Ser. 1, **1–5**, 1935–1939; titled *A M*[*agyar*] *Kir*[*ályi*] *Kertészeti Akademiei Közlémenyei.* / *Mitteilungen der kgl. ungarischen Gartenbau-Akademie.* / *Bulletin de l'Académie royale hongroise d'Horticulture.* / *Bulletin of the royal Hungarian horticultural College* (*new Series*), **6–8**, 1940–1942 (also called Ser. 2, **1–3**); titled *M*[*agyar*] *kir*[*ályi*] *Kertészeti és Szőlészeti Főiskola Közleményei.* / *Mitteilungen der* [*köni*]*gl*[*ichen*] *ungarischen Hochschule für Garten- und Weinbau.* / *Bulletin de l'École supérieur royale hongroise d'Horticulture et Viticulture.* / *Bulletin of the royal Hungarian College for Horticulture and Viniculture*, **9–10**, 1943–1944 (also called Ser 2, **4–5**); titled ***Ann. Sect. Horti-Viticult. Univ. Sci. Agr. (Budapest)***, *Agrártudományi egyetem kert és szőlőgazdaságtudomány. karanák évkönyve.* / *Annales Sectionis Horti- et Viticulturae Universitatis Scientiae Agriculturae.* / *Mitteilungen des Fakultät für Garten- und Weinbau der Universität für Agrarwissenschaften.* / *Bulletin de la Faculté horticole et viticole de l'Université des Sciences agricoles.* / *Bulletin of the Faculty of Horticulture and Viticulture of the University of Agriculture.* / Известия Факультета садоводственных и виноградственных Наук Венгерского Университета аграрных Наук [*Izvestija Fakul'teta sadovodstvennykh i vinogradstvennykh Nauk Vengerskogo Universiteta agrarnykh Nauk*], **11–13**, 1945–1949; titled *Agrártudományi egyetem kert- és szőlőgazdaságtudományi karának évkönyve.* / *Annales Sectionis Horti- et Viticulturae Universitatis Scientiae Agriculturae*, **1–3** (also numbered **14–16**), 1950–1952; titled *A Kertészeti és Szőlészeti Főiskola évkönyve.* / *Annales Academiae Horti -et Viticulturae*, **17–27**, 1953–1963; titled *A Kertészeti és Szőlészeti Főiskola Közleményei.* / *Publicationes Academiae Horti- et Viticulturae*, **28**→, 1964→.

Mitt. Naturw. Ver. Steierm.
Mitteilungen des naturwissenschaftlichen Vereins für Steiermark. Graz. **1**→, 1863→.

Mitt. Thür. Bot. Ges.
Mitteilungen des thüringischen botanischen Gesellschaft. Weimar. **1–2**, 1949–1960. Also ***Beih.***, *Beihefte* **1–3**, 1949–1952. Continued as *Haussknechtia. Mitteilungen der thüringen botanischen Gesellschaft*. Jena. **1**→, 1984→.

Mitt. Thür Bot. Ver.
Mittheilungen des thüringischen botanischen Vereins. Weimar. Ser. 1, **1–9**, 1882–1890 (incorporated in *Mitteilungen der geographischen Gesellschaft «für Thüringen» zu Jena.*) Nov. ser., **1–51**, 1891–1944. (From Nov. ser. **18**, 1903, titled *Mitteilungen ...*). From **10**, *Mitteilungen der geographischen Gesellschaft «für Thüringen» zu Jena* and *Mitt. Thür. Bot. Ver.* became separate journals.

Monatsber. Koenigl. Akad. (Berlin)
Monatsberichte der königlich preussischen Akademie der Wissenschaften zu Berlin. Berlin. 1856–1881. Year is used as vol. no.

Monde Pl.
Le Monde des Plantes. Revue mensuelle de Botanique. Organe de l'Académie internationale de Géographie botanique. Le Mans. **1–8**: p. 56, 1891–1898. Continued as *Le Monde des Plantes. Revue trimestrielle et internationale de Bibliographie*, **1**→, 1899→ (later published at Toulouse; from no. 5, subtitle changes to *Intermédiare des Botanistes ...*). Also continued as ***Bull. Acad. Int. Géogr. Bot. (Le Mans)***, *Bulletin de l'Académie internationale de Géographie botanique*, **8**: p. 47 [57]–**19**, 1899–1910; titled *Bulletin de*

Géographie botanique. Organe mensuel de l'Académie internationale de Botanique, **21–27**, 1911–1919. (**16-20** at Paris.)

Monit. Jard. Bot. Tiflis

Вѣстникъ Тифлисскаго ботаническаго Сада [*Věstnik Tiflisskago botaničeskago Sada*]. / *Moniteur du Jardin botanique de Tiflis.* / **ტფილისის ბოტანიკურ ბაღის მოამბე** [*Tp'hilisis botanik'our baghis moamb*]. Tiflis. Ser. 1, **1–51**, 1905–1921. Nov. ser., **1–5**, 1925–1931.

Naturaliste (Paris)

Le Naturaliste. Journal des Échanges et des Nouvelles. Paris. **1–53** (= année **1–9**), 1879–1889. Ser. 2, **1–548** (= année **9–32**), 1887–1910.

Nat. Sicil.

Il Naturalista siciliano. Giornale di Scienze naturali. Palermo. **1–14**, 1881–1895. Nov. ser., **1–20**, 1896–1908. Nov. ser., **1–27**, 1909–1930. Ser. 3 [*sic*], **1–3**, 1946–1948. Ser. 4, **1→**, 1978→. (Cf. *Cat. Périod. Bibl. Cons. Bot. Genève, Suppl. 1980–1987*: 68 (1988).)

Nederl. Kruidk. Arch.

Nederlandsch kruidkundig Archief. Verslagen en mededelingen der nederlandsche botanische Vereeniging. Amsterdam, later Leyden. Ser. 1, **1–5**, 1846–1870. Ser. 2, **1–6**, 1871–1895. Ser. 3, **1–2**, 1896–1900/1904. Years are then used as vol. nos., **1904–1932**, 1904–1932. Vol. nos. then revert to ser. 1, **43–58**, 1933–1951. (Ser. 2, ser. 3 & **1904–1913**(**1**) at Nijmegen; **1913**(**2**)**–1919** at Groningen; **1920–1921** at Utrecht; **1922–1932** and **43–58** at Amsterdam.)

Neue Denkschr. Schweiz. Ges. Naturw.

Neue Denkschriften der allgemeinen schweizerischen Gesellschaft für die gesammten Naturwissenschaften. / Nouveaux Mémoires de la Société helvétique des Sciences naturelles. Neuchâtel. **1–40**, 1837–1906; titled ***Neue Denkschr. Schweiz. Naturf. Ges.***, *Neue Denkschriften der schweizerischen naturforschenden Gesellschaft*, same French title, **41–54**, 1906–1918; titled ***Denkschr. Schweiz. Naturf. Ges.***, *Denkschriften der schweizerischen naturforschenden Gesellschaft. / Mémoires de la Société helvétique des Sciences naturelles*, **55→**, 1920→. (**8–9**, 1847, at Neuenberg; **10**, 1849, at Neuchâtel; **11→**, 1850→, at Zürich.)

Neue Denkschr. Schweiz. Naturf. Ges.

Cf. *Neue Denkschr. Schweiz. Ges. Naturw.*

Neues Hannover Mag.

Cf. *Hannover Mag.*

New Phytol.

The new Phytologist; a British botanical Journal. (Subtitle omitted from **31→**, 1932→.) London. **1→**, 1902→. (**6–10**, 1907–1912, at Cambridge; **11–28**, 1912–1929, at London; **29–54**, 1930–1955, at Cambridge; **55–81**, 1956–1978, at Oxford; **82→**, 1979→, at London.)

Nordic Jour. Bot.

Nordic Journal of Botany. Copenhagen. **1→**, 1981→. (Continuation of *Bot. Not.*, *Bot. Tidsskr.*, *Dansk Bot. Arkiv.*, *Friesia* and *Nytt Mag. Bot.*)

Norges Svalb.-Ishavs-Undersøk. Skr.

Skrifter om Svalbard og Ishavet. Oslo. **1–81**, 1929–1940; titled *Norges Svalbard- og Ishavs-Undersøkelser Skrifter*, **82–89**, 1941–1947; titled *Skrifter Norsk Polarinstitutt*, **90→**, 1948→.

Not. Syst. (Leningrad)

Ботанические Материалы Гербария главного ботанического Сада РСФСР [*Botaničeskie Materialy Gerbarija glavnogo botaničeskogo Sada RSFSR*]. / *Notulae systematicae ex Herbario Horti botanici petropolitani.* Petrograd. **1–4**, 1919–1923; titled the same in Russian and *Notulae systematicae ex Herbario Horti botanici Reipublicae rossicae*, **5**, 1924; titled **Ботанические Материалы Гербария главного ботанического Сада СССР** [*Botaničeskie Materialy Gerbarija glavnogo botaničeskogo Sada SSSR*]. / *Notulae systematicae ex Herbario Horti botanici URSS*, Leningrad, **6**, 1926; titled **Ботанические Материалы Гербария ботанического Института Академии Наук СССР** [*Botaničeskie Materialy Gerbarija botaničeskogo Instituta Akademii Nauk SSSR*]. / *Notulae systematicae ex Herbario Instituti botanici Academiae Scientiarum URSS*, **7–8**(**3**), 1937–1938; titled **Ботанические Материалы Гербария ботанического Института Имени В. Л. Комарова Академии Наук СССР** [*Botaničeskie Materialy Gerbarija botaničeskogo Instituta Imeni V. L. Komarova Akademii Nauk SSSR*]. / *Notulae systematicae ex Herbario Instituti botanice nomine V. L. Komarovii Academiae Scientiarum URSS*, **8**(**4**)**–22**, 1940–1963.

Notes Roy. Bot. Gard. Edinb.

Notes from the Royal Botanic Garden, Edinburgh. Edinburgh. **1–46**, 1900–1990; titled *Edinburgh Journal of Botany*, **47→**, 1990→.

Notizbl. Bot. Gart. Berlin

Notizblatt des königlichen botanischen Gartens und Museums zu Berlin. Leipzig. **1–15**, 1895–1944; with minor changes of title, becoming *Notizblatt des botanisches Gartens und Museums zu Berlin-Dahlem*, **7**(**68**)**–15**(**7**), 1920–1944. (**6–7**, 1913–1920, at Leipzig & Berlin; **8–15**, 1921–1944, at Berlin-Dahlem.)

Nouv. Dict. Hist. Nat.

Nouveau Dictionnaire d'Histoire naturelle, appliquée aux Arts, principalement à l'Agriculture et à l'Économie rurale et domestique: par une Société de Naturalistes et d'Agriculteurs: avec Figures tirées des trois Règnes de la Nature. Paris. 1802–1804. (**1–3** in 1802; **4–21** in 1803; **22–24** in 1804.) Ed. 2. Paris. 1816–1819. (**1–6** in 1816; **7–18** in 1817; **19–27** in 1818; **28–36** in 1819.)

Növ. Közl.

Növénytani Közlemények. Budapest. **1–7**, 1902–1908; titled ***Bot. Közl.***, *Botanikai Közlemények*, **8→**, 1909→.

Nov. Syst. Pl. Vasc. (Leningrad)

Новости систематики Высших Растений [*Novosti sistematiki vysšikh Rastenij*] / *Novitates systematicae Plantarum vascularium.* Moskva & Leningrad. 1964→. (In first 5 vols., year serves as volume number; from **6→**, 1968→, volumes are numbered on title page.)

Nova Acta Acad. Leop.-Carol.

Nova Acta physico-medica Academiae Caesareae Leopoldino-Carolinae Naturae curiosorum exhibentia Ephemerides sive Observationes Historias et Experimenta. (From **9**, 1818, titled *Nova Acta . . . curiosorum.*) Norimbergae. Ser. 1, **1–19**(**1**), 1757–1839; titled *Novorum Actorum Academiae caesareae Leopoldinae-Carolinae Naturae curiosorum. Verhandlungen der kaiserlichen Leopoldinisch-Carolinischen Akademie der Naturforscher*, **19**(**2**)**–26**, 1842–1858; titled *Novorum Actorum Academia caesareae Leopoldinae-Carolinae germanicae Naturae curiosorum. Verhandlungen der kaiserlichen Leopoldinisch-Carolinischen deutschen Akademie der Naturforscher*, **27–35**, 1860–1870; titled *Nova Acta Academiae Caesareae Leopoldino-Carolinae germanicae Naturae curiosorum. Verhandlungen der kaiserlich Leopoldinisch-Carolinischen deutschen Akademie der Naturforscher*, **36–106**, 1873–1922; titled *Nova Acta Abhandlungen der Leopoldinisch-Carolinischen deutschen Akademie der Naturforscher*, **107–108**, 1923–1926; titled *Nova Acta Abhandlungen der kaiserlich Leopoldinischen deutschen Akademie der Naturforscher*, **109–110**, 1928. Nov. ser., titled *Nova Acta Leopoldina. Abhandlungen der kaiserlich Leopoldinisch-Carolinisch deutschen Akademie der Naturforscher* (German title varies), **1→**, 1932→. (**19**(**2**)**–26**, 1842–1858, at Breslau; **27–35**, 1860–1870, at Jena; **36–106**, 1873–1922, at Dresden; **107–110**, 1923–1926, and Nov. ser. **1→**, 1932→, at Halle.)

Nova Acta Acad. Petrop.
Nova Acta Academiae Scientiarum imperialis petropolitanae. Praecedit Historia ejusdem Academiae. St. Petersburg. **1–15**, 1787–1806.

Novi Comment. Acad. Sci. Petrop.
Novi Commentarii Academiae Scientiarum imperialis petropolitanae. Petropoli. **1–20**, 1750–1776.

Novi Comment. Gotting.
Cf. *Comment. Gotting.*

Novit. Bot. Inst. Bot. Univ. Carol. Prag.
Novitates botanicae et Delectus Seminum, Fructuum, Sporarumque Anno 1960 collectorum, quae Praefectus Horti botanici Universitatis carolinae pragensis libentissime pro mutua Commutatione offert. Praga. 1960. Titled *Novitates botanicae (cum Delectu Seminum, Fructuum, Sporarum Plantarumque), quas scient. Praefectus Horti botanici Universitatis carolinae pragensis in Anno 1961* [*1962, 1963*] *libentissime pro mutua Commutatione offert*, Praga, 1961–1963; titled *Novitates botanicae cum Delectu Seminum, Fructuum, Sporarum Plantarumque quas Institutum botanicum et Hortus botanicus Universitatis carolinae pragensis in Anno 1964* [*... 1970*] *libentissime pro mutua Commutatione offert*, 1964–1970. (Title of 1967 issue, *Novitates botanicae quas Institutum botanicum Universitatis carolinae pragensis in Anno 1967 ... offert.*) Titled as *Novitates botanicae ex Instituto et Horto botanico Universitatis carolinae pragensis 1971* [1972, 1973–1975]. *Litterae taxonomicae, nomenclatoricae aut phytogeographicae nec non inventis Plantarum Florae Reipublicae socialisticae Čechoslovacae novarum imprimi breviores*, Praga, 1971, 1973 and 1976; titled *Novitates botanicae ex Universitate Carolinae*, Praha, **1**, 1982.

Nucleus (Calcutta)
Nucleus. International Journal of Cytology and allied Topics. Calcutta. **1**→, 1958→.

Nuovo Gior. Bot. Ital.
Cf. *Gior. Bot. Ital.*

Nyt. Mag. Naturvid. (Christiania)
Nyt Magazin for Naturvidenskaberne. Christiania. **1–74**, 1836–1934. (**21–26**, 1875–1881, also as ser. 2, **1–6**; **27–32**, 1882–1892, also as ser. 3, **1–6**; **33-37**, 1893–1900, also as ser. 4, **1-5**.) Titled *Nytt Magazin for Naturvidenskaperne*, Oslo, **75–88**, 1935–1951.

Nytt Mag. Bot.
Nytt Magasin för Botanikk. Oslo. **1–17**, 1952–1970; titled *Norwegian Journal of Botany*, **18–27**, 1971–1980(–1981? **27**(**4**), 1980, not received at E until 14 January 1981.) Continued as ***Nordic Jour. Bot.***, *Nordic Journal of Botany*. Copenhagen. **1**→, 1981→.

Oč. Fitosoc. Fitogeog.
Оцерки по Фитосоциологий и Фитогеографий [*Očerki po Fitosociologij i Fitogeografij*]. Moscow. 1929. (A collection of papers dedicated to V. N. Sukaev. Abstracted in *Bot. Centr.* **17**: 174, etc. (1930) under the alternative German title *Skizzen zur Phytosoziologie und Phytogeographie.*)

Op. Bot.
Cf. *Op. Bot.* (*Lund*).

Op. Bot. (Lund)
Opera botanica. A Societate botanica lundensi in Supplementum Seriei Botaniska Notiser *edita.* Stockholm. **1–60**, 1953–1980. (**13–60**, 1967–1980, at Lund.) Continued as ***Op. Bot.***, *Opera botanica.* Copenhagen. **61**→, 1981→.

Ostenia
Ostenia. Colección de Trabajos botánicos dedicados a Don Cornelio Osten. Montevideo. 1933.

Österr. Bot. Wochenbl.
Österreichisches botanisches Wochenblatt. Wien. **1–7**, 1851–1857; titled ***Österr. Bot. Zeitschr.***, *Österreichische* [*Oesterreichische*] *botanische Zeitschrift*, **8–91** & **94–122**, 1858–1942, 1947–1973 (**92–93**, 1943–1944, titled *Wiener botanische Zeitschrift*); titled ***Pl. Syst. Evol.***, *Plant Systematics and Evolution. / Entwicklungsgeschichte und Systematik der Pflanzen*, **123**→, 1975→. (German subtitle discontinued from **155**→, 1987→.)

Österr. Bot. Zeitschr.
Cf. *Österr. Bot. Wochenbl.*

Österr. Monatschr. Forstwes.
Österreichische Monatschrift für Forstwesen. Wien. **15–37**, 1863–1882.

Pallas Nagy Lexikona
A Pallas Nagy Lexikona. Budapest. (**10** in 1875. It has not been possible to ascertain whether all vols. of this encyclopaedia were published simultaneously.)

Penny Cycl.
The Penny Cyclopaedia of the Society for the Diffusion of useful Knowledge. London. 1833–1858.

Period. Soc. Méd. Cádiz
Periodico de la Sociedad médico-quirúrgica de Cádiz. Cádiz. **1–4**, 1820–1824.

Philipp. Jour. Sci. (Bot.)
The Philippine Journal of Science. Section C. Botany. Manila. **1–13**, 1906–1918. (Sectional title only in **2–13**.) Titled *The Philippine Journal of Science*, **14**→, 1919→.

Philos. Trans. Roy. Soc. Lond.
Philosophical Transactions: giving some Account of the present Undertakings, Studies, and Labours of the Ingenious in many Parts of the World. London. **1–65**, 1665–1775; titled *Philosophical Transactions of the Royal Society of London*, **66–177**, 1776–1886; titled *Philosophical Transactions of the Royal Society of London. Series B, containing Papers of a biological Character*, **178–223**, 1887–1933; titled *Philosophical Transactions ... Series B. Biological Sciences*, **224**→, 1934/1935→.)

Phytologist (Newman)
The Phytologist: a popular botanical Miscellany. Conducted by George Luxford. London. Ser. 1, edit. G. Luxford & E. Newman, **1** & **2**: pp. 1–372, 1841–1845; ... *Conducted by Edward Newman*, edit. E. Newman (only), **2**: pp. 373–end, **3–5**, 1846–1854. Ser. 2, by A. Irvine and subtitled *A botanical Journal*, **1–6**, 1855–1863. (Cf. F. A. Stafleu & R. S. Cowan, *Taxonomic Literature* ed. 2, **3**: 199 & 737 (1981).)

Phyton (Austria)
Phyton. Annales Rei botanicae. Horn. **1**→, 1948→.

Pl. Polon. Exsicc.
Rośliny Polskie. / Plantae Poloniae exsiccatae. Cracoviae.

Pl. Syst. Evol.
Cf. *Österr. Bot. Wochenbl.*

Planta
Planta. Archiv für wissenschaftliche Botanik. Zeitschrift für wissenschaftliche Biologie. Abt. E. Berlin-Dahlem. **1**→, 1925→.

Pollen et Spores
Pollen et Spores. Paris. **1**→, 1959→.

Polsk. Towarz. Bot. (Monogr. Bot.)
Polskie Towarzystwo botaniczne. Monographae botanicae. Warszawa. **1**→, 1953→.

Preslia
Preslia. Věstnik české botanické Společnosti. Praha. **1**, 1914; titled *Preslia. Věstnik československé botanické Společnosti. / Bulletin de la Société botanique tchécoslovaque à Prague. / Reports of the Czechoslovak botanical Society of Prague*, **2–15**, 1923–1936; titled *Preslia. Věstnik čs. botanické Společnosti. / Bulletin de la Société botanique tchéque à*

Prague. / Reports of the Czech botanical Society of Prague, **16–17**, 1939; titled *Preslia. Věstník české botanické Společnosti*, **18–21**, 1940–1942; titled *Preslia. Věstník československé botanické Společnosti v Praze*. **22–23**, 1948; titled *Preslia. Časopis československé botanické Společnosti*, **24**→, 1952→.

Proc. Amer. Acad. Arts Sci.
Proceedings of the American Academy of Arts and Sciences. Boston. **1–85**, 1846–1958. Continued as (?amalgamated with) *Daedalus. Proceedings*... Boston. (**1–85**, 1846–1954) **86**→, 1955→.

Proc. Roy. Irish Acad.
Proceedings of the Royal Irish Academy. Dublin. **1**→, 1930→.

Proc. Study Fauna Fl. USSR, N.S. Sect. Bot.
Материалы к Познанию Фауны и Флоры СССР, Издаваемые Московский Обществом Испытателей Природы, Новая Серия, Отдел ботанический [*Materialy k Poznaniu Fauny i Flory SSSR, Izdavaemye Moskovskii Obščestvom Ispytatelej Prirody, Novaya Seriya, Otdel botaniceskij*]. / *Proceedings of the Study of the Fauna and Flora of the USSR. New Series. Section of Botany*. Moskva. [**15(23)** in 1968. Full details of volumation unavailable.]

Publ. Cairo Univ. Herb.
Publications from the Cairo University Herbarium. Cairo. **1–7/8**, 1969–1977. (**1–6** at Cairo; **7/8** [single issue] at Koenigstein. **1** in 1970; **2** in 1969; **3** in 1970; **4** in 1971; **5** in 1972; **6** in 1975; **7/8** in 1977.) Continued as *Taeckholmia*. Koenigstein. **9**→, 1980→. (**9**, for 1978, in 1980; cf. *Taxon* **29**: 555 (1980). **10**→, 1987→, at Cairo.) *Additional Series* (*Flora of Egypt*), Cairo, **1**→, 1980→.

Publ. Inst. Biol. Apl. (Barcelona)
Publicaciones del Instituto de Biológia aplicada. Barcelona. **1**→, 1946→.

Quart. Bull. Alp. Gard. Soc.
Cf. *Bull. Alp. Gard. Soc.*

Rad. Polj. Šum. Fakult. Sarajevu
Radovi poljoprivredno-šumarskog Fakulteta, Univerziteta u Sarajevu. Sarajevo. **1**→, 1956→.

Razpr. Mat.-Prir. Akad. Ljubljani
Razprave. Matematično-Prirodoslovnega Razreda Akademije Znanosti in Umetnosti v Ljubljani. Ljubljana. **1–3**, 1940–1942; titled *Razprave. Slovenska Akademija Znanosti in Umetnosti. Ljubljana. Razred Mat., Prirod. Med. Tech. Prirod. Odsek.*, **4**, 1949; titled *Razprave. Slovenska Akademija Znanosti in Umetnosti, Razred za Prirodoslovne in Medicinske vede. / Dissertationes. Academia Scientiarum et Artium slovenica, Classis 4, Historia naturalis et Medecina*, **1–17**, 1951–1974 (**5**, for 1959, published in 1960; from **6–17**, 1961–1974, *Oddelek za Prirodoslovne vede. / Pars historico-naturalis* is added to the title.) Cf. *Cat. Périod. Bibl. Cons. Jard. Bot. Genève, Suppl. 1980–1987*: 88 (1988).

Rendic. Accad. Sci. Fis. Mat. (Napoli)
Rendiconti dell'Accademia delle Scienze fisiche e matematiche. Napoli. **1–20**, 1842–1861; titled ***Rendic. Reale Accad. Sci. (Napoli)***, *Rendiconti della Reale Accademia* ..., Ser. 1, **1–25**, 1862–1886; ser. 2, **1-8**, 1887–1894; ser. 3, **1–36**, 1895–1930; ser. 4, **1**→, 1931→.

Rep. Bot. Exch. Club Brit. Is.
The London botanical Exchange Club. Report of the Curators for 1866 (... *–1868*). London. **1866–1868**, 1867–1869 (**1867–1868**, 1868–1869, titled *Report* ... *and List of Desiderata for 1869* [*1870*]); titled *Report of the botanical Exchange Club for 1869* (... *–1877/78*), **1869–1877/1878**, 1870–1879; titled *Report of the botanical Exchange Club of the British Isles*, **1(1879–1900)**, 1880–1901; titled *Report of the botanical Exchange Club and Society of the British Isles*, **2(1901)–3(6)**, 1902–1914; titled *Report of the botanical Society and Exchange Club of the British Isles*, **4–13**, 1914/1916–1947. (**1(1879–1900)** & **2(1901–1902)**, 1880–1903, at Manchester; **2(1903)–3(4)**, 1904–1913, at Oxford; **3(5)–13**, 1913–1947, at Arbroath.)

Revista Biol.
Revista de Biologia. Lisboa, Lunda, Lourenço Marques & Rio de Janeiro. **1**→. 1956→. (**12**→ published at Lisboa and Rio de Janeiro only.)

Revue Hort. (Paris)
Revue horticole. Paris. **1**, 1829–1832. Ser. 1, **1-3**, 1832–1841. Ser. 2, **1-5**, 1841–1846. Ser. 3, **1–5**, 1847–1851. Ser. 4, **1–5**, 1852–1856. Then **1857–1865**, 1857–1865. Then **37**→, 1866→.

Rhodora
Rhodora. Journal of the New England botanical Club. Boston, Massachusetts & Providence, Rhode I. **1**→, 1891→. (**31–44**, 1929–1942, at Lancaster, Pennsylvania & Boston; **45–60**, 1943–1958, at Lancaster, Pennsylvania & Cambridge, Massachusetts; **61**→, 1959→, at Cambridge, Massachusetts.)

Rozpr. Akad. Um. (Mat.-Przyr.)
Rozprawy i Sprawozdania z Posiedzeń Wydziału matematyczno-przyrodniczego Akademii Umiejętności. Kraków. **1–20**, 1874–1890. Ser. 2, titled *Rozprawy Akademii Umiejętności. Wydział matematyczno-przyrodniczy*, **1–19**, 1891–1902. Ser. 3, titled *Rozprawy Wydziału matematyczno-przyrodniczego Akademii Umiejętności. Dział B. Nauki biologiczne*, **1–18**, 1902–1918.

Rozpr. Wydz. Mat.-Przyr. Polsk. Akad. Um. (Biol.)
Rozprawy Wydziału matematyczno-przyrodniczego Polskiej Akademii Umiejętności. Dział A/B. Nauki matematyczno-fizyczne oraz biologiczne. Kraków. Ser. 3, **19–25/26**, 1920–1928; titled *Polska Akademja Umiejętności. Rozprawy Wydziału matematyczno-przyrodniczego. Dział B. Nauki biologiczne*, 1928–1954.

Sborn. Bălg. Akad. Nauk.
Сборникъ на Българската Академия на Науките [*Sbornik na Bălgarskata Akademija na Naukitě*]. Sofija. 1911→.

Sched. Fl. Hung. Exsicc.
Jegyzék Magyarország Növényeinek Gyüjteményéhez, kiadja a Magyar Nemzeti Múzeum Növénytani Osztálya. / Schedae ad Floram hungaricam exsiccatam a Sectione botanica Musei nationalis hungarici editam. Budapest. **1-10**, 1912–1932. (**1** in 1912; **2** in 1914; **3** in 1914; **4** in 1916; **5** in 1919; **6** in 1923; **7** in 1925; **8** in 1927; **9** in 1932; **10** in 1932.)

Sched. Herb. Fl. Ross.
Schedae ad Herbarium Florae rossicae, a Sectione botanica Societatis imp. petropolitanae Naturae Curiosorum editum. / Список Растений Гербария Русской Флоры [*Spisok Rastenij Gerbarija Russkoj Flory*]. Peterburg. **1–8**, 1898–1922. (**1** in 1898; **2** in 1900; **3** in 1901; **4** in 1902; **5** in 1905; **6** in 1908; **7** in 1911; **8** in 1922.) Contributors include B. M. Koso-Poliansky, D. I. Litvinov and J. N. Woronow. Titled ***Sched. Herb. Fl. URSS.***, *Schedae ad Herbarium Florae URSS* / Список Растений Гербария Флоры СССР [*Spisok Rastenij Gerbarija Flory SSSR*]. **9**→, 1932→. (**9** in 1932; **10** in 1936; **11** in 1949; **12** in 1953; **13** in 1955; **14** in 1957; **15** in 1963; **16** in 1966; **17** in 1967; **18** in 1970; **19** in 1972.) **12** & **13** lack Latin title and are abbreviated as ***Spisok Rast. Gerb. Fl. SSSR***.

Schriftenreihe Österr. Ges. Holzforsch.
Schriftenreihe der Österreischischen Gesellschaft für Holzforschung. Vienna. **1–6**, 1950–1956.

Sci. Mag. Biol. (Kharkov)
Наукові Записки Біології [*Naukovi Zapysky Biologiji*]. / *Scientific Magazine of Biology*. Kharkov. **1**→, 1927→.

Scient. Pharmac.
Scientia pharmaceutica. Wien. 1930→.

Séances Publiq. Acad. Sci. Besançon
Séances publiques de l'Académie des Sciences, Belles-lettres et Arts de Besançon. Besançon. 1754–1813?; titled *Académie des Sciences, Belles-lettres et Arts, de Besançon. Séance[s] publique[s]*, 1814?–1858, 1866–1875. (1859–1865, titled *Mémoires et Documents de l'Académie des Sciences, Belles-lettres et Arts, de Besançon.*)

Sitz.-Ber. Böhm. Ges. Wiss.
Sitzungsberichte der königl. böhmischen Gesellschaft der Wissenschaften in Prag. Prag. 1859–1873; titled *Zprávy o Zasedáni královské české Společnosti Nauk v Praze. / Sitzungsberichte der königl. böhmischen Gesellschaft der Wissenschaften in Prag*, 1874–1885; titled *Zprávy o Zasedáni královské české Společnosti Nauk. Třida mathematicko-přirodovědecká. / Sitzungsberichte der königl. böhmischen Gesellschaft der Wissenschaften. Mathematisch-naturwissenschaftliche Classe*, 1886; titled ***Vestn. Král. České Společn. Nauk, Tř. Math.-Přir.***, *Věstnik královské české Společnosti Nauk. Třida mathematicko-přirodovědecká. / Sitzungsberichte der königl. böhmischen Gesellschaft der Wissenschaften. Mathematisch-naturwissenschaftliche Classe*, 1887–1915; titled *Věstnik královské české Společnosti Nauk. Třida mathematicko-přirodovědecká. / Mémoires de la Société royale des Sciences de Bohême. Classe des Sciences*, 1919–1935; titled *Věstnik královské české Společnosti Nauk. Třida mathematicko-přirodovědecká. / Mémoires de la Société royale des Lettres et des Sciences de Bohême. Classe des Sciences*, 1936–1953. (1940–1944 lacks French title.)

Skr. Kiøbenhavnske Selsk. Lærd. Vid.
Skrifter som udi det Kiøbenhavnske Selskab af Lærdoms og Videnskabers Elskere ere fremlagte og oplæste. København. **1–10**, 1745–1770; titled *Skrifter som udi det kongelige Videnskabers Selskab ere fremlagte, og nu til Trykken befordrede*, **11–12**, 1777–1779. Continued as ***Kong. Danske Vid. Selsk.***, *Nye Samling af det kongelige danske Videnskabers Selskabs Skrifter*, **1–5**, 1781–1799. Continued as ***Danske Vid. Selsk.***, *Det kongelige danske Videnskabers-Selskabs Skrivter.* Ser. 3, **1-7**, 1800–1818. Several later series, with minor title variations, until 1938. Continued as *Biologiske Skrifter. Kongelige danske Videnskabernes Selskab*, **1→**, 1939→.

Skr. Vid.-Selsk. Christ.
Skrifter udgivne af Videnskabsselskabet i Christiania. Matematisk-naturvidenskabelig Klasse. Christiania. 1894–1925. (With minor changes of title, becoming *Skrifter utgit av Videnskapsselskapet i Kristiania*, Kristiania, 1911–1925.)

Sommerfeltia
Sommerfeltia. Oslo. **1→**, 1985→.

Specch. Sci.
Specchio delle Scienze o Giornale enciclopedico di Sicilia. Palermo. **1–2**, 1814.

Spraw. Kom. Fizyogr. Krakow.
Sprawozdanie Komisyi fizyjograficznej c. k. Towarzystwa naukowego krakowskiego. Kraków. **1–6**, 1867–1872; titled *Akademia Umiejętności w Krakowie. Sprawozdanie Komisyi fizyjograficznej*, **7–52**, 1873–1918 (from **27**, 1892, *...fizyograficznej...*); titled *Polska Akademia Umiejętności. Sprawozdania Komisji fizjograficznej*, **53–73**, 1920–1939.

Svensk Bot. Tidskr.
Svensk botanisk Tidskrift. Stockholm. **1→**, 1907→. (**16→**, 1922→, at Uppsala.)

Syll. Pl. Nov. Ratisbon. (Königl. Baier. Bot. Ges.)
Sylloge Plantarum novarum itemque minus cognitarum a praestantissimis Botanicis adhuc viventibus collecta et a Societate regia botanica ratisbonensi edita. (Königlich-baierische botanische Gesellschaft in Regensburg.) Ratisbonae. **1–2**, 1824–1828. (**1** in 1824; **2** in 1828.)

Symb. Bot. Upsal.
Symbolae botanicae upsalienses. Arbeten från botaniska Institutionen Uppsala. Uppsala. **1→**, 1932→.

Syst. Bot.
Systematic Botany. Quarterly Journal of the American Society of Plant Taxonomists. Tallahassee, Florida. **1→**, 1976→. (**3(4)–7(1)** 1978–1982, at Ann Arbor, Michigan; **7(2)–10(4)**, 1982–1985, at Kent, Ohio; **11→**, 1986→, at New York.)

Taxon
Taxon. Official News Bulletin of the international Association for Plant Taxonomy. Utrecht. **1–16**, 1951–1967; titled *Taxon. Journal of the International Association for Plant Taxonomy*, **17→**, 1968→. (**37→**, 1988→, at Berlin.)

Term. Füz.
Természetrajzi Füzetek. Budapest. **1–25**, 1877–1902.

Term.-Tud. Közl.
Természettudományi Közlöny. Budapest. **1–76**, 1869–1944; **88** (= Nov. ser. **1**)→, 1957→. (Titled *Természettudományi*, **1–3**, 1946–1948; titled *Természet és Technika*, **108–112**, 1949–1953; titled *Természet és Társadalam*, **113–115**, 1954–1956.) ***Pótfüz.***, *Pótfüzetek a Természettudományi Közlönyhoz.* **1–88**, 1888–1907.

Trab. Dep. Bot. Fisiol. Veg.
Trabajos del Departamento de Botanica y Fisiologia vegetal, Universidad de Madrid. Madrid. **1→**, 1968→.

Trans. Acad. Sci. St. Louis
Transactions of the Academy of Science of St. Louis. St. Louis, Missouri. **1→**, 1856→.

Trans. Amer. Philos. Soc.
Transactions of the American philosophical Society, held at Philadelphia for promoting useful Knowledge. Philadelphia, Pennsylvania. **1-6**, 1771–1809. Nov. ser. **1→**, 1818→.

Trans. Cambr. Philos. Soc.
Transactions of the Cambridge philosophical Society. Cambridge. **1–23**, 1821 or 1822–1928. (Cf. *British Library KIST* microfiche catalogue, which gives different starting dates for four different holdings.)

Trans. Linn. Soc. London
Transactions of the Linnean Society. London. Ser. 1, **1–6**, 1791–1802; titled *Transactions of the Linnean Society of London*, **7–30**, 1804–1875; Ser. 2, titled ***Trans. Linn. Soc. London (Bot.)***, *Transactions of the Linnean Society of London. Botany*, **1–9**, 1875–1922. Ser. 3, titled *Transactions of the Linnean Society of London*, **1**, 1939–1955. (**1(1)** in 1939; **1(2)** in 1940; **1(3)** in 1955; *Botany* and *Zoology* are combined.)

Trans. Linn. Soc. London (Bot.)
Cf. *Trans. Linn. Soc. London.*

Trans. Proc. Bot. Soc. Edinb.
Transactions of the botanical Society. Edinburgh. **1–11**, 1844–1873; titled *Transactions and Proceedings of the botanical Society*, **12–19**, 1873–1893 ('*and Proceedings*' not in title of **16–18**, 1886–1891); titled *Transactions and Proceedings of the botanical Society of Edinburgh*, **20–40**, 1894–1969; titled ***Trans. Bot. Soc. Edinb.***, *Botanical Society of Edinburgh. Transactions*, **41–45**, 1970–1991; titled *Botanical Journal of Scotland*, **46(1)** (for 1991), 1992→.

Trans. Proc. N.Z. Inst.
Transactions and Proceedings of the New Zealand Institute. Wellington. **1–63**, 1868–1934 (**63**, 1933/1934, at Dunedin.) Titled *Transactions and Proceedings of the Royal Society of New Zealand*, Dunedin, **64–79**, 1934–1952; titled *Transactions of the Royal Society of New Zealand*, Dunedin, **80–88**, 1952–1966. Nov. ser., titled *Transactions of the Royal Society of New Zealand. Botany*, **1–3**, 1961–1969; titled *Transactions of the Royal Society of New Zealand.*

Biological Sciences, **11–12**, 1968–1970 (with volume numbering continuing that of *Zoology*); titled *Journal of the Royal Society of New Zealand*, **1→**, 1971→.

Trav. Inst. Bot. (Charkov)

Travaux de l'Institut botanique. / Труди Н.-Д. Інстіїуту ботаніки [*Trudy N.-D. Ynstytutu botanyki*]. / *Traveaux de l'Institut botanique.* Charkov. **1–3**, 1936–1938. (**3**, 1938, also titled Учені Записки Харівського державного Університету Ім. О. М. Горького **14** [*Učeny Zapiski Kharivs'kogo deržavnogo Ynyversitety Im. O. M. Gor'kogo* **14**.] / *Kharkov A Gorky State University, Book* **14**, *Proceedings of the Botanical Institute.*)

Trav. Mus. Bot. Acad. Pétersb.

Труды ботаническаго Музея императорской Академии Наук [*Trudy botaničeskago Muzeja imperatorskoj Akademii Nauk*]. / *Travaux du Musée botanique de l'Académie impériale des Sciences de Saint Pétersbourg* [*de Petrograd*]. St. Petersburg (Petrograd). **1–16**, 1902–1916; titled Труды ботаническаго[ого]) Музея Российской Алкадемии Наук [*Trudy botaničeskago*[*ogo*] *Muzeja Rossijskoj Akademii Nauk*]. / *Travaux du Musée botanique de l'Académie des Sciences de Russie.* Petrograd. **17–18**, 1918–1920; titled Труды ботанического Музея [*Trudy botaničeskogo Muzeja*]. / *Travaux du Musée botanique.* Leningrad. **19–25**, 1928–1932.

Trav. Mus. Géorgie

საქართველოს მუზეუმის შრომები.
[*Sakarthk'elos Mouzoumis šromebi*]. / *Travaux du Musée de Géorgie.* Tph'lissi. 1920–1933.

Treb. Inst. Catalana Hist. Nat.

Treballs de la Institució catalana d'Historia natural. Barcelona. **1–7**, 1915–1924.

Tromsø Mus. Aarshefter

Tromsø Museums Aarshefter. Tromsø. **1–70**, 1878–1947.

Trudy Kommiss. Kiev. Učebn. Okr.

Труды Коммиссии, высочайше учрежденной при императорском Университете Св. Владимира, для Описания Губернии Киевскаго Учебнаго Округа. [*Trudy Kommissii, vysočajše učreždennoj pri imperatorskom Universitete Sv. Vladimira, dlja Opisanija Gubernii Kievskago Učebnago Okruga*]. Kiev.

Trudy Nauč.-Issled. Inst. Biol. (Har'kov)

Труди научно-исследовательского Института Біології. Праці Науково-дослідного Институту Біології [*Trudy naučno-issledovatel'skogo Instituta Biologii. Praci naukovo-doslidnogo Instytutu Biologiji*]. Later titled Труді ... Віології і біологического Факульътета, Праці ... Біології і біологічного Факультету [*Trudy ... Biologii i biologičeskogo Fakul'teta. Praci ... Biologiji i biologičnogo Fakul'tetu*]. Har'kov. **12–35**, 1947–1963; **37→**. 1963→ (**1**, 1936, and **3–11**, 193?–1941, titled Праці науковогдослідного зоолого-біологічного Институту [*Praci naukovogdoslidnogo zoologo-biologičnogo Institutu*] / Трауді научноисследовательского зоолого-біологического Института. [*Traudi naučnoissledovtel'skogo zoologo-biologičeskogo Instituta*]. / *Proceedings of the zoological-biological Institute for scientific Research*; **2**, 1934, titled Праці зообіологічного Институту. Труді зообіологического Института [*Praci zoobiologičnogo Instytutu. Trudy zoobiologičeskogo Instituta*]; **36**, 1963, titled Труди биологического Факультета по Генетіке і Зоології [*Trudy biologičeskogo Fakul'teta po Genetike i Zoologii*].

Trudy Odessk. Obšč. Sad.

Труды Одесскаго Общества садтсва [*Trudy Odesskago Obščestva sadstva*]. / *Travaux de la Société d'Horticulture d'Odessa.*

Trudy Petergof. Est.-Nauč. Inst.

Труды Петергофского естественно-научного Института [*Trudy Petergofskogo estestvenno-naučnogo Instituta*]. / *Travaux de l'Institut des Sciences naturelles de Peterhoff.* Peterhoff. **1–8**, 1925–1952; titled Труды Петергофского биологического Института [*Trudy Petergofskogo biologičeskogo Instituta*]. / *Travaux de l'Institut biologique de Péterhof.* Moscow. **9–13/14**, 1932–1935; **16–17**, 1938–1939. (**15**, 1935, at Leningrad and titled Труды Петергофского биологического Института Ленинградского государственного Университета [*Trudy Petergofskogo biologičeskogo Instituta Leningradskogo gosudarstvennogo Universiteta*].)

Trudy Tifliss. Bot. Sada

Труды Тифлисскаго ботаническаго Сада [*Trudy Tiflisskago botaničeskago Sada*]. / *Travaux du Jardin botanique de Tiflis* (from **22**) Труды Тифлисского ботанического Сада [*Trudy Tiflisskogo botaničeskogo Sada*]. / ტფილისის ბოტანიკურ ბაღის შრომები [*Thbilisis botanikuri bagir šromebi*]. / *Travaux du Jardin botanique de Tiflis.* Tiflis. **1–20**, 1895–1917; **21–26** (= ser. 2, **1–6**), 1920–1933/1934. Continued as ტფილისის ბოტანიკურ შრომები [*Thbilisis botanikuri šromebi*]. / Труды Тбилиского ботанического Сада [*Trudy Tbilisskogo botaničeskogo Sada*]. / *Travaux du Jardin botanique du Tibilissi*, Tiflis (Tbilisi), **27–38** (= ser. 2, **7–10** & **6–13**, nos. **6–10** being used twice), 1938–1949.

Ukr. Bot. Žur.

Українськиї ботанічниї Журнал [*Ukrajinskyji botaničnyji Žurnal*]. / *The Ukrainian botanical Review.* Kiev. **1–5**, 1922–1929. Continued as ***Zur. Inst. Bot. URSR,*** Журнал Інституту ботаніки АН УРСР [*Žurnal Instytutu botaniky AN URSR*]. / *Journal de l'Institut botanique de l'Académie des Sciences de la RSS d'Ukraine.* **1–23**, 1934–1940. Continued as ***Ukr. Bot. Žur.,*** Ботанічниї Журнал [*Botaničnyj Žurnal*]. / *Journal botanique de l'Académie des Sciences de la RSS d'Ukraine.* Nov. ser., **1–12**, 1940–1955. Titled Український ботанічний Журнал [*Ukrajinskyji botaničnyji Žurnal*], **13→**, 1956→.

Univ. Izv. (Kiev)

Университетскія Извѣстія [*Unyversytetskija Yzvěstija*]. Kiev. **1–59**, 1861–1919.

Verh. Bot. Ver. Brandenb.

Verhandlungen des botanischen Vereins für die Provinz Brandenburg und die angrenzenden Länder. Berlin. **1–115**, 1859–1980. (From **12**, 1870, ... *und die angrenzenden Länder* is dropped; from **14**, 1872, titled *Verhandlungen des botanisches Vereins der Provinz Brandenburg.*) Continued as *Verhandlungen des Berliner botanischen Vereins*, **1→**, 1982→.

Verh. Ges. Vaterl. Mus. Böhm.

Verhandlungen der Gesellschaft des vaterländischen Museums in Böhmen. / *Časopis Českeho Museum.* Praha. **1–4**, 1823–1826; **10–34**, 1832–1856. (Vols. **5–9** not used; titled *Monatsschrift der Gesellschaft des vaterländischen Museums in Böhmen*, 1827–1829; titled *Jahrbücher des böhmischer Museums für Natur- und Landerkunde, Geschichte, Kunst, und Literatur*, **1–2**, 1830–1831.)

Verh. Mitt. Siebenb. Ver. Naturw.

Verhandlungen und Mittheilungen [*Mitteilungen*] *des siebenbürgischen Vereins für Naturwissenschaften zu Hermannstadt.* Hermannstadt. **1–90**, 1850–1940; titled *Mitteilungen der Arbeitsgemeinschaft für Naturwissenschaften Sibiu-Hermannstadt*, **91–92**, 1941–1942.

Verh. Naturf. Ges. Basel

Verhandlungen der naturforschenden Gesellschaft in Basel. Basel. **1→**, 1854→.

Verh. Naturf. Ver. Brünn

Verhandlungen des naturforschenden Vereins in Brünn. Brünn. **1–74**, 1862–1944.

Verh. Schweiz. Naturf. Ges.

Eröffnungsrede der Jahresversammlung der allgemeinen schweizerischen Gesellschaft für die gesammten Wissenschaften. Zurich. 1817 &

1819; titled *Discours ..., en ouvrant la première Séance de la Réunion périodique de la Sociéte helvétique des Sciences naturelles*. Lausanne. 1818; titled *Discours d'Ouverture de la Session de la Société helvétique des Sciences naturelles*, Lausanne, 1820; titled *Eröffnungs-Rede der Jahresversammlung der allgem. schweizerischen Gesellschaft für gesammte Naturwissenschaften*, Zurich. **7**, 1821; titled *Eröffnungs-Rede der schweizerischen Gesellschaft für die gesammten Naturwissenschaften*, Zurich, **8**, 1823; titled *Kurze Übersicht der Verhandlungen der allgemeinen schweizerischen Gesellschaft für die gesammten Naturwissenschaften*, Aarau, **9–10**, 1823–1824; titled *Verhandlungen der allgemeinen schweizerischen Gesellschaft für die gesammten Naturwissenschaften*, Solothurn, **11–13**, 1825–1827, **16**, 1831, **19**, 1835, **21**, 1837; titled *Actes de la Société helvétique des Sciences naturelles*, Lausanne, **14–15**, 1829–1830, **17**, 1832, **22**, 1837, **25**, 1841, **28**, 1845; titled *Atti della Società elvetica delle Scienze naturali*, Lugano, **18**, 1833 & **44**, 1861; titled *Verhandlungen der schweizerischen naturforschenden Gesellschaft*, Basel, **23–24**, 1838–1839, **26–27**, 1842–1843, **29**, 1845, **31–43**, 1847–1860 & **46–140**, 1863–1960 (with changes of main title and sometimes alternative titles in French and/or Italian; **45**, Lausanne, 1862, titled *Actes de la Société suisse des Sciences naturelles*); titled *Verhandlungen der schweizerischen naturforschenden Gesellschaft, Wissenschaftlicher Teil*, Basel, **141–157**, 1961–1977; titled *Jahrbuch der schweizerischen naturforschenden Gesellschaft. Wissenschaftlicher Teil*, Bern, **158→**, 1978→.

Verh. Zool.-Bot. Ges. Wien
Verhandlungen des zoologisch-botanischen Vereins in Wien. Wien. **1–7**, 1852–1857; titled *Verhandlungen der kaiserlich-königlichen zoologisch-botanischen Gesellschaft in Wien*, **8–67**, 1858–1918; titled *Verhandlungen der zoologisch-botanischen Gesellschaft in Wien*, **68→**, 1918→.

Veröff. Überseemus. Bremen A
Veröffentlichungen aus dem Übersee-Museum Bremen. Reihe A. Bremen. **1→**, 1952→.

Verz. Tausch Säm. Früchte Zürich
Verzeichnis im Tausch abgebbarer Sämereien und Früchte des botanischen Gartens der Universität Zürich. Zürich. 1826→. Year is used for vol. no.

Věstn. Král. České Společn. Nauk. Tř. Mat.-Přír.
Cf. *Sitz.-Ber. Böhm. Ges. Wiss.*

Vid. Meddel. Dansk Naturh. Foren. Kjøbenhavn
Videnskabelige Meddelelser fra den naturhistoriske Forening i Kjöbenhavn. Kjöbenhavn. **1–63**, 1849–1912; titled *Videnskabelige Meddelelser fra dansk naturhistorisk Forening i Kjøbenhavn*, **64→**, 1913→. Years used as vol. nos. until **63**, 1912.

Vidensk. Selsk. Skr.
Skrifter udgivne af Videnskabs-Selskabet: Christiana. Mathematisk-naturvidenskabelig Klasse. Christiania. 1894–1924.

Viert. Naturf. Ges. Zürich
Vierteljahrsschrift der naturforschenden Gesellschaft in Zürich. Zürich. **1→**, 1856→.

Watsonia
Watsonia. Journal of the botanical Society of the British Isles. Arbroath. **1→**, 1949→. (**3→**, 1953→, at London; **8–16**, 1970–1987, subtitled *Journal and Proceedings of the botanical Society of the British Isles*.)

Webbia
Webbia. Raccolta di Scritti botanici pubblicati in Occasione del 50° Anniversario della Morte di Filippi Barker Webb. Firenze. **1–5**, 1905–1923 ('*pubblicati ... Webb*' only in **1**, 1905); **6→**, 1948→. (Publication suspended 1924–1947; from **39→**, 1985→, subtitled *Rivista internazionale di Sistematica e Fitogeografia. / International Journal of Plant Taxonomy and Geography*.)

Wentia
Wentia. Amsterdam. **1–17**, 1959–1966.

Willdenowia
Willdenowia. Mitteilungen aus dem botanischen Garten und Museum Berlin-Dahlem. Berlin-Dahlem. **1(2)→**, 1953→. (**1(1)** with above title, omitting *Willdenowia*.)

Wiss. Zeitschr. Friedrich-Schiller Univ., Math.-Nat. Reihe
Wissenschaftliche Zeitschrift der Friedrich-Schiller-Universität Jena/Thüringen. Jena. **1**, 1951–1952; titled *Wissenschaftliche Zeitschrift der Friedrich-Schiller Universität Jena/Thüringen. Mathematisch-naturwissenschaftliche Reihe*, **2→**, 1952→. (*Thüringen* not in all vols.; series title becomes *Naturwissenschaftliche Reihe* from **33→**, 1984.)

Wiss. Zeitschr. Univ. Halle
Wissenschaftliche Zeitschrift der Martin-Luther-Universität Halle-Wittenberg. Halle. **1→**, 1951→. (***Math.-Nat. Reihe,*** *Mathematisch-naturwissenschaftliche Reihe*, is added to title from **3→**, 1953→.)

Wochenschr. Gartn. Pflanzenk.
Verhandlungen des Vereins zur Beförderung des Gartenbaues in den königlich preussischen Staaten. Berlin. **1–20**, 1824–1853. Nov. ser. 1854–1860. (Also numbered **1–28**.) Continued as *Wochenschrift für Gärtnerei und Pflanzenkunde*, **1–2**, 1858–1859; titled *Wochenschrift des Vereines zur Beförderung des Gartenbaues in den königlich preussischen Staaten für Gärtnerei und Pflanzenkunde*, **3–15**, 1860–1872; titled *Monatsschrift des Vereins zur Beförderung des Gartenbaues in den königlich preussischen Staaten für Gärtnerei und Pflanzenkunde*, **16–21**, 1873–1878; titled *Monatsschrift des Vereins zur Beförderung des Gartenbaues in den königlich preussischen Staaten und der Gesellschaft der Gartenfreunde Berlins*, **22–24**, 1879–1881; titled *Garten-Zeitung. Monatsschrift für Gärtner und Gartenfreunde*, **1–4**, 1882–1885; titled *Garten-Zeitung. Deutsche Wochenschrift für Gärtner und Gartenfreunde*, **1**, 1886; titled *Verhandlungen des Vereins zur Beförderung des Gartenbaues und der Gesellschaft der Gartenfreunde Berlins*, 1887–1893.

Zeitschr. Ferdinand. Tirol (Innsbruck)
Zeitschr. für Tirol und Vorarlberg. Innsbruck. Ser. 1, **1–8**, 1825–1834. Ser. 2, titled *Neue Zeitschrift des Ferdinandeums für Tirol und Vorarlberg*, **1–12**, 1835–1846. Ser. 3, titled *Zeitschrift des Ferdinandeums für Tirol und Vorarlberg*, **1–60**, 1852–1920.

Zeitschr. Indukt. Abstamm. Vererbungslehre
Zeitschrift für induktive Abstammungs- und Vererbungslehre. Berlin. **1–88,** 1908–1957; titled *Zeitschrift für Vererbungslehre*. **89→**, 1958→.

Zeitschr. Sukkulentenk.
Zeitschrift für Sukkulentenkunde. Berlin. **1-3**, 1923–1928.

APPENDIX IV

GLOSSARY OF TECHNICAL TERMS

The number of technical terms used in *Flora Europaea* has been kept as low as is consistent with a reasonable standard of accuracy and brevity. Most of them are used in well-established traditional senses, and their meanings may be ascertained by reference to glossaries such as H. I. Featherly, *Taxonomic Terminology of the Higher Plants* (Ames, Iowa, U.S.A., 1954). No term is used in a sense inconsistent with that given by Featherly.

Experience has shown, however, that some useful terms are liable to misinterpretation, and others, which can be used in a wider sense, are used in a restricted sense in *Flora Europaea*. This glossary is intended simply to indicate without ambiguity the sense in which these potentially ambiguous terms are employed.

Certain technical terms, which are restricted to descriptions in particular families or genera, are explained under the family or genus concerned.

ABOVE Used to indicate both the upper surface of a normally horizontal organ and the upper part of an organ or of the whole plant.

ACHENE A small, dry, 1-seeded, indehiscent fruit, whether derived from a superior or from an inferior ovary.

ALTERNATE Arising singly at a node; includes regularly spiral, as well as distichous arrangements.

ANNUAL Completing its life-cycle from seed to seed in less than 12 months; includes 'overwintering' annuals, which germinate in autumn and flower the following year.

BELOW Used to indicate the basal part of a plant, stem or inflorescence; cf. *beneath*.

BENEATH Used to indicate the lower surface of a normally horizontal organ; cf. *below*.

BIDENTATE With two teeth

BISERRATE Serrate, with the teeth themselves serrate.

CADUCOUS Falling unusually early.

CILIATE With hairs on the margin.

DECIDUOUS Of leaves: falling in autumn; of other organs: falling before the majority of adjacent or associated organs.

ERECTO-PATENT Diverging at an angle of 15–45° from the axis on which the structure is borne.

FLOCCOSE Clothed with woolly hairs, which are disposed in tufts or tend to rub off and adhere in small masses.

GLABRESCENT Becoming glabrous with increasing age or maturity. For structures very slightly but persistently hairy the term *subglabrous* is used.

HIRSUTE Covered with long, moderately stiff and not interwoven hairs.

HISPID Covered with stiff hairs or bristles.

LANATE Covered with soft, flexuous, intertwined hairs.

PELTATE Denotes an organ of which the stalk is attached to a more or less flat surface, and not to the margin; the attachment is not, however, necessarily central.

PUBERULENT With very short hairs.

PUBESCENT With soft, short hairs.

PYRENE A small stone, consisting of one or few seeds with a hard covering, enclosed in fleshy tissue, e.g. *Crataegus*, *Ilex*.

SEMI-PATENT Between patent and appressed.

SERICEOUS With silky, appressed hairs.

SETOSE Covered with stout, rigid bristles.

SIMPLE HAIR Indicates an unbranched hair; it may or may not bear a gland.

STOCK The persistent, usually somewhat woody base of an otherwise herbaceous perennial.

STOLON A short-lived, horizontal stem, either above or below the surface of the ground, rooting at one or more nodes.

STRIGOSE With stiff, appressed, straight hairs.

TERETE More or less cylindrical, without grooves or ridges.

TOMENTOSE With hairs compacted into a felty mass.

TUBERCULATE Covered with smooth, knob-like elevations.

VELUTINOUS With a dense indumentum of fine, soft, straight hairs.

VERRUCOSE Covered with rough, wart-like elevations.

VILLOUS Covered with long, soft, straight hairs.

APPENDIX V

VOCABULARIUM ANGLO-LATINUM

IN USUM LECTORUM LINGUAE ANGLICAE MINUS PERITORUM CONFECTUM

N.B. Plurimi termini ad descriptionem botanicam in lingua anglica usurpati aequipollentibus latinis persimiles sunt, e.g. *ovate* (ovatus), *inflorescence* (inflorescentia). Talia verba omnia sunt omissa.

above insuper, supra, super
all omnes
almost fere, paene
always semper
arable fields arva
around circum
arranged dispositus
awn arista
back dorsum
backward(s) retro
bank ripa
barbed pilis hamatis obsitus
bare nudus
bark cortex
basin-shaped pelviformis
beak rostrum
bearded barbatus
become fieri
below infra, sub
beneath infra, subtus
bent inflexus
berry bacca
between inter
bind colligare, firmare
bitter amarus
black niger, ater
blue caeruleus
bloom pruina
boat navicula
border margo
borne prolatus
branch ramus
breadth latitudo
bright laete
bristle seta
broad latus
brown fuscus, brunneus
bud gemma
bushy spisse et iteratim ramosus
catkin amentum
chaffy paleaceus
chamber loculus
chequered cancellatus
chestnut castaneus
chief principalis
claw unguis
cliff rupes
climbing scandens
close propinquus, affinis
closed clausus
clothed vestitus
cluster glomerulus
coarse crassus, grossus
coast litus, ora
coat tunica
common vulgaris
completely omnino, ex toto
compound compositus
cone strobilus
corner angulus
cornfield seges
covered obtectus
cream ochroleucus, albido-flavescens
crevice fissura
crimson kermesinus, sanguineus; ut flos *Papaveris rhoeadis* coloratus
crowded confertus
cultivated cultus, sativus
curled crispus
cushion pulvinus
damp humidus
dark obscurus
dead emortuus
decay dissolutio
deep profundus; intense
developed evolutus
die mori
docks navalia
downwards deorsum
downy lanuginosus
dry siccus
dull opacus; impolitus
dwarf nanus
early prius, mox, praecoce
eastern orientalis
eastwards orientem versus
edge margo
edible edulis
either ... or aut ... aut
end pars terminalis
enlarge crescere, augere
entire integer
entirely omnino
equal aequalis, aequans
established subspontaneus
evening vesper
evergreen sempervirens
exceeding superans
face facies
fan-shaped flabellatus
female femineus, pistillatus
feebly debiliter, perleviter
few pauci
finely subtiliter
first primus
flap valva, ligula
flat planus
flattened compressus, applanatus
flax *Linum usitatissimum*
flesh-coloured carneus, pallide et opace roseus
fleshy carnosus
floating natans
flooded inundatus
flower flos
fodder bestiarum pabulum
fold plica
following sequens
forest silva magna
forwards porro
free liber
fringe fimbriae
fruit fructus
furnished munitus
furrow sulcus
garden hortus
glossy nitidus
grassy graminosus
gravelly glareosus
green viridis
grey cinereus
grooved canaliculatus, sulcatus
ground solum
grow crescere, habitare
hair pilum
hairy pilis munitus
half dimidium
hard durus
head caput, capitulum
heath ericetum, callunetum
hedge saepes
helmet galea
hill collis
hoary incanus
hollow fistulosus, cavus; cavum, excavatio
hood cucullus

hooked uncinatus
inner interior, internus
inside intus, intra; pagina vel pars interior
introduced inquilinus, allatus
jagged argutus
jointed articulatus
juice succus
keel carina
lake lacus
late sero
later postea
leaf folium
leafless foliis carens
leaflet foliolum
length longitudo
less minus
level altitudo
light clare
limestone calx
lip labium
locally hic inde
low humilis, pusillus
lower inferior
lowland campestris, planitiem incolens
main principalis
male masculus, stamineus
many multi
marbled marmoratus
marsh palus
mat stratum e ramulis procumbentibus intertextis compositum
meadow pratum
mealy farinosus
medicinal officinalis
middle pars centralis; medius
midrib costa, folii nervus principalis
milky lacteus
mistake error
more plus, magis
most plerique, pars major
mountain mons
mouth os
much multo, multum
naked nudus
narrow angustus
native indigenus
naturalized subspontaneus
near prope
nearly paene, fere
neither... nor nec... nec
net reticulum
never numquam
nodding nutans, cernuus
none nulli
northern borealis
northwards septentrionem versus
notch incisio
nut nux
often saepe
open apertus
orange aurantiacus
ornament decus
other alius, alter
otherwise aliter
outer exterior, externus
outside extra; pagina vel pars exterior
overlapping imbricatus
pale pallidus
papery chartaceus
pasture pascuum
patch macula
peat-bog turbarium
pink roseus
pitted foveolatus
planted cultus
point acumen
pond stagnum
pool stagnum
prickle aculeus
purple purpureus
quarter pars quarta
rank ordo
rarely raro
ray radius
red ruber
related affinis
remains reliquiae
rest ceteri
rib costa
rice-field oryzetum
ridge carina
ring annulus
river flumen
road via
rock saxum, rupes
root radix
rosette rosula
rough asper
rounded rotundatus
rust-coloured ferrugineus
salt-marsh palus salsa
sand arena
scale squama
scanty exiguus
scar cicatrix
scarcely vix
scarlet laete et clare ruber, paullulo aurantiaco affectus; ut flos *Salviae splendentis* coloratus
scattered sparsus
scented fragrans
scree clivus alpestris, saxis deorsum conjectis copertus
sea mare
seed semen
seldom raro
several nonnulli, complures
shady umbrosus
shallow haud profundus
shape forma
sharply acute
sheath vagina
shelter tegmen contra ventum
shingle glarea maritima vel fluviatilis
shiny nitidus
shoot caudiculus, surculus
shore litus, ora
short brevis
shoulder angulus obtusus
shrub frutex
side latus, pagina
silky sericeus
silvery argenteus
slender tenuis, gracilis
slightly leviter, paullo
slipper calceolus
slit rima, foramen longum sed angustum
slope clivus, declivitas
small parvus
smell odor
smooth laevis
snow-patch locus in montibus ubi nix sero perdurat
soft mollis
soil solum
sometimes interdum
southern australis
southwards meridiem versus
spikelet spicula
spot punctum, macula
spreading patens, divaricatus
spring ver
spur calcar
square quadratus
stalk stipes
standard vexillum
stem caulis
stiff rigidus
stock caudex
stony lapidosus
stout crassus, robustus
straight rectus
streak linea
stripe vitta
strong robustus, validus
suddenly abrupte
summer aestas
sunk immersus
surface superficies, pagina
sweet dulcis
swollen tumidus, inflatus
tall altus
tawny fulvus
teeth dentes
thick crassus, densus, spissus
thicket dumetum
thin tenuis
third pars tertia
timber materia; lignum ad usum hominum aptum
tinged suffusus
tip apex
tipped ad apicem munitus vel tinctus
tooth dens
top vertex
tough lentus

tree arbor
tufted in fasciculos dispositus, caespitosus
twice bis
twig ramulus, virga
twining volubilis
twisted contortus
uncertain incertus, dubius
undivided indivisus
unequal inaequalis
united conjunctus, connatus
upper superior
uppermost supremus
upwards sursum
usually plerumque
veil velum
vein nervus
velvety velutinus
violet violaceus
wart verruca
waste incultus
weak debilis, flaccidus
well bene
western occidentalis
westwards occidentem versus
wet madidus
white albus, candidus
whorled verticillatus
wide latus
widespread late diffusus
width latitudo
wing ala
winter hiems
wiry filo ferreo similis
withered marcidus
without sine
wood silva; lignum
woody lignosus
woolly lanatus
wrinkled rugosus
yellow flavus, luteus

INDEX

INDEX

Generic names adopted in *Flora Europaea* are printed in **bold-face** type; specific and subspecific epithets adopted are printed in ordinary type. All synonyms are printed in *italic* type.

MAPS

MAP I

To illustrate the boundaries of Europe for the purposes of *Flora Europaea*, and its division into 'territories' which are indicated by two-letter abbreviations after the summary of geographical distribution for each species. These abbreviations are derived from the Latin name of the territory concerned.

Al Albania
Au Austria, with Liechtenstein
Az Açores
Be Belgium, with Luxembourg
Bl Islas Baleares
Br Britain, including Orkney, Zetland and Isle of Man; excluding Channel Islands and Northern Ireland
Bu Bulgaria
Co Corse
Cr Kriti (*Creta*), with Karpathos, Kasos and Gavdhos
Cz Czechoslovakia
Da Denmark (*Dania*), including Bornholm
Fa Færöer
Fe Finland (*Fennia*), including Ahvenanmaa (Åland Islands)
Ga France (*Gallia*), with the Channel Islands (Îles Normandes) and Monaco; excluding Corse
Ge Germany
Gr Greece, excluding those islands included under Kriti (*supra*) and those which are outside Europe as defined for *Flora Europaea*
Hb Ireland (*Hibernia*); both the Republic of Ireland and Northern Ireland
He Switzerland (*Helvetia*)
Ho Netherlands (*Hollandia*)
Hs Spain (*Hispania*), with Gibraltar and Andorra; excluding Islas Baleares
Hu Hungary
Is Iceland (*Islandia*)
It Italy, including the Arcipelago Toscano; excluding Sardegna and Sicilia as defined *infra*
Ju Jugoslavia (see note on p. xlv)
Lu Portugal (*Lusitania*)
No Norway
Po Poland
Rm Romania
Rs Territories of the former U.S.S.R. (*Rossia*). This has been subdivided as follows, using the floristic divisions of Komarov's *Flora U.R.S.S.*; in a few places, however, our boundaries deviate slightly from those of Komarov.*
Rs(N) *Northern division*: Arctic Europe, Karelo-Lapland, Dvina-Pečora
Rs(B) *Baltic division*: Estonia, Latvia, Lithuania, Kaliningradskaja Oblast'
Rs(C) *Central division*: Ladoga-Ilmen, Upper Volga, Volga-Kama, Upper Dnepr, Volga-Don, Ural
Rs(W) *South-western division*: Moldavia, Middle Dnepr, Black Sea, Upper Dnestr
Rs(K) Krym (*Crimea*)
Rs(E) *South-eastern division*: Lower Don, Lower Volga, Transvolga
White Russia (Bjelorussija) is entirely in Rs(C). Ukraine is largely in Rs(W), but small parts are in Rs(C), Rs(E) and Rs(K). The European part of Kazakhstan is in Rs(E).
Sa Sardegna
Sb Svalbard, comprising Spitsbergen, Björnöya (Bear Island) and Jan Mayen
Si Sicilia, with Pantelleria, Isole Pelagie, Isole Lipari and Ustica; also the Malta archipelago
Su Sweden (*Suecia*), including Öland and Gotland
Tu Turkey (European part), including Gökçeada (Imroz)

* For the relation of these subdivisions to those used in the *Flora partis Europaeae URSS* (edited originally by An. A. Fedorov, and continued by N. N. Tzvelev) see p. xxii.

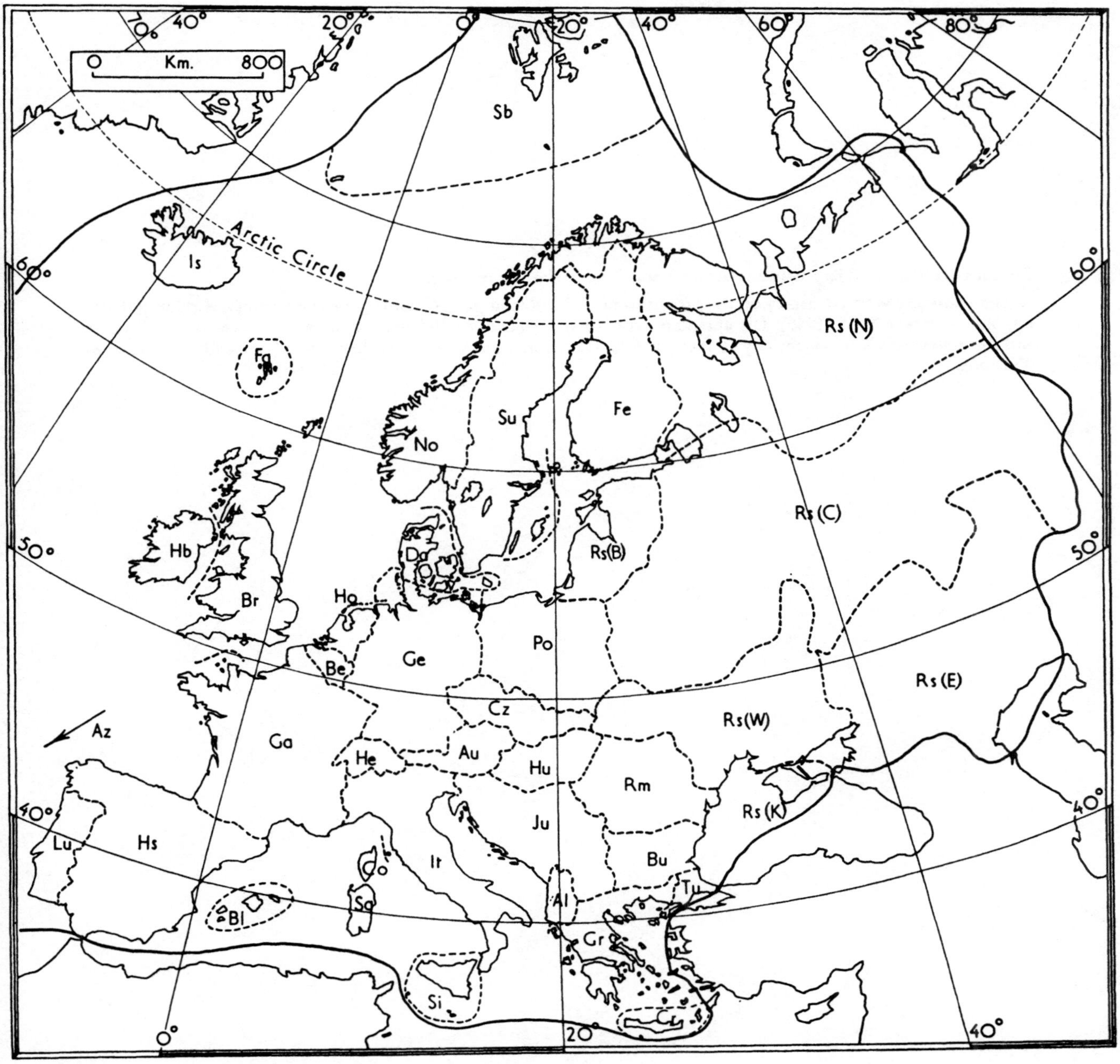

0 Km. 800
70°
40°
20°
0°
20°
40°
60°
80°
Sb
Arctic Circle
60°
Is
Fa
Rs (N)
Su
Fe
No
Hb
Rs (C)
Da
Rs(B)
50°
Br
Ho
Po
Ge
Be
Rs (E)
Cz
Rs(W)
Az
Ga
Au
He
Hu
Rm
40°
Ju
Rs (K)
Lu
Hs
Co
It
Bu
Al
Tu
Bl
Sa
Gr
Si
Cr
0°
20°
40°

MAP I

MAP II

To illustrate the boundary between Europe and Asia in the Aegean region.

The boundary is based largely on the proposals of K. H. Rechinger, 'Grundzüge der Pflanzenverbreitung in der Aegäis', *Vegetatio* **2**: 55 (1949). His northern, western and Kikladhes divisions are regarded as entirely in Europe and his eastern division as entirely in Asia; it was, however, necessary to divide his southern and north-eastern divisions.

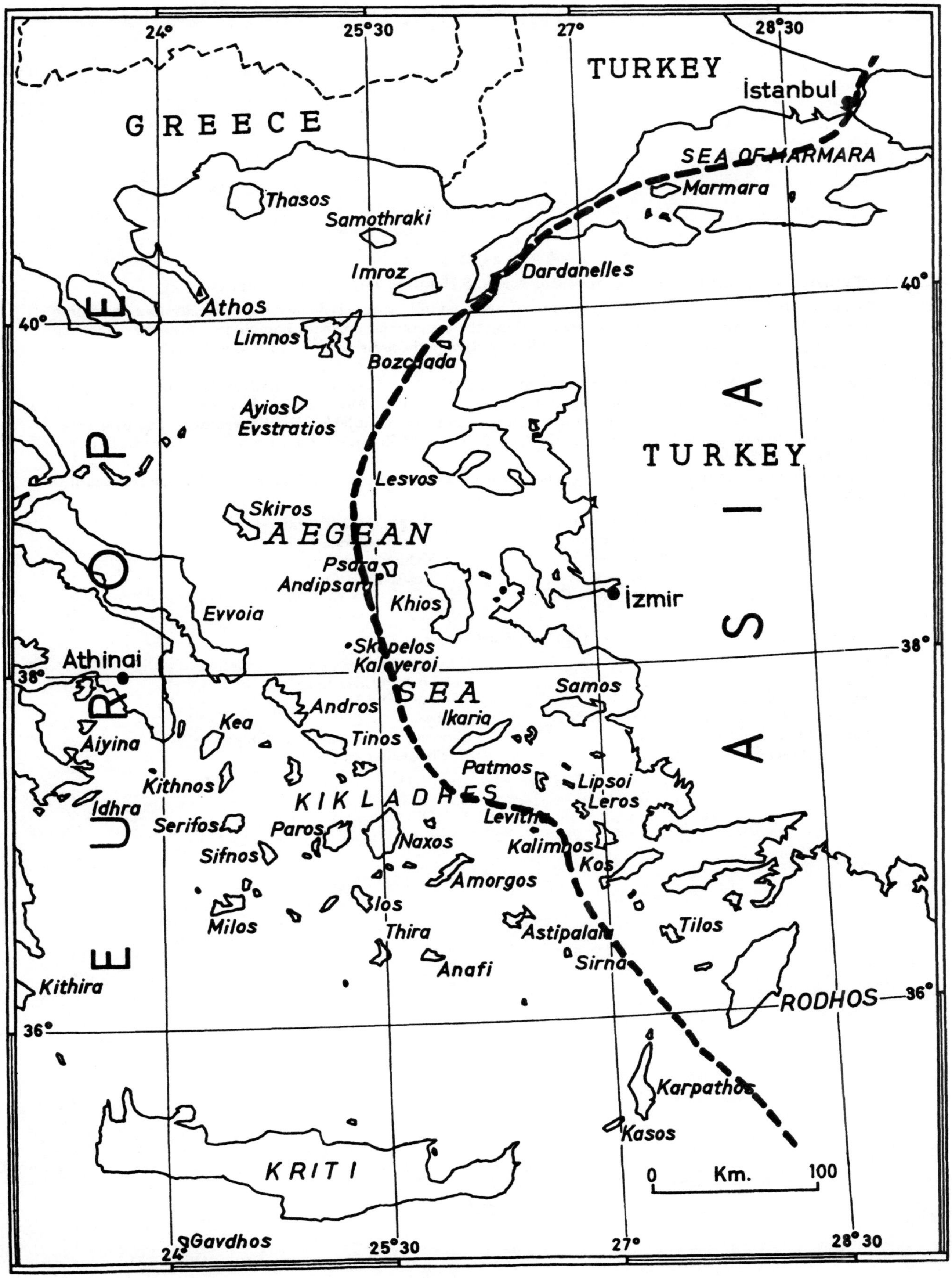

24°
25°30
27°
28°30
TURKEY
İstanbul
GREECE
SEA OF MARMARA
Marmara
Thasos
Samothraki
Imroz
Dardanelles
40°
Athos
Limnos
Bozcaada
Ayios
Evstratios
TURKEY
Lesvos
Skiros
AEGEAN
Psara
Andipsara
Khios
İzmir
Evvoia
Skopelos
Kaloyeroi
38°
Athinai
SEA
Samos
Andros
Ikaria
Kea
Tinos
Aiyina
Patmos
Lipsoi
Kithnos
KIKLADHES
Leros
Idhra
Levitha
Serifos
Paros
Naxos
Kalimnos
Sifnos
Kos
Amorgos
Ios
Milos
Tilos
Thira
Astipalaia
Anafi
Sirna
Kithira
RODHOS
36°
Karpathos
Kasos
KRITI
0
Km.
100
Gavdhos
EUROPE
ASIA

MAP II

MAP III

To illustrate the boundary between Europe and Asia in S.E. Russia and adjacent territories.

The southern boundary of Europe between the Caspian and Black Seas is defined for *Flora Europaea* as running up the Terek River westwards to 45° E.; thence along the eastern and northern boundaries of the Stavropol'skij Kraj (as marked in *The Times Atlas*) to meet the Kuban River a short distance east of Kropotkin; thence down the Kuban River to its more southerly mouth.

The eastern boundary of Europe is defined as running in the Arctic Ocean between Novaja Zemlja and Vajgač; up the Kara River to 68° N.; thence along the crest of the Ural Mountains (following the administrative boundaries) to 58° 30′ N.; thence by an arbitrary straight line to a point 50 km E. of Sverdlovsk, and by another arbitrary straight line to the head-waters of the Ural River (S. of Zlatoust); thence along the Ural River to the Caspian Sea.

The following administrative districts of Russia near the eastern or southern boundary of Europe are regarded as entirely in Europe:

Arkhangel'skaja Obl.
Komi Resp.
Permskaja Obl.
Tatarskaja Resp.
Saratovskaja Obl.
Volgogradskaja Obl.
Astrakhanskaja Obl.
Kalmytskaja Resp.
Rostovskaja Obl.

The following are regarded as partly in Europe, partly in Asia:

Russia
Sverdlovskaja Obl.
Čeljabinskaja Obl.
Baškirskaja A.S.S.R. (only the extreme N.E. corner being in Asia)
Orenburgskaja Obl.
Dagestanskaja Resp.
Čečeno-Inguškaja Resp.
Krasnodarskij Kraj

Kazakhstan
Zapadno-Kazakhstanskaja Obl.
Gur'jevskaja Obl.

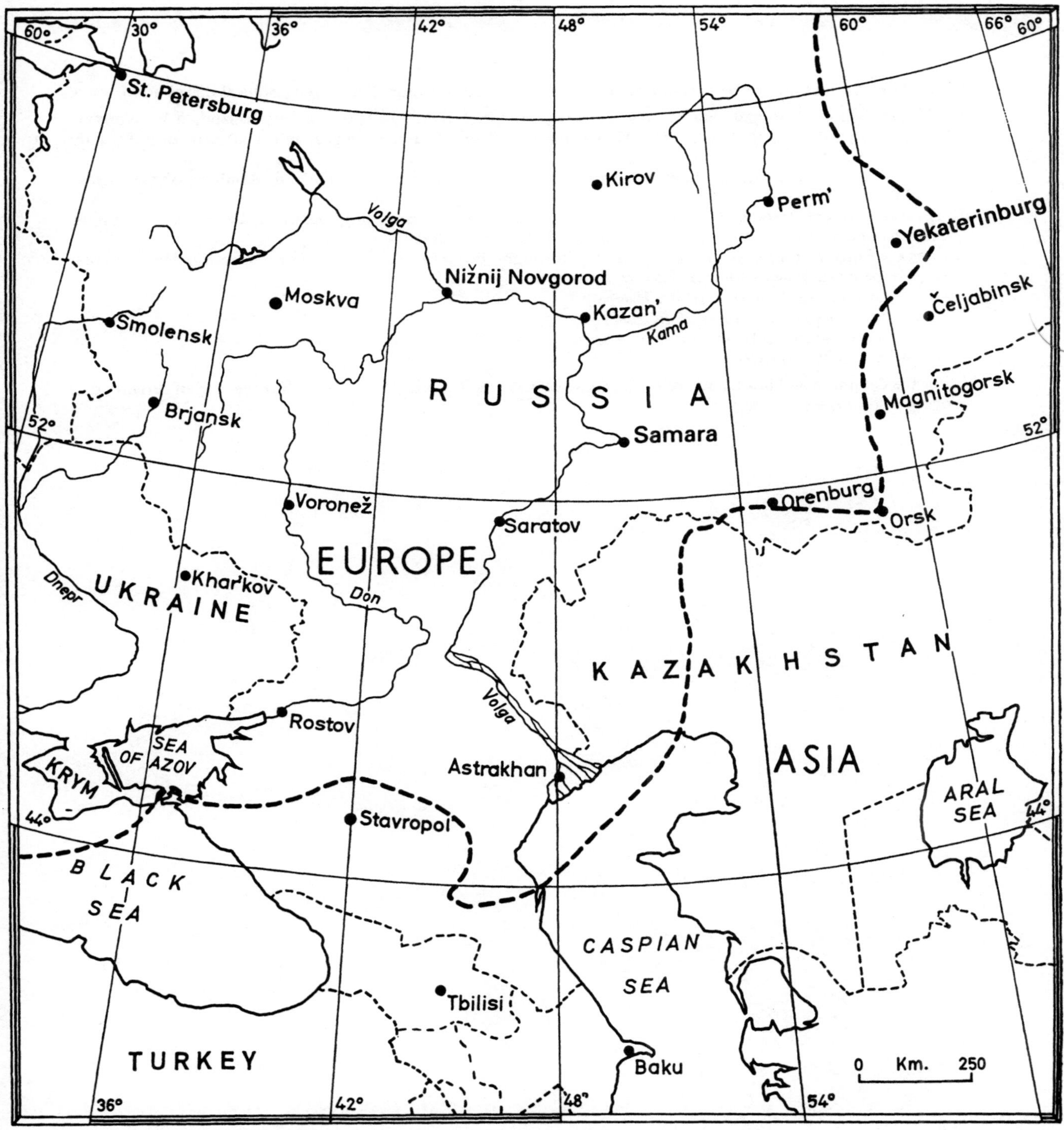
60°
30°
36°
42°
48°
54°
60°
66°
60°
St. Petersburg
Kirov
Perm'
Volga
Yekaterinburg
Nižnij Novgorod
Moskva
Čeljabinsk
Smolensk
Kazan'
Kama
Magnitogorsk
R U S S I A
Brjansk
52°
Samara
52°
Voronež
Orenburg
Orsk
Saratov
EUROPE
Khar'kov
Dnepr
UKRAINE
Don
K A Z A K H S T A N
Volga
Rostov
SEA OF AZOV
Astrakhan
ASIA
KRYM
ARAL SEA
44°
Stavropol
44°
B L A C K
S E A
CASPIAN
SEA
Tbilisi
TURKEY
Baku
0 Km. 250
36°
42°
48°
54°

MAP III

MAP IV

To illustrate the meaning to be attached to certain phrases used in summaries of geographical distribution.

W. Europe: Açores, Portugal, Spain, Islas Baleares, France, Ireland, Britain, Færöer, Iceland, S.W. Norway, Netherlands, Belgium, N.W. Germany, W. Denmark (Jylland), Corse, Sardegna, and small parts of N.W. Italy and W. Switzerland

E. Europe: N.E. Greece and the Aegean islands, Bulgaria, S. & E. Romania, Finland, republics of the former U.S.S.R.

N. Europe: Svalbard, Iceland, Færöer, Ireland, Britain (excluding S. England), Denmark, Fennoscandia, former U.S.S.R. north of a line running through Minsk–Tula–Penza–Orsk

S. Europe: Europe south of a line running through Bordeaux–Chambéry–Aosta–Locarno–Riva–Udine–Zagreb–Beograd–Ploesti–Odessa–Rostov–Astrakhan'.

-------- eastern boundary of *W. Europe*

oooooooo western boundary of *E. Europe*

—·—·—·— southern boundary of *N. Europe*

× × × × × × × northern boundary of *S. Europe*

For the definition and illustration of the meaning of S.W., N.W., S.E., N.E. and C. Europe, and of certain other geographical phrases, see map v.

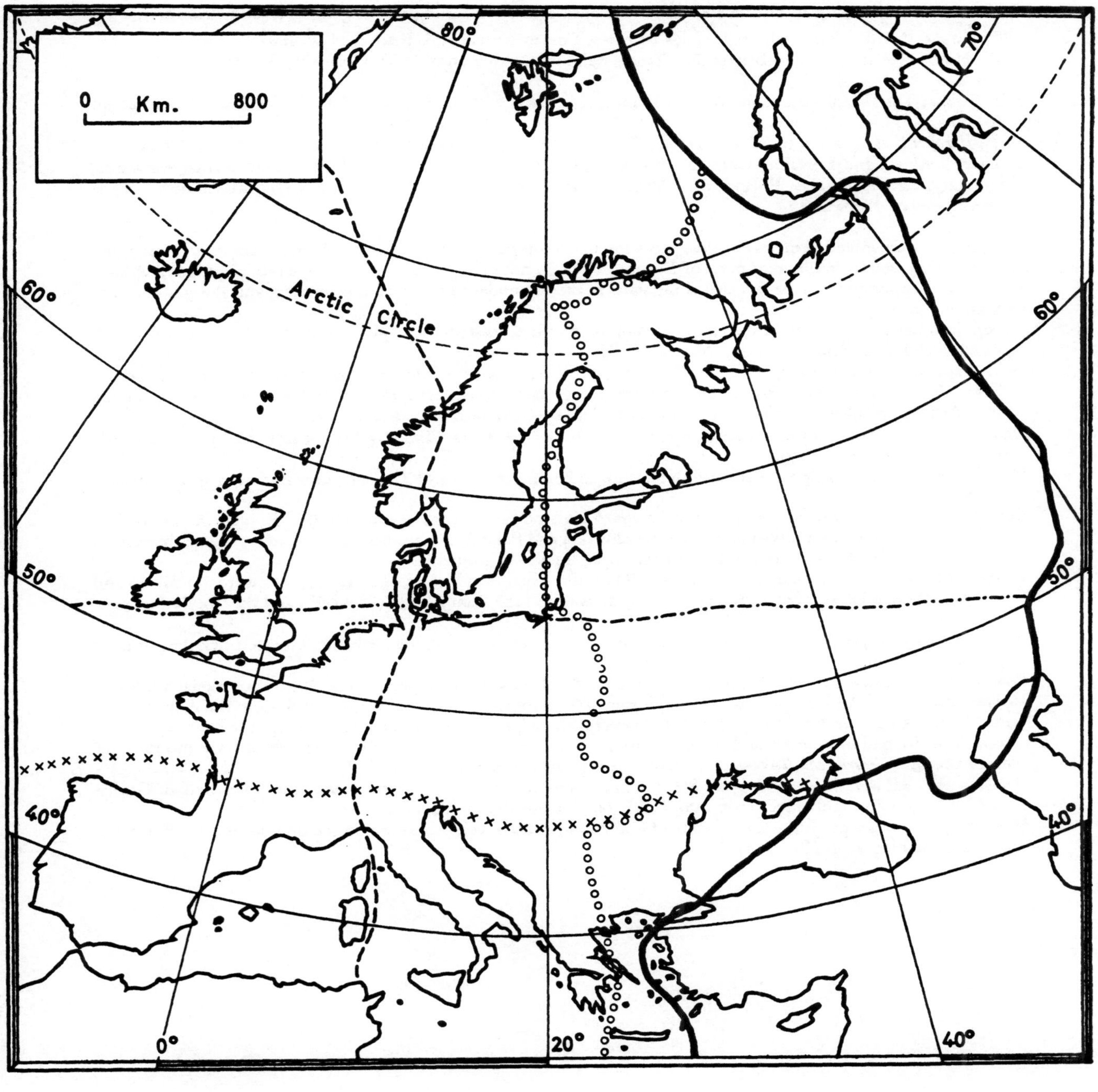

MAP IV

MAP V

To illustrate the meaning to be attached to certain phrases used in summaries of geographical distribution.

S.W. Europe: Açores, Portugal, Spain, Islas Baleares, Corse, Sardegna, S. France, N.W. Italy.

N.W. Europe: Iceland, Færöer, Britain, N. France, Belgium, Netherlands, N.W. Germany, W. Denmark (Jylland), Norway.

S.E. Europe: The Balkan peninsula, Aegean islands, S.E. Italy, S. & E. Romania, former U.S.S.R. south of about 48° N.

N.E. Europe: Russia north of 56° N., Latvia, Estonia, Finland, E. Sweden, and a small part of N.E. Norway.

C. Europe: Alsace and Lorraine, Germany, Switzerland, Austria, the Italian Alps from Monte Bianco eastwards, Hungary, Czechoslovakia, Poland, the Ukrainian Carpathians, N., W. & C. Romania, Jugoslavia north of the Danube–Sava–Kupa line.

Maps IV and V are intended merely to give precision to certain geographical phrases which are commonly used, but used in various senses in different parts of Europe. They do not purport to divide Europe into phytogeographical regions, as is apparent from the fact that along parts of their boundaries these regions overlap, and along other parts they are not contiguous.

Certain other phrases used in the summaries of geographical distribution, but not illustrated in the maps, may be briefly defined as follows:

Alps: Separated from the Appennini at 8° 15′ E. (above Savona); bounded on the east by the line Semmering–Graz–Maribor–Ljubljana–Trieste. Divided into three major divisions: *eastern*, *central*, and *south-western*, by the lines Arlberg–St Moritz–Chiavenna–Como and Genève–Chamonix–Aosta–Ivrea.

Arctic: This term is used to designate all territories north of the Arctic Circle, and is not restricted to those which have only 'arctic' vegetation.

Extreme north: Arctic Europe, together with Iceland, and Fennoscandia and Russia northwards from approximately 64° N.

Carpathians: Divided into *western*, *eastern* and *southern* divisions at the pass of Łupków (22° E.) and the Oituz Pass (46° 05′ N.). The western division is in Czechoslovakia and Poland, the southern entirely in Romania, the eastern extends from Czechoslovakia and Poland through Ukraine to Romania.

Pyrenees: Includes the subsidiary chains within 50 km of the main watershed, and extends westwards to Bilbao and Vitoria. Divided into *eastern*, *central* and *western* divisions at the Pont du Roi (0° 45′ E). and the Col du Somport (0° 30′ W.).

Balkan peninsula: Jugoslavia south of the Danube–Sava–Kupa line, Bulgaria, Albania, Greece (including islands close to the mainland) and Turkey-in-Europe.

Fennoscandia: Norway, Sweden, Finland and part of N.W. Russia (Murmanskaja Oblast' and Karelskaja A.S.S.R.).

Mediterranean region: All European territories within 100 km of the Mediterranean Sea (including the Adriatic, but not the Black Sea), and including also all Italy except the Alpine region and all Spain except the west and north-west. It is divided into *eastern* and *western* divisions by a line following the main watershed of Italy and running east of Sicilia. *Central Mediterranean* indicates the region between 8° E. and 20° E.

Aegean region: All islands in the Aegean Sea which come within the scope of the *Flora*, and those parts of Greece and Turkey-in-Europe which drain into the Aegean Sea or the Dardanelles.

Macedonia: Comprises the Jugoslav republic of Makedonija, the Greek province of Makedhonia, and the Bulgarian province of Blagoevgrad.

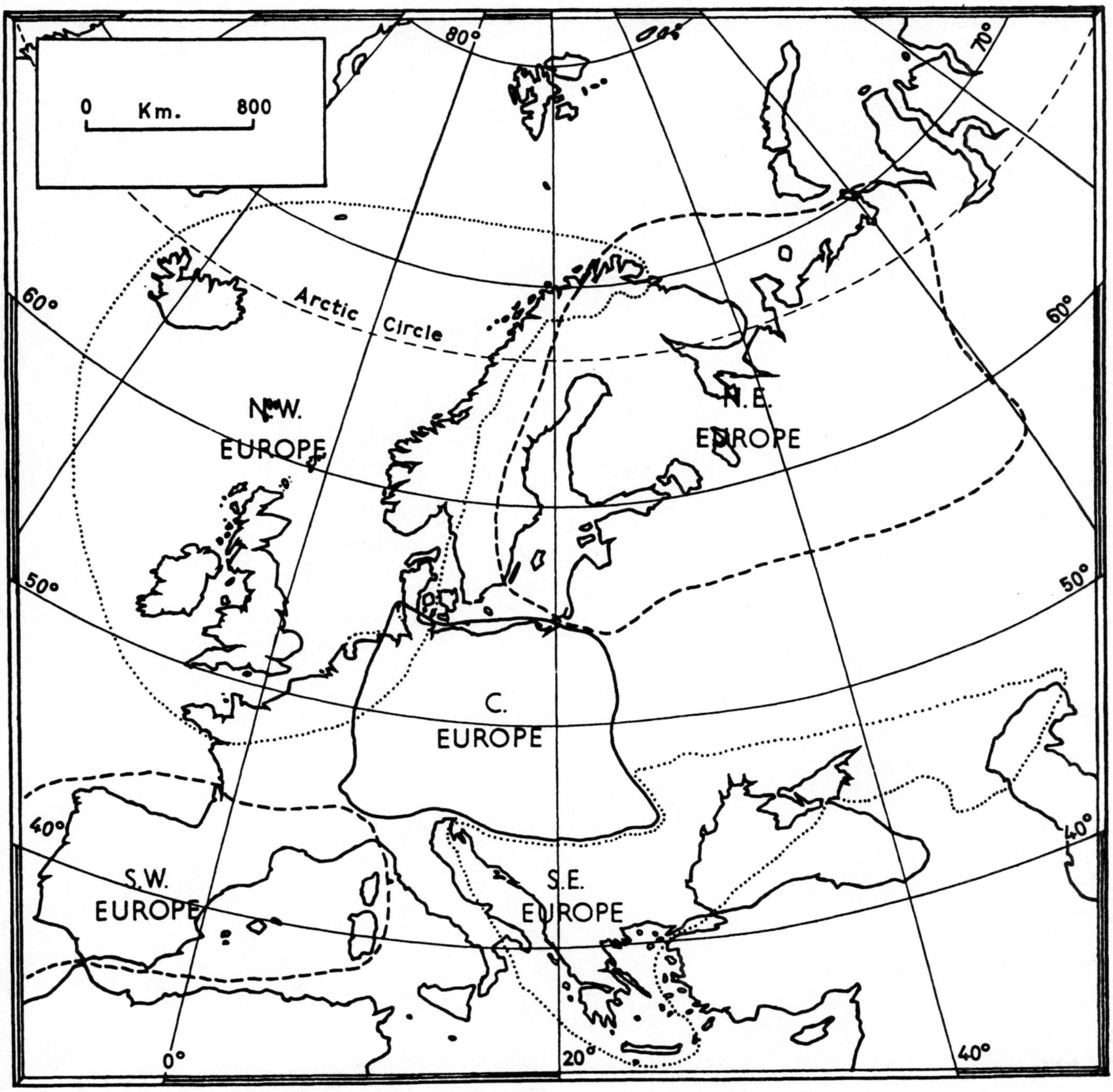
0 Km. 800
80°
70°
60°
60°
Arctic Circle
N.W.
EUROPE
N.E.
EUROPE
50°
50°
C.
EUROPE
40°
40°
S.W.
EUROPE
S.E.
EUROPE
0°
20°
40°

MAP V

Lightning Source UK Ltd.
Milton Keynes UK
15 December 2010

164447UK00001B/14/P

9 780521 153669